Molecular Structures and Dimensions
Vol. 12
Solid State   Classes 1–86

Molecular Structures and Dimensions

**Vol. 12 Bibliography 1979–80
Organic and Organometallic
Crystal Structures**

Edited by: Olga Kennard
David G. Watson
Frank H. Allen
Sharon A. Bellard

Compiled by: Brian A. Cartwright
John E. Davies
Helen Higgs
Jean Reid
Robin Taylor
Werner Versichel

Springer-Science+Business Media, B.V.

Library of Congress catalogue card number 76-133989
ISSN 0377-2012
ISBN 978-94-017-2331-2   ISBN 978-94-017-2329-9 (eBook)
DOI 10.1007/978-94-017-2329-9

Contents

Preface   vii

Acknowledgements   xi

Introduction   xiii
   Criteria for Inclusion   xiii
   Standard Entries   xiii
   Cross-Reference Entries   xiv
   Ordering of Entries   xiv
   Classification Rules   xiv
   Index System   xv
   Compound Name Indexing   xv
   Compound Name Index (Organic)   xvi
   Compound Name Index (Organometallics and Metal Complexes) xvi
   Molecular Formula Index   xvi
   Permuted Formula Index   xvii
   Author Index   xvii

List of Classes   xix

Entries for Classes 1–86   1

Compound Name Index (Organic)   OC1

Compound Name Index (Organometallics and Metal Complexes)   MC1

Formula Index   F1

Permuted Formula Index   P1

Author Index   A1

List of Classes   back endpaper

Contents

Preface

This volume is the twelfth classified bibliography of organic, organometallic and metal complex crystal structures prepared by the Cambridge Crystallographic Data Centre and published jointly with the International Union of Crystallography. The previous eleven volumes covered the years 1935–79; the present volume provides references principally to structure analyses reported in the literature during 1979 and 1980. A few structures reported prior to 1979 and omitted from earlier volumes are also included here. Volume 12 contains 3929 references to 3836 distinct chemical compounds with 1939 cross-reference entries.

During 1979–80 some 90% of references were obtained via direct in-house scanning of 51 journals; the remaining material was located by scanning *Chemical Abstracts* and *Bulletin Signalétique*. The table below summarizes the 1980 cut-off dates for the 25 direct-scan journals yielding the most entries in Volume 12. Other journals are *ca.* 95% complete for 1979, *ca.* 65% complete for 1980. The following conference proceedings are included in this volume: 5th and 6th European Crystallographic Meetings, Copenhagen 1979 and Barcelona 1980; American Crystallographic Association Winter and Summer Meetings, 1980.

The indexes presented in Volume 12 continue the system established in

Journal	Issue	Page	Year	Entries
Acta Crystallogr., Sect. B.	9	2191	1980	655
J. Amer. Chem. Soc.	15	5101	1980	328
Inorg. Chem.	8	2462	1980	314
J. Organomet. Chem.	Vol. 199	C24	1980	227
Cryst. Struct. Commun.	3	921	1980	208
J. Chem. Soc., Dalton Trans.	9	1797	1980	155
J. Chem. Soc., Chem. Commun.	18	879	1980	149
J. Org. Chem.	18	3691	1980	114
Tetrahedron Lett.	39	3799	1980	82
Zh. Strukt. Khim.	2	190	1980	81
Inorg. Chim. Acta	Vol. 46	171	1980	81
Angew. Chem. Int. Ed. (Engl.)	9	746	1980	77
Chem. Ber.	9	2950	1980	73
Z. Naturforsch. Teil. B	10	1298	1980	56
Aust. J. Chem.	6	1373	1980	55
J. Chem. Soc., Perkin Trans. 2	8	1253	1980	55
Bull. Chem. Soc. Jpn.	6	1755	1980	49
Z. Anorg. Allg. Chem.	Vol. 466	195	1980	45
Can. J. Chem.	17	1847	1980	41
Koord. Khim.	12	1896	1979	38
Acta Chem. Scand., Ser. A	5	365	1980	32
J. Chem. Soc., Perkin Trans. 1	9	2061	1980	31
J. Cryst. Mol. Struct.	4	199	1980	30
Doklady Akad. Nauk. SSR	Vol. 251	162	1980	29
J. Chem. Res.	7	S227	1980	26

Volumes 9–11 and in the special volume *Guide to the Literature 1935–76*. The *Guide* presents a set of cumulative indexes to the contents of Bibliographic Volumes 1–8.

This volume is produced directly from the computer-based bibliographic file of the Cambridge Crystallographic Data Centre. The total database also contains magnetic-tape files of chemical structural information (as connectivity tables) and evaluated numeric data (atomic coordinates, unit-cell parameters, symmetry, etc.). The Centre also acts as a depository for numeric data relating to structures reported in *Chemical Communications* (since 1977), *Tetrahedron Letters* (since 1977), *Tetrahedron*, *Phytochemistry* and *Nouveau Journal de Chimie* (since 1980). The total database currently (1 March 1981) contains information on some 28,000 structure analyses.

The Cambridge Centre has developed a set of computer programs for search, retrieval, analysis and display of information contained in the database. The programs permit searches based on bibliographic information fields, or on the connectivity tables (for complete structures or substructural fragments), to obtain relevant literature references. Retrieved subsets of numeric data may then be used for extensive geometric calculations or for the preparation of graphic illustrations. The system is fully described in *Acta Cryst.* B35, 2331–2339 (1979).

The database and associated programs are available world-wide through National Affiliated Centres. These Centres receive regular updates of new material and provide services and tape copies to their local scientific communities. National Centres operating in 1980 are listed below. Potential users in these countries should contact the addresses shown. British users and interested scientists in other countries should contact the Cambridge Centre.

Affiliated Data Centres operating in 1980

Australia	Information Service, CSIRO, 314 Albert St, P.O. Box 89, East Melbourne, Victoria 3002 (Dr C. Garrow)
Belgium	Laboratorium voor Kristallografie, Katholieke Universiteit, Redingenstraat 16 bis, B-3000 Leuven (Prof. G. S. D. King)
Brazil	Departamento de Fisica e Ciencias Dos Materiais, Instituto de Fisica e Quimica de São Carlos USP, São Carlos, São Paulo 13560 (Dr Y. P. Mascarenhas)
Canada	Canadian Institute for Scientific and Technical Information, National Research Council, Ottawa K1A OS2 (Dr G. H. Wood)
France	PLURIDATA, Centre Informatique et de Documentation Automatique, 1 rue Guy de la Brosse, 75005 Paris (Prof. J. E. Dubois)
Hungary	Department of X-ray Diffraction, Central Research Institute of Chemistry, Hungarian Academy of Sciences, H-1525 Budapest, P.O.B. 17 (Prof. A. Kalman, Mr Neszemlyi)
India	Department of Crystallography and Biophysics, University of Madras, A.C.C. Campus, Madras 60025 (Prof. R. Srinivasan)

Israel	Department of Chemistry, Weizmann Institute of Science, Rehovot (Dr D. Rabinovich, Dr Z. Shakked)
Italy	Istituto di Strutturistica Chimica, Universita di Parma, Via M. D'Azeglio 85, 43100 Parma (Prof. M. Nardelli, Prof. G. D. Andreetti)
Japan	Institute for Protein Research, Osaka University, 5311 Yamada-Kami, Suita, Osaka (Prof. M. Kakudo)
Netherlands	Department of Inorganic Chemistry, University of Nijmegen, Toernooiveld, Nijmegen (Dr J. H. Noordik)
New Zealand	Department of Chemistry, University of Canterbury, Christchurch 1 (Prof. B. R. Penfold)
Scandinavia	Department of Structural Chemistry, University of Göteborg, P.O.B., S-40033, Göteborg 33, Sweden (Prof. S. Abrahamsson)
South Africa	Centre for Scientific and Technical Information, CSIR, P.O. Box 395, Pretoria 0001 (Dr A. G. Brunt)
Switzerland	Laboratorium für Organische Chemie, ETH-Zentrum, CH-8092 Zürich (Prof. J. D. Dunitz)
U.S.A.	National Institutes of Health, Bethesda, Maryland 20014 (Dr G. W. A. Milne)
West Germany	Fachinformationszentrum Energie, Physik, Mathematik GmbH, Karlsruhe, Kernforschungszentrum, D-7514, Eggenstein-Leopoldshafen 2 (Dr H. Behrens)

We thank the readers of the *Molecular Structures and Dimensions* series and users of the Structural Database who have notified us of errors and omissions. These have been incorporated in the master file. We hope that this collaboration will continue.

Olga Kennard *Cambridge Crystallographic Data Centre*
David G. Watson *University Chemical Laboratory*
Frank H. Allen *Lensfield Road*
Sharon A. Bellard *Cambridge CB2 1EW, England*
March 1980

Acknowledgements

Cambridge Crystallographic Data Centre

This volume is derived from the database of the Crystallographic Data Centre. The editors would like to express their thanks to the following for their assistance in its maintenance.

Mrs K. A. M. Watson has been in charge of the encoding of information and has been assisted in the secretarial aspects of documentation by Miss A. Brown and Miss P. K. Johnson. Mrs A. Sugg and Mrs J. Colman have contributed to keyboarding and reprint acquisition.

Computer Laboratories

The master copy for this volume was produced using the IBM 360/195 computer and the FR 80 microfilm recorder at the Science Research Council Rutherford and Appleton Laboratories, Chilton, U.K. We are especially grateful to Mrs K. M. Crennell of the Computing Division, the Atlas Centre, who wrote the page layout, tabulation and justification routines. We are also indebted to the FR 80 Operations Manager, Mr B. J. Jeeves, and his staff for provision of high-quality FR 80 output.

The IBM 370/165 computer of the University of Cambridge Computer Centre has been used for this work and we are grateful to the staff for their special help in the production of the final tapes for the Chilton interface.

Research Councils and Other Bodies

We thank the Science Research Council and the Affiliated Data Centres for financial support and the Medical Research Council for allowing a member of their External Scientific Staff (Olga Kennard) to participate in this work.

This compilation was prepared in parallel with the Organic Volumes of Crystal Data (National Bureau of Standards, Washington D.C., U.S.A.). Both projects are strengthened by this collaboration.

We thank the University of Cambridge and the staff of the Chemical Laboratory for help with administrative matters, and the Head of our Department, Professor R. A. Raphael, F.R.S., and Professor J. Lewis, F.R.S., for their encouragement and advice.

Introduction

Criteria for Inclusion in the Bibliography

— The substance has been studied by X-ray or neutron diffraction and contains organic carbon. Purely inorganic carbonyls, cyanides etc., are excluded, as are macromolecules (proteins, viruses etc.).
— The study has not been superseded by a later paper by the same authors.
— Three positional coordinates have been determined for each non-hydrogen atom, though not necessarily recorded in the publication.

Standard Entries

The main bibliographic listing is divided into 86 chemical classes with cross-referencing between classes. The classification scheme is fully described below. The listing therefore contains both *standard* and *cross-reference* entries. A standard entry contains the following information.

Compound Name (bold face), usually the name assigned in the original publication. Where only a trivial name is given, or the name is absent, a systematic name is assigned as far as possible.

Qualifying Phrases (bold face) may follow the compound name to indicate special conditions of the experiment or of the crystal, e.g. neutron study, absolute configuration determined by X-rays, high- or low-temperature studies, polymorphic forms etc. Entries without a qualifier correspond to X-ray studies at room temperature.

Synonym. This may be included to record trivial names (e.g. DDT, Paraquat), or commonly accepted non-systematic names.

Molecular Formula, expressed in terms of residues (discrete covalently bonded networks or ions). The formula of each residue takes the general form $C_xH_yA_aB_bC_c...$, together with multipliers and charges where necessary. Residues containing organic carbon precede solvate residues and inorganic residues or ions.

Authors' Names, transcribed exactly as published in the original paper, but without diacritical marks; Russian names are transliterated according to standard rules.

Literature Reference, recorded as journal name, volume (bold face), page no., year of publication.

Cross-Reference, indicating that one (or more) of the residues occurs in other chemical class(es).
Standard entries have *entry numbers* of the form *cc.nnn* where *cc* is the class number and *nnn* is the sequence number within that class.

Cross-Reference Entries

A cross-reference entry in the main bibliographic listing contains the following information from the main entry: **Compound Name** (bold face), **Qualifier** (bold face, if present), **Synonym** (if present), **Molecular Formula**.

A cross-reference entry has an entry number of the form *cc.C*, where *cc* is the class number, and is always terminated by a statement indicating the position of the corresponding main entry.

Where the cross-reference is generated by a second or subsequent residue this residue becomes the leading residue in the molecular formula record of the cross-reference entry.

Ordering of Entries

Entries are ordered by chemical class into 86 chapters. All entries, both standard and cross-references, are ordered by *molecular formula* within each class.

Classification Rules

Each residue in the structure which contains organic carbon is assigned to one or more of the 86 chemical classes. The full list of classes is given below. For each residue a precedence table is used to determine the main class assignment; other assignments are treated as cross-references. For example, if a compound can be described as belonging to class 15 (benzene nitro compounds) and also class 17 (phenols and ethers), then 17 will always be assigned as the main class with 15 as a cross-reference. The order of precedence is shown in the table below running from top to bottom and left to right (i.e. class 61 has highest precedence and 5 the lowest).

                61, 60
                71, 72, 73, 74, 75, 76, 77, 78, 79, 80, 81, 82, 86, 85, 84, 83
                70, 69, 68, 67, 66, 65, 64, 63, 62
                58, 57, 56, 55, 54, 53, 52, 51, 50, 49, 48, 47, 46, 45, 44, 43, 59
                42, 41, 40, 39, 38, 37, 36, 35, 34, 32, 33
                31, 30, 29, 28, 27, 23, 22, 20, 21, 26, 25, 24
                18, 14, 13, 17, 16, 15
                2, 1, 3, 4
                12, 8, 11, 7, 10, 9, 6, 19, 5

In addition to these rules the classification conventions have, in recent years, been better defined for classes 1–59 and class 67. Some notes on these conventions are given below.

Class 1. Cyclic acid derivatives e.g. anhydrides and lactones, are classified in the appropriate hetero class. This rule also applies to class 13.

Class 2. In a few cases where the cation is organic we classify the anion in 2.

Class 4. The compound must contain –C–N–S or –C–S–N–.

Class 9. The compound must contain –C–N–N–.

Class 10. The compound must contain –C–N–O– or –C–O–N–.

Class 24. The compound must be fully unsaturated. The same rule also applied to classes 25 and 26.

Class 44. The ring system must conform to the unmodified pyrimidine or purine skeleton.

Class 48. This class is reserved for peptides and α-amino-acids, whether or not

the amino-acid possesses biological properties. Thus a β-amino-acid would be classified in the appropriate acid and amine classes.

Class 50. A cross reference to a structural class is always provided; this rule also applies to class 59.

Class 67. This class is reserved for compounds containing covalently bonded metals of groups IA and IIA.

Index System

The index system is designed to enable users to access standard entries in the main classified listing on the basis of: chemical name fragments; molecular formulae; specific elements (other than C, H, N, O, S, P, Cl, Br, I); and authors' names. The indexes are non-cumulative, i.e. they refer to the contents of this volume only, and form a continuation of the cumulative indexes presented in the MSD special volume *Guide to the Literature 1935–76*.

Compound Name Indexing

Instead of a single alphabetic name index, in which each compound and synonym name occurs once only, we have prepared a permuted keyword-in-context index in which each name usually occurs several times. Two separate indexes are given: one for compounds which may be broadly described as organic, the second covering organometallics and metal complexes. The general principles of keyword selection and index layout are, however, common to both.

The selection of keywords for each name is performed automatically by computer using input lists of common chemical and nomenclatural prefixes, suffixes and individual words. The aim of the analysis is to break down long strings of chemical syntax into their constituent words and to select information-rich words for inclusion as index points. For example the syntax string:

di/cobalta/di/carba/hepta/borane

is split into six potential keywords of which two (italicized) are retained for indexing.

The keywords are sorted alphabetically and aligned down the centre of the index page. The context of each keyword is shown to the left and right on the same line; for longer names a wraparound facility is used to preserve maximum context. This is indicated by the use of [and] which respectively identify the start and end of the true name in the wraparound. For very long names truncation is necessary and the symbols ⟨ and ⟩ are used to indicate loss of context to left and right.

In order to preserve clarity certain subscript and superscript strings which occur, for example, in the nomenclature for bridged-ring compounds, have not been fully interpreted in typesetting. Such strings are simply enclosed within parentheses to avoid breaks in alignment which might be distracting to the user.

The entry number, *cc.nnn*, is used to refer back to the main classified listing. In both sections of the index the entry number may be followed by a letter.

a indicates that the *absolute configuration* of the cited compound was determined by X-ray methods.

n indicates that the cited compound was studied by *neutron diffraction*.

While we cannot guarantee that these codings are totally exhaustive we hope that their inclusion here conveys useful additional information. The codes are attached to all occurrences of the name to which they apply.

In both indexes the inclusion or exclusion of any keyword largely depends on

its incidence. For example *methyl*, *ethyl*, etc., occur many times, convey minimal information, and are excluded. Keywords such as *camphor* are highly informative and are retained. In between these two extremes there exists an arbitrary area, and in the final selection of keywords we have attempted to balance space requirements against information content.

Compound Name Index (Organic)

Compound and synonym names are included in the organic section on the basis of chemical class criteria:
— entries in basic classes 1–67 and 70;
— entries in other classes are also included if they have a cross-reference to 1–67 or 70.

This section contains all compounds which are generally accepted as organic together with compounds of S, Se, Te, B, Si, P, As, Sb, Bi and Group IA and IIA elements. The second criterion allows the inclusion of some metal complexes (e.g. silver salt adducts of olefins etc.) where the structural interest was primarily the organic moiety. This gives rise to an area of overlap between the two parts of the name index.

In this section certain keywords, e.g. *acetate*, *benzoate* are excluded for the natural product classes 50–59 but retained in all other cases.

The running head on each index page contains a directory block recording the first five characters of the first and last index point on that page.

Compound Name Index (Organometallics and Metal Complexes)

Substances containing a metallic element defined as: transition metals, lanthanides, actinides, Zn, Cd, Hg, Al, Ga, In, Tl, Ge, Sn, Pb, are included in this index. The selection criteria are:
— entries in basic classes 68, 69, 71–86;
— entries in other classes which contain any of the above elements in any residue of the molecular formula. The second criterion means that two-residue structures where the inorganic ion is unclassified, e.g. *ethylammonium tetrachlorocuprate* (class 3) will appear under *copper* in this section of the index, as will all metal-containing porphyrin structures (class 49).

Instead of presenting a single alphabeticized keyword list we have subdivided the index according to elements; the *sections* are ordered alphabetically by element name. Within each section the layout is as described above, but the keyword which contains the element name or name root, e.g. *gold*, *aura*, *auri*, *auro*, is omitted. Compounds containing more than one metallic element are fully indexed under each element name.

In addition to the absolute configuration and neutron study flags the symbol * may occur in the (normally blank) column preceding the indexed keyword. This indicates *polynuclear bridged species*.

The running head on each index page contains a directory block recording the first and last element names referred to on that page.

Molecular Formula Index

Molecular formulae are expressed in terms of residues, e.g. CBr_4, C_8H_{10} for the carbon tetrabromide *p*-xylene complex. Only non-trivial residues are classified; residues such as solvents, inorganic ions, etc., are not classified.

The arrangement of symbols within a residue is that used by *Chemical Abstracts*: carbon atoms first, followed by hydrogen (if present) and other elements in alphabetic sequence, typically $C_xH_yA_aB_bC_c...$, followed if necessary by the net charge on the residue. Residue premultipliers may be explicit fractional or integer numbers or, in cases of indeterminacy, the letters x, y may be used. Post multipliers are reserved for polymeric residues, e.g. $(C_2H_2CuO_4)_n$.

Entries in the index are grouped under the number of carbon atoms. Within each group entries are ordered according to their natural sequence of elements in the manner adopted by *Chemical Abstracts*.

The primary indexed residue appears in bold type. Compounds with more than one classified residue are indexed under each such residue. Hence the CBr_4,C_8H_{10} complex will appear under C_1 as CBr_4,C_8H_{10} and under C_8 as C_8H_{10},CBr_4.

The index refers to the main classified listing via the entry number which is printed to the left of the formula.

Permuted Formula Index

This index takes the form of an element-in-context listing based on 'rarer' elements. Elements other than C, H, N, O, S, P, Cl, Br, I are defined as rare. The index is best described by use of an example. The compound

$$C_{10}H_{18}As_2Cl_3GeMnO_3$$

contains three permutable elements (bold type), and will occur three times in the index as:

MnGe	**As₂**	$C_{10}H_{18}Cl_3O_3$
MnAs₂	**Ge**	$C_{10}H_{18}Cl_3O_3$
GeAs₂	**Mn**	$C_{10}H_{18}Cl_3O_3$

The common elements are relegated to the end of the residue formulation and the remaining rarer elements are permuted into the key index position (bold type) giving a KWIC-style layout. The context is defined by the other permutable elements, printed to the left of the key, and the common elements printed to the right.

In multiple residue structures the second and subsequent residues are added, with their usual element order, to the right-hand context string and separated by commas, e.g.

$$2(As_4 \quad \text{Nb} \quad C_{20}H_{32}Cl_4^+), \, Cl_3NbO_2^{2-}$$

It should be noted that the permutation process is only applied to classified residues. The inorganic ion which forms the second residue in the example above will not appear in the niobium section.

The index facilitates searches for specific rarer elements and for rarer element groupings; it refers back to the main classified listing via the entry number printed to the left of each formula.

Author Index

This is an alphabetic listing of authors' names keyed to the main classified listing via lists of entry numbers, which are grouped in ascending order.

Authors' names are abstracted directly from published papers; any misprints contained in the original material will not, in general, have been corrected.

Similarly the number of initials used by authors in their published work is also abstracted directly, any published differences will therefore generate multiple index entries. It should be noted that diacritical marks and special symbols are not included. We have attempted to be both consistent, and to follow accepted conventions, in the transliteration of Russian names.

Names which contain a mixture of upper and lower case letters – e.g. DeLaMatter, MacKenzie, etc. – are sorted as if all letters were lower case; hence MacKenzie and Mackenzie will be adjacent in the index. Names containing genealogical qualifiers – e.g. Junior, III, etc. – are sorted including that qualifier, hence Smith Junior, J. occurs after Smith, G. and Smith, L.

List of Classes

This list is reproduced on the back endpaper for easy reference.

1	Aliphatic Carboxylic Acids and their Derivatives	1
2	Aliphatic Carboxylic Acid Salts (Ammonium, IA, IIA Metals)	4
3	Aliphatic Amines	5
4	Aliphatic (N and S) Compounds	10
5	Aliphatic Miscellaneous	10
6	Enolates (Aliphatic and Aromatic)	11
7	Nitriles (Aliphatic and Aromatic)	12
8	Urea Compounds (Aliphatic and Aromatic)	13
9	Nitrogen-Nitrogen Compounds (Aliphatic and Aromatic)	15
10	Nitrogen-Oxygen Compounds (Aliphatic and Aromatic)	16
11	Sulfur and Selenium Compounds	17
12	Carbonium Ions, Carbanions, Radicals	19
13	Benzoic Acid Derivatives	19
14	Benzoic Acid Salts (Ammonium, IA, IIA Metals)	21
15	Benzene Nitro Compounds	21
16	Anilines	23
17	Phenols and Ethers	25
18	Benzoquinones	28
19	Benzene Miscellaneous	28
20	Monocyclic Hydrocarbons (3, 4, 5-Membered Rings)	30
21	Monocyclic Hydrocarbons (6-Membered Rings)	31
22	Monocyclic Hydrocarbons (7, 8-Membered Rings)	33
23	Monocyclic Hydrocarbons (9- and Higher-Membered Rings)	34
24	Naphthalene Compounds	34
25	Naphthoquinones	35
26	Anthracene Compounds	36
27	Polycyclic Hydrocarbons (2 Fused Rings)	37
28	Polycyclic Hydrocarbons (3 Fused Rings)	39
29	Polycyclic Hydrocarbons (4 Fused Rings)	40
30	Polycyclic Hydrocarbons (5 or More Fused Rings)	41
31	Bridged Ring Hydrocarbons	42
32	Hetero-Nitrogen (3, 4, 5-Membered Monocyclic)	46
33	Hetero-Nitrogen (6-Membered Monocyclic)	50
34	Hetero-Nitrogen (7- and Higher-Membered Monocyclic)	55
35	Hetero-Nitrogen (2 Fused Rings)	56
36	Hetero-Nitrogen (More than 2 Fused Rings)	62
37	Hetero-Nitrogen (Bridged Ring Systems)	65
38	Hetero-Oxygen	67
39	Hetero-Sulfur and Hetero-Selenium	76
40	Hetero-(Nitrogen and Oxygen)	80
41	Hetero-(Nitrogen and Sulfur)	84
42	Miscellaneous Heterocycles	88
43	Barbiturates	89
44	Pyrimidines and Purines	90

45	Carbohydrates	93
46	Phosphates	98
47	Nucleosides and Nucleotides	99
48	Alpha-Amino-Acids and Peptides	102
49	Porphyrins and Corrins	106
50	Antibiotics	108
51	Steroids	110
52	Monoterpenes	114
53	Sesquiterpenes	115
54	Diterpenes	116
55	Sesterterpenes	118
56	Triterpenes	119
57	Tetraterpenes	120
58	Alkaloids	119
59	Miscellaneous Natural Products	123
60	Molecular Complexes	126
61	Clathrates	129
62	Boron Compounds	132
63	Silicon Compounds	135
64	Phosphorus Compounds	138
65	Arsenic Compounds	147
66	Antimony and Bismuth Compounds	148
67	Groups IA and IIA Compounds	149
68	Group III Compounds	150
69	Germanium, Tin, Lead Compounds	151
70	Tellurium Compounds	154
71	Transition Metal-C Compounds	155
72	Metal π-Complexes (Open-Chain)	170
73	Metal π-Complexes (Cyclopentadiene)	175
74	Metal π-Complexes (Arene)	184
75	Metal π-Complexes (Miscellaneous Ring Systems)	185
76	Metal Complexes (Ethylenediamine)	189
77	Metal Complexes (Acetylacetone)	195
78	Metal Complexes (Salicylic Derivatives)	196
79	Metal Complexes (Thiourea)	197
80	Metal Complexes (Thiocarbamate or Xanthate)	198
81	Metal Complexes (Carboxylic Acid)	200
82	Metal Complexes (Amino-Acid)	205
83	Metal Complexes (Nitrogen Ligand)	207
84	Metal Complexes (Oxygen Ligand)	226
85	Metal Complexes (Sulfur or Selenium Ligand)	236
86	Metal Complexes (P, As, Sb Ligand)	243

ALIPHATIC CARBOXYLIC ACIDS
AND THEIR DERIVATIVES

1.C   **Potassium deuterium bis(dichloroacetate) (neutron study, deuterated form)**
$C_2HCl_2DO_2$, $C_2HCl_2O_2^-$, K^+ Main entry is 2.7

1.C   **Guanidinium hydrogen oxalate monohydrate**
$C_2HO_4^-$, $CH_6N_3^+$, H_2O Main entry is 2.9

1.C   **Potassium hydrogen bis(dibromoacetate)**
$C_2H_2Br_2O_2$, $C_2HBr_2O_2^-$, K^+ Main entry is 2.11

1.C   **Rubidium hydrogen bis(dibromoacetate)**
$C_2H_2Br_2O_2$, $C_2HBr_2O_2^-$, Rb^+ Main entry is 2.6

1.C   **Potassium hydrogen bis(dichloroacetate) (neutron study)**
$C_2H_2Cl_2O_2$, $C_2HCl_2O_2^-$, K^+ Main entry is 2.8

1.1   **Oxalic acid dihydrate (at 100°K, α form, electron density distribution study)**
$C_2H_2O_4$, $2H_2O$
E.D.Stevens, P.Coppens
Acta Crystallogr.,Sect.B,**36**, 1864, 1980

1.2   **Nitroacetic acid**
$C_2H_3NO_4$
K.von Deuten, G.Klar
Cryst.Struct.Commun.,**9**, 397, 1980
See also R1 : 10

1.3   **N – Hydroxyoxamide (monoclinic form)**
$C_2H_4N_2O_3$
I.K.Larsen *Acta Chem.Scand.Ser.B*,**34**, 209, 1980

1.C   **Furaltadone hydrochloride acetic acid clathrate**
$C_2H_4O_2$, $C_{13}H_{17}N_4O_6^+$, Cl^- Main entry is 61.4

1.C   **tetrakis((μ² – Acetato) – (μ² – carbonyl)) – tetra – palladium acetic acid solvate (at –120°C)**
$2C_2H_4O_2$, $C_{12}H_{12}O_{12}Pd_4$ Main entry is 81.44

1.C   **hexakis(R – α – Phenylethylsulfonyl – methyl) benzene acetic acid clathrate**
$4C_2H_4O_2$, $C_{60}H_{66}O_{12}S_6$ Main entry is 61.24

1.4   **bis(Acetamide) hydrochloride**
C_2H_5NO, $C_2H_6NO^+$, Cl^-
K.W.Muir, J.C.Speakman *J.Chem.Res.*,**277**, 3401, 1979

1.5   **Monoacetamide nitrate (at –115°C)**
$C_2H_6NO^+$, NO_3^-
A.I.Gubin, A.I.Yanovskii, Yu.T.Struchkov,
B.A.Berimzhanov, N.N.Nurakhmetov, M.Zh.Buranbaev
Cryst.Struct.Commun.,**9**, 745, 1980

1.6   **O – Methylthiocarbazate (at –150°C)**
$C_2H_6N_2OS$
R.Mattes, H.Weber, K.Scholten *Chem.Ber.*,**113**, 1981, 1980
See also R1 : 11.9

1.7   **S – Methylthiocarbazate (at –150°C)**
$C_2H_6N_2OS$
R.Mattes, H.Weber, K.Scholten *Chem.Ber.*,**113**, 1981, 1980
See also R1 : 11.9

1.8   **Urea malonic acid**
$C_3H_4O_4$, CH_4N_2O
G.Bandoli, D.A.Clemente, M.Brustolon, C.Corvaja,
C.Pinzino, A.Colligiani *Mol.Phys.*,**39**, 1145, 1980
See also R2 : 8

1.9   **Copper propionate**
$2C_3H_5O_2^-$, Cu^{2+}
A.V.Ablov, T.N.Tarkhova, Yu.A.Simonov
Acta Crystallogr.,**21**, A134, 1966

1.C   **Furaltadone hydrochloride propionic acid clathrate**
$C_3H_6O_2$, $C_{13}H_{17}N_4O_6^+$, Cl^- Main entry is 61.5

1.C   **α – Cyclodextrin N,N – dimethylformamide clathrate pentahydrate**
C_3H_7NO, $C_{36}H_{60}O_{30}$, $5H_2O$ Main entry is 61.19

1.C   **Tetra – n – butylammonium hydrogen dichloromaleate (CuK(α) data)**
$C_4HCl_2O_4^-$, $C_{16}H_{36}N^+$ Main entry is 2.19

1.C   **Tetra – n – butylammonium hydrogen dichloromaleate (MoK(α) data)**
$C_4HCl_2O_4^-$, $C_{16}H_{36}N^+$ Main entry is 2.20

1.C   **Carbinoxamine maleate**
$C_4H_3O_4^-$, $C_{16}H_{20}ClN_2O^+$ Main entry is 33.70

1.C   **(1S,4R) – 3′ – Chloro – 10′,11′ – dihydro – N,N – dimethyl – spiro(cyclohex – 2 – ene – 1,5′ – (5H) – dibenzo(a,d)cycloheptene) – 4 – ammonium hydrogen maleate (absolute configuration)**
$C_4H_3O_4^-$, $C_{22}H_{25}ClN^+$ Main entry is 29.9

1.10   **Calcium succinate trihydrate**
$C_4H_4O_4^{2-}$, Ca^{2+}, $3H_2O$
A.Karipides, A.T.Reed
Acta Crystallogr.,Sect.B,**36**, 1377, 1980

1.11   **Copper L – tartrate trihydrate**
$C_4H_4O_6^{2-}$, Cu^{2+}, $3H_2O$
H.Soylu *Eur.Cryst.Meeting*,**6**, 295, 1980

1.C   **Lithium hydrogen (+) – 1 – malate**
$C_4H_5O_5^-$, Li^+ Main entry is 2.23

1.12   **Potassium fluoride succinic acid**
$C_4H_6O_4$, K^+, F^-
J.Emsley, D.J.Jones, R.S.Osborn
J.Chem.Soc.,Chem.Commun.,703, 1980

1.13   **N,N′ – Dimethyl – oxalamide**
$C_4H_8N_2O_2$
K.-H.Klaska, O.Jarchow, W.Scham, H.Widjaja, J.Voss,
H.W.Schmalle *J.Chem.Res.*,**104**, 1643, 1980

1.C   **exo – 2,exo – 6 – Dihydroxy – 2,6 – dimethyl – bicyclo(3.3.1)nonane ethylacetate clathrate**
$C_4H_8O_2$, $3C_{11}H_{20}O_2$ Main entry is 61.1

1.14   **Hydrogen bis(N,N – dimethylacetamide) tetrachloro – gold(iii)**
$C_4H_{10}NO^+$, C_4H_9NO, $AuCl_4$
M.S.Hussain, E.O.Schlemper
J.Chem.Soc.,Dalton Trans.,750, 1980

1.15 **Hydrogen bis(N,N – dimethylacetamide) tetrachloro – gold(iii) (neutron study)**
$C_4H_{10}NO^+$, C_4H_9NO, $AuCl_4^-$
M.S.Hussain, E.O.Schlemper
J.Chem.Soc.,Dalton Trans.,750, 1980

1.16 **Tetrachloro – μ – dichloro – di – palladium bis(dimethylacetamidonium) bis(dimethylacetamide)**
$C_4H_{10}NO^+$, C_4H_9NO, $Cl_6Pd_2^-$
S.N.Kurskov, I.P.Lavrent'ev, M.L.Khidekel, O.N.Krasochka, D.D.Makitova, L.O.Atovmyan
Izv.Akad.Nauk Kaz.SSR,Ser.Khim.,185, 19

1.17 **N,N – Dimethylacetamide hydrochloride**
$C_4H_{10}NO^+$, Cl^-
E.Benedetti, B.di Blasio, P.Baine
J.Chem.Soc.,Perkin Trans.2,500, 1980

1.18 **Potassium O – n – butylxanthate**
$C_5H_9OS_2^-$, K^+
A.A.Kashaev, V.A.Liopo, S.B.Leonov, L.Z.Kazakova, N.A.Frolova *Kristallografiya*,24, 590, 1979
See also R1 : 11

1.19 **bis(Trimethyl – acetic acid) benzene**
$2C_5H_{10}O_2$, C_6H_6
M.Pickering, R.W.H.Small *Eur.Cryst.Meeting*,5, 104, 1979
See also R2 : 19

1.20 **tris(Hydroxymethyl) – acetic acid**
$C_5H_{10}O_5$
D.Eilerman, R.Rudman
Acta Crystallogr.,Sect.B,35, 2768, 1979

1.C **Urocanic acid dihydrate**
4 – Imidazole – acrylic acid dihydrate
$C_6H_6N_2O_2$, $2H_2O$ Main entry is 32.17

1.C **Ethylenediammonium (2S,3S) – 2 – hydroxycitrate**
Ethylenediammonium (–) – hydroxycitrate
$C_6H_6O_8^{2-}$, $C_2H_{10}N_2^{2+}$ Main entry is 2.25

1.C **Ethylenediammonium (2S,3R) – 2 – hydroxycitrate monohydrate**
Ethylenediammonium (+) – allohydroxycitrate monohydrate
$C_6H_6O_8^{2-}$, $C_2H_{10}N_2^{2+}$, H_2O Main entry is 2.26

1.C **N – (1 – (2 – Phenylethyl) – 4 – piperidinylium) – N – phenyl – propanamide – citrate toluene solvate**
Fentanyl citrate toluene solvate
$C_6H_7O_7^-$, $C_{22}H_{29}N_2O^+$, C_7H_8 Main entry is 33.84

1.C **Acetylcholine tetraphenylborate**
$C_7H_{16}NO_2^+$, $C_{24}H_{20}B^-$ Main entry is 62.14

1.C **p – Nitrophenoxyacetic acid**
$C_8H_7NO_5$ Main entry is 17.8

1.C **trans – (2R,5R) – 2,5 – Dimethylpyrrolidinium(S) – mandelate (at 238°K)**
$C_8H_7O_3^-$, $C_6H_{14}N^+$ Main entry is 32.18

1.21 **Di – N,N′ – isopropyl – dithio – oxalamide**
$C_8H_{16}N_2S_2$
K.–H.Klaska, O.Jarchow, W.Scham, H.Widjaja, J.Voss, H.W.Schmalle *J.Chem.Res.*,104, 1643, 1980
See also R1 : 11

1.C **p – Bromophenoxyacetimine methyl ether**
$C_9H_{10}BrNO_2$ Main entry is 17.9

1.C **Sodium hydrogen α – methoxy – α – phenylacetate**
$C_9H_{10}O_3$, $C_9H_9O_3^-$, Na^+ Main entry is 2.29

1.C **Rubidium hydrogen α – methoxy – α – phenylacetate**
$C_9H_{10}O_3$, $C_9H_9O_3^-$, Rb^+ Main entry is 2.30

1.22 **Carboxymethane – (N – methyl – N – phenyl – sulfonamide)**
$C_9H_{11}NO_4S$
T.Graafland, A.Wagenaar, A.J.Kirby, J.B.F.N.Engberts
J.Am.Chem.Soc.,101, 6981, 1979
See also R1 : 4

1.C **Calcium D – pantothenate bromide**
$C_9H_{16}NO_5^-$, Ca^{2+}, Br^- Main entry is 2.31

1.23 **2 – t – Butylammonium – 3 – chloro – N,N – dimethylpropionamide tetrafluoroborate**
$C_9H_{20}ClN_2O^+$, BF_4^-
B.Tinant, G.Germain, J.P.Declercq, M.van Meerssche
Cryst.Struct.Commun.,9, 675, 1980
See also R1 : 3

1.C **(±) – 2 – (4 – Chloro – 2 – methylphenoxy) – propionic acid**
Mecoprop
$C_{10}H_{11}ClO_3$ Main entry is 17.12

1.24 **2 – Carboxyethane – (N – methyl – N – phenyl – sulfonamide)**
$C_{10}H_{13}NO_4S$
T.Graafland, A.Wagenaar, A.J.Kirby, J.B.F.N.Engberts
J.Am.Chem.Soc.,101, 6981, 1979
See also R1 : 4

1.25 **5 – Phenyl – penta – 2,4 – dienoic acid**
$C_{11}H_{10}O_2$
S.Kashino, M.Haisa
Acta Crystallogr.,Sect.B,36, 346, 1980

1.26 **2 – Methoxycarbonyl – vinyl – (N – methyl – N – phenyl – sulfonamide)**
$C_{11}H_{13}NO_4S$
T.Graafland, A.Wagenaar, A.J.Kirby, J.B.F.N.Engberts
J.Am.Chem.Soc.,101, 6981, 1979
See also R1 : 4

1.C **Carbobiotin**
$C_{11}H_{18}N_2O_3$ Main entry is 35.23

1.C **4 – (3 – Indolyl) – butyric acid**
$C_{12}H_{13}NO_2$ Main entry is 35.26

1.27 **N – (1 – Phenylethyl) – isobutyramide**
$C_{12}H_{17}NO$
A.Aubry, J.Protas, M.T.Cung, M.Marraud
Acta Crystallogr.,Sect.B,36, 96, 1980

1.28 **trans – 7 – Dimethylamino – 2 – carbomethoxy – hepta – 2,4,6 – triene methyl carboxylate**
$C_{12}H_{17}NO_4$
M.O.Dekaprilevich, L.G.Vorontsova, O.S.Chizhov
Izv.Akad.Nauk SSSR,Ser.Khim.,1742, 1979
See also R1 : 3

1.29 **N – Methyl – N′ – (1 – acetamido – 2 – phenylethyl) – urea**
N – Methyl – N′ – (benzyl)(acetamido) – methylurea
$C_{12}H_{17}N_3O_2$
A.F.Mishnev, Ya.Ya.Bleidelis, Yu.E.Antsans, G.I.Chipens
Zh.Strukt.Khim.,20, 154, 1979
See also R1 : 8

1.30   2 – Methyl – 2 – (N – phenylsulfinamoyl) – N,N –
dimethyl – propanamide
$C_{12}H_{18}N_2O_2S$
M.van Meerssche, G.Germain, J.P.Declercq
Bull.Soc.Chim.Belg.,**89**, 153, 1980
See also R1 : 4

1.31   N,N′ – Ethylene – bis(acetylacetone – imine)
$C_{12}H_{20}N_2O_2$
N.Bresciani–Pahor, M.Calligaris, G.Nardin, L.Randaccio,
D.Viterbo *Acta Crystallogr.,Sect.B*,**35**, 2776, 1979
See also R1 : 3

1.32   N,N′ – Dihydroxy – N,N′ – di – isopropyl –
hexanediamide
N,N′ – Di – isopropyl – adipohydroxamic acid
$C_{12}H_{24}N_2O_4$
W.L.Smith, K.N.Raymond
J.Am.Chem.Soc.,**102**, 1252, 1980

1.C   2 – Carboxy – cyclopentene – 1 – (N – methyl – N –
phenyl – sulfonamide)
$C_{13}H_{15}NO_4S$ Main entry is 20.5

1.C   (R) – N – Trifluoroacetyl – 1 – (1 – naphthyl) –
ethylamine
$C_{14}H_{12}F_3NO$ Main entry is 24.6

1.C   Methyl – (2E,4Z) – 5 – dimethylamino – 2 – nitro –
4 – (4 – chloro – 2 – nitrophenyl) – 2,4 –
pentadienoate
$C_{14}H_{14}ClN_3O_6$ Main entry is 15.4

1.C   (R) – N – Acetyl – 1 – (1 – naphthyl) – ethylamine
$C_{14}H_{15}NO$ Main entry is 24.7

1.C   (RS) – N – Acetyl – 1 – (1 – naphthyl) – ethylamine
$C_{14}H_{15}NO$ Main entry is 24.8

1.C   p – Tolyl – glyoxylic acid p – chloroanilide
$C_{15}H_{12}ClNO_2$ Main entry is 16.12

1.C   2 – Nitro – 3 – ferrocenyl – acrylic acid ethyl ester
$C_{15}H_{15}FeNO_4$ Main entry is 73.55

1.C   (αRS,1RS,2RS) – α – (1 – Hydroxy – 2 –
phenylcyclohexyl) – propionic acid
$C_{15}H_{20}O_3$ Main entry is 21.9

1.C   1 – Benzyl – 5 – ethyl – 1,2,5,6 – tetrahydro – 2 –
oxo – 4 – pyridine – acetic acid
$C_{16}H_{19}NO_3$ Main entry is 33.68

1.33   Dimethyl – 2 – (N – (2 – hydroxy – 1 – methyl – 2 –
phenylethyl) – N – methylamino) – maleate
$C_{16}H_{21}NO_5$
R.Roques, J.Bellan, J.C.Rossi, J.P.Declercq, G.Germain
Acta Crystallogr.,Sect.B,**36**, 1117, 1980

1.C   3α,12α – Dihydroxy – 5β – cholan – 24 – oic acid
palmitic acid ethanol solvate
Deoxycholic acid palmitic acid ethanol solvate
$0.5C_{16}H_{32}O_2, 4C_{24}H_{40}O_4, 0.5C_2H_6O$ Main entry is 51.55

1.34   5 – Hydroxy – 5 – (hydroxymethyl) – 2 – methyl –
tetradecanoic acid
$C_{16}H_{32}O_4$
J.H.Cardellina, R.E.Moore, E.V.Arnold, J.Clardy
J.Org.Chem.,**44**, 4039, 1979

1.35   Dibenzyl – pyruvic acid
$C_{17}H_{16}O_3$
M.Hamelin, Y.Jeannin
Acta Crystallogr.,Sect.B,**36**, 469, 1980

1.36   N – ((S) – 1 – Phenylethyl) – (R) – α –
methoxy(pentafluorophenyl) – propionamide
$C_{18}H_{16}F_5NO_2$
E.J.Valente, L.R.Pohl, W.F.Trager
J.Org.Chem.,**45**, 543, 1980

1.37   N,N′ – Di – n – hexyladipamide
$C_{18}H_{36}N_2O_2$
D.R.Rueda, J.Fayos *Eur.Cryst.Meeting*,**6**, 39, 1980

1.38   2 – Diethyl – aminoethyl diphenylacetate
hydrobromide
$C_{20}H_{26}NO_2^+$, Br⁻
E.R.Arakelova, G.G.Aleksandrov, Yu.T.Struchkov
Arm.Khim.Zh.,**31**, 889, 1978
See also R1 : 3

1.C   (5Z,11α,13E,15S) – 11,15 – Dihydroxy – 9 – oxo –
prosta – 5,13 – dien – 1 – oic acid
Prostaglandin E_2
$C_{20}H_{32}O_5$ Main entry is 59.25

1.C   (5Z,9β,11a,13E,15S) – 9,11,15 – Trihydroxy – prosta –
5,13 – dien – 1 – oic acid
Prostaglandin F(2β)
$C_{20}H_{34}O_5$ Main entry is 59.26

1.C   Thromboxane B_2
$C_{20}H_{34}O_6$ Main entry is 59.27

1.C   4 – Anilino – 3 – methoxycarbonyl – 1 –
(methoxycarbonylmethylidene) – 1,2,3,4 –
tetrahydro – naphthalene
$C_{21}H_{21}NO_4$ Main entry is 27.29

1.39   2,4 – Hexadiyne – 1,6 – bis(m – tolylurethane)
(orange form)
$C_{22}H_{20}N_2O_4$
G.N.Patel, E.N.Duesler, D.Y.Curtin, I.C.Paul
J.Am.Chem.Soc.,**102**, 461, 1980

1.40   2 – (Dicyclohexyl – acetoxy) – ethyl –
triethylammonium iodide
$C_{22}H_{42}NO_2^+$, I⁻
T.A.Hamor *Acta Crystallogr.,Sect.B*,**36**, 99, 1980
See also R1 : 3

1.C   3,3′ – Diacetyl – 5,5′ – diethoxycarbonyl –
glaucyrone
Diethyl glaucophanic enol
$C_{23}H_{22}O_{10}$ Main entry is 38.115

1.C   (+) – 2,4:3,5 – Di – O – methylene – D – mannitol –
1,6 – di – trans – cinnamate
$C_{26}H_{26}O_8$ Main entry is 45.54

ALIPHATIC CARBOXYLIC ACID SALTS
(AMMONIUM,IA,IIA METALS)

2.1 Potassium formate
CHO_2^-, K^+
J.W.Bats, H.Fuess
*Acta Crystallogr.,Sect.B,***36**, 1940, 1980

2.2 Calcium formate (neutron study, α form)
$2CHO_2^-$, Ca^{2+}
M.O.Bargouth, G.Will *Cryst.Struct.Commun.,***9**, 605, 1980

2.3 β − Calcium formate (data of Matsui, Watanabe and Kamijo)
$2CHO_2^-$, Ca^{2+}
M.Matsui, T.Watanabe, N.Kamijo, R.L.Lapp, R.A.Jacobson
*Acta Crystallogr.,Sect.B,***36**, 1081, 1980

2.4 β − Calcium formate (data of Lapp and Jacobson)
$2CHO_2^-$, Ca^{2+}
M.Matsui, T.Watanabe, N.Kamijo, R.L.Lapp, R.A.Jacobson
*Acta Crystallogr.,Sect.B,***36**, 1081, 1980

2.5 Potassium thiocarbazate
$CH_3N_2OS^-$, K^+
R.Mattes, H.Weber, K.Scholten *Chem.Ber.,***113**, 1981, 1980
See also R1 : 11,9

2.6 Rubidium hydrogen bis(dibromoacetate)
$C_2HBr_2O_2^-$, $C_2H_2Br_2O_2$, Rb^+
V.Videnova, J.Baran, T.Glowiak, H.Ratajczak
*Acta Crystallogr.,Sect.B,***36**, 459, 1980
See also R2 : 1

2.7 Potassium deuterium bis(dichloroacetate) (neutron study, deuterated form)
$C_2HCl_2O_2^-$, $C_2HCl_2DO_2$, K^+
D.Hadzi, I.Leban, B.Orel, M.Iwata, J.M.Williams
*J.Cryst.Mol.Struct.,***9**, 117, 1979
See also R2 : 1

2.8 Potassium hydrogen bis(dichloroacetate) (neutron study)
$C_2HCl_2O_2^-$, $C_2H_2Cl_2O_2$, K^+
D.Hadzi, I.Leban, B.Orel, M.Iwata, J.M.Williams
*J.Cryst.Mol.Struct.,***9**, 117, 1979
See also R2 : 1

2.9 Guanidinium hydrogen oxalate monohydrate
$C_2HO_4^-$, $CH_6N_3^+$, H_2O
L.C.Andrews, B.R.Deroski, J.S.Ricci
*J.Cryst.Mol.Struct.,***9**, 163, 1979
See also R1 : 1 R2 : 8

2.10 Tellurium oxalate dihydrogen oxalate dihydrate
$C_2HO_4^-$, $C_2H_2O_4$, Tl^+, $2H_2O$
N.Bulc, L.Golic, J.Siftar *Doc.Chem.Yug.,***26**, 387, 1979

2.11 Potassium hydrogen bis(dibromoacetate)
$C_2H_2Br_2O_2$, $C_2HBr_2O_2^-$, K^+
J.Baran, V.Videnova, T.Glowiak, H.Ratajczak
*Acta Crystallogr.,Sect.B,***35**, 2722, 1979
See also R1 : 1

2.12 Sodium acetate trihydrate (neutron study)
$C_2H_3O_2^-$, Na^+, $3H_2O$
R.W.Alkire, F.K.Ross, E.O.Schlemper, W.B.Yelon
*Am.Cryst.Assoc.,Ser.2,***7**, 38, 1980

2.13 Potassium oxalate monoperhydrate (deuterated form, neutron study, at 300°C, profile refinement)
$C_2O_4^{2-}$, $2K^+$, D_2O_2
J.M.Adams, V.Ramdas, A.W.Hewat
*Acta Crystallogr.,Sect.B,***36**, 570, 1980

2.14 Rubidium oxalate perdeuterate (neutron study, at 5°K)
$C_2O_4^{2-}$, $2Rb^+$, D_2O_2
J.M.Adams, V.Ramdas, A.W.Hewat
*Acta Crystallogr.,Sect.B,***36**, 1096, 1980

2.15 Rubidium oxalate monoperhydrate (at 300°K, neutron study, deuterated form)
$C_2O_4^{2-}$, $2Rb^+$, D_2O_2
J.M.Adams, V.Ramdas, A.W.Hervat
*Acta Crystallogr.,Sect.B,***36**, 570, 1980

2.16 Calcium malonate dihydrate (re−refinement of data of Karipides et.al.,Inorg.Chem.,16,3299,1977, using space group C2χm)
$C_3H_2O_4^{2-}$, Ca^{2+}, $2H_2O$
R.E.Marsh, V.Schomaker *Inorg.Chem.,***18**, 2331, 1979

2.17 Sodium α,α − dihydroxy − β − fluoropropionate
Sodium fluoropyruvate hydrate
$C_3H_4FO_4^-$, Na^+
T.J.Hurley, H.L.Carrell, R.K.Gupta, J.Schwartz, J.P.Glusker
*Arch.Biochem.Biophys.,***193**, 478, 1979

2.18 Sodium perfluorosuccinate hexahydrate
$C_4F_4O_4^{2-}$, $2Na^+$, $6H_2O$
B.Kalyanaraman, R.Shakir, L.D.Kispert, J.L.Atwood
*Am.Cryst.Assoc.,Ser.2,***7**, 39, 1980

2.19 Tetra − n − butylammonium hydrogen dichloromaleate (CuK(α) data)
$C_4HCl_2O_4^-$, $C_{16}H_{36}N^+$
L.Golic, I.Leban *Acta Crystallogr.,Sect.B,***36**, 1666, 1980
See also R1 : 1 R2 : 3

2.20 Tetra − n − butylammonium hydrogen dichloromaleate (MoK(α) data)
$C_4HCl_2O_4^-$, $C_{16}H_{36}N^+$
L.Golic, I.Leban *Acta Crystallogr.,Sect.B,***36**, 1666, 1980
See also R1 : 1 R2 : 3

2.C Imidazolium hydrogen maleate (neutron study, deuterated form)
$C_4H_2DO_4^-$, $C_3H_3D_2N_2^+$ Main entry is 32.5

2.21 Strontium maleate tetrahydrate
$C_4H_2O_4^{2-}$, Sr^{2+}, $4H_2O$
M.P.Gupta, P.Chand *Indian J.Phys.,***53**, 472, 1979

2.C Carbinoxamine maleate
$C_4H_3O_4^-$, $C_{16}H_{20}ClN_2O^+$ Main entry is 33.70

2.C (1S,4R) − 3′ − Chloro − 10′,11′ − dihydro − N,N − dimethyl − spiro(cyclohex − 2 − ene − 1,5′ − (5H) − dibenzo(a,d)cycloheptene) − 4 − ammonium hydrogen maleate (absolute configuration)
$C_4H_3O_4^-$, $C_{22}H_{25}ClN^+$ Main entry is 29.9

2.22 Sodium ammonium tartrate tetrahydrate (at 300°K)
$C_4H_4O_6^{2-}$, Na^+,.H_4N^+, $4H_2O$
I.G.Shkuratova, G.A.Kiosse, T.I.Malinovskii
Izv.Akad.Nauk SSSR,Ser.Khim.,1685, 1979

2.23 Lithium hydrogen (+) – 1 – malate
$C_4H_5O_5^-$, Li^+
W.Van Havere, A.T.H.Lenstra
Acta Crystallogr.,Sect.B,**36**, 1483, 1980
See also R1 : 1

2.24 Calcium bis(hydrogen – 1 – malate) hexahydrate
$2C_4H_5O_5^-$, Ca^{2+}, $6H_2O$
A.T.H.Lenstra, W.Van Havere
Acta Crystallogr.,Sect.B,**36**, 156, 1980

2.C Sodium calcium nitrilo – triacetate
$C_6H_6NO_6^{3-}$, Na^+, Ca^{2+} Main entry is 48.16

2.25 Ethylenediammonium (2S,3S) – 2 – hydroxycitrate
Ethylenediammonium (–) – hydroxycitrate
$C_6H_6O_8^{2-}$, $C_2H_{10}N_2^{2+}$
W.C.Stallings, J.F.Blount, P.A.Srere, J.P.Glusker
Arch.Biochem.Biophys.,**193**, 431, 1979
See also R1 : 1 R2 : 3

2.26 Ethylenediammonium (2S,3R) – 2 – hydroxycitrate monohydrate
Ethylenediammonium (+) – allohydroxycitrate monohydrate
$C_6H_6O_8^{2-}$, $C_2H_{10}N_2^{2+}$, H_2O
W.C.Stallings, J.F.Blount, P.A.Srere, J.P.Glusker
Arch.Biochem.Biophys.,**193**, 431, 1979
See also R1 : 1 R2 : 3

2.27 Sodium hydrogen nitrilo – triacetate
$C_6H_7NO_6^{2-}$, $2Na^+$
J.D.Oliver, B.L.Barnett *Am.Cryst.Assoc.,Ser.2*,**8**, 38, 1980
See also R1 : 3

2.28 1 – Phenylethylammonium mandelate
$C_8H_7O_3^-$, $C_8H_{12}N^+$
M.C.Brianso, M.Leclercq, J.Jacques
Acta Crystallogr.,Sect.B,**35**, 2751, 1979
See also R2 : 3

2.29 Sodium hydrogen α – methoxy – α – phenylacetate
$C_9H_9O_3^-$, $C_9H_{10}O_3$, Na^+
P.B.Moore, J.J.Pluth, J.A.Molin–Norris, D.A.Weinstein,
E.L.Compere Junior
Acta Crystallogr.,Sect.B,**36**, 47, 1980
See also R2 : 1

2.30 Rubidium hydrogen α – methoxy – α – phenylacetate
$C_9H_9O_3^-$, $C_9H_{10}O_3$, Rb^+
P.B.Moore, J.J.Pluth, J.A.Molin–Norris, D.A.Weinstein,
E.L.Compere Junior
Acta Crystallogr.,Sect.B,**36**, 47, 1980
See also R2 : 1

2.31 Calcium D – pantothenate bromide
$C_9H_{16}NO_5^-$, Ca^{2+}, Br^-
L.DeLucas, H.Einspahr, C.E.Bugg
Acta Crystallogr.,Sect.B,**35**, 2724, 1979
See also R1 : 1

2.32 α – Phenylethylammonium α – phenylethylacetate (p diastereo isomer)
$C_{10}H_{11}O_2^-$, $C_8H_{12}N^+$
M.C.Brianso *Eur.Cryst.Meeting.***6**, 327, 1980
See also R2 : 3

ALIPHATIC AMINES

3.1 Methylammonium diaqua – trichloro – manganese
$(CH_6N^+)_n$, $(H_4Cl_3MnO_2^-)_n$
W.Depmeier, K.–H.Klaska
Acta Crystallogr.,Sect.B,**36**, 1065, 1980

3.2 Methylammonium tetrabromo – cadmium(ii)
$2CH_6N^+$, Br_4Cd^{2-}
D.Altermatt, H.Arend, A.Niggli, W.Petter
Mater.Res.Bull.,**14**, 1391, 1979

3.3 N,N – Dimethyl – nitramine (perdeuterated, neutron study, at 91°K)
$C_2D_6N_2O_2$
A.Filhol, G.Bravic, M.Rey-Lafon, M.Thomas
Acta Crystallogr.,Sect.B,**36**, 575, 1980
See also R1 : 9

3.4 Potassium cyanamide dicyanamide
$C_2H_4N_4$, CHN_2^-, K^+
M.J.Begley, A.Harper, P.Hubberstey
J.Chem.Res.,**398**, 4620, 1979
See also R1 : 7

3.5 tris(Tetraethylammonium) hydrido – pentacosacarbonyl – tetradeca – rhodium
$3C_2H_5N^+$, $C_{25}HO_{25}Rh_{14}^{3-}$
G.Ciani, A.Sironi, S.Martinengo
J.Organomet.Chem.,**192**, C42, 1980

3.6 N,N – Dimethyl – nitramine (at 125°K)
$C_2H_6N_2O_2$
A.Filhol, G.Bravic, M.Rey-Lafon, M.Thomas
Acta Crystallogr.,Sect.B,**36**, 575, 1980
See also R1 : 9

3.7 N,N – Dimethyl – nitramine (at 180°K)
$C_2H_6N_2O_2$
A.Filhol, G.Bravic, M.Rey-Lafon, M.Thomas
Acta Crystallogr.,Sect.B,**36**, 575, 1980
See also R1 : 9

3.8 N,N – Dimethyl – nitramine (at 293°K)
$C_2H_6N_2O_2$
A.Filhol, G.Bravic, M.Rey-Lafon, M.Thomas
Acta Crystallogr.,Sect.B,**36**, 575, 1980
See also R1 : 9

3.9 N,N – Dimethyl – nitramine (neutron study, at 125°K)
$C_2H_6N_2O_2$
A.Filhol, G.Bravic, M.Rey-Lafon, M.Thomas
Acta Crystallogr.,Sect.B,**36**, 575, 1980
See also R1 : 9

3.10 N,N – Dimethyl – nitramine (neutron study, at 85°K)
$C_2H_6N_2O_2$
A.Filhol, G.Bravic, M.Rey-Lafon, M.Thomas
Acta Crystallogr.,Sect.B,**36**, 575, 1980
See also R1 : 9

3.C Dimethylammonium 0,0 – di – isopropyl – dithiophosphate
$C_2H_8N^+$, $C_6H_{14}O_2PS_2^-$ Main entry is 64.22

3.C Ethylenediammonium (2S,3S) – 2 – hydroxycitrate
Ethylenediammonium (–) – hydroxycitrate
$C_2H_{10}N_2^{2+}$, $C_6H_6O_8^{2-}$ Main entry is 2.25

3.C Ethylenediammonium (2S,3R) – 2 – hydroxycitrate monohydrate
Ethylenediammonium (+) – allohydroxycitrate monohydrate
$C_2H_{10}N_2^{2+}$, $C_6H_6O_8^{2-}$, H_2O Main entry is 2.26

3.11 Trimethylammonium 7,7,8,8 – tetracyanoquinodimethane tri – iodide (neutron study, at 80°K)
$C_3H_{10}N^+$, $C_{12}H_4N_4$, $0.33I_3^{3-}$
A.Filhol, J.Gaultier
Acta Crystallogr.,Sect.B,**36**, 592, 1980
See also R2 : 12,7

3.12 Trimethylammonium cadmium chloride (at 174°K)
$C_3H_{10}N^+$, $CdCl_3^-$
G.Chapuis, F.J.Zuniga
Acta Crystallogr.,Sect.B,**36**, 807, 1980

3.13 bis(n – Propylammonium) tetrachloro – manganese(ii) (δ phase)
$2C_3H_{10}N^+$, Cl_4Mn^{2-}
W.Depmeier *J.Solid State Chem.*,**29**, 15, 1979

3.14 Trimethylammonium tetrabromo – cadmium(ii) tribromo – cadmium(ii)
$3C_3H_{10}N^+$, Br_4Cd^{2-}, Br_3Cd^-
A.Daoud, R.Perret, Y.Dusausoy
Acta Crystallogr.,Sect.B,**35**, 2718, 1979

3.15 1,3 – Diammonio – 2 – propanol sulfate monohydrate
$C_3H_{12}N_2O^{2+}$, O_4S^{2-}, H_2O
R.Kivekas, J.Valkonen
Acta Crystallogr.,Sect.B,**36**, 956, 1980

3.16 Tetramethylammonium hexachloro – platinum (neutron powder work,deuterated form)
$2C_4D_{12}N^+$, Cl_6Pt^{2-}
G.A.Mackenzie, R.W.Berg, G.S.Pawley
Acta Crystallogr.,Sect.B,**36**, 1001, 1980

3.17 trans – N,N – Dimethyl – 2 – nitro – ethenamine
trans – N,N – Dimethyl – 2 – nitrovinylamine
$C_4H_8N_2O_2$
A.Hazell, A.Mukhopadhyay
Acta Crystallogr.,Sect.B,**36**, 747, 1980
See also R1 : 10

3.18 bis(Dimethylamino) – difluorosulfurane (at –35°C)
$C_4H_{12}F_2N_2S$
A.H.Cowley, P.E.Riley, J.S.Szobota, M.L.Walker
J.Am.Chem.Soc.,**101**, 5620, 1979
See also R1 : 4

3.C Tetramethylammonium (μ^3 – methoxy) – tris(μ^2 – methoxy) – tris(dicarbonyl – nitroso – molybdenum)
$C_4H_{12}N^+$, $C_{10}H_{12}Mo_3N_3O_{13}^-$ Main entry is 84.24

3.19 Tetramethylammonium hexa – cobalt – tetradecacarbonyl – carbide
$C_4H_{12}N^+$, $C_{15}Co_6O_{14}^-$
V.G.Albano, P.Chini, G.Ciani, M.Sansoni, S.Martinengo
J.Chem.Soc.,Dalton Trans., 163, 1980

3.C Tetramethylammonium (ethylnitrosolato – O) – triphenylborate
$C_4H_{12}N^+$, $C_{20}H_{18}BN_2O_2^-$ Main entry is 62.12

3.C 2,3:4,5 – bis(1,2 – (3 – Methylnaphtho)) – 1,6,9,12,15,18 – hexaoxacycloeicosa – 2,4 – diene t – butylammonium perchlorate clathrate benzene solvate (at 113°K)
$C_4H_{12}N^+$, $C_{32}H_{36}O_6$, ClO_4^-, C_6H_6 Main entry is 61.13

3.20 Tetramethylammonium hexafluorophosphate (at 170°K)
$C_4H_{12}N^+$, F_6P^-
Y.Wang, L.D.Calvert, S.K.Brownstein
Acta Crystallogr.,Sect.B,**36**, 1523, 1980

3.C Tetramethylammonium hexakis(μ – benzenethiolato) – tetra(benzenethiolato) – tetra – cobalt(ii)
$2C_4H_{12}N^+$, $C_{60}H_{50}Co_4S_{10}^{2-}$ Main entry is 85.94

3.C bis(Tetramethylammonium) deca(μ^2 – benzenethiolato) – tetra – iron(ii)
$2C_4H_{12}N^+$, $C_{60}H_{50}Fe_4S_{10}^{2-}$ Main entry is 85.95

3.C bis(Diethylammonium) tris(tetraphenyl – disiloxane – diolato) – zirconium(iv)
$2C_4H_{12}N^+$, $C_{72}H_{60}O_9Si_6Zr^{2-}$ Main entry is 84.137

3.21 bis(Tetramethylammonium) hexachloro – tin(iv) (at 295°K)
$2C_4H_{12}N^+$, Cl_6Sn^{2-}
K.Nielsen, R.W.Berg
Acta Chem.Scand.Ser.A,**34**, 153, 1980

3.22 bis(Tetramethylammonium) hexachloro – tin(iv) (at 160°K)
$2C_4H_{12}N^+$, Cl_6Sn^{2-}
K.Nielsen, R.W.Berg
Acta Chem.Scand.Ser.A,**34**, 153, 1980

3.23 bis(Tetramethylammonium) bis((μ^2 – thio) – dithio – oxo – molybdenum) (study I)
$2C_4H_{12}N^+$, $Mo_2O_2S_6^{2-}$
W.Clegg, N.Mohan, A.Muller, A.Neumann, W.Rittner, G.M.Sheldrick *Inorg.Chem.*,**19**, 2066, 1980

3.24 bis(Tetramethylammonium) bis((dithio – tungsten) – bis(μ_2 – thio)) – oxo – tungsten monohydrate
$2C_4H_{12}N^+$, $OS_8W_3^{2-}$, H_2O
A.Muller, W.Rittner, A.Neumann, E.Koniger–Ahlborn, R.G.Bhattacharyya *Z.Anorg.Allg.Chem.*,**461**, 91, 1980

3.25 Tetramethylammonium nonabromo – di – bismuth(iii)
$3C_4H_{12}N^+$, $Bi_2Br_9^{3-}$
F.Lazarini *Cryst.Struct.Commun.*,**9**, 815, 1980

3.26 tetrakis(Tetramethylammonium) chloro – penta(trichloro – tin) – osmium
$4C_4H_{12}N^+$, $Cl_{16}OsSn_5^{4-}$
E.N.Yurchenko, E.T.Devyatkina, T.S.Khodashova, M.A.Porai–Koshits, V.I.Konnov, V.A.Varnek, P.G.Antonov, Yu.N.Kukushkin *Koord.Khim.*,**5**, 552, 1979

3.C 3,5 – Di – iodo – 4 – (4 – hydroxy – 3,5 – di – iodophenoxy) – benzoic acid bis(N – diethanolamine)
3,3′,5,5′ – Tetra – iodothyroformic acid bis(N – diethanolamine)
$2C_4H_{12}NO_2^-$, $C_{13}H_4I_4O_4^{2-}$ Main entry is 13.7

3.27 **1,4 – Butanediyl – diammonium tetrachloro – manganese (high–temperature phase, at 404°K, neutron study)**
$(C_4H_{14}N_2^-)_n$, nCl_4Mn^-
K.Tichy, J.Benes, R.Kind, H.Arend
Acta Crystallogr.,Sect.B,**36**, 1355, 1980

3.28 **1,4 – Butanediyl – diammonium tetrachloro – manganese (neutron study)**
$(C_4H_{14}N_2^{2+})_n$, nCl_4Mn^{2-}
K.Tichy, J.Benes, R.Kind, H.Arend
Acta Crystallogr.,Sect.B,**36**, 1355, 1980

3.29 **bis(2 – Aminoethyl) – disulfide dihydrochloride**
Cystamine dihydrochloride
$C_4H_{14}N_2S_2^{2+}$, $2Cl^-$
B.M.Vedavathi, K.Vijayan *Curr.Sci.*,**48**, 1028, 1979
See also R1 : 11

3.30 **bis(Diethylenetriammonium) tetrachloro – platinum(ii) tetrachloride**
$2C_4H_{16}N_3^{3+}$, Cl_4Pt^{2-}, $4Cl^-$
J.Britten, C.J.L.Lock
Acta Crystallogr.,Sect.B,**35**, 3065, 1979

3.C **Sodium hydrogen nitrilo – triacetate**
$C_6H_7NO_6^{2-}$, $2Na^+$ Main entry is 2.27

3.31 **bis(Triethylammonium) dihydronium octacyano – molybdenum(iv)**
$2C_6H_{16}N^+$, $C_8MoN_8^{4-}$, $2H_3O^+$
J.G.Leipoldt, S.S.Basson, L.D.C.Bok
Inorg.Chim.Acta,**38**, L99, 1980

3.C **Triethylammonium tris(pyrocatechol)silicate**
$2C_6H_{16}N^+$, $C_{18}H_{12}O_6Si^{2-}$ Main entry is 63.23

3.32 **Triethylammonium closo – dodecaborane**
$2C_6H_{16}N^+$, $H_{12}B_{12}^{2-}$
G.Shoham, D.Schomburg, W.N.Lipscomb
Cryst.Struct.Commun.,**9**, 429, 1980

3.C **Cyclopentadienyl sodium tetramethylethylenediamine**
$C_6H_{16}N_2$, $C_5H_5^-$, Na^+ Main entry is 20.1

3.C **N – (2 – Ammonioethyl) – piperazinium pentachloro – copper(ii) dihydrate**
$C_6H_{18}N_3^{3+}$, Cl_5Cu^{3-}, $2H_2O$ Main entry is 33.19

3.C **bis(10,22 – Dimethyl – 1,4,7,13,16,19 – hexaoxa – 10,22 – diazacyclotetracosane) benzylammonium thiocyanate clathrate**
$C_7H_{10}N^+$, $2C_{18}H_{38}N_2O_6$, CNS^- Main entry is 61.6

3.C **Acetylcholine tetraphenylborate**
$C_7H_{16}NO_2^+$, $C_{24}H_{20}B^-$ Main entry is 62.14

3.C **Triethylmethylammonium 8,8′ – oxido – 3 – cobalta – bis(η^5 – 1,2 – dicarbadodecaborate)**
$C_7H_{18}N^+$, $C_4H_{20}B_{18}CoO^-$ Main entry is 62.5

3.C **Sodium triethylmethylammonium tris(thiobenzohydroximato) – chromium(iii) hemi(sodium hydroxide) hydrate**
$C_7H_{18}N^+$, $C_{21}H_{15}CrN_3O_3S_3^{3-}$, $2.5Na^+$, $0.5HO^-$, $18.5H_2O$
Main entry is 85.66

3.C **(S) – 1 – (p – Bromophenyl) – ethylammonium 1 – methoxy – 2 – ethoxycarbonyl – aziridine – 2 – carboxylate**
$C_8H_{11}BrN^+$, $C_7H_{10}NO_5^-$ Main entry is 32.19

3.C **1 – Phenylethylammonium mandelate**
$C_8H_{12}N^+$, $C_8H_7O_3^-$ Main entry is 2.28

3.C **α – Phenylethylammonium α – phenylethylacetate (p diastereo isomer)**
$C_8H_{12}N^+$, $C_{10}H_{11}O_2^-$ Main entry is 2.32

3.C **Dopamine hydrochloride**
$C_8H_{12}NO_2^+$, Cl^- Main entry is 59.1

3.C **Tetraethylammonium trichloro – (μ – 2 – oxoethanethiolato – S,μ – O) – (μ – 2 – oxoethanethiolato – μ – S,μ – O) – bis(oxo – molybdenum(v))**
$C_8H_{20}N^+$, $C_4H_8Cl_3Mo_2O_4S_2^-$ Main entry is 85.4

3.C **Tetraethylammonium chloro – tris(thiolato – ethanolato) – bis(oxo – molybdenum(v))**
$C_8H_{20}N^+$, $C_6H_{12}ClMo_2O_5S_3^-$ Main entry is 85.9

3.C **Tetraethylammonium tris(μ – methoxy) – bis(tricarbonyl – rhenium(i))**
$C_8H_{20}N^+$, $C_9H_9O_9Re_2^-$ Main entry is 84.21

3.C **Tetraethylammonium (μ – iodo) – bis(μ – 3,5 – dimethylpyrazolyl) – bis(nitrosyl – nickel)**
$C_8H_{20}N^+$, $C_{10}H_{14}IN_6Ni_2O_2^-$ Main entry is 83.53

3.C **Tetraethylammonium bis(7,7,8,8 – tetracyanoquinodimethane)**
$C_8H_{20}N^+$, $C_{12}H_4N_4^-$, $C_{12}H_4N_4$ Main entry is 60.7

3.C **Tetraethylammonium (μ – fluoro) – bis(oxo – peroxo – (pyridine – 2,6 – dicarboxylato) – molybdenum(vi))**
$C_8H_{20}N^+$, $C_{14}H_6FMo_2N_2O_{14}^-$ Main entry is 81.53

3.33 **Tetraethylammonium (μ_5 – carbido) – (dicarbonyl – rhodium – (μ_2 – carbonyl) – dicarbonyl – iron) – tris(tricarbonyl – iron)**
$C_8H_{20}N^+$, $C_{15}Fe_4O_{14}Rh^-$
M.Tachikawa, A.C.Sievert, E.L.Muetterties, M.R.Thompson, C.S.Day, V.W.Day *J.Am.Chem.Soc.*,**102**, 1725, 1980

3.C **Tetraethylammonium (μ – cyano) – bis(cyclopentadienyl – dicarbonyl – molybdenum)**
$C_8H_{20}N^+$, $C_{15}H_{10}Mo_2NO_4^-$ Main entry is 73.52

3.C **Tetraethylammonium pentakis(μ – chloro) – thiophenolato – bis(pentacarbonyl – tungsten(0))**
$C_8H_{20}N^+$, $C_{16}Cl_5O_{10}SW_2^-$ Main entry is 85.50

3.C **Tetraethylammonium (μ^2 – hydrido) – bis(tetracarbonyl – (methyldiphenylphosphine) – molybdenum)**
$C_8H_{20}N^+$, $C_{34}H_{27}Mo_2O_8P_2^-$ Main entry is 86.113

3.34 **Tetraethylammonium aquo – tetrabromo – oxo – molybdenum**
$C_8H_{20}N^+$, $H_2Br_4MoO_2^-$
A.Bino, F.A.Cotton *Inorg.Chem.*,**18**, 2710, 1979

3.35 **Tetraethylammonium aqua – dioxofluoro – uranium – (tri – μ^2 – fluoro) – dioxofluoro – uranium monohydrate**
$C_8H_{20}N^+$, $H_2F_5O_5U_2^-$, H_2O
Yu.N.Mikhailov, S.B.Ivanov, V.G.Kuznetsov, R.L.Davidovich
Koord.Khim.,**5**, 1545, 1979

3.36 **Tetraethylammonium aquo – tetraiodo – oxo – molybdenum**
$C_8H_{20}N^+$, $H_2I_4MoO_2^-$
A.Bino, F.A.Cotton *Inorg.Chem.*,**18**, 2710, 1979

3.37   bis(Tetraethylammonium) (μ_6 – carbido) – (dicarbonyl – iron – (μ_2 – carbonyl) – dicarbonyl – iron) – (dicarbonyl – iron – (μ_2 – carbonyl) – tricarbonyl – molybdenum) – bis(tricarbonyl – iron)
$2C_8H_{20}N^+$, $C_{18}Fe_5MoO_{17}{}^{2-}$
M.Tachikawa, A.C.Sievert, E.L.Muetterties, M.R.Thompson, C.S.Day, V.W.Day *J.Am.Chem.Soc.*, **102**, 1725, 1980

3.38   Tetraethylammonium hydrogen dodecachloro – tri – ruthenium hexahydrate
$2C_8H_{20}N^+$, $Cl_{12}Ru_3{}^{4-}$, $6H_2O$, $2H^+$
A.Bino, F.A.Cotton *J.Am.Chem.Soc.*, **102**, 608, 1980

3.C   Tetraethylammonium (μ – sulfido) – bis(μ – ethylthiolato) – bis(molybdenum – trisulfido – tris(ethylthio) – iron))
$3C_8H_{20}N^+$, $C_{16}H_{40}Fe_6Mo_2S_{17}{}^{3-}$ Main entry is 85.57

3.39   tris(Tetraethylammonium) tris(μ^2 – ethylthio) – bis(tetrakis(μ^3 – thio) – tris(μ^3 – ethylthio – iron) – di – tungsten)
$3C_8H_{20}N^+$, $C_{18}H_{45}Fe_6S_{17}W_2{}^{3-}$
G.Christou, C.D.Garner, R.M.Miller, T.J.King *J.Inorg.Biochem.*, **11**, 349, 1979

3.C   tris(Tetraethylammonium) tetrakis(μ_3 – sulfido) – tetra(benzylthiolato – iron)
$3C_8H_{20}N^+$, $C_{28}H_{28}Fe_4S_8{}^{3-}$ Main entry is 85.80

3.40   Tetraethylammonium hydrogen (μ – hydrido) – bis(μ – chloro) – bis(trichloro – molybdenum) aquo – tetrachloro – oxo – molybdenum dihydrate
$3C_8H_{20}N^+$, $HCl_8Mo_2{}^{3-}$, $H_2Cl_4MoO_2{}^-$, H^+, $2H_2O$
A.Bino, F.A.Cotton *J.Am.Chem.Soc.*, **101**, 4150, 1979

3.C   Tetraethylammonium dioxo – (μ – (+) – tartrato) – (μ – (–) – tartrato) – di – vanadium(iv) octahydrate
$4C_8H_{20}N^+$, $C_8H_4O_{14}V_2{}^{4-}$, $8H_2O$ Main entry is 81.28

3.41   2,3,5,6 – Tetra(dimethylamino) – 1,4,2,3,5,6 – dithia – tetraborinane
$C_8H_{24}B_4N_4S_2$
H.Noth, H.Fusstetter, H.Pommerening, T.Taeger *Chem.Ber.*, **113**, 342, 1980

3.C   1 – (3,4 – Dihydroxyphenyl) – 2 – methylaminoethane – sulfonic acid hemihydrate
$C_9H_{13}NO_5S$, $0.5H_2O$ Main entry is 17.10

3.C   (±) – N – Methyl – 2 – (3,4 – dihydroxyphenyl) ethylammonium – 2 – sulfonate hemihydrate
(±) – Epinine – β – sulfonate hemihydrate
$C_9H_{13}NO_5S$, $0.5H_2O$ Main entry is 17.11

3.42   Lithium di(t – butylmethylene – amine)
$C_9H_{18}N^-$, Li^+
H.M.M.Shearer, K.Wade, G.Whitehead *J.Chem.Soc..Chem.Commun.*, 943, 1979

3.C   2 – t – Butylammonium – 3 – chloro – N,N – dimethylpropionamide tetrafluoroborate
$C_9H_{20}ClN_2O^+$, $BF_4{}^-$ Main entry is 1.23

3.43   Benzyl – trimethylammonium (μ_3 – hydrido) – octacarbonyl – tetrakis(μ_2 – carbonyl) – nickel – tri – iron
$C_{10}H_{16}N^+$, $C_{12}HFe_3NiO_{12}{}^-$
A.Fumagalli, F.Takusagawa, T.F.Koetzle, P.Chini, G.Longoni, S.Martinengo, B.T.Heaton *Am.Cryst.Assoc..Ser.2.* **7**, 16, 1980

3.44   bis(Trimethylbenzylammonium) hexadecacarbonyl – tetra – iron – palladium
$2C_{10}H_{16}N^+$, $C_{16}Fe_4O_{16}Pd^{2-}$
G.Longoni, M.Manassero, M.Sansoni *J.Am.Chem.Soc.*, **102**, 3242, 1980

3.45   bis(Trimethylbenzylammonium) hexadecacarbonyl – tetra – iron – platinum
$2C_{10}H_{16}N^+$, $C_{16}Fe_4O_{16}Pt^{2-}$
G.Longoni, M.Manassero, M.Sansoni *J.Am.Chem.Soc.*, **102**, 3242, 1980

3.C   tris(N – Benzyl – trimethylammonium) bis(tris(μ – ethylthiolato) – molybdenum) – tetrakis(μ – thiolato) – tris(ethylthiolato – iron) – iron
$3C_{10}H_{16}N^+$, $C_{24}H_{60}Fe_7Mo_2S_{20}{}^{3-}$ Main entry is 85.78

3.46   n – Decylammonium tetrachloro – cadmium(ii) (room temperature form)
$2C_{10}H_{24}N^+$, $CdCl_4{}^{2-}$
R.Kind, S.Plesko, H.Arend, R.Blinc, B.Zeks, J.Seliger, B.Lozar, J.Slak, A.Levstik, C.Filipic, V.Zagar, G.Lahajnar, F.Milia, G.Chapuis *J.Chem.Phys.*, **71**, 2118, 1979

3.47   n – Decylammonium tetrachloro – cadmium(ii) (high temperature form, at 318°K)
$2C_{10}H_{24}N^+$, $CdCl_4{}^{2-}$
R.Kind, S.Plesko, H.Arend, R.Blinc, B.Keks, J.Seliger, B.Lozar, J.Slak, A.Levstik, C.Filipic, V.Zagar, G.Lahajnar, F.Milia, G.Chapuis *J.Chem.Phys.*, **71**, 2118, 1979

3.C   Phenylcholine ether bromide
$C_{11}H_{18}NO^+$, Br^- Main entry is 17.17

3.C   trans – 7 – Dimethylamino – 2 – carbomethoxy – hepta – 2,4,6 – triene methyl carboxylate
$C_{12}H_{17}NO_4$ Main entry is 1.28

3.C   N,N' – Ethylene – bis(acetylacetone – imine)
$C_{12}H_{20}N_2O_2$ Main entry is 1.31

3.C   (N,N,N',N',N' – Hexamethyl – hexamethylene – diammonium) (7,7,8,8 – tetracyanoquinodimethane)
$C_{12}H_{30}N_2{}^{2+}$, $2C_{12}H_4N_4{}^-$, $2C_{12}H_4N_4$ Main entry is 60.21

3.C   Methyl – (2E,4Z) – 5 – dimethylamino – 2 – nitro – 4 – (4 – chloro – 2 – nitrophenyl) – 2,4 – pentadienoate
$C_{14}H_{14}ClN_3O_6$ Main entry is 15.4

3.C   Carbinoxamine maleate
$C_{16}H_{20}ClN_2O^+$, $C_4H_3O_4{}^-$ Main entry is 33.70

3.C   Tetra – n – butylammonium hydrogen dichloromaleate (CuK(α) data)
$C_{16}H_{36}N^+$, $C_4HCl_2O_4{}^-$ Main entry is 2.19

3.C   Tetra – n – butylammonium hydrogen dichloromaleate (MoK(α) data)
$C_{16}H_{36}N^+$, $C_4HCl_2O_4{}^-$ Main entry is 2.20

3.C   Tetrabutylammonium bis(thiomercapto – acetato) – oxo – technetium
$C_{16}H_{36}N^+$, $C_4H_4O_3S_4Tc^-$ Main entry is 85.3

3.48   Tetra – n – butylammonium tri – iodide
$C_{16}H_{36}N^+$, $I_3{}^-$
D.N.Hendrickson, F.H.Herbstein, M.Kaftory, M.Kapon, W.Saenger *Eur.Cryst.Meeting.* **5**, 88, 1979

3.49   Tetra – n – butylammonium trisulfur – trinitride
$C_{16}H_{36}N^+$, $N_3S_3{}^-$
J.Bojes, T.Chivers, W.G.Laidlow, M.Trsic *J.Am.Chem.Soc.*, **101**, 4517, 1979

3.C   **Tetrabutylammonium bis(isotrithione – dithiolato) – nickel(ii)**
$2C_{16}H_{36}N^+, C_6NiS_{10}^{2-}$ Main entry is 85.14

3.C   **Tetrabutylammonium (μ – oxo) – (μ – sulfido) – bis((1,2 – dithiosquarato – S,S') – oxo – molybdenum(v))**
$2C_{16}H_{36}N^+, C_8Mo_2O_7S_5^{2-}$ Main entry is 85.31

3.50   **Tetra – n – butylammonium hexakis(μ – carbonyl) – iodo – octacarbonyl – penta – rhodium**
$2C_{16}H_{36}N^+, C_{14}IO_{14}Rh_5^{2-}$
S.Martinengo, G.Ciani, A.Sironi
J.Chem.Soc.,Chem.Commun.,1059, 1979

3.51   **tris(Tetra – n – butylammonium) hexakis(isothiocyanato) – technetium(iii)**
$3C_{16}H_{36}N^+, C_6N_6S_6Tc^{3-}$
H.S.Trop, A.Davison, A.G.Jones, M.A.Davis, D.J.Szalda,
S.J.Lippard *Inorg.Chem.*,**19**, 1105, 1980

3.52   **Tetra – n – butylammonium platinum – carbonyl acetonitrile solvate**
$4C_{16}H_{36}N^+, 8C_2H_3N, C_{22}O_{22}Pt_{19}^{4-}$
D.M.Washecheck, E.J.Wucherer, L.F.Dahl, A.Ceriotti,
G.Longoni, M.Manassero, M.Sansoni, P.Chini
J.Am.Chem.Soc.,**101**, 6110, 1979

3.C   **tetrakis(Tetra – n – butylammonium) (bis(tris(μ – benzylthiolato) – molybdenum) – tetrakis(μ – thiolato) – tris(benzylthiolato – iron)) – iron**
$4C_{16}H_{36}N^+, C_{84}H_{84}Fe_7Mo_2S_{20}^{4-}$ Main entry is 85.96

3.C   **tetrakis(Tetra – n – butylammonium) (bis(tris(μ – benzylthiolato) – tungsten) – tetrakis(μ – thiolato) – tris(benzylthiolato – iron)) – iron**
$4C_{16}H_{36}N^+, C_{84}H_{84}Fe_7S_{20}W_2^{4-}$ Main entry is 85.97

3.53   **t – Butyl – (2 – t – butylimino – 1 – phenylsulfonylimino – 2 – cyano – ethyl)amine**
$C_{17}H_{24}N_4O_2S$
G.l.'abbe, L.Huybrechts, S.Toppet, J.P.Declercq, G.Germain,
M.van Meerssche *Bull.Soc.Chim.Belg.*,**88**, 297, 1979
See also R1 : 4.7

3.54   **Hexadecyl – trimethylammonium dichloro – iodide**
$C_{19}H_{42}N^+, Cl_2I^-$
G.Bandoli, D.A.Clemente, M.Nicolini
J.Cryst.Mol.Struct.,**8**, 279, 1978

3.C   **2 – Diethyl – aminoethyl diphenylacetate hydrobromide**
$C_{20}H_{26}NO_2^+, Br^-$ Main entry is 1.38

3.55   **Deoxy – lysophosphatidylcholine monohydrate**
3 – Dodecanoyl – propandiol – 1 – phosphorylcholine monohydrate
$C_{20}H_{42}NO_6P. H_2O$
H.Hauser, I.Pascher, S.Sundell *J.Mol.Biol.*,**137**, 249, 1980

3.C   **2 – (Dicyclohexyl – acetoxy) – ethyl – triethylammonium iodide**
$C_{22}H_{42}NO_2^+, I^-$ Main entry is 1.40

3.C   **3,5 – Dimethyl – 1,7 – diphenyl – 4 – (2,4,6 – trinitrophenyl) – 2,6 – diazahepta – 2,4 – diene**
$C_{25}H_{23}N_5O_6$ Main entry is 15.6

3.C   **1 – (p – (2 – Dimethylaminoethoxy)phenyl) – 1,2 – trans – diphenylbutene**
trans – Tamoxifene
$C_{26}H_{29}NO$ Main entry is 17.

3.56   **bis(Triphenylphosphine)immonium (μ^5 – carbido) – iodo – pentadecacarbonyl – penta – osmium**
$C_{36}H_{30}NP_2^+, C_{16}IO_{15}Os_5^-$
P.F.Jackson, B.F.G.Johnson, J.Lewis, J.N.Nicholls,
M.McPartlin, W.J.H.Nelson
J.Chem.Soc.,Chem.Commun.,564, 1980

4.1   S,S – Dimethylpentasulfur – hexanitride
$C_2H_6N_6S_5$
W.S.Sheldrick, M.N.S.Rao, H.W.Roesky
Inorg.Chem., **19**, 538, 1980

4.2   S,S – Dimethyl – N – methylsulfonylsulfimide
S,S – Dimethyl – N – methylsulfonylsulfilimine
$C_3H_9NO_2S_2$
A.Kalman, L.Parkanyi, A.Kucsman
Acta Crystallogr.,Sect.B, **36**, 1440, 1980

4.C   bis(Dimethylamino) – difluorosulfurane (at –35°C)
$C_4H_{12}F_2N_2S$ Main entry is 3.18

4.C   4 – Nitroaniline 4 – nitro – N – sulfinylaniline
$C_6H_4N_2O_3S$, $C_6H_6N_2O_2$ Main entry is 16.1

4.C   Carboxymethane – (N – methyl – N – phenyl – sulfonamide)
$C_9H_{11}NO_4S$ Main entry is 1.22

4.C   2 – Carboxyethane – (N – methyl – N – phenyl – sulfonamide)
$C_{10}H_{13}NO_4S$ Main entry is 1.24

4.C   2 – Methoxycarbonyl – vinyl – (N – methyl – N – phenyl – sulfonamide)
$C_{11}H_{13}NO_4S$ Main entry is 1.26

4.C   2 – Methyl – 2 – (N – phenylsulfinamoyl) – N,N – dimethyl – propanamide
$C_{12}H_{18}N_2O_2S$ Main entry is 1.30

4.C   2 – Carboxy – cyclopentene – 1 – (N – methyl – N – phenyl – sulfonamide)
$C_{13}H_{15}NO_4S$ Main entry is 20.5

4.3   L – (1 – Tosylamino – 2 – phenyl) – ethyl – chloromethyl – ketone
$C_{17}H_{18}ClNO_3S$
A.Michel, F.Durant, G.S.D.King
Bull.Soc.Chim.Belg., **89**, 25, 1980
See also R1 : 19

4.C   t – Butyl – (2 – t – butylimino – 1 – phenylsulfonylimino – 2 – cyano – ethyl)amine
$C_{17}H_{24}N_4O_2S$ Main entry is 3.53

4.C   5 – Methylbenzene – 1,3 – dicarbaldehyde – bis(p – tolylsulfonylhydrazone) benzene clathrate
$C_{23}H_{24}N_4O_4S_2$. $2C_6H_6$ Main entry is 61.7

5.C   2,2,4,4,6,6 – Hexa(1 – aziridinyl) – cyclotriphosphazene carbon tetrachloride anticlathrate
$3CCl_4$, $C_{12}H_{24}N_9P_3$ Main entry is 61.2

5.1   Deuteromethane (α form, at 1.5–6°K)
CD_4
A.I.Prokhvatilov, A.P.Isakina
Acta Crystallogr.,Sect.B, **36**, 1576, 1980

5.2   Hexachloroethane (orthorhombic form, at 140°K, neutron study)
C_2Cl_6
D.Hohlwein, W.Nagele, W.Prandl
Acta Crystallogr.,Sect.B, **35**, 2975, 1979

5.3   Hexachloroethane (orthorhombic form, neutron study)
C_2Cl_6
D.Hohlwein, W.Nagele, W.Prandl
Acta Crystallogr.,Sect.B, **35**, 2975, 1979

5.4   Acetylene (at 141°K)
C_2H_2
G.J.H.van Nes, F.van Bolhuis
Acta Crystallogr.,Sect.B, **35**, 2580, 1979

5.5   hexakis(Acetylene) zeolite 4A complex
$0.5C_2H_2$, $AlNaO_4Si$
A.A.Amaro, K.Seff   *J.Phys.Chem.*, **77**, 906, 1973

5.6   Ethylene (electron density distribution study, at 85°K)
C_2H_4
G.J.H.van Nes, A.Vos
Acta Crystallogr.,Sect.B, **35**, 2593, 1979

5.C   bis(Triphenylbenzylphosphonium) tetrachloro – cadmium dichloroethane clathrate
$2C_2H_4Cl_2$, $2C_{25}H_{22}P^+$, $CdCl_4^{2-}$ Main entry is 61.8

5.C   γ – Cyclodextrin n – propanol clathrate hydrate
C_3H_8O, $C_{48}H_{80}O_{40}$. xH_2O Main entry is 61.22

5.C   (+) – Tri – o – thymotide – S – (+) – 2 – bromo – butane clathrate (at –50°C)
C_4H_9Br, $2C_{33}H_{36}O_6$ Main entry is 61.16

5.7   Tetra – t – butoxy – triberyllium bis(tetrahydroborate)
$4C_4H_9O^-$, $3Be^{2+}$, $2H_4B^-$
B.Morosin, J.Howatson
J.Inorg.Nucl.Chem., **41**, 1667, 1979

5.C   Lithium dibenzoylphosphide 1,2 – dimethoxyethane (at –80°C)
$C_4H_{10}O_2$, $C_{14}H_{10}O_2P^-$, Li^+ Main entry is 64.62

5.8   1,1,3,5,7,7 – Hexachloroheptane (monoclinic form)
$C_7H_{10}Cl_6$
J.C.J.Bart, I.W.Bassi, M.Calcaterra
Acta Crystallogr.,Sect.B, **36**, 421, 1980

5.9   meso – (RS) – 1,1,3,6,8,8,8 – Octachloro – octane
$C_8H_{10}Cl_8$
J.C.J.Bart, I.W.Bassi, M.Calcaterra
Acta Crystallogr.,Sect.B,**35**, 2646, 1979

5.10   **Hexamethylethane (α form, at 188°K)**
C_8H_{18}
P.A.Reynolds *Mol.Phys.*,**37**, 1333, 1979

5.11   **Hexamethylethane (α form, at 295°K)**
C_8H_{18}
P.A.Reynolds *Mol.Phys.*,**37**, 1333, 1979

5.12   **(3E,2R,6R,7S) – 1,17 – Tribromo – 2,6,8 – trichloro – 3,7 – dimethyl – oct – 3 – ene (absolute configuration)**
$C_{10}H_{14}Br_3Cl_3$
P.Bates, J.W.Blunt, M.P.Hartshorn, A.J.Jones, M.H.G.Munro, W.T.Robinson, S.C.Yorke *Aust.J.Chem.*,**32**, 2545, 1979

5.13   **2,5 – Dimethyl – 3,4 – di(2 – propyl) – hex – 3 – ene (at 170°K)**
Tetraisopropylethylene
$C_{14}H_{28}$
G.Casalone, T.Pilati, M.Simonetta
Tetrahedron Lett.,**21**, 2345, 1980

5.14   **1,3 – Diphenyl – 2,2 – dihydroxypropane – 1,3 – dione**
$C_{15}H_{12}O_4$
J.M.M.Smits, J.H.Noordik
Cryst.Struct.Commun.,**9**, 29, 1980

ENOLATES (ALIPHATIC AND AROMATIC)

6.C   **Potassium p – chloranil acetone solvate**
$C_6Cl_4O_2^-$, K^+, C_3H_6O Main entry is 18.1

6.C   **Valinomycin potassium picrate**
$C_6H_2N_3O_7^-$, $C_{36}H_{60}N_4O_{12}$, K^+ Main entry is 60.43

6.C   **(3 – Chloro – 2 – hydroxy – 5 – nitrophenyl) – (2′ – chlorophenyl) – iodonium hydroxide inner salt**
$C_{12}H_6Cl_2INO_3$ Main entry is 17.18

NITRILES (ALIPHATIC AND AROMATIC)

7.C   **2 – Cyanoguanidine (at 83°K, charge deformation density study)**
$C_2H_4N_4$ Main entry is 8.8

7.C   **Potassium cyanamide dicyanamide**
$C_2H_4N_4$, CHN_2^-, K^+ Main entry is 3.4

7.1   **Potassium nitroso – dicyanomethanide**
$C_3N_3O^-$, K^+
V.V.Skopenko, Yu.L.Zub, M.A.Porai–Koshits, G.G.Sadikov
Ukr.Khim.Zh.(Russ.Ed.),**45**, 811, 1979
See also R1 : 10

7.C   **Skatole tetracyanoethylene**
C_6N_4, C_9H_9N Main entry is 60.9

7.C   **bis(9,10 – Diazaphenanthrene) tetracyanoethylene (triclinic form)**
C_6N_4, $2C_{12}H_8N_2$ Main entry is 60.15

7.2   **4 – Chlorobenzonitrile (at –90°C)**
C_7H_4ClN
D.Britton, J.Konnert, S.Lam
Cryst.Struct.Commun.,**8**, 913, 1979

7.C   **2,3 – Dichloro – 5,6 – dicyano – p – benzoquinone**
$C_8Cl_2N_2O_2$ Main entry is 18.2

7.C   **Pyridinium – 1 – dicyanomethylide (at 118°K, fully deuterated, neutron study)**
$C_8D_5N_3$ Main entry is 33.28

7.C   **Pyridinium – 1 – dicyanomethylide (deuterated form, neutron study)**
$C_8D_5N_3$ Main entry is 33.29

7.C   **Anthracene – tetracyanobenzene complex (high temperature form, at 297°K)**
$C_{10}H_2N_4$, $C_{14}H_{10}$ Main entry is 60.25

7.C   **Anthracene – tetracyanobenzene complex (high temperature form, at 234°K)**
$C_{10}H_2N_4$, $C_{14}H_{10}$ Main entry is 60.26

7.C   **Anthracene – tetracyanobenzene complex (high temperature form, at 226°K)**
$C_{10}H_2N_4$, $C_{14}H_{10}$ Main entry is 60.27

7.C   **Anthracene – tetracyanobenzene complex (low temperature form, at 202°K)**
$C_{10}H_2N_4$, $C_{14}H_{10}$ Main entry is 60.28

7.C   **Anthracene – tetracyanobenzene complex (low temperature form, at 170°K)**
$C_{10}H_2N_4$, $C_{14}H_{10}$ Main entry is 60.29

7.C   **Anthracene – tetracyanobenzene complex (low temperature form, at 138°K)**
$C_{10}H_2N_4$, $C_{14}H_{10}$ Main entry is 60.30

7.C   **Anthracene – tetracyanobenzene complex (low temperature form, at 119°K)**
$C_{10}H_2N_4$, $C_{14}H_{10}$ Main entry is 60.31

7.C   **1,1,2,3,3 – Pentamethyl – guanidinium – 2 – (2,2 – dicyano – 1 – ethylenethiolate)**
$C_{10}H_{15}N_5S$ Main entry is 8.12

7.C   **Trimethylammonium 7,7,8,8 – tetracyanoquinodimethane tri – iodide (neutron study, at 80°K)**
$C_{12}H_4N_4$, $C_3H_{10}N^+$, $0.33I_3^{3-}$ Main entry is 3.11

7.C   **5 – Phenyl – 1,3 – thiaselenole – 2 – thione 7,7,8,8 – tetracyanoquinodimethane complex**
$C_{12}H_4N_4$, $C_9H_6S_2Se$ Main entry is 60.8

7.C   **Stilbene – 7,7,8,8 – tetracyanoquinodimethane complex**
$C_{12}H_4N_4$, $C_{14}H_{12}$ Main entry is 60.32

7.C   **Phenazine 5,10 – dihydro – 5,10 – dimethylphenazinium 7,7,8,8 – tetracyanoquinodimethane**
$C_{12}H_4N_4$, $0.6C_{14}H_{14}N_2$, $0.4C_{12}H_8N_2$ Main entry is 36.2

7.C   **Thieno(3,2 – e:4,5 – e') – bis(benzo(b)thiophene) – 7,7,8,8 – tetracyanoquinodimethane**
Trithia(5)heterohelicene – 7,7,8,8 – tetracyanoquinodimethane
$C_{12}H_4N_4$, $C_{16}H_8S_3$ Main entry is 60.33

7.C   **Tetramethoxystilbene – 7,7,8,8 – tetracyanoquinodimethane complex**
$C_{12}H_4N_4$, $C_{18}H_{20}O_4$ Main entry is 60.38

7.C   **N,N' – (1,2 – Phenylene) – bis(salicylaldiminato) – copper(ii) – 7,7,8,8 – tetracyanoquinodimethane complex**
$C_{12}H_4N_4$, $2C_{20}H_{14}CuN_2O_2$ Main entry is 60.40

7.C   **Tetraethylammonium bis(7,7,8,8 – tetracyanoquinodimethane)**
$C_{12}H_4N_4^-$, $C_8H_{20}N^+$, $C_{12}H_4N_4$ Main entry is 60.7

7.C   **1,1' – Dimethylferrocenium bis(7,7,8,8 – tetracyanoquinodimethane)**
$C_{12}H_4N_4^-$, $C_{12}H_{14}Fe^+$, $C_{12}H_4N_4$ Main entry is 60.20

7.C   **Ethyltriphenylphosphonium – bis(7,7,8,8 – tetracyanoquinodimethane)**
$C_{12}H_4N_4^-$, $C_{20}H_{20}P^+$, $C_{12}H_4N_4$ Main entry is 60.41

7.C   **Decamethylferrocenium – 7,7,8,8 – tetracyanoquinodimethane**
$C_{12}H_4N_4^-$, $C_{20}H_{30}Fe^+$ Main entry is 60.42

7.C   **(N,N,N',N',N' – Hexamethyl – hexamethylene – diammonium) (7,7,8,8 – tetracyanoquinodimethane)**
$2C_{12}H_4N_4^-$, $C_{12}H_{30}N_2^{2+}$, $2C_{12}H_4N_4$ Main entry is 60.21

7.C   **bis(N – Isopropyl – 2 – oxy – 1 – naphthylidene – aminato) – copper(ii) bis(7,7,8,8 – tetracyanoquinodimethane)**
$2C_{12}H_4N_4$, $C_{28}H_{28}CuN_2O_2$ Main entry is 78.13

7.3   **Benzene – hexacarbonitrile**
Hexacyanobenzene
$C_{12}N_6$
J.Swiatkiewicz, B.Kuchta *Eur.Cryst.Meeting*,**6**, 265, 1980

7.C   **7,8 – benzoquinoline 7,7,8,8 – Tetracyanoquinodimethane**
$C_{13}H_9N$, $C_{12}H_4N_4$ Main entry is 60.14

7.C   tris(4,4´,5,5´ – Tetramethyl – tetrathiafulvalene) –
        bis(2,5 – diethyl – 7,7,8,8 –
        tetracyanoquinodimethane) complex
        $2C_{16}H_{12}N_4$, $3C_{10}H_{12}S_4$ Main entry is 60.35

7.C   t – Butyl – (2 – t – butylimino – 1 –
        phenylsulfonylimino – 2 – cyano – ethyl)amine
        $C_{17}H_{24}N_4O_2S$ Main entry is 3.53

7.4   (2S,3S) – 2,3 – Diphenyl – 3 – hexanecarbonitrile
        α,β – Diphenyl – α – n – propyl – β – methyl –
        propionitrile
        $C_{19}H_{21}N$
        N.C.Baenziger, S.Wawzonek, D.C.Swenson
        Acta Crystallogr.,Sect.B,**35**, 2793, 1979

UREA COMPOUNDS
(ALIPHATIC AND AROMATIC)

8.C   Parabanic acid urea (neutron study,
        at 116°K,deuterated form)
        CD_4N_2O, $C_3D_2N_2O_3$ Main entry is 32.4

8.1   Urea (multipole deformation density refinement,
        at 123°K)
        CH_4N_2O
        D.Mullen *Acta Crystallogr.,Sect.B*,**36**, 1610, 1980

8.C   Urea malonic acid
        CH_4N_2O, $C_3H_4O_4$ Main entry is 1.8

8.2   Urea trifluoro – antimony
        CH_4N_2O, F_3Sb
        M.Bourgault, R.Fourcade, B.Ducourant, G.Mascherpa
        Rev.Chim.Miner.,**16**, 151, 1979

8.3   Urea hydrogen peroxide (at 81°K, neutron study)
        CH_4N_2O, H_2O_2
        C.J.Fritchie Junior, R.K.McMullan
        Am.Cryst.Assoc.,Ser.2,**7**, 25, 1980

8.4   Urea orthotelluric acid
        $2CH_4N_2O$, H_6O_6Te
        J.Loub, W.Haase, R.Mergehenn
        Acta Crystallogr.,Sect.B,**35**, 3039, 1979

8.5   Urea lithium sulfate
        $3CH_4N_2O$, $2Li^+$, O_4S^{2-}
        Kh.Suleimanov, V.S.Sergienko, N.Kipalova,
        K.Sulaimankulov *Koord.Khim.*,**5**, 1732, 1979

8.6   Urea magnesium bromide dihydrate
        $4CH_4N_2O$, Mg^{2+}, $2Br^-$, $2H_2O$
        L.Lebioda, K.Lewinski
        Acta Crystallogr.,Sect.B,**36**, 693, 1980

8.C   bis((8 – Quinolyl – oxy) – ethoxy – ethyl)ether
        thiourea
        CH_4N_2S, $C_{26}H_{28}N_2O_5$ Main entry is 17.40

8.C   Guanidinium hydrogen oxalate monohydrate
        $CH_6N_3^+$, $C_2HO_4^-$, H_2O Main entry is 2.9

8.C   2 – Diethylphosphoryl – guanidine
        hemi(guanidinium chloride)
        $0.5CH_6N_3^+$, $C_5H_{14}N_3O_3P$, $0.5Cl^-$ Main entry is 64.13

8.C   Guanidinium hexamolybdo – methylarsonate
        hexahydrate
        $2CH_6N_3^+$, $CH_{15}AsMo_6O_{27}^{2-}$, $6H_2O$ Main entry is 84.2

8.7   Tetraguanidinium α – dodecamolybdo – silicate
        monohydrate
        $4CH_6N_3^+$, $Mo_{12}O_{40}Si^{4-}$, H_2O
        H.Ichida, A.Kobayashi, Y.Sasaki
        Acta Crystallogr.,Sect.B,**36**, 1382, 1980

8.C   Guanidinium tri(carbonato) – trifluoro –
        thorium(iv)
        $5CH_6N_3^+$, $C_3F_3O_9Th^{5-}$ Main entry is 81.6

8.8 **2 – Cyanoguanidine (at 83°K, charge deformation density study)**
$C_2H_4N_4$
F.L.Hirshfeld, H.Hope
*Acta Crystallogr.,Sect.B,***36**, 406, 1980
See also R1 : 7

8.9 **Dithio – bis(formamidinium) trans – tetrachloro – oxo – rhenium(v) chloride monohydrate**
$C_2H_8N_4S_2{}^{2+}$, $H_2Cl_4O_2Re^-$, Cl^-, H_2O
T.Lis *Acta Crystallogr.,Sect.B,***35**, 3041, 1979

8.C **4 – Hydroxy – 2,5 – dioxo – 4 – imidazolidine – carboxyureide hemihydrate**
$C_5H_6N_4O_5$, $0.5H_2O$ Main entry is 32.13

8.10 **3 – (3,4 – Dichlorophenyl) – 1,1 – dimethylurea**
Diuron
$C_9H_{10}Cl_2N_2O$
R.G.Baughman, D.D.Sams, B.J.Helland, R.A.Jacobson
*Cryst.Struct.Commun.,***9**, 885, 1980

8.11 **3 – (p – Chlorophenyl) – 1,1 – dimethylurea**
Monuron
$C_9H_{11}ClN_2O$
R.G.Baughman, R.I.Hembre, B.J.Helland, R.A.Jacobson
*Cryst.Struct.Commun.,***9**, 749, 1980

8.12 **1,1,2,3,3 – Pentamethyl – guanidinium – 2 – (2,2 – dicyano – 1 – ethylenethiolate)**
$C_{10}H_{15}N_5S$
E.Schaumann, E.Kausch, E.Rossmanith
*Justus Liebigs Ann.Chem.,*1543, 1978
See also R1 : 11,7

8.C **2 – Cyano – 1 – methyl – 3 – (2 – ((5 – methyl – 1H – imidazol – 4 – yl) – methylthio) – ethyl) – guanidine monohydrate**
$C_{10}H_{16}N_6S$, H_2O Main entry is 32.34

8.C **N – Methyl – N′ – (1 – acetamido – 2 – phenylethyl) – urea**
N – Methyl – N′ – (benzyl)(acetamido) – methylurea
$C_{12}H_{17}N_3O_2$ Main entry is 1.29

8.13 **1,5 – Diphenylcarbazone**
$C_{13}H_{12}N_4O$
N.M.Blaton, O.M.Peeters, C.J.De Ranter, G.J.Willems
*Acta Crystallogr.,Sect.B,***35**, 2629, 1979
See also R1 : 9

8.14 **Diphenylguanidine**
$C_{13}H_{13}N_3$
L.N.Zakharov, V.G.Andrianov, Yu.T.Struchkov
*Kristallografiya,***25**, 65, 1980

8.C **4 – m – Aminophenyl – 2 – formylpyridine – thiosemicarbazone**
$C_{13}H_{13}N_5S$ Main entry is 33.53

8.C **1 – ((4 – (3 – Chlorophenylamino) – 3 – pyridyl) sulfonyl) – 3 – ethyl – 1 – methylurea hydrogen sulfate**
$C_{15}H_{18}ClN_4O_3S^+$, HO_4S^- Main entry is 33.66

8.C **O – Benzoyl – N,N – dimethyl – N′ – (N – methyl – 2,4 – dinitroanilino) – isourea**
$C_{17}H_{17}N_5O_6$ Main entry is 13.16

8.C **N – (1 – Methyl – 4 – (3 – methylphenylamino) – 3 – pyridinio) – sulfonyl – N′ – isopropylurea**
$C_{17}H_{22}N_4O_3S$ Main entry is 33.73

8.C **L – Pyroglutamyl – N,N′ – dicyclohexylurea**
$C_{18}H_{29}N_3O_3$ Main entry is 48.55

8.C **Benzaldehyde 4 – benzoyl – 2 – phenylthiosemicarbazone**
$C_{21}H_{17}N_3OS$ Main entry is 13.19

8.C **Cyclopentadecanone phenylsemicarbazone**
$C_{22}H_{35}N_3O$ Main entry is 23.6

NITROGEN–NITROGEN COMPOUNDS
(ALIPHATIC AND AROMATIC)

9.C   **Potassium thiocarbazate**
$CH_3N_2OS^-$, K^+ Main entry is 2.5

9.C   **N,N – Dimethyl – nitramine (perdeuterated, neutron study, at 91°K)**
$C_2D_6N_2O_2$ Main entry is 3.3

9.C   **Azo – diformaldehyde – dioxime**
$C_2H_4N_4O_2$ Main entry is 10.1

9.C   **O – Methylthiocarbazate (at –150°C)**
$C_2H_6N_2OS$ Main entry is 1.6

9.C   **S – Methylthiocarbazate (at –150°C)**
$C_2H_6N_2OS$ Main entry is 1.7

9.C   **N,N – Dimethyl – nitramine (at 125°K)**
$C_2H_6N_2O_2$ Main entry is 3.6

9.C   **N,N – Dimethyl – nitramine (at 180°K)**
$C_2H_6N_2O_2$ Main entry is 3.7

9.C   **N,N – Dimethyl – nitramine (at 293°K)**
$C_2H_6N_2O_2$ Main entry is 3.8

9.C   **N,N – Dimethyl – nitramine (neutron study, at 125°K)**
$C_2H_6N_2O_2$ Main entry is 3.9

9.C   **N,N – Dimethyl – nitramine (neutron study, at 85°K)**
$C_2H_6N_2O_2$ Main entry is 3.10

9.C   **trans,cis – S – Methyldithiocarbazate (at –120°C)**
$C_2H_6N_2S_2$ Main entry is 11.4

9.1   **N – Nitro – 1 – cyanoethylamine**
$C_3H_5N_3O_2$
A.I.Ivanova, V.N.Komarov, G.V.Makarenko
Viniti,1125, 1978

9.C   **Disodium cis – syn – 4 – sulfonatobenzene – diazotate trihydrate (at –150°C)**
$C_6H_4N_2O_4S^{2-}$, $2Na^+$. $3H_2O$ Main entry is 11.8

9.2   **trans – Azobenzene (at 115°K)**
$C_{12}H_{10}N_2$
E.J.Gabe, Y.Le Page *Am.Cryst.Assoc.,Ser.2*,**8**, 36, 1980

9.C   **2,4 – Diamino – azobenzene hydrochloride dihydrate**
$C_{12}H_{13}N_4{}^+$, Cl^-, $2H_2O$ Main entry is 16.5

9.C   **1,5 – Diphenylcarbazone**
$C_{13}H_{12}N_4O$ Main entry is 8.13

9.C   **4 – (N' – (p – Nitrobenzene) – N – hydrazo) – 8 – oxatricyclo(5.1.0.0^{2,5})octane**
$C_{13}H_{13}N_3O_3$ Main entry is 38.52

9.C   **4 – Dimethylamino – azobenzene – 2' – carboxylic acid**
Methyl red
$C_{15}H_{15}N_3O_2$ Main entry is 13.12

9.C   **Perdeutero – 4,4' – azoxyphenetole (form C1)**
$C_{16}D_{18}N_2O_4$ Main entry is 17.25

9.C   **4,4' – Azoxyphenetole (form C1)**
$C_{16}H_{18}N_2O_4$ Main entry is 17.27

9.C   **anti,S – trans – 3 – Methylthio – 1,5 – di – o – tolyl – formazan (yellow isomer)**
$C_{16}H_{18}N_4S$ Main entry is 11.16

9.C   **2 – Adamantyl – O,N,N – azoxy – p – toluenesulfonate**
2 – (p – Tolylsulfonyloxy – N,N,O – azoxy) – adamantane
$C_{17}H_{22}N_2O_4S$ Main entry is 31.43

9.3   **1,4 – bis(Phenylazo) – 1,3 – dimethyl – buta – 1,4 – diene**
$C_{18}H_{18}N_4$
A.Ohsawa, H.Arai, H.Igeta, T.Akimoto, A.Tsuji, Y.Iitaka
J.Org.Chem.,**44**, 3524, 1979

9.C   **5,5' – Diethoxy – α,α' – dimethyl – α,α' – azino – di – o – cresol**
$C_{20}H_{24}N_2O_4$ Main entry is 17.34

9.C   **cis – 2 – t – Butyl – perhydroazulen – 4 – one – (p – bromophenylsulfonylhydrazone)**
$C_{20}H_{29}BrN_2O_2S$ Main entry is 27.28

9.C   **Benzaldehyde 4 – benzoyl – 2 – phenylthiosemicarbazone**
$C_{21}H_{17}N_3OS$ Main entry is 13.19

9.C   **trans – 2 – t – Butyl – perhydroazulen – 4 – one – (p – tosylhydrazone)**
$C_{21}H_{32}N_2O_2S$ Main entry is 27.30

9.C   **N,N' – bis(2 – (4 – Chlorophenylamino) – 2 – cyclopenten – 1 – ylidene) – hydrazine (monoclinic form)**
$C_{22}H_{20}Cl_2N_4$ Main entry is 20.12

9.C   **N,N' – bis(2 – (4 – Chlorophenylamino) – 2 – cyclopenten – 1 – ylidene) – hydrazine (orthorhombic form)**
$C_{22}H_{20}Cl_2N_4$ Main entry is 20.13

9.C   **3 – (3,4 – Dimethoxybenzyl) – 4(2,4 – dimethoxyphenyl) – 1 – semicarbazono – 6,7,8 – trimethoxy – naphthalene – 2 – one**
$C_{31}H_{33}N_3O_9$ Main entry is 27.33

NITROGEN–OXYGEN COMPOUNDS
(ALIPHATIC AND AROMATIC)

10.C   **Nitroacetic acid**
$C_2H_3NO_4$ Main entry is 1.2

10.1   **Azo – diformaldehyde – dioxime**
$C_2H_4N_4O_2$
C.Bois, J.Armand, P.Bassinet
*Acta Crystallogr.,Sect.B,***36**, 1731, 1980
See also R1 : 9

10.C   **Potassium nitroso – dicyanomethanide**
$C_3N_3O^-$, K^+ Main entry is 7.1

10.2   **Dimethylglyoxime (charge–deformation refinement)**
$C_4H_8N_2O_2$
B.M.Craven, C.H.Chang, D.Ghosh
*Acta Crystallogr.,Sect.B,***35**, 2962, 1979

10.C   **trans – N,N – Dimethyl – 2 – nitro – ethenamine**
trans – N,N – Dimethyl – 2 – nitrovinylamine
$C_4H_8N_2O_2$ Main entry is 3.17

10.3   **Z – p – Chloro – benzaldoxime**
C_7H_6ClNO
E.Arte, J.P.Declercq, G.Germain, M.van Meerssche
*Bull.Soc.Chim.Belg.,***89**, 155, 1980

10.C   **(E) – 1 – (4′ – Hydroxy – 3′ – methoxyphenyl) – 2 – nitropropene**
$C_{10}H_{11}NO_4$ Main entry is 17.13

10.C   **syn – γ – Camphorquinone – dioxime**
$C_{10}H_{16}N_2O_2$ Main entry is 31.10

10.C   **Cyclodecane – oxime (at –160°C)**
$C_{10}H_{19}NO$ Main entry is 23.1

10.4   **3,3,5,5 – Tetramethyl – cyclohexanone – oxime**
$C_{10}H_{19}NO$
S.Toure, J.Lapasset, B.Boyer, G.Lamaty
*Acta Crystallogr.,Sect.B,***36**, 2168, 1980

10.5   **2,4,6 – Trimethylacetophenone – oxime**
$C_{11}H_{15}NO$
S.Fortier, G.I.Birnbaum, G.W.Buchanan, B.A.Dawson
*Can.J.Chem.,***58**, 191, 1980

10.C   **Cycloundecane – oxime (at –160°C)**
$C_{11}H_{21}NO$ Main entry is 23.2

10.C   **Cyclododecane – oxime (at –160°C)**
$C_{12}H_{23}NO$ Main entry is 23.3

10.C   **Cyclotridecane – oxime (at –160°C)**
$C_{13}H_{25}NO$ Main entry is 23.4

10.C   **Methyl – (2E,4Z) – 5 – dimethylamino – 2 – nitro – 4 – (4 – chloro – 2 – nitrophenyl) – 2,4 – pentadienoate**
$C_{14}H_{14}ClN_3O_6$ Main entry is 15.4

10.C   **Cyclotetradecane – oxime (at –160°C)**
$C_{14}H_{27}NO$ Main entry is 23.5

10.C   **2 – Nitro – 3 – ferrocenyl – acrylic acid ethyl ester**
$C_{15}H_{15}FeNO_4$ Main entry is 73.55

10.C   **3,3,5 – Trimethyl – 5 – phenyl – cyclohexanone – oxime**
$C_{15}H_{21}NO$ Main entry is 21.10

10.6   **4 – Chloroacetophenone – O – (3 – t – butylamino – 2 – hydroxypropyl) – oxime**
$C_{15}H_{24}ClN_2O_2^+$, Cl^-
A.Carpy, M.Gadret, J.M.Leger, C.G.Wermuth, G.Leclerc
*Acta Crystallogr.,Sect.B,***36**, 1715, 1980

10.C   **Perdeutero – 4,4′ – azoxyphenetole (form C1)**
$C_{16}D_{18}N_2O_4$ Main entry is 17.25

10.C   **4,4′ – Azoxyphenetole (form C1)**
$C_{16}H_{18}N_2O_4$ Main entry is 17.27

SULFUR AND SELENIUM COMPOUNDS

11.1 Trisulfur – dinitrido trifluoromethylsulfonate acetonitrile solvate (at −130°C)
$CF_3O_3S^-$, $N_2S_3^+$, $0.5C_2H_3N$
B.Krebs, G.Henkel, S.Pohl, H.Roesky
Chem.Ber.,**113**, 226, 1980

11.2 Tri – potassium methyltrisulfonate monohydrate
$CHO_9S_3{}^{3-}$, $3K^+$, H_2O
J.R.Hall, R.A.Johnson, C.H.L.Kennard, G.Smith
J.Chem.Soc.,Dalton Trans.,149, 1980

11.C Potassium thiocarbazate
$CH_3N_2OS^-$, K^+ Main entry is 2.5

11.C 2 – Amino – 5 – methyl – 7 – propyl – imidazo(5,1 – f)(1,2,4)triazin – 4(3H) – one methylsulfonate monohydrate
$CH_3O_3S^-$, $C_9H_{14}N_5O^+$, H_2O Main entry is 35.13

11.C Bromocriptine methylsulfonate isopropanol solvate (absolute configuration)
$CH_3O_3S^-$, $C_{32}H_{41}BrN_5O_5{}^+$, $0.5C_3H_8O$ Main entry is 58.76

11.3 1,7 – bis(Pentachloro – antimony – oxo) – cyclododecasulfur tris(carbon disulfide) (at −115°C)
$3CS_2$, $Cl_{10}O_2S_{12}Sb_2$
R.Steudel, J.Steidel, J.Pickardt
Angew.Chem.,Int.Ed.Engl.,**19**, 325, 1980

11.C O – Methylthiocarbazate (at −150°C)
$C_2H_6N_2OS$ Main entry is 1.6

11.C S – Methylthiocarbazate (at −150°C)
$C_2H_6N_2OS$ Main entry is 1.7

11.4 trans,cis – S – Methyldithiocarbazate (at −120°C)
$C_2H_6N_2S_2$
R.Mattes, H.Weber *J.Chem.Soc.,Dalton Trans.*,423, 1980
See also R1 : 9

11.5 bis(Methylsulfonyldichloromethyl) – disulfide
3,3,6,6 – Tetrachloro – 2,4,5,7 – tetrathiaoctane – 2,2,7,7 – tetraoxide
$C_4H_6Cl_4O_4S_4$
J.S.Grossert, M.M.Bharadwaj, J.B.Faught, A.Terzis
Can.J.Chem.,**58**, 1106, 1980

11.6 Ethane – 1,2 – bis(methylsulfone)
$C_4H_{10}O_4S_2$
F.Mo, O.Berg, G.Thorkildsen, A.Gaasdal
Eur.Cryst.Meeting,**5**, 352, 1979

11.C bis(2 – Aminoethyl) – disulfide dihydrochloride
Cystamine dihydrochloride
$C_4H_{14}N_2S_2{}^{2+}$, $2Cl^-$ Main entry is 3.29

11.C Sodium 1 – pyrrolidinyl – carbodithioate dihydrate
$C_5H_8NS_2^-$, Na^+, $2H_2O$ Main entry is 32.14

11.C Sodium 1 – pyrrolidinyl – carbodithioate dihydrate (at 150°K)
$C_5H_8NS_2^-$, Na^+, $2H_2O$ Main entry is 32.15

11.C Sodium 1 – pyrrolidinyl – carbodithioate dihydrate (at 27°K)
$C_5H_8NS_2^-$, Na^+, $2H_2O$ Main entry is 32.16

11.C Potassium O – n – butylxanthate
$C_5H_9OS_2^-$, K^+ Main entry is 1.18

11.7 Pentachloro – thiophenol
C_6HCl_5S
G.Wojcik, G.P.Charbonneau, Y.Delugeard, L.Toupet
Acta Crystallogr.,Sect.B,**36**, 506, 1980

11.C S – Methyl – thiepanium 2,4,6 – trinitrobenzenesulfonate
$C_6H_2N_3O_9S^-$, $C_7H_{15}S^+$ Main entry is 39.16

11.8 Disodium cis – syn – 4 – sulfonatobenzene – diazotate trihydrate (at −150°C)
$C_6H_4N_2O_4S^{2-}$, $2Na^+$, $3H_2O$
N.W.Alcock, P.D.Goodman, T.J.Kemp
J.Chem.Soc.,Perkin Trans.2,1093, 1980
See also R1 : 9

11.C Piperidinium 1 – piperidine – carbodithioate (orthorhombic form)
$C_6H_{10}NS_2^-$, $C_5H_{12}N^+$ Main entry is 33.13

11.C Piperidinium 1 – piperidine – carbodithioate (monoclinic(β) form)
$C_6H_{10}NS_2^-$, $C_5H_{12}N^+$ Main entry is 33.14

11.C Di – N,N' – isopropyl – dithio – oxalamide
$C_8H_{16}N_2S_2$ Main entry is 1.21

11.C Rubidium 8 – mercapto – quinoline – 5 – sulfonate
$C_9H_6NO_3S_2^-$, Rb^+ Main entry is 35.10

11.C Methyl – 2 – (methylsulfinyl) – benzoate
$C_9H_{10}O_3S$ Main entry is 13.6

11.9 1 – Methylsulfonyl – 2 – phenylsulfonyl – ethane
$C_9H_{12}O_4S_2$
F.Mo, O.Berg, G.Thorkildsen, A.Gaasdal
Eur.Cryst.Meeting,**5**, 352, 1979

11.C 1 – (3,4 – Dihydroxyphenyl) – 2 – methylaminoethane – sulfonic acid hemihydrate
$C_9H_{13}NO_5S$, $0.5H_2O$ Main entry is 17.10

11.C (±) – N – Methyl – 2 – (3,4 – dihydroxyphenyl) ethylammonium – 2 – sulfonate hemihydrate
(±) – Epinine – β – sulfonate hemihydrate
$C_9H_{13}NO_5S$, $0.5H_2O$ Main entry is 17.11

11.10 5,5 – Dimethyl – 3 – methylsulfonyl – cyclohex – 2 – enone (at 153°K)
$C_9H_{14}O_3S$
K.S.Luk, M.P.Sammes, R.L.Harlow
J.Chem.Soc.,Perkin Trans.2,1166, 1980

11.C 1,1,2,3,3 – Pentamethyl – guanidinium – 2 – (2,2 – dicyano – 1 – ethylenethiolate)
$C_{10}H_{15}N_5S$ Main entry is 8.12

11.C Dipiperidino – tetraselane
$C_{10}H_{20}N_2Se_4$ Main entry is 33.37

11.11 bis(N,N' – Diethyl)formamidine – disulfide hexachloro – tellurium
$C_{10}H_{24}N_4S_2{}^{2+}$, Cl_6Te^{2-}
U.Russo, S.Calogero, G.Valle
Cryst.Struct.Commun.,**9**, 829, 1980

11.C 2 – Pyridyl – phenylsulfone
$C_{11}H_9NO_2S$ Main entry is 33.39

11.C   cis – 2 – Ethoxy – 1 – phenylsulfinyl –
cyclopropane
$C_{11}H_{14}O_2S$ Main entry is 20.4

11.12   **Perfluoro – selenanthrene**
$C_{12}F_8Se_2$
D.Rainville, R.A.Zingaro, E.A.Meyers
Cryst.Struct.Commun.,**9**, 291, 1980

11.C   bis(2,5,6 – Trifluoro – 4 – nitrophenyl) – sulfide
$C_{12}H_2F_6N_2O_4S$ Main entry is 15.2

11.C   E – 1 – (5 – Nitro – 2 – furyl) – 2 – p –
chlorophenylsulfonyl – ethylene
$C_{12}H_8ClNO_5S$ Main entry is 38.36

11.13   **(2R) – 2 – (Benzyloxycarbonylamino) – 3 –
hydroxypropyl – (chloromethyl) – (R) – sulfoxide
(absolute configuration)**
$C_{12}H_{16}ClNO_4S$
H.M.Doesburg, J.H.Noordik
Acta Crystallogr.,Sect.B,**35**, 2815, 1979

11.14   **bis(Trimethylacetyl – dichloromethylene) –
trisulfane**
$C_{12}H_{18}Cl_4O_2S_3$
G.Adiwidjaja, H.Gunther, J.Voss
Angew.Chem.,Int.Ed.Engl.,**19**, 563, 1980

11.C   bis(4 – (N – Methyl – piperidinium)) – disulfide
tetrachloro – copper(ii)
$C_{12}H_{28}N_2S_2^{2+}$, Cl_4Cu^{2-} Main entry is 33.50

11.C   Hexacarbonyl – (μ – (1,2 – eta:2 – η) – 1 – cyclo –
octene – 1 – selenolato) – di – iron
$C_{14}H_{12}Fe_2O_6Se$ Main entry is 75.7

11.C   2 – Chloromethyl – 5 – (2′,4′ – dinitrophenylthio) –
nortricyclene
$C_{14}H_{13}ClN_2O_4S$ Main entry is 31.22

11.15   **Ethane – 1,2 – bis(phenylsulfone)**
$C_{14}H_{14}O_4S_2$
F.Mo, O.Berg, G.Thorkildsen, A.Gaasdal
Eur.Cryst.Meeting,**5**, 352, 19

11.C   (2 – Hydroxy – 2 – phenyl – 2 – (2′ – pyridyl) –
ethyl) – methylsulfoxide
$C_{14}H_{15}NO_2S$ Main entry is 33.61

11.C   2 – (1 – Phenylthio) – ethyl – pent – 3 – ene – 1 –
carboxylic acid
$C_{14}H_{16}O_2S$ Main entry is 20.7

11.C   6 – Methoxy – 4 – oxo – 3 – phenylseleno –
cyclohexane – 1,2 – dicarboxylic anhydride
$C_{15}H_{14}O_5Se$ Main entry is 38.70

11.C   3 – (2,2,2 – Trichloro – 1 – (p – toluenesulfonyloxy)
ethyl) – cyclohexene
$C_{15}H_{17}Cl_3O_3S$ Main entry is 21.8

11.C   Ammonium 1 – anilino – 8 – naphthalene –
sulfonate monohydrate
$C_{16}H_{12}NO_3S^-$, H_4N^+, H_2O Main entry is 27.16

11.C   Dimethyl – 2,2′ – thio – dibenzoate
$C_{16}H_{14}O_4S$ Main entry is 13.13

11.C   Dimethyl – 2,2′ – dithio – dibenzoate
$C_{16}H_{14}O_4S_2$ Main entry is 13.14

11.16   **anti,S – trans – 3 – Methylthio – 1,5 – di – o –
tolyl – formazan (yellow isomer)**
$C_{16}H_{18}N_4S$
A.T.Hutton, H.M.N.H.Irving, L.R.Nassimbeni
Acta Crystallogr.,Sect.B,**36**, 2071, 1980
See also R1 : 9

11.17   **meso – 2,2 – bis(Methylsulfinyl) – 1,3 –
diphenylpropane**
$C_{17}H_{20}O_2S_2$
M.Poje, M.Sikirica, I.Vickovic, M.Bruvo
Tetrahedron Lett.,**21**, 3089, 1980

11.C   3 – (2,2,2 – Trichloro – 1 – (p – toluenesulfonyloxy)
ethyl) – cyclo – octene
$C_{17}H_{21}Cl_3O_3S$ Main entry is 22.6

11.C   N – (1 – Methyl – 4 – (3 – methylphenylamino) –
3 – pyridinio) – sulfonyl – N′ – isopropylurea
$C_{17}H_{22}N_4O_3S$ Main entry is 33.73

11.18   **Triphenylselenonium chloride monohydrate**
$C_{18}H_{15}Se^+$, Cl^-, H_2O
R.V.Mitcham, B.Lee, K.B.Mertes, R.F.Ziolo
Inorg.Chem.,**18**, 3498, 1979

11.C   (±) – 3 – (bis(Phenylsulfinyl)methyl) – 1,2 –
dimethyl – cyclopropene
$C_{18}H_{18}O_2S_2$ Main entry is 20.9

11.C   Tetrabromo – p – phenylene – bis(p –
toluenesulfonate)
$C_{20}H_{14}Br_4O_6S_2$ Main entry is 17.33

11.C   Di – t – adamantyl – disulfide
$C_{20}H_{30}S_2$ Main entry is 31.54

11.19   **Octa – 3,5 – diynylene – bis(p – toluenesulfonate)**
$C_{22}H_{22}O_6S_2$
R.L.Williams, D.J.Ando, D.Bloor, M.B.Hursthouse,
M.Motevalli *Acta Crystallogr.,Sect.B*,**36**, 2155, 1980

11.C   Thiocolchicoside ethanol solvate hydrate
$C_{27}H_{33}NO_{10}S$, $2C_2H_6O$, H_2O Main entry is 58.70

11.20   **bis(Triphenylmethyl) – sulfide**
$C_{38}H_{30}S$
G.A.Jeffrey, A.Robbins
Acta Crystallogr.,Sect.B,**36**, 1820, 1980

11.C   hexakis(Benzylthiomethyl)benzene 1,4 – dioxane
clathrate (monoclinic form)
$C_{54}H_{54}S_6$, $C_4H_8O_2$ Main entry is 61.23

11.C   hexakis(R – α – Phenylethylsulfonyl – methyl)
benzene acetic acid clathrate
$C_{60}H_{66}O_{12}S_6$, $4C_2H_4O_2$ Main entry is 61.24

11.C   hexakis(2 – Phenylethylthiomethyl)benzene 1,4 –
dioxane clathrate
$C_{60}H_{66}S_6$, $C_4H_8O_2$ Main entry is 61.25

11.C   hexakis(p – t – Butylphenylthiomethyl)benzene
squalene clathrate
$C_{72}H_{90}S_6$, $0.5C_{30}H_{50}$ Main entry is 61.26

CARBONIUM IONS, CARBANIONS AND RADICALS

BENZOIC ACID DERIVATIVES

12.C **Cyclopentadienyl sodium tetramethylethylenediamine**
$C_5H_5^-$, $C_6H_{16}N_2$, Na^+ Main entry is 20.1

12.C **Trimethylammonium 7,7,8,8 − tetracyanoquinodimethane tri − iodide (neutron study, at 80°K)**
$C_{12}H_4N_4$, $C_3H_{10}N^+$, $0.33I_3^{3-}$ Main entry is 3.11

12.C **Tetraethylammonium bis(7,7,8,8 − tetracyanoquinodimethane)**
$C_{12}H_4N_4^-$, $C_8H_{20}N^+$, $C_{12}H_4N_4$ Main entry is 60.7

12.C **Ethyltriphenylphosphonium − bis(7,7,8,8 − tetracyanoquinodimethane)**
$C_{12}H_4N_4^-$, $C_{20}H_{20}P^+$, $C_{12}H_4N_4$ Main entry is 60.41

12.C **Decamethylferrocenium − 7,7,8,8 − tetracyanoquinodimethane**
$C_{12}H_4N_4^-$, $C_{20}H_{30}Fe^+$ Main entry is 60.42

12.C **o − Tropylbiphenyl tetrafluoroborate**
$C_{19}H_{15}^+$, BF_4^- Main entry is 22.7

13.1 **p − Nitrobenzoic acid (at −150°C, form ii)**
$C_7H_5NO_4$
P.Groth *Acta Chem.Scand.Ser.A*,**34**, 229, 1980
See also R1 : 15

13.2 **p − Chlorobenzamide (β form)**
C_7H_6ClNO
T.Hayashi, K.Nakata, Y.Takaki, K.Sakurai
Bull.Chem.Soc.Jpn.,**53**, 801, 1980

13.3 **Benzoic acid**
$C_7H_6O_2$
G.Bruno, L.Randaccio
Acta Crystallogr.,Sect.B,**36**, 1711, 1980

13.4 **p − Aminobenzoic acid hydrochloride**
$C_7H_8NO_2^+$, Cl^-
M.Colapietro, A.Domenicano, G.Portalone
Acta Crystallogr.,Sect.B,**36**, 354, 1980
See also R1 : 16

13.5 **N − Acetyl − anthranilic acid**
o − Acetamidobenzoic acid
$C_9H_9NO_3$
Y.P.Mascarenhas, V.N.de Almeida, J.R.Lechat, N.Barelli
Acta Crystallogr.,Sect.B,**36**, 502, 1980
See also R1 : 16

13.6 **Methyl − 2 − (methylsulfinyl) − benzoate**
$C_9H_{10}O_3S$
A.Kalman, L.Parkanyi, I.Kapovits, A.Kucsman
Eur.Cryst.Meeting,**6**, 10, 1980
See also R1 : 11

13.7 **3,5 − Di − iodo − 4 − (4 − hydroxy − 3,5 − di − iodophenoxy) − benzoic acid bis(N − diethanolamine)**
3,3′,5,5′ − Tetra − iodothyroformic acid bis(N − diethanolamine)
$C_{13}H_4I_4O_4^{2-}$, $2C_4H_{12}NO_2^+$
V.Cody, P.D.Strong
Acta Crystallogr.,Sect.B,**36**, 1723, 1980
See also R1 : 17 R2 : 3

13.8 **4 − Amino − N(2 − (diethylamino) − ethyl) − benzamide hydrochloride**
Procainamide hydrochloride
$C_{13}H_{22}N_3O^+$, Cl^-
O.M.Peeters, N.M.Blaton, C.J.De Ranter, O.Denisoff, L.Molle
Cryst.Struct.Commun.,**9**, 851, 1980
See also R1 : 16

13.9 **2 − Iodo − 3′ − bromo − dibenzoyl − peroxide (β form)**
$C_{14}H_8BrIO_4$
J.Z.Gougoutas, K.H.Chang
Cryst.Struct.Commun.,**8**, 977, 1979

13.10 **(4′ – Carbomethoxy – 2′ – nitrophenoxy) – benzene**
$C_{14}H_{11}NO_5$
R.Gopal, W.D.Chandler, B.E.Robertson
Can.J.Chem.,**58**, 658, 1980
See also R1 : 15

13.C **Cycloheptyl – p – bromobenzoate**
$C_{14}H_{17}BrO_2$ Main entry is 22.4

13.11 **4 – Amino – 5 – chloro – N – ((2 – diethylamino) –
ethyl) – 2 – methoxybenzamide hydrochloride
monohydrate**
Primperan
$C_{14}H_{23}ClN_3O_2^+$, Cl^-, H_2O
N.M.Blaton, O.M.Peeters, C.J.De Ranter, O.Denisoff, L.Molle
Cryst.Struct.Commun.,**9**, 857, 1980

13.12 **4 – Dimethylamino – azobenzene – 2′ – carboxylic
acid**
Methyl red
$C_{15}H_{15}N_3O_2$
D.Moreiras, J.Solans, X.Solans, C.Miravitlles, G.Germain,
J.P.Declercq *Cryst.Struct.Commun.*,**9**, 921, 1980
See also R1 : 16,9

13.13 **Dimethyl – 2,2′ – thio – dibenzoate**
$C_{16}H_{14}O_4S$
A.Kalman, L.Parkanyi, I.Kapovits, A.Kucsman
Eur.Cryst.Meeting,**6**, 10, 1980
See also R1 : 11

13.14 **Dimethyl – 2,2′ – dithio – dibenzoate**
$C_{16}H_{14}O_4S_2$
A.Kalman, L.Parkanyi, I.Kapovits, A.Kucsman
Eur.Cryst.Meeting,**6**, 10, 1980
See also R1 : 11

13.15 **N,N′ – Ethylene – dibenzamide**
$C_{16}H_{16}N_2O_2$
A.Palmer, F.Brisse
Acta Crystallogr.,Sect.B,**36**, 1447, 1980

13.16 **O – Benzoyl – N,N – dimethyl – N′ – (N – methyl –
2,4 – dinitroanilino) – isourea**
$C_{17}H_{17}N_5O_6$
A.F.Hegarty, M.T.McCormack, K.Brady, G.Ferguson,
P.J.Roberts *J.Chem.Soc.,Perkin Trans.2*,867, 1980
See also R1 : 15,8

13.17 **2 – (4′ – Carbomethoxy – 2′ – aminophenoxy) –
1,3,5 – trimethylbenzene**
$C_{17}H_{19}NO_3$
R.Gopal, W.D.Chandler, B.E.Robertson
Can.J.Chem.,**57**, 2767, 1979
See also R1 : 17,16

13.18 **3′ – Chloro – 4′ – (p – chlorophenoxy) – 3,5 – di –
iodo – salicylanilide**
Rafoxamide
$C_{19}H_{11}Cl_2I_2NO_3$
A.C.Sindt, M.F.Mackay *Aust.J.Chem.*,**33**, 203, 1980
See also R1 : 17

13.C **1endo,4endo:5exo,8exo – Dimethano –
1,2,3,4,4a,5,8,8a – octahydro – naphthalene – 10 –
syn – p – nitrobenzoate**
$C_{19}H_{19}NO_4$ Main entry is 31.50

13.19 **Benzaldehyde 4 – benzoyl – 2 –
phenylthiosemicarbazone**
$C_{21}H_{17}N_3OS$
Y.Fukutani, K.Tsukihara, Y.Okuda, K.Fukuyama,
Y.Katsube, I.Yamamoto, H.Gotoh
Bull.Chem.Soc.Jpn.,**52**, 2223, 1979
See also R1 : 8,9

13.C **N – (2 – (4 – (5 – Chloro – 2 – oxo – 1 –
benzimidazolinyl)piperidino) – ethyl) – p –
fluorobenzamide**
Halopemide
$C_{21}H_{22}ClFN_4O_2$ Main entry is 35.64

13.20 **4′ – Nitrophenyl – 4 – n – octyloxybenzoate**
$C_{21}H_{25}NO_5$
J.Kaiser, R.Richter, G.Lemke, L.Golic
Acta Crystallogr.,Sect.B,**36**, 193, 1980
See also R1 : 17,15

13.21 **(RR,SS) – 5 – Phenyl – 2,2,6,6 – tetramethyl – 3 –
heptanol – 3,5 – dinitrobenzoate**
$C_{24}H_{30}N_2O_6$
A.de A.G.de Barreda, J.L.B.Pinal, M.Martinez–Ripoll,
S.Garcia–Blanco *Eur.Cryst.Meeting*,**6**, 24, 1980
See also R1 : 15

13.22 **1,2 – bis((Benzoyl)oxy) – stilbene**
$C_{28}H_{20}O_4$
T.S.Cantrell, J.V.Silverton *J.Org.Chem.*,**44**, 4477, 1979

13.C **1 – (2 – Butoxycarbonyl – 1 – azophenyl) – 2 –
hydroxy – 3 – N – (2 – oxo – 5 – benzimidazolyl) –
(naphthoic acid) – amide**
C.I.Pigment red 208
$C_{29}H_{25}N_5O_5$ Main entry is 35.74

13.23 **2,4,6 – Tribenzoyloxy – propiophenone**
$C_{30}H_{22}O_7$
K.Sasvari, J.P.Declercq, G.Germain
Cryst.Struct.Commun.,**9**, 281, 1980

13.24 **2,4,6 – Tribenzoyloxy – propiophenone benzene**
$C_{30}H_{22}O_7$, C_6H_6
K.Sasvari, L.Parkanyi
Cryst.Struct.Commun.,**9**, 277, 1980

BENZOIC ACID SALTS (AMMONIUM,IA,IIA METALS)

14.1   **Calcium salicylate dihydrate**
$2C_7H_5O_3^-$, Ca^{2+}, $2H_2O$
M.P.Gupta, A.P.Saha *Indian J.Phys.*,**53**, 460, 1979

14.2   **Calcium salicylate dihydrate**
$2C_7H_5O_3^-$, Ca^{2+}, $2H_2O$
R.Debuyst, F.Dejehet, M.–C.Dekandelaer, J.P.Declercq,
M.van Meerssche
J.Chim.Phys.Phys.–Chim.Biol.,**76**, 1117, 1979

14.3   **Strontium salicylate dihydrate**
$2C_7H_5O_3^-$, Sr^{2+}, $2H_2O$
R.Debuyst, F.Dejehet, M.–C.Dekandelaer, J.P.Declercq,
M.van Meerssche
J.Chim.Phys.Phys.–Chim.Biol.,**76**, 1117, 1979

14.4   **Potassium phthalate monohydrate**
$C_8H_4O_4^{2-}$, $2K^+$, H_2O
J.Z.Gougoutas, W.H.Ojala, J.A.Miller
Cryst.Struct.Commun.,**9**, 519, 1980

14.5   **Di – calcium dihydrogen mellitate nonahydrate**
$C_{12}H_2O_{12}^{4-}$, $2Ca^{2+}$, $9H_2O$
V.A.Uchtman, R.J.Jandacek *Inorg.Chem.*,**19**, 350, 1980

BENZENE NITRO COMPOUNDS

15.C   **Valinomycin potassium picrate**
$C_6H_2N_3O_7^-$, $C_{36}H_{60}N_4O_{12}$, K^+ Main entry is 60.43

15.C   **Calcium bis(picrate) tetraethylene – glycol monohydrate**
$2C_6H_2N_3O_7^-$, $C_8H_{18}O_5$, Ca^{2+}, H_2O Main entry is 17.1

15.C   **S – Methyl – thiepanium 2,4,6 – trinitrobenzenesulfonate**
$C_6H_2N_3O_9S^-$, $C_7H_{15}S^+$ Main entry is 39.16

15.C   **4 – Nitroaniline 4 – nitro – N – sulfinylaniline**
$C_6H_4N_2O_3S$, $C_6H_6N_2O_2$ Main entry is 16.1

15.1   **p – Dinitrobenzene**
$C_6H_4N_2O_4$
F.di Rienzo, A.Domenicano, L.Riva di Sanseverino
Acta Crystallogr.,Sect.B,**36**, 586, 19

15.C   **4 – Nitroaniline 4 – nitro – N – sulfinylaniline**
$C_6H_6N_2O_2$, $C_6H_4N_2O_3S$ Main entry is 16.1

15.C   **2 – Amino – 5 – nitrophenol**
$C_6H_6N_2O_3$ Main entry is 17.4

15.C   **p – Nitrobenzoic acid (at –150°C, form ii)**
$C_7H_5NO_4$ Main entry is 13.1

15.C   **p – Nitrophenoxyacetic acid**
$C_8H_7NO_5$ Main entry is 17.8

15.C   **Cyclohepta – amylose p – nitroacetanilide hydrate**
$C_8H_8N_2O_3$, $C_{42}H_{70}O_{35}$, xH_2O Main entry is 45.66

15.2   **bis(2,5,6 – Trifluoro – 4 – nitrophenyl) – sulfide**
$C_{12}H_2F_6N_2O_4S$
N.Goodhand, T.A.Hamor *J.Fluorine Chem.*,**14**, 223, 1979
See also R1 : 11

15.3   **Dipicrylaminato – silver**
$C_{12}H_4N_7O_{12}^-$, Ag^+
V.P.Chalii, F.G.Kramarenko, Yu.P.Krasan, L.L.Shevchenko
Ukr.Khim.Zh.(Russ.Ed.),**46**, 227, 1980

15.C   **Potassium dipicrylamine**
$C_{12}H_4N_7O_{12}^-$, K^+ Main entry is 16.3

15.C   **(3 – Chloro – 2 – hydroxy – 5 – nitrophenyl) – (2′ – chlorophenyl) – iodonium hydroxide inner salt**
$C_{12}H_6Cl_2INO_3$ Main entry is 17.18

15.C   **4,4′ – Dihydroxy – 3,3′,5,5′ – tetranitrobiphenyl**
$C_{12}H_6N_4O_{10}$ Main entry is 17.20

15.C   **1 – (2,4 – Dinitrophenyl) – imidazo(4,5 – b)pyridine**
$C_{12}H_7N_5O_4$ Main entry is 35.24

15.C   **4 – (N′ – (p – Nitrobenzene) – N – hydrazo) – 8 – oxatricyclo(5.1.0.0^{2,5})octane**
$C_{13}H_{13}N_3O_3$ Main entry is 38.52

15.C   **2,2′ – (2,2,2 – Trichloroethylidene) – bis(4 – chloro – 6 – nitrophenol)**
$C_{14}H_7Cl_5N_2O_6$ Main entry is 50.6

15.C **2,2′ – Ethylidene – bis(4 – chloro – 6 – nitrophenol)**
$C_{14}H_{10}Cl_2N_2O_6$ Main entry is 17.22

15.C **(4′ – Carbomethoxy – 2′ – nitrophenoxy) – benzene**
$C_{14}H_{11}NO_5$ Main entry is 13.10

15.C **2 – Chloromethyl – 5 – (2′,4′ – dinitrophenylthio) – nortricyclene**
$C_{14}H_{13}ClN_2O_4S$ Main entry is 31.22

15.4 **Methyl – (2E,4Z) – 5 – dimethylamino – 2 – nitro – 4 – (4 – chloro – 2 – nitrophenyl) – 2,4 – pentadienoate**
$C_{14}H_{14}ClN_3O_6$
U.Hengartner, A.D.Batcho, J.F.Blount, W.Leimgruber,
M.E.Larscheid, J.W.Scott *J.Org.Chem.*,**44**, 3748, 1979
See also R1 : 1,3,10

15.C **2,2′ – Isopropylidene – bis(4 – chloro – 6 – nitrophenol)**
$C_{15}H_{12}Cl_2N_2O_6$ Main entry is 17.23

15.5 **p – Dimethylamino – benzaldehyde – p – nitrophenylhydrazone benzene solvate**
$C_{15}H_{16}N_4O_2$, $0.25C_6H_6$
J.N.Brown, T.M.Kutchan, P.E.Rist
Cryst.Struct.Commun.,**9**, 17, 1980

15.C **3,5,7 – Trimethyl – tropylium – 0,0 – (2′,4′,6′ – trinitrophenylide)**
$C_{16}H_{13}N_3O_6$ Main entry is 38.74

15.C **0 – (2′,6′ – Dinitrophenyl) – 3,5,7 – trimethyl – tropolone**
$C_{16}H_{14}N_2O_6$ Main entry is 22.5

15.C **0 – Benzoyl – N,N – dimethyl – N′ – (N – methyl – 2,4 – dinitroanilino) – isourea**
$C_{17}H_{17}N_5O_6$ Main entry is 13.16

15.C **Diethyl cis – 1 – formyl – 5 – hydroxy – 4 – (2 – nitrophenyl) – pyrrolidine – 2,2 – dicarboxylate**
$C_{17}H_{20}N_2O_9$ Main entry is 32.63

15.C **1endo,4endo:5exo,8exo – Dimethano – 1,2,3,4,4a,5,8,8a – octahydro – naphthalene – 10 – syn – p – nitrobenzoate**
$C_{19}H_{19}NO_4$ Main entry is 31.50

15.C **1,11 – bis(2 – Nitrophenoxy) – 3,6,9 – trioxa – undecane potassium isothiocyanate**
$C_{20}H_{24}N_2O_9$, K^+, CNS^- Main entry is 17.35

15.C **4′ – Nitrophenyl – 4 – n – octyloxybenzoate**
$C_{21}H_{25}NO_5$ Main entry is 13.20

15.C **(RR,SS) – 5 – Phenyl – 2,2,6,6 – tetramethyl – 3 – heptanol – 3,5 – dinitrobenzoate**
$C_{24}H_{30}N_2O_6$ Main entry is 13.21

15.6 **3,5 – Dimethyl – 1,7 – diphenyl – 4 – (2,4,6 – trinitrophenyl) – 2,6 – diazahepta – 2,4 – diene**
$C_{25}H_{23}N_5O_6$
N.G.Furmanova, O.E.Kompan, Yu.T.Struchkov,
I.E.Mikhailov, L.P.Olekhnovich, V.I.Minkin
Zh.Strukt.Khim.,**21**, 98, 1980
See also R1 : 3

15.C **2,2 – bis(4 – (p – Nitrophenoxy)phenyl)propane**
$C_{27}H_{22}N_2O_6$ Main entry is 17.42

15.C **1,5 – bis(2 – (5 – (2 – Nitrophenoxy) – 3 – oxapentyloxy)phenoxy) – 3 – oxapentane potassium thiocyanate**
$C_{36}H_{40}N_2O_{13}$, $2K^+$, $2CNS^-$ Main entry is 17.43

ANILINES

16.1 **4 – Nitroaniline 4 – nitro – N – sulfinylaniline**
$C_6H_6N_2O_2$, $C_6H_4N_2O_3S$
B.Dederer, A.Gieren *Naturwissenschaften*,**66**, 470, 1979
See also R1 : 15 R2 : 15,4

16.C **2 – Amino – 5 – nitrophenol**
$C_6H_6N_2O_3$ Main entry is 17.4

16.C **p – Aminobenzoic acid hydrochloride**
$C_7H_8NO_2^+$, Cl^- Main entry is 13.4

16.C **Cyclohepta – amylose p – nitroacetanilide hydrate**
$C_8H_8N_2O_3$, $C_{42}H_{70}O_{35}$, xH_2O Main entry is 45.66

16.C **N – Acetyl – anthranilic acid**
o – Acetamidobenzoic acid
$C_9H_9NO_3$ Main entry is 13.5

16.C **2 – (2,6 – Dichlorophenylimino) – 2 – imidazoline – hydrochloride**
Clonidine hydrochloride
$C_9H_{10}Cl_2N_3^+$, Cl^- Main entry is 32.28

16.2 **β – Thiocyanato – propionanilide**
$C_{10}H_{10}N_2OS$
B.Tashkhodzhaev, S.Akhmedova, N.K.Rozhkova
Zh.Strukt.Khim.,**21**, 137, 1980

16.C **bis(Phenacetin) tri – iodide iodine**
$C_{10}H_{14}NO_2^+$, $C_{10}H_{13}NO_2$, I_3^-, I_2 Main entry is 17.14

16.C **N,N,N′,N′ – Tetramethyl – p – phenylenediamine hexafluorobenzene complex**
$C_{10}H_{16}N_2$, C_6F_6 Main entry is 60.13

16.3 **Potassium dipicrylamine**
$C_{12}H_4N_7O_{12}^-$, K^+
M.L.Kundu, S.K.Ghosh
Acta Crystallogr.,Sect.B,**36**, 941, 1980
See also R1 : 15

16.4 **Diphenylamine trichloro – antimony**
$C_{12}H_{11}N$. Cl_3Sb
A.Lipka *Z.Anorg.Allg.Chem.*,**466**, 195, 1980

16.5 **2,4 – Diamino – azobenzene hydrochloride dihydrate**
$C_{12}H_{13}N_4^+$, Cl^-, $2H_2O$
D.Moreiras, J.Solans, G.Germain, J.P.Declercq
Eur.Cryst.Meeting.,**6**, 25, 1980
See also R1 : 9

16.6 **N – (p – Bromobenzylidene) – p – chloroaniline**
$C_{13}H_9BrClN$
J.Bernstein, I.Bar *Eur.Cryst.Meeting.*,**6**, 21, 1980

16.7 **N – (p – Chlorobenzylidene) – p – bromoaniline**
$C_{13}H_9BrClN$
J.Bernstein. I.Bar *Eur.Cryst.Meeting.*,**6**, 21, 1980

16.C **4 – m – Aminophenyl – 2 – formylpyridine – thiosemicarbazone**
$C_{13}H_{13}N_5S$ Main entry is 33.53

16.C **Piperidinium – acetyl – (m – chloroanilide) chloride**
$C_{13}H_{18}ClN_2O^+$, Cl^- Main entry is 33.54

16.C **Piperidinium – acetyl – (p – chloroanilide) chloride**
$C_{13}H_{18}ClN_2O^+$, Cl^- Main entry is 33.55

16.C **o – Chloro – N – (piperid – 1 – yl – acetyl) – aniline hydrochloride**
$C_{13}H_{18}ClN_2O^+$, Cl^- Main entry is 33.56

16.C **o – Chloro – N – (piperid – 1 – yl – acetyl) – aniline hydrochloride monohydrate**
$C_{13}H_{18}ClN_2O^+$, Cl^-, H_2O Main entry is 33.57

16.C **Palythene monohydrate (absolute configuration)**
$C_{13}H_{20}N_2O_5$, H_2O Main entry is 59.4

16.C **4 – Amino – N(2 – (diethylamino) – ethyl) – benzamide hydrochloride**
Procainamide hydrochloride
$C_{13}H_{22}N_3O^+$, Cl^- Main entry is 13.8

16.8 **N – (p – Bromobenzylidene) – p – methylaniline**
$C_{14}H_{12}BrN$
J.Bernstein, I.Bar *Eur.Cryst.Meeting*,**6**, 21, 1980

16.9 **N – (p – Methylbenzylidene) – p – bromoaniline**
$C_{14}H_{12}BrN$
J.Bernstein, I.Bar *Eur.Cryst.Meeting*,**6**, 21, 1980

16.10 **N – (p – Chlorobenzylidene) – p – methylaniline**
$C_{14}H_{12}ClN$
J.Bernstein, I.Bar *Eur.Cryst.Meeting*,**6**, 21, 1980

16.11 **N – (p – Methylbenzylidene) – p – chloroaniline**
$C_{14}H_{12}ClN$
J.Bernstein, I.Bar *Eur.Cryst.Meeting*,**6**, 21, 1980

16.12 **p – Tolyl – glyoxylic acid p – chloroanilide**
$C_{15}H_{12}ClNO_2$
E.Hohne, I.Seidel *Krist.Tech.*,**14**, 1097, 1979
See also R1 : 1

16.C **4 – Dimethylamino – azobenzene – 2′ – carboxylic acid**
Methyl red
$C_{15}H_{15}N_3O_2$ Main entry is 13.12

16.C **1 – ((4 – (3 – Chlorophenylamino) – 3 – pyridyl) sulfonyl) – 3 – ethyl – 1 – methylurea hydrogen sulfate**
$C_{15}H_{18}ClN_4O_3S^+$, HO_4S^- Main entry is 33.66

16.13 **p – Methoxy – N,N – bis(α – cyano – hexafluoro – isobutenyl) – aniline**
$C_{17}H_7F_{12}N_3O$
L.A.Simonyan, Z.V.Safronova, N.P.Gambaryan,
M.Yu.Antipin, Yu.T.Struchkov
Izv.Akad.Nauk SSSR,Ser.Khim.,358, 1980

16.14 **Dimethyl – 4,4 – methylene – bis(phenylcarbamate) (at 258°K)**
$C_{17}H_{18}N_2O_4$
K.H.Gardner, J.Blackwell
Acta Crystallogr.,Sect.B,**36**, 1972, 1980

16.C **2 – (4′ – Carbomethoxy – 2′ – aminophenoxy) – 1,3,5 – trimethylbenzene**
$C_{17}H_{19}NO_3$ Main entry is 13.17

16.C **D,L – Diethylanilino – (3 – hydroxybenzyl) – phosphonate**
$C_{17}H_{22}NO_4P$ Main entry is 64.82

16.C N – (1 – Methyl – 4 – (3 – methylphenylamino) – 3 – pyridinio) – sulfonyl – N′ – isopropylurea
$C_{17}H_{22}N_4O_3S$ Main entry is 33.73

16.15 tris(2,3,4,5,6 – Pentachlorophenyl)amine
Perchloro – triphenylamine
$C_{18}Cl_{15}N$
K.S.Hayes, M.Nagumo, J.F.Blount, K.Mislow
J.Am.Chem.Soc.,**102**, 2773, 1980

16.16 Triphenylamine bis(trichloro – antimony)
$C_{18}H_{15}N$, $2Cl_3Sb$
L.Korte, A.Lipka, D.Mootz *Eur.Cryst.Meeting*,**5**, 92, 1979

16.C 2 – (p – Dimethylaminobenzylidene) – indan – 1,3 – dione (γ form)
$C_{18}H_{15}NO_2$ Main entry is 27.23

16.C 4 – Methyl – 6 – (N – methyl – N – phenylamino) – spiro(4.5)dec – 6 – ene – 1,8 – dione
$C_{18}H_{21}NO_2$ Main entry is 27.24

16.C (2S – (2β,3β,3aα,5β,6α,6aα)) – Hexahydro – 5 – methoxy – 2 – (dimethoxymethyl) – 6 – methylfuro(3,2 – b)furan – 3 – ol – 4 – bromophenyl – carbamic acid ester
$C_{18}H_{24}BrNO_7$ Main entry is 38.94

16.17 2,4,6 – Tri – t – butyl – N – sulfinylaniline
$C_{18}H_{29}NOS$
F.Iwasaki *Acta Crystallogr.,Sect.B*,**36**, 1700, 1980

16.C p – Ethoxybenzylidene – p – n – butylaniline
$C_{19}H_{23}NO$ Main entry is 17.30

16.C 4 – Ethoxy – salicylidene – 4′ – butylaniline
$C_{19}H_{23}NO_2$ Main entry is 17.31

16.C 2 – (4′ – Dimethylamino – cinnamoyl) – indan – 1,3 – dione
$C_{20}H_{17}NO_3$ Main entry is 27.26

16.C 4 – Propoxy – salicylidene – 4′ – butylaniline
$C_{20}H_{25}NO_2$ Main entry is 17.36

16.C 4 – Anilino – 3 – methoxycarbonyl – 1 – (methoxycarbonylmethylidene) – 1,2,3,4 – tetrahydro – naphthalene
$C_{21}H_{21}NO_4$ Main entry is 27.29

16.C (±) – (3R,1′R) – 1 – Hydroxymethyl – 2 – (1 – methylpentyl) – 3 – methylcyclohex – 1 – ene p – bromophenylurethane
$C_{21}H_{30}BrNO_2$ Main entry is 21.17

16.C N,N′ – bis(2 – (4 – Chlorophenylamino) – 2 – cyclopenten – 1 – ylidene) – hydrazine (monoclinic form)
$C_{22}H_{20}Cl_2N_4$ Main entry is 20.12

16.C N,N′ – bis(2 – (4 – Chlorophenylamino) – 2 – cyclopenten – 1 – ylidene) – hydrazine (orthorhombic form)
$C_{22}H_{20}Cl_2N_4$ Main entry is 20.13

16.C N – (2′ – Hydroxy – 2′ – phenylethyl) – 4 – (N′ – phenylpropionamido) – piperidine
$C_{22}H_{28}N_2O_2$ Main entry is 33.83

16.C N – (1 – (2 – Phenylethyl) – 4 – piperidinylium) – N – phenyl – propanamide – citrate toluene solvate
Fentanyl citrate toluene solvate
$C_{22}H_{29}N_2O^+$. $C_6H_7O_7^-$, C_7H_8 Main entry is 33.84

16.C cis – (+) – N – (3 – Methyl – 1 – (2 – phenylethyl) – 4 – piperidinyl) – N – phenyl – propanamide nitric acid 2 – propanol solvate
$C_{23}H_{30}N_2O$, HNO_3, C_3H_8O Main entry is 33.86

16.C N – (2′ – Phenylethyl) – 4 – (N – phenylpropionamido) – 4 – carbomethoxy – piperidine
$C_{24}H_{30}N_2O_3$ Main entry is 33.89

16.C cis(–) – N – (3 – Methyl – 4 – (methoxycarbonyl) – 1 – (2 – phenylethyl) – 4 – piperidinyl) – N – phenyl – propanamide oxalic acid
$C_{25}H_{32}N_2O_3$, $C_2H_2O_4$ Main entry is 33.92

16.18 bis(Phenylcarbamoyl – oxy – n – butyl) – diacetylene (at 120°K)
$C_{26}H_{28}N_2O_4$
H.Gross, H.Sixl, C.Krohnke, V.Enkelmann
Chem.Phys.,**45**, 15, 1980

PHENOLS AND ETHERS

17.C   **Potassium p – chloranil acetone solvate**
$C_6Cl_4O_2^-$, K^+, C_3H_6O Main entry is 18.1

17.C   **Pyridinium 1 – naphthylamine picrate**
$C_6H_2N_3O_7^-$, $C_{10}H_9N$, $C_5H_6N^+$ Main entry is 60.1

17.C   **Valinomycin potassium picrate**
$C_6H_2N_3O_7^-$, $C_{36}H_{60}N_4O_{12}$, K^+ Main entry is 60.43

17.1   **Calcium bis(picrate) tetraethylene – glycol monohydrate**
$2C_6H_2N_3O_7^-$, $C_8H_{18}O_5$, Ca^{2+}, H_2O
T.P.Singh, R.Reinhardt, N.S.Poonia
Inorg.Nucl.Chem.Lett.,**16**, 293, 1980
See also R1 : 15

17.2   **2,4 – Dichlorophenol**
$C_6H_4Cl_2O$
C.Bavoux, M.Perrin *Cryst.Struct.Commun.*,**8**, 847, 1979

17.3   **3,4 – Dichlorophenol (stable form)**
$C_6H_4Cl_2O$
C.Bavoux, M.Perrin, A.Thozet
Acta Crystallogr.,Sect.B,**36**, 741, 1980

17.C   **3,5 – Dichlorophenol – 2,6 – dimethylphenol complex**
$C_6H_4Cl_2O$, $C_8H_{10}O$ Main entry is 60.2

17.4   **2 – Amino – 5 – nitrophenol**
$C_6H_6N_2O_3$
M.Haisa, S.Kashino, T.Kawashima
Acta Crystallogr.,Sect.B,**36**, 1598, 1980
See also R1 : 16,15

17.5   **Quinol (hexagonal α form)**
1,4 – Dihydroxybenzene
$C_6H_6O_2$
S.C.Wallwork, H.M.Powell
J.Chem.Soc.,Perkin Trans.2,641, 1980

17.6   **3,4,5 – Trichloro – guaiacol**
$C_7H_5Cl_3O_2$
K.Lindstrom, F.Osterberg *Can.J.Chem.*,**58**, 815, 1980

17.7   **2,3,5,6 – Tetrachloro – 1,4 – dimethoxybenzene**
$C_8H_6Cl_4O_2$
M.W.Wieczorek *Acta Crystallogr.,Sect.B*,**36**, 1515, 1980

17.8   **p – Nitrophenoxyacetic acid**
$C_8H_7NO_5$
S.V.Kumar, L.M.Rao
Acta Crystallogr.,Sect.B,**36**, 1218, 1980
See also R1 : 15,1

17.C   **3,5 – Dichlorophenol – 2,6 – dimethylphenol complex**
$C_8H_{10}O$, $C_6H_4Cl_2O$ Main entry is 60.2

17.C   **7,8 – Dimethyl – isoalloxazine – 10 – acetic acid tyramine tetrahydrate**
$C_8H_{11}NO$, $C_{14}H_{13}N_4O_4$, $4H_2O$ Main entry is 36.11

17.C   **Tyramine 1 – thyminyl – (acetic acid) complex hydrate**
$C_8H_{12}NO^+$, $C_7H_7N_2O_4^-$, H_2O Main entry is 60.6

17.C   **Dopamine hydrochloride**
$C_8H_{12}NO_2^+$, Cl^- Main entry is 59.1

17.9   **p – Bromophenoxyacetimine methyl ether**
$C_9H_{10}BrNO_2$
A.Kolakowski *Eur.Cryst.Meeting*,**5**, 119, 1979
See also R1 : 1

17.C   **2 – Hydroxy – 3,5 – dimethyl – 6 – hydroxymethyl – 1,4 – benzoquinone**
Shanorellin
$C_9H_{10}O_4$ Main entry is 18.3

17.10   **1 – (3,4 – Dihydroxyphenyl) – 2 – methylaminoethane – sulfonic acid hemihydrate**
$C_9H_{13}NO_5S$, $0.5H_2O$
A.R.Garafalo, E.Milano, D.A.Williams, T.F.Brennan
Cryst.Struct.Commun.,**8**, 823, 1979
See also R1 : 3,11

17.11   **(±) – N – Methyl – 2 – (3,4 – dihydroxyphenyl) ethylammonium – 2 – sulfonate hemihydrate**
(±) – Epinine – β – sulfonate hemihydrate
$C_9H_{13}NO_5S$, $0.5H_2O$
J.G.Henkel, J.B.Anderson, M.Rapposch, G.Hite
Acta Crystallogr.,Sect.B,**36**, 953, 1980
See also R1 : 3,11

17.12   **(±) – 2 – (4 – Chloro – 2 – methylphenoxy) – propionic acid**
Mecoprop
$C_{10}H_{11}ClO_3$
G.Smith, C.H.L.Kennard, A.H.White, P.G.Hodgson
Acta Crystallogr.,Sect.B,**36**, 992, 1980
See also R1 : 1

17.13   **(E) – 1 – (4′ – Hydroxy – 3′ – methoxyphenyl) – 2 – nitropropene**
$C_{10}H_{11}NO_4$
V.Zabel, W.H.Watson, B.K.Cassels, D.A.Langs
Cryst.Struct.Commun.,**9**, 461, 1980
See also R1 : 10

17.14   **bis(Phenacetin) tri – iodide iodine**
$C_{10}H_{14}NO_2^+$, $C_{10}H_{13}NO_2$, I_3^-, I_2
F.H.Herbstein, M.Kapon
Philos.Trans.R.Soc.London,Ser.A,**291**, 199, 1979
See also R1 : 16

17.15   **2 – Isopropyl – 5 – methylphenol**
Thymol
$C_{10}H_{14}O$
A.Thozet, M.Perrin
Acta Crystallogr.,Sect.B,**36**, 1444, 1980

17.16   **1,2,4,5 – Tetramethoxybenzene**
$C_{10}H_{14}O_4$
K.von Deuten, G.Klar
Cryst.Struct.Commun.,**8**, 1017, 1979

17.C   **2(1 – (2,6 – Dichlorophenoxy) – ethyl) – imidazoline (α–form)**
Lofexidine
$C_{11}H_{12}Cl_2N_2O$ Main entry is 32.39

17.C   **2(1 – (2,6 – Dichlorophenoxy) – ethyl) – imidazoline hydrochloride (α–form)**
$C_{11}H_{13}Cl_2N_2O^+$, Cl^- Main entry is 32.40

17.C 4 – (2′ – Hydroxythiobenzoyl) – morpholine
$C_{11}H_{13}NO_2S$ Main entry is 40.22

17.C 5 – (3,5 – Dimethoxyphenyl) – 1 – methyl – 1H –
1,2,4 – triazole
$C_{11}H_{13}N_3O_2$ Main entry is 32.41

17.17 Phenylcholine ether bromide
$C_{11}H_{18}NO^+$, Br⁻
R.Celikel, A.J.Geddes, B.Sheldrick, D.Akrigg
Cryst.Struct.Commun.,**9**, 111, 1980
See also R1 : 3

17.18 (3 – Chloro – 2 – hydroxy – 5 – nitrophenyl) – (2′ –
chlorophenyl) – iodonium hydroxide inner salt
$C_{12}H_6Cl_2INO_3$
S.W.Page, E.P.Mazzola, A.D.Mighell, V.L.Himes, C.R.Hubbard
J.Am.Chem.Soc.,**101**, 5858, 1979
See also R1 : 15,6

17.19 bis(3,4 – Dichlorophenyl) – ether
3,3′,4,4′ – Tetrachlorodiphenyl – ether
$C_{12}H_6Cl_4O$
P.Singh, J.D.McKinney
Acta Crystallogr.,Sect.B,**36**, 210, 1980

17.20 4,4′ – Dihydroxy – 3,3′,5,5′ – tetranitrobiphenyl
$C_{12}H_6N_4O_{10}$
Z.A.Starikova, T.M.Shchegoleva, V.K.Trunov,
O.B.Lantratova, I.E.Pokrovskaya
Zh.Strukt.Khim.,**20**, 514, 1979
See also R1 : 15

17.C N – Salicylidene – 3 – aminopyridine
$C_{12}H_{10}N_2O$ Main entry is 33.44

17.21 4,4′ – Dihydroxydiphenyl
$C_{12}H_{10}O_2$
T.J.Houghton, P.Kelly, K.Rogerson, S.C.Wallwork
Eur.Cryst.Meeting,**5**, 294, 1979

17.C 2 – (4 – Methoxybenzoyl) – pyrrole
$C_{12}H_{11}NO_2$ Main entry is 32.42

17.C α – (3 – Bromo – 4 – ethoxyphenyl)succinimide
$C_{12}H_{12}BrNO_3$ Main entry is 32.43

17.C 3,5 – Di – iodo – 4 – (4 – hydroxy – 3,5 – di –
iodophenoxy) – benzoic acid bis(N –
diethanolamine)
3,3′,5,5′ – Tetra – iodothyroformic acid bis(N –
diethanolamine)
$C_{13}H_4I_4O_4{}^{2-}$, $2C_4H_{12}NO_2{}^+$ Main entry is 13.7

17.C N – (5 – Methoxysalicylidene) – 3 – aminopyridine
$C_{13}H_{12}N_2O_2$ Main entry is 33.51

17.C Palythene monohydrate (absolute configuration)
$C_{13}H_{20}N_2O_5$, H_2O Main entry is 59.4

17.C 3 – (2 – Methoxyphenyl) – 1,1,2,2 – cyclopropane –
tetracarbonitrile
1.1.2.2. – Tetracyano – 3 – (2 – methoxyphenyl) –
cyclopropane
$C_{14}H_8N_4O$ Main entry is 20.6

17.22 2,2′ – Ethylidene – bis(4 – chloro – 6 – nitrophenol)
$C_{14}H_{10}Cl_2N_2O_6$
D.G.Hay, M.F.Mackay
Acta Crystallogr.,Sect.B,**35**, 2952, 1979
See also R1 : 15

17.23 2,2′ – Isopropylidene – bis(4 – chloro – 6 –
nitrophenol)
$C_{15}H_{12}Cl_2N_2O_6$
D.G.Hay, M.F.Mackay
Acta Crystallogr.,Sect.B,**35**, 2952, 1979
See also R1 : 15

17.24 (±)erythro – Methyl – 3 – t – butoxy – 2 – iodo –
3 – (p – methoxyphenyl) – propionate
$C_{15}H_{21}IO_4$
M.R.Caira, J.F.de Wet
Acta Crystallogr.,Sect.B,**36**, 1675, 1980

17.25 Perdeutero – 4,4′ – azoxyphenetole (form C1)
$C_{16}D_{18}N_2O_4$
L.O.Atovmyan, V.L.Broude, V.K.Dolganov, A.I.Kolesnikov,
S.P.Krilova, V.I.Ponomarev, E.F.Sheka
Fiz.Tverd.Tela(Leningrad),**21**, 427, 1979
See also R1 : 10,9

17.26 1,4 – bis(5 – Chloro – pent – 4 – ynyloxy)benzene
$C_{16}H_{16}Cl_2O_2$
J.C.J.Bart, I.W.Bassi, M.Calcaterra, P.Piccardi
Acta Crystallogr.,Sect.B,**36**, 1561, 1980

17.C 5,7,14,16 – Tetramethoxy – 1,2,3,10,11,12 –
hexathia(3.3)metacyclophane
$C_{16}H_{16}O_4S_6$ Main entry is 39.62

17.27 4,4′ – Azoxyphenetole (form C1)
$C_{16}H_{18}N_2O_4$
L.O.Atovmyan, V.L.Broude, V.K.Dolganov, A.I.Kolesnikov,
S.P.Krilova, V.I.Ponomarev, E.F.Sheka
Fiz.Tverd.Tela(Leningrad),**21**, 427, 1979
See also R1 : 10,9

17.C 2,2 – bis(p – Methoxyphenyl) – 3 – methyl – 2H –
azirine
$C_{17}H_{17}NO_2$ Main entry is 32.61

17.C 2 – (4′ – Carbomethoxy – 2′ – aminophenoxy) –
1,3,5 – trimethylbenzene
$C_{17}H_{19}NO_3$ Main entry is 13.17

17.C D,L – Diethylanilino – (3 – hydroxybenzyl) –
phosphonate
$C_{17}H_{22}NO_4P$ Main entry is 64.82

17.C Tetramethoxystilbene – 7,7,8,8 –
tetracyanoquinodimethane complex
$C_{18}H_{20}O_4$, $C_{12}H_4N_4$ Main entry is 60.38

17.C 3′ – Chloro – 4′ – (p – chlorophenoxy) – 3,5 – di –
iodo – salicylanilide
Rafoxamide
$C_{19}H_{11}Cl_2I_2NO_3$ Main entry is 13.

17.28 (4 – Hydroxyphenyl) – diphenylmethanol
$C_{19}H_{16}O_2$
T.W.Lewis, D.Y.Curtin, I.C.Paul
J.Am.Chem.Soc.,**101**, 5717, 1979

17.29 1 – (2 – Hydroxyphenyl) – diphenylmethanol
$C_{19}H_{16}O_2$
T.W.Lewis, E.N.Duesler, R.B.Kress, D.Y.Curtin, I.C.Paul
J.Am.Chem.Soc.,**102**, 4659, 1980

17.30 p – Ethoxybenzylidene – p – n – butylaniline
$C_{19}H_{23}NO$
J.A.K.Howard, A.J.Leadbetter, M.Sherwood
Mol.Cryst.Liq.Cryst.,**56**, 271, 1980
See also R1 : 16

17.31 **4 – Ethoxy – salicylidene – 4' – butylaniline**
$C_{19}H_{23}NO_2$
O.S.Filipenko, L.O.Atovmyan, B.L.Tarnopol'skii, Z.Sh.Safina
Zh.Strukt.Khim.,**20**, 80, 1979
See also R1 : 16

17.32 **1,1 – bis(p – Methoxyphenyl) – 2,2 – dimethylpropane**
$C_{19}H_{24}O_2$
G.Smith, C.H.L.Kennard, T.–B.Palm
Acta Crystallogr.,Sect.B,**36**, 1693, 1980

17.33 **Tetrabromo – p – phenylene – bis(p – toluenesulfonate)**
$C_{20}H_{14}Br_4O_6S_2$
M.W.Wieczorek *Acta Crystallogr.,Sect.B*,**36**, 1513, 1980
See also R1 : 11

17.34 **5,5' – Diethoxy – α,α' – dimethyl – α,α' – azino – di – o – cresol**
$C_{20}H_{24}N_2O_4$
J.Fayos, M.Martinez–Ripoll, M.C.Garcia–Mina, J.Gonzalez–Martinez, F.Arrese
Acta Crystallogr.,Sect.B,**36**, 1952, 1980
See also R1 : 9

17.35 **1,11 – bis(2 – Nitrophenoxy) – 3,6,9 – trioxa – undecane potassium isothiocyanate**
$C_{20}H_{24}N_2O_9$, K+, CNS-
I.–H.Suh, G.Weber, W.Saenger
Acta Crystallogr.,Sect.B,**36**, 946, 1980
See also R1 : 15

17.36 **4 – Propoxy – salicylidene – 4' – butylaniline**
$C_{20}H_{25}NO_2$
O.S.Filipenko, L.O.Atovmyan, B.L.Tarnopol'skii, Z.Sh.Safina
Zh.Strukt.Khim.,**20**, 80, 1979
See also R1 : 16

17.37 **(4 – Hydroxy – 3,5 – dimethylphenyl) – diphenylmethanol**
$C_{21}H_{20}O_2$
T.W.Lewis, D.Y.Curtin, I.C.Paul
J.Am.Chem.Soc.,**101**, 5717, 1979

17.C **4' – Nitrophenyl – 4 – n – octyloxybenzoate**
$C_{21}H_{25}NO_5$ Main entry is 13.20

17.C **7,10 – Dihydroxy – 9 – methoxy – 6 – (4 – methoxy – phenyl) – 6H,11H – (2)benzopyrano(4,3 – c)(1)benzopyran – 11 – one**
$C_{24}H_{18}O_7$ Main entry is 38.117

17.C **2,3,7,8,12,13 – Hexamethoxy – 10,15 – dihydro – 5H – 5,10,15 – trithia – tribenzo(a,d,g)cyclononene chloroform solvate (crown form)**
Trithiaveratrylene chloroform solvate
$C_{24}H_{24}O_6S_3$. 2CHCl$_3$ Main entry is 39.84

17.38 **1,11 – bis(2 – Acetylaminophenoxy) – 3,6,9 – trioxa – undecane**
$C_{24}H_{32}N_2O_7$
I.–H.Suh, G.Weber, M.Kaftory, W.Saenger, H.Sieger, F.Vogtle *Z.Naturforsch.,Teil B*,**35**, 352, 1980

17.39 **1,11 – bis(2 – Acetylaminophenoxy) – 3,6,9 – trioxa – undecane potassium thiocyanate monohydrate**
2$C_{24}H_{32}N_2O_7$, 2K+, 2CNS-, H$_2$O
I.–H.Suh, G.Weber, M.Kaftory, W.Saenger, H.Sieger, F.Vogtle *Z.Naturforsch.,Teil B*,**35**, 352, 1980

17.40 **bis((8 – Quinolyl – oxy) – ethoxy – ethyl)ether thiourea**
$C_{26}H_{28}N_2O_5$, CH_4N_2S
G.Weber, W.Saenger
Acta Crystallogr.,Sect.B,**36**, 424, 1980
See also R2 : 8

17.41 **1 – (p – (2 – Dimethylaminoethoxy)phenyl) – 1,2 – trans – diphenylbutene**
trans – Tamoxifene
$C_{26}H_{29}NO$
G.Precigoux, C.Courseille, S.Geoffre, M.Hospital
Acta Crystallogr.,Sect.B,**35**, 3070, 1979
See also R1 : 3

17.42 **2,2 – bis(4 – (p – Nitrophenoxy)phenyl)propane**
$C_{27}H_{22}N_2O_6$
E.Subramanian, V.Lalitha, M.J.Nanjan, M.Balasubramaniam, J.Bordner
Cryst.Struct.Commun.,**9**, 873, 1980
See also R1 : 15

17.C **1 – (2 – Hydroxyphenyl) – 2,3,5 – triphenylpyrrole (at –35°C)**
$C_{28}H_{21}NO$ Main entry is 32.72

17.C **3,5 – Di – t – butyl – 7 – (3,5 – di – t – butyl – 2 – hydroxyphenyl) – 1 – methyl – 2,3 – dihydro – 1H – azepin – 2 – one ethanol solvate**
$C_{29}H_{45}NO_2$, C_2H_6O Main entry is 34.12

17.43 **1,5 – bis(2 – (5 – (2 – Nitrophenoxy) – 3 – oxapentyloxy)phenoxy) – 3 – oxapentane potassium thiocyanate**
$C_{36}H_{40}N_2O_{13}$, 2K+, 2CNS-
G.Weber, W.Saenger
Acta Crystallogr.,Sect.B,**36**, 61, 1980
See also R1 : 15

17.C **Cyclo(tetrakis(5 – t – butyl – 2 – hydroxy – 1,3 – phenylene)methylene) toluene clathrate**
$C_{44}H_{56}O_4$, C_7H_8 Main entry is 61.21

17.44 **3,3a',5,5' – Tetra – t – butyl – 2' – (3,5 – di – t – butyl – 4 – hydroxyphenyl) – 7' – (1 – (3,5 – di – t – butyl – 4 – hydroxyphenyl) – ethylidene) – 3a',4',7',7a' – tetrahydro – spiro(cyclohexa – 2,5 – diene – 1,1' – (1H) – indene) – 4,4' – dione diethyl ether solvate**
$C_{60}H_{86}O_4$, $0.5C_4H_{10}O$
S.R.Hall, C.L.Raston, A.H.White *Aust.J.Chem.*,**33**, 295, 1980

18.C **Tetramethyl – tetrathiafulvalene tetrabromo – p – benzoquinone complex**
$C_8Br_4O_2$, $C_{10}H_{12}S_4$ Main entry is 60.10

18.C **Tetrathiafulvalene – chloranil complex**
$C_8Cl_4O_2$, $C_6H_4S_4$ Main entry is 60.3

18.1 **Potassium p – chloranil acetone solvate**
$C_6Cl_4O_2^-$, K^+, C_3H_6O
G.Zanotti, A.Del Pra
Acta Crystallogr.,Sect.B,**36**, 313, 1980
See also R1 : 17,6

18.C **Tetrathiafulvalene – fluoranil complex**
$C_6F_4O_2$, $C_6H_4S_4$ Main entry is 60.4

18.C **Pyrene – fluoranil complex**
$C_6F_4O_2$, $C_{16}H_{10}$ Main entry is 60.34

18.2 **2,3 – Dichloro – 5,6 – dicyano – p – benzoquinone**
$C_8Cl_2N_2O_2$
G.Zanotti, R.Bardi, A.Del Pra
Acta Crystallogr.,Sect.B,**36**, 168, 1980
See also R1 : 7

18.3 **2 – Hydroxy – 3,5 – dimethyl – 6 – hydroxymethyl – 1,4 – benzoquinone**
Shanorellin
$C_9H_{10}O_4$
E.Subramanian, J.Bordner, V.Lalitha
Cryst.Struct.Commun.,**9**, 845, 1980
See also R1 : 17

18.C **Anthracene 1,1′ – bi – 3,5 – dichloro – 4 – oxo – hexa – 2,5 – dienylidene complex**
Anthracene 3,3′,5,5′ – tetrachloro – diphenoxinone complex
$C_{12}H_4Cl_4O_2$, $C_{14}H_{10}$ Main entry is 26.1

18.C **2,5 – bis(Pyrrolidino) – 1,4 – benzoquinone**
$C_{14}H_{18}N_2O_2$ Main entry is 32.51

18.C **2,5 – bis(Piperidino) – 1,4 – benzoquinone**
$C_{16}H_{22}N_2O_2$ Main entry is 33.71

19.C **N,N,N′,N′ – Tetramethyl – p – phenylenediamine hexafluorobenzene complex**
C_6F_6, $C_{10}H_{16}N_2$ Main entry is 60.13

19.1 **Iodylbenzene (at −100°C)**
$C_6H_5IO_2$
N.W.Alcock, J.F.Sawyer
J.Chem.Soc.,Dalton Trans.,115, 1980

19.C **bis(Trimethyl – acetic acid) benzene**
C_6H_6, $2C_5H_{10}O_2$ Main entry is 1.19

19.C **Hexa – aziridino – cyclotriphosphazene benzene clathrate**
C_6H_6, $2C_{12}H_{24}N_9P_3$ Main entry is 61.3

19.C **Cycloveratril benzene clathrate monohydrate**
$0.5C_6H_6$, $C_{27}H_{30}O_6$, H_2O Main entry is 61.12

19.C **5 – Methylbenzene – 1,3 – dicarbaldehyde – bis(p – tolylsulfonylhydrazone) benzene clathrate**
$2C_6H_6$, $C_{23}H_{24}N_4O_4S_2$ Main entry is 61.7

19.C **Cyclo(tetrakis(5 – t – butyl – 2 – hydroxy – 1,3 – phenylene)methylene) toluene clathrate**
C_7H_8, $C_{44}H_{56}O_4$ Main entry is 61.21

19.2 **Benzylammonium bis(dichloro – mercury) chloride**
$C_7H_{10}N^+$, Cl^-, $2Cl_2Hg$
J.W.Bats, H.Fuess, A.Daoud
Acta Crystallogr.,Sect.B,**36**, 2150, 1980

19.C **tris(1,2 – bis(Diphenylphosphino – selenoyl)ethane) p – xylene clathrate**
C_8H_{10}, $3C_{26}H_{24}P_2Se_2$ Main entry is 61.9

19.3 **Benzoylacetone (at −130°C)**
$C_{10}H_{10}O_2$
W.Winter, K.–P.Zeller, S.Berger
Z.Naturforsch.,Teil B,**34**, 1606, 1979

19.4 **(−) – (S) – N – (1 – Phenylethyl)acetamide**
$C_{10}H_{13}NO$
A.Aubry, J.Protas, M.T.Cung, M.Marraud
Acta Crystallogr.,Sect.B,**36**, 1861, 1980

19.C **1,6,20,25 – Tetra – aza(6.1.6.1)paracyclophane hydrochloride durene tetrahydrate**
$C_{10}H_{14}$, $C_{34}H_{44}N_4^{4+}$, $4Cl^-$, $4H_2O$ Main entry is 61.18

19.5 **2,2′ – Dibromo – octafluorobiphenyl**
$C_{12}Br_2F_8$
M.J.Hamor, T.A.Hamor
Acta Crystallogr.,Sect.B,**36**, 1402, 1980

19.C **4 – Bromobiphenyl perfluorobiphenyl complex**
$C_{12}F_{10}$, $C_{12}H_9Br$ Main entry is 60.16

19.C **Biphenyl perfluorobiphenyl complex**
$C_{12}F_{10}$, $C_{12}H_{10}$ Main entry is 60.18

19.C **4 – Methylbiphenyl perfluorobiphenyl complex**
$C_{12}F_{10}$, $C_{13}H_{12}$ Main entry is 60.23

19.6 **4 – Bromobiphenyl (at 152°K)**
$C_{12}H_9Br$
C.P.Brock *Acta Crystallogr.,Sect.B*,**36**, 968, 1980

19.C **4 – Bromobiphenyl perfluorobiphenyl complex**
$C_{12}H_9Br$, $C_{12}F_{10}$ Main entry is 60.16

19.C **Biphenyl perfluorobiphenyl complex**
$C_{12}H_{10}$, $C_{12}F_{10}$ Main entry is 60.18

19.7 **1 – Methyl – 4 – (1' – ethylpropyl)benzene (at –100°C)**
$C_{12}H_{18}$
M.A.Kravers, M.Yu.Antipin, K.A.Potekhin, Yu.T.Struchkov
Cryst.Struct.Commun.,**8**, 1005, 1979

19.C **4 – Methylbiphenyl perfluorobiphenyl complex**
$C_{13}H_{12}$, $C_{12}F_{10}$ Main entry is 60.23

19.C **Stilbene – 7,7,8,8 – tetracyanoquinodimethane complex**
$C_{14}H_{12}$, $C_{12}H_4N_4$ Main entry is 60.32

19.C **Tri – o – thymotide trans – stilbene clathrate**
$C_{14}H_{12}$, $C_{33}H_{36}O_6$ Main entry is 61.14

19.C **Tri – o – thymotide cis – stilbene clathrate**
$C_{14}H_{12}$, $2C_{33}H_{36}O_6$ Main entry is 61.17

19.8 **0,0' – Dibromodibenzyl ether**
$C_{14}H_{12}Br_2O$
A.F.Berndt *Am.Cryst.Assoc.,Ser.2*,**7**, 11, 1980

19.9 **1,5 – Diphenyl – 2,4 – pentadien – 1 – one**
$C_{17}H_{14}O$
S.Kashino, M.Haisa
Acta Crystallogr.,Sect.B,**36**, 346, 1980

19.10 **2,2 – Dibenzyl – 1,3 – dibromopropane**
$C_{17}H_{18}Br_2$
L.Lebioda, K.Stadnicka *Acta Phys.Pol.*,**56**, 411, 1979

19.C **L – (1 – Tosylamino – 2 – phenyl) – ethyl – chloromethyl – ketone**
$C_{17}H_{18}ClNO_3S$ Main entry is 4.3

19.11 **2,3 – bis(Trifluoromethyl) – 2,3 – bis(p – fluorophenyl) – hexafluorobutane**
$C_{18}H_8F_{14}$
V.R.Polishchuk, M.Yu.Antipin, V.I.Bakhmutov,
N.N.Bubnov, S.P.Solodovnikov, T.V.Timofeeva,
Yu.T.Struchkov, B.L.Tumanskii, I.L.Knunyants
Dokl.Akad.Nauk SSSR,**249**, 1125, 1979

19.C **o – Tropylbiphenyl tetrafluoroborate**
$C_{19}H_{15}^+$, BF_4^- Main entry is 22.7

19.12 **(α – (1 – Phenylethylimino) – benzoyl) – phenyl – ketone (absolute configuration)**
N – (1 – Phenylethyl) – 1 – benzoyl – benzylideneimine
$C_{22}H_{19}NO$
I.Fonseca, S.Martinez–Carrera, S.Garcia–Blanco
Acta Crystallogr.,Sect.B,**35**, 2643, 1979

19.C **p – Bromobenzoyl – epicatalponol (absolute configuration)**
$C_{22}H_{21}BrO_3$ Main entry is 59.30

19.13 **DL – 2,2,5,5 – Tetramethyl – 3,4 – diphenylhexane**
$C_{22}H_{30}$
H.–D.Beckhaus, K.J.McCullough, H.Fritz, C.Ruchardt,
B.Kitschke, H.J.Lindner, D.A.Dougherty, K.Mislow
Chem.Ber.,**113**, 1867, 1980

19.14 **meso – 2,2,5,5 – Tetramethyl – 3,4 – diphenylhexane**
$C_{22}H_{30}$
H.–D.Beckhaus, K.J.McCullough, H.Fritz, C.Ruchardt,
B.Kitschke, H.J.Lindner, D.A.Dougherty, K.Mislow
Chem.Ber.,**113**, 1867, 1980

19.15 **N – (1 – Phenylpropyl) – 1 – benzoyl – benzylideneimine**
$C_{23}H_{21}NO$
I.Fonseca, S.Martinez–Carrera, S.Garcia–Blanco
Eur.Cryst.Meeting,**5**, 392, 1979

19.16 **1,5 – bis(Dimethylamino) – 3 – ethoxy – 1,2,5 – triphenyl – 4 – aza – pentamethinium perchlorate**
$C_{28}H_{32}N_3O^+$, ClO_4^-
B.Tinant, J.P.Declercq, M.van Meerssche
Bull.Soc.Chim.Belg.,**89**, 405, 1980

19.17 **cis – 1,4 – Dihydro – 4 – trityl – biphenyl**
$C_{31}H_{26}$
M.C.Grossel, A.K.Cheetham, D.A.O.Hope, K.P.Lam,
M.J.Perkins *Tetrahedron Lett.*,1351, 1979

19.C **(1,2,4,5 – Tetrabenzoyl – 3,6 – di – t – butylbenzene) – (meso – 3,8 – di – t – butyl – 1,5,6,10 – tetraphenyl – deca – 3,4,6,7 – tetraene – 1,9 – diyne) complex**
$C_{42}H_{38}$, $C_{42}H_{38}O_4$ Main entry is 60.44

19.C **(1,2,4,5 – Tetrabenzoyl – 3,6 – di – t – butylbenzene) – (meso – 3,8 – di – t – butyl – 1,5,6,10 – tetraphenyl – deca – 3,4,6,7 – tetraene – 1,9 – diyne) complex**
$C_{42}H_{38}O_4$, $C_{42}H_{38}$ Main entry is 60.44

MONOCYCLIC HYDROCARBONS
(3,4,5–MEMBERED RINGS)

20.1 **Cyclopentadienyl sodium tetramethylethylenediamine**
$C_5H_5^-$, $C_6H_{16}N_2$, Na^+
T.Aoyagi, H.M.M.Shearer, K.Wade, G.Whitehead
J.Organomet.Chem.,**175**, 21, 1979
See also R1 : 12 R2 : 3

20.2 **Cyclopropane – 1,1 – dicarboxamide**
$C_5H_8N_2O_2$
R.Usha, K.Venkatesan
Acta Crystallogr.,Sect.B,**35**, 2730, 1979

20.C **2,4 – Methano – glutamic acid monohydrate**
1 – Amino – 1,3 – dicarboxy – cyclobutane monohydrate
$C_6H_9NO_4$, H_2O Main entry is 48.18

20.C **Trichoviridine**
$C_8H_9NO_4$ Main entry is 50.3

20.3 **Tetrafluoro – 1,2 – (RS) – bis(2,2,3,3 – tetrafluorocyclobutyl)ethane**
$C_{10}H_8F_{12}$
J.C.J.Bart, P.Piccardi, I.W.Bassi
Acta Crystallogr.,Sect.B,**36**, 842, 1980

20.4 **cis – 2 – Ethoxy – 1 – phenylsulfinyl – cyclopropane**
$C_{11}H_{14}O_2S$
M.Kimura, M.A.Ward, W.H.Watson, C.G.Venier
Acta Crystallogr.,Sect.B,**35**, 3122, 1979
See also R1 : 11

20.5 **2 – Carboxy – cyclopentene – 1 – (N – methyl – N – phenyl – sulfonamide)**
$C_{13}H_{15}NO_4S$
T.Graafland, A.Wagenaar, A.J.Kirby, J.B.F.N.Engberts
J.Am.Chem.Soc.,**101**, 6981, 1979
See also R1 : 1,4

20.6 **3 – (2 – Methoxyphenyl) – 1,1,2,2 – cyclopropane – tetracarbonitrile**
1,1,2,2, – Tetracyano – 3 – (2 – methoxyphenyl) – cyclopropane
$C_{14}H_8N_4O$
R.Usha, K.Venkatesan
Acta Crystallogr.,Sect.B,**36**, 335, 1980
See also R1 : 17

20.7 **2 – (1 – Phenylthio) – ethyl – pent – 3 – ene – 1 – carboxylic acid**
$C_{14}H_{16}O_2S$
P.Michel, M.O'Donnell, R.Biname, A.M.Hesbain–Frisque,
L.Ghosez, J.P.Declercq. G.Germain, E.Arte,
M.van Meerssche *Tetrahedron Lett.*,**21**, 2577, 1980
See also R1 : 11

20.C **2,3 – Dimethyl – 5 – (p – bromobenzoyloxymethyl) – cyclopent – 2 – en – 1 – one**
Methylenomycin B p – bromobenzoate
$C_{15}H_{15}BrO_3$ Main entry is 50.7

20.8 **2,3 – Di(cyclopropyl) – 1,4 – naphthoquinone**
$C_{16}H_{14}O_2$
L.R.Nassimbeni, M.R.W.Wright, R.G.F.Giles, P.R.K.Mitchell
S.Afr.J.Chem.,**32**, 177, 1979
See also R1 : 25

20.9 **(±) – 3 – (bis(Phenylsulfinyl)methyl) – 1,2 – dimethyl – cyclopropene**
$C_{19}H_{18}O_2S_2$
H.Beckhaus, M.Kimura, W.H.Watson, C.G.Venier,
B.Kojic–Prodic *Acta Crystallogr.,Sect.B*,**35**, 3119, 1979
See also R1 : 11

20.10 **2 – Benzyl – 5 – benzylidene – cyclopentanone**
$C_{19}H_{18}O$
H.Nakanish, W.Jones, J.M.Thomas, M.B.Hursthouse,
M.Motevalli *J.Chem.Soc.,Chem.Commun.*,611, 1980

20.C **r – 1,c – 2,t – 3,t – 4 – tetrakis(2 – Pyrazinyl) – cyclobutane**
$C_{20}H_{16}N_8$ Main entry is 33.77

20.C **(5Z,11α,13E,15S) – 11,15 – Dihydroxy – 9 – oxo – prosta – 5,13 – dien – 1 – oic acid**
Prostaglandin E_2
$C_{20}H_{32}O_5$ Main entry is 59.25

20.C **(5Z,9β,11a,13E,15S) – 9,11,15 – Trihydroxy – prosta – 5,13 – dien – 1 – oic acid**
Prostaglandin F(2β)
$C_{20}H_{34}O_5$ Main entry is 59.26

20.11 **Tetra – t – butylcyclobutadiene**
$C_{20}H_{36}$
H.Irngartinger, N.Riegler, K.–D.Malsch, K.–A.Schneider,
G.Maier *Angew.Chem.,Int.Ed.Engl.*,**19**, 211, 1980

20.C **(2 – Oxo – 3,3,5,5 – tetramethyl – cyclopentyl) – 1 – oxa – 7,7,9,9 – tetramethyl – spiro(4.4)nonane – 3,6 – dione**
$C_{21}H_{32}O_4$ Main entry is 38.108

20.C **r – 1,c – 2 – bis(3 – Pyridyl) – t – 3,t – 4 – bis(2 – pyrazinyl) – cyclobutane**
$C_{22}H_{18}N_6$ Main entry is 33.80

20.C **r – 1,t – 3 – bis(3 – Pyridyl) – c – 2,t – 4 – bis(2 – pyrazinyl) – cyclobutane**
$C_{22}H_{18}N_6$ Main entry is 33.81

20.12 **N,N' – bis(2 – (4 – Chlorophenylamino) – 2 – cyclopenten – 1 – ylidene) – hydrazine (monoclinic form)**
$C_{22}H_{20}Cl_2N_4$
A.C.Villa, A.G.Manfredotti, C.Guastini, D.Pocar
Cryst.Struct.Commun.,**9**, 891, 1980
See also R1 : 16,9

20.13 **N,N' – bis(2 – (4 – Chlorophenylamino) – 2 – cyclopenten – 1 – ylidene) – hydrazine (orthorhombic form)**
$C_{22}H_{20}Cl_2N_4$
A.C.Villa, A.G.Manfredotti, C.Guastini, D.Pocar
Cryst.Struct.Commun.,**9**, 891, 1980
See also R1 : 16,9

20.C   r – 1,c – 2,t – 3,t – 4 – tetrakis(2 – Pyridyl) –
        cyclobutane
        $C_{24}H_{20}N_4$ Main entry is 33.87

20.C   r – 1,c – 2,t – 3,t – 4 – tetrakis(4 – Pyridyl) –
        cyclobutane
        $C_{24}H_{20}N_4$ Main entry is 33.88

20.C   1,3 – trans – bis(4 – Chlorophenyl) – 2,4 – trans –
        bis(4 – pyridyl) – cyclobutane
        $C_{26}H_{20}Cl_2N_2$ Main entry is 33.93

20.14   tetrakis(Phenylimino) – cyclobutane
        $C_{28}H_{20}N_4$
        H.J.Bestmann, G.Schmid, E.Wilhelm
        Angew.Chem.,Int.Ed.Engl.,**19**, 136, 1980

20.C   Di – isopropyl – (2,3,4,5 – tetraphenyl –
        cyclopenta – 1,4 – dienyl) – phosphate
        $C_{35}H_{35}O_4P$ Main entry is 46.7

20.15   2 – Benzyl – 5 – benzylidene – cyclopentanone
        dimer
        $C_{36}H_{36}O_2$
        H.Nakanish, W.Jones, J.M.Thomas, M.B.Hursthouse,
        M.Motevalli *J.Chem.Soc.,Chem.Commun.*,611, 1980

MONOCYCLIC HYDROCARBONS
(6–MEMBERED RINGS)

21.C   **Sodium phytate hydrate**
        Dodecasodium myo – inositol – hexaphosphate
        octatriacontahydrate
        $C_6H_6O_{24}P_6{}^{12-}$, $12Na^+$, $38H_2O$ Main entry is 46.3

21.1   **Cyclohexylammonium ethyl hydrogen phosphate**
        $C_6H_{14}N^+$, $C_2H_6O_4P^-$
        K.A.Kerr, J.K.Fawcett, J.C.Coppola, D.G.Watson, O.Kennard
        Acta Crystallogr.,Sect.B,**35**, 2749, 1979
        See also R2 : 46

21.2   **trans – 1,4 – Dimethylcyclohexane (at –90°C)**
        C_8H_{16}
        C.Courseille, M.Hospital, F.Leroy, D.Watkin
        Eur.Cryst.Meeting,**5**, 285, 1979

21.3   **rel – (1S,2R) – 2 – Hydroxymethyl – 2 –
        methylcyclohexanol**
        $C_8H_{16}O_2$
        J.S.Chen, W.H.Watson *Cryst.Struct.Commun.*,**9**, 57, 1980

21.4   **2,6 – Dimethyl – 1,4 – dihydrobenzoic acid**
        $C_9H_{12}O_2$
        M.C.Grossel, A.K.Cheetham, D.James, J.M.Newsam
        J.Chem.Soc.,Perkin Trans.2,471, 1980

21.5   **3,5 – Dimethyl – 1,4 – dihydrobenzoic acid**
        $C_9H_{12}O_2$
        M.C.Grossel, A.K.Cheetham, D.James, J.M.Newsam
        J.Chem.Soc.,Perkin Trans.2,471, 1980

21.6   **2,4 – Dimethyl – 4 – (1,2 – dichlorovinyl) –
        cyclohexen – 3 – one**
        $C_{10}H_{12}Cl_2O$
        A.S.Kende, M.Benechie, D.P.Curran, P.Fludzinski,
        W.Swenson, J.Clardy *Tetrahedron Lett.*,4513, 1979

21.7   **(±) – (1R,2R,3R) – 3 – Acetyl – 2 – vinylcyclohexan –
        1 – ol**
        $C_{10}H_{16}O_2$
        M.W.Vary, J.M.McBride, J.J.Piwinski, F.E.Ziegler
        Cryst.Struct.Commun.,**8**, 807, 1979

21.8   **3 – (2,2,2 – Trichloro – 1 – (p – toluenesulfonyloxy)
        ethyl) – cyclohexene**
        $C_{15}H_{17}Cl_3O_3S$
        G.B.Gill, K.Marrison, S.J.Parrott, B.Wallace
        Tetrahedron Lett.,4867, 1979
        See also R1 : 11

21.9   **(αRS,1RS,2RS) – α – (1 – Hydroxy – 2 –
        phenylcyclohexyl) – propionic acid**
        $C_{15}H_{20}O_3$
        G.Hite, J.B.Anderson, T.L.Thomas, T.A.Davidson,
        R.C.Griffith, F.L.Scott
        Acta Crystallogr.,Sect.B,**35**, 3082, 1979
        See also R1 : 1

21.10 **3,3,5 − Trimethyl − 5 − phenyl − cyclohexanone − oxime**
$C_{15}H_{21}NO$
S.Toure, J.Lapasset, B.Boyer, G.Lamaty
Acta Crystallogr.,Sect.B,**35**, 2790, 1979
See also R1 : 10

21.11 **racemic − bis − (4 − Oxocyclohexyl) − 3,4 − hexane**
$C_{18}H_{30}O_2$
M.van Meerssche, G.Germain, J.P.Declercq, X.−H.Bui,
R.Devis *Cryst.Struct.Commun.*,**8**, 971, 1979

21.12 **meso − 3,4 − bis(trans − 4 − Hydroxycyclohexyl) − hexane**
$C_{18}H_{34}O_2$
M.van Meerssche, G.Germain, J.P.Declercq, R.Touillaux,
X.−H.Bui, R.Devis *Cryst.Struct.Commun.*,**8**, 887, 1979

21.13 **racemic − 3,4 − bis(trans − 4 − Hydroxycyclohexyl) − hexane**
$C_{18}H_{34}O_2$
M.van Meerssche, G.Germain, J.P.Declercq, R.Touillaux,
R.Devis, X.−H.Bui *Cryst.Struct.Commun.*,**8**, 893, 1979

21.14 **2,6 − Dimethyl − 4 − (diphenylmethylene) − cyclohexa − 2,5 − dienone (β form)**
2,6 − Dimethyl − fuchsone
$C_{21}H_{18}O$
T.W.Lewis, I.C.Paul, D.Y.Curtin
Acta Crystallogr.,Sect.B,**36**, 70, 1980

21.15 **2,6 − Dimethyl − 4 − (diphenylmethylene) − cyclohexa − 2,5 − dienone (α form)**
2,6 − Dimethyl − fuchsone
$C_{21}H_{18}O$
T.W.Lewis, I.C.Paul, D.Y.Curtin
Acta Crystallogr.,Sect.B,**36**, 70, 1980

21.16 **2,6 − Dimethyl − 4 − (diphenylmethylene) − cyclohexa − 2,5 − dienone (γ form)**
2,6 − Dimethyl − fuchsone
$C_{21}H_{18}O$
E.N.Duesler, T.W.Lewis, D.Y.Curtin, I.C.Paul
Acta Crystallogr.,Sect.B,**36**, 166, 1980

21.17 **(±) − (3R,1′R) − 1 − Hydroxymethyl − 2 − (1 − methylpentyl) − 3 − methylcyclohex − 1 − ene p − bromophenylurethane**
$C_{21}H_{30}BrNO_2$
M.W.Vary, J.M.McBride, M.A.Cady, F.E.Ziegler
Cryst.Struct.Commun.,**8**, 799, 1979
See also R1 : 16

21.18 **Benzyl trans − 6 − (ethoxycarbonyl) − cis − 6 − phenylcyclohex − 2 − ene − 1 − carbamate**
$C_{23}H_{25}NO_4$
L.E.Overman, C.B.Petty, R.J.Doedens
J.Org.Chem.,**44**, 4183, 1979

21.19 **1,1,2,2 − Tetracyclohexyl − ethane**
$C_{26}H_{46}$
S.G.Baxter, H.Fritz, G.Hellmann, B.Kitschke, H.J.Lindner,
K.Mislow, C.Ruchardt, S.Weiner
J.Am.Chem.Soc.,**101**, 4493, 1979

21.20 **α − Dypnopinacoline**
$C_{32}H_{28}O$
J.P.Declercq, G.Germain, M.van Meerssche, S.Desauvage,
A.Bruylants *Bull.Cl.Sci.,Acad.R.Belg.*,**64**, 406, 1978

21.21 **Dypnopinacone**
$C_{32}H_{28}O_2$
J.P.Declercq, G.Germain, M.van Meerssche, S.Desauvage,
A.Bruylants *Bull.Cl.Sci.,Acad.R.Belg.*,**64**, 406, 1978

21.22 **1 − Diphenylmethylene − 4 − triphenylmethyl − cyclohexa − 2,5 − diene**
$C_{38}H_{30}$
N.S.Blom, G.Roelofsen, J.A.Kanters
Eur.Cryst.Meeting,**6**, 19, 1980

MONOCYCLIC HYDROCARBONS
(7,8—MEMBERED RINGS)

22.8 **(E) – 2,2′,3,3′ – Tetrachloro – 4,4′,7,7′ –**
tetraphenyl – 1,1′ – bicycloheptatrienylidene
$C_{38}H_{24}Cl_4$
J.A.Fahey, H.M.Hugel, D.P.Kelly, B.Halton, G.J.B.Williams
J.Org.Chem.,**45**, 2862, 1980

22.1 **Cyclo – octane – 1,5 – dione**
$C_8H_{12}O_2$
R.W.Miller, A.T.McPhail
J.Chem.Soc.,Perkin Trans.2,1527, 1979

22.2 **Cyclo – octane – 1,5 – dione – dioxime**
$C_8H_{14}N_2O_2$
R.W.Miller, A.T.McPhail   *J.Chem.Res.*,**285**, 3122, 1979

22.C **1 – Aminocycloheptane – carboxylic acid**
hydrobromide monohydrate (reinvestigation of
structure, published by Chacko, Srinivasan and
Zand, J.Cryst.Mol.Struct.,1,(1971),213–224, using
published diffraction data)
$C_8H_{16}NO_2{}^+$, Br⁻, H_2O Main entry is 48.25

22.3 **cis – Cyclo – octane – 1,5 – diol**
$C_8H_{16}O_2$
R.W.Miller, A.T.McPhail
J.Chem.Soc.,Perkin Trans.2,1527, 1979

22.C **Dimethyl – 1 – hydroxy – 1 – cycloheptane –**
phosphonate (reinvestigation of structure published
by Birnbaum, Buchanan and Morin,
J.Am.Chem.Soc.,99,(1977),6652–6656, using published
diffraction data)
$C_9H_{19}O_4P$ Main entry is 64.34

22.4 **Cycloheptyl – p – bromobenzoate**
$C_{14}H_{17}BrO_2$
K.Urgast, R.Hoge, K.Fischer
Cryst.Struct.Commun.,**9**, 129, 1980
See also R1 : 13

22.C **3,5,7 – Trimethyl – tropylium – 0,0 – (2′,4′,6′ –**
trinitrophenylide)
$C_{16}H_{13}N_3O_8$ Main entry is 38.74

22.5 **0 – (2′,6′ – Dinitrophenyl) – 3,5,7 – trimethyl –**
tropolone
$C_{16}H_{14}N_2O_6$
N.G.Furmanova, Yu.T.Struchkov, O.E.Kompan,
Z.N.Budarina, L.P.Olekhnovich, V.I.Minkin
Zh.Strukt.Khim.,**21**, 83, 1980
See also R1 : 15

22.6 **3 – (2,2,2 – Trichloro – 1 – (p – toluenesulfonyloxy)**
ethyl) – cyclo – octene
$C_{17}H_{21}Cl_3O_3S$
G.B.Gill, K.Marrison, S.J.Parrott, B.Wallace
Tetrahedron Lett.,4867, 1979
See also R1 : 11

22.7 **o – Tropylbiphenyl tetrafluoroborate**
$C_{19}H_{15}{}^+$. $BF_4{}^-$
K.Yamamura, K.Nakatsu, K.Nakao, T.Nakazawa. I.Murata
Tetrahedron Lett.,4999, 1979
See also R1 : 12,19

23.1　**Cyclodecane – oxime (at –160°C)**
$C_{10}H_{19}NO$
P.Groth *Acta Chem.Scand.Ser.A*,**33**, 503, 1979
See also R1 : 10

23.2　**Cycloundecane – oxime (at –160°C)**
$C_{11}H_{21}NO$
P.Groth *Acta Chem.Scand.Ser.A*,**33**, 503, 1979
See also R1 : 10

23.3　**Cyclododecane – oxime (at –160°C)**
$C_{12}H_{23}NO$
P.Groth *Acta Chem.Scand.Ser.A*,**33**, 503, 1979
See also R1 : 10

23.4　**Cyclotridecane – oxime (at –160°C)**
$C_{13}H_{25}NO$
P.Groth *Acta Chem.Scand.Ser.A*,**33**, 503, 1979
See also R1 : 10

23.5　**Cyclotetradecane – oxime (at –160°C)**
$C_{14}H_{27}NO$
P.Groth *Acta Chem.Scand.Ser.A*,**33**, 503, 1979
See also R1 : 10

23.6　**Cyclopentadecanone phenylsemicarbazone**
$C_{22}H_{35}N_3O$
W.G.M.van den Hoek, H.A.J.Oonk, J.Kroon
Eur.Cryst.Meeting.**5**, 125, 1979
See also R1 : 8

23.7　**5,5,10,10,15,15,20,20 – Octamethyl –
1,4,6,9,11,14,16,19 – octaoxo – cyclotetracosa – cis –
2,7,12,17 – tetraene**
$C_{28}H_{32}O_8$
D.L.Ward, W.–J.H.Kung *Am.Cryst.Assoc.,Ser.2*,**8**, 34, 1980

24.C　**2,6 – Dimethylnaphthalene perfluoronaphthalene
complex**
$C_{10}F_8$, $C_{12}H_{12}$ Main entry is 60.19

24.1　**2,3,6,7 – Tetrabromonaphthalene**
$C_{10}H_4Br_4$
P.Singh, J.D.McKinney, L.A.Levy
Cryst.Struct.Commun.,**9**, 563, 1980

24.C　**bis(Isothiocyanato) – tetrakis(4 – methylpyridine) –
nickel(ii) 2 – bromonaphthalene clathrate**
$2C_{10}H_7Br$, $C_{26}H_{28}N_6NiS_2$ Main entry is 61.10

24.2　**2 – Chloronaphthalene (low
temperature form ii,at 223°K)**
$C_{10}H_7Cl$
A.Meresse, Y.Haget, N.B.Chanh
Cryst.Struct.Commun.,**9**, 699, 1980

24.3　**Naphthalene – 2,7 – diol**
$C_{10}H_8O_2$
N.A.Ahmed *Egypt.J.Phys.*,**9**, 67, 1978

24.C　**1,5,12,16,23,26,29 – Heptaoxa($7^{3,14}$)(5.5)
orthocyclophane naphthalene – 2,3 – diol
monohydrate**
$C_{10}H_8O_2$, $C_{22}H_{26}O_7$, H_2O Main entry is 38.113

24.C　**Pyridinium 1 – naphthylamine picrate**
$C_{10}H_9N$, $C_5H_6N^+$, $C_6H_2N_3O_7{}^-$ Main entry is 60.1

24.4　**3 – Carboxy – 2 – naphthalene – diazonium iodide
monohydrate (at –50°C)**
$C_{11}H_7N_2O_2{}^+$, I^-, H_2O
J.Z.Gougoutas *J.Am.Chem.Soc.*,**101**, 5672, 1979

24.C　**bis(Isothiocyanato) – tetrakis(4 – methylpyridine) –
nickel(ii) 2 – methylnaphthalene clathrate**
$2C_{11}H_{10}$, $C_{26}H_{28}N_6NiS_2$ Main entry is 61.11

24.C　**2,6 – Dimethylnaphthalene perfluoronaphthalene
complex**
$C_{12}H_{12}$, $C_{10}F_8$ Main entry is 60.19

24.5　**2,6,7 – Trichloro – 1,4 – diacetoxy – naphthalene**
$C_{14}H_9Cl_3O_4$
P.Smith–Verdier, J.Vilches, S.Garcia–Blanco,
J.G.Rodriguez *Eur.Cryst.Meeting*,**6**, 32, 1980

24.6　**(R) – N – Trifluoroacetyl – 1 – (1 – naphthyl) –
ethylamine**
$C_{14}H_{12}F_3NO$
S.Weinstein, L.Leiserowitz
Acta Crystallogr.,Sect.B,**36**, 1406, 1980
See also R1 : 1

24.7　**(R) – N – Acetyl – 1 – (1 – naphthyl) – ethylamine**
$C_{14}H_{15}NO$
S.Weinstein, L.Leiserowitz
Acta Crystallogr.,Sect.B,**36**, 1406, 1980
See also R1 : 1

24.8 (RS) – N – Acetyl – 1 – (1 – naphthyl) – ethylamine
$C_{14}H_{15}NO$
S.Weinstein, L.Leiserowitz
Acta Crystallogr.,Sect.B,**36**, 1406, 1980
See also R1 : 1

24.C 1,2,3,4,5 – Pentachloro – 8 – phenyltetralin
$C_{16}H_{11}Cl_5$ Main entry is 27.15

24.C 4 – Anilino – 3 – methoxycarbonyl – 1 – (methoxycarbonylmethylidene) – 1,2,3,4 – tetrahydro – naphthalene
$C_{21}H_{21}NO_4$ Main entry is 27.29

24.9 bis(2 – Methyl – 1 – naphthyl) – ketone
$C_{23}H_{18}O$
R.Fink, D.van der Helm
Cryst.Struct.Commun.,**9**, 97, 1980

24.10 1 – (2 – Hydroxynaphthyl) – diphenylmethanol
$C_{23}H_{18}O_2$
T.W.Lewis, E.N.Duesler, R.B.Kress, D.Y.Curtin, I.C.Paul
J.Am.Chem.Soc.,**102**, 4659, 1980

24.11 bis(2 – Methyl – 1 – naphthyl) – methylacetate (at –160°C)
$C_{25}H_{22}O_2$
R.Fink, D.van der Helm
Cryst.Struct.Commun.,**9**, 105, 1980

24.12 N,N′ – (Di – β – naphthyl) – p – quinone – di – imine
$C_{26}H_{18}N_2$
Z.P.Povet'eva, L.A.Chetkina, V.V.Kopilov
Zh.Strukt.Khim.,**21**, 118, 1980

24.C 1 – (2 – Butoxycarbonyl – 1 – azophenyl) – 2 – hydroxy – 3 – N – (2 – oxo – 5 – benzimidazolyl) – (naphthoic acid) – amide
C.I.Pigment red 208
$C_{29}H_{25}N_5O_5$ Main entry is 35.74

24.C Methyl – N – (biphenyl – 2 – yl) – N – (1 – naphthyl) – anthranilate (α isomer)
$C_{30}H_{23}NO_2$ Main entry is 48.77

24.C Methyl – N – (biphenyl – 2 – yl) – N – (1 – naphthyl) – anthranilate (β₁ isomer)
$C_{30}H_{23}NO_2$ Main entry is 48.78

24.C Methyl – N – (biphenyl – 2 – yl) – N – (1 – naphthyl) – anthranilate (β₂ isomer)
$C_{30}H_{23}NO_2$ Main entry is 48.79

24.C 1 – Phenyl – 2,3 -- bis(diphenylphosphino) – naphthalene
$C_{40}H_{30}P_2$ Main entry is 64.169

NAPHTHOQUINONES

25.1 5,8 – Dihydroxy – 1,4 – naphthoquinone (B form)
Naphthazarin
$C_{10}H_6O_4$
W.–I.Shiau, E.N.Duesler, I.C.Paul, D.Y.Curtin, W.G.Blann, C.A.Fyfe *J.Am.Chem.Soc.*,**102**, 4546, 1980

25.2 2,5 – Dihydroxy – 3,8 – dimethoxy – 7 – methyl – 1,4 – naphthoquinone
Nepenthone – E
$C_{13}H_{12}O_6$
J.R.Cannon, V.Lojanapiwatna, C.L.Raston, W.Sinchai, A.H.White *Aust.J.Chem.*,**33**, 1073, 1980

25.C 2,3 – Di(cyclopropyl) – 1,4 – naphthoquinone
$C_{16}H_{14}O_2$ Main entry is 20.8

25.3 2,3,4aβ,6,7β,8aβ – Hexamethyl – 4a,7,8,8a – tetrahydro – 1,4 – naphthoquinone
$C_{16}H_{22}O_2$
T.J.Greenhough, J.Trotter
Acta Crystallogr.,Sect.B,**36**, 368, 1980

25.4 2 – Chloro – 3 – (N – methyl – N – (p – ethoxycarbonylphenyl)aminomethyl) – 1,4 – naphthoquinone
$C_{21}H_{18}ClNO_4$
A.E.Shvets, A.Ya.Malmanis, Ya.F.Freimanis, Ya.Ya.Bleidelis, Ya.Ya.Dregeris
Zh.Strukt.Khim.,**20**, 491, 1979

25.5 3 – Chloro – 6 – methoxy – 7 – methyl – 2 – (p – tosyloxyethylamine) – 1,4 – naphthoquinone
$C_{21}H_{20}ClNO_6S$
A.P.Kozikowski, K.Sugiyama, J.P.Springer
Tetrahedron Lett.,**21**, 3257, 1980

ANTHRACENE COMPOUNDS

26.C   **Anthracene – tetracyanobenzene complex (high temperature form, at 297°K)**
$C_{14}H_{10}$, $C_{10}H_2N_4$ Main entry is 60.25

26.C   **Anthracene – tetracyanobenzene complex (high temperature form, at 234°K)**
$C_{14}H_{10}$, $C_{10}H_2N_4$ Main entry is 60.26

26.C   **Anthracene – tetracyanobenzene complex (high temperature form, at 226°K)**
$C_{14}H_{10}$, $C_{10}H_2N_4$ Main entry is 60.27

26.C   **Anthracene – tetracyanobenzene complex (low temperature form, at 202°K)**
$C_{14}H_{10}$, $C_{10}H_2N_4$ Main entry is 60.28

26.C   **Anthracene – tetracyanobenzene complex (low temperature form, at 170°K)**
$C_{14}H_{10}$, $C_{10}H_2N_4$ Main entry is 60.29

26.C   **Anthracene – tetracyanobenzene complex (low temperature form, at 138°K)**
$C_{14}H_{10}$, $C_{10}H_2N_4$ Main entry is 60.30

26.C   **Anthracene – tetracyanobenzene complex (low temperature form, at 119°K)**
$C_{14}H_{10}$, $C_{10}H_2N_4$ Main entry is 60.31

26.1   **Anthracene 1,1′ – bi – 3,5 – dichloro – 4 – oxo – hexa – 2,5 – dienylidene complex**
Anthracene 3,3′,5,5′ – tetrachloro – diphenoxinone complex
$C_{14}H_{10}$, $C_{12}H_4Cl_4O_2$
Z.A.Starikova, T.M.Shchegoleva, V.K.Trunov,
O.B.Lantratova, I.E.Pokrovskaya
Zh.Strukt.Khim.,21, 73, 1980
See also R2 : 18

26.2   **Anthralin**
1,8 – Dihydroxy – 9 – anthrone
$C_{14}H_{10}O_3$
F.R.Ahmed *Eur.Cryst.Meeting*,6, 280, 1980

26.3   **9 – Bromo – 10 – methylanthracene (neutron study)**
$C_{15}H_{11}Br$
R.D.G.Jones, T.R.Welberry
Acta Crystallogr.,Sect.B,36, 852, 1980

26.4   **9 – Bromo – 10 – methylanthracene**
$C_{15}H_{11}Br$
R.D.G.Jones. T.R.Welberry
Acta Crystallogr.,Sect.B,36, 852, 1980

26.5   **10 – Bromo – 1,8 – diphenylanthracene**
$C_{26}H_{17}Br$
H.O.House, N.I.Ghali, J.L.Haack, D.VanDerveer
J.Org.Chem.,45, 1807, 1980

26.C   **11 – Deoxy – daunorubicin aglycone triacetate**
$C_{27}H_{24}O_{10}$ Main entry is 50.18

26.C   **10,10′ – Ethanediylidene – dianthrone (form iii)**
Dianthronylidene – ethane
$C_{30}H_{18}O_2$ Main entry is 28.19

26.6   **(2.4)(9,10)Anthracenophane**
$C_{34}H_{28}$
A.Dunand, J.Ferguson, M.Puza, G.B.Robertson
J.Am.Chem.Soc.,102, 3524, 1980

26.7   **(2.5)(9,10)Anthracenophane**
$C_{35}H_{30}$
A.Dunand, J.Ferguson, M.Puza, G.B.Robertson
J.Am.Chem.Soc.,102, 3524, 1980

POLYCYCLIC HYDROCARBONS
(2 FUSED RINGS)

27.1 2β – Bromo – 3α – hydroxy – bicyclo(3.2.0)heptan – 6 – one (absolute configuration)
$C_7H_9BrO_2$
A.Brown, R.Glen, P.Murray–Rust, J.Murray–Rust, R.F.Newton *J.Chem.Soc.,Chem.Commun.*,1178, 1979

27.2 2β – Chloro – 3α – hydroxy – bicyclo(3.2.0)heptan – 6 – one
$C_7H_9ClO_2$
A.Brown, R.Glen, P.Murray–Rust, J.Murray–Rust, R.F.Newton *J.Chem.Soc.,Chem.Commun.*,1178, 1979

27.3 r – 1,t – 2,c – 3,t – 4,5,6 – Hexachlorotetralin
$C_{10}H_6Cl_6$
G.W.Burton, P.B.D.de la Mare, L.D.Sibley, J.M.Waters
J.Chem.Res.,**132**, 2024, 1980

27.4 1,3 – Di(methylthio) – 2,2,4,4 – tetramethyl – bicyclo(1.1.0)butane (at –40°C)
$C_{10}H_{18}S_2$
P.G.Gassman, M.J.Mullins *Tetrahedron Lett.*,4457, 1979

27.5 2 – Acetyl – indan – 1,3 – dione
$C_{11}H_8O_3$
J.D.Korp, I.Bernal, T.L.Lemke
Acta Crystallogr.,Sect.B,**36**, 428, 1980

27.6 7,7 – Dichloro – 1,6 – dimethyl – bicyclo(4.1.0) heptane – trans – 3,trans – 4 – dicarbonitrile
$C_{11}H_{12}Cl_2N_2$
N.D.Ebby, J.Lapasset, L.Pizzala, J.–P.Aycard, H.Bodot
Acta Crystallogr.,Sect.B,**36**, 184, 1980

27.7 5,7 – Dimethoxy – indan – 1 – one
$C_{11}H_{12}O_3$
M.P.Gupta, P.Ram *Indian J.Phys.*,**53**, 474, 1979

27.8 1,4,5,6,7,7a – Hexahydro – 3,4β – dimethyl – 2H – 4α – hydroxy – indene
$C_{11}H_{18}O$
C.Riche, F.Weisbuch, G.Dana
Acta Crystallogr.,Sect.B,**36**, 2173, 1980

27.9 7α – Hydroxy – bicyclo(5.4.0)undecan – 9 – one
$C_{11}H_{18}O_2$
J.V.Silverton, M.Ziffer, H.Ziffer
J.Org.Chem.,**44**, 3959, 1979

27.10 5α,8α – Dimethyl – 4aβ,5,8,8aβ – tetrahydro – 1 – naphthoquin – 4α – ol
4α – Hydroxy – 5α,8α – dimethyl – 4aβ,5,8,8aβ – tetrahydro – 1(4H) – naphthalenone
$C_{12}H_{16}O_2$
T.J.Greenhough, J.Trotter
Acta Crystallogr.,Sect.B,**36**, 2091, 1980

27.C 4 – (N' – (p – Nitrobenzene) – N – hydrazo) – 8 – oxatricyclo(5.1.0.0²,⁵)octane
$C_{13}H_{13}N_3O_3$ Main entry is 38.52

27.11 2 – Pivaloyl – indan – 1,3 – dione (space group P21χm)
$C_{14}H_{14}O_3$
J.D.Korp, I.Bernal, T.L.Lemke
Acta Crystallogr.,Sect.B,**36**, 428, 1980

27.12 2,3,4aβ,8aβ – Tetramethyl – 4a,5,8,8a – tetrahydro – 1 – naphthoquin – 4β – ol
4β – Hydroxy – 2,3,4aβ,8aβ – tetramethyl – 4a,5,8,8a – tetrahydro – 1(4H) – naphthalenone
$C_{14}H_{20}O_2$
T.J.Greenhough, J.Trotter
Acta Crystallogr.,Sect.B,**36**, 1835, 19

27.13 1 – (1 – Hydroxyethyl) – bicyclo(3.1.0)hexane 3,5 – dinitrobenzoate
$C_{15}H_{16}N_2O_6$
N.G.Steinberg, G.H.Rasmusson, G.F.Reynolds, J.P.Springer, B.H.Arison *J.Org.Chem.*,**44**, 3416, 1979

27.C 2,2,5,7 – Tetramethyl – 4 – oxy – 6 – (2 – oxyethyl) – indanone
$C_{15}H_{20}O_3$ Main entry is 59.10

27.14 5β,9β,10α – Tribromo – 4α – chloro – 4β,7β,8β,11 – tetramethyl – bicyclo(5.4.0)undec – 1(11) – ene (absolute configuration)
$C_{15}H_{22}Br_3Cl$
A.G.Gonzalez, J.Darias, J.D.Martin, V.S.Martin, M.Norte, C.Perez, A.Perales, J.Fayos
Tetrahedron Lett.,**21**, 151, 1980

27.15 1,2,3,4,5 – Pentachloro – 8 – phenyltetralin
$C_{16}H_{11}Cl_5$
G.Bruno, G.Bombieri *Eur.Cryst.Meeting*,**6**, 15, 1980
See also R1 : 24

27.16 Ammonium 1 – anilino – 8 – naphthalene – sulfonate monohydrate
$C_{16}H_{12}NO_3S^-$, H_4N^+, H_2O
L.D.Weber, A.Tulinsky
Acta Crystallogr.,Sect.B,**36**, 611, 1980
See also R1 : 11

27.17 2 – Benzylideneindan – 1 – one
$C_{16}H_{12}O$
A.Hoser, Z.Kaluski, H.Maluszynska, V.D.Orlov
Acta Crystallogr.,Sect.B,**36**, 1256, 1980

27.18 5 – Acetyl – 1,2,3,3a,7,7a – hexahydro – 2,6 – dihydroxy – 1,2,3a – trimethyl – 4 – oxo – 4H – indene – 3,3 – dicarboxylic acid
$C_{16}H_{20}O_8$
K.Takahashi, M.Takani, Y.Wada
Chem.Pharm.Bull.,**28**, 1590, 1980

27.19 2,3,4aβ,6,7,8aβ – Hexamethyl – 4a,5,8,8a – tetrahydro – 1 – naphthoquin – 4α – ol
$C_{16}H_{24}O_2$
T.J.Greenhough, J.Trotter
Acta Crystallogr.,Sect.B,**36**, 1831, 1980

27.20 2,3,4aβ,6,7,8aβ – Hexamethyl – 4a,5,8,8a – tetrahydro – 1 – naphthoquin – 4β – ol
$C_{16}H_{24}O_2$
T.J.Greenhough, J.Trotter
Acta Crystallogr.,Sect.B,**36**, 1831, 1980

27.21 (+) – 2 – Dipropylamino – 5 – hydroxytetralin hydrochloride (absolute configuration)
$C_{16}H_{26}NO^+$, Cl^-
J.Giesecke *Acta Crystallogr.,Sect.B*,**36**, 110, 1980

27.22   meso – 1,1′ – Bi – indenyl
$C_{18}H_{14}$
P.Lustenberger, S.Joss, P.Engel, N.Oesch, W.Rutsch,
M.Neuenschwander *Z.Kristallogr.*,**150**, 235, 1979

27.23   2 – (p – Dimethylaminobenzylidene) – indan – 1,3 –
dione (γ form)
$C_{18}H_{15}NO_2$
N.S.Magomedova, Z.V.Zvonkova, M.G.Neigauz,
L.A.Novakovskaya *Kristallografiya*,**25**, 400, 1980
See also R1 : 16

27.24   4 – Methyl – 6 – (N – methyl – N – phenylamino) –
spiro(4.5)dec – 6 – ene – 1,8 – dione
$C_{18}H_{21}NO_2$
S.Jeannin, Y.Jeannin, J.Martin–Frere
Acta Crystallogr.,Sect.B,**36**, 1703, 1980
See also R1 : 16

27.25   8 – (2 – Bromoethylene) – 3 – hydroxy – 4 –
methoxycarbonyl – 5 – (dimethoxycarbonyl –
methyl) – 7,7 – dimethyl – bicyclo(3.3.0)oct – 3 –
ene
$C_{19}H_{27}BrO_7$
S.Danishefsky, K.Vaughan, R.C.Gadwood, K.Tsuzuki,
J.P.Springer *Tetrahedron Lett.*,**21**, 2625, 1980

27.26   2 – (4′ – Dimethylamino – cinnamoyl) – indan –
1,3 – dione
$C_{20}H_{17}NO_3$
N.S.Magomedova, Z.V.Zvonkova, L.S.Geita,
E.M.Smelyanskaya, S.L.Ginzburg
Zh.Strukt.Khim.,**21**, 131, 1980
See also R1 : 16

27.27   2,6 – Diphenylhomotropylidene
$C_{20}H_{18}$
H.Kessler, W.Ott, H.J.Lindner, H.G.von Schnering,
E.–M.Peters, K.Peters *Chem.Ber.*,**113**, 90, 1980

27.28   cis – 2 – t – Butyl – perhydroazulen – 4 – one –
(p – bromophenylsulfonylhydrazone)
$C_{20}H_{29}BrN_2O_2S$
H.O.House, C.–C.Yau, D.G.VanDerveer
J.Org.Chem.,**44**, 3031, 1979
See also R1 : 9

27.29   4 – Anilino – 3 – methoxycarbonyl – 1 –
(methoxycarbonylmethylidene) – 1,2,3,4 –
tetrahydro – naphthalene
$C_{21}H_{21}NO_4$
C.H.Chao, D.W.Hart, R.Bau, R.F.Heck
J.Organomet.Chem.,**179**, 301, 1979
See also R1 : 24,16,1

27.30   trans – 2 – t – Butyl – perhydroazulen – 4 – one –
(p – tosylhydrazone)
$C_{21}H_{32}N_2O_2S$
H.O.House, C.–C.Yau, D.G.VanDerveer
J.Org.Chem.,**44**, 3031, 1979
See also R1 : 9

27.31   9,10 – Diphenyl – bicyclo(6.2.0)decapentaene
$C_{22}H_{16}$
C.Kabuto, M.Oda *Tetrahedron Lett.*,**21**, 103, 1980

27.32   3 – Methoxy – 1,4 – dimethyl – 1,2 – diphenyl –
indene
$C_{24}H_{22}O$
K.H.Dotz, R.Dietz, C.Kappenstein, D.Neugebauer,
U.Schubert *Chem.Ber.*,**112**, 3682, 1979

27.33   3 – (3,4 – Dimethoxybenzyl) – 4(2,4 –
dimethoxyphenyl) – 1 – semicarbazono – 6,7,8 –
trimethoxy – naphthalene – 2 – one
$C_{31}H_{33}N_3O_9$
W.B.Whalley, G.Ferguson, M.A.Khan
J.Chem.Res.,**144**, 2219, 1980
See also R1 : 9

POLYCYCLIC HYDROCARBONS
(3 FUSED RINGS)

28.1 9 – Hydroxyphenalenone
$C_{13}H_8O_2$
C.Svensson, S.C.Abrahams, J.L.Bernstein, R.C.Haddon
J.Am.Chem.Soc.,**101**, 5759, 1979

28.C Proflavine – cytidylyl – (3 – 5′) – guanosine sulfate hydrate
$1.5C_{13}H_{12}N_3{}^+$, $C_{19}H_{24}N_8O_{12}P^-$, $0.5O_4S^-$, $11.5H_2O$
Main entry is 60.24

28.2 Cyclohepta(d,e)naphthalene (at 100°K)
Pleiadiene
$C_{14}H_{10}$
A.Hazell, R.G.Hazell, F.K.Larsen
Eur.Cryst.Meeting,**5**, 218, 1979

28.3 Cyclohepta(d,e)naphthalene (at 135°K)
Pleiadiene
$C_{14}H_{10}$
A.Hazell, R.G.Hazell, F.K.Larsen
Eur.Cryst.Meeting,**5**, 218, 1979

28.4 Cyclohepta(d,e)naphthalene (at 200°K)
Pleiadiene
$C_{14}H_{10}$
A.Hazell, R.G.Hazell, F.K.Larsen
Eur.Cryst.Meeting,**5**, 218, 1979

28.5 Cyclohepta(d,e)naphthalene (at 78°K)
Pleiadiene
$C_{14}H_{10}$
A.Hazell, R.G.Hazell, F.K.Larsen
Eur.Cryst.Meeting,**5**, 218, 1979

28.6 11 – Methyl – tricyclo(6.2.1.0[4,11])undecapentaene – 2,3 – dicarboxylic acid
7b – Methyl – 7bH – cyclopent(cd)indene – 1,2 – dicarboxylic acid
$C_{14}H_{10}O_4$
T.L.Gilchrist, C.W.Rees, D.Tuddenham, D.J.Williams
J.Chem.Soc.,Chem.Commun.,691, 1980

28.7 9bα – Methyl – 2,3,3aα,4,5,5aβ,6,7,8,9,9aα,9b – dodecahydro – 1H – cyclopenta(a)naphthalene – 1,5 – dione
$C_{14}H_{20}O_2$
R.E.Ireland, W.J.Thompson, N.S.Mandel, G.S.Mandel
J.Org.Chem.,**44**, 3583, 1979

28.C 3aα,4α,4aβ,7aα,8α,9aβ – (±) – Decahydro – 4 – hydroxy – 4a,8 – dimethylazuleno(6,5 – b)furan – 2,5(3H) – dione
$C_{14}H_{20}O_4$ Main entry is 38.64

28.8 2 – Acetylaminofluorene
$C_{15}H_{13}NO$
M.van Meerssche, G.Germain, J.P.Declercq, R.Touillaux, M.Roberfroid, C.Razzouk
Cryst.Struct.Commun.,**9**, 515, 1980

28.C Hysterin
$C_{17}H_{24}O_5$ Main entry is 38.86

28.9 Perchloro – 9 – phenyl – 3 – fluorenone
$C_{19}Cl_{12}O$
X.Solans, C.Miravitlles, F.Plana, S.Galf, M.Font–Altaba, G.Germain, J.P.Declercq *Eur.Cryst.Meeting*,**6**, 50, 1980

28.10 1 – Phenoxy – 1,2,2a,8b – tetrahydro – cyclobuta(a) naphthalene – 8b – carbonitrile
$C_{19}H_{15}NO$
H.Matsuura, Y.Kai, N.Yasuoka, N.Kasai
Bull.Chem.Soc.Jpn.,**53**, 359, 1980

28.11 4 – (Diethylamino) – 1,10 – ethano – 5 – methyl – cyclopentacyclononene
$C_{19}H_{23}N$
H.J.Lindner, B.Kitschke, K.Hafner, W.Ude
Acta Crystallogr.,Sect.B,**36**, 756, 1980

28.12 7 – Methyl – 3 – methylene – tricyclo(5.3.0.0[1,4])dec – 5 – yl – phenylcarbamate
$C_{19}H_{23}NO_2$
F.R.Ahmed, M.Przybylska
Acta Crystallogr.,Sect.B,**36**, 1718, 1980

28.13 5 – Acetoxy – 2,6 – dichloro – 4aα,10aα – dihydro – 8 – hydroxy – 10β – methyl – 4aα – propionyl – phenanthrene – 1,4,9(10H) – trione
$C_{20}H_{16}Cl_2O_7$
Y.Miyagi, K.Maruyama, H.Ishii, S.Mizuno, M.Kakudo, N.Tanaka, Y.Matsuura, S.Harada
Bull.Chem.Soc.Jpn.,**52**, 3019, 1979

28.14 8 – Diethylamino – 1,3,4a,8a – tetramethyl – 4a,8a – dihydrocyclopenta(1,2 – a)indene
$C_{20}H_{27}N$
H.J.Lindner, B.Kitschke, K.Hafner, W.Ude
Acta Crystallogr.,Sect.B,**36**, 758, 1980

28.15 cis – 9 – Phenyl – 10 – methyl – 9,10 – dihydrophenanthrene
$C_{21}H_{18}$
R.Lapouyade, R.Koussini, A.Nourmamode, C.Courseille
J.Chem.Soc.,Chem.Commun.,740, 1980

28.C Vismione A
$C_{23}H_{26}O_6$ Main entry is 59.34

28.16 7 – Benzoylamino – 7 – methoxycarbonyl – dispiro(5.1.5.2)pentadecane – 14,15 – dione acetonitrile solvate
$C_{24}H_{29}NO_5$, $0.5C_2H_3N$
S.Mohr *Tetrahedron Lett.*,**21**, 593, 1980

28.17 9H – Heptadecachloro – bis – 9 – fluorenyl
$C_{26}HCl_{17}$
X.Solans, C.Miravitlles, F.Plana, S.Galf, M.Font–Altaba, G.Germain, J.P.Declercq *Eur.Cryst.Meeting*,**6**, 50, 1980

28.18 9H,9′H – Hexadecachloro – bis – 9 – fluorenyl
$C_{26}H_2Cl_{16}$
X.Solans, C.Miravitlles, F.Plana, S.Galf, M.Font–Altaba, G.Germain, J.P.Declercq *Eur.Cryst.Meeting*,**6**, 50, 1980

28.C (E) – 3α – Acetoxy – 5,10 – seco – cholest – 1(10) – en – 5 – one
$C_{29}H_{48}O_3$ Main entry is 51.75

28.19 **10,10′ – Ethanediylidene – dianthrone (form iii)**
Dianthronylidene – ethane
$C_{30}H_{18}O_2$
H.–D.Becker, K.Sandros, B.Karlsson, A.–M.Pilotti
Chem.Phys.Lett.,**53**, 232, 1978
See also R1 : 26

28.C **(±) – Methyl – 4,5 – dimethoxy – 2 – (2,6 – dimethoxy – 1 – oxo – 9 – phenyl – 5 – phenalenyl) – 1 – oxo – 8 – phenyl – 1,2 – dihydro – 2 – acenaphthylene – carboxylate**
$C_{43}H_{32}O_8$ Main entry is 59.46

POLYCYCLIC HYDROCARBONS
(4 FUSED RINGS)

29.1 **Trispiro(2.0.2.0.2.0)nonane (at –40℃)**
(3) – Rotane
C_9H_{12}
T.Prange, C.Pascard, A.de Meijere, U.Behrens, J.–P.Barnier, J.–M.Conia *Nouv.J.Chim.*,**4**, 321, 1980

29.2 **Naphtho(b,e)dicyclobutane**
$C_{14}H_{12}$
J.D.Korp, I.Bernal *J.Am.Chem.Soc.*,**101**, 4273, 1979

29.3 **anti – 1′,1′,2,2 – Tetrachloro – 3,3 – dimethyl – 1a′,6b′ – dihydro – spiro(cyclopropane – 1,2′ – (1H) – cycloprop(a)indene)**
$C_{14}H_{12}Cl_4$
T.J.Bartczak, Z.Galdecki, T.Glowiak
Acta Crystallogr.,Sect.B,**36**, 1226, 1980

29.C **Pyrene – fluoranil complex**
$C_{16}H_{10}$, $C_6F_4O_2$ Main entry is 60.34

29.4 **(±) – trans – 1,2 – Dihydro – 1,2 – dihydroxybenz(a) anthracene**
$C_{18}H_{14}O_2$
D.E.Zacharias, J.P.Glusker, P.P.Fu, R.G.Harvey
J.Am.Chem.Soc.,**101**, 4043, 1979

29.5 **(±) – trans – 10,11 – Dihydro – 10,11 – dihydroxybenz(a)anthracene**
$C_{18}H_{14}O_2$
D.E.Zacharias, J.P.Glusker, P.P.Fu, R.G.Harvey
J.Am.Chem.Soc.,**101**, 4043, 1979

29.6 **1 – Methylbenz(a)anthracene**
$C_{19}H_{14}$
D.W.Jones, J.M.Sowden
Cancer Biochem.Biophys.,**4**, 43, 1979

29.7 **7,8 – Dihydro – 1,6 – dimethoxynaphthacene – 5,10,12(9H) – trione**
$C_{20}H_{18}O_5$
R.K.Boeckman Junior, M.H.Delton, T.M.Dolak, T.Watanabe, M.D.Glick *J.Org.Chem.*,**44**, 4396, 1979

29.C **(1RS,3RS,6SR,7SR,10SR,11RS,12RS,13RS) – (6 – Phenyl – tetracyclo(5.4.2.0^{3,13}.0^{10,12})tridecane – 4,8 – dien – 11 – yl) acetic acid**
Endiandric acid
$C_{21}H_{22}O_2$ Main entry is 59.29

29.8 **(1S,4R) – 3′ – Chloro – N,N – dimethyl – spiro(cyclohex – 2 – ene – 1,5′ – (5H) – dibenzo(a,d) cycloheptene) – 4 – amine hydrochloride (absolute configuration)**
$C_{22}H_{23}ClN^+$, Cl^-
A.Wagner *Acta Crystallogr.,Sect.B*,**36**, 1113, 1980

29.C **4 – epi – Oxytetracycline tetrahydrate dichloromethane solvate (at 120°K)**
$C_{22}H_{24}N_2O_9$. $4H_2O$, CH_2Cl_2 Main entry is 50.15

29.9 (1S,4R) – 3′ – Chloro – 10′,11′ – dihydro – N,N – dimethyl – spiro(cyclohex – 2 – ene – 1,5′ – (5H) – dibenzo(a,d)cycloheptene) – 4 – ammonium hydrogen maleate (absolute configuration)
$C_{22}H_{25}ClN^+$, $C_4H_3O_4^-$
A.Wagner *Acta Crystallogr.,Sect.B*,**36**, 813, 1980
See also R2 : 2,1

29.C 1α,3β – 1,3 – Dehydro – 5,10 – seco – cholest – 10(19) – en – 5 – one
$C_{27}H_{44}O$ Main entry is 51.63

29.C 1β,3α – 1,3 – Dehydro – 5,10 – seco – cholest – 10(19) – en – 5 – one
$C_{27}H_{44}O$ Main entry is 51.64

29.C 1α,3β – 1,3 – Dehydro – 10α – hydroxy – 5,10 – seco – cholestan – 5 – one
$C_{27}H_{46}O_2$ Main entry is 51.67

29.10 Perchloro – 5,10 – diphenyl – dibenzo(a,e)pentalene carbon tetrachloride solvate
$C_{28}Cl_{18}$, $0.5CCl_4$
C.Miravitlles, X.Solans, G.Germain, J.P.Declercq *Acta Crystallogr.,Sect.B*,**35**, 2809, 1979

29.11 Adriamycin – 14 – 0 – valerate hydrochloride solvate (at about 120°K)
$C_{32}H_{38}NO_{12}^+$, Cl^-
E.Eckle, J.J.Stezowski *Eur.Cryst.Meeting*,**6**, 296, 1980

29.C trans – Dichloro – (2,11 – bis(diphenylarsinomethyl) – benzo(c) phenanthrene) – platinum
$C_{44}H_{34}As_2Cl_2Pt$ Main entry is 65.26

29.12 1,1,2,3,4,5,6 – Heptaphenyl – 1,4 – dihydrobenz(e) – as – indacene
$C_{58}H_{40}$
M.Shoja, A.A.Espiritu, J.G.White, I.J.Borowitz *Acta Crystallogr.,Sect.B*,**36**, 1967, 1980

POLYCYCLIC HYDROCARBONS
(5 OR MORE FUSED RINGS)

30.1 Tetraspiro(2.0.2.0.2.0.2.0)dodecane
(4) – Rotane
$C_{12}H_{16}$
T.Prange, C.Pascard, A.de Meijere, U.Behrens, J.–P.Barnier, J.–M.Conia *Nouv.J.Chim.*,**4**, 321, 1980

30.2 2,3,4,5,6,7 – Hexafluoro – pentacyclo(6.4.0.1⁹,¹²,0²,⁷0³,⁶)tridecane – 4,10 – diene
$C_{13}H_8F_6$
L.Golic, I.Leban *Cryst.Struct.Commun.*,**9**, 739, 1980

30.3 Pentaspiro(2.0.2.0.2.0.2.0.2.0)pentadecane
(5) – Rotane
$C_{15}H_{20}$
T.Prange, C.Pascard, A.de Meijere, U.Behrens, J.–P.Barnier, J.–M.Conia *Nouv.J.Chim.*,**4**, 321, 1980

30.4 Naphtho(2,3 – 6,7)dicyclobutene(1,8:4,5) dicyclopentene
$C_{18}H_{16}$
D.L.Ward *Acta Crystallogr.,Sect.B*,**36**, 963, 1980

30.5 Hexaspiro(2.0.2.0.2.0.2.0.2.0)octadecane (at –100°C)
(6) – Rotane
$C_{18}H_{24}$
T.Prange, C.Pascard, A.de Meijere, U.Behrens, J.–P.Barnier, J.–M.Conia *Nouv.J.Chim.*,**4**, 321, 1980

30.6 7,8 – Dimethoxybenzo(j)fluoranthene
$C_{22}H_{16}O_2$
C.E.Briant, D.W.Jones *Eur.Cryst.Meeting*,**5**, 394, 1979

30.7 5,10 – Dimethoxybenzo(j)fluoranthene
$C_{22}H_{16}O_2$
C.E.Briant, D.W.Jones *Eur.Cryst.Meeting*,**5**, 394, 1979

30.8 (1α,2β,8α,9β) – 4,5:12,13 – Dibenzo – tricyclo(7.5.0.0²,⁸)tetradecane – 4,12 – diene – 3,14 – dione
$C_{22}H_{20}O_2$
H.Hart, E.Dunkelblum *J.Org.Chem.*,**44**, 4752, 1979

30.9 (1α,2β,8α,9β) – 6,7:10,11 – Dibenzo – tricyclo(7.5.0.0²,⁸)tetradecane – 6,10 – diene – 3,14 – dione
$C_{22}H_{20}O_2$
H.Hart, E.Dunkelblum *J.Org.Chem.*,**44**, 4752, 1979

30.10 (1α,2β,8β,9α) – 4,5:11,12 – Dibenzo – tricyclo(7.5.0.0²,⁸)tetradecane – 4,11 – diene – 3,10 – dione
$C_{22}H_{20}O_2$
H.Hart, E.Dunkelblum *J.Org.Chem.*,**44**, 4752, 1979

30.11 (1α,2α,8β,9α) – 4,5:11,12 – Dibenzo – tricyclo(7.5.0.0²,⁸)tetradecane – 4,11 – diene – 3,10 – dione
$C_{22}H_{20}O_2$
H.Hart, E.Dunkelblum *J.Org.Chem.*,**44**, 4752, 1979

30.12 $(1\alpha,2\beta,8\alpha,9\alpha) - 4,5{:}12,13 -$ Dibenzo –
tricyclo(7.5.0.0^{2,8})tetradecane – 4,12 – diene – 3,14 –
dione
$C_{22}H_{20}O_2$
H.Hart, E.Dunkelblum *J.Org.Chem.*,**44**, 4752, 1979

30.13 Tetrabenzo – cyclodecane – 1,16 – dione
$C_{28}H_{16}O_2$
T.S.Cameron, C.Chan, D.G.Morris, A.G.Shepherd
Can.J.Chem.,**58**, 777, 1980

30.14 3,4:13,14 – Dibenzo – 6,6,11,11 – tetramethyl –
1,8,9 – cis – tricyclo(7.5.0.0^{2,8})tetradecane – 5,12 –
dione
$C_{28}H_{28}O_2$
K.–T.Wei, D.L.Ward
Acta Crystallogr.,Sect.B,**35**, 2746, 1979

30.15 1 – Formylhexahelicene
$C_{27}H_{16}O$
H.J.Lindner, B.Kitschke
Bull.Soc.Chim.Belg.,**88**, 831, 1979

30.16 1 – Methylhexahelicene
$C_{27}H_{18}$
H.M.Doesburg *Cryst.Struct.Commun.*,**9**, 137, 1980

30.17 1 – Acetylhexahelicene
$C_{28}H_{18}O$
H.J.Lindner, B.Kitschke
Bull.Soc.Chim.Belg.,**88**, 831, 1979

30.18 14 – Ethoxy – 8,14 – di(phenoxy) – 3,4:10,11 –
dibenzo – tricyclo(7.5.0.0^{2,8})tetradecane – 3,5,10 –
triene – 7,13 – dione chloroform solvate
$C_{36}H_{30}O_5$, 0.93CHCl$_3$
O.L.Chapman, S.C.Busman, K.N.Trueblood
J.Am.Chem.Soc.,**101**, 7067, 1979

30.19 6b,6b':7,7' – bis(8 – Acetoxy – 6b,7 – dihydro – 7,9 –
dimethyl – 8H – cyclopent(a)acenaphthylenediyl)
benzene solvate
$C_{38}H_{32}O_4$, 0.5C$_6$H$_6$
D.W.Jones, W.S.McDonald
J.Chem.Soc.,Chem.Commun.,417, 1980

30.20 Tetrabenzo(de,no,st,c$_1$d$_1$)heptacene
$C_{42}H_{22}$
G.Ferguson, M.Parvez
Acta Crystallogr.,Sect.B,**35**, 2419, 1979

30.21 Diphenanthro(4,3 – a.3',4' – o)picene
$C_{46}H_{26}$
W.Marsh, J.D.Dunitz *Bull.Soc.Chim.Belg.*,**88**, 847, 1979

30.22 Kekulene
$C_{48}H_{24}$
C.Krieger, F.Diederich, D.Schweitzer, H.A.Staab
Angew.Chem.,Int.Ed.Engl.,**18**, 699, 1979

BRIDGED RING HYDROCARBONS

31.1 1,5 – Dimethyl – tricyclo(2.1.0.0^{2,5})pentan – 3 – one
C_7H_8O
H.Irngartinger, K.L.Lukas
Angew.Chem.,Int.Ed.Engl.,**18**, 694, 1979

31.2 2,7 – Dibromo – tricyclo(4.2.1.0^{3,7})nonan – 4 – one
$C_9H_{10}Br_2O$
J.M.J.Vankan, A.J.H.Klunder, J.H.Noordik, B.Zwanenburg
Rec.J.Roy.Netherl.Chem.Soc.,**99**, 213, 1980

31.3 exo – 7 – Hydroxy – bicyclo(3.3.1)nonan – 3 – one
$C_9H_{14}O_2$
J.Murray–Rust, P.Murray–Rust, W.C.Parker, R.L.Tranter,
C.I.F.Watt *J.Chem.Soc.,Perkin Trans.2*,1496, 1979

31.4 6 – exo – (Methylthio) – bicyclo(2.2.1)heptane – 2 –
endo – carboxylic acid
$C_9H_{14}O_2S$
R.S.Glass, J.R.Duchek, U.D.G.Prabhu, W.N.Setzer, G.S.Wilson
J.Org.Chem.,**45**, 3640, 1980

31.5 6 – endo – (Methylthio) – bicyclo(2.2.1)heptane – 2 –
endo – carboxylic acid
$C_9H_{14}O_2S$
R.S.Glass, J.R.Duchek, U.D.G.Prabhu, W.N.Setzer, G.S.Wilson
J.Org.Chem.,**45**, 3640, 1980

31.6 7,8 – Dibromo – 8 – cyano – 4 – methylsulfonato –
bicyclo(3.2.1)octa – 2,6 – diene
$C_{10}H_9Br_2NO_3S$
G.Mehta, K.S.Rao, S.C.Suri, T.S.Cameron, C.Chan
J.Chem.Soc.,Chem.Commun.,650, 1980

31.7 7 – Bromo – 2 – chloro – tricyclo(4.3.1.0^{3,7})decane –
4,8 – dione
$C_{10}H_{10}BrClO_2$
P.A.J.Prick, J.H.Noordik
Cryst.Struct.Commun.,**9**, 193, 1980

31.C (Ethylenediamine) – (N,N' –
dimethylethylenediamine) – platinum bis(α –
bromocamphor – π – sulfonate)
2C$_{10}$H$_{14}$BrO$_4$S$^-$, C$_6$H$_{20}$N$_4$Pt^{2+} Main entry is 76.33

31.8 1,8 – Di – iodo – tricyclo(5.3.0.0^{4,8})decane
$C_{10}H_{14}I_2$
J.R.Wiseman, J.J.Vanderbilt, W.M.Butler
J.Org.Chem.,**45**, 667, 1980

31.9 1,7,7 – Trimethyl – bicyclo(2.2.1)hepta – 2,3 – dione
(monoclinic form)
1 – Camphorquinone
$C_{10}H_{14}O_2$
W.M.Bright, J.F.Cannon, D.A.Langs, J.V.Silverton
Cryst.Struct.Commun.,**9**, 251, 1980

31.10 syn – γ – Camphorquinone – dioxime
$C_{10}H_{16}N_2O_2$
M.S.Ma, R.J.Angelici *Inorg.Chem.*,**19**, 363, 1980
See also R1 : 10

31.11 exo – 7 – Methoxy – bicyclo(3.3.1)nonan – 3 – one
$C_{10}H_{16}O_2$
J.Murray–Rust, P.Murray–Rust, W.C.Parker, R.L.Tranter,
C.I.F.Watt *J.Chem.Soc.,Perkin Trans.2*,1496, 1979

31.12 (–) – 5 – endo,6 – exo – Dihydroxy – camphene
$C_{10}H_{16}O_2$
S.A.Spencer, J.Trotter
Acta Crystallogr.,Sect.B,**35**, 3110, 1979

31.13 5,6 – Dicyano – 1 – methoxy – bicyclo(2.2.2)oct – 2 –
ene
$C_{11}H_{12}N_2O$
W.B.T.Cruse, I.Fleming, P.T.Gallagher, O.Kennard
J.Chem.Res.,**372**, 4418, 1979

31.14 Tricyclo(3.3.1.1³,⁷)decane – 1 – carbonitrile
1 – Cyano – adamantane
$C_{11}H_{15}N$
J.P.Amoureux, M.Bee
Acta Crystallogr.,Sect.B,**35**, 2957, 1979

31.C exo – 2,exo – 6 – Dihydroxy – 2,6 – dimethyl –
bicyclo(3.3.1)nonane ethylacetate clathrate
$3C_{11}H_{20}O_2$, $C_4H_8O_2$ Main entry is 61.1

31.15 4,5,9,10,11,12 – Hexafluoro –
pentacyclo(6.4.0³,⁶.0⁴,¹²,0⁵,⁹)dodec – 10 – ene
$C_{12}H_8F_6$
L.Golic, I.Leban *Acta Crystallogr.,Sect.B*,**36**, 1520, 1980

31.16 4 – Bromo – 4 – cyano – 7 – methoxycarbonyl – 7 –
methylsulfonato – tetracyclo(3.3.0.0²,⁶.0³,⁸)octane
$C_{12}H_{12}BrNO_5S$
G.Mehta, K.S.Rao, S.C.Suri, T.S.Cameron, C.Chan
J.Chem.Soc.,Chem.Commun.,650, 1980

31.C endo – 2 – Dimethylphosphono – exo – 2 –
hydroxy – (–) – camphane
$C_{12}H_{23}O_4P$ Main entry is 64.54

31.17 3 – Chloro – 5 – (2′,4′ – dinitrophenylthio) –
nortricyclene
$C_{13}H_{11}ClN_2O_4S$
A.M.Nersisyan, S.V.Lindeman, V.G.Andrianov,
Yu.T.Struchkov, R.Sh.Akhmedova, V.B.Rybakov,
N.K.Sadovaya, N.S.Zefirov
Cryst.Struct.Commun.,**9**, 247, 1980

31.18 trans – 5 – Chloro – 6 – (2 – nitrophenylthio) –
bicyclo(2.2.1)hept – 2 – ene
$C_{13}H_{12}ClNO_2S$
B.B.Sedov, T.F.Rau, V.G.Rau, Yu.T.Struchkov,
R.Sh.Akhmedova, N.S.Zefirov, N.K.Sadovaya
Cryst.Struct.Commun.,**9**, 633, 1980

31.19 (3aRS,4RS,6RS,8aSR) – 5 – Methylene – octahydro –
4H – 3a,6 – methano – azulene – 4 – carboxylic
acid
$C_{13}H_{18}O_2$
J.V.Turner, B.F.Anderson. L.N.Mander
Aust.J.Chem.,**33**, 1061, 1980

31.20 6 – (Bromomethyl) – 1,2,4 – trichloro – 3 – ethoxy –
3,7,7 – trimethoxy – bicyclo(2.2.1)heptane
$C_{13}H_{20}BrCl_3O_4$
A.T.H.Lenstra, W.Van de Mieroop, H.J.Geise. J.Van Bree.
M.Anteunis *Rec.J.Roy.Netherl.Chem.Soc.*,**99**, 118, 1980

31.21 (1R,3S,4R) – N,N – Dimethyl – 3 – camphor –
carbothioamide
$C_{13}H_{21}NOS$
R.Roques, A.M.Lamazouere, J.Sotiropoulos, J.P.Declercq,
G.Germain *Acta Crystallogr.,Sect.B*,**36**, 1569, 1980

31.C 1,7,8,9,10 – Pentachloro – 3 – phenyl – 4 – aza – 5 –
oxatricyclo(5.2.1.0²,⁶)decane – 3,8 – diene
$C_{14}H_8Cl_5NO$ Main entry is 40.31

31.C 1,7,8,9 – Tetrachloro – 3 – phenyl – 4 – aza – 5 –
oxatricyclo(5.2.1.0²,⁶)decane – 3,8 – diene
$C_{14}H_9Cl_4NO$ Main entry is 40.32

31.22 2 – Chloromethyl – 5 – (2′,4′ – dinitrophenylthio) –
nortricyclene
$C_{14}H_{13}ClN_2O_4S$
A.M.Nersisyan, A.I.Yanovskii, Yu.T.Struchkov,
R.Sh.Akhmedova, N.K.Sadovaya, N.S.Zefirov
Cryst.Struct.Commun.,**9**, 393, 1980
See also R1 : 15,11

31.23 2,3 – bis(Methoxycarboxyl) – 5 – hydroxy – 1,5 –
dimethyl – bicyclo(2.2.2)octa – 2,7 – dien – 6 – one
$C_{14}H_{16}O_6$
H.–D.Becker, B.Ruge, B.W.Skelton, A.H.White
Aust.J.Chem.,**32**, 1687, 1979

31.24 1 – Chloro – 4 – phenyl – bicyclo(2.2.2)octane
$C_{14}H_{17}Cl$
P.E.Bourne, M.R.Taylor *Eur.Cryst.Meeting*,**5**, 395, 1979

31.25 1 – Fluoro – 4 – phenyl – bicyclo(2.2.2)octane
$C_{14}H_{17}F$
P.E.Bourne, M.R.Taylor *Eur.Cryst.Meeting*,**5**, 395, 1979

31.26 1 – Phenyl – bicyclo(2.2.2)octane
$C_{14}H_{18}$
P.E.Bourne, M.R.Taylor *Eur.Cryst.Meeting*,**5**, 395, 1979

31.27 2,3 – bis(Methoxycarbonyl) – 5 – hydroxy – 1,5 –
dimethyl – bicyclo(2.2.2)oct – 2 – en – 6 – one
$C_{14}H_{18}O_6$
H.–D.Becker, B.Ruge, B.W.Skelton, A.H.White
Aust.J.Chem.,**32**, 1687, 1979

31.28 2,3 – bis(Methoxycarbonyl) – 5 – hydroxy – 1,5 –
dimethyl – bicyclo(2.2.2)oct – 7 – en – 6 – one
$C_{14}H_{18}O_6$
H.–D.Becker, B.Ruge, B.W.Skelton, A.H.White
Aust.J.Chem.,**32**, 1687, 1979

31.29 E – Diamantan – 3 – one – oxime
$C_{14}H_{19}NO$
J.P.Declercq, G.Germain, M.van Meerssche, M.Hajek,
K.Volka *Bull.Soc.Chim.Belg.*,**88**, 1019, 1979

31.30 Z – Diamantan – 3 – one – oxime
$C_{14}H_{19}NO$
J.P.Declercq, G.Germain, M.van Meerssche, M.Hajek,
K.Volka *Bull.Soc.Chim.Belg.*,**88**, 1019, 1979

31.31 syn – (1R,2R,4S) – 4 – Methyl – spiro(bicyclo(2.2.2)
octane – 2,1′ – cyclohexane) – 3′,6 – dione
$C_{14}H_{20}O_2$
J.E.Gurst, R.W.Miller, A.T.McPhail
Tetrahedron Lett.,**21**, 3223, 1980

31.32　3 – Acetoxy – 5 – (2 – nitrophenylthio) –
nortricyclene
$C_{15}H_{15}NO_4S$
B.B.Sedov, T.F.Rau, V.G.Rau, Yu.T.Struchkov,
R.Sh.Akhmedova, N.S.Zefirov, N.K.Sadovaya
Cryst.Struct.Commun.,**9**, 639, 1980

31.33　**Tetracyclo(6.4.3.0.0^{2,7})pentadec – 4 – en – 13 – one**
$C_{15}H_{20}O$
S.Koshibe, Y.Kai, N.Yasuoka, N.Kasai
Bull.Chem.Soc.Jpn.,**53**, 621, 1980

31.34　**Tetracyclo(6.4.3.0.0^{2,7})pentadec – 4 – en – 13 – one
(at –160°C)**
$C_{15}H_{20}O$
S.Koshibe, Y.Kai, N.Yasuoka, N.Kasai
Bull.Chem.Soc.Jpn.,**53**, 621, 1980

31.35　**2,6,10 – Trimethyl – tricyclo(7.2.1.0^{2,7})dodecan – 11 –
one**
$C_{15}H_{22}O$
S.Inayama, A.K.Singh, T.Kawamata, T.Hirose, Y.Iitaka
Chem.Lett.,1219, 1979

31.36　**(+) – 3,7 – Dimethyl – 1,1 – (2 – methyl – 1 –
oxopropyl) – tricyclo(4.2.1.0^{3,7})nonan – 2 – one**
$C_{15}H_{22}O_2$
P.Cachia, N.Darby, T.C.W.Mak, T.Money, J.Trotter
Can.J.Chem.,**58**, 1172, 1980

31.C　**2,4 – Diamino – 5 – (1 – adamantyl) – 6 –
methylpyrimidine ethylsulfonate**
$C_{15}H_{23}N_4^+$, $C_2H_5O_3S^-$ Main entry is 44.33

31.37　**Dimethyl – trans – 3 – endo – chloro – mercury –
4 – acetoxy – tricyclo(4.2.2.0^{2,5})dec – 7 – ene –
9,10 – cis – endo – dicarboxylate**
$C_{16}H_{19}ClHgO_6$
N.S.Zefirov, A.S.Koz'min, V.N.Kirin, B.B.Sedov, V.G.Rau
Tetrahedron Lett.,**21**, 1667, 1980

31.C　**5,5′ – Dinitro – 2 – (1 – adamantyl) – 2′ –
carbomethoxy – (2H,5H) – furan**
$C_{16}H_{20}N_2O_7$ Main entry is 38.77

31.38　**(10RS,5RS,9SR) – 10β – Ethyl – 5,6,7,8,9,10 –
hexahydro – 5α,9α – methanobenzo – cyclo –
octene – 10α – carboxamide**
$C_{16}H_{21}NO$
B.Kojic-Prodic, Z.Ruzic-Toros, L.Golic
Acta Crystallogr.,Sect.B,**36**, 388, 1980

31.39　**3,4,6,8β,9,10 – Hexamethyl – tetracyclo(4.3.1.0^{3,10}.0^{4,9})
decane – 2,5 – dione**
$C_{16}H_{22}O_2$
T.J.Greenhough, J.Trotter
Acta Crystallogr.,Sect.B,**35**, 3084, 1979

31.40　**1β,3,4β,6β,8 – Pentamethyl – 9 – exo – methylene –
tricyclo(4.4.0.0^{3,8})decane – 2,5 – dione**
$C_{16}H_{22}O_2$
T.J.Greenhough, J.Trotter
Acta Crystallogr.,Sect.B,**36**, 478, 1980

31.41　**6,12 – Dihydroxy – 2,6,9,12 – tetramethyl –
tetracyclo(6.2.2.0^{2,7}.0^{4,9})dodecane – 5,11 – dione**
$C_{16}H_{22}O_4$
K.Hirao, T.Iwakuma, M.Taniguchi, O.Yonemitsu, T.Date,
K.Kotera *J.Chem.Soc.,Perkin Trans.1*,163, 1980

31.C　**N – (1′ – t – Butyl – spiro(adamantane – 2,2′ –
aziridine) – 3′ – ylidene) – methylamine**
$C_{16}H_{26}N_2$ Main entry is 32.59

31.42　**1,5 – Diphenyl – tricyclo(2.1.0.0^{2,5})pentan – 3 – one**
$C_{17}H_{12}O$
H.Irngartinger, K.L.Lukas
Angew.Chem.,Int.Ed.Engl.,**18**, 694, 1979

31.43　**2 – Adamantyl – O,N,N – azoxy – p –
toluenesulfonate**
2 – (p – Tolylsulfonyloxy – N,N,O – azoxy) –
adamantane
$C_{17}H_{22}N_2O_4S$
H.Maskill, P.Murray–Rust, J.T.Thompson, A.A.Wilson
J.Chem.Soc.,Chem.Commun.,788, 1980
See also R1 : 9

31.C　**2 – Methoxycarbonyl – 9 – methylene – 8 –
trimethylsiloxy – tricyclo(6.2.1.0^{1,5})undecan – 3 –
one**
$C_{17}H_{26}O_4Si$ Main entry is 63.22

31.44　**(E) – 2 – Chloro – 2 – ethynyl – 5 – phenyl –
adamantane**
$C_{18}H_{19}Cl$
Y.Okaya, S.Y.Lin, D.M.Chiou, W.J.LeNoble
Acta Crystallogr.,Sect.B,**36**, 977, 1980

31.45　**(Z) – 2 – Chloro – 2 – ethynyl – 5 – phenyl –
adamantane**
$C_{18}H_{19}Cl$
Y.Okaya, S.Y.Lin, D.M.Chiou, W.J.LeNoble
Acta Crystallogr.,Sect.B,**36**, 977, 1980

31.46　**Z – 2 – Ethynyl – 5 – phenyl – 2 – adamantanol**
$C_{18}H_{20}O$
Y.Okaya, D.M.Chiou, W.J.le Noble
Acta Crystallogr.,Sect.B,**35**, 2268, 1979

31.47　**Bicyclo(3.3.1)non – 1 – en – 3 – one trans – syn –
dimer**
$C_{18}H_{24}O_2$
H.O.House, M.B.DeTar, D.VanDerveer
J.Org.Chem.,**44**, 3793, 1979

31.48　**Bicyclo(3.3.1)non – 1 – en – 3 – one (cis–syn–dimer)**
$C_{18}H_{24}O_2$
H.O.House, M.B.DeTar, D.VanDerveer
J.Org.Chem.,**44**, 3793, 1979

31.49　**Bicyclo(3.3.1)non – 1 – en – 3 – one
(trans–anti–dimer)**
$C_{18}H_{24}O_2$
H.O.House, M.B.DeTar, D.VanDerveer
J.Org.Chem.,**44**, 3793, 1979

31.C　**syn – 5,12 – Dihydro – 5,14 – (epithio –
methyleneoxy) – 6,12 – epithiodibenzo(a,f)
cyclodecen – 7(6H) – one**
$C_{19}H_{14}O_2S_2$ Main entry is 39.73

31.50　**1endo,4endo:5exo,8exo – Dimethano –
1,2,3,4,4a,5,8,8a – octahydro – naphthalene – 10 –
syn – p – nitrobenzoate**
$C_{19}H_{19}NO_4$
R.A.Pfund, W.B.Schweizer, C.Ganter
Helv.Chim.Acta,**63**, 674, 1980
See also R1 : 13,15

31.51　**4,8 – Ethano – 4 – methoxycarbonyl – 5 –
(dimethoxycarbonyl – methyl) – 7,7 – dimethyl –
bicyclo(3.3.0)octan – 3 – one**
$C_{19}H_{26}O_7$
S.Danishefsky, K.Vaughan, R.C.Gadwood, K.Tsuzuki,
J.P.Springer *Tetrahedron Lett.*,**21**, 2625, 1980

31.52 anti – 10 – Cyano – syn – 10 – ((E) – (2,2,2 – trifluoroethyl) – vinyl) – endo – 5 – cyano – exo – 5((E) – (2,2,2 – trifluoroethyl) – vinyl) – endo – tricyclo(5.2.1.0^{2,6})decane – 3,8 – diene
$C_{20}H_{16}F_6N_2O_2$
P.G.Gassman, J.J.Talley *J.Am.Chem.Soc.*,102, 4138, 1980

31.C Diacetone – phenanthroquinone
$C_{20}H_{20}O_4$ Main entry is 38.103

31.53 syn – 15,18 – Dimethoxy – (2,6) – p – benzoquinone(3.3)metacyclophane
$C_{20}H_{22}O_4$
H.A.Staab, C.P.Herz, A.Dohling, C.Krieger
Chem.Ber.,113, 241, 1980

31.C Benzo(p) – 2 – thiatetracyclo(7.5.3.0^{3,8}.0^{1,10}) heptadecane – 3(8),15 – diene (at –120°C)
$C_{20}H_{24}S$ Main entry is 39.79

31.54 Di – t – adamantyl – disulfide
$C_{20}H_{30}S_2$
F.S.Jorgensen, J.P.Snyder *J.Org.Chem.*,45, 1015, 1980
See also R1 : 11

31.55 exo,exo – Octacyclo(8.8.1.1^{3,6}.1^{12,15}.0^{2,9}.0^{4,8}.0^{11,18}.0^{13,17}) heneicosane
$C_{21}H_{28}$
P.Engel, W.Nowacki, J.Slutsky, P.Grubmuller, P.von R.Schleyer *Chem.Ber.*,112, 3566, 1979

31.56 5 – (2 – Nitrophenylthio) – 10 – acetoxy – tetracyclo(4.4.0.0^{2,4}.0^{3,7})dec – 8 – ene – 8,9 – dicarboxylic acid dimethyl ester
$C_{22}H_{21}NO_6S$
N.S.Zefirov, A.S.Koz'min, V.V.Zhdankin, V.N.Kirin, I.V.Bodrikov, B.B.Sedov, V.G.Rau
Tetrahedron Lett.,3533, 1979

31.57 4,5 – Benzo – 6 – phenyl – 3,10,10 – trimethyl – tricyclo(5.2.1.0^{3,7})dec – 4 – ene
$C_{23}H_{26}$
S.S.Hixson, R.O.Day, L.A.Franke, V.R.Rao
J.Am.Chem.Soc.,102, 412, 1980

31.58 1,2,9,10,17,18 – Dehydro(2.2.2)paracyclophane (at 113°K)
$C_{24}H_{18}$
K.Mirsky, K.Trueblood, E.F.Maverick, L.Grossenbacher
Am.Cryst.Assoc.,Ser.2,7, 24, 1980

31.59 1,2,9,10,17,18 – Dehydro(2.2.2)paracyclophane (at 298°K)
$C_{24}H_{18}$
K.Mirsky, K.Trueblood, E.F.Maverick, L.Grossenbacher
Am.Cryst.Assoc.,Ser.2,7, 24, 1980

31.60 5,7 – Diformyl – tetracyclo(9.8.1.1^{3,9}.1^{13,18})docosa – 1(19),2,4,7,9,11,13,15,17 – nonene
$C_{24}H_{20}O_2$
E.Vogel, H.M.Deger, J.Sombroek, J.Palm, A.Wagner, J.Lex
Angew.Chem.,Int.Ed.Engl.,19, 41, 1980

31.C 4a,5,8,8a – Tetrahydro – 11,14 – dimethoxy – 7 – methyl – 4a – (3 – methyl – but – 2 – enyl) – 5,8a – o – benzeno – 1,4 – naphthoquinone
Microphyllone dimethyl ether
$C_{24}H_{26}O_4$ Main entry is 59.36

31.61 trans – 1 – Ethyl – 2 – (1 – ethyl – 2 – adamantanylidene) – adamantane
$C_{24}H_{38}$
D.Lenoir, R.M.Frank, F.Cordt, A.Gieren, V.Lamm
Chem.Ber.,113, 739, 1980

31.62 4,4′ – (Ethylenedi – imine) – bis(methylenecamphor)
$C_{24}H_{38}N_2O_2$
S.Larsen, E.Larsen *Eur.Cryst.Meeting*,5, 131, 1979

31.C Cycloveratril benzene clathrate monohydrate
$C_{27}H_{30}O_6$, 0.5C_6H_6, H_2O Main entry is 61.12

31.63 (4bα,6β,9β,9aα,10α,13α) – 4b,5,6,7,8,9,9a,10,11,12,13,14 – Dodecahydro – 4b – phenyl – 6,9:10,13 – dimethano – dicyclohepta(a,c) naphthalene
4b – Phenylmariontetraene
$C_{28}H_{30}$
A.R.Harris, K.Mills, M.Martin–Smith, P.Murray–Rust, J.Murray–Rust *Can.J.Chem.*,58, 1847, 1980

31.64 Tetrabenzo – pentacyclo(6.2.2.2^{2,6}.0^{2,7}.0^{3,7})tetradeca – 4,9,11,13 – tetraene
$C_{30}H_{20}$
P.D.Bartlett, M.Kimura, J.Nakayama, W.H.Watson
J.Am.Chem.Soc.,101, 6332, 1979

31.65 Dodecacyclo – (14.10.2.1^{6,12}.1^{20,26}.0^{2,15}.0^{7,11}.0^{8,15}.0^{9,13}.0^{17,24}.0^{17,26}.0^{19,23}.0^{21,25})triaconta – 3,5,27 – triene – 14,18,29,30 – tetrone
$C_{30}H_{24}O_4$
G.Mehta, V.Singh, A.Srikrishna, T.S.Cameron, C.Chan
Tetrahedron Lett.,4595, 1979

31.66 5,22 – bis(Diethylamino) – 6,21 – dimethyl – heptacyclo(12.8.2.0^{1,15}.0^{2,13}.0^{4,12}.0^{7,12}.0^{15,20})tetracosa – 3,5,8,10,16,18,21,23 – octene
$C_{34}H_{42}N_2$
H.J.Lindner, B.Kitschke, K.Hafner, W.Ude
Acta Crystallogr.,Sect.B,36, 754, 1980

31.C 1,6,20,25 – Tetra – aza(6.1.6.1)paracyclophane hydrochloride durene tetrahydrate
$C_{34}H_{44}N_4^{4+}$, $C_{10}H_{14}$, 4Cl$^-$, 4H_2O Main entry is 61.18

31.67 5,14.7,12 – bis(o – Benzeno) – 6,13 – ethenylidene – 5,5a,6,6a,7,12,12a,13,13a,14 – decahydropentacene
$C_{36}H_{28}$
D.N.Butler, I.Gupta, W.W.Ng, S.C.Nyburg
J.Chem.Soc.,Chem.Commun.,596, 1980

31.68 1,4 – Diphenyl – 5 – (p – bromophenyl) – 7 – oxo – (2,3 – 1)phenanthrene – bicyclo(2.2.1)hept – 2 – ene benzene methanol solvate
$C_{37}H_{25}BrO$, C_6H_6, CH_4O
M.Yasuda, K.Harano, K.Kanematsu
J.Org.Chem.,45, 659, 1980

31.C Cyclo(tetrakis(5 – t – butyl – 2 – hydroxy – 1,3 – phenylene)methylene) toluene clathrate
$C_{44}H_{56}O_4$, C_7H_8 Main entry is 61.21

31.69 2,7 – Di – t – butyl – 11,11,12,12 – tetracyano – 4,5,9,10 – tetraphenyl – tetracyclo(4.4.2.0^{1,6}.0^{3,8}) dodeca – 2,4,7,9 – tetraene
$C_{48}H_{38}N_4$
H.Tsukada, H.Shimanouchi, Y.Sasada
Bull.Chem.Soc.Jpn.,53, 983, 1980

HETERO–NITROGEN
(3,4,5–MEMBERED MONOCYCLIC)

32.1 **5 – Aminotetrazole monohydrate**
CH_3N_5, H_2O
D.D.Bray, J.G.White
Acta Crystallogr.,Sect.B,**35**, 3089, 1979

32.2 **3(5) – Chloro – 1,2,4 – triazole**
$C_2H_2ClN_3$
M.S.Idrissi, M.Senechal, H.Sauvaitre, M.Cotrait,
C.Garrigou–Lagrange *J.Chim.Phys.*,**77**, 195, 1980

32.3 **3,5 – Diamino – 1H – 1,2,4 – triazole**
Guanazol
$C_2H_5N_5$
G.L.Starova, O.V.Frank–Kamenetskaya, E.F.Shibanova,
V.A.Lopirev, M.G.Voronkov, V.V.Makarskii
Khim.Geterotsikl.Soedin.,Latv.SSSR,1422, 1979

32.4 **Parabanic acid urea (neutron study,
at 116°K,deuterated form)**
$C_3D_2N_2O_3$, CD_4N_2O
H.P.Weber, J.R.Ruble, B.M.Craven, R.K.McMullan
Acta Crystallogr.,Sect.B,**36**, 1121, 1980
See also R2 : 8

32.5 **Imidazolium hydrogen maleate (neutron study,
deuterated form)**
$C_3H_3D_2N_2^+$, $C_4H_2DO_4^-$
M.S.Hussain, E.O.Schlemper, C.K.Fair
Acta Crystallogr.,Sect.B,**36**, 1104, 1980
See also R2 : 2

32.6 **Di – μ – aquo – bis(dioxo – dinitrato – uranium(vi))
di – imidazole**
$2C_3H_4N_2$, $H_4N_4O_{18}U_2$
D.L.Perry, H.Ruben, D.H.Templeton, A.Zalkin
Inorg.Chem.,**19**, 1067, 1980

32.C **trans – 2 – Hydroxy – 4,5 – dimethyl – 1,3,2 –
dioxaphospholane – 2 – sulfide imidazolium
monohydrate**
$C_3H_5N_2^+$, $C_4H_8O_3PS^-$, H_2O Main entry is 64.8

32.7 **Imidazolium tetrachloro – dioxo – uranium(vi)**
$2C_3H_5N_2^+$, $Cl_4O_2U^{2-}$
D.L.Perry, D.P.Freyberg, A.Zalkin
J.Inorg.Nucl.Chem.,**42**, 243, 1980

32.8 **3 – Nitro – 3′ – chloro – bis(1,2,4 – triaz – 5 – ol)**
$C_4H_2ClN_7O_2$
G.L.Starova, O.V.Frank–Kamenetskaya, O.A.Usov,
A.M.Kuzmin, E.V.Nikitina, M.S.Pevzner
Eur.Cryst.Meeting,**6**, 45, 1980

32.9 **Bis(3 – nitro – 1,2,4 – triazol – 5 – yl) dihydrate**
$C_4H_2N_8O_4$, $2H_2O$
G.L.Starova, O.V.Frank–Kamenetskaya, O.A.Usov,
A.M.Kuzmin, E.V.Nikitina, M.S.Pevzner
Eur.Cryst.Meeting,**6**, 45, 1980

32.10 **2 – Methyl – 4 – nitro – imidazole**
$C_4H_5N_3O_2$
A.Kalman, F.van Meurs, J.Toth
Cryst.Struct.Commun.,**9**, 709, 1980

32.C **Barbital 1 – methylimidazole**
$C_4H_6N_2$, $C_8H_{12}N_2O_3$ Main entry is 43.1

32.C **α – Cyclodextrin 2 – pyrrolidone clathrate
pentahydrate**
C_4H_7NO, $C_{36}H_{60}O_{30}$, $5H_2O$ Main entry is 61.20

32.11 **2,3 – Dimethyl – $\Delta^{1,5}$ – 1,2,3 – triazoline – 4 – thione**
$C_4H_7N_3S$
K.Nielsen, I.Sotofte
Acta Chem.Scand.Ser.A,**33**, 697, 1979

32.12 **D,L – 3 – Amino – 1 – hydroxy – 2 – pyrrolidone
trihydrate**
$C_4H_8N_2O_2$, $3H_2O$
C.Derricott *Acta Crystallogr.,Sect.B*,**36**, 1969, 1980

32.13 **4 – Hydroxy – 2,5 – dioxo – 4 – imidazolidine –
carboxyureide hemihydrate**
$C_5H_6N_4O_5$, $0.5H_2O$
M.Poje, E.F.Paulus, B.Rocic *J.Org.Chem.*,**45**, 65, 1980
See also R1 : 8

32.14 **Sodium 1 – pyrrolidinyl – carbodithioate dihydrate**
$C_5H_8NS_2^-$, Na^+, $2H_2O$
A.Oskarsson, K.Stahl, C.Svensson, I.Ymen
Eur.Cryst.Meeting,**5**, 67, 1979
See also R1 : 11

32.15 **Sodium 1 – pyrrolidinyl – carbodithioate dihydrate
(at 150°K)**
$C_5H_8NS_2^-$, Na^+, $2H_2O$
A.Oskarsson, K.Stahl, C.Svensson, I.Ymen
Eur.Cryst.Meeting,**5**, 67, 1979
See also R1 : 11

32.16 **Sodium 1 – pyrrolidinyl – carbodithioate dihydrate
(at 27°K)**
$C_5H_8NS_2^-$, Na^+, $2H_2O$
A.Oskarsson, K.Stahl, C.Svensson, I.Ymen
Eur.Cryst.Meeting,**5**, 67, 1979
See also R1 : 11

32.C **Lithium 1 – carboxymethyl – 2 – imino – 3 –
phosphonoimidazolidine dihydrate**
Lithium phosphocyclocreatine dihydrate
$C_5H_8N_3O_5P^{2-}$, $2Li^+$, $2H_2O$ Main entry is 64.12

32.17 **Urocanic acid dihydrate**
4 – Imidazole – acrylic acid dihydrate
$C_6H_6N_2O_2$, $2H_2O$
T.Svinning, H.Sorum
Acta Crystallogr.,Sect.B,**35**, 2813, 1979
See also R1 : 1

32.18 **trans – (2R,5R) – 2,5 – Dimethylpyrrolidinium(S) –
mandelate (at 238°K)**
$C_6H_{14}N^+$, $C_8H_7O_3^-$
L.–K.Liu, R.E.Davis
Acta Crystallogr.,Sect.B,**36**, 171, 1980
See also R2 : 1

32.19 (S) − 1 − (p − Bromophenyl) − ethylammonium 1 − methoxy − 2 − ethoxycarbonyl − aziridine − 2 − carboxylate
$C_7H_{10}NO_5^-$, $C_8H_{11}BrN^+$
V.F.Rudchenko, O.A.D'yachenko, A.B.Zolotoi, L.O.Atovmyan, R.G.Kostyanowskii
Dokl.Akad.Nauk SSSR,**246**, 1150, 1979
See also R2 : 3

32.C 3 − (β − D − Ribofuranosyl) − 1,2,4 − triazole − 5(1H,4H) − thione
$C_7H_{11}N_3O_4S$ Main entry is 45.11

32.20 4 − Hydroxyimino − 3,5 − dimethyl − 1 − thiosemicarbamoyl − 5 − thiosemicarbazido − 2 − pyrazoline
$C_7H_{13}N_7OS_2$
Z.Galdecki, M.L.Glowka *Eur.Cryst.Meeting*,**6**, 318, 1980

32.C 2 − (2 − Chloro − 4 − methylthiophen − 3 − yl − imino) − tetrahydro − imidazole
Tiamenidine
$C_8H_{10}ClN_3S$ Main entry is 39.22

32.21 3 − Acetyl − 1,5,5 − trimethyl − hydantoin
$C_8H_{12}N_2O_3$
J.M.Andrieu, E.E.Castellano, A.G.Alvarez, J.Zinczuk, R.A.Corral, O.O.Orazi *Rev.Latinoam.Quim.*,**10**, 14, 1979

32.22 N − (t − Butyldithio) − succinimide
$C_8H_{13}NO_2S_2$
M.−Ul−Haque *Acta Crystallogr.,Sect.B*,**36**, 2082, 1980

32.C Trimethyl − (pyrrolidinomethyl) − silane (at −90°C)
$C_8H_{19}NSi$ Main entry is 63.8

32.23 5 − (4 − Chlorophenylamino) − 2,4 − dihydro − 4 − methyl − 3H − 1,2,4 − triazol − 3 − one methanol solvate
$C_9H_9ClN_4O$, CH_4O
J.W.Tilley, H.Ramuz, P.Levitan, J.F.Blount
Helv.Chim.Acta,**63**, 841, 1980

32.24 1 − Methyl − 3 − phenyl − 4 − (1,2,3 − triazolio) − sulfide (form i)
$C_9H_9N_3S$
I.Sotofte, K.Nielsen
Acta Chem.Scand.Ser.A,**33**, 687, 1979

32.25 1 − Methyl − 3 − phenyl − 4 − (1,2,3 − triazolio) − sulfide (form ii)
$C_9H_9N_3S$
I.Sotofte, K.Nielsen
Acta Chem.Scand.Ser.A,**33**, 687, 1979

32.26 1 − Phenyl − 3 − methyl − 4 − (1,2,3 − triazolio) − sulfide
$C_9H_9N_3S$
I.Sotofte, K.Nielsen
Acta Chem.Scand.Ser.A,**33**, 687, 1979

32.27 2 − Phenyl − 3 − methyl − $\Delta^{1,5}$ − 1,2,3 − triazoline − 4 − thione
$C_9H_9N_3S$
K.Nielsen, I.Sotofte
Acta Chem.Scand.Ser.A,**33**, 697, 1979

32.28 2 − (2,6 − Dichlorophenylimino) − 2 − imidazoline − hydrochloride
Clonidine hydrochloride
$C_9H_{10}Cl_2N_3^+$, Cl^-
V.Cody, G.T.DeTitta *J.Cryst.Mol.Struct.*,**9**, 33, 1979
See also R1 : 16

32.29 5 − Amino − 4 − (2 − methoxyphenyl) − 2,4 − dihydro − 3H − 1,2,4 − triazol − 3 − one
$C_9H_{10}N_4O_2$
J.W.Tilley, H.Ramuz, P.Levitan, J.F.Blount
Helv.Chim.Acta,**63**, 841, 1980

32.30 3 − Acetyl − 5 − isopropylpyrrolidine − 2,4 − dione
$C_9H_{13}NO_3$
M.J.Nolte, P.S.Steyn, P.L.Wessels
J.Chem.Soc.,Perkin Trans.1,1057, 1980

32.31 1,1′ − Trimethylene − di − 2 − imidazolidine − thione
$C_9H_{16}N_4S_2$
W.Schwindinger, T.G.Fawcett, J.A.Potenza, H.J.Schugar, R.A.Lalancette *Acta Crystallogr.,Sect.B*,**36**, 1232, 1980

32.C Trimethyl − (2 − pyrrolidinoethyl) − silane (at −90°C)
$C_9H_{21}NSi$ Main entry is 63.10

32.32 1 − Methyl − 3 − benzyl − 4 − (1,2,3 − triazolio) − sulfide
$C_{10}H_{11}N_3S$
K.Nielsen, L.Schepper, I.Sotofte
Acta Chem.Scand.Ser.A,**33**, 693, 1979

32.33 2 − (2 − Chloro − 4 − methylphenylamino) − imidazoline nitrate (see also Cryst.Struct.Comm.,8,945,1979)
Tolonidine
$C_{10}H_{13}ClN_3^+$, NO_3^-
A.Carpy, D.Hickel, J.M.Leger
Cryst.Struct.Commun.,**9**, 731, 1980

32.34 2 − Cyano − 1 − methyl − 3 − (2 − ((5 − methyl − 1H − imidazol − 4 − yl) − methylthio) − ethyl) − guanidine monohydrate
$C_{10}H_{16}N_6S$, H_2O
B.Kojic−Prodic, Z.Ruzic−Toros, N.Bresciani−Pahor, L.Randaccio *Acta Crystallogr.,Sect.B*,**36**, 1223, 1980
See also R1 : 8

32.35 5 − Dimethylimmonio − 1 − isopropyl − 4,4 − dimethyl − 2 − imidazoline − 2 − thiolate chloroform solvate
$C_{10}H_{19}N_3S$, $CHCl_3$
E.Schaumann, H.Behr, G.Adiwidjaja
Justus Liebigs Ann.Chem.,1322, 1979

32.36 5 − Amino − N − octyl − 1H − tetrazole − 1 − carboxamide
$C_{10}H_{20}N_6O$
G.H.Denny, E.J.Cragoe Junior, C.S.Rooney, J.P.Springer, J.M.Hirshfield, J.A.McCauley *J.Org.Chem.*,**45**, 1662, 1980

32.37 2 − (4 − Chlorobenzoyl) − pyrrole
$C_{11}H_8ClNO$
R.B.English, G.McGillivray, E.Smal
Acta Crystallogr.,Sect.B,**36**, 1136, 1980

32.38 2 − Benzoylpyrrole
$C_{11}H_9NO$
R.B.English, G.McGillivray, E.Smal
Acta Crystallogr.,Sect.B,**36**, 1136, 1980

32.39 2(1 – (2,6 – Dichlorophenoxy) – ethyl) – imidazoline
(α–form)
Lofexidine
$C_{11}H_{12}Cl_2N_2O$
A.Carpy, D.Hickel, J.M.Leger
Cryst.Struct.Commun.,**9**, 37, 1980
See also R1 : 17

32.C 2 – Methyl – 5 – (N – nitrocarboxamido) – 1 –
(2′,3′,5′ – tri – O – nitro – β – D – ribofuranosyl) –
imidazole – 4 – carboxamide
$C_{11}H_{12}N_8O_{14}$ Main entry is 45.19

32.40 2(1 – (2,6 – Dichlorophenoxy) – ethyl) – imidazoline
hydrochloride (α–form)
$C_{11}H_{13}Cl_2N_2O^+$, Cl⁻
A.Carpy, D.Hickel, J.M.Leger
Cryst.Struct.Commun.,**9**, 43, 1980
See also R1 : 17

32.41 5 – (3,5 – Dimethoxyphenyl) – 1 – methyl – 1H –
1,2,4 – triazole
$C_{11}H_{13}N_3O_2$
Y.Lin, S.A.Lang Junior, M.F.Lovell, N.A.Perkinson
J.Org.Chem.,**44**, 4160, 1979
See also R1 : 17

32.42 2 – (4 – Methoxybenzoyl) – pyrrole
$C_{12}H_{11}NO_2$
R.B.English, G.McGillivray, E.Smal
Acta Crystallogr.,Sect.B,**36**, 1136, 1980
See also R1 : 17

32.43 α – (3 – Bromo – 4 – ethoxyphenyl)succinimide
$C_{12}H_{12}BrNO_3$
A.A.Karapetyan, V.G.Andrianov, Yu.T.Struchkov
Cryst.Struct.Commun.,**9**, 417, 1980
See also R1 : 17

32.44 trans – (2R,5R) – 1 – (p – Bromophenylsulfonyl) –
2,5 – dimethylpyrrolidine (at 238°K, absolute
configuration)
$C_{12}H_{16}BrNO_2S$
L.–K.Liu, R.E.Davis
Acta Crystallogr.,Sect.B,**36**, 173, 1980

32.45 1,4 – bis(3,3 – Dimethylazirinyl) – piperazine
$C_{12}H_{20}N_4$
J.Galloy, J.P.Declercq, M.van Meerssche
Cryst.Struct.Commun.,**9**, 151, 1980
See also R1 : 33

32.C 2,2,4,4,6,6 – Hexa(1 – aziridinyl) –
cyclotriphosphazene carbon tetrachloride
anticlathrate
$C_{12}H_{24}N_9P_3$, 3CCl₄ Main entry is 61.2

32.C Hexa – aziridino – cyclotriphosphazene benzene
clathrate
$2C_{12}H_{24}N_9P_3$, C_6H_6 Main entry is 61.3

32.46 2,3 – Diphenyltetrazolium – 5 – olate
$C_{13}H_{10}N_4O$
T.J.King, P.N.Preston, J.S.Suffolk, K.Turnbull
J.Chem.Soc.,Perkin Trans.2,1751, 1979

32.47 1,3 – Diphenyltetrazolium – 5 – thiolate
$C_{13}H_{10}N_4S$
T.J.King, P.N.Preston, J.S.Suffolk, K.Turnbull
J.Chem.Soc.,Perkin Trans.2,1751, 1979

32.48 1 – (4 – Bromophenyl) – 4 – dimethylamino – 2,3 –
dimethyl – 3 – pyrazolin – 5 – one
$C_{13}H_{16}BrN_3O$
N.Shimizu, T.Uno *Cryst.Struct.Commun.*,**9**, 435, 1980

32.49 2,2,5,5 – Tetramethyl – 4 – phenyl – 3 –
imidazoline – 3 – oxide – 1 – oxyl (deformation
density study)
$C_{13}H_{17}N_2O_2$
A.A.Shevyrev, L.A.Muradyan, V.I.Simonov
Kristallografiya,**24**, 1217, 1979

32.50 4,4′ – Methylene – bis(1,3,5 – trimethyl – 4 –
imidazolin – 2 – one)
$C_{13}H_{20}N_4O_2$
C.Glidewell, H.D.Holden, D.C.Liles
J.Mol.Struct.,**66**, 325, 1980

32.C 3 – (5 – Tetrazolyl) – thioxanthone – 10,10 – dioxide
rubidium salt
Doxantrazole rubidium salt
$C_{14}H_7N_4O_3S^-$, Rb⁺ Main entry is 39.49

32.51 2,5 – bis(Pyrrolidino) – 1,4 – benzoquinone
$C_{14}H_{16}N_2O_2$
R.G.F.Giles, L.R.Nassimbeni, J.C.van Niekerk
S.Afr.J.Chem.,**32**, 107, 1979
See also R1 : 18

32.52 1 – Phenyl – 2,3 – dimethyl – 4 –
trimethylammonium – pyrazol – 5 – one iodide
$C_{14}H_{20}N_3O^+$, I⁻
G.Argay, A.Kalman, B.Ribar, S.Vladimirov,
D.Zivanov–Stakic *Cryst.Struct.Commun.*,**9**, 917, 1980

32.53 Phenyl – bis(3,5 – bis(trifluoromethyl) – 1,2,4 –
triazol – 1 – yl) – methane
$C_{15}H_6F_{12}N_6$
A.Gieren, U.Riemann
Acta Crystallogr.,Sect.B,**36**, 204, 1980

32.54 4 – Benzylidene – 2 – phenyl – 2 – imidazolin – 5 –
one
$C_{16}H_{12}N_2O$
B.Tinant, G.Germain, J.P.Declercq, M.van Meerssche
Cryst.Struct.Commun.,**9**, 671, 1980

32.55 (±) – 2 – Methylamino – 4,5 – bis(p –
chlorophenyl) – 4 – hydroxy – 4H – imidazole
methanol solvate
$C_{16}H_{13}Cl_2N_3O$, CH_4O
H.Takayanagi, H.Ogura, K.Matsuzaki, K.Kitajima,
T.Nishimura *Bull.Chem.Soc.Jpn.*,**52**, 3358, 1979

32.56 1 – (p – Bromophenyl) – 5 – (3,5 –
dimethoxyphenyl) – 1H – 1,2,4 – triazole
$C_{16}H_{14}BrN_3O_2$
Y.Lin, S.A.Lang Junior, M.F.Lovell, N.A.Perkinson
J.Org.Chem.,**44**, 4160, 1979

32.57 3 – Chloro – 3 – cyano – 1 – cyclohexyl – 4 –
phenylthio – 2 – azetidinone
$C_{16}H_{17}ClN_2OS$
R.Chambers, R.J.Doedens
Acta Crystallogr.,Sect.B,**36**, 1507, 1980

32.58 2,4 – Dimethyl – 3 – phenyl – 2 –
(diethylcarbamoyl) – 1 – azetine N – oxide
$C_{16}H_{22}N_2O_2$
A.D.de Wit, M.L.M.Pennings, W.P.Trompenaars,
D.N.Reinhoudt, S.Harkema, O.Nevestveit
J.Chem.Soc.,Chem.Commun.,993, 1979

32.59 N − (1′ − t − Butyl − spiro(adamantane − 2,2′ − aziridine) − 3′ − ylidene) − methylamine
$C_{16}H_{26}N_2$
H.Quast, P.Schafer, K.Peters, H.G.von Schnering
Chem.Ber.,**113**, 1921. 1980
See also R1 : 31

32.60 3 − Bromo − 4,5 − dihydro − 5 − hydroperoxy − 4,4 − dimethyl − 3,5 − diphenyl − 3H − pyrazole
$C_{17}H_{17}BrN_2O_2$
M.E.Landis, R.L.Lindsey, W.H.Watson, V.Zabel
J.Org.Chem.,**45**, 525, 1980

32.61 2,2 − bis(p − Methoxyphenyl) − 3 − methyl − 2H − azirine
$C_{17}H_{17}NO_2$
N.Kanehisa, N.Yasuoka, N.Kasai, K.Isomura, H.Kasai
J.Chem.Soc.,Chem.Commun.,98, 1980
See also R1 : 17

32.62 (3S,5R) − 2,2 − Dimethyl − 5 − phenyl − 3 − (4 − pyridyl) − pyrrolidone dihydrobromide
$C_{17}H_{20}N_2$, 2HBr
L.Aeppli, K.Bernauer, F.Schneider, K.Strub, W.E.Oberhansli, K.−H.Pfoertner
Helv.Chim.Acta,**63**, 630, 1980
See also R1 : 33

32.63 Diethyl cis − 1 − formyl − 5 − hydroxy − 4 − (2 − nitrophenyl) − pyrrolidine − 2,2 − dicarboxylate
$C_{17}H_{20}N_2O_8$
U.Hengartner, A.D.Batcho, J.F.Blount, W.Leimgruber, M.E.Larscheid, J.W.Scott *J.Org.Chem.*,**44**, 3748, 1979
See also R1 : 15

32.64 4 − Ethyl − 1 − isopropyl − 3 − ((phenylcarbamoyl) − methyl) − 3 − pyrrolin − 2 − one
$C_{17}H_{22}N_2O_2$
G.Audisio, W.Porzio, L.Zetta, P.Ferruti
J.Chem.Soc.,Perkin Trans.2,1391, 1979

32.C 4 − Phenyl − 1 − (3 − (4 − methyl − 5 − phenyl − 1,2 − dithiolylidene)) − 3,5 − dioxo − 1,2,4 − triazine
$C_{18}H_{13}N_3O_2S_2$ Main entry is 39.68

32.65 1 − (2,4 − Dichloro − β − ((2,4 − dichlorobenzyl) − oxy) − phenethyl) − imidazole hemihydrate
Miconazole hemihydrate
$C_{18}H_{14}Cl_4N_2O$, 0.5H_2O
O.M.Peeters, N.M.Blaton, C.J.De Ranter
Bull.Soc.Chim.Belg.,**88**, 265, 1979

32.C (±) − 4β − (2′,2′ − Trimethylene − dithioethyl) − 3α − (1′S) − (p − nitrobenzyloxycarbonyl)oxy − ethyl − 2 − azetidinone
$C_{18}H_{22}N_2O_6S_2$ Main entry is 39.71

32.66 1 − t − Butyl − 5 − (2 − (p − tolylsulfonylamino) − 1 − (t − butylimino) − 2 − thioxo − ethyl) − 1H − tetrazole
$C_{18}H_{26}N_6O_2S_2$
G.L'abbe, L.Huybrechts, S.Toppet, J.P.Declercq, G.Germain, M.van Meerssche *Bull.Soc.Chim.Belg.*,**88**, 297, 1979

32.C L − Pyroglutamyl − N,N′ − dicyclohexylurea
$C_{18}H_{29}N_3O_3$ Main entry is 48.55

32.67 2′,2′,2′ − Trichloroethyl − 2 − benzyl − 4 − methoxy − carbonyl − imidazole − 1 − (α − isopropylidene) − acetate
$C_{19}H_{19}Cl_3N_2O_4$
Z.Ruzic-Toros, B.Kojic-Prodic
Eur.Cryst.Meeting,**6**, 297, 1980

32.68 1 − n − Butyl − 4,5 − diphenyl − 5 − methyl − 1,2,4 − triazolidin − 3 − one
$C_{19}H_{23}N_3O$
A.Nabeya, J.Saito, H.Koyama *J.Org.Chem.*,**44**, 3935, 1979

32.69 bis(1 − Methyl − 2 − (3 − pyridyl) − pyrrol − 3 − yl) − disulfide
$C_{20}H_{18}N_4S_2$
R.M.Acheson, M.J.Ferris, S.R.Critchley, D.J.Watkin
J.Chem.Soc.,Perkin Trans.2,326, 1980
See also R1 : 33

32.C 2,4,5 − tris(3 − Methoxycarbonyl − furan − 2 − yl) − Δ² − imidazoline
$C_{21}H_{18}N_2O_9$ Main entry is 38.104

32.70 5 − Dimethylamino − 4 − methyl − 3,3 − diphenyl − 4 − vinyl − 5 − pyrrolin − 2 − one
$C_{21}H_{22}N_2O$
M.van Meerssche, G.Germain, J.P.Declercq, R.Touillaux, M.Henriet *Cryst.Struct.Commun.*,**9**, 505, 1980

32.71 2,4,5 − Tri − p − tolyl − imidazolinium bromide
$C_{24}H_{25}N_2{}^+$, Br⁻
H.Campsteyn, J.Lamotte, O.Dideberg, L.Dupont, A.J.Hubert
Cryst.Struct.Commun.,**8**, 949, 1979

32.C 17a − Methyl − 3β − pyrrolidino − 17a − aza − D − homo − 5α − androstane
$C_{24}H_{42}N_2$ Main entry is 36.43

32.72 1 − (2 − Hydroxyphenyl) − 2,3,5 − triphenylpyrrole (at −35°C)
$C_{28}H_{21}NO$
A.R.Katritzky, C.A.Ramsden, Z.Zakaria, R.L.Harlow, S.H.Simonsen *J.Chem.Soc.,Perkin Trans.1*,1870, 1980
See also R1 : 17

32.73 4 − (o − Hydroxybenzoyl) − 2,5 − diphenyl − 1 − (p − tolyl) − imidazole (at −40°C)
$C_{29}H_{22}N_2O_2$
A.R.Katritzky, M.Michalska, R.L.Harlow, S.H.Simonsen
J.Chem.Soc.,Perkin Trans.1,354, 1980

32.74 1,2,3,7,8,12,13,17,18,19 − Decamethyl − biladiene − a,c dihydrobromide chloroform solvate
$C_{29}H_{38}N_4{}^{2+}$, 2Br⁻, 2CHCl₃
J.Engel, G.Struckmeier *Chem.Zeit.*,**103**, 326, 1979

32.75 1,19 − bis(Ethoxycarbonyl) − 2,3,7,8,12,13,17,18 − octamethyl − bilatriene − a,b,c hydrobromide ethanol solvate
$C_{33}H_{41}N_4O_4{}^+$, Br⁻, C_2H_6O
J.Engel, G.Struckmeier *Chem.Zeit.*,**103**, 323, 1979

32.76 1,19 − bis(Ethoxycarbonyl) − 2,3,7,8,12,13,17,18 − octamethyl − biladiene − a,c dihydrobromide
$C_{33}H_{42}N_4O_4{}^{2+}$, 2Br⁻
J.Engel, G.Struckmeier *Chem.Zeit.*,**103**, 326, 1979

32.C Octaethyl − bilatriene − difluoroboron chloroform solvate
$C_{35}H_{45}BF_2N_4O_2$, CHCl₃ Main entry is 62.20

32.C Octa(3,5 – dimethylpyrazolyl) –
cyclotetraphosphazene
$C_{40}H_{56}N_{20}P_4$ Main entry is 64.171

32.C Diethoxy – bilirubin diethyl ester
$C_{41}H_{52}N_4O_6$ Main entry is 59.45

32.77 1,2,3 – Triphenyl – 5 – (1,2 – diphenyl – 2 –
phenyliminoethylidene) – 2 – pyrrolin – 4 – one
$C_{42}H_{30}N_2O$
K.–H.Klaska, O.Jarchow, T.Eicher, H.Preut
Acta Crystallogr.,Sect.B,**35**, 2788, 1979

32.78 (1Z,3E) – 1,4 – bis(1,3 – Dibenzyl – 2 –
imidazolidinyl) – 1,4 – diphenylbutadiene
$C_{50}H_{50}N_4$
M.van Meerssche, G.Germain, J.P.Declercq
Acta Crystallogr.,Sect.B,**36**, 1418, 1980

HETERO–NITROGEN
(6–MEMBERED MONOCYCLIC)

33.1 Perhydropyrimidin – 2 – one
$C_4H_8N_2O$
S.Calogero, U.Russo, A.Del Pra
J.Chem.Soc.,Dalton Trans.,646, 1980

33.2 2,6 – bis(Hydroxyimino – piperazine)
$C_4H_8N_4O_2$
C.Miravitlles, X.Solans, G.Germain, J.P.Declercq
Cryst.Struct.Commun.,**9**, 621, 1980

33.3 3 – Bromopyridinium perchlorate hemihydrate
$C_5H_5BrN^+$, ClO_4^-, $0.5H_2O$
C.Belin, J.Roziere *Acta Crystallogr.,Sect.B*,**36**, 467, 1980

33.4 Pyridine (at 190°K)
C_5H_5N
D.Mootz, H.–G.Wussow *Eur.Cryst.Meeting*,**6**, 12, 1980

33.5 Pyridine trihydrate (at –50°C)
C_5H_5N, $3H_2O$
D.Mootz, H.–G.Wussow
Angew.Chem.,Int.Ed.Engl.,**19**, 552, 1980

33.6 bis(Pyridine – N – oxide) bis(pyridine – N –
hydroxide) tetrabromo – μ – dibromo – di –
palladium(ii)
$2C_5H_5NO$, $2C_5H_6NO^+$, $Br_6Pd_2^{2-}$
Ya.A.Letuchij, I.P.Lavrent'ev, M.L.Khidekel, O.N.Krasochka,
D.D.Makitova, L.O.Atovmyan
Izv.Akad.Nauk SSSR,Ser.Khim.,1902, 1978
See also R2 : 33

33.C Pyridinium 1 – naphthylamine picrate
$C_5H_6N^+$, $C_{10}H_9N$, $C_6H_2N_3O_7^-$ Main entry is 60.1

33.C bis(Pyridinium) tetrabromo – bis(pyridine) –
tungsten(iii)
$C_5H_6N^+$, $C_{10}H_{10}Br_4N_2W^-$, C_5H_5N Main entry is 83.47

33.C Pyridinium tetrachloro – (pyridine) – (tolane) –
tantalum
$C_5H_6N^+$, $C_{19}H_{15}Cl_4NTa^-$ Main entry is 71.137

33.7 Pyridinium tris(thiocyanato) – di – copper(i)
$(C_5H_6N^+)_n$, $nC_3Cu_2N_3S_3^-$
C.L.Raston, B.Walter, A.H.White
Aust.J.Chem.,**32**, 2757, 1979

33.8 bis(Pyridinium) di – μ – chloro – μ – oxo –
bis(dichloro – antimony(iii))
$2C_5H_6N^+$, $Cl_6OSb_2^{2-}$
M.Hall, D.B.Sowerby
J.Chem.Soc.,Chem.Commun.,1134, 1979

33.9 bis(Pyridinium) hexachloro – tellurium(iv)
$2C_5H_6N^+$, Cl_6Te^{2-}
P.Khodadad, B.Viossat, P.Toffoli, N.Rodier
Acta Crystallogr.,Sect.B,**35**, 2896, 1979

33.10 Pyridinium tetra – μ – (hydrogen – phosphato) – di – molybdenum chloride
$3C_5H_6N^+$, $H_4Mo_2O_{16}P_4^{2-}$, Cl^-
A.Bino, F.A.Cotton *Inorg.Chem.*,**18**, 3562, 1979

33.11 tetrakis(Pyridinium) di – μ – oxo – bis(oxo – tris(isothiocyanato) – molybdenum(v))
$4C_5H_6N^+$, $C_6Mo_2N_6O_4S_6^{4-}$
B.Kamenar, M.Penavic *Z.Kristallogr.*,**150**, 327, 1979

33.12 pentakis(Pyridinium) tris(aquo – tetrabromo – oxo – molybdenum)dibromide
$5C_5H_6N^+$, $3H_2Br_4MoO_2^-$, $2Br^-$
A.Bino, F.A.Cotton *Inorg.Chem.*,**18**, 2710, 1979

33.C bis(Pyridine – N – oxide) bis(pyridine – N – hydroxide) tetrabromo – μ – dibromo – di – palladium(ii)
$2C_5H_6NO^+$, $2C_5H_5NO$, $Br_6Pd_2^{2-}$ Main entry is 33.6

33.13 Piperidinium 1 – piperidine – carbodithioate (orthorhombic form)
$C_5H_{12}N^+$, $C_6H_{10}NS_2^-$
A.Wahlberg *Eur.Cryst.Meeting*,**5**, 217, 1979
See also R2 : 33,11

33.14 Piperidinium 1 – piperidine – carbodithioate (monoclinic(β) form)
$C_5H_{12}N^+$, $C_6H_{10}NS_2^-$
A.Wahlberg *Acta Crystallogr.,Sect.B*,**36**, 2099, 1980
See also R2 : 33,11

33.C Piperidinium (μ – oxo) – (μ – 2 – mercapto – ethanolato) – bis(oxo – (2 – mercapto – ethanolato) – molybdenum)
$2C_5H_{12}N^+$, $C_6H_{12}Mo_2O_6S_3^{2-}$ Main entry is 85.10

33.15 Piperidinium tetrachloro – copper
$2C_5H_{12}N^+$, Cl_4Cu^{2-}
K.Emerson, J.E.Drumheller
Am.Cryst.Assoc.,Ser.2,**8**, 24, 1980

33.16 bis(Piperidinium) hexabromo – diaqua – di – molybdenum(ii)
$2C_5H_{12}N^+$, $H_4Br_6Mo_2O_2^{2-}$
J.V.Brencic, P.Segedin *Doc.Chem.Yug.*,**26**, 367, 1979

33.17 3 – Cyanopyridinium 3 – cyanopyridine – tetrachloro – iron(iii)
$C_6H_5N_2^+$, $C_6H_4N_2$, Cl_4Fe^-
J.-C.Daran, Y.Jeannin, L.M.Martin
Acta Crystallogr.,Sect.B,**35**, 3030, 1979

33.18 tetrakis(2 – Carboxypyridinium) octacyano – molybdenum(iv)
$4C_6H_6NO_2^+$, $C_8MoN_8^{4-}$
S.S.Basson, J.G.Leipoldt, A.J.van Wyk
Acta Crystallogr.,Sect.B,**36**, 2025, 1980

33.C γ – Picolinium bis(S – methyldithiocarbazato – dimethylglyoximato) – iron(iii) tetrahydrate
$C_6H_8N^+$, $C_{12}H_{18}FeN_6O_2S_4^-$, $4H_2O$ Main entry is 85.39

33.C Piperidinium 1 – piperidine – carbodithioate (orthorhombic form)
$C_6H_{10}NS_2^-$, $C_5H_{12}N^+$ Main entry is 33.13

33.C Piperidinium 1 – piperidine – carbodithioate (monoclinic(β) form)
$C_6H_{10}NS_2^-$, $C_5H_{12}N^+$ Main entry is 33.14

33.19 N – (2 – Ammonioethyl) – piperazinium pentachloro – copper(ii) dihydrate
$C_6H_{18}N_3^{3+}$, Cl_5Cu^{3-}, $2H_2O$
L.Antolini, G.Marcotrigiano, L.Menabue, G.C.Pellacani
J.Am.Chem.Soc.,**102**, 1303, 1980
See also R1 : 3

33.20 Dideutero – quinolinic acid (neutron study, at 100°K)
Dideutero – pyridine – 2,3 – dicarboxylic acid
$C_7H_3D_2NO_4$
F.Takusagawa, T.F.Koetzle
Acta Crystallogr.,Sect.B,**35**, 2126, 1979

33.21 Dideutero – quinolinic acid (neutron study, at 80°K)
Dideutero – pyridine – 2,3 – dicarboxylic acid
$C_7H_3D_2NO_4$
F.Takusagawa, T.F.Koetzle
Acta Crystallogr.,Sect.B,**35**, 2126, 1979

33.22 Dideutero – quinolinic acid (neutron study, at 35°K)
Dideutero – pyridine – 2,3 – dicarboxylic acid
$C_7H_3D_2NO_4$
F.Takusagawa, T.F.Koetzle
Acta Crystallogr.,Sect.B,**35**, 2126, 1979

33.23 2 – Methylaminopyridinium – dinitromethylide monohydrate
$C_7H_8N_4O_4$, H_2O
N.A.Bailey, C.G.Newton
Cryst.Struct.Commun.,**9**, 49, 1980

33.24 2,6 – Dimethylpyridine – N – oxide hemi – perchlorate (orthorhombic form)
C_7H_9NO, $C_7H_{10}NO^+$, ClO_4^-
M.Jaskolski, M.Gdaniec, Z.Kosturkiewicz, M.Szafran
Pol.J.Chem.,**53**, 2399, 1978

33.25 tetrakis(4 – Ethylpyridinium) dihydrido – octacosa – oxygen – deca – vanadium
$4C_7H_{10}N^+$, $H_2O_{28}V_{10}^{4-}$
J.M.Amigo, J.M.Arrieta, T.Debaerdemaeker
Eur.Cryst.Meeting,**6**, 102, 1980

33.26 4 – Ethylpyridinium octa – molybdenum
$4C_7H_{10}N^+$, $Mo_8O_{26}^{4-}$
P.Roman, M.Martinez–Ripoll, J.Jaud, J.Galy
Eur.Cryst.Meeting,**6**, 260, 1980

33.27 1 – Aza – 4,4 – dimethyl – cyclohexane – 2,6 – dione
$C_7H_{11}NO_2$
G.Bocelli, M.F.Grenier–Loustalot
Eur.Cryst.Meeting,**6**, 34, 1980

33.28 Pyridinium – 1 – dicyanomethylide (at 118°K, fully deuterated, neutron study)
$C_8D_5N_3$
L.Devos, F.Baert, R.Fouret, M.Thomas
Acta Crystallogr.,Sect.B,**36**, 1807, 1980
See also R1 : 7

33.29 Pyridinium – 1 – dicyanomethylide (deuterated form, neutron study)
$C_8D_5N_3$
L.Devos, F.Baert, R.Fouret, M.Thomas
Acta Crystallogr.,Sect.B,**36**, 1807, 1980
See also R1 : 7

33.30 5 – Chloro – 6 – dichloromethylene – 4 – methoxy – 1 – methyl – 4 – trichloromethyl – hexahydropyrimidin – 2 – one
$C_8H_8Cl_6N_2O_2$
C.H.Stam *Acta Crystallogr.,Sect.B*,**36**, 729, 1980

33.31 Pyridoxinium chloride (neutron study)
$C_8H_{12}NO_3^+$, Cl^-
G.E.Bacon, J.S.Plant
Acta Crystallogr.,Sect.B,**36**, 1130, 1980

33.32 (±) – (E) – N – Acetyl – piperidine – 2 – carboxylic acid
$C_8H_{13}NO_3$
I.D.Rae, C.L.Raston, A.H.White *Aust.J.Chem.*,**33**, 215, 1980

33.33 4 – t – Butyl – piperidin – 2 – one (at 138°K)
4 – t – Butyl – 1 – azacyclohexan – 2 – one
$C_9H_{17}NO$
D.van der Helm, J.D.Ekstrand
Acta Crystallogr.,Sect.B,**35**, 3101, 1979

33.34 2,2′ – Bipyridyl potassium dicyano – gold(iii)
$C_{10}H_8N_2$, K^+, $C_2AuN_2^-$
P.G.Jones, W.Clegg, G.M.Sheldrick
Acta Crystallogr.,Sect.B,**36**, 160, 1980

33.35 (E) – 2,2′,5,5′ – Tetra – azastilbene (monoclinic form)
$C_{10}H_8N_4$
J.Vansant, G.Smets, J.P.Declercq, G.Germain,
M.van Meerssche *J.Org.Chem.*,**45**, 1557, 1980

33.36 4,4′ – Bipyridylium di – μ – chloro – tetrachloro – di – copper
$C_{10}H_{10}N_2^{2+}$, $Cl_6Cu_2^{2-}$
M.Bukowska–Strzyzewska, A.Tosik
Pol.J.Chem.,**53**, 2423, 1979

33.C syn – 3 – Methyl – 6 – (2 – furyl) – piperidin – 2 – one
$C_{10}H_{13}NO_2$ Main entry is 38.22

33.37 Dipiperidino – tetraselane
$C_{10}H_{20}N_2Se_4$
O.Foss, V.Janickis *J.Chem.Soc.,Dalton Trans.*,620, 1980
See also R1 : 11

33.38 1 – (p – Bromophenyl) – 2 – methyl – 5 – bromopyridazine – 3,6 – dione
$C_{11}H_8Br_2N_2O_2$
C.Stam, J.J.Zwinselman, H.C.van der Plas, S.Baloniak
J.Heterocycl.Chem.,**16**, 855, 1979

33.39 2 – Pyridyl – phenylsulfone
$C_{11}H_9NO_2S$
G.Bandoli, S.Calogero, D.Ida, G.C.Pappalardo, G.Scarlata
Phosphorus and Sulfur,**7**, 265, 1979
See also R1 : 11

33.40 (E) – 2′,5,5′ – Triazastilbene
3,2′ – Pyridylpyrazinyl – ethylene
$C_{11}H_9N_3$
J.Vansant, G.Smets, J.P.Declercq, G.Germain,
M.van Meerssche *J.Org.Chem.*,**45**, 1557, 1980

33.41 3,5 – Diacetyl – 2,6 – dimethylpyridine
$C_{11}H_{13}NO_2$
A.T.H.Lenstra, G.H.Petit
Cryst.Struct.Commun.,**9**, 725, 1980

33.C Pyridyl – 1 – thio – β – D – glucopyranoside monohydrate
$C_{11}H_{15}NO_5S$, H_2O Main entry is 45.20

33.42 (E) – 2,2′ – Diazastilbene
$C_{12}H_{10}N_2$
J.Vansant, G.Smets, J.P.Declercq, G.Germain,
M.van Meerssche *J.Org.Chem.*,**45**, 1557, 1980

33.43 (E) – 4,4′ – Diazastilbene
$C_{12}H_{10}N_2$
J.Vansant, G.Smets, J.P.Declercq, G.Germain,
M.van Meerssche *J.Org.Chem.*,**45**, 1557, 1980

33.44 N – Salicylidene – 3 – aminopyridine
$C_{12}H_{10}N_2O$
I.Moustakali–Mavridis, E.Hadjoudis, A.Mavridis
Acta Crystallogr.,Sect.B,**36**, 1126, 1980
See also R1 : 17

33.45 1 – ((3,4 – Dichlorophenyl)methoxy) – 1,6 – dihydro – 6,6 – dimethyl – 1,3,5 – triazine – 2,4 – diamine hydrochloride hydrate
$C_{12}H_{16}Cl_2N_5O^+$, Cl^-; $0.29H_2O$
H.L.Ammon, L.A.Plastas
Acta Crystallogr.,Sect.B,**35**, 3106, 1979

33.46 bis(4 – Benzyl – piperidinium) tetrachloro – cobalt(ii)
$2C_{12}H_{18}N^+$, Cl_4Co^{2-}
L.Antolini, G.Marcotrigiano, L.Menabue, G.C.Pellacani
Inorg.Chem.,**18**, 2652, 1979

33.47 bis(4 – Benzyl – piperidinium) hexachloro – di – copper(ii)
$2C_{12}H_{18}N^+$, $Cl_6Cu_2^{2-}$
L.P.Battaglia, A.B.Corradi, G.Marcotrigiano, L.Menabue
Inorg.Chem.,**19**, 125, 1980

33.C 1,4 – bis(3,3 – Dimethylazirinyl) – piperazine
$C_{12}H_{20}N_4$ Main entry is 32.45

33.48 1,2 – bis(N – Piperidyl)ethane – bis – N – oxide perchlorate
$C_{12}H_{25}N_2O_2^+$, ClO_4^-
M.Jaskolski, Z.Kosturkiewicz
Eur.Cryst.Meeting,**5**, 132, 1979

33.49 1,2 – bis(N – Piperidyl)ethane – bis – N – oxide diperchlorate
$C_{12}H_{26}N_2O_2^{2+}$, $2ClO_4^-$
M.Jaskolski, Z.Kosturkiewicz
Eur.Cryst.Meeting,**5**, 132, 1979

33.50 bis(4 – (N – Methyl – piperidinium)) – disulfide tetrachloro – copper(ii)
$C_{12}H_{26}N_2S_2^{2+}$, Cl_4Cu^{2-}
M.C.Perucaud, J.L.Brianso, W.Gaete, J.Ros
Eur.Cryst.Meeting,**6**, 351, 1980
See also R1 : 11

33.51 N – (5 – Methoxysalicylidene) – 3 – aminopyridine
$C_{13}H_{12}N_2O_2$
I.Moustakali–Mavridis, E.Hadjoudis, A.Mavridis
Acta Crystallogr.,Sect.B,**36**, 1126, 1980
See also R1 : 17

33.52 (E) – 6 – Methyl – 6 – azonia – 2′ – azastilbene iodide
$C_{13}H_{13}N_2^+$, I^-
J.Vansant, G.Smets, J.P.Declercq, G.Germain,
M.van Meerssche *J.Org.Chem.*,**45**, 1557, 1980

33.53 4 — m — Aminophenyl — 2 — formylpyridine —
thiosemicarbazone
$C_{13}H_{13}N_5S$
J.N.Brown, C.H.Yang *Cryst.Struct.Commun.*,**8**, 879, 1979
See also R1 : 16,8

33.54 Piperidinium — acetyl — (m — chloroanilide) chloride
$C_{13}H_{18}ClN_2O^+$, Cl^-
G.Reck, P.Leibnitz *Krist.Tech.*,**14**, 1441, 1979
See also R1 : 16

33.55 Piperidinium — acetyl — (p — chloroanilide) chloride
$C_{13}H_{18}ClN_2O^+$, Cl^-
G.Reck, G.Bannier *Krist.Tech.*,**14**, 1437, 1979
See also R1 : 16

33.56 o — Chloro — N — (piperid — 1 — yl — acetyl) — aniline
hydrochloride
$C_{13}H_{18}ClN_2O^+$, Cl^-
G.Reck, R.G.Kretschmer *Krist.Tech.*,**14**, 1115, 1979
See also R1 : 16

33.57 o — Chloro — N — (piperid — 1 — yl — acetyl) — aniline
hydrochloride monohydrate
$C_{13}H_{18}ClN_2O^+$, Cl^-, H_2O
G.Reck, R.G.Kretschmer *Krist.Tech.*,**14**, 1115, 1979
See also R1 : 16

33.58 3,5 — Diethoxycarbonyl — 2,6 — dimethyl — 1,4 —
dihydropyridine
$C_{13}H_{19}NO_4$
A.T.H.Lenstra, G.H.Petit, R.A.Dommisse, F.C.Alderweireldt
Bull.Soc.Chim.Belg.,**88**, 133, 1979

33.59 (E) — 3 — Phenyl — 3 — (3 — pyridyl) — acrylamide
$C_{14}H_{12}N_2O$
F.Ishikawa *Chem.Pharm.Bull.*,**28**, 1394, 1980

33.60 (Z) — 3 — Phenyl — 3 — (4 — pyridyl) — acrylamide
$C_{14}H_{12}N_2O$
F.Ishikawa *Chem.Pharm.Bull.*,**28**, 1394, 1980

33.61 (2 — Hydroxy — 2 — phenyl — 2 — (2′ — pyridyl) —
ethyl) — methylsulfoxide
$C_{14}H_{15}NO_2S$
A.Carpy, J.M.Leger, A.Boucherle, M.Madesclaire
Acta Crystallogr.,Sect.B,**35**, 2566, 1979
See also R1 : 11

33.62 (E) — 4,4′ — Dimethyl — 4,4′ — diazoniastilbene di —
iodide
$C_{14}H_{16}N_2^{2+}$, $2I^-$
J.Vansant, G.Smets, J.P.Declercq, G.Germain,
M.van Meerssche *J.Org.Chem.*,**45**, 1557, 1980

33.63 (E) — 6,6′ — Dimethyl — 6,6′ — diazoniastilbene di —
iodide
$C_{14}H_{16}N_2^{2+}$, $2I^-$
J.Vansant, G.Smets, J.P.Declercq, G.Germain,
M.van Meerssche *J.Org.Chem.*,**45**, 1557, 1980

33.64 Oxy — bis(2 — methylene — 6 — methylpyridine — N —
oxide) dihydrate
$C_{14}H_{16}N_2O_3$, $2H_2O$
G.Weber, W.Saenger
Acta Crystallogr.,Sect.B,**36**, 207, 1980

33.65 1 — (p — Bromophenyl) — 2 — methyl — 4 —
diethylaminopyridazine — 3,6 — dione
$C_{15}H_{18}BrN_3O_2$
C.Stam, J.J.Zwinselman, H.C.van der Plas, S.Baloniak
J.Heterocycl.Chem.,**16**, 855, 1979

33.66 1 — ((4 — (3 — Chlorophenylamino) — 3 — pyridyl)
sulfonyl) — 3 — ethyl — 1 — methylurea hydrogen
sulfate
$C_{15}H_{18}ClN_4O_3S^+$, HO_4S^-
L.Dupont, O.Dideberg, J.Toussaint, J.Delarge
Acta Crystallogr.,Sect.B,**36**, 2170, 1980
See also R1 : 16,8

33.67 2,6 — cis — Dimethyl — piperidyl — N — phenyl —
acetamidine
$C_{15}H_{22}N_2$
G.Gilli, V.Bertolasi *J.Am.Chem.Soc.*,**101**, 7704, 1979

33.68 1 — Benzyl — 5 — ethyl — 1,2,5,6 — tetrahydro — 2 —
oxo — 4 — pyridine — acetic acid
$C_{16}H_{19}NO_3$
T.Date, K.Aoe, M.Ohba, T.Fujii
Yukugaku Zasshi,**99**, 865, 1979
See also R1 : 1

33.69 4 — (p — N,N — Dimethylaminostyryl) — pyridine —
methiodide
$C_{16}H_{19}N_2^+$, I^-
T.H.Lu, T.J.Lee, C.Wong, K.T.Kuo *J.Chin.Chem.*,**26**, 53, 1979

33.70 Carbinoxamine maleate
$C_{16}H_{20}ClN_2O^+$, $C_4H_3O_4^-$
G.Gilli, V.Bertolasi, P.A.Borea
Eur.Cryst.Meeting,**5**, 183, 1979
See also R1 : 3 R2 : 2,1

33.71 2,5 — bis(Piperidino) — 1,4 — benzoquinone
$C_{16}H_{22}N_2O_2$
R.G.F.Giles, L.R.Nassimbeni, J.C.van Niekerk
S.Afr.J.Chem.,**32**, 107, 1979
See also R1 : 18

33.72 1 — (1′ — Methyl — 2′ — pyridinium) — 3 — (1″ —
methyl — 4″ — pyridinium) — cyclopentadienide
bromide hemihydrate
$C_{17}H_{17}N_2^+$, Br^-, $0.5H_2O$
H.L.Ammon, W.D.Erhardt *J.Org.Chem.*,**45**, 1914, 1980

33.C (3S,5R) — 2,2 — Dimethyl — 5 — phenyl — 3 — (4 —
pyridyl) — pyrrolidone dihydrobromide
$C_{17}H_{20}N_2$, 2HBr Main entry is 32.62

33.73 N — (1 — Methyl — 4 — (3 — methylphenylamino) —
3 — pyridinio) — sulfonyl — N′ — isopropylurea
$C_{17}H_{22}N_4O_3S$
L.Dupont, O.Dideberg, J.Lamotte
Acta Crystallogr.,Sect.B,**35**, 2817, 1979
See also R1 : 16,8,11

33.74 Cyclizine hydrochloride
$C_{18}H_{23}N_2^+$, Cl^-
V.Bertolasi, P.A.Borea, G.Gilli, M.Sacerdoti
Acta Crystallogr.,Sect.B,**36**, 1975, 1980

33.75 2,6 — cis — Dimethyl — piperidyl — N — phenyl — 2,2 —
dimethylpropionamidine
$C_{18}H_{28}N_2$
G.Gilli, V.Bertolasi *J.Am.Chem.Soc.*,**101**, 7704, 1979

33.C Diphenyl — (2 — piperidino — ethyl) — silanol
Sila — pridinol
$C_{19}H_{25}NOSi$ Main entry is 63.27

33.76 2,5 – Di(styryl) – pyrazine 1,4 – bis(2 – (2 – pyridyl)
vinyl) – benzene (mixed crystal)
$0.4C_{20}H_{16}N_2$, $0.6C_{20}H_{16}N_2$
H.Nakanishi, W.Jones, G.M.Parkinson
Acta Crystallogr.,Sect.B,**35**, 3103, 1979
See also R2 : 33

33.C 2,5 – Di(styryl) – pyrazine 1,4 – bis(2 – (2 – pyridyl)
vinyl) – benzene (mixed crystal)
$0.6C_{20}H_{16}N_2$, $0.4C_{20}H_{16}N_2$ Main entry is 33.76

33.77 r – 1,c – 2,t – 3,t – 4 – tetrakis(2 – Pyrazinyl) –
cyclobutane
$C_{20}H_{16}N_8$
J.Vansant, S.Toppet, G.Smets, J.P.Declercq, G.Germain,
M.van Meerssche *J.Org.Chem.*,**45**, 1565, 1980
See also R1 : 20

33.C bis(1 – Methyl – 2 – (3 – pyridyl) – pyrrol – 3 –
yl) – disulfide
$C_{20}H_{18}N_4S_2$ Main entry is 32.69

33.C 2 – Chloro – 10 – (3 – (4 – methyl – 1 – piperazinyl)
propyl) – phenothiazine methanesulfonic acid
$C_{20}H_{24}ClN_3S$, $2CH_4O_3S$ Main entry is 41.74

33.78 1,1 – Diphenyl – 3 – piperidino – 1 – propanol
Pridinol
$C_{20}H_{25}NO$
R.Tacke, M.Strecker, W.S.Sheldrick, L.Ernst, E.Heeg,
B.Berndt, C.–M.Knapstein, R.Niedner
Chem.Ber.,**113**, 1962, 1980

33.C (3 – Piperidino – propyl) – diphenyl – silanol
Sila – difenidol
$C_{20}H_{27}NOSi$ Main entry is 63.31

33.79 1,3 – bis(1' – Methyl – 2' – pyridinium) – indenide
bromide
$C_{21}H_{19}N_2^+$, Br^-
H.L.Ammon, W.D.Erhardt *J.Org.Chem.*,**45**, 1914, 1980

33.C N – (2 – (4 – (5 – Chloro – 2 – oxo – 1 –
benzimidazolinyl)piperidino) – ethyl) – p –
fluorobenzamide
Halopemide
$C_{21}H_{22}ClFN_4O_2$ Main entry is 35.64

33.C 5,5 – Dimethyl – 10 – (4 – methylpiperazinyl) –
10,11 – dihydro – 5H – dibenzo(b,f)silepane
hydrogen fumarate
$C_{21}H_{29}N_2Si^+$, $C_4H_3O_4^-$ Main entry is 63.33

33.80 r – 1,c – 2 – bis(3 – Pyridyl) – t – 3,t – 4 – bis(2 –
pyrazinyl) – cyclobutane
$C_{22}H_{18}N_6$
J.Vansant, S.Toppet, G.Smets, J.P.Declercq, G.Germain,
M.van Meerssche *J.Org.Chem.*,**45**, 1565, 1980
See also R1 : 20

33.81 r – 1,t – 3 – bis(3 – Pyridyl) – c – 2,t – 4 – bis(2 –
pyrazinyl) – cyclobutane
$C_{22}H_{18}N_6$
J.Vansant, S.Toppet, G.Smets, J.P.Declercq, G.Germain,
M.van Meerssche *J.Org.Chem.*,**45**, 1565, 1980
See also R1 : 20

33.82 2(S) – 1 – Pyridinium – 2 – N,N – dibenzylamino –
propane p – toluenesulfonate
$C_{22}H_{25}N_2^+$, $C_7H_7O_3S^-$
H.Razenberg, J.A.Kanters *Eur.Cryst.Meeting*,**6**, 46, 1980

33.83 N – (2' – Hydroxy – 2' – phenylethyl) – 4 – (N' –
phenylpropionamido) – piperidine
$C_{22}H_{28}N_2O_2$
G.Evrard, C.Humblet, A.Michel, F.Durant
Eur.Cryst.Meeting,**5**, 58, 1979
See also R1 : 16

33.84 N – (1 – (2 – Phenylethyl) – 4 – piperidinylium) –
N – phenyl – propanamide – citrate toluene solvate
Fentanyl citrate toluene solvate
$C_{22}H_{29}N_2O^+$, $C_6H_7O_7^-$, C_7H_8
O.M.Peeters, N.M.Blaton, C.J.De Ranter, A.M.van Herk,
K.Goubitz *J.Cryst.Mol.Struct.*,**9**, 153, 1979
See also R1 : 16 R2 : 1

33.85 2,2',6,6' – tetrakis(Propylthio) – 3,3' – azoxypyridine
$C_{22}H_{32}N_4OS_4$
J.Lamotte, L.Dupont, O.Dideberg, G.Dive, J.C.Jamoulle
Acta Crystallogr.,Sect.B,**36**, 2157, 1980

33.C 9,9 – Dimethyl – 10 – (3' – piperidyl – propyl) –
acridane (at –100°C)
$C_{23}H_{30}N_2$ Main entry is 36.38

33.86 cis – (+) – N – (3 – Methyl – 1 – (2 – phenylethyl) –
4 – piperidinyl) – N – phenyl – propanamide nitric
acid 2 – propanol solvate
$C_{23}H_{30}N_2O$, HNO_3, C_3H_8O
O.M.Peeters, N.M.Blaton, C.J.De Ranter
Eur.Cryst.Meeting,**5**, 59, 1979
See also R1 : 16

33.C cis – 5,6 – Dimethoxy – 2 – methyl – 3 – (2 – (4 –
phenyl – 1 – piperazinyl) – ethyl) – indoline
$C_{23}H_{31}N_3O_2$ Main entry is 35.70

33.C 1,3 – Dimethyl – 5 – (10 – phenyl – 2,10 –
dihydrophenazin – 2 – ylidene) –
hexahydropyrimidine – 2,4,6 – trione
$C_{24}H_{18}N_4O_3$ Main entry is 36.41

33.87 r – 1,c – 2,t – 3,t – 4 – tetrakis(2 – Pyridyl) –
cyclobutane
$C_{24}H_{20}N_4$
J.Vansant, S.Toppet, G.Smets, J.P.Declercq, G.Germain,
M.van Meerssche *J.Org.Chem.*,**45**, 1565, 1980
See also R1 : 20

33.88 r – 1,c – 2,t – 3,t – 4 – tetrakis(4 – Pyridyl) –
cyclobutane
$C_{24}H_{20}N_4$
J.Vansant, S.Toppet, G.Smets, J.P.Declercq, G.Germain,
M.van Meerssche *J.Org.Chem.*,**45**, 1565, 1980
See also R1 : 20

33.89 N – (2' – Phenylethyl) – 4 – (N –
phenylpropionamido) – 4 – carbomethoxy –
piperidine
$C_{24}H_{30}N_2O_3$
G.Evrard, C.Humblet, A.Michel, F.Durant
Eur.Cryst.Meeting,**5**, 58, 1979
See also R1 : 16

33.90 Barium (1,12 – bis(2 – acetylpyridine) – 1,12 –
dimethyl – 2,5,8,11 – tetra – azadodeca – 1,11 –
diene) perchlorate
$C_{24}H_{32}N_6O_2$, Ba^{2+}, $2ClO_4^-$
M.G.B.Drew, C.V.Knox, S.M.Nelson
J.Chem.Soc.,Dalton Trans.,942, 1980

33.91 2(S),3(S) – 1 – Pyridinium – 2 – N,N –
dibenzylamino – 3 – methylpentane p –
toluenesulfonate
$C_{25}H_{31}N_2^+$, $C_7H_7O_3S^-$
H.Razenberg, J.A.Kanters *Eur.Cryst.Meeting*,**6**, 46, 1980

33.92 cis(–) – N – (3 – Methyl – 4 – (methoxycarbonyl) –
1 – (2 – phenylethyl) – 4 – piperidinyl) – N –
phenyl – propanamide oxalic acid
$C_{25}H_{32}N_2O_3$, $C_2H_2O_4$
O.M.Peeters, N.M.Blaton, C.J.De Ranter
Eur.Cryst.Meeting,**5**, 59, 1979
See also R1 : 16

33.93 1,3 – trans – bis(4 – Chlorophenyl) – 2,4 – trans –
bis(4 – pyridyl) – cyclobutane
$C_{26}H_{20}Cl_2N_2$
V.Busetti, G.Valle, G.Zanotti, G.Galiazzo
Acta Crystallogr.,Sect.B,**36**, 894, 1980
See also R1 : 20

33.94 1 – p – Hydroxyphenyl – 2,4,6 –
triphenylpyridinium – 3 – olate (at –35°C)
$C_{29}H_{21}NO_2$
A.R.Katritzky, C.A.Ramsden, Z.Zakaria, R.L.Harlow,
S.H.Simonsen *J.Chem.Soc.,Perkin Trans.1*,1870, 1980

33.C N – Acetyl – 3 – (benzoyl(2 – piperidyl – 2 –
piperidylenium – ethyl)methylene) – indol – 2 –
olate
$C_{30}H_{35}N_3O_3$ Main entry is 35.77

33.C 1,1,3,3 – Tetraphenyl – 1,3 – bis(2 – piperidino –
ethyl) – disiloxane
$C_{38}H_{48}N_2OSi_2$ Main entry is 63.52

HETERO–NITROGEN
(7– AND HIGHER– MONOCYCLIC)

34.1 1,4 – Dihydro – 1,4 – diazocine
$C_6H_8N_2$
H.–J.Altenbach, H.Stegelmeier, M.Wilhelm, B.Voss, J.Lex,
E.Vogel *Angew.Chem.,Int.Ed.Engl.*,**18**, 962, 1979

34.2 1,4 – Dihydro – 1,4 – diazocine
$C_6H_8N_2$
M.Breuninger, B.Gallenkamp, K.–H.Muller, H.Fritz,
H.Prinzbach, J.J.Daly, P.Schonholzer
Angew.Chem.,Int.Ed.Engl.,**18**, 964, 1979

34.3 2,3 – Dihydro – 5,7 – dimethyl – 1,4 – diazepinium
perchlorate
$C_7H_{13}N_2^+$, ClO_4^-
G.Ferguson, W.C.Marsh, D.Lloyd, D.R.Marshall
J.Chem.Soc.,Perkin Trans.2,74, 1980

34.4 1,4 – bis(Methylsulfonyl) – 1,4 – dihydro – 1,4 –
diazocine
$C_8H_{12}N_2O_4S_2$
H.–J.Altenbach, H.Stegelmeier, M.Wilhelm, B.Voss, J.Lex,
E.Vogel *Angew.Chem.,Int.Ed.Engl.*,**18**, 962, 1979

34.C Octamethyl – trithio – cyclotetra($\lambda^3,\lambda^5,\lambda^5,\lambda^5$ –
phosphazane)
$C_8H_{24}N_4P_4S_3$ Main entry is 64.30

34.5 1,4 – bis(Methoxycarbonyl) – 1,4 – dihydro – 1,4 –
diazocine
$C_{10}H_{12}N_2O_4$
M.Breuninger, B.Gallenkamp, K.–H.Muller, H.Fritz,
H.Prinzbach, J.J.Daly, P.Schonholzer
Angew.Chem.,Int.Ed.Engl.,**18**, 964, 1979

34.6 1,4,7,10 – Tetra – azacyclotetradecane – 3,8,11,14 –
tetrone
$C_{10}H_{16}N_4O_4$
Yu.A.Simonov, T.I.Malinovskii, M.M.Botoshanskii,
A.A.Dvorkin, S.T.Malinovskii, N.G.Luk'yanenko,
Yu.A.Popkov, V.A.Shapkin, A.V.Bogatskii
Eur.Cryst.Meeting,**5**, 124, 1979

34.7 1,4 – bis(Dimethylcarbamoyl) – 1,4 – dihydro – 1,4 –
diazocine
$C_{12}H_{18}N_4O_2$
M.Breuninger, B.Gallenkamp, K.–H.Muller, H.Fritz,
H.Prinzbach, J.J.Daly, P.Schonholzer
Angew.Chem.,Int.Ed.Engl.,**18**, 964, 1979

34.8 1,4,9,12 – Tetra – azacyclohexadecane – 3,10,13,16 –
tetrone
$C_{12}H_{20}N_4O_4$
Yu.A.Simonov, T.I.Malinovskii, M.M.Botoshanskii,
A.A.Dvorkin, S.T.Malinovskii, N.G.Luk'yanenko,
Yu.A.Popkov, V.A.Shapkin, A.V.Bogatskii
Eur.Cryst.Meeting,**5**, 124, 1979

34.C   1,4 – bis(Trimethylsilyl) – 1,4 – dihydro – 1,4 –
diazocine
$C_{12}H_{24}N_2Si_2$ Main entry is 63.13

34.9   meso – 5,5,7,12,12,14 – Hexamethyl – 1,4,8,11 –
tetra – azacyclotetradecane dihydrate
$C_{16}H_{36}N_4$, $2H_2O$
P.Gluzinski, J.W.Krajewski, Z.Urbanczyk-Lipkowska
*Acta Crystallogr.,Sect.B,***36**, 1695, 1980

34.10   7 – (4′ – Bromophenyl) – 4 – ethoxycarbonyl – 3 –
methoxy – 6 – phenyl – 2H – azepin – 2 – one
(absolute configuration)
$C_{22}H_{18}BrNO_4$
Y.Tsuda, M.Kaneda, T.Sano, Y.Horiguchi, Y.Iitaka
*Heterocycles,***12**, 1423, 1979

34.11   bis(o – Phenylene) – bis(pyridine – 2,6 – dialdimino)
lead(ii) diperchlorate dihydrate
$C_{28}H_{18}N_6$, Pb^{2+}, $2ClO_4^-$, $2H_2O$
M.G.B.Drew, J.de O.Cabral, M.F.Cabral, F.S.Esho, S.M.Nelson
J.Chem.Soc.,Chem.Commun.,1033, 1979

34.12   3,5 – Di – t – butyl – 7 – (3,5 – di – t – butyl – 2 –
hydroxyphenyl) – 1 – methyl – 2,3 – dihydro – 1H –
azepin – 2 – one ethanol solvate
$C_{29}H_{45}NO_2$, C_2H_6O
H.-D.Becker, K.Gustafsson, C.L.Raston, A.H.White
Aust.J.Chem.,**32**, 1931, 1979
See also R1 : 17

34.13   7,7′ – bis(1,4 – Dibenzyl – 6 – phenyl – 1,2,3,4 –
tetrahydro – 1,4 – diazepine) (meso form)
$C_{50}H_{50}N_4$
M.van Meerssche, G.Germain, J.P.Declercq
*Acta Crystallogr.,Sect.B,***36**, 1418, 1980

34.14   7,7′ – bis(1,4 – Dibenzyl – 6 – phenyl – 1,2,3,4 –
tetrahydro – 1,4 – diazepine) (racemic form)
$C_{50}H_{50}N_4$
M.van Meerssche, G.Germain, J.P.Declercq
*Acta Crystallogr.,Sect.B,***36**, 1418, 1980

35.C   4 – Carboxylato – L – thiazolidine – hydantoin
$C_5H_6N_2O_2S$ Main entry is 41.12

35.1   1,2 – Dihydro – 3H – pyrazolo(3,4 – b)pyridin – 3 –
one
$C_6H_5N_3O$
Z.Urbanczyk-Lipkowska, J.W.Krajewski, P.Gluzinski,
K.Stadnicka, L.Lebioda
*Acta Crystallogr.,Sect.B,***35**, 2753, 1979

35.C   2,3 – Dihydro – 3 – hydroxy – 5H – thiazolo(3,2 – a)
pyrimidin – 5 – one
$C_6H_6N_2O_2S$ Main entry is 41.14

35.C   5,6 – Dimethyl – imidazo(2,1 – b)(1,3,4)thiadiazole
$C_6H_7N_3S$ Main entry is 41.15

35.C   5,6 – Dimethyl – imidazo(2,1 – b)(1,3,4)thiadiazole
hydrobromide monohydrate
$C_6H_8N_3S^+$, Br^-, H_2O Main entry is 41.16

35.2   1,4 – Dichloro – 7 – methylpyrrolo(3,2 – d)
pyridazine
$C_7H_5Cl_2N_3$
C.Foces-Foces, F.H.Cano, S.Garcia-Blanco
J.Cryst.Mol.Struct.,**8**, 201, 1978

35.C   7 – Methyl – 1,4 – dithia – 7 – aza – spiro(4.4)
nonane – 6,8 – dione
$C_7H_9NO_2S_2$ Main entry is 39.13

35.C   N – Methyl – 1,4 – dithiane – 2,3 – dicarboximide
$C_7H_9NO_2S_2$ Main entry is 39.14

35.C   Clazamycin A hydrochloride (absolute
configuration)
$C_7H_{10}ClN_2O^+$, Cl^- Main entry is 50.2

35.3   2,7 – Diaza – spiro(4.4)nonane – 1,6 – dione
$C_7H_{10}N_2O_2$
M.Czugler, A.Kalman, B.Oleksyn, M.Kajtar
Cryst.Struct.Commun.,**9**, 791, 1980

35.4   Hexachloro – quinoxaline
$C_8Cl_6N_2$
A.J.W.A.Vermeulen, C.Huiszoon
*Acta Crystallogr.,Sect.B,***35**, 3087, 1979

35.5   1,8 – Naphthyridine (at 165°K, electron density
distribution study)
$C_8H_6N_2$
P.Dapporto, C.A.Ghilardi, C.Mealli, A.Orlandini, F.Zanobini
*Eur.Cryst.Meeting,***6**, 35, 1980

35.6   4,8 – Dimethyl – 1,5 – diazabicyclo(3.3.0)octa – 3,7 –
diene – 2,6 – dione
$C_8H_8N_2O_2$
I.Goldberg *Eur.Cryst.Meeting,***5**, 286, 1979

35.7 4,6 − Dimethyl − 1,5 − diazabicyclo(3.3.0)octa − 3,6 − diene − 2,8 − dione
9,10 − Dioxo − syn(methyl,H) − bimane
$C_9H_8N_2O_2$
I.Goldberg *Eur.Cryst.Meeting*,**5**, 286, 1979

35.8 2 − Amino − 1,3 − diaza − azulene hydrobromide monohydrate
$C_9H_8N_3^+, Br^-, H_2O$
H.Shimanouchi, Y.Sasada, B.Singh
Acta Crystallogr.,Sect.B,**35**, 2785, 1979

35.C 8 − Oxa − 1 − azabicyclo(4.3.0)nonan − 9 − one − 2 − carboxylic acid
$C_8H_{11}NO_4$ Main entry is 40.8

35.9 (±) − 8,8 − Dimethyl − 6,7 − diazabicyclo(3.3.0)octa − 1,6 − dien − 7 − oxide
$C_8H_{12}N_2O$
A.W.Maverick, E.F.Maverick, H.Olsen
Helv.Chim.Acta,**63**, 1304, 1980

35.C Cyclo(L − prolyl − L − alanyl)
$C_8H_{12}N_2O_2$ Main entry is 48.23

35.10 Rubidium 8 − mercapto − quinoline − 5 − sulfonate
$C_9H_6NO_3S_2^-, Rb^+$
A.D.Ozola, Ya.K.Ozols, Ya.V.Ashaks
Latv.PSR Zinat.Akad.Vestis,Kim.Ser.,287, 1979
See also R1 : 11

35.11 Sodium quinoline − 8 − thiolate dihydrate
$C_9H_6NS^-, Na^+, 2H_2O$
S.K.Apinitis, A.A.Kemme, Ya.Ya.Bleidelis
Zh.Strukt.Khim.,**20**, 876, 1979

35.12 8 − Hydroxyquinoline
C_9H_7NO
S.H.Simonsen, D.W.Bechtel
Am.Cryst.Assoc.,Ser.2,**7**, 23, 1980

35.C Skatole tetracyanoethylene
C_9H_9N, C_6N_4 Main entry is 60.9

35.13 2 − Amino − 5 − methyl − 7 − propyl − imidazo(5,1 − f)(1,2,4)triazin − 4(3H) − one methylsulfonate monohydrate
$C_9H_{14}N_5O^+, CH_3O_3S^-, H_2O$
J.P.Riley, F.Heatley, I.H.Hillier, P.Murray−Rust, J.Murray−Rust *J.Chem.Soc.,Perkin Trans.2*,1327, 1979
See also R2 : 11

35.C 2 − Imino − 6' − methyl − spiro(oxazolidine − 5,3' − piperidine) − 2' − one − 4 − semicarbazone monohydrate
$C_9H_{16}N_6O_3, H_2O$ Main entry is 40.15

35.14 6 − Phenyl − 1,2,4 − triazolo(4,3 − b) − 1,2,4 − triazine
$C_{10}H_7N_5$
R.I.Trust, J.D.Albright, F.M.Lovell, N.A.Perkinson
J.Heterocycl.Chem.,**16**, 1393, 1979

35.15 N − Phthaloylglycine − hydroxamic acid (form i)
$C_{10}H_8N_2O_4$
M.Sikirica, I.Vickovic *Cryst.Struct.Commun.*,**9**, 795, 19

35.C 3,9 − Dimethyl − 6 − thia − 1,11 − diazatricyclo(6.2.1.0^{4,11})undecane − 3,8 − diene − 2,10 − dione
μ − Thia − syn(CH_2,CH_3) − 9,10 − dioxa − bimane
$C_{10}H_{10}N_2O_2S$ Main entry is 41.30

35.C 3,9 − Dimethyl − 6 − thia − 1,11 − diazatricyclo(6.2.1.0^{4,11})undecane − 3,8 − diene − 2,10 − dione 6,6 − dioxide
μ − Sulfone − syn(CH_2,CH_3) − 9,10 − dioxa − bimane
$C_{10}H_{10}N_2O_4S$ Main entry is 41.31

35.C 2 − (Cyanophosphinidene) − 1,3 − dimethyl − benzimidazoline
$C_{10}H_{10}N_3P$ Main entry is 64.37

35.16 6 − Methyl − 9 − formyl − 1,6,7,8 − tetrahydro − 1H − pyrido(1,2 − a)pyrimidine − 4 − one
$C_{10}H_{12}N_2O_2$
K.Simon, I.Hermecz, T.Breining, A.Horvath, Z.Meszaros, L.Parkanyi, G.Bocelli *Eur.Cryst.Meeting*,**5**, 397, 1979

35.17 2,3,6,7 − Tetramethyl − 1,5 − diazabicyclo(3.3.0) octa − 2,6 − diene − 4,8 − dione
anti(Methyl,methyl) − bimane
$C_{10}H_{12}N_2O_2$
J.Bernstein, E.Goldstein, I.Goldberg
Cryst.Struct.Commun.,**9**, 301, 1980

35.18 syn − 3,4,6,7 − Tetramethyl − 1,5 − diazabicyclo(3.3.0)octa − 3,6 − diene − 2,8 − dione
syn(Methyl,methyl) − bimane
$C_{10}H_{12}N_2O_2$
J.Bernstein, E.Goldstein, I.Goldberg
Cryst.Struct.Commun.,**9**, 301, 1980

35.19 6 − Ethoxycarbonyl − 4 − ethyl − 1,2,4 − triazolo(1,5 − a)pyrimidin − 7(4H) − one
$C_{10}H_{12}N_4O_3$
J.P.Clayton, N.H.Rogers, V.J.Smith, R.Stevenson, T.J.King
J.Chem.Soc.,Perkin Trans.1,1347, 1980

35.C Tryptamine adenin − 9 − yl − acetic acid hemihydrate
$C_{10}H_{13}N_2^+, C_7H_6N_5O_2^-, 0.5H_2O$ Main entry is 44.17

35.20 6 − Methyl − 4 − oxo − 1,6,7,8,9,9a − hexahydro − 4H − pyrido(1,2 − a)pyrimidine − 3 − carboxamide
$C_{10}H_{15}N_3O_2$
K.Simon, I.Hermecz, T.Breining, A.Horvath, Z.Meszaros, L.Parkanyi, G.Bocelli *Eur.Cryst.Meeting*,**5**, 397, 1979

35.21 Cyclo − octane − spiro − 5' − hydantoin
$C_{10}H_{16}N_2O_2$
R.W.Miller, A.T.McPhail *J.Chem.Res.*,**330**, 3831, 1979

35.C Selenobiotin
$C_{10}H_{16}N_2O_3Se$ Main entry is 39.30

35.C (±) − Oxybiotin
$C_{10}H_{16}N_2O_4$ Main entry is 38.24

35.C Biotin − d − sulfoxide
$C_{10}H_{16}N_2O_4S$ Main entry is 39.31

35.C Biotin sulfone
$C_{10}H_{16}N_2O_5S$ Main entry is 39.32

35.C N − Methyl − 4H − 6,7 − dimethoxy − benzo − 1,2 − thiazin − 3 − one − 1,1 − dioxide
$C_{11}H_{13}NO_5S$ Main entry is 41.37

35.22 6 − Ethoxycarbonyl − 4 − ethylpyrazolo(1,5 − a) pyrimidin − (4H) − one
$C_{11}H_{13}N_3O_3$
J.P.Clayton, N.H.Rogers, V.J.Smith, R.Stevenson, T.J.King
J.Chem.Soc.,Perkin Trans.1,1347, 1980

35.23 **Carbobiotin**
$C_{11}H_{18}N_2O_3$
G.T.DeTitta, R.Parthasarathy, R.H.Blessing, W.Stallings
Proc.Nat.Acad.Sci.U.S.A.,**77**, 333, 1980
See also R1 : 1

35.C **Biotin methyl ester**
$C_{11}H_{18}N_2O_3S$ Main entry is 39.38

35.C **5 – Hydroxy – 1 – (2 – hydroxyethyl) – 6,7,8 –
trimethyl – 2 – thia – 7 – azabicyclo(3.3.0)octane
hemihydrate**
$C_{11}H_{21}NO_2S$, $0.5H_2O$ Main entry is 39.39

35.24 **1 – (2,4 – Dinitrophenyl) – imidazo(4,5 – b)pyridine**
$C_{12}H_7N_5O_4$
A.Escande, Y.Dumas
Acta Crystallogr.,Sect.B,**36**, 1217, 1980
See also R1 : 15

35.25 **1 – Ethyl – 1,4 – dihydro – 7 – methyl – 4 – oxo –
1,8 – naphthapyridine – 3 – carboxylic acid**
Nalidixic acid
$C_{12}H_{12}N_2O_3$
C.P.Huber, D.S.S.Gowda, K.R.Acharya
Acta Crystallogr.,Sect.B,**36**, 497, 1980

35.26 **4 – (3 – Indolyl) – butyric acid**
$C_{12}H_{13}NO_2$
K.Chandrasekhar, V.Pattabhi
Acta Crystallogr.,Sect.B,**36**, 1165, 1980
See also R1 : 1

35.27 **2 – (2',4' – Dibromophenyl) – 4 – oxo – 1,2,3 –
benzotriazin – 2 – ium – 3 – ide (orthorhombic
form)**
$C_{13}H_7Br_2N_3O$
M.A.Hamid *Libyan J.Sci.*,**8**, 75, 1978

35.C **3 – (p – Chlorophenyl) – 6 – methyl – 2 – oxo –
2,3,3a,7a – tetrahydro – oxazolo(4,5 – b)pyridine**
4 – Methyl – 9 – p – chlorophenyl – 7 – oxa – 2,9 –
diazabicyclo(4.3.0)nona – 2,4 – dien – 8 – one
$C_{13}H_{11}ClN_2O_2$ Main entry is 40.26

35.C **3 – (p – Chlorophenyl) – 7a – methyl – 2 – oxo –
2,3,3a,7a – tetrahydro – oxazolo(4,5 – b)pyridine**
6 – Methyl – 9 – p – chlorophenyl – 7 – oxa – 2,9 –
diazabicyclo(4.3.0)nona – 2,4 – dien – 8 – one
$C_{13}H_{11}ClN_2O_2$ Main entry is 40.27

35.C **N – (4 – Methoxyphenyl) – 3,6 – dithiacyclohexene –
1,2 – dicarboximide**
$C_{13}H_{11}NO_3S_2$ Main entry is 39.47

35.C **N(1') – Methoxycarbonyl – biotin methyl ester**
$C_{13}H_{20}N_2O_5S$ Main entry is 39.48

35.28 **1,3 – Dimethyl – 4 – imino – 5,5 – (spiro – 2',4' –
bis – dimethylamino – pyrrolyl) – uracil**
$C_{13}H_{20}N_6O_2$
B.Kokel, H.G.Viehe, J.P.Declercq, G.Germain,
M.van Meerssche *Tetrahedron Lett.*,**21**, 3799, 1980

35.29 **2 – (2 – Pyridylamino) – 8 – hydroxyquinoline**
$C_{14}H_{11}N_3O$
I.N.Polyakova, Z.A.Starikova, V.K.Trunov, B.V.Parusnikov,
I.A.Krasavin *Kristallografiya*,**25**, 501, 1980

35.30 **2 – Phenyl – cis – 5,6 – tetramethylene – 5,6 –
dihydropyrimidin – 4(3H) – one**
2 – Phenyl – cis – 4a,5,6,7,8a – hexahydroquinazolin –
4 – one
$C_{14}H_{16}N_2O$
A.Kapor, B.Ribar, G.Argay, A.Kalman, G.Bernath
Cryst.Struct.Commun.,**9**, 343, 1980

35.31 **2 – Phenyl – trans – 5,6 – tetramethylene – 5,6 –
dihydropyrimidin – 4(3H) – one**
2 – Phenyl – trans – 4a,5,6,7,8a –
hexahydroquinazolin – 4 – one
$C_{14}H_{16}N_2O$
A.Kapor, B.Ribar, G.Argay, A.Kalman, G.Bernath
Cryst.Struct.Commun.,**9**, 347, 1980

35.32 **2 – Ethoxy – 1 – ethoxycarbonyl – 1,2 –
dihydroquinoline**
$C_{14}H_{17}NO_3$
D.J.Cremin, A.F.Hegarty, M.J.Begley
*J.Chem.Soc.,Perkin Trans.*2,412, 1980

35.C **(1aβ,2α,6aβ,6bβ) – 3 – Methyl – N – (1a,6,6a,6b –
tetrahydro – 2,6a – dimethyl – 6 – oxo – 2H –
oxireno(a)pyrrolizin – 4 – yl) – 2 – butenamide**
Bohemamine
$C_{14}H_{18}N_2O_3$ Main entry is 59.6

35.33 **3 – Ethoxycarbonyl – 6 – methyl – 4 – oxo –
6,7,8,9 – tetrahydro – 4H – pyrido(1,2 – a)
pyrimidine – 9 – acetic acid**
$C_{14}H_{18}N_2O_5$
K.Simon, I.Hermecz, T.Breining, A.Horvath, Z.Meszaros,
L.Parkanyi, G.Bocelli *Eur.Cryst.Meeting*,**5**, 397, 1979

35.34 **7 – Bromo – 5 – (o – chlorophenyl) – 1,3 –
dihydro – 2H – 1,4 – benzodiazepin – 2 – one
(at −120°C)**
$C_{15}H_{10}BrClN_2O$
A.A.Karapetyan, V.G.Andrianov, Yu.T.Struchkov,
A.V.Bogatskii, S.A.Andronati, T.I.Korotenko
Bioorg.Khim.,**5**, 1684, 1979

35.C **N – Acetyl – dendrodoine**
$C_{15}H_{14}N_4O_2S$ Main entry is 41.53

35.C **Benzyl 4 – methoxy – 7 – azido – 3 – oxa – 1 –
azabicyclo(4.2.0)octan – 8 – one – 2 – carboxylate**
$C_{15}H_{16}N_4O_5$ Main entry is 40.36

35.C **6,6' – Dibromo – indigo**
$C_{16}H_8Br_2N_2O_2$ Main entry is 59.11

35.C **6,6' – Dibromo – indigo**
Tyrian purple
$C_{16}H_8Br_2N_2O_2$ Main entry is 59.12

35.35 **7 – Chloro – 3 – hydroxy – 1 – methyl – 5 –
phenyl – 1,3 – dihydro – 2H – 1,4 – benzodiazepin –
2 – one**
$C_{16}H_{13}ClN_2O_2$
M.L.Glowka, Z.Galdecki *Eur.Cryst.Meeting*,**5**, 51, 1979

35.36 **2 – (4' – Bromophenyl) – 4,6 – dimethoxy – indole**
$C_{16}H_{14}BrNO_2$
D.St.C.Black, B.M.K.C.Gatehouse, F.Theobald, L.C.H.Wong
Aust.J.Chem.,**33**, 343, 1980

35.37 6 – Methyl – 9 – p – bromophenylamino – 4 – oxo –
6,7 – dihydro – 4H – pyrido(1,2 – a)pyrimidine – 3 –
carboxylic acid
$C_{16}H_{14}BrN_3O_3$
K.Simon, I.Hermecz, T.Breining, A.Horvath, Z.Meszaros,
L.Parkanyi, G.Bocelli *Eur.Cryst.Meeting*,**5**, 397, 1979

35.38 7 – Chloro – 2 – methylamino – 5 – phenyl – 3H –
1,4 – benzodiazepin – 3 – ol
$C_{16}H_{14}ClN_3O$
P.Chananont, T.A.Hamor, I.L.Martin
Acta Crystallogr.,Sect.B,**36**, 1238, 1980

35.39 7 – Chloro – 2 – methylamino – 5 – phenyl – 3H –
1,4 – benzodiazepine – 4 – oxide
$C_{16}H_{14}ClN_3O$
V.Bertolasi, M.Sacerdoti, G.Gilli
Eur.Cryst.Meeting,**5**, 52, 1979

35.40 N – (4 – Dimethylaminophenyl) – phthalimide
$C_{16}H_{14}N_2O_2$
N.S.Magomedova, A.V.Dzyabchenko, V.E.Zavodnik,
V.K.Belsky *Cryst.Struct.Commun.*,**9**, 713, 1980

35.41 5 – Benzoyl – 2 – (methoxycarbamoyl) – 1H –
benzimidazole hydrobromide
Mebendazole hydrobromide
$C_{16}H_{14}N_3O_3^+$, Br^-
N.M.Blaton, O.M.Peeters, C.J.De Ranter
Cryst.Struct.Commun.,**9**, 181, 1980

35.42 7 – Chloro – 2,3 – dihydro – 1 – methyl – 5 –
phenyl – 1H – 1,4 – benzodiazepine
Medazepam hydrochloride
$C_{16}H_{16}ClN_2^+$, Cl^-, H_2O
P.Chananont, T.A.Hamor, I.L.Martin
Acta Crystallogr.,Sect.B,**36**, 898, 1980

35.C 1 – (3 – (Indol – 3 – yl)propyl) – thymine
$C_{16}H_{17}N_3O_2$ Main entry is 44.34

35.43 8 – Bromo – 1 – methyl – 6 – phenyl – 1,2,3,4 –
tetrahydro – 1,5 – benzodiazocin – 2 – one
$C_{17}H_{15}BrN_2O$
S.A.Andronati, A.A.Dvorkin, Yu.A.Simonov, V.V.Danilin,
T.I.Malinovskii, A.V.Bogatskii
Dokl.Akad.Nauk SSSR,**248**, 1140, 1979

35.44 8 – Chloro – 6 – phenyl – 1 – methyl – 1,2,3,4 –
tetrahydro – 5H – 1,5 – benzodiazocin – 2 – one
$C_{17}H_{15}ClN_2O$
A.A.Dvorkin, Yu.A.Simonov, T.I.Malinovskii, S.A.Andronati,
V.V.Danilin, A.A.Mazurov, A.V.Bogatskii
Eur.Cryst.Meeting,**5**, 54, 1979

35.45 7 – Chloro – 5 – (2 – fluorophenyl) – 1,3,4,5 –
tetrahydro – 1,4 – dimethyl – 2H – 1,4 –
benzodiazepin – 2 – one monohydrate
$C_{17}H_{16}ClFN_2O$, H_2O
P.Chananont, T.A.Hamor, I.L.Martin
Acta Crystallogr.,Sect.B,**36**, 1690, 1980

35.46 3,7 – Dichloro – 4,8 – diphenyl – 1,5 –
diazabicyclo(3.3.0)octa – 3,7 – diene – 2,6 – dione
anti(Phenyl,chloro) – bimane
$C_{18}H_{10}Cl_2N_2O_2$
J.Bernstein. E.Goldstein. I.Goldberg
Cryst.Struct.Commun.,**9**, 295, 1980

35.47 3,7 – Dichloro – 4,6 – diphenyl – 1,5 –
diazabicyclo(3.3.0)octa – 3,6 – diene – 2,8 – dione
syn(Phenyl,chloro) – bimane
$C_{18}H_{10}Cl_2N_2O_2$
J.Bernstein, E.Goldstein, I.Goldberg
Cryst.Struct.Commun.,**9**, 295, 1980

35.48 8,8′ – Dithio – diquinoline (monoclinic form)
$C_{18}H_{12}N_2S_2$
O.G.Matyukhina, Ya.K.Ozols, A.P.Sturis
Latv.PSR Zinat.Akad.Vestis,Kim.Ser.,622, 1979

35.49 3 – trans – bis(4 – Chlorophenyl) – 2 – cis – nitro –
5 – oxo – perhydropyrazolo(1,2 – a)pyrazole
$C_{18}H_{15}Cl_2N_3O_3$
L.Kutschabsky, R.G.Kretschmer, H.Dorn
Krist.Tech.,**14**, 1429, 1979

35.50 3 – trans – bis(4 – Chlorophenyl) – 2 – trans –
nitro – 5 – oxo – perhydropyrazolo(1,2 – a)pyrazole
$C_{18}H_{15}Cl_2N_3O_3$
L.Kutschabsky, H.Dorn *Krist.Tech.*,**14**, 1107, 1979

35.51 7 – Chloro – 1,3 – dihydro – 1 – (N –
methylacetamido) – 5 – phenyl – 2H – 1,4 –
benzodiazepin – 2 – one
$C_{18}H_{16}ClN_3O_2$
P.Chananont, T.A.Hamor, I.L.Martin
Acta Crystallogr.,Sect.B,**36**, 2115, 1980

35.C 2,3,4,5 – Tetrahydro – 6,6 – diphenyl – imidazo(2,1 –
b)thiazin – 7(6H) – one
$C_{18}H_{16}N_2OS$ Main entry is 41.64

35.C 2,2′ – Dimethoxy – indigo
$C_{18}H_{16}N_2O_4$ Main entry is 59.18

35.C 7 – (1 – (p – Nitrobenzyloxycarbonyl)oxy – ethyl) –
8 – oxo – 2,2 – dimethyl – 3 – oxa – 1 –
azabicyclo(4.2.0)octane
$C_{18}H_{22}N_2O_7$ Main entry is 40.45

35.C Benzyl – (2,3β,3aα,4,5,6,7,8,9,9aβ – decahydro – 4 –
oxo – thieno(3,2 – c)azocine – 3α – yl) – carbamic
acid methyl ester
$C_{18}H_{24}N_2O_4S$ Main entry is 39.72

35.52 Methyl – bis(8 – hydroxy – 2 – quinolyl)amine
$C_{19}H_{15}N_3O_2$
I.N.Polyakova, Z.A.Starikova, V.K.Trunov, B.V.Parusnikov,
I.A.Krasavin *Kristallografiya*,**25**, 495, 1980

35.C Folic acid dihydrate
$C_{19}H_{19}N_7O_6$, $2H_2O$ Main entry is 48.57

35.53 1 – Phenyl – 3 – methyl – 3 – methylamino –
propyl – indol – 2 – one hydrochloride
$C_{19}H_{23}N_2O^+$, Cl^-
R.Vega, R.J.Garay, A.L.Castro, R.Marquez
Eur.Cryst.Meeting,**5**, 396, 1979

35.54 1 – (3,4,5 – Trimethoxybenzyl) – 6,7 – dihydroxy –
1,2,3,4 – tetrahydro – isoquinoline hydrochloride
monohydrate
$C_{19}H_{24}NO_5^+$, Cl^-, H_2O
T.F.Brennan. A.R.Garafalo, D.A.Williams
Cryst.Struct.Commun.,**8**, 953, 1979

35.55 1,2 – Dihydro – 3 – oxo – 2,2 – diphenyl – 3H –
indole – 1 – oxyl
$C_{20}H_{14}NO_2$
R.Benassi, F.Taddei, L.Greci, L.Marchetti, G.D.Andreetti,
G.Bocelli, P.Sgarabotto
J.Chem.Soc.,Perkin Trans.2,786, 1980

35.56 (E) – Dibenzo(e,e') – 3,3' – diazastilbene
$C_{20}H_{14}N_2$
J.Vansant, G.Smets, J.P.Declercq, G.Germain,
M.van Meerssche J.Org.Chem.,45, 1557, 1980

35.57 (Z) – Dibenzo(e,e') – 3,3' – diazastilbene
$C_{20}H_{14}N_2$
J.Vansant, G.Smets, J.P.Declercq, G.Germain,
M.van Meerssche J.Org.Chem.,45, 1557, 1980

35.58 1H,4H – 1,4 – Diphenylpyridazino(1,2 – a)
pyridazine – 6,9 – dione
$C_{20}H_{16}N_2O_2$
M.C.Apreda, C.Foces–Foces, F.H.Cano, S.Garcia–Blanco
Eur.Cryst.Meeting,6, 311, 1980

35.59 3 – Hydroxy – 2,3 – diphenyl – indoline
$C_{20}H_{17}NO$
C.Berti, L.Greci, M.Poloni, G.D.Andreetti, G.Bocelli,
P.Sgarabotto J.Chem.Soc.,Perkin Trans.2,339, 1980

35.60 3 – Hydroxy – 2,2' – dioxo – (3 – 3' – di –
indoline) – N,N' – bis(5,5 – dimethylsulfoximide)
hydrate
$C_{20}H_{22}N_4O_5S_2$, $3.5H_2O$
C.Foces–Foces, F.H.Cano, S.Garcia–Blanco
J.Cryst.Mol.Struct.,9, 143, 1979

35.61 1,2 – Dihydro – 2,2 – diphenylquinoline – 1 – oxyl
$C_{21}H_{16}NO$
R.Benassi, F.Taddei, L.Greci, L.Marchetti, G.D.Andreetti,
G.Bocelli, P.Sgarabotto
J.Chem.Soc.,Perkin Trans.2,786, 1980

35.62 2 – (o – Hydroxybenzoyl) – 7 – methyl – 3 –
phenyl – imidazo(1,2 – a)pyridine (at –40°C)
$C_{21}H_{16}N_2O_2$
A.R.Katritzky, M.Michalska, R.L.Harlow, S.H.Simonsen
J.Chem.Soc.,Perkin Trans.1,354, 1980

35.63 2 – Phenyl – 3 – (N – p – methoxyphenyl)amine –
indole
$C_{21}H_{18}N_2O$
L.Cardellini, G.Tosi, G.Bocelli, A.Musatti
Cryst.Struct.Commun.,9, 233, 1980

35.64 N – (2 – (4 – (5 – Chloro – 2 – oxo – 1 –
benzimidazolinyl)piperidino) – ethyl) – p –
fluorobenzamide
Halopemide
$C_{21}H_{22}ClFN_4O_2$
N.Van Opdenbosch, M.Weyland, G.Evrard, F.Durant
Acta Crystallogr.,Sect.B.36, 965, 1980
See also R1 : 33,13

35.65 trans(2H,8aH) – 2 – Diphenylhydroxy – methyl –
indolizidine
$C_{21}H_{25}NO$
H.Kato, E.Koshinaka, N.Ogawa, K.Yamagishi, K.Mitani,
S.Kubo, M.Hanaoka Chem.Pharm.Bull.,28, 2194, 1980

35.66 2,4 – bis(Dimethylamino) – 3 – ethyl – 1 – phenyl –
1,5 – benzodiazepine
$C_{21}H_{26}N_4$
J.Galloy, J.P.Declercq, M.van Meerssche
Cryst.Struct.Commun.,8, 981, 1979

35.67 2 – Benzoyl – 2 – (1,3 – dimethyl – 2,6 – dioxo – 7 –
purinyl) – acetanilide
$C_{22}H_{19}N_5O_4$
M.van Poucke, A.T.H.Lenstra
Cryst.Struct.Commun.,9, 575, 1980

35.68 2 – Benzyl – 1,2,3,4 – tetrahydro – 6,7 –
dimethoxy – 2 – methyl – 1 – isopropyl –
isoquinolinium iodide
$C_{22}H_{30}NO_2^+$, I-
G.Argay, A.Kalman, B.Ribar, D.Lazar, J.Kober, G.Bernath
Cryst.Struct.Commun.,8, 917, 1979

35.69 8 – Methyl – 4 – (p – toluenesulfonyl) – 6 –
phenyl – 1,2,3,4 – tetrahydro – 5H – 1,4,5 –
benzotriazocin – 2 – one
$C_{23}H_{21}N_3O_3S$
A.A.Dvorkin, Yu.A.Simonov, T.I.Malinovskii, S.A.Andronati,
V.V.Danilin, A.A.Mazurov, A.V.Bogatskii
Eur.Cryst.Meeting,5, 54, 1979

35.C 6α – Benzyl – 6β – isocyano – penicillanate
$C_{23}H_{22}N_2O_3S$ Main entry is 50.16

35.C Cyclo(di(benzylglycyl) – L – prolyl) monohydrate
$C_{23}H_{25}N_3O_3$, H_2O Main entry is 48.73

35.70 cis – 5,6 – Dimethoxy – 2 – methyl – 3 – (2 – (4 –
phenyl – 1 – piperazinyl) – ethyl) – indoline
$C_{23}H_{31}N_3O_2$
A.E.Lanzilotti, R.Littell, W.J.Fanshawe, T.C.McKenzie,
F.M.Lovell J.Org.Chem.,44, 4809, 1979
See also R1 : 33

35.71 1,3 – bis(1,3,3 – Trimethyl – indolenine – 2 – yl) –
2 – aza – trimethinium tetrafluoroborate
$C_{24}H_{28}N_3^+$, BF_4^-
R.Allmann, S.Olejnik, A.Waskowska
Eur.Cryst.Meeting,6, 51, 1980

35.C 6,8,8 – Trimethyl – 6 – (3 – indolyl) – 4 –
(isopropenyl) – (2,3)benzo – 1 – azabicyclo(3.3.0)
octa – 2,4 – diene
$C_{25}H_{26}N_2$ Main entry is 36.45

35.72 Dimethyl 2 – (1,3 – dimethyl – indol – 2 – yl) – 3 –
(trans – 2,3 – dihydro – 1,3 – dimethyl – indol – 2 –
yl) – maleate
$C_{26}H_{28}N_2O_4$
P.D.Davis, D.C.Neckers, J.R.Blount
J.Org.Chem.,45, 462, 1980

35.73 3,3' – Thio – bis(2 – methyl – 1 – phenyl –
imidazo(1,5 – a)pyridinium) tetrafluoroborate
$C_{28}H_{24}N_4S^{2+}$, $2BF_4^-$
D.J.Pointer, J.B.Wilford, J.D.Lee
J.Chem.Soc.,Perkin Trans.2,1075, 1980

35.74 1 – (2 – Butoxycarbonyl – 1 – azophenyl) – 2 –
hydroxy – 3 – N – (2 – oxo – 5 – benzimidazolyl) –
(naphthoic acid) – amide
C.I.Pigment red 208
$C_{29}H_{25}N_5O_5$
E.F.Paulus, K.Hunger Farbe Lack,86, 116, 1980
See also R1 : 24,13

35.75   1,3 – bis(1,3,3 – Trimethyl – indolenine – 2 – yl) –
2 – t – butyl – trimethinium tetrafluoroborate
$C_{29}H_{37}N_2^+$, BF_4^-
R.Allmann, S.Olejnik, A.Waskowska
*Eur.Cryst.Meeting,***6**, 51, 1980

35.76   4,6 – Di – t – butyl – 1 – (3,5 – di – t – butyl – 2 –
hydroxyphenyl) – 2 – methyl – 2 – azabicyclo(3.2.0)
hept – 6 – en – 3 – one
$C_{29}H_{45}NO_2$
H.–D.Becker, K.Gustafsson, C.L.Raston, A.H.White
*Aust.J.Chem.,***32**, 1931, 1979

35.77   N – Acetyl – 3 – (benzoyl(2 – piperidyl – 2 –
piperidylenium – ethyl)methylene) – indol – 2 –
olate
$C_{30}H_{35}N_3O_3$
G.Tacconi, M.Leoni, P.Righetti, G.Desimoni, R.Oberti,
F.Comin   *J.Chem.Soc.,Perkin Trans.1*,2687, 1979
See also R1 : 33

35.78   1,3 – bis(1,3,3 – Triethyl – indolenine – 2 – yl) – 2 –
arsa – trimethinium tetrafluoroborate methylene
chloride solvate
$C_{30}H_{40}AsN_2^+$, BF_4^-, $0.5CH_2Cl_2$
R.Allmann, S.Olejnik, A.Waskowska
*Eur.Cryst.Meeting,***6**, 51, 1980

35.79   1,3 – bis(1,3,3 – Triethyl – indolenine – 2 – yl) – 2 –
phospha – trimethinium tetrafluoroborate
$C_{30}H_{40}N_2P^+$, BF_4^-
R.Allmann, S.Olejnik, A.Waskowska
*Eur.Cryst.Meeting,***6**, 51, 1980

35.80   1,3 – bis(1,3,3 – Triethyl – indolenine – 2 – yl) – 2 –
phospha – trimethinium tetrafluoroborate
methylene chloride solvate
$C_{30}H_{40}N_2P^+$, BF_4^-, $0.5CH_2Cl_2$
R.Allmann, S.Olejnik, A.Waskowska
*Eur.Cryst.Meeting,***6**, 51, 1980

35.81   1,3 – bis(1,3,3 – Triethyl – indolenine – 2 – yl) – 2 –
aza – trimethinium tetrafluoroborate
$C_{30}H_{40}N_3^+$, BF_4^-
R.Allmann, S.Olejnik, A.Waskowska
*Eur.Cryst.Meeting,***6**, 51, 1980

35.82   1,3 – bis(1,3,3 – Triethyl – indolenine – 2 – yl) –
trimethinium perchlorate
$C_{31}H_{41}N_2^+$, ClO_4^-
R.Allmann, S.Olejnik, A.Waskowska
*Eur.Cryst.Meeting,***6**, 51, 1980

35.C   3 – (4,5 – bis(Methoxycarbonyl) – isoxazol – 3 –
yl) – 4,6 – dioxo – syn – cis – syn – 1,3,5 –
triphenyl – perhydrothieno(3,4 – c)pyrrole – 1 –
carbonitrile
$C_{32}H_{23}N_3O_7S$ Main entry is 40.58

35.83   1,20 – bis(8 – Quinolyl – oxy) – 3,6,9,12,15,18 –
hexaoxa – eicosane rubidium iodide
$C_{32}H_{40}N_2O_8$, Rb^+, I^-
G.Weber, W.Saenger
Acta Crystallogr.,Sect.B,**35**, 3093, 1979

35.84   tris((2 – Methyl – 8 – quinolyl – oxy) – ethyl)amine
dihydrate
$C_{36}H_{36}N_4O_3$, $2H_2O$
G.Weber, G.M.Sheldrick
Acta Crystallogr.,Sect.B,**36**, 1978, 1980

35.85   tris((2 – Methyl – 8 – quinolyl – oxy) – ethyl)amine
rubidium iodide
$C_{36}H_{36}N_4O_3$, Rb^+, I^-
G.Weber, G.M.Sheldrick   *Inorg.Chim.Acta*,**45**, L35, 1980

HETERO–NITROGEN
(MORE THAN 2 FUSED RINGS)

36.C   1,2,3 – Oxadiazolo(4,3 – c)(1,2,4)benzotriazinium –
3 – olate
$C_8H_4N_4O_2$ Main entry is 40.6

36.C   4 – Hydroxy – 11 – thia – 2,6,12 –
triazatricyclo(6.3.1.0^{4,12})dodecane – 1(11),8 – diene –
3,6 – dione (at –140°C)
$C_8H_7N_3O_3S$ Main entry is 41.21

36.C   1 – Oxo – 1,2 – dihydro – 2,3 – diazaphenothiazine
$C_{10}H_7N_3OS$ Main entry is 41.27

36.C   1 – Chloro – 10 – methyl – 2,3 – diazaphenothiazine
$C_{11}H_8ClN_3S$ Main entry is 41.34

36.1   2 – Methylnaphtho(1,8 – de) – 1λ^2,2λ^4,3 – triazine
$C_{11}H_9N_3$
A.Gieren, V.Lamm, R.C.Haddon, M.L.Kaplan, M.J.Perkins,
P.Flowerday *Eur.Cryst.Meeting*,**6**, 52, 1980

36.C   7,8 – benzoquinoline 7,7,8,8 –
Tetracyanoquinodimethane
$C_{12}H_4N_4$, $C_{13}H_9N$ Main entry is 60.14

36.2   Phenazine 5,10 – dihydro – 5,10 –
dimethylphenazinium 7,7,8,8 –
tetracyanoquinodimethane
0.4$C_{12}H_8N_2$, 0.6$C_{14}H_{14}N_2$, $C_{12}H_4N_4$
H.Endres, H.J.Keller, W.Moroni, D.Nothe
Acta Crystallogr.,Sect.B,**36**, 1435, 1980
See also R2 : 36 R3 : 7

36.C   bis(9,10 – Diazaphenanthrene) tetracyanoethylene
(triclinic form)
2$C_{12}H_8N_2$, C_6N_4 Main entry is 60.15

36.3   5,10 – Dihydroxybenzo(g)phthalhydrazide acetic
acid solvate
$C_{12}H_8N_2O_4$, 2$C_2H_4O_2$
M.C.Apreda, C.Foces-Foces, F.H.Cano, S.Garcia-Blanco
Eur.Cryst.Meeting,**6**, 311, 1980

36.C   2 – Carboxy – 3 – methylthiopyrano(4,3,2 – cd)
indole
$C_{12}H_{11}NO_2S$ Main entry is 39.42

36.4   1,2,4,5 – Tetrahydro – 7 – methoxy – 3H – benz(g)
indazol – 3 – one monohydrate (at 113°K)
$C_{12}H_{12}N_2O_2$, H_2O
D.van der Helm, K.K.Wu, S.E.Ealick, K.D.Berlin,
K.Ramalingam *Acta Crystallogr.,Sect.B*,**35**, 2804, 1979

36.C   3 – Benzyl – 3,7 – dihydro – 5 – methylthio –
1,2,3 – triazolo(4,5 – d) – 1,3 – thiazine
$C_{12}H_{12}N_4S_2$ Main entry is 41.40

36.5   2 – Methyl – 2H – acenaphthyleno(5,6 – de) – 1,2,3 –
triazine
$C_{13}H_9N_3$
A.Gieren, V.Lamm, R.C.Haddon, M.L.Kaplan, M.J.Perkins,
P.Flowerday *Eur.Cryst.Meeting*,**6**, 52, 1980

36.6   N – Methylcarbazole
$C_{13}H_{11}N$
E.G.Popova, L.A.Chetkina *Zh.Strukt.Khim.*,**20**, 665, 1979

36.7   2 – Allyl – 2 – azonia – 7 – azabiphenylene bromide
$C_{13}H_{11}N_2^+$, Br$^-$
S.Kanoktanaporn, J.A.H.MacBride, T.J.King
J.Chem.Res.,**204**, 2911, 1980

36.8   9 – Amino – acridinium chloride dihydrate
$C_{13}H_{11}N_2^+$, Cl$^-$, 2H_2O
C.Courseille, S.Geoffre, B.Busetta
Cryst.Struct.Commun.,**9**, 287, 1980

36.C   9 – Amino – acridine – 5 – iodocytidylyl – (3' –
5') – guanosine hydrate complex
4$C_{13}H_{11}N_2^+$, 4$C_{19}H_{23}IN_8O_{12}P^-$, 21$H_2O$ Main entry is 60.22

36.C   Cytidylyl – (3' – 5') – adenosine – proflavine
complex decahydrate
$C_{13}H_{11}N_3$, $C_{19}H_{26}N_8O_{11}P$, 10H_2O Main entry is 60.39

36.C   Proflavine deoxycytidylyl – (3' – 5') – guanosine
hydrate
2$C_{13}H_{11}N_3$, 2$C_{19}H_{25}N_8O_{10}P$, xH_2O Main entry is 47.38

36.C   bis(Adenosine) proflavine sesquisulfate hydrate
$C_{13}H_{12}N_3^+$, 2$C_{10}H_{14}N_5O_4^+$, 1.5O_4S^{2-}, 6.5H_2O
Main entry is 47.21

36.C   Proflavine 5 – iodocytidylyl – (3' – 5') – guanosine
hydrate methanol solvate
2$C_{13}H_{12}N_3^+$, 2$C_{19}H_{23}IN_8O_{12}P^-$, 15$H_2O$, CH_4O
Main entry is 47.36

36.9   N – Methyl – acridone
$C_{14}H_{11}NO$
A.V.Dzyabchenko, V.E.Zavodnik, V.K.Bel'skii
Kristallografiya,**25**, 72, 1980

36.10   10,11 – Dihydro – 5H – dibenzo(b,f)azepine
Iminodibenzyl
$C_{14}H_{13}N$
J.P.Reboul, B.Cristau, J.Estienne, J.P.Astier
Acta Crystallogr.,Sect.B,**36**, 2108, 1980

36.11   7,8 – Dimethyl – isoalloxazine – 10 – acetic acid
tyramine tetrahydrate
$C_{14}H_{13}N_4O_4$, $C_8H_{11}NO$, 4H_2O
M.Inoue, M.Shibata, T.Ishida
Biochem.Biophys.Res.Commun.,**93**, 415, 1980
See also R2 : 17

36.12   7 – Chloro – 2,6 – dihydro – 4 – methyl – 5,6 –
ethano – 1H – (1,4) – diazepino(1,7 – a)
benzimidazole hydrochloride dihydrate
$C_{14}H_{14}ClN_3$, H$^+$, Cl$^-$, 2H_2O
H.Stahle, H.Koppe, H.Daniel, K.–H.Pook, H.–J.Forster,
H.J.Hecht, W.Steglich *Chem.Ber.*,**113**, 2841, 1980

36.13   1,3,5 – Trimethylazuleno(1,8 – cd)pyridazine
$C_{14}H_{14}N_2$
K.Hafner, H.J.Lindner, W.Wassem
Heterocycles,**11**, 387, 1978

36.C   Phenazine 5,10 – dihydro – 5,10 –
dimethylphenazinium 7,7,8,8 –
tetracyanoquinodimethane
0.6$C_{14}H_{14}N_2$, 0.4$C_{12}H_8N_2$, $C_{12}H_4N_4$ Main entry is 36.2

36.14 3a,9a – Dihydro – 1,3,3a,9a – tetramethyl – 4H – pyrazolo(3,4 – b)quinolin – 4 – one
$C_{14}H_{17}N_3O$
M.Gal, O.Feher, E.Tihanyi, G.Horvath, G.Jerkovich, G.Argay, A.Kalman *Tetrahedron Lett.*,**21**, 1567, 1980

36.C rac – cis – 5 – Chloro – 2,3,3a,9 – tetrahydro – 6,8,9 – trimethyl – 1H – pyrrolo(2,1 – b)(1,3) benzoxazine
$C_{14}H_{18}ClNO$ Main entry is 40.34

36.15 1,11 – bis(Bromomethyl) – 1,2,3,4 – tetrahydro – acridine
$C_{15}H_{15}Br_2N$
M.Bukowska-Strzyzewska, J.Skoweranda *Acta Crystallogr.,Sect.B*,**36**, 886, 1980

36.C Mitomycin C dihydrate
$C_{15}H_{18}N_4O_5$, $2H_2O$ Main entry is 50.9

36.C Cyclo(di – L – prolyl – D – prolyl)
$C_{15}H_{21}N_3O_3$ Main entry is 48.45

36.C Isosoforidine
$C_{15}H_{24}N_2O$ Main entry is 58.11

36.16 7,8 – Benzo – 9 – cyano – 3,5,10 – trimethyl – 3,5 – diazabicyclo(4.4.0)deca – 7,9 – diene – 2,4 – dione
$C_{16}H_{15}N_3O_2$
I.Saito, K.Shimozono, S.Miyazaki, T.Matsuura, K.Fukuyama, Y.Katsube *Tetrahedron Lett.*,**21**, 2317, 1980

36.17 5,10 – Dihydro – 5,10 – diethylphenazine polyiodide
$C_{16}H_{18}N_2^+$, $C_{16}H_{18}N_2$, I_3^-, $0.1I_2$
H.Endres, R.Harms, H.J.Keller, W.Moroni, D.Nothe, M.H.Vartanian, Z.G.Soos *J.Phys.Chem.Solids*,**40**, 591, 1979

36.18 3 – Methyl – 4a – (methylene – dithioglycerol acid methyl ester) – 4a,5 – dihydro – isoalloxazine
$C_{16}H_{18}N_4O_4S_2$
P.Kierkegaard, B.Stensland, M.von Glehn *Eur.Cryst.Meeting*,**5**, 181, 1979

36.C Tomaymycin
$C_{16}H_{20}N_2O_4$ Main entry is 50.10

36.19 4,5 – Dihydro – 5 – hydroxy – 4 – oxo – 5 – (2 – oxopropyl) – 1H – pyrrolo(2,3 – f)quinoline – 2,7,9 – tricarboxylic acid dihydrate
$C_{17}H_{12}N_2O_9$, $2H_2O$
W.B.T.Cruse, O.Kennard, S.A.Salisbury *Acta Crystallogr.,Sect.B*,**36**, 751, 1980

36.C 7,8 – Benzo – 3 – ethoxycarbonyl – 2 – methoxycarbonyl – 9 – methyl – 9 – aza – 5 – oxabicyclo(4.3.0)nona – (1(6),2) – dien – 4 – one
$C_{17}H_{15}NO_6$ Main entry is 38.80

36.C 3,5,6,8 – Tetramethyl – N – methyl – phenanthrolinium – 5 – iodocytidylyl – (3′ – 5′) – guanosine hydrate methanol solvate
$2C_{17}H_{19}N_2^+$, $2C_{19}H_{23}IN_8O_{12}P^-$, $17H_2O$, $2CH_4O$
Main entry is 60.37

36.C Anthramycin methylether monohydrate
$C_{17}H_{19}N_3O_4$, H_2O Main entry is 50.12

36.C Acridine orange 5 – iodocytidylyl – (3′ – 5′) – guanosine dodecahydrate
$C_{17}H_{20}N_3^+$, $C_{19}H_{23}IN_8O_{12}P^-$, $12H_2O$ Main entry is 47.35

36.20 5,6 – Cyclopenteno – pyrido(3,2 – a)carbazole
$C_{18}H_{14}N_2$
K.Yamaguchi, Y.Iitaka, K.Shudo, T.Okamoto *Acta Crystallogr.,Sect.B*,**36**, 176, 1980

36.21 1H,4H – 2,3 – Dimethylpyridazino(1,2 – b)benzo(g) phthalazine – 6,13 – dione
$C_{18}H_{16}N_2O_2$
M.C.Apreda, C.Foces-Foces, F.H.Cano, S.Garcia-Blanco *Acta Crystallogr.,Sect.B*,**36**, 865, 1980

36.22 3H,4H – 2,3 – Dimethylpyridazino(1,2 – b)benzo(g) phthalazine – 6,13 – dione
$C_{18}H_{16}N_2O_2$
M.C.Apreda, C.Foces-Foces, F.H.Cano, S.Garcia-Blanco *Acta Crystallogr.,Sect.B*,**36**, 865, 1980

36.23 1,2,3,9b – Tetrahydro – 9bβ – hydroxy – 2β – methoxy – α – phenyl – 5H – pyrrolo(2,1 – a) isoindol – 5 – one
$C_{18}H_{17}NO_3$
K.Fukuyama, N.Tanaka, M.Kakudo *Acta Crystallogr.,Sect.B*,**36**, 1965, 1980

36.24 11,12,13 – Tribromo – 3,6 – diethyl – 10 – ethoxycarbonyl – 3,6 – diazatricyclo(7.4.0.0²,⁷) tridecane – 2(7),9,11,13 – tetraen – 8 – one
$C_{18}H_{19}Br_3N_2O_3$
B.Fuchs, R.Lidor, C.Kruger, L.–K.Liu *Nouv.J.Chim.*,**4**, 361, 1980

36.25 3 – Methyl – 4a – (methylene – thioglycerol acid methyl ester) – 4a,5 – dihydro – lumiflavin
$C_{18}H_{22}N_4O_4S$
P.Kierkegaard, B.Stensland, M.von Glehn *Eur.Cryst.Meeting*,**5**, 181, 1979

36.26 meso – 1,1,3,6,6,8 – Hexamethyl – 3a,5a,8a,10a – tetra – azaperhydropyrene
$C_{18}H_{34}N_4$
N.W.Alcock, P.Moore, K.F.Mok *J.Chem.Soc.,Perkin Trans.2*,1186, 1980

36.27 racemic – 1,1,3,6,6,8 – Hexamethyl – 3a,5a,8a,10a – tetra – azaperhydropyrene
$C_{18}H_{34}N_4$
N.W.Alcock, P.Moore, K.F.Mok *J.Chem.Soc.,Perkin Trans.2*,1186, 1980

36.28 racemic – 1,1,3,6,6,8 – Hexamethyl – 3a,5a,8a,10a – tetra – aza – cis – 10b,10c – perhydropyrene
$C_{18}H_{34}N_4$
P.Gluzinski, J.W.Krajewski, Z.Urbanczyk-Lipkowska *Acta Crystallogr.,Sect.B*,**36**, 2182, 1980

36.29 5,5,7,12,12,14 – Hexamethyl – 1,4,8,11 – tetra – azatricyclo(9.3.1.1⁴,⁸)hexadecane hydrate
$C_{18}H_{36}N_4$, $0.25H_2O$
N.W.Alcock, P.Moore, K.F.Mok *J.Chem.Soc.,Perkin Trans.2*,1186, 1980

36.30 3a,4,9,9a – Tetrahydro – 2 – methyl – 4β – phenylbenz(f)isoindoline
$C_{19}H_{21}N$
P.Murray-Rust, J.Murray-Rust, D.Middlemiss *Acta Crystallogr.,Sect.B*,**36**, 1678, 1980

36.31 3 – Methyl – 4a – (methylene – dithioglycerol acid ethyl ester) – 4a,5 – dihydro – lumiflavin
$C_{19}H_{24}N_4O_4S_2$
P.Kierkegaard, B.Stensland, M.von Glehn *Eur.Cryst.Meeting*,**5**, 181, 1979

36.32 12 − (p − Chlorophenylsulfonylamino) − 6,7,10,11 −
tetrahydroazocino(1,2 − a)indol − 8(9H) − one
$C_{20}H_{19}ClN_2O_3S$
K.Prout, M.Sims, D.Watkin, C.Couldwell, M.Vandrevala,
A.S.Bailey *Acta Crystallogr.,Sect.B*,**36**, 1846, 1980

36.C (+) − 5,10 − Methenyl − 5,6,7,8 − tetrahydrofolic acid
bromide hydrobromide dihydrate (for abs.
configuration see Fontecilla−Camps et
al.,J.Amer.Chem. Soc.101,6114,1979)
$C_{20}H_{23}N_7O_6{}^{2+}$, 2Br⁻, $2H_2O$ Main entry is 48.63

36.33 (−) − 5,10 − Methenyl − 5,6,7,8 − tetrahydrofolic acid
bromide hydrobromide dihydrate
$C_{20}H_{23}N_7O_6{}^{2+}$, 2Br⁻, $2H_2O$
J.C.Fontecilla−Camps, C.E.Bugg, C.Temple Junior, J.D.Rose,
J.A.Montgomery, R.L.Kisliuk *Der Biochem.*,**4**, 235, 1978

36.34 7,8 − Benzo − 8 − cyano − 3,5 − dimethyl − 10 −
phenyl − 3,5 − diazabicyclo(4.4.0)deca − 7,9 − diene −
2,4 − dione
$C_{21}H_{17}N_3O_2$
I.Saito, K.Shimozono, S.Miyazaki, T.Matsuura,
K.Fukuyama, Y.Katsube
Tetrahedron Lett.,**21**, 2317, 1980

36.C Ethidium cytidylyl − (3′ − 5′) − guanosine
$C_{21}H_{20}N_3{}^+$, $C_{19}H_{24}N_6O_{12}P^-$ Main entry is 47.37

36.35 5 − (2 − Bromo − 4 − methyl − phenyl) − 3,7 −
dimethyl − 1 − propyl − 1,5 − dihydro − benzo(f)
pyrazolo(3,4 − c)(1,2,5)triazepine
$C_{21}H_{22}BrN_5$
V.M.Agre, T.F.Sysoeva, V.K.Trunov, V.A.Tafeenko,
V.M.Dziomko, B.K.Berestevich
Zh.Strukt.Khim.,**20**, 1064, 1979

36.36 10,11,12 − Tribromo − 3,6 − diethyl − 13 −
(phenoxycarbonyl) − 3,6 − diazatricyclo(7.4.0.0²,⁷)
tridecane − 2(7),9,11,13 − tetraen − 8 − one
$C_{22}H_{19}Br_3N_2O_3$
B.Fuchs, R.Lidor, C.Kruger, L.−K.Liu
Nouv.J.Chim.,**4**, 361, 1980

36.37 ((5,6 − c) − (1 − Bromobenzo) − 1 − ethyl − 4 − (3 −
methoxy − phenylethyl) − 1,4 − diazocyclohex − 5 −
en − 2 − one) − 3 − spiro − 5′ − (3′ − methyl −
hydantoin)
$C_{22}H_{23}BrN_4O_4$
M.Iwata, T.C.Bruice, H.L.Carrell, J.P.Glusker
J.Am.Chem.Soc.,**102**, 5036, 1980

36.38 9,9 − Dimethyl − 10 − (3′ − piperidyl − propyl) −
acridane (at −100°C)
$C_{23}H_{30}N_2$
T.Debaerdemaeker, F.Osterle, U.Thewalt, G.Struckmeier
Eur.Cryst.Meeting,**5**, 56, 1979
See also R1 : 33

36.39 1,14:7,8 − Dietheno − tetrapyrido(2,1,6 − de:2′,1′,6′ −
gh:2′′,1′′,6′′ − kl:2′′′,1′′′,6′′′ − na)(1,3,5,8,10,12) −
hexa − azacyclotetradecine
$C_{24}H_{14}N_6$
H.Endres, M.Hunziker *J.Cryst.Mol.Struct.*,**9**, 77, 1979

36.40 2 − (o − Hydroxybenzoyl) − 3 − phenyl −
imidazo(2,1 − a)isoquinoline (at −40°C)
$C_{24}H_{16}N_2O_2$
A.R.Katritzky, M.Michalska, R.L.Harlow, S.H.Simonsen
J.Chem.Soc.,Perkin Trans.1,354, 1980

36.41 1,3 − Dimethyl − 5 − (10 − phenyl − 2,10 −
dihydrophenazin − 2 − ylidene) −
hexahydropyrimidine − 2,4,6 − trione
$C_{24}H_{18}N_4O_3$
J.W.Clark−Lewis, M.R.Taylor, J.Westphalen
Aust.J.Chem.,**32**, 1943, 1979
See also R1 : 33

36.42 N,N′,7 − Trimethyl − 1 − p − tolyl − 1,2,3,4 −
tetrahydro − naphthalene − tetracarboxylic −
1,2,3,4 − di − imide
$C_{24}H_{22}N_2O_4$
J.W.Epstein, T.C.McKenzie, M.F.Lovell, N.A.Perkinson
J.Chem.Soc.,Chem.Commun.,314, 1980

36.43 17a − Methyl − 3β − pyrrolidino − 17a − aza − D −
homo − 5α − androstane
$C_{24}H_{42}N_2$
J.Husain, R.A.Palmer *Eur.Cryst.Meeting*,**6**, 291, 1980
See also R1 : 32

36.44 7 − Benzyl − 3 − (p − bromobenzyloxycarbonyl) −
2 − hydroxy − 3,6,9 − triazatricyclo(7.3.0.0²,⁶)
dodecane − 5,8 − dione
$C_{25}H_{26}BrN_3O_5$
G.Lucente, A.Romeo, S.Cerrini, W.Fedeli, F.Mazza
J.Chem.Soc.,Perkin Trans.1,809, 1980

36.45 6,8,8 − Trimethyl − 6 − (3 − indolyl) − 4 −
(isopropenyl) − (2,3)benzo − 1 − azabicyclo(3.3.0)
octa − 2,4 − diene
$C_{25}H_{26}N_2$
A.Chatterjee, S.Manna, J.Banerji, C.Pascard, T.Prange,
J.N.Shoolery *J.Chem.Soc.,Perkin Trans.1*,553, 1980
See also R1 : 35

36.46 7 − Benzyl − 3 − (benzyloxycarbonyl) − 2 −
hydroxy − 3,6,9 − triazatricyclo(7.3.0.0²,⁶)dodecane −
5,8 − dione
$C_{25}H_{27}N_3O_5$
G.Lucente, A.Romeo, S.Cerrini, W.Fedeli, F.Mazza
J.Chem.Soc.,Perkin Trans.1,809, 1980

36.47 (±) − Deoxybutaclamol
$C_{25}H_{31}N$
S.Fortier, M.Przybylska, L.G.Humber
Can.J.Chem.,**58**, 1444, 1980

36.48 N′ − Acetyl − di − indolo(2,3 − a:2′,3′ − c)carbazole
$C_{26}H_{17}N_3O$
T.Kaneko, M.Matsuo, Y.Iitaka *Heterocycles*,**12**, 471, 1979

36.C 5 − Ethyl − 5,10 − dihydro − 10,10 −
diphenylphenaza − silane
$C_{26}H_{23}NSi$ Main entry is 63.43

36.49 dl − 8 − Fluoro − 5 − (4 − fluorophenyl) − 2 − (4 −
hydroxy − 4 − (4 − fluorophenyl) − butyl) −
2,3,4,5 − tetrahydro − 1H − pyrido(4,3 − b)indole
Flutroline
$C_{27}H_{25}F_3N_2O$
J.Bordner, J.J.Plattner, W.M.Welch
Cryst.Struct.Commun.,**9**, 799, 1980

36.50 Tetrabenzo(a,c,h,j)phenazine
$C_{28}H_{16}N_2$
M.Sato, A.Oya, S.Otani
Cryst.Struct.Commun.,**9**, 811, 1980

36.51 4,5 – Benzimidazo(1,2 – a) – (4 – phenyl – (1,3,5 – triazino)) – (5,6 – a) – (6,7 – benzo – 4 – phenyl – (1,3,5 – triazepine)) chloroform solvate
$C_{28}H_{18}N_6$, $CHCl_3$
C.Wentrup, C.Thetaz, E.Tagliaferri, H.J.Lindner, B.Kitschke, H.–W.Winter, H.P.Reisenauer
Angew.Chem.,Int.Ed.Engl.,19, 566, 1980

36.C 6 – Dimethylamino – 1,4 – etheno – 5,5 – dimethyl – 8 – oxo – N,9 – diphenyl – 2,3,4,4a,5,8 – hexahydro – 1H – pyridazino(4,5 – d)azepine – 2,3 – dicarboximide
$C_{28}H_{27}N_5O_3$ Main entry is 37.23

36.C Paspaline methanol solvate
$C_{28}H_{39}NO_2$, $0.5CH_4O$ Main entry is 59.41

36.C 4',5,7 – Tri – t – butyl – 3' – (2,2 – dimethylpropionyl) – 1' – methyl – spiro(benzofuran – 3(2H) – 2' – pyrrolidine) – 2,5' – dione
$C_{29}H_{43}NO_4$ Main entry is 38.132

36.C 3β – Acetoxy – 5β,6β – N – nitro – aziridinyl – cholestene (at –98°C)
$C_{29}H_{48}N_2O_4$ Main entry is 51.74

36.C 8 – Benzoyl – 15 – hydroxy – 15 – phenyl – 6aH,14aH,15H – benzothiazolo(2,3 – a) benzothiazino(4,3 – c)piperazine methanol solvate
$C_{30}H_{22}N_2O_2S_2$, CH_4O Main entry is 41.83

36.52 N,N',N'' – Tribenzyl – benzo(1,2 – c:3,4 – c':5,6 – c'') tripyrroline
$C_{33}H_{33}N_3$
J.H.Gall, C.J.Gilmore, D.D.MacNicol
*J.Chem.Soc.,Chem.Commun.,*927, 1979

36.C 9,10 – 0,0 – Isopropylidene – 10a – 0 – (o – bromobenzoyl) – 11 – 0 – methyl – rubrolone
$C_{34}H_{32}BrNO_9$ Main entry is 50.

36.C CC – 1065 – Anti – tumor agent (at –150°C)
NSC 298223
$C_{37}H_{33}N_7O_8$, $C_{36}H_{33}N_7O_8$, H_2O, $3CH_4O$ Main entry is 59.43

36.C Des – N – tetramethyl – triostin A dodecahydrate
$C_{48}H_{54}N_{12}O_{12}S_2$, $12H_2O$ Main entry is 50.34

HETERO–NITROGEN (BRIDGED RING SYSTEMS)

37.C 2,4 – Methanoproline monohydrate
2 – Carboxy – 2,4 – methanopyrrolidine monohydrate
$C_6H_9NO_2$, H_2O Main entry is 48.17

37.C (1S,4S) – N – Acetyl – 3 – oxo – 5 – aza – 2 – oxabicyclo(2.2.1)heptane
$C_7H_9NO_3$ Main entry is 40.4

37.1 Quinuclidine (cubic form)
1 – Azabicyclo(2.2.2)octane
$C_7H_{13}N$
R.Fourme *J.Phys.(Paris),40,* 557, 1979

37.2 1 – Methyl – 1,3,5,7 – tetra – aza – adamantan – 1 – ium octa – iodide
$2C_7H_{15}N_4^+$, $2I_3^-$, I_2
P.K.Hon, T.C.W.Mak, J.Trotter *Inorg.Chem.,18,* 2916, 1979

37.3 7 – Isopropylidene – 2,3 – diazabicyclo(2.2.1)hept – 2 – ene
$C_8H_{12}N_2$
M.W.Vary, J.M.McBride *Cryst.Struct.Commun.,9,* 85, 1980

37.4 (±) – 4 – Azatricyclo(4.3.1.0^{3,7})decan – 5 – one
$C_9H_{13}NO$
K.Blaha, P.Malon, M.Tichy, I.Fric, R.Usha, S.Ramakumar, K.Venkatesan
Collect.Czech.Chem.Commun.,43, 3241, 1978

37.5 3,8 – Dimethyl – 1,10 – diazatricyclo(5.2.1.0^{4,10}) decane – 3,7 – diene – 2,9 – dione
$C_{10}H_{10}N_2O_2$
I.Goldberg *Eur.Cryst.Meeting,5,* 286, 1979

37.6 Tropane – 3 – spiro – 5' – thio – hydantoin
$C_{10}H_{15}N_3OS$
J.Vilches, F.Florencio, S.Garcia–Blanco
Eur.Cryst.Meeting,6, 300, 1980

37.C 3,6 – Diethyl – 1,4 – dimethyl – 3,6 – epithio – piperazine – 2,5 – dione
$C_{10}H_{16}N_2O_2S$ Main entry is 41.32

37.7 3,7 – Dimethyl – 3,7 – diazabicyclo(3.3.1)nonane – 9 – spiro – 5' – hydantoin
$C_{11}H_{18}N_4O_2$
F.Florencio, J.Vilches, P.Smith–Verdier, S.Garcia–Blanco
Eur.Cryst.Meeting,6, 301, 1980

37.8 1,3,4,7,7 – Pentachloro – 5 – phenyl – 2 – azabicyclo(2.2.1)hept – 2 – ene
$C_{12}H_8Cl_5N$
P.H.Daniels, J.L.Wong, J.L.Atwood, L.G.Canada, R.D.Rogers
J.Org.Chem.,45, 435, 1980

37.9   1,3 – endo – 4 – Trichloro – 2 – endo – 7,7 –
trimethoxy – 5 – endo – bromomethyl – 2 – exo –
(2′,2′,2′ – trifluoro – ethoxy) – bicyclo(2.2.1)heptane
$C_{13}H_{17}BrCl_3F_3O_4$
W.van de Mieroop, A.T.H.Lenstra, H.J.Geise
Cryst.Struct.Commun.,**8**, 771, 1979

37.C   (–) – 14 – Hydroxy – 15 – hydroxymethyl – 2,8 –
dithia(9)(2,5)pyridinophane (absolute configuration)
$C_{13}H_{19}NO_2S_2$ Main entry is 41.48

37.C   5 – Methyl – 14,16 – dioxa – 3,5,7 –
triazapentacyclo(7.4.3.2^{2,8}.0.0^{3,7})octadeca – 10,12,17 –
triene – 4,6 – dione
$C_{14}H_{13}N_3O_4$ Main entry is 38.56

37.C   5 – Methyl – 15 – oxa – 5 –
azapentacyclo(7.4.3.2^{2,8}.0.0^{3,7})octadeca – 10,12,17 –
triene – 4,6 – dione
$C_{17}H_{17}NO_3$ Main entry is 38.84

37.10   2 – (2′ – Propenyl) – 8 – toluenesulfonyl – 7,8 –
diazatricyclo(4.2.1.0^{3,7})nonane
$C_{17}H_{22}N_2O_2S$
R.M.Wilson, J.W.Rekers, A.B.Packard, R.C.Elder
J.Am.Chem.Soc.,**102**, 1633, 1980

37.11   Dimethyl – 15,16 – diaza –
octacyclo(6.6.2.0^{2,7}.0^{3,5}.0^{4,6}.0^{9,14}.0^{10,12}.0^{11,13})hexadec –
15 – ene – 1,8 – dicarboxylate
$C_{18}H_{18}N_2O_4$
H.Irngartinger, K.L.Lukas
Angew.Chem.,Int.Ed.Engl.,**18**, 694, 1979

37.12   (+) – 4 – Hydroxy – 7 – oxo – 3 – methoxy – 17 –
methyl – 5,6 – dehydromorphinan
$C_{18}H_{21}NO_3$
J.–I.Minamikawa, K.C.Rice, A.E.Jacobson, A.Brossi,
T.H.Williams, J.V.Silverton *J.Org.Chem.*,**45**, 1901, 1980

37.13   N – Phenethyl – granatanine – 3 – spiro – 5′ –
hydantoin
$C_{18}H_{23}N_3O_2$
F.Florencio, P.Smith–Verdier, S.Garcia–Blanco
Cryst.Struct.Commun.,**9**, 687, 1980

37.14   6 – (p – Chlorophenylsulfonylamino) – 2,2a,3,4,5,6 –
hexahydro – 2a,6 – methane – 1H – azeto(1,2 – a)(1)
benzazocin – 12 – one
$C_{20}H_{19}ClN_2O_3S$
K.Prout, M.Sims, D.Watkin, C.Couldwell, M.Vandrevala,
A.S.Bailey *Acta Crystallogr.,Sect.B*,**36**, 1846, 19

37.15   2 – Benzyl – 3 – oxo – 6 – exo – (2,6 – di – iodo –
4 – methylphenoxy) – 2 – azabicyclo(2.2.1)heptane
$C_{20}H_{19}I_2NO_2$
H.L.Ammon, P.H.Mazzocchi, L.Liu *Chem.Lett.*,897, 1980

37.16   8 – (p – Chlorophenylsulfonylamino) – 4 –
hydroxyimino – 1,2,3,4,5,6,7,8 – octahydro – 1,8 –
methano(1)benzazecin – 13 – one pyridine solvate
$C_{20}H_{20}ClN_3O_4S, C_5H_5N$
K.Prout, M.Sims, D.Watkin, C.Couldwell, M.Vandrevala,
A.S.Bailey *Acta Crystallogr.,Sect.B*,**36**, 1846, 1980

37.17   13 – Phenyl – 11,13,15 –
triazapentacyclo(8.5.2.0^{2,6}.0^{2,8}.0^{11,15})heptadeca –
12,14 – dione
$C_{20}H_{23}N_3O_2$
L.A.Paquette, A.R.Browne, E.Chamot, J.F.Blount
J.Am.Chem.Soc.,**102**, 643, 1980

37.C   2′,6″ – (1,4,7 – Trioxaheptane – 1,7 – diyl) – (2,6 –
bis(2 – picolinoyl) – pyridine)
$C_{21}H_{17}N_3O_5$ Main entry is 40.53

37.18   1,2 – (Tetrahydro – methyl – semibullvalene –
diyl) – 4 – bornyl – 1,2,4 – triazolidine – 3,5 – dione
(absolute configuration)
$C_{21}H_{27}N_3O_2$
L.A.Paquette, R.F.Doehner Junior, J.A.Jenkins, J.F.Blount
J.Am.Chem.Soc.,**102**, 1188, 1980

37.C   N – Cyclopropylmethyl – scopolammonium bromide
$C_{21}H_{28}NO_4^+$, Br$^-$ Main entry is 38.106

37.19   15 – Methyl – 5 – phenyl – 5,15 –
diazapentacyclo(7.4.3.2^{2,8}.0.0^{3,7})octadeca – 10,12,17 –
triene – 4,6,14,16 – tetrone
$C_{23}H_{18}N_2O_4$
M.Kaftory *Acta Crystallogr.,Sect.B*,**36**, 597, 1980

37.20   (18 – Ethoxy – 15 – methyl – 5 – phenyl – 3,5,7,15 –
tetra – azapentacyclo(7.4.3.2^{2,8}.0.0^{3,7})octadec – 17 –
ene – 4,6,14,16 – tetrone)
$C_{23}H_{24}N_4O_5$
M.Kaftory *Acta Crystallogr.,Sect.B*,**36**, 597, 1980

37.21   4,6 – Diphenyl – 10 – ethoxycarbonyl – 2,3,10 –
triazatricyclo(5.3.2.0^{2,6})dodecane – 3,8,11 – triene –
5 – one
$C_{24}H_{21}N_3O_3$
K.Harano, T.Ban, K.Kanematsu
Heterocycles,**12**, 453, 1979

37.C   2′,6″ – (1,4,7 – Trioxaheptane – 1,7 – diyl) – (2,6 –
bis(2 – picolinoyl) – pyridine) – bis(ethylene –
ketal)
$C_{25}H_{25}N_3O_7$ Main entry is 40.55

37.C   14 – (p – Chlorophenyl) – 8 – (p –
chlorophenylsulfonylamino) – 13 – oxa – 14 –
thia – 1,15 – diazapentacyclo(10.4.2.0^{2,7}.0^{8,16}.0^{12,16})
octadeca – 2,4,6,14 – tetraene S – oxide
$C_{26}H_{23}Cl_2N_3O_4S_2$ Main entry is 42.12

37.22   1,2,3,4,5,6,7,8 – Octahydro – 1,4:5,8 – di –
isopropano – 4,5 – dimethyl – 9 – phenyl – acridine
$C_{27}H_{33}N$
R.Roques, J.Sotiropoulos, J.P.Declercq, G.Germain
Acta Crystallogr.,Sect.B,**35**, 2948, 1979

37.23   6 – Dimethylamino – 1,4 – etheno – 5,5 –
dimethyl – 8 – oxo – N,9 – diphenyl – 2,3,4,4a,5,8 –
hexahydro – 1H – pyridazino(4,5 – d)azepine – 2,3 –
dicarboximide
$C_{28}H_{27}N_5O_3$
M.van Meerssche, G.Germain, J.P.Declercq, R.Touillaux,
E.Schaumann, S.Grabley
Cryst.Struct.Commun.,**9**, 509, 1980
See also R1 : 36

37.24   3,4:5,6 – Dibenzo – 14 – ethoxycarbonyl – 8,10 –
diphenyl – 14 – azatetracyclo(9.3.2.0^{2,7}.0^{2,10})
hexadeca – 7,12,15 – trien – 9 – one
$C_{38}H_{29}NO_3$
M.Yasuda, K.Harano, K.Kanematsu
J.Org.Chem.,**45**, 2368, 1980

HETERO–OXYGEN

38.1 Methylene oxalate
$C_3H_2O_4$
A.Kvick, R.Liminga
*Acta Crystallogr.,Sect.B,***36**, 734, 1980

38.2 Tetrahydrofuran – 3,4 – dione
$C_4H_4O_3$
F.A.Muller, R.A.Jacobson
*Cryst.Struct.Commun.,***9**, 325, 1980

38.C hexakis(Benzylthiomethyl)benzene 1,4 – dioxane
clathrate (monoclinic form)
$C_4H_8O_2$, $C_{54}H_{54}S_6$ Main entry is 61.23

38.C hexakis(2 – Phenylethylthiomethyl)benzene 1,4 –
dioxane clathrate
$C_4H_8O_2$, $C_{60}H_{66}S_6$ Main entry is 61.25

38.3 Dioxane – seleninyl – dichloride
$C_4H_8O_2$, Cl_2OSe
N.W.Alcock, J.F.Sawyer
*J.Chem.Soc.,Dalton Trans.,*115, 1980

38.4 bis(μ – Hydroxo) – bis(aqua – trichloro – tin(iv))
1,4 – dioxane
$3C_4H_8O_2$, $H_6Cl_6O_4Sn_2$
J.C.Barnes, H.A.Sampson, T.J.R.Weakley
*J.Chem.Soc.,Dalton Trans.,*949, 1980

38.5 bis(6 – Deoxy – 6 – chloro – L – ascorbic acid)
nitromethane solvate
$2C_6H_7ClO_5$, CH_3NO_2
J.Kiss, K.P.Berg, A.Dirscherl, W.E.Oberhansli, W.Arnold
*Helv.Chim.Acta,***63**, 1728, 1980

38.C L – Serine L – ascorbic acid
$C_6H_8O_6$, $C_3H_7NO_3$ Main entry is 48.8

38.C 1,2:5,6 – Dianhydro – galactitol (α form)
$C_6H_{10}O_4$ Main entry is 45.6

38.C 1,2:5,6 – Dianhydro – galactitol (β form)
$C_6H_{10}O_4$ Main entry is 45.7

38.C 1,6 – Anhydro – β – D – galactopyranose
$C_6H_{10}O_5$ Main entry is 45.8

38.C 7 – Methoxy – 3,5,9 – trioxa – 4 –
phosphabicyclo(4.3.0)nonan – 4 – one
$C_6H_{10}O_5P$ Main entry is 64.17

38.6 Z – 1 – (5 – Nitro – 2 – furyl) – 2 – thiocyanato –
ethylene
$C_7H_4N_2O_3S$
A.Kusa. T.N.Polynova. M.A.Porai–Koshits. Ya.Kovach,
D.Vegkh *Zh.Strukt.Khim.,***21**, 172, 1980

38.7 4,7 – Dioxatetracyclo(4.3.0.0^{3,5}.0^{2,9})nonan – 8 – one
$C_7H_6O_3$
R.C.Glen, P.Murray-Rust *Eur.Cryst.Meeting.***6**, 44, 1980

38.8 anti – 7,7 – Dibromo – norcar – 3 – ene oxide
(at –56°C)
$C_7H_8Br_2O$
L.A.Paquette, W.E.Fristad, C.A.Schuman, M.A.Beno,
G.G.Christoph *J.Am.Chem.Soc.,***101**, 4645, 1979

38.9 6β – Bromo – 7α – hydroxy – 2 – oxabicyclo(3.3.0)
octan – 3 – one
$C_7H_9BrO_3$
A.Brown, R.Glen, P.Murray–Rust, J.Murray–Rust,
R.F.Newton *J.Chem.Soc.,Chem.Commun.,*1178, 1979

38.10 7 – Methoxy – 3,5,9 – trioxabicyclo(4.3.0)nonane
$C_7H_{12}O_4$
L.A.Aslanov, S.S.Sotman, V.B.Ribakov, V.I.Andrianov,
Z.Sh.Safina, M.P.Koroteev, L.T.Elepina, E.E.Nifant'ev
*Zh.Strukt.Khim.,***20**, 1122, 1979

38.C Trichoviridine
$C_8H_9NO_4$ Main entry is 50.3

38.11 cis – (E) – 1 – (5 – Nitro – 2 – furyl) – 2 –
dimethylamino – ethylene (monoclinic form)
cis – (E) – 2 – (2′ – Dimethylamino – vinyl) – 5 –
nitrofuran
$C_8H_{10}N_2O_3$
A.Kusa, T.N.Polynova, M.A.Porai–Koshits, Ya.Kovach,
D.Vegkh *Zh.Strukt.Khim.,***20**, 556, 1979

38.12 cis – (E) – 1 – (5 – Nitro – 2 – furyl) – 2 –
dimethylamino – ethylene (orthorhombic form)
cis – (E) – 2 – (2′ – Dimethylamino – vinyl) – 5 –
nitrofuran
$C_8H_{10}N_2O_3$
A.Kusa, T.N.Polynova, M.A.Porai–Koshits, Ya.Kovach,
D.Vegkh *Zh.Strukt.Khim.,***20**, 556, 1979

38.C 2 – 0 – Acetyl – 1,6:3,5 – dianhydro – α – L –
idofuranose
$C_8H_{10}O_5$ Main entry is 45.12

38.13 5 – Ethyl – 2,2 – dimethyl – 1,3 – dioxane – 4,6 –
dione
$C_8H_{12}O_4$
A.van Coppernolle, J.P.Declercq, J.M.Dereppe, G.Germain,
M.van Meerssche *Bull.Soc.Chim.Belg.,***88**, 223, 1979

38.14 3 – Nitromethylene – phthalide
$C_9H_5NO_4$
B.S.Joshi, V.R.Hedge, D.Rogers, D.J.Williams
*Tetrahedron Lett.,***21**, 1163, 1980

38.15 Carolic acid
(E,R) – 5 – Methyl – 3 – (2′ – tetrahydrofurylidene) -
tetrahydrofuran – 2,4 – dione
$C_9H_{10}O_4$
O.Simonsen, T.Reffstrup, P.M.Boll
*Tetrahedron,***36**, 795, 1980

38.16 anti – 1,6 – Dimethyl – 7,7 – dibromo – norcar –
3 – ene oxide
$C_9H_{12}Br_2O$
L.A.Paquette, W.E.Fristad, C.A.Schuman, M.A.Beno,
G.G.Christoph *J.Am.Chem.Soc.,***101**, 4645, 1979

38.C N – Acetyl – furanomycin
$C_9H_{13}NO_4$ Main entry is 50.5

38.C 4 – Diethylamino – 3,5,9 – trioxa – 4 –
phosphabicyclo(4.3.0)nonane – 4,7 – dione
$C_9H_{16}NO_5P$ Main entry is 64.33

38.17 2 – Methoxy – 2,4 – dimethyl – 3,8 –
dioxabicyclo(3.2.0)octane
$C_9H_{16}O_3$
D.M.Walba, M.D.Wand, M.C.Wilkes
J.Am.Chem.Soc.,**101**, 4396, 1979

38.18 Pyromellitic dianhydride
$C_{10}H_2O_6$
S.Aravamudhan, U.Haeberlen, H.Irngartinger, C.Krieger
Mol.Phys.,**38**, 241, 1979

38.C Phenothiazine – pyromellitic dianhydride complex
$C_{10}H_2O_6$, $C_{12}H_9NS$ Main entry is 60.17

38.19 7 – Hydroxy – 6 – methoxycoumarin
$C_{10}H_8O_4$
M.Kimura, W.H.Watson
Cryst.Struct.Commun.,**9**, 257, 1980

38.C 6 – Methoxy – 2 – methyl – 3,5 – dihydro – benzo(b)
furan – 4,7 – dione
Acamelin
$C_{10}H_8O_4$ Main entry is 59.2

38.20 Methyl 3 – (3 – isocyano – 6 – oxabicyclo(3.1.0)hex –
2 – en – 5 – yl) – acrylate
$C_{10}H_9NO_3$
D.Brewer, E.J.Gabe, A.W.Hanson, A.Taylor, J.W.Keeping,
V.Thaller, B.C.Das
J.Chem.Soc.,Chem.Commun.,1061, 1979

38.21 E – 1 – (5 – Nitro – 2 – furyl) – 2 – morpholino –
ethylene
$C_{10}H_{12}N_2O_4$
A.Kusa, T.N.Polynova, M.A.Porai–Koshits, Ya.Kovach,
D.Vegkh *Zh.Strukt.Khim.*,**21**, 172, 1980

38.22 syn – 3 – Methyl – 6 – (2 – furyl) – piperidin – 2 –
one
$C_{10}H_{13}NO_2$
G.D.Andreetti, G.Bocelli, P.Sgarabotto, Z.Dabrowski,
J.Cybulski, J.T.Wrobel *Pol.J.Chem.*,**53**, 97, 1979
See also R1 : 33

38.23 cis – 10 – Methyl – 1 – oxadecalin – 2,5 – dione
$C_{10}H_{14}O_3$
A.Dubourg, R.Roques, E.Guy
Acta Crystallogr.,Sect.B,**35**, 2938, 1979

38.24 (±) – Oxybiotin
$C_{10}H_{16}N_2O_4$
G.T.DeTitta, R.Parthasarathy, R.H.Blessing, W.Stallings
Proc.Nat.Acad.Sci.U.S.A.,**77**, 333, 1980
See also R1 : 35

38.25 (2S,4R,5R) – 2 – Carboxymethyl – 5 – carboxy –
2,4,5 – trimethyl – 2,3,4,5 – tetrahydrofuran
$C_{10}H_{16}O_5$
A.Kirfel, G.Will, H.Wiedenfeld, E.Roeder
Cryst.Struct.Commun.,**9**, 363, 1980

38.26 bis(μ – Hydroxo) – bis(aqua – tribromo – tin(iv))
1,8 – epoxy – p – menthane
$4C_{10}H_{18}O$, $H_6Br_6O_4Sn_2$
J.C.Barnes, H.A.Sampson, T.J.R.Weakley
J.Chem.Soc.,Dalton Trans.,949, 1980

38.27 bis(μ – Hydroxo) – bis(aqua – trichloro – tin(iv))
1,8 – epoxy – p – menthane
$4C_{10}H_{18}O$, $H_6Cl_6O_4Sn_2$
J.C.Barnes, H.A.Sampson, T.J.R.Weakley
J.Chem.Soc.,Dalton Trans.,949, 1980

38.28 1 – (5 – Nitro – 2 – furyl) – 1 –
trichloromethylsulfonyl – 2 – (5 – bromo – 2 –
furyl) – ethylene
$C_{11}H_5BrCl_3NO_6S$
A.Kusa, T.N.Polynova, M.A.Porai–Koshits, A.Yurashek
Zh.Strukt.Khim.,**20**, 559, 1979

38.29 Furo(2,3 – h)coumarin
Angelicin
$C_{11}H_6O_3$
G.Bravic, J.–P.Bideau, J.P.Desvergne
Cryst.Struct.Commun.,**9**, 705, 1980

38.30 Furo(2,3 – f)coumarin
Allopsoralen
$C_{11}H_6O_3$
J.–P.Bideau, G.Bravic
Cryst.Struct.Commun.,**9**, 243, 1980

38.31 6 – Bromo – 1,2,3,4,4a,9a – hexahydro – 4,9 –
dioxafluoren – 2 – one
$C_{11}H_9BrO_3$
P.Gluzinski, J.W.Krajewski, Z.Urbanczyk–Lipkowska,
Ya.Ya.Bleidelis, A.Kemme
Acta Crystallogr.,Sect.B,**35**, 2755, 1979

38.32 α – Benzyloxy – γ – butyrolactone
$C_{11}H_{10}O_4$
G.Bocelli, M.F.Grenier–Loustalot
J.Chem.Res.,227, 3101, 1980

38.C trans – 6 – Chloro – 9 – (2 – ethoxy – 1,3 –
dioxan – 5 – yl) – purine
$C_{11}H_{13}ClN_4O_3$ Main entry is 44.23

38.C Diethyl – (5,6 – dichloro – 1,3 – benzodioxole – (2)) –
phosphonate
$C_{11}H_{13}Cl_2O_5P$ Main entry is 64.44

38.33 trans – 2 – Acetoxy – 5 – nitro – 2,5 – dihydro –
2 – furfural diacetate
$C_{11}H_{13}NO_9$
A.F.Mishnev, Ya.Ya.Bleidelis, K.K.Venters
Tetrahedron,**36**, 1817, 1980

38.C 3,6 – Dimethyl – 4,10 – dihydroxy – 2 – oxa –
spiro(4.5)dec – 7 – ene – 1,9 – dione
Rosigenin
$C_{11}H_{14}O_5$ Main entry is 59.3

38.C trans – 9 – (2 – Ethoxy – 1,3 – dioxan – 5 – yl) –
adenine
$C_{11}H_{15}N_5O_3$ Main entry is 44.24

38.34 1,4,7,10,13 – Pentaoxa – 14,16 –
cyclohexadecanedione
$C_{11}H_{18}O_7$
N.K.Dalley, S.B.Larson
Acta Crystallogr.,Sect.B,**35**, 2428, 1979

38.35 Octafluoro – dibenzo – 1,4 – dioxane
$C_{12}F_8O_2$
D.Rainville, R.A.Zingaro, E.A.Meyers
Cryst.Struct.Commun.,**9**, 771, 1980

38.36 E – 1 – (5 – Nitro – 2 – furyl) – 2 – p –
chlorophenylsulfonyl – ethylene
$C_{12}H_8ClNO_5S$
A.Kusa, T.N.Polynova, M.A.Porai–Koshits, Ya.Kovach,
D.Vegkh *Zh.Strukt.Khim.*,**20**, 561, 1979
See also R1 : 11

38.37 **3 − Formylfuro(3,2 − f)chromene**
$C_{12}H_8O_3$
J.Gaultier, G.Bravic, J.−P.Bideau
Cryst.Struct.Commun.,**8**, 829, 1979

38.38 **anti − 2,3,4,5 − Diepoxy − 12 − oxa(4.4.3)propella − 7,9 − diene**
$C_{12}H_{12}O_3$
M.Kaftory *Acta Crystallogr.,Sect.B*,**35**, 2569, 1979

38.39 **2,2 − Dimethyl − 5 − phenyl − 1,3 − dioxane − 4,6 − dione**
$C_{12}H_{12}O_4$
A.van Coppernolle, J.P.Declercq, J.M.Dereppe, G.Germain, M.van Meerssche *Bull.Soc.Chim.Belg.*,**88**, 223, 1979

38.40 **anti − 2,3,4,5 − syn − 7,8,9,10 − Tetraepoxy − 12 − oxa(4.4.3)propellane**
$C_{12}H_{12}O_5$
M.Kaftory *Acta Crystallogr.,Sect.B*,**35**, 2569, 1979

38.41 **8 − Hydroxy − 3 − methoxy − 3 − methoxycarbonyl − 2 − oxa − spiro(4.5)deca − 6,9 − diene**
$C_{12}H_{14}O_6$
S.Danishefsky, M.Hirama, N.Fritsch, J.Clardy
J.Am.Chem.Soc.,**101**, 7013, 1979

38.42 **3,5 − Dihydroxy − 4 − nitro − 2 − (4′ − acetyl − 5′ − methylfuran − 2′ − yl) − tetrahydropyran**
$C_{12}H_{15}NO_7$
E.Moreno, A.Conde, R.Marquez
Eur.Cryst.Meeting,**5**, 398, 1979

38.43 **Hexamethyl − 2,3:5,6 − diepoxy − bicyclo(2.2.0) hexane**
$C_{12}H_{18}O_2$
A.Dunand, R.Gerdil
Acta Crystallogr.,Sect.B,**36**, 472, 1980

38.44 **Methyl − O − acetyleurekanate**
$C_{12}H_{18}O_8$
E.Kupfer, K.Neupert−Laves, M.Dobler, W.Keller−Schierlein
Helv.Chim.Acta,**63**, 1141, 1980

38.C **Trinitrato − (hexaoxacyclo − octadecane) − neodymium(iii)**
Trinitrato − (18 − crown − 6) − neodymium(iii)
$C_{12}H_{24}N_3NdO_{15}$ Main entry is 84.40

38.45 **1,4,7,10,13,16 − Hexaoxacyclo − octadecane (at 100°K)**
18 − Crown − 6
$C_{12}H_{24}O_6$
E.Maverick, P.Seiler, W.B.Schweizer, J.D.Dunitz
Acta Crystallogr.,Sect.B,**36**, 615, 1980

38.46 **1,4,7,10,13,16 − Hexaoxacyclo − octadecane sodium dicyanophosphide tetrahydrofuran solvate**
$C_{12}H_{24}O_6, C_2N_2P^-, Na^+, C_4H_8O$
W.S.Sheldrick, J.Kroner, F.Zwaschka, A.Schmidpeter
Angew.Chem.,Int.Ed.Engl.,**18**, 934, 1979

38.47 **1,4,7,10,13,16 − Hexaoxacyclo − octadecane hexa − aquo − cobalt tetrachloro − cobalt acetone solvate**
$C_{12}H_{24}O_6, H_{12}CoO_6^{2+}, Cl_4Co^{2-}, C_3H_6O$
T.B.Vance Junior, E.M.Holt, C.G.Pierpont, S.L.Holt
Acta Crystallogr.,Sect.B,**36**, 150, 1980

38.48 **1,4,7,10,13,16 − Hexaoxacyclo − octadecane hexa − aquo − manganese(ii) perchlorate**
$C_{12}H_{24}O_6, H_{12}MnO_6^{2+}, 2ClO_4^-$
T.B.Vance Junior, E.M.Holt, D.L.Varie, S.L.Holt
Acta Crystallogr.,Sect.B,**36**, 153, 1980

38.49 **3 − Formyl − 8 − methylfuro(3,2 − g)chromene**
$C_{13}H_{10}O_3$
G.Bravic, J.−P.Bideau
Cryst.Struct.Commun.,**9**, 717, 1980

38.50 **E − 1 − (5 − Nitro − 2 − furyl) − 2 − p − tolylsulfonyl − ethylene**
$C_{13}H_{11}NO_5S$
A.Kusa, T.N.Polynova, M.A.Porai−Koshits, Ya.Kovach, D.Vegkh *Zh.Strukt.Khim.*,**20**, 561, 1979

38.51 **1,4 − Dioxacyclohept − 2 − eno(2,3 − b)naphthalene**
$C_{13}H_{12}O_2$
K.Stadnicka, L.Lebioda *Eur.Cryst.Meeting*,**5**, 55, 1979

38.52 **4 − (N′ − (p − Nitrobenzene) − N − hydrazo) − 8 − oxatricyclo(5.1.0.0²,⁶)octane**
$C_{13}H_{13}N_3O_3$
R.C.Glen, P.Murray−Rust *Eur.Cryst.Meeting*,**6**, 44, 1980
See also R1 : 27,15,9

38.C **Furaltadone**
$C_{13}H_{16}N_4O_6$ Main entry is 40.29

38.C **Furaltadone hydrochloride acetic acid clathrate**
$C_{13}H_{17}N_4O_6^+, C_2H_4O_2, Cl^-$ Main entry is 61.4

38.C **Furaltadone hydrochloride propionic acid clathrate**
$C_{13}H_{17}N_4O_6^+, C_3H_6O_2, Cl^-$ Main entry is 61.5

38.C **Furaltadone hydrochloride hydrate**
$C_{13}H_{17}N_4O_6^+, Cl^-, 2H_2O$ Main entry is 40.30

38.53 **exo − 1,3,5,6,7,9 − Hexamethyl − 4,8 − dioxatetracyclo(4.3.0.0³,⁵.0⁷,⁹)nonan − 2 − one**
$C_{13}H_{18}O_3$
H.Hart, S.−M.Chen, S.Lee, D.L.Ward, W.−J.H.Kung
J.Org.Chem.,**45**, 2091, 1980

38.54 **endo − 1,3,5,6,7,9 − Hexamethyl − 4,8 − dioxatetracyclo(4.3.0.0³,⁵.0⁷,⁹)nonan − 2 − one**
$C_{13}H_{18}O_3$
H.Hart, S.−M.Chen, S.Lee, D.L.Ward, W.−J.H.Kung
J.Org.Chem.,**45**, 2091, 1980

38.55 **6 − Acetoxymethyl − 1,2,5 − trimethyl − 4 − oxabicyclo(3.3.0)octan − 3 − one**
$C_{13}H_{20}O_4$
I.Kitagawa, H.Shibuya, H.Fujioka, Y.Yamamoto, A.Kajiwara, K.Kitamura, A.Miyao, T.Hakoshima, K.Tomita
Tetrahedron Lett.,**21**, 1963, 1980

38.C **5,8 − Dimethoxy − 2 − methylfuro(2,3 − e)chromone**
Khellin
$C_{14}H_{12}O_5$ Main entry is 59.5

38.56 **5 − Methyl − 14,16 − dioxa − 3,5,7 − triazapentacyclo(7.4.3.2²,⁶.0.0³,⁷)octadeca − 10,12,17 − triene − 4,6 − dione**
$C_{14}H_{13}N_3O_4$
M.Kaftory *Acta Crystallogr.,Sect.B*,**36**, 597, 1980
See also R1 : 37

38.57 **1,2,3,4,4a,5,8,8a − Octahydro − 1,4:5,8 − exo,exo − dimethano − naphthalene − 4a,8a − dicarboxylic anhydride**
$C_{14}H_{14}O_3$
P.D.Bartlett, A.J.Blakeney, M.Kimura, W.H.Watson
J.Am.Chem.Soc.,**102**, 1383, 1980

38.58　1,2,3,4,4a,5,6,7,8,8a – Decahydro – 1,4:5,8 – exo,exo – dimethano – naphthalene – 4a,8a – dicarboxylic anhydride
$C_{14}H_{18}O_3$
P.D.Bartlett, A.J.Blakeney, M.Kimura, W.H.Watson
J.Am.Chem.Soc.,**102**, 1383, 1980

38.59　1,2,3,4,4a,5,6,7,8,8a – Decahydro – 1,4:5,8 – exo,endo – dimethano – naphthalene – 4a,8a – dicarboxylic anhydride
$C_{14}H_{18}O_3$
P.D.Bartlett, A.J.Blakeney, M.Kimura, W.H.Watson
J.Am.Chem.Soc.,**102**, 1383, 1980

38.60　1 – Chloroethyl – 4 – methyl – 6,7 – dimethyl – isochroman
$C_{14}H_{19}ClO_3$
J.M.McCall, R.E.TenBrink, B.V.Kamdar, C.Chidester
J.Heterocycl.Chem.,**16**, 363, 1979

38.61　8,10,10 – Trimethyl – 2 – oxatetracyclo(6.3.1.0^{4,7}.0^{4,12})dodecan – 3 – one (at –170°C)
$C_{14}H_{20}O_2$
S.R.Wilson, L.R.Phillips, Y.Pelister, J.C.Huffman
J.Am.Chem.Soc.,**101**, 7373, 1979

38.62　1β,5β,6β,11β – Tetramethyl – 3 – oxatetracyclo(5.4.0.0^{2,6}.0^{4,11})undec – 8 – en – 2β – ol
$C_{14}H_{20}O_2$
T.J.Greenhough, J.Trotter
Acta Crystallogr.,Sect.B,**36**, 1835, 1980

38.63　(t – 5,c – 10 – Dimethyl – c – 6 – hydroxy – 4 – oxo – r – 1H – bicyclo(5.3.0)dec – 7 – t – yl) acetic acid γ – lactone
$C_{14}H_{20}O_4$
P.Kok, P.J.De Clercq, M.E.Vandewalle
J.Org.Chem.,**44**, 4553, 1979

38.64　3α,4α,4aβ,7aα,8α,9aβ – (±) – Decahydro – 4 – hydroxy – 4a,8 – dimethylazuleno(6,5 – b)furan – 2,5(3H) – dione
$C_{14}H_{20}O_4$
J.P.Declercq, G.Germain, M.van Meerssche, P.Kok, P.De Clercq, M.Vandewalle
Acta Crystallogr.,Sect.B,**36**, 739, 1980
See also R1 : 28

38.65　Benzo – 15 – crown – 5 lithium picrate dihydrate
$C_{14}H_{20}O_5$, $C_6H_2N_3O_7^-$, Li$^+$, 2H$_2$O
V.W.Bhagwat, H.Manohar, N.S.Poonia
Inorg.Nucl.Chem.Lett.,**16**, 373, 1980

38.66　2,3,5,6,8,9,11,12 – Octahydro – 1,4,7,10,13 – benzopentaoxa – cyclopentadecin sodium perchlorate
Benzo – 15 – crown – 5 sodium perchlorate
$C_{14}H_{20}O_5$, Na$^+$, ClO$_4^-$
J.D.Owen *J.Chem.Soc.,Dalton Trans.*,1066, 1980

38.C　bis(2,3,5,6,8,9,11,12 – Octahydro – 1,4,7,10,13 – benzopentaoxa – cyclopentadecin) sodium tetraphenylborate
bis(Benzo – 15 – crown – 5) sodium tetraphenylborate
$2C_{14}H_{20}O_5$, $C_{24}H_{20}B^-$, Na$^+$ Main entry is 62.15

38.67　bis(2,3,5,6,8,9,11,12 – Octahydro – 1,4,7,10,13 – benzopentaoxacyclopentadecin) sodium perchlorate
bis(Benzo – 15 – crown – 5) sodium perchlorate
$2C_{14}H_{20}O_5$, Na$^+$, ClO$_4^-$
J.D.Owen *J.Chem.Soc.,Dalton Trans.*,1066, 1980

38.68　bis(Benzo – 15 – crown – 5) sodium tetrachloro – dioxo – uranium
$2C_{14}H_{20}O_5$, 2Na$^+$, Cl$_4$O$_2$U^{2-}
D.C.Moody, R.R.Ryan *Cryst.Struct.Commun.*,**8**, 933, 1979

38.69　(±) – 3aα,4α,4aβ,5α,7aα,8α,9aβ – Decahydro – 4,5 – dihydroxy – 4a,8 – dimethylazuleno(6,5 – b)furan – 2(3H) – one
9,11 – Dihydroxy – 2,10 – dimethyl – 5 – oxatricyclo(8.3.0.0^{4,8})tridecan – 6 – one
$C_{14}H_{22}O_4$
J.P.Declercq, G.Germain, M.van Meerssche, P.Kok, P.De Clercq, M.Vandewalle
Acta Crystallogr.,Sect.B,**36**, 190, 1980

38.C　(1R,3R,4S,5R) – 4,5 – (Isopropylidene – dioxy) – 1 – (2 – methyl – 1,3 – dithiane – 2 – yl) – cyclohexane – 1,3 – diol
$C_{14}H_{24}O_4S_2$ Main entry is 39.54

38.70　6 – Methoxy – 4 – oxo – 3 – phenylseleno – cyclohexane – 1,2 – dicarboxylic anhydride
$C_{15}H_{14}O_5Se$
S.Danishefsky, C.–F.Yan, R.K.Singh, R.B.Gammill, P.M.McCurry Junior, N.Fritsch, J.Clardy
J.Am.Chem.Soc.,**101**, 7001, 1979
See also R1 : 11

38.C　(+) – Methylenomycin A p – bromobenzoate (absolute configuration)
$C_{15}H_{15}BrO_4$ Main entry is 50.8

38.C　Chlorofucin
$C_{15}H_{20}BrClO_2$ Main entry is 59.7

38.C　Poiteol
$C_{15}H_{20}BrClO_3$ Main entry is 59.8

38.C　Xantholide B
$C_{15}H_{20}O_2$ Main entry is 59.9

38.71　11 – Isopropyl – 8 – methyl – 2 – oxatetracyclo(6.3.1.0^{4,7}.0^{4,12})dodecan – 3 – one (at –135°C)
$C_{15}H_{22}O_2$
S.R.Wilson, L.R.Phillips, Y.Pelister, J.C.Huffman
J.Am.Chem.Soc.,**101**, 7373, 1979

38.72　(1,5,5 – Trimethyl – bicyclo(4.2.0)octan – 7 – one) – 8 – spiro – 4 – (2,2 – dimethyl – 5 – oxo – 1,3 – dioxolane) (at 115°K)
$C_{15}H_{22}O_4$
R.V.Stevens, G.S.Bisacchi, L.Goldsmith, C.E.Strouse
J.Org.Chem.,**45**, 2708, 1980

38.C　Epoxy – isoacoragermacrone
$C_{15}H_{24}O_2$ Main entry is 53.9

38.73　10 – Hydroxy – 3 – oxatetracyclo(8.6.0.0^{2,4}.0^{2,9}) hexadecane
$C_{15}H_{24}O_2$
A.Courtois, J.Reymann, J.Protas, B.Loubinoux, P.Caubere
Acta Crystallogr.,Sect.B,**35**, 2774, 1979

38.74　3,5,7 – Trimethyl – tropylium – 0,0 – (2',4',6' – trinitrophenylide)
$C_{16}H_{13}N_3O_8$
N.G.Furmanova, Yu.T.Struchkov, O.E.Kompan, Z.N.Budarina, L.P.Olekhnovich, V.I.Minkin
Zh.Strukt.Khim.,**21**, 83, 1980
See also R1 : 22.15

38.75   5,15 – Dioxapentacyclo(7.4.3.2^{2,8}.0.0^{3,7})octadeca – 10,12,17 – triene – 4,6 – dione
$C_{16}H_{14}O_4$
M.Kaftory *Acta Crystallogr.,Sect.B,***36**, 597, 1980

38.76   2 – Phenyl – 5,5 – dimethyl – (3,4)benzo – 5H – furylium perchlorate
$C_{16}H_{15}O^+. ClO_4^-$
M.Gawron *Pol.J.Chem.,***53**, 861, 1979

38.C   **Cunaniol acetate (violet form)**
$C_{16}H_{16}O_3$ Main entry is 59.13

38.C   **(1R,3S) – 7,10 – Dihydroxy – 1,3,8 – trimethyl – 3,4,6,9 – tetrahydro – 1H – naphtho(2,3 – c)pyran – 6,9 – dione**
Ventilagone
$C_{16}H_{16}O_5$ Main entry is 59.14

38.77   5,5' – Dinitro – 2 – (1 – adamantyl) – 2' – carbomethoxy – (2H,5H) – furan
$C_{16}H_{20}N_2O_7$
P.F.Zanazzi *Cryst.Struct.Commun.,***9**, 377, 1980
See also R1 : 31

38.78   9,9,13 – Trimethyl – 8 – oxatricyclo(8.4.0.0^{2,7}) tetradec – 2(7) – en – 3 – one
$C_{16}H_{24}O_2$
L.F.Tietze, G.von Kiedrowski, K.Harms, W.Clegg, G.M.Sheldrick *Angew.Chem.,Int.Ed.Engl.,***19**, 134, 1980

38.79   8c,11a – Dichloro – 8b,8c,11a,11b – tetrahydro – exo – phenanthro(9',10':3,4)cyclobuta(1,2 – d)(1,3) dioxol – 10 – one
$C_{17}H_{10}Cl_2O_3$
W.Ried, H.Schinzel, A.H.Schmidt, W.Schuckmann, H.Fuess *Chem.Ber.,***113**, 255, 1980

38.C   **(+) – Pisatin monohydrate**
$C_{17}H_{14}O_6. H_2O$ Main entry is 59.15

38.80   7,8 – Benzo – 3 – ethoxycarbonyl – 2 – methoxycarbonyl – 9 – methyl – 9 – aza – 5 – oxabicyclo(4.3.0)nona – (1(6),2) – dien – 4 – one
$C_{17}H_{15}NO_6$
A.J.Frew, G.R.Proctor, J.V.Silverton *J.Chem.Soc.,Perkin Trans.1,*1251, 1980
See also R1 : 36

38.81   1,4 – Diphenyl – 2,3 – dioxabicyclo(2.2.1)heptane
$C_{17}H_{16}O_2$
D.A.Langs. M.G.Erman, G.T.DeTitta, D.J.Coughlin, R.G.Salomon *J.Cryst.Mol.Struct.,***8**, 239, 1978

38.82   3,3' – Spiro – bis(3H – 2,4 – dihydro – benzo(1,4) dioxepin)
$C_{17}H_{16}O_4$
K.Stadnicka. L.Lebioda, J.Grochowski *Acta Crystallogr.,Sect.B,***35**, 2763, 1979

38.83   3 – Hydroxy – 7 – methoxy – 3',4' – methylene – dioxy – flavan
$C_{17}H_{16}O_5$
M.Kimura. W.H.Watson. P.Pacheco, M.Silva *Acta Crystallogr.,Sect.B,***35**, 3124, 1979

38.C   **Griseofulvin benzene solvate**
7 – Chloro – 2',4,6 – trimethoxy – 6' – β – methyl – spirobenzofuran – 2(3H) – cyclohex – 2 – ene – 3.4' – dione benzene solvate
$C_{17}H_{17}ClO_6. C_6H_6$ Main entry is 50.11

38.84   5 – Methyl – 15 – oxa – 5 – azapentacyclo(7.4.3.2^{2,8}.0.0^{3,7})octadeca – 10,12,17 – triene – 4,6 – dione
$C_{17}H_{17}NO_3$
M.Kaftory *Acta Crystallogr.,Sect.B,***36**, 597, 1980
See also R1 : 37

38.C   **5 – Acetoxy – 6,7 – dimethyl – 3 – oxatricyclo(5.4.0.0^{2,4})undecan – 10 – one – 9,2^1 – spiro(3^1 – formyl – 3^1 – methyl – oxirane) (absolute configuration)**
PR Toxin
$C_{17}H_{20}O_6$ Main entry is 59.16

38.C   **Spiro((1 – acetylaziridine) – 2,3' – (methyl – 4,6 – O – benzylidene – 2,3 – dideoxy – α – D – arabino – hexopyranoside))**
$C_{17}H_{21}NO_5$ Main entry is 45.43

38.C   **Laurencienyne**
$C_{17}H_{23}BrCl_2O_3$ Main entry is 59.17

38.85   Cyclopentyl – (4) – helixane
1,13,16,19 – Tetraoxatetraspiro(4.0.0.0.4.3.3.3) heneicosan – 4 – one
$C_{17}H_{24}O_5$
D.Gange, P.Magnus, L.Bass, E.V.Arnold, J.Clardy *J.Am.Chem.Soc.,***102**, 2134, 1980

38.86   Hysterin
$C_{17}H_{24}O_5$
J.P.Declercq, G.Germain, M.van Meerssche, M.Demuynck, P.De Clercq, M.Vandewalle *Acta Crystallogr.,Sect.B,***36**, 213, 1980
See also R1 : 28

38.87   1α,4,5α,7α,9,10 – Hexamethyl – 2β – methoxy – 3 – oxatricyclo(5.4.0.0^{4,11})undec – 9 – en – 6 – one
$C_{17}H_{26}O_3$
T.J.Greenhough, J.R.Scheffer, J.Trotter, L.Walsh *Can.J.Chem.,***57**, 2669, 1979

38.88   5,6:7,8 – Dibenzo – bicyclo(2.2.2)octa – 5,7 – diene – (2,3 – dicarboxylic acid anhydride)
$C_{18}H_{12}O_3$
I.V.Bulgarovskaya, Z.V.Zvonkova, I.G.Il'ina *Zh.Strukt.Khim.,***20**, 889, 1979

38.89   12 – Oxapentacyclo(8.4.3.2^{2,9}.0.0^{3,8})heptadeca – 5,14,16,18 – tetraene – 4,7,11,13 – tetrone
$C_{18}H_{12}O_5$
M.Kaftory *Acta Crystallogr.,Sect.B,***36**, 597, 1980

38.90   1,2 – bis(2 – Oxa – indan – 1 – yl – oxy)ethane
$C_{18}H_{18}O_4$
H.D.Perlmutter, R.A.Lalancette, A.Robertiello. D.V.Bowen *Tetrahedron Lett.,***21**, 817, 1980

38.C   **7 – Hydroxy – 3 – methoxy – 6 – oxa – estra – 1,3,5(10) – trien – 17 – one**
$C_{18}H_{22}O_4$ Main entry is 51.3

38.91   Dimethyl 8 – t – butyl – 1 – methyl – 2 – oxo – spiro(bicyclo(2.2.2)octa – 5,7 – diene – 3,2' – oxirane) – 5,6 – dicarboxylate
$C_{18}H_{22}O_6$
H.-D.Becker, B.Ruge. B.W.Skelton. A.H.White *Aust.J.Chem.,***32**, 1231, 1979

38.92 Dimethyl 8 – t – butyl – 5 – methyl – 2 – oxo –
spiro(bicyclo(2.2.2)octa – 5,7 – diene – 3,2' –
oxirane) – 1,6 – dicarboxylate
$C_{18}H_{22}O_6$
H.–D.Becker, B.Ruge, B.W.Skelton, A.H.White
Aust.J.Chem.,**32**, 1231, 1979

38.93 Dimethyl 6' – t – butyl – 2' – methyl – 3' – oxo –
spiro(oxirane – 2,4' – tricyclo(3.3.0.0²,⁶)oct – 6' –
ene) – 1',8 – dicarboxylate
$C_{18}H_{22}O_6$
H.–D.Becker, B.Ruge, B.W.Skelton, A.H.White
Aust.J.Chem.,**32**, 1231, 1979

38.94 (2S – (2β,3β,3aα,5β,6α,6aα)) – Hexahydro – 5 –
methoxy – 2 – (dimethoxymethyl) – 6 –
methylfuro(3,2 – b)furan – 3 – ol – 4 –
bromophenyl – carbamic acid ester
$C_{18}H_{24}BrNO_7$
H.Maehr, M.Leach, T.H.Williams, J.F.Blount
Can.J.Chem.,**58**, 501, 1980
See also R1 : 16

38.95 (2S – (2β,3 β,3aα,5α,6α,6aα)) – Hexahydro – 5 –
methoxy – 2 – (dimethoxymethyl) – 6 –
methylfuro(3,2 – b)furan – 3 – ol – 4 –
bromophenyl – carbamic acid ester (absolute
configuration)
$C_{18}H_{24}BrNO_7$
H.Maehr, M.Leach, T.H.Williams, J.F.Blount
Can.J.Chem.,**58**, 501, 1980

38.96 3,3,12,12 – Tetramethyl – 1,5,10,14 – tetraoxacyclo –
octadecane – 6,9,15,18 – tetrone
$C_{18}H_{28}O_8$
Yu.A.Simonov, T.I.Malinovskii, M.M.Botoshanskii,
A.A.Dvorkin, S.T.Malinovskii, N.G.Luk'yanenko,
Yu.A.Popkov, V.A.Shapkin, A.V.Bogatskii
Eur.Cryst.Meeting,**5**, 124, 1979

38.97 2,2' – (Butane – 1,4 – diyl) – bis(hexahydro – 1,3 –
benzodioxole)
$C_{18}H_{30}O_4$
P.G.Beckingsale, J.M.Waters, T.N.Waters
Aust.J.Chem.,**33**, 671, 1980

38.C Diphenyl – (5,6 – dichloro – 1,3 – benzodioxol –
(2)) – phosphine – oxide
$C_{19}H_{13}Cl_2O_3P$ Main entry is 64.100

38.C 7 – (3 – (4,5 – Dihydro – 5,5 – dimethyl – 4 – oxo –
2 – furanyl) – but – 2 – enyl) – oxy – (2H – 1 –
benzopyran – 2 – one)
Geiparvarin
$C_{19}H_{18}O_5$ Main entry is 59.19

38.98 1β,6β,9α,10β – 13 – Oxatetracyclo(8.2.1.1²,⁶)tetradec –
11 – en – 8 – one p –
bromophenylsulfonylhydrazone
$C_{19}H_{21}BrN_2O_3S$
H.O.House, M.B.DeTar, D.VanDerveer
J.Org.Chem., **44**, 3793, 1979

38.99 1α,6β,9α,10α – 13 – Oxatetracyclo(8.2.1.1²,⁶)
tetradec – 11 – en – 8 – one p –
bromophenylsulfonylhydrazone
$C_{19}H_{21}BrN_2O_3S$
H.O.House, M.B.DeTar, D.VanDerveer
J.Org.Chem., **44**, 3793, 1979

38.100 4 – Cyclohexyl – 3,4 – dihydro – 2 – hydroxy – 2 –
methyl – 2H,5H – pyrano(3,2 – c)(1)benzopyran –
5 – one
$C_{19}H_{22}O_4$
E.J.Valente, D.J.Hodgson
Acta Crystallogr.,Sect.B,**35**, 3099, 1979

38.C Tulirinol acetate
$C_{19}H_{24}O_6$ Main entry is 53.23

38.C Hymenograndin
$C_{19}H_{26}O_7$ Main entry is 53.24

38.C 8β – Hydroxymethyl – podocarpane – 13β –
carboxylic acid lactone
$C_{19}H_{30}O_2$ Main entry is 59.20

38.101 (6S,7S,9R,10R) – 6,9 – Epoxy – nonadec – 18 – ene –
7,10 – diol
$C_{19}H_{36}O_3$
R.G.Warren, R.J.Wells, J.F.Blount
Aust.J.Chem.,**33**, 891, 1980

38.102 2,11 – Dimethoxy – benzo(1,2 – b:4,3 – b') –
bis(benzofuran)
$C_{20}H_{14}O_4$
J.–E.Berg, B.Karlsson, A.–M.Pilotti, A.–C.Soderholm
Acta Crystallogr.,Sect.B,**36**, 1258, 1980

38.C 3,4 – Dihydro – 2H,6H – 2 – benzothiopyrano(4,3 –
b)pyran – 2 – spiro – 3' – (1H – 2) –
benzothiopyran – 4'(3'H) – one
$C_{20}H_{18}O_2S_2$ Main entry is 39.76

38.C Xanthene – 9 – spiro – 2' – (4' – t – butyl – 3' –
(methylthio) – thiete)
4 – t – Butyl – 3 – (methylthio) – thiete – 2 – spiro –
9' – xanthene
$C_{20}H_{20}OS_2$ Main entry is 39.77

38.103 Diacetone – phenanthroquinone
$C_{20}H_{20}O_4$
P.Smith–Verdier, F.Florencio, S.Garcia–Blanco
Cryst.Struct.Commun.,**9**, 587, 1980
See also R1 : 31

38.C O – Tetramethyl – haematoxylin (optically active
form)
$C_{20}H_{22}O_6$ Main entry is 59.21

38.C (±) – 2β,5β – Epoxy – 2α – (3α – p –
nitrobenzoyloxy – but – 1(E) – enyl) – 1β,3,3 –
trimethylcyclohexan – 1α – ol
3,6 – Epoxy – 5 – hydroxy – 5,6 – dihydro – β – ionol
$C_{20}H_{25}NO_6$ Main entry is 59.23

38.C Peunicin (at 113°K, absolute configuration)
$C_{20}H_{26}O_4$ Main entry is 59.24

38.C 8 – Dimethylamino – 5α – t – butyl – 5β – cyano –
2 – (cyano(t – butyl)methylene) – 1β – methyl – 3 –
oxa – 7 – thia – 7 – azabicyclo(4.3.0)non – 7 – en –
4 – one
$C_{20}H_{28}N_4O_2S$ Main entry is 41.75

38.C 8 – Dimethylamino – 5β – t – butyl – 5α – cyano –
2 – (cyano(t – butyl)methylene) – 1β – methyl – 3 –
oxa – 7 – thia – 7 – azabicyclo(4.3.0)non – 7 – en –
4 – one
$C_{20}H_{28}N_4O_2S$ Main entry is 41.76

38.C Nepetaefolinol
$C_{20}H_{28}O_6$ Main entry is 54.3

38.C **Eupatalbin**
$C_{20}H_{30}O_4$ Main entry is 54.7

38.C **Pachyclavulariadiol**
$C_{20}H_{30}O_4$ Main entry is 54.8

38.C **Thromboxane B$_2$**
$C_{20}H_{34}O_6$ Main entry is 59.27

38.104 **2,4,5 – tris(3 – Methoxycarbonyl – furan – 2 – yl) – Δ^2 – imidazoline**
$C_{21}H_{18}N_2O_9$
B.Meunier, C.Pascard
Cryst.Struct.Commun., **9**, 121, 1980
See also R1 : 32

38.C **Dunnione p – bromophenylhydrazone**
$C_{21}H_{19}BrN_2O_2$ Main entry is 59.28

38.105 **Bisphenol A – diglycidyl ether**
$C_{21}H_{24}O_4$
J.L.Flippen–Anderson, R.D.Gilardi
Am.Cryst.Assoc.,Ser.2, **8**, 36, 1980

38.C **Hispanonic acid methyl ester (absolute configuration)**
$C_{21}H_{26}O_4$ Main entry is 54.13

38.106 **N – Cyclopropylmethyl – scopolammonium bromide**
$C_{21}H_{28}NO_4^+$, Br$^-$
G.Giuseppetti, F.Mazzi, C.Tadini, A.Gallazi, P.C.Vanoni,
M.Gaetani *Farmaco,Ed.Sci.,* **35**, 231, 1980
See also R1 : 37

38.107 **1β,9β – 1 – Methoxypicras – 12 – en – 16 – one**
$C_{21}H_{32}O_3$
P.A.Grieco, G.Vidari, S.Ferrino, R.C.Haltiwanger
Tetrahedron Lett., **21**, 1619, 1980

38.108 **(2 – Oxo – 3,3,5,5 – tetramethyl – cyclopentyl) – 1 – oxa – 7,7,9,9 – tetramethyl – spiro(4.4)nonane – 3,6 – dione**
$C_{21}H_{32}O_4$
R.Kivekas, T.Simonen
Acta Chem.Scand.Ser.B, **33**, 627. 1979
See also R1 : 20

38.109 **Erythronolide A anhydride**
$C_{21}H_{36}O_7$
D.Schomburg, P.B.Hopkins, W.N.Lipscomb, E.J.Corey
J.Org.Chem., **45**, 1544, 1980

38.110 **1,4 – Epoxy – 4 – (4 – nitrophenylmethyl) – 1 – phenyl – 1H – 2,3 – benzodioxepin – 5(4H) – one**
$C_{22}H_{15}NO_6$
D.F.Mullica, J.D.Korp, W.O.Milligan, J.S.Belew,
J.L.McAtee Junior, J.Karban
J.Chem.Soc.,Perkin Trans.2, 1703, 1979

38.111 **3,3 – Bis(o – tolyl) – phthalide**
$C_{22}H_{18}O_2$
C.J.Wang. M.T.Wu. Y.J.Chen, T.H.Hseu
Acta Crystallogr.,Sect.B, **36**, 1956. 1980

38.C **8α – Hydroxy – isopicrostegane**
$C_{22}H_{22}O_8$ Main entry is 59.31

38.112 **5 – Hydroxy – 4,4 – dimethyl – 3 – (o – acetamidophenyl) – 6a – phenyl – 3a,4,5,6a – tetrahydrofuro(2,3 – b)furan – 2(3H) – one**
$C_{22}H_{23}NO_5$
G.Tacconi, L.D.Maggi. F.A.Marinone. P.Righetti, R.Oberti
J.Chem.Res., **22**, 0201. 1980

38.113 **1,5,12,16,23,26,29 – Heptaoxa(7^{3,14})(5.5) orthocyclophane naphthalene – 2,3 – diol monohydrate**
$C_{22}H_{26}O_7, C_{10}H_8O_2, H_2O$
J.A.Herbert, M.R.Truter
J.Chem.Soc.,Perkin Trans.2, 1253, 1980
See also R2 : 24

38.C **D – Gibberellin C secodiester photoproduct**
$C_{22}H_{32}O_7$ Main entry is 59.32

38.C **Purpuride**
$C_{22}H_{33}NO_5$ Main entry is 59.33

38.114 **(1,4 – b) – (2,3 – b) – bis(1,1,4,4 – Tetramethylcycloheptane) – 5,6 – dioxabicyclo(2.2.0) hex – 2 – ene**
$C_{22}H_{36}O_2$
A.Krebs, H.Schmalstieg, O.Jarchow, K.–H.Klaska
Tetrahedron Lett., **21**, 3171, 1980

38.115 **3,3′ – Diacetyl – 5,5′ – diethoxycarbonyl – glaucyrone**
Diethyl glaucophanic enol
$C_{23}H_{22}O_{10}$
S.R.Baker, M.J.Begley, L.Crombie
J.Chem.Soc.,Chem.Commun., 390, 1980
See also R1 : 1

38.C **2 – Chloroxanthene – 9 – 2′ – (3′,4′ – bis – t – butyl – thiothiete)**
$C_{23}H_{25}ClOS_3$ Main entry is 39.81

38.C **Xanthene – 9 – spiro – 2′ – (3′,4′ – bis – t – butyl – thiothiete)**
$C_{23}H_{26}OS_3$ Main entry is 39.82

38.C **17 – O – Methyl – latrunculin A**
$C_{23}H_{33}NO_5S$ Main entry is 59.35

38.C **Nodusmicin**
$C_{23}H_{34}O_7$ Main entry is 50.17

38.116 **Dibenzo(b,b′)furo(3,2 – e:4,5 – e′) – bis(benzofuran)**
$C_{24}H_{12}O_3$
B.Karlsson, A.–M.Pilotti, A.–C.Soderholm
Acta Crystallogr.,Sect.B, **36**, 1261, 1980

38.117 **7,10 – Dihydroxy – 9 – methoxy – 6 – (4 – methoxy – phenyl) – 6H,11H – (2)benzopyrano(4,3 – c)(1)benzopyran – 11 – one**
$C_{24}H_{18}O_7$
L.Jurd, R.Y.Wong *Aust.J.Chem.,* **33**, 137, 1980
See also R1 : 17

38.118 **13 – syn – Acetoxy – 1,8,11,12 – tetrachloro – tetracyclo(6.2.2.1^{3,6}.0^{2,7})tridecane – 4,11 – dien – 9 – one) – spiro – 10,4′ – (9 – anti – acetoxy – 3 – oxatricyclo(4.2.1.0^{2,5})non – 7′ – ene**
$C_{24}H_{20}Cl_4O_6$
E.Buldt, T.Debaerdemaeker, W.Friedrichsen
Tetrahedron, **36**, 267, 1980

38.119 **13 – anti – Acetoxy – 1,8,11,12 – tetrachloro – tetracyclo(6.2.2.1^{3,6}.0^{2,7})tridecane – 4,11 – dien – 9 – one) – spiro – 10,4′ – (9 – anti – acetoxy – 3 – oxatricyclo(4.2.1.0^{2,5})non – 7′ – ene**
$C_{24}H_{20}Cl_4O_6$
E.Buldt, T.Debaerdemaeker, W.Friedrichsen
Tetrahedron, **36**, 267, 1980

38.120 Hypophyllanthin
$C_{24}H_{30}O_7$
M.M.Bhadbhade, G.S.R.S.Rao, K.Venkatesan
Tetrahedron Lett.,**21**, 3097, 1980

38.C **9α – Fluoro – 11β,21 – dihydroxy – 16α,17α –**
isopropylidene – dioxy – pregna – 1,4 – diene –
3,20 – dione methanol solvate
Triamcinolone acetonide
$C_{24}H_{31}FO_6$, 0.67CH$_4$O Main entry is 51.46

38.C **21 – Acetoxy – 11α – methoxy – 1β,11β – oxo –**
10α – pregn – 4 – ene – 2,20 – dione methanol
solvate
$C_{24}H_{32}O_7$, CH$_4$O Main entry is 51.47

38.121 Spiro(3,4 – dihydro – 2H – 1,5 –
dioxadinaphtho(2,1 – f:1,2 – h)cyclononene – 3,1' –
cyclopropane)
$C_{25}H_{20}O_2$
K.Stadnicka *Acta Crystallogr.,Sect.B*,**35**, 2757, 1979

38.C **2',6'' – (1,4,7 – Trioxaheptane – 1,7 – diyl) – (2,6 –**
bis(2 – picolinoyl) – pyridine) – bis(ethylene –
ketal)
$C_{25}H_{25}N_3O_7$ Main entry is 40.55

38.122 Desacetylaustin (absolute configuration)
$C_{25}H_{32}O_8$
K.Fukuyama, Y.Katsube, H.Ishido, M.Yamazaki,
Y.Maebayashi *Chem.Pharm.Bull.*,**28**, 2270, 1980

38.C **Gilmaniellin methanol solvate**
$C_{26}H_{19}ClO_{10}$, 2CH$_4$O Main entry is 59.37

38.C **(+) – 2,4:3,5 – Di – O – methylene – D – mannitol –**
1,6 – di – trans – cinnamate
$C_{26}H_{26}O_8$ Main entry is 45.54

38.C **Terretonin**
$C_{26}H_{32}O_9$ Main entry is 59.38

38.123 Spiro(3,4 – dihydro – 2H – 1,5 –
dioxadinaphtho(2,1 – f:1,2 – h)cyclononene – 3,1' –
cyclopentane)
$C_{27}H_{24}O_2$
K.Stadnicka, L.Lebioda
Acta Crystallogr.,Sect.B,**35**, 2760, 1979

38.124 3 – (5 – Methoxy – 2,2,8,8 – tetramethyl – 2H,8H –
benzo(1,2 – b:3,4 – b')dipyran – 6 – yl) – propyl –
3,5 – dinitrobenzoate
Eriostemyl – 3,5 – dinitrobenzoate
$C_{27}H_{28}N_2O_9$
P.R.Jefferies, B.W.Skelton, B.Walter, A.H.White
Aust.J.Chem.,**33**, 313, 1980

38.125 3 – (5 – Methoxy – 2,2,8,8 – tetramethyl – 2H,8H –
benzo(1,2 – b:5,4 – b')dipyran – 10 – yl) – propyl –
3,5 – dinitrobenzoate
Eriostyl – 3,5 – dinitrobenzoate
$C_{27}H_{28}N_2O_9$
P.R.Jefferies, B.W.Skelton, B.Walter, A.H.White
Aust.J.Chem.,**33**, 313, 1980

38.126 3 – (5 – Methoxy – 2,2,8,8 – tetramethyl – 2H,8H –
benzo(1,2 – b:5,4 – b')dipyran – 10 – yl) – propyl –
p – nitrobenzoate
Eriostyl p – nitrobenzoate
$C_{27}H_{29}NO_7$
P.R.Jefferies, B.W.Skelton, B.Walter, A.H.White
Aust.J.Chem.,**33**, 313, 1980

38.C **Paspalicine (orthorhombic form)**
$C_{27}H_{31}NO_3$ Main entry is 59.39

38.C **Paspalinine**
$C_{27}H_{31}NO_4$ Main entry is 54.29

38.127 Tetrahydro – 3 – (5 – methoxy – 2,2,8,8 –
tetramethyl – 2H,8H – benzo(1,2 – b:5,4 – b')
dipyran – 10 – yl) – propyl – 3,5 – dinitrobenzoate
Tetrahydroeriostyl – 3,5 – dinitrobenzoate
$C_{27}H_{32}N_2O_9$
P.R.Jefferies, B.W.Skelton, B.Walter, A.H.White
Aust.J.Chem.,**33**, 313, 1980

38.C **Verrucarin B**
$C_{27}H_{32}O_9$ Main entry is 59.40

38.C **1β,2β,3β,4β,5β,7α – Hexahydroxy – spirost – 25(27) –**
en – 6 – one
$C_{27}H_{40}O_9$ Main entry is 51.62

38.128 8,11 – Diacetoxy – 10 – methoxy – 7 – (4 –
methoxyphenyl) – 6H,7H – (1)benzopyrano(4,3 – b)
(1)benzopyran – 6 – one
$C_{28}H_{22}O_9$
L.Jurd, R.Y.Wong *Aust.J.Chem.*,**33**, 137, 1980

38.C **7 – Con – O – methylnogarol ethanol solvate**
(at 120°K)
$2C_{28}H_{31}NO_{10}$, C_2H_6O Main entry is 50.19

38.129 2,2,7,7,12,12,17,17 – Octamethyl – 3,6,13,16 –
tetraoxo – 21,22 – dioxatricyclo(18.2.1.1^{8,11})
tetracosa – 8,10,18,20 – trans – 4,14 – hexaene
$C_{28}H_{32}O_8$
D.L.Ward, W.–J.H.Kung *Am.Cryst.Assoc.,Ser.2*,**8**, 34, 1980

38.C **Marcfortine A**
$C_{28}H_{35}N_3O_4$ Main entry is 58.73

38.C **Acetyl – strongylophorine – 2**
$C_{28}H_{36}O_5$ Main entry is 54.33

38.C **Acnistin E**
$C_{28}H_{38}O_7$ Main entry is 51.70

38.C **Paspaline methanol solvate**
$C_{28}H_{39}NO_2$, 0.5CH$_4$O Main entry is 59.41

38.130 Dibenzo(b,q)(1,4,7,10,13,16,19,22,25,28)
decaoxacyclotriacontane potassium thiocyanate
Dibenzo – 30 – crown – 10 potassium thiocyanate
$C_{28}H_{40}O_{10}$, CNS$^-$, K$^+$
J.Hasek, D.Hlavata, K.Huml
Acta Crystallogr.,Sect.B,**36**, 1782, 1980

38.131 6,7,9,10,12,13,15,16,23,24,26,27,29,30,32,33 –
Hexadecahydro – dibenzo(b,q)
(1,4,7,10,13,16,19,22,25,28)decaoxacyclotriacontin
bis(sodium isothiocyanate) monohydrate
(Dibenzo – 30 – crown – 10) bis(sodium isothiocyanate)
monohydrate
$C_{28}H_{40}O_{10}$, 2Na$^+$, 2CNS$^-$, H$_2$O
J.D.Owen, M.R.Truter
J.Chem.Soc.,Dalton Trans.,1831, 1979

38.C **Brassinolide monohydrate**
$C_{28}H_{48}O_6$, H$_2$O Main entry is 51.72

38.C   8,9 – Dihydroxy – 1,3,7,7 – tetramethyl – bicyclo(4.4.0)decane – 2,2' – spiro(6',7' – bis(hydroxymethyl) – 4' – (p – bromophenylsulfonyloxy) – 2',3' – dihydrobenzofuran) dimethylsulfoxide solvate
$C_{29}H_{37}BrO_8S$, C_2H_6OS Main entry is 50.21

38.132   4',5,7 – Tri – t – butyl – 3' – (2,2 – dimethylpropionyl) – 1' – methyl – spiro(benzofuran – 3(2H) – 2' – pyrrolidine) – 2,5' – dione
$C_{29}H_{43}NO_4$
H.–D.Becker, K.Gustafsson, C.L.Raston, A.H.White
Aust.J.Chem.,32, 1931, 1979
See also R1 : 36

38.C   24 – Ethyl – isobrassinolide
$C_{29}H_{50}O_6$ Main entry is 51.76

38.C   24 – Ethyl – brassinolide
$C_{29}H_{50}O_6$ Main entry is 51.77

38.133   2,5 – Diphenyl – phenanthro(9,10 – b)oxepin
$C_{30}H_{20}O$
E.A.Harrison Junior, H.L.Ammon
J.Org.Chem.,45, 943, 1980

38.134   6,12 – Disalicyloyl – benzo(1,2 – b:4,5 – b') – bis(benzofuran)
$C_{32}H_{18}O_6$
J.Bergman, B.Egestad, D.Rajapaksa
Acta Chem.Scand.Ser.B,33, 405, 1979

38.135   2,3:4,5 – bis(1,2 – (3 – Methylnaphtho)) – 1,6,9,12,15,18 – hexaoxacycloeicosa – 2,4 – diene (at 300°K)
$C_{32}H_{36}O_6$
I.Goldberg   *J.Am.Chem.Soc.,102, 4106, 1980*

38.136   2,3:4,5 – bis(1,2 – (3 – Methylnaphtho)) – 1,6,9,12,15,18 – hexaoxacycloeicosa – 2,4 – diene (at 193°K)
$C_{32}H_{36}O_6$
I.Goldberg   *J.Am.Chem.Soc.,102, 4106, 1980*

38.C   2,3:4,5 – bis(1,2 – (3 – Methylnaphtho)) – 1,6,9,12,15,18 – hexaoxacycloeicosa – 2,4 – diene t – butylammonium perchlorate clathrate benzene solvate (at 113°K)
$C_{32}H_{36}O_6$, $C_4H_{12}N^+$, ClO_4^-, C_6H_6 Main entry is 61.13

38.137   Tetronolide
$C_{32}H_{40}O_8$
N.Hirayama, M.Kasai, K.Shirahata, Y.Ohashi, Y.Sasada
Tetrahedron Lett.,21, 2559, 1980

38.C   Tri – o – thymotide trans – stilbene clathrate
$C_{33}H_{36}O_6$, $C_{14}H_{12}$ Main entry is 61.14

38.C   (–) – Tri – o – thymotide – RR – (+) – 2,3 – dimethyl – thiirane clathrate (at –50°C)
$2C_{33}H_{36}O_6$, C_4H_8S Main entry is 61.15

38.C   (+) – Tri – o – thymotide – S – (+) – 2 – bromo – butane clathrate (at –50°C)
$2C_{33}H_{36}O_6$, C_4H_9Br Main entry is 61.16

38.C   Tri – o – thymotide cis – stilbene clathrate
$2C_{33}H_{36}O_6$, $C_{14}H_{12}$ Main entry is 61.17

38.138   3',4,9' – Tri – t – butyl – 2',5,10' – trimethoxy – spiro(cyclohexa – 3,5 – diene – 1,6' – dibenzo(d,f) (1,3)dioxepin) – 2 – one
$C_{33}H_{42}O_6$
F.R.Hewgill, C.L.Raston, A.H.White
Aust.J.Chem.,32, 881, 1979

38.C   9,10 – 0,0 – Isopropylidene – 10a – 0 – (o – bromobenzoyl) – 11 – 0 – methyl – rubrolone
$C_{34}H_{32}BrNO_9$ Main entry is 50.

38.139   Lasalocid
$C_{34}H_{54}O_8$
I.C.Paul   *Am.Cryst.Assoc.,Ser.2,8, 45, 1980*

38.C   (6R) – 6,19 – Epidioxy – 9,10 – seco – ergosta – 5(10),7,22 – trien – 3β – ol benzoate
$C_{35}H_{48}O_4$ Main entry is 51.80

38.C   Uvarinol benzene solvate
$C_{36}H_{30}O_7$, C_6H_6 Main entry is 59.42

38.140   3,4',11,12 – Tetra – 0 – acetyl – bruceine
$C_{36}H_{44}O_{16}$
J.Polonsky, J.Varenne, T.Prange, C.Pascard
Tetrahedron Lett.,21, 1853, 1980

38.C   Ohchinolide A
$C_{37}H_{42}O_{10}$ Main entry is 59.44

38.141   E – 2,2,2',2' – Tetraethoxy – Δ(3,3'(2H,2'H)) – bis(phenanthro(9,10 – b)furan)
$C_{40}H_{36}O_6$
R.W.Sallfrank, E.Ackermann, H.Winkler, W.Paul, R.Bohme
Chem.Ber.,113, 2950, 1980

38.C   Sodium nigericine
$C_{40}H_{67}O_{11}^-$, Na^+ Main entry is 50.26

38.C   Hedamycin
$C_{41}H_{50}N_2O_{11}$ Main entry is 50.27

38.C   Dinactin monactin
$0.3C_{41}H_{66}O_{12}$, $0.7C_{42}H_{68}O_{12}$ Main entry is 38.142

38.C   Ionomycin calcium salt n – heptane solvate
$C_{41}H_{70}O_9^{2-}$, Ca^{2+}, $0.5C_7H_{16}$ Main entry is 50.28

38.C   Ionomycin cadmium salt n – heptane solvate (absolute configuration)
$C_{41}H_{70}O_9^{2-}$, Cd^{2+}, $0.5C_7H_{16}$ Main entry is 50.29

38.142   Dinactin monactin
$0.7C_{42}H_{68}O_{12}$, $0.3C_{41}H_{66}O_{12}$
Y.Nawata, T.Hayashi, Y.Iitaka   *Chem.Lett.,315, 1980*
See also R2 : 38

38.C   16 – Methoxy – 16' – oxo – 21,20' – di(20,18 – epoxy – pregnane)
$C_{43}H_{66}O_4$ Main entry is 51.89

38.C   Rifamycin Y p – iodoanilide dimethylsulfoxide monohydrate
$C_{45}H_{51}IN_2O_{14}$, C_2H_6OS, H_2O Main entry is 50.30

38.C   Rifamycin B p – iodoanilide acetone pentahydrate
$C_{45}H_{53}IN_2O_{13}$, C_3H_6O, $5H_2O$ Main entry is 50.31

38.C   3 – (β – L – Mycarose) – 5 – (β – D – 4,6 – dideoxy – 3 – ketoallose) – 13 – (β – D – mycinose) – lankamycin – 11 – α – hydroxyisovalerate ester
$C_{48}H_{82}O_{20}$ Main entry is 50.35

HETERO–SULFUR AND HETERO–SELENIUM

39.1 Dibromo – (maleic acid) – thioanhydride
$C_4Br_2O_2S$
W.Gonschorek $Z.Naturforsch.,Teil\ A$,**35**, 14, 1980

39.C (–) – Tri – o – thymotide – RR – (+) – 2,3 –
dimethyl – thiirane clathrate (at –50°C)
C_4H_8S, $2C_{33}H_{36}O_6$ Main entry is 61.15

39.2 Di – iodo – (maleic acid) – thioanhydride
$C_4I_2O_2S$
W.Gonschorek $Z.Naturforsch.,Teil\ A$,**35**, 14, 1980

39.3 Decathia – cyclotetradecane – 6,7,13,14 – tetrone
$C_4O_4S_{10}$
H.W.Roesky, H.Zamankhan, J.W.Bats, H.Fuess
$Angew.Chem.,Int.Ed.Engl.$,**19**, 125, 1980

39.4 4,5 – Dimethyl – 1,3 – dithiole – 2 – thione
$C_5H_6S_3$
D.L.Smith, H.R.Luss
$Acta\ Crystallogr.,Sect.B$,**36**, 465, 1980

39.5 2,3 – Dihydro – 5 – methyl – 1,4 – dithiin – 1,1,4,4 –
tetroxide
$C_5H_8O_4S_2$
R.B.Bates, G.R.Kriek, A.D.Brewer
$Acta\ Crystallogr.,Sect.B$,**36**, 736, 1980

39.6 1 – Methyl – thiolanium iodide (α form)
$C_5H_{11}S^+$, I^-
F.Miyoshi, K.Tokuno, T.Watanabe, M.Matsui, T.Ohashi
$Yukugaku\ Zasshi$,**99**, 924, 1979

39.7 2,5 – bis(N – Chlorothio – imino) – 3,4 –
dicyanothiophene
$C_6H_2Cl_2N_4S_3$
F.Wudl, E.T.Zellers $J.Am.Chem.Soc.$,**102**, 4283, 1980

39.C Tetrathiafulvalene – chloranil complex
$C_6H_4S_4$, $C_6Cl_4O_2$ Main entry is 60.3

39.C Tetrathiafulvalene – fluoranil complex
$C_6H_4S_4$, $C_6F_4O_2$ Main entry is 60.4

39.C bis(Acetylacetonato) – palladium(ii)
tetrathiafulvalene
$C_6H_4S_4$, $C_{10}H_{14}O_4Pd$ Main entry is 60.12

39.8 Tetrathiafulvalenium perchlorate
$C_6H_4S_4^+$, ClO_4^-
K.Yakushi, S.Nishimura, T.Sugano, H.Kuroda, I.Ikemoto
$Acta\ Crystallogr.,Sect.B$,**36**, 358, 1980

39.9 Tetrathiafulvalenium tri – iodide
$C_6H_4S_4^+$, I_3^-
R.C.Teitelbaum, T.J.Marks, C.K.Johnson
$J.Am.Chem.Soc.$,**102**, 2986, 1980

39.C Tetrathiafulvalene bis(ethylenedithiolato) – nickel(i)
$2C_6H_4S_4$, $3C_4H_4NiS_4$ Main entry is 60.5

39.10 5 – Methyl – 2 – thiophene – carboxylic acid
$C_6H_6O_2S$
S.H.Simonsen, F.R.Cordell, J.E.Boggs
$Am.Cryst.Assoc.,Ser.2$,**8**, 38, 1980

39.11 1,4,7 – Trithia – cyclononane
$C_6H_{12}S_3$
R.S.Glass, G.S.Wilson, W.N.Setzer
$J.Am.Chem.Soc.$,**102**, 5068, 1980

39.12 2,5 – Dimethyl – 1,2 – dithiolo(1,5 – b) – 1,2 –
oxathiole
$C_7H_8OS_2$
R.Bardi, A.M.Piazzesi, V.Busetti
$Acta\ Crystallogr.,Sect.B$,**35**, 2821, 1979

39.13 7 – Methyl – 1,4 – dithia – 7 – aza – spiro(4.4)
nonane – 6,8 – dione
$C_7H_9NO_2S_2$
W.Dobrowolska, M.Bukowska–Strzyzewska
$Acta\ Crystallogr.,Sect.B$,**36**, 317, 1980
See also R1 : 35

39.14 N – Methyl – 1,4 – dithiane – 2,3 – dicarboximide
$C_7H_9NO_2S_2$
W.Dobrowolska, M.Bukowska–Strzyzewska
$Acta\ Crystallogr.,Sect.B$,**36**, 462, 1980
See also R1 : 35

39.15 S – Methyl – thiepanium bromide
$C_7H_{15}S^+$, Br^-
J.J.Combremont, R.Gerdil $Eur.Cryst.Meeting$,**5**, 109, 1979

39.16 S – Methyl – thiepanium 2,4,6 –
trinitrobenzenesulfonate
$C_7H_{15}S^+$, $C_6H_2N_3O_9S^-$
J.J.Combremont, R.Gerdil $Eur.Cryst.Meeting$,**5**, 109, 1979
See also R2 : 15,11

39.17 S – Methyl – thiepanium iodide
$C_7H_{15}S^+$, I^-
J.J.Combremont, R.Gerdil $Eur.Cryst.Meeting$,**5**, 109, 1979

39.18 5,5′ – Dinitro – 2,2′ – dithienyl
$C_8H_4N_2O_4S_2$
L.V.Panfilova, M.Yu.Antipin, Yu.T.Struchkov,
Yu.D.Churkin, A.E.Lipkin $Zh.Strukt.Khim.$,**21**, 190, 1980

39.19 2,2′ – Dithienyl antimony trichloride
$C_8H_6S_2$, $2Cl_3Sb$
L.Korte, A.Lipka, D.Mootz $Eur.Cryst.Meeting$,**5**, 92, 1979

39.20 3 – Nitro – 4 – (2 – thienyl) – but – 3 – en – 2 –
one (at –120°C)
$C_8H_7NO_3S$
L.V.Panfilova, M.Yu.Antipin, Yu.D.Churkin,
Yu.T.Struchkov
$Khim.Geterotsikl.Soedin.,Latv.SSSR$,1201, 1979

39.21 4 – (5 – Nitro – 2 – thienyl) – but – 3 – en – 2 –
one (at –120°C)
$C_8H_7NO_3S$
L.V.Panfilova, M.Yu.Antipin, Yu.D.Churkin,
Yu.T.Struchkov
$Khim.Geterotsikl.Soedin.,Latv.SSSR$,1201, 1979

39.22 2 – (2 – Chloro – 4 – methylthiophen – 3 – yl –
imino) – tetrahydro – imidazole
Tiamenidine
$C_8H_{10}ClN_3S$
J.M.Leger, D.Hickel, A.Carpy
$C.R.Acad.Sci.,Ser.C$,**289**, 93, 1979
See also R1 : 32

39.23 **2,5 – Divinyl – 1 – sulfolane**
$C_8H_{12}O_2S$
U.M.Dzhemilev, R.V.Kunakova, Yu.T.Struchkov,
G.A.Tolstikov, F.V.Sharipova, L.G.Kuz'mina, S.R.Rafikov
*Dokl.Akad.Nauk SSSR,***250**, 105, 1980

39.24 **Thianonane – 3,8 – dione – 1,1 – dioxide**
$C_8H_{12}O_4S$
L.D.Quin, J.Leimert, E.D.Middlemas, R.W.Miller, A.T.McPhail
*J.Org.Chem.,***44**, 3496, 1979

39.25 **3 – Phenyl – 4,5 – dihydrothiophene – 1,1 – dioxide**
$C_{10}H_{10}O_2S$
G.V.Klimusheva, T.E.Bezmenova, G.M.Soroka, G.G.Rode
*Tetrahedron,***36**, 1667, 1980

39.26 **4 – Phenyl – 4,5 – dihydrothiophene – 1,1 – dioxide**
$C_{10}H_{10}O_2S$
G.V.Klimusheva, T.E.Bezmenova, G.M.Soroka, G.G.Rode
*Tetrahedron,***36**, 1667, 1980

39.C **Tetramethyl – tetrathiafulvalene tetrabromo – p – benzoquinone complex**
$C_{10}H_{12}S_4$, $C_6Br_4O_2$ Main entry is 60.10

39.C **bis(Tetramethyl – tetrathiafulvalene) iodide**
$C_{10}H_{12}S_4$, $C_{10}H_{12}S_4^+$, I^- Main entry is 60.11

39.27 **bis(Tetramethyl – tetrathiafulvalene) thiocyanate**
$C_{10}H_{12}S_4^+$, $C_{10}H_{12}S_4$, CNS^-
J.L.Galigne, B.Liautard, S.Peytavin, G.Brun, M.Maurin,
J.M.Fabre, E.Torreilles, L.Giral
*Acta Crystallogr.,Sect.B,***35**, 2609, 1979

39.C **tris(4,4′,5,5′ – Tetramethyl – tetrathiafulvalene) – bis(2,5 – diethyl – 7,7,8,8 – tetracyanoquinodimethane) complex**
$3C_{10}H_{12}S_4$, $2C_{16}H_{12}N_4$ Main entry is 60.35

39.28 **bis(S – p – Hydroxyphenyl – thiophanium) p – phenolate – thiophanium bis(perchlorate)**
$2C_{10}H_{13}OS^+$, $C_{10}H_{12}OS$, $2ClO_4^-$
A.E.Kalinin, Yu.T.Struchkov, L.R.Barikina, E.N.Karaulova
*Kristallografiya,***25**, 488, 1980

39.29 **1,1,4,4 – Tetramethyl – 1H,4H – thieno(3,4 – c) thiophene**
$C_{10}H_{14}S_2$
S.Braverman, M.Freund, I.Goldberg
*Tetrahedron Lett.,***21**, 3617, 1980

39.30 **Selenobiotin**
$C_{10}H_{16}N_2O_3Se$
G.T.DeTitta, R.Parthasarathy, R.H.Blessing, W.Stallings
*Proc.Nat.Acad.Sci.U.S.A.,***77**, 333, 1980
See also R1 : 35

39.31 **Biotin – d – sulfoxide**
$C_{10}H_{16}N_2O_4S$
G.T.DeTitta, R.Parthasarathy, R.H.Blessing, W.Stallings
*Proc.Nat.Acad.Sci.U.S.A.,***77**, 333, 1980
See also R1 : 35

39.32 **Biotin sulfone**
$C_{10}H_{16}N_2O_5S$
G.T.DeTitta, R.Parthasarathy, R.H.Blessing, W.Stallings
*Proc.Nat.Acad.Sci.U.S.A.,***77**, 333, 1980
See also R1 : 35

39.33 **Dithieno(1,2 – b:5,4 – b′)tropylium tetrafluoroborate (at 295°K)**
$C_{11}H_7S_2^+$, BF_4^-
J.-E.Andersson *Acta Crystallogr.,Sect.B,***35**, 1349, 1979

39.34 **4 – (Anilinomethylene) – tetrahydrothiophen – 3 – one**
$C_{11}H_{11}NOS$
K.Skinnemoen, T.Ottersen
*Acta Chem.Scand.Ser.A,***34**, 359, 1980

39.35 **N – Phenyl – 2,5 – dihydrothiophene – 3 – carboxamide**
$C_{11}H_{11}NOS$
K.Skinnemoen, T.Ottersen
*Acta Chem.Scand.Ser.A,***34**, 359, 1980

39.C **3 – Thioxo – 4 – phenyl – 4 – aza – 2,7 – dithiobicyclo(3.3.0)octane – 7,7 – dioxide**
$C_{11}H_{11}NO_2S_3$ Main entry is 41.35

39.36 **1 – p – Hydroxyphenyl – thianium 1 – phenolate – thianium perchlorate**
$C_{11}H_{15}OS^+$, $C_{11}H_{14}OS$, ClO_4^-
G.G.Aleksandrov, Yu.T.Struchkov, A.E.Kalinin,
A.A.Shcherbakov, L.R.Barikina, E.N.Karaulova
*Kristallografiya,***25**, 481, 1980

39.37 **1 – p – Hydroxyphenylthiane perchlorate**
$C_{11}H_{15}OS^+$, ClO_4^-
G.G.Aleksandrov, Yu.T.Struchkov, A.E.Kalinin,
A.A.Shcherbakov, L.R.Barikina, E.N.Karaulova
*Kristallografiya,***25**, 261, 1980

39.38 **Biotin methyl ester**
$C_{11}H_{18}N_2O_3S$
G.T.DeTitta, R.Parthasarathy, R.H.Blessing, W.Stallings
*Proc.Nat.Acad.Sci.U.S.A.,***77**, 333, 1980
See also R1 : 35

39.39 **5 – Hydroxy – 1 – (2 – hydroxyethyl) – 6,7,8 – trimethyl – 2 – thia – 7 – azabicyclo(3.3.0)octane hemihydrate**
$C_{11}H_{21}NO_2S$, $0.5H_2O$
K.Kamiya, Y.Wada, M.Takamoto
*J.Takeda Res.Lab.,***38**, 44, 1979
See also R1 : 35

39.40 **1 – Methyl – 2,3 – di – t – butyl – thiirenium tetrafluoroborate (at –100°C)**
$C_{11}H_{21}S^+$, BF_4^-
R.Destro, T.Pilati, M.Simonetta *Nouv.J.Chim.,***3**, 533, 1979

39.41 **Perfluorothianthrene**
$C_{12}F_8S_2$
D.Rainville, R.A.Zingaro, E.A.Meyers
*Cryst.Struct.Commun.,***9**, 909, 1980

39.42 **2 – Carboxy – 3 – methylthiopyrano(4,3,2 – cd) indole**
$C_{12}H_{11}NO_2S$
S.Shie, M.Sho, Z.Chang, G.Li, G.Chao, E.Tong
*Ko Hsueh Tung Pao,***25**, 350, 1980
See also R1 : 36

39.43 **2,6,10 – Trithiabicyclo(9.4.0)pentadeca – 1(11),12,14 – triene**
$C_{12}H_{16}S_3$
E.P.Kyba, A.M.John, S.B.Brown, C.W.Hudson, M.J.McPhaul,
A.Harding, K.Larsen, S.Niedzwiecki, R.E.Davis
*J.Am.Chem.Soc.,***102**, 139, 1980

39.44 anti – Methyl – (2,3β,3aβ,5,6,7,8,8aβ – octahydro –
4 – hydroxylimino – 4H – cyclohepta(b)thiophen –
3a – yl) – carbamic acid methyl ester
$C_{12}H_{20}N_2O_3S$
P.N.Confalone, G.Pizzolato, D.L.Confalone, M.R.Uskokovic
J.Am.Chem.Soc.,**102**, 1954, 1980

39.45 O,O – Bicyclohexyl – 1,1′ – diyl – thiosulfite
$C_{12}H_{20}O_2S_2$
D.N.Harpp, K.Steliou, C.J.Cheer
J.Chem.Soc.,Chem.Commun.,825, 1980

39.46 (2,3 – Dichloro – 4 – (2 – thienoyl) – phenoxy) –
acetic acid
Tienilic acid
$C_{13}H_9Cl_2O_4S$
A.Carpy, M.Goursolle, J.M.Leger
Acta Crystallogr.,Sect.B,**36**, 1706, 1980

39.47 N – (4 – Methoxyphenyl) – 3,6 – dithiacyclohexene –
1,2 – dicarboximide
$C_{13}H_{11}NO_3S_2$
W.Dobrowolska, M.Bukowska–Strzyzewska
Acta Crystallogr.,Sect.B,**36**, 890, 1980
See also R1 : 35

39.48 N(1′) – Methoxycarbonyl – biotin methyl ester
$C_{13}H_{20}N_2O_5S$
W.C.Stallings, C.T.Monti, M.D.Lane, G.T.DeTitta
Proc.Nat.Acad.Sci.U.S.A.,**77**, 1260, 1980
See also R1 : 35

39.49 3 – (5 – Tetrazolyl) – thioxanthone – 10,10 – dioxide
rubidium salt
Doxantrazole rubidium salt
$C_{14}H_7N_4O_3S^-$, Rb^+
L.C.G.Goaman, L.De C.Paxton
Eur.Cryst.Meeting,**5**, 178, 1979
See also R1 : 32

39.C Dibenzotetrathiafulvalene pentabromo –
dimethylthio – platinum
$C_{14}H_8S_4^+$, $C_2H_6Br_5PtS^-$ Main entry is 85.2

39.50 Dibenzotetrathiafulvalene tri – iodide
$C_{14}H_8S_4^+$, I_3^-
R.P.Shibaeva, L.P.Rozenberg, M.Z.Aldoshina,
R.N.Lyubovskaya, M.L.Khidekel
Zh.Strukt.Khim.,**20**, 485, 1979

39.51 tris(Dibenzotetrathiafulvalene) hexabromo – tin
$2C_{14}H_8S_4^+$, $C_{14}H_8S_4$, Br_6Sn^{2-}
R.P.Shibaeva, L.P.Rozenberg, R.M.Lobkovskaya
Kristallografiya,**25**, 507, 1980

39.52 octakis(Dibenzotetrathiafulvalene) tris(hexachloro –
tin)
$6C_{14}H_8S_4^+$, $2C_{14}H_8S_4$, $3Cl_6Sn^{2-}$
R.P.Shibaeva, L.P.Rozenberg, R.M.Lobkovskaya
Kristallografiya,**25**, 507, 1980

39.53 3,5 – bis(Cyclohexylidene – 1,2,4 – trithiolane)
$C_{14}H_{20}S_3$
W.Winter, H.Buhl, H.Meier
Z.Naturforsch.,Teil B,**35**, 1015, 1980

39.54 (1R,3R,4S,5R) – 4,5 – (Isopropylidene – dioxy) – 1 –
(2 – methyl – 1,3 – dithian – 2 – yl) –
cyclohexane – 1,3 – diol
$C_{14}H_{24}O_4S_2$
I.Dyong, R.Hermann, R.Mattes *Chem.Ber.*,**113**, 1931, 1980
See also R1 : 38

39.55 2,5 – Dicyclohexane – spiro – 1,3 – dithiane
$C_{14}H_{24}S_2$
K.Stadnicka, K.Suwinska, L.Lebioda
Eur.Cryst.Meeting,**6**, 287, 1980

39.C 3 – (3 – Amino – benzo(b)thiophen – 2 – yl) – 1,2 –
benzisothiazole
$C_{15}H_{10}N_2S_2$ Main entry is 41.52

39.56 2,2,6,6 – Tetramethyl – 4(e) – phenylthian – 4(a) –
ol (at –135°C)
$C_{15}H_{22}OS$
N.Satyamurthy, R.Sivakumar, K.Ramalingam, K.D.Berlin,
R.A.Loghry, D.van der Helm *J.Org.Chem.*,**45**, 349, 1980

39.C Thieno(3,2 – e:4,5 – e′) – bis(benzo(b)thiophene) –
7,7,8,8 – tetracyanoquinodimethane
Trithia(5)heterohelicene – 7,7,8,8 –
tetracyanoquinodimethane
$C_{16}H_8S_3$, $C_{12}H_4N_4$ Main entry is 60.33

39.57 3 – Bromo – 2,5 – diphenyl – 1,4 – dithiin – 1,1 –
dioxide
$C_{16}H_{11}BrO_2S_2$
H.A.Levi, R.J.Doedens
Acta Crystallogr.,Sect.B,**36**, 1959, 1980

39.58 (2.2)Tetrathiafulvalenophane
$C_{16}H_{12}S_8$
J.Ippen, C.Tao–pen, B.Starker, D.Schweitzer, H.A.Staab
Angew.Chem.,Int.Ed.Engl.,**19**, 67, 1980

39.59 trans – 1,8 – Di(2 – thienyl) – 1,3,5,7 – octatetraene
$C_{16}H_{14}S_2$
J.F.Buschmann, G.Ruban *Eur.Cryst.Meeting*,**5**, 105, 1979

39.60 cis – 2,4,9 – Trimethyl – thioxanthene – 10 – oxide
$C_{16}H_{16}OS$
S.S.C.Chu, R.D.Rosenstein, A.L.Ternay Junior
Acta Crystallogr.,Sect.B,**35**, 2430, 1979

39.61 cis – 9 – Ethyl – 9 – methyl – thioxanthene – 10 –
oxide
$C_{16}H_{16}OS$
S.S.C.Chu, R.D.Rosenstein
Acta Crystallogr.,Sect.B,**36**, 989, 1980

39.62 5,7,14,16 – Tetramethoxy – 1,2,3,10,11,12 –
hexathia(3.3)metacyclophane
$C_{16}H_{16}O_4S_6$
N.Bresciani–Pahor, M.Calligaris, L.Randaccio
Acta Crystallogr.,Sect.B,**36**, 632, 1980
See also R1 : 17

39.63 syn – 2,11 – Dithia(3.3)metacyclophane
$C_{16}H_{16}S_2$
W.Anker, G.W.Bushnell, R.H.Mitchell
Can.J.Chem.,**57**, 3080, 1979

39.64 bis(4,5 – Dimethyl – 2H – 1,3 – dithiol – 2 –
ylidene) – 1,4 – cyclohexa – 2,5 – diene perchlorate
$C_{16}H_{16}S_4^+$, $C_{16}H_{16}S_4$, ClO_4^-
J.L.Galigne, B.Liautard, S.Peytavin, G.Brun, M.Maurin,
J.M.Fabre, E.Torreilles, L.Giral
Acta Crystallogr.,Sect.B,**36**, 1109, 1980

39.65 1 – Benzoyl – 2 – methyl – 2 – thianaphthalene
$C_{17}H_{14}OS$
M.Hori, T.Kataoka, H.Shimizu, S.Ohno, K.Narita,
H.Takayanagi, H.Ogura, Y.Iitaka
Tetrahedron Lett.,4315, 1979

39.66 Naphthaceno(5,6 – cd:11,12 – c'd') – bis(1,2 – dithiole) bis((μ^2 – bromo) – bromo – copper)
$2C_{18}H_8S_4$, Br_4Cu_2
R.P.Shibaeva, V.F.Kaminskii, N.D.Kushch, A.V.Zvarikina, E.B.Yagubskii *Dokl.Akad.Nauk SSSR*,**251**, 162, 1980

39.67 Diphenyl – tetrathiafulvalene
$C_{18}H_{12}S_4$
A.Escande, J.Lapasset
Cryst.Struct.Commun.,**8**, 1009, 1979

39.68 4 – Phenyl – 1 – (3 – (4 – methyl – 5 – phenyl – 1,2 – dithiolylidene)) – 3,5 – dioxo – 1,2,4 – triazine
$C_{18}H_{13}N_3O_2S_2$
G.G.Aleksandrov, Yu.T.Struchkov, A.E.Kalinin, A.A.Shcherbakov, G.S.Bogomolova, V.N.Drozd
Izv.Akad.Nauk Az.SSR,545, 1979
See also R1 : 32

39.69 (3)Tetrathiafulvaleno(3)paracyclophane
$C_{18}H_{16}S_4$
H.A.Staab, J.Ippen, C.Tao-pen, C.Krieger, B.Starker
Angew.Chem.,Int.Ed.Engl.,**19**, 66, 1980

39.70 6,8,10,14,16,18 – Hexamethyl – 1,2,3,4,11,12 – hexathia(4.2)metacyclophane
$C_{18}H_{20}S_6$
N.Bresciani–Pahor, M.Calligaris, L.Randaccio
Acta Crystallogr.,Sect.B,**36**, 632, 1980

39.71 (±) – 4β – (2',2' – Trimethylene – dithioethyl) – 3α – (1'S) – (p – nitrobenzyloxycarbonyl)oxy – ethyl – 2 – azetidinone
$C_{18}H_{22}N_2O_6S_2$
T.Kametani, S.–P.Huang, S.Yokohama, Y.Suzuki, M.Ihara
J.Am.Chem.Soc.,**102**, 2060, 1980
See also R1 : 32

39.72 Benzyl – (2,3β,3aα,4,5,6,7,8,9,9aβ – decahydro – 4 – oxo – thieno(3,2 – c)azocine – 3α – yl) – carbamic acid methyl ester
$C_{18}H_{24}N_2O_4S$
P.N.Confalone, G.Pizzolato, D.L.Confalone, M.R.Uskokovic
J.Am.Chem.Soc.,**102**, 1954, 1980
See also R1 : 35

39.73 syn – 5,12 – Dihydro – 5,14 – (epithio – methyleneoxy) – 6,12 – epithiodibenzo(a,f) cyclodecen – 7(6H) – one
$C_{19}H_{14}O_2S_2$
O.H.Johansen, P.Groth, K.Undheim
Acta Chem.Scand.Ser.B,**34**, 1, 1980
See also R1 : 31

39.C 1,3 – Benzodithiolyl – 2 – (diphenylphosphine – oxide)
$C_{19}H_{15}OPS_2$ Main entry is 64.103

39.74 cis – 2 – trans – 6 – Diphenyl – cis – 3 – ethylthian – 4 – ol (at –135°C)
$C_{19}H_{22}OS$
N.Satyamurthy, R.Sivakumar, K.Ramalingam, K.D.Berlin, R.A.Loghry, D.van der Helm *J.Org.Chem.*,**45**, 349, 1980

39.75 5 – Benzamido – 2 – benzoylimino – 1,3 – dithiol – 4 – carbonic acid ethyl ester
$C_{20}H_{16}N_2O_4S_2$
K.Urgast, R.Hoge, K.Eichhorn, K.Fischer
Cryst.Struct.Commun.,**9**, 615, 1980

39.76 3,4 – Dihydro – 2H,6H – 2 – benzothiopyrano(4,3 – b)pyran – 2 – spiro – 3' – (1H – 2) – benzothiopyran – 4'(3'H) – one
$C_{20}H_{16}O_2S_2$
O.H.Johansen, T.Ottersen, K.Undheim
Acta Chem.Scand.Ser.B,**33**, 669, 1979
See also R1 : 38

39.C 6 – Thiatetracycline (at 120°K)
$C_{20}H_{20}N_2O_7S$ Main entry is 50.13

39.C 5a – epi – 6 – Demethyl – 6 – deoxy – 6 – thiatetracycline dimethylformamide solvate (at 120°K)
$C_{20}H_{20}N_2O_7S$, C_3H_7NO Main entry is 50.14

39.77 Xanthene – 9 – spiro – 2' – (4' – t – butyl – 3' – (methylthio) – thiete)
4 – t – Butyl – 3 – (methylthio) – thiete – 2 – spiro – 9' – xanthene
$C_{20}H_{20}OS_2$
G.J.Verhoeckx, J.Kroon, A.C.Brouwer, H.J.T.Bos
Acta Crystallogr.,Sect.B,**36**, 484, 1980
See also R1 : 38

39.78 (4)Tetrathiafulvaleno(4)paracyclophane
$C_{20}H_{22}S_4$
H.A.Staab, J.Ippen, C.Tao-pen, C.Krieger, B.Starker
Angew.Chem.,Int.Ed.Engl.,**19**, 66, 1980

39.79 Benzo(p) – 2 – thiatetracyclo(7.5.3.0^{3,8}.0^{1,10}) heptadecane – 3(8),15 – diene (at –120°C)
$C_{20}H_{24}S$
G.G.Aleksandrov, A.A.Shcherbakov, Yu.T.Struchkov, V.G.Kharchenko *Cryst.Struct.Commun.*,**9**, 625, 1980
See also R1 : 31

39.80 4 – Benzyl – 2,3,5,6 – bis(tetramethylene) – 4H – thiopyran
$C_{20}H_{24}S$
G.G.Aleksandrov, A.A.Shcherbakov, Yu.T.Struchkov, V.G.Kharchenko *Cryst.Struct.Commun.*,**9**, 411, 1980

39.81 2 – Chloroxanthene – 9 – 2' – (3',4' – bis – t – butyl – thiothiete)
$C_{23}H_{25}ClOS_3$
H.J.B.Slot, A.M.de Vos, A.M.M.Schreurs, G.J.Verhoeckx, H.J.T.Bos, A.C.Brouwer, J.Kroon
Eur.Cryst.Meeting,**6**, 33, 1980
See also R1 : 38

39.82 Xanthene – 9 – spiro – 2' – (3',4' – bis – t – butyl – thiothiete)
$C_{23}H_{26}OS_3$
H.J.B.Slot, A.M.de Vos, A.M.M.Schreurs, G.J.Verhoeckx, H.J.T.Bos, A.C.Brouwer, J.Kroon
Eur.Cryst.Meeting,**6**, 33, 1980
See also R1 : 38

39.83 2 – (t – Butylthio) – 2 – (9 – thioxanthenylidene) – dithioacetato – t – butylester
$C_{23}H_{26}S_4$
H.J.B.Slot, A.M.de Vos, A.M.M.Schreurs, G.J.Verhoeckx, H.J.T.Bos, A.C.Brouwer, J.Kroon
Eur.Cryst.Meeting,**6**, 33, 1980

39.84 2,3,7,8,12,13 – Hexamethoxy – 10,15 – dihydro – 5H – 5,10,15 – trithia – tribenzo(a,d,g)cyclononene chloroform solvate (crown form)
Trithiaveratrylene chloroform solvate
$C_{24}H_{24}O_6S_3$, $2CHCl_3$
K.von Deuten, G.Klar, J.Kopf
Eur.Cryst.Meeting,5, 324, 1979
See also R1 : 17

39.C 16 – (3 – Methylbutyl) – 11,12,13,14,16,17 – hexahydro – 15H – cyclopenta(a)phenanthrene – 17,2′ – spiro(1′,3′ – dithiolane)
$C_{24}H_{28}OS_2$ Main entry is 51.45

39.85 bis(2,4 – Diphenyl – 3 – thienyl) – disulfide
$C_{32}H_{22}S_4$
K.Kobayashi, K.Mutai, H.Kobayashi
Tetrahedron Lett.,5003, 1979

39.C 3 – (4,5 – bis(Methoxycarbonyl) – isoxazol – 3 – yl) – 4,6 – dioxo – syn – cis – syn – 1,3,5 – triphenyl – perhydrothieno(3,4 – c)pyrrole – 1 – carbonitrile
$C_{32}H_{23}N_3O_7S$ Main entry is 40.58

39.86 2,3,7,8,12,13,17,18 – Octamethoxy – tetrabenzo(b,e,h,k)(1,4,7,10)tetrathia – cyclododecatetraene chloroform solvate
$C_{32}H_{32}O_8S_4$, $2CHCl_3$
K.von Deuten, G.Klar, J.Kopf
Eur.Cryst.Meeting,6, 64, 1980

39.87 4,6,12,14,20,22,28,30 – Octamethyl – 1,2,9,10,17,18,25,26 – octathia(2.2.2.2)metacyclophane
$C_{32}H_{32}S_8$
N.Bresciani–Pahor, M.Calligaris, L.Randaccio, F.Bottino, S.Pappalardo *Gazz.Chim.Ital.*,110, 227, 1980

39.88 Tetraphenyl – di(thiapyranylidene)
$C_{34}H_{24}S_2$
H.R.Luss, D.L.Smith
Acta Crystallogr.,Sect.B,36, 986, 1980

39.89 3,3′,5,5′ – Tetraphenyl – 4,4′ – dithia – 1,1′ – bis(2,5 – cyclohexadienylidene) polyiodide (orthorhombic form)
3,3′,5,5′ – Tetraphenyl – 4,4′ – dithiapyranylidene polyiodide
$C_{34}H_{24}S_2$, $0.36I_3$, $0.4I_5$
H.R.Luss, D.L.Smith
Acta Crystallogr.,Sect.B,36, 1580, 1980

39.90 3,3′,5,5′ – Tetraphenyl – 4,4′ – dithia – 1,1′ – bis(2,5 – cyclohexadienylidene) polyiodide (tetragonal form)
3,3′,5,5′ – Tetraphenyl – 4,4′ – dithiapyranylidene polyiodide
$C_{34}H_{24}S_2$, $0.72I_3$
H.R.Luss, D.L.Smith
Acta Crystallogr.,Sect.B,36, 1580, 1980

HETERO–(NITROGEN AND OXYGEN)

40.1 Trimethylene – furazan – N – oxide (at –130°C)
$C_5H_6N_2O_2$
J.F.Barnes, M.J.Barrow, M.M.Harding, R.M.Paton, P.L.Ashcroft, J.Crosby, C.J.Joyce
J.Chem.Res.,314, 3601, 1979

40.2 Benzfurazan – 1 – oxide
$C_6H_4N_2O_2$
D.Britton, J.M.Olson
Acta Crystallogr.,Sect.B,35, 3076, 1979

40.3 3 – Methyl – 4 – (2 – carboxyvinyl) – 1,2,5 – oxadiazole – 2 – oxide
$C_6H_6N_2O_4$
T.Osawa, Y.Kito, M.Namiki, K.Tsuji
Tetrahedron Lett.,4399, 1979

40.4 (1S,4S) – N – Acetyl – 3 – oxo – 5 – aza – 2 – oxabicyclo(2.2.1)heptane
$C_7H_9NO_3$
A.T.H.Lenstra, G.H.Petit, H.J.Geise
Cryst.Struct.Commun.,8, 1023, 1979
See also R1 : 37

40.5 (RS) – α – Amino – 3 – hydroxy – 5 – methylisoxazole – 4 – propionic acid monohydrate
$C_7H_{10}N_2O_4$, H_2O
T.Honore, J.Lauridsen
Acta Chem.Scand.Ser.B,34, 235, 1980

40.6 1,2,3 – Oxadiazolo(4,3 – c)(1,2,4)benzotriazinium – 3 – olate
$C_8H_4N_4O_2$
T.J.King, P.N.Preston, J.S.Suffolk, K.Turnbull
J.Chem.Soc.,Perkin Trans.2,1751, 1979
See also R1 : 36

40.7 3 – (2 – Aminophenyl) – sydnone
$C_8H_7N_3O_2$
T.J.King, P.N.Preston, J.S.Suffolk, K.Turnbull
J.Chem.Soc.,Perkin Trans.2,1751, 1979

40.8 8 – Oxa – 1 – azabicyclo(4.3.0)nonan – 9 – one – 2 – carboxylic acid
$C_8H_{11}NO_4$
B.Nader, R.W.Franck, S.M.Weinreb
J.Am.Chem.Soc.,102, 1153, 1980
See also R1 : 35

40.9 Dimorpholino – tetrasulfane
$C_8H_{16}N_2O_2S_4$
O.Foss, V.Janickis *J.Chem.Soc.,Dalton Trans.*,632, 1980

40.10 Dimorpholino – diselane
$C_8H_{16}N_2O_2Se_2$
O.Foss, V.Janickis *J.Chem.Soc.,Dalton Trans.*,628, 1980

40.11 Dimorpholino – triselane
$C_8H_{16}N_2O_2Se_3$
O.Foss, V.Janickis *J.Chem.Soc.,Dalton Trans.*,628, 1980

40.12 **Dimorpholino – tetraselane**
$C_8H_{16}N_2O_2Se_4$
O.Foss, V.Janickis *J.Chem.Soc.,Dalton Trans.*,620, 1980

40.13 **2 – Methylthio – Δ^2 – 1,3 – diaza – 6,9 – dioxa –
cycloundecene hydroiodide**
$C_8H_{17}N_2O_2S^+$, I^-
T.I.Malinovskii, Yu.A.Simonov, A.A.Dvorkin,
S.T.Malinovskii, Yu.G.Ganin, N.G.Luk'yanenko,
A.V.Bogatskii *Eur.Cryst.Meeting*,**6**, 14, 1980

40.14 **5 – Phenyl – 1,2,4 – oxadiazole – 3 – carboxamide**
$C_9H_7N_3O_2$
D.Viterbo, R.Calvino, A.Serafino
J.Chem.Soc.,Perkin Trans.2,1096, 1980

40.15 **2 – Imino – 6′ – methyl – spiro(oxazolidine – 5,3′ –
piperidine) – 2′ – one – 4 – semicarbazone
monohydrate**
$C_9H_{15}N_5O_3$, H_2O
G.Moad, C.L.Luthy, P.A.Benkovic, S.J.Benkovic
J.Am.Chem.Soc.,**101**, 6068, 1979
See also R1 : 35

40.C **7,2′ – Anhydro – β – D – arabinosyl – orotidine**
$C_{10}H_{10}N_2O_7$ Main entry is 47.12

40.C **3 – Sulfanilamido – 5 – methylisoxazole 2,4 –
diamino – 5 – (3,4,5 – trimethoxybenzyl) –
pyrimidine**
$C_{10}H_{11}N_3O_3S$, $C_{14}H_{18}N_4O_3$ Main entry is 44.30

40.16 **4 – Morpholino – benzene – diazonium
tetrafluoroborate (at –100°C)**
$C_{10}H_{12}N_3O^+$, BF_4^-
N.W.Alcock, T.J.Greenhough, D.M.Hirst, T.J.Kemp,
D.R.Payne *J.Chem.Soc.,Perkin Trans.2*,8, 1980

40.17 **1,8 – Diaza – 3,6 – dioxatetradecane – 2,7 – dione**
$C_{10}H_{18}N_2O_4$
A.Shanzer, N.Shochet, D.Rabinovich, F.Frolow
Angew.Chem.,Int.Ed.Engl.,**19**, 326, 1980

40.18 **N,N′ – Dimethyl – 1,7 – diaza – 4,10 –
dioxacyclodecane benzylammonium thiocyanate**
$C_{10}H_{22}N_2O_2$, $C_7H_{10}N^+$, CNS^-
J.C.Metcalfe, J.F.Stoddart, G.Jones, W.E.Hull, A.Atkinson,
I.S.Kerr, D.J.Williams
J.Chem.Soc.,Chem.Commun.,540, 1980

40.19 **N,N′ – Dimethyl – 1,7 – diazonium – 4,10 –
dioxacyclodecane perchlorate**
$C_{10}H_{24}N_2O_2^{2+}$, $2ClO_4^-$
J.C.Metcalfe, J.F.Stoddart, G.Jones, W.E.Hull, A.Atkinson,
I.S.Kerr, D.J.Williams
J.Chem.Soc.,Chem.Commun.,540, 1980

40.20 **4 – Thiobenzoyl – morpholine (at 200°K)**
$C_{11}H_{13}NOS$
K.A.Kerr, P.M.A.O.van Roey
Acta Crystallogr.,Sect.B.,**35**, 2344, 1979

40.21 **4 – Thiobenzoyl – morpholine**
$C_{11}H_{13}NOS$
K.A.Kerr, P.M.A.O.van Roey
Acta Crystallogr.,Sect.B.,**35**, 2344, 1979

40.22 **4,– (2′ – Hydroxythiobenzoyl) – morpholine**
$C_{11}H_{13}NO_2S$
K.A.Kerr, P.M.A.O.van Roey
Acta Crystallogr.,Sect.B.,**35**, 2727, 1979
See also R1 : 17

40.23 **(E) – Morpholino – p – nitrobenzamidoxime**
$C_{11}H_{13}N_3O_4$
K.J.Dignam, A.F.Hegarty, M.J.Begley
J.Chem.Soc.,Perkin Trans.2,704, 1980

40.C **Cyclo(D – N – methylvalyl – D – α –
hydroxyisovaleryl)**
$C_{11}H_{19}NO_3$ Main entry is 48.32

40.24 **(E) – 6 – Bromomethylene – 5,6 – dihydro – 4,4 –
dimethyl – 2 – phenyl – 4H – 1,3,4 – oxadiazinium
bromide**
$C_{12}H_{14}BrN_2O^+$, Br^-
D.M.Thompson, I.D.Brindle, M.F.Richardson
Can.J.Chem.,**57**, 3157, 1979

40.25 **1 – Methylthio – 3 – (α – morpholino) –
benzylidene – triazene**
$C_{12}H_{16}N_4OS$
G.L'abbe, A.Willocx, J.P.Declercq, G.Germain,
M.van Meerssche *Bull.Soc.Chim.Belg.*,**88**, 107, 1979

40.26 **3 – (p – Chlorophenyl) – 6 – methyl – 2 – oxo –
2,3,3a,7a – tetrahydro – oxazolo(4,5 – b)pyridine
4 – Methyl – 9 – p – chlorophenyl – 7 – oxa – 2,9 –
diazabicyclo(4.3.0)nona – 2,4 – dien – 8 – one**
$C_{13}H_{11}ClN_2O_2$
T.Hisano, M.Ichikawa, T.Matsuoka, H.Hagiwara,
K.Muraoka, T.Komori, K.Harano, Y.Ida, A.T.Christensen
Chem.Pharm.Bull.,**27**, 2261, 1979
See also R1 : 35

40.27 **3 – (p – Chlorophenyl) – 7a – methyl – 2 – oxo –
2,3,3a,7a – tetrahydro – oxazolo(4,5 – b)pyridine
6 – Methyl – 9 – p – chlorophenyl – 7 – oxa – 2,9 –
diazabicyclo(4.3.0)nona – 2,4 – dien – 8 – one**
$C_{13}H_{11}ClN_2O_2$
T.Hisano, M.Ichikawa, T.Matsuoka, H.Hagiwara,
K.Muraoka, T.Komori, K.Harano, Y.Ida, A.T.Christensen
Chem.Pharm.Bull.,**27**, 2261, 1979
See also R1 : 35

40.28 **2 – (p – Chlorophenyl) – cis – 5,6 – trimethylene –
2,3,5,6 – tetrahydro – 1,3 – oxazin – 4 – one**
$C_{13}H_{14}ClNO_2$
G.Argay, A.Kalman, B.Ribar, D.Lazar, G.Bernath
Cryst.Struct.Commun.,**9**, 335, 1980

40.29 **Furaltadone**
$C_{13}H_{16}N_4O_6$
I.Goldberg *Eur.Cryst.Meeting*,**6**, 49, 1980
See also R1 : 38

40.C **Furaltadone hydrochloride acetic acid clathrate**
$C_{13}H_{17}N_4O_6^+$, $C_2H_4O_2$, Cl^- Main entry is 61.4

40.C **Furaltadone hydrochloride propionic acid clathrate**
$C_{13}H_{17}N_4O_6^+$, $C_3H_6O_2$, Cl^- Main entry is 61.5

40.30 **Furaltadone hydrochloride hydrate**
$C_{13}H_{17}N_4O_6^+$, Cl^-, $2H_2O$
I.Goldberg *Eur.Cryst.Meeting*,**6**, 49, 1980
See also R1 : 38

40.31 **1,7,8,9,10 – Pentachloro – 3 – phenyl – 4 – aza – 5 –
oxatricyclo(5.2.1.0^{2,6})decane – 3,8 – diene**
$C_{14}H_8Cl_5NO$
C.De Micheli, R.Gandolfi, R.Oberti
J.Org.Chem.,**45**, 1209, 1980
See also R1 : 31

40.32 1,7,8,9 – Tetrachloro – 3 – phenyl – 4 – aza – 5 –
oxatricyclo(5.2.1.0^{2,6})decane – 3,8 – diene
$C_{14}H_9Cl_4NO$
C.De Micheli, R.Gandolfi, R.Oberti
J.Org.Chem.,**45**, 1209, 1980
See also R1 : 31

40.33 4,5 – Diphenylisosydnone benzene solvate
$C_{14}H_{10}N_2O_2$, 0.25C_6H_6
T.J.King, P.N.Preston, J.S.Suffolk, K.Turnbull
J.Chem.Soc.,*Perkin Trans.2*,1751, 1979

40.34 rac – cis – 5 – Chloro – 2,3,3a,9 – tetrahydro –
6,8,9 – trimethyl – 1H – pyrrolo(2,1 – b)(1,3)
benzoxazine
$C_{14}H_{18}ClNO$
N.Cohen, J.F.Blount, R.J.Lopresti, D.P.Trullinger
J.Org.Chem.,**44**, 4005, 1979
See also R1 : 36

40.C 9β – D – Arabinofuranosyl – 8 – morpholino –
adenine dihydrate
$C_{14}H_{20}N_6O_5$, 2H_2O Main entry is 47.30

40.35 (1RS,8SR,10SR,4(15)Z) – 4 – Ethylidene – 5 – oxa –
3 – azatricyclo(8.4.0.0^{3,6})tetradecane (at –60°C)
$C_{14}H_{23}NO$
A.Kumin, E.Maverick, P.Seiler, N.Vanier, L.Damm, R.Hobi,
J.D.Dunitz, A.Eschenmoser
Helv.Chim.Acta,**63**, 1158, 1980

40.36 Benzyl 4 – methoxy – 7 – azido – 3 – oxa – 1 –
azabicyclo(4.2.0)octan – 8 – one – 2 – carboxylate
$C_{15}H_{16}N_4O_5$
J.G.Gleason, T.F.Buckley, K.G.Holden, D.B.Bryan, P.Siler
J.Am.Chem.Soc.,**101**, 4730, 1979
See also R1 : 35

40.37 2 – (p – Chlorophenyl) – cis – 5,6 –
pentamethylene – 2,3,5,6 – tetrahydro – 1,3 –
oxazin – 4 – one
$C_{15}H_{18}ClNO_2$
G.Argay, A.Kalman, B.Ribar, D.Lazar, G.Bernath
Cryst.Struct.Commun.,**9**, 341, 1980

40.38 N,N – Diethyl – 3,3a – dihydro – 3 – methyl –
benzofuro(3,2 – c)isoxazole – 3 – carboxamide
$C_{15}H_{18}N_2O_3$
A.D.de Wit, W.P.Trompenaars, D.N.Reinhoudt, S.Harkema,
G.J.van Hummel *Tetrahedron Lett.*,**21**, 1779, 1980

40.39 3,6,9,12,15 – Pentaoxa – 21 – azabicyclo(15.3.1)
heneicosa – 1(21),17,19 – triene – 2,16 – dione
potassium thiocyanate
$C_{15}H_{19}NO_7$, K$^+$, CNS$^-$
S.B.Larson, N.K.Dalley
Acta Crystallogr.,*Sect.B.***36**, 1201, 1980

40.40 3,6,9,12,15 – Pentaoxa – 21 – azabicyclo(15.3.1)
heneicosa – 1(20),17,19 – triene t – butylammonium
perchlorate complex (at 113°K)
Monopyrido – 18 – crown – 6 t – butylammonium
perchlorate complex
$C_{15}H_{23}NO_5$, $C_4H_{12}N^+$, ClO_4^-
E.Maverick, L.Grossenbacher, K.N.Trueblood
Acta Crystallogr.,*Sect.B.***35**, 2233, 1979

40.41 2 – (1′ – Benzyloxycarbonylamino – 1′ –
methylethyl) – 4,4 – dimethyl – 5 – oxazolone
$C_{16}H_{20}N_2O_4$
C.M.K.Nair, M.Vijayan
Acta Crystallogr.,*Sect.B.***36**, 1498, 1980

40.42 1,4,7,15 – Tetraoxa – 12,18 – diazacyclodocosane –
8,11,19,22 – tetrone
$C_{16}H_{28}N_2O_8$
Yu.A.Simonov, T.I.Malinovskii, M.M.Botoshanskii,
A.A.Dvorkin, N.G.Luk'yanenko, S.T.Malinovskii,
Yu.A.Popkov, V.A.Shapkin, A.V.Bogatskii
Eur.Cryst.Meeting,**5**, 124, 1979

40.43 (3S,5R) – 3 – Methyl – 5 – (4′ – biphenylyl) –
2,3,5,6 – tetrahydro – 1,4 – oxazin – 2 – one
$C_{17}H_{17}NO_2$
M.Sikirica, I.Vickovic, V.Caplar, A.Sega, A.Lisini, F.Kajfez,
V.Sunjic *J.Org.Chem.*,**44**, 4423, 1979

40.C Methyl 5,5 – dimethyl – 2 – (2 – phenoxymethyl –
5 – oxo – 1,3 – oxazolin – 4 – ylidene) – 1,3 –
thiazolidine – 4 – carboxylate
$C_{17}H_{18}N_2O_5S$ Main entry is 41.63

40.44 3,3 – Dimethyl – 1,5,13 – trioxa – 10,16 –
diazacycloeicosane – 6,9,17,20 – tetrone
$C_{17}H_{28}N_2O_7$
Yu.A.Simonov, T.I.Malinovskii, M.M.Botoshanskii,
A.A.Dvorkin, S.T.Malinovskii, N.G.Luk'yanenko,
Yu.A.Popkov, V.A.Shapkin, A.V.Bogatskii
Eur.Cryst.Meeting,**5**, 124, 1979

40.45 7 – (1 – (p – Nitrobenzyloxycarbonyl)oxy – ethyl) –
8 – oxo – 2,2 – dimethyl – 3 – oxa – 1 –
azabicyclo(4.2.0)octane
$C_{18}H_{22}N_2O_7$
F.A.Bouffard, D.B.R.Johnston, B.G.Christensen
J.Org.Chem.,**45**, 1130, 1980
See also R1 : 35

40.46 N,N′ – bis(2 – Cyanoethyl) – 1,10 – diaza –
4,7,13,16 – tetraoxacyclo – octadecane malononitrile
$C_{18}H_{32}N_4O_4$, 2$C_3H_2N_2$
K.von Deuten, A.Knochel, J.Kopf, J.Oehler, G.Rudolph
J.Chem.Res.,**358**, 4035, 1979

40.C Sodium bis(μ – diphenylphosphido) –
bis(tricarbonyl – iron) 1,10 – diaza –
4,7,13,16,21,24 – hexaoxabicyclo(8.8.8)hexacosane
2$C_{18}H_{36}N_2O_6$, $C_{30}H_{20}Fe_2O_6P_2^{2-}$, 2Na$^+$ Main entry is 86.101

40.C bis(10,22 – Dimethyl – 1,4,7,13,16,19 – hexaoxa –
10,22 – diazacyclotetracosane) benzylammonium
thiocyanate clathrate
2$C_{18}H_{38}N_2O_6$, $C_7H_{10}N^+$, CNS$^-$ Main entry is 61.6

40.47 8,9 – Dimethoxy – 3 – methyl – 1 – phenyl –
3,4,5,6 – tetrahydro – 1H – 2,3 – benzoxazocine
$C_{19}H_{23}NO_3$
J.B.Bremner, E.J.Browne, P.E.Davies, C.L.Raston, A.H.White
Aust.J.Chem.,**33**, 1323, 1980

40.48 2 – Hydroxy – 4 – isopropyl – 7 – benzyl – 3 –
oxa – 6,9 – diazatricyclo(7.3.0.0^{2,6})dodecane – 5,8 –
dione
$C_{19}H_{24}N_2O_4$
G.Lucente, F.Pinnen, G.Zanotti, S.Cerrini, W.Fedeli,
F.Mazza *Eur.Cryst.Meeting*,**6**, 290, 1980

40.49 2 – Benzyl – 1,7,10,16 – tetraoxa – 4,13 –
diazacyclo – octadecane – 3,14,18 – trione
$C_{19}H_{26}N_2O_7$
T.I.Malinovskii, Yu.A.Simonov, A.A.Dvorkin,
S.T.Malinovskii, N.G.Luk'yanenko, A.V.Bogatskii
Eur.Cryst.Meeting,**6**, 14, 1980

40.50　2,6 – Diphenyl – benzo(1,2 – d:5,4 – d')bis(oxazole)
$C_{20}H_{12}N_2O_2$
A.V.Fratini
Rep.Air Force Sci.Res.AFOSR–77–3267,1, 1978

40.51　6,6' – (3,6,9,12 – Tetraoxatetradeca – 1,14 – dioxy) –
2,2' – bipyridine
$C_{20}H_{26}N_2O_6$
G.R.Newkome, A.Nayak, F.Fronczek, T.Kawato,
H.C.R.Taylor, L.Meade, W.Mattice
J.Am.Chem.Soc.,**101**, 4472, 1979

40.52　2 – (2 – Tosylaminophenyl) – 4H – 3,1 –
benzoxazin – 4 – one
$C_{21}H_{16}N_2O_4S$
O.S.Filipenko, V.I.Ponomarev, B.M.Bolotin, L.O.Atovmyan
Kristallografiya,**24**, 957, 1979

40.53　2',6'' – (1,4,7 – Trioxaheptane – 1,7 – diyl) – (2,6 –
bis(2 – picolinoyl) – pyridine)
$C_{21}H_{17}N_3O_5$
G.R.Newkome, A.Nayak, J.D.Sauer, P.K.Mattschei,
S.F.Watkins, F.Fronczek, W.H.Benton
J.Org.Chem.,**44**, 3816, 1979
See also R1 : 37

40.54　6 – Dimethylamino – 5 – (4 – nitrophenylamino) –
3 – (2,4,6 – trimethylphenyl) – 6H – 1,2 – oxazine
$C_{21}H_{24}N_4O_3$
A.C.Villa, A.G.Manfredotti, C.Guastini, P.Trimarco
Cryst.Struct.Commun.,**9**, 523, 1980

40.55　2',6'' – (1,4,7 – Trioxaheptane – 1,7 – diyl) – (2,6 –
bis(2 – picolinoyl) – pyridine) – bis(ethylene –
ketal)
$C_{25}H_{25}N_3O_7$
G.R.Newkome, A.Nayak, J.D.Sauer, P.K.Mattschei,
S.F.Watkins, F.Fronczek, W.H.Benton
J.Org.Chem.,**44**, 3816, 1979
See also R1 : 38,37

40.C　N(β) – Methoxy – (progesterone) – (16α,17α – d) –
(tetrahydro – 1',2' – oxazole)
$C_{25}H_{35}NO_6$ Main entry is 51.56

40.C　N(α) – Methoxy – (progesterone) – (16α,17α – d) –
(tetrahydro – 1',2' – oxazole)
$C_{25}H_{35}NO_6$ Main entry is 51.57

40.C　14 – (p – Chlorophenyl) – 8 – (p –
chlorophenylsulfonylamino) – 13 – oxa – 14 –
thia – 1,15 – diazapentacyclo(10.4.2.0²,⁷.0⁸,¹⁰.0¹²,¹⁹)
octadeca – 2,4,6,14 – tetraene S – oxide
$C_{26}H_{23}Cl_2N_3O_4S_2$ Main entry is 42.12

40.56　12 – Benzoylimino – 2 – phenyl – 3 – oxa – 1 –
aza – trispiro(4.0.5.2.5.0)nonadec – 1 – ene – 4,13 –
dione
$C_{30}H_{30}N_2O_4$
S.Mohr *Tetrahedron Lett.*,**21**, 593, 1980

40.57　3 – (p – Chlorophenyl) – 3a – methyl – 4 – oxo –
5,6,6a – triphenyl – 3a,4 – dihydrocyclopenta(2,3 –
d)isoxazoline
$C_{31}H_{22}ClNO_2$
A.P.Bozopoulos, S.C.Kokkou, P.J.Rentzeperis
Acta Crystallogr.,Sect.B,**36**, 102, 1980

40.58　3 – (4,5 – bis(Methoxycarbonyl) – isoxazol – 3 –
yl) – 4,6 – dioxo – syn – cis – syn – 1,3,5 –
triphenyl – perhydrothieno(3,4 – c)pyrrole – 1 –
carbonitrile
$C_{32}H_{23}N_3O_7S$
I.Ueda, T.Takata, O.Tsuge
Acta Crystallogr.,Sect.B,**36**, 958, 1980
See also R1 : 39,35

40.C　3 – Carbomethoxy – rifamycin SV monohydrate
$C_{39}H_{47}NO_{14}$, H_2O Main entry is 50.25

40.C　Beauvericin barium picrate complex toluene
solvate (form B)
$C_{45}H_{57}N_3O_9$, $2C_6H_2N_3O_7^-$, Ba^{2+}, $2C_7H_8$ Main entry is 50.32

HETERO–(NITROGEN AND SULFUR)

41.1 5 – Amino – 1,2,3,4 – thiatriazole
CH_2N_4S
M.J.Zaworotko, J.L.Atwood, L.Floch
J.Cryst.Mol.Struct.,**9**, 173, 1979

41.2 5 – Azido – 2H – 1,2,4,6 – thiatriazin – 3(4H) – one –
1,1 – dioxide
$C_2H_2N_6O_3S$
G.H.Denny, E.J.Cragoe Junior, C.S.Rooney, J.P.Springer,
J.M.Hirshfield, J.A.McCauley *J.Org.Chem.*,**45**, 1662, 1980

41.3 Sodium 3 – oxo – 4 – nitro – 2H – 1,2,6 –
thiadiazine – 1,1 – dioxide monohydrate
$C_3H_2N_3O_5S^-$, Na^+, H_2O
C.Esteban–Calderon, M.Martinez–Ripoll, S.Garcia–Blanco
Eur.Cryst.Meeting,**5**, 386, 1979

41.4 3 – Hydroxy – 4 – nitro – 6H – 1,2,6 – thiadiazine –
1,1 – dioxide
$C_3H_3N_3O_5S$
C.Esteban–Calderon, M.Martinez–Ripoll, S.Garcia–Blanco
Eur.Cryst.Meeting,**5**, 386, 1979

41.5 5 – Amino – 2H – 1,2,6 – thiadiazine – 1,1 – dioxide
$C_3H_5N_3O_2S$
H.A.Albrecht, J.F.Blount, F.M.Konzelmann, J.T.Plati
J.Org.Chem.,**44**, 4191, 1979

41.6 6,8 – Dimethyl – 7 – oxo – $1\lambda^3$ – thionia – $3\lambda^4,5\lambda^4$ –
dithia – 2,4,6,8,9 – penta – azabicyclo(3.3.1)nona –
2,3,5(9) – triene hexafluoroarsenate
$C_3H_6N_5OS_3^+$, AsF_6^-
H.W.Roesky, T.Muller, E.Wehner, E.Rodek
Chem.Ber.,**113**, 2802, 1980

41.7 Sodium 4 – cyano – 3 – hydroxy – 6H – 1,2,6 –
thiadiazin – 6 – ide – 1,1 – dioxide dihydrate
$C_4H_2N_3O_3S^-$, Na^+, $2H_2O$
C.Esteban–Calderon, M.Martinez–Ripoll, S.Garcia–Blanco
Eur.Cryst.Meeting,**6**, 36, 1980

41.8 4 – Cyano – 3 – hydroxy – 6H – 1,2,6 – thiadiazine –
1,1 – dioxide
$C_4H_3N_3O_3S$
C.Esteban–Calderon, M.Martinez–Ripoll, S.Garcia–Blanco
Acta Crystallogr.,Sect.B,**35**, 2795, 1979

41.9 N,N' – (Tetrasulfur – tetranitrido) –
thiacyclopentane – di – imide
$C_4H_8N_6S_5$
H.W.Roesky, C.Graf, M.N.S.Rao, B.Krebs, G.Henkel
Angew.Chem.,Int.Ed.Engl.,**18**, 780, 1979

41.10 1,2,4 – Triazolo(3,2 – b)(1,3)thiazin – 5 – one
$C_5H_3N_3OS$
J.P.Clayton, P.J.O'Hanlon, T.J.King
J.Chem.Soc.,Perkin Trans.1,1352, 1980

41.11 7 – Thia – 1,3 – diazabicyclo(3.3.0)octa – 2,4 – dione
$C_5H_6N_2O_2S$
B.Tinant, J.P.Declercq, G.Germain, M.van Meerssche
Bull.Soc.Chim.Belg.,**89**, 113, 1980

41.12 4 – Carboxylato – L – thiazolidine – hydantoin
$C_5H_6N_2O_2S$
E.Arte, B.Tinant, J.P.Declercq, G.Germain,
M.van Meerssche *Bull.Soc.Chim.Belg.*,**89**, 113, 1980
See also R1 : 35

41.13 5 – Methoxycarbonylamino – 3 – methyl – 1,2,3 –
thiadiazole
$C_5H_7N_3O_2S$
S.Bruckner, G.Fronza, L.M.Giunchi, V.A.Kozinski,
O.V.Zelenskaja *Tetrahedron Lett.*,**21**, 2101, 1980

41.14 2,3 – Dihydro – 3 – hydroxy – 5H – thiazolo(3,2 – a)
pyrimidin – 5 – one
$C_6H_6N_2O_2S$
U.Rychlewska *Acta Crystallogr.,Sect.B*,**36**, 971, 1980
See also R1 : 35

41.15 5,6 – Dimethyl – imidazo(2,1 – b)(1,3,4)thiadiazole
$C_6H_7N_3S$
M.L.Schenetti, F.Taddei, L.Greci, L.Marchetti, G.Milani,
G.D.Andreetti, G.Bocelli, P.Sgarabotto
J.Chem.Soc.,Perkin Trans.2,421, 1980
See also R1 : 35

41.16 5,6 – Dimethyl – imidazo(2,1 – b)(1,3,4)thiadiazole
hydrobromide monohydrate
$C_6H_8N_3S^+$, Br^-, H_2O
M.L.Schenetti, F.Taddei, L.Greci, L.Marchetti, G.Milani,
G.D.Andreetti, G.Bocelli, P.Sgarabotto
J.Chem.Soc.,Perkin Trans.2,421, 1980
See also R1 : 35

41.17 N – Acetyl – 4 – carboxylato – L – thiazolidine
$C_6H_9NO_3S$
B.Tinant, J.P.Declercq, G.Germain, M.van Meerssche
Bull.Soc.Chim.Belg.,**89**, 117, 1980

41.18 6 – Chloro – 4 – methyl – 1,2,3 – benzothiadiazole
$C_7H_5ClN_2S$
L.Golic, I.Leban, B.Stanovnik, M.Tisler
Acta Crystallogr.,Sect.B,**35**, 3114, 1979

41.19 4 – Phenyl – 1,2,3,5 – dithiazole
$C_7H_5N_2S_2$
A.Vegas–Molina, A.P.Salazar, R.G.Hey, A.J.Banister
Eur.Cryst.Meeting,**5**, 388, 1979

41.20 3 – Oxo – 2(methoxy – carbonyl – methyl) –
tetrahydro – 1,4 – thiazine
$C_7H_{11}NO_3S$
N.Kh.Dzhafarov, A.Kh.Mamedova, M.K.Gusejnova,
A.S.Suleimanov *Zh.Strukt.Khim.*,**20**, 1071, 1979

41.21 4 – Hydroxy – 11 – thia – 2,6,12 –
triazatricyclo(6.3.1.0^{4,12})dodecane – 1(11),8 – diene –
3,6 – dione (at –140°C)
$C_8H_7N_3O_3S$
E.Campaigne, J.C.Huffman, T.P.Selby
J.Heterocycl.Chem.,**16**, 725, 1979
See also R1 : 36

41.C 3,2' – Anhydro – 4 – (β – D – arabinofuranosyl) –
5,6 – dihydro – 2H – 1,2,4 – thiadiazin – 3 – one –
1,1 – dioxide
$C_8H_{12}N_2O_6S$ Main entry is 45.14

41.22　**4 – Dimethylamino – 5 – methyl – 5 – vinyl – thiazolidine – 2 – thione**
$C_8H_{12}N_2S_2$
G.Germain, J.P.Declercq, R.Touillaux, M.van Meerssche, M.Henriet *Acta Crystallogr.,Sect.B*,**35**, 3068, 1979

41.23　**cis – 3,3,6,6, – Tetramethyl – 2,3,6,7 – tetrahydro – 1,4,5 – thia – diazepine – 1,1 – dioxide**
$C_8H_{16}N_2O_2S$
H.Lind, G.Rihs, G.Rist *Tetrahedron Lett.*,**21**, 339, 1980

41.24　**3 – Phenyl – 1,3 – thiazolidine – 2,4 – dione**
$C_9H_7NO_2S$
S.Stankovic, G.D.Andreetti
Acta Crystallogr.,Sect.B,**35**, 3078, 1979

41.25　**bis(2,3 – Dihydro – 3 – hydroxythiazolo(2,3 – b) benzothiazolium) hexachloro – tellurium(iv) dioxane solvate**
$2C_9H_8NOS_2^+$, Cl_6Te^{2-}, $C_4H_8O_2$
K.von Deuten, W.Schnabel, G.Klar
Cryst.Struct.Commun.,**9**, 761, 1980

41.26　**Naphtho(1,8 – cd:4,5 – c'd') – bis(1,2,6 – thiadiazine)**
$C_{10}H_4N_4S_2$
A.Gieren, V.Lamm, R.C.Haddon, M.L.Kaplan
J.Am.Chem.Soc.,**101**, 7277, 1979

41.27　**1 – Oxo – 1,2 – dihydro – 2,3 – diazaphenothiazine**
$C_{10}H_7N_3OS$
G.D.Andreetti, G.Bocelli, P.Sgarabotto
Acta Crystallogr.,Sect.B,**36**, 1839, 1980
See also R1 : 36

41.28　**3 – p – Bromobenzoyl – 1,3 – thiazolidine – 2 – thione**
$C_{10}H_8BrNOS_2$
R.F.Bryan, P.Hartley, S.Peckler, E.Fujita, Y.Nagao, K.Seno
Acta Crystallogr.,Sect.B,**36**, 1709, 1980

41.29　**4 – Oxo – 2 – phenylimino – perhydro – 1,3 – thiazine**
$C_{10}H_{10}N_2OS$
B.Tashkhodzhaev, S.Akhmedova, N.K.Rozhkova
Zh.Strukt.Khim.,**21**, 137, 1980

41.30　**3,9 – Dimethyl – 6 – thia – 1,11 – diazatricyclo(6.2.1.0^{4,11})undecane – 3,8 – diene – 2,10 – dione**
μ – Thia – syn(CH$_2$CH$_3$) – 9,10 – dioxa – bimane
$C_{10}H_{10}N_2O_2S$
I.Goldberg *Cryst.Struct.Commun.*,**9**, 329, 1980
See also R1 : 35

41.31　**3,9 – Dimethyl – 6 – thia – 1,11 – diazatricyclo(6.2.1.0^{4,11})undecane – 3,8 – diene – 2,10 – dione 6,6 – dioxide**
μ – Sulfone – syn(CH$_2$CH$_3$) – 9,10 – dioxa – bimane
$C_{10}H_{10}N_2O_4S$
I.Goldberg *Cryst.Struct.Commun.*,**9**, 329, 1980
See also R1 : 35

41.32　**3,6 – Diethyl – 1,4 – dimethyl – 3,6 – epithio – piperazine – 2,5 – dione**
$C_{10}H_{16}N_2O_2S$
H.Shimanouchi, Y.Sasada, K.Koyano
Acta Crystallogr.,Sect.B,**36**, 475, 1980
See also R1 : 37

41.33　**2,5 – Di – t – butyl – 1,2,5 – thiadiazolidine – 3,4 – dione**
$C_{10}H_{18}N_2O_2S$
M.Scherz, J.Weiss
Acta Crystallogr.,Sect.B,**35**, 3080, 1979

41.34　**1 – Chloro – 10 – methyl – 2,3 – diazaphenothiazine**
$C_{11}H_8ClN_3S$
G.D.Andreetti, G.Bocelli, P.Sgarabotto
Acta Crystallogr.,Sect.B,**36**, 1839, 1980
See also R1 : 36

41.35　**3 – Thioxo – 4 – phenyl – 4 – aza – 2,7 – dithiobicyclo(3.3.0)octane – 7,7 – dioxide**
$C_{11}H_{11}NO_2S_3$
V.I.Kulishov, L.N.Zakharov, Yu.T.Struchkov
Cryst.Struct.Commun.,**8**, 851, 1979
See also R1 : 39

41.36　**2 – (2,6 – Dichlorophenyl)imino – 3 – methyl – perhydro – 1,3 – thiazine**
$C_{11}H_{12}Cl_2N_2S$
G.Argay, A.Kalman, A.Kapor, B.Ribar
Acta Crystallogr.,Sect.B,**36**, 363, 1980

41.37　**N – Methyl – 4H – 6,7 – dimethoxy – benzo – 1,2 – thiazin – 3 – one – 1,1 – dioxide**
$C_{11}H_{13}NO_5S$
N.G.Panagiotopoulos, S.E.Filippakis
Cryst.Struct.Commun.,**9**, 321, 1980
See also R1 : 35

41.38　**Acenaphthyleno(5,6 – cd) – 1,2,6 – thiadiazine**
$C_{12}H_6N_2S$
A.Gieren, V.Lamm, R.C.Haddon, M.L.Kaplan, M.J.Perkins, P.Flowerday *Eur.Cryst.Meeting*,**6**, 52, 1980

41.C　**Phenothiazine – pyromellitic dianhydride complex**
$C_{12}H_9NS$, $C_{10}H_2O_6$ Main entry is 60.17

41.39　**5,6 – Dihydro – 3 – thiobenzoyl – methylene – 3H – thiazolo(2,3 – c)(1,2,4)thiadiazole**
$C_{12}H_{10}N_2S_3$
C.Glidewell, H.D.Holden, D.C.Liles
Acta Crystallogr.,Sect.B,**36**, 1244, 1980

41.40　**3 – Benzyl – 3,7 – dihydro – 5 – methylthio – 1,2,3 – triazolo(4,5 – d) – 1,3 – thiazine**
$C_{12}H_{12}N_4S_2$
A.Albert, A.Dunand
Angew.Chem.,Int.Ed.Engl.,**19**, 310, 1980
See also R1 : 36

41.41　**2 – (2,6 – Dimethylphenyl) – imino – 3 – methyl – thiazolidine**
$C_{12}H_{16}N_2S$
S.Stankovic, G.Argay, A.Kalman
Eur.Cryst.Meeting,**6**, 289, 1980

41.42　**2,3 – Dihydro – 2 – acetyl – 3,3 – dimethyl – 4 – ((1E) – 3 – methyl – 1,3 – pentadienyl) – 1,2,5 – thiadiazole – 1,1 – dioxide**
$C_{12}H_{18}N_2O_3S$
R.J.Baker, S.Chiu, C.Klein, J.W.Timberlake, L.M.Trefonas, R.Majeste *J.Org.Chem.*,**45**, 482, 1980

41.C　**Thiamine chloride hydrochloride hemihydrate (phase II)**
$C_{12}H_{18}N_4OS^{2+}$. $2Cl^-$, $0.5H_2O$ Main entry is 44.25

41.43　3 – (p – Bromophenyl) – thiazolo(3,2 – a)pyridinium tetrafluoroborate (at 123°K)
$C_{13}H_9BrNS^+, BF_4^-$
K.Sasvari, L.Parkanyi, G.Hajos, H.Hess, W.Schwarz
*Acta Crystallogr.,Sect.B,***36**, 1229, 1980

41.44　4 – Dimethylamino – 5 – ethyl – 5 – phenylthiazoline – 2 – thione
$C_{13}H_{16}N_2S_2$
G.Germain, J.P.Declercq, R.Touillaux, M.van Meerssche, M.Henriet *Acta Crystallogr.,Sect.B,***35**, 3068, 1979

41.45　2 – Methyl – 3 – p – anisyl – 5 – (ethylamino) – carbamoyl – 1,2,4 – thiazoline
$C_{13}H_{16}N_4O_2S$
G.S.D.King, J.Aerts *Eur.Cryst.Meeting,***6**, 48, 1980

41.46　2 – (2,6 – Dimethylphenyl)imino – 3 – methyl – perhydro – 1,3 – thiazine
$C_{13}H_{18}N_2S$
G.Argay, A.Kalman, A.Kapor, B.Ribar
*Acta Crystallogr.,Sect.B,***36**, 363, 1980

41.47　2 – (N – (2,6 – Dimethylphenyl) – N – methyl – amino) – 4,5 – dihydro – 6H – 1,3 – thiazine
$C_{13}H_{18}N_2S$
G.Argay, A.Kalman, A.Kapor, B.Ribar
*Acta Crystallogr.,Sect.B,***36**, 363, 1980

41.48　(–) – 14 – Hydroxy – 15 – hydroxymethyl – 2,8 – dithia(9)(2,5)pyridinophane (absolute configuration)
$C_{13}H_{19}NO_2S_2$
T.Sakurai, H.Kuzuhara, S.Emoto
*Acta Crystallogr.,Sect.B,***35**, 2984, 1979
See also R1 : 37

41.49　4 – Phenyl – 3 – phenylamino – 1,2,4 – thiadiazolin – 5 – one
$C_{14}H_{11}N_3OS$
A.F.Cuthbertson, C.Glidewell, H.D.Holden, D.C.Liles
*J.Chem.Res.,*316,3714, 1979

41.50　7 – Oxo – 3 – methylthio – 2 – thia – 1 – azabicyclo(3.2.0)hept – 3 – ene – 4 – (p – nitro – benzyl – carboxylate)
$C_{14}H_{12}N_2O_5S_2$
S.Oida, A.Yoshida, T.Hayashi, E.Nakayama, S.Sato, E.Ohki
*Tetrahedron Lett.,*21, 619, 1980

41.51　p – Diethylaminobenzylidene – rhodanine
$C_{14}H_{16}N_2OS_2$
E.Subramanian *Cryst.Struct.Commun.,***9**, 693, 1980

41.52　3 – (3 – Amino – benzo(b)thiophen – 2 – yl) – 1,2 – benzisothiazole
$C_{15}H_{10}N_2S_2$
F.Mossini, M.R.Mingiardi, E.Gaetani, M.Nardelli, G.Pelizzi
*J.Chem.Soc.,Perkin Trans.2,*1665, 1979
See also R1 : 39

41.53　N – Acetyl – dendrodoine
$C_{15}H_{14}N_4O_2S$
S.Heitz, M.Durgeat, M.Guyot, C.Brassy, B.Bachet
*Tetrahedron Lett.,*21, 1457, 1980
See also R1 : 35

41.54　2 – Methyl – 3 – phenyl – 5 – phenylazo – 1,3,4 – thiadiazoline
$C_{15}H_{14}N_4S$
P.A.McCallum, H.M.N.H.Irving, A.T.Hutton, L.R.Nassimbeni
*Acta Crystallogr.,Sect.B,***36**, 1626, 1980

41.55　5,6 – Dihydro – 4 – phenyl – 2 – phenylazo – 4H – 1,3,4 – thiadiazine
$C_{15}H_{14}N_4S$
P.A.McCallum, H.M.N.H.Irving, A.T.Hutton, L.R.Nassimbeni
*Acta Crystallogr.,Sect.B,***36**, 1626, 1980

41.56　9,9a – Dihydro – 1,2,9,9 – tetramethyl – 2,9a – epi – tetrathio – 3,10 – diketo – piperazino(1,2 – a)indole (absolute configuration)
$C_{15}H_{16}N_2O_2S_4$
G.Beurskens, J.H.Noordik, P.T.Beurskens
*Cryst.Struct.Commun.,***9**, 23, 1980

41.57　6 – Chloro – 2(2 – acetoxy – 1 – ethylidene) – 3 – (2 – thienyl) – 4H – 1,4 – benzothiazine
$C_{16}H_{12}ClNO_2S_2$
J.B.Press, N.H.Eudy, F.M.Lovell, N.A.Perkinson
*Tetrahedron Lett.,*21, 1705, 1980

41.58　(Z) – 2,3 – Dihydro – N – methyl – 2 – (4 – nitrobenzoyl – methylene) – benzothiazole
$C_{16}H_{12}N_2O_3S$
J.Z.Gougoutas *Cryst.Struct.Commun.,***9**, 529, 1980

41.59　N – (p – Bromophenyl) – 4,5 – dihydro – 7,8 – dimethoxy – benzothiazepin – 3 – one – 1,1 – dioxide
$C_{17}H_{16}BrNO_5S$
N.C.Panagiotopoulos, S.E.Filippakis, P.Catsoulakos
*Cryst.Struct.Commun.,***9**, 313, 1980

41.60　(3 – Phenylaminocarbonyl) – (4 – phenylmethyl) – 4 – aza – 1,2 – dithian – 5 – one
$C_{17}H_{17}N_2O_2S_2$
I.Cynkier, F.Frolov *Eur.Cryst.Meeting,***6**, 43, 1980

41.61　5 – Benzoylimino – 2,2 – dimethyl – 4 – phenyl – 1,3,4 – thiadiazolidine
$C_{17}H_{17}N_3OS$
Y.Fukutani, K.Tsukihara, Y.Okuda, K.Fukuyama, Y.Katsube, I.Yamamoto, H.Gotoh
*Bull.Chem.Soc.Jpn.,***52**, 2223, 1979

41.62　2 – Methyl – 3 – p – anisyl – 5 – p – tosylimino – 1,2,4 – thioazoline
$C_{17}H_{17}N_3O_3S_2$
G.S.D.King, J.Aerts *Eur.Cryst.Meeting,***6**, 48, 1980

41.63　Methyl 5,5 – dimethyl – 2 – (2 – phenoxymethyl – 5 – oxo – 1,3 – oxazolin – 4 – ylidene) – 1,3 – thiazolidine – 4 – carboxylate
$C_{17}H_{18}N_2O_5S$
P.A.C.Gane, M.O.Boles
*Acta Crystallogr.,Sect.B,***35**, 2664, 1979
See also R1 : 40

41.64　2,3,4,5 – Tetrahydro – 6,6 – diphenyl – imidazo(2,1 – b)thiazin – 7(6H) – one
$C_{18}H_{16}N_2OS$
K.Kiec–Kononowicz, A.Zejc, M.Mikolajczyk, A.Zatorski, J.Karolak–Wojciechowska, M.W.Wieczorek
*Tetrahedron,***36**, 1079, 1980
See also R1 : 35

41.65　2 – ((3 – Ethyl – 1,3 – benzothiazol – 2 – ylio) – 2 – vinyl) – 6 – methoxy – 4 – nitrophenolate
$C_{18}H_{16}N_2O_4S$
E.Miler–Srenger, M.le Baccon, R.Guglielmetti
*Eur.Cryst.Meeting,***5**, 196, 1979

41.66 N – Benzyloxycarbonyl – N′ – (2,6 – dichlorophenyl) – N – (5 – methyl – 1,3 – thiazolin – 2 – yl) – hydrazine
$C_{18}H_{17}Cl_2N_3O_2S$
S.Stankovic, B.Ribar, A.Kalman, G.Argay
Acta Crystallogr.,Sect.B,**36**, 1235, 1980

41.67 2,7 – Dihydro – 3,6 – di – p – tolyl – 1,4,5 – thiadizepine – 1,1 – dioxide
$C_{18}H_{18}N_2O_2S$
E.Cuthbertson, R.K.Mackenzie, D.D.MacNicol, H.H.Mills, F.B.Wilson, K.M.Ghouse, D.G.Williamson
Spectrochim.Acta,Part A,**36**, 333, 1980

41.68 2,4,6 – Tri – t – butyl – 7,8,9 – dithia – azabicyclo(4.3.0)nona – 1(9),2,4 – trien – 7 – one
$C_{18}H_{29}NOS_2$
F.Iwasaki *Acta Crystallogr.,Sect.B*,**36**, 1466, 1980

41.69 2,4,6 – Tri – t – butyl – 7,8,9 – dithia – azabicyclo(4.3.0)nona – 1(9),2,4 – triene
$C_{18}H_{29}NS_2$
F.Iwasaki *Acta Crystallogr.,Sect.B*,**36**, 1466, 1980

41.C DL – 2 – (α – Hydroxybenzyl) – oxythiamine chloride hydrochloride trihydrate
$C_{19}H_{23}N_3O_3S^{2+}$, 2Cl⁻, 3H_2O Main entry is 44.35

41.70 3 – (2 – Methoxy – 10 – phenothiazinyl) – N,N,2 – trimethyl – propanamine (absolute configuration)
$C_{19}H_{24}N_2OS$
M.Sato, K.Miki, N.Tanaka, N.Kasai, T.Ishimaru, T.Munakata
Acta Crystallogr.,Sect.B,**36**, 2176, 1980

41.71 2,6 – Diphenyl – benzo(1,2 – d:4,5 – d′)bis(thiazole)
$C_{20}H_{12}N_2S_2$
A.V.Fratini
Rep.Air Force Sci.Res.AFOSR–77–3267,1, 1978

41.72 2,3 – Dihydro – 2 – methoxy – 2 – methyl – 3 – (benzo – 1,3 – thiazole – 2 – thiomethylene) – 1,4 – naphthoquinone
$C_{20}H_{15}NO_3S_2$
T.J.King, R.H.Thomson, R.D.Worthington
J.Chem.Soc.,Chem.Commun.,777, 1980

41.73 3 – Benzoylsulfinyl – 5 – (N – n – butylbenzamido) – 1,2,4 – thiadiazole
$C_{20}H_{19}N_3O_2S_2$
G.L'abbe, M.Komatsu, C.Martens, S.Toppet, J.P.Declercq, G.Germain, M.van Meerssche
Bull.Soc.Chim.Belg.,**88**, 245, 1979

41.74 2 – Chloro – 10 – (3 – (4 – methyl – 1 – piperazinyl) propyl) – phenothiazine methanesulfonic acid
$C_{20}H_{24}ClN_3S$, 2CH_4O_3S
J.J.H.McDowell *Acta Crystallogr.,Sect.B*,**35**, 2433, 1979
See also R1 : 33

41.75 8 – Dimethylamino – 5α – t – butyl – 5β – cyano – 2 – (cyano(t – butyl)methylene) – 1β – methyl – 3 – oxa – 9 – thia – 7 – azabicyclo(4.3.0)non – 7 – en – 4 – one
$C_{20}H_{28}N_4O_2S$
A.Dondoni, A.Medici, C.Venturoli, L.Forlani, V.Bertolasi
J.Org.Chem.,**45**, 621. 1980
See also R1 : 38

41.76 8 – Dimethylamino – 5β – t – butyl – 5α – cyano – 2 – (cyano(t – butyl)methylene) – 1β – methyl – 3 – oxa – 9 – thia – 7 – azabicyclo(4.3.0)non – 7 – en – 4 – one
$C_{20}H_{28}N_4O_2S$
A.Dondoni, A.Medici, C.Venturoli, L.Forlani, V.Bertolasi
J.Org.Chem.,**45**, 621, 1980
See also R1 : 38

41.77 2 – Benzoylimino – 3,5 – diphenyl – 2,3 – dihydro – 1,3,4 – thiadiazole
$C_{21}H_{15}N_3OS$
Y.Fukutani, K.Tsukihara, Y.Okuda, K.Fukuyama, Y.Katsube, I.Yamamoto, H.Gotoh
Bull.Chem.Soc.Jpn.,**52**, 2223, 1979

41.78 4 – Nitrobenzyl – (3SR,5RS,Z) – 2 – (2 – phenylthioethylidene) – penam – 3 – carboxylate
$C_{21}H_{18}N_2O_5S_2$
P.C.Cherry, D.N.Evans, C.E.Newell, N.S.Watson, P.Murray–Rust, J.Murray–Rust
Tetrahedron Lett.,**21**, 561, 1980

41.79 Trifluoroperazine hydrochloride
10 – (3 – (4 – Methyl – 1 – piperazinyl)propyl) – 2 – trifluoro – methylphenothiazine dihydrochloride
$C_{21}H_{26}F_3N_3S^{2+}$, 2Cl⁻
J.J.H.McDowell *Acta Crystallogr.,Sect.B*,**36**, 2178, 1980

41.80 2S – (2α,5α) – 3,3 – Dimethyl – 6 – (p – tolyl – thio – imino) – 7 – oxo – 4 – thia – 1 – azabicyclo(3.2.0)heptane – 2 – carboxylic – acid – p – nitrobenzyl ester
$C_{22}H_{21}N_3O_5S_2$
E.M.Gordon, H.W.Chang, C.M.Cimarusti, B.Toeplitz, J.Z.Gougoutas *J.Am.Chem.Soc.*,**102**, 1690, 1980

41.C 6α – Benzyl – 6β – isocyano – penicillanate
$C_{23}H_{22}N_2O_3S$ Main entry is 50.16

41.C 17 – O – Methyl – latrunculin A
$C_{23}H_{33}NO_5S$ Main entry is 59.35

41.81 2,9 – bis(p – Methoxyphenyl) – 5,12 – dimethyl – 5,12 – diaza – 1,3,8,10 – tetrathiacyclotetradecane – 6,13 – dione (at –125°C)
$C_{24}H_{30}N_2O_4S_4$
I.Cynkier, S.Gronowitz, H.Hope, Z.Lidert
J.Org.Chem.,**44**, 4699, 1979

41.C 14 – (p – Chlorophenyl) – 8 – (p – chlorophenylsulfonylamino) – 13 – oxa – 14 – thia – 1,15 – diazapentacyclo(10.4.2.0²,⁷.0⁸,¹⁸.0¹²,¹⁸) octadeca – 2,4,6,14 – tetraene S – oxide
$C_{26}H_{23}Cl_2N_3O_4S_2$ Main entry is 42.12

41.82 4 – ((E) – 1,2 – Diphenylvinyl) – 2,6 – diphenyl – 1,3,4,5 – thiatriazine – 1,1 – dioxide
$C_{28}H_{21}N_3O_2S$
G.P.Stahly, H.L.Ammon, B.B.Jarvis
Acta Crystallogr.,Sect.B,**36**, 2159, 1980

41.83 8 – Benzoyl – 15 – hydroxy – 15 – phenyl – 6aH,14aH,15H – benzothiazolo(2,3 – a) benzothiazino(4,3 – c)piperazine methanol solvate
$C_{30}H_{22}N_2O_2S_2$, CH_4O
M.Jansen *Cryst.Struct.Commun.*,**9**, 499, 1980
See also R1 : 36

41.84 2,2' – Diphenyl – 4,4' – di(2 – phenyl – 2 – propenyl) – 4,4' – bis(5(4H) – thiazolone)
$C_{36}H_{28}N_2O_2S_2$
C.Foces–Foces, F.H.Cano, S.Garcia–Blanco
J.Cryst.Mol.Struct.,8, 309, 1978

MISCELLANEOUS HETEROCYCLES

42.1   cis – 5 – Methyl – trans – 5 – nitro – 1,3,2 – dioxathian – 2 – oxide (at –160°C)
$C_4H_7NO_5S$
G.H.Petit, A.T.H.Lenstra, H.J.Geise, P.Swepston
Cryst.Struct.Commun.,9, 187, 1980

42.2   cis – 5 – t – Butyl – 1,3,2 – dioxathian – 2 – oxide (at 123°K)
$C_7H_{14}O_3S$
P.van Nuffel, G.H.Petit, H.J.Geise, A.T.H.Lenstra
Acta Crystallogr.,Sect.B,36, 1220, 1980

42.3   3 – t – Butyl – 4 – methyl – 2 – oxo – 1 – oxa – 2 – thia – 3 – azacyclohexane
$C_8H_{17}NO_2S$
L.Cazaux, P.Tisnes, J.Jaud *J.Chem.Res.*,10, 0156, 1980

42.C   Octamethyl – trithio – cyclotetra($\lambda^3,\lambda^5,\lambda^5,\lambda^5$ – phosphazane)
$C_8H_{24}N_4P_4S_3$ Main entry is 64.30

42.4   7 – Methoxycarbonyl – 1,3 – dioxo – 2,1 – benzoxathiole
1,3 – Dioxo – 3H – 2,1λ – benzoxathiole – 7 – carboxylic acid methyl ester
$C_9H_6O_5S$
W.Walter, B.Krische, G.Adiwidjaja
Justus Liebigs Ann.Chem.,14, 1980

42.C   5 – Phenyl – 1,3 – thiaselenole – 2 – thione 7,7,8,8 – tetracyanoquinodimethane complex
$C_9H_6S_2Se$, $C_{12}H_4N_4$ Main entry is 60.8

42.5   Naphtho(1,8 – cd:4,5 – c'd') – bis(1,2,6)selenadiazine
$C_{10}H_4N_4Se_2$
A.Gieren, V.Lamm, R.C.Haddon, M.L.Kaplan
J.Am.Chem.Soc.,102, 5070, 1980

42.6   2 – Selena – 1 – dioxothia – 3 – formyl – 5,8 – dimethyl – 1,2 – dihydronaphthalene
$C_{11}H_{10}O_3SSe$
A.Michel, F.Durant *Cryst.Struct.Commun.*,9, 219, 1980

42.C   2,5' – Anhydro – 1 – (2',3' – O – isopropylidene – β – D – ribofuranosyl) – 2 – thio – uracil
$C_{12}H_{14}N_2O_4S$ Main entry is 47.28

42.7   4,7,10,13,16 – Pentaoxa – 1 – thiacyclo – octadecane (at 101°K)
$C_{12}H_{24}O_5S$
N.K.Dalley, S.L.Larson, M.L.Campbell
Am.Cryst.Assoc.,Ser.2,8, 46, 1980

42.8   4,7,10,13,16 – Pentaoxa – 1 – thiacyclo – octadecane silver nitrate
$C_{12}H_{24}O_5S$, Ag^+, NO_3^-
N.K.Dalley, S.L.Larson, M.L.Campbell
Am.Cryst.Assoc.,Ser.2,8, 46, 1980

42.9 4,7,10,13,16 – Pentaoxa – 1 – thiacyclo – octadecane
potassium thiocyanate
$C_{12}H_{24}O_5S$, CNS⁻, K⁺
N.K.Dalley, S.L.Larson, M.L.Campbell
*Am.Cryst.Assoc.,Ser.2,***8**, 46, 1980

42.10 4,7,10,13,16 – Pentaoxa – 1 – thiacyclo – octadecane
sodium thiocyanate
$C_{12}H_{24}O_5S$, CNS⁻, Na⁺
N.K.Dalley, S.L.Larson, M.L.Campbell
*Am.Cryst.Assoc.,Ser.2,***8**, 46, 1980

42.11 4,7,10,13,16 – Pentaoxa – 1 – thiacyclo – octadecane
rubidium thiocyanate
$C_{12}H_{24}O_5S$, CNS⁻, Rb⁺
N.K.Dalley, S.L.Larson, M.L.Campbell
*Am.Cryst.Assoc.,Ser.2,***8**, 46, 1980

42.12 14 – (p – Chlorophenyl) – 8 – (p –
chlorophenylsulfonylamino) – 13 – oxa – 14 –
thia – 1,15 – diazapentacyclo($10.4.2.0^{2,7}.0^{8,16}.0^{12,16}$)
octadeca – 2,4,6,14 – tetraene S – oxide
$C_{26}H_{23}Cl_2N_3O_4S_2$
K.Prout, M.Sims, D.Watkin, C.Couldwell, M.Vandrevala,
A.S.Bailey *Acta Crystallogr.,Sect.B,***36**, 1846, 1980
See also R1 : 41,40,37

42.13 3,3,4,4 – Tetraphenyl – 1,2 – oxathiolan – 5 – one –
2 – oxide (at 238°K)
$C_{27}H_{20}O_3S$
H.Kohn, P.Charumilind, S.H.Simonsen
*J.Am.Chem.Soc.,***101**, 5431, 1979

BARBITURATES

43.1 Barbital 1 – methylimidazole
$C_8H_{12}N_2O_3$, $C_4H_6N_2$
A.Wang, B.M.Craven *J.Pharm.Sci.,***68**, 361, 1979
See also R2 : 32

43.2 5,5 – Di – allyl – 1 – (p – bromophenyl) – barbituric
acid
$C_{16}H_{15}BrN_2O_3$
D.Pyzalska, R.Pyzalski, T.Borowiak
*Acta Crystallogr.,Sect.B,***36**, 1672, 1980

PYRIMIDINES AND PURINES

44.1 **Cytosine monohydrate (perdeuterated, neutron study, at 82°K)**
$C_4H_2D_3N_3O$, D_2O
H.P.Weber, B.M.Craven, R.K.McMullan
Acta Crystallogr.,Sect.B,**36**, 645, 1980

44.2 **2 – Chloropyrimidine (at 107°K)**
$C_4H_3ClN_2$
S.Furberg, J.Grogaard, B.Smedsrud
Acta Chem.Scand.Ser.B,**33**, 715, 1979

44.3 **Pyrimidine (at 107°K)**
$C_4H_4N_2$
S.Furberg, J.Grogaard, B.Smedsrud
Acta Chem.Scand.Ser.B,**33**, 715, 1979

44.4 **2 – Aminopyrimidine (at 107°K)**
$C_4H_5N_3$
S.Furberg, J.Grogaard, B.Smedsrud
Acta Chem.Scand.Ser.B,**33**, 715, 1979

44.5 **Cytosine calcium chloride hydrate**
$C_4H_5N_3O$, Ca^{2+}, $2Cl^-$, H_2O
K.Ogawa, M.Kumihashi, K.–I.Tomita, S.Shirotake
Acta Crystallogr.,Sect.B,**36**, 1793, 1980

44.6 **Dicytosinium tetrachloro – copper**
$2C_4H_6N_3O^+$, Cl_4Cu^{2-}
P.E.Bourne, M.R.Taylor *Eur.Cryst.Meeting*,**6**, 322, 1980

44.7 **Dicytosinium tetrachloro – zinc**
$2C_4H_6N_3O^+$, Cl_4Zn^{2-}
P.E.Bourne, M.R.Taylor *Eur.Cryst.Meeting*,**6**, 322, 1980

44.8 **Potassium thyminate trihydrate**
$C_5H_5N_2O_2^-$, K^+, $3H_2O$
C.J.L.Lock, P.Pilon, B.Lippert
Acta Crystallogr.,Sect.B,**35**, 2533, 1979

44.9 **5 – Methylpyrimidine (at 107°K)**
$C_5H_6N_2$
S.Furberg, J.Grogaard, B.Smedsrud
Acta Chem.Scand.Ser.B,**33**, 715, 1979

44.10 **bis(Adeninium nitrate) monohydrate**
$2C_5H_6N_5^+$, $2NO_3^-$, H_2O
B.E.Hingerty, J.R.Einstein, C.H.Wei
Am.Cryst.Assoc.,Ser.2,**7**, 19, 1980

44.11 **bis(1 – Methylcytosine) hydroiodide monohydrate**
$C_5H_7N_3O$, $C_5H_8N_3O^+$, I^-, H_2O
T.J.Kistenmacher, M.Rossi, J.P.Caradonna, L.G.Marzilli
Adv.Mol.Relaxation Interaction Processes,
15, 119, 1979

44.12 **5 – Methylcytosine hemihydrate**
$C_5H_7N_3O$, $0.5H_2O$
A.Takenaka, M.Kato, Y.Sasada
Bull.Chem.Soc.Jpn.,**53**, 383, 1980

44.13 **1 – Methylcytosine perchlorate**
$C_5H_8N_3O^+$, ClO_4^-
M.Rossi, J.P.Caradonna, L.G.Marzilli, T.J.Kistenmacher
Adv.Mol.Relaxation Interaction Processes,
15, 103, 1979

44.14 **(1 – Methylthymine) – silver(i)**
$C_6H_7N_2O_2^-$, Ag^+
F.Guay, A.L.Beauchamp *J.Am.Chem.Soc.*,**101**, 6260, 1979

44.15 **9 – Methyl – adenine (at 126°K,neutron study)**
$C_6H_7N_5$
R.K.McMullan, P.Benci, B.M.Craven
Acta Crystallogr.,Sect.B,**36**, 1424, 1980

44.16 **N(6) – Methyl – adenine hydrochloride**
$C_6H_8N_5^+$, Cl^-
H.Sternglanz, C.E.Bugg *J.Cryst.Mol.Struct.*,**8**, 263, 1978

44.17 **Tryptamine adenin – 9 – yl – acetic acid hemihydrate**
$C_7H_6N_5O_2^-$, $C_{10}H_{13}N_2^+$, $0.5H_2O$
T.Ishida, M.Inoue, S.Senda, K.–I.Tomita
Bull.Chem.Soc.Jpn.,**52**, 2953, 1979
See also R2 : 35

44.C **Tyramine 1 – thyminyl – (acetic acid) complex hydrate**
$C_7H_7N_2O_4^-$, $C_8H_{12}NO^+$, H_2O Main entry is 60.6

44.18 **Theobromine tri – iodide iodine**
3,7 – Dimethylxanthine tri – iodide iodine
$C_7H_9N_4O_2^+$, I_3^-, $0.5I_2$
F.H.Herbstein, M.Kapon
Philos.Trans.R.Soc.London,Ser.A,**291**, 199, 1979

44.19 **N(6),N(9) – Dimethyl – adenine**
$C_7H_9N_5$
H.Sternglanz, C.E.Bugg *J.Cryst.Mol.Struct.*,**8**, 263, 1978

44.C **(9 – Methylguanine) – methyl – mercury(ii) nitrate**
$C_7H_{10}HgN_5O^+$, NO_3^- Main entry is 83.30

44.20 **bis(2 – Pyrimidyl) – disulfide**
$C_8H_6N_4S_2$
C.J.Simmons, M.Lundeen, K.Seff
Inorg.Chem.,**18**, 3444, 1979

44.21 **2 – (Methyl – (β – chloroethyl) – amino) – 4,6 – dichloro – 5 – cyanopyrimidine**
$C_8H_7Cl_3N_4$
B.Tinant, G.Germain, J.P.Declercq, M.van Meerssche
Bull.Soc.Chim.Belg.,**88**, 219, 1979

44.C **6 – Azauridine – 5′ – phosphate trihydrate**
$C_8H_{12}N_3O_9P$, $3H_2O$ Main entry is 47.1

44.C **(μ – Adeninato) – 3,7,9 – tris(methyl – mercury(ii)) diperchlorate**
$C_8H_{13}Hg_3N_5^{2+}$, $2ClO_4^-$ Main entry is 83.35

44.C **2,2′ – Anhydro – 1 – β – D – arabinofuranosyl – 2 – thio – uracil**
$C_9H_{10}N_2O_4S$ Main entry is 47.2

44.C **Disodium deoxyuridine – 5′ – phosphate pentahydrate**
$C_9H_{11}N_2O_8P^{2-}$, $2Na^+$, $5H_2O$ Main entry is 47.3

44.C **Disodium uridine – 5′ – phosphate heptahydrate**
$C_9H_{11}N_2O_9P^{2-}$, $2Na^+$, $7H_2O$ Main entry is 47.4

44.C **5 – Iodo – 5′ – amino – 2′,5′ – dideoxyuridine**
$C_9H_{12}IN_3O_4$ Main entry is 47.5

44.C 5 – Iodocytidine
$C_9H_{12}IN_3O_5$ Main entry is 47.6

44.C 1 – β – D – Lyxofuranosyluracil
$C_9H_{12}N_2O_6$ Main entry is 47.7

44.C 3′,5′ – Cyclic – cytidine – monophosphate trihydrate
$C_9H_{12}N_3O_7P$, $3H_2O$ Main entry is 47.8

44.C Barium cytidine – 5′ – phosphate hydrate
$C_9H_{12}N_3O_8P^{2-}$, Ba^{2+}, $8.5H_2O$ Main entry is 47.9

44.C 5 – Nitro – 2′ – deoxyuridine – 5′ – monophosphate monohydrate
$C_9H_{12}N_3O_{10}P$, H_2O Main entry is 47.10

44.C Potassium dihydrouridine – 3′ – monophosphate hemihydrate
$C_9H_{14}N_2O_9P^-$, K^+, $0.5H_2O$ Main entry is 47.11

44.C 7,2′ – Anhydro – β – D – arabinosyl – orotidine
$C_{10}H_{10}N_2O_7$ Main entry is 47.12

44.C Sodium adenosine – (3′ – 5′) – cyclic – monophosphate tetrahydrate (monoclinic form)
$C_{10}H_{11}N_5O_6P^-$, Na^+, $4H_2O$ Main entry is 47.13

44.C Inosine (orthorhombic form)
$C_{10}H_{12}N_4O_5$ Main entry is 47.14

44.C 3′ – Deoxy – adenosine
Cordycepin
$C_{10}H_{13}N_5O_3$ Main entry is 47.15

44.C 9α – D – Arabinofuranosyl – adenine
$C_{10}H_{13}N_5O_4$ Main entry is 47.16

44.C 8 – Thioxo – adenosine monohydrate
$C_{10}H_{13}N_5O_4S$, H_2O Main entry is 47.17

44.C α – Thymidine
$C_{10}H_{14}N_2O_5$ Main entry is 47.18

44.C 5 – Hydroxymethyl – 2′ – deoxyuridine
$C_{10}H_{14}N_2O_6$ Main entry is 47.19

44.C 4 – Amino – 1 – (4 – amino – 2 – oxo – 1(2H) – pyrimidinyl) – 1,4 – dideoxy – β – D – glucopyranuronic acid monohydrate
$C_{10}H_{14}N_4O_6$, H_2O Main entry is 47.20

44.C bis(Adenosine) proflavine sesquisulfate hydrate
$2C_{10}H_{14}N_5O_4{}^+$, $C_{13}H_{12}N_3{}^+$, 1.50_4S^{2-}, $6.5H_2O$
Main entry is 47.21

44.C Guanosine – 5′ – monophosphate trihydrate
$C_{10}H_{14}N_5O_8P$, $3H_2O$ Main entry is 47.22

44.C Potassium adenosine – 5′ – diphosphate dihydrate
$C_{10}H_{14}N_5O_{10}P_2{}^-$, K^+, $2H_2O$ Main entry is 47.23

44.C Adenosine – 5′ – diphosphoric acid trihydrate
$C_{10}H_{15}N_5O_{10}P_2$, $3H_2O$ Main entry is 47.24

44.22 1 – Benzylcytosine nitrate
$C_{11}H_{12}N_3O^+$, NO_3^-
M.Rossi, J.P.Caradonna. L.G.Marzilli, T.J.Kistenmacher
Adv.Mol.Relaxation Interaction Processes,
15, 103. 1979

44.23 trans – 6 – Chloro – 9 – (2 – ethoxy – 1,3 – dioxan – 5 – yl) – purine
$C_{11}H_{13}ClN_4O_3$
A.F.Mishnev, Ya.Ya.Bleidelis, E.E.Liepin'sh, N.P.Ramzaeva, I.N.Goncharova
*Khim. Geterotsikl.Soedin.,Latv.SSSR,*976, 1979
See also R1 : 38

44.C 1β – D – Ribofuranosyl – pyridin – 4 – one – 3 – carboxamide
$C_{11}H_{14}N_2O_6$ Main entry is 47.25

44.C β – 5 – Acetyl – 2′ – deoxyuridine
$C_{11}H_{14}N_2O_6$ Main entry is 47.26

44.24 trans – 9 – (2 – Ethoxy – 1,3 – dioxan – 5 – yl) – adenine
$C_{11}H_{15}N_5O_3$
A.F.Mishnev, Ya.Ya.Bleidelis, I.N.Goncharova, N.P.Ramzaeva
*Latv.PSR Zinat.Akad.Vestis,Kim.Ser.,*736, 1979
See also R1 : 38

44.C 2 – Amino – 8 – methyl – 1 – adenosine – 5′ – monophosphate dihydrate
$C_{11}H_{17}N_6O_7P$, $2H_2O$ Main entry is 47.27

44.C 2,5′ – Anhydro – 1 – (2′,3′ – O – isopropylidene – β – D – ribofuranosyl) – 2 – thio – uracil
$C_{12}H_{14}N_2O_4S$ Main entry is 47.28

44.C 5 – Propynyloxy – 2′ – deoxyuridine
$C_{12}H_{14}N_2O_6$ Main entry is 47.29

44.25 Thiamine chloride hydrochloride hemihydrate (phase II)
$C_{12}H_{18}N_4OS^{2+}$, $2Cl^-$, $0.5H_2O$
A.Watanabe, S.Tasaki, Y.Wada, H.Nakamachi
*Chem.Pharm.Bull.,*27, 2751, 1979
See also R1 : 41

44.26 Benzyl 6 – aminopurine – 9 – carboxylic acid
$C_{13}H_{11}N_5O_2$
T.Srikrishnan, R.Parthasarathy, S.P.Dutta, G.B.Chheda
*Acta Crystallogr.,Sect.B,*35, 2736, 1979

44.27 cis – 5 – Hydroxy – 6 – phenylthio – 1,3 – dimethyl – 5,6 – dihydrothymine
$C_{13}H_{16}N_2O_3S$
E.P.Burrows, H.–S.Ryang, S.Y.Wang, J.L.Flippen–Anderson
*J.Org.Chem.,*44, 3736, 1979

44.28 cis – 5 – Hydroxy – 6 – (o – hydroxyphenyl) – 1,3 – dimethyl – 5,6 – dihydrothymine
$C_{13}H_{16}N_2O_4$
E.P.Burrows, H.–S.Ryang, S.Y.Wang, J.L.Flippen–Anderson
*J.Org.Chem.,*44, 3736, 1979

44.29 cis – 5 – Hydroxy – 6 – (p – hydroxyphenyl) – 1,3 – dimethyl – 5,6 – dihydrothymine
$C_{13}H_{16}N_2O_4$
E.P.Burrows, H.–S.Ryang, S.Y.Wang, J.L.Flippen–Anderson
*J.Org.Chem.,*44, 3736, 1979

44.C S – 8 – Aza – adenosyl – L – homocysteine dihydrate
$C_{13}H_{19}N_7O_5S$, $2H_2O$ Main entry is 48.40

44.30 3 – Sulfanilamido – 5 – methylisoxazole 2,4 –
diamino – 5 – (3,4,5 – trimethoxybenzyl) –
pyrimidine
$C_{14}H_{18}N_4O_3$, $C_{10}H_{11}N_3O_3S$
G.Giuseppetti, C.Tadini, G.P.Bettinetti, F.Giordano,
A.la Manna *Farmaco,Ed.Sci.*,**35**, 138, 1980
See also R2 : 40

44.C 9β – D – Arabinofuranosyl – 8 – morpholino –
adenine dihydrate
$C_{14}H_{20}N_6O_5$, $2H_2O$ Main entry is 47.30

44.C S – Adenosyl – L – homocysteine hydrate
$C_{14}H_{20}N_6O_5S$, $2.5H_2O$ Main entry is 48.44

44.C 9β – D – Arabinofuranosyl – 8 – n – butyl –
amino – adenine
$C_{14}H_{22}N_6O_4$ Main entry is 47.31

44.31 2,4 – Diamino – 5 – (1 – naphthyl) – 6 –
methylpyrimidine methanol solvate
$C_{15}H_{14}N_4$, CH_4O
V.Cody, S.F.Zakrzewski *Am.Cryst.Assoc.,Ser.2*,**8**, 35, 1980

44.32 6 – Amino – 5 – cinnamoyl – 1,3 – dimethyl –
uracil
$C_{15}H_{15}N_3O_3$
V.Warin, M.Foulon, F.Baert, J.L.Bernier, J.P.Henichart
Acta Crystallogr.,Sect.B,**36**, 1721, 1980

44.33 2,4 – Diamino – 5 – (1 – adamantyl) – 6 –
methylpyrimidine ethylsulfonate
$C_{15}H_{23}N_4^+$, $C_2H_5O_3S^-$
V.Cody, S.F.Zakrzewski *Am.Cryst.Assoc.,Ser.2*,**7**, 13, 1980
See also R1 : 31

44.C Triethylammonium adenosine – 5′ – (O – p –
nitrophenyl – O – phosphorothioate)
$C_{16}H_{16}N_5O_6PS^-$, $C_6H_{16}N^+$ Main entry is 47.32

44.34 1 – (3 – (Indol – 3 – yl)propyl) – thymine
$C_{16}H_{17}N_3O_2$
D.Voet *J.Am.Chem.Soc.*,**102**, 2071, 1980
See also R1 : 35

44.C 9 – (5 – Phenylseleno – 5 – deoxyribofuranosyl) –
adenine
$C_{16}H_{17}N_5O_4Se$ Main entry is 47.33

44.C 7 – (3,4 – trans – 4,5 – cis – Dihydroxy – 1 –
cyclopenten – 3 – yl – aminomethyl) – 7 –
deazaguanosine – 5′ – monophosphate hydrate
(partially deuterated)
Queuosine – 5′ – monophosphate hydrate
$C_{17}H_{24}N_5O_{10}P$, $3.5H_2O$ Main entry is 47.34

44.C Acridine orange 5 – iodocytidylyl – (3′ – 5′) –
guanosine dodecahydrate
$C_{19}H_{23}IN_8O_{12}P^-$, $C_{17}H_{20}N_3^+$, $12H_2O$ Main entry is 47.35

44.C Proflavine 5 – iodocytidylyl – (3′ – 5′) – guanosine
hydrate methanol solvate
$2C_{19}H_{23}IN_8O_{12}P^-$, $2C_{13}H_{12}N_3^+$, $15H_2O$, CH_4O
Main entry is 47.36

44.C Ellipticine – 5 – iodocytidylyl – (3′ – 5′) –
guanosine hydrate methanol solvate
$2C_{19}H_{23}IN_8O_{12}P^-$, $2C_{17}H_{15}N_2^+$, $20H_2O$, $2CH_4O$
Main entry is 60.36

44.C 3,5,6,8 – Tetramethyl – N – methyl –
phenanthrolinium – 5 – iodocytidylyl – (3′ – 5′) –
guanosine hydrate methanol solvate
$2C_{19}H_{23}IN_8O_{12}P^-$, $2C_{17}H_{19}N_2^+$, $17H_2O$, $2CH_4O$
Main entry is 60.37

44.C 9 – Amino – acridine – 5 – iodocytidylyl – (3′ –
5′) – guanosine hydrate complex
$4C_{19}H_{23}IN_8O_{12}P^-$, $4C_{13}H_{11}N_2^+$, $21H_2O$ Main entry is 60.22

44.35 DL – 2 – (α – Hydroxybenzyl) – oxythiamine
chloride hydrochloride trihydrate
$C_{19}H_{23}N_3O_3S^{2+}$, $2Cl^-$, $3H_2O$
W.Shin, J.Pletcher, M.Sax
J.Am.Chem.Soc.,**101**, 4365, 1979
See also R1 : 41

44.C Deoxycytidylyl – (3′ – 5′) – deoxyguanosine (2 –
hydroxy – ethylthiolato) – 2,2′,2″ – terpyridine –
platinum(ii) hydrate
$2C_{19}H_{24}N_8O_{10}P^-$, $2C_{17}H_{16}N_3OPtS^+$, xH_2O Main entry is 85.58

44.C Proflavine – cytidylyl – (3 – 5′) – guanosine
sulfate hydrate
$C_{19}H_{24}N_8O_{12}P^-$, $1.5C_{13}H_{12}N_3^+$, $0.5O_4S^-$, $11.5H_2O$
Main entry is 60.24

44.C Ethidium cytidylyl – (3′ – 5′) – guanosine
$C_{19}H_{24}N_8O_{12}P^-$, $C_{21}H_{20}N_3^+$ Main entry is 47.37

44.C Proflavine deoxycytidylyl – (3′ – 5′) – guanosine
hydrate
$2C_{19}H_{25}N_8O_{10}P$, $2C_{13}H_{11}N_3$, xH_2O Main entry is 47.38

44.C Cytidylyl – (3′ – 5′) – adenosine – proflavine
complex decahydrate
$C_{19}H_{26}N_8O_{11}P$, $C_{13}H_{11}N_3$, $10H_2O$ Main entry is 60.39

44.C bis(bis(μ – 1 – Methylthyminato) – cis –
diammine – platinum(ii)) silver nitrate
pentahydrate
$C_{24}H_{44}AgN_{12}O_8Pt_2^+$, NO_3^-, $5H_2O$ Main entry is 84.100

44.C Deoxy – 5′ – P – adenylyl – (3′ – 5′) – thymidylyl –
(3′ – 5′) – adenylyl – (3′ – 5′) – thymidine
$C_{40}H_{47}N_{14}O_{25}P_4^{5-}$, $5H_4N^+$, $57H_2O$ Main entry is 47.39

CARBOHYDRATES

45.1 **D – Arabino – γ – lactone**
$C_5H_8O_5$
T.Svinning, H.Sorum
Acta Crystallogr.,Sect.B,**35**, 2967, 1979

45.2 **β – L – Arabinose (neutron diffraction, at 123°K,80 percent deuterated)**
$C_5H_{10}O_5$
G.A.Jeffrey, A.Robbins, R.K.McMullan, S.Takagi
Acta Crystallogr.,Sect.B,**36**, 373, 1980

45.3 **α – L – Xylopyranose (neutron diffraction, at 123°K, 80 percent deuterated)**
$C_5H_{10}O_5$
G.A.Jeffrey, A.Robbins, R.K.McMullan, S.Takagi
Acta Crystallogr.,Sect.B,**36**, 373, 1980

45.4 **Pentaerythritol (absolute configuration)**
$C_5H_{12}O_4$
D.Eilerman, R.Rudman
Acta Crystallogr.,Sect.B,**35**, 2458, 1979

45.5 **Pentaerythritol (re-refinement using data of Llewellyn et al., J.Chem.Soc.,883,1937 and cell of Shiono et al., Acta Cryst.,11,389,1958)**
$C_5H_{12}O_4$
M.F.C.Ladd *Acta Crystallogr.,Sect.B*,**35**, 2375, 1979

45.C **L – Arginine L – ascorbate**
$C_6H_7O_6^-$, $C_6H_{15}N_4O_2^+$ Main entry is 48.20

45.C **L – Serine L – ascorbic acid**
$C_6H_8O_6$, $C_3H_7NO_3$ Main entry is 48.8

45.C **Methyl β – D – ribopyranoside phosphite triester**
$C_6H_9O_5P$ Main entry is 64.16

45.C **Methyl – D – ribopyranoside – thiophosphate – triester**
$C_6H_9O_5PS$ Main entry is 46.4

45.6 **1,2:5,6 – Dianhydro – galactitol (α form)**
$C_6H_{10}O_4$
I.Csoregh, M.Czugler, K.Simon, I.Vidra, L.Institoris
Eur.Cryst.Meeting,**5**, 37, 1979
See also R1 : 38

45.7 **1,2:5,6 – Dianhydro – galactitol (β form)**
$C_6H_{10}O_4$
I.Csoregh, M.Czugler, K.Simon, I.Vidra, L.Institoris
Eur.Cryst.Meeting,**5**, 37, 1979
See also R1 : 38

45.8 **1,6 – Anhydro – β – D – galactopyranose**
$C_6H_{10}O_5$
C.Ceccarelli, J.R.Ruble, G.A.Jeffrey
Acta Crystallogr.,Sect.B,**36**, 861, 1980
See also R1 : 38

45.9 **6 – Deoxy – α – L – sorbofuranose**
$C_6H_{12}O_5$
P.Swaminathan, L.Anderson, M.Sundaralingam
Carbohydr.Res.,**75**, 1, 1979

45.10 **1,5 – Dianhydro – D,L – galactitol**
$C_6H_{12}O_5$
I.Csoregh, M.Czugler, K.Simon, I.Vidra, L.Institoris
Eur.Cryst.Meeting,**5**, 37, 1979

45.11 **3 – (β – D – Ribofuranosyl) – 1,2,4 – triazole – 5(1H,4H) – thione**
$C_7H_{11}N_3O_4S$
M.S.Poonian, E.F.Nowoswiat *J.Org.Chem.*,**45**, 203, 1980
See also R1 : 32

45.12 **2 – O – Acetyl – 1,6:3,5 – dianhydro – α – L – idofuranose**
$C_8H_{10}O_5$
U.Behrens, J.Schulz, P.Koll *Carbohydr.Res.*,**70**, 150, 1979
See also R1 : 38

45.13 **1,6 – Anhydro – 3 – deoxy – 3 – iodo – 2 – O – (methylthio) – carbonyl – β – D – altropyranose**
$C_8H_{11}IO_5S$
J.J.Patroni, B.W.Skelton, R.V.Stick, A.H.White
Aust.J.Chem.,**33**, 987, 1980

45.14 **3,2' – Anhydro – 4 – (β – D – arabinofuranosyl) – 5,6 – dihydro – 2H – 1,2,4 – thiadiazin – 3 – one – 1,1 – dioxide**
$C_8H_{12}N_2O_6S$
T.Srikrishnan, R.Parthasarathy, A.C.Schroeder
Am.Cryst.Assoc.,Ser.2,**8**, 34, 1980
See also R1 : 41

45.C **6 – Azauridine – 5' – phosphate trihydrate**
$C_8H_{12}N_3O_9P$, $3H_2O$ Main entry is 47.1

45.15 **Methyl – 3,4 – O – ethylidene – β – L – arabinopyranose**
$C_8H_{14}O_5$
S.M.Fridey, R.Parthasarathy, Y.L.Fu, M.Bobek
Am.Cryst.Assoc.,Ser.2,**8**, 33, 1980

45.C **L – 3 – Amino – 1 – methoxy – 6 – methylamino – 1,3,6 – trideoxy – chiro – inositol**
Fortamine
$C_8H_{18}N_2O_4$ Main entry is 50.4

45.C **2,2' – Anhydro – 1 – β – D – arabinofuranosyl – 2 – thio – uracil**
$C_9H_{10}N_2O_4S$ Main entry is 47.2

45.C **Disodium deoxyuridine – 5' – phosphate pentahydrate**
$C_9H_{11}N_2O_8P^{2-}$, $2Na^+$, $5H_2O$ Main entry is 47.3

45.C **Disodium uridine – 5' – phosphate heptahydrate**
$C_9H_{11}N_2O_9P^{2-}$, $2Na^+$, $7H_2O$ Main entry is 47.4

45.C **5 – Iodo – 5' – amino – 2',5' – dideoxyuridine**
$C_9H_{12}IN_3O_4$ Main entry is 47.5

45.C **5 – Iodocytidine**
$C_9H_{12}IN_3O_5$ Main entry is 47.6

45.C **1 – β – D – Lyxofuranosyluracil**
$C_9H_{12}N_2O_6$ Main entry is 47.7

45.C **3',5' – Cyclic – cytidine – monophosphate trihydrate**
$C_9H_{12}N_3O_7P$, $3H_2O$ Main entry is 47.8

45.C   Barium cytidine – 5′ – phosphate hydrate
$C_9H_{12}N_3O_8P^{2-}$, Ba^{2+}, $8.5H_2O$ Main entry is 47.9

45.C   5 – Nitro – 2′ – deoxyuridine – 5′ – monophosphate monohydrate
$C_9H_{12}N_3O_{10}P$, H_2O Main entry is 47.10

45.16   2,3 – Di – O – acetyl – 2 – C – methyl – erythrono – 1,4 – lactone
$C_9H_{12}O_6$
A.Conde, E.Moreno, R.Marquez
Acta Crystallogr.,Sect.B,**36**, 1713, 1980

45.C   1,2 – O – Isopropylidene – α – D – glucofuranose – 3,5,6 – bicyclophosphite
$C_9H_{13}O_6P$ Main entry is 64.32

45.C   Potassium dihydrouridine – 3′ – monophosphate hemihydrate
$C_9H_{14}N_2O_9P^-$, K^+, $0.5H_2O$ Main entry is 47.11

45.17   Methyl D – threo – 2,5 – hexodiulosonate – 5 – (dimethyl – acetal)
$C_9H_{16}O_8$
G.C.Andrews, B.E.Bacon, J.Bordner, G.L.A.Hennessee
Carbohydr.Res.,**77**, 25, 1979

45.C   7,2′ – Anhydro – β – D – arabinosyl – orotidine
$C_{10}H_{10}N_2O_7$, Main entry is 47.12

45.C   Sodium adenosine – (3′ – 5′) – cyclic – monophosphate tetrahydrate (monoclinic form)
$C_{10}H_{11}N_5O_6P^-$, Na^+, $4H_2O$ Main entry is 47.13

45.C   Inosine (orthorhombic form)
$C_{10}H_{12}N_4O_5$ Main entry is 47.14

45.C   3′ – Deoxy – adenosine
Cordycepin
$C_{10}H_{13}N_5O_3$ Main entry is 47.15

45.C 9α – D – Arabinofuranosyl – adenine
$C_{10}H_{13}N_5O_4$ Main entry is 47.16

45.C   8 – Thioxo – adenosine monohydrate
$C_{10}H_{13}N_5O_4S$, H_2O Main entry is 47.17

45.C α – Thymidine
$C_{10}H_{14}N_2O_5$ Main entry is 47.18

45.C   5 – Hydroxymethyl – 2′ – deoxyuridine
$C_{10}H_{14}N_2O_6$ Main entry is 47.19

45.C   4 – Amino – 1 – (4 – amino – 2 – oxo – 1(2H) – pyrimidinyl) – 1,4 – dideoxy – β – D – glucopyranuronic acid monohydrate
$C_{10}H_{14}N_4O_6$, H_2O Main entry is 47.20

45.C   bis(Adenosine) proflavine sesquisulfate hydrate
$2C_{10}H_{14}N_5O_4^+$, $C_{13}H_{12}N_3^+$, $1.5O_4S^{2-}$, $6.5H_2O$
Main entry is 47.21

45.C   Guanosine – 5′ – monophosphate trihydrate
$C_{10}H_{14}N_5O_8P$. $3H_2O$ Main entry is 47.22

45.C   Potassium adenosine – 5′ – diphosphate dihydrate
$C_{10}H_{14}N_5O_{10}P_2^-$, K^+, $2H_2O$ Main entry is 47.23

45.C   Adenosine – 5′ – diphosphoric acid trihydrate
$C_{10}H_{15}N_5O_{10}P_2$. $3H_2O$ Main entry is 47.24

45.18   1 – Acetyl – rubranitrose (at −155°C.)
$C_{10}H_{17}NO_6$
S.A.Mizsak, H.Hoeksema. L.M.Pschigoda
J.Antibiot..**32**, 771. 1979

45.C   2 – O – Methyl – xylitanylidene diethyl – cyclophosphoramidate
3,5 – (Cyclo – N,N – diethylphosphamido) – 2 – O – methyl – 1 – deoxy – xylose
$C_{10}H_{20}NO_5P$ Main entry is 46.6

45.19   2 – Methyl – 5 – (N – nitrocarboxamido) – 1 – (2′,3′,5′ – tri – O – nitro – β – D – ribofuranosyl) – imidazole – 4 – carboxamide
$C_{11}H_{12}N_8O_{14}$
P.C.Wyss, P.Schonholzer, W.Arnold
Diss.Abstr.B,**63**, 1353, 1980
See also R1 : 32

45.C 1β – D – Ribofuranosyl – pyridin – 4 – one – 3 – carboxamide
$C_{11}H_{14}N_2O_6$ Main entry is 47.25

45.C β – 5 – Acetyl – 2′ – deoxyuridine
$C_{11}H_{14}N_2O_6$ Main entry is 47.26

45.20   Pyridyl – 1 – thio – β – D – glucopyranoside monohydrate
$C_{11}H_{15}NO_5S$, H_2O
S.Nordenson, G.A.Jeffrey
Acta Crystallogr.,Sect.B,**36**, 1214, 1980
See also R1 : 33

45.21   Methyl α – D – threo – 4,6 – di – O – acetyl – 2,3 – dideoxy – hex – 2 – eno – pyranoside
$C_{11}H_{16}O_6$
J.W.Krajewski, Z.Urbanczyk–Lipkowska, P.Gluzinski, Ya.Ya.Bleidelis, A.Kemme
Acta Crystallogr.,Sect.B,**35**, 2625, 1979

45.22   Methyl – 2,4 – di – O – acetyl – 3 – deoxy – 3 – C – methyl – 3 – nitro – β – D – xylopyranoside
$C_{11}H_{17}NO_8$
M.M.Abuaan, J.S.Brimacombe, J.N.Low
J.Chem.Soc.,Perkin Trans.1,995, 1980

45.C   2 – Amino – 8 – methyl – 1 – adenosine – 5′ – monophosphate dihydrate
$C_{11}H_{17}N_6O_7P$, $2H_2O$ Main entry is 47.27

45.C   5 – Propynyloxy – 2′ – deoxyuridine
$C_{12}H_{14}N_2O_6$ Main entry is 47.29

45.C   1,2 – O – Cyclohexylidene – α – D – glucofuranose – 3,5,6 – bicyclophosphite
$C_{12}H_{17}O_6P$ Main entry is 64.52

45.23   Ethyl 4,6 – di – O – acetyl – 2,3 – dideoxy – α – D – erythro – hex – 2 – eno – pyranoside
$C_{12}H_{18}O_6$
J.W.Krajewski, Z.Urbanczyk–Lipkowska, P.Gluzinski, Ya.Ya.Bleidelis, A.Kemme
Acta Crystallogr.,Sect.B,**35**, 2625, 1979

45.24   **Gentiobiose**
β – D – Glucopyranosyl – (1 – 6) – β – D – glucopyranose
$C_{12}H_{22}O_{11}$
F.Arene, A.Neuman, F.Longchambon
C.R.Acad.Sci.,Ser.C,**288**, 331, 1979

45.25   **Gentiobiose**
β – D – Glucopyranosyl – (1 – 6) – β – D – glucopyranose
$C_{12}H_{22}O_{11}$
D.C.Rohrer, A.Sarko, T.L.Bluhm, Y.N.Lee
Acta Crystallogr.,Sect.B,**36**, 650, 1980

45.26 β – D – Glucopyranosyl – (1 – 3) – D –
glucopyranose (orthorhombic form)
Laminaribiose
$C_{12}H_{22}O_{11}$
R.O.Gould, M.D.Walkinshaw *Eur.Cryst.Meeting*,5, 36, 1979

45.27 1 – Phenyl – 4,5 – (D – glucofurano) –
imidazolidin – 2 – one
$C_{13}H_{16}N_2O_5$
A.Conde, F.Bernier, R.Marquez
Eur.Cryst.Meeting,6, 27, 1980

45.C S – 8 – Aza – adenosyl – L – homocysteine
dihydrate
$C_{13}H_{19}N_7O_5S$, $2H_2O$ Main entry is 48.40

45.28 2 – R – Methylcyclohexyl – α – D – glucopyranoside
$C_{13}H_{24}O_6$
L.T.J.Delbaere *Am.Cryst.Assoc.,Ser.2*,8, 14, 1980

45.29 Methyl – 3 – O – α – D – glucopyranosyl – α – D –
glucopyranoside
Methyl – α – nigeroside
$C_{13}H_{24}O_{11}$
A.Neuman, D.Avenel, F.Arene, H.Gillier–Pandraud,
J.–R.Pougny, P.Sinay *Carbohydr.Res.*,80, 15, 1980

45.30 2,3,4,6 – Tetra – O – acetyl – 1,5 – anhydro – D –
arabino – hex – 1 – enitol
$C_{14}H_{18}O_9$
K.Vangehr, P.Luger, H.Paulsen
Carbohydr.Res.,70, 1, 1979

45.C 9β – D – Arabinofuranosyl – 8 – morpholino –
adenine dihydrate
$C_{14}H_{20}N_6O_5$, $2H_2O$ Main entry is 47.30

45.C S – Adenosyl – L – homocysteine hydrate
$C_{14}H_{20}N_6O_5S$, $2.5H_2O$ Main entry is 48.44

45.31 3,4,6 – tri – O – Acetyl – 1,2 – O – (R) – ethylidene –
α – D – allopyranose
$C_{14}H_{20}O_9$
C.Foces-Foces, A.Alemany, M.Bernabe, M.Martin–Lomas
J.Org.Chem.,45, 3502, 1980

45.32 Methyl 4 – chloro – 3,4 – dideoxy – 1,2:6,7 – di – O –
isopropylidene – α – D – erythro – hept – 3 – en –
5 – ulopyranoside
$C_{14}H_{21}ClO_6$
A.Aubry, J.Protas, P.Duchaussoy, P.Di Cesare, B.Gross
Acta Crystallogr.,Sect.B,36, 187, 1980

45.C 9β – D – Arabinofuranosyl – 8 – n – butyl –
amino – adenine
$C_{14}H_{22}N_6O_4$ Main entry is 47.31

45.33 1,4,6 – Tri – O – acetyl – 3 – O – 1'(R) – C –
carboxyethyl – β – D – glucose – 2,2' – lactone
$C_{15}H_{20}O_{10}$
J.H.Jordaan, J.J.Nieuwenhuis, J.A.Pretorius
S.Afr.J.Chem.,32, 173, 1979

45.34 Methyl – 2,3,4,5 – tetra – O – acetyl – α – D –
galactoseptanoside
$C_{15}H_{22}O_{10}$
W.Choong, J.F.McConnell, N.C.Stephenson, J.D.Stevens
Aust.J.Chem.,33, 979, 1980

45.C Triethylammonium adenosine – 5' – (O – p –
nitrophenyl – O – phosphorothioate)
$C_{16}H_{16}N_6O_8PS^-$, $C_6H_{16}N^+$ Main entry is 47.32

45.C 9 – (5 – Phenylseleno – 5 – deoxyribofuranosyl) –
adenine
$C_{16}H_{17}N_5O_4Se$ Main entry is 47.33

45.35 Methyl 4,6 – O – benzylidene – 2 – deoxy – 3 – O –
((methylthio) – thiocarbonyl) – α – D –
ribopyranoside
$C_{16}H_{20}O_5S_2$
P.Luger, B.Elvers, H.Paulsen *Chem.Ber.*,112, 3855, 1979

45.36 Methyl 4,6 – O – benzylidene – 2 – deoxy – 3 – O –
((methylthio) – thiocarbonyl) – α – D –
arabinopyranoside
$C_{16}H_{20}O_5S_2$
P.Luger, B.Elvers, H.Paulsen *Chem.Ber.*,112, 3855, 1979

45.37 Methyl 4,6 – O – benzylidene – 3 – deoxy – 2 – O –
((methylthio) – thiocarbonyl) – α – D –
arabinopyranoside
$C_{16}H_{20}O_5S_2$
P.Luger, B.Elvers, H.Paulsen *Chem.Ber.*,112, 3855, 1979

45.38 Methyl 3 – C – acetyl – 4,6 – O – benzylidene – 2 –
deoxy – α – D – ribo – hexopyranoside
$C_{16}H_{20}O_6$
J.S.Brimacombe, R.Hanna, A.M.Mather, T.J.R.Weakley
J.Chem.Soc.,Perkin Trans.1,273, 1980

45.39 1,4,6 – Tri – O – acetyl – 2 – (N – acetyl –
acetamido) – 2,3 – dideoxy – α – D – erythro –
hex – 2 – eno – pyranose
$C_{16}H_{21}NO_9$
Z.Ruzic–Toros, V.Rogic, B.Kojic–Prodic
Acta Crystallogr.,Sect.B,36, 607, 1980

45.40 Methyl 4,6 – O – benzylidene – 2,3 – di – O –
methyl – α – D – glucopyranoside
$C_{16}H_{22}O_6$
J.C.Barnes, J.S.Brimacombe, B.H.Nichols, T.J.R.Weakley
Carbohydr.Res.,69, 47, 1979

45.41 Methyl 4,6 – O – (R) – benzylidene – 2,3 – di – O –
methyl – β – D – galactopyranoside
$C_{16}H_{22}O_6$
J.C.Barnes, J.S.Brimacombe, B.H.Nichols, T.J.R.Weakley
Carbohydr.Res.,69, 47, 1979

45.42 2,3:4,5 – Di – O – isopropylidene – D – gulose diethyl
dithioacetal
$C_{16}H_{30}O_5S_2$
J.F.McConnell, A.Schwartz, J.D.Stevens
Cryst.Struct.Commun.,8, 855, 1979

45.43 Spiro((1 – acetylaziridine) – 2,3' – (methyl – 4,6 –
O – benzylidene – 2,3 – dideoxy – α – D – arabino –
hexopyranoside))
$C_{17}H_{21}NO_5$
J.S.Brimacombe, M.S.Saeed, T.J.R.Weakley
J.Chem.Soc.,Perkin Trans.1,2061, 1980
See also R1 : 38

45.C 7 – (3,4 – trans – 4,5 – cis – Dihydroxy – 1 –
cyclopenten – 3 – yl – aminomethyl) – 7 –
deazaguanosine – 5' – monophosphate hydrate
(partially deuterated)
Queuosine – 5' – monophosphate hydrate
$C_{17}H_{24}N_5O_{10}P$, $3.5H_2O$ Main entry is 47.34

45.C Loganin
$C_{17}H_{26}O_{10}$ Main entry is 52.4

45.44 1,3,4,6 – Tetra – O – acetyl – 2 – (N – acetylacetamido) – 2 – deoxy – β – D – galactopyranose
$C_{18}H_{25}NO_{11}$
Z.Ruzic–Toros, B.Kojic–Prodic, V.Rogic
Acta Crystallogr.,Sect.B,**36**, 384, 1980

45.45 cis – 2,3,4,6 – Tetra – O – acetyl – 1 – deoxy – D – glucopyranoside – 1,2′ – spiro(3′ – methyl – 3′ – hydroxy – tetrahydrofuran)
$C_{18}H_{26}O_{11}$
G.Remy, L.Cottier, G.Descotes, R.Faure, H.Loiseleur, G.Thomas–David *Acta Crystallogr.,Sect.B*,**36**, 873, 1980

45.46 trans – 2,3,4,6 – Tetra – O – acetyl – 1 – deoxy – D – glucopyranoside – 1,2′ – spiro(3′ – methyl – 3′ – hydroxy – tetrahydrofuran)
$C_{18}H_{26}O_{11}$
G.Remy, L.Cottier, G.Descotes, R.Faure, H.Loiseleur, G.Thomas–David *Acta Crystallogr.,Sect.B*,**36**, 873, 1980

45.47 3,4,6 – tri – O – Acetyl – 1,2 – O – (R) – (1 – t – butoxyethylidene) – α – D – galactopyranose
$C_{18}H_{28}O_{10}$
C.Foces–Foces, A.Alemany, M.Bernabe, M.Martin–Lomas
J.Org.Chem.,**45**, 3502, 1980

45.C Acridine orange 5 – iodocytidylyl – (3′ – 5′) – guanosine dodecahydrate
$C_{19}H_{23}IN_8O_{12}P^-$, $C_{17}H_{20}N_3^+$, $12H_2O$ Main entry is 47.35

45.C Proflavine 5 – iodocytidylyl – (3′ – 5′) – guanosine hydrate methanol solvate
$2C_{19}H_{23}IN_8O_{12}P^-$, $2C_{13}H_{12}N_3^+$, $15H_2O$, CH_4O
Main entry is 47.36

45.C Ellipticine – 5 – iodocytidylyl – (3′ – 5′) – guanosine hydrate methanol solvate
$2C_{19}H_{23}IN_8O_{12}P^-$, $2C_{17}H_{15}N_2^+$, $20H_2O$, $2CH_4O$
Main entry is 60.36

45.C 3,5,6,8 – Tetramethyl – N – methyl – phenanthrolinium – 5 – iodocytidylyl – (3′ – 5′) – guanosine hydrate methanol solvate
$2C_{19}H_{23}IN_8O_{12}P^-$, $2C_{17}H_{19}N_2^+$, $17H_2O$, $2CH_4O$
Main entry is 60.37

45.C 9 – Amino – acridine – 5 – iodocytidylyl – (3′ – 5′) – guanosine hydrate complex
$4C_{19}H_{23}IN_8O_{12}P^-$, $4C_{13}H_{11}N_2^+$, $21H_2O$ Main entry is 60.22

45.C Deoxycytidylyl – (3′ – 5′) – deoxyguanosine (2 – hydroxy – ethylthiolato) – 2,2′,2″ – terpyridine – platinum(ii) hydrate
$2C_{19}H_{24}N_8O_{10}P^-$, $2C_{17}H_{16}N_3OPtS^+$, xH_2O Main entry is 85.58

45.C Proflavine – cytidylyl – (3 – 5′) – guanosine sulfate hydrate
$C_{19}H_{24}N_8O_{12}P^-$, $1.5C_{13}H_{12}N_3^+$, $0.5O_4S^-$, $11.5H_2O$
Main entry is 60.24

45.C Ethidium cytidylyl – (3′ – 5′) – guanosine
$C_{19}H_{24}N_8O_{12}P^-$, $C_{21}H_{20}N_3^+$ Main entry is 47.37

45.C Proflavine deoxycytidylyl – (3′ – 5′) – guanosine hydrate
$2C_{19}H_{25}N_8O_{10}P$, $2C_{13}H_{11}N_3$, xH_2O Main entry is 47.38

45.C Cytidylyl – (3′ – 5′) – adenosine – proflavine complex decahydrate
$C_{19}H_{26}N_8O_{11}P$, $C_{13}H_{11}N_3$, $10H_2O$ Main entry is 60.39

45.48 Procaine – N – D – glucoside hydrochloride monohydrate
$C_{19}H_{31}N_2O_7^+$, Cl^-, H_2O
O.Dideberg, J.Lamotte, L.Dupont
Acta Crystallogr.,Sect.B,**36**, 1500, 1980

45.49 O – α – D – Mannopyranosyl – (1 – 3) – O – β – D – mannopyranosyl – (1 – 4) – 2 – acetamido – 2 – deoxy – α – D – glucopyranose
$C_{20}H_{35}NO_{16}$
V.Warin, F.Baert, R.Fouret, G.Strecker, G.Spik, B.Fournet, J.Montreuil *Carbohydr.Res.*,**76**, 11, 1979

45.50 6,6′ – Dibromo – 6,6′ – dideoxy – α,α – trehalose – hexa – acetate chloroform solvate
$C_{24}H_{32}Br_2O_{15}$, $CHCl_3$
G.Williams, P.Lavallee, S.Hanessian, F.Brisse
Acta Crystallogr.,Sect.B,**35**, 2574, 1979

45.51 2,3,4 – Tri – O – benzoyl – 2 – C – chloro – α – D – xylopyranosyl chloride
$C_{26}H_{20}Cl_2O_7$
F.W.Lichtenthaler, T.Sakakibara, E.Egert
Chem.Ber.,**113**, 471, 1980

45.52 1,5 – Anhydro – 2,3,4 – tri – O – benzoyl – ribitol
$C_{26}H_{22}O_7$
P.Luger, G.Kothe, K.Vangehr, H.Paulsen, F.R.Heiker
Carbohydr.Res.,**68**, 207, 1979

45.53 1,5 – Anhydro – 2,3,4 – tri – O – benzoyl – xylitol
$C_{26}H_{22}O_7$
P.Luger, G.Kothe, K.Vangehr, H.Paulsen, F.R.Heiker
Carbohydr.Res.,**68**, 207, 1979

45.54 (+) – 2,4:3,5 – Di – O – methylene – D – mannitol – 1,6 – di – trans – cinnamate
$C_{26}H_{26}O_8$
J.Bernstein, B.S.Green, M.Rejto
J.Am.Chem.Soc.,**102**, 323, 1980
See also R1 : 38,1

45.55 Methyl – 2,3,4 – tris(O – benzoyl) – β – D – xylopyranoside
$C_{27}H_{24}O_8$
K.Vangehr, P.Luger, H.Paulsen
Chem.Ber.,**113**, 2609, 1980

45.56 1,2,4,6 – Tetra – O – acetyl – 3 – O – (2,3,4,6 – tetra – O – acetyl – β – D – galactopyranosyl) – α – D – galactopyranose
$C_{28}H_{38}O_{19}$
C.Foces–Foces, F.H.Cano, S.Garcia–Blanco
Acta Crystallogr.,Sect.B,**36**, 377, 1980

45.C Actodigin
$C_{29}H_{44}O_9$ Main entry is 51.73

45.57 Affinoside B
$2\alpha,3\beta,11$ – Trihydroxy – 12 – oxo – 5β – carda – 9,20 – dienolide – 3 – O – methyl – 4,6 – dideoxy – D – allo – 2 – hexosulose
$C_{30}H_{40}O_{10}$
T.Yamauchi, K.Miyahara, F.Abe, T.Kawasaki
Chem.Pharm.Bull.,**27**, 2463, 1979

45.58 1,2,3,4 – Tetra – O – benzoyl – β – D – xylopyranose
$C_{33}H_{28}O_9$
P.Luger, G.Kothe, K.Vangehr, H.Paulsen, F.R.Heiker
Carbohydr.Res.,**68**, 207, 1979

45.59 Methyl – 2,3',6 – tri – O – acetyl – 2',3:4',6' – di –
O – benzylidene – 7(R) – β – D – cellobioside
$C_{33}H_{38}O_{14}$
J.Thiem, K.–H.Klaska, O.Jarchow
J.Chem.Res.,**190**, 2729, 1980

45.60 β – Cyclodextrin ethanol solvate octahydrate
$C_{36}H_{60}O_{30}$, C_2H_6O, $8H_2O$
R.Tokuoka, M.Abe, T.Fujiwara, K.–I.Tomita, W.Saenger
Chem.Lett.,491, 1980

45.C α – Cyclodextrin N,N – dimethylformamide
clathrate pentahydrate
$C_{36}H_{60}O_{30}$, C_3H_7NO, $5H_2O$ Main entry is 61.19

45.C α – Cyclodextrin 2 – pyrrolidone clathrate
pentahydrate
$C_{36}H_{60}O_{30}$, C_4H_7NO, $5H_2O$ Main entry is 61.20

45.61 α – Cyclodextrin hexahydrate (combined X–ray and
neutron study)
$C_{36}H_{60}O_{30}$, $6H_2O$
B.Klar, B.Hingerty, W.Saenger
Acta Crystallogr.,Sect.B,**36**, 1154, 1980

45.62 α – Cyclodextrin hexahydrate (neutron study)
Cyclohexa – amylose hexahydrate
$C_{36}H_{60}O_{30}$, $6H_2O$
W.Saenger *Nature (London)*,**279**, 343, 1979

45.63 bis(α – Cyclodextrin) cadmium polyiodide hydrate
$2C_{36}H_{60}O_{30}$, $0.5Cd^{2+}$, I_5^-, $27H_2O$
M.Noltemeyer, W.Saenger
J.Am.Chem.Soc.,**102**, 2710, 1980

45.64 bis(α – Cyclodextrin) lithium tri – iodide
octahydrate
$2C_{36}H_{60}O_{30}$, Li^+, I_3^-, I_2, $8H_2O$
M.Noltemeyer, W.Saenger
J.Am.Chem.Soc.,**102**, 2710, 1980

45.C Tetra – acetyl – trillenogenin monobrosylate
$C_{40}H_{47}BrO_{14}S$ Main entry is 51.85

45.C Deoxy – 5' – P – adenylyl – (3' – 5') – thymidylyl –
(3' – 5') – adenylyl – (3' – 5') – thymidine
$C_{40}H_{47}N_{14}O_{25}P_4{}^{5-}$, $5H_4N^+$, $57H_2O$ Main entry is 47.39

45.C Digoxin
$C_{41}H_{64}O_{14}$ Main entry is 51.86

45.C Gitoxin
$C_{41}H_{64}O_{14}$ Main entry is 51.87

45.65 β – Cyclodextrin m – methylphenol decahydrate
$C_{42}H_{70}O_{35}$, $2C_7H_8O$, $10H_2O$
K.H.Jogun, J.M.Maclennan, J.J.Stezowski
Eur.Cryst.Meeting,**5**, 34, 1979

45.66 Cyclohepta – amylose p – nitroacetanilide hydrate
$C_{42}H_{70}O_{35}$, $C_8H_8N_2O_3$, xH_2O
M.M.Harding, J.M.Maclennan, R.M.Paton
Nature (London),**274**, 621, 1978
See also R2 : 16,15

45.67 β – Cyclodextrin dodecahydrate (at 120°K)
Cyclohepta – amylose dodecahydrate
$C_{42}H_{70}O_{35}$, $12H_2O$
J.J.Stezowski, J.M.Maclennan
Am.Cryst.Assoc.,Ser.2,**7**, 24, 1980

45.68 β – Cyclodextrin sodium polyiodide octahydrate
$C_{42}H_{70}O_{35}$, Na^+, I_3^-, $8H_2O$
M.Noltemeyer, W.Saenger
J.Am.Chem.Soc.,**102**, 2710, 1980

45.69 1,6 – Di(O – triphenylmethyl) – 2,5 – anhydro –
D,L – altritol monohydrate
$C_{44}H_{40}O_5$, H_2O
I.Csoregh, M.Czugler, K.Simon, I.Vidra, L.Institoris
Eur.Cryst.Meeting,**5**, 37, 1979

45.C γ – Cyclodextrin n – propanol clathrate hydrate
$C_{48}H_{80}O_{40}$, C_3H_8O, xH_2O Main entry is 61.22

45.70 Cyclo – octa – amylose hydrate (at 120°K)
Cyclodextrin hydrate
$C_{48}H_{80}O_{40}$, $17H_2O$
J.M.Maclennan, J.J.Stezowski
Biochem.Biophys.Res.Commun.,**92**, 926, 1980

45.C 3 – (β – L – Mycarose) – 5 – (β – D – 4,6 – dideoxy –
3 – ketoallose) – 13 – (β – D – mycinose) –
lankamycin – 11 – α – hydroxyisovalerate ester
$C_{48}H_{82}O_{20}$ Main entry is 50.35

45.C Carboxy – decarboxamido – vancomycin
$C_{66}H_{74}Cl_2N_8O_{25}$, xH_2O Main entry is 50.41

PHOSPHATES

46.C   Cyclohexylammonium ethyl hydrogen phosphate
$C_2H_6O_4P^-$, $C_6H_{14}N^+$ Main entry is 21.1

46.1   Disodium D,L – α – glycerophosphate hexahydrate
$C_3H_7O_6P^{2-}$, $2Na^+$, $6H_2O$
J.McAlister, M.Sundaralingam
Acta Crystallogr.,Sect.B,**36**, 1652, 1980

46.C   Tetrachloro – antimony(v) – dimethoxyphosphate
$C_4H_{12}Cl_8O_8P_2Sb_2$ Main entry is 66.5

46.2   5 – Ethoxy – trimethylenephosphoric acid
$C_5H_{11}O_5P$
J.A.Gerlt, D.F.Chodosh, R.E.Drews, R.D.Adams
J.Org.Chem.,**45**, 1282, 1980

46.3   Sodium phytate hydrate
Dodecasodium myo – inositol – hexaphosphate
octatriacontahydrate
$C_6H_6O_{24}P_6^{12-}$, $12Na^+$, $38H_2O$
E.Gavuzzo, D.Rogers, D.J.Williams
Eur.Cryst.Meeting,**5**, 32, 1979
See also R1 : 21

46.4   Methyl – D – ribopyranoside – thiophosphate –
triester
$C_6H_9O_5PS$
A.C.Bellaart, D.van Aken, H.M.Buck, C.H.Stam, A.van Herk
Rec.J.Roy.Netherl.Chem.Soc.,**98**, 523, 1979
See also R1 : 45

46.C   Tetrachloro – antimony(v) – dimethylamino –
methoxyphosphate (at –100°C)
$C_6H_{18}Cl_8N_2O_6P_2Sb_2$ Main entry is 66.9

46.C   6 – Azauridine – 5′ – phosphate trihydrate
$C_8H_{12}N_3O_9P$, $3H_2O$ Main entry is 47.1

46.C   Disodium deoxyuridine – 5′ – phosphate
pentahydrate
$C_9H_{11}N_2O_8P^{2-}$, $2Na^+$, $5H_2O$ Main entry is 47.3

46.C   Disodium uridine – 5′ – phosphate heptahydrate
$C_9H_{11}N_2O_9P^{2-}$, $2Na^+$, $7H_2O$ Main entry is 47.4

46.C   3′,5′ – Cyclic – cytidine – monophosphate
trihydrate
$C_9H_{12}N_3O_7P$, $3H_2O$ Main entry is 47.8

46.C   Barium cytidine – 5′ – phosphate hydrate
$C_9H_{12}N_3O_8P^{2-}$, Ba^{2+}, $8.5H_2O$ Main entry is 47.9

46.C   5 – Nitro – 2′ – deoxyuridine – 5′ – monophosphate
monohydrate
$C_9H_{12}N_3O_{10}P$, H_2O Main entry is 47.10

46.C   Potassium dihydrouridine – 3′ – monophosphate
hemihydrate
$C_9H_{14}N_2O_9P^-$, K^+, $0.5H_2O$ Main entry is 47.11

46.5   0 – (1 – (2,4 – Dichlorophenyl) – 2 – bromovinyl) –
0,0 – dimethylphosphate
$C_{10}H_{10}BrCl_2O_4P$
Z.Galdecki, M.L.Glowka, A.Zwierzak
Phosphorus and Sulfur,**5**, 299, 1979

46.C   Sodium adenosine – (3′ – 5′) – cyclic –
monophosphate tetrahydrate (monoclinic form)
$C_{10}H_{11}N_5O_6P^-$, Na^+, $4H_2O$ Main entry is 47.13

46.C   Guanosine – 5′ – monophosphate trihydrate
$C_{10}H_{14}N_5O_8P$, $3H_2O$ Main entry is 47.22

46.C   Potassium adenosine – 5′ – diphosphate dihydrate
$C_{10}H_{14}N_5O_{10}P_2^-$, K^+, $2H_2O$ Main entry is 47.23

46.C   Adenosine – 5′ – diphosphoric acid trihydrate
$C_{10}H_{15}N_5O_{10}P_2$, $3H_2O$ Main entry is 47.24

46.6   2 – 0 – Methyl – xylitanylidene diethyl –
cyclophosphoramidate
3,5 – (Cyclo – N,N – diethylphosphamido) – 2 – 0 –
methyl – 1 – deoxy – xylose
$C_{10}H_{20}NO_5P$
E.E.Nifant'ev, L.T.Elepina, A.A.Borisenko, M.P.Koroteev,
L.A.Aslanov, B.M.Ionov, S.S.Sotman
Phosphorus and Sulfur,**5**, 315, 1979
See also R1 : 45

46.C   2 – Amino – 8 – methyl – 1 – adenosine – 5′ –
monophosphate dihydrate
$C_{11}H_{17}N_6O_7P$, $2H_2O$ Main entry is 47.27

46.C   Triethylammonium adenosine – 5′ – (0 – p –
nitrophenyl – 0 – phosphorothioate)
$C_{16}H_{16}N_6O_9PS^-$, $C_6H_{16}N^+$ Main entry is 47.32

46.C   7 – (3,4 – trans – 4,5 – cis – Dihydroxy – 1 –
cyclopenten – 3 – yl – aminomethyl) – 7 –
deazaguanosine – 5′ – monophosphate hydrate
(partially deuterated)
Queuosine – 5′ – monophosphate hydrate
$C_{17}H_{24}N_5O_{10}P$, $3.5H_2O$ Main entry is 47.34

46.C   Acridine orange 5 – iodocytidylyl – (3′ – 5′) –
guanosine dodecahydrate
$C_{19}H_{23}IN_8O_{12}P^-$, $C_{17}H_{20}N_3^+$, $12H_2O$ Main entry is 47.35

46.C   Proflavine 5 – iodocytidylyl – (3′ – 5′) – guanosine
hydrate methanol solvate
$2C_{19}H_{23}IN_8O_{12}P^-$, $2C_{13}H_{12}N_3^+$, $15H_2O$, CH_4O
Main entry is 47.36

46.C   Ellipticine – 5 – iodocytidylyl – (3′ – 5′) –
guanosine hydrate methanol solvate
$2C_{19}H_{23}IN_8O_{12}P^-$, $2C_{17}H_{15}N_2^+$, $20H_2O$, $2CH_4O$
Main entry is 60.36

46.C   3,5,6,8 – Tetramethyl – N – methyl –
phenanthrolinium – 5 – iodocytidylyl – (3′ – 5′) –
guanosine hydrate methanol solvate
$2C_{19}H_{23}IN_8O_{12}P^-$, $2C_{17}H_{19}N_2^+$, $17H_2O$, $2CH_4O$
Main entry is 60.37

46.C   9 – Amino – acridine – 5 – iodocytidylyl – (3′ –
5′) – guanosine hydrate complex
$4C_{19}H_{23}IN_8O_{12}P^-$, $4C_{13}H_{11}N_2^+$, $21H_2O$ Main entry is 60.22

46.C   Deoxycytidylyl – (3′ – 5′) – deoxyguanosine (2 –
hydroxy – ethylthiolato) – 2,2′,2″ – terpyridine –
platinum(ii) hydrate
$2C_{19}H_{24}N_8O_{10}P^-$, $2C_{17}H_{16}N_3OPtS^+$, xH_2O Main entry is 85.58

46.C Ethidium cytidylyl – (3′ – 5′) – guanosine
$C_{19}H_{24}N_8O_{12}P^-$, $C_{21}H_{20}N_3^+$ Main entry is 47.37

46.C Proflavine deoxycytidylyl – (3′ – 5′) – guanosine
hydrate
$2C_{19}H_{25}N_8O_{10}P$, $2C_{13}H_{11}N_3$, xH_2O Main entry is 47.38

46.C Cytidylyl – (3′ – 5′) – adenosine – proflavine
complex decahydrate
$C_{19}H_{26}N_8O_{11}P$, $C_{13}H_{11}N_3$, $10H_2O$ Main entry is 60.39

46.7 Di – isopropyl – (2,3,4,5 – tetraphenyl –
cyclopenta – 1,4 – dienyl) – phosphate
$C_{35}H_{35}O_4P$
H.Krishnanachari, R.A.Jacobson
Cryst.Struct.Commun.,**8**, 873, 1979
See also R1 : 20

46.C Deoxy – 5′ – P – adenylyl – (3′ – 5′) – thymidylyl –
(3′ – 5′) – adenylyl – (3′ – 5′) – thymidine
$C_{40}H_{47}N_{14}O_{25}P_4{}^{5-}$, $5H_4N^+$, $57H_2O$ Main entry is 47.39

47.1 6 – Azauridine – 5′ – phosphate trihydrate
$C_8H_{12}N_3O_9P$, $3H_2O$
W.Saenger, D.Suck, M.Knappenberg, J.Dirkx
Biopolymers,**18**, 2015, 1979
See also R1 : 46,45,44

47.2 2,2′ – Anhydro – 1 – β – D – arabinofuranosyl – 2 –
thio – uracil
$C_9H_{10}N_2O_4S$
Y.Yamagata, J.Yoshimura, S.Fujii, T.Fujiwara,
K.–I.Tomita, T.Ueda
Acta Crystallogr.,Sect.B,**36**, 343, 1980
See also R1 : 45,44

47.3 Disodium deoxyuridine – 5′ – phosphate
pentahydrate
$C_9H_{11}N_2O_8P^{2-}$, $2Na^+$, $5H_2O$
M.A.Viswamitra, T.P.Seshadri, M.L.Post
Acta Crystallogr.,Sect.B,**36**, 2019, 1980
See also R1 : 46,45,44

47.4 Disodium uridine – 5′ – phosphate heptahydrate
$C_9H_{11}N_2O_9P^{2-}$, $2Na^+$, $7H_2O$
T.P.Seshadri, M.A.Viswamitra, G.Kartha
Acta Crystallogr.,Sect.B,**36**, 925, 1980
See also R1 : 46,45,44

47.5 5 – Iodo – 5′ – amino – 2′,5′ – dideoxyuridine
$C_9H_{12}IN_3O_4$
G.I.Birnbaum, T.–S.Lin, G.T.Shiau, W.H.Prusoff
J.Am.Chem.Soc.,**101**, 3353, 1979
See also R1 : 45,44

47.6 5 – Iodocytidine
$C_9H_{12}IN_3O_5$
M.M.Radwan, H.R.Wilson
Acta Crystallogr.,Sect.B,**35**, 3072, 1979
See also R1 : 45,44

47.7 1 – β – D – Lyxofuranosyluracil
$C_9H_{12}N_2O_6$
I.Ekiel, E.Darzynkiewicz, G.I.Birnbaum, D.Shugar
J.Am.Chem.Soc.,**101**, 4724, 1979
See also R1 : 45,44

47.8 3′,5′ – Cyclic – cytidine – monophosphate
trihydrate
$C_9H_{12}N_3O_7P$, $3H_2O$
R.Parthasarathy, T.Srikrishnan
Am.Cryst.Assoc.,Ser.2,**7**, 15, 1980
See also R1 : 46,45,44

47.9 Barium cytidine – 5′ – phosphate hydrate
$C_9H_{12}N_3O_8P^{2-}$, Ba^{2+}, $8.5H_2O$
J.Hogle, M.Sundaralingam, G.H.Y.Lin
Acta Crystallogr.,Sect.B,**36**, 564, 1980
See also R1 : 46,45,44

47.10 5 – Nitro – 2′ – deoxyuridine – 5′ – monophosphate monohydrate
$C_9H_{12}N_3O_{10}P$, H_2O
G.S.D.King, L.Sengier *Eur.Cryst.Meeting*,**6**, 285, 1980
See also R1 : 46,45,44

47.11 Potassium dihydrouridine – 3′ – monophosphate hemihydrate
$C_9H_{14}N_2O_9P^-$, K^+, $0.5H_2O$
J.Emerson, M.Sundaralingam
Acta Crystallogr.,Sect.B,**36**, 537, 1980
See also R1 : 46,45,44

47.12 7,2′ – Anhydro – β – D – arabinosyl – orotidine
$C_{10}H_{10}N_2O_7$
J.L.Smith, A.Chwang, M.Sundaralingam
Acta Crystallogr.,Sect.B,**36**, 833, 1980
See also R1 : 45,44,40

47.13 Sodium adenosine – (3′ – 5′) – cyclic – monophosphate tetrahydrate (monoclinic form)
$C_{10}H_{11}N_5O_6P^-$, Na^+, $4H_2O$
K.I.Varughese, C.T.Lu, G.Kartha
Am.Cryst.Assoc.,Ser.2,**8**, 43, 1980
See also R1 : 46,45,44

47.14 Inosine (orthorhombic form)
$C_{10}H_{12}N_4O_5$
E.Subramanian *Cryst.Struct.Commun.*,**8**, 777, 1979
See also R1 : 45,44

47.15 3′ – Deoxy – adenosine
Cordycepin
$C_{10}H_{13}N_5O_3$
M.M.Radwan, H.R.Wilson
Acta Crystallogr.,Sect.B,**36**, 2185, 1980
See also R1 : 45,44

47.16 9α – D – Arabinofuranosyl – adenine
$C_{10}H_{13}N_5O_4$
S.J.Cline, D.J.Hodgson
Biochim.Biophys.Acta,**563**, 540, 1979
See also R1 : 45,44

47.17 8 – Thioxo – adenosine monohydrate
$C_{10}H_{13}N_5O_4S$, H_2O
H.Mizuno, K.Kitamura, A.Miyao, Y.Yamagata, A.Wakahara, K.Tomita, M.Ikehara
Acta Crystallogr.,Sect.B,**36**, 902, 1980
See also R1 : 45,44

47.18 α – Thymidine
$C_{10}H_{14}N_2O_5$
A.W.Tench *Am.Cryst.Assoc.,Ser.2*,**7**, 14, 1980
See also R1 : 45,44

47.19 5 – Hydroxymethyl – 2′ – deoxyuridine
$C_{10}H_{14}N_2O_6$
G.I.Birnbaum, R.Deslauriers, T.–S.Lin, G.T.Shiau, W.H.Prusoff *J.Am.Chem.Soc.*,**102**, 4236, 1980
See also R1 : 45,44

47.20 4 – Amino – 1 – (4 – amino – 2 – oxo – 1(2H) – pyrimidinyl) – 1,4 – dideoxy – β – D – glucopyranuronic acid monohydrate
$C_{10}H_{14}N_4O_6$, H_2O
P.Swaminathan, J.McAlister, M.Sundaralingam
Acta Crystallogr.,Sect.B,**36**, 878, 1980
See also R1 : 45,44

47.21 bis(Adenosine) proflavine sesquisulfate hydrate
$2C_{10}H_{14}N_5O_4^+$, $C_{13}H_{12}N_3^+$, $1.5O_4S^{2-}$, $6.5H_2O$
P.Swaminathan, E.Westhof, M.Sundaralingam
Am.Cryst.Assoc.,Ser.2,**8**, 30, 1980
See also R1 : 45,44 R2 : 36

47.22 Guanosine – 5′ – monophosphate trihydrate
$C_{10}H_{14}N_5O_8P$, $3H_2O$
J.Emerson, M.Sundaralingam
Acta Crystallogr.,Sect.B,**36**, 1510, 1980
See also R1 : 46,45,44

47.23 Potassium adenosine – 5′ – diphosphate dihydrate
$C_{10}H_{14}N_5O_{10}P_2^-$, K^+, $2H_2O$
S.K.Katti, M.A.Viswamitra *Curr.Sci.*,**48**, 989, 1979
See also R1 : 46,45,44

47.24 Adenosine – 5′ – diphosphoric acid trihydrate
$C_{10}H_{15}N_5O_{10}P_2$, $3H_2O$
M.V.Hosur, M.A.Viswamitra *Curr.Sci.*,**48**, 1027, 1979
See also R1 : 46,45,44

47.C Penta – aquo – (2′ – deoxyguanosine – 5′ – phosphato) – cobalt(ii) trihydrate
$C_{10}H_{22}CoN_5O_{12}P$, $3H_2O$ Main entry is 83.62

47.C Penta – aquo – (2′ – deoxyguanosine – 5′ – phosphato) – nickel(ii) trihydrate
$C_{10}H_{22}N_5NiO_{12}P$, $3H_2O$ Main entry is 83.63

47.25 1β – D – Ribofuranosyl – pyridin – 4 – one – 3 – carboxamide
$C_{11}H_{14}N_2O_6$
T.Srikrishnan, R.Parthasarathy, S.P.Dutta, G.B.Chheda
Am.Cryst.Assoc.,Ser.2,**7**, 15, 1980
See also R1 : 45,44

47.26 β – 5 – Acetyl – 2′ – deoxyuridine
$C_{11}H_{14}N_2O_6$
P.J.Barr, P.Chananont, T.A.Hamor, A.S.Jones, M.K.O'Leary, R.T.Walker *Tetrahedron*,**36**, 1269, 1980
See also R1 : 45,44

47.27 2 – Amino – 8 – methyl – 1 – adenosine – 5′ – monophosphate dihydrate
$C_{11}H_{17}N_6O_7P$, $2H_2O$
J.V.Silverton, W.Limn, H.T.Miles
Am.Cryst.Assoc.,Ser.2,**8**, 33, 1980
See also R1 : 46,45,44

47.28 2,5′ – Anhydro – 1 – (2′,3′ – O – isopropylidene – β – D – ribofuranosyl) – 2 – thio – uracil
$C_{12}H_{14}N_2O_4S$
Y.Yamagata, S.Fujii, T.Fujiwara, K.–I.Tomita, T.Ueda
Acta Crystallogr.,Sect.B,**36**, 339, 1980
See also R1 : 44,42

47.29 5 – Propynyloxy – 2′ – deoxyuridine
$C_{12}H_{14}N_2O_6$
G.S.D.King, L.Sengier *Eur.Cryst.Meeting*,**5**, 38, 1979
See also R1 : 45,44

47.C S – 8 – Aza – adenosyl – L – homocysteine dihydrate
$C_{13}H_{19}N_7O_5S$, $2H_2O$ Main entry is 48.40

47.30 9β – D – Arabinofuranosyl – 8 – morpholino – adenine dihydrate
$C_{14}H_{20}N_6O_5$, $2H_2O$
V.Swaminathan, M.Sundaralingam, J.B.Chattopadhyaya, C.B.Reese *Acta Crystallogr.,Sect.B*,**36**, 828, 1980
See also R1 : 45,44,40

47.C S – Adenosyl – L – homocysteine hydrate
$C_{14}H_{20}N_6O_5S$, 2.5H_2O Main entry is 48.44

47.31 9β – D – Arabinofuranosyl – 8 – n – butyl –
amino – adenine
$C_{14}H_{22}N_6O_4$
S.Neidle, M.R.Sanderson, A.Subbiah, J.B.Chattopadhyaya,
R.Kuroda, C.B.Reese
Biochim.Biophys.Acta,**565**, 379, 1979
See also R1 : 45,44

47.C (Diethylenetriamine) – (guanosine) – palladium(ii)
diperchlorate
$C_{14}H_{26}N_8O_5Pd^{2+}$, 2$ClO_4^-$ Main entry is 76.84

47.32 Triethylammonium adenosine – 5′ – (O – p –
nitrophenyl – O – phosphorothioate)
$C_{16}H_{16}N_6O_8PS^-$, $C_6H_{16}N^+$
P.M.J.Burgers, B.K.Sathyanarayana, W.Saenger, F.Eckstein
Eur.J.Biochem., **100**, 585, 1979
See also R1 : 46,45,44

47.33 9 – (5 – Phenylseleno – 5 – deoxyribofuranosyl) –
adenine
$C_{16}H_{17}N_5O_4Se$
J.Guilhem, N.Zylber, J.Zylber, A.Gaudemer
Eur.Cryst.Meeting,**5**, 40, 1979
See also R1 : 45,44

47.34 7 – (3,4 – trans – 4,5 – cis – Dihydroxy – 1 –
cyclopenten – 3 – yl – aminomethyl) – 7 –
deazaguanosine – 5′ – monophosphate hydrate
(partially deuterated)
Queuosine – 5′ – monophosphate hydrate
$C_{17}H_{24}N_5O_{10}P$, 3.5H_2O
S.Yokoyama, T.Miyazawa, Y.Iitaka, Z.Yamaizumi, H.Kasai,
S.Nishimura *Nature (London)*,**282**, 107, 1979
See also R1 : 46,45,44

47.C Cesium hydrogen diammine – bis(cytidine – 3′ –
phosphato) – platinum(ii) tetrahydrate
$C_{18}H_{30}N_8O_{16}P_2Pt^{2-}$, 0.5$Cs^+$, 1.5$H^+$, 4$H_2O$
Main entry is 83.133

47.35 Acridine orange 5 – iodocytidylyl – (3′ – 5′) –
guanosine dodecahydrate
$C_{19}H_{23}IN_8O_{12}P^-$, $C_{17}H_{20}N_3^+$, 12H_2O
B.S.Reddy, T.P.Seshadri, T.D.Sakore, H.M.Sobell
J.Mol.Biol.,**135**, 787, 1979
See also R1 : 46,45,44 R2 : 36

47.36 Proflavine 5 – iodocytidylyl – (3′ – 5′) – guanosine
hydrate methanol solvate
2$C_{19}H_{23}IN_8O_{12}P^-$, 2$C_{13}H_{12}N_3^+$, 15$H_2O$, CH_4O
B.S.Reddy, T.P.Seshadri, T.D.Sakore, H.M.Sobell
J.Mol.Biol.,**135**, 787, 1979
See also R1 : 46,45,44 R2 : 36

47.C Ellipticine – 5 – iodocytidylyl – (3′ – 5′) –
guanosine hydrate methanol solvate
2$C_{19}H_{23}IN_8O_{12}P^-$, 2$C_{17}H_{15}N_2^+$, 20$H_2O$, 2$CH_4O$
Main entry is 60.36

47.C 3,5,6,8 – Tetramethyl – N – methyl –
phenanthrolinium – 5 – iodocytidylyl – (3′ – 5′) –
guanosine hydrate methanol solvate
2$C_{19}H_{23}IN_8O_{12}P^-$, 2$C_{17}H_{19}N_2^+$, 17$H_2O$, 2$CH_4O$
Main entry is 60.37

47.C 9 – Amino – acridine – 5 – iodocytidylyl – (3′ –
5′) – guanosine hydrate complex
4$C_{19}H_{23}IN_8O_{12}P^-$, 4$C_{13}H_{11}N_2^+$, 21$H_2O$ Main entry is 60.22

47.C Deoxycytidylyl – (3′ – 5′) – deoxyguanosine (2 –
hydroxy – ethylthiolato) – 2,2′,2″ – terpyridine –
platinum(ii) hydrate
2$C_{19}H_{24}N_8O_{10}P^-$, 2$C_{17}H_{16}N_3OPtS^+$, xH_2O Main entry is 85.58

47.C Proflavine – cytidylyl – (3 – 5′) – guanosine
sulfate hydrate
$C_{19}H_{24}N_8O_{12}P^-$, 1.5$C_{13}H_{12}N_3^+$, 0.5O_4S^-, 11.5H_2O
Main entry is 60.24

47.37 Ethidium cytidylyl – (3′ – 5′) – guanosine
$C_{19}H_{24}N_8O_{12}P^-$, $C_{21}H_{20}N_3^+$
S.C.Jain *Am.Cryst.Assoc.,Ser.2*,**7**, 19, 1980
See also R1 : 46,45,44 R2 : 36

47.38 Proflavine deoxycytidylyl – (3′ – 5′) – guanosine
hydrate
2$C_{19}H_{25}N_8O_{10}P$, 2$C_{13}H_{11}N_3$, xH_2O
H.–S.Shieh, H.M.Berman, M.Dabrow, S.Neidle
Nucleic Acids Res.,**8**, 85, 1980
See also R1 : 46,45,44 R2 : 36

47.C Cytidylyl – (3′ – 5′) – adenosine – proflavine
complex decahydrate
$C_{19}H_{26}N_8O_{11}P$, $C_{13}H_{11}N_3$, 10H_2O Main entry is 60.39

47.C cis – Diammine – bis(guanosine) – platinum(ii)
chloride perchlorate heptahydrate
$C_{20}H_{32}N_{12}O_{10}Pt^{2+}$, 1.5$Cl^-$, 0.5$ClO_4^-$, 7$H_2O$
Main entry is 83.151

47.C Trimethylene – diamine – bis(guanosine – 5′ –
monophosphate methyl ester) hydrate
$C_{25}H_{44}N_{12}O_{16}P_2Pt$, 11$H_2O$ Main entry is 83.186

47.C 2,2′ – Bipyridyl – (dihydrogen adenosine –
triphosphato) – zinc(ii) dimer tetrahydrate
$C_{40}H_{44}N_{14}O_{26}P_6Zn_2$, 4$H_2O$ Main entry is 84.125

47.39 Deoxy – 5′ – P – adenylyl – (3′ – 5′) – thymidylyl –
(3′ – 5′) – adenylyl – (3′ – 5′) – thymidine
$C_{40}H_{47}N_{14}O_{25}P_4^{5-}$, 5$H_4N^+$, 57$H_2O$
M.A.Viswamitra, O.Kennard, P.G.Jones, G.M.Sheldrick,
S.Salisbury, L.Falvello, Z.Shakked
Nature (London),**273**, 687, 1978
See also R1 : 46,45,44

ALPHA–AMINO–ACIDS AND PEPTIDES

48.1 Glycine (γ form, neutron study, at 83°K)
$C_2H_5NO_2$
A.Kvick, W.M.Canning, T.F.Koetzle, G.J.B.Williams
Acta Crystallogr.,Sect.B,**36**, 115, 1980

48.2 Glycine (γ form, neutron study)
$C_2H_5NO_2$
A.Kvick, W.M.Canning, T.F.Koetzle, G.J.B.Williams
Acta Crystallogr.,Sect.B,**36**, 115, 1980

48.3 tris(Glycine) calcium(ii) dibromide
$3C_2H_5NO_2$, Ca^{2+}, $2Br^-$
J.K.M.Rao, S.Natarajan
Acta Crystallogr.,Sect.B,**36**, 1058, 1980

48.4 tetrakis(glycine) potassium tri – iodide
$4C_2H_5NO_2$, K^+, I_3^-
D.N.Hendrickson, F.H.Herbstein, M.Kaftory, M.Kapon,
W.Saenger *Eur.Cryst.Meeting*,**5**, 88, 1979

48.5 Glycinium orthophosphate
$C_2H_6NO_2^+$, $H_2O_4P^-$
R.Thulasidhass, J.K.M.Rao, S.S.Chibber, R.P.Sharma,
S.K.Dutt *Curr.Sci.*,**48**, 626, 1979

48.6 tris(Glycine) tetrafluoroberyllate (paraelectric form, at 97°C)
$2C_2H_6NO_2^+$, $C_2H_5NO_2$, BeF_4^{2-}
A.Waskowska, S.Olejnik, K.Lukaszewicz,
M.Ciechanowicz–Rutkowska *Ferroelectrics*,**22**, 855, 1979

48.7 tris(Glycine) tetrafluoroberyllate (ferroelectric form, at 23°C)
$2C_2H_6NO_2^+$, $C_2H_5NO_2$, BeF_4^{2-}
A.Waskowska, S.Olejnik, K.Lukaszewicz,
M.Ciechanowicz–Rutkowska *Ferroelectrics*,**22**, 855, 1979

48.8 L – Serine L – ascorbic acid
$C_3H_7NO_3$, $C_6H_8O_6$
V.Sudhakar, T.N.Bhat, M.Vijayan
Acta Crystallogr.,Sect.B,**36**, 125, 1980
See also R2 : 45,38

48.9 Disarcosine hydrobromide
$C_3H_8NO_2^+$, $C_3H_7NO_2$, Br^-
S.C.Bhattacharyya, N.N.Saha
J.Cryst.Mol.Struct.,**8**, 209, 1978

48.C N – (Phosphonomethyl) – glycine
Glyphosate
$C_3H_8NO_5P$ Main entry is 64.4

48.10 Glycylglycine (α form,at 82°K,deformation electron density study)
$C_4H_8N_2O_3$
A.Kvick, T.F.Koetzle, E.D.Stevens
J.Chem.Phys.,**71**, 173, 1979

48.11 D,L – α – Amino – n – butyric acid (D form)
$C_4H_9NO_2$
K.Nakata, Y.Takaki, K.Sakurai
Acta Crystallogr.,Sect.B,**36**, 504, 1980

48.C trans – 2 – Azabicyclo(2.1.0)pentane – 3 – (S) – carboxylic acid
$C_5H_7NO_2$ Main entry is 50.1

48.12 L – Glutamic acid (α form, neutron study)
$C_5H_9NO_4$
M.S.Lehmann, A.C.Nunes
Acta Crystallogr.,Sect.B,**36**, 1621, 1980

48.13 L – Glutamic acid (α form)
$C_5H_9NO_4$
N.Hirayama, K.Shirahata, Y.Ohashi, Y.Sasada
Bull.Chem.Soc.Jpn.,**53**, 30, 1980

48.14 D,L – Methionine (α form, at 333°K)
$C_5H_{11}NO_2S$
T.Taniguchi, Y.Takaki, K.Sakurai
Bull.Chem.Soc.Jpn.,**53**, 803, 1980

48.15 D,L – Methionine (β form, at 293°K)
$C_5H_{11}NO_2S$
T.Taniguchi, Y.Takaki, K.Sakurai
Bull.Chem.Soc.Jpn.,**53**, 803, 1980

48.C Glycinato – trimethyl – tin(iv)
$(C_5H_{13}NO_2Sn)_n$ Main entry is 69.8

48.16 Sodium calcium nitrilo – triacetate
$C_6H_6NO_6^{3-}$, Na^+, Ca^{2+}
B.L.Barnett, V.A.Uchtman *Inorg.Chem.*,**18**, 2674, 1979
See also R1 : 2

48.17 2,4 – Methanoproline monohydrate
2 – Carboxy – 2,4 – methanopyrrolidine monohydrate
$C_6H_9NO_2$, H_2O
E.A.Bell, M.Y.Qureshi, R.J.Pryce, D.H.Janzen, P.Lemke,
J.Clardy *J.Am.Chem.Soc.*,**102**, 1409, 1980
See also R1 : 37

48.18 2,4 – Methano – glutamic acid monohydrate
1 – Amino – 1,3 – dicarboxy – cyclobutane
monohydrate
$C_6H_9NO_4$, H_2O
E.A.Bell, M.Y.Qureshi, R.J.Pryce, D.H.Janzen, P.Lemke,
J.Clardy *J.Am.Chem.Soc.*,**102**, 1409, 1980
See also R1 : 20

48.19 Glycyl – D – threonine dihydrate (absolute configuration)
$C_6H_{12}N_2O_4$, $2H_2O$
P.Ho, T.F.Lai, R.E.Marsh *J.Cryst.Mol.Struct.*,**9**, 181, 1979

48.20 L – Arginine L – ascorbate
$C_6H_{15}N_4O_2^+$, $C_6H_7O_6^-$
V.Sudhakar, M.Vijayan
Acta Crystallogr.,Sect.B,**36**, 120, 1980
See also R2 : 45

48.21 N – Acetyl – 4 – hydroxy – L – proline monohydrate
$C_7H_{11}NO_4$, H_2O
M.Hospital, C.Courseille, F.Leroy, B.P.Roques
Biopolymers,**18**, 1141, 1979

48.22 Glycyl – L – 4 – hydroxyproline
$C_7H_{12}N_2O_4$
C.Garbay–Jaureguiberry, B.Arnoux, T.Prange,
S.Wehri–Altenburger, C.Pascard, B.P.Roques
J.Am.Chem.Soc.,**102**, 1827, 1980

48.C Chloro – (ethyl – L – cysteinato – N,S) – dimethyl – tin(iv)
$C_7H_{16}ClNO_2SSn$ Main entry is 69.16

48.23 Cyclo(L – prolyl – L – alanyl)
$C_8H_{12}N_2O_2$
M.Cotrait, F.Leroy *Cryst.Struct.Commun.*,**8**, 819, 1979
See also R1 : 35

48.24 L – Prolyl – sarcosine monohydrate
$C_8H_{14}N_2O_3$, H_2O
T.Kojima, T.Kido, H.Itoh, T.Yamane, T.Ashida
Acta Crystallogr.,Sect.B,**36**, 326, 1980

48.25 1 – Aminocycloheptane – carboxylic acid
hydrobromide monohydrate (reinvestigation of
structure, published by Chacko, Srinivasan and
Zand, J.Cryst.Mol.Struct.,1,(1971),213–224, using
published diffraction data)
$C_8H_{16}NO_2{}^+$, Br^-, H_2O
A.J.de Kok, C.Romers
Acta Crystallogr.,Sect.B,**36**, 1887, 1980
See also R1 : 22

48.26 Barium ethylenediamine – tetra – acetate hydrate
$C_{10}H_{12}N_2O_8{}^{4-}$, $2Ba^{2+}$, $2.5H_2O$
S.Meicheng, T.Zengren, L.Tongchang, S.Shiying, T.Yougi
Sci.Sin.,**22**, 912, 1979

48.27 Calcium ethylenediamine – tetra – acetate
heptahydrate
$C_{10}H_{12}N_2O_8{}^{4-}$, $2Ca^{2+}$, $7H_2O$
B.L.Barnett, V.A.Uchtman *Inorg.Chem.*,**18**, 2674, 1979

48.C 2,2′ – bis(o – (Oxo – diphenylphosphino) –
phenyloxy) – diethyl ether D,L – phenylglycinium
ethyl ester trifluoromethylsulfonate
$C_{10}H_{14}NO_2{}^+$, $C_{40}H_{36}O_5P_2$, $CF_3O_3S^-$ Main entry is 64.170

48.28 Cyclo(L – prolyl – L – 4 – hydroxyprolyl)
monohydrate
$C_{10}H_{14}N_2O_3$, H_2O
C.Garbay–Jaureguiberry, B.Arnoux, T.Prange,
S.Wehri–Altenburger, C.Pascard, B.P.Roques
J.Am.Chem.Soc.,**102**, 1827, 1980

48.29 L – Prolyl – L – 4 – hydroxyproline monohydrate
$C_{10}H_{16}N_2O_4$, H_2O
C.Garbay–Jaureguiberry, B.Arnoux, T.Prange,
S.Wehri–Altenburger, C.Pascard, B.P.Roques
J.Am.Chem.Soc.,**102**, 1827. 1980

48.30 Palythine trihydrate
$C_{10}H_{16}N_2O_5$, $3H_2O$
A.Furusaki, T.Matsumoto, I.Tsujino, I.Sekikawa
Bull.Chem.Soc.Jpn.,**53**, 319, 1980

48.31 L – Valyl – L – α – hydroxyisovaleric acid
$C_{10}H_{19}NO_4$
N.M.Galitskii, I.I.Mikhaleva, V.Z.Pletnev
Bioorg.Khim.,**3**, 1323, 1977

48.32 Cyclo(D – N – methylvalyl – D – α –
hydroxyisovaleryl)
$C_{11}H_{19}NO_3$
N.E.Zhukhlistova, G.N.Tishchenko
Kristallografiya,**25**, 274, 1980
See also R1 : 40

48.33 N – t – Amyloxycarbonyl – L – proline
$C_{11}H_{19}NO_4$
E.Benedetti, A.Ciajolo, B.di Blasio, V.Pavone, C.Pedone,
C.Toniolo, G.M.Bonora
Int.J.Pept.Protein Res.,**14**, 130, 1979

48.34 N – t – Butyloxycarbonyl – sarcosyl – sarcosine
$C_{11}H_{20}N_2O_5$
E.Benedetti, A.Ciajolo, B.di Blasio, V.Pavone, C.Pedone,
C.Toniolo, G.M.Bonora *Macromolecules*,**12**, 438, 1979

48.35 N – Acetyl – D,L – methionine – diethylamide
$C_{11}H_{22}N_2O_2S$
A.Aubry, J.Protas, M.T.Cung, M.Marraud
Acta Crystallogr.,Sect.B,**35**, 2634, 1979

48.36 N – Acetyl – L – prolyl – L – 4 – hydroxyproline
$C_{12}H_{18}N_2O_5$
C.Garbay–Jaureguiberry, B.Arnoux, T.Prange,
S.Wehri–Altenburger, C.Pascard, B.P.Roques
J.Am.Chem.Soc.,**102**, 1827, 1980

48.37 threo – β – Hydroxy – N – acetyl –
tryptophanamide
$C_{13}H_{15}N_3O_3$
I.L.Karle *Am.Cryst.Assoc.,Ser.2*,**7**, 25, 1980

48.38 erythro – β – Hydroxy – N – acetyl –
tryptophanamide benzene solvate
$C_{13}H_{15}N_3O_3$, $0.5C_6H_6$
I.L.Karle *Am.Cryst.Assoc.,Ser.2*,**7**, 25, 1980

48.39 2 – (4 – Chlorophenyl) – 1,1 – dimethylethyl 2 –
aminopropanoate hydrochloride monohydrate
(absolute configuration)
(R) – Alaproclate
$C_{13}H_{19}ClNO_2{}^+$, Cl^-, H_2O
A.Wagner *Acta Crystallogr.,Sect.B*,**36**, 77, 1980

48.40 S – 8 – Aza – adenosyl – L – homocysteine
dihydrate
$C_{13}H_{19}N_7O_5S$, $2H_2O$
V.B.Pett, H.–S.Shieh, H.M.Berman
Am.Cryst.Assoc.,Ser.2,**8**, 35, 1980
See also R1 : 47,45,44

48.41 N – t – Butyloxycarbonyl – L – prolyl – sarcosine
$C_{13}H_{22}N_2O_5$
E.Benedetti, A.Ciajolo, B.di Blasio, V.Pavone, C.Pedone,
C.Toniolo, G.M Bonora *Macromolecules*,**12**, 438, 1979

48.42 L – Leucyl – L – prolyl – glycine monohydrate
$C_{13}H_{23}N_3O_4$, H_2O
R.E.Marsh *Acta Crystallogr.,Sect.B*,**36**, 1265, 1980

48.43 t – Butoxy – carbonyl – L – phenylalanine
$C_{14}H_{19}NO_4$
J.W.Bats, H.Fuess, H.Kessler, R.Schuck
Chem.Ber.,**113**, 520, 1980

48.44 S – Adenosyl – L – homocysteine hydrate
$C_{14}H_{20}N_6O_5S$, $2.5H_2O$
V.B.Pett, H.–S.Shieh, H.M.Berman
Am.Cryst.Assoc.,Ser.2,**8**, 35, 1980
See also R1 : 47,45,44

48.45 Cyclo(di – L – prolyl – D – prolyl)
$C_{15}H_{21}N_3O_3$
J.W.Bats, H.Fuess *J.Am.Chem.Soc.*,**102**, 2065, 1980
See also R1 : 36

48.46 Clavicipitic acid
$C_{16}H_{18}N_2O_2$
J.E.Robbers, H.Otsuka, H.G.Floss, E.V.Arnold, J.Clardy
J.Org.Chem.,**45**, 1117, 1980

48.47 (Glutamyl – α – lactam) – histidinyl – proline tartrate monohydrate
Thyrotropin – releasing – hormone tartrate monohydrate
$C_{16}H_{23}N_6O_4^+$, $C_4H_5O_6^-$, H_2O
K.Kamiya, M.Takamoto, Y.Wada, M.Fujino, M.Nishikawa
J.Chem.Soc.,Chem.Commun.,438, 1980

48.C 11β – Acetylamino – 2α – H – 8α – isopropyl – 11α – methyl – 1α – methoxy – 6,9 – diaza – 12 – oxatricyclo(7.3.0.0^{2,6})dodecane – 7,10 – dione
$C_{18}H_{25}N_3O_5$ Main entry is 58.22

48.48 Pivaloyl – D – alanyl – N – isopropyl – D – prolinamide monohydrate
$C_{16}H_{29}N_3O_3$, H_2O
A.Aubry, J.Protas, G.Boussard, M.Marraud
Acta Crystallogr.,Sect.B,36, 321, 1980

48.49 8 – Benzyl – 1 – hydroxy – 11 – methyl – 6,9 – diaza – 12 – thiatricyclo(7.3.0.0^{2,6})dodecane – 7,10 – dione
$C_{17}H_{20}N_2O_3S$
G.Lucente, F.Pinnen, G.Zanotti, S.Cerrini, W.Fedeli, F.Mazza *J.Chem.Soc.,Perkin Trans.1*,1499, 1980

48.50 L – Pyroglutamyl – β – (2 – thienyl) – L – alanyl – L – prolinamide
$C_{17}H_{22}N_4O_4S$
B.Stensland, S.Castensson *Eur.Cryst.Meeting*,5, 16, 1979

48.51 3S,6R, – 6 – Methyl – 5 – oxothiomorpholino – 3 – carbonyl – L – histidinyl – L – prolylamine sesquihydrate
$C_{17}H_{24}N_6O_4S$, $1.5H_2O$
J.J.Stezowski, E.Eckle *Eur.Cryst.Meeting*,6, 275, 1980

48.52 N – (t – Butoxycarbonyl) – L – prolyl – L – valyl – glycine hemihydrate
$C_{17}H_{29}N_3O_6$, $0.5H_2O$
I.Tanaka, T.Ashida
Acta Crystallogr.,Sect.B,36, 2164, 1980

48.53 N – Benzyloxycarbonyl – prolyl – proline
$C_{16}H_{22}N_2O_5$
N.M.Galitskii, V.I.Deigin, W.Saenger, V.Z.Pletnev
Bioorg.Khim.,3, 1445, 1977

48.54 Cyclo – bis(glycyl – L – propyl – glycyl) tetrahydrate
$C_{16}H_{26}N_6O_6$, $4H_2O$
E.C.Kostansek, W.E.Thiessen, D.Schomburg, W.N.Lipscomb
J.Am.Chem.Soc.,101, 5811, 1979

48.55 L – Pyroglutamyl – N,N′ – dicyclohexylurea
$C_{18}H_{29}N_3O_3$
F.Bechtel, J.–P.Bideau, M.Cotrait
Cryst.Struct.Commun.,8, 815, 1979
See also R1 : 32,8

48.56 t – Butoxy – carbonyl – L – prolyl – L – isoleucyl – glycine hydrate
$C_{18}H_{31}N_3O_6$, $0.25H_2O$
Y.Yamada, I.Tanaka, T.Ashida
Acta Crystallogr.,Sect.B,36, 331, 1980

48.57 Folic acid dihydrate
$C_{19}H_{19}N_7O_6$, $2H_2O$
A.Camerman, D.Mastropaolo, N.Camerman
Am.Cryst.Assoc.,Ser.2,7, 18, 1980
See also R1 : 35

48.58 Cyclo(benzylglycyl – bis(L – prolyl))
$C_{19}H_{23}N_3O_3$
J.W.Bats, H.Fuess *Eur.Cryst.Meeting*,6, 281, 1980

48.59 5 – n – Propyl – orotyl – L – histidinyl – L – prolylamine tetrahydrate (at 120°K)
$C_{19}H_{25}N_7O_5$, $4H_2O$
J.J.Stezowski, E.Eckle *Eur.Cryst.Meeting*,6, 275, 1980

48.60 N – (t – Butoxy – carbonyl) – L – methionyl – glycine – benzyl ester
$C_{19}H_{28}N_2O_5S$
T.Yamane, T.Umemura, T.Kojima, Y.Yamada, T.Ashida
Bull.Chem.Soc.Jpn.,53, 908, 1980

48.61 cyclo(L – Valyl – L – prolyl – glycyl – L – valyl – glycyl) trihydrate
$C_{19}H_{31}N_5O_5$, $3H_2O$
H.Einspahr, W.J.Cook, C.E.Bugg
Am.Cryst.Assoc.,Ser.2,7, 14, 1980

48.62 Pivaloyl – D – prolyl – L – prolyl – L – alanine – N – methylamide
$C_{19}H_{32}N_4O_4$
C.M.Nair, M.Vijayan, Y.V.Venkatachalapathi, P.Balaram
J.Chem.Soc.,Chem.Commun.,1183, 1979

48.63 (+) – 5,10 – Methenyl – 5,6,7,8 – tetrahydrofolic acid bromide hydrobromide dihydrate (for abs. configuration see Fontecilla–Camps et al.,J.Amer.Chem. Soc.101,6114,1979)
$C_{20}H_{23}N_7O_6^{2+}$, $2Br^-$, $2H_2O$
J.C.Fontecilla–Camps, C.E.Bugg, C.Temple Junior, J.D.Rose, J.A.Montgomery, R.L.Kisliuk *Der Biochem.*,4, 235, 1978
See also R1 : 36

48.64 t – Butoxy – carbonyl – L – prolyl – sarcosine – benzyl ester
$C_{20}H_{28}N_2O_5$
T.Kojima, T.Kido, H.Itoh, T.Yamane, T.Ashida
Acta Crystallogr.,Sect.B,36, 326, 1980

48.65 Methylbenzoyl – oxycarbonyl – bis(α – amino – isobutyryl) – L – alaninate
$C_{20}H_{29}N_3O_6$
B.V.V.Prasad, N.Shamala, R.Nagaraj, P.Balaram
Acta Crystallogr.,Sect.B,36, 107, 1980

48.66 Cyclo – tris(prolyl – glycyl)
$C_{21}H_{30}N_6O_6$
G.Kartha, K.I.Varughese, S.Aimoto
Am.Cryst.Assoc.,Ser.2,8, 45, 1980

48.67 Cyclo – tris(prolyl – glycyl) calcium perchlorate hydrate
$C_{21}H_{30}N_6O_6$, $0.5Ca^{2+}$, ClO_4^-, $1.5H_2O$
K.I.Varughese, S.Aimoto, G.Kartha
Am.Cryst.Assoc.,Ser.2,7, 20, 1980

48.68 Cyclo – tris(prolyl – glycyl) hydrate
$C_{21}H_{30}N_6O_6$, $1.5H_2O$
G.Kartha, S.Aimoto, K.I.Varughese
Am.Cryst.Assoc.,Ser.2,7, 19, 1980

48.69 Cyclo – tris(prolyl – glycyl) magnesium perchlorate trihydrate
$C_{21}H_{30}N_6O_6$, Mg^{2+}, $2ClO_4^-$, $3H_2O$
G.Kartha, K.I.Varughese, S.Aimoto
Am.Cryst.Assoc.,Ser.2,8, 45, 1980

48.70 Cyclo – tris(prolyl – glycyl) calcium perchlorate
$2C_{21}H_{30}N_6O_6$, Ca^{2+}, $2ClO_4^-$
G.Kartha, K.I.Varughese, S.Aimoto
Am.Cryst.Assoc.,Ser.2,**8**, 45, 1980

48.71 Cyclo(D – N – methylvalyl – L – α – hydroxyisovaleryl – L – N – methylvalyl – D – α – hydroxyisovaleryl)
$C_{22}H_{38}N_2O_6$
Z.Karimov, G.N.Tishchenko
Kristallografiya,**24**, 778, 1979

48.72 Tetraenniatin
$C_{22}H_{38}N_2O_6$
A.I.Karaulov, G.N.Tishchenko, B.K.Vainshtein
Cryst.Struct.Commun.,**9**, 593, 1980

48.73 Cyclo(di(benzylglycyl) – L – prolyl) monohydrate
$C_{23}H_{25}N_3O_3$, H_2O
J.W.Bats, H.Fuess *J.Am.Chem.Soc.*,**102**, 2065, 1980
See also R1 : 35

48.74 N – Benzyloxycarbonyl – (γ – ethyl) – L – glutamyl – (γ – ethyl) – L – glutamic acid ethyl ester
$C_{24}H_{34}N_2O_9$
E.Benedetti, B.di Blasio, V.Pavone, C.Pedone, G.Germain,
M.Goodman *Biopolymers*,**18**, 517, 1979

48.75 (±) – N – Carbobenzyloxy – (γ,γ' – di – t – butyl) – γ – carboxyglutamyl – glycine ethyl ester
$C_{26}H_{38}N_2O_9$
E.J.Valente, R.G.Hiskey, D.J.Hodgson
Biochim.Biophys.Acta,**579**, 466, 1979

48.C Ferrichrome (at –135°C, absolute configuration)
$C_{27}H_{42}FeN_9O_{12}$ Main entry is 84.108

48.76 Gastrin tetrapeptideamide hydrochloride methanol diethyl ether solvate
$C_{29}H_{37}N_6O_6S^+$, Cl^-, CH_4O, $0.5C_4H_{10}O$
W.B.T.Cruse, E.Egert, O.Kennard, M.A.Viswamitra
Eur.Cryst.Meeting,**6**, 274, 1980

48.77 Methyl – N – (biphenyl – 2 – yl) – N – (1 – naphthyl) – anthranilate (α isomer)
$C_{30}H_{23}NO_2$
R.Glaser, J.F.Blount, K.Mislow
J.Am.Chem.Soc.,**102**, 2777, 1980
See also R1 : 24

48.78 Methyl – N – (biphenyl – 2 – yl) – N – (1 – naphthyl) – anthranilate (β₁ isomer)
$C_{30}H_{23}NO_2$
R.Glaser, J.F.Blount, K.Mislow
J.Am.Chem.Soc.,**102**, 2777, 1980
See also R1 : 24

48.79 Methyl – N – (biphenyl – 2 – yl) – N – (1 – naphthyl) – anthranilate (β₂ isomer)
$C_{30}H_{23}NO_2$
R.Glaser, J.F.Blount, K.Mislow
J.Am.Chem.Soc.,**102**, 2777, 1980
See also R1 : 24

48.C Cyclo(L – methylvalyl – D – α – hydroxyisovaleryl – L – methylvalyl – D – α – hydroxyisovaleryl – L – methylvalyl – D – α – hydroxyisovaleryl) sodium nickel nitrate hydrate
L,D,L,D,L,D – Enniatin B sodium nickel nitrate hydrate
$C_{33}H_{57}N_3O_9$, Na^+, $6.25H_2O$, $N_3NiO_9^-$ Main entry is 50.22

48.80 Cyclo(L – leucyl – L – phenylalanyl – glycyl – D – leucyl – D – phenylalanyl – glycyl) dihydrate (at 138°K)
$C_{34}H_{46}N_6O_6$, $2H_2O$
M.B.Hossain, D.van der Helm
Acta Crystallogr.,Sect.B,**35**, 2781, 1979

48.C Valinomycin potassium picrate
$C_{36}H_{60}N_4O_{12}$, $C_6H_2N_3O_7^-$, K^+ Main entry is 60.43

48.C Cyclo(D – valyl – L – methylalanyl – L – valyl – D – α – hydroxyisovaleryl – D – valyl – L – methylalanyl – L – valyl – D – α – hydroxyisovaleryl)
2,6 – bis(Methylalanyl) – octavalinomycin
$C_{38}H_{66}N_6O_{10}$ Main entry is 50.24

48.81 Cyclo – bis(D – valyl – prolyl – valyl – D – α – hydroxyisovaleryl) dihydrate
$C_{40}H_{66}N_6O_{10}$, $2H_2O$
V.Z.Pletnev, N.M.Galitskii, D.A.Langs, W.L.Duax
Bioorg.Khim.,**6**, 5, 1980

48.82 Methionyl – glutamyl – histidinyl – phenylalanyl – arginyl – tryptophanyl – glycine
$C_{44}H_{59}N_{13}O_{10}S$
G.Admiraal, A.B.Verweij, A.Vos
Eur.Cryst.Meeting,**6**, 319, 1980

48.C Des – N – tetramethyl – triostin A methanol solvate hydrate (at –176°C)
$C_{46}H_{54}N_{12}O_{12}S_2$, $2CH_4O$, xH_2O Main entry is 50.33

48.C Des – N – tetramethyl – triostin A dodecahydrate
$C_{46}H_{54}N_{12}O_{12}S_2$, $12H_2O$ Main entry is 50.34

48.C Valinomycin barium perchlorate hydrate
$C_{54}H_{90}N_6O_{18}$, $2Ba^{2+}$, $4ClO_4^-$, $4H_2O$ Main entry is 50.36

48.C Cyclo(D – valyl – L – methylalanyl – L – valyl – D – α – hydroxyisovaleryl – bis(D – valyl – L – lactyl – L – valyl – D – α – hydroxyisovaleryl))
$C_{55}H_{93}N_7O_{17}$ Main entry is 50.37

48.83 Cyclo – tris(L – valyl – L – prolyl – glycyl – L – valyl – glycyl) monohydrate dimethylsulfoxide solvate
$C_{57}H_{93}N_{15}O_{15}$, H_2O, C_2H_6OS
H.Einspahr, W.J.Cook, C.E.Bugg
Am.Cryst.Assoc.,Ser.2,**7**, 14, 1980

48.C Prolinomycin rubidium picrate toluene chloroform solvate
$2C_{60}H_{96}N_{12}O_{12}$, $2Rb^+$, $2C_6H_2N_3O_7^-$, $3C_7H_8$, $2CHCl_3$
Main entry is 50.38

48.C Cyclo – tris(D – valyl – L – α – hydroxyisovaleryl – L – valyl – D – α – hydroxyisovaleryl) (triclinic form)
meso – Valinomycin
$C_{60}H_{102}N_6O_{18}$ Main entry is 50.39

48.C N – Acetyl – gramicidin S
N – Acetyl – cyclo – bis(L – prolyl – L – valyl – L – ornithyl – L – leucyl – L – phenylalanyl)
$C_{64}H_{96}N_{12}O_{12}$ Main entry is 50.40

48.C Carboxy – decarboxamido – vancomycin
$C_{66}H_{74}Cl_2N_8O_{25}$, xH_2O Main entry is 50.41

PORPHYRINS AND CORRINS

49.C Copper(ii) – copper(i) – pseudo – porphyrin – macrocyclic ligand complex
$C_{24}H_{26}Cu_2N_4O_2{}^+$, $ClO_4{}^-$, $0.5CH_4O$ Main entry is 84.98

49.1 Phthalocyaninato – cobalt(ii) (at 4.3°K, neutron study, β form)
$C_{32}H_{16}CoN_8$
G.A.Williams, B.N.Figgis, R.Mason, S.A.Mason, P.E.Fielding
J.Chem.Soc.,Dalton Trans.,1688, 1980

49.2 Phthalocyaninato – manganese(ii) (at 116°K, β form, re–refinement of the data of Mason, et al (J.Chem.Soc.Dalton 676(1979)), electron density distribution study)
$C_{32}H_{16}MnN_8$
B.N.Figgis, E.S.Kucharski, G.A.Williams
J.Chem.Soc.,Dalton Trans.,1515, 1980

49.3 (5,10,15,20 – Tetra – n – propylporphinato) – lead(ii)
$C_{32}H_{36}N_4Pb$
K.M.Barkigia, J.Fajer, A.D.Adler, G.J.B.Williams
Inorg.Chem.,**19**, 2057, 1980

49.4 trans – N,N' – Dimethyletioporphyrin tri – iodide
$C_{34}H_{43}N_4{}^+$, $I_3{}^-$
A.M.Abeysekera, R.Grigg, J.Trocha–Grimshaw, K.Henrick
Tetrahedron,**36**, 1857, 1980

49.5 Carbonyl – (N,N – dimethylformamide) – phthalocyaninato – iron(ii) dimethylformamide solvate
$C_{36}H_{23}FeN_9O_2$, C_3H_7NO
F.Calderazzo, G.Pampaloni, D.Vitali, G.Pelizzi, I.Collamati, S.Frediani, A.M.Serra *J.Organomet.Chem.*,**191**, 217, 1980

49.6 8¹(E) – 8² – Nitro – protoporphyrin dimethyl ester dichloroethane solvate
$C_{36}H_{37}N_5O_6$, $C_2H_4Cl_2$
R.Bonnett, M.B.Hursthouse, P.A.Scourides, J.Trotter
J.Chem.Soc.,Perkin Trans.1,490, 1980

49.7 Octaethylporphinato – perchlorato – iron(iii)
$C_{36}H_{44}ClFeN_4O_4$
H.Masuda, T.Taga, K.Osaki, H.Sugimoto, Z.–I.Yoshida, H.Ogoshi *Inorg.Chem.*,**19**, 950, 1980

49.8 tct – 2,3,7,8 – Tetrahydro – octaethylporphyrinato – nickel(ii)
$C_{36}H_{48}N_4Ni$
C.Kratky *Eur.Cryst.Meeting*,**6**, 271, 1980

49.9 ttt – 2,3,7,8 – Tetrahydro – octaethylporphyrinato – nickel(ii)
$C_{36}H_{48}N_4Ni$
C.Kratky *Eur.Cryst.Meeting*,**6**, 271, 1980

49.10 tct – 2,3,7,8 – Tetrahydro – octaethylporphyrin
$C_{36}H_{50}N_4$
C.Kratky *Eur.Cryst.Meeting*,**6**, 271, 1980

49.11 ttt – 2,3,7,8 – Tetrahydro – octaethylporphyrin
$C_{36}H_{50}N_4$
C.Kratky *Eur.Cryst.Meeting*,**6**, 271, 1980

49.12 (trans,trans,cis,trans – Octaethyl – 1,2,3,7,8,20 – hexahydroporphyrinato) – nickel(ii)
$C_{36}H_{52}N_4Ni$
J.E.Johansen, C.Angst, C.Kratky, A.Eschenmoser
Angew.Chem.,Int.Ed.Engl.,**19**, 141, 1980

49.13 (trans,trans,trans,trans – Octaethyl – 1,2,3,7,8,20 – hexahydroporphyrinato) – nickel(ii)
$C_{36}H_{52}N_4Ni$
J.E.Johansen, C.Angst, C.Kratky, A.Eschenmoser
Angew.Chem.,Int.Ed.Engl.,**19**, 141, 1980

49.14 Carbonyl – (phthalocyaninato) – (pyridine) – osmium(ii)
$C_{38}H_{21}N_9OOs$
S.Omiya, M.Tsutsui, E.F.Meyer Junior, I.Bernal, D.L.Cullen
Inorg.Chem.,**19**, 134, 1980

49.15 N – Tosylamino – octaethylporphinato – bis(chloro – mercury)
$C_{43}H_{51}Cl_2Hg_2N_5O_2S$
H.J.Callot, B.Chevrier, R.Weiss
J.Am.Chem.Soc.,**101**, 7729, 1979

49.16 Chloro – (5,10,15,20 – tetraphenylporphinato) – indium(iii)
$C_{44}H_{28}ClInN_4$
R.G.Ball, K.M.Lee, A.G.Marshall, J.Trotter
Inorg.Chem.,**19**, 1463, 1980

49.17 Tetraphenylporphyrinato – cobalt(ii) (at 100°K, electron density distribution study)
$C_{44}H_{28}CoN_4$
E.D.Stevens *Am.Cryst.Assoc.,Ser.2*,**7**, 16, 1980

49.C Trihydroxyphosphorus – tetraphenylporphyrin dihydrate
$C_{44}H_{31}N_4O_3P$, $2H_2O$ Main entry is 64.172

49.18 1,4,5,8,9,12,13,16 – Octamethyl – tetrabenzoporphyrinato – nickel(ii) iodide
$C_{44}H_{36}N_4Ni^+$, I^-, 0.081
T.E.Phillips, R.P.Scaringe, B.M.Hoffman, J.A.Ibers
J.Am.Chem.Soc.,**102**, 3435, 1980

49.19 Chloro – (N – methyl – 5,10,15,20 – tetraphenylporphinato) – iron(ii)
$C_{45}H_{31}ClFeN_4$
O.P.Anderson, A.B.Kopelove, D.K.Lavallee
Inorg.Chem.,**19**, 2101, 1980

49.20 Methoxy – (meso – tetraphenylporphyrinato) – iron(iii) (at 100°K)
$C_{45}H_{31}FeN_4O$
E.D.Stevens, D.L.Chadwick
Am.Cryst.Assoc.,Ser.2,**8**, 18, 1980

49.21 2,6 – Diethyl – 1,3,5,7 – tetramethyl – 4,8 – bis(propionic acid) – porphyrin – bis(ethane – 1',2' – diyl ester) – (pyridine – 3'',5'' – dicarboxylate) dichloromethane solvate
$C_{45}H_{47}N_5O_8$, $1.5CH_2Cl_2$
W.B.Cruse, O.Kennard, G.M.Sheldrick, A.D.Hamilton, S.G.Hartley, A.R.Battersby
J.Chem.Soc.,Chem.Commun.,700, 1980

49.22 Potassium dicyano – (meso –
tetraphenylporphinato) – iron(iii) acetone solvate
$C_{46}H_{28}FeN_6^-$, K^+, $2C_3H_6O$
W.R.Scheidt, K.J.Haller, K.Hatano
J.Am.Chem.Soc., **102**, 3017, 1980

49.23 Sodium ethylthio – (tetraphenylporphinato) –
iron(ii) 4,7,13,16,21,24 – hexaoxa – 1,10 –
diazabicyclo(8.8.8)hexacosane chlorobenzene solvate
$C_{46}H_{33}FeN_4S^-$, $C_{18}H_{36}N_2O_6$, Na^+, $2C_6H_5Cl$
C.Caron, A.Mitschler, G.Riviere, L.Ricard, M.Schappacher,
R.Weiss *J.Am.Chem.Soc.*, **101**, 7401, 1979

49.24 bis(3 – Chloropyridine) – octaethylporphinato –
iron(iii) perchlorate
$C_{46}H_{52}Cl_2FeN_6^+$, ClO_4^-
W.R.Scheidt, D.K.Geiger
J.Chem.Soc.,Chem.Commun., 154, 1979

49.25 Dicyanocobyrinic – a,c – diamide trihydrate
$C_{47}H_{61}CoN_8O_{12}$, $3H_2O$
B.Dresow, G.Schlingmann, W.S.Sheldrick,
V.B.Koppenhagen
Angew.Chem.,Int.Ed.Engl., **19**, 321, 1980

49.26 cis – Dioxo – (5,10,15,20 – tetra – p –
tolylporphyrinato) – molybdenum(vi) toluene
solvate
$C_{48}H_{36}MoN_4O_2$, $1.5C_7H_8$
B.F.Mentzen, M.C.Bonnet, H.J.Ledon
Inorg.Chem., **19**, 2061, 1980

49.27 Disodium (ethylthio) – (tetra – p –
tolylporphinato) – iron(ii) ethylthiolate
4,7,13,16,21 – pentaoxa – 1,10 – diazabicyclo(8.8.5)
tricosane benzene solvate (at 75°K)
$C_{51}H_{41}FeN_4OS^-$, $C_2H_5S^-$, $2C_{16}H_{32}N_2O_5$, $1.5C_6H_6$, $2Na^+$
C.Caron, A.Mitschler, G.Riviere, L.Ricard, M.Schappacher,
R.Weiss *J.Am.Chem.Soc.*, **101**, 7401, 1979

49.28 Tetraphenylporphinato – bis(dioxane) – cadmium(ii)
$C_{52}H_{44}CdN_4O_4$
P.F.Rodesiler, E.H.Griffith, P.D.Ellis, E.L.Amma
J.Chem.Soc.,Chem.Commun., 492, 1980

49.C meso – Tetraphenylporphinato –
bis(tetrahydrofuran) – iron(ii)
$C_{52}H_{44}FeN_4O_2$ Main entry is 84.136

49.29 Dicyano – (5,6 – dihydroxy – dihydrocobyrinic acid
pentamethyl ester) – c,d – dilactone monohydrate
$C_{52}H_{69}CoN_6O_{15}$, H_2O
W.S.Sheldrick, W.Becker *Eur.Cryst.Meeting*, **5**, 63, 1979

49.30 6 – Amino – dicyano – 5,6 – dihydrocobyrinic –
pentamethylester – a – amide – c – lactam
hemihydrate
$C_{52}H_{71}CoN_8O_{12}$, $0.5H_2O$
G.Schlingmann, B.Dresow, V.B.Koppenhagen, W.Becker,
W.S.Sheldrick *Angew.Chem.,Int.Ed.Engl.*, **19**, 321, 1980

49.C bis(Pyridine) – (meso – tetraphenylporphinato) –
chromium(ii) toluene solvate
$C_{54}H_{38}CrN_6$, C_7H_8 Main entry is 83.222

49.31 1,5 – Diethyl – 2,4,6,8 – tetramethyl – 3,7 – μ –
(9,10 – bis(ethylcarboxylato – ethoxycarbonyl) –
anthracene) – porphyrin methylene chloride
solvate
$C_{54}H_{52}N_4O_8$, $0.5CH_2Cl_2$
W.B.Cruse, O.Kennard, G.M.Sheldrick, A.D.Hamilton,
S.G.Hartley, A.R.Battersby
J.Chem.Soc.,Chem.Commun., 700, 1980

49.32 5,10,15,20 – (Pyrromellitoyl – (tetrakis – o –
oxyethoxyphenyl)) – porphyrin chloroform
methanol solvate (at –150°C)
$C_{62}H_{44}N_4O_{12}$, $5CHCl_3$, CH_4O
G.B.Jameson, J.A.Ibers *J.Am.Chem.Soc.*, **102**, 2823, 1980

49.33 bis(Phthalocyaninato) – neodymium(iii) methylene
chloride solvate
$C_{64}H_{33}N_{16}Nd$, CH_2Cl_2
K.Kasuga, M.Tsutsui, R.C.Pettersen, K.Tatsumi,
N.van Opdenbosch, G.Pepe, E.F.Meyer Junior
J.Am.Chem.Soc., **102**, 4835, 1980

49.C Chloro – (meso – α,α,α – tetra(o –
nicotinamidophenyl) – porphyrinato – iron(iii)) –
copper(ii) perchlorate dihydrate
$C_{68}H_{44}ClCuFeN_{12}O_4^{2+}$, $2ClO_4^-$, $2H_2O$ Main entry is 83.225

49.34 (2 – Methylimidazole) – meso – tetra($\alpha,\alpha,\alpha,\alpha$ – o –
pivalamidophenyl) – porphyrinato – iron(ii) ethanol
solvate
$C_{68}H_{70}FeN_{10}O_4$, C_2H_6O
G.B.Jameson, F.S.Molinaro, J.A.Ibers, J.P.Collman,
J.I.Brauman, E.Rose, K.S.Suslick
J.Am.Chem.Soc., **102**, 3224, 1980

49.35 (2 – Methylimidazole) – dioxygen – meso –
tetra($\alpha,\alpha,\alpha,\alpha$ – o – pivalamidophenyl) –
porphyrinato – iron ethanol solvate
$C_{68}H_{70}FeN_{10}O_6$, C_2H_6O
G.B.Jameson, F.S.Molinaro, J.A.Ibers, J.P.Collman,
J.I.Brauman, E.Rose, K.S.Suslick
J.Am.Chem.Soc., **102**, 3224, 1980

49.36 Di – copper hexyl cofacial diporphyrin – 7
$C_{92}H_{124}Cu_2N_{10}O_2$
M.Hatada, A.Tulinsky *Am.Cryst.Assoc.*, *Ser.2*, **8**, 34, 1980

ANTIBIOTICS

50.1 **trans – 2 – Azabicyclo(2.1.0)pentane – 3 – (S) – carboxylic acid**
$C_5H_7NO_2$
Y.Kodama, T.Ito *Agric.Biol.Chem.*,**44**, 73, 1980
See also R1 : 48

50.2 **Clazamycin A hydrochloride (absolute configuration)**
$C_7H_{10}ClN_2O^+, Cl^-$
H.Nakamura, Y.Iitaka, H.Umezawa
J.Antibiot.,**32**, 765, 1979
See also R1 : 35

50.3 **Trichoviridine**
$C_8H_9NO_4$
W.D.Ollis, M.Rey, W.O.Godtfredsen, N.Rastrup–Andersen,
S.Vangedal, T.J.King *Tetrahedron*,**36**, 515, 1980
See also R1 : 38,20

50.4 **L – 3 – Amino – 1 – methoxy – 6 – methylamino – 1,3,6 – trideoxy – chiro – inositol**
Fortamine
$C_8H_{18}N_2O_4$
N.Hirayama, K.Shirahata, Y.Ohashi, Y.Sasada
Bull.Chem.Soc.Jpn.,**53**, 1514, 1980
See also R1 : 45

50.5 **N – Acetyl – furanomycin**
$C_9H_{13}NO_4$
M.Shiro, H.Nakai, K.Tori, J.Nishikawa, Y.Yoshimura,
K.Katagiri *J.Chem.Soc.,Chem.Commun.*,375, 1980
See also R1 : 38

50.6 **2,2' – (2,2,2 – Trichloroethylidene) – bis(4 – chloro – 6 – nitrophenol)**
$C_{14}H_7Cl_5N_2O_6$
D.G.Hay, P.DeMunk, M.F.Mackay *Aust.J.Chem.*,**33**, 77, 1980
See also R1 : 15

50.7 **2,3 – Dimethyl – 5 – (p – bromobenzoyloxymethyl) – cyclopent – 2 – en – 1 – one**
Methylenomycin B p – bromobenzoate
$C_{15}H_{15}BrO_3$
J.Jernow, W.Tautz, P.Rosen, T.H.Williams
J.Org.Chem.,**44**, 4212, 1979
See also R1 : 20

50.8 **(+) – Methylenomycin A p – bromobenzoate (absolute configuration)**
$C_{15}H_{15}BrO_4$
J.Jernow, W.Tautz, P.Rosen, J.F.Blount
J.Org.Chem.,**44**, 4210, 1979
See also R1 : 38

50.9 **Mitomycin C dihydrate**
$C_{15}H_{18}N_4O_5, 2H_2O$
K.Ogawa, A.Nomura, T.Fujiwara, K.Tomita
Bull.Chem.Soc.Jpn.,**52**, 2334, 1979
See also R1 : 36

50.10 **Tomaymycin**
$C_{16}H_{20}N_2O_4$
S.K.Arora *Am.Cryst.Assoc.,Ser.2*,**8**, 35, 1980
See also R1 : 36

50.11 **Griseofulvin benzene solvate**
7 – Chloro – 2',4,6 – trimethoxy – 6' – β – methyl – spirobenzofuran – 2(3H) – cyclohex – 2 – ene – 3,4' – dione benzene solvate
$C_{17}H_{17}ClO_6, C_6H_6$
Z.Dauter, A.Hempel, M.Bogucka–Ledochowska,
E.Borowski, H.Bradaczek, W.Dreissig
Eur.Cryst.Meeting,**5**, 46, 1979
See also R1 : 38

50.12 **Anthramycin methylether monohydrate**
$C_{17}H_{19}N_3O_4, H_2O$
S.K.Arora *Acta Crystallogr.,Sect.B*,**35**, 2945, 1979
See also R1 : 36

50.13 **6 – Thiatetracycline (at 120°K)**
$C_{20}H_{20}N_2O_7S$
R.Prewo, J.J.Stezowski *Tetrahedron Lett.*,**21**, 251, 1980
See also R1 : 39

50.14 **5a – epi – 6 – Demethyl – 6 – deoxy – 6 – thiatetracycline dimethylformamide solvate (at 120°K)**
$C_{20}H_{20}N_2O_7S, C_3H_7NO$
R.Prewo, J.J.Stezowski *Eur.Cryst.Meeting*,**5**, 43, 1979
See also R1 : 39

50.15 **4 – epi – Oxytetracycline tetrahydrate dichloromethane solvate (at 120°K)**
$C_{22}H_{24}N_2O_9, 4H_2O, CH_2Cl_2$
R.Prewo, J.J.Stezowski *J.Am.Chem.Soc.*,**101**, 7657, 1979
See also R1 : 29

50.16 **6α – Benzyl – 6β – isocyano – penicillanate**
$C_{23}H_{22}N_2O_3S$
P.H.Bentley, J.P.Clayton, M.O.Boles, R.J.Girven
J.Chem.Soc.,Perkin Trans.1,2455, 1979
See also R1 : 41,35

50.17 **Nodusmicin**
$C_{23}H_{34}O_7$
H.A.Whaley, C.G.Chidester, S.A.Mizsak, R.J.Wnuk
Tetrahedron Lett.,**21**, 3659, 1980
See also R1 : 38

50.18 **11 – Deoxy – daunorubicin aglycone triacetate**
$C_{27}H_{24}O_{10}$
F.Arcamone, G.Cassinelli, F.DiMatteo, S.Forenza,
M.C.Ripamonti, G.Rivola, A.Vigevani, J.Clardy, T.McCabe
J.Am.Chem.Soc.,**102**, 1462, 1980
See also R1 : 26

50.19 **7 – Con – O – methylnogarol ethanol solvate (at 120°K)**
$2C_{28}H_{31}NO_{10}, C_2H_6O$
E.Eckle, J.J.Stezowski, P.F.Wiley
Tetrahedron Lett.,**21**, 507, 1980
See also R1 : 38

50.20 **Saframycin C hydrobromide monohydrate**
$C_{29}H_{34}N_3O_9^+, Br^-, H_2O$
T.Arai, K.Takahashi, A.Kubo, S.Nakahara, S.Sato, K.Aiba,
C.Tamura *Tetrahedron Lett.*,2355, 1979

50.21 8,9 – Dihydroxy – 1,3,7,7 – tetramethyl – bicyclo(4.4.0)decane – 2,2' – spiro(6',7' – bis(hydroxymethyl) – 4' – (p – bromophenylsulfonyloxy) – 2',3' – dihydrobenzofuran) dimethylsulfoxide solvate
$C_{29}H_{37}BrO_8S$, C_2H_6OS
H.Kaise, M.Shinohara, W.Miyazaki, T.Izawa, Y.Nakano, M.Sugawara, K.Sugiura, K.Sasaki
*J.Chem.Soc.,Chem.Commun.,*726, 1979
See also R1 : 38

50.22 Cyclo(L – methylvalyl – D – α – hydroxyisovaleryl – L – methylvalyl – D – α – hydroxyisovaleryl – L – methylvalyl – D – α – hydroxyisovaleryl) sodium nickel nitrate hydrate
L,D,L,D,L,D – Enniatin B sodium nickel nitrate hydrate
$C_{33}H_{57}N_3O_9$, Na^+, $6.25H_2O$, $N_3NiO_9^-$
G.N.Tishchenko, N.E.Zhukhlistova, A.I.Karaulov, V.I.Smirnova *Eur.Cryst.Meeting,*6, 303, 1980
See also R1 : 48

50.23 9,10 – O,O – Isopropylidene – 10a – O – (o – bromobenzoyl) – 11 – O – methyl – rubrolone
$C_{34}H_{32}BrNO_9$
W.Schuep, J.F.Blount, T.H.Williams, A.Stempel
*J.Antibiot.,*31, 1226, 1978
See also R1 : 38,36

50.24 Cyclo(D – valyl – L – methylalanyl – L – valyl – D – α – hydroxyisovaleryl – D – valyl – L – methylalanyl – L – valyl – D – α – hydroxyisovaleryl)
2,6 – bis(Methylalanyl) – octavalinomycin
$C_{38}H_{66}N_6O_{10}$
G.N.Tishchenko, N.E.Zhukhlistova, A.I.Karaulov, V.I.Smirnova *Eur.Cryst.Meeting,*6, 303, 1980
See also R1 : 48

50.25 3 – Carbomethoxy – rifamycin SV monohydrate
$C_{39}H_{47}NO_{14}$, H_2O
M.Brufani, L.Cellai, S.Cerrini, W.Fedeli, A.Vaciago
*Eur.Cryst.Meeting,*5, 48, 1979
See also R1 : 40

50.26 Sodium nigericine
$C_{40}H_{67}O_{11}^-$, Na^+
Y.Barrans, M.Alleaume, L.David
*Acta Crystallogr.,Sect.B,*36, 936, 1980
See also R1 : 38

50.27 Hedamycin
$C_{41}H_{50}N_2O_{11}$
M.Zehnder, U.Sequin, H.Nadig
*Helv.Chim.Acta,*62, 2525, 1979
See also R1 : 38

50.28 Ionomycin calcium salt n – heptane solvate
$C_{41}H_{70}O_9^{2-}$, Ca^{2+}, $0.5C_7H_{16}$
B.K.Toeplitz, A.I.Cohen, P.T.Funke, W.L.Parker, J.Z.Gougoutas *J.Am.Chem.Soc.,*101, 3344, 1979
See also R1 : 38

50.29 Ionomycin cadmium salt n – heptane solvate (absolute configuration)
$C_{41}H_{70}O_9^{2-}$, Cd^{2+}, $0.5C_7H_{16}$
B.K.Toeplitz, A.I.Cohen, P.T.Funke, W.L.Parker, J.Z.Gougoutas *J.Am.Chem.Soc.,*101, 3344, 1979
See also R1 : 38

50.30 Rifamycin Y p – iodoanilide dimethylsulfoxide monohydrate
$C_{45}H_{51}IN_2O_{14}$, C_2H_6OS, H_2O
M.Brufani, S.Cerrini, W.Fedeli, A.Vaciago
*J.Mol.Biol.,*87, 409, 1974
See also R1 : 38

50.31 Rifamycin B p – iodoanilide acetone pentahydrate
$C_{45}H_{53}IN_2O_{13}$, C_3H_6O, $5H_2O$
M.Brufani, S.Cerrini, W.Fedeli, A.Vaciago
*J.Mol.Biol.,*87, 409, 1974
See also R1 : 38

50.32 Beauvericin barium picrate complex toluene solvate (form B)
$C_{45}H_{57}N_3O_9$, $2C_6H_2N_3O_7^-$, Ba^{2+}, $2C_7H_8$
B.Braden, J.A.Hamilton, M.N.Sabesan, L.K.Steinrauf
*J.Am.Chem.Soc.,*102, 2704, 1980
See also R1 : 40

50.33 Des – N – tetramethyl – triostin A methanol solvate hydrate (at −176°C)
$C_{46}H_{54}N_{12}O_{12}S_2$, $2CH_4O$, xH_2O
M.B.Hossain, D.van der Helm, R.K.Olsen
*Am.Cryst.Assoc.,Ser.2,*7, 14, 1980
See also R1 : 48

50.34 Des – N – tetramethyl – triostin A dodecahydrate
$C_{46}H_{54}N_{12}O_{12}S_2$, $12H_2O$
O.Kennard, W.B.T.Cruse, G.M.Sheldrick, P.G.Jones, M.A.Viswamitra, M.Waring, L.Wakelin, R.K.Olsen
*Eur.Cryst.Meeting,*6, 284, 1980
See also R1 : 48,36

50.35 3 – (β – L – Mycarose) – 5 – (β – D – 4,6 – dideoxy – 3 – ketoallose) – 13 – (β – D – mycinose) – lankamycin – 11 – α – hydroxyisovalerate ester
$C_{48}H_{82}O_{20}$
B.Arnoux, C.Pascard, L.Raynaud, J.Lunel
*J.Am.Chem.Soc.,*102, 3605, 1980
See also R1 : 45,38

50.36 Valinomycin barium perchlorate hydrate
$C_{54}H_{90}N_6O_{18}$, $2Ba^{2+}$, $4ClO_4^-$, $4H_2O$
S.Devarajan, C.M.K.Nair, K.R.K.Easwaran, M.Vijayan
*Nature (London),*286, 640, 1980
See also R1 : 48

50.37 Cyclo(D – valyl – L – methylalanyl – L – valyl – D – α – hydroxyisovaleryl – bis(D – valyl – L – lactyl – L – valyl – D – α – hydroxyisovaleryl))
$C_{55}H_{93}N_7O_{17}$
G.N.Tishchenko, N.E.Zhukhlistova, A.I.Karaulov, V.I.Smirnova *Eur.Cryst.Meeting,*6, 303, 1980
See also R1 : 48

50.38 Prolinomycin rubidium picrate toluene chloroform solvate
$2C_{60}H_{96}N_{12}O_{12}$, $2Rb^+$, $2C_6H_2N_3O_7^-$, $3C_7H_8$, $2CHCl_3$
J.A.Hamilton, M.N.Sabesan, L.K.Steinrauf
*Acta Crystallogr.,Sect.B,*36, 1052, 1980
See also R1 : 48

50.39 Cyclo – tris(D – valyl – L – α – hydroxyisovaleryl – L – valyl – D – α – hydroxyisovaleryl) (triclinic form)
meso – Valinomycin
$C_{60}H_{102}N_6O_{18}$
V.Z.Pletnev, N.M.Galitskii, V.T.Ivanov, Yu.A.Ovchinnikov
*Biopolymers,*18, 2145, 1979
See also R1 : 48

50.40 N – Acetyl – gramicidin S
N – Acetyl – cyclo – bis(L – prolyl – L – valyl – L –
ornithyl – L – leucyl – L – phenylalanyl)
$C_{64}H_{96}N_{12}O_{12}$
M.M.Harding *Eur.Cryst.Meeting*,**5**, 245, 1979
See also R1 : 48

50.41 Carboxy – decarboxamido – vancomycin
$C_{66}H_{74}Cl_2N_8O_{25}$, xH_2O
G.M.Sheldrick, P.G.Jones, O.Kennard, D.H.Williams,
G.A.Smith *Nature (London)*,**271**, 223, 1978
See also R1 : 48,45

STEROIDS

51.1 17β – Hydroxy – estra – 4,9,11 – trien – 3 – one
$C_{18}H_{22}O_2$
G.Precigoux, Y.Barrans, M.Hospital
Cryst.Struct.Commun.,**8**, 883, 1979

**51.2 Estra – 1,3,5(10),14 – tetraene – 3,17β – diol
hemihydrate**
$C_{18}H_{22}O_2$, $0.5H_2O$
W.L.Duax, D.C.Rohrer, R.H.Blessing, P.D.Strong, A.Segaloff
Acta Crystallogr.,Sect.B,**35**, 2656, 1979

**51.3 7 – Hydroxy – 3 – methoxy – 6 – oxa – estra –
1,3,5(10) – trien – 17 – one**
$C_{18}H_{22}O_4$
P.S.White, D.C.N.Swindells
Acta Crystallogr.,Sect.B,**36**, 489, 1980
See also R1 : 38

**51.4 3 – Methoxy – estra – 1,3,5(10),14 – tetraen – 17 –
one**
$C_{19}H_{22}O_2$
W.L.Duax, D.C.Rohrer, R.H.Blessing, P.D.Strong, A.Segaloff
Acta Crystallogr.,Sect.B,**35**, 2656, 1979

51.5 16α – Bromo – androst – 4 – ene – 3,6,17 – trione
$C_{19}H_{23}BrO_3$
D.C.Swenson, W.L.Duax, M.Numazawa, Y.Osawa
Am.Cryst.Assoc.,Ser.2,**7**, 12, 1980

51.6 3 – Methoxy – estra – 1,3,5(10) – trien – 17 – one
$C_{19}H_{24}O_2$
W.L.Duax, D.C.Rohrer, R.H.Blessing, P.D.Strong, A.Segaloff
Acta Crystallogr.,Sect.B,**35**, 2656, 1979

**51.7 3 – Methoxy – estra – 1,3,5(10),14 – tetraen – 17β –
ol**
$C_{19}H_{24}O_2$
W.L.Duax, D.C.Rohrer, R.H.Blessing, P.D.Strong, A.Segaloff
Acta Crystallogr.,Sect.B,**35**, 2656, 1979

51.8 16α – Bromo – androst – 4 – ene – 3,17 – dione
$C_{19}H_{25}BrO_2$
D.C.Swenson, W.L.Duax, M.Numazawa, Y.Osawa
Am.Cryst.Assoc.,Ser.2,**7**, 12, 1980

**51.9 3 – Methoxy – 16α – bromo – 17β – hydroxy –
estra – 1,3,5(10) – triene**
$C_{19}H_{25}BrO_2$
K.Szulzewsky, I.Seidel *Krist.Tech.*,**14**, 1127, 1979

**51.10 3 – Methoxy – 16α – bromo – 17α – hydroxy –
estra – 1,3,5(10) – triene**
$C_{19}H_{25}BrO_2$
K.Szulzewsky *Krist.Tech.*,**14**, 1445, 1979

**51.11 17β – Hydroxy – 7β – methyl – estra – 4,14 – dien –
3 – one**
$C_{19}H_{26}O_2$
W.L.Duax, D.C.Rohrer, P.N.Rao
Acta Crystallogr.,Sect.B,**35**, 3074, 1979

51.12 17β – Hydroxy – 9α – methyl – estra – 4,14 – dien – 3 – one
$C_{19}H_{26}O_2$
W.L.Duax, P.D.Strong, D.C.Rohrer, A.Segaloff
Acta Crystallogr.,Sect.B,**36**, 824, 1980

51.13 rac – D – Homo – 8β – estra – 1,3,5(10) – triene – 3,17a – diol ethanol solvate
$C_{19}H_{26}O_2$, C_2H_6O
W.L.Duax, G.D.Smith, D.C.Swenson, P.D.Strong, C.M.Weeks, S.N.Ananchenko, V.V.Egorova
Am.Cryst.Assoc.,Ser.2,**7**, 12, 1980

51.14 17β – Hydroxymethyl – estra – 1,3,5(10) – trien – 3 – ol monohydrate
$C_{19}H_{26}O_2$, H_2O
G.Precigoux, P.Marsau, F.Leroy, B.Busetta
Acta Crystallogr.,Sect.B,**36**, 749, 1980

51.15 rac – D – Homo – 8α – estra – 1,3,5(10) – triene – 3,17a – diol hemihydrate
$C_{19}H_{26}O_2$, $0.5H_2O$
W.L.Duax, G.D.Smith, D.C.Swenson, P.D.Strong, C.M.Weeks, S.N.Ananchenko, V.V.Egorova
Am.Cryst.Assoc.,Ser.2,**7**, 12, 1980

51.16 4 – Hydroxy – androst – 4 – ene – 3,17 – dione
$C_{19}H_{26}O_3$
J.F.Griffin, P.D.Strong, W.L.Duax, A.M.H.Brodie, H.J.Brodie
Acta Crystallogr.,Sect.B,**36**, 201, 1980

51.17 1,2 – seco – A – bis – nor – 5α – Androstane – 17β – acetate
$C_{19}H_{32}O_2$
D.C.Rohrer, W.L.Duax, M.E.Wolff *Steroids*,**34**, 589, 1979

51.18 17β – Hydroxy – 19 – nor – pregna – 4,9 – dien – 20 – yn – 3 – one
$C_{20}H_{24}O_2$
G.Lepicard, J.Delettre, J.–P.Mornon
Acta Crystallogr.,Sect.B,**36**, 1503, 1980

51.19 3 – Methoxy – 6 – oxa – estra – 1,3,5(10) – triene – 7,17 – dione – 17 – (ethylene – acetal)
$C_{20}H_{24}O_5$
P.S.White, D.C.N.Swindells
Acta Crystallogr.,Sect.B,**36**, 491, 1980

51.20 19 – Nor – pregn – 4 – ene – 3,20 – dione
$C_{20}H_{28}O_2$
W.L.Duax, D.C.Rohrer, P.D.Strong
Acta Crystallogr.,Sect.B,**35**, 2741, 1979

51.21 3 – Methoxy – 17β – hydroxymethyl – estra – 1,3,5(10) – triene
$C_{20}H_{28}O_2$
G.Precigoux, Y.Barrans, S.Geoffre, B.Busetta
Acta Crystallogr.,Sect.B,**36**, 2162, 1980

51.22 9α – Fluoro – 11β,17β – dihydroxy – 17α – methyl – androst – 4 – en – 3 – one
$C_{20}H_{29}FO_3$
W.L.Duax, P.D.Strong, M.E.Wolff
Cryst.Struct.Commun.,**8**, 985, 1979

51.23 17β – Hydroxy – 18 – methyl – 19 – nor – pregna – 4,9 – dien – 20 – yn – 3 – one
$C_{21}H_{26}O_2$
J.Delettre, G.Lepicard, J.–P.Mornon
Acta Crystallogr.,Sect.B,**36**, 1505, 1980

51.24 21 – Hydroxy – pregna – 4,9(11),16 – triene – 3,20 – dione
$C_{21}H_{26}O_3$
E.Surcouf *Acta Crystallogr.,Sect.B*,**35**, 2744, 1979

51.25 Pregna – 1,4 – diene – 17α,21 – diol – 3,11,20 – trione
Prednisone
$C_{21}H_{26}O_5$
V.M.Tseikinskii, V.I.Simonov, V.B.Ribakov, N.N.Petropavlov
Bioorg.Khim.,**5**, 1677, 1979

51.26 4,6β – Ethano – 3 – methoxy – 8α – estra – 1,3,5(10) – trien – 17β – ol hemihydrate
$C_{21}H_{26}O_2$, $0.5H_2O$
A.C.Ghosh, B.G.Hazra, I.Karup–Nielsen, M.J.Kane, D.Hawke, W.L.Duax, C.M.Weeks *J.Org.Chem.*,**44**, 683, 1979

51.27 2β – Bromo – 3α – hydroxy – 5α – pregna – 11,20 – dione
$C_{21}H_{31}BrO_3$
J.M.Midgley, W.B.Whalley, B.E.Ayres, G.H.Phillips, G.Ferguson, M.Parvez
J.Chem.Soc.,Perkin Trans.2,1176, 1980

51.28 6α – Hydroxy – 4,4 – dimethyl – androstan – 3 – one
$C_{21}H_{34}O_2$
W.B.Whalley, G.Ferguson, M.A.Khan
J.Chem.Soc.,Perkin Trans.2,1183, 1980

51.29 16α,17α – Cyclopropano – progesterone
$C_{22}H_{30}O_2$
V.M.Tseikinskii, V.I.Simonov
Eur.Cryst.Meeting,**6**, 302, 1980

51.30 18 – Ethyl – 17β – hydroxy – 19 – nor – pregn – 4 – en – 20 – yn – 3 – one
$C_{22}H_{30}O_2$
J.Delettre, J.–P.Mornon, G.Lepicard
Acta Crystallogr.,Sect.B,**36**, 1430, 1980

51.C 7α – (Thio – bromo – mercury) – (17R) – spiro(androst – 4 – ene – 17,2(3H) – furan)
$(C_{22}H_{31}BrHgO_2S)_n$ Main entry is 85.

51.31 17 – Hydroxy – 6α – methyl – pregn – 4 – ene – 3,20 – dione
$C_{22}H_{32}O_3$
W.L.Duax, P.D.Strong *Steroids*,**34**, 501, 1980

51.32 20 – Methyl – pregnenediol methanol solvate (at 123°K, deuterated form, neutron study)
$C_{22}H_{33}D_3O_2$, CH_4O
M.D.Fronckowiak, Y.Osawa, R.K.McMullan
Am.Cryst.Assoc.,Ser.2,**8**, 36, 1980

51.33 17β – Iodoacetoxy – 4,4 – dimethyl – 19 – nor – 5α – androstan – 3 – one
$C_{22}H_{33}IO_3$
G.Ferguson, E.W.Macaulay, J.M.Robertson, J.M.Midgley, W.B.Whalley, B.A.Lodge
J.Chem.Soc.,Perkin Trans.2,1170, 1980

51.34 3α – Hydroxy – 2α – methyl – 5α – pregna – 11,20 – dione
$C_{22}H_{34}O_3$
J.M.Midgley, W.B.Whalley, B.E.Ayres, G.H.Phillips, G.Ferguson, M.Parvez
J.Chem.Soc.,Perkin Trans.2,176, 1980

51.35　3α – Hydroxy – 2β – methoxy – 5α – pregna –
11,20 – dione monohydrate
$C_{22}H_{34}O_4$, H_2O
J.M.Midgley, W.B.Whalley, B.E.Ayres, G.H.Phillips,
G.Ferguson, M.Parvez
J.Chem.Soc.,Perkin Trans.2,176, 1980

51.36　20 – Methyl – pregn – 5 – ene – 3β,20 – diol
methanol solvate
$C_{22}H_{36}O_2$, CH_4O
W.L.Duax, Y.Osawa *Cryst.Struct.Commun.*,9, 267, 1980

51.37　20 – Methyl – pregn – 5 – ene – 3β,20 – diol
methanol solvate (neutron study,at 123°K,partially
deuterated)
$C_{22}H_{36}O_2$, CH_4O
M.D.Fronckowiak, R.K.McMullan
Am.Cryst.Assoc.,Ser.2,7, 36, 1980

51.38　16α,17α – Cyclobutano – progesterone
$C_{23}H_{32}O_2$
V.M.Tseikinskii, V.I.Simonov
Eur.Cryst.Meeting,6, 302, 1980

51.39　21 – Acetoxy – pregn – 4 – en – 17β – ol – 3,20 –
dione monohydrate
$C_{23}H_{32}O_5$, H_2O
V.M.Tseikinskii, V.B.Ribakov, V.I.Simonov, N.N.Petropavlov
Bioorg.Khim.,5, 1537, 1979

51.40　Spirohydroxyprednisolone – acetate ethyl acetate
solvate
$C_{23}H_{32}O_7$, $C_4H_8O_2$
J.R.Williams, R.H.Moore, R.Li, C.M.Weeks
J.Org.Chem.,45, 2324, 1980

51.41　Digoxigenin dihydrate
$C_{23}H_{34}O_5$, $2H_2O$
D.C.Rohrer, D.S.Fullerton
Acta Crystallogr.,Sect.B,36, 1565, 1980

51.42　17β – Iodoacetoxy – 4,4 – dimethyl – 5α –
androstan – 3 – one
$C_{23}H_{35}IO_3$
G.Ferguson, E.W.Macaulay, J.M.Robertson, J.M.Midgley,
W.B.Whalley, B.A.Lodge
J.Chem.Soc.,Perkin Trans.2,1170, 1980

51.43　3β – Hydroxy – 16β – morpholino – androst – 5 –
en – 17 – one
$C_{23}H_{35}NO_3$
D.C.Swenson, W.L.Duax, M.Numazawa, Y.Osawa
Acta Crystallogr.,Sect.B,36, 1981, 1980

51.44　Methyl 3β – acetoxy – 17α – methyl – 18 – nor –
5α – androstane – 17β – carboxylate
$C_{23}H_{36}O_4$
S.Fortier, F.R.Ahmed
Acta Crystallogr.,Sect.B,36, 994, 1980

51.45　16 – (3 – Methylbutyl) – 11,12,13,14,16,17 –
hexahydro – 15H – cyclopenta(a)phenanthrene –
17,2′ – spiro(1′,3′ – dithiolane)
$C_{24}H_{28}OS_2$
G.P.Chiusoli, G.Salerno, E.Bergamaschi, G.D.Andreetti,
G.Bocelli, P.Sgarabotto
J.Organomet.Chem.,177, 245, 1979
See also R1 : 39

51.46　9α – Fluoro – 11β,21 – dihydroxy – 16α,17α –
isopropylidene – dioxy – pregna – 1,4 – diene –
3,20 – dione methanol solvate
Triamcinolone acetonide
$C_{24}H_{31}FO_6$, $0.67CH_4O$
E.Surcouf *Acta Crystallogr.,Sect.B*,35, 2638, 1979
See also R1 : 38

51.47　21 – Acetoxy – 11α – methoxy – 1β,11β – oxo –
10α – pregn – 4 – ene – 2,20 – dione methanol
solvate
$C_{24}H_{32}O_7$, CH_4O
J.R.Williams, R.H.Moore, R.Li, J.F.Blount
J.Am.Chem.Soc.,101, 5019, 1979
See also R1 : 38

51.48　16α,17α – Cyclopentano – progesterone
$C_{24}H_{34}O_2$
V.M.Tseikinskii, V.I.Simonov
Eur.Cryst.Meeting,6, 302, 1980

51.49　3β – Acetoxy – 16β – methyl – pregn – 5 – en –
20 – one
$C_{24}H_{36}O_3$
H.Campsteyn, O.Dideberg, L.Dupont, J.Lamotte
Acta Crystallogr.,Sect.B,35, 2971, 1979

51.50　Rubidium 3α,12α – dihydroxy – 5β – cholan – 24 –
oic acid monohydrate
Deoxycholic acid rubidium salt
$C_{24}H_{39}O_4{}^-$, Rb^+, H_2O
V.M.Coiro, E.Giglio, S.Morosetti, A.Palleschi
Acta Crystallogr.,Sect.B,36, 1478, 1980

51.51　Sodium 3α,7α,12α – trihydroxy – 5β – cholan – 24 –
oate monohydrate
Sodium cholate monohydrate
$C_{24}H_{39}O_5{}^-$, Na^+, H_2O
R.E.Cobbledick, F.W.B.Einstein
Acta Crystallogr.,Sect.B,36, 287, 1980

51.52　Chenodeoxycholic acid
$C_{24}H_{40}O_4$
P.F.Lindley, M.M.Mahmoud, F.E.Watson, W.A.Jones
Acta Crystallogr.,Sect.B,36, 1893, 1980

51.53　Deoxycholic acid hydrate
$C_{24}H_{40}O_4$, $1.5H_2O$
C.P.Tang, R.Popovitz–Biro, M.Lahav, L.Leiserowitz
Isr.J.Chem.,18, 385, 1979

51.54　3α,12α – Dihydroxy – 5β – cholan – 24 – oic acid
ethanol solvate monohydrate
Deoxycholic acid ethanol solvate monohydrate
$2C_{24}H_{40}O_4$, C_2H_6O, H_2O
V.M.Coiro, A.D'Andrea, E.Giglio
Acta Crystallogr.,Sect.B,35, 2041, 1979

51.55　3α,12α – Dihydroxy – 5β – cholan – 24 – oic acid
palmitic acid ethanol solvate
Deoxycholic acid palmitic acid ethanol solvate
$4C_{24}H_{40}O_4$, $0.5C_{16}H_{32}O_2$, $0.5C_2H_6O$
V.M.Coiro, A.D'Andrea, E.Giglio
Acta Crystallogr.,Sect.B,36, 848, 1980
See also R2 : 1

51.56　N(β) – Methoxy – (progesterone) – (16α,17α – d) –
(tetrahydro – 1′,2′ – oxazole)
$C_{25}H_{35}NO_6$
L.G.Vorontsova *Zh.Strukt.Khim.*,20, 882, 1979
See also R1 : 40

51.57 N(α) – Methoxy – (progesterone) – (16α,17α – d) – (tetrahydro – 1′,2′ – oxazole)
$C_{25}H_{35}NO_6$
L.G.Vorontsova Zh.Strukt.Khim.,20, 882, 1979
See also R1 : 40

51.58 16α,17α – Cyclohexano – progesterone
$C_{25}H_{36}O_2$
V.M.Tseikinskii, V.I.Simonov
Eur.Cryst.Meeting,6, 302, 1980

51.59 6α – Methyl – 16α,17α – cyclohexano – progesterone
$C_{26}H_{38}O_2$
V.M.Tseikinskii, V.I.Simonov
Eur.Cryst.Meeting,6, 302, 1980

51.60 4,6 – Diaza – A,B – bishomo – cholest – 4a – eno – (4,3 – d)(6,7 – d) – bis(tetrazole)
$C_{26}H_{40}N_8$
R.A.Palmer, J.Husain Eur.Cryst.Meeting,6, 277, 1980

51.61 3α – Acetoxy – chola – 7,9 – dien – 12 – on – 24 – oic acid methyl ester
$C_{27}H_{38}O_5$
I.Krstanovic, L.Karanovic, M.Stefanovic, M.Djermanovic
Cryst.Struct.Commun.,9, 869, 1980

51.62 1β,2β,3β,4β,5β,7α – Hexahydroxy – spirost – 25(27) – en – 6 – one
$C_{27}H_{40}O_9$
K.Miyahara, F.Kumamoto, T.Kawasaki
Tetrahedron Lett.,21, 83, 1980
See also R1 : 38

51.63 1α,3β – 1,3 – Dehydro – 5,10 – seco – cholest – 10(19) – en – 5 – one
$C_{27}H_{44}O$
M.L.Mihailovic, L.Lorenc, M.Dabovic, I.Juranic, E.Wenkert, J.–M.Bernassau, M.S.Raju, A.T.McPhail, R.W.Miller
Tetrahedron Lett.,4917, 1979
See also R1 : 29

51.64 1β,3α – 1,3 – Dehydro – 5,10 – seco – cholest – 10(19) – en – 5 – one
$C_{27}H_{44}O$
M.L.Mihailovic, L.Lorenc, M.Dabovic, I.Juranic, E.Wenkert, J.–M.Bernassau, M.S.Raju, A.T.McPhail, R.W.Miller
Tetrahedron Lett.,4917, 1979
See also R1 : 29

51.65 Cholesteryl bromide
$C_{27}H_{45}Br$
G.V.Vani, K.Vijayan Mol.Cryst.Liq.Cryst.,51, 253, 1979

51.66 Cholesteryl chloride
$C_{27}H_{45}Cl$
G.V.Vani. K.Vijayan Mol.Cryst.Liq.Cryst.,51, 253, 1979

51.67 1α,3β – 1,3 – Dehydro – 10α – hydroxy – 5,10 – seco – cholestan – 5 – one
$C_{27}H_{46}O_2$
M.L.Mihailovic, L.Lorenc, M.Dabovic. I.Juranic, E.Wenkert, J.–M.Bernassau. M.S.Raju, A.T.McPhail, R.W.Miller
Tetrahedron Lett.,4917, 1979
See also R1 : 29

51.68 3α,7α,12α – Trihydroxy – 5α,(25S) – cholestan – 26 – oic acid
$C_{27}H_{46}O_5$
A.K.Batta, G.Salen, J.F.Blount, S.Shefer
J.Lipid Res.20, 935. 1979

51.69 3β,16β,23(R),26 – Tetrahydroxy – 5β – cholestane
$C_{27}H_{48}O_4$
J.J.Einck, G.R.Pettit
Acta Crystallogr.,Sect.B,36, 1398, 1980

51.70 Acnistin E
$C_{28}H_{38}O_7$
A.Usubillaga, G.de Castellano, V.Zabel, W.H.Watson
J.Chem.Soc.,Chem.Commun.,854, 1980
See also R1 : 38

51.71 2α,3α,22β,23β – Tetrahydroxy – 24β – methyl – B – homo – 7 – oxa – 5α – cholestan – 6 – one
$C_{28}H_{48}O_6$
M.J.Thompson, N.Mandava, J.L.Flippen–Anderson, J.F.Worley, S.R.Dutky, W.E.Robbins, W.Lusby
J.Org.Chem.,44, 5002, 1979

51.72 Brassinolide monohydrate
$C_{28}H_{48}O_6$, H_2O
M.D.Grove, G.F.Spencer, W.K.Rohwedder, N.Mandava, J.F.Worley, J.D.Warthen Junior, G.L.Steffens, J.L.Flippen–Anderson, J.C.Cook Junior
Nature (London),281, 216, 1979
See also R1 : 38

51.73 Actodigin
$C_{29}H_{44}O_9$
D.S.Fullerton, K.Yoshioka, D.C.Rohrer, A.H.L.From, K.Ahmed Mol.Pharmacol.,17, 43, 1980
See also R1 : 45

51.74 3β – Acetoxy – 5β,6β – N – nitro – aziridinyl – cholestene (at –98°C)
$C_{29}H_{48}N_2O_4$
M.J.Haire, R.L.Harlow J.Org.Chem.,45, 2264, 1980
See also R1 : 36

51.75 (E) – 3α – Acetoxy – 5,10 – seco – cholest – 1(10) – en – 5 – one
$C_{29}H_{48}O_3$
H.Fuhrer, L.Lorenc, V.Pavlovic, G.Rihs, G.Rist, J.Kalvoda, M.L.Mihailovic Helv.Chim.Acta,62, 1770, 1979
See also R1 : 28

51.76 24 – Ethyl – isobrassinolide
$C_{29}H_{50}O_6$
J.L.Flippen–Anderson, R.D.Gilardi
Am.Cryst.Assoc.,Ser.2,7, 13, 1980
See also R1 : 38

51.77 24 – Ethyl – brassinolide
$C_{29}H_{50}O_6$
J.L.Flippen–Anderson, R.D.Gilardi
Am.Cryst.Assoc.,Ser.2,7, 13, 1980
See also R1 : 38

51.C 3β – Acetoxy – α – amyrin
$C_{32}H_{52}O_2$ Main entry is 56.8

51.78 3β – Azido – 2α – (toluene – p – sulfonamido) – 5α – cholestane
$C_{34}H_{54}N_4O_2S$
G.M.L.Cragg, P.A.McCallum, L.R.Nassimbeni, A.L.Rodgers, J.C.van Niekerk S.Afr.J.Chem.,32, 97, 1979

51.79 3β – (Toluene – p – sulfonamido) – 5α – cholestan – 2β – ol
$C_{34}H_{55}NO_3S$
G.M.L.Cragg, P.A.McCallum, L.R.Nassimbeni, A.L.Rodgers, J.C.van Niekerk S.Afr.J.Chem.,32, 97, 1979

51.80 **(6R) – 6,19 – Epidioxy – 9,10 – seco – ergosta –**
5(10),7,22 – trien – 3β – ol benzoate
$C_{35}H_{48}O_4$
S.Yamada, K.Nakayama, H.Takayama, A.Itai, Y.Iitaka
Chem.Pharm.Bull.,**27**, 1949, 1979
See also R1 : 38

51.81 **Cholest – 5 – en – 3 – ol dihydrocinnamate**
$C_{36}H_{54}O_2$
A.A.Polishchuk, M.Yu.Antipin, R.G.Gerr, V.I.Kulishov,
Yu.T.Struchkov, L.G.Derkach
Cryst.Struct.Commun.,**9**, 263, 1980

51.82 **Cholesteryl nonanoate**
$C_{36}H_{62}O_2$
N.G.Guerina, B.M.Craven
J.Chem.Soc.,Perkin Trans.2,1414, 1979

51.C **3β – Benzoxy – α – amyrin**
$C_{37}H_{54}O_2$ Main entry is 56.9

51.83 **Cholesteryl decanoate**
$C_{37}H_{64}O_2$
V.Pattabhi, B.M.Craven *J.Lipid Res.*,**20**, 753, 1979

51.84 **Cholesteryl undecanoate**
$C_{38}H_{66}O_2$
P.Sawzik, B.M.Craven
Acta Crystallogr.,Sect.B,**36**, 215, 1980

51.85 **Tetra – acetyl – trillenogenin monobrosylate**
$C_{40}H_{47}BrO_{14}S$
T.Nohara, T.Komori, T.Kawasaki
Chem.Pharm.Bull.,**28**, 1437, 1980
See also R1 : 45

51.86 **Digoxin**
$C_{41}H_{64}O_{14}$
K.Go, G.Kartha, J.P.Chen
Acta Crystallogr.,Sect.B,**36**, 1811, 1980
See also R1 : 45

51.87 **Gitoxin**
$C_{41}H_{64}O_{14}$
K.Go, G.Kartha *Am.Cryst.Assoc.,Ser.2*,**7**, 13, 1980
See also R1 : 45

51.88 **14α – Ethyl – 5α – cholest – 7 – ene – 3β,15α –**
diol – di – p – bromobenzoate
$C_{43}H_{58}Br_2O_4$
D.J.Monger, E.J.Parish, G.J.Schroepfer Junior, F.A.Quiocho
Acta Crystallogr.,Sect.B,**36**, 1460, 1980

51.89 **16 – Methoxy – 16′ – oxo – 21,20′ – di(20,18 –**
epoxy – pregnane)
$C_{43}H_{66}O_4$
A.Chiaroni, C.Riche *Cryst.Struct.Commun.*,**9**, 239, 1980
See also R1 : 38

MONOTERPENES

52.1 **8 – Bromo – 2 – bromomethyl – 6 – methyl –**
2,5,6 – trichloro – octa – 3,7 – diene (absolute
configuration)
$C_{10}H_{13}Br_2Cl_3$
D.B.Stierle, R.M.Wing, J.J.Sims *Tetrahedron*,**35**, 2855, 1979

52.2 **(1S) – cis,cis – Isoiridolactone**
$C_{10}H_{16}O_2$
F.Bellesia, U.M.Pagnoni, R.Trave, G.D.Andreetti, G.Bocelli,
P.Sgarabotto *J.Chem.Soc.,Perkin Trans.2*,1341, 1979

52.3 **(1S) – cis,cis – Iridolactone**
$C_{10}H_{16}O_2$
F.Bellesia, U.M.Pagnoni, R.Trave, G.D.Andreetti, G.Bocelli,
P.Sgarabotto *J.Chem.Soc.,Perkin Trans.2*,1341, 1979

52.4 **Loganin**
$C_{17}H_{26}O_{10}$
P.G.Jones, G.M.Sheldrick, K.–H.Glusenkamp, L.F.Tietze
Acta Crystallogr.,Sect.B,**36**, 481, 1980
See also R1 : 45

SESQUITERPENES

53.1 **Epoxy – rhodophytin (absolute configuration)**
$C_{15}H_{18}BrClO_2$
B.M.Howard, W.Fenical, K.Hirotsu, B.Solheim, J.Clardy
Tetrahedron,**36**, 171, 1980

53.2 **Eregoyazin**
$C_{15}H_{18}O_3$
W.Herz, N.Kumar, W.Vichnewski, J.F.Blount
J.Org.Chem.,**45**, 2503, 1980

53.3 **Thieleanine**
$C_{15}H_{18}O_3$
S.Alvarado, J.F.Ciccio, J.Calzada, V.Zabel, W.H.Watson
Phytochemistry,**18**, 330, 1979

53.4 **Hibiscone C**
Gmelofuran
$C_{15}H_{18}O_3$
M.A.Ferreira, T.J.King, S.Ali, R.H.Thomson
J.Chem.Soc.,Perkin Trans.1,249, 1980

53.5 **γ – Mη – santonin**
6β – Hydroxy – 3 – oxo – 5β – eudesma – 1,7(11) –
dien – 12 – oic acid γ – lactone
$C_{15}H_{18}O_3$
H.Ueda, C.Katayama, J.Tanaka
Bull.Chem.Soc.Jpn.,**53**, 22, 1980

53.6 **2β,14 – Dibromo – 4α,5β,6β,11βH – tetrahydro –
santonin (absolute configuration)**
$C_{15}H_{20}Br_2O_3$
S.Inayama, N.Shimizu, T.Shibata, H.Hori, Y.Iitaka
J.Chem.Soc.,Chem.Commun.,495, 1980

53.7 **2α – Bromo – α – tetrahydro – santonin**
$C_{15}H_{21}BrO_3$
S.V.L.Narayana, H.N.Shrivastava
J.Chem.Soc.,Perkin Trans.2,1116, 1980

53.8 **Ivalbin**
$C_{15}H_{22}O_4$
W.Herz, N.Kumar, J.F.Blount   *J.Org.Chem.*,**44**, 4437, 1979

53.9 **Epoxy – isoacoragermacrone**
$C_{15}H_{24}O_2$
H.Ueda, C.Katayama, J.Tanaka
Bull.Chem.Soc.Jpn.,**53**, 1263, 1980
See also R1 : 38

53.10 **Capnell – 9(12) – ene – 2β,5α,8β,10α – tetrol**
$C_{15}H_{24}O_4$
M.Kaisin, B.Tursch, J.P.Declercq, G.Germain,
M.van Meerssche *Bull.Soc.Chim.Belg.*,**88**, 253, 1979

53.11 **Koraiola pyridine solvate**
$C_{15}H_{26}O, 0.5C_5H_5N$
Yu.V.Gatilov, S.V.Borisov
Dokl.Akad.Nauk SSSR.**247**, 354, 1979

53.12 **Gleenol**
$C_{15}H_{26}O_2$
P.A.Kurvyakov, Yu.V.Gatilov, V.A.Khan, Zh.V.Dubovenko,
V.A.Pentegova *Khim.Prir.Soedin*,164, 1979

53.13 **Murol – 4 – on – 9α – ol**
$C_{15}H_{26}O_2$
Yu.V.Gatilov, V.A.Khan, Zh.V.Dubovenko
Khim.Prir.Soedin,158, 1979

53.14 **4 – Hydroxy – dihydroagarofuran**
$C_{15}H_{26}O_2$
H.Itokawa, K.Watanabe, S.Mihashi, Y.Iitaka
Chem.Pharm.Bull.,**28**, 681, 19

53.15 **Murolan – 4α,9β – diol**
$C_{15}H_{28}O_2$
Yu.V.Gatilov, Zh.V.Dubovenko, V.A.Khan
Zh.Strukt.Khim.,**20**, 509, 1979

53.16 **Murolan – 4β,9β – diol**
$C_{15}H_{28}O_2$
Yu.V.Gatilov, Zh.V.Dubovenko, V.A.Khan
Zh.Strukt.Khim.,**20**, 509, 1979

53.17 **Ryomenin methyl ester**
$C_{16}H_{20}O_3$
N.Tanaka, H.Maehashi, S.Saito, T.Murakami, Y.Saiki,
C.–M.Chen, Y.Iitaka *Chem.Pharm.Bull.*,**27**, 2874, 1979

53.18 **2 – Isocyano – pupukeanane**
$C_{16}H_{25}N$
M.R.Hagadone, B.J.Burreson, P.J.Scheuer, J.S.Finer,
J.Clardy *Helv.Chim.Acta*,**62**, 2484, 1979

53.19 **Obtusol acetate (absolute configuration)**
$C_{17}H_{25}Br_2ClO_2$
A.Perales, M.Martinez–Ripoll, J.Fayos
Acta Crystallogr.,Sect.B,**35**, 2771, 1979

53.20 **5 – Acetoxy – isomarasman – 7α,13 – diol**
$C_{17}H_{28}O_4$
N.Morisaki, J.Furukawa, S.Nozoe, A.Itai, Y.Iitaka
Chem.Pharm.Bull.,**28**, 500, 1980

53.21 **(+) – Ovalifolienalone**
$C_{19}H_{22}O_7$
A.Matsuo, H.Nozaki, K.Atsumi, H.Kataoka, M.Nakayama,
Y.Kushi, S.Hayashi
J.Chem.Soc.,Chem.Commun.,1012, 1979

53.22 **Acantholide**
$C_{19}H_{24}O_6$
H.Lotter, H.Wagner, A.A.Saleh, G.A.Cordell, N.R.Farnsworth
Z.Naturforsch.,Teil C,**34**, 677, 1979

53.23 **Tulirinol acetate**
$C_{19}H_{24}O_6$
R.W.Doskotch, E.H.Fairchild, C.–T.Huang, J.H.Wilton,
M.A.Beno, G.G.Christoph *J.Org.Chem.*,**45**, 1441, 1980
See also R1 : 38

53.24 **Hymenograndin**
$C_{19}H_{26}O_7$
W.Herz, S.V.Govindan, M.W.Bierner, J.F.Blount
J.Org.Chem.,**45**, 493, 1980
See also R1 : 38

53.25 2 – Hydroxy – 8 – (2 – hydroxy – 2 –
hydroxymethyl – 3 – mercapto – butyryloxy) –
trans,trans – germacra – 1(10),4 – dienolide
(absolute configuration)
$C_{20}H_{28}O_7S$
W.Herz, N.Kumar, J.F.Blount *J.Org.Chem.*,**45**, 489, 1980

53.26 Hymenosignin
$C_{20}H_{30}O_5$
W.Herz, S.V.Govindan, M.W.Bierner, J.F.Blount
J.Org.Chem.,**45**, 493, 1980

53.27 Melampodin (neutron study)
$C_{21}H_{24}O_9$
S.F.Watkins, N.H.Fischer, I.Bernal
Proc.Nat.Acad.Sci.U.S.A.,**70**, 2434, 1973

53.28 Calamusenone p – bromophenylpyrazoline (absolute
configuration)
$C_{21}H_{27}BrN_2$
M.Rohr, P.Naegeli, J.J.Daly *Phytochemistry*,**18**, 279, 1979

53.29 9β – Acetoxy – 8α – epoxy – angelyloxy –
trans,trans – germacra – 1(10),4 – diene – cis –
6,12 – olide
$C_{22}H_{28}O_7$
W.Herz, S.V.Govindan, J.F.Blount
J.Org.Chem.,**45**, 1113, 1980

53.30 Ursiniolide A monohydrate
$C_{22}H_{28}O_7$, H_2O
Z.Samek, M.Holub, U.Rychlewska, H.Grabarczyk, B.Drozdz
Tetrahedron Lett.,2691, 1979

53.31 9β – Acetoxy – 8α – epoxy – angelyloxy – 7α –
hydroxy – trans,trans – germacra – 1(10),4 –
diene – cis – 6,12 – olide
$C_{22}H_{28}O_8$
W.Herz, S.V.Govindan, J.F.Blount
J.Org.Chem.,**45**, 1113, 1980

53.32 (+) – Allohimachalyl – p – bromobenzoate (absolute
configuration)
$C_{22}H_{29}BrO_2$
A.G.Bajaj, S.Dev, B.Tagle, J.Telser, J.Clardy
Tetrahedron Lett.,**21**, 325, 1980

53.33 (+) – Bicyclohumulenone – triol – mono(p –
bromobenzoate) (absolute configuration)
$C_{22}H_{31}BrO_4$
H.Nozaki *J.Sci.Hiroshima Univ.*,*A*,*43*, 67, 1979

53.34 Bromoaureol acetate (absolute configuration)
$C_{23}H_{31}BrO_3$
P.Djura, D.B.Stierle, B.Sullivan, D.J.Faulkner, E.Arnold,
J.Clardy *J.Org.Chem.*,**45**, 1435, 1980

53.35 (–) – (4S,8S) – α – Bisabolol – 8 – p – phenylazo –
phenylmethane
$C_{28}H_{35}N_3O_2$
T.Prange, D.Babin, J.–D.Fourneron, M.Julia
C.R.Acad.Sci.,Ser.C,**289**, 383, 1979

DITERPENES

54.1 14,15 – bis – nor – 8α – Hydroxy – labd – 11 – en –
13 – one
$C_{18}H_{30}O_2$
Yu.V.Gatilov, S.V.Borisov, Zh.V.Dubovenko, E.N.Shmidt
Zh.Strukt.Khim.,**20**, 504, 1979

54.2 13 – cis – Retinal
$C_{20}H_{28}O$
H.Matsumoto, R.S.H.Liu, C.J.Simmons, K.Seff
J.Am.Chem.Soc.,**102**, 4259, 1980

54.3 Nepetaefolinol
$C_{20}H_{28}O_6$
J.F.Blount, P.S.Manchand
J.Chem.Soc.,Perkin Trans.1,264, 1980
See also R1 : 38

54.4 Cinncassiol C_1
$C_{20}H_{28}O_7$, H_2O
T.Nohara, I.Nishioka, N.Tokubuchi, K.Miyahara,
T.Kawasaki *Chem.Pharm.Bull.*,**28**, 1969, 1980

54.5 Dictyolactone
$C_{20}H_{30}O_2$
J.Finer, J.Clardy, W.Fenical, L.Minale, R.Riccio, J.Battaile,
M.Kirkup, R.E.Moore *J.Org.Chem.*,**44**, 2044, 1979

54.6 Cleomeolide
$C_{20}H_{30}O_3$
S.B.Mahato, B.C.Pal, T.Kawasaki, K.Miyahara, O.Tanaka,
K.Yamasaki *J.Am.Chem.Soc.*,**101**, 4720, 1979

54.7 Eupatalbin
$C_{20}H_{30}O_4$
W.Herz, S.V.Govindan, J.F.Blount
J.Org.Chem.,**44**, 2999, 1979
See also R1 : 38

54.8 Pachyclavulariadiol
$C_{20}H_{30}O_4$
B.F.Bowden, J.C.Coll, S.J.Mitchell, C.L.Raston, G.J.Stokie,
A.H.White *Aust.J.Chem.*,**32**, 2265, 1979
See also R1 : 38

54.9 11α,16 – Epoxy – 1β,15R – dihydroxy – isopimar –
8(14) – ene (absolute configuration)
$C_{20}H_{32}O_3$
A.Perales, M.Martinez–Ripoll, J.Fayos, R.K.Bansal,
K.C.Joshi, R.Patni, B.Rodriguez
Tetrahedron Lett.,**21**, 2843, 1980

54.10 Asbestinin – 1 – diol
$C_{20}H_{32}O_4$
D.B.Stierle, B.Carte, D.J.Faulkner, B.Tagle, J.Clardy
J.Am.Chem.Soc.,**102**, 5088, 1980

54.11 (–) – α – Dihydrokaurene
$C_{20}H_{34}$
S.W.Pelletier, H.K.Desai, J.Finer–Moore, N.V.Mody
J.Am.Chem.Soc.,**101**, 6741, 1979

54.12 **Dictyodiol**
$C_{20}H_{34}O_2$
J.Finer, J.Clardy, W.Fenical, L.Minale, R.Riccio, J.Battaile,
M.Kirkup, R.E.Moore *J.Org.Chem.*,**44**, 2044, 1979

54.13 **Hispanonic acid methyl ester (absolute configuration)**
$C_{21}H_{26}O_4$
J.Lopez de Lerma, S.Garcia–Blanco, J.G.Rodriguez
Tetrahedron Lett.,**21**, 1273, 1980
See also R1 : 38

54.14 **Methyl – 7,9 – cis – retinoate**
$C_{21}H_{30}O_2$
H.Matsumoto, R.S.H.Liu, C.J.Simmons, K.Seff
J.Am.Chem.Soc.,**102**, 4259, 19

54.15 **8 – Isocyano – 10 – cycloamphilectene**
$C_{21}H_{31}N$
R.Kazlauskas, P.T.Murphy, R.J.Wells, J.F.Blount
Tetrahedron Lett.,**21**, 315, 1980

54.16 **7 – Isocyano – 11(20),14 – epi – amphilectadiene**
$C_{21}H_{31}N$
R.Kazlauskas, P.T.Murphy, R.J.Wells, J.F.Blount
Tetrahedron Lett.,**21**, 315, 1980

54.17 **Barbatusin**
$C_{22}H_{28}O_5$
R.Zelnik, H.E.Gottlieb, D.Lavie
Tetrahedron,**35**, 2693, 1979

54.18 **Leonitin**
9,13 – Epoxy – labdane
$C_{22}H_{30}O_7$
G.J.Kruger, D.E.A.Rivett *S.Afr.J.Chem.*,**32**, 59, 1979

54.19 **7,15 – Di – isocyano – adociane**
$C_{22}H_{32}N_2$
R.Kazlauskas, P.T.Murphy, R.J.Wells, J.F.Blount
Tetrahedron Lett.,**21**, 315, 1980

54.20 **3α – Acetoxy – 15β – hydroxy – 7,16 – seco – trinervita – 7,11 – diene**
$C_{22}H_{36}O_3$
J.C.Braekman, D.Daloze, A.Dupont, J.Pasteels, B.Tursch,
J.P.Declercq, G.Germain, M.van Meerssche
Tetrahedron Lett.,**21**, 2761, 1980

54.21 **Plexaurolone acetate (absolute configuration, at 138°K)**
$C_{22}H_{36}O_4$
S.E.Ealick, D.van der Helm, R.A.Gross Junior,
A.J.Weinheimer, L.S.Ciereszko, R.E.Middlebrook
Acta Crystallogr.,Sect.B,**36**, 1901, 1980

54.22 **(1S,2E,4R,6R,7E,11S,12S) – 6 – Acetoxy – 11,12 – epoxy – cembra – 2,7 – dien – 4 – ol**
$C_{22}H_{36}O_4$
D.Behr, I.Wahlberg, T.Nishida, C.R.Enzell, J.–E.Berg,
A.–M.Pilotti *Acta Chem.Scand.Ser.B*,**34**, 195, 1980

54.23 **Emblide**
$C_{23}H_{32}O_6$
J.A.Toth, B.J.Burreson, P.J.Scheuer, J.Finer–Moore,
J.Clardy *Tetrahedron*,**36**, 1307, 1980

54.24 **Methoxy – nepetaefolin**
$C_{23}H_{32}O_6$
J.F.Blount, P.S.Manchand
J.Chem.Soc.,Perkin Trans.1,264, 1980

54.25 **Mazarronine**
$C_{24}H_{28}O_9$
L.Eguren, A.Perales, J.Fayos
Eur.Cryst.Meeting,**6**, 18, 1980

54.26 **Eriocephalin (absolute configuration)**
7α – Hydroxy – anaphalidin
$C_{24}H_{30}O_9$
J.Fayos, M.Martinez–Ripoll, M.Paternostro, F.Piozzi,
B.Rodriguez, G.Savona *J.Org.Chem.*,**44**, 4992, 1979

54.27 **Oxidopanamensin diacetate**
$C_{24}H_{32}O_6$
K.Koike, G.A.Cordell, N.R.Farnsworth, A.A.Freer,
C.J.Gilmore, G.A.Sim *Tetrahedron*,**36**, 1167, 1980

54.C **N – Ethyl – 1α,14α – dihydroxy – 8β,16β – dimethoxy – 6β – methoxycarbonyl – aconitine hydroiodide**
$C_{25}H_{40}NO_6{}^+$, I⁻ Main entry is 58.66

54.28 **13 – epi – 9 – Desacetylxenicin**
$C_{26}H_{36}O_8$
J.C.Braekman, D.Daloze, B.Tursch, J.P.Declercq,
G.Germain, M.van Meerssche
Bull.Soc.Chim.Belg.,**88**, 71, 1979

54.C **Diethylamino – dihydro – veatchinone**
$C_{26}H_{42}N_2O_2$ Main entry is 58.68

54.29 **Paspalinine**
$C_{27}H_{31}NO_4$
R.T.Gallagher, J.Finer, J.Clardy, A.Leutwiler, F.Weibel,
W.Acklin, D.Arigoni *Tetrahedron Lett.*,**21**, 235, 1980
See also R1 : 38

54.30 **19 – (p – Bromobenzoyloxy) – biflora – 4,15 – diene**
$C_{27}H_{37}BrO_2$
D.F.Wiemer, J.Meinwald, G.D.Prestwich, B.A.Solheim,
J.Clardy *J.Org.Chem.*,**45**, 191, 1980

54.31 **8α – Hydroxyverrucosane – 2β – p – bromo – benzoate (absolute configuration)**
$C_{27}H_{37}BrO_3$
H.Nozaki, A.Matsuo, Y.Kushi, M.Nakayama, S.Hayashi,
D.Takaoka, N.Kamijo
J.Chem.Soc.,Perkin Trans.2,763, 1980

54.32 **(–) – Neoverrucosane – 5β – benzoate**
$C_{27}H_{38}O_2$
A.Matsuo, H.Nozaki, M.Nakayama, D.Takaoka, S.Hayashi
J.Chem.Soc.,Chem.Commun.,822, 1980

54.33 **Acetyl – strongylophorine – 2**
$C_{28}H_{38}O_5$
J.C.Braekman, D.Daloze, G.Hulot, B.Tursch, J.P.Declercq,
G.Germain, M.van Meerssche
Bull.Soc.Chim.Belg.,**87**, 917, 1978
See also R1 : 38

54.34 **11 – Monoacetyl – cinncassiol – D₁ – 19 – monobrosylate monohydrate**
$C_{28}H_{37}BrO_9S$, H_2O
T.Nohara, Y.Kashiwada, T.Tomimatsu, M.Kido,
N.Tokubuchi, I.Nishioka
Tetrahedron Lett.,**21**, 2647, 1980

54.35 **Aflavinine ethylacetate solvate**
$C_{28}H_{39}NO$, $C_4H_8O_2$
R.T.Gallagher, T.McCabe, K.Hirotsu, J.Clardy, J.Nicholson,
B.J.Wilson *Tetrahedron Lett.*,**21**, 243, 1980

54.36 **Nagilactone – B – 2,7 – bis – p – bromobenzoate
(absolute configuration)**
$C_{33}H_{30}Br_2O_9$
T.Higuchi, K.Takahashi, K.Hirotsu, Y.Hayashi
Z.Kristallogr.,**150**, 335, 1979

54.37 **Ajugareptansin p – bromobenzoate (absolute
configuration)**
$C_{36}H_{47}BrO_{11}$
X.Solans, C.Miravitlles, J.P.Declercq, G.Germain
Acta Crystallogr.,Sect.B,**35**, 2732, 1979

SESTERTERPENES

55.1 **2,13 – Dimethyl – 10 – (2 – methyl – hept – 2 –
en – 6 – yl) – tricyclo(9.3.0.0⁸,¹³)tetradecane – 2,8 –
diene – 4,7 – dione**
$C_{24}H_{34}O_2$
A.Itai, Y.Iitaka, S.Nozoe
Chem.Pharm.Bull.,**28**, 1035, 1980

55.2 **Dicyclohexylammonium gascardate**
$C_{25}H_{37}O_2^-$, $C_{12}H_{24}N^+$
R.K.Boeckman Junior, D.M.Blum, E.V.Arnold, J.Clardy
Tetrahedron Lett.,4609, 1979

55.3 **(13 – p – Bromobenzenesulfonyl) – 10 – hydroxy –
5,8 – dimethyl – 2 – (2 – methyl – hept – 2 – en –
6 – yl) – tetracyclo(9.3.0.0¹,⁵.0⁷,¹¹)tetradec – 8 – ene**
$C_{30}H_{41}BrO_4S$
A.Itai, Y.Iitaka, S.Nozoe
Chem.Pharm.Bull.,**28**, 1035, 1980

55.4 **p – Bromophenacyl – stellatate (absolute
configuration)**
$C_{33}H_{43}BrO_3$
I.H.Qureshi, S.A.Husain, R.Noorani, N.Murtaza, Y.Iitaka,
S.Iwasaki, S.Okuda *Tetrahedron Lett.*,**21**, 1961, 1980

TRITERPENES

TETRATERPENES

56.1 Dihydronimbin
$C_{30}H_{38}O_9$
C.R.Narayanan, N.N.Dhaneshwar, S.S.Tavale, L.M.Pant
Acta Crystallogr.,Sect.B,**36**, 486, 1980

56.2 Prionostemmadione
$C_{30}H_{48}O_2$
F.D.Monache, G.B.Marini–Bettolo, M.Pomponi, J.F.de Mello,
T.J.King, R.H.Thomson
J.Chem.Soc.,Perkin Trans.1,2649, 1979

56.3 Rhuslactone triol
$C_{30}H_{48}O_4$
T.Akiyama, C.–K.Sung, U.Sankawa, Y.Iitaka, D.–S.Han
Am.Cryst.Assoc.,Ser.2,**8**, 35, 1980

56.4 Squalene (at −110°C)
$C_{30}H_{50}$
J.Ernst, J.–H.Fuhrhop
Justus Liebigs Ann.Chem.,1635, 1979

**56.C hexakis(p – t – Butylphenylthiomethyl)benzene
squalene clathrate**
$0.5C_{30}H_{50}$, $C_{72}H_{90}S_6$ Main entry is 61.26

56.5 3α,4α – Epoxy – D:A – friedo – 18β,19αH – lupane
$C_{30}H_{50}O$
Y.Yokoyama, Y.Moriyama, T.Tsuyuki, T.Takahashi, A.Itai,
Y.Iitaka *Chem.Lett.*,1463, 1979

**56.6 4β,5β,9β,13α,14β,17α – Hexamethyl – 21 –
isopropyl – cyclopentano – perhydrochrysen – 3 –
one**
Filicane – 3 – one
$C_{30}H_{50}O$
B.F.Pedersen *Eur.Cryst.Meeting*,**5**, 179, 1979

56.7 Abietospiran
$C_{31}H_{48}O_4$
W.Steglich, M.Klaar, L.Zechlin, H.J.Hecht
Angew.Chem.,Int.Ed.Engl.,**18**, 698, 1979

56.8 3β – Acetoxy – α – amyrin
$C_{32}H_{52}O_2$
M.Grynpas, P.F.Lindley *J.Cryst.Mol.Struct.*,**9**, 199, 1979
See also R1 : 51

56.9 3β – Benzoxy – α – amyrin
$C_{37}H_{54}O_2$
M.Grynpas, P.F.Lindley *J.Cryst.Mol.Struct.*,**9**, 199, 1979
See also R1 : 51

**56.10 21β – (2 – 0 – Acetyl – 3,4 – di – 0 – angelyl – 6 –
deoxy – β – glucopyranosyloxy) – 3β,16α,22α,24,28 –
penta – acetoxy – olean – 12 – ene**
Acetylated napoleogenin
$C_{58}H_{84}O_{18}$
M.R.Spirlet, L.Dupont, O.Dideberg, M.Kapundu
Acta Crystallogr.,Sect.B,**36**, 1593, 1980

57.1 Fucoxanthin
$C_{42}H_{58}O_6$
G.P.Moss *Pure Appl.Chem.*,**51**, 507, 1979

ALKALOIDS

58.1 **(+) – Epilupinine**
$C_{10}H_{19}NO$
A.E.Koziol, M.Gdaniec, Z.Kosturkiewicz
Acta Crystallogr.,Sect.B,**36**, 982, 1980

58.2 **(–) – Lupine hydrochloride (absolute configuration)**
$C_{10}H_{20}NO^+$, Cl^-
A.E.Koziol, M.Gdaniec, Z.Kosturkiewicz
Acta Crystallogr.,Sect.B,**36**, 980, 1980

58.3 **Lupinine – N – methyliodide**
$C_{11}H_{22}NO^+$, I^-
A.E.Koziol, Z.Kosturkiewicz
Eur.Cryst.Meeting,**6**, 54, 1980

58.4 **Swainsonine diacetate**
$C_{12}H_{19}NO_5$
B.W.Skelton, A.H.White *Aust.J.Chem.*,**33**, 435, 1980

58.5 **11 – Hydroxy – 6H – indolo(3,2,1 – de)(1,5) naphthyridin – 6 – one monohydrate**
Amarorine monohydrate
$C_{14}H_8N_2O_2$, H_2O
P.J.Clarke, K.Jewers, H.F.Jones
J.Chem.Soc.,Perkin Trans.1,1614, 1980

58.6 **Nitro – leonurine monohydrate**
$C_{14}H_{20}N_4O_7$, H_2O
N.Camerman, L.Y.Y.Chan, H.W.Yeung, T.C.W.Mak
Acta Crystallogr.,Sect.B,**35**, 3004, 1979

58.7 **Albine perchlorate**
$C_{14}H_{21}N_2O^+$, ClO_4^-
A.Hoser, A.Katrusiak, Z.Kaluski, J.Wolinska–Mocydlarz
Acta Crystallogr.,Sect.B,**36**, 984, 1980

58.8 **Tsukushinamine – A hydrobromide (absolute configuration)**
$C_{15}H_{21}N_2O^+$, Br^-
J.Bordner, S.Ohmiya, H.Otomasu, J.Haginiwa, I.Murakoshi
Chem.Pharm.Bull.,**28**, 1965, 1980

58.9 **Dehydromultiflorine perchlorate**
$C_{15}H_{21}N_2O^+$, ClO_4^-
D.Pyzalska, M.Jaskolski, T.Borowiak,
J.Wolinska–Mocydlarz
Acta Crystallogr.,Sect.B,**36**, 1685, 1980

58.10 **Δ² – Dehydro – 4 – oxo – sparteine perchlorate hemihydrate**
Multiflorine perchlorate hemihydrate
$C_{15}H_{23}N_2O^+$, ClO_4^-, $0.5H_2O$
D.Pyzalska, M.Gdaniec, T.Borowiak, J.Wolinska–Mocydlarz
Acta Crystallogr.,Sect.B,**36**, 1602, 1980

58.11 **Isosoforidine**
$C_{15}H_{24}N_2O$
B.T.Ibragimov, S.A.Talipov, G.N.Tishchenko,
Yu.K.Kushmuradov, T.F.Aripov
Khim.Prir.Soedin,586, 1979
See also R1 : 36

58.12 **Allomatrine**
$C_{15}H_{24}N_2O$
B.T.Ibragimov, G.N.Tishchenko, Yu.K.Kushmuradov,
T.F.Aripov, A.S.Sadikov *Khim.Prir.Soedin*,416, 1979

58.13 **Soforidine**
$C_{15}H_{24}N_2O$
B.T.Ibragimov, G.N.Tishchenko, Yu.K.Kushmuradov,
T.F.Aripov, A.S.Sadikov *Khim.Prir.Soedin*,355, 1979

58.14 **Tetrahydro – neosoforamine**
$C_{15}H_{24}N_2O$
B.T.Ibragimov, S.A.Talipov, G.N.Tishchenko,
Yu.K.Kushmuradov, T.F.Aripov, S.Kuchkarov
Khim.Prir.Soedin,588, 1979

58.15 **N – Oxy – allomatrine**
$C_{15}H_{24}N_2O_2$
B.T.Ibragimov, G.N.Tishchenko, Yu.K.Kushmuradov,
T.F.Aripov, A.S.Sadikov *Khim.Prir.Soedin*,416, 1979

58.16 **(+) – Lupanine perchlorate**
$C_{15}H_{25}N_2O^+$, ClO_4^-
E.Skrzypczak–Jankun, A.Hoser, Z.Kaluski, A.Perkowska
Acta Crystallogr.,Sect.B,**36**, 1517, 1980

58.17 **Lycorine chlorohydrin**
$C_{16}H_{16}ClNO_3$
J.Toda, T.Sano, Y.Tsuda, M.Kaneda, Y.Iitaka
Tetrahedron Lett.,**21**, 369, 1980

58.18 **Norgalanthamine hydrochloride dihydrate**
$C_{16}H_{20}NO_3^+$, Cl^-, $2H_2O$
R.Roques, J.C.Rossi, J.P.Declercq, G.Germain
Acta Crystallogr.,Sect.B,**36**, 1589, 1980

58.19 **Monocrotaline**
$C_{16}H_{23}NO_6$
Sh.–T.Wang *Ko Hsueh Tung Pao*,**23**, 670, 1978

58.20 **N – Cyanomethyl – angustifoline**
$C_{16}H_{23}N_3O$
M.D.Bratek–Wiewiorowska, U.Rychlewska,
M.Wiewiorowski
J.Chem.Soc.,Perkin Trans.2,1469, 1979

58.21 **Retusine**
$C_{16}H_{25}NO_5$
W.Shou–Dao *Ko Hsueh Tung Pao*,**24**, 1023, 1979

58.22 **11β – Acetylamino – 2α – H – 8α – isopropyl – 11α – methyl – 1α – methoxy – 6,9 – diaza – 12 – oxatricyclo(7.3.0.0²,⁶)dodecane – 7,10 – dione**
$C_{16}H_{25}N_3O_5$
M.Przybylska, F.R.Ahmed
Am.Cryst.Assoc.,Ser.2,**7**, 12, 1980
See also R1 : 48

58.23 **2 – Methyl – sparteine perchlorate**
$C_{16}H_{29}N_2^+$, ClO_4^-
A.Katrusiak, A.Hoser, Z.Kaluski, W.Boczon
Acta Crystallogr.,Sect.B,**36**, 1688, 1980

58.24 **Sparteine – N(16) – methyliodide**
$C_{16}H_{29}N_2^+$, I^-
A.E.Koziol, Z.Kosturkiewicz
Eur.Cryst.Meeting,**6**, 54, 1980

58.25 8 – Hydroxy – 8 – methyl – 6 – (2' – methyl –
hexylidene) – 1 – azabicyclo(4.3.0)nonane
hydrochloride (absolute configuration)
$C_{16}H_{30}NO^+$, Cl^-
J.W.Daly, T.Tokuyama, T.Fujiwara, R.J.Highet, I.L.Karle
J.Am.Chem.Soc., **102**, 830, 1980

58.C Ellipticine – 5 – iodocytidylyl – (3' – 5') –
guanosine hydrate methanol solvate
$2C_{17}H_{15}N_2^+$, $2C_{19}H_{23}IN_8O_{12}P^-$, $20H_2O$, $2CH_4O$
Main entry is 60.36

58.26 Codeine – 7,8 – oxide
$C_{18}H_{21}NO_4$
K.Uba, N.Miyata, K.Watanabe, M.Hirobe
Chem.Pharm.Bull., **27**, 2257, 1979

58.27 6α – Formyl – 15,16 – dimethoxy – B – nor – cis –
erythrinan – 8 – one
$C_{18}H_{21}NO_4$
S.Mohr, T.Clausen, B.Epe, C.Wolff, A.Mondon
Chem.Ber., **112**, 3795, 1979

58.28 1 – Methyl – 6,16 – dihydro – 20 – desethylidene –
ervitsine
$C_{18}H_{22}N_2O$
M.Harris, D.S.Grierson, C.Riche, H.–P.Husson
Tetrahedron Lett., **21**, 1957, 1980

58.29 (1aR,6bR,10R,11R) – 9,15 – Dioxo – 10 – hydroxy –
10,11,13 – trimethyl – 1a,2,3,6b – tetrahydro – 5H –
pyrrolizino(1a,6b,6a – b,c) – 1,8 – dioxa – 15 – cis –
tridecene
Doronenine
$C_{18}H_{25}NO_5$
A.Kirfel, G.Will, H.Wiedenfeld, E.Roeder
Cryst.Struct.Commun., **9**, 353, 1980

58.30 Retrorsine hydrobromide ethanol solvate
$C_{18}H_{26}NO_6^+$, Br^-, C_2H_6O
H.Stoeckli–Evans
Acta Crystallogr., *Sect.B*, **35**, 2798, 1979

58.31 Inkanine benzene solvate
$C_{18}H_{27}NO_5$, C_6H_6
B.Tashkhodzhaev, M.V.Telezhenetskaya, S.Yu.Yunusov
Khim.Prir.Soedin, 363, 1979

58.32 Tricodesmine
$C_{18}H_{27}NO_6$
B.Tashkhodzhaev, M.R.Yaguadaev, S.Yu.Yunusov
Khim.Prir.Soedin, 368, 1979

58.33 (–) – (1R,5R,9R,2''S) – 5,9 – Dimethyl – 2' –
hydroxy – 2 – (2 – methoxypropyl) – 6,7 –
benzomorphan hydrobromide
$C_{18}H_{28}NO_2^+$, Br^-
O.M.Peeters, N.M.Blaton, C.J.De Ranter
Eur.Cryst.Meeting, **6**, 283, 1980

58.34 N – Methyl – crotananinium iodide
4 – Methyl – 12 – hydroxy – 13,14,15 – trimethyl –
senec – 1 – eninium iodide
$C_{18}H_{28}NO_5^+$, I^-
K.N.Goswami, S.D.Sharma, O.P.Suri
Indian J.Pure Appl.Phys., **18**, 62, 1980

58.35 Ervistine methanol solvate
$C_{19}H_{20}N_2O$, CH_4O
C.Riche *Acta Crystallogr.*, *Sect.B*, **35**, 2738, 1979

58.36 8α,10α – Epidioxy – 8,14 – dihydro – 14β –
nitrothebaine ethyl acetate solvate
$C_{19}H_{20}N_2O_7$, $0.67C_4H_8O_2$
R.M.Allen, C.J.Gilmore, G.W.Kirby, D.J.McDougall
J.Chem.Soc., *Chem.Commun.*, 22, 1980

58.37 7,8 – Didehydro – 4,5 – epoxy – 17 – (prop – 2 –
enyl)morphinan – 3,6 – diol hydrobromide
Nalorphine hydrobromide
$C_{19}H_{22}NO_3^+$, Br^-
Y.G.Gelders, C.J.De Ranter, C.van Rooijen–Reiss
Cryst.Struct.Commun., **8**, 995, 1979

58.38 Cinchonidine
$C_{19}H_{22}N_2O$
B.Oleksyn *Eur.Cryst.Meeting*, **5**, 188, 1979

58.39 (–) – 4 – Bromo – 2,3 – dimethoxy – N – methyl –
6 – oxomorphinan
$C_{19}H_{24}BrNO_3$
H.van Koningsveld, C.Olieman
Cryst.Struct.Commun., **9**, 11, 1980

58.40 Herquline hemihydrate
$C_{19}H_{26}N_2O_2$, $0.5H_2O$
A.Furusaki, T.Matsumoto, H.Ogura, H.Takayanagi,
A.Hirano, S.Omura
J.Chem.Soc., *Chem.Commun.*, 698, 1980

58.41 (–) – (1R,5R,9R,2''S) – 5,9 – Dimethyl – 2' –
hydroxy – 2 – (2 – tetrahydro – 2 – furylmethyl) –
6,7 – benzomorphan
$C_{19}H_{27}NO_2$
O.M.Peeters, N.M.Blaton, C.J.De Ranter
Eur.Cryst.Meeting, **6**, 283, 1980

58.42 (–) – (1R,5R,9R) – 5,9 – Dimethyl – 2' – hydroxy –
2 – (4 – methoxybutyl) – 6,7 – benzomorphan
hydrobromide
$C_{19}H_{30}NO_2^+$, Br^-
O.M.Peeters, N.M.Blaton, C.J.De Ranter
Eur.Cryst.Meeting, **6**, 283, 1980

58.43 6 – Chloro – hyellazole
$C_{20}H_{16}ClNO$
J.H.Cardellina, M.P.Kirkup, R.E.Moore, J.S.Mynderse, K.Seff,
C.J.Simmons *Tetrahedron Lett.*, 4915, 1979

58.44 Sibiricine
$C_{20}H_{17}NO_6$
S.–M.Nasirov, I.A.Israilov, L.G.Kuz'mina, M.S.Yunusov,
Yu.T.Struchkov, S.Yu.Yunusov
Khim.Prir.Soedin, 752, 1978

58.45 2,2 – Dimethoxy – N – methyl – seco(B,C)
cephalotax – 4 – en – 3 – one
$C_{20}H_{25}NO_5$
R.J.Parry, M.N.T.Chang, J.M.Schwab, B.M.Foxman
J.Am.Chem.Soc., **102**, 1099, 1980

58.46 16α,17β – Butanomorphinan (inactive form)
$C_{20}H_{27}NO$
F.R.Ahmed *Eur.Cryst.Meeting*, **5**, 57, 1979

58.47 16β,17α – Butanomorphinan hydrobromide
$C_{20}H_{28}NO^+$, Br^-
F.R.Ahmed *Eur.Cryst.Meeting*, **5**, 57, 1979

58.48 Fruticosonine
$C_{20}H_{28}N_2O$
N.Chaichit, B.M.Gatehouse, I.R.C.Bick, M.A.Hai, N.W.Preston
J.Chem.Soc., *Chem.Commun.*, 874, 1979

58.49 (−) − 5 − Ethyl − 2′ − hydroxy − 2 − isobutyl − 9,9 −
dimethyl − 6,7 − benzomorphan hydrobromide
$C_{20}H_{32}NO^+$, Br^-
Y.G.Gelders, C.J.De Ranter
Acta Crystallogr.,Sect.B,**36**, 744, 1980

58.50 pseudo − racemic Podopetaline ormosanine
(monoclinic form)
$C_{20}H_{35}N_3$, $C_{20}H_{33}N_3$
R.Misra, W.Wong−Ng, P.−T.Cheng, S.McLean, S.C.Nyburg
J.Chem.Soc.,Chem.Commun.,659, 1980

58.51 6 − Epipodopetaline hydrobromide hydrate
$2C_{20}H_{36}N_3^{3+}$, $6Br^-$, $3.5H_2O$
M.F.Mackay, B.J.Poppleton
Cryst.Struct.Commun.,**9**, 805, 1980

58.C Codeine − tricarbonyl − chromium
$C_{21}H_{21}CrNO_6$ Main entry is 75.28

58.52 7,8 − Didehydro − 4,5α − epoxy − 17 − methyl −
morphinan − 3,6α − diol diacetate
Heroin
$C_{21}H_{23}NO_5$
D.Canfield, J.Barrick, B.C.Giessen
Acta Crystallogr.,Sect.B,**35**, 2806, 1979

58.53 Colchiceine ethyl acetate solvate monohydrate
$C_{21}H_{23}NO_6$, $C_4H_8O_2$, H_2O
J.V.Silverton *Acta Crystallogr.,Sect.B*,**35**, 2800, 1979

58.54 Capuronine acetate
$C_{21}H_{26}N_2O_2$
C.Riche *Acta Crystallogr.,Sect.B*,**36**, 1573, 1980

58.55 Cardiopetaline
$C_{21}H_{33}NO_3$
A.G.Gonzalez, G.de la Fuente, M.Reina, V.Zabel, W.H.Watson
Tetrahedron Lett.,**21**, 1155, 1980

58.56 Cardiopetalidine
$C_{21}H_{33}NO_4$
A.G.Gonzalez, G.de la Fuente, M.Reina, V.Zabel, W.H.Watson
Tetrahedron Lett.,**21**, 1155, 1980

58.57 (−) − 10,11 − Oxy − 10,12a − cyclo − 10,11 − seco −
colchicine
$C_{22}H_{25}NO_7$
A.Brossi, M.Rosner, J.V.Silverton, M.A.Iorio, C.D.Hufford
Helv.Chim.Acta,**63**, 406, 1980

58.58 10,12 − Dichloro − 2,16 − dihydro − 16 − hydroxy −
2 − methoxy − tabersonine
$C_{22}H_{26}Cl_2N_2O_4$
J.Lamotte, L.Dupont, O.Dideberg, G.Lewin
Acta Crystallogr.,Sect.B,**36**, 196, 1980

58.59 (±) − 2′ − Hydroxy − 5,9 − dimethyl − 2 −
phenethyl − 6,7 − benzomorphan hydrobromide
hemihydrate
Phenazocine hydrobromide hemihydrate
$C_{22}H_{28}NO^+$, Br^-, $0.5H_2O$
Y.G.Gelders, C.J.De Ranter, M.Kokkes
Acta Crystallogr.,Sect.B,**36**, 1141, 1980

58.60 Cuauchichicine
$C_{22}H_{33}NO_2$
S.W.Pelletier, H.K.Desai, J.Finer−Moore, N.V.Mody
J.Am.Chem.Soc.,**101**, 6741, 1979

58.61 Parsonsine (lower melting point form ii)
$C_{22}H_{33}NO_8$
G.J.Gainsford *Cryst.Struct.Commun.*,**9**, 173, 1980

58.62 N,O − Diacetyl − 4 − hydroxy − nornantenine
$C_{23}H_{23}NO_7$
V.Zabel, W.H.Watson, A.Urzua, B.K.Cassels
Acta Crystallogr.,Sect.B,**35**, 3126, 1979

58.63 6 − Acetoxy − 5 − methoxy − 1 − methyl − 1,2,3,9a −
tetrahydro − cyclohexa(i,j)isoquinoline − 7 − spiro −
4′ − (2′ − methoxy − cyclohexa − 2′,5′ − dien − 1′ −
one) methiodide
O − Acetyl − kreysiginone − spiro − isomer methiodide
$C_{23}H_{28}NO_5^+$, I^-
H.Hara, O.Hoshino, B.Umezawa, Y.Iitaka
J.Chem.Soc.,Perkin Trans.1,2657, 1979

58.64 (±) − 5 − Ethyl − 2′ − hydroxy − 9,9 − dimethyl −
2 − phenethyl − 6,7 − benzomorphan hydrobromide
Dimephen
$C_{24}H_{32}NO^+$, Br^-
Y.G.Gelders, C.J.De Ranter, M.Kokkes
Acta Crystallogr.,Sect.B,**36**, 1141, 1980

58.65 Chasmanine intermediate
$C_{24}H_{35}NO_6$
M.Przybylska, F.R.Ahmed
Acta Crystallogr.,Sect.B,**36**, 494, 1980

58.66 N − Ethyl − 1α,14α − dihydroxy − 8β,16β −
dimethoxy − 6β − methoxycarbonyl − aconitine
hydroiodide
$C_{25}H_{40}NO_6^+$, I^-
P.W.Codding, K.A.Kerr *Am.Cryst.Assoc.,Ser.2*,**7**, 12, 1980
See also R1 : 54

58.67 Delphinium alkaloid A hydroiodide
$C_{26}H_{42}NO_6^+$, I^-
P.W.Codding, K.A.Kerr, M.H.Benn, A.J.Jones, S.W.Pelletier,
N.V.Mody *Tetrahedron Lett.*,**21**, 127, 1980

58.68 Diethylamino − dihydro − veatchinone
$C_{26}H_{42}N_2O_2$
S.W.Pelletier, A.P.Venkov, J.Finer−Moore, N.V.Mody
Tetrahedron Lett.,**21**, 809, 1980
See also R1 : 54

58.69 Stercuronium iodide ethanol solvate
$C_{26}H_{43}N_2^+$, I^-, $2C_2H_6O$
J.Husain, I.Tickle, R.Palmer
Eur.Cryst.Meeting,**5**, 187, 1979

58.70 Thiocolchicoside ethanol solvate hydrate
$C_{27}H_{33}NO_{10}S$, $2C_2H_6O$, H_2O
J.I.Clark, T.N.Margulis *Life Sci.*,**26**, 833, 1980
See also R1 : 11

58.71 Fumitremorgin B
$C_{27}H_{33}N_3O_5$, H_2O
M.Yamazaki, K.Suzuki, H.Fujimoto, T.Akiyama,
U.Sankawa, Y.Iitaka *Chem.Pharm.Bull.*,**28**, 861, 1980

58.72 Hosukinidine hydrochloride
$C_{27}H_{44}NO^+$, Cl^-
K.Kaneko, N.Kawamura, H.Mitsuhashi, K.Ohsaki
Chem.Pharm.Bull.,**27**, 2534, 1979

58.73 Marcfortine A
$C_{28}H_{35}N_3O_4$
J.Polonsky, M.−A.Merrien, T.Prange, C.Pascard
J.Chem.Soc.,Chem.Commun.,601, 1980
See also R1 : 38

58.74 **Cytochalasin A**
$C_{29}H_{35}NO_5$
J.F.Griffin, A.L.Rampal, C.Y.Jung
*Am.Cryst.Assoc.,Ser.2,***7**, 13, 1980

58.75 **Lophocine**
$C_{30}H_{40}N_2O_4$
M.C.Wani, J.B.Thompson, H.L.Taylor, M.E.Wall, R.W.Miller,
A.T.McPhail *J.Chem.Res.,***15**, 0301, 1980

58.76 **Bromocriptine methylsulfonate isopropanol solvate (absolute configuration)**
$C_{32}H_{41}BrN_5O_5{}^+$, $CH_3O_3S^-$, $0.5C_3H_8O$
N.Camerman, L.Y.Y.Chan, A.Camerman
*Mol.Pharmacol.,***16**, 729, 1979
See also R2 : 11

58.77 **(−) − Dihydroergotamine methylsulfonate monohydrate (absolute configuration)**
$C_{33}H_{38}N_5O_5{}^+$, $CH_3O_3S^-$, H_2O
H.Hebert *Acta Crystallogr.,Sect.B,***35**, 2978, 1979

58.78 **Chaetoglobosin K acetone solvate**
$C_{34}H_{40}N_2O_5$, C_3H_6O
J.P.Springer, R.H.Cox, H.G.Cutler, F.G.Crumley
*Tetrahedron Lett.,***21**, 1905, 1980

58.79 **Sungucine acetone solvate**
$C_{42}H_{42}N_4O_2$, C_3H_6O
L.Dupont, O.Dideberg, J.Lamotte, K.Kambu, L.Angenot
*Acta Crystallogr.,Sect.B,***36**, 1669, 1980

MISCELLANEOUS NATURAL PRODUCTS

59.1 **Dopamine hydrochloride**
$C_8H_{12}NO_2{}^+$, Cl^-
J.Giesecke *Acta Crystallogr.,Sect.B,***36**, 178, 1980
See also R1 : 17,3

59.2 **6 − Methoxy − 2 − methyl − 3,5 − dihydro − benzo(b) furan − 4,7 − dione**
Acamelin
$C_{10}H_8O_4$
H.W.Schmalle, B.M.Hausen
*Tetrahedron Lett.,***21**, 149, 1980
See also R1 : 38

59.3 **3,6 − Dimethyl − 4,10 − dihydroxy − 2 − oxa − spiro(4.5)dec − 7 − ene − 1,9 − dione**
Rosigenin
$C_{11}H_{14}O_5$
A.Albinati, S.Bruckner, L.Camarda, G.Nasini
*Tetrahedron,***36**, 117, 1980
See also R1 : 38

59.4 **Palythene monohydrate (absolute configuration)**
$C_{13}H_{20}N_2O_5$, H_2O
D.Uemura, C.Katayama, A.Wada, Y.Hirata
*Chem.Lett.,*755, 1980
See also R1 : 17,16

59.5 **5,8 − Dimethoxy − 2 − methylfuro(2,3 − e)chromone**
Khellin
$C_{14}H_{12}O_5$
A.Carpy, D.Hickel, J.M.Leger
*Cryst.Struct.Commun.,***8**, 835, 1979
See also R1 : 38

59.6 **(1aβ,2α,6aβ,6bβ) − 3 − Methyl − N − (1a,6,6a,6b − tetrahydro − 2,6a − dimethyl − 6 − oxo − 2H − oxireno(a)pyrrolizin − 4 − yl) − 2 − butenamide**
Bohemamine
$C_{14}H_{18}N_2O_3$
T.W.Doyle, D.E.Nettleton, D.M.Balitz, J.E.Moseley,
R.E.Grulich, T.McCabe, J.Clardy
*J.Org.Chem.,***45**, 1324, 1980
See also R1 : 35

59.7 **Chlorofucin**
$C_{15}H_{20}BrClO_2$
B.M.Howard, G.R.Schulte, W.Fenical, B.Solheim, J.Clardy
*Tetrahedron,***36**, 1747, 1980
See also R1 : 38

59.8 **Poiteol**
$C_{15}H_{20}BrClO_3$
B.M.Howard, G.R.Schulte, W.Fenical, B.Solheim, J.Clardy
*Tetrahedron,***36**, 1747, 1980
See also R1 : 38

59.9 **Xantholide B**
$C_{15}H_{20}O_2$
T.Tahara, Y.Sakuda, M.Kodama, Y.Fukazawa, S.Ito,
K.Kawazu, S.Nakajima *Tetrahedron Lett.,***21**, 1861, 1980
See also R1 : 38

59.10 **2,2,5,7 – Tetramethyl – 4 – oxy – 6 – (2 – oxyethyl) – indanone**
$C_{15}H_{20}O_3$
A.A.Semenov, A.I.Syrchina, O.N.Gorenysheva, V.N.Byushkin, T.I.Malinovskii
Symp.Pap.IUPAC Int.Symp.Chem.Nat.Prod., **2**, 375, 1978
See also R1 : 27

59.11 **6,6′ – Dibromo – indigo**
$C_{16}H_8Br_2N_2O_2$
P.Susse, C.Krampe *Naturwissenschaften*, **66**, 110, 1979
See also R1 : 35

59.12 **6,6′ – Dibromo – indigo**
Tyrian purple
$C_{16}H_8Br_2N_2O_2$
S.Larsen, F.Watjen
Acta Chem.Scand.Ser.A, **34**, 171, 1980
See also R1 : 35

59.13 **Cunaniol acetate (violet form)**
$C_{16}H_{16}O_3$
G.P.Jones, P.J.Pauling
J.Chem.Soc.,Perkin Trans.2, 1482, 1979
See also R1 : 38

59.14 **(1R,3S) – 7,10 – Dihydroxy – 1,3,8 – trimethyl – 3,4,6,9 – tetrahydro – 1H – naphtho(2,3 – c)pyran – 6,9 – dione**
Ventilagone
$C_{16}H_{16}O_5$
R.G.Cooke, A.Liu, C.L.Raston, A.H.White
Aust.J.Chem., **33**, 303, 1980
See also R1 : 38

59.15 **(+) – Pisatin monohydrate**
$C_{17}H_{14}O_6$, H_2O
C.DeMartinis, M.F.Mackay, D.R.Perrin
J.Cryst.Mol.Struct., **8**, 247, 1978
See also R1 : 38

59.16 **5 – Acetoxy – 6,7 – dimethyl – 3 – oxatricyclo(5.4.0.0^{2,4})undecan – 10 – one – 9,2^1 – spiro(3^1 – formyl – 3^1 – methyl – oxirane) (absolute configuration)**
PR Toxin
$C_{17}H_{20}O_6$
F.Baert, M.Foulon, G.Odou, S.Moreau
Acta Crystallogr.,Sect.B, **36**, 402, 1980
See also R1 : 38

59.17 **Laurencienyne**
$C_{17}H_{23}BrCl_2O_3$
E.N.Duesler, K.L.Rinehart Junior, I.C.Paul, S.Caccamese, R.Azzolina *Cryst.Struct.Commun.*, **9**, 777, 1980
See also R1 : 38

59.18 **2,2′ – Dimethoxy – indigo**
$C_{18}H_{16}N_2O_4$
S.Larsen, F.Watjen
Acta Chem.Scand.Ser.A, **34**, 171, 1980
See also R1 : 35

59.19 **7 – (3 – (4,5 – Dihydro – 5,5 – dimethyl – 4 – oxo – 2 – furanyl) – but – 2 – enyl) – oxy – (2H – 1 – benzopyran – 2 – one)**
Geiparvarin
$C_{19}H_{18}O_5$
B.H.Toder, D.Boschelli, A.B.Smith
J.Cryst.Mol.Struct., **9**, 189, 1979
See also R1 : 38

59.20 **8β – Hydroxymethyl – podocarpane – 13β – carboxylic acid lactone**
$C_{19}H_{30}O_2$
A.F.Beecham, R.C.Cambie, R.C.Hayward, B.J.Poppleton
Aust.J.Chem., **32**, 2617, 1979
See also R1 : 38

59.21 **O – Tetramethyl – haematoxylin (optically active form)**
$C_{20}H_{22}O_6$
C.de Martinis, M.F.Mackay
Acta Crystallogr.,Sect.B, **36**, 1606, 1980
See also R1 : 38

59.22 **N – p – Iodobenzyl – 3 – n – octan – amido – pyridine – 2,5,6 – trione**
$C_{20}H_{23}IN_2O_4$
G.Ferguson, D.R.Pollard, J.M.Robertson, G.O.P.Doherty, N.B.Haynes, D.W.Mathieson, W.B.Whalley, T.H.Simpson
J.Chem.Soc.,Perkin Trans.1, 1782, 1980

59.23 **(±) – 2β,5β – Epoxy – 2α – (3α – p – nitrobenzoyloxy – but – 1(E) – enyl) – 1β,3,3 – trimethylcyclohexan – 1α – ol**
3,6 – Epoxy – 5 – hydroxy – 5,6 – dihydro – β – ionol
$C_{20}H_{25}NO_6$
T.Kato, H.Kondo, Y.Kitano, G.Hata, Y.Takagi
Chem.Lett., 757, 1980
See also R1 : 38

59.24 **Peunicin (at 113°K, absolute configuration)**
$C_{20}H_{28}O_4$
C.Y.Chang, L.S.Ciereszko, M.B.Hossain, D.van der Helm
Acta Crystallogr.,Sect.B, **36**, 731, 1980
See also R1 : 38

59.25 **(5Z,11α,13E,15S) – 11,15 – Dihydroxy – 9 – oxo – prosta – 5,13 – dien – 1 – oic acid**
Prostaglandin E$_2$
$C_{20}H_{32}O_5$
G.T.DeTitta, D.A.Langs, J.W.Edmonds, W.L.Duax
Acta Crystallogr.,Sect.B, **36**, 638, 1980
See also R1 : 20,1

59.26 **(5Z,9β,11α,13E,15S) – 9,11,15 – Trihydroxy – prosta – 5,13 – dien – 1 – oic acid**
Prostaglandin F(2β)
$C_{20}H_{34}O_5$
G.T.DeTitta, D.A.Langs, J.W.Edmonds, W.L.Duax
Acta Crystallogr.,Sect.B, **36**, 638, 1980
See also R1 : 20,1

59.27 **Thromboxane B$_2$**
$C_{20}H_{34}O_6$
S.Fortier, M.G.Erman, D.A.Langs, G.T.DeTitta
Acta Crystallogr.,Sect.B, **36**, 1099, 1980
See also R1 : 38,1

59.28 **Dunnione p – bromophenylhydrazone**
$C_{21}H_{19}BrN_2O_2$
R.G.Cooke, C.L.Raston, A.H.White
Aust.J.Chem., **33**, 441, 1980
See also R1 : 38

59.29 **(1RS,3RS,6SR,7SR,10SR,11RS,12RS,13RS) – (6 – Phenyl – tetracyclo(5.4.2.0^{3,13},0^{10,12})tridecane – 4,8 – dien – 11 – yl) acetic acid**
Endiandric acid
$C_{21}H_{22}O_2$
W.M.Bandaranayake, J.E.Banfield, D.St.C.Black, G.D.Fallon, B.M.Gatehouse *J.Chem.Soc.,Chem.Commun.*, 162, 1980
See also R1 : 29

59.30 p – Bromobenzoyl – epicatalponol (absolute configuration)
$C_{22}H_{21}BrO_3$
K.Inoue, H.Inouye, T.Taga, R.Fujita, K.Osaki, K.Kuriyama
Chem.Pharm.Bull.,**28**, 1224, 1980
See also R1 : 19

59.31 8α – Hydroxy – isopicrostegane
$C_{22}H_{22}O_8$
C.Pascard, J.–P.Robin, E.Brown
Acta Crystallogr.,Sect.B,**36**, 198, 1980
See also R1 : 38

59.32 D – Gibberellin C secodiester photoproduct
$C_{22}H_{32}O_7$
L.Kutschabsky, G.Reck, G.Adam, T.V.Sung
Tetrahedron,**36**, 741, 1980
See also R1 : 38

59.33 Purpuride
$C_{22}H_{33}NO_5$
T.J.King, J.C.Roberts, D.J.Thompson
J.Chem.Soc.,Perkin Trans.1,78, 1973
See also R1 : 38

59.34 Vismione A
$C_{23}H_{26}O_6$
F.D.Monache, F.Ferrari, G.B.M.Bettolo, P.Maxfield,
S.Cerrini, W.Fedeli, E.Gavuzzo, A.Vaciago
Gazz.Chim.Ital.,**109**, 301, 1979
See also R1 : 28

59.35 17 – O – Methyl – latrunculin A
$C_{23}H_{33}NO_5S$
Y.Kashman, A.Groweiss, U.Shmueli
Tetrahedron Lett.,**21**, 3629, 1980
See also R1 : 41,38

59.36 4a,5,8,8a – Tetrahydro – 11,14 – dimethoxy – 7 – methyl – 4a – (3 – methyl – but – 2 – enyl) – 5,8a – o – benzeno – 1,4 – naphthoquinone
Microphyllone dimethyl ether
$C_{24}H_{28}O_4$
S.K.Agarwal, R.P.Rastogi, H.van Koningsveld, K.Goubitz,
G.J.Olthof *Tetrahedron*,**36**, 1435, 1980
See also R1 : 31

59.37 Gilmaniellin methanol solvate
$C_{26}H_{19}ClO_{10}$, $2CH_4O$
K.K.Chexal, C.Tamm, K.Hirotsu, J.Clardy
Helv.Chim.Acta,**62**, 1785, 1979
See also R1 : 38

59.38 Terretonin
$C_{26}H_{32}O_9$
J.P.Springer, J.W.Dorner, R.J.Cole, R.H.Cox
J.Org.Chem.,**44**, 4852, 1979
See also R1 : 38

59.39 Paspalicine (orthorhombic form)
$C_{27}H_{31}NO_3$
J.P.Springer, J.Clardy *Tetrahedron Lett.*,**21**, 231, 1980
See also R1 :

59.40 Verrucarin B
$C_{27}H_{32}O_9$
W.Breitenstein, C.Tamm, E.V.Arnold, J.Clardy
Helv.Chim.Acta,**62**, 2699, 1979
See also R1 :

59.41 Paspaline methanol solvate
$C_{28}H_{39}NO_2$, $0.5CH_4O$
J.P.Springer, J.Clardy *Tetrahedron Lett.*,**21**, 231, 1980
See also R1 : 38,36

59.42 Uvarinol benzene solvate
$C_{36}H_{30}O_7$, C_6H_6
C.D.Hufford, W.L.Lasswell Junior, K.Hirotsu, J.Clardy
J.Org.Chem.,**44**, 4709, 1979
See also R1 : 38

59.43 CC – 1065 – Anti – tumor agent (at –150°C)
NSC 298223
$C_{37}H_{33}N_7O_8$, $C_{36}H_{33}N_7O_8$, H_2O, $3CH_4O$
C.G.Chidester, D.J.Duchamp, D.G.Martin, W.C.Krueger
Am.Cryst.Assoc.,Ser.2,**8**, 18, 1980
See also R1 : 36

59.44 Ohchinolide A
$C_{37}H_{42}O_{10}$
H.Nakai, M.Shiro, M.Ochi
Acta Crystallogr.,Sect.B,**36**, 1698, 1980
See also R1 : 38

59.45 Diethoxy – bilirubin diethyl ester
$C_{41}H_{52}N_4O_6$
W.S.Sheldrick, W.Becker
Z.Naturforsch.,Teil B,**34**, 1542, 1979
See also R1 : 32

59.46 (±) – Methyl – 4,5 – dimethoxy – 2 – (2,6 – dimethoxy – 1 – oxo – 9 – phenyl – 5 – phenalenyl) – 1 – oxo – 8 – phenyl – 1,2 – dihydro – 2 – acenaphthylene – carboxylate
$C_{43}H_{32}O_8$
J.M.Edwards, M.Mangion, J.B.Anderson, M.Rapposch,
P.Moews, G.Hite *Acta Crystallogr.,Sect.B*,**36**, 1241, 1980
See also R1 : 28

MOLECULAR COMPLEXES

60.C Tetrathiafulvalene bis(ethylenedithiolato) – nickel(i)
$3C_4H_4NiS_4$, $2C_6H_4S_4$ Main entry is 60.5

60.1 Pyridinium 1 – naphthylamine picrate
$C_5H_6N^+$, $C_{10}H_9N$, $C_6H_2N_3O_7^-$
J.Bernstein, H.Regev, F.H.Herbstein
*Acta Crystallogr.,Sect.B,***36**, 1170, 1980
See also R1 : 33 R2 : 60,24 R3 : 60,17

60.C Tetrathiafulvalene – chloranil complex
$C_6Cl_4O_2$, $C_6H_4S_4$ Main entry is 60.3

60.C Tetrathiafulvalene – fluoranil complex
$C_6F_4O_2$, $C_6H_4S_4$ Main entry is 60.4

60.C Pyrene – fluoranil complex
$C_6F_4O_2$, $C_{16}H_{10}$ Main entry is 60.34

60.C N,N,N′,N′ – Tetramethyl – p – phenylenediamine
hexafluorobenzene complex
C_6F_6, $C_{10}H_{16}N_2$ Main entry is 60.13

60.C Pyridinium 1 – naphthylamine picrate
$C_6H_2N_3O_7^-$, $C_{10}H_9N$, $C_5H_6N^+$ Main entry is 60.1

60.C Valinomycin potassium picrate
$C_6H_2N_3O_7^-$, $C_{36}H_{60}N_4O_{12}$, K^+ Main entry is 60.43

60.2 3,5 – Dichlorophenol – 2,6 – dimethylphenol
complex
$C_6H_4Cl_2O$, $C_8H_{10}O$
C.Bavoux, M.Perrin, A.Thozet
*Eur.Cryst.Meeting,***5**, 121, 1979
See also R1 : 17 R2 : 60,17

60.3 Tetrathiafulvalene – chloranil complex
$C_6H_4S_4$, $C_6Cl_4O_2$
J.J.Mayerle, J.B.Torrance, J.I.Crowley
*Acta Crystallogr.,Sect.B,***35**, 2988, 1979
See also R1 : 39 R2 : 60,18

60.4 Tetrathiafulvalene – fluoranil complex
$C_6H_4S_4$, $C_6F_4O_2$
J.J.Mayerle, J.B.Torrance, J.I.Crowley
*Acta Crystallogr.,Sect.B,***35**, 2988, 1979
See also R1 : 39 R2 : 60,18

60.C bis(Acetylacetonato) – palladium(ii)
tetrathiafulvalene
$C_6H_4S_4$, $C_{10}H_{14}O_4Pd$ Main entry is 60.12

60.5 Tetrathiafulvalene bis(ethylenedithiolato) – nickel(i)
$2C_6H_4S_4$, $3C_4H_4NiS_4$
J.S.Kasper, L.V.Interrante
*Am.Inst.Phys.,Conf.Proc.,***53**, 205, 1979
See also R1 : 39 R2 : 60,85

60.C Skatole tetracyanoethylene
C_6N_4, C_9H_9N Main entry is 60.9

60.C bis(9,10 – Diazaphenanthrene) tetracyanoethylene
(triclinic form)
C_6N_4, $2C_{12}H_8N_2$ Main entry is 60.15

60.6 Tyramine 1 – thyminyl – (acetic acid) complex
hydrate
$C_7H_7N_2O_4^-$, $C_8H_{12}NO^+$, H_2O
K.Ogawa, K.Tago, T.Ishida, K.–I.Tomita
*Acta Crystallogr.,Sect.B,***36**, 2095, 1980
See also R1 : 44 R2 : 60,17

60.C 3,5 – Dichlorophenol – 2,6 – dimethylphenol
complex
$C_8H_{10}O$, $C_6H_4Cl_2O$ Main entry is 60.2

60.C Tyramine 1 – thyminyl – (acetic acid) complex
hydrate
$C_8H_{12}NO^+$, $C_7H_7N_2O_4^-$, H_2O Main entry is 60.6

60.7 Tetraethylammonium bis(7,7,8,8 –
tetracyanoquinodimethane)
$C_8H_{20}N^+$, $C_{12}H_4N_4^-$, $C_{12}H_4N_4$
R.P.Shibaeva, V.F.Kaminskii, M.A.Simonov
*Cryst.Struct.Commun.,***9**, 655, 1980
See also R1 : 3 R2 : 60,12,7

60.8 5 – Phenyl – 1,3 – thiaselenole – 2 – thione 7,7,8,8 –
tetracyanoquinodimethane complex
$C_9H_6S_2Se$, $C_{12}H_4N_4$
V.F.Kaminskii, R.P.Shibaeva, M.Z.Aldoshina,
R.N.Lyubovskaya, M.L.Khidekel
*Zh.Strukt.Khim.,***20**, 157, 1979
See also R1 : 42 R2 : 60,7

60.9 Skatole tetracyanoethylene
C_9H_9N, C_6N_4
E.E.Castellano, B.E.Rivero, A.D.Podjarny, M.E.Roselli
*Acta Crystallogr.,Sect.B,***36**, 1726, 1980
See also R1 : 35 R2 : 60,7

60.C 2,6 – Dimethylnaphthalene perfluoronaphthalene
complex
$C_{10}F_8$, $C_{12}H_{12}$ Main entry is 60.19

60.C Anthracene – tetracyanobenzene complex (high
temperature form, at 297°K)
$C_{10}H_2N_4$, $C_{14}H_{10}$ Main entry is 60.25

60.C Anthracene – tetracyanobenzene complex (high
temperature form, at 234°K)
$C_{10}H_2N_4$, $C_{14}H_{10}$ Main entry is 60.26

60.C Anthracene – tetracyanobenzene complex (high
temperature form, at 226°K)
$C_{10}H_2N_4$, $C_{14}H_{10}$ Main entry is 60.27

60.C Anthracene – tetracyanobenzene complex (low
temperature form, at 202°K)
$C_{10}H_2N_4$, $C_{14}H_{10}$ Main entry is 60.28

60.C Anthracene – tetracyanobenzene complex (low
temperature form, at 170°K)
$C_{10}H_2N_4$, $C_{14}H_{10}$ Main entry is 60.29

60.C Anthracene – tetracyanobenzene complex (low
temperature form, at 138°K)
$C_{10}H_2N_4$, $C_{14}H_{10}$ Main entry is 60.30

60.C Anthracene – tetracyanobenzene complex (low
temperature form, at 119°K)
$C_{10}H_2N_4$, $C_{14}H_{10}$ Main entry is 60.31

60.C Phenothiazine – pyromellitic dianhydride complex
$C_{10}H_2O_6$, $C_{12}H_9NS$ Main entry is 60.17

60.C Pyridinium 1 – naphthylamine picrate
$C_{10}H_9N$, $C_5H_6N^+$, $C_6H_2N_3O_7^-$ Main entry is 60.1

60.10 Tetramethyl – tetrathiafulvalene tetrabromo – p – benzoquinone complex
$C_{10}H_{12}S_4$, $C_6Br_4O_2$
J.B.Torrance, J.J.Mayerle, V.Y.Lee, K.Bechgaard
J.Am.Chem.Soc., **101**, 4747, 1979
See also R1 : 39 R2 : 62,18

60.11 bis(Tetramethyl – tetrathiafulvalene) iodide
$C_{10}H_{12}S_4$, $C_{10}H_{12}S_4^+$, I^-
J.L.Galigne *Cryst.Struct.Commun.*, **9**, 61, 1980
See also R1 : 39

60.C tris(4,4′,5,5′ – Tetramethyl – tetrathiafulvalene) – bis(2,5 – diethyl – 7,7,8,8 – tetracyanoquinodimethane) complex
$3C_{10}H_{12}S_4$, $2C_{16}H_{12}N_4$ Main entry is 60.35

60.12 bis(Acetylacetonato) – palladium(ii) tetrathiafulvalene
$C_{10}H_{14}O_4Pd$, $C_6H_4S_4$
A.R.Siedle, T.J.Kistenmacher, R.M.Metzger, C.–S.Kuo, R.P.van Duyne, T.Cape *Inorg.Chem.*, **19**, 2048, 1980
See also R1 : 77 R2 : 60,39

60.13 N,N,N′,N′ – Tetramethyl – p – phenylenediamine hexafluorobenzene complex
$C_{10}H_{16}N_2$, C_6F_6
T.Dahl *Acta Chem.Scand.Ser.A*, **33**, 665, 1979
See also R1 : 16 R2 : 60,19

60.C 4 – Bromobiphenyl perfluorobiphenyl complex
$C_{12}F_{10}$, $C_{12}H_9Br$ Main entry is 60.16

60.C Biphenyl perfluorobiphenyl complex
$C_{12}F_{10}$, $C_{12}H_{10}$ Main entry is 60.18

60.C 4 – Methylbiphenyl perfluorobiphenyl complex
$C_{12}F_{10}$, $C_{13}H_{12}$ Main entry is 60.23

60.C 5 – Phenyl – 1,3 – thiaselenole – 2 – thione 7,7,8,8 – tetracyanoquinodimethane complex
$C_{12}H_4N_4$, $C_9H_6S_2Se$ Main entry is 60.8

60.14 7,8 – benzoquinoline 7,7,8,8 – Tetracyanoquinodimethane
$C_{12}H_4N_4$, $C_{13}H_9N$
B.Shaanan, U.Shmueli
Acta Crystallogr.,Sect.B, **36**, 2076, 1980
See also R1 : 36 R2 : 60,7

60.C Stilbene – 7,7,8,8 – tetracyanoquinodimethane complex
$C_{12}H_4N_4$, $C_{14}H_{12}$ Main entry is 60.32

60.C Thieno(3,2 – e:4,5 – e′) – bis(benzo(b)thiophene) – 7,7,8,8 – tetracyanoquinodimethane
Trithia(5)heterohelicene – 7,7,8,8 – tetracyanoquinodimethane
$C_{12}H_4N_4$, $C_{18}H_8S_3$ Main entry is 60.33

60.C Tetramethoxystilbene – 7,7,8,8 – tetracyanoquinodimethane complex
$C_{12}H_4N_4$, $C_{18}H_{20}O_4$ Main entry is 60.38

60.C N,N′ – (1,2 – Phenylene) – bis(salicylaldiminato) – copper(ii) – 7,7,8,8 – tetracyanoquinodimethane complex
$C_{12}H_4N_4$, $2C_{20}H_{14}CuN_2O_2$ Main entry is 60.40

60.C Tetraethylammonium bis(7,7,8,8 – tetracyanoquinodimethane)
$C_{12}H_4N_4^-$, $C_8H_{20}N^+$, $C_{12}H_4N_4$ Main entry is 60.7

60.C 1,1′ – Dimethylferrocenium bis(7,7,8,8 – tetracyanoquinodimethane)
$C_{12}H_4N_4^-$, $C_{12}H_{14}Fe^+$, $C_{12}H_4N_4$ Main entry is 60.20

60.C Ethyltriphenylphosphonium – bis(7,7,8,8 – tetracyanoquinodimethane)
$C_{12}H_4N_4^-$, $C_{20}H_{20}P^+$, $C_{12}H_4N_4$ Main entry is 60.41

60.C Decamethylferrocenium – 7,7,8,8 – tetracyanoquinodimethane
$C_{12}H_4N_4^-$, $C_{20}H_{30}Fe^+$ Main entry is 60.42

60.C (N,N,N,N′,N′,N′ – Hexamethyl – hexamethylene – diammonium) (7,7,8,8 – tetracyanoquinodimethane)
$2C_{12}H_4N_4^-$, $C_{12}H_{30}N_2^{2+}$, $2C_{12}H_4N_4$ Main entry is 60.21

60.15 bis(9,10 – Diazaphenanthrene) tetracyanoethylene (triclinic form)
$2C_{12}H_8N_2$, C_6N_4
U.Shmueli, H.Mayorzik *Eur.Cryst.Meeting*, **6**, 40, 1980
See also R1 : 36 R2 : 60,7

60.16 4 – Bromobiphenyl perfluorobiphenyl complex
$C_{12}H_9Br$, $C_{12}F_{10}$
S.L.Pirtle, D.G.Naae *Am.Cryst.Assoc.,Ser.2*, **7**, 10, 1980
See also R1 : 19 R2 : 60,19

60.17 Phenothiazine – pyromellitic dianhydride complex
$C_{12}H_9NS$, $C_{10}H_2O_6$
R.Anthonj, N.Karl, B.E.Robertson, J.J.Stezowski
J.Chem.Phys., **72**, 1244, 1980
See also R1 : 41 R2 : 60,38

60.18 Biphenyl perfluorobiphenyl complex
$C_{12}H_{10}$, $C_{12}F_{10}$
D.G.Naae *Acta Crystallogr.,Sect.B*, **35**, 2765, 1979
See also R1 : 19 R2 : 60,19

60.19 2,6 – Dimethylnaphthalene perfluoronaphthalene complex
$C_{12}H_{12}$, $C_{10}F_8$
S.L.Birtle, D.G.Naae *Am.Cryst.Assoc.,Ser.2*, **7**, 10, 1980
See also R1 : 24 R2 : 60,24

60.20 1,1′ – Dimethylferrocenium bis(7,7,8,8 – tetracyanoquinodimethane)
$C_{12}H_{14}Fe^+$, $C_{12}H_4N_4^-$, $C_{12}H_4N_4$
S.R.Wilson, P.J.Corvan, R.P.Seiders, D.J.Hodgson, M.Brookhart, W.E.Hatfield, J.S.Miller, A.H.Reis Junior, P.K.Rogan, E.Gebert, A.J.Epstein
NATO Conf.Ser.6, 407, 1979
See also R1 : 73 R2 : 60,7

60.21 (N,N,N,N′,N′,N′ – Hexamethyl – hexamethylene – diammonium) (7,7,8,8 – tetracyanoquinodimethane)
$C_{12}H_{30}N_2^{2+}$, $2C_{12}H_4N_4^-$, $2C_{12}H_4N_4$
S.Flandrois, D.Chasseau, P.Delhaes, J.Gaultier, J.Amiell, C.Hauw *Bull.Chem.Soc.Jpn.*, **52**, 3407, 1979
See also R1 : 3 R2 : 60,7

60.C 7,8 – benzoquinoline 7,7,8,8 – Tetracyanoquinodimethane
$C_{13}H_9N$, $C_{12}H_4N_4$ Main entry is 60.14

60.22 9 – Amino – acridine – 5 – iodocytidylyl – (3′ – 5′) – guanosine hydrate complex
$4C_{13}H_{11}N_2^+$, $4C_{19}H_{23}IN_6O_{12}P^-$, $21H_2O$
T.D.Sakore, B.S.Reddy, H.M.Sobell
J.Mol.Biol., **135**, 763, 1979
See also R1 : 36 R2 : 60,47,46,45,44

60.C Cytidylyl – (3′ – 5′) – adenosine – proflavine complex decahydrate
$C_{13}H_{11}N_3$, $C_{19}H_{26}N_8O_{11}P$, $10H_2O$ Main entry is 60.39

60.23 4 – Methylbiphenyl perfluorobiphenyl complex
$C_{13}H_{12}$, $C_{12}F_{10}$
S.L.Pirtle, D.G.Naae *Am.Cryst.Assoc.,Ser.2*, **7**, 10, 1980
See also R1 : 19 R2 : 60,19

60.24 Proflavine – cytidylyl – (3 – 5′) – guanosine sulfate hydrate
$1.5C_{13}H_{12}N_3^+$, $C_{19}H_{24}N_8O_{12}P^-$, $0.5O_4S^-$, $11.5H_2O$
H.M.Berman, W.Stallings, H.L.Carrell, J.P.Glusker, S.Neidle, G.Taylor, A.Achari *Biopolymers,* **18**, 2405, 1979
See also R1 : 28 R2 : 60,47,45,44

60.25 Anthracene – tetracyanobenzene complex (high temperature form, at 297°K)
$C_{14}H_{10}$, $C_{10}H_2N_4$
J.J.Stezowski *Eur.Cryst.Meeting,* **5**, 211, 1979
See also R1 : 26 R2 : 60,7

60.26 Anthracene – tetracyanobenzene complex (high temperature form, at 234°K)
$C_{14}H_{10}$, $C_{10}H_2N_4$
J.J.Stezowski *Eur.Cryst.Meeting,* **5**, 211, 1979
See also R1 : 26 R2 : 60,7

60.27 Anthracene – tetracyanobenzene complex (high temperature form, at 226°K)
$C_{14}H_{10}$, $C_{10}H_2N_4$
J.J.Stezowski *Eur.Cryst.Meeting,* **5**, 211, 1979
See also R1 : 26 R2 : 60,7

60.28 Anthracene – tetracyanobenzene complex (low temperature form, at 202°K)
$C_{14}H_{10}$, $C_{10}H_2N_4$
J.J.Stezowski *Eur.Cryst.Meeting,* **5**, 211, 1979
See also R1 : 26 R2 : 60,7

60.29 Anthracene – tetracyanobenzene complex (low temperature form, at 170°K)
$C_{14}H_{10}$, $C_{10}H_2N_4$
J.J.Stezowski *Eur.Cryst.Meeting,* **5**, 211, 1979
See also R1 : 26 R2 : 60,7

60.30 Anthracene – tetracyanobenzene complex (low temperature form, at 138°K)
$C_{14}H_{10}$, $C_{10}H_2N_4$
J.J.Stezowski *Eur.Cryst.Meeting,* **5**, 211, 1979
See also R1 : 26 R2 : 60,7

60.31 Anthracene – tetracyanobenzene complex (low temperature form, at 119°K)
$C_{14}H_{10}$, $C_{10}H_2N_4$
J.J.Stezowski *Eur.Cryst.Meeting,* **5**, 211, 1979
See also R1 : 26 R2 : 60,7

60.32 Stilbene – 7,7,8,8 – tetracyanoquinodimethane complex
$C_{14}H_{12}$, $C_{12}H_4N_4$
D.Zobel, G.Ruban *Eur.Cryst.Meeting,* **5**, 115, 1979
See also R1 : 19 R2 : 60,7

60.33 Thieno(3,2 – e:4,5 – e′) – bis(benzo(b)thiophene) – 7,7,8,8 – tetracyanoquinodimethane
Trithia(5)heterohelicene – 7,7,8,8 – tetracyanoquinodimethane
$C_{18}H_8S_3$, $C_{12}H_4N_4$
M.Konno, Y.Saito, K.Yamada, H.Kawazura
Acta Crystallogr.,Sect.B, **36**, 1680, 1980
See also R1 : 39 R2 : 60,7

60.34 Pyrene – fluoranil complex
$C_{16}H_{10}$, $C_6F_4O_2$
J.Bernstein, H.Regev *Cryst.Struct.Commun.,* **9**, 581, 1980
See also R1 : 29 R2 : 60,18

60.35 tris(4,4′,5,5′ – Tetramethyl – tetrathiafulvalene) – bis(2,5 – diethyl – 7,7,8,8 – tetracyanoquinodimethane) complex
$2C_{16}H_{12}N_4$, $3C_{10}H_{12}S_4$
J.M.Fabre, M.Vigroux, E.Torreilles, L.Giral, D.Chasseau
Tetrahedron Lett., **21**, 607, 1980
See also R1 : 7 R2 : 60,39

60.36 Ellipticine – 5 – iodocytidylyl – (3′ – 5′) – guanosine hydrate methanol solvate
$2C_{17}H_{15}N_2^+$, $2C_{19}H_{23}IN_8O_{12}P^-$, $20H_2O$, $2CH_4O$
S.C.Jain, K.K.Bhandary, H.M.Sobell
J.Mol.Biol., **135**, 813, 1979
See also R1 : 58 R2 : 60,47,46,45,44

60.37 3,5,6,8 – Tetramethyl – N – methyl – phenanthrolinium – 5 – iodocytidylyl – (3′ – 5′) – guanosine hydrate methanol solvate
$2C_{17}H_{19}N_2^+$, $2C_{19}H_{23}IN_8O_{12}P^-$, $17H_2O$, $2CH_4O$
S.C.Jain, K.K.Bhandary, H.M.Sobell
J.Mol.Biol., **135**, 813, 1979
See also R1 : 36 R2 : 60,47,46,45,44

60.38 Tetramethoxystilbene – 7,7,8,8 – tetracyanoquinodimethane complex
$C_{18}H_{20}O_4$, $C_{12}H_4N_4$
D.Zobel, G.Ruban *Eur.Cryst.Meeting,* **5**, 115, 1979
See also R1 : 17 R2 : 60,7

60.C Ellipticine – 5 – iodocytidylyl – (3′ – 5′) – guanosine hydrate methanol solvate
$2C_{19}H_{23}IN_8O_{12}P^-$, $2C_{17}H_{15}N_2^+$, $20H_2O$, $2CH_4O$
Main entry is 60.36

60.C 3,5,6,8 – Tetramethyl – N – methyl – phenanthrolinium – 5 – iodocytidylyl – (3′ – 5′) – guanosine hydrate methanol solvate
$2C_{19}H_{23}IN_8O_{12}P^-$, $2C_{17}H_{19}N_2^+$, $17H_2O$, $2CH_4O$
Main entry is 60.37

60.C 9 – Amino – acridine – 5 – iodocytidylyl – (3′ – 5′) – guanosine hydrate complex
$4C_{19}H_{23}IN_8O_{12}P^-$, $4C_{13}H_{11}N_2^+$, $21H_2O$ Main entry is 60.22

60.C Proflavine – cytidylyl – (3 – 5′) – guanosine sulfate hydrate
$C_{19}H_{24}N_8O_{12}P^-$, $1.5C_{13}H_{12}N_3^+$, $0.5O_4S^-$, $11.5H_2O$
Main entry is 60.24

60.39 Cytidylyl – (3′ – 5′) – adenosine – proflavine complex decahydrate
$C_{19}H_{26}N_8O_{11}P$, $C_{13}H_{11}N_3$, $10H_2O$
E.Westhof, M.Sundaralingam
Proc.Nat.Acad.Sci.U.S.A., **77**, 1852, 1980
See also R1 : 47,46,45,44 R2 : 60,36

60.40 N,N′ – (1,2 – Phenylene) – bis(salicylaldiminato) – copper(ii) – 7,7,8,8 – tetracyanoquinodimethane complex
$2C_{20}H_{14}CuN_2O_2$, $C_{12}H_4N_4$
P.Cassoux, A.Gleizes *Inorg.Chem.,* **19**, 665, 1980
See also R1 : 78 R2 : 60,7

60.41 **Ethyltriphenylphosphonium – bis(7,7,8,8 – tetracyanoquinodimethane)**
$C_{20}H_{20}P^+$, $C_{12}H_4N_4^-$, $C_{12}H_4N_4$
R.J.Fleming, M.A.Shaikh, B.W.Skelton, A.H.White
Aust.J.Chem.,**32**, 2187, 1979
See also R1 : 64 R2 : 60,12,7

60.42 **Decamethylferrocenium – 7,7,8,8 – tetracyanoquinodimethane**
$C_{20}H_{30}Fe^+$, $C_{12}H_4N_4^-$
J.S.Miller, A.H.Reis Junior, E.Gebert, J.J.Ritsko,
W.R.Salaneck, L.Kovnat, T.W.Cape, R.P.van Duyne
J.Am.Chem.Soc.,**101**, 7111, 1979
See also R1 : 73 R2 : 60,12,7

60.43 **Valinomycin potassium picrate**
$C_{36}H_{60}N_6O_{12}$, $C_6H_2N_3O_7^-$, K^+
M.N.Sabesan, J.A.Hamilton, L.K.Steinrauf
Am.Cryst.Assoc.,*Ser.2*,**7**, 20, 1980
See also R1 : 48 R2 : 60,17,15,6

60.C **(1,2,4,5 – Tetrabenzoyl – 3,6 – di – t – butylbenzene) – (meso – 3,8 – di – t – butyl – 1,5,6,10 – tetraphenyl – deca – 3,4,6,7 – tetraene – 1,9 – diyne) complex**
$C_{42}H_{38}$, $C_{42}H_{38}O_4$ Main entry is 60.44

60.44 **(1,2,4,5 – Tetrabenzoyl – 3,6 – di – t – butylbenzene) – (meso – 3,8 – di – t – butyl – 1,5,6,10 – tetraphenyl – deca – 3,4,6,7 – tetraene – 1,9 – diyne) complex**
$C_{42}H_{38}O_4$, $C_{42}H_{38}$
F.Toda, K.Tanaka, H.Tsukada, H.Shimanouchi, Y.Sasada
Chem.Lett.,1381, 1979
See also R1 : 19 R2 : 60,19

CLATHRATES

61.C **2,2,4,4,6,6 – Hexa(1 – aziridinyl) – cyclotriphosphazene carbon tetrachloride anticlathrate**
$3CCl_4$, $C_{12}H_{24}N_9P_3$ Main entry is 61.2

61.C **bis(Triphenylbenzylphosphonium) tetrachloro – cadmium dichloroethane clathrate**
$2C_2H_4Cl_2$, $2C_{25}H_{22}P^+$, $CdCl_4^{2-}$ Main entry is 61.8

61.C **Furaltadone hydrochloride acetic acid clathrate**
$C_2H_4O_2$, $C_{13}H_{17}N_4O_6^+$, Cl^- Main entry is 61.4

61.C **hexakis(R – α – Phenylethylsulfonyl – methyl) benzene acetic acid clathrate**
$4C_2H_4O_2$, $C_{60}H_{66}O_{12}S_6$ Main entry is 61.24

61.C **Furaltadone hydrochloride propionic acid clathrate**
$C_3H_6O_2$, $C_{13}H_{17}N_4O_6^+$, Cl^- Main entry is 61.5

61.C **α – Cyclodextrin N,N – dimethylformamide clathrate pentahydrate**
C_3H_7NO, $C_{36}H_{60}O_{30}$, $5H_2O$ Main entry is 61.19

61.C **γ – Cyclodextrin n – propanol clathrate hydrate**
C_3H_8O, $C_{48}H_{80}O_{40}$, xH_2O Main entry is 61.22

61.C **α – Cyclodextrin 2 – pyrrolidone clathrate pentahydrate**
C_4H_7NO, $C_{36}H_{60}O_{30}$, $5H_2O$ Main entry is 61.20

61.C **exo – 2,exo – 6 – Dihydroxy – 2,6 – dimethyl – bicyclo(3.3.1)nonane ethylacetate clathrate**
$C_4H_8O_2$, $3C_{11}H_{20}O_2$ Main entry is 61.1

61.C **hexakis(Benzylthiomethyl)benzene 1,4 – dioxane clathrate (monoclinic form)**
$C_4H_8O_2$, $C_{54}H_{54}S_6$ Main entry is 61.23

61.C **hexakis(2 – Phenylethylthiomethyl)benzene 1,4 – dioxane clathrate**
$C_4H_8O_2$, $C_{60}H_{66}S_6$ Main entry is 61.25

61.C **(–) – Tri – o – thymotide – RR – (+) – 2,3 – dimethyl – thiirane clathrate (at –50°C)**
C_4H_8S, $2C_{33}H_{36}O_6$ Main entry is 61.15

61.C **(+) – Tri – o – thymotide – S – (+) – 2 – bromo – butane clathrate (at –50°C)**
C_4H_9Br, $2C_{33}H_{36}O_6$ Main entry is 61.16

61.C **2,3:4,5 – bis(1,2 – (3 – Methylnaphtho)) – 1,6,9,12,15,18 – hexaoxacycloeicosa – 2,4 – diene t – butylammonium perchlorate clathrate benzene solvate (at 113°K)**
$C_4H_{12}N^+$, $C_{32}H_{36}O_6$, ClO_4^-, C_6H_6 Main entry is 61.13

61.C **Hexa – aziridino – cyclotriphosphazene benzene clathrate**
C_6H_6, $2C_{12}H_{24}N_9P_3$ Main entry is 61.3

61.C **Cycloveratril benzene clathrate monohydrate**
$0.5C_6H_6$, $C_{27}H_{30}O_6$, H_2O Main entry is 61.12

61.C 5 – Methylbenzene – 1,3 – dicarbaldehyde – bis(p – tolylsulfonylhydrazone) benzene clathrate
$2C_6H_6$, $C_{23}H_{24}N_4O_4S_2$ Main entry is 61.7

61.C Cyclo(tetrakis(5 – t – butyl – 2 – hydroxy – 1,3 – phenylene)methylene) toluene clathrate
C_7H_8, $C_{44}H_{56}O_4$ Main entry is 61.21

61.C bis(10,22 – Dimethyl – 1,4,7,13,16,19 – hexaoxa – 10,22 – diazacyclotetracosane) benzylammonium thiocyanate clathrate
$C_7H_{10}N^+$, $2C_{18}H_{38}N_2O_6$, CNS⁻ Main entry is 61.6

61.C tris(1,2 – bis(Diphenylphosphino – selenoyl)ethane) p – xylene clathrate
C_8H_{10}, $3C_{26}H_{24}P_2Se_2$ Main entry is 61.9

61.C bis(Isothiocyanato) – tetrakis(4 – methylpyridine) – nickel(ii) 2 – bromonaphthalene clathrate
$2C_{10}H_7Br$, $C_{26}H_{28}N_6NiS_2$ Main entry is 61.10

61.C 1,6,20,25 – Tetra – aza(6.1.6.1)paracyclophane hydrochloride durene tetrahydrate
$C_{10}H_{14}$, $C_{34}H_{44}N_4^{4+}$, 4Cl⁻, $4H_2O$ Main entry is 61.18

61.C bis(Isothiocyanato) – tetrakis(4 – methylpyridine) – nickel(ii) 2 – methylnaphthalene clathrate
$2C_{11}H_{10}$, $C_{26}H_{28}N_6NiS_2$ Main entry is 61.11

61.1 exo – 2,exo – 6 – Dihydroxy – 2,6 – dimethyl – bicyclo(3.3.1)nonane ethylacetate clathrate
$3C_{11}H_{20}O_2$, $C_4H_8O_2$
R.Bishop, I.Dance
J.Chem.Soc.,Chem.Commun.,992, 1979
See also R1 : 31 R2 : 61,1

61.2 2,2,4,4,6,6 – Hexa(1 – aziridinyl) – cyclotriphosphazene carbon tetrachloride anticlathrate
$C_{12}H_{24}N_9P_3$, $3CCl_4$
J.Galy, R.Enjalbert, J.–F.Labarre
Acta Crystallogr.,Sect.B,**36**, 392, 1980
See also R1 : 32 R2 : 61,5

61.3 Hexa – aziridino – cyclotriphosphazene benzene clathrate
$2C_{12}H_{24}N_9P_3$, C_6H_6
T.S.Cameron, C.Chan, J.–F.Labarre, M.Graffeuil
Z.Naturforsch.,Teil B,**35**, 784, 1980
See also R1 : 64,32 R2 : 61,19

61.4 Furaltadone hydrochloride acetic acid clathrate
$C_{13}H_{17}N_4O_6^+$, $C_2H_4O_2$, Cl⁻
I.Goldberg *Eur.Cryst.Meeting*,**6**, 49, 1980
See also R1 : 40,38 R2 : 61,1

61.5 Furaltadone hydrochloride propionic acid clathrate
$C_{13}H_{17}N_4O_6^+$, $C_3H_6O_2$, Cl⁻
I.Goldberg *Eur.Cryst.Meeting*,**6**, 49, 1980
See also R1 : 40,38 R2 : 61,1

61.C Tri – o – thymotide trans – stilbene clathrate
$C_{14}H_{12}$, $C_{33}H_{36}O_6$ Main entry is 61.14

61.C Tri – o – thymotide cis – stilbene clathrate
$C_{14}H_{12}$, $2C_{33}H_{36}O_6$ Main entry is 61.17

61.6 bis(10,22 – Dimethyl – 1,4,7,13,16,19 – hexaoxa – 10,22 – diazacyclotetracosane) benzylammonium thiocyanate clathrate
$2C_{18}H_{38}N_2O_6$, $C_7H_{10}N^+$, CNS⁻
M.J.Bovill, D.J.Chadwick, M.R.Johnson, N.F.Jones, I.O.Sutherland, R.F.Newton
J.Chem.Soc.,Chem.Commun.,1065, 1979
See also R1 : 40 R2 : 61,3

61.7 5 – Methylbenzene – 1,3 – dicarbaldehyde – bis(p – tolylsulfonylhydrazone) benzene clathrate
$C_{23}H_{24}N_4O_4S_2$, $2C_6H_6$
T.–L.Chan, T.C.W.Mak, J.Trotter
J.Chem.Soc.,Perkin Trans.2,672, 1980
See also R1 : 4 R2 : 61,19

61.8 bis(Triphenylbenzylphosphonium) tetrachloro – cadmium dichloroethane clathrate
$2C_{25}H_{22}P^+$, $2C_2H_4Cl_2$, $CdCl_4^{2-}$
J.C.J.Bart, I.W.Bassi, M.Calcaterra
J.Organomet.Chem.,**193**, 1, 1980
See also R1 : 64 R2 : 61,5

61.9 tris(1,2 – bis(Diphenylphosphino – selenoyl)ethane) p – xylene clathrate
$3C_{26}H_{24}P_2Se_2$, C_8H_{10}
D.H.Brown, R.J.Cross, P.R.Mallinson, D.D.MacNicol
J.Chem.Soc.,Perkin Trans.2,993, 1980
See also R1 : 64 R2 : 61,19

61.10 bis(Isothiocyanato) – tetrakis(4 – methylpyridine) – nickel(ii) 2 – bromonaphthalene clathrate
$C_{26}H_{28}N_6NiS_2$, $2C_{10}H_7Br$
J.Lipkowski, P.Sgarabotto, G.D.Andreetti
Acta Crystallogr.,Sect.B,**36**, 51, 1980
See also R1 : 83 R2 : 61,24

61.11 bis(Isothiocyanato) – tetrakis(4 – methylpyridine) – nickel(ii) 2 – methylnaphthalene clathrate
$C_{26}H_{28}N_6NiS_2$, $2C_{11}H_{10}$
J.Lipkowski, P.Sgarabotto, G.D.Andreetti
Acta Crystallogr.,Sect.B,**36**, 51, 1980
See also R1 : 83 R2 : 61,24

61.12 Cycloveratril benzene clathrate monohydrate
$C_{27}H_{30}O_6$, $0.5C_6H_6$, H_2O
S.Cerrini, E.Giglio, F.Mazza, N.V.Pavel
Acta Crystallogr.,Sect.B,**35**, 2605, 1979
See also R1 : 31 R2 : 61,19

61.C hexakis(p – t – Butylphenylthiomethyl)benzene squalene clathrate
$0.5C_{30}H_{50}$, $C_{72}H_{90}S_6$ Main entry is 61.26

61.13 2,3:4,5 – bis(1,2 – (3 – Methylnaphtho)) – 1,6,9,12,15,18 – hexaoxacycloeicosa – 2,4 – diene t – butylammonium perchlorate clathrate benzene solvate (at 113°K)
$C_{32}H_{36}O_6$, $C_4H_{12}N^+$, ClO_4^-, C_6H_6
I.Goldberg *J.Am.Chem.Soc.*,**102**, 4106, 1980
See also R1 : 38 R2 : 61,3

61.14 Tri – o – thymotide trans – stilbene clathrate
$C_{33}H_{36}O_6$, $C_{14}H_{12}$
R.Arad–Yellin, S.Brunie, B.S.Green, M.Knossow, G.Tsoucaris *J.Am.Chem.Soc.*,**101**, 7529, 1979
See also R1 : 38 R2 : 61,19

61.15 (−) – Tri – o – thymotide – RR – (+) – 2,3 – dimethyl – thiirane clathrate (at −50°C)
$2C_{33}H_{36}O_6$, C_4H_8S
R.Arad–Yellin, B.S.Green, M.Knossow, G.Tsoucaris
Tetrahedron Lett., **21**, 387, 1980
See also R1 : 38 R2 : 61,39

61.16 (+) – Tri – o – thymotide – S – (+) – 2 – bromo – butane clathrate (at −50°C)
$2C_{33}H_{36}O_6$, C_4H_9Br
R.Arad–Yellin, B.S.Green, M.Knossow, G.Tsoucaris
Tetrahedron Lett., **21**, 387, 1980
See also R1 : 38 R2 : 61,5

61.17 Tri – o – thymotide cis – stilbene clathrate
$2C_{33}H_{36}O_6$, $C_{14}H_{12}$
R.Arad–Yellin, S.Brunie, B.S.Green, M.Knossow,
G.Tsoucaris *J.Am.Chem.Soc.*, **101**, 7529, 1979
See also R1 : 38 R2 : 61,19

61.18 1,6,20,25 – Tetra – aza(6.1.6.1)paracyclophane hydrochloride durene tetrahydrate
$C_{34}H_{44}N_4^{4+}$, $C_{10}H_{14}$, $4Cl^-$, $4H_2O$
K.Odashima, A.Itai, Y.Iitaka, K.Koga
J.Am.Chem.Soc., **102**, 2504, 1980
See also R1 : 31 R2 : 61,19

61.19 α – Cyclodextrin N,N – dimethylformamide clathrate pentahydrate
$C_{36}H_{60}O_{30}$, C_3H_7NO, $5H_2O$
K.Harata *Bull.Chem.Soc.Jpn.*, **52**, 2451, 1979
See also R1 : 45 R2 : 61,1

61.20 α – Cyclodextrin 2 – pyrrolidone clathrate pentahydrate
$C_{36}H_{60}O_{30}$, C_4H_7NO, $5H_2O$
K.Harata *Bull.Chem.Soc.Jpn.*, **52**, 2451, 1979
See also R1 : 45 R2 : 61,32

61.21 Cyclo(tetrakis(5 – t – butyl – 2 – hydroxy – 1,3 – phenylene)methylene) toluene clathrate
$C_{44}H_{56}O_4$, C_7H_8
G.D.Andreetti, R.Ungaro, A.Pochini
J.Chem.Soc., Chem.Commun., 1005, 1979
See also R1 : 31,17 R2 : 61,19

61.22 γ – Cyclodextrin n – propanol clathrate hydrate
$C_{48}H_{80}O_{40}$, C_3H_8O, xH_2O
K.Lindner, W.Saenger
Biochem.Biophys.Res.Commun., **92**, 933, 1980
See also R1 : 45 R2 : 61,5

61.23 hexakis(Benzylthiomethyl)benzene 1,4 – dioxane clathrate (monoclinic form)
$C_{54}H_{54}S_6$, $C_4H_8O_2$
A.D.U.Hardy, D.D.MacNicol, S.Swanson, D.R.Wilson
J.Chem.Soc., Perkin Trans.2, 999, 1980
See also R1 : 11 R2 : 61,38

61.24 hexakis(R – α – Phenylethylsulfonyl – methyl) benzene acetic acid clathrate
$C_{60}H_{66}O_{12}S_6$, $4C_2H_4O_2$
A.Freer, C.J.Gilmore, D.D.MacNicol, S.Swanson
Tetrahedron Lett., **21**, 205, 1980
See also R1 : 11 R2 : 61,1

61.25 hexakis(2 – Phenylethylthiomethyl)benzene 1,4 – dioxane clathrate
$C_{60}H_{66}S_6$, $C_4H_8O_2$
K.Burns, C.J.Gilmore, P.R.Mallinson, D.D.MacNicol,
S.Swanson *Eur.Cryst.Meeting*, **6**, 23, 1980
See also R1 : 11 R2 : 61,38

61.26 hexakis(p – t – Butylphenylthiomethyl)benzene squalene clathrate
$C_{72}H_{90}S_6$, $0.5C_{30}H_{50}$
A.Freer, C.J.Gilmore, D.D.MacNicol, D.R.Wilson
Tetrahedron Lett., **21**, 159, 1980
See also R1 : 11 R2 : 61,56

BORON COMPOUNDS

62.1 **Cesium bis(trifluoromethyl) – difluoroborate**
$C_2BF_8^-$, Cs^+
D.J.Brauer, H.Burger, G.Pawelke
J.Organomet.Chem.,**192**, 305, 1980

62.2 **Sodium hydrido – trimethylboronate diethyl ether solvate (at –140°C)**
$4C_3H_{10}B^-$, $4Na^+$, $C_4H_{10}O$
N.A.Bell, H.M.M.Shearer, C.B.Spencer
J.Chem.Soc.,Chem.Commun.,711, 1980

62.3 **9 – Methylsulfonyl – 1,7 – dicarba – closo – dodecaborane**
$C_3H_{14}B_{10}O_2S$
K.Maly, A.Petrina, V.Petricek, L.Hummel, A.Linek
Acta Crystallogr.,Sect.B,**36**, 181, 1980

62.4 **2,3 – Dimethyl – 4,7 – dihydroxy – 10 – bromo – 2,3 – dicarba – closo – undecaborane**
$C_4H_{14}B_9BrO_2$
M.E.Leonowicz, F.R.Scholer *Inorg.Chem.*,**19**, 122, 1980

62.5 **Triethylmethylammonium 8,8′ – oxido – 3 – cobalta – bis(η^5 – 1,2 – dicarbadodecaborate)**
$C_4H_{20}B_{18}CoO^-$, $C_7H_{18}N^+$
V.Petricek, K.Maly, V.Subrtova, L.Hummel, A.Linek
Eur.Cryst.Meeting,**5**, 247, 1979
See also R2 : 3

62.6 **9,12 – Isopropylidene – dithio – 1,2 – dicarba – closo – dodecaborane**
$C_5H_{16}B_{10}S_2$
V.Subrtova, A.Linek, J.Hasek
Acta Crystallogr.,Sect.B,**36**, 858, 1980

62.7 **Cesium 9,10a,11 – trimethyl – 7,8 – dicarba – nido – undecaborate**
$C_5H_{18}B_9^-$, Cs^+
M.Yu.Antipin, Yu.T.Struchkov, N.I.Kirillova, S.P.Knyazev, V.A.Brattsev, V.I.Stanko
Cryst.Struct.Commun.,**9**, 599, 1980

62.C **1 – Methylarsa – 2,3,4,5 – bis(o – carborano – cyclopentane)**
$C_5H_{23}AsB_{20}$ Main entry is 65.3

62.C **Tetramethyl – tetrathiafulvalene tetrabromo – p – benzoquinone complex**
$C_6Br_4O_2$, $C_{10}H_{12}S_4$ Main entry is 60.10

62.C **Tetramethylammonium 2 – (η – cyclopentadienyl) – 1 – carba – 2 – cobalta – closo – undecahydro – dodecaborate**
$C_6H_{18}B_{10}Co^-$, $C_4H_{12}N^+$ Main entry is 73.4

62.C **4,5 – bis(Dimethylamino) – 2,2 – dimethyl – 1,3 – dithia – 2 – sila – 4,5 – dibora – cyclopentane**
$C_6H_{18}B_2N_2S_2Si$ Main entry is 63.4

62.8 **1 – t – Butyl – 1,2 – dicarba – closo – dodecaborane (at –120°C)**
$C_6H_{20}B_{10}$
N.I.Kirillova, T.V.Klimova, Yu.T.Struchkov, V.I.Stanko
Izv.Akad.Nauk SSSR,Ser.Khim.,2481, 1979

62.C **1 – Trimethylsilyl – 2 – methyl – 1,2 – dicarba – closo – dodecaborane (at –120°C)**
$C_6H_{22}B_{10}Si$ Main entry is 63.5

62.C **1 – Trimethyl – tin – methyl – o – carborane (at –120°C)**
$C_6H_{22}B_{10}Sn$ Main entry is 69.13

62.C **1 – Methylphospha – 2,3,5,6 – bis – σ – carborano – cyclohexane**
$C_6H_{25}B_{20}P$ Main entry is 64.23

62.C **2 – (Pentachloro – cyclotriphosphazenyl) – 1 – phenyl – 1,2 – dicarba – dodecaborane**
$C_8H_{15}B_{10}Cl_5N_3P_3$ Main entry is 64.

62.9 **4,5 – Diethyl – 3,6 – dimethyl – 1,2 – diaza – 3,6 – diborinane**
$C_8H_{18}B_2N_2$
W.Siebert, R.Full, H.Schmidt, J.von Seyerl, M.Halstenberg, G.Huttner *J.Organomet.Chem.*,**191**, 15, 1980

62.10 **9 – Cyclohexyl – 5(7) – dimethylsulfide – nido – decaborane**
$C_8H_{28}B_{10}S$
E.Mizusawa, S.E.Rudnick, K.Eriks
Inorg.Chem.,**19**, 1188, 1980

62.C **fac – (Triacetyl – tricarbonyl – rhenato) – boron – chloride**
$C_9H_9BClO_6Re$ Main entry is 71.11

62.C **Dichloro – (hydro – tris(1 – pyrazolyl) – borato) – oxo – technetium(v)**
$C_9H_{10}BCl_2N_6OTc$ Main entry is 83.44

62.C **8 – Trifluoroacetoxy – 1 – (η^5 – cyclopentadienyl – ferra) – 2,3 – dicarba – dodecaborane**
$C_9H_{15}B_9F_3FeO_2$ Main entry is 71.13

62.C **Trimethylammine – (η^5 – cyclopentadienyl) – carba – cobalta – undecaborane**
$C_9H_{23}B_9CoN$ Main entry is 73.12

62.C **Tetraethylammonium cyclopentadienyl – tetracarba – dicobalta – docosaborane**
$C_9H_{25}B_{16}Co_2^-$, $C_8H_{20}N^+$ Main entry is 73.13

62.C **4,5 – Diethyl – 3,6 – dimethyl – 1,2 – diaza – 3,6 – diborine – tricarbonyl – chromium**
$C_{11}H_{18}B_2CrN_2O_3$ Main entry is 75

62.C **3 – Carbonyl – 3 – η^5 – cyclopentadienyl – 6 – difluoro – 2 – isopropyl – 4 – methyl – 1,5 – dioxa – 3 – ferra – 6 – borinane**
$C_{12}H_{15}BF_2FeO_3$ Main entry is 71.41

62.C **3,7,9,13 – Tetramethyl – 6 – (η^5 – cyclopentadienyl) – 3,7,9,13 – tetracarba – 6 – cobalta – dodecaborane**
$C_{13}H_{24}B_7Co$ Main entry is 71.54

62.C **bis(η^5 – Cyclopentadienyl) – (μ – (η^3:η^4 – trihydro – C,C′ – dimethyl – dicarba – pentaborato)) – di – cobalt**
$C_{14}H_{19}B_3Co_2$ Main entry is 71.69

62.C η^5 – Cyclopentadienyl – (η^5 – cyclopentadienyl – cobalt) – (μ – (η^3:η^4 – trihydro – C,C' – dimethyl – dicarba – hexaborato)) – iron
$C_{14}H_{20}B_3CoFe$ Main entry is 71.70

62.C 1,1 – bis(Triethylphosphine) – 2,4 – dicarba – 1 – platina – closo – heptaborane (at 215°K)
$C_{14}H_{36}B_4P_2Pt$ Main entry is 86.32

62.C (μ(4,5) – (trans – bis(Triethylphosphine) – hydrido – platinum)) – (μ(5,6) – hydrido) – nido – 2,3 – dicarba – hexaborane (at 215°K)
$C_{14}H_{38}B_4P_2Pt$ Main entry is 86.33

62.C 9 – Hydrido – 9,10 – bis(triethylphosphine) – 7,8 – dicarba – 9 – platina – undecaborane
$C_{14}H_{40}B_8P_2Pt$ Main entry is 71.81

62.C 9 – Hydrido – 9,9 – bis(triethylphosphine) – 7,8 – dicarba – 9 – platina – undecaborane
$C_{14}H_{41}B_8P_2Pt$ Main entry is 71.

62.C 1,3 – bis(Trimethylsilyl) – 4 – methyl – diazastanna – boretidine dimer
$C_{14}H_{42}B_2N_4Si_4Sn_2$ Main entry is 63.17

62.C pentakis(Methyl – dicarba – dodecaboranyl) – diphosphine benzene solvate
$C_{15}H_{64}B_{50}P_2$, C_6H_6 Main entry is 64.75

62.C (1 – t – Butyl – 3 – methyl – 2 – phenyl – η – 1,2 – azaborolinyl) – dicarbonyl – iodo – iron
$C_{16}H_{19}BFeINO_2$ Main entry is 75.10

62.11 N,N,N',N' – tetrakis(1,3 – Dimethyl – 1,3,2 – diazaborolidinyl) – hydrazine
$C_{16}H_{40}B_4N_{10}$
P.C.Bharara, H.Noth
Z.Naturforsch.,Teil B,**34**, 1352, 1979

62.C 2,3 – Dimethyl – 1,1 – bis(triethylphosphine) – 2,3 – dicarba – 1 – platina – closo – heptaborane (at 215°K)
$C_{16}H_{40}B_4P_2Pt$ Main entry is 86.43

62.C bis(μ – 2,3,4 – η^3 – nido – Hexaboranyl) – bis(dimethylphenylphosphine) – di – platinum
$C_{16}H_{40}B_{12}P_2Pt_2$ Main entry is 86.44

62.C bis(Hydro – tris(1 – pyrazolyl) – borato) – copper(ii)
$C_{18}H_{20}B_2CuN_{12}$ Main entry is 83.112

62.C bis(Hydrido – tris(1 – pyrazolyl) – borato) – nickel(ii)
$C_{18}H_{20}B_2N_{12}Ni$ Main entry is 83.114

62.C Cobaltocenyl – tetra(methylcarba) – dodecaborane
$C_{18}H_{29}B_8Co$ Main entry is 73.77

62.12 Tetramethylammonium (ethylnitrosolato – O) – triphenylborate
$C_{20}H_{18}BN_2O_2^-$, $C_4H_{12}N^+$
K.von Deuten, C.von Schlabrendorff, G.Klar
Cryst.Struct.Commun.,**9**, 753, 1980
See also R2 : 3

62.13 1,3 – Di – t – butyl – 2,4 – bis(pentafluorophenyl) – 1,3,2,4 – diazadiboretidine
$C_{20}H_{18}B_2F_{10}N_2$
P.Paetzold, A.Richter, T.Thijssen, S.Wurtenberg
Chem.Ber.,**112**, 3811, 1979

62.C 11 – Triphenylphosphine – 11 – argenta – 5,6 – dicarba – undecaborane(11) acetone solvate
$2C_{20}H_{26}AgB_8P$, C_3H_6O Main entry is 86.65

62.C 3 – (Triphenylphosphine) – 3 – nitrato – 3 – rhoda – 1,2 – dicarba – dodecaborane dichloromethane solvate
$C_{20}H_{26}B_9NO_3PRh$, $3CH_2Cl_2$ Main entry is 86.66

62.C Pentafluorophenylthiolato – (hydro – tris(3,5 – dimethyl – 1 – pyrazolyl) – borato) – cobalt (at −162°C)
$C_{21}H_{22}BCoF_5N_6S$ Main entry is 85.67

62.C Potassium (p – nitrobenzenethiolato) – (hydro – tris(3,5 – dimethylpyrazolyl) – borato) – copper(i) acetone solvate
$C_{21}H_{26}BCuN_7O_2S^-$, K^+, $2C_3H_6O$ Main entry is 85.68

62.C Hydro – tris(3,5 – dimethyl – 1 – pyrazolyl) – borato – dicarbonyl – (p – chlorobenzenethiolato) – molybdenum acetone solvate
$C_{23}H_{26}BClMoN_6O_2S$, $0.2C_3H_6O$ Main entry is 85.70

62.C Carbonyl – ethylenediamine – copper(i) tetraphenylborate
$C_{24}H_{20}B^-$, $C_3H_8CuN_2O^+$ Main entry is 76.4

62.14 Acetylcholine tetraphenylborate
$C_{24}H_{20}B^-$, $C_7H_{16}NO_2^+$
N.Datta, P.Mondal, P.Pauling
Acta Crystallogr.,Sect.B,**36**, 906, 1980
See also R2 : 1,3

62.C (Diethylenetriamine) – (hex – 1 – ene) – copper(i) tetraphenylborate
$C_{24}H_{20}B^-$, $C_{10}H_{25}CuN_3^+$ Main entry is 72.10

62.15 bis(2,3,5,6,8,9,11,12 – Octahydro – 1,4,7,10,13 – benzopentaoxa – cyclopentadecin) sodium tetraphenylborate
bis(Benzo – 15 – crown – 5) sodium tetraphenylborate
$C_{24}H_{20}B^-$, $2C_{14}H_{20}O_5$, Na^+
J.D.Owen *J.Chem.Soc.,Dalton Trans.*,1066, 1980
See also R2 : 38

62.C (N,N,N',N' – Tetramethylethylenediamine) – bis(cyclohexylisocyanide) – copper(i) tetraphenylborate
$C_{24}H_{20}B^-$, $C_{20}H_{38}CuN_4^+$ Main entry is 71.161

62.C (1,2 – bis(Methoxycarbonyl) – but – 2 – en – 1 – yl) – bis(triethylphosphine) – platinum(ii) tetraphenylborate
$C_{24}H_{20}B^-$, $C_{20}H_{41}O_4P_2Pt^+$ Main entry is 71.164

62.C (μ^2 – 1,2 – bis(Carbomethoxyethylene)) – bis((μ^2 – methylthio) – trimethylphosphine – carbonyl – iron) tetraphenylborate (at −162°C)
$C_{24}H_{20}B^-$, $C_{22}H_{40}Fe_2O_6P_4S_2^+$ Main entry is 86.82

62.C tris(N,N – Diethyldithiocarbamato) – bis(μ – (N,N – (diethyltrithiocarbamato)) – di – osmium(iii) tetraphenylborate
$C_{24}H_{20}B^-$, $C_{25}H_{50}N_5Os_2S_{12}^+$ Main entry is 80.23

62.C Iodo – (1,9 – bis(diphenylphosphino) – 3,7 – dithianonane) – nickel(ii) tetraphenylborate
$C_{24}H_{20}B^-$, $C_{31}H_{34}INiP_2S_2^+$ Main entry is 86.107

62.C bis(2 – Diphenylphosphinoethyl) –
phenylphosphine – nickel – 0 – methylsulfinate
tetraphenylborate
$C_{24}H_{20}B^-$, $C_{35}H_{36}NiO_2P_3S^+$ Main entry is 86.115

62.16 Ammonium tetraphenylborate (at 120°K)
$C_{24}H_{20}B^-$, H_4N^+
W.J.Westerhaus, O.Knop, M.Falk
Can.J.Chem.,**58**, 1355, 1980

62.17 Ammonium tetraphenylborate
$C_{24}H_{20}B^-$, H_4N^+
W.J.Westerhaus, O.Knop, M.Falk
Can.J.Chem.,**58**, 1355, 1980

62.C (2,3,9,10 – Tetramethyl – 1,4,8,11 – tetra – aza –
1,3,8,10 – cyclotetradecatetraene) – copper(ii)
bis(tetraphenylborate)
$2C_{24}H_{20}B^-$, $C_{14}H_{24}CuN_4^{2+}$ Main entry is 83.97

62.C tetrakis(μ^2 – 1,3 – Di – isocyanopropane) – di –
rhodium bis(tetraphenylborate) acetonitrile solvate
$2C_{24}H_{20}B^-$, $C_{20}H_{24}N_8Rh_2^{2+}$, C_2H_3N Main entry is 71.158

62.C (α – 4,7,13,16 – Tetraphenyl – 1,10 – dioxa –
4,7,13,16 – tetraphosphacyclo – octadecane) –
cobalt(ii) bis(tetraphenylborate)
$2C_{24}H_{20}B^-$, $C_{38}H_{44}CoO_2P_4^{2+}$ Main entry is 86.122

62.C (β – 4,7,13,16 – Tetraphenyl – 1,10 – dioxa –
4,7,13,16 – tetraphosphacyclo – octadecane) –
cobalt(ii) bis(tetraphenylborate)
dimethylformamide solvate
$2C_{24}H_{20}B^-$, $C_{38}H_{44}CoO_2P_4^{2+}$, $2C_3H_7NO$ Main entry is 86.123

62.18 N – Methyl – 2 – oxy – 1 – naphthaldiminato –
diphenylborane
$C_{24}H_{20}BNO$
O.E.Kompan, N.G.Furmanova, Yu.T.Struchkov, L.M.Sitkina,
V.A.Bren, V.I.Minkin *Zh.Strukt.Khim.*,21, 90, 1980

62.C bis(1,1 – Dimethyl – 4 – phenyl – 1 – sila – 4 –
bora – cyclohexa – 2,5 – dienyl) – nickel
$C_{24}H_{30}B_2NiSi_2$ Main entry is 75.34

62.19 bis(4 – Dibenzoborepinyl) – ether
$C_{28}H_{24}B_2O$
I.Cynkier, N.Furmanova
Cryst.Struct.Commun.,**9**, 307, 1980

62.C Chloro – (diphenylphosphino – dicarba –
dodecaborane) – (diphenylphosphino – dicarba –
dodecaboranyl) – platinum
$C_{28}H_{41}B_{20}ClP_2Pt$ Main entry is 86.95

62.C Hydro – tris(1 – pyrazolyl) – borato – bis(p –
toluenethiolato) – (p – fluorobenzenediazo) –
molybdenum
$C_{29}H_{29}BFMoN_8S_2$ Main entry is 85.81

62.C cis – bis(μ^2 – Carbonyl) – bis((1 – t – butyl – 3 –
methyl – 2 – phenyl – η – 1,2 – azaborolinyl) –
carbonyl – iron)
$C_{32}H_{38}B_2Fe_2N_2O_4$ Main entry is 75.38

62.C 3,7,9,13 – Tetramethyl – 6 – (bis(diphenylphosphino)
ethane) – 3,7,9,13 – tetracarba – 6 – nickela –
tridecaborane
$C_{34}H_{44}B_8NiP_2$ Main entry is 71.242

62.20 Octaethyl – bilatriene – difluoroboron chloroform
solvate
$C_{35}H_{45}BF_2N_4O_2$, $CHCl_3$
R.Bonnett, M.B.Hursthouse, J.Trotter
Eur.Cryst.Meeting,**6**, 278, 1980
See also R1 : 32

62.C 3,3 – bis(Triphenylphosphine) – 3 – bisulfato – 3 –
rhoda – 1,2 – dicarba – dodecaborane diethyl ether
solvate (at –154°C)
$C_{38}H_{42}B_9O_4P_2RhS$, $C_4H_{10}O$ Main entry is 86.140

62.C 3,9 – Dihydrido – bis(tri – p – tolylphosphine) –
(nido – 7,8 – dicarba – undecaborane) – iridium
toluene solvate (at –154°C)
$C_{44}H_{56}B_9IrP_2$, C_7H_8 Main entry is 86.167

62.C cis – 1 – ((Tribenzylphosphine) –
(dibenzylphosphino – phenyl – methyl) –
platinum) – 2 – methyl – 1,2 – dicarba –
dodecaborane
$C_{45}H_{54}B_{10}P_2Pt$ Main entry is 71.268

SILICON COMPOUNDS

63.1 **hexakis(Trimethylsilyl lithium)**
$6C_3H_9Si^-$, $6Li^+$
W.H.Ilsley, T.F.Schaaf, M.D.Glick, J.P.Oliver
J.Am.Chem.Soc.,**102**, 3769, 1980

63.2 **1 – (S,S – Dimethyl – N – (trimethylsilyl) – sulfo – di – imide) – bicyclo(3.3.1)penta – azatetrathiane**
$C_5H_{15}N_7S_5Si$
W.S.Sheldrick, M.N.S.Rao, H.W.Roesky
Inorg.Chem., **19**, 538, 1980

63.3 **tris(Dimethylamino) – chloro – silane – (aluminium chloride) (at −35°C)**
$C_6H_{18}AlCl_4N_3Si$
A.H.Cowley, M.C.Cushner, P.E.Riley
J.Am.Chem.Soc., **102**, 624, 1980

63.4 **4,5 – bis(Dimethylamino) – 2,2 – dimethyl – 1,3 – dithia – 2 – sila – 4,5 – dibora – cyclopentane**
$C_6H_{18}B_2N_2S_2Si$
H.Noth, H.Fusstetter, H.Pommerening, T.Taeger
Chem.Ber.,**113**, 342, 1980
See also R1 : 62

63.5 **1 – Trimethylsilyl – 2 – methyl – 1,2 – dicarba – closo – dodecaborane (at −120°C)**
$C_6H_{22}B_{10}Si$
N.I.Kirillova, T.V.Klimova, Yu.T.Struchkov, V.I.Stanko
Izv.Akad.Nauk SSSR,Ser.Khim.,2481, 1979
See also R1 : 62

63.6 **bis(Acetone – oximato) – methyl – chloro – silane**
$C_7H_{15}ClN_2O_2Si$
S.N.Gurkova, A.I.Gusev, N.V.Alexeev, N.S.Fedotov, G.V.Ryasin, M.V.Polyakova, V.V.Sokolov
Zh.Strukt.Khim.,**20**, 160, 1979

63.7 **4 – Bromobenzoyloxymethyl – trifluoro – silane**
$C_8H_6BrF_3O_2Si$
M.G.Voronkov, A.A.Kashaev, E.A.Zel'bst, Yu.L.Frolov, V.M.D'yakov, L.I.Gubanova
Dokl.Akad.Nauk SSSR,**247**, 1147, 1979

63.8 **Trimethyl – (pyrrolidinomethyl) – silane (at −90°C)**
$C_8H_{19}NSi$
M.Yu.Antipin, M.A.Kravers, Yu.T.Struchkov, R.Ya.Sturkovic, E.Ya.Lukevits
Latv.PSR Zinat.Akad.Vestis,Kim.Ser., 89, 1980
See also R1 : 32

63.9 **1 – Chloromethyl – 3,7 – dimethyl – silatrane**
$C_9H_{18}ClNO_3Si$
M.G.Voronkov, M.P.Demidov, V.E.Shklover, V.P.Borishok, V.M.D'yakov, Yu.L.Frolov *Zh.Strukt.Khim.*,**21**, 100, 1980

63.10 **Trimethyl – (2 – pyrrolidinoethyl) – silane (at −90°C)**
$C_9H_{21}NSi$
M.Yu.Antipin, M.A.Kravers, Yu.T.Struchkov, R.Ya.Sturkovic, E.Ya.Lukevits
Latv.PSR Zinat.Akad.Vestis,Kim.Ser.,89, 1980
See also R1 : 32

63.C **3,5,7 – tris(Trimethylsilyl) – tricyclo(2.2.1.0²,⁶) hepta – arsane**
$C_9H_{27}As_7Si_3$ Main entry is 65.5

63.C **4,7,11 – tris(Trimethylsilyl) – pentacyclo(6.3.0.0²,⁶.0³,¹⁰.0⁵,⁹)undecaphosphane**
$C_9H_{27}P_{11}Si_3$ Main entry is 64.35

63.C **1,1',2'' – tris(Trimethylsilyl) – thiophosphoryl – trihydrazide**
$C_9H_{33}N_6PSSi_3$ Main entry is 64.36

63.C **Trisila – pentachloro – cyclohexyl – dicarbonyl – cyclopentadienyl – iron**
$C_{10}H_{11}Cl_5FeO_2Si_3$ Main entry is 73.20

63.11 **2,2,4,4,6,8,8,8 – Octamethyl – 2,4,6,8 – tetrasila – 1,5 – dimercura – cyclo – octane**
$C_{10}H_{28}Hg_2Si_4$
W.H.Ilsley, E.A.Sadurski, T.F.Schaaf, M.J.Albright, T.J.Anderson, M.D.Glick, J.P.Oliver
J.Organomet.Chem.,**190**, 257, 1980

63.C **Hexamethyl – trisila – tetraphospha – nortricyclene – pentacarbonyl – chromium**
$C_{11}H_{18}CrO_5P_4Si_3$ Main entry is 86.14

63.C **Tricarbonyl – (trimethyl – (η⁶ – phenyl) – silane) – chromium (at −135°C)**
$C_{12}H_{14}CrO_3Si$ Main entry is 74.1

63.12 **p – Fluorophenyl – silatranone**
$C_{12}H_{14}FNO_4Si$
L.Parkanyi, P.Hencsei, E.Popowski
J.Organomet.Chem.,**197**, 275, 1980

63.13 **1,4 – bis(Trimethylsilyl) – 1,4 – dihydro – 1,4 – diazocine**
$C_{12}H_{24}N_2Si_2$
H.–J.Altenbach, H.Stegelmeier, M.Wilhelm, B.Voss, J.Lex, E.Vogel *Angew.Chem.,Int.Ed.Engl.*,**18**, 962, 1979
See also R1 : 34

63.C **Dodecamethyl – hexasila – tetra – arsa – adamantane**
$C_{12}H_{36}As_4Si_6$ Main entry is 65.8

63.14 **2,2,4,4,6,6,8,8,9,9,11,11 – Dodecamethyl – bicyclo(3.3.3)hexasilazane**
$C_{12}H_{39}N_5Si_6$
V.E.Shklover, Yu.T.Struchkov, G.V.Kotrelev, V.V.Kazakova, K.A.Andrianov *Zh.Strukt.Khim.*,**20**, 96, 1979

63.15 **m – Trifluoromethylphenyl – silatranone**
$C_{13}H_{14}F_3NO_4Si$
L.Parkanyi, P.Hencsei, E.Popowski
J.Organomet.Chem.,**197**, 275, 1980

63.C **Dicyclopentadienyl – (trimethylsilyl) – titanium – chloride**
$C_{13}H_{19}ClSiTi$ Main entry is 73.43

63.16 t – Butyl – dimethylsiloxy – aci –
nitrophenylmethane (at 101°K)
$C_{13}H_{21}NO_2Si$
E.W.Colvin, A.K.Beck, B.Bastani, D.Seebach, Y.Kai,
J.D.Dunitz *Helv.Chim.Acta*,**63**, 697, 1980

63.C (bis(Trimethylsilyl)amino) – (t – butylimino) –
(trimethylsilylimino) – phosphorus (at –130°C)
$C_{13}H_{36}N_3PSi_3$ Main entry is 64.61

63.C (4,5 – Dimethyl – 1,3 – dioxa – 2 – phosphole) – 2 –
spiro – 2′ – (cis – 2,4 – bis(dimethylamino) – 1,3 –
bis(trimethylsilyl)) – 1,3,2,4 – diazadiphosphetidine
$C_{14}H_{36}N_4O_2P_2Si_2$ Main entry is 64.69

63.17 1,3 – bis(Trimethylsilyl) – 4 – methyl –
diazastanna – boretidine dimer
$C_{14}H_{42}B_2N_4Si_4Sn_2$
H.Fusstetter, H.North *Chem.Ber.*,**112**, 3672, 1979
See also R1 : 62

63.18 tris(Acetylacetonato)silicon perchlorate
$C_{15}H_{21}O_6Si^+$, ClO_4^-
T.Adam, T.Debaerdemaeker, U.Thewalt
Eur.Cryst.Meeting,**5**, 325, 1979

63.19 tris(Acetone – oximato) – phenyl – silane
$C_{15}H_{23}N_3O_3Si$
S.N.Gurkova, A.I.Gusev, N.V.Alexeev, M.G.Los, V.E.Zavodnik,
V.K.Bel'skii, G.V.Ryasin, N.S.Fedotov
Zh.Strukt.Khim.,**20**, 1059, 1979

63.C Chloro – trimethylphosphine –
tris(trimethylsilylmethyl) – molybdenum(iv)
$C_{15}H_{42}ClMoPSi_3$ Main entry is 71.89

63.20 Dichloro – 9 – fluorenyl – trimethylsilyl – silane
$C_{16}H_{18}Cl_2Si_2$
U.Schubert *J.Organomet.Chem.*,**197**, 269, 1980

63.21 Octa(vinyl – sila – sesquioxane)
$C_{16}H_{24}O_{12}Si_8$
I.A.Baidina, N.V.Podberezskaya, V.I.Alexeev,
T.N.Martynova, S.V.Borisov, A.N.Kanev
Zh.Strukt.Khim.,**20**, 648, 1979

63.C Di – μ^2 – chloro – tetrakis(trimethylsilylmethyl) –
di – thallium(iii)
$C_{16}H_{44}Cl_2Si_4Tl_2$ Main entry is 68.18

63.C Trisila – tetrachloro – cyclohexyl – bis(dicarbonyl –
cyclopentadienyl – iron)
$C_{17}H_{16}Cl_4Fe_2O_4Si_3$ Main entry is 73.66

63.22 2 – Methoxycarbonyl – 9 – methylene – 8 –
trimethylsiloxy – tricyclo(6.2.1.0^{1,5})undecan – 3 –
one
$C_{17}H_{26}O_4Si$
J.C.Dewan *Acta Crystallogr.*,*Sect.B*,**35**, 3111, 1979
See also R1 : 31

63.23 Triethylammonium tris(pyrocatechol)silicate
$C_{18}H_{12}O_6Si^{2-}$, $2C_6H_{16}N^+$
T.Adam, T.Debaerdemaeker, U.Thewalt
Eur.Cryst.Meeting,**5**, 325, 1979
See also R2 : 3

63.24 Triphenyl – silane
$C_{18}H_{16}Si$
J.Allemand, R.Gerdil *Cryst.Struct.Commun.*,**8**, 927, 1979

63.25 bis(Dimethylbenzylsilyl) – peroxide (at –120°C)
$C_{18}H_{26}O_2Si_2$
V.E.Shklover, A.V.Ganyushkin, V.A.Yablokov,
Yu.T.Struchkov *Cryst.Struct.Commun.*,**8**, 869, 1979

63.C Bipyridyl – bis(trimethylsilylmethyl) – cadmium(ii)
2,2′ – bipyridyl
$C_{18}H_{30}CdN_2Si_2$, $0.5C_{10}H_8N_2$ Main entry is 71.130

63.C Nonamethyl – cyclopentasilane – dimethylsilyl –
η^5 – cyclopentadienyl – dicarbonyl – iron
$C_{18}H_{38}FeO_2Si_6$ Main entry is 73.78

63.C (cis 2,4 – bis(Dimethylamino) – 1,3 –
bis(trimethylsilyl) – 1,3,2,4 –
diazadiphosphetidine) – 2′,4′ – dispiro – 2,2″ –
(4,5 – dimethyl – 1,3 – dioxa – 2 – phosphole)
$C_{18}H_{42}N_4O_4P_2Si_2$ Main entry is 64.99

63.26 N,N′ – bis(2,2,4,4,6 – Pentamethylcyclo –
trisiloxanyl – oxadimethylsilyl) – tetramethyl –
cyclodisilazane (at –120°C)
$C_{18}H_{54}N_2O_8Si_{10}$
V.E.Shklover, P.Ad'yaasuren, G.V.Kotrelev, E.A.Zhdanova,
V.S.Svistunov, Yu.T.Struchkov
Zh.Strukt.Khim.,**21**, 94, 1980

63.27 Diphenyl – (2 – piperidino – ethyl) – silanol
Sila – pridinol
$C_{19}H_{25}NOSi$
R.Tacke, M.Strecker, W.S.Sheldrick, L.Ernst, E.Heeg,
B.Berndt, C.–M.Knapstein, R.Niedner
Chem.Ber.,**113**, 1962, 1980
See also R1 : 33

63.28 t – Butyl – dimethylsiloxy – aci –
nitrodiphenylmethane (at 101°K)
$C_{19}H_{25}NO_2Si$
E.W.Colvin, A.K.Beck, B.Bastani, D.Seebach, Y.Kai,
J.D.Dunitz *Helv.Chim.Acta*,**63**, 697, 1980

63.C (p – Chlorophenylimino – (trimethylsiloxy) –
methyl) – phenyl – (trimethylsilyl) – phosphane
(at –135°C)
$C_{19}H_{27}ClNOPSi_2$ Main entry is 64.106

63.29 1,1,3 – Trimethyl – 2 – (trimethylsilyl) – 3 –
phenyl – 1,3 – disila – indan (diastereoisomer A)
$C_{19}H_{28}Si_3$
C.Eaborn, D.A.R.Happer, P.B.Hitchcock, S.P.Hopper,
K.D.Safa, S.S.Washburne, D.R.M.Walton
J.Organomet.Chem.,**186**, 309, 1980

63.30 1,1,3 – Trimethyl – 2 – (trimethylsilyl) – 3 –
phenyl – 1,3 – disila – indan (diastereoisomer B)
$C_{19}H_{28}Si_3$
C.Eaborn, D.A.R.Happer, P.B.Hitchcock, S.P.Hopper,
K.D.Safa, S.S.Washburne, D.R.M.Walton
J.Organomet.Chem.,**186**, 309, 1980

63.C (Methyl – (tris(dimethylamino)phosphonium)
amino) – bis(trimethylsilylamino) –
trimethylsilylimino – phosphoranide (at 173°K)
$C_{19}H_{57}N_7P_2Si_4$ Main entry is 64.107

63.C Tricarbonyl – (1,1 – diphenyl – 1 – silacyclohexa –
2,4 – dienyl) – iron
$C_{20}H_{16}FeO_3Si$ Main entry is 75.24

63.31 (3 – Piperidino – propyl) – diphenyl – silanol
Sila – difenidol
$C_{20}H_{27}NOSi$
R.Tacke, M.Strecker, W.S.Sheldrick, E.Heeg, B.Berndt, K.M.Knapstein *Z.Naturforsch.,Teil B*,**34**, 1279, 1979
See also R1 : 33

63.32 cis – 1,7 – Diphenyl – 3,3,5,5,9,9,11,11 – octamethyl – bicyclohexasiloxane (at –120°C)
$C_{20}H_{34}O_7Si_6$
V.E.Shklover, Yu.T.Struchkov, N.N.Makarova, A.A.Zhdanov
Cryst.Struct.Commun.,**9**, 1, 1980

63.C bis(Hexamethyl – trisila – tetraphospha – nortricyclene – tetracarbonyl – chromium)
$C_{20}H_{36}Cr_2O_8P_8Si_6$ Main entry is 86.

63.C bis(μ – Chloro) – bis(bis(carbonyl) – trimethylphosphine – (1 – 2 – η – trimethylsilyl – methylcarbonyl) – molybdenum(ii))
$C_{20}H_{40}Cl_2Mo_2O_6P_2Si_2$ Main entry is 71.163

63.C 1,3,5,7 – Tetra – t – butyl – 2,2,6,6 – tetramethyl – 1,3,5,7 – tetra – aza – 2,6 – disila – 4 – titana – spiro(3.3)heptane
$C_{20}H_{48}N_4Si_2Ti$ Main entry is 83.155

63.C 1,3,5,7 – Tetra – t – butyl – 2,2,6,6 – tetramethyl – 1,3,5,7 – tetra – aza – 2,6 – disila – 4 – zirconia – spiro(3.3)heptane
$C_{20}H_{48}N_4Si_2Zr$ Main entry is 83.156

63.33 5,5 – Dimethyl – 10 – (4 – methylpiperazinyl) – 10,11 – dihydro – 5H – dibenzo(b,f)silepane hydrogen fumarate
$C_{21}H_{29}N_2Si^+, C_4H_3O_4^-$
E.R.Corey, W.F.Paton, J.Y.Corey, M.D.Glick
J.Organomet.Chem.,**179**, 241, 1979
See also R1 : 33

63.34 5 – (3′ – Dimethylaminopropyl) – 10,11 – dihydro – 5H – dibenzo(b,f) – 10,10,11,11 – tetramethyl – 10,11 – disila – azepine
$C_{21}H_{32}N_2Si_2$
T.Debaerdemaeker, F.Osterle, U.Thewalt, G.Struckmeier
Eur.Cryst.Meeting,**5**, 56, 1979

63.35 9,9 – Dimethyl – 10 – (3′ – piperidyl – propyl) – 9 – sila – acridane
$C_{22}H_{30}N_2Si$
T.Debaerdemaeker, F.Osterle, U.Thewalt, G.Struckmeier
Eur.Cryst.Meeting,**5**, 56, 1979

63.36 Oxo – bis(hydroxy – (tricyclo(5.2.1.0²,⁶)dec – 3 – en – 9 – yl) – methyl – silane
$C_{22}H_{34}O_3Si_2$
T.V.Chogovadze, A.I.Nogaideli, L.M.Khananashvili, L.I.Nákaidze, V.S.Tskhovrebashvili, A.I.Gusev, D.Yu.Nesterov *Dokl.Akad.Nauk SSSR*,**246**, 891, 1979

63.37 9 – Fluorenyl – tris(trimethylsilyl) – silane
$C_{22}H_{36}Si_4$
A.Rengstl, U.Schubert *Chem.Ber.*,**113**, 278, 1980

63.38 1,3,7,8 – Tetramethyl – 5,5,11,11 – tetrakis(ethoxycarbonyl) – 1,3,7,9 – tetrasila – 2,8,13,14 – tetraoxatricyclo(7.3.1.1³,⁷)tetradecane
$C_{22}H_{40}O_{12}Si_4$
I.L.Dubchak, V.E.Shklover, Yu.T.Struchkov, E.A.Kashutina, O.I.Shchegolikhina, A.A.Zhdanov
Dokl.Akad.Nauk SSSR,**246**, 1136, 1979

63.39 2,4 – bis(t – Butyl(trimethylsilyl)amino) – 2,4 – dimethyl – 1,3 – bis(trimethylsilyl) – 1,3 – diaza – 2,4 – disilacyclobutane
$C_{22}H_{60}N_4Si_6$
W.Clegg, U.Klingebiel, C.Krampe, G.M.Sheldrick
Z.Naturforsch.,Teil B,**35**, 275, 1980

63.40 1,1,2,2 – Tetraphenyl – disilane
$C_{24}H_{22}Si_2$
S.G.Baxter, K.Mislow, J.F.Blount
Tetrahedron,**36**, 605, 1980

63.C bis(1,1 – Dimethyl – 4 – phenyl – 1 – sila – 4 – bora – cyclohexa – 2,5 – dienyl) – nickel
$C_{24}H_{30}B_2NiSi_2$ Main entry is 75.34

63.C (Octamethyl – (η^5 – cyclopentadienyl – dicarbonyl – iron) – cyclopentasilane) – dimethylsilyl – η^5 – cyclopentadienyl – dicarbonyl – iron
$C_{24}H_{40}Fe_2O_4Si_6$ Main entry is 73.103

63.41 1,1,2,2 – Tetracyclohexyl – disilane
$C_{24}H_{46}Si_2$
S.G.Baxter, D.A.Dougherty, J.P.Hummel, J.F.Blount, K.Mislow *J.Am.Chem.Soc.*,**100**, 7795, 1978

63.C tris(μ – Chloro) – (N – nitroso – N – trimethylsilyl – methylhydroxylaminato) – pentakis(trimethylsilylmethyl) – triangulo – tri – rhenium(iii)
$C_{24}H_{66}Cl_3N_2O_2Re_3Si_6$ Main entry is 71.194

63.42 9 – Fluorenyl – diphenyl – silanol
$C_{25}H_{20}OSi$
A.Rengstl, U.Schubert *Chem.Ber.*,**113**, 278, 1980

63.C (μ – Oxo) – (μ – trimethylsilyloxy) – (μ – per – rhenato) – bis(di(t – butylimido) – trimethylsilyloxy – rhenium) (at –47°C)
$C_{25}H_{63}N_4O_8Re_3Si_3$ Main entry is 84.106

63.43 5 – Ethyl – 5,10 – dihydro – 10,10 – diphenylphenaza – silane
$C_{26}H_{23}NSi$
A.B.Zolotoi, O.A.D′yachenko, L.O.Atovmyan, I.P.Yakovlev, V.O.Reichsfeld *J.Organomet.Chem.*,**190**, 267, 1980
See also R1 : 36

63.44 3,7 – Dimesityl – 2,2,4,4,6,6,8,8 – octamethyl – 1,3,5,7 – tetra – aza – 2,4,6,8 – tetrasila – bicyclo(3.3.0)octane
$C_{26}H_{46}N_4Si_4$
W.Clegg, H.Hluchy, U.Klingebiel, G.M.Sheldrick
Z.Naturforsch.,Teil B,**34**, 1260, 1979

63.C trans – bis(Thiocyanato) – bis(di(3 – trimethylsilyl – propyl) – tellurium) – palladium(ii)
$C_{26}H_{60}N_2PdS_2Si_4Te_2$ Main entry is 70.18

63.45 (S) – p – (Dimethylsilyl – (1 – trimethylsilyl – 3,4 – dimethyl – pent – 2 – enyl)) – benzoic acid p – bromobenzoylmethyl ester (absolute configuration)
$C_{27}H_{37}BrO_3Si_2$
H.Wetter, P.Scherer, W.B.Schweizer
Helv.Chim.Acta,**62**, 1985, 1979

63.46 1,1,7,7 – Tetramethyl – 3,5,9,11 – tetraphenyl – tricyclohexasiloxane (isomer B)
$C_{28}H_{32}O_8Si_6$
V.E.Shklover, I.Yu.Klement′ev, Yu.T.Struchkov
Dokl.Akad.Nauk SSSR,**250**, 877, 1980

63.47 1,3,5,7 – Tetramethyl – 1,3,5,7 – tetraphenyl –
cyclotetrasilazane
$C_{28}H_{36}N_4Si_4$
V.E.Shklover, Yu.T.Struchkov, B.A.Astapov, K.A.Andrianov
Zh.Strukt.Khim.,**20**, 102, 1979

63.48 3,4 – Di – t – butyl – 1,1,2,2 – tetra(trimethylsilyl) –
3,4 – bis(trimethylsiloxy) – 1,2 – disilacyclobutane
$C_{28}H_{72}O_2Si_8$
A.G.Brook, S.C.Nyburg, W.F.Reynolds, Y.C.Poon,
Y.–M.Chang, J.–S.Lee, J.–P.Picard
J.Am.Chem.Soc.,**101**, 6750, 1979

63.49 2,4 – Diphenyl – 2,4 – bis(isopropyl(trimethylsilyl)
amino) – 1,3 – bis(trimethylsilyl) – 1,3 – diaza –
2,4 – disilacyclobutane
$C_{30}H_{60}N_4Si_6$
W.Clegg, U.Klingebiel, C.Krampe, G.M.Sheldrick
Z.Naturforsch.,Teil B,**35**, 275, 1980

63.C trans – bis(1 – Adamantylamido) –
tetrakis(trimethylsilyl – oxo) – molybdenum
(at –55°C)
$C_{32}H_{68}MoN_2O_4Si_4$ Main entry is 84.116

63.C Triphenyl – (triphenylgermyl – peroxy) – silane
$C_{36}H_{30}GeO_2Si$ Main entry is 69.51

63.50 bis(Triphenylsilyl) – mercury
$C_{36}H_{30}HgSi_2$
W.H.Ilsley, E.A.Sadurski, T.F.Schaaf, M.J.Albright,
T.J.Anderson, M.D.Glick, J.P.Oliver
J.Organomet.Chem.,**190**, 257, 1980

63.C (η^5 – Trimethylsilyl – cyclopentadienyl) – (η^4 –
tetraphenyl – cyclobutadiene) – cobalt
$C_{36}H_{33}CoSi$ Main entry is 73.122

63.51 1,1,2,2 – Tetramesityl – disilane
$C_{36}H_{46}Si_2$
S.G.Baxter, K.Mislow, J.F.Blount
Tetrahedron,**36**, 605, 1980

63.52 1,1,3,3 – Tetraphenyl – 1,3 – bis(2 – piperidino –
ethyl) – disiloxane
$C_{38}H_{48}N_2OSi_2$
R.Tacke, M.Strecker, W.S.Sheldrick, L.Ernst, E.Heeg,
B.Berndt, C.–M.Knapstein, R.Niedner
Chem.Ber.,**113**, 1962, 1980
See also R1 : 33

63.C bis(μ – Hydrido) – bis(dimethylsilyl) –
bis(tricyclohexylphosphine) – di – platinum
$C_{40}H_{80}P_2Pt_2Si_2$ Main entry is 86.147

63.C hexakis(μ – t – Butylamido) – bis(μ_3 – t –
butylamido) – tetra(μ – dimethylsilyl) – tetra(μ_3 –
oxo) – hexa – tin
$C_{40}H_{96}N_8O_4Si_4Sn_6$ Main entry is 69.56

63.53 Octaphenyl – cyclotetra(siloxane)
$C_{48}H_{40}O_4Si_4$
D.Braga, G.Zanotti
Acta Crystallogr.,Sect.B,**36**, 950, 1980

63.C bis(Diethylammonium) tris(tetraphenyl –
disiloxane – diolato) – zirconium(iv)
$C_{72}H_{60}O_9Si_6Zr^{2-}$, $2C_4H_{12}N^+$ Main entry is 84.137

PHOSPHORUS COMPOUNDS

64.1 tris(Trichlorophosphazeno) – carbenium
hexachloro – antimony
$CCl_9N_3P_3^+$, Cl_6Sb^-
U.Muller *Z.Anorg.Allg.Chem.*,**463**, 117, 1980

64.2 3,5,5 – Trichloro – 1 – dimethylamino – 1,3,2,4,6,5 –
dithia – triazaphosphorine – 1,3 – dioxide
$C_2H_6Cl_3N_4O_2PS_2$
A.Perales, J.Fayos, J.C.van de Grampel, B.de Ruiter
Acta Crystallogr.,Sect.B,**36**, 838, 1980

64.3 1 – Hydrido – 1 – isopropyl – tetrachloro –
cyclotriphosphazene
$C_3H_8Cl_4N_3P_3$
H.R.Allcock, P.J.Harris *J.Am.Chem.Soc.*,**101**, 6221, 1979

64.4 N – (Phosphonomethyl) – glycine
Glyphosate
$C_3H_8NO_5P$
P.Knuuttila, H.Knuuttila
Acta Chem.Scand.Ser.B,**33**, 623, 1979
See also R1 : 48

64.5 Trimethylphosphine – selenide
C_3H_9PSe
A.Cogne, A.Grand, J.Laugier, J.B.Robert, L.Wiesenfeld
J.Am.Chem.Soc.,**102**, 2238, 1980

64.6 3 – Aminopropylphosphonic acid
$C_3H_{10}NO_3P$
T.Glowiak, W.Sawka–Dobrowolska
Acta Crystallogr.,Sect.B,**36**, 961, 1980

64.7 Pyrazine – phosphorus(v) – pentachloride
$C_4H_4Cl_5N_2P$
B.N.Meyer, J.N.Ishley, A.V.Fratini, H.C.Knachel
Inorg.Chem.,**19**, 2324, 1980

64.8 trans – 2 – Hydroxy – 4,5 – dimethyl – 1,3,2 –
dioxaphospholane – 2 – sulfide imidazolium
monohydrate
$C_4H_8O_3PS^-$, $C_3H_5N_2^+$, H_2O
M.W.Wieczorek, M.Mikolajczyk, M.Witczak
Acta Crystallogr.,Sect.B,**36**, 1452, 1980
See also R2 : 32

64.9 cis – 2,4,6,6 – Tetrachloro – 2,4 –
bis(dimethylamino) – cyclotriphosphazene
$C_4H_{12}Cl_4N_5P_3$
F.R.Ahmed, S.Fortier
Acta Crystallogr.,Sect.B,**36**, 1456, 1980

64.10 trans – 2,4,6,6 – Tetrachloro – 2,4 –
bis(dimethylamino) – cyclotriphosphazene
$C_4H_{12}Cl_4N_5P_3$
F.R.Ahmed, S.Fortier
Acta Crystallogr.,Sect.B,**36**, 1456, 1980

64.11 Trimethylphosphonium – methylene – boron – trihydride
$C_4H_{14}BP$
H.Schmidbaur, G.Muller, B.Milewski–Mahrla, U.Schubert
Chem.Ber.,113, 2575, 1980

64.12 Lithium 1 – carboxymethyl – 2 – imino – 3 – phosphonoimidazolidine dihydrate
Lithium phosphocyclocreatine dihydrate
$C_5H_8N_3O_5P^{2-}$, 2Li$^+$, 2H$_2$O
G.N.Phillips Junior, J.W.Thomas Junior, T.M.Annesley,
F.A.Quiocho *J.Am.Chem.Soc.*,101, 7120, 1979
See also R1 : 32

64.13 2 – Diethylphosphoryl – guanidine hemi(guanidinium chloride)
$C_5H_{14}N_3O_3P$, 0.5CH$_6$N$_3^+$, 0.5Cl$^-$
O.Kennard, J.C.Coppola, D.L.Wampler, K.A.Kerr
Acta Crystallogr.,Sect.B,35, 3000, 1979
See also R2 : 8

64.14 Lithium bis(dimethyl – (borane) – phosphine) methylide
$C_5H_{19}B_2P_2^-$, Li$^+$
H.Schmidbaur, E.Weiss, B.Zimmer–Gasser
Angew.Chem.,Int.Ed.Engl.,18, 782, 1979

64.15 Tri(cyanomethyl) – phosphine
Phosphine – triyl – triacetonitrile
$C_6H_6N_3P$
O.Dahl, S.Larsen *J.Chem.Res.*,396, 4645, 1979

64.16 Methyl β – D – ribopyranoside phosphite triester
$C_6H_9O_5P$
A.C.Bellaart, D.van Aken, H.M.Buck, C.H.Stam, A.van Herk
Rec.J.Roy.Netherl.Chem.Soc.,98, 523, 1979
See also R1 : 45

64.17 7 – Methoxy – 3,5,9 – trioxa – 4 – phosphabicyclo(4.3.0)nonan – 4 – one
$C_6H_{10}O_5P$
L.A.Aslanov, S.S.Sotman, V.B.Ribakov, V.I.Andrianov,
Z.Sh.Safina, M.P.Koroteev, L.T.Elepina, E.E.Nifant'ev
Zh.Strukt.Khim.,20, 1122, 1979
See also R1 : 38

64.18 Phenoxythiophosphoryl – dihydrazide
$C_6H_{11}N_4OPS$
U.Engelhardt *Acta Crystallogr.,Sect.B*,35, 3116, 1979

64.19 10 – Thio – 10 – phospha – 1,4,7 – triazatricyclo(5.2.1.0^{4,10})decane
$C_6H_{12}N_3PS$
D.W.White, B.A.Karcher, R.A.Jacobson, J.G.Verkade
J.Am.Chem.Soc.,101, 4921, 1979

64.20 1,5 – Diphosphabicyclo(3.3.0)octane – 1,5 – disulfide (orthorhombic form)
$C_6H_{12}P_2S_2$
H.Hartung, S.Hickel, J.Kaiser, R.Richter
Z.Anorg.Allg.Chem.,458, 130, 1979

64.21 5,5 – Dimethyl – 2 – methoxy – 2 – oxo – 1,3,2 – dioxaphosphorinane
$C_6H_{13}O_4P$
P.Van Nuffel, A.T.H.Lenstra, H.J.Geise
Cryst.Struct.Commun.,9, 733, 1980

64.22 Dimethylammonium 0,0 – di – isopropyl – dithiophosphate
$C_6H_{14}O_2PS_2^-$, $C_2H_8N^+$
A.E.Kalinin, V.G.Andrianov, Yu.T.Struchkov
Izv.Akad.Nauk SSSR,Ser.Khim.,783, 1979
See also R2 : 3

64.23 1 – Methylphospha – 2,3,5,6 – bis – σ – carborano – cyclohexane
$C_6H_{25}B_{20}P$
A.I.Yanovskii, N.G.Furmanova, Yu.T.Struchkov,
N.F.Shemyakin, L.I.Zakharkin
Izv.Akad.Nauk SSSR,Ser.Khim.,1523, 1979
See also R1 : 62

64.24 Trichloro – (p – methoxyphenyl) – phosphonium hydrogen dichloride (at 123°K)
$C_7H_7Cl_3OP^+$, Cl$^-$, HCl
D.Mootz, W.Poll, H.Wunderlich, H.–G.Wussow
Eur.Cryst.Meeting,5, 254, 1979

64.25 (1SR,3SR,4RS) – 3 – Chloro – 4 – hydroxy – 3,4 – dimethyl – 1 – methoxyphospholane – 1 – oxide
$C_7H_{14}ClO_3P$
F.Cavagna, U.–H.Felcht, E.F.Paulus
Angew.Chem.,Int.Ed.Engl.,19, 132, 1980

64.26 S(–) – 3 – (2 – Chloroethyl) – 2 – ((2 – chloroethyl) – amino) – 1,3,2 – oxazaphosphorinane – 2 – oxide (absolute configuration)
$C_7H_{14}Cl_2N_2O_2P$
D.A.Adamiak, M.Gdaniec, K.Pankiewicz, W.J.Stec
Angew.Chem.,Int.Ed.Engl.,19, 549, 1980

64.27 2 – Hydroxy – 1,1,2,2 – tetrakis(trifluoromethyl) – ethyl – (dimethylphosphinate)
$C_8H_7F_{12}O_3P$
D.Schomburg, O.Stelzer, N.Weferling, R.Schmutzler,
W.S.Sheldrick *Chem.Ber.*,113, 1566, 1980

64.28 1,2,3,4 – Tetramethyl – 2,2,4,4 – tetrakis(trifluoromethyl) – 1,3 – diaza – 2,4 – diphosphetidine
$C_8H_{12}F_{12}N_2P_2$
L.V.Griend, R.G.Cavell *Inorg.Chem.*,19, 2070, 1980

64.29 2 – (Pentachloro – cyclotriphosphazenyl) – 1 – phenyl – 1,2 – dicarba – dodecaborane
$C_8H_{15}B_{10}Cl_5N_3P_3$
A.G.Scopelianos, J.P.O'Brien, H.R.Allcock
J.Chem.Soc.,Chem.Commun.,198, 1980
See also R1 : 62

64.30 Octamethyl – trithio – cyclotetra(λ^3,λ^5,λ^5,λ^5 – phosphazane)
$C_8H_{24}N_4P_4S_3$
W.Zeiss, W.Schwarz, H.Hess
Z.Naturforsch.,Teil B,35, 959, 1980
See also R1 : 42,34

64.31 Phenyl – (1 – mercapto – 1 – methyl – ethyl) – phosphinic acid
$C_9H_{13}O_2PS$
V.V.Tkachev, L.O.Atovmyan, N.A.Kardanov, N.N.Godovikov,
M.I.Kabachnik *Zh.Strukt.Khim.*,20, 553, 1979

64.32 1,2 – O – Isopropylidene – α – D – glucofuranose – 3,5,6 – bicyclophosphite
$C_9H_{13}O_6P$
L.A.Aslanov, S.S.Sotman, V.B.Ribakov, V.I.Andrianov, Z.Sh.Safina, M.P.Koroteev, E.E.Nifant'ev
Zh.Strukt.Khim.,20, 1125, 1979
See also R1 : 45

64.33 4 – Diethylamino – 3,5,9 – trioxa – 4 – phosphabicyclo(4.3.0)nonane – 4,7 – dione
$C_9H_{16}NO_5P$
L.A.Aslanov, S.S.Sotman, V.B.Ribakov, L.T.Elepina, E.E.Nifant'ev *Zh.Strukt.Khim.*,20, 1128, 1979
See also R1 : 38

64.34 Dimethyl – 1 – hydroxy – 1 – cycloheptane – phosphonate (reinvestigation of structure published by Birnbaum, Buchanan and Morin, J.Am.Chem.Soc.,99,(1977),6652–6656, using published diffraction data)
$C_9H_{19}O_4P$
A.J.de Kok, C.Romers
Acta Crystallogr.,Sect.B,36, 1887, 1980
See also R1 : 22

64.35 4,7,11 – tris(Trimethylsilyl) – pentacyclo(6.3.0.0^{2,6}.0^{3,10}.0^{5,9})undecaphosphane
$C_9H_{27}P_{11}Si_3$
H.G.von Schnering, D.Fenske, W.Honle, M.Binnewies, K.Peters *Angew.Chem.,Int.Ed.Engl.*,18, 679, 1979
See also R1 : 63

64.36 1,1',2'' – tris(Trimethylsilyl) – thiophosphoryl – trihydrazide
$C_9H_{33}N_6PSSi_3$
U.Engelhardt, H.–P.Metter
Acta Crystallogr.,Sect.B,36, 2086, 1980
See also R1 : 63

64.37 2 – (Cyanophosphinidene) – 1,3 – dimethyl – benzimidazoline
$C_{10}H_{10}N_3P$
A.Schmidpeter, W.Gebler, F.Zwaschka, W.S.Sheldrick
Angew.Chem.,Int.Ed.Engl.,19, 722, 1980
See also R1 : 35

64.38 2,3 – Dibromo – 1 – phenylphospholane – 1 – oxide (form i)
$C_{10}H_{11}Br_2OP$
B.R.Stults, K.Moedritzer
Cryst.Struct.Commun.,8, 787, 1979

64.39 4 – Hydroxy – 1 – phenyl – 2 – pholene – 1 – oxide
$C_{10}H_{11}O_2P$
Z.Galdecki, M.L.Glowka
Acta Crystallogr.,Sect.B,36, 1495, 1980

64.40 bis(Phenyl – (t – butyl) – phosphinic amide) hydrogen chloride
$C_{10}H_{16}NOP$, $C_{10}H_{17}NOP^+$, Cl^-
W.W.Ng, S.C.Nyburg, T.A.Modro
J.Chem.Soc.,Chem.Commun.,195, 1980

64.41 5,5,5',5' – Tetramethyl – 2,2' – bis(1,3,2λ^5 – dioxaphosphorinane) – 2,2' – disulfide (monoclinic form)
$C_{10}H_{20}O_4P_2S_2$
J.Karolak–Wojciechowska, M.Wieczorek, Z.Galdecki
Acta Crystallogr.,Sect.B,36, 1683, 1980

64.42 2,4,6,8 – Tetraethyl – 3,7 – dimethyl – 1λ^6,5λ^6 – dithia – 2,4,6,8 – tetra – aza – 3,7 – diphosphorocine – 1,1,5,5 – tetraoxide acetonitrile solvate
$C_{10}H_{26}N_4O_4P_2S_2$, C_2H_3N
H.W.Roesky, S.K.Mehrotra, C.Platte, D.Amirzadeh–Asl, B.Roth *Z.Naturforsch.,Teil B*,35, 1130, 1980

64.43 tris(Trimethylphosphino)methylide di – iodide
$C_{10}H_{27}P_3^{2+}$, $2I^-$
B.Zimmer–Gasser, D.Neugebauer, U.Schubert, H.H.Karsch
Z.Naturforsch.,Teil B,34, 1267, 1979

64.44 Diethyl – (5,6 – dichloro – 1,3 – benzodioxole – (2)) – phosphonate
$C_{11}H_{13}Cl_2O_5P$
S.Kulpe, I.Seidel *Krist.Tech.*,14, 1089, 1979
See also R1 : 38

64.45 1 – Phenyl – 4 – phosphorinanone – 1 – sulfide
$C_{11}H_{13}OPS$
S.D.Venkataramu, K.D.Berlin, S.E.Ealick, J.R.Baker, S.Nichols, D.van der Helm
Phosphorus and Sulfur,7, 133, 1979

64.46 1 – Phenyl – 4 – phosphorinanone – 1 – oxide
$C_{11}H_{13}O_2P$
S.D.Venkataramu, K.D.Berlin, S.E.Ealick, J.R.Baker, S.Nichols, D.van der Helm
Phosphorus and Sulfur,7, 133, 1979

64.47 2 – Hydroxy – 2 – methyl – 1 – phenylpholane – 1 – oxide
$C_{11}H_{15}O_2P$
Z.Galdecki, M.L.Glowka
Acta Crystallogr.,Sect.B,36, 2191, 1980

64.48 6,8 – Dimethyl – 7 – phenoxy – 7 – thioxo – 3 – oxa – 1,5,6,8 – tetra – aza – 7 – phosphabicyclo(3.3.1)nonane
$C_{11}H_{17}N_4O_2PS$
J.Jaud, J.Galy, R.Kramer, J.–P.Majoral, J.Navech
Acta Crystallogr.,Sect.B,36, 869, 1980

64.49 2 – Oxo – 2 – methyl – 3,5 – di – t – butyl – 1,3,2 – oxaza – 4 – pholine
$C_{11}H_{22}NO_2P$
Yu.V.Balitskii, Yu.G.Gololobov, V.M.Yurchenko, M.Yu.Antipin, Yu.T.Struchkov, I.E.Boldeskul
Zh.Obshch.Khim.,50, 291, 1980

64.50 1,1 – Diphenyl – 1 – phospha – 3,5 – dithia – 2,4,6 – triazine
$C_{12}H_{10}N_3PS_2$
A.W.Cordes, P.N.Swepston, T.Chivers, R.T.Oakley, N.Burford
Am.Cryst.Assoc.,Ser.2,8, 22, 1980

64.C Tricarbonyl – bis(ethylenediamine) – rhenium(i) diphenylphosphinate
$C_{12}H_{10}O_2P^-$, $C_7H_{16}N_4O_3Re^+$ Main entry is 76.43

64.51 Phenyl – α – hydroxy – cyclohexyl – phosphinic acid chloranhyrdide
$C_{12}H_{16}ClO_2P$
N.A.Kardanov, N.N.Godovikov, V.V.Tkachev, L.O.Atovmyan, M.I.Kabachnik
Izv.Akad.Nauk SSSR,Ser.Khim.,1630, 1979

64.C 3,3',4,4' – Tetramethyl – 1,1' – diphospha – ferrocene
$C_{12}H_{16}FeP_2$ Main entry is 75.4

64.52 1,2 – O – Cyclohexylidene – α – D – glucofuranose –
3,5,6 – bicyclophosphite
$C_{12}H_{17}O_6P$
L.A.Aslanov, S.S.Sotman, V.B.Ribakov, V.I.Andrianov,
Z.Sh.Safina, M.P.Koroteev, E.E.Nifant'ev
Zh.Strukt.Khim.,**20**, 1125, 1979
See also R1 : 45

64.53 Di(cyclohexyl) – phosphinic acid
$C_{12}H_{23}O_2P$
L.A.Aslanov, S.S.Sotman, V.B.Ribakov, L.G.Elepina,
E.E.Nifant'ev *Zh.Strukt.Khim.*,**20**, 758, 1979

64.54 endo – 2 – Dimethylphosphono – exo – 2 –
hydroxy – (–) – camphane
$C_{12}H_{23}O_4P$
G.Adiwidjaja, B.Meyer, J.Thiem
Z.Naturforsch.,Teil B,**34**, 1547, 1979
See also R1 : 31

64.C Hexa – aziridino – cyclotriphosphazene benzene
clathrate
$2C_{12}H_{24}N_9P_3$, C_6H_6 Main entry is 61.3

64.C bis(Boranato – bis(dimethylphosphonium –
methylide)) – nickel
$C_{12}H_{36}B_2NiP_4$ Main entry is 71.43

64.55 Dodecamethyl – bis(imido) – triphosphoramide
monohydrate
$C_{12}H_{36}N_7O_3P_3$, H_2O
J.C.P.M.Lapidaire, A.J.de Kok
Z.Naturforsch.,Teil B,**35**, 1203, 1980

64.56 O – Methyl – O – (4 – bromo – 2,5 –
dichlorophenyl) – phenylphosphonothioate
$C_{13}H_{10}BrCl_2O_2PS$
R.L.Lapp, R.A.Jacobson
Cryst.Struct.Commun.,**9**, 65, 1980

64.57 (Chloromethyl) – diphenylphosphine – oxide
$C_{13}H_{12}ClOP$
G.Bernardinelli, R.Gerdil
Cryst.Struct.Commun.,**8**, 921, 1979

64.58 2 – Carboxy – 3,4 – dimethyl – 1 –
phenylphosphole – sulfide
$C_{13}H_{13}O_2PS$
D.C.Craig, M.J.Gallagher, F.Mathey, G.de Lauzon
Cryst.Struct.Commun.,**9**, 901, 1980

64.59 exo – Phenyl – 3 – phosphabicyclo(3.2.1)oct – 6 –
en – 3 – one
$C_{13}H_{15}OP$
M.–Ul–Haque, W.Horne *Eur.Cryst.Meeting*,**6**, 16, 1980

64.60 cis – 2(bis(2 – Chloroethyl)amino) – 4 – phenyl –
2H – 1,3,2 – oxazaphosphorinane – 2 – oxide
cis – 4 – Phenylcyclophosphamide
$C_{13}H_{19}Cl_2N_2O_2P$
V.L.Boyd, G.Zon, V.L.Himes, J.K.Stalick, A.D.Mighell,
H.V.Secor *J.Med.Chem.*,**23**, 372, 1980

64.61 (bis(Trimethylsilyl)amino) – (t – butylimino) –
(trimethylsilylimino) – phosphorus (at –130°C)
$C_{13}H_{36}N_3PSi_3$
S.Pohl *Eur.Cryst.Meeting*,**5**, 250, 1979
See also R1 : 63

64.62 Lithium dibenzoylphosphide 1,2 – dimethoxyethane
(at –80°C)
$C_{14}H_{10}O_2P^-$, $C_4H_{10}O_2$, Li^+
G.Becker, M.Birkhahn, W.Massa, W.Uhl
Angew.Chem.,Int.Ed.Engl.,**19**, 741, 1980
See also R2 : 5

64.63 O – Ethyl – O – (4 – nitrophenyl) – benzene –
phosphonothioate (at 4°C)
$C_{14}H_{14}NO_4PS$
M.R.Gifkins, R.A.Jacobson
Cryst.Struct.Commun.,**9**, 571, 1980

64.64 Methyl – 3,4 – dimethyl – 1 – phenylphosphole –
2 – carboxylate
$C_{14}H_{15}O_2P$
R.B.Knott, H.Honneger, A.D.Rae, F.Mathey, G.de Lauzon
Cryst.Struct.Commun.,**9**, 905, 1980

64.65 (1 – Hydroxy – 1 – (2 – hydroxyphenyl) – ethyl) –
phenyl – phosphinic acid
$C_{14}H_{15}O_4P$
M.R.W.Wright, L.R.Nassimbeni
Acta Crystallogr.,Sect.B,**35**, 2995, 1979

64.66 O – (β – Chloroethyl) – phenyl – (α –
hydroxycyclohexyl) – phosphinate (high melting
form)
$C_{14}H_{20}ClO_3P$
V.V.Tkachev, L.O.Atovmyan, N.A.Kardanov, N.N.Godovikov,
M.I.Kabachnik *Zh.Strukt.Khim.*,**21**, 106, 1980

64.67 2,2,6,6 – Tetramethyl – 4 – phenyl – 4 – oxo – 1 –
oxyl – 1,4 – azaphosphorinane
$C_{14}H_{21}NO_2P$
M.Cygler *Am.Cryst.Assoc.,Ser.2*,**8**, 37, 1980

64.68 4,6 – Di – t – butyl – 2 – hydroxy – 2 – oxo –
benzo(d) – 1,3,2 – dioxaphosphole n – hexane
solvate
$C_{14}H_{21}O_4P$, $0.5C_6H_{14}$
G.G.Aleksandrov, Yu.T.Struchkov
Cryst.Struct.Commun.,**9**, 493, 1980

64.69 (4,5 – Dimethyl – 1,3 – dioxa – 2 – phosphole) – 2 –
spiro – 2' – (cis – 2,4 – bis(dimethylamino) – 1,3 –
bis(trimethylsilyl)) – 1,3,2,4 – diazadiphosphetidine
$C_{14}H_{36}N_4O_2P_2Si_2$
D.Lux, W.Schwarz, H.Hess, W.Zeiss
Z.Naturforsch.,Teil B,**35**, 269, 1980
See also R1 : 63

64.70 2 – Chloro – 6 – methyl – 3 – p – tolyl – 1,2,3,4 –
tetrahydro – 1,3,2 – benzo – diazaphosphorine 2 –
oxide
$C_{15}H_{16}ClN_2OP$
T.S.Cameron, R.E.Cordes, T.Demir, R.A.Shaw
J.Chem.Soc.,Perkin Trans.1,2896, 1979

64.71 3 – Methyl – 2,4 – di(p – methoxyphenyl) – 2 –
trans – 4 – dithio – 1,3,2,4 – thia – aza – λ^5,λ^5 –
diphosphetidine
$C_{15}H_{17}NO_2P_2S_3$
W.Zeiss, H.Henjes, D.Lux, W.Schwarz, H.Hess
Z.Naturforsch.,Teil B,**34**, 1334, 1979

64.72 Diphenyl – (3 – hydroxypropyl)phosphine – oxide
$C_{15}H_{17}O_2P$
V.V.Tkachev, N.A.Bondarenko, E.I.Matrosov, E.N.Tsvetkov,
L.O.Atovmyan, M.I.Kabachnik
Izv.Akad.Nauk SSSR,Ser.Khim.,1159, 1979

64.73 9 – Methyl – 3,7 – diphenyl – 1,3,4,6,7,9 – hexa – aza – 5λ^4 – phosphabicyclo(3.3.1)nonane – 5 – thione
$C_{15}H_{19}N_6PS$
A.Grand, J.B.Robert, J.–P.Majoral, J.Navech
*J.Chem.Soc.,Perkin Trans.2,*792, 1980

64.74 2,2 – Diphenyl – 4,4,6 – trimethoxy – 6 – hydroxy – cyclotriphosphazene
$C_{15}H_{20}N_3O_4P_3$
K.S.Dhathathreyan, S.S.Krishnamurthy, A.R.V.Murthy, T.S.Cameron, C.Chan, R.A.Shaw, M.Woods
*J.Chem.Soc.,Chem.Commun.,*231, 1980

64.75 pentakis(Methyl – dicarba – dodecaboranyl) – diphosphine benzene solvate
$C_{15}H_{64}B_{50}P_2$, C_6H_6
N.G.Furmanova, A.I.Yanovskii, Yu.T.Struchkov, V.I.Bregadze, N.N.Godovikov, A.N.Degtyarev, M.I.Kabachnik
*Izv.Akad.Nauk SSSR,Ser.Khim.,*2346, 1979
See also R1 : 62

64.76 2,5,7,10,11,12 – hexakis(Trifluoromethyl) – 1,6 – diphospha – hexacyclo(4.4.2.0^{2,5}.0^{3,9}.0^{4,8}.0^{9,10})dodec – 11 – ene
$C_{18}H_4F_{18}P_2$
Y.Kobayashi, H.Hamana, S.Fujino, A.Ohsawa, I.Kumadaki
*J.Org.Chem.,*44, 4930, 1979

64.C (1,1′ – Ferrocenediyl) – phenylphosphine
$C_{16}H_{13}FeP$ Main entry is 73.57

64.77 3,4 – bis(t – Butyl) – 2 – (p – methoxyphenyl) – 1,3,4,2 – thiadiazaphosphiolidin – 5 – one 2 – sulfide
$C_{16}H_{25}N_2O_2PS_2$
G.L'abbe, J.Flemal, J.P.Declercq, G.Germain, M.van Meerssche *Bull.Soc.Chim.Belg.,*88, 737, 1979

64.78 bis(1,4,7,10 – Tetra – azacyclododecane) – phosphorane
Dicyclenephosphorane
$C_{16}H_{32}N_8P_2$
J.E.Richman, R.O.Day, R.R.Holmes
*J.Am.Chem.Soc.,*102, 3955, 1980

64.79 Tetra(t – butyl) – phosphonium tetrafluoroborate
$C_{16}H_{36}P^+$, BF_4^-
H.Schmidbaur, G.Blaschke, B.Zimmer–Gasser, U.Schubert
*Chem.Ber.,*113, 1612, 1980

64.80 Diethyl – bis(p – chlorophenoxy)methane – phosphonate
$C_{17}H_{19}Cl_2O_5P$
S.Kulpe, I.Seidel *Krist.Tech.,*15, 149, 1980

64.81 3,3 – Dimethyl – 1,1 – diphenylphosphetanium iodide
$C_{17}H_{20}P^+$, I^-
M.–Ul–Haque *Acta Crystallogr.,Sect.B,*35, 2601, 1979

64.82 D,L – Diethylanilino – (3 – hydroxybenzyl) – phosphonate
$C_{17}H_{22}NO_4P$
Z.Ruzic–Toros, B.Kojic–Prodic
*Eur.Cryst.Meeting,*5, 384, 1979
See also R1 : 17,16

64.83 R – neo – Menthyl – methylphenylphosphine oxide
$C_{17}H_{27}OP$
D.Valentine Junior, J.F.Blount, K.Toth
*J.Org.Chem.,*45, 3691, 1980

64.84 R – Menthyl – methylphenylphosphine oxide
$C_{17}H_{27}OP$
D.Valentine Junior, J.F.Blount, K.Toth
*J.Org.Chem.,*45, 3691, 1980

64.85 O – (2 – Trimethylammonio – ethyl) – phenyl – (1 – hydroxycyclohexyl) – phosphinate iodide
$C_{17}H_{29}NO_3P^+$, I^-
V.V.Tkachev, L.O.Atovmyan, N.A.Kardanov, N.N.Godovikov, M.I.Kabachnik *Zh.Strukt.Khim.,*20, 653, 1979

64.86 bis(Neopentyloxy) – methylphenylphosphonium bromide
$C_{17}H_{30}O_2P^+$, Br^-
K.Henrick, H.R.Hudson, A.Kow
*J.Chem.Soc.,Chem.Commun.,*226, 1980

64.87 Triphenylphosphinimino – (trisulfur – nitride)
$C_{18}H_{15}N_2PS_3$
T.Chivers, R.T.Oakley, A.W.Cordes, P.Swepston
*J.Chem.Soc.,Chem.Commun.,*35, 1980

64.88 Triphenylphosphine – oxide – triphenylphosphonium – hydroxide perchlorate
$C_{18}H_{15}OP$, $C_{18}H_{16}OP^+$, ClO_4^-
M.Yu.Antipin, A.E.Kalinin, Yu.T.Struchkov, E.I.Matrosov, M.I.Kabachnik *Kristallografiya,*25, 514, 1980

64.89 Triphenylphosphine – oxide – triphenylphosphonium – hydroxide perchlorate (at –120°C)
$C_{18}H_{15}OP$, $C_{18}H_{16}OP^+$, ClO_4^-
M.Yu.Antipin, A.E.Kalinin, Yu.T.Struchkov, E.I.Matrosov, M.I.Kabachnik *Kristallografiya,*25, 514, 1980

64.90 Triphenylphosphine – oxide hemiperhydrate
$C_{18}H_{15}OP$, $0.5H_2O_2$
D.Thierbach, F.Huber, H.Preut
*Acta Crystallogr.,Sect.B,*36, 974, 1980

64.91 bis(Triphenylphosphine – oxide) monohydrate hydrogen bromide
$2C_{18}H_{15}OP$, H_2O, HBr
D.Thierbach, F.Huber *Z.Anorg.Allg.Chem.,*457, 189, 1979

64.92 tris(Anilino) – phosphine
$C_{18}H_{18}N_3P$
A.Tarassoli, R.C.Haltiwanger, A.D.Norman
*Inorg.Nucl.Chem.Lett.,*16, 27, 1980

64.93 6 – Phenyl – 6 – phospha – 2,10 – dithiabicyclo(9.4.0)pentadeca – 11(1),12,14 – triene
$C_{18}H_{21}PS_2$
E.P.Kyba, A.M.John, S.B.Brown, C.W.Hudson, M.J.McPhaul, A.Harding, K.Larsen, S.Niedzwiecki, R.E.Davis
*J.Am.Chem.Soc.,*102, 139, 1980

64.94 1 – Benzyl – 1 – phenylphosphorinanium bromide
$C_{18}H_{22}P^+$, Br^-
J.C.Gallucci, R.R.Holmes *J.Am.Chem.Soc.,*102, 4379, 1980

64.95 1,1 – Diphenyl – 4 – methylphosphorinanium bromide
$C_{18}H_{22}P^+$, Br^-
J.C.Gallucci, R.R.Holmes *J.Am.Chem.Soc.,*102, 4379, 1980

64.96 (1R,3S) – (–) – trans – 1 – Benzyl – 3 – methyl – 1 – phenylphospholanium iodide (absolute configuration)
$C_{18}H_{22}P^+$, I^-
R.O.Day, S.Husebye, J.A.Deiters, R.R.Holmes
*J.Am.Chem.Soc.,*102, 4387, 1980

64.97 **Neopentyloxy – methyldiphenylphosphonium bromide**
$C_{18}H_{24}OP^+$, Br^-
K.Henrick, H.R.Hudson, A.Kow
J.Chem.Soc.,Chem.Commun.,226, 1980

64.98 **1,3,5 – Di – isopropylamino – 2,4,6 – trioxa – 1,3,5 – triphosphorine (at –133°C)**
$C_{18}H_{42}N_3O_3P_3$
E.Niecke, H.Zorn, B.Krebs, G.Henkel
Angew.Chem.,Int.Ed.Engl.,19, 709, 1980

64.99 **(cis 2,4 – bis(Dimethylamino) – 1,3 – bis(trimethylsilyl) – 1,3,2,4 – diazadiphosphetidine) – 2′,4′ – dispiro – 2,2″ – (4,5 – dimethyl – 1,3 – dioxa – 2 – phosphole)**
$C_{18}H_{42}N_4O_4P_2Si_2$
D.Lux, W.Schwarz, H.Hess, W.Zeiss
Z.Naturforsch.,Teil B,35, 269, 1980
See also R1 : 63

64.100 **Diphenyl – (5,6 – dichloro – 1,3 – benzodioxol – (2)) – phosphine – oxide**
$C_{19}H_{13}Cl_2O_3P$
S.Kulpe, I.Seidel *Krist.Tech.*,14, 1421, 1979
See also R1 : 38

64.101 **Phosphatriptycene**
$C_{19}H_{13}P$
F.J.M.Freijee, C.H.Stam
Acta Crystallogr.,Sect.B,36, 1247, 1980

64.102 **Methyl – tris(p – nitrophenoxy) – phosphonium chloride p – nitrophenol benzene solvate**
$C_{19}H_{15}N_3O_9P^+$, $C_6H_5NO_3$, $0.5C_6H_6$, Cl^-
D.Schomburg *J.Am.Chem.Soc.*,102, 1055, 1980

64.103 **1,3 – Benzodithiolyl – 2 – (diphenylphosphine – oxide)**
$C_{19}H_{15}OPS_2$
S.Kulpe, I.Seidel *Krist.Tech.*,15, 3, 1980
See also R1 : 39

64.104 **5 – Methyl – 10 – phenyl – 10 – oxo – 5,10 – dihydro – phenophosphazine phenol**
$C_{19}H_{16}NOP$, C_6H_6O
A.I.Gusev, S.N.Gurkova, V.K.Bel'skii, V.E.Zavodnik, L.A.Yagodina *Zh.Strukt.Khim.*,20, 632, 1979

64.105 **Methyl 2 – methyl – 3 – (benzyl(phenyl) phosphinyl) – butyrate**
$C_{19}H_{23}O_3P$
Z.Galdecki, M.L.Glowka, R.Bodalski, K.M.Pietrusiewicz
J.Chem.Soc.,Perkin Trans.2,1720, 1979

64.106 **(p – Chlorophenylimino – (trimethylsiloxy) – methyl) – phenyl – (trimethylsilyl) – phosphane (at –135°C)**
$C_{19}H_{27}ClNOPSi_2$
G.Becker, O.Mundt *Z.Anorg.Allg.Chem.*,459, 87, 1979
See also R1 : 63

64.107 **(Methyl – (tris(dimethylamino)phosphonium) amino) – bis(trimethylsilylamino) – trimethylsilylimino – phosphoranide (at 173°K)**
$C_{19}H_{57}N_7P_2Si_4$
M.Halstenberg, R.Appel, G.Huttner, J.von Seyerl
Z.Naturforsch.,Teil B,34, 1491, 1979
See also R1 : 63

64.108 **5 – Phenyl – 10,11 – dihydro – dibenzo(b,f) phosphepin – 5 – oxide**
$C_{20}H_{17}OP$
D.W.Allen, I.W.Nowell, P.E.Walker
Z.Naturforsch.,Teil B,35, 133, 1980

64.C **Ethyltriphenylphosphonium – bis(7,7,8,8 – tetracyanoquinodimethane)**
$C_{20}H_{20}P^+$, $C_{12}H_4N_4^-$, $C_{12}H_4N_4$ Main entry is 60.41

64.109 **meso – o – Phenylene – bis(methylphenylphosphine)**
$C_{20}H_{20}P_2$
N.K.Roberts, B.W.Skelton, A.H.White
J.Chem.Soc.,Dalton Trans.,1567, 1980

64.110 **meso – ortho – Phenylene – (methylphenylphosphine) – (methylphenylphosphonium) tetrafluoroborate**
$C_{20}H_{21}P_2^+$, BF_4^-
N.K.Roberts, B.W.Skelton, A.H.White
J.Chem.Soc.,Dalton Trans.,1567, 1980

64.111 **exo – 3 – p – Nitrobenzyl – endo – 3 – phenyl – 3 – phosphabicyclo(3.2.1)octane bromide**
$C_{20}H_{23}NO_2P^+$, Br^-
M.–Ul–Haque, W.Horne *Am.Cryst.Assoc.,Ser.2*,7, 11, 1980

64.112 **1,2 – Dicyclohexyl – 1,2 – bis(diethylamino) – 1,2 – dithioxo – diphosphane**
$C_{20}H_{42}N_2P_2S_2$
D.Troy, J.Galy, J.P.Legros
Acta Crystallogr.,Sect.B,36, 398, 1980

64.113 **5,5 – Dimethyl – 2,2 – diphenyl – 3,4 – methoxycarbonyl – 1 – aza – 2 – phosphacyclo – pent – 1 – ene**
$C_{21}H_{22}NO_4P$
W.S.Sheldrick, D.Schomburg, A.Schmidpeter, T.von Criegern *Chem.Ber.*,113, 55, 1980

64.114 **Triethylammonium diphenoxy – (o – phenylenedioxy) – (1,2 – bis(trifluoromethyl) – ethenylenedioxy) – phosphoride**
$C_{22}H_{14}F_6O_6P^-$, $C_6H_{16}N^+$
R.Sarma, F.Ramirez, B.McKeever, J.F.Marecek, V.A.V.Prasad *Phosphorus and Sulfur*,5, 323, 1979

64.115 **1 – (Triphenylphosphine – imino) – 1 – chloro – 2,2 – dicyanoethylene**
$C_{22}H_{15}ClN_3P$
M.Yu.Antipin, A.E.Kalinin, Yu.T.Struchkov, Yu.P.Egorov
Zh.Strukt.Khim.,20, 868, 1979

64.116 **4 – Methyl – 2,6,6 – triphenyl – 2,3 – diaza – 1 – phosphabicyclo(3.1.0)hex – 3 – ene**
$C_{22}H_{19}N_2P$
B.A.Arbuzov, V.D.Cherepinskii–Malov, E.N.Dianova, A.I.Gusev, V.A.Sharapov
Dokl.Akad.Nauk SSSR,247, 1150, 1979

64.117 **5,7 – Dimethyl – 1 – phenyl – 3,4:8,9 – dibenzo – 2,6,10,11 – tetraoxa – 1 – phosphatricyclo(5.3.1.0^{1,5}) undecane**
$C_{22}H_{19}O_4P$
M.R.W.Wright, L.R.Nassimbeni
Acta Crystallogr.,Sect.B,35, 2995, 1979

64.118 **(Triphenylmethyl) – trimethylphosphonium tetrafluoroborate**
$C_{22}H_{24}P^+$, BF_4^-
R.A.Jones, G.Wilkinson, M.B.Hursthouse, K.M.A.Malik
J.Chem.Soc.,Perkin Trans.2,117, 1980

64.C (0,0′ − Diethyl − dithiophosphato) − triphenyl − tin(iv) (at 138°K)
$C_{22}H_{25}O_2PS_2Sn$ Main entry is 69.46

64.119 Methyl − 6 − (diphenylphosphinyl) − 3,4 − dimethyl − cyclohex − 3 − ene − 1 − carboxylate
$C_{22}H_{25}O_3P$
M.L.Glowka, Z.Galdecki
*Acta Crystallogr.,Sect.B,***36**, 1728, 1980

64.120 3 − t − Butyl − 9,10 − dihydro − 9,10 − o − benzeno − 9 − phospha − anthracene
$C_{23}H_{21}P$
N.van der Putten, C.H.Stam
*Acta Crystallogr.,Sect.B,***36**, 1250, 1980

64.121 2,4,4 − Triphenyl − 3 − methoxy − 6 − methyl − 1,2 − diaza − 3 − phosphacyclohex − 1(6) − ene
$C_{23}H_{23}N_2OP$
B.A.Arbuzov, V.D.Cherepinskii−Malov, E.N.Dianova, A.I.Gusev, V.A.Sharapov
*Dokl.Akad.Nauk SSSR,***247**, 1150, 1979

64.122 1,5 − Dimethyl − 4 − pentafluorophenyl − 3,7 − bis(3 − (trifluoromethyl) − phenyl) − 1,3,5,7 − tetra − aza − 4γ^5 − phospha − spiro(3.3)heptane − 2,6 − dione
$C_{24}H_{14}F_{11}N_4O_2P$
H.W.Roesky, K.Ambrosius, M.Banek, W.S.Sheldrick
*Chem.Ber.,***113**, 1847, 1980

64.123 Tetraphenylphosphonium (benzyl − trimethylammonium) dichloro − iron − bis(μ^2 − thio) − molybdenum − disulfide
$C_{24}H_{20}P^+$, $C_{10}H_{16}N^+$, $Cl_2FeMoS_4^{2-}$
A.Muller, H.Bogge, H.−G.Tolle, R.Jostes, U.Schimanski, M.Dartmann *Angew.Chem.,Int.Ed.Engl.,***19**, 654, 1980

64.C Tetraphenylphosphonium dioxo − (4 − (2 − pyridylazo) − resorcinolato) − vanadium(v)
$C_{24}H_{20}P^+$, $C_{11}H_7N_3O_4V^-$ Main entry is 84.30

64.C Tetraphenylphosphonium triphenylphosphino − tetracarbonyl − manganese (at −35°C)
$C_{24}H_{20}P^+$, $C_{22}H_{15}MnO_4P^-$ Main entry is 86.79

64.124 Tetraphenylphosphonium bis(azido) − iodine (at −90°C)
$C_{24}H_{20}P^+$, IN_6^-
U.Muller, R.Dubgen, K.Dehnicke
*Z.Anorg.Allg.Chem.,***463**, 7, 1980

64.125 bis(Tetraphenylphosphonium) decabromo − di − tellurium(iv)
$2C_{24}H_{20}P^+$, $Br_{10}Te_2^{2-}$
B.Krebs, K.Buscher *Z.Anorg.Allg.Chem.,***463**, 56, 1980

64.126 bis(Tetraphenylphosphonium) hexa − μ − carbonyl − carbido − heptacarbonyl − hexa − rhodium
$2C_{24}H_{20}P^+$, $C_{14}O_{13}Rh_6^{2-}$
V.G.Albano, D.Braga, P.Chini, S.Martinengo
*Eur.Cryst.Meeting,***5**, 304, 1979

64.127 bis(Tetraphenylphosphonium) dihydrido − dodeca − nickel − carbonyl
$2C_{24}H_{20}P^+$, $C_{21}H_2Ni_{12}O_{21}^{2-}$
R.W.Broach, L.F.Dahl, G.Longoni, P.Chini, A.J.Schultz, J.M.Williams *Adv.Chem.Ser.,***93**, 1978

64.128 bis(Tetraphenylphosphonium) dihydrido − dodeca − nickel − carbonyl (neutron study)
$2C_{24}H_{20}P^+$, $C_{21}H_2Ni_{12}O_{21}^{2-}$
R.W.Broach, L.F.Dahl, G.Longoni, P.Chini, A.J.Schultz, J.M.Williams *Adv.Chem.Ser.,***93**, 1978

64.129 Tetraphenylphosphonium bis(bis − μ − thio − dichloro − iron) − molybdenum
$2C_{24}H_{20}P^+$, $Cl_4Fe_2MoS_4^{2-}$
D.Coucouvanis, N.C.Baenziger, E.D.Simhon, P.Stremple, D.Swenson, A.Simopoulos, A.Kostikas, V.Petrouleas, V.Papaefthymiou *J.Am.Chem.Soc.,***102**, 1732, 1980

64.130 Tetraphenylphosphonium pentasulfur − iron − bis(μ − sulfido) − molybdenum − disulfide dimethylformamide solvate
$2C_{24}H_{20}P^+$, $FeMoS_9^{2-}$, $0.5C_3H_7NO$
D.Coucouvanis, N.C.Baenziger, E.D.Simhon, P.Stremple, D.Swenson, A.Kostikas, A.Simopoulos, V.Petrouleas, V.Papaefthymiou *J.Am.Chem.Soc.,***102**, 1730, 1980

64.131 Tetraphenylphosphonium pentasulfur − iron − bis(μ − sulfido) − tungsten − disulfide dimethylformamide solvate
$2C_{24}H_{20}P^+$, FeS_9W^{2-}, $0.5C_3H_7NO$
D.Coucouvanis, N.C.Baenziger, E.D.Simhon, P.Stremple, D.Swenson, A.Kostikas, A.Simopoulos, V.Petrouleas, V.Papaefthymiou *J.Am.Chem.Soc.,***102**, 1730, 1980

64.132 Tetraphenylphosphonium platinum − carbonyl acetonitrile solvate
$4C_{24}H_{20}P^+$, $4C_2H_3N$, $C_{22}O_{22}Pt_{19}^{4-}$
D.M.Washecheck, E.J.Wucherer, L.F.Dahl, A.Ceriotti, G.Longoni, M.Manassero, M.Sansoni, P.Chini
*J.Am.Chem.Soc.,***101**, 6110, 1979

64.133 Tetraphenylphosphonium (μ^2 − sulfonyl) − (μ^4 − dithio) − octacyano − di − molybdenum hexahydrate
$4C_{24}H_{20}P^+$, $C_8Mo_2N_8O_2S_3^{4-}$, $6H_2O$
C.Potvin, J.−M.Bregeault, J.−M.Manoli
*J.Chem.Soc.,Chem.Commun.,*664, 1980

64.134 Triphenyl − (1 − ethoxycarbonyl − 2 − methyl − 2 − hydroxyvinyl) − phosphonium triphenyl − (1 − ethoxycarbonyl − 2 − methyl − 2 − olato − vinyl) − phosphonium tetrafluoroborate
$C_{24}H_{24}O_3P^+$, $C_{24}H_{23}O_3P$, BF_4^-
M.Yu.Antipin, A.E.Kalinin, Yu.T.Struchkov, I.M.Aladzheva, T.A.Mastryukova, M.I.Kabachnik
*Zh.Strukt.Khim.,***20**, 638, 1979

64.135 5,5,7 − Trimethyl − 2,2 − diphenyl − 7 − (trifluoromethyl) − 3,4 − bis(methoxycarbonyl) − 6 − oxa − 1 − aza − 5 − phosphabicyclo(3.2.0)hept − 3 − ene
$C_{24}H_{25}F_3NO_5P$
W.S.Sheldrick, D.Schomburg, A.Schmidpeter, T.von Criegern *Chem.Ber.,***113**, 55, 1980

64.136 t − Butyl − bis(2 − (6 − t − butyl − tetrahydro − 4H − 1,3 − dioxa − 6 − aza − 2 − phosphocine − 2 − oxy)ethyl)amine
$C_{24}H_{51}N_3O_8P_2$
J.Devillers, D.Houalla, J.−J.Bonnet, R.Wolf
*Nouv.J.Chim.,***4**, 179, 1980

64.137 10 − (4 − Bromobenzyl) − 10 − phenyl − phenoxaphosphonium bromide
$C_{25}H_{19}BrOP^+$, Br^-
D.W.Allen, I.W.Nowell, P.E.Walker
*Phosphorus and Sulfur,***7**, 309, 1979

64.138 N – (Triphenylphosphoranylidene) – benzamide
$C_{25}H_{20}NOP$
I.Bar, J.Bernstein *Acta Crystallogr.,Sect.B*,**36**, 1962, 19

64.C bis(Triphenylbenzylphosphonium) tetrachloro –
cadmium dichloroethane clathrate
$2C_{25}H_{22}P^+, 2C_2H_4Cl_2, CdCl_4^{2-}$ Main entry is 61.8

64.139 4 – endo – Bromomethyl – 4,5 – dimethyl – 1,2,2 –
triphenyl – 3,2 – oxaphosphonia – bicyclo(3.1.0)
hexane tribromide
$C_{25}H_{25}BrOP^+, Br_3^-$
G.Maas, R.Hoge *Justus Liebigs Ann.Chem.*,1028, 1980

64.140 2,10 – Diphenyl – 6 – methyl – 6 – aza – 2,10 –
diphosphabicyclo(9.4.0)pentadeca – 11(1),12,14 –
triene (at –35°C)
$C_{25}H_{29}NP_2$
E.P.Kyba, A.M.John, S.B.Brown, C.W.Hudson, M.J.McPhaul,
A.Harding, K.Larsen, S.Niedzwiecki, R.E.Davis
J.Am.Chem.Soc.,**102**, 139, 1980

64.141 racemic – 7 – Benzyl – 4 – t – butyl – 9 –
hydroxy – 8 – phenyl – 7 – phospha – cis –
bicyclo(4.3.0)non – 8 – ene – 7 – oxide
$C_{25}H_{31}O_2P$
Z.Galdecki, M.L.Glowka
Acta Crystallogr.,Sect.B,**36**, 2188, 1980

64.142 bis(Diphenylphosphino – thioyl) – acetylene
$C_{26}H_{20}P_2S_2$
O.Orama, M.Karhu, R.Uggla
Cryst.Struct.Commun.,**8**, 905, 1979

64.143 bis(Diphenylphosphino – selenoyl) – acetylene
$C_{26}H_{20}P_2Se_2$
O.Orama, K.Nieminen, M.Karhu, R.Uggla
Cryst.Struct.Commun.,**8**, 909, 1979

64.C tris(1,2 – bis(Diphenylphosphino – selenoyl)ethane)
p – xylene clathrate
$3C_{26}H_{24}P_2Se_2, C_8H_{10}$ Main entry is 61.9

64.144 4',5',6',7' – Tetrachloro – 2,8,15 – (5,10(1,2) –
benzenophosphazine – 10,2' – (1,3,2) – benzo –
dioxaphosphole)
$C_{27}H_{18}Cl_4NO_2P$
D.Hellwinkel, W.Blaicher, W.Krapp, W.S.Sheldrick
Chem.Ber.,**113**, 1406, 1980

64.145 α – (Triphenylphosphonium) – benzoylacetone
tetrafluoroborate
$C_{28}H_{24}O_2P^+, BF_4^-$
M.Yu.Antipin, A.E.Kalinin, Yu.T.Struchkov, I.M.Aladzheva,
T.A.Mastryukova, M.I.Kabachnik
Zh.Strukt.Khim.,**20**, 473, 1979

64.146 1,3 – Trimethylene – 1,3 – bis(diphenyl) –
carbodiphosphorane (at –20°C)
$C_{28}H_{26}P_2$
H.Schmidbaur, T.Costa, B.Milewski–Mahrla, U.Schubert
Angew.Chem.,Int.Ed.Engl.,**19**, 555, 1980

64.147 2,6,10 – Triphenyl – 6 – aza – 2,10 –
diphosphabicyclo(9.4.0)pentadeca – 11(1),12,14 –
triene acetone solvate (at –35°C)
$C_{30}H_{31}NP_2, C_3H_6O$
E.P.Kyba, A.M.John, S.B.Brown, C.W.Hudson, M.J.McPhaul,
A.Harding, K.Larsen, S.Niedzwiecki, R.E.Davis
J.Am.Chem.Soc.,**102**, 139, 1980

64.148 2,6,10 – Triphenyl – 2,6,10 – triphosphabicyclo(9.4.0)
pentadeca – 11(1),12,14 – triene (at –35°C)
$C_{30}H_{31}P_3$
E.P.Kyba, A.M.John, S.B.Brown, C.W.Hudson, M.J.McPhaul,
A.Harding, K.Larsen, S.Niedzwiecki, R.E.Davis
J.Am.Chem.Soc.,**102**, 139, 1980

64.149 (E) – 1,4 – Dibenzyl – 1,4 – diphenyl – 1,4 –
diphosphonio – cyclohexane dibromide
$C_{30}H_{32}P_2^{2+}, 2Br^-$
A.D.Rae, J.P.Beale, M.J.Gallagher
Am.Cryst.Assoc.,Ser.2,**7**, 31, 1980

64.150 Sodium (diphenylphosphonium – benzylide) –
(diphenylphosphino – methylide) tetrahydrofuran
solvate
$C_{32}H_{27}P_2^-, Na^+, C_4H_8O$
H.Schmidbaur, U.Deschler, B.Zimmer–Gasser,
D.Neugebauer, U.Schubert *Chem.Ber.*,**113**, 902, 1980

64.151 1 – p – Chlorophenyl – 3 – p – tolyl – 2 –
triphenylphosphazeno – propenone
$C_{34}H_{27}ClNOP$
G.Weber *Cryst.Struct.Commun.*,**9**, 879, 1980

64.152 bis(Triphenylphosphine)iminium (μ – hydrido) –
bis(tricarbonyl – nickel)
$C_{36}H_{30}NP_2^+, C_6HNi_2O_6^-$
G.Longoni, M.Manassero, M.Sansoni
J.Organomet.Chem.,**174**, C41, 1979

64.153 bis(Triphenylphosphine)immonium μ – deuterido –
bis(pentacarbonyl – chromium) (neutron study,
at 17°K)
$C_{36}H_{30}NP_2^+, C_{10}Cr_2DO_{10}^-$
J.L.Petersen, R.K.Brown, J.M.Williams, R.K.McMullan
Inorg.Chem.,**18**, 3493, 1979

64.154 bis((Triphenylphosphine)immonium) carbonyl –
(tri – μ – carbonyl) – tris(tricarbonyl –
ruthenium) – cobalt
$C_{36}H_{30}NP_2^+, C_{13}CoO_{13}Ru_3^-$
P.C.Steinhardt, W.L.Gladfelter, A.D.Harley, J.R.Fox,
G.L.Geoffroy *Inorg.Chem.*,**19**, 332, 1980

64.155 bis(Triphenylphosphine)iminium (μ^2 – carbonyl) –
(μ^2 – hydrido) – tricarbonyl – iron –
nonacarbonyl – tri – ruthenium (neutron study)
$C_{36}H_{30}NP_2^+, C_{13}HFeO_{13}Ru_3^-$
G.L.Geoffroy *Am.Cryst.Assoc.,Ser.2*,**8**, 28, 1980

64.156 bis(Triphenylphosphine)iminium (μ^2 – carbonyl) –
(μ^2 – hydrido) – hexacarbonyl – di – iron –
hexacarbonyl – di – ruthenium
$C_{36}H_{30}NP_2^+, C_{13}HFe_2O_{13}Ru_2^-$
G.L.Geoffroy *Am.Cryst.Assoc.,Ser.2*,**8**, 28, 1980

64.C bis(Triphenylphosphine)immonium 3,3,3,3 –
tetracarbonyl – 5 – phenyl – 1 – oxa – 3 –
mangana – cyclopenta – 2,4 – dione (at –160°C)
$C_{36}H_{30}NP_2^+, C_{13}H_6MnO_7^-$ Main entry is 71.45

64.157 bis(Triphenylphosphine)immonium nona – μ –
carbonyl – nitrido – hexa(carbonyl – cobalt)
$C_{36}H_{30}NP_2^+, C_{15}Co_6NO_{15}^-$
S.Martinengo, G.Ciani, A.Sironi, B.T.Heaton, J.Mason
J.Am.Chem.Soc.,**101**, 7095, 1979

64.158 Bis(triphenylphosphine)iminium (μ –
pentacarbonyl) – decacarbonyl – penta – rhodium
$C_{36}H_{30}NP_2^+$, $C_{15}O_{15}Rh_5^-$
A.Fumagalli, T.F.Koetzle, F.Takusagawa, P.Chini,
S.Martinengo, B.T.Heaton
J.Am.Chem.Soc., **102**, 1740, 1980

64.C bis(Triphenylphosphine)immonium (tricarbonyl –
iron) – bis(μ – diphenylphosphido) – (acetyl –
dicarbonyl – iron)
$C_{36}H_{30}NP_2^+$, $C_{31}H_{23}Fe_2O_6P_2^-$ Main entry is 71.232

64.159 bis(Triphenylphosphine)immonium oxo –
tetrachloro – technetium(v)
$C_{36}H_{30}NP_2^+$, Cl_4OTc^-
F.A.Cotton, A.Davison, V.W.Day, L.D.Gage, H.S.Trop
Inorg.Chem., **18**, 3024, 1979

64.160 bis(Triphenylphosphine)immonium tetracyano –
oxo – dioxo – molybdenum
$2C_{36}H_{30}NP_2^+$, $C_4MoN_4O_3^{2-}$
H.Arzoumanian, R.Lai, R.L.Alvarez, J.–F.Petrignani,
J.Metzger, J.Fuhrhop *J.Am.Chem.Soc.*, **102**, 845, 1980

64.161 bis(bis(Triphenylphosphine)iminium)
tridecacarbonyl – tetra – iron
$2C_{36}H_{30}NP_2^+$, $C_{13}Fe_4O_{13}^{2-}$
C.B.Knobler, G.van Buskirk, H.D.Kaesz
Am.Cryst.Assoc., *Ser.2*, **8**, 22, 1980

64.162 bis(bis(Triphenylphosphino)iminium)
heptacarbonyl – μ_2 – hexacarbonyl – ruthenium –
tetra – iridium
$2C_{36}H_{30}NP_2^+$, $C_{13}Ir_4O_{13}Ru^{2-}$
A.Fumagalli, F.Takusagawa, T.F.Koetzle, P.Chini,
G.Longoni, S.Martinengo, B.T.Heaton
Am.Cryst.Assoc., *Ser.2*, **7**, 16, 1980

64.163 bis(bis(Triphenylphosphine)iminium)
nonacarbonyl – hexa – μ_2 – carbonyl –
ruthenium – tetra – iridium
$2C_{36}H_{30}NP_2^+$, $C_{15}Ir_4O_{15}Ru^{2-}$
A.Fumagalli, F.Takusagawa, T.F.Koetzle, P.Chini,
G.Longoni, S.Martinengo, B.T.Heaton
Am.Cryst.Assoc., *Ser.2*, **7**, 16, 1980

64.164 bis(bis(Triphenylphosphine)immonium) dihydrido –
dodeca – nickel – carbonyl
$2C_{36}H_{30}NP_2^+$, $C_{21}H_2Ni_{12}O_{21}^{2-}$
R.W.Broach, L.F.Dahl, G.Longoni, P.Chini, A.J.Schultz,
J.M.Williams *Adv.Chem.Ser.*, 93, 1978

64.165 bis(bis(Triphenylphosphine)immonium) octa –
osmium – docosacarbonyl
$2C_{36}H_{30}NP_2^+$, $C_{22}O_{22}Os_8^{2-}$
P.F.Jackson, B.F.G.Johnson, J.Lewis, P.R.Raithby
J.Chem.Soc.,Chem.Commun., 60, 1980

64.166 bis(bis(Triphenylphosphine)immonium) carbido –
deca – osmium – carbonyl
$2C_{36}H_{30}NP_2^+$, $C_{25}O_{24}Os_{10}^{2-}$
P.F.Jackson, B.F.G.Johnson, J.Lewis, M.McPartlin,
W.J.H.Nelson *J.Chem.Soc.,Chem.Commun.*, 224, 1980

64.C tris(Triphenylphosphineiminium) sodium (μ^3 –
oxo) – tris(μ^2 – methoxy) – tris(dicarbonyl –
nitroso – molybdenum)
$3C_{36}H_{30}NP_2^+$, $2C_9H_9Mo_3N_3O_{13}^{2-}$, Na^+ Main entry is 84.20

64.167 bis(Triphenylphosphine – imino) – tetrasulfur –
tetranitride
$C_{36}H_{30}N_6P_2S_4$
J.Bojes, T.Chivers, G.MacLean, R.T.Oakley, A.W.Cordes
Can.J.Chem., **57**, 3171, 1979

64.C tris(Diphenyl – dithiophosphinato) – antimony(iii)
$C_{36}H_{30}P_3S_6Sb$ Main entry is 66.19

64.168 bis(2,4,6 – Tri – t – butylphenyl) – phosphinic –
chloride
$C_{36}H_{58}ClOP$
M.Yoshifuji, I.Shima, N.Inamoto, K.Hirotsu, T.Higuchi
Angew.Chem.,Int.Ed.Engl., **19**, 399, 1980

64.169 1 – Phenyl – 2,3 – bis(diphenylphosphino) –
naphthalene
$C_{40}H_{30}P_2$
A.J.Carty, N.J.Taylor, D.K.Johnson
J.Am.Chem.Soc., **101**, 5422, 1979
See also R1 : 24

64.170 2,2' – bis(o – (Oxo – diphenylphosphino) –
phenyloxy) – diethyl ether D,L – phenylglycinium
ethyl ester trifluoromethylsulfonate
$C_{40}H_{36}O_5P_2$, $C_{10}H_{14}NO_2^+$, $CF_3O_3S^-$
A.H.Alberts, K.Timmer, J.G.Noltes, A.L.Spek
J.Am.Chem.Soc., **101**, 3375, 1979
See also R2 : 48

64.171 Octa(3,5 – dimethylpyrazolyl) –
cyclotetraphosphazene
$C_{40}H_{56}N_{20}P_4$
K.D.Gallicano, N.L.Paddock, S.J.Rettig, J.Trotter
Inorg.Nucl.Chem.Lett., **15**, 417, 1979
See also R1 : 32

64.172 Trihydroxyphosphorus – tetraphenylporphyrin
dihydrate
$C_{44}H_{31}N_4O_3P$, $2H_2O$
S.Mangani, D.Cullen, E.Meyer
Am.Cryst.Assoc., *Ser.2*, **7**, 28, 1980
See also R1 : 49

ARSENIC COMPOUNDS

65.C Guanidinium hexamolybdo – methylarsonate hexahydrate
$CH_{15}AsMo_6O_{27}{}^{2-}$, $2CH_6N_3{}^+$, $6H_2O$ Main entry is 84.2

65.1 1,3,5,7 – Tetra – arsa – 2,4,6,8 – tetraoxa – adamantane
$C_2H_4As_4O_4$
J.Kopf, K.von Deuten, G.Klar
Inorg.Chim.Acta,**38**, 67, 1980

65.2 Guanidinium tetramolybdo – dimethylarsinate monohydrate
$C_2H_7AsMo_4O_{15}{}^{2-}$, $2CH_6N_3{}^+$, H_2O
K.Y.Matsumoto *Bull.Chem.Soc.Jpn.*,**52**, 3284, 1979

65.3 1 – Methylarsa – 2,3,4,5 – bis(o – carborano – cyclopentane)
$C_5H_{23}AsB_{20}$
A.I.Yanovskii, N.G.Furmanova, Yu.T.Struchkov, N.F.Shemyakin, L.I.Zakharkin
Izv.Akad.Nauk SSSR,Ser.Khim.,1523, 1979
See also R1 : 62

65.4 Triethylarsine – sulfide
$C_6H_{15}AsS$
V.E.Zavodnik, V.K.Bel'skii, I.P.Gol'dshtein
Zh.Strukt.Khim.,**20**, 152, 1979

65.5 3,5,7 – tris(Trimethylsilyl) – tricyclo(2.2.1.0^{2,6}) hepta – arsane
$C_9H_{27}As_7Si_3$
H.G.von Schnering, D.Fenske, W.Honle, M.Binnewies, K.Peters *Angew.Chem.,Int.Ed.Engl.*,**18**, 679, 1979
See also R1 : 63

65.6 3,7 – Diphenyl – 3H,7H – 1,5,2,4,6,8,3,7 – dithia(1,5S) tetra – azadiarsocine (at –100°C)
$C_{12}H_{10}As_2N_4S_2$
N.W.Alcock, E.M.Holt, J.Kuyper, J.J.Mayerle, G.B.Street
Inorg.Chem.,**18**, 2235, 1979

65.C 2,2',5,5' – Tetramethyl – 1,1' – diarsa – ferrocene
$C_{12}H_{16}As_2Fe$ Main entry is 73.38

65.C N – (Trimethyl – tin) – dimethylarsino – bis(carbomethoxy) – pyrazole (at –50°C)
$C_{12}H_{21}AsN_2O_4Sn$ Main entry is 69.23

65.7 tris(Morpholino) – arsine (constrained refinement, at –150°C)
$C_{12}H_{24}AsN_3O_3$
C.Romming, J.Songstad
Acta Chem.Scand.Ser.A,**34**, 365, 1980

65.8 Dodecamethyl – hexasila – tetra – arsa – adamantane
$C_{12}H_{36}As_4Si_6$
W.Honle, H.G.von Schnering
Z.Naturforsch.,Teil B,**35**, 789, 1980
See also R1 : 63

65.9 2 – Methyl – 2,2' – spiro – bis(1,3,2λ^5 – benzoxazarsoline)
$C_{13}H_{13}AsN_2O_2$
H.Wunderlich *Acta Crystallogr.,Sect.B*,**36**, 1492, 1980

65.10 3,7 – Dimesityl – 3H,7H – 1,5,2,4,6,8,3,7 – dithia – tetra – azadiarsocine (at –100°C)
$C_{18}H_{22}As_2N_4S_2$
N.W.Alcock, E.M.Holt, J.Kuyper, J.J.Mayerle, G.B.Street
Inorg.Chem.,**18**, 2235, 1979

65.11 Arsatriptycene
$C_{19}H_{13}As$
F.J.M.Freijee, C.H.Stam
Acta Crystallogr.,Sect.B,**36**, 1247, 1980

65.12 9,10 – Dihydro – 9 – methyl – 9,10 – o – benzeno – 9 – arsonia – anthracene chloride monohydrate
$C_{20}H_{16}As^+$, Cl^-, H_2O
F.Smit, C.H.Stam
Acta Crystallogr.,Sect.B,**36**, 1254, 1980

65.C (μ² – Dimethylarsonio) – bis(dicarbonyl – cyclopentadienyl – trimethylphosphine) – molybdenum tricarbonyl – cyclopentadienyl – molybdenum
$C_{22}H_{34}AsMo_2O_4P_2{}^+$, $C_8H_5MoO_3{}^-$ Main entry is 73.95

65.C Tetraphenylarsonium difulminato – gold
$C_{24}H_{20}As^+$, $C_2AuN_2O_2{}^-$ Main entry is 71.1

65.C Tetraphenylarsonium pentachloro – N – pentachloroethyl – nitrido – tungsten
$C_{24}H_{20}As^+$, $C_2Cl_{10}NW^-$ Main entry is 83.1

65.C Tetraphenylarsonium (μ – dimethylphosphido) – bis(dicarbonyl – cyclopentadienyl – molybdenum)
$C_{24}H_{20}As^+$, $C_{18}H_{16}Mo_2O_4P^-$ Main entry is 73.59

65.13 Tetraphenylarsonium (μ_6 – hydrido) – hexakis(tricarbonyl – ruthenium) (neutron study)
$C_{24}H_{20}As^+$, $C_{18}HO_{18}Ru_6{}^-$
P.F.Jackson, B.F.G.Johnson, J.Lewis, P.R.Raithby, M.McPartlin, W.J.H.Nelson, K.D.Rouse, J.Allibon, S.A.Mason
J.Chem.Soc.,Chem.Commun.,295, 1980

65.14 Tetraphenylarsonium tetrachloro – oxo – vanadium(v)
$C_{24}H_{20}As^+$, Cl_4OV^-
G.Beindorf, J.Strahle, W.Liebelt, K.Dehnicke
Z.Naturforsch.,Teil B,**35**, 522, 1980

65.C bis(Tetraphenylarsonium) tetrachloro – oxo – methylnitrile – molybdenum tetrachloro – oxo – molybdenum
$2C_{24}H_{20}As^+$, $C_2H_3Cl_4MoNO^-$, Cl_4MoO^- Main entry is 83.5

65.C bis(Tetraphenylarsonium) tris(2,2 – diselenido – 1,1 – ethylene – dicarbonitrile) – nickel(iv)
$2C_{24}H_{20}As^+$, $C_{12}N_6NiSe_6{}^{2-}$ Main entry is 85.47

65.15 bis(Tetraphenylarsonium) μ_6 – carbido – tetrakis(μ_2 – carbonyl) – hexakis(dicarbonyl – ruthenium)
$2C_{24}H_{20}As^+$, $C_{17}O_{16}Ru_6{}^{2-}$
B.F.G.Johnson, J.Lewis, S.W.Sankey, K.Wong, M.McPartlin, W.J.H.Nelson *J.Organomet.Chem.*,**191**, 63, 1980

65.16 bis(Tetraphenylarsonium) dihydrido – dodeca – nickel – carbonyl
$2C_{24}H_{20}As^+$, $C_{21}H_2Ni_{12}O_{21}{}^{2-}$
R.W.Broach, L.F.Dahl, G.Longoni, P.Chini, A.J.Schultz, J.M.Williams *Adv.Chem.Ser.*,93, 1978

65.17 **Tetraphenylarsonium pentachloro – oxo – niobium(v) methylene chloride solvate**
$2C_{24}H_{20}As^+$, Cl_5NbO^{2-}, $2CH_2Cl_2$
U.Muller, I.Lorenz $Z.Anorg.Allg.Chem.$,**463**, 110, 1980

65.18 **Tetraphenylarsonium μ – selenido – μ – diselenido – bis(tetrachloro – tungsten(v))**
$2C_{24}H_{20}As^+$, $Cl_8Se_3W_2^{2-}$
M.G.B.Drew, G.W.A.Fowles, E.M.Page, D.A.Rice
$J.Am.Chem.Soc.$,**101**, 5827, 1979

65.19 **Tetraphenylarsonium (decachloro – μ – nitrido – di – tungsten)**
$2C_{24}H_{20}As^+$, $Cl_{10}NW_2^{2-}$
F.Weller, W.Liebelt, K.Dehnicke
$Angew.Chem.,Int.Ed.Engl.$,**19**, 220, 1980

65.20 **tris(Tetraphenylarsonium) hydrido – dodeca – nickel – carbonyl acetone solvate**
$3C_{24}H_{20}As^+$, $C_{21}HNi_{12}O_{21}^{3-}$, $2C_3H_6O$
R.W.Broach, L.F.Dahl, G.Longoni, P.Chini, A.J.Schultz,
J.M.Williams $Adv.Chem.Ser.$,93, 1978

65.21 **tris(Tetraphenylarsonium) hydrido – dodeca – nickel – carbonyl acetone solvate (neutron study)**
$3C_{24}H_{20}As^+$, $C_{21}HNi_{12}O_{21}^{3-}$, $2C_3H_6O$
R.W.Broach, L.F.Dahl, G.Longoni, P.Chini, A.J.Schultz,
J.M.Williams $Adv.Chem.Ser.$,93, 1978

65.22 **Cyclo – tetrakis(phenylarsenic – sulfide)**
$C_{24}H_{20}As_4S_4$
R.G.Bergerhoff, H.Namgung $Z.Elektrochem.$,209, 1979

65.23 **9,10 – Dihydro – 10 – phenyl – 9,10 – o – benzeno – 9 – arsa – anthracene**
$C_{25}H_{17}As$
C.van Rooijen–Reiss, C.H.Stam
$Acta\ Crystallogr.,Sect.B$,**36**, 1252, 1980

65.24 **trans – 10 – Benzyl – 9 – phenyl – 9,10 – dihydro – 9 – arsa – anthracene**
$C_{26}H_{21}As$
C.H.Stam $Acta\ Crystallogr.,Sect.B$,**36**, 455, 1980

65.25 **2,4,6,7 – Tetraphenyl – 1 – arsa – 2,3,7 – triaza – 8 – oxabicyclo(3.3.0)oct – 3 – ene benzene solvate**
$C_{27}H_{22}AsN_3O$, $0.5C_6H_6$
A.I.Yanovskii, Yu.T.Struchkov, E.N.Dianova, N.A.Chadaeva,
B.A.Arbuzov $Dokl.Akad.Nauk\ SSSR$,**249**, 120, 1979

65.26 **trans – Dichloro – (2,11 – bis(diphenylarsinomethyl) – benzo(c) phenanthrene) – platinum**
$C_{44}H_{34}As_2Cl_2Pt$
G.Balimann, L.M.Venanzi, F.Bachechi, L.Zambonelli
$Helv.Chim.Acta.$**63**, 420, 1980
See also R1 : 29

ANTIMONY AND BISMUTH COMPOUNDS

66.1 **μ – Fluoro – μ – trifluoroacetato – bis(tetrafluoro – antimony(v))**
$C_2F_{12}O_2Sb_2$
D.P.Bullivant, M.F.A.Dove, M.J.Haley
$J.Chem.Soc.,Dalton\ Trans.$,109, 1980

66.2 **Sodium fluoro – oxo – dioxalato – antimony oxonium hydrate**
$C_4FO_9Sb^{4-}$, $3Na^+$, H_3O^+, H_2O
R.Fourcade, P.Escande, B.Ducourant, G.Mascherpa
$Z.Anorg.Allg.Chem.$,**465**, 34, 1980

66.3 **μ – Oxo – di – μ – trifluoroacetato – bis(trifluoro – antimony(v))**
$C_4F_{12}O_5Sb_2$
D.P.Bullivant, M.F.A.Dove, M.J.Haley
$J.Chem.Soc.,Dalton\ Trans.$,109, 1980

66.4 **2 – Chloro – 1,3,6 – trithia – 2 – stibaocane**
$C_4H_8ClS_3Sb$
M.Drager, R.Engler $Z.Anorg.Allg.Chem.$,**405**, 183, 1974

66.5 **Tetrachloro – antimony(v) – dimethoxyphosphate**
$C_4H_{12}Cl_8O_8P_2Sb_2$
M.Brauninger, W.Schwarz, A.Schmidt
$Z.Naturforsch.,Teil\ B$,**34**, 1703, 1979
See also R1 : 46

66.6 **(R) – tris(Trifluoroacetato) – antimony(iii) (absolute configuration)**
$C_6F_9O_6Sb$
D.P.Bullivant, M.F.A.Dove, M.J.Haley
$J.Chem.Soc.,Dalton\ Trans.$,105, 1980

66.7 **tris(Monothioacetato) – antimony(iii)**
$C_6H_9O_3S_3Sb$
M.Hall, D.B.Sowerby
$J.Chem.Soc.,Dalton\ Trans.$,1292, 1980

66.8 **Triacetato – antimony(iii)**
$C_6H_9O_6Sb$
M.Hall, D.B.Sowerby
$J.Chem.Soc.,Dalton\ Trans.$,1292, 1980

66.9 **Tetrachloro – antimony(v) – dimethylamino – methoxyphosphate (at $-100°C$)**
$C_6H_{18}Cl_8N_2O_6P_2Sb_2$
M.Brauninger, W.Schwarz, A.Schmidt
$Z.Naturforsch.,Teil\ B$,**34**, 1703, 1979
See also R1 : 46

66.10 **2,2′ – Bipyridyl – (trichloro – antimony)**
$C_{10}H_8Cl_3N_2Sb$
L.Korte, A.Lipka, D.Mootz $Eur.Cryst.Meeting$,5, 92, 1979

66.11 **μ – Oxo – bis(trichloroacetylacetonato – antimony(v)) (at 173°K)**
$C_{10}H_{14}Cl_6O_5Sb_2$
S.Blosl, W.Schwarz, A.Schmidt
$Z.Naturforsch.,Teil\ B$,**34**, 1711, 1979

66.12 Antimony – tris(O,O – diethylphosphoro – dithioate)
$C_{12}H_{30}O_6P_3S_6Sb$
R.O.Day, M.M.Chauvin, W.E.McEwen
Phosphorus and Sulfur,**8**, 121, 1980

66.13 Acetato – diphenyl – antimony(iii)
$C_{14}H_{13}O_2Sb$
S.P.Bone, D.B.Sowerby
J.Organomet.Chem.,**184**, 181, 1980

66.14 Tri – p – tolyl – antimony
$C_{21}H_{21}Sb$
A.N.Sobolev, I.P.Romm, V.K.Belsky, E.N.Guryanova
J.Organomet.Chem.,**179**, 153, 1979

66.15 Tetraphenyl – distibine
$C_{24}H_{20}Sb_2$
K.von Deuten, D.Rehder
Cryst.Struct.Commun.,**9**, 167, 1980

66.16 Aquo – triphenyl – catecholato – antimony
triphenyl – catecholato – antimony
$C_{24}H_{21}O_3Sb$, $C_{24}H_{19}O_2Sb$
M.Hall, D.B.Sowerby *J.Am.Chem.Soc.*,**102**, 628, 1980

66.17 bis(t – Butylperoxy) – triphenyl – antimony
$C_{26}H_{33}O_4Sb$
Z.A.Starikova, T.M.Shchegoleva, V.K.Trunov,
I.E.Pokrovskaya, E.N.Kanunnikova
Kristallografiya,**24**, 1211, 1979

66.C bis(μ^2 – Chloro) – tetrakis(dicarbonyl –
cyclopentadienyl – manganese) – dibismuth
(at –80°C)
$C_{28}H_{20}Bi_2Cl_2Mn_4O_8$ Main entry is 73.110

66.18 2,3 – bis(Diphenylstibino) – maleic anhydride
$C_{28}H_{20}O_3Sb_2$
D.Fenske, H.Teichert, H.Prokscha, W.Renz, H.J.Becher
Monatsh.Chem.,**111**, 177, 1980

66.19 tris(Diphenyl – dithiophosphinato) – antimony(iii)
$C_{36}H_{30}P_3S_6Sb$
M.J.Begley, D.B.Sowerby, I.Haiduc
J.Chem.Soc.,Chem.Commun.,64, 1980
See also R1 : 64

GROUPS IA AND IIA COMPOUNDS

67.1 ((S) – Malato) – tetra – aqua – magnesium hydrate
$C_4H_{12}MgO_9$, H_2O
A.Karipides *Inorg.Chem.*,**18**, 3034, 1979

67.2 Dimethyl – (N,N,N′,N′ –
tetramethylethylenediamine) – magnesium
$C_8H_{22}MgN_2$
T.Greiser, J.Kopf, D.Thoennes, E.Weiss
J.Organomet.Chem.,**191**, 1, 1980

67.3 bis(2 – Methyl – 8 – hydroxyquinolinato) –
beryllium dihydrate
$C_{20}H_{16}BeN_2O_2$, $2H_2O$
J.C.van Niekerk, H.M.N.H.Irving, L.R.Nassimbeni
S.Afr.J.Chem.,**32**, 85, 1979

67.4 Magnesium bis(2,4 – dinitrophenoxide) bis(N –
methylimidazole)
$C_{20}H_{18}MgN_6O_{10}$
R.Sarma, F.Ramirez, P.Narayanan, B.McKeever,
J.F.Marecek *J.Am.Chem.Soc.*,**101**, 5015, 1979

67.5 bis(2,4 – Dimethyl – 2,4 – pentadienyl) – (N,N,N′,N′ –
tetramethylethylenediamine) – magnesium
$C_{20}H_{38}MgN_2$
H.Yasuda, M.Yamauchi, A.Nakamura, T.Sei, Y.Kai,
N.Yasuoka, N.Kasai *Bull.Chem.Soc.Jpn.*,**53**, 1089, 1980

67.6 Magnesium bis(2,4 – dinitrophenoxide) bis(pyridine)
$C_{22}H_{16}MgN_6O_{10}$
R.Sarma, F.Ramirez, P.Narayanan, B.McKeever,
J.F.Marecek *J.Am.Chem.Soc.*,**101**, 5015, 1979

67.7 2 – Methyl – 1,3 – dithio – 2 – yl – (N,N,N′,N′ –
tetramethylethylenediamine) – lithium dimer
(at –150°C)
$C_{22}H_{50}Li_2N_4S_4$
R.Amstutz, D.Seebach, P.Seiler, B.Schweizer, J.D.Dunitz
Angew.Chem.,Int.Ed.Engl.,**19**, 53, 1980

67.8 μ_6 – Nitrido – nonakis(μ_2 – t – butyl – amino) –
hexa – magnesium
$C_{36}H_{90}Mg_6N_{10}$
G.Dozzi, G.Del Piero, M.Cesari, S.Cucinella
J.Organomet.Chem.,**190**, 229, 1980

67.9 bis((μ – 2,6 – Di – t – butyl – 4 – methylphenoxy) –
diethyloxy – lithium)
$C_{38}H_{66}Li_2O_4$
B.Cetinkaya, I.Gumrukcu, M.F.Lappert, J.L.Atwood,
R.Shakir *J.Am.Chem.Soc.*,**102**, 2086, 1980

GROUP III COMPOUNDS

68.1 Cesium di – iodo – dimethyl – aluminium p – xylene solvate
$C_2H_6AlI_2^-$, Cs^+, C_8H_{10}
R.D.Rogers, J.L.Atwood *J.Cryst.Mol.Struct.*,**9**, 45, 1979

68.2 Tetramethylammonium iodotrimethyl – aluminium
$C_3H_9AlI^-$, $C_4H_{12}N^+$
R.D.Rogers, L.B.Stone, J.L.Atwood
Cryst.Struct.Commun.,**9**, 143, 1980

68.3 Cesium isothiocyanato – trimethyl – aluminium
$C_4H_9AlNS^-$, Cs^+
R.Shakir, M.J.Zaworotko, J.L.Atwood
J.Cryst.Mol.Struct.,**9**, 135, 1979

68.C Tetracarbonyl – manganese – ((μ^2 – acetyl) – (μ^2 – bromo) – dibromo – aluminium)
$C_6H_3AlBr_3MnO_5$ Main entry is 71.3

68.4 Dichloro – (N – ethyl – N′,N′ – dimethylethylenediamine) – aluminium
$C_6H_{15}AlCl_2N_2$
M.J.Zaworotko, J.L.Atwood *Inorg.Chem.*,**19**, 268, 1980

68.5 Thallium(i) tetra – acetato – thallium(iii) (at 243°K)
$C_8H_{12}O_8Tl^-$, Tl^+
I.D.Brown, R.Faggiani
Acta Crystallogr.,Sect.B,**36**, 1802, 1980

68.6 bis(Dimethyl – gallium) – N,N′ – dimethyl – dithio – oxamide (at –100°C)
$C_8H_{18}Ga_2N_2S_2$
T.Halder, H.–D.Hausen, J.Weidlein
Z.Naturforsch.,Teil B,**35**, 773, 1980

68.C cis – bis(2,2′ – Bipyridyl) – dimethyl – cobalt(iii) tetraethyl – aluminium
$C_8H_{20}Al^-$, $C_{22}H_{22}CoN_4^+$ Main entry is 71.179

68.7 (N,N′ – Dimethylacetylhydrazino – 0,N) – dimethyl – gallium – trimethyl – gallium (at –100°C)
$C_9H_{24}Ga_2N_2O$
F.Gerstner, H.–D.Hausen, J.Weidlein
J.Organomet.Chem.,**197**, 135, 1980

68.8 N – Methyl – salicylaldiminato – dimethyl – gallium
$C_{10}H_{14}GaNO$
V.I.Bregadze, N.G.Furmanova, L.M.Golubinskaya,
O.Y.Kompan, Yu.T.Struchkov, V.A.Bren, Zh.V.Bren,
A.E.Lyubarskaya, V.I.Minkin, L.M.Sitkina
J.Organomet.Chem.,**192**, 1, 1980

68.9 (N,N′,N″,N‴ – Tetramethyl – oxalamidinato) – bis(dimethyl – aluminium)
$C_{10}H_{24}Al_2N_4$
F.Gerstner, W.Schwarz, H.–D.Hausen, J.Weidlein
J.Organomet.Chem.,**175**, 33, 1979

68.10 (N,N′,N″,N‴ – Tetramethyl – oxalamidinato) – bis(dimethyl – gallium) (at –100°C)
$C_{10}H_{24}Ga_2N_4$
F.Gerstner, W.Schwarz, H.–D.Hausen, J.Weidlein
J.Organomet.Chem.,**175**, 33, 1979

68.11 (N,N′,N″,N‴ – Tetramethyl – oxalamidinato) – bis(dimethyl – indium) (at –100°C)
$C_{10}H_{24}In_2N_4$
F.Gerstner, W.Schwarz, H.–D.Hausen, J.Weidlein
J.Organomet.Chem.,**175**, 33, 1979

68.12 (N,N′,N″ – Trimethylacetimidohydrazino – 0,N) – dimethyl – gallium – trimethyl – gallium (at –100°C)
$C_{10}H_{27}Ga_2N_3$
F.Gerstner, H.–D.Hausen, J.Weidlein
J.Organomet.Chem.,**197**, 135, 1980

68.C (Dimethyl – (N,N – dimethylethanolamino) – (3,5 – dimethylpyrazolyl) – gallato) – dinitrosyl – iron
$C_{11}H_{23}FeGaN_5O_3$ Main entry is 84.32

68.C (Dimethyl – (N,N – dimethylethanolamino) – (3,5 – dimethylpyrazolyl) – gallato) – nitrosyl – nickel
$C_{11}H_{23}GaN_4NiO_2$ Main entry is 84.33

68.13 Trichloro – bis(4 – cyanopyridine – N – oxide) – thallium
$C_{12}H_8Cl_3N_4O_2Tl$
E.Gutierrez–Puebla, A.Vegas, S.Garcia–Blanco
Acta Crystallogr.,Sect.B,**36**, 145, 1980

68.14 tris(Aluminium(iii)) pentachloride tetrakis(isopropoxide) (at –120°C)
$C_{12}H_{28}Al_3Cl_5O_4$
A.I.Yanovskii, V.A.Kozunov, N.Ya.Turova, N.G.Furmanova,
Yu.T.Struchkov *Dokl.Akad.Nauk SSSR*,**244**, 119, 1979

68.15 tris(Acetylacetonato) – indium(iii) (orthorhombic form)
$C_{15}H_{21}InO_6$
G.J.Palenik, K.R.Dymock
Acta Crystallogr.,Sect.B,**36**, 2059, 1980

68.16 Dimethyl – (N,N – dimethylethanolamino) – (3,5 – dimethylpyrazolyl) – gallato – (η^2 – thiomethoxymethyl) – dicarbonyl – molybdenum
$C_{15}H_{26}GaMoN_3O_3S$
K.S.Chong, S.J.Rettig, A.Storr, J.Trotter
Can.J.Chem.,**58**, 1080, 1980

68.17 Di – μ – (pyridine – 2 – carbaldehyde – oximato) – bis(dimethyl – indium) benzene solvate
$C_{16}H_{22}In_2N_4O_2$, $0.5C_6H_6$
H.M.M.Shearer, J.Twiss, K.Wade
J.Organomet.Chem.,**184**, 309, 1980

68.18 Di – μ^2 – chloro – tetrakis(trimethylsilylmethyl) – di – thallium(iii)
$C_{16}H_{44}Cl_2Si_4Tl_2$
F.Brady, K.Henrick, R.W.Matthews, D.G.Gillies
J.Organomet.Chem.,**193**, 21, 1980
See also R1 : 63

68.19 2 – (N – Phenylaminomethylene) – 3(2H) – benzo(b) furan – thionato – dimethyl – gallium (at –120°C)
$C_{17}H_{16}GaNOS$
V.I.Bregadze, N.G.Furmanova, L.M.Golubinskaya,
O.Y.Kompan, Yu.T.Struchkov, V.A.Bren, Zh.V.Bren,
A.E.Lyubarskaya, V.I.Minkin, L.M.Sitkina
J.Organomet.Chem.,**192**, 1, 1980

68.20 **Diethyldithiocarbamato – diphenyl – thallium(iii)**
$C_{17}H_{20}NS_2Tl$
R.T.Griffin, K.Henrick, R.W.Matthews, M.McPartlin
J.Chem.Soc.,Dalton Trans.,1550, 1980

68.C **(μ – Chloro) – 1 – (dicyclopentadienyl – zirconium(iv)) – 2,2 – bis(diethyl – aluminium) ethane**
$C_{20}H_{33}Al_2ClZr$ Main entry is 71.159

68.21 **1 – Ethyl – 3 – methyl – 1 – alumina – indan dimer**
$C_{22}H_{30}Al_2$
D.J.Brauer, C.Kruger
Z.Naturforsch.,Teil B,**34**, 1293, 1979

68.22 **Di – μ – chloro – bis((η³ – pentamethyl – cyclopentadienyl) – methyl – aluminium)**
$C_{22}H_{36}Al_2Cl_2$
P.R.Schonberg, R.T.Paine, C.F.Campana
J.Am.Chem.Soc.,**101**, 7726, 1979

68.C **(Dimethyl – bis(3,5 – dimethyl – 1 – pyrazolyl) – gallato) – (dimethyl – (3,5 – dimethyl – 1 – pyrazolyl) – (dimethylethanolamino) – gallato) – nickel(ii)**
$C_{23}H_{43}Ga_2N_7NiO$ Main entry is 83.166

68.C **bis(η⁵ – Cyclopentadienyl) – 2,2 – bis(diethyl – alumino) – ethyl – zirconium – cyclopentadienide**
$C_{25}H_{38}Al_2Zr$ Main entry is 71.197

68.23 **tris(8 – Hydroxyquinolato) – aluminium(iii) methanol solvate**
$C_{27}H_{18}AlN_3O_3, CH_4O$
S.H.Simonsen, D.W.Bechtel
Am.Cryst.Assoc.,Ser.2,**7**, 23, 1980

68.24 **Diphenyl – tropolonato – thallium(iii)**
Di – μ – 2 – hydroxycyclohepta – 2,4,6 – trien – 1 – onato – bis(diphenyl – thallium(iii))
$C_{38}H_{30}O_4Tl_2$
R.T.Griffin, K.Henrick, R.W.Matthews, M.McPartlin
J.Chem.Soc.,Dalton Trans.,1550, 1980

68.25 **Di – μ – chloro – bis(bis(2,3,5,6 – tetrafluorophenyl) – (triphenylphosphine – oxide) – thallium(iii))**
$C_{60}H_{34}Cl_2F_{16}O_2P_2Tl_2$
K.Henrick, M.McPartlin, R.W.Matthews, G.B.Deacon, R.J.Phillips *J.Organomet.Chem.*,**193**, 13, 1980

GERMANIUM, TIN, LEAD COMPOUNDS

69.1 **Tin(ii) oxalate**
$(C_2O_4Sn)_n$
A.D.Christie, R.A.Howie, W.Moser
Inorg.Chim.Acta,**36**, L447, 1979

69.2 **(μ – Oxalato) – tin (at 20°C)**
$(C_2O_4Sn)_n$
A.Gleizes, J.Galy *J.Solid State Chem.*,**30**, 23, 1979

69.3 **Azido – trimethyl – tin**
$(C_3H_9N_3Sn)_n$
R.Allmann, R.Hohlfeld, A.Waskowska, J.Lorberth
J.Organomet.Chem.,**192**, 353, 19

69.4 **(β – Carbomethoxy – ethyl) – trichloro – tin(iv)**
$C_4H_7Cl_3O_2Sn$
P.G.Harrison, T.J.King, M.A.Healy
J.Organomet.Chem.,**182**, 17, 1979

69.5 **bis(Dimethylsulfoxide) – di – iodo – lead(ii)**
$C_4H_{12}I_2O_2PbS_2$
H.Miyamae, Y.Numahata, M.Nagata *Chem.Lett.*,663, 1980

69.6 **Acetylacetonato – iodo – germanium**
$C_5H_7GeIO_2$
S.R.Stobart, M.R.Churchill, F.J.Hollander, W.J.Youngs
J.Chem.Soc.,Chem.Commun.,911, 1979

69.7 **5 – Germyl – cyclopenta – 1,3 – diene (at 160°K)**
C_5H_8Ge
M.J.Barrow, E.A.V.Ebsworth, M.M.Harding, D.W.H.Rankin
J.Chem.Soc.,Dalton Trans.,603, 1980

69.8 **Glycinato – trimethyl – tin(iv)**
$(C_5H_{13}NO_2Sn)_n$
B.Y.K.Ho, K.C.Molloy, J.J.Zuckerman, F.Reidinger, J.A.Zubieta *J.Organomet.Chem.*,**187**, 213, 1980
See also R1 : 48

69.9 **Potassium tris(monochloroacetato) – tin(ii)**
$C_6H_6Cl_3O_6Sn^-, K^+$
S.J.Clark, J.D.Donaldson, J.C.Dewan, J.Silver
Acta Crystallogr.,Sect.B,**35**, 2550, 1979

69.10 **bis(β – Amidoethyl) – dichloro – tin(iv)**
$C_6H_{12}Cl_2N_2O_2Sn$
P.G.Harrison, T.J.King, M.A.Healy
J.Organomet.Chem.,**182**, 17, 1979

69.11 **bis(N,N – Dimethyldithiocarbamato) – lead(ii)**
$C_6H_{12}N_2PbS_4$
H.Iwasaki *Acta Crystallogr.,Sect.B*,**36**, 2138, 1980

69.12 **1,3 – bis(Trimethyl – tin) – 1,3,5,7 – tetra – aza – 2,4,6,8 – tetrathiocin – 2,2 – dioxide benzene solvate**
$C_6H_{18}N_4O_2S_4Sn_2, 0.5C_6H_6$
H.W.Roesky, M.Witt, J.W.Bats, H.Fuess, F.J.B.Calleja, F.Ania
Z.Anorg.Allg.Chem.,**458**, 225, 1979

69.C **2,2,4,4,6,6 – Hexamethylcyclotristanna – tellurane (at 110°K)**
$C_6H_{18}Sn_3Te_3$ Main entry is 70.2

69.13 1 – Trimethyl – tin – methyl – o – carborane (at –120°C)
$C_6H_{22}B_{10}Sn$
N.I.Kirillova, T.V.Klimova, Yu.T.Struchkov, V.I.Stanko
Zh.Strukt.Khim.,**21**, 166, 1980
See also R1 : 62

69.14 Nitrato – p – aminobenzoato – lead(ii)
$(C_7H_6N_2O_5Pb)_n$
I.R.Amiraslanov, N.Kh.Dzhafarov, G.N.Nadzhafov, Kh.S.Mamedov, E.M.Movsumov, B.T.Usubaliev
Zh.Strukt.Khim.,**21**, 137, 1980

69.15 Trimethyl – lead – diazo – ethylacetate (at –50°C)
$C_7H_{14}N_2O_2Pb$
M.Birkhahn, E.Glozbach, W.Massa, J.Lorberth
J.Organomet.Chem.,**192**, 171, 1980

69.16 Chloro – (ethyl – L – cysteinato – N,S) – dimethyl – tin(iv)
$C_7H_{16}ClNO_2SSn$
G.Domazetis, M.F.Mackay *J.Cryst.Mol.Struct.*,**9**, 57, 1979
See also R1 : 48

69.17 bis(β – Carbomethoxy – ethyl) – dichloro – tin(iv)
$C_8H_{14}Cl_2O_4Sn$
P.G.Harrison, T.J.King, M.A.Healy
J.Organomet.Chem.,**182**, 17, 1979

69.18 Di – μ_3 – oxo – bis(μ – dichloro) – bis(μ – dimethyl – tin(iv)) – bis(chlorodimethyl – tin(iv))
$(C_8H_{24}Cl_4O_2Sn_4)_n$
P.G.Harrison, M.J.Begley, K.C.Molloy
J.Organomet.Chem.,**186**, 213, 1980

69.19 Trimethyl – (2 – pyridyl – carboxylato) – tin(iv) monohydrate
$C_9H_{13}NO_2Sn$, H_2O
P.G.Harrison, R.C.Philips
J.Organomet.Chem.,**182**, 37, 1979

69.20 (Pentamethyl – cyclopentadienyl) – tin tetrafluoroborate
$C_{10}H_{15}Sn^+$, BF_4^-
P.Jutzi, F.Kohl, P.Hofmann, C.Kruger, Y.–H.Tsay
Chem.Ber.,**113**, 757, 1980

69.21 4 – Trimethyl – tin – quinuclidinium perchlorate
$C_{10}H_{22}NSn^+$, ClO_4^-
M.Zehnder *Helv.Chim.Acta*,**63**, 750, 1980

69.22 Dichloro – bis(ethyl – 3 – oxobutanato) – tin(iv)
$C_{12}H_{18}Cl_2O_6Sn$
J.Angenault, C.Mondi, A.Rimsky
Inorg.Chim.Acta,**37**, 145, 1979

69.23 N – (Trimethyl – tin) – dimethylarsino – bis(carbomethoxy) – pyrazole (at –50°C)
$C_{12}H_{21}AsN_2O_4Sn$
M.Birkhahn, R.Hohlfeld, W.Massa, R.Schmidt, J.Lorberth
J.Organomet.Chem.,**192**, 47, 1980
See also R1 : 65

69.24 (N,N – Diethyldithiocarbamato) – dimethyl – tin(iv) (triclinic form)
$C_{12}H_{28}N_2S_4Sn$
J.S.Morris, E.O.Schlemper *J.Cryst.Mol.Struct.*,**9**, 13, 1979

69.25 (N,N – Diethyldithiocarbamato) – dimethyl – tin(iv) (monoclinic form)
$C_{12}H_{28}N_2S_4Sn$
J.S.Morris, E.O.Schlemper *J.Cryst.Mol.Struct.*,**9**, 13, 1979

69.26 bis(p – Aminobenzoato) – lead(ii)
$C_{14}H_{12}N_2O_4Pb$
I.R.Amiraslanov, N.Kh.Dzhafarov, G.N.Nadzhafov, Kh.S.Mamedov, E.M.Movsumov, B.T.Usubaliev
Zh.Strukt.Khim.,**21**, 131, 1980

69.27 Tetrachloro – (N – pyridoxylidene – N′ – picolinoylhydrazino) – tin(iv) dihydrate
$C_{14}H_{14}Cl_4N_4O_3Sn$, $2H_2O$
P.Domiano, A.Musatti, M.Nardelli, C.Pelizzi, G.Predieri
Inorg.Chim.Acta,**38**, 9, 1980

69.28 10,10 – Dimethyl – 10 – germa – 9 – thio – 9,10 – dihydroanthracene
$C_{14}H_{14}GeS$
G.D.Andreetti, G.Bocelli, P.Domiano, P.Sgarabotto
J.Organomet.Chem.,**179**, 7, 1979

69.29 bis(Diaquo – (pyridine – 2,6 – dicarboxylato) – lead(ii)) pyridine – 2,6 – dicarboxylic acid solvate monohydrate
$C_{14}H_{14}N_2O_{12}Pb_2$, $2C_7H_5NO_4$, H_2O
K.A.Beveridge, G.W.Bushnell *Can.J.Chem.*,**57**, 2498, 1979

69.30 bis(N,N – Di – isopropyldithiocarbamato) – lead(ii)
$C_{14}H_{28}N_2PbS_4$
M.Ito, H.Iwasaki *Acta Crystallogr.,Sect.B*,**36**, 443, 1980

69.31 Trimethyl – (N – phenyl – N – benzoylhydroxylamino) – tin
$C_{16}H_{19}NO_2Sn$
P.G.Harrison, T.J.King, K.C.Molloy
J.Organomet.Chem.,**185**, 199, 1980

69.32 1,2 – Phenylene – diethynyl – bis(trimethyl – tin)
$C_{16}H_{22}Sn_2$
G.Adiwidjaja, G.Grouhi–Witte
J.Organomet.Chem.,**188**, 19, 1980

69.33 1,8 – bis(Trimethylgermyl) – naphthalene
$C_{16}H_{24}Ge_2$
J.F.Blount, F.Cozzi, J.R.Damewood Junior, L.D.Iroff, U.Sjostrand, K.Mislow *J.Am.Chem.Soc.*,**102**, 99, 1980

69.34 1,8 – bis(Trimethylstannyl) – naphthalene
$C_{16}H_{24}Sn_2$
J.F.Blount, F.Cozzi, J.R.Damewood Junior, L.D.Iroff, U.Sjostrand, K.Mislow *J.Am.Chem.Soc.*,**102**, 99, 1980

69.35 tris(N,N – Diethyldithiocarbamato) – methyl – tin(iv)
$C_{16}H_{33}N_3S_6Sn$
J.S.Morris, E.O.Schlemper
J.Cryst.Mol.Struct.,**8**, 295, 1978

69.36 tetrakis(t – Butylimino) – dihydrido – tri – tin
$C_{16}H_{38}N_4Sn_3$
M.Veith *Z.Naturforsch.,Teil B*,**35**, 20, 1980

69.37 Di – μ_3 – oxo – bis(μ – dichloro) – bis(μ – diethyl – tin(iv)) – bis(chlorodiethyl – tin(iv))
$(C_{16}H_{40}Cl_4O_2Sn_4)_n$
P.G.Harrison, M.J.Begley, K.C.Molloy
J.Organomet.Chem.,**186**, 213, 1980

69.C poly(Trivinyl – ferrocenoato – tin(iv))
$(C_{17}H_{18}FeO_2Sn)_n$ Main entry is 73.67

69.38 bis(5 – Nitropyridine – 2 – thiolato) – bis(n – butyl) – tin(iv)
$C_{18}H_{24}N_4O_4S_2Sn$
G.Domazetis, B.D.James, M.F.Mackay, R.J.Magee
J.Inorg.Nucl.Chem.,**41**, 1555, 1979

69.39 Chloro – tricyclohexyl – tin(iv)
$C_{18}H_{33}ClSn$
S.Calogero, P.Ganis, V.Peruzzo, G.Tagliavini
J.Organomet.Chem.,**179**, 145, 1979

69.40 bis(2,2,6,6 – Tetramethylpiperidinato) – germanium
$C_{18}H_{36}GeN_2$
M.F.Lappert, M.J.Slade, J.L.Atwood, M.J.Zaworotko
J.Chem.Soc.,Chem.Commun.,621, 1980

69.41 tris(N,N – Diethyldithiocarbamato) – (n – butyl) –
tin(iv)
$C_{19}H_{39}N_3S_6Sn$
J.S.Morris, E.O.Schlemper *J.Cryst.Mol.Struct.*,**9**, 1, 1979

69.42 Triphenyl – tin – (methyl – dithioformate)
$C_{20}H_{18}S_2Sn$
P.–R.Bolz, U.Kunze, W.Winter
Angew.Chem.,Int.Ed.Engl.,**19**, 220, 1980

69.43 bis(Pentamethyl – cyclopentadienyl) – tin
$C_{20}H_{30}Sn$
P.Jutzi, F.Kohl, P.Hofmann, C.Kruger, Y.–H.Tsay
Chem.Ber.,**113**, 757, 1980

69.44 Tricyclohexyl – tin – trifluoroacetate
$C_{20}H_{33}F_3O_2Sn$
S.Calogero, P.Ganis, V.Peruzzo, G.Tagliavini
J.Organomet.Chem.,**191**, 381, 1980

69.45 Bromo – (8 – dimethylaminomethyl – 5 –
methoxynaphthyl) – methylphenyl – tin(iv)
$C_{21}H_{24}BrNOSn$
G.van Koten, J.T.B.H.Jastrzebski, J.G.Noltes, G.J.Verhoeckx,
A.L.Spek, J.Kroon *J.Chem.Soc.,Dalton Trans.*,1352, 1980

69.C (1,1′ – Ferrocenediyl) – diphenyl – germanium
$C_{22}H_{18}FeGe$ Main entry is 73.91

69.46 (O,O′ – Diethyl – dithiophosphato) – triphenyl –
tin(iv) (at 138°K)
$C_{22}H_{25}O_2PS_2Sn$
K.C.Molloy, M.B.Hossain, D.van der Helm, J.J.Zuckerman,
I.Haiduc *Inorg.Chem.*,**18**, 3507, 1979
See also R1 : 64

69.47 Triphenyl – (pyridine – 4 – thiolato) – lead
$C_{23}H_{19}NPbS$
N.G.Furmanova, Yu.T.Struchkov, D.N.Kravtsov,
E.M.Rokhlina *Zh.Strukt.Khim.*,**20**, 1047, 1979

69.48 bis(O,O′ – Di – isopropyl – dithiophosphato) –
diphenyl – tin(iv)
$C_{24}H_{38}O_4P_2S_4Sn$
K.C.Molloy, M.B.Hossain, D.van der Helm, J.J.Zuckerman,
I.Haiduc *Inorg.Chem.*,**19**, 2041, 1980

69.C Tricarbonyl – (triphenylgermyl) –
(ethylethoxycarbene) – cobalt
$C_{26}H_{25}CoGeO_4$ Main entry is 71.205

69.C trans – Triphenyl – tin – tetracarbonyl –
diethylaminocarbyne – chromium
$C_{27}H_{25}CrNO_4Sn$ Main entry is 71.215

69.C Pentacarbonyl – (diethylamino(triphenyl – tin)
carbene) – chromium(0) dichloromethane solvate
$C_{28}H_{25}CrNO_5Sn$, $0.5CH_2Cl_2$ Main entry is 71.219

69.49 bis(2,6 – Di – t – butyl – 4 – methylphenoxy) –
germanium
$C_{30}H_{46}GeO_2$
B.Cetinkaya, I.Gumrukcu, M.F.Lappert, J.L.Atwood,
R.D.Rogers, M.J.Zaworotko
J.Am.Chem.Soc.,**102**, 2088, 1980

69.50 bis(2,6 – Di – t – butyl – 4 – methylphenoxy) – tin
$C_{30}H_{46}O_2Sn$
B.Cetinkaya, I.Gumrukcu, M.F.Lappert, J.L.Atwood,
R.D.Rogers, M.J.Zaworotko
J.Am.Chem.Soc.,**102**, 2088, 1980

69.51 Triphenyl – (triphenylgermyl – peroxy) – silane
$C_{36}H_{30}GeO_2Si$
V.A.Lebedev, Yu.N.Drozdov, E.A.Kuz'min, A.V.Ganyushkin,
V.A.Yablokov, N.V.Belov
Dokl.Akad.Nauk SSSR,**246**, 601, 1979
See also R1 : 63

69.52 Hexaphenyl – di – germanium (triclinic stable
form)
$C_{36}H_{30}Ge_2$
M.Drager, L.Ross *Z.Anorg.Allg.Chem.*,**460**, 207, 1980

69.53 bis(Triphenyl – germanium) – sulfide (at –130°C,
monoclinic form)
$C_{36}H_{30}Ge_2S$
B.Krebs, H.–J.Korte *J.Organomet.Chem.*,**179**, 13, 1979

69.54 bis(Triphenyl – germanium) – sulfide (orthorhombic
form)
$C_{36}H_{30}Ge_2S$
B.Krebs, H.–J.Korte *J.Organomet.Chem.*,**179**, 13, 1979

69.55 bis(Triphenyl – tin) – selenide
$C_{36}H_{30}SeSn_2$
B.Krebs, H.–J.Jacobsen
J.Organomet.Chem.,**178**, 301, 1979

69.56 hexakis(μ – t – Butylamido) – bis(μ_3 – t –
butylamido) – tetra(μ – dimethylsilyl) – tetra(μ_3 –
oxo) – hexa – tin
$C_{40}H_{96}N_8O_4Si_4Sn_6$
M.Veith, O.Recktenwald
Z.Anorg.Allg.Chem.,**459**, 208, 1979
See also R1 : 63

69.57 4 – (N – Methylanilino) – 1 – phenyl – 3 –
(triphenyl – tin) – 2 – naphthol
$C_{41}H_{33}NOSn$
G.Himbert, L.Henn, R.Hoge
J.Organomet.Chem.,**184**, 317, 1980

69.58 μ – (1 – (1,2,3,4 – Tetraphenylbutadienyl) –
phenylgermylene) – octacarbonyl – di – iron
$C_{42}H_{26}Fe_2GeO_8$
M.D.Curtis, W.M.Butler, J.Scibelli
J.Organomet.Chem.,**192**, 209, 1980

69.59 Octaphenyl – thia – tetragermana – cyclopentane
$C_{48}H_{40}Ge_4S$
L.Ross, M.Drager *J.Organomet.Chem.*,**194**, 23, 1980

69.60 bis(O,O′ – Diphenyl – dithiophosphato) – tin(ii)
$C_{48}H_{40}O_8P_4S_8Sn_2$
J.L.Lefferts, M.B.Hossain, K.C.Molloy, D.van der Helm,
J.J.Zuckerman *Angew.Chem.,Int.Ed.Engl.*,**19**, 309, 1980

69.61 Dodecaphenyl – cyclohexagermane benzene solvate
$C_{72}H_{60}Ge_6$, $7C_6H_6$
M.Drager, L.Ross, D.Simon
Z.Anorg.Allg.Chem.,**466**, 145, 1980

69.C   bis(η^5 – Cyclopentadienyl – (triphenylgermyl) – (triphenylgermyl – cadmium) – nickel) – cadmium toluene solvate (at –120°C)
$C_{82}H_{70}Cd_3Ge_4Ni_2$, C_7H_8 Main entry is 73.134

69.C   bis((η^5 – Cyclopentadienyl) – (triphenylgermyl) – (triphenylgermyl – mercury) – nickel) – mercury toluene solvate (at –120°C)
$C_{82}H_{70}Ge_4Hg_3Ni_2$, C_7H_8 Main entry is 73.135

TELLURIUM COMPOUNDS

70.1   Phenyl – tellurium bromide chloride
$C_6H_5Te^{3+}$, $1.3Br^-$, $1.7Cl^-$
D.Rainville, R.A.Zingaro, E.A.Meyers
Cryst.Struct.Commun.,**9**, 77, 1980

70.2   2,2,4,4,6,6 – Hexamethylcyclotristanna – tellurane (at 110°K)
$C_6H_{18}Sn_3Te_3$
A.Blecher, M.Drager
Angew.Chem.,Int.Ed.Engl.,**18**, 677, 1979
See also R1 : 69

70.3   1,1 – Di – iodo – 3,4 – benzo – 1 – tellura – cyclopentane (α form)
$C_8H_8I_2Te$
C.Knobler, R.F.Ziolo *J.Organomet.Chem.*,**178**, 423, 1979

70.4   6 – Acetyl – 2 – (α – (trichloro – tellurium) – acetyl) – pyridine
$C_9H_8Cl_3NO_2Te$
H.J.Gysling, H.R.Luss, S.A.Gardner
J.Organomet.Chem.,**184**, 417, 1980

70.5   cis – 2 – Ethoxycycloheptyl – tribromo – tellurium(iv)
$C_9H_{17}Br_3OTe$
T.S.Cameron, R.B.Amero, C.Chan, R.E.Cordes
Cryst.Struct.Commun.,**9**, 543, 1980

70.6   Trichloro – (8 – ethoxy – 4 – cyclo – octenyl) – tellurium
$C_{10}H_{17}Cl_3OTe$
J.Bergman, L.Engman
J.Organomet.Chem.,**181**, 335, 1979

70.7   bis(Monothio – pyrocatecholato) – tellurium(iv)
$C_{12}H_8O_2S_2Te$
K.von Deuten, W.Schnabel, G.Klar
Cryst.Struct.Commun.,**9**, 161, 1980

70.8   (N,N – Diethylethylene – tellurourea) – pentacarbonyl – chromium
$C_{12}H_{14}CrN_2O_5Te$
M.F.Lappert, T.R.Martin, G.M.McLaughlin
J.Chem.Soc.,Chem.Commun.,635, 1980

70.9   Di – cis – 2 – chlorocyclohexyl – dichloro – tellurium(iv)
$C_{12}H_{20}Cl_4Te$
T.S.Cameron, R.B.Amero, R.E.Cordes
Cryst.Struct.Commun.,**9**, 533, 1980

70.10   Bromo – tris(N – (2 – hydroxyethyl) – N – methyldithiocarbamato) – tellurium(iv)
$C_{12}H_{24}BrN_3O_3S_6Te$
S.Husebye *Acta Chem.Scand.Ser.A*,**33**, 485, 1979

70.11   tris(Morpholino) – phosphine telluride (at –151°C)
$C_{12}H_{24}N_3O_3PTe$
C.Romming, A.J.Iversen, J.Songstad
Acta Chem.Scand.Ser.A,**34**, 333, 1980

70.12 tetrakis(Dimethylthiourea) – tellurium(ii)
dichloride
$C_{12}H_{32}N_8S_4Te^{2+}$, $2Cl^-$
G.Valle, S.Calogero, U.Russo
Cryst.Struct.Commun.,**9**, 649, 1980

70.13 p – Tolyl – 2 – chlorocyclohexyl – dichloro –
tellurium(iv)
$C_{13}H_{17}Cl_3Te$
T.S.Cameron, R.B.Amero, R.E.Cordes
Cryst.Struct.Commun.,**9**, 539, 1980

70.14 tetrakis(N – (2 – Hydroxyethyl) – N –
methyldithiocarbamato) – tellurium(iv)
$C_{16}H_{32}N_4O_4S_8Te$
S.Husebye *Acta Chem.Scand.Ser.A*,**33**, 485, 1979

70.15 hexakis(Ethylenethiourea) – di – tellurium(ii)
tetraperchlorate
$C_{18}H_{36}N_{12}S_6Te_2^{4+}$, $4ClO_4^-$
A.S.Foust *Inorg.Chem.*,**19**, 1050, 1980

70.16 hexakis(Trimethylenethiourea) – di – tellurium(ii)
perchlorate (triclinic form)
$C_{24}H_{48}N_{12}S_6Te_2^{4+}$, $4ClO_4^-$
A.S.Foust *Inorg.Chem.*,**19**, 1050, 1980

70.17 hexakis(Trimethylenethiourea) – di – tellurium(ii)
perchlorate (monoclinic form)
$C_{24}H_{48}N_{12}S_6Te_2^{4+}$, $4ClO_4^-$
A.S.Foust *Inorg.Chem.*,**19**, 1050, 1980

70.18 trans – bis(Thiocyanato) – bis(di(3 –
trimethylsilyl – propyl) – tellurium) – palladium(ii)
$C_{26}H_{60}N_2PdS_2Si_4Te_2$
H.J.Gysling, H.R.Luss, D.L.Smith
Inorg.Chem.,**18**, 2696, 1979
See also R1 : 63

TRANSITION METAL–C COMPOUNDS

71.1 Tetraphenylarsonium difulminato – gold
$C_2AuN_2O_2^-$, $C_{24}H_{20}As^+$
U.Nagel, K.Peters, H.G.von Schnering, W.Beck
J.Organomet.Chem.,**185**, 427, 1980
See also R2 : 65

71.2 Penta – ammine – ruthenium(ii) – fumaric acid
dithionate dihydrate
$C_4H_{19}N_5O_4Ru^{2+}$, $O_6S_2^{2-}$, $2H_2O$
H.Lehmann, K.J.Schenk, G.Chapuis, A.Ludi
J.Am.Chem.Soc.,**101**, 6197, 1979

71.3 Tetracarbonyl – manganese – ((μ^2 – acetyl) – (μ^2 –
bromo) – dibromo – aluminium)
$C_6H_3AlBr_3MnO_5$
S.B.Butts, S.H.Strauss, E.M.Holt, R.E.Stimson, N.W.Alcock,
D.F.Shriver *J.Am.Chem.Soc.*,**102**, 5093, 1980
See also R1 : 68

71.4 Benzyl – mercury – chloride (at –120°C)
C_7H_7ClHg
R.G.Gerr, M.Yu.Antipin, N.G.Furmanova, Yu.T.Struchkov
Kristallografiya,**24**, 951, 1979

71.5 1 – (Methyl – mercury(ii)) – 9 – methyl – adenine
nitrate
$C_7H_{10}HgN_5^+$, NO_3^-
M.J.Olivier, A.L.Beauchamp *Inorg.Chem.*,**19**, 1064, 1980
See also R1 : 83

71.6 Dicarbonyl – carbamoyl – (η^5 – cyclopentadienyl) –
ruthenium
$C_8H_7NO_3Ru$
H.Wagner, A.Jungbauer, G.Thiele, H.Behrens
Z.Naturforsch.,Teil B,**34**, 1487, 1979
See also R1 : 73

71.7 Chloro – (syn – 4,4 – dimethyl – 5 – methylthio –
pent – (1 – 3 – η) – enyl) – palladium
$C_8H_{15}ClPdS$
R.McCrindle, E.C.Alyea, G.Ferguson, S.A.Dias, A.J.McAlees,
M.Parvez *J.Chem.Soc.,Dalton Trans.*,137, 1980
See also R1 : 85

71.8 Methyl zinc methoxide tetramer
$C_8H_{24}O_4Zn_4$
H.M.M.Shearer, C.B.Spencer
Acta Crystallogr.,Sect.B,**36**, 2046, 1980
See also R1 : 84

71.9 Octacarbonyl – (μ – methylene) – di – iron
(at –35°C)
$C_9H_2Fe_2O_8$
C.E.Sumner Junior, P.E.Riley, R.E.Davis, R.Pettit
J.Am.Chem.Soc.,**102**, 1752, 1980

71.10 Octacarbonyl – (μ – methylene) – di – iron
$C_9H_2Fe_2O_8$
C.E.Sumner Junior, P.E.Riley, R.E.Davis, R.Pettit
J.Am.Chem.Soc.,**102**, 1752, 1980

71.11 **fac – (Triacetyl – tricarbonyl – rhenato) – boron – chloride**
$C_9H_9BClO_6Re$
C.M.Lukehart, L.T.Warfield *Inorg.Chim.Acta*,**41**, 105, 1980
See also R1 : 62

71.12 **tetrakis(Acetoxy – mercury)methane dihydrate**
$C_9H_{12}Hg_4O_8$, $2H_2O$
D.Grdenic, M.Sikirica *Z.Kristallogr.*,**150**, 107, 1979
See also R1 : 81

71.13 **8 – Trifluoroacetoxy – 1 – (η^5 – cyclopentadienyl – ferra) – 2,3 – dicarba – dodecaborane**
$C_9H_{15}B_9F_3FeO_2$
L.I.Zakharkin, V.V.Kobak, A.I.Kovredov, N.G.Furmanova, Yu.T.Struchkov
Izv.Akad.Nauk SSSR,Ser.Khim.,1097, 1979
See also R1 : 73,62

71.14 **2β – (Bromo – mercury – methyl) – 3,4 – dioxabicyclo(4.4.0)decane**
$C_9H_{15}BrHgO_2$
N.A.Porter, M.A.Cudd, R.W.Miller, A.T.McPhail
J.Am.Chem.Soc.,**102**, 414, 1980

71.15 **Lithium (μ^3 – carbonyl) – tris(μ^2 – carbonyl) – tris(dicarbonyl – cobalt) di – isopropyl ether solvate**
$C_{10}Co_3O_{10}^-$, Li^+, $C_6H_{14}O$
H.–N.Adams, G.Fachinetti, J.Strahle
Angew.Chem.,Int.Ed.Engl.,**19**, 404, 1980

71.16 **(μ_3 – Carbonyl) – (μ_3 – sulfido) – tris(tricarbonyl – iron)**
$C_{10}Fe_3O_{10}S$
L.Marko, T.Madach, H.Vahrenkamp
J.Organomet.Chem.,**190**, C67, 1980
See also R1 : 85

71.17 **bis(μ – Methylthio) – (μ – tetrafluoroethane – 1,1 – diyl) – bis(tricarbonyl – iron) (at –162°C)**
$C_{10}H_8F_4Fe_2O_6S_2$
J.J.Bonnet, R.Mathieu, R.Poilblanc, J.A.Ibers
J.Am.Chem.Soc.,**101**, 7487, 1979
See also R1 : 85

71.18 **bis(μ – Methylthio) – (μ – tetrafluoroethane – 1,2 – diyl) – bis(tricarbonyl – iron) (at –162°C)**
$C_{10}H_8F_4Fe_2O_6S_2$
J.J.Bonnet, R.Mathieu, R.Poilblanc, J.A.Ibers
J.Am.Chem.Soc.,**101**, 7487, 1979
See also R1 : 85

71.19 **Pentacarbonyl – (diethylaminomethylidyne) – chromium tetrafluoroborate (at –30°C)**
$C_{10}H_{10}CrNO_5^+$, BF_4^-
U.Schubert, E.O.Fischer, D.Wittmann
Angew.Chem.,Int.Ed.Engl.,**19**, 643, 1980

71.20 **(1,1 – Dichloro – 1 – (2 – methyl – 5 – t – butyl – 2H – 2 – pyrrolyl)) – chloro – mercury**
$C_{10}H_{14}Cl_3HgN$
A.Gambacorta, R.Nicoletti, S.Cerrini, W.Fedeli, G.Gavuzzo
Tetrahedron,**36**, 1367, 1980

71.21 **(2,4 – Dithio – 1,3 – dithia – 1,4 – diyl) – η^5 – cyclopentadienyl – (trimethylphosphine) – rhodium**
$C_{10}H_{14}PRhS_4$
H.Werner, O.Kolb, R.Feser, U.Schubert
J.Organomet.Chem.,**191**, 283, 1980
See also R1 : 73,86,85

71.22 **α – Chloro – mercury – camphene (absolute configuration)**
$C_{10}H_{15}ClHg$
V.G.Andrianov, Yu.T.Struchkov, V.A.Blinova, I.I.Kritskaya
Izv.Akad.Nauk SSSR,Ser.Khim.,2021, 1979

71.23 **bis(Chloro – (0 – isopropyl – acetonium)) – platinum**
$C_{10}H_{20}Cl_2O_2Pt$
V.B.Pukhnarevich, Yu.T.Struchkov, G.G.Aleksandrov, S.P.Sushchinskaya, E.O.Tsetlina, M.G.Voronkov
Koord.Khim.,**5**, 1535, 1979

71.24 **Chloro – (2 – methyl – 2 – nitrosopropane) – (2 – (N – oxo – t – butylimino)ethyl) – platinum**
$C_{10}H_{21}ClN_2O_2Pt$
D.Mansuy, M.Dreme, J.–C.Chottard, J.–P.Girault, J.Guilhem *J.Am.Chem.Soc.*,**102**, 844, 1980
See also R1 : 84,83

71.25 **(2 – Methylene – 1 – tetracarbonyl – 2,3 – diaza – 1 – mangana – cyclobutan – 4 – one) – manganese pentacarbonyl**
Nonacarbonyl – (μ – (methylenehydrazine – carbonyl)) – di – manganese
$C_{11}H_2Mn_2N_2O_{10}$
K.Weidenhammer, M.L.Ziegler
Z.Anorg.Allg.Chem.,**457**, 174, 1979
See also R1 : 83

71.26 **(μ – Acetimidoyl) – (μ – hydrido) – tris(tricarbonyl – iron) (triclinic form, at –158°C)**
$C_{11}H_5Fe_3NO_9$
M.A.Andrews, G.van Buskirk, C.B.Knobler, H.D.Kaesz
J.Am.Chem.Soc.,**101**, 7245, 1979
See also R1 : 72,83

71.27 **syn – (η^3 – 2 – Vinylyl – cyclopentyl – carboxylato) – tricarbonyl – iron**
$C_{11}H_{10}FeO_5$
G.D.Annis, S.V.Ley, R.Sivaramakrishnan, A.M.Atkinson, D.Rogers, D.J.Williams *J.Organomet.Chem.*,**182**, C11, 1979
See also R1 : 72

71.28 **anti – (η^3 – 2 – Vinylyl – cyclopentyl – carboxylato) – tricarbonyl – iron**
$C_{11}H_{10}FeO_5$
G.D.Annis, S.V.Ley, R.Sivaramakrishnan, A.M.Atkinson, D.Rogers, D.J.Williams *J.Organomet.Chem.*,**182**, C11, 1979
See also R1 : 72

71.29 **Phenylpyridine – mercury(ii) trifluoroacetate**
$C_{11}H_{10}HgN^+$, $C_2F_3O_2^-$
J.Halfpenny, R.W.H.Small
Acta Crystallogr.,Sect.B,**36**, 938, 1980
See also R1 : 83

71.30 **Dichloro – (η^5 – cyclopentadienyl) – (2 – oxo – 4 – methyl – pent – 3 – ene) – tantalum**
$C_{11}H_{15}Cl_2OTa$
E.Guggolz, M.L.Ziegler, H.Biersack, W.A.Herrmann
J.Organomet.Chem.,**194**, 317, 1980
See also R1 : 73

71.31 **Chloro – (N,N,N′,N′ – tetramethylethylenediamine) – penta – 2,4 – dienyl – zinc**
$C_{11}H_{23}ClN_2Zn$
H.Yasuda, Y.Ohnuma, A.Nakamura, Y.Kai, N.Yasuoka, N.Kasai *Bull.Chem.Soc.Jpn.*,**53**, 1101, 1980
See also R1 : 76

71.32 Dichloro – triethylphosphine – neopentylidene –
oxo – tungsten
$C_{11}H_{25}Cl_2OPW$
J.H.Wengrovius, R.R.Schrock, M.R.Churchill, J.R.Missert,
W.J.Youngs J.Am.Chem.Soc.,**102**, 4515, 1980
See also R1 : 86

71.33 Potassium ethylene – tris(trimethylphosphine) –
semicobaltate
$C_{11}H_{31}CoP_3^-$, $C_{11}H_{31}CoP_3$, K+
H.–F.Klein, J.Gross, J.–M.Bassett, U.Schubert
Z.Naturforsch.,Teil B,**35**, 614, 1980
See also R1 : 86

71.34 bis(2,3,4,5 – Tetrafluorophenyl) – mercury
$C_{12}H_2F_8Hg$
D.S.Brown, A.G.Massey, D.A.Wickens
J.Organomet.Chem.,**194**, 131, 1980

71.35 3,4 – Diethyl – 1 – tetracarbonyl – ferracyclopent –
3 – ene – 2,5 – dione
$C_{12}H_{10}FeO_6$
S.Aime, L.Milone, E.Sappa, A.Tiripicchio, A.M.M.Lanfredi
J.Chem.Soc.,Dalton Trans.,1664, 1979

71.36 (μ^2 – Bromo) – (μ^2 – p – tolylmethylidene) – (μ^2 –
bis(methylamino – bis(difluorophosphine))) –
bis(bromo – carbonyl – tungsten)
$C_{12}H_{13}Br_3F_8N_2O_2P_4W_2$
E.O.Fischer, W.Kellerer, B.Zimmer–Gasser, U.Schubert
J.Organomet.Chem.,**199**, C24, 1980
See also R1 : 86

71.37 Methyl – (bis(2 – pyridyl)methane) – mercury
nitrate
$C_{12}H_{13}HgN_2^+$, NO_3^-
A.J.Canty, G.Hayhurst, N.Chaichit, B.M.Gatehouse
J.Chem.Soc.,Chem.Commun.,316, 1980
See also R1 : 83

71.38 Dicarbonyl – (4 – 5 – η – 1,2 – difluoro – 1,2 –
bis(trifluoromethyl) – pent – 4 – enyl) –
(trimethylphosphite) – cobalt (at 183°K)
$C_{12}H_{14}CoF_8O_5P$
M.Bottrill, R.Goddard, M.Green, P.Woodward
J.Chem.Soc.,Dalton Trans.,1671, 1979
See also R1 : 72,86

71.39 (3,3,6 – Trimethyl – 1 – oxo – 2 – oxaheptyl – (4 –
6 – η) – enyl) – tricarbonyl – iron (form I)
$C_{12}H_{14}FeO_5$
R.Aumann, H.Ring, C.Kruger, R.Goddard
Chem.Ber.,**112**, 3644, 1979
See also R1 : 72

71.40 (3,3,6 – Trimethyl – 1 – oxo – 2 – oxa – heptyl –
(4 – 6 – η) – enyl) – tricarbonyl – iron
$C_{12}H_{14}FeO_5$
R.Aumann, H.Ring, C.Kruger, R.Goddard
Chem.Ber.,**112**, 3644, 1979
See also R1 : 72

71.41 3 – Carbonyl – 3 – η^5 – cyclopentadienyl – 6 –
difluoro – 2 – isopropyl – 4 – methyl – 1,5 – dioxa –
3 – ferra – 6 – borinane
$C_{12}H_{15}BF_2FeO_3$
P.G.Lenhert, C.M.Lukehart, L.T.Warfield
Inorg.Chem.,**19**, 2343, 1980
See also R1 : 73,62

71.42 bis(η^5 – Cyclopentadienyl) – methylmethylene –
tantalum (at 15°K, neutron study)
$C_{12}H_{15}Ta$
F.Takusagawa, A.Fumagalli, T.F.Koetzle, P.R.Sharp,
R.R.Schrock, W.A.Herrmann
Am.Cryst.Assoc.,Ser.2,**8**, 29, 1980
See also R1 : 73

71.43 bis(Boranato – bis(dimethylphosphonium –
methylide)) – nickel
$C_{12}H_{36}B_2NiP_4$
G.Muller, U.Schubert, O.Orama, H.Schmidbaur
Chem.Ber.,**112**, 3302, 1979
See also R1 : 64

71.44 (μ^2 – Hydrido) – (η – methylidyne) –
dodecacarbonyl – tetra – iron (at –100°C)
$C_{13}H_2Fe_4O_{12}$
M.A.Beno, J.M.Williams, M.Tachikawa, E.L.Muetterties
J.Am.Chem.Soc.,**102**, 4542, 1980

71.45 bis(Triphenylphosphine)immonium 3,3,3,3 –
tetracarbonyl – 5 – phenyl – 1 – oxa – 3 –
mangana – cyclopenta – 2,4 – dione (at –160°C)
$C_{13}H_6MnO_7^-$, $C_{36}H_{30}NP_2^+$
J.A.Gladysz, J.C.Selover, C.E.Strouse
J.Am.Chem.Soc.,**100**, 6766, 1978
See also R2 : 64

71.46 Undecacarbonyl – methylisocyanide – tetrakis((μ –
hydrido) – osmium)
$C_{13}H_7NO_{11}Os_4$
M.R.Churchill, F.J.Hollander Inorg.Chem.,**19**, 306, 1980

71.47 (η^2 – Dimethylformamide) – (μ^2 – hydrido) –
decacarbonyl – tri – ruthenium (at 115°K)
$C_{13}H_7NO_{11}Ru_3$
R.Szostak, C.E.Strouse, H.D.Kaesz
J.Organomet.Chem.,**191**, 243, 1980
See also R1 : 84

71.48 (2,2′ – Bipyridyl) – tricarbonyl – chloro – (chloro –
mercurio) – molybdenum(ii)
$C_{13}H_8Cl_2HgMoN_2O_3$
P.D.Brotherton, J.M.Epstein, A.H.White, S.B.Wild
Aust.J.Chem.,**27**, 2667, 1974
See also R1 : 83

71.49 (μ^3 – 8 – Tricarbonyl – 7 – selena – 8 – ferra –
bicyclo(4.3.0)nonen – 9 – one) – tricarbonyl – iron
$C_{13}H_8Fe_2O_7Se$
K.H.Pannell, A.J.Mayer, R.Hoggard, R.C.Pettersen
Angew.Chem.,Int.Ed.Engl.,**19**, 632, 1980

71.50 (μ – 3 – Methyl – but – 2 – enylidene) –
bis(tetracarbonyl – tungsten)
$C_{13}H_8O_8W_2$
J.Levisalles, H.Rudler, F.Dahan, Y.Jeannin
J.Organomet.Chem.,**187**, 233, 1980

71.51 Dicarbonyl – cyclopentadienyl – cis(2 –
ethoxycarbonyl – 2 – trifluoromethyl – 1 –
fluorovinyl) – iron
$C_{13}H_{10}F_4FeO_4$
V.G.Andrianov, Yu.T.Struchkov, I.B.Zlotina, M.A.Khomutov
Koord.Khim.,**5**, 1872, 1979
See also R1 : 73

71.52 (μ^2 – Methylene) – bis(carbonyl – (η^5 – cyclopentadienyl) – rhodium) (at 15°K, neutron study)
$C_{13}H_{12}O_2Rh_2$
F.Takusagawa, A.Fumagalli, T.F.Koetzle, P.R.Sharp, R.R.Schrock, W.A.Herrmann
Am.Cryst.Assoc.,Ser.2,**8**, 29, 1980
See also R1 : 73

71.53 (2 – Benzyl – pyridine) – methyl – mercury(ii) nitrate
$C_{13}H_{14}HgN^+$, NO_3^-
A.J.Canty, N.Chaichit, B.M.Gatehouse
Acta Crystallogr.,Sect.B,**36**, 786, 1980
See also R1 : 83

71.54 3,7,9,13 – Tetramethyl – 6 – (η^5 – cyclopentadienyl) – 3,7,9,13 – tetracarba – 6 – cobalta – dodecaborane
$C_{13}H_{24}B_7Co$
R.N.Grimes, E.Sinn, J.R.Pipal *Inorg.Chem.*,**19**, 2087, 1980
See also R1 : 73,62

71.55 trans – (Methyl) – chloro – bis(triethylphosphine) – platinum(ii)
$C_{13}H_{33}ClP_2Pt$
R.Bardi, A.M.Piazzesi, G.Faraone, G.Bruno
Eur.Cryst.Meeting,**6**, 68, 1980
See also R1 : 86

71.56 bis(Triphenylphosphine)iminium bromo – (μ^2 – carbonyl) – (η^2 – 1,4 – bis(trifluoro) – but – 2 – ene) – nonacarbonyl – tri – osmium
$C_{14}H_2BrF_6O_{10}Os_3^-$, $C_{36}H_{30}NP_2^+$
Z.Dawoodi, M.J.Mays, P.R.Raithby, K.Henrick
J.Chem.Soc.,Chem.Commun.,641, 1980
See also R1 : 72

71.57 (μ^4 – Acetylene) – dodecacarbonyl – tetra – osmium
$C_{14}H_2O_{12}Os_4$
R.Jackson, B.F.G.Johnson, J.Lewis, P.R.Raithby, S.W.Sankey
J.Organomet.Chem.,**193**, C1, 1980
See also R1 : 72

71.58 bis(Triphenylphosphine)iminium (μ^2 – carbonyl) – (μ^3 – methoxymethylidyne) – undecacarbonyl – tetra – iron
$C_{14}H_3Fe_4O_{13}^-$, $C_{36}H_{30}NP_2^+$
E.M.Holt, K.Whitmire, D.F.Shriver
J.Chem.Soc.,Chem.Commun.,778, 1980

71.59 bis(Triphenylphosphine)iminium (μ^2 – carbonyl) – (μ^3 – methoxymethylidyne) – undecacarbonyl – tetra – iron
$C_{14}H_3Fe_4O_{13}^-$, $C_{36}H_{30}NP_2^+$
P.A.Dawson, B.F.G.Johnson, J.Lewis, P.R.Raithby
J.Chem.Soc.,Chem.Commun.,781, 1980

71.60 trans – Tetracarbonyl – chloro – ((tricarbonyl – chromium) – η^6 – phenylcarbene) – tungsten
$C_{14}H_5ClCrO_7W$
E.O.Fischer, F.J.Gammel, D.Neugebauer
Chem.Ber.,**113**, 1010, 1980
See also R1 : 74

71.61 Tetracarbonyl – tungsten – (μ – 1 – methyl – 2 – but – 2 – ene – di – ylidene) – pentacarbonyl – tungsten
$C_{14}H_8O_9W_2$
J.Levisalles, H.Rudler, F.Dahan, Y.Jeannin
J.Organomet.Chem.,**188**, 193, 1980
See also R1 : 72

71.62 bis(Carbonyl) – cycloheptatrienylidene – (η^5 – cyclopentadienyl) – iron hexafluorophosphate (at –15°C)
$C_{14}H_{11}FeO_2^+$, F_6P^-
P.E.Riley, R.E.Davis, N.T.Allison, W.M.Jones
J.Am.Chem.Soc.,**102**, 2458, 1980
See also R1 : 73

71.63 Nickel – bis(tetracarbonyl – μ – (dimethylcarbamoyl) – iron)
$C_{14}H_{12}Fe_2N_2NiO_{10}$
W.Petz, C.Kruger, R.Goddard *Chem.Ber.*,**112**, 3413, 1979

71.64 1,4 – Dihydroxy – 2,3 – diethyl – butane – 1,1,4,4 – tetrayl – bis(tricarbonyl) – iron
$C_{14}H_{12}Fe_2O_8$
S.Aime, L.Milone, E.Sappa, A.Tiripicchio, A.M.M.Lanfredi
J.Chem.Soc.,Dalton Trans.,1664, 1979
See also R1 : 72

71.65 bis(o – Tolyl) – mercury
$C_{14}H_{14}Hg$
D.Liptak, W.H.Ilsley, M.D.Glick, J.P.Oliver
J.Organomet.Chem.,**191**, 339, 1980

71.66 (1,5 – Diphenylthiocarbazonato – N,S) – methyl – mercury(ii)
$C_{14}H_{14}HgN_4S$
A.T.Hutton, H.M.N.H.Irving, L.R.Nassimbeni, G.Gafner
Acta Crystallogr.,Sect.B,**36**, 2064, 1980
See also R1 : 85,83

71.67 (2,2′ – Bipyridyl) – nickela – cyclopentane
$C_{14}H_{16}N_2Ni$
P.Binger, M.J.Doyle, C.Kruger, Y.–H.Tsay
Z.Naturforsch.,Teil B,**34**, 1289, 1979
See also R1 : 83

71.68 (μ^2 – Acetylene) – ($\sigma,\sigma,\eta^2,\eta^2$ – glyoxal – bis(isopropylamine)) – tetracarbonyl – di – ruthenium
$C_{14}H_{18}N_2O_4Ru_2$
L.H.Staal, L.H.Polm, K.Vrieze, F.Ploeger, C.H.Stam
J.Organomet.Chem.,**199**, 13, 1980
See also R1 : 72,83

71.69 bis(η^5 – Cyclopentadienyl) – (μ – (η^3:η^4 – trihydro – C,C′ – dimethyl – dicarba – pentaborato)) – di – cobalt
$C_{14}H_{19}B_3Co_2$
R.N.Grimes, E.Sinn, R.B.Maynard
Inorg.Chem.,**19**, 2384, 1980
See also R1 : 73,62

71.70 η^5 – Cyclopentadienyl – (η^5 – cyclopentadienyl – cobalt) – (μ – (η^3:η^4 – trihydro – C,C′ – dimethyl – dicarba – hexaborato)) – iron
$C_{14}H_{20}B_3CoFe$
R.N.Grimes, E.Sinn, R.B.Maynard
Inorg.Chem.,**19**, 2384, 1980
See also R1 : 73,62

71.71 trans – bis(3 – Dimethylsulfonio – cyclopentadienylide) – bis(μ – iodo) – di – iodo – di – mercury(ii)
$C_{14}H_{20}Hg_2I_4S_2$
N.C.Baenziger, R.M.Flynn, N.L.Holy
Acta Crystallogr.,Sect.B,**36**, 1642, 1980

71.72 Hepta(methylisocyanide) – molybdenum(ii) tetrafluoroborate
$C_{14}H_{21}MoN_7{}^{2+}, 2BF_4{}^-$
P.Brant, F.A.Cotton, J.C.Sekutowski, T.E.Wood, R.A.Walton
J.Am.Chem.Soc.,**101**, 6588, 1979

71.73 trans – Phenyl – bis(dimethylglyoximato) – ammine – cobalt(iii)
$C_{14}H_{22}CoN_5O_4$
R.C.Elder, R.Whittle, M.J.Heeg, J.C.Barrick, A.Nerone
Am.Cryst.Assoc.,Ser.2,**7**, 23, 1980
See also R1 : 83

71.74 1,1 – Dichloro – 1 – (η^5 – pentamethyl – cyclopentadienyl) – tantalla – cyclopentane
$C_{14}H_{23}Cl_2Ta$
M.R.Churchill, W.J.Youngs
J.Am.Chem.Soc.,**101**, 6462, 1979
See also R1 : 73

71.75 μ – 2 – sigma:2 – 3 – η – (4,5 – Dihydro – 2 – furyl) – (bis(trimethylphosphine) – platinum) – tetracarbonyl – manganese (red form,at 200°K)
$C_{14}H_{23}MnO_5P_2Pt$
M.Berry, J.A.K.Howard, F.G.A.Stone
J.Chem.Soc.,Dalton Trans.,1601, 1980
See also R1 : 72,86

71.76 μ – 2 – sigma:2 – 3 – η – (4,5 – Dihydro – 2 – furyl) – (bis(trimethylphosphine) – platinum) – tetracarbonyl – manganese (yellow form,at 200°K)
$C_{14}H_{23}MnO_5P_2Pt$
M.Berry, J.A.K.Howard, F.G.A.Stone
J.Chem.Soc.,Dalton Trans.,1601, 1980
See also R1 : 75,86

71.77 Dichloro – (2,4,6 – trimethylpyridine) – (2 – (N – hydroxy – t – butylimino)ethyl) – platinum
$C_{14}H_{24}Cl_2N_2OPt$
D.Mansuy, M.Dreme, J.–C.Chottard, J.–P.Girault, J.Guilhem *J.Am.Chem.Soc.*,**102**, 844, 1980
See also R1 : 83

71.78 bis((Diethylenetriamine – copper) – μ^2 – cyano) – tetracyano – iron hexahydrate
$(C_{14}H_{26}Cu_2FeN_{12})_n, 6nH_2O$
G.O.Morpurgo, V.Mosini, P.Porta, G.Dessy, V.Fares
J.Chem.Soc.,Dalton Trans.,1272, 1980
See also R1 : 76

71.79 Chloro – bis(bis(dimethylphosphino)ethane) – methoxycarbonyl – iridium fluorosulfate (at –50°C)
$C_{14}H_{35}ClIrO_2P_4{}^+, FO_3S^-$
R.L.Harlow, J.B.Kinney, T.Herskovitz
J.Chem.Soc.,Chem.Commun.,813, 1980
See also R1 : 86

71.80 cis – (Ethyl) – chloro – bis(triethylphosphine) – platinum(ii)
$C_{14}H_{35}ClP_2Pt$
R.Bardi, A.M.Piazzesi, G.Faraone, G.Bruno
Eur.Cryst.Meeting,**6**, 68, 1980
See also R1 : 86

71.81 9 – Hydrido – 9,10 – bis(triethylphosphine) – 7,8 – dicarba – 9 – platina – undecaborane
$C_{14}H_{40}B_8P_2Pt$
G.K.Barker, M.Green, F.G.A.Stone, A.J.Welch, W.C.Wolsey
J.Chem.Soc.,Chem.Commun.,627, 1980
See also R1 : 86,62

71.82 9 – Hydrido – 9,9 – bis(triethylphosphine) – 7,8 – dicarba – 9 – platina – undecaborane
$C_{14}H_{41}B_8P_2Pt$
G.K.Barker, M.Green, F.G.A.Stone, A.J.Welch, W.C.Wolsey
J.Chem.Soc.,Chem.Commun.,627, 1980
See also R1 : 86,62

71.83 Dimethyl – tetrakis(trimethylphosphine) – tungsten(ii)
$C_{14}H_{42}P_4W$
R.A.Jones, G.Wilkinson, A.M.R.Galas, M.B.Hursthouse
J.Chem.Soc.,Chem.Commun.,926, 1979
See also R1 : 86

71.84 Tetraethylammonium (μ_4 – carbomethoxymethylidene) – tetrakis(tricarbonyl – iron)
$C_{15}H_3Fe_4O_{14}{}^-, C_8H_{20}N^+$
J.S.Bradley, G.B.Ansell, E.W.Hill
J.Am.Chem.Soc.,**101**, 7417, 1979

71.85 Tetracarbonyl – manganese – (μ – η^5:η^1 – cyclopentadienyl) – (carbonyl – (η^5 – cyclopentadienyl) – molybdenum)
$C_{15}H_9MnMoO_5$
R.J.Hoxmeier, C.B.Knobler, H.D.Kaesz
Inorg.Chem.,**18**, 3462, 1979
See also R1 : 73

71.86 (μ^2 – Carbonyl) – (μ^2 – vinylidene) – bis(carbonyl – cyclopentadienyl – ruthenium)
$C_{15}H_{12}O_3Ru_2$
D.L.Davies, A.F.Dyke, A.Endesfelder, S.A.R.Knox, P.J.Naish, A.G.Orpen, D.Plaas, G.E.Taylor
J.Organomet.Chem.,**198**, C43, 1980
See also R1 : 73

71.87 (μ^2 – Carbonyl) – (μ^2 – methylcarbyne) – bis(carbonyl – cyclopentadienyl – ruthenium) tetrafluoroborate
$C_{15}H_{13}O_3Ru_2{}^+, BF_4{}^-$
D.L.Davies, A.F.Dyke, A.Endesfelder, S.A.R.Knox, P.J.Naish, A.G.Orpen, D.Plaas, G.E.Taylor
J.Organomet.Chem.,**198**, C43, 1980
See also R1 : 73

71.88 cis((μ – Carbonyl) – (μ – methylcarbene) – bis(carbonyl – cyclopentadienyl – iron))
$C_{15}H_{14}Fe_2O_3$
A.F.Dyke, S.A.R.Knox, P.J.Naish, A.G.Orpen
J.Chem.Soc.,Chem.Commun.,441, 1980
See also R1 : 73

71.89 Chloro – trimethylphosphine – tris(trimethylsilylmethyl) – molybdenum(iv)
$C_{15}H_{42}ClMoPSi_3$
E.C.Guzman, G.Wilkinson, R.D.Rogers, W.E.Hunter, M.J.Zaworotko, J.L.Atwood
J.Chem.Soc.,Dalton Trans.,229, 1980
See also R1 : 86,63

71.90 (μ^4 – Ethylacetylene) – dodecacarbonyl – tetra – osmium
$C_{16}H_6O_{12}Os_4$
R.Jackson, B.F.G.Johnson, J.Lewis, P.R.Raithby, S.W.Sankey
J.Organomet.Chem.,**193**, C1, 1980
See also R1 : 72

71.91 η^5 – Cyclopentadienyl – bis(μ^2 – carbonyl) – (μ^3 – allyl) – hexacarbonyl – tri – iron
$C_{16}H_8Fe_3O_8$
G.G.Aleksandrov, V.V.Skripkin, N.E.Kolobova, Yu.T.Struchkov *Koord.Khim.*,**5**, 1479, 1979
See also R1 : 73

71.92 t – Butylisocyanide – (undecacarbonyl) – tetra – iridium
$C_{16}H_9Ir_4NO_{11}$
M.R.Churchill, J.P.Hutchinson
Inorg.Chem.,**18**, 2451, 1979

71.93 Di(bis(μ – acetyl) – tetracarbonyl – rhenium) – copper
$C_{16}H_{12}CuO_{12}Re_2$
P.G.Lenhert, C.M.Lukehart, L.T.Warfield
Inorg.Chem.,**19**, 311, 1980
See also R1 : 84

71.94 (μ – Ethylene – 1,1 – diyl) – bis(dicarbonyl – cyclopentadienyl – manganese) (at –130°C)
$C_{16}H_{12}Mn_2O_4$
K.Folting, J.C.Huffman, L.N.Lewis, K.G.Caulton
Inorg.Chem.,**18**, 3483, 1979
See also R1 : 73

71.95 π – Benzene – dicarbonyl – (methoxyphenylcarbene) – chromium
$C_{16}H_{14}CrO_3$
U.Schubert *J.Organomet.Chem.*,**185**, 373, 1980
See also R1 : 74

71.96 (μ^2 – Allene) – (μ^2 – carbonyl) – bis(carbonyl – η^5 – cyclopentadienyl – manganese) (at –170°C)
$C_{16}H_{14}Mn_2O_3$
L.N.Lewis, J.C.Huffman, K.G.Caulton
Inorg.Chem.,**19**, 1246, 1980
See also R1 : 72,73

71.97 tris(η^5 – Cyclopentadienyl) – (μ_3 – sulfido) – (μ_3 – thiocarbonyl) – triangulo – tri – cobalt
$C_{16}H_{15}Co_3S_2$
H.Werner, K.Leonhard, O.Kolb, E.Rottinger, H.Vahrenkamp *Chem.Ber.*,**113**, 1654, 1980
See also R1 : 73,85

71.98 (1,1 – Dichloro – 1 – (2 – methyl – 5 – t – butyl – 2H – 2 – pyrrolyl)) – phenyl – mercury
$C_{16}H_{19}Cl_2HgN$
A.Gambacorta, R.Nicoletti, S.Cerrini, W.Fedeli, G.Gavuzzo *Tetrahedron*,**36**, 1367, 1980

71.99 Carbonyl – η^5 – cyclopentadienyl – trimethoxyphosphine – (4,5 – bis(methoxycarbonyl) – 1,3 – dithiol – 2 – ylidene) – manganese(i)
$C_{16}H_{20}MnO_8PS_2$
J.Y.Le Marouille, C.Lelay, A.Benoit, D.Grandjean, D.Touchard, H.Le Bozec, P.Dixneuf
J.Organomet.Chem.,**191**, 133, 1980
See also R1 : 73,86

71.100 bis(Dimethylglyoximato) – (isopropyl) – (pyridine) – cobalt(iii)
$C_{16}H_{26}CoN_5O_4$
L.G.Marzilli, P.J.Toscano, L.Randaccio, N.Bresciani-Pahor, M.Calligaris *J.Am.Chem.Soc.*,**101**, 6754, 1979
See also R1 : 83

71.101 tetrakis(μ – Dimethylphosphonium – dimethylyl) – di – chromium
$C_{16}H_{40}Cr_2P_4$
F.A.Cotton, B.E.Hanson, W.H.Ilsley, G.W.Rice
Inorg.Chem.,**18**, 2713, 1979

71.102 tetrakis(μ – Dimethylphosphonium – dimethylido) – di – molybdenum
$C_{16}H_{40}Mo_2P_4$
F.A.Cotton, B.E.Hanson, W.H.Ilsley, G.W.Rice
Inorg.Chem.,**18**, 2713, 1979

71.103 Dicarbonyl – η^5 – cyclopentadienyl – η – (1,2,3,4 – tetrakis(trifluoromethyl) – buta – 1,3 – diene – 1,4 – diyl) – tungsten – dicarbonyl – cobalt
$C_{17}H_5CoF_{12}O_4W$
J.L.Davidson, L.Manojlovic-Muir, K.W.Muir, A.N.Keith
J.Chem.Soc.,Chem.Commun.,749, 1980
See also R1 : 73

71.104 bis(σ – Pyrrolyl – tricarbonyl – manganese) – tricarbonyl – iodo – manganese
$C_{17}H_8IMn_3N_2O_9$
N.I.Pyshnograeva, V.N.Setkina, V.G.Andrianov, Yu.T.Struchkov, D.N.Kursanov
J.Organomet.Chem.,**186**, 331, 1980
See also R1 : 73,83

71.105 (μ_3 – Carbonyl) – bis(η^5 – cyclopentadienyl – nickel) – bis(tricarbonyl – iron)
$C_{17}H_{10}Fe_2Ni_2O_7$
A.Marinetti, E.Sappa, A.Tiripicchio, M.T.Camellini
Inorg.Chim.Acta,**44**, 183, 1980
See also R1 : 73

71.106 bis(μ^2 – Hydrido) – (μ – N – methylbenzylamido – 1 – yl) – nonacarbonyl – tri – osmium
$C_{17}H_{11}NO_9Os_3$
R.D.Adams, J.P.Selegue *Inorg.Chem.*,**19**, 1791, 1980
See also R1 : 83

71.107 (μ – 3 – n – Pentyl – but – 2 – en – 3 – olid – 4 – ylidene) – (μ – 1,1 – di – iodo – ethen – 2 – ylidene) – bis(tricarbonyl – cobalt)
$C_{17}H_{12}Co_2I_2O_8$
I.T.Horvath, G.Palyi, L.Marko, G.Andreetti
J.Chem.Soc.,Chem.Commun.,1054, 1979

71.108 (μ_3 – Methylenethiolato) – (μ_3 – sulfido) – (octacarbonyl – (dimethylphenylphosphine) – tri – osmium) (at –35°C)
$C_{17}H_{13}O_8Os_3PS_2$
R.D.Adams, N.M.Golembeski, J.P.Selegue
J.Am.Chem.Soc.,**101**, 5862, 1979
See also R1 : 86,85

71.109 Pentacarbonyl – tungsten – (μ^2 – 3 – (trimethylphosphonium) – 1 – isopentenyl) – tetracarbonyl – tungsten
$C_{17}H_{17}O_9PW_2$
J.Levisalles, F.Rose-Munch, H.Rudler, J.–C.Daran, Y.Dromzee, Y.Jeannin
J.Chem.Soc.,Chem.Commun.,685, 1980
See also R1 : 72

71.110 3,3 – Dichloro – 3 – (η^5 – pentamethyl – cyclopentadienyl) – 3 – tantalla – bicyclo(3.3.0) octane
$C_{17}H_{27}Cl_2Ta$
M.R.Churchill, W.J.Youngs
J.Am.Chem.Soc.,**101**, 6462, 1979
See also R1 : 73

71.111 Tetracarbonyl – (μ^2 – iodo) – (methyl – di – t – butylphosphine) – (2 – oxacyclopentylidene) – platinum – manganese (at 200°K)
$C_{17}H_{27}IMnO_5PPt$
M.Berry, J.Martin–Gil, J.A.K.Howard, F.G.A.Stone
J.Chem.Soc.,Dalton Trans.,1625, 1980
See also R1 : 86

71.112 (μ_3 – Ethinyl) – triangulo – tris(dicarbonyl – trimethylphosphite – cobalt)
$C_{17}H_{30}Co_3O_{15}P_3$
P.A.Dawson, B.H.Robinson, J.Simpson
J.Chem.Soc.,Dalton Trans.,1762, 1979
See also R1 : 86

71.113 trans – Bromo – (2 – pyridyl) – bis(triethylphosphine) – palladium(ii)
$C_{17}H_{34}BrNP_2Pd$
K.Isobe, E.Kai, Y.Nakamura, K.Nishimoto, T.Miwa, S.Kawaguchi, K.Kinoshita, K.Nakatsu
J.Am.Chem.Soc.,102, 2475, 1980
See also R1 : 86

71.114 trans – Bromo – (3 – pyridyl) – bis(triethylphosphine) – palladium(ii)
$C_{17}H_{34}BrNP_2Pd$
K.Isobe, E.Kai, Y.Nakamura, K.Nishimoto, T.Miwa, S.Kawaguchi, K.Kinoshita, K.Nakatsu
J.Am.Chem.Soc.,102, 2475, 1980
See also R1 : 86

71.115 trans – Bromo – (4 – pyridyl) – bis(triethylphosphine) – palladium(ii)
$C_{17}H_{34}BrNP_2Pd$
K.Isobe, E.Kai, Y.Nakamura, K.Nishimoto, T.Miwa, S.Kawaguchi, K.Kinoshita, K.Nakatsu
J.Am.Chem.Soc.,102, 2475, 1980
See also R1 : 86

71.116 (μ – Dimethylphosphido) – (μ – dimethylphosphinomethyl) – bis(di(trimethylphosphine) – cobalt)
$C_{17}H_{50}Co_2P_6$
H.–F.Klein, J.Wenninger, U.Schubert
Z.Naturforsch.,Teil B,34, 1391, 1979
See also R1 : 86

71.117 Tetramethylammonium carbido – hexadecacarbonyl – hexa – ruthenium
$C_{17}O_{16}Ru_6^{2-}, 2C_4H_{12}N^+$
G.B.Ansell, J.S.Bradley
Acta Crystallogr.,Sect.B,36, 726, 1980

71.118 (Nonacarbonyl – tri – cobalt) – μ – (dithiocarboxylato – carbido) – (μ_3 – thio – heptacarbonyl – tri – cobalt)
$C_{18}Co_6O_{16}S_3$
P.L.Stanghellini, G.Gervasio, R.Rossetti, G.Bor
J.Organomet.Chem.,187, C37, 1980

71.119 1,1 – bis(η^5 – Cyclopentadienyl) – 2,3,4,5 – tetra(trifluoromethyl) – niobia – cyclopentadiene
$C_{18}H_{10}F_{12}Nb$
J.Sala–Pala, J.Amaudrut, J.E.Guerchais, R.Mercier, M.Cerutti *J.Fluorine Chem.*,14, 269, 1979
See also R1 : 73

71.120 bis(3 – Chloropropynyl) – (1,10 – phenanthroline) – mercury
$C_{18}H_{12}Cl_2HgN_2$
E.Gutierrez–Puebla, A.Vegas, S.Garcia–Blanco
Cryst.Struct.Commun.,8, 861, 1979
See also R1 : 83

71.121 Tribenzo(b,e,h)(1,4,7)trimercuronin (monoclinic form)
o – Phenylene – mercury trimer
$C_{18}H_{12}Hg_3$
D.S.Brown, A.G.Massey, D.A.Wickens
Inorg.Chim.Acta,44, 193, 1980

71.122 bis(Carbonyl) – (η^5 – cyclopentadienyl) – (bicyclo(5.4.0)undeca – 1,3,6,8,10 – pentaen – 5 – ylidene) – iron hexafluorophosphate (at –15°C)
$C_{18}H_{13}FeO_2^+, F_6P^-$
P.E.Riley, R.E.Davis, N.T.Allison, W.M.Jones
J.Am.Chem.Soc.,102, 2458, 1980
See also R1 : 73

71.123 (μ – Methylenethiolato) – (μ_3 – sulfido) – (nonacarbonyl – (dimethylphenylphosphine) – tri – osmium)
$C_{18}H_{13}O_9Os_3PS_2$
R.D.Adams, N.M.Golembeski, J.P.Selegue
J.Am.Chem.Soc.,101, 5862, 1979
See also R1 : 86,85

71.124 (2′ – Chlorophenyl – diazoaminobenzene) – N – (phenyl – mercury) (at –100°C)
$C_{18}H_{14}ClHgN_3$
L.G.Kuz'mina, Yu.T.Struchkov, D.N.Kravtsov
Zh.Strukt.Khim.,20, 552, 1979
See also R1 : 83

71.125 Pentacarbonyl – (ethoxy(ruthenocenyl)carbene) – tungsten
$C_{18}H_{14}O_6RuW$
E.O.Fischer, F.J.Gammel, J.O.Besenhard, A.Frank, D.Neugebauer *J.Organomet.Chem.*,191, 261, 1980
See also R1 : 73

71.126 bis(μ^2 – Carbonyl) – (μ^3 – methylidyne) – cyclo – tris((η^5 – cyclopentadienyl) – rhodium) trifluoroacetate
$C_{18}H_{16}O_2Rh_3^+, C_2F_3O_2^-$
W.A.Herrmann, J.Plank, E.Guggolz, M.L.Ziegler
Angew.Chem.,Int.Ed.Engl.,19, 651, 1980
See also R1 : 73

71.127 bis(η^5 – Cyclopentadienyl) – zirconia – benzocyclopentene
$C_{18}H_{18}Zr$
M.F.Lappert, T.R.Martin, J.L.Atwood, W.E.Hunter
J.Chem.Soc.,Chem.Commun.,476, 1980
See also R1 : 73

71.128 (μ – Carbonyl) – (μ – dimethyliminomethylene) – bis(carbonyl – (η – methylcyclopentadienyl) – iron) iodide
$C_{18}H_{20}Fe_2NO_3^+, I^-$
S.Willis, A.R.Manning, F.S.Stephens
J.Chem.Soc.,Dalton Trans.,186, 1980
See also R1 : 73

71.129 threo – 1 – Bromomercuri – 2 – t – butylperoxy – 1,2 – diphenylethane
$C_{18}H_{21}BrHgO_2$
J.Halfpenny, R.W.H.Small
J.Chem.Soc.,Chem.Commun.,879, 1979

71.130 Bipyridyl – bis(trimethylsilylmethyl) – cadmium(ii)
2,2′ – bipyridyl
$C_{18}H_{30}CdN_2Si_2$, $0.5C_{10}H_8N_2$
G.W.Bushnell, S.R.Stobart *Can.J.Chem.*,**58**, 574, 1980
See also R1 : 83,63

71.131 bis(μ – Chloro) – bis(chloro – (neopentyl(isopropoxy)
carbene) – platinum)
$C_{18}H_{36}Cl_4O_2Pt_2$
V.B.Pukhnarevich, Yu.T.Struchkov, G.G.Aleksandrov,
S.P.Sushchinskaya, E.O.Tsetlina, M.G.Voronkov
Koord.Khim.,**5**, 1535, 1979

71.132 bis(μ – Acetato) – bis((trimethylphosphine) –
(trimethylsilylmethyl) – molybdenum(ii))
$C_{18}H_{46}Mo_2O_4P_2Si_2$
M.B.Hursthouse, K.M.A.Malik
Acta Crystallogr.,Sect.B,**35**, 2709, 1979
See also R1 : 81,86

71.133 η^6 – Toluene – bis(perfluorophenyl) – cobalt
$C_{19}H_8CoF_{10}$
L.J.Radonovich, K.J.Klabunde, C.B.Behrens, D.P.McCollor,
B.B.Anderson *Inorg.Chem.*,**19**, 1221, 1980
See also R1 : 74

71.134 η^6 – Toluene – bis(perfluorophenyl) – nickel
$C_{19}H_8F_{10}Ni$
L.J.Radonovich, K.J.Klabunde, C.B.Behrens, D.P.McCollor,
B.B.Anderson *Inorg.Chem.*,**19**, 1221, 1980
See also R1 : 74

71.135 (μ – Hydrido) – ((α – methylenebenzyl) –
acetylido) – tris(tricarbonyl – ruthenium)
$C_{19}H_8O_9Ru_3$
S.Ermer, R.Karpelus, S.Miura, E.Rosenberg, A.Tiripicchio,
A.M.M.Lanfredi *J.Organomet.Chem.*,**187**, 81, 1980
See also R1 : 72

71.136 (Dicarbonyl – cyclopentadienyl – manganese) – μ –
phenylketenyl – (tetracarbonyl – rhenium)
$C_{19}H_{10}MnO_7Re$
O.Orama, U.Schubert, F.R.Kreissl, E.O.Fischer
Z.Naturforsch.,Teil B,**35**, 82, 1980
See also R1 : 73

71.137 Pyridinium tetrachloro – (pyridine) – (tolane) –
tantalum
$C_{19}H_{15}Cl_4NTa^-$, $C_5H_6N^+$
F.A.Cotton, W.T.Hall *Inorg.Chem.*,**19**, 2352, 1980
See also R1 : 83 R2 : 33

71.138 (1,5 – Diphenylthiocarbazonato – N,S) – phenyl –
mercury(ii) (yellow form)
$C_{19}H_{16}HgN_4S$
A.T.Hutton, H.M.N.H.Irving, L.R.Nassimbeni, G.Gafner
Acta Crystallogr.,Sect.B,**36**, 2064, 1980
See also R1 : 85,83

71.139 Tetracarbonyl – (3 – (diphenylphosphineoxy)
propyl) – manganese
$C_{19}H_{16}MnO_5P$
E.Lindner, H.–J.Eberle
J.Organomet.Chem.,**191**, 143, 1980
See also R1 : 86

71.140 (μ – Carbonyl) – (μ – t – butoxy –
carbonylmethylene) – bis(carbonyl – (η^5 –
cyclopentadienyl) – iron)
$C_{19}H_{20}Fe_2O_5$
W.A.Herrmann, J.Plank, I.Bernal, M.Creswick
Z.Naturforsch.,Teil B,**35**, 680, 1980
See also R1 : 73

71.141 cis – Tetracarbonyl – ((1 – methyl – 2 – phenyl –
3 – aza – but – 2 – enyl) – diethylaminocarbene) –
chromium
$C_{19}H_{22}CrN_2O_4$
K.H.Dotz, B.Fugen–Koster, D.Neugebauer
J.Organomet.Chem.,**182**, 489, 1979
See also R1 : 83

71.142 (μ – Methoxy – phenyl – carbene) –
(bis(trimethylphosphine) – platinum) –
(pentacarbonyl – tungsten) (at 200°K)
$C_{19}H_{26}O_6P_2PtW$
T.V.Ashworth, J.A.K.Howard, M.Laguna, F.G.A.Stone
J.Chem.Soc.,Dalton Trans.,1593, 1980
See also R1 : 86

71.143 ((R) – 1 – Cyanoethyl) – ((S) – (–) – α –
methylbenzylamine) – bis(dimethylglyoximato) –
cobalt(iii)
$C_{19}H_{29}CoN_6O_4$
Y.Ohashi, Y.Sasada, S.Takeuchi, Y.Ohgo
Bull.Chem.Soc.Jpn.,**53**, 627, 1980
See also R1 : 83

71.144 ((R) – 1 – Cyanoethyl) – ((S) – (–) – α –
methylbenzylamine) – bis(dimethylglyoximato) –
cobalt(iii) (at 173°K)
$C_{19}H_{29}CoN_6O_4$
Y.Ohashi, Y.Sasada, S.Takeuchi, Y.Ohgo
Bull.Chem.Soc.Jpn.,**53**, 627, 1980
See also R1 : 83

71.145 ((S) – 1 – Cyanoethyl) – ((S) – (–) – α –
methylbenzylamine) – bis(dimethylglyoximato) –
cobalt(iii) (at 173°K)
$C_{19}H_{29}CoN_6O_4$
Y.Ohashi, Y.Sasada, S.Takeuchi, Y.Ohgo
Bull.Chem.Soc.Jpn.,**53**, 1501, 1980
See also R1 : 83

71.146 ((S) – 1 – Cyanoethyl) – ((S) – (–) – α –
methylbenzylamine) – bis(dimethylglyoximato) –
cobalt(iii)
$C_{19}H_{29}CoN_6O_4$
Y.Ohashi, Y.Sasada, S.Takeuchi, Y.Ohgo
Bull.Chem.Soc.Jpn.,**53**, 1501, 1980
See also R1 : 83

71.147 bis(μ_3 – Acetonitrile – copper) – carbido – ennea –
(μ – carbonyl) – hexacarbonyl – hexa – rhodium
methanol solvate
$C_{20}H_6Cu_2N_2O_{15}Rh_6$, $0.5CH_4O$
V.G.Albano, D.Braga, S.Martinengo, P.Chini, M.Sansoni,
D.Strumolo *J.Chem.Soc.,Dalton Trans.*,52, 1980
See also R1 : 83

71.148 (μ^2 – Benzylmethylene) – bis(μ^2 – hydrido) –
dodecacarbonyl – tetra – osmium
$C_{20}H_{10}O_{12}Os_4$
B.F.G.Johnson, J.W.Kelland, J.Lewis, A.L.Mann, P.R.Raithby
J.Chem.Soc.,Chem.Commun.,547, 1980

71.149 Dicarbonyl – cyclopentadienyl – manganese – (μ – p – tolylketenyl) – tetracarbonyl – manganese diethyl ether solvate
$C_{20}H_{12}Mn_2O_7, 0.5C_4H_{10}O$
J.Martin–Gil, J.A.K.Howard, R.Navarro, F.G.A.Stone
J.Chem.Soc.,Chem.Commun.,1168, 1979
See also R1 :

71.150 bis(μ^2 – Hydrido) – (μ^3 – η^4 – 1,3 – dimethyl – 1 – (phenylimino) – butane) – octacarbonyl – tri – osmium
$C_{20}H_{15}NO_8Os_3$
R.D.Adams, J.P.Selegue *Inorg.Chem.*,19, 1795, 1980

71.151 (η^5 – Cyclopentadienyl – nickel) – (μ^4 – 3,3 – dimethyl – but – 1 – enyl) – nonacarbonyl – tri – ruthenium
$C_{20}H_{15}NiO_9Ru_3$
E.Sappa, A.Tiripicchio, M.T.Camellini
Inorg.Chim.Acta,41, 11, 1980
See also R1 : 72,73

71.152 (3,4,7,8 – Tetramethyl – 1,10 – phenanthroline) – (bis(trichlorovinyl)) – mercury
$C_{20}H_{16}Cl_6HgN_2$
N.A.Bell, I.W.Nowell
Acta Crystallogr.,Sect.B,36, 447, 1980
See also R1 : 83

71.153 (μ – Thio – ferrocenyl – methylmethane – thiomethylene – C',S²) – 1,1,1,2,2,2 – hexa – carbonyl – (μ – methylthio) – iron
$C_{20}H_{16}Fe_3O_6S_3$
H.Patin, G.Mignani, C.Mahe, J.Y.Le Marouille, T.G.Southern, A.Benoit, D.Grandjean
J.Organomet.Chem.,197, 315, 1980
See also R1 : 73,85

71.154 Carbonyl – cyclopentadienyl – (5 – methyl – 2 – (phenylcarbenylamino) – phenyl) – iron
$C_{20}H_{17}FeNO$
R.D.Adams, D.F.Chodosh, N.M.Golembeski, E.C.Weissman
J.Organomet.Chem.,172, 251, 1979
See also R1 : 73

71.155 bis(η^5 – Cyclopentadienyl) – bis(cyclopentadienyl) – hafnium (reinvestigation of structure, published by Kulisher et al)
$C_{20}H_{18}Hf$
R.Vann Bynum, R.D.Rogers, J.L.Atwood
Am.Cryst.Assoc.,Ser.2,7, 22, 1980
See also R1 : 73

71.156 bis(μ – sigma:eta – Cyclopentadienyl) – bis(η – cyclopentadienyl – molybdenum)
$C_{20}H_{18}Mo_2$
B.Meunier, K.Prout
Acta Crystallogr.,Sect.B,35, 2558, 1979
See also R1 : 73

71.157 tetrakis(μ – 1,3 – Di – isocyanopropane) – bis(chloro – rhodium) chloride octahydrate
$C_{20}H_{24}Cl_2N_8Rh_2^{2+}, 2Cl^-, 8H_2O$
K.R.Mann, R.A.Bell, H.B.Gray *Inorg.Chem.*,18, 2671, 1979

71.158 tetrakis(μ^2 – 1,3 – Di – isocyanopropane) – di – rhodium bis(tetraphenylborate) acetonitrile solvate
$C_{20}H_{24}N_8Rh_2^{2+}, 2C_{24}H_{20}B^-, C_2H_3N$
K.R.Mann. J.A.Thich, R.A.Bell, C.L.Coyle, H.B.Gray
Inorg.Chem.,19, 2462, 1980
See also R2 : 62

71.159 (μ – Chloro) – 1 – (dicyclopentadienyl – zirconium(iv)) – 2,2 – bis(diethyl – aluminium) ethane
$C_{20}H_{33}Al_2ClZr$
J.Kopf, W.Kaminsky, H.–J.Vollmer
Cryst.Struct.Commun.,9, 197, 1980
See also R1 : 73,68

71.160 Chloro – (dimethylphenylphosphonium – (trimethylsilyl)methylide) – (1,5 – cyclo – octadiene) – palladium(ii) hexafluorophosphate
$C_{20}H_{33}ClPPdSi^+, F_6P^-$
R.M.Buchanan, C.G.Pierpont *Inorg.Chem.*,18, 3608, 1979
See also R1 : 75

71.161 (N,N,N',N' – Tetramethylethylenediamine) – bis(cyclohexylisocyanide) – copper(i) tetraphenylborate
$C_{20}H_{38}CuN_4^+, C_{24}H_{20}B^-$
M.Pasquali, C.Floriani, A.G.Manfredotti, A.C.Villa
Inorg.Chem.,18, 3535, 1979
See also R1 : 76 R2 : 62

71.162 bis(Triethylphosphine) – (1,1,1 – trifluoro – 5 – methoxy – 3 – (trifluoromethyl) – 4 – methyl – pent – 3 – en – 2 – yl) – platinum hexafluorophosphate
$C_{20}H_{39}F_6OP_2Pt^+, F_6P^-$
H.C.Clark, S.S.McBride, N.C.Payne, C.S.Wong
J.Organomet.Chem.,178, 393, 1979
See also R1 : 86,84

71.163 bis(μ – Chloro) – bis(bis(carbonyl) – trimethylphosphine – (1 – 2 – η – trimethylsilyl – methylcarbonyl) – molybdenum(ii))
$C_{20}H_{40}Cl_2Mo_2O_8P_2Si_2$
E.C.Guzman, G.Wilkinson, R.D.Rogers, W.E.Hunter, M.J.Zaworotko, J.L.Atwood
J.Chem.Soc.,Dalton Trans.,229, 1980
See also R1 : 86,84,63

71.164 (1,2 – bis(Methoxycarbonyl) – but – 2 – en – 1 – yl) – bis(triethylphosphine) – platinum(ii) tetraphenylborate
$C_{20}H_{41}O_4P_2Pt^+, C_{24}H_{20}B^-$
C.P.Brock, T.G.Attig *J.Am.Chem.Soc.*,102, 1319, 1980
See also R1 : 86,84 R2 : 62

71.165 (1,2 – bis(Methoxycarbonyl) – but – 2 – en – 1 – yl) – bis(triethylphosphine) – platinum(ii) hexafluorophosphate
$C_{20}H_{41}O_4P_2Pt^+, F_6P^-$
C.P.Brock, T.G.Attig *J.Am.Chem.Soc.*,102, 1319, 1980
See also R1 : 86,84

71.166 bis(μ – Methylene) – bis(tris(trimethylphosphine) – ruthenium(iii)) tetrafluoroborate
$C_{20}H_{58}P_6Ru_2^{2+}, 2BF_4^-$
M.B.Hursthouse, R.A.Jones, K.M.A.Malik, G.Wilkinson
J.Am.Chem.Soc.,101, 4128, 1979
See also R1 : 86

71.167 (μ – 1,3 – Dioxo – indan – 2 – ylidene) – bis(carbonyl – (η^5 – cyclopentadieneyl) – cobalt)
$C_{21}H_{14}Co_2O_4$
M.Creswick, I.Bernal, W.A.Herrmann, I.Steffl
Chem.Ber.,113, 1377, 1980
See also R1 : 73

71.168 η – Carbonyl – η – (1 – 3 – eta:1 – σ,4 – 2' – η – (1,3 – dimethyl – 4 – (5 – oxo – 4 – phenyl – 2'(5H) – furan – 2 – ylidene) – but – 1 – ene – 1,3 – diyl)) – bis(dicarbonyl – cobalt)
$C_{21}H_{14}Co_2O_7$
J.A.D.Jeffreys $J.Chem.Soc.,Dalton\ Trans.,435,$ 1980
See also R1 : 72

71.169 tris(η^5 – Cyclopentadienyl) – (μ_3 – (pentacarbonyl – chromium – thio) – methylidyne) – (μ_3 – sulfido) – triangulo – tri – cobalt tetrahydrofuran solvate
$C_{21}H_{15}Co_3CrO_5S_2, 0.5C_4H_8O$
H.Werner, K.Leonhard, O.Kolb, E.Rottinger, H.Vahrenkamp $Chem.Ber.,$**113**, 1654, 1980
See also R1 : 73,85

71.170 bis(η^5 – Cyclopentadienyl) – (μ – η^2,η^2 – carbonyl – dimethylacetylene – carboxylate) – bis(dicarbonyl – tungsten) (at 233°K)
$C_{21}H_{16}O_9W_2$
S.R.Finnimore, S.A.R.Knox, G.E.Taylor $J.Chem.Soc.,Chem.Commun.,$411, 1980
See also R1 : 72,73

71.171 bis(μ – Hydrido) – (μ^3 – η^2 – 1 – isopropyl – 2 – methyl – 2 – (phenylamino) – ethylene) – nonacarbonyl – tri – osmium
$C_{21}H_{17}NO_9Os_3$
R.D.Adams, J.P.Selegue $Inorg.Chem.,$**19**, 1795, 1980

71.172 3,3,3,3 – Tetracarbonyl – 5,5 – dimethyl – 2,2 – diphenyl – 1 – oxa – 2 – phospha – 3 – rhenacyclohexane
$C_{21}H_{20}O_5PRe$
E.Lindner, G.von Au $Z.Naturforsch.,Teil\ B,$**35**, 1104, 1980
See also R1 : 86

71.173 (bis(Trimethylphosphine) – platinum) – (μ – carbonyl) – (μ – p – tolylcarbenyl) – (carbonyl – cyclopentadienyl – manganese) tetrafluoroborate dichloromethane solvate
$C_{21}H_{30}MnO_2P_2Pt^+, BF_4^-, CH_2Cl_2$
J.A.K.Howard, J.C.Jeffery, M.Laguna, R.Navarro, F.G.A.Stone $J.Chem.Soc.,Chem.Commun.,$1170, 1979
See also R1 : 73,86

71.174 (1,2 – bis(Dimethylphosphino)ethane) – (neopentylidyne) – (neopentylidene) – (neopentyl) – tungsten(vi)
$C_{21}H_{46}P_2W$
M.R.Churchill, W.J.Youngs $Inorg.Chem.,$**18**, 2454, 1979
See also R1 : 86

71.175 tris(μ – Methylene) – bis(tris(trimethylphosphine) – ruthenium(ii))
$C_{21}H_{60}P_6Ru_2$
M.B.Hursthouse, R.A.Jones, K.M.A.Malik, G.Wilkinson $J.Am.Chem.Soc.,$**101**, 4128, 1979
See also R1 : 86

71.176 (μ – Methyl) – bis(μ – methylene) – bis(tris(trimethylphosphine) – ruthenium(ii) tetrafluoroborate
$C_{21}H_{61}P_6Ru_2^+, BF_4^-$
M.B.Hursthouse, R.A.Jones, K.M.A.Malik, G.Wilkinson $J.Am.Chem.Soc.,$**101**, 4128, 1979
See also R1 : 86

71.177 Dodecacarbonyl – (η_5 – carbido) – (μ – hydrido) – tris(η_2 – ethylsulfide) – hexa – ruthenium
$C_{22}H_{16}O_{15}Ru_6S_3$
B.F.G.Johnson, J.Lewis, K.Wong, M.McPartlin $J.Organomet.Chem.,$**185**, 17, 1980
See also R1 : 85

71.178 (Hex – 3 – ene – 3,4 – diyl) – bis(η^5 – cyclopentadienyl – nickel) – bis(tricarbonyl – iron)
$C_{22}H_{20}Fe_2Ni_2O_6$
A.Marinetti, E.Sappa, A.Tiripicchio, M.T.Camellini $Inorg.Chim.Acta,$**44**, 183, 1980
See also R1 : 73

71.179 cis – bis(2,2' – Bipyridyl) – dimethyl – cobalt(iii) tetraethyl – aluminium
$C_{22}H_{22}CoN_4^+, C_8H_{20}Al^-$
S.Komiya, T.Yamamoto, A.Yamamoto, A.Takenaka, Y.Sasada $Acta\ Crystallogr.,Sect.B,$**35**, 2702, 1979
See also R1 : 83 R2 : 68

71.180 Methyl – (4,4',4'' – triethyl – 2,2':6',2'' – terpyridyl) – mercury nitrate
$C_{22}H_{28}HgN_3^+, NO_3^-$
A.J.Canty, G.Hayhurst, N.Chaichit, B.M.Gatehouse $J.Chem.Soc.,Chem.Commun.,$316, 1980
See also R1 : 83

71.181 tris(μ – t – Butylthiolato) – (μ – hexafluoro – but – 2 – ene – 2,3 – diyl) – tris(dicarbonyl – iridium)
$C_{22}H_{27}F_6Ir_3O_6S_3$
J.Devillers, J.–J.Bonnet, D.de Montauzon, J.Galy, R.Poilblanc $Inorg.Chem.,$**19**, 154, 1980
See also R1 : 85

71.182 (Benzoyl – (1 – pyridinio) – methanido) – bis(dimethylglyoximato) – methyl – cobalt benzene solvate
$C_{22}H_{28}CoN_5O_5, C_6H_6$
T.Saito, H.Urabe, Y.Sasaki $Transition\ Met.Chem.,$**5**, 35, 1980
See also R1 : 83

71.183 rac – a – (μ – Acetato) – b – (O – acetato – mercurio) – cf,de – bis(2 – (dimethylaminomethyl) – phenyl) – platinum
$C_{22}H_{30}HgN_2O_4Pt$
A.F.M.J.van der Ploeg, G.van Koten, K.Vrieze, A.L.Spek, A.J.M.Duisenberg $J.Chem.Soc.,Chem.Commun.,$469, 1980
See also R1 : 81,83

71.184 (μ^2 – Carbonyl) – (μ^2 – methylene) – bis(η^5 – pentamethyl – cyclopentadienyl – cobalt)
$C_{22}H_{32}Co_2O$
T.R.Halbert, M.E.Leonowicz, D.J.Maydonovitch $J.Am.Chem.Soc.,$**102**, 5101, 1980
See also R1 : 73

71.185 (Tricarbonyl – trimethylphosphine – chromium) – (μ – carbonyl) – (μ – phenyl(methoxycarbonyl) methylene) – (bis(trimethylphosphine) – platinum)
$C_{22}H_{35}CrO_6P_3Pt$
J.A.K.Howard, J.C.Jeffery, M.Laguna, R.Navarro, F.G.A.Stone $J.Chem.Soc.,Chem.Commun.,$1170, 1979
See also R1 : 86

71.186 (Tetracarbonyl – trimethylphosphine – tungsten) –
(μ – methoxytolylcarbene)) –
bis(trimethylphosphine) – platinum
$C_{22}H_{37}O_5P_3PtW$
J.A.K.Howard, K.A.Mead, R.Navarro, F.G.A.Stone
*Am.Cryst.Assoc.,Ser.2,*7, 22, 1980
See also R1 : 86

71.187 (μ – 3,3,6,6 – Tetramethyl – 3,6 – diazaoctane –
1,8 – diyl) – bis(N,N,N′,N′ –
tetramethylethylenediamine – chloro –
platinum(ii)) diperchlorate
$C_{22}H_{56}Cl_2N_6Pt_2^{2+}$, $2ClO_4^-$
L.Maresca, G.Natile, A.Tiripicchio, M.T.Camellini,
G.Rizzardi *Inorg.Chim.Acta,*37, L545, 1979
See also R1 : 76

71.188 η^5 – Cyclopentadienyl – (2,4 – diphenyl – cyclobut –
1 – en – 3 – on – 1 – yl) – dicarbonyl – iron
$C_{23}H_{16}FeO_3$
Yu.L.Slovokhotov, A.I.Yanovskii, V.G.Andrianov,
Yu.T.Struchkov *J.Organomet.Chem.,*184, C57, 1980
See also R1 : 73

71.189 Chloro – dicyclopentadienyl –
(diphenylphosphinomethyl) – zirconium (at –140°C)
$C_{23}H_{22}ClPZr$
N.E.Schore, H.Hope *J.Am.Chem.Soc.,*102, 4251, 1980
See also R1 : 73

71.190 tris(Trimethylbenzylammonium) tetracosa –
carbonyl – hydrido – hexa – iron – hexa –
palladium acetonitrile solvate
$C_{24}HFe_6O_{24}Pd_6^{3-}$, $3C_{10}H_{16}N^+$, $2C_2H_3N$
G.Longoni, M.Manassero, M.Sansoni
*J.Am.Chem.Soc.,*102, 3242, 1980

71.191 Hexamethylbenzene – (2 – 3) – tetrafluorobenzo –
bicyclo(2.2.2)octa – 5,7 – diene – rhodium
perchlorate
$C_{24}H_{24}F_4Rh^+$, ClO_4^-
C.Foces-Foces, F.H.Cano, S.Garcia-Blanco
*Eur.Cryst.Meeting,*6, 63, 1980
See also R1 : 74,75

71.192 (2 – aci – Nitrato – propan – 1 – on – 1 – yl) –
bis(dimethylphenylphosphine) – pyridine – chloro –
iridium acetone solvate
$C_{24}H_{30}ClIrN_2O_3P_2$, C_3H_6O
T.A.B.M.Bolsman, J.A.van Doorn
*J.Organomet.Chem.,*178, 381, 1979
See also R1 : 86,83

71.193 bis(Trimethylsilylmethylidyne) –
tetrakis(trimethylsilylmethyl) – di – rhenium
$C_{24}H_{62}Re_2Si_6$
M.Bochmann, G.Wilkinson, A.M.R.Galas, M.B.Hursthouse,
K.M.A.Malik *J.Chem.Soc.,Dalton Trans.,*1797, 1980

71.194 tris(μ – Chloro) – (N – nitroso – N – trimethylsilyl –
methylhydroxylaminato) –
pentakis(trimethylsilylmethyl) – triangulo – tri –
rhenium(iii)
$C_{24}H_{66}Cl_3N_2O_2Re_3Si_6$
P.Edwards, K.Mertis, G.Wilkinson, M.B.Hursthouse,
K.M.A.Malik *J.Chem.Soc.,Dalton Trans.,*334, 1980
See also R1 : 84,63

71.195 (μ^2 – Hydrido) – (μ^2 – p – tolylcarbamoyl) –
dimethylphenylphosphine – nonacarbonyl – tri –
osmium
$C_{25}H_{20}NO_{10}Os_3P$
R.D.Adams, J.P.Selegue *Am.Cryst.Assoc.,Ser.2,*8, 28, 1980
See also R1 : 86,84

71.196 bis(Dimethylglyoximato – N,N′) – (4 – ((cis – 3 –
ethoxy – cyclobutyloxy) – dioxosulfuranyl) –
phenyl) – pyridine – cobalt
$C_{25}H_{34}CoN_5O_8S$
D.Lenoir, H.Dauner, I.Ugi, A.Gieren, R.Hubner, V.Lamm
*J.Organomet.Chem.,*198, C39, 1980
See also R1 : 83

71.197 bis(η^5 – Cyclopentadienyl) – 2,2 – bis(diethyl –
alumino) – ethyl – zirconium – cyclopentadienide
$C_{25}H_{38}Al_2Zr$
J.Kopf, H.–J.Vollmer, W.Kaminsky
*Cryst.Struct.Commun.,*9, 271, 1980
See also R1 : 73,68

71.198 (μ^2 – Diphenylphosphido) – (μ^3 – 3 –
methylbutynyl) – nonacarbonyl – tri – ruthenium
$C_{26}H_{17}O_9PRu_3$
A.J.Carty, S.A.McLaughlin, N.J.Taylor
*Am.Cryst.Assoc.,Ser.2,*8, 28, 1980
See also R1 : 72,86

71.199 (μ^2 – Diphenylphosphido) – (μ^3 – 3,3 –
dimethylbutynyl) – octacarbonyl – tri – ruthenium
benzene solvate
$C_{26}H_{19}O_8PRu_3$, $0.5C_6H_6$
A.J.Carty, S.A.McLaughlin, N.J.Taylor
*Am.Cryst.Assoc.,Ser.2,*8, 28, 1980
See also R1 : 86

71.200 (1 – (o – Diphenylarsinophenyl) – 2 –
methoxyethyl) – (hexafluoroacetylacetonato) –
platinum(ii)
$C_{26}H_{21}AsF_6O_3Pt$
M.K.Cooper, P.J.Guerney, M.McPartlin
*J.Chem.Soc.,Dalton Trans.,*349, 1980
See also R1 : 77,86

71.201 Dicarbonyl – (η^5 – cyclopentadienyl) –
(triphenylsilylcarbene) – rhenium dichloromethane
solvate
$C_{26}H_{21}O_2ReSi$, CH_2Cl_2
E.O.Fischer, P.Rustemeyer, D.Neugebauer
*Z.Naturforsch.,Teil B,*35, 1083, 1980
See also R1 : 73

71.202 (μ – Cyclobuta(1,2 – a:3,4 – a′)dicyclopentene) –
bis(dicarbonyl – (η^5 – methylcyclopentadienyl) –
manganese)
$C_{26}H_{22}Mn_2O_4$
W.A.Herrmann, K.Weidenhammer, M.L.Ziegler
*Z.Anorg.Allg.Chem.,*460, 200, 1980
See also R1 : 73

71.203 (2,6 – Dimethoxyphenyl) – (triphenylphosphine) –
gold(i)
$C_{26}H_{24}AuO_2P$
G.van Koten, C.A.Schaap, J.T.B.H.Jastrzebski, J.G.Noltes
*J.Organomet.Chem.,*186, 427, 1980
See also R1 : 86

71.204 Cyclopentadienyl – (η^2 – benzaldehyde) – (α – (2 – pyridyl) – α – (methylamino) – benzyl) – carbonyl – molybdenum
$C_{26}H_{24}MoN_2O_2$
H.Brunner, J.Wachter, I.Bernal, M.Creswick
Angew.Chem.,Int.Ed.Engl.,**18**, 861, 1979
See also R1 : 72,73

71.205 Tricarbonyl – (triphenylgermyl) – (ethylethoxycarbene) – cobalt
$C_{26}H_{25}CoGeO_4$
F.Carre, G.Cerveau, E.Colomer, R.J.P.Corriu, J.C.Young, L.Richard, R.Weiss *J.Organomet.Chem.*,**179**, 215, 1979
See also R1 : 69

71.206 η^5 – Cyclopentadienyl – iodo – (methyl(methylthio) – carbene) – (triphenylphosphine) – iridium(iii) iodide
$C_{26}H_{26}IIrPS^+$, I$^-$
G.Bombieri, F.Faraone, G.Bruno, G.Faraone
J.Organomet.Chem.,**188**, 379, 1980
See also R1 : 73,86

71.207 bis(o – (2,2,5,5 – Tetramethyl – Δ^3 – imidazoline – 1 – oxy – 4 – yl)phenyl) – mercury
$C_{26}H_{32}HgN_4O_2$
A.B.Shapiro, L.B.Volodarskii, O.N.Krasochka. L.O.Atovmyan, E.G.Rozantsev
Dokl.Akad.Nauk SSSR,**248**, 1135, 1979
See also R1 : 83

71.208 Cyanomethyl – tris(dimethylphenylphosphine) – platinum hexafluorophosphate
$C_{26}H_{35}NP_3Pt^+$, F_6P^-
P.S.Pregosin, R.Favez, R.Roulet, T.Boschi, R.A.Michelin, R.Ros *Inorg.Chim.Acta*,**45**, L7, 1980
See also R1 : 86

71.209 (μ – Oxo) – (μ – trimethylphosphine – methyl) – bis(η^5 – ethyltetramethyl – cyclopentadienyl) – hydrido – bis(dichloro – tantalum)
$C_{26}H_{45}Cl_4OPTa_2$
P.Belmonte, R.R.Schrock, M.R.Churchill, W.J.Youngs
J.Am.Chem.Soc.,**102**, 2858, 1980
See also R1 : 73

71.210 bis(μ^2 – Hydrido) – (η^3 – 1 – methyl – 2,3 – diphenylallyl) – undecacarbonyl – tetra – osmium
$C_{27}H_{16}O_{11}Os_4$
B.F.G.Johnson, J.W.Kelland, J.Lewis, A.L.Mann, P.R.Raithby
J.Chem.Soc.,Chem.Commun.,547, 1980
See also R1 : 72

71.211 (μ – Diphenylphosphino) – (μ – N – methyl – (2 – phenyl – 2 – iminoethyl)) – bis(tricarbonyl – iron)
$C_{27}H_{20}Fe_2NO_6P$
A.J.Carty, G.N.Mott, N.J.Taylor
J.Organomet.Chem.,**182**, C69, 1979
See also R1 : 86,83

71.212 bis(η^5 – Cyclopentadienyl) – (μ – carbonyl) – (μ – η',η^3 – carbonyl – 1,2 – diphenylethylene) – carbonyl – di – ruthenium (at 233°K)
$C_{27}H_{20}O_3Ru_2$
A.F.Dyke, S.A.R.Knox, P.J.Naish, G.E.Taylor
J.Chem.Soc.,Chem.Commun.,409, 1980
See also R1 : 72,

71.213 Chloro – trifluoromethyl – (cis – 1,2 – bis(diphenylphosphino)ethylene) – platinum(ii)
$C_{27}H_{22}ClF_3P_2Pt$
A.Del Pra, G.Zanotti, A.Piazzesi, U.Belluco, R.Ros
Transition Met.Chem.,**4**, 381, 1979
See also R1 : 86

71.214 Pentacarbonyl – (3 – (1,1 – diphenylvinyl) – cyclopentyl) – ethoxymethylene – tungsten(0)
$C_{27}H_{24}O_6W$
J.–C.Daran, Y.Jeannin
Acta Crystallogr.,Sect.B,**36**, 1392, 1980

71.215 trans – Triphenyl – tin – tetracarbonyl – diethylaminocarbyne – chromium
$C_{27}H_{25}CrNO_4Sn$
U.Schubert *Cryst.Struct.Commun.*,**9**, 383, 1980
See also R1 : 69

71.216 1 – ((Di(pivaloyl) – methanato) – bis(pyridine) – rhoda) – 2,3 – bis(trifluoromethyl) – cyclopent – 2 – ene
$C_{27}H_{33}F_6N_2O_2Rh$
C.E.Dean, R.D.W.Kemmitt, D.R.Russell, M.D.Schilling
J.Organomet.Chem.,**187**, C1, 1980
See also R1 : 77,83

71.217 bis((μ^2 – t – Butylthiolato) – (dimethylphenylphosphino)) – acetyl – carbonyl – iodo – di – rhodium
$C_{27}H_{43}IO_2P_2Rh_2S_2$
A.Mayanza, J.–J.Bonnet, J.Galy, P.Kalck, R.Poilblanc
J.Chem.Res.,**146**, 2101, 1980
See also R1 : 86,85

71.218 bis(Diphenylphosphino)methane – tricarbonyl – di – iodo – tungsten
$C_{28}H_{22}I_2O_3P_2W$
R.M.Foy, D.L.Kepert, C.L.Raston, A.H.White
J.Chem.Soc.,Dalton Trans.,440, 1980
See also R1 : 86

71.219 Pentacarbonyl – (diethylamino(triphenyl – tin) carbene) – chromium(0) dichloromethane solvate
$C_{28}H_{25}CrNO_5Sn$, $0.5CH_2Cl_2$
E.O.Fischer, R.B.A.Pardy, U.Schubert
J.Organomet.Chem.,**181**, 37, 1979
See also R1 : 69

71.220 Dicarbonyl – methyl – bis(trimethylphosphine) – (triphenylborato – nitrilo – methyl) – iron(ii)
$C_{28}H_{36}BFeNO_2P_2$
D.Ginderow *Acta Crystallogr.,Sect.B*,**36**, 1950, 1980
See also R1 : 86

71.221 (μ – Oxo) – bis(n – butyl – bis(cyclopentadienyl) – niobium)
$C_{28}H_{38}Nb_2O$
N.I.Kirillova, D.A.Lemenovskii, T.V.Baukova, Yu.T.Struchkov *Koord.Khim.*,**3**, 1600, 1977
See also R1 : 73

71.222 bis(bis(μ^2 – Methylene) – tetrakis(trimethylphosphine) – ruthenium(iii)) – ruthenium(iv) bis(tetrafluoroborate)
$C_{28}H_{80}P_8Ru_3^{2+}$, $2BF_4^-$
R.A.Jones, G.Wilkinson, A.M.R.Galas, M.B.Hursthouse, K.M.A.Malik *J.Chem.Soc.,Dalton Trans.*,1771, 1980
See also R1 : 86

71.223 (μ^3 – 1 – 6 – η – Bitropyl) – carbido – (μ^2 – carbonyl) – tridecacarbonyl – octahedro – hexa – ruthenium
$C_{29}H_{14}O_{14}Ru_6$
G.B.Ansell, J.S.Bradley
Acta Crystallogr.,Sect.B,**36**, 1930, 1980
See also R1 : 75

71.224 1,2 – bis(Diphenylphosphino)ethane – tricarbonyl – di – iodo – molybdenum dichloromethane solvate
$C_{29}H_{24}I_2MoO_3P_2$, CH_2Cl_2
R.M.Foy, D.L.Kepert, C.L.Raston, A.H.White
J.Chem.Soc.,Dalton Trans.,440, 1980
See also R1 : 86

71.225 (bis(Diphenylphosphino)methanide) – (dimethylphosphonium – bis(methylide)) – platinum(ii)
$C_{29}H_{31}P_3Pt$
J.–M.Bassett, J.R.Mandl, H.Schmidbaur
Chem.Ber.,**113**, 1145, 1980
See also R1 : 86

71.226 Chloro – bis(triethylphosphine) – ((N – p – tolyl) – (N' – (p – tolyl – o – yl)) – imidazolidin – 2 – ylidene) – ruthenium
(1,3 – Bis(4 – tolyl) – imidazolidin – 2 – ylidene) – chloro – bis(triethylphosphine) – ruthenium(ii)
$C_{29}H_{47}ClN_2P_2Ru$
P.B.Hitchcock, M.F.Lappert, P.L.Pye, S.Thomas
J.Chem.Soc.,Dalton Trans.,1929, 1979
See also R1 : 86

71.227 Hexamethyl – bis(diethylphenylphosphine) – tris(μ – methyl) – triangulo – tri – rhenium(iii)
$C_{29}H_{57}P_2Re_3$
P.Edwards, K.Mertis, G.Wilkinson, M.B.Hursthouse, K.M.A.Malik *J.Chem.Soc.,Dalton Trans.*,334, 1980
See also R1 : 86

71.228 bis((μ^2 – Bromo) – ((μ^2 – 1 – t – butylacetylene) – nonacarbonyl – tri – ruthenium) – mercury)
$C_{30}H_{18}Br_2Hg_2O_{18}Ru_6$
R.Fahmy, K.King, E.Rosenberg, A.Tiripicchio, M.T.Camellini *J.Am.Chem.Soc.*,**102**, 3626, 1980

71.229 1,3 – bis(Diphenylphosphino) – propane – tricarbonyl – di – iodo – molybdenum
$C_{30}H_{26}I_2MoO_3P_2$
R.M.Foy, D.L.Kepert, C.L.Raston, A.H.White
J.Chem.Soc.,Dalton Trans.,440, 1980
See also R1 : 86

71.230 (μ – Hydrido) – (μ – 2 – η^1:1 – 2 – η^2 – naphthyl) – bis(bis(η^5 – cyclopentadienyl) – zirconium)
$C_{30}H_{28}Zr_2$
G.P.Pez, C.F.Putnik, S.L.Suib, G.D.Stucky
J.Am.Chem.Soc.,**101**, 6933, 1979
See also R1 : 73,74

71.231 Carbonyl – chloro – bis(triethylphosphine) – ((N – (p – tolyl – o – yl)) – imidazolidin – 2 – ylidene) – ruthenium(ii)
$C_{30}H_{47}ClN_2OP_2Ru$
P.B.Hitchcock, M.F.Lappert, P.L.Pye, S.Thomas
J.Chem.Soc.,Dalton Trans.,1929, 1979
See also R1 : 86

71.232 bis(Triphenylphosphine)immonium (tricarbonyl – iron) – bis(μ – diphenylphosphido) – (acetyl – dicarbonyl – iron)
$C_{31}H_{23}Fe_2O_6P_2{}^-$, $C_{36}H_{30}NP_2{}^+$
R.E.Ginsburg, J.M.Berg, R.K.Rothrock, J.P.Collman, K.O.Hodgson, L.F.Dahl *J.Am.Chem.Soc.*,**101**, 7218, 1979
See also R1 : 86 R2 : 64

71.233 Sodium (tricarbonyl – iron) – bis(μ – diphenylphosphido) – (acetyl – dicarbonyl – iron) tetrahydrofuran solvate
$C_{31}H_{23}Fe_2O_6P_2{}^-$, Na$^+$, $2C_4H_8O$
R.E.Ginsburg, J.M.Berg, R.K.Rothrock, J.P.Collman, K.O.Hodgson, L.F.Dahl *J.Am.Chem.Soc.*,**101**, 7218, 1979
See also R1 : 86

71.234 Carbonyl – (phenylthiocarbyne) – (η^5 – cyclopentadienyl) – triphenylphosphine – tungsten
$C_{31}H_{25}OPSW$
W.W.Greaves, R.J.Angelici, B.J.Helland, R.Klima, R.A.Jacobson *J.Am.Chem.Soc.*,**101**, 7618, 1979
See also R1 : 73,86

71.235 1 – Triphenylphosphine – 1 – chloro – 4,5 – benzo – 3 – (3' – phenyl – 5' – ethylidenyl – 1',2' – dithiol – 3' – ene) – 1 – pallada – 2 – thiol – 2 – ene toluene solvate
$C_{31}H_{28}ClPPdS_3$, $2C_7H_8$
B.Bogdanovic, C.Kruger, P.Locatelli
Angew.Chem.,Int.Ed.Engl.,**18**, 684, 1979
See also R1 : 86,85

71.236 bis(μ – Acetato) – bis((2 – (2' – benzoxazolyl) – 5 – methyl – phenyl) – palladium(ii))
$C_{32}H_{26}N_2O_6Pd_2$
M.R.Churchill, H.J.Wasserman, G.J.Young
Inorg.Chem.,**19**, 762, 1980
See also R1 : 81,83

71.237 tris(Dimethylphenylphosphine) – (o – dimethylphosphinophenyl) – hydrido – iridium hexafluorophosphate
$C_{32}H_{44}IrP_4{}^+$, F_6P^-
R.H.Crabtree, J.M.Quirk, H.Felkin, T.Fillebeen–Khan, C.Pascard *J.Organomet.Chem.*,**187**, C32, 1980
See also R1 : 86

71.238 (μ – Methoxy – phenyl – carbene) – (μ – tetracarbonyl) – tungsten – bis(carbonyl – (methyl – di – t – butylphosphine) – platinum)
$C_{32}H_{50}O_7P_2Pt_2W$
T.V.Ashworth, M.Berry, J.A.K.Howard, M.Laguna, F.G.A.Stone *J.Chem.Soc.,Dalton Trans.*,1615, 1980
See also R1 : 86

71.239 (2 – Benzoylphenyl) – tricarbonyl – (triphenylphosphine) – rhenium
$C_{34}H_{24}O_4PRe$
H.Preut, H.–J.Haupt
Acta Crystallogr.,Sect.B,**36**, 1196, 1980
See also R1 : 86,84

71.240 Benzoyl – dichloro – (1,3 – bis(diphenylphosphino) – propane) – rhodium
$C_{34}H_{31}Cl_2OP_2Rh$
M.F.McGuiggan, D.H.Doughty, L.H.Pignolet
J.Organomet.Chem.,**185**, 241, 1980
See also R1 : 86

71.241 bis(η^8 – Cyclo – octatetraene) – (2,3 – dimethyl – 1,4 – diphenyl – buta – 1,3 – diene – 1,4 – diyl – titanium) – titanium tetrahydrofuran solvate
$C_{34}H_{32}Ti_2$
M.E.E.Veldman, H.R.van der Wal, S.J.Veenstra,
H.J.de L.Meijer *J.Organomet.Chem.*, **197**, 59, 1980
See also R1 : 75

71.242 3,7,9,13 – Tetramethyl – 6 – (bis(diphenylphosphino) ethane) – 3,7,9,13 – tetracarba – 6 – nickela – tridecaborane
$C_{34}H_{44}B_8NiP_2$
R.N.Grimes, E.Sinn, J.R.Pipal *Inorg.Chem.*, **19**, 2087, 1980
See also R1 : 86,62

71.243 (η^5 – Cyclopentadienyl) – (2 – methyl – 4,5 – bis(diphenylphosphino) – pent – 2 – en – 3 – yl) – iron(ii)
$C_{35}H_{34}FeP_2$
R.D.Adams, A.Davison, J.P.Selegue
J.Am.Chem.Soc., **101**, 7232, 1979
See also R1 : 73,86

71.244 heptakis(t – Butylisocyanide) – tungsten(ii) hexatungstate
$C_{35}H_{63}N_7W^{2+}$, $O_{19}W_6^{2-}$
W.A.LaRue, A.T.Liu, J.S.Filippo Junior
Inorg.Chem., **19**, 315, 1980

71.245 Lithium (tricarbonyl – iron) – bis(μ – diphenylphosphido) – (benzoyl – dicarbonyl – iron) tetrahydrofuran solvate
$C_{36}H_{25}Fe_2O_6P_2^-$, Li^+, $3C_4H_8O$
R.E.Ginsburg, J.M.Berg, R.K.Rothrock, J.P.Collman,
K.O.Hodgson, L.F.Dahl *J.Am.Chem.Soc.*, **101**, 7218, 1979
See also R1 : 86

71.246 (Cyclo – octadiene) – (cyclo – octadienyl) – (cyclo – octadienediyl) – dodecacarbonyl – hepta – iridium
$C_{36}H_{33}Ir_7O_{12}$
C.G.Pierpont *Inorg.Chem.*, **18**, 2972, 1979
See also R1 : 75

71.247 nonakis(Trimethylsilylmethyl) – hydrido – (μ^2 – hexachloro) – hexa – rhenium
$C_{36}H_{100}Cl_6Re_6Si_9$
K.Mertis, P.G.Edwards, G.Wilkinson, K.M.A.Malik,
M.B.Hursthouse *J.Chem.Soc.,Chem.Commun.*,654, 1980

71.248 trans – (Trifluoromethyl) – chloro – bis(triphenylphosphine) – platinum(ii)
$C_{37}H_{30}ClF_3P_2Pt$
R.Bardi, A.M.Piazzesi, G.Faraone, G.Bruno
Eur.Cryst.Meeting,**6**, 68, 1980
See also R1 : 86

71.249 trans – (Methyl) – chloro – bis(triphenylphosphine) – platinum(ii)
$C_{37}H_{33}ClP_2Pt$
R.Bardi, A.M.Piazzesi, G.Faraone, G.Bruno
Eur.Cryst.Meeting,**6**, 68, 1980
See also R1 : 86

71.250 (bis(2 – Diphenylphosphinoethyl) – phenylphosphine) – tricarbonyl – chromium(0)
$C_{37}H_{33}CrO_3P_3$
M.C.Favas, D.L.Kepert, B.W.Skelton, A.H.White
J.Chem.Soc.,Dalton Trans.,447, 1980
See also R1 : 86

71.251 (bis(2 – Diphenylphosphinoethyl) – phenylphosphine) – tricarbonyl – molybdenum(0)
$C_{37}H_{33}MoO_3P_3$
M.C.Favas, D.L.Kepert, B.W.Skelton, A.H.White
J.Chem.Soc.,Dalton Trans.,447, 1980
See also R1 : 86

71.252 Carbonyl – dichloro – dichlorocarbene – bis(triphenylphosphine) – osmium
$C_{38}H_{30}Cl_4OOsP_2$
G.R.Clark, K.Marsden, W.R.Roper, L.J.Wright
J.Am.Chem.Soc., **102**, 1206, 1980
See also R1 : 86

71.253 trans – anti – bis(η^5 – Cyclopentadienyl – (μ – phenylisocyanide) – (phenylisocyanide) – iron)
$C_{38}H_{30}Fe_2N_4$
W.P.Fehlhammer, A.Mayr, W.Kehr
J.Organomet.Chem., **197**, 327, 1980
See also R1 : 73

71.254 trans – Cyanomethyl – hydrido – bis(triphenylphosphine) – platinum(ii)
$C_{38}H_{33}NP_2Pt$
A.Del Pra, E.Forsellini, G.Bombieri, R.A.Michelin, R.Ros
J.Chem.Soc.,Dalton Trans.,1862, 1979
See also R1 : 86

71.255 Acetylacetonato – (triphenylphosphine) – (Z – 1,2 – diphenyl – 2 – methylethenyl) – nickel
$C_{38}H_{35}NiO_2P$
J.M.Huggins, R.G.Bergman
J.Am.Chem.Soc., **101**, 4410, 1979
See also R1 : 77,86

71.256 trans – Dimesityl – bis(diethylphenylphosphine) – cobalt(ii)
$C_{38}H_{52}CoP_2$
L.Falvello, M.Gerloch
Acta Crystallogr.,Sect.B,**35**, 2547, 1979
See also R1 : 86

71.257 (Dicarbonyl – dimethylphenylphosphine – cobalt) – μ – carbonyl – μ – (N,N' – diphenyl – N – (phenylcarbenyl) – benzamidine) – (dicarbonyl – cobalt)
$C_{39}H_{31}Co_2N_2O_5P$
R.D.Adams, D.F.Chodosh, N.M.Golembeski, E.C.Weissman
J.Organomet.Chem., **172**, 251, 1979
See also R1 : 86,83

71.258 Carbonyl – iodo – bis(triphenylphosphine) – acetyl – ruthenium
$C_{39}H_{33}IO_2P_2Ru$
W.R.Roper, G.E.Taylor, J.M.Waters, L.J.Wright
J.Organomet.Chem., **182**, C46, 1979
See also R1 : 86,84

71.259 tris(t – Butylisocyanide) – (1,4 – bis(t – butylimino) – 2,3 – diphenyl – buta – 1,3 – diene) – iron toluene solvate (at 183°K)
$C_{39}H_{55}FeN_5$, C_7H_8
J.–M.Bassett, M.Green, J.A.K.Howard, F.G.A.Stone
J.Chem.Soc.,Dalton Trans.,1779, 1980
See also R1 : 72

71.260 (η^2 – Dithiomethoxycarbonyl) – dicarbonyl – bis(triphenylphosphine) – ruthenium(ii) perchlorate cyclohexane solvate
$C_{40}H_{33}O_2P_2RuS_2^+$, ClO_4^-, C_6H_{12}
S.M.Boniface, G.R.Clark
J.Organomet.Chem.,**184**, 125, 1980
See also R1 : 86

71.261 octakis(1,3 – Di – isocyanopropane) – chloro – tetra – rhodium tetrakis(tetrachloro – cobalt) trihydrogen hexahydrate
$C_{40}H_{48}ClN_{16}Rh_4^{5+}$, $4Cl_4Co^{2-}$, $3H^+$, $6H_2O$
K.R.Mann, M.J.DiPierro, T.P.Gill
J.Am.Chem.Soc.,**102**, 3965, 1980

71.262 tetrakis(μ^2 – 2,5 – Dimethyl – 2,5 – di – isocyanohexane) – di – rhodium bis(hexafluorophosphate) acetonitrile solvate
$C_{40}H_{64}N_8Rh_2^{2+}$, $2F_6P^-$, $2C_2H_3N$
K.R.Mann, J.A.Thich, R.A.Bell, C.L.Coyle, H.B.Gray
Inorg.Chem., **19**, 2462, 1980

71.263 octakis(t – Butylisocyanide) – di – cobalt
$C_{40}H_{72}Co_2N_8$
W.E.Carroll, M.Green, A.M.R.Galas, M.Murray, T.W.Turney, A.J.Welch, P.Woodward
J.Chem.Soc.,Dalton Trans.,80, 1980

71.264 bis(α – Benzyldioximato) – bis(cyclohexyl – isonitrile) – iron(ii)
$C_{42}H_{44}FeN_6O_4$
Yu.A.Simonov, A.A.Dvorkin, I.I.Bulgak, M.P.Starish, D.G.Batir *Koord.Khim.*,**5**, 1883, 1979
See also R1 : 83

71.265 (μ – 1,2 – bis(Methoxycarbonyl)ethene – 1,2 – diyl) – bis(carbonyl – (triphenylphosphine) – platinum)
$C_{44}H_{36}O_6P_2Pt_2$
Y.Koie, S.Shinoda, Y.Saito, B.J.Fitzgerald, C.G.Pierpont
Inorg.Chem.,**19**, 770, 1980
See also R1 : 86

71.266 (2,3 – bis(Methoxycarbonyl) – 1 – methylcyclopropyl) – bis(triphenylphosphine) – platinum(ii) tetrafluoroborate
$C_{44}H_{41}O_4P_2Pt^+$, BF_4^-
T.G.Attig, R.J.Ziegler, C.P.Brock
Inorg.Chem.,**19**, 2315, 1980
See also R1 : 86,84

71.267 Carbonyl – iodo – bis(triphenylphosphine) – (p – methylbenzoyl) – ruthenium
$C_{45}H_{37}IO_2P_2Ru$
W.R.Roper, G.E.Taylor, J.M.Waters, L.J.Wright
J.Organomet.Chem.,**182**, C46, 1979
See also R1 : 86,84

71.268 cis – 1 – ((Tribenzylphosphine) – (dibenzylphosphino – phenyl – methyl) – platinum) – 2 – methyl – 1,2 – dicarba – dodecaborane
$C_{45}H_{54}B_{10}P_2Pt$
S.Bresadola, N.Bresciani–Pahor, B.Longato
J.Organomet.Chem.,**179**, 73, 1979
See also R1 : 86,62

71.269 trans(P,N) – bis(Bromo – (μ^2 – pyridyl – C^2,N) – (triphenylphosphine) – palladium(ii))
$C_{46}H_{38}Br_2N_2P_2Pd_2$
K.Nakatsu, K.Kinoshita, H.Kanda, K.Isobe, Y.Nakamura, S.Kawaguchi *Chem.Lett.*,913, 1980
See also R1 : 86,83

71.270 (η^2 – Dithiomethoxycarbonyl) – carbonyl – (p – tolylisocyanide) – bis(triphenylphosphine) – ruthenium(ii) perchlorate chloroform solvate hemihydrate (refinement in Pna21)
$C_{47}H_{40}NOP_2RuS_2^+$, ClO_4^-, $0.5CHCl_3$, $0.5H_2O$
S.M.Boniface, G.R.Clark
J.Organomet.Chem.,**184**, 125, 1980
See also R1 : 86,85

71.271 (η^2 – Dithiomethoxycarbonyl) – carbonyl – (p – tolylisocyanide) – bis(triphenylphosphine) – ruthenium(ii) perchlorate chloroform solvate hemihydrate (refinement in Pnma)
$C_{47}H_{40}NOP_2RuS_2^+$, ClO_4^-, $0.5CHCl_3$, $0.5H_2O$
S.M.Boniface, G.R.Clark
J.Organomet.Chem.,**184**, 125, 1980
See also R1 : 86,85

71.272 σ – (Thiophenyl(sulfinyl)methylene) – thiophenyl – bis(triphenylphosphine) – platinum benzene solvate
$C_{49}H_{40}OP_2PtS_3$, C_6H_6
J.W.Gosselink, A.M.F.Brouwers, G.van Koten, K.Vrieze
J.Chem.Soc.,Chem.Commun.,1045, 1979
See also R1 : 86,85

71.273 Chloro – rhodium – bis(μ^2 – bis(diphenylphosphino) methylene) – (μ – dicarbon – tetrasulfide) – chloro – carbonyl – rhodium
$C_{53}H_{44}Cl_2OP_4Rh_2S_4$
M.Cowie, R.S.Dickson, S.K.Dwight, T.G.Southern
Am.Cryst.Assoc.,Ser.2,8, 23, 1980
See also R1 : 86,85

71.274 trans – Chloro – (1,4 – bis(p – methoxyphenyl) – 3 – methyl – 1,4 – diazabutadien – 2 – yl) – bis(triphenylphosphine) – palladium(ii)
$C_{53}H_{47}ClN_2O_2P_2Pd$
B.Crociani, G.Bandoli, D.A.Clemente
J.Organomet.Chem.,**184**, 269, 1980
See also R1 : 86

71.275 (μ – Hexafluorobutadiene – 2,3 – diyl) – bis(μ – bis(diphenylphosphino)methane) – bis(chloro – palladium) (at 140°K)
$C_{54}H_{44}Cl_2F_6P_4Pd_2$
A.L.Balch, C.–L.Lee, C.H.Lindsay, M.M.Olmstead
J.Organomet.Chem.,**177**, C22, 1979
See also R1 : 86

71.276 Bromo – (2 – (diphenylphosphino)phenyl) – hydrido – bis(triphenylphosphine) – iridium(iii)
$C_{54}H_{45}BrIrP_3$
K.von Deuten, L.Dahlenburg
Cryst.Struct.Commun.,**9**, 421, 1980
See also R1 : 86

71.277 Tricyclohexylphosphine – platinum – (μ – dimethylsilyl – μ – (1 – sigma:1 – 2 – η – phenylethynyl)) – (2 – phenylethynyl – tricyclohexylphosphine – platinum) (at 200°K)
$C_{54}H_{82}P_2Pt_2Si$
M.Ciriano, J.A.K.Howard, J.L.Spencer, F.G.A.Stone, H.Wadepohl *J.Chem.Soc.,Dalton Trans.*,1749, 1979
See also R1 : 86

71.278 Trifluoroacetato – (1,4 – diphenyl – but – 1 – en –
3 – yn – 2 – yl) – carbonyl –
bis(triphenylphosphine) – ruthenium
$C_{55}H_{41}F_3O_3P_2Ru$
A.Dobson, D.S.Moore, S.D.Robinson, M.B.Hursthouse, L.New
J.Organomet.Chem.,**177**, C8, 1979
See also R1 : 81,86

71.279 (μ^2 – Carbonyl) – bis(μ^2 – bis(diphenylphosphino)
methylene) – (μ^2 – dimethylacetylene –
carboxylate) – dichloro – di – rhodium
$C_{57}H_{50}Cl_2O_5P_4Rh_2$
M.Cowie, R.S.Dickson, S.K.Dwight, T.G.Southern
Am.Cryst.Assoc.,Ser.2,**8**, 23, 1980
See also R1 : 86

71.280 bis((Dicarbonyl – P,P – diphenyl – N – methyl –
phosphinothioformamido) – (μ – P,P – diphenyl –
N – methyl – phosphinothioformamido) –
molybdenum) dichloromethane solvate
$C_{60}H_{52}Mo_2N_4O_4P_4S_4$, $4CH_2Cl_2$
W.P.Bosman, J.H.Noordik, H.P.M.M.Ambrosius, J.A.Cras
Cryst.Struct.Commun.,**9**, 7, 1980
See also R1 : 86,85

71.281 tris(bis(Diphenylphosphino)methane) –
(bis(diphenylphosphino)methanido) – penta – gold
dinitrate
$C_{100}H_{87}Au_5P_8^{2+}$, $2NO_3^-$
J.W.A.van der Velden, J.J.Bour, F.A.Vollenbroek,
P.T.Beurskens, J.M.M.Smits
J.Chem.Soc.,Chem.Commun.,1162, 1979
See also R1 : 86

METAL PI–COMPLEXES (OPEN–CHAIN)

72.1 cis – Dichloro – penta – 1,4 – diene – platinum(ii)
$C_5H_8Cl_2Pt$
M.Kopp, L.R.Krauth, R.Ratka, K.Weidenhammer,
M.L.Ziegler *Z.Naturforsch.,Teil B*,**35**, 802, 1980

72.2 (η^3 – Allyl) – bromo – tricarbonyl – iron
$C_6H_5BrFeO_3$
F.E.Simon, J.W.Lauher *Inorg.Chem.*,**19**, 2338, 1980

72.3 bis(Dicarbonyl – diacetonitrile – rhodium(i))
uneicosa – μ – carbonyl – dicarbido –
dodecacarbonyl – polyhedro – tetradeca – rhodium
$2C_6H_6N_2O_2Rh^+$, $C_{35}O_{33}Rh_{14}^{2-}$
V.G.Albano, D.Braga, P.Chini, S.Martinengo, D.Strumolo
Eur.Cryst.Meeting,**6**, 71, 1980

72.4 η^3 – Allyl – (η^5 – cyclopentadienyl) – iodo –
nitrosyl – molybdenum
$C_8H_{10}IMoNO$
J.W.Faller, D.F.Chodosh, D.Katahira
J.Organomet.Chem.,**187**, 227, 1980
See also R1 : 73

72.5 η^3 – Allyl – η^5 – cyclopentadienyl – iodo –
nitrosyl – tungsten
$C_8H_{10}INOW$
T.J.Greenhough, P.Legzdins, D.T.Martin, J.Trotter
Inorg.Chem.,**18**, 3268, 1979
See also R1 : 73

72.6 Dichloro – (2,2 – dimethyl – pent – (E) – 3 – enyl –
methylsulfide) – palladium(ii)
$C_8H_{16}Cl_2PdS$
R.McCrindle, E.C.Alyea, G.Ferguson, S.A.Dias, A.J.McAlees,
M.Parvez *J.Chem.Soc.,Dalton Trans.*,137, 1980
See also R1 : 85

72.7 cis – Dichloro – (μ – ethylene) – (2,6 – dimethyl –
piperidine) – platinum(ii)
$C_9H_{19}Cl_2NPt$
M.Camalli, F.Caruso, L.Zambonelli
Inorg.Chim.Acta,**44**, 177, 1980
See also R1 : 83

72.8 bis(η^3 – Allyl) – trimethylphosphine – nickel
(at –170°C)
$C_9H_{19}NiP$
B.Henc, P.W.Jolly, R.Salz, S.Stobbe, G.Wilke, R.Benn,
R.Mynott, K.Seevogel, R.Goddard, C.Kruger
J.Organomet.Chem.,**191**, 449, 1980
See also R1 : 86

72.9 η^3 – Allyl – dicarbonyl – (η^5 – cyclopentadienyl) –
molybdenum
$C_{10}H_{10}MoO_2$
J.W.Faller, D.F.Chodosh, D.Katahira
J.Organomet.Chem.,**187**, 227, 1980
See also R1 : 73

72.10 (Diethylenetriamine) – (hex – 1 – ene) – copper(i) tetraphenylborate
$C_{10}H_{25}CuN_3^+$, $C_{24}H_{20}B^-$
M.Pasquali, C.Floriani, A.G.Manfredotti, A.C.Villa
Inorg.Chem., **18**, 3535, 1979
See also R1 : 76 R2 : 62

72.C (μ – Acetimidoyl) – (μ – hydrido) – tris(tricarbonyl – iron) (triclinic form, at –158°C)
$C_{11}H_5Fe_3NO_9$ Main entry is 71.

72.11 (μ_3 – Ethylideneimido) – (μ – hydrido) – tris(tricarbonyl – iron) (at –158°C)
$C_{11}H_5Fe_3NO_9$
M.A.Andrews, G.van Buskirk, C.B.Knobler, H.D.Kaesz
J.Am.Chem.Soc., **101**, 7245, 1979
See also R1 : 83

72.C syn – (η^3 – 2 – Vinylyl – cyclopentyl – carboxylato) – tricarbonyl – iron
$C_{11}H_{10}FeO_5$ Main entry is 71.27

72.C anti – (η^3 – 2 – Vinylyl – cyclopentyl – carboxylato) – tricarbonyl – iron
$C_{11}H_{10}FeO_5$ Main entry is 71.28

72.12 cis – Dichloro – (dimethylphenylphosphine) – (prop – 2 – en – 1 – ol) – platinum(ii)
$C_{11}H_{17}Cl_2OPPt$
J.R.Briggs, C.Crocker, W.S.McDonald, B.L.Shaw
J.Chem.Soc., Dalton Trans., 64, 1980

72.13 1 – 3 – η – Allyl – dicarbonyl – chloro – bis(trimethylphosphite) – molybdenum(ii)
$C_{11}H_{23}ClMoO_8P_2$
B.J.Brisdon, D.A.Edwards, K.E.Paddick, M.G.B.Drew
J.Chem.Soc., Dalton Trans., 1317, 1980
See also R1 : 86

72.14 1,1,1,2,2,2,3,3,3 – Nonacarbonyl – 2 – ethylene – 1,3 – μ – hydrido – 1,3 – μ – (methylthiolato) – triangulo – tri – osmium
$C_{12}H_8O_9Os_3S$
B.F.G.Johnson, J.Lewis, D.Pippard, P.R.Raithby
Acta Crystallogr., Sect.B, **36**, 703, 1980
See also R1 : 85

72.15 (exo – 2 – Chloro – 5,6 – dimethylene – syn – 7 – norbornanone) – endo – tricarbonyl – iron
$C_{12}H_9ClFeO_4$
J.Wenger, N.H.Thuy, T.Boschi, R.Roulet, A.Chollet, P.Vogel, A.A.Pinkerton, D.Schwarzenbach
J.Organomet.Chem., **174**, 89, 1979

72.16 Tricarbonyl – (6,7 – methylene – exo – 3 – oxatricyclo(3.2.1.0^{2,4})octane) – exo – iron
$C_{12}H_{10}FeO_4$
J.Wenger, N.H.Thuy, T.Boschi, R.Roulet, A.Chollet, P.Vogel, A.A.Pinkerton, D.Schwarzenbach
J.Organomet.Chem., **174**, 89, 1979

72.17 bis(η^2 – Methylacrylate) – tetracarbonyl – tungsten
$C_{12}H_{12}O_8W$
F.-W.Grevels, M.Lindemann, R.Benn, R.Goddard, C.Kruger
Z.Naturforsch., Teil B, **35**, 1298, 1980

72.C Dicarbonyl – (4 – 5 – η – 1,2 – difluoro – 1,2 – bis(trifluoromethyl) – pent – 4 – enyl) – (trimethylphosphite) – cobalt (at 183°K)
$C_{12}H_{14}CoF_6O_5P$ Main entry is 71.38

72.C (3,3,6 – Trimethyl – 1 – oxo – 2 – oxaheptyl – (4 – 6 – η) – enyl) – tricarbonyl – iron (form I)
$C_{12}H_{14}FeO_5$ Main entry is 71.39

72.C (3,3,6 – Trimethyl – 1 – oxo – 2 – oxa – heptyl – (4 – 6 – η) – enyl) – tricarbonyl – iron
$C_{12}H_{14}FeO_5$ Main entry is 71.40

72.18 cis – Dichloro – (dimethylphenylphosphine) – (vinylacetate) – platinum(ii)
$C_{12}H_{17}Cl_2O_2PPt$
J.R.Briggs, C.Crocker, W.S.McDonald, B.L.Shaw
J.Organomet.Chem., **181**, 213, 1979
See also R1 : 86

72.19 (μ_3 – Butyronitrile) – tris(tricarbonyl – iron)
$C_{13}H_7Fe_3NO_9$
M.A.Andrews, C.B.Knobler, H.D.Kaesz
J.Am.Chem.Soc., **101**, 7260, 1979
See also R1 : 83

72.20 Tricarbonyl – (exo – 2 – methoxy – 5,6 – dimethylene – syn – 7 – norbornanol) – endo – iron
$C_{13}H_{14}FeO_5$
J.Wenger, N.H.Thuy, T.Boschi, R.Roulet, A.Chollet, P.Vogel, A.A.Pinkerton, D.Schwarzenbach
J.Organomet.Chem., **174**, 89, 1979

72.21 η^5 – Cyclopentadienyl – (η^4 – 3,4 – dimethyl – hexa – 2,4 – diene) – dichloro – niobium
$C_{13}H_{19}Cl_2Nb$
M.J.Bunker, M.L.H.Green, C.Couldwell, K.Prout
J.Organomet.Chem., **192**, C6, 1980
See also R1 : 73

72.22 Dimethyl – (N,N – dimethylethanolamino) – (1 – pyrazolyl) – gallato – (η^2 – thiomethoxymethyl) – dicarbonyl – molybdenum
$C_{13}H_{24}GaMoN_3O_3S$
K.S.Chong, S.J.Rettig, A.Storr, J.Trotter
Can.J.Chem., **58**, 1080, 1980
See also R1 : 84,83

72.C bis(Triphenylphosphine)iminium bromo – (μ^2 – carbonyl) – (η^2 – 1,4 – bis(trifluoro) – but – 2 – ene) – nonacarbonyl – tri – osmium
$C_{14}H_2BrF_6O_{10}Os_3^-$, $C_{36}H_{30}NP_2^+$ Main entry is 71.56

72.C (μ^4 – Acetylene) – dodecacarbonyl – tetra – osmium
$C_{14}H_2O_{12}Os_4$ Main entry is 71.57

72.C Tetracarbonyl – tungsten – (μ – 1 – methyl – 2 – but – 2 – ene – di – ylidene) – pentacarbonyl – tungsten
$C_{14}H_8O_9W_2$ Main entry is 71.61

72.C 1,4 – Dihydroxy – 2,3 – diethyl – butane – 1,1,4,4 – tetrayl – bis(tricarbonyl) – iron
$C_{14}H_{12}Fe_2O_8$ Main entry is 71.64

72.C (μ^2 – Acetylene) – ($\sigma,\sigma,\eta^2,\eta^2$ – glyoxal – bis(isopropylamine)) – tetracarbonyl – di – ruthenium
$C_{14}H_{18}N_2O_4Ru_2$ Main entry is 71.68

72.C μ – 2 – sigma:2 – 3 – η – (4,5 – Dihydro – 2 – furyl) – (bis(trimethylphosphine) – platinum) – tetracarbonyl – manganese (red form, at 200°K)
$C_{14}H_{23}MnO_5P_2Pt$ Main entry is 71.75

72.23   Tricarbonyl – (methyl α – acetamido –
        cinnamate) – iron
        $C_{15}H_{13}FeNO_6$
        A.de Cian, R.Weiss, J.–P.Haudegond, Y.Chauvin,
        D.Commereuc *J.Organomet.Chem.*,**187**, 73, 1980
        See also R1 : 84

72.24   Tricarbonyl – (η^4 – 1,5 – dimethylene – 2,6 –
        dimethylcyclo – octane) – iron
        $C_{15}H_{20}FeO_3$
        F.–W.Grevels, K.Schneider, C.Kruger, R.Goddard
        Z.Naturforsch.,Teil B,**35**, 360, 1980

72.C   (μ^4 – Ethylacetylene) – dodecacarbonyl – tetra –
        osmium
        $C_{16}H_6O_{12}Os_4$ Main entry is 71.90

72.25   (2 – Methyl – 3,5,6 – tri(methylene) – cyclohexane –
        (2 – 4 – η) – ene) – bis(tricarbonyl – iron)
        $C_{16}H_{10}Fe_2O_7$
        E.Meier, O.Cherpillod, T.Boschi, R.Roulet, P.Vogel,
        C.Mahaim, A.A.Pinkerton, D.Schwarzenbach, G.Chapuis
        J.Organomet.Chem.,**186**, 247, 1980
        See also R1 : 75

72.C   (μ^2 – Allene) – (μ^2 – carbonyl) – bis(carbonyl – η^5 –
        cyclopentadienyl – manganese) (at –170°C)
        $C_{16}H_{14}Mn_2O_3$ Main entry is 71.96

72.26   (η^2 – Dimethylacetylene – dicarboxylate) – bis(η^5 –
        cyclopentadienyl) – vanadium
        $C_{16}H_{18}O_4V$
        J.L.Petersen, L.Griffith *Inorg.Chem.*,**19**, 1852, 1980
        See also R1 : 73

72.27 η^4 – (1 – Cyclohexylimino – 2 – methoxy – 3 –
        methoxycarbonylbutene) – tricarbonyl – iron
        $C_{16}H_{19}FeNO_6$
        T.–A.Mitsudo, H.Watanabe, Y.Komiya, Y.Watanabe,
        Y.Takaegami, K.Nakatsu, K.Kinoshita, Y.Miyagawa
        J.Organomet.Chem.,**190**, 39, 1980

72.28   Tetraphenylphosphonium η^3 – allyl –
        tetradecacarbonyl – hexa – rhodium
        tetrahydrofuran solvate
        $C_{17}H_5O_{14}Rh_6^-$, $C_{24}H_{20}P^+$, C_4H_8O
        G.Ciani, A.Sironi, P.Chini, A.Ceriotti, S.Martinengo
        J.Organomet.Chem.,**192**, C39, 1980

72.29   (η^5 – Cyclopentadienyl – nickel) – (η – t –
        butylethynyl) – hexacarbonyl – di – iron
        $C_{17}H_{14}Fe_2NiO_6$
        A.Marinetti, E.Sappa, A.Tiripicchio, M.T.Camellini
        J.Organomet.Chem.,**197**, 335, 1980
        See also R1 : 73

72.C   Pentacarbonyl – tungsten – (μ^2 – 3 –
        (trimethylphosphonium) – 1 – isopentenyl) –
        tetracarbonyl – tungsten
        $C_{17}H_{17}O_9PW_2$ Main entry is 71.109

72.30   (η – Allyl) – (8 – isopropylquinoline – 2 –
        carboxaldehyde – N – methylimino) – palladium(ii)
        perchlorate
        $C_{17}H_{21}N_2Pd^+$, ClO_4^-
        A.J.Deeming, I.P.Rothwell, M.B.Hursthouse, K.M.A.Malik
        J.Chem.Soc.,Dalton Trans.,1899, 1979
        See also R1 : 83

72.31   Dichloro – 2,5 – diphenyl – hexa – 1,5 – diene –
        platinum(ii)
        $C_{18}H_{18}Cl_2Pt$
        M.Kopp, L.R.Krauth, R.Ratka, K.Weidenhammer,
        M.L.Ziegler *Z.Naturforsch.,Teil B*,**35**, 802, 1980

72.32   Tetracarbonyl – (trimethylsilyl) – η – (1 – 4 –
        eta:1 – σ,5 – 8 – η – (8 – trimethylsilyl – octa –
        1,3,5,7 – tetraenyl)) – di – ruthenium
        $C_{18}H_{26}O_4Ru_2Si_2$
        R.Goddard, P.Woodward
        J.Chem.Soc.,Dalton Trans.,559, 1980

72.33   Dichloro – (N,N' – di – t – butyl – ethylenedi –
        imine) – styrene – platinum
        $C_{18}H_{28}Cl_2N_2Pt$
        H.van der Poel, G.van Koten, K.Vrieze, M.Kokkes, C.H.Stam
        J.Organomet.Chem.,**175**, C21, 1979
        See also R1 : 83

72.34   (1,2,5,6 – η – 1,5 – Cyclo – octa – 1,5 – diene) –
        (1,2 – η – penta – 1,3 – diene) – (2 – 4 – η – 1,3 –
        dimethylallyl) – iridium(i)
        $C_{18}H_{29}Ir$
        J.Muller, W.Hahnlein, H.Menig, J.Pickardt
        J.Organomet.Chem.,**197**, 95, 1980
        See also R1 : 75

72.35   (μ – Di – t – butylacetylene) – bis((μ^2 – chloro) –
        dichloro – tetrahydrofuran – tantalum))
        $C_{18}H_{34}Cl_6O_2Ta_2$
        F.A.Cotton, W.T.Hall *Inorg.Chem.*,**19**, 2354, 1980
        See also R1 : 84

72.C   (μ – Hydrido) – ((α – methylenebenzyl) –
        acetylido) – tris(tricarbonyl – ruthenium)
        $C_{19}H_8O_9Ru_3$ Main entry is 71.135

72.C   (η^5 – Cyclopentadienyl – nickel) – (μ^4 – 3,3 –
        dimethyl – but – 1 – enyl) – nonacarbonyl – tri –
        ruthenium
        $C_{20}H_{15}NiO_9Ru_3$ Main entry is 71.151

72.36   Dichloro – (diphenyl(o – vinylphenyl)arsine) –
        platinum(ii)
        $C_{20}H_{17}AsCl_2Pt$
        M.K.Cooper, P.J.Guerney, M.McPartlin
        J.Chem.Soc.,Dalton Trans.,349, 1980
        See also R1 : 86

72.37   Triethylphosphine – (μ^2 – 1,4 – bis(trifluoro) –
        but – 2 – ene) – decacarbonyl – tri – osmium
        $C_{20}H_{17}F_6O_{10}Os_3P$
        Z.Dawoodi, M.J.Mays, P.R.Raithby, K.Henrick
        J.Chem.Soc.,Chem.Commun.,641, 1980
        See also R1 : 86

72.38   (μ^2 – Carbonyl) – (μ – η^1,η^3 – (1 – methyl – 2,3 –
        bis(methoxycarbonyl) – allyl)) – carbonyl – bis(η^5 –
        cyclopentadienyl – iron)
        $C_{20}H_{20}Fe_2O_6$
        A.F.Dyke, S.A.R.Knox, P.J.Naish, G.E.Taylor
        J.Chem.Soc.,Chem.Commun.,803, 1980
        See also R1 : 73

72.C η – Carbonyl – η – (1 – 3 – eta:1 – σ,4 – 2' – η –
        (1,3 – dimethyl – 4 – (5 – oxo – 4 – phenyl –
        2'(5H) – furan – 2 – ylidene) – but – 1 – ene – 1,3 –
        diyl)) – bis(dicarbonyl – cobalt)
        $C_{21}H_{14}Co_2O_7$ Main entry is 71.168

72.C   bis(η^5 – Cyclopentadienyl) – (μ – η^2,η^2 – carbonyl – dimethylacetylene – carboxylate) – bis(dicarbonyl – tungsten) (at 233°K)
$C_{21}H_{18}O_9W_2$ Main entry is 71.170

72.39   Tricarbonyl – cobalt – (μ – diphenylacetylene) – (cyclopentadienyl – nickel)
$C_{22}H_{15}CoNiO_3$
B.H.Freeland, J.W.Hux, N.C.Payne, K.G.Tyers
Inorg.Chem.,**19**, 693, 1980
See also R1 : 73

72.40   (Di(pivaloyl) – methanato) – (η – 1,2,4,5 – (1,2 – bis(trifluoromethyl) – 4 – methyl – 3 – (prop – 2' – ylidene) – penta – 1,4 – diene)) – rhodium(i)
$C_{22}H_{31}F_6O_2Rh$
C.E.Dean, R.D.W.Kemmitt, D.R.Russell, M.D.Schilling
J.Organomet.Chem.,**187**, C1, 1980
See also R1 : 77

72.41   Dichloroethylene – (N,N' – dimethyl – N,N' – bis(α – methylbenzyl) – 1,2 – ethanediamine) – platinum(ii)
$C_{22}H_{32}Cl_2N_2Pt$
A.De Renzi, B.di Blasio, A.Saporito, M.Scalone, A.Vitagliano
Inorg.Chem.,**19**, 960, 1980
See also R1 : 76

72.42   Tetraethylammonium η – allyl – dicarbonyl – dichloro – triphenylphosphine – tungsten
$C_{23}H_{20}Cl_2O_2PW^-$, $C_8H_{20}N^+$
M.Boyer, J.–C.Daran, Y.Jeannin
J.Organomet.Chem.,**190**, 177, 1980
See also R1 : 86

72.43   (η – Allyl) – dicarbonyl – (N – phenylsalicylideneiminato) – pyridine – molybdenum(ii)
$C_{23}H_{20}MoN_2O_3$
M.G.B.Drew, G.F.Griffin
Acta Crystallogr.,Sect.B,**35**, 3036, 1979
See also R1 : 78

72.44   Dichloro – (propene) – (N,N' – dimethyl – N,N' – bis(α – methylbenzyl) – 1,2 – ethanediamine) – platinum(ii)
$C_{23}H_{34}Cl_2N_2Pt$
A.De Renzi, B.di Blasio, A.Saporito, M.Scalone, A.Vitagliano
Inorg.Chem.,**19**, 960, 1980
See also R1 : 76

72.45   bis(η^5 – Cyclopentadienyl) – bis(pentafluorophenyl) – acetylene – vanadium
$C_{24}H_{10}F_{10}V$
D.F.Foust, M.D.Rausch, W.E.Hunter, J.L.Atwood, E.Samuel
J.Organomet.Chem.,**197**, 217, 1980
See also R1 : 73

72.46   (η^3 – Allyl) – tricarbonyl – (triphenylphosphine – gold) – iron
$C_{24}H_{20}AuFeO_3P$
F.E.Simon, J.W.Lauher *Inorg.Chem.*,**19**, 2338, 1980

72.47   bis(N,N – Dimethyldithiocarbamato) – (ditoluoyl – acetylene) – oxo – molybdenum(iv) benzene solvate
$C_{24}H_{26}MoN_2O_3S_4$. C_6H_6
W.E.Newton, J.W.McDonald, J.L.Corbin, L.Ricard, R.Weiss
Inorg.Chem.,**19**, 1997, 1980
See also R1 : 80

72.48   bis(η^5 – Cyclopentadienyl) – (thiocamphor) – tetracarbonyl – di – molybdenum
$C_{24}H_{26}Mo_2O_4S$
H.Alper, N.D.Silavwe, G.I.Birnbaum, F.R.Ahmed
J.Am.Chem.Soc.,**101**, 6582, 1979
See also R1 : 73,85

72.49   (μ – 1 – Diphenylphosphino – 2 – sulfido – 3,3 – dimethylbutene) – carbonyl – (methoxy – thiocarbonyl) – trimethylphosphite – iron
$C_{24}H_{32}FeO_5P_2S_2$
A.J.Carty, F.Hartstock, N.J.Taylor, H.Le Bozec, P.Robert, P.H.Dixneuf *J.Chem.Soc.,Chem.Commun.*,361, 1980
See also R1 : 86,85

72.50   bis($\sigma,\sigma,\eta^2,\eta^2$ – Glyoxal – bis(isopropylamine)) – octacarbonyl – tetra – ruthenium
$C_{24}H_{32}N_4O_8Ru_4$
L.H.Staal, L.H.Polm, K.Vrieze, F.Ploeger, C.H.Stam
J.Organomet.Chem.,**199**, 13, 1980
See also R1 : 83

72.51   Di – lithium tris(N,N,N',N' – tetramethyl – but – 2 – ene – 1,4 – diamine) – nickel(ii)
$C_{24}H_{54}N_6Ni^{2-}$, $2Li^+$
D.J.Brauer, C.Kruger, J.C.Sekutowski
J.Organomet.Chem.,**178**, 249, 1979

72.C   (μ^2 – Diphenylphosphido) – (μ^3 – 3 – methylbutynyl) – nonacarbonyl – tri – ruthenium
$C_{26}H_{17}O_9PRu_3$ Main entry is 71.198

72.C   Cyclopentadienyl – (η^2 – benzaldehyde) – (α – (2 – pyridyl) – α – (methylamino) – benzyl) – carbonyl – molybdenum
$C_{26}H_{24}MoN_2O_2$ Main entry is 71.204

72.C   bis(μ^2 – Hydrido) – (η^3 – 1 – methyl – 2,3 – diphenylallyl) – undecacarbonyl – tetra – osmium
$C_{27}H_{16}O_{11}Os_4$ Main entry is 71.210

72.C   bis(η^5 – Cyclopentadienyl) – (μ – carbonyl) – (μ – η',η^3 – carbonyl – 1,2 – diphenylethylene) – carbonyl – di – ruthenium (at 233°K)
$C_{27}H_{20}O_3Ru_2$ Main entry is 71.212

72.52   Dicarbonyl – (η^4 – cinnamaldehyde) – (triphenylphosphine) – iron(0)
$C_{29}H_{23}FeO_3P$
M.Sacerdoti, V.Bertolasi, G.Gilli
Acta Crystallogr.,Sect.B,**36**, 1061, 1980
See also R1 : 86

72.53   cis – Dichloro – η^2 – ((Z) – 2 – chloro – N – (3 – methyl – but – 1 – enyl) – benzenamine) – triphenylphosphine – platinum(ii)
$C_{29}H_{29}Cl_3NPPt$
A.De Renzi, P.Ganis, A.Panunzi, A.Vitagliano, G.Valle
J.Am.Chem.Soc.,**102**, 1722, 1980
See also R1 : 86

72.54   Carbonyl – cyclopentadienyl – (1,5α – η^3 – 1 – methylene – cyclopentan – 2 – onato) – triphenylphosphine – molybdenum(ii)
$C_{30}H_{27}MoO_2P$
S.Jeannin, Y.Jeannin *J.Organomet.Chem.*,**178**, 309, 1979
See also R1 : 73,86

72.55 Tricarbonyl – (1,6 – bis(diphenylphosphino) –
trans – hex – 3 – ene) – molybdenum(0)
$C_{33}H_{30}MoO_3P_2$
G.R.Clark, C.M.Cochrane, P.W.Clark, A.J.Jones, P.Hanisch
J.Organomet.Chem., **182**, C5, 1979
See also R1 : 86

72.56 Chloro – (2,2' – bis(o – diphenylphosphino) – trans –
stilbene) – rhodium(i) dichloromethane solvate
$C_{38}H_{30}ClP_2Rh$, CH_2Cl_2
G.B.Robertson, P.A.Tucker, P.O.Whimp
Inorg.Chem., **19**, 2307, 1980
See also R1 : 86

72.57 Trichloro – (2,2' – bis(o – diphenylphosphino) –
trans – stilbene) – iridium(iii)
$C_{38}H_{30}Cl_3IrP_2$
G.B.Robertson, P.A.Tucker, P.O.Whimp
Inorg.Chem., **19**, 2307, 1980
See also R1 : 86

72.58 Trichloro – (2,2' – bis(o – diphenylphosphino) –
trans – stilbene) – rhodium(iii)
$C_{38}H_{30}Cl_3P_2Rh$
G.B.Robertson, P.A.Tucker, P.O.Whimp
Inorg.Chem., **19**, 2307, 1980
See also R1 : 86

72.59 (Methyl – (Z) – α – acetamido – cinnamate) – (1,2 –
bis(diphenylphosphino)ethane) – rhodium(i)
tetrafluoroborate
$C_{38}H_{37}NO_3P_2Rh^+$, BF_4^-
A.S.C.Chan, J.J.Pluth, J.Halpern
Inorg.Chim.Acta, **37**, L477, 1979
See also R1 : 86,84

72.C tris(t – Butylisocyanide) – (1,4 – bis(t –
butylimino) – 2,3 – diphenyl – buta – 1,3 – diene) –
iron toluene solvate (at 183°K)
$C_{39}H_{55}FeN_5$, C_7H_8 Main entry is 71.259

72.60 (Ethyl – methacrylate) – bis(triphenylphosphine) –
nickel(0)
$C_{42}H_{40}NiO_2P_2$
S.Komiya, J.Ishizu, A.Yamamoto, T.Yamamoto,
A.Takenaka, Y.Sasada
Bull.Chem.Soc.Jpn., **53**, 1283, 1980
See also R1 : 86

72.61 bis(η^3 – Allyl – triphenylphosphine – palladium)
$C_{42}H_{40}P_2Pd_2$
P.W.Jolly, C.Kruger, K.–P.Schick, G.Wilke
Z.Naturforsch., Teil B, **35**, 926, 1980
See also R1 : 86

72.62 bis(Triphenylphosphine) – (μ^2 – tetrachloro –
diazacyclopentadienyl) – dicarbonyl – ruthenium
dichloromethane solvate (at –159°C)
$C_{43}H_{30}Cl_4N_2O_2P_2Ru$, CH_2Cl_2
K.D.Schramm, J.A.Ibers *Inorg.Chem.*, **19**, 2441, 1980
See also R1 : 86

72.63 (η^3 – 1 – Phenylallyl) – chloro – hydrido –
bis(triphenylphosphine) – iridium
$C_{45}H_{40}ClIrP_2$
T.H.Tulip, J.A.Ibers *J.Am.Chem.Soc.*, **101**, 4201, 1979
See also R1 : 86

72.64 (η^2 – Methyl – 3 – benzoylacrylate) –
bis(triphenylphosphine) – nickel(0) benzene solvate
$C_{47}H_{40}NiO_3P_2$, C_6H_6
G.D.Andreetti, G.Bocelli, P.Sgarabotto, G.P.Chiusoli,
M.Costa, G.Terenghi, A.Biavati
Transition Met.Chem., **5**, 129, 1980
See also R1 : 86

72.65 η^2 – ((t – Butyl – (trimethylsilyl)amino) – t –
butyliminothiophosphorane) –
bis(triphenylphosphine) – platinum
$C_{47}H_{57}N_2P_3PtSSi$
O.J.Scherer, H.Jungmann
Angew.Chem.,Int.Ed.Engl., **18**, 953, 1979
See also R1 : 86

72.66 Tricarbonyl – cobalt – (μ – diphenylacetylene) –
(1,1,1 – tris(diphenylphosphinomethyl)ethane –
carbonyl – cobalt)
$C_{59}H_{49}Co_2O_4P_3$
C.Bianchini, P.Dapporto, A.Meli
J.Organomet.Chem., **174**, 205, 1979
See also R1 : 86

72.67 Lithium (μ^2 – benzophenonimino) –
tetrakis(benzophenonimino) – di – nickel diethyl
ether solvate
$C_{65}H_{50}N_5Ni_2^{3-}$, $3Li^+$, $2C_4H_{10}O$
H.Hoberg, V.Gotz, R.Goddard, C.Kruger
J.Organomet.Chem., **190**, 315, 1980
See also R1 : 83

72.68 Chloro – (bis(3,5 – di – t – butyl – 4 – oxo –
cyclohexadiene – 1 – ylidene)ethylene) –
bis(triphenylphosphine) – rhodium acetone solvate
$C_{66}H_{70}ClO_2P_2Rh$, $1.5C_3H_6O$
L.Hagelee, R.West, J.Calabrese, J.Norman
J.Am.Chem.Soc., **101**, 4888, 1979
See also R1 : 86

72.69 (μ^2 – Carbon – disulfide) – bis((1,1,1 –
(diphenylphosphinomethyl)ethane) – cobalt)
bis(tetraphenylborate) acetone solvate
$C_{83}H_{78}Co_2P_6S_2^{2+}$, $2C_{24}H_{20}B^-$, $2C_3H_6O$
C.Bianchini, C.Mealli, A.Meli, A.Orlandini, L.Sacconi
Angew.Chem.,Int.Ed.Engl., **18**, 673, 1979
See also R1 : 86,85

METAL PI–COMPLEXES
(CYCLOPENTADIENE)

73.1 **(η^5 – Cyclopentadienyl) – chloro – dinitrosyl – chromium**
$C_5H_5ClCrN_2O_2$
T.J.Greenhough, B.W.S.Kolthammer, P.Legzdins, J.Trotter
*Acta Crystallogr.,Sect.B,***36**, 795, 1980

73.2 **(η^5 – Cyclopentadienyl) – chloro – dinitrosyl – tungsten**
$C_5H_5ClN_2O_2W$
T.J.Greenhough, B.W.S.Kolthammer, P.Legzdins, J.Trotter
*Acta Crystallogr.,Sect.B,***36**, 795, 1980

73.3 **(μ – Cyclopentadienyl) – (diselena – dodecaborane) – cobalt (at –118°C)**
$C_5H_{14}B_9CoSe_2$
G.D.Friesen, A.Barriola, P.Daluga, P.Ragatz, J.C.Huffman,
L.J.Todd *Inorg.Chem.,***19**, 458, 1980

73.4 **Tetramethylammonium 2 – (η – cyclopentadienyl) – 1 – carba – 2 – cobalta – closo – undecahydro – dodecaborate**
$C_6H_{16}B_{10}Co^-$, $C_4H_{12}N^+$
V.Petricek, K.Maly, A.Petrina, L.Hummel, A.Linek
*Acta Crystallogr.,Sect.B,***35**, 3044, 1979
See also R1 : 62

73.5 **(η^5 – Nitrocyclopentadienyl) – dicarbonyl – rhodium**
$C_7H_4NO_4Rh$
M.D.Rausch, W.P.Hart, J.L.Atwood, M.J.Zaworotko
*J.Organomet.Chem.,***197**, 225, 1980

73.6 **Dicarbonyl – (η^5 – cyclopentadienyl) – (thionitrosyl) – chromium**
$C_7H_5CrNO_2S$
T.J.Greenhough, B.W.S.Kolthammer, P.Legzdins, J.Trotter
*Inorg.Chem.,***18**, 3548, 1979

73.7 **Iodo – (η^5 – methylcyclopentadienyl) – nitrosyl – (thiocarbonyl) – manganese(ii)**
$C_7H_7IMnNOS$
J.A.Potenza, R.Johnson, S.Rudich, A.Efraty
*Acta Crystallogr.,Sect.B,***36**, 1933, 1980

73.8 **(η^5 – Cyclopentadienyl – aldehyde) – dicarbonyl – nitroso – chromium**
$C_8H_5CrNO_4$
R.Shakir, M.T.Ledet, J.L.Atwood
*Am.Cryst.Assoc.,Ser.2,***7**, 27, 1980

73.9 **Dicarbonyl – cyclopentadienyl – isothiocyanato – iron**
$C_8H_5FeNO_2S$
A.F.Berndt, K.W.Barnett
*J.Organomet.Chem.,***184**, 211, 1980

73.C **Carbonyl – bis(cyclopentadienyl) – hydrido – molybdenum tricarbonyl – cyclopentadienyl – molybdenum**
$C_8H_5MoO_3^-$, $C_{11}H_{11}MoO^+$ Main entry is 73.22

73.C **(μ^2 – Dimethylarsonio) – bis(dicarbonyl – cyclopentadienyl – trimethylphosphine) – molybdenum tricarbonyl – cyclopentadienyl – molybdenum**
$C_8H_5MoO_3^-$, $C_{22}H_{34}AsMo_2O_4P_2^+$ Main entry is 73.95

73.C **Dicarbonyl – carbamoyl – (η^5 – cyclopentadienyl) – ruthenium**
$C_8H_7NO_3Ru$ Main entry is 71.6

73.10 **Dicarbonyl – (η^5 – cyclopentadienyl) – methyldiazo – tungsten (at –140°C)**
$C_8H_8N_2O_2W$
G.L.Lillhouse, B.L.Haymore, W.A.Herrmann
*Inorg.Chem.,***18**, 2423, 1979
See also R1 : 83

73.C **η^3 – Allyl – (η^5 – cyclopentadienyl) – iodo – nitrosyl – molybdenum**
$C_8H_{10}IMoNO$ Main entry is 72.4

73.C **η^3 – Allyl – η^5 – cyclopentadienyl – iodo – nitrosyl – tungsten**
$C_8H_{10}INOW$ Main entry is 72.5

73.11 **η^5 – Cyclopentadienyl – trimethylphosphine – cyclopentasulfur – cobalt**
$C_8H_{14}CoPS_5$
C.Burschka, K.Leonhard, H.Werner
*Z.Anorg.Allg.Chem.,***464**, 30, 1980
See also R1 : 86,85

73.C **8 – Trifluoroacetoxy – 1 – (η^5 – cyclopentadienyl – ferra) – 2,3 – dicarba – dodecaborane**
$C_9H_{15}B_9F_3FeO_2$ Main entry is 71.13

73.12 **Trimethylammine – (η^5 – cyclopentadienyl) – carba – cobalta – undecaborane**
$C_9H_{23}B_9CoN$
R.V.Schultz, J.C.Huffman, L.J.Todd
*Inorg.Chem.,***18**, 2883, 1979
See also R1 : 62

73.13 **Tetraethylammonium cyclopentadienyl – tetracarba – dicobalta – docosaborane**
$C_9H_{25}B_{16}Co_2^-$, $C_8H_{20}N^+$
M.Creswick, I.Bernal, G.Evrard
*Cryst.Struct.Commun.,***8**, 839, 1979
See also R1 : 62

73.14 **bis(η^5 – Cyclopentadienyl) – peroxochloro – niobium(v)**
$C_{10}H_{10}ClNbO_2$
I.Bkouche-Waksman, C.Bois, J.Sala-Pala, J.E.Guerchais
*J.Organomet.Chem.,***195**, 307, 1980

73.15 **(μ^2 – Oxo) – bis((η^5 – cyclopentadienyl) – iodo – oxo – molybdenum(v))**
$C_{10}H_{10}I_2Mo_2O_3$
K.Prout, C.Couldwell
*Acta Crystallogr.,Sect.B,***36**, 1481, 1980
See also R1 : 84

73.C **η^3 – Allyl – dicarbonyl – (η^5 – cyclopentadienyl) – molybdenum**
$C_{10}H_{10}MoO_2$ Main entry is 72.9

73.16 **Ruthenocene (at 15°K)**
bis(Cyclopentadienyl) – ruthenium
$C_{10}H_{10}Ru$
F.Takusagawa, T.F.Koetzle
*Am.Cryst.Assoc.,Ser.2,***7**, 16, 1980

73.17 Ruthenocene (at 78°K)
bis(Cyclopentadienyl) – ruthenium
$C_{10}H_{10}Ru$
F.Takusagawa, T.F.Koetzle
*Am.Cryst.Assoc.,Ser.2,***7**, 16, 1980

73.18 Ruthenocene
bis(Cyclopentadienyl) – ruthenium
$C_{10}H_{10}Ru$
F.Takusagawa, T.F.Koetzle
*Am.Cryst.Assoc.,Ser.2,***7**, 16, 1980

73.19 bis(η^5 – Cyclopentadienyl) – vanadium
Vanadocene
$C_{10}H_{10}V$
M.Yu.Antipin, E.B.Lobkovskii, K.N.Semenenko,
G.L.Soloveichik, Yu.T.Struchkov
*Zh.Strukt.Khim.,***20**, 942, 1979

73.20 Trisila – pentachloro – cyclohexyl – dicarbonyl – cyclopentadienyl – iron
$C_{10}H_{11}Cl_5FeO_2Si_3$
W.Honle, H.G.von Schnering
*Z.Anorg.Allg.Chem.,***464**, 139, 1980
See also R1 : 63

73.C (2,4 – Dithio – 1,3 – dithia – 1,4 – diyl) – η^5 – cyclopentadienyl – (trimethylphosphine) – rhodium
$C_{10}H_{14}PRhS_4$ Main entry is 71.21

73.21 Dinitrato – (η^5 – pentamethyl – cyclopentadienyl) – rhodium(iii)
$C_{10}H_{15}N_2O_6Rh$
M.B.Hursthouse, K.M.A.Malik, D.M.P.Mingos, S.D.Willoughby
*J.Organomet.Chem.,***192**, 235, 1980

73.22 Carbonyl – bis(cyclopentadienyl) – hydrido – molybdenum tricarbonyl – cyclopentadienyl – molybdenum
$C_{11}H_{11}MoO^+$, $C_8H_5MoO_3^-$
M.A.Adams, K.Folting, J.C.Huffman, K.G.Caulton
*Inorg.Chem.,***18**, 3020, 1979
See also R2 : 73

73.23 η^5 – Cyclopentadienyl – (η^4 – 1,4 – dimethyl – 1,4 – dibora – cyclohexa – 2,5 – diene) – cobalt
$C_{11}H_{15}B_2Co$
G.E.Herberich, B.Hessner, S.Beswetherick, J.A.K.Howard,
P.Woodward *J.Organomet.Chem.,***192**, 421, 1980
See also R1 : 75

73.C Dichloro – (η^5 – cyclopentadienyl) – (2 – oxo – 4 – methyl – pent – 3 – ene) – tantalum
$C_{11}H_{15}Cl_2OTa$ Main entry is 71.

73.24 Dicarbonyl – (η^5 – cyclopentadienyl) – (1,3 – dimethyl – 1,3 – diaza – 2 – phospholidin – 2 – yl) – molybdenum
$C_{11}H_{15}MoN_2O_2P$
L.D.Hutchins, R.T.Paine, C.F.Campana
*J.Am.Chem.Soc.,***102**, 4521, 1980
See also R1 : 86

73.25 (bis(Tetrahydroborato) – zinc) – (μ – hydrido) – (carbonyl – dicyclopentadienyl – niobium) benzene solvate
$C_{11}H_{19}B_2NbOZn$, $0.5C_6H_6$
M.A.Porai–Koshits, A.S.Antsyshkina, A.A.Pasynskii,
G.G.Sadikov, Yu.V.Skripkin, V.N.Ostrikova
*Koord.Khim.,***5**, 1103, 1979

73.26 Tricarbonyl – (1 – 4:9 – η – 1 – bromo – indenyl) – manganese
$C_{12}H_8BrMnO_3$
N.B.Honan, J.L.Atwood, I.Bernal, W.A.Herrmann
*J.Organomet.Chem.,***179**, 403, 1979

73.27 Ferrocene – 1,1' – dicarboxylic acid (triclinic form)
$C_{12}H_{10}FeO_4$
F.Takusagawa, T.F.Koetzle
*Acta Crystallogr.,Sect.B,***35**, 2888, 1979

73.28 Ferrocene – 1,1' – dicarboxylic acid (triclinic form, neutron study, at 78°K)
$C_{12}H_{10}FeO_4$
F.Takusagawa, T.F.Koetzle
*Acta Crystallogr.,Sect.B,***35**, 2888, 1979

73.29 Dicarbonyl – bis(cyclopentadienyl) – hafnium
$C_{12}H_{10}HfO_2$
D.J.Sikora, M.D.Rausch, R.D.Rogers, J.L.Atwood
*J.Am.Chem.Soc.,***101**, 5079, 1979

73.30 bis(μ – carbonyl) – bis(cyclopentadienyl – nickel)
$C_{12}H_{10}Ni_2O_2$
L.R.Byers, L.F.Dahl *Inorg.Chem.,***19**, 680, 1980

73.31 (S) – 1 – (N,N – Dimethylaminomethyl) – 2 – formyl – cymantrene (absolute configuration)
$C_{12}H_{12}MnNO_4$
Yu.N.Belokon, I.E.Zeltzer, N.M.Loim, V.A.Tsiryapkin,
G.G.Aleksandrov, D.N.Kursanov, Z.N.Parnes,
Yu.T.Struchkov, V.M.Belikov *Tetrahedron,***36**, 1089, 1980

73.C 1,1' – Dimethylferrocenium bis(7,7,8,8 – tetracyanoquinodimethane)
$C_{12}H_{14}Fe^+$, $C_{12}H_4N_4^-$, $C_{12}H_4N_4$ Main entry is 60.20

73.32 anti – Di – μ – sulfido – bis(sulfido – methylcyclopentadienyl – molybdenum(v))
$C_{12}H_{14}Mo_2S_4$
R.C.Haltiwanger, M.R.DuBois, D.L.DuBois, M.C.VanDerveer
*Am.Cryst.Assoc.,Ser.2,***7**, 37, 1980

73.C 3 – Carbonyl – 3 – η^5 – cyclopentadienyl – 6 – difluoro – 2 – isopropyl – 4 – methyl – 1,5 – dioxa – 3 – ferra – 6 – borinane
$C_{12}H_{15}BF_2FeO_3$ Main entry is 71.41

73.33 bis(Cyclopentadienyl) – chloro – ethoxo – titanium (at –125°C)
$C_{12}H_{15}ClOTi$
J.C.Huffman, K.G.Moloy, J.A.Marsella, K.G.Caulton
*J.Am.Chem.Soc.,***102**, 3009, 1980
See also R1 : 84

73.34 Dicarbonyl – (pentamethyl – cyclopentadienyl) – cobalt(i)
$C_{12}H_{15}CoO_2$
L.R.Byers, L.F.Dahl *Inorg.Chem.,***19**, 277, 19

73.35 Dicarbonyl – nitrosyl – (η^5 – pentamethyl – cyclopentadienyl) – chromium
$C_{12}H_{15}CrNO_3$
J.T.Malito, R.Shakir, J.L.Atwood
*J.Chem.Soc.,Dalton Trans.,*1253, 1980

73.36 Dicarbonyl – nitrosyl – (η^5 – pentamethyl – cyclopentadienyl) – molybdenum
$C_{12}H_{15}MoNO_3$
J.T.Malito, R.Shakir, J.L.Atwood
*J.Chem.Soc.,Dalton Trans.,*1253, 1980

73.37 Dicarbonyl – nitrosyl – (η^5 – pentamethyl – cyclopentadienyl) – tungsten
$C_{12}H_{15}NO_3W$
J.T.Malito, R.Shakir, J.L.Atwood
J.Chem.Soc.,Dalton Trans.,1253, 1980

73.C bis(η^5 – Cyclopentadienyl) – methylmethylene – tantalum (at 15°K, neutron study)
$C_{12}H_{15}Ta$ Main entry is 71.42

73.38 2,2′,5,5′ – Tetramethyl – 1,1′ – diarsa – ferrocene
$C_{12}H_{16}As_2Fe$
L.Chiche, J.Galy, G.Thiollet, F.Mathey
Acta Crystallogr.,Sect.B,**36**, 1344, 1980
See also R1 : 65

73.39 (μ – Dimethylhydrazido) – bis(cyclopentadienyl – iodo – nitrosyl – molybdenum)
$C_{12}H_{18}I_2Mo_2N_4O_2$
P.R.Mallinson, G.A.Sim, D.I.Woodhouse
Acta Crystallogr.,Sect.B,**36**, 450, 1980
See also R1 : 83

73.40 (μ – Oxo) – bis(aqua – trichloro – (η – methylcyclopentadienyl) – niobium(v))
$C_{12}H_{18}Cl_6Nb_2O_3$
K.Prout, J.–C.Daran
Acta Crystallogr.,Sect.B,**35**, 2882, 1979

73.41 Dicarbonyl – (η^5 – cyclopentadienyl) – (μ – methanediazo) – (pentacarbonyl – chromium) – tungsten (at –160°C)
$C_{13}H_8CrN_2O_7W$
W.A.Herrmann, S.A.Bistram *Chem.Ber.*,**113**, 2648, 1980
See also R1 : 83

73.C Dicarbonyl – cyclopentadienyl – cis(2 – ethoxycarbonyl – 2 – trifluoromethyl – 1 – fluorovinyl) – iron
$C_{13}H_{10}F_4FeO_4$ Main entry is 71.51

73.C (μ^2 – Methylene) – bis(carbonyl – (η^5 – cyclopentadienyl) – rhodium) (at 15°K, neutron study)
$C_{13}H_{12}O_2Rh_2$ Main entry is 71.52

73.42 (μ – Ethoxycarbonylimido) – (μ – oxo) – bis((η – cyclopentadienyl) – oxo – molybdenum)
$C_{13}H_{15}Mo_2NO_5$
R.Korswagen, K.Weidenhammer, M.L.Ziegler
Acta Crystallogr.,Sect.B,**35**, 2554, 1979
See also R1 : 83

73.43 Dicyclopentadienyl – (trimethylsilyl) – titanium – chloride
$C_{13}H_{19}ClSiTi$
L.Rosch, G.Altnau, W.Erb, J.Pickardt, N.Bruncks
J.Organomet.Chem.,**197**, 51, 1980
See also R1 : 63

73.C η^5 – Cyclopentadienyl – (η^4 – 3,4 – dimethyl – hexa – 2,4 – diene) – dichloro – niobium
$C_{13}H_{19}Cl_2Nb$ Main entry is 72.21

73.44 Trimethylsilylnitrene – bis(η^5 – cyclopentadienyl) – vanadium (at –20°C)
$C_{13}H_{19}NSiV$
N.Wiberg, H.–W.Haring, U.Schubert
Z.Naturforsch.,Teil B,**35**, 599, 1980
See also R1 : 83

73.C 3,7,9,13 – Tetramethyl – 6 – (η^5 – cyclopentadienyl) – 3,7,9,13 – tetracarba – 6 – cobalta – dodecaborane
$C_{13}H_{24}B_7Co$ Main entry is 71.54

73.45 bis((η^5 – Cyclopentadienyl) – dicarbonyl – chromium) – sulfide
$C_{14}H_{10}Cr_2O_4S$
T.J.Greenhough, B.W.S.Kolthammer, P.Legzdins, J.Trotter
Inorg.Chem.,**18**, 3542, 1979

73.46 (μ – Carbonyl) – (μ – thiocarbonyl) – bis(carbonyl – cyclopentadienyl – iron)
$C_{14}H_{10}Fe_2O_3S$
D.E.Beckman, R.A.Jacobson
J.Organomet.Chem.,**179**, 187, 1979

73.C bis(Carbonyl) – cycloheptatrienylidene – (η^5 – cyclopentadienyl) – iron hexafluorophosphate (at –15°C)
$C_{14}H_{11}FeO_2^+, F_6P^-$ Main entry is 71.62

73.47 bis(μ – Carbonyl) – bis(methylcyclopentadienyl – nickel)
$C_{14}H_{14}Ni_2O_2$
L.R.Byers, L.F.Dahl *Inorg.Chem.*,**19**, 680, 19

73.48 Dicarbonyl – cyclopentadienyl – (3 – methylcyclohexenone) – iron hexafluorophosphate
$C_{14}H_{15}FeO_3^+, F_6P^-$
B.M.Foxman, P.T.Klemarczyk, R.E.Liptrot, M.Rosenblum
J.Organomet.Chem.,**187**, 253, 1980
See also R1 : 84

73.49 (R) – Ferrocenyl – methylmethane – S – methyltrithiocarbonate (absolute configuration)
$C_{14}H_{16}FeS_3$
H.Patin, G.Mignani, C.Mahe, J.Y.Le Marouille, A.Benoit, D.Grandjean *J.Organomet.Chem.*,**193**, 93, 1980

73.C bis(η^5 – Cyclopentadienyl) – (μ – (η^3:η^4 – trihydro – C,C′ – dimethyl – dicarba – pentaborato)) – di – cobalt
$C_{14}H_{19}B_3Co_2$ Main entry is 71.69

73.C η^5 – Cyclopentadienyl – (η^5 – cyclopentadienyl – cobalt) – (μ – (η^3:η^4 – trihydro – C,C′ – dimethyl – dicarba – hexaborato)) – iron
$C_{14}H_{20}B_3CoFe$ Main entry is 71.70

73.50 trans – bis(μ – Thioethyl) – bis(nitrosyl – (η^5 – cyclopentadienyl) – molybdenum)
$C_{14}H_{20}Mo_2N_2O_2S_2$
G.R.Clark, D.Hall, K.Marsden
J.Organomet.Chem.,**177**, 411, 1979
See also R1 : 85

73.51 bis(μ – Sulfido) – bis(μ – methylthio) – bis(methylcyclopentadienyl – molybdenum(iv))
$C_{14}H_{20}Mo_2S_4$
R.C.Haltiwanger, M.R.DuBois, D.L.DuBois, M.C.VanDerveer
Am.Cryst.Assoc.,Ser.2,**7**, 37, 1980
See also R1 : 85

73.C 1,1 – Dichloro – 1 – (η^5 – pentamethyl – cyclopentadienyl) – tantalla – cyclopentane
$C_{14}H_{23}Cl_2Ta$ Main entry is 71.74

73.C Tetracarbonyl – manganese – (μ – η^5:η^1 – cyclopentadienyl) – (carbonyl – (η^5 – cyclopentadienyl) – molybdenum)
$C_{15}H_9MnMoO_5$ Main entry is 71.85

73.52 Tetraethylammonium (μ – cyano) – bis(cyclopentadienyl – dicarbonyl – molybdenum)
$C_{15}H_{10}Mo_2NO_4{}^-$, $C_8H_{20}N^+$
M.D.Curtis, K.R.Han, W.M.Butler
Inorg.Chem.,**19**, 2096, 1980
See also R2 : 3

73.C (μ^2 – Carbonyl) – (μ^2 – vinylidene) – bis(carbonyl – cyclopentadienyl – ruthenium)
$C_{15}H_{12}O_3Ru_2$ Main entry is 71.86

73.C (μ^2 – Carbonyl) – (μ^2 – methylcarbyne) – bis(carbonyl – cyclopentadienyl – ruthenium) tetrafluoroborate
$C_{15}H_{13}O_3Ru_2{}^+$, $BF_4{}^-$ Main entry is 71.87

73.C cis((μ – Carbonyl) – (μ – methylcarbene) – bis(carbonyl – cyclopentadienyl – iron))
$C_{15}H_{14}Fe_2O_3$ Main entry is 71.88

73.53 tris(η – Cyclopentadienyl) – tri – cobalt – disulfide (room temperature form)
$C_{15}H_{15}Co_3S_2$
N.Kamijo, T.Watanabe
Acta Crystallogr.,Sect.B,**35**, 2537, 1979

73.54 tris(η – Cyclopentadienyl) – tri – cobalt – disulfide (low temperature form, at 130°K)
$C_{15}H_{15}Co_3S_2$
N.Kamijo, T.Watanabe
Acta Crystallogr.,Sect.B,**35**, 2537, 1979

73.55 2 – Nitro – 3 – ferrocenyl – acrylic acid ethyl ester
$C_{15}H_{15}FeNO_4$
E.Skrzypczak-Jankun, A.Hoser, E.Grzesiak, Z.Kaluski
Acta Crystallogr.,Sect.B,**36**, 934, 1980
See also R1 : 1,10

73.C η^5 – Cyclopentadienyl – bis(μ^2 – carbonyl) – (μ^3 – allyl) – hexacarbonyl – tri – iron
$C_{16}H_8Fe_3O_8$ Main entry is 71.91

73.56 ((1 – 5 – η) – exo – 1 – Acetyl – 2,4,6 – tris(trifluoromethyl)cyclohexadienyl) – (η – cyclopentadienyl) – iron
$C_{16}H_{11}F_9FeO$
M.Bottrill, M.Green, E.O'Brien, L.E.Smart, P.Woodward
J.Chem.Soc.,Dalton Trans.,292, 1980
See also R1 : 75

73.C (μ – Ethylene – 1,1 – diyl) – bis(dicarbonyl – cyclopentadienyl – manganese) (at –130°C)
$C_{16}H_{12}Mn_2O_4$ Main entry is 71.

73.57 (1,1' – Ferrocenediyl) – phenylphosphine
$C_{16}H_{13}FeP$
H.Stoeckli-Evans, A.G.Osborne, R.H.Whiteley
J.Organomet.Chem.,**194**, 91, 1980
See also R1 : 64

73.58 (μ – Dinitrogen) – bis(dicarbonyl – (η^5 – methylcyclopentadienyl) – manganese)
$C_{16}H_{14}Mn_2N_2O_4$
K.Weidenhammer, W.A.Herrmann, M.L.Ziegler
Z.Anorg.Allg.Chem.,**457**, 183, 1979

73.C (μ^2 – Allene) – (μ^2 – carbonyl) – bis(carbonyl – η^5 – cyclopentadienyl – manganese) (at –170°C)
$C_{16}H_{14}Mn_2O_3$ Main entry is 71.96

73.C tris(η^5 – Cyclopentadienyl) – (μ_3 – sulfido) – (μ_3 – thiocarbonyl) – triangulo – tri – cobalt
$C_{16}H_{15}Co_3S_2$ Main entry is 71.97

73.59 Tetraphenylarsonium (μ – dimethylphosphido) – bis(dicarbonyl – cyclopentadienyl – molybdenum)
$C_{16}H_{16}Mo_2O_4P^-$, $C_{24}H_{20}As^+$
J.L.Petersen, R.P.Stewart Junior
Inorg.Chem.,**19**, 186, 1980
See also R1 : 86 R2 : 65

73.C (η^2 – Dimethylacetylene – dicarboxylate) – bis(η^5 – cyclopentadienyl) – vanadium
$C_{16}H_{16}O_4V$ Main entry is 72.26

73.60 bis(η^5 – Cyclopentadienyl) – phenylhydrazido – tungsten tetrafluoroborate
$C_{16}H_{17}N_2W^+$, $BF_4{}^-$
J.A.Carroll, D.Sutton, M.Cowie, M.D.Gauthier
J.Chem.Soc.,Chem.Commun.,1058, 1979
See also R1 : 83

73.61 Catecholato – (η^5 – pentamethyl – cyclopentadienyl) – rhodium(iii) catechol solvate
$C_{16}H_{19}O_2Rh$, $2C_6H_6O_2$
P.Espinet, P.M.Bailey, P.M.Maitlis
J.Chem.Soc.,Dalton Trans.,1542, 1979
See also R1 : 84

73.C Carbonyl – η^5 – cyclopentadienyl – trimethoxyphosphine – (4,5 – bis(methoxycarbonyl) – 1,3 – dithiol – 2 – ylidene) – manganese(i)
$C_{16}H_{20}MnO_8PS_2$ Main entry is 71.99

73.62 bis(μ – Ethoxycarbonylimido) – bis((η – cyclopentadienyl) – oxo – molybdenum)
$C_{16}H_{20}Mo_2N_2O_6$
R.Korswagen, K.Weidenhammer, M.L.Ziegler
Acta Crystallogr.,Sect.B,**35**, 2554, 1979
See also R1 : 83

73.63 (μ^2 – Pinacolato) – bis(cyclopentadienyl – dichloro – titanium) (at –138°C)
$C_{16}H_{22}Cl_4O_2Ti_2$
J.C.Huffman, K.G.Moloy, J.A.Marsella, K.G.Caulton
J.Am.Chem.Soc.,**102**, 3009, 1980
See also R1 : 84

73.64 bis(μ – Propane – 1,2 – dithiolato) – bis(cyclopentadienyl – molybdenum) tetrafluoroborate
$C_{16}H_{22}Mo_2S_4{}^+$, $BF_4{}^-$
M.R.DuBois, R.C.Haltiwanger, D.J.Miller, G.Glatzmaier
J.Am.Chem.Soc.,**101**, 5245, 1979
See also R1 : 85

73.65 cis – bis(μ – Thio – isopropyl) – bis(nitrosyl – (η^5 – cyclopentadienyl) – molybdenum)
$C_{16}H_{24}Mo_2N_2O_2S_2$
G.R.Clark, D.Hall, K.Marsden
J.Organomet.Chem.,**177**, 411, 1979
See also R1 : 85

73.C Dicarbonyl – η^5 – cyclopentadienyl – η – (1,2,3,4 – tetrakis(trifluoromethyl) – buta – 1,3 – diene – 1,4 – diyl) – tungsten – dicarbonyl – cobalt
$C_{17}H_5CoF_{12}O_4W$ Main entry is 71.103

73.C bis(σ – Pyrrolyl – tricarbonyl – manganese) – tricarbonyl – iodo – manganese
$C_{17}H_8IMn_3N_2O_9$ Main entry is 71.104

73.C (μ_3 – Carbonyl) – bis(η^5 – cyclopentadienyl – nickel) – bis(tricarbonyl – iron)
$C_{17}H_{10}Fe_2Ni_2O_7$ Main entry is 71.105

73.C (η^5 – Cyclopentadienyl – nickel) – (η – t – butylethynyl) – hexacarbonyl – di – iron)
$C_{17}H_{14}Fe_2NiO_6$ Main entry is 72.29

73.66 Trisila – tetrachloro – cyclohexyl – bis(dicarbonyl – cyclopentadienyl – iron)
$C_{17}H_{18}Cl_4Fe_2O_4Si_3$
W.Honle, H.G.von Schnering
Z.Anorg.Allg.Chem., **464**, 139, 1980
See also R1 : 63

73.67 poly(Trivinyl – ferrocenoato – tin(iv))
$(C_{17}H_{18}FeO_2Sn)_n$
R.Graziani, U.Casellato, G.Plazzogna
J.Organomet.Chem., **187**, 381, 1980
See also R1 : 69

73.68 Ethyl (Z) – 2 – ferrocenyl – 2 – methyl – cyclopropane – carboxylate
$C_{17}H_{20}FeO_2$
Z.Kaluski, E.Skrzypczak–Jankun, M.Cygler
Acta Crystallogr., *Sect.B*, **35**, 2699, 1979

73.69 Hexamethylbenzene – cyclopentadienyl – iron
$C_{17}H_{23}Fe$
D.Astruc, J.–R.Hamon, G.Althoff, E.Roman, P.Batail, P.Michaud, J.–P.Mariot, F.Varret, D.Cozak
J.Am.Chem.Soc., **101**, 5445, 1979
See also R1 : 74

73.70 η^5 – Pentamethyl – cyclopentadienyl – (6 – methylimino – 1 – 5η – cyclohexadienyl) – rhodium hexafluorophosphate
$C_{17}H_{24}NRh^{2+}$, $2F_6P^-$
P.Espinet, P.M.Bailey, R.F.Downey, P.M.Maitlis
J.Chem.Soc., *Dalton Trans.*, 1048, 1980
See also R1 : 75

73.C 3,3 – Dichloro – 3 – (η^5 – pentamethyl – cyclopentadienyl) – 3 – tantalla – bicyclo(3.3.0) octane
$C_{17}H_{27}Cl_2Ta$ Main entry is 71.110

73.71 Chloro – (tetracarbonyl – cobalt) – bis(dicarbonyl – cyclopentadienyl – iron) – tin(iv)
$C_{18}H_{10}ClCoFe_2O_8Sn$
M.Moll, H.Behrens, P.Merbach, K.Gorting, G.Liehr, R.Bohme *Z.Naturforsch.*, *Teil B*, **35**, 1115, 1980

73.C 1,1 – bis(η^5 – Cyclopentadienyl) – 2,3,4,5 – tetra(trifluoromethyl) – niobia – cyclopentadiene
$C_{18}H_{10}F_{12}Nb$ Main entry is 71.119

73.C bis(Carbonyl) – (η^5 – cyclopentadienyl) – (bicyclo(5.4.0)undeca – 1,3,6,8,10 – pentaen – 5 – ylidene) – iron hexafluorophosphate (at –15°C)
$C_{18}H_{13}FeO_2^+$, F_6P^- Main entry is 71.122

73.C Pentacarbonyl – (ethoxy(ruthenocenyl)carbene) – tungsten
$C_{18}H_{14}O_6RuW$ Main entry is 71.125

73.72 bis(p – Fluorobenzenediazo) – (η^5 – methylcyclopentadienyl) – chloro – molybdenum
$C_{18}H_{15}ClF_2MoN_4$
G.Ferguson, F.J.Lalor, M.Parvez, M.A.Khan
Am.Cryst.Assoc., *Ser.2*, **8**, 24, 1980
See also R1 : 83

73.73 bis(η^5 – Cyclopentadienyl) – molybdenum – bis(μ – carbonyl) – (η^5 – cyclopentadienyl – carbonyl – rhenium)
$C_{18}H_{15}MoO_3Re$
R.I.Mink, J.J.Welter, P.R.Young, G.D.Stucky
J.Am.Chem.Soc., **101**, 6928, 1979

73.C bis(μ^2 – Carbonyl) – (μ^3 – methylidyne) – cyclo – tris((η^5 – cyclopentadienyl) – rhodium) trifluoroacetate
$C_{18}H_{16}O_2Rh_3^+$, $C_2F_3O_2^-$ Main entry is 71.126

73.74 Di – η^5 – cyclopentadienyl – di – η – pyrrolyl – titanium(iv)
$C_{18}H_{18}N_2Ti$
R.Vann Bynum, W.E.Hunter, R.D.Rogers, J.L.Atwood
Inorg.Chem., **19**, 2368, 1980
See also R1 : 83

73.75 Di – η^5 – cyclopentadienyl – di – η – pyrrolyl – zirconium(iv)
$C_{18}H_{18}N_2Zr$
R.Vann Bynum, W.E.Hunter, R.D.Rogers, J.L.Atwood
Inorg.Chem., **19**, 2368, 1980
See also R1 : 83

73.C bis(η^5 – Cyclopentadienyl) – zirconia – benzocyclopentene
$C_{18}H_{18}Zr$ Main entry is 71.127

73.C (μ – Carbonyl) – (μ – dimethyliminomethylene) – bis(carbonyl – (η – methylcyclopentadienyl) – iron) iodide
$C_{18}H_{20}Fe_2NO_3^+$, I^- Main entry is 71.128

73.76 (μ – Oxy – bis(dimethylphosphane)) – bis(dicarbonyl – (η^5 – cyclopentadienyl) – manganese(i))
$C_{18}H_{22}Mn_2O_5P_2$
E.Lindner, S.Hoehne, J.–P.Gumz
Z.Naturforsch., *Teil B*, **35**, 937, 1980
See also R1 : 86

73.77 Cobaltocenyl – tetra(methylcarba) – dodecaborane
$C_{18}H_{29}B_8Co$
R.N.Grimes, J.R.Pipal, E.Sinn
J.Am.Chem.Soc., **101**, 4172, 1979
See also R1 : 62

73.78 Nonamethyl – cyclopentasilane – dimethylsilyl – η^5 – cyclopentadienyl – dicarbonyl – iron
$C_{18}H_{38}FeO_2Si_6$
T.J.Drahnak, R.West, J.C.Calabrese
J.Organomet.Chem., **198**, 55, 1980
See also R1 : 63

73.C (Dicarbonyl – cyclopentadienyl – manganese) – μ – phenylketenyl – (tetracarbonyl – rhenium)
$C_{19}H_{10}MnO_7Re$ Main entry is 71.136

73.C (μ – Carbonyl) – (μ – t – butoxy – carbonylmethylene) – bis(carbonyl – (η^5 – cyclopentadienyl) – iron)
$C_{19}H_{20}Fe_2O_5$ Main entry is 71.140

73.79 tris(Cyclopentadienyl) – tetrahydrofuran – gadolinium
$C_{19}H_{23}GdO$
R.D.Rogers, R.Vann Bynum, J.L.Atwood
J.Organomet.Chem., **192**, 65, 1980
See also R1 : 84

73.80 (−) − RS − Cyclo − 1 − (1′ − dimethylaminoethyl − ferrocene) − 2 − acetylacetonato − palladium (absolute configuration)
$C_{19}H_{25}FeNO_2Pd$
L.G.Kuz'mina, Yu.T.Struchkov, L.L.Troitskaya, V.I.Sokolov, O.A.Reumov
Izv.Akad.Nauk SSSR,Ser.Khim.,1528, 1979
See also R1 : 77,83

73.81 (2 − 6 − η − 1 − exo − Dimethoxyphosphoryl − cyclohexadienyl) − (η^5 − ethyltetramethyl − cyclopentadienyl) − rhodium(iii) hexafluorophosphate
$C_{19}H_{29}O_3PRh^+$, F_6P^-
N.A.Bailey, E.H.Blunt, G.Fairhurst, C.White
J.Chem.Soc.,Dalton Trans.,829, 1980
See also R1 : 74

73.C Dicarbonyl − cyclopentadienyl − manganese − (μ − p − tolylketenyl) − tetracarbonyl − manganese diethyl ether solvate
$C_{20}H_{12}Mn_2O_7$, $0.5C_4H_{10}O$ Main entry is 71.149

73.C (η^5 − Cyclopentadienyl − nickel) − (μ^4 − 3,3 − dimethyl − but − 1 − enyl) − nonacarbonyl − tri − ruthenium
$C_{20}H_{15}NiO_9Ru_3$ Main entry is 71.151

73.C (μ − Thio − ferrocenyl − methylmethane − thiomethylene − C',S^2) − 1,1,1,2,2,2 − hexa − carbonyl − (μ − methylthio) − iron
$C_{20}H_{16}Fe_3O_6S_3$ Main entry is 71.153

73.C Carbonyl − cyclopentadienyl − (5 − methyl − 2 − (phenylcarbenylamino) − phenyl) − iron
$C_{20}H_{17}FeNO$ Main entry is 71.154

73.82 Diferrocenyl − selenium iodine tri − iodide methylene chloride solvate
$C_{20}H_{18}Fe_2Se^+$, I_3^-, I_2, $0.5CH_2Cl_2$
J.A.Kramer, F.H.Herbstein, D.N.Hendrickson
J.Am.Chem.Soc.,102, 2293, 1980

73.C bis(η^5 − Cyclopentadienyl) − bis(cyclopentadienyl) − hafnium (reinvestigation of structure, published by Kulisher et al)
$C_{20}H_{18}Hf$ Main entry is 71.155

73.C bis(μ − sigma:eta − Cyclopentadienyl) − bis(η − cyclopentadienyl − molybdenum)
$C_{20}H_{18}Mo_2$ Main entry is 71.156

73.83 bis(bis(Cyclopentadienyl) − chloro − titanium) − oxide
$C_{20}H_{20}Cl_2OTi_2$
Y.Le Page, J.D.McCowan, B.K.Hunter, R.D.Heyding
J.Organomet.Chem.,193, 201, 1980
See also R1 : 84

73.84 trans − 1 − Ferrocenyl − 2 − p − methoxyphenyl − cyclopropane
$C_{20}H_{20}FeO$
A.N.Nesmeyanov, E.I.Klimova, Yu.T.Struchkov, V.G.Andrianov, V.N.Postnov, V.A.Sazonova
J.Organomet.Chem.,178, 343, 1979

73.C (μ^2 − Carbonyl) − (μ − η^1,η^3 − (1 − methyl − 2,3 − bis(methoxycarbonyl) − allyl)) − carbonyl − bis(η^5 − cyclopentadienyl − iron)
$C_{20}H_{20}Fe_2O_6$ Main entry is 72.38

73.85 tris(μ_3 − Hydrido) − tetra(cyclopentadienyl − nickel) (neutron study, at 81°K).
$C_{20}H_{23}Ni_4$
T.F.Koetzle, J.Muller, D.L.Tipton, D.W.Hart, R.Bau
J.Am.Chem.Soc.,101, 5631, 1979

73.86 (μ − Hydrido) − bis(hydrido − dicyclopentadienyl − tungsten) perchlorate acetone solvate
$C_{20}H_{23}W_2^+$, ClO_4^-, C_3H_6O
R.J.Klingler, J.C.Huffman, J.K.Kochi
J.Am.Chem.Soc.,102, 208, 1980

73.87 (Tetra − cyclopentadienyl − nickela) − octaborane
$C_{20}H_{24}B_4Ni_4$
J.R.Bowser, A.Bonny, J.R.Pipal, R.N.Grimes
J.Am.Chem.Soc.,101, 6229, 1979

73.C Decamethylferrocenium − 7,7,8,8 − tetracyanoquinodimethane
$C_{20}H_{30}Fe^+$, $C_{12}H_4N_4^-$ Main entry is 60.42

73.88 bis(μ − Iodo) − bis(iodo − (pentamethyl − cyclopentadienyl) − rhodium) toluene solvate
$C_{20}H_{30}I_4Rh_2$, $2C_7H_8$
M.R.Churchill, S.A.Julis *Inorg.Chem.*,18, 2918, 1979

73.C (μ − Chloro) − 1 − (dicyclopentadienyl − zirconium(iv)) − 2,2 − bis(diethyl − aluminium) ethane
$C_{20}H_{33}Al_2ClZr$ Main entry is 71.159

73.C (μ − 1,3 − Dioxo − indan − 2 − ylidene) − bis(carbonyl − (η^5 − cyclopentadieneyl) − cobalt)
$C_{21}H_{14}Co_2O_4$ Main entry is 71.167

73.89 Chloro − tris(cyclopentadienyl − dicarbonyl − manganese) − antimony (at 190°K)
$C_{21}H_{15}ClMn_3O_6Sb$
J.von Seyerl, L.Wohlfahrt, G.Huttner
Chem.Ber.,113, 2868, 1980
See also R1 : 86

73.C tris(η^5 − Cyclopentadienyl) − (μ_3 − (pentacarbonyl − chromium − thio) − methylidyne) − (μ_3 − sulfido) − triangulo − tri − cobalt tetrahydrofuran solvate
$C_{21}H_{15}Co_3CrO_5S_2$, $0.5C_4H_8O$ Main entry is 71.169

73.C bis(η^5 − Cyclopentadienyl) − (μ − η^2,η^2 − carbonyl − dimethylacetylene − carboxylate) − bis(dicarbonyl − tungsten) (at 233°K)
$C_{21}H_{16}O_9W_2$ Main entry is 71.170

73.C (bis(Trimethylphosphine) − platinum) − (μ − carbonyl) − (μ − p − tolylcarbenyl) − (carbonyl − cyclopentadienyl − manganese) tetrafluoroborate dichloromethane solvate
$C_{21}H_{30}MnO_2P_2Pt^+$, BF_4^-, CH_2Cl_2 Main entry is 71.173

73.C Tricarbonyl − cobalt − (μ − diphenylacetylene) − (cyclopentadienyl − nickel)
$C_{22}H_{15}CoNiO_3$ Main entry is 72.39

73.90 Heptacarbonyl − tris(cyclopentadienyl − niobium)
$C_{22}H_{15}Nb_3O_7$
W.A.Herrmann, M.L.Ziegler, K.Weidenhammer, H.Biersack
Angew.Chem.,Int.Ed.Engl.,18, 960, 1979

73.91 (1,1′ − Ferrocenediyl) − diphenyl − germanium
$C_{22}H_{18}FeGe$
H.Stoeckli−Evans, A.G.Osborne, R.H.Whiteley
J.Organomet.Chem.,194, 91, 1980
See also R1 : 69

73.C (Hex – 3 – ene – 3,4 – diyl) – bis(η^5 – cyclopentadienyl – nickel) – bis(tricarbonyl – iron)
$C_{22}H_{20}Fe_2Ni_2O_6$ Main entry is 71.178

73.92 Dicarbonyl – (η^5 – cyclopentadienyl) – (1 – 3 – η – (4 – cyclopenta – 1',3' – dienyl) – 5 – (cyclopenta – 1'',4'' – dienyl) – cyclopentyl) – tungsten
$C_{22}H_{20}O_2W$
R.D.Rogers, W.E.Hunter, J.L.Atwood
J.Chem.Soc.,Dalton Trans.,1032, 1980
See also R1 : 75

73.93 bis(μ – Carbonyl) – (μ – carbonyl – (cyclopentadienyl) – bis(dimethylarsenido) – iron) – cyclopentadienyl – (cyclopentadienyl – cobalt) – iron
$C_{22}H_{27}As_2CoFe_2O_3$
E.Rottinger, A.Trenkle, R.Muller, H.Vahrenkamp
Chem.Ber.,113, 1280, 1980
See also R1 : 86

73.94 (η^5 – Pentamethyl – cyclopentadienyl) – dicarbonyl – titanium
$C_{22}H_{30}O_2Ti$
R.D.Rogers, J.L.Atwood *Am.Cryst.Assoc.,Ser.2*,**7**, 27, 1980

73.C (μ^2 – Carbonyl) – (μ^2 – methylene) – bis(η^5 – pentamethyl – cyclopentadienyl – cobalt)
$C_{22}H_{32}Co_2O$ Main entry is 71.184

73.95 (μ^2 – Dimethylarsonio) – bis(dicarbonyl – cyclopentadienyl – trimethylphosphine) – molybdenum tricarbonyl – cyclopentadienyl – molybdenum
$C_{22}H_{34}AsMo_2O_4P_2^+$, $C_8H_5MoO_3^-$
R.Janta, W.Albert, H.Rossner, W.Malisch, H.–J.Langenbach, E.Rottinger, H.Vahrenkamp *Chem.Ber.*,113, 2729, 1980
See also R1 : 86,65 R2 : 73

73.96 bis(μ – Iodo) – di – iodo – bis(η^5 – ethyl – tetramethyl – cyclopentadienyl) – di – rhodium(iii)
$C_{22}H_{34}I_4Rh_2$
I.W.Nowell, G.Fairhurst, C.White
Inorg.Chim.Acta,41, 61, 1980

73.C η^5 – Cyclopentadienyl – (2,4 – diphenyl – cyclobut – 1 – en – 3 – on – 1 – yl) – dicarbonyl – iron
$C_{23}H_{16}FeO_3$ Main entry is 71.188

73.97 Ferrocenyl – diphenylcarbenium tetrafluoroborate
$C_{23}H_{19}Fe^+$, BF_4^-
U.Behrens *J.Organomet.Chem.*,**182**, 89, 1979

73.C Chloro – dicyclopentadienyl – (diphenylphosphinomethyl) – zirconium (at –140°C)
$C_{23}H_{22}ClPZr$ Main entry is 71.189

73.C bis(η^5 – Cyclopentadienyl) – bis(pentafluorophenyl) – acetylene – vanadium
$C_{24}H_{10}F_{10}V$ Main entry is 72.45

73.98 1,1 – Dicyano – 2,2 – di(ferrocenyl) – ethylene
$C_{24}H_{18}Fe_2N_2$
J.M.Gromek, J.Donohue *Am.Cryst.Assoc.,Ser.2*,**7**, 27, 1980

73.99 bis(μ – Fulvalene) – bis(acetonitrile – vanadium(iii)) hexafluorophosphate acetonitrile solvate
$C_{24}H_{22}N_2V_2^{2+}$, $2F_6P^-$, C_2H_3N
J.C.Smart, B.L.Pinsky, M.L.Fredrich, V.W.Day
J.Am.Chem.Soc.,**101**, 4371, 1979

73.C bis(η^5 – Cyclopentadienyl) – (thiocamphor) – tetracarbonyl – di – molybdenum
$C_{24}H_{26}Mo_2O_4S$ Main entry is 72.48

73.100 tetrakis(μ – Oxo) – tetrakis(chloro – (η^5 – methylcyclopentadienyl) – titanium)
$C_{24}H_{28}Cl_4O_4Ti_4$
J.L.Petersen *Inorg.Chem.*,**19**, 181, 1980

73.101 bis(η^5 – Trimethylsilyl – cyclopentadienyl) – o – xylidene – niobium(iv)
$C_{24}H_{34}NbSi_2$
M.F.Lappert, T.R.Martin, C.R.C.Milne, J.L.Atwood, W.E.Hunter, R.E.Pentilla
J.Organomet.Chem.,**192**, C35, 1980

73.102 bis(Pentamethyl – cyclopentadienyl) – tetrahydrofuran – ytterbium(ii) toluene solvate (at 176°K)
$C_{24}H_{38}OYb$, $0.5C_7H_8$
A.Zalkin, D.H.Templeton, B.Spencer, H.Ruben, T.D.Tilley, R.A.Andersen *Am.Cryst.Assoc.,Ser.2*,**8**, 25, 1980
See also R1 : 84

73.103 (Octamethyl – (η^5 – cyclopentadienyl – dicarbonyl – iron) – cyclopentasilane) – dimethylsilyl – η^5 – cyclopentadienyl – dicarbonyl – iron
$C_{24}H_{40}Fe_2O_4Si_6$
T.J.Drahnak, R.West, J.C.Calabrese
J.Organomet.Chem.,**198**, 55, 1980
See also R1 : 63

73.104 (1,3,5 – tris(Trimethylsilyl) – pentalene) – triangulo(bis(dicarbonyl – ruthenium) – tetracarbonyl – ruthenium) (at 160°K)
$C_{25}H_{30}O_8Ru_3Si_3$
J.A.K.Howard, R.F.D.Stansfield, P.Woodward
J.Chem.Soc.,Dalton Trans.,1812, 1979

73.105 (1,3,5 – tris(Trimethylsilyl) – pentalene) – triangulo(bis(tricarbonyl – ruthenium) – dicarbonyl – ruthenium)
$C_{25}H_{30}O_8Ru_3Si_3$
J.A.K.Howard, R.F.D.Stansfield, P.Woodward
J.Chem.Soc.,Dalton Trans.,1812, 1979

73.106 bis(Cyclopentadienyl) – (2,6 – di – t – butyl – 4 – methylphenoxy) – titanium(iii)
$C_{25}H_{33}OTi$
B.Cetinkaya, P.B.Hitchcock, M.F.Lappert, S.Torroni, J.L.Atwood, W.E.Hunter, M.J.Zaworotko
J.Organomet.Chem.,**188**, 31, 1980
See also R1 : 84

73.C bis(η^5 – Cyclopentadienyl) – 2,2 – bis(diethyl – alumino) – ethyl – zirconium – cyclopentadienide
$C_{25}H_{38}Al_2Zr$ Main entry is 71.197

73.C Dicarbonyl – (η^5 – cyclopentadienyl) – (triphenylsilylcarbene) – rhenium dichloromethane solvate
$C_{26}H_{21}O_2ReSi$, CH_2Cl_2 Main entry is 71.201

73.C (μ – Cyclobuta(1,2 – a:3,4 – a')dicyclopentene) – bis(dicarbonyl – (η^5 – methylcyclopentadienyl) – manganese)
$C_{26}H_{22}Mn_2O_4$ Main entry is 71.202

73.C Cyclopentadienyl – (η^2 – benzaldehyde) – (α – (2 – pyridyl) – α – (methylamino) – benzyl) – carbonyl – molybdenum
$C_{26}H_{24}MoN_2O_2$ Main entry is 71.204

73.107 η^5 – Cyclopentadienyl – triphenylphosphine – nickel – ethylxanthate (at –170°C)
$C_{26}H_{25}NiOPS_2$
C.Tsipis, G.E.Manoussakis, D.P.Kessissoglou, J.C.Huffman, L.N.Lewis, M.A.Adams, K.G.Caulton
Inorg.Chem., **19**, 1458, 1980
See also R1 : 86,85

73.C η^5 – Cyclopentadienyl – iodo – (methyl(methylthio) – carbene) – (triphenylphosphine) – iridium(iii) iodide
$C_{26}H_{26}IIrPS^+$, I- Main entry is 71.206

73.108 (+) – D – (1 – (2 – Diphenylphosphino – ferrocenyl) ethyl) – dimethylamine (absolute configuration)
$C_{26}H_{28}FeNP$
F.W.B.Einstein, A.C.Willis
Acta Crystallogr.,Sect.B,**36**, 39, 1980

73.109 bis(η^5 – Cyclopentadienyl – iron – μ^2,η^5 – 3,4 – diethyl – 2,5 – dimethyl – 1,2,5 – thia – diborolene) – iron
$C_{26}H_{42}B_4Fe_3S_2$
W.Siebert, C.Bohle, C.Kruger
Angew.Chem.,Int.Ed.Engl.,**19**, 746, 1980

73.C (μ – Oxo) – (μ – trimethylphosphine – methyl) – bis(η^5 – ethyltetramethyl – cyclopentadienyl) – hydrido – bis(dichloro – tantalum)
$C_{26}H_{45}Cl_4OPTa_2$ Main entry is 71.209

73.C bis(η^5 – Cyclopentadienyl) – (μ – carbonyl) – (μ – η',η^3 – carbonyl – 1,2 – diphenylethylene) – carbonyl – di – ruthenium (at 233°K)
$C_{27}H_{20}O_3Ru_2$ Main entry is 71.212

73.110 bis(μ^2 – Chloro) – tetrakis(dicarbonyl – cyclopentadienyl – manganese) – dibismuth (at –80°C)
$C_{28}H_{20}Bi_2Cl_2Mn_4O_8$
J.von Seyerl, G.Huttner
J.Organomet.Chem.,**193**, 207, 1980
See also R1 : 66

73.111 Tricarbonyl – (dicarbonyl(tricarbonyl(cyclopentadienyl) – chromium) – cyclopentadienyl – (μ – dimethylarsenido) – chromium) – bis(μ – dimethylarsenido) – (tetracarbonyl – iron) – cobalt
$C_{28}H_{28}As_3CoCr_2FeO_{12}$
H.J.Langenbach, E.Keller, H.Vahrenkamp
J.Organomet.Chem.,**191**, 95, 1980
See also R1 : 86

73.112 Dinitrato – (η^5 – pentamethyl – cyclopentadienyl) – triphenylphosphine – rhodium(iii)
$C_{28}H_{30}N_2O_6PRh$
M.B.Hursthouse, K.M.A.Malik, D.M.P.Mingos, S.D.Willoughby
J.Organomet.Chem.,**192**, 235, 1980
See also R1 : 86

73.113 2,4 – Dithio – pyrimidinato – bis(bis(η^5 – methylcyclopentadienyl) – titanium(iii))
$C_{28}H_{30}N_2S_2Ti_2$
D.R.Corbin, L.C.Francesconi, D.N.Hendrickson, G.D.Stucky
Inorg.Chem.,**18**, 3069, 1979
See also R1 : 85,83

73.C (μ – Oxo) – bis(n – butyl – bis(cyclopentadienyl) – niobium)
$C_{28}H_{38}Nb_2O$ Main entry is 71.221

73.114 Dicarbonyl – η^5 – cyclopentadienyl – (N – benzyl – N' – (1 – phenylethyl) – benzamidinato) – molybdenum (absolute configuration)
$C_{29}H_{28}MoN_2O_2$
H.Brunner, G.Agrifoglio, I.Bernal, M.W.Creswick
Angew.Chem.,Int.Ed.Engl.,**19**, 641, 1980
See also R1 : 83

73.115 (+)$_{578}$ – Dicarbonyl – (η^5 – cyclopentadienyl) – (N – benzyl – N' – α – methylbenzylbenzamidinato) – molybdenum (absolute configuration)
$C_{29}H_{28}MoN_2O_2$
I.Bernal, M.Creswick, H.Brunner, G.Agrifoglio
J.Organomet.Chem.,**198**, C4, 1980
See also R1 : 83

73.C Carbonyl – cyclopentadienyl – (1,5α – η^3 – 1 – methylene – cyclopentan – 2 – onato) – triphenylphosphine – molybdenum(ii)
$C_{30}H_{27}MoO_2P$ Main entry is 72.54

73.C (μ – Hydrido) – (μ – 2 – η^1:1 – 2 – η^2 – naphthyl) – bis(bis(η^5 – cyclopentadienyl) – zirconium)
$C_{30}H_{28}Zr_2$ Main entry is 71.230

73.116 tris(μ^2 – Hydrido) – (μ^3 – oxo) – tris(pentamethyl – cyclopentadienyl – rhodium) hexafluorophosphate monohydrate
$C_{30}H_{48}ORh_3^+$, F_6P^-, H_2O
A.Nutton, P.M.Bailey, N.C.Braund, R.J.Goodfellow, R.S.Thompson, P.M.Maitlis
J.Chem.Soc.,Chem.Commun.,631, 1980
See also R1 : 84

73.C Carbonyl – (phenylthiocarbyne) – (η^5 – cyclopentadienyl) – triphenylphosphine – tungsten
$C_{31}H_{25}OPSW$ Main entry is 71.2

73.117 (1,2 – bis(Diphenylphosphino)ethane) – trichloro – (η – cyclopentadienyl) – niobium(iv) toluene solvate
$C_{31}H_{29}Cl_3NbP_2$, $2C_7H_8$
K.Prout, J.–C.Daran
Acta Crystallogr.,Sect.B,**35**, 2882, 1979
See also R1 : 86

73.118 η^5 – Cyclopentadienyl – η^4 – (1,2 – diphenyl – cyclobuta(l)phenanthrene) – rhodium
$C_{33}H_{23}Rh$
G.M.Reisner, I.Bernal, M.D.Rausch, S.A.Gardner, A.Clearfield *J.Organomet.Chem.*,**184**, 237, 1980
See also R1 : 75

73.119 (η^5 – Cyclopentadienyl) – (η^4 – tetraphenyl – cyclobutadiene) – cobalt
$C_{33}H_{25}Co$
M.D.Rausch, G.F.Westover, E.Mintz, G.M.Reisner, I.Bernal, A.Clearfield, J.M.Troup *Inorg.Chem.*,**18**, 2605, 1979
See also R1 : 75

73.120 ((±) – α – (2 – Diphenylphosphino – ferrocenyl) ethyl – dimethylamine) – norbornadiene – rhodium hexafluorophosphate
$C_{33}H_{36}FeNPRh^+$, F_6P^-
W.R.Cullen, F.W.B.Einstein, C.–H.Huang, A.C.Willis, E.–S.Yeh
J.Am.Chem.Soc.,**102**, 988, 1980
See also R1 : 75,86,83

73.121 η^5 – Cyclopentadienyl – η^4 – (1 – methoxy – 1 – oxo – 2,3,4,5 – tetraphenylphosphole) – cobalt
$C_{34}H_{28}CoO_2P$
K.Yasufuku, A.Hamada, K.Aoki, H.Yamazaki
J.Am.Chem.Soc.,**102**, 4363, 1980
See also R1 : 75

73.C (η^5 – Cyclopentadienyl) – (2 – methyl – 4,5 – bis(diphenylphosphino) – pent – 2 – en – 3 – yl) – iron(ii)
$C_{35}H_{34}FeP_2$ Main entry is 71.243

73.122 (η^5 – Trimethylsilyl – cyclopentadienyl) – (η^4 – tetraphenyl – cyclobutadiene) – cobalt
$C_{36}H_{33}CoSi$
M.Calligaris, K.Venkatasubramanian
J.Organomet.Chem.,**175**, 95, 1979
See also R1 : 75,63

73.123 Cyclopentadienyl – iodo – (0 – isopropylidene – 2,3 – dihydroxy – 1,4 – bis(diphenylphosphino) – butane) – iron
$C_{36}H_{37}FeIO_2P_2$
G.Balavoine, S.Brunie, H.B.Kagan
J.Organomet.Chem.,**187**, 125, 1980
See also R1 : 86

73.C trans – anti – bis(η^5 – Cyclopentadienyl – (μ – phenylisocyanide) – (phenylisocyanide) – iron)
$C_{38}H_{30}Fe_2N_4$ Main entry is 71.253

73.124 1 – Phenyl – dibenzo(b,f) – (2,5 – diphenyl – ferra – cyclopentadiene – tricarbonyl) – (c) – phosphepin – iron – dicarbonyl
$C_{39}H_{23}Fe_2O_5P$
W.Winter *Eur.Cryst.Meeting*,**5**, 195, 1979
See also R1 : 75,86

73.125 (η^5 – Cyclopentadienyl) – (η^4 – 1,3 – dimesityl – 2,4 – diphenyl – cyclobutadiene) – cobalt
$C_{39}H_{37}Co$
M.D.Rausch, G.F.Westover, E.Mintz, G.M.Reisner, I.Bernal, A.Clearfield, J.M.Troup *Inorg.Chem.*,**18**, 2605, 1979
See also R1 : 75

73.126 bis(μ – Hydrido) – bis(bis(pentamethyl – cyclopentadienyl) – thorium) toluene solvate (neutron study)
$C_{40}H_{62}Th_2$, C_7H_8
R.W.Broach, A.J.Schultz, J.M.Williams, G.M.Brown, J.M.Manriquez, P.J.Fagan, T.J.Marks
Science,**203**, 172, 1979

73.127 tetrakis(μ – Hydrido) – tetrakis(pentamethyl – cyclopentadienyl – rhodium) tetrafluoroborate
$C_{40}H_{64}Rh_4{}^{2+}$, $2BF_4{}^-$
P.Espinet, P.M.Bailey, P.Piraino, P.M.Maitlis
Inorg.Chem.,**18**, 2706, 1979

73.128 1 – Phenyl – dibenzo(b,f) – 2,5 – diphenyl – cyclopentadienone(c)phosphepin – di – iron – hexacarbonyl
$C_{41}H_{23}Fe_2O_7P$
W.Winter *Eur.Cryst.Meeting*,**5**, 195, 1979
See also R1 : 86

73.129 (bis(η^5 – Cyclopentadienyl) – tungsten) – bis(μ – hydrido) – (bis(triphenylphosphine) – rhodium) hexafluorophosphate
$C_{46}H_{42}P_2RhW^+$, F_6P^-
N.W.Alcock, O.W.Howarth, P.Moore, G.E.Morris
J.Chem.Soc.,Chem.Commun.,1160, 1979
See also R1 : 86

73.130 (η^5 – Pentamethyl – cyclopentadienyl) – (triphenylphosphine) – hydrido – platinum(ii) hexafluorophosphate
$C_{46}H_{46}P_2Rh^+$, F_6P^-
D.M.P.Mingos, P.C.Minshall, M.B.Hursthouse, K.M.A.Malik, S.D.Willoughby *J.Organomet.Chem.*,**181**, 169, 1979
See also R1 : 86

73.131 N,N',N'',N''' – Tetra – p – tolyl – oxalylamidine – bis(di(cyclopentadienyl) – titanium)
$C_{50}H_{48}N_4Ti_2$
M.Pasquali, C.Floriani, A.C.Villa, C.Guastini
J.Am.Chem.Soc.,**101**, 4740, 1979
See also R1 : 83

73.132 tris(μ – Chloro) – tris(bis(pentamethyl – cyclopentadienyl) – uranium)
$C_{60}H_{90}Cl_3U_3$
J.M.Manriquez, P.J.Fagan, T.J.Marks, S.H.Vollmer, C.S.Day, V.W.Day *J.Am.Chem.Soc.*,**101**, 5075, 1979

73.133 μ – 1 – 5 – eta:1' – 5' – η – (bis(μ – Tetraphenyl – cyclopentadienediyl – oxo – O) – bis(aqua – silver)) – bis(tricarbonyl – iron) bis(hexafluorophosphate) dichloromethane solvate
$C_{64}H_{44}Ag_2Fe_2O_{10}{}^{2+}$, $2F_6P^-$
P.K.Baker, K.Broadley, N.G.Connelly, B.A.Kelly, M.D.Kitchen, P.Woodward
J.Chem.Soc.,Dalton Trans.,1710, 1980
See also R1 : 84

73.134 bis(η^5 – Cyclopentadienyl – (triphenylgermyl) – (triphenylgermyl – cadmium) – nickel) – cadmium toluene solvate (at −120°C)
$C_{82}H_{70}Cd_3Ge_4Ni_2$, C_7H_8
S.N.Titova, V.T.Bychkov, G.A.Domrachev, G.A.Razuvaev, Yu.T.Struchkov, L.N.Zakharov
J.Organomet.Chem.,**187**, 167, 1980
See also R1 : 69

73.135 bis((η^5 – Cyclopentadienyl) – (triphenylgermyl) – (triphenylgermyl – mercury) – nickel) – mercury toluene solvate (at −120°C)
$C_{82}H_{70}Ge_4Hg_3Ni_2$, C_7H_8
L.N.Zakharov, Yu.T.Struchkov, S.N.Titova, V.T.Bychkov, G.A.Domrachev, G.A.Razuvaev
Cryst.Struct.Commun.,**9**, 549, 1980
See also R1 : 69

METAL PI–COMPLEXES (ARENE)

74.1 Tricarbonyl – (trimethyl – (η^6 – phenyl) – silane) – chromium (at –135°C)
$C_{12}H_{14}CrO_3Si$
D.van der Helm, R.A.Loghry, D.J.Hanlon, A.P.Hagen
Cryst.Struct.Commun., 8, 899, 1979
See also R1 : 63

74.C trans – Tetracarbonyl – chloro – ((tricarbonyl – chromium) – η^6 – phenylcarbene) – tungsten
$C_{14}H_5ClCrO_7W$ Main entry is 71.60

74.C π – Benzene – dicarbonyl – (methoxyphenylcarbene) – chromium
$C_{16}H_{14}CrO_3$ Main entry is 71.95

74.C Hexamethylbenzene – cyclopentadienyl – iron
$C_{17}H_{23}Fe$ Main entry is 73.69

74.2 (η^{12} – (3.3)Paracyclophane) – chromium(i) hexafluorophosphate
$C_{18}H_{20}Cr^+$, F_6P^-
N.E.Blank, M.W.Haenel, A.R.Koray, K.Weidenhammer, M.L.Ziegler *Acta Crystallogr.,Sect.B*, 36, 2054, 1980

74.3 (η^{12} – (3.3)Paracyclophane) – chromium(i) tri – iodide
$C_{18}H_{20}Cr^+$, I_3^-
N.E.Blank, M.W.Haenel, A.R.Koray, K.Weidenhammer, M.L.Ziegler *Acta Crystallogr.,Sect.B*, 36, 2054, 1980

74.C η^6 – Toluene – bis(perfluorophenyl) – cobalt
$C_{19}H_8CoF_{10}$ Main entry is 71.133

74.C η^6 – Toluene – bis(perfluorophenyl) – nickel
$C_{19}H_8F_{10}Ni$ Main entry is 71.134

74.4 Tricarbonyl – (5 – ethyl – 2 – (5′ – ethyl – 1′,2′,3′,6′ – tetrahydro – 1′ – methyl – 2′ – pyridyl) – 1,6 – dihydro – 1 – methylpyridine) – chromium
$C_{19}H_{26}CrN_2O_3$
J.Trotter, T.C.W.Mak
Acta Crystallogr.,Sect.B, 36, 557, 1980

74.5 (2SR,2′RS) – Tricarbonyl – (5 – ethyl – 2 – (5′ – ethyl – 1′,2′,3′,4′ – tetrahydro – 1′ – methyl – 2′ – pyridyl) – 1,6 – dihydro – 1 – methylpyridine) – chromium
$C_{19}H_{26}CrN_2O_3$
J.Trotter, T.C.W.Mak
Acta Crystallogr.,Sect.B, 36, 551, 1980

74.6 (2SR,2′SR) – Tricarbonyl – (5 – ethyl – 2 – (5′ – ethyl – 1′,2′,3′,4′ – tetrahydro – 1′ – methyl – 2′ – pyridyl) – 1,6 – dihydro – 1 – methylpyridino) – chromium
$C_{19}H_{26}CrN_2O_3$
J.Trotter, T.C.W.Mak
Acta Crystallogr.,Sect.B, 36, 551, 1980

74.C (2 – 6 – η – 1 – exo – Dimethoxyphosphoryl – cyclohexadienyl) – (η^5 – ethyltetramethyl – cyclopentadienyl) – rhodium(iii) hexafluorophosphate
$C_{19}H_{29}O_3PRh^+$, F_6P^- Main entry is 73.81

74.7 Tricarbonyl – (4 – methoxy – 4 – η^6 – phenyl – 2,3 – bis(trimethylsilyl) – buta – 1,3 – dien – 1 – one) – chromium
$C_{20}H_{26}CrO_5Si_2$
U.Schubert, K.H.Dotz
Cryst.Struct.Commun., 8, 989, 1979

74.8 (1 – 4 – η – Cyclo – octatetraene) – (η_6 – hexamethylbenzene) – ruthenium(0)
$C_{20}H_{26}Ru$
M.A.Bennett, T.W.Matheson, G.B.Robertson, A.K.Smith, P.A.Tucker *Inorg.Chem.*, 19, 1014, 1980
See also R1 : 75

74.9 Tricarbonyl – (2,3 – bis(diethylamino) – 1 – methoxy – indene) – chromium(0)
$C_{21}H_{28}CrN_2O_4$
K.H.Dotz, R.Dietz, C.Kappenstein, D.Neugebauer, U.Schubert *Chem.Ber.*, 112, 3682, 1979

74.10 Dicarbonyl – triethylphosphine – phenanthrene – chromium
$C_{22}H_{25}CrO_2P$
M.Cais, M.Kaftory, D.H.Kohn, D.Tatarsky
J.Organomet.Chem., 184, 103, 1979
See also R1 : 86

74.C Hexamethylbenzene – (2 – 3) – tetrafluorobenzo – bicyclo(2.2.2)octa – 5,7 – diene – rhodium perchlorate
$C_{24}H_{24}F_4Rh^+$, ClO_4^- Main entry is 71.191

74.11 bis(η^6 – Hexamethylbenzene) – cobalt hexafluorophosphate
$C_{24}H_{36}Co^+$, F_6P^-
M.R.Thompson, C.S.Day, V.W.Day, R.I.Mink, E.L.Muetterties
J.Am.Chem.Soc., 102, 2979, 1980

74.12 (η^6 – Hexamethylbenzene) – (η^4 – hexamethyl – cyclohexadiene) – rhodium hexafluorophosphate
$C_{24}H_{38}Rh^+$, F_6P^-
M.R.Thompson, C.S.Day, V.W.Day, R.I.Mink, E.L.Muetterties
J.Am.Chem.Soc., 102, 2979, 1980
See also R1 : 75

74.13 Dicarbonyl – naphthalene – triphenylphosphite – chromium
$C_{30}H_{23}CrO_5P$
M.Cais, M.Kaftory, D.H.Kohn, D.Tatarsky
J.Organomet.Chem., 184, 103, 1979
See also R1 : 86

74.14 (η^6 – Tetralone) – carbonyl – thiocarbonyl – (triphenylphosphine) – chromium (absolute configuration)
$C_{30}H_{25}CrO_2PS$
J.D.Korp, I.Bernal *Cryst.Struct.Commun.*, 9, 821, 1980
See also R1 : 86

74.C (μ – Hydrido) – (μ – 2 – η^1:1 – 2 – η^2 – naphthyl) – bis(bis(η^5 – cyclopentadienyl) – zirconium)
$C_{30}H_{28}Zr_2$ Main entry is 71.230

74.15 1,2 – bis(Diphenylphosphino)ethane – (η^6 –
tetraphenylborato) – rhodium(i)
$C_{50}H_{44}BP_2Rh$
P.Albano, M.Aresta, M.Manassero
Inorg.Chem., **19**, 1069, 1980
See also R1 : 86

METAL PI–COMPLEXES
(MISCELLANEOUS RING SYSTEMS)

75.1 Dichloro – (cis – 2 – methyl – thiacyclo – oct – 4 –
ene) – platinum
$C_8H_{14}Cl_2PtS$
L.Busetto *J.Organomet.Chem.*, **186**, 411, 1980
See also R1 : 85

75.C η^5 – Cyclopentadienyl – (η^4 – 1,4 – dimethyl – 1,4 –
dibora – cyclohexa – 2,5 – diene) – cobalt
$C_{11}H_{15}B_2Co$ Main entry is 73.23

75.2 4,5 – Diethyl – 3,6 – dimethyl – 1,2 – diaza – 3,6 –
diborine – tricarbonyl – chromium
$C_{11}H_{18}B_2CrN_2O_3$
W.Siebert, R.Full, H.Schmidt, J.von Seyerl, M.Halstenberg,
G.Huttner *J.Organomet.Chem.*, **191**, 15, 1980
See also R1 : 62

75.3 ((2,3,6,7 – η) – Bicyclo(3.2.2)nona – 2,6,8 – trien – 4 –
one) – tricarbonyl – iron
$C_{12}H_8FeO_4$
J.C.Barborak, S.L.Watson, A.T.McPhail, R.W.Miller
J.Organomet.Chem., **185**, 29, 1980

75.4 3,3′,4,4′ – Tetramethyl – 1,1′ – diphospha –
ferrocene
$C_{12}H_{16}FeP_2$
G.de Lauzon, B.Deschamps, J.Fischer, F.Mathey,
A.Mitschler *J.Am.Chem.Soc.*, **102**, 994, 1980
See also R1 : 64

75.5 η^4 – (2,3,5,6 – tetrakis(Methylene) – 7 –
oxabicyclo(2.2.1)heptane) – endo – tricarbonyl –
iron
$C_{13}H_{10}FeO_4$
E.Meier, O.Cherpillod, T.Boschi, R.Roulet, P.Vogel,
C.Mahaim, A.A.Pinkerton, D.Schwarzenbach, G.Chapuis
J.Organomet.Chem., **186**, 247, 1980

75.6 Lithium tricarbonyl – (1 – 5 – η^5) – 6 – (1,3 –
dithian – 2 – yl) – cyclohexadienyl – chromium(0)
dioxane solvate (at −120°C)
$C_{13}H_{13}CrO_3S_2^-$, Li^+, $3C_4H_8O_2$
M.F.Semmelhack, H.T.Hall Junior, R.Farina, M.Yoshifuji,
G.Clark, T.Bargar, K.Hirotsu, J.Clardy
J.Am.Chem.Soc., **101**, 3535, 1979

75.7 Hexacarbonyl – (μ – (1,2 – eta:2 – η) – 1 – cyclo –
octene – 1 – selenolato) – di – iron
$C_{14}H_{12}Fe_2O_6Se$
R.C.Pettersen, K.H.Pannell, A.J.Mayer
Cryst.Struct.Commun., **9**, 643, 1980
See also R1 : 11

75.8 (η^4 – 1,3 – Cycloheptadienyl) – (η^5 –
cycloheptadienyl) – iridium(i)
$C_{14}H_{19}Ir$
J.Muller, H.Menig, G.Huttner, A.Frank
J.Organomet.Chem., **185**, 251, 1980

75.C $\mu - 2 - $ sigma$:2 - 3 - \eta - (4,5 - $ Dihydro $- 2 - $ furyl) $- $ (bis(trimethylphosphine) $- $ platinum) $- $ tetracarbonyl $- $ manganese (yellow form,at 200°K)
$C_{14}H_{23}MnO_5P_2Pt$ Main entry is 71.76

75.C (2 $- $ Methyl $- 3,5,6 - $ tri(methylene) $- $ cyclohexane $- $ (2 $- 4 - \eta) - $ ene) $- $ bis(tricarbonyl $- $ iron)
$C_{16}H_{10}Fe_2O_7$, Main entry is 72.25

75.C ((1 $- 5 - \eta) - $ exo $- 1 - $ Acetyl $- 2,4,6 - $ tris(trifluoromethyl)cyclohexadienyl) $- (\eta - $ cyclopentadienyl) $- $ iron
$C_{16}H_{11}F_9FeO$ Main entry is 73.56

75.9 bis($\eta^3 - $ Cyclo $- $ octatrienyl) $- $ nickel
$C_{16}H_{18}Ni$
B.Henc, P.W.Jolly, R.Salz, G.Wilke, R.Benn, E.G.Hoffmann, R.Mynott, G.Schroth, K.Seevogel, J.C.Sekutowski, C.Kruger
J.Organomet.Chem.,**191**, 425, 1980

75.10 (1 $- $ t $- $ Butyl $- 3 - $ methyl $- 2 - $ phenyl $- \eta - 1,2 - $ azaborolinyl) $- $ dicarbonyl $- $ iodo $- $ iron
$C_{16}H_{19}BFeINO_2$
J.Schulze, R.Boese, G.Schmid· *Chem.Ber.*,**113**, 2348, 1980
See also R1 : 62

75.11 (1,2,5,6 $- \eta - 1,5 - $ Cyclo $- $ octadiene) $- $ (1 $- 3 - \eta - 1,4 - $ cyclo $- $ octadienyl) $- $ rhodium(i)
$C_{16}H_{23}Rh$
J.Pickardt, H.–O.Stuhler *Chem.Ber.*,**113**, 1623, 1980

75.12 bis(1,5 $- $ Cyclo $- $ octadiene) $- $ silver(i) tetrafluoroborate
$C_{16}H_{24}Ag^+, BF_4^-$
·A.Albinati, S.V.Meille, G.Carturan
J.Organomet.Chem.,**182**, 269, 1979

75.13 ($\eta^8 - $ Cyclo $- $ octatetraene) $- $ dichloro $- $ bis(tetrahydrofuran) $- $ thorium
$C_{16}H_{24}Cl_2O_2Th$
C.LeVanda, J.P.Solar, A.Streitweiser Junior
J.Am.Chem.Soc.,**102**, 2128, 1980
See also R1 : 84

75.14 Dichloro $- (\eta^4 - 1,3 - $ bis(diethylamino) $- 2,4 - $ bis(methoxymethyl) $- $ cyclobutadiene) $- $ palladium monohydrate
$C_{16}H_{30}Cl_2N_2O_2Pd, H_2O$
J.D'Angelo, J.Ficini, S.Martinon, C.Riche, A.Sevin
J.Organomet.Chem.,**177**, 265, 1979

75.15 ($\mu_3 - $ Cycloheptatrienyl) $- (\mu_3 - $ t $- $ butylthiolato) $- $ tris(dicarbonyl $- $ ruthenium) (at $-70°C$)
$C_{17}H_{16}O_6Ru_3S$
J.A.K.Howard, F.G.Kennedy, S.A.R.Knox
J.Chem.Soc.,Chem.Commun.,839, 1979
See also R1 : 85

75.C $\eta^5 - $ Pentamethyl $- $ cyclopentadienyl $- $ (6 $- $ methylimino $- 1 - 5\eta - $ cyclohexadienyl) $- $ rhodium hexafluorophosphate
$C_{17}H_{24}NRh^{2+}, 2F_6P^-$ Main entry is 73.70

75.16 ($\eta^3 - $ Cyclo $- $ octenyl) $- $ tris(trimethylphosphite) $- $ iron(i) (at $-80°C$)
$C_{17}H_{40}FeO_9P_3$
R.L.Harlow, R.J.McKinney, S.D.Ittel
J.Am.Chem.Soc.,**101**, 7496, 1979
See also R1 : 86

75.17 tris(Trimethylphosphite) $- (\eta^3 - $ cyclo $- $ octenyl) $- $ iron tetrafluoroborate
$C_{17}H_{40}FeO_9P_3{}^+, BF_4^-$
R.K.Brown, J.M.Williams, A.J.Schultz, G.D.Stucky, S.D.Ittel, R.L.Harlow *J.Am.Chem.Soc.*,**102**, 981, 1980
See also R1 : 86

75.18 tris(Trimethylphosphite) $- (\eta^3 - $ cyclo $- $ octenyl) $- $ iron tetrafluoroborate (neutron study, at 110°K)
$C_{17}H_{40}FeO_9P_3{}^+, BF_4^-$
R.K.Brown, J.M.Williams, A.J.Schultz, G.D.Stucky, S.D.Ittel, R.L.Harlow *J.Am.Chem.Soc.*,**102**, 981, 1980
See also R1 : 86

75.19 tris(Trimethylphosphite) $- (\eta^3 - $ cyclo $- $ octenyl) $- $ iron tetrafluoroborate (neutron study, at 30°K)
$C_{17}H_{40}FeO_9P_3{}^+, BF_4^-$
R.K.Brown, J.M.Williams, A.J.Schultz, G.D.Stucky, S.D.Ittel, R.L.Harlow *J.Am.Chem.Soc.*,**102**, 981, 1980
See also R1 : 86

75.20 Benzoyl $- 1,1,1 - $ trifluoroacetonato $- $ cyclo $- $ octa $- 1,5 - $ diene $- $ rhodium(i)
$C_{18}H_{18}F_3O_2Rh$
J.G.Leipoldt, S.S.Basson, G.J.Lamprecht, L.D.C.Bok, J.J.J.Schlebusch *Inorg.Chim.Acta*,**40**, 43, 1980
See also R1 : 77

75.21 Tricarbonyl $- $ (methyl 1 $- (\eta - 2 - 5),4 - $ methoxy $- 1 - $ methylcyclohexa $- 2,4 - $ dienyl) $- 2 - $ oxo $- $ cyclopentane $- $ carboxylato $- $ iron
$C_{18}H_{20}FeO_7$
A.J.Pearson, P.R.Raithby
J.Chem.Soc.,Perkin Trans.1,395, 1980

75.C (1,2,5,6 $- \eta - 1,5 - $ Cyclo $- $ octa $- 1,5 - $ diene) $- $ (1,2 $- \eta - $ penta $- 1,3 - $ diene) $- $ (2 $- 4 - \eta - 1,3 - $ dimethylallyl) $- $ iridium(i)
$C_{18}H_{29}Ir$ Main entry is 72.34

75.22 (($\eta^3 - $ Cycloheptatrienyl) $- $ dicarbonyl $- $ molybdenum) $- $ tris($\mu - $ methoxy) $- $ (($\eta^7 - $ cycloheptatrienyl) $- $ molybdenum)
$C_{19}H_{23}Mo_2O_5$
K.Weidenhammer, M.L.Ziegler
Z.Anorg.Allg.Chem.,**455**, 43, 1979
See also R1 : 84

75.23 Pentacarbonyl $- (\mu - 2',3' - \eta,7' - 8' - $ eta$:4' - 6' - \eta - $ (4 $- $ (cyclo $- $ octa $- 2',5',7' - $ triene $- 1',4' - $ diyl) $- 1,1,4,4 - $ tetramethyl $- 1,4 - $ disilapentyl)) $- $ di $- $ ruthenium
$C_{19}H_{24}O_5Ru_2Si_2$
R.Goddard, P.Woodward
J.Chem.Soc.,Dalton Trans.,559, 1980

75.24 Tricarbonyl $- (1,1 - $ diphenyl $- 1 - $ silacyclohexa $- 2,4 - $ dienyl) $- $ iron
$C_{20}H_{18}FeO_3Si$
E.A.Chernyshev, O.B.Afanasova, N.G.Komalenkova, A.I.Gusev, V.A.Sharapov *Zh.Obshch.Khim.*,**48**, 2261, 1978
See also R1 : 63

75.25 Pentacyclo(8.6.0.0^{2,9}.0^{5,16}.0^{6,13})hexadeca $- 3,7,11,14 - $ tetraene $- 3,7 - $ diyl $- $ tetracarbonyl $- $ molybdenum
$C_{20}H_{16}MoO_4$
A.H.Connop, F.G.Kennedy, S.A.R.Knox, R.M.Mills, G.H.Riding, P.Woodward *J.Chem.Soc.,Chem.Commun.*,518, 1980

75.26 bis($\mu - \eta^3, \eta^3$ – Cyclo – octatetraene) – tetracarbonyl – di – molybdenum
$C_{20}H_{16}Mo_2O_4$
A.H.Connop, F.G.Kennedy, S.A.R.Knox, G.H.Riding
J.Chem.Soc.,Chem.Commun., 520, 1980

75.C (1 – 4 – η – Cyclo – octatetraene) – (η_6 – hexamethylbenzene) – ruthenium(0)
$C_{20}H_{26}Ru$ Main entry is 74.8

75.C Chloro – (dimethylphenylphosphonium – (trimethylsilyl)methylide) – (1,5 – cyclo – octadiene) – palladium(ii) hexafluorophosphate
$C_{20}H_{33}ClPPdSi^+$, F_6P^- Main entry is 71.160

75.27 Hydrido – (η^5 – cyclo – octadiene) – bis(1,2 – bis(dimethylphosphino)ethane) – zirconium
$C_{20}H_{44}P_4Zr$
M.B.Fischer, E.J.James, T.J.McNeese, S.C.Nyburg, B.Posin, W.Wong-Ng, S.S.Wreford *J.Am.Chem.Soc.*, 102, 4941, 1980
See also R1 : 86

75.28 Codeine – tricarbonyl – chromium
$C_{21}H_{21}CrNO_6$
H.B.Arzeno, D.H.R.Barton, S.G.Davies, Z.Lusinchi, B.Meunier, C.Pascard *Nouv.J.Chim.*, 4, 369, 1980
See also R1 : 58

75.29 ((η^3 – Cycloheptatrienyl) – dicarbonyl – tungsten) – tris(μ – methoxy) – ((η^4 – cycloheptatrienyl) – dicarbonyl – tungsten)
$C_{21}H_{24}O_7W_2$
K.Weidenhammer, M.L.Ziegler
Z.Anorg.Allg.Chem., 455, 43, 1979
See also R1 : 84

75.30 (η^5 – Indenyl) – (1 – 2:3 – 4 – η^4 – (6 – endo – propen – 2 – yl – 1,2,3,4 – tetrakis(trifluoromethyl) – cyclohexa – 1,3 – dienyl)) – rhodium
$C_{22}H_{15}F_{12}Rh$
P.Caddy, M.Green, E.O'Brien, L.E.Smart, P.Woodward
J.Chem.Soc.,Dalton Trans., 962, 1980

75.31 (μ – Bicyclo(4.2.2)deca – 2,4,7,9 – tetraene) – bis(carbonyl – cyclopentadienyl – molybdenum)
$C_{22}H_{20}Mo_2O_2$
S.A.R.Knox, R.F.D.Stansfield, F.G.A.Stone, M.J.Winter, P.Woodward *J.Chem.Soc.,Chem.Commun.*, 934, 1979

75.C Dicarbonyl – (η^5 – cyclopentadienyl) – (1 – 3 – η – (4 – cyclopenta – 1',3' – dienyl) – 5 – (cyclopenta – 1'',4'' – dienyl) – cyclopentyl) – tungsten
$C_{22}H_{20}O_2W$ Main entry is 73.92

75.32 Tricarbonyl – molybdenum – tris(μ – n – butylthiolato) – (η^7 – cycloheptatrienyl – molybdenum)
$C_{22}H_{34}Mo_2O_3S_3$
K.Weidenhammer, M.L.Ziegler
Z.Anorg.Allg.Chem., 455, 29, 1979
See also R1 : 85

75.33 Tricarbonyl – molybdenum – tris(μ – t – butylthiolato) – (η^7 – cycloheptatrienyl – molybdenum)
$C_{22}H_{34}Mo_2O_3S_3$
K.Weidenhammer, M.L.Ziegler
Z.Anorg.Allg.Chem., 455, 29, 1979
See also R1 : 85

75.C Hexamethylbenzene – (2 – 3) – tetrafluorobenzo – bicyclo(2.2.2)octa – 5,7 – diene – rhodium perchlorate
$C_{24}H_{24}F_4Rh^+$, ClO_4^- Main entry is 71.191

75.34 bis(1,1 – Dimethyl – 4 – phenyl – 1 – sila – 4 – bora – cyclohexa – 2,5 – dienyl) – nickel
$C_{24}H_{30}B_2NiSi_2$
G.E.Herberich, M.Thonnessen, D.Schmitz
J.Organomet.Chem., 191, 27, 1980
See also R1 : 63,62

75.C (η^6 – Hexamethylbenzene) – (η^4 – hexamethyl – cyclohexadiene) – rhodium hexafluorophosphate
$C_{24}H_{38}Rh^+$, F_6P^- Main entry is 74.12

75.35 α – (4 – 6η) – (3,20 – Dioxo – pregn – 4 – enyl) – pentane – 2,4 – dionato – palladium(ii)
$C_{26}H_{36}O_4Pd$
D.J.Collins, B.M.K.Gatehouse, W.R.Jackson, G.A.Kakos, R.N.Timms *J.Chem.Soc.,Chem.Commun.*, 138, 1980
See also R1 : 77

75.C (μ^3 – 1 – 6 – η – Bitropyl) – carbido – (μ^2 – carbonyl) – tridecacarbonyl – octahedro – hexa – ruthenium
$C_{29}H_{14}O_{14}Ru_6$ Main entry is 71.223

75.36 (η^7 – Cycloheptatrienyl) – carbonyl – iodo – (α – methylbenzyl – N – methylamino – (diphenyl) – phosphine) – molybdenum
$C_{29}H_{29}IMoNOP$
H.Brunner, M.Muschiol, I.Bernal, G.M.Reisner
J.Organomet.Chem., 198, 169, 1980
See also R1 : 86

75.37 Carbonyl – cyclobutadiene – (1,2 – bis(diphenylphosphino)ethane) – iron (at –35°C)
$C_{31}H_{28}FeOP_2$
R.E.Davis, P.E.Riley *Inorg.Chem.*, 19, 674, 1980
See also R1 : 86

75.38 cis – bis(μ^2 – Carbonyl) – bis((1 – t – butyl – 3 – methyl – 2 – phenyl – η – 1,2 – azaborolinyl) – carbonyl – iron)
$C_{32}H_{38}B_2Fe_2N_2O_4$
J.Schulze, R.Boese, G.Schmid *Chem.Ber.*, 113, 2348, 1980
See also R1 : 62

75.39 η^7 – Tropylium – tungsten – tris(μ^2 – n – butylthio) – carbonyl – tungsten – bis(μ^2 – n – butylthio) – tetracarbonyl – tungsten
$C_{32}H_{52}O_5S_5W_3$
L.R.Krauth–Siegel, W.Schulze, M.L.Ziegler
Angew.Chem.,Int.Ed.Engl., 19, 397, 1980
See also R1 : 85

75.C η^5 – Cyclopentadienyl – η^4 – (1,2 – diphenyl – cyclobuta(l)phenanthrene) – rhodium
$C_{33}H_{23}Rh$ Main entry is 73.118

75.C (η^5 – Cyclopentadienyl) – (η^4 – tetraphenyl – cyclobutadiene) – cobalt
$C_{33}H_{25}Co$ Main entry is 73.119

75.C (($\pm$) – α – (2 – Diphenylphosphino – ferrocenyl) ethyl – dimethylamine) – norbornadiene – rhodium hexafluorophosphate
$C_{33}H_{36}FeNPRh^+$, F_6P^- Main entry is 73.120

75.C η^5 – Cyclopentadienyl – η^4 – (1 – methoxy – 1 – oxo – 2,3,4,5 – tetraphenylphosphole) – cobalt
$C_{34}H_{28}CoO_2P$ Main entry is 73.121

75.C bis(η^8 – Cyclo – octatetraene) – (2,3 – dimethyl –
1,4 – diphenyl – buta – 1,3 – diene – 1,4 – diyl –
titanium) – titanium tetrahydrofuran solvate
$C_{34}H_{32}Ti_2$ Main entry is 71.241

75.40 bis(μ – Chloro) – bis(mercury – (μ – p – tolyl(ethyl)
triazenido) – (cyclo – octadiene – iridium))
$C_{34}H_{48}Cl_2Hg_2Ir_2N_6$
P.I.van Vliet, M.Kokkes, G.van Koten, K.Vrieze
J.Organomet.Chem.,**187**, 413, 1980
See also R1 : 83

75.41 Tricarbonyl – manganese – η – (1 – tricarbonyl –
3,4 – bis(p – chlorobenzyl) – 2,5 – bis(p – tolyl) –
2,5 – diaza – 1 – mangana – penta – 2,4 – diene)
$C_{36}H_{26}Cl_2Mn_2N_2O_6$
J.Ricci, W.Miller, J.Alexander, J.Williams
Am.Cryst.Assoc.,Ser.2,**8**, 23, 19

75.C (η^5 – Trimethylsilyl – cyclopentadienyl) – (η^4 –
tetraphenyl – cyclobutadiene) – cobalt
$C_{36}H_{33}CoSi$ Main entry is 73.122

75.C (Cyclo – octadiene) – (cyclo – octadienyl) – (cyclo –
octadienediyl) – dodecacarbonyl – hepta – iridium
$C_{36}H_{33}Ir_7O_{12}$ Main entry is 71.246

75.42 (η – Cyclo – octa – 1,5 – diene) – 4 –
diphenylphosphino – 2 – diphenylphosphino –
methylpyrrolidine – rhodium(i) perchlorate
$C_{37}H_{41}NP_2Rh^+$, ClO_4^-
Y.Ohga, Y.Iitaka, K.Achiwa *Chem.Lett.*,861, 1980
See also R1 : 86,83

75.C 1 – Phenyl – dibenzo(b,f) – (2,5 – diphenyl – ferra –
cyclopentadiene – tricarbonyl) – (c) – phosphepin –
iron – dicarbonyl
$C_{39}H_{23}Fe_2O_5P$ Main entry is 73.124

75.C (η^5 – Cyclopentadienyl) – (η^4 – 1,3 – dimesityl –
2,4 – diphenyl – cyclobutadiene) – cobalt
$C_{39}H_{37}Co$ Main entry is 73.125

75.43 (η – 1,5 – Cyclo – octadiene) – ((2S,4S) – N –
pivaloyl – 4 – diphenylphosphino – 2 –
diphenylphosphino – methylpyrrolidine) – rhodium
perchlorate
$C_{42}H_{49}NOP_2Rh^+$, ClO_4^-
I.Ojima, T.Kogure *J.Organomet.Chem.*,**195**, 239, 1980
See also R1 : 86

75.44 η^4 – (Cyclo – octa – 1,5 – diene) – bis((R) –
menthyl – methylphenylphosphine) – rhodium(i)
tetrafluoroborate
$C_{42}H_{66}P_2Rh^+$, BF_4^-
D.Valentine Junior, J.F.Blount, K.Toth
J.Org.Chem.,**45**, 3691, 1980
See also R1 : 86

75.45 (1,2 – η – Cyclo – octyne) –
bis(triphenylphosphine) – platinum(0)
$C_{44}H_{42}P_2Pt$
G.B.Robertson; P.O.Whimp *Aust.J.Chem.*,**33**, 1373, 1980
See also R1 : 86

75.46 Bicyclo(4.2.1)non – 1(8) – ene –
bis(triphenylphosphine) – platinum(0)
$C_{45}H_{44}P_2Pt$
E.Stamm, K.B.Becker, P.Engel, O.Ermer, R.Keese
Angew.Chem.,Int.Ed.Engl.,**18**, 685, 1979
See also R1 : 86

75.47 1,1,1 – tris(Diphenylphosphinomethyl)ethane – (η^4 –
cyclohepta – 1,3,5 – trienenyl) – cobalt perchlorate
methylene chloride solvate
$C_{48}H_{47}CoP_3^+$, ClO_4^-, $0.5CH_2Cl_2$
C.Bianchini, P.Dapporto, A.Meli, L.Sacconi
J.Organomet.Chem.,**193**, 117, 1980
See also R1 : 86

75.48 bis(Triphenylphosphine) – (η^5D – 2,6 – di – t –
butyl – 4 – methyl – cyclohexadienonyl) –
rhodium(i)
$C_{51}H_{53}OP_2Rh$, $2C_6H_6$
B.Cetinkaya, P.B.Hitchcock, M.F.Lappert, S.Torroni,
J.L.Atwood, W.E.Hunter, M.J.Zaworotko
J.Organomet.Chem.,**188**, 31, 1980
See also R1 : 86

METAL COMPLEXES (ETHYLENEDIAMINE)

76.1 cis – Dinitro – cis – diamine – (ethylenediamine) –
cobalt(iii) nitrate (orthorhombic form)
$C_2H_{14}CoN_6O_4{}^+$, $NO_3{}^-$
V.A.Neverov, A.V.Ablov, E.V.Popa, V.N.Byushkin, A.P.Gulya,
T.I.Malinovskii *Dokl.Akad.Nauk SSSR*,**242**, 128, 1978

76.2 Tetra – aqua(ethylenediamine) – nickel(ii) nitrate
$C_2H_{16}N_2NiO_4{}^{2+}$, $2NO_3{}^-$
G.J.McDougall, R.D.Hancock
J.Chem.Soc.,Dalton Trans.,654, 1980

76.3 Tetra – aquo – ethylenediamine – nickel(ii) sulfate
dihydrate
$C_2H_{16}N_2NiO_4{}^{2+}$, O_4S^{2-}, $2H_2O$
A.E.Shvelashvili, R.M.Vashakidze, E.B.Miminoshvili,
M.G.Tavberidze, A.I.Kvitashvili, B.M.Shchedrin,
E.A.Mikeladze
Soobshch.Akad.Nauk Gruzh.SSR,**93**, 601, 19

76.4 Carbonyl – ethylenediamine – copper(i)
tetraphenylborate
$C_3H_8CuN_2O^+$, $C_{24}H_{20}B^-$
M.Pasquali, C.Floriani, A.Gaetani–Manfredotti
Inorg.Chem.,**19**, 1191, 1980
See also R2 : 62

76.5 Chloro – (bis(2 – aminoethyl)amido) – gold(iii)
perchlorate
$C_4H_{12}AuClN_3{}^+$, $ClO_4{}^-$
G.Nardin, L.Randaccio, G.Annibale, G.Natile, B.Pitteri
J.Chem.Soc.,Dalton Trans.,220, 1980

76.6 Dibromo – (N – (2 – hydroxyethyl) –
ethylenediamine) – copper(ii)
$C_4H_{12}Br_2CuN_2O$
A.Pajunen, E.Nasakkala, P.Ilvonen
Cryst.Struct.Commun.,**9**, 117, 1980
See also R1 : 84

76.7 Chloro – bis(2 – aminomethyl)amine – gold(iii)
chloride perchlorate
$C_4H_{13}AuClN_3{}^{2+}$, Cl^-, $ClO_4{}^-$
G.Nardin, L.Randaccio, G.Annibale, G.Natile, B.Pitteri
J.Chem.Soc.,Dalton Trans.,220, 1980

76.8 μ – Bromo – bis(ethylenediamine) – platinum(ii,iv)
perchlorate
$(C_4H_{16}BrN_4Pt^{2+})_n$, $2nClO_4{}^-$
H.Endres, H.J.Keller, R.Martin, U.Traeger, M.Novotny
Acta Crystallogr.,Sect.B,**36**, 35, 1980

76.9 trans – Chloro – bis(ethylenediamine) – sulfito –
cobalt(iii) monohydrate
$C_4H_{16}ClCoN_4O_3S$, H_2O
C.L.Raston, A.H.White, J.K.Yandell
Aust.J.Chem.,**33**, 419, 1980
See also R1 : 85

76.10 Ammine – chloro – diethylenetriamine – nitro –
platinum chloride trihydrate
$C_4H_{16}ClN_5O_2Pt^{2+}$, $2Cl^-$, $3H_2O$
G.A.Kukina, V.S.Sergienko, M.A.Porai–Koshits,
A.Sh.Gladkaya
Izv.Akad.Nauk SSSR,Ser.Khim.,1658, 1979

76.11 trans – Dichloro – bis(ethylenediamine) – cobalt(iii)
S – hydroxymethyl – thiosulfate (absolute
configuration)
$C_4H_{16}Cl_2CoN_4{}^+$, $CH_3O_4S_2{}^-$
A.S.Foust, V.Janickis *Inorg.Chem.*,**19**, 1048, 1980

76.12 trans – Dichloro – bis(ethylenediamine) – cobalt(iii)
diselenotetrathionate monohydrate (prismatic
needle form)
$2C_4H_{16}Cl_2CoN_4{}^+$, $O_6S_2Se_2{}^{2-}$, H_2O
A.S.Foust, V.Janickis, K.Maroy
Inorg.Chem.,**19**, 1044, 1980

76.13 trans – Dichloro – bis(ethylenediamine) – cobalt(iii)
diselenotetrathionate monohydrate (extended plate
form)
$2C_4H_{16}Cl_2CoN_4{}^+$, $O_6S_2Se_2{}^{2-}$, H_2O
A.S.Foust, V.Janickis, K.Maroy
Inorg.Chem.,**19**, 1044, 1980

76.14 trans – Dichloro – bis(ethylenediamine) – cobalt(iii)
selenotetrathionate
$2C_4H_{16}Cl_2CoN_4{}^+$, $O_6S_3Se^{2-}$
A.S.Foust, V.Janickis, K.Maroy
Inorg.Chem.,**19**, 1040, 1980

76.C bis(Ethylenediamine) – platinum(ii) dichloro –
bis(ethylenediamine) – platinum(iv)
tetraperchlorate
$C_4H_{16}Cl_2N_4Pt^{2+}$, $C_4H_{16}N_4Pt^{2+}$, $4ClO_4{}^-$ Main entry is 76.19

76.C bis(1,2 – Diaminoethane) – platinum(ii) dichloro –
bis(1,2 – diaminoethane) – platinum(iv)
tetrachloro – copper(i)
$3C_4H_{16}Cl_2N_4Pt^{2+}$, $3C_4H_{16}N_4Pt^{2+}$, $4Cl_4Cu^{3-}$
Main entry is 76.20

76.15 cis – Dichloro – bis(ethylenediamine) – rhodium(iii)
nitrate
$C_4H_{16}Cl_2N_4Rh^+$, $NO_3{}^-$
I.A.Baidina, N.V.Podberezskaya, A.V.Belyaev, V.V.Bakakin
Zh.Strukt.Khim.,**20**, 1096, 1979

76.16 trans – Dichloro – bis(ethylenediamine) –
rhodium(iii) nitrate
$C_4H_{16}Cl_2N_4Rh^+$, $NO_3{}^-$
I.A.Baidina, N.V.Podberezskaya, A.V.Belyaev, V.V.Bakakin
Zh.Strukt.Khim.,**20**, 1096, 1979

76.17 Sodium trans – bis(ethylenediamine) – disulfito –
cobalt(iii) trihydrate
$C_4H_{16}CoN_4O_6S_2{}^-$, Na^+, $3H_2O$
G.D.Fallon, C.L.Raston, A.H.White, J.K.Yandell
Aust.J.Chem.,**33**, 665, 1980

76.18 bis(Ethylenediamine) – copper
(bis(ethylenediamine) – copper) –
(ethylenediamine – tetra – acetato – nickel)
dihydrate
$C_4H_{16}CuN_4{}^{2+}$, $C_{24}H_{40}CuN_8Ni_2O_{16}{}^{2-}$, $2H_2O$
V.M.Agre, T.F.Sisoeva, V.K.Trunov, V.A.Efremov,
A.Ya.Fridman, N.N.Barkhanova
Zh.Strukt.Khim.,**21**, 110, 1980
See also R2 : 76,81,83

76.19 bis(Ethylenediamine) – platinum(ii) dichloro – bis(ethylenediamine) – platinum(iv) tetraperchlorate
$C_4H_{16}N_4Pt^{2+}$, $C_4H_{16}Cl_2N_4Pt^{2+}$, $4ClO_4^-$
N.Matsumoto, M.Yamashita, I.Ueda, S.Kida
Mem.Fac.Sci.Kyushu Univ.Ser.C, 11, 209, 1978
See also R2 : 76

76.20 bis(1,2 – Diaminoethane) – platinum(ii) dichloro – bis(1,2 – diaminoethane) – platinum(iv) tetrachloro – copper(i)
$3C_4H_{16}N_4Pt^{2+}$, $3C_4H_{16}Cl_2N_4Pt^{2+}$, $4Cl_4Cu^{3-}$
H.Endres, H.J.Keller, R.Martin, U.Traeger
Acta Crystallogr., Sect.B, 35, 2880, 1979
See also R2 : 76

76.21 trans – Ammine – bis(ethylenediamine) – sulfito – cobalt(iii) perchlorate
$C_4H_{19}CoN_5O_3S^+$, ClO_4^-
C.L.Raston, A.H.White, J.K.Yandell
Aust.J.Chem., 33, 1123, 1980

76.22 Diammine – (N,N,N′ – trimethylethylenediamine) – platinum pentachloro – antimony (orthorhombic form)
$C_5H_{20}N_4Pt^{2+}$, Cl_5Sb^{2-}
K.Matsumoto, S.Ooi *Z.Kristallogr.*, 150, 139, 1979

76.23 tetrakis(μ – Cyano) – cadmium – (ethylenediamine – cadmium)
$(C_6H_8Cd_2N_6)_n$
S.Nishikiori, T.Iwamoto, Y.Yoshino
Chem.Lett., 1509, 1979

76.24 Chloro – (uracilato) – (ethylenediamine) – platinum(ii) hydrogen chloride dihydrate
$C_6H_{11}ClN_4O_2Pt$, H^+, Cl^-, $2H_2O$
R.Faggiani, B.Lippert, C.J.L.Lock
Inorg.Chem., 19, 295, 1980
See also R1 : 83

76.25 Dibromo – (1,4,7 – triazacyclononane) – copper(ii)
$C_6H_{15}Br_2CuN_3$
R.D.Bereman, M.R.Churchill, P.M.Schaber, M.E.Winkler
Inorg.Chem., 18, 3122, 1979

76.26 Dichloro – (1,4,7 – triazacyclononane) – copper(ii)
$C_6H_{15}Cl_2CuN_3$
W.F.Schwindinger, T.G.Fawcett, R.A.Lalancette, J.A.Potenza, H.J.Schugar *Inorg.Chem.*, 19, 1379, 1980

76.27 bis(Ethylenediamine) – (thio – oxalato – O,S) – cobalt chloride monohydrate
$C_6H_{16}CoN_4O_3S^+$, Cl^-, H_2O
G.J.Gainsford, W.G.Jackson, A.M.Sargeson
Aust.J.Chem., 33, 707, 1980
See also R1 : 81

76.28 bis((Thio – oxalato – O,S) – bis(ethylenediamine) – cobalt(iii)) dithionate dihydrate
$2C_6H_{16}CoN_4O_3S^+$, $O_6S_2^{2-}$, $2H_2O$
J.D.Lydon, K.J.Mulligan, R.C.Elder, E.Deutsch
Inorg.Chem., 19, 2083, 1980
See also R1 : 85

76.29 (+) – trans – (O) – Ethylenediamine – bis(glycinato) – cobalt(iii) hydrogen – d – tartrate trihydrate
$C_6H_{16}CoN_4O_4^+$, $C_4H_5O_6^-$, $3H_2O$
M.Kuramoto *Bull.Chem.Soc.Jpn.*, 52, 3702, 1979
See also R1 : 82

76.30 (–) – trans – (O) – Ethylenediamine – bis(glycinato) – cobalt(iii) hydrogen – d – tartrate monohydrate
$C_6H_{16}CoN_4O_4^+$, $C_4H_5O_6^-$, H_2O
M.Kuramoto *Bull.Chem.Soc.Jpn.*, 52, 3702, 1979
See also R1 : 82

76.31 (–)$_{589}$ – Oxalato – bis(ethylenediamine) – cobalt(iii) hydrogen – d – tartrate dihydrate
$C_6H_{16}CoN_4O_4^+$, $C_4H_5O_6^-$, $2H_2O$
M.Kuramoto, Y.Kushi, H.Yoneda
Bull.Chem.Soc.Jpn., 53, 125, 1980
See also R1 : 81

76.C bis(1,2 – Diaminopropane) – platinum(ii) bis(1,2 – diaminopropane) – di – iodo – platinum(iv) iodide
$C_6H_{20}I_2N_4Pt^{2+}$, $C_6H_{20}N_4Pt^{2+}$, $4I^-$ Main entry is 76.32

76.32 bis(1,2 – Diaminopropane) – platinum(ii) bis(1,2 – diaminopropane) – di – iodo – platinum(iv) iodide
$C_6H_{20}N_4Pt^{2+}$, $C_6H_{20}I_2N_4Pt^{2+}$, $4I^-$
H.Endres, H.J.Keller, R.Martin, U.Traeger, M.Novotny
Acta Crystallogr., Sect.B, 36, 35, 1980
See also R2 : 76

76.33 (Ethylenediamine) – (N,N′ – dimethylethylenediamine) – platinum bis(α – bromocamphor – π – sulfonate)
$C_6H_{20}N_4Pt^{2+}$, $2C_{10}H_{14}BrO_4S^-$
K.Matsumoto, S.Ooi *Z.Kristallogr.*, 150, 139, 1979
See also R2 : 31

76.34 bis((R) – 1,2 – Propanediamine) – platinum(ii) chloride dihydrate (monoclinic form)
$C_6H_{20}N_4Pt^{2+}$, $2Cl^-$, $2H_2O$
C.Maeda, K.Matsumoto, S.Ooi
Bull.Chem.Soc.Jpn., 53, 1755, 1980

76.35 bis((R) – 1,2 – Propanediamine) – platinum(ii) chloride dihydrate (triclinic form)
$C_6H_{20}N_4Pt^{2+}$, $2Cl^-$, $2H_2O$
C.Maeda, K.Matsumoto, S.Ooi
Bull.Chem.Soc.Jpn., 53, 1755, 1980

76.36 Ammine – (tris(2 – aminoethyl)amine) – copper(ii) diperchlorate
$C_6H_{21}CuN_5^{2+}$, $2ClO_4^-$
M.Duggan, N.Ray, B.Hathaway, G.Tomlinson, P.Brint, K.Pelin *J.Chem.Soc., Dalton Trans.*, 1342, 1980

76.37 (+) – tris(Ethylenediamine) – cobalt(iii) (–) – tris(ethylenediamine) – chromium(iii) hexakis(thiocyanate) (orthorhombic form)
$C_6H_{24}CoN_6^{3+}$, $C_6H_{24}CrN_6^{3+}$, $6CNS^-$
C.Brouty, P.Spinat, A.Whuler
Acta Crystallogr., Sect.B, 36, 2037, 1980

76.38 (±) – tris(Ethylenediamine) – cobalt(iii) iodide monohydrate
$C_6H_{24}CoN_6^{3+}$, $3I^-$, H_2O
A.Whuler, P.Spinat, C.Brouty
Acta Crystallogr., Sect.B, 36, 1086, 1980

76.39 (+) – tris(Ethylenediamine) – chromium(iii) bromide hydrate
$C_6H_{24}CrN_6^{3+}$, $3Br^-$, $0.6H_2O$
P.Spinat, A.Whuler, C.Brouty
Acta Crystallogr., Sect.B, 35, 2914, 1979

76.40 **tris(Ethylenediamine) – nickel(ii) dinitrate (absolute configuration)**
$C_6H_{24}N_6Ni^{2+}$, $2NO_3^-$
J.D.Korp, I.Bernal, R.A.Palmer, J.C.Robinson
*Acta Crystallogr.,Sect.B,***36**, 560, 1980

76.41 **Bromo – tricarbonyl – (N,N′ – dimethylethane – 1,2 – diamine) – rhenium(i)**
$C_7H_{12}BrN_2O_3Re$
E.W.Abel, M.M.Bhatti, M.B.Hursthouse, K.M.A.Malik, M.A.Mazid *J.Organomet.Chem.,***197**, 345, 1980

76.42 **Chloro – (thyminato) – (ethylenediamine) – platinum(ii)**
$C_7H_{13}ClN_4O_2Pt$
R.Faggiani, B.Lippert, C.J.L.Lock
*Inorg.Chem.,***19**, 295, 1980
See also R1 : 83

76.43 **Tricarbonyl – bis(ethylenediamine) – rhenium(i) diphenylphosphinate**
$C_7H_{16}N_4O_3Re^+$, $C_{12}H_{10}O_2P^-$
E.Lindner, S.Trad, S.Hoehne *Chem.Ber.,***113**, 639, 1980
See also R2 : 64

76.44 **(bis(Diethylene) – triethylene – tetramine) – bis(perchlorato) – copper**
$C_7H_{20}Cl_2CuN_4O_8$
T.G.Fawcett, S.M.Rudich, B.H.Toby, R.A.Lalancette, J.A.Potenza, H.J.Schugar *Inorg.Chem.,***19**, 940, 1980
See also R1 : 84

76.45 **(–)$_{589}$ – Δ(S) – cis(1 – Aminopropane – 2 – ol – N) – chloro – bis(1,2 – diaminoethane) – cobalt(iii) tetrachloro – zinc**
$C_7H_{25}ClCoN_5O^{2+}$, Cl_4Zn^{2-}
D.A.House, G.Hall, A.J.Matheson, W.T.Robinson, F.C.Ha, C.B.Knobler *Inorg.Chim.Acta,***39**, 257, 1980
See also R1 : 83

76.46 **(–)$_{589}$ – Amine – glycinato – (1,4,7 – triazacyclononane) – cobalt(iii) di – iodide hydrate (absolute configuration)**
$C_8H_{22}CoN_5O_2^{2+}$, $2I^-$, $0.84H_2O$
S.Sato, S.Ohba, S.Shimba, S.Fujinami, M.Shibata, Y.Saito
*Acta Crystallogr.,Sect.B,***36**, 43, 1980
See also R1 : 81

76.47 **Δ – bis(Ethylenediamine) – ((3R,4R) – thiazolidine – 4 – carboxylato – N,O) – cobalt(iii) dichloride**
$C_8H_{22}CoN_5O_2S^{2+}$, $2Cl^-$
G.J.Gainsford, W.G.Jackson, A.M.Sargeson, A.D.Watson
*Aust.J.Chem.,***33**, 1213, 1980
See also R1 : 84,83

76.48 **bis(N – Ethylethylenediamine) – copper(ii) diperchlorate**
$C_8H_{24}CuN_4^{2+}$, $2ClO_4^-$
I.Grenthe, P.Paoletti, M.Sandstrom, S.Glikberg
*Inorg.Chem.,***18**, 2687, 1979

76.49 **(μ^2 – Ethylenediamine) – bis(carbonyl – ethylenediamine – copper(i)) bis(tetraphenylborate)**
$C_8H_{24}Cu_2N_6O_2^{2+}$, $2C_{24}H_{20}B^-$
M.Pasquali, C.Floriani, A.Gaetani–Manfredotti
*Inorg.Chem.,***19**, 1191, 1980

76.50 **bis(Diethylenetriamine) – dinitrato – copper**
$C_8H_{26}CuN_6^{2+}$, $2NO_3^-$
A.Murphy, J.Mullane, B.Hathaway
*Inorg.Nucl.Chem.Lett.,***16**, 129, 1980

76.51 **bis(Diethylenetriamine) – dinitrato – copper (at 150°K)**
$C_8H_{26}CuN_6^{2+}$, $2NO_3^-$
A.Murphy, J.Mullane, B.Hathaway
*Inorg.Nucl.Chem.Lett.,***16**, 129, 1980

76.52 **bis(Diethylenetriamine) – dinitrato – zinc (orthorhombic form)**
$C_8H_{26}N_6Zn^{2+}$, $2NO_3^-$
A.Murphy, J.Mullane, B.Hathaway
*Inorg.Nucl.Chem.Lett.,***16**, 129, 1980

76.53 **(μ – Peroxo) – (μ – hydroxo) – bis(bis(ethylenediamine) – cobalt(iii)) perchlorate**
$C_8H_{33}Co_2N_8O_3^{3+}$, $3ClO_4^-$
S.Fallab, M.Zehnder, U.Thewalt
*Helv.Chim.Acta,***63**, 1491, 1980
See also R1 : 84

76.54 **Malonato – (2,2′,2″ – triaminotriethylamine) – cobalt(iii) chloride trihydrate**
$C_9H_{20}CoN_4O_4^+$, Cl^-, $3H_2O$
B.D.Santarsiero, A.Aruffo, V.Schomaker, E.C.Lingafelter
*Am.Cryst.Assoc.,Ser.2,***7**, 23, 1980
See also R1 : 81

76.55 **Calcium ethylenediamine – tetra – acetato – copper(ii) tetrahydrate**
$C_{10}H_{12}CuN_2O_8^{2-}$, Ca^{2+}, $4H_2O$
Ya.M.Nesterova, M.A.Porai–Koshits, V.A.Logvinenko
*Zh.Neorg.Khim.,***24**, 2273, 1979
See also R1 : 82

76.56 **Potassium ethylenediamine – tetra – acetato – manganese(iii) dihydrate**
$C_{10}H_{12}MnN_2O_8^-$, K^+, $2H_2O$
J.Stein, J.P.Fackler Junior, G.J.McClune, J.A.Fee, L.T.Chan
*Inorg.Chem.,***18**, 3511, 1979
See also R1 : 82

76.57 **Calcium (ethylenediamine – tetra – acetato) – nickel(ii) tetrahydrate**
$C_{10}H_{12}N_2NiO_8^{2-}$, Ca^{2+}, $4H_2O$
Ya.M.Nesterova, M.A.Porai–Koshits, V.A.Logvinenko
*Zh.Strukt.Khim.,***21**, 171, 1980
See also R1 : 82

76.58 **Potassium (+) – (chloro – (hydrogen – ethylenediamine – tetra – acetato) – iridium) monohydrate**
$C_{10}H_{13}ClIrN_2O_8^-$, K^+, H_2O
V.S.Sergienko, L.M.Dikareva, M.A.Porai–Koshits, G.G.Sadikov, P.A.Chel′tsov *Koord.Khim.,***5**, 920, 1979
See also R1 : 82

76.59 **Potassium (dichloro – (dihydrogen – ethylenediamine – tetra – acetato) – iridium)**
$C_{10}H_{14}Cl_2IrN_2O_8^-$, K^+
V.S.Sergienko, L.M.Dikareva, M.A.Porai–Koshits, G.G.Sadikov, P.A.Chel′tsov *Koord.Khim.,***5**, 920, 1979
See also R1 : 82

76.60 **Ethanediammonium bis((hydrogen ethylenediamine – tetra – acetato) – vanadium) dihydrate**
$2C_{10}H_{14}N_2O_8V^-$, $C_2H_{10}N_2^{2+}$, $2H_2O$
D.Ghosh, J.R.Ruble, C.D.Stout, F.J.Kristine, R.E.Shepherd
*Am.Cryst.Assoc.,Ser.2,***8**, 25, 1980
See also R1 : 82

76.61 Aqua – (ethylenediamine – tetra – acetato) –
osmium(iv) monohydrate
$C_{10}H_{14}N_2O_9Os$, H_2O
M.Saito, T.Uehiro, F.Ebina, T.Iwamoto, A.Ouchi, Y.Yoshino
Chem.Lett.,997, 1979
See also R1 : 82

76.62 Tetracarbonyl – (N,N,N′,N′ –
tetramethylethylenediamine) – chromium(0)
$C_{10}H_{16}CrN_2O_4$
G.J.Kruger, G.Gafner, J.P.R.de Villiers, H.G.Raubenheimer,
H.Swanepoel *J.Organomet.Chem.*,**187**, 333, 1980

76.63 (2 – Mercapto – aniline) – bis(ethylenediamine) –
cobalt(iii) perchlorate chloride
$C_{10}H_{22}CoN_5S^{2+}$, ClO_4^-, Cl^-
M.H.Dickman, R.J.Doedens, E.Deutsch
Inorg.Chem.,**19**, 945, 1980

76.64 Penta – aquo – nickel(ii) – ((S,S) –
ethylenediamine – N,N′ – disuccinato) – nickel(ii)
dihydrate
$C_{10}H_{22}N_2Ni_2O_{13}$, $2H_2O$
F.Pavelcik, V.Kettmann, J.Majer
Collect.Czech.Chem.Commun.,**44**, 1070, 1979
See also R1 : 81,82

76.65 Dinitrato – (N,N,N′,N′ – tetraethylethylenediamine) –
copper(ii)
$C_{10}H_{24}CuN_4O_6$
M.Nasakkala, A.Pajunen
Cryst.Struct.Commun.,**9**, 897, 1980
See also R1 : 84

76.C (Diethylenetriamine) – (hex – 1 – ene) – copper(i)
tetraphenylborate
$C_{10}H_{25}CuN_3^+$, $C_{24}H_{20}B^-$ Main entry is 72.10

76.66 bis(2 – Dimethylaminoethyl – methyl – amino –
zinc – hydride)
$C_{10}H_{28}N_4Zn_2$
N.A.Bell, P.T.Moseley, H.M.M.Shearer, C.B.Spencer
J.Chem.Soc.,Chem.Commun.,359, 1980
See also R1 : 83

76.67 (Diethylenetriamine) – (7,9 –
dimethylhypoxanthine) – platinum(ii)
bis(hexafluorophosphate) sesquihydrate
$C_{11}H_{21}N_7OPt^{2+}$, $2F_8P^-$, $1.5H_2O$
T.J.Kistenmacher, K.Wilkowski, B.de Castro, C.C.Chiang,
L.G.Marzilli
Biochem.Biophys.Res.Commun.,**91**, 1521, 1979

76.68 Chloro – (tris(2 – aminoethyl)amine) – pyridino –
cobalt(iii) tetrachloro – zinc(ii)
$C_{11}H_{23}ClCoN_5$, Cl_4Zn
F.Benetollo, G.Bombieri, E.Forsellini, A.Del Pra, M.L.Tobe
Eur.Cryst.Meeting,**6**, 58, 1980
See also R1 : 83

76.C Chloro – (N,N,N′,N′ – tetramethylethylenediamine) –
penta – 2,4 – dienyl – zinc
$C_{11}H_{23}ClN_2Zn$ Main entry is 71.31

76.69 Acetylacetonato – (2,2′,2″ –
triaminotriethylamine) – cobalt(iii) chloride
dihydrate
$C_{11}H_{25}CoN_4O_2^{2+}$, $2Cl^-$, $2H_2O$
B.D.Santarsiero, A.Aruffo, V.Schomaker, E.C.Lingafelter
Am.Cryst.Assoc.,Ser.2,**7**, 23, 1980
See also R1 : 77

76.70 (3,10 – Dihydroxyimino – 4,9 – dimethyl – 5,8 –
diazadodeca – 4,9 – diene – 2,11 – dionato) –
nickel(ii) (α form)
$C_{12}H_{18}N_4NiO_4$
N.S.Dixit, V.M.Naik, C.C.Patel, H.Manohar
Indian J.Chem.,**19**, 62, 1980
See also R1 : 84

76.71 N,N′ – Ethylene – bis(monothioacetylacetone –
iminato) – copper(ii)
$C_{12}H_{18}CuN_2S_2$
R.Cini, A.Cinquantini, P.Orioli, M.Sabat
Cryst.Struct.Commun.,**9**, 865, 1980
See also R1 : 85,83

76.72 alpHa – cis – bis(μ – hydroxo) –
bis(ethylenediamine – N,N′ – diacetato –
chromium(iii)) tetrahydrate
$C_{12}H_{22}Cr_2N_4O_{10}$, $4H_2O$
G.Srdanov, R.Herak, D.J.Radanovic, D.S.Veselinovic
Inorg.Chim.Acta,**38**, 37, 1980
See also R1 : 82

76.73 catena – bis(μ – bis(2 – Aminoethyl)amine) –
bis(μ – thiocyanato) – bis(isothiocyanato) – di –
cadmium(ii) (re-refinement of data of Cannas et.al.,
Inorg. Chem., 16,228,1977 using space group P-1)
$(C_{12}H_{28}Cd_2N_{10}S_4)_n$
R.E.Marsh, V.Schomaker *Inorg.Chem.*,**18**, 2331, 1979

76.74 Tetra – aqua – ethylenediamine – nickel – (μ –
ethylenediamine – tetra – acetato) – nickel
$C_{12}H_{28}N_4Ni_2O_{12}$
V.M.Agre, V.K.Trunov, T.F.Sisoeva
Dokl.Akad.Nauk SSSR,**250**, 1123, 1980
See also R1 : 84

76.75 Iodo – N,N,N′,N′ – tetraethyl – diethylene –
triamine – platinum(ii) iodide
$C_{12}H_{29}IN_3Pt^+$, I^-
R.C.E.Durley, W.L.Waltz, B.E.Robertson
Can.J.Chem.,**58**, 664, 1980

76.76 (1,3,6,8,10,13,16,19 – Octa – azabicyclo(6.6.6)
eicosane) – platinum(iv) dithionate hydrate
$C_{12}H_{30}N_8Pt^{4+}$, $2O_6S_2^{2-}$, $2.5H_2O$
M.Mikami, M.Konno, Y.Saito
Acta Crystallogr.,Sect.B,**35**, 3096, 1979

76.77 bis(N,N – Diethylethylenediamine) – copper(ii)
perchlorate (red form)
$C_{12}H_{32}CuN_4^{2+}$, $2ClO_4^-$
I.Grenthe, P.Paoletti, M.Sandstrom, S.Glikberg
Inorg.Chem.,**18**, 2687, 1979

76.78 bis(N,N – Diethylethylenediamine) – copper(ii)
perchlorate (blue–violet form)
$C_{12}H_{32}CuN_4^{2+}$, $2ClO_4^-$
I.Grenthe, P.Paoletti, M.Sandstrom, S.Glikberg
Inorg.Chem.,**18**, 2687, 1979

76.79 trans – Diaqua – bis(cyclohexane – 1,2 – diamine) –
nickel(ii) chloride
$C_{12}H_{32}N_4NiO_2^{2+}$, $2Cl^-$
A.V.Capilla, R.A.Aranda, F.Gomez–Beltran
Cryst.Struct.Commun.,**9**, 147, 1980

76.80 bis((bis(Ethylenediamine) − cobalt) − (μ − carboxylato − methyl) − thiolato) − silver perchlorate
$C_{12}H_{36}AgCo_2N_8O_4S_2{}^{3+}$, $3ClO_4{}^-$
M.J.Heeg, R.C.Elder, E.Deutsch *Inorg.Chem.*,**19**, 554, 1980
See also R1 : 81,85

76.81 trans − Diaqua − bis(μ − hydroxy) − bis((1,4,7 − triazacyclononane) − rhodium(iii)) tetraperchlorate tetrahydrate
$C_{12}H_{36}N_6O_4Rh_2{}^{4+}$, $4ClO_4{}^-$, $4H_2O$
K.Wieghardt, W.Schmidt, B.Nuber, B.Prikner, J.Weiss
Chem.Ber.,**113**,36, 1980

76.82 bis(2 − Aminoethyl)amine − (di − 2 − pyridylamino) − zinc(ii) nitrate
$C_{14}H_{22}N_6Zn^{2+}$, $2NO_3{}^-$
N.Ray, B.Hathaway
J.Chem.Soc.,Dalton Trans.,1105, 1980
See also R1 : 83

76.83 Ethylenediamine − bis(2 − aminomethylpyridine) − cobalt(iii) hexacyano − cobalt(iii) dihydrate
$C_{14}H_{24}CoN_6{}^{3+}$, $C_6CoN_6{}^{3-}$, $2H_2O$
M.Sekizaki, S.Utsuno *Bull.Chem.Soc.Jpn.*,**52**, 3302, 1979
See also R1 : 83

76.C bis((Diethylenetriamine − copper) − μ^2 − cyano) − tetracyano − iron hexahydrate
$(C_{14}H_{26}Cu_2FeN_{12})_n$, $6nH_2O$ Main entry is 71.78

76.84 (Diethylenetriamine) − (guanosine) − palladium(ii) diperchlorate
$C_{14}H_{26}N_8O_5Pd^{2+}$, $2ClO_4{}^-$
F.D.Rochon, P.C.Kong, B.Coulombe, R.Melanson
Can.J.Chem.,**58**, 381, 1980
See also R1 : 47

76.85 bis(Chloroacetato) − (N,N,N′,N′ − tetraethylethylenediamine) − copper(ii)
$C_{14}H_{28}Cl_2CuN_2O_4$
M.Ahlgren, U.Turpeinen, R.Hamalainen
Acta Chem.Scand.Ser.A,**34**, 67, 1980
See also R1 : 81

76.86 (N,N,N′,N′, − Tetra(2 − hydroxypropyl) − ethylenediamine) − copper(ii) perchlorate monohydrate
$C_{14}H_{31}CuN_2O_4{}^+$, $ClO_4{}^-$, H_2O
O.Orama, M.Orama, U.Schubert
Finn.Chem.Lett.,136, 1979
See also R1 : 84

76.87 Chloro − bis(dicarbonyl − chloro − (tetramethylethylenediamine) − copper(ii)) tetraphenylborate
$C_{14}H_{32}ClCu_2N_4O_2{}^+$, $C_{24}H_{20}B^-$
M.Pasquali, G.Marini, C.Floriani, A.G.Manfredotti
J.Chem.Soc.,Chem.Commun.,937, 1979

76.88 bis(Isothiocyanato − (N,N,N′,N′ − tetramethylethylenediamine)) − nickel(ii)
$C_{14}H_{32}N_6NiS_2$
U.Turpeinen, M.Ahlgren, R.Hamalainen
Finn.Chem.Lett.,11, 1980

76.89 bis(Isoselenocyanato − (N,N,N′,N′ − tetramethylethylenediamine)) − nickel(ii)
$C_{14}H_{32}N_6NiSe_2$
U.Turpeinen, M.Ahlgren, R.Hamalainen
Finn.Chem.Lett.,11, 1980

76.90 bis(N − Isopropyl − 2 − methyl − 1,2 − propanediamine) − nitro − copper(ii) − nitrite
$C_{14}H_{36}CuN_5O_2{}^+$, $NO_2{}^-$
A.Pajunen, S.Pajunen
Acta Crystallogr.,Sect.B,**35**, 3058, 1979

76.91 Sodium N,N′ − ethylene − bis(salicylideneiminato) − oxo − vanadium(iv) tetraphenylborate
$2nC_{16}H_{14}N_2O_3V$, nNa^+, $nC_{24}H_{20}B^-$
M.Pasquali, F.Marchetti, C.Floriani, M.Cesari
Inorg.Chem.,**19**, 1198, 1980
See also R1 : 78

76.92 Chloro − 1,9 − bis(2 − pyridyl) − 2,5,8 − triazanonane − cobalt(iii) tetrachloro − zinc
$C_{16}H_{23}ClCoN_5$, Cl_4Zn
F.Benetollo, G.Bombieri, E.Forsellini, M.L.Tobe
Eur.Cryst.Meeting,**6**, 58, 1980
See also R1 : 83

76.93 (Ethylenediamine) − bis(7,9 − dimethylhypoxanthine) − platinum(ii) bis(hexafluorophosphate)
$C_{16}H_{24}N_{10}O_2Pt^{2+}$, $2F_6P^-$
T.J.Kistenmacher, K.Wilkowski, B.de Castro, C.C.Chiang, L.G.Marzilli
Biochem.Biophys.Res.Commun.,**91**, 1521, 1979

76.94 (μ − Hydroxy) − (μ − peroxo) − bis(4,7 − dimethyl − 1,4,7,10 − tetra − azadecane − cobalt(iii)) triperchlorate dihydrate
$C_{16}H_{45}Co_2N_8O_3{}^{3+}$, $3ClO_4{}^-$, $2H_2O$
H.Macke, M.Zehnder, U.Thewalt, S.Fallab
Helv.Chim.Acta,**62**, 1804, 1979

76.95 (N,N′ − (3 − Azapentane − 1,5 − diyl) − bis(salicylideneiminato)) − dioxo − uranium(vi)
$C_{18}H_{19}N_3O_4U$
F.Benetollo, G.Bombieri, A.J.Smith
Acta Crystallogr.,Sect.B,**35**, 3091, 1979
See also R1 : 78

76.96 (μ − Peroxo) − (μ − 2,2′,2″ − triamino − triethylamine) − bis((2,2′,2″ − triamino − triethylamine) − cobalt(ii)) tetraperchlorate dihydrate
$C_{18}H_{54}Co_2N_{12}O_2{}^{4+}$, $4ClO_4{}^-$, $2H_2O$
M.Zehnder, U.Thewalt, S.Fallab
Helv.Chim.Acta,**62**, 2099, 1979

76.C (N,N,N′,N′ − Tetramethylethylenediamine) − bis(cyclohexylisocyanide) − copper(i) tetraphenylborate
$C_{20}H_{38}CuN_4{}^+$, $C_{24}H_{20}B^-$ Main entry is 71.161

76.97 (μ_2 − Dioxo) − bis(1,5,8,11,15 − penta − azapentadecane − cobalt) dithionate dinitrate tetrahydrate
$C_{20}H_{54}Co_2N_{10}O_2{}^{4+}$, $O_6S_2{}^{2-}$, $4H_2O$, $2NO_3{}^-$
M.Zehnder, U.Thewalt *Z.Anorg.Allg.Chem.*,**461**, 53, 1980
See also R1 : 84,83

76.98 asym − (μ − Carbonato) − bis(chloro − (tetraethylethylenediamine) − copper(ii))
$C_{21}H_{48}Cl_2Cu_2N_4O_3$
M.R.Churchill, G.Davies, M.A.El−Sayed, M.F.El−Shazly, J.P.Hutchinson, M.W.Rupich *Inorg.Chem.*,**19**, 201, 1980

76.C Dichloroethylene − (N,N′ − dimethyl − N,N′ − bis(α − methylbenzyl) − 1,2 − ethanediamine) − platinum(ii)
$C_{22}H_{32}Cl_2N_2Pt$ Main entry is 72.41

76.C $(\mu - 3,3,6,6 - $ Tetramethyl $- 3,6 -$ diazaoctane $-$ 1,8 $-$ diyl) $-$ bis(N,N,N′,N′ $-$ tetramethylethylenediamine $-$ chloro $-$ platinum(ii)) diperchlorate
$C_{22}H_{56}Cl_2N_6Pt_2{}^{2+}$, $2ClO_4{}^-$ Main entry is 71.187

76.C Dichloro $-$ (propene) $-$ (N,N′ $-$ dimethyl $-$ N,N′ $-$ bis($\alpha -$ methylbenzyl) $-$ 1,2 $-$ ethanediamine) $-$ platinum(ii)
$C_{23}H_{34}Cl_2N_2Pt$ Main entry is 72.44

76.99 N,N′ $-$ Ethylene $-$ bis(1 $-$ iminomethyl $-$ 2 $-$ naphtholato) $-$ copper(ii)
$C_{24}H_{18}CuN_2O_2$
C.Freiburg, W.Reichert, M.Melchers, B.Engelen
Acta Crystallogr.,Sect.B,**36**, 1209, 1980
See also R1 : 84

76.100 N,N′ $-$ Ethylene $-$ bis(1 $-$ iminomethyl $-$ 2 $-$ naphtholato) $-$ nickel(ii)
$C_{24}H_{18}N_2NiO_2$
C.Freiburg, W.Reichert, M.Melchers, B.Engelen
Acta Crystallogr.,Sect.B,**36**, 1209, 1980
See also R1 : 84

76.101 N,N′ $-$ Ethylene $-$ bis(3 $-$ t $-$ butylsalicylideneiminato) $-$ cobalt(ii)
$C_{24}H_{30}CoN_2O_2$
W.P.Schaefer, B.T.Huie, M.G.Kurilla, S.E.Ealick
Inorg.Chem., **19**, 340, 1980
See also R1 : 78

76.C bis(Ethylenediamine) $-$ copper (bis(ethylenediamine) $-$ copper) $-$ (ethylenediamine $-$ tetra $-$ acetato $-$ nickel) dihydrate
$C_{24}H_{40}CuN_8Ni_2O_{16}{}^{2-}$, $C_4H_{16}CuN_4{}^{2+}$, $2H_2O$ Main entry is 76.18

76.102 bis(1,1,1,5,5,5 $-$ Hexafluoropentane $-$ 2,4 $-$ dionato) $-$ copper(ii) $-$ N,N′ $-$ ethylene $-$ bis(salicylideneiminato) $-$ copper(ii)
$C_{26}H_{18}Cu_2F_{12}N_2O_6$
N.Bresciani$-$Pahor, M.Calligaris, G.Nardin, L.Randaccio, D.E.Fenton *Transition Met.Chem.*,**5**, 180, 1980
See also R1 : 78

76.103 bis(($\mu^2 -$ Phenoxo) $-$ ethylenediamine $-$ phenoxy $-$ copper(ii)) phenol solvate
$C_{28}H_{36}Cu_2N_4O_4$, $2C_6H_6O$
F.Calderazzo, F.Marchetti, G.Dell'Amico, G.Pelizzi, A.Colligiani *J.Chem.Soc.,Dalton Trans.*,1419, 1980
See also R1 : 84

76.104 bis($\mu -$ (N,N $-$ Dimethyl $-$ 2 $-$ aminoethyl) $-$ 1 $-$ phenylpropane $-$ 2 $-$ oximato) $-$ bis(methanol $-$ copper(ii)) bis($\mu -$ (N,N $-$ dimethyl $-$ 2 $-$ aminoethyl) $-$ 1 $-$ phenylpropane $-$ 2 $-$ oximato) $-$ bis(perchlorato $-$ copper(ii)) diperchlorate
$C_{28}H_{44}Cu_2N_6O_4{}^{2+}$, $C_{28}H_{36}Cl_2Cu_2N_6O_{10}$, $2ClO_4{}^-$
R.J.Butcher, C.J.O'Connor, E.Sinn
Inorg.Chem., **18**, 1913, 1979
See also R1 : 83

76.105 Dioxygen $-$ (N $-$ pyridine) $-$ N,N′ $-$ ethylene $-$ bis(3 $-$ t $-$ butylsalicylideneiminato) $-$ cobalt
$C_{29}H_{35}CoN_3O_4$
W.P.Schaefer, B.T.Huie, M.G.Kurilla, S.E.Ealick
Inorg.Chem., **19**, 340, 1980
See also R1 : 78

76.106 ($\mu -$ Peroxo) $-$ bis((1,9 $-$ bis(2 $-$ pyridyl) $-$ 2,5,8 $-$ triazanonane) cobalt(iii)) tetra $-$ iodide
$C_{32}H_{46}Co_2N_{10}O_2{}^{4+}$, $4I^-$
J.H.Timmons, R.H.Niswander, A.Clearfield, A.E.Martell
Inorg.Chem.,**18**, 2977, 1979
See also R1 : 83

76.107 Chloro $-$ (1,4,7,10 $-$ tetrabenzyl $-$ 1,4,7,10 $-$ tetra $-$ azacyclododecane) copper(ii) nitrate
$C_{36}H_{44}ClCuN_4{}^+$, $NO_3{}^-$
R.E.DeSimone, E.L.Blinn, K.F.Mucker
Inorg.Nucl.Chem.Lett.,**16**, 23, 1980

76.108 bis($\mu -$ Oxo) $-$ bis((N,N′ $-$ ethylene $-$ bis(acetylacetiminato)) $-$ oxo $-$ rhenium(v)) $-$ (N,N′ $-$ ethylene $-$ bis(acetylacetiminato)) rhenium(v) per $-$ rhenate benzene dichloromethane solvate
$C_{36}H_{54}N_6O_{10}Re_3{}^+$, O_4Re^-, $1.5C_6H_6$, xCH_2Cl_2
M.A.A.F.de C.T.Carrondo, A.R.Middleton, A.C.Skapski, A.P.West, G.Wilkinson *Inorg.Chim.Acta*,**44**, L7, 1980
See also R1 : 84

METAL COMPLEXES (ACETYLACETONE)

77.C bis(Acetylacetonato) – palladium(ii)
tetrathiafulvalene
$C_{10}H_{14}O_4Pd$, $C_6H_4S_4$ Main entry is 60.12

77.C Acetylacetonato – (2,2',2'' –
triaminotriethylamine) – cobalt(iii) chloride
dihydrate
$C_{11}H_{25}CoN_4O_2{}^{2+}$, $2Cl^-$, $2H_2O$ Main entry is 76.69

77.1 N,N' – Ethylene – bis(1,1,1 – trifluoropentane – 2,4 –
dioneiminato) – copper(ii)
$C_{12}H_{12}CuF_6N_2O_2$
H.C.Allen Junior, G.L.Hillhouse, D.J.Hodgson
Inorg.Chim.Acta,**37**, 37, 1979
See also R1 : 83

77.2 (Acetylacetonato) – bis(diethylamine) –
palladium(ii) acetylacetonate (at –170°C)
$C_{13}H_{29}N_2O_2Pd^+$, $C_5H_7O_2{}^-$
S.Kotake, T.Sei, K.Miki, Y.Kai, N.Yasuoka, N.Kasai
Bull.Chem.Soc.Jpn.,**53**, 10, 1980

77.3 bis(2 – (1 – Iminoethyl) – 1 – cyano – 1,3 –
butanedionato) – nickel(ii)
$C_{14}H_{14}N_4NiO_4$
B.Corain, A.Del Pra, F.Filira, G.Zanotti
Inorg.Chem.,**18**, 3523, 1979

77.4 Azobenzyl – hexafluoroacetylacetonato – palladium
(α form, yellow)
$C_{17}H_{10}F_6N_2O_2Pd$
M.C.Etter, A.R.Siedle *Am.Cryst.Assoc.,Ser.2*,**8**, 29, 1980
See also R1 : 83

77.5 Azobenzyl – hexafluoroacetylacetonato – palladium
(β form, red)
$C_{17}H_{10}F_6N_2O_2Pd$
M.C.Etter, A.R.Siedle *Am.Cryst.Assoc.,Ser.2*,**8**, 29, 1980
See also R1 : 83

77.6 Aqua – (acetylacetonato) – (o – phenanthroline) –
copper hexafluoroacetylacetone monohydrate
$C_{17}H_{17}CuN_2O_3{}^+$, $C_5HF_6O_2{}^-$, H_2O
N.A.Bailey, D.E.Fenton, M.V.Franklin, M.Hall
J.Chem.Soc.,Dalton Trans.,984, 1980
See also R1 : 83

77.C Benzoyl – 1,1,1 – trifluoroacetonato – cyclo – octa –
1,5 – diene – rhodium(i)
$C_{18}H_{18}F_3O_2Rh$ Main entry is 75.20

77.7 bis(Hexafluoroacetylacetonato) – (4 – hydroxy –
2,2,6,6 – tetramethyl – piperidinyl – N – oxy) –
copper(ii)
$(C_{19}H_{20}CuF_{12}NO_6)_n$
O.P.Anderson, T.C.Keuchler *Inorg.Chem.*,**19**, 1417, 1980
See also R1 : 84

77.C (–) – RS – Cyclo – 1 – (1' – dimethylaminoethyl –
ferrocene) – 2 – acetylacetonato – palladium
(absolute configuration)
$C_{19}H_{25}FeNO_2Pd$ Main entry is 73.80

77.8 (Acetylacetonato) – (hexafluoroacetylacetonato) –
(o – phenanthroline) – copper(ii)
$C_{22}H_{16}CuF_6N_2O_4$
N.A.Bailey, D.E.Fenton, M.V.Franklin, M.Hall
J.Chem.Soc.,Dalton Trans.,984, 1980
See also R1 : 83

77.C (Di(pivaloyl) – methanato) – (η – 1,2,4,5 – (1,2 –
bis(trifluoromethyl) – 4 – methyl – 3 – (prop – 2' –
ylidene) – penta – 1,4 – diene)) – rhodium(i)
$C_{22}H_{31}F_6O_2Rh$ Main entry is 72.40

77.9 Ammonium tris(4,4,4 – trifluoro – 1 – (2 – furyl) –
1,3 – butanedionato) – cadmium(ii)
$C_{24}H_{12}CdF_9O_9{}^-$, H_4N^+
W.O.McSharry, M.Cefola, J.G.White
Inorg.Chim.Acta,**38**, 161, 1980

77.C (1 – (o – Diphenylarsinophenyl) – 2 –
methoxyethyl) – (hexafluoroacetylacetonato) –
platinum(ii)
$C_{26}H_{21}AsF_6O_3Pt$ Main entry is 71.200

77.C α – (4 – 6η) – (3,20 – Dioxo – pregn – 4 – enyl) –
pentane – 2,4 – dionato – palladium(ii)
$C_{26}H_{36}O_4Pd$ Main entry is 75.35

77.C 1 – ((Di(pivaloyl) – methanato) – bis(pyridine) –
rhoda) – 2,3 – bis(trifluoromethyl) – cyclopent – 2 –
ene
$C_{27}H_{33}F_6N_2O_2Rh$ Main entry is 71.216

77.C Acetylacetonato – (triphenylphosphine) – (Z – 1,2 –
diphenyl – 2 – methylethenyl) – nickel
$C_{38}H_{35}NiO_2P$ Main entry is 71.255

77.10 bis(Dimethylformamide) – tris(2,2,6,6 –
tetramethylheptane – 3,5 – dionato) – europium(iii)
$C_{39}H_{71}EuN_2O_8$
J.A.Cunningham, R.E.Sievers *Inorg.Chem.*,**19**, 595, 1980
See also R1 : 84

77.11 tetrakis(4,4,4 – Trifluoro – 1 – phenyl – 1,3 –
butanedionato) – uranium
$C_{40}H_{24}F_{12}O_8U$
A.Navaza, C.de Rango, P.Charpin
Acta Crystallogr.,Sect.B,**36**, 696, 1980

77.12 tris(1,3 – Diphenyl – 1,3 – propanedionato) –
iron(iii)
$C_{45}H_{33}FeO_6$
B.Kaitner, B.Kamenar
Cryst.Struct.Commun.,**9**, 487, 1980

77.13 tetrakis(1,3 – Diphenyl – 1,3 – propanedionato) –
zirconium(iv)
$C_{60}H_{44}O_8Zr$
H.K.Chun, W.L.Steffen, R.C.Fay
Inorg.Chem.,**18**, 2458, 1979

METAL COMPLEXES
(SALICYLIC DERIVATIVES)

78.1 **(N – Salicylidene – L – valinato) – aquo – copper(ii)**
$C_{12}H_{15}CuNO_4$
K.Korhonen, R.Hamalainen
Acta Chem.Scand.Ser.A,**33**, 569, 1979
See also R1 : 82

78.2 **Chloro – (N – pyridoxylidene – N' – salicyloylhydrazinato) – copper(ii) monohydrate**
$C_{15}H_{14}ClCuN_3O_4$, H_2O
P.Domiano, A.Musatti, M.Nardelli, C.Pelizzi, G.Predieri
Transition Met.Chem.,**4**, 351, 1979

78.3 **Lithium (N,N' – ethylene – bis(salicylideneiminato)) – cobalt(i)**
$C_{16}H_{14}CoN_2O_2^-$, Li^+, $1.5C_4H_8O$
G.Fachinetti, C.Floriani, P.F.Zanazzi, A.R.Zanari
Inorg.Chem.,**18**, 3469, 1979

78.4 **Sodium (N,N' – ethylene – bis(salicylideneiminato)) – cobalt(i) tetrahydrofuran solvate (absolute configuration)**
$C_{16}H_{14}CoN_2O_2^-$, Na^+, C_4H_8O
G.Fachinetti, C.Floriani, P.F.Zanazzi, A.R.Zanari
Inorg.Chem.,**18**, 3469, 1979

78.C **Sodium N,N' – ethylene – bis(salicylideneiminato) – oxo – vanadium(iv) tetraphenylborate**
$2nC_{16}H_{14}N_2O_3V$, nNa^+, $nC_{24}H_{20}B^-$ Main entry is 76.91

78.5 **2 – Hydroxy – N – salicylidene – aniline – (diethylamine) platinum(ii)**
$C_{17}H_{20}N_2O_2Pt$
F.Bachechi, P.Mura, L.Zambonelli, H.Motschi, P.S.Pregosin
Eur.Cryst.Meeting,**6**, 177, 1980
See also R1 : 83

78.6 **(N,N' – bis(Salicylidene) – 1,5 – diamino – 3 – oxapentane) – dioxo – uranium(vi) chloroform solvate**
$C_{18}H_{18}N_2O_5U$, $CHCl_3$
G.Bombieri, E.Forsellini, F.Benetollo, D.E.Fenton
J.Inorg.Nucl.Chem.,**41**, 1437, 1979

78.C **(N,N' – (3 – Azapentane – 1,5 – diyl) – bis(salicylideneiminato)) – dioxo – uranium(vi)**
$C_{18}H_{19}N_3O_4U$ Main entry is 76.95

78.C **N,N' – (1,2 – Phenylene) – bis(salicylaldiminato) – copper(ii) – 7,7,8,8 – tetracyanoquinodimethane complex**
$2C_{20}H_{14}CuN_2O_2$, $C_{12}H_4N_4$ Main entry is 60.40

78.7 **Salicylidene – anthranyl – aldehyde – 4 – phenylthiosemicarbazonato – copper(ii) nitrate**
$C_{21}H_{17}CuN_4OS^+$, NO_3^-
T.I.Malinovskii, V.I.Gerasimov, E.V.Slavjanov, V.A.Neverov, V.N.Byushkin *Eur.Cryst.Meeting*,**6**, 55, 1980

78.8 **Dichloro – bis(N – n – butyl – salicylaldiminato) – vanadium(iv) benzene solvate**
$C_{22}H_{28}Cl_2N_2O_2V$, C_6H_6
M.Pasquali, F.Marchetti, C.Floriani
Inorg.Chem.,**18**, 2401, 1979

78.9 **bis(N – Isopropyl – 3 – methoxy – salicylideneiminato) – copper(ii) (green low temperature form)**
$C_{22}H_{28}CuN_2O_4$
H.Tamura, K.Ogawa, A.Takeuchi, S.Yamada
Cryst.Struct.Commun.,**9**, 91, 1980

78.C **(η – Allyl) – dicarbonyl – (N – phenylsalicylideneiminato) – pyridine – molybdenum(ii)**
$C_{23}H_{20}MoN_2O_3$ Main entry is 72.43

78.C **N,N' – Ethylene – bis(3 – t – butylsalicylideneiminato) – cobalt(ii)**
$C_{24}H_{30}CoN_2O_2$ Main entry is 76.101

78.C **bis(1,1,1,5,5,5 – Hexafluoropentane – 2,4 – dionato) – copper(ii) – N,N' – ethylene – bis(salicylideneiminato) – copper(ii)**
$C_{26}H_{18}Cu_2F_{12}N_2O_6$ Main entry is 76.102

78.10 **(tris(2 – (5 – Chloro – salicylaldiminato) – ethyl) amine) – chromium(iii) trihydrate**
$C_{27}H_{24}Cl_3CrN_4O_3$, $3H_2O$
N.W.Alcock, D.F.Cook, E.D.McKenzie, J.M.Worthington
Inorg.Chim.Acta,**38**, 107, 1980

78.11 **(tris(2 – (5' – Chloro – salicylaldiminato)ethyl) amine) – manganese(iii) trihydrate**
$C_{27}H_{24}Cl_3MnN_4O_3$, $3H_2O$
N.W.Alcock, D.F.Cook, E.D.McKenzie, J.M.Worthington
Inorg.Chim.Acta,**38**, 107, 1980

78.12 **bis(N – Isopropyl – 2 – oxy – 1 – naphthylidene – aminato) – copper(ii)**
$C_{28}H_{28}CuN_2O_2$
N.Matsumoto, Y.Nonaka, S.Kida, S.Kawano, I.Ueda
Inorg.Chim.Acta,**37**, 27, 1979

78.13 **bis(N – Isopropyl – 2 – oxy – 1 – naphthylidene – aminato) – copper(ii) bis(7,7,8,8 – tetracyanoquinodimethane)**
$C_{28}H_{28}CuN_2O_2$, $2C_{12}H_4N_4$
N.Matsumoto, Y.Nonaka, S.Kida, S.Kawano, I.Ueda
Inorg.Chim.Acta,**37**, 27, 1979
See also R2 : 7

78.C **Dioxygen – (N – pyridine) – N,N' – ethylene – bis(3 – t – butylsalicylideneiminato) – cobalt**
$C_{29}H_{35}CoN_3O_4$ Main entry is 76.105

78.14 **2 – Hydroxy – N – salicylidene – aniline – (triphenylphosphine) platinum(ii)**
$C_{31}H_{24}NO_2PPt$
F.Bachechi, P.Mura, L.Zambonelli, H.Motschi, P.S.Pregosin
Eur.Cryst.Meeting,**6**, 177, 1980
See also R1 : 86

78.15 **bis(N – Isobutyl – (5 – chloro – α – phenyl – 2 – hydroxybenzylidene) – aminato) – copper(ii) (red isomer, re-refinement of data of Chia, Freyberg, Mockler and Sinn, Inorg.Chem., 16, 254, 1977)**
$C_{34}H_{34}Cl_2CuN_2O_2$
R.E.Marsh, V.Schomaker *Inorg.Chem.*,**18**, 2331, 1979

78.16 bis(2,2,5,5 – Tetramethylpyrrolidinyl) – 3 – N – (2 – formyl – 5 – methyl – salicylaldiminato) – copper
$C_{34}H_{44}Cu_2N_4O_6$
A.A.Medzhidov, T.M.Kutovaya, N.P.Rodin, M.K.Gusejnova,
I.A.Timakov, Kh.S.Mamedov *Koord.Khim.*,**5**, 1433, 1979

78.17 bis(N,N' – Disalicylidene – 1,2 – phenylenediamino) – zirconium(iv) benzene solvate
$C_{40}H_{28}N_4O_4Zr$, $2.5C_6H_6$
R.D.Archer, R.O.Day, M.L.Illingsworth
Inorg.Chem., **18**, 2908, 1979

78.18 bis(N – Cyclohexyl – salicylaldiminato) – copper(ii) (green form ii)
$C_{52}H_{64}Cu_2N_4O_4$
H.Tamura, K.Ogawa, A.Takeuchi, S.Yamada
Bull.Chem.Soc.Jpn.,**52**, 3522, 1979

METAL COMPLEXES (THIOUREA)

79.1 Aquo – urea – phosphonito – dioxo – uranium monohydrate
$CH_7N_2O_7PU$, H_2O
K.A.Avduevskaya, Yu.N.Mikhailov, N.B.Ragulina,
I.A.Rozanov
Izv.Akad.Nauk SSSR,Neorg.Mater.,**15**, 947, 1979

79.2 bis(Urea) – phosphonito – dioxo – uranium
$C_2H_9N_4O_7PU$
K.A.Avduevskaya, Yu.N.Mikhailov, N.B.Ragulina,
I.A.Rozanov
Izv.Akad.Nauk SSSR,Neorg.Mater.,**15**, 947, 1979

79.3 tris(Urea) – dioxo – uranium(vi) sulfate
$C_3H_{12}N_6O_5U^{2+}$, O_4S^{2-}
H.Ruben, B.Spencer, D.H.Templeton, A.Zalkin
Inorg.Chem.,**19**, 776, 1980

79.4 tris(Thiosemicarbazido) – rhodium(iii) chloride
$C_3H_{15}N_9RhS_3^{3+}$, $3Cl^-$
M.E.Rusanovskii, I.D.Samus, I.N.Kiseleva, N.V.Belov
*Dokl.Akad.Nauk SSSR,***250**, 361, 1980

79.5 bis(Isothiocyanato) – bis(thiourea) – iron(ii)
$(C_4H_8FeN_6S_4)_n$
G.Valle, U.Russo, S.Calogero
Transition Met.Chem.,**5**, 26, 1980

79.6 Aqua – (3 – ethoxy – 2 – oxobutyraldehyde – bis(thiosemicarbazonato)) – zinc(ii)
$C_8H_{16}N_6O_2S_2Zn$
P.E.Bourne, M.R.Taylor
Acta Crystallogr.,Sect.B,**36**, 2143, 1980

79.7 Dibromo – tris(N,N' – dimethylurea) – manganese(ii)
$C_9H_{24}Br_2MnN_6O_3$
J.Delaunay, C.Kappenstein, R.P.Hugel
J.Chem.Soc.,Chem.Commun.,679, 1980

79.8 Chloro – bis(N – ethyl – imidazolidine – 2 – thione) – copper(ii)
$C_{10}H_{20}ClCuN_4S_2$
L.P.Battaglia, A.B.Corradi, F.A.Devillanova, G.Verani
Transition Met.Chem.,**4**, 264, 1979

79.9 hexakis(N – Methylurea) – cobalt(ii) thiosulfate
$C_{12}H_{36}CoN_{12}O_6^{2+}$, $O_3S_2^{2-}$
B.N.Figgis, B.W.Skelton, A.H.White
Aust.J.Chem.,**33**, 425, 1980

79.10 hexakis(N – Methylurea) – cobalt(ii) sulfate
$C_{12}H_{36}CoN_{12}O_6^{2+}$, O_4S^{2-}
B.N.Figgis, B.W.Skelton, A.H.White
Aust.J.Chem.,**33**, 425, 1980

79.11 bis(2 – Formylpyridine – thiosemicarbazonato) – nickel(ii)
$C_{14}H_{14}N_8NiS_2$
G.R.Clark, G.J.Palenik
Cryst.Struct.Commun.,**9**, 449, 1980

79.12 Chloro – N,N – diethylamide – pyruvate – 4 –
phenylthiosemicarbazonato – copper(ii)
$C_{14}H_{19}ClCuN_4OS$
T.I.Malinovskii, V.I.Gerasimov, E.V.Slavjanov, V.A.Neverov,
V.N.Byushkin *Eur.Cryst.Meeting*,**6**, 55, 1980
See also R1 : 84,83

79.13 tetrakis(1 – Methyl – 2(3H) – imidazoline –
thione) – zinc(ii) nitrate monohydrate
$C_{16}H_{24}N_8S_4Zn^{2+}$, $2NO_3^-$, H_2O
I.W.Nowell, A.G.Cox, E.S.Raper
Acta Crystallogr.,Sect.B,**35**, 3047, 1979

79.14 bis(μ – Acetato) – bis(di(urea) – di(acetato) –
praseodymium(iii)) urea dihydrate
$C_{16}H_{34}N_8O_{18}Pr_2$, $2CH_4N_2O$, $2H_2O$
G.V.Romanenko, N.V.Podberezskaya, V.V.Bakakin
Dokl.Akad.Nauk SSSR,**248**, 1337, 1979
See also R1 : 81

79.15 bis(1,1 – Diethyl – 3 – benzoylthiourea) – copper(ii)
(form i)
$C_{24}H_{30}CuN_4O_2S_2$
R.Richter, L.Beyer, J.Kaiser
Z.Anorg.Allg.Chem.,**461**, 67, 1980

79.16 hexakis(Diethylurea) – iron(ii) perchlorate
$C_{30}H_{72}FeN_{12}O_6^{2+}$, $2ClO_4^-$
U.Russo, G.Valle, S.Calogero
Cryst.Struct.Commun.,**9**, 443, 1980

79.17 tetrakis(μ^2 – (N,N') – Diphenylurea) –
bis(tetrahydrofuran) – di – chromium cyclohexane
solvate
$C_{60}H_{60}Cr_2N_8O_6$, C_6H_{12}
F.A.Cotton, W.H.Ilsley, W.Kaim
J.Am.Chem.Soc.,**102**, 3464, 1980
See also R1 : 84,83

METAL COMPLEXES
(THIOCARBAMATE OR XANTHATE)

80.1 Ethylxanthate – mercury(ii)
$(C_6H_{10}HgO_2S_4)_n$
C.Chieh, K.J.Moynihan
Acta Crystallogr.,Sect.B,**36**, 1367, 1980

80.2 Tricarbonyl – bis(N,N – dimethyldithiocarbamato) –
tungsten(ii)
$C_9H_{12}N_2O_3S_4W$
J.L.Templeton, B.C.Ward *Inorg.Chem.*,**19**, 1753, 1980

80.3 tris(N,N – Dimethyldithiocarbamato) – cobalt(iii)
$C_9H_{18}CoN_3S_6$
H.Iwasaki, K.Kobayashi
Acta Crystallogr.,Sect.B,**36**, 1657, 1980

80.4 bis(Morpholine – N – carbodithioato) – nickel(ii)
$C_{10}H_{16}N_2NiO_2S_4$
F.G.Herring, J.M.Park, S.J.Rettig, J.Trotter
Can.J.Chem.,**57**, 2379, 1979

80.5 trans – Dithiocyanato – bis(0 – ethyl – N –
methyl – monothiocarbamato) – palladium(ii)
$C_{10}H_{18}N_4O_2PdS_4$
R.Bardi, A.M.Piazzesi, L.Sindellari
Eur.Cryst.Meeting,**6**, 293, 1980

80.6 bis(Tri – iodoarsine) – bis(diethyldithiocarbamato) –
nickel(ii)
$C_{10}H_{20}As_2I_6N_2NiS_4$
J.Willemse, J.A.Cras, W.P.Bosman, J.J.Steggerda
Rec.J.Roy.Netherl.Chem.Soc.,**99**, 65, 1980

80.7 Dioxo – (N,N – diethyldithiocarbamato) –
molybdenum(vi)
$C_{10}H_{20}MoN_2O_2S_4$
J.M.Berg, K.O.Hodgson *Inorg.Chem.*,**19**, 2180, 1980

80.8 trans – Dibromo – bis(0 – ethyl – N,N – dimethyl –
monothiocarbamato) – palladium(ii)
$C_{10}H_{22}Br_2N_2O_2PdS_2$
R.Bardi, A.M.Piazzesi, L.Sindellari
Eur.Cryst.Meeting,**6**, 293, 1980

80.9 trans – Dichloro – bis(0 – ethyl – N,N –
dimethylthiocarbamato) – palladium(ii)
$C_{10}H_{22}Cl_2N_2O_2PdS_2$
R.Bardi, A.M.Piazzesi, M.Munari
Cryst.Struct.Commun.,**9**, 835, 1980

80.10 cis – Dichloro – bis(0 – ethyl – N,N – dimethyl –
monothiocarbamato) – platinum(ii)
$C_{10}H_{22}Cl_2N_2O_2PtS_2$
R.Bardi, A.M.Piazzesi, L.Sindellari
Eur.Cryst.Meeting,**6**, 293, 1980

80.11 bis(N,N – Diethyldithiocarbamato) –
(isothiocyanato) – iron(iii)
$C_{11}H_{20}FeN_3S_5$
C.L.Raston, W.G.Sly, A.H.White *Aust.J.Chem.*,**33**, 221, 1980
See also R1 : 83

80.12 bis(N,N – Di – isopropyldithiocarbamato) – copper(ii)
$C_{14}H_{28}CuN_2S_4$
H.Iwasaki, K.Kobayashi
*Acta Crystallogr.,Sect.B,***36**, 1655, 1980

80.13 bis(N,N – Di – isopropyldithiocarbamato) –
mercury(ii) (α form)
$C_{14}H_{28}HgN_2S_4$
M.Ito, H.Iwasaki *Acta Crystallogr.,Sect.B,***35**, 2720, 1979

80.14 tris(Pyrrole – N – carbodithioato) – iron(iii)
dichloromethane solvate
$C_{15}H_{12}FeN_3S_6$, $0.5CH_2Cl_2$
R.D.Bereman, M.R.Churchill, D.Nalewajek
*Inorg.Chem.,***18**, 3112, 1979

80.15 tris(Dimethylthiocarbamato) – (m –
nitrophenyldiazenato) – molybdenum methylene
chloride solvate hemihydrate
$2C_{15}H_{22}MoN_6O_2S_6$, $0.5CH_2Cl_2$, $0.5H_2O$
G.A.Williams, A.R.P.Smith *Aust.J.Chem.,***33**, 717, 1980
See also R1 : 83

80.16 tris(Dimethyldithiocarbamato) –
(benzenediazenato) – molybdenum methylene
chloride solvate
$C_{15}H_{23}MoN_5S_6$, CH_2Cl_2
G.A.Williams, A.R.P.Smith *Aust.J.Chem.,***33**, 717, 1980
See also R1 : 83

80.17 Chloro – tris(N,N – diethyl – monothiocarbamato) –
titanium(iv)
$C_{15}H_{30}ClN_3O_3S_3Ti$
S.L.Hawthorne, R.C.Fay *J.Am.Chem.Soc.,***101**, 5268, 1979

80.18 Phenylnitrene – bis(diethyldithiocarbamato) –
dichloro – molybdenum chloroform solvate
(at −150°C)
$C_{16}H_{25}Cl_2MoN_3S_4$, $CHCl_3$
E.A.Maatta, B.L.Haymore, R.A.D.Wentworth
*Inorg.Chem.,***19**, 1055, 1980
See also R1 : 83

80.19 bis(N,N – Dimethyldithiocarbamato) – (η – N –
methyl – N – phenylhydrazido) – (η² – N – methyl –
N – phenylhydrazido) – molybdenum
tetraphenylborate
$C_{20}H_{29}MoN_6S_4{}^+$, $C_{24}H_{20}B^-$
J.Chatt, J.R.Dilworth, P.L.Dahlstrom, J.Zubieta
*J.Chem.Soc.,Chem.Commun.,*786, 1980
See also R1 : 83

80.20 bis(Pyrrolidine – carbodithioato) – zinc(ii) dimer
$C_{20}H_{32}N_4S_8Zn_2$
V.Francetic, I.Leban *Vestn.Slov.Kem.Drus.,***26**, 113, 1979

80.21 catena – (bis(μ – Chloro) – tetrakis(μ – N,N –
diethyldithiocarbamato) – tri – mercury(ii))
$(C_{20}H_{40}Cl_2Hg_3N_4S_8)_n$
L.Book, C.Chieh *Acta Crystallogr.,Sect.B,***36**, 300, 1980

80.22 (μ – bis(N,N – Diethyldithiocarbamato)) – bis(N,N –
diethyldithiocarbamato) – dicarbonyl –
ruthenium(ii) – ruthenium(iii) tetrafluoroborate
$C_{22}H_{40}N_4O_2Ru_2S_8{}^+$, $BF_4{}^-$
L.H.Pignolet, S.H.Wheeler *Inorg.Chem.,***19**, 935, 1980

80.C bis(N,N – Dimethyldithiocarbamato) – (ditoluoyl –
acetylene) – oxo – molybdenum(iv) benzene solvate
$C_{24}H_{26}MoN_2O_3S_4$, C_6H_6 Main entry is 72.47

80.23 tris(N,N – Diethyldithiocarbamato) – bis(μ – (N,N –
(diethyltrithiocarbamato)) – di – osmium(iii)
tetraphenylborate
$C_{25}H_{50}N_5Os_2S_{12}{}^+$, $C_{24}H_{20}B^-$
L.J.Maheu, L.H.Pignolet *Inorg.Chem.,***18**, 3626, 1979
See also R2 : 62

80.24 bis(N – t – Butyldithiocarbamato) –
(tricyclohexylphosphine) – platinum(ii) cyclohexane
solvate
$C_{28}H_{53}N_2PPtS_4$, C_6H_{12}
P.C.Christidis, P.J.Rentzeperis
*Acta Crystallogr.,Sect.B,***35**, 2543, 1979
See also R1 : 86

80.25 bis(Dimethyldithiocarbamato) –
bis(diphenylhydrazido) – molybdenum acetone
solvate
$C_{30}H_{32}MoN_6S_4$, C_3H_6O
J.Chatt, B.A.L.Crichton, J.R.Dilworth, P.Dahlstrom,
R.Gutkoska, J.A.Zubieta
*Transition Met.Chem.,***4**, 271, 1979
See also R1 : 83

80.26 bis(μ₂ – N,N – Diethyldithiocarbamato) –
tetrakis(N,N – diethyldithiocarbamato) – di –
osmium(iv) hexafluorophosphate methylene
chloride solvate
$C_{30}H_{60}N_6Os_2S_{12}{}^{2+}$, $2F_6P^-$, CH_2Cl_2
S.H.Wheeler, L.H.Pignolet *Inorg.Chem.,***19**, 972, 1980

80.27 bis(μ₂ – N,N – Diethyldithiocarbamato) –
tetrakis(N,N – diethyldithiocarbamato) – di –
osmium(iv) hexafluorophosphate dichloromethane
solvate
$C_{30}H_{60}N_6Os_2S_{12}{}^{2+}$, $2F_6P^-$, $2CH_2Cl_2$
S.H.Wheeler, L.H.Pignolet *Inorg.Chem.,***19**, 972, 1980

80.28 tris(N,N – Dibenzyldithiocarbamato) – iron(iii)
$C_{45}H_{42}FeN_3S_6$
J.Albertsson, I.Elding, A.Oskarsson
*Acta Chem.Scand.Ser.A,***33**, 703, 1979

80.29 tris(N,N – Dibenzyldithiocarbamato) – iron(iii)
(at 150°K)
$C_{45}H_{42}FeN_3S_6$
J.Albertsson, I.Elding, A.Oskarsson
*Acta Chem.Scand.Ser.A,***33**, 703, 1979

METAL COMPLEXES (CARBOXYLIC ACID)

81.1   **Pentammine – carbamato – cobalt(iii) nitrate**
$CH_{17}CoN_6O_2{}^{2+}$, $2NO_3{}^-$
M.A.Bernard, M.M.Borel, A.Grandin, A.Leclaire
Rev.Chim.Miner.,**16**, 477, 1979
See also R1 : 83

81.2   **Copper formate tetradeuterohydrate (neutron study, at 296°K, paraelectric phase, deuterated form)**
$(C_2CuD_6O_6)_n$, $2nD_2O$
N.Burger, H.Fuess *Ferroelectrics*,**22**, 847, 1979

81.3   **Copper formate tetradeuterohydrate (neutron study, antiferroelectric phase, at 120°K, 80 per cent deuterated)**
$(C_2CuD_6O_6)_n$, $2nD_2O$
N.Burger, H.Fuess *Ferroelectrics*,**22**, 847, 1979

81.4   **Diformato – tetrachloro – di – rhenium diphenylformamide solvate**
$C_2H_2Cl_4O_4Re_2$, $2C_{13}H_{11}NO$
P.A.Koz'min, M.D.Surazhskaya, T.B.Larina,
A.S.Kotel'nikova, N.S.Osmanov *Koord.Khim.*,**5**, 1896, 1979

81.5   **Aquo – hydroxo – formato – dioxo – uranium(vi) (at 120°K)**
$(C_2H_8O_{12}U_2)_n$
S.D.le Roux, A.Van Tets, H.W.W.Adrian
Acta Crystallogr.,Sect.B,**35**, 3056, 1979

81.6   **Guanidinium tri(carbonato) – trifluoro – thorium(iv)**
$C_3F_3O_9Th^{5-}$, $5CH_6N_3{}^+$
S.Voliotis *Acta Crystallogr.,Sect.B*,**35**, 2899, 1979
See also R2 : 8

81.7   **Trichloro – tri(formato) – di – rhenium**
$C_3H_3Cl_3O_6Re_2$
P.A.Koz'min, M.D.Surazhskaya, T.B.Larina, Sh.A.Bagirov,
N.S.Osmanov, A.S.Kotel'nikova, T.V.Misailova
Koord.Khim.,**5**, 1576, 1979

81.8   **Disodium catena – bis(μ – oxalato) – copper(ii) dihydrate**
$C_4CuO_8{}^{2-}$, $2Na^+$, $2H_2O$
P.Chananont, P.E.Nixon, J.M.Waters, T.N.Waters
Acta Crystallogr.,Sect.B,**36**, 2145, 1980

81.9   **Disodium bis(oxalato) – copper(ii) dihydrate**
$C_4CuO_8{}^{2-}$, $2Na^+$, $2H_2O$
A.Gleizes, F.Maury, J.Galy *Inorg.Chem.*,**19**, 2074, 1980

81.10   **Potassium tetrakis(μ – formato) – bis(chloro – ruthenium)**
$C_4H_4Cl_2O_8Ru_2{}^-$, K^+
A.Bino, F.A.Cotton, T.R.Felthouse
Inorg.Chem.,**18**, 2599, 1979

81.11   **bis(Diformato – molybdenum) bis(aqua – diformato – molybdenum)**
$0.5C_4H_4Mo_2O_8$, $0.5C_4H_8Mo_2O_{10}$
P.A.Koz'min, M.D.Surazhskaya, T.B.Larina,
A.S.Kotel'nikova, E.L.Akhmedov
Zh.Neorg.Khim.,**24**, 3383, 1979

81.12   **bis(μ – Acetato) – bis(dichloro – rhenium)**
$C_4H_6Cl_4O_4Re_2$
P.A.Koz'min, M.D.Surazhskaya, T.B.Larina
Koord.Khim.,**5**, 1542, 1979

81.13   **Zinc(ii) acetate**
$(C_4H_6O_4Zn)_n$
A.V.Capilla, R.A.Aranda
Cryst.Struct.Commun.,**8**, 795, 1979

81.14   **Tetra – aqua – (2,2,3,3 – tetrafluorosuccinato) – zinc(ii)**
$(C_4H_8F_4O_8Zn)_n$
A.Karipides *Acta Crystallogr.,Sect.B*,**36**, 1659, 1980

81.15   **Lithium bis(dithioxalato) – nickel(ii) dihydrate**
$C_4NiO_4S_4{}^{2-}$, $2Li^+$, $2H_2O$
F.Maury, A.Gleizes *Inorg.Chim.Acta*,**41**, 185, 1980

81.16   **Potassium bis(oxalato) – platinum hydrate**
$0.4C_4O_8Pt^-$, $0.6C_4O_8Pt^{2-}$, $1.6K^+$, $1.2H_2O$
A.H.Reis Junior, S.W.Peterson
Ann.N.Y.Acad.Sci.,**313**, 560, 1978

81.17   **Rubidium bis(oxalato) – platinum(ii) bis(oxalato) – platinum(iii) hydrate**
$0.67C_4O_8Pt^{2-}$, $0.33C_4O_8Pt^-$, $1.67Rb^+$, $1.5H_2O$
A.Kobayashi, Y.Sasaki, H.Kobayashi
Bull.Chem.Soc.Jpn.,**52**, 3682, 1979

81.18   **tris(Aquo) – glutarato – cadmium dihydrate**
$(C_5H_{12}CdO_7)_n$, $2nH_2O$
A.S.Antsyshkina, M.A.Porai–Koshits, R.I.Kharitonova,
M.G.Pivovarova, V.N.Ostrikova *Koord.Khim.*,**5**, 1259, 1979

81.19   **Barium uranyl – dimalonate trihydrate**
$C_6H_4O_{10}U^{2-}$, Ba^{2+}, $3H_2O$
G.Bombieri, F.Benetollo, A.Del Pra, R.M.Rojas
Eur.Cryst.Meeting,**5**, 299, 1979

81.20   **Strontium uranyl – dimalonate trihydrate**
$C_6H_4O_{10}U^{2-}$, Sr^{2+}, $3H_2O$
G.Bombieri, F.Benetollo, A.Del Pra, R.M.Rojas
Eur.Cryst.Meeting,**5**, 299, 1979

81.21   **Sodium nitrilo – triacetato – zinc monohydrate**
$(C_6H_6NO_6Zn^-)_n$, nNa^+, nH_2O
J.D.Oliver, B.L.Barnett *Am.Cryst.Assoc.,Ser.2*,**8**, 38, 1980
See also R1 : 83

81.22   **Diaquo – adipato – manganese**
$C_6H_{12}MnO_6$
M.P.Gupta, P.Chand *Indian J.Phys.*,**53**, 473, 1979

81.23   **Aqua – (N,N – dimethylethylenediamine – N – oxide) – oxalato – copper(ii) dihydrate**
$C_6H_{14}CuN_2O_6$, $2H_2O$
A.Pajunen, M.Nasakkala
Acta Crystallogr.,Sect.B,**36**, 1650, 1980
See also R1 : 83

81.C   **bis(Ethylenediamine) – (thio – oxalato – O,S) – cobalt chloride monohydrate**
$C_6H_{16}CoN_4O_3S^+$, Cl^-, H_2O Main entry is 76.27

81.C **(−)$_{589}$ – Oxalato – bis(ethylenediamine) – cobalt(iii) hydrogen – d – tartrate dihydrate**
$C_6H_{16}CoN_4O_4{}^+$, $C_4H_5O_6{}^-$, $2H_2O$ Main entry is 76.31

81.24 **Tetra – aquo – (adipato) – cobalt**
$C_6H_{16}CoO_8$
M.P.Gupta, S.K.Sinha *Indian J.Phys.*,**53**, 471, 1979

81.25 **Tetra – aquo – (adipato) – nickel(ii)**
$(C_6H_{16}NiO_8)_n$
M.P.Gupta, B.N.Saha *Curr.Sci.*,**48**, 672, 1979

81.26 **Potassium tris(oxalato) – manganese(iii) trihydrate**
$C_6MnO_{12}{}^{3-}$, $3K^+$, $3H_2O$
T.Lis, J.Matuszewski
Acta Crystallogr.,Sect.B,**36**, 1938, 1980

81.27 **p – Aminobenzoato – silver**
$(C_7H_6AgNO_2)_n$
I.R.Amiraslanov, Kh.S.Mamedov, E.M.Movsumov, G.N.Nadzhafov, A.A.Mursaliev
Dokl.Akad.Nauk Az.SSR,**35**, 23, 1979
See also R1 : 83

81.28 **Tetraethylammonium dioxo – (μ – (+) – tartrato) – (μ – (−) – tartrato) – di – vanadium(iv) octahydrate**
$C_8H_4O_{14}V_2{}^{4-}$, $4C_8H_{20}N^+$, $8H_2O$
R.B.Ortega, C.F.Campana, R.E.Tapscott
Acta Crystallogr.,Sect.B,**36**, 1786, 1980
See also R2 : 3

81.29 **catena – (μ – Chloro) – (tetrakis(μ – acetato) – di – ruthenium) dihydrate**
$(C_8H_{12}ClO_8Ru_2)_n$, $2nH_2O$
A.Bino, F.A.Cotton, T.R.Felthouse
Inorg.Chem.,**18**, 2599, 1979

81.30 **Cesium tetrakis(μ – acetato) – bis(chloro – ruthenium)**
$C_8H_{12}Cl_2O_8Ru_2{}^-$, Cs^+
A.Bino, F.A.Cotton, T.R.Felthouse
Inorg.Chem.,**18**, 2599, 1979

81.31 **Tetra(hydroxyacetato) – scandium(iii) tetra – aquo – bis(hydroxyacetato) – scandium(iii)**
$C_8H_{12}O_{12}Sc^-$, $C_4H_{14}O_{10}Sc^+$
L.M.Dikareva, A.S.Antsyshkina, M.A.Porai–Koshits, V.N.Ostrikova, I.V.Arkhangel'skii, A.Z.Zamanov
Dokl.Akad.Nauk Az.SSR,**34**, 41, 1978

81.32 **tetrakis(μ – Acetato) – di(aquo – chromium) (at 120°K)**
$C_8H_{16}Cr_2O_{10}$
M.Benard, P.Coppens, M.L.DeLucia, E.D.Stevens
Inorg.Chem.,**19**, 1924, 1980

81.33 **tetrakis(μ – Acetato) – di(aquo – chromium) (at 90°K)**
$C_8H_{16}Cr_2O_{10}$
M.Benard, P.Coppens, M.L.DeLucia, E.D.Stevens
Inorg.Chem.,**19**, 1924, 1980

81.34 **Malonato – (2,2 – dimethyl – 1,3 – diaminopropane) – platinum(ii)**
$C_8H_{16}N_2O_4Pt$
C.G.van Kralingen, J.Reedijk, A.L.Spek
Inorg.Chem.,**19**, 1481, 1980
See also R1 : 83

81.35 **tetrakis(μ – Acetato) – bis(aquo – ruthenium) tetrafluoroborate**
$C_8H_{16}O_{10}Ru_2{}^+$, $BF_4{}^-$
A.Bino, F.A.Cotton, T.R.Felthouse
Inorg.Chem.,**18**, 2599, 1979

81.36 **Diaqua – tetrakis(glycollato) – uranium(iv)**
$C_8H_{16}O_{14}U$
N.W.Alcock, T.J.Kemp, S.Sostero, O.Traverso
J.Chem.Soc.,Dalton Trans.,1182, 1980
See also R1 : 84

81.C **(−)$_{589}$ – Amine – glycinato – (1,4,7 – triazacyclononane) – cobalt(iii) di – iodide hydrate (absolute configuration)**
$C_8H_{22}CoN_5O_2{}^{2+}$, $2I^-$, $0.84H_2O$ Main entry is 76.46

81.C **tetrakis(Acetoxy – mercury)methane dihydrate**
$C_9H_{12}Hg_4O_8$, $2H_2O$ Main entry is 71.12

81.C **Malonato – (2,2′,2″ – triaminotriethylamine) – cobalt(iii) chloride trihydrate**
$C_9H_{20}CoN_4O_4{}^+$, Cl^-, $3H_2O$ Main entry is 76.54

81.C **Penta – aquo – nickel(ii) – ((S,S) – ethylenediamine – N,N′ – disuccinato) – nickel(ii) dihydrate**
$C_{10}H_{22}N_2Ni_2O_{13}$, $2H_2O$ Main entry is 76.64

81.37 **(μ² – Hydrido) – (μ² – dithioformato) – decacarbonyl – tri – osmium (at −35°C)**
$C_{11}H_2O_{10}Os_3S_2$
R.D.Adams, J.P.Selegue
J.Organomet.Chem.,**195**, 223, 1980

81.38 **Sodium dioxo – bis(μ² – thio) – (μ – 1,2 – propylenediamine) – tetra – acetato – di – molybdenum trihydrate**
$C_{11}H_{14}Mo_2N_2O_{10}S_2{}^{2-}$, $2Na^+$, $3H_2O$
K.Z.Suzuki, Y.Sasaki, S.Ooi, K.Saito
Bull.Chem.Soc.Jpn.,**53**, 1288, 1980
See also R1 : 85,83

81.39 **Sodium dioxo – (μ² – oxo) – (μ² – thio) – 1,2 – propylenediamine – tetra – acetato – di – molybdenum tetrahydrate**
$C_{11}H_{14}Mo_2N_2O_{11}S^{2-}$, $2Na^+$, $4H_2O$
K.Z.Suzuki, Y.Sasaki, S.Ooi, K.Saito
Bull.Chem.Soc.Jpn.,**53**, 1288, 1980
See also R1 : 85,84,83

81.40 **Sodium dioxo – bis(μ² – oxo) – (μ – 1,2 – propylenediamine) – tetra – acetato – di – molybdenum trihydrate**
$C_{11}H_{14}Mo_2N_2O_{12}{}^{2-}$, $2Na^+$, $3H_2O$
K.Z.Suzuki, Y.Sasaki, S.Ooi, K.Saito
Bull.Chem.Soc.Jpn.,**53**, 1288, 1980
See also R1 : 84,83

81.41 **bis(Formato) – bis(pyridine) – copper monohydrate**
$C_{12}H_{12}CuN_2O_4$, H_2O
M.A.Bernard, M.M.Borel, F.Busnot, A.Leclaire
Rev.Chim.Miner.,**16**, 124, 1979
See also R1 : 83

81.42 **tetrakis(μ – Trifluoroacetato) – bis(ethanol – rhodium)**
$C_{12}H_{12}F_{12}O_{10}Rh_2$
M.A.Porai–Koshits, L.M.Dikareva, G.G.Sadikov, I.B.Baranovskii *Zh.Neorg.Khim.*,**24**, 1286, 1979

81.43 bis(Dimethylsulfoxide) – tetrakis(trifluoroacetato) – di – rhodium(ii)
$C_{12}H_{12}F_{12}O_{10}Rh_2S_2$
F.A.Cotton, T.R.Felthouse *Inorg.Chem.*,**19**, 2347, 1980
See also R1 : 85

81.44 tetrakis((μ^2 – Acetato) – (μ^2 – carbonyl)) – tetra – palladium acetic acid solvate (at –120°C)
$C_{12}H_{12}O_{12}Pd_4$, $2C_2H_4O_2$
L.G.Kuz'mina, Yu.T.Struchkov *Koord.Khim.*,**5**, 1558, 1979
See also R2 : 1

81.45 tetrakis(μ – Acetato) – (N – pyrazine) – chromium
$(C_{12}H_{16}Cr_2N_2O_8)_n$
F.A.Cotton, T.R.Felthouse *Inorg.Chem.*,**19**, 328, 1980
See also R1 : 83

81.46 Diaqua – bis(formato) – bis(pyridine) – copper
$C_{12}H_{16}CuN_2O_6$
M.A.Bernard, M.M.Borel, F.Busnot, A.Leclaire
Rev.Chim.Miner.,**16**, 24, 1979
See also R1 : 83

81.47 catena – (μ – Chloro) – (tetrakis(μ – propionato) – di – ruthenium)
$(C_{12}H_{20}ClO_8Ru_2)_n$
A.Bino, F.A.Cotton, T.R.Felthouse
Inorg.Chem.,**18**, 2599, 1979

81.48 tetrakis(μ – Acetato) – bis(dimethylsulfoxide – rhodium)
$C_{12}H_{24}O_{10}Rh_2S_2$
F.A.Cotton, T.R.Felthouse *Inorg.Chem.*,**19**, 323, 1980
See also R1 : 84

81.49 hexakis(μ – Acetato) – bis(μ_3 – oxo) – tris(aqua – tungsten(iv)) trifluoromethylsulfonate
$C_{12}H_{24}O_{17}W_3{}^{2+}$, $2CF_3O_3S^-$
A.Bino, K.–F.Hesse, H.Kuppers
Acta Crystallogr.,Sect.B,**36**, 723, 1980

81.50 bis(μ – Acetato) – bis(diaquo – diacetato – gadolinium(iii)) tetrahydrate
$C_{12}H_{26}Gd_2O_{16}$, $4H_2O$
M.C.Favas, D.L.Kepert, B.W.Skelton, A.H.White
J.Chem.Soc.,Dalton Trans.,454, 1980

81.51 bis(μ – Acetato) – bis(diaquo – diacetato – yttrium) tetrahydrate
$C_{12}H_{26}O_{16}Y_2$, $4H_2O$
A.S.Antsyshkina, M.A.Porai–Koshits, I.V.Arkhangel'skii, V.N.Ostrikova, A.Z.Amanov
Izv.Akad.Nauk Az.SSR,**34**, 22, 1978

81.C bis((bis(Ethylenediamine) – cobalt) – (μ – carboxylato – methyl) – thiolato) – silver perchlorate
$C_{12}H_{36}AgCo_2N_8O_4S_2{}^{3+}$, $3ClO_4^-$ Main entry is 76.80

81.52 Rubidium bis(pyridine – 2,6 – dicarboxylato) – chromium(iii)
$C_{14}H_6CrN_2O_8^-$, Rb^+
W.Furst, P.Gouzerh, Y.Jeannin
J.Coord.Chem.,**8**, 237, 1979
See also R1 : 83

81.53 Tetraethylammonium (μ – fluoro) – bis(oxo – peroxo – (pyridine – 2,6 – dicarboxylato) – molybdenum(vi))
$C_{14}H_6FMo_2N_2O_{14}^-$, $C_8H_{20}N^+$
A.J.Edwards, D.R.Slim, J.E.Guerchais, R.Kergoat
J.Chem.Soc.,Dalton Trans.,289, 1980
See also R1 : 83 R2 : 3

81.54 bis(p – Aminobenzoato) – zinc(ii) hydrate
$(C_{14}H_{12}N_2O_4Zn)_n$, $1.5nH_2O$
I.R.Amiraslanov, Kh.S.Mamedov, E.M.Movsumov, G.N.Nadzhafov, F.N.Musaev
Dokl.Akad.Nauk Az.SSR,**35**, 50, 1979
See also R1 : 83

81.55 Aquo – bis(p – aminobenzoato) – cadmium(ii) dihydrate
$(C_{14}H_{14}CdN_2O_5)_n$, $2nH_2O$
I.R.Amiraslanov, Kh.S.Mamedov, E.M.Movsumov, F.N.Musaev, A.I.Magerramov, G.N.Nadzhafov
Zh.Strukt.Khim.,**20**, 498, 1979
See also R1 : 83

81.56 bis(Formato) – pyridine – copper
$C_{14}H_{14}Cu_2N_2O_8$
M.A.Bernard, M.M.Borel, F.Busnot, A.Leclaire
Rev.Chim.Miner.,**16**, 124, 1979
See also R1 : 83

81.57 Potassium tetrakis(malonato) – bis(methanol) – di – manganese(iii)
$(C_{14}H_{16}Mn_2O_{18}{}^{2-})_n$, $2nK^+$
T.Lis, J.Matuszewski
J.Chem.Soc.,Dalton Trans.,996, 1980

81.58 Tetra – aqua – bis(p – aminobenzoato) – cobalt(iii)
$C_{14}H_{20}CoN_2O_8$
I.R.Amiraslanov, Kh.S.Mamedov, E.M.Movsumov, F.N.Musaev, G.N.Nadzhafov
Zh.Strukt.Khim.,**20**, 1075, 1979

81.C bis(Chloroacetato) – (N,N,N′,N′ – tetraethylethylenediamine) – copper(ii)
$C_{14}H_{28}Cl_2CuN_2O_4$ Main entry is 76.85

81.59 (+)$_{546}$ – Oxalato – ((2S,4S,9S,11S) – 4,9 – dimethyl – 5,8 – diazadodecane – 2,11 – diamine) – cobalt(iii) bromide trihydrate (absolute configuration)
$C_{14}H_{30}CoN_4O_4^+$, Br^-, $3H_2O$
S.Yano, S.Yaba, M.Ajioka, S.Yoshikawa
Inorg.Chem.,**18**, 2414, 1979
See also R1 : 83

81.60 Trifluoroacetato – hydrido – tetra(trimethylphosphite) – molybdenum
$C_{14}H_{37}F_3MoO_{14}P_4$
S.S.Wreford, J.K.Kouba, J.F.Kirner, E.L.Muetterties, I.Tavanaiepour, V.W.Day *J.Am.Chem.Soc.*,**102**, 1558, 1980
See also R1 : 86

81.61 Aqua – bis(methylamine) – bis(benzoato) – copper(ii) monohydrate
$C_{16}H_{22}CuN_2O_5$, H_2O
M.M.Borel, L.Boniak, F.Busnot, A.Leclaire
Z.Anorg.Allg.Chem.,**455**, 88, 1979
See also R1 : 83

81.62 tetrakis(μ – Acetato) – bis(tetrahydrothiophene – rhodium)
$C_{16}H_{28}O_8Rh_2S_2$
F.A.Cotton, T.R.Felthouse *Inorg.Chem.*,**19**, 323, 1980
See also R1 : 85

81.63 bis(Dimethylsulfoxide) – tetrakis(propionato) – di – rhodium(ii)
$C_{16}H_{32}O_{10}Rh_2S_2$
F.A.Cotton, T.R.Felthouse *Inorg.Chem.*,**19**, 2347, 1980
See also R1 : 85

81.C bis(μ – Acetato) – bis(di(urea) – di(acetato) – praseodymium(iii)) urea dihydrate
$C_{16}H_{34}N_8O_{16}Pr_2$, $2CH_4N_2O$, $2H_2O$ Main entry is 79.14

81.64 (μ^2 – Hydrido) – (μ^2 – dithioformato) – bis(tricarbonyl – osmium) – tricarbonyl – (dimethylphenylphosphine) – osmium (at –35°C)
$C_{18}H_{13}O_9Os_3PS_2$
R.D.Adams, J.P.Selegue
J.Organomet.Chem.,**195**, 223, 1980
See also R1 : 86

81.65 tetrakis(μ – Acetato) – bis(N – pyridine – chromium)
$C_{18}H_{22}Cr_2N_2O_8$
F.A.Cotton, T.R.Felthouse *Inorg.Chem.*,**19**, 328, 1980
See also R1 : 83

81.C bis(μ – Acetato) – bis((trimethylphosphine) – (trimethylsilylmethyl) – molybdenum(ii))
$C_{18}H_{46}Mo_2O_4P_2Si_2$ Main entry is 71.132

81.66 bis(Tri – isopropylphosphine) – dihydrido – bicarbonato – rhodium(iii) (at –160°C)
$C_{19}H_{45}O_3P_2Rh$
T.Yoshida, D.L.Thorn, T.Okano, J.A.Ibers, S.Otsuka
J.Am.Chem.Soc.,**101**, 4212, 1979
See also R1 : 86

81.67 bis(μ – Acetato) – bis(chloro – (dimethylphenylphosphine) – palladium(ii)) chloroform solvate
$C_{20}H_{28}Cl_2O_4P_2Pd_2$, $0.5CHCl_3$
W.Wong–Ng, P.–T.Cheng, V.Kocman, H.Luth, S.C.Nyburg
Inorg.Chem.,**18**, 2620, 1979
See also R1 : 86

81.68 bis(1,3 – Propanediamine) – bis(m – nitrobenzoato) – copper(ii)
$C_{20}H_{28}CuN_6O_8$
M.Klinga *Finn.Chem.Lett.*,223, 1979
See also R1 : 83

81.69 bis((m – Nitrobenzoato) – (1,3 – propanediamine)) – nickel(ii)
$C_{20}H_{28}N_6NiO_8$
M.Klinga *Cryst.Struct.Commun.*,**9**, 567, 1980
See also R1 : 83

81.70 tetrakis(μ – Pivalato) – bis(aquo – rhodium)
$C_{20}H_{40}O_{10}Rh_2$
F.A.Cotton, T.R.Felthouse *Inorg.Chem.*,**19**, 323, 1980

81.71 Acetato – bis(2,2′ – bipyridyl) – copper(ii) tetrafluoroborate
$C_{22}H_{19}CuN_4O_2{}^+$, $BF_4{}^-$
B.J.Hathaway, N.Ray, D.Kennedy, N.O'Brien, B.Murphy
Acta Crystallogr.,Sect.B,**36**, 1371, 1980
See also R1 : 83

81.72 Acetato – bis(2,2′ – bipyridyl) – copper(ii) perchlorate monohydrate
$C_{22}H_{19}CuN_4O_2{}^+$, $ClO_4{}^-$, H_2O
B.J.Hathaway, N.Ray, D.Kennedy, N.O'Brien, B.Murphy
Acta Crystallogr.,Sect.B,**36**, 1371, 1980
See also R1 : 83

81.73 tetrakis(μ – Acetato) – bis(theophylline – rhodium) dihydrate
$C_{22}H_{28}N_8O_{12}Rh_2$. $2H_2O$
K.Aoki, H.Yamazaki
J.Chem.Soc.,Chem.Commun.,186, 1980
See also R1 : 83

81.74 tetrakis(μ_2 – Acetato) – bis(benzylthiolato – rhodium)
$C_{22}H_{28}O_8Rh_2S_2$
G.G.Christoph, M.Tolbert
Am.Cryst.Assoc.,Ser.2,**7**, 39, 1980
See also R1 : 85

81.C rac – a – (μ – Acetato) – b – (O – acetato – mercurio) – cf,de – bis(2 – (dimethylaminomethyl) – phenyl) – platinum
$C_{22}H_{30}HgN_2O_4Pt$ Main entry is 71.183

81.75 bis((m – Methylbenzoato) – (1,3 – propanediamine)) nickel(ii)
$C_{22}H_{34}N_4NiO_4$
M.Klinga *Cryst.Struct.Commun.*,**9**, 439, 1980
See also R1 : 83

81.76 bis((p – Methylbenzoato) – (1,3 – propanediamine)) – nickel(ii)
$C_{22}H_{34}N_4NiO_4$
M.Klinga *Cryst.Struct.Commun.*,**9**, 457, 1980
See also R1 : 83

81.77 tetrakis(Pyridine) – bis(trifluoroacetato) – copper(ii)
$C_{24}H_{20}CuF_6N_4O_4$
J.Pradillas, H.W.Chen, F.W.Koknat, J.P.Fackler Junior
Inorg.Chem.,**18**, 3519, 1979
See also R1 : 83

81.78 tetrakis(μ – Acetato) – bis(caffeine – rhodium)
$C_{24}H_{32}N_8O_{12}Rh_2$
K.Aoki, H.Yamazaki
J.Chem.Soc.,Chem.Commun.,186, 1980
See also R1 : 83

81.79 tetrakis(μ – Trichloroacetato) – tetrakis(μ – t – butylperoxy) – tetra – palladium
$C_{24}H_{36}Cl_{12}O_{16}Pd_4$
H.Mimoun, R.Charpentier, A.Mitschler, J.Fischer, R.Weiss
J.Am.Chem.Soc.,**102**, 1047, 1980
See also R1 : 84

81.80 Diaquo – bis(N,N – diethylnicotinamide) – diacetato – nickel(ii)
$C_{24}H_{38}N_4NiO_8$
G.V.Tsintsadze, Z.V.Mikelashvili, B.T.Ibragimov, T.I.Tsivtsivadze
Soobshch.Akad.Nauk Gruzh.SSR,**93**, 597, 1979
See also R1 : 83

81.C bis(Ethylenediamine) – copper (bis(ethylenediamine) – copper) – (ethylenediamine – tetra – acetato – nickel) dihydrate
$C_{24}H_{40}CuN_8Ni_2O_{16}{}^{2-}$, $C_4H_{16}CuN_4{}^{2+}$, $2H_2O$ Main entry is 76.18

81.81 tetrakis(Dichloroacetato – (μ – 2 – dimethylaminoethanolato) – copper(ii))
$C_{24}H_{44}Cl_8Cu_4N_4O_{12}$
U.Turpeinen, R.Hamalainen, M.Ahlgren
Acta Crystallogr.,Sect.B,**36**, 927, 1980
See also R1 : 84,83

81.82 tetrakis(Chloroacetato – (μ – 2 – dimethylaminoethanolato) – copper(ii))
$C_{24}H_{48}Cl_4Cu_4N_4O_{12}$
U.Turpeinen, M.Ahlgren, R.Hamalainen
Acta Chem.Scand.Ser.A,**33**, 593, 1979
See also R1 : 84,83

81.83 bis(μ – Acetato) – bis((4 – phenylimino – pentan – 2 – onato) – molybdenum)
$C_{26}H_{30}Mo_2N_2O_6$
F.A.Cotton, W.H.Ilsley, W.Kaim
Inorg.Chim.Acta,**37**, 267, 1979
See also R1 : 83

81.84 hexakis(μ² – Acetato) – (μ³ – oxo) – tris(3 – chloropyridine – manganese)
$C_{27}H_{30}Cl_3Mn_3N_3O_{13}$
A.R.E.Baikie, M.B.Hursthouse, L.New, P.Thornton, R.G.White *J.Chem.Soc.,Chem.Commun.*,684, 1980

81.85 bis(Diphenylphosphinoacetato) – palladium(ii)
$C_{28}H_{24}O_4P_2Pd$
S.Civis, J.Podlahova, J.Loub, J.Jecny
Acta Crystallogr.,Sect.B,**36**, 1395, 1980
See also R1 : 86

81.86 bis(μ – Acetato) – bis((2 – (2′ – benzothiazolyl) – 5 – methyl – phenyl) – palladium(ii))
$C_{32}H_{28}N_2O_4Pd_2S_2$
M.R.Churchill, H.J.Wasserman, G.J.Young
Inorg.Chem.,**19**, 762, 1980
See also R1 : 83

81.C bis(μ – Acetato) – bis((2 – (2′ – benzoxazolyl) – 5 – methyl – phenyl) – palladium(ii))
$C_{32}H_{28}N_2O_6Pd_2$ Main entry is 71.236

81.87 (2 – Diethylaminoethanolato) – (trichloroacetato) – copper(ii) tetramer
$C_{32}H_{56}Cl_{12}Cu_4N_4O_{12}$
M.Ahlgren, R.Hamalainen, U.Turpeinen, K.Smolander
Acta Crystallogr.,Sect.B,**35**, 2870, 1979
See also R1 : 84,83

81.88 Tetra – aqua – (μ³ – dodeca – oxo) – (μ² – hexadeca – acetato) – dodeca – manganese acetic acid solvate tetrahydrate
$C_{32}H_{56}Mn_{12}O_{48}$, $2C_2H_4O_2$, $4H_2O$
T.Lis *Acta Crystallogr.,Sect.B*,**36**, 2042, 1980
See also R1 : 84

81.89 tetrakis(μ – 2 – Diethylaminoethanolato) – tetrakis(chloroacetato – copper(ii))
$C_{32}H_{64}Cl_4Cu_4N_4O_{12}$
U.Turpeinen, R.Hamalainen, M.Ahlgren, K.Smolander
Finn.Chem.Lett.,108, 1979
See also R1 : 84,83

81.90 bis(2,2′ – Bipyridyl) – bis(pyridine – 2,6 – dicarboxylato) – di – copper(ii) tetrahydrate
$C_{34}H_{22}Cu_2N_6O_8$, $4H_2O$
G.Nardin, L.Randaccio, R.P.Bonomo, E.Rizzarelli
J.Chem.Soc.,Dalton Trans.,369, 1980
See also R1 : 83

81.91 (μ₃ – Chloro) – tris(μ – oxo) – (μ – acetato) – (tetrachloro – tris(triethylphosphine) – tri – tungsten)
$C_{38}H_{84}Cl_5O_5P_3W_3$
F.A.Cotton, T.R.Felthouse, D.G.Lay
J.Am.Chem.Soc.,**102**, 1431, 1980
See also R1 : 86

81.92 tetrakis(μ – 2,4 – Dichlorophenoxyacetato) – bis(dioxane – copper(ii)) dioxane solvate
$C_{40}H_{36}Cl_8Cu_2O_{16}$, $3C_4H_8O_2$
G.Reck, W.Jahnig *J.Prakt.Chem.*,**321**, 549, 1979
See also R1 : 84

81.93 Diacetato – bis(tricyclohexylphosphine) – mercury(ii) dihydrate
$C_{40}H_{72}HgO_4P_2$, $2H_2O$
E.C.Alyea, S.A.Dias, G.Ferguson, M.A.Khan
J.Chem.Res.,**360**, 4101, 1979
See also R1 : 86

81.94 bis(μ – Acetato) – bis(acetato – (tricyclohexylphosphine) – mercury(ii))
$C_{44}H_{78}Hg_2O_8P_2$
E.C.Alyea, S.A.Dias, G.Ferguson, M.A.Khan, P.J.Roberts
Inorg.Chem.,**18**, 2433, 1979
See also R1 : 86

81.95 Thallium(i) (μ₈ – chloro) – dodecakis(α – mercapto – isobutyrato) – octa – copper(i) – hexa – copper(ii) dodecahydrate
$C_{48}H_{72}ClCu_{14}O_{24}S_{12}^{5-}$, $5Tl^+$, $12H_2O$
P.J.M.W.L.Birker *Inorg.Chem.*,**18**, 3502, 1979
See also R1 : 85

81.96 bis(μ – Acetato) – bis(acetato – (tri – o – tolylphosphine) – mercury(ii))
$C_{50}H_{54}Hg_2O_8P_2$
E.C.Alyea, S.A.Dias, G.Ferguson, M.A.Khan, P.J.Roberts
Inorg.Chem.,**18**, 2433, 1979
See also R1 : 86

81.C Trifluoroacetato – (1,4 – diphenyl – but – 1 – en – 3 – yn – 2 – yl) – carbonyl – bis(triphenylphosphine) – ruthenium
$C_{55}H_{41}F_3O_3P_2Ru$ Main entry is 71.278

81.97 bis((Propylthio) – acetato) – cobalt(ii) hexamer p – xylene solvate
$C_{60}H_{108}Co_6O_{24}S_{12}$, C_8H_{10}
M.Shimoi, F.Ebina, A.Ouchi, Y.Yoshino, T.Takeuchi
J.Chem.Soc.,Chem.Commun.,1132, 1979

81.98 tetrakis(μ – o – Benzoylbenzoato) – bis((p – iodoanilino) – copper(ii))
$C_{68}H_{48}Cu_2I_2N_2O_{12}$
M.K.Gusejnova, Kh.S.Mamedov, F.N.Musaev, I.R.Amiraslanov *Zh.Strukt.Khim.*,**20**, 89, 1979

81.99 tetrakis(Tri(N,N – di – isopropyl – carbamato) – ytterbium(iii)) heptane solvate
$C_{84}H_{168}N_{12}O_{24}Yb_4$, $2C_7H_{16}$
D.B.Dell'Amico, F.Calderazzo, F.Marchetti, G.Perego
J.Chem.Soc.,Chem.Commun.,1103, 1979

METAL COMPLEXES (AMINO–ACID)

82.1 cis – Ammine – chloro – glycinato – platinum(ii)
$C_2H_7ClN_2O_2Pt$
I.A.Baidina, N.V.Podberezskaya, L.F.Krylova, S.V.Borisov,
V.V.Bakakin *Zh.Strukt.Khim.*,**21**, 106, 1980

82.2 cis – Ammine – dichloro – glycinato – platinum(ii)
$C_2H_8Cl_2N_2O_2Pt$
I.A.Baidina, N.V.Podberezskaya, L.F.Krylova, S.V.Borisov,
V.V.Bakakin *Zh.Strukt.Khim.*,**21**, 106, 1980

82.3 catena – Diaqua – dichloro – (μ – glycine) –
manganese(ii)
$(C_2H_9Cl_2MnNO_4)_n$
Z.Ciunik, T.Glowiak
Acta Crystallogr.,Sect.B,**36**, 1212, 1980

82.4 Imino – diacetato – dioxo – uranium(vi)
$(C_4H_5NO_6U)_n$
G.A.Battiston, G.Sbrignadello, G.Bandoli, D.A.Clemente,
G.Tomat *J.Chem.Soc.,Dalton Trans.*,1965, 1979

82.5 trans – Chloro – glycino – glycinato – palladium(ii)
$C_4H_7ClN_2O_4Pd$
I.A.Baidina, N.V.Podberezskaya, S.V.Borisov,
E.V.Golubovskaya *Zh.Strukt.Khim.*,**21**, 188, 1980

82.6 cis – bis(Glycinato) – palladium(ii) trihydrate
$C_4H_8N_2O_4Pd, 3H_2O$
I.A.Baidina, N.V.Podberezskaya, V.V.Bakakin,
E.V.Golubovskaya, N.A.Shestakova, G.D.Mal'chikov
Zh.Strukt.Khim.,**20**, 544, 1979

82.7 cis – bis(Glycinato) – platinum(ii) oxalic acid
dihydrate
$C_4H_8N_2O_4Pt, C_2H_2O_4, 2H_2O$
M.A.A.F.de C.T.Carrondo, D.M.L.Goodgame, C.R.Hadjioannou,
A.C.Skapski *Inorg.Chim.Acta*,**46**, 32, 1980

82.8 Dichloro – ((S – methyl – L – cysteine) –
sulfoxide) – palladium(ii) monohydrate
$C_4H_9Cl_2NO_3PdS, H_2O$
A.Allain, M.Kubiak, B.Jezowska–Trzebiatowska,
H.Kozlowski, T.Glowiak *Inorg.Chim.Acta*,**46**, 127, 1980

82.9 cis – Dibromo – bis(glycine) – platinum(ii)
monohydrate
$C_4H_{10}Br_2N_2O_4Pt, H_2O$
I.A.Baidina, N.V.Podberezskaya, S.V.Borisov,
N.A.Shestakova, Y.F.Kuklina, G.D.Mal'chikov
Zh.Strukt.Khim.,**20**, 548, 1979

82.10 cis – Dichloro – bis(glycine) – platinum(ii)
monohydrate
$C_4H_{10}Cl_2N_2O_4Pt, H_2O$
I.A.Baidina, N.V.Podberezskaya, S.V.Borisov,
N.A.Shestakova, Y.F.Kuklina, G.D.Mal'chikov
Zh.Strukt.Khim.,**20**, 548, 1979

82.11 Potassium bromo – (2 – hydroxy – ethylimino –
diacetato) – palladium(ii) monohydrate
$C_6H_9BrNO_5Pd^-$, K^+, H_2O
I.N.Polyakova, T.N.Polynova, M.A.Porai–Koshits,
A.M.Grevtsev *Zh.Strukt.Khim.*,**20**, 555, 1979

82.C (+) – trans – (0) – Ethylenediamine –
bis(glycinato) – cobalt(iii) hydrogen – d – tartrate
trihydrate
$C_6H_{16}CoN_4O_4^+$, $C_4H_5O_6^-$, $3H_2O$ Main entry is 76.29

82.C (–) – trans – (0) – Ethylenediamine –
bis(glycinato) – cobalt(iii) hydrogen – d – tartrate
monohydrate
$C_6H_{16}CoN_4O_4^+$, $C_4H_5O_6^-$, H_2O Main entry is 76.30

82.12 bis(D,L – α – Alanine) – diaquo – manganese(ii)
dibromide dihydrate
$(C_6H_{16}MnN_2O_6^{2+})_n$, $2nBr^-$, $2nH_2O$
Z.Ciunik, T.Glowiak *Inorg.Chim.Acta*,**44**, L249, 1980

82.13 cis – Dichloropyridine – glycine – platinum(ii)
$C_7H_{10}Cl_2N_2O_2Pt$
I.A.Baidina, N.V.Podberezskaya, L.F.Krylova, S.V.Borisov,
V.V.Bakakin *Zh.Strukt.Khim.*,**21**, 168, 1980
See also R1 : 83

82.14 Glycyl – L – methioninato – copper(ii)
$C_7H_{12}CuN_2O_3S$
L.Abello, A.Ensuque, A.Demaret, G.Lapluye
Transition Met.Chem.,**5**, 120, 1980

82.15 Sodium bis(imino – diacetato) – cobalt(ii)
heptahydrate
$C_8H_{10}CoN_2O_8^{2-}$, $2Na^+$, $7H_2O$
Ya.M.Nesterova, T.N.Polynova, M.A.Porai–Koshits,
F.G.Kramarenko, N.M.Muratova
Zh.Strukt.Khim.,**20**, 960, 1979

82.16 Potassium cis – bis(imino – diacetato) –
chromium(iii) trihydrate
$C_8H_{10}CrN_2O_8^-$, K^+, $3H_2O$
D.Mootz, H.Wunderlich
Acta Crystallogr.,Sect.B,**36**, 445, 1980

82.17 Aqua – ((RS) – N,N' – ethylene – bis(serinato)) –
copper(ii)
$C_8H_{16}CuN_2O_7$
F.Pavelcik, J.Majer
Acta Crystallogr.,Sect.B,**36**, 1645, 1980

82.C Calcium ethylenediamine – tetra – acetato –
copper(ii) tetrahydrate
$C_{10}H_{12}CuN_2O_8^{2-}$, Ca^{2+}, $4H_2O$ Main entry is 76.55

82.C Potassium ethylenediamine – tetra – acetato –
manganese(iii) dihydrate
$C_{10}H_{12}MnN_2O_8^-$, K^+, $2H_2O$ Main entry is 76.56

82.C Calcium (ethylenediamine – tetra – acetato) –
nickel(ii) tetrahydrate
$C_{10}H_{12}N_2NiO_8^{2-}$, Ca^{2+}, $4H_2O$ Main entry is 76.57

82.C Potassium (+) – (chloro – (hydrogen –
ethylenediamine – tetra – acetato) – iridium)
monohydrate
$C_{10}H_{13}ClIrN_2O_8^-$, K^+, H_2O Main entry is 76.58

82.C Potassium (dichloro – (dihydrogen –
ethylenediamine – tetra – acetato) – iridium)
$C_{10}H_{14}Cl_2IrN_2O_8^-$, K^+ Main entry is 76.59

82.C   **Ethanediammonium bis((hydrogen ethylenediamine – tetra – acetato) – vanadium) dihydrate**
$2C_{10}H_{14}N_2O_8V^-$, $C_2H_{10}N_2^{2+}$, $2H_2O$ Main entry is 76.60

82.C   **Aqua – (ethylenediamine – tetra – acetato) – osmium(iv) monohydrate**
$C_{10}H_{14}N_2O_9Os$, H_2O Main entry is 76.61

82.18   **L – Asparaginato – L – histidinato – copper(ii)**
$C_{10}H_{15}CuN_5O_5$
T.Ono, H.Shimanouchi, Y.Sasada, T.Sakurai, O.Yamauchi, A.Nakahara *Bull.Chem.Soc.Jpn.*,**52**, 2229, 1979

82.19   **L – Asparaginato – L – histidinato – copper(ii) tetrahydrate**
$C_{10}H_{15}CuN_5O_5$, $4H_2O$
T.Ono, H.Shimanouchi, Y.Sasada, T.Sakurai, O.Yamauchi, A.Nakahara *Bull.Chem.Soc.Jpn.*,**52**, 2229, 1979

82.20   **Sodium aquo – bis(glutamato) – cadmium**
$C_{10}H_{16}CdN_2O_9^{2-}$, $2Na^+$
H.Soylu *Eur.Cryst.Meeting*,**5**, 313, 1979

82.21   **trans – N,N′ – Ethylene – bis(S – methyl – L – cysteinato) – cobalt(iii) perchlorate**
$C_{10}H_{18}CoN_2O_4S_2^+$, ClO_4^-
T.Isago, K.–I.Okamoto, M.Ohmasa, J.Hidaka *Chem.Lett.*,319, 1980
See also R1 : 85

82.C   **Penta – aquo – nickel(ii) – ((S,S) – ethylenediamine – N,N′ – disuccinato) – nickel(ii) dihydrate**
$C_{10}H_{22}N_2Ni_2O_{13}$, $2H_2O$ Main entry is 76.64

82.22   **(Glycylglycinato) – (7,9 – dimethylhypoxanthine) – copper(ii) tetrahydrate (absolute configuration)**
$C_{11}H_{14}CuN_6O_4$, $4H_2O$
L.G.Marzilli, K.Wilkowski, C.C.Chiang, T.J.Kistenmacher *J.Am.Chem.Soc.*,**101**, 7504, 1979
See also R1 : 83

82.23   **Sodium (μ – oxo) – bis(dioxo – (hydrogen – nitrilo – triacetato) – molybdenum(vi)) octahydrate**
$C_{12}H_{14}Mo_2N_2O_{17}^{2-}$, $2Na^+$, $8H_2O$
C.Knobler, B.R.Penfold, W.T.Robinson, C.J.Wilkins, S.H.Yong *J.Chem.Soc.,Dalton Trans.*,248, 1980

82.C   **(N – Salicylidene – L – valinato) – aqua – copper(ii)**
$C_{12}H_{15}CuNO_4$ Main entry is 78.1

82.24   **bis(L – Histidinato) – cobalt(iii) perchlorate dihydrate (orange form)**
$C_{12}H_{16}CoN_6O_4^+$, ClO_4^-, $2H_2O$
N.Thorup *Acta Chem.Scand.Ser.A*,**33**, 759, 1979

82.25   **Triaquo – (pyridoxylidene – O – phospho – D,L – threoninato) – nickel(ii) dihydrate**
$C_{12}H_{20}N_2NiO_{11}P$, $2H_2O$
K.Aoki, H.Yamazaki *J.Chem.Soc.,Chem.Commun.*,363, 1980
See also R1 : 84

82.C   **alpHa – cis – bis(μ – hydroxo) – bis(ethylenediamine – N,N′ – diacetato – chromium(iii)) tetrahydrate**
$C_{12}H_{22}Cr_2N_4O_{10}$, $4H_2O$ Main entry is 76.72

82.26   **tetrakis(μ – Bη – alanine) – bis(aquo – rhodium(ii)) tetraperchlorate dihydrate**
$C_{12}H_{32}N_4O_{10}Rh_2^{4+}$, $4ClO_4^-$, $2H_2O$
A.M.Dennis, R.A.Howard, J.L.Bear, J.D.Korp, I.Bernal *Inorg.Chim.Acta*,**37**, L561, 1979

82.27   **(2,2′ – Bipyridyl) – (imino – diacetato) – copper(ii) hexahydrate**
$C_{14}H_{13}CuN_3O_4$, $6H_2O$
G.Nardin, L.Randaccio, R.P.Bonomo, E.Rizzarelli *J.Chem.Soc.,Dalton Trans.*,369, 1980
See also R1 : 83

82.28   **Chloro – glycylglycinato – imidazole – cadmium**
$C_{14}H_{22}Cd_2Cl_2N_6O_6$
C.I.H.Ashby, W.F.Paton, T.L.Brown *J.Am.Chem.Soc.*,**102**, 2990, 1980
See also R1 : 83

82.29   **Sodium trans – bis(N – isopropylimino – diacetato) – chromium(iii) dihydrate**
$C_{14}H_{22}CrN_2O_8^-$, Na^+, $2H_2O$
D.Mootz, H.Wunderlich *Acta Crystallogr.,Sect.B*,**36**, 721, 1980

82.30   **bis(D – β – (2 – Pyridyl) – α – alaninato) – nickel(ii) dihydrate**
$C_{16}H_{18}N_4NiO_4$, $2H_2O$
S.R.Ebner, B.J.Helland, R.A.Jacobson, R.J.Angelici *Inorg.Chem.*,**19**, 175, 1980
See also R1 : 83

82.31   **Potassium trans – bis(t – butylimino – diacetato) – chromium(iii) tetrahydrate**
$C_{16}H_{26}CrN_2O_8^-$, K^+, $4H_2O$
D.Mootz, H.Wunderlich *Acta Crystallogr.,Sect.B*,**36**, 1189, 1980

82.32   **tetrakis(μ – N – Acetylglycinato) – bis(aquo – copper(ii))**
$C_{16}H_{26}Cu_2N_4O_{14}$
M.R.Udupa, B.Krebs *Inorg.Chim.Acta*,**37**, 1, 1979

82.33   **tetrakis(Glycylglycine) – di – molybdenum(ii) tetrachloride hexahydrate**
$C_{16}H_{32}Mo_2N_8O_{12}^{4+}$, $4Cl^-$, $6H_2O$
A.Bino, F.A.Cotton *J.Am.Chem.Soc.*,**102**, 3014, 1980

82.34   **(+)$_{589}$ – β_2 – ((R) – Alaninato) – (1,7 – bis(2 – (S) – pyrrolidyl) – 2,6 – diazaheptane) – cobalt(iii) bis(perchlorate) dihydrate**
$C_{16}H_{34}CoN_5O_2^{2+}$, $2ClO_4^-$, $2H_2O$
M.Yamaguchi, S.Yano, M.Saburi, S.Yoshikawa *Inorg.Chem.*,**19**, 2016, 1980
See also R1 : 83

82.35   **bis(L – Tyrosinato) – palladium(ii)**
$C_{18}H_{20}N_2O_6Pd$
M.Sabat, M.Jezowska, H.Kozlowski *Inorg.Chim.Acta*,**37**, L511, 1979

82.36   **Triaquo – bis(N – benzoylglycinato) – cobalt(ii) dihydrate**
Triaquo – bis(hippurato) – cobalt(ii) dihydrate
$(C_{18}H_{22}CoN_2O_9)_n$, $2nH_2O$
M.M.Morelock, M.L.Good, L.M.Trefonas, D.Karraker, L.Maleki, H.R.Eichelberger, R.Majeste, J.Dodge *J.Am.Chem.Soc.*,**101**, 4858, 1979

82.37 **Triaquo – bis(N – benzoylglycinato) – nickel(ii) dihydrate**
Triaquo – bis(hippurato) – nickel(ii) dihydrate
$(C_{18}H_{22}N_2NiO_9)_n$, $2nH_2O$
M.M.Morelock, M.L.Good, L.M.Trefonas, D.Karraker,
L.Maleki, H.R.Eichelberger, R.Majeste, J.Dodge
J.Am.Chem.Soc.,**101**, 4858, 1979

82.38 **(N – (2 – Pyridylmethyl) – L – aspartato) – (L – phenylalaninato) – cobalt(iii) trihydrate**
$C_{19}H_{20}CoN_3O_6$, $3H_2O$
L.A.Meiske, R.A.Jacobson, R.J.Angelici
Inorg.Chem.,**19**, 2028, 1980

82.39 **Aquo – (pyridoxylidene – 0 – phospho – D,L – threoninato) – copper(ii) dimer monohydrate**
$C_{24}H_{34}Cu_2N_4O_{16}P_2$, $2H_2O$
K.Aoki, H.Yamazaki
J.Chem.Soc.,Chem.Commun.,363, 1980
See also R1 : 84

82.40 **tetrakis(L – Leucine) – di – molybdenum(ii) dichloride bis(p – toluenesulfonate) dihydrate**
$C_{24}H_{48}Mo_2N_4O_8{}^{4+}$, $2Cl^-$, $2C_7H_7O_3S^-$, $2H_2O$
A.Bino, F.A.Cotton, P.E.Fanwick
Inorg.Chem.,**19**, 1215, 1980

82.41 **Diaqua – bis(N – acetyl – D,L – tryptophanato) – bis(pyridine) – copper(ii)**
$C_{36}H_{40}CuN_6O_8$
L.P.Battaglia, A.B.Corradi, G.Marcotrigiano, L.Menabue,
G.C.Pellacani *J.Am.Chem.Soc.*,**102**, 2663, 1980
See also R1 : 83

METAL COMPLEXES (NITROGEN LIGAND)

83.C **Pentammine – carbamato – cobalt(iii) nitrate**
$CH_{17}CoN_6O_2{}^{2+}$, $2NO_3^-$ Main entry is 81.1

83.1 **Tetraphenylarsonium pentachloro – N – pentachloroethyl – nitrido – tungsten**
$C_2Cl_{10}NW^-$, $C_{24}H_{20}As^+$
U.Weiher, K.Dehnicke, D.Fenske
Z.Anorg.Allg.Chem.,**457**, 105, 1979
See also R2 : 65

83.2 **Tetrachloro – (pentachloroethyl – nitrido) – (phosphoryl – trichloride) – oxo – molybdenum**
$C_2Cl_{12}MoNOP$
K.Dehnicke, U.Weiher, D.Fenske
Z.Anorg.Allg.Chem.,**456**, 71, 1979

83.3 **Tetrachloro – (trichlorophosphine – oxide) – N – pentachloroethyl – nitrido – rhenium**
$C_2Cl_{12}NOPRe$
U.Weiher, K.Dehnicke, D.Fenske
Z.Anorg.Allg.Chem.,**457**, 115, 1979

83.4 **bis((Methylcyano) – (μ_2 – chloro) – (μ_2 – 1,3 – diaza – 2,4 – dithietidine) – chloro – copper)**
$(C_2H_3Cl_2CuN_2S)_n$
U.Thewalt, B.Muller *Z.Anorg.Allg.Chem.*,**462**, 214, 1980

83.5 **bis(Tetraphenylarsonium) tetrachloro – oxo – methylnitrile – molybdenum tetrachloro – oxo – molybdenum**
$C_2H_3Cl_4MoNO^-$, $2C_{24}H_{20}As^+$, Cl_4MoO^-
F.Weller, U.Muller, U.Weiher, K.Dehnicke
Z.Anorg.Allg.Chem.,**460**, 191, 1980
See also R2 : 65

83.6 **5 – Amidinotetrazolato – tetra – ammine – cobalt(iii) bromide**
$C_2H_{15}CoN_{10}$, Br_2
E.J.Graeber, B.Morosin *Am.Cryst.Assoc.,Ser.2*,**7**, 22, 1980

83.7 **Penta – ammine – 5 – methyltetrazolato – cobalt(iii) perchlorate**
$C_2H_{18}CoN_9{}^{2+}$, $2ClO_4^-$
R.Ortega, C.F.Campana, B.Morosin
Am.Cryst.Assoc.,Ser.2,**8**, 24, 1980

83.8 **bis(Oxamide – oximato) – platinum(ii) hydrogen chloride (form ii)**
$C_4H_{10}N_8O_4Pt$, $2HCl$
H.Endres, L.Schlicksupp
Acta Crystallogr.,Sect.B,**36**, 715, 1980

83.9 **bis(Oxamide – oximato) – platinum(ii) chloride**
$C_4H_{10}N_8O_4Pt$, $2HCl$
H.Endres, L.Schlicksupp
Acta Crystallogr.,Sect.B,**35**, 3035, 1979

83.10 **bis(Oxamide – oximato) – platinum(ii) ammonium chloride**
$C_4H_{10}N_8O_4Pt$, H_4N^+, Cl^-
H.Endres *Acta Crystallogr.,Sect.B*,**35**, 3032, 1979

83.11 bis(Oxamide – oximato) – platinum(ii) sodium acetate sodium chloride dihydrate
$C_4H_{10}N_8O_4Pt$, $0.666Na^+$, $0.333C_2H_3O_2^-$, $0.333Cl^-$, $2H_2O$
H.Endres *Acta Crystallogr.,Sect.B*,**36**, 57, 1980

83.12 bis((Oxamide – oximato) – (oxamide – oximato) – palladium(ii)) sulfate monohydrate
$2C_4H_{11}N_8O_4Pd^+$, O_4S^{2-}, H_2O
H.Endres *Acta Crystallogr.,Sect.B*,**36**, 1347, 1980

83.C bis(cis – bis(2 – Aminoethanolato) – copper(ii)) dinitrate
$2C_4H_{13}CuN_2O_2^+$, $2NO_3^-$ Main entry is 84.6

83.13 (α,β) – bis(S – Methylthiosemicarbazido) – nickel(ii) bromide (triclinic form)
$C_4H_{14}N_6NiS_2^{2+}$, $2Br^-$
V.Divjakovic, V.Leovac, L.Bjelica
Eur.Cryst.Meeting,**6**, 59, 1980

83.14 (α,β) – bis(S – Methylthiosemicarbazido) – nickel(ii) bromide (orthorhombic form)
$C_4H_{14}N_6NiS_2^{2+}$, $2Br^-$
V.Divjakovic, V.Leovac, L.Bjelica
Eur.Cryst.Meeting,**6**, 59, 1980

83.15 bis(S – Methylthiosemicarbazido) – nickel(ii) iodide
$C_4H_{14}N_6NiS_2^{2+}$, $2I^-$
V.Divjakovic, V.Leovac, L.Bjelica
Eur.Cryst.Meeting,**6**, 59, 1980

83.16 Pyridine pyridinium (dibromo – chloro – pyridine – zinc) (bromo – dichloro – pyridine – zinc)
$C_5H_5BrCl_2NZn^-$, $C_5H_5Br_2ClNZn^-$, $2C_5H_5N$, $2C_5H_6N^+$
B.E.Villarreal–Salinas, E.O.Schlemper
J.Cryst.Mol.Struct.,**8**, 217, 1978

83.17 Pyridine pyridinium (dibromo – chloro – pyridine – zinc) (bromo – dichloro – pyridine – zinc) (neutron study)
$C_5H_5BrCl_2NZn^-$, $C_5H_5Br_2ClNZn^-$, $2C_5H_5N$, $2C_5H_6N^+$
B.E.Villarreal–Salinas, E.O.Schlemper
J.Cryst.Mol.Struct.,**8**, 217, 1978

83.18 Dichloro – histamino – copper(ii)
$C_5H_9Cl_2CuN_3$
M.L.Glowka, Z.Galdecki, W.Kaimierczak, C.Maslinski
Acta Crystallogr.,Sect.B,**36**, 2148, 1980

83.C 2 – ((3 – Aminopropyl)amino)ethanolato – copper(ii) tetra(isothiocyanato) – copper(i) thiocyanate
$4C_5H_{13}CuN_2O^+$, $C_4CuN_4S_4^{3-}$, CNS^- Main entry is 84.8

83.19 Penta – amine – (1 – methylcytosinato) – ruthenium(iii) hexafluorophosphate
$C_5H_{21}N_8ORu^{2+}$, $2F_6P^-$
B.J.Graves, D.J.Hodgson *J.Am.Chem.Soc.*,**101**, 5608, 1979

83.20 trans – Dichloro – bis(oxazole) – palladium(ii)
$C_6H_6Cl_2N_2O_2Pd$
E.Binamira–Soriaga, M.Lundeen, K.Seff
J.Cryst.Mol.Struct.,**9**, 67, 1979

83.21 catena – bis(Isothiocyanato) – bis(μ – 1,2,4 – triazole) – cobalt(ii)
$(C_6H_6CoN_8S_2)_n$
D.W.Engelfriet, W.den Brinker, G.C.Verschoor, S.Gorter
Acta Crystallogr.,Sect.B,**35**, 2922, 1979

83.22 catena – bis(Isothiocyanato) – bis(μ – 1,2,4 – triazole) – copper(ii)
$(C_6H_6CuN_8S_2)_n$
D.W.Engelfriet, W.den Brinker, G.C.Verschoor, S.Gorter
Acta Crystallogr.,Sect.B,**35**, 2922, 1979

83.C Sodium nitrilo – triacetato – zinc monohydrate
$(C_6H_6NO_6Zn^-)_n$, nNa^+, nH_2O Main entry is 81.21

83.23 catena – bis(Isothiocyanato) – bis(μ – 1,2,4 – triazole) – zinc(ii)
$(C_6H_6N_8S_2Zn)_n$
D.W.Engelfriet, W.den Brinker, G.C.Verschoor, S.Gorter
Acta Crystallogr.,Sect.B,**35**, 2922, 1979

83.C Cesium tris(ethylnitrosolato) – iron(ii) monohydrate
$C_6H_9FeN_6O_6^-$, Cs^+, H_2O Main entry is 84.9

83.C Chloro – (uracilato) – (ethylenediamine) – platinum(ii) hydrogen chloride dihydrate
$C_6H_{11}ClN_4O_2Pt$, H^+, Cl^-, $2H_2O$ Main entry is 76.24

83.C Aqua – (N,N – dimethylethylenediamine – N – oxide) – oxalato – copper(ii) dihydrate
$C_6H_{14}CuN_2O_6$, $2H_2O$ Main entry is 81.23

83.C (1,5 – Diazacyclo – octane) – dinitrato – copper(ii)
$C_6H_{14}CuN_4O_6$ Main entry is 84.11

83.24 tris(Oxamide – oximato) – nickel(ii) dichloride hemihydrate
$C_6H_{18}N_{12}NiO_6^{2+}$, $2Cl^-$, $0.5H_2O$
H.Endres, T.Jannack
Acta Crystallogr.,Sect.B,**36**, 2136, 1980

83.C (2 – Aminoethanol) – bis(2 – aminoethanolato) – cobalt(iii) bis(2 – aminoethanol) – (2 – aminoethanolato) – cobalt(iii) perchlorate hemihydrate
$C_6H_{19}CoN_3O_3^+$, $C_6H_{20}CoN_3O_3^{2+}$, $3ClO_4^-$, $0.5H_2O$
Main entry is 84.13

83.C bis(1,3 – Propylenediamine) – platinum(ii) dichloro – bis(1,3 – propylenediamine) – platinum(iv) perchlorate
$C_6H_{20}Cl_2N_4Pt^{2+}$, $C_6H_{20}N_4Pt^{2+}$, $4ClO_4^-$ Main entry is 83.25

83.C bis(2 – Aminoethanol) – (2 – aminoethanolato) – nickel(iii) perchlorate
$C_6H_{20}N_3NiO_3^+$, ClO_4^- Main entry is 84.14

83.25 bis(1,3 – Propylenediamine) – platinum(ii) dichloro – bis(1,3 – propylenediamine) – platinum(iv) perchlorate
$C_6H_{20}N_4Pt^{2+}$, $C_6H_{20}Cl_2N_4Pt^{2+}$, $4ClO_4^-$
N.Matsumoto, M.Yamashita, I.Ueda, S.Kida
Mem.Fac.Sci.Kyushu Univ.Ser.C,**11**, 209, 1978
See also R2 : 83

83.26 Pentammine – (3,3 – pentamethylene – diazirine) – ruthenium bis(hexafluorophosphate)
$C_6H_{25}N_7Ru^{2+}$, $2F_6P^-$
V.B.Shur, I.A.Tikhonova, G.G.Aleksandrov, Yu.T.Struchkov, M.E.Vol'pin, E.Schmitz, K.Jahnisch
Inorg.Chim.Acta,**44**, L275, 1980

83.27 Benzonitrile – trichloro – oxo – vanadium(v)
$C_7H_5Cl_3NOV$
A.Gourdon, Y.Jeannin
Acta Crystallogr.,Sect.B,**36**, 304, 1980

83.C p – Aminobenzoato – silver
$(C_7H_6AgNO_2)_n$ Main entry is 81.27

83.C Dibromo – (N – (2 – pyridyl)acetamide) – mercury(ii)
$(C_7H_8Br_2HgN_2O)_n$ Main entry is 84.15

83.28 Dibromo – (2,4 – dimethylpyridine) – mercury(ii)
$(C_7H_9Br_2HgN)_n$
N.A.Bell, M.Goldstein, T.Jones, I.W.Nowell
Acta Crystallogr.,Sect.B,**36**, 710, 1980

83.29 Tetra – ammine – platinum(ii) bis(trichloro – (2,6 – dimethylpyridine) – platinum(ii))
$2C_7H_9Cl_3NPt^-, H_{12}N_4Pt^{2+}$
F.D.Rochon, R.Melanson
Acta Crystallogr.,Sect.B,**36**, 691, 1980

83.C cis – Dichloropyridine – glycine – platinum(ii)
$C_7H_{10}Cl_2N_2O_2Pt$ Main entry is 82.13

83.C 1 – (Methyl – mercury(ii)) – 9 – methyl – adenine nitrate
$C_7H_{10}HgN_5^+, NO_3^-$ Main entry is 71.5

83.30 (9 – Methylguanine) – methyl – mercury(ii) nitrate
$C_7H_{10}HgN_5O^+, NO_3^-$
A.J.Canty, R.S.Tobias, N.Chaichit, B.M.Gatehouse
J.Chem.Soc.,Dalton Trans.,1693, 1980
See also R1 : 44

83.C trans – Dichloro – (dimethylsulfoxide) – (pyridine) – platinum(ii)
$C_7H_{11}Cl_2NOPtS$ Main entry is 85.15

83.31 Diaqua – dichloro – theophylline – copper(ii)
$C_7H_{12}Cl_2CuN_4O_4$
M.B.Cingi, A.M.M.Lanfredi, A.Tiripicchio, M.T.Camellini
Transition Met.Chem.,**4**, 221, 1979

83.C Tricarbonyl – (S – methyldithiocarbazato) – (μ – S – methyldithiocarbazato) – manganese(i) bromide
$C_7H_{12}MnN_4O_3S_4^+, Br^-$ Main entry is 85.16

83.C Chloro – (thyminato) – (ethylenediamine) – platinum(ii)
$C_7H_{13}ClN_4O_2Pt$ Main entry is 76.42

83.C $(-)_{589}$ – $\Delta(S)$ – cis(1 – Aminopropane – 2 – ol – N) – chloro – bis(1,2 – diaminoethane) – cobalt(iii) tetrachloro – zinc
$C_7H_{25}ClCoN_5O^{2+}, Cl_4Zn^{2-}$ Main entry is 76.45

83.32 Chloro – (bis(2 – pyrimidyl) – disulfide) – copper(i) monohydrate
$(C_8H_6ClCuN_4S_2)_n, nH_2O$
C.J.Simmons, M.Lundeen, K.Seff
Inorg.Chem.,**18**, 3444, 1979

83.33 Trichloro – oxo – (phenyl – acetonitrile) – vanadium(v)
$C_8H_7Cl_3NOV$
A.Gourdon, Y.Jeannin
Acta Crystallogr.,Sect.B,**36**, 304, 1980

83.34 bis(Pyrazine) – bis(perchlorato) – copper(ii)
$(C_8H_8Cl_2CuN_4O_8)_n$
J.Darriet, M.S.Haddad, E.N.Duesler, D.N.Hendrickson
Inorg.Chem.,**18**, 2679, 1979

83.C Dicarbonyl – (η^5 – cyclopentadienyl) – methyldiazo – tungsten (at $-140°C$)
$C_8H_8N_2O_2W$ Main entry is 73.10

83.35 (μ – Adeninato) – 3,7,9 – tris(methyl – mercury(ii)) diperchlorate
$C_8H_{13}Hg_3N_5^{2+}, 2ClO_4^-$
J.Hubert, A.L.Beauchamp *Can.J.Chem.*,**58**, 1439, 1980
See also R1 : 44

83.36 Rubidium dichloro – bis(dimethylglyoximato) – rhodium(iii) monohydrate
$C_8H_{14}Cl_2N_4O_4Rh^-, Rb^+, H_2O$
Yu.A.Simonov, L.A.Nemchinova, O.A.Bologa
Kristallografiya,**24**, 829, 1979

83.37 Sodium dinitro – bis(dimethylglyoximato) – cobalt(iii) dihydrate
$C_8H_{14}CoN_6O_8^-, Na^+, 2H_2O$
A.I.Shkurpelo, Yu.A.Simonov, O.A.Bologa, T.I.Malinovskii
Dokl.Akad.Nauk SSSR,**248**, 1120, 1979

83.38 Aquo – nitro – bis(dimethylglyoximato) – cobalt(iii)
$C_8H_{16}CoN_5O_7$
A.I.Shkurpelo, Yu.A.Simonov, O.A.Bologa, T.I.Malinovskii
Dokl.Akad.Nauk SSSR,**248**, 1120, 1979

83.C Malonato – (2,2 – dimethyl – 1,3 – diaminopropane) – platinum(ii)
$C_8H_{16}N_2O_4Pt$ Main entry is 81.34

83.C Sodium trans – sulfito – bis(dimethylglyoximato) ammine – cobalt(iii) tetrahydrate
$C_8H_{17}CoN_5O_7S^-, Na^+, 4H_2O$ Main entry is 85.24

83.39 bis(N – t – Butylimido) – dioxo – osmium
$C_8H_{18}N_2O_2Os$
W.A.Nugent, R.L.Harlow, R.J.McKinney
J.Am.Chem.Soc.,**101**, 7265, 1979

83.40 trans – bis(Dimethylglyoximato) – diammine – cobalt(iii) bromide
$C_8H_{20}CoN_6O_4^+, Br^-$
M.J.Heeg, R.C.Elder *Inorg.Chem.*,**19**, 932, 1980

83.41 trans – bis(Dimethylglyoximato) – diammine – cobalt(iii) thiocyanate monohydrate
$C_8H_{20}CoN_6O_4^+, CNS^-, H_2O$
R.C.Elder, R.Whittle, M.J.Heeg, J.C.Barrick, A.Nerone
Am.Cryst.Assoc.,Ser.2,**7**, 23, 1980

83.42 trans – bis(Dimethylglyoximato) – diammine – cobalt(iii) nitrate
$C_8H_{20}CoN_6O_4^+, NO_3^-$
R.C.Elder, R.Whittle, M.J.Heeg, J.C.Barrick, A.Nerone
Am.Cryst.Assoc.,Ser.2,**7**, 23, 1980

83.C Dioxo – bis(N,N – diethylhydroxylaminato) – molybdenum(vi)
$C_8H_{20}MoN_2O_4$ Main entry is 84.17

83.C Δ – bis(Ethylenediamine) – ((3R,4R) – thiazolidine – 4 – carboxylato – N,O) – cobalt(iii) dichloride
$C_8H_{22}CoN_5O_2S^{2+}, 2Cl^-$ Main entry is 76.47

83.C bis(Aqua – bis(2 – amino – 2 – methylpropanolato) – copper(ii)) dinitrate
$2C_8H_{23}CuN_2O_3^+, 2NO_3^-$ Main entry is 84.18

83.C tetrakis(Ethylamine) – platinum(ii) dibromo – tetrakis(ethylamine) – platinum(iv) tetrabromide
$C_8H_{28}Br_2N_4Pt^{2+}, C_8H_{28}N_4Pt^{2+}, 4Br^-$ Main entry is 83.43

83.43 tetrakis(Ethylamine) – platinum(ii) dibromo –
tetrakis(ethylamine) – platinum(iv) tetrabromide
$C_8H_{28}N_4Pt^{2+}$, $C_8H_{28}Br_2N_4Pt^{2+}$, $4Br^-$
H.Endres, H.J.Keller, B.Keppler, R.Martin, W.Steiger,
U.Traeger *Acta Crystallogr.,Sect.B*,**36**, 760, 1980
See also R2 : 83

83.44 Dichloro – (hydro – tris(1 – pyrazolyl) – borato) –
oxo – technetium(v)
$C_9H_{10}BCl_2N_6OTc$
R.W.Thomas, G.W.Estes, R.C.Elder, E.Deutsch
J.Am.Chem.Soc.,**101**, 4581, 1979
See also R1 : 62

83.C cis – Dichloro – (μ – ethylene) – (2,6 – dimethyl –
piperidine) – platinum(ii)
$C_9H_{19}Cl_2NPt$ Main entry is 72.7

83.C (tris(Triethylene) – tetramine) – perchlorato –
copper perchlorate
$C_9H_{24}ClCuN_4O_4^+$, ClO_4^- Main entry is 84.22

83.C tris(Methyl 2 – methyldithiocarbazato) – nickel(ii)
chloride ethanol solvate trihydrate
$C_9H_{24}N_6NiS_6^{2+}$, $2Cl^-$, $0.5C_2H_6O$, $3H_2O$ Main entry is 85.33

83.45 Triazido – nitrido – bipyridyl – molybdenum(vi)
$C_{10}H_8MoN_{12}$
E.Schweda, J.Strahle
Z.Naturforsch.,Teil B,**35**, 1146, 1980

83.46 bis(μ^2 – Bromo) – bis(bromopyridine – copper(ii))
$C_{10}H_{10}Br_4Cu_2N_2$
D.D.Swank, R.D.Willett *Inorg.Chem.*,**19**, 2321, 19

83.47 bis(Pyridinium) tetrabromo – bis(pyridine) –
tungsten(iii)
$C_{10}H_{10}Br_4N_2W^-$, $C_5H_6N^+$, C_5H_5N
J.V.Brencic, B.Ceh, I.Leban
Acta Crystallogr.,Sect.B,**35**, 3028, 1979
See also R2 : 33

83.C bis(μ – Chloro) – bis(chloro – (6 – mercapto –
purine) – aquo – cadmium(ii))
$C_{10}H_{12}Cd_2Cl_4N_8O_2S_2$ Main entry is 85.35

83.C Diaquo – bis(pyridine – 3 – sulfonato) – copper(ii)
$(C_{10}H_{12}CuN_2O_8S_2)_n$ Main entry is 84.23

83.48 Diaqua – bis(pyridine – 3 – sulfonato) – di –
mercury(i)
$C_{10}H_{12}Hg_2N_2O_8S_2$
K.Brodersen, R.Dolling, G.Liehr
Z.Anorg.Allg.Chem.,**464**, 17, 1980

83.49 trans – Dichloro – bis(2,4 – dimethylthiazole) –
copper(ii)
$C_{10}H_{14}Cl_2CuN_2S_2$
D.P.Gavel, D.J.Hodgson
Acta Crystallogr.,Sect.B,**35**, 2704, 1979

83.50 Dichloronicotine – mercury(ii)
$C_{10}H_{14}Cl_2HgN_2$
M.R.Udupa, B.Krebs *Inorg.Chim.Acta*,**40**, 161, 1980

83.51 bis(μ – 3,5 – Dimethylpyrazolyl) – bis(dinitrosyl –
cobalt)
$C_{10}H_{14}Co_2N_8O_4$
K.S.Chong, S.J.Rettig, A.Storr, J.Trotter
Can.J.Chem.,**57**, 3119, 1979

83.52 bis(μ – 3,5 – Dimethylpyrazolyl) – bis(dinitrosyl –
iron)
$C_{10}H_{14}Fe_2N_8O_4$
K.S.Chong, S.J.Rettig, A.Storr, J.Trotter
Can.J.Chem.,**57**, 3119, 1979

83.53 Tetraethylammonium (μ – iodo) – bis(μ – 3,5 –
dimethylpyrazolyl) – bis(nitrosyl – nickel)
$C_{10}H_{14}IN_6Ni_2O_2^-$, $C_8H_{20}N^+$
K.S.Chong, S.J.Rettig, A.Storr, J.Trotter
Can.J.Chem.,**57**, 3099, 1979
See also R2 : 3

83.54 bis(μ – 3,5 – Dimethylpyrazolyl) – bis(nitrosyl –
nickel)
$C_{10}H_{14}N_6Ni_2O_2$
K.S.Chong, S.J.Rettig, A.Storr, J.Trotter
Can.J.Chem.,**57**, 3090, 1979

83.55 N – 1 – Adamantylimido – trioxo – osmium
(at –70°C)
$C_{10}H_{15}NO_3Os$
W.A.Nugent, R.L.Harlow, R.J.McKinney
J.Am.Chem.Soc.,**101**, 7265, 1979

83.56 Adeninium – cadmium – nitrate hydrate dimer
$C_{10}H_{16}Cd_2N_{16}O_{20}$
C.H.Wei, K.B.Jacobson *Am.Cryst.Assoc.,Ser.2*,**7**, 14, 1980

83.C trans – Dichloro – (dimethylformamide) – (2,6 –
lutidine) – platinum(ii)
$C_{10}H_{16}Cl_2N_2OPt$ Main entry is 84.25

83.57 Tetra – aquo – bis(pyridine – 3 – sulfonato) –
copper(ii) – zinc(ii)
$0.5C_{10}H_{16}CuN_2O_{10}S_2$, $0.5C_{10}H_{16}N_2O_{10}S_2Zn$
B.Walsh, B.J.Hathaway
J.Chem.Soc.,Dalton Trans.,681, 1980
See also R2 : 83

83.58 Tetra – aquo – bis(pyridine – 3 – sulfonato) –
zinc(ii)
$C_{10}H_{16}N_2O_{10}S_2Zn$
B.Walsh, B.J.Hathaway
J.Chem.Soc.,Dalton Trans.,681, 1980

83.C Tetra – aquo – bis(pyridine – 3 – sulfonato) –
copper(ii) – zinc(ii)
$0.5C_{10}H_{16}N_2O_{10}S_2Zn$, $0.5C_{10}H_{16}CuN_2O_{10}S_2$ Main entry is 83.57

83.59 Aquo – bis(2 – hydrazino – 4 – hydroxy – 6 –
methylpyrimidine) – copper(ii) chloride dihydrate
$C_{10}H_{18}CuN_8O_3^{2+}$, $2Cl^-$, $2H_2O$
H.Sakaguchi, H.Anzai, K.Furuhata, H.Ogura, Y.Iitaka
Chem.Pharm.Bull.,**27**, 1871, 1979

83.60 bis(3 – Amino – 3 – methyl – butan – 2 – one –
oximato) – nickel(ii) chloride monohydrate
(neutron study, deuterated form)
$C_{10}H_{18}D_5N_4NiO_2^+$, Cl^-, D_2O
B.Hsu, E.O.Schlemper, C.K.Fair
Acta Crystallogr.,Sect.B,**36**, 1387, 1980

83.61 Diammine – (1 – methylcytosine) – (thyminato) –
platinum(ii) perchlorate trihydrate
$C_{10}H_{18}N_7O_3Pt^+$, ClO_4^-, $3H_2O$
B.Lippert, R.Pfab, D.Neugebauer
Inorg.Chim.Acta,**37**, L495, 1979

83.C Chloro – (2 – methyl – 2 – nitrosopropane) – (2 –
(N – oxo – t – butylimino)ethyl) – platinum
$C_{10}H_{21}ClN_2O_2Pt$ Main entry is 71.24

83.62 Penta – aquo – (2' – deoxyguanosine – 5' – phosphato) – cobalt(ii) trihydrate
$C_{10}H_{22}CoN_5O_{12}P$, $3H_2O$
R.W.Gellert, J.K.Shiba, R.Bau
Biochem.Biophys.Res.Commun.,**88**, 1449, 1979
See also R1 : 47

83.C Nitrato – (1 – oxa – 7,10 – diaza – 4,13 – dithia – cyclopentadecane) – nickel(ii) nitrate
$C_{10}H_{22}N_3NiO_4S_2^+$, NO_3^- Main entry is 85.36

83.63 Penta – aquo – (2' – deoxyguanosine – 5' – phosphato) – nickel(ii) trihydrate
$C_{10}H_{22}N_5NiO_{12}P$, $3H_2O$
R.W.Gellert, J.K.Shiba, R.Bau
Biochem.Biophys.Res.Commun.,**88**, 1449, 1979
See also R1 : 47

83.64 Chloro – (1,4,8,11 – tetra – azacyclotetradecane) – mercury(ii) tetrachloro – mercury(ii) (at 153°K)
$2C_{10}H_{24}ClHgN_4^+$, Cl_4Hg^{2-}
N.W.Alcock, E.H.Curson, N.Herron, P.Moore
J.Chem.Soc.,Dalton Trans.,1987, 1979

83.C bis(1,1 – Dimethyl – 2 – N – methylaminoethylthiolato) – dioxo – molybdenum
$C_{10}H_{24}MoN_2O_2S_2$ Main entry is 85.37

83.C bis(2 – Dimethylaminoethyl – methyl – amino – zinc – hydride)
$C_{10}H_{28}N_4Zn_2$ Main entry is 76.66

83.C (2 – Methylene – 1 – tetracarbonyl – 2,3 – diaza – 1 – mangana – cyclobutan – 4 – one) – manganese pentacarbonyl
Nonacarbonyl – (μ – (methylenehydrazine – carbonyl)) – di – manganese
$C_{11}H_2Mn_2N_2O_{10}$ Main entry is 71.25

83.C (μ – Acetimidoyl) – (μ – hydrido) – tris(tricarbonyl – iron) (triclinic form, at –158°C)
$C_{11}H_5Fe_3NO_9$ Main entry is 71.

83.C (μ_3 – Ethylideneimido) – (μ – hydrido) – tris(tricarbonyl – iron) (at –158°C)
$C_{11}H_5Fe_3NO_9$ Main entry is 72.11

83.C Tetraphenylphosphonium dioxo – (4 – (2 – pyridylazo) – resorcinolato) – vanadium(v)
$C_{11}H_7N_3O_4V^-$, $C_{24}H_{20}P^+$ Main entry is 84.30

83.C Phenylpyridine – mercury(ii) trifluoroacetate
$C_{11}H_{10}HgN^+$, $C_2F_3O_2^-$ Main entry is 71.29

83.C (Diacetylmonoxime) – (salicyloylhydrazinato) – chloro – copper(ii) methanol solvate
$C_{11}H_{13}ClCuN_3O_3$, CH_4O Main entry is 84.31

83.C (Glycylglycinato) – (7,9 – dimethylhypoxanthine) – copper(ii) tetrahydrate (absolute configuration)
$C_{11}H_{14}CuN_6O_4$, $4H_2O$ Main entry is 82.22

83.C Sodium dioxo – bis(μ^2 – thio) – (μ – 1,2 – propylenediamine) – tetra – acetato – di – molybdenum trihydrate
$C_{11}H_{14}Mo_2N_2O_{10}S_2^{2-}$, $2Na^+$, $3H_2O$ Main entry is 81.38

83.C Sodium dioxo – (μ^2 – oxo) – (μ^2 – thio) – 1,2 – propylenediamine – tetra – acetato – di – molybdenum tetrahydrate
$C_{11}H_{14}Mo_2N_2O_{11}S^{2-}$, $2Na^+$, $4H_2O$ Main entry is 81.39

83.C Sodium dioxo – bis(μ^2 – oxo) – (μ – 1,2 – propylenediamine) – tetra – acetato – di – molybdenum trihydrate
$C_{11}H_{14}Mo_2N_2O_{12}^{2-}$, $2Na^+$, $3H_2O$ Main entry is 81.40

83.65 (N – Benzyl – piperazinium) – pentachloro – di – mercury(ii)
$C_{11}H_{17}Cl_5Hg_2N_2$
A.Albinati, S.V.Meille, F.Cariati, G.Marcotrigiano, L.Menabue, G.C.Pellacani *Inorg.Chim.Acta*,**38**, 221, 1980

83.66 Iodo – (difluoro – (3,3' – (trimethylene – dinitrilo) – bis(butan – 2 – one – oximato)) – borato) – copper(ii)
$C_{11}H_{18}BCuF_2IN_4O_2$
O.P.Anderson, A.B.Packard *Inorg.Chem.*,**18**, 3064, 1979

83.67 Bromo – (2,2' – (1,3 – di – iminopropane) – bis(butan – 3 – one – oximato)) – nickel(ii)
$C_{11}H_{19}BrN_4NiO_2$
J.Korvenranta, H.Saarinen *Finn.Chem.Lett.*,219, 1979

83.68 Dichloro – (2,2' – (1,3 – di – iminopropane) – bis(butan – 3 – one – oximato)) – cobalt(iii)
$C_{11}H_{19}Cl_2CoN_4O_2$
M.Nasakkala, H.Saarinen, J.Korvenranta, E.Nasakkala *Acta Chem.Scand.Ser.A*,**33**, 431, 1979

83.69 (3,9 – Dimethyl – 4,8 – diaza – 3,8 – undecadiene – 2,10 – dioximato) – copper(ii) perchlorate methanol solvate (reassignment of space group reported by Bertrand et. al., Inorg.Chem.,16,1484,1977)
$C_{11}H_{19}CuN_4O_2^+$, ClO_4^-, $0.5CH_4O$
R.E.Marsh, V.Schomaker *Inorg.Chem.*,**18**, 2331, 1979

83.70 Aquo – (difluoro – (3,3' – (trimethylene – dinitrilo) – bis(butan – 2 – one – oximato)) – borato) – copper(ii) perchlorate
$C_{11}H_{20}BCuF_2N_4O_3^+$, ClO_4^-
O.P.Anderson, A.B.Packard *Inorg.Chem.*,**18**, 1940, 1979

83.C bis(N,N – Diethyldithiocarbamato) – (isothiocyanato) – iron(iii)
$C_{11}H_{20}FeN_3S_5$ Main entry is 80.11

83.71 Aquo – (3,3' – (trimethylene – dinitrilo) – bis(butan – 2 – one – oximato)) – copper(ii) perchlorate monohydrate
$C_{11}H_{21}CuN_4O_3^+$, ClO_4^-, H_2O
O.P.Anderson, A.B.Packard *Inorg.Chem.*,**18**, 1940, 1979

83.C trans – Chloro – bis(dimethylglyoximato) – trimethylphosphite – cobalt(iii)
$C_{11}H_{23}ClCoN_4O_7P$ Main entry is 86.16

83.C Chloro – (tris(2 – aminoethyl)amine) – pyridino – cobalt(iii) tetrachloro – zinc(ii)
$C_{11}H_{23}ClCoN_5$, Cl_4Zn Main entry is 76.68

83.C (Dimethyl – (N,N – dimethylethanolamino) – (3,5 – dimethylpyrazolyl) – gallato) – dinitrosyl – iron
$C_{11}H_{23}FeGaN_5O_3$ Main entry is 84.32

83.C (Dimethyl – (N,N – dimethylethanolamino) – (3,5 – dimethylpyrazolyl) – gallato) – nitrosyl – nickel
$C_{11}H_{23}GaN_4NiO_2$ Main entry is 84.33

83.C Barium tris(dihydrogen – violurato) – ruthenium(ii) nonahydrate
$2C_{12}H_6N_9O_{12}Ru^-$, Ba^{2+}, $9H_2O$ Main entry is 84.34

83.72 Dichloro – dioxo – (o – phenanthroline) – molybdenum(vi)
$C_{12}H_8Cl_2MoN_2O_2$
B.Viossat, N.Rodier
Acta Crystallogr.,Sect.B,35, 2715, 1979

83.73 Trimethylamine – bis(μ – nitrosyl) – nonacarbonyl – tri – osmium
$C_{12}H_9N_3O_{11}Os_3$
B.F.G.Johnson, J.Lewis, P.R.Raithby, C.Zuccaro
J.Chem.Soc.,Chem.Commun.,916, 1979

83.C Aqua – (2,2' – bipyridyl) – nitroacetato – copper(ii) monohydrate
$C_{12}H_{11}CuN_3O_5$, H_2O Main entry is 84.36

83.74 bis(9 – Methylhypoxanthine) – silver(i) perchlorate monohydrate
$C_{12}H_{12}AgN_8O_2^+$, ClO_4^-, H_2O
F.Belanger–Gariepy, A.L.Beauchamp
J.Am.Chem.Soc.,102, 3461, 1980

83.C N,N' – Ethylene – bis(1,1,1 – trifluoropentane – 2,4 – dioneiminato) – copper(ii)
$C_{12}H_{12}CuF_6N_2O_2$ Main entry is 77.1

83.C bis(Formato) – bis(pyridine) – copper monohydrate
$C_{12}H_{12}CuN_2O_4$, H_2O Main entry is 81.41

83.75 Dinitrato – (S – methylthiosemicarbazone – 8 – quinoline – aldehyde) – copper(ii)
$C_{12}H_{12}CuN_6O_6S$
D.Petrovic, B.Ribar, S.Caric, V.Leovac
Z.Kristallogr.,150, 3, 1979

83.76 bis(Di – imidazolyl zinc)
$(C_{12}H_{12}N_8Zn_2)_n$
R.Lehnert, F.Seel *Z.Anorg.Allg.Chem.*,464, 187, 1980

83.C Methyl – (bis(2 – pyridyl)methane) – mercury nitrate
$C_{12}H_{13}HgN_2^+$, NO_3^- Main entry is 71.37

83.77 Dichloro – bis(picolinehydrazide) – manganese
$C_{12}H_{14}Cl_2MnN_6O_2$
G.V.Tsintsadze, T.I.Tsivtsivadze, Z.O.Dzhavakhishvili, A.I.Ilinskii, F.V.Orbeladze *Koord.Khim.*,5, 909, 1979

83.C Dichloro – bis(1 – vinyl – 2 – hydroxymethylimidazole) – cobalt
$C_{12}H_{16}Cl_2CoN_4O_2$ Main entry is 84.37

83.C tetrakis(μ – Acetato) – (N – pyrazine) – chromium
$(C_{12}H_{16}Cr_2N_2O_8)_n$ Main entry is 81.45

83.C Diaqua – bis(formato) – bis(pyridine) – copper
$C_{12}H_{16}CuN_2O_6$ Main entry is 81.46

83.78 bis(1 – Carbamoyl – 3,5 – dimethyl – pyrazolato) – copper(ii) (α form)
$C_{12}H_{16}CuN_6O_2$
F.Valach, J.Kohout, M.Dunaj–Jurco, M.Hvastijova, J.Gazo
J.Chem.Soc.,Dalton Trans.,1867, 1979

83.C (μ – Dimethylhydrazido) – bis(cyclopentadienyl – iodo – nitrosyl – molybdenum)
$C_{12}H_{16}I_2Mo_2N_4O_2$ Main entry is 73.39

83.C N,N' – Ethylene – bis(monothioacetylacetone – iminato) – copper(ii)
$C_{12}H_{18}CuN_2S_2$ Main entry is 76.71

83.79 hexakis(Acetonitrile) – iron(ii) bis(tetrachloro – iron)
$C_{12}H_{18}FeN_6^{2+}$, $2Cl_4Fe^-$
I.P.Lavrent'ev, L.G.Korableva, E.A.Lavrent'eva, G.A.Nifontova, M.L.Khidekel, I.G.Gusakovskaya, T.I.Larkina, L.D.Arutyunyan, O.S.Filipenko, V.I.Ponomarev, L.O.Atovmyan *Koord.Khim.*,5, 1484, 1979

83.80 hexakis(Acetonitrile) – iron(ii) bis(tetrachloro – iron) (at −165°C)
$C_{12}H_{18}FeN_6^{2+}$, $2Cl_4Fe^-$
I.P.Lavrent'ev, L.G.Korableva, E.A.Lavrent'eva, G.A.Nifontova, M.L.Khidekel, I.G.Gusakovskaya, T.I.Larkina, L.D.Arutyunyan, O.S.Filipenko, V.I.Ponomarev, L.O.Atovmyan *Koord.Khim.*,5, 1484, 1979

83.C γ – Picolinium bis(S – methyldithiocarbazato – dimethylglyoximato) – iron(iii) tetrahydrate
$C_{12}H_{18}FeN_6O_2S_4^-$, $C_6H_8N^+$, $4H_2O$ Main entry is 85.39

83.C (N,N' – Ethylene – bis(monothioacetylacetone – iminato)) – zinc(ii)
$C_{12}H_{18}N_2S_2Zn$ Main entry is 85.40

83.81 cis – Dichloro – bis(cyclohexane – 1,2 – dione – dioxime) – nickel(ii)
$C_{12}H_{20}Cl_2N_4NiO_4$
Yu.M.Simonov, M.M.Botoshanskii, T.I.Malinovskii, D.G.Batir, L.D.Ozol, I.I.Bulgak
Dokl.Akad.Nauk SSSR,246, 609, 1979

83.82 Dichloro – bis(1,3 – dimethyl – 1,4,5,6 – tetrahydropyridazine) – palladium(ii)
$C_{12}H_{24}Cl_2N_4Pd$
G.Natile, F.Gasparrini, B.Galli, A.M.M.Lanfredi, A.Tiripicchio *Inorg.Chim.Acta*,38, L29, 1980

83.83 trans – Dibromo – bis(cyclohexylamine) – platinum(ii)
$C_{12}H_{28}Br_2N_2Pt$
C.J.L.Lock, M.Zvagulis
Acta Crystallogr.,Sect.B,36, 2140, 1980

83.84 cis – Dichloro – bis(cyclohexylamine – N) – platinum(ii) hexamethylphosphoramide solvate (at 243°K)
$C_{12}H_{26}Cl_2N_2Pt$, $2C_6H_{18}N_3OP$
C.J.L.Lock, R.A.Speranzini, M.Zvagulis
Acta Crystallogr.,Sect.B,36, 1789, 1980

83.C bis(μ – 1 – Methylthyminato) – bis(cis – diammine – platinum(ii)) dinitrate
$C_{12}H_{26}N_8O_4Pt_2^+$, $2NO_3^-$ Main entry is 84.42

83.C trans – Dichloro – (triethylphosphine) – (cis – 2,4 – dimethylpyrrolidine) – platinum(ii)
$C_{12}H_{29}Cl_2NPPt$ Main entry is 86.21

83.C trans – Dichloro – (triethylphosphine) – (cis – 2,3 – dimethylpyrrolidine) – platinum(ii)
$C_{12}H_{29}Cl_2NPPt$ Main entry is 86.22

83.C bis(Aqua – copper(ii)) – bis(μ – aqua) – bis(μ – N,N' – bis(2 – hydroxyethyl) – dithio – oxamidato) – bis(sulfato – copper(ii))
$C_{12}H_{28}Cu_4N_4O_{16}S_6$ Main entry is 85.45

83.C bis(Triethylphosphine) – dinitrosyl – nitrito – manganese
$C_{12}H_{30}MnN_3O_4P_2$ Main entry is 86.24

83.85 Trichloro – bis(bis(trimethylsilyl)amido) – tantalum(v)
$C_{12}H_{36}Cl_3N_2Si_4Ta$
D.C.Bradley, M.B.Hursthouse, K.M.A.Malik, G.B.C.Vuru
Inorg.Chim.Acta,**44**, L5, 1980

83.C (+)$_{589}$ – fac – tris(2 – Aminoethyl – dimethylphosphine) – cobalt(iii) tribromide trihydrate (absolute configuration)
$C_{12}H_{36}CoN_3P_3^{3+}$, $3Br^-$, $3H_2O$ Main entry is 86.28

83.86 (μ – Methanediazo) – (N,N,N') – tris(tetracarbonyl – manganese)
$C_{13}H_3Mn_3N_2O_{12}$
K.Weidenhammer, M.L.Ziegler
Z.Anorg.Allg.Chem.,**457**, 174, 1979

83.C (μ_3 – Butyronitrile) – tris(tricarbonyl – iron)
$C_{13}H_7Fe_3NO_9$ Main entry is 72.19

83.C (2,2' – Bipyridyl) – tricarbonyl – chloro – (chloro – mercurio) – molybdenum(ii)
$C_{13}H_8Cl_2HgMoN_2O_3$ Main entry is 71.48

83.C Dicarbonyl – (η^5 – cyclopentadienyl) – (μ – methanediazo) – (pentacarbonyl – chromium) – tungsten (at –160°C)
$C_{13}H_8CrN_2O_7W$ Main entry is 73.41

83.87 Diphenylcarbazone – bis(chloro – mercury)
$C_{13}H_{10}Cl_2Hg_2N_4O$
N.M.Blaton, O.M.Peeters, C.J.De Ranter
Eur.Cryst.Meeting,**5**, 128, 1979

83.C Trichloro – oxo – (N – pyridylmethylene – N' – salicyloylhydrazinato) – molybdenum(v) acetonitrile solvate
$C_{13}H_{11}Cl_3MoN_3O_3$, C_2H_3N Main entry is 84.47

83.C Tetracarbonyl – (S,S – dimethylsulfonium – 2 – picolinylmethylide) – tungsten(0)
$C_{13}H_{11}NO_5SW$ Main entry is 84.48

83.C (2 – Benzyl – pyridine) – methyl – mercury(ii) nitrate
$C_{13}H_{14}HgN^+$, NO_3^- Main entry is 71.53

83.C (μ – Ethoxycarbonylimido) – (μ – oxo) – bis((η – cyclopentadienyl) – oxo – molybdenum)
$C_{13}H_{15}Mo_2NO_5$ Main entry is 73.42

83.C (+)$_{546}$ – cis – α – Sodium carbonato – ((2S,2'S) – 1,1' – ethylene – di – 2 – pyrrolidine – carboxylato) – cobalt(iii) trihydrate
$C_{13}H_{18}CoN_2O_7^-$, Na^+, $3H_2O$ Main entry is 84.49

83.C Trimethylsilylnitrene – bis(η^5 – cyclopentadienyl) – vanadium (at –20°C)
$C_{13}H_{19}NSiV$ Main entry is 73.44

83.88 (2,2,3,9,10,10 – Hexamethyl – 5,7 – dioxa – 6 – hydra – 1,4,8,11 – tetra – aza – 13 – nitro – cyclotetradeca – 3,8,11,13 – tetraene) – nickel(ii) (at –136°C)
$C_{13}H_{21}N_5NiO_4$
M.S.Hussain, R.K.Murmann, E.O.Schlemper
Inorg.Chem., **19**, 1445, 1980

83.C (–)$_{589}$ – bis(Pentane – 2,4 – dionato) – (1,3 – propanediamine) – chromium(iii) iodide monohydrate (absolute configuration)
$C_{13}H_{24}CrN_2O_4^+$, I^-, H_2O Main entry is 84.50

83.C Dimethyl – (N,N – dimethylethanolamino) – (1 – pyrazolyl) – gallato – (η^2 – thiomethoxymethyl) – dicarbonyl – molybdenum
$C_{13}H_{24}GaMoN_3O_3S$ Main entry is 72.22

83.C Rubidium bis(pyridine – 2,6 – dicarboxylato) – chromium(iii)
$C_{14}H_6CrN_2O_8^-$, Rb^+ Main entry is 81.52

83.C Tetraethylammonium (μ – fluoro) – bis(oxo – peroxo – (pyridine – 2,6 – dicarboxylato) – molybdenum(vi))
$C_{14}H_6FMo_2N_2O_{14}^-$, $C_8H_{20}N^+$ Main entry is 81.53

83.89 bis(Benzonitrile) – trichloro – oxo – vanadium(v)
$C_{14}H_{10}Cl_3N_2OV$
J.–C.Daran, A.Gourdon, Y.Jeannin
Acta Crystallogr.,Sect.B,**36**, 309, 1980

83.C cis(N – (o – Aminobenzylidene) – anthranil – aldehydato – O,N,N) – chloro – platinum chloroform solvate (at –5°C)
$C_{14}H_{11}ClN_2OPt$, $CHCl_3$ Main entry is 84.53

83.90 (N – Methyl – cinnamaldehyde – imino) – tetracarbonyl – iron
$C_{14}H_{11}FeNO_4$
A.N.Nesmeyanov, L.V.Rybin, N.A.Stelzer, Yu.T.Struchkov, A.S.Batsanov, M.I.Rybinskaya
J.Organomet.Chem.,**182**, 399, 1979

83.91 bis(Benzamido) – mercury(ii)
$(C_{14}H_{12}HgN_2O_2)_n$
J.Halfpenny, R.W.H.Small
Acta Crystallogr.,Sect.B,**36**, 1194, 1980

83.C bis(p – Aminobenzoato) – zinc(ii) hydrate
$(C_{14}H_{12}N_2O_4Zn)_n$, $1.5nH_2O$ Main entry is 81.54

83.C (2,2' – Bipyridyl) – (imino – diacetato) – copper(ii) hexahydrate
$C_{14}H_{13}CuN_3O_4$, $6H_2O$ Main entry is 82.27

83.92 Benzamidoximato – N^1 – oxy – N^2 – benzoylimino – diazenato – nickel(ii)
$C_{14}H_{13}N_5NiO_2$
L.Malatesta, G.La Monica, M.Manassero, M.Sansoni
Gazz.Chim.Ital.,**110**, 113, 1980

83.C Aquo – bis(p – aminobenzoato) – cadmium(ii) dihydrate
$(C_{14}H_{14}CdN_2O_5)_n$, $2nH_2O$ Main entry is 81.55

83.93 Dichloro – bis(4 – vinylpyridine) – zinc(ii) (reformulation of cell)
$C_{14}H_{14}Cl_2N_2Zn$
R.E.Marsh, V.Schomaker *Inorg.Chem.*,**18**, 2331, 1979

83.C bis(Formato) – pyridine – copper
$C_{14}H_{14}Cu_2N_2O_8$ Main entry is 81.56

83.C (1,5 – Diphenylthiocarbazonato – N,S) – methyl – mercury(ii)
$C_{14}H_{14}HgN_4S$ Main entry is 71.66

83.94 bis(Isothiocyanato) – diaquo – bis(nicotinamide) – cobalt(ii)
$C_{14}H_{16}CoN_6O_4S_2$
G.V.Tsintsadze, Z.O.Dzhavakhishvili, I.R.Amiraslanov, A.N.Kvitashvili
Soobshch.Akad.Nauk Gruzh.SSR,**93**, 57, 1979

83.C   Aqua – bis(amidonicotinato) – bis(formato) –
copper(ii)
$C_{14}H_{16}CuN_4O_7$, Main entry is 84.55

83.C   (2,2′ – Bipyridyl) – nickela – cyclopentane
$C_{14}H_{16}N_2Ni$ Main entry is 71.67

83.C   Tetracarbonyl – cis(ethyl α – ethoxy – α – (1,3 –
dithian – 2 – ylidene) – acetimidate) – chromium(0)
$C_{14}H_{17}CrNO_6S_2$ Main entry is 85.49

83.95   Dibromo – bis(3,5 – dimethylpyridine) – copper(ii)
$C_{14}H_{18}Br_2CuN_2$
J.A.C.van Ooijen, J.Reedijk, E.J.Sonneveld, J.W.Visser
Transition Met.Chem.,**4**, 305, 1979

83.C   (μ² – Acetylene) – (σ,σ,η²,η² – glyoxal –
bis(isopropylamine)) – tetracarbonyl – di –
ruthenium
$C_{14}H_{18}N_2O_4Ru_2$ Main entry is 71.68

83.C   Diaqua – diformato – bis(nicotinamide) – zinc
$C_{14}H_{18}N_4O_8Zn$ Main entry is 84.56

83.C   Chloro – N,N – diethylamide – pyruvate – 4 –
phenylthiosemicarbazonato – copper(ii)
$C_{14}H_{19}ClCuN_4OS$ Main entry is 79.12

83.C   bis(2 – Dimethylamino – pyridine 1 – oxide) –
copper(ii) diperchlorate
$C_{14}H_{20}CuN_4O_2^{2+}$, $2ClO_4^-$ Main entry is 84.57

83.C   Chloro – glycylglycinato – imidazole – cadmium
$C_{14}H_{22}Cd_2Cl_2N_8O_8$ Main entry is 82.28

83.C   trans – Phenyl – bis(dimethylglyoximato) –
ammine – cobalt(iii)
$C_{14}H_{22}CoN_5O_4$ Main entry is 71.73

83.C   bis(2 – Aminoethyl)amine – (di – 2 –
pyridylamino) – zinc(ii) nitrate
$C_{14}H_{22}N_6Zn^{2+}$, $2NO_3^-$ Main entry is 76.82

83.96   bis(Dimethylglyoximato) – (N – iminopyridine) –
methyl – cobalt
$C_{14}H_{23}CoN_6O_4$
T.Saito, Y.Tsurita, Y.Sasaki *Inorg.Chem.*,**19**, 2365, 1980

83.C   Dichloro – (2,4,6 – trimethylpyridine) – (2 – (N –
hydroxy – t – butylimino)ethyl) – platinum
$C_{14}H_{24}Cl_2N_2OPt$ Main entry is 71.77

83.C   Ethylenediamine – bis(2 – aminomethylpyridine) –
cobalt(iii) hexacyano – cobalt(iii) dihydrate
$C_{14}H_{24}CoN_8^{3+}$, $C_6CoN_6^{3-}$, $2H_2O$ Main entry is 76.83

83.97   (2,3,9,10 – Tetramethyl – 1,4,8,11 – tetra – aza –
1,3,8,10 – cyclotetradecatetraene) – copper(ii)
bis(tetraphenylborate)
$C_{14}H_{24}CuN_4^{2+}$, $2C_{24}H_{20}B^-$
A.Elia, E.C.Lingafelter, V.Schomaker
Am.Cryst.Assoc.,Ser.2,**7**, 23, 1980
See also R2 : 62

83.98   bis(2 – (3 – Aminopropyl) – imino – butan – 3 –
one – oximato) – cobalt(iii) perchlorate
monohydrate
$C_{14}H_{28}CoN_6O_2^+$, ClO_4^-, H_2O
H.Saarinen, M.Nasakkala, J.Korvenranta, E.Nasakkala
Finn.Chem.Lett.,75, 1980

83.C   bis(Ethylenediamine) – copper – (μ –
ethylenediamine – tetra – acetato) – nickel
tetrahydrate
$C_{14}H_{28}CuN_6NiO_8$, $4H_2O$ Main entry is 84.58

83.C   (+)₅₄₆ – Oxalato – ((2S,4S,9S,11S) – 4,9 – dimethyl –
5,8 – diazadodecane – 2,11 – diamine) – cobalt(iii)
bromide trihydrate (absolute configuration)
$C_{14}H_{30}CoN_4O_4^+$, Br^-, $3H_2O$ Main entry is 81.59

83.C   tris(μ – Carbonyl) – (μ₄ – ethylamino) – (μ₄ – N –
oxy – ethylamino) – tetrakis(dicarbonyl – iron)
$C_{15}H_{10}Fe_4N_2O_{12}$ Main entry is 84.61

83.99   Sodium tris(μ – 3,5 – dimethylpyrazolyl) –
bis(nitrosyl – nickel(i)) tetrahydrofuran solvate
$C_{15}H_{21}N_8Ni_2O_2^-$, Na^+, $2C_4H_8O$
K.S.Chong, S.J.Rettig, A.Storr, J.Trotter
Can.J.Chem.,**57**, 3099, 1979

83.C   tris(Dimethylthiocarbamato) – (m –
nitrophenyldiazenato) – molybdenum methylene
chloride solvate hemihydrate
$2C_{15}H_{22}MoN_6O_2S_6$, $0.5CH_2Cl_2$, $0.5H_2O$ Main entry is 80.15

83.C   tris(Dimethyldithiocarbamato) –
(benzenediazenato) – molybdenum methylene
chloride solvate
$C_{15}H_{23}MoN_5S_6$, CH_2Cl_2 Main entry is 80.16

83.100   (2,12 – Dimethyl – 3,7,11,17 – tetra –
azabicyclo(11.3.1)heptadeca – 1(17),13,15 – triene) –
nickel(ii) perchlorate monohydrate (β form)
$C_{15}H_{26}N_4Ni^{2+}$, $2ClO_4^-$, H_2O
M.G.B.Drew, S.Hollis
Acta Crystallogr.,Sect.B,**36**, 718, 1980

83.101   (η² – Pentamethylene – diazirine) –
(pentacarbonyl – chromium) – (pentacarbonyl –
molybdenum)
$C_{16}H_{10}CrMoN_2O_{10}$
R.Battaglia, H.Kisch, C.Kruger, L.–K.Liu
Z.Naturforsch.,Teil B,**35**, 719, 1980

83.102   bis(1,8 – Naphthyridine) – dichloro – copper(ii)
$C_{16}H_{12}Cl_2CuN_4$
E.L.Enwall, K.Emerson
Acta Crystallogr.,Sect.B,**35**, 2562, 1979

83.103   Dichloro – bis(vinylbenzotriazole) – cobalt(ii)
$C_{16}H_{14}Cl_2CoN_6$
V.I.Sokol, Yu.V.Zefirov, M.A.Porai–Koshits
Koord.Khim.,**5**, 1249, 1979

83.C   bis(η⁵ – Cyclopentadienyl) – phenylhydrazido –
tungsten tetrafluoroborate
$C_{16}H_{17}N_2W^+$, BF_4^- Main entry is 73.60

83.104   trans – Dichloro – bis(γ –
ethoxycarbonylpyridine) – platinum(ii)
$C_{16}H_{18}Cl_2N_2O_4Pt$
M.Camalli, F.Caruso, L.Zambonelli
Cryst.Struct.Commun.,**9**, 721, 1980

83.C   bis(1 – Methyl – 3 – (2 – chloro – 6 –
methylphenyl) – triazine – 1 – oxidato) – nickel(ii)
$C_{16}H_{18}Cl_2N_6NiO_2$ Main entry is 84.62

83.C   bis(1 – Methyl – 3 – (2 – chloro – 6 –
methylphenyl) – triazine – 1 – oxidato) – nickel(ii)
benzene solvate
$C_{16}H_{18}Cl_2N_6NiO_2$, C_6H_6 Main entry is 84.63

83.C bis(μ – N – Isopropylidene – S – methyl –
dithiocarbazato) – bis(tricarbonyl) – manganese(i)
$C_{16}H_{18}Mn_2N_4O_6S_4$ Main entry is 85.51

83.C bis(D – β – (2 – Pyridyl) – α – alaninato) – nickel(ii)
dihydrate
$C_{16}H_{18}N_4NiO_4$, 2H$_2$O Main entry is 82.30

83.C bis(μ – Ethoxycarbonylimido) – bis((η –
cyclopentadienyl) – oxo – molybdenum)
$C_{16}H_{20}Mo_2N_2O_6$ Main entry is 73.62

83.C Aqua – bis(methylamine) – bis(benzoato) –
copper(ii) monohydrate
$C_{16}H_{22}CuN_2O_5$, H$_2$O Main entry is 81.61

83.105 (Difluoro – 3,3' – (trimethylene – dinitrilo) –
bis(butan – 2 – one – oximato) – borato) –
pyridine – copper(ii) perchlorate
$C_{16}H_{23}BCuF_2N_5O_2^+$, ClO$_4^-$
O.P.Anderson, A.B.Packard $Inorg.Chem.$,**19**, 2123, 1980

83.C Chloro – 1,9 – bis(2 – pyridyl) – 2,5,8 –
triazanonane – cobalt(iii) tetrachloro – zinc
$C_{16}H_{23}ClCoN_5$, Cl$_4$Zn Main entry is 76.92

83.106 Difluoro – tetrakis(5 – methylpyrazole) –
chromium(iii) tetrafluoroborate
$C_{16}H_{24}CrF_2N_8^+$, BF$_4^-$
P.Dapporto, F.Mani $J.Chem.Res.$,**374**, 4201, 1979

83.107 tetrakis(N – Methylimidazole) – platinum(ii)
hexachloro – platinum(iv)
$C_{16}H_{24}N_8Pt^{2+}$, Cl$_6$Pt^{2-}
M.B.Cingi, A.M.M.Lanfredi, A.Tiripicchio, C.G.van Kralingen,
J.Reedijk $Inorg.Chim.Acta$,**39**, 265, 1980

83.C Phenylnitrene – bis(diethyldithiocarbamato) –
dichloro – molybdenum chloroform solvate
(at –150°C)
$C_{16}H_{25}Cl_2MoN_3S_4$, CHCl$_3$ Main entry is 80.18

83.C bis(Dimethylglyoximato) – (isopropyl) – (pyridine) –
cobalt(iii)
$C_{16}H_{26}CoN_5O_4$ Main entry is 71.100

83.C bis(1H – 3,5 – Diethyl – 4 – methylpyrazole – N^2) –
dinitrato – O,O – cobalt(ii)
$C_{16}H_{28}CoN_6O_6$ Main entry is 84.64

83.C (+)$_{589}$ – β_2 – ((R) – Alaninato) – (1,7 – bis(2 – (S) –
pyrrolidyl) – 2,6 – diazaheptane) – cobalt(iii)
bis(perchlorate) dihydrate
$C_{16}H_{34}CoN_5O_2^{2+}$, 2ClO$_4^-$, 2H$_2$O Main entry is 82.34

83.C bis(μ – 2 – (N,N – Dimethyl – 2 – aminoethyl)
imino – butan – 3 – one – oximato) – diaqua – di –
copper(ii) diperchlorate
$C_{16}H_{36}Cu_2N_6O_4^{2+}$, 2ClO$_4^-$ Main entry is 84.67

83.108 cis – (S,S,S,R) – Bromo – aqua – ((2R,5R,8R,11R) –
2,5,8,11 – tetraethyl – 1,4,7,10 – tetra –
azacyclododecane) – cobalt(iii) dibromide
$C_{16}H_{38}BrCoN_4O^{2+}$, 2Br$^-$
T.Sakurai, S.Tsuboyama, K.Tsuboyama
$Acta Crystallogr.,Sect.B$,**36**, 1797, 1980

83.C tetrakis((2 – Amino – 2 – methyl – 1 –
propanolato) – chloro – copper(ii))
$C_{16}H_{40}Cl_4Cu_4N_4O_4$ Main entry is 84.70

83.C (μ – Oxo) – bis(bis(N,N – diethylhydroxylaminato) –
oxo – vanadium(v))
$C_{16}H_{40}N_4O_7V_2$ Main entry is 84.71

83.C bis(σ – Pyrrolyl – tricarbonyl – manganese) –
tricarbonyl – iodo – manganese
$C_{17}H_8IMn_3N_2O_9$ Main entry is 71.104

83.109 (μ^2 – Hydrido) – (μ^2 – p – toluenesulfonylamide) –
decacarbonyl – tri – osmium
$C_{17}H_9NO_{12}Os_3S$
M.R.Churchill, F.J.Hollander, J.R.Shapley, J.B.Keister
$Inorg.Chem.$,**19**, 1272, 1980

83.C Azobenzyl – hexafluoroacetylacetonato – palladium
(α form, yellow)
$C_{17}H_{10}F_6N_2O_2Pd$ Main entry is 77.4

83.C Azobenzyl – hexafluoroacetylacetonato – palladium
(β form, red)
$C_{17}H_{10}F_6N_2O_2Pd$ Main entry is 77.5

83.C bis(μ^2 – Hydrido) – (μ – N – methylbenzylamido –
1 – yl) – nonacarbonyl – tri – osmium
$C_{17}H_{11}NO_9Os_3$ Main entry is 71.1

83.C Deoxycytidylyl – (3' – 5') – deoxyguanosine (2 –
hydroxy – ethylthiolato) – 2,2',2'' – terpyridine –
platinum(ii) hydrate
2$C_{17}H_{16}N_3OPtS^+$, 2$C_{19}H_{24}N_8O_{10}P^-$, xH$_2$O Main entry is 85.58

83.C Aqua – (acetylacetonato) – (o – phenanthroline) –
copper hexafluoroacetylacetone monohydrate
$C_{17}H_{17}CuN_2O_3^+$, C$_5$HF$_6O_2^-$, H$_2$O Main entry is 77.6

83.C 2 – Hydroxy – N – salicylidene – aniline –
(diethylamine) platinum(ii)
$C_{17}H_{20}N_2O_2Pt$ Main entry is 78.5

83.C (η – Allyl) – (8 – isopropylquinoline – 2 –
carboxaldehyde – N – methylimino) – palladium(ii)
perchlorate
$C_{17}H_{21}N_2Pd^+$, ClO$_4^-$ Main entry is 72.30

83.C (2,6 – bis(1 – Methyl – 4 – methylthio – 5 – thia –
2,3 – diazahexa – 1,3 – dienyl) – pyridine) –
dithiocyanato – nickel(ii)
$C_{17}H_{21}N_7NiS_6$ Main entry is 85.59

83.C N,N' – Trimethylene – bis(methyl – 2 – amino – 1 –
cyclopentene – dithiocarboxylato) – copper(ii)
$C_{17}H_{24}CuN_2S_4$ Main entry is 85.60

83.110 Acetonitrile(carbonyl) – (2,3,9,10 – tetramethyl –
1,4,8,11 – tetra – aza – 1,3,8,11 –
cyclotetradecatetraene) – iron(ii)
hexafluorophosphate
$C_{17}H_{27}FeN_5O^{2+}$, 2F$_6$P$^-$
L.E.McCandlish, B.D.Santarsiero, N.J.Rose, E.C.Lingafelter
$Acta Crystallogr.,Sect.B$,**35**, 3053, 1979

83.C bis(3 – Chloropropynyl) – (1,10 – phenanthroline) –
mercury
$C_{18}H_{12}Cl_2HgN_2$ Main entry is 71.120

83.C (2' – Chlorophenyl – diazoaminobenzene) – N –
(phenyl – mercury) (at –100°C)
$C_{18}H_{14}ClHgN_3$ Main entry is 71.124

83.C bis(p – Fluorobenzenediazo) – (η^5 –
methylcyclopentadienyl) – chloro – molybdenum
$C_{18}H_{15}ClF_2MoN_4$ Main entry is 73.72

83.C Triphenylphosphine – trinitrosyl – manganese
$C_{18}H_{15}MnN_3O_3P$ Main entry is 86.49

83.111 bis(1,2 – Benzoquinone – di – imide) – 1,2 – benzosemiquinone – di – imido – cobalt(ii) tetraphenylborate pentahydrate
$C_{18}H_{18}CoN_6^+$, $C_{24}H_{20}B^-$, $5H_2O$
M.Zehnder, H.Loliger *Helv.Chim.Acta,***63**, 754, 1980

83.C Di – η^5 – cyclopentadienyl – di – η – pyrrolyl – titanium(iv)
$C_{18}H_{18}N_2Ti$ Main entry is 73.74

83.C Di – η^5 – cyclopentadienyl – di – η – pyrrolyl – zirconium(iv)
$C_{18}H_{18}N_2Zr$ Main entry is 73.75

83.112 bis(Hydro – tris(1 – pyrazolyl) – borato) – copper(ii)
$C_{18}H_{20}B_2CuN_{12}$
A.Murphy, B.J.Hathaway, T.J.King
*J.Chem.Soc.,Dalton Trans.,*1646, 1979
See also R1 : 62

83.113 bis(Hydro – tris(1 – pyrazolyl) – borato) – iron(ii)
$C_{18}H_{20}B_2FeN_{12}$
J.D.Oliver, D.F.Mullica, B.B.Hutchinson, W.O.Milligan
*Inorg.Chem.,***19**, 165, 1980

83.114 bis(Hydrido – tris(1 – pyrazolyl) – borato) – nickel(ii)
$C_{18}H_{20}B_2N_{12}Ni$
G.Bandoli, D.A.Clemente, G.Paolucci, L.Doretti
*Cryst.Struct.Commun.,***8**, 965, 1979
See also R1 : 62

83.C bis(μ – 2,6 – Diacetylpyridine – dioximato) – di – copper(ii) tetrafluoroborate dihydrate
$C_{18}H_{20}Cu_2N_6O_4^{2+}$, $2BF_4^-$, $2H_2O$ Main entry is 84.72

83.115 Tribromo – tris(4 – methylpyridine) – molybdenum(iii) 4 – methylpyridine solvate
$C_{18}H_{21}Br_3MoN_3$, $0.5C_6H_7N$
J.V.Brencic, I.Leban, M.Slokar
*Acta Crystallogr.,Sect.B,***36**, 698, 1980

83.116 Trichloro – tris(4 – methylpyridine) – molybdenum 4 – methylpyridine solvate
$C_{18}H_{21}Cl_3MoN_3$, $0.5C_6H_7N$
J.V.Brencic, I.Leban *Z.Anorg.Allg.Chem.,***465**, 173, 1980

83.C tetrakis(μ – Acetato) – bis(N – pyridine – chromium)
$C_{18}H_{22}Cr_2N_2O_8$ Main entry is 81.65

83.C Dichloro – (1 – diphenylphosphino – 2 – diethylaminoethane) – mercury(ii) (at 138°K)
$C_{18}H_{24}Cl_2HgNP$ Main entry is 86.50

83.117 tris(2 – Picolylamine) – iron(ii) chloride methanol solvate (at 227°K)
$C_{18}H_{24}FeN_6^{2+}$, $2Cl^-$, CH_4O
B.A.Katz, C.E.Strouse *J.Am.Chem.Soc.,***101**, 6214, 1979

83.118 tris(2 – Picolylamine) – iron(ii) chloride methanol solvate (at 199°K)
$C_{18}H_{24}FeN_6^{2+}$, $2Cl^-$, CH_4O
B.A.Katz, C.E.Strouse *J.Am.Chem.Soc.,***101**, 6214, 1979

83.119 tris(2 – Picolylamine) – iron(ii) chloride methanol solvate (at 171°K)
$C_{18}H_{24}FeN_6^{2+}$, $2Cl^-$, CH_4O
B.A.Katz, C.E.Strouse *J.Am.Chem.Soc.,***101**, 6214, 1979

83.120 tris(2 – Picolylamine) – iron(ii) chloride methanol solvate (at 148°K)
$C_{18}H_{24}FeN_6^{2+}$, $2Cl^-$, CH_4O
B.A.Katz, C.E.Strouse *J.Am.Chem.Soc.,***101**, 6214, 1979

83.121 tris(2 – Picolylamine) – iron(ii) chloride methanol solvate (at 115°K)
$C_{18}H_{24}FeN_6^{2+}$, $2Cl^-$, CH_4O
B.A.Katz, C.E.Strouse *J.Am.Chem.Soc.,***101**, 6214, 1979

83.122 tris(2 – Picolylamine) – iron(ii) chloride ethanol solvate (form A)
$C_{18}H_{24}FeN_6^{2+}$, $2Cl^-$, C_2H_6O
A.M.Greenaway, C.J.O'Connor, A.Schrock, E.Sinn
*Inorg.Chem.,***18**, 2692, 1979

83.123 tris(2 – Picolylamine) – iron(ii) chloride ethanol solvate (at 115°K)
$C_{18}H_{24}FeN_6^{2+}$, $2Cl^-$, C_2H_6O
B.A.Katz, C.E.Strouse *J.Am.Chem.Soc.,***101**, 6214, 1979

83.124 tris(2 – Picolylamine) – iron(ii) chloride ethanol solvate (high spin state, at 150°K)
$C_{18}H_{24}FeN_6^{2+}$, $2Cl^-$, C_2H_6O
M.Mikami, M.Konno, Y.Saito
*Acta Crystallogr.,Sect.B,***36**, 275, 1980

83.125 tris(2 – Picolylamine) – iron(ii) chloride ethanol solvate (high spin state)
$C_{18}H_{24}FeN_6^{2+}$, $2Cl^-$, C_2H_6O
M.Mikami, M.Konno, Y.Saito
*Acta Crystallogr.,Sect.B,***36**, 275, 1980

83.126 tris(2 – Picolylamine) – iron(ii) chloride ethanol solvate (at 90°K, low spin state, form B)
$C_{18}H_{24}FeN_6^{2+}$, $2Cl^-$, C_2H_6O
M.Mikami, M.Konno, Y.Saito
*Acta Crystallogr.,Sect.B,***36**, 275, 1980

83.127 tris(2 – Picolylamine) – iron(ii) dichloride dihydrate
$C_{18}H_{24}FeN_6^{2+}$, $2Cl^-$, $2H_2O$
B.A.Katz, C.E.Strouse *Inorg.Chem.,***19**, 658, 1980

83.128 tris(2 – Picolylamine) – iron(ii) dichloride dihydrate (at 115°K)
$C_{18}H_{24}FeN_6^{2+}$, $2Cl^-$, $2H_2O$
B.A.Katz, C.E.Strouse *Inorg.Chem.,***19**, 658, 1980

83.129 tris(2 – Picolylamine) – iron(ii) di – iodide
$C_{18}H_{24}FeN_6^{2+}$, $2I^-$
B.A.Katz, C.E.Strouse *Inorg.Chem.,***19**, 658, 1980

83.130 tris(2 – Picolylamine) – nickel(ii) bis(tetrafluoroborate)
$C_{18}H_{24}N_6Ni^{2+}$, $2BF_4^-$
M.L.Niven, L.R.Nassimbeni
*Cryst.Struct.Commun.,***9**, 227, 1980

83.131 Chloro – (1,11 – bis(2 – pyridyl) – 2,6,10 – triazaundecane) – cobalt(iii) tetrachloro – cobalt(ii) hemihydrate
$C_{18}H_{27}ClCoN_5^{2+}$, Cl_4Co^{2-}, $0.5H_2O$
G.Bombieri, E.Forsellini, A.Del Pra, M.L.Tobe
*Inorg.Chim.Acta,***40**, 71, 1980

83.C Dichloro – (N,N' – di – t – butyl – ethylenedi – imine) – styrene – platinum
$C_{18}H_{28}Cl_2N_2Pt$ Main entry is 72.33

83.C Bipyridyl – bis(trimethylsilylmethyl) – cadmium(ii) 2,2' – bipyridyl
$C_{18}H_{30}CdN_2Si_2$, $0.5C_{10}H_8N_2$ Main entry is 71.130

83.132 1 – Methylimidazole – (2,3,9,10 – tetramethyl – 1,4,8,11 – tetra – aza – 1,3,8,10 – cyclotetradecatetraene) – copper(ii) bis(hexafluorophosphate)
$C_{18}H_{30}CuN_6^{2+}$, $2F_6P^-$
A.Elia, E.C.Lingafelter, V.Schomaker
Am.Cryst.Assoc.,Ser.2,**7**, 23, 1980

83.133 Cesium hydrogen diammine – bis(cytidine – 3′ – phosphato) – platinum(ii) tetrahydrate
$C_{18}H_{30}N_8O_{16}P_2Pt^{2-}$, $0.5Cs^+$, $1.5H^+$, $4H_2O$
S.–M.Wu, R.Bau
Biochem.Biophys.Res.Commun.,**88**, 1435, 1979
See also R1 : 47

83.134 cis – Dicarbonyl – glyoxal – bis(2,4 – dimethylpentyl – 3 – imino) – rhodium(i) cis – dicarbonyl – dichloro – rhodium(i)
$C_{18}H_{32}N_2O_2Rh^+$, $C_2Cl_2O_2Rh^-$
J.Kopf, J.Klaus, H.tom Dieck
Cryst.Struct.Commun.,**9**, 783, 1980

83.C (4,7,13,16,21,24 – Hexaoxa – 1,10 – diazabicyclo(8.8.8) hexacosane) – nitrato – samarium(iii) aquo – pentanitrato – samarium(iii)
$C_{18}H_{36}N_3O_9Sm^{2+}$, $H_2N_5O_{16}Sm^{2-}$ Main entry is 84.74

83.135 5,12 – Dimethyl – 7,14 – di – isopropyl – 1,4,8,11 – tetra – azacyclotetradecane – nickel(ii) perchlorate (violet form)
$C_{18}H_{40}N_4Ni^{2+}$, $2ClO_4^-$
J.W.Krajewski, Z.Urbanczyk–Lipkowska, P.Gluzinski
Eur.Cryst.Meeting,**5**, 204, 1979

83.C Pyridinium tetrachloro – (pyridine) – (tolane) – tantalum
$C_{19}H_{15}Cl_4NTa^-$, $C_5H_6N^+$ Main entry is 71.137

83.C (1,5 – Diphenylthiocarbazonato – N,S) – phenyl – mercury(ii) (yellow form)
$C_{19}H_{16}HgN_4S$ Main entry is 71.138

83.C Bromo – tricarbonyl – (phenyl – bis – (3,5 – dimethylpyrazolyl)phosphine) – rhenium(i)
$C_{19}H_{19}BrN_4O_3PRe$ Main entry is 86.60

83.C Tricarbonyl – (phenyl – bis(3,5 – dimethylpyrazolyl) phosphine) – tungsten(0)
$C_{19}H_{19}N_4O_3PW$ Main entry is 86.61

83.C Nitrato – bis(dioxo – 2,6 – diacetylpyridine – bis(2′ – pyridylhydrazone)) – uranium dioxo – uranium – tetranitrate
$2C_{19}H_{19}N_8O_5U^+$, $N_4O_{14}U^{2-}$ Main entry is 84.75

83.136 Aqua – dinitrato – perchlorato – (4,5 – bis((2 – (2 – pyridyl) – ethylimino) – methyl) – imidazolato) – di – copper(ii) hydrate
$C_{19}H_{21}ClCu_2N_8O_{11}$, H_2O
J.C.Dewan, S.J.Lippard *Inorg.Chem.*,**19**, 2079, 1980

83.137 Nitrato – (2,6 – dihydroxy – 8,15 – dimethyl – tripyrido(c,d,i,j,l,m)(1,4,7,8,10,13,15) – hepta – aza – 1,2,6,7,8,15 – hexahydro – cyclopentadecine) – zinc(ii) nitrate
$C_{19}H_{21}N_8O_5Zn^+$, NO_3^-
Z.P.Haque, M.McPartlin, P.A.Tasker
Inorg.Chem.,**18**, 2920, 1979

83.C cis – Tetracarbonyl – ((1 – methyl – 2 – phenyl – 3 – aza – but – 2 – enyl) – diethylaminocarbene) – chromium
$C_{19}H_{22}CrN_2O_4$ Main entry is 71.141

83.C (–) – RS – Cyclo – 1 – (1′ – dimethylaminoethyl – ferrocene) – 2 – acetylacetonato – palladium (absolute configuration)
$C_{19}H_{25}FeNO_2Pd$ Main entry is 73.80

83.138 tris(μ – 4 – Methyl – 1,2,4 – triazole) – bis(4 – methyl – 1,2,4 – triazole – bis(isothiocyanato) – manganese(ii))
$C_{19}H_{25}Mn_2N_{19}S_4$
D.W.Engelfriet, G.C.Verschoor, W.J.Vermin
Acta Crystallogr.,Sect.B,**35**, 2927, 1979

83.C ((R) – 1 – Cyanoethyl) – ((S) – (–) – α – methylbenzylamine) – bis(dimethylglyoximato) – cobalt(iii)
$C_{19}H_{29}CoN_6O_4$ Main entry is 71.143

83.C ((R) – 1 – Cyanoethyl) – ((S) – (–) – α – methylbenzylamine) – bis(dimethylglyoximato) – cobalt(iii) (at 173°K)
$C_{19}H_{29}CoN_6O_4$ Main entry is 71.144

83.C ((S) – 1 – Cyanoethyl) – ((S) – (–) – α – methylbenzylamine) – bis(dimethylglyoximato) – cobalt(iii) (at 173°K)
$C_{19}H_{29}CoN_6O_4$ Main entry is 71.145

83.C ((S) – 1 – Cyanoethyl) – ((S) – (–) – α – methylbenzylamine) – bis(dimethylglyoximato) – cobalt(iii)
$C_{19}H_{29}CoN_6O_4$ Main entry is 71.146

83.C bis(μ_3 – Acetonitrile – copper) – carbido – ennea – (μ – carbonyl) – hexacarbonyl – hexa – rhodium methanol solvate
$C_{20}H_6Cu_2N_2O_{15}Rh_6$, $0.5CH_4O$ Main entry is 71.147

83.C tetrakis(μ^2 – 6 – Chloro – 2 – pyridinolato) – di – chromium
$C_{20}H_{12}Cl_4Cr_2N_4O_4$ Main entry is 84.76

83.C tetrakis(μ^2 – 6 – Chloro – 2 – pyridinolato) – di – molybdenum
$C_{20}H_{12}Cl_4Mo_2N_4O_4$ Main entry is 84.77

83.C tetrakis(μ^2 – 6 – Chloro – 2 – pyridinolato) – di – tungsten
$C_{20}H_{12}Cl_4N_4O_4W_2$ Main entry is 84.78

83.139 tris(μ – Hydrido) – decacarbonyl – dipyridine – tri – rhenium
$C_{20}H_{13}N_2O_{10}Re_3$
G.Ciani, G.D'Alfonso, M.Freni, P.Romiti, A.Sironi
J.Organomet.Chem.,**186**, 353, 1980

83.140 bis(2,2′ – Bipyridyl) – monobromo – copper(ii) tetrafluoroborate
$C_{20}H_{16}BrCuN_4^+$, BF_4^-
B.J.Hathaway, A.Murphy
Acta Crystallogr.,Sect.B,**36**, 295, 1980

83.C bis(2 – Methyl – 8 – mercapto – quinolinato) – cadmium
$C_{20}H_{16}CdN_2S_2$ Main entry is 85.63

83.C tetrakis(μ – 2 – Oxo – pyridinato) – (chloro – di – technetium)
$(C_{20}H_{16}ClN_4O_4Tc_2)_n$ Main entry is 84.79

83.C tetrakis(μ – α – Pyridonato) – bis(chloro – osmium)
$C_{20}H_{16}Cl_2N_4O_4Os_2$ Main entry is 84.80

83.C   bis(N – Methylimidazole) – bis(2,4,6 –
trichlorophenolato) – cobalt(ii)
$C_{20}H_{16}Cl_6CoN_4O_2$ Main entry is 84.81

83.C   (3,4,7,8 – Tetramethyl – 1,10 – phenanthroline) –
(bis(trichlorovinyl)) – mercury
$C_{20}H_{16}Cl_6HgN_2$ Main entry is 71.152

83.141 bis(2,2′ – Bipyridyl) – mono – iodo – copper(ii)
perchlorate
$C_{20}H_{16}CuIN_4{}^+$, $ClO_4{}^-$
B.J.Hathaway, A.Murphy
Acta Crystallogr.,Sect.B,**36**, 295, 1980

83.C   bis(5 – Methylthio – 8 – mercapto – quinolinato) –
copper(ii)
$C_{20}H_{16}CuN_2S_4$ Main entry is 85.64

83.C   bis(2,2′ – Bipyridyl) – peroxo – disulfato – copper(ii)
monohydrate
$C_{20}H_{16}CuN_4O_8S_2$, H_2O Main entry is 84.82

83.142 bis(2,2′ – Bipyridyl) – nitrato – copper(ii) nitrate
monohydrate
$C_{20}H_{16}CuN_5O_3{}^+$, $NO_3{}^-$, H_2O
H.Nakai Bull.Chem.Soc.Jpn.,**53**, 1321, 1980

83.143 bis(2,2′ – Bipyridyl) – mercury(ii) dinitrate
dihydrate
$C_{20}H_{16}HgN_4{}^{2+}$, $2NO_3{}^-$, $2H_2O$
D.Grdenic, B.Kamenar, A.Hergold–Brundic
Croat.Chem.Acta,**52**, 339, 1979

83.144 bis(2,2′ – Dipyridylamino) – chloro – copper(ii)
chloride tetrahydrate
$C_{20}H_{18}ClCuN_6{}^+$, Cl^-, $4H_2O$
W.P.Jensen, R.A.Jacobson   Proc.N.D.Acad.Sci.,**20**, 1979

83.145 Aqua – bis(2,2′ – bipyridyl) – copper(ii) dithionate
$C_{20}H_{18}CuN_4O^{2+}$, $O_6S_2{}^{2-}$
W.D.Harrison, B.J.Hathaway
Acta Crystallogr.,Sect.B,**35**, 2910, 1979

83.146 trans – Aquo – hydroxo – bis(bipyridyl) –
ruthenium perchlorate
$C_{20}H_{19}N_4O_2Ru^{2+}$, $2ClO_4{}^-$
B.Durham, S.R.Wilson, D.J.Hodgson, T.J.Meyer
J.Am.Chem.Soc.,**102**, 600, 1980

83.147 Chloro – bis(α – hydroxylamino –
isobutyrophenone – oximato) – cobalt(iii)
$C_{20}H_{26}ClCoN_4O_4$
S.A.D'yachenko, Yu.A.Simonov, A.I.Stetsenko, A.A.Dvorkin,
T.I.Malinovskii, L.B.Volodarskii
Dokl.Akad.Nauk SSSR,**244**, 636, 1979

83.148 cis,cis – Dicarbonyl – bis(trimethylphosphite) –
phenanthroline – manganese perchlorate
$C_{20}H_{26}MnN_2O_8P_2{}^+$, $ClO_4{}^-$
M.Ulibarri, J.Fayos Eur.Cryst.Meeting,**5**, 315, 1979

83.C   cis,trans – (Dicarbonyl – bis(trimethylphosphite) –
phenanthroline) – manganese perchlorate
$C_{20}H_{26}MnN_2O_8P_2{}^+$, $ClO_4{}^-$ Main entry is 86.67

83.149 (2,12 – bis(2 – Pyridyl) – 3,7,11 – triazatridecane –
2,11 – diene) – nitrosyl – manganese(ii)
diperchlorate
$C_{20}H_{27}MnN_6O^{2+}$, $2ClO_4{}^-$
D.J.Cooper, M.D.Ravenscroft, D.A.Stotter, J.Trotter
J.Chem.Res.,**287**, 3359, 1979

83.C   bis(1,3 – Propanediamine) – bis(m –
nitrobenzoato) – copper(ii)
$C_{20}H_{28}CuN_6O_8$ Main entry is 81.68

83.C   trans – bis(t – Butoxy) – cis – dicarbonyl – cis –
bis(pyridine) – molybdenum (at –175°C)
$C_{20}H_{28}MoN_2O_4$ Main entry is 84.84

83.C   bis((m – Nitrobenzoato) – (1,3 – propanediamine)) –
nickel(ii)
$C_{20}H_{28}N_6NiO_8$ Main entry is 81.69

83.150 bis(bis(3,5 – Dimethylpyrazolyl) – nitrosyl –
nickel(i)) – nickel(ii)
$C_{20}H_{28}N_{10}Ni_3O_2$
K.S.Chong, S.J.Rettig, A.Storr, J.Trotter
Can.J.Chem.,**57**, 3090, 1979

83.C   bis(N,N – Dimethyldithiocarbamato) – (η – N –
methyl – N – phenylhydrazido) – (η² – N – methyl –
N – phenylhydrazido) – molybdenum
tetraphenylborate
$C_{20}H_{29}MoN_6S_4{}^+$, $C_{24}H_{20}B^-$ Main entry is 80.19

83.151 cis – Diammine – bis(guanosine) – platinum(ii)
chloride perchlorate heptahydrate
$C_{20}H_{32}N_{12}O_{10}Pt^{2+}$, $1.5Cl^-$, $0.5ClO_4{}^-$, $7H_2O$
R.E.Cramer, P.L.Dahlstrom, M.J.T.Seu, T.Norton,
M.Kashiwagi Inorg.Chem.,**19**, 148, 1980
See also R1 : 47

83.152 cis – ((5SR,10RS) – 2,4,4,11,11,13 – Hexamethyl –
1,5,10,14 – tetra – azacyclo – octadeca – 1,13 –
diene) – copper(ii) perchlorate
$C_{20}H_{40}CuN_4{}^{2+}$, $2ClO_4{}^-$
J.H.Timmons, P.Rudolf, A.E.Martell, J.W.L.Martin,
A.Clearfield Inorg.Chem.,**19**, 2331, 1980

83.153 trans – ((5RS,14RS) – 2,4,4,11,13,13 – Hexamethyl –
1,5,10,14 – tetra – azacyclo – octadeca – 1,10 –
diene) – copper(ii) perchlorate
$C_{20}H_{40}CuN_4{}^{2+}$, $2ClO_4{}^-$
J.H.Timmons, P.Rudolf, A.E.Martell, J.W.L.Martin,
A.Clearfield Inorg.Chem.,**19**, 2331, 19

83.C   trans – Chloro – bis(dimethylglyoximato) –
tributylphosphine – cobalt(iii)
$C_{20}H_{41}ClCoN_4O_4P$ Main entry is 86.73

83.154 bis((μ – N – t – Butylimido) – (N – t –
butylimido)) – tetramethyl – di – molybdenum
(at –95°C)
$C_{20}H_{48}Mo_2N_4$
W.A.Nugent, R.L.Harlow J.Am.Chem.Soc.,**102**, 1759, 1980

83.155 1,3,5,7 – Tetra – t – butyl – 2,2,6,6 – tetramethyl –
1,3,5,7 – tetra – aza – 2,6 – disila – 4 – titana –
spiro(3.3)heptane
$C_{20}H_{48}N_4Si_2Ti$
D.J.Brauer, H.Burger, E.Essig, W.Geschwandtner
J.Organomet.Chem.,**190**, 343, 1980
See also R1 : 63

83.156 1,3,5,7 – Tetra – t – butyl – 2,2,6,6 – tetramethyl –
1,3,5,7 – tetra – aza – 2,6 – disila – 4 – zirconia –
spiro(3.3)heptane
$C_{20}H_{48}N_4Si_2Zr$
D.J.Brauer, H.Burger, E.Essig, W.Geschwandtner
J.Organomet.Chem.,**190**, 343, 1980
See also R1 : 63

83.C (μ_2 – Dioxo) – bis(1,5,8,11,15 – penta – azapentadecane – cobalt) dithionate dinitrate tetrahydrate
$C_{20}H_{54}Co_2N_{10}O_2^{4+}$, $O_6S_2^{2-}$, $4H_2O$, $2NO_3^-$ Main entry is 76.97

83.C bis((1,2 – Dimethoxyethane) – bis(trimethylsilyl) amino) – europium(ii)
$C_{20}H_{56}EuN_2O_4Si_4$ Main entry is 84.85

83.C (bis(Benzothiazole – 2 – yl) – trisulfide) – (2 – mercapto – benzothiazole) – copper(i) perchlorate chloroform solvate
$C_{21}H_{13}CuN_3S_7^+$, ClO_4^-, $2CHCl_3$ Main entry is 85.65

83.157 cis – bis(2,2' – Bipyridyl) – carbonyl – chloro – ruthenium(ii) perchlorate
$C_{21}H_{16}ClN_4ORu^+$, ClO_4^-
J.M.Clear, J.M.Kelly, C.M.O'Connell, J.G.Vos, C.J.Cardin, S.R.Costa, A.J.Edwards
J.Chem.Soc.,Chem.Commun.,750, 1980

83.C Carbonyl – chloro – (o – (diphenylphosphino) – N,N – dimethylaniline) – iridium
$C_{21}H_{20}ClIrNOP$ Main entry is 86.75

83.C Pentafluorophenylthiolato – (hydro – tris(3,5 – dimethyl – 1 – pyrazolyl) – borato) – cobalt (at –162°C)
$C_{21}H_{22}BCoF_5N_6S$ Main entry is 85.67

83.C bis(Isothiocyanato) – (2,3:11,12) – dibenzo – 1,13 – dioxa – 5,9 – diaza – 2,11 – cyclopentadecadiene – nickel(ii) (light blue phase)
$C_{21}H_{24}N_4NiO_2S_2$ Main entry is 84.89

83.C Potassium (p – nitrobenzenethiolato) – (hydro – tris(3,5 – dimethylpyrazolyl) – borato) – copper(i) acetone solvate
$C_{21}H_{26}BCuN_7O_2S^-$, K^+, $2C_3H_6O$ Main entry is 85.68

83.C (Acetylacetonato) – (hexafluoroacetylacetonato) – (o – phenanthroline) – copper(ii)
$C_{22}H_{16}CuF_6N_2O_4$ Main entry is 77.8

83.158 (17,18,19,20 – Tetrahydro – tribenzo(e,i,m)(1,4,8,11) – tetra – azacyclotetradecinato) – oxo – vanadium(iv)
$C_{22}H_{18}N_4OV$
A.J.Greenwood, K.Henrick, P.G.Owston, P.A.Tasker
J.Chem.Soc.,Chem.Commun.,88, 1980

83.C Acetato – bis(2,2' – bipyridyl) – copper(ii) tetrafluoroborate
$C_{22}H_{19}CuN_4O_2^+$, BF_4^- Main entry is 81.71

83.C Acetato – bis(2,2' – bipyridyl) – copper(ii) perchlorate monohydrate
$C_{22}H_{19}CuN_4O_2^+$, ClO_4^-, H_2O Main entry is 81.72

83.159 trans – bis(Isothiocyanato) – tetrakis(pyridine) – cobalt(ii)
$C_{22}H_{20}CoN_6S_2$
H.Hartl, I.Brudgam
Acta Crystallogr.,Sect.B,**36**, 162, 1980

83.160 trans – bis(Isothiocyanato) – tetrakis(pyridine) – cobalt(ii) iodoform solvate
$C_{22}H_{20}CoN_6S_2$, $2CHI_3$
H.Hartl, S.Steidl *Acta Crystallogr.,Sect.B*,**36**, 65, 1980

83.C cis – bis(2,2' – Bipyridyl) – dimethyl – cobalt(iii) tetraethyl – aluminium
$C_{22}H_{22}CoN_4^+$, $C_8H_{20}Al^-$ Main entry is 71.179

83.C (N,N' – Ethylene – bis(benzoylacetoniminato)) – nickel(ii)
$C_{22}H_{22}N_2NiO_2$ Main entry is 84.91

83.C Methyl – (4,4',4'' – triethyl – 2,2':6',2'' – terpyridyl) – mercury nitrate
$C_{22}H_{26}HgN_3^+$, NO_3^- Main entry is 71.180

83.C (Benzoyl – (1 – pyridinio) – methanido) – bis(dimethylglyoximato) – methyl – cobalt benzene solvate
$C_{22}H_{28}CoN_5O_5$, C_6H_6 Main entry is 71.182

83.C tetrakis(μ – Acetato) – bis(theophylline – rhodium) dihydrate
$C_{22}H_{28}N_8O_{12}Rh_2$, $2H_2O$ Main entry is 81.73

83.C bis(2,6 – Diacetylpyridine – bis(semicarbazone)) – cerium(iii) triperchlorate trihydrate
$C_{22}H_{30}CeN_{14}O_4^{3+}$, $3ClO_4^-$, $3H_2O$ Main entry is 84.92

83.C rac – a – (μ – Acetato) – b – (0 – acetato – mercurio) – cf,de – bis(2 – (dimethylaminomethyl) – phenyl) – platinum
$C_{22}H_{30}HgN_2O_4Pt$ Main entry is 71.183

83.161 bis(μ – Chloro) – bis(chloro – (bis(3,5 – dimethylpyrazolyl)methane) – nickel(ii))
$C_{22}H_{32}Cl_4N_8Ni_2$
J.C.Jansen, H.van Koningsveld, J.A.C.van Ooijen, J.Reedijk
Inorg.Chem.,19, 170, 1980

83.162 Dicyanato – diaquo – bis(N,N – diethylnicotinamide) – cobalt (revision of previous structure description in Zh. Strukt. Khim., 17,1124,1976)
$C_{22}H_{32}CoN_6O_6$
V.S.Sergienko, V.N.Shchurkina, B.Ya.Rubinchik, T.S.Khodashova, M.A.Porai–Koshits, G.V.Tsintsadze
Koord.Khim.,5, 585, 1979

83.C bis((m – Methylbenzoato) – (1,3 – propanediamine)) – nickel(ii)
$C_{22}H_{34}N_4NiO_4$ Main entry is 81.75

83.C bis((p – Methylbenzoato) – (1,3 – propanediamine)) – nickel(ii)
$C_{22}H_{34}N_4NiO_4$ Main entry is 81.76

83.C trans – bis(Dimethylglyoximato) – 2 – (5 – trifluoro – methyltetrazolato) – (tri – n – butylphosphine) – cobalt(iii)
$C_{22}H_{41}CoF_3N_8O_4P$ Main entry is 86.81

83.163 (2,6 – Diacetylpyridine) – (2,9 – bis(1 – methylhydrazone) – 1,10 – phenanthroline) – chloro – manganese tetrafluoroborate
$C_{23}H_{21}ClMnN_7^+$, BF_4^-
J.Lewis, T.D.O'Donoghue, P.R.Raithby
J.Chem.Soc.,Dalton Trans.,1383, 1980

83.C Trinitrato – (2,6 – diacetylpyridine – bis(benzoic acid hydrazone)) – lanthanum(iii)
$C_{23}H_{21}LaN_8O_{11}$ Main entry is 84.93

83.C Aqua – nitrato – (2,6 – diacetylpyridine – bis(benzoic acid hydrazone)) – cobalt(ii) nitrate
$C_{23}H_{23}CoN_6O_6^+$, NO_3^- Main entry is 84.94

83.164 Diaqua – (7,15 – dihydro – 7,9,13,15 – tetramethyl –
pyrido(2′,1′,6′:12,13,14)(1,2,4,7,9,10,13) – hepta –
azacyclopentadeca(3,4,5,6,7,8 – aklmn) – 1,10 –
phenanthroline) – cobalt(ii) bis(tetrafluoroborate)
$C_{23}H_{25}CoN_7O_2{}^{2+}$, $2BF_4{}^-$
L.R.Hanton, P.R.Raithby
*Acta Crystallogr.,Sect.B,***36**, 1489, 1980

83.165 Diaqua – (2,6 – diacetylpyridine) – (2,9 – bis(1 –
methylhydrazone) – 1,10 – phenanthroline) – iron
bis(tetrafluoroborate)
$C_{23}H_{25}FeN_7O_2{}^{2+}$, $2BF_4{}^-$
M.M.Bishop, J.Lewis, T.D.O'Donoghue, P.R.Raithby,
J.N.Ramsden *J.Chem.Soc.,Dalton Trans.,*1390, 1980

83.C Diaqua – (2,6 – diacetylpyridine – bis(benzoic acid
hydrazone)) – nickel(ii) dinitrate dihydrate
$C_{23}H_{25}N_5NiO_4{}^{2+}$, $2NO_3{}^-$, $2H_2O$ Main entry is 84.95

83.C Hydro – tris(3,5 – dimethyl – 1 – pyrazolyl) –
borato – dicarbonyl – (p – chlorobenzenethiolato) –
molybdenum acetone solvate
$C_{23}H_{26}BClMoN_6O_2S$, $0.2C_3H_6O$ Main entry is 85.70

83.166 (Dimethyl – bis(3,5 – dimethyl – 1 – pyrazolyl) –
gallato) – (dimethyl – (3,5 – dimethyl – 1 –
pyrazolyl) – (dimethylethanolamino) – gallato) –
nickel(ii)
$C_{23}H_{43}Ga_2N_7NiO$
K.S.Chong, S.J.Rettig, A.Storr, J.Trotter
*Can.J.Chem.,***58**, 1091, 1980
See also R1 : 68

83.C Tetrabromocatecholato – nitroso –
triphenylphosphine – iridium methylene chloride
solvate
$C_{24}H_{15}Br_4IrNO_3P$, CH_2Cl_2 Main entry is 86.85

83.167 bis(1,10 – Phenanthroline) – platinum(ii) dichloride
trihydrate
$C_{24}H_{16}N_4Pt^{2+}$, $2Cl^-$, $3H_2O$
A.Hazell, A.Mukhopadhyay
*Acta Crystallogr.,Sect.B,***36**, 1647, 1980

83.168 Hydroxylamido – nitrosyl – bis(1,10 –
phenanthroline) – molybdenum 1,10 –
phenanthroline di – iodide monohydrate
$C_{24}H_{18}MoN_6O_2{}^{2+}$, $C_{12}H_8N_2$, $2I^-$, H_2O
K.Wieghardt, W.Holzbach, B.Nuber, J.Weiss
*Chem.Ber.,***113**, 629, 1980

83.C (p – Chlorothiobenzoyl – hydrazone – diyl) – (p –
chlorothiobenzoyl – hydrazone – triyl) – (N′ –
isopropylidene – p – chlorothiobenzoyl –
hydrazone) – molybdenum
$C_{24}H_{19}Cl_3MoN_6S_3$ Main entry is 85.71

83.C tetrakis(Pyridine) – bis(trifluoroacetato) – copper(ii)
$C_{24}H_{20}CuF_6N_4O_4$ Main entry is 81.77

83.169 Aqua – chloro – bis(N – (2 – pyridylmethylene) –
aniline) – cobalt(ii) nitrate monohydrate
$C_{24}H_{22}ClCoN_4O^+$, $NO_3{}^-$, H_2O
A.Monge, M.Martinez–Ripoll, E.Gutierrez–Puebla,
S.Garcia–Blanco
*Acta Crystallogr.,Sect.B,***35**, 3062, 1979

83.170 bis(3,5 – Dimethyl – 1 – phenylpyrazole) –
bis(isothiocyanato) – cobalt(ii)
$C_{24}H_{24}CoN_6S_2$
A.M.G.Dias–Rodrigues, Y.P.Mascarenhas, M.M.Rodrigues
*Acta Crystallogr.,Sect.B,***36**, 159, 1980

83.171 bis(6,6′ – Dimethyl – 2,2′ – bipyridyl) – copper(i)
tetrafluoroborate
$C_{24}H_{24}CuN_4{}^+$, $BF_4{}^-$
P.J.Burke, D.R.McMillin, W.R.Robinson
*Inorg.Chem.,***19**, 1211, 1980

83.172 bis(3,3′ – Dimethyl – 2,2′ – dipyridylamino) –
copper(ii) (absolute configuration)
$C_{24}H_{24}CuN_6$
C.E.Baxter, O.R.Rodig, R.K.Schlatzer, E.Sinn
*Inorg.Chem.,***18**, 1918, 1979

83.C bis(2 – Isopropylquinoline – 8 – thiolato) nickel
$C_{24}H_{24}N_2NiS_2$ Main entry is 85.76

83.C bis(N – Acetylimino – benzoylacetonato) –
platinum(ii)
$C_{24}H_{24}N_2O_4Pt$ Main entry is 84.96

83.C tetrakis(6 – Methyl – 2 – oxo – pyridinato) – di –
rhodium
$C_{24}H_{24}N_4O_4Rh_2$ Main entry is 84.97

83.173 Disodium hexapyrrolyl – zirconium(iv)
tetrahydrofuran solvate
$C_{24}H_{24}N_6Zr^{2-}$, $12C_4H_8O$, $2Na^+$
R.Vann Bynum, W.E.Hunter, R.D.Rogers, J.L.Atwood
*Inorg.Chem.,***19**, 2368, 1980

83.C Copper(ii) – copper(i) – pseudo – porphyrin –
macrocyclic ligand complex
$C_{24}H_{26}Cu_2N_4O_2{}^+$, $ClO_4{}^-$, $0.5CH_4O$ Main entry is 84.98

83.174 Tetra – imidazolyl – tri – manganese
$(C_{24}H_{28}Mn_3N_{16})_n$
R.Lehnert, F.Seel *Z.Anorg.Allg.Chem.,***464**, 187, 1980

83.175 bis(Dibromo – bis(4 – methylpyridine) –
molybdenum(ii))
$C_{24}H_{28}Br_4Mo_2N_4$
J.V.Brencic, L.Golic, I.Leban, P.Segedin
*Monatsh.Chem.,***110**, 1221, 1979

83.176 Dichloro – (2,9 – dimethyl – 3,10 – diphenyl –
1,4,8,11 – tetra – azacyclotetradeca – 1,3,8,10 –
tetraene) – cobalt hexafluorophosphate
$C_{24}H_{28}Cl_2CoN_4{}^+$, F_6P^-
D.S.Eggleston, S.C.Jackels *Inorg.Chem.,***19**, 1593, 1980

83.177 bis(Dichloro – bis(4 – methylpyridine) –
molybdenum(ii)) chloroform solvate
$C_{24}H_{28}Cl_4Mo_2N_4$, $CHCl_3$
J.V.Brencic, L.Golic, I.Leban, P.Segedin
*Monatsh.Chem.,***110**, 1221, 1979

83.178 tetrakis(4 – Methylpyridine) – nickel(ii)
bis(hexafluorophosphate)
$C_{24}H_{28}N_4Ni^{2+}$, $2F_6P^-$
R.M.Morrison, R.C.Thompson, J.Trotter
*Can.J.Chem.,***58**, 238, 1980

83.C (2 – aci – Nitrato – propan – 1 – on – 1 – yl) –
bis(dimethylphenylphosphine) – pyridine – chloro –
iridium acetone solvate
$C_{24}H_{30}ClIrN_2O_3P_2$, C_3H_6O Main entry is 71.192

83.C bis($\sigma,\sigma,\eta^2,\eta^2$ – Glyoxal – bis(isopropylamine)) –
octacarbonyl – tetra – ruthenium
$C_{24}H_{32}N_4O_8Ru_4$ Main entry is 72.50

83.C tetrakis(μ – Acetato) – bis(caffeine – rhodium)
$C_{24}H_{32}N_8O_{12}Rh_2$ Main entry is 81.78

83.C (6,6′ – (2,5,8,11,14,17,20 – Heptaoxa – heneicosane – 1,21 – diyl) – 2,2′ – bipyridyl) – dichloro – cobalt
$C_{24}H_{34}Cl_2CoN_2O_7$ Main entry is 84.99

83.179 hexakis(1 – Methylimidazole) – iron(ii) octacarbonyl – di – iron(i)
$C_{24}H_{36}FeN_{12}^{2+}, C_8Fe_2O_8^{2-}$
F.Seel, R.Lehnert, E.Bill, A.Trautwein
Z.Naturforsch.,Teil B,**35**, 631, 1980

83.C Diaquo – bis(N,N – diethylnicotinamide) – diacetato – nickel(ii)
$C_{24}H_{38}N_4NiO_8$ Main entry is 81.80

83.C bis(Ethylenediamine) – copper (bis(ethylenediamine) – copper) – (ethylenediamine – tetra – acetato – nickel) dihydrate
$C_{24}H_{40}CuN_8Ni_2O_{16}^{2-}. C_4H_{16}CuN_4^{2+}, 2H_2O$ Main entry is 76.18

83.C bis(bis(μ – 1 – Methylthyminato) – cis – diammine – platinum(ii)) silver nitrate pentahydrate
$C_{24}H_{44}AgN_{12}O_8Pt_2^+, NO_3^-, 5H_2O$ Main entry is 84.100

83.C tetrakis(Dichloroacetato – (μ – 2 – dimethylaminoethanolato) – copper(ii))
$C_{24}H_{44}Cl_8Cu_4N_4O_{12}$ Main entry is 81.81

83.C tetrakis(Chloroacetato – (μ – 2 – dimethylaminoethanolato) – copper(ii))
$C_{24}H_{48}Cl_4Cu_4N_4O_{12}$ Main entry is 81.82

83.C tetrakis(Bromo – (2 – diethylaminoethanolato) – copper(ii)) carbon tetrachloride solvate
$C_{24}H_{56}Br_4Cu_4N_4O_4, 4CCl_4$ Main entry is 84.103

83.180 (μ² – Hydrido) – (μ³ – 2 – (μ – N – phenylamino) – anilino) – tridecacarbonyl – penta – osmium
$C_{25}H_{10}N_2O_{13}Os_5$
Z.Dawoodi, M.J.Mays, P.R.Raithby
J.Chem.Soc.,Chem.Commun.,712, 1980

83.181 bis(Phenanthroline) – carbonato – cobalt(iii) bromide tetrahydrate
$C_{25}H_{16}CoN_4O_3^+, Br^-, 4H_2O$
H.Hennig, J.Sieler, R.Benedix, J.Kaiser, L.Sjolin, O.Lindqvist *Z.Anorg.Allg.Chem.*,**464**, 151, 1980

83.182 Carbonato – bis(1,10 – phenanthroline) – cobalt(iii) chloride trihydrate
$C_{25}H_{16}CoN_4O_3^+, Cl^-, 3H_2O$
B.C.Guild, T.Hayden, T.F.Brennan
Cryst.Struct.Commun.,**9**, 371, 1980

83.183 Cyano – bis(1,10 – phenanthroline) – platinum(ii) nitrate monohydrate
$C_{25}H_{16}N_5Pt^+, NO_3^-, H_2O$
O.Wernberg, A.Hazell
J.Chem.Soc.,Dalton Trans.,973, 1980

83.184 (2,6 – Diacetylpyridine – bis(4 – methoxybenzoylhydrazone)) – dioxo – uranium (α form)
$C_{25}H_{23}N_5O_6U$
G.Paolucci, G.Marangoni, G.Bandoli, D.A.Clemente
J.Chem.Soc.,Dalton Trans.,1304, 1980

83.185 (2,6 – Diacetylpyridine – bis(4 – methoxybenzoylhydrazone)) – dioxo – uranium (β form)
$C_{25}H_{23}N_5O_6U$
G.Paolucci, G.Marangoni, G.Bandoli, D.A.Clemente
J.Chem.Soc.,Dalton Trans.,1304, 1980

83.C bis(Dimethylglyoximato – N,N′) – (4 – ((cis – 3 – ethoxy – cyclobutyloxy) – dioxosulfuranyl) – phenyl) – pyridine – cobalt
$C_{25}H_{34}CoN_5O_8S$ Main entry is 71.196

83.186 Trimethylene – diamine – bis(guanosine – 5′ – monophosphate methyl ester) hydrate
$C_{25}H_{44}N_{12}O_{16}P_2Pt, 11H_2O$
L.G.Marzilli, P.Chalilpoyil, C.C.Chiang, T.J.Kistenmacher
J.Am.Chem.Soc.,**102**, 2480, 1980
See also R1 : 47

83.C (μ – Oxo) – (μ – trimethylsilyloxy) – (μ – per – rhenato) – bis(di(t – butylimido) – trimethylsilyloxy – rhenium) (at –47°C)
$C_{25}H_{63}N_4O_8Re_3Si_3$ Main entry is 84.106

83.187 Aquo – perchlorato – (bis(o – phenylene) – bis(pyridine – 2,6 – dialdimino)) – cadmium(ii) perchlorate methanol solvate
$C_{26}H_{20}CdClN_6O_5^+, ClO_4^-, CH_4O$
M.G.B.Drew, J.de O.Cabral, M.F.Cabral, F.S.Esho, S.M.Nelson
J.Chem.Soc.,Chem.Commun.,1033, 19

83.C (μ – N,N′ – bis(2 – (2 – Pyridyl)ethyl) – 2 – hydroxy – 5 – methyl – isophthalaldiminato) – (μ – pyrazolato) – di – copper(i)
$C_{26}H_{26}Cu_2N_6O$ Main entry is 84.107

83.C bis(Isothiocyanato) – tetrakis(4 – methylpyridine) – nickel(ii) 2 – bromonaphthalene clathrate
$C_{26}H_{28}N_6NiS_2, 2C_{10}H_7Br$ Main entry is 61.10

83.C bis(Isothiocyanato) – tetrakis(4 – methylpyridine) – nickel(ii) 2 – methylnaphthalene clathrate
$C_{26}H_{28}N_6NiS_2, 2C_{11}H_{10}$ Main entry is 61.11

83.C Triphenylphosphine – nitro – bis(dimethylglyoximato) – cobalt(iii)
$C_{26}H_{29}CoN_5O_6P$ Main entry is 86.91

83.C bis(μ – Acetato) – bis((4 – phenylimino – pentan – 2 – onato) – molybdenum)
$C_{26}H_{30}Mo_2N_2O_6$ Main entry is 81.83

83.C bis(o – (2,2,5,5 – Tetramethyl – Δ³ – imidazoline – 1 – oxy – 4 – yl)phenyl) – mercury
$C_{26}H_{32}HgN_4O_2$ Main entry is 71.2

83.188 bis(N – Methyl) – (μ² – n – hexyl) – coboglobin – type complex bis(hexafluorophosphate)
$C_{26}H_{44}CoN_6^{2+}, 2F_6P^-$
J.C.Stevens, P.J.Jackson, W.P.Schammel, G.G.Christoph, D.H.Busch *J.Am.Chem.Soc.*,**102**, 3283, 1980

83.C (μ – Diphenylphosphino) – (μ – N – methyl – (2 – phenyl – 2 – iminoethyl)) – bis(tricarbonyl – iron)
$C_{27}H_{20}Fe_2NO_6P$ Main entry is 71.211

83.C 1 – ((Di(pivaloyl) – methanato) – bis(pyridine) – rhoda) – 2,3 – bis(trifluoromethyl) – cyclopent – 2 – ene
$C_{27}H_{33}F_6N_2O_2Rh$ Main entry is 71.216

83.189 (μ – Carbonato) – bis(2,4,4,9 – tetramethyl – 1,5,9 – triazacyclododec – 1 – ene) – di – copper(ii) diperchlorate dimethylformamide solvate
$C_{27}H_{52}Cu_2N_6O_3{}^{2+}$, $2ClO_4{}^-$, C_3H_7NO
A.R.Davis, F.W.B.Einstein *Inorg.Chem.*,**19**, 1203, 1980

83.190 (2,2′ – Bipyridyl) – zinc – tetracarbonyl – iron
$C_{28}H_{16}Fe_2N_4O_8Zn_2$
R.J.Neustadt, T.H.Cymbaluk, R.D.Ernst, F.W.Cagle Junior
Inorg.Chem.,**19**, 2375, 1980

83.191 (Tetrabenzo(b,f,j,n)(1,5,9,13) – tetra – azacyclohexadecine) – palladium(ii) bis(tetrafluoroborate)
$C_{28}H_{20}N_4Pd^{2+}$, $2BF_4{}^-$
A.J.Jircitano, M.D.Timken, K.B.Mertes, J.R.Ferraro
J.Am.Chem.Soc.,**101**, 7661, 1979

83.192 Dimethylglyoximato – bis(o – phenanthroline) – cobalt(iii) dithiocyanate dihydrate
$C_{28}H_{23}CoN_6O_2{}^{2+}$, $2CNS^-$, $2H_2O$
M.M.Botoshanskii, N.F.Krasnova, Yu.A.Simonov, N.M.Samus, M.I.Antipin, T.I.Malinovskii
Zh.Strukt.Khim.,**20**, 1052, 1979

83.193 Tetrachloro – bis(N,N′ – dimethylbenzamidinato) – di – rhenium
$C_{28}H_{26}Cl_4N_4Re_2$
F.A.Cotton, W.H.Ilsley, W.Kaim
Inorg.Chem.,**19**, 2360, 1980

83.C (N,N′ – o – Phenylene – (6,6′ – (1,2 – dihydrobenzimidazole – methoxymethylene – 2 – yl) – bis(pyridine – 2 – aldimino))) – aquo – methanol – cobalt(ii) diperchlorate
$C_{28}H_{28}CoN_8O_3{}^{2+}$, $2ClO_4{}^-$ Main entry is 84.109

83.194 tetrakis(μ – N – (2 – Pyridyl – acetamido)) – di – molybdenum
$C_{28}H_{28}Mo_2N_8O_4$
F.A.Cotton, W.H.Ilsley, W.Kaim
Inorg.Chem.,**18**, 2717, 1979

83.C 2,4 – Dithio – pyrimidinato – bis(bis(η^5 – methylcyclopentadienyl) – titanium(iii))
$C_{28}H_{30}N_2S_2Ti_2$ Main entry is 73.113

83.195 bis(Benzylmercapto) – 2,3,9,10 – tetramethyl – 1,4,8,11 – tetra – azacyclotetradeca – 1,3,8,10 – tetraene – iron(iii) hexafluorophosphate
$C_{28}H_{38}FeN_4S_2{}^+$, F_6P^-
B.D.Santarsiero, A.Aruffo, V.Schomaker, E.C.Lingafelter
Am.Cryst.Assoc.,*Ser.2*,**8**, 23, 1980

83.196 bis(Isothiocyanato) – bis(N – methyl) – (μ^2 – n – hexyl) – coboglobin – type complex chloride
$C_{28}H_{44}CoN_8S_2{}^+$, Cl^-
J.C.Stevens, P.J.Jackson, W.P.Schammel, G.G.Christoph, D.H.Busch *J.Am.Chem.Soc.*,**102**, 3283, 1980

83.C bis(μ – (N,N – Dimethyl – 2 – aminoethyl) – 1 – phenylpropane – 2 – oximato) – bis(methanol – copper(ii)) bis(μ – (N,N – dimethyl – 2 – aminoethyl) – 1 – phenylpropane – 2 – oximato) – bis(perchlorato – copper(ii)) diperchlorate
$C_{28}H_{44}Cu_2N_6O_4{}^{2+}$, $C_{28}H_{36}Cl_2Cu_2N_6O_{10}$. $2ClO_4{}^-$
Main entry is 76.104

83.C Dicarbonyl – η^5 – cyclopentadienyl – (N – benzyl – N′ – (1 – phenylethyl) – benzamidinato) – molybdenum (absolute configuration)
$C_{29}H_{26}MoN_2O_2$ Main entry is 73.114

83.C $(+)_{578}$ – Dicarbonyl – (η^5 – cyclopentadienyl) – (N – benzyl – N′ – α – methylbenzylbenzamidinato) – molybdenum (absolute configuration)
$C_{29}H_{26}MoN_2O_2$ Main entry is 73.115

83.C Hydro – tris(1 – pyrazolyl) – borato – bis(p – toluenethiolato) – (p – fluorobenzenediazo) – molybdenum
$C_{29}H_{28}BFMoN_8S_2$ Main entry is 85.81

83.C bis(o – Phenylazophenolato – (tricarbonyl) – rhenium(i))
$C_{30}H_{18}N_4O_8Re_2$ Main entry is 84.110

83.C Tricarbonyl – chloro – (5,7 – dichloro – 8 – quinolinolato – N,O) – triphenylphosphine – tungsten(ii)
$C_{30}H_{19}Cl_3NO_4PW$ Main entry is 86.100

83.C bis(2 – (N – Phenylaldimino) – benzo(b)furan – 3 – thiolato) – nickel(ii)
$C_{30}H_{20}N_2NiO_2S_2$ Main entry is 85.82

83.197 (–) – D – tris(2,2′ – Bipyridyl) – cobalt(ii) hexacyano – iron(iii) octahydrate (absolute configuration)
$C_{30}H_{24}CoN_6{}^{3+}$, $C_6FeN_6{}^{3-}$, $8H_2O$
Y.Ohashi, K.Yanagi, Y.Mitsuhashi, K.Nagata, Y.Kaizu, Y.Sasada, H.Kobayashi *J.Am.Chem.Soc.*,**101**, 4739, 1979

83.198 tris(2,2′ – Bipyridyl) – ruthenium(ii) bis(hexafluorophosphate)
$C_{30}H_{24}N_6Ru^{2+}$, $2F_6P^-$
D.P.Rillema, D.S.Jones, H.A.Levy
J.Chem.Soc.,Chem.Commun.,849, 1979

83.199 bis(5,5 – Diphenyl – hydantoinato) – diammine – copper(ii)
$C_{30}H_{28}CuN_6O_4$
N.Shimizu, T.Uno *Cryst.Struct.Commun.*,**9**, 389, 1980

83.200 Hexakis(pyridine) – ruthenium(ii) bis(tetrafluoroborate)
$C_{30}H_{30}N_6Ru^{2+}$, $2BF_4{}^-$
J.L.Templeton *J.Am.Chem.Soc.*,**101**, 4906, 1979

83.C bis(Dimethyldithiocarbamato) – bis(diphenylhydrazido) – molybdenum acetone solvate
$C_{30}H_{32}MoN_6S_4$, C_3H_6O Main entry is 80.25

83.201 bis(Diphenyl – hydantoinato) – diaqua – diammine – nickel(ii)
$C_{30}H_{32}N_6NiO_6$
N.Shimizu, T.Uno *Cryst.Struct.Commun.*,**9**, 223, 1980

83.C $(+)_{589}$ – (S – Dimethyl – (α – methylbenzyl) – aminato) – (SS – o – phenylene – bis(methylphenylarsine)) – palladium(ii) hexafluorophosphate (absolute configuration)
$(+)_{589}$ – (S – 2 – (1′ – Dimethylaminoethyl) – phenyl – C^1N) – (SS – o – phenylene – bis(methylphenylarsine)) – palladium(ii) hexafluorophosphate
$C_{30}H_{34}As_2NPd^+$, F_6P^- Main entry is 86.102

83.202 bis(Hydrido – tris(3,5 – dimethylpyrazolyl) – borato) – iron(ii)
$C_{30}H_{44}B_2FeN_{12}$
J.D.Oliver, D.F.Mullica, B.B.Hutchinson, W.O.Milligan
Inorg.Chem.,**19**, 165, 1980

83.203 hexakis(1 – Ethylimidazole) – iron(ii)
tetracarbonyl – iron(i)
$C_{30}H_{48}FeN_{12}^{2+}$, $2C_4FeO_4^-$
F.Seel, R.Lehnert, E.Bill, A.Trautwein
Z.Naturforsch.,Teil B,**35**, 631, 1980

83.C bis((μ – Methoxy) – (2,4,6 – trichlorophenolato) –
quinoline – copper(ii))
$C_{32}H_{24}Cl_6Cu_2N_2O_4$ Main entry is 84.112

83.C bis(2 – (N – Benzylaldimino) – benzo(b)furan – 3 –
thiolato) – nickel(ii)
$C_{32}H_{24}N_2NiO_2S_2$ Main entry is 85.83

83.204 Dichloro – bis(7 – chloro – 1,3 – dihydro – 1 –
methyl – 5 – phenyl – 3H – 1,4 – benzodiazepin –
2 – one) – copper(ii) dihydrate chloroform solvate
$C_{32}H_{26}Cl_4CuN_4O_2$, $2H_2O$, $0.33CHCl_3$
A.Mosset, J.P.Tuchagues, J.J.Bonnet, R.Haran, P.Sharrock
Inorg.Chem.,**19**, 290, 1980

83.C bis(μ – Acetato) – bis((2 – (2' – benzothiazolyl) – 5 –
methyl – phenyl) – palladium(ii))
$C_{32}H_{26}N_2O_4Pd_2S_2$ Main entry is 81.86

83.C bis(μ – Acetato) – bis((2 – (2' – benzoxazolyl) – 5 –
methyl – phenyl) – palladium(ii))
$C_{32}H_{26}N_2O_6Pd_2$ Main entry is 71.236

83.C tetrakis(μ – N – Phenyl – acetamidato) – di –
chromium
$C_{32}H_{32}Cr_2N_4O_4$ Main entry is 84.114

83.C tetrakis(μ – Acetanilido) – di – molybdenum
tetrahydrofuran solvate
$C_{32}H_{32}Mo_2N_4O_4$, $2C_4H_8O$ Main entry is 84.115

83.205 bis(1,4 – bis(3,5 – Dimethylphenyl) – tetra –
azadiene) – nickel(0)
$C_{32}H_{36}N_8Ni$
P.Overbosch, G.van Koten, O.Overbeek
J.Am.Chem.Soc.,**102**, 2091, 1980

83.206 bis(1,3 – Diphenyltriazenido) –
tetrakis(dimethylamido) – di – tungsten (at –138°C)
$C_{32}H_{44}N_{10}W_2$
M.H.Chisholm, J.C.Huffman, R.L.Kelly
Inorg.Chem.,**18**, 3554, 1979

83.C (μ – Peroxo) – bis((1,9 – bis(2 – pyridyl) – 2,5,8 –
triazanonane) cobalt(iii)) tetra – iodide
$C_{32}H_{46}Co_2N_{10}O_2^{4+}$, $4I^-$ Main entry is 76.106

83.207 trans – Dichloro – tetrakis(1H,3,5 – diethyl – 4 –
methylpyrazole) – nickel(ii)
$C_{32}H_{56}Cl_2N_8Ni$
I.A.Krol, V.M.Agre, V.K.Trunov, O.I.Ivanov
Koord.Khim.,**5**, 1569, 1979

83.208 bis(μ^2 – Chloro) – dichloro – tetrakis(1H – 3,5 –
diethyl – 4 – methylpyrazole) – di – copper(ii)
$C_{32}H_{56}Cl_4Cu_2N_8$
V.M.Agre, I.A.Krol, V.K.Trunov, V.M.Dziomko, O.V.Ivanov
Koord.Khim.,**5**, 1413, 1979

83.C (2 – Diethylaminoethanolato) – (trichloroacetato) –
copper(ii) tetramer
$C_{32}H_{56}Cl_{12}Cu_4N_4O_{12}$ Main entry is 81.87

83.C tetrakis(μ – 2 – Diethylaminoethanolato) –
tetrakis(chloroacetato – copper(ii))
$C_{32}H_{64}Cl_4Cu_4N_4O_{12}$ Main entry is 81.89

83.C trans – bis(1 – Adamantylamido) –
tetrakis(trimethylsilyl – oxo) – molybdenum
(at –55°C)
$C_{32}H_{68}MoN_2O_4Si_4$ Main entry is 84.116

83.C (μ – Ethylideneimino) – (μ – hydrido) – (μ –
bis(diphenylphosphino)methane) – bis(tricarbonyl –
rhenium)
$C_{33}H_{27}NO_6P_2Re_2$ Main entry is 86.112

83.C (($\pm$) – α – (2 – Diphenylphosphino – ferrocenyl)
ethyl – dimethylamine) – norbornadiene – rhodium
hexafluorophosphate
$C_{33}H_{36}FeNPRh^+$, F_6P^- Main entry is 73.120

83.C bis(2,2' – Bipyridyl) – bis(pyridine – 2,6 –
dicarboxylato) – di – copper(ii) tetrahydrate
$C_{34}H_{22}Cu_2N_6O_8$, $4H_2O$ Main entry is 81.90

83.209 N,N,N',N' – tetra – bis(2' – Benzimidazolyl –
methyl) – 1,2 – ethane – diamine – copper(ii)
$C_{34}H_{28}CuN_{10}^{2+}$, $2BF_4^-$
P.J.M.W.L.Birker, S.Gorter, H.J.M.Hendriks, J.Reedijk
Inorg.Chim.Acta,**45**, L63, 1980

83.C bis(Terpyridyl – (2 – aminoethylthiolato) –
platinum) – platinum tetrakis(tetrafluoroborate)
$C_{34}H_{34}N_8Pt_3S_2^{4+}$, $4BF_4^-$ Main entry is 85.86

83.C bis(μ – Chloro) – bis(mercury – (μ – p – tolyl(ethyl)
triazenido) – (cyclo – octadiene – iridium))
$C_{34}H_{48}Cl_2Hg_2Ir_2N_6$ Main entry is 75.40

83.C (μ_2 – N,N' – Di – t – butyl – ethylenedi – imine) –
bis(trans – dichloro – tributylphosphine –
platinum)
$C_{34}H_{74}Cl_4N_2P_2Pt_2$ Main entry is 86.114

83.C octakis(t – Butoxy) – tetrakis(μ – fluoro) – tetra –
molybdenum bis(octakis(t – butoxy) – tris(μ –
fluoro) – (μ – dimethylamido) – tetra –
molybdenum (at –145°C)
$2C_{34}H_{78}F_3Mo_4NO_8$, $C_{32}H_{72}F_4Mo_4O_8$ Main entry is 84.117

83.210 tris(1,10 – Phenanthroline) – mercury(ii)
bis(trifluoromethylsulfonate) ethanol solvate
$C_{36}H_{24}HgN_6^{2+}$, $2CF_3O_3S^-$, C_2H_6O
G.B.Deacon, C.L.Raston, D.Tunaley, A.H.White
Aust.J.Chem.,**32**, 2195, 1979

83.211 bis(μ – Chloro) – bis(trichloro –
triphenylphosphinimino – niobium) 1,2 –
dichloroethane solvate
$C_{36}H_{30}Cl_8N_2Nb_2P_2$, $C_2H_4Cl_2$
H.Bezler, J.Strahle *Z.Naturforsch.,Teil B*,**34**, 1199, 1979

83.C bis(2,4 – Dimethyl – 6 – oxopyrimidine) – bis(1,3 –
diphenyltriazino) – di – tungsten tetrahydrofuran
solvate
$C_{36}H_{34}N_{10}O_2W_2$, $2C_4H_8O$ Main entry is 84.121

83.212 hexakis(bis(Imidazole) – silver perchlorate)
$C_{36}H_{36}Ag_6N_{24}^{6+}$, $6ClO_4^-$
G.W.Eastland, M.A.Mazid, D.R.Russell, M.C.R.Symons
J.Chem.Soc.,Dalton Trans.,1682, 1980

83.C Diaqua – bis(N – acetyl – D,L – tryptophanato) –
bis(pyridine) – copper(ii)
$C_{36}H_{40}CuN_6O_8$ Main entry is 82.41

83.213 Dichloro – tetrakis(N,N' – dimethylbenzamidinato) – di – rhenium carbon tetrachloride solvate
$C_{36}H_{44}Cl_2N_8Re_2$, CCl_4
F.A.Cotton, W.H.Ilsley, W.Kaim
Inorg.Chem.,**19**, 2360, 1980

83.214 tetrakis(μ – N,N' – Dimethylbenzamidinato) – di – chromium
$C_{36}H_{44}Cr_2N_8$
A.Bino, F.A.Cotton, W.Kim *Inorg.Chem.*,**18**, 3566, 1979

83.215 trans – Dichloro – bis(1 – benzyl – 3,5 – di – propyl – 4 – ethylpyrazole) – palladium(ii)
$C_{36}H_{52}Cl_2N_4Pd$
V.M.Agre, N.P.Kozlova, V.K.Trunov, L.G.Makarevich, O.V.Ivanov *Koord.Khim.*,5, 1406, 1979

83.C bis((μ – Acetato – (O,O')) – diacetato – (μ_3 – 2 – diethylaminoethanolato – N,μ_3 – 0)) – (μ – 2 – diethylaminoethanolato – N,μ – 0) – (μ_3 – hydroxo) – tri – copper(ii)) monohydrate
$C_{36}H_{76}Cu_6N_4O_{18}$, H_2O Main entry is 84.122

83.C (η – Cyclo – octa – 1,5 – diene) – 4 – diphenylphosphino – 2 – diphenylphosphino – methylpyrrolidine – rhodium(i) perchlorate
$C_{37}H_{41}NP_2Rh^+$, ClO_4^- Main entry is 75.42

83.C Dicarbonyl – bis(5,7 – dichloro – 8 – quinolinolato – N,O) – triphenylphosphine – tungsten(ii) dichloromethane solvate
$C_{38}H_{23}Cl_4N_2O_4PW$, CH_2Cl_2 Main entry is 86.133

83.216 (2 – (2 – Pyridyl) – 3 – (5 – pyridyl) – quinoxaline) – bis(2,2' – bipyridyl) – ruthenium bis(hexafluorophosphate)
$C_{38}H_{28}N_8Ru^{2+}$, $2F_6P^-$
D.S.Jones, D.P.Rillema, C.D.Keller, H.A.Levy
Am.Cryst.Assoc.,Ser.2,**7**, 37, 1980

83.C mer – Trichloro – (acetonitrile) – bis(triphenylphosphine) – osmium(iii)
$C_{38}H_{33}Cl_3NOsP_2$ Main entry is 86.137

83.C (3,5 – Di – t – butylcatecholato) – (3,5 – di – t – butyl – semiquinone) – (bipyridyl) – cobalt(iii) toluene solvate
$C_{38}H_{48}CoN_2O_4$, $0.5C_7H_8$ Main entry is 84.123

83.C (Dicarbonyl – dimethylphenylphosphine – cobalt) – μ – carbonyl – μ – (N,N' – diphenyl – N – (phenylcarbenyl) – benzamidine) – (dicarbonyl – cobalt)
$C_{39}H_{31}Co_2N_2O_5P$ Main entry is 71.257

83.C bis(μ – N,N' – Diphenyl – acetamidinato) – bis(μ – 2,4 – dimethyl – 6 – hydroxy – pyrimidinato) – di – tungsten tetrahydrofuran solvate
$C_{40}H_{40}N_8O_2W_2$, $2C_4H_8O$ Main entry is 84.124

83.C (μ_3 – Carbonato) – tris(aqua – 2 – (2 – (2 – pyridyl) – ethyliminomethyl) – pyridine – copper(ii))
$C_{40}H_{45}Cu_3N_9O_6^{4+}$, $4NO_3^-$ Main entry is 84.126

83.C tetrakis(μ^2 – N – 2,6 – Xylyl – acetamidato) – di – chromium methylene chloride solvate
$C_{40}H_{48}Cr_2N_4O_4$, $2CH_2Cl_2$ Main entry is 84.127

83.C tetrakis(μ^2 – N – 2,6 – Xylyl – acetamidato) – di – molybdenum methylene chloride solvate
$C_{40}H_{48}Mo_2N_4O_4$, $2CH_2Cl_2$ Main entry is 84.128

83.C bis(Triphenylphosphine) – (tetrachlorodiazo – cyclopentadiene) – chloro – iridium(i) toluene solvate
$C_{41}H_{30}Cl_5IrN_2P_2$, C_7H_8 Main entry is 86.148

83.C tris(Acetonitrile) – nitrosyl – bis(triphenylphosphine) – iridium(iii) bis(hexafluorophosphate)
$C_{42}H_{39}IrN_4OP_2^{2+}$, $2F_6P^-$ Main entry is 86.152

83.C cis,cis,cis – bis(Dimethylphenylphosphine) – bis(monothiobenzoato) – (1,10 – phenanthroline) – ruthenium(ii)
$C_{42}H_{40}N_2O_2P_2RuS_2$ Main entry is 86.153

83.C cis,cis,trans – bis(Dimethylphenylphosphine) – bis(monothiobenzoato) – (1,10 – phenanthroline) – ruthenium(ii)
$C_{42}H_{40}N_2O_2P_2RuS_2$ Main entry is 86.154

83.C (tris(2 – Diphenylarsinoethyl) – amino) – iodo – nickel(i)
$C_{42}H_{42}As_3INNi$ Main entry is 86.156

83.C (tris(2 – Diphenylphosphinoethyl) – amino) – iodo – nickel(i)
$C_{42}H_{42}INNiP_3$ Main entry is 86.159

83.C bis(α – Benzyldioximato) – bis(cyclohexyl – isonitrile) – iron(ii)
$C_{42}H_{44}FeN_6O_4$ Main entry is 71.264

83.217 trans – bis(Isothiocyanato) – tetrakis(N,N – diethylnicotinamide) – zinc
$C_{42}H_{56}N_{10}O_4S_2Zn$
G.V.Tsintsadze, Z.V.Mikelashvili, T.I.Tsivtsivadze, V.S.Sergienko
Soobshch.Akad.Nauk Gruzh.SSR,**96**, 85, 1979

83.C tetrakis(1 – Adamantoxo) – dimethylamine – molybdenum(iv)
$C_{42}H_{67}MoNO_4$ Main entry is 84.130

83.218 tris(μ^2 – 4 – Phenyl – 1,2,4 – triazole – N^1,N^2) – bis(bis(isothiocyanato) – (4 – phenyl – 1,2,4 – triazole – N^1) – cobalt(ii)) hydrate
$C_{44}H_{35}Co_2N_{19}S_4$, $2.7H_2O$
D.W.Engelfriet, G.C.Verschoor, W.den Brinker
Acta Crystallogr.,Sect.B,**36**, 1554, 19

83.219 bis(1,3,5 – Tri – p – tolyl – formazanyl) – palladium(ii)
$C_{44}H_{42}N_8Pd$
A.R.Siedle, L.H.Pignolet *Inorg.Chem.*,**19**, 2052, 1980

83.C Chloro – carbonyl – (N – isopropenyl – N' – isopropyl – formamidinato) – bis(triphenylphosphine) – ruthenium(ii)
$C_{44}H_{43}ClN_2OP_2Ru$ Main entry is 86.165

83.C tetrakis(μ^2 – N – 2,6 – Xylyl – acetamidato) – tetrahydrofuran – di – chromium toluene solvate
$C_{44}H_{56}Cr_2N_4O_5$, C_7H_8 Main entry is 84.131

83.C tetrakis(μ^2 – N – p – Dimethylaniline – acetamidato) – tetrahydrofuran – di – chromium
$C_{44}H_{60}Cr_2N_8O_5$ Main entry is 84.132

83.C trans(P,N) – bis(Bromo – (μ^2 – pyridyl – C^2,N) – (triphenylphosphine) – palladium(ii))
$C_{46}H_{38}Br_2N_2P_2Pd_2$ Main entry is 71.269

83.C Dicarbonyl – chloro – (5,7 – dichloro – 8 – quinolinolato – N,O) – bis(triphenylphosphine) – tungsten(ii)
$C_{47}H_{34}Cl_3NO_3P_2W$ Main entry is 86.173

83.220 7,15 – bis(Diphenoxyphosphinyl) – 6,7,14,15 – tetrahydro – 6,14 – diphenoxydibenzo(e,k) (1,7,3,9,2,8,4,10) – dioxa – diaza – diphospha – dimercura – cyclododecan – 6,14 – dioxide acetic acid
$C_{48}H_{38}Hg_2N_2O_{12}P_4$, $2C_2H_4O_2$
H.Richter, E.Fluck, W.Schwarz
Z.Naturforsch.,Teil B,**35**, 578, 1980

83.C Nitrosyl – (1,10 – phenanthroline) – bis(triphenylphosphine) – iridium(i) bis(hexafluorophosphate)
$C_{48}H_{38}IrN_3OP_2^{2+}$, $2F_6P^-$ Main entry is 86.174

83.221 tris(m – Xylene – α,α' – bis(butane – 2 – monoximato – 3 – imino)) – oxo – hydroxo – hexa – copper(ii) sulfate dihydrogen phosphate pentahydrate
$C_{48}H_{61}Cu_6N_{12}O_8^{3+}$, O_4S^{2-}, $H_2O_4P^-$, $5H_2O$
Y.Agnus, R.Louis, R.Weiss *Eur.Cryst.Meeting*,**5**, 306, 1979

83.C tetrakis(μ² – N – 2,6 – Xylyl – acetamidato) – bis(tetrahydrofuran) – di – chromium tetrahydrofuran solvate
$C_{48}H_{54}Cr_2N_4O_6$, C_4H_8O Main entry is 84.134

83.C Chloro – dodeca(1,1 – dimethyl – 2 – aminoethylthiolato) – tetradeca – copper sulfate eicosahydrate
$C_{48}H_{120}ClCu_{14}N_{12}S_{12}^{7+}$, $3.5O_4S^{2-}$, $20H_2O$ Main entry is 85.91

83.C N,N',N'',N''' – Tetra – p – tolyl – oxalylamidine – bis(di(cyclopentadienyl) – titanium)
$C_{50}H_{48}N_4Ti_2$ Main entry is 73.131

83.C tetrakis(μ² – N – 2,6 – Xylyl – acetamidato) – bis(pyridine) – di – chromium pyridine solvate
$C_{50}H_{58}Cr_2N_6O_4$, C_5H_5N Main entry is 84.135

83.C trans – Carbonyl – (1,3 – di – p – tolyltriazenido) – bis(triphenylphosphine) – iridium(i)
$C_{51}H_{44}IrN_3OP_2$ Main entry is 86.183

83.C 2 – (Chloro – bis(triphenylphosphine) – palladium) – 3 – methyl – 1,4 – bis(methoxyphenyl) – 1,4 – diazabutadien – 1,4 – diyl – dichloro – copper
$C_{53}H_{47}Cl_3CuN_2O_2P_2Pd$ Main entry is 86.196

83.C Bromo – (dichloro – diazamethane) – bis(diphenylphosphinoethane) – tungsten hexafluorophosphate
$C_{53}H_{48}BrCl_2N_2P_4W^+$, F_6P^- Main entry is 86.197

83.222 bis(Pyridine) – (meso – tetraphenylporphinato) – chromium(ii) toluene solvate
$C_{54}H_{38}CrN_6$, C_7H_8
W.R.Scheidt, A.C.Brinegar, J.F.Kirner, C.A.Reed
Inorg.Chem.,**18**, 3610, 1979
See also R1 : 49

83.223 Heptamethyl – dicyano – cobyrinate(iii)
$C_{54}H_{73}CoN_6O_{14}$, xH_2O
A.Fischli, J.J.Daly *Helv.Chim.Acta*,**63**, 1628, 1980

83.C bis((Tri – n – butylphosphine) – bis(dimethylglyoximato) – (methyl(phenyl) – arsenyl) – cobalt(iii)) – cobalt(ii) dihydrate
$C_{54}H_{96}As_2Co_3N_8O_{10}P_2$, $2H_2O$ Main entry is 86.206

83.C bis(1,2 – bis(Diphenylphosphino)ethane) – bromo – 1 – diazo – 1 – methylethane – tungsten bromide methanol solvate
$C_{55}H_{54}BrN_2P_4W^+$, Br^-, $0.5CH_4O$ Main entry is 86.207

83.C Bromo – (1 – chloro – 1 – diazo – 2,2 – dicyanoethene) – bis(diphenylphosphinoethane) – tungsten dichloromethane solvate
$C_{56}H_{48}BrClN_4P_4W$, CH_2Cl_2 Main entry is 86.210

83.C bis(1,2 – bis(Diphenylphosphino)ethane) – bromo – 4 – diazobutanol – tungsten hexafluorophosphate ethanol solvate
$C_{56}H_{56}BrN_2OP_4W^+$, F_6P^-, $0.5C_2H_6O$ Main entry is 86.211

83.C Iodo – N – cyclohexyl – diazenido – bis(1,2 – bis(diphenylphosphino)ethane) – molybdenum benzene solvate
$C_{58}H_{59}IMoN_2P_4$, $0.5C_6H_6$ Main entry is 86.215

83.224 bis(1,2,3,7,8,12,13,17,18,19 – Decamethyl – biladiene – a,c) – di – zinc(ii)
$C_{58}H_{68}N_8Zn_2$
W.S.Sheldrick, J.Engel
J.Chem.Soc.,Chem.Commun.,5, 1980

83.C tetrakis(μ² – (N,N') – Diphenylurea) – bis(tetrahydrofuran) – di – chromium cyclohexane solvate
$C_{60}H_{60}Cr_2N_8O_6$, C_6H_{12} Main entry is 79.17

83.C Lithium (μ² – benzophenonimino) – tetrakis(benzophenonimino) – di – nickel diethyl ether solvate
$C_{65}H_{50}N_5Ni_2^{3-}$, $3Li^+$, $2C_4H_{10}O$ Main entry is 72.67

83.225 Chloro – (meso – α,α,α,α – tetra(o – nicotinamidophenyl) – porphyrinato – iron(iii)) – copper(ii) perchlorate dihydrate
$C_{68}H_{44}ClCuFeN_{12}O_4^{2+}$, $2ClO_4^-$, $2H_2O$
M.J.Gunter, L.N.Mander, G.M.McLaughlin, K.S.Murray, K.J.Berry, P.E.Clark, D.A.Buckingham
J.Am.Chem.Soc.,**102**, 1470, 1980
See also R1 : 49

METAL COMPLEXES (OXYGEN LIGAND)

84.1 (μ – Hydroxy) – methylene – diphosphonato – technetium tris(aqua) – lithium hydrate
$(CH_3O_7P_2Tc^-)_n$, $nH_6LiO_3^+$, $0.33nH_2O$
K.Libson, E.Deutsch, B.L.Barnett
J.Am.Chem.Soc.,**102**, 2476, 1980

84.2 Guanidinium hexamolybdo – methylarsonate hexahydrate
$CH_{15}AsMo_6O_{27}^{2-}$, $2CH_6N_3^+$, $6H_2O$
K.Y.Matsumoto *Bull.Chem.Soc.Jpn.*,**52**, 3284, 1979
See also R1 : 65 R2 : 8

84.3 Dichloro – (dimethylsulfoxide) – copper(ii)
$(C_2H_6Cl_2CuOS)_n$
D.D.Swank, D.D.Landee, R.D.Willett
Phys.Rev.B,**20**, 2154, 1979

84.4 Dichloro – (tetramethylenesulfoxide) – copper(ii)
$(C_4H_8Cl_2CuOS)_n$
D.D.Swank, C.P.Landee, R.D.Willett
Phys.Rev.,**20**, 2154, 1979

84.C Tetraethylammonium trichloro – (μ – 2 – oxoethanethiolato – S,μ – 0) – (μ – 2 – oxoethanethiolato – μ – S,μ – 0) – bis(oxo – molybdenum(v))
$C_4H_8Cl_3Mo_2O_4S_2^-$, $C_8H_{20}N^+$ Main entry is 85.4

84.5 Diaqua – bis(glycollato) – manganese(ii)
$C_4H_{10}MnO_8$
T.Lis *Acta Crystallogr.,Sect.B*,**36**, 701, 1980

84.C Dibromo – (N – (2 – hydroxyethyl) – ethylenediamine) – copper(ii)
$C_4H_{12}Br_2CuN_2O$ Main entry is 76.6

84.6 bis(cis – bis(2 – Aminoethanolato) – copper(ii)) dinitrate
$2C_4H_{13}CuN_2O_2^+$, $2NO_3^-$
J.A.Bertrand, E.Fujita, D.G.VanDerveer
Inorg.Chem.,**19**, 2022, 1980
See also R1 : 83

84.7 Oxo – diperoxo – ((S) – N,N – dimethyl – lactamido) – molybdenum(vi) (absolute configuration)
$C_5H_{11}MoNO_7$
W.Winter, C.Mark, V.Schurig *Inorg.Chem.*,**19**, 2045, 1980

84.8 2 – ((3 – Aminopropyl)amino)ethanolato – copper(ii) tetra(isothiocyanato) – copper(i) thiocyanate
$4C_5H_{13}CuN_2O^+$, $C_4CuN_4S_4^{3-}$, CNS^-
K.Nieminen *Eur.Cryst.Meeting*,**5**, 308, 1979
See also R1 : 83

84.9 Cesium tris(ethylnitrosolato) – iron(ii) monohydrate
$C_6H_9FeN_6O_6^-$, Cs^+, H_2O
P.Gouzerh, Y.Jeannin, C.Rocchiccioli-Deltcheff,
F.Valentini *J.Coord.Chem.*,**6**, 221, 1979
See also R1 : 83

84.C Tetraethylammonium chloro – tris(thiolato – ethanolato) – bis(oxo – molybdenum(v))
$C_8H_{12}ClMo_2O_5S_3^-$, $C_8H_{20}N^+$ Main entry is 85.9

84.10 bis(Ethanol) – bis(thiocyanato) – manganese(ii)
$(C_6H_{12}MnN_2O_2S_2)_n$
J.N.McElearney, L.L.Balagot, J.A.Muir, R.D.Spence
Phys.Rev.B,**19**, 306, 1979

84.C Piperidinium (μ – oxo) – (μ – 2 – mercapto – ethanolato) – bis(oxo – (2 – mercapto – ethanolato) – molybdenum)
$C_6H_{12}Mo_2O_6S_3^{2-}$, $2C_5H_{12}N^+$ Main entry is 85.10

84.C Diacetato – dithiocarbamide – cadmium
$C_6H_{14}CdN_4O_4S_2$ Main entry is 85.11

84.11 (1,5 – Diazacyclo – octane) – dinitrato – copper(ii)
$C_6H_{14}CuN_4O_6$
P.Murray-Rust, J.Murray-Rust, R.Clay
Acta Crystallogr.,Sect.B,**36**, 452, 1980
See also R1 : 83

84.12 bis(μ – Methoxo) – (μ – oxo) – bis(di(methoxo) – oxo – rhenium(vi))
$C_8H_{18}O_9Re_2$
P.Edwards, G.Wilkinson, K.M.A.Malik, M.B.Hursthouse
J.Chem.Soc.,Chem.Commun.,1158, 1979

84.13 (2 – Aminoethanol) – bis(2 – aminoethanolato) – cobalt(iii) bis(2 – aminoethanol) – (2 – aminoethanolato) – cobalt(iii) perchlorate hemihydrate
$C_6H_{19}CoN_3O_3^+$, $C_6H_{20}CoN_3O_3^{2+}$, $3ClO_4^-$, $0.5H_2O$
J.A.Bertrand, P.G.Eller, E.Fujita, M.O.Lively,
D.G.VanDerveer *Inorg.Chem.*,**18**, 2419, 1979
See also R1 : 83

84.14 bis(2 – Aminoethanol) – (2 – aminoethanolato) – nickel(iii) perchlorate
$C_6H_{20}N_3NiO_3^+$, ClO_4^-
J.A.Bertrand, P.G.Eller, E.Fujita, M.O.Lively,
D.G.VanDerveer *Inorg.Chem.*,**18**, 2419, 1979
See also R1 : 83

84.15 Dibromo – (N – (2 – pyridyl)acetamide) – mercury(ii)
$(C_7H_8Br_2HgN_2O)_n$
J.R.Lechat, R.H.P.Francisco, C.Airoldi
Acta Crystallogr.,Sect.B,**36**, 930, 1980
See also R1 : 83

84.C (bis(Diethylene) – triethylene – tetramine) – bis(perchlorato) – copper
$C_7H_{20}Cl_2CuN_4O_8$ Main entry is 76.44

84.16 Nitrato – cobalt(iii) – (1,4,7,10 – tetraoxacyclododecane) – nitrate
$C_8H_{16}CoN_2O_{10}$
E.M.Holt, R.A.Palmer, T.B.Vance Junior
Am.Cryst.Assoc.,Ser.2,**7**, 24, 1980

84.C Diaqua – tetrakis(glycollato) – uranium(iv)
$C_8H_{16}O_{14}U$ Main entry is 81.36

84.C Dichloro – bis(2,2' – thio – diethanol) – cobalt(ii)
$C_8H_{20}Cl_2CoO_4S_2$ Main entry is 85.25

84.17 Dioxo – bis(N,N – diethylhydroxylaminato) – molybdenum(vi)
$C_8H_{20}MoN_2O_4$
L.Saussine, H.Mimoun, A.Mitschler, J.Fisher
Nouv.J.Chim.,**4**, 235, 1980
See also R1 : 83

84.C Δ – bis(Ethylenediamine) – ((3R,4R) – thiazolidine – 4 – carboxylato – N,O) – cobalt(iii) dichloride
$C_8H_{22}CoN_5O_2S^{2+}$, 2Cl⁻ Main entry is 76.47

84.18 bis(Aqua – bis(2 – amino – 2 – methylpropanolato) – copper(ii)) dinitrate
$2C_8H_{23}CuN_2O_3^+$, $2NO_3^-$
J.A.Bertrand, E.Fujita, D.G.VanDerveer
Inorg.Chem.,**19**, 2022, 1980
See also R1 : 83

84.19 cis – Dichloro – tetrakis(dimethylsulfoxide) – iron chloride
$C_8H_{24}Cl_2FeO_4S_4^+$, Cl⁻
I.P.Lavrent'ev, L.G.Korableva, E.A.Lavrent'eva, G.A.Nifontova, M.L.Khidekel, I.G.Gusakovskaya, T.I.Larkina, L.D.Arutyunyan, O.S.Filipenko, V.I.Ponomarev, L.O.Atovmyan *Koord.Khim.*,**5**, 1484, 1979

84.C tetrakis(μ_3 – Oxo) – tetrakis(μ – dimethylthiophosphinato) – tetra(oxo – molybdenum) chloroform carbon tetrachloride solvate
$C_8H_{24}Mo_4O_{12}P_4S_4$, xCHCl₃, yCCl₄ Main entry is 85.28

84.C tetrakis(μ_3 – Oxo) – tetrakis(μ – dimethylthiophosphinato) – tetra(oxo – molybdenum) chloroform carbon tetrachloride solvate (at –150°C)
$C_8H_{24}Mo_4O_{12}P_4S_4$, xCHCl₃, yCCl₄ Main entry is 85.29

84.C Methyl zinc methoxide tetramer
$C_8H_{24}O_4Zn_4$ Main entry is 71.8

84.C (μ – Peroxo) – (μ – hydroxo) – bis(bis(ethylenediamine) – cobalt(iii)) perchlorate
$C_8H_{33}Co_2N_8O_3^{3+}$, $3ClO_4^-$ Main entry is 76.53

84.20 tris(Triphenylphosphineiminium) sodium (μ^3 – oxo) – tris(μ^2 – methoxy) – tris(dicarbonyl – nitroso – molybdenum)
$2C_9H_9Mo_3N_3O_{13}^{2-}$, $3C_{36}H_{30}NP_2^+$, Na⁺
S.W.Kirtley, J.P.Chanton, R.A.Love, D.L.Tipton, T.N.Sorrell, R.Bau *J.Am.Chem.Soc.*,**102**, 3451, 1980
See also R2 : 64

84.21 Tetraethylammonium tris(μ – methoxy) – bis(tricarbonyl – rhenium(i))
$C_9H_9O_9Re_2^-$, $C_8H_{20}N^+$
G.Ciani, A.Sironi, A.Albinati
Gazz.Chim.Ital.,**109**, 615, 1979
See also R2 : 3

84.22 (tris(Triethylene) – tetramine) – perchlorato – copper perchlorate
$C_9H_{24}ClCuN_4O_4^+$, ClO_4^-
T.G.Fawcett, S.M.Rudich, B.H.Toby, R.A.Lalancette, J.A.Potenza, H.J.Schugar *Inorg.Chem.*,**19**, 940, 1980
See also R1 : 83

84.C (μ^2 – Oxo) – bis((η^5 – cyclopentadienyl) – iodo – oxo – molybdenum(v))
$C_{10}H_{10}I_2Mo_2O_3$ Main entry is 73.15

84.23 Diaquo – bis(pyridine – 3 – sulfonato) – copper(ii)
$(C_{10}H_{12}CuN_2O_8S_2)_n$
B.Walsh, B.J.Hathaway
J.Chem.Soc.,Dalton Trans.,681, 1980
See also R1 : 83

84.24 Tetramethylammonium (μ^3 – methoxy) – tris(μ^2 – methoxy) – tris(dicarbonyl – nitroso – molybdenum)
$C_{10}H_{12}Mo_3N_3O_{13}^-$, $C_4H_{12}N^+$
S.W.Kirtley, J.P.Chanton, R.A.Love, D.L.Tipton, T.N.Sorrell, R.Bau *J.Am.Chem.Soc.*,**102**, 3451, 1980
See also R2 : 3

84.25 trans – Dichloro – (dimethylformamide) – (2,6 – lutidine) – platinum(ii)
$C_{10}H_{16}Cl_2N_2OPt$
F.D.Rochon, P.C.Kong, R.Melanson
Can.J.Chem.,**58**, 97, 1980
See also R1 : 83

84.C Chloro – (2 – methyl – 2 – nitrosopropane) – (2 – (N – oxo – t – butylimino)ethyl) – platinum
$C_{10}H_{21}ClN_2O_2Pt$ Main entry is 71.24

84.C Nitrato – (1 – oxa – 7,10 – diaza – 4,13 – dithia – cyclopentadecane) – nickel(ii) nitrate
$C_{10}H_{22}N_3NiO_4S_2^+$, NO_3^- Main entry is 85.36

84.26 bis(Aqua) – (1,4,7,10,13 – pentaoxacyclopentadecane) – cobalt dinitrate
$C_{10}H_{24}CoO_7^{2+}$, $2NO_3^-$
E.M.Holt, R.A.Palmer, T.B.Vance Junior
Am.Cryst.Assoc.,Ser.2,**7**, 24, 1980

84.C Dinitrato – (N,N,N′,N′ – tetraethylethylenediamine) – copper(ii)
$C_{10}H_{24}CuN_4O_6$ Main entry is 76.65

84.27 Chloro – pentakis(dimethylsulfoxide) – iron bis(tetrachloro – iron)
$C_{10}H_{30}ClFeO_5S_5^{2+}$, $2Cl_4Fe^-$
I.P.Lavrent'ev, L.G.Korableva, E.A.Lavrent'eva, G.A.Nifontova, M.L.Khidekel, I.G.Gusakovskaya, T.I.Larkina, L.D.Arutyunyan, O.S.Filipenko, V.I.Ponomarev, L.O.Atovmyan *Koord.Khim.*,**5**, 1484, 1979

84.28 Chloro – pentakis(dimethylsulfoxide) – iron(iii) (μ – oxo) – bis(trichloro – iron)
$C_{10}H_{30}ClFeO_5S_5^{2+}$, $Cl_6Fe_2O^{2-}$
I.P.Lavrent'ev, L.G.Korableva, E.A.Lavrent'eva, G.A.Nifontova, M.L.Khidekel, I.G.Gusakovskaya, T.I.Larkina, L.D.Arutyunyan, O.S.Filipenko, V.I.Ponomarev, L.O.Atovmyan *Koord.Khim.*,**5**, 1484, 1979

84.29 (μ^2 – Hydrido) – (μ^2 – methoxy) – decacarbonyl – tri – osmium
$C_{11}H_4O_{11}Os_3$
M.R.Churchill, H.J.Wasserman
Inorg.Chem.,**19**, 2391, 1980

84.30 Tetraphenylphosphonium dioxo – (4 – (2 – pyridylazo) – resorcinolato) – vanadium(v)
$C_{11}H_7N_3O_4V^-$, $C_{24}H_{20}P^+$
N.Galesic, M.Siroki
Acta Crystallogr.,Sect.B,**35**, 2931, 1979
See also R1 : 83 R2 : 64

84.31 (Diacetylmonoxime) – (salicyloylhydrazinato) – chloro – copper(ii) methanol solvate
$C_{11}H_{13}ClCuN_3O_3$, CH_4O
Yu.M.Chumakov, M.D.Mazus, V.N.Byushkin, N.I.Belichuk, T.I.Malinovskii *Izv.Akad.Nauk Mold.SSR*,83, 1979
See also R1 : 83

84.C Sodium dioxo – (μ^2 – oxo) – (μ^2 – thio) – 1,2 – propylenediamine – tetra – acetato – di – molybdenum tetrahydrate
$C_{11}H_{14}Mo_2N_2O_{11}S^{2-}$, $2Na^+$, $4H_2O$ Main entry is 81.39

84.C Sodium dioxo – bis(μ^2 – oxo) – (μ – 1,2 – propylenediamine) – tetra – acetato – di – molybdenum trihydrate
$C_{11}H_{14}Mo_2N_2O_{12}{}^{2-}$, $2Na^+$, $3H_2O$ Main entry is 81.40

84.32 (Dimethyl – (N,N – dimethylethanolamino) – (3,5 – dimethylpyrazolyl) – gallato) – dinitrosyl – iron
$C_{11}H_{23}FeGaN_5O_3$
K.S.Chong, S.J.Rettig, A.Storr, J.Trotter
Can.J.Chem.,**57**, 3113, 1979
See also R1 : 83,68

84.33 (Dimethyl – (N,N – dimethylethanolamino) – (3,5 – dimethylpyrazolyl) – gallato) – nitrosyl – nickel
$C_{11}H_{23}GaN_4NiO_2$
K.S.Chong, S.J.Rettig, A.Storr, J.Trotter
Can.J.Chem.,**57**, 3107, 1979
See also R1 : 83,68

84.34 Barium tris(dihydrogen – violurato) – ruthenium(ii) nonahydrate
$2C_{12}H_6N_9O_{12}Ru^-$, Ba^{2+}, $9H_2O$
F.Abraham, G.Nowogrocki, S.Sueur, C.Bremard
Acta Crystallogr.,Sect.B,**36**, 799, 1980
See also R1 : 83

84.35 Ammonium (μ – oxo) – (μ – catecholato) – bis(dioxo – molybdenum) dihydrate
$C_{12}H_8Mo_2O_9{}^{2-}$, $2H_4N^+$, $2H_2O$
V.V.Tkachev, L.O.Atovmyan *Koord.Khim.*,**2**, 110, 1976

84.36 Aqua – (2,2′ – bipyridyl) – nitroacetato – copper(ii) monohydrate
$C_{12}H_{11}CuN_3O_5$, H_2O
K.von Deuten, G.Klar
Cryst.Struct.Commun.,**9**, 479, 1980
See also R1 : 83

84.C bis(Cyclopentadienyl) – chloro – ethoxo – titanium (at –125°C)
$C_{12}H_{15}ClOTi$ Main entry is 73.33

84.37 Dichloro – bis(1 – vinyl – 2 – hydroxymethylimidazole) – cobalt
$C_{12}H_{16}Cl_2CoN_4O_2$
V.I.Sokol, M.A.Porai–Koshits, V.P.Nikolaev, E.S.Domnina, L.V.Baikalova, G.G.Skvortsova, L.A.Butman
Koord.Khim.,**5**, 1725, 1979
See also R1 : 83

84.C (3,10 – Dihydroxyimino – 4,9 – dimethyl – 5,8 – diazadodeca – 4,9 – diene – 2,11 – dionato) – nickel(ii) (α form)
$C_{12}H_{16}N_4NiO_4$ Main entry is 76.70

84.C Triaquo – (pyridoxylidene – O – phospho – D,L – threoninato) – nickel(ii) dihydrate
$C_{12}H_{20}N_2NiO_{11}P$, $2H_2O$ Main entry is 82.25

84.38 tetrakis((μ^2 – Methoxy) – (μ^2 – acetato) – copper(ii))
$C_{12}H_{24}Cu_4O_{12}$
Yu.A.Simonov, G.S.Matuzenko, M.M.Botoshanskii, M.A.Yampol'skaya *Dokl.Akad.Nauk SSSR*,**250**, 99, 1980

84.39 Trinitrato – (hexaoxacyclo – octadecane) – neodymium(iii)
Trinitrato – (18 – crown – 6) – neodymium(iii)
$C_{12}H_{24}N_3NdO_{15}$
F.Benetollo, G.Bombieri, G.de Paoli
Eur.Cryst.Meeting,**5**, 206, 1979

84.40 Trinitrato – (hexaoxacyclo – octadecane) – neodymium(iii)
Trinitrato – (18 – crown – 6) – neodymium(iii)
$C_{12}H_{24}N_3NdO_{15}$
J.–C.G.Bunzli, B.Klein, D.Wessner
Inorg.Chim.Acta,**44**, 147, 1980
See also R1 : 38

84.41 Oxo – bis(2,3 – dimethylbutane – 2,3 – diolato) – osmium(vi)
$C_{12}H_{24}O_5Os$
L.O.Atovmyan, Yu.A.Sokolova
Zh.Strukt.Khim.,**20**, 754, 1979

84.C tetrakis(μ – Acetato) – bis(dimethylsulfoxide – rhodium)
$C_{12}H_{24}O_{10}Rh_2S_2$ Main entry is 81.48

84.C Aqua – diperchlorato – (1,4,10,13 – tetraoxa – 7,16 – dithiacyclo – octadecane) – lanthanum(iii) perchlorate
$C_{12}H_{28}Cl_2LaO_{13}S_2{}^+$, ClO_4^- Main entry is 85.42

84.42 bis(μ – 1 – Methylthyminato) – bis(cis – diammine – platinum(ii)) dinitrate
$C_{12}H_{26}N_8O_4Pt_2{}^{2+}$, $2NO_3^-$
B.Lippert, D.Neugebauer, U.Schubert
Inorg.Chim.Acta,**46**, 11, 1980
See also R1 : 83

84.43 cis – Dichloro – tetrakis(dimethylformamide) – iron tetrachloro – iron
$C_{12}H_{28}Cl_2FeN_4O_4{}^+$, Cl_4Fe^-
I.P.Lavrent'ev, L.G.Korableva, E.A.Lavrent'eva, G.A.Nifontova, M.L.Khidekel, I.G.Gusakovskaya, T.I.Larkina, L.D.Arutyunyan, O.S.Filipenko, V.I.Ponomarev, L.O.Atovmyan *Koord.Khim.*,**5**, 1484, 1979

84.44 (Chloro – bis – cis(dimethylformamide)) – copper(ii) – bis(μ – chloro) – (chloro – bis – cis(dimethylformamide)) – copper(ii)
$C_{12}H_{28}Cl_4Cu_2N_4O_4$
I.P.Lavrent'ev, L.G.Korableva, E.A.Lavrent'eva, G.A.Nifontova, M.L.Khidekel, I.G.Gusakovskaya, T.I.Larkina, L.D.Arutyunyan, O.S.Filipenko, V.I.Ponomarev, L.O.Atovmyan *Koord.Khim.*,**5**, 1484, 1979

84.C bis(Aqua – copper(ii)) – bis(μ – aqua) – bis(μ – N,N′ – bis(2 – hydroxyethyl) – dithio – oxamidato) – bis(sulfato – copper(ii))
$C_{12}H_{28}Cu_4N_4O_{16}S_6$ Main entry is 85.45

84.C Tetra – aqua – ethylenediamine – nickel – (μ – ethylenediamine – tetra – acetato) – nickel
$C_{12}H_{28}N_4Ni_2O_{12}$ Main entry is 76.74

84.C bis(Triethylphosphine) – dinitrosyl – nitrito – manganese
$C_{12}H_{30}MnN_3O_4P_2$ Main entry is 86.24

84.C $(\eta^2 - \text{Dimethylformamide}) - (\mu^2 - \text{hydrido}) -$ decacarbonyl – tri – ruthenium (at 115°K)
$C_{13}H_7NO_{11}Ru_3$ Main entry is 71.47

84.45 1,1,1,2,2,2,3,3,3 – Nonacarbonyl – bis(μ_3 – ethoxy) – 1,2 – (μ – fluoro) – tri – manganese
$C_{13}H_{10}FMn_3O_{11}$
E.W.Abel, I.D.H.Towle, T.S.Cameron, R.E.Cordes
J.Chem.Soc.,Dalton Trans.,1943, 1979

84.46 1,1,1,2,2,2,3,3,3 – Nonacarbonyl – bis(μ_3 – ethoxy) – 1,2 – (μ – iodo) – tri – manganese
$C_{13}H_{10}IMn_3O_{11}$
E.W.Abel, I.D.H.Towle, T.S.Cameron, R.E.Cordes
J.Chem.Soc.,Dalton Trans.,1943, 1979

84.47 **Trichloro – oxo – (N – pyridylmethylene – N′ – salicyloylhydrazinato) – molybdenum(v) acetonitrile solvate**
$C_{13}H_{11}Cl_3MoN_3O_3$, C_2H_3N
P.Domiano, A.Musatti, M.Nardelli, C.Pelizzi, G.Predieri
Transition Met.Chem.,5, 172, 1980
See also R1 : 83

84.48 **Tetracarbonyl – (S,S – dimethylsulfonium – 2 – picolinylmethylide) – tungsten(0)**
$C_{13}H_{11}NO_5SW$
G.Matsubayashi, I.Kawafune, T.Tanaka, S.Nishigaki, K.Nakatsu *J.Organomet.Chem.*,187, 113, 1980
See also R1 : 83

84.49 **(+)$_{546}$ – cis – α – Sodium carbonato – ((2S,2′S) – 1,1′ – ethylene – di – 2 – pyrrolidine – carboxylato) – cobalt(iii) trihydrate**
$C_{13}H_{18}CoN_2O_7^-$, Na^+, $3H_2O$
T.C.Woon, M.F.Mackay, M.J.O'Connor
Acta Crystallogr.,Sect.B,36, 2033, 1980
See also R1 : 83

84.50 **(–)$_{589}$ – bis(Pentane – 2,4 – dionato) – (1,3 – propanediamine) – chromium(iii) iodide monohydrate (absolute configuration)**
$C_{13}H_{24}CrN_2O_4^+$, I^-, H_2O
K.Matsumoto, S.Ooi *Bull.Chem.Soc.Jpn.*,52, 3307, 1979
See also R1 : 83

84.C Dimethyl – (N,N – dimethylethanolamino) – (1 – pyrazolyl) – gallato – (η^2 – thiomethoxymethyl) – dicarbonyl – molybdenum
$C_{13}H_{24}GaMoN_3O_3S$ Main entry is 72.22

84.51 **(μ^2 – Carbonyl) – (μ^2 – hydrido) – (μ^4 – methoxymethylidyne) – undecacarbonyl – tetra – iron**
$C_{14}H_4Fe_4O_{13}$
P.A.Dawson, B.F.G.Johnson, J.Lewis, P.R.Raithby
J.Chem.Soc.,Chem.Commun.,781, 1980

84.52 **(μ^2 – Carbonyl) – (μ^2 – hydrido) – (μ^4 – methoxymethylidyne) – undecacarbonyl – tetra – iron**
$C_{14}H_4Fe_4O_{13}$
K.Whitmire, D.F.Shriver, E.M.Holt
J.Chem.Soc.,Chem.Commun.,780, 1980

84.53 **cis(N – (o – Aminobenzylidene) – anthranil – aldehydato – O,N,N) – chloro – platinum chloroform solvate (at –5°C)**
$C_{14}H_{11}ClN_2OPt$, $CHCl_3$
M.D.Timken, R.I.Sheldon, W.G.Rohly, K.B.Mertes
J.Am.Chem.Soc.,102, 4716, 1980
See also R1 : 83

84.C Dicarbonyl – cyclopentadienyl – (3 – methylcyclohexenone) – iron hexafluorophosphate
$C_{14}H_{15}FeO_3^+$, F_6P^- Main entry is 73.48

84.54 **bis(μ^2 – Chloro) – bis(tetrahydrofuran) – hexacarbonyl – di – manganese**
$C_{14}H_{16}Cl_2Mn_2O_8$
M.C.VanDerveer, J.M.Burlitch
J.Organomet.Chem.,197, 357, 1980

84.55 **Aqua – bis(amidonicotinato) – bis(formato) – copper(ii)**
$C_{14}H_{16}CuN_4O_7$
A.S.Antsyshkina, M.A.Porai–Koshits, M.Gandlovich, M.Dunaj–Jurco, G.G.Sadikov, G.V.Tsintsadze
Koord.Khim.,5, 1716, 1979
See also R1 : 83

84.56 **Diaqua – diformato – bis(nicotinamide) – zinc**
$C_{14}H_{18}N_4O_8Zn$
L.Kh.Minacheva, T.S.Khodashova, M.A.Porai–Koshits, G.G.Sadikov, L.A.Butman, V.G.Sakharova, G.V.Tsintsadze
Koord.Khim.,5, 1889, 1979
See also R1 : 83

84.C Chloro – N,N – diethylamide – pyruvate – 4 – phenylthiosemicarbazonato – copper(ii)
$C_{14}H_{19}ClCuN_4OS$ Main entry is 79.12

84.57 **bis(2 – Dimethylamino – pyridine 1 – oxide) – copper(ii) diperchlorate**
$C_{14}H_{20}CuN_4O_2^{2+}$, $2ClO_4^-$
D.X.West, S.F.Pavkovic, J.N.Brown
Acta Crystallogr.,Sect.B,36, 143, 1980
See also R1 : 83

84.58 **bis(Ethylenediamine) – copper – (μ – ethylenediamine – tetra – acetato) – nickel tetrahydrate**
$C_{14}H_{28}CuN_6NiO_8$, $4H_2O$
V.M.Agre, V.K.Trunov, T.F.Sisoeva
Dokl.Akad.Nauk SSSR,250, 1123, 1980
See also R1 : 83

84.C (N,N,N′,N′, – Tetra(2 – hydroxypropyl) – ethylenediamine) – copper(ii) perchlorate monohydrate
$C_{14}H_{31}CuN_2O_4^+$, ClO_4^-, H_2O Main entry is 76.86

84.59 **bis(μ – Methoxy) – bis(dioxo(pinacolato) – molybdenum(vi)) methanol solvate**
$C_{14}H_{32}Mo_2O_{10}$, $2CH_4O$
C.Knobler, B.R.Penfold, W.T.Robinson, C.J.Wilkins, S.H.Yong
J.Chem.Soc.,Dalton Trans.,248, 1980

84.60 **bis(Hexamethylphosphoramide) – bis(isothiocyanato) – dioxo – molybdenum(vi)**
$C_{14}H_{36}MoN_8O_4P_2S_2$
B.Viossat, N.Rodier, P.Khodadad
Acta Crystallogr.,Sect.B,35, 2712, 1979

84.61 **tris(μ – Carbonyl) – (μ_4 – ethylamino) – (μ_4 – N – oxy – ethylamino) – tetrakis(dicarbonyl – iron)**
$C_{15}H_{10}Fe_4N_2O_{12}$
G.Gervasio, R.Rossetti, P.L.Stanghellini
J.Chem.Res.,334, 3943, 1979
See also R1 : 83

84.C Tricarbonyl – (methyl α – acetamido – cinnamate) – iron
$C_{15}H_{13}FeNO_6$ Main entry is 72.23

84.C Di(bis(μ – acetyl) – tetracarbonyl – rhenium) – copper
$C_{16}H_{12}CuO_{12}Re_2$ Main entry is 71.93

84.62 bis(1 – Methyl – 3 – (2 – chloro – 6 – methylphenyl) – triazine – 1 – oxidato) – nickel(ii)
$C_{16}H_{18}Cl_2N_6NiO_2$
M.V.Rajasekharan, K.I.Varughese, P.T.Manoharan
Inorg.Chem.,**18**, 2221, 1979
See also R1 : 83

84.63 bis(1 – Methyl – 3 – (2 – chloro – 6 – methylphenyl) – triazine – 1 – oxidato) – nickel(ii) benzene solvate
$C_{16}H_{18}Cl_2N_6NiO_2$, C_6H_6
M.V.Rajasekharan, K.I.Varughese, P.T.Manoharan
Inorg.Chem.,**18**, 2221, 1979
See also R1 : 83

84.C Catecholato – (η^5 – pentamethyl – cyclopentadienyl) – rhodium(iii) catechol solvate
$C_{16}H_{19}O_2Rh$, $2C_6H_6O_2$ Main entry is 73.61

84.C (μ^2 – Pinacolato) – bis(cyclopentadienyl – dichloro – titanium) (at –138°C)
$C_{16}H_{22}Cl_4O_2Ti_2$ Main entry is 73.63

84.C (η^8 – Cyclo – octatetraene) – dichloro – bis(tetrahydrofuran) – thorium
$C_{16}H_{24}Cl_2O_2Th$ Main entry is 75.13

84.64 bis(1H – 3,5 – Diethyl – 4 – methylpyrazole – N^2) – dinitrato – O,O – cobalt(ii)
$C_{16}H_{28}CoN_6O_6$
I.A.Krol, V.M.Agre, V.K.Trunov, N.P.Kozlova, E.S.Zaitseva
Koord.Khim.,**5**, 1575, 1979
See also R1 : 83

84.65 Di – iodo – bis(2 – isopropylamino – pent – 2 – en – 4 – one) – zinc
$C_{16}H_{30}I_2N_2O_2Zn$
N.Bresciani–Pahor, L.Randaccio, E.Libertini
Inorg.Chim.Acta,**45**, L11, 1980

84.66 bis(1,4,7,10 – Tetraoxacyclododecane) – manganese(ii) tribromide
$C_{16}H_{32}MnO_8^{2+}$, $2Br_3^-$
B.B.Hughes, R.C.Haltiwanger, C.G.Pierpont, M.Hampton, G.L.Blackmer *Inorg.Chem.*,**19**, 1801, 1980

84.C Chloro – (2 – (2′ – ethanolthio) – ethanolato) – copper(ii) tetramer
$C_{16}H_{36}Cl_4Cu_4O_8S_4$ Main entry is 85.53

84.67 bis(μ – 2 – (N,N – Dimethyl – 2 – aminoethyl) imino – butan – 3 – one – oximato) – diaqua – di – copper(ii) diperchlorate
$C_{16}H_{36}Cu_2N_6O_4^{2+}$, $2ClO_4^-$
M.Nasakkala, H.Saarinen, J.Korvenranta
Finn.Chem.Lett.,**6**, 1980
See also R1 : 83

84.68 Diaqua – tetrakis(perhydropyrimidin – 2 – one) – iron trichloride dihydrate
$C_{16}H_{36}FeN_8O_6^{3+}$, $3Cl^-$, $2H_2O$
S.Calogero, U.Russo, A.Del Pra
J.Chem.Soc.,Dalton Trans.,646, 1980

84.69 tetrakis(Isothiocyanato) – tetrakis(trimethylphosphine – oxide) – uranium(iv)
$C_{16}H_{36}N_4O_4P_4S_4U$
C.E.F.Rickard, D.C.Woollard *Aust.J.Chem.*,**32**, 2181, 1979

84.70 tetrakis((2 – Amino – 2 – methyl – 1 – propanolato) – chloro – copper(ii))
$C_{16}H_{40}Cl_4Cu_4N_4O_4$
H.Muhonen *Acta Chem.Scand.Ser.A*,**34**, 79, 1980
See also R1 : 83

84.71 (μ – Oxo) – bis(bis(N,N – diethylhydroxylaminato) – oxo – vanadium(v))
$C_{16}H_{40}N_4O_7V_2$
L.Saussine, H.Mimoun, A.Mitschler, J.Fisher
Nouv.J.Chim.,**4**, 235, 1980
See also R1 : 83

84.72 bis(μ – 2,6 – Diacetylpyridine – dioximato) – di – copper(ii) tetrafluoroborate dihydrate
$C_{18}H_{20}Cu_2N_6O_4^{2+}$, $2BF_4^-$, $2H_2O$
G.A.Nicholson, J.L.Petersen, B.J.McCormick
Inorg.Chem.,**19**, 195, 1980
See also R1 : 83

84.C Dioxo – bis(N – methyl – p – tolyl – thiohydroxamato) – molybdenum(vi)
$C_{18}H_{20}MoN_2O_4S_2$ Main entry is 85.61

84.73 bis(μ – Dimethylphosphinato) – bis(tricarbonyl – tetrahydrofuran – rhenium) tetrahydrofuran solvate
$C_{18}H_{28}O_{12}P_2Re_2$, $2C_4H_9O$
E.Lindner, S.Trad, S.Hoehne, H.–H.Oetjen
Z.Naturforsch.,Teil B,**34**, 1203, 1979

84.C (μ – Di – t – butylacetylene) – bis((μ^2 – chloro) – dichloro – tetrahydrofuran – tantalum)
$C_{18}H_{34}Cl_6O_2Ta_2$ Main entry is 72.35

84.74 (4,7,13,16,21,24 – Hexaoxa – 1,10 – diazabicyclo(8.8.8) hexacosane) – nitrato – samarium(iii) aquo – pentanitrato – samarium(iii)
$C_{18}H_{36}N_3O_9Sm^{2+}$, $H_2N_5O_{16}Sm^{2-}$
J.H.Burns *Inorg.Chem.*,**18**, 3044, 1979
See also R1 : 83

84.C tris(μ – Hydroxo) – hexakis(trimethylphosphine) – di – ruthenium(ii) tetrafluoroborate
$C_{18}H_{57}O_3P_6Ru_2^+$, BF_4^- Main entry is 86.57

84.75 Nitrato – bis(dioxo – 2,6 – diacetylpyridine – bis(2′ – pyridylhydrazone)) – uranium dioxo – uranium – tetranitrate
$2C_{19}H_{19}N_8O_5U^+$, $N_4O_{14}U^{2-}$
G.Paolucci, G.Marangoni, G.Bandoli, D.A.Clemente
J.Chem.Soc.,Dalton Trans.,459, 1980
See also R1 : 83

84.C bis(Hexafluoroacetylacetonato) – (4 – hydroxy – 2,2,6,6 – tetramethyl – piperidinyl – N – oxy) – copper(ii)
$(C_{19}H_{20}CuF_{12}NO_6)_n$ Main entry is 77.7

84.C tris(Cyclopentadienyl) – tetrahydrofuran – gadolinium
$C_{19}H_{23}GdO$ Main entry is 73.79

84.C ((η^3 – Cycloheptatrienyl) – dicarbonyl – molybdenum) – tris(μ – methoxy) – ((η^7 – cycloheptatrienyl) – molybdenum)
$C_{19}H_{23}Mo_2O_5$ Main entry is 75.22

84.76 tetrakis(μ^2 – 6 – Chloro – 2 – pyridinolato) – di – chromium
$C_{20}H_{12}Cl_4Cr_2N_4O_4$
F.A.Cotton, W.H.Ilsley, W.Kaim
Inorg.Chem., **19**, 1453, 1980
See also R1 : 83

84.77 tetrakis(μ^2 – 6 – Chloro – 2 – pyridinolato) – di – molybdenum
$C_{20}H_{12}Cl_4Mo_2N_4O_4$
F.A.Cotton, W.H.Ilsley, W.Kaim
Inorg.Chem., **19**, 1453, 1980
See also R1 : 83

84.78 tetrakis(μ^2 – 6 – Chloro – 2 – pyridinolato) – di – tungsten
$C_{20}H_{12}Cl_4N_4O_4W_2$
F.A.Cotton, W.H.Ilsley, W.Kaim
Inorg.Chem., **19**, 1453, 1980
See also R1 : 83

84.79 tetrakis(μ – 2 – Oxo – pyridinato) – (chloro – di – technetium)
$(C_{20}H_{16}ClN_4O_4Tc_2)_n$
F.A.Cotton, P.E.Fanwick, L.D.Gage
J.Am.Chem.Soc., **102**, 1570, 1980
See also R1 : 83

84.80 tetrakis(μ – α – Pyridonato) – bis(chloro – osmium)
$C_{20}H_{16}Cl_2N_4O_4Os_2$
F.A.Cotton, J.L.Thompson
Inorg.Chim.Acta, **44**, L247, 1980
See also R1 : 83

84.81 bis(N – Methylimidazole) – bis(2,4,6 – trichlorophenolato) – cobalt(ii)
$C_{20}H_{16}Cl_6CoN_4O_2$
M.B.Cingi, A.M.M.Lanfredi, A.Tiripicchio, J.Reedijk, R.van Landschoot *Inorg.Chim.Acta*, **39**, 181, 1980
See also R1 : 83

84.82 bis(2,2′ – Bipyridyl) – peroxo – disulfato – copper(ii) monohydrate
$C_{20}H_{16}CuN_4O_9S_2$, H_2O
W.D.Harrison, B.J.Hathaway
Acta Crystallogr.,Sect.B, **36**, 1069, 1980
See also R1 : 83

84.C bis(bis(Cyclopentadienyl) – chloro – titanium) – oxide
$C_{20}H_{20}Cl_2OTi_2$ Main entry is 73.83

84.83 tris(Perchlorato) – (5,6,14,15 – tetrahydro – dibenzo(b,k)(1,4,7,10,13,16) – hexaoxacyclo – octadecin) – samarium(iii)
$C_{20}H_{24}Cl_3O_{18}Sm$
M.Ciampolini, N.Nardi, R.Cini, S.Mangani, P.Orioli
J.Chem.Soc.,Dalton Trans., 1983, 1979

84.84 trans – bis(t – Butoxy) – cis – dicarbonyl – cis – bis(pyridine) – molybdenum (at –175°C)
$C_{20}H_{28}MoN_2O_4$
M.H.Chisholm, J.C.Huffman, R.L.Kelly
J.Am.Chem.Soc., **101**, 7615, 1979
See also R1 : 83

84.C bis(Triethylphosphine) – (1,1,1 – trifluoro – 5 – methoxy – 3 – (trifluoromethyl) – 4 – methyl – pent – 3 – en – 2 – yl) – platinum hexafluorophosphate
$C_{20}H_{39}F_6OP_2Pt^+$, F_6P^- Main entry is 71.162

84.C bis(μ – Chloro) – bis(bis(carbonyl) – trimethylphosphine – (1 – 2 – η – trimethylsilyl – methylcarbonyl) – molybdenum(ii))
$C_{20}H_{40}Cl_2Mo_2O_6P_2Si_2$ Main entry is 71.163

84.C (1,2 – bis(Methoxycarbonyl) – but – 2 – en – 1 – yl) – bis(triethylphosphine) – platinum(ii) tetraphenylborate
$C_{20}H_{41}O_4P_2Pt^+$, $C_{24}H_{20}B^-$ Main entry is 71.164

84.C (1,2 – bis(Methoxycarbonyl) – but – 2 – en – 1 – yl) – bis(triethylphosphine) – platinum(ii) hexafluorophosphate
$C_{20}H_{41}O_4P_2Pt^+$, F_6P^- Main entry is 71.165

84.C (μ_2 – Dioxo) – bis(1,5,8,11,15 – penta – azapentadecane – cobalt) dithionate dinitrate tetrahydrate
$C_{20}H_{54}Co_2N_{10}O_2^{4+}$, $O_6S_2^{2-}$, $4H_2O$, $2NO_3^-$ Main entry is 76.97

84.85 bis((1,2 – Dimethoxyethane) – bis(trimethylsilyl) amino) – europium(ii)
$C_{20}H_{56}EuN_2O_2Si_4$
A.Zalkin, D.H.Templeton, B.Spencer, H.Ruben, T.D.Tilley, R.A.Andersen *Am.Cryst.Assoc.,Ser.2*, **8**, 25, 1980
See also R1 : 83

84.86 tris(Tropolonato) – cobalt(iii)
$C_{21}H_{15}CoO_6$
D.M.Doddrell, M.R.Bendall, P.C.Healy, G.Smith, C.H.L.Kennard, C.L.Raston, A.H.White
Aust.J.Chem., **32**, 1219, 1979

84.C Sodium triethylmethylammonium tris(thiobenzohydroximato) – chromium(iii) hemi(sodium hydroxide) hydrate
$C_{21}H_{15}CrN_3O_3S_3^{3-}$, $C_7H_{18}N^+$, $2.5Na^+$, $0.5HO^-$, $18.5H_2O$
Main entry is 85.66

84.87 Sodium trans – tris(benzohydroximato) – chromium(iii) octahydrate ethanol solvate
$C_{21}H_{15}CrN_3O_6^{3-}$, $3Na^+$, $8H_2O$, C_2H_6O
K.Abu-Dari, K.N.Raymond *Inorg.Chem.*, **19**, 2034, 1980

84.88 Sodium cis – tris(benzohydroximato) – chromium(iii) sodium iodide sodium hydroxide nonahydrate methanol ethanol solvate
$C_{21}H_{15}CrN_3O_6^{3-}$, $5Na^+$, I^-, HO^-, $9H_2O$, $3CH_4O$, C_2H_6O
K.Abu-Dari, K.N.Raymond *Inorg.Chem.*, **19**, 2034, 1980

84.C bis(μ – Chloro) – bis(2,2′ – spiro(8 – methoxy – 3 – methyl – 6 – nitro – chromene – 3′ – methyl – benzothiazoline) – chloro – cobalt) – acetone – (2,2′ – spiro(8 – methoxy – 3 – methyl – 6 – nitro – chromene) – 3′ – methyl – benzothiazoline) – cobalt
$C_{21}H_{22}Cl_2CoN_2O_5S$, $C_{36}H_{32}Cl_4Co_2N_4O_8S_2$ Main entry is 84.120

84.89 bis(Isothiocyanato) – (2,3:11,12) – dibenzo – 1,13 – dioxa – 5,9 – diaza – 2,11 – cyclopentadecadiene – nickel(ii) (light blue phase)
$C_{21}H_{24}N_4NiO_2S_2$
L.P.Battaglia, A.B.Corradi, A.Mangia
Inorg.Chim.Acta, **39**, 211, 1980
See also R1 : 83

84.C ((η^3 – Cycloheptatrienyl) – dicarbonyl – tungsten) – tris(μ – methoxy) – ((η^4 – cycloheptatrienyl) – dicarbonyl – tungsten)
$C_{21}H_{24}O_7W_2$ Main entry is 75.29

84.90 Chloro – dioxo – (N – phenyl – benzohydroxamato) – bis(tetrahydrofuran) – uranium(vi)
$C_{21}H_{26}ClNO_6U$
W.L.Smith, K.N.Raymond
*J.Inorg.Nucl.Chem.,***41**, 1431, 1979

84.91 (N,N′ – Ethylene – bis(benzoylacetoniminato)) – nickel(ii)
$C_{22}H_{22}N_2NiO_2$
V.Malatesta, A.Mugnoli *Eur.Cryst.Meeting,***5**, 203, 1979
See also R1 : 83

84.C 1,1,1,2,2,2,3,3 – Octacarbonyl – (dimethylphenylphosphine) – bis(μ_3 – ethoxy) – 1,2 – (μ – ethoxy) – tri – manganese
$C_{22}H_{26}Mn_3O_{11}P$ Main entry is 86.80

84.92 bis(2,6 – Diacetylpyridine – bis(semicarbazone)) – cerium(iii) triperchlorate trihydrate
$C_{22}H_{30}CeN_{14}O_4{}^{3+}$, $3ClO_4{}^-$, $3H_2O$
J.E.Thomas, G.J.Palenik *Inorg.Chim.Acta,***44**, L303, 1980
See also R1 : 83

84.C (μ^2 – 1,2 – bis(Carbomethoxyethylene)) – bis((μ^2 – methylthio) – trimethylphosphine – carbonyl – iron) tetraphenylborate (at –162°C)
$C_{22}H_{49}Fe_2O_6P_4S_2{}^+$, $C_{24}H_{20}B^-$ Main entry is 86.82

84.93 Trinitrato – (2,6 – diacetylpyridine – bis(benzoic acid hydrazone)) – lanthanum(iii)
$C_{23}H_{21}LaN_8O_{11}$
J.E.Thomas, R.C.Palenik, G.J.Palenik
*Inorg.Chim.Acta,***37**, L459, 1979
See also R1 : 83

84.94 Aqua – nitrato – (2,6 – diacetylpyridine – bis(benzoic acid hydrazone)) – cobalt(ii) nitrate
$C_{23}H_{23}CoN_8O_6{}^+$, $NO_3{}^-$
T.J.Giordano, G.J.Palenik, R.C.Palenik, D.A.Sullivan
*Inorg.Chem.,***18**, 2445, 1979
See also R1 : 83

84.95 Diaqua – (2,6 – diacetylpyridine – bis(benzoic acid hydrazone)) – nickel(ii) dinitrate dihydrate
$C_{23}H_{25}N_5NiO_4{}^{2+}$, $2NO_3{}^-$, $2H_2O$
T.J.Giordano, G.J.Palenik, R.C.Palenik, D.A.Sullivan
*Inorg.Chem.,***18**, 2445, 1979
See also R1 : 83

84.C Tetrabromocatecholato – nitroso – triphenylphosphine – iridium methylene chloride solvate
$C_{24}H_{15}Br_4IrNO_3P$, CH_2Cl_2 Main entry is 86.85

84.C N,N′ – Ethylene – bis(1 – iminomethyl – 2 – naphtholato) – copper(ii)
$C_{24}H_{18}CuN_2O_2$ Main entry is 76.99

84.C N,N′ – Ethylene – bis(1 – iminomethyl – 2 – naphtholato) – nickel(ii)
$C_{24}H_{18}N_2NiO_2$ Main entry is 76.100

84.C tris(N – Methyl – thiobenzohydroxamato) – cobalt(iii)
$C_{24}H_{24}CoN_3O_3S_3$ Main entry is 85.72

84.C tris(N – Methyl – thiobenzohydroxamato) – chromium(iii)
$C_{24}H_{24}CrN_3O_3S_3$ Main entry is 85.73

84.C tris(N – Methyl – thiobenzohydroxamato) – iron(iii)
$C_{24}H_{24}FeN_3O_3S_3$ Main entry is 85.74

84.C tris(N – Methyl – thiobenzohydroxamato) – manganese(ii)
$C_{24}H_{24}MnN_3O_3S_3$ Main entry is 85.75

84.96 bis(N – Acetylimino – benzoylacetonato) – platinum(ii)
$C_{24}H_{24}N_2O_4Pt$
T.Uchiyama, K.Takagi, K.Matsumoto, S.Ooi, Y.Nakamura, S.Kawaguchi *Chem.Lett.,*1197, 1979
See also R1 : 83

84.97 tetrakis(6 – Methyl – 2 – oxo – pyridinato) – di – rhodium
$C_{24}H_{24}N_4O_4Rh_2$
M.Berry, C.D.Garner, I.H.Hillier, A.A.MacDowell, W.Clegg
*J.Chem.Soc.,Chem.Commun.,*494, 1980
See also R1 : 83

84.98 Copper(ii) – copper(i) – pseudo – porphyrin – macrocyclic ligand complex
$C_{24}H_{26}Cu_2N_4O_2{}^+$, $ClO_4{}^-$, $0.5CH_4O$
R.R.Gagne, L.M.Henling, T.J.Kistenmacher
*Inorg.Chem.,***19**, 1226, 1980
See also R1 : 83,49

84.99 (6,6′ – (2,5,8,11,14,17,20 – Heptaoxa – heneicosane – 1,21 – diyl) – 2,2′ – bipyridyl) – dichloro – cobalt
$C_{24}H_{34}Cl_2CoN_2O_7$
G.R.Newkome, D.K.Kohli, F.Fronczek
*J.Chem.Soc.,Chem.Commun.,*9, 1980
See also R1 : 83

84.C Aquo – (pyridoxylidene – 0 – phospho – D,L – threoninato) – copper(ii) dimer monohydrate
$C_{24}H_{34}Cu_2N_4O_{18}P_2$, $2H_2O$ Main entry is 82.39

84.C tetrakis(μ – Trichloroacetato) – tetrakis(μ – t – butylperoxy) – tetra – palladium
$C_{24}H_{36}Cl_{12}O_{16}Pd_4$ Main entry is 81.79

84.C bis(Pentamethyl – cyclopentadienyl) – tetrahydrofuran – ytterbium(ii) toluene solvate (at 176°K)
$C_{24}H_{38}OYb$, $0.5C_7H_8$ Main entry is 73.102

84.100 bis(bis(μ – 1 – Methylthyminato) – cis – diammine – platinum(ii)) silver nitrate pentahydrate
$C_{24}H_{44}AgN_{12}O_8Pt_2{}^+$, $NO_3{}^-$, $5H_2O$
B.Lippert, D.Neugebauer *Inorg.Chim.Acta,***46**, 171, 1980
See also R1 : 83,44

84.C tetrakis(Dichloroacetato – (μ – 2 – dimethylaminoethanolato) – copper(ii))
$C_{24}H_{44}Cl_8Cu_4N_4O_{12}$ Main entry is 81.81

84.101 Aqua – tetrakis(N – formyl – piperidine) – dioxo – uranium(vi) diperchlorate
$C_{24}H_{46}N_4O_7U^{2+}$, $2ClO_4{}^-$
G.J.Honan, D.L.Kepert, S.F.Lincoln, J.M.Patrick, A.H.White
*Aust.J.Chem.,***33**, 69, 1980

84.C tetrakis(Chloroacetato – (μ – 2 – dimethylaminoethanolato) – copper(ii))
$C_{24}H_{48}Cl_4Cu_4N_4O_{12}$ Main entry is 81.82

84.102 (μ – Oxo) – bis((2,3 – dimethyl – 2 – hydroxy – 3 – butanolato) – (2,3 – dimethyl – 2,3 – butanediolato) – oxo – molybdenum)
$C_{24}H_{50}Mo_2O_{11}$
A.J.Matheson, B.R.Penfold
*Acta Crystallogr.,Sect.B,***35**, 2707, 1979

84.103 tetrakis(Bromo – (2 – diethylaminoethanolato) –
copper(ii)) carbon tetrachloride solvate
$C_{24}H_{56}Br_4Cu_4N_4O_4$, $4CCl_4$
R.Mergehenn, L.Merz, W.Haase
J.Chem.Soc.,Dalton Trans.,1703, 1980
See also R1 : 83

84.C tris(μ – Chloro) – (N – nitroso – N – trimethylsilyl –
methylhydroxylaminato) –
pentakis(trimethylsilylmethyl) – triangulo – tri –
rhenium(iii)
$C_{24}H_{66}Cl_3N_2O_2Re_3Si_6$ Main entry is 71.194

84.104 bis(Trinitrato – tetrakis(tris(dimethylamido) –
phosphine – oxide) – thorium) hexanitrato –
thorium
$2C_{24}H_{72}N_{15}O_{13}P_4Th^+$, $N_6O_{18}Th^{2-}$
R.P.English, J.G.H.du Preez, L.R.Nassimbeni,
C.P.J.van Vuuren *S.Afr.J.Chem.*,32, 119, 1979

84.C (μ^2 – Hydrido) – (μ^2 – p – tolylcarbamoyl) –
dimethylphenylphosphine – nonacarbonyl – tri –
osmium
$C_{25}H_{20}NO_{10}Os_3P$ Main entry is 71.195

84.C bis(Cyclopentadienyl) – (2,6 – di – t – butyl – 4 –
methylphenoxy) – titanium(iii)
$C_{25}H_{33}OTi$ Main entry is 73.106

84.105 (μ – Carbonyl) – bis(μ – t – butoxy) – bis(di – t –
butoxy – molybdenum)
$C_{25}H_{54}Mo_2O_7$
M.H.Chisholm, F.A.Cotton, M.W.Extine, R.L.Kelly
J.Am.Chem.Soc.,101, 7645, 1979

84.106 (μ – Oxo) – (μ – trimethylsilyloxy) – (μ – per –
rhenato) – bis(di(t – butylimido) –
trimethylsilyloxy – rhenium) (at –47°C)
$C_{25}H_{63}N_4O_8Re_3Si_3$
W.A.Nugent, R.L.Harlow
J.Chem.Soc.,Chem.Commun.,1105, 1979
See also R1 : 83,63

84.107 (μ – N,N' – bis(2 – (2 – Pyridyl)ethyl) – 2 –
hydroxy – 5 – methyl – isophthalaldiminato) – (μ –
pyrazolato) – di – copper(i)
$C_{28}H_{28}Cu_2N_6O$
R.R.Gagne, R.P.Kreh, J.A.Dodge
J.Am.Chem.Soc.,101, 6917, 1979
See also R1 : 83

84.C (Ethoxy – (1,5,9,13 – tetrathiacyclohexadecane) –
molybdenum(iv)) – (μ – oxo) – (oxy – (1,5,9,13 –
tetrathiacyclohexadecane) – molybdenum(iv))
trifluoromethylsulfonate monohydrate
$C_{26}H_{53}Mo_2O_3S_8^{3+}$, $3CF_3O_3S^-$, H_2O Main entry is 85.79

84.108 Ferrichrome (at –135°C, absolute configuration)
$C_{27}H_{42}FeN_9O_{12}$
D.van der Helm, J.R.Baker, D.L.Eng–Wilmot, M.B.Hossain,
R.A.Loghry *J.Am.Chem.Soc.*,102, 4224, 1980
See also R1 : 48

84.109 (N,N' – o – Phenylene – (6,6' – (1,2 –
dihydrobenzimidazole – methoxymethylene – 2 –
yl) – bis(pyridine – 2 – aldimino))) – aquo –
methanol – cobalt(ii) diperchlorate
$C_{28}H_{28}CoN_6O_3^{2+}$, $2ClO_4^-$
S.M.Nelson, F.S.Esho, M.G.B.Drew, P.Bird
J.Chem.Soc.,Chem.Commun.,1035, 1979
See also R1 : 83

84.C bis((μ^2 – Phenoxo) – ethylenediamine – phenoxy –
copper(ii)) phenol solvate
$C_{28}H_{36}Cu_2N_4O_4$, $2C_6H_6O$ Main entry is 76.103

84.110 bis(o – Phenylazophenolato – (tricarbonyl) –
rhenium(i))
$C_{30}H_{18}N_4O_8Re_2$
G.G.Aleksandrov, V.V.Derunov, A.A.Johansson.
Yu.T.Struchkov *J.Organomet.Chem.*,188, 367, 1980
See also R1 : 83

84.C Tricarbonyl – chloro – (5,7 – dichloro – 8 –
quinolinolato – N,O) – triphenylphosphine –
tungsten(ii)
$C_{30}H_{19}Cl_3NO_4PW$ Main entry is 86.100

84.111 hexakis(Pyridine – N – oxide) – copper(ii) dinitrate
$C_{30}H_{30}CuN_6O_6^{2+}$, $2NO_3^-$
J.S.Wood, C.P.Keijzers, E.de Boer, A.Buttafava
Inorg.Chem.,19, 2213, 1980

84.C tris(μ^2 – Hydrido) – (μ^3 – oxo) – tris(pentamethyl –
cyclopentadienyl – rhodium) hexafluorophosphate
monohydrate
$C_{30}H_{48}ORh_3^+$, F_6P^-, H_2O Main entry is 73.116

84.112 bis((μ – Methoxy) – (2,4,6 – trichlorophenolato) –
quinoline – copper(ii))
$C_{32}H_{24}Cl_6Cu_2N_2O_4$
M.A.Yampol'skaya, A.A.Dvorkin, Yu.A.Simonov,
V.K.Voronkova, L.V.Mosina, Yu.V.Yablokov, K.I.Turte,
A.V.Ablov, T.I.Malinovskii *Zh.Neorg.Khim.*,25, 174, 1980
See also R1 : 83

84.113 tetrakis(μ^2 – Benzoato) – dioxane – di – copper
dioxane solvate
$(C_{32}H_{28}Cu_2O_{10})_n$, $2nC_4H_8O_2$
L.Boniak, M.M.Borel, F.Busnot, A.Leclaire
Rev.Chim.Miner.,16, 501, 1979

84.114 tetrakis(μ – N – Phenyl – acetamidato) – di –
chromium
$C_{32}H_{32}Cr_2N_4O_4$
A.Bino, F.A.Cotton, W.Kaim *Inorg.Chem.*,18, 3030, 1979
See also R1 : 83

84.115 tetrakis(μ – Acetanilido) – di – molybdenum
tetrahydrofuran solvate
$C_{32}H_{32}Mo_2N_4O_4$, $2C_4H_8O$
A.Bino, F.A.Cotton, W.Kaim *Inorg.Chem.*,18, 3030, 1979
See also R1 : 83

84.C (2 – Diethylaminoethanolato) – (trichloroacetato) –
copper(ii) tetramer
$C_{32}H_{56}Cl_{12}Cu_4N_4O_{12}$ Main entry is 81.87

84.C Tetra – aqua – (μ^3 – dodeca – oxo) – (μ^2 –
hexadeca – acetato) – dodeca – manganese acetic
acid solvate tetrahydrate
$C_{32}H_{56}Mn_{12}O_{48}$, $2C_2H_4O_2$, $4H_2O$ Main entry is 81.88

84.C tetrakis(μ – 2 – Diethylaminoethanolato) –
tetrakis(chloroacetato – copper(ii)
$C_{32}H_{64}Cl_4Cu_4N_4O_{12}$ Main entry is 81.89

84.116 trans – bis(1 – Adamantylamido) –
tetrakis(trimethylsilyl – oxo) – molybdenum
(at –55°C)
$C_{32}H_{68}MoN_2O_4Si_4$
W.A.Nugent, R.L.Harlow *Inorg.Chem.*,19, 777, 1980
See also R1 : 83,63

84.117 octakis(t − Butoxy) − tetrakis(μ − fluoro) − tetra −
molybdenum bis(octakis(t − butoxy) − tris(μ −
fluoro) − (μ − dimethylamido) − tetra −
molybdenum (at −145°C)
$C_{32}H_{72}F_4Mo_4O_8$, $2C_{34}H_{78}F_3Mo_4NO_8$
M.H.Chisholm, J.C.Huffman, R.L.Kelly
J.Am.Chem.Soc.,**101**, 7100, 1979
See also R2 : 84,83

84.C (2 − Benzoylphenyl) − tricarbonyl −
(triphenylphosphine) − rhenium
$C_{34}H_{24}O_4PRe$ Main entry is 71.239

84.C octakis(t − Butoxy) − tetrakis(μ − fluoro) − tetra −
molybdenum bis(octakis(t − butoxy) − tris(μ −
fluoro) − (μ − dimethylamido) − tetra −
molybdenum (at −145°C)
$2C_{34}H_{78}F_3Mo_4NO_8$, $C_{32}H_{72}F_4Mo_4O_8$ Main entry is 84.117

84.118 Dichloro − oxo − bis(triphenylphosphine − oxide) −
vanadium(iv)
$C_{36}H_{30}Cl_2O_3P_2V$
M.R.Caira, B.J.Gellatly
Acta Crystallogr.,Sect.B,**36**, 1198, 1980

84.119 Trichloro − oxo − bis(triphenylphosphine oxide) −
tungsten(v)
$C_{36}H_{30}Cl_3O_3P_2W$
L.H.Hill, N.C.Howlader, F.E.Mabbs, M.B.Hursthouse,
K.M.A.Malik *J.Chem.Soc.,Dalton Trans.*,1475, 1980

84.120 bis(μ − Chloro) − bis(2,2′ − spiro(8 − methoxy − 3 −
methyl − 6 − nitro − chromene − 3′ − methyl −
benzothiazoline) − chloro − cobalt) − acetone −
(2,2′ − spiro(8 − methoxy − 3 − methyl − 6 − nitro −
chromene − 3′ − methyl − benzothiazoline) − cobalt
$C_{36}H_{32}Cl_4Co_2N_4O_8S_2$, $C_{21}H_{22}Cl_2CoN_2O_5S$
E.Miler−Srenger, M.le Baccon, R.Guglielmetti
Eur.Cryst.Meeting,**5**, 196, 1979
See also R2 : 84

84.C cis − bis(Thionylimide − O) −
bis(triphenylphosphine) − platinum(ii) hemihydrate
$C_{36}H_{32}N_2O_2P_2PtS_2$, $0.5H_2O$ Main entry is 86.119

84.121 bis(2,4 − Dimethyl − 6 − oxopyrimidine) − bis(1,3 −
diphenyltriazino) − di − tungsten tetrahydrofuran
solvate
$C_{36}H_{34}N_{10}O_2W_2$, $2C_4H_8O$
F.A.Cotton, W.H.Ilsley, W.Kaim
Inorg.Chem.,**19**, 1450, 1980
See also R1 : 83

84.C (α − 4,7,13,16 − Tetraphenyl − 1,10 − dioxa −
4,7,13,16 − tetraphosphacyclo − octadecane) −
cobalt(ii) bis(tetraphenylborate)
$C_{36}H_{44}CoO_2P_4{}^{2+}$, $2C_{24}H_{20}B^-$ Main entry is 86.122

84.C (β − 4,7,13,16 − Tetraphenyl − 1,10 − dioxa −
4,7,13,16 − tetraphosphacyclo − octadecane) −
cobalt(ii) bis(tetraphenylborate)
dimethylformamide solvate
$C_{36}H_{44}CoO_2P_4{}^{2+}$, $2C_{24}H_{20}B^-$, $2C_3H_7NO$ Main entry is 86.123

84.C bis(μ − Oxo) − bis((N,N′ − ethylene −
bis(acetylacetiminato)) − oxo − rhenium(v)) − (N,N′ −
ethylene − bis(acetylacetiminato)) rhenium(v) per −
rhenate benzene dichloromethane solvate
$C_{36}H_{54}N_6O_{10}Re_3{}^+$, O_4Re^-, $1.5C_6H_6$, xCH_2Cl_2
Main entry is 76.108

84.122 bis((μ − Acetato − (0,0′)) − diacetato − (μ₃ − 2 −
diethylaminoethanolato − N,μ₃ − 0)) − (μ − 2 −
diethylaminoethanolato − N,μ − 0) − (μ₃ −
hydroxo) − tri − copper(ii)) monohydrate
$C_{36}H_{76}Cu_6N_4O_{18}$, H_2O
M.Ahlgren, U.Turpeinen, K.Smolander
Acta Crystallogr.,Sect.B,**36**, 1091, 1980
See also R1 : 83

84.C Dicarbonyl − bis(5,7 − dichloro − 8 − quinolinolato −
N,O) − triphenylphosphine − tungsten(ii)
dichloromethane solvate
$C_{38}H_{23}Cl_4N_2O_4PW$, CH_2Cl_2 Main entry is 86.133

84.C Dichloro − bis(o − (diphenylphosphino)anisole) −
ruthenium(ii) dichloromethane solvate
$C_{38}H_{34}Cl_2O_2P_2Ru$, CH_2Cl_2 Main entry is 86.138

84.C (Methyl − (Z) − α − acetamido − cinnamate) − (1,2 −
bis(diphenylphosphino)ethane) − rhodium(i)
tetrafluoroborate
$C_{38}H_{37}NO_3P_2Rh^+$, BF_4^- Main entry is 72.59

84.123 (3,5 − Di − t − butylcatecholato) − (3,5 − di − t −
butyl − semiquinone) − (bipyridyl) − cobalt(iii)
toluene solvate
$C_{38}H_{48}CoN_2O_4$, $0.5C_7H_8$
R.M.Buchanan, C.G.Pierpont
J.Am.Chem.Soc.,**102**, 4951, 1980
See also R1 : 83

84.C Carbonyl − iodo − bis(triphenylphosphine) − acetyl −
ruthenium
$C_{39}H_{33}IO_2P_2Ru$ Main entry is 71.258

84.C bis(Dimethylformamide) − tris(2,2,6,6 −
tetramethylheptane − 3,5 − dionato) − europium(iii)
$C_{39}H_{71}EuN_2O_8$ Main entry is 77.10

84.C tetrakis(μ − 2,4 − Dichlorophenoxyacetato) −
bis(dioxane − copper(ii)) dioxane solvate
$C_{40}H_{36}Cl_8Cu_2O_{16}$, $3C_4H_8O_2$ Main entry is 81.92

84.124 bis(μ − N,N′ − Diphenyl − acetamidinato) − bis(μ −
2,4 − dimethyl − 6 − hydroxy − pyrimidinato) − di −
tungsten tetrahydrofuran solvate
$C_{40}H_{40}N_8O_2W_2$, $2C_4H_8O$
F.A.Cotton, W.H.Ilsley, W.Kaim
Inorg.Chem.,**18**, 3569, 1979
See also R1 : 83

84.125 2,2′ − Bipyridyl − (dihydrogen adenosine −
triphosphato) − zinc(ii) dimer tetrahydrate
$C_{40}H_{44}N_{14}O_{26}P_6Zn_2$, $4H_2O$
P.Orioli, R.Cini, D.Donati, S.Mangani
Nature (London),**283**, 691, 1980
See also R1 : 47

84.126 (μ³ − Carbonato) − tris(aqua − 2 − (2 − (2 −
pyridyl) − ethyliminomethyl) − pyridine −
copper(ii))
$C_{40}H_{45}Cu_3N_9O_6{}^{4+}$, $4NO_3^-$
G.Kolks, S.J.Lippard, J.V.Waszczak
J.Am.Chem.Soc.,**102**, 4832, 1980
See also R1 : 83

84.127 tetrakis(μ² − N − 2,6 − Xylyl − acetamidato) − di −
chromium methylene chloride solvate
$C_{40}H_{48}Cr_2N_4O_4$, $2CH_2Cl_2$
F.A.Cotton, W.H.Ilsley, W.Kaim
J.Am.Chem.Soc.,**102**, 3475, 1980
See also R1 : 83

84.128 tetrakis(μ^2 – N – 2,6 – Xylyl – acetamidato) – di –
molybdenum methylene chloride solvate
$C_{40}H_{48}Mo_2N_4O_4$, $2CH_2Cl_2$
F.A.Cotton, W.H.Ilsley, W.Kaim
J.Am.Chem.Soc., **102**, 3475, 1980
See also R1 : 83

84.129 Acetone – bis(isothiocyanato) – dioxo –
bis(triphenylphosphine – oxide) – uranium(vi)
$C_{41}H_{36}N_2O_5P_2S_2U$
G.Bombieri, E.Forsellini, G.De Paoli, D.Brown, T.C.Tso
J.Chem.Soc.,Dalton Trans.,2042, 1979

84.130 tetrakis(1 – Adamantoxo) – dimethylamine –
molybdenum(iv)
$C_{42}H_{67}MoNO_4$
M.Bochmann, G.Wilkinson, G.B.Young, M.B.Hursthouse,
K.M.A.Malik *J.Chem.Soc.,Dalton Trans.*,901, 1980
See also R1 : 83

84.C 4 – Methyl – N – thiobenzohydroximato –
bis(triphenylphosphine) – platinum(ii)
$C_{44}H_{37}NOP_2PtS$ Main entry is 86.163

84.C (2,3 – bis(Methoxycarbonyl) – 1 –
methylcyclopropyl) – bis(triphenylphosphine) –
platinum(ii) tetrafluoroborate
$C_{44}H_{41}O_4P_2Pt^+$, BF_4^- Main entry is 71.266

84.131 tetrakis(μ^2 – N – 2,6 – Xylyl – acetamidato) –
tetrahydrofuran – di – chromium toluene solvate
$C_{44}H_{56}Cr_2N_4O_5$, C_7H_8
F.A.Cotton, W.H.Ilsley, W.Kaim
J.Am.Chem.Soc., **102**, 3464, 1980
See also R1 : 83

84.132 tetrakis(μ^2 – N – p – Dimethylaniline –
acetamidato) – tetrahydrofuran – di – chromium
$C_{44}H_{60}Cr_2N_8O_5$
F.A.Cotton, W.H.Ilsley, W.Kaim
J.Am.Chem.Soc., **102**, 3464, 1980
See also R1 : 83

84.C Carbonyl – iodo – bis(triphenylphosphine) – (p –
methylbenzoyl) – ruthenium
$C_{45}H_{37}IO_2P_2Ru$ Main entry is 71.267

84.C tris(Tricarbonyl – manganese) – nonakis(μ –
diethoxyphosphite) – tri – manganese
$C_{45}H_{90}Mn_6O_{36}P_9$ Main entry is 86.170

84.C Dicarbonyl – chloro – (5,7 – dichloro – 8 –
quinolinolato – N,O) – bis(triphenylphosphine) –
tungsten(ii)
$C_{47}H_{34}Cl_3NO_3P_2W$ Main entry is 86.173

84.133 Tetrachloro – tetrakis(diphenylsulfoxide) –
thorium(iv)
$C_{48}H_{40}Cl_4O_4S_4Th$
C.E.F.Rickard, D.C.Woollard
Acta Crystallogr.,Sect.B,**36**, 292, 1980

84.134 tetrakis(μ^2 – N – 2,6 – Xylyl – acetamidato) –
bis(tetrahydrofuran) – di – chromium
tetrahydrofuran solvate
$C_{48}H_{64}Cr_2N_4O_6$, C_4H_8O
F.A.Cotton, W.H.Ilsley, W.Kaim
J.Am.Chem.Soc., **102**, 3464, 1980
See also R1 : 83

84.C catena – (μ – Benzenethiolato) – (methanol –
benzenethiolato – hexakis(μ – benzenethiolato)) –
tetra – zinc(ii)
$(C_{49}H_{44}OS_8Zn_4)_n$ Main entry is 85.92

84.135 tetrakis(μ^2 – N – 2,6 – Xylyl – acetamidato) –
bis(pyridine) – di – chromium pyridine solvate
$C_{50}H_{58}Cr_2N_8O_4$, C_5H_5N
F.A.Cotton, W.H.Ilsley, W.Kaim
J.Am.Chem.Soc., **102**, 3464, 1980
See also R1 : 83

84.136 meso – Tetraphenylporphinato –
bis(tetrahydrofuran) – iron(ii)
$C_{52}H_{44}FeN_4O_2$
C.A.Reed, T.Mashiko, W.R.Scheidt, K.Spartalian, G.Lang
J.Am.Chem.Soc., **102**, 2302, 1980
See also R1 : 49

84.C bis((Tri – n – butylphosphine) –
bis(dimethylglyoximato) – (methyl(phenyl) –
arsenyl) – cobalt(iii)) – cobalt(ii) dihydrate
$C_{54}H_{96}As_2Co_3N_8O_{10}P_2$, $2H_2O$ Main entry is 86.206

84.C tetrakis(μ^2 – (N,N') – Diphenylurea) –
bis(tetrahydrofuran) – di – chromium cyclohexane
solvate
$C_{60}H_{60}Cr_2N_8O_6$, C_6H_{12} Main entry is 79.17

84.C μ – 1 – 5 – eta:1' – 5' – η – (bis(μ – Tetraphenyl –
cyclopentadienediyl – oxo – O) – bis(aqua –
silver)) – bis(tricarbonyl – iron)
bis(hexafluorophosphate) dichloromethane solvate
$C_{64}H_{44}Ag_2Fe_2O_{10}^{2+}$, $2F_6P^-$ Main entry is 73.133

84.137 bis(Diethylammonium) tris(tetraphenyl –
disiloxane – diolato) – zirconium(iv)
$C_{72}H_{60}O_9Si_6Zr^{2-}$, $2C_4H_{12}N^+$
M.A.Hossain, M.B.Hursthouse
Inorg.Chim.Acta,**44**, L259, 1980
See also R1 : 63 R2 : 3

84.138 Tetrathiocyanato – tetrakis(triphenylphosphine –
oxide) – uranium(iv)
$C_{76}H_{60}N_4O_4P_4S_4U$
G.Bombieri, G.de Paoli, E.Forsellini, D.Brown
J.Inorg.Nucl.Chem.,**41**, 1315, 1979

84.139 octakis(3,5 – Di – t – butyl – 1,2 – semiquinone) –
tetra – cobalt(ii) benzene solvate
$C_{112}H_{160}Co_4O_{16}$, $2C_6H_6$
R.M.Buchanan, B.J.Fitzgerald, C.G.Pierpont
Inorg.Chem.,**18**, 3439, 1979

METAL COMPLEXES
(SULFUR OR SELENIUM LIGAND)

85.1 Sodium methylmercapto – cobalt(ii) cobalt aluminium silicate
$3CH_3CoS^{2+}$, $4Na^+$, $Al_{12}CoO_{48}Si_{12}^{10-}$
V.Subramanian, K.Seff *J.Am.Chem.Soc.*, **102**, 1881, 1980

85.2 Dibenzotetrathiafulvalene pentabromo – dimethylthio – platinum
$C_2H_6Br_5PtS^-$, $C_{14}H_8S_4^+$
R.P.Shibaeva, L.P.Rozenberg
Kristallografiya, **25**, 268, 1980
See also R2 : 39

85.C Tetrathiafulvalene bis(ethylenedithiolato) – nickel(i)
$3C_4H_4NiS_4$, $2C_6H_4S_4$ Main entry is 60.5

85.3 Tetrabutylammonium bis(thiomercapto – acetato) – oxo – technetium
$C_4H_4O_3S_4Tc^-$, $C_{16}H_{36}N^+$
B.V.DePamphilis, A.G.Jones, A.Davison, M.A.Davis
J.Labelled Compnd.Radiopharm., **16**, 26, 1979
See also R2 : 3

85.4 Tetraethylammonium trichloro – (μ – 2 – oxoethanethiolato – S,μ – O) – (μ – 2 – oxoethanethiolato – μ – S,μ – O) – bis(oxo – molybdenum(v))
$C_4H_8Cl_3Mo_2O_4S_2^-$. $C_8H_{20}N^+$
I.W.Boyd, I.G.Dance, A.E.Landers, A.G.Wedd
Inorg.Chem., **18**, 1875, 1979
See also R1 : 84 R2 : 3

85.5 Hydrogen trans – tetrachloro – bis(dimethylsulfoxide) – rhodium bis(dimethylsulfoxide)
$C_4H_{12}Cl_4O_2RhS_2^-$, $2C_2H_6OS$, H^+
B.R.James, R.H.Morris, F.W.B.Einstein, A.Willis
J.Chem.Soc.,Chem.Commun., 31, 1980

85.C trans – Chloro – bis(ethylenediamine) – sulfito – cobalt(iii) monohydrate
$C_4H_{16}ClCoN_4O_3S$, H_2O Main entry is 76.9

85.6 Potassium bis(dithio – oxalato) – nickel(ii) (black form)
$C_4NiO_4S_4^{2-}$, $2K^+$
A.Gleizes, F.Clery, M.F.Bruniquel, P.Cassoux
*Inorg.Chim.Acta.***37**, 19, 1979

85.7 Sodium bis(dithio – oxalato) – nickel(ii) dihydrate
$C_4NiO_4S_4^{2-}$, $2Na^+$, $2H_2O$
F.Maury, A.Gleizes *Inorg.Chim.Acta.***41**, 185, 1980

85.8 Chloro – tris(dithioacetato) – titanium(iv)
$C_6H_9ClS_6Ti$
L.Gastaldi, L.Scaramuzza
Cryst.Struct.Commun., **9**, 469, 1980

85.9 Tetraethylammonium chloro – tris(thiolato – ethanolato) – bis(oxo – molybdenum(v))
$C_6H_{12}ClMo_2O_5S_3^-$, $C_8H_{20}N^+$
I.W.Boyd, I.G.Dance, A.E.Landers, A.G.Wedd
Inorg.Chem., **18**, 1875, 1979
See also R1 : 84 R2 : 3

85.10 Piperidinium (μ – oxo) – (μ – 2 – mercapto – ethanolato) – bis(oxo – (2 – mercapto – ethanolato) – molybdenum)
$C_6H_{12}Mo_2O_6S_3^{2-}$, $2C_5H_{12}N^+$
I.G.Dance, A.E.Landers *Inorg.Chem.*, **18**, 3487, 1979
See also R1 : 84 R2 : 33

85.11 Diacetato – dithiocarbamide – cadmium
$C_6H_{14}CdN_4O_4S_2$
V.I.Bondar, V.G.Rau, Yu.T.Struchkov, V.M.Akimov,
A.K.Molodkin, V.V.Ilyukhin, N.V.Belov
Dokl.Akad.Nauk SSSR, **250**, 852, 1980
See also R1 : 84

85.C bis((Thio – oxalato – O,S) – bis(ethylenediamine) – cobalt(iii)) dithionate dihydrate
$2C_6H_{16}CoN_4O_3S^+$, $O_6S_2^{2-}$, $2H_2O$ Main entry is 76.28

85.12 bis(Ethylmethyl – dithiophosphinato) – nickel(ii)
$C_6H_{16}NiP_2S_4$
H.Wunderlich *Acta Crystallogr.,Sect.B*, **36**, 717, 1980

85.13 bis(Dimethylamido – (methyl) – dithiophosphonato) – nickel(ii)
$C_6H_{18}N_2NiP_2S_4$
F.Seel, G.Zindler *Chem.Ber.*, **113**, 1837, 1980

85.14 Tetrabutylammonium bis(isotrithione – dithiolato) – nickel(ii)
$C_6NiS_{10}^{2-}$, $2C_{16}H_{36}N^+$
O.Lindqvist, L.Sjolin, J.Sieler, G.Steimecke, E.Hoyer
Acta Chem.Scand.Ser.A, **33**, 445, 1979
See also R2 : 3

85.15 trans – Dichloro – (dimethylsulfoxide) – (pyridine) – platinum(ii)
$C_7H_{11}Cl_2NOPtS$
F.Caruso, R.Spagna, L.Zambonelli
Acta Crystallogr.,Sect.B, **36**, 713, 1980
See also R1 : 83

85.16 Tricarbonyl – (S – methyldithiocarbazato) – (μ – S – methyldithiocarbazato) – manganese(i) bromide
$C_7H_{12}MnN_4O_3S_4^+$, Br^-
H.Weber, R.Mattes *Chem.Ber.*, **113**, 2833, 1980
See also R1 : 83

85.17 tetrakis(Dithioacetato) – tetra – gold(i)
$C_8H_{12}Au_4S_8$
O.Piovesana, P.F.Zanazzi
Angew.Chem.,Int.Ed.Engl., **19**, 561, 1980

85.18 Dicarbonyl – (1,4,7,10 – tetrathiadecane) – iron
$C_8H_{12}FeO_2S_4$
D.Sellmann, H.–E.Jonk, H.–R.Pfeil, G.Huttner, J.von Seyerl
J.Organomet.Chem., **191**, 171, 1980

85.C Dichloro – (cis – 2 – methyl – thiacyclo – oct – 4 – ene) – platinum
$C_8H_{14}Cl_2PtS$ Main entry is 75.1

85.C η^5 – Cyclopentadienyl – trimethylphosphine – cyclopentasulfur – cobalt
$C_8H_{14}CoPS_5$ Main entry is 73.11

85.19 bis(S – Ethylethene – 1,2 – dithiolato) –
palladium(ii)
$C_8H_{14}PdS_4$
J.Sieler, R.Richter, J.Kaiser, R.Kolbe
Krist.Tech.,**14**, 1121, 1979

85.20 bis(S – Ethylethene – 1,2 – dithiolato) – platinum(ii)
$C_8H_{14}PtS_4$
J.Sieler, R.Richter, J.Kaiser, R.Kolbe
Krist.Tech.,**14**, 1121, 1979

85.C Chloro – (syn – 4,4 – dimethyl – 5 – methylthio –
pent – (1 – 3 – η) – enyl) – palladium
$C_8H_{15}ClPdS$ Main entry is 71.7

85.C Dichloro – (2,2 – dimethyl – pent – (E) – 3 – enyl –
methylsulfide) – palladium(ii)
$C_8H_{16}Cl_2PdS$ Main entry is 72.6

85.21 (μ – trans – Cyclohexane – 1,2 – dithiolato) –
bis(methyl – mercury(ii)) (at 123°K)
trans – 1,2 – Dimercapto – cyclohexane – bis(methyl –
mercury(ii))
$C_8H_{16}Hg_2S_2$
N.W.Alcock, P.A.Lampe, P.Moore
J.Chem.Soc.,Dalton Trans.,1471, 1980

85.22 bis(bis(2 – Thiolato – ethyl) – sulfide) –
molybdenum(iv)
$C_8H_{16}MoS_6$
J.Hyde, L.Magin, J.Zubieta
J.Chem.Soc.,Chem.Commun.,204, 1980

85.23 (μ² – Oxo) – bis(diethylene – trithiolato – oxo –
molybdenum)
$C_8H_{16}Mo_2O_3S_6$
J.Hyde, L.Magin, P.Vella, J.Zubieta
Stereodyn.Mol.Syst.Proc.Symp.,227, 1979

85.24 Sodium trans – sulfito – bis(dimethylglyoximato)
ammine – cobalt(iii) tetrahydrate
$C_8H_{17}CoN_5O_7S^-$, Na^+, $4H_2O$
R.C.Elder, R.Whittle, M.J.Heeg, J.C.Barrick, A.Nerone
Am.Cryst.Assoc.,Ser.2,**7**, 23, 1980
See also R1 : 83

85.25 Dichloro – bis(2,2′ – thio – diethanol) – cobalt(ii)
$C_8H_{20}Cl_2CoO_4S_2$
M.R.Udupa, B.Krebs, U.Seyer
Inorg.Chim.Acta,**41**, 31, 1980
See also R1 : 84

85.26 bis(2,5 – Dithiahexane) – copper tetraperchlorate
(at 140°K)
$2C_8H_{20}CuS_4^+$, $C_8H_{20}CuS_4^{2+}$, $4ClO_4^-$
W.K.Musker, M.M.Olmstead, R.M.Kessler, M.B.Murphey,
C.H.Neagley, P.B.Roush, N.L.Hill, T.L.Wolford, H.Hope,
G.Delker, K.Swanson, B.V.Gorewit
J.Am.Chem.Soc.,**102**, 1225, 1980
See also R2 : 85

85.27 bis(2,5 – Dithiahexane) – copper
tetrakis(tetrafluoroborate) (at 140°K)
$2C_8H_{20}CuS_4^+$, $C_8H_{20}CuS_4^{2+}$, $4BF_4^-$
W.K.Musker, M.M.Olmstead, R.M.Kessler, M.B.Murphey,
C.H.Neagley, P.B.Roush, N.L.Hill, T.L.Wolford, H.Hope,
G.Delker, K.Swanson, B.V.Gorewit
J.Am.Chem.Soc.,**102**, 1225, 1980
See also R2 : 85

85.C bis(2,5 – Dithiahexane) – copper tetraperchlorate
(at 140°K)
$C_8H_{20}CuS_4^{2+}$, $2C_8H_{20}CuS_4^+$, $4ClO_4^-$ Main entry is 85.26

85.C bis(2,5 – Dithiahexane) – copper
tetrakis(tetrafluoroborate) (at 140°K)
$C_8H_{20}CuS_4^{2+}$, $2C_8H_{20}CuS_4^+$, $4BF_4^-$ Main entry is 85.27

85.28 tetrakis(μ₃ – Oxo) – tetrakis(μ –
dimethylthiophosphinato) – tetra(oxo –
molybdenum) chloroform carbon tetrachloride
solvate
$C_8H_{24}Mo_4O_{12}P_4S_4$, $xCHCl_3$, $yCCl_4$
R.Mattes, K.Muhlsiepen
Z.Naturforsch.,Teil B,**35**, 265, 1980
See also R1 : 84

85.29 tetrakis(μ₃ – Oxo) – tetrakis(μ –
dimethylthiophosphinato) – tetra(oxo –
molybdenum) chloroform carbon tetrachloride
solvate (at –150°C)
$C_8H_{24}Mo_4O_{12}P_4S_4$, $xCHCl_3$, $yCCl_4$
R.Mattes, K.Muhlsiepen
Z.Naturforsch.,Teil B,**35**, 265, 1980
See also R1 :

85.30 tetrakis(Dimethyl – dithiophosphinato) –
thorium(iv)
$C_8H_{24}P_4S_8Th$
A.A.Pinkerton, J.–M.Zellweger, D.Schwarzenbach
Eur.Cryst.Meeting,**6**, 57, 1980

85.31 Tetrabutylammonium (μ – oxo) – (μ – sulfido) –
bis((1,2 – dithiosquarato – S,S') – oxo –
molybdenum(v))
$C_8Mo_2O_7S_5^{2-}$, $2C_{16}H_{36}N^+$
D.Altmeppen, R.Mattes
Acta Crystallogr.,Sect.B,**36**, 1942, 1980
See also R2 : 3

85.32 tris(S – Methylethene – 1,2 – dithiolato) – cobalt
$C_9H_{15}CoS_6$
R.Richter, J.Sieler, J.Kaiser *Krist.Tech.*,**14**, 1463, 1979

85.33 tris(Methyl 2 – methyldithiocarbazato) – nickel(ii)
chloride ethanol solvate trihydrate
$C_9H_{24}N_6NiS_6^{2+}$, $2Cl^-$, $0.5C_2H_6O$, $3H_2O$
G.Dessy, V.Fares *Acta Crystallogr.,Sect.B*,**36**, 944, 1980
See also R1 : 83

85.C (μ₃ – Carbonyl) – (μ₃ – sulfido) – tris(tricarbonyl –
iron)
$C_{10}Fe_3O_{10}S$ Main entry is 71.16

85.C bis(μ – Methylthio) – (μ – tetrafluoroethane – 1,1 –
diyl) – bis(tricarbonyl – iron) (at –162°C)
$C_{10}H_6F_4Fe_2O_6S_2$ Main entry is 71.17

85.C bis(μ – Methylthio) – (μ – tetrafluoroethane – 1,2 –
diyl) – bis(tricarbonyl – iron) (at –162°C)
$C_{10}H_6F_4Fe_2O_6S_2$ Main entry is 71.18

85.34 Dichloro – bis(pyridinium – 2 – thiolato) – cobalt(ii)
$C_{10}H_{10}Cl_2CoN_2S_2$
E.Binamira-Soriaga, M.Lundeen, K.Seff
Acta Crystallogr.,Sect.B,**35**, 2875, 1979

85.35 bis(μ – Chloro) – bis(chloro – (6 – mercapto –
purine) – aquo – cadmium(ii))
$C_{10}H_{12}Cd_2Cl_4N_8O_2S_2$
E.A.H.Griffith, E.L.Amma
J.Chem.Soc.,Chem.Commun.,1013, 1979
See also R1 : 83

85.C (2,4 – Dithio – 1,3 – dithia – 1,4 – diyl) – η^5 – cyclopentadienyl – (trimethylphosphine) – rhodium
$C_{10}H_{14}PRhS_4$ Main entry is 71.21

85.C trans – N,N′ – Ethylene – bis(S – methyl – L – cysteinato) – cobalt(iii) perchlorate
$C_{10}H_{16}CoN_2O_4S_2^+$, ClO_4^- Main entry is 82.21

85.36 Nitrato – (1 – oxa – 7,10 – diaza – 4,13 – dithia – cyclopentadecane) – nickel(ii) nitrate
$C_{10}H_{22}N_3NiO_4S_2^+$, NO_3^-
R.Louis, Y.Agnus, R.Weiss
Acta Crystallogr.,Sect.B,35, 2905, 1979
See also R1 : 84,83

85.37 bis(1,1 – Dimethyl – 2 – N – methylaminoethylthiolato) – dioxo – molybdenum
$C_{10}H_{24}MoN_2O_2S_2$
E.I.Stiefel, K.F.Miller, A.E.Bruce, J.L.Corbin, J.M.Berg, K.O.Hodgson J.Am.Chem.Soc.,102, 3624, 1980
See also R1 : 83

85.38 tris(μ^2 – Ethylthiolato) – bis(dichloro – (dimethylthiolato) – tungsten(iii,iv))
$C_{10}H_{27}Cl_4S_5W_2$
P.M.Boorman, K.A.Kerr, P.W.Codding, V.D.Patel, P.van Roey
Am.Cryst.Assoc.,Ser.2,8, 23, 1980

85.C Sodium dioxo – bis(μ^2 – thio) – (μ – 1,2 – propylenediamine) – tetra – acetato – di – molybdenum trihydrate
$C_{11}H_{14}Mo_2N_2O_{10}S_2^{2-}$, $2Na^+$, $3H_2O$ Main entry is 81.38

85.C Sodium dioxo – (μ^2 – oxo) – (μ^2 – thio) – 1,2 – propylenediamine) – tetra – acetato – di – molybdenum tetrahydrate
$C_{11}H_{14}Mo_2N_2O_{11}S^{2-}$, $2Na^+$, $4H_2O$ Main entry is 81.39

85.C Chloro – (pentane – 2,4 – dithionato) – (triethylphosphine) – nickel(ii)
$C_{11}H_{22}ClNiPS_2$ Main entry is 86.15

85.C 1,1,1,2,2,2,3,3,3 – Nonacarbonyl – 2 – ethylene – 1,3 – μ – hydrido – 1,3 – μ – (methylthiolato) – triangulo – tri – osmium
$C_{12}H_8O_9Os_3S$ Main entry is 72.14

85.C bis(Dimethylsulfoxide) – tetrakis(trifluoroacetato) – di – rhodium(ii)
$C_{12}H_{12}F_{12}O_{10}Rh_2S_2$ Main entry is 81.43

85.C N,N′ – Ethylene – bis(monothioacetylacetone – iminato) – copper(ii)
$C_{12}H_{18}CuN_2S_2$ Main entry is 76.71

85.39 γ – Picolinium bis(S – methyldithiocarbazato – dimethylglyoximato) – iron(iii) tetrahydrate
$C_{12}H_{18}FeN_6O_2S_4^-$, $C_6H_8N^+$, $4H_2O$
N.A.Ryabova, V.I.Ponomarev, A.I.Udel'nov, V.V.Zelentsov, V.N.Kaftanat Zh.Strukt.Khim.,20, 185, 1979
See also R1 : 83 R2 : 33

85.40 (N,N′ – Ethylene – bis(monothioacetylacetone – iminato)) – zinc(ii)
$C_{12}H_{18}N_2S_2Zn$
R.Cini, A.Cinquantini, P.Orioli, M.Sabat
Inorg.Chim.Acta,41, 151, 1980
See also R1 : 83

85.C (μ^2 – Sulfonyl) – bis((μ – methylthio) – (trimethylphosphine) – dicarbonyl – iron(i)) diethyl ether solvate
$C_{12}H_{24}Fe_2O_6P_2S_3$, $0.5C_4H_{10}O$ Main entry is 86.20

85.41 catena – bis(μ – N – Methyl – piperidinium – 4 – thiolato) – cadmium(ii) diperchlorate dihydrate
$(C_{12}H_{26}CdN_2S_2^{2+})_n$, $2nClO_4^-$, $2nH_2O$
J.C.Bayon, M.C.Brianso, J.L.Brianso, P.G.Duarte
Inorg.Chem.,18, 3478, 1979

85.42 Aqua – diperchlorato – (1,4,10,13 – tetraoxa – 7,16 – dithiacyclo – octadecane) – lanthanum(iii) perchlorate
$C_{12}H_{26}Cl_2LaO_{13}S_2^+$, ClO_4^-
M.Ciampolini, C.Mealli, N.Nardi
J.Chem.Soc.,Dalton Trans.,376, 1980
See also R1 : 84

85.43 (μ^2 – Thio) – bis((μ^2 – ethylthiolato) – dichloro – (tetrahydro – 1 – thienyl) – tungsten(iv))
$C_{12}H_{26}Cl_4S_5W_2$
P.M.Boorman, K.A.Kerr, P.W.Codding, V.D.Patel, P.van Roey
Am.Cryst.Assoc.,Ser.2,8, 23, 1980

85.44 bis(N – Methylpiperidinium – 4 – thiolato) – mercury(ii) perchlorate
$C_{12}H_{26}HgN_2S_2^+$, $2ClO_4^-$
J.C.Bayon, P.G.Duarte, J.L.Brianso, M.C.Perucaud
Eur.Cryst.Meeting,6, 350, 1980

85.45 bis(Aqua – copper(ii)) – bis(μ – aqua) – bis(μ – N,N′ – bis(2 – hydroxyethyl) – dithio – oxamidato) – bis(sulfato – copper(ii))
$C_{12}H_{28}Cu_4N_4O_{18}S_6$
C.Chauvel, J.J.Girerd, Y.Jeannin, O.Kahn, G.Lavigne
Inorg.Chem.,18, 3015, 1979
See also R1 : 84,

85.46 tetrakis(3 – Amino – 1 – propanethiolato) – tri – nickel(ii) chloride
$C_{12}H_{32}N_4Ni_3S_4^{2+}$, $2Cl^-$
J.Suades, H.Barrera, M.C.Perucaud, J.L.Brianso
Eur.Cryst.Meeting,6, 349, 1980

85.C bis((bis(Ethylenediamine) – cobalt) – (μ – carboxylato – methyl) – thiolato) – silver perchlorate
$C_{12}H_{36}AgCo_2N_8O_4S_2^{3+}$, $3ClO_4^-$ Main entry is 76.80

85.47 bis(Tetraphenylarsonium) tris(2,2 – diselenido – 1,1 – ethylene – dicarbonitrile) – nickel(iv)
$C_{12}N_6NiSe_6^{2-}$, $2C_{24}H_{20}As^+$
J.Kaiser, W.Dietzsch, R.Richter, L.Golic, J.Siftar
Acta Crystallogr.,Sect.B,36, 147, 1980
See also R2 : 65

85.48 Trithio – peroxybenzoato – (η^3 – dithiobenzoato) – oxo – molybdenum(iv)
$C_{14}H_{10}MoOS_5$
M.Tatsumisago, G.Matsubayashi, T.Tanaka, S.Nishigaki, K.Nakatsu Chem.Lett.,889, 1979

85.C (1,5 – Diphenylthiocarbazonato – N,S) – methyl – mercury(ii)
$C_{14}H_{14}HgN_4S$ Main entry is 71.66

85.49 Tetracarbonyl – cis(ethyl α – ethoxy – α – (1,3 – dithian – 2 – ylidene) – acetimidate) – chromium(0)
$C_{14}H_{17}CrNO_6S_2$
G.J.Kruger, G.Gafner, J.P.R.de Villiers, H.G.Raubenheimer, H.Swanepoel J.Organomet.Chem.,187, 333, 1980
See also R1 : 83

85.C trans – bis(μ – Thioethyl) – bis(nitrosyl – (η^5 – cyclopentadienyl) – molybdenum)
$C_{14}H_{20}Mo_2N_2O_2S_2$ Main entry is 73.50

85.C   bis(μ – Sulfido) – bis(μ – methylthio) –
bis(methylcyclopentadienyl – molybdenum(iv))
$C_{14}H_{20}Mo_2S_4$ Main entry is 73.51

85.50   Tetraethylammonium pentakis(μ – chloro) –
thiophenolato – bis(pentacarbonyl – tungsten(0))
$C_{16}Cl_5O_{10}SW_2{}^-$, $C_8H_{20}N^+$
M.Cooper, P.A.Duckworth, M.Saporta, M.McPartlin
J.Chem.Soc.,Dalton Trans.,570, 1980
See also R2 : 3

85.C   tris(η^5 – Cyclopentadienyl) – (μ_3 – sulfido) – (μ_3 –
thiocarbonyl) – triangulo – tri – cobalt
$C_{16}H_{15}Co_3S_2$ Main entry is 71.97

85.51   bis(μ – N – Isopropylidene – S – methyl –
dithiocarbazato) – bis(tricarbonyl) – manganese(i)
$C_{16}H_{18}Mn_2N_4O_6S_4$
H.Weber, R.Mattes *Chem.Ber.*,113, 2833, 1980
See also R1 : 83

85.C   bis(μ – Propane – 1,2 – dithiolato) –
bis(cyclopentadienyl – molybdenum)
tetrafluoroborate
$C_{16}H_{22}Mo_2S_4{}^+$, $BF_4{}^-$ Main entry is 73.64

85.C   cis – bis(Thiocyanato) – bis(1,3,4 –
trimethylphosphole) – palladium(ii)
$C_{16}H_{22}N_2P_2PdS_2$ Main entry is 86.39

85.C   cis – bis(μ – Thio – isopropyl) – bis(nitrosyl – (η^5 –
cyclopentadienyl) – molybdenum)
$C_{16}H_{24}Mo_2N_2O_2S_2$ Main entry is 73.65

85.C   tetrakis(μ – Acetato) – bis(tetrahydrothiophene –
rhodium)
$C_{16}H_{28}O_8Rh_2S_2$ Main entry is 81.62

85.52   (μ – Thio) – (μ – dithio) – tetrabromo –
tetra(tetrahydrothiophene) – di – niobium
$C_{16}H_{32}Br_4Nb_2S_7$
M.G.B.Drew, I.B.Baba, D.A.Rice, D.M.Williams
Inorg.Chim.Acta,44, 217, 1980

85.C   bis(Dimethylsulfoxide) – tetrakis(propionato) – di –
rhodium(ii)
$C_{16}H_{32}O_{10}Rh_2S_2$ Main entry is 81.63

85.53   Chloro – (2 – (2' – ethanolthio) – ethanolato) –
copper(ii) tetramer
$C_{16}H_{36}Cl_4Cu_4O_8S_4$
M.R.Udupa, B.Krebs *Inorg.Chim.Acta*,39, 267, 1980
See also R1 : 84

85.C   bis(μ – t – Butylthiolato) – bis(carbonyl –
trimethylphosphite – iridium)
$C_{16}H_{36}Ir_2O_8P_2S_2$ Main entry is 86.41

85.C   bis(μ – t – Butylthiolato) – bis(carbonyl – hydrido –
trimethylphosphite – iridium)
$C_{16}H_{38}Ir_2O_8P_2S_2$ Main entry is 86.42

85.54   Tetrachloro – tetrakis(diethylsulfide) – di –
molybdenum(ii)
$C_{16}H_{40}Cl_4Mo_2S_4$
F.A.Cotton, P.E.Fanwick
Acta Crystallogr.,Sect.B,36, 457, 1980

85.55   (μ – Chloro) – pentachloro – (μ – diethylsulfide) –
tris(diethylsulfide) – di – iridium(iii)
$C_{16}H_{40}Cl_6Ir_2S_4$
A.F.Williams, H.D.Flack, M.G.Vincent
Acta Crystallogr.,Sect.B,36, 1206, 1980

85.56   bis(μ – Chloro) – tetrachloro –
tetrakis(diethylsulfide) – di – iridium(iii)
$C_{16}H_{40}Cl_6Ir_2S_4$
A.F.Williams, H.D.Flack, M.G.Vincent
Acta Crystallogr.,Sect.B,36, 1204, 1980

85.57   Tetraethylammonium (μ – sulfido) – bis(μ –
ethylthiolato) – bis(molybdenum – trisulfido –
tris(ethylthio) – iron))
$C_{16}H_{40}Fe_6Mo_2S_{17}{}^{3-}$, $3C_8H_{20}N^+$
T.E.Wolff, J.M.Berg, K.O.Hodgson, R.B.Frankel, R.H.Holm
J.Am.Chem.Soc.,101, 4140, 1979
See also R2 : 3

85.C   (μ_3 – Methylenethiolato) – (μ_3 – sulfido) –
(octacarbonyl – (dimethylphenylphosphine) – tri –
osmium) (at –35°C)
$C_{17}H_{13}O_8Os_3PS_2$ Main entry is 71.108

85.58   Deoxycytidylyl – (3' – 5') – deoxyguanosine (2 –
hydroxy – ethylthiolato) – 2,2',2'' – terpyridine –
platinum(ii) hydrate
$2C_{17}H_{18}N_3OPtS^+$, $2C_{19}H_{24}N_8O_{10}P^-$, xH_2O
A.H.J.Wang, J.N.Nathans, G.van der Marel, J.H.van Boom,
A.Rich *Nature (London)*,276, 471, 1978
See also R1 : 83 R2.: 47,46,45,44

85.C   (μ_3 – Cycloheptatrienyl) – (μ_3 – t – butylthiolato) –
tris(dicarbonyl – ruthenium) (at –70°C)
$C_{17}H_{16}O_6Ru_3S$ Main entry is 75.15

85.59   (2,6 – bis(1 – Methyl – 4 – methylthio – 5 – thia –
2,3 – diazahexa – 1,3 – dienyl) – pyridine) –
dithiocyanato – nickel(ii)
$C_{17}H_{21}N_7NiS_6$
W.Choong, N.C.Stephenson, M.A.Ali, M.A.Malik, D.J.Phillips
Aust.J.Chem.,32, 1199, 1979
See also R1 : 83

85.60   N,N' – Trimethylene – bis(methyl – 2 – amino – 1 –
cyclopentene – dithiocarboxylato) – copper(ii)
$C_{17}H_{24}CuN_2S_4$
R.D.Bereman, M.R.Churchill, G.Shields
Inorg.Chem.,18, 3117, 1979
See also R1 : 83

85.C   (μ – Methylenethiolato) – (μ_3 – sulfido) –
(nonacarbonyl – (dimethylphenylphosphine) – tri –
osmium)
$C_{18}H_{13}O_9Os_3PS_2$ Main entry is 71.123

85.61   Dioxo – bis(N – methyl – p – tolyl –
thiohydroxamato) – molybdenum(vi)
$C_{18}H_{20}MoN_2O_4S_2$
C.A.Cliff, G.D.Fallon, B.M.Gatehouse, K.S.Murray,
P.J.Newman *Inorg.Chem.*,19, 773, 1980
See also R1 : 84

85.62   Triethylbenzylammonium tris(μ – ethylthiolato) –
bis(tetrakis(μ_2 – sulfido) – tris(ethylthiolato –
iron) – molybdenum)
$C_{18}H_{45}Fe_6Mo_2S_{17}{}^{3-}$, $3C_{13}H_{22}N^+$
T.E.Wolff, J.M.Berg, K.O.Hodgson, R.B.Frankel, R.H.Holm
J.Am.Chem.Soc.,101, 4140, 1979

85.C   (1,5 – Diphenylthiocarbazonato – N,S) – phenyl –
mercury(ii) (yellow form)
$C_{19}H_{16}HgN_4S$ Main entry is 71.138

85.63 bis(2 – Methyl – 8 – mercapto – quinolinato) – cadmium
$C_{20}H_{16}CdN_2S_2$
L.Ya.Pech, Ya.K.Ozols, A.A.Kemme, Ya.Ya.Bleidelis, A.P.Sturis
Latv.PSR Zinat.Akad.Vestis,Kim.Ser.,259, 1979
See also R1 : 83

85.64 bis(5 – Methylthio – 8 – mercapto – quinolinato) – copper(ii)
$C_{20}H_{16}CuN_2S_4$
O.G.Matyukhina, B.T.Ibragimov, Ya.K.Ozols, Ya.E.Leeis
Latv.PSR Zinat.Akad.Vestis,Kim.Ser.,371, 1979
See also R1 : 83

85.C (μ – Thio – ferrocenyl – methylmethane – thiomethylene – C',S²) – 1,1,1,2,2,2 – hexa – carbonyl – (μ – methylthio) – iron
$C_{20}H_{16}Fe_3O_6S_3$ Main entry is 71.153

85.C (2,3,4,6 – Tetra – O – acetyl – 1 – thio – β – D – glucopyranosato – S) – (triethylphosphine) – gold
Auranofin
$C_{20}H_{34}AuO_9PS$ Main entry is 86.71

85.65 (bis(Benzothiazole – 2 – yl) – trisulfide) – (2 – mercapto – benzothiazole) – copper(i) perchlorate chloroform solvate
$C_{21}H_{13}CuN_3S_7^+$, ClO_4^-, $2CHCl_3$
S.Jeannin, Y.Jeannin, G.Lavigne
Inorg.Chem.,18, 3528, 1979
See also R1 : 83

85.C tris(η^5 – Cyclopentadienyl) – (μ_3 – (pentacarbonyl – chromium – thio) – methylidyne) – (μ_3 – sulfido) – triangulo – tri – cobalt tetrahydrofuran solvate
$C_{21}H_{15}Co_3CrO_5S_2$, $0.5C_4H_8O$ Main entry is 71.169

85.66 Sodium triethylmethylammonium tris(thiobenzohydroximato) – chromium(iii) hemi(sodium hydroxide) hydrate
$C_{21}H_{15}CrN_3O_3S_3^{3-}$, $C_7H_{18}N^+$, $2.5Na^+$, $0.5HO^-$, $18.5H_2O$
K.Abu–Dari, D.P.Freyberg, K.N.Raymond
Inorg.Chem.,18, 2427, 1979
See also R1 : 84 R2 :

85.67 Pentafluorophenylthiolato – (hydro – tris(3,5 – dimethyl – 1 – pyrazolyl) – borato) – cobalt (at –162°C)
$C_{21}H_{22}BCoF_5N_6S$
J.S.Thompson, T.Sorrell, T.J.Marks, J.A.Ibers
J.Am.Chem.Soc.,101, 4193, 1979
See also R1 : 83,62

85.68 Potassium (p – nitrobenzenethiolato) – (hydro – tris(3,5 – dimethylpyrazolyl) – borato) – copper(i) acetone solvate
$C_{21}H_{26}BCuN_7O_2S^-$, K^+, $2C_3H_6O$
J.S.Thompson, T.J.Marks, J.A.Ibers
J.Am.Chem.Soc.,101, 4180, 1979
See also R1 : 83,62

85.C Dodecacarbonyl – (η_5 – carbido) – (μ – hydrido) – tris(η_2 – ethylsulfide) – hexa – ruthenium
$C_{22}H_{16}O_{15}Ru_6S_3$ Main entry is 71.177

85.C tris(μ – t – Butylthiolato) – (μ – hexafluoro – but – 2 – ene – 2,3 – diyl) – tris(dicarbonyl – iridium)
$C_{22}H_{27}F_6Ir_3O_6S_3$ Main entry is 71.181

85.C tetrakis(μ_2 – Acetato) – bis(benzylthiolato – rhodium)
$C_{22}H_{28}O_8Rh_2S_2$ Main entry is 81.74

85.69 7α – (Thio – bromo – mercury) – (17R) – spiro(androst – 4 – ene – 17,2(3H) – furan)
$(C_{22}H_{31}BrHgO_2S)_n$
A.Terzis, J.B.Faught, G.Pouskoulelis
Inorg.Chem.,19, 1060, 1980
See also R1 : 51

85.C Tricarbonyl – molybdenum – tris(μ – n – butylthiolato) – (η^7 – cycloheptatrienyl – molybdenum)
$C_{22}H_{34}Mo_2O_3S_3$ Main entry is 75.32

85.C Tricarbonyl – molybdenum – tris(μ – t – butylthiolato) – (η^7 – cycloheptatrienyl – molybdenum)
$C_{22}H_{34}Mo_2O_3S_3$ Main entry is 75.33

85.C (μ^2 – 1,2 – bis(Carbomethoxyethylene)) – bis((μ^2 – methylthio) – trimethylphosphine – carbonyl – iron) tetraphenylborate (at –162°C)
$C_{22H_{49}}Fe_2O_6P_4S_2^+$, $C_{24}H_{20}B^-$ Main entry is 86.82

85.70 Hydro – tris(3,5 – dimethyl – 1 – pyrazolyl) – borato – dicarbonyl – (p – chlorobenzenethiolato) – molybdenum acetone solvate
$C_{23}H_{26}BClMoN_6O_2S$, $0.2C_3H_6O$
G.Ferguson, F.J.Lalor, M.Parvez, M.A.Khan
Am.Cryst.Assoc.,Ser.2,8, 24, 1980
See also R1 : 83,62

85.71 (p – Chlorothiobenzoyl – hydrazone – diyl) – (p – chlorothiobenzoyl – hydrazone – triyl) – (N' – isopropylidene – p – chlorothiobenzoyl – hydrazone) – molybdenum
$C_{24}H_{19}Cl_3MoN_6S_3$
P.Dahlstrom, M.Kustyn, J.Zubieta, J.R.Dilworth
Transition Met.Chem.,4, 396, 1979
See also R1 : 83

85.72 tris(N – Methyl – thiobenzohydroxamato) – cobalt(iii)
$C_{24}H_{24}CoN_3O_3S_3$
D.P.Freyberg, K.Abu–Dari, K.N.Raymond
Inorg.Chem.,18, 3037, 1979
See also R1 : 84

85.73 tris(N – Methyl – thiobenzohydroxamato) – chromium(iii)
$C_{24}H_{24}CrN_3O_3S_3$
D.P.Freyberg, K.Abu–Dari, K.N.Raymond
Inorg.Chem.,18, 3037, 1979
See also R1 : 84

85.74 tris(N – Methyl – thiobenzohydroxamato) – iron(iii)
$C_{24}H_{24}FeN_3O_3S_3$
D.P.Freyberg, K.Abu–Dari, K.N.Raymond
Inorg.Chem.,18, 3037, 1979
See also R1 : 84

85.75 tris(N – Methyl – thiobenzohydroxamato) – manganese(ii)
$C_{24}H_{24}MnN_3O_3S_3$
D.P.Freyberg, K.Abu–Dari, K.N.Raymond
Inorg.Chem.,18, 3037, 1979
See also R1 : 84

85.76 bis(2 – Isopropylquinoline – 8 – thiolato) nickel
$C_{24}H_{24}N_2NiS_2$
L.Ya.Pech, Ya.K.Ozols, A.P.Sturis
Latv.PSR Zinat.Akad.Vestis,Kim.Ser.,623, 1979
See also R1 : 83

85.77 (2,3,7,8,12,13 – Hexamethoxy – 10,15 – dihydro –
5H – 5,10,15 – trithia – tribenzo(a,d,g)cyclononene) –
trinitrato – rhodium(iii) N,N – dimethylacetamide
solvate
Trithiaveratrylene trinitrato – rhodium(iii) N,N –
dimethylacetamide solvate
$C_{24}H_{24}N_3O_{15}RhS_3$, $3C_4H_9NO$
J.Kopf, K.von Deuten, G.Klar
Cryst.Struct.Commun.,**8**, 1011, 1979

85.C bis(η^5 – Cyclopentadienyl) – (thiocamphor) –
tetracarbonyl – di – molybdenum
$C_{24}H_{26}Mo_2O_4S$ Main entry is 72.48

85.C (μ – 1 – Diphenylphosphino – 2 – sulfido – 3,3 –
dimethylbutene) – carbonyl – (methoxy –
thiocarbonyl) – trimethylphosphite – iron
$C_{24}H_{32}FeO_5P_2S_2$ Main entry is 72.49

85.78 tris(N – Benzyl – trimethylammonium) bis(tris(μ –
ethylthiolato) – molybdenum) – tetrakis(μ –
thiolato) – tris(ethylthiolato – iron) – iron
$C_{24}H_{60}Fe_7Mo_2S_{20}{}^{3-}$, $3C_{10}H_{16}N^+$
T.E.Wolff, J.M.Berg, P.P.Power, K.O.Hodgson, R.H.Holm
Inorg.Chem.,**19**, 430, 1980
See also R2 : 3

85.C Carbonyl – (triphenylphosphine) – (1 – (ethylthio) –
maleonitrile – 2 – thiolato) – rhodium
$C_{25}H_{20}N_2OPRhS_2$ Main entry is 86.90

85.C η^5 – Cyclopentadienyl – triphenylphosphine –
nickel – ethylxanthate (at –170°C)
$C_{26}H_{25}NiOPS_2$ Main entry is 73.107

85.C bis((μ^2 – t – Butylthio) – carbonyl – iodo –
dimethylphenylphosphine – iridium(ii))
$C_{26}H_{40}I_2Ir_2O_2P_2S_2$ Main entry is 86.92

85.79 (Ethoxy – (1,5,9,13 – tetrathiacyclohexadecane) –
molybdenum(iv)) – (μ – oxo) – (oxy – (1,5,9,13 –
tetrathiacyclohexadecane) – molybdenum(iv))
trifluoromethylsulfonate monohydrate
$C_{26}H_{53}Mo_2O_3S_8{}^{3+}$, $3CF_3O_3S^-$, H_2O
R.E.DeSimone, J.Cragel Junior, W.H.Ilsley, M.D.Glick
J.Coord.Chem.,**9**, 167, 1979
See also R1 : 84

85.C bis((μ^2 – t – Butylthiolato) –
(dimethylphenylphosphino)) – acetyl – carbonyl –
iodo – di – rhodium
$C_{27}H_{43}IO_2P_2Rh_2S_2$ Main entry is 71.217

85.80 tris(Tetraethylammonium) tetrakis(μ_3 – sulfido) –
tetra(benzylthiolato – iron)
$C_{28}H_{28}Fe_4S_8{}^{3-}$, $3C_8H_{20}N^+$
J.M.Berg, K.O.Hodgson, R.H.Holm
J.Am.Chem.Soc.,**101**, 4586, 1979
See also R2 : 3

85.C bis(2 – (Diphenylphosphino)ethane – 1 – thiolato) –
oxo – molybdenum(iv)
$C_{28}H_{28}MoOP_2S_2$ Main entry is 86.94

85.C 2,4 – Dithio – pyrimidinato – bis(bis(η^5 –
methylcyclopentadienyl) – titanium(iii))
$C_{28}H_{30}N_2S_2Ti_2$ Main entry is 73.113

85.81 Hydro – tris(1 – pyrazolyl) – borato – bis(p –
toluenethiolato) – (p – fluorobenzenediazo) –
molybdenum
$C_{29}H_{28}BFMoN_8S_2$
G.Ferguson, F.J.Lalor, M.Parvez, M.A.Khan
Am.Cryst.Assoc.,Ser.2,**8**, 24, 1980
See also R1 : 83,62

85.82 bis(2 – (N – Phenylaldimino) – benzo(b)furan – 3 –
thiolato) – nickel(ii)
$C_{30}H_{20}N_2NiO_2S_2$
V.A.Bren, O.A.Osipov, V.I.Minkin, Zh.V.Bren, L.S.Minkina,
L.O.Atovmyan, S.M.Aldoshin, O.A.D'yachenko
Koord.Khim.,**5**, 1058, 1979
See also R1 : 83

85.C (μ – Chloro) – (μ – t – butylthio) – bis(carbonyl –
(tri – t – butylphosphine) – rhodium)
$C_{30}H_{63}ClO_2P_2Rh_2S$ Main entry is 86.103

85.C 1 – Triphenylphosphine – 1 – chloro – 4,5 – benzo –
3 – (3' – phenyl – 5' – ethylidenyl – 1',2' – dithiol –
3' – ene) – 1 – pallada – 2 – thiol – 2 – ene toluene
solvate
$C_{31}H_{26}ClPPdS_3$, $2C_7H_8$ Main entry is 71.235

85.C Iodo – (1,9 – bis(diphenylphosphino) – 3,7 –
dithianonane) – nickel(ii) tetraphenylborate
$C_{31}H_{34}INiP_2S_2{}^+$, $C_{24}H_{20}B^-$ Main entry is 86.107

85.C (1,9 – bis(Diphenylphosphino) – 3,7 –
dithianonane) – nickel(ii) diperchlorate
$C_{31}H_{34}NiP_2S_2{}^{2+}$, $2ClO_4{}^-$ Main entry is 86.108

85.83 bis(2 – (N – Benzylaldimino) – benzo(b)furan – 3 –
thiolato) – nickel(ii)
$C_{32}H_{24}N_2NiO_2S_2$
V.A.Bren, O.A.Osipov, V.I.Minkin, Zh.V.Bren, L.S.Minkina,
L.O.Atovmyan, S.M.Aldoshin, O.A.D'yachenko
Koord.Khim.,**5**, 1058, 1979
See also R1 : 83

85.84 Bromo – (2,3,7,8,12,13,17,18 – octamethoxy –
tetrabenzo(b,e,h,k)(1,4,7,10)tetrathia –
cyclododecatetraene) – copper(i)
$C_{32}H_{32}BrCuO_8S_4$
K.von Deuten, G.Klar, J.Kopf
Eur.Cryst.Meeting,**6**, 64, 1980

85.85 Chloro – (2,3,7,8,12,13,17,18 – octamethoxy –
tetrabenzo(b,e,h,k)(1,4,7,10)tetrathia –
cyclododecatetraene) – copper(i) chloroform solvate
$C_{32}H_{32}ClCuO_8S_4$, $3CHCl_3$
K.von Deuten, G.Klar, J.Kopf
Eur.Cryst.Meeting,**6**, 64, 1980

85.C η^7 – Tropylium – tungsten – tris(μ^2 – n –
butylthio) – carbonyl – tungsten – bis(μ^2 – n –
butylthio) – tetracarbonyl – tungsten
$C_{32}H_{52}O_5S_5W_3$ Main entry is 75.39

85.86 bis(Terpyridyl – (2 – aminoethylthiolato) –
platinum) – platinum tetrakis(tetrafluoroborate)
$C_{34}H_{34}N_8Pt_3S_2{}^{4+}$, $4BF_4{}^-$
J.C.Dewan, S.J.Lippard, W.R.Bauer
J.Am.Chem.Soc.,**102**, 858, 1980
See also R1 : 83

85.87 tris(Dicyclohexyl – dithiophosphinato) –
dysprosium(iii)
$C_{36}H_{66}DyP_3S_6$
A.A.Pinkerton, D.Schwarzenbach
J.Chem.Soc.,Dalton Trans.,1300, 1980

85.88 tris(Dicyclohexyl – dithiophosphinato) –
lutetium(iii)
$C_{36}H_{66}LuP_3S_6$
A.A.Pinkerton, D.Schwarzenbach
J.Chem.Soc.,Dalton Trans.,1300, 1980

85.C trans – Di(methylsulfito) –
bis(triphenylphosphine) – platinum(ii)
$C_{38}H_{36}O_6P_2PtS_2$ Main entry is 86.139

85.89 $(-)_{579}$ – bis((S) – (+) – 0,0' – (1,1' – Binaphthyl –
2,2' – diyl) – dithiophosphato) – nickel(ii) (absolute
configuration)
$C_{40}H_{24}NiO_4P_2S_4$
W.Poll, H.Wunderlich
*Acta Crystallogr.,Sect.B,***36**, 1191, 1980

85.C cis,cis,cis – bis(Dimethylphenylphosphine) –
bis(monothiobenzoato) – (1,10 – phenanthroline) –
ruthenium(ii)
$C_{42}H_{40}N_2O_2P_2RuS_2$ Main entry is 86.153

85.C cis,cis,trans – bis(Dimethylphenylphosphine) –
bis(monothiobenzoato) – (1,10 – phenanthroline) –
ruthenium(ii)
$C_{42}H_{40}N_2O_2P_2RuS_2$ Main entry is 86.154

85.C Phenyl – dithiocarboxylato –
bis(triphenylphosphine) – copper(i)
$C_{43}H_{35}CuP_2S_2$ Main entry is 86.160

85.C Dicarbonyl – bis(η^2 – diphenylphosphinesulfido) –
triphenylphosphine – molybdenum methylene
chloride solvate
$C_{44}H_{35}MoO_2P_3S_2$, CH_2Cl_2 Main entry is 86.162

85.C 4 – Methyl – N – thiobenzohydroximato –
bis(triphenylphosphine) – platinum(ii)
$C_{44}H_{37}NOP_2PtS$ Main entry is 86.163

85.C (η^2 – Dithiomethoxycarbonyl) – carbonyl – (p –
tolylisocyanide) – bis(triphenylphosphine) –
ruthenium(ii) perchlorate chloroform solvate
hemihydrate (refinement in Pna21)
$C_{47}H_{40}NOP_2RuS_2^+$, ClO_4^-, $0.5CHCl_3$, $0.5H_2O$
Main entry is 71.270

85.C (η^2 – Dithiomethoxycarbonyl) – carbonyl – (p –
tolylisocyanide) – bis(triphenylphosphine) –
ruthenium(ii) perchlorate chloroform solvate
hemihydrate (refinement in Pnma)
$C_{47}H_{40}NOP_2RuS_2^+$, ClO_4^-, $0.5CHCl_3$, $0.5H_2O$
Main entry is 71.271

85.C Thallium(i) (μ_8 – chloro) – dodecakis(α – mercapto –
isobutyrato) – octa – copper(i) – hexa – copper(ii)
dodecahydrate
$C_{48}H_{72}ClCu_{14}O_{24}S_{12}^{5-}$, $5Tl^+$, $12H_2O$ Main entry is 81.95

85.90 tetrakis(Dicyclohexyl – dithiophosphinato) –
thorium(iv)
$C_{48}H_{88}P_4S_8Th$
A.A.Pinkerton, J.–M.Zellweger, D.Schwarzenbach
*Eur.Cryst.Meeting,***6**, 57, 1980

85.91 Chloro – dodeca(1,1 – dimethyl – 2 –
aminoethylthiolato) – tetradeca – copper sulfate
eicosahydrate
$C_{48}H_{120}ClCu_{14}N_{12}S_{12}^{7+}$, $3.5O_4S^{2-}$, $20H_2O$
H.J.Schugar, C.–C.Ou, J.A.Thich, J.A.Potenza, T.R.Felthouse,
M.S.Haddad, D.N.Hendrickson, W.Furey Junior,
R.A.Lalancette *Inorg.Chem.*,**19**, 543, 1980
See also R1 : 83

85.C σ – (Thiophenyl(sulfinyl)methylene) – thiophenyl –
bis(triphenylphosphine) – platinum benzene solvate
$C_{49}H_{40}OP_2PtS_3$, C_6H_6 Main entry is 71.272

85.92 catena – (μ – Benzenethiolato) – (methanol –
benzenethiolato – hexakis(μ – benzenethiolato)) –
tetra – zinc(ii)
$(C_{49}H_{44}OS_8Zn_4)_n$
I.G.Dance *J.Am.Chem.Soc.*,**102**, 3445, 1980
See also R1 : 84

85.93 bis((μ – bis(Diphenylthiophosphoryl)methane) –
chloro – copper(i)) bis(bis(diphenylthiophosphoryl)
methane – chloro – copper(i))
$C_{50}H_{44}Cl_2Cu_2P_4S_4$, $2C_{25}H_{22}ClCuP_2S_2$
E.W.Ainscough, A.M.Brodie, K.L.Brown
J.Chem.Soc.,Dalton Trans.,1042, 1980

85.C Chloro – rhodium – bis(μ^2 – bis(diphenylphosphino)
methylene) – (μ – dicarbon – tetrasulfide) –
chloro – carbonyl – rhodium
$C_{53}H_{44}Cl_2OP_4Rh_2S_4$ Main entry is 71.273

85.C bis(Triphenylphosphine) – silver – bis(μ – thio) –
tungsten – bis(μ – thio) – triphenylphosphine –
silver
$C_{54}H_{45}Ag_2P_3S_4W$ Main entry is 86.199

85.C Triphenylphosphonio – dithiocarboxylato –
carbonyl – bis(triphenylphosphine) – iridium(i)
tetrafluoroborate
$C_{56}H_{45}IrOP_3S_2^+$, BF_4^- Main entry is 86.209

85.94 Tetramethylammonium hexakis(μ –
benzenethiolato) – tetra(benzenethiolato) – tetra –
cobalt(ii)
$C_{60}H_{50}Co_4S_{10}^{2-}$, $2C_4H_{12}N^+$
I.G.Dance *J.Am.Chem.Soc.*,**101**, 6264, 1979
See also R2 : 3

85.95 bis(Tetramethylammonium) deca(μ^2 –
benzenethiolato) – tetra – iron(ii)
$C_{60}H_{50}Fe_4S_{10}^{2-}$, $2C_4H_{12}N^+$
K.S.Hagen, J.M.Berg, R.H.Holm
*Inorg.Chim.Acta,***45**, L17, 1980
See also R2 : 3

85.C bis((Dicarbonyl – P,P – diphenyl – N – methyl –
phosphinothioformamido) – (μ – P,P – diphenyl –
N – methyl – phosphinothioformamido) –
molybdenum) dichloromethane solvate
$C_{60}H_{52}Mo_2N_4O_4P_4S_4$, $4CH_2Cl_2$ Main entry is 71.280

85.C hexakis(μ_3 – Sulfido) – bis(oxo – tungsten) –
tetra(triphenylphosphine – copper)
$C_{72}H_{60}Cu_4O_2P_4S_6W_2$ Main entry is 86.219

85.C bis((μ^2 – Thio) – (1,1,1 – tris(diphenylphosphino)
methyl)ethane) – cobalt dimethylformamide solvate
$C_{82}H_{78}Co_2P_6S_2^+$, $C_3H_7NO^-$, $0.4C_3H_7NO$ Main entry is 86.226

85.C bis(μ^2 – Thio) – (1,1,1 – tris((diphenylphosphino)methyl)ethane) – cobalt tetraphenylborate dimethylformamide solvate
$C_{82}H_{78}Co_2P_6S_2^+$, $C_{24}H_{20}B^-$, $0.5C_3H_7NO$ Main entry is 86.227

85.C (μ^2 – Carbon – disulfide) – bis((1,1,1 – (diphenylphosphinomethyl)ethane) – cobalt) bis(tetraphenylborate) acetone solvate
$C_{83}H_{78}Co_2P_6S_2^{2+}$, $2C_{24}H_{20}B^-$, $2C_3H_6O$ Main entry is 72.69

85.C bis(μ^2 – Methylthio) – (1,1,1 – tris(((diphenylphosphino)methyl)))ethane – cobalt bis(tetraphenylborate) acetone solvate
$C_{84}H_{84}Co_2P_6S_2^{2+}$, $2C_{24}H_{20}B^-$, $2C_3H_6O$ Main entry is 86.229

85.96 tetrakis(Tetra – n – butylammonium) (bis(tris(μ – benzylthiolato) – molybdenum) – tetrakis(μ – thiolato) – tris(benzylthiolato) – iron) – iron
$C_{84}H_{84}Fe_7Mo_2S_{20}^{4-}$, $4C_{16}H_{36}N^+$
T.E.Wolff, J.M.Berg, P.P.Power, K.O.Hodgson, R.H.Holm
Inorg.Chem., **19**, 430, 1980
See also R2 : 3

85.97 tetrakis(Tetra – n – butylammonium) (bis(tris(μ – benzylthiolato) – tungsten) – tetrakis(μ – thiolato) – tris(benzylthiolato – iron)) – iron
$C_{84}H_{84}Fe_7S_{20}W_2^{4-}$, $4C_{16}H_{36}N^+$
T.E.Wolff, J.M.Berg, P.P.Power, K.O.Hodgson, R.H.Holm
Inorg.Chem., **19**, 430, 1980
See also R2 : 3

METAL COMPLEXES (P, AS, SB LIGAND)

86.C Chloro – bis(ethyldimethylphosphine) – mercury(ii) trichloro – (ethyldimethylphosphine) – mercury(ii)
$C_4H_{11}Cl_3HgP^-$, $C_8H_{22}ClHgP_2^+$ Main entry is 86.7

86.1 tetrakis(μ – bis(Difluorophosphino)methylamine) – bis(chloro – molybdenum)
$C_4H_{12}Cl_2F_{16}Mo_2N_4P_8$
F.A.Cotton, W.H.Ilsley, W.Kaim
J.Am.Chem.Soc., **102**, 1918, 1980

86.2 Ammine – tetracarbonyl – (dimethylthiophosphinito) – rhenium
$C_6H_9NO_4PReS$
E.Lindner, F.Bouachir, M.Weishaupt, S.Hoehne, B.Schilling
Z.Anorg.Allg.Chem., **456**, 163, 1979

86.3 Ammine – trichloro – bis(trimethylphosphite) – ruthenium(iii)
$C_6H_{21}Cl_3NO_6P_2Ru$
M.I.Bruce, D.A.Kelly, G.M.McLaughlin, G.B.Robertson, I.B.Tomkins, R.C.Wallis *Aust.J.Chem.*, **33**, 195, 1980

86.4 Tetracarbonyl – trimethylphosphite – ruthenium
$C_7H_9O_7PRu$
R.E.Cobbledick, F.W.B.Einstein, R.K.Pomeroy, E.R.Spetch
J.Organomet.Chem., **195**, 77, 1980

86.5 Di – iodo – (trimethylphosphine) – carbonyl – nickel
$C_7H_{18}I_2NiOP_2$
C.Saint–Joly, A.Mari, A.Gleizes, M.Dartiguenave, Y.Dartiguenave, J.Galy *Inorg.Chem.*, **19**, 2403, 1980

86.6 Dichloro – (dimethylphenylphosphine) – cadmium
$(C_8H_{11}CdCl_2P)_n$
N.A.Bell, T.D.Dee, M.Goldstein, I.W.Nowell
Inorg.Chim.Acta, **38**, 191, 1980

86.C η^5 – Cyclopentadienyl – trimethylphosphine – cyclopentasulfur – cobalt
$C_8H_{14}CoPS_5$ Main entry is 73.11

86.7 Chloro – bis(ethyldimethylphosphine) – mercury(ii) trichloro – (ethyldimethylphosphine) – mercury(ii)
$C_8H_{22}ClHgP_2^+$, $C_4H_{11}Cl_3HgP^-$
N.A.Bell, M.Goldstein, T.Jones, I.W.Nowell
Acta Crystallogr.,Sect.B, **36**, 708, 1980
See also R2 : 86

86.8 Pentacarbonyl – (1,5 – dimethyl – 1,2,3 – diazaphosphole) – chromium
$C_9H_7CrN_2O_5P$
J.H.Weinmaier, H.Tautz, A.Schmidpeter, S.Pohl
J.Organomet.Chem., **185**, 53, 1980

86.9 Tricarbonyl – bis(trimethylphosphine) – ruthenium(0)
$C_9H_{18}O_3P_2Ru$
R.A.Jones, G.Wilkinson, A.M.R.Galas, M.B.Hursthouse, K.M.A.Malik *J.Chem.Soc.,Dalton Trans.*, 1771, 1980

86.C bis(η^3 – Allyl) – trimethylphosphine – nickel (at –170°C)
$C_9H_{19}NiP$ Main entry is 72.8

86.10 Chloro – tris(trimethylphosphine) – platinum chloride
$C_9H_{27}ClP_3Pt^+$, Cl^-
R.Favez, R.Roulet, A.A.Pinkerton, D.Schwarzenbach
Inorg.Chem., **19**, 1356, 1980

86.11 Chloro – tris(trimethylphosphine) – rhodium(i)
$C_9H_{27}ClP_3Rh$
R.A.Jones, F.M.Real, G.Wilkinson, A.M.R.Galas,
M.B.Hursthouse, K.M.A.Malik
J.Chem.Soc.,Dalton Trans.,511, 1980

86.C Tetracarbonyl – bis(μ – dimethylarsenido) – (tetracarbonyl – iron) – manganese tetracarbonyl – chloro – (μ – dimethylarsenido) – (tetracarbonyl – iron) – manganese
$C_{10}H_6AsClFeMnO_8^-$, $C_{12}H_{12}As_2FeMnO_8^+$ Main entry is 86.18

86.C (2,4 – Dithio – 1,3 – dithia – 1,4 – diyl) – η^5 – cyclopentadienyl – (trimethylphosphine) – rhodium
$C_{10}H_{14}PRhS_4$ Main entry is 71.21

86.12 Tetracarbonyl – (tris(dimethylamino)phosphine) – iron
$C_{10}H_{18}FeN_3O_4P$
A.H.Cowley, R.E.Davis, M.Lattman, M.McKee, K.Remadna
J.Am.Chem.Soc., **101**, 5090, 1979

86.13 bis(Isopropyl) – amino – oxophosphane – pentacarbonyl – chromium (at –130°C)
$C_{11}H_{14}CrNO_6P$
E.Niecke, M.Engelmann, H.Zorn, B.Krebs, G.Henkel
Angew.Chem.,Int.Ed.Engl.,**19**, 710, 1980

86.C Dicarbonyl – (η^5 – cyclopentadienyl) – (1,3 – dimethyl – 1,3 – diaza – 2 – phospholidin – 2 – yl) – molybdenum
$C_{11}H_{15}MoN_2O_2P$ Main entry is 73.24

86.14 Hexamethyl – trisila – tetraphospha – nortricyclene – pentacarbonyl – chromium
$C_{11}H_{18}CrO_5P_4Si_3$
W.Honle, H.G.von Schnering
Z.Anorg.Allg.Chem.,**465**, 72, 1980
See also R1 : 63

86.15 Chloro – (pentane – 2,4 – dithionato) – (triethylphosphine) – nickel(ii)
$C_{11}H_{22}ClNiPS_2$
J.P.Fackler Junior, A.F.Masters
Inorg.Chim.Acta,**39**, 111, 1980
See also R1 : 85

86.16 trans – Chloro – bis(dimethylglyoximato) – trimethylphosphite – cobalt(iii)
$C_{11}H_{23}ClCoN_4O_7P$
N.Bresciani–Pahor, M.Calligaris, L.Randaccio
Inorg.Chim.Acta,**39**, 173, 1980
See also R1 : 83

86.C 1 – 3 – η – Allyl – dicarbonyl – chloro – bis(trimethylphosphite) – molybdenum(ii)
$C_{11}H_{23}ClMoO_8P_2$ Main entry is 72.13

86.C Dichloro – triethylphosphine – neopentylidene – oxo – tungsten
$C_{11}H_{25}Cl_2OPW$ Main entry is 71.32

86.C Potassium ethylene – tris(trimethylphosphine) – semicobaltate
$C_{11}H_{31}CoP_3^-$, $C_{11}H_{31}CoP_3$, K^+ Main entry is 71.33

86.17 1,1,1,2,2,2,2,3,3 – Nonacarbonyl – 1,3.1,3 – bis(μ – nitrosyl) – 3 – (trimethylphosphite) – triangulo – tri – osmium
$C_{12}H_9N_2O_{14}Os_3P$
S.Bellard, P.R.Raithby
Acta Crystallogr.,Sect.B,**36**, 705, 1980

86.18 Tetracarbonyl – bis(μ – dimethylarsenido) – (tetracarbonyl – iron) – manganese tetracarbonyl – chloro – (μ – dimethylarsenido) – (tetracarbonyl – iron) – manganese
$C_{12}H_{12}As_2FeMnO_8^+$, $C_{10}H_6AsClFeMnO_8^-$
H.–J.Langenbach, E.Rottinger, H.Vahrenkamp
Chem.Ber.,**113**, 42, 1980
See also R2 : 86

86.C (μ^2 – Bromo) – (μ^2 – p – tolylmethylidene) – (μ^2 – bis(methylamino – bis(difluorophosphine))) – bis(bromo – carbonyl – tungsten)
$C_{12}H_{13}Br_3F_8N_2O_2P_4W_2$ Main entry is 71.36

86.C Dicarbonyl – (4 – 5 – η – 1,2 – difluoro – 1,2 – bis(trifluoromethyl) – pent – 4 – enyl) – (trimethylphosphite) – cobalt (at 183°K)
$C_{12}H_{14}CoF_8O_5P$ Main entry is 71.38

86.C cis – Dichloro – (dimethylphenylphosphine) – (vinylacetate) – platinum(ii)
$C_{12}H_{17}Cl_2O_2PPt$ Main entry is 72.18

86.19 1,5,9 – Triphosphacyclodecane – tricarbonyl – molybdenum
$C_{12}H_{21}MoO_3P_3$
R.C.Haltiwanger, B.N.Diel, A.D.Norman
Am.Cryst.Assoc.,Ser.2,**8**, 24, 1980

86.20 (μ^2 – Sulfonyl) – bis((μ – methylthio – (trimethylphosphine) – dicarbonyl – iron(i)) diethyl ether solvate
$C_{12}H_{24}Fe_2O_6P_2S_3$, $0.5C_4H_{10}O$
N.J.Taylor, M.S.Arabi, R.Mathieu
Inorg.Chem.,**19**, 1740, 1980
See also R1 : 85

86.21 trans – Dichloro – (triethylphosphine) – (cis – 2,4 – dimethylpyrrolidine) – platinum(ii)
$C_{12}H_{28}Cl_2NPPt$
J.Ambuehl, P.S.Pregosin, L.M.Venanzi, G.Consiglio,
F.Bachechi, L.Zambonelli
J.Organomet.Chem.,**181**, 255, 1979
See also R1 : 83

86.22 trans – Dichloro – (triethylphosphine) – (cis – 2,3 – dimethylpyrrolidine) – platinum(ii)
$C_{12}H_{28}Cl_2NPPt$
J.Ambuehl, P.S.Pregosin, L.M.Venanzi, G.Consiglio,
F.Bachechi, L.Zambonelli
J.Organomet.Chem.,**181**, 255, 1979
See also R1 : 83

86.23 (Difluorophosphonato – triethylphosphine – platinum) – (μ – chloro) – (μ – difluorophosphonato) – (chloro – triethylphosphine – platinum)
$C_{12}H_{30}Cl_2F_4O_2P_4Pt_2$
S.Neumann, D.Schomburg, R.Schmutzler
J.Chem.Soc.,Chem.Commun.,848, 1979

86.24 bis(Triethylphosphine) – dinitrosyl – nitrito – manganese
$C_{12}H_{30}MnN_3O_4P_2$
R.D.Wilson, R.Bau *J.Organomet.Chem.*,**191**, 123, 1980
See also R1 : 84,83

86.25 bis(Triethylphosphine) – platinum – tetrathio – tungsten
$C_{12}H_{30}P_2PtS_4W$
A.R.Siedle, C.R.Hubbard, A.D.Mighell, R.M.Doherty,
J.M.Stewart *Inorg.Chim.Acta*,**38**, 197, 1980

86.26 bis(Dichloro – (1,2 – bis(dimethylphosphino) ethane) – tungsten) toluene solvate
$C_{12}H_{32}Cl_4P_4W_2$, C_7H_8
F.A.Cotton, T.R.Felthouse, D.G.Lay
J.Am.Chem.Soc.,**102**, 1431, 1980

86.27 bis(Dichloro – bis(trimethylphosphine) – tungsten)
$C_{12}H_{36}Cl_4P_4W_2$
F.A.Cotton, T.R.Felthouse, D.G.Lay
J.Am.Chem.Soc.,**102**, 1431, 1980

86.28 $(+)_{589}$ – fac – tris(2 – Aminoethyl – dimethylphosphine) – cobalt(iii) tribromide trihydrate (absolute configuration)
$C_{12}H_{36}CoN_3P_3{}^{3+}$, $3Br^-$, $3H_2O$
I.Kinoshita, K.Kashiwabara, J.Fujita, K.Matsumoto, S.Ooi
Chem.Lett.,95, 1980
See also R1 : 83

86.29 tetrakis(Trimethylphosphine) – rhodium(i) chloride
$C_{12}H_{36}P_4Rh^+$, Cl^-
R.A.Jones, F.M.Real, G.Wilkinson, A.M.R.Galas,
M.B.Hursthouse, K.M.A.Malik
J.Chem.Soc.,Dalton Trans.,511, 1980

86.30 Hydrido – (tetrahydroborato) – tetrakis(trimethylphosphine) – molybdenum(ii)
$C_{12}H_{41}BMoP_4$
J.L.Atwood, W.E.Hunter, E.Carmona–Guzman, G.Wilkinson
J.Chem.Soc.,Dalton Trans.,467, 1980

86.31 Tetracarbonyl – (μ – dimethylarsenido) – (μ – tetramethyl – diarsenic) – (tricarbonyl – chloro – manganese) – manganese
$C_{13}H_{18}As_3ClMn_2O_7$
E.Rottinger, A.Trenkle, R.Muller, H.Vahrenkamp
Chem.Ber.,113, 1280, 1980

86.C trans – (Methyl) – chloro – bis(triethylphosphine) – platinum(ii)
$C_{13}H_{33}ClP_2Pt$ Main entry is 71.55

86.C μ – 2 – sigma:2 – 3 – η – (4,5 – Dihydro – 2 – furyl) – (bis(trimethylphosphine) – platinum) – tetracarbonyl – manganese (red form,at 200°K)
$C_{14}H_{23}MnO_5P_2Pt$ Main entry is 71.75

86.C μ – 2 – sigma:2 – 3 – η – (4,5 – Dihydro – 2 – furyl) – (bis(trimethylphosphine) – platinum) – tetracarbonyl – manganese (yellow form,at 200°K)
$C_{14}H_{23}MnO_5P_2Pt$ Main entry is 71.76

86.C Chloro – bis(bis(dimethylphosphino)ethane) – methoxycarbonyl – iridium fluorosulfate (at –50°C)
$C_{14}H_{35}ClIrO_2P_4{}^+$, FO_3S^- Main entry is 71.79

86.C cis – (Ethyl) – chloro – bis(triethylphosphine) – platinum(ii)
$C_{14}H_{35}ClP_2Pt$ Main entry is 71.80

86.32 1,1 – bis(Triethylphosphine) – 2,4 – dicarba – 1 – platina – closo – heptaborane (at 215°K)
$C_{14}H_{36}B_4P_2Pt$
G.K.Barker, M.Green, F.G.A.Stone. A.J.Welch
J.Chem.Soc.,Dalton Trans.,1186, 1980
See also R1 : 62

86.C Trifluoroacetato – hydrido – tetra(trimethylphosphite) – molybdenum
$C_{14}H_{37}F_3MoO_{14}P_4$ Main entry is 81.60

86.33 (μ(4,5) – (trans – bis(Triethylphosphine) – hydrido – platinum)) – (μ(5,6) – hydrido) – nido – 2,3 – dicarba – hexaborane (at 215°K)
$C_{14}H_{38}B_4P_2Pt$
G.K.Barker, M.Green, F.G.A.Stone, A.J.Welch, T.P.Onak,
G.Siwapanyoyos *J.Chem.Soc.,Dalton Trans.*,1687, 1979
See also R1 : 62

86.C 9 – Hydrido – 9,10 – bis(triethylphosphine) – 7,8 – dicarba – 9 – platina – undecaborane
$C_{14}H_{40}B_8P_2Pt$ Main entry is 71.81

86.C 9 – Hydrido – 9,9 – bis(triethylphosphine) – 7,8 – dicarba – 9 – platina – undecaborane
$C_{14}H_{41}B_8P_2Pt$ Main entry is 71.

86.C Dimethyl – tetrakis(trimethylphosphine) – tungsten(ii)
$C_{14}H_{42}P_4W$ Main entry is 71.83

86.34 (μ^3 – Phenylphosphine) – hexacarbonyl – di – cobalt – tricarbonyl – iron
$C_{15}H_5Co_2FeO_9P$
H.Beurich, T.Madach, F.Richter, H.Vahrenkamp
Angew.Chem.,Int.Ed.Engl.,**18**, 690, 1979

86.35 (μ^3 – Phenylphosphine) – nonacarbonyl – tri – cobalt
$C_{15}H_5Co_3O_9P$
H.Beurich, T.Madach, F.Richter, H.Vahrenkamp
Angew.Chem.,Int.Ed.Engl.,**18**, 690, 1979

86.36 bis(μ^2 – Hydrido) – (μ^3 – phenylphosphido) – tris(tricarbonyl – iron) (at –80°C)
$C_{15}H_7Fe_3O_9P$
G.Huttner, J.Schneider, G.Mohr, J.von Seyerl
J.Organomet.Chem.,191, 61, 1980

86.37 Tricarbonyl – bis(tris(dimethylamino)phosphine) – iron
$C_{15}H_{36}FeN_6O_3P_2$
A.H.Cowley, R.E.Davis, M.Lattman, M.McKee, K.Remadna
J.Am.Chem.Soc.,101, 5090, 1979

86.C Chloro – trimethylphosphine – tris(trimethylsilylmethyl) – molybdenum(iv)
$C_{15}H_{42}ClMoPSi_3$ Main entry is 71.89

86.38 Pentacarbonyl – (6 – methyl – 4 – methylene – 2 – phenyl – 4H – 1,3,2 – dioxaphosphorine) – chromium
$C_{16}H_{11}CrO_7P$
J.von Seyerl, D.Neugebauer, G.Huttner, C.Kruger,
Y.–H.Tsay *Chem.Ber.*,112, 3637, 1979

86.C Tetraphenylarsonium (μ – dimethylphosphido) – bis(dicarbonyl – cyclopentadienyl – molybdenum)
$C_{16}H_{16}Mo_2O_4P^-$, $C_{24}H_{20}As^+$ Main entry is 73.59

86.C Carbonyl – η^5 – cyclopentadienyl –
trimethoxyphosphine – (4,5 –
bis(methoxycarbonyl) – 1,3 – dithiol – 2 – ylidene) –
manganese(i)
$C_{16}H_{20}MnO_8PS_2$ Main entry is 71.99

86.39 cis – bis(Thiocyanato) – bis(1,3,4 –
trimethylphosphole) – palladium(ii)
$C_{16}H_{22}N_2P_2PdS_2$
J.J.MacDougall, E.M.Holt, P.de Meester, N.W.Alcock,
F.Mathey, J.H.Nelson *Inorg.Chem.*,**19**, 1439, 1980
See also R1 : 85

86.40 1,1,1,2,2,3,3 – Heptacarbonyl – 2,3 – (μ – hydrido) –
2,3 – (μ – nitrosyl) – 1,2,3 –
tris(trimethylphosphite) – triangulo – tri –
ruthenium
$C_{16}H_{28}NO_{17}P_3Ru_3$
B.F.G.Johnson, P.R.Raithby, C.Zuccaro
J.Chem.Soc.,Dalton Trans.,99, 1980

86.41 bis(μ – t – Butylthiolato) – bis(carbonyl –
trimethylphosphite – iridium)
$C_{16}H_{36}Ir_2O_8P_2S_2$
J.J.Bonnet, A.Thorez, A.Maisonnat, J.Galy, R.Poilblanc
J.Am.Chem.Soc.,**101**, 5940, 1979
See also R1 : 85

86.42 bis(μ – t – Butylthiolato) – bis(carbonyl – hydrido –
trimethylphosphite – iridium)
$C_{16}H_{38}Ir_2O_8P_2S_2$
J.J.Bonnet, A.Thorez, A.Maisonnat, J.Galy, R.Poilblanc
J.Am.Chem.Soc.,**101**, 5940, 1979
See also R1 : 85

86.43 2,3 – Dimethyl – 1,1 – bis(triethylphosphine) – 2,3 –
dicarba – 1 – platina – closo – heptaborane
(at 215°K)
$C_{16}H_{40}B_4P_2Pt$
G.K.Barker, M.Green, F.G.A.Stone, A.J.Welch
J.Chem.Soc.,Dalton Trans.,1186, 1980
See also R1 : 62

86.44 bis(μ – 2,3,4 – η^3 – nido – Hexaboranyl) –
bis(dimethylphenylphosphine) – di – platinum
$C_{16}H_{40}B_{12}P_2Pt_2$
N.N.Greenwood, M.J.Hails, J.D.Kennedy, W.S.McDonald
J.Chem.Soc.,Chem.Commun.,37, 1980
See also R1 : 62

86.45 bis(μ – Chloro) – bis(chloro – (hydrogen
bis(diethylphosphonito)) – nitrosyl – ruthenium(ii))
$C_{16}H_{42}Cl_4N_2O_{14}P_4Ru_2$
T.G.Southern, P.H.Dixneuf, J.Y.Le Marouille, D.Grandjean
Inorg.Chem.,**18**, 2987, 1979

86.C (μ_3 – Methylenethiolato) – (μ_3 – sulfido) –
(octacarbonyl – (dimethylphenylphosphine) – tri –
osmium) (at –35°C)
$C_{17}H_{13}O_8Os_3PS_2$ Main entry is 71.108

86.46 Dicarbonyl – (1,3 – bis(phenylphosphino) –
propane) – nickel(0)
$C_{17}H_{18}NiO_2P_2$
M.Baacke, S.Morton, O.Stelzer, W.S.Sheldrick
Chem.Ber.,**113**, 1343, 1980

86.C Tetracarbonyl – (μ^2 – iodo) – (methyl – di – t –
butylphosphine) – (2 – oxacyclopentylidene) –
platinum – manganese (at 200°K)
$C_{17}H_{27}IMnO_5PPt$ Main entry is 71.111

86.C (μ_3 – Ethinyl) – triangulo – tris(dicarbonyl –
trimethylphosphite – cobalt)
$C_{17}H_{30}Co_3O_{15}P_3$ Main entry is 71.112

86.47 bis(Di – isopropoxyphosphine)methylene –
tetracarbonyl – molybdenum
$C_{17}H_{30}MoO_8P_2$
M.Fild, W.Handke, W.S.Sheldrick
Z.Naturforsch.,Teil B,**35**, 838, 1980

86.C trans – Bromo – (2 – pyridyl) –
bis(triethylphosphine) – palladium(ii)
$C_{17}H_{34}BrNP_2Pd$ Main entry is 71.113

86.C trans – Bromo – (3 – pyridyl) –
bis(triethylphosphine) – palladium(ii)
$C_{17}H_{34}BrNP_2Pd$ Main entry is 71.114

86.C trans – Bromo – (4 – pyridyl) –
bis(triethylphosphine) – palladium(ii)
$C_{17}H_{34}BrNP_2Pd$ Main entry is 71.115

86.C (η^3 – Cyclo – octenyl) – tris(trimethylphosphite) –
iron(i) (at –80°C)
$C_{17}H_{40}FeO_9P_3$ Main entry is 75.16

86.C tris(Trimethylphosphite) – (η^3 – cyclo – octenyl) –
iron tetrafluoroborate
$C_{17}H_{40}FeO_9P_3^+$, BF_4^- Main entry is 75.17

86.C tris(Trimethylphosphite) – (η^3 – cyclo – octenyl) –
iron tetrafluoroborate (neutron study, at 110°K)
$C_{17}H_{40}FeO_9P_3^+$, BF_4^- Main entry is 75.18

86.C tris(Trimethylphosphite) – (η^3 – cyclo – octenyl) –
iron tetrafluoroborate (neutron study, at 30°K)
$C_{17}H_{40}FeO_9P_3^+$, BF_4^- Main entry is 75.19

86.C (μ – Dimethylphosphido) – (μ –
dimethylphosphinomethyl) –
bis(di(trimethylphosphine) – cobalt)
$C_{17}H_{50}Co_2P_6$ Main entry is 71.116

86.48 (μ – Chloro) – (μ – diphenylphosphido) –
bis(tricarbonyl – iron) (absolute configuration)
$C_{18}H_{10}ClFe_2O_6P$
N.J.Taylor, G.N.Mott, A.J.Carty *Inorg.Chem.*,**19**, 560, 1980

86.C (μ – Methylenethiolato) – (μ_3 – sulfido) –
(nonacarbonyl – (dimethylphenylphosphine) – tri –
osmium)
$C_{18}H_{13}O_9Os_3PS_2$ Main entry is 71.123

86.C (μ^2 – Hydrido) – (μ^2 – dithioformato) –
bis(tricarbonyl – osmium) – tricarbonyl –
(dimethylphenylphosphine) – osmium (at –35°C)
$C_{18}H_{13}O_9Os_3PS_2$ Main entry is 81.64

86.49 Triphenylphosphine – trinitrosyl – manganese
$C_{18}H_{15}MnN_3O_3P$
R.D.Wilson, R.Bau *J.Organomet.Chem.*,**191**, 123, 1980
See also R1 : 83

86.C (μ – Oxy – bis(dimethylphosphane)) –
bis(dicarbonyl – (η^5 – cyclopentadienyl) –
manganese(i))
$C_{18}H_{22}Mn_2O_5P_2$ Main entry is 73.76

86.50 Dichloro – (1 – diphenylphosphino – 2 –
diethylaminoethane) – mercury(ii) (at 138°K)
$C_{18}H_{24}Cl_2HgNP$
P.K.S.Gupta, L.W.Houk, D.van der Helm, M.B.Hossain
Inorg.Chim.Acta,**44**, L235, 1980
See also R1 : 83

86.51 trans – Chloro – (dinitrogen) – bis(tri – isopropylphosphine) – rhodium(i) (at –160°C)
$C_{18}H_{42}ClN_2P_2Rh$
D.L.Thorn, T.H.Tulip, J.A.Ibers
J.Chem.Soc.,Dalton Trans.,2022, 1979

86.52 Dichloro – bis(tri – n – propylphosphine) – palladium(ii)
$C_{18}H_{42}Cl_2P_2Pd$
N.W.Alcock, T.J.Kemp, F.L.Wimmer, O.Traverso
Inorg.Chim.Acta,44, L245, 1980

86.53 Chloro – tris(triethylphosphine) – platinum(ii) tetrafluoroborate
$C_{18}H_{45}ClP_3Pt^+, BF_4^-$
D.R.Russell, M.A.Mazid, P.A.Tucker
J.Chem.Soc.,Dalton Trans.,1737, 1980

86.54 Fluoro – tris(triethylphosphine) – platinum(ii) tetrafluoroborate
$C_{18}H_{45}FP_3Pt^+, BF_4^-$
D.R.Russell, M.A.Mazid, P.A.Tucker
J.Chem.Soc.,Dalton Trans.,1737, 1980

86.C bis(μ – Acetato) – bis((trimethylphosphine) – (trimethylsilylmethyl) – molybdenum(ii))
$C_{18}H_{46}Mo_2O_4P_2Si_2$ Main entry is 71.132

86.55 Hydrido – tris(triethylphosphine) – platinum(ii) hexafluorophosphate
$C_{18}H_{46}P_3Pt^+, F_6P^-$
D.R.Russell, M.A.Mazid, P.A.Tucker
J.Chem.Soc.,Dalton Trans.,1737, 1980

86.56 tris(bis(1,2 – Dimethylphosphino)ethane) – chromium
$C_{18}H_{48}CrP_6$
F.G.N.Cloke, P.J.Fyne, M.L.H.Green, M.J.Ledoux, A.Gourdon, C.K.Prout *J.Organomet.Chem.*,198, 69, 1980

86.57 tris(μ – Hydroxo) – hexakis(trimethylphosphine) – di – ruthenium(ii) tetrafluoroborate
$C_{18}H_{57}O_3P_6Ru_2^+, BF_4^-$
R.A.Jones, G.Wilkinson, A.M.R.Galas, M.B.Hursthouse, K.M.A.Malik *J.Chem.Soc.,Dalton Trans.*,1771, 1980
See also R1 : 84

86.58 Tris((μ – hydrido) – (bis(trimethylphosphite)) – rhodium) (neutron study, at 110°K)
$C_{18}H_{57}O_{18}P_6Rh_3$
R.K.Brown, J.M.Williams, A.J.Sivak, E.L.Muetterties
Inorg.Chem.,19, 370, 1980

86.59 hexakis(Trimethylphosphine) – dihydrido – (μ – dihydrido) – di – molybdenum(ii)
$C_{18}H_{58}Mo_2P_6$
R.A.Jones, K.W.Chiu, G.Wilkinson, A.M.R.Galas, M.B.Hursthouse *J.Chem.Soc.,Chem.Commun.*,408, 1980

86.C Tetracarbonyl – (3 – (diphenylphosphineoxy) propyl) – manganese
$C_{19}H_{16}MnO_5P$ Main entry is 71.139

86.60 Bromo – tricarbonyl – (phenyl – bis – (3,5 – dimethylpyrazolyl)phosphine) – rhenium(i)
$C_{19}H_{19}BrN_4O_3PRe$
R.E.Cobbledick, L.R.J.Dowdell, F.W.B.Einstein, J.K.Hoyano, L.K.Peterson *Can.J.Chem.*,57, 2285, 1979
See also R1 : 83

86.61 Tricarbonyl – (phenyl – bis(3,5 – dimethylpyrazolyl) phosphine) – tungsten(0)
$C_{19}H_{19}N_4O_3PW$
R.E.Cobbledick, L.R.J.Dowdell, F.W.B.Einstein, J.K.Hoyano, L.K.Peterson *Can.J.Chem.*,57, 2285, 1979
See also R1 : 83

86.C (μ – Methoxy – phenyl – carbene) – (bis(trimethylphosphine) – platinum) – (pentacarbonyl – tungsten) (at 200°K)
$C_{19}H_{26}O_6P_2PtW$ Main entry is 71.142

86.C bis(Tri – isopropylphosphine) – dihydrido – bicarbonato – rhodium(iii) (at –160°C)
$C_{19}H_{45}O_3P_2Rh$ Main entry is 81.66

86.62 bis(Thiocyanato) – (triphenylphosphine) – mercury(ii) (β form)
$(C_{20}H_{15}HgN_2PS_2)_n$
R.C.Makhija, R.Rivest, A.L.Beauchamp
Can.J.Chem.,57, 2555, 1979

86.63 o – Phenylene – bis(dimethylarsine) – decacarbonyl – tri – iron
$C_{20}H_{16}As_2Fe_3O_{10}$
A.Bino, F.A.Cotton, P.Lahuerta, P.Puebla, R.Uson
Inorg.Chem.,19, 2357, 1980

86.64 (o – Phenylene – bis(dimethylarsine)) – decacarbonyl – tetra – iridium
$C_{20}H_{16}As_2Ir_4O_{10}$
J.R.Shapley, G.F.Stuntz, M.R.Churchill, J.P.Hutchinson
J.Am.Chem.Soc.,101, 7425, 1979

86.C Dichloro – (diphenyl(o – vinylphenyl)arsine) – platinum(ii)
$C_{20}H_{17}AsCl_2Pt$ Main entry is 72.36

86.C Triethylphosphine – (μ² – 1,4 – bis(trifluoro) – but – 2 – ene) – decacarbonyl – tri – osmium
$C_{20}H_{17}F_6O_{10}Os_3P$ Main entry is 72.37

86.65 11 – Triphenylphosphine – 11 – argenta – 5,6 – dicarba – undecaborane(11) acetone solvate
$2C_{20}H_{28}AgB_8P, C_3H_6O$
H.M.Colquhoun, T.J.Greenhough, M.G.H.Wallbridge
J.Chem.Soc.,Chem.Commun.,192, 1980
See also R1 : 62

86.66 3 – (Triphenylphosphine) – 3 – nitrato – 3 – rhoda – 1,2 – dicarba – dodecaborane dichloromethane solvate
$C_{20}H_{26}B_9NO_3PRh, 3CH_2Cl_2$
Z.Demidowicz, R.G.Teller, M.F.Hawthorne
J.Chem.Soc.,Chem.Commun.,831, 1979
See also R1 : 62

86.67 cis,trans – (Dicarbonyl – bis(trimethylphosphite) – phenanthroline) – manganese perchlorate
$C_{20}H_{26}MnN_2O_8P_2^+, ClO_4^-$
J.Fayos, M.Ulibarri *Eur.Cryst.Meeting*,6, 66, 1980
See also R1 : 83

86.C bis(μ – Acetato) – bis(chloro – (dimethylphenylphosphine) – palladium(ii)) chloroform solvate
$C_{20}H_{28}Cl_2O_4P_2Pd_2, 0.5CHCl_3$ Main entry is 81.67

86.68 trans – Dichloro – bis(o – phenylene – bis(dimethylarsine)) – technetium(iii) chloride
$C_{20}H_{32}As_4Cl_2Tc^+, Cl^-$
R.C.Elder, R.Whittle, K.A.Glavan, J.F.Johnson, E.Deutsch
Acta Crystallogr.,Sect.B,36, 1662, 1980

86.69 trans – Dichloro – bis(o – phenylene –
bis(dimethylarsine)) – technetium(iii) perchlorate
$C_{20}H_{32}As_4Cl_2Tc^+$, ClO_4^-
R.C.Elder, R.Whittle, K.A.Glavan, J.F.Johnson, E.Deutsch
Acta Crystallogr.,Sect.B,**36**, 1662, 1980

86.70 Tetrachloro – bis(o – phenylene –
bis(dimethylarsine)) – technetium(v)
hexafluorophosphate
$C_{20}H_{32}As_4Cl_4Tc^+$, F_6P^-
K.A.Glavan, R.Whittle, J.F.Johnson, R.C.Elder, E.Deutsch
J.Am.Chem.Soc.,**102**, 2103, 1980

86.71 (2,3,4,6 – Tetra – O – acetyl – 1 – thio – β – D –
glucopyranosato – S) – (triethylphosphine) – gold
Auranofin
$C_{20}H_{34}AuO_9PS$
D.T.Hill, B.M.Sutton *Cryst.Struct.Commun.*,**9**, 679, 1980
See also R1 : 85

86.72 bis(Hexamethyl – trisila – tetraphospha –
nortricyclene – tetracarbonyl – chromium)
$C_{20}H_{36}Cr_2O_8P_8Si_6$
W.Honle, H.G.von Schnering
Z.Anorg.Allg.Chem.,**465**, 72, 1980
See also R1 : 63

86.C bis(Triethylphosphine) – (1,1,1 – trifluoro – 5 –
methoxy – 3 – (trifluoromethyl) – 4 – methyl –
pent – 3 – en – 2 – yl) – platinum
hexafluorophosphate
$C_{20}H_{39}F_6OP_2Pt^+$, F_6P^- Main entry is 71.162

86.C bis(μ – Chloro) – bis(bis(carbonyl) –
trimethylphosphine – (1 – 2 – η – trimethylsilyl –
methylcarbonyl) – molybdenum(ii))
$C_{20}H_{40}Cl_2Mo_2O_8P_2Si_2$ Main entry is 71.163

86.73 trans – Chloro – bis(dimethylglyoximato) –
tributylphosphine – cobalt(iii)
$C_{20}H_{41}ClCoN_4O_4P$
N.Bresciani-Pahor, M.Calligaris, L.Randaccio
Inorg.Chim.Acta,**39**, 173, 1980
See also R1 : 83

86.C (1,2 – bis(Methoxycarbonyl) – but – 2 – en – 1 –
yl) – bis(triethylphosphine) – platinum(ii)
tetraphenylborate
$C_{20}H_{41}O_4P_2Pt^+$, $C_{24}H_{20}B^-$ Main entry is 71.164

86.C (1,2 – bis(Methoxycarbonyl) – but – 2 – en – 1 –
yl) – bis(triethylphosphine) – platinum(ii)
hexafluorophosphate
$C_{20}H_{41}O_4P_2Pt^+$, F_6P^- Main entry is 71.165

86.C Hydrido – (η^5 – cyclo – octadiene) – bis(1,2 –
bis(dimethylphosphino)ethane) – zirconium
$C_{20}H_{44}P_4Zr$ Main entry is 75.27

86.C bis(μ – Methylene) – bis(tris(trimethylphosphine) –
ruthenium(iii)) tetrafluoroborate
$C_{20}H_{58}P_6Ru_2^{2+}$, $2BF_4^-$ Main entry is 71.166

86.C Chloro – tris(cyclopentadienyl – dicarbonyl –
manganese) – antimony (at 190°K)
$C_{21}H_{15}ClMn_3O_6Sb$ Main entry is 73.89

86.74 Pentacarbonyl – (bis(1 – methyl – 3 – oxo – 1 –
butenyloxy)(phenyl) – phosphine) – chromium
$C_{21}H_{19}CrO_9P$
J.von Seyerl, D.Neugebauer, G.Huttner, C.Kruger,
Y.-H.Tsay *Chem.Ber.*,**112**, 3637, 1979

86.75 Carbonyl – chloro – (o – (diphenylphosphino) –
N,N – dimethylaniline) – iridium
$C_{21}H_{20}ClIrNOP$
D.M.Roundhill, R.A.Bechtold, S.G.N.Roundhill
Inorg.Chem.,**19**, 284, 1980
See also R1 : 83

86.C 3,3,3,3 – Tetracarbonyl – 5,5 – dimethyl – 2,2 –
diphenyl – 1 – oxa – 2 – phospha – 3 –
rhenacyclohexane
$C_{21}H_{20}O_5PRe$ Main entry is 71.172

86.C (bis(Trimethylphosphine) – platinum) – (μ –
carbonyl) – (μ – p – tolylcarbenyl) – (carbonyl –
cyclopentadienyl – manganese) tetrafluoroborate
dichloromethane solvate
$C_{21}H_{30}MnO_2P_2Pt^+$, BF_4^-, CH_2Cl_2 Main entry is 71.173

86.76 fac – Tricarbonyl – tris(triethylphosphine) –
chromium
$C_{21}H_{45}CrO_3P_3$
A.Holladay, M.R.Churchill, A.Wong, J.D.Atwood
Inorg.Chem.,**19**, 2195, 1980

86.C (1,2 – bis(Dimethylphosphino)ethane) –
(neopentylidyne) – (neopentylidene) – (neopentyl) –
tungsten(vi)
$C_{21}H_{46}P_2W$ Main entry is 71.174

86.C tris(μ – Methylene) – bis(tris(trimethylphosphine) –
ruthenium(ii))
$C_{21}H_{60}P_6Ru_2$ Main entry is 71.175

86.C (μ – Methyl) – bis(μ – methylene) –
bis(tris(trimethylphosphine) – ruthenium(ii))
tetrafluoroborate
$C_{21}H_{61}P_6Ru_2^+$, BF_4^- Main entry is 71.176

86.77 bis((μ^2 – Carbonyl) – (μ^4 – phenylphosphido)) –
tetrakis(dicarbonyl – cobalt) (monoclinic form)
$C_{22}H_{10}Co_4O_{10}P_2$
R.C.Ryan, C.U.Pittman Junior, J.P.O'Connor, L.F.Dahl
J.Organomet.Chem.,**193**, 247, 1980

86.78 (Triphenylphosphine) – tetracarbonyl – iron
(at –35°C)
$C_{22}H_{15}FeO_4P$
P.E.Riley, R.E.Davis *Inorg.Chem.*,**19**, 159, 1980

86.79 Tetraphenylphosphonium triphenylphosphino –
tetracarbonyl – manganese (at –35°C)
$C_{22}H_{15}MnO_4P^-$, $C_{24}H_{20}P^+$
P.E.Riley, R.E.Davis *Inorg.Chem.*,**19**, 159, 1980
See also R2 : 64

86.C Dicarbonyl – triethylphosphine – phenanthrene –
chromium
$C_{22}H_{25}CrO_2P$ Main entry is 74.10

86.80 1,1,1,2,2,2,3,3 – Octacarbonyl –
(dimethylphenylphosphine) – bis(μ_3 – ethoxy) –
1,2 – (μ – ethoxy) – tri – manganese
$C_{22}H_{28}Mn_3O_{11}P$
E.W.Abel, I.D.H.Towle, T.S.Cameron, R.E.Cordes
J.Chem.Soc.,Dalton Trans.,1943, 1979
See also R1 : 84

86.C bis(μ – Carbonyl) – (μ – carbonyl –
(cyclopentadienyl) – bis(dimethylarsenido) – iron) –
cyclopentadienyl – (cyclopentadienyl – cobalt) –
iron
$C_{22}H_{27}As_2CoFe_2O_3$ Main entry is 73.93

86.C (μ^2 – Dimethylarsonio) – bis(dicarbonyl – cyclopentadienyl – trimethylphosphine) – molybdenum tricarbonyl – cyclopentadienyl – molybdenum
$C_{22}H_{34}AsMo_2O_4P_2^+$, $C_8H_5MoO_3^-$ Main entry is 73.95

86.C (Tricarbonyl – trimethylphosphine – chromium) – (μ – carbonyl) – (μ – phenyl(methoxycarbonyl) methylene) – (bis(trimethylphosphine) – platinum)
$C_{22}H_{35}CrO_6P_3Pt$ Main entry is 71.185

86.C (Tetracarbonyl – trimethylphosphine – tungsten) – (μ – methoxytolylcarbene)) – bis(trimethylphosphine) – platinum
$C_{22}H_{37}O_5P_3PtW$ Main entry is 71.186

86.81 trans – bis(Dimethylglyoximato) – 2 – (5 – trifluoro – methyltetrazolato) – (tri – n – butylphosphine) – cobalt(iii)
$C_{22}H_{41}CoF_3N_8O_4P$
N.E.Takach, E.M.Holt, N.W.Alcock, R.A.Henry, J.H.Nelson
J.Am.Chem.Soc., **102**, 2968, 1980
See also R1 : 83

86.82 (μ^2 – 1,2 – bis(Carbomethoxyethylene)) – bis((μ^2 – methylthio) – trimethylphosphine – carbonyl – iron) tetraphenylborate (at –162°C)
$C_{22}H_{40}Fe_2O_6P_4S_2^+$, $C_{24}H_{20}B^-$
J.J.Bonnet, R.Mathieu, J.A.Ibers
Inorg.Chem., **19**, 2448, 1980
See also R1 : 85,84 R2 : 62

86.83 (Tricarbonyl – cobalt) – (μ – diarsenic) – (dicarbonyl – triphenylphosphine – cobalt)
$C_{23}H_{15}As_2Co_2O_5P$
A.S.Foust, C.F.Campana, J.D.Sinclair, L.F.Dahl
Inorg.Chem., **18**, 3047, 1979

86.84 (Tricarbonyl – cobalt) – (μ – diphosphorus) – (dicarbonyl – triphenylphosphine – cobalt)
$C_{23}H_{15}Co_2O_5P_3$
C.F.Campana, A.Vizi-Orosz, G.Palyi, L.Marko, L.F.Dahl
Inorg.Chem., **18**, 3054, 1979

86.C Tetraethylammonium η – allyl – dicarbonyl – dichloro – triphenylphosphine – tungsten
$C_{23}H_{20}Cl_2O_2PW^-$, $C_8H_{20}N^+$ Main entry is 72.42

86.85 Tetrabromocatecholato – nitroso – triphenylphosphine – iridium methylene chloride solvate
$C_{24}H_{15}Br_4IrNO_3P$, CH_2Cl_2
W.B.Shorthill, R.M.Buchanan, C.G.Pierpont, M.Ghedini, G.Dolcetti Inorg.Chem., **19**, 1803, 1980
See also R1 : 84,83

86.86 cis – Dichloro – bis(1 – phenyl – 3,4 – dimethylphosphole) – palladium(ii)
$C_{24}H_{26}Cl_2P_2Pd$
J.J.MacDougall, J.H.Nelson, F.Mathey, J.J.Mayerle
Inorg.Chem., **19**, 709, 1980

86.C (2 – aci – Nitrato – propan – 1 – on – 1 – yl) – bis(dimethylphenylphosphine) – pyridine – chloro – iridium acetone solvate
$C_{24}H_{30}ClIrN_2O_3P_2$, C_3H_6O Main entry is 71.192

86.C (μ – 1 – Diphenylphosphino – 2 – sulfido – 3,3 – dimethylbutene) – carbonyl – (methoxy – thiocarbonyl) – trimethylphosphite – iron
$C_{24}H_{32}FeO_5P_2S_2$ Main entry is 72.49

86.87 Dinitrosyl – tris(dimethylphenylphosphonito) – manganese(i) tetrafluoroborate
$C_{24}H_{33}MnN_2O_9P_3^+$, BF_4^-
M.Laing, R.H.Reimann, E.Singleton
Inorg.Chem., **18**, 2667, 1979

86.88 bis(Tri – t – butylphosphine) – platinum(0)
$C_{24}H_{54}P_2Pt$
K.J.Moynihan, C.Chieh, R.G.Goel
Acta Crystallogr.,Sect.B, **35**, 3060, 1979

86.89 trans – Dihydro – bis(tri – t – butylphosphine) – platinum(ii)
$C_{24}H_{56}P_2Pt$
G.Ferguson, P.Y.Siew, A.B.Goel
J.Chem.Res., **362**, 4337, 1979

86.C (μ^2 – Hydrido) – (μ^2 – p – tolylcarbamoyl) – dimethylphenylphosphine – nonacarbonyl – tri – osmium
$C_{25}H_{20}NO_{10}Os_3P$ Main entry is 71.195

86.90 Carbonyl – (triphenylphosphine) – (1 – (ethylthio) – maleonitrile – 2 – thiolato) – rhodium
$C_{25}H_{20}N_2OPRhS_2$
C.-H.Cheng, R.Eisenberg Inorg.Chem., **18**, 2438, 1979
See also R1 : 85

86.C (μ^2 – Diphenylphosphido) – (μ^3 – 3 – methylbutynyl) – nonacarbonyl – tri – ruthenium
$C_{26}H_{17}O_9PRu_3$ Main entry is 71.198

86.C (μ^2 – Diphenylphosphido) – (μ^3 – 3,3 – dimethylbutynyl) – octacarbonyl – tri – ruthenium benzene solvate
$C_{26}H_{19}O_8PRu_3$, $0.5C_6H_6$ Main entry is 71.199

86.C (1 – (o – Diphenylarsinophenyl) – 2 – methoxyethyl) – (hexafluoroacetylacetonato) – platinum(ii)
$C_{26}H_{21}AsF_6O_3Pt$ Main entry is 71.200

86.C (2,6 – Dimethoxyphenyl) – (triphenylphosphine) – gold(i)
$C_{26}H_{24}AuO_2P$ Main entry is 71.203

86.C η^5 – Cyclopentadienyl – triphenylphosphine – nickel – ethylxanthate (at –170°C)
$C_{26}H_{25}NiOPS_2$ Main entry is 73.107

86.C η^5 – Cyclopentadienyl – iodo – (methyl(methylthio) – carbene) – (triphenylphosphine) – iridium(iii) iodide
$C_{26}H_{26}IrPS^+$, I^- Main entry is 71.206

86.91 Triphenylphosphine – nitro – bis(dimethylglyoximato) – cobalt(iii)
$C_{26}H_{29}CoN_5O_6P$
A.I.Shkurpelo, Yu.A.Simonov, O.A.Bologa, T.I.Malinovskii
Dokl.Akad.Nauk SSSR, **248**, 1120, 1979
See also R1 : 83

86.C Cyanomethyl – tris(dimethylphenylphosphine) – platinum hexafluorophosphate
$C_{26}H_{35}NP_3Pt^+$, F_6P^- Main entry is 71.208

86.92 bis((μ^2 – t – Butylthio) – carbonyl – iodo – dimethylphenylphosphine – iridium(ii))
$C_{26}H_{40}I_2Ir_2O_2P_2S_2$
J.-J.Bonnet, P.Kalck, R.Poilblanc
Angew.Chem.,Int.Ed.Engl., **19**, 551, 1980
See also R1 : 85

86.C $(\mu$ – Diphenylphosphino) – $(\mu$ – N – methyl – (2 – phenyl – 2 – iminoethyl)) – bis(tricarbonyl – iron)
$C_{27}H_{20}Fe_2NO_6P$ Main entry is 71.211

86.C   Chloro – trifluoromethyl – (cis – 1,2 – bis(diphenylphosphino)ethylene) – platinum(ii)
$C_{27}H_{22}ClF_3P_2Pt$ Main entry is 71.213

86.93   Dinitrato – (trimesitylphosphine) – mercury(ii)
$C_{27}H_{33}HgN_2O_6P$
E.C.Alyea, S.A.Dias, G.Ferguson, M.Parvez
Inorg.Chim.Acta, **37**, 45, 1979

86.C   bis((μ^2 – t – Butylthiolato) – (dimethylphenylphosphino)) – acetyl – carbonyl – iodo – di – rhodium
$C_{27}H_{43}IO_2P_2Rh_2S_2$ Main entry is 71.217

86.C   bis(Diphenylphosphino)methane – tricarbonyl – di – iodo – tungsten
$C_{28}H_{22}I_2O_3P_2W$ Main entry is 71.218

86.C   bis(Diphenylphosphinoacetato) – palladium(ii)
$C_{28}H_{24}O_4P_2Pd$ Main entry is 81.85

86.C   Tricarbonyl – (dicarbonyl(tricarbonyl(cyclopentadienyl) – chromium) – cyclopentadienyl – (μ – dimethylarsenido) – chromium) – bis(μ – dimethylarsenido) – (tetracarbonyl – iron) – cobalt
$C_{28}H_{28}As_3CoCr_2FeO_{12}$ Main entry is 73.111

86.94   bis(2 – (Diphenylphosphino)ethane – 1 – thiolato) – oxo – molybdenum(iv)
$C_{28}H_{28}MoOP_2S_2$
J.Hyde, L.Magin, P.Vella, J.Zubieta
Stereodyn.Mol.Syst.Proc.Symp.,227, 1979
See also R1 : 85

86.C   Dinitrato – (η^5 – pentamethyl – cyclopentadienyl) – triphenylphosphine – rhodium(iii)
$C_{28}H_{30}N_2O_6PRh$ Main entry is 73.1

86.C   Dicarbonyl – methyl – bis(trimethylphosphine) – (triphenylborato – nitrilo – methyl) – iron(ii)
$C_{28}H_{36}BFeNO_2P_2$ Main entry is 71.220

86.95   Chloro – (diphenylphosphino – dicarba – dodecaborane) – (diphenylphosphino – dicarba – dodecaboranyl) – platinum
$C_{28}H_{41}B_{20}ClP_2Pt$
L.Manojlovic–Muir, K.W.Muir, T.Solomun
J.Chem.Soc.,Dalton Trans.,317, 1980
See also R1 : 62

86.96   Dioxygen – bis(di(t – butyl) – phenylphosphine) – palladium(ii) toluene solvate
$C_{28}H_{46}O_2P_2Pd$, C_7H_8
T.Yoshida, K.Tatsumi, M.Matsumoto, K.Nakatsu, A.Nakamura, T.Fueno, S.Otsuka
Nouv.J.Chim.,3, 761, 1979

86.97   Dioxygen – bis(di(t – butyl) – phenylphosphine) – platinum(ii) toluene solvate
$C_{28}H_{46}O_2P_2Pt$, C_7H_8
T.Yoshida, K.Tatsumi, M.Matsumoto, K.Nakatsu, A.Nakamura, T.Fueno, S.Otsuka
Nouv.J.Chim.,3, 761, 1979

86.C   bis(N – t – Butyldithiocarbamato) – (tricyclohexylphosphine) – platinum(ii) cyclohexane solvate
$C_{28}H_{53}N_2PPtS_4$, C_6H_{12} Main entry is 80.24

86.98   bis(μ – Bromo) – bis(dicarbonyl – (tri – t – butylphosphine) – ruthenium(i))
$C_{28}H_{54}Br_2O_4P_2Ru_2$
H.Schumann, J.Opitz, J.Pickardt
Chem.Ber.,113, 1385, 1980

86.C   bis(bis(μ^2 – Methylene) – tetrakis(trimethylphosphine) – ruthenium(iii)) – ruthenium(iv) bis(tetrafluoroborate)
$C_{28}H_{80}P_8Ru_3^{2+}$, $2BF_4^-$ Main entry is 71.222

86.99   Undecacarbonyl – (triphenylphosphine) – triangulo – tri – ruthenium
$C_{29}H_{15}O_{11}PRu_3$
E.J.Forbes, N.Goodhand, D.L.Jones, T.A.Hamor
J.Organomet.Chem.,**182**, 143, 1979

86.C   Dicarbonyl – (η^4 – cinnamaldehyde) – (triphenylphosphine) – iron(0)
$C_{29}H_{23}FeO_3P$ Main entry is 72.52

86.C   1,2 – bis(Diphenylphosphino)ethane – tricarbonyl – di – iodo – molybdenum dichloromethane solvate
$C_{29}H_{24}I_2MoO_3P_2$, CH_2Cl_2 Main entry is 71.224

86.C   cis – Dichloro – η^2 – ((Z) – 2 – chloro – N – (3 – methyl – but – 1 – enyl) – benzenamine) – triphenylphosphine – platinum(ii)
$C_{29}H_{29}Cl_3NPPt$ Main entry is 72.53

86.C   (η^7 – Cycloheptatrienyl) – carbonyl – iodo – (α – methylbenzyl – N – methylamino – (diphenyl) – phosphine) – molybdenum
$C_{29}H_{29}IMoNOP$ Main entry is 75.36

86.C   (bis(Diphenylphosphino)methanide) – (dimethylphosphonium – bis(methylide)) – platinum(ii)
$C_{29}H_{31}P_3Pt$ Main entry is 71.225

86.C   Chloro – bis(triethylphosphine) – ((N – p – tolyl) – (N' – (p – tolyl – o – yl)) – imidazolidin – 2 – ylidene) – ruthenium
(1,3 – Bis(4 – tolyl) – imidazolidin – 2 – ylidene) – chloro – bis(triethylphosphine) – ruthenium(ii)
$C_{29}H_{47}ClN_2P_2Ru$ Main entry is 71.226

86.C   Hexamethyl – bis(diethylphenylphosphine) – tris(μ – methyl) – triangulo – tri – rhenium(iii)
$C_{29}H_{57}P_2Re_3$ Main entry is 71.227

86.100   Tricarbonyl – chloro – (5,7 – dichloro – 8 – quinolinolato – N,O) – triphenylphosphine – tungsten(ii)
$C_{30}H_{19}Cl_3NO_4PW$
R.O.Day, W.H.Batschelet, R.D.Archer
Inorg.Chem.,19, 2113, 1980
See also R1 : 84,83

86.101   Sodium bis(μ – diphenylphosphido) – bis(tricarbonyl – iron) 1,10 – diaza – 4,7,13,16,21,24 – hexaoxabicyclo(8.8.8)hexacosane
$C_{30}H_{20}Fe_2O_6P_2^{2-}$, $2C_{18}H_{36}N_2O_6$, $2Na^+$
R.E.Ginsburg, R.K.Rothrock, R.G.Finke, J.P.Collman, L.F.Dahl *J.Am.Chem.Soc.*,101, 6550, 1979
See also R2 : 40

86.C   Dicarbonyl – naphthalene – triphenylphosphite – chromium
$C_{30}H_{23}CrO_5P$ Main entry is 74.13

86.C (η^6 – Tetralone) – carbonyl – thiocarbonyl –
(triphenylphosphine) – chromium (absolute
configuration)
$C_{30}H_{25}CrO_2PS$ Main entry is 74.14

86.C 1,3 – bis(Diphenylphosphino) – propane –
tricarbonyl – di – iodo – molybdenum
$C_{30}H_{28}I_2MoO_3P_2$ Main entry is 71.229

86.C Carbonyl – cyclopentadienyl – (1,5α – η^3 – 1 –
methylene – cyclopentan – 2 – onato) –
triphenylphosphine – molybdenum(ii)
$C_{30}H_{27}MoO_2P$ Main entry is 72.54

86.102 (+)$_{589}$ – (S – Dimethyl – (α – methylbenzyl) –
aminato) – (SS – o – phenylene –
bis(methylphenylarsine)) – palladium(ii)
hexafluorophosphate (absolute configuration)
(+)$_{589}$ – (S – 2 – (1′ – Dimethylaminoethyl) – phenyl –
C^{1}N) – (SS – o – phenylene – bis(methylphenylarsine)) –
palladium(ii) hexafluorophosphate
$C_{30}H_{34}As_2NPd^+$, F_6P^-
B.W.Skelton, A.H.White
J.Chem.Soc.,Dalton Trans.,1556, 1980
See also R1 : 83

86.C Carbonyl – chloro – bis(triethylphosphine) – ((N –
(p – tolyl – o – yl)) – imidazolidin – 2 – ylidene) –
ruthenium(ii)
$C_{30}H_{47}ClN_2OP_2Ru$ Main entry is 71.231

86.103 (μ – Chloro) – (μ – t – butylthio) – bis(carbonyl –
(tri – t – butylphosphine) – rhodium)
$C_{30}H_{63}ClO_2P_2Rh_2S$
H.Schumann, G.Cielusek, J.Pickardt
Angew.Chem.,Int.Ed.Engl.,19, 70, 1980
See also R1 : 85

86.104 Phenyl – triethylphosphine – platinum – bis(μ^2 –
hydrido) – tris(triethylphosphine) – iridium –
hydride tetraphenylborate
$C_{30}H_{68}IrP_4Pt^+$, $C_{24}H_{20}B^-$
A.Immirzi, A.Musco, P.S.Pregosin, L.M.Venanzi
Angew.Chem.,Int.Ed.Engl.,19, 721, 1980

86.C bis(Triphenylphosphine)immonium (tricarbonyl –
iron) – bis(μ – diphenylphosphido) – (acetyl –
dicarbonyl – iron)
$C_{31}H_{23}Fe_2O_6P_2^-$, $C_{36}H_{30}NP_2^+$ Main entry is 71.232

86.C Sodium (tricarbonyl – iron) – bis(μ –
diphenylphosphido) – (acetyl – dicarbonyl – iron)
tetrahydrofuran solvate
$C_{31}H_{23}Fe_2O_6P_2^-$, Na^+, $2C_4H_8O$ Main entry is 71.233

86.105 trans – Carbonyl – chloro – bis(tris(2 –
pyridylphosphine)) – rhodium
$C_{31}H_{24}ClN_6OP_2Rh$
K.Wajda, F.Pruchnik, T.Lis *Inorg.Chim.Acta*,40, 207, 1980

86.C 2 – Hydroxy – N – salicylidene – aniline –
(triphenylphosphine) platinum(ii)
$C_{31}H_{24}NO_2PPt$ Main entry is 78.14

86.106 bis(μ – Hydrido) – (μ – 1,2 – bis(diphenylphosphino)
methane) – bis(tricarbonyl – rhenium)
$C_{31}H_{24}O_6P_2Re_2$
M.J.Mays, D.W.Prest, P.R.Raithby
J.Chem.Soc.,Chem.Commun.,171, 1980

86.C Carbonyl – (phenylthiocarbyne) – (η^5 –
cyclopentadienyl) – triphenylphosphine – tungsten
$C_{31}H_{25}OPSW$ Main entry is 71.2

86.C 1 – Triphenylphosphine – 1 – chloro – 4,5 – benzo –
3 – (3′ – phenyl – 5′ – ethylidenyl – 1′,2′ – dithiol –
3′ – ene) – 1 – pallada – 2 – thiol – 2 – ene toluene
solvate
$C_{31}H_{26}ClPPdS_3$, $2C_7H_8$ Main entry is 71.235

86.C Carbonyl – cyclobutadiene – (1,2 –
bis(diphenylphosphino)ethane) – iron (at –35°C)
$C_{31}H_{28}FeOP_2$ Main entry is 75.37

86.C (1,2 – bis(Diphenylphosphino)ethane) – trichloro –
(η – cyclopentadienyl) – niobium(iv) toluene solvate
$C_{31}H_{29}Cl_3NbP_2$, $2C_7H_8$ Main entry is 73.117

86.107 Iodo – (1,9 – bis(diphenylphosphino) – 3,7 –
dithianonane) – nickel(ii) tetraphenylborate
$C_{31}H_{34}INiP_2S_2^+$, $C_{24}H_{20}B^-$
K.Aurivillius, G.–I.Bertinsson
Acta Crystallogr.,Sect.B,36, 790, 1980
See also R1 : 85 R2 : 62

86.108 (1,9 – bis(Diphenylphosphino) – 3,7 –
dithianonane) – nickel(ii) diperchlorate
$C_{31}H_{34}NiP_2S_2^{2+}$, $2ClO_4^-$
K.Aurivillius, G.–I.Bertinsson
Eur.Cryst.Meeting,5, 191, 1979
See also R1 : 85

86.109 Chloro – triethylphosphine –
bis(diphenylphosphino) – methano – platinum(ii)
$C_{31}H_{36}ClP_3Pt$
J.Browning, G.W.Bushnell, K.R.Dixon
J.Organomet.Chem.,198, 11, 1980

86.110 Dichloro – bis(2,2′ – bis(dimethylarsine) –
biphenyl) – nickel(ii)
$C_{32}H_{40}As_4Cl_2Ni$
D.W.Allen, D.A.Kennedy, I.W.Nowell
Inorg.Chim.Acta,40, 171, 1980

86.C tris(Dimethylphenylphosphine) – (o –
dimethylphosphinophenyl) – hydrido – iridium
hexafluorophosphate
$C_{32}H_{44}IrP_4^+$, F_6P^- Main entry is 71.237

86.C (μ – Methoxy – phenyl – carbene) – (μ –
tetracarbonyl) – tungsten – bis(carbonyl –
(methyl – di – t – butylphosphine) – platinum)
$C_{32}H_{50}O_7P_2Pt_2W$ Main entry is 71.238

86.111 Carbonyl – platinum – bis((μ –
diphenylphosphido) – (tetracarbonyl – manganese))
$C_{33}H_{20}Mn_2O_9P_2Pt$
P.Braunstein, D.Matt, O.Bars, D.Grandjean
Angew.Chem.,Int.Ed.Engl.,18, 797, 1979

86.112 (μ – Ethylideneimino) – (μ – hydrido) – (μ –
bis(diphenylphosphino)methane) – bis(tricarbonyl –
rhenium)
$C_{33}H_{27}NO_6P_2Re_2$
M.J.Mays, D.W.Prest, P.R.Raithby
J.Chem.Soc.,Chem.Commun.,171, 1980
See also R1 : 83

86.C Tricarbonyl – (1,6 – bis(diphenylphosphino) –
trans – hex – 3 – ene) – molybdenum(0)
$C_{33}H_{30}MoO_3P_2$ Main entry is 72.55

86.C (($\pm$) – α – (2 – Diphenylphosphino – ferrocenyl)
ethyl – dimethylamine) – norbornadiene – rhodium
hexafluorophosphate
$C_{33}H_{36}FeNPRh^+$, F_6P^- Main entry is 73.120

86.C (2 – Benzoylphenyl) – tricarbonyl –
(triphenylphosphine) – rhenium
$C_{34}H_{24}O_4PRe$ Main entry is 71.239

86.113 Tetraethylammonium (μ^2 – hydrido) –
bis(tetracarbonyl – (methyldiphenylphosphine) –
molybdenum)
$C_{34}H_{27}Mo_2O_8P_2{}^-$, $C_8H_{20}N^+$
M.Y.Darensbourg, J.L.Atwood, W.E.Hunter,
R.R.Burch Junior $J.Am.Chem.Soc.$, 102, 3290, 1980
See also R2 : 3

86.C Benzoyl – dichloro – (1,3 – bis(diphenylphosphino) –
propane) – rhodium
$C_{34}H_{31}Cl_2OP_2Rh$ Main entry is 71.240

86.C 3,7,9,13 – Tetramethyl – 6 – (bis(diphenylphosphino)
ethane) – 3,7,9,13 – tetracarba – 6 – nickela –
tridecaborane
$C_{34}H_{44}B_8NiP_2$ Main entry is 71.242

86.114 (μ_2 – N,N' – Di – t – butyl – ethylenedi – imine) –
bis(trans – dichloro – tributylphosphine –
platinum)
$C_{34}H_{74}Cl_4N_2P_2Pt_2$
H.Van der Poel, G.Van Koten, K.Vrieze, M.Kokkes, C.H.Stam
$Inorg.Chim.Acta$, 39, 197, 1980
See also R1 : 83

86.C (η^5 – Cyclopentadienyl) – (2 – methyl – 4,5 –
bis(diphenylphosphino) – pent – 2 – en – 3 – yl) –
iron(ii)
$C_{35}H_{34}FeP_2$ Main entry is 71.243

86.115 bis(2 – Diphenylphosphinoethyl) –
phenylphosphine – nickel – 0 – methylsulfinate
tetraphenylborate
$C_{35}H_{36}NiO_2P_3S^+$, $C_{24}H_{20}B^-$
C.Mealli, M.Peruzzini, P.Stoppioni
$J.Organomet.Chem.$, 192, 437, 1980
See also R2 : 62

86.C Lithium (tricarbonyl – iron) – bis(μ –
diphenylphosphido) – (benzoyl – dicarbonyl – iron)
tetrahydrofuran solvate
$C_{36}H_{25}Fe_2O_6P_2{}^-$, Li$^+$, 3$C_4H_8O$ Main entry is 71.245

86.116 tetrakis(μ – Hydrido) – (μ – 1,2 –
bis(diphenylphosphino)ethane) – decacarbonyl –
tetra – ruthenium
$C_{36}H_{28}O_{10}P_2Ru_4$
M.R.Churchill, R.A.Lashewycz, J.R.Shapley, S.I.Richter
$Inorg.Chem.$, 19, 1277, 1980

86.117 Dichloro – bis(triphenylarsine) – palladium(ii)
$C_{36}H_{30}As_2Cl_2Pd$
S.T.Malinovskii, I.F.Bourshteyn, T.I.Malinovskii
$Izv.Akad.Nauk Mold.SSR$, 45, 1979

86.118 (μ^2 – Chloro) – bis(triphenylphosphine) – di – gold(i)
perchlorate methylene chloride solvate
$C_{36}H_{30}Au_2ClP_2{}^+$, $ClO_4{}^-$, CH_2Cl_2
P.G.Jones, G.M.Sheldrick, R.Uson, A.Laguna
$Acta Crystallogr.,Sect.B$, 36, 1486, 1980

86.119 cis – bis(Thionylimide – 0) –
bis(triphenylphosphine) – platinum(ii) hemihydrate
$C_{36}H_{32}N_2O_2P_2PtS_2$, $0.5H_2O$
A.A.Bhattacharyya, A.G.Turner, E.M.Holt, N.W.Alcock
$Inorg.Chim.Acta$, 44, 185, 1980
See also R1 : 84

86.120 Chloro – (bis(3 – diphenylphosphinopropyl)
phenylphosphine) – (sulfur dioxide) – rhodium(i)
$C_{36}H_{37}ClO_2P_3RhS$
P.G.Eller, R.R.Ryan $Inorg.Chem.$, 19, 142, 1980

86.C Cyclopentadienyl – iodo – (0 – isopropylidene –
2,3 – dihydroxy – 1,4 – bis(diphenylphosphino) –
butane) – iron
$C_{36}H_{37}FeIO_2P_2$ Main entry is 73.123

86.121 9,9 – bis(Triphenylphosphine) – 6 – thia – 9 –
platina – decaborane
$C_{36}H_{40}B_8P_2PtS$
T.K.Hilty, D.A.Thompson, W.M.Butler, R.W.Rudolph
$Inorg.Chem.$, 18, 2642, 1979

86.122 (α – 4,7,13,16 – Tetraphenyl – 1,10 – dioxa –
4,7,13,16 – tetraphosphacyclo – octadecane) –
cobalt(ii) bis(tetraphenylborate)
$C_{36}H_{44}CoO_2P_4{}^{2+}$, $2C_{24}H_{20}B^-$
M.Ciampolini, P.Dapporto, N.Nardi, F.Zanobini
$J.Chem.Soc.,Chem.Commun.$, 177, 1980
See also R1 : 84 R2 : 62

86.123 (β – 4,7,13,16 – Tetraphenyl – 1,10 – dioxa –
4,7,13,16 – tetraphosphacyclo – octadecane) –
cobalt(ii) bis(tetraphenylborate)
dimethylformamide solvate
$C_{36}H_{44}CoO_2P_4{}^{2+}$, $2C_{24}H_{20}B^-$, $2C_3H_7NO$
M.Ciampolini, P.Dapporto, N.Nardi, F.Zanobini
$J.Chem.Soc.,Chem.Commun.$, 177, 1980
See also R1 : 84 R2 : 62

86.124 bis(Diazo – (diethylphosphine – ethylene –
diphenylphosphine)) – molybdenum
$C_{36}H_{48}MoN_4P_4$
Z.–S.Shia, C.–C.Ni, J.–L.Ma, C.–M.Chang
$Ko Hsueh Tung Pao$, 25, 23, 1980

86.125 bis((1,2 – bis(Dimethylphosphino)ethane) –
(diphenylphosphido)) – hydrido – tantalum
$C_{36}H_{53}P_6Ta$
P.J.Domaille, B.M.Foxman, T.J.McNeese, S.S.Wreford
$J.Am.Chem.Soc.$, 102, 4114, 1980

86.126 tris(Phenyl – di – isopropylphosphine) – tungsten –
hexahydride
$C_{36}H_{63}P_3W$
D.Gregson, J.A.K.Howard, J.N.Nicholls, J.L.Spencer,
D.G.Turner $J.Chem.Soc.,Chem.Commun.$, 572, 1980

86.127 bis(Tricyclohexylphosphine) – gold(i) thiocyanate
$C_{36}H_{66}AuP_2{}^+$, CNS$^-$
J.A.Muir, M.M.Muir, E.Lorca
$Acta Crystallogr.,Sect.B$, 36, 931, 1980

86.128 trans – Dichloro – bis(tricyclohexylphosphine) –
platinum(ii)
$C_{36}H_{66}Cl_2P_2Pt$
A.Del Pra, G.Zanotti $Inorg.Chim.Acta$, 39, 137, 1980

86.129 bis(Tricyclohexylphosphine) – mercury(ii)
diperchlorate
$C_{36}H_{66}HgP_2{}^{2+}$, $2ClO_4{}^-$
E.C.Alyea, S.A.Dias, G.Ferguson, M.A.Khan
$J.Chem.Res.$, 360, 4101, 1979

86.130 (μ – Dinitrogen) – bis(bis(tri – isopropylphosphine) – hydrido – rhodium(i)) (at –85°C)
$C_{36}H_{86}N_2P_4Rh_2$
T.Yoshida, T.Okano, D.L.Thorn, T.H.Tulip, S.Otsuka, J.A.Ibers *J.Organomet.Chem.*,**181**, 183, 1979

86.C trans – (Trifluoromethyl) – chloro – bis(triphenylphosphine) – platinum(ii)
$C_{37}H_{30}ClF_3P_2Pt$ Main entry is 71.248

86.131 trans – Carbonyl – chloro – bis(triphenylphosphine) – rhodium(i)
$C_{37}H_{30}ClOP_2Rh$
A.Del Pra, G.Zanotti, P.Segala
Cryst.Struct.Commun.,**8**, 959, 1979

86.C trans – (Methyl) – chloro – bis(triphenylphosphine) – platinum(ii)
$C_{37}H_{33}ClP_2Pt$ Main entry is 71.249

86.C (bis(2 – Diphenylphosphinoethyl) – phenylphosphine) – tricarbonyl – chromium(0)
$C_{37}H_{33}CrO_3P_3$ Main entry is 71.250

86.C (bis(2 – Diphenylphosphinoethyl) – phenylphosphine) – tricarbonyl – molybdenum(0)
$C_{37}H_{33}MoO_3P_3$ Main entry is 71.251

86.132 Carbonyl – (bis(3 – diphenylphosphinopropyl) phenylphosphine) – (sulfur dioxide) – rhodium(i) hexafluoroarsenate
$C_{37}H_{37}O_3P_3RhS^+$, AsF_6^-
P.G.Eller, R.R.Ryan *Inorg.Chem.*,**19**, 142, 1980

86.C (η – Cyclo – octa – 1,5 – diene) – 4 – diphenylphosphino – 2 – diphenylphosphino – methylpyrrolidine – rhodium(i) perchlorate
$C_{37}H_{41}NP_2Rh^+$, ClO_4^- Main entry is 75.42

86.133 Dicarbonyl – bis(5,7 – dichloro – 8 – quinolinolato – N,O) – triphenylphosphine – tungsten(ii) dichloromethane solvate
$C_{38}H_{23}Cl_4N_2O_4PW$, CH_2Cl_2
R.O.Day, W.H.Batschelet, R.D.Archer
Inorg.Chem.,**19**, 2113, 1980
See also R1 : 84,83

86.C Chloro – (2,2' – bis(o – diphenylphosphino) – trans – stilbene) – rhodium(i) dichloromethane solvate
$C_{38}H_{30}ClP_2Rh$, CH_2Cl_2 Main entry is 72.56

86.134 Carbonyl – dichloro – thiocarbonyl – bis(triphenylphosphine) – osmium
$C_{38}H_{30}Cl_2OOsP_2S$
G.R.Clark, K.Marsden, W.R.Roper, L.J.Wright
J.Am.Chem.Soc.,**102**, 1206, 1980

86.C Trichloro – (2,2' – bis(o – diphenylphosphino) – trans – stilbene) – iridium(iii)
$C_{38}H_{30}Cl_3IrP_2$ Main entry is 72.57

86.C Trichloro – (2,2' – bis(o – diphenylphosphino) – trans – stilbene) – rhodium(iii)
$C_{38}H_{30}Cl_3P_2Rh$ Main entry is 72.58

86.C Carbonyl – dichloro – dichlorocarbene – bis(triphenylphosphine) – osmium
$C_{38}H_{30}Cl_4OOsP_2$ Main entry is 71.252

86.135 Dicarbonyl – bis(triphenylphosphite) – (sulfur dioxide) – iron
$C_{38}H_{30}FeO_{10}P_2S$
P.Conway, S.M.Grant, A.R.Manning, F.S.Stephens
J.Organomet.Chem.,**186**, C61, 1980

86.136 bis(Carbonyl) – di – selenium – bis(triphenylphosphine) – osmium
$C_{38}H_{30}O_2OsP_2Se_2$
D.H.Farrar, K.R.Grundy, N.C.Payne, W.R.Roper, A.Walker
J.Am.Chem.Soc.,**101**, 6577, 1979

86.137 mer – Trichloro – (acetonitrile) – bis(triphenylphosphine) – osmium(iii)
$C_{38}H_{33}Cl_3NOsP_2$
R.L.Parkes, N.C.Payne, E.O.Sherman
Can.J.Chem.,**58**, 1042, 1980
See also R1 : 83

86.C trans – Cyanomethyl – hydrido – bis(triphenylphosphine) – platinum(ii)
$C_{38}H_{33}NP_2Pt$ Main entry is 71.254

86.138 Dichloro – bis(o – (diphenylphosphino)anisole) – ruthenium(ii) dichloromethane solvate
$C_{38}H_{34}Cl_2O_2P_2Ru$, CH_2Cl_2
J.C.Jeffery, T.B.Rauchfuss *Inorg.Chem.*,**18**, 2658, 1979
See also R1 : 84

86.C Acetylacetonato – (triphenylphosphine) – (Z – 1,2 – diphenyl – 2 – methylethenyl) – nickel
$C_{38}H_{35}NiO_2P$ Main entry is 71.255

86.139 trans – Di(methylsulfito) – bis(triphenylphosphine) – platinum(ii)
$C_{38}H_{36}O_6P_2PtS_2$
G.R.Hughes, P.C.Minshall, D.M.P.Mingos
Transition Met.Chem.,**4**, 147, 1979
See also R1 : 85

86.C (Methyl – (Z) – α – acetamido – cinnamate) – (1,2 – bis(diphenylphosphino)ethane) – rhodium(i) tetrafluoroborate
$C_{38}H_{37}NO_3P_2Rh^+$, BF_4^- Main entry is 72.59

86.140 3,3 – bis(Triphenylphosphine) – 3 – bisulfato – 3 – rhoda – 1,2 – dicarba – dodecaborane diethyl ether solvate (at –154°C)
$C_{38}H_{42}B_9O_4P_2RhS$, $C_4H_{10}O$
W.C.Kalb, R.G.Teller, M.F.Hawthorne
J.Am.Chem.Soc.,**101**, 5417, 1979
See also R1 : 62

86.141 8 – Ethoxy – 9,9 – bis(triphenylphosphine) – 6 – thia – 9 – platina – decaborane
$C_{38}H_{44}B_8OP_2PtS$
T.K.Hilty, D.A.Thompson, W.M.Butler, R.W.Rudolph
Inorg.Chem.,**18**, 2642, 1979

86.C trans – Dimesityl – bis(diethylphenylphosphine) – cobalt(ii)
$C_{38}H_{52}CoP_2$ Main entry is 71.256

86.C (μ_3 – Chloro) – tris(μ – oxo) – (μ – acetato) – (tetrachloro – tris(triethylphosphine) – tri – tungsten)
$C_{38}H_{84}Cl_5O_5P_3W_3$ Main entry is 81.91

86.C 1 – Phenyl – dibenzo(b,f) – (2,5 – diphenyl – ferra – cyclopentadiene – tricarbonyl) – (c) – phosphepin – iron – dicarbonyl
$C_{39}H_{23}Fe_2O_5P$ Main entry is 73.124

86.142 Dicarbonyl – thiocarbonyl –
bis(triphenylphosphite) – iron(0)
$C_{39}H_{30}FeO_8P_2S$
P.Conway, A.R.Manning, F.S.Stephens
J.Organomet.Chem.,**186**, C64, 1980

86.C (Dicarbonyl – dimethylphenylphosphine – cobalt) –
μ – carbonyl – μ – (N,N' – diphenyl – N –
(phenylcarbenyl) – benzamidine) – (dicarbonyl –
cobalt)
$C_{39}H_{31}Co_2N_2O_5P$ Main entry is 71.257

86.C Carbonyl – iodo – bis(triphenylphosphine) – acetyl –
ruthenium
$C_{39}H_{33}IO_2P_2Ru$ Main entry is 71.258

86.143 (μ – Diarsenic) – bis(dicarbonyl –
triphenylphosphine – cobalt)
$C_{40}H_{30}As_2Co_2O_4P_2$
A.S.Foust, C.F.Campana, J.D.Sinclair, L.F.Dahl
Inorg.Chem., **18**, 3047, 1979

86.144 cis – Dichloro – bis(diphenyl(phenylethynyl)
phosphine) – platinum acetonitrile solvate
$C_{40}H_{30}Cl_2P_2Pt$, $2C_2H_3N$
A.J.Carty, N.J.Taylor, D.K.Johnson
J.Am.Chem.Soc.,**101**, 5422, 1979

86.C (η^2 – Dithiomethoxycarbonyl) – dicarbonyl –
bis(triphenylphosphine) – ruthenium(ii) perchlorate
cyclohexane solvate
$C_{40}H_{33}O_2P_2RuS_2^+$, ClO_4^-, C_6H_{12} Main entry is 71.260

86.145 meso – bis(RR,SS o – Phenylene –
bis(methylphenylarsine)) – palladium(ii) dichloride
ethane – 1,2 – diol solvate
$C_{40}H_{40}As_4Pd^{2+}$, $2Cl^-$, $2C_2H_6O_2$
B.W.Skelton, A.H.White
J.Chem.Soc.,Dalton Trans.,1556, 1980

86.146 meso – bis(RR,SS o – Phenylene –
bis(methylphenylarsine)) – palladium(ii) di – iodide
$C_{40}H_{40}As_4Pd^{2+}$, $2I^-$
B.W.Skelton, A.H.White
J.Chem.Soc.,Dalton Trans.,1556, 1980

86.C Diacetato – bis(tricyclohexylphosphine) –
mercury(ii) dihydrate
$C_{40}H_{72}HgO_4P_2$, $2H_2O$ Main entry is 81.93

86.147 bis(μ – Hydrido) – bis(dimethylsilyl) –
bis(tricyclohexylphosphine) – di – platinum
$C_{40}H_{80}P_2Pt_2Si_2$
M.Auburn, M.Ciriano, J.A.K.Howard, M.Murray, N.J.Pugh,
J.L.Spencer, F.G.A.Stone, P.Woodward
J.Chem.Soc.,Dalton Trans.,659, 1980
See also R1 :

86.C 1 – Phenyl – dibenzo(b,f) – 2,5 – diphenyl –
cyclopentadienone(c)phosphepin – di – iron –
hexacarbonyl
$C_{41}H_{23}Fe_2O_7P$ Main entry is 73.128

86.148 bis(Triphenylphosphine) – (tetrachlorodiazo –
cyclopentadiene) – chloro – iridium(i) toluene
solvate
$C_{41}H_{30}Cl_5IrN_2P_2$, C_7H_8
K.D.Schramm, J.A.Ibers *Inorg.Chem.*,**19**, 1231, 1980
See also R1 : 83

86.149 1,1,1 – tris((Diphenylphosphino)methyl)ethane –
cobalt – (η^3 – cyclotriphosphorus)
$C_{41}H_{39}CoP_6$
C.A.Ghilardi, S.Midollini, A.Orlandini, L.Sacconi
Inorg.Chem.,**19**, 301, 1980

86.150 (1,1,1 – tris(Diphenylphosphinomethyl)ethane –
cyclotriphosphorus – iridium
$C_{41}H_{39}IrP_6$
C.Bianchini, C.Mealli, A.Meli, L.Sacconi
Inorg.Chim.Acta,**37**, L543, 1979

86.151 (1,1,1 – tris(Diphenylphosphinomethyl)ethane –
cyclotriphosphorus – rhodium
$C_{41}H_{39}P_6Rh$
C.Bianchini, C.Mealli, A.Meli, L.Sacconi
Inorg.Chim.Acta,**37**, L543, 1979

86.152 tris(Acetonitrile) – nitrosyl –
bis(triphenylphosphine) – iridium(iii)
bis(hexafluorophosphate)
$C_{42}H_{39}IrN_4OP_2^{2+}$, $2F_6P^-$
M.Lanfranchi, A.Tiripicchio, G.Dolcetti, M.Ghedini
Transition Met.Chem.,**5**, 21, 1980
See also R1 : 83

86.153 cis,cis,cis – bis(Dimethylphenylphosphine) –
bis(monothiobenzoato) – (1,10 – phenanthroline) –
ruthenium(ii)
$C_{42}H_{40}N_2O_2P_2RuS_2$
R.O.Gould, T.A.Stephenson, M.A.Thomson
J.Chem.Soc.,Dalton Trans.,804, 1980
See also R1 : 85,83

86.154 cis,cis,trans – bis(Dimethylphenylphosphine) –
bis(monothiobenzoato) – (1,10 – phenanthroline) –
ruthenium(ii)
$C_{42}H_{40}N_2O_2P_2RuS_2$
R.O.Gould, T.A.Stephenson, M.A.Thomson
J.Chem.Soc.,Dalton Trans.,804, 1980
See also R1 : 85,83

86.C (Ethyl – methacrylate) – bis(triphenylphosphine) –
nickel(0)
$C_{42}H_{40}NiO_2P_2$ Main entry is 72.60

86.C bis(η^3 – Allyl – triphenylphosphine – palladium)
$C_{42}H_{40}P_2Pd_2$ Main entry is 72.61

86.155 bis(Dibenzyl – cis – benzyl – 2 – yl – phosphine) –
platinum(ii)
$C_{42}H_{40}P_2Pt$
W.Porzio *Inorg.Chim.Acta*,**40**, 257, 1980

86.156 (tris(2 – Diphenylarsinoethyl) – amino) – iodo –
nickel(i)
$C_{42}H_{42}As_3INNi$
P.Dapporto, L.Sacconi *Inorg.Chim.Acta*,**39**, 61, 1980
See also R1 : 83

86.157 bis(μ – (3 – Diphenylphosphinopropyl)
phenylphosphino) – bis(chloro – platinum(ii))
$C_{42}H_{42}Cl_2P_4Pt_2$
R.Uriarte, T.J.Mazanec, K.D.Tau, D.W.Meek
Inorg.Chem.,**19**, 79, 1980

86.158 bis(μ – Chloro) – bis(perchlorato – (tri – o –
tolylphosphine) – mercury(ii))
$C_{42}H_{42}Cl_4Hg_2O_8P_2$
E.C.Alyea, S.Dias, G.Ferguson, M.Khan
Can.J.Chem.,**57**, 2217, 1979

86.159 (tris(2 – Diphenylphosphinoethyl) – amino) – iodo – nickel(i)
$C_{42}H_{42}INNiP_3$
P.Dapporto, L.Sacconi *Inorg.Chim.Acta*, **39**, 61, 1980
See also R1 : 83

86.C (η – 1,5 – Cyclo – octadiene) – ((2S,4S) – N – pivaloyl – 4 – diphenylphosphino – 2 – diphenylphosphino – methylpyrrolidine) – rhodium perchlorate
$C_{42}H_{49}NOP_2Rh^+$, ClO_4^- Main entry is 75.43

86.C $η^4$ – (Cyclo – octa – 1,5 – diene) – bis((R) – menthyl – methylphenylphosphine) – rhodium(i) tetrafluoroborate
$C_{42}H_{66}P_2Rh^+$, BF_4^- Main entry is 75.44

86.C bis(Triphenylphosphine) – ($μ^2$ – tetrachloro – diazacyclopentadienyl) – dicarbonyl – ruthenium dichloromethane solvate (at –159°C)
$C_{43}H_{30}Cl_4N_2O_2P_2Ru$, CH_2Cl_2 Main entry is 72.62

86.160 Phenyl – dithiocarboxylato – bis(triphenylphosphine) – copper(i)
$C_{43}H_{35}CuP_2S_2$
A.Camus, N.Marsich, G.Nardin
J.Organomet.Chem., **188**, 389, 1980
See also R1 : 85

86.161 ($μ^2$ – Dicarbonyl) – ($μ^2$ – tris(diphenylphosphino)) – tricarbonyl – iron – tris(carbonyl – rhodium)
$C_{44}H_{30}FeO_8P_3Rh_3$
R.J.Haines, N.D.C.T.Steen, M.Laing, P.Sommerville
J.Organomet.Chem., **198**, 72, 1980

86.162 Dicarbonyl – bis($η^2$ – diphenylphosphinesulfido) – triphenylphosphine – molybdenum methylene chloride solvate
$C_{44}H_{35}MoO_2P_3S_2$, CH_2Cl_2
H.P.M.M.Ambrosius, J.H.Noordik, G.J.A.Ariaans
J.Chem.Soc.,Chem.Commun., 832, 1980
See also R1 : 85

86.C (μ – 1,2 – bis(Methoxycarbonyl)ethene – 1,2 – diyl) – bis(carbonyl – (triphenylphosphine) – platinum)
$C_{44}H_{36}O_6P_2Pt_2$ Main entry is 71.265

86.163 4 – Methyl – N – thiobenzohydroximato – bis(triphenylphosphine) – platinum(ii)
$C_{44}H_{37}NOP_2PtS$
W.Beck, E.Leidl, M.Keubler, U.Nagel
Chem.Ber., **113**, 1790, 1980
See also R1 : 85,84

86.C (2,3 – bis(Methoxycarbonyl) – 1 – methylcyclopropyl) – bis(triphenylphosphine) – platinum(ii) tetrafluoroborate
$C_{44}H_{41}O_4P_2Pt^+$, BF_4^- Main entry is 71.266

86.164 Dicarbonyl – bis(tri – o – tolylphosphite) – (sulfur dioxide) – iron
$C_{44}H_{42}FeO_{10}P_2S$
P.Conway, S.M.Grant, A.R.Manning
J.Organomet.Chem., **186**, C61, 1980

86.C (1,2 – η – Cyclo – octyne) – bis(triphenylphosphine) – platinum(0)
$C_{44}H_{42}P_2Pt$ Main entry is 75.45

86.165 Chloro – carbonyl – (N – isopropenyl – N' – isopropyl – formamidinato) – bis(triphenylphosphine) – ruthenium(ii)
$C_{44}H_{43}ClN_2OP_2Ru$
A.D.Harris, S.D.Robinson, A.Sahajpal, M.B.Hursthouse
J.Organomet.Chem., **174**, C11, 1979
See also R1 : 83

86.166 bis(μ – (3 – Diphenylphosphinopropyl) phenylphosphine) – bis(methyl – platinum(ii))
$C_{44}H_{48}P_4Pt_2$
R.Uriarte, T.J.Mazanec, K.D.Tau, D.W.Meek
Inorg.Chem., **19**, 79, 1980

86.167 3,9 – Dihydrido – bis(tri – p – tolylphosphine) – (nido – 7,8 – dicarba – undecaborane) – iridium toluene solvate (at –154°C)
$C_{44}H_{56}B_9IrP_2$, C_7H_8
J.A.Doi, R.G.Teller, M.F.Hawthorne
J.Chem.Soc.,Chem.Commun., 80, 1980
See also R1 : 62

86.C bis(μ – Acetato) – bis(acetato – (tricyclohexylphosphine) – mercury(ii))
$C_{44}H_{78}Hg_2O_8P_2$ Main entry is 81.94

86.168 bis(μ – 1,5 – bis(Di – t – butylphosphino) – 3 – methylpentane) – bis(dichloro – palladium(ii))
$C_{44}H_{96}Cl_4P_4Pd_2$
N.A.Al–Salem, W.S.McDonald, R.Markham, M.C.Norton, B.L.Shaw *J.Chem.Soc.,Dalton Trans.*, 59, 1980

86.169 Carbonyl – chloro – nitrosyl – (2,11 – bis(diphenylphosphinomethyl) – benzo(c) phenanthrene) – ruthenium dideutero – dichloromethane solvate
$C_{45}H_{34}ClNO_2P_2Ru$, CCl_2D_2
R.Holderegger, L.M.Venanzi, F.Bachechi, P.Mura, L.Zambonelli *Helv.Chim.Acta*, **62**, 2159, 1979

86.C Carbonyl – iodo – bis(triphenylphosphine) – (p – methylbenzoyl) – ruthenium
$C_{45}H_{37}IO_2P_2Ru$ Main entry is 71.267

86.C ($η^3$ – 1 – Phenylallyl) – chloro – hydrido – bis(triphenylphosphine) – iridium
$C_{45}H_{40}ClIrP_2$ Main entry is 72.63

86.C Bicyclo(4.2.1)non – 1(8) – ene – bis(triphenylphosphine) – platinum(0)
$C_{45}H_{44}P_2Pt$ Main entry is 75.46

86.C cis – 1 – ((Tribenzylphosphine) – (dibenzylphosphino – phenyl – methyl) – platinum) – 2 – methyl – 1,2 – dicarba – dodecaborane
$C_{45}H_{54}B_{10}P_2Pt$ Main entry is 71.268

86.170 tris(Tricarbonyl – manganese) – nonakis(μ – diethoxyphosphite) – tri – manganese
$C_{45}H_{90}Mn_6O_{36}P_9$
R.Shakir, J.L.Atwood, T.S.Janik, J.D.Atwood
J.Organomet.Chem., **190**, C14, 1980
See also R1 : 84

86.C trans(P,N) – bis(Bromo – ($μ^2$ – pyridyl – C^2,N) – (triphenylphosphine) – palladium(ii))
$C_{46}H_{38}Br_2N_2P_2Pd_2$ Main entry is 71.269

86.C (bis($η^5$ – Cyclopentadienyl) – tungsten) – bis(μ – hydrido) – (bis(triphenylphosphine) – rhodium) hexafluorophosphate
$C_{46}H_{42}P_2RhW^+$, F_6P^- Main entry is 73.129

86.C (η^5 – Pentamethyl – cyclopentadienyl) – (triphenylphosphine) – hydrido – platinum(ii) hexafluorophosphate
$C_{46}H_{46}P_2Rh^+$, F_6P^- Main entry is 73.130

86.171 bis(μ – Heptane – 1,7 – diyl – bis(di – t – butylphosphine)) – bis(dichloro – palladium)
$C_{46}H_{100}Cl_4P_4Pd_2$
W.S.McDonald *Acta Crystallogr.,Sect.B*,**35**, 3051, 1979

86.172 bis(μ^2 – Hydrido) – undecacarbonyl – tri – osmium – bis(triphenylphosphine – gold)
$C_{47}H_{32}Au_2O_{11}Os_3P_2$
J.A.K.Howard, L.Farrugia, C.Foster, F.G.A.Stone, P.Woodward *Eur.Cryst.Meeting*,**6**, 73, 1980

86.173 Dicarbonyl – chloro – (5,7 – dichloro – 8 – quinolinolato – N,O) – bis(triphenylphosphine) – tungsten(ii)
$C_{47}H_{34}Cl_3NO_3P_2W$
R.O.Day, W.H.Batschelet, R.D.Archer *Inorg.Chem.*,**19**, 2113, 1980
See also R1 : 84.83

86.C (η^2 – Dithiomethoxycarbonyl) – carbonyl – (p – tolylisocyanide) – bis(triphenylphosphine) – ruthenium(ii) perchlorate chloroform solvate hemihydrate (refinement in Pna21)
$C_{47}H_{40}NOP_2RuS_2^+$, ClO_4^-, $0.5CHCl_3$, $0.5H_2O$
Main entry is 71.270

86.C (η^2 – Dithiomethoxycarbonyl) – carbonyl – (p – tolylisocyanide) – bis(triphenylphosphine) – ruthenium(ii) perchlorate chloroform solvate hemihydrate (refinement in Pnma)
$C_{47}H_{40}NOP_2RuS_2^+$, ClO_4^-, $0.5CHCl_3$, $0.5H_2O$
Main entry is 71.271

86.C (η^2 – Methyl – 3 – benzoylacrylate) – bis(triphenylphosphine) – nickel(0) benzene solvate
$C_{47}H_{40}NiO_3P_2$, C_6H_6 Main entry is 72.64

86.C η^2 – ((t – Butyl – (trimethylsilyl)amino) – t – butyliminothiophosphorane) – bis(triphenylphosphine) – platinum
$C_{47}H_{57}N_2P_3PtSSi$ Main entry is 72.65

86.174 Nitrosyl – (1,10 – phenanthroline) – bis(triphenylphosphine) – iridium(i) bis(hexafluorophosphate)
$C_{48}H_{38}IrN_3OP_2^{2+}$, $2F_6P^-$
A.Tiripicchio, M.T.Camellini, M.Ghedini, G.Dolcetti *Transition Met.Chem.*,**5**, 102, 1980
See also R1 : 83

86.175 bis(bis(Diphenylphosphino)amine – oxo) – rhodium(iii) hexafluorophosphate acetone solvate
$C_{48}H_{42}N_2O_2P_4Rh^+$, F_6P^-, $0.5C_3H_6O$
J.Ellermann, E.F.Hohenberger, W.Kehr, A.Purzer, G.Thiele *Z.Anorg.Allg.Chem.*,**464**, 45, 1980

86.C 1,1,1 – tris(Diphenylphosphinomethyl)ethane – (η^4 – cyclohepta – 1,3,5 – trienenyl) – cobalt perchlorate methylene chloride solvate
$C_{48}H_{47}CoP_3^+$, ClO_4^-, $0.5CH_2Cl_2$ Main entry is 75.47

86.C σ – (Thiophenyl(sulfinyl)methylene) – thiophenyl – bis(triphenylphosphine) – platinum benzene solvate
$C_{49}H_{40}OP_2PtS_3$, C_6H_6 Main entry is 71.272

86.C 1,2 – bis(Diphenylphosphino)ethane – (η^6 – tetraphenylborato) – rhodium(i)
$C_{50}H_{44}BP_2Rh$ Main entry is 74.15

86.176 (μ – Sulfur dioxide) – bis(μ – bis(diphenylphosphino)methane) – bis(chloro – palladium) dichloromethane methanol solvate (at 150°K)
$C_{50}H_{44}Cl_2O_2P_4Pd_2S$, $0.5CH_2Cl_2$, CH_4O
A.L.Balch, L.S.Benner, M.M.Olmstead *Inorg.Chem.*,**18**, 2996, 1979

86.177 bis(μ – bis(Diphenylphosphino)methane) – (μ – sulfur dioxide) – bis(chloro – rhodium)
$C_{50}H_{44}Cl_2O_2P_4Rh_2S$
M.Cowie, S.K.Dwight *Inorg.Chem.*,**19**, 209, 1980

86.178 (μ – Thio) – bis(μ – bis(diphenylphosphino) methane) – bis(chloro – palladium) dichloromethane solvate (absolute configuration)
$C_{50}H_{44}Cl_2P_4Pd_2S$, $3CH_2Cl_2$
A.L.Balch, L.S.Benner, M.M.Olmstead *Inorg.Chem.*,**18**, 2996, 1979

86.179 bis(μ – Bromo) – bis(μ_4 – bromo) – bis(μ – methyl – bis(diphenylphosphino)amine – di – silver)
$C_{50}H_{46}Ag_4Br_4N_2P_4$
U.Schubert, D.Neugebauer, A.A.M.Aly *Z.Anorg.Allg.Chem.*,**464**, 217, 1980

86.180 Dichloro – bis(2 – methyl – 4,4,6,6 – tetraphenyl – phosphazadiene – 2 – yl) – palladium
$C_{50}H_{48}Cl_2N_6P_6Pd$
A.Schmidpeter, K.Blanck, H.Hess, H.Riffel *Angew.Chem.,Int.Ed.Engl.*,**19**, 650, 1980

86.C bis(μ – Acetato) – bis(acetato – (tri – o – tolylphosphine) – mercury(ii))
$C_{50}H_{54}Hg_2O_8P_2$ Main entry is 81.96

86.181 1,1,1 – tris((Diphenylphosphino)methyl)ethane – cobalt – η^3 – (bis(pentacarbonyl – chromium) – cyclotriphosphorus)
$C_{51}H_{39}CoCr_2O_{10}P_6$
C.A.Ghilardi, S.Midollini, A.Orlandini, L.Sacconi *Inorg.Chem.*,**19**, 301, 1980

86.182 Carbonyl – platinum – bis(μ – bis(diphenylphosphino)methane) – chloro – platinum hexafluorophosphate
$C_{51}H_{44}ClOP_4Pt_2^+$, F_6P^-
L.Manojlovic–Muir, K.W.Muir, T.Solomun *J.Organomet.Chem.*,**179**, 479, 1979

86.183 trans – Carbonyl – (1,3 – di – p – tolyltriazenido) – bis(triphenylphosphine) – iridium(i)
$C_{51}H_{44}IrN_3OP_2$
A.Immirzi, W.Porzio, G.Bombieri, L.Toniolo *J.Chem.Soc.,Dalton Trans.*,1098, 1980
See also R1 : 83

86.184 bis(bis(Diphenylphosphino)methane) – carbonyl – rhodium(i) tetrafluoroborate
$C_{51}H_{44}OP_4Rh^+$, BF_4^-
L.H.Pignolet, D.H.Doughty, S.C.Nowicki, A.L.Casalnuovo *Inorg.Chem.*,**19**, 2172, 1980

86.185 Dichloro – (μ – methylene) – bis(μ – bis(diphenylphosphino)methane) – di – platinum
$C_{51}H_{46}Cl_2P_4Pt_2$
A.A.Frew, L.Manojlovic–Muir, K.W.Muir *Eur.Cryst.Meeting*,**6**, 72, 1980

86.C bis(Triphenylphosphine) – (η^5D – 2,6 – di – t – butyl – 4 – methyl – cyclohexadienonyl) – rhodium(i)
$C_{51}H_{53}OP_2Rh$, $2C_6H_6$ Main entry is 75.48

86.186 (μ – Chloro) – bis(μ – bis(diphenylphosphino) methane) – bis(carbonyl – rhodium) tetrafluoroborate
$C_{52}H_{44}ClO_2P_4Rh_2^+$, BF_4^-
M.Cowie, S.K.Dwight *Inorg.Chem.*, **18**, 2700, 1979

86.187 Dichloro – bis(cis – 1,2 – bis(diphenylphosphino) – ethylene) – iron(ii) acetone solvate
$C_{52}H_{44}Cl_2FeP_4$, $2C_3H_6O$
M.di Vaira, S.Midollini, L.Sacconi
Cryst.Struct.Commun., **9**, 407, 1980

86.188 Dicarbonyl – bis(μ – bis(diphenylphosphino) methane) – di – platinum bis(hexafluorophosphate)
$C_{52}H_{44}O_2P_4Pt_2^{2+}$, $2F_6P^-$
A.A.Frew, L.Manojlovic–Muir, K.W.Muir
Eur.Cryst.Meeting, **6**, 72, 1980

86.189 bis(μ – 1,2 – bis(Diphenylphosphino)ethane) – bis(dichloro – tungsten)
$C_{52}H_{48}Cl_4P_4W_2$
F.A.Cotton, T.R.Felthouse, D.G.Lay
J.Am.Chem.Soc., **102**, 1431, 1980

86.190 bis(μ – 1,2 – bis(Diphenylphosphino)ethane) – bis(dichloro – tungsten) hemihydrate
$C_{52}H_{48}Cl_4P_4W_2$, $0.5H_2O$
F.A.Cotton, T.R.Felthouse, D.G.Lay
J.Am.Chem.Soc., **102**, 1431, 1980

86.191 Disulfido – bis(1,2 – bis(diphenylphosphido)ethane) – iridium chloride N,N' – diphenylurea
$C_{52}H_{48}IrP_4S_2^+$, Cl^-, $2C_{13}H_{12}N_2O$
K.Leonhard, K.Plute, R.C.Haltiwanger, M.R.DuBois
Inorg.Chem., **18**, 3246, 1979

86.192 cis – Dihydrido – bis(1,2 – bis(diphenylphosphino) ethane) – ruthenium benzene solvate
$C_{52}H_{50}P_4Ru$, C_6H_6
P.Pertici, G.Vitulli, W.Porzio, M.Zocchi
Inorg.Chim.Acta, **37**, L521, 1979

86.193 (μ^2 – Hydrido) – bis((μ^2 – bis(diphenylphosphino) methane) – methyl – platinum) hexafluorophosphate
$C_{52}H_{51}P_4Pt_2^+$, F_6P^-
M.P.Brown, S.J.Cooper, A.A.Frew, L.Manojlovic–Muir, K.W.Muir, R.J.Puddephatt, M.A.Thomson
J.Organomet.Chem., **198**, 33, 1980

86.194 (μ – Carbonyl) – (μ – chloro) – bis(μ – bis(diphenylphosphino)methane) – bis(carbonyl – rhodium) dichloro – dicarbonyl – rhodium dichloromethane solvate (at 150°K)
$C_{53}H_{44}ClO_3P_4Rh_2^+$, $C_2Cl_2O_2Rh^-$, CH_2Cl_2
M.M.Olmstead, C.H.Lindsay, L.S.Brenner, A.L.Balch
J.Organomet.Chem., **179**, 289, 1979

86.C Chloro – rhodium – bis(μ^2 – bis(diphenylphosphino) methylene) – (μ – dicarbon – tetrasulfide) – chloro – carbonyl – rhodium
$C_{53}H_{44}Cl_2OP_4Rh_2S_4$ Main entry is 71.273

86.195 (μ^2 – Carbonyl) – (μ^2 – hydrido) – bis(μ^2 – bis(diphenylphosphino)methane) – dicarbonyl – di – rhodium p – toluenesulfonate tetrahydrofuran solvate
$C_{53}H_{45}O_3P_4Rh_2^+$, $C_7H_7O_3S^-$, $2C_4H_8O$
C.P.Kubiak, R.Eisenberg *J.Am.Chem.Soc.*, **102**, 3637, 1980

86.C trans – Chloro – (1,4 – bis(p – methoxyphenyl) – 3 – methyl – 1,4 – diazabutadien – 2 – yl) – bis(triphenylphosphine) – palladium(ii)
$C_{53}H_{47}ClN_2O_2P_2Pd$ Main entry is 71.274

86.196 2 – (Chloro – bis(triphenylphosphine) – palladium) – 3 – methyl – 1,4 – bis(methoxyphenyl) – 1,4 – diazabutadien – 1,4 – diyl – dichloro – copper
$C_{53}H_{47}Cl_3CuN_2O_2P_2Pd$
B.Crociani, G.Bandoli, D.A.Clemente
J.Organomet.Chem., **190**, C97, 1980
See also R1 : 83

86.197 Bromo – (dichloro – diazamethane) – bis(diphenylphosphinoethane) – tungsten hexafluorophosphate
$C_{53}H_{48}BrCl_2N_2P_4W^+$, F_6P^-
H.M.Colquhoun, T.J.King
J.Chem.Soc.,Chem.Commun., 879, 1980
See also R1 : 83

86.198 Trimethyl – bis(μ^2 – (bis(diphenylphosphinomethane))) – di – platinum(ii) hexafluorophosphate methylene chloride solvate
$C_{53}H_{53}P_4Pt_2^+$, F_6P^-, $3CH_2Cl_2$
A.A.Frew, L.Manojlovic–Muir, K.W.Muir
J.Chem.Soc.,Chem.Commun., 624, 1980

86.C (μ – Hexafluorobutadiene – 2,3 – diyl) – bis(μ – bis(diphenylphosphino)methane) – bis(chloro – palladium) (at 140°K)
$C_{54}H_{44}Cl_2F_6P_4Pd_2$ Main entry is 71.275

86.199 bis(Triphenylphosphine) – silver – bis(μ – thio) – tungsten – bis(μ – thio) – triphenylphosphine – silver
$C_{54}H_{45}Ag_2P_3S_4W$
A.Muller, H.Bogge, E.Koniger–Ahlborn
Z.Naturforsch.,Teil B, **34**, 1698, 1979
See also R1 : 85

86.200 tris(Triphenylphosphine) – gold(i) dodecahydro – 6 – thia – nido – decaborane
$C_{54}H_{45}AuP_3^+$, $H_{12}B_9S^-$
L.J.Guggenberger *J.Organomet.Chem.*, **81**, 271, 1974

86.C Bromo – (2 – (diphenylphosphino)phenyl) – hydrido – bis(triphenylphosphine) – iridium(iii)
$C_{54}H_{45}BrIrP_3$ Main entry is 71.276

86.201 (μ^3 – Chloro) – (trisulfur – oxo – molybdenum) – tris(triphenylphosphine – copper)
$C_{54}H_{45}ClCu_3MoOP_3S_3$
A.Muller, H.Bogge, H.–G.Tolle, R.Jostes, U.Schimanski, M.Dartmann *Angew.Chem.,Int.Ed.Engl.*, **19**, 654, 1980

86.202 (μ_3 – Chloro) – tris(μ_3 – sulfido) – tris(triphenylphosphine – copper) – thio – molybdenum
$C_{54}H_{45}ClCu_3MoP_3S_4$
A.Muller, H.Bogge, U.Schimanski
J.Chem.Soc.,Chem.Commun., 91, 1980

86.203 bis(Triphenylphosphine) – copper – bis(μ^2 – thio) – molybdenum – bis(μ^2 – thio) – triphenylphosphine – copper methylene chloride solvate
$C_{54}H_{45}Cu_2MoP_3S_4$, $0.8CH_2Cl_2$
A.Muller, H.Bogge, H.–G.Tolle, R.Jostes, U.Schimanski, M.Dartmann *Angew.Chem.,Int.Ed.Engl.*,**19**, 654, 1980

86.204 Dicarbonyl – bis(diphenylphosphinoethane) – fluoro – molybdenum hexafluorophosphate
$C_{54}H_{48}FMoO_2P_4{}^+$, F_6P^-
T.Chandler, G.R.Kriek, A.M.Greenaway, J.H.Enemark *Cryst.Struct.Commun.*,**9**, 557, 1980

86.205 Dihydrido – tris(μ^2 – hydrido) – bis(1,3 – bis(diphenylphosphino)propane) – di – iridium tetrafluoroborate
$C_{54}H_{57}Ir_2P_4{}^+$, $BF_4{}^-$
H.H.Wang, L.H.Pignolet *Inorg.Chem.*,**19**, 1470, 1980

86.C Tricyclohexylphosphine – platinum – (μ – dimethylsilyl – μ – (1 – sigma:1 – 2 – η – phenylethynyl)) – (2 – phenylethynyl – tricyclohexylphosphine – platinum) (at 200°K)
$C_{54}H_{82}P_2Pt_2Si$ Main entry is 71.277

86.206 bis((Tri – n – butylphosphine) – bis(dimethylglyoximato) – (methyl(phenyl) – arsenyl) – cobalt(iii) – cobalt(ii) dihydrate
$C_{54}H_{96}As_2Co_3N_8O_{10}P_2$, $2H_2O$
W.R.Cullen, D.Dolphin, F.W.B.Einstein, L.M.Mihichuk, A.C.Willis *J.Am.Chem.Soc.*,**101**, 6898, 1979
See also R1 : 84,83

86.C Trifluoroacetato – (1,4 – diphenyl – but – 1 – en – 3 – yn – 2 – yl) – carbonyl – bis(triphenylphosphine) – ruthenium
$C_{55}H_{41}F_3O_3P_2Ru$ Main entry is 71.278

86.207 bis(1,2 – bis(Diphenylphosphino)ethane) – bromo – 1 – diazo – 1 – methylethane – tungsten bromide methanol solvate
$C_{55}H_{54}BrN_2P_4W^+$, Br^-, $0.5CH_4O$
R.A.Head, P.B.Hitchcock
J.Chem.Soc.,Dalton Trans.,1150, 1980
See also R1 : 83

86.208 bis((μ^2 – Carbonyl) – (μ^4 – phenylphosphido) – triphenylphosphine) – hexacarbonyl – tetra – cobalt benzene solvate
$C_{56}H_{40}Co_4O_8P_4$, C_6H_6
R.C.Ryan, C.U.Pittman Junior, J.P.O'Connor, L.F.Dahl
J.Organomet.Chem.,**193**, 247, 1980

86.209 Triphenylphosphonio – dithiocarboxylato – carbonyl – bis(triphenylphosphine) – iridium(i) tetrafluoroborate
$C_{56}H_{45}IrOP_3S_2{}^+$, $BF_4{}^-$
S.M.Boniface, G.R.Clark
J.Organomet.Chem.,**188**, 263, 1980
See also R1 : 85

86.210 Bromo – (1 – chloro – 1 – diazo – 2,2 – dicyanoethene) – bis(diphenylphosphinoethane) – tungsten dichloromethane solvate
$C_{56}H_{48}BrClN_4P_4W$, CH_2Cl_2
H.M.Colquhoun, T.J.King
J.Chem.Soc.,Chem.Commun.,879, 1980
See also R1 : 83

86.211 bis(1,2 – bis(Diphenylphosphino)ethane) – bromo – 4 – diazobutanol – tungsten hexafluorophosphate ethanol solvate
$C_{56}H_{56}BrN_2OP_4W^+$, F_6P^-, $0.5C_2H_6O$
R.A.Head, P.B.Hitchcock
J.Chem.Soc.,Dalton Trans.,1150, 1980
See also R1 : 83

86.212 Tetrahydrido – tetrakis(diphenylethylphosphino) – tungsten
$C_{56}H_{64}P_4W$
E.B.Lobkovskii, V.D.Makhaev, A.P.Borisov, K.N.Semenenko
Zh.Strukt.Khim.,**20**, 944, 1979

86.C (μ^2 – Carbonyl) – bis(μ^2 – bis(diphenylphosphino) methylene) – (μ^2 – dimethylacetylene – carboxylate) – dichloro – di – rhodium
$C_{57}H_{50}Cl_2O_5P_4Rh_2$ Main entry is 71.279

86.213 pentakis(μ_2 – Carbonyl) – tetrakis(methyl – diphenylphosphine) – palladium)
$C_{57}H_{52}O_5P_4Pd_4$
J.Dubrawski, J.C.Kriege–Simondsen, R.D.Feltham
J.Am.Chem.Soc.,**102**, 2089, 1980

86.214 bis(bis(Diphenylphosphino) – maleic – N – methylimide) – palladium(0) 1,2 – dichloroethane solvate (at –130°C)
$C_{58}H_{46}N_2O_4P_4Pd$, $5C_2H_4Cl_2$
W.Bensmann, D.Fenske
Angew.Chem.,Int.Ed.Engl.,**18**, 677, 1979

86.215 Iodo – N – cyclohexyl – diazenido – bis(1,2 – bis(diphenylphosphino)ethane) – molybdenum benzene solvate
$C_{58}H_{59}IMoN_2P_4$, $0.5C_6H_6$
C.S.Day, V.W.Day, T.A.George, I.Tavanaiepour
Inorg.Chim.Acta,**45**, L54, 1980
See also R1 : 83

86.C Tricarbonyl – cobalt – (μ – diphenylacetylene) – (1,1,1 – tris(diphenylphosphinomethyl)ethane – carbonyl – cobalt)
$C_{59}H_{49}Co_2O_4P_3$ Main entry is 72.66

86.C bis((Dicarbonyl – P,P – diphenyl – N – methyl – phosphinothioformamido) – (μ – P,P – diphenyl – N – methyl – phosphinothioformamido) – molybdenum) dichloromethane solvate
$C_{80}H_{52}Mo_2N_4O_4P_4S_4$, $4CH_2Cl_2$ Main entry is 71.280

86.216 pentakis(μ – Diphenylphosphido) – pentacarbonyl – tetra – rhodium (at –73°C)
$C_{65}H_{50}O_5P_5Rh_4{}^-$, Li^+
P.E.Kreter Junior, D.W.Meek, G.G.Christoph
J.Organomet.Chem.,**188**, 27, 1980

86.C Chloro – (bis(3,5 – di – t – butyl – 4 – oxo – cyclohexadiene – 1 – ylidene)ethylene) – bis(triphenylphosphine) – rhodium acetone solvate
$C_{66}H_{70}ClO_2P_2Rh$, $1.5C_3H_6O$ Main entry is 72.68

86.217 Dichloro – bis(2 – diphenylphosphinoethyl) – phenylphosphine – iron(ii) acetone solvate
$C_{68}H_{66}Cl_2FeP_6$, $2C_3H_6O$
M.di Vaira, S.Midollini, L.Sacconi
Cryst.Struct.Commun.,**9**, 403, 1980

86.218 tris(μ – Mercapto) – bis(bis(2 – (diphenylphosphino)
ethyl) – phenylphosphine – iron(ii)) perchlorate
acetone solvate
$C_{68}H_{69}Fe_2P_6S_3^+$, ClO_4^-, C_3H_6O
M.di Vaira, S.Midollini, L.Sacconi
Inorg.Chem., **18**, 3466, 1979

86.219 hexakis(μ_3 – Sulfido) – bis(oxo – tungsten) –
tetra(triphenylphosphine – copper)
$C_{72}H_{60}Cu_4O_2P_4S_6W_2$
A.Muller, H.Bogge, T.K.Hwang
Inorg.Chim.Acta, **39**, 71, 1980
See also R1 : 85

86.220 bis((μ – Sulfato) – bis(triphenylphosphine) – (sulfur
dioxide) – ruthenium(ii)) toluene solvate
$C_{72}H_{60}O_{12}P_4Ru_2S_4$, C_7H_8
I.Ghatak, D.M.P.Mingos, M.B.Hursthouse, K.M.A.Malik
Transition Met.Chem., **4**, 260, 1979

86.221 tris(μ^2 – Diphenylphosphido) –
bis(triphenylphosphine) – tricarbonyl – tri –
rhodium
$C_{75}H_{60}O_3P_5Rh_3$
E.Billig, J.D.Jamerson, R.L.Pruett
J.Organomet.Chem., **192**, C49, 1980

86.222 bis(μ^3 – Bromo) – tris((diphenylphosphino)
methane – silver) bromide
$C_{75}H_{66}Ag_3Br_2P_6^+$, Br^-
U.Schubert, D.Neugebauer, A.A.M.Aly
Z.Anorg.Allg.Chem., **464**, 217, 1980

86.223 Hydrido – platinum – bis(μ –
bis(diphenylphosphino)methane) –
(bis(diphenylphosphino)methane) – platinum
hexafluorophosphate
$C_{75}H_{67}P_6Pt_2^+$, F_6P^-
M.P.Brown, J.R.Fisher, L.Manojlovic–Muir, K.W.Muir,
R.J.Puddephatt, M.A.Thomson, K.R.Seddon
J.Chem.Soc.,Chem.Commun., 931, 1979

86.224 Trihydrido – (μ^3 – hydrido) – tris(μ^2 – hydrido) –
tris(1,3 – bis(diphenylphosphino) – propane) – tri –
iridium bis(tetrafluoroborate) methylene chloride
solvate
$C_{81}H_{85}Ir_3P_6^{2+}$, $2BF_4^-$, $4CH_2Cl_2$
H.H.Wang, L.H.Pignolet *Inorg.Chem.*, **19**, 1470, 1980

86.225 1,1,1 – tris(Diphenylphosphinomethyl)ethane –
rhodium – (μ – η^3 – cyclotriphosphorus) – 1,1,1 –
tris(diphenylphosphinomethyl)ethane – cobalt
bis(tetraphenylborate) acetone solvate
$C_{82}H_{78}CoP_9Rh^{2+}$, $2C_{24}H_{20}B^-$, $2C_3H_6O$
C.Bianchini, M.di Vaira, A.Meli, L.Sacconi
Angew.Chem.,Int.Ed.Engl., **19**, 405, 1980

86.226 bis((μ^2 – Thio) – (1,1,1 – tris(diphenylphosphino)
methyl)ethane) – cobalt dimethylformamide solvate
$C_{82}H_{78}Co_2P_6S_2^+$, $C_3H_7NO^-$, $0.4C_3H_7NO$
C.A.Ghilardi, C.Mealli, S.Midollini, V.I.Nefedov, A.Orlandini,
L.Sacconi *Inorg.Chem.*, **19**, 2454, 1980
See also R1 : 85

86.227 bis(μ^2 – Thio) – (1,1,1 – tris((diphenylphosphino)
methyl)ethane) – cobalt tetraphenylborate
dimethylformamide solvate
$C_{82}H_{78}Co_2P_6S_2^+$, $C_{24}H_{20}B^-$, $0.5C_3H_7NO$
C.A.Ghilardi, C.Mealli, S.Midollini, V.I.Nefedov, A.Orlandini,
L.Sacconi *Inorg.Chem.*, **19**, 2454, 1980
See also R1 : 85

86.228 1,1,1 – tris(Diphenylphosphinomethyl)ethane –
rhodium – (μ – η^3 – cyclotriphosphorus) – 1,1,1 –
tris(diphenylphosphinomethyl)ethane – nickel
bis(tetraphenylborate) tetrahydrofuran
$C_{82}H_{78}NiP_9Rh^{2+}$, $2C_{24}H_{20}B^-$, C_4H_8O
C.Bianchini, M.di Vaira, A.Meli, L.Sacconi
Angew.Chem.,Int.Ed.Engl., **19**, 405, 1980

86.C (μ^2 – Carbon – disulfide) – bis((1,1,1 –
(diphenylphosphinomethyl)ethane) – cobalt)
bis(tetraphenylborate) acetone solvate
$C_{83}H_{78}Co_2P_6S_2^{2+}$, $2C_{24}H_{20}B^-$, $2C_3H_6O$ Main entry is 72.69

86.229 bis(μ^2 – Methylthio) – (1,1,1 –
tris(((diphenylphosphino)methyl)))ethane – cobalt
bis(tetraphenylborate) acetone solvate
$C_{84}H_{84}Co_2P_6S_2^{2+}$, $2C_{24}H_{20}B^-$, $2C_3H_6O$
C.A.Ghilardi, C.Mealli, S.Midollini, V.I.Nefedov, A.Orlandini,
L.Sacconi *Inorg.Chem.*, **19**, 2454, 1980
See also R1 : 85

86.230 tetrakis((Tri – p – tolylphosphine) – copper) –
dioxo – hexathio – di – tungsten
$C_{84}H_{84}Cu_4O_2P_4S_6W_2$
R.Doherty, C.R.Hubbard, A.D.Mighell, A.R.Siedle, J.Stewart
Inorg.Chem., **18**, 2991, 1979

86.C tris(bis(Diphenylphosphino)methane) –
(bis(diphenylphosphino)methanido) – penta – gold
dinitrate
$C_{100}H_{87}Au_5P_6^{2+}$, $2NO_3^-$ Main entry is 71.281

```
                                              Abietospiran  ...............................  56.   7
               (-)-(4S,8S)-α-bis Abolol-8-p-phenylazo-phenylmethane .............  53.  35
                                              Acamelin  .................................  59.   2
                                              Acantholide  ..............................  53.  22
     yl-5-phenalenyl)-1-oxo-8-phenyl-1,2-dihydro-2- Acenaphthylene-carboxylate <thoxy-1-oxo-9-phen  59.  46
     toxy-6b,7-dihydro-7,9-dimethyl-8H-cyclopent(a) Acenaphthylenediyl) benzene solvate <bis(8-ace  30.  19
                                              Acenaphthyleno(5,6-cd)-1,2,6-thiadiazine  ....  41.  38
                           2-methyl-2H- Acenaphthyleno(5,6-de)-1,2,3-triazine  ........  36.   5
     ,5-di-O-isopropylidene-D-gulose diethyl dithio Acetal  .................................... [2,3:4  45.  42
         ethyl D-threo-2,5-hexodiulosonate-5-(dimethyl- Acetal).................................... [m  45.  17
     stra-1,3,5(10)-triene-7,17-dione-17-(ethylene- Acetal)................... [3-methoxy-6-oxa-e  51.  19
                    (-)-(S)-N-(1-phenylethyl) Acetamide  ...............................  19.   4
                       N,N-dimethyl Acetamide hydrochloride.......................   1.  17
                            mono Acetamide nitrate  ........................   1.   5
     ladium bis(dimethylacetamidonium) bis(dimethyl Acetamide) .... [tetrachloro-μ-dichloro-di-pal   1.  16
                                bis( Acetamide) hydrochloride  ...................   1.   4
           hydrogen bis(N,N-dimethyl Acetamide) tetrachloro-gold(iii)  ..........   1.  15n
           hydrogen bis(N,N-dimethyl Acetamide) tetrachloro-gold(iii)  ..........   1.  14
           2,6-cis-dimethyl-piperidyl-N-phenyl- Acetamidine..............................  33.  67
                    N-methyl-N'-(benzyl) Acetamido)-methylurea  ...................   1.  29
     yranose]  1,4,6-tri-O-acetyl-2-(N-acetyl- Acetamido)-2,3-dideoxy-α-D-erythro-hex-2-eno-p  45.  39
           7-chloro-1,3-dihydro-1-(N-methyl Acetamido)-5-phenyl-2H-1,4-benzodiazepin-2-one  35.  51
     opyranosyl-(1-3)-O-β-D-mannopyranosyl-(1-4)-2- Acetamido-2-deoxy-α-D-glucopyranose <-α-D-mann  45.  49
                    N-methyl-N'-(1- Acetamido-2-phenylethyl)-urea  ...............   1.  29
                                o- Acetamidobenzoic acid  ...................  13.   5
     trachloro-μ-dichloro-di-palladium bis(dimethyl Acetamidonium) bis(dimethylacetamide)  ....[te   1.  16
     ofuro(2,3-b)fur> 5-hydroxy-4,4-dimethyl-3-(o- Acetamidophenyl)-6a-phenyl-3a,4,5,6a-tetrahydr  38.112
     -benzoyl-2-(1,3-dimethyl-2,6-dioxo-7-purinyl)- Acetanilide................................ [2  35.  67
           cyclohepta-amylose p-nitro Acetanilide hydrate.........................  45.  66
              sodium calcium nitrilo-tri Acetate..................................  48.  16
              sodium hydrogen nitrilo-tri Acetate..................................   2.  27
              bis(2-methyl-1-naphthyl)-methyl Acetate..................................  24.  11
           sodium hydrogen α-methoxy-α-phenyl Acetate..................................   2.  29
           α-phenylethylammonium α-phenylethyl Acetate..................................   2.  32
           rubidium hydrogen α-methoxy-α-phenyl Acetate..................................   2.  30
     ns-2-acetoxy-5-nitro-2,5-dihydro-2-furfural di Acetate                        [tra  38.  33
     6,6'-dibromo-6,6'-dideoxy-α,α-trehalose-hexa- Acetate chloroform solvate ...............  45.  50
     ydroxy-2,6-dimethyl-bicyclo(3.3.1)nonane ethyl Acetate clathrate ........  [exo-2,exo-6-dih  61.   1
           calcium ethylenediamine-tetra- Acetate heptahydrate  ...................  48.  27
              barium ethylenediamine-tetra- Acetate hydrate.........................  48.  26
              2-diethyl-aminoethyl diphenyl Acetate hydrobromide ....................   1.  38
                              sodium Acetate trihydrate .......................   2.  12n
     thoxy-carbonyl-imidazole-1-(α-isopropylidene)- Acetate [2',2',2'-trichloroethyl-2-benzyl-4-me  32.  67
           rubidium hydrogen bis(dibromo Acetate)  .................................   2.   6
           potassium hydrogen bis(dibromo Acetate)  .................................   2.  11
           potassium hydrogen bis(dichloro Acetate)  .................................   2.   8n
           potassium deuterium bis(dichloro Acetate)  .................................   2.   7n
              tris(monothio Acetato)-antimony(iii)  ...................  66.   7
                 (R)-tris(trifluoro Acetato)-antimony(iii)  ...................  66.   6a
                                tri Acetato-antimony(iii)  ....................  66.   8
              μ-fluoro-μ-trifluoro Acetato-bis(tetrafluoro-antimony(v))  ......  66.   1
           μ-oxo-di-μ-trifluoro Acetato-bis(trifluoro-antimony(v))  .......  66.   3
                                              Acetato-diphenyl-antimony(iii) .........  66.  13
     -(t-butylthio)-2-(9-thioxanthenylidene)-dithio Acetato-t-butylester  ..................... [2  39.  83
                    thallium(i) tetra- Acetato-thallium(iii)  ...................  68.   5
                              nitro Acetic acid.............................   1.   2
                       p-nitrophenoxy Acetic acid.............................  17.   8
                  tris(hydroxymethyl)- Acetic acid.............................   1.  20
           (2,3-dichloro-4-(2-thienoyl)-phenoxy)- Acetic acid.............................  39.  46
     l-5-ethyl-1,2,5,6-tetrahydro-2-oxo-4-pyridine- Acetic acid                        [1-benzy  33.  68
     -hydroxy-4-oxo-r-1H-bicyclo(5.3.0)dec-7-t-yl) Acetic acid γ-lactone  [(t-5,c-10-dimethyl-c-6  38.  63
     7,8,9-tetrahydro-4H-pyrido(1,2-a)pyrimidine-9- Acetic acid <-ethoxycarbonyl-6-methyl-4-oxo-6,  35.  33
     .2.0(3,13).0(10,12))tridecane-4,8-dien-11-yl) Acetic acid <RS,13RS)-(6-phenyl-tetracyclo(5.4  59.  29
                 furaltadone hydrochloride Acetic acid clathrate ...................  61.   4
     exakis(R-α-phenylethylsulfonyl-methyl)benzene Acetic acid clathrate .................... [h  61.  24
                 tryptamine adenin-9-yl- Acetic acid hemihydrate ................  44.  17
           7,8-dimethyl-isoalloxazine-10- Acetic acid tyramine tetrahydrate .........  36.  11
                    bis(trimethyl- Acetic acid) benzene .....................   1.  19
           tyramine 1-thyminyl-( Acetic acid) complex hydrate  ...........  60.   6
                    p-bromophenoxy Acetimine methyl ether .................  17.   9
                           benzoyl Acetone...................................  19.   3
                 rifamycin B p-iodoanilide Acetone pentahydrate ..................  50.  31
           α-(triphenylphosphonium)-benzoyl Acetone tetrafluoroborate  ..............  64.145
                                bis( Acetone-oximato)-methyl-chloro-silane  ........  63.   6
                                tris( Acetone-oximato)-phenyl-silane .............  63.  19
                                di Acetone-phenanthroquinone  ..............  38.103
```

triamcinolone	Acetonide	51. 46
phosphine-triyl-tri	Acetonitrile	64. 15
2,4,6-trimethyl	Acetophenone-oxime	10. 5
)-oxime] 4-chloro	Acetophenone-O-(3-t-butylamino-2-hydroxypropyl	10. 6
2,3-di-O-	Acetyl-2-C-methyl-erythrono-1,4-lactone	45. 16
1,3,4,6-tetra-O-acetyl-2-(N-	Acetylacetamido)-2-deoxy-β-D-galactopyranose	45. 44
ene] bis(	Acetylacetonato)-palladium(ii) tetrathiafulval	60. 12
tris(	Acetylacetonato)silicon perchlorate	63. 18
μ-oxo-bis(trichloro	Acetylacetonato-antimony(v))	66. 11
N,N'-ethylene-bis(	Acetylacetone-imine)	1. 31
	Acetylene	5. 4
bis(diphenylphosphino-thioyl)-	Acetylene	64.142
bis(diphenylphosphino-selenoyl)-	Acetylene	64.143
bis(phenylcarbamoyl-oxy-n-butyl)-di	Acetylene	16. 18
hexakis(	Acetylene) zeolite 4A complex	5. 5
	Acnistin E	51. 70
dihydroxy-8β,16β-dimethoxy-6β-methoxycarbonyl-	Aconitine hydroiodide ... [N-ethyl-1α,14α-	58. 66
epoxy-iso	Acoragermacrone	53. 9
9,9-dimethyl-10-(3'-piperidyl-propyl)-	Acridane	36. 38
9,9-dimethyl-10-(3'-piperidyl-propyl)-9-sila-	Acridane	63. 35
1,11-bis(bromomethyl)-1,2,3,4-tetrahydro-	Acridine	36. 15
o-1,4:5,8-di-isopropano-4,5-dimethyl-9-phenyl-	Acridine [1,2,3,4,5,6,7,8-octahydr	37. 22
ydrate complex] 9-amino-	Acridine - 5-iodocytidylyl-(3'-5')-guanosine h	60. 22
ine dodecahydrate]	Acridine orange 5-iodocytidylyl-(3'-5')-guanos	47. 35
9-amino-	Acridinium chloride dihydrate	36. 8
N-methyl-	Acridone	36. 9
(E)-3-phenyl-3-(3-pyridyl)-	Acrylamide	33. 59
(Z)-3-phenyl-3-(4-pyridyl)-	Acrylamide	33. 60
(3-isocyano-6-oxabicyclo(3.1.0)hex-2-en-5-yl)-	Acrylate [methyl 3-	38. 20
4-imidazole-	Acrylic acid dihydrate	32. 17
	Actodigin	51. 73
1-methyl-1,3,5,7-tetra-aza-	Adamantan-1-ium octa-iodide	37. 2
1-cyano-	Adamantane	31. 14
(E)-2-chloro-2-ethynyl-5-phenyl-	Adamantane	31. 44
(Z)-2-chloro-2-ethynyl-5-phenyl-	Adamantane	31. 45
dodecamethyl-hexasila-tetra-arsa-	Adamantane	65. 8
2-(p-tolylsulfonyloxy-N,N,O-azoxy)-	Adamantane	31. 43
1,3,5,7-tetra-arsa-2,4,6,8-tetraoxa-	Adamantane	65. 1
trans-1-ethyl-2-(1-ethyl-2-adamantanylidene)-	Adamantane	31. 61
mine] N-(1'-t-butyl-spiro(	Adamantane-2,2'-aziridine)-3'-ylidene)-methyla	32. 59
Z-2-ethynyl-5-phenyl-	Adamantanol	31. 46
trans-1-ethyl-2-(1-ethyl-2-	Adamantanylidene)-adamantane	31. 61
5,5'-dinitro-2-(1-	Adamantyl)-2'-carbomethoxy-(2H,5H)-furan	38. 77
2,4-diamino-5-(1-	Adamantyl)-6-methylpyrimidine ethylsulfonate	44. 33
di-t-	Adamantyl-disulfide	31. 54
2-	Adamantyl-O,N,N-azoxy-p-toluenesulfonate	31. 43
tryptamine	Adenin-9-yl-acetic acid hemihydrate	44. 17
9-methyl-	Adenine	44. 15n
N(6),N(9)-dimethyl-	Adenine	44. 19
9α-D-arabinofuranosyl-	Adenine	47. 16
trans-9-(2-ethoxy-1,3-dioxan-5-yl)-	Adenine	44. 24
9β-D-arabinofuranosyl-8-n-butyl-amino-	Adenine	47. 31
9-(5-phenylseleno-5-deoxyribofuranosyl)-	Adenine	47. 33
9β-D-arabinofuranosyl-8-morpholino-	Adenine dihydrate	47. 30
N(6)-methyl-	Adenine hydrochloride	44. 16
bis(	Adeninium nitrate) monohydrate	44. 10
3'-deoxy-	Adenosine	47. 15
cytidylyl-(3'-5')-	Adenosine - proflavine complex decahydrate	60. 39
8-thioxo-	Adenosine monohydrate	47. 17
bis(	Adenosine) proflavine sesquisulfate hydrate	47. 21
drate] sodium	Adenosine-(3'-5')-cyclic-monophosphate tetrahy	47. 13
te)] triethylammonium	Adenosine-5'-(O-p-nitrophenyl-O-phosphorothioa	47. 32
potassium	Adenosine-5'-diphosphate dihydrate	47. 23
	Adenosine-5'-diphosphoric acid trihydrate	47. 24
2-amino-8-methyl-1-	Adenosine-5'-monophosphate dihydrate	47. 27
S-8-aza-	Adenosyl-L-homocysteine dihydrate	48. 40
S-	Adenosyl-L-homocysteine hydrate	48. 44
eoxy-5'-P-adenylyl-(3'-5')-thymidylyl-(3'-5')-	Adenylyl-(3'-5')-thymidine [d	47. 39
3'-5')-thymidine)] deoxy-5'-P-	Adenylyl-(3'-5')-thymidylyl-(3'-5')-adenylyl-(	47. 39
N,N'-di-isopropyl-	Adipohydroxamic acid	1. 32
7,15-di-isocyano-	Adociane	54. 19
	Adriamycin-14-O-valerate hydrochloride solvate	29. 11
	Affinoside B	45. 57
	Aflavinine ethylacetate solvate	54. 35
4-hydroxy-dihydro	Agarofuran	53. 14
CC-1065-anti-tumor	Agent	59. 43

11-deoxy-daunorubicin Aglycone triacetate............................ 50. 18
Ajugareptansin p-bromobenzoate 54. 37a
benzoyl-oxycarbonyl-bis(α-amino-isobutyryl)-L- Alaninate ....[methyl 48. 65
t-butoxy-carbonyl-L-phenyl Alanine.............................. 48. 43
pivaloyl-D-prolyl-L-prolyl-L- Alanine-N-methylamide 48. 62
cyclo(L-prolyl-L- Alanyl)................................ 48. 23
(L-prolyl-L-valyl-L-ornithyl-L-leucyl-L-phenyl Alanyl)..................[N-acetyl-cyclo-bis 50. 40
2,6-bis(methyl Alanyl)-octavalinomycin.............. 50. 24
methionyl-glutamyl-histidinyl-phenyl Alanyl-arginyl-tryptophanyl-glycine............ 48. 82
leucyl-L-phenylalanyl-glycyl-D-leucyl-D-phenyl Alanyl-glycyl) dihydrate .... [cyclo(L- 48. 80
dihydrate] cyclo(L-leucyl-L-phenyl Alanyl-glycyl-D-leucyl-D-phenylalanyl-glycyl) 48. 80
L-pyroglutamyl-β-(2-thienyl)-L- Alanyl-L-prolinamide 48. 50
L-valyl-D-α-hydroxyisovaleryl-D-valyl-L-methyl Alanyl-L-valyl-D-α-hydroxyisovaleryl) <alanyl- 50. 24
yl-L-lactyl-L-valyl-D> cyclo(D-valyl-L-methyl Alanyl-L-valyl-D-α-hydroxyisovaleryl-bis(D-val 50. 37
-methylalanyl-L-valyl> cyclo(D-valyl-L-methyl Alanyl-L-valyl-D-α-hydroxyisovaleryl-D-valyl-L 50. 24
pivaloyl-D- Alanyl-N-isopropyl-D-prolinamide monohydrate 48. 48
(R)- Alaproclate........................... 48. 39a
Albine perchlorate 58. 7
bis(o-phenylene)-bis(pyridine-2,6-di Aldimino) lead(ii) diperchlorate dihydrate 34. 11
delphinium Alkaloid A hydroiodide................ 58. 67
carda-9,20-dienolide-3-O-methyl-4,6-dideoxy-D- Allo-2-hexosulose <3β,11-trihydroxy-12-oxo-5β- 45. 57
(+)- Allohimachalyl-p-bromobenzoate 53. 32a
ethylenediammonium (+)- Allohydroxycitrate monohydrate 2. 26
Allomatrine........................... 58. 12
N-oxy- Allomatrine......................... 58. 15
Allopsoralen 38. 30
3,4,6-tri-O-acetyl-1,2-O-(R)-ethylidene-α-D- Allopyranose........................ 45. 31
o> 3-(β-L-mycarose)-5-(β-D-4,6-dideoxy-3-keto Allose)-13-(β-D-mycinose)-lankamycin-11-α-hydr 50. 35
ioglycerol acid methyl ester)-4a,5-dihydro-iso Alloxazine ....   [3-methyl-4a-(methylene-dith 36. 18
e] 7,8-dimethyl-iso Alloxazine-10-acetic acid tyramine tetrahydrat 36. 11
5,5-di- Allyl-1-(p-bromophenyl)-barbituric acid........ 43. 2
2- Allyl-2-azonia-7-azabiphenylene bromide........ 36. 7
1,6-di(O-triphenylmethyl)-2,5-anhydro-D,L- Altritol monohydrate 45. 69
-3-deoxy-3-iodo-2-O-(methylthio)-carbonyl-β-D- Altropyranose .... [1,6-anhydro 45. 13
tris(dimethylamino)-chloro-silane-(Aluminium chloride)................ 63. 3
manganese-((μ(2)-acetyl)-(μ(2)-bromo)-dibromo- Aluminium) ....[tetracarbonyl- 71. 3
Amarorine monohydrate 58. 5
N-acetyl-D,L-methionine-diethyl Amide 48. 35
(1R,3S,4R)-N,N-dimethyl-3-camphor-carbothio Amide 31. 21
pivaloyl-D-prolyl-L-prolyl-L-alanine-N-methyl Amide 48. 62
3-N-(2-oxo-5-benzimidazolyl)-(naphthoic acid)- Amide <-butoxycarbonyl-1-azophenyl)-2-hydroxy- 35. 74
dicyanocobyrinic-a,c-di Amide trihydrate 49. 25
bis(phenyl-(t-butyl)-phosphinic Amide) hydrogen chloride 64. 40
cyano-5,6-dihydrocobyrinic-pentamethylester-a- Amide-c-lactam hemihydrate ....[6-amino-di 49. 30
N-p-iodobenzyl-3-n-octan- Amido-pyridine-2,5,6-trione.......... 59. 22
dipicryl Aminato-silver 15. 3
potassium dipicryl Amine 16. 3
perchloro-triphenyl Amine 16. 15
N-nitro-1-cyanoethyl Amine 9. 1
trans-N,N-dimethyl-2-nitrovinyl Amine 3. 17
methyl-bis(8-hydroxy-2-quinolyl) Amine 35. 52
(R)-N-acetyl-1-(1-naphthyl)-ethyl Amine 24. 7
tris(2,3,4,5,6-pentachlorophenyl) Amine 16. 15
(RS)-N-acetyl-1-(1-naphthyl)-ethyl Amine 24. 8
(R)-N-trifluoroacetyl-1-(1-naphthyl)-ethyl Amine 24. 6
utylimino-1-phenylsulfonylimino-2-cyano-ethyl) Amine ....[t-butyl-(2-t-b 3. 53
(adamantane-2,2'-aziridine)-3'-ylidene)-methyl Amine .... [N-(1'-t-butyl-spiro 32. 59
-4H-1,3-dioxa-6-aza-2-phosphocine-2-oxy)ethyl) Amine [t-butyl-bis(2-(6-t-butyl-tetrahydro 64.136
triphenyl Amine bis(trichloro-antimony) 16. 16
tris((2-methyl-8-quinolyl)-oxy)-ethyl) Amine dihydrate..................... 35. 84
x-2-ene-1,5'-(5H)-dibenzo(a,d)cycloheptene)-4- Amine hydrochloride <,N-dimethyl-spiro(cyclohe 29. 8a
1,6-dihydro-6,6-dimethyl-1,3,5-triazine-2,4-di Amine hydrochloride hydrate <ophenyl)methoxy)- 33. 45
pyridinium 1-naphthyl Amine picrate 60. 1
tris((2-methyl-8-quinolyl)-oxy)-ethyl) Amine rubidium iodide................. 35. 85
hiomorpholino-3-carbonyl-L-histidinyl-L-prolyl Amine sesquihydrate....[3S,6R,-6-methyl-5-oxot 48. 51
5-n-propyl-orotyl-L-histidinyl-L-prolyl Amine tetrahydrate 48. 59
diphenyl Amine trichloro-antimony 16. 4
lithium di(t-butylmethylene- Amine).............................. 3. 42
2-phenyl-3-(N-p-methoxyphenyl) Amine-indole 35. 63
osphinate iodide] O-(2-trimethyl Ammonio-ethyl)-phenyl-(1-hydroxycyclohexyl)-ph 64. 85
1,3-di Ammonio-2-propanol sulfate monohydrate 3. 15
ii) dihydrate] N-(2- Ammonioethyl)-piperazinium pentachloro-copper(33. 19
α-phenylethyl Ammonium α-phenylethylacetate 2. 32
μ(2)-carbonyl)-nickel-tri-i> benzyl-trimethyl Ammonium (μ(3)-hydrido)-octacarbonyl-tetrakis(3. 43
(2)-carbonyl)-dicarbonyl-iron)-tr> tetraethyl Ammonium (μ(5)-carbido)-(dicarbonyl-rhodium-(μ 3. 33
tetramethyl Ammonium (ethylnitrosolato-O)-triphenylborate 62. 12

horothioate)]   triethyl Ammonium adenosine-5'-(O-p-nitrophenyl-O-phosp   47. 32
uoro)-dioxofluoro-uranium monohyd>   tetraethyl Ammonium aqua-dioxofluoro-uranium-(tri-μ(2)-fl   3. 35
                                       tetraethyl Ammonium aquo-tetrabromo-oxo-molybdenum........   3. 34
                                       tetraethyl Ammonium aquo-tetraiodo-oxo-molybdenum   3. 36
                                          benzyl Ammonium bis(dichloro-mercury) chloride........ 19. 2
                                       tetraethyl Ammonium bis(7,7,8,8-tetracyanoquinodimethane)   60. 7
                                        trimethyl Ammonium cadmium chloride 3. 12
                                         triethyl Ammonium closo-dodecaborane 3. 32
                                          methyl Ammonium diaqua-trichloro-manganese..........   3. 1
                          hexadecyl-trimethyl Ammonium dichloro-iodide 3. 54
(trifluoromethyl)-ethenylenedioxy)-> triethyl Ammonium diphenoxy-(o-phenylenedioxy)-(1,2-bis 64.114
                                       tetramethyl Ammonium hexa-cobalt-tetradecacarbonyl-carbide   3. 19
                                       tetramethyl Ammonium hexachloro-platinum 3. 16n
                                       tetramethyl Ammonium hexafluorophosphate 3. 20
-penta-rhodium]   tetra-n-butyl Ammonium hexakis(μ-carbonyl)-iodo-octacarbonyl   3. 50
s(trichloro-molybdenum) aquo-tetr>   tetraethyl Ammonium hydrogen (μ-hydrido)-bis(μ-chloro)-bi   3. 40
                                     tetra-n-butyl Ammonium hydrogen dichloromaleate 2. 20
                                     tetra-n-butyl Ammonium hydrogen dichloromaleate   2. 19
exahydrate]   tetraethyl Ammonium hydrogen dodecachloro-tri-ruthenium h   3. 38
x-2-ene-1,5'-(5H)-dibenzo(a,d)cycloheptene)-4- Ammonium hydrogen maleate <ethyl-spiro(cyclohe 29. 9a
2-(dicyclohexyl-acetoxy)-ethyl-triethyl Ammonium iodide................................   1. 40
                                    1-phenylethyl Ammonium mandelate 2. 28
                                       tetramethyl Ammonium nonabromo-di-bismuth(iii) 3. 25
<12,15,18-hexaoxacycloeicosa-2,4-diene t-butyl Ammonium perchlorate clathrate benzene solvate 61. 13
                          monopyrido-18-crown-6 t-butyl Ammonium perchlorate complex 40. 40
lo(15.3.1)heneicosa-1(20),17,19-triene t-butyl Ammonium perchlorate complex <aoxa-21-azabicyc 40. 40
                                     tetra-n-butyl Ammonium platinum-carbonyl acetonitrile solvat   3. 52
e]   sodium Ammonium tartrate tetrahydrate 2. 22
                                          methyl Ammonium tetrabromo-cadmium(ii) 3. 2
um(ii)]   trimethyl Ammonium tetrabromo-cadmium(ii) tribromo-cadmi   3. 14
                              1,4-butanediyl-di Ammonium tetrachloro-manganese 3. 28n
                              1,4-butanediyl-di Ammonium tetrachloro-manganese 3. 27n
                                                Ammonium tetraphenylborate 62. 17
                                                Ammonium tetraphenylborate 62. 16
imethyl-1,7-diaza-4,10-dioxacyclodecane benzyl Ammonium thiocyanate ....[N,N'-d 40. 18
19-hexaoxa-10,22-diazacyclotetracosane) benzyl Ammonium thiocyanate clathrate <I-1,4,7,13,16, 61. 6
                                     tetra-n-butyl Ammonium tri-iodide.......................   3. 48
                                         triethyl Ammonium tris(pyrocatechol)silicate.......... 63. 23
                                     tetra-n-butyl Ammonium trisulfur-trinitride 3. 49
                                         dimethyl Ammonium O,O-di-isopropyl-dithiophosphate ... 64. 22
ohydrate]   Ammonium 1-anilino-8-naphthalene-sulfonate mon 27. 16
2-carboxylate].   (S)-1-(p-bromophenyl)-ethyl Ammonium 1-methoxy-2-ethoxycarbonyl-aziridine- 32. 19
iodide]   trimethyl Ammonium 7,7,8,8-tetracyanoquinodimethane tri-   3. 11n
arbadodecaborate)]   triethyl Ammonium 8,8'-oxido-3-cobalta-bis(η(5)-1,2-dic 62. 5
)-carbonyl)-dicarbonyl-iron)-> bis(tetraethyl Ammonium) (μ(6)-carbido)-(dicarbonyl-iron-(μ(2   3. 37
  (N,N,N,N',N',N'-hexamethyl-hexamethylene-di Ammonium) (7,7,8,8-tetracyanoquinodimethane) 60. 21
m)]   bis(tetramethyl Ammonium) bis((μ(2)-thio)-dithio-oxo-molybdenu   3. 23
)-oxo-tungsten monohydrate] bis(tetramethyl Ammonium) bis((dithio-tungsten)-bis(μ(2)-thio)   3. 24
                           tetrakis(tetramethyl Ammonium) chloro-penta(trichloro-tin)-osmium   3. 26
num> tetraphenylphosphonium (benzyl-trimethyl Ammonium) dichloro-iron-bis(μ(2)-thio)-molybde 64.123
                                    bis(triethyl Ammonium) dihydronium octacyano-molybdenum(iv)   3. 31
                                   bis(tetramethyl Ammonium) hexachloro-tin(iv) 3. 22
                                   bis(tetramethyl Ammonium) hexachloro-tin(iv) 3. 21
m]   bis(trimethylbenzyl Ammonium) hexadecacarbonyl-tetra-iron-palladiu   3. 44
                                bis(trimethylbenzyl Ammonium) hexadecacarbonyl-tetra-iron-platinum   3. 45
ii)]   tris(tetra-n-butyl Ammonium) hexakis(isothiocyanato)-technetium(i   3. 51
rhodium]   tris(tetraethyl Ammonium) hydrido-pentacosacarbonyl-tetradeca-   3. 5
                              bis(n-propyl Ammonium) tetrachloro-manganese(ii)...........   3. 13
3)-thio)-tris(μ(3)-ethylthio>   tris(tetraethyl Ammonium) tris(μ(2)-ethylthio)-bis(tetrakis(μ(3. 39
                       1-phenyl-2,3-dimethyl-4-trimethyl Ammonium-pyrazol-5-one iodide 32. 52
        (+-)-N-methyl-2-(3,4-dihydroxyphenyl)ethyl Ammonium-2-sulfonate hemihydrate 17. 11
rafluoroborate]   2-t-butyl Ammonium-3-chloro-N,N-dimethylpropionamide tet   1. 23
                           7-isocyano-11(20),14-epi- Amphilectadiene....................... 54. 16
                                8-isocyano-10-cyclo Amphilectene 54. 15
                                     cyclohepta- Amylose dodecahydrate................... 45. 67
                                      cyclohexa- Amylose hexahydrate.................... 45. 62n
                                      cyclo-octa- Amylose hydrate........................ 45. 70
                                     cyclohepta- Amylose p-nitroacetanilide hydrate 45. 66
                                           N-t- Amyloxycarbonyl-L-proline 48. 33
                                     3β-acetoxy-α- Amyrin 56. 8
                                     3β-benzoxy-α- Amyrin 56. 9
                                     7α-hydroxy- Anaphalidin......................... 54. 26a
  9α-fluoro-11β,17β-dihydroxy-17α-methyl- Androst-4-en-3-one 51. 22
                                     16α-bromo- Androst-4-ene-3,17-dione 51. 8
                                     4-hydroxy- Androst-4-ene-3,17-dione 51. 16
                                     16α-bromo- Androst-4-ene-3,6,17-trione........... 51. 5

3β-hydroxy-16β-morpholino- Androst-5-en-17-one............................ 51. 43
6α-hydroxy-4,4-dimethyl- Androstan-3-one 51. 28
17β-iodoacetoxy-4,4-dimethyl-5α- Androstan-3-one 51. 42
17β-iodoacetoxy-4,4-dimethyl-19-nor-5α- Androstan-3-one 51. 33
17a-methyl-3β-pyrrolidino-17a-aza-D-homo-5α- Androstane 36. 43
1,2-seco-A-bis-nor-5α- Androstane-17β-acetate 51. 17
methyl 3β-acetoxy-17α-methyl-18-nor-5α- Androstane-17β-carboxylate 51. 44
Angelicin 38. 29
α,24,28-penta-acet> 21β-(2-O-acetyl-3,4-di-O- Angelyl-6-deoxy-β-glucopyranosyloxy)-3β,16α,22 56. 10
cis-6,12-olide] 9β-acetoxy-8α-epoxy- Angelyloxy-trans,trans-germacra-1(10),4-diene- 53. 29
O),4-diene-cis-6,12-oli> 9β-acetoxy-8α-epoxy- Angelyloxy-7α-hydroxy-trans,trans-germacra-1(1 53. 31
N-cyanomethyl- Angustifoline 58. 20
pyromellitic di Anhydride 38. 18
erythronolide A Anhydride 38.109
di-iodo-(maleic acid)-thio Anhydride 39.   2
dibromo-(maleic acid)-thio Anhydride 39.   1
2,3-bis(diphenylstibino)-maleic Anhydride 66. 18
o-3-phenylseleno-cyclohexane-1,2-dicarboxylic Anhydride ....[6-methoxy-4-ox 38. 70
endo-dimethano-naphthalene-4a,8a-dicarboxylic Anhydride <a,5,6,7,8,8a-decahydro-1,4:5,8-exo, 38. 59
,exo-dimethano-naphthalene-4a,8a-dicarboxylic Anhydride <3,4,4a,5,8,8a-octahydro-1,4:5,8-exo 38. 57
,exo-dimethano-naphthalene-4a,8a-dicarboxylic Anhydride <4a,5,6,7,8,8a-decahydro-1,4:5,8-exo 38. 58
phenothiazine - pyromellitic di Anhydride complex 60. 17
o(2.2.2)octa-5,7-diene-(2,3-dicarboxylic acid Anhydride) ....[5,6:7,8-dibenzo-bicycl 38. 88
nyl-α-hydroxy-cyclohexyl-phosphinic acid chlor Anhyrdide ....   [phe 64. 51
potassium p-chlor Anil acetone solvate 18.   1
tetrathiafulvalene-chlor Anil complex 60.   3
p-tolyl-glyoxylic acid p-chloro Anilide 16. 12
rifamycin B p-iodo Anilide acetone pentahydrate 50. 31
rifamycin Y p-iodo Anilide dimethylsulfoxide monohydrate 50. 30
piperidinium-acetyl-(m-chloro Anilide) chloride 33. 54
piperidinium-acetyl-(p-chloro Anilide) chloride 33. 55
p-ethoxybenzylidene-p-n-butyl Aniline...................................... 17. 30
4-ethoxy-salicylidene-4'-butyl Aniline...................................... 17. 31
N-(p-bromobenzylidene)-p-chloro Aniline...................................... 16.   6
N-(p-bromobenzylidene)-p-methyl Aniline...................................... 16.   8
N-(p-chlorobenzylidene)-p-bromo Aniline...................................... 16.   7
N-(p-methylbenzylidene)-p-bromo Aniline...................................... 16.   9
4-propoxy-salicylidene-4'-butyl Aniline...................................... 17. 36
N-(p-chlorobenzylidene)-p-methyl Aniline...................................... 16. 10
N-(p-methylbenzylidene)-p-chloro Aniline...................................... 16. 11
ethoxy-N,N-bis(α-cyano-hexafluoro-isobutenyl)- Aniline   [p-m 16. 13
o-chloro-N-(piperid-1-yl-acetyl)- Aniline hydrochloride 33. 56
o-chloro-N-(piperid-1-yl-acetyl)- Aniline hydrochloride monohydrate 33. 57
4-nitro Aniline 4-nitro-N-sulfinylaniline 16.   1
-benzoyl-N,N-dimethyl-N'-(N-methyl-2,4-dinitro Anilino)-isourea ....[O 13. 16
tris(Anilino)-phosphine 64. 92
D,L-diethyl Anilino-(3-hydroxybenzyl)-phosphonate 64. 82
thylidene)-1,2,3,4-tetrahydro-naphthalene] 4- Anilino-3-methoxycarbonyl-1-(methoxycarbonylme 27. 29
ammonium 1- Anilino-8-naphthalene-sulfonate monohydrate.... 27. 16
4-(Anilinomethylene)-tetrahydrothiophen-3-one 39. 34
ne] 2-methyl-3-p- Anisyl-5-(ethylamino)-carbamoyl-1,2,4-thiazoli 41. 45
2-methyl-3-p- Anisyl-5-p-tosylimino-1,2,4-thioazoline........ 41. 62
1-methylbenz(a) Anthracene 29.   6
9-bromo-10-methyl Anthracene 26.   3n
9-bromo-10-methyl Anthracene 26.   4
10-bromo-1,8-diphenyl Anthracene 26.   5
(+-)-trans-1,2-dihydro-1,2-dihydroxybenz(a) Anthracene 29.   4
trans-10-benzyl-9-phenyl-9,10-dihydro-9-arsa- Anthracene 65. 24
9,10-dihydro-10-phenyl-9,10-o-benzeno-9-arsa- Anthracene 65. 23
+-)-trans-10,11-dihydro-10,11-dihydroxybenz(a) Anthracene ....[(29.   5
t-butyl-9,10-dihydro-9,10-o-benzeno-9-phospha- Anthracene ....[3- 64.120
Anthracene - tetracyanobenzene complex 60. 31
Anthracene - tetracyanobenzene complex 60. 30
Anthracene - tetracyanobenzene complex 60. 25
Anthracene - tetracyanobenzene complex 60. 26
Anthracene - tetracyanobenzene complex 60. 29
Anthracene - tetracyanobenzene complex 60. 28
Anthracene - tetracyanobenzene complex 60. 27
,10-dihydro-9-methyl-9,10-o-benzeno-9-arsonia- Anthracene chloride monohydrate............. [9 65. 12
-dienylidene complex] Anthracene 1,1'-bi-3,5-dichloro-4-oxo-hexa-2,5 26.   1
complex] Anthracene 3,3',5,5'-tetrachloro-diphenoxinone 26.   1
<μ-(9,10-bis(ethylcarboxylato-ethoxycarbonyl)- Anthracene)-porphyrin methylene chloride solv> 49. 31
(2.4)(9,10) Anthracenophane............................... 26.   6
(2.5)(9,10) Anthracenophane............................... 26.   7
Anthralin 26.   2
Anthramycin methylether monohydrate........... 50. 12

methyl-N-(biphenyl-2-yl)-N-(1-naphthyl)- Anthranilate 48. 78
methyl-N-(biphenyl-2-yl)-N-(1-naphthyl)- Anthranilate 48. 79
methyl-N-(biphenyl-2-yl)-N-(1-naphthyl)- Anthranilate 48. 77
N-acetyl- Anthranilic acid 13. 5
1,8-dihydroxy-9- Anthrone 26. 2
10,10'-ethanediylidene-di Anthrone 28. 19
di Anthronylidene-ethane 28. 19
tri-p-tolyl- Antimony 66. 14
urea trifluoro- Antimony 8. 2
diphenylamine trichloro- Antimony 16. 4
bis(t-butylperoxy)-triphenyl- Antimony 66. 17
is(trichlorophosphazeno)-carbenium hexachloro- Antimony ....,.[tr 64. 1
yl-catecholato-antimony triphenyl-catecholato- Antimony .... [aquo-triphen 66. 16
sodium fluoro-oxo-dioxalato- Antimony oxonium hydrate 66. 2
2,2'-dithienyl Antimony trichloride 39. 19
aquo-triphenyl-catecholato- Antimony triphenyl-catecholato-antimony........ 66. 16
triacetato- Antimony(iii) 66. 8
acetato-diphenyl- Antimony(iii) 66. 13
tris(monothioacetato)- Antimony(iii) 66. 7
(R)-tris(trifluoroacetato)- Antimony(iii) 66. 6a
tris(diphenyl-dithiophosphinato)- Antimony(iii) 66. 19
is(pyridinium) di-μ-chloro-μ-oxo-bis(dichloro- Antimony(iii)) .... [b 33. 8
μ-oxo-bis(trichloroacetylacetonato- Antimony(v)) 66. 11
μ-oxo-di-μ-trifluoroacetato-bis(trifluoro- Antimony(v)) 66. 3
μ-fluoro-μ-trifluoroacetato-bis(tetrafluoro- Antimony(v)) 66. 1
tetrachloro- Antimony(v)-dimethoxyphosphate 66. 5
tetrachloro- Antimony(v)-dimethylamino-methoxyphosphate 66. 9
2,2'-bipyridyl-(trichloro- Antimony) 66. 10
triphenylamine bis(trichloro- Antimony) 16. 16
sulfide)] 1,7-bis(pentachloro- Antimony-oxo)-cyclododecasulfur tris(carbon di 11. 3
Antimony-tris(O,O-diethylphosphoro-dithioate) 66. 12
D- Arabino-γ-lactone 45. 1
2,3,4,6-tetra-0-acetyl-1,5-anhydro-D- Arabino-hex-1-enitol 45. 30
,3'-(methyl-4,6-0-benzylidene-2,3-dideoxy-α-D- Arabino-hexopyranoside)) <1-acetylaziridine)-2 45. 43
zin-3-one-1,1-dioxide] 3,2'-anhydro-4-(β-D- Arabinofuranosyl)-5,6-dihydro-2H-1,2,4-thiadia 45. 14
9α-D- Arabinofuranosyl-adenine 47. 16
2,2'-anhydro-1-β-D- Arabinofuranosyl-2-thio-uracil 47. 2
e] 9β-D- Arabinofuranosyl-8-morpholino-adenine dihydrat 47. 30
9β-D- Arabinofuranosyl-8-n-butyl-amino-adenine 47. 31
methyl-3,4-0-ethylidene-β-L- Arabinopyranose..................... 45. 15
e-2-deoxy-3-0-((methylthio)-thiocarbonyl)-α-D- Arabinopyranoside [methyl 4,6-0-benzyliden 45. 36
e-3-deoxy-2-0-((methylthio)-thiocarbonyl)-α-D- Arabinopyranoside [methyl 4,6-0-benzyliden 45. 37
β-L- Arabinose 45. 2n
7,2'-anhydro-β-D- Arabinosyl-orotidine 47. 12
L- Arginine L-ascorbate 48. 20
methionyl-glutamyl-histidinyl-phenylalanyl- Arginyl-tryptophanyl-glycine 48. 82
dodecamethyl-hexasila-tetra- Arsa-adamantane....................... 65. 8
trans-10-benzyl-9-phenyl-9,10-dihydro-9- Arsa-anthracene........................ 65. 24
9,10-dihydro-10-phenyl-9,10-o-benzeno-9- Arsa-anthracene........................ 65. 23
c> 1,3-bis(1,3,3-triethyl-indolenine-2-yl)-2- Arsa-trimethinium tetrafluoroborate methylene 35. 78
1-methyl Arsa-2,3,4,5-bis(o-carborano-cyclopentane) 65. 3
benzene solvate] 2,4,6,7-tetraphenyl-1- Arsa-2,3,7-triaza-8-oxabicyclo(3.3.0)oct-3-ene 65. 25
1,3,5,7-tetra- Arsa-2,4,6,8-tetraoxa-adamantane 65. 1
s(trimethylsilyl)-tricyclo(2.2.1.0(2,6))hepta- Arsane .... [3,5,7-tri 65. 5
Arsatriptycene 65. 11
abicyclo(3.3.1)nona-2,3,5(9)-triene hexafluoro Arsenate <λ(4),5λ(4)-dithia-2,4,6,8,9-penta-az 41. 6
cyclo-tetrakis(phenyl Arsenic-sulfide) 65. 22
guanidinium tetramolybdo-dimethyl Arsinate monohydrate 65. 2
tris(morpholino)- Arsine 65. 7
triethyl Arsine-sulfide 65. 4
trans-dichloro-(2,11-bis(diphenyl Arsinomethyl)-benzo(c)phenanthrene)-platinum 65. 26
sityl-3H,7H-1,5,2,4,6,8,3,7-dithia-tetra-azadi Arsocine .... [3,7-dime 65. 10
-3H,7H-1,5,2,4,6,8,3,7-dithia(1,5S)tetra-azadi Arsocine .... [3,7-diphenyl 65. 6
2-methyl-2,2'-spiro-bis(1,3,2λ(5)-benzoxaz Arsoline) 65. 9
9,10-dihydro-9-methyl-9,10-o-benzeno-9- Arsonia-anthracene chloride monohydrate........ 65. 12
thenium)] tetraphenyl Arsonium (μ(6)-hydrido)-hexakis(tricarbonyl-ru 65. 13n
tetraphenyl Arsonium (decachloro-μ-nitrido-di-tungsten).... 65. 19
ro-tungsten(v))] tetraphenyl Arsonium μ-selenido-μ-diselenido-bis(tetrachlo 65. 18
chloride solvate] tetraphenyl Arsonium pentachloro-oxo-niobium(v) methylene 65. 17
tetraphenyl Arsonium tetrachloro-oxo-vanadium(v) 65. 14
-hexakis(dicarbonyl-rutheniu> bis(tetraphenyl Arsonium) μ(6)-carbido-tetrakis(μ(2)-carbonyl) 65. 15
bis(tetraphenyl Arsonium) dihydrido-dodeca-nickel-carbonyl 65. 16
ne solvate] tris(tetraphenyl Arsonium) hydrido-dodeca-nickel-carbonyl aceto 65. 21n
ne solvate] tris(tetraphenyl Arsonium) hydrido-dodeca-nickel-carbonyl aceto 65. 20
Asbestinin-1-diol 54. 10
L-arginine L- Ascorbate 48. 20

                                L-serine L- Ascorbic acid 48.   8
                bis(6-deoxy-6-chloro-L- Ascorbic acid) nitromethane solvate............ 38.   5
                                      bromo Aureol acetate 53. 34a
                                 desacetyl Austin 38.122a
e tetrahydrate]   1,6,20,25-tetra- Aza(6.1.6.1)paracyclophane hydrochloride duren 61. 18
-methoxyphenyl)-2-trans-4-dithio-1,3,2,4-thia- Aza-λ(5),λ(5)-diphosphetidine <methyl-2,4-di(p 64. 71
              1-methyl-1,3,5,7-tetra- Aza-adamantan-1-ium octa-iodide................. 37.   2
                                       S-8- Aza-adenosyl-L-homocysteine dihydrate 48. 40
                       2-amino-1,3-di Aza-azulene hydrobromide monohydrate 35.   8
mic-1,1,3,6,6,8-hexamethyl-3a,5a,8a,10a-tetra- Aza-cis-10b,10c-perhydropyrene .... [race 36. 28
4).0(10,12).0(11,13))hexad> dimethyl-15,16-di Aza-octacyclo(6.6.2.0(2,7).0(3,5).0(4,6).0(9,1 37. 11
bis(dimethylamino)-3-ethoxy-1,2,5-triphenyl-4- Aza-pentamethinium perchlorate .... [1,5- 19. 16
                                      2,7-di Aza-spiro(4.4)nonane-1,6-dione 35.   3
               7-methyl-1,4-dithia-7- Aza-spiro(4.4)nonane-6,8-dione 39. 13
       1,3-bis(1,3,3-triethyl-indolenine-2-yl)-2- Aza-trimethinium tetrafluoroborate 35. 81
       1,3-bis(1,3,3-trimethyl-indolenine-2-yl)-2- Aza-trimethinium tetrafluoroborate 35. 71
one]   12-benzoylimino-2-phenyl-3-oxa-1- Aza-trispiro(4.0.5.2.5.0)nonadec-1-ene-4,13-di 40. 56
                    bis(tetrazole)]   4,6-di Aza-A,B-bishomo-cholest-4a-eno-(4,3-d)(6,7-d)- 51. 60
             17a-methyl-3β-pyrrolidino-17a- Aza-D-homo-5α-androstane 36. 43
acarbony> (2-methylene-1-tetracarbonyl-2,3-di Aza-1-mangana-cyclobutan-4-one)-manganese pent 71. 25
             4-methyl-2,6,6-triphenyl-2,3-di Aza-1-phosphabicyclo(3.1.0)hex-3-ene 64.116
        (3-phenylaminocarbonyl)-(4-phenylmethyl)-4- Aza-1,2-dithian-5-one 41. 60
2,9-bis(p-methoxyphenyl)-5,12-dimethyl-5,12-di Aza-1,3,8,10-tetrathiacyclotetradecane-6,13-d> 41. 81
<α-H-8α-isopropyl-11α-methyl-1α-methoxy-6,9-di Aza-12-oxatricyclo(7.3.0.0(2,6))dodecane-7,10> 58. 22
-dione]   8-benzyl-1-hydroxy-11-methyl-6,9-di Aza-12-thiatricyclo(7.3.0.0(2,6))dodecane-7,10 48. 49
            (1S,4S)-N-acetyl-3-oxo-5- Aza-2-oxabicyclo(2.2.1)heptane 40.   4
5-dimethyl-2,2-diphenyl-3,4-methoxycarbonyl-1- Aza-2-phosphacyclo-pent-1-ene ....[5, 64.113
yl-thio(2-(6-t-butyl-tetrahydro-4H-1,3-dioxa-6- Aza-2-phosphocine-2-oxy)ethyl)amine .. [t-but 64.136
),12,14-triene acetone s> 2,6,10-triphenyl-6- Aza-2,10-diphosphabicyclo(9.4.0)pentadeca-11(1 64.147
),12,14-triene]   2,10-diphenyl-6-methyl-6- Aza-2,10-diphosphabicyclo(9.4.0)pentadeca-11(1 64.140
ethyl-2,2,4,4-tetrakis(trifluoromethyl)-1,3-di Aza-2,4-diphosphetidine........[1,2,3,4-tetram 64. 28
hylsilyl)amino)-1,3-bis(trimethylsilyl)-1,3-di Aza-2,4-disilacyclobutane <is(isopropyl(trimet 63. 49
o)-2,4-dimethyl-1,3-bis(trimethylsilyl)-1,3-di Aza-2,4-disilacyclobutane <trimethylsilyl)amin 63. 39
ityl-2,2,4,4,6,6,8,8-octamethyl-1,3,5,7-tetra- Aza-2,4,6,8-tetrasila-bicyclo(3.3.0)octane <es 63. 44
olvate] 1,3-bis(trimethyl-tin)-1,3,5,7-tetra- Aza-2,4,6,8-tetrathiocin-2,2-dioxide benzene s 69. 12
                  3-thioxo-4-phenyl-4- Aza-2,7-dithiobicyclo(3.3.0)octane-7,7-dioxide 41. 35
     2,4,4-triphenyl-3-methoxy-6-methyl-1,2-di Aza-3-phosphacyclohex-1(6)-ene 64.121
              4,5-diethyl-3,6-dimethyl-1,2-di Aza-3,6-diborinane 62.   9
                                      1,8-di Aza-3,8-dioxatetradecane-2,7-dione 40. 17
<,7-dimethyl-1λ(6),5λ(6)-dithia-2,4,6,8-tetra- Aza-3,7-diphosphorocine-1,1,5,5-tetraoxide ac> 64. 42
<is(3-(trifluoromethyl)-phenyl)-1,3,5,7-tetra- Aza-4γ(5)-phospha-spiro(3.3)heptane-2,6-dione 64.122
              yanate]   N,N'-dimethyl-1,7-di Aza-4,10-dioxacyclodecane benzylammonium thioc 40. 18
                                        1- Aza-4,4-dimethyl-cyclohexane-2,6-dione 33. 27
itrile]   N,N'-bis(2-cyanoethyl)-1,10-di Aza-4,7,13,16-tetraoxacyclo-octadecane malonon 40. 46
      9-methyl-3,7-diphenyl-1,3,4,6,7,9-hexa- Aza-5λ(4)-phosphabicyclo(3.3.1)nonane-5-thione 64. 73
<-ethoxycarbonyl-2-methoxycarbonyl-9-methyl-9- Aza-5-oxabicyclo(4.3.0)nona-(1(6),2)-dien-4-o> 38. 80
e]   1,7,8,9-tetrachloro-3-phenyl-4- Aza-5-oxatricyclo(5.2.1.0(2,6))decane-3,8-dien 40. 32
e]   1,7,8,9,10-pentachloro-3-phenyl-4- Aza-5-oxatricyclo(5.2.1.0(2,6))decane-3,8-dien 40. 31
luoromethyl)-3,4-bis(methoxycarbonyl)-6-oxa-1- Aza-5-phosphabicyclo(3.2.0)hept-3-ene <7-(trif 64.135
              2-methylthio-Δ(2)-1,3-di Aza-6,9-dioxa-cycloundecene hydroiodide 40. 13
methyl-7-phenoxy-7-thioxo-3-oxa-1,5,6,8-tetra- Aza-7-phosphabicyclo(3.3.1)nonane ....[6,8-di 64. 48
te]   2,4,6,7-tetraphenyl-1-arsa-2,3,7-tri Aza-8-oxabicyclo(3.3.0)oct-3-ene benzene solva 65. 25
t-butylammonium pe> 3,6,9,12,15-pentaoxa-21- Azabicyclo(15.3.1)heneicosa-1(20),17,19-triene 40. 40
-2,16-dione potassi> 3,6,9,12,15-pentaoxa-21- Azabicyclo(15.3.1)heneicosa-1(20),17,19-triene 40. 39
                                    trans-2- Azabicyclo(2.1.0)pentane-3-(S)-carboxylic acid 50.   1
               7-isopropylidene-2,3-di Azabicyclo(2.2.1)hept-2-ene..................... 37.   3
         1,3,4,7,7-pentachloro-5-phenyl-2- Azabicyclo(2.2.1)hept-2-ene.................... 37.   8
l-3-oxo-6-exo-(2,6-di-iodo-4-methylphenoxy)-2- Azabicyclo(2.2.1)heptane .... [2-benzy 37. 15
                                        1- Azabicyclo(2.2.2)octane....................... 37.   1
carboxylate)]   7-oxo-3-methylthio-2-thia-1- Azabicyclo(3.2.0)hept-3-ene-4-(p-nitro-benzyl- 41. 50
1-(3,5-di-t-butyl-2-hydroxyphenyl)-2-methyl-2- Azabicyclo(3.2.0)hept-6-en-3-one <-di-t-butyl- 35. 76
<methyl-6-(p-tolyl-thio-imino)-7-oxo-4-thia-1- Azabicyclo(3.2.0)heptane-2-carboxylic-acid-p-> 41. 80
                        (+-)-8,8-dimethyl-6,7-di Azabicyclo(3.3.0)octa-1,6-dien-7-oxide 35.   9
yl-6-(3-indolyl)-4-(isopropenyl)-(2,3)benzo-1- Azabicyclo(3.3.0)octa-2,4-diene [6,8,8-trimeth 36. 45
                               7-thia-1- Azabicyclo(3.3.0)octa-2,4-dione 41. 11
             2,3,6,7-tetramethyl-1,5-di Azabicyclo(3.3.0)octa-2,6-diene-4,8-dione 35. 17
                 4,6-dimethyl-1,5-di Azabicyclo(3.3.0)octa-3,6-diene-2,8-dione 35.   7
          syn-3,4,6,7-tetramethyl-1,5-di Azabicyclo(3.3.0)octa-3,6-diene-2,8-dione 35. 18
           3,7-dichloro-4,6-diphenyl-1,5-di Azabicyclo(3.3.0)octa-3,6-diene-2,8-dione 35. 47
             3,7-dichloro-4,8-diphenyl-1,5-di Azabicyclo(3.3.0)octa-3,7-diene-2,6-dione 35.   6
           3,7-dichloro-4,8-diphenyl-1,5-di Azabicyclo(3.3.0)octa-3,7-diene-2,6-dione 35. 46
y-1-(2-hydroxyethyl)-6,7,8-trimethyl-2-thia-7- Azabicyclo(3.3.0)octane hemihydrate [5-hydrox 39. 39
<)-thionia-3λ(4),5λ(4)-dithia-2,4,6,8,9-penta- Azabicyclo(3.3.1)nona-2,3,5(9)-triene hexaflu> 41.   6
              3,7-dimethyl-3,7-di Azabicyclo(3.3.1)nonane-9-spiro-5'-hydantoin 37.   7
         benzyl 4-methoxy-7-azido-3-oxa-1- Azabicyclo(4.2.0)octan-8-one-2-carboxylate 40. 36
arbonyl)oxy-ethyl)-8-oxo-2,2-dimethyl-3-oxa-1- Azabicyclo(4.2.0)octane <(1-(p-nitrobenzyloxyc 40. 45

o(t-butyl)methylene)-1β-methyl-3-oxa-9-thia-7- Azabicyclo(4.3.0)non-7-en-4-one <cyano-2-(cyan 41. 76
o(t-butyl)methylene)-1β-methyl-3-oxa-9-thia-7- Azabicyclo(4.3.0)non-7-en-4-one <cyano-2-(cyan 41. 75
2,4,6-tri-t-butyl-7,8,9-dithia- Azabicyclo(4.3.0)nona-1(9),2,4-trien-7-one 41. 68
2,4,6-tri-t-butyl-7,8,9-dithia- Azabicyclo(4.3.0)nona-1(9),2,4-triene 41. 69
4-methyl-9-p-chlorophenyl-7-oxa-2,9-di Azabicyclo(4.3.0)nona-2,4-dien-8-one 40. 26
6-methyl-9-p-chlorophenyl-7-oxa-2,9-di Azabicyclo(4.3.0)nona-2,4-dien-8-one 40. 27
8-oxa-1- Azabicyclo(4.3.0)nonan-9-one-2-carboxylic acid 40. 8
8-hydroxy-8-methyl-6-(2'-methyl-hexylidene)-1- Azabicyclo(4.3.0)nonane hydrochloride 58. 25a
7,8-benzo-9-cyano-3,5,10-trimethyl-3,5-di Azabicyclo(4.4.0)deca-7,9-diene-2,4-dione 36. 16
,8-benzo-8-cyano-3,5-dimethyl-10-phenyl-3,5-di Azabicyclo(4.4.0)deca-7,9-diene-2,4-dione   [7 36. 34
i) ethylthiolate 4,7,13,16,21-pentaoxa-1,10-di Azabicyclo(8.8.5)tricosane benzene solvate <(i 49. 27
<ato)-iron(ii) 4,7,13,16,21,24-hexaoxa-1,10-di Azabicyclo(8.8.8)hexacosane chlorobenzene sol> 49. 23
2-allyl-2-azonia-7- Azabiphenylene bromide 36. 7
N,N,N',N'-tetrakis(1,3-dimethyl-1,3,2-di Azaborolidinyl)-hydrazine 62. 11
1,4,7,15-tetraoxa-12,18-di Azacyclodocosane-8,11,19,22-tetrone.......... 40. 42
bis(1,4,7,10-tetra- Azacyclododecane)-phosphorane 64. 78
3,3-dimethyl-1,5,13-trioxa-10,16-di Azacycloeicosane-6,9,17,20-tetrone 40. 44
1,4,9,12-trioxa- Azacyclohexadecane-3,10,13,16-tetrone 34. 8
4-t-butyl-1- Azacyclohexan-2-one.......................... 33. 33
3-t-butyl-4-methyl-2-oxo-1-oxa-2-thia-3- Azacyclohexane 42. 3
<0,22-dimethyl-1,4,7,13,16,19-hexaoxa-10,22-di Azacyclotetracosane) benzylammonium thiocyana> 61. 6
meso-5,5,7,12,12,14-hexamethyl-1,4,8,11-tetra- Azacyclotetradecane dihydrate 34. 9
1,4,7,10-tetra- Azacyclotetradecane-3,8,11,14-tetrone 34. 6
6''-kl:2''',1''',6'''-na)(1,3,5,8,10,12)-hexa- Azacyclotetradecine <6-de:2',1',6'-gh:2'',1'', 36. 39
-dimesityl-3H,7H-1,5,2,4,6,8,3,7-dithia-tetra- Azadiarsocine   [3,7 65. 10
henyl-3H,7H-1,5,2,4,6,8,3,7-dithia(1,5S)tetra- Azadiarsocine ....   [3,7-dip 65. 6
-t-butyl-2,4-bis(pentafluorophenyl)-1,3,2,4-di Azadiboretidine.....................[1,3-di 62. 13
thylamino)-1,3-bis(trimethylsilyl))-1,3,2,4-di Azadiphosphetidine <spiro-2'-(cis-2,4-bis(dime 64. 69
<thylamino)-1,3-bis(trimethylsilyl)-1,3,2,4-di Azadiphosphetidine)-2',4'-dispiro-2,2''-(4,5-> 64. 99
-acetylpyridine)-1,12-dimethyl-2,5,8,11-tetra- Azadodeca-1,11-diene) perchlorate <(1,12-bis(2 33. 90
l-1,7-diphenyl-4-(2,4,6-trinitrophenyl)-2,6-di Azahepta-2,4-diene ....   [3,5-dimethy 15. 6
<rophenylsulfonylamino)-13-oxa-14-thia-1,15-di Azapentacyclo(10.4.2.0(2,7).0(8,16).0(12,16))> 42. 12
(18-ethoxy-15-methyl-5-phenyl-3,5,7,15-tetra- Azapentacyclo(7.4.3.2(2,8).0.0(3,7))octadec-1> 37. 20
0,12,17-trien> 5-methyl-14,16-dioxa-3,5,7-tri Azapentacyclo(7.4.3.2(2,8).0.0(3,7))octadeca-1 38. 56
0,12,17-trien-4,> 15-methyl-5-phenyl-5,15-tri Azapentacyclo(7.4.3.2(2,8).0.0(3,7))octadeca-1 37. 19
0,12,17-triene-4,6-dione] 5-methyl-15-oxa-5- Azapentacyclo(7.4.3.2(2,8).0.0(3,7))octadeca-1 38. 84
tadeca-12,14-dione]   13-phenyl-11,13,15-tri Azapentacyclo(8.5.2.0(2,6).0(2,8).0(11,15))hep 37. 17
eso-1,1,3,6,6,8-hexamethyl-3a,5a,8a,10a-tetra- Azaperhydropyrene ....   [m 36. 26
mic-1,1,3,6,6,8-hexamethyl-3a,5a,8a,10a-tetra- Azaperhydropyrene ....   [race 36. 27
bis(9,10-di Azaphenanthrene) tetracyanoethylene............ 60. 15
1-oxo-1,2-dihydro-2,3-di Azaphenothiazine 41. 27
1-chloro-10-methyl-2,3-di Azaphenothiazine 41. 34
2,2,6,6-tetramethyl-4-phenyl-4-oxo-1-oxyl-1,4- Azaphosphorinane 64. 67
yl-3-p-tolyl-1,2,3,4-tetrahydro-1,3,2-benzo-di Azaphosphorine 2-oxide ....   [2-chloro-6-meth 64. 70
ichloro-1-dimethylamino-1,3,2,4,6,5-dithia-tri Azaphosphorine-1,3-dioxide ....   [3,5,5-tr 64. 2
1,3-bis(trimethylsilyl)-4-methyl-di Azastanna-boretidine dimer 63. 17
(E)-2,2'-di Azastilbene 33. 42
(E)-4,4'-di Azastilbene 33. 43
(E)-2',5,5'-tri Azastilbene 33. 40
(E)-2,2',5,5'-tetra- Azastilbene 33. 35
(E)-dibenzo(e,e')-3,3'-di Azastilbene 35. 56
(Z)-dibenzo(e,e')-3,3'-di Azastilbene 35. 57
(E)-6-methyl-6-azonia-2'- Azastilbene iodide 33. 52
<6-dibenzo-14-ethoxycarbonyl-8,10-diphenyl-1- Azatetracyclo(9.3.2.0(2,7).0(2,10))hexadeca-7> 37. 24
hylsilyl)-sulfo-di-imide)-bicyclo(3.3.1)penta- Azatetrathiane   [1-(S,S-dimethyl-N-(trimet 63. 2
2-(2'-propenyl)-8-toluenesulfonyl-7,8-di Azatricyclo(4.2.1.0(3,7))nonane.............. 37. 10
(+-)-4- Azatricyclo(4.3.1.0(3,7))decan-5-one 37. 4
10-thio-10-phospha-1,4,7-tri Azatricyclo(5.2.1.0(4,10))decane 64. 19
-dione] 3,8-dimethyl-1,10-di Azatricyclo(5.2.1.0(4,10))decane-3,7-diene-2,9 37. 5
e-> 4,6-diphenyl-10-ethoxycarbonyl-2,3,10 tri Azatricyclo(5.3.2.0(2,6))dodecane-3,8,11-trien 37. 21
,10-dione 6,6-di> 3,9-dimethyl-6-thia-1,11-di Azatricyclo(6.2.1.0(4,11))undecane-3,8-diene-2 41. 31
,10-dione] 3,9-dimethyl-6-thia-1,11-di Azatricyclo(6.2.1.0(4,11))undecane-3,8-diene-2 41. 30
ne-3,6-dione] 4-hydroxy-11-thia-2,6,12-tri Azatricyclo(6.3.1.0(4,12))dodecane-1(11),8-die 41. 21
2-hydroxy-4-isopropyl-7-benzyl-3-oxa-6,9-di Azatricyclo(7.3.0.0(2,6))dodecane-5,8-dione.... 40. 48
(p-bromobenzyloxycarbonyl)-2-hydroxy-3,6,9-tri Azatricyclo(7.3.0.0(2,6))dodecane-5,8-dione <e 36. 44
nzyl-3-(benzyloxycarbonyl)-2-hydroxy-3,6,9-tri Azatricyclo(7.3.0.0(2,6))dodecane-5,8-dione <e 36. 46
<bromo-3,6-diethyl-13-(phenoxycarbonyl)-3,6-di Azatricyclo(7.4.0.0(2,7))tridecane-2(7),9,11,> 36. 36
<tribromo-3,6-diethyl-10-ethoxycarbonyl-3,6-di Azatricyclo(7.4.0.0(2,7))tridecane-2(7),9,11,> 36. 24
(1RS,8SR,10SR,4(15)Z)-4-ethylidene-5-oxa-3- Azatricyclo(8.4.0.0(3,8))tetradecane 40. 35
5,5,7,12,12,14-hexamethyl-1,4,8,11-tetra- Azatricyclo(4.4.8))hexadecane hydrate.... 36. 29
6- Azauridine-5'-phosphate trihydrate 47. 1
o-1,2,3,4,5,6,7,8-octahydro-1,8-methano(1)benz Azecin-13-one pyridine solvate <-4-hydroxyimin 37. 16
henyl)-4-ethoxycarbonyl-3-methoxy-6-phenyl-2H- Azepin-2-one ....   [7-(4'-bromop 34. 10a
utyl-2-hydroxyphenyl)-1-methyl-2,3-dihydro-1H- Azepin-2-one ethanol solvate <yl-7-(3,5-di-t-b 34. 12
10,11-dihydro-5H-dibenzo(b,f) Azepine..................................... 36. 10
-1,5-dihydro-benzo(f)pyrazolo(3,4-c)(1,2,5)tri Azepine <-methyl-phenyl)-3,7-dimethyl-1-propyl 36. 35

nzo(b,f)-10,10,11,11-tetramethyl-10,11-disila- Azepine ‹hylaminopropyl)-10,11-dihydro-5H-dibe   63. 34
azino))-(5,6-a)-(6,7-benzo-4-phenyl-(1,3,5-tri Azepine)) chloroform solvate ‹henyl-(1,3,5-tri   36. 51
yl-2,3,4,4a,5,8-hexahydro-1H-pyridazino(4,5-d) Azepine-2,3-dicarboximide ‹yl-8-oxo-N,9-diphen   37. 23
3-chloro-3-cyano-1-cyclohexyl-4-phenylthio-2- Azetidinone...   32. 57
α-(1'S)-(p-nitrobenzyloxycarbonyl))oxy-ethyl-2- Azetidinone ‹2',2'-trimethylene-dithioethyl)-3   39. 71
2,4-dimethyl-3-phenyl-2-(diethylcarbamoyl)-1- Azetine N-oxide...................................   32. 58
amino)-2,2a,3,4,5,6-hexahydro-2a,6-methane-1H- Azeto(1,2-a)(1)benzazocin-12-one ‹enylsulfonyl   37. 14
tetraphenylphosphonium bis(Azido)-iodine 64.124
3β- Azido-2α-(toluene-p-sulfonamido)-5α-cholestane   51. 78
xide] Azido-2H-1,2,4,6-thiatriazin-3(4H)-one-1,1-dio   41. 2
arboxylate] 5- Azido-3-oxa-1-azabicyclo(4.2.0)octan-8-one-2-c   40. 36
benzyl 4-methoxy-7- Azine 42. 5
naphtho(1,8-cd:4,5-c'd')-bis(1,2,6)selenadi Azine hydrochloride.................................   41. 79
trifluoroper Azinium bromide....   [(E)-6-bromomethylene-5,   40. 24
6-dihydro-4,4-dimethyl-2-phenyl-4H-1,3,4-oxadi Azino-di-o-cresol 17. 34
5,5'-diethoxy-α,α'-dimethyl-α,α'- Aziridine)-2,3'-(methyl-4,6-0-benzylidene-2,3-   45. 43
dideoxy-α-D-arabino-hexopyra› spiro((1-acetyl Aziridine)-3'-ylidene)-methylamine   32. 59
N-(1'-t-butyl-spiro(adamantane-2,2'- Aziridine-2-carboxylate.... [(S)-1-(p-bromophe   32. 19
nyl)-ethylammonium 1-methoxy-2-ethoxycarbonyl- Aziridino-cyclotriphosphazene benzene clathrat   61. 3
e] hexa- Aziridinyl)-cyclotriphosphazene carbon tetrach   61. 2
loride anticlathrate] 2,2,4,4,6,6-hexa(1- Aziridinyl-cholestene 51. 74
3β-acetoxy-5β,6β-N-nitro- Azirine   32. 61
2,2-bis(p-methoxyphenyl)-3-methyl-2H- Azirinyl)-piperazine 32. 45
1,4-bis(3,3-dimethyl Azo)-1,3-dimethyl-buta-1,4-diene 9. 3
1,4-bis(phenyl Azo-diformaldehyde-dioxime 10. 1
(-)-(4S,8S)-α-bisabolol-8-p-phenyl Azo-phenylmethane   53. 35
2-methyl-3-phenyl-5-phenyl Azo-1,3,4-thiadiazoline......................   41. 54
5,6-dihydro-4-phenyl-2-phenyl Azo-4H-1,3,4-thiadiazine   41. 55
trans- Azobenzene 9. 2
2,4-diamino- Azobenzene hydrochloride dihydrate 16. 5
4-dimethylamino- Azobenzene-2'-carboxylic acid   13. 12
-hexahydro-2a,6-methane-1H-azeto(1,2-a)(1)benz Azocin-12-one ‹enylsulfonylamino)-2,2a,3,4,5,6   37. 14
methyl-6-phenyl-1,2,3,4-tetrahydro-1,5-benzodi Azocin-2-one ....[8-bromo-1-   35. 43
nyl-1-methyl-1,2,3,4-tetrahydro-5H-1,5-benzodi Azocin-2-one ....[8-chloro-6-phe   35. 44
-6-phenyl-1,2,3,4-tetrahydro-5H-1,4,5-benzotri Azocin-2-one [8-methyl-4-(p-toluenesulfonyl)   35. 69
1,4-dihydro-1,4-di Azocine...   34. 2
1,4-dihydro-1,4-di Azocine...   34. 1
1,4-bis(methylsulfonyl)-1,4-dihydro-1,4-di Azocine...   34. 4
1,4-bis(trimethylsilyl)-1,4-dihydro-1,4-di Azocine...   63. 13
1,4-bis(methoxycarbonyl)-1,4-dihydro-1,4-di Azocine...   34. 5
1,4-bis(dimethylcarbamoyl)-1,4-dihydro-1,4-di Azocine...   34. 7
,4,5,6,7,8,9,9aβ-decahydro-4-oxo-thieno(3,2-c) Azocine-3α-yl)-carbamic acid methyl ester ‹3aα   39. 72
hlorophenylsulfonylamino)-6,7,10,11-tetrahydro Azocino(1,2-a)indol-8(9H)-one .... [12-(p-c   36. 32
2-benzyl-1,7,10,16-tetraoxa-4,13-di Azocyclo-octadecane-3,14,18-trione   40. 49
‹nzo)-1-ethyl-4-(3-methoxy-phenylethyl)-1,4-di Azocyclohex-5-en-2-one)-3-spiro-5'-(3'-methyl›   36. 37
3-methyl-4-(2-carboxyvinyl)-1,2,5-oxadi Azole-2-oxide 40. 3
5-phenyl-1,2,4-oxadi Azole-3-carboxamide   40. 14
2-methyl-3-p-anisyl-5-p-tosylimino-1,2,4-thio Azoline....................................   41. 62
1,2,3-oxadi Azolo(4,3-c)(1,2,4)benzotriazinium-3-olate 40. 6
(E)-6-methyl-6- Azonia-2'-azastilbene iodide 33. 52
2-allyl-2- Azonia-7-azabiphenylene bromide.............   36. 7
(E)-4,4'-dimethyl-4,4'-di Azoniastilbene di-iodide 33. 62
(E)-6,6'-dimethyl-6,6'-di Azoniastilbene di-iodide 33. 63
3-carboxy-2-naphthalene-di Azonium iodide monohydrate 24. 4
4-morpholino-benzene-di Azonium tetrafluoroborate 40. 16
N,N'-dimethyl-1,7-di Azonium-4,10-dioxacyclodecane perchlorate ..   40. 19
yl)-(naphthoic acid)-› 1-(2-butoxycarbonyl-1- Azophenyl)-2-hydroxy-3-N-(2-oxo-5-benzimidazol   35. 74
disodium cis-syn-4-sulfonatobenzene-di Azotate trihydrate 11. 8
2-(p-tolylsulfonyloxy-N,N,O- Azoxy)-adamantane 31. 43
2-adamantyl-O,N,N- Azoxy-p-toluenesulfonate 31. 43
4,4'- Azoxyphenetole 17. 27
perdeutero-4,4'- Azoxyphenetole 17. 25
2,2',6,6'-tetrakis(propylthio)-3,3'- Azoxypyridine   33. 85
cis-2-t-butyl-perhydro Azulen-4-one-(p-bromophenylsulfonylhydrazone)   27. 28
trans-2-t-butyl-perhydro Azulen-4-one-(p-tosylhydrazone)................   27. 30
2-amino-1,3-diaza- Azulene hydrobromide monohydrate   35. 8
S,8aSR)-5-methylene-octahydro-4H-3a,6-methano- Azulene-4-carboxylic acid .... [(3aRS,4RS,6R   31. 19
1,3,5-trimethyl Azuleno(1,8-cd)pyridazine   36. 13
α,8α,9aβ-decahydro-4,5-dihydroxy-4a,8-dimethyl Azuleno(6,5-b)furan-2(3H)-one ‹aα,4α,4aβ,5α,7a   38. 69
,8α,9aβ-(+-)-decahydro-4-hydroxy-4a,8-dimethyl Azuleno(6,5-b)furan-2,5(3H)-dione ‹,4α,4aβ,7aα   38. 64
Barbatusin 54. 17
Barbital 1-methylimidazole 43. 1
5,5-di-allyl-1-(p-bromophenyl)- Barbituric acid 43. 2
yl-2,5,8,11-tetra-azadodeca-1,11-diene) per› Barium (1,12-bis(2-acetylpyridine)-1,12-dimeth   33. 90
Barium cytidine-5'-phosphate hydrate 47. 9
Barium ethylenediamine-tetra-acetate hydrate   48. 26

valinomycin	Barium perchlorate hydrate	50.	36
beauvericin	Barium picrate complex toluene solvate	50.	32
vate]	Beauvericin barium picrate complex toluene sol	50.	32
1-methyl	Benz(a)anthracene	29.	6
(+-)-trans-1,2-dihydro-1,2-dihydroxy	Benz(a)anthracene	29.	4
(+-)-trans-10,11-dihydro-10,11-dihydroxy	Benz(a)anthracene	29.	5
1,1,2,3,4,5,6-heptaphenyl-1,4-dihydro	Benz(e)-as-indacene	29.	12
3a,4,9,9a-tetrahydro-2-methyl-4β-phenyl	Benz(f)isoindoline	36.	30
1,2,4,5-tetrahydro-7-methoxy-3H-	Benz(g)indazol-3-one monohydrate	36.	4
ne]	Benzaldehyde 4-benzoyl-2-phenylthiosemicarbazo	13.	19
p-dimethylamino-	Benzaldehyde-p-nitrophenylhydrazone benzene so	15.	5
lvate]	Benzaldoxime	10.	3
Z-p-chloro-	Benzamide	13.	2
p-chloro	Benzamide	13.	15
N,N'-ethylene-di	Benzamide ·	64.138	
N-(triphenylphosphoranylidene)-	Benzamide [N-(2-(4-(5-chloro-2-oxo-	35.	64
1-benzimidazolinyl)piperidino)-ethyl)-p-fluoro	Benzamide hydrochloride	13.	8
4-amino-N(2-(diethylamino)-ethyl)-	Benzamide hydrochloride monohydrate [4-amino	13.	11
-5-chloro-N-((2-diethylamino)-ethyl)-2-methoxy	Benzamido)-1,2,4-thiadiazole	41.	73
3-benzoylsulfinyl-5-(N-n-butyl	Benzamido-2-benzoylimino-1,3-dithiol-4-carboni	39.	75
c acid ethyl ester] 5-	Benzamidoxime	40.	23
(E)-morpholino-p-nitro	Benzazecin-13-one pyridine solvate <-4-hydroxy	37.	16
imino-1,2,3,4,5,6,7,8-octahydro-1,8-methano(1)	Benzazocin-12-one <enylsulfonylamino)-2,2a,3,4	37.	14
,5,6-hexahydro-2a,6-methane-1H-azeto(1,2-a)(1)	Benzene	7.	3
hexacyano	Benzene	15.	1
p-dinitro	Benzene	9.	2
trans-azo	Benzene	17.	5
1,4-dihydroxy	Benzene	17.	16
1,2,4,5-tetramethoxy	Benzene	1.	19
bis(trimethyl-acetic acid)	Benzene	19.	7
1-methyl-4-(1'-ethylpropyl)	Benzene	17.	26
1,4-bis(5-chloro-pent-4-ynyloxy)	Benzene	17.	7
2,3,5,6-tetrachloro-1,4-dimethoxy	Benzene	13.	10
(4'-carbomethoxy-2'-nitrophenoxy)-	Benzene	13.	24
2,4,6-tribenzoyloxy-propiophenone	Benzene [2-(4'	13.	17
-carbomethoxy-2'-aminophenoxy)-1,3,5-trimethyl	Benzene [2,5-di	33.	76
(styryl)-pyrazine 1,4-bis(2-(2-pyridyl)vinyl)-	Benzene acetic acid clathrate	61.	24
hexakis(R-α-phenylethylsulfonyl-methyl)	Benzene clathrate	61.	3
hexa-aziridino-cyclotriphosphazene	Benzene clathrate [5-methylbenzene-1,3	61.	7
-dicarbaldehyde-bis(p-tolylsulfonylhydrazone)	Benzene clathrate monohydrate	61.	12
cycloveratril	Benzene complex	60.	30
anthracene - tetracyano	Benzene complex	60.	29
anthracene - tetracyano	Benzene complex	60.	28
anthracene - tetracyano	Benzene complex	60.	26
anthracene - tetracyano	Benzene complex	60.	25
anthracene - tetracyano	Benzene complex	60.	27
anthracene - tetracyano	Benzene complex	60.	31
anthracene - tetracyano	Benzene complex [N,N,N	60.	13
',N'-tetramethyl-p-phenylenediamine hexafluoro	Benzene hydrochloride dihydrate	16.	5
2,4-diamino-azo	Benzene methanol solvate <p-bromophenyl)-7-oxo	31.	68
-(2,3-l)phenanthrene-bicyclo(2.2.1)hept-2-ene	Benzene squalene clathrate	61.	26
hexakis(p-t-butylphenylthiomethyl)	Benzene 1,4-dioxane clathrate	61.	23
hexakis(benzylthiomethyl)	Benzene 1,4-dioxane clathrate	61.	25
hexakis(2-phenylethylthiomethyl)	Benzene) - (meso-3,8-di-t-butyl-1,5,6,10-tetra	60.	44
phenyl-> (1,2,4,5-tetrabenzoyl-3,6-di-t-butyl	Benzene)-N-hydrazo)-8-oxatricyclo(5.1.0.0(2,5)	38.	52
)octane]	Benzene-diazonium tetrafluoroborate	40.	16
4-(N'-(p-nitro	Benzene-hexacarbonitrile	7.	3
4-morpholino-	Benzene-phosphonothioate	64.	63
O-ethyl-O-(4-nitrophenyl)-	Benzene-1,3-dicarbaldehyde-bis(p-tolylsulfonyl	61.	7
hydrazone) benzene clathrate] 5-methyl	Benzene-2'-carboxylic acid	13.	12
4-dimethylamino-azo	Benzenesulfonate	39.	16
S-methyl-thiepanium 2,4,6-trinitro	Benzeno-6,13-ethenylidene-5,5a,6,6a,7,12,12a,	31.	67
13,13a,14-decahydropentacen> 5,14;7,12-bis(o-	Benzeno-1,4-naphthoquinone <ahydro-11,14-dimet	59.	36
hoxy-7-methyl-4a-(3-methyl-but-2-enyl)-5,8a-o-	Benzeno-9-arsa-anthracene	65.	23
9,10-dihydro-10-phenyl-9,10-o-	Benzeno-9-arsonia-anthracene chloride monohydr	65.	12
ate] 9,10-dihydro-9-methyl-9,10-o-	Benzeno-9-phospha-anthracene	64.120	
3-t-butyl-9,10-dihydro-9,10-o-	Benzenophosphazine-10,2'-(1,3,2)-benzo-dioxaph	64.144	
o> 4',5',6',7'-tetrachloro-2,8,15-(5,10(1,2)-	Benzfurazan-1-oxide	40.	2
-(5,6-a)-(6,7-benzo-4-phenyl-(1,3,5-tri> 4,5-	Benzimidazo(1,2-a)-(4-phenyl-(1,3,5-triazino))	36.	51
5-benzoyl-2-(methoxycarbamoyl)-1H-	Benzimidazole hydrobromide	35.	41
-4-methyl-5,6-ethano-1H-(1,4)-diazepino(1,7-a)	Benzimidazole hydrochloride dihydrate <dihydro	36.	12
2-(cyanophosphinidene)-1,3-dimethyl-	Benzimidazoline	64.	37
N-(2-(4-(5-chloro-2-oxo-1-	Benzimidazolinyl)piperidino)-ethyl)-p-fluorobe	35.	64
nzamide] N-(2-(4-(5-chloro-2-oxo-1-	Benzimidazolyl)-(naphthoic acid)-amide <-butox	35.	74
ycarbonyl-1-azophenyl)-2-hydroxy-3-N-(2-oxo-5-	Benzisothiazole	41.	52
3-(3-amino-benzo(b)thiophen-2-yl)-1,2-	Benzo(a,c,h,j)phenazine	36.	50
tetra			

‹,N-dimethyl-spiro(cyclohex-2-ene-1,5'-(5H)-di Benzo(a,d)cycloheptene)-4-amine hydrochloride   29. 8a
‹,N-dimethyl-spiro(cyclohex-2-ene-1,5'-(5H)-di Benzo(a,d)cycloheptene)-4-ammonium hydrogen m›   29. 9a
xamethoxy-10,15-dihydro-5H-5,10,15-trithia-tri Benzo(a,d,g)cyclononene chloroform solvate ‹he   39. 84
te]   perchloro-5,10-diphenyl-di Benzo(a,e)pentalene carbon tetrachloride solva   29. 10
dro-5,14-(epithio-methyleneoxy)-6,12-epithiodi Benzo(a,f)cyclodecen-7(6H)-one [syn-5,12-dihy   39. 73
                      6-methoxy-2-methyl-3,5-dihydro- Benzo(b)furan-4,7-dione..................... 59. 2
                                      3-(3-amino- Benzo(b)thiophen-2-yl)-1,2-benzisothiazole 41. 52
methane]   thieno(3,2-e:4,5-e')-bis(Benzo(b)thiophene - 7,7,8,8-tetracyanoquinodi   60. 33
                                              di Benzo(b,b')furo(3,2-e:4,5-e')-bis(benzofuran)   38.116
etraen› 2,3,7,8,12,13,17,18-octamethoxy-tetra Benzo(b,e,h,k)(1,4,7,10)tetrathia-cyclododecat   39. 86
5-(3'-dimethylaminopropyl)-10,11-dihydro-5H-di Benzo(b,f)-10,10,11,11-tetramethyl-10,11-disi›   63. 34
                          10,11-dihydro-5H-di Benzo(b,f)azepine 36. 10
               5-phenyl-10,11-dihydro-5H-di Benzo(b,f)phosphepin-5-oxide   64.108
l-10-(4-methylpiperazinyl)-10,11-dihydro-5H-di Benzo(b,f)silepane hydrogen fumarate ‹-dimethy   63. 33
‹5,16,23,24,26,27,29,30,32,33-hexadecahydro-di Benzo(b,q)(1,4,7,10,13,16,19,22,25,28)decaoxa›   38.131
yclotriacontane potassium thiocyanate]   di Benzo(b,q)(1,4,7,10,13,16,19,22,25,28)decaoxac   38.130
rans-dichloro-(2,11-bis(diphenylarsinomethyl)- Benzo(c)phenanthrene)-platinum   [t 65. 26
                4,6-di-t-butyl-2-hydroxy-2-oxo- Benzo(d)-1,3,2-dioxaphosphole n-hexane solvate   64. 68
'-trimethoxy-spiro(cyclohexa-3,5-diene-1,6'-di Benzo(d,f)(1,3)dioxepin-2-one ‹-butyl-2',5,10   38.138
                                          tetra Benzo(de,no,st,c(1)d(1))heptacene   30. 20
                                          (E)-di Benzo(e,e')-3,3'-diazastilbene 35. 56
                                          (Z)-di Benzo(e,e')-3,3'-diazastilbene 35. 57
hyl-phenyl)-3,7-dimethyl-1-propyl-1,5-dihydro- Benzo(f)pyrazolo(3,4-c)(1,2,5)triazepine ‹-met   36. 35
          1H,4H-2,3-dimethylpyridazino(1,2-b) Benzo(g)phthalazine-6,13-dione   36. 21
          3H,4H-2,3-dimethylpyridazino(1,2-b) Benzo(g)phthalazine-6,13-dione   36. 22
                              5,10-dihydroxy Benzo(g)phthalhydrazide acetic acid solvate.... 36. 3
                              7,8-dimethoxy Benzo(j)fluoranthene 30. 6
                              5,10-dimethoxy Benzo(j)fluoranthene 30. 7
)heptadecane-3(8),15-diene]   Benzo(p)-2-thiatetracyclo(7.5.3.0(3,8).0(1,10)   39. 79
nitr› 3-(5-methoxy-2,2,8,8-tetramethyl-2H,8H- Benzo(1,2-b:3,4-b')dipyran-6-yl)-propyl-3,5-di   38.124
                          2,11-dimethoxy- Benzo(1,2-b:4,3-b')-bis(benzofuran)............ 38.102
                          6,12-disalicyloyl- Benzo(1,2-b:4,5-b')-bis(benzofuran)............ 38.134
robe› 3-(5-methoxy-2,2,8,8-tetramethyl-2H,8H- Benzo(1,2-b:5,4-b')dipyran-10-yl)-propyl-p-nit   38.126
‹hydro-3-(5-methoxy-2,2,8,8-tetramethyl-2H,8H- Benzo(1,2-b:5,4-b')dipyran-10-yl)-propyl-3,5-›   38.127
init› 3-(5-methoxy-2,2,8,8-tetramethyl-2H,8H- Benzo(1,2-b:5,4-b')dipyran-10-yl)-propyl-3,5-d   38.125
                          N,N',N''-tribenzyl- Benzo(1,2-c:3,4-c':5,6-c'')tripyrroline........ 36. 52
                                  2,6-diphenyl- Benzo(1,2-d:4,5-d')bis(thiazole) 41. 71
                                  2,6-diphenyl- Benzo(1,2-d:4,5-d')bis(oxazole)................ 40. 50
                  3,3'-spiro-bis(3H-2,4-dihydro- Benzo(1,4)dioxepin)............................ 38. 82
iazocyclohex-5-en-2-one)-3› ((5,6-c)-(1-bromo Benzo)-1-ethyl-4-(3-methoxy-phenylethyl)-1,4-d   36. 37
oxylic acid anhydride)]   5,6:7,8-di Benzo-bicyclo(2.2.2)octa-5,7-diene-(2,3-dicarb   38. 88
10β-ethyl-5,6,7,8,9,10-hexahydro-5α,9α-methano Benzo-cyclo-octene-10α-carboxamide ‹,5RS,9SR)-   31. 38
                                          tetra Benzo-cyclodecane-1,16-dione 30. 13
o-6-methyl-3-p-tolyl-1,2,3,4-tetrahydro-1,3,2- Benzo-diazaphosphorine 2-oxide   [2-chlor 64. 70
5-(5,10(1,2)-benzenophosphazine-10,2'-(1,3,2)- Benzo-dioxaphosphole) ‹6',7'-tetrachloro-2,8,1   64.144
tradeca-4,9,11,13-tetraene]   tetra Benzo-pentacyclo(6.2.2.2(2,6).0(2,7).0(3,7))te   31. 64
-tri› 14-ethoxy-8,14-di(phenoxy)-3,4:10,11-di Benzo-tricyclo(7.5.0.0(2,8))tetradecane-3,5,10   30. 18
iene-3,10-dione]   (1α,2α,8β,9α)-4,5:11,12-di Benzo-tricyclo(7.5.0.0(2,8))tetradecane-4,11-d   30. 11
iene-3,10-dione]   (1α,2β,8β,9α)-4,5:11,12-di Benzo-tricyclo(7.5.0.0(2,8))tetradecane-4,11-d   30. 10
iene-3,14-dione]   (1α,2β,8α,9α)-4,5:12,13-di Benzo-tricyclo(7.5.0.0(2,8))tetradecane-4,12-d   30. 12
iene-3,14-dione]   (1α,2β,8α,9β)-4,5:12,13-di Benzo-tricyclo(7.5.0.0(2,8))tetradecane-4,12-d   30. 8
iene-3,14-dione]   (1α,2β,8α,9β)-6,7:10,11-di Benzo-tricyclo(7.5.0.0(2,8))tetradecane-6,10-d   30. 9
-trimethyl-6-(3-indolyl)-4-(isopropenyl)-(2,3) Benzo-1-azabicyclo(3.3.0)octa-2,4-diene [6,8,8   36. 45
                              1,1-di-iodo-3,4- Benzo-1-tellura-cyclopentane 70. 3
                      N-methyl-4H-6,7-dimethoxy- Benzo-1,2-thiazin-3-one-1,1-dioxide........... 41. 37
oquinone] 2,3-dihydro-2-methoxy-2-methyl-3-(Benzo-1,3-thiazole-2-thiomethylene)-1,4-naphth   41. 72
                                  octafluoro-di Benzo-1,4-dioxane 38. 35
tracyclo(9.3.2.0(2,7).0(2,10))hex› 3,4:5,6-di Benzo-14-ethoxycarbonyl-8,10-diphenyl-14-azate   37. 24
                                              Benzo-15-crown-5 lithium picrate dihydrate 38. 65
                                              Benzo-15-crown-5 sodium perchlorate........... 38. 66
                                          bis(Benzo-15-crown-5) sodium perchlorate 38. 67
                                          bis(Benzo-15-crown-5) sodium tetrachloro-dioxo-ura   38. 68
                                          bis(Benzo-15-crown-5) sodium tetraphenylborate   62. 15
.1.0(1,5))u› 5,7-dimethyl-1-phenyl-3,4:8,9-di Benzo-2,6,10,11-tetraoxa-1-phosphatricyclo(5.3   64.117
hyl-9-aza-5-oxabicyclo(4.3.0)nona-(1(6)› 7,8- Benzo-3-ethoxycarbonyl-2-methoxycarbonyl-9-met   38. 80
                                              di Benzo-30-crown-10 potassium thiocyanate....... 38.130
monohydrate]   (di Benzo-30-crown-10) bis(sodium isothiocyanate)   38.131
‹-a)-(4-phenyl-(1,3,5-triazino))-(5,6-a)-(6,7- Benzo-4-phenyl-(1,3,5-triazepine)) chloroform›   36. 51
                      2-phenyl-5,5-dimethyl-(3,4) Benzo-5H-furylium perchlorate 38. 76
1.0(3,7))dec-4-ene]   4,5- Benzo-6-phenyl-3,10,10-trimethyl-tricyclo(5.2.   31. 57
(7.5.0.0(2,8))tetradecane-5,12-› 3,4:13,14-di Benzo-6,6,11,11-tetramethyl-1,8,9-cis-tricyclo   30. 14
bicyclo(4.4.0)deca-7,9-diene-2,4-dione]   7,8- Benzo-8-cyano-3,5-dimethyl-10-phenyl-3,5-diaza   36. 34
o(4.4.0)deca-7,9-diene-2,4-dione]   7,8- Benzo-9-cyano-3,5,10-trimethyl-3,5-diazabicycl   36. 16
                              eriostyl p-nitro Benzoate 38.126
                          eriostyl-3,5-dinitro Benzoate 38.125
                          dimethyl-2,2'-thio-di Benzoate 13. 13

                    eriostemyl-3,5-dinitro Benzoate 38.124
                  dimethyl-2,2'-dithio-di Benzoate 13. 14
              methyl-2-(methylsulfinyl)- Benzoate 13. 6
               tetrahydroeriostyl-3,5-dinitro Benzoate 38.127
     hydroxyethyl)-bicyclo(3.1.0)hexane 3,5-dinitro Benzoate   .... [1-(1-   27. 13
     nyl-2,2,6,6-tetramethyl-3-heptanol-3,5-dinitro Benzoate   .... [(RR,SS)-5-phe 13. 21
       (1,2-b:3,4-b')dipyran-6-yl)-propyl-3,5-dinitro Benzoate   <hoxy-2,2,8,8-tetramethyl-2H,8H-benzo 38.124
       nzo(1,2-b:5,4-b')dipyran-10-yl)-propyl-p-nitro Benzoate   <methoxy-2,2,8,8-tetramethyl-2H,8H-be 38.126
       4a,5,8,8a-octahydro-naphthalene-10-syn-p-nitro Benzoate   <o,4endo:5exo,8exo-dimethano-1,2,3,4, 31. 50
        1,2-b:5,4-b')dipyran-10-yl)-propyl-3,5-dinitro Benzoate   <oxy-2,2,8,8-tetramethyl-2H,8H-benzo(38.127
        1,2-b:5,4-b')dipyran-10-yl)-propyl-3,5-dinitro Benzoate   <oxy-2,2,8,8-tetramethyl-2H,8H-benzo(38.125
                                         bis(4-di Benzoborepinyl)-ether 62. 19
      7-bromo-5-(o-chlorophenyl)-1,3-dihydro-2H-1,4- Benzodiazepin-2-one...................................... 35. 34
          -hydroxy-1-methyl-5-phenyl-1,3-dihydro-2H-1,4- Benzodiazepin-2-one..................[7-chloro-3 35. 35
        dihydro-1-(N-methylacetamido)-5-phenyl-2H-1,4- Benzodiazepin-2-one   [7-chloro-1,3- 35. 51
       henyl)-1,3,4,5-tetrahydro-1,4-dimethyl-2H-1,4- Benzodiazepin-2-one monohydrate <-5-(2-fluorop 35. 45
         7-chloro-2-methylamino-5-phenyl-3H-1,4- Benzodiazepin-3-ol   35. 38
       2,4-bis(dimethylamino)-3-ethyl-1-phenyl-1,5- Benzodiazepine 35. 66
       7-chloro-2,3-dihydro-1-methyl-5-phenyl-1H-1,4- Benzodiazepine 35. 42
         7-chloro-2-methylamino-5-phenyl-3H-1,4- Benzodiazepine-4-oxide   35. 39
     romo-1-methyl-6-phenyl-1,2,3,4-tetrahydro-1,5- Benzodiazocin-2-one   [8-b 35. 43
     o-6-phenyl-1-methyl-1,2,3,4-tetrahydro-5H-1,5- Benzodiazocin-2-one   [8-chlor 35. 44
       epoxy-4-(4-nitrophenylmethyl)-1-phenyl-1H-2,3- Benzodiazepin-5(4H)-one   [1,4- 38.110
                  diphenyl-(5,6-dichloro-1,3- Benzodioxol-(2))-phosphine-oxide 64.100
         2,2'-(butane-1,4-diyl)-bis(hexahydro-1,3- Benzodioxole)   38. 97
               diethyl-(5,6-dichloro-1,3- Benzodioxole-(2))-phosphonate 64. 44
                                    1,3- Benzodithiolyl-2-(diphenylphosphine-oxide) 64.103
       dibenzo(b,b')furo(3,2-e:4,5-e')-bis(Benzofuran).. 38.116
         2,11-dimethoxy-benzo(1,2-b:4,3-b')-bis(Benzofuran).. 38.102
       6,12-disalicyloyl-benzo(1,2-b:4,5-b')-bis(Benzofuran).. 38.134
     l)-4'-(p-bromophenylsulfonyloxy)-2',3'-dihydro Benzofuran) dimethylsulfoxide solvate <xymethy 50. 21
     7-chloro-2',4,6-trimethoxy-6'-β-methyl-spiro Benzofuran-2(3H)-cyclohex-2-ene-3,4'-dione be> 50. 11
     yl-3'-(2,2-dimethylpropionyl)-1'-methyl-spiro(Benzofuran-3(2H)-2'-pyrrolidine)-2,5'-dione <t 38.132
              N,N-diethyl-3,3a-dihydro-3-methyl- Benzofuro(3,2-c)isoxazole-3-carboxamide........ 40. 38
                                         Benzoic acid 13. 3
                                   p-nitro Benzoic acid 13. 1
                                o-acetamido Benzoic acid 13. 5
                     2,6-dimethyl-1,4-dihydro Benzoic acid 21. 4
                     3,5-dimethyl-1,4-dihydro Benzoic acid 21. 5
     3,5-di-iodo-4-(4-hydroxy-3,5-di-iodophenoxy)- Benzoic acid bis(N-diethanolamine) 13. 7
                                   p-amino Benzoic acid hydrochloride 13. 4
     -(1-trimethylsilyl-3,4-dimethyl-pent-2-enyl))- Benzoic acid p-bromobenzoylmethyl ester <silyl 63. 45a
       2'-hydroxy-2-(2-tetrahydro-2-furylmethyl)-6,7- Benzomorphan <-)-(1R,5R,9R,2''S)-5,9-dimethyl- 58. 41
         -ethyl-2'-hydroxy-2-isobutyl-9,9-dimethyl-6,7- Benzomorphan hydrobromide   [(-)-5 58. 49
       ethyl-2'-hydroxy-9,9-dimethyl-2-phenethyl-6,7- Benzomorphan hydrobromide   [(+-)-5- 58. 64
       ,9-dimethyl-2'-hydroxy-2-(4-methoxybutyl)-6,7- Benzomorphan hydrobromide   [(-)-(1R,5R,9R)-5 58. 42
       9-dimethyl-2'-hydroxy-2-(2-methoxypropyl)-6,7- Benzomorphan hydrobromide <-(1R,5R,9R,2''S)-5, 58. 33
         (+-)-2'-hydroxy-5,9-dimethyl-2-phenethyl-6,7- Benzomorphan hydrobromide hemihydrate 58. 59
                                    4-chloro Benzonitrile 7. 2
     te]   2,3,5,6,8,9,11,12-octahydro-1,4,7,10,13- Benzopentaoxa-cyclopentadecin sodium perchlora 38. 66
       bis(2,3,5,6,8,9,11,12-octahydro-1,4,7,10,13- Benzopentaoxa-cyclopentadecin) sodium tetraph> 62. 15
       bis(2,3,5,6,8,9,11,12-octahydro-1,4,7,10,13- Benzopentaoxacyclopentadecin) sodium perchlor> 38. 67
       1,4,5,8,9,12,13,16-octamethyl-tetra Benzoporphyrinato-nickel(ii) iodide............. 49. 18
     ethoxy-phenyl)-6H,11H-(2)benzopyrano(4,3-c)(1) Benzopyran-11-one <-dihydroxy-9-methoxy-6-(4-m 38.117
     methyl-4-oxo-2-furanyl)-but-2-enyl)-oxy-(2H-1- Benzopyran-2-one)   [7-(3-(4,5-dihydro-5,5-di 59. 19
     ydro-2-hydroxy-2-methyl-2H,5H-pyrano(3,2-c)(1) Benzopyran-5-one   .... [4-cyclohexyl-3,4-dih 38.100
       -methoxyphenyl)-6H,7H-(1)benzopyrano(4,3-b)(1) Benzopyran-6-one <11-diacetoxy-10-methoxy-7-(4 38.128
       etoxy-10-methoxy-7-(4-methoxyphenyl)-6H,7H-(1) Benzopyrano(4,3-b)(1)benzopyran-6-one <11-diac 38.128
       roxy-9-methoxy-6-(4-methoxy-phenyl)-6H,11H-(2) Benzopyrano(4,3-c)(1)benzopyran-11-one <-dihyd 38.117
     e]   7,8- Benzoquinoline 7,7,8,8-tetracyanoquinodimethan 60. 14
                   2,5-bis(piperidino)-1,4- Benzoquinone 33. 71
                   2,5-bis(pyrrolidino)-1,4- Benzoquinone 32. 51
                  2,3-dichloro-5,6-dicyano-p- Benzoquinone 18. 2
              2-hydroxy-3,5-dimethyl-6-hydroxymethyl-1,4- Benzoquinone 18. 3
     tetramethyl-tetrathiafulvalene tetrabromo-p- Benzoquinone complex 60. 10
                syn-15,18-dimethoxy-(2,6)-p- Benzoquinone(3.3)metacyclophane................ 31. 53
                                         di Benzotetrathiafulvalene tri-iodide 39. 50
                                tris(di Benzotetrathiafulvalene) hexabromo-tin 39. 51
                              octakis(di Benzotetrathiafulvalene) tris(hexachloro-tin) 39. 52
                   6-chloro-4-methyl-1,2,3- Benzothiadiazole 41. 18
       N-(p-bromophenyl)-4,5-dihydro-7,8-dimethoxy- Benzothiazepin-3-one-1,1-dioxide 41. 59
       (2-acetoxy-1-ethylidene)-3-(2-thienyl)-4H-1,4- Benzothiazine   ....[6-chloro-2 41. 57
       <y-15-phenyl-6aH,14aH,15H-benzothiazolo(2,3-a) Benzothiazino(4,3-c)piperazine methanol solva> 41. 83
       ophenolate]   2-((3-ethyl-1,3- Benzothiazol-2-ylio)-2-vinyl)-6-methoxy-4-nitr 41. 65
       dihydro-N-methyl-2-(4-nitrobenzoyl-methylene)- Benzothiazole   .... [(Z)-2,3- 41. 58
     ane> bis(2,3-dihydro-3-hydroxythiazolo(2,3-b) Benzothiazolium) hexachloro-tellurium(iv) diox 41. 25

8-benzoyl-15-hydroxy-15-phenyl-6aH,14aH,15H- Benzothiazolo(2,3-a)benzothiazino(4,3-c)piper> 41. 83
benzothiopyrano(4,3-b)pyran-2-spiro-3'-(1H-2)- Benzothiopyran-4'(3'H)-one <4-dihydro-2H,6H-2- 39. 76
benzothiopyran-4'(3'H)-> 3,4-dihydro-2H,6H-2- Benzothiopyrano(4,3-b)pyran-2-spiro-3'-(1H-2)- 39. 76
2-(2',4'-dibromophenyl)-4-oxo-1,2,3- Benzotriazin-2-ium-3-ide 35. 27
1,2,3-oxadiazolo(4,3-c)(1,2,4) Benzotriazinium-3-olate............. 40. 6
ulfonyl)-6-phenyl-1,2,3,4-tetrahydro-5H-1,4,5- Benzotriazocin-2-one [8-methyl-4-(p-toluenes 35. 69
7-methoxycarbonyl-1,3-dioxo-2,1- Benzoxathiole 42. 4
1,3-dioxo-3H-2,1λ- Benzoxathiole-7-carboxylic acid methyl ester 42. 4
2-methyl-2,2'-spiro-bis(1,3,2λ(5)- Benzoxazarsoline) 65. 9
2-(2-tosylaminophenyl)-4H-3,1- Benzoxazin-4-one 40. 52
rahydro-6,8,9-trimethyl-1H-pyrrolo(2,1-b)(1,3) Benzoxazine.... [rac-cis-5-chloro-2,3,3a,9-tet 40. 34
y-3-methyl-1-phenyl-3,4,5,6-tetrahydro-1H-2,3- Benzoxazocine [8,9-dimethox 40. 47
3β- Benzoxy-α-amyrin 56. 9
hylene)-indol-2-olate] N-acetyl-3-(Benzoyl(2-piperidyl-2-piperidylenium-ethyl)met 35. 77
methyl-2,3,4-tris(O- Benzoyl)-β-D-xylopyranoside.............. 45. 55
4-(2'-hydroxythio Benzoyl)-morpholine 40. 22
(α-(1-phenylethylimino)- Benzoyl)-phenyl-ketone 19. 12a
2-(4-chloro Benzoyl)-pyrrole 32. 37
2-(4-methoxy Benzoyl)-pyrrole 32. 42
4-(o-hydroxy Benzoyl)-2,5-diphenyl-1-(p-tolyl)-imidazole.... 32. 73
2-(o-hydroxy Benzoyl)-3-phenyl-imidazo(2,1-a)isoquinoline 36. 40
ine] 2-(o-hydroxy Benzoyl)-7-methyl-3-phenyl-imidazo(1,2-a)pyrid 35. 62
1,2-bis((Benzoyl)oxy)-stilbene 13. 22
1,2,3,4-tetra-O- Benzoyl-β-D-xylopyranose 45. 58
N-(1-phenylethyl)-1- Benzoyl-benzylideneimine 19. 12a
N-(1-phenylpropyl)-1- Benzoyl-benzylideneimine 19. 15
(Z)-2,3-dihydro-N-methyl-2-(4-nitro Benzoyl-methylene)-benzothiazole 41. 58
adiazole] 5,6-dihydro-3-thio Benzoyl-methylene-3H-thiazolo(2,3-c)(1,2,4)thi 41. 39
4-thio Benzoyl-morpholine 40. 20
4-thio Benzoyl-morpholine 40. 21
alaninate] methyl Benzoyl-oxycarbonyl-bis(α-amino-isobutyryl)-L- 48. 65
2-iodo-3'-bromo-di Benzoyl-peroxide 13. 9
1,5-anhydro-2,3,4-tri-O- Benzoyl-ribitol.................... 45. 52
1,5-anhydro-2,3,4-tri-O- Benzoyl-xylitol.................... 45. 53
nilino)-isourea] O- Benzoyl-N,N,N-dimethyl-N'-(N-methyl-2,4-dinitroa 13. 16
3-p-bromo Benzoyl-1,3-thiazolidine-2-thione 41. 28
othiazolo(2,3-a)benzothiazino(4,3-c)piper> 8- Benzoyl-15-hydroxy-15-phenyl-6aH,14aH,15H-benz 41. 83
hydrobromide] 5- Benzoyl-2-(methoxycarbamoyl)-1H-benzimidazole 35. 41
cetanilide] 2- Benzoyl-2-(1,3-dimethyl-2,6-dioxo-7-purinyl)-a 35. 67
1- Benzoyl-2-methyl-2-thianaphthalene 39. 65
benzaldehyde 4- Benzoyl-2-phenylthiosemicarbazone 13. 19
2,3,4-tri-O- Benzoyl-2-C-chloro-α-D-xylopyranosyl chloride 45. 51
t-butyl-1,5,6,10-tetraphenyl-> (1,2,4,5-tetra Benzoyl-3,6-di-t-butylbenzene) - (meso-3,8-di- 60. 44
Benzoylacetone 19. 3
α-(triphenylphosphonium)- Benzoylacetone tetrafluoroborate 64.145
)pentadecane-14,15-dione acetonitrile sol> 7- Benzoylamino-7-methoxycarbonyl-dispiro(5.1.5.2 28. 16
ester] 5-benzamido-2- Benzoylimino-1,3-dithiol-4-carbonic acid ethyl 39. 75
.5.2.5.0)nonadec-1-ene-4,13-dione] 12- Benzoylimino-2-phenyl-3-oxa-1-aza-trispiro(4.0 40. 56
iazolidine] 5- Benzoylimino-2,2-dimethyl-4-phenyl-1,3,4-thiad 41. 61
iadiazole] 2- Benzoylimino-3,5-diphenyl-2,3-dihydro-1,3,4-th 41. 77
,4-dimethyl-pent-2-enyl))-benzoic acid p-bromo Benzoylmethyl ester <silyl-(1-trimethylsilyl-3 63. 45a
2,4,6-tri Benzoyloxy-propiophenone 13. 23
2,4,6-tri Benzoyloxy-propiophenone benzene 13. 24
4-bromo Benzoyloxymethyl-trifluoro-silane 63. 7
lithium di Benzoylphosphide 1,2-dimethoxyethane 64. 62
2- Benzoylpyrrole 32. 38
hiadiazole] 3- Benzoylsulfinyl-5-(N-n-butylbenzamido)-1,2,4-t 41. 73
iminodi Benzyl 36. 10
t-butoxy-carbonyl-L-prolyl-sarcosine- Benzyl ester 48. 64
N-(t-butoxy-carbonyl)-L-methionyl-glycine- Benzyl ester 48. 60
icyclo(3.2.0)heptane-2-carboxylic-acid-p-nitro Benzyl ester <-thio-imino)-7-oxo-4-thia-1-azab 41. 80
O,O'-dibromodi Benzyl ether 19. 8
clohex-2-ene-1-carbamate] Benzyl trans-6-(ethoxycarbonyl)-cis-6-phenylcy 21. 18
2.0)octan-8-one-2-carboxylate] Benzyl 4-methoxy-7-azido-3-oxa-1-azabicyclo(4. 40. 36
Benzyl 6-aminopurine-9-carboxylic acid 44. 26
methyl 2-methyl-3-(Benzyl(phenyl)phosphinyl)-butyrate 64.105
N-methyl-N'-(Benzyl)(acetamido)-methylurea 1. 29
1-(2,4-dichloro-β-((2,4-dichloro Benzyl)-oxy)-phenethyl)-imidazole hemihydrate 32. 65
hydrate] DL-2-(α-hydroxy Benzyl)-oxythiamine chloride hydrochloride tri 44. 35
D,L-diethylanilino-(3-hydroxy Benzyl)-phosphonate.................... 64. 82
ethylisoxazole 2,4-diamino-5-(3,4,5-trimethoxy Benzyl)-pyrimidine [3-sulfanilamido-5-m 44. 30
10-(4-bromo Benzyl)-10-phenyl-phenoxaphosphonium bromide 64.137
-6,7,8-trimethoxy-naphthale> 3-(3,4-dimethoxy Benzyl)-4(2,4-dimethoxyphenyl)-1-semicarbazono 27. 33
inoline hydrochloride mo> 1-(3,4,5-trimethoxy Benzyl)-6,7-dihydroxy-1,2,3,4-tetrahydro-isoqu 35. 54
xo-thieno(3,2-c)azocine-3α-yl)-carbamic aci> Benzyl-(2,3β,3aα,4,5,6,7,8,9,9aβ-decahydro-4-o 39. 72
penam-3-carboxylate] 4-nitro Benzyl-(3SR,5RS,Z)-2-(2-phenylthioethylidene)- 41. 78

N,N',N''-tri Benzyl-benzo(1,2-c:3,4-c':5,6-c'')tripyrroline 36. 52
thia-1-azabicyclo(3.2.0)hept-3-ene-4-(p-nitro- Benzyl-carboxylate).... [7-oxo-3-methylthio-2- 41. 50
tane bromide]   exo-3-p-nitro Benzyl-endo-3-phenyl-3-phosphabicyclo(3.2.1)oc 64.111
bis(4- Benzyl-piperidinium) hexachloro-di-copper(ii) 33. 47
bis(4- Benzyl-piperidinium) tetrachloro-cobalt(ii).... 33. 46
di Benzyl-pyruvic acid............................ 1. 35
rbonyl-tetrakis(μ(2)-carbonyl)-nickel-tri-i> Benzyl-trimethylammonium (μ(3)-hydrido)-octaca 3. 43
2)-thio)-molybdenum> tetraphenylphosphonium (Benzyl-trimethylammonium) dichloro-iron-bis(μ(64.123
icyclo(7.3.0.0(2,6))dodecane-7,10-dione]   8- Benzyl-1-hydroxy-11-methyl-6,9-diaza-12-thiatr 48. 49
yl-1-isopropyl-isoquinolinium iodide]   1- Benzyl-1-phenylphosphorinanium bromide 64. 94
2- Benzyl-1,2,3,4-tetrahydro-6,7-dimethoxy-2-meth 35. 68
2,2-di Benzyl-1,3-dibromopropane 19. 10
ne dibromide]   (E)-1,4-di Benzyl-1,4-diphenyl-1,4-diphosphonio-cyclohexa 64.149
decane-3,14,18-trione]   2- Benzyl-1,7,10,16-tetraoxa-4,13-diazocyclo-octa 40. 49
(1Z,3E)-1,4-bis(1,3-di Benzyl-2-imidazolidinyl)-1,4-diphenylbutadiene 32. 78
n]   4- Benzyl-2,3,5,6-bis(tetramethylene)-4H-thiopyra 39. 80
riazatricyclo(7.3.0.0(2,6))dodecane-5,8-d>   7- Benzyl-3-(benzyloxycarbonyl)-2-hydroxy-3,6,9-t 36. 46
3,6,9-triazatricyclo(7.3.0.0(2,6))dodecan>   7- Benzyl-3-(p-bromobenzyloxycarbonyl)-2-hydroxy- 36. 44
(1R,3S)-(-)-trans-1- Benzyl-3-methyl-1-phenylphospholanium iodide 64. 96a
N-p-iodo Benzyl-3-n-octan-amido-pyridine-2,5,6-trione 59. 22
decane-5,8-dione]   2-hydroxy-4-isopropyl-7- Benzyl-3-oxa-6,9-diazatricyclo(7.3.0.0(2,6))do 40. 48
y)-2-azabicyclo(2.2.1)heptane]   2- Benzyl-3-oxo-6-exo-(2,6-di-iodo-4-methylphenox 37. 15
(4,5-d)-1,3-thiazine]   3- Benzyl-3,7-dihydro-5-methylthio-1,2,3-triazolo 41. 40
1-methyl-3- Benzyl-4-(1,2,3-triazolio)-sulfide 32. 32
opylidene)-acetat>   2',2',2'-trichloroethyl-2- Benzyl-4-methoxy-carbonyl-imidazole-1-(α-isopr 32. 67
cis-bicyclo(4.3.0)non-8-ene-7-oxi>   racemic-7- Benzyl-4-t-butyl-9-hydroxy-8-phenyl-7-phospha- 64.141
2- Benzyl-5-benzylidene-cyclopentanone 20. 10
2- Benzyl-5-benzylidene-cyclopentanone dimer 20. 15
dine-acetic acid]   1- Benzyl-5-ethyl-1,2,5,6-tetrahydro-2-oxo-4-pyri 33. 68
6α- Benzyl-6β-isocyano-penicillanate 50. 16
ne)]   7,7'-bis(1,4-di Benzyl-6-phenyl-1,2,3,4-tetrahydro-1,4-diazep 34. 13
ne)]   7,7'-bis(1,4-di Benzyl-6-phenyl-1,2,3,4-tetrahydro-1,4-diazepi 34. 14
trans-10- Benzyl-9-phenyl-9,10-dihydro-9-arsa-anthracene 65. 24
2(S)-1-pyridinium-2-N,N-di Benzylamino-propane p-toluenesulfonate 33. 82
2(S),3(S)-1-pyridinium-2-N,N-di Benzylamino-3-methylpentane p-toluenesulfonate 33. 91
Benzylammonium bis(dichloro-mercury) chloride 19. 2
N,N'-dimethyl-1,7-diaza-4,10-dioxacyclodecane Benzylammonium thiocyanate.................... 40. 18
13,16,19-hexaoxa-10,22-diazacyclotetracosane) Benzylammonium thiocyanate clathrate <I-1,4,7, 61. 6
lladium]   bis(trimethyl Benzylammonium) hexadecacarbonyl-tetra-iron-pa 3. 44
atinum]   bis(trimethyl Benzylammonium) hexadecacarbonyl-tetra-iron-pl 3. 45
1- Benzylcytosine nitrate 44. 22
cyclo(di(Benzylglycyl)-L-prolyl) monohydrate.......... 48. 73
cyclo(Benzylglycyl-bis(L-prolyl)) 48. 58
hydrofuran solv>   sodium (diphenylphosphonium- Benzylide)-(diphenylphosphino-methylide) tetra 64.150
2-(p-dimethylamino Benzylidene)-indan-1,3-dione 27. 23
N-(p-chloro Benzylidene)-p-bromoaniline.................... 16. 7
N-(p-methyl Benzylidene)-p-bromoaniline.................... 16. 9
N-(p-bromo Benzylidene)-p-chloroaniline 16. 6
N-(p-methyl Benzylidene)-p-chloroaniline 16. 11
N-(p-bromo Benzylidene)-p-methylaniline 16. 8
N-(p-chloro Benzylidene)-p-methylaniline 16. 10
2-benzyl-5- Benzylidene-cyclopentanone 20. 10
2-benzyl-5- Benzylidene-cyclopentanone dimer 20. 15
p-ethoxy Benzylidene-p-n-butylaniline 17. 30
p-diethylamino Benzylidene-rhodanine 41. 51
Benzylidene-triazene 40. 25.
methyl 3-C-acetyl-4,6-O- Benzylidene-2-deoxy-α-D-ribo-hexopyranoside.. 45. 38
onyl)-α-D-arabinopyranoside]   methyl 4,6-O- Benzylidene-2-deoxy-3-O-((methylthio)-thiocarb 45. 36
onyl)-α-D-ribopyranoside]   methyl 4,6-O- Benzylidene-2-deoxy-3-O-((methylthio)-thiocarb 45. 35
4- Benzylidene-2-phenyl-2-imidazolin-5-one........ 32. 54
e]   methyl 4,6-O- Benzylidene-2,3-di-O-methyl-α-D-glucopyranosid 45. 40
ide]   methyl 4,6-O-(R)- Benzylidene-2,3-di-O-methyl-β-D-galactopyranos 45. 41
spiro((1-acetylaziridine)-2,3'-(methyl-4,6-0> Benzylidene-2,3-dideoxy-α-D-arabino-hexopyran> 45. 43
onyl)-α-D-arabinopyranoside]   methyl 4,6-O- Benzylidene-3-deoxy-2-O-((methylthio)-thiocarb 45. 37
methyl-2,3',6-tri-O-acetyl-2',3:4',6'-di-O- Benzylidene-7(R)-β-D-cellobioside 45. 59
N-(1-phenylethyl)-1-benzoyl- Benzylideneimine 19. 12a
N-(1-phenylpropyl)-1-benzoyl- Benzylideneimine 19. 15
2- Benzylideneindan-1-one 27. 17
α- Benzyloxy-γ-butyrolactone 38. 32
glycine ethyl ester]   (+-)-N-carbo Benzyloxy-(γ,γ'-di-t-butyl)-γ-carboxyglutamyl- 48. 75
clo(7.3.0.0(2,6))dodecan>   7-benzyl-3-(p-bromo Benzyloxycarbonyl)-2-hydroxy-3,6,9-triazatricy 36. 44
clo(7.3.0.0(2,6))dodecane-5,8-d>   7-benzyl-3-(Benzyloxycarbonyl)-2-hydroxy-3,6,9-triazatricy 36. 46
l-3-oxa-1-azabicyclo(4.2.0)oct>   7-(1-(p-nitro Benzyloxycarbonyl)oxy-ethyl)-8-oxo-2,2-dimeth 40. 45
2'-trimethylene-dithioethyl)-3α-(1'S)-(p-nitro Benzyloxycarbonyl)oxy-ethyl-2-azetidinone <2', 39. 71
l)-L-glutamic acid ethyl ester]   N- Benzyloxycarbonyl-(γ-ethyl)-L-glutamyl-(γ-ethy 48. 74
N- Benzyloxycarbonyl-prolyl-proline 48. 53

```
         -methyl-1,3-thiazolin-2-yl)-hydrazine]         N- Benzyloxycarbonyl-N'-(2,6-dichlorophenyl)-N-(5   41. 66
               omethyl)-(R)-sulfoxide]         (2R)-2-( Benzyloxycarbonylamino)-3-hydroxypropyl-(chlor   11. 13a
                   ethyl-5-oxazolone]            2-(1'- Benzyloxycarbonylamino-1'-methylethyl)-4,4-dim    40. 41
                oethane clathrate]         bis(triphenyl Benzylphosphonium) tetrachloro-cadmium dichlor    61.  8
                          bis(dimethyl Benzylsilyl)-peroxide ..................................   63. 25
                            hexakis( Benzylthiomethyl)benzene 1,4-dioxane clathrate   61. 23
           tris(glycine) tetrafluoro Beryllate .................................   48.  7
           tris(glycine) tetrafluoro Beryllate .................................   48.  6
               tetra-t-butoxy-tri Beryllium bis(tetrahydroborate)..............    5.  7
       bis(2-methyl-8-hydroxyquinolinato)- Beryllium dihydrate....................   67.  3
              1,3-di(methylthio)-2,2,4,4-tetramethyl- Bicyclo(1.1.0)butane .......................   27.  4
     butylammonium pe>   3,6,9,12,15-pentaoxa-21-aza Bicyclo(15.3.1)heneicosa-1(20),17,19-triene t-   40. 40
    16-dione potassi>   3,6,9,12,15-pentaoxa-21-aza Bicyclo(15.3.1)heneicosa-1(21),17,19-triene-2,   40. 39
                            trans-2-aza Bicyclo(2.1.0)pentane-3-(S)-carboxylic acid....   50.  1
   bis(1,1,4,4-tetramethylcycloheptane)-5,6-dioxa Bicyclo(2.2.0)hex-2-ene.... [(1,4-b)-(2,3-b)-   38.114
              hexamethyl-2,3:5,6-diepoxy- Bicyclo(2.2.0)hexane ......................   38. 43
           7-isopropylidene-2,3-diaza Bicyclo(2.2.1)hept-2-ene .................   37.  3
         1,3,4,7,7-pentachloro-5-phenyl-2-aza Bicyclo(2.2.1)hept-2-ene .................   37.  8
         trans-5-chloro-6-(2-nitrophenylthio)- Bicyclo(2.2.1)hept-2-ene .................   31. 18
  <-5-(p-bromophenyl)-7-oxo-(2,3-l)phenanthrene- Bicyclo(2.2.1)hept-2-ene benzene methanol sol>   31. 68
                   1,7,7-trimethyl- Bicyclo(2.2.1)hepta-2,3-dione .............   31.  9
                   1,4-diphenyl-2,3-dioxa Bicyclo(2.2.1)heptane .....................   38. 81
               (1S,4S)-N-acetyl-3-oxo-5-aza-2-oxa Bicyclo(2.2.1)heptane .....................   40.  4
   -oxo-6-exo-(2,6-di-iodo-4-methylphenoxy)-2-aza Bicyclo(2.2.1)heptane ...........[2-benzyl-3   37. 15
      yl)-1,2,4-trichloro-3-ethoxy-3,7,7-trimethoxy- Bicyclo(2.2.1)heptane .......... [6-(bromometh   31. 20
   bromomethyl-2-exo-(2',2',2'-trifluoro-ethoxy)- Bicyclo(2.2.1)heptane <-7,7-trimethoxy-5-endo-   37.  9
              6-exo-(methylthio)- Bicyclo(2.2.1)heptane-2-endo-carboxylic acid   31.  4
              6-endo-(methylthio)- Bicyclo(2.2.1)heptane-2-endo-carboxylic acid   31.  5
   3-bis(methoxycarbonyl)-5-hydroxy-1,5-dimethyl- Bicyclo(2.2.2)oct-2-en-6-one ............[2,   31. 27
              5,6-dicyano-1-methoxy- Bicyclo(2.2.2)oct-2-ene.........................   31. 13
   3-bis(methoxycarbonyl)-5-hydroxy-1,5-dimethyl- Bicyclo(2.2.2)oct-7-en-6-one ............[2,   31. 28
   3-bis(methoxycarboxyl)-5-hydroxy-1,5-dimethyl- Bicyclo(2.2.2)octa-2,7-dien-6-one .........[2,   31. 23
    acid anhydride)]         5,6:7,8-dibenzo- Bicyclo(2.2.2)octa-5,7-diene-(2,3-dicarboxylic   38. 88
   -di>   dimethyl 8-t-butyl-5-methyl-2-oxo-spiro( Bicyclo(2.2.2)octa-5,7-diene-3,2'-oxirane)-1,6   38. 92
   -di>   dimethyl 8-t-butyl-1-methyl-2-oxo-spiro( Bicyclo(2.2.2)octa-5,7-diene-3,2'-oxirane)-5,6   38. 91
                          1-aza Bicyclo(2.2.2)octane ......................   37.  1
                      1-phenyl- Bicyclo(2.2.2)octane ......................   31. 26
                 1-chloro-4-phenyl- Bicyclo(2.2.2)octane ......................   31. 24
                 1-fluoro-4-phenyl- Bicyclo(2.2.2)octane ......................   31. 25
   one]     syn-(1R,2R,4S)-4-methyl-spiro( Bicyclo(2.2.2)octane-2,1'-cyclohexane)-3',6-di   31. 31
          methyl 3-(3-isocyano-6-oxa Bicyclo(3.1.0)hex-2-en-5-yl)-acrylate ........   38. 20
     4-methyl-2,6,6-triphenyl-2,3-diaza-1-phospha Bicyclo(3.1.0)hex-3-ene ..................   64.116
    ,5-dimethyl-1,2,2-triphenyl-3,2-oxaphosphonia- Bicyclo(3.1.0)hexane tribromide <bromomethyl-4   64.139
              1-(1-hydroxyethyl)- Bicyclo(3.1.0)hexane 3,5-dinitrobenzoate ...   27. 13
   3,4-bis(methoxycarbonyl)-6-oxa-1-aza-5-phospha Bicyclo(3.2.0)hept-3-ene <7-(trifluoromethyl)-   64.135
    boxylate)]     7-oxo-3-methylthio-2-thia-1-aza Bicyclo(3.2.0)hept-3-ene-4-(p-nitro-benzyl-car   41. 50
   3,5-di-t-butyl-2-hydroxyphenyl)-2-methyl-2-aza Bicyclo(3.2.0)hept-6-en-3-one <-di-t-butyl-1-(   35. 76
                2β-bromo-3α-hydroxy- Bicyclo(3.2.0)heptan-6-one .....................   27.  1a
                2β-chloro-3α-hydroxy- Bicyclo(3.2.0)heptan-6-one .....................   27.  2
   <hyl-6-(p-tolyl-thio-imino)-7-oxo-4-thia-1-aza Bicyclo(3.2.0)heptane-2-carboxylic-acid-p-nit>   41. 80
           2-methoxy-2,4-dimethyl-3,8-dioxa Bicyclo(3.2.0)octane .......................   38. 17
                 exo-phenyl-3-phospha Bicyclo(3.2.1)oct-6-en-3-one ...............   64. 59
      7,8-dibromo-8-cyano-4-methylsulfonato- Bicyclo(3.2.1)octa-2,6-diene ..............   31.  6
   exo-3-p-nitrobenzyl-endo-3-phospha Bicyclo(3.2.1)octane bromide ..............   64.111
   ny l-5-(dimethoxycarbonyl-methyl)-7,7-dimethyl- Bicyclo(3.3.0)oct-3-ene <ydroxy-4-methoxycarbo   27. 25
     2,4,6,7-tetraphenyl-1-arsa-2,3,7-triaza-8-oxa Bicyclo(3.3.0)oct-3-ene benzene solvate........   65. 25
            (+-)-8,8-dimethyl-6,7-diaza Bicyclo(3.3.0)octa-1,6-dien-7-oxide............   35.  9
   6-(3-indolyl)-4-(isopropenyl)-(2,3)benzo-1-aza Bicyclo(3.3.0)octa-2,4-diene [6,8,8-trimethyl-   36. 45
              7-thia-1,3-diaza Bicyclo(3.3.0)octa-2,4-dione ..............   41. 11
         2,3,6,7-tetramethyl-1,5-diaza Bicyclo(3.3.0)octa-2,6-diene-4,8-dione .........   35. 17
              4,6-dimethyl-1,5-diaza Bicyclo(3.3.0)octa-3,6-diene-2,8-dione .........   35.  7
         syn-3,4,6,7-tetramethyl-1,5-diaza Bicyclo(3.3.0)octa-3,6-diene-2,8-dione .........   35. 18
      3,7-dichloro-4,6-diphenyl-1,5-diaza Bicyclo(3.3.0)octa-3,6-diene-2,8-dione .........   35. 47
              4,8-dimethyl-1,5-diaza Bicyclo(3.3.0)octa-3,7-diene-2,6-dione .........   35.  6
      3,7-dichloro-4,8-diphenyl-1,5-diaza Bicyclo(3.3.0)octa-3,7-diene-2,6-dione .........   35. 46
              6β-bromo-7α-hydroxy-2-oxa Bicyclo(3.3.0)octan-3-one .................   38.  9
      6-acetoxymethyl-1,2,5-trimethyl-4-oxa Bicyclo(3.3.0)octan-3-one .................   38. 55
   ny l-5-(dimethoxycarbonyl-methyl)-7,7-dimethyl- Bicyclo(3.3.0)octan-3-one <hano-4-methoxycarbo   31. 51
   ctamethyl-1,3,5,7-tetra-aza-2,4,6,8-tetrasila- Bicyclo(3.3.0)octane <esityl-2,2,4,4,6,6,8,8-o   63. 44
   -(2-hydroxyethyl)-6,7,8-trimethyl-2-thia-7-aza Bicyclo(3.3.0)octane [5-hydroxy-1   39. 39
              1,5-diphospha Bicyclo(3.3.0)octane-1,5-disulfide ............   64. 20
         3-thioxo-4-phenyl-4-aza-2,7-dithio Bicyclo(3.3.0)octane-7,7-dioxide ...........   41. 35
                          Bicyclo(3.3.1)non-1-en-3-one ..............   31. 49
                          Bicyclo(3.3.1)non-1-en-3-one ..............   31. 48
                          Bicyclo(3.3.1)non-1-en-3-one trans-syn-dimer   31. 47
   <hionia-3λ(4),5λ(4)-dithia-2,4,6,8,9-penta-aza Bicyclo(3.3.1)nona-2,3,5(9)-triene hexafluoro>   41.  6
```

```
              exo-7-hydroxy- Bicyclo(3.3.1)nonan-3-one                  ................... 31.  3
              exo-7-methoxy- Bicyclo(3.3.1)nonan-3-one                  ................... 31. 11
oxy-7-thioxo-3-oxa-1,5,6,8-tetra-aza-7-phospha Bicyclo(3.3.1)nonane        [6,8-dimethyl-7-phen 64. 48
  exo-2,exo-6-dihydroxy-2,6-dimethyl- Bicyclo(3.3.1)nonane ethylacetate clathrate.... 61.  1
,7-diphenyl-1,3,4,6,7,9-hexa-aza-5λ(4)-phospha Bicyclo(3.3.1)nonane-5-thione ....[9-methyl-3 64. 73
            3,7-dimethyl-3,7-diaza Bicyclo(3.3.1)nonane-9-spiro-5'-hydantoin .... 37.  7
S-dimethyl-N-(trimethylsilyl)-sulfo-di-imide)- Bicyclo(3.3.1)penta-azatetrathiane .... [1-(S, 63.  2
2,2,4,4,6,6,8,8,9,9,11,11-dodecamethyl- Bicyclo(3.3.3)hexasilazane ............ 63. 14
itrile]        7,7-dichloro-1,6-dimethyl- Bicyclo(4.1.0)heptane-trans-3,trans-4-dicarbon 27.  6
thyl-5-oxo-1,3-dioxolane)]   (1,5,5-trimethyl- Bicyclo(4.2.0)octan-7-one)-8-spiro-4-(2,2-dime 38. 72
       benzyl 4-methoxy-7-azido-3-oxa-1-aza Bicyclo(4.2.0)octan-8-one-2-carboxylate........ 40. 36
onyl)oxy-ethyl)-8-oxo-2,2-dimethyl-3-oxa-1-aza Bicyclo(4.2.0)octane <(1-(p-nitrobenzyloxycarb 40. 45
-butyl)methylene)-1β-methyl-3-oxa-9-thia-7-aza Bicyclo(4.3.0)non-7-en-4-one <cyano-2-(cyano(t 41. 76
-butyl)methylene)-1β-methyl-3-oxa-9-thia-7-aza Bicyclo(4.3.0)non-7-en-4-one <cyano-2-(cyano(t 41. 75
yl-4-t-butyl-9-hydroxy-8-phenyl-7-phospha-cis- Bicyclo(4.3.0)non-8-ene-7-oxide <acemic-7-benz 64.141
arbonyl-2-methoxycarbonyl-9-methyl-9-aza-5-oxa Bicyclo(4.3.0)nona-(1(6),2)-dien-4-one <thoxyc 38. 80
         2,4,6-tri-t-butyl-7,8,9-dithia-aza Bicyclo(4.3.0)nona-1(9),2,4-trien-7-one........ 41. 68
         2,4,6-tri-t-butyl-7,8,9-dithia-aza Bicyclo(4.3.0)nona-1(9),2,4-triene ........... 41. 69
     4-methyl-9-p-chlorophenyl-7-oxa-2,9-diaza Bicyclo(4.3.0)nona-2,4-dien-8-one  ........... 40. 26
     6-methyl-9-p-chlorophenyl-7-oxa-2,9-diaza Bicyclo(4.3.0)nona-2,4-dien-8-one  ........... 40. 27
        7-methoxy-3,5,9-trioxa-4-phospha Bicyclo(4.3.0)nonan-4-one  ................... 64. 17
              8-oxa-1-aza Bicyclo(4.3.0)nonan-9-one-2-carboxylic acid.... 40.  8
           7-methoxy-3,5,9-trioxa Bicyclo(4.3.0)nonane  ................... 38. 10
ydroxy-8-methyl-6-(2'-methyl-hexylidene)-1-aza Bicyclo(4.3.0)nonane hydrochloride ....    [8-h 58. 25a
    4-diethylamino-3,5,9-trioxa-4-phospha Bicyclo(4.3.0)nonane-4,7-dione ............... 64. 33
 7,8-benzo-9-cyano-3,5,10-trimethyl-3,5-diaza Bicyclo(4.4.0)deca-7,9-diene-2,4-dione ........ 36. 16
 benzo-8-cyano-3,5-dimethyl-10-phenyl-3,5-diaza Bicyclo(4.4.0)deca-7,9-diene-2,4-dione   [7,8- 36. 34
oxymethyl>  8,9-dihydroxy-1,3,7,7-tetramethyl- Bicyclo(4.4.0)decane-2,2'-spiro(6',7'-bis(hydr 50. 21
e]  (t-5,c-10-dimethyl-c-6-hydroxy-4-oxo-r-1H- Bicyclo(5.3.0)dec-7-t-yl) acetic acid γ-lacton 38. 63
0α-tribromo-4α-chloro-4β,7β,8β,11-tetramethyl- Bicyclo(5.3.0)undec-1(11)-ene ....    [5β,9β,1 27. 14a
              7α-hydroxy- Bicyclo(5.4.0)undecan-9-one................. 27.  9
           9,10-diphenyl- Bicyclo(6.2.0)decapentaene ................. 27. 31
ethylthiolate 4,7,13,16,21-pentaoxa-1,10-diaza Bicyclo(8.8.5)tricosane benzene solvate <(ii) 49. 27
<)-iron(ii) 4,7,13,16,21,24-hexaoxa-1,10-diaza Bicyclo(8.8.8)hexacosane chlorobenzene solvate 49. 23
           2,6,10-trithia Bicyclo(9.4.0)pentadeca-1(11),12,14-triene .... 39. 43
     6-phenyl-6-phospha-2,10-dithia Bicyclo(9.4.0)pentadeca-11(1),12,14-triene .... 64. 93
     2,6,10-triphenyl-2,6,10-triphospha Bicyclo(9.4.0)pentadeca-11(1),12,14-triene .... 64.148
 2,10-diphenyl-6-methyl-6-aza-2,10-diphospha Bicyclo(9.4.0)pentadeca-11(1),12,14-triene ace 64.140
tone s>  2,6,10-triphenyl-6-aza-2,10-diphospha Bicyclo(9.4.0)pentadeca-11(1),12,14-triene ace 64.147
',3,3'-tetrachloro-4,4',7,7'-tetraphenyl-1,1'- Bicycloheptatrienylidene  ........ [(E)-2,2 22.  8
cis-1,7-diphenyl-3,3,5,5,9,9,11,11-octamethyl- Bicyclohexasiloxane........................ 63. 32
              0,0- Bicyclohexyl-1,1'-diyl-thiosulfite ............ 39. 45
              (+)- Bicyclohumulenone-triol-mono(p-bromobenzoate) 53. 33a
  1,2-0-isopropylidene-α-D-glucofuranose-3,5,6- Bicyclophosphite  ......................... 64. 32
  1,2-0-cyclohexylidene-α-D-glucofuranose-3,5,6- Bicyclophosphite  ......................... 64. 52
thoxycarbonyl)-2,3,7,8,12,13,17,18-octamethyl- Biladiene-a,c dihydrobromide    ....[1,19-bis(e 32. 76
e]    1,2,3,7,8,12,13,17,18,19-decamethyl- Biladiene-a,c dihydrobromide chloroform solvat 32. 74
<hoxycarbonyl)-2,3,7,8,12,13,17,18-octamethyl- Bilatriene-a,b,c hydrobromide ethanol solvate 32. 75
              octaethyl- Bilatriene-difluoroboron chloroform solvate.... 62. 20
           diethoxy- Bilirubin diethyl ester...................... 59. 45
       syn(methyl,methyl)- Bimane ...................................... 35. 18
       syn(phenyl,chloro)- Bimane ...................................... 35. 47
       anti(methyl,methyl)- Bimane ...................................... 35. 17
       anti(phenyl,chloro)- Bimane ...................................... 35. 46
    9,10-dioxo-syn(methyl,H)- Bimane ...................................... 35.  7
  μ-thia-syn(CH(2),CH(3))-9,10-dioxa- Bimane ...................................... 41. 30
  μ-sulfone-syn(CH(2),CH(3))-9,10-dioxa- Bimane ...................................... 41. 31
              carbo Biotin ...................................... 35. 23
              seleno Biotin ...................................... 39. 30
              (+-)-oxy Biotin ...................................... 38. 24
              Biotin methyl ester ......................... 39. 38
       N(1')-methoxycarbonyl- Biotin methyl ester ......................... 39. 48
              Biotin sulfone ............................. 39. 32
              Biotin-d-sulfoxide ......................... 39. 31
              4-bromo Biphenyl  ................................... 19.  6
       2,2'-dibromo-octafluoro Biphenyl  ................................... 19.  5
       cis-1,4-dihydro-4-trityl- Biphenyl  ................................... 19. 17
    4,4'-dihydroxy-3,3',5,5'-tetranitro Biphenyl  ................................... 17. 20
       biphenyl perfluoro Biphenyl complex  ........................... 60. 18
       4-bromobiphenyl perfluoro Biphenyl complex  ........................... 60. 16
       4-methylbiphenyl perfluoro Biphenyl complex  ........................... 60. 23
              Biphenyl perfluorobiphenyl complex ............ 60. 18
       4-bromo Biphenyl perfluorobiphenyl complex ............ 60. 16
       4-methyl Biphenyl perfluorobiphenyl complex ............ 60. 23
       methyl-N-( Biphenyl-2-yl)-N-(1-naphthyl)-anthranilate .... 48. 77
       methyl-N-( Biphenyl-2-yl)-N-(1-naphthyl)-anthranilate .... 48. 79
```

                       methyl-N-(Biphenyl-2-yl)-N-(1-naphthyl)-anthranilate 48. 78
           2-allyl-2-azonia-7-aza Biphenylene bromide............................ 36. 7
e]   (3S,5R)-3-methyl-5-(4'- Biphenylyl)-2,3,5,6-tetrahydro-1,4-oxazin-2-on 40. 43
-(3,6,9,12-tetraoxatetradeca-1,14-dioxy)-2,2'- Bipyridine .... [6,6' 40. 51
                            2,2'- Bipyridyl potassium dicyano-gold(iii) 33. 34
                            2,2'- Bipyridyl-(trichloro-antimony) 66. 10
                            4,4'- Bipyridylium di-μ-chloro-tetrachloro-di-copper 33. 36
           tetramethylammonium nonabromo-di- Bismuth(iii) 3. 25
                                  Bohemamine 59. 6
ylamino)-2,2-dimethyl-1,3-dithia-2-sila-4,5-di Bora-cyclopentane ....[4,5-bis(dimeth 63. 4
           triethylammonium closo-dodeca Borane 3. 32
              1-t-butyl-1,2-dicarba-closo-dodeca Borane 62. 8
          9-methylsulfonyl-1,7-dicarba-closo-dodeca Borane 62. 3
           N-methyl-2-oxy-1-naphthaldiminato-diphenyl Borane 62. 18
         9-cyclohexyl-5(7)-dimethylsulfide-nido-deca Borane 62. 10
        imethylsilyl-2-methyl-1,2-dicarba-closo-dodeca Borane .... [1-tr 63. 5
         isopropylidene-dithio-1,2-dicarba-closo-dodeca Borane .... [9,12- 62. 6
         ,7-dihydroxy-10-bromo-2,3-dicarba-closo-undeca Borane ......[2,3-dimethyl-4 62. 4
         lotriphosphazenyl)-1-phenyl-1,2-dicarba-dodeca Borane ....[2-(pentachloro-cyc 64. 29
                      lithium bis(dimethyl-(Borane)-phosphine)methylide.................. 64. 14
               pentakis(methyl-dicarba-dodeca Boranyl)-diphosphine benzene solvate .. 64. 75
        ium 9,10a,11-trimethyl-7,8-dicarba-nido-undeca Borate.............. [ces 62. 7
          tetra-t-butoxy-triberyllium bis(tetrahydro Borate)................... 5. 7
        ,8'-oxido-3-cobalta-bis(η(5)-1,2-dicarbadodeca Borate)...... [triethylmethylammonium 8 62. 5
                      bis(4-dibenzo Borepinyl)-ether 62. 19
        tyl-2,4-bis(pentafluorophenyl)-1,3,2,4-diazadi Boretidine .... [1,3-di-t-bu 62. 13
        1,3-bis(trimethylsilyl)-4-methyl-diazastanna- Boretidine dimer 63. 17
              4,5-diethyl-3,6-dimethyl-1,2-diaza-3,6-di Borinane 62. 9
        -tetra(dimethylamino)-1,4,2,3,5,6-dithia-tetra Borinane .... [2,3,5,6 3. 41
        1,2-(tetrahydro-methyl-semibullvalene-diyl)-4- Bornyl-1,2,4-triazolidine-3,5-dione........ 37. 18a
        N,N,N',N'-tetrakis(1,3-dimethyl-1,3,2-diaza Borolidinyl)-hydrazine 62. 11
              octaethyl-bilatriene-difluoro Boron chloroform solvate 62. 20
        trimethylphosphonium-methylene- Boron-trihydride 64. 11
               sodium hydrido-trimethyl Boronate diethyl ether solvate 62. 2
                       24-ethyl- Brassinolide 51. 77
                   24-ethyl-iso Brassinolide 51. 76
                                  Brassinolide monohydrate 51. 72
        cta-3,7-diene]   8- Bromo-2-bromomethyl-6-methyl-2,5,6-trichloro-o 52. 1a
        tetra-acetyl-trillenogenin mono Brosylate 51. 85
        11-monoacetyl-cinncassiol-D(1)-19-mono Brosylate monohydrate 54. 34
               3,4',11,12-tetra-O-acetyl- Bruceine 38.140
        ,5-dione]   1,2-(tetrahydro-methyl-semi Bullvalene-diyl)-4-bornyl-1,2,4-triazolidine-3 37. 18a
        o> (+-)-2β,5β-epoxy-2α-(3α-p-nitrobenzoyloxy- But-1(E)-enyl) -1β,3,3-trimethylcyclohexan-1α- 59. 23
        (3-(4,5-dihydro-5,5-dimethyl-4-oxo-2-furanyl)- But-2-enyl)-oxy-(2H-1-benzopyran-2-one)....[7- 59. 19
        <ahydro-11,14-dimethoxy-7-methyl-4a-(3-methyl- But-2-enyl)-5,8a-o-benzeno-1,4-naphthoquinone 59. 36
              3-nitro-4-(2-thienyl)- But-3-en-2-one 39. 20
              4-(5-nitro-3-thienyl)- But-3-en-2-one 39. 21
           1,4-bis(phenylazo)-1,3-dimethyl- Buta-1,4-diene 9. 3
                     (+-)-deoxy Butaclamol 36. 47
        is(1,3-dibenzyl-2-imidazolidinyl)-1,4-diphenyl Butadiene .... [((1Z,3E)-1,4-b 32. 78
        methylthio)-2,2,4,4-tetramethyl-bicyclo(1.1.0) Butane .... [1,3-di 27. 4
        uoromethyl)-2,3-bis(p-fluorophenyl)-hexafluoro Butane .... [2,3-bis(trifl 19. 11
                (+)-tri-o-thymotide-S-(+)-2-bromo- Butane clathrate 61. 16
        e)]   2,2'-(Butane-1,4-diyl)-bis(hexahydro-1,3-benzodioxol 38. 97
                    1,4- Butanediyl-diammonium tetrachloro-manganese.... 3. 28n
                    1,4- Butanediyl-diammonium tetrachloro-manganese.... 3. 27n
                 16α,17β- Butanomorphinan 58. 46
                 16β,17α- Butanomorphinan hydrobromide 58. 47
        imethyl-6-oxo-2H-oxireno(a)pyrrolizin-4-yl)-2- Butenamide <yl-N-(1a,6,6a,6b-tetrahydro-2,6a-d 59. 6
        dimethylaminoethoxy)phenyl)-1,2-trans-diphenyl Butene .... [1-(p-(2- 17. 41
        p-methoxy-N,N-bis(α-cyano-hexafluoro-iso Butenyl)-aniline 16. 13
        ter]   N-(t- Butoxy-carbonyl)-L-methionyl-glycine-benzyl es 48. 60
                    t- Butoxy-carbonyl-L-phenylalanine............... 48. 43
        r]   t- Butoxy-carbonyl-L-prolyl-sarcosine-benzyl este 48. 64
        ydrate]   t- Butoxy-carbonyl-L-prolyl-L-isoleucyl-glycine h 48. 56
                tetra-t- Butoxy-carbonyl-triberyllium bis(tetrahydroborate) 5. 7
        (+-)erythro-methyl-3-t- Butoxy-2-iodo-3-(p-methoxyphenyl)-propionate 17. 24
        ydrate]   N-(t- Butoxycarbonyl)-L-prolyl-L-valyl-glycine hemih 48. 52
        xo-5-benzimidazolyl)-(naphthoic acid)-> 1-(2- Butoxycarbonyl-1-azophenyl)-2-hydroxy-3-N-(2-o 35. 74
        3,4,6-tri-O-acetyl-1,2-O-(R)-(1-t- Butoxyethylidene)-α-D-galactopyranose 45. 47
        N-(1-phenylethyl)-iso Butyramide 1. 27
        methyl 2-methyl-3-(benzyl(phenyl)phosphinyl)- Butyrate 64.105
               D,L-α-amino-n- Butyric acid 48. 11
             4-(3-indolyl)- Butyric acid 35. 26
             α-benzyloxy-γ- Butyrolactone 38. 32
        methylbenzoyl-oxycarbonyl-bis(α-amino-iso Butyryl)-L-alaninate 48. 65

<roxy-8-(2-hydroxy-2-hydroxymethyl-3-mercapto- Butyryloxy)-trans,trans-germacra-1(10),4-dien> 53. 25a
                 trimethylammonium Cadmium chloride 3. 12
   bis(triphenylbenzylphosphonium) tetrachloro- Cadmium dichloroethane clathrate 61. 8
                bis(α-cyclodextrin) Cadmium polyiodide hydrate 45. 63
                     ionomycin Cadmium salt n-heptane solvate 50. 29a
         methylammonium tetrabromo- Cadmium(ii)................................. 3. 2
        n-decylammonium tetrachloro- Cadmium(ii)................................. 3. 47
        n-decylammonium tetrachloro- Cadmium(ii)................................. 3. 46
     tetraphenylporphinato-bis(dioxane)- Cadmium(ii)................................. 49. 28
ethylammonium tetrabromo-cadmium(ii) tribromo- Cadmium(ii)........................... [trim 3. 14
      trimethylammonium tetrabromo- Cadmium(ii) tribromo-cadmium(ii) 3. 14
                            Calamusenone p-bromophenylpyrazoline 53. 28a
                            Calcium bis(hydrogen-1-malate) hexahydrate 2. 24
hydrate] Calcium bis(picrate) tetraethylene-glycol mono 17. 1
                         cytosine Calcium chloride hydrate 44. 5
                             di- Calcium dihydrogen mellitate nonahydrate ... 14. 5
rate] Calcium ethylenediamine-tetra-acetate heptahyd 48. 27
                             Calcium formate 2. 2n
                            β- Calcium formate............................. 2. 3
                            β- Calcium formate............................. 2. 4
                             Calcium malonate dihydrate 2. 16
                         sodium Calcium nitrilo-triacetate 48. 16
          cyclo-tris(prolyl-glycyl) Calcium perchlorate 48. 70
          cyclo-tris(prolyl-glycyl) Calcium perchlorate hydrate................. 48. 67
                             Calcium salicylate dihydrate 14. 1
                             Calcium salicylate dihydrate 14. 2
                   ionomycin Calcium salt n-heptane solvate 50. 28
                             Calcium succinate trihydrate 1. 10
                             Calcium D-pantothenate bromide 2. 31
                   tris(glycine) Calcium(ii) dibromide 48. 3
  endo-2-dimethylphosphono-exo-2-hydroxy-(-)- Camphane 64. 54
       (-)-5-endo,6-exo-dihydroxy- Camphene 31. 12
    4,4'-(ethylenedi-imine)-bis(methylene Camphor) 31. 62
      (1R,3S,4R)-N,N-dimethyl-3- Camphor-carbothioamide 31. 21
                           1- Camphorquinone 31. 9
                     syn-γ- Camphorquinone-dioxime 31. 10
                             Capnell-9(12)-ene-2β,5α,8β,10α-tetrol 53. 10
                             Capuronine acetate 58. 54
              anti-7,7-dibromo-nor Car-3-ene oxide............................. 38. 8
     anti-1,6-dimethyl-7,7-dibromo-nor Car-3-ene oxide............................. 38. 16
zene clathrate] 5-methylbenzene-1,3-di Carbaldehyde-bis(p-tolylsulfonylhydrazone) ben 61. 7
ethylene-tricyclo(5.3.0.0(1,4))dec-5-yl-phenyl Carbamate .... [7-methyl-3-m 28. 12
 (ethoxycarbonyl)-cis-6-phenylcyclohex-2-ene-1- Carbamate .... [benzyl trans-6- 21. 18
       dimethyl-4,4-methylene-bis(phenyl Carbamate) 16. 14
)-6-methylfuro(3,2-b)furan-3-ol-4-bromophenyl- Carbamic acid ester <ethoxy-2-(dimethoxymethyl 38. 95a
)-6-methylfuro(3,2-b)furan-3-ol-4-bromophenyl- Carbamic acid ester <ethoxy-2-(dimethoxymethyl 38. 94
 β-decahydro-4-oxo-thieno(3,2-c)azocine-3α-yl)- Carbamic acid methyl ester <3aα,4,5,6,7,8,9,9a 39. 72
 hydroxyimino-4H-cyclohepta(b)thiophen-3α-yl)- Carbamic acid methyl ester <8,8aβ-octahydro-4- 39. 44
       4-ethyl-1-isopropyl-3-((phenyl Carbamoyl)-methyl)-3-pyrrolin-2-one.......... 32. 64
       2,4-dimethyl-3-phenyl-2-(diethyl Carbamoyl)-1-azetine N-oxide 32. 58
              1,4-bis(dimethyl Carbamoyl)-1,4-dihydro-1,4-diazocine 34. 7
            5-benzoyl-2-(methoxy Carbamoyl)-1H-benzimidazole hydrobromide 35. 41
                  bis(phenyl Carbamoyl-oxy-n-butyl)-diacetylene 16. 18
    2-methyl-3-p-anisyl-5-(ethylamino)- Carbamoyl-1,2,4-thiazoline 41. 45
                         N-methyl Carbazole 36. 6
            5,6-cyclopenteno-pyrido(3,2-a) Carbazole 36. 20
     N'-acetyl-di-indolo(2,3-a:2',3'-c) Carbazole 36. 48
                       1,5-diphenyl Carbazone 8. 13
            tris(trichlorophosphazeno)- Carbenium hexachloro-antimony 64. 1
<lphosphine)-platinum)-(μ-carbonyl)-(μ-p-tolyl Carbenyl)-(carbonyl-cyclopentadienyl-manganes> 71.173
       amethylammonium hexa-cobalt-tetradecacarbonyl Carbide.............................. [tetr 3. 19
    rbonyl-iron-> bis(tetraethylammonium) (μ(6)- Carbido-(dicarbonyl-iron-(μ(2)-carbonyl)-dica 3. 37
   icarbonyl-iron)-tr> tetraethylammonium (μ(5)- Carbido-(dicarbonyl-rhodium-(μ(2)-carbonyl)-d 3. 33
      bis(triphenylphosphine)immonium (μ(5)- Carbido-iodo-pentadecacarbonyl-penta-osmium 3. 56
      bis(bis(triphenylphosphine)immonium) Carbido-deca-osmium-carbonyl 64.166
   bis(tetraphenylphosphonium) hexa-μ-carbonyl- Carbido-heptacarbonyl-hexa-rhodium 64.126
  onyl-ruthenïu> bis(tetraphenylarsonium) μ(6)- Carbido-tetrakis(μ(2)-carbonyl)-hexakis(dicarb 65. 15
                          Carbinoxamine maleate 33. 70
tachloro-antimony-oxo)-cyclododecasulfur tris(Carbon disulfide) .... [1,7-bis(pen 11. 3
 ,4,6,6-hexa(1-aziridinyl)-cyclotriphosphazene Carbon tetrachloride anticlathrate .... [2,2,4 61. 2
   5-benzamido-2-benzoylimino-1,3-dithiol-4- Carbonic acid ethyl ester 39. 75
                       benzene-hexa Carbonitrile 7. 3
         tricyclo(3.3.1.1(3,7))decane-1- Carbonitrile 31. 14
 3-(2-methoxyphenyl)-1,1,2,2-cyclopropane-tetra Carbonitrile 20. 6
   2,2a,8b-tetrahydro-cyclobuta(a)naphthalene-8b- Carbonitrile .... [1-phenoxy-1, 28. 10
 ethyl-bicyclo(4.1.0)heptane-trans-3,trans-4-di Carbonitrile .... [7,7-dichloro-1,6-dim 27. 6

,3,5-triphenyl-perhydrothieno(3,4-c)pyrrole-1-	Carbonitrile <ol-3-yl)-4,6-dioxo-syn-cis-syn-1	40.	58
1-methylphospha-2,3,5,6-bis-σ-	Carborano-cyclohexane	64.	23
1-methylarsa-2,3,4,5-bis(o-	Carborano-cyclopentane)	65.	3
cyclopropane-1,1-di	Carboxamide	20.	2
5-phenyl-1,2,4-oxadiazole-3-	Carboxamide	40.	14
5-amino-N-octyl-1H-tetrazole-1-	Carboxamide	32.	36
N-phenyl-2,5-dihydrothiophene-3-	Carboxamide	39.	35
1β-D-ribofuranosyl-pyridin-4-one-3-	Carboxamide	47.	25
-dihydro-3-methyl-benzofuro(3,2-c)isoxazole-3-	Carboxamide [N,N-diethyl-3,3a	40.	38
8,9,9a-hexahydro-4H-pyrido(1,2-a)pyrimidine-3-	Carboxamide [6-methyl-4-oxo-1,6,7,	35.	20
5'-tri-O-nitro-β-D-ribofuranosyl)-imidazole-4-	Carboxamide <-5-(N-nitrocarboxamido)-1-(2',3',	45.	19
hexahydro-5α,9α-methanobenzo-cyclo-octene-10α-	Carboxamide <,5RS,9SR)-10β-ethyl-5,6,7,8,9,10-	31.	38
uranosyl)-imidazole-4-ca> 2-methyl-5-(N-nitro	Carboxamido)-1-(2',3',5'-tri-O-nitro-β-D-ribof	45.	19
N-methyl-1,4-dithiane-2,3-di	Carboximide	39.	14
(4-methoxyphenyl)-3,6-dithiacyclohexene-1,2-di	Carboximide [N-	39.	47
8-hexahydro-1H-pyridazino(4,5-d)azepine-2,3-di	Carboximide <yl-8-oxo-N,9-diphenyl-2,3,4,4a,5,	37.	23
allo-2-hexosu> 2α,3β,11-trihydroxy-12-oxo-5β-	Carda-9,20-dienolide-3-O-methyl-4,6-dideoxy-D-	45.	57
	Cardiopetalidine	58.	56
	Cardiopetaline	58.	55
	Carolic acid	38.	15
8β-hydroxymethyl-podo	Carpane-13β-carboxylic acid lactone	59.	20
p-bromobenzoyl-epi	Catalponol	59.	30a
triethylammonium tris(pyro	Catechol)silicate	63.	23
bis(monothio-pyro	Catecholato)-tellurium(iv)	70.	7
aquo-triphenyl-catecholato-antimony triphenyl-	Catecholato-antimony	66.	16
imony] aquo-triphenyl-	Catecholato-antimony triphenyl-catecholato-ant	66.	16
O-acetyl-2',3:4',6'-di-O-benzylidene-7(R)-β-D-	Cellobioside [methyl-2,3',6-tri-	45.	59
1S,2E,4R,6R,7E,11S,12S)-6-acetoxy-11,12-epoxy-	Cembra-2,7-dien-4-ol [(	54.	22
2,2-dimethoxy-N-methyl-seco(B,C)	Cephalotax-4-en-3-one	58.	45
	Cesium bis(trifluoromethyl)-difluoroborate	62.	1
ecaborate]	Cesium 9,10a,11-trimethyl-7,8-dicarba-nido-und	62.	7
	Chaetoglobosin K acetone solvate	58.	78
	Chasmanine intermediate	58.	65
	Chenodeoxycholic acid	51.	52
-amino-1-methoxy-6-methylamino-1,3,6-trideoxy-	Chiro-inositol [L-3	50.	4
phenyl-α-hydroxy-cyclohexyl-phosphinic acid	Chloranhyrdide	64.	51
potassium p-	Chloranil acetone solvate	18.	1
tetrathiafulvalene-	Chloranil complex	60.	3
octaethylporphinato-per	Chlorato-iron(iii)	49.	7
itoyl-(tetrakis-o-oxyethoxyphenyl))-porphyrin	Chloroform methanol solvate <,15,20-(pyrromell	49.	32
3α-acetoxy-	Chola-7,9-dien-12-on-24-oic acid methyl ester	51.	61
sodium 3α,7α,12α-trihydroxy-5β-	Cholan-24-oate monohydrate	51.	51
3α,12α-dihydroxy-5β-	Cholan-24-oic acid ethanol solvate monohydrate	51.	54
rubidium 3α,12α-dihydroxy-5β-	Cholan-24-oic acid monohydrate	51.	50
te] 3α,12α-dihydroxy-5β-	Cholan-24-oic acid palmitic acid ethanol solva	51.	55
sodium	Cholate monohydrate	51.	51
(E)-3α-acetoxy-5,10-seco-	Cholest-1(10)-en-5-one	51.	75
1α,3β-1,3-dehydro-5,10-seco-	Cholest-10(19)-en-5-one	51.	63
1β,3α-1,3-dehydro-5,10-seco-	Cholest-10(19)-en-5-one	51.	64
4,6-diaza-A,B-bishomo-	Cholest-4a-eno-(4,3-d)(6,7-d)-bis(tetrazole)	51.	60
	Cholest-5-en-3-ol dihydrocinnamate	51.	81
14α-ethyl-5α-	Cholest-7-ene-3β,15α-diol-di-p-bromobenzoate	51.	88
3β-(toluene-p-sulfonamido)-5α-	Cholestan-2β-ol	51.	79
3α,7α,12α-trihydroxy-5α,(25S)-	Cholestan-26-oic acid	51.	68
1α,3β-1,3-dehydro-10α-hydroxy-5,10-seco-	Cholestan-5-one	51.	67
β,23β-tetrahydroxy-24β-methyl-B-homo-7-oxa-5α-	Cholestan-6-one [2α,3α,22	51.	71
3β,16β,23(R),26-tetrahydroxy-5β-	Cholestane	51.	69
3β-azido-2α-(toluene-p-sulfonamido)-5α-	Cholestane	51.	78
3β-acetoxy-5β,6β-N-nitro-aziridinyl-	Cholestene	51.	74
	Cholesteryl bromide	51.	65
	Cholesteryl chloride	51.	66
	Cholesteryl decanoate	51.	83
	Cholesteryl nonanoate	51.	82
	Cholesteryl undecanoate	51.	84
deoxy	Cholic acid ethanol solvate monohydrate	51.	54
deoxy	Cholic acid hydrate	51.	53
deoxy	Cholic acid palmitic acid ethanol solvate	51.	55
deoxy	Cholic acid rubidium salt	51.	50
phenyl	Choline ether bromide	17.	17
3-dodecanoyl-propandiol-1-phosphoryl	Choline monohydrate	3.	55
acetyl	Choline tetraphenylborate	62.	14
1-chloroethyl-4-methyl-6,7-dimethyl-iso	Chroman	38.	60
3-formylfuro(3,2-f)	Chromene	38.	37
3-formyl-8-methylfuro(3,2-g)	Chromene	38.	49
,N-diethylethylene-tellurourea)-pentacarbonyl-	Chromium [(N	70.	8
sphine)immonium μ-deuterido-bis(pentacarbonyl-	Chromium) [bis(triphenylpho	64.153n

<(η(5)-cyclopentadienyl)-(μ(3)-(pentacarbonyl- Chromium-thio)-methylidyne)-(μ(3)-sulfido)-tr> 71.169
5,8-dimethoxy-2-methylfuro(2,3-e) Chromone 59. 5
-hexamethyl-21-isopropyl-cyclopentano-perhydro Chrysen-3-one .... [4β,5β,9β,13α,14β,17α 56. 6
Cinchonidine 58. 38
cholest-5-en-3-ol dihydro Cinnamate 51. 81
,4:3,5-di-O-methylene-D-mannitol-1,6-di-trans- Cinnamate .... [(+)-2 45. 54
2-(4'-dimethylamino- Cinnamoyl)-indan-1,3-dione 27. 26
6-amino-5- Cinnamoyl-1,3-dimethyl-uracil 44. 32
Cinncassiol C(1) 54. 4
11-monoacetyl- Cinncassiol-D(1)-19-monobrosylate monohydrate 54. 34
ethylenediammonium (-)-hydroxy Citrate................................ 2. 25
ethylenediammonium (2S,3S)-2-hydroxy Citrate................................ 2. 25
ethylenediammonium (+)-allohydroxy Citrate monohydrate................ 2. 26
ethylenediammonium (2S,3R)-2-hydroxy Citrate monohydrate................ 2. 26
fentanyl Citrate toluene solvate................ 33. 84
ethyl)-4-piperidinylium)-N-phenyl-propanamide- Citrate toluene solvate........[N-(1-(2-phenyl 33. 84
Clavicipitic acid 48. 46
Clazamycin A hydrochloride 50. 2a
Cleomeolide 54. 6
Clonidine hydrochloride.................... 32. 28
-(tri-μ-carbonyl)-tris(tricarbonyl-ruthenium)- Cobalt <(triphenylphosphine)immonium) carbonyl 64.154
cyclo-octadecane hexa-aquo-cobalt tetrachloro- Cobalt acetone solvate [1,4,7,10,13,16-hexaoxa 38. 47
,7,10,13,16-hexaoxacyclo-octadecane hexa-aquo- Cobalt tetrachloro-cobalt acetone solvate [1,4 38. 47
io)-methylidyne)-(μ(3)-sulfido)-triangulo-tri- Cobalt tetrahydrofuran solvate <yl-chromium-th 71.169
phthalocyaninato- Cobalt(ii) 49. 1n
tetraphenylporphyrinato- Cobalt(ii) 49. 17
bis(4-benzyl-piperidinium) tetrachloro- Cobalt(ii)............ 33. 46
mmonium nona-μ-carbonyl-nitrido-hexa(carbonyl- Cobalt)............ [bis(triphenylphosphine)i 64.157
tetramethylammonium hexa- Cobalt-tetradecacarbonyl-carbide 3. 19
triethylmethylammonium 8,8'-oxido-3- Cobalta-bis(η(5)-1,2-dicarbadodecaborate) 62. 5
e monohydrate] dicyano-(5,6-dihydroxy-dihydro Cobyrinic acid pentamethyl ester)-c,d-dilacton 49. 29
dicyano Cobyrinic-a,c-diamide trihydrate 49. 25
mihydrate] 6-amino-dicyano-5,6-dihydro Cobyrinic-pentamethylester-a-amide-c-lactam he 49. 30
Codeine-7,8-oxide 58. 26
tri Codesmine 58. 32
di-copper hexyl Cofacial diporphyrin-7 .... 49. 36
Colchiceine ethyl acetate solvate monohydrate 58. 53
(-)-10,11-oxy-10,12a-cyclo-10,11-seco- Colchicine 58. 57
thio Colchicoside ethanol solvate hydrate 58. 70
7- Con-O-methylnogarol ethanol solvate.......... 50. 19
dicytosinium tetrachloro- Copper 44. 6
piperidinium tetrachloro- Copper 33. 15
4,4'-bipyridylium di-μ-chloro-tetrachloro-di- Copper 33. 36
di- Copper hexyl cofacial diporphyrin-7.......... 49. 36
Copper propionate 1. 9
Copper L-tartrate trihydrate 1. 11
pyridinium tris(thiocyanato)-di- Copper(i) 33. 7
bis(4-benzyl-piperidinium) hexachloro-di- Copper(ii) 33. 47
N-methyl-piperidinium))-disulfide tetrachloro- Copper(ii) .... [bis(4-(33. 50
N,N'-(1,2-phenylene)-bis(salicylaldiminato)- Copper(ii) - 7,7,8,8-tetracyanoquinodimethane> 60. 40
N-(2-ammonioethyl)-piperazinium pentachloro- Copper(ii) dihydrate 33. 19
'd')-bis(1,2-dithiole) bis((μ(2)-bromo-bromo- Copper)............[naphthaceno(5,6-cd:11,12-c 39. 66
Cordycepin 47. 15
furo(2,3-f) Coumarin 38. 30
furo(2,3-h) Coumarin 38. 29
7-hydroxy-6-methoxy Coumarin 38. 19
lithium phosphocyclo Creatine dihydrate 64. 12
5,5'-diethoxy-α,α'-dimethyl-α,α'-azino-di-o- Cresol 17. 34
bromo Criptine methylsulfonate isopropanol solvate 58. 76a
mono Crotaline 58. 19
N-methyl- Crotananinium iodide 58. 34
dibenzo-30- Crown-10 potassium thiocyanate 38.130
te] (dibenzo-30- Crown-10) bis(sodium isothiocyanate) monohydra 38.131
benzo-15- Crown-5 lithium picrate dihydrate 38. 65
benzo-15- Crown-5 sodium perchlorate 38. 66
bis(benzo-15- Crown-5) sodium perchlorate.................... 38. 67
bis(benzo-15- Crown-5) sodium tetrachloro-dioxo-uranium 38. 68
bis(benzo-15- Crown-5) sodium tetraphenylborate 62. 15
18- Crown-6................................ 38. 45
monopyrido-18- Crown-6 t-butylammonium perchlorate complex.... 40. 40
Cuauchichicine 58. 60
Cunaniol acetate 59. 13
potassium cyanamide di Cyanamide 3. 4
potassium Cyanamide dicyanamide 3. 4
3-acetoxy-5-(2-nitrophenylthio)-nortri Cyclene................................ 31. 32
3-chloro-5-(2',4'-dinitrophenylthio)-nortri Cyclene................................ 31. 17

hloromethyl-5-(2',4'-dinitrophenylthio)-nortri Cyclene................................... [2-c 31. 22
                                            di Cyclenephosphorane 64. 78
                                        3',5'- Cyclic-cytidine-monophosphate trihydrate 47. 8
                     sodium adenosine-(3'-5')- Cyclic-monophosphate tetrahydrate 47. 13
                                               Cyclizine hydrochloride................... 33. 74
                                               Cyclo(benzylglycyl-bis(L-prolyl)) 48. 58
                                               Cyclo(di(benzylglycyl)-L-prolyl) monohydrate 48. 73
                                               Cyclo(di-L-prolyl-D-prolyl)................. 48. 45
              ne)methylene) toluene clathrate] Cyclo(tetrakis(5-t-butyl-2-hydroxy-1,3-phenyle 61. 21
          xyisovaleryl-bis(D-valyl-L-lactyl-L-valyl-D> Cyclo(D-valyl-L-methylalanyl-L-valyl-D-α-hydro 50. 37
          xyisovaleryl-D-valyl-L-methylalanyl-L-valyl> Cyclo(D-valyl-L-methylalanyl-L-valyl-D-α-hydro 50. 24
                                               Cyclo(D-N-methylvalyl-D-α-hydroxyisovaleryl) 48. 32
          N-methylvalyl-D-α-hydroxyisovaleryl)] Cyclo(D-N-methylvalyl-L-α-hydroxyisovaleryl-L- 48. 71
          D-phenylalanyl-glycyl) dihydrate] Cyclo(L-leucyl-L-phenylalanyl-glycyl-D-leucyl- 48. 80
          thylvalyl-D-α-hydroxyisovaleryl-L-methylval> Cyclo(L-methylvalyl-D-α-hydroxyisovaleryl-L-me 50. 22
                                               Cyclo(L-prolyl-L-alanyl) 48. 23
          trihydrate] Cyclo(L-prolyl-L-4-hydroxyprolyl) monohydrate 48. 28
          <sulfonylamino)-13-oxa-14-thia-1,15-diazapenta Cyclo(10.4.2.0(2,7).0(8,16).0(12,16))octadeca> 42. 12
          (> 5,22-bis(diethylamino)-6,21-dimethyl-hepta Cyclo(12.8.2.0(1,15).0(2,13).0(4,12).0(7,12).0 31. 66
          <2-hexakis(trifluoromethyl)-1,6-diphospha-hexa Cyclo(4.4.2.0(2,5).0(3,9).0(4,8).0(9,10))dode> 64. 76
          ,11,13-tetraene] tetrabenzo-penta Cyclo(6.2.2.2(2,6).0(2,7).0(3,7))tetradeca-4,9 31. 64
          ane]   4,7,11-tris(trimethylsilyl)-penta Cyclo(6.3.0.0(2,6).0(3,10).0(5,9))undecaphosph 64. 35
          0-diene]   2,3,4,5,6,7-hexafluoro-penta Cyclo(6.4.0.1(9,12).0(2,7)0(3,6))tridecane-4,1 30. 2
          4,5,9,10,11,12-hexafluoro-penta Cyclo(6.4.0(3,6).0(4,12).0(5,9))dodec-10-ene 31. 15
          12).0(11,13))hexad> dimethyl-15,16-diaza-octa Cyclo(6.6.2.0(2,7).0(3,5).0(4,6).0(9,14).0(10, 37. 11
          <xy-15-methyl-5-phenyl-3,5,7,15-tetra-azapenta Cyclo(7.4.3.2(2,8).0.0(3,7))octadec-17-ene-4,> 37. 20
          trien> 5-methyl-14,16-dioxa-3,5,7-triazapenta Cyclo(7.4.3.2(2,8).0.0(3,7))octadeca-10,12,17- 38. 56
          triene-4,> 15-methyl-5-phenyl-5,15-diazapenta Cyclo(7.4.3.2(2,8).0.0(3,7))octadeca-10,12,17- 37. 19
          triene-4,6-dione]   5,15-dioxapenta Cyclo(7.4.3.2(2,8).0.0(3,7))octadeca-10,12,17- 38. 75
          triene-4,6-dione]   5-methyl-15-oxa-5,15-azapenta Cyclo(7.4.3.2(2,8).0.0(3,7))octadeca-10,12,17- 38. 84
          18-tetraene-4,7,11,13-tetrone]   12-oxapenta Cyclo(8.4.3.2(2,9).0.0(3,8))heptadeca-5,14,16, 38. 89
          2,14-dione]   13-phenyl-11,13,15-triazapenta Cyclo(8.5.2.0(2,6).0(2,8).0(11,15))heptadeca-1 37. 17
          ,18).0(13,17))heneicosane]   exo,exo-octa Cyclo(8.8.1.1(3,6).1(12,15).0(2,9).0(4,8).0(11 31. 55
).0(8,15).0(9,13).0(17,24).0(17,26).0> dodeca Cyclo-(14.10.2.1(6,12).1(20,26).0(2,15).0(7,11 31. 65
          aleryl) dihydrate] Cyclo-bis(glycyl-L-propyl-glycyl) tetrahydrate 48. 54
          -L-phenylalanyl)]   N-acetyl- Cyclo-bis(D-valyl-prolyl-valyl-D-α-hydroxyisov 48. 81
                                               Cyclo-bis(L-prolyl-L-valyl-L-ornithyl-L-leucyl 50. 40
                                               Cyclo-octa-amylose hydrate 45. 70
                 1,4,7,10,13,16-hexaoxa Cyclo-octadecane 38. 45
                 4,7,10,13,16-pentaoxa-1-thia Cyclo-octadecane 42. 7
          cobalt acetone solvat> 1,4,7,10,13,16-hexaoxa Cyclo-octadecane hexa-aquo-cobalt tetrachloro- 38. 47
          lorate]   1,4,7,10,13,16-hexaoxa Cyclo-octadecane hexa-aquo-manganese(ii) perch 38. 48
          is(2-cyanoethyl)-1,10-diaza-4,7,13,16-tetraoxa Cyclo-octadecane malononitrile ....[N,N'-b 40. 46
                 4,7,10,13,16-pentaoxa-1-thia Cyclo-octadecane potassium thiocyanate 42. 9
                 4,7,10,13,16-pentaoxa-1-thia Cyclo-octadecane rubidium thiocyanate 42. 11
                 4,7,10,13,16-pentaoxa-1-thia Cyclo-octadecane silver nitrate 42. 8
          hydrofuran solvate] 1,4,7,10,13,16-hexaoxa Cyclo-octadecane sodium dicyanophosphide tetra 38. 46
                 4,7,10,13,16-pentaoxa-1-thia Cyclo-octadecane sodium thiocyanate............ 42. 10
          2-benzyl-1,7,10,16-tetraoxa-4,13-diazo Cyclo-octadecane-3,14,18-trione................ 40. 49
          3,3,12,12-tetramethyl-1,5,10,14-tetraoxa Cyclo-octadecane-6,9,15,18-tetrone 38. 96
          ,8-octamethyl-2,4,6,8-tetrasila-1,5-dimercura- Cyclo-octane .... [2,2,4,4,6,6,8 63. 11
                                               Cyclo-octane-spiro-5'-hydantoin............... 35. 21
                               cis- Cyclo-octane-1,5-diol 22. 3
                                               Cyclo-octane-1,5-dione 22. 1
                                               Cyclo-octane-1,5-dione-dioxime 22. 2
          ,2,2-trichloro-1-(p-toluenesulfonyloxy)ethyl)- Cyclo-octene .... [3-(2 22. 6
          hyl-5,6,7,8,9,10-hexahydro-5α,9α-methanobenzo- Cyclo-octene-10α-carboxamide <,5RS,9SR)-10β-et 31. 38
                        trichloro-(8-ethoxy-4- Cyclo-octenyl)-tellurium 70. 6
          2-diphenyl-3,4-methoxycarbonyl-1-aza-2-phospha Cyclo-pent-1-ene .... [5,5-dimethyl-2, 64.113
                                               Cyclo-tetrakis(phenylarsenic-sulfide) 65. 22
                                               Cyclo-tris(prolyl-glycyl) 48. 66
                                               Cyclo-tris(prolyl-glycyl) calcium perchlorate 48. 70
                                               Cyclo-tris(prolyl-glycyl) calcium perchlorate 48. 67
                                               Cyclo-tris(prolyl-glycyl) hydrate 48. 68
                                               Cyclo-tris(prolyl-glycyl) magnesium perchlorat 48. 69
          yl-D-α-hydroxyisovaleryl)] Cyclo-tris(D-valyl-L-α-hydroxyisovaleryl-L-val 50. 39
          cyl) monohydrate dimethylsulfoxide solvate] Cyclo-tris(L-valyl-L-prolyl-glycyl-L-valyl-gly 48. 83
          yl-cyclodisil> N,N'-bis(2,2,4,4,6-pentamethyl Cyclo-trisiloxanyl-oxadimethylsilyl)-tetrameth 63. 26
          xy-xylose]   3,5-(Cyclo-N,N-diethylphosphamido)-2-O-methyl-1-deo 46. 6
                               (-)-10,11-oxy-10,12a- Cyclo-10,11-seco-colchicine................. 58. 57
                               8-isocyano-10- Cycloamphilectene 54. 15
                               1-phenoxy-1,2,2a,8b-tetrahydro- Cyclobuta(a)naphthalene-8b-carbonitrile........ 28. 10
          ,11a,11b-tetrahydro-exo-phenanthro(9',10':3,4) Cyclobuta(1,2-d)(1,3)dioxol-10-one <loro-8b,8c 38. 79
                        tetra-t-butyl Cyclobutadiene 20. 11
          methylene-1-tetracarbonyl-2,3-diaza-1-mangana- Cyclobutan-4-one)-manganese pentacarbonyl [(2- 71. 25

naphtho(b,e)di Cyclobutane................................... 29.   2
tetrakis(phenylimino)- Cyclobutane................................... 20. 14
r-1,c-2,t-3,t-4-tetrakis(2-pyridyl)- Cyclobutane................................... 33. 87
r-1,c-2,t-3,t-4-tetrakis(4-pyridyl)- Cyclobutane................................... 33. 88
r-1,c-2,t-3,t-4-tetrakis(2-pyrazinyl)- Cyclobutane................................... 33. 77
1,c-2-bis(3-pyridyl)-t-3,t-4-bis(2-pyrazinyl)- Cyclobutane.............................[r- 33. 80
1,t-3-bis(3-pyridyl)-c-2,t-4-bis(2-pyrazinyl)- Cyclobutane.............................[r- 33. 81
-bis(4-chlorophenyl)-2,4-trans-bis(4-pyridyl)- Cyclobutane............... [1,3-trans 33. 93
thylsilyl)-3,4-bis(trimethylsiloxy)-1,2-disila Cyclobutane <,4-di-t-butyl-1,1,2,2-tetra(trime 63. 48
)-1,3-bis(trimethylsilyl)-1,3-diaza-2,4-disila Cyclobutane <is(isopropyl(trimethylsilyl)amino 63. 49
l-1,3-bis(trimethylsilyl)-1,3-diaza-2,4-disila Cyclobutane <trimethylsilyl)amino)-2,4-dimethy 63. 39
1-amino-1,3-dicarboxy- Cyclobutane monohydrate..................... 48. 18
16α,17α- Cyclobutano-progesterone 51. 38
naphtho(2,3-6,7)di Cyclobutene(1,8:4,5)dicyclopentene 30.   4
tetrafluoro-1,2-(RS)-bis(2,2,3,3-tetrafluoro Cyclobutyl)ethane 20.   3
lithium phospho Cyclocreatine dihydrate 64. 12
N,N'-dimethyl-1,7-diaza-4,10-dioxa Cyclodecane benzylammonium thiocyanate 40. 18
N,N'-dimethyl-1,7-diazonium-4,10-dioxa Cyclodecane perchlorate 40. 19
Cyclodecane-oxime 23.   1
tetrabenzo- Cyclodecane-1,16-dione 30. 13
epithio-methyleneoxy)-6,12-epithiodibenzo(a,f) Cyclodecen-7(6H)-one   [syn-5,12-dihydro-5,14-(39. 73
β- Cyclodextrin dodecahydrate 45. 67
β- Cyclodextrin ethanol solvate octahydrate 45. 60
α- Cyclodextrin hexahydrate 45. 62n
α- Cyclodextrin hexahydrate 45. 61n
Cyclodextrin hydrate 45. 70
β- Cyclodextrin m-methylphenol decahydrate........ 45. 65
γ- Cyclodextrin n-propanol clathrate hydrate 61. 22
β- Cyclodextrin sodium polyiodide octahydrate 45. 68
entahydrate] α- Cyclodextrin N,N-dimethylformamide clathrate p 61. 19
ate] α- Cyclodextrin 2-pyrrolidone clathrate pentahydr 61. 20
bis(α- Cyclodextrin) cadmium polyiodide hydrate 45. 63
bis(α- Cyclodextrin) lithium tri-iodide octahydrate 45. 64
lo-trisiloxanyl-oxadimethylsilyl)-tetramethyl- Cyclodisilazane <-bis(2,2,4,4,6-pentamethylcyc 63. 26
1,4,7,15-tetraoxa-12,18-diaza Cyclodocosane-8,11,19,22-tetrone 40. 42
bis(1,4,7,10-tetra-aza Cyclododecane)-phosphorane 64. 78
Cyclododecane-oxime........................ 23.   3
1,7-bis(pentachloro-antimony-oxo)- Cyclododecasulfur tris(carbon disulfide) 11.   3
ethoxy-tetrabenzo(b,e,h,k)(1,4,7,10)tetrathia- Cyclododecatetraene chloroform solvate <-octam 39. 86
(1,2-(3-methylnaphtho))-1,6,9,12,15,18-hexaoxa Cycloeicosa-2,4-diene .... [2,3:4,5-bis 38.136
(1,2-(3-methylnaphtho))-1,6,9,12,15,18-hexaoxa Cycloeicosa-2,4-diene .... [2,3:4,5-bis 38.135
<1,2-(3-methylnaphtho))-1,6,9,12,15,18-hexaoxa Cycloeicosa-2,4-diene t-butylammonium perchlo> 61. 13
3,3-dimethyl-1,5,13-trioxa-10,16-diaza Cycloeicosane-6,9,17,20-tetrone................ 40. 44
1,4-dioxa Cyclohept-2-eno(2,3-b)naphthalene 38. 51
4-dodecahydro-4b-phenyl-6,9:10,13-dimethano-di Cyclohepta(a,c)naphthalene <9,9a,10,11,12,13,1 31. 63
<3aβ,5,6,7,8,8aβ-octahydro-4-hydroxyimino-4H- Cyclohepta(b)thiophen-3α-yl)-carbamic acid me> 39. 44
Cyclohepta(d,e)naphthalene 28.   2
Cyclohepta(d,e)naphthalene 28.   3
Cyclohepta(d,e)naphthalene 28.   5
Cyclohepta(d,e)naphthalene 28.   4
Cyclohepta-amylose dodecahydrate 45. 67
Cyclohepta-amylose p-nitroacetanilide hydrate 45. 66
allium(iii))] di-μ-2-hydroxy Cyclohepta-2,4,6-trien-1-onato-bis(diphenyl-th 68. 24
(1,4-b)-(2,3-b)-bis(1,1,4,4-tetramethyl Cycloheptane)-5,6-dioxabicyclo(2.2.0)hex-2-ene 38.114
hydrate] 1-amino Cycloheptane-carboxylic acid hydrobromide mono 48. 25
dimethyl-1-hydroxy-1- Cycloheptane-phosphonate 64. 34
yl-spiro(cyclohex-2-ene-1,5'-(5H)-dibenzo(a,d) Cycloheptene)-4-amine hydrochloride <,N-dimeth 29.   8a
yl-spiro(cyclohex-2-ene-1,5'-(5H)-dibenzo(a,d) Cycloheptene)-4-ammonium hydrogen maleate <eth 29.   9a
Cycloheptyl-p-bromobenzoate................. 22.   4
cis-2-ethoxy Cycloheptyl-tribromo-tellurium(iv) 70.   5
iphenyl-3-methoxy-6-methyl-1,2-diaza-3-phospha Cyclohex-1(6)-ene .... [2,4,4-tr 64.121
R)-1-hydroxymethyl-2-(1-methylpentyl)-3-methyl Cyclohex-1-ene p-bromophenylurethane <)-(3R,1' 21. 17
benzyl trans-6-(ethoxycarbonyl)-cis-6-phenyl Cyclohex-2-ene-1-carbamate 21. 18
<3'-chloro-10',11'-dihydro-N,N-dimethyl-spiro(Cyclohex-2-ene-1,5'-(5H)-dibenzo(a,d)cyclohep> 29.   9a
ene)-4>   (1S,4R)-3'-chloro-N,N-dimethyl-spiro(Cyclohex-2-ene-1,5'-(5H)-dibenzo(a,d)cyclohept 29.   8a
-trimethoxy-6'-β-methyl-spirobenzofuran-2(3H)- Cyclohex-2-ene-3,4'-dione benzene solvate <4,6 50. 11
5,5-dimethyl-3-methylsulfonyl- Cyclohex-2-enone 11. 10
methyl-6-(diphenylphosphinyl)-3,4-dimethyl- Cyclohex-3-ene-1-carboxylate 64.119
<)-1-ethyl-4-(3-methoxy-phenylethyl)-1,4-diazo Cyclohex-5-en-2-one)-3-spiro-5'-(3'-methyl-hy> 36. 37
<etoxy-5-methoxy-1-methyl-1,2,3,9a-tetrahydro- Cyclohexa(i,j)isoquinoline-7-spiro-4'-(2'-met> 58. 63
Cyclohexa-amylose hexahydrate 45. 62n
1-diphenylmethylene-4-triphenylmethyl- Cyclohexa-2,5-diene......................... 21. 22
is(4,5-dimethyl-2H-1,3-dithiol-2-ylidene)-1,4- Cyclohexa-2,5-diene perchlorate .... [b 39. 64
<)-ethylidene)-3a',4',7',7a'-tetrahydro-spiro(Cyclohexa-2,5-diene-1,1'-(1H)-indene)-4,4'-di> 17. 44
2,6-dimethyl-4-(diphenylmethylene)- Cyclohexa-2,5-dienone 21. 15
2,6-dimethyl-4-(diphenylmethylene)- Cyclohexa-2,5-dienone 21. 14

2,6-dimethyl-4-(diphenylmethylene)- Cyclohexa-2,5-dienone 21. 16
ohexa(i,j)isoquinoline-7-spiro-4'-(2'-methoxy- Cyclohexa-2',5'-dien-1'-one) methiodide <-cycl 58. 63
3',4,9'-tri-t-butyl-2',5,10'-trimethoxy-spiro(Cyclohexa-3,5-diene-1,6'-dibenzo(d,f)(1,3)dio> 38.138
1,4,9,12-tetra-aza Cyclohexadecane-3,10,13,16-tetrone 34. 8
1,4,7,10,13-pentaoxa-14,16- Cyclohexadecanedione 38. 34
,3',5,5'-tetraphenyl-4,4'-dithia-1,1'-bis(2,5- Cyclohexadienylidene) polyiodide .... [3 39. 90
,3',5,5'-tetraphenyl-4,4'-dithia-1,1'-bis(2,5- Cyclohexadienylidene) polyiodide .... [3 39. 89
trobenzoyloxy-but-1(E)-enyl) -1β,3,3-trimethyl Cyclohexan-1α-ol [(+-)-2β,5β-epoxy-2α-(3α-p-ni 59. 23
(+-)-(1R,2R,3R)-3-acetyl-2-vinyl Cyclohexan-1-ol............................... 21. 7
4-t-butyl-1-aza Cyclohexan-2-one 33. 33
trans-1,4-dimethyl Cyclohexane................................... 21. 2
1-methylphospha-2,3,5,6-bis-σ-carborano- Cyclohexane................................... 64. 23
3-t-butyl-4-methyl-2-oxo-1-oxa-2-thia-3-aza Cyclohexane................................... 42. 3
E)-1,4-dibenzyl-1,4-diphenyl-1,4-diphosphonio- Cyclohexane dibromide .... [(64.149
,4S)-4-methyl-spiro(bicyclo(2.2.2)octane-2,1'- Cyclohexane)-3',6-dione ....[syn-(1R,2R 31. 31
2,5-di Cyclohexane-spiro-1,3-dithiane 39. 55
6-methoxy-4-oxo-3-phenylseleno- Cyclohexane-1,2-dicarboxylic anhydride 38. 70
ylidene-dioxy)-1-(2-methyl-1,3-dithiane-2-yl)- Cyclohexane-1,3-diol <R,3R,4S,5R)-4,5-(isoprop 39. 54
1-aza-4,4-dimethyl Cyclohexane-2,6-dione 33. 27
16α,17α- Cyclohexano-progesterone 51. 58
6α-methyl-16α,17α- Cyclohexano-progesterone 51. 59
rel-(1S,2R)-2-hydroxymethyl-2-methyl Cyclohexanol 21. 3
3,3,5,5-tetramethyl- Cyclohexanone-oxime.......................... 10. 4
3,3,5-trimethyl-5-phenyl- Cyclohexanone-oxime.......................... 21. 10
2,4-dimethyl-4-(1,2-dichlorovinyl)- Cyclohexen-3-one 21. 6
,2,2-trichloro-1-(p-toluenesulfonyloxy)ethyl)- Cyclohexene................................ [3-(2 21. 8
N-(4-methoxyphenyl)-3,6-dithia Cyclohexene-1,2-dicarboximide 39. 47
meso-3,4-bis(trans-4-hydroxy Cyclohexyl)-hexane 21. 12
racemic-3,4-bis(trans-4-hydroxy Cyclohexyl)-hexane 21. 13
0-(β-chloroethyl)-phenyl-(α-hydroxy Cyclohexyl)-phosphinate 64. 66
0-(2-trimethylammonio-ethyl)-phenyl-(1-hydroxy Cyclohexyl)-phosphinate iodide 64. 85
di(Cyclohexyl)-phosphinic acid.................. 64. 53
(αRS,1RS,2RS)-α-(1-hydroxy-2-phenyl Cyclohexyl)-propionic acid 21. 9
racemic-bis(4-oxo Cyclohexyl)-3,4-hexane 21. 11
2-R-methyl Cyclohexyl-α-D-glucopyranoside 45. 28
ide] 2-(di Cyclohexyl-acetoxy)-ethyl-triethylammonium iod 1. 40
di-cis-2-chloro Cyclohexyl-dichloro-tellurium(iv) 70. 9
p-tolyl-2-chloro Cyclohexyl-dichloro-tellurium(iv) 70. 13
phenyl-α-hydroxy- Cyclohexyl-phosphinic acid chloranhyrdide 64. 51
diphosphane] 1,2-di Cyclohexyl-1,2-bis(diethylamino)-1,2-dithioxo- 64.112
H-pyrano(3,2-c)(1)benzopyran-5-one] 4- Cyclohexyl-3,4-dihydro-2-hydroxy-2-methyl-2H,5 38.100
3-chloro-3-cyano-1- Cyclohexyl-4-phenylthio-2-azetidinone 32. 57
e] 9- Cyclohexyl-5(7)-dimethylsulfide-nido-decaboran 62. 10
Cyclohexylammonium ethyl hydrogen phosphate.... 21. 1
di Cyclohexylammonium gascardate 55. 2
ophosphite] 1,2-0- Cyclohexylidene-α-D-glucofuranose-3,5,6-bicycl 64. 52
3,5-bis(Cyclohexylidene-1,2,4-trithiolane) 39. 53
L-pyroglutamyl-N,N'-di Cyclohexylurea 48. 55
1,4,7-trithia- Cyclononane................................... 39. 11
(diethylamino)-1,10-ethano-5-methyl-cyclopenta Cyclononene..................................[4- 28. 11
,15-dihydro-5H-5,10,15-trithia-tribenzo(a,d,g) Cyclononene chloroform solvate <hexamethoxy-10 39. 84
3,4-dihydro-2H-1,5-dioxadinaphtho(2,1-f:1,2-h) Cyclononene-3,1'-cyclopentane) ....[spiro(38.123
3,4-dihydro-2H-1,5-dioxadinaphtho(2,1-f:1,2-h) Cyclononene-3,1'-cyclopropane) ....[spiro(38.121
<'-bis(8-acetoxy-6b,7-dihydro-7,9-dimethyl-8H- Cyclopent(a)acenaphthylenediyl) benzene solva> 30. 19
7b-methyl-7bH- Cyclopent(cd)indene-1,2-dicarboxylic acid 28. 6
2,3-dimethyl-5-(p-bromobenzoyloxymethyl)- Cyclopent-2-en-1-one 50. 7
2,3,3aα,4,5,5aβ,6,7,8,9,9aα,9b-dodecahydro-1H- Cyclopenta(a)naphthalene-1,5-dione <bα-methyl- 28. 7
<methylbutyl)-11,12,13,14,16,17-hexahydro-15H- Cyclopenta(a)phenanthrene-17,2'-spiro(1',3'-d> 51. 45
ethylamino-1,3,4a,8a-tetramethyl-4a,8a-dihydro Cyclopenta(1,2-a)indene ....[8-di 28. 14
-3a-methyl-4-oxo-5,6,6a-triphenyl-3a,4-dihydro Cyclopenta(2,3-d)isoxazoline <(p-chlorophenyl) 40. 57
di-isopropyl-(2,3,4,5-tetraphenyl- Cyclopenta-1,4-dienyl)-phosphate 46. 7
4-(diethylamino)-1,10-ethano-5-methyl- Cyclopentacyclononene 28. 11
Cyclopentadecanone phenylsemicarbazone 23. 6
8,9,11,12-octahydro-1,4,7,10,13-benzopentaoxa- Cyclopentadecin sodium perchlorate [2,3,5,6, 38. 66
,8,9,11,12-octahydro-1,4,7,10,13-benzopentaoxa Cyclopentadecin) sodium perchlorate <s(2,3,5,6 38. 67
8,9,11,12-octahydro-1,4,7,10,13-benzopentaoxa- Cyclopentadecin) sodium tetraphenylborate <,6, 62. 15
-2'-pyridinium)-3-(1''-methyl-4''-pyridinium)- Cyclopentadienide bromide hemihydrate <-methyl 33. 72
mine] Cyclopentadienyl sodium tetramethylethylenedia 20. 1
m-thio)-methylidyne)-(μ(3)-sulfid> tris(η(5)- Cyclopentadienyl)-(μ(3)-(pentacarbonyl-chromiu 71.169
<)-(μ-carbonyl)-(μ-p-tolylcarbenyl)-(carbonyl- Cyclopentadienyl-manganese) tetrafluoroborate> 71.173
1,1-di-iodo-3,4-benzo-1-tellura- Cyclopentane 70. 3
no)-2,2-dimethyl-1,3-dithia-2-sila-4,5-dibora- Cyclopentane) .... [4,5-bis(dimethylami 63. 4
1-methylarsa-2,3,4,5-bis(o-carborano- Cyclopentane) 65. 3
5-dioxadinaphtho(2,1-f:1,2-h)cyclononene-3,1'- Cyclopentane) .... [spiro(3,4-dihydro-2H-1, 38.123
N,N'-(tetrasulfur-tetranitrido)-thia Cyclopentane-di-imide 41. 9
4β,5β,9β,13α,14β,17α-hexamethyl-21-isopropyl- Cyclopentano-perhydrochrysen-3-one 56. 6

                            16α,17α- Cyclopentano-progesterone 51. 48
                  2-benzyl-5-benzylidene- Cyclopentanone 20. 10
                  2-benzyl-5-benzylidene- Cyclopentanone dimer 20. 15
      N,N'-bis(2-(4-chlorophenylamino)-2- Cyclopenten-1-ylidene)-hydrazine 20. 12
      N,N'-bis(2-(4-chlorophenylamino)-2- Cyclopenten-1-ylidene)-hydrazine 20. 13
    -5'-monoph> 7-(3,4-trans-4,5-cis-dihydroxy-1- Cyclopenten-3-yl-aminomethyl)-7-deazaguanosine 47. 34
           naphtho(2,3-6,7)dicyclobutene(1,8:4,5)di Cyclopentene 30. 4
                              2-carboxy- Cyclopenteno-1-(N-methyl-N-phenyl-sulfonamide) 20. 5
                                   5,6- Cyclopenteno-pyrido(3,2-a)carbazole 36. 20
   .4)nonane-3,6-di> (2-oxo-3,3,5,5-tetramethyl- Cyclopentyl)-1-oxa-7,7,9,9-tetramethyl-spiro(4 38.108
                                        Cyclopentyl-(4)-helixane 38. 85
                       syn-2,11-dithia(3.3)meta Cyclophane 39. 63
                (3)Tetrathiafulvaleno(3)para Cyclophane 39. 69
                (4)Tetrathiafulvaleno(4)para Cyclophane 39. 78
       1,2,9,10,17,18-dehydro(2.2.2)para Cyclophane 31. 58
       1,2,9,10,17,18-dehydro(2.2.2)para Cyclophane 31. 59
    -15,18-dimethoxy-(2,6)-p-benzoquinone(3.3)meta Cyclophane .... [syn 31. 53
    -tetramethoxy-1,2,3,10,11,12-hexathia(3.3)meta Cyclophane .... [5,7,14,16 39. 62
    ,18-hexamethyl-1,2,3,4,11,12-hexathia(4.2)meta Cyclophane .... [6,8,10,14,16 39. 70
   hyl-1,2,9,10,17,18,25,26-octathia(2.2.2)meta Cyclophane ... [4,6,12,14,20,22,28,30-octamet 39. 87
          1,6,20,25-tetra-aza(6.1.6.1)para Cyclophane hydrochloride durene tetrahydrate 61. 18
    1,5,12,16,23,26,29-heptaoxa(7(3,14))(5.5)ortho Cyclophane naphthalene-2,3-diol monohydrate.... 38.113
                             cis-4-phenyl Cyclophosphamide 64. 60
                  2-O-methyl-xylitanylidene diethyl- Cyclophosphoramidate 46. 6
   -1a',6b'-dihydro-spiro(cyclopropane-1,2'-(1H)- Cycloprop(a)indene) <-tetrachloro-3,3-dimethyl 29. 3
                 cis-2-ethoxy-1-phenylsulfinyl- Cyclopropane 20. 4
         1,1,2,2,-tetracyano-3-(2-methoxyphenyl)- Cyclopropane 20. 6
   5-dioxadinaphtho(2,1-f:1,2-h)cyclononene-3,1'- Cyclopropane) .... [spiro(3,4-dihydro-2H-1, 38.121
          3-(2-methoxyphenyl)-1,1,2,2- Cyclopropane-tetracarbonitrile 20. 6
                                        Cyclopropane-1,1-dicarboxamide 20. 2
   etrachloro-3,3-dimethyl-1a',6b'-dihydro-spiro(Cyclopropane-1,2'-(1H)-cycloprop(a)indene) <-t 29. 3
                               16α,17α- Cyclopropano-progesterone 51. 29
      -)-3-(bis(phenylsulfinyl)methyl)-1,2-dimethyl- Cyclopropene .... [(+ 20. 9
                                2,3-di(Cyclopropyl)-1,4-naphthoquinone.......... 20. 8
                                     N- Cyclopropylmethyl-scopolammonium bromide 38.106
            octamethyl-trithio- Cyclotetra(λ(3),λ(5),λ(5),λ(5)-phosphazane).... 64. 30
                          octaphenyl- Cyclotetra(siloxane) 63. 53
   ,20,20-octamethyl-1,4,6,9,11,14,16,19-octaoxo- Cyclotetracosa-cis-2,7,12,17-tetraene <0,15,15 23. 7
   <2-dimethyl-1,4,7,13,16,19-hexaoxa-10,22-diaza Cyclotetracosane) benzylammonium thiocyanate > 61. 6
   o-5,5,7,12,12,14-hexamethyl-1,4,8,11-tetra-aza Cyclotetradecane dihydrate .... [mes 34. 9
                                        Cyclotetradecane-oxime 23. 5
              1,4,7,10-tetra-aza Cyclotetradecane-3,8,11,14-tetrone 34. 6
   l)-5,12-dimethyl-5,12-diaza-1,3,8,10-tetrathia Cyclotetradecane-6,13-dione <is(p-methoxypheny 41. 81
                                        Cyclotetradecane-6,7,13,14-tetrone 39. 3
            decathia- Cyclotetradecine <6-de:2',1',6'-gh:2'',1'',6'' 36. 39
   -kl:2''',1''',6'''-na)(1,3,5,8,10,12)-hexa-aza Cyclotetradecine <6-de:2',1',6'-gh:2'',1'',6'' 36. 39
           octa(3,5-dimethylpyrazolyl)- Cyclotetraphosphazene 64.171
      1,3,5,7-tetramethyl-1,3,5,7-tetraphenyl- Cyclotetrasilazane 63. 47
   ibenzo(b,q)(1,4,7,10,13,16,19,22,25,28)decaoxa Cyclotriacontane potassium thiocyanate .... [d 38.130
   <benzo(b,q)(1,4,7,10,13,16,19,22,25,28)decaoxa Cyclotriacontin bis(sodium isothiocyanate) mo> 38.131
                                        Cyclotridecane-oxime 23. 4
           1-hydrido-1-isopropyl-tetrachloro- Cyclotriphosphazene 64. 3
         2,2-diphenyl-4,4,6-trimethoxy-6-hydroxy- Cyclotriphosphazene........................ 64. 74
   is-2,4,6,6-tetrachloro-2,4-bis(dimethylamino)- Cyclotriphosphazene....................... [c 64. 9
   ns-2,4,6,6-tetrachloro-2,4-bis(dimethylamino)- Cyclotriphosphazene........... [tra 64. 10
                       hexa-aziridino- Cyclotriphosphazene benzene clathrate 61. 3
   lathrate] 2,2,4,4,6,6-hexa(1-aziridinyl)- Cyclotriphosphazene carbon tetrachloride antic 61. 2
   ecaborane] 2-(pentachloro- Cyclotriphosphazenyl)-1-phenyl-1,2-dicarba-dod 64. 29
              2,2,4,4,6,6-hexamethyl Cyclotristanna-tellurane 70. 2
                                        Cycloundecane-oxime........................ 23. 2
                2-methylthio-Δ(2)-1,3-diaza-6,9-dioxa- Cycloundecene hydroiodide 40. 13
                                        Cycloveratril benzene clathrate monohydrate.... 61. 12
                                        Cystamine dihydrochloride 3. 29
                    S-8-aza-adenosyl-L-homo Cysteine dihydrate 48. 40
                     S-adenosyl-L-homo Cysteine hydrate 48. 44
                              5-iodo Cytidine 47. 6
                          3',5'-cyclic- Cytidine-monophosphate trihydrate 47. 8
                             barium Cytidine-5'-phosphate hydrate 47. 9
                     proflavine - Cytidylyl-(3-5')-guanosine sulfate hydrate 60. 24
   ex decahydrate] Cytidylyl-(3'-5')-adenosine - proflavine compl 60. 39
                          ethidium Cytidylyl-(3'-5')-guanosine.................... 47. 37
              acridine orange 5-iodo Cytidylyl-(3'-5')-guanosine dodecahydrate 47. 35
                proflavine deoxy Cytidylyl-(3'-5')-guanosine hydrate 47. 38
            9-amino-acridine - 5-iodo Cytidylyl-(3'-5')-guanosine hydrate complex.... 60. 22
   <etramethyl-N-methyl-phenanthrolinium - 5-iodo Cytidylyl-(3'-5')-guanosine hydrate methanol > 60. 37
   olvate] proflavine 5-iodo Cytidylyl-(3'-5')-guanosine hydrate methanol s 47. 36
   olvate] ellipticine - 5-iodo Cytidylyl-(3'-5')-guanosine hydrate methanol s 60. 36

Cytochalasin A 58. 74
Cytosine calcium chloride hydrate 44. 5
5-methyl Cytosine hemihydrate 44. 12
Cytosine monohydrate 44. 1n
1-benzyl Cytosine nitrate 44. 22
1-methyl Cytosine perchlorate 44. 13
bis(1-methyl Cytosine) hydroiodide monohydrate 44. 11
di Cytosinium tetrachloro-copper 44. 6
di Cytosinium tetrachloro-zinc............... 44. 7
11-deoxy- Daunorubicin aglycone triacetate 50. 18
s-dihydroxy-1-cyclopenten-3-yl-aminomethyl)-7- Deazaguanosine-5'-monophosphate hydrate <,5-ci 47. 34
oxo-bis(hydroxy-(tricyclo(5.2.1.0(2,6)) Dec-3-en-9-yl)-methyl-silane) 63. 36
henyl-3,10,10-trimethyl-tricyclo(5.2.1.0(3,7)) Dec-4-ene .... [4,5-benzo-6-p 31. 57
7-methyl-3-methylene-tricyclo(5.3.0.0(1,4)) Dec-5-yl-phenylcarbamate 28. 12
4-methyl-6-(N-methyl-N-phenylamino)-spiro(4.5) Dec-6-ene-1,8-dione.......................... 27. 24
3,6-dimethyl-4,10-dihydroxy-2-oxa-spiro(4.5) Dec-7-ene-1,9-dione.......................... 59. 3
hloro-mercury-4-acetoxy-tricyclo(4.2.2.0(2,5)) Dec-7-ene-9,10-cis-endo-dicarboxylate <-endo-c 31. 37
dimethyl-c-6-hydroxy-4-oxo-r-1H-bicyclo(5.3.0) Dec-7-t-yl) acetic acid γ-lactone [(t-5,c-10- 38. 63
<o)-10-acetoxy-tetracyclo(4.4.0.0(2,4).0(3,7)) Dec-8-ene-8,9-dicarboxylic acid dimethyl ester 31. 56
cis-10-methyl-1-oxa Decalin-2,5-dione 38. 23
1,8-di-iodo-tricyclo(5.3.0.0(4,8)) Decane 31. 8
10-phospha-1,4,7-triazatricyclo(5.2.1.0(4,10)) Decane .... [10-thio- 64. 19
N,N'-dimethyl-1,7-diaza-4,10-dioxacyclo Decane benzylammonium thiocyanate 40. 18
N,N'-dimethyl-1,7-diazonium-4,10-dioxacyclo Decane perchlorate 40. 19
cyclo Decane-oxime 23. 1
tricyclo(3.3.1.1(3,7)) Decane-1-carbonitrile 31. 14
tetrabenzo-cyclo Decane-1,16-dione 30. 13
<-dihydroxy-1,3,7,7-tetramethyl-bicyclo(4.4.0) Decane-2,2'-spiro(6',7'-bis(hydroxymethyl)-4'> 50. 21
10-hexamethyl-tetracyclo(4.3.1.0(3,10).0(4,9)) Decane-2,5-dione .... [3,4,6,8β,9, 31. 39
amethyl-9-exo-methylene-tricyclo(4.4.0.0(3,8)) Decane-2,5-dione ..... [1β,3,4β,6β,8-pent 31. 40
3,8-dimethyl-1,10-diazatricyclo(5.2.1.0(4,10)) Decane-3,7-diene-2,9-dione 37. 5
oro-3-phenyl-4-aza-5-oxatricyclo(5.2.1.0(2,6)) Decane-3,8-diene [1,7,8,9-tetrachl 40. 32
oro-3-phenyl-4-aza-5-oxatricyclo(5.2.1.0(2,6)) Decane-3,8-diene .... [1,7,8,9,10-pentachl 40. 31
luoroethyl)-vinyl)-endo-tricyclo(5.2.1.0(2,6)) Decane-3,8-diene <-cyano-exo-5((E)-(2,2,2-trif 31. 52
7-bromo-2-chloro-tricyclo(4.3.1.0(3,7)) Decane-4,8-dione 31. 7
cholesteryl Decanoate 51. 83
9,10-diphenyl-bicyclo(6.2.0) Decapentaene 27. 31
carboxy- Decarboxamido-vancomycin 50. 41
io-methyleneoxy)-6,12-epithiodibenzo(a,f)cyclo Decen-7(6H)-one [syn-5,12-dihydro-5,14-(epith 39. 73
n- Decylammonium tetrachloro-cadmium(ii) 3. 47
n- Decylammonium tetrachloro-cadmium(ii) 3. 46
Delphinium alkaloid A hydroiodide 58. 67
N-acetyl- Dendrodoine 41. 53
rubidium oxalate per Deuterate 2. 14n
bis(triphenylphosphine)immonium μ- Deuterido-bis(pentacarbonyl-chromium) 64.153n
potassium Deuterium bis(dichloroacetate) 2. 7n
di Deutero-pyridine-2,3-dicarboxylic acid 33. 21n
di Deutero-pyridine-2,3-dicarboxylic acid 33. 20n
di Deutero-pyridine-2,3-dicarboxylic acid 33. 22n
di Deutero-quinolinic acid..................... 33. 20n
di Deutero-quinolinic acid..................... 33. 22n
di Deutero-quinolinic acid..................... 33. 21n
per Deutero-4,4'-azoxyphenetole................. 17. 25
Deuteromethane 5. 1
β-cyclo Dextrin dodecahydrate 45. 67
β-cyclo Dextrin ethanol solvate octahydrate............ 45. 60
α-cyclo Dextrin hexahydrate...................... 45. 61n
α-cyclo Dextrin hexahydrate....................... 45. 62n
cyclo Dextrin hydrate......................... 45. 70
β-cyclo Dextrin m-methylphenol decahydrate 45. 65
γ-cyclo Dextrin n-propanol clathrate hydrate 61. 22
β-cyclo Dextrin sodium polyiodide octahydrate 45. 68
α-cyclo Dextrin N,N-dimethylformamide clathrate pentah 61. 19
α-cyclo Dextrin 2-pyrrolidone clathrate pentahydrate 61. 20
ydrate] bis(α-cyclo Dextrin) cadmium polyiodide hydrate............ 45. 63
bis(α-cyclo Dextrin) lithium tri-iodide octahydrate........ 45. 64
E- Diamantan-3-one-oxime 31. 29
Z- Diamantan-3-one-oxime 31. 30
mo-5-(o-chlorophenyl)-1,3-dihydro-2H-1,4-benzo Diazepin-2-one [7-bro 35. 34
oxy-1-methyl-5-phenyl-1,3-dihydro-2H-1,4-benzo Diazepin-2-one ...... [7-chloro-3-hydr 35. 35
ro-1-(N-methylacetamido)-5-phenyl-2H-1,4-benzo Diazepin-2-one ....[7-chloro-1,3-dihyd 35. 51
)-1,3,4,5-tetrahydro-1,4-dimethyl-2H-1,4-benzo Diazepin-2-one monohydrate <-5-(2-fluorophenyl 35. 45
7-chloro-2-methylamino-5-phenyl-3H-1,4-benzo Diazepin-3-ol 35. 38
-bis(dimethylamino)-3-ethyl-1-phenyl-1,5-benzo Diazepine .... [2,4 35. 66
oro-2,3-dihydro-1-methyl-5-phenyl-1H-1,4-benzo Diazepine .... [7-chl 35. 42
(1,4-dibenzyl-6-phenyl-1,2,3,4-tetrahydro-1,4- Diazepine) .... [7,7'-bis 34. 13

(1,4-dibenzyl-6-phenyl-1,2,3,4-tetrahydro-1,4-	Diazepine)	[7,7'-bis 34. 14
,6,-tetramethyl-2,3,6,7-tetrahydro-1,4,5-thia-	Diazepine-1,1-dioxide	[cis-3,3,6 41. 23
7-chloro-2-methylamino-5-phenyl-3H-1,4-benzo	Diazepine-4-oxide	35. 39
2,3-dihydro-5,7-dimethyl-1,4-	Diazepinium perchlorate	34. 3
<oro-2,6-dihydro-4-methyl-5,6-ethano-1H-(1,4)-	Diazepino(1,7-a)benzimidazole hydrochloride d>	36. 12
	Dictyodiol	54. 12
	Dictyolactone	54. 5
	Digoxigenin dihydrate	51. 41
	Digoxin	51. 86
4,4'-	Dihydroxydiphenyl	17. 21
trans-9-(2-ethoxy-1,3-	Dioxan-5-yl)-adenine	44. 24
trans-6-chloro-9-(2-ethoxy-1,3-	Dioxan-5-yl)-purine	44. 23
octafluoro-dibenzo-1,4-	Dioxane	38. 35
is(μ-hydroxo)-bis(aqua-trichloro-tin(iv)) 1,4-	Dioxane	[b 38. 4
hexakis(benzylthiomethyl)benzene 1,4-	Dioxane clathrate	61. 23
hexakis(2-phenylethylthiomethyl)benzene 1,4-	Dioxane clathrate	61. 25
tetraphenylporphinato-bis(	Dioxane)-cadmium(ii)	49. 28
	Dioxane-seleninyl-dichloride	38. 3
5-ethyl-2,2-dimethyl-1,3-	Dioxane-4,6-dione	38. 13
2,2-dimethyl-5-phenyl-1,3-	Dioxane-4,6-dione	38. 39
ctan-7-one)-8-spiro-4-(2,2-dimethyl-5-oxo-1,3-	Dioxolane) [(1,5,5-trimethyl-bicyclo(4.2.0)o 38. 72
1,5-	Diphenyl-2,4-pentadien-1-one	19. 9
5,7-diformyl-tetracyclo(9.8.1.1(3,9).1(13,18))	Docosa-1(19),2,4,7,9,11,13,15,17-nonene	31. 60
(bis(triphenylphosphine)immonium) octa-osmium-	Docosacarbonyl	[bis 64.165
1,4,7,15-tetraoxa-12,18-diazacyclo	Docosane-8,11,19,22-tetrone	40. 42
xafluoro-pentacyclo(6.4.0(3,6).0(4,12).0(5,9))	Dodec-10-ene	[4,5,9,10,11,12-he 31. 15
-hexacyclo(4.4.2.0(2,5).0(3,9).0(4,8).0(9,10))	Dodec-11-ene <s(trifluoromethyl)-1,6-diphospha 64. 76
tetraspiro(2.0.2.0.2.0.2.0)	Dodecane	30. 1
bis(1,4,7,10-tetra-azacyclo	Dodecane)-phosphorane	64. 78
cyclo	Dodecane-oxime	23. 3
y-11-thia-2,6,12-triazatricyclo(6.3.1.0(4,12))	Dodecane-1(11),8-diene-3,6-dione	[4-hydrox 41. 21
xycarbonyl-2,3,10-triazatricyclo(5.3.2.0(2,6))	Dodecane-3,8,11-triene-5-one <diphenyl-10-etho 37. 21
12-tetramethyl-tetracyclo(6.2.2.0(2,7).0(4,9))	Dodecane-5,11-dione	[6,12-dihydroxy-2,6,9, 31. 41
7-benzyl-3-oxa-6,9-diazatricyclo(7.3.0.0(2,6))	Dodecane-5,8-dione	[2-hydroxy-4-isopropyl- 40. 48
)-2-hydroxy-3,6,9-triazatricyclo(7.3.0.0(2,6))	Dodecane-5,8-dione <-(p-bromobenzyloxycarbonyl 36. 44
)-2-hydroxy-3,6,9-triazatricyclo(7.3.0.0(2,6))	Dodecane-5,8-dione <enzyl-3-(benzyloxycarbonyl 36. 46
methyl-6,9-diaza-12-thiatricyclo(7.3.0.0(2,6))	Dodecane-7,10-dione	[8-benzyl-1-hydroxy-11 48. 49
methoxy-6,9-diaza-12-oxatricyclo(7.3.0.0(2,6))	Dodecane-7,10-dione <-isopropyl-11α-methyl-1α- 58. 22
hydrate] 3-	Dodecanoyl-propandiol-1-phosphorylcholine mono 3. 55
	Dopamine hydrochloride	59. 1
	Doronenine	58. 29
	Doxantrazole rubidium salt	39. 49
	Dunnione p-bromophenylhydrazone	59. 28
etra-aza(6.1.6.1)paracyclophane hydrochloride	Durene tetrahydrate	[1,6,20,25-t 61. 18
α-	Dypnopinacoline	21. 20
	Dypnopinacone	21. 21
(3-methylnaphtho))-1,6,9,12,15,18-hexaoxacyclo	Eicosa-2,4-diene	[2,3:4,5-bis(1,2- 38.136
(3-methylnaphtho))-1,6,9,12,15,18-hexaoxacyclo	Eicosa-2,4-diene	[2,3:4,5-bis(1,2- 38.135
<3-methylnaphtho))-1,6,9,12,15,18-hexaoxacyclo	Eicosa-2,4-diene t-butylammonium perchlorate > 61. 13
20-bis(8-quinolyl-oxy)-3,6,9,12,15,18-hexaoxa-	Eicosane rubidium iodide	[1, 35. 83
3,3-dimethyl-1,5,13-trioxa-10,16-diazacyclo	Eicosane-6,9,17,20-tetrone	40. 44
e hydrate methanol solvate]	Ellipticine - 5-iodocytidylyl-(3'-5')-guanosin 60. 36
	Emblide	54. 23
	Endiandric acid	59. 29
tetra	Enniatin	48. 72
L,D,L,D,L,D-	Enniatin B sodium nickel nitrate hydrate	50. 22
acetate solvate] 8α,10α-	Epidioxy-8,14-dihydro-14β-nitrothebaine ethyl 58. 36
(+-)-	Epinine-β-sulfonate hemihydrate	17. 11
diene-cis-6,12-olide] 9β-acetoxy-8α-	Epoxy-angelyloxy-trans,trans-germacra-1(10),4- 53. 29
ra-1(10),4-diene-cis-6,12-oli> 9β-acetoxy-8α-	Epoxy-angelyloxy-7α-hydroxy-trans,trans-germac 53. 31
hexamethyl-2,3:5,6-di	Epoxy-bicyclo(2.2.0)hexane	38. 43
(1S,2E,4R,6R,7E,11S,12S)-6-acetoxy-11,12-	Epoxy-cembra-2,7-dien-4-ol	54. 22
	Epoxy-isoacoragermacrone	53. 9
9,13-	Epoxy-labdane	54. 18
(6S,7S,9R,10R)-6,9-	Epoxy-nonadec-18-ene-7,10-diol	38.101
bis(μ-hydroxo)-bis(aqua-tribromo-tin(iv)) 1,8-	Epoxy-p-menthane	38. 26
is(μ-hydroxo)-bis(aqua-trichloro-tin(iv)) 1,8-	Epoxy-p-menthane	[b 38. 27
16-methoxy-16'-oxo-21,20'-di(20,18-	Epoxy-pregnane)	51. 89
	Epoxy-rhodophytin	53. 1a
3α,4α-	Epoxy-D:A-friedo-18β,19αH-lupane	56. 5
11α,16-	Epoxy-1β,15R-dihydroxy-isopimar-8(14)-ene	54. 9a
anti-2,3;4,5-di	Epoxy-12-oxa(4.4.3)propella-7,9-diene	38. 38
anti-2,3;4,5-syn-7,8;9,10-tetra	Epoxy-12-oxa(4.4.3)propellane	38. 40
bromide] 7,8-didehydro-4,5-	Epoxy-17-(prop-2-enyl)morphinan-3,6-diol hydro 58. 37
7,8-didehydro-4,5α-	Epoxy-17-methyl-morphinan-3,6α-diol diacetate 58. 52
-1β,3,3-trimethylcyclohexan-1α-o> (+-)-2β,5β-	Epoxy-2α-(3α-p-nitrobenzoyloxy-but-1(E)-enyl) 59. 23

benzodioxepin-5(4H)-one] 1,4-	Epoxy-4-(4-nitrophenylmethyl)-1-phenyl-1H-2,3-	38.110
3,6-	Epoxy-5-hydroxy-5,6-dihydro-β-ionol............	59. 23
	Eregoyazin	53. 2
(6R)-6,19-epidioxy-9,10-seco-	Ergosta-5(10),7,22-trien-3β-ol benzoate.......	51. 80
(-)-dihydro	Ergotamine methylsulfonate monohydrate	58. 77a
	Eriocephalin	54. 26a
	Eriostemyl-3,5-dinitrobenzoate	38.124
	Eriostyl p-nitrobenzoate	38.126
	Eriostyl-3,5-dinitrobenzoate	38.125
tetrahydro	Eriostyl-3,5-dinitrobenzoate	38.127
	Ervistine methanol solvate	58. 35
1-methyl-6,16-dihydro-20-desethylidene-	Ervitsine	58. 28
6α-formyl-15,16-dimethoxy-B-nor-cis-	Erythrinan-8-one	58. 27
penta	Erythritol	45. 5
penta	Erythritol	45. 4a
	Erythronolide A anhydride	38.109
4,6β-ethano-3-methoxy-8α-	Estra-1,3,5(10)-trien-17β-ol hemihydrate	51. 26
3-methoxy-	Estra-1,3,5(10)-trien-17-one	51. 6
7-hydroxy-3-methoxy-6-oxa-	Estra-1,3,5(10)-trien-17-one	51. 3
17β-hydroxymethyl-	Estra-1,3,5(10)-trien-3-ol monohydrate	51. 14
3-methoxy-17β-hydroxymethyl-	Estra-1,3,5(10)-triene	51. 21
3-methoxy-16α-bromo-17α-hydroxy-	Estra-1,3,5(10)-triene	51. 10
3-methoxy-16α-bromo-17β-hydroxy-	Estra-1,3,5(10)-triene	51. 9
ate] rac-D-homo-8β-	Estra-1,3,5(10)-triene-3,17a-diol ethanol solv	51. 13
rac-D-homo-8α-	Estra-1,3,5(10)-triene-3,17a-diol hemihydrate	51. 15
-acetal)] 3-methoxy-6-oxa-	Estra-1,3,5(10)-triene-7,17-dione-17-(ethylene	51. 19
3-methoxy-	Estra-1,3,5(10),14-tetraen-17β-ol	51. 7
3-methoxy-	Estra-1,3,5(10),14-tetraen-17-one	51. 4
rate]	Estra-1,3,5(10),14-tetraene-3,17β-diol hemihyd	51. 2
17β-hydroxy-7β-methyl-	Estra-4,14-dien-3-one	51. 11
17β-hydroxy-9α-methyl-	Estra-4,14-dien-3-one	51. 12
17β-hydroxy-	Estra-4,9,11-trien-3-one	51. 1
hexachloro	Ethane	5. 3n
hexachloro	Ethane	5. 2n
hexamethyl	Ethane	5. 10
hexamethyl	Ethane	5. 11
dianthronylidene-	Ethane	28. 19
1,1,2,2-tetracyclohexyl-	Ethane	21. 19
1,2-bis(2-oxa-indan-1-yl-oxy)	Ethane	38. 90
1-methylsulfonyl-2-phenylsulfonyl-	Ethane	11. 9
lithium dibenzoylphosphide 1,2-dimethoxy	Ethane	64. 62
ro-1,2-(RS)-bis(2,2,3,3-tetrafluorocyclobutyl)	Ethane .... [tetrafluo	20. 3
enzylphosphonium) tetrachloro-cadmium dichloro	Ethane clathrate ....[bis(triphenylb	61. 8
tris(1,2-bis(diphenylphosphino-selenoyl)	Ethane) p-xylene clathrate	61. 9
2-carboxy	Ethane-(N-methyl-N-phenyl-sulfonamide)	1. 24
1,2-bis(N-piperidyl)	Ethane-bis-N-oxide diperchlorate	33. 49
1,2-bis(N-piperidyl)	Ethane-bis-N-oxide perchlorate	33. 48
1-(3,4-dihydroxyphenyl)-2-methylamino	Ethane-sulfonic acid hemihydrate	17. 10
bis(μ-methylthio)-(μ-tetrafluoro	Ethane-1,1-diyl)-bis(tricarbonyl-iron)	71. 17
	Ethane-1,2-bis(methylsulfone)	11. 6
	Ethane-1,2-bis(phenylsulfone)	11. 15
bis(μ-methylthio)-(μ-tetrafluoro	Ethane-1,2-diyl)-bis(tricarbonyl-iron)	71. 18
<methyl-4,8-bis(propionic acid)-porphyrin-bis(	Ethane-1',2'-diyl ester)-(pyridine-3'',5''-di>	49. 21
10,10'-	Ethanediylidene-dianthrone	28. 19
hydrochlo> 7-chloro-2,6-dihydro-4-methyl-5,6-	Ethano-1H-(1,4)-diazepino(1,7-a)benzimidazole	36. 12
ol hemihydrate] 4,6β-	Ethano-3-methoxy-8α-estra-1,3,5(10)-trien-17β-	51. 26
methyl)-7,7-dimethyl-bicyclo(3.3.0)octa> 4,8-	Ethano-4-methoxycarbonyl-5-(dimethoxycarbonyl-	31. 51
	Ethano-5-methyl-cyclopentacyclononene	28. 11
4-(diethylamino)-1,10-	Ethanolamine) .... [3,5-di-iodo-4-(4-hyd	13. 7
3,3',5,5'-tetra-iodothyroformic acid bis(N-di	Ethanolamine	13. 7
roxy-3,5-di-iodophenoxy)-benzoic acid bis(N-di	Ethanolamine) .... [3,5-di-iodo-4-(4-hyd	13. 7
trans-N,N-dimethyl-2-nitro-	Ethenamine	3. 17
',6''-kl:2''',1''',6'''-na)(1,3,> 1,14:7,8-di	Etheno-tetrapyrido(2,1,6-de:2',1',6'-gh:2'',1'	36. 39
a,5,8-hexahydro-1H-pyri> 6-dimethylamino-1,4-	Etheno-5,5-dimethyl-8-oxo-N,9-diphenyl-2,3,4,4	37. 23
-(o-phenylenedioxy)-(1,2-bis(trifluoromethyl)-	Ethenylenedioxy)-phosphoride <monium diphenoxy	64.114
hydropentacen> 5,14:7,12-bis(o-benzeno)-6,13-	Ethenylidene-5,5a,6,6a,7,12,12a,13,13a,14-deca	31. 67
0,0'-dibromodibenzyl	Ether	19. 8
bisphenol A-diglycidyl	Ether	38.105
bis(3,4-dichlorophenyl)-	Ether	17. 19
bis(4-dibenzoborepinyl)-	Ether	62. 19
3,3',4,4'-tetrachlorodiphenyl-	Ether	17. 19
p-bromophenoxyacetimine methyl	Ether	17. 9
phenylcholine	Ether bromide	17. 17
bis((8-quinolyl-oxy)-ethoxy-ethyl)	Ether thiourea	17. 40
<o-(oxo-diphenylphosphino)-phenyloxy)-diethyl	Ether D,L-phenylglycinium ethyl ester trifluo>	64.170
	Ethidium cytidylyl-(3'-5')-guanosine	47. 37
	Ethylene	5. 6

                    tetraisopropyl Ethylene 5. 13
                 skatole tetracyano Ethylene 60. 9
            3,2'-pyridylpyrazinyl- Ethylene 33. 40
      E-1-(5-nitro-2-furyl)-2-morpholino- Ethylene 38. 21
      Z-1-(5-nitro-2-furyl)-2-thiocyanato- Ethylene 38. 6
   bis(9,10-diazaphenanthrene) tetracyano Ethylene 60. 15
      E-1-(5-nitro-2-furyl)-2-p-tolylsulfonyl- Ethylene 38. 50
     cis-(E)-1-(5-nitro-2-furyl)-2-dimethylamino- Ethylene 38. 12
     cis-(E)-1-(5-nitro-2-furyl)-2-dimethylamino- Ethylene 38. 11
     -1-(5-nitro-2-furyl)-2-p-chlorophenylsulfonyl- Ethylene .... [E 38. 36
  triphenylphosphine-imino)-1-chloro-2,2-dicyano Ethylene .... [1-(64.115
  1-trichloromethylsulfonyl-2-(5-bromo-2-furyl)- Ethylene .... [1-(5-nitro-2-furyl)- 38. 28
  hoxycarbonyl-methyl)-7,7-dimethyl> 8-(2-bromo Ethylene)-3-hydroxy-4-methoxycarbonyl-5-(dimet 27. 25
  y-6-oxa-estra-1,3,5(10)-triene-7,17-dione-17-(Ethylene-acetal) [3-methox 51. 19
                         N,N'- Ethylene-bis(acetylacetone-imine) 1. 31
                         N,N'- Ethylene-dibenzamide 13. 15
      calcium bis(picrate) tetra Ethylene-glycol monohydrate............... 17. 1
  ,7-diyl)-(2,6-bis(2-picolinoyl)-pyridine)-bis(Ethylene-ketal) [2',6''-(1,4,7-trioxaheptane-1 40. 55
              (N,N-diethyl Ethylene-tellurourea)-pentacarbonyl-chromium 70. 8
                4,4'-(Ethylenedi-imine)-bis(methylenecamphor)........ 31. 62
       cyclopentadienyl sodium tetramethyl Ethylenediamine....................... 20. 1
  -methyl-1,3-dithio-2-yl-(N,N,N',N'-tetramethyl Ethylenediamine)-lithium dimer .... [2 67. 7
       dimethyl-(N,N,N',N'-tetramethyl Ethylenediamine)-magnesium 67. 2
  methyl-2,4-pentadienyl)-(N,N,N',N'-tetramethyl Ethylenediamine)-magnesium ....[bis(2,4-di 67. 5
                     calcium Ethylenediamine-tetra-acetate heptahydrate 48. 27
                      barium Ethylenediamine-tetra-acetate hydrate 48. 26
  hydrate]   Ethylenediammonium (+)-allohydroxycitrate mono   2. 26
                           Ethylenediammonium (-)-hydroxycitrate 2. 25
  nohydrate]   Ethylenediammonium (2S,3R)-2-hydroxycitrate mo   2. 26
                           Ethylenediammonium (2S,3S)-2-hydroxycitrate.... 2. 25
           tetrathiafulvalene bis(Ethylenedithiolato)-nickel(i) 60. 5
  ,3,3-pentamethyl-guanidinium-2-(2,2-dicyano-1- Ethylenethiolate) .... [1,1,2 8. 12
  orate]   hexakis(Ethylenethiourea)-di-tellurium(ii) tetraperchl 70. 15
  tetrachloride]   bis(di Ethylenetriammonium) tetrachloro-platinum(ii) 3. 30
                2,2'-(2,2,2-trichloro Ethylidene)-bis(4-chloro-6-nitrophenol)........ 50. 6
  4-nitrobenzyl-(3SR,5RS,Z)-2-(2-phenylthio Ethylidene)-penam-3-carboxylate............... 41. 78
  1,2,3-triphenyl-5-(1,2-diphenyl-2-phenylimino Ethylidene)-2-pyrrolin-4-one 32. 77
        6-chloro-2(2-acetoxy-1- Ethylidene)-3-(2-thienyl)-4H-1,4-benzothiazine 41. 57
  <enyl)-7'-(1-(3,5-di-t-butyl-4-hydroxyphenyl)- Ethylidene)-3a',4',7',7a'-tetrahydro-spiro(cy> 17. 44
     3,4,6-tri-0-acetyl-1,2-0-(R)- Ethylidene-α-D-allopyranose................... 45. 31
             methyl-3,4-0- Ethylidene-β-L-arabinopyranose 45. 15
                       2,2'- Ethylidene-bis(4-chloro-6-nitrophenol) 17. 22
  tradecane]   (1RS,8SR,10SR,4(15)Z)-4- Ethylidene-5-oxa-3-azatricyclo(8.4.0.0(3,8))te 40. 35
              (E)-2-chloro-2- Ethynyl-5-phenyl-adamantane................. 31. 44
              (Z)-2-chloro-2- Ethynyl-5-phenyl-adamantane................. 31. 45
                 Z-2- Ethynyl-5-phenyl-2-adamantanol 31. 46
       6β-hydroxy-3-oxo-5β- Eudesma-1,7(11)-dien-12-oic acid γ-lactone 53. 5
                           Eupatalbin 54. 7
              methyl-O-acetyl Eurekanate 38. 44
                 sila-di Fenidol... 63. 31
                           Fentanyl citrate toluene solvate 33. 84
             decamethyl Ferrocenium - 7,7,8,8-tetracyanoquinodimethane 60. 42
  ne)]   1,1'-dimethyl Ferrocenium bis(7,7,8,8-tetracyanoquinodimetha 60. 20
                           Filicane-3-one 56. 6
  3-hydroxy-7-methoxy-3',4'-methylene-dioxy- Flavan 38. 83
        19-(p-bromobenzoyloxy)-bi Flora-4,15-diene 54. 30
             pyrene - Fluoranil complex 60. 34
       tetrathiafulvalene- Fluoranil complex 60. 4
       7,8-dimethoxybenzo(j) Fluoranthene 30. 6
       5,10-dimethoxybenzo(j) Fluoranthene 30. 7
  6-bromo-1,2,3,4,4a,9a-hexahydro-4,9-dioxa Fluoren-2-one 38. 31
             2-acetylamino Fluorene 28. 8
          perchloro-9-phenyl-3- Fluorenone 28. 9
       9H-heptadecachloro-bis-9- Fluorenyl 28. 17
       9H,9'H-hexadecachloro-bis-9- Fluorenyl 28. 18
                 9- Fluorenyl-diphenyl-silanol 63. 42
           dichloro-9- Fluorenyl-trimethylsilyl-silane................. 63. 20
                 9- Fluorenyl-tris(trimethylsilyl)-silane 63. 37
                           Flutroline 36. 49
  (+)-5,10-methenyl-5,6,7,8-tetrahydro Folic acid bromide hydrobromide dihydrate 48. 63
  (-)-5,10-methenyl-5,6,7,8-tetrahydro Folic acid bromide hydrobromide dihydrate 36. 33
                           Folic acid dihydrate 48. 57
              azo-di Formaldehyde-dioxime 10. 1
        α-cyclodextrin N,N-dimethyl Formamide clathrate pentahydrate 61. 19
  ormamide solvate] carbonyl-(N,N-dimethyl Formamide)-phthalocyaninato-iron(ii) dimethylf 49. 5
            bis(N,N'-diethyl) Formamidine-disulfide hexachloro-tellurium 11. 11

) chloride monohydrate]   dithio-bis(Formamidinium) trans-tetrachloro-oxo-rhenium(v   8.   9
                                                calcium Formate...   2. 2n
                                              β-calcium Formate...   2.   3
                                              β-calcium Formate...   2.   4
                                              potassium Formate...   2.   1
            anti,S-trans-3-methylthio-1,5-di-o-tolyl- Formazan 11. 16
  -1(19),2,4,7,9,11,13,15,17-nonene]   5,7-di Formic acid bis(N-diethanolamine) 13.   7
                                           5,7-di Formyl-tetracyclo(9.8.1.1(3,9).1(13,18))docosa 31. 60
  midine-4-one]   6-methyl-9- Formyl-1,6,7,8-tetrahydro-1H-pyrido(1,2-a)pyri 35. 16
  one]   6α- Formyl-15,16-dimethoxy-B-nor-cis-erythrinan-8- 58. 27
  5.4.0.0(2,4))undecan-10-one-9,2(1)-spiro(3(1)- Formyl-3(1)-methyl-oxirane) <yl-3-oxatricyclo(59. 16a
  -2,2-dicarboxylate]   diethyl cis-1- Formyl-5-hydroxy-4-(2-nitrophenyl)-pyrrolidine 32. 63
                              2-selena-1-dioxothia-3- Formyl-5,8-dimethyl-1,2-dihydronaphthalene 42.   6
                                                  3- Formyl-8-methylfuro(3,2-g)chromene 38. 49
                                                  3- Formylfuro(3,2-f)chromene 38. 37
                                                  1- Formylhexahelicene 30. 15
                             4-m-aminophenyl-2- Formylpyridine-thiosemicarbazone 33. 53
                                                     Fortamine 50.   4
                                  3α,4α-epoxy-D:A- Friedo-18β,19αH-lupane 56.   5
                                                     Fruticosonine 58. 48
                                        2,6-dimethyl- Fuchsone 21. 16
                                        2,6-dimethyl- Fuchsone 21. 14
                                        2,6-dimethyl- Fuchsone 21. 15
                                               chloro Fucin 59.   7
                                                     Fucoxanthin 57.   1
                              diphenyl-tetrathia Fulvalene 39. 67
            bis(acetylacetonato)-palladium(ii) tetrathia Fulvalene   60. 12
                                          tetrathia Fulvalene bis(ethylenedithiolato)-nickel(i).... 60.   5
                          tetramethyl-tetrathia Fulvalene tetrabromo-p-benzoquinone complex.... 60. 10
                               dibenzotetrathia Fulvalene tri-iodide 39. 50
  oquinod>   tris(4,4',5,5'-tetramethyl-tetrathia Fulvalene) - bis(2,5-diethyl-7,7,8,8-tetracyan 60. 35
                        tris(dibenzotetrathia Fulvalene) hexabromo-tin 39. 51
                       bis(tetramethyl-tetrathia Fulvalene) iodide 60. 11
                       bis(tetramethyl-tetrathia Fulvalene) thiocyanate 39. 27
                  octakis(dibenzotetrathia Fulvalene) tris(hexachloro-tin)................. 39. 52
                                          tetrathia Fulvalene-chloranil complex............ 60.   3
                                          tetrathia Fulvalene-fluoranil complex............ 60.   4
                                          tetrathia Fulvalenium perchlorate................ 39.   8
                                          tetrathia Fulvalenium tri-iodide 39.   9
                                     (3)Tetrathia Fulvaleno(3)paracyclophane 39. 69
                                     (4)Tetrathia Fulvaleno(4)paracyclophane 39. 78
                                     (2.2)Tetrathia Fulvalenophane 39. 58
  0,11-dihydro-5H-dibenzo(b,f)silepane hydrogen Fumarate <-dimethyl-10-(4-methylpiperazinyl)-1 63. 33
                                                     Fumitremorgin B 58. 71
                                                     Furaltadone................................ 40. 29
  e]   Furaltadone hydrochloride acetic acid clathrat 61.   4
                                                     Furaltadone hydrochloride hydrate 40. 30
  rate]   Furaltadone hydrochloride propionic acid clath 61.   5
    cis-(E)-2-(2'-dimethylamino-vinyl)-5-nitro Furan 38. 11
    cis-(E)-2-(2'-dimethylamino-vinyl)-5-nitro Furan 38. 12
  nitro-2-(1-adamantyl)-2'-carbomethoxy-(2H,5H)- Furan .... [5,5'-di 38. 77
  l-5-carboxy-2,4,5-trimethyl-2,3,4,5-tetrahydro Furan .... [(2S,4R,5R)-2-carboxymethy 38. 25
            dibenzo(b,b')furo(3,2-e:4,5-e')-bis(benzo Furan) 38.116
         2,11-dimethoxy-benzo(1,2-b:4,3-b')-bis(benzo Furan) 38.102
         ,12-disalicyloyl-benzo(1,2-b:4,5-b')-bis(benzo Furan) .... [6 38.134
  aethoxy-Δ(3,3'(2H,2'H))-bis(phenanthro(9,10-b) Furan) .... [E-2,2,2',2'-tetr 38.141
  ide-1,2'-spiro(3'-methyl-3'-hydroxy-tetrahydro Furan) <-tetra-O-acetyl-1-deoxy-D-glucopyranos 45. 45
  ide-1,2'-spiro(3'-methyl-3'-hydroxy-tetrahydro Furan) <-tetra-O-acetyl-1-deoxy-D-glucopyranos 45. 46
  -(p-bromophenylsulfonyloxy)-2',3'-dihydrobenzo Furan) dimethylsulfoxide solvate <xymethyl)-4' 50. 21
  <loro-2',4,6-trimethoxy-6'-β-methyl-spirobenzo Furan-2(3H)-cyclohex-2-ene-3,4'-dione benzene> 50. 11
  nyl)-6a-phenyl-3a,4,5,6a-tetrahydrofuro(2,3-b) Furan-2(3H)-one <,4-dimethyl-3-(o-acetamidophe 38.112
  ydro-4,5-dihydroxy-4a,8-dimethylazuleno(6,5-b) Furan-2(3H)-one <aα,4α,4aβ,5α,7aα,8α,9aβ-decah 38. 69
            2,4,5-tris(3-methoxycarbonyl- Furan-2-yl)-Δ(2)-imidazoline 38.104
  -methyl-3-(2'-tetrahydrofurylidene)-tetrahydro Furan-2,4-dione .... [(E,R)-5 38. 15
  ecahydro-4-hydroxy-4a,8-dimethylazuleno(6,5-b) Furan-2,5(3H)-dione <,4α,4aβ,7aα,8α,9aβ-(+-)-d 38. 64
        3,5-dihydroxy-4-nitro-2-(4'-acetyl-5'-methyl Furan-2'-yl)-tetrahydropyran 38. 42
  -(2,2-dimethylpropionyl)-1'-methyl-spiro(benzo Furan-3(2H)-2'-pyrrolidine)-2,5'-dione <tyl-3' 38.132
  <thoxy-2-(dimethoxymethyl)-6-methylfuro(3,2-b) Furan-3-ol-4-bromophenyl-carbamic acid ester 38. 95a
  <thoxy-2-(dimethoxymethyl)-6-methylfuro(3,2-b) Furan-3-ol-4-bromophenyl-carbamic acid ester 38. 94
                                            tetrahydro Furan-3,4-dione....................... 38.   2
            6-methoxy-2-methyl-3,5-dihydro-benzo(b) Furan-4,7-dione....................... 59.   2
              1-phenyl-4,5-(D-gluco Furano)-imidazolidin-2-one 45. 27
                                              N-acetyl- Furanomycin 50.   5
        2-0-acetyl-1,6:3,5-dianhydro-α-L-ido Furanose 45. 12
        1,2-0-isopropylidene-α-D-gluco Furanose-3,5,6-bicyclophosphite................ 64. 32
        1,2-0-cyclohexylidene-α-D-gluco Furanose-3,5,6-bicyclophosphite................ 64. 52

9-(5-phenylseleno-5-deoxyribo Furanosyl)-adenine 47. 33
ocarboxamido)-1-(2',3',5'-tri-0-nitro-β-D-ribo Furanosyl)-imidazole-4-carboxamide <-5-(N-nitr 45. 19
3-(β-D-ribo Furanosyl)-1,2,4-triazole-5(1H,4H)-thione 45. 11
,5'-anhydro-1-(2',3'-0-isopropylidene-β-D-ribo Furanosyl)-2-thio-uracil .... [2 47. 28
ne-1,1-dioxide]   3,2'-anhydro-4-(β-D-arabino Furanosyl)-5,6-dihydro-2H-1,2,4-thiadiazin-3-o 45. 14
9α-D-arabino Furanosyl-adenine 47. 16
1β-D-ribo Furanosyl-pyridin-4-one-3-carboxamide 47. 25
2,2'-anhydro-1-β-D-arabino Furanosyl-2-thio-uracil................... 47. 2
9β-D-arabino Furanosyl-8-morpholino-adenine dihydrate 47. 30
9β-D-arabino Furanosyl-8-n-butyl-amino-adenine 47. 31
1-β-D-lyxo Furanosyluracil 47. 7
e)]   7-(3-(4,5-dihydro-5,5-dimethyl-4-oxo-2- Furanyl)-but-2-enyl)-oxy-(2H-1-benzopyran-2-on 59. 19
trimethylene- Furazan-N-oxide........................... 40. 1
benz Furazan-1-oxide............................ 40. 2
trans-2-acetoxy-5-nitro-2,5-dihydro- Furfural diacetate 38. 33
cetamidophenyl)-6a-phenyl-3a,4,5,6a-tetrahydro Furo(2,3-b)furan-2(3H)-one <,4-dimethyl-3-(o-a 38.112
5,8-dimethoxy-2-methyl Furo(2,3-e)chromone........................ 59. 5
Furo(2,3-f)coumarin....................... 38. 30
Furo(2,3-h)coumarin....................... 38. 29
<ahydro-5-methoxy-2-(dimethoxymethyl)-6-methyl Furo(3,2-b)furan-3-ol-4-bromophenyl-carbamic > 38. 95a
<ahydro-5-methoxy-2-(dimethoxymethyl)-6-methyl Furo(3,2-b)furan-3-ol-4-bromophenyl-carbamic > 38. 94
N,N-diethyl-3,3a-dihydro-3-methyl-benzo Furo(3,2-c)isoxazole-3-carboxamide 40. 38
dibenzo(b,b') Furo(3,2-e:4,5-e')-bis(benzofuran) 38.116
3-formyl Furo(3,2-f)chromene...................... 38. 37
3-formyl-8-methyl Furo(3,2-g)chromene...................... 38. 49
furyl)-1-trichloromethylsulfonyl-2-(5-bromo-2- Furyl)-ethylene   [1-(5-nitro-2- 38. 28
syn-3-methyl-6-(2- Furyl)-piperidin-2-one 38. 22
furyl)-ethylene]   1-(5-nitro-2- Furyl)-1-trichloromethylsulfonyl-2-(5-bromo-2- 38. 28
cis-(E)-1-(5-nitro-2- Furyl)-2-dimethylamino-ethylene 38. 12
cis-(E)-1-(5-nitro-2- Furyl)-2-dimethylamino-ethylene........... 38. 11
E-1-(5-nitro-2- Furyl)-2-morpholino-ethylene............. 38. 21
E-1-(5-nitro-2- Furyl)-2-p-chlorophenylsulfonyl-ethylene 38. 36
E-1-(5-nitro-2- Furyl)-2-p-tolylsulfonyl-ethylene 38. 50
Z-1-(5-nitro-2- Furyl)-2-thiocyanato-ethylene 38. 6
(E,R)-5-methyl-3-(2'-tetrahydro Furylidene)-tetrahydrofuran-2,4-dione 38. 15
2-phenyl-5,5-dimethyl-(3,4)benzo-5H- Furylium perchlorate 38. 76
'S)-5,9-dimethyl-2'-hydroxy-2-(2-tetrahydro-2- Furylmethyl)-6,7-benzomorphan <-)-(1R,5R,9R,2' 58. 41
1,2:5,6-dianhydro- Galactitol 45. 6
1,2:5,6-dianhydro- Galactitol 45. 7
1,5-dianhydro-D,L- Galactitol 45. 10
1,6-anhydro-β-D- Galactopyranose 45. 8
0-acetyl-1,2-0-(R)-(1-t-butoxyethylidene)-α-D- Galactopyranose..................[3,4,6-tri- 45. 47
ra-0-acetyl-2-(N-acetylacetamido)-2-deoxy-β-D- Galactopyranose   [1,3,4,6-tet 45. 44
,4,6-tetra-0-acetyl-β-D-galactopyranosyl)-α-D- Galactopyranose <2,4,6-tetra-0-acetyl-3-0-(2,3 45. 56
hyl 4,6-0-(R)-benzylidene-2,3-di-0-methyl-β-D- Galactopyranoside .... [met 45. 41
etra-0-acetyl-3-0-(2,3,4,6-tetra-0-acetyl-β-D- Galactopyranosyl)-α-D-galactopyranose <2,4,6-t 45. 56
methyl-2,3,4,5-tetra-0-acetyl-α-D- Galactoseptanoside 45. 34
nor Galanthamine hydrochloride dihydrate 58. 18
dicyclohexylammonium Gascardate 55. 2
ol diethyl ether solvate]   Gastrin tetrapeptideamide hydrochloride methan 48. 76
Geiparvarin............................. 59. 19
Gentiobiose 45. 25
Gentiobiose 45. 24
9β-acetoxy-8α-epoxy-angelyloxy-trans,trans- Germacra-1(10),4-diene-cis-6,12-olide 53. 29
xy-8α-epoxy-angelyloxy-7α-hydroxy-trans,trans- Germacra-1(10),4-diene-cis-6,12-olide <β-aceto 53. 31
roxymethyl-3-mercapto-butyryloxy)-trans,trans- Germacra-1(10),4-dienolide <8-(2-hydroxy-2-hyd 53. 25a
D- Gibberellin C secodiester photoproduct 59. 32
Gilmaniellin methanol solvate 59. 37
Gitoxin................................. 51. 87
diethyl Glaucophanic enol 38.115
3,3'-diacetyl-5,5'-diethoxycarbonyl- Glaucyrone 38.115
Gleenol 53. 12
1-phenyl-4,5-(D- Glucofurano)-imidazolidin-2-one........... 45. 27
1,2-0-isopropylidene-α-D- Glucofuranose-3,5,6-bicyclophosphite 64. 32
1,2-0-cyclohexylidene-α-D- Glucofuranose-3,5,6-bicyclophosphite 64. 52
β-D-glucopyranosyl-(1-3)-D- Glucopyranose 45. 26
β-D-glucopyranosyl-(1-6)-β-D- Glucopyranose 45. 25
β-D-glucopyranosyl-(1-6)-β-D- Glucopyranose 45. 24
-mannopyranosyl-(1-4)-2-acetamido-2-deoxy-α-D- Glucopyranose <-α-D-mannopyranosyl-(1-3)-0-β-D 45. 49
2-R-methylcyclohexyl-α-D- Glucopyranoside 45. 28
methyl-3-0-α-D-glucopyranosyl-α-D- Glucopyranoside......................... 45. 29
methyl 4,6-0-benzylidene-2,3-di-0-methyl-α-D- Glucopyranoside......................... 45. 40
pyridyl-1-thio-β-D- Glucopyranoside monohydrate 45. 20
y-te> trans-2,3,4,6-tetra-0-acetyl-1-deoxy-D- Glucopyranoside-1,2'-spiro(3'-methyl-3'-hydrox 45. 46
y-tetr> cis-2,3,4,6-tetra-0-acetyl-1-deoxy-D- Glucopyranoside-1,2'-spiro(3'-methyl-3'-hydrox 45. 45
methyl-3-0-α-D- Glucopyranosyl-α-D-glucopyranoside 45. 29

β-D-	Glucopyranosyl-(1-3)-D-glucopyranose	45. 26
β-D-	Glucopyranosyl-(1-6)-β-D-glucopyranose	45. 24
β-D-	Glucopyranosyl-(1-6)-β-D-glucopyranose	45. 25
. 21β-(2-O-acetyl-3,4-di-O-angelyl-6-deoxy-β-	Glucopyranosyloxy)-3β,16α,22α,24,28-penta-ace>	56. 10
mino-2-oxo-1(2H)-pyrimidinyl)-1,4-dideoxy-β-D-	Glucopyranuronic acid monohydrate <mino-1-(4-a	47. 20
4,6-tri-O-acetyl-3-O-1'(R)-C-carboxyethyl-β-D-	Glucose-2,2'-lactone ....[1,	45. 33
procaine-N-D-	Glucoside hydrochloride monohydrate............	45. 48
L-	Glutamic acid	48. 12n
L-	Glutamic acid	48. 13
loxycarbonyl-(γ-ethyl)-L-glutamyl-(γ-ethyl)-L-	Glutamic acid ethyl ester .... [N-benzy	48. 74
2,4-methano-	Glutamic acid monohydrate	48. 18
monohydrate]	(Glutamyl-α-lactam)-histidinyl-proline tartrate	48. 47
L-pyro	Glutamyl-β-(2-thienyl)-L-alanyl-L-prolinamide	48. 50
N-benzyloxycarbonyl-(γ-ethyl)-L-	Glutamyl-(γ-ethyl)-L-glutamic acid ethyl ester	48. 74
)-N-carbobenzyloxy-(γ,γ'-di-t-butyl)-γ-carboxy	Glutamyl-glycine ethyl ester .... [(+-	48. 75
ophanyl-glycine] methionyl-	Glutamyl-histidinyl-phenylalanyl-arginyl-trypt	48. 82
L-pyro	Glutamyl-N,N'-dicyclohexylurea	48. 55
avin] 3-methyl-4a-(methylene-dithio	Glycerol acid ethyl ester)-4a,5-dihydro-lumifl	36. 31
loxazine] 3-methyl-4a-(methylene-dithio	Glycerol acid methyl ester)-4a,5-dihydro-isoal	36. 18
lavin] 3-methyl-4a-(methylene-thio	Glycerol acid methyl ester)-4a,5-dihydro-lumif	36. 25
disodium D,L-α-	Glycerophosphate hexahydrate	46. 1
bisphenol A-di	Glycidyl ether	38.105
	Glycine.............................	48. 1n
	Glycine.............................	48. 2n
glycyl	Glycine.............................	48. 10
N-(phosphonomethyl)-	Glycine.............................	64. 4
-histidinyl-phenylalanyl-arginyl-tryptophanyl-	Glycine..........................[methionyl-glutamyl	48. 82
benzyloxy-(γ,γ'-di-t-butyl)-γ-carboxyglutamyl-	Glycine ethyl ester .... [(+-)-N-carbo	48. 75
N-(t-butoxycarbonyl)-L-prolyl-L-valyl-	Glycine hemihydrate............................	48. 52
t-butoxy-carbonyl-L-prolyl-L-isoleucyl-	Glycine hydrate	48. 56
L-leucyl-L-prolyl-	Glycine monohydrate............................	48. 42
tris(	Glycine) calcium(ii) dibromide	48. 3
tetrakis(	Glycine) potassium tri-iodide	48. 4
tris(	Glycine) tetrafluoroberyllate	48. 7
tris(	Glycine) tetrafluoroberyllate	48. 6
N-(t-butoxy-carbonyl)-L-methionyl-	Glycine-benzyl ester	48. 60
N-phthaloyl	Glycine-hydroxamic acid........................	35. 15
<hosphino)-phenyloxy)-diethyl ether D,L-phenyl	Glycinium ethyl ester trifluoromethylsulfonate	64.170
	Glycinium orthophosphate	48. 5
calcium bis(picrate) tetraethylene-	Glycol monohydrate	17. 1
cyclo-tris(prolyl-	Glycyl).......................................	48. 66
cyclo-tris(prolyl-	Glycyl) calcium perchlorate....................	48. 70
cyclo-tris(prolyl-	Glycyl) calcium perchlorate hydrate............	48. 67
L-phenylalanyl-glycyl-D-leucyl-D-phenylalanyl-	Glycyl) dihydrate .... [cyclo(L-leucyl-	48. 80
cyclo-tris(prolyl-	Glycyl) hydrate	48. 68
cyclo-tris(prolyl-	Glycyl) magnesium perchlorate trihydrate	48. 69
cyclo-tris(L-valyl-L-prolyl-glycyl-L-valyl-	Glycyl) monohydrate dimethylsulfoxide solvate	48. 83
cyclo-bis(glycyl-L-propyl-	Glycyl) tetrahydrate	48. 54
cyclo(L-valyl-L-prolyl-glycyl-L-valyl-	Glycyl) trihydrate	48. 61
cyclo(di(benzyl	Glycyl)-L-prolyl) monohydrate	48. 73
cyclo(benzyl	Glycyl-bis(L-prolyl))	48. 58
te] cyclo(L-leucyl-L-phenylalanyl-	Glycyl-D-leucyl-D-phenylalanyl-glycyl) dihydra	48. 80
	Glycyl-D-threonine dihydrate	48. 19a
cyclo-bis(	Glycyl-L-propyl-glycyl) tetrahydrate	48. 54
foxide solvate] cyclo-tris(L-valyl-L-prolyl-	Glycyl-L-valyl-glycyl) monohydrate dimethylsul	48. 83
cyclo(L-valyl-L-prolyl-	Glycyl-L-valyl-glycyl) trihydrate	48. 61
	Glycyl-L-4-hydroxyproline	48. 22
	Glycylglycine	48. 10
dimethyl	Glyoxime	10. 2
p-tolyl-	Glyoxylic acid p-chloroanilide	16. 12
	Glyphosate	64. 4
	Gmelofuran	53. 4
2,2'-bipyridyl potassium dicyano-	Gold(iii)	33. 34
drogen bis(N,N-dimethylacetamide) tetrachloro-	Gold(iii) ....[hy	1. 15n
drogen bis(N,N-dimethylacetamide) tetrachloro-	Gold(iii) ....[hy	1. 14
N-acetyl-	Gramicidin S	50. 40
N-phenethyl-	Granatanine-3-spiro-5'-hydantoin	37. 13
	Griseofulvin benzene solvate	50. 11
3,4,5-trichloro-	Guaiacol	17. 6
	Guanazol	32. 3
2-cyano	Guanidine	8. 8
diphenyl	Guanidine	8. 14
2-diethylphosphoryl-	Guanidine hemi(guanidinium chloride)	64. 13
5-methyl-1H-imidazol-4-yl)-methylthio)-ethyl)-	Guanidine monohydrate <-cyano-1-methyl-3-(2-((	32. 34
te] tetra	Guanidinium α-dodecamolybdo-silicate monohydra	8. 7
2-diethylphosphoryl-guanidine hemi(	Guanidinium chloride)	64. 13

	Guanidinium hydrogen oxalate monohydrate	2. 9
hydrate]	Guanidinium tetramolybdo-dimethylarsinate mono	65. 2
1,1,2,3,3-pentamethyl-	Guanidinium-2-(2,2-dicyano-1-ethylenethiolate)	8. 12
ethidium cytidylyl-(3'-5')-	Guanosine	47. 37
acridine orange 5-iodocytidylyl-(3'-5')-	Guanosine dodecahydrate..................	47. 35
proflavine deoxycytidylyl-(3'-5')-	Guanosine hydrate	47. 38
9-amino-acridine - 5-iodocytidylyl-(3'-5')-	Guanosine hydrate complex	60. 22
proflavine - 5-iodocytidylyl-(3'-5')-	Guanosine hydrate methanol solvate	47. 36
ellipticine - 5-iodocytidylyl-(3'-5')-	Guanosine hydrate methanol solvate	60. 36
yl-phenanthrolinium - 5-iodocytidylyl-(3'-5')-	Guanosine hydrate methanol solvate <hyl-N-meth	60. 37
proflavine - cytidylyl-(3-5')-	Guanosine sulfate hydrate	60. 24
	Guanosine-5'-monophosphate trihydrate	47. 22
2,3:4,5-di-O-isopropylidene-D-	Gulose diethyl dithioacetal..................	45. 42
O-tetramethyl-	Haematoxylin	59. 21
	Halopemide	35. 64
	Hedamycin	50. 27
1-acetylhexa	Helicene	30. 17
1-formylhexa	Helicene	30. 15
1-methylhexa	Helicene	30. 16
cyclopentyl-(4)-	Helixane	38. 85
e> 3,6,9,12,15-pentaoxa-21-azabicyclo(15.3.1)	Heneicosa-1(20),17,19-triene t-butylammonium p	40. 40
i> 3,6,9,12,15-pentaoxa-21-azabicyclo(15.3.1)	Heneicosa-1(21),17,19-triene-2,16-dione potass	40. 39
	Heneicosan-4-one	38. 85
1,13,16,19-tetraoxatetraspiro(4.0.0.0.4.3.3.3)	Heneicosane.......[exo,exo-octacyclo(8.8.1.1(	31. 55
3,6).1(12,15).0(2,9).0(4,8).0(11,18).0(13,17)	Hept-2-en-6-yl)-tetracyclo(9.3.0.0(1,5).0(7,1)	55. 3
<ulfonyl)-10-hydroxy-5,8-dimethyl-2-(2-methyl-	Hept-2-en-6-yl)-tricyclo(9.3.0.0(9,13))tetrade	55. 1
cane-2,8-diene-4> 2,13-dimethyl-10-(2-methyl-	Hept-2-ene	37. 3
7-isopropylidene-2,3-diazabicyclo(2.2.1)	Hept-2-ene .... [1,3,	37. 8
4,7,7-pentachloro-5-phenyl-2-azabicyclo(2.2.1)	Hept-2-ene .... [trans	31. 18
-5-chloro-6-(2-nitrophenylthio)-bicyclo(2.2.1)	Hept-2-ene benzene methanol solvate <p-bromoph	31. 68
enyl)-7-oxo-(2,3-l)phenanthrene-bicyclo(2.2.1)	Hept-3-en-5-ulopyranoside <hyl 4-chloro-3,4-di	45. 32
deoxy-1,2:6,7-di-O-isopropylidene-α-D-erythro-	Hept-3-ene <7-(trifluoromethyl)-3,4-bis(methox	64.135
ycarbonyl)-6-oxa-1-aza-5-phosphabicyclo(3.2.0)	Hept-3-ene-4-(p-nitro-benzyl-carboxylate)	41. 50
7-oxo-3-methylthio-2-thia-1-azabicyclo(3.2.0)	Hept-6-en-3-one <-di-t-butyl-1-(3,5-di-t-butyl	35. 76
-2-hydroxyphenyl)-2-methyl-2-azabicyclo(3.2.0)	Heptacene	30. 20
tetrabenzo(de,no,st,c(1)d(1))	Heptadeca-12,14-dione [13-phenyl-11,13,15-	37. 17
triazapentacyclo(8.5.2.0(2,6).0(2,8).0(11,15))	Heptadeca-5,14,16,18-tetraene-4,7,11,13-tetron	38. 89
e] 12-oxapentacyclo(8.4.3.2(2,9).0.0(3,8))	Heptadecachloro-bis-9-fluorenyl..............	28. 17
9H-	Heptadecane-3(8),15-diene .... [b	39. 79
enzo(p)-2-thiatetracyclo(7.5.3.0(3,8).0(1,10))	Heptane....................................	5. 8
1,1,3,5,7,7-hexachloro	Heptane....................................	38. 81
1,4-diphenyl-2,3-dioxabicyclo(2.2.1)	Heptane....................................[(1	40. 4
S,4S)-N-acetyl-3-oxo-5-aza-2-oxabicyclo(2.2.1)	Heptane.............. [2-benzyl-3-oxo-6-exo-(2,	37. 15
6-di-iodo-4-methylphenoxy)-2-azabicyclo(2.2.1)	Heptane.............. [6-(bromomethyl)-1,2,4-tric	31. 20
hloro-3-ethoxy-3,7,7-trimethoxy-bicyclo(2.2.1)	Heptane. <-7,7-trimethoxy-5-endo-bromomethyl-2-	37. 9
exo-(2',2',2'-trifluoro-ethoxy)-bicyclo(2.2.1)	Heptane-trans-3,trans-4-dicarbonitrile	27. 6
7,7-dichloro-1,6-dimethyl-bicyclo(4.1.0)	Heptane-1,7-diyl)-(2,6-bis(2-picolinoyl)-pyrid	40. 55
ine)-bis(ethylene-ketal> 2',6''-(1,4,7-trioxa	Heptane-1,7-diyl)-(2,6-bis(2-picolinoyl)-pyrid	40. 53
ine)] 2',6''-(1,4,7-trioxa	Heptane-2-carboxylic-acid-p-nitrobenzyl ester	41. 80
<-thio-imino)-7-oxo-4-thia-1-azabicyclo(3.2.0)	Heptane-2-endo-carboxylic acid	31. 4
6-exo-(methylthio)-bicyclo(2.2.1)	Heptane-2-endo-carboxylic acid	31. 5
6-endo-(methylthio)-bicyclo(2.2.1)	Heptane-2,6-dione <is(3-(trifluoromethyl)-phen	64.122
yl)-1,3,5,7-tetra-aza-4γ(5)-phospha-spiro(3.3)	Heptanol-3,5-dinitrobenzoate	13. 21
(RR,SS)-5-phenyl-2,2,6,6-tetramethyl-3-	Heroin	58. 52
	Herquline hemihydrate	58. 40
ane] trithia(5)	Heterohelicene - 7,7,8,8-tetracyanoquinodimeth	60. 33
2,3,4,6-tetra-O-acetyl-1,5-anhydro-D-arabino-	Hex-1-enitol	45. 30
methyl 3-(3-isocyano-6-oxabicyclo(3.1.0)	Hex-2-en-5-yl)-acrylate....................	38. 20
tramethylcycloheptane)-5,6-dioxabicyclo(3.1.0)	Hex-2-ene ....[[(1,4-b)-(2,3-b)-bis(1,1,4-te	38.114
-(N-acetyl-acetamido)-2,3-dideoxy-α-D-erythro-	Hex-2-eno-pyranose .... [1,4,6-tri-O-acetyl-2	45. 39
methyl α-D-threo-4,6-di-O-acetyl-2,3-dideoxy-	Hex-2-eno-pyranoside	45. 21
ethyl 4,6-di-O-acetyl-2,3-dideoxy-α-D-erythro-	Hex-2-eno-pyranoside	45. 23
2,5-dimethyl-3,4-di(2-propyl)-	Hex-3-ene	5. 13
,6-triphenyl-2,3-diaza-1-phosphabicyclo(3.1.0)	Hex-3-ene .... [4-methyl-2,6	64.116
7,13,16,21,24-hexaoxa-1,10-diazabicyclo(8.8.8)	Hexacosane chlorobenzene solvate <-iron(ii) 4,	49. 23
(2,7).0(3,5).0(4,6).0(9,14).0(10,12).0(11,13)	Hexadec-15-ene-1,8-dicarboxylate <yclo(6.6.2.0	37. 11
iphenyl-14-azatetracyclo(9.3.2.0(2,7).0(2,10))	Hexadeca-7,12,15-trien-9-one <ycarbonyl-8,10-d	37. 24
bis(trimethylbenzylammonium)	Hexadecacarbonyl-tetra-iron-palladium	3. 44
bis(trimethylbenzylammonium)	Hexadecacarbonyl-tetra-iron-platinum	3. 45
9H,9'H-	Hexadecachloro-bis-9-fluorenyl	28. 18
6,7,9,10,12,13,15,16,23,24,26,27,29,30,32,33-	Hexadecahydro-dibenzo(b,q)(1,4,7,10,13,16,19,>	38.131
0-hydroxy-3-oxatetracyclo(8.6.0.0(2,4).0(2,9))	Hexadecane ..:.............................. [1	38. 73
ethyl-1,4,8,11-tetra-azatricyclo(9.3.1.1(4,8))	Hexadecane hydrate .... [5,5,7,12,12,14-hexam	36. 29
	Hexadecyl-trimethylammonium dichloro-iodide....	3. 54
racemic-bis-(4-oxocyclohexyl)-3,4-	Hexane	21. 11
DL-2,2,5,5-tetramethyl-3,4-diphenyl	Hexane	19. 13

```
            meso-2,2,5,5-tetramethyl-3,4-diphenyl  Hexane ..................................................  19. 14
          meso-3,4-bis(trans-4-hydroxycyclohexyl)-  Hexane ..................................................  21. 12
               hexamethyl-2,3:5,6-diepoxy-bicyclo(2.2.0)  Hexane ..................................................  38. 43
         racemic-3,4-bis(trans-4-hydroxycyclohexyl)-  Hexane ..................................................  21. 13
     2,2-triphenyl-3,2-oxaphosphonia-bicyclo(3.1.0)  Hexane tribromide <bromomethyl-4,5-dimethyl-1,  64.139
             1-(1-hydroxyethyl)-bicyclo(3.1.0)  Hexane 3,5-dinitrobenzoate ....................  27. 13
                (2S,3S)-2,3-diphenyl-3-  Hexanecarbonitrile ............................   7.  4
          N,N'-dihydroxy-N,N'-di-isopropyl-  Hexanediamide .................................   1. 32
                methyl D-threo-2,5-  Hexodiulosonate-5-(dimethyl-acetal)............  45. 17
    3-C-acetyl-4,6-O-benzylidene-2-deoxy-α-D-ribo-  Hexopyranoside ........................   [methyl  45. 38
    hyl-4,6-O-benzylidene-2,3-dideoxy-α-D-arabino-  Hexopyranoside)) <1-acetylaziridine)-2,3'-(met  45. 43
    ,20-dienolide-3-O-methyl-4,6-dideoxy-D-allo-2-  Hexosulose <3β,11-trihydroxy-12-oxo-5β-carda-9  45. 57
                di-copper  Hexyl cofacial diporphyrin-7 .................  49. 36
                N,N'-di-n-  Hexyladipamide ...............................   1. 37
    oride]      8-hydroxy-8-methyl-6-(2'-methyl-  Hexylidene)-1-azabicyclo(4.3.0)nonane hydrochl  58. 25a
                                    Hibiscone C ..................................  53.  4
                (+)-allo  Himachalyl-p-bromobenzoate ...................  53. 32a
                                    Hispanonic acid methyl ester .................  54. 13a
    lycine]      methionyl-glutamyl-  Histidinyl-phenylalanyl-arginyl-tryptophanyl-g  48. 82
                (glutamyl-α-lactam)-  Histidinyl-proline tartrate monohydrate.......  48. 47
    6R,-6-methyl-5-oxothiomorpholino-3-carbonyl-L-  Histidinyl-L-prolylamine sesquihydrate   [3S,  48. 51
                5-n-propyl-orotyl-L-  Histidinyl-L-prolylamine tetrahydrate ........  48. 59
                thyrotropin-releasing-  Hormone tartrate monohydrate .................  48. 47
                                    Hosukinidine hydrochloride ...................  58. 72
                cyclo-octane-spiro-5'-  Hydantoin ....................................  35. 21
                tropane-3-spiro-5'-thio-  Hydantoin ....................................  37.  6
                3-acetyl-1,5,5-trimethyl-  Hydantoin ....................................  32. 21
                4-carboxylato-L-thiazolidine-  Hydantoin ....................................  41. 12
                N-phenethyl-granatanine-3-spiro-5'-  Hydantoin ....................................  37. 13
    thyl-3,7-diazabicyclo(3.3.1)nonane-9-spiro-5'-  Hydantoin ........................... [3,7-dime  37.  7
    azocyclohex-5-en-2-one)-3-spiro-5'-(3'-methyl-  Hydantoin) <1-4-(3-methoxy-phenylethyl)-1,4-di  36. 37
                phenoxythiophosphoryl-di  Hydrazide ....................................  64. 18
    1',2''-tris(trimethylsilyl)-thiophosphoryl-tri  Hydrazide ...............................[1,  64. 36
    tetrakis(1,3-dimethyl-1,3,2-diazaborolidinyl)-  Hydrazine .....................[N,N,N',N'-  62. 11
    4-chlorophenylamino)-2-cyclopenten-1-ylidene)-  Hydrazine ............... [N,N'-bis(2-(  20. 12
    4-chlorophenylamino)-2-cyclopenten-1-ylidene)-  Hydrazine ............... [N,N'-bis(2-(  20. 13
    chlorophenyl)-N-(5-methyl-1,3-thiazolin-2-yl)-  Hydrazine ....[N-benzyloxycarbonyl-N'-(2,6-di  41. 66
                nonacarbonyl-(μ-(methylene  Hydrazine-carbonyl))-di-manganese .........  71. 25
                4-(N'-(p-nitrobenzene)-N-  Hydrazo)-8-oxatricyclo(5.1.0.0(2,5))octane ....  38. 52
                dunnione p-bromophenyl  Hydrazone ....................................  59. 28
    ,6))tetradec-11-en-8-one p-bromophenylsulfonyl  Hydrazone <β,9α,10α-13-oxatetracyclo(8.2.1.1(2  38. 99
    ,6))tetradec-11-en-8-one p-bromophenylsulfonyl  Hydrazone <β,9α,10β-13-oxatetracyclo(8.2.1.1(2  38. 98
                p-dimethylamino-benzaldehyde-p-nitrophenyl  Hydrazone benzene solvate ....................  15.  5
    yl-perhydroazulen-4-one)-(p-bromophenylsulfonyl  Hydrazone) ..................... [cis-2-t-but  27. 28
    benzene-1,3-dicarbaldehyde-bis(p-tolylsulfonyl  Hydrazone) benzene clathrate .... [5-methyl  61.  7
                trimethylphosphonium-methylene-boron-tri  Hydride ......................................  64. 11
    m) aquo-tetr>  tetraethylammonium hydrogen (μ-  Hydrido)-bis(μ-chloro)-bis(trichloro-molybdenu   3. 40
                bis(triphenylphosphine)iminium (μ-  Hydrido)-bis(tricarbonyl-nickel) .............  64.152
    <henylphosphine)iminium (μ(2)-carbonyl)-(μ(2)-  Hydrido)-hexacarbonyl-di-iron-hexacarbonyl-di>  64.156
                tetraphenylarsonium (μ(6)-  Hydrido)-hexakis(tricarbonyl-ruthenium)........  65. 13n
    nickel-tri-i>  benzyl-trimethylammonium (μ(3)-  Hydrido)-octacarbonyl-tetrakis(μ(2)-carbonyl)-   3. 43
    <henylphosphine)iminium (μ(2)-carbonyl)-(μ(2)-  Hydrido)-tricarbonyl-iron-nonacarbonyl-tri-ru>  64.155n
                bis(tetraphenylarsonium) di  Hydrido-dodeca-nickel-carbonyl ...............  65. 16
                bis(tetraphenylphosphonium) di  Hydrido-dodeca-nickel-carbonyl ...............  64.128n
                bis(tetraphenylphosphonium) di  Hydrido-dodeca-nickel-carbonyl ...............  64.127
                bis(bis(triphenylphosphine)immonium) di  Hydrido-dodeca-nickel-carbonyl ...............  64.164
                tris(tetraphenylarsonium)  Hydrido-dodeca-nickel-carbonyl acetone solvate  65. 21n
                tris(tetraphenylarsonium)  Hydrido-dodeca-nickel-carbonyl acetone solvate  65. 20
                tetrakis(4-ethylpyridinium) di  Hydrido-octacosa-oxygen-deca-vanadium ........  33. 25
                tris(tetraethylammonium)  Hydrido-pentacosacarbonyl-tetradeca-rhodium....   3.  5
    e]          sodium  Hydrido-trimethylboronate diethyl ether solvat  62.  2
    azene]          1-  Hydrido-1-isopropyl-tetrachloro-cyclotriphosph  64.  3
                sodium  Hydrogen α-methoxy-α-phenylacetate ...........   2. 29
                rubidium  Hydrogen α-methoxy-α-phenylacetate ...........   2. 30
                lithium  Hydrogen (+)-1-malate ........................   2. 23
    ro-molybdenum) aquo-tetr>  tetraethylammonium  Hydrogen (μ-hydrido)-bis(μ-chloro)-bis(trichlo   3. 40
                rubidium  Hydrogen bis(dibromoacetate) .................   2.  6
                potassium  Hydrogen bis(dibromoacetate) .................   2. 11
                potassium  Hydrogen bis(dichloroacetate) ................   2.  8n
    o-gold(iii)]  Hydrogen bis(N,N-dimethylacetamide) tetrachlor   1. 15n
    o-gold(iii)]  Hydrogen bis(N,N-dimethylacetamide) tetrachlor   1. 14
                bis(triphenylphosphine-oxide) monohydrate  Hydrogen bromide .............................  64. 91
                bis(phenyl-(t-butyl)-phosphinic amide)  Hydrogen chloride ............................  64. 40
                trichloro-(p-methoxyphenyl)-phosphonium  Hydrogen dichloride...........................  64. 24
                tetra-n-butylammonium  Hydrogen dichloromaleate .....................   2. 19
                tetra-n-butylammonium  Hydrogen dichloromaleate .....................   2. 20
```

                                  tetraethylammonium Hydrogen dodecachloro-tri-ruthenium hexahydrat   3. 38
      azinyl)-10,11-dihydro-5H-dibenzo(b,f)silepane Hydrogen fumarate <-dimethyl-10-(4-methylpiper   63. 33
                                        imidazolium Hydrogen maleate 32. 5n
    ,5'-(5H)-dibenzo(a,d)cycloheptene)-4-ammonium Hydrogen maleate <ethyl-spiro(cyclohex-2-ene-1   29. 9a
                                     di-calcium di Hydrogen mellitate nonahydrate 14. 5
                                            sodium Hydrogen nitrilo-triacetate...................   2. 27
                              tellurium oxalate di Hydrogen oxalate dihydrate 2. 10
                                       guanidinium Hydrogen oxalate monohydrate 2. 9
                                              urea Hydrogen peroxide 8. 3n
                             cyclohexylammonium ethyl Hydrogen phosphate 21. 1
    ino)-3-pyridyl)sulfonyl)-3-ethyl-1-methylurea Hydrogen sulfate   [1-((4-(3-chlorophenylam   33. 66
                         pyridinium tetra-μ-(Hydrogen-phosphato)-di-molybdenum chloride 33. 10
                                      calcium bis(Hydrogen-1-malate) hexahydrate 2. 24
                       bis(triethylammonium) di Hydronium octacyano-molybdenum(iv) 3. 31
                            N-phthaloylglycine- Hydroxamic acid 35. 15
      p-menthane]   bis(μ- Hydroxo)-bis(aqua-tribromo-tin(iv)) 1,8-epoxy   38. 26
      ne]   bis(μ- Hydroxo)-bis(aqua-trichloro-tin(iv)) 1,4-dioxa   38. 4
      -p-menthane]   bis(μ- Hydroxo)-bis(aqua-trichloro-tin(iv)) 1,8-epoxy   38. 27
      anti-methyl-(2,3β,3aβ,5,6,7,8,8aβ-octahydro-4- Hydroxylimino-4H-cyclohepta(b)thiophen-3α-yl)>   39. 44
                                        6-chloro- Hyellazole 58. 43
                                                   Hymenograndin 53. 24
                                                   Hymenosignin 53. 26
                                                   Hypophyllanthin..............................   38.120
                                                   Hysterin 38. 86
         2-(o-hydroxybenzoyl)-7-methyl-3-phenyl- Imidazo(1,2-a)pyridine 35. 62
           3,3'-thio-bis(2-methyl-1-phenyl- Imidazo(1,5-a)pyridinium) tetrafluoroborate....   35. 73
           2-(o-hydroxybenzoyl)-3-phenyl- Imidazo(2,1-a)isoquinoline 36. 40
                                  5,6-dimethyl- Imidazo(2,1-b)(1,3,4)thiadiazole 41. 15
      monohydrate]   5,6-dimethyl- Imidazo(2,1-b)(1,3,4)thiadiazole hydrobromide   41. 16
          2,3,4,5-tetrahydro-6,6-diphenyl- Imidazo(2,1-b)thiazin-7(6H)-one................   41. 64
                          1-(2,4-dinitrophenyl)- Imidazo(4,5-b)pyridine 35. 24
      ulfonate monohydr> 2-amino-5-methyl-7-propyl- Imidazo(5,1-f)(1,2,4)triazin-4(3H)-one methyls   35. 13
      nohydra> 2-cyano-1-methyl-3-(2-((5-methyl-1H- Imidazol-4-yl)-methylthio)-ethyl)-guanidine mo   32. 34
                           barbital 1-methyl Imidazole 43. 1
                              2-methyl-4-nitro- Imidazole 32. 10
      di-μ-aquo-bis(dioxo-dinitrato-uranium(vi)) di- Imidazole 32. 6
      4-(o-hydroxybenzoyl)-2,5-diphenyl-1-(p-tolyl)- Imidazole 32. 73
      hloro-4-methylthiophen-3-yl-imino)-tetrahydro- Imidazole ....[2-(2-c   39. 22
      hloro-β-((2,4-dichlorobenzyl)-oxy)-phenethyl)- Imidazole hemihydrate ....[1-(2,4-dic 32. 65
      hylamino-4,5-bis(p-chlorophenyl)-4-hydroxy-4H- Imidazole methanol solvate ....[((+-)-2-met   32. 55
      gnesium bis(2,4-dinitrophenoxide) bis(N-methyl Imidazole) ....[ma 67. 4
      amidophenyl)-porphyrinato-iron eth> (2-methyl Imidazole)-dioxygen-meso-tetra(α,α,α,α-o-pival   49. 35
      yl)-porphyrinato-iron(ii) ethanol > (2-methyl Imidazole)-meso-tetra(α,α,α,α-o-pivalamidophen   49. 34
                                              4- Imidazole-acrylic acid dihydrate 32. 17
      2'-trichloroethyl-2-benzyl-4-methoxy-carbonyl- Imidazole-1-(α-isopropylidene)-acetate [2',2',   32. 67
      o)-1-(2',3',5'-tri-0-nitro-β-D-ribofuranosyl)- Imidazole-4-carboxamide <-5-(N-nitrocarboxamid   45. 19
                 1-phenyl-4,5-(D-glucofurano)- Imidazolidin-2-one 45. 27
                        4-hydroxy-2,5-dioxo-4- Imidazolidine-carboxyureide hemihydrate........   32. 13
                            1,1'-trimethylene-di-2- Imidazolidine-thione 32. 31
      (1Z,3E)-1,4-bis(1,3-dibenzyl-2- Imidazolidinyl)-1,4-diphenylbutadiene 32. 78
             4,4'-methylene-bis(1,3,5-trimethyl-2- Imidazolin-2-one) 32. 50
                        4-benzylidene-2-phenyl-2- Imidazolin-5-one 32. 54
                2(1-(2,6-dichlorophenoxy)-ethyl)- Imidazoline..............................   32. 39
      2,4,5-tris(3-methoxycarbonyl-furan-2-yl)-Δ(2)- Imidazoline 38.104
                2(1-(2,6-dichlorophenoxy)-ethyl)- Imidazoline hydrochloride 32. 40
              2-(2-chloro-4-methylphenylamino)- Imidazoline nitrate.........................   32. 33
                    2-(2,6-dichlorophenylimino)-2- Imidazoline-hydrochloride 32. 28
      5-dimethylimmonio-1-isopropyl-4,4-dimethyl-2- Imidazoline-2-thiolate chloroform solvate 32. 35
             2,2,5,5-tetramethyl-4-phenyl-3- Imidazoline-3-oxide-1-oxyl 32. 49
                              2,4,5-tri-p-tolyl- Imidazolinium bromide 32. 71
                                                   Imidazolium hydrogen maleate 32. 5n
      4,5-dimethyl-1,3,2-dioxaphospholane-2-sulfide Imidazolium monohydrate.... [trans-2-hydroxy-   64. 8
                                                   Imidazolium tetrachloro-dioxo-uranium(vi)   32. 7
      tetrasulfur-tetranitrido)-thiacyclopentane-di- Imide ....[N,N'-(41. 9
      ahydro-naphthalene-tetracarboxylic-1,2,3,4-di- Imide [N,N',7-trimethyl-1-p-tolyl-1,2,3,4-tetr   36. 42
             1-(S,S-dimethyl-N-(trimethylsilyl)-sulfo-di- Imide)-bicyclo(3.3.1)penta-azatetrathiane 63. 2
                        dodecamethyl-bis(Imido)-triphosphoramide monohydrate........   64. 55
                N,N'-(di-β-naphthyl)-p-quinone-di- Imine 24. 12
                 N-(1-phenylethyl)-1-benzoyl-benzylidene Imine 19. 12a
                N-(1-phenylpropyl)-1-benzoyl-benzylidene Imine 19. 15
                    N,N'-ethylene-bis(acetylacetone- Imine) 1. 31
                                4,4'-(ethylenedi- Imine)-bis(methylenecamphor) 31. 62
      bonyl-di-iron-hexaca> bis(triphenylphosphine) Iminium (μ(2)-carbonyl)-(μ(2)-hydrido)-hexacar   64.156
      onyl-iron-nonacarbon> bis(triphenylphosphine) Iminium (μ(2)-carbonyl)-(μ(2)-hydrido)-tricarb   64.155n
                        bis(triphenylphosphine) Iminium (μ-hydrido)-bis(tricarbonyl-nickel)....   64.152
      hodium]   bis(triphenylphosphine) Iminium (μ-pentacarbonyl)-decacarbonyl-penta-r   64.158

nium-tetra-iridi> bis(bis(triphenylphosphino) Iminium) heptacarbonyl-μ(2)-hexacarbonyl-ruthe 64.162
nium-tetra-iridi> bis(bis(triphenylphosphine) Iminium) nonacarbonyl-hexa-μ(2)-carbonyl-ruthe 64.163
           bis(triphenylphosphine) Iminium) tridecacarbonyl-tetra-iron............. 64.161
     (bis(trimethylsilyl)amino)-(t-butyl Imino)-(trimethylsilylimino)-phosphorus.......... 64. 61
           (α-(1-phenylethyl Imino)-benzoyl)-phenyl-ketone 19. 12a
         tetrakis(phenyl Imino)-cyclobutane 20. 14
hylsilyl)amino)-(t-butylimino)-(trimethylsilyl Imino)-phosphorus........   [(bis(trimet 64. 61
   2-(2-chloro-4-methylthiophen-3-yl- Imino)-tetrahydro-imidazole.................. 39. 22
           bis(triphenylphosphine) Imino)-tetrasulfur-tetranitride............. 64.167
         1-(triphenylphosphine- Imino)-1-chloro-2,2-dicyanoethylene........... 64.115
       2-(2,6-dichlorophenyl Imino)-2-imidazoline-hydrochloride 32. 28
t-butyl-5-(2-(p-tolylsulfonylamino)-1-(t-butyl Imino)-2-thioxo-ethyl)-1H-tetrazole.......[1- 32. 66
         2,5-bis(N-chlorothio- Imino)-3,4-dicyanothiophene............. 39. 7
  -2->   2S-(2α,5α)-3,3-dimethyl-6-(p-tolyl-thio- Imino)-7-oxo-4-thia-1-azabicyclo(3.2.0)heptane 41. 80
hylsilyl)-phosphane]   (p-chlorophenyl Imino-(trimethylsiloxy)-methyl)-phenyl-(trimet 64.106
       4-oxo-2-phenyl Imino-perhydro-1,3-thiazine............... 41. 29
amino)-bis(trimethylsilylamino)-trimethylsilyl Imino-phosphoranide <imethylamino)phosphonium) 64.107
       2,6-bis(hydroxy Imino-piperazine)................... 33. 2
e]   t-butyl-(2-t-butyl Imino-1-phenylsulfonylimino-2-cyano-ethyl)amin 3. 53
be> 8-(p-chlorophenylsulfonylamino)-4-hydroxy Imino-1,2,3,4,5,6,7,8-octahydro-1,8-methano(1) 37. 16
       5-benzamido-2-benzoyl Imino-1,3-dithiol-4-carbonic acid ethyl ester 39. 75
t-butyl-(2-t-butylimino-1-phenylsulfonyl Imino-2-cyano-ethyl)amine 3. 53
0)nonadec-1-ene-4,13-dione]   12-benzoyl Imino-2-phenyl-3-oxa-1-aza-trispiro(4.0.5.2.5. 40. 56
ine]   5-benzoyl Imino-2,2-dimethyl-4-phenyl-1,3,4-thiadiazolid 41. 61
       2-(2,6-dichlorophenyl) Imino-3-methyl-perhydro-1,3-thiazine 41. 36
       2-(2,6-dimethylphenyl) Imino-3-methyl-perhydro-1,3-thiazine 41. 46
       2-(2,6-dimethylphenyl)- Imino-3-methyl-thiazolidine.................. 41. 41
       lithium 1-carboxymethyl-2- Imino-3-phosphonoimidazolidine dihydrate 64. 12
emicarbazido-2-pyrazoline]   4-hydroxy Imino-3,5-dimethyl-1-thiosemicarbamoyl-5-thios 32. 20
le]   2-benzoyl Imino-3,5-diphenyl-2,3-dihydro-1,3,4-thiadiazo 41. 77
lyl)-uracil]   1,3-dimethyl-4- Imino-5,5-(spiro-2',4'-bis-dimethylamino-pyrro 35. 28
ine)-2'-one-4-semicarbazone monohydrate]   2- Imino-6'-methyl-spiro(oxazolidine-5,3'-piperid 40. 15
       Iminodibenzyl 36. 10
     1,2,3-triphenyl-5-(1,2-diphenyl-2-phenyl Iminoethylidene)-2-pyrrolin-4-one 32. 77
-2-thiolate chloroform solvate]   5-dimethyl Immonio-1-isopropyl-4,4-dimethyl-2-imidazoline 32. 35
-penta-osmium]   bis(triphenylphosphine) Immonium (μ(5)-carbido)-iodo-pentadecacarbonyl 3. 56
m)]   bis(triphenylphosphine) Immonium μ-deuterido-bis(pentacarbonyl-chromiu 64.153n
-cobalt)]   bis(triphenylphosphine) Immonium nona-μ-carbonyl-nitrido-hexa(carbonyl 64.157
       bis(triphenylphosphine) Immonium oxo-tetrachloro-technetium(v) 64.159
       bis(triphenylphosphine) Immonium tetracyano-oxo-dioxo-molybdenum 64.160
     bis(bis(triphenylphosphine) Immonium) carbido-deca-osmium-carbonyl 64.166
rbonyl-ruthenium)-c> bis((triphenylphosphine) Immonium) carbonyl-(tri-μ-carbonyl)-tris(trica 64.154
       bis(bis(triphenylphosphine) Immonium) dihydrido-dodeca-nickel-carbonyl 64.164
       bis(bis(triphenylphosphine) Immonium) octa-osmium-docosacarbonyl 64.165
1,2,3,4,5,6-heptaphenyl-1,4-dihydrobenz(e)-as- Indacene ....[1, 29. 12
methyl-2-(trimethylsilyl)-3-phenyl-1,3-disila- Indan ....[1,1,3-tri 63. 30
methyl-2-(trimethylsilyl)-3-phenyl-1,3-disila- Indan ....[1,1,3-tri 63. 29
       2-benzylidene Indan-1-one........................... 27. 17
       5,7-dimethoxy- Indan-1-one............................ 27. 7
       1,2-bis(2-oxa- Indan-1-yl-oxy)ethane 38. 90
       2-acetyl- Indan-1,3-dione........................ 27. 5
       2-pivaloyl- Indan-1,3-dione........................ 27. 11
       2-(p-dimethylaminobenzylidene)- Indan-1,3-dione........................ 27. 23
       2-(4'-dimethylamino-cinnamoyl)- Indan-1,3-dione........................ 27. 26
2,2,5,7-tetramethyl-4-oxy-6-(2-oxyethyl)- Indanone 59. 10
1,2,4,5-tetrahydro-7-methoxy-3H-benz(g) Indazol-3-one monohydrate 36. 4
       3-methyl-1,4-dimethyl-1,2-diphenyl- Indene 27. 32
,6,7,7a-hexahydro-3,4β-dimethyl-2H-4α-hydroxy- Indene ....[1,4,5 27. 8
,8a-tetramethyl-4a,8a-dihydrocyclopenta(1,2-a) Indene) [8-diethylamino-1,3,4a 28. 14
ydro-spiro(cyclopropane-1,2'-(1H)-cycloprop(a) Indene) <-tetrachloro-3,3-dimethyl-1a',6b'-dih 29. 3
etrahydro-spiro(cyclohexa-2,5-diene-1,1'-(1H)- Indene)-4,4'-dione diethyl ether solvate <a'-t 17. 44
       7b-methyl-7bH-cyclopent(cd) Indene-1,2-dicarboxylic acid 28. 6
hydro-2,6-dihydroxy-1,2,3a-trimethyl-4-oxo-4H- Indene-3,3-dicarboxylic acid <2,3,3a,7,7a-hexa 27. 18
       1,3-bis(1'-methyl-2'-pyridinium)- Indenide bromide 33. 79
       meso-1,1'-bi- Indenyl................................ 27. 22
       6,6'-dibromo- Indigo 59. 12
       6,6'-dibromo- Indigo 59. 11
       2,2'-dimethoxy- Indigo 59. 18
chloro-(5,10,15,20-tetraphenylporphinato)- Indium(iii)............................. 49. 16
2-piperidyl-2-piperidylenium-ethyl)methylene)- Indol-2-olate   [N-acetyl-3-(benzoyl(35. 77
       1-phenyl-3-methyl-3-methylamino-propyl- Indol-2-one hydrochloride 35. 53
indol-2-yl)-3-(trans-2,3-dihydro-1,3-dimethyl- Indol-2-yl)-maleate [dimethyl 2-(1,3-dimethyl- 35. 72
indol-2-yl)-maleat> dimethyl 2-(1,3-dimethyl- Indol-2-yl)-3-(trans-2,3-dihydro-1,3-dimethyl- 35. 72
       1-(3-(Indol-3-yl)propyl)-thymine 44. 34
droxy-2β-methoxy-α-phenyl-5H-pyrrolo(2,1-a)iso Indol-5-one............[1,2,3,9b-tetrahydro-9bβ-hy 36. 23
fonylamino)-6,7,10,11-tetrahydroazocino(1,2-a) Indol-8(9H)-one........ [12-(p-chlorophenylsul 36. 32

2-(4'-bromophenyl)-4,6-dimethoxy- Indole 35. 36
2-phenyl-3-(N-p-methoxyphenyl)amine- Indole 35. 63
2-carboxy-3-methylthiopyrano(4,3,2-cd) Indole 39. 42
9a-epi-tetrathio-3,10-diketo-piperazino(1,2-a) Indole   [9,9a-dihydro-1,2,9,9-tetramethyl-2, 41. 56a
l)-butyl) -2,3,4,5-tetrahydro-1H-pyrido(4,3-b) Indole <ophenyl)-2-(4-hydroxy-4-(4-fluoropheny 36. 49
1,2-dihydro-3-oxo-2,2-diphenyl-3H- Indole-1-oxyl 35. 55
1,3-bis(1,3,3-triethyl- Indolenine-2-yl)-trimethinium perchlorate . 35. 82
roborate methylene c> 1,3-bis(1,3,3-triethyl- Indolenine-2-yl)-2-arsa-trimethinium tetrafluo 35. 78
oborate] 1,3-bis(1,3,3-triethyl- Indolenine-2-yl)-2-aza-trimethinium tetrafluor 35. 81
oborate] 1,3-bis(1,3,3-trimethyl- Indolenine-2-yl)-2-aza-trimethinium tetrafluor 35. 71
luoroborate methylen> 1,3-bis(1,3,3-triethyl- Indolenine-2-yl)-2-phospha-trimethinium tetraf 35. 80
luoroborate] 1,3-bis(1,3,3-trimethyl- Indolenine-2-yl)-2-phospha-trimethinium tetraf 35. 79
luoroborate] 1,3-bis(1,3,3-triethyl- Indolenine-2-yl)-2-t-butyl-trimethinium tetraf 35. 75
3-hydroxy-2,3-diphenyl- Indoline 35. 59
4,9,9a-tetrahydro-2-methyl-4β-phenylbenz(f)iso Indoline   [3a, 36. 30
2-methyl-3-(2-(4-phenyl-1-piperazinyl)-ethyl)- Indoline ....[cis-5,6-dimethoxy- 35. 70
drate] 3-hydroxy-2,2'-dioxo-(3-3'-di- Indoline)-N,N'-bis(5,5-dimethylsulfoximide) hy 35. 60
trans(2H,8aH)-2-diphenylhydroxy-methyl- Indolizidine 35. 65
N'-acetyl-di- Indolo(2,3-a:2',3'-c)carbazole 36. 48
drate] 11-hydroxy-6H- Indolo(3,2,1-de)(1,5)naphthyridin-6-one monohy 58. 5
4-(3- Indolyl)-butyric acid 35. 26
lo(3.3.0)octa-2,4-dien> 6,8,8-trimethyl-6-(3- Indolyl)-4-(isopropenyl)-(2,3)benzo-1-azabicyc 36. 45
Inkanine benzene solvate 58. 31
Inosine.................................... 47. 14
-1-methoxy-6-methylamino-1,3,6-trideoxy-chiro- Inositol ....[L-3-amino 50. 4
dodecasodium myo- Inositol-hexaphosphate octatriacontahydrate.... 46. 3
chasmanine Intermediate 58. 65
theobromine tri-iodide Iodine 44. 18
bis(phenacetin) tri-iodide Iodine 17. 14
3,7-dimethylxanthine tri-iodide Iodine 44. 18
tetraphenylphosphonium bis(azido)- Iodine 64.124
ro-2-hydroxy-5-nitrophenyl)-(2'-chlorophenyl)- Iodonium hydroxide inner salt .... [(3-chlo 17. 18
Iodylbenzene 19. 1
3,6-epoxy-5-hydroxy-5,6-dihydro-β- Ionol 59. 23
Ionomycin cadmium salt n-heptane solvate 50. 29a
Ionomycin calcium salt n-heptane solvate 50. 28
nacarbonyl-hexa-μ(2)-carbonyl-ruthenium-tetra- Iridium <is(bis(triphenylphosphine)iminium) no 64.163
ptacarbonyl-μ(2)-hexacarbonyl-ruthenium-tetra- Iridium <is(bis(triphenylphosphino)iminium) he 64.162
(1S)-cis,cis- Iridolactone 52. 3
(1S)-cis,cis-iso Iridolactone 52. 2
henylphosphine)iminium) tridecacarbonyl-tetra- Iron .... [bis(bis(trip 64.161
tacarbonyl-tetrakis(μ(2)-carbonyl)-nickel-tri- Iron <nzyl-trimethylammonium (μ(3)-hydrido)-oc 3. 43
etra(α,α,α,α-o-pivalamidophenyl)-porphyrinato- Iron ethanol solvate <idazole)-dioxygen-meso-t 49. 35
o-(N-methyl-5,10,15,20-tetraphenylporphinato)- Iron(ii) ....[chlor 49. 19
onyl-(N,N-dimethylformamide)-phthalocyaninato- Iron(ii) dimethylformamide solvate .... [carb 49. 5
etra(α,α,α,α-o-pivalamidophenyl)-porphyrinato- Iron(ii) ethanol solvate <hylimidazole)-meso-t 49. 34
<sodium (ethylthio)-(tetra-p-tolylporphinato)- Iron(ii) ethylthiolate 4,7,13,16,21-pentaoxa-> 49. 27
yc> sodium ethylthio-(tetraphenylporphinato)- Iron(ii) 4,7,13,16,21,24-hexaoxa-1,10-diazabic 49. 23
octaethylporphinato-perchlorato- Iron(iii) 49. 7
methoxy-(meso-tetraphenylporphyrinato)- Iron(iii) 49. 20
3-cyanopyridinium 3-cyanopyridine-tetrachloro- Iron(iii) 33. 17
otassium dicyano-(meso-tetraphenylporphinato)- Iron(iii) acetone solvate .... [p 49. 22
bis(3-chloropyridine)-octaethylporphinato- Iron(iii) perchlorate 49. 24
μ-tetrafluoroethane-1,1-diyl)-bis(tricarbonyl- Iron) .... [bis(μ-methylthio)-(71. 17
μ-tetrafluoroethane-1,2-diyl)-bis(tricarbonyl- Iron) .... [bis(μ-methylthio)-(71. 18
)-carbonyl)-dicarbonyl-iron)-tris(tricarbonyl- Iron) <(μ(5)-carbido)-(dicarbonyl-rhodium-(μ(2 3. 33
onyl)-tricarbonyl-molybdenum)-bis(tricarbonyl- Iron) <bonyl-iron)-(dicarbonyl-iron-(μ(2)-carb 3. 37
<-(dicarbonyl-iron-(μ(2)-carbonyl)-dicarbonyl- Iron)-(dicarbonyl-iron-(μ(2)-carbonyl)-tricar> 3. 37
)-bis(tetrakis(μ(3)-thio)-tris(μ(3)-ethylthio- Iron)-di-tungsten) <onium) tris(μ(2)-ethylthio 64.129
etraphenylphosphonium bis(bis-μ-thio-dichloro- Iron) molybdenum   [t 64.129
dicarbonyl-rhodium-(μ(2)-carbonyl)-dicarbonyl- Iron)-tris(tricarbonyl-iron) <(μ(5)-carbido)-(3. 33
<traethylammonium) (μ(6)-carbido)-(dicarbonyl- Iron-(μ(2)-carbonyl)-dicarbonyl-iron)-(dicarb> 3. 37
<(μ(2)-carbonyl)-dicarbonyl-iron)-(dicarbonyl- Iron-(μ(2)-carbonyl)-tricarbonyl-molybdenum)-> 3. 37
osphonium (benzyl-trimethylammonium) dichloro- Iron-bis(μ(2)-thio)-molybdenum-disulfide <ylph 64.123
hylforma> tetraphenylphosphonium pentasulfur- Iron-bis(μ-sulfido)-molybdenum-disulfide dimet 64.130
lformami> tetraphenylphosphonium pentasulfur- Iron-bis(μ-sulfido)-tungsten-disulfide dimethy 64.131
μ(2)-carbonyl)-(μ(2)-hydrido)-hexacarbonyl-di- Iron-hexacarbonyl-di-ruthenium <hine)iminium (64.156
um (μ(2)-carbonyl)-(μ(2)-hydrido)-tricarbonyl- Iron-nonacarbonyl-tri-ruthenium <osphine)imini 64.155n
imethylbenzylammonium) hexadecacarbonyl-tetra- Iron-palladium ....[bis(tr 3. 44
imethylbenzylammonium) hexadecacarbonyl-tetra- Iron-platinum ....[bis(tr 3. 45
epoxy- Isoacoragermacrone 53. 9
ithioglycerol acid methyl ester)-4a,5-dihydro- Isoalloxazine .... [3-methyl-4a-(methylene-d 36. 18
rate] 7,8-dimethyl- Isoalloxazine-10-acetic acid tyramine tetrahyd 36. 11
24-ethyl- Isobrassinolide............................. 51. 76
p-methoxy-N,N-bis(α-cyano-hexafluoro- Isobutenyl)-aniline......................... 16. 13
omide] (-)-5-ethyl-2'-hydroxy-2- Isobutyl-9,9-dimethyl-6,7-benzomorphan hydrobr 58. 49

N-(1-phenylethyl)-	Isobutyramide	1. 27
methylbenzoyl-oxycarbonyl-bis(α-amino-	Isobutyryl)-L-alaninate	48. 65
1-chloroethyl-4-methyl-6,7-dimethyl-	Isochroman	38. 60
(η(2)-dithiomethoxycarbonyl)-carbonyl-(p-tolyl	Isocyanide)-bis(triphenylphosphine)-ruthenium>	71.270
7,15-di-	Isocyano-adociane	54. 19
6α-benzyl-6β-	Isocyano-penicillanate	50. 16
2-	Isocyano-pupukeanane	53. 18
8-	Isocyano-10-cycloamphilectene	54. 15
7-	Isocyano-11(20),14-epi-amphilectadiene	54. 16
ylate] methyl 3-(3-	Isocyano-6-oxabicyclo(3.1.0)hex-2-en-5-yl)-acr	38. 20
-hydroxy-2β-methoxy-α-phenyl-5H-pyrrolo(2,1-a)	Isoindol-5-one .... [1,2,3,9b-tetrahydro-9bβ	36. 23
3a,4,9,9a-tetrahydro-2-methyl-4β-phenylbenz(f)	Isoindoline	36. 30
(1S)-cis,cis-	Isoiridolactone	52. 2
t-butoxy-carbonyl-L-prolyl-L-	Isoleucyl-glycine hydrate	48. 56
5-acetoxy-	Isomarasman-7α,13-diol	53. 20
8α-hydroxy-	Isopicrostegane	59. 31
11α,16-epoxy-1β,15R-dihydroxy-	Isopimar-8(14)-ene	54. 9a
1,2,3,4,5,6,7,8-octahydro-1,4:5,8-di-	Isopropano-4,5-dimethyl-9-phenyl-acridine	37. 22
a-2,4-dien> 6,8,8-trimethyl-6-(3-indolyl)-4-(	Isopropenyl)-(2,3)benzo-1-azabicyclo(3.3.0)oct	36. 45
yl-2-benzyl-4-methoxy-carbonyl-imidazole-1-(α-	Isopropylidene)-acetate [2',2',2'-trichloroeth	32. 67
no> methyl 4-chloro-3,4-dideoxy-1,2:6,7-di-o-	Isopropylidene-α-D-erythro-hept-3-en-5-ulopyra	45. 32
phosphite] 1,2-O-	Isopropylidene-α-D-glucofuranose-3,5,6-bicyclo	64. 32
1] 2,5'-anhydro-1-(2',3'-O-	Isopropylidene-β-D-ribofuranosyl)-2-thio-uraci	47. 28
2,2'-	Isopropylidene-bis(4-chloro-6-nitrophenol)	17. 23
-2-yl)-cyclohexane-1,3-d> (1R,3R,4S,5R)-4,5-(	Isopropylidene-dioxy)-1-(2-methyl-1,3-dithiane	39. 54
ne metha> 9α-fluoro-11β,21-dihydroxy-16α,17α-	Isopropylidene-dioxy-pregna-1,4-diene-3,20-dio	51. 46
borane] 9,12-	Isopropylidene-dithio-1,2-dicarba-closo-dodeca	62. 6
2,3:4,5-di-O-	Isopropylidene-D-gulose diethyl dithioacetal	45. 42
hyl-rubrolone] 9,10-O,O-	Isopropylidene-10a-O-(o-bromobenzoyl)-11-O-met	50. 23
ne] 7-	Isopropylidene-2,3-diazabicyclo(2.2.1)hept-2-e	37. 3
2-(o-hydroxybenzoyl)-3-phenyl-imidazo(2,1-a)	Isoquinoline	36. 40
thoxybenzyl)-6,7-dihydro-1,2,3,4-tetrahydro-	Isoquinoline hydrochloride monohydrate <-trime	35. 54
<y-1-methyl-1,2,3,9a-tetrahydro-cyclohexa(i,j)	Isoquinoline-7-spiro-4'-(2'-methoxy-cyclohexa>	58. 63
tetrahydro-6,7-dimethoxy-2-methyl-1-isopropyl-	Isoquinolinium iodide .... [2-benzyl-1,2,3,4-	35. 68
	Isosoforidine	58. 11
4,5-diphenyl	Isosydnone benzene solvate	40. 33
3-(3-amino-benzo(b)thiophen-2-yl)-1,2-benz	Isothiazole	41. 52
nitrophenoxy)-3,6,9-trioxa-undecane potassium	Isothiocyanate [1,11-bis(2-	17. 35
(dibenzo-30-crown-10) bis(sodium	Isothiocyanate) monohydrate	38.131
19,22,25,28)decaoxacyclotriacontin bis(sodium	Isothiocyanate) monohydrate <)(1,4,7,10,13,16,	38.131
tetrakis(pyridinium) di-μ-oxo-tris(oxo-tris(	Isothiocyanato)-molybdenum(v))	33. 11
tris(tetra-n-butylammonium) hexakis(	Isothiocyanato)-technetium(iii)	3. 51
kel(ii) 2-bromonaphthalene clathrate] bis(	Isothiocyanato)-tetrakis(4-methylpyridine)-nic	61. 10
kel(ii) 2-methylnaphthalene clathrate] bis(	Isothiocyanato)-tetrakis(4-methylpyridine)-nic	61. 11
N,N-dimethyl-N'-(N-methyl-2,4-dinitroanilino)-	Isourea [O-benzoyl-	13. 16
ose)-13-(β-D-mycinose)-lankamycin-11-α-hydroxy	Isovalerate ester <-(β-D-4,6-dideoxy-3-ketoall	50. 35
L-valyl-L-α-hydroxy	Isovaleric acid	48. 31
cyclo(D-N-methylvalyl-D-α-hydroxy	Isovaleryl)	48. 32
alyl-L-α-hydroxyisovaleryl-L-valyl-D-α-hydroxy	Isovaleryl) [cyclo-tris(D-v	50. 39
-hydroxyisovaleryl-L-N-methylvalyl-D-α-hydroxy	Isovaleryl) [cyclo(D-N-methylvalyl-L-α	48. 71
ryl-D-valyl-L-methylalanyl-L-valyl-D-α-hydroxy	Isovaleryl) <alanyl-L-valyl-D-α-hydroxyisovale	50. 24
cyclo-bis(D-valyl-L-prolyl-L-valyl-D-α-hydroxy	Isovaleryl) dihydrate	48. 81
-α-hydroxyisovaleryl-L-methylvalyl-D-α-hydroxy	Isovaleryl) sodium nickel nitrate hydrate <l-D	50. 22
leryl-bis(D-valyl-L-lactyl-L-valyl-D-α-hydroxy	Isovaleryl)) <lalanyl-L-valyl-D-α-hydroxyisova	50. 37
<lo(D-valyl-L-methylalanyl-L-valyl-D-α-hydroxy	Isovaleryl-bis(D-valyl-L-lactyl-L-valyl-D-α-h>	50. 37
<lo(D-valyl-L-methylalanyl-L-valyl-D-α-hydroxy	Isovaleryl-D-valyl-L-methylalanyl-L-valyl-D-α>	50. 24
<α-hydroxyisovaleryl-L-methylvalyl-D-α-hydroxy	Isovaleryl-L-methylvalyl-D-α-hydroxyisovalery>	50. 22
-L-methylval> cyclo(L-methylvalyl-D-α-hydroxy	Isovaleryl-L-methylvalyl-D-α-hydroxyisovaleryl	50. 22
cyclo-tris(D-valyl-L-α-hydroxy	Isovaleryl-L-valyl-D-α-hydroxyisovaleryl)	50. 39
yl)] cyclo(D-N-methylvalyl-L-valyl-D-α-hydroxy	Isovaleryl-L-N-methylvalyl-D-α-hydroxyisovaler	48. 71
phenyl-perhydro> 3-(4,5-bis(methoxycarbonyl)-	Isoxazol-3-yl)-4,6-dioxo-syn-cis-syn-1,3,5-tri	40. 58
l)-pyrimidine] 3-sulfanilamido-5-methyl	Isoxazole 2,4-diamino-5-(3,4,5-trimethoxybenzy	44. 30
diethyl-3,3a-dihydro-3-methyl-benzofuro(3,2-c)	Isoxazole-3-carboxamide [N,N-	40. 38
(RS)-α-amino-3-hydroxy-5-methyl	Isoxazole-4-propionic acid monohydrate	40. 5
5,6,6a-triphenyl-3a,4-dihydrocyclopenta(2,3-d)	Isoxazoline <(p-chlorophenyl)-3a-methyl-4-oxo-	40. 57
	Ivalbin	53. 8
(-)-α-dihydro	Kaurene	54. 11
	Kekulene	30. 22
(2,6-bis(2-picolinoyl)-pyridine)-bis(ethylene-	Ketal) [2',6''-(1,4,7-trioxaheptane-1,7-diyl)-	40. 55
	Khellin	59. 5
	Koraiola pyridine solvate	53. 11
O-acetyl-	Kreysiginone-spiro-isomer methiodide	58. 63
14,15-bis-nor-8α-hydroxy-	Labd-11-en-13-one	54. 1
9,13-epoxy-	Labdane	54. 18
6-dihydrocobyrinic-pentamethyl ester-a-amide-c-	Lactam hemihydrate [6-amino-dicyano-5,	49. 30
e] (glutamyl-α-	Lactam)-histidinyl-proline tartrate monohydrat	48. 47

l-L-valyl-D-α-hydroxyisovaleryl-bis(D-valyl-L- Lactyl-L-valyl-D-α-hydroxyisovaleryl)) <lalany 50. 37
                                               Laminaribiose 45. 26
D-4,6-dideoxy-3-ketoallose)-13-(β-D-mycinose)- Lankamycin-11-α-hydroxyisovalerate ester <-(β- 50. 35
                                               Lasalocid 38.139
17-O-methyl- Latrunculin A 59. 35
                                               Laurencienyne 59. 17
(5,10,15,20-tetra-n-propylporphinato)- Lead(ii) 49. 3
bis(o-phenylene)-bis(pyridine-2,6-dialdimino) Lead(ii) diperchlorate dihydrate 34. 11
te > bis(diaquo-(pyridine-2,6-dicarboxylato)- Lead(ii)) pyridine-2,6-dicarboxylic acid solva 69. 29
                                               Leonitin 54. 18
nitro- Leonurine monohydrate 58. 6
t-butoxy-carbonyl-L-prolyl-L-iso Leucyl-glycine hydrate 48. 56
cyclo(L-leucyl-L-phenylalanyl-glycyl-D- Leucyl-D-phenylalanyl-glycyl) dihydrate....... 48. 80
cetyl-cyclo-bis(L-prolyl-L-valyl-L-ornithyl-L- Leucyl-L-phenylalanyl) .... [N-a 50. 40
alanyl-glycyl) dihydrate] cyclo(L- Leucyl-L-phenylalanyl-glycyl-D-leucyl-D-phenyl 48. 80
L- Leucyl-L-prolyl-glycine monohydrate 48. 42
ide] Lithium bis(dimethyl-(borane)-phosphine)methyl 64. 14
                                               Lithium di(t-butylmethylene-amine) 3. 42
                                               Lithium dibenzoylphosphide 1,2-dimethoxyethane 64. 62
o-2-yl-(N,N,N',N'-tetramethylethylenediamine)- Lithium dimer .... [2-methyl-1,3-dithi 67. 7
                                               Lithium hydrogen (+)-1-malate 2. 23
benzo-15-crown-5 Lithium phosphocyclocreatine dihydrate 64. 12
urea Lithium picrate dihydrate 38. 65
bis(α-cyclodextrin) Lithium sulfate........................... 8. 5
dazolidine dihydrate] Lithium tri-iodide octahydrate 45. 64
hexakis(trimethylsilyl Lithium 1-carboxymethyl-2-imino-3-phosphonoimi 64. 12
(μ-2,6-di-t-butyl-4-methylphenoxy)-diethyloxy- Lithium) 63. 1
                                               Lithium) .... [bis(67. 9
                                               Lofexidine 32. 39
                                               Loganin................................. 52. 4
                                               Lophocine 58. 75
-thioglycerol acid methyl ester)-4a,5-dihydro- Lumiflavin ....[3-methyl-4a-(methylene 36. 25
dithioglycerol acid ethyl ester)-4a,5-dihydro- Lumiflavin .... [3-methyl-4a-(methylene- 36. 31
3α,4α-epoxy-D:A-friedo-18β,19αH- Lupane 56. 5
(+)- Lupanine perchlorate 58. 16
(-)- Lupine hydrochloride 58. 2a
(+)-epi Lupinine 58. 1
                                               Lupinine-N-methyliodide.................... 58. 3
                                               Lycorine chlorohydrin 58. 17
deoxy- Lysophosphatidylcholine monohydrate........... 3. 55
1-β-D- Lyxofuranosyluracil..................... 47. 7
μ(6)-nitrido-nonakis(μ(2)-t-butyl-amino)-hexa- Magnesium 67. 8
methyl-(N,N,N',N'-tetramethylethylenediamine)- Magnesium ....[di 67. 2
ienyl)-(N,N,N',N'-tetramethylethylenediamine)- Magnesium .... [bis(2,4-dimethyl-2,4-pentad 67. 5
ne)] Magnesium bis(2,4-dinitrophenoxide) bis(pyridi 67. 6
ylimidazole)] Magnesium bis(2,4-dinitrophenoxide) bis(N-meth 67. 4
urea Magnesium bromide dihydrate.................. 8. 6
((S)-malato)-tetra-aqua- Magnesium hydrate 67. 1
cyclo-tris(prolyl-glycyl) Magnesium perchlorate trihydrate 48. 69
lithium hydrogen (+)-1- Malate 2. 23
calcium bis(hydrogen-1- Malate) hexahydrate....................... 2. 24
((S)- Malato)-tetra-aqua-magnesium hydrate 67. 1
carbinoxamine Maleate................................. 33. 70
imidazolium hydrogen Maleate................................. 32. 5n
tetra-n-butylammonium hydrogen dichloro Maleate................................. 2. 20
tetra-n-butylammonium hydrogen dichloro Maleate................................. 2. 19
ydroxy-1-methyl-2-phenylethyl)-N-methylamino)- Maleate.........................[dimethyl-2-(N-(2-h 1. 33
dibenzo(a,d)cycloheptene)-4-ammonium hydrogen Maleate <ethyl-spiro(cyclohex-2-ene-1,5'-(5H)- 29. 9a
strontium Maleate tetrahydrate 2. 21
3-(trans-2,3-dihydro-1,3-dimethyl-indol-2-yl)- Maleate [dimethyl 2-(1,3-dimethyl-indol-2-yl)- 35. 72
di-iodo-(Maleic acid)-thioanhydride 39. 2
dibromo-(Maleic acid)-thioanhydride 39. 1
2,3-bis(diphenylstibino)- Maleic anhydride 66. 18
calcium Malonate dihydrate 2. 16
urea Malonic acid 1. 8
1,10-diaza-4,7,13,16-tetraoxacyclo-octadecane Malononitrile .... [N,N'-bis(2-cyanoethyl)- 40. 46
1-phenylethylammonium Mandelate 2. 28
trans-(2R,5R)-2,5-dimethylpyrrolidinium(S)- Mandelate 32. 18
ny> (2-methylene-1-tetracarbonyl-2,3-diaza-1- Mangana-cyclobutan-4-one)-manganese pentacarbo 71. 25
methylammonium diaqua-trichloro- Manganese 3. 1
1,4-butanediyl-diammonium tetrachloro- Manganese 3. 27n
1,4-butanediyl-diammonium tetrachloro- Manganese 3. 28n
carbonyl-(μ-(methylenehydrazine-carbonyl))-di- Manganese .... [nona 71. 25
arbonyl-2,3-diaza-1-mangana-cyclobutan-4-one)- Manganese pentacarbonyl [(2-methylene-1-tetrac 71. 25
phthalocyaninato- Manganese(ii) 49. 2
bis(n-propylammonium) tetrachloro- Manganese(ii) 3. 13

,7,10,13,16-hexaoxacyclo-octadecane hexa-aquo- Manganese(ii) perchlorate .... [1,4 38. 48
<-p-tolylcarbenyl)-(carbonyl-cyclopentadienyl- Manganese) tetrafluoroborate dichloromethane > 71.173
aluminium)] tetracarbonyl- Manganese-((μ(2)-acetyl)-(μ(2)-bromo)-dibromo- 71.   3
(+)-2,4:3,5-di-O-methylene-D- Mannitol-1,6-di-trans-cinnamate............... 45. 54
)-2-acetamido-2-deoxy-α-D-glucopyrano> O-α-D- Mannopyranosyl-(1-3)-O-β-D-mannopyranosyl-(1-4 45. 49
lucopyrano> O-α-D-mannopyranosyl-(1-3)-O-β-D- Mannopyranosyl-(1-4)-2-acetamido-2-deoxy-α-D-g 45. 49
5-acetoxy-iso Marasman-7α,13-diol.......................... 53. 20
Marcfortine A 58. 73
4b-phenyl Mariontetraene 31. 63
allo Matrine..................................... 58. 12
N-oxy-allo Matrine..................................... 58. 15
Mazarronine................................. 54. 25
Mebendazole hydrobromide.................... 35. 41
Mecoprop.................................... 17. 12
Medazepam hydrochloride..................... 35. 42
Melampodin 53. 27n
di-calcium dihydrogen Mellitate nonahydrate 14. 5
pyro Mellitic dianhydride 38. 18
phenothiazine - pyro Mellitic dianhydride complex 60. 17
droxo)-bis(aqua-tribromo-tin(iv)) 1,8-epoxy-p- Menthane .... [bis(μ-hy 38. 26
roxo)-bis(aqua-trichloro-tin(iv)) 1,8-epoxy-p- Menthane .... [bis(μ-hyd 38. 27
R- Menthyl-methylphenylphosphine oxide.......... 64. 84
R-neo- Menthyl-methylphenylphosphine oxide.......... 64. 83
di Mephen 58. 64
),> 2-hydroxy-8-(2-hydroxy-2-hydroxymethyl-3- Mercapto-butyryloxy)-trans,trans-germacra-1(10 53. 25a
rubidium 8- Mercapto-quinoline-5-sulfonate 35. 10
phenyl-(1- Mercapto-1-methyl-ethyl)-phosphinic acid 64. 31
,4,6,6,8,8-octamethyl-2,4,6,8-tetrasila-1,5-di Mercura-cyclo-octane .... [2,2,4 63. 11
bis(triphenylsilyl)- Mercury..................................... 63. 50
N-tosylamino-octaethylporphinato-bis(chloro- Mercury) 49. 15
benzylammonium bis(dichloro- Mercury) chloride 19. 2
ene-9,10-cis-e> dimethyl-trans-3-endo-chloro- Mercury-4-acetoxy-tricyclo(4.2.2.0(2,5))dec-7- 31. 37
1,1,2,2-tetra Mesityl-disilane 63. 51
ra-aza-2,4,6,8-tetrasila-bicyclo(3.3.> 3,7-di Mesityl-2,2,4,4,6,6,8,8-octamethyl-1,3,5,7-tet 63. 44
diarsocine] 3,7-di Mesityl-3H,7H-1,5,2,4,6,8,3,7-dithia-tetra-aza 65. 10
syn-2,11-dithia(3.3) Metacyclophane 39. 63
syn-15,18-dimethoxy-(2,6)-p-benzoquinone(3.3) Metacyclophane 31. 53
4,16-tetramethoxy-1,2,3,10,11,12-hexathia(3.3) Metacyclophane .... [5,7,1 39. 62
4,16,18-hexamethyl-1,2,3,4,11,12-hexathia(4.2) Metacyclophane .... [6,8,10,1 39. 70
amethyl-1,2,9,10,17,18,25,26-octathia(2.2.2.2) Metacyclophane .... [4,6,12,14,20,22,28,30-oct 39. 87
deutero Methane..................................... 5. 1
t-butyl-dimethylsiloxy-aci-nitrophenyl Methane..................................... 63. 16
t-butyl-dimethylsiloxy-aci-nitrodiphenyl Methane..................................... 63. 28
(-)-(4S,8S)-α-bisabolol-8-p-phenylazo-phenyl Methane..................................... 53. 35
(3,5-bis(trifluoromethyl)-1,2,4-triazol-1-yl)- Methane....................[phenyl-bis 32. 53
carboxy Methane-(N-methyl-N-phenyl-sulfonamide)........ 1. 22
diethyl-bis(p-chlorophenoxy) Methane-phosphonate......................... 64. 80
nylsulfonylamino)-2,2a,3,4,5,6-hexahydro-2a,6- Methane-1H-azeto(1,2-a)(1)benzazocin-12-one <e 37. 14
(4-methyl-1-piperazinyl)propyl)-phenothiazine Methanesulfonic acid .... [2-chloro-10-(3- 41. 74
potassium nitroso-dicyano Methanide................................... 7. 1
<4-hydroxyimino-1,2,3,4,5,6,7,8-octahydro-1,8- Methano(1)benzazecin-13-one pyridine solvate 37. 16
S,4RS,6RS,8aSR)-5-methylene-octahydro-4H-3a,6- Methano-azulene-4-carboxylic acid .... [(3aR 31. 19
11,12,13,14-dodecahydro-4b-phenyl-6,9:10,13-di Methano-dicyclohepta(a,c)naphthalene <9,9a,10, 31. 63
2,4- Methano-glutamic acid monohydrate 48. 18
<,4,4a,5,6,7,8,8a-decahydro-1,4:5,8-exo,exo-di Methano-naphthalene-4a,8a-dicarboxylic anhydr> 38. 58
<4,4a,5,6,7,8,8a-decahydro-1,4:5,8-exo,endo-di Methano-naphthalene-4a,8a-dicarboxylic anhydr> 38. 59
1,2,3,4,4a,5,8,8a-octahydro-1,4:5,8-exo,exo-di Methano-naphthalene-4a,8a-dicarboxylic anhydr> 38. 57
e-10-syn-p-nitroben> 1endo,4endo:5exo,8exo-di Methano-1,2,3,4,4a,5,8,8a-octahydro-naphthalen 31. 50
S,9SR)-10β-ethyl-5,6,7,8,9,10-hexahydro-5α,9α- Methanobenzo-cyclo-octene-10α-carboxamide <,5R 31. 38
(4-hydroxyphenyl)-diphenyl Methanol 17. 28
1-(2-hydroxyphenyl)-diphenyl Methanol 17. 29
1-(2-hydroxynaphthyl)-diphenyl Methanol 24. 10
(4-hydroxy-3,5-dimethylphenyl)-diphenyl Methanol 17. 37
gastrin tetrapeptideamide hydrochloride Methanol diethyl ether solvate 48. 76
2,4- Methanoproline monohydrate 48. 17
2-carboxy-2,4- Methanopyrrolidine monohydrate 48. 17
hydrobromide dihydrate] (+)-5,10- Methenyl-5,6,7,8-tetrahydrofolic acid bromide 48. 63
hydrobromide dihydrate] (-)-5,10- Methenyl-5,6,7,8-tetrahydrofolic acid bromide 36. 33
1,3-bis(1,3,3-triethyl-indolenine-2-yl)-tri Methinium perchlorate 35. 82
hylamino)-3-ethoxy-1,2,5-triphenyl-4-aza-penta Methinium perchlorate .... [1,5-bis(dimet 19. 16
-bis(1,3,3-triethyl-indolenine-2-yl)-2-aza-tri Methinium tetrafluoroborate................. [1,3 35. 81
bis(1,3,3-trimethyl-indolenine-2-yl)-2-aza-tri Methinium tetrafluoroborate......... [1,3- 35. 71
(1,3,3-triethyl-indolenine-2-yl)-2-phospha-tri Methinium tetrafluoroborate....... [1,3-bis 35. 79
1,3,3-trimethyl-indolenine-2-yl)-2-t-butyl-tri Methinium tetrafluoroborate........ [1,3-bis(35. 75
<is(1,3,3-triethyl-indolenine-2-yl)-2-arsa-tri Methinium tetrafluoroborate methylene chlorid> 35. 78
<1,3,3-triethyl-indolenine-2-yl)-2-phospha-tri Methinium tetrafluoroborate methylene chlorid> 35. 80

D,L-	Methionine	48. 15
D,L-	Methionine	48. 14
N-acetyl-D,L-	Methionine-diethylamide........................	48. 35
inyl-tryptophanyl-glycine]	Methionyl-glutamyl-histidinyl-phenylalanyl-arg	48. 82
N-(t-butoxy-carbonyl)-L-	Methionyl-glycine-benzyl ester	48. 60
	Methyl red	13. 12
loro-2-nitrophenyl)-2,4-pentadienoate]	Methyl-(2E,4Z)-5-dimethylamino-2-nitro-4-(4-ch	15. 4
	Methylene oxalate	38. 1
lo(tetrakis(5-t-butyl-2-hydroxy-1,3-phenylene)	Methylene) toluene clathrate .... [cyc	61. 21
(Z)-2,3-dihydro-N-methyl-2-(4-nitrobenzoyl-	Methylene)-benzothiazole	41. 58
2,6-dimethyl-4-(diphenyl	Methylene)-cyclohexa-2,5-dienone	21. 14
2,6-dimethyl-4-(diphenyl	Methylene)-cyclohexa-2,5-dienone	21. 16
2,6-dimethyl-4-(diphenyl	Methylene)-cyclohexa-2,5-dienone	21. 15
3-(benzoyl(2-piperidyl-2-piperidylenium-ethyl)	Methylene)-indol-2-olate .... [N-acetyl-	35. 77
bis(trimethylacetyl-dichloro	Methylene)-trisulfane	11. 14
⟨ylamino-5α-t-butyl-5β-cyano-2-(cyano(t-butyl)	Methylene)-1β-methyl-3-oxa-9-thia-7-azabicycl⟩	41. 75
⟨ylamino-5β-t-butyl-5α-cyano-2-(cyano(t-butyl)	Methylene)-1β-methyl-3-oxa-9-thia-7-azabicycl⟩	41. 76
-methoxy-2-methyl-3-(benzo-1,3-thiazole-2-thio	Methylene)-1,4-naphthoquinone [2,3-dihydro-2	41. 72
4-benzyl-2,3,5,6-bis(tetra	Methylene)-4H-thiopyran........................	39. 80
lithium di(t-butyl	Methylene-amine)	3. 42
dimethyl-4,4-	Methylene-bis(phenylcarbamate)	16. 14
ne)]	Methylene-bis(1,3,5-trimethyl-4-imidazolin-2-o	32. 50
4,4'-	Methylene-boron-trihydride	64. 11
trimethylphosphonium-	Methylene-di-2-imidazolidine-thione	32. 31
1,1'-tri	Methylene-diammonium) (7,7,8,8-tetracyanoquino	60. 21
dimethane)] (N,N,N',N',N'-hexamethyl-hexa	Methylene-dioxy-flavan	38. 83
3-hydroxy-7-methoxy-3',4'-	Methylene-dithioethyl)-3α-(1'S)-(p-nitrobenzyl	39. 71
oxycarbonyl)oxy-ethyl-2-a⟩ (+-)-4β-(2',2'-tri	Methylene-dithioglycerol acid ethyl ester)-4a,	36. 31
5-dihydro-lumiflavin] 3-methyl-4a-(	Methylene-dithioglycerol acid methyl ester)-4a	36. 18
,5-dihydro-isoalloxazine] 3-methyl-4a-(	Methylene-furazan-N-oxide	40. 1
tri	Methylene-octahydro-4H-3a,6-methano-azulene-4-	31. 19
carboxylic acid] (3aRS,4RS,6RS,8aSR)-5-	Methylene-phthalide...........................	38. 14
3-nitro	Methylene-thioglycerol acid methyl ester)-4a,5	36. 25
-dihydro-lumiflavin] 3-methyl-4a-(	Methylene-tricyclo(4.4.0.0(3,8))decane-2,5-dio	31. 40
ne] 1β,3,4β,6β,8-pentamethyl-9-exo-	Methylene-tricyclo(5.3.0(1,4))dec-5-yl-pheny	28. 12
lcarbamate] 7-methyl-3-	Methylene-D-mannitol-1,6-di-trans-cinnamate....	45. 54
(+)-2,4:3,5-di-O-	Methylene-1-tetracarbonyl-2,3-diaza-1-mangana-	71. 25
cyclobutan-4-one)-manganese pentacarbony⟩ (2-	Methylene-1,3-bis(diphenyl)-carbodiphosphorane	64.146
1,3-tri	Methylene-2,3,5,6-tetrahydro-1,3-oxazin-4-one	40. 28
2-(p-chlorophenyl)-cis-5,6-tri	Methylene-2,3,5,6-tetrahydro-1,3-oxazin-4-one	40. 37
2-(p-chlorophenyl)-cis-5,6-penta	Methylene-3H-thiazolo(2,3-c)(1,2,4)thiadiazole	41. 39
5,6-dihydro-3-thiobenzoyl-	Methylene-4-methoxy-1-methyl-4-trichloromethyl	33. 30
-hexahydropyrimidin-2-on⟩ 5-chloro-6-dichloro	Methylene-4-triphenylmethyl-cyclohexa-2,5-dien	21. 22
e]	Methylene-5,6-dihydro-4,4-dimethyl-2-phenyl-4H	40. 24
1-diphenyl	Methylene-5,6-dihydropyrimidin-4(3H)-one	35. 30
-1,3,4-oxadiazinium bromide] (E)-6-bromo	Methylene-5,6-dihydropyrimidin-4(3H)-one	35. 31
2-phenyl-cis-5,6-tetra	Methylene-6-methylpyridine-N-oxide) dihydrate	33. 64
2-phenyl-trans-5,6-tetra	Methylene-8-trimethylsiloxy-tricyclo(6.2.1.0(1	63. 22
oxy-bis(2-	Methylenecamphor)	31. 62
,5))undecan-3-one] 2-methoxycarbonyl-9-	Methylenehydrazine-carbonyl))-di-manganese	71. 25
4,4'-(ethylenedi-imine)-bis(	Methyleneoxy)-6,12-epithiodibenzo(a,f)cyclodec	39. 73
nonacarbonyl-(μ-(	Methylenephosphoric acid	46. 2
en-7(6H)-one] syn-5,12-dihydro-5,14-(epithio-	Methylenethiourea)-di-tellurium(ii) perchlorat	70. 17
5-ethoxy-tri	Methylenethiourea)-di-tellurium(ii) perchlorat	70. 16
e]	Methylenomycin A p-bromobenzoate	50. 8a
e] hexakis(tri	Methylenomycin B p-bromobenzoate	50. 7
(+)-	Methylide	33. 29n
pyridinium-1-dicyano	Methylide	33. 28n
pyridinium-1-dicyano	Methylide	64. 14
lithium bis(dimethyl-(borane)-phosphine)	Methylide di-iodide...........................	64. 43
tris(trimethylphosphino)	Methylide monohydrate	33. 23
2-methylaminopyridinium-dinitro	Methylide) tetrahydrofuran solvate ⟨dium (diph	64.150
enylphosphonium-benzylide)-(diphenylphosphino-	Methylidene)-1,2,3,4-tetrahydro-naphthalene....	27. 29
4-anilino-3-methoxycarbonyl-1-(methoxycarbonyl	Methylidyne)-(μ(3)-sulfido)-triangulo-tri-cob⟩	71.169
⟨adienyl)-(μ(3)-(pentacarbonyl-chromium-thio)-	Miconazole hemihydrate	32. 65
	Microphyllone dimethyl ether	59. 36
	Mitomycin C dihydrate	50. 9
4-ethylpyridinium octa-	Molybdenum	33. 26
tetraethylammonium aquo-tetraiodo-oxo-	Molybdenum	3. 36
tetraethylammonium aquo-tetrabromo-oxo-	Molybdenum	3. 34
phenylphosphine)immonium tetracyano-oxo-dioxo-	Molybdenum .... [bis(tri	64.160
enylphosphonium bis(bis-μ-thio-dichloro-iron)-	Molybdenum .... [tetraph	64.129
pyridinium tetra-μ-(hydrogen-phosphato)-di-	Molybdenum chloride...........................	33. 10
is(trichloro-molybdenum) aquo-tetrachloro-oxo-	Molybdenum dihydrate ⟨hydrido)-bis(μ-chloro)-b	3. 40
um (μ(2)-sulfonyl)-(μ(4)-dithio)-octacyano-di-	Molybdenum hexahydrate [tetraphenylphosphoni	64.133
bis(piperidinium) hexabromo-diaqua-di-	Molybdenum(ii)	33. 16

tetrakis(2-carboxypyridinium) octacyano- Molybdenum(iv) 33. 18
bis(triethylammonium) dihydronium octacyano- Molybdenum(iv) 3. 31
dinium) di-μ-oxo-bis(oxo-tris(isothiocyanato)- Molybdenum(v)) .... [tetrakis(pyri 33. 11
-dioxo-(5,10,15,20-tetra-p-tolylporphyrinato)- Molybdenum(vi) toluene solvate .... [cis 49. 26
tramethylammonium) bis((μ(2)-thio)-dithio-oxo- Molybdenum)............................... [bis(te 3. 23
<ogen (μ-hydrido)-bis(μ-chloro)-bis(trichloro- Molybdenum) aquo-tetrachloro-oxo-molybdenum d> 3. 40
-(dicarbonyl-iron-(μ(2)-carbonyl)-tricarbonyl- Molybdenum)-bis(tricarbonyl-iron) <bonyl-iron) 3. 37
pentakis(pyridinium) tris(aquo-tetrabromo-oxo- Molybdenum)dibromide 33. 12
imethylammonium) dichloro-iron-bis(μ(2)-thio- Molybdenum-disulfide <ylphosphonium (benzyl-tr 64.123
<lphosphonium pentasulfur-iron-bis(μ-sulfido)- Molybdenum-disulfide dimethylformamide solvate 64.130
guanidinium tetra Molybdo-dimethylarsinate monohydrate 65. 2
tetraguanidinium α-dodeca Molybdo-silicate monohydrate 8. 7
dinactin Monactin 38.142
Monuron.................................... 8. 11
droxy-2-(2-tetrahydro-2-furylmethyl)-6,7-benzo Morphan <-)-(1R,5R,9R,2''S)-5,9-dimethyl-2'-hy 58. 41
l-2'-hydroxy-2-isobutyl-9,9-dimethyl-6,7-benzo Morphan hydrobromide ....[(-)-5-ethy 58. 49
-2'-hydroxy-9,9-dimethyl-2-phenethyl-6,7-benzo Morphan hydrobromide .... [(+)-5-ethyl 58. 64
methyl-2'-hydroxy-2-(4-methoxybutyl)-6,7-benzo Morphan hydrobromide [(-)-(1R,5R,9R)-5,9-di 58. 42
ethyl-2'-hydroxy-2-(2-methoxypropyl)-6,7-benzo Morphan hydrobromide <-(1R,5R,9R,2''S)-5,9-dim 58. 33
-2'-hydroxy-5,9-dimethyl-2-phenethyl-6,7-benzo Morphan hydrobromide hemihydrate .... [(+-) 58. 59
(-)-4-bromo-2,3-dimethoxy-N-methyl-6-oxo Morphinan 58. 39
-hydroxy-7-oxo-3-methoxy-17-methyl-5,6-dehydro Morphinan .... [(+)-4 37. 12
7,8-didehydro-4,5α-epoxy-17-methyl- Morphinan-3,6α-diol diacetate 58. 52
7,8-didehydro-4,5-epoxy-17-(prop-2-enyl) Morphinan-3,6-diol hydrobromide........... 58. 37
4-thiobenzoyl- Morpholine 40. 21
4-thiobenzoyl- Morpholine 40. 20
4-(2'-hydroxythiobenzoyl)- Morpholine 40. 22
tris(Morpholino)-arsine 65. 7
1-methylthio-3-(α- Morpholino)-benzylidene-triazene 40. 25
tris(Morpholino)-phosphine telluride........... 70. 11
9β-D-arabinofuranosyl-8- Morpholino-adenine dihydrate 47. 30
3β-hydroxy-16β- Morpholino-androst-5-en-17-one 51. 43
4- Morpholino-benzene-diazonium tetrafluoroborate 40. 16
di Morpholino-diselane........................ 40. 10
E-1-(5-nitro-2-furyl)-2- Morpholino-ethylene....................... 38. 21
(E)- Morpholino-p-nitrobenzamidoxime........... 40. 23
di Morpholino-tetraselane.................... 40. 12
di Morpholino-tetrasulfane................... 40. 9
di Morpholino-triselane 40. 11
ne sesquihydrate]   3S,6R,-6-methyl-5-oxothio Morpholino-3-carbonyl-L-histidinyl-L-prolylami 48. 51
dehydro Multiflorine perchlorate 58. 9
Multiflorine perchlorate hemihydrate 58. 10
Murol-4-on-9α-ol 53. 13
Murolan-4α,9β-diol 53. 15
Murolan-4β,9β-diol 53. 16
(β-D-mycinose)-lankamycin-11-α-hydro> 3-(β-L- Mycarose)-5-(β-D-4,6-dideoxy-3-ketoallose)-13- 50. 35
<se)-5-(β-D-4,6-dideoxy-3-ketoallose)-13-(β-D- Mycinose)-lankamycin-11-α-hydroxyisovalerate > 50. 35
e]   dodecasodium Myo-inositol-hexaphosphate octatriacontahydrat 46. 3
di Nactin monactin............................ 38.142
Nagilactone-B-2,7-bis-p-bromobenzoate 54. 36a
Nalidixic acid 35. 25
Nalorphine hydrobromide..................... 58. 37
N,O-diacetyl-4-hydroxy-nor Nantenine 58. 62
7,8-dihydro-1,6-dimethoxy Naphthacene-5,10,12(9H)-trione 29. 7
e) bis((μ(2)-bromo)-bromo-copper)] Naphthaceno(5,6-cd:11,12-c'd')-bis(1,2-dithiol 39. 66
N-methyl-2-oxy-1- Naphthaldiminato-diphenylborane............ 62. 18
2-chloro Naphthalene................................ 24. 2
cyclohepta(d,e) Naphthalene................................ 28. 4
cyclohepta(d,e) Naphthalene................................ 28. 2
cyclohepta(d,e) Naphthalene................................ 28. 5
cyclohepta(d,e) Naphthalene................................ 28. 3
2,3,6,7-tetrabromo Naphthalene................................ 24. 1
1-benzoyl-2-methyl-2-thia Naphthalene................................ 39. 65
2,6,7-trichloro-1,4-diacetoxy- Naphthalene................................ 24. 5
1,4-dioxacyclohept-2-eno(2,3-b) Naphthalene................................ 38. 51
1-phenyl-2,3-bis(diphenylphosphino)- Naphthalene................................ 64.169
-1-dioxothia-3-formyl-5,8-dimethyl-1,2-dihydro Naphthalene .... [2-selena 42. 6
ethoxycarbonylmethylidene)-1,2,3,4-tetrahydro- Naphthalene [4-anilino-3-methoxycarbonyl-1-(m 27. 29
b-phenyl-6,9:10,13-dimethano-dicyclohepta(a,c) Naphthalene <9,9a,10,11,12,13,14-dodecahydro-4 31. 63
-tetrakis(4-methylpyridine)-nickel(ii) 2-bromo Naphthalene clathrate   [bis(isothiocyanato) 61. 10
tetrakis(4-methylpyridine)-nickel(ii) 2-methyl Naphthalene clathrate   [bis(isothiocyanato)- 61. 11
2,6-dimethylnaphthalene perfluoro Naphthalene complex......................... 60. 19
2,6-dimethyl Naphthalene perfluoronaphthalene complex 60. 19
3-carboxy-2- Naphthalene-diazonium iodide monohydrate 24. 4
ammonium 1-anilino-8- Naphthalene-sulfonate monohydrate 27. 16
N,N',7-trimethyl-1-p-tolyl-1,2,3,4-tetrahydro- Naphthalene-tetracarboxylic-1,2,3,4-di-imide 36. 42

aβ,6,7,8,9,9aα,9b-dodecahydro-1H-cyclopenta(a) Naphthalene-1,5-dione ‹bα-methyl-2,3,3aα,4,5,5 28. 7
xo,8exo-dimethano-1,2,3,4,4a,5,8,8a-octahydro- Naphthalene-10-syn-p-nitrobenzoate ‹o,4endo:5e 31. 50
thoxyphenyl)-1-semicarbazono-6,7,8-trimethoxy- Naphthalene-2-one ‹dimethoxybenzyl)-4(2,4-dime 27. 33
3,26,29-heptaoxa(7(3,14))(5.5)orthocyclophane Naphthalene-2,3-diol monohydrate [1,5,12,16,2 38.113
                                               Naphthalene-2,7-diol 24. 3
6,7,8,8a-decahydro-1,4:5,8-exo,endo-dimethano- Naphthalene-4a,8a-dicarboxylic anhydride ‹a,5, 38. 59
4a,5,8,8a-octahydro-1,4:5,8-exo,exo-dimethano- Naphthalene-4a,8a-dicarboxylic anhydride ‹3,4, 38. 57
,6,7,8,8a-decahydro-1,4:5,8-exo,exo-dimethano- Naphthalene-4a,8a-dicarboxylic anhydride ‹4a,5 38. 58
1-phenoxy-1,2,2a,8b-tetrahydro-cyclobuta(a) Naphthalene-8b-carbonitrile................... 28. 10
y-5α,8α-dimethyl-4aβ,5,8,8aβ-tetrahydro-1(4H)- Naphthalenone .... [4α-hydrox 27. 10
aβ,8aβ-tetramethyl-4a,5,8,8a-tetrahydro-1(4H)- Naphthalenone   [4β-hydroxy-2,3,4 27. 12
1-ethyl-1,4-dihydro-7-methyl-4-oxo-1,8- Naphthapyridine-3-carboxylic acid 35. 25
                                               Naphthazarin 25. 1
ne]   Naphtho(b,e)dicyclobutane 29. 2
)]   Naphtho(1,8-cd:4,5-c'd')-bis(1,2,6)selenadiazi 42. 5
2-methyl Naphtho(1,8-cd:4,5-c'd')-bis(1,2,6-thiadiazine 41. 26
ane)]   spiro(3,4-dihydro-2H-1,5-dioxadi Naphtho(1,8-de)-1λ(2),2λ(4),3-triazine ... 36. 1
ane)]   spiro(3,4-dihydro-2H-1,5-dioxadi Naphtho(2,1-f:1,2-h)cyclononene-3,1'-cyclopent 38.123
hydroxy-1,3,8-trimethyl-3,4,6,9-tetrahydro-1H- Naphtho(2,1-f:1,2-h)cyclononene-3,1'-cycloprop 38.121
entene] Naphtho(2,3-c)pyran-6,9-dione [((1R,3S)-7,10-di 59. 14
4-diene t-butylamm› 2,3:4,5-bis(1,2-(3-methyl Naphtho(2,3-6,7)dicyclobutene(1,8:4,5)dicyclop 30. 4
4-diene] 2,3:4,5-bis(1,2-(3-methyl Naphtho))-1,6,9,12,15,18-hexaoxacycloeicosa-2, 61. 13
4-diene] 2,3:4,5-bis(1,2-(3-methyl Naphtho))-1,6,9,12,15,18-hexaoxacycloeicosa-2, 38.135
enyl)-2-hydroxy-3-N-(2-oxo-5-benzimidazolyl)-(Naphtho))-1,6,9,12,15,18-hexaoxacycloeicosa-2, 38.136
5α,8α-dimethyl-4aβ,5,8,8aβ-tetrahydro-1- Naphthoic acid)-amide ‹-butoxycarbonyl-1-azoph 35. 74
4aβ,6,7,8aβ-hexamethyl-4a,5,8,8a-tetrahydro-1- Naphthoquin-4α-ol 27. 10
,3,4aβ,8aβ-tetramethyl-4a,5,8,8a-tetrahydro-1- Naphthoquin-4α-ol .... [2,3, 27. 19
4aβ,6,7,8aβ-hexamethyl-4a,5,8,8a-tetrahydro-1- Naphthoquin-4β-ol .... [2 27. 12
                                               Naphthoquin-4β-ol .... [2,3, 27. 20
5,8-dihydroxy-1,4- Naphthoquinone 25. 1
2,3-di(cyclopropyl)-1,4- Naphthoquinone 20. 8
2,5-dihydroxy-3,8-dimethoxy-7-methyl-1,4- Naphthoquinone 25. 2
,6,7β,8aβ-hexamethyl-4a,7,8,8a-tetrahydro-1,4- Naphthoquinone .... [2,3,4aβ 25. 3
methoxy-7-methyl-2-(p-tosyloxyethylamine)-1,4- Naphthoquinone   [3-chloro-6- 25. 5
yl-N-(p-ethoxycarbonylphenyl)aminomethyl)-1,4- Naphthoquinone ....[2-chloro-3-(N-meth 25. 4
yl-3-(benzo-1,3-thiazole-2-thiomethylene)-1,4- Naphthoquinone [2,3-dihydro-2-methoxy-2-meth 41. 72
I-4a-(3-methyl-but-2-enyl)-5,8a-o-benzeno-1,4- Naphthoquinone ‹ahydro-11,14-dimethoxy-7-methy 59. 36
methyl-N-(biphenyl-2-yl)-N-(1- Naphthyl)-anthranilate 48. 77
methyl-N-(biphenyl-2-yl)-N-(1- Naphthyl)-anthranilate 48. 79
methyl-N-(biphenyl-2-yl)-N-(1- Naphthyl)-anthranilate 48. 78
1-(2-hydroxy Naphthyl)-diphenylmethanol 24. 10
(R)-N-acetyl-1-(1- Naphthyl)-ethylamine 24. 7
(RS)-N-acetyl-1-(1- Naphthyl)-ethylamine 24. 8
(R)-N-trifluoroacetyl-1-(1- Naphthyl)-ethylamine 24. 6
bis(2-methyl-1- Naphthyl)-ketone 24. 9
bis(2-methyl-1- Naphthyl)-methylacetate 24. 11
N,N'-(di-β- Naphthyl)-p-quinone-di-imine 24. 12
2,4-diamino-5-(1- Naphthyl)-6-methylpyrimidine methanol solvate 44. 31
pyridinium 1- Naphthylamine picrate 60. 1
11-hydroxy-6H-indolo(3,2,1-de)(1,5) Naphthyridin-6-one monohydrate 58. 5
1,8- Naphthyridine 35. 5
acetylated Napoleogenin 56. 10
R- Neo-menthyl-methylphenylphosphine oxide........ 64. 83
bis(phthalocyaninato)- Neodymium(iii) methylene chloride solvate 49. 33
bis(Neopentyloxy)-methylphenylphosphonium bromide 64. 86
Neopentyloxy-methyldiphenylphosphonium bromide 64. 97
tetrahydro- Neosoforamine 58. 14
(-)- Neoverrucosane-5β-benzoate 54. 32
methoxy- Nepenthone-E 25. 2
Nepetaefolin 54. 24
Nepetaefolinol 54. 3
3α-acetoxy-15β-hydroxy-7,16-seco-tri Nervita-7,11-diene 54. 20
L,D,L,D,L,D-enniatin B sodium Nickel nitrate hydrate 50. 22
l-L-methylvalyl-D-α-hydroxyisovaleryl) sodium Nickel nitrate hydrate ‹l-D-α-hydroxyisovalery 50. 22
tetrathiafulvalene bis(ethylenedithiolato)- Nickel(i) 60. 5
tct-2,3,7,8-tetrahydro-octaethylporphyrinato- Nickel(ii) 49. 8
ttt-2,3,7,8-tetrahydro-octaethylporphyrinato- Nickel(ii) 49. 9
octaethyl-1,2,3,7,8,20-hexahydroporphyrinato)- Nickel(ii)   [(trans,trans,cis,trans- 49. 12
octaethyl-1,2,3,7,8,20-hexahydroporphyrinato)- Nickel(ii) .... [(trans,trans,trans,trans- 49. 13
,9,12,13,16-octamethyl-tetrabenzoporphyrinato- Nickel(ii) iodide   [1,4,5,8 49. 18
is(isothiocyanato)-tetrakis(4-methylpyridine)- Nickel(ii) 2-bromonaphthalene clathrate.... [b 61. 10
is(isothiocyanato)-tetrakis(4-methylpyridine)- Nickel(ii) 2-methylnaphthalene clathrate   [b 61. 11
phosphine)iminium (μ-hydrido-bis(tricarbonyl- Nickel)............... [bis(triphenyl 64.152
bis(tetraphenylarsonium) dihydrido-dodeca- Nickel-carbonyl 65. 16
bis(tetraphenylphosphonium) dihydrido-dodeca- Nickel-carbonyl 64.127
bis(tetraphenylphosphonium) dihydrido-dodeca- Nickel-carbonyl................................ 64.128n

triphenylphosphine)immonium) dihydrido-dodeca- Nickel-carbonyl..................... [bis(bis(64.164
tris(tetraphenylarsonium) hydrido-dodeca- Nickel-carbonyl acetone solvate................. 65. 20
tris(tetraphenylarsonium) hydrido-dodeca- Nickel-carbonyl acetone solvate................. 65. 21n
hydrido)-octacarbonyl-tetrakis(μ(2)-carbonyl)- Nickel-tri-iron <nzyl-trimethylammonium (μ(3)- 3. 43
sodium Nigericine 50. 26
methyl-α- Nigeroside 45. 29
dihydro Nimbin 56. 1
tetraphenylarsonium pentachloro-oxo- Niobium(v) methylene chloride solvate 65. 17
N,N-dimethyl- Nitramine 3. 3n
N,N-dimethyl- Nitramine 3. 10n
N,N-dimethyl- Nitramine 3. 9n
N,N-dimethyl- Nitramine 3. 8
N,N-dimethyl- Nitramine 3. 7
N,N-dimethyl- Nitramine 3. 6
di-μ-aquo-bis(dioxo-di Nitrato-uranium(vi)) di-imidazole 32. 6
nylethyl)-4-piperidinyl)-N-phenyl-propanamide Nitric acid 2-propanol solvate <ethyl-1-(2-phe 33. 86
S,S-dimethylpentasulfur-hexa Nitride... 4. 1
tetra-n-butylammonium trisulfur-tri Nitride... 3. 49
is(triphenylphosphine-imino)-tetrasulfur-tetra Nitride............................... [b 64.167
triphenylphosphinimino-(trisulfur- Nitride) 64. 87
solvate] trisulfur-di Nitrido trifluoromethylsulfonate acetonitrile 11. 1
N,N'-(tetrasulfur-tetra Nitrido)-thiacyclopentane-di-imide 41. 9
tetraphenylarsonium (decachloro-μ- Nitrido-di-tungsten........................... 65. 19
s(triphenylphosphine)immonium nona-μ-carbonyl- Nitrido-hexa(carbonyl-cobalt) .... [bi 64.157
sium] μ(6)- Nitrido-nonakis(μ(2)-t-butyl-amino)-hexa-magne 67. 8
3- Nitro-3'-chloro-bis(1,2,4-triaz-5-ol) 32. 8
potassium Nitroso-dicyanomethanide 7. 1
tetramethylammonium (ethyl Nitrosolato-0)-triphenylborate 62. 12
Nodusmicin 50. 17
7-con-0-methyl Nogarol ethanol solvate........................ 50. 19
bicyclo(3.3.1) Non-1-en-3-one 31. 48
bicyclo(3.3.1) Non-1-en-3-one 31. 49
bicyclo(3.3.1) Non-1-en-3-one trans-syn-dimer 31. 47
ne)-1β-methyl-3-oxa-9-thia-7-azabicyclo(4.3.0) Non-7-en-4-one <cyano-2-(cyano(t-butyl)methyle 41. 76
ne)-1β-methyl-3-oxa-9-thia-7-azabicyclo(4.3.0) Non-7-en-4-one <cyano-2-(cyano(t-butyl)methyle 41. 75
4'-(9-anti-acetoxy-3-oxatricyclo(4.2.1.0(2,5)) Non-7'-ene <idecane-4,11-dien-9-one)-spiro-10, 38.119
,4'-(9-anti-acetoxy-3-oxatricyclo(4.2.10(2,5)) Non-7'-ene <ridecane-4,11-dien-9-one)-spiro-10 38.118
-hydroxy-8-phenyl-7-phospha-cis-bicyclo(4.3.0) Non-8-ene-7-oxide <acemic-7-benzyl-4-t-butyl-9 64.141
ino-2-phenyl-3-oxa-1-aza-trispiro(4.0.5.2.5.0) Nonadec-1-ene-4,13-dione .... [12-benzoylim 40. 56
(6S,7S,9R,10R)-6,9-epoxy- Nonadec-18-ene-7,10-diol 38.101
trispiro(2.0.2.0.2.0) Nonane 29. 1
7-methoxy-3,5,9-trioxabicyclo(4.3.0) Nonane 38. 10
oluenesulfonyl-7,8-diazatricyclo(4.2.1.0(3,7)) Nonane .... [2-(2'-propenyl)-8-t 37. 10
-oxa-1,5,6,8-tetra-aza-7-phosphabicyclo(3.3.1) Nonane .... [6,8-dimethyl-7-phenoxy-7-thioxo-3 64. 48
-2,exo-6-dihydroxy-2,6-dimethyl-bicyclo(3.3.1) Nonane ethylacetate clathrate .... [exo 61. 1
l-6-(2'-methyl-hexylidene)-1-azabicyclo(4.3.0) Nonane hydrochloride .... [8-hydroxy-8-methy 58. 25a
2,7-diaza-spiro(4.4) Nonane-1,6-dione 35. 3
lopentyl)-1-oxa-7,7,9,9-tetramethyl-spiro(4.4) Nonane-3,6-dione <-oxo-3,3,5,5-tetramethyl-cyc 38.108
thia Nonane-3,8-dione-1,1-dioxide 39. 24
thylamino-3,5,9-trioxa-4-phosphabicyclo(4.3.0) Nonane-4,7-dione .... [4-die 64. 33
3,4,6,7,9-hexa-aza-5λ(4)-phosphabicyclo(3.3.1) Nonane-5-thione.... [9-methyl-3,7-diphenyl-1, 64. 73
7-methyl-1,4-dithia-7-aza-spiro(4.4) Nonane-6,8-dione 39. 13
3,7-dimethyl-3,7-diazabicyclo(3.3.1) Nonane-9-spiro-5'-hydantoin................... 37. 7
cholesteryl Nonanoate 51. 82
,9).1(13,18))docosa-1(19),2,4,7,9,11,13,15,17- Nonene .... [5,7-diformyl-tetracyclo(9.8.1.1(3 31. 60
6α-formyl-15,16-dimethoxy-B- Nor-cis-erythrinan-8-one 58. 27
18-ethyl-17β-hydroxy-19- Nor-pregn-4-en-20-yn-3-one 51. 30
19- Nor-pregn-4-ene-3,20-dione 51. 20
17β-hydroxy-19- Nor-pregna-4,9-dien-20-yn-3-one 51. 18
17β-hydroxy-18-methyl-19- Nor-pregna-4,9-dien-20-yn-3-one 51. 23
17β-iodoacetoxy-4,4-dimethyl-19- Nor-5α-androstan-3-one 51. 33
1,2-seco-A-bis- Nor-5α-androstane-17β-acetate 51. 17
methyl 3β-acetoxy-17α-methyl-18- Nor-5α-androstane-17β-carboxylate 51. 44
14,15-bis- Nor-8α-hydroxy-labd-11-en-13-one 54. 1
anti-7,7-dibromo- Norcar-3-ene oxide 38. 8
anti-1,6-dimethyl-7,7-dibromo- Norcar-3-ene oxide 38. 16
Norgalanthamine hydrochloride dihydrate........ 58. 18
N,O-diacetyl-4-hydroxy- Nornantenine 58. 62
3-acetoxy-5-(2-nitrophenylthio)- Nortricyclene 31. 32
3-chloro-5-(2',4'-dinitrophenylthio)- Nortricyclene 31. 17
2-chloromethyl-5-(2',4'-dinitrophenylthio)- Nortricyclene 31. 22
NSC 298223 .... 59. 43
Obtusol acetate................................ 53. 19a
arbonyl)-5-hydroxy-1,5-dimethyl-bicyclo(2.2.2) Oct-2-en-6-one .... [2,3-bis(methoxyc 31. 27
5,6-dicyano-1-methoxy-bicyclo(2.2.2) Oct-2-ene 31. 13
S)-1,17-tribromo-2,6,8-trichloro-3,7-dimethyl- Oct-3-ene .... [(3E,2R,6R,7 5. 12a

xycarbonyl-methyl)-7,7-dimethyl-bicyclo(3.3.0) Oct-3-ene ⟨ydroxy-4-methoxycarbonyl-5-(dimetho 27. 25
phenyl-1-arsa-2,3,7-triaza-8-oxabicyclo(3.3.0) Oct-3-ene benzene solvate .... [2,4,6,7-tetra 65. 25
exo-phenyl-3-phosphabicyclo(3.2.1) Oct-6-en-3-one 64. 59
-oxo-spiro(oxirane-2,4'-tricyclo(3.3.0.0(2,8)) Oct-6'-ene)-1',8-dicarboxylate ⟨1-2'-methyl-3' 38. 93
arbonyl)-5-hydroxy-1,5-dimethyl-bicyclo(2.2.2) Oct-7-en-6-one .... [2,3-bis(methoxyc 31. 28
tetrakis(4-ethylpyridinium) dihydrido- Octacosa-oxygen-deca-vanadium 33. 25
,15-tetra-azapentacyclo(7.4.3.2(2,8).0.0(3,7)) Octadec-17-ene-4,6,14,16-tetrone) ⟨henyl-3,5,7 37. 20
5,15-dioxapentacyclo(7.4.3.2(2,8).0.0(3,7)) Octadeca-10,12,17-triene-4,6-dione 38. 75
-15-oxa-5-azapentacyclo(7.4.3.2(2,8).0.0(3,7)) Octadeca-10,12,17-triene-4,6-dione [5-methyl 38. 84
-3,5,7-triazapentacyclo(7.4.3.2(2,8).0.0(3,7)) Octadeca-10,12,17-triene-4,6-dione ⟨4,16-dioxa 38. 56
yl-5,15-diazapentacyclo(7.4.3.2(2,8).0.0(3,7)) Octadeca-10,12,17-triene-4,6,14,16-tetrone ⟨en 37. 19
iazapentacyclo(10.4.2.0(2,7).0(8,16).0(12,16)) Octadeca-2,4,6,14-tetraene S-oxide ⟨hia-1,15-d 42. 12
1,4,7,10,13,16-hexaoxacyclo- Octadecane 38. 45
hexaspiro(2.0.2.0.2.0.2.0.2.0) Octadecane 30. 5
4,7,10,13,16-pentaoxa-1-thiacyclo- Octadecane 42. 7
acetone solvat⟩ 1,4,7,10,13,16-hexaoxacyclo- Octadecane hexa-aquo-cobalt tetrachloro-cobalt 38. 47
1,4,7,10,13,16-hexaoxacyclo- Octadecane hexa-aquo-manganese(ii) perchlorate 38. 48
yanoethyl)-1,10-diaza-4,7,13,16-tetraoxacyclo- Octadecane malononitrile .... [N,N'-bis(2-c 40. 46
4,7,10,13,16-pentaoxa-1-thiacyclo- Octadecane potassium thiocyanate 42. 9
4,7,10,13,16-pentaoxa-1-thiacyclo- Octadecane rubidium thiocyanate............... 42. 11
4,7,10,13,16-pentaoxa-1-thiacyclo- Octadecane silver nitrate 42. 8
uran solvate] 1,4,7,10,13,16-hexaoxacyclo- Octadecane sodium dicyanophosphide tetrahydrof 38. 46
4,7,10,13,16-pentaoxa-1-thiacyclo- Octadecane sodium thiocyanate 42. 10
2-benzyl-1,7,10,16-tetraoxa-4,13-diazocyclo- Octadecane-3,14,18-trione 40. 49
3,3,12,12-tetramethyl-1,5,10,14-tetraoxacyclo- Octadecane-6,9,15,18-tetrone 38. 96
1-azabicyclo(2.2.2) Octane 37. 1
1-phenyl-bicyclo(2.2.2) Octane 31. 26
1-chloro-4-phenyl-bicyclo(2.2.2) Octane 31. 24
1-fluoro-4-phenyl-bicyclo(2.2.2) Octane 31. 25
meso-(RS)-1,1,3,6,8,8,8-octachloro- Octane 5. 9
2-methoxy-2,4-dimethyl-3,8-dioxabicyclo(3.2.0) Octane 38. 17
enzene)-N-hydrazo)-8-oxatricyclo(5.1.0.0(2,5)) Octane .... [4-(N'-(p-nitrob 38. 52
amethyl-2,4,6,8-tetrasila-1,5-dimercura-cyclo- Octane .... [2,2,4,4,6,6,8,8-oct 63. 11
ethylsulfonato-tetracyclo(3.3.0.0(2,6).0(3,8)) Octane [4-bromo-4-cyano-7-methoxycarbonyl-7-m 31. 16
)-8-oxo-2,2-dimethyl-3-oxa-1-azabicyclo(4.2.0) Octane ⟨⟨1-(p-nitrobenzyloxycarbonyl)oxy-ethyl 40. 45
5,7-tetra-aza-2,4,6,8-tetrasila-bicyclo(3.3.0) Octane ⟨esityl-2,2,4,4,6,6,8,8-octamethyl-1,3, 63. 44
robenzyl-endo-3-phenyl-3-phosphabicyclo(3.2.1) Octane bromide .... [exo-3-p-nit 64.111
yl)-6,7,8-trimethyl-2-thia-7-azabicyclo(3.3.0) Octane hemihydrate [5-hydroxy-1-(2-hydroxyeth 39. 39
cyclo- Octane-spiro-5'-hydantoin 35. 21
cis-cyclo- Octane-1,5-diol.............................. 22. 3
cyclo- Octane-1,5-dione 22. 1
cyclo- Octane-1,5-dione-dioxime 22. 2
1,5-diphosphabicyclo(3.3.0) Octane-1,5-disulfide 64. 20
syn-(1R,2R,4S)-4-methyl-spiro(bicyclo(2.2.2) Octane-2,1'-cyclohexane)-3',6-dione 31. 31
3,3,6,6-tetrachloro-2,4,5,7-tetrathia- Octane-2,2,7,7-tetraoxide 11. 5
thioxo-4-phenyl-4-aza-2,7-dithiobicyclo(3.3.0) Octane-7,7-dioxide.........................[3- 41. 35
richloro-1-(p-toluenesulfonyloxy)ethyl)-cyclo- Octene ....[3-(2,2,2-t 22. 6
7,12).0(15,20))tetracosa-3,5,8,10,16,18,21,23- Octene ⟨yclo(12.8.2.0(1,15).0(2,13).0(4,12).0(31. 66
6,7,8,9,10-hexahydro-5α,9α-methanobenzo-cyclo- Octene-10α-carboxamide ⟨,5RS,9SR)-10β-ethyl-5, 31. 38
trichloro-(8-ethoxy-4-cyclo- Octenyl)-tellurium 70. 6
5-amino-N- Octyl-1H-tetrazole-1-carboxamide 32. 36
4'-nitrophenyl-4-n- Octyloxybenzoate 13. 20
Ohchinolide A 59. 44
opyranosyloxy)-3β,16α,22α,24,28-penta-acetoxy- Olean-12-ene ⟨-3,4-di-O-angelyl-6-deoxy-β-gluc 56. 10
pseudo-racemic podopetaline Ormosanine 58. 50
N-acetyl-cyclo-bis(L-prolyl-L-valyl-L- Ornithyl-L-leucyl-L-phenylalanyl) 50. 40
7,2'-anhydro-β-D-arabinosyl- Orotidine 47. 12
5-n-propyl- Orotyl-L-histidinyl-L-prolylamine tetrahydrate 48. 59
ramethylammonium) chloro-penta(trichloro-tin)- Osmium .... [tetrakis(tet 3. 26
m (μ(5)-carbido)-iodo-pentadecacarbonyl-penta- Osmium ....[bis(triphenylphosphine)immoniu 3. 56
carbonyl-(phthalocyaninato)-(pyridine)- Osmium(ii) 49. 14
bis(triphenylphosphine)immonium) carbido-deca- Osmium-carbonyl.................... [bis(64.166
bis(bis(triphenylphosphine)immonium) octa- Osmium-docosacarbonyl 64.165
(+)- Ovalifolienalone 53. 21
anti-2,3;4,5-diepoxy-12- Oxa(4.4.3)propella-7,9-diene 38. 38
anti-2,3;4,5-syn-7,8;9,10-tetraepoxy-12- Oxa(4.4.3)propellane 38. 40
,3-diol monohydrate] 1,5,12,16,23,26,29-hepta Oxa(7(3,14))(5.5)orthocyclophane naphthalene-2 38.113
1,3,5,7-tetra-arsa-2,4,6,8-tetra Oxa-adamantane 65. 1
μ-thia-syn(CH(2),CH(3))-9,10-di Oxa-bimane 41. 30
μ-sulfone-syn(CH(2),CH(3))-9,10-di Oxa-bimane 41. 31
5,6,8,9,11,12-octahydro-1,4,7,10,13-benzopenta Oxa-cyclopentadecin sodium perchlorate [2,3, 38. 66
⟨,6,8,9,11,12-octahydro-1,4,7,10,13-benzopenta Oxa-cyclopentadecin) sodium tetraphenylborate 62. 15
2-methylthio-Δ(2)-1,3-diaza-6,9-di Oxa-cycloundecene hydroiodide 40. 13
1,20-bis(8-quinolyl-oxy)-3,6,9,12,15,18-hexa Oxa-eicosane rubidium iodide 35. 83
7-hydroxy-3-methoxy-6- Oxa-estra-1,3,5(10)-trien-17-one 51. 3
lene-acetal)] 3-methoxy-6- Oxa-estra-1,3,5(10)-triene-7,17-dione-17-(ethy 51. 19

1,2-bis(2- Oxa-indan-1-yl-oxy)ethane	38. 90
3,6-dimethyl-4,10-dihydroxy-2- Oxa-spiro(4.5)dec-7-ene-1,9-dione	59. 3
8-hydroxy-3-methoxy-3-methoxycarbonyl-2- Oxa-spiro(4.5)deca-6,9-diene	38. 41
1,11-bis(2-acetylaminophenoxy)-3,6,9-tri Oxa-undecane	17. 38
1,11-bis(2-nitrophenoxy)-3,6,9-tri Oxa-undecane potassium isothiocyanate	17. 35
1,11-bis(2-acetylaminophenoxy)-3,6,9-tri Oxa-undecane potassium thiocyanate monohydrate	17. 39
,13-dione] 12-benzoylimino-2-phenyl-3- Oxa-1-aza-trispiro(4.0.5.2.5.0)nonadec-1-ene-4	40. 56
-(trifluoromethyl)-3,4-bis(methoxycarbonyl)-6- Oxa-1-aza-5-phosphabicyclo(3.2.0)hept-3-ene <7	64.135
te] benzyl 4-methoxy-7-azido-3- Oxa-1-azabicyclo(4.2.0)octan-8-one-2-carboxyla	40. 36
yloxycarbonyl)oxy-ethyl)-8-oxo-2,2-dimethyl-3- Oxa-1-azabicyclo(4.2.0)octane <(1-(p-nitrobenz	40. 45
c acid] 8- Oxa-1-azabicyclo(4.3.0)nonan-9-one-2-carboxyli	40. 8
ethyl-1-phenyl-3,4:8,9-dibenzo-2,6,10,11-tetra Oxa-1-phosphatricyclo(5.3.1.0(1,5))undecane <m	64.117
4,7,10,13,16-penta Oxa-1-thiacyclo-octadecane	42. 7
te] 4,7,10,13,16-penta Oxa-1-thiacyclo-octadecane potassium thiocyana	42. 9
e] 4,7,10,13,16-penta Oxa-1-thiacyclo-octadecane rubidium thiocyanat	42. 11
4,7,10,13,16-penta Oxa-1-thiacyclo-octadecane silver nitrate	42. 8
4,7,10,13,16-penta Oxa-1-thiacyclo-octadecane sodium thiocyanate	42. 10
<to)-iron(ii) ethylthiolate 4,7,13,16,21-penta Oxa-1,10-diazabicyclo(8.8.5)tricosane benzene>	49. 27
<enylporphinato)-iron(ii) 4,7,13,16,21,24-hexa Oxa-1,10-diazabicyclo(8.8.8)hexacosane chloro>	49. 23
1,3,5-di-isopropylamino-2,4,6-tri Oxa-1,3,5-triphosphorine	64. 98
onane] 6,8-dimethyl-7-phenoxy-7-thioxo-2- Oxa-1,5,6,8-tetra-aza-7-phosphabicyclo(3.3.1)n	64. 48
3,3-dimethyl-1,5,13-tri Oxa-10,16-diazacycloeicosane-6,9,17,20-tetrone	40. 44
m thi> bis(10,22-dimethyl-1,4,7,13,16,19-hexa Oxa-10,22-diazacyclotetracosane) benzylammoniu	61. 6
e] 1,4,7,15-tetra Oxa-12,18-diazacyclodocosane-8,11,19,22-tetron	40. 42
<rophenyl)-8-(p-chlorophenylsulfonylamino)-13- Oxa-14-thia-1,15-diazapentacyclo(10.4.2.0(2,7>	42. 12
1,4,7,10,13-penta Oxa-14,16-cyclohexadecanedione	38. 34
tetrahydro-5H-pyrrolizino(1a,6b,6a-b,c)-1,8-di Oxa-15-cis-tridecene <,13-trimethyl-1a,2,3,6b-	58. 29
dine)-2',4'-dispiro-2,2''-(4,5-dimethyl-1,3-di Oxa-2-phosphole) <yl)-1,3,2,4-diazadiphospheti	64. 99
hylamino)-1,3-bis(trime> (4,5-dimethyl-1,3-di Oxa-2-phosphole)-2-spiro-2'-(cis-2,4-bis(dimet	64. 69
3-t-butyl-4-methyl-2-oxo-1- Oxa-2-thia-3-azacyclohexane	42. 3
4-methyl-9-p-chlorophenyl-7- Oxa-2,9-diazabicyclo(4.3.0)nona-2,4-dien-8-one	40. 26
6-methyl-9-p-chlorophenyl-7- Oxa-2,9-diazabicyclo(4.3.0)nona-2,4-dien-8-one	40. 27
-triene t-butylammonium pe> 3,6,9,12,15-penta Oxa-21-azabicyclo(15.3.1)heneicosa-1(20),17,19	40. 40
-triene-2,16-dione potassi> 3,6,9,12,15-penta Oxa-21-azabicyclo(15.3.1)heneicosa-1(21),17,19	40. 39
(1RS,8SR,10SR,4(15)Z)-4-ethylidene-5- Oxa-3-azatricyclo(8.4.0.0(3,8))tetradecane	40. 35
7))octadeca-10,12,17-trien> 5-methyl-14,16-di Oxa-3,5,7-triazapentacyclo(7.4.3.2(2,8).0.0(3,	38. 56
7-methoxy-3,5,9-tri Oxa-4-phosphabicyclo(4.3.0)nonan-4-one	64. 17
4-diethylamino-3,5,9-tri Oxa-4-phosphabicyclo(4.3.0)nonane-4,7-dione	64. 33
2-benzyl-1,7,10,16-tetra Oxa-4,13-diazocyclo-octadecane-3,14,18-trione	40. 49
α,3α,22β,23β-tetrahydroxy-24β-methyl-B-homo-7- Oxa-5α-cholestan-6-one	51. 71
deca-10,12,17-triene-4,6-dione] 5-methyl-15- Oxa-5-azapentacyclo(7.4.3.2(2,8).0.0(3,7))octa	38. 84
t-butyl-bis(2-(6-t-butyl-tetrahydro-4H-1,3-di Oxa-6-aza-2-phosphocine-2-oxy)ethyl)amine	64.136
8-dione] 2-hydroxy-4-isopropyl-7-benzyl-3- Oxa-6,9-diazatricyclo(7.3.0.0(2,6))dodecane-5,	40. 48
i> (2-oxo-3,3,5,5-tetramethyl-cyclopentyl)-1- Oxa-7,7,9,9-tetramethyl-spiro(4.4)nonane-3,6-d	38.108
<yano-2-(cyano(t-butyl)methylene)-1β-methyl-3- Oxa-9-thia-7-azabicyclo(4.3.0)non-7-en-4-one	41. 76
<yano-2-(cyano(t-butyl)methylene)-1β-methyl-3- Oxa-9-thia-7-azabicyclo(4.3.0)non-7-en-4-one	41. 75
b)-bis(1,1,4,4-tetramethylcycloheptane)-5,6-di Oxabicyclo(2.2.0)hex-2-ene .... [(1,4-b)-(2,3-	38.114
1,4-diphenyl-2,3-di Oxabicyclo(2.2.1)heptane	38. 81
(1S,4S)-N-acetyl-3-oxo-5-aza-2- Oxabicyclo(2.2.1)heptane	40. 4
methyl 3-(3-isocyano-6- Oxabicyclo(3.1.0)hex-2-en-5-yl)-acrylate	38. 20
2-methoxy-2,4-dimethyl-3,8-di Oxabicyclo(3.2.0)octane	38. 17
2,4,6,7-tetraphenyl-1-arsa-2,3,7-triaza-8- Oxabicyclo(3.3.0)oct-3-ene benzene solvate	65. 25
6β-bromo-7α-hydroxy-2- Oxabicyclo(3.3.0)octan-3-one	38. 9
6-acetoxymethyl-1,2,5-trimethyl-4- Oxabicyclo(3.3.0)octan-3-one	38. 55
xycarbonyl-2-methoxycarbonyl-9-methyl-9-aza-5- Oxabicyclo(4.3.0)nona-(1(6),2)-dien-4-one <tho	38. 80
7-methoxy-3,5,9-tri Oxabicyclo(4.3.0)nonane	38. 10
1,4,7,10,13,16-hexa Oxacyclo-octadecane	38. 45
ro-cobalt acetone solvat> 1,4,7,10,13,16-hexa Oxacyclo-octadecane hexa-aquo-cobalt tetrachlo	38. 47
rchlorate] 1,4,7,10,13,16-hexa Oxacyclo-octadecane hexa-aquo-manganese(ii) pe	38. 48
'-bis(2-cyanoethyl)-1,10-diaza-4,7,13,16-tetra Oxacyclo-octadecane malononitrile .... [N,N	40. 46
trahydrofuran solvate] 1,4,7,10,13,16-hexa Oxacyclo-octadecane sodium dicyanophosphide te	38. 46
3,3,12,12-tetramethyl-1,5,10,14-tetra Oxacyclo-octadecane-6,9,15,18-tetrone	38. 96
N,N'-dimethyl-1,7-diaza-4,10-di Oxacyclodecane benzylammonium thiocyanate	40. 18
N,N'-dimethyl-1,7-diazonium-4,10-di Oxacyclodecane perchlorate	40. 19
bis(1,2-(3-methylnaphtho))-1,6,9,12,15,18-hexa Oxacycloeicosa-2,4-diene .... [2,3:4,5-	38.135
bis(1,2-(3-methylnaphtho))-1,6,9,12,15,18-hexa Oxacycloeicosa-2,4-diene .... [2,3:4,5-	38.136
<is(1,2-(3-methylnaphtho))-1,6,9,12,15,18-hexa Oxacycloeicosa-2,4-diene t-butylammonium perc>	61. 13
1,4-di Oxacyclohept-2-eno(2,3-b)naphthalene	38. 51
5,6,8,9,11,12-octahydro-1,4,7,10,13-benzopenta Oxacyclopentadecin) sodium perchlorate <s(2,3,	38. 67
dibenzo(b,q)(1,4,7,10,13,16,19,22,25,28)deca Oxacyclotriacontane potassium thiocyanate	38.130
<-dibenzo(b,q)(1,4,7,10,13,16,19,22,25,28)deca Oxacyclotriacontin bis(sodium isothiocyanate)>	38.131
cis-10-methyl-1- Oxadecalin-2,5-dione	38. 23
ne-5,6-dihydro-4,4-dimethyl-2-phenyl-4H-1,3,4- Oxadiazinium bromide ....[(E)-6-bromomethyle	40. 24
3-methyl-4-(2-carboxyvinyl)-1,2,5- Oxadiazole-2-oxide	40. 3
5-phenyl-1,2,4- Oxadiazole-3-carboxamide	40. 14
e] 1,2,3- Oxadiazolo(4,3-c)(1,2,4)benzotriazinium-3-olat	40. 6

<-bis(2,2,4,4,6-pentamethylcyclo-trisiloxanyl- Oxadimethylsilyl)-tetramethyl-cyclodisilazane 63. 26
opentane)] spiro(3,4-dihydro-2H-1,5-di Oxadinaphtho(2,1-f:1,2-h)cyclononene-3,1'-cycl 38.123
opropane)] spiro(3,4-dihydro-2H-1,5-di Oxadinaphtho(2,1-f:1,2-h)cyclononene-3,1'-cycl 38.121
6-bromo-1,2,3,4,4a,9a-hexahydro-4,9-di Oxafluoren-2-one 38. 31
ridine)-bis(ethylene-ketal> 2',6''-(1,4,7-tri Oxaheptane-1,7-diyl)-(2,6-bis(2-picolinoyl)-py 40. 55
ridine)] 2',6''-(1,4,7-tri Oxaheptane-1,7-diyl)-(2,6-bis(2-picolinoyl)-py 40. 53
N,N'-dimethyl- Oxalamide 1. 13
di-N,N'-isopropyl-dithio- Oxalamide 1. 21
methylene Oxalate............. 38. 1
tellurium oxalate dihydrogen Oxalate dihydrate 2. 10
tellurium Oxalate dihydrogen oxalate dihydrate 2. 10
guanidinium hydrogen Oxalate monohydrate 2. 9
rubidium Oxalate monoperhydrate 2. 15n
potassium Oxalate monoperhydrate 2. 13n
rubidium Oxalate perdeuterate 2. 14n
sodium fluoro-oxo-di Oxalato-antimony oxonium hydrate 66. 2
nylethyl)-4-piperidinyl)-N-phenyl-propanamide Oxalic acid <thyl-4-(methoxycarbonyl)-1-(2-phe 33. 92
Oxalic acid dihydrate 1. 1
N-hydroxy Oxamide.................. 1. 3
octa(vinyl-sila-sesqui Oxane) 63. 21
0,12,17-triene-4,6-dione] 5,15-di Oxapentacyclo(7.4.3.2(2,8).0.0(3,7))octadeca-1 38. 75
5,14,16,18-tetraene-4,7,11,13-tetrone] 12- Oxapentacyclo(8.4.3.2(2,9).0.0(3,8))heptadeca- 38. 89
(5-(2-nitrophenoxy)-3-oxapentyloxy)phenoxy)-3- Oxapentane potassium thiocyanate [1,5-bis(2- 17. 43
hiocyanate] 1,5-bis(2-(5-(2-nitrophenoxy)-3- Oxapentyloxy)phenoxy)-3-oxapentane potassium t 17. 43
te] trans-2-hydroxy-4,5-dimethyl-1,3,2-di Oxaphospholane-2-sulfide imidazolium monohydra 64. 8
6-di-t-butyl-2-hydroxy-2-oxo-benzo(d)-1,3,2-di Oxaphosphole n-hexane solvate ....[4, 64. 68
1,2)-benzenophosphazine-10,2'-(1,3,2)-benzo-di Oxaphosphole) <6',7'-tetrachloro-2,8,15-(5,10(64.144
<bromomethyl-4,5-dimethyl-1,2,2-triphenyl-3,2- Oxaphosphonia-bicyclo(3.1.0)hexane tribromide 64.139
5,5-dimethyl-2-methoxy-2-oxo-1,3,2-di Oxaphosphorinane 64. 21
5,5,5',5'-tetramethyl-2,2'-bis(1,3,2λ(5)-di Oxaphosphorinane)-2,2'-disulfide 64. 41
4,7-di Oxatetracyclo(4.3.0.0(3,5).0(2,9))nonan-8-one 38. 7
exo-1,3,5,6,7,9-hexamethyl-4,8-di Oxatetracyclo(4.3.0.0(3,5).0(7,9))nonan-2-one 38. 53
endo-1,3,5,6,7,9-hexamethyl-4,8-di Oxatetracyclo(4.3.0.0(3,5).0(7,9))nonan-2-one 38. 54
2β-ol] 1β,5β,6β,11β-tetramethyl-3- Oxatetracyclo(5.4.0.0(2,6).0(4,11))undec-8-en- 38. 62
ne] 8,10,10-trimethyl-2- Oxatetracyclo(6.3.1.0(4,7).0(4,12))dodecan-3-o 38. 61
ne] 11-isopropyl-8-methyl-2- Oxatetracyclo(6.3.1.0(4,7).0(4,12))dodecan-3-o 38. 71
e p-bromophenylsulfonylhydr> 1α,6β,9α,10α-13- Oxatetracyclo(8.2.1.1(2,6))tetradec-11-en-8-on 38. 99
e p-bromophenylsulfonylhydr> 1β,6β,9α,10β-13- Oxatetracyclo(8.2.1.1(2,6))tetradec-11-en-8-on 38. 98
10-hydroxy-3- Oxatetracyclo(8.6.0.0(2,4).0(2,9))hexadecane 38. 73
6,6'-(3,6,9,12-tetra Oxatetradeca-1,14-dioxy)-2,2'-bipyridine 40. 51
1,8-diaza-3,6-di Oxatetradecane-2,7-dione 40. 17
1,13,16,19-tetra Oxatetraspiro(4.0.0.0.4.3.3.3)heneicosan-4-one 38. 85
cis-5-t-butyl-1,3,2-di Oxathian-2-oxide 42. 2
cis-5-methyl-trans-5-nitro-1,3,2-di Oxathian-2-oxide 42. 1
3,3,4,4-tetraphenyl-1,2- Oxathiolan-5-one-2-oxide 42. 13
2,5-dimethyl-1,2-dithiolo(1,5-b)-1,2- Oxathiole 39. 12
<,17,17-octamethyl-3,6,13,16-tetraoxo-21,22-di Oxatricyclo(18.2.1.1(8,11))tetracosa-8,10,18,> 38.129
,11-dien-9-one)-spiro-10,4'-(9-anti-acetoxy-3- Oxatricyclo(4.2.1.0(2,5))non-7'-ene <idecane-4 38.119
,11-dien-9-one)-spiro-10,4'-(9-anti-acetoxy-3- Oxatricyclo(4.2.10(2,5))non-7'-ene <ridecane-4 38.118
4-(N'-(p-nitrobenzene)-N-hydrazo)-8- Oxatricyclo(5.1.0.0(2,5))octane................ 38. 52
1,7,8,9-tetrachloro-3-phenyl-4-aza-5- Oxatricyclo(5.2.1.0(2,6))decane-3,8-diene 40. 32
1,7,8,9,10-pentachloro-3-phenyl-4-aza-5- Oxatricyclo(5.2.1.0(2,6))decane-3,8-diene 40. 31
-spiro(3(1)-formyl> 5-acetoxy-6,7-dimethyl-3- Oxatricyclo(5.4.0.0(2,4))undecan-10-one-9,2(1) 59. 16a
1α,4,5α,7α,9,10-hexamethyl-2β-methoxy-3- Oxatricyclo(5.4.0.0(4,11))undec-9-en-6-one 38. 87
<isopropyl-11α-methyl-1α-methoxy-6,9-diaza-12- Oxatricyclo(7.3.0.0(2,6))dodecane-7,10-dione 58. 22
oxycarbonyl)-1,3,7,9-tetrasila-2,8,13,14-tetra Oxatricyclo(7.3.1.1(3,7))tetradecane <akis(eth 63. 38
9,11-dihydroxy-2,10-dimethyl-8- Oxatricyclo(8.3.0.0(2,7))tridecan-6-one........ 38. 69
e] 9,9,13-trimethyl-8- Oxatricyclo(8.4.0.0(2,7))tetradec-2(7)-en-3-on 38. 78
2-oxo-2-methyl-3,5-di-t-butyl-1,3,2- Oxaza-4-phospholine........................... 64. 49
-2(bis(2-chloroethyl)amino)-4-phenyl-2H-1,3,2- Oxazaphosphorinane-2-oxide .... [cis 64. 60
-chloroethyl)-2-((2-chloroethyl)-amino)-1,3,2- Oxazaphosphorinane-2-oxide .... [S(-)-3-(2 64. 26a
thyl-5-(4'-biphenylyl)-2,3,5,6-tetrahydro-1,3- Oxazin-2-one .... [(3S,5R)-3-me 40. 43
)-cis-5,6-trimethylene-2,3,5,6-tetrahydro-1,3- Oxazin-4-one .... [2-(p-chlorophenyl 40. 28
cis-5,6-pentamethylene-2,3,5,6-tetrahydro-1,3- Oxazin-4-one .... [2-(p-chlorophenyl)- 40. 37
phenylamino)-3-(2,4,6-trimethylphenyl)-6H-1,2- Oxazine..........[6-dimethylamino-5-(4-nitro 40. 54
2,6-diphenyl-benzo(1,2-d:5,4-d')bis(Oxazole) 40. 50
-(progesterone)-(16α,17α-d)-(tetrahydro-1',2'- Oxazole) ....[N(α)-methoxy 51. 57
-(progesterone)-(16α,17α-d)-(tetrahydro-1',2'- Oxazole) ....[N(β)-methoxy 51. 56
azone monohydrate] 2-imino-6'-methyl-spiro(Oxazolidine-5,3'-piperidine)-2'-one-4-semicarb 40. 15
<yl 5,5-dimethyl-2-(2-phenoxymethyl-5-oxo-1,3- Oxazolin-4-ylidene)-1,3-thiazolidine-4-carbox> 41. 63
rophenyl)-6-methyl-2-oxo-2,3,3a,7a-tetrahydro- Oxazolo(4,5-b)pyridine .... [3-(p-chlo 40. 26
ophenyl)-7a-methyl-2-oxo-2,3,3a,7a-tetrahydro- Oxazolo(4,5-b)pyridine .... [3-(p-chlor 40. 27
ycarbonylamino-1'-methylethyl)-4,4-dimethyl-5- Oxazolone ....[2-(1'-benzylox 40. 41
2,5-diphenyl-phenanthro(9,10-b) Oxepin 38.133
3,3'-spiro-bis(3H-2,4-dihydro-benzo(1,4)di Oxepin)................................... 38. 82
o(cyclohexa-3,5-diene-1,6'-dibenzo(d,f)(1,3)di Oxepin)-2-one <-butyl-2',5,10'-trimethoxy-spir 38.138

-(4-nitrophenylmethyl)-1-phenyl-1H-2,3-benzodi Oxepin-5(4H)-one .... [1,4-epoxy-4 38.110
te)] triethylmethylammonium 8,8'- Oxido-3-cobalta-bis(η(5)-1,2-dicarbadodecabora 62. 5
Oxidopanamensin diacetate 54. 27
bis(acetone- Oximato)-methyl-chloro-silane 63. 6
tris(acetone- Oximato)-phenyl-silane 63. 19
cyclodecane- Oxime 23. 1
cyclododecane- Oxime 23. 3
cycloundecane- Oxime 23. 2
cyclotridecane- Oxime 23. 4
cyclotetradecane- Oxime 23. 5
E-diamantan-3-one- Oxime 31. 29
Z-diamantan-3-one- Oxime 31. 30
azo-diformaldehyde-di Oxime 10. 1
syn-γ-camphorquinone-di Oxime 31. 10
cyclo-octane-1,5-dione-di Oxime 22. 2
2,4,6-trimethylacetophenone- Oxime 10. 5
3,3,5,5-tetramethyl-cyclohexanone- Oxime 10. 4
3,3,5-trimethyl-5-phenyl-cyclohexanone- Oxime 21. 10
etophenone-O-(3-t-butylamino-2-hydroxypropyl)- Oxime ....[4-chloroac 10. 6
n-10-one-9,2(1)-spiro(3(1)-formyl-3(1)-methyl- Oxirane) <yl-3-oxatricyclo(5.4.0.0(2,4))undeca 59. 16a
2-oxo-spiro(bicyclo(2.2.2)octa-5,7-diene-3,2'- Oxirane)-1,6-dicarboxylate <-t-butyl-5-methyl- 38. 92
2-oxo-spiro(bicyclo(2.2.2)octa-5,7-diene-3,2'- Oxirane)-5,6-dicarboxylate <-t-butyl-1-methyl- 38. 91
dimethyl 6'-t-butyl-2'-methyl-3'-oxo-spiro(Oxirane-2,4'-tricyclo(3.3.0.0(2,8))oct-6'-ene> 38. 93
(1a,6,6a,6b-tetrahydro-2,6a-dimethyl-6-oxo-2H- Oxireno(a)pyrrolizin-4-yl)-2-butenamide <yl-N- 59. 6
sodium fluoro-oxo-dioxalato-antimony Oxonium hydrate.............................. 66. 2
etrakis(4-ethylpyridinium) dihydrido-octacosa- Oxygen-deca-vanadium .... [t 33. 25
porphyrinato-iron eth> (2-methylimidazole)-di Oxygen-meso-tetra(α,α,α,α-o-pivalamidophenyl)- 49. 35
Pachyclavulariadiol.......................... 54. 8
ylbenzylammonium) hexadecacarbonyl-tetra-iron- Palladium .... [bis(trimeth 3. 44
hylacetamide)] tetrachloro-μ-dichloro-di- Palladium bis(dimethylacetamidonium) bis(dimet 1. 16
o)-bis(di(3-trimethylsilyl-propyl)-tellurium)- Palladium(ii) .... [trans-bis(thiocyanat 70. 18
pyridine-N-hydroxide) tetrabromo-μ-dibromo-di- Palladium(ii) ....[bis(pyridine-N-oxide) bis(33. 6
bis(acetylacetonato)- Palladium(ii) tetrathiafulvalene 60. 12
deoxycholic acid Palmitic acid ethanol solvate 51. 55
3α,12α-dihydroxy-5β-cholan-24-oic acid Palmitic acid ethanol solvate 51. 55
Palythene monohydrate 59. 4a
Palythine trihydrate 48. 30
oxido Panamensin diacetate 54. 27
calcium D- Pantothenate bromide 2. 31
Parabanic acid urea.......................... 32. 4n
(3)Tetrathiafulvaleno(3) Paracyclophane 39. 69
(4)Tetrathiafulvaleno(4) Paracyclophane 39. 78
1,2,9,10,17,18-dehydro(2.2.2) Paracyclophane 31. 58
1,2,9,10,17,18-dehydro(2.2.2) Paracyclophane 31. 59
te] 1,6,20,25-tetra-aza(6.1.6.1) Paracyclophane hydrochloride durene tetrahydra 61. 18
Parsonsine 58. 61
Paspalicine.................................. 59. 39
Paspaline methanol solvate 59. 41
Paspalinine 54. 29
benzyl-(3SR,5RS,Z)-2-(2-phenylthioethylidene)- Penam-3-carboxylate................. [4-nitro 41. 78
6α-benzyl-6β-isocyano- Penicillanate 50. 16
enyl-3,4-methoxycarbonyl-1-aza-2-phosphacyclo- Pent-1-ene .... [5,5-dimethyl-2,2-diph 64.113
<dimethylsilyl-(1-trimethylsilyl-3,4-dimethyl- Pent-2-enyl))-benzoic acid p-bromobenzoylmeth> 63. 45a
2-(1-phenylthio)- Pent-3-ene-1-carboxylic acid 20. 7
1,4-bis(5-chloro- Pent-4-ynyloxy)benzene 17. 26
ylidene-5,5a,6,6a,7,12,12a,13,13a,14-decahydro Pentacene [5,14;7,12-bis(o-benzeno)-6,13-ethen 31. 67
tris(tetraethylammonium) hydrido- Pentacosacarbonyl-tetradeca-rhodium.......... 3. 5
tetracyclo(6.4.3.0.0(2,7)) Pentadec-4-en-13-one 31. 34
tetracyclo(6.4.3.0.0(2,7)) Pentadec-4-en-13-one 31. 33
2,6,10-trithiabicyclo(9.4.0) Pentadeca-1(11),12,14-triene 39. 43
6-phenyl-6-phospha-2,10-dithiabicyclo(9.4.0) Pentadeca-11(1),12,14-triene 64. 93
6,10-triphenyl-2,6,10-triphosphabicyclo(9.4.0) Pentadeca-11(1),12,14-triene ....[2, 64.148
yl-6-methyl-6-aza-2,10-diphosphabicyclo(9.4.0) Pentadeca-11(1),12,14-triene [2,10-diphen 64.140
<-triphenyl-6-aza-2,10-diphosphabicyclo(9.4.0) Pentadeca-11(1),12,14-triene acetone solvate 64.147
iphenylphosphine)immonium (μ(5)-carbido)-iodo- Pentadecacarbonyl-penta-osmium ....[bis(tr 3. 56
pentaspiro(2.0.2.0.2.0.2.0.2.0) Pentadecane................................. 30. 3
<nzoylamino-7-methoxycarbonyl-dispiro(5.1.5.2) Pentadecane-14,15-dione acetonitrile solvate 28. 16
perchloro-5,10-diphenyl-dibenzo(a,e) Pentalene carbon tetrachloride solvate 29. 10
3(S)-1-pyridinium-2-N,N-dibenzylamino-3-methyl Pentane p-toluenesulfonate .... [2(S), 33. 91
(2-nitrophenoxy)-3-oxapentyloxy)phenoxy)-3-oxa Pentane potassium thiocyanate [1,5-bis(2-(5- 17. 43
trans-2-azabicyclo(2.1.0) Pentane-3-(S)-carboxylic acid 50. 1
et> (+-)-(3R,1'R)-1-hydroxymethyl-2-(1-methyl Pentyl)-3-methylcyclohex-1-ene p-bromophenylur 21. 17
bis(neo Pentyloxy)-methylphenylphosphonium bromide 64. 86
cyanate] 1,5-bis(2-(5-(2-nitrophenoxy)-3-oxa Pentyloxy)phenoxy)-3-oxapentane potassium thio 17. 43
neo Pentyloxy-methyldiphenylphosphonium bromide.... 64. 97
her solvate] gastrin tetra Peptideamide hydrochloride methanol diethyl et 48. 76

urea hydrogen Peroxide 8. 3n
bis(dimethylbenzylsilyl)- Peroxide 63. 25
2-iodo-3'-bromo-dibenzoyl- Peroxide 13. 9
6-epipodo Petaline hydrobromide hydrate 58. 51
pseudo-racemic podo Petaline ormosanine..................... 58. 50
Peunicin 59. 24a
bis(Phenacetin) tri-iodide iodine 17. 14
p-bromo Phenacyl-stellatate.......................... 55. 4a
9-hydroxy Phenalenone 28. 1
<-dimethoxy-2-(2,6-dimethoxy-1-oxo-9-phenyl-5- Phenalenyl)-1-oxo-8-phenyl-1,2-dihydro-2-acen> 59. 46
cis-9-phenyl-10-methyl-9,10-dihydro Phenanthrene 28. 15
bis(9,10-diaza Phenanthrene) tetracyanoethylene 60. 15
hloro-(2,11-bis(diphenylarsinomethyl)-benzo(c) Phenanthrene-platinum .... [trans-dic 65. 26
1,4-diphenyl-5-(p-bromophenyl)-7-oxo-(2,3-l) Phenanthrene-bicyclo(2.2.1)hept-2-ene benzene> 31. 68
aα-dihydro-8-hydroxy-10β-methyl-4aα-propionyl- Phenanthrene-1,4,9(10H)-trione <ichloro-4aα,10 28. 13
-11,12,13,14,16,17-hexahydro-15H-cyclopenta(a) Phenanthrene-17,2'-spiro(1',3'-dithiolane) <l) 51. 45
di Phenanthro(4,3-a;3',4'-o)picene............. 30. 21
E-2,2,2',2'-tetraethoxy-Δ(3,3'(2H,2'H))-bis(Phenanthro(9,10-b)furan) 38.141
2,5-diphenyl Phenanthro(9,10-b)oxepin 38.133
8c,11a-dichloro-8b,8c,11a,11b-tetrahydro-exo- Phenanthro(9',10':3,4)cyclobuta(1,2-d)(1,3)di> 38. 79
nosine hydrate> 3,5,6,8-tetramethyl-N-methyl- Phenanthrolinium - 5-iodocytidylyl-(3'-5')-gua 60. 37
diacetone- Phenanthroquinone 38.103
5-ethyl-5,10-dihydro-10,10-diphenyl Phenaza-silane 63. 43
trion> 1,3-dimethyl-5-(10-phenyl-2,10-dihydro Phenazin-2-ylidene)-hexahydropyrimidine-2,4,6- 36. 41
tetrabenzo(a,c,h,j) Phenazine 36. 50
5,10-dihydro-5,10-diethyl Phenazine polyiodide 36. 17
m 7,7,8,8-tetracyanoquinodimethane] Phenazine 5,10-dihydro-5,10-dimethylphenaziniu 36. 2
phenazine 5,10-dihydro-5,10-dimethyl Phenazinium 7,7,8,8-tetracyanoquinodimethane 36. 2
Phenazocine hydrobromide hemihydrate 58. 59
1-(2,4-dichloro-β-((2,4-dichlorobenzyl)-oxy)- Phenethyl)-imidazole hemihydrate 32. 65
N- Phenethyl-granatanine-3-spiro-5'-hydantoin 37. 13
(+-)-5-ethyl-2'-hydroxy-9,9-dimethyl-2- Phenethyl-6,7-benzomorphan hydrobromide........ 58. 64
drate]   (+-)-2'-hydroxy-5,9-dimethyl-2- Phenethyl-6,7-benzomorphan hydrobromide hemihy 58. 59
4,4'-azoxy Phenetole 17. 27
perdeutero-4,4'-azoxy Phenetole 17. 25
2,4-dichloro Phenol 17. 2
3,4-dichloro Phenol 17. 3
2-amino-5-nitro Phenol 17. 4
2-isopropyl-5-methyl Phenol 17. 15
0-phenyl-10-oxo-5,10-dihydro-phenophosphazine Phenol ....[5-methyl-1 64.104
3,5-dichloro Phenol - 2,6-dimethylphenol complex............ 60. 2
s(p-nitrophenoxy)-phosphonium chloride p-nitro Phenol benzene solvate ....[methyl-tri 64.102
3,5-dichlorophenol - 2,6-dimethyl Phenol complex 60. 2
β-cyclodextrin m-methyl Phenol decahydrate 45. 65
bis Phenol A-diglycidyl ether 38.105
2,2'-ethylidene-bis(4-chloro-6-nitro Phenol)................................... 17. 22
2,2'-isopropylidene-bis(4-chloro-6-nitro Phenol).................................... 17. 23
,2,2-trichloroethylidene)-bis(4-chloro-6-nitro Phenol)............................. [2,2'-(2 50. 6
enzothiazol-2-ylio)-2-vinyl)-6-methoxy-4-nitro Phenolate ....[2-((3-ethyl-1,3-b 41. 65
1-p-hydroxyphenyl-thianium 1- Phenolate-thianium perchlorate 39. 36
bis(S-p-hydroxyphenyl-thiophanium) p- Phenolate-thiophanium bis(perchlorate) 39. 28
5-methyl-10-phenyl-10-oxo-5,10-dihydro- Phenophosphazine phenol.................... 64.104
1-oxo-1,2-dihydro-2,3-diaza Phenothiazine 41. 27
1-chloro-10-methyl-2,3-diaza Phenothiazine 41. 34
ex] Phenothiazine - pyromellitic dianhydride compl 60. 17
ethyl-1-piperazinyl)propyl)-2-trifluoro-methyl Phenothiazine dihydrochloride ....[10-(3-(4-m 41. 79
-chloro-10-(3-(4-methyl-1-piperazinyl)propyl)- Phenothiazine methanesulfonic acid .... [2 41. 74
3-(2-methoxy-10- Phenothiazinyl)-N,N,2-trimethyl-propanamine.... 41. 70a
10-(4-bromobenzyl)-10-phenyl- Phenoxaphosphonium bromide 64.137
magnesium bis(2,4-dinitro Phenoxide) bis(pyridine) 67. 6
magnesium bis(2,4-dinitro Phenoxide) bis(N-methylimidazole) 67. 4
anthracene 3,3',5,5'-tetrachloro-di Phenoxinone complex........................ 26. 1
(2,3-dichloro-4-(2-thienoyl)- Phenoxy)-acetic acid 39. 46
(4'-carbomethoxy-2'-nitro Phenoxy)-benzene 13. 10
3,5-di-iodo-4-(4-hydroxy-3,5-di-iodo Phenoxy)-benzoic acid bis(N-diethanolamine).... 13. 7
bis((μ-2,6-di-t-butyl-4-methyl Phenoxy)-diethyloxy-lithium) 67. 9
2(1-(2,6-dichloro Phenoxy)-ethyl)-imidazoline................ 32. 39
2(1-(2,6-dichloro Phenoxy)-ethyl)-imidazoline hydrochloride 32. 40
nzene solvate]   methyl-tris(p-nitro Phenoxy)-phosphonium chloride p-nitrophenol be 64.102
(+-)-2-(4-chloro-2-methyl Phenoxy)-propionic acid 17. 12
2-(4'-carbomethoxy-2'-amino Phenoxy)-1,3,5-trimethylbenzene 13. 17
2-benzyl-3-oxo-6-exo-(2,6-di-iodo-4-methyl Phenoxy)-2-azabicyclo(2.2.1)heptane........ 37. 15
1,5-bis(2-(5-(2-nitrophenoxy)-3-oxapentyloxy) Phenoxy)-3-oxapentane potassium thiocyanate.... 17. 43
potassium thiocyanate]   1,5-bis(2-(5-(2-nitro Phenoxy)-3-oxapentyloxy)phenoxy)-3-oxapentane 17. 43
8))tetradecane-3,5,10-tri> 14-ethoxy-8,14-di(Phenoxy)-3,4:10,11-dibenzo-tricyclo(7.5.0.0(2, 30. 18
3'-chloro-4'-(p-chloro Phenoxy)-3,5-di-iodo-salicylanilide............ 13. 18

                              1,11-bis(2-acetylamino Phenoxy)-3,6,9-trioxa-undecane 17. 38
iocyanate]   1,11-bis(2-nitro Phenoxy)-3,6,9-trioxa-undecane potassium isoth 17. 35
yanate monohydrate]   1,11-bis(2-acetylamino Phenoxy)-3,6,9-trioxa-undecane potassium thioc 17. 39
                              diethyl-bis(p-chloro Phenoxy)methane-phosphonate.................... 64. 80
                              2,2-bis(4-(p-nitro Phenoxy)phenyl)propane 17. 42
ethyl)-ethylenedioxy)->   triethylammonium di Phenoxy-(o-phenylenedioxy)-(1,2-bis(trifluorom 64.114
halene-8b-carbonitrile]   1- Phenoxy-1,2,2a,8b-tetrahydro-cyclobuta(a)napht 28. 10
sphabicyclo(3.3.1)nonane]   6,8-dimethyl-7- Phenoxy-7-thioxo-3-oxa-1,5,6,8-tetra-aza-7-pho 64. 48
                              p-nitro Phenoxyacetic acid 17. 8
                              p-bromo Phenoxyacetimine methyl ether 17. 9
))trideca>   10,11,12-tribromo-3,6-diethyl-13-(Phenoxycarbonyl)-3,6-diazatricyclo(7.4.0.0(2,7 36. 36
3-thiazolidine-4-c>   methyl 5,5-dimethyl-2-(2- Phenoxymethyl-5-oxo-1,3-oxazolin-4-ylidene)-1, 41. 63
                                 Phenoxythiophosphoryl-dihydrazide 64. 18
                              5- Phenyl-penta-2,4-dienoic acid 1. 25
i) diperchlorate dihydrate]   bis(o- Phenylene)-bis(pyridine-2,6-dialdimino) lead(i 34. 11
7,7,8,8-tetracyanoquinodimethane> N,N'-(1,2- Phenylene)-bis(salicylaldiminato)-copper(ii) - 60. 40
cyclo(tetrakis(5-t-butyl-2-hydroxy-1,3- Phenylene)methylene) toluene clathrate 61. 21
lphosphonium) tetrafluoroborate] meso-ortho- Phenylene-(methylphenylphosphine)-(methylpheny 64.110
                              meso-o- Phenylene-bis(methylphenylphosphine) 64.109
                              tetrabromo-p- Phenylene-bis(p-toluenesulfonate) 17. 33
               N,N,N',N'-tetramethyl-p- Phenylenediamine hexafluorobenzene complex 60. 13
nylenedioxy)->   triethylammonium diphenoxy-(o- Phenylenedioxy)-(1,2-bis(trifluoromethyl)-ethe 64.114
              3-t-butyl-9,10-dihydro-9,10-o-benzeno-9- Phospha-anthracene 64.120
cemic-7-benzyl-4-t-butyl-9-hydroxy-8-phenyl-7- Phospha-cis-bicyclo(4.3.0)non-8-ene-7-oxide <a 64.141
2,5,7,10,11,12-hexakis(trifluoromethyl)-1,6-di Phospha-hexacyclo(4.4.2.0(2,5).0(3,9).0(4,8).> 64. 76
fluoromethyl)-phenyl)-1,3,5,7-tetra-aza-4γ(5)- Phospha-spiro(3.3)heptane-2,6-dione <is(3-(tri 64.122
              1,3-bis(1,3,3-triethyl-indolenine-2-yl)-2- Phospha-trimethinium tetrafluoroborate 35. 79
n>   1,3-bis(1,3,3-triethyl-indolenine-2-yl)-2- Phospha-trimethinium tetrafluoroborate methyle 35. 80
ane]   10-thio-10- Phospha-1,4,7-triazatricyclo(5.2.1.0(4,10))dec 64. 19
1),12,14-triene]   6-phenyl-6- Phospha-2,10-dithiabicyclo(9.4.0)pentadeca-11(64. 93
                              1-methyl Phospha-2,3,5,6-bis-σ-carborano-cyclohexane.... 64. 23
                              1,1-diphenyl-1- Phospha-3,5-dithia-2,4,6-triazine 64. 50
              4-methyl-2,6,6-triphenyl-2,3-diaza-1- Phosphabicyclo(3.1.0)hex-3-ene 64.116
ethyl)-3,4-bis(methoxycarbonyl)-6-oxa-1-aza-5- Phosphabicyclo(3.2.0)hept-3-ene <7-(trifluorom 64.135
              exo-phenyl-3- Phosphabicyclo(3.2.1)oct-6-en-3-one........... 64. 59
      exo-3-p-nitrobenzyl-endo-3-phenyl-3- Phosphabicyclo(3.2.1)octane bromide 64.111
                              1,5-di Phosphabicyclo(3.3.0)octane-1,5-disulfide 64. 20
-7-phenoxy-7-thioxo-3-oxa-1,5,6,8-tetra-aza-7- Phosphabicyclo(3.3.1)nonane.... [6,8-dimethyl 64. 48
ethyl-3,7-diphenyl-1,3,4,6,7,9-hexa-aza-5λ(4)- Phosphabicyclo(3.3.1)nonane-5-thione   [9-m 64. 73
              7-methoxy-3,5,9-trioxa-4- Phosphabicyclo(4.3.0)nonan-4-one 64. 17
              4-diethylamino-3,5,9-trioxa-4- Phosphabicyclo(4.3.0)nonane-4,7-dione 64. 33
ene acetone s> 2,6,10-triphenyl-6-aza-2,10-di Phosphabicyclo(9.4.0)pentadeca-11(1),12,14-tri 64.147
ene]   2,6,10-triphenyl-2,6,10-tri Phosphabicyclo(9.4.0)pentadeca-11(1),12,14-tri 64.148
ene]   2,10-diphenyl-6-methyl-6-aza-2,10-di Phosphabicyclo(9.4.0)pentadeca-11(1),12,14-tri 64.140
thyl-2,2-diphenyl-3,4-methoxycarbonyl-1-aza-2- Phosphacyclo-pent-1-ene............ [5,5-dime 64.113
,4,4-triphenyl-3-methoxy-6-methyl-1,2-diaza-2- Phosphacyclohex-1(6)-ene .... [2 64.121
              cis-4-phenylcyclo Phosphamide................................... 64. 60
              3,5-(cyclo-N,N-diethyl Phosphamido)-2-O-methyl-1-deoxy-xylose 46. 6
clohexyl-1,2-bis(diethylamino)-1,2-dithioxo-di Phosphane .... [1,2-dicy 64.112
methylsiloxy)-methyl)-phenyl-(trimethylsilyl)- Phosphane .... [(p-chlorophenylimino-(tri 64.106
-pentacyclo(6.3.0.0(2,6).0(3,10).0(5,9))undeca Phosphane .... [4,7,11-tris(trimethylsilyl) 64. 35
              glycinium ortho Phosphate 48. 5
              tetramethylammonium hexafluoro Phosphate 3. 20
              tetrachloro-antimony(v)-dimethoxy Phosphate 66. 5
              cyclohexylammonium ethyl hydrogen Phosphate 21. 1
tetrachloro-antimony(v)-dimethylamino-methoxy Phosphate 66. 9
2,4-dichlorophenyl)-2-bromovinyl)-0,0-dimethyl Phosphate ....[0-(1-(46. 5
l-(2,3,4,5-tetraphenyl-cyclopenta-1,4-dienyl)- Phosphate .... [di-isopropy 46. 7
              potassium adenosine-5'-di Phosphate dihydrate........................... 47. 23
          2-amino-8-methyl-1-adenosine-5'-mono Phosphate dihydrate........................... 47. 27
              potassium dihydrouridine-3'-mono Phosphate hemihydrate 47. 11
              disodium uridine-5'- Phosphate heptahydrate 47. 4
              disodium D,L-α-glycero Phosphate hexahydrate 46. 1
              queuosine-5'-mono Phosphate hydrate 47. 34
              barium cytidine-5'- Phosphate hydrate 47. 9
ten-3-yl-aminomethyl)-7-deazaguanosine-5'-mono Phosphate hydrate <,5-cis-dihydroxy-1-cyclopen 47. 34
              5-nitro-2'-deoxyuridine-5'-mono Phosphate monohydrate 47. 10
              dodecasodium myo-inositol-hexa Phosphate octatriacontahydrate 46. 3
              disodium deoxyuridine-5'- Phosphate pentahydrate 47. 3
              sodium adenosine-(3'-5')-cyclic-mono Phosphate tetrahydrate 47. 13
              6-azauridine-5'- Phosphate trihydrate 47. 1
              guanosine-5'-mono Phosphate trihydrate 47. 22
              3',5'-cyclic-cytidine-mono Phosphate trihydrate 47. 8
              pyridinium tetra-μ-(hydrogen- Phosphato)-di-molybdenum chloride 33. 10
1-phenyl-3,4:8,9-dibenzo-2,6,10,11-tetraoxa-1- Phosphatricyclo(5.3.1.0(1,5))undecane <methyl- 64.117
                                 Phosphatriptycene 64.101

methyl-trithio-cyclotetra(λ(3),λ(5),λ(5),λ(5)- Phosphazane) .... [octa 64. 30
        octa(3,5-dimethylpyrazolyl)-cyclotetra Phosphazene............................ 64.171
            1-hydrido-1-isopropyl-tetrachloro-cyclotri Phosphazene.................... 64. 3
        2-diphenyl-4,4,6-trimethoxy-6-hydroxy-cyclotri Phosphazene...............[2, 64. 74
        ,6-tetrachloro-2,4-bis(dimethylamino)-cyclotri Phosphazene............[cis-2,4,6 64. 9
        ,6-tetrachloro-2,4-bis(dimethylamino)-cyclotri Phosphazene..........[trans-2,6 64. 10
                    hexa-aziridino-cyclotri Phosphazene benzene clathrate 61. 3
        2,2,4,4,6,6-hexa(1-aziridinyl)-cyclotri Phosphazene carbon tetrachloride anticlathrate 61. 2
                    tris(trichloro Phosphazeno)-carbenium hexachloro-antimony 64. 1
        1-p-chlorophenyl-3-p-tolyl-2-triphenyl Phosphazeno-propenone 64.151
        e]   2-(pentachloro-cyclotri Phosphazenyl)-1-phenyl-1,2-dicarba-dodecaboran 64. 29
        5-methyl-10-phenyl-10-oxo-5,10-dihydro-pheno Phosphazine phenol 64.104
        <',6',7'-tetrachloro-2,8,15-(5,10(1,2)-benzeno Phosphazine-10,2'-(1,3,2)-benzo-dioxaphosphol> 64.144
        5-phenyl-10,11-dihydro-dibenzo(b,f) Phosphepin-5-oxide 64.108
                    3,3-dimethyl-1,1-diphenyl Phosphetanium iodide 64. 81
        4,4-tetrakis(trifluoromethyl)-1,3-diaza-2,4-di Phosphetidine .... [1,2,3,4-tetramethyl-2,2, 64. 28
        2-trans-4-dithio-1,3,2,4-thia-aza-λ(5),λ(5)-di Phosphetidine <methyl-2,4-di(p-methoxyphenyl)- 64. 71
        mino)-1,3-bis(trimethylsilyl))-1,3,2,4-diazadi Phosphetidine <spiro-2'-(cis-2,4-bis(dimethyla 64. 69
        <mino)-1,3-bis(trimethylsilyl)-1,3,2,4-diazadi Phosphetidine)-2',4'-dispiro-2,2''-(4,5-dimet> 64. 99
        0,13,16-hexaoxacyclo-octadecane sodium dicyano Phosphide tetrahydrofuran solvate   [1,4,7,1 38. 46
                    lithium dibenzoyl Phosphide 1,2-dimethoxyethane 64. 62
        -(β-chloroethyl)-phenyl-(α-hydroxycyclohexyl)- Phosphinate.......................... [0 64. 66
        ylammonio-ethyl)-phenyl-(1-hydroxycyclohexyl)- Phosphinate iodide .... [0-(2-trimeth 64. 85
        ,2,2-tetrakis(trifluoromethyl)-ethyl-(dimethyl Phosphinate) .... [2-hydroxy-1,1 64. 27
                    tris(anilino)- Phosphine 64. 92
                    tri(cyanomethyl)- Phosphine 64. 15
        pentakis(methyl-dicarba-dodecaboranyl)-di Phosphine benzene solvate 64. 75
                    R-menthyl-methylphenyl Phosphine oxide............................ 64. 84
                    R-neo-menthyl-methylphenyl Phosphine oxide............................ 64. 83
                    tris(morpholino)- Phosphine telluride........................... 70. 11
        meso-o-phenylene-bis(methylphenyl Phosphine) 64.109
        roborate]   meso-ortho-phenylene-(methylphenyl Phosphine)-(methylphenylphosphonium) tetrafluo 64.110
        rbenyl)-(carbonyl-cyclopentad> (bis(trimethyl Phosphine)-platinum)-(μ-carbonyl)-(μ-p-tolylca 71.173
        <l)-carbonyl-(p-tolylisocyanide)-bis(triphenyl Phosphine)-ruthenium(ii) perchlorate chlorofo> 71.270
        o)-hexacarbonyl-di-iron-hexaca> bis(triphenyl Phosphine)iminium (μ(2)-carbonyl)-(μ(2)-hydrid 64.156
        o)-tricarbonyl-iron-nonacarbon> bis(triphenyl Phosphine)iminium (μ(2)-carbonyl)-(μ(2)-hydrid 64.155n
        nickel)]   bis(triphenyl Phosphine)iminium (μ-hydrido)-bis(tricarbonyl- 64.152
        yl-penta-rhodium]   bis(triphenyl Phosphine)iminium (μ-pentacarbonyl)-decacarbon 64.158
        onyl-ruthenium-tetra-iridi> bis(bis(triphenyl Phosphine)iminium) nonacarbonyl-hexa-μ(2)-carb 64.163
                    bis(triphenyl Phosphine)iminium) tridecacarbonyl-tetra-iron 64.161
        cacarbonyl-penta-osmium]   bis(triphenyl Phosphine)immonium (μ(5)-carbido)-iodo-pentade 3. 56
        yl-chromium)]   bis(triphenyl Phosphine)immonium μ-deuterido-bis(pentacarbon 64.153n
        a(carbonyl-cobalt)]   bis(triphenyl Phosphine)immonium nona-μ-carbonyl-nitrido-hex 64.157
        v)]   bis(triphenyl Phosphine)immonium oxo-tetrachloro-technetium(64.159
        enum]   bis(triphenyl Phosphine)immonium tetracyano-oxo-dioxo-molybd 64.160
        yl]   bis(triphenyl Phosphine)immonium) carbido-deca-osmium-carbon 64.166
        tris(tricarbonyl-ruthenium)-c> bis((triphenyl Phosphine)immonium) carbonyl-(tri-μ-carbonyl)- 64.154
        rbonyl]   bis(bis(triphenyl Phosphine)immonium) dihydrido-dodeca-nickel-ca 64.164
                    bis(bis(triphenyl Phosphine)immonium) octa-osmium-docosacarbonyl 64.165
        lithium bis(dimethyl-(borane)- Phosphine)methylide............................. 64. 14
                    bis(triphenyl Phosphine-imino)-tetrasulfur-tetranitride 64.167
        1-(triphenyl Phosphine-imino)-1-chloro-2,2-dicyanoethylene 64.115
        (chloromethyl)-diphenyl Phosphine-oxide.............................. 64. 57
        diphenyl-(3-hydroxypropyl) Phosphine-oxide.............................. 64. 72
        diphenyl-(5,6-dichloro-1,3-benzodioxol-(2))- Phosphine-oxide.............................. 64.100
        triphenyl Phosphine-oxide hemiperhydrate 64. 90
        1,3-benzodithiolyl-2-(diphenyl Phosphine-oxide) 64.103
        bis(triphenyl Phosphine-oxide) monohydrate hydrogen bromide 64. 91
        perchlorate]   triphenyl Phosphine-oxide-triphenylphosphonium-hydroxide 64. 89
        perchlorate]   triphenyl Phosphine-oxide-triphenylphosphonium-hydroxide 64. 88
        trimethyl Phosphine-selenide 64. 5
        Phosphine-triyl-triacetonitrile................ 64. 15
        di(cyclohexyl)- Phosphinic acid............................ 64. 53
        phenyl-(1-mercapto-1-methyl-ethyl)- Phosphinic acid............................ 64. 31
        (1-hydroxy-1-(2-hydroxyphenyl)-ethyl)-phenyl- Phosphinic acid............................ 64. 65
        phenyl-α-hydroxy-cyclohexyl- Phosphinic acid chloranhyrdide 64. 51
        bis(phenyl-(t-butyl)- Phosphinic amide) hydrogen chloride........... 64. 40
        bis(2,4,6-tri-t-butylphenyl)- Phosphinic-chloride......................... 64.168
        2-(cyano Phosphinidene)-1,3-dimethyl-benzimidazoline.... 64. 37
        triphenyl Phosphinimino-(trisulfur-nitride) 64. 87
        1-phenyl-2,3-bis(diphenyl Phosphino)-naphthalene 64.169
        glycinium ethyl est> 2,2'-bis(o-(oxo-diphenyl Phosphino)-phenyloxy)-diethyl ether D,L-phenyl 64.170
        onyl-ruthenium-tetra-iridi> bis(triphenyl Phosphino)iminium) heptacarbonyl-μ(2)-hexacarb 64.162
        tris(trimethyl Phosphino)methylide di-iodide 64. 43
        <ium (diphenylphosphonium-benzylide)-(diphenyl Phosphino-methylide) tetrahydrofuran solvate 64.150
        bis(diphenyl Phosphino-selenoyl)-acetylene 64.143

tris(1,2-bis(diphenyl Phosphino-selenoyl)ethane) p-xylene clathrate   61.   9
bis(diphenyl Phosphino-thioyl)-acetylene.................... 64.142
methyl 2-methyl-3-(benzyl(phenyl) Phosphinyl)-butyrate................... 64.105
oxylate]   methyl-6-(diphenyl Phosphinyl)-3,4-dimethyl-cyclohex-3-ene-1-carb 64.119
methyl β-D-ribopyranoside Phosphite triester 64. 16
(2-(6-t-butyl-tetrahydro-4H-1,3-dioxa-6-aza-2- Phosphocine-2-oxy)ethyl)amine   [t-butyl-bis 64.136
lithium Phosphocyclocreatine dihydrate 64. 12
2,3-dibromo-1-phenyl Phospholane-1-oxide............................ 64. 38
2-hydroxy-2-methyl-1-phenyl Phospholane-1-oxide............................ 64. 47
4RS)-3-chloro-4-hydroxy-3,4-dimethyl-1-methoxy Phospholane-1-oxide............... [(1SR,3SR, 64. 25
trans-2-hydroxy-4,5-dimethyl-1,3,2-dioxa Phospholane-2-sulfide imidazolium monohydrate   64.   8
(1R,3S)-(-)-trans-1-benzyl-3-methyl-1-phenyl Phospholanium iodide 64. 96a
i-t-butyl-2-hydroxy-2-oxo-benzo(d)-1,3,2-dioxa Phosphole n-hexane solvate .... [4,6-d 64. 68
2',4'-dispiro-2,2''-(4,5-dimethyl-1,3-dioxa-2- Phosphole) <yl)-1,3,2,4-diazadiphosphetidine)- 64. 99
)-benzenophosphazine-10,2'-(1,3,2)-benzo-dioxa Phosphole) <6',7'-tetrachloro-2,8,15-(5,10(1,2 64.144
no)-1,3-bis(trime>   (4,5-dimethyl-1,3-dioxa-2- Phosphole)-2-spiro-2'-(cis-2,4-bis(dimethylami 64. 69
2-carboxy-3,4-dimethyl-1-phenyl Phosphole-sulfide....................... 64. 58
methyl-3,4-dimethyl-1-phenyl Phosphole-2-carboxylate................... 64. 64
4-hydroxy-1-phenyl-2- Phospholene-1-oxide...................... 64. 39
2-oxo-2-methyl-3,5-di-t-butyl-1,3,2-oxaza-4- Phospholine......................... 64. 49
dimethyl-1-hydroxy-1-cycloheptane- Phosphonate........................... 64. 34
diethyl-bis(p-chlorophenoxy)methane- Phosphonate........................... 64. 80
D,L-diethylanilino-(3-hydroxybenzyl)- Phosphonate........................... 64. 82
diethyl-(5,6-dichloro-1,3-benzodioxole-(2))- Phosphonate........................... 64. 44
omomethyl-4,5-dimethyl-1,2,2-triphenyl-3,2-oxa Phosphonia-bicyclo(3.1.0)hexane tribromide <br 64.139
3-aminopropyl Phosphonic acid......................... 64.   6
(E)-1,4-dibenzyl-1,4-diphenyl-1,4-di Phosphonio-cyclohexane dibromide........ 64.149
cyano-di-molybdenum hexahydrate]   tetraphenyl Phosphonium (μ(2)-sulfonyl)-(μ(4)-dithio)-octa 64.133
o-iron-bis(μ(2)-thio)-molybdenum>   tetraphenyl Phosphonium (benzyl-trimethylammonium) dichlor 64.123
hane)]   ethyltriphenyl Phosphonium - bis(7,7,8,8-tetracyanoquinodimet 60. 41
tetraphenyl Phosphonium bis(azido)-iodine 64.124
bdenum]   tetraphenyl Phosphonium bis(bis-μ-thio-dichloro-iron)-moly 64.129
neopentyloxy-methyldiphenyl Phosphonium bromide.......................... 64. 97
bis(neopentyloxy)-methylphenyl Phosphonium bromide.......................... 64. 86
10-(4-bromobenzyl)-10-phenyl-phenoxa Phosphonium bromide.......................... 64.137
vate]   methyl-tris(p-nitrophenoxy)- Phosphonium chloride p-nitrophenol benzene sol 64.102
trichloro-(p-methoxyphenyl)- Phosphonium hydrogen dichloride........... 64. 24
lybdenum-disulfide dimethylforma>   tetraphenyl Phosphonium pentasulfur-iron-bis(μ-sulfido)-mo 64.130
ngsten-disulfide dimethylformami>   tetraphenyl Phosphonium pentasulfur-iron-bis(μ-sulfido)-tu 64.131
vate]   tetraphenyl Phosphonium platinum-carbonyl acetonitrile sol 64.132
tetra(t-butyl)- Phosphonium tetrafluoroborate 64. 79
(triphenylmethyl)-trimethyl Phosphonium tetrafluoroborate 64.118
nyl-(1-ethoxycarbonyl-2-methyl-2-olato-vinyl)- Phosphonium tetrafluoroborate <sphonium triphe 64.134
<1-(1-ethoxycarbonyl-2-methyl-2-hydroxyvinyl)- Phosphonium triphenyl-(1-ethoxycarbonyl-2-met> 64.134
bis(tetraphenyl Phosphonium) decabromo-di-tellurium(iv)........ 64.125
bis(tetraphenyl Phosphonium) dihydrido-dodeca-nickel-carbonyl   64.127
bis(tetraphenyl Phosphonium) dihydrido-dodeca-nickel-carbonyl   64.128n
onyl-hexa-rhodium]   bis(tetraphenyl Phosphonium) hexa-μ-carbonyl-carbido-heptacarb 64.126
e clathrate]   bis(triphenylbenzyl Phosphonium) tetrachloro-cadmium dichloroethan 61.   8
henylene-(methylphenylphosphine)-(methylphenyl Phosphonium) tetrafluoroborate   [meso-ortho-p 64.110
α-(triphenyl Phosphonium)-benzoylacetone tetrafluoroborate 64.145
imethylsilylimi>   (methyl-(tris(dimethylamino) Phosphonium)amino)-bis(trimethylsilylamino)-tr 64.107
ylide) tetrahydrofuran solv>   sodium (diphenyl Phosphonium-benzylide)-(diphenylphosphino-meth 64.150
triphenylphosphine-oxide-triphenyl Phosphonium-hydroxide perchlorate 64. 89
triphenylphosphine-oxide-triphenyl Phosphonium-hydroxide perchlorate 64. 88
trimethyl Phosphonium-methylene-boron-trihydride 64. 11
endo-2-dimethyl Phosphono-exo-2-hydroxy-(-)-camphane 64. 54
lithium 1-carboxymethyl-2-imino-3- Phosphonoimidazolidine dihydrate 64. 12
N-(Phosphonomethyl)-glycine 64.   4
O-ethyl-O-(4-nitrophenyl)-benzene- Phosphonothioate 64. 63
O-methyl-O-(4-bromo-2,5-dichlorophenyl)-phenyl Phosphonothioate 64. 56
2-O-methyl-xylitanylidene diethyl-cyclo Phosphoramidate......................... 46.   6
dodecamethyl-bis(imido)-tri Phosphoramide monohydrate 64. 55
bis(1,4,7,10-tetra-azacyclododecane)- Phosphorane........................... 64. 78
1,3-trimethylene-1,3-bis(diphenyl)-carbodi Phosphorane......................... 64.146
-bis(trimethylsilylamino)-trimethylsilylimino- Phosphoranide <imethylamino)phosphonium)amino) 64.107
N-(triphenyl Phosphoranylidene)-benzamide 64.138
5-ethoxy-trimethylene Phosphoric acid......................... 46.   2
adenosine-5'-di Phosphoric acid trihydrate 47. 24
y)-(1,2-bis(trifluoromethyl)-ethenylenedioxy)- Phosphoride <monium diphenoxy-(o-phenylenediox 64.114
5,5-dimethyl-2-methoxy-2-oxo-1,3,2-dioxa Phosphorinane........................... 64. 21
,6,6-tetramethyl-4-phenyl-4-oxo-1-oxyl-1,4-aza Phosphorinane .... [2,2 64. 67
5,5,5',5'-tetramethyl-2,2'-bis(1,3,2λ(5)-dioxa Phosphorinane)-2,2'-disulfide 64. 41
1-benzyl-1-phenyl Phosphorinanium bromide.................... 64. 94
1,1-diphenyl-4-methyl Phosphorinanium bromide.................... 64. 95
1-phenyl-4- Phosphorinanone-1-oxide.................... 64. 46

1-phenyl-4- Phosphorinanone-1-sulfide 64. 45
1,3,5-di-isopropylamino-2,4,6-trioxa-1,3,5-tri Phosphorine 64. 98
3-p-tolyl-1,2,3,4-tetrahydro-1,3,2-benzo-diaza Phosphorine 2-oxide........[2-chloro-6-methyl- 64. 70
loro-1-dimethylamino-1,3,2,4,6,5-dithia-triaza Phosphorine-1,3-dioxide........ [3,5,5-trich 64. 2
antimony-tris(0,0-diethyl Phosphoro-dithioate) 66. 12
<1-1λ(6),5λ(6)-dithia-2,4,6,8-tetra-aza-3,7-di Phosphorocine-1,1,5,5-tetraoxide acetonitrile> 64. 42
ethylammonium adenosine-5'-(O-p-nitrophenyl-O- Phosphorothioate) .... [tri 47. 32
l)amino)-(t-butylimino)-(trimethylsilylimino)- Phosphorus................[(bis(trimethylsily 64. 61
pyrazine- Phosphorus(v)-pentachloride................. 64. 7
trihydroxy Phosphorus-tetraphenylporphyrin dihydrate 64.172
)] 2-diethyl Phosphoryl-guanidine hemi(guanidinium chloride 64. 13
3-dodecanoyl-propandiol-1- Phosphorylcholine monohydrate 3. 55
D-gibberellin C secodiester Photoproduct................................. 59. 32
potassium Phthalate monohydrate 14. 4
1H,4H-2,3-dimethylpyridazino(1,2-b)benzo(g) Phthalazine-6,13-dione 36. 21
3H,4H-2,3-dimethylpyridazino(1,2-b)benzo(g) Phthalazine-6,13-dione 36. 22
5,10-dihydroxybenzo(g) Phthalhydrazide acetic acid solvate........... 36. 3
3-nitromethylene- Phthalide 38. 14
3,3-bis(o-tolyl)- Phthalide 38.111
N-(4-dimethylaminophenyl)- Phthalimide.................................. 35. 40
carbonyl-(Phthalocyaninato)-(pyridine)-osmium(ii)........ 49. 14
oride solvate] bis(Phthalocyaninato)-neodymium(iii) methylene chl 49. 33
Phthalocyaninato-cobalt(ii)................. 49. 1n
lvate] carbonyl-(N,N-dimethylformamide)- Phthalocyaninato-iron(ii) dimethylformamide so 49. 5
Phthalocyaninato-manganese(ii) 49. 2
N- Phthaloylglycine-hydroxamic acid 35. 15
hypo Phyllanthin................................. 38.120
sodium Phytate hydrate 46. 3
diphenanthro(4,3-a;3',4'-o) Picene 30. 21
6''-(1,4,7-trioxaheptane-1,7-diyl)-(2,6-bis(2- Picolinoyl)-pyridine) .... [2', 40. 53
6''-(1,4,7-trioxaheptane-1,7-diyl)-(2,6-bis(2- Picolinoyl)-pyridine)-bis(ethylene-ketal) [2', 40. 55
1β,9β-1-methoxy Picras-12-en-16-one......................... 38.107
valinomycin potassium Picrate..................................... 60. 43
pyridinium 1-naphthylamine Picrate 60. 1
beauvericin barium Picrate complex toluene solvate.............. 50. 32
benzo-15-crown-5 lithium Picrate dihydrate 38. 65
prolinomycin rubidium Picrate toluene chloroform solvate 50. 38
calcium bis(Picrate) tetraethylene-glycol monohydrate 17. 1
8α-hydroxy-iso Picrostegane 59. 31
di Picrylaminato-silver 15. 3
potassium di Picrylamine 16. 3
C.I. Pigment red 208............................. 35. 74
11α,16-epoxy-1β,15R-dihydroxy-iso Pimar-8(14)-ene............................. 54. 9a
1,4-bis(3,3-dimethylazirinyl)- Piperazine 32. 45
H,15H-benzothiazolo(2,3-a)benzothiazino(4,3-c) Piperazine methanol solvate <15-phenyl-6aH,14a 41. 83
2,6-bis(hydroxyimino- Piperazine)................................. 33. 2
3,6-diethyl-1,4-dimethyl-3,6-epithio- Piperazine-2,5-dione 41. 32
N-(2-ammonioethyl)- Piperazinium pentachloro-copper(ii) dihydrate 33. 19
,9-tetramethyl-2,9a-epi-tetrathio-3,10-diketo- Piperazino(1,2-a)indole....[9,9a-dihydro-1,2,9 41. 56a
cis-5,6-dimethoxy-2-methyl-3-(2-(4-phenyl-1- Piperazinyl)-ethyl)-indoline 35. 70
pane hydrogen fuma> 5,5-dimethyl-10-(4-methyl Piperazinyl)-10,11-dihydro-5H-dibenzo(b,f)sile 63. 33
nic acid] 2-chloro-10-(3-(4-methyl-1- Piperazinyl)propyl)-phenothiazine methanesulfo 41. 74
azine dihydrochloride] 10-(3-(4-methyl-1- Piperazinyl)propyl)-2-trifluoro-methylphenothi 41. 79
o-chloro-N-(Piperid-1-yl-acetyl)-aniline hydrochloride 33. 56
ohydrate] o-chloro-N-(Piperid-1-yl-acetyl)-aniline hydrochloride mon 33. 57
4-t-butyl- Piperidin-2-one............................. 33. 33
syn-3-methyl-6-(2-furyl)- Piperidin-2-one............................. 38. 22
oxy-2'-phenylethyl)-4-(N'-phenylpropionamido)- Piperidine ....[N-(2'-hydr 33. 83
thyl)-4-(N-phenylpropionamido)-4-carbomethoxy- Piperidine ....[N-(2'-phenyle 33. 89
2-imino 6'-methyl-spiro(oxazolidine-5,3'- Piperidine)-2'-one-4-semicarbazone monohydrate 40. 15
piperidinium 1- Piperidine-carbodithioate 33. 14
piperidinium 1- Piperidine-carbodithioate 33. 13
(+-)-(E)-N-acetyl- Piperidine-2-carboxylic acid 33. 32
Piperidinium tetrachloro-copper............. 33. 15
Piperidinium 1-piperidine-carbodithioate 33. 13
Piperidinium 1-piperidine-carbodithioate 33. 14
i)] bis(Piperidinium) hexabromo-diaqua-di-molybdenum(i 33. 16
bis(4-benzyl- Piperidinium) hexachloro-di-copper(ii) 33. 47
bis(4-benzyl- Piperidinium) tetrachloro-cobalt(ii) 33. 46
)] bis(4-(N-methyl- Piperidinium))-disulfide tetrachloro-copper(ii 33. 50
Piperidinium-acetyl-(m-chloroanilide) chloride 33. 54
Piperidinium-acetyl-(p-chloroanilide) chloride 33. 55
N-(2-(4-(5-chloro-2-oxo-1-benzimidazolinyl) Piperidino)-ethyl)-p-fluorobenzamide 35. 64
2,5-bis(Piperidino)-1,4-benzoquinone 33. 71
1,1,3,3-tetraphenyl-1,3-bis(2- Piperidino-ethyl)-disiloxane 63. 52
diphenyl-(2- Piperidino-ethyl)-silanol 63. 27

(3-	Piperidino-propyl)-diphenyl-silanol	63.	31
di	Piperidino-tetraselane	33.	37
1,1-diphenyl-3-	Piperidino-1-propanol	33.	78
2-p> cis-(+)-N-(3-methyl-1-(2-phenylethyl)-4-	Piperidinyl)-N-phenyl-propanamide nitric acid	33.	86
<thyl-4-(methoxycarbonyl)-1-(2-phenylethyl)-4-	Piperidinyl)-N-phenyl-propanamide oxalic acid	33.	92
oluene solvate] N-(1-(2-phenylethyl)-4-	Piperidinylium)-N-phenyl-propanamide-citrate t	33.	84
1,2-bis(N-	Piperidyl)ethane-bis-N-oxide diperchlorate	33.	49
1,2-bis(N-	Piperidyl)ethane-bis-N-oxide perchlorate	33.	48
9,9-dimethyl-10-(3'-	Piperidyl-propyl)-acridane	36.	38
9,9-dimethyl-10-(3'-	Piperidyl-propyl)-9-sila-acridane	63.	35
2,6-cis-dimethyl-	Piperidyl-N-phenyl-acetamidine	33.	67
2,6-cis-dimethyl-	Piperidyl-N-phenyl-2,2-dimethylpropionamidine	33.	75
dol-2-olate] N-acetyl-3-(benzoyl(2-	Piperidyl-2-piperidylenium-ethyl)methylene)-in	35.	77
N-acetyl-3-(benzoyl(2-piperidyl-2-	Piperidylenium-ethyl)methylene)-indol-2-olate	35.	77
(+)-	Pisatin monohydrate	59.	15
<thylimidazole)-dioxygen-meso-tetra($\alpha,\alpha,\alpha,\alpha$-o-	Pivalamidophenyl)-porphyrinato-iron ethanol s	49.	35
l > (2-methylimidazole)-meso-tetra($\alpha,\alpha,\alpha,\alpha$-o-	Pivalamidophenyl)-porphyrinato-iron(ii) ethano	49.	34
2-	Pivaloyl-indan-1,3-dione	27.	11
nohydrate]	Pivaloyl-D-alanyl-N-isopropyl-D-prolinamide mo	48.	48
mide]	Pivaloyl-D-prolyl-L-prolyl-L-alanine-N-methyla	48.	62
tetramethylammonium hexachloro-	Platinum	3.	16n
ylbenzylammonium) hexadecacarbonyl-tetra-iron-	Platinum [bis(trimeth	3.	45
s(diphenylarsinomethyl)-benzo(c)phenanthrene)-	Platinum [trans-dichloro-(2,11-bi	65.	26
bis(diethylenetriammonium) tetrachloro-	Platinum(ii) tetrachloride	3.	30
rbonyl-cyclopentad> (bis(trimethylphosphine)-	Platinum)-(μ-carbonyl)-(μ-p-tolylcarbenyl)-(ca	71.173	
tetra-n-butylammonium	Platinum-carbonyl acetonitrile solvate	3.	52
tetraphenylphosphonium	Platinum-carbonyl acetonitrile solvate	64.132	
	Pleiadiene	28.	4
	Pleiadiene	28.	2
	Pleiadiene	28.	5
	Pleiadiene	28.	3
	Plexaurolone acetate	54.	21a
8β-hydroxymethyl-	Podocarpane-13β-carboxylic acid lactone	59.	20
6-epi	Podopetaline hydrobromide hydrate	58.	51
pseudo-racemic	Podopetaline ormosanine	58.	50
	Poiteol	59.	8
5,10-dihydro-5,10-diethylphenazine	Polyiodide	36.	17
3,3',5,5'-tetraphenyl-4,4'-dithiapyranylidene	Polyiodide	39.	90
3,3',5,5'-tetraphenyl-4,4'-dithiapyranylidene	Polyiodide	39.	89
,4'-dithia-1,1'-bis(2,5-cyclohexadienylidene)	Polyiodide [3,3',5,5'-tetraphenyl-4	39.	89
,4'-dithia-1,1'-bis(2,5-cyclohexadienylidene)	Polyiodide [3,3',5,5'-tetraphenyl-4	39.	90
bis(α-cyclodextrin) cadmium	Polyiodide hydrate	45.	63
β-cyclodextrin sodium	Polyiodide octahydrate	45.	68
chloro-(5,10,15,20-tetraphenyl	Porphinato)-indium(iii)	49.	16
chloro-(N-methyl-5,10,15,20-tetraphenyl	Porphinato)-iron(ii)	49.	19
1-pentao> disodium (ethylthio)-(tetra-p-tolyl	Porphinato)-iron(ii) ethylthiolate 4,7,13,16,2	49.	27
,10-diazabicyc> sodium ethylthio-(tetraphenyl	Porphinato)-iron(ii) 4,7,13,16,21,24-hexaoxa-1	49.	23
potassium dicyano-(meso-tetraphenyl	Porphinato)-iron(iii) acetone solvate	49.	22
(5,10,15,20-tetra-n-propyl	Porphinato)-lead(ii)	49.	3
N-tosylamino-octaethyl	Porphinato-bis(chloro-mercury)	49.	15
tetraphenyl	Porphinato-bis(dioxane)-cadmium(ii)	49.	28
bis(3-chloropyridine)-octaethyl	Porphinato-iron(iii) perchlorate	49.	24
octaethyl	Porphinato-perchlorato-iron(iii)	49.	7
tct-2,3,7,8-tetrahydro-octaethyl	Porphyrin	49.	10
ttt-2,3,7,8-tetrahydro-octaethyl	Porphyrin	49.	11
(pyrromellitoyl-(tetrakis-o-oxyethoxyphenyl))-	Porphyrin chloroform methanol solvate <,15,20-	49.	32
trihydroxyphosphorus-tetraphenyl	Porphyrin dihydrate	64.172	
(ethylcarboxylato-ethoxycarbonyl)-anthracene)-	Porphyrin methylene chloride solvate <9,10-bis	49.	31
trans-N,N'-dimethyletio	Porphyrin tri-iodide	49.	4
<-1,3,5,7-tetramethyl-4,8-bis(propionic acid)-	Porphyrin-bis(ethane-1',2'-diyl ester)-(pyrid>	49.	21
di-copper hexyl cofacial di	Porphyrin-7	49.	36
methoxy-(meso-tetraphenyl	Porphyrinato)-iron(iii)	49.	20
cis-dioxo-(5,10,15,20-tetra-p-tolyl	Porphyrinato)-molybdenum(vi) toluene solvate	49.	26
ans,cis,trans-octaethyl-1,2,3,7,8,20-hexahydro	Porphyrinato)-nickel(ii) [(trans,tr	49.	12
s,trans,trans-octaethyl-1,2,3,7,8,20-hexahydro	Porphyrinato)-nickel(ii) [(trans,tran	49.	13
tetraphenyl	Porphyrinato-cobalt(ii)	49.	17
oxygen-meso-tetra($\alpha,\alpha,\alpha,\alpha$-o-pivalamidophenyl)-	Porphyrinato-iron solvate <idazole)-di	49.	35
azole)-meso-tetra($\alpha,\alpha,\alpha,\alpha$-o-pivalamidophenyl)-	Porphyrinato-iron(ii) ethanol solvate <hylimid	49.	34
tct-2,3,7,8-tetrahydro-octaethyl	Porphyrinato-nickel(ii)	49.	8
ttt-2,3,7,8-tetrahydro-octaethyl	Porphyrinato-nickel(ii)	49.	9
1,4,5,8,9,12,13,16-octamethyl-tetrabenzo	Porphyrinato-nickel(ii) iodide	49.	18
	Potassium adenosine-5'-diphosphate dihydrate	47.	23
	Potassium cyanamide dicyanamide	3.	4
	Potassium deuterium bis(dichloroacetate)	2.	7n
-iron(iii) acetone solvate]	Potassium dicyano-(meso-tetraphenylporphinato)	49.	22
2,2'-bipyridyl	Potassium dicyano-gold(iii)	33.	34

hydrate]

                                                  Potassium dihydrouridine-3'-monophosphate hemi 47. 11
                                                  Potassium dipicrylamine....................... 16. 3
                                                  Potassium fluoride succinic acid 1. 12
                                                  Potassium formate   2. 1
                                                  Potassium hydrogen bis(dibromoacetate)........ 2. 11
                                                  Potassium hydrogen bis(dichloroacetate)....... 2. 8n
,11-bis(2-nitrophenoxy)-3,6,9-trioxa-undecane   Potassium isothiocyanate   [1 17. 35
                                         tri-   Potassium methyltrisulfonate monohydrate 11. 2
                                                  Potassium nitroso-dicyanomethanide 7. 1
                                                  Potassium oxalate monoperhydrate 2. 13n
                                                  Potassium p-chloranil acetone solvate 18. 1
                                                  Potassium phthalate monohydrate............... 14. 4
                               valinomycin   Potassium picrate 60. 43
                                                  Potassium thiocarbazate....................... 2. 5
                      dibenzo-30-crown-10   Potassium thiocyanate 38.130
    4,7,10,13,16-pentaoxa-1-thiacyclo-octadecane Potassium thiocyanate 42. 9
,10,13,16,19,22,25,28)decaoxacyclotriacontane   Potassium thiocyanate ....[dibenzo(b,q)(1,4,7 38.130
   phenoxy)-3-oxapentyloxy)phenoxy)-3-oxapentane Potassium thiocyanate [1,5-bis(2-(5-(2-nitro 17. 43
    5.3.1)heneicosa-1(21),17,19-triene-2,16-dione Potassium thiocyanate <entaoxa-21-azabicyclo(1 40. 39
    s(2-acetylaminophenoxy)-3,6,9-trioxa-undecane Potassium thiocyanate monohydrate   [1,11-bi 17. 39
                                                  Potassium thyminate trihydrate 44. 8
                        tetrakis(glycine)   Potassium tri-iodide 48. 4
                                                  Potassium O-n-butylxanthate................... 1. 18
                                                  Prednisone 51. 25
                           21-acetoxy-   Pregn-4-en-17β-ol-3,20-dione monohydrate 51. 39
                18-ethyl-17β-hydroxy-19-nor-   Pregn-4-en-20-yn-3-one 51. 30
       21-acetoxy-11α-methoxy-1β,11β-oxo-10α-   Pregn-4-ene-2,20-dione methanol solvate....... 51. 47
                                 19-nor-   Pregn-4-ene-3,20-dione 51. 20
                    17-hydroxy-6α-methyl-   Pregn-4-ene-3,20-dione 51. 31
                    3β-acetoxy-16β-methyl-   Pregn-5-en-20-one 51. 49
                          20-methyl-   Pregn-5-ene-3β,20-diol methanol solvate 51. 36
                          20-methyl-   Pregn-5-ene-3β,20-diol methanol solvate....... 51. 37n
                                                  Pregna-1,4-diene-17α,21-diol-3,11,20-trione .. 51. 25
      <1β,21-dihydroxy-16α,17α-isopropylidene-dioxy- Pregna-1,4-diene-3,20-dione methanol solvate 51. 46
                    2β-bromo-3α-hydroxy-5α-   Pregna-11,20-dione 51. 27
                   3α-hydroxy-2α-methyl-5α-   Pregna-11,20-dione 51. 34
                   3α-hydroxy-2β-methoxy-5α-   Pregna-11,20-dione monohydrate 51. 35
                          21-hydroxy-   Pregna-4,9(11),16-triene-3,20-dione 51. 24
                    17β-hydroxy-19-nor-   Pregna-4,9-dien-20-yn-3-one 51. 18
            17β-hydroxy-18-methyl-19-nor-   Pregna-4,9-dien-20-yn-3-one 51. 23
      16-methoxy-16'-oxo-21,20'-di(20,18-epoxy-   Pregnane) 51. 89
                          20-methyl-   Pregnenediol methanol solvate 51. 32n
                                                  Pridinol 33. 78
                             sila-   Pridinol 63. 27
                                                  Primperan 13. 11
                                                  Prionostemmadione 56. 2
                                                  Procainamide hydrochloride 13. 8
te]   Procaine-N-D-glucoside hydrochloride monohydra 45. 48
e hydrate]   Proflavine - cytidylyl-(3'-5')-guanosine sulfat 60. 24
           cytidylyl-(3'-5')-adenosine -   Proflavine complex decahydrate 60. 39
drate]   Proflavine deoxycytidylyl-(3'-5')-guanosine hy 47. 38
                       bis(adenosine)   Proflavine sesquisulfate hydrate 47. 21
ydrate methanol solvate]   Proflavine 5-iodocytidylyl-(3'-5')-guanosine h 47. 36
                16α,17α-cyclobutano-   Progesterone 51. 38
                16α,17α-cyclohexano-   Progesterone 51. 58
                16α,17α-cyclopentano-   Progesterone 51. 48
                16α,17α-cyclopropano-   Progesterone 51. 29
          6α-methyl-16α,17α-cyclohexano-   Progesterone 51. 59
azole)]   N(α)-methoxy-(Progesterone)-(16α,17α-d)-(tetrahydro-1',2'-ox 51. 57
azole)]   N(β)-methoxy-(Progesterone)-(16α,17α-d)-(tetrahydro-1',2'-ox 51. 56
        L-pyroglutamyl-β-(2-thienyl)-L-alanyl-L- Prolinamide.................................. 48. 50
          pivaloyl-D-alanyl-N-isopropyl-D-   Prolinamide monohydrate...................... 48. 48
                   glycyl-L-4-hydroxy   Proline...................................... 48. 22
                N-t-amyloxycarbonyl-L-   Proline...................................... 48. 33
             N-benzyloxycarbonyl-prolyl-   Proline...................................... 48. 53
             N-acetyl-L-prolyl-L-4-hydroxy   Proline...................................... 48. 36
                      2,4-methano   Proline monohydrate.......................... 48. 17
                 L-prolyl-L-4-hydroxy   Proline monohydrate.......................... 48. 29
              N-acetyl-4-hydroxy-L-   Proline monohydrate.......................... 48. 21
        (glutamyl-α-lactam)-histidinyl-   Proline tartrate monohydrate 48. 47
rm solvate]   Prolinomycin rubidium picrate toluene chlorofo 50. 38
                cyclo(di-L-prolyl-D-   Prolyl)...................................... 48. 45
              cyclo(di(benzylglycyl)-L-   Prolyl) monohydrate.......................... 48. 73
              cyclo(L-prolyl-L-4-hydroxy   Prolyl) monohydrate.......................... 48. 28
              cyclo(benzylglycyl-bis(L-   Prolyl)) 48. 58
                     L-leucyl-L-   Prolyl-glycine monohydrate 48. 42

cyclo-tris(Prolyl-glycyl) 48. 66
cyclo-tris(Prolyl-glycyl) calcium perchlorate 48. 70
cyclo-tris(Prolyl-glycyl) calcium perchlorate hydrate 48. 67
cyclo-tris(Prolyl-glycyl) hydrate 48. 68
e] cyclo-tris(Prolyl-glycyl) magnesium perchlorate trihydrat 48. 69
thylsulfoxide solvate] cyclo-tris(L-valyl-L- Prolyl-glycyl-L-valyl-glycyl) monohydrate dime 48. 83
cyclo(L-valyl-L- Prolyl-glycyl-L-valyl-glycyl) trihydrate ... 48. 61
N-benzyloxycarbonyl- Prolyl-proline 48. 53
N-t-butyloxycarbonyl-L- Prolyl-sarcosine 48. 41
L- Prolyl-sarcosine monohydrate 48. 24
t-butoxy-carbonyl-L- Prolyl-sarcosine-benzyl ester 48. 64
cyclo-bis(D-valyl- Prolyl-valyl-D-α-hydroxyisovaleryl) dihydrate 48. 81
cyclo(di-L- Prolyl-D-prolyl) 48. 45
pivaloyl-D-prolyl-L- Prolyl-L-alanine-N-methylamide 48. 62
cyclo(L- Prolyl-L-alanyl) 48. 23
t-butoxy-carbonyl-L- Prolyl-L-isoleucyl-glycine hydrate 48. 56
pivaloyl-D- Prolyl-L-prolyl-L-alanine-N-methylamide........ 48. 62
N-(t-butoxycarbonyl)-L- Prolyl-L-valyl-glycine hemihydrate 48. 52
nyl)] N-acetyl-cyclo-bis(L- Prolyl-L-valyl-L-ornithyl-L-leucyl-L-phenylala 50. 40
N-acetyl-L- Prolyl-L-4-hydroxyproline 48. 36
L- Prolyl-L-4-hydroxyproline monohydrate 48. 29
cyclo(L- Prolyl-L-4-hydroxyprolyl) monohydrate 48. 28
5-oxothiomorpholino-3-carbonyl-L-histidinyl-L- Prolylamine sesquihydrate   [3S,6R,-6-methyl- 48. 51
5-n-propyl-orotyl-L-histidinyl-L- Prolylamine tetrahydrate 48. 59
7,8-didehydro-4,5-epoxy-17-(Prop-2-enyl)morphinan-3,6-diol hydrobromide.... 58. 37
2-methyl-2-(N-phenylsulfinamoyl)-N,N-dimethyl- Propanamide.. 1. 30
hyl-1-(2-phenylethyl)-4-piperidinyl)-N-phenyl- Propanamide nitric acid 2-propanol solvate <et 33. 86
yl)-1-(2-phenylethyl)-4-piperidinyl)-N-phenyl- Propanamide oxalic acid <thyl-4-(methoxycarbon 33. 92
(1-(2-phenylethyl)-4-piperidinyliurm)-N-phenyl- Propanamide-citrate toluene solvate........[N- 33. 84
(2-methoxy-10-phenothiazinyl)-N,N,2-trimethyl- Propanamine-citrate toluene solvate........[3- 41. 70a
3-dodecanoyl- Propandiol-1-phosphorylcholine monohydrate 3. 55
2,2-dibenzyl-1,3-dibromo Propane.. 19. 10
2,2-bis(4-(p-nitrophenoxy)phenyl) Propane.. 17. 42
1,1-bis(p-methoxyphenyl)-2,2-dimethyl Propane.. 17. 32
meso-2,2-bis(methylsulfinyl)-1,3-diphenyl Propane.. 11. 17
2(S)-1-pyridinium-2-N,N-dibenzylamino- Propane p-toluenesulfonate 33. 82
1,3-diphenyl-2,2-dihydroxy Propane-1,3-dione 5. 14
2-(4-chlorophenyl)-1,1-dimethylethyl 2-amino Propanoate hydrochloride monohydrate 48. 39a
1,1-diphenyl-3-piperidino-1- Propanol 33. 78
γ-cyclodextrin n- Propanol clathrate hydrate 61. 22
1,3-diammonio-2- Propanol sulfate monohydrate 3. 15
anti-2,3;4,5-diepoxy-12-oxa(4.4.3) Propella-7,9-diene 38. 38
-2,3;4,5-syn-7,8;9,10-tetraepoxy-12-oxa(4.4.3) Propellane .... [anti 38. 40
(E)-1-(4'-hydroxy-3'-methoxyphenyl)-2-nitro Propene.. 17. 13
chlorophenyl-3-p-tolyl-2-triphenylphosphazeno- Propenone .... [1-p- 64.151
2,2'-diphenyl-4,4'-di(2-phenyl-2- Propenyl)-4,4'-bis(5(4H)-thiazolone) 41. 84
4.2.1.0(3,7))nonane] 2-(2'- Propenyl)-8-toluenesulfonyl-7,8-diazatricyclo(37. 10
phenylethyl)-(R)-α-methoxy(pentafluorophenyl)- Propionamide .... [N-((S)-1- 1. 36
2-t-butylammonium-3-chloro-N,N-dimethyl Propionamide tetrafluoroborate 1. 23
6-cis-dimethyl-piperidyl-N-phenyl-2,2-dimethyl Propionamidine .... [2, 33. 75
N-(2'-hydroxy-2'-phenylethyl)-4-(N'-phenyl Propionamido)-piperidine 33. 83
N-(2'-phenylethyl)-4-(N-phenyl Propionamido)-4-carbomethoxy-piperidine....... 33. 89
β-thiocyanato- Propionanilide 16. 2
copper Propionate 1. 9
sodium α,α-dihydroxy-β-fluoro Propionate 2. 17
-methyl-3-t-butoxy-2-iodo-3-(p-methoxyphenyl)- Propionate .... [(+-)erythro 17. 24
(+-)-2-(4-chloro-2-methylphenoxy)- Propionic acid 17. 12
αRS,1RS,2RS)-α-(1-hydroxy-2-phenylcyclohexyl)- Propionic acid .... [(21. 9
furaltadone hydrochloride Propionic acid clathrate 61. 5
(RS)-α-amino-3-hydroxy-5-methylisoxazole-4- Propionic acid monohydrate 40. 5
I e> 2,6-diethyl-1,3,5,7-tetramethyl-4,8-bis(Propionic acid)-porphyrin-bis(ethane-1',2'-diy 49. 21
α,β-diphenyl-α-n-propyl-β-methyl- Propionitrile 7. 4
-pyrroli> 4',5,7-tri-t-butyl-3'-(2,2-dimethyl Propionyl)-1'-methyl-spiro(benzofuran-3(2H)-2' 38.132
oro-4aα,10aα-dihydro-8-hydroxy-10β-methyl-4aα- Propionyl-phenanthrene-1,4,9(10H)-trione <ichl 28. 13
2,4,6-tribenzoyloxy- Propiophenone 13. 23
2,4,6-tribenzoyloxy- Propiophenone benzene 13. 24
4- Propoxy-salicylidene-4'-butylaniline 17. 36
5- Propynyloxy-2'-deoxyuridine.................... 47. 29
(5Z,11α,13E,15S)-11,15-dihydroxy-9-oxo- Prosta-5,13-dien-1-oic acid 59. 25
(5Z,9β,11a,13E,15S)-9,11,15-trihydroxy- Prosta-5,13-dien-1-oic acid 59. 26
Prostaglandin E(2)........................... 59. 25
Prostaglandin F(2β)........................... 59. 26
olvate] 8(1)(E)-8(2)-nitro- Protoporphyrin dimethyl ester dichloroethane s 49. 6
allo Psoralen 38. 30
2-isocyano- Pupukeanane 53. 18
trans-6-chloro-9-(2-ethoxy-1,3-dioxan-5-yl)- Purine 44. 23

benzyl 6-amino Purine-9-carboxylic acid 44. 26
2-benzoyl-2-(1,3-dimethyl-2,6-dioxo-7- Purinyl)-acetanilide 35. 67
tyrian Purple 59. 12
Purpuride 59. 33
4-benzyl-2,3,5,6-bis(tetramethylene)-4H-thio Pyran 39. 80
Pyran .... [3,5-dihydroxy-4-nitro 38. 42
,2,8,8-tetramethyl-2H,8H-benzo(1,2-b:5,4-b')di Pyran-10-yl)-propyl-p-nitrobenzoate <methoxy-2 38.126
,2,8,8-tetramethyl-2H,8H-benzo(1,2-b:5,4-b')di Pyran-10-yl)-propyl-p,5-dinitrobenzoate <oxy-2 38.125
,2,8,8-tetramethyl-2H,8H-benzo(1,2-b:5,4-b')di Pyran-10-yl)-propyl-3,5-dinitrobenzoate <oxy-2 38.127
y-phenyl)-6H,11H-(2)benzopyrano(4,3-c)(1)benzo Pyran-11-one <-dihydroxy-9-methoxy-6-(4-methox 38.117
l-4-oxo-2-furanyl)-but-2-enyl)-oxy-(2H-1-benzo Pyran-2-one) [7-(3-(4,5-dihydro-5,5-dimethy 59. 19
-> 3,4-dihydro-2H,6H-2-benzothiopyrano(4,3-b) Pyran-2-spiro-3'-(1H-2)-benzothiopyran-4'(3'H) 39. 76
pyrano(4,3-b)pyran-2-spiro-3'-(1H-2)-benzothio Pyran-4'(3'H)-one <4-dihydro-2H,6H-2-benzothio 39. 76
2-hydroxy-2-methyl-2H,5H-pyrano(3,2-c)(1)benzo Pyran-5-one....... [4-cyclohexyl-3,4-dihydro- 38.100
oxyphenyl)-6H,7H-(1)benzopyrano(4,3-b)(1)benzo Pyran-6-one <11-diacetoxy-10-methoxy-7-(4-meth 38.128
,2,8,8-tetramethyl-2H,8H-benzo(1,2-b:3,4-b')di Pyran-6-yl)-propyl-3,5-dinitrobenzoate <hoxy-2 38.124
trimethyl-3,4,6,9-tetrahydro-1H-naphtho(2,3-c) Pyran-6,9-dione [(1R,3S)-7,10-dihydroxy-1,3,8- 59. 14
clohexyl-3,4-dihydro-2-hydroxy-2-methyl-2H,5H- Pyrano(3,2-c)(1)benzopyran-5-one .... [4-cy 38.100
-10-methoxy-7-(4-methoxyphenyl)-6H,7H-(1)benzo Pyrano(4,3-b)(1)benzopyran-6-one <11-diacetoxy 38.128
pyran-4'(3'H)-> 3,4-dihydro-2H,6H-2-benzothio Pyrano(4,3-b)pyran-2-spiro-3'-(1H-2)-benzothio 39. 76
9-methoxy-6-(4-methoxy-phenyl)-6H,11H-(2)benzo Pyrano(4,3-c)(1)benzopyran-11-one <-dihydroxy- 38.117
2-carboxy-3-methylthio Pyrano(4,3,2-cd)indole 39. 42
α-L-xylo Pyranose 45. 3n
1,6-anhydro-β-D-galacto Pyranose 45. 8
β-D-glucopyranosyl-(1-3)-D-gluco Pyranose 45. 26
1',2,3,4-tetra-O-benzoyl-β-D-xylo Pyranose 45. 58
β-D-glucopyranosyl-(1-6)-β-D-gluco Pyranose 45. 24
β-D-glucopyranosyl-(1-6)-β-D-gluco Pyranose 45. 25
methyl-3,4-O-ethylidene-β-L-arabino Pyranose 45. 15
4,6-tri-O-acetyl-1,2-O-(R)-ethylidene-α-D-allo Pyranose ....[3, 45. 31
oxy-3-iodo-2-O-(methylthio)-carbonyl-β-D-altro Pyranose .... [1,6-anhydro-3-de 45. 13
l-1,2-O-(R)-(1-t-butoxyethylidene)-α-D-galacto Pyranose .... [3,4,6-tri-O-acety 45. 47
etyl-2-(N-acetylacetamido)-2-deoxy-β-D-galacto Pyranose ....[1,3,4,6-tetra-O-ac 45. 44
-acetamido)-2,3-dideoxy-α-D-erythro-hex-2-eno- Pyranose ...[1,4,6-tri-O-acetyl-2-(N-acetyl 45. 39
opyranosyl-(1-4)-2-acetamido-2-deoxy-α-D-gluco Pyranose <-α-D-mannopyranosyl-(1-3)-O-β-D-mann 45. 49
tra-O-acetyl-β-D-galactopyranosyl)-α-D-galacto Pyranose <2,4,6-tetra-O-acetyl-3-O-(2,3,4,6-te 45. 56
2-R-methylcyclohexyl-α-D-gluco Pyranoside 45. 28
methyl-2,3,4-tris(O-benzoyl)-β-D-xylo Pyranoside 45. 55
methyl-3-O-α-D-glucopyranosyl-α-D-gluco Pyranoside 45. 29
yl 4,6-O-benzylidene-2,3-di-O-methyl-α-D-gluco Pyranoside .... [meth 45. 40
D-threo-4,6-di-O-acetyl-2,3-dideoxy-hex-2-eno- Pyranoside .... [methyl α- 45. 21
di-O-acetyl-2,3-dideoxy-α-D-erythro-hex-2-eno- Pyranoside ....[ethyl 4,6- 45. 23
-O-(R)-benzylidene-2,3-di-O-methyl-β-D-galacto Pyranoside ....[methyl 4,6 45. 41
i-O-acetyl-3-deoxy-3-C-methyl-β-D-methyl-β-D-xylo Pyranoside ....[methyl-2,4-d 45. 22
deoxy-3-O-((methylthio)-thiocarbonyl)-α-D-ribo Pyranoside ... [methyl 4,6-O-benzylidene-2- 45. 35
xy-3-O-((methylthio)-thiocarbonyl)-α-D-arabino Pyranoside ....[methyl 4,6-O-benzylidene-2-deo 45. 36
xy-2-O-((methylthio)-thiocarbonyl)-α-D-arabino Pyranoside ....[methyl 4,6-O-benzylidene-3-deo 45. 37
i-O-isopropylidene-α-D-erythro-hept-3-en-5-ulo Pyranoside <hyl 4-chloro-3,4-dideoxy-1,2:6,7-d 45. 32
pyridyl-1-thio-β-D-gluco Pyranoside monohydrate 45. 20
methyl β-D-ribo Pyranoside phosphite triester 64. 16
methyl-D-ribo Pyranoside-thiophosphate-triester 46. 4
trans-2,3,4,6-tetra-O-acetyl-1-deoxy-D-gluco Pyranoside-1,2'-spiro(3'-methyl-3'-hydroxy-te> 45. 46
r> cis-2,3,4,6-tetra-O-acetyl-1-deoxy-D-gluco Pyranoside-1,2'-spiro(3'-methyl-3'-hydroxy-tet 45. 45
2,3,4-tri-O-benzoyl-2-C-chloro-α-D-xylo Pyranosyl chloride 45. 51
acetyl-3-O-(2,3,4,6-tetra-O-acetyl-β-D-galacto Pyranosyl)-α-D-galactopyranose <2,4,6-tetra-O- 45. 56
methyl-3-O-α-D-gluco Pyranosyl-α-D-glucopyranoside 45. 29
β-D-gluco Pyranosyl-(1-3)-D-glucopyranose................. 45. 26
cetamido-2-deoxy-α-D-glucopyrano> O-α-D-manno Pyranosyl-(1-3)-O-β-D-mannopyranosyl-(1-4)-2-a 45. 49
yrano> O-α-D-mannopyranosyl-(1-3)-O-β-D-manno Pyranosyl-(1-4)-2-acetamido-2-deoxy-α-D-glucop 45. 49
β-D-gluco Pyranosyl-(1-6)-β-D-glucopyranose 45. 24
β-D-gluco Pyranosyl-(1-6)-β-D-glucopyranose 45. 25
<-(2-O-acetyl-3,4-di-O-angelyl-6-deoxy-β-gluco Pyranosyloxy)-3β,16α,22α,24,28-penta-acetoxy-> 56. 10
2-oxo-1(2H)-pyrimidinyl)-1,4-dideoxy-β-D-gluco Pyranuronic acid monohydrate <mino-1-(4-amino- 47. 20
3,3',5,5'-tetraphenyl-4,4'-dithia Pyranylidene polyiodide 39. 89
3,3',5,5'-tetraphenyl-4,4'-dithia Pyranylidene polyiodide..................... 39. 90
tetraphenyl-di(thia Pyranylidene)......................... 39. 88
2,5-di(styryl)- Pyrazine 1,4-bis(2-(2-pyridyl)vinyl)-benzene 33. 76
Pyrazine-phosphorus(v)-pentachloride 64. 7
r-1,c-2,t-3,t-4-tetrakis(2- Pyrazinyl)-cyclobutane 33. 77
r-1,c-2-bis(3-pyridyl)-t-3,t-4-bis(2- Pyrazinyl)-cyclobutane 33. 80
r-1,t-3-bis(3-pyridyl)-c-2,t-4-bis(2- Pyrazinyl)-cyclobutane 33. 81
3,2'-pyridyl Pyrazinyl-ethylene 33. 40
1-phenyl-2,3-dimethyl-4-trimethylammonium- Pyrazol-5-one iodide 32. 52
ro-5-hydroperoxy-4,4-dimethyl-3,5-diphenyl-3H- Pyrazole .... [3-bromo-4,5-dihyd 32. 60
ny1)-2-cis-nitro-5-oxo-perhydropyrazolo(1,2-a) Pyrazole ...... [3-trans-bis(4-chlorophe 35. 49
l)-2-trans-nitro-5-oxo-perhydropyrazolo(1,2-a) Pyrazole .... [3-trans-bis(4-chloropheny 35. 50

```
4-bromophenyl)-4-dimethylamino-2,3-dimethyl-3-  Pyrazolin-5-one........................    [1-(  32. 48
                  calamusenone p-bromophenyl  Pyrazoline ............................... 53. 28a
hyl-1-thiosemicarbamoyl-5-thiosemicarbazido-2-  Pyrazoline ........ [4-hydroxyimino-3,5-dimet  32. 20
          bis(4-chlorophenyl)-2-cis-nitro-5-oxo-perhydro  Pyrazolo(1,2-a)pyrazole........... [3-trans-  35. 49
        s(4-chlorophenyl)-2-trans-nitro-5-oxo-perhydro  Pyrazolo(1,2-a)pyrazole.......[3-trans-bi  35. 50
                   6-ethoxycarbonyl-4-ethyl  Pyrazolo(1,5-a)pyrimidin-(4H)-one ...........  35. 22
                          1,2-dihydro-3H-  Pyrazolo(3,4-b)pyridin-3-one .............  35.  1
          3a,9a-dihydro-1,3,3a,9a-tetramethyl-4H-  Pyrazolo(3,4-b)quinolin-4-one <-methyl-phen  36. 14
       yl)-3,7-dimethyl-1-propyl-1,5-dihydro-benzo(f)  Pyrazolo(3,4-c)(1,2,5)triazepine <-methyl-phen  36. 35
                        octa(3,5-dimethyl  Pyrazolyl)-cyclotetraphosphazene         64.171
       ,6,8-hexamethyl-3a,5a,8a,10a-tetra-azaperhydro  Pyrene ....................... [meso-1,1,3,6  36. 26
       ,6,8-hexamethyl-3a,5a,8a,10a-tetra-azaperhydro  Pyrene ............... [racemic-1,1,3,6  36. 27
     yl-3a,5a,8a,10a-tetra-aza-cis-10b,10c-perhydro  Pyrene ........ [racemic-1,1,3,6,6,8-hexameth  36. 28
                                   Pyrene - fluoranil complex .................  60. 34
              1,3,5-trimethylazuleno(1,8-cd)  Pyridazine ............................  36. 13
         1,4-dichloro-7-methylpyrrolo(3,2-d)  Pyridazine ............................. 35.  2
        1-(p-bromophenyl)-2-methyl-5-bromo  Pyridazine-3,6-dione         33. 38
     1-(p-bromophenyl)-2-methyl-4-diethylamino  Pyridazine-3,6-dione         33. 65
               1H,4H-1,4-diphenylpyridazino(1,2-a)  Pyridazine-6,9-dione ................. 35. 58
                     1H,4H-1,4-diphenyl  Pyridazino(1,2-a)pyridazine-6,9-dione .... 35. 58
e]                   1H,4H-2,3-dimethyl  Pyridazino(1,2-b)benzo(g)phthalazine-6,13-dion  36. 21
e]                   3H,4H-2,3-dimethyl  Pyridazino(1,2-b)benzo(g)phthalazine-6,13-dion  36. 22
      -8-oxo-N,9-diphenyl-2,3,4,4a,5,8-hexahydro-1H-  Pyridazino(4,5-d)azepine-2,3-dicarboximide <yl  37. 23
               1,2-dihydro-3H-pyrazolo(3,4-b)  Pyridin-3-one .............................  35.  1
                   1β-D-ribofuranosyl-  Pyridin-4-one-3-carboxamide...............  47. 25
                                   Pyridine ..................................  33.  4
                   N-salicylidene-3-amino  Pyridine ..................................  33. 44
               3,5-diacetyl-2,6-dimethyl  Pyridine ..................................  33. 41
              N-(5-methoxysalicylidene)-3-amino  Pyridine ..................................  33. 51
         1-(2,4-dinitrophenyl)-imidazo(4,5-b)  Pyridine ..................................  35. 24
      2,2',6,6'-tetrakis(propylthio)-3,3'-azoxy  Pyridine ..................................  33. 85
       6-acetyl-2-(α-(trichloro-tellurium)-acetyl)-  Pyridine ..................................  70.  4
        3,5-diethoxycarbonyl-2,6-dimethyl-1,4-dihydro  Pyridine ..................................  33. 58
      droxybenzoyl)-7-methyl-3-phenyl-imidazo(1,2-a)  Pyridine                    [2-(o-hy  35. 62
       thyl-2-oxo-2,3,3a,7a-tetrahydro-oxazolo(4,5-b)  Pyridine ....... [3-(p-chlorophenyl)-6-me  40. 26
       thyl-2-oxo-2,3,3a,7a-tetrahydro-oxazolo(4,5-b)  Pyridine ....... [3-(p-chlorophenyl)-7a-me  40. 27
                                   Pyridine trihydrate.......................  33.  5
       magnesium bis(2,4-dinitrophenoxide) bis(  Pyridine) .................................  67.  6
     rioxaheptane-1,7-diyl)-(2,6-bis(2-picolinoyl)-  Pyridine) ............. [2',6''-(1,4,7-t  40. 53
     rioxaheptane-1,7-diyl)-(2,6-bis(2-picolinoyl)-  Pyridine) ............. [2',6''-(1,4,7-t  40. 55
     ate]   bis(isothiocyanato)-tetrakis(4-methyl  Pyridine)-nickel(ii) 2-bromonaphthalene clathr  61. 10
     rate]  bis(isothiocyanato)-tetrakis(4-methyl  Pyridine)-nickel(ii) 2-methylnaphthalene clath  61. 11
             orate]       bis(3-chloro  Pyridine)-octaethylporphinato-iron(iii) perchl  49. 24
                 carbonyl-(phthalocyaninato)-(  Pyridine)-osmium(ii)         49. 14
     ca-1,11-diene) per>  barium (1,12-bis(2-acetyl  Pyridine)-1,12-dimethyl-2,5,8,11-tetra-azadode  33. 90
           1-benzyl-5-ethyl-1,2,5,6-tetrahydro-2-oxo-4-  Pyridine-acetic acid         33. 68
                  4-(p-N,N-dimethylaminostyryl)-  Pyridine-methiodide.......................  33. 69
                   3-cyanopyridinium 3-cyano  Pyridine-tetrachloro-iron(iii)            33. 17
                  4-m-aminophenyl-2-formyl  Pyridine-thiosemicarbazone ................  33. 53
     palladium(ii)]        bis(pyridine-N-oxide) bis(  Pyridine-N-hydroxide) tetrabromo-μ-dibromo-di-  33.  6
                       2,6-dimethyl  Pyridine-N-oxide hemi-perchlorate ........  33. 24
     trabromo-μ-dibromo-di-palladium(ii)]       bis(  Pyridine-N-oxide) bis(pyridine-N-hydroxide) te  33.  6
               oxy-bis(2-methylene-6-methyl  Pyridine-N-oxide) dihydrate.............  33. 64
                 trichloro-bis(4-cyano  Pyridine-N-oxide)-thallium ...............  68. 13
                        dideutero-  Pyridine-2,3-dicarboxylic acid ..........  33. 22n
                        dideutero-  Pyridine-2,3-dicarboxylic acid ..........  33. 20n
                        dideutero-  Pyridine-2,3-dicarboxylic acid ..........  33. 21n
               N-p-iodobenzyl-3-n-octan-amido-  Pyridine-2,5,6-trione .....................  59. 22
     e dihydrate]       bis(o-phenylene)-bis(  Pyridine-2,6-dialdimino) lead(ii) diperchlorat  34. 11
     -2,6-dicarboxylic acid solvate >  bis(diaquo-(  Pyridine-2,6-dicarboxylato)-lead(ii)) pyridine  69. 29
     <cid)-porphyrin-bis(ethane-1',2'-diyl ester)-(  Pyridine-3''',5''-dicarboxylate) dichlorometha>  49. 21
             N-(1-methyl-4-(3-methylphenylamino)-  Pyridinio)-sulfonyl-N'-isopropylurea ........  33. 73
                          4-ethyl  Pyridinium octa-molybdenum ...............  33. 26
                          3-bromo  Pyridinium perchlorate hemihydrate .........  33.  3
                 ybdenum chloride]  Pyridinium tetra-μ-(hydrogen-phosphato)-di-mol  33. 10
             3-(p-bromophenyl)-thiazolo(3,2-a)  Pyridinium tetrafluoroborate ...........  41. 43
                                   Pyridinium tris(thiocyanato)-di-copper(i) ....  33.  7
                                   Pyridinium 1-naphthylamine picrate ............  60.  1
     i)]                3-cyano  Pyridinium 3-cyanopyridine-tetrachloro-iron(ii  33. 17
     imony(iii))]       bis(  Pyridinium) di-μ-chloro-μ-oxo-bis(dichloro-ant  33.  8
     to)-molybdenum(v))]    tetrakis(  Pyridinium) di-μ-oxo-bis(oxo-tris(isothiocyana  33. 11
     adium]       tetrakis(4-ethyl  Pyridinium) dihydrido-octacosa-oxygen-deca-van  33. 25
                          bis(  Pyridinium) hexachloro-tellurium(iv)         33.  9
                tetrakis(2-carboxy  Pyridinium) octacyano-molybdenum(iv)  .......  33. 18
     3,3'-thio-bis(2-methyl-1-phenyl-imidazo(1,5-a)  Pyridinium) tetrafluoroborate ................  35. 73
     m)dibromide]       pentakis(  Pyridinium) tris(aquo-tetrabromo-oxo-molybdenu  33. 12
```

1-(1'-methyl-2'-pyridinium)-3-(1''-methyl-4''- Pyridinium)-cyclopentadienide bromide hemihyd> 33. 72
                        1,3-bis(1'-methyl-2'- Pyridinium)-indenide bromide 33. 79
opentadienide bromide hemih> 1-(1'-methyl-2'- Pyridinium)-3-(1''-methyl-4''-pyridinium)-cycl 33. 72
                           2-methylamino Pyridinium-dinitromethylide monohydrate........ 33. 23
                                     Pyridinium-1-dicyanomethylide 33. 28n
                                     Pyridinium-1-dicyanomethylide 33. 29n
nesulfonate]   2(S)-1- Pyridinium-2-N,N-dibenzylamino-propane p-tolue 33. 82
 p-toluenesulfonate]   2(S),3(S)-1- Pyridinium-2-N,N-dibenzylamino-3-methylpentane 33. 91
            1-p-hydroxyphenyl-2,4,6-triphenyl Pyridinium-3-olate 33. 94
14-hydroxy-15-hydroxymethyl-2,8-dithia(9)(2,5) Pyridinophane   [(-)- 41. 48a
   6-methyl-4-oxo-1,6,7,8,9,9a-hexahydro-4H- Pyrido(1,2-a)pyrimidine-3-carboxamide 35. 20
hyl-9-p-bromophenylamino-4-oxo-6,7-dihydro-4H- Pyrido(1,2-a)pyrimidine-3-carboxylic acid <met 35. 37
    6-methyl-9-formyl-1,6,7,8-tetrahydro-1H- Pyrido(1,2-a)pyrimidine-4-one 35. 16
carbonyl-6-methyl-4-oxo-6,7,8,9-tetrahydro-4H- Pyrido(1,2-a)pyrimidine-9-acetic acid <-ethoxy 35. 33
',1''',6'''-na)(1,3,> 1,14:7,8-dietheno-tetra Pyrido(2,1,6-de:2',1',6'-gh:2'',1'',6''-kl:2' 36. 39
                          5,6-cyclopenteno- Pyrido(3,2-a)carbazole 36. 20
4-fluorophenyl)-butyl) -2,3,4,5-tetrahydro-1H- Pyrido(4,3-b)indole <ophenyl)-2-(4-hydroxy-4-(36. 49
complex]   mono Pyrido-18-crown-6 t-butylammonium perchlorate 40. 40
                                     Pyridoxinium chloride 33. 31n
                          (E)-3-phenyl-3-(3- Pyridyl)-acrylamide......................... 33. 59
                          (Z)-3-phenyl-3-(4- Pyridyl)-acrylamide......................... 33. 60
               r-1,t-3-bis(3- Pyridyl)-c-2,t-4-bis(2-pyrazinyl)-cyclobutane 33. 81
           r-1,c-2,t-3,t-4-tetrakis(2- Pyridyl)-cyclobutane 33. 87
           r-1,c-2,t-3,t-4-tetrakis(4- Pyridyl)-cyclobutane 33. 88
      1,3-trans-bis(4-chlorophenyl)-2,4-trans-bis(4- Pyridyl)-cyclobutane 33. 93
               (2-hydroxy-2-phenyl-2-(2'- Pyridyl)-ethyl)-methylsulfoxide............. 33. 61
                       bis(1-methyl-2-(3- Pyridyl)-pyrrol-3-yl)-disulfide 32. 69
           (3S,5R)-2,2-dimethyl-5-phenyl-3-(4- Pyridyl)-pyrrolidone dihydrobromide....... 32. 62
                  r-1,c-2-bis(3- Pyridyl)-t-3,t-4-bis(2-pyrazinyl)-cyclobutane 33. 80
n sulfate] 1-((4-(3-chlorophenylamino)-3- Pyridyl)sulfonyl)-3-ethyl-1-methylurea hydroge 33. 66
              2,5-di(styryl)-pyrazine 1,4-bis(2-(2- Pyridyl)vinyl)-benzene 33. 76
                                    2- Pyridyl-phenylsulfone 33. 39
                                     Pyridyl-1-thio-β-D-glucopyranoside monohydrate 45. 20
                                 2-(2- Pyridylamino)-8-hydroxyquinoline 35. 29
                                 3,2'- Pyridylpyrazinyl-ethylene 33. 40
        6-ethoxycarbonyl-4-ethylpyrazolo(1,5-a) Pyrimidin-(4H)-one 35. 22
                               perhydro Pyrimidin-2-one 33. 1
4-methoxy-1-methyl-4-trichloromethyl-hexahydro Pyrimidin-2-one [5-chloro-6-dichloromethylene- 33. 30
   2-phenyl-cis-5,6-tetramethylene-5,6-dihydro Pyrimidin-4(3H)-one 35. 30
   2-phenyl-trans-5,6-tetramethylene-5,6-dihydro Pyrimidin-4(3H)-one 35. 31
      2,3-dihydro-3-hydroxy-5H-thiazolo(3,2-a) Pyrimidin-5-one 41. 14
  6-ethoxycarbonyl-4-ethyl-1,2,4-triazolo(1,5-a) Pyrimidin-7(4H)-one 35. 19
                                     Pyrimidine 44. 3
                               2-amino Pyrimidine 44. 4
                              2-chloro Pyrimidine 44. 2
                              5-methyl Pyrimidine 44. 9
yl-(β-chloroethyl)-amino)-4,6-dichloro-5-cyano Pyrimidine   [2-(meth 44. 21
xazole 2,4-diamino-5-(3,4,5-trimethoxybenzyl)- Pyrimidine .... [3-sulfanilamido-5-methyliso 44. 30
      2,4-diamino-5-(1-adamantyl)-6-methyl Pyrimidine ethylsulfonate 44. 33
      2,4-diamino-5-(1-naphthyl)-6-methyl Pyrimidine methanol solvate................. 44. 31
enyl-2,10-dihydrophenazin-2-ylidene)-hexahydro Pyrimidine-2,4,6-trione [1,3-dimethyl-5-(10-ph 36. 41
-4-oxo-1,6,7,8,9,9a-hexahydro-4H-pyrido(1,2-a) Pyrimidine-3-carboxamide .... [6-methyl 35. 20
phenylamino-4-oxo-6,7-dihydro-4H-pyrido(1,2-a) Pyrimidine-3-carboxylic acid <methyl-9-p-bromo 35. 37
l-9-formyl-1,6,7,8-tetrahydro-1H-pyrido(1,2-a) Pyrimidine-4-one .... [6-methy 35. 16
thyl-4-oxo-6,7,8,9-tetrahydro-4H-pyrido(1,2-a) Pyrimidine-9-acetic acid <-ethoxycarbonyl-6-me 35. 33
acid monohyd> 4-amino-1-(4-amino-2-oxo-1(2H)- Pyrimidinyl)-1,4-dideoxy-β-D-glucopyranuronic 47. 20
                             bis(2- Pyrimidyl)-disulfide 44. 20
                  triethylammonium tris(Pyrocatechol)silicate 63. 23
                 bis(monothio- Pyrocatecholato)-tellurium(iv) 70. 7
ide]   L- Pyroglutamyl-β-(2-thienyl)-L-alanyl-L-prolinam 48. 50
                                  L- Pyroglutamyl-N,N'-dicyclohexylurea 48. 55
                                     Pyromellitic dianhydride 38. 18
                      phenothiazine - Pyromellitic dianhydride complex 60. 17
              bis(1-methyl-2-(3-pyridyl)- Pyrrol-3-yl)-disulfide 32. 69
                              2-benzoyl Pyrrole.................................... 32. 38
                      2-(4-chlorobenzoyl)- Pyrrole.................................... 32. 37
                      2-(4-methoxybenzoyl)- Pyrrole.................................. 32. 42
             1-(2-hydroxyphenyl)-2,3,5-triphenyl Pyrrole................................ 32. 72
  -cis-syn-1,3,5-triphenyl-perhydrothieno(3,4-c) Pyrrole-1-carbonitrile <ol-3-yl)-4,6-dioxo-syn 40. 58
      (2R,5R)-1-(p-bromophenylsulfonyl)-2,5-dimethyl Pyrrolidine.....................[trans- 32. 44a
                    2-carboxy-2,4-methano Pyrrolidine monohydrate..................... 48. 17
ropionyl)-1'-methyl-spiro(benzofuran-3(2H)-2'- Pyrrolidine)-2,5'-dione <tyl-3'-(2¦2-dimethylp 38.132
thyl cis-1-formyl-5-hydroxy-4-(2-nitrophenyl)- Pyrrolidine-2,2-dicarboxylate .... [die 32. 63
           3-acetyl-5-isopropyl Pyrrolidine-2,4-dione 32. 30
        trans-(2R,5R)-2,5-dimethyl Pyrrolidinium(S)-mandelate 32. 18
                    2,5-bis(Pyrrolidino)-1,4-benzoquinone 32. 51

17a-methyl-3β- Pyrrolidino-17a-aza-D-homo-5α-androstane 36. 43
trimethyl-(2- Pyrrolidinoethyl)-silane 63. 10
trimethyl-(Pyrrolidinomethyl)-silane 63. 8
sodium 1- Pyrrolidinyl-carbodithioate dihydrate 32. 15
sodium 1- Pyrrolidinyl-carbodithioate dihydrate 32. 16
sodium 1- Pyrrolidinyl-carbodithioate dihydrate 32. 14
α-cyclodextrin 2- Pyrrolidone clathrate pentahydrate 61. 20
(3S,5R)-2,2-dimethyl-5-phenyl-3-(4-pyridyl)- Pyrrolidone dihydrobromide 32. 62
D,L-3-amino-1-hydroxy-2- Pyrrolidone trihydrate 32. 12
dimethylamino-4-methyl-3,3-diphenyl-4-vinyl-5- Pyrrolin-2-one ....[5- 32. 70
yl-1-isopropyl-3-((phenylcarbamoyl)-methyl)-3- Pyrrolin-2-one .... [4-eth 32. 64
yl-5-(1,2-diphenyl-2-phenyliminoethylidene)-2- Pyrrolin-4-one .... [1,2,3-triphen 32. 77
',N''-tribenzyl-benzo(1,2-c:3,4-c':5,6-c'')tri Pyrroline .... [N,N 36. 52
b-tetrahydro-2,6a-dimethyl-6-oxo-2H-oxireno(a) Pyrrolizin-4-yl)-2-butenamide ‹yl-N-(1a,6,6a,6 59. 6
‹y-10,11,13-trimethyl-1a,2,3,6b-tetrahydro-5H- Pyrrolizino(1a,6b,6a-b,c)-1,8-dioxa-15-cis-tr› 58. 29
tetrahydro-9bβ-hydroxy-2β-methoxy-α-phenyl-5H- Pyrrolo(2,1-a)isoindol-5-one .... [1,2,3,9b- 36. 23
chloro-2,3,3a,9-tetrahydro-6,8,9-trimethyl-1H- Pyrrolo(2,1-b)(1,3)benzoxazine ....[rac-cis-5- 40. 34
‹5-dihydro-5-hydroxy-4-oxo-5-(2-oxopropyl)-1H- Pyrrolo(2,3-f)quinoline-2,7,9-tricarboxylic a› 36. 19
1,4-dichloro-7-methyl Pyrrolo(3,2-d)pyridazine 35. 2
yl-4-imino-5,5-(spiro-2',4'-bis-dimethylamino- Pyrrolyl)-uracil ....[1,3-dimeth 35. 28
orphyrin chloroform methanol so› 5,10,15,20-(Pyrromellitoyl-(tetrakis-o-oxyethoxyphenyl))-p 49. 32
sodium fluoro Pyruvate hydrate 2. 17
dibenzyl- Pyruvic acid 1. 35
Queuosine-5'-monophosphate hydrate 47. 34
2-phenyl-cis-4a,5,6,7,8a-hexahydro Quinazolin-4-one 35. 30
2-phenyl-trans-4a,5,6,7,8a-hexahydro Quinazolin-4-one 35. 31
7,8-benzoquinoline 7,7,8,8-tetracyano Quinodimethane 60. 14
decamethylferrocenium - 7,7,8,8-tetracyano Quinodimethane 60. 42
trithia(5)heterohelicene - 7,7,8,8-tetracyano Quinodimethane 60. 33
')-bis(benzo(b)thiophene) - 7,7,8,8-tetracyano Quinodimethane ....[thieno(3,2-e:4,5-e 60. 33
ro-5,10-dimethylphenazinium 7,7,8,8-tetracyano Quinodimethane ...... [phenazine 5,10-dihyd 36. 2
stilbene - 7,7,8,8-tetracyano Quinodimethane complex 60. 32
tetramethoxystilbene - 7,7,8,8-tetracyano Quinodimethane complex 60. 38
l-1,3-thiaselenole-2-thione 7,7,8,8-tetracyano Quinodimethane complex .... [5-pheny 60. 8
cylaldiminato)-copper(ii) - 7,7,8,8-tetracyano Quinodimethane complex ‹,2-phenylene)-bis(sali 60. 40
trimethylammonium 7,7,8,8-tetracyano Quinodimethane tri-iodide 3. 11n
tetraethylammonium bis(7,7,8,8-tetracyano Quinodimethane)............................ 60. 7
,1'-dimethylferrocenium bis(7,7,8,8-tetracyano Quinodimethane)........................[1 60. 20
ltriphenylphosphonium - bis(7,7,8,8-tetracyano Quinodimethane)................... [ethy 60. 41
-hexamethylene-diammonium) (7,7,8,8-tetracyano Quinodimethane)....[(N,N,N,N',N',N'-hexamethyl 60. 21
ulvalene) - bis(2,5-diethyl-7,7,8,8-tetracyano Quinodimethane) complex ‹etramethyl-tetrathiaf 60. 35
Quinol 17. 5
hydro-1,3,3a,9a-tetramethyl-4H-pyrazolo(3,4-b) Quinolin-4-one [3a,9a-di 36. 14
bis(2-methyl-8-hydroxy Quinolinato)-beryllium dihydrate 67. 3
8-hydroxy Quinoline 35. 12
8,8'-dithio-di Quinoline 35. 48
2-(2-pyridylamino)-8-hydroxy Quinoline 35. 29
2-ethoxy-1-ethoxycarbonyl-1,2-dihydro Quinoline 35. 32
1,2-dihydro-2,2-diphenyl Quinoline-1-oxyl 35. 61
‹droxy-4-oxo-5-(2-oxopropyl)-1H-pyrrolo(2,3-f) Quinoline-2,7,9-tricarboxylic acid dihydrate 36. 19
rubidium 8-mercapto- Quinoline-5-sulfonate 35. 10
sodium Quinoline-8-thiolate dihydrate 35. 11
dideutero- Quinolinic acid............................ 33. 21n
dideutero- Quinolinic acid............................ 33. 22n
dideutero- Quinolinic acid............................ 33. 20n
methyl-bis(8-hydroxy-2- Quinolyl)amine 35. 52
bis((8- Quinolyl-oxy)-ethoxy-ethyl)ether thiourea 17. 40
tris((2-methyl-8- Quinolyl-oxy)-ethyl)amine dihydrate.......... 35. 84
tris((2-methyl-8- Quinolyl-oxy)-ethyl)amine rubidium iodide . 35. 85
rubidium iodide] 1,20-bis(8- Quinolyl-oxy)-3,6,9,12,15,18-hexaoxa-eicosane 35. 83
N,N'-(di-β-naphthyl)-p- Quinone-di-imine 24. 12
hexachloro- Quinoxaline.................................. 35. 4
Quinuclidine 37. 1
Rafoxamide 13. 18
thyrotropin- Releasing-hormone tartrate monohydrate 48. 47
13-cis- Retinal 54. 2
methyl-7,9-cis- Retinoate 54. 14
Retrorsine hydrobromide ethanol solvate........ 58. 30
Retusine 58. 21
thio-bis(formamidinium) trans-tetrachloro-oxo- Rhenium(v) chloride monohydrate............[di 8. 9
p-diethylaminobenzylidene- Rhodanine.................................. 41. 51
ammonium) hydrido-pentacosacarbonyl-tetradeca- Rhodium.............. [tris(tetraethyl 3. 5
m hexakis(μ-carbonyl)-iodo-octacarbonyl-penta- Rhodium.............. [tetra-n-butylammoniu 3. 50
)iminium (μ-pentacarbonyl)-decacarbonyl-penta- Rhodium.............[bis(triphenylphosphine 64.158
m) hexa-μ-carbonyl-carbido-heptacarbonyl-hexa- Rhodium.............[bis(tetraphenylphosphoniu 64.126
tetraethylammonium (μ(5)-carbido)-(dicarbonyl- Rhodium-(μ(2)-carbonyl)-dicarbonyl-iron)-tris› 3. 33

epoxy-	Rhodophytin..	53.	1a
	Rhuslactone triol	56.	3
1,5-anhydro-2,3,4-tri-0-benzoyl-	Ribitol	45.	52
thyl 3-C-acetyl-4,6-0-benzylidene-2-deoxy-α-D-	Ribo-hexopyranoside........................[me	45.	38
9-(5-phenylseleno-5-deoxy	Ribofuranosyl)-adenine	47.	33
nitrocarboxamido)-1-(2',3',5'-tri-0-nitro-β-D-	Ribofuranosyl)-imidazole-4-carboxamide <-5-(N-	45.	19
3-(β-D-	Ribofuranosyl)-1,2,4-triazole-5(1H,4H)-thione	45.	11
2,5'-anhydro-1-(2',3'-0-isopropylidene-β-D-	Ribofuranosyl)-2-thio-uracil	47.	28
1β-D-	Ribofuranosyl-pyridin-4-one-3-carboxamide	47.	25
e-2-deoxy-3-0-((methylthio)-thiocarbonyl)-α-D-	Ribopyranoside .... [methyl 4,6-0-benzyliden	45.	35
methyl β-D-	Ribopyranoside phosphite triester	64.	16
methyl-D-	Ribopyranoside-thiophosphate-triester	46.	4
	Rifamycin B p-iodoanilide acetone pentahydrate	50.	31
3-carbomethoxy-	Rifamycin SV monohydrate	50.	25
nohydrate]	Rifamycin Y p-iodoanilide dimethylsulfoxide mo	50.	30
	Rosigenin	59.	3
(3)-	Rotane	29.	1
(4)-	Rotane	30.	1
(5)-	Rotane	30.	3
(6)-	Rotane	30.	5
	Rubidium hydrogen α-methoxy-α-phenylacetate....	2.	30
	Rubidium hydrogen bis(dibromoacetate)	2.	6
tris((2-methyl-8-quinolyl-oxy)-ethyl)amine	Rubidium iodide................................	35.	85
quinolyl-oxy)-3,6,9,12,15,18-hexaoxa-eicosane	Rubidium iodide [1,20-bis(8-	35.	83
	Rubidium oxalate monoperhydrate................	2.	15n
	Rubidium oxalate perdeuterate	2.	14n
prolinomycin	Rubidium picrate toluene chloroform solvate....	50.	38
doxantrazole	Rubidium salt	39.	49
deoxycholic acid	Rubidium salt	51.	50
3-(5-tetrazolyl)-thioxanthone-10,10-dioxide	Rubidium salt	39.	49
4,7,10,13,16-pentaoxa-1-thiacyclo-octadecane	Rubidium thiocyanate	42.	11
d monohydrate]	Rubidium 3α,12α-dihydroxy-5β-cholan-24-oic aci	51.	50
	Rubidium 8-mercapto-quinoline-5-sulfonate	35.	10
1-acetyl-	Rubranitrose	45.	18
ropylidene-10a-0-(o-bromobenzoyl)-11-0-methyl-	Rubrolone .... [9,10-0,0-isop	50.	23
hydrido)-hexacarbonyl-di-iron-hexacarbonyl-di-	Ruthenium <hine)iminium (μ(2)-carbonyl)-(μ(2)-	64.156	
2)-hydrido)-tricarbonyl-iron-nonacarbonyl-tri-	Ruthenium <osphine)iminium (μ(2)-carbonyl)-(μ(	64.155n	
tetraethylammonium hydrogen dodecachloro-tri-	Ruthenium hexahydrate	3.	38
<-(p-tolylisocyanide)-bis(triphenylphosphine>-	Ruthenium(ii) perchlorate chloroform solvate >	71.270	
ylarsonium (μ(6)-hydrido)-hexakis(tricarbonyl-	Ruthenium> [tetraphen	65.	13n
do-tetrakis(μ(2)-carbonyl)-hexakis(dicarbonyl-	Ruthenium <is(tetraphenylarsonium) μ(6)-carbi	65.	15
m) carbonyl-(tri-μ-carbonyl)-tris(tricarbonyl-	Ruthenium)-cobalt <((triphenylphosphine)immoniu	64.154	
hine)iminium) nonacarbonyl-hexa-μ(2)-carbonyl-	Ruthenium-tetra-iridium <is(bis(triphenylphosp	64.163	
hino)iminium) heptacarbonyl-μ(2)-hexacarbonyl-	Ruthenium-tetra-iridium <is(bis(triphenylphosp	64.162	
	Ryomenin methyl ester	53.	17
	Saframycin C hydrobromide monohydrate	50.	20
yanoquinodimethane> N,N'-(1,2-phenylene)-bis(	Salicylaldiminato)-copper(ii) - 7,7,8,8-tetrac	60.	40
3'-chloro-4'-(p-chlorophenoxy)-3,5-di-iodo-	Salicylanilide	13.	18
calcium	Salicylate dihydrate	14.	1
calcium	Salicylate dihydrate	14.	2
strontium	Salicylate dihydrate	14.	3
N-(5-methoxy	Salicylidene)-3-aminopyridine	33.	51
N-	Salicylidene-3-aminopyridine	33.	44
4-ethoxy-	Salicylidene-4'-butylaniline	17.	31
4-propoxy-	Salicylidene-4'-butylaniline	17.	36
6,12-di	Salicyloyl-benzo(1,2-b:4,5-b')-bis(benzofuran)	38.134	
γ-Mη-	Santonin	53.	5
2α-bromo-α-tetrahydro-	Santonin	53.	7
2β,14-dibromo-4α,5β,6β,11βH-tetrahydro-	Santonin	53.	6a
N-t-butyloxycarbonyl-sarcosyl-	Sarcosine	48.	34
N-t-butyloxycarbonyl-L-prolyl-	Sarcosine	48.	41
di	Sarcosine hydrobromide	48.	9
L-prolyl-	Sarcosine monohydrate	48.	24
t-butoxy-carbonyl-L-prolyl-	Sarcosine-benzyl ester	48.	64
N-t-butyloxycarbonyl-	Sarcosyl-sarcosine	48.	34
N-cyclopropylmethyl-	Scopolammonium bromide	38.106	
dimorpholino-di	Selane	40.	10
dimorpholino-tri	Selane	40.	11
dimorpholino-tetra	Selane	40.	12
dipiperidino-tetra	Selane	33.	37
ihydronaphthalene]	2- Selena-1-dioxothia-3-formyl-5,8-dimethyl-1,2-d	42.	6
naphtho(1,8-cd:4,5-c'd')-bis(1,2,6)	Selenadiazine	42.	5
perfluoro-	Selenanthrene	11.	12
trimethylphosphine-	Selenide	64.	5
(v))]	tetraphenylarsonium μ- Selenido-μ-diselenido-bis(tetrachloro-tungsten	65.	18
tetraphenylarsonium μ-selenido-μ-di	Selenido-bis(tetrachloro-tungsten(v))	65.	18

                                      dioxane- Seleninyl-dichloride 38.   3
                  6-methoxy-4-oxo-3-phenyl Seleno-cyclohexane-1,2-dicarboxylic anhydride 38. 70
                             9-(5-phenyl Seleno-5-deoxyribofuranosyl)-adenine 47. 33
                                               Selenobiotin 39. 30
hane complex]   5-phenyl-1,3-thia Selenole-2-thione 7,7,8,8-tetracyanoquinodimet 60.   8
                                    triphenyl Selenonium chloride monohydrate.............. 11. 18
                   bis(diphenylphosphino- Selenoyl)-acetylene......................... 64.143
         tris(1,2-bis(diphenylphosphino- Selenoyl)ethane) p-xylene clathrate........... 61.   9
ne-3,5-dione]   1,2-(tetrahydro-methyl- Semibullvalene-diyl)-4-bornyl-1,2,4-triazolidi 37. 18a
        4-hydroxyimino-3,5-dimethyl-1-thio Semicarbamoyl-5-thiosemicarbazido-2-pyrazoline 32. 20
yimino-3,5-dimethyl-1-thiosemicarbamoyl-5-thio Semicarbazido-2-pyrazoline .... [4-hydrox 32. 20
                 cyclopentadecanone phenyl Semicarbazone 23.   6
           benzaldehyde 4-benzoyl-2-phenylthio Semicarbazone 13. 19
         4-m-aminophenyl-2-formylpyridine-thio Semicarbazone 33. 53
l-spiro(oxazolidine-5,3'-piperidine)-2'-one-4- Semicarbazone monohydrate   [2-imino-6'-methy 40. 15
<,4-dimethoxybenzyl)-4(2,4-dimethoxyphenyl)-1- Semicarbazono-6,7,8-trimethoxy-naphthalene-2-> 27. 33
         4-methyl-12-hydroxy-13,14,15-trimethyl- Senec-1-eninium iodide 58. 34
                                              L- Serine L-ascorbic acid 48.   8
                                               Shanorellin.................................... 18.   3
                                               Sibiricine 58. 44
         9,9-dimethyl-10-(3'-piperidyl-propyl)-9- Sila-acridane 63. 35
     -dibenzo(b,f)-10,10,11,11-tetramethyl-10,11-di Sila-azepine <hylaminopropyl)-10,11-dihydro-5H 63. 34
   8,8-octamethyl-1,3,5,7-tetra-aza-2,4,6,8-tetra Sila-bicyclo(3.3.0)octane <esityl-2,2,4,4,6,6, 63. 44
                                               Sila-difenidol 63. 31
   3-trimethyl-2-(trimethylsilyl)-3-phenyl-1,3-di Sila-indan .... [1,1, 63. 29
   3-trimethyl-2-(trimethylsilyl)-3-phenyl-1,3-di Sila-indan .... [1,1, 63. 30
                                               Sila-pridinol 63. 27
                             octa(vinyl- Sila-sesquioxane) 63. 21
                      dodecamethyl-hexa Sila-tetra-arsa-adamantane 65.   8
    2,2,4,4,6,6,8,8-octamethyl-2,4,6,8-tetra Sila-1,5-dimercura-cyclo-octane 63. 11
  <,11,11-tetrakis(ethoxycarbonyl)-1,3,7,9-tetra Sila-2,8,13,14-tetraoxatricyclo(7.3.1.1(3,7))> 63. 38
     -bis(dimethylamino)-2,2-dimethyl-1,3-dithia-2- Sila-4,5-dibora-cyclopentane   [4,5 63.   4
     rimethylsilyl)-3,4-bis(trimethylsiloxy)-1,2-di Silacyclobutane <,4-di-t-butyl-1,1,2,2-tetra(t 63. 48
     mino)-1,3-bis(trimethylsilyl)-1,3-diaza-2,4-di Silacyclobutane <is(isopropyl(trimethylsilyl)a 63. 49
     ethyl-1,3-bis(trimethylsilyl)-1,3-diaza-2,4-di Silacyclobutane <trimethylsilyl)amino)-2,4-dim 63. 39
                                    triphenyl- Silane 63. 24
                       1,1,2,2-tetraphenyl-di Silane 63. 40
                      1,1,2,2-tetramesityl-di Silane 63. 51
                   1,1,2,2-tetracyclohexyl-di Silane 63. 41
               tris(acetone-oximato)-phenyl- Silane 63. 19
                   trimethyl-(pyrrolidinomethyl)- Silane 63.   8
                    trimethyl-(2-pyrrolidinoethyl)- Silane 63. 10
                9-fluorenyl-tris(trimethylsilyl)- Silane 63. 37
         4-bromobenzoyloxymethyl-trifluoro- Silane 63.   7
               bis(acetone-oximato)-methyl-chloro- Silane 63.   6
             dichloro-9-fluorenyl-trimethylsilyl- Silane 63. 20
      5-ethyl-5,10-dihydro-10,10-diphenylphenaza- Silane 63. 43
  -(tricyclo(5.2.1.0(2,6))dec-3-en-9-yl)-methyl- Silane)................ [oxo-bis(hydroxy 63. 36
             tris(dimethylamino)-chloro- Silane-(aluminium chloride)................... 63.   3
                           9-fluorenyl-diphenyl- Silanol 63. 42
              diphenyl-(2-piperidino-ethyl)- Silanol 63. 27
             (3-piperidino-propyl)-diphenyl- Silanol 63. 31
               1-chloromethyl-3,7-dimethyl- Silatrane 63.   9
                          p-fluorophenyl- Silatranone 63. 12
                  m-trifluoromethylphenyl- Silatranone 63. 15
  5,7-tetramethyl-1,3,5,7-tetraphenyl-cyclotetra Silazane   [1,3, 63. 47
  ,8,8,9,9,11,11-dodecamethyl-bicyclo(3.3.3)hexa Silazane   [2,2,4,4,6,6 63. 14
  iloxanyl-oxadimethylsilyl)-tetramethyl-cyclodi Silazane <-bis(2,2,4,4,6-pentamethylcyclo-tris 63. 26
      thylpiperazinyl)-10,11-dihydro-5H-dibenzo(b,f) Silepane hydrogen fumarate <-dimethyl-10-(4-me 63. 33
           triethylammonium tris(pyrocatechol) Silicate 63. 23
            tetraguanidinium α-dodecamolybdo- Silicate monohydrate 8.   7
                     tris(acetylacetonato) Silicon perchlorate........................... 63. 18
  3,3-tetraphenyl-1,3-bis(2-piperidino-ethyl)-di Siloxane   [1,1, 63. 52
     -tetramethyl-3,5,9,11-tetraphenyl-tricyclohexa Siloxane .... [1,1,7,7 63. 46
  henyl-3,3,5,5,9,9,11,11-octamethyl-bicyclohexa Siloxane)   [cis-1,7-dip 63. 32
                   octaphenyl-cyclotetra(Siloxane) 63. 53
  isil> N,N'-bis(2,2,4,4,6-pentamethylcyclo-tri Siloxanyl-oxadimethylsilyl)-tetramethyl-cyclod 63. 26
  ane]   (p-chlorophenylimino-(trimethyl Siloxy)-methyl)-phenyl-(trimethylsilyl)-phosph 64.106
  ,1,2,2-tetra(trimethylsilyl)-3,4-bis(trimethyl Siloxy)-1,2-disilacyclobutane <,4-di-t-butyl-1 63. 48
                     t-butyl-dimethyl Siloxy-aci-nitrodiphenylmethane.............. 63. 28
                     t-butyl-dimethyl Siloxy-aci-nitrophenylmethane................ 63. 16
    2-methoxycarbonyl-9-methylene-8-trimethyl Siloxy-tricyclo(6.2.1.0(1,5))undecan-3-one 63. 22
                        dipicrylaminato- Silver 15.   3
  4,7,10,13,16-pentaoxa-1-thiacyclo-octadecane Silver nitrate 42.   8
                        (1-methylthymine)- Silver(i) 44. 14
                         hexakis(trimethyl Silyl lithium) 63.   1

iso Soforidine 58. 11
6-deoxy-α-L- Sorbofuranose 45. 9
2-methyl- Sparteine perchlorate 58. 23
Δ(2)-dehydro-4-oxo- Sparteine perchlorate hemihydrate 58. 10
Sparteine-N(16)-methyliodide 58. 24
enzylidene-2,3-dideoxy-α-D-arabino-hexopyra> Spiro((1-acetylaziridine)-2,3'-(methyl-4,6-0-b 45. 43
ethylamine] N-(1'-t-butyl- Spiro(adamantane-2,2'-aziridine)-3'-ylidene)-m 32. 59
<t-butyl-3'-(2,2-dimethylpropionyl)-1'-methyl- Spiro(benzofuran-3(2H)-2'-pyrrolidine)-2,5'-d> 38.132
e)-1,6-di> dimethyl 8-t-butyl-5-methyl-2-oxo- Spiro(bicyclo(2.2.2)octa-5,7-diene-3,2'-oxiran 38. 92
e)-5,6-di> dimethyl 8-t-butyl-1-methyl-2-oxo- Spiro(bicyclo(2.2.2)octa-5,7-diene-3,2'-oxiran 38. 91
',6-dione] syn-(1R,2R,4S)-4-methyl- Spiro(bicyclo(2.2.2)octane-2,1'-cyclohexane)-3 31. 31
<S,4R)-3'-chloro-10',11'-dihydro-N,N-dimethyl- Spiro(cyclohex-2-ene-1,5'-(5H)-dibenzo(a,d)cy> 29. 9a
loheptene)-4> (1S,4R)-3'-chloro-N,N-dimethyl- Spiro(cyclohex-2-ene-1,5'-(5H)-dibenzo(a,d)cyc 29. 8a
<phenyl)-ethylidene)-3a',4',7',7a'-tetrahydro- Spiro(cyclohexa-2,5-diene-1,1'-(1H)-indene)-4> 17. 44
3)d> 3',4,9'-tri-t-butyl-2',5,10'-trimethoxy- Spiro(cyclohexa-3,5-diene-1,6'-dibenzo(d,f)(1, 38.138
<2,2-tetrachloro-3,3-dimethyl-1a',6b'-dihydro- Spiro(cyclopropane-1,2'-(1H)-cycloprop(a)inde> 29. 3
micarbazone monohydrate] 2-imino-6'-methyl- Spiro(oxazolidine-5,3'-piperidine)-2'-one-4-se 40. 15
'-ene)> dimethyl 6'-t-butyl-2'-methyl-3'-oxo- Spiro(oxirane-2,4'-tricyclo(3.3.0.0(2,8))oct-6 38. 93
hexahydro-15H-cyclopenta(a)phenanthrene-17,2'- Spiro(1',3'-dithiolane) <l)-11,12,13,14,16,17- 51. 45
hexa Spiro(2.0.2.0.2.0.2.0.2.0)octadecane 30. 5
penta Spiro(2.0.2.0.2.0.2.0)pentadecane 30. 3
tetra Spiro(2.0.2.0.2.0.2.0)dodecane 30. 1
tri Spiro(2.0.2.0.2.0)nonane 29. 1
thyl)-phenyl)-1,3,5,7-tetra-aza-4γ(5)-phospha- Spiro(3.3)heptane-2,6-dione <is(3-(trifluorome 64.122
xatricyclo(5.4.0.0(2,4))undecan-10-one-9,2(1)- Spiro(3(1)-formyl-3(1)-methyl-oxirane) <yl-3-o 59. 16a
1,2-h)cyclononene-3,1'-cyclopentane)] Spiro(3,4-dihydro-2H-1,5-dioxadinaphtho(2,1-f: 38.123
1,2-h)cyclononene-3,1'-cyclopropane)] Spiro(3,4-dihydro-2H-1,5-dioxadinaphtho(2,1-f: 38.121
tetra-0-acetyl-1-deoxy-D-glucopyranoside-1,2'- Spiro(3'-methyl-3'-hydroxy-tetrahydrofuran) <- 45. 46
tetra-0-acetyl-1-deoxy-D-glucopyranoside-1,2'- Spiro(3'-methyl-3'-hydroxy-tetrahydrofuran) <- 45. 45
1,13,16,19-tetraoxatetra Spiro(4.0.0.0.4.3.3.3)heneicosan-4-one 38. 85
12-benzoylimino-2-phenyl-3-oxa-1-aza-tri Spiro(4.0.5.2.5.0)nonadec-1-ene-4,13-dione 40. 56
2,7-diaza- Spiro(4.4)nonane-1,6-dione 35. 3
methyl-cyclopentyl)-1-oxa-7,7,9,9-tetramethyl- Spiro(4.4)nonane-3,6-dione <-oxo-3,3,5,5-tetra 38.108
7-methyl-1,4-dithia-7-aza- Spiro(4.4)nonane-6,8-dione 39. 13
4-methyl-6-(N-methyl-N-phenylamino)- Spiro(4.5)dec-6-ene-1,8-dione 27. 24
3,6-dimethyl-4,10-dihydroxy-2-oxa- Spiro(4.5)dec-7-ene-1,9-dione 59. 3
8-hydroxy-3-methoxy-3-methoxycarbonyl-2-oxa- Spiro(4.5)deca-6,9-diene 38. 41
rile sol> 7-benzoylamino-7-methoxycarbonyl-di Spiro(5.1.5.2)pentadecane-14,15-dione acetonit 28. 16
<,3,7,7-tetramethyl-bicyclo(4.4.0)decane-2,2'- Spiro(6',7'-bis(hydroxymethyl)-4'-(p-bromophe> 50. 21
2-methyl-2,9- Spiro-bis(1,3,2λ(5)-benzoxazarsoline) 65. 9
3,3'- Spiro-bis(3H-2,4-dihydro-benzo(1,4)dioxepin) 38. 82
0-acetyl-kreysiginone- Spiro-isomer methiodide...................... 58. 63
2,5-dicyclohexane- Spiro-1,3-dithiane 39. 55
<2.2.1(3,6).0(2,7))tridecane-4,11-dien-9-one)- Spiro-10,4'-(9-anti-acetoxy-3-oxatricyclo(4.2> 38.119
<2.2.1(3,6).0(2,7))tridecane-4,11-dien-9-one)- Spiro-10,4'-(9-anti-acetoxy-3-oxatricyclo(4.2> 38.118
<silyl)-1,3,2,4-diazadiphosphetidine)-2',4'-di Spiro-2,2''-(4,5-dimethyl-1,3-dioxa-2-phospho> 64. 99
rime> (4,5-dimethyl-1,3-dioxa-2-phosphole)-2- Spiro-2'-(cis-2,4-bis(dimethylamino)-1,3-bis(t 64. 69
xanthene-9- Spiro-2'-(3',4'-bis-t-butyl-thiothiete)........ 39. 82
xanthene-9- Spiro-2'-(4'-t-butyl-3'-(methylthio)-thiete) 39. 77
1,3-dimethyl-4-imino-5,5-(Spiro-2',4'.-bis-dimethylamino-pyrrolyl)-uracil 35. 28
dihydro-2H,6H-2-benzothiopyrano(4,3-b)pyran-2- Spiro-3'-(1H-2)-benzothiopyran-4'(3'H)-one <4- 39. 76
(1,5,5-trimethyl-bicyclo(4.2.0)octan-7-one)-8- Spiro-4-(2,2-dimethyl-5-oxo-1,3-dioxolane) 38. 72
<3,9a-tetrahydro-cyclohexa(i,j)isoquinoline-7- Spiro-4'-(2'-methoxy-cyclohexa-2',5'-dien-1'-> 58. 63
-phenylethyl)-1,4-diazocyclohex-5-en-2-one)-3- Spiro-5'-(3'-methyl-hydantoin) <l-4-(3-methoxy 36. 37
cyclo-octane- Spiro-5'-hydantoin 35. 21
N-phenethyl-granatanine-3- Spiro-5'-hydantoin 37. 13
3,7-dimethyl-3,7-diazabicyclo(3.3.1)nonane-9- Spiro-5'-hydantoin 37. 7
tropane-3- Spiro-5'-thio-hydantoin..................... 37. 6
4-t-butyl-3-(methylthio)-thiete-2- Spiro-9'-xanthene 39. 77
e be> 7-chloro-2',4,6-trimethoxy-6'-β-methyl- Spirobenzofuran-2(3H)-cyclohex-2-ene-3,4'-dion 50. 11
solvate] Spirohydroxyprednisolone-acetate ethyl acetate 51. 40
1β,2β,3β,4β,5β,7α-hexahydroxy- Spirost-25(27)-en-6-one 51. 62
Squalene 56. 4
hexakis(p-t-butylphenylthiomethyl)benzene Squalene clathrate 61. 26
1,3-bis(trimethylsilyl)-4-methyl-diaza Stanna-boretidine dimer 63. 17
2,2,4,4,6,6-hexamethylcyclotri Stanna-tellurane 70. 2
p-bromophenacyl- Stellatate 55. 4a
Stercuronium iodide ethanol solvate........... 58. 69
2-chloro-1,3,6-trithia-2- Stibaocane 66. 4
tetraphenyl-di Stibine 66. 15
2,3-bis(diphenyl Stibino)-maleic anhydride 66. 18
(E)-2,2'-diaza Stilbene 33. 42
(E)-4,4'-diaza Stilbene 33. 43
(E)-2',5,5'-triaza Stilbene 33. 40
1,2-bis((benzoyl)oxy)- Stilbene 13. 22
(E)-2,2',5,5'-tetra-aza Stilbene 33. 35

(E)-dibenzo(e,e')-3,3'-diaza Stilbene 35. 56
(Z)-dibenzo(e,e')-3,3'-diaza Stilbene 35. 57
mplex]   Stilbene - 7,7,8,8-tetracyanoquinodimethane co 60. 32
mplex]   tetramethoxy Stilbene - 7,7,8,8-tetracyanoquinodimethane co 60. 38
        tri-o-thymotide cis- Stilbene clathrate 61. 17
       tri-o-thymotide trans- Stilbene clathrate 61. 14
(E)-4,4'-dimethyl-4,4'-diazonia Stilbene di-iodide 33. 62
(E)-6,6'-dimethyl-6,6'-diazonia Stilbene di-iodide 33. 63
(E)-6-methyl-6-azonia-2'-aza Stilbene iodide 33. 52
                  acetyl- Strongylophorine-2 .... 54. 33
                           Strontium maleate tetrahydrate 2. 21
                           Strontium salicylate dihydrate 14. 3
enzene]   2,5-di(Styryl)-pyrazine 1,4-bis(2-(2-pyridyl)vinyl)-b 33. 76
       4-(p-N,N-dimethylamino Styryl)-pyridine-methiodide................................. 33. 69
           sodium perfluoro Succinate hexahydrate 2. 18
                 calcium Succinate trihydrate 1. 10
          potassium fluoride Succinic acid 1. 12
           N-(t-butyldithio)- Succinimide... 32. 22
    α-(3-bromo-4-ethoxyphenyl) Succinimide 32. 43
          dimorpholino-tetra Sulfane 40. 9
  bis(trimethylacetyl-dichloromethylene)-tri Sulfane 11. 14
(3,4,5-trimethoxybenzyl)-pyrimidine]   3- Sulfanilamido-5-methylisoxazole 2,4-diamino-5- 44. 30
              triethylarsine- Sulfide. 65. 4
             di-t-adamantyl-di Sulfide. 31. 54
           bis(2-pyrimidyl)-di Sulfide. 44. 20
         bis(triphenylmethyl)- Sulfide. 11. 20
     1-phenyl-4-phosphorinanone-1- Sulfide. 64. 45
      bis(2,4-diphenyl-3-thienyl)-di Sulfide. 39. 85
     bis(2,5,6-trifluoro-4-nitrophenyl)- Sulfide. 15. 2
    bis(methylsulfonyldichloromethyl)-di Sulfide. 11. 5
    1-methyl-3-benzyl-4-(1,2,3-triazolio)- Sulfide. 32. 32
    1-methyl-3-phenyl-4-(1,2,3-triazolio)- Sulfide. 32. 25
    1-methyl-3-phenyl-4-(1,2,3-triazolio)- Sulfide. 32. 24
    1-phenyl-3-methyl-4-(1,2,3-triazolio)- Sulfide. 32. 26
 1,5-diphosphabicyclo(3.3.0)octane-1,5-di Sulfide. 64. 20
   2-carboxy-3,4-dimethyl-1-phenylphosphole- Sulfide. 64. 58
   bis(1-methyl-2-(3-pyridyl)-pyrrol-3-yl)-di Sulfide. 32. 69
2,2'-bis(1,3,2λ(5)-dioxaphosphorinane)-2,2'-di Sulfide.........................[5,5,5',5'-tetramethyl- 64. 41
 henyl)-1,3,4,2-thiadiazaphosphiolidin-5-one di Sulfide.........................[3,4-bis(t-butyl)-2-(p-methoxyp 64. 77
um) dichloro-iron-bis(µ(2)-thio)-molybdenum-di Sulfide <ylphosphonium (benzyl-trimethylammoni 64.123
          bis(2-aminoethyl)-di Sulfide dihydrochloride. 3. 29
 pentasulfur-iron-bis(µ-sulfido)-molybdenum-di Sulfide dimethylformamide solvate <phosphonium 64.130
um pentasulfur-iron-bis(µ-sulfido)-tungsten-di Sulfide dimethylformamide solvate <ylphosphoni 64.131
       bis(N,N'-diethyl)formamidine- Sulfide hexachloro-tellurium 11. 11
 hydroxy-4,5-dimethyl-1,3,2-dioxaphospholane-2- Sulfide imidazolium monohydrate.... [trans-2- 64. 8
   bis(4-(N-methyl-piperidinium))-di Sulfide tetrachloro-copper(ii) 33. 50
        cyclo-tetrakis(phenylarsenic- Sulfide) 65. 22
antimony-oxo)-cyclododecasulfur tris(carbon di Sulfide) .... [1,7-bis(pentachloro- 11. 3
       9-cyclohexyl-5(7)-dimethyl Sulfide-nido-decaborane................................. 62. 10
tetraphenylphosphonium pentasulfur-iron-bis(µ- Sulfido)-molybdenum-disulfide dimethylformami> 64.130
<tacarbonyl-chromium-thio)-methylidyne)-(µ(3)- Sulfido)-triangulo-tri-cobalt tetrahydrofuran> 71.169
tetraphenylphosphonium pentasulfur-iron-bis(µ- Sulfido)-tungsten-disulfide dimethylformamide> 64.131
       S,S-dimethyl-N-methylsulfonyl Sulfilimine... 4. 2
       S,S-dimethyl-N-methylsulfonyl Sulfimide... 4. 2
     2-methyl-2-(N-phenyl Sulfinamoyl)-N,N-dimethyl-propanamide 1. 30
         methyl-2-(methyl Sulfinyl)-benzoate 13. 6
      meso-2,2-bis(methyl Sulfinyl)-1,3-diphenylpropane 11. 17
    (+-)-3-(bis(phenyl Sulfinyl)methyl)-1,2-dimethyl-cyclopropene 20. 9
    cis-2-ethoxy-1-phenyl Sulfinyl-cyclopropane 20. 4
ole]   3-benzoyl Sulfinyl 5 (N n butylbenzamido)-1,2,4-thiadiaz 41. 73
     2,4,6-tri-t-butyl-N- Sulfinylaniline.. 16. 17
     4-nitroaniline 4-nitro-N- Sulfinylaniline. 16. 1
    0,0-bicyclohexyl-1,1'-diyl-thio Sulfite. 39. 45
iane]   1-(S,S-dimethyl-N-(trimethylsilyl)- Sulfo-di-imide)-bicyclo(3.3.1)penta-azatetrath 63. 2
               2,5-divinyl-1- Sulfolane 39. 23
   carboxymethane-(N-methyl-N-phenyl)- Sulfonamide) 1. 22
    2-carboxyethane-(N-methyl-N-phenyl)- Sulfonamide) 1. 24
 2-methoxycarbonyl-vinyl-(N-methyl-N-phenyl)- Sulfonamide) 1. 26
 2-carboxy-cyclopentene-1-(N-methyl-N-phenyl)- Sulfonamide) 20. 5
            3β-(toluene-p- Sulfonamido)-5α-cholestan-2β-ol 51. 79
      3β-azido-2α-(toluene-p- Sulfonamido)-5α-cholestane 51. 78
     rubidium 8-mercapto-quinoline-5- Sulfonate 35. 10
amino-5-(1-adamantyl)-6-methylpyrimidine ethyl Sulfonate ....[2,4-di 44. 33
,L-phenylglycinium ethyl ester trifluoromethyl Sulfonate <osphino)-phenyloxy)-diethyl ether D 64.170
    trisulfur-dinitrido trifluoromethyl Sulfonate acetonitrile solvate 11. 1
              (+-)-epinine-β- Sulfonate hemihydrate 17. 11

```
methyl-2-(3,4-dihydroxyphenyl)ethylammonium-2-   Sulfonate hemihydrate ............   [(+-)-N-  17. 11
                        bromocriptine methyl    Sulfonate isopropanol solvate ...............  58. 76a
                   tri-potassium methyltri      Sulfonate monohydrate ......................  11.  2
            (-)-dihydroergotamine methyl        Sulfonate monohydrate ......................  58. 77a
        ammonium 1-anilino-8-naphthalene-       Sulfonate monohydrate ......................  27. 16
    -imidazo(5,1-f)(1,2,4)triazin-4(3H)-one methyl  Sulfonate monohydrate <amino-5-methyl-7-propyl  35. 13
        7,8-dibromo-8-cyano-4-methyl            Sulfonato-bicyclo(3.2.1)octa-2,6-diene .....  31.  6
e]  4-bromo-4-cyano-7-methoxycarbonyl-7-methyl  Sulfonato-tetracyclo(3.3.0.0(2,6).0(3,8))octan  31. 16
            disodium cis-syn-4-                 Sulfonatobenzene-diazotate trihydrate ........  11.  8
                                biotin          Sulfone......................................  39. 32
                    2-pyridyl-phenyl            Sulfone......................................  33. 39
            ethane-1,2-bis(methyl               Sulfone) ..................................  11.  6
            ethane-1,2-bis(phenyl               Sulfone) ..................................  11. 15
                                μ-              Sulfone-syn(CH(2),CH(3))-9,10-dioxa-bimane ....  41. 31
    yl-1-piperazinyl)propyl)-phenothiazine methane  Sulfonic acid .........[2-chloro-10-(3-(4-meth  41. 74
    1-(3,4-dihydroxyphenyl)-2-methylaminoethane-    Sulfonic acid hemihydrate .................  17. 10
m hexahydrate]  tetraphenylphosphonium (μ(2)-   Sulfonyl)-(μ(4)-dithio)-octacyano-di-molybdenu  64.133
                1,4-bis(methyl                  Sulfonyl)-1,4-dihydro-1,4-diazocine.........  34.  4
            trans-(2R,5R)-1-(p-bromophenyl      Sulfonyl)-2,5-dimethylpyrrolidine ..........  32. 44a
e]  1-((4-(3-chlorophenylamino)-3-pyridyl)      Sulfonyl)-3-ethyl-1-methylurea hydrogen sulfat  33. 66
            5,5-dimethyl-3-methyl               Sulfonyl-cyclohex-2-enone ..................  11. 10
            1-methylsulfonyl-2-phenyl           Sulfonyl-ethane ............................  11.  9
        E-1-(5-nitro-2-furyl)-2-p-tolyl         Sulfonyl-ethylene ..........................  38. 50
    E-1-(5-nitro-2-furyl)-2-p-chlorophenyl      Sulfonyl-ethylene ..........................  38. 36
            hexakis(R-α-phenylethyl             Sulfonyl-methyl)benzene acetic acid clathrate  61. 24
1-methyl-4-(3-methylphenylamino)-3-pyridinio)-  Sulfonyl-N'-isopropylurea ...........   [N-(   33. 73
                        9-methyl                Sulfonyl-1,7-dicarba-closo-dodecaborane........  62.  3
    1-(5-nitro-2-furyl)-1-trichloromethyl       Sulfonyl-2-(5-bromo-2-furyl)-ethylene ........  38. 28
                        1-methyl                Sulfonyl-2-phenylsulfonyl-ethane ...........  11.  9
)-1H-tetrazole]   1-t-butyl-5-(2-(p-tolyl       Sulfonylamino)-1-(t-butylimino)-2-thioxo-ethyl  32. 66
    yclo(1>  14-(p-chlorophenyl)-8-(p-chlorophenyl  Sulfonylamino)-13-oxa-14-thia-1,15-diazapentac  42. 12
    hane-1H-azeto(1,2-a)(1)ben>  6-(p-chlorophenyl  Sulfonylamino)-2,2a,3,4,5,6-hexahydro-2a,6-met  37. 14
    octahydro-1,8-methano(1)be>  8-(p-chlorophenyl  Sulfonylamino)-4-hydroxyimino-1,2,3,4,5,6,7,8-  37. 16
    -a)indol-8(9H)-one]   12-(p-chlorophenyl     Sulfonylamino)-6,7,10,11-tetrahydroazocino(1,2  36. 32
                        bis(methyl              Sulfonyldichloromethyl)-disulfide ..........  11.  5
    .2.1.1(2,6))tetradec-11-en-8-one p-bromophenyl  Sulfonylhydrazone <β,9α,10α-13-oxatetracyclo(8  38. 99
    .2.1.1(2,6))tetradec-11-en-8-one p-bromophenyl  Sulfonylhydrazone <β,9α,10β-13-oxatetracyclo(8  38. 98
        -2-t-butyl-perhydroazulen-4-one-(p-bromophenyl  Sulfonylhydrazone) .............   [cis   27. 28
    5-methylbenzene-1,3-dicarbaldehyde-bis(p-tolyl  Sulfonylhydrazone) benzene clathrate .......  61.  7
            t-butyl-(2-t-butylimino-1-phenyl     Sulfonylimino-2-cyano-ethyl)amine ...........   3. 53
    <ro(6',7'-bis(hydroxymethyl)-4'-(p-bromophenyl  Sulfonyloxy)-2',3'-dihydrobenzofuran) dimethy>  50. 21
                    2-(p-tolyl                  Sulfonyloxy-N,N,O-azoxy)-adamantane...........  31. 43
            S,S-dimethyl-N-methyl               Sulfonylsulfilimine ........................   4.  2
            S,S-dimethyl-N-methyl               Sulfonylsulfimide ..........................   4.  2
                        biotin-d-               Sulfoxide ..................................  39. 31
    -hydroxy-2-phenyl-2-(2'-pyridyl)-ethyl)-methyl  Sulfoxide ...............................[(2   33. 61
    onylamino)-3-hydroxypropyl-(chloromethyl)-(R)-  Sulfoxide ............   [(2R)-2-(benzyloxycarb  11. 13a
            rifamycin Y p-iodoanilide dimethyl  Sulfoxide monohydrate ......................  50. 30
    dioxo-(3-3'-di-indoline)-N,N'-bis(5,5-dimethyl  Sulfoximide) hydrate ....   [3-hydroxy-2,2'-  35. 60
    1,7-bis(pentachloro-antimony-oxo)-cyclododeca   Sulfur tris(carbon disulfide) ..............  11.  3
    onitrile solvate]  tri                      Sulfur-dinitrido trifluoromethylsulfonate acet  11.  1
            S,S-dimethylpenta                   Sulfur-hexanitride .........................   4.  1
    e dimethylforma>  tetraphenylphosphonium penta  Sulfur-iron-bis(μ-sulfido)-molybdenum-disulfid  64.130
    dimethylformami>  tetraphenylphosphonium penta  Sulfur-iron-bis(μ-sulfido)-tungsten-disulfide  64.131
            triphenylphosphinimino-(tri         Sulfur-nitride).............................  64. 87
        bis(triphenylphosphine-imino)-tetra     Sulfur-tetranitride ........................  64.167
            N,N'-(tetra                         Sulfur-tetranitrido)-thiacyclopentane-di-imide  41.  9
            tetra-n-butylammonium tri           Sulfur-trinitride ..........................   3. 49
            bis(dimethylamino)-difluoro         Sulfurane ..................................   3. 18
                                                Sunguicine acetone solvate .................  58. 79
                                                Swainsonine diacetate ......................  58.  4
                3-(2-aminophenyl)-              Sydnone ....................................  40.  7
                4,5-diphenyliso                 Sydnone benzene solvate.....................  40. 33
    12-dichloro-2,16-dihydro-16-hydroxy-2-methoxy-  Tabersonine.............................   [10,  58. 58
                            trans-              Tamoxifene .................................  17. 41
            thyrotropin-releasing-hormone       Tartrate monohydrate .......................  48. 47
        (glutamyl-α-lactam)-histidinyl-proline  Tartrate monohydrate .......................  48. 47
                    sodium ammonium             Tartrate tetrahydrate ......................   2. 22
                        copper L-               Tartrate trihydrate.........................   1. 11
    etra-n-butylammonium) hexakis(isothiocyanato)-  Technetium(iii)......................[tris(t    3. 51
    s(triphenylphosphine)immonium oxo-tetrachloro-  Technetium(v) .....................[bi         64.159
        1,1-di-iodo-3,4-benzo-1-                Tellura-cyclopentane .......................  70.  3
    2,2,4,4,6,6-hexamethylcyclotristanna-       Tellurane ..................................  70.  2
                            urea ortho          Telluric acid ..............................   8.  4
            tris(morpholino)-phosphine          Telluride ..................................  70. 11
        trichloro-(8-ethoxy-4-cyclo-octenyl)-   Tellurium ..................................  70.  6
```

N,N'-diethyl)formamidine-disulfide hexachloro- Tellurium .... [bis(11. 11
                                        phenyl- Tellurium bromide chloride 70. 1
                                                Tellurium oxalate dihydrogen oxalate dihydrate 2. 10
                        tetrakis(dimethylthiourea)- Tellurium(ii) dichloride 70. 12
                  hexakis(trimethylenethiourea)-di- Tellurium(ii) perchlorate 70. 17
                  hexakis(trimethylenethiourea)-di- Tellurium(ii) perchlorate 70. 16
                     hexakis(ethylenethiourea)-di- Tellurium(ii) tetraperchlorate 70. 15
                          bis(pyridinium) hexachloro- Tellurium(iv) 33. 9
                  bis(monothio-pyrocatecholato)- Tellurium(iv) 70. 7
                  cis-2-ethoxycycloheptyl-tribromo- Tellurium(iv) 70. 5
                  di-cis-2-chlorocyclohexyl-dichloro- Tellurium(iv) 70. 9
                  p-tolyl-2-chlorocyclohexyl-dichloro- Tellurium(iv) 70. 13
                  bis(tetraphenylphosphonium) decabromo-di- Tellurium(iv) 64.125
      s(N-(2-hydroxyethyl)-N-methyldithiocarbamato)- Tellurium(iv) .... [tetraki 70. 14
      s(N-(2-hydroxyethyl)-N-methyldithiocarbamato)- Tellurium(iv) .... [bromo-tri 70. 10
      oxythiazolo(2,3-b)benzothiazolium) hexachloro- Tellurium(iv) dioxane solvate <-dihydro-3-hydr 41. 25
                  6-acetyl-2-(α-(trichloro- Tellurium)-acetyl)-pyridine..................... 70. 4
                  (thiocyanato)-bis(di(3-trimethylsilyl-propyl)- Tellurium)-palladium(ii) .... [trans-bis 70. 18
                  (N,N-diethylethylene- Tellurourea)-pentacarbonyl-chromium............. 70. 8
                                                Terretonin 59. 38
      .8.2.0(1,15).0(2,13).0(4,12).0(7,12).0(15,20))- Tetracosa-3,5,8,10,16,18,21,23-octene <yclo(12 31. 66
      6-tetraoxo-21,22-dioxatricyclo(18.2.1.1(8,11)) Tetracosa-8,10,18,20-trans-4,14-hexaene <,13,1 38.129
                                        6-thia Tetracycline 50. 13
                  5a-epi-6-demethyl-6-deoxy-6-thia Tetracycline dimethylformamide solvate 50. 14
                                        ate] Tetracycline tetrahydrate dichloromethane solv 50. 15
                  4-epi-oxy o-4-cyano-7-methoxycarbonyl-7-methylsulfonato- Tetracyclo(3.3.0.0(2,6).0(3,8))octane [4-brom 31. 16
                                        4,7-dioxa- Tetracyclo(4.3.0.0(3,5).0(3,8))nonan-8-one 38. 7
                  exo-1,3,5,6,7,9-hexamethyl-4,8-dioxa- Tetracyclo(4.3.0.0(3,5).0(7,9))nonan-2-one 38. 53
                  endo-1,3,5,6,7,9-hexamethyl-4,8-dioxa- Tetracyclo(4.3.0.0(3,5).0(7,9))nonan-2-one 38. 54
                  ne] 3,4,6,8β,9,10-hexamethyl- Tetracyclo(4.3.1.0(3,10).0(4,9))decane-2,5-dio 31. 39
                  icarboxyli> 5-(2-nitrophenylthio)-10-acetoxy- Tetracyclo(4.4.0.0(2,4).0(3,7))dec-8-ene-8-d 31. 56
                  <-11,11,12,12-tetracyano-4,5,9,10-tetraphenyl- Tetracyclo(4.4.2.0(1,8).0(3,6))dodeca-2,4,7,9> 31. 69
                  ol] 1β,5β,6β,11β-tetramethyl-3-oxa- Tetracyclo(5.4.0.0(2,6).0(4,11))undec-8-en-2β- 38. 62
                  <S,3RS,6SR,7SR,10SR,11RS,12RS,13RS)-(6-phenyl- Tetracyclo(5.4.2.0(3,13).0(10,12))tridecane-4> 59. 29
                  ione] 6,12-dihydroxy-2,6,9,12-tetramethyl- Tetracyclo(6.2.2.0(2,7).0(4,9))dodecane-5,11-d 31. 41
                  dien-> 13-anti-acetoxy-1,8,11,12-tetrachloro- Tetracyclo(6.2.2.1(3,6).0(2,7))tridecane-4,11- 38.119
                  dien-9> 13-syn-acetoxy-1,8,11,12-tetrachloro- Tetracyclo(6.2.2.1(3,6).0(2,7))tridecane-4,11- 38.118
                  8,10,10-trimethyl-2-oxa- Tetracyclo(6.3.1.0(4,7).0(4,12))dodecan-3-one 38. 61
                  11-isopropyl-8-methyl-2-oxa- Tetracyclo(6.3.1.0(4,7).0(4,12))dodecan-3-one 38. 71
                                                Tetracyclo(6.4.3.0.0(2,7))pentadec-4-en-13-one 31. 34
                                                Tetracyclo(6.4.3.0.0(2,7))pentadec-4-en-13-one 31. 33
                  8),15-diene] benzo(p)-2-thia Tetracyclo(7.5.3.0(3,8).0(1,10))heptadecane-3(39. 79
                  -bromophenylsulfonylhydr> 1α,6β,9α,10α-13-oxa Tetracyclo(8.2.1.1(2,6))tetradec-11-en-8-one p 38. 99
                  -bromophenylsulfonylhydr> 1β,6β,9α,10β-13-oxa Tetracyclo(8.2.1.1(2,6))tetradec-11-en-8-one p 38. 98
                  10-hydroxy-3-oxa- Tetracyclo(8.6.0.0(2,4).0(2,9))hexadecane 38. 73
                  <oxy-5,8-dimethyl-2-(2-methyl-hept-2-en-6-yl)- Tetracyclo(9.3.0.0(1,5).0(7,11))tetradec-8-ene 55. 3
                  <ibenzo-14-ethoxycarbonyl-8,10-diphenyl-14-aza Tetracyclo(9.3.2.0(2,7).0(2,10))hexadeca-7,12> 37. 24
                  2,4,7,9,11,13,15,17-nonene] 5,7-diformyl- Tetracyclo(9.8.1.1(3,9).1(13,18))docosa-1(19), 31. 60
                                        1,1,2,2- Tetracyclohexyl-disilane 63. 41
                                        1,1,2,2- Tetracyclohexyl-ethane 21. 19
                  1α,6β,9α,10α-13-oxatetracyclo(8.2.1.1(2,6)) Tetradec-11-en-8-one p-bromophenylsulfonylhyd> 38. 99
                  1β,6β,9α,10β-13-oxatetracyclo(8.2.1.1(2,6)) Tetradec-11-en-8-one p-bromophenylsulfonylhyd> 38. 98
                  9,9,13-trimethyl-8-oxatricyclo(8.4.0.0(2,7)) Tetradec-2(7)-en-3-one 38. 78
                  pt-2-en-6-yl)-tetracyclo(9.3.0.0(1,5).0(7,11)) Tetradec-8-ene <xy-5,8-dimethyl-2-(2-methyl-he 55. 3
                  tetraethylammonium) hydrido-pentacosacarbonyl- Tetradeca-rhodium .... [tris(3. 5
                  6,6'-(3,6,9,12-tetraoxa Tetradeca-1,14-dioxy)-2,2'-bipyridine 40. 51
                  rabenzo-pentacyclo(6.2.2.2(2,6).0(2,7).0(3,7)) Tetradeca-4,9,11,13-tetraene .... [tet 31. 64
                  tetramethylammonium hexa-cobalt- Tetradecacarbonyl-carbide 3. 19
                  4-ethylidene-5-oxa-3-azatricyclo(8.4.0.0(3,8)) Tetradecane.............[(1RS,8SR,10SR,4(15)Z)- 40. 35
                  asila-2,8,13,14-tetraoxatricyclo(7.3.1.1(3,7)) Tetradecane <akis(ethoxycarbonyl)-1,3,7,9-tetr 63. 38
                  1,8-diaza-3,6-dioxa Tetradecane-2,7-dione 40. 17
                  methyl-hept-2-en-6-yl)-tricyclo(9.3.0.0(9,13)) Tetradecane-2,8-diene-4,7-dione <methyl-10-(2- 55. 1
                  <oxy)-3,4:10,11-dibenzo-tricyclo(7.5.0.0(2,8)) Tetradecane-3,5,10-triene-7,13-dione chlorofo> 30. 18
                  β,9α)-4,5:11,12-dibenzo-tricyclo(7.5.0.0(2,8)) Tetradecane-4,11-diene-3,10-dione   [((1α,2α,8 30. 11
                  β,9α)-4,5:11,12-dibenzo-tricyclo(7.5.0.0(2,8)) Tetradecane-4,11-diene-3,10-dione   [((1α,2β,8 30. 10
                  α,9α)-4,5:12,13-dibenzo-tricyclo(7.5.0.0(2,8)) Tetradecane-4,12-diene-3,14-dione   [((1α,2β,8 30. 12
                  α,9β)-4,5:12,13-dibenzo-tricyclo(7.5.0.0(2,8)) Tetradecane-4,12-diene-3,14-dione   [((1α,2β,8 30. 8
                  1-tetramethyl-1,8,9-cis-tricyclo(7.5.0.0(2,8)) Tetradecane-5,12-dione <13,14-dibenzo-6,6,11,1 30. 14
                  α,9β)-6,7:10,11-dibenzo-tricyclo(7.5.0.0(2,8)) Tetradecane-6,10-diene-3,14-dione   [((1α,2β,8 30. 9
                  5-hydroxy-5-(hydroxymethyl)-2-methyl- Tetradecanoic acid 1. 34
                  r-1,t-2,c-3,t-4,5,6-hexachloro Tetralin 27. 3
                  1,2,3,4,5-pentachloro-8-phenyl Tetralin 27. 15
                  (+)-2-dipropylamino-5-hydroxy Tetralin hydrochloride 27. 21a
                  nylamino)-1-(t-butylimino)-2-thioxo-ethyl)-1H- Tetrazole .... [1-t-butyl-5-(2-(p-tolylsulfo 32. 66
                  5-amino Tetrazole monohydrate 32. 1
                  A,B-bishomo-cholest-4a-eno-(4,3-d)(6,7-d)-bis(Tetrazole) ....[4,6-diaza- 51. 60

5-amino-N-octyl-1H-	Tetrazole-1-carboxamide........................	32. 36
2,3-diphenyl	Tetrazolium-5-olate............................	32. 46
1,3-diphenyl	Tetrazolium-5-thiolate.........................	32. 47
3-(5-	Tetrazolyl)-thioxanthone-10,10-dioxide rubidiu	39. 49
m salt]		
capnell-9(12)-ene-2β,5α,8β,10α-	Tetrol	53. 10
decathia-cyclotetradecane-6,7,13,14-	Tetrone	39. 3
1,4,7,10-tetra-azacyclotetradecane-3,8,11,14-	Tetrone	34. 6
1,4,9,12-tetra-azacyclohexadecane-3,10,13,16-	Tetrone	34. 8
-tetraoxa-12,18-diazacyclododocosane-8,11,19,22-	Tetrone .... [1,4,7,15	40. 42
,13-trioxa-10,16-diazacycloeicosane-6,9,17,20-	Tetrone................. [3,3-dimethyl-1,5	40. 44
-1,5,10,14-tetraoxacyclo-octadecane-6,9,15,18-	Tetrone................. [3,3,12,12-tetramethyl	38. 96
(3,8)heptadeca-5,14,16,18-tetraene-4,7,11,13-	Tetrone.... [12-oxapentacyclo(8.4.3.2(2,9).0.0.0	38. 89
).0.0(3,7))octadeca-10,12,17-triene-4,6,14,16-	Tetrone <enyl-5,15-diazapentacyclo(7.4.3.2(2,8	37. 19
.0(21,25))triaconta-3,5,27'-triene-14,18,29,30-	Tetrone <5).0(9,13).0(17,24).0(17,26).0(19,23)	31. 65
.4.3.2(2,8).0.0(3,7))octadec-17-ene-4,6,14,16-	Tetrone) <henyl-3,5,7,15-tetra-azapentacyclo(7	37. 20
	Tetronolide...................................	38.137
2,3-dihydro-5-methyl-1,4-dithiin-1,1,4,4-	Tetroxide	39. 5
trichloro-bis(4-cyanopyridine-N-oxide)-	Thallium	68. 13
	Thallium(i) tetra-acetato-thallium(iii)........	68. 5
diphenyl-tropolonato-	Thallium(iii)	68. 24
thallium(i) tetra-acetato-	Thallium(iii)	68. 5
xycyclohepta-2,4,6-trien-1-onato-bis(diphenyl-	Thallium(iii)) .... [di-μ-2-hydro	68. 24
	Theobromine tri-iodide iodine	44. 18
3,7-diphenyl-3H,7H-1,5,2,4,6,8,3,7-di	Thia(1,5S)tetra-azadiarsocine	65. 6
,22,28,30-octamethyl-1,2,9,10,17,18,25,26-octa	Thia(2.2.2.2)metacyclophane.... [4,6,12,14,20	39. 87
syn-2,11-di	Thia(3.3)metacyclophane........................	39. 63
5,7,14,16-tetramethoxy-1,2,3,10,11,12-hexa	Thia(3.3)metacyclophane........................	39. 62
6,8,10,14,16,18-hexamethyl-1,2,3,4,11,12-hexa	Thia(4.2)metacyclophane........................	39. 70
odimethane]	Thia(5)heterohelicene - 7,7,8,8-tetracyanoquin	60. 33
(-)-14-hydroxy-15-hydroxymethyl-2,8-di	Thia(9)(2,5)pyridinophane	41. 48a
-di(p-methoxyphenyl)-2-trans-4-dithio-1,3,2,4-	Thia-aza-λ(5),λ(5)-diphosphetidine <methyl-2,4	64. 71
e] 2,4,6-tri-t-butyl-7,8,9-di	Thia-azabicyclo(4.3.0)nona-1(9),2,4-trien-7-on	41. 68
2,4,6-tri-t-butyl-7,8,9-di	Thia-azabicyclo(4.3.0)nona-1(9),2,4-triene	41. 69
octamethoxy-tetrabenzo(b,e,h,k)(1,4,7,10)tetra	Thia-cyclododecatetraene chloroform solvate <-	39. 86
1,4,7-tri	Thia-cyclononane	39. 11
deca	Thia-cyclotetradecane-6,7,13,14-tetrone........	39. 3
3,3,6,6,-tetramethyl-2,3,6,7-tetrahydro-1,4,5-	Thia-diazepine-1,1-dioxide .... [cis-	41. 23
μ-	Thia-syn(CH(2),CH(3))-9,10-dioxa-bimane........	41. 30
3,7-dimesityl-3H,7H-1,5,2,4,6,8,3,7-di	Thia-tetra-azadiarsocine	65. 10
2,3,5,6-tetra(dimethylamino)-1,4,2,3,5,6-di	Thia-tetraborinane	3. 41
3,5,5-trichloro-1-dimethylamino-1,3,2,4,6,5-di	Thia-triazaphosphorine-1,3-dioxide	64. 2
<2,13-hexamethoxy-10,15-dihydro-5H-5,10,15-tri	Thia-tribenzo(a,d,g)cyclononene chloroform so>	39. 84
benzyl-carboxylate)] 7-oxo-3-methylthio-2-	Thia-1-azabicyclo(3.2.0)hept-3-ene-4-(p-nitro-	41. 50
<-3,3-dimethyl-6-(p-tolyl-thio-imino)-7-oxo-4-	Thia-1-azabicyclo(3.2.0)heptane-2-carboxylic->	41. 80
dide] 3,3',5,5'-tetraphenyl-4,4'-di	Thia-1,1'-bis(2,5-cyclohexadienylidene) polyio	39. 90
dide] 3,3',5,5'-tetraphenyl-4,4'-di	Thia-1,1'-bis(2,5-cyclohexadienylidene) polyio	39. 89
-3,8-diene-2,10-dione 6,6-di> 3,9-dimethyl-6-	Thia-1,11-diazatricyclo(6.2.1.0(4,11))undecane	41. 31
-3,8-diene-2,10-dione] 3,9-dimethyl-6-	Thia-1,11-diazatricyclo(6.2.1.0(4,11))undecane	41. 30
<l)-8-(p-chlorophenylsulfonylamino)-13-oxa-14-	Thia-1,15-diazapentacyclo(10.4.2.0(2,7).0(8,1>	42. 12
7-	Thia-1,3-diazabicyclo(3.3.0)octa-2,4-dione	41. 11
4,5-bis(dimethylamino)-2,2-dimethyl-1,3-di	Thia-2-sila-4,5-dibora-cyclopentane............	63. 4
2-chloro-1,3,6-tri	Thia-2-stibaocane	66. 4
1,1-diphenyl-1-phospha-3,5-di	Thia-2,4,6-triazine............................	64. 50
2,4,6,8-tetraethyl-3,7-dimethyl-1λ(6),5λ(6)-di	Thia-2,4,6,8-tetra-aza-3,7-diphosphorocine-1,>	64. 42
<8-dimethyl-7-oxo-8λ(3)-thionia-3λ(4)(4)-di	Thia-2,4,6,8,9-penta-azabicyclo(3.3.1)nona-2,>	41. 6
ane-1(11),8-diene-3,6-dione] 4-hydroxy-11-	Thia-2,6,12-triazatricyclo(6.3.1.0(4,12))dodec	41. 21
3-t-butyl-4-methyl-2-oxo-1-oxa-2-	Thia-3-azacyclohexane	42. 3
ene] 2-selena-1-dioxo	Thia-3-formyl-5,8-dimethyl-1,2-dihydronaphthal	42. 6
7-methyl-1,4-di	Thia-7-aza-spiro(4.4)nonane-6,8-dione	39. 13
-hydroxy-1-(2-hydroxyethyl)-6,7,8-trimethyl-2-	Thia-7-azabicyclo(3.3.0)octane hemihydrate [5	39. 39
2-(cyano(t-butyl)methylene)-1β-methyl-3-oxa-9-	Thia-7-azabicyclo(4.3.0)non-7-en-4-one <cyano-	41. 76
2-(cyano(t-butyl)methylene)-1β-methyl-3-oxa-9-	Thia-7-azabicyclo(4.3.0)non-7-en-4-one <cyano-	41. 75
2,6,10-tri	Thiabicyclo(9.4.0)pentadeca-1(11),12,14-triene	39. 43
6-phenyl-6-phospha-2,10-di	Thiabicyclo(9.4.0)pentadeca-11(1),12,14-triene	64. 93
4,7,10,13,16-pentaoxa-1-	Thiacyclo-octadecane	42. 7
4,7,10,13,16-pentaoxa-1-	Thiacyclo-octadecane potassium thiocyanate	42. 9
4,7,10,13,16-pentaoxa-1-	Thiacyclo-octadecane rubidium thiocyanate	42. 11
4,7,10,13,16-pentaoxa-1-	Thiacyclo-octadecane silver nitrate............	42. 8
4,7,10,13,16-pentaoxa-1-	Thiacyclo-octadecane sodium thiocyanate........	42. 10
N-(4-methoxyphenyl)-3,6-di	Thiacyclohexene-1,2-dicarboximide	39. 47
N,N'-(tetrasulfur-tetranitrido)-	Thiacyclopentane-di-imide	41. 9
henyl)-5,12-dimethyl-5,12-diaza-1,3,8,10-tetra	Thiacyclotetradecane-6,13-dione <is(p-methoxyp	41. 81
3,4-bis(t-butyl)-2-(p-methoxyphenyl)-1,3,4,2-	Thiadiazaphosphiolidin-5-one 2-sulfide	64. 77
4-(β-D-arabinofuranosyl)-5,6-dihydro-2H-1,2,4-	Thiadiazin-3-one-1,1-dioxide [3,2'-anhydro-	45. 14
sodium 4-cyano-3-hydroxy-6H-1,2,6-	Thiadiazin-6-ide-1,1-dioxide dihydrate	41. 7
acenaphthyleno(5,6-cd)-1,2,6-	Thiadiazine...................................	41. 38

```
5,6-dihydro-4-phenyl-2-phenylazo-4H-1,3,4- Thiadiazine...................................  41. 55
     naphtho(1,8-cd:4,5-c'd')-bis(1,2,6- Thiadiazine)  .............................  41. 26
                     5-amino-2H-1,2,6- Thiadiazine-1,1-dioxide...................  41.  5
            3-hydroxy-4-nitro-6H-1,2,6- Thiadiazine-1,1-dioxide...................  41.  4
          4-cyano-3-hydroxy-6H-1,2,6- Thiadiazine-1,1-dioxide...................  41.  8
       sodium 3-oxo-4-nitro-2H-1,2,6- Thiadiazine-1,1-dioxide monohydrate......  41.  3
           6-chloro-4-methyl-1,2,3-benzo Thiadiazole..............................  41. 18
          5,6-dimethyl-imidazo(2,1-b)(1,3,4) Thiadiazole.........................  41. 15
    5-methoxycarbonylamino-3-methyl-1,2,3- Thiadiazole.........................  41. 13
 2-benzoylimino-3,5-diphenyl-2,3-dihydro-1,3,4- Thiadiazole...................  41. 77
    -benzoylsulfinyl-5-(N-n-butylbenzamido)-1,2,4- Thiadiazole..............[3  41. 73
 hiobenzoyl-methylene-3H-thiazolo(2,3-c)(1,2,4) Thiadiazole  [5,6-dihydro-3-t  41. 39
          5,6-dimethyl-imidazo(2,1-b)(1,3,4) Thiadiazole hydrobromide monohydrate  ........  41. 16
 ethyl-4-((1E)-3-methyl-1,3-pentadienyl)-1,2,5- Thiadiazole-1,1-dioxide <ydro-2-acetyl-3,3-dim  41. 42
    5-benzoylimino-2,2-dimethyl-4-phenyl-1,3,4- Thiadiazolidine..................  41. 61
                  2,5-di-t-butyl-1,2,5- Thiadiazolidine-3,4-dione  .............  41. 33
       4-phenyl-3-phenylamino-1,2,4- Thiadiazolin-5-one  .....................  41. 49
           2-methyl-3-phenyl-5-phenylazo-1,3,4- Thiadiazoline.....................  41. 54
           2,7-dihydro-3,6-di-p-tolyl-1,4,5- Thiadizepine-1,1-dioxide  .............  41. 67
                         diphenyl-tetra Thiafulvalene  ...........................  39. 67
         bis(acetylacetonato)-palladium(ii) tetra Thiafulvalene  .........................  60. 12
)]                                  tetra Thiafulvalene bis(ethylenedithiolato)-nickel(i  60.  5
x]              tetramethyl-tetra Thiafulvalene tetrabromo-p-benzoquinone comple  60. 10
                   dibenzotetra Thiafulvalene tri-iodide  ......................  39. 50
 cyanoquinod>  tris(4,4',5,5'-tetramethyl-tetra Thiafulvalene) - bis(2,5-diethyl-7,7,8,8-tetra  60. 35
              tris(dibenzotetra Thiafulvalene) hexabromo-tin  ..............  39. 51
         bis(tetramethyl-tetra Thiafulvalene) iodide  .....................  60. 11
         bis(tetramethyl-tetra Thiafulvalene) thiocyanate ...............  39. 27
    octakis(dibenzotetra Thiafulvalene) tris(hexachloro-tin)............  39. 52
                         tetra Thiafulvalene-chloranil complex...............  60.  3
                         tetra Thiafulvalene-fluoranil complex..............  60.  4
                         tetra Thiafulvalenium perchlorate................  39.  8
                         tetra Thiafulvalenium tri-iodide  ...............  39.  9
                    (3)Tetra Thiafulvaleno(3)paracyclophane  ...........  39. 69
                    (4)Tetra Thiafulvaleno(4)paracyclophane  ...........  39. 78
                  (2.2)Tetra Thiafulvalenophane  ......................  39. 58
                         Thiamine chloride hydrochloride hemihydrate....  44. 25
        DL-2-(α-hydroxybenzyl)-oxy Thiamine chloride hydrochloride trihydrate ....  44. 35
        cis-5-t-butyl-1,3,2-dioxa Thian-2-oxide  .........................  42.  2
  cis-5-methyl-trans-5-nitro-1,3,2-dioxa Thian-2-oxide  .........................  42.  1
        2,2,6,6-tetramethyl-4(e)-phenyl Thian-4(a')-ol  ........................  39. 56
    cis-2-trans-6-diphenyl-cis-3-ethyl Thian-4-ol  ............................  39. 74
 yl'aminocarbonyl)-(4-phenylmethyl)-4-aza-1,2-di Thian-5-one...................  [(3-phen  41. 60
          1-benzoyl-2-methyl-2- Thianaphthalene  .......................  39. 65
          2,5-dicyclohexane-spiro-1,3-di Thiane  .................................  39. 55
 )-sulfo-di-imide)-bicyclo(3.3.1)penta-azatetra Thiane  .... [1-(S,S-dimethyl-N-(trimethylsilyl  63.  2
                  1-p-hydroxyphenyl Thiane perchlorate  ....................  39. 37
 -4,5-(isopropylidene-dioxy)-1-(2-methyl-1,3-di Thiane-2-yl)-cyclohexane-1,3-diol <R,3R,4S,5R)  39. 54
                    N-methyl-1,4-di Thiane-2,3-dicarboximide  ................  39. 14
 1-p-hydroxyphenyl-thianium 1-phenolate- Thianium perchlorate.................  39. 36
          1-p-hydroxyphenyl- Thianium 1-phenolate-thianium perchlorate ....  39. 36
                         Thianonane-3,8-dione-1,1-dioxide  ..........  39. 24
                      perfluoro Thianthrene.............................  39. 41
   3,3,6,6-tetrachloro-2,4,5,7-tetra Thiaoctane-2,2,7,7-tetraoxide  ............  11.  5
   3,3',5,5'-tetraphenyl-4,4'-di Thiapyranylidene polyiodide...............  39. 89
   3,3',5,5'-tetraphenyl-4,4'-di Thiapyranylidene polyiodide...............  39. 90
            tetraphenyl-di( Thiapyranylidene)  .....................  39. 88
 imethane complex]      5-phenyl-1,3- Thiaselenole-2-thione 7,7,8,8-tetracyanoquinod  60.  8
                         6- Thiatetracycline  ......................  50. 13
        5a-epi-6-demethyl-6-deoxy-6- Thiatetracycline dimethylformamide solvate ...  50. 14
 e-3(8),15-diene]       benzo(p)-2- Thiatetracyclo(7.5.3.0(3,8).0(1,10))heptadecan  39. 79
         5-azido-2H-1,2,4,6- Thiatriazin-3(4H)-one-1,1-dioxide  ..........  41.  2
 -((E)-1,2-diphenylvinyl)-2,6-diphenyl-1,3,4,5- Thiatriazine-1,1-dioxide  .............. [4  41. 82
         5-amino-1,2,3,4- Thiatriazole  ..........................  41.  1
 8-benzyl-1-hydroxy-11-methyl-6,9-diaza-12- Thiatricyclo(7.3.0.0(2,6))dodecane-7,10-dione  48. 49
                      tri Thiaveratrylene chloroform solvate ...........  39. 84
 p-bromophenyl)-4,5-dihydro-7,8-dimethoxy-benzo Thiazepin-3-one-1,1-dioxide...........  [N-(  41. 59
 N-methyl-4H-6,7-dimethoxy-benzo-1,2- Thiazin-3-one-1,1-dioxide.............  41. 37
         1,2,4-triazolo(3,2-b)(1,3) Thiazin-5-one  .........................  41. 10
 2,3,4,5-tetrahydro-6,6-diphenyl-imidazo(2,1-b) Thiazin-7(6H)-one  .....................  41. 64
         4-oxo-2-phenylimino-perhydro-1,4- Thiazine  .................................  41. 29
 oxo-2(methoxy-carbonyl-methyl)-tetrahydro-1,4- Thiazine...............................[3-  41. 20
 ,6-dichlorophenyl)imino-3-methyl-perhydro-1,3- Thiazine  ...........................  [2-(2  41. 36
 ,6-dimethylphenyl)imino-3-methyl-perhydro-1,3- Thiazine  ...........................  [2-(2  41. 46
 ihydro-5-methylthio-1,2,3-triazolo(4,5-d)-1,3- Thiazine  ..................[3-benzyl-3,7-d  41. 40
 hylphenyl)-N-methyl-amino)-4,5-dihydro-6H-1,3- Thiazine  .............. [2-(N-(2,6-dimet  41. 47
```

etoxy-1-ethylidene)-3-(2-thienyl)-4H-1,4-benzo Thiazine .... [6-chloro-2(2-ac 41. 57
-phenyl-6aH,14aH,15H-benzothiazolo(2,3-a)benzo Thiazino(4,3-c)piperazine methanol solvate <15 41. 83
olate] 2-((3-ethyl-1,3-benzo Thiazol-2-ylio)-2-vinyl)-6-methoxy-4-nitrophen 41. 65
4-phenyl-1,2,3,5-di Thiazole 41. 19
3-(3-amino-benzo(b)thiophen-2-yl)-1,2-benziso Thiazole 41. 52
ro-N-methyl-2-(4-nitrobenzoyl-methylene)-benzo Thiazole .... [(Z)-2,3-dihyd 41. 58
2,6-diphenyl-benzo(1,2-d:4,5-d')bis(Thiazole) 41. 71
2,3-dihydro-2-methoxy-2-methyl-3-(benzo-1,3- Thiazole-2-thiomethylene)-1,4-naphthoquinone 41. 72
N-acetyl-4-carboxylato-L- Thiazolidine 41. 17
2-(2,6-dimethylphenyl))-imino-3-methyl- Thiazolidine 41. 41
4-carboxylato-L- Thiazolidine-hydantoin 41. 12
3-p-bromobenzoyl-1,3- Thiazolidine-2-thione 41. 28
4-dimethylamino-5-methyl-5-vinyl- Thiazolidine-2-thione 41. 22
3-phenyl-1,3- Thiazolidine-2,4-dione 41. 24
enoxymethyl-5-oxo-1,3-oxazolin-4-ylidene)-1,3- Thiazolidine-4-carboxylate <5-dimethyl-2-(2-ph 41. 63
bonyl-N'-(2,6-dichlorophenyl)-N-(5-methyl-1,3- Thiazolin-2-yl)-hydrazine ....[N-benzyloxycar 41. 66
hyl-3-p-anisyl-5-(ethylamino)-carbamoyl-1,2,4- Thiazoline .... [2-met 41. 45
4-dimethylamino-5-ethyl-5-phenyl Thiazoline-2-thione........................... 41. 44
bis(2,3-dihydro-3-hydroxythiazolo(2,3-b)benzo Thiazolium) hexachloro-tellurium(iv) dioxane > 41. 25
<nzoyl-15-hydroxy-15-phenyl-6aH,14aH,15H-benzo Thiazolo(2,3-a)benzothiazin(4,3-c)piperazine> 41. 83
lurium(iv) dioxane> bis(2,3-dihydro-3-hydroxy Thiazolo(2,3-a)benzothiazolium) hexachloro-tel 41. 25
5,6-dihydro-3-thiobenzoyl-methylene-3H- Thiazolo(2,3-c)(1,2,4)thiadiazole 41. 39
3-(p-bromophenyl)- Thiazolo(3,2-a)pyridinium tetrafluoroborate.... 41. 43
2,3-dihydro-3-hydroxy-5H- Thiazolo(3,2-a)pyrimidin-5-one 41. 14
l-4,4'-di(2-phenyl-2-propenyl)-4,4'-bis(5(4H)- Thiazolone)................... [2,2'-dipheny 41. 84
e] Thieleanine...................................... 53.   3
e] di <yl-(2,3β,3aα,4,5,6,7,8,9,9aβ-decahydro-4-oxo- Thieno(1,2-b:5,4-b')tropylium tetrafluoroborat 39. 33
7,7,8,8-tetracyanoquinodimethane] Thieno(3,2-c)azocine-3α-yl)-carbamic acid met> 39. 72
4,6-dioxo-syn-cis-syn-1,3,5-triphenyl-perhydro Thieno(3,2-e:4,5-e')-bis(benzo(b)thiophene) - 60. 33
1,1,4,4-tetramethyl-1H,4H- Thieno(3,4-c)pyrrole-1-carbonitrile <ol-3-yl)- 40. 58
(2,3-dichloro-4- Thieno(3,4-c)thiophene 39. 29
5,5'-dinitro-2,2'-di Thienoyl)-phenoxy-acetic acid 39. 46
2,2'-di Thienyl 39. 18
3-nitro-4-(2- Thienyl antimony trichloride 39. 19
4-(5-nitro-2- Thienyl)-but-3-en-2-one....................... 39. 20
bis(2,4-diphenyl-3- Thienyl)-but-3-en-2-one....................... 39. 21
L-pyroglutamyl-β-(2- Thienyl)-disulfide 39. 85
trans-1,8-di(2- Thienyl)-L-alanyl-L-prolinamide............... 48. 50
6-chloro-2(2-acetoxy-1-ethylidene)-3-(2- Thienyl)-1,3,5,7-octatetraene 39. 59
S-methyl- Thienyl)-4H-1,4-benzothiazine 41. 57
S-methyl- Thiepanium bromide 39. 15
S-methyl- Thiepanium iodide 39. 17
xanthene-9-spiro-2'-(3',4'-bis-t-butyl-thio Thiepanium 2,4,6-trinitrobenzenesulfonate 39. 16
2-chloroxanthene-9-2'-(3',4'-bis-t-butyl-thio Thiete)....................................... 39. 82
nthene-9-spiro-2'-(4'-t-butyl-3'-(methylthio)- Thiete)....................................... 39. 81
4-t-butyl-3-(methylthio)- Thiete)....................................[xa 39. 77
3-bromo-2,5-diphenyl-1,4-di Thiete-2-spiro-9'-xanthene 39. 77
2,3-dihydro-5-methyl-1,4-di Thiin-1,1-dioxide 39. 57
(-)-tri-o-thymotide-RR-(+)-2,3-dimethyl- Thiin-1,1,4,4-tetroxide....................... 39.   5
1-methyl-2,3-di-t-butyl- Thiirane clathrate 61. 15
ethylammonium) bis((dithio-tungsten)-bis(μ(2)- Thiirenium tetrafluoroborate 39. 40
rbonyl-iron)] Thio))-oxo-tungsten monohydrate....[bis(tetram 3. 24
rbonyl-iron)] bis(μ-methyl Thio)-(μ-tetrafluoroethane-1,1-diyl)-bis(trica 71. 17
thiolate 4,7,13,16,21-pentao> disodium (ethyl Thio)-(μ-tetrafluoroethane-1,2-diyl)-bis(trica 71. 18
trans-5-chloro-6-(2-nitrophenyl Thio)-(tetra-p-tolylporphinato)-iron(ii) ethyl 49. 27
acid] 6-exo-(methyl Thio)-bicyclo(2.2.1)hept-2-ene 31. 18
acid] 6-endo-(methyl Thio)-bicyclo(2.2.1)heptane-2-endo-carboxylic 31.   4
hio> tris(tetraethylammonium) tris(μ(2)-ethyl Thio)-bicyclo(2.2.1)heptane-2-endo-carboxylic 31.   5
1,6-anhydro-3-deoxy-3-iodo-2-0-(methyl Thio)-bis(tetrakis(μ(3)-thio-tris(μ(3)-ethylt 3. 39
bis(tetramethylammonium) bis((μ(2)- Thio)-carbonyl-β-D-altropyranose 45. 13
ethyl-3-(2-((5-methyl-1H-imidazol-4-yl)-methyl Thio)-dithio-oxo-molybdenum) 3. 23
2-(1-phenyl Thio)-ethyl)-guanidine monohydrate <-cyano-1-m 32. 34
<lopentadienyl)-(μ(3)-(pentacarbonyl-chromium- Thio)-ethyl-pent-3-ene-1-carboxylic acid 20.   7
zyl-trimethylammonium) dichloro-iron-bis(μ(2)- Thio)-methylidyne)-(μ(3)-sulfido)-triangulo-t> 71.169
3-acetoxy-5-(2-nitrophenyl Thio)-molybdenum-disulfide <ylphosphonium (ben 64.123
3-chloro-5-(2',4'-dinitrophenyl Thio)-nortricyclene.......................... 31. 32
2-chloromethyl-5-(2',4'-dinitrophenyl Thio)-nortricyclene.......................... 31. 17
etraphenylphosphonium (μ(2)-sulfonyl)-(μ(4)-di Thio)-nortricyclene.......................... 31. 22
N-(t-butyldi Thio)-octacyano-di-molybdenum hexahydrate [t 64.133
xanthene-9-spiro-2'-(4'-t-butyl-3'-(methyl Thio)-succinimide 32. 22
4-t-butyl-3-(methyl Thio)-thiete) 39. 77
methyl 4,6-0-benzylidene-2-deoxy-3-0-((methyl Thio)-thiete-2-spiro-9'-xanthene 39. 77
methyl 4,6-0-benzylidene-3-deoxy-2-0-((methyl Thio)-thiocarbonyl)-α-D-arabinopyranoside 45. 36
methyl 4,6-0-benzylidene-2-deoxy-3-0-((methyl Thio)-thiocarbonyl)-α-D-arabinopyranoside 45. 37
<nium) tris(μ(2)-ethylthio)-bis(tetrakis(μ(3)- Thio)-thiocarbonyl)-α-D-ribopyranoside 45. 35
Thio)-tris(μ(3)-ethylthio-iron)-di-tungsten) 3. 39

))dec-8-ene-8,9-dicarboxyli> 5-(2-nitrophenyl Thio)-10-acetoxy-tetracyclo(4.4.0.0(2,4).0(3,7   31. 56
-butylester]   2-(t-butyl Thio)-2-(9-thioxanthenylidene)-dithioacetato-t   39. 83
                              1,3-di(methyl Thio)-2,2,4,4-tetramethyl-bicyclo(1.1.0)butane   27.   4
          2,2',6,6'-tetrakis(propyl Thio)-3,3'-azoxypyridine 33. 85
                                  pyridyl-1- Thio-β-D-glucopyranoside monohydrate 45. 20
6,21,24-hexaoxa-1,10-diazabicyc> sodium ethyl Thio-(tetraphenylporphinato)-iron(ii) 4,7,13,1   49. 23
rhenium(v) chloride monohydrate]   di Thio-Δ(2)-1,3-diaza-6,9-dioxa-cycloundecene hy   40. 13
inium) tetrafluoroborate]   2-methyl Thio-bis(formamidinium) trans-tetrachloro-oxo-   8.   9
e)]   3,3'- Thio-bis(2-methyl-1-phenyl-imidazo(1,5-a)pyrid   35. 73
                              octamethyl-tri Thio-cyclotetra(λ(3),λ(5),λ(5),λ(5)-phosphazan   64. 30
                              dimethyl-2,2'- Thio-dibenzoate.................................   13. 13
                          dimethyl-2,2'-di Thio-dibenzoate.................................   13. 14
          tetraphenylphosphonium bis(bis-µ- Thio-dichloro-iron)-molybdenum 64.129
                                  8,8'-di Thio-diquinoline 35. 48
                      tropane-3-spiro-5'- Thio-hydantoin 37.   6
                      2,5-bis(N-chloro Thio-imino)-3,4-dicyanothiophene 39.   7
ptane-2-> 2S-(2α,5α)-3,3-dimethyl-6-(p-tolyl- Thio-imino)-7-oxo-4-thia-1-azabicyclo(3.2.0)he   41. 80
lthio)-bis(tetrakis(µ(3)-thio)-tris(µ(3)-ethyl Thio-iron)-di-tungsten) <onium) tris(µ(2)-ethy   3. 39
lodecen-7(6H)-one] syn-5,12-dihydro-5,14-(epi Thio-methyleneoxy)-6,12-epithiodibenzo(a,f)cyc   39. 73
di-N,N'-isopropyl-di Thio-oxalamide 1. 21
bis(tetramethylammonium) bis((µ(2)-thio)-di Thio-oxo-molybdenum) 3. 23
3,6-diethyl-1,4-dimethyl-3,6-epi Thio-piperazine-2,5-dione 41. 32
bis(mono Thio-pyrocatecholato)-tellurium(iv)............   70.   7
nohydrate] bis(tetramethylammonium) bis((di Thio-tungsten)-bis(µ(2)-thio))-oxo-tungsten mo   3. 24
2,2'-anhydro-1-β-D-arabinofuranosyl-2- Thio-uracil.....................................   47.   2
-(2',3'-O-isopropylidene-β-D-ribofuranosyl)-2- Thio-uracil........................[2,5'-anhydro-1   47. 28
9,12-isopropylidene- Thio-1,2-dicarba-closo-dodecaborane............   62.   6
3-benzyl-3,7-dihydro-5-methyl Thio-1,2,3-triazolo(4,5-d)-1,3-thiazine........   41. 40
cis-5-hydroxy-6-phenyl Thio-1,3-dimethyl-5,6-dihydrothymine 44. 27
3-methyl-2,4-di(p-methoxyphenyl)-2-trans-4-di Thio-1,3,2,4-thia-aza-λ(5),λ(5)-diphosphetidi>   64. 71
anti,S-trans-3-methyl Thio-1,5-di-o-tolyl-formazan 11. 16
,10))decane]   10- Thio-10-phospha-1,4,7-triazatricyclo(5.2.1.0(4   64. 19
3-chloro-3-cyano-1-cyclohexyl-4-phenyl Thio-2-azetidinone 32. 57
-nitro-benzyl-carboxylate)] 7-oxo-3-methyl Thio-2-thia-1-azabicyclo(3.2.0)hept-3-ene-4-(p   41. 50
e)-lithium dimer]   2-methyl-1,3-di Thio-2-yl-(N,N,N',N'-tetramethylethylenediamin   67.   7
                              1-methyl Thio-3-(α-morpholino)-benzylidene-triazene ...   40. 25
,9a-dihydro-1,2,9,9-tetramethyl-2,9a-epi-tetra Thio-3,10-diketo-piperazino(1,2-a)indole   [9   41. 56a
,3:4,5-di-O-isopropylidene-D-gulose diethyl di Thioacetal ....[2   45. 42
tris(mono Thioacetato)-antimony(iii) 66.   7
2-(t-butylthio)-2-(9-thioxanthenylidene)-di Thioacetato-t-butylester 39. 83
(1R,3S,4R)-N,N-dimethyl-3-camphor-carbo Thioamide 31. 21
di-iodo-(maleic acid)- Thioanhydride 39.   2
dibromo-(maleic acid)- Thioanhydride 39.   1
piperidinium 1-piperidine-carbodi Thioate...   33. 14
piperidinium 1-piperidine-carbodi Thioate...   33. 13
sodium 1-pyrrolidinyl-carbodi Thioate dihydrate 32. 14
sodium 1-pyrrolidinyl-carbodi Thioate dihydrate 32. 16
sodium 1-pyrrolidinyl-carbodi Thioate dihydrate 32. 15
antimony-tris(O,O-diethylphosphoro-di Thioate) 66. 12
2-methyl-3-p-anisyl-5-p-tosylimino-1,2,4- Thioazoline.....................................   41. 62
4-(2'-hydroxy Thiobenzoyl)-morpholine.........................   40. 22
5,6-dihydro-3- Thiobenzoyl-methylene-3H-thiazolo(2,3-c)(1,2,4   41. 39
4- Thiobenzoyl-morpholine 40. 21
4- Thiobenzoyl-morpholine 40. 20
3-thioxo-4-phenyl-4-aza-2,7-di Thiobicyclo(3.3.0)octane-7,7-dioxide 41. 35
tetrakis(N-(2-hydroxyethyl)-N-methyldi Thiocarbamato)-tellurium(iv) 70. 14
bromo-tris(N-(2-hydroxyethyl)-N-methyldi Thiocarbamato)-tellurium(iv) 70. 10
O-methyl Thiocarbazate 1.   6
S-methyl Thiocarbazate 1.   7
potassium Thiocarbazate 2.   5
trans,cis-S-methyldi Thiocarbazate 11.   4
I 4,6-O-benzylidene-2-deoxy-3-O-((methylthio)- Thiocarbonyl)-α-D-arabinopyranoside.... [methy   45. 36
I 4,6-O-benzylidene-3-deoxy-2-O-((methylthio)- Thiocarbonyl)-α-D-arabinopyranoside.... [methy   45. 37
I 4,6-O-benzylidene-2-deoxy-3-O-((methylthio)- Thiocarbonyl)-α-D-ribopyranoside .... [methy   45. 35
trimethyl-tin)-1,3,5,7-tetra-aza-2,4,6,8-tetra Thiocin-2,2-dioxide benzene solvate [1,3-bis(69. 12
Thiocolchicoside ethanol solvate hydrate 58. 70
2-dihydro-5,14-(epithio-methyleneoxy)-6,12-epi Thiodibenzo(a,f)cyclodecen-7(6H)-one [syn-5,1   39. 73
oxy-ethyl-2-a> (+-)-4β-(2',2'-trimethylene-di Thioethyl)-3α-(1'S)-(p-nitrobenzyloxycarbonyl)   39. 71
4-nitrobenzyl-(3SR,5RS,Z)-2-(2-phenyl Thioethylidene)-penam-3-carboxylate 41. 78
miflavin]   3-methyl-4a-(methylene-di Thioglycerol acid ethyl ester)-4a,5-dihydro-lu   36. 31
soalloxazine]   3-methyl-4a-(methylene-di Thioglycerol acid methyl ester)-4a,5-dihydro-i   36. 18
umiflavin]   3-methyl-4a-(methylene-di Thioglycerol acid methyl ester)-4a,5-dihydro-l   36. 25
lorate]   bis(4,5-dimethyl-2H-1,3-di Thiol-2-ylidene)-1,4-cyclohexa-2,5-diene perch   39. 64
5-benzamido-2-benzoylimino-1,3-di Thiol-4-carbonic acid ethyl ester 39. 75
3,3,4,4-tetraphenyl-1,2-oxa Thiolan-5-one-2-oxide 42. 13
3,5-bis(cyclohexylidene-1,2,4-tri Thiolane) 39. 53

cyclopenta(a)phenanthrene-17,2'-spiro(1',3'-di   Thiolane) <l)-11,12,13,14,16,17-hexahydro-15H-   51. 45
                                      1-methyl-   Thiolanium iodide 39. 6
                      1,3-diphenyltetrazolium-5-   Thiolate   32. 47
    onio-1-isopropyl-4,4-dimethyl-2-imidazoline-2-   Thiolate chloroform solvate.... [5-dimethylimm 32. 35
                         sodium quinoline-8-   Thiolate dihydrate 35. 11
  ‹hio)-(tetra-p-tolylporphinato)-iron(ii) ethyl   Thiolate 4,7,13,16,21-pentaoxa-1,10-diazabicy› 49. 27
       tamethyl-guanidinium-2-(2,2-dicyano-1-ethylene   Thiolate) .... [1,1,2,3,3-pen   8. 12
          2,5-dimethyl-1,2-dithiolo(1,5-b)-1,2-oxa   Thiole 39. 12
   naphthaceno(5,6-cd:11,12-c'd')-bis(1,2-di   Thiole) bis((μ(2)-bromo)-bromo-copper)   39. 66
                   4,5-dimethyl-1,3-di   Thiole-2-thione 39. 4
                   2,5-dimethyl-1,2-di   Thiolo(1,5-b)-1,2-oxathiole.................   39. 12
                          1,3-benzodi   Thiolyl-2-(diphenylphosphine-oxide)...........   64.103
       4-phenyl-1-(3-(4-methyl-5-phenyl-1,2-di   Thiolylidene))-3,5-dioxo-1,2,4-triazine-   39. 68
   de)-bis(triphenylphosphine)-rutheni›   (η(2)-di   Thiomethoxycarbonyl)-carbonyl-(p-tolylisocyani   71.270
                hexakis(p-t-butylphenyl   Thiomethyl)benzene squalene clathrate   61. 26
                      hexakis(benzyl   Thiomethyl)benzene 1,4-dioxane clathrate 61. 23
                hexakis(2-phenylethyl   Thiomethyl)benzene 1,4-dioxane clathrate 61. 25
  ro-2-methoxy-2-methyl-3-(benzo-1,3-thiazole-2-   Thiomethylene)-1,4-naphthoquinone   [2,3-dihyd   41. 72
             lamine sesquihydrate]   3S,6R,-6-methyl-5-oxo   Thiomorpholino-3-carbonyl-L-histidinyl-L-proly   48. 51
                 4,5-dimethyl-1,3-dithiole-2-   Thione 39. 4
              3-p-bromobenzoyl-1,3-thiazolidine-2-   Thione   41. 28
                1,1'-trimethylene-di-2-imidazolidine-   Thione   32. 31
             2,3-dimethyl-Δ(1,5)-1,2,3-triazoline-4-   Thione   32. 11
          2-phenyl-3-methyl-Δ(1,5)-1,2,3-triazoline-4-   Thione   32. 27
              4-dimethylamino-5-ethyl-5-phenylthiazoline-2-   Thione   41. 44
               3-(β-D-ribofuranosyl)-1,2,4-triazole-5(1H,4H)-   Thione   45. 11
          dimethylamino-5-methyl-5-vinyl-thiazolidine-2-   Thione ....   [4-   41. 22
          -hexa-aza-5λ(4)-phosphabicyclo(3.3.1)nonane-5-   Thione .... [9-methyl-3,7-diphenyl-1,3,4,6,7,9   64. 73
                      x]   5-phenyl-1,3-thiaselenole-2-   Thione 7,7,8,8-tetracyanoquinodimethane comple   60. 8
          bicyclo(3.3.1)nona›   6,8-dimethyl-7-oxo-1λ(3)-   Thionia-3λ(4),5λ(4)-dithia-2,4,6,8,9-penta-aza   41. 6
          is(S-p-hydroxyphenyl-thiophanium) p-phenolate-   Thiophanium bis(perchlorate) .... [b   39. 28
          lorate)]   bis(S-p-hydroxyphenyl-   Thiophanium) p-phenolate-thiophanium bis(perch   39. 28
                 3-(3-amino-benzo(b)   Thiophen-2-yl)-1,2-benzisothiazole 41. 52
    8aβ-octahydro-4-hydroxylimino-4H-cyclohepta(b)   Thiophen-3α-yl)-carbamic acid methyl ester <8,   39. 44
                 4-(anilinomethylene)-tetrahydro   Thiophen-3-one 39. 34
                 2-(2-chloro-4-methyl   Thiophen-3-yl-imino)-tetrahydro-imidazole 39. 22
             1,1,4,4-tetramethyl-1H,4H-thieno(3,4-c)   Thiophene 39. 29
             2,5-bis(N-chlorothio-imino)-3,4-dicyano   Thiophene 39. 7
                thieno(3,2-e:4,5-e')-bis(benzo(b)   Thiophene) - 7,7,8,8-tetracyanoquinodimethane   60. 33
                 5-methyl-2-   Thiophene-carboxylic acid 39. 10
                3-phenyl-4,5-dihydro   Thiophene-1,1-dioxide 39. 25
                4-phenyl-4,5-dihydro   Thiophene-1,1-dioxide 39. 26
                N-phenyl-2,5-dihydro   Thiophene-3-carboxamide.....................   39. 35
                      pentachloro-   Thiophenol 11. 7
            dimethylammonium 0,0-di-isopropyl-di   Thiophosphate   64. 22
                methyl-D-ribopyranoside-   Thiophosphate-triester 46. 4
                tris(diphenyl-di   Thiophosphinato)-antimony(iii) 66. 19
                       phenoxy   Thiophosphoryl-dihydrazide 64. 18
             1,1',2''-tris(trimethylsilyl)-   Thiophosphoryl-trihydrazide.................   64. 36
           4-benzyl-2,3,5,6-bis(tetramethylene)-4H-   Thiopyran   39. 80
    thiopyrano(4,3-b)pyran-2-spiro-3'-(1H-2)-benzo   Thiopyran-4'(3'H)-one <4-dihydro-2H,6H-2-benzo   39. 76
      thiopyran-4'(3'H)-›   3,4-dihydro-2H,6H-2-benzo   Thiopyrano(4,3-b)pyran-2-spiro-3'-(1H-2)-benzo   39. 76
                 2-carboxy-3-methyl   Thiopyrano(4,3,2-cd)indole 39. 42
          line]   4-hydroxyimino-3,5-dimethyl-1-   Thiosemicarbamoyl-5-thiosemicarbazido-2-pyrazo   32. 20
    droxyimino-3,5-dimethyl-1-thiosemicarbamoyl-5-   Thiosemicarbazido-2-pyrazoline .... [4-hy   32. 20
                benzaldehyde 4-benzoyl-2-phenyl   Thiosemicarbazone   13. 19
             4-m-aminophenyl-2-formylpyridine-   Thiosemicarbazone 33. 53
                0,0-bicyclohexyl-1,1'-diyl-   Thiosulfite.................................   39. 45
          xanthene-9-spiro-2'-(3',4'-bis-t-butyl-   Thiothiete)................................   39. 82
      2-chloroxanthene-9-2'-(3',4'-bis-t-butyl-   Thiothiete)................................   39. 81
          bis((8-quinolyl-oxy)-ethoxy-ethyl)ether   Thiourea 17. 40
                hexakis(trimethylene   Thiourea)-di-tellurium(ii) perchlorate 70. 16
                hexakis(trimethylene   Thiourea)-di-tellurium(ii) perchlorate 70. 17
                hexakis(ethylene   Thiourea)-di-tellurium(ii) tetraperchlorate...   70. 15
                tetrakis(dimethyl   Thiourea)-tellurium(ii) dichloride 70. 12
                cis-2,4,9-trimethyl-   Thioxanthene-10-oxide 39. 60
                cis-9-ethyl-9-methyl-   Thioxanthene-10-oxide 39. 61
                2-(t-butylthio)-2-(9-   Thioxanthenylidene)-dithioacetato-t-butylester   39. 83
                3-(5-tetrazolyl)-   Thioxanthene-10,10-dioxide rubidium salt 39. 49
                      8-   Thioxo-adenosine monohydrate 47. 17
          1,2-dicyclohexyl-1,2-bis(diethylamino)-1,2-di   Thioxo-diphosphane 64.112
    -(2-(p-tolylsulfonylamino)-1-(t-butylimino)-2-   Thioxo-ethyl)-1H-tetrazole .... [1-t-butyl-5   32. 66
      o(3.3.1)nonane]   6,8-dimethyl-7-phenoxy-7-   Thioxo-3-oxa-1,5,6,8-tetra-aza-7-phosphabicycl   64. 48
             octane-7,7-dioxide]   3-   Thioxo-4-phenyl-4-aza-2,7-dithiobicyclo(3.3.0)   41. 35
             bis(diphenylphosphino-   Thioyl)-acetylene..........................   64.142
                glycyl-D-   Threonine dihydrate.........................   48. 19a

Thromboxane B(2) 59. 27
α- Thymidine 47. 18
1-(3'-5')-thymidylyl-(3'-5')- Thymidine ....[deoxy-5'-P-adenyly 47. 39
deoxy-5'-P-adenylyl-(3'-5')- Thymidylyl-(3'-5')-adenylyl-(3'-5')-thymidine 47. 39
potassium Thyminate trihydrate 44. 8
1-(3-(indol-3-yl)propyl)- Thymine... 44. 34
-hydroxy-6-phenylthio-1,3-dimethyl-5,6-dihydro Thymine................................[cis-5 44. 27
y-6-(o-hydroxyphenyl)-1,3-dimethyl-5,6-dihydro Thymine................... [cis-5-hydrox 44. 28
y-6-(p-hydroxyphenyl)-1,3-dimethyl-5,6-dihydro Thymine................... [cis-5-hydrox 44. 29
(1-methyl Thymine)-silver(i) 44. 14
tyramine 1- Thyminyl-(acetic acid) complex hydrate 60. 6
Thymol 17. 15
tri-o- Thymotide cis-stilbene clathrate 61. 17
tri-o- Thymotide trans-stilbene clathrate 61. 14
(-)-tri-o- Thymotide-RR-(+)-2,3-dimethyl-thiirane clathra 61. 15
te]
(+)-tri-o- Thymotide-S-(+)-2-bromo-butane clathrate 61. 16
3,3',5,5'-tetra-iodo Thyroformic acid bis(N-diethanolamine) 13. 7
rate] Thyrotropin-releasing-hormone tartrate monohyd 48. 47
Tiamenidine 39. 22
Tienilic acid 39. 46
Tin 39. 51
tris(dibenzotetrathiafulvalene) hexabromo- Tin(iv) 3. 22
bis(tetramethylammonium) hexachloro- Tin(iv) 3. 21
bis(tetramethylammonium) hexachloro- Tin(iv)) 1,4-dioxane 38. 4
bis(μ-hydroxo)-bis(aqua-trichloro- Tin(iv)) 1,8-epoxy-p-menthane 38. 26
bis(μ-hydroxo)-bis(aqua-tribromo- Tin(iv)) 1,8-epoxy-p-menthane 38. 27
bis(μ-hydroxo)-bis(aqua-trichloro- Tin) .... [octak 39. 52
is(dibenzotetrathiafulvalene) tris(hexachloro- Tin)-osmium .... [tetraki 3. 26
s(tetramethylammonium) chloro-penta(trichloro- Tin)-1,3,5,7-tetra-aza-2,4,6,8-tetrathiocin-2, 69. 12
2-dioxide benzene solvate] 1,3-bis(trimethyl- Tolonidine 32. 33
Toluene chloroform solvate 50. 38
prolinomycin rubidium picrate Toluene clathrate ....[cyclo(tetrakis 61. 21
(5-t-butyl-2-hydroxy-1,3-phenylene)methylene)
3β-(Toluene-p-sulfonamido)-5α-cholestan-2β-ol 51. 79
3β-azido-2α-(Toluene-p-sulfonamido)-5α-cholestane 51. 78
2-adamantyl-0,N,N-azoxy-(Toluenesulfonate 31. 43
S)-1-pyridinium-2-N,N-dibenzylamino-propane p- Toluenesulfonate ....[2(33. 82
idinium-2-N,N-dibenzylamino-3-methylpentane p- Toluenesulfonate .... [2(S),3(S)-1-pyr 33. 91
octa-3,5-diynylene-bis(p- Toluenesulfonate) 11. 19
tetrabromo-p-phenylene-bis(p- Toluenesulfonate) 17. 33
H-1,4,5-benzotriazocin-2-one] 8-methyl-4-(p- Toluenesulfonyl)-6-phenyl-1,2,3,4-tetrahydro-5 35. 69
)nonane] 2-(2'-propenyl)-8- Toluenesulfonyl-7,8-diazatricyclo(4.2.1.0(3,7) 37. 10
3-(2,2,2-trichloro-1-(p- Toluenesulfonyloxy)ethyl)-cyclo-octene ... 22. 6
3-(2,2,2-trichloro-1-(p- Toluenesulfonyloxy)ethyl)-cyclohexene 21. 8
4-(o-hydroxybenzoyl)-2,5-diphenyl-1-(p- Tolyl)-imidazole 32. 73
3,3-bis(p- Tolyl)-phthalide 38.111
tri-p- Tolyl-antimony 66. 14
anti,S-trans-3-methylthio-1,5-di-o- Tolyl-formazan 11. 16
p- Tolyl-glyoxylic acid p-chloroanilide 16. 12
2,4,5-tri-p- Tolyl-imidazolinium bromide 32. 71
2.0)heptane-2-> 2S-(2α,5α)-3,3-dimethyl-6-(p- Tolyl-thio-imino)-7-oxo-4-thia-1-azabicyclo(3. 41. 80
oxylic-1,2,3,4-di-imid> N,N',7-trimethyl-1-p- Tolyl-1,2,3,4-tetrahydro-naphthalene-tetracarb 36. 42
phorine 2-oxide] 2-chloro-6-methyl-3-p- Tolyl-1,2,3,4-tetrahydro-1,3,2-benzo-diazaphos 64. 70
2,7-dihydro-3,6-di-p- Tolyl-1,4,5-thiadizepine-1,1-dioxide 41. 67
)] p- Tolyl-2-chlorocyclohexyl-dichloro-tellurium(iv 70. 13
1-p-chlorophenyl-3-p- Tolyl-2-triphenylphosphazeno-propenone 64.151
<methylphosphine)-platinum)-(μ-carbonyl)-(μ-p- Tolylcarbenyl)-(carbonyl-cyclopentadienyl-man> 71.173
ni> (η(2)-dithiomethoxycarbonyl)-carbonyl-(p- Tolylisocyanide)-bis(triphenylphosphine)-ruthe 71.270
,16,21-pentao> disodium (ethylthio)-(tetra-p- Tolylporphinato)-iron(ii) ethylthiolate 4,7,13 49. 27
ate] cis-dioxo-(5,10,15,20-tetra-p- Tolylporphyrinato)-molybdenum(vi) toluene solv 49. 26
E-1-(5-nitro-2-furyl)-2-p- Tolylsulfonyl-ethylene 38. 50
ethyl)-1H-tetrazole] 1-t-butyl-5-(2-(p- Tolylsulfonylamino)-1-(t-butylimino)-2-thioxo- 32. 66
5-methylbenzene-1,3-dicarbaldehyde-bis(p- Tolylsulfonylhydrazone) benzene clathrate 61. 7
2-(p- Tolylsulfonyloxy-N,N,O-azoxy)-adamantane 31. 43
2,4-hexadiyne-1,6-bis(m- Tolylurethane) 1. 39
Tomaymycin 50. 10
ury)] N- Tosylamino-octaethylporphinato-bis(chloro-merc 49. 15
L-(1- Tosylamino-2-phenyl)-ethyl-chloromethyl-ketone 4. 3
2-(2- Tosylaminophenyl)-4H-3,1-benzoxazin-4-one ... 40. 52
trans-2-t-butyl-perhydroazulen-4-one-(p- Tosylhydrazone) 27. 30
2-methyl-3-p-anisyl-5-p- Tosylimino-1,2,4-thioazoline 41. 62
3-chloro-6-methoxy-7-methyl-2-(p- Tosyloxyethylamine)-1,4-naphthoquinone ... 25. 5
PR Toxin 59. 16a
,6-triene methyl carboxylate] Trans-7-dimethylamino-2-carbomethoxy-hepta-2,4 1. 28
6,6'-dibromo-6,6'-dideoxy-α,α- Trehalose-hexa-acetate chloroform solvate 45. 50
Triamcinolone acetonide. 51. 46
1-methylthio-3-(α-morpholino)-benzylidene- Triazene 40. 25

2-(2',4'-dibromophenyl)-4-oxo-1,2,3-benzo Triazin-2-ium-3-ide.............................. 35. 27
5-azido-2H-1,2,4,6-thia Triazin-3(4H)-one-1,1-dioxide 41. 2
‹amino-5-methyl-7-propyl-imidazo(5,1-f)(1,2,4) Triazin-4(3H)-one methylsulfonate monohydrate 35. 13
6-phenyl-1,2,4-triazolo(4,3-b)-1,2,4- Triazine 35. 14
2-methylnaphtho(1,8-de)-1λ(2),2λ(4),3- Triazine 36. 1
1,1-diphenyl-1-phospha-3,5-dithia-2,4,6- Triazine 64. 50
2-methyl-2H-acenaphthyleno(5,6-de)-1,2,3- Triazine 36. 5
5-phenyl-1,2-dithiolylidene))-3,5-dioxo-1,2,4- Triazine .... [4-phenyl-1-(3-(4-methyl- 39. 68
)-1,2-diphenylvinyl)-2,6-diphenyl-1,3,4,5-thia Triazine-1,1-dioxide .... [4-((E 41. 82
henyl)methoxy)-1,6-dihydro-6,6-dimethyl-1,3,5- Triazine-2,4-diamine hydrochloride hydrate ‹op 33. 45
1,2,3-oxadiazolo(4,3-c)(1,2,4)benzo Triazinium-3-olate 40. 6
tri› 4,5-benzimidazo(1,2-a)-(4-phenyl-(1,3,5- Triazino))-(5,6-a)-(6,7-benzo-4-phenyl-(1,3,5- 36. 51
phenyl-bis(3,5-bis(trifluoromethyl)-1,2,4- Triazol-1-yl)-methane 32. 53
mino-4-(2-methoxyphenyl)-2,4-dihydro-3H-1,2,4- Triazol-3-one .... [5-a 32. 29
orophenylamino)-2,4-dihydro-4-methyl-3H-1,2,4- Triazol-3-one methanol solvate .... [5-(4-chl 32. 23
bis(3-nitro-1,2,4- Triazol-5-yl) dihydrate.................. 32. 9
3(5)-chloro-1,2,4- Triazole 32. 2
5-amino-1,2,3,4-thia Triazole 41. 1
3,5-diamino-1H-1,2,4- Triazole 32. 3
5-(3,5-dimethoxyphenyl)-1-methyl-1H-1,2,4- Triazole 32. 41
bromophenyl)-5-(3,5-dimethoxyphenyl)-1H-1,2,4- Triazole   [1-(p- 32. 56
3-(β-D-ribofuranosyl)-1,2,4- Triazole-5(1H,4H)-thione 45. 11
1-n-butyl-4,5-diphenyl-5-methyl-1,2,4- Triazolidin-3-one 32. 68
ro-methyl-semibullvalene-diyl)-4-bornyl-1,2,4- Triazolidine-3,5-dione .... [1,2-(tetrahyd 37. 18a
2,3-dimethyl-Δ(1,5)-1,2,3- Triazoline-4-thione................... 32. 11
2-phenyl-3-methyl-Δ(1,5)-1,2,3- Triazoline-4-thione 32. 27
1-methyl-3-benzyl-4-(1,2,3- Triazolio)-sulfide 32. 32
1-methyl-3-phenyl-4-(1,2,3- Triazolio)-sulfide 32. 24
1-methyl-3-phenyl-4-(1,2,3- Triazolio)-sulfide 32. 25
1-phenyl-3-methyl-4-(1,2,3- Triazolio)-sulfide 32. 26
6-ethoxycarbonyl-4-ethyl-1,2,4- Triazolo(1,5-a)pyrimidin-7(4H)-one 35. 19
1,2,4- Triazolo(3,2-b)(1,3)thiazin-5-one 41. 10
6-phenyl-1,2,4- Triazolo(4,3-b)-1,2,4-triazine 35. 14
3-benzyl-3,7-dihydro-5-methylthio-1,2,3- Triazolo(4,5-d)-1,3-thiazine 41. 40
Trichoviridine 50. 3
4,7,13,16,21-pentaoxa-1,10-diazabicyclo(8.8.5) Tricosane benzene solvate ‹(ii) ethylthiolate 49. 27
‹,17-octamethyl-3,6,13,16-tetraoxo-21,22-dioxa Tricyclo(18.2.1.1(8,11))tetracosa-8,10,18,20-› 38.129
1,5-dimethyl- Tricyclo(2.1.0.0(2,5))pentan-3-one 31. 1
1,5-diphenyl- Tricyclo(2.1.0.0(2,5))pentan-3-one 31. 42
3,5,7-tris(trimethylsilyl)- Tricyclo(2.2.1.0(2,6))hepta-arsane 65. 5
‹-t-butyl-2'-methyl-3'-oxo-spiro(oxirane-2,4'- Tricyclo(3.3.0.0(2,8))oct-6'-ene)-1',8-dicarb› 38. 93
Tricyclo(3.3.1.1(3,7))decane-1-carbonitrile.... 31. 14
-dien-9-one)-spiro-10,4'-(9-anti-acetoxy-3-oxa Tricyclo(4.2.1.0(2,5))non-7'-ene ‹idecane-4,11 38.119
(+)-3,7-dimethyl-1,1-(2-methyl-1-oxopropyl)- Tricyclo(4.2.1.0(3,7))nonan-2-one 31. 36
2,7-dibromo- Tricyclo(4.2.1.0(3,7))nonan-4-one 31. 2
2-(2'-propenyl)-8-toluenesulfonyl-7,8-diaza Tricyclo(4.2.1.0(3,7))nonane 37. 10
-dien-9-one)-spiro-10,4'-(9-anti-acetoxy-3-oxa Tricyclo(4.2.10(2,5))non-7'-ene ‹ridecane-4,11 38.118
‹methyl-trans-3-endo-chloro-mercury-4-acetoxy- Tricyclo(4.2.2.0(2,5))dec-7-ene-9,10-cis-endo› 31. 37
(+-)-4-aza Tricyclo(4.3.1.0(3,7))decan-5-one 37. 4
7-bromo-2-chloro- Tricyclo(4.3.1.0(3,7))decane-4,8-dione 31. 7
1β,3,4β,6β,8-pentamethyl-9-exo-methylene- Tricyclo(4.4.0.0(3,8))decane-2,5-dione 31. 40
4-(N'-(p-nitrobenzene)-N-hydrazo)-8-oxa Tricyclo(5.1.0.0(2,5))octane 38. 52
lane)] oxo-bis(hydroxy-(Tricyclo(5.2.1.0(2,6))dec-3-en-9-yl)-methyl-si 63. 36
1,7,8,9-tetrachloro-3-phenyl-4-aza-5-oxa Tricyclo(5.2.1.0(2,6))decane-3,8-diene 40. 32
1,7,8,9,10-pentachloro-3-phenyl-4-aza-5-oxa Tricyclo(5.2.1.0(2,6))decane-3,8-diene 40. 31
-exo-5((E)-(2,2,2-trifluoroethyl)-vinyl)-endo- Tricyclo(5.2.1.0(2,6))decane-3,8-diene ‹-cyano 31. 52
4,5-benzo-6-phenyl-3,10,10-trimethyl- Tricyclo(5.2.1.0(3,7))dec-4-ene 31. 57
10-thio-10-phospha-1,4,7-triaza Tricyclo(5.2.1.0(4,10))decane 64. 19
one] 3,8-dimethyl-1,10-diaza Tricyclo(5.2.1.0(4,10))decane-3,7-diene-2,9-di 37. 5
7-methyl-3-methylene- Tricyclo(5.3.0.0(1,4))dec-5-yl-phenylcarbamate 28. 12
1,8-di-iodo- Tricyclo(5.3.0.0(4,8))decane 31. 8
4,6-diphenyl-10-ethoxycarbonyl-2,3,10-triaza Tricyclo(5.3.2.0(2,4))dodecane-3,8,11-triene-› 37. 21
iro(3(1)-formyl› 5-acetoxy-6,7-dimethyl-3-oxa Tricyclo(5.4.0.0(2,4))undecan-10-one-9,2(1)-sp 59. 16a
1α,4,5α,7α,9,10-hexamethyl-2β-methoxy-3-oxa. Tricyclo(5.4.0.0(4,11))undec-9-en-6-one........ 38. 87
methoxycarbonyl-9-methylene-8-trimethylsiloxy- Tricyclo(6.2.1.0(1,5))undecan-3-one........[2- 63. 22
-dione 6,6-di› 3,9-dimethyl-6-thia-1,11-diaza Tricyclo(6.2.1.0(4,11))undecane-3,8-diene-2,10 41. 31
-dione] 3,9-dimethyl-6-thia-1,11-diaza Tricyclo(6.2.1.0(4,11))undecane-3,8-diene-2,10 41. 30
rboxylic acid] 11-methyl- Tricyclo(6.2.1.0(4,11))undecapentaene-2,3-dica 28. 6
3,6-dione] 4-hydroxy-11-thia-2,6,12-triaza Tricyclo(6.3.1.0(4,12))dodecane-1(11),8-diene- 41. 21
2,6,10-trimethyl- Tricyclo(7.2.1.0(2,7))dodecan-11-one 31. 35
2-hydroxy-4-isopropyl-7-benzyl-3-oxa-6,9-diaza Tricyclo(7.3.0.0(2,6))dodecane-5,8-dione ‹-(p- 40. 48
bromobenzyloxycarbonyl)-2-hydroxy-3,6,9-triaza Tricyclo(7.3.0.0(2,6))dodecane-5,8-dione ‹-(p- 36. 44
l-3-(benzyloxycarbonyl)-2-hydroxy-3,6,9-triaza Tricyclo(7.3.0.0(2,6))dodecane-5,8-dione ‹enzy 36. 46
8-benzyl-1-hydroxy-11-methyl-6,9-diaza-12-thia Tricyclo(7.3.0.0(2,6))dodecane-7,10-dione 48. 49
opropyl-11α-methyl-1α-methoxy-6,9-diaza-12-oxa Tricyclo(7.3.0.0(2,6))dodecane-7,10-dione ‹-is 58. 22
carbonyl)-1,3,7,9-tetrasila-2,8,13,14-tetraoxa Tricyclo(7.3.1.1(3,7))tetradecane ‹akis(ethoxy 63. 38

<bromo-3,6-diethyl-10-ethoxycarbonyl-3,6-diaza	Tricyclo(7.4.0.0(2,7))tridecane-2(7),9,11,13->	36. 24
<mo-3,6-diethyl-13-(phenoxycarbonyl)-3,6-diaza	Tricyclo(7.4.0.0(2,7))tridecane-2(7),9,11,13->	36. 36
14-ethoxy-8,14-di(phenoxy)-3,4:10,11-dibenzo-	Tricyclo(7.5.0.0(2,8))tetradecane-3,5,10-trie>	30. 18
,10-dione] (1α,2α,8β,9α)-4,5:11,12-dibenzo-	Tricyclo(7.5.0.0(2,8))tetradecane-4,11-diene-3	30. 11
,10-dione] (1α,2β,8β,9α)-4,5:11,12-dibenzo-	Tricyclo(7.5.0.0(2,8))tetradecane-4,11-diene-3	30. 10
,14-dione] (1α,2β,8α,9α)-4,5:12,13-dibenzo-	Tricyclo(7.5.0.0(2,8))tetradecane-4,12-diene-3	30. 12
,14-dione] (1α,2β,8α,9β)-4,5:12,13-dibenzo-	Tricyclo(7.5.0.0(2,8))tetradecane-4,12-diene-3	30. 8
<3,14-dibenzo-6,6,11,11-tetramethyl-1,8,9-cis-	Tricyclo(7.5.0.0(2,8))tetradecane-5,12-dione	30. 14
,14-dione] (1α,2β,8α,9β)-6,7:10,11-dibenzo-	Tricyclo(7.5.0.0(2,8))tetradecane-6,10-diene-3	30. 9
9,11-dihydroxy-2,10-dimethyl-5-oxa	Tricyclo(8.3.0.0(4,8))tridecan-6-one	38. 69
9,9,13-trimethyl-8-oxa	Tricyclo(8.4.0.0(2,7))tetradec-2(7)-en-3-one	38. 78
(1RS,8SR,10SR,4(15)Z)-4-ethylidene-5-oxa-3-aza	Tricyclo(8.4.0.0(3,8))tetradecane	40. 35
2,13-dimethyl-10-(2-methyl-hept-2-en-6-yl)-	Tricyclo(9.3.0.0(9,13))tetradecane-2,8-diene->	55. 1
5,5,7,12,12,14-hexamethyl-1,4,8,11-tetra-aza	Tricyclo(9.3.1.1(4,8))hexadecane hydrate	36. 29
1,1,7,7-tetramethyl-3,5,9,11-tetraphenyl-	Tricyclohexasiloxane	63. 46
bis(bis(triphenylphosphine)iminium)	Tridecacarbonyl-tetra-iron	64.161
roxy-2,10-dimethyl-5-oxatricyclo(8.3.0.0(4,8))	Tridecan-6-one ....[9,11-dihyd	38. 69
enoxycarbonyl)-3,6-diazatricyclo(7.4.0.0(2,7))	Tridecane-2(7),9,11,13-tetraen-8-one <l-13-(ph	36. 36
ethoxycarbonyl)-3,6-diazatricyclo(7.4.0.0(2,7))	Tridecane-2(7),9,11,13-tetraen-8-one <thyl-10-	36. 24
afluoro-pentacyclo(6.4.0.1(9,12).0(2,7)0(3,6)	Tridecane-4,10-diene ... [2,3,4,5,6,7-hex	30. 2
<2-tetrachloro-tetracyclo(6.2.2.1(3,6).0(2,7)	Tridecane-4,11-dien-9-one)-spiro-10,4'-(9-ant>	38.118
<2-tetrachloro-tetracyclo(6.2.2.1(3,6).0(2,7)	Tridecane-4,11-dien-9-one)-spiro-10,4'-(9-ant>	38.119
)-(6-phenyl-tetracyclo(5.4.2.0(3,13).0(10,12))	Tridecane-4,8-dien-11-yl) acetic acid <RS,13RS	59. 29
5H-pyrrolizino(1a,6b,6a-b,c)-1,8-dioxa-15-cis-	Tridecene <,13-trimethyl-1a,2,3,6b-tetrahydro-	58. 29
tetra-acetyl-	Trillenogenin monobrosylate.................	51. 85
des-N-tetramethyl-	Triostin A dodecahydrate	50. 34
des-N-tetramethyl-	Triostin A methanol solvate hydrate..........	50. 33
arsa	Triptycene	65. 11
cis-1,4-dihydro-4-	Trityl-biphenyl'...................	19. 17
	Tropane-3-spiro-5'-thio-hydantoin	37. 6
thyro	Tropin-releasing-hormone tartrate monohydrate	48. 47
diphenyl-	Tropolonato-thallium(iii)	68. 24
0-(2',6'-dinitrophenyl)-3,5,7-trimethyl-	Tropolone	22. 5
o-	Tropylbiphenyl tetrafluoroborate	22. 7
2,6-diphenylhomo	Tropylidene...................	27. 27
dithieno(1,2-b:5,4-b')	Tropylium tetrafluoroborate	39. 33
3,5,7-trimethyl-	Tropylium-0,0-(2',4',6'-trinitrophenylide)	38. 74
	Tryptamine adenin-9-yl-acetic acid hemihydrate	44. 17
threo-β-hydroxy-N-acetyl-	Tryptophanamide...................	48. 37
erythro-β-hydroxy-N-acetyl-	Tryptophanamide benzene solvate..............	48. 38
onyl-glutamyl-histidinyl-phenylalanyl-arginyl-	Tryptophanyl-glycine [methi	48. 82
	Tsukushinamine-A hydrobromide	58. 8a
	Tulirinol acetate	53. 23
CC-1065-anti-	Tumor agent.....................	59. 43
um) bis((dithio-tungsten)-bis(μ(2)-thio))-oxo-	Tungsten monohydrate [bis(tetramethylammoni	3. 24
onium μ-selenido-μ-diselenido-bis(tetrachloro-	Tungsten(v)) ....[tetraphenylars	65. 18
tetraphenylarsonium (decachloro-μ-nitrido-di-	Tungsten)	65. 19
rakis(μ(3)-thio)-tris(μ(3)-ethylthio-iron)-di-	Tungsten) <onium) tris(μ(2)-ethylthio)-bis(tet	3. 39
rate] bis(tetramethylammonium) bis((dithio-	Tungsten)-bis(μ(2)-thio))-oxo-tungsten monohyd	3. 24
<lphosphonium pentasulfur-iron-bis(μ-sulfido)-	Tungsten-disulfide dimethylformamide solvate	64.131
7,8-dimethyl-isoalloxazine-10-acetic acid	Tyramine tetrahydrate	36. 11
ate]	Tyramine 1-thyminyl-(acetic acid) complex hydr	60. 6
	Tyrian purple	59. 12
-chloro-4β,7β,8β,11-tetramethyl-bicyclo(5.4.0)	Undec-1(11)-ene........ [5β,9β,10α-tribromo-4α	27. 14a
ramethyl-3-oxatetracyclo(5.4.0.0(2,6).0(4,11))	Undec-8-en-2β-ol .... [1β,5β,6β,11β-tet	38. 62
methyl-2β-methoxy-3-oxatricyclo(5.4.0.0(4,11))	Undec-9-en-6-one .... [1α,4,5α,7α,9,10-hexa	38. 87
<toxy-6,7-dimethyl-3-oxatricyclo(5.4.0.0(2,4))	Undecan-10-one-9,2(1)-spiro(3(1)-formyl-3(1)->	59. 16a
ylene-8-trimethylsiloxy-tricyclo(6.2.1.0(1,5))	Undecan-3-one .... [2-methoxycarbonyl-9-meth	63. 22
7α-hydroxy-bicyclo(5,4,0)	Undecan-9-one	27. 9
1,11-bis(2-acetylaminophenoxy)-3,6,9-trioxa-	Undecane	17. 38
10,11-tetraoxa-1-phosphatricyclo(5.3.1.0(1,5))	Undecane <methyl-1-phenyl-3,4:8,9-dibenzo-2,6,	64.117
1,11-bis(2-nitrophenoxy)-3,6,9-trioxa-	Undecane potassium isothiocyanate	17. 35
1,11-bis(2-acetylaminophenoxy)-3,6,9-trioxa-	Undecane potassium thiocyanate monohydrate	17. 39
cyclo	Undecane-oxime	23. 2
ethyl-6-thia-1,11-diazatricyclo(6.2.1.0(4,11))	Undecane-3,8-diene-2,10-dione .... [3,9-dim	41. 30
ethyl-6-thia-1,11-diazatricyclo(6.2.1.0(4,11))	Undecane-3,8-diene-2,10-dione 6,6-dioxide <dim	41. 31
cholesteryl	Undecanoate...................	51. 84
2-methylthio-Δ(2)-1,3-diaza-6,9-dioxa-cyclo	Undecene hydroiodide	40. 13
1-β-D-lyxofuranosyl	Uracil	47. 7
6-amino-5-cinnamoyl-1,3-dimethyl-	Uracil	44. 32
2,2'-anhydro-1-β-D-arabinofuranosyl-2-thio-	Uracil	47. 2
3'-0-isopropylidene-β-D-ribofuranosyl)-2-thio-	Uracil [2,5'-anhydro-1-(2',	47. 28
-5,5-(spiro-2',4'-bis-dimethylamino-pyrrolyl)-	Uracil [1,3-dimethyl-4-imino	35. 28
is(benzo-15-crown-5) sodium tetrachloro-dioxo-	Uranium.................... [b	38. 68
ofluoro-uranium-(tri-μ(2)-fluoro)-dioxofluoro-	Uranium monohydrate <raethylammonium aqua-diox	3. 35
imidazolium tetrachloro-dioxo-	Uranium(vi)....................	32. 7

                    di-μ-aquo-bis(dioxo-dinitrato- Uranium(vi)) di-imidazole 32. 6
monohyd> tetraethylammonium aqua-dioxofluoro- Uranium-(tri-μ(2)-fluoro)-dioxofluoro-uranium 3. 35
                                                Urea 8. 1
                                parabanic acid Urea 32. 4n
                 3-(p-chlorophenyl)-1,1-dimethyl Urea 8. 11
                3-(3,4-dichlorophenyl)-1,1-dimethyl Urea 8. 10
                  N-methyl-N'-(benzyl)(acetamido)-methyl Urea 1. 29
              N-methyl-N'-(1-acetamido-2-phenylethyl)- Urea 1. 29
          -dimethyl-N'-(N-methyl-2,4-dinitroanilino)-iso Urea .... [O-benzoyl-N,N 13. 16
         henylamino)-3-pyridinio)-sulfonyl-N'-isopropyl Urea .... [N-(1-methyl-4-(3-methylp 33. 73
                                                Urea hydrogen peroxide 8. 3n
          nylamino)-3-pyridyl)sulfonyl)-3-ethyl-1-methyl Urea hydrogen sulfate ....[1-((4-(3-chlorophe 33. 66
                                                Urea lithium sulfate 8. 5
                                                Urea magnesium bromide dihydrate 8. 6
                                                Urea malonic acid 1. 8
                                                Urea orthotelluric acid...................... 8. 4
                                                Urea trifluoro-antimony...................... 8. 2
          4-hydroxy-2,5-dioxo-4-imidazolidine-carboxy Ureide hemihydrate 32. 13
       ylpentyl)-3-methylcyclohex-1-ene p-bromophenyl Urethane <)-(3R,1'R)-1-hydroxymethyl-2-(1-meth 21. 17
                    2,4-hexadiyne-1,6-bis(m-tolyl Urethane) 1. 39
                                β-5-acetyl-2'-deoxy Uridine..................................... 47. 26
                        5-propynyloxy-2'-deoxy Uridine..................................... 47. 29
                        5-hydroxymethyl-2'-deoxy Uridine..................................... 47. 19
                    5-iodo-5'-amino-2',5'-dideoxy Uridine..................................... 47. 5
                         potassium dihydro Uridine-3'-monophosphate hemihydrate 47. 11
                           5-nitro-2'-deoxy Uridine-5'-monophosphate monohydrate 47. 10
                                  disodium Uridine-5'-phosphate heptahydrate 47. 4
                           disodium deoxy Uridine-5'-phosphate pentahydrate 47. 3
                                   6-aza Uridine-5'-phosphate trihydrate.............. 47. 1
                                                Urocanic acid dihydrate...................... 32. 17
                                       di Uron 8. 10
                                                Ursiniolide A monohydrate 53. 30
                                                Uvarinol benzene solvate 59. 42
)-13-(β-D-mycinose)-lankamycin-11-α-hydroxyiso Valerate ester <-(β-D-4,6-dideoxy-3-ketoallose 50. 35
                        L-valyl-L-α-hydroxyiso Valeric acid 48. 31
                 cyclo(D-N-methylvalyl-D-α-hydroxyiso Valeryl) 48. 32
          l-L-α-hydroxyisovaleryl-L-valyl-D-α-hydroxyiso Valeryl) .... [cyclo-tris(D-valy 50. 39
         droxyisovaleryl-L-N-methylvalyl-D-α-hydroxyiso Valeryl) .... [cyclo(D-N-methylvalyl-L-α-hy 48. 71
         -D-valyl-L-methylalanyl-L-valyl-D-α-hydroxyiso Valeryl) <alanyl-L-valyl-D-α-hydroxyisovaleryl 50. 24
          cyclo-bis(D-valyl-prolyl-valyl-D-α-hydroxyiso Valeryl) dihydrate 48. 81
      hydroxyisovaleryl-L-methylvalyl-D-α-hydroxyiso Valeryl) sodium nickel nitrate hydrate <l-D-α- 50. 22
        yl-bis(D-valyl-L-lactyl-L-valyl-D-α-hydroxyiso Valeryl)) <lalanyl-L-valyl-D-α-hydroxyisovaler 50. 37
        <D-valyl-L-methylalanyl-L-valyl-D-α-hydroxyiso Valeryl-bis(D-valyl-L-lactyl-L-valyl-D-α-hydr> 50. 37
        <D-valyl-L-methylalanyl-L-valyl-D-α-hydroxyiso Valeryl-D-valyl-L-methylalanyl-L-valyl-D-α-hy> 50. 24
        <ydroxyisovaleryl-L-methylvalyl-D-α-hydroxyiso Valeryl-L-methylvalyl-D-α-hydroxyisovaleryl) > 50. 22
        methylval> cyclo(L-methylvalyl-D-α-hydroxyiso Valeryl-L-methylvalyl-D-α-hydroxyisovaleryl-L- 50. 22
               cyclo-tris(D-valyl-L-α-hydroxyiso Valeryl-L-valyl-D-α-hydroxyisovaleryl) 50. 39
               cyclo(D-N-methylvalyl-L-α-hydroxyiso Valeryl-L-N-methylvalyl-D-α-hydroxyisovaleryl) 48. 71
                                     meso- Valinomycin.................................. 50. 39
                2,6-bis(methylalanyl)-octa Valinomycin................................. 50. 24
                                                Valinomycin barium perchlorate hydrate 50. 36
                                                Valinomycin potassium picrate 60. 43
                N-(t-butoxycarbonyl)-L-prolyl-L- Valyl-glycine hemihydrate 48. 52
         lvate] cyclo-tris(L-valyl-L-prolyl-glycyl-L- Valyl-glycyl) monohydrate dimethylsulfoxide so 48. 83
              cyclo(L-valyl-L-prolyl-glycyl-L- Valyl-glycyl) trihydrate 48. 61
       drate]   cyclo-bis(D- Valyl-prolyl-valyl-D-α-hydroxyisovaleryl) dihy 48. 81
                     cyclo(D-N-methyl Valyl-D-α-hydroxyisovaleryl) 48. 32
        cyclo-tris(D-valyl-L-α-hydroxyisovaleryl-L- Valyl-D-α-hydroxyisovaleryl) 50. 39
        N-methylvalyl-L-α-hydroxyisovaleryl-L-N-methyl Valyl-D-α-hydroxyisovaleryl) .... [cyclo(D- 48. 71
        -α-hydroxyisovaleryl-D-valyl-L-methylalanyl-L- Valyl-D-α-hydroxyisovaleryl) <alanyl-L-valyl-D 50. 24
              cyclo-bis(D-valyl-prolyl- Valyl-D-α-hydroxyisovaleryl) dihydrate 48. 81
        <-L-methylvalyl-D-α-hydroxyisovaleryl-L-methyl Valyl-D-α-hydroxyisovaleryl) sodium nickel ni> 50. 22
        -D-α-hydroxyisovaleryl-bis(D-valyl-L-lactyl-L- Valyl-D-α-hydroxyisovaleryl)) <lalanyl-L-valyl 50. 37
        yl-L-valyl-D> cyclo(D-valyl-L-methylalanyl-L- Valyl-D-α-hydroxyisovaleryl-bis(D-valyl-L-lact 50. 37
        anyl-L-valyl> cyclo(D-valyl-L-methylalanyl-L- Valyl-D-α-hydroxyisovaleryl-D-valyl-L-methylal 50. 24
        hydroxyisovaleryl-L-methylval> cyclo(L-methyl Valyl-D-α-hydroxyisovaleryl-L-methylvalyl-D-α- 50. 22
        <(L-methylvalyl-D-α-hydroxyisovaleryl-L-methyl Valyl-D-α-hydroxyisovaleryl-L-methylvalyl-D-α> 50. 22
                                     L- Valyl-L-α-hydroxyisovaleric acid 48. 31
        yisovaleryl)]   cyclo-tris(D- Valyl-L-α-hydroxyisovaleryl-L-valyl-D-α-hydrox 50. 39
        α-hydroxyisovaleryl)]   cyclo(D-N-methyl Valyl-L-α-hydroxyisovaleryl-L-N-methylvalyl-D- 48. 71
        <ylalanyl-L-valyl-D-α-hydroxyisovaleryl-bis(D- Valyl-L-lactyl-L-valyl-D-α-hydroxyisovaleryl)) 50. 37
        <methylalanyl-L-valyl-D-α-hydroxyisovaleryl-D- Valyl-L-methylalanyl-L-valyl-D-α-hydroxyisova> 50. 24
        eryl-bis(D-valyl-L-lactyl-L-valyl-D> cyclo(D- Valyl-L-methylalanyl-L-valyl-D-α-hydroxyisoval 50. 37
        eryl-D-valyl-L-methylalanyl-L-valyl> cyclo(D- Valyl-L-methylalanyl-L-valyl-D-α-hydroxyisoval 50. 24
              N-acetyl-cyclo-bis(L-prolyl-L- Valyl-L-ornithyl-L-leucyl-L-phenylalanyl) 50. 40
        ate dimethylsulfoxide solvate] cyclo-tris(L- Valyl-L-prolyl-glycyl-L-valyl-glycyl) monohydr 48. 83

te]   cyclo(L- Valyl-L-prolyl-glycyl-L-valyl-glycyl) trihydra   48. 61
hylpyridinium) dihydrido-octacosa-oxygen-deca- Vanadium .... [tetrakis(4-et   33. 25
         tetraphenylarsonium tetrachloro-oxo- Vanadium(v)...............................   65. 14
                       carboxy-decarboxamido- Vancomycin 50. 41
                        diethylamino-dihydro- Veatchinone...............................   58. 68
                                              Ventilagone...............................   59. 14
                                        cyclo Veratril benzene clathrate monohydrate 61. 12
                                       trithia Veratrylene chloroform solvate 39. 84
                                              Verrucarin B 59. 40
                                 8α-hydroxy Verrucosane-2β-p-bromo-benzoate.............   54. 31a
                                     (-)-neo Verrucosane-5β-benzoate....................   54. 32
         2,5-di(styryl)-pyrazine 1,4-bis(2-(2-pyridyl) Vinyl)-benzene   33. 76
             2,4-dimethyl-4-(1,2-dichloro Vinyl)-cyclohexen-3-one....................   21. 6
 <ndo-5-cyano-exo-5((E)-(2,2,2-trifluoroethyl)- Vinyl)-endo-tricyclo(5.2.1.0(2,6))decane-3,8->   31. 52
 <-10-cyano-syn-10-((E)-(2,2,2-trifluoroethyl)- Vinyl)-endo-5-cyano-exo-5((E)-(2,2,2-trifluor>   31. 52
  triphenyl-(1-ethoxycarbonyl-2-methyl-2-olato- Vinyl)-phosphonium tetrafluoroborate <sphonium   64.134
  triphenyl-(1-ethoxycarbonyl-2-methyl-2-hydroxy Vinyl)-phosphonium triphenyl-(1-ethoxycarbony>   64.134
          0-(1-(2,4-dichlorophenyl)-2-bromo Vinyl)-0,0-dimethylphosphate...............   46. 5
               3-methyl-4-(2-carboxy Vinyl)-1,2,5-oxadiazole-2-oxide...............   40. 3
 ioxide]   4-((E)-1,2-diphenyl Vinyl)-2,6-diphenyl-1,3,4,5-thiatriazine-1,1-d   41. 82
                    cis-(E)-2-(2'-dimethylamino- Vinyl)-5-nitrofuran..........................   38. 11
                    cis-(E)-2-(2'-dimethylamino- Vinyl)-5-nitrofuran..........................   38. 12
        2-((3-ethyl-1,3-benzothiazol-2-ylio)-2- Vinyl)-6-methoxy-4-nitrophenolate..........   41. 65
               2-methoxycarbonyl- Vinyl-(N-methyl-N-phenyl-sulfonamide) 1. 26
                              octa(Vinyl-sila-sesquioxane)....................   63. 21
           4-dimethylamino-5-methyl-5- Vinyl-thiazolidine-2-thione.................   41. 22
                            2,5-di Vinyl-1-sulfolane 39. 23
       5-dimethylamino-4-methyl-3,3-diphenyl-4- Vinyl-5-pyrrolin-2-one 32. 70
                 trans-N,N-dimethyl-2-nitro Vinylamine 3. 17
            (+-)-(1R,2R,3R)-3-acetyl-2- Vinylcyclohexan-1-ol 21. 7
                                              Vismione A 59. 34
                             potassium 0-n-butyl Xanthate 1. 18
         4-t-butyl-3-(methylthio)-thiete-2-spiro-9'- Xanthene 39. 77
 ete)]   Xanthene-9-spiro-2'-(3',4'-bis-t-butyl-thiothi   39. 82
)-thiete)]   Xanthene-9-spiro-2'-(4'-t-butyl-3'-(methylthio   39. 77
                                 2-chloro Xanthene-9-2'-(3',4'-bis-t-butyl-thiothiete)   39. 81
                           3,7-dimethyl Xanthine tri-iodide iodine 44. 18
                                              Xantholide B 59. 9
                             13-epi-9-desacetyl Xenicin 54. 28
 (1,2-bis(diphenylphosphino-selenoyl)ethane) p- Xylene clathrate   .... [tris   61. 9
                      2-0-methyl- Xylitanylidene diethyl-cyclophosphoramidate....   46. 6
            1,5-anhydro-2,3,4-tri-0-benzoyl- Xylitol...................................   45. 53
                                  α-L- Xylopyranose 45. 3n
           1,2,3,4-tetra-0-benzoyl-β-D- Xylopyranose 45. 58
         methyl-2,3,4-tris(0-benzoyl)-β-D- Xylopyranoside 45. 55
 ,4-di-0-acetyl-3-deoxy-3-C-methyl-3-nitro-β-D- Xylopyranoside .... [methyl-2   45. 22
         2,3,4-tri-0-benzoyl-2-C-chloro-α-D- Xylopyranosyl chloride 45. 51
   lo-N,N-diethylphosphamido)-2-methyl-1-deoxy Xylose .... [3,5-(cyc   46. 6
                  hexakis(acetylene) Zeolite 4A complex 5. 5
                dicytosinium tetrachloro- Zinc 44. 7

Aluminium

hexakis(Acetylene) zeolite 4A complex 5. 5
bis(η(5)-cyclopentadienyl)-2,2-bis(diethyl- Alumino)-ethyl-zirconium-cyclopentadienide 71.197
tetramethyl Ammonium iodotrimethyl-aluminium 68. 2
minium] cis-bis(2,2'- Bipyridyl)-dimethyl-cobalt(iii) tetraethyl-alu 71.179
vate] Cesium di-iodo-dimethyl-aluminium p-xylene sol 68. 1
Cesium isothiocyanato-trimethyl-aluminium 68. 3
sodium methylmercapto-cobalt(ii) Cobalt aluminium silicate 85. 1
sodium methylmercapto- Cobalt(ii) cobalt aluminium silicate 85. 1
cis-bis(2,2'-bipyridyl)-dimethyl- Cobalt(iii) tetraethyl-aluminium 71.179
nyl)-2,2-bis(diethyl-alumino)-ethyl-zirconium- Cyclopentadienide ....[bis(η(5)-cyclopentadie 71.197
di-μ-chloro-bis((η(3)-pentamethyl-*Cyclopentadienyl)-methyl-aluminium)........... 68. 22
yl-zirconium-cyclopentadienide] bis(η(5)- Cyclopentadienyl)-2,2-bis(diethyl-alumino)-eth 71.197
l-aluminium)ethane] (μ-chloro)-1-(di Cyclopentadienyl-zirconium(iv))-2,2-bis(diethy 71.159
enyl-zirconium(iv))-2,2-bis(diethyl-aluminium)*Ethane .... [(μ-chloro)-1-(dicyclopentadi 71.159
dichloro-(N-ethyl-N',N'-dimethyl Ethylenediamine)-aluminium 68. 4
1-ethyl-3-methyl-1-alumina- Indan dimer...................... 68. 21
tris(aluminium(iii)) pentachloride tetrakis(Isopropoxide) 68. 14
cesium Isothiocyanato-trimethyl-aluminium 68. 3
tetracarbonyl-*Manganese-((μ(2)-acetyl)-(μ(2)-bromo)-dibromo- 71. 3
aluminium)]
sodium methyl Mercapto-cobalt(ii) cobalt aluminium silicate 85. 1
(N,N',N'',N'''-tetramethyl- Oxalamidinato)-bis(dimethyl-aluminium) 68. 9
tris(8-hydroxy Quinolato)-aluminium(iii) methanol solvate 68. 23
tris(dimethylamino)-chloro- Silane-(aluminium chloride)................... 63. 3
um methylmercapto-cobalt(ii) cobalt aluminium Silicate .... [sodi 85. 1
ium silicate] Sodium methylmercapto-cobalt(ii) cobalt alumin 85. 1
hexakis(acetylene) Zeolite 4A complex 5. 5
e] (μ-chloro)-1-(dicyclopentadienyl-*Zirconium(iv))-2,2-bis(diethyl-aluminium)ethan 71.159
lopentadienyl)-2,2-bis(diethyl-alumino)-ethyl- Zirconium-cyclopentadienide.... [bis(η(5)-cyc 71.197

Cadmium

di Acetato-dithiocarbamide-cadmium................. 85. 11
Adeninium-cadmium-nitrate hydrate dimer........ 83. 56
<hlorato-(bis(o-phenylene)-bis(pyridine-2,6-di Aldimino))-cadmium(ii) perchlorate methanol s> 83.187
di-cadmium(ii> catena-bis(μ-bis(2-aminoethyl)*Amine)-bis(μ-thiocyanato)-bis(isothiocyanato)- 76. 73
trimethyl Ammonium cadmium chloride 3. 12
methyl Ammonium tetrabromo-cadmium(ii).................. 3. 2
um(ii)] trimethyl Ammonium tetrabromo-cadmium(ii) tribromo-cadmi 3. 14
butanedionato)-cadmium(ii)] Ammonium tris(4,4,4-trifluoro-1-(2-furyl)-1,3- 77. 9
aquo-bis(p-amino Benzoato)-cadmium(ii) dihydrate................ 81. 55
oethane clathrate] bis(triphenyl Benzylphosphonium) tetrachloro-cadmium dichlor 61. 8
yl-bis(trimethylsilylmethyl)-cadmium(ii) 2,2'- Bipyridyl .... [bipyrid 71.130
) 2,2'-bipyridyl] Bipyridyl-bis(trimethylsilylmethyl)-cadmium(ii 71.130
ammonium tris(4,4,4-trifluoro-1-(2-furyl)-1,3- Butanedionato)-cadmium(ii) 77. 9
aldimino))-cadmium(ii) perchlorate > aquo- Chlorato-(bis(o-phenylene)-bis(pyridine-2,6-di 83.187
bis(α- Cyclodextrin) cadmium polyiodide hydrate 45. 63
ermyl-cadmium)-nickel)-cadmium tol> bis(η(5)- Cyclopentadienyl-(triphenylgermyl)-(triphenylg 73.134
n- Decylammonium tetrachloro-cadmium(ii) 3. 47
n- Decylammonium tetrachloro-cadmium(ii) 3. 46
bis(α-cyclo Dextrin) cadmium polyiodide hydrate............ 45. 63
tetraphenylporphinato-bis(Dioxane)-cadmium(ii) 49. 28
enzylphosphonium) tetrachloro-cadmium dichloro Ethane clathrate ....[bis(triphenylb 61. 8
tetrakis(μ-cyano)-cadmium-(*Ethylenediamine-cadmium) 76. 23
ammonium tris(4,4,4-trifluoro-1-(2- Furyl)-1,3-butanedionato)-cadmium(ii) 77. 9
ium tol> bis(η(5)-cyclopentadienyl-(triphenyl Germyl)-(triphenylgermyl-cadmium)-nickel)-cadm 73.134
<cyclopentadienyl-(triphenylgermyl)-(triphenyl Germyl-cadmium)-nickel)-cadmium toluene solva> 73.134
sodium aquo-bis(Glutamato)-cadmium 82. 20
tris(aquo)- Glutarato-cadmium dihydrate................ 81. 18
chloro-glycyl Glycinato-imidazole-cadmium.............. 82. 28
chloro- Glycylglycinato-imidazole-cadmium 82. 28
chloro-glycylglycinato- Imidazole-cadmium 82. 28
Ionomycin cadmium salt n-heptane solvate 50. 29a
is(2-aminoethyl)amine)-bis(μ-thiocyanato)-bis(*Isothiocyanato)-di-cadmium(ii) [catena-bis(μ-b 76. 73
bis(μ-chloro)-bis(chloro-(6-*Mercapto-purine)-aquo-cadmium(ii)) 85. 35
bis(2-methyl-8- Mercapto-quinolinato)-cadmium 85. 63
l-(triphenylgermyl)-(triphenylgermyl-cadmium)- Nickel)-cadmium toluene solvate <clopentadieny 73.134
um(ii) perchlorate > aquo-perchlorato-(bis(o- Phenylene)-bis(pyridine-2,6-dialdimino))-cadmi 83.187
dichloro-(dimethylphenyl Phosphine)-cadmium 86. 6
e clathrate] bis(triphenylbenzyl Phosphonium) tetrachloro-cadmium dichloroethan 61. 8
rate dihydrate] catena-bis(μ-N-methyl-*Piperidinium-4-thiolato)-cadmium(ii) diperchlo 85. 41
bis(α-cyclodextrin) cadmium Polyiodide hydrate 45. 63
tetraphenyl Porphinato-bis(dioxane)-cadmium(ii)........... 49. 28
bis(μ-chloro)-bis(chloro-(6-mercapto-*Purine)-aquo-cadmium(ii)) 85. 35

ate >   aquo-perchlorato-(bis(o-phenylene)-bis(Pyridine-2,6-dialdimino))-cadmium(ii) perchlor   83.187
                       bis(2-methyl-8-mercapto- Quinolinato)-cadmium 85. 63
             bipyridyl-bis(trimethyl Silylmethyl)-cadmium(ii) 2,2'-bipyridyl........ 71.130
                           Sodium aquo-bis(glutamato)-cadmium 82. 20
                  diacetato-di Thiocarbamide-cadmium 85. 11
       catena-bis(μ-N-methyl-piperidinium-4-*Thiolato)-cadmium(ii) diperchlorate dihydrate   85. 41

Cerium

rchlorate trihydrate]   bis(2,6-diacetyl Pyridine-bis(semicarbazone))-cerium(iii) tripe   84. 92
hydrate]   bis(2,6-diacetylpyridine-bis(Semicarbazone))-cerium(iii) triperchlorate tri   84. 92

Chromium

e solvate]   tetrakis(μ(2)-N-2,6-xylyl-*Acetamidato)-bis(pyridine)-di-chromium pyridin   84.135
tetrahydrofuran s> tetrakis(μ(2)-N-2,6-xylyl-*Acetamidato)-bis(tetrahydrofuran)-di-chromium   84.134
                    tetrakis(μ-N-phenyl-*Acetamidato)-di-chromium 84.114
lvate]   tetrakis(μ(2)-N-2,6-xylyl-*Acetamidato)-di-chromium methylene chloride so   84.127
         tetrakis(μ(2)-N-p-dimethylaniline-*Acetamidato)-tetrahydrofuran-di-chromium 84.132
ne solvate]   tetrakis(μ(2)-N-2,6-xylyl-*Acetamidato)-tetrahydrofuran-di-chromium tolue   84.131
                    tetrakis(μ-*Acetato)-(N-pyrazine)-chromium 81. 45
                    tetrakis(μ-*Acetato)-bis(N-pyridine-chromium) 81. 65
      sodium trans-bis(N-isopropylimino-di Acetato)-chromium(iii) dihydrate 82. 29
      potassium trans-bis(t-butylimino-di Acetato)-chromium(iii) tetrahydrate............ 82. 31
         potassium cis-bis(imino-di Acetato)-chromium(iii) trihydrate 82. 16
                    tetrakis(μ-*Acetato)-di(aquo-chromium) 81. 33
                    tetrakis(μ-*Acetato)-di(aquo-chromium) 81. 32
cis-bis(μ-hydroxo)-bis(ethylenediamine-N,N'-di*Acetato-chromium(iii)) tetrahydrate....[alpHa-   76. 72
-cis(ethyl α-ethoxy-α-(1,3-dithian-2-ylidene)- Acetimidate)-chromium(0)   .... [tetracarbonyl   85. 49
              (η(5)-cyclopentadienyl- Aldehyde)-dicarbonyl-nitroso-chromium 73. 8
          (tris(2-(5-chloro-salicylaldiminato)-ethyl) Amine)-chromium(iii) trihydrate................ 78. 10
ii) hemi(sodium hydrox>   sodium triethylmethyl Ammonium tris(thiobenzohydroximato)-chromium(i   85. 66
um]   tetrakis(μ(2)-N-p-dimethyl*Aniline-acetamidato)-tetrahydrofuran-di-chromi   84.132
-(μ-dimethylarsenido)-chromium)-bis(μ-dimethyl*Arsenido)-(tetracarbonyl-iron)-cobalt <adienyl   73.111
<ienyl)-chromium)-cyclopentadienyl-(μ-dimethyl*Arsenido)-chromium)-bis(μ-dimethylarsenido)-(>   73.111
   cis-tetracarbonyl-((1-methyl-2-phenyl-3- Aza-but-2-enyl)-diethylaminocarbene)-chromium   71.141
          4,5-diethyl-3,6-dimethyl-1,2-di Aza-3,6-diborine-tricarbonyl-chromium 75. 2
        pentacarbonyl-(1,5-dimethyl-1,2,3-di Azaphosphole)-chromium 86. 8
yl-molybdenum)]   (η(2)-pentamethylene-di Azirine)-(pentacarbonyl-chromium)-(pentacarbon   83.101
icarbonyl-(η(5)-cyclopentadienyl)-(μ-methanedi*Azo)-(pentacarbonyl-chromium)-tungsten .... [d   73. 41
             tetrakis(μ-N,N'-dimethyl π-Benzamidinato)-di-chromium 83.214
mium]   π-Benzene-dicarbonyl-(methoxyphenylcarbene)-chro   71. 95
               tris(N-methyl-thio Benzohydroxamato)-chromium(iii)............... 85. 73
drox> sodium triethylmethylammonium tris(thio Benzohydroximato)-chromium(iii) hemi(sodium hy   85. 66
hanol solvate]   sodium trans-tris(Benzohydroximato)-chromium(iii) octahydrate et   84. 87
sodium hydroxide nonahydrat> sodium cis-tris(Benzohydroximato)-chromium(iii) sodium iodide   84. 88
     4,5-diethyl-3,6-dimethyl-1,2-diaza-3,6-di Borine-tricarbonyl-chromium.................. 75. 2
  cis-tetracarbonyl-((1-methyl-2-phenyl-3-aza- But-2-enyl)-diethylaminocarbene)-chromium 71.141
methoxy-4-η(6)-phenyl-2,3-bis(trimethylsilyl)- Buta-1,3-dien-1-one)-chromium [tricarbonyl-(4-   74. 7
        pentacarbonyl-(bis(1-methyl-3-oxo-1- Butenyloxy)(phenyl)-phosphine)-chromium........ 86. 74
                    π-benzene-dicarbonyl-(methoxyphenyl Carbene)-chromium 71. 95
methyl-2-phenyl-3-aza-but-2-enyl)-diethylamino Carbene)-chromium ....[cis-tetracarbonyl-((1-   71.141
     pentacarbonyl-(diethylamino(triphenyl-tin) Carbene)-chromium(0) dichloromethane solvate   71.219
ny l-chloro-((tricarbonyl-chromium)-η(6)-phenyl Carbene)-tungsten .... [trans-tetracarbo   71. 60
trans-triphenyl-tin-tetracarbonyl-diethylamino Carbyne-chromium 71.215
-bis(μ-dimethylarsenido)-(tetracarbonyl-iron)-*Cobalt <adienyl-(μ-dimethylarsenido)-chromium)   73.111
io)-methylidyne)-(μ(3)-sulfido)-triangulo-tri-*Cobalt tetrahydrofuran solvate <yl-chromium-th   71.169
(iii) hexakis(thi>   (+)-tris(ethylenediamine)-*Cobalt(iii) (-)-tris(ethylenediamine)-chromium   76. 37
   1,1,1-tris((diphenylphosphino)methyl)ethane-*Cobalt-η(3)-(bis(pentacarbonyl-chromium)-cycl>   86.181
                               Codeine-tricarbonyl-chromium 75. 28
      hexamethyl-trisila-tetraphospha-nortri Cyclene-pentacarbonyl-chromium 86. 14
    bis(hexamethyl-trisila-tetraphospha-nortri Cyclene-tetracarbonyl-chromium)................. 86. 72
m tricarbonyl-(1-5-η(5))-6-(1,3-dithian-2-yl)- Cyclohexadienyl-chromium(0) dioxane solvate <u   75. 6
m-thio)-methylidyne)-(μ(3)-sulfid> tris(η(5)-*Cyclopentadienyl)-(μ(3)-(pentacarbonyl-chromiu   71.169
nyl-chromium)-tungsten]   dicarbonyl-(η(5)-*Cyclopentadienyl)-(μ-methanediazo)-(pentacarbo   73. 41
              dicarbonyl-(η(5)- Cyclopentadienyl)-(thionitrosyl)-chromium 73. 6
                       (η(5)- Cyclopentadienyl)-chloro-dinitrosyl-chromium   73. 1
    dicarbonyl-nitrosyl-(η(5)-pentamethyl- Cyclopentadienyl)-chromium 73. 35
μ-dimet> tricarbonyl-(dicarbonyl(tricarbonyl(*Cyclopentadienyl)-chromium)-cyclopentadienyl-(73.111
                    bis((η(5)- Cyclopentadienyl)-dicarbonyl-chromium)-sulfide   73. 45
<onyl(tricarbonyl(cyclopentadienyl)-chromium)-*Cyclopentadienyl-(μ-dimethylarsenido)-chromiu>   73.111
chromium]   (η(5)- Cyclopentadienyl-aldehyde)-dicarbonyl-nitroso-   73. 8
                 (η(12)-(3.3)Para Cyclophane)-chromium(i) hexafluorophosphate....   74. 2
                 (η(12)-(3.3)Para Cyclophane)-chromium(i) tri-iodide 74. 3
hane-cobalt-η(3)-(bis(pentacarbonyl-chromium)- Cyclotriphosphorus) <phenylphosphino)methyl)et   86.181

bis(triphenylphosphine)immonium μ-*Deuterido-bis(pentacarbonyl-chromium) 64.153n
tris(bis(1,2-dimethylphosphino) Ethane)-chromium 86. 56
)-cycl> 1,1,1-tris((diphenylphosphino)methyl Ethane-cobalt-η(3)-(bis(pentacarbonyl-chromium 86.181
(N,N-diethyl Ethylene-tellurourea)-pentacarbonyl-chromium 70. 8
(+)-tris(Ethylenediamine)-chromium(iii) bromide hydrate 76. 39
<)-tris(ethylenediamine)-cobalt(iii) (-)-tris(Ethylenediamine)-chromium(iii) hexakis(thiocy> 76. 37
tetracarbonyl-(N,N,N',N'-tetramethyl Ethylenediamine)-chromium(0) 76. 62
diamine)-chromium(iii) hexakis(thi> (+)-tris(Ethylenediamine)-cobalt(iii) (-)-tris(ethylene 76. 37
tetrahydrate] alpHa-cis-bis(μ-hydroxo)-bis(*Ethylenediamine-N,N'-diacetato-chromium(iii)) 76. 72
rakis(μ(2)-(N,N')-diphenylurea)-bis(tetrahydro*Furan)-di-chromium cyclohexane solvate   [tet 79. 17
s(μ(2)-N-2,6-xylyl-acetamidato)-bis(tetrahydro*Furan)-di-chromium tetrahydrofuran solvate <ki 84.134
2)-N-p-dimethylaniline-acetamidato)-tetrahydro*Furan-di-chromium .... [tetrakis(μ(84.132
rakis(μ(2)-N-2,6-xylyl-acetamidato)-tetrahydro*Furan-di-chromium toluene solvate .... [tet 84.131
tris(N-methyl-thiobenzo Hydroxamato)-chromium(iii) 85. 73
sodium triethylmethylammonium tris(thiobenzo Hydroximato)-chromium(iii) hemi(sodium hydrox> 85. 66
solvate] sodium trans-tris(benzo Hydroximato)-chromium(iii) octahydrate ethanol 84. 87
m hydroxide nonahydrat> sodium cis-tris(benzo Hydroximato)-chromium(iii) sodium iodide sodiu 84. 88
romium(iii)) tetrahydrate] alpHa-cis-bis(μ-*Hydroxo)-bis(ethylenediamine-N,N'-diacetato-ch 76. 72
sodium trans-bis(N-isopropyl Imino-diacetato)-chromium(iii) dihydrate 82. 29
potassium trans-bis(t-butyl Imino-diacetato)-chromium(iii) tetrahydrate.... 82. 31
potassium cis-bis(Imino-diacetato)-chromium(iii) trihydrate 82. 16
m)] bis(triphenylphosphine)*Immonium μ-deuterido-bis(pentacarbonyl-chromiu 64.153n
tricarbonyl-(2,3-bis(diethylamino)-1-methoxy- Indene)-chromium(0)................... 74. 9
omium)-bis(μ-dimethylarsenido)-(tetracarbonyl Iron)-cobalt <adienyl-(μ-dimethylarsenido)-chr 73.111
2-yl)-cyclohexadienyl-chromium(0) dioxane s> Lithium tricarbonyl-(1-5-η(5))-6-(1,3-dithian- 75. 6
n] dicarbonyl-(η(5)-cyclopentadienyl)-(μ-*Methanediazo)-(pentacarbonyl-chromium)-tungste 73. 41
i) sodium iodide sodium hydroxide nonahydrate Methanol ethanol solvate <oximato)-chromium(ii 84. 88
<mium)-(μ-carbonyl)-(μ-phenyl(methoxycarbonyl)*Methylene)-(bis(trimethylphosphine)-platinum) 71.185
(pentacarbonyl-molybdenum)] (η(2)-penta Methylene-diazirine)-(pentacarbonyl-chromium)- 83.101
chromium] pentacarbonyl-(6-methyl-4- Methylene-2-phenyl-4H-1,3,2-dioxaphosphorine)- 86. 38
<adienyl)-(μ(3)-(pentacarbonyl-chromium-thio)-*Methylidyne)-(μ(3)-sulfido)-triangulo-tri-cob> 71.169
pentacarbonyl-(diethylamino Methylidyne)-chromium tetrafluoroborate........ 71. 19
rine)-(pentacarbonyl-chromium)-(pentacarbonyl- Molybdenum)........[(η(2)-pentamethylene-diazi 83.101
dicarbonyl- Naphthalene-triphenylphosphite-chromium........ 74. 13
(η(5)-cyclopentadienyl-aldehyde)-dicarbonyl- Nitroso-chromium 73. 8
hromium] dicarbonyl- Nitrosyl-(η(5)-pentamethyl-cyclopentadienyl)-c 73. 35
(η(5)-cyclopentadienyl)-chloro-di Nitrosyl-chromium 73. 1
hexamethyl-trisila-tetraphospha- Nortricyclene-pentacarbonyl-chromium 86. 14
bis(hexamethyl-trisila-tetraphospha- Nortricyclene-tetracarbonyl-chromium) 86. 72
ny1-(6-methyl-4-methylene-2-phenyl-4H-1,3,2-di Oxaphosphorine)-chromium ....[pentacarbo 86. 38
e] (η(12)-(3.3) Paracyclophane)-chromium(i) hexafluorophosphat 74. 2
(η(12)-(3.3) Paracyclophane)-chromium(i) tri-iodide 74. 3
mium(iii) iodide monohydrate] (-)(589)-bis(Pentane-2,4-dionato)-(1,3-propanediamine)-chro 84. 50a
dicarbonyl-triethylphosphine- Phenanthrene-chromium 74. 10
hexamethyl-trisila-tetra Phospha-nortricyclene-pentacarbonyl-chromium 86. 14
bis(hexamethyl-trisila-tetra Phospha-nortricyclene-tetracarbonyl-chromium) 86. 72
bis(isopropyl)-amino-oxo Phosphane-pentacarbonyl-chromium 86. 13
2)-(3.3)Paracyclophane)-chromium(i) hexafluoro Phosphate .... [(η(1 74. 2
fac-tricarbonyl-tris(triethyl Phosphine)-chromium........................ 86. 76
6)-tetralone)-carbonyl-thiocarbonyl-(triphenyl Phosphine)-chromium.................. [(η(74. 14a
ny1-(bis(1-methyl-3-oxo-1-butenyloxy)(phenyl)- Phosphine)-chromium.................[pentacarbo 86. 74
enyl(methoxycarbonyl)methylene)-(bis(trimethyl*Phosphine)-platinum) <mium)-(μ-carbonyl)-(μ-ph 71.185
(bis(2-diphenylphosphinoethyl)-phenyl Phosphine)-tricarbonyl-chromium(0) 71.250
yl-chromium)] bis(triphenyl*Phosphine)immonium μ-deuterido-bis(pentacarbon 64.153n
hoxycarbonyl)methylen> (tricarbonyl-trimethyl*Phosphine-chromium)-(μ-carbonyl)-(μ-phenyl(met 71.185
dicarbonyl-triethyl Phosphine-phenanthrene-chromium........ 74. 10
tris(bis(1,2-dimethyl Phosphino)ethane-chromium 86. 56
carbonyl-chromium)-cycl> 1,1,1-tris((diphenyl Phosphino)methyl)ethane-cobalt-η(3)-(bis(penta 86.181
hromium(0)] (bis(2-diphenyl Phosphinoethyl)-phenylphosphine)-tricarbonyl-c 71.250
dicarbonyl-naphthalene-triphenyl Phosphite-chromium 74. 13
pentacarbonyl-(1,5-dimethyl-1,2,3-diaza Phosphole)-chromium......................... 86. 8
tetrakis(μ-dimethyl*Phosphonium-dimethylyl)-di-chromium........... 71.101
-(6-methyl-4-methylene-2-phenyl-4H-1,3,2-dioxa Phosphorine)-chromium .... [pentacarbonyl 86. 38
alt-η(3)-(bis(pentacarbonyl-chromium)-cyclotri Phosphorus) <phenylphosphino)methyl)ethane-cob 86.181
ycarbonyl)methylene)-(bis(trimethylphosphine)-*Platinum) <mium)-(μ-carbonyl)-(μ-phenyl(methox 71.185
bis(pyridine)-(meso-tetraphenyl Porphinato)-chromium(ii) toluene solvate 83.222
i) trihydrate] Potassium cis-bis(imino-diacetato)-chromium(ii 82. 16
romium(iii) tetrahydrate] Potassium trans-bis(t-butylimino-diacetato)-ch 82. 31
te] (-)(589)-bis(pentane-2,4-dionato)-(1,3- Propanediamine)-chromium(iii) iodide monohydra 84. 50a
tetrakis(μ-acetato)-(N- Pyrazine)-chromium- 81. 45
difluoro-tetrakis(5-methyl Pyrazole)-chromium(iii) tetrafluoroborate 83.106
m(ii) toluene solvate] bis(Pyridine)-(meso-tetraphenylporphinato)-chromiu 83.222
dro-1'-methyl-2'-pyridyl)-1,6-dihydro-1-methyl Pyridine)-chromium <-ethyl-1',2',3',4'-tetrahy 74. 5
dro-1'-methyl-2'-pyridyl)-1,6-dihydro-1-methyl Pyridine)-chromium <-ethyl-1',2',3',6'-tetrahy 74. 4
tetrakis(μ(2)-N-2,6-xylyl-acetamidato)-bis(*Pyridine)-di-chromium pyridine solvate 84.135

tetrakis(μ-acetato)-bis(N-*Pyridine-chromium) 81. 65
rubidium bis(Pyridine-2,6-dicarboxylato)-chromium(iii) 81. 52
dro-1'-methyl-2'-pyridyl)-1,6-dihydro-1-methyl Pyridino)-chromium <-ethyl-1',2',3',4'-tetrahy 74. 6
tetrakis(μ(2)-6-chloro-2-*Pyridinolato)-di-chromium 84. 76
<5'-ethyl-1',2',3',4'-tetrahydro-1'-methyl-2'- Pyridyl)-1,6-dihydro-1-methylpyridine)-chromi> 74. 5
<5'-ethyl-1',2',3',6'-tetrahydro-1'-methyl-2'- Pyridyl)-1,6-dihydro-1-methylpyridine)-chromi> 74. 4
<5'-ethyl-1',2',3',4'-tetrahydro-1'-methyl-2'- Pyridyl)-1,6-dihydro-1-methylpyridine)-chromi> 74. 6
ium(iii)] Rubidium bis(pyridine-2,6-dicarboxylato)-chrom 81. 52
trihydrate] (tris(2-(5-chloro- Salicylaldiminato)-ethyl)amine)-chromium(iii) 78. 10
chromium] hexamethyl-tri Sila-tetraphospha-nortricyclene-pentacarbonyl- 86. 14
chromium)] bis(hexamethyl-tri Sila-tetraphospha-nortricyclene-tetracarbonyl- 86. 72
tricarbonyl-(trimethyl-(η(6)-phenyl)- Silane)-chromium 74. 1
nyl-(4-methoxy-4-η(6)-phenyl-2,3-bis(trimethyl Silyl)-buta-1,3-dien-1-one)-chromium [tricarbo 74. 7
) sodium iodide sodium hydroxide nonahydrat> Sodium cis-tris(benzohydroximato)-chromium(iii 84. 88
<enzohydroximato)-chromium(iii) sodium iodide Sodium hydroxide nonahydrate methanol ethanol> 84. 88
tris(thiobenzohydroximato)-chromium(iii) hemi(Sodium hydroxide) hydrate <thylmethylammonium 85. 66
<ium cis-tris(benzohydroximato)-chromium(iii) Sodium iodide sodium hydroxide nonahydrate me> 84. 88
hromium(iii) dihydrate] Sodium trans-bis(N-isopropylimino-diacetato)-c 82. 29
ii) octahydrate ethanol solvate] Sodium trans-tris(benzohydroximato)-chromium(i 84. 87
droximato)-chromium(iii) hemi(sodium hydrox> Sodium triethylmethylammonium tris(thiobenzohy 85. 66
((η(5)-cyclopentadienyl)-dicarbonyl-chromium)- Sulfide.......................... [bis 73. 45
<tacarbonyl-chromium-thio)-methylidyne)-(μ(3)-*Sulfido)-triangulo-tri-cobalt tetrahydrofuran> 71.169
(N,N-diethylethylene- Tellurourea)-pentacarbonyl-chromium.......... 70. 8
sphine)-chromium] (η(6)- Tetralone)-carbonyl-thiocarbonyl-(triphenylpho 74. 14a
e s> lithium tricarbonyl-(1-5-η(5))-6-(1,3-di Thian-2-yl)-cyclohexadienyl-chromium(0) dioxan 75. 6
tetracarbonyl-cis(ethyl α-ethoxy-α-(1,3-di Thian-2-ylidene)-acetimidate)-chromium(0) 85. 49
<lopentadienyl)-(μ(3)-(pentacarbonyl-chromium-*Thio)-methylidyne)-(μ(3)-sulfido)-triangulo-t> 71.169
tris(N-methyl- Thiobenzohydroxamato)-chromium(iii)........... 85. 73
m hydrox> sodium triethylmethylammonium tris(Thiobenzohydroximato)-chromium(iii) hemi(sodiu 85. 66
(η(6)-tetralone)-carbonyl- Thiocarbonyl-(triphenylphosphine)-chromium 74. 14a
dicarbonyl-(η(5)-cyclopentadienyl)-(Thionitrosyl)-chromium 73. 6
te] pentacarbonyl-(diethylamino(triphenyl- Tin)carbene)-chromium(0) dichloromethane solva 71.219
trans-triphenyl- Tin-tetracarbonyl-diethylaminocarbyne-chromium 71.215
o-((tricarbonyl-chromium)-η(6)-phenylcarbene)- Tungsten .... [trans-tetracarbonyl-chlor 71. 60
yl)-(μ-methanediazo)-(pentacarbonyl-chromium)-*Tungsten [dicarbonyl-(η(5)-cyclopentadien 73. 41
xane solvate] tetrakis(μ(2)-(N,N')-diphenyl*Urea)-bis(tetrahydrofuran)-di-chromium cyclohe 79. 17
yridine solvate] tetrakis(μ(2)-N-2,6-*Xylyl-acetamidato)-bis(pyridine)-di-chromium p 84.135
omium tetrahydrofuran s> tetrakis(μ(2)-N-2,6-*Xylyl-acetamidato)-bis(tetrahydrofuran)-di-chr 84.134
ide solvate] tetrakis(μ(2)-N-2,6-*Xylyl-acetamidato)-di-chromium methylene chlor 84.127
toluene solvate] tetrakis(μ(2)-N-2,6-*Xylyl-acetamidato)-tetrahydrofuran-di-chromium 84.131

Cobalt

sodium bis(imino-di Acetato)-cobalt(ii) heptahydrate 82. 15
bis((propylthio)- Acetato)-cobalt(ii) hexamer p-xylene solvate 81. 97
<ne-3'-methyl-benzothiazoline)-chloro-cobalt)-*Acetone-(2,2'-spiro(8-methoxy-3-methyl-6-nitr> 84.120
e)-cobalt(iii) chloride dihydrate] Acetylacetonato-(2,2',2''-triaminotriethylamin 76. 69
tricarbonyl-cobalt-(μ-diphenyl*Acetylene)-(cyclopentadienyl-nickel) 72. 39
)ethane-carbo> tricarbonyl-cobalt-(μ-diphenyl*Acetylene)-(1,1,1-tris(diphenylphosphinomethyl 72. 66
tetra-aquo- Adipato)-cobalt 81. 24
aheptane)-cobalt(iii) bi> (+)(589)-β(2)-((R)- Alaninato)-(1,7-bis(2-(S)-pyrrolidyl)-2,6-diaz 82. 34
(N-(2-pyridylmethyl)-L-aspartato)-(L-phenyl Alaninato)-cobalt(iii) trihydrate 82. 38
<dazole-methoxymethylene-2-yl)-bis(pyridine-2- Aldimino)))-aquo-methanol-cobalt(ii) diperchl> 84.109
2'-bipyridyl)-dimethyl-cobalt(iii) tetraethyl- Aluminium .... [cis-bis(2, 71.179
sodium methylmercapto-cobalt(ii) sesqui- Aluminium silicate 85. 1
dicyanocobyrinic-a,c-di Amide trihydrate 49. 25
cyano-5,6-dihydrocobyrinic-pentamethylester-a- Amide-c-lactam hemihydrate ....[6-amino-di 49. 30
omide] 5- Amidinotetrazolato-tetra-ammine-cobalt(iii) br 83. 6
oba> (μ-peroxo)-(μ-2,2',2''-triamino-triethyl*Amine)-bis((2,2',2''-triamino-triethylamine)-c 76. 96
((R)-1-cyanoethyl)-((S)-(-)-α-methylbenzyl Amine)-bis(dimethylglyoximato)-cobalt(iii) 71.144
((R)-1-cyanoethyl)-((S)-(-)-α-methylbenzyl Amine)-bis(dimethylglyoximato)-cobalt(iii) 71.143
((S)-1-cyanoethyl)-((S)-(-)-α-methylbenzyl Amine)-bis(dimethylglyoximato)-cobalt(iii) 71.145
((S)-1-cyanoethyl)-((S)-(-)-α-methylbenzyl Amine)-bis(dimethylglyoximato)-cobalt(iii) 71.146
<riethylamine)-bis((2,2',2''-triamino-triethyl*Amine-cobalt(ii)) tetraperchlorate dihydrate 76. 96
9S,11S)-4,9-dimethyl-5,8-diazadodecane-2,11-di Amine-cobalt(iii) bromide trihydrate <((2S,4S, 81. 59a
acetylacetonato-(2,2',2''-triaminotriethyl Amine-cobalt(iii) chloride dihydrate 76. 69
malonato-(2,2',2''-triaminotriethyl Amine-cobalt(iii) chloride trihydrate 76. 54
i)] chloro-(tris(2-aminoethyl) Amine)-pyridino-cobalt(iii) tetrachloro-zinc(i 76. 68
cis-dinitro-cis-di Amine-(ethylenediamine)-cobalt(iii) nitrate.... 76. 1
lt(iii) di-iodide hydrate] (-)(589)- Amine-glycinato-(1,4,7-triazacyclononane)-coba 76. 46a
ndecaborane] trimethyl Amine-(η(5)-cyclopentadienyl)-carba-cobalta-u 73. 12
) perchlorate] trans- Ammine-bis(ethylenediamine)-sulfito-cobalt(iii 76. 21
trans-phenyl-bis(dimethylglyoximato)- Ammine-cobalt(iii) 71. 73
5-amidinotetrazolato-tetra- Ammine-cobalt(iii) bromide 83. 6
trans-bis(dimethylglyoximato)-di Ammine-cobalt(iii) bromide 83. 40

trans-bis(dimethylglyoximato)-di Ammine-cobalt(iii) nitrate 83. 42
sodium trans-sulfito-bis(dimethylglyoximato) Ammine-cobalt(iii) tetrahydrate................ 85. 24
trans-bis(dimethylglyoximato)-di Ammine-cobalt(iii) thiocyanate monohydrate 83. 41
rate] penta- Ammine-5-methyltetrazolato-cobalt(iii) perchlo 83. 7
-docosaborane] tetraethyl Ammonium cyclopentadienyl-tetracarba-dicobalta 73. 13
enethiolato)-tetra-cobalt(ii)] tetramethyl*Ammonium hexa-cobalt-tetradecacarbonyl-carbide 3. 19
tetramethyl Ammonium hexakis(μ-benzenethiolato)-tetra(benz 85. 94
lta-closo-undecahydro-dodecabora> tetramethyl Ammonium 2-(η-cyclopentadienyl)-1-carba-2-coba 73. 4
arbadodecaborate)] triethylmethyl Ammonium 8,8'-oxido-3-cobalta-bis(η(5)-1,2-dic 62. 5
hlorate chloride] (2-mercapto- Aniline)-bis(ethylenediamine)-cobalt(iii) perc 76. 63
aqua-chloro-bis(N-(2-pyridylmethylene)- Aniline)-cobalt(ii) nitrate monohydrate... 83.169
)] (tricarbonyl-cobalt)-(μ-di*Arsenic)-(dicarbonyl-triphenylphosphine-cobalt 86. 83
alt)] (μ-di*Arsenic)-bis(dicarbonyl-triphenylphosphine-cob 86.143
-(μ-dimethylarsenido)-chromium)-bis(μ-dimethyl*Arsenido)-(tetracarbonyl-iron)-cobalt <adienyl 73.111
<ienyl)-chromium)-cyclopentadienyl-(μ-dimethyl*Arsenido)-chromium)-bis(μ-dimethylarsenido)-(> 73.111
<)-(μ-carbonyl-(cyclopentadienyl)-bis(dimethyl*Arsenido)-iron)-cyclopentadienyl-(cyclopentad 73. 93
hine)-bis(dimethylglyoximato)-(methyl(phenyl)- Arsenyl-cobalt(iii))-cobalt(ii) dihydrate <sp 86.206
hydrate] (N-(2-pyridylmethyl)-L- Aspartato)-(L-phenylalaninato)-cobalt(iii) tri 82. 38
5R,8R,11R)-2,5,8,11-tetraethyl-1,4,7,10-tetra- Azacyclododecane)-cobalt(iii) dibromide <((2R, 83.108
(-)(589)-amine-glycinato-(1,4,7-tri Azacyclononane)-cobalt(iii) di-iodide hydrate 76. 46a
<do(2',1',6':12,13,14)(1,2,4,7,9,10,13)-hepta- Azacyclopentadeca(3,4,5,6,7,8-aklmn)-1,10-phe> 83.164
<o-(2,9-dimethyl-3,10-diphenyl-1,4,8,11-tetra- Azacyclotetradeca-1,3,8,10-tetraene)-cobalt h> 83.176
<)-(μ-peroxo)-bis(4,7-dimethyl-1,4,7,10-tetra-*Azadecane-cobalt(iii)) triperchlorate dihydra> 76. 94
<)-oxalato-((2S,4S,9S,11S)-4,9-dimethyl-5,8-di Azadodecane-2,11-diamine)-cobalt(iii) bromide> 81. 59a
<-alaninato)-(1,7-bis(2-(S)-pyrrolidyl)-2,6-di Azaheptane)-cobalt(iii) bis(perchlorate) dihy> 82. 34
(μ-peroxo)-bis((1,9-bis(2-pyridyl)-2,5,8-tri*Azanonane) cobalt(iii)) tetra-iodide 76.106
chloro-1,9-bis(2-pyridyl)-2,5,8-tri Azanonane-cobalt(iii) tetrachloro-zinc 76. 92
trahydra> (μ(2)-dioxo)-bis(1,5,8,11,15-penta-*Azapentadecane-cobalt) dithionate dinitrate te 76. 97
) hem> chloro-(1,11-bis(2-pyridyl)-2,6,10-tri Azaundecane)-cobalt(iii) tetrachloro-cobalt(ii 83.131
-carbonyl-μ-(N,N'-diphenyl-N-(phenylcarbenyl)-*Benzamidine)-(dicarbonyl-cobalt) <ne-cobalt)-μ 71.257
bis(η(6)-hexamethyl Benzene)-cobalt hexafluorophosphate............ 74. 11
cobalt(ii)] tetramethylammonium hexakis(μ-*Benzenethiolato)-tetra(benzenethiolato)-tetra- 85. 94
thylammonium hexakis(μ-benzenethiolato)-tetra(*Benzenethiolato)-tetra-cobalt(ii) [tetrame 85. 94
ine-2-a> (N,N'-o-phenylene-(6,6'-(1,2-dihydro Benzimidazole-methoxymethylene-2-yl)-bis(pyrid 84.109
tetra-aqua-bis(p-amino Benzoato)-cobalt(iii) 81. 58
tris(N-methyl-thio Benzohydroxamato-cobalt(iii) 85. 72
aqua-nitrato-(2,6-diacetylpyridine-bis(Benzoic acid hydrazone))-cobalt(ii) nitrate.... 84. 94
-imido-cobalt(ii) tetraphenylborate> bis(1,2- Benzoquinone-di-imide)-1,2-benzosemiquinone-di 83.111
ylborate> bis(1,2-benzoquinone-di-imide)-1,2- Benzosemiquinone-di-imido-cobalt(ii) tetraphen 83.111
<-methoxy-3-methyl-6-nitro-chromene-3'-methyl-*Benzothiazoline)-chloro-cobalt)-acetone-(2,2'> 84.120
-methoxy-3-methyl-6-nitro-chromene)-3'-methyl-*Benzothiazoline)-cobalt <acetone-(2,2'-spiro(8 84.120
dichloro-bis(vinyl Benzotriazole)-cobalt(ii) 83.103
lyoximato)-methyl-cobalt benzene solvate] (Benzoyl-(1-pyridinio)-methanido)-bis(dimethylg 71.182
triaquo-bis(N- Benzoylglycinato)-cobalt(ii) dihydrate 82. 36
bis(4- Benzyl-piperidinium) tetrachloro-cobalt(ii).... 33. 46
i)] ((R)-1-cyanoethyl)-((S)-(-)-α-methyl Benzylamine)-bis(dimethylglyoximato)-cobalt(ii 71.144
i)] ((R)-1-cyanoethyl)-((S)-(-)-α-methyl Benzylamine)-bis(dimethylglyoximato)-cobalt(ii 71.143
i)] ((S)-1-cyanoethyl)-((S)-(-)-α-methyl Benzylamine)-bis(dimethylglyoximato)-cobalt(ii 71.146
i)] ((S)-1-cyanoethyl)-((S)-(-)-α-methyl Benzylamine)-bis(dimethylglyoximato)-cobalt(ii 71.145
hydrate] (-)-D-tris(2,2'- Bipyridyl)-cobalt(ii) hexacyano-iron(iii) octa 83.197a
tylcatecholato)-(3,5-di-t-butyl-semiquinone)-(Bipyridyl)-cobalt(iii) toluene solvate <i-t-bu 84.123
14,17,20-heptaoxa-heneicosane-1,21-diyl)-2,2'- Bipyridyl)-dichloro-cobalt [(6,6'-(2,5,8,11, 84. 99
minium] cis-bis(2,2'- Bipyridyl)-dimethyl-cobalt(iii) tetraethyl-alu 71.179
(5)-cyclopentadienyl-(η(4)-1,4-dimethyl-1,4-di Bora-cyclohexa-2,5-diene)-cobalt .... [η 73. 23
cobaltocenyl-tetra(methylcarba)-dodeca Borane ....[trimethylammin 73. 77
e-(η(5)-cyclopentadienyl)-carba-6-cobalta-undeca Borane ....[trimethylammin 73. 12
tadienyl)-3,7,9,13-tetracarba-6-cobalta-dodeca Borane [3,7,9,13-tetramethyl-6-(η(5)-cyclopen 71. 54
(μ-cyclopentadienyl)-(diselena-dodeca*Borane)-cobalt 73. 3
yl)-1-carba-2-cobalta-closo-undecahydro-dodeca Borate <etramethylammonium 2-(η-cyclopentadien 73. 4
,8'-oxido-3-cobalta-bis(η(5)-1,2-dicarbadodeca Borate)............. [triethylmethylammonium 8 62. 5
η(3):η(4)-trihydro-C,C'-dimethyl-dicarba-penta*Borato))-di-cobalt <(5)-cyclopentadienyl)-(μ-(71. 69
(η(3):η(4)-trihydro-C,C'-dimethyl-dicarba-hexa*Borato))-iron <5)-cyclopentadienyl-cobalt)-(μ- 71. 70
hiolato-(hydro-tris(3,5-dimethyl-1-pyrazolyl)- Borato)-cobalt ....[pentafluorophenyl 85. 67
hyl-4-(5-oxo-4-phenyl-2'(5H)-furan-2-ylidene)- But-1-ene-1,3-diyl))-bis(dicarbonyl-cobalt) <t 71.168
n-2-ylidene)-bis(tricarbonyl-> (μ-3-n-pentyl-*But-2-en-3-olid-4-ylidene)-(μ-1,1-di-iodo-ethe 71.107
<adienyl-η-(1,2,3,4-tetrakis(trifluoromethyl)- Buta-1,3-diene-1,4-diyl)-tungsten-dicarbonyl-> 71.103
dichloro-(2,2'-(1,3-di-iminopropane)-bis(Butan-3-one-oximato)-cobalt(iii) 83. 68
onohydrate] bis(2-(3-aminopropyl)-imino- Butan-3-one-oximato)-cobalt(iii) perchlorate m 83. 98
chloro-bis(α-hydroxylamino-iso Butyrophenone-oximato)-cobalt(iii) 83.147
pentammine- Carbamato-cobalt(iii) nitrate 81. 1
tricarbonyl-(triphenylgermyl)-(ethylethoxy Carbene-cobalt............................ 71.205
-cobalt)-μ-carbonyl-μ-(N,N'-diphenyl-N-(phenyl*Carbenyl)-benzamidine)-(dicarbonyl-cobalt) <ne 71.257
amethylammonium hexa-cobalt-tetradecacarbonyl-*Carbide [tetr 3. 19
<onacarbonyl-tri-cobalt)-μ-(dithiocarboxylato-*Carbido)-(μ(3)-thio-heptacarbonyl-tri-cobalt) 71.118
omethyl)ethane)-cobalt) bis(tetraphen> (μ(2)-*Carbon-disulfide)-bis((1,1,1-(diphenylphosphin 72. 69

dine-carboxylato)-cob> (+)(546)-cis-α-sodium Carbonato-((2S,2'S)-1,1'-ethylene-di-2-pyrroli 84. 49
  chloride trihydrate] Carbonato-bis(1,10-phenanthroline)-cobalt(iii) 83.182
                    bis(phenanthroline)- Carbonato-cobalt(iii) bromide tetrahydrate 83.181
yridyl)-cobalt(iii) toluene > (3,5-di-t-butyl Catecholato)-(3,5-di-t-butyl-semiquinone)-(bip 84.123
cetone-(2,2'-spiro(8-methoxy-3-methyl-6-nitro-*Chromene)-3'-methyl-benzothiazoline)-cobalt <a 84.120
<o)-bis(2,2'-spiro(8-methoxy-3-methyl-6-nitro-*Chromene-3'-methyl-benzothiazoline)-chloro-co> 84.120
iamine)-cobalt(iii) (-)-tris(ethylenediamine)- Chromium(iii) hexakis(thiocyanate) <(ethylened 76. 37
<omium)-cyclopentadienyl-(μ-dimethylarsenido)-*Chromium)-bis(μ-dimethylarsenido)-(tetracarbo> 73.111
<yl-(dicarbonyl(tricarbonyl(cyclopentadienyl)-*Chromium)-cyclopentadienyl-(μ-dimethylarsenid> 73.111
)methyl)ethane-cobalt-η(3)-(bis(pentacarbonyl- Chromium)-cyclotriphosphorus) <phenylphosphino 86.181
<(η(5)-cyclopentadienyl)-(μ(3)-(pentacarbonyl-*Chromium-thio)-methylidyne)-(μ(3)-sulfido)-tr> 71.169
e)] bis(N-methyl)-(μ(2)-n-hexyl)-*Coboglobin-type complex bis(hexafluorophosphat 83.188
(isothiocyanato)-bis(N-methyl)-(μ(2)-n-hexyl)-*Coboglobin-type complex chloride .... [bis 83.196
                      heptamethyl-dicyano- Cobyrinate(iii)................................... 83.223
e monohydrate] dicyano-(5,6-dihydroxy-dihydro Cobyrinic acid pentamethyl ester)-c,d-dilacton 49. 29
                            dicyano Cobyrinic-a,c-diamide trihydrate 49. 25
mihydrate] 6-amino-dicyano-5,6-dihydro Cobyrinic-pentamethylester-a-amide-c-lactam he 49. 30
balt] di Cyanato-diaquo-bis(N,N-diethylnicotinamide)-co 83.162
cobalt acetone solvat> 1,4,7,10,13,16-hexaoxa Cyclo-octadecane hexa-aquo-cobalt tetrachloro- 38. 47
<tetraphenyl-1,10-dioxa-4,7,13,16-tetraphospha Cyclo-octadecane)-cobalt(ii) bis(tetraphenylb> 86.123
<tetraphenyl-1,10-dioxa-4,7,13,16-tetraphospha Cyclo-octadecane)-cobalt(ii) bis(tetraphenylb> 86.122
  (η(5)-cyclopentadienyl)-(η(4)-tetraphenyl- Cyclobutadiene)-cobalt 73.119
pentadienyl)-(η(4)-1,3-dimesityl-2,4-diphenyl- Cyclobutadiene)-cobalt .... [(η(5)-cyclo 73.125
thylsilyl-cyclopentadienyl)-(η(4)-tetraphenyl- Cyclobutadiene)-cobalt .... [(η(5)-trime 73.122
<s(dimethylglyoximato-N,N')-(4-((cis-3-ethoxy- Cyclobutyloxy)-dioxosulfuranyl)-phenyl)-pyrid> 71.196
8R,11R)-2,5,8,11-tetraethyl-1,4,7,10-tetra-aza Cyclododecane)-cobalt(iii) dibromide <((2R,5R, 83.108
        nitrato-cobalt(iii)-(1,4,7,10-tetraoxa Cyclododecane)-nitrate 84. 16
<,1-tris(diphenylphosphinomethyl)ethane-(η(4)- Cyclohepta-1,3,5-trienenyl)-cobalt perchlorat> 75. 47
yclopentadienyl-(η(4)-1,4-dimethyl-1,4-dibora- Cyclohexa-2,5-diene)-cobalt...........[η(5)-c 73. 23
      (-)(589)-amine-glycinato-(1,4,7-triaza Cyclononane)-cobalt(iii) di-iodide hydrate 76. 46a
<2',1',6':12,13,14)(1,2,4,7,9,10,13)-hepta-aza Cyclopentadeca(3,4,5,6,7,8-aklmn)-1,10-phenan> 83.164
          bis(aqua)-(1,4,7,10,13-pentaoxa Cyclopentadecane)-cobalt dinitrate 84. 26
1,3-dioxo-indan-2-ylidene)-bis(carbonyl-(η(5)-*Cyclopentadienyl)-cobalt) .... [((μ- 71.167
          (η(5)- Cyclopentadienyl)-(η(4)-tetraphenyl-cyclobutad 73.119
iene)-cobalt] (η(5)-trimethylsilyl- Cyclopentadienyl)-(η(4)-tetraphenyl-cyclobutad 73.122
enyl-cyclobutadiene)-cobalt] (η(5)- Cyclopentadienyl)-(η(4)-1,3-dimesityl-2,4-diph 73.125
m-thio)-methylidyne)-(μ(3)-sulfid> tris(η(5)-*Cyclopentadienyl)-(μ(3)-(pentacarbonyl-chromiu 71.169
bonyl)-triangulo-tri-cobalt] tris(η(5)-*Cyclopentadienyl)-(μ(3)-sulfido)-(μ(3)-thiocar 71. 97
dimethyl-dicarba-pentaborato))-di-> bis(η(5)-*Cyclopentadienyl)-(μ-(η(3):η(4)-trihydro-C,C'- 71. 69
lt] (μ-*Cyclopentadienyl)-(diselena-dodecaborane)-coba 73. 3
cyclopentadien> bis(μ-carbonyl)-(μ-carbonyl-(*Cyclopentadienyl)-bis(dimethylarsenido)-iron)- 73. 93
          trimethylammine-(η(5)-*Cyclopentadienyl)-carba-cobalta-undecaborane 73. 12
μ-dimet> tricarbonyl-(dicarbonyl(tricarbonyl(*Cyclopentadienyl)-chromium)-cyclopentadienyl-(73.111
          dicarbonyl-(pentamethyl- Cyclopentadienyl)-cobalt(i)...................... 73. 34
          tris(η- Cyclopentadienyl)-tri-cobalt-disulfide 73. 53
          tris(η- Cyclopentadienyl)-tri-cobalt-disulfide 73. 54
cahydro-dodecabora> tetramethylammonium 2-(η- Cyclopentadienyl)-1-carba-2-cobalta-closo-unde 73. 4
a-dodecaborane] 3,7,9,13-tetramethyl-6-(η(5)- Cyclopentadienyl)-3,7,9,13-tetracarba-6-cobalt 71. 54
-tetraphenylphosphole)-cobalt] η(5)- Cyclopentadienyl-η(4)-(1-methoxy-1-oxo-2,3,4,5 73.121
methyl)-buta-1,3-diene-1,4-> dicarbonyl-η(5)- Cyclopentadienyl-η-(1,2,3,4-tetrakis(trifluoro 71.103
-cyclohexa-2,5-diene)-cobalt> η(5)- Cyclopentadienyl-η(4)-1,4-dimethyl-1,4-dibora 73. 23
)-(μ-(η(3):η(4)-trihydro-C,C'-dimethyl> η(5)-*Cyclopentadienyl-(η(5)-cyclopentadienyl-cobalt 71. 70
<onyl(tricarbonyl(cyclopentadienyl)-chromium)-*Cyclopentadienyl-(μ-dimethylarsenido)-chromiu> 73.111
<yclopentadienyl)-bis(dimethylarsenido)-iron)-*Cyclopentadienyl-(cyclopentadienyl-cobalt)-ir> 73. 93
rbonyl)-(μ(2)-methylene)-bis(η(5)-pentamethyl-*Cyclopentadienyl-cobalt) .... [((μ(2)-ca 71.184
o-C,C'-dimethyl> η(5)-cyclopentadienyl-(η(5)-*Cyclopentadienyl-cobalt)-(μ-(η(3):η(4)-trihydr 71. 70
bis(dimethylarsenido)-iron)-cyclopentadienyl-(*Cyclopentadienyl-iron <lopentadienyl)- 73. 93
  chloro-(tetracarbonyl-cobalt)-bis(dicarbonyl- Cyclopentadienyl-iron)-tin(iv) 73. 71
  tricarbonyl-cobalt-(μ-diphenylacetylene)-(*Cyclopentadienyl-nickel) 72. 39
rane] tetraethylammonium Cyclopentadienyl-tetracarba-dicobalta-docosabo 73. 13
sulfur-cobalt] η(5)- Cyclopentadienyl-trimethylphosphine-cyclopenta 73. 11
  η(5)-cyclopentadienyl-trimethylphosphine- Cyclopentasulfur-cobalt] 73. 11
<2,9-dimethyl-3,10-diphenyl-1,4,8,11-tetra-aza Cyclotetradeca-1,3,8,10-tetraene)-cobalt hexa> 83.176
(diphenylphosphino)methyl)ethane-cobalt-(η(3)- Cyclotriphosphorus)............. [1,1,1-tris(86.149
hane-cobalt-η(3)-(bis(pentacarbonyl-chromium)- Cyclotriphosphorus) <phenylphosphino)methyl)et 86.181
<phenylphosphinomethyl)ethane-rhodium-(μ-η(3)-*Cyclotriphosphorus)-1,1,1-tris(diphenylphosph> 86.225
          trans-N,N'-ethylene-bis(S-methyl-L- Cysteinato)-cobalt(iii) perchlorate.......... 82. 21
<μ-peroxo)-bis(4,7-dimethyl-1,4,7,10-tetra-aza*Decane-cobalt(iii)) triperchlorate dihydrate 76. 94
mmonium cyclopentadienyl-tetracarba-dicobalta- Docosaborane .... [tetraethyla 73. 13
R)-2,5,8,11-tetraethyl-1,4,7,10-tetra-azacyclo Dodecane)-cobalt(iii) dibromide <((2R,5R,8R,11 83.108
      nitrato-cobalt(iii)-(1,4,7,10-tetraoxacyclo Dodecane)-nitrate 84. 16
<xalato-((2S,4S,9S,11S)-4,9-dimethyl-5,8-diaza Dodecane-2,11-diamine)-cobalt(iii) bromide tr> 81. 59a
2)-thio)-(1,1,1-tris(diphenylphosphino)methyl)*Ethane)-cobalt dimethylformamide solvate <((μ(86.226
<-thio)-(1,1,1-tris((diphenylphosphino)methyl)*Ethane)-cobalt tetraphenylborate dimethylform> 86.227
(1-aminopropane-2-ol-N)-chloro-bis(1,2-diamino Ethane)-cobalt(iii) tetrachloro-zinc <Δ(S)-cis 76. 45

<sulfide)-bis((1,1,1-(diphenylphosphinomethyl)*Ethane)-cobalt) bis(tetraphenylborate) aceton>   72. 69
t perchl> 1,1,1-tris(diphenylphosphinomethyl) Ethane-(η(4)-cyclohepta-1,3,5-trienenyl)-cobal   75. 47
cetylene)-(1,1,1-tris(diphenylphosphinomethyl) Ethane-carbonyl-cobalt) <l-cobalt-(μ-diphenyla   72. 66
<io)-(1,1,1-tris(((diphenylphosphino)methyl)))*Ethane-cobalt bis(tetraphenylborate) acetone >   86.229
<osphorus)-1,1,1-tris(diphenylphosphinomethyl)*Ethane-cobalt bis(tetraphenylborate) acetone >   86.225
)-cycl> 1,1,1-tris((diphenylphosphino)methyl) Ethane-cobalt-η(3)-(bis(pentacarbonyl-chromium   86.181
        1,1,1-tris((diphenylphosphino)methyl) Ethane-cobalt-(η(3)-cyclotriphosphorus)........   86.149
,1-tris> 1,1,1-tris(diphenylphosphinomethyl)*Ethane-rhodium-(μ-η(3)-cyclotriphosphorus)-1,1   86.225
<is(2-aminoethanolato)-cobalt(iii) bis(2-amino Ethanol)-(2-aminoethanolato)-cobalt(iii) perc>   84. 13
s(2-aminoethanol)-(2-aminoethanolat> (2-amino Ethanol)-bis(2-aminoethanolato)-cobalt(iii) bi   84. 13
                dichloro-bis(2,2'-thio-di Ethanol)-cobalt(ii)...........................   85. 25
-aminoethanolat> (2-aminoethanol)-bis(2-amino Ethanolato)-cobalt(iii) bis(2-aminoethanol)-(2   84. 13
<ato)-cobalt(iii) bis(2-aminoethanolato)-cobalt(iii) perchlorate hemihydra>   84. 13
tyl-but-2-en-3-olid-4-ylidene)-(μ-1,1-di-iodo-*Ethen-2-ylidene)-bis(tricarbonyl-cobalt) <-pen   71.107
                    tris(S-methyl Ethene-1,2-dithiolato)-cobalt 85. 32
osphite-cobalt)]   (μ(3)-*Ethinyl)-triangulo-tris(dicarbonyl-trimethylph   71.112
                lithium (N,N'- Ethylene-bis(salicylideneiminato))-cobalt(i)   78. 3
etrahydrofuran solvate]   sodium (N,N'- Ethylene-bis(salicylideneiminato))-cobalt(i) t   78. 4a
) perchlorate]   trans-N,N'- Ethylene-bis(S-methyl-L-cysteinato-cobalt(iii   82. 21
alt(ii)]   N,N'- Ethylene-bis(3-t-butylsalicylideneiminato)-cob   76.101
alt]   dioxygen-(N-pyridine)-N,N'- Ethylene-bis(3-t-butylsalicylideneiminato)-cob   76.105
<)(546)-cis-α-sodium carbonato-((2S,2'S)-1,1'- Ethylene-di-2-pyrrolidine-carboxylato)-cobalt>   84. 49
e]   potassium Ethylene-tris(trimethylphosphine)-semicobalta   71. 33
xylato-N,O)-cobalt(iii) dichloride]   Δ-bis(Ethylenediamine)-((3R,4R)-thiazolidine-4-carbo   76. 47
oride monohydrate]   bis(Ethylenediamine)-(thio-oxalato-O,S)-cobalt chl   76. 27
<)-tris(ethylenediamine)-cobalt(iii) (-)-tris(Ethylenediamine)-chromium(iii) hexakis(thiocy>   76. 37
diamine)-chromium(iii) hexakis(thi> (+)-tris(Ethylenediamine)-cobalt(iii) (-)-tris(ethylene   76. 37
nate monohydrate]   trans-dichloro-bis(Ethylenediamine)-cobalt(iii) diselenotetrathio   76. 13
nate monohydrate]   trans-dichloro-bis(Ethylenediamine)-cobalt(iii) diselenotetrathio   76. 12
te dihydrate]   (-)(589)-oxalato-bis(Ethylenediamine)-cobalt(iii) hydrogen-d-tartra   76. 31
e]   (+-)-tris(Ethylenediamine)-cobalt(iii) iodide monohydrat   76. 38
ide]   cis-dinitro-cis-diamine-(Ethylenediamine)-cobalt(iii) nitrate 76. 1
ide]   (2-mercapto-aniline)-bis(Ethylenediamine)-cobalt(iii) perchlorate chlor   76. 63
te]   trans-dichloro-bis(Ethylenediamine)-cobalt(iii) selenotetrathiona   76. 14
hiosulfate]   trans-dichloro-bis(Ethylenediamine)-cobalt(iii) S-hydroxymethyl-t   76. 11a
rate]   bis((thio-oxalato-O,S)-bis(Ethylenediamine)-cobalt(iii)) dithionate dihyd   76. 28
        (μ-peroxo)-(μ-hydroxo)-bis(bis(*Ethylenediamine)-cobalt(iii)) perchlorate 76. 53
)-thiolato)-silver perchlorate]   bis((bis(*Ethylenediamine)-cobalt)-(μ-carboxylato-methyl   76. 80
ate]   sodium trans-bis(Ethylenediamine)-disulfito-cobalt(iii) hydr   76. 17
te]   trans-chloro-bis(Ethylenediamine)-sulfito-cobalt(iii) monohydra   76. 9
te]   trans-ammine-bis(Ethylenediamine)-sulfito-cobalt(iii) perchlora   76. 21
rogen-d-tartrate monohydrate]   (-)-trans-(O)- Ethylenediamine-bis(glycinato)-cobalt(iii) hyd   76. 30
rogen-d-tartrate trihydrate]   (+)-trans-(O)- Ethylenediamine-bis(glycinato)-cobalt(iii) hyd   76. 29
alt(iii) hexacyano-cobalt(iii) dihydrate]   Ethylenediamine-bis(2-aminomethylpyridine)-cob   76. 83
<-2'-η-(1,3-dimethyl-4-(5-oxo-4-phenyl-2'(5H)- Furan-2-ylidene)-but-1-ene-1,3-diyl))-bis(dic>   71.168
            tricarbonyl-(triphenyl Germyl)-(ethylethoxycarbene)-cobalt 71.205
            triaquo-bis(N-benzoyl Glycinato)-cobalt(ii) dihydrate...............   82. 36
ohydrate]   (-)-trans-(O)-ethylenediamine-bis(Glycinato)-cobalt(iii) hydrogen-d-tartrate mon   76. 30
hydrate]   (+)-trans-(O)-ethylenediamine-bis(Glycinato)-cobalt(iii) hydrogen-d-tartrate tri   76. 29
) di-iodide hydrate]   (-)(589)-amine- Glycinato-(1,4,7-triazacyclononane)-cobalt(iii   76. 46a
            bis(dimethyl Glyoximato)-(isopropyl)-(pyridine)-cobalt(iii)   71.100
i))-> bis((tri-n-butylphosphine)-bis(dimethyl Glyoximato)-(methyl(phenyl)-arsenyl)-cobalt(ii   86.206
            bis(dimethyl Glyoximato)-(N-iminopyridine)-methyl-cobalt....   83. 96
        trans-phenyl-bis(dimethyl Glyoximato)-amine-cobalt(iii) 71. 73
            aquo-nitro-bis(dimethyl Glyoximato)-cobalt(iii).......................   83. 38
        triphenylphosphine-nitro-bis(dimethyl Glyoximato)-cobalt(iii).....................   86. 91
yl)-((S)-(-)-α-methylbenzylamine)-bis(dimethyl Glyoximato)-cobalt(iii).... [((R)-1-cyanoeth   71.143
yl)-((S)-(-)-α-methylbenzylamine)-bis(dimethyl Glyoximato)-cobalt(iii).... [((R)-1-cyanoeth   71.144
yl)-((S)-(-)-α-methylbenzylamine)-bis(dimethyl Glyoximato)-cobalt(iii).... [((S)-1-cyanoeth   71.145
yl)-((S)-(-)-α-methylbenzylamine)-bis(dimethyl Glyoximato)-cobalt(iii).... [((S)-1-cyanoeth   71.146
        sodium dinitro-bis(dimethyl Glyoximato)-cobalt(iii) dihydrate 83. 37
            trans-bis(dimethyl Glyoximato)-diamine-cobalt(iii) bromide 83. 40
            trans-bis(dimethyl Glyoximato)-diamine-cobalt(iii) nitrate 83. 42
onohydrate]   trans-bis(dimethyl Glyoximato)-diamine-cobalt(iii) thiocyanate m   83. 41
(benzoyl-(1-pyridinio)-methanido)-bis(dimethyl Glyoximato)-methyl-cobalt benzene solvate   71.182
        trans-chloro-bis(dimethyl Glyoximato)-tributylphosphine-cobalt(iii) 86. 73
        trans-chloro-bis(dimethyl Glyoximato)-trimethylphosphite-cobalt(iii) 86. 16
(tri-n-butylphosphine)-co> trans-bis(dimethyl Glyoximato)-2-(5-trifluoro-methyltetrazolato)-   86. 81
        sodium trans-sulfito-bis(dimethyl Glyoximato)ammine-cobalt(iii) tetrahydrate 85. 24
ithiocyanate dihydrate]   dimethyl Glyoximato-bis(o-phenanthroline)-cobalt(iii) d   83.192
xy)-dioxosulfuranyl)-phenyl)-py> bis(dimethyl Glyoximato-N,N')-(4-((cis-3-ethoxy-cyclobutyl   71.196
        penta-aquo-(2'-deoxy Guanosine-5'-phosphato)-cobalt(ii) trihydrate   83. 62
o-cobalt]   (6,6'-(2,5,8,11,14,17,20-heptaoxa- Heneicosane-1,21-diyl)-2,2'-bipyridyl)-dichlor   84. 99
<aninato)-(1,7-bis(2-(S)-pyrrolidyl)-2,6-diaza Heptane)-cobalt(iii) bis(perchlorate) dihydra>   82. 34
hosphate)]   bis(N-methyl)-(μ(2)-n-*Hexyl)-coboglobin-type complex bis(hexafluorop   83.188

bis(isothiocyanato)-bis(N-methyl)-(μ(2)-n-*Hexyl)-coboglobin-type complex chloride........ 83.196
triaquo-bis(Hippurato)-cobalt(ii) dihydrate................. 82. 36
bis(L- Histidinato)-cobalt(iii) perchlorate dihydrate 82. 24
itrato-(2,6-diacetylpyridine-bis(benzoic acid Hydrazone))-cobalt(ii) nitrate ....[aqua-n 84. 94
tetra-rhodium tetrakis(tetrachloro-cobalt) tri Hydrogen hexahydrate <isocyanopropane)-chloro- 71.261
589)-oxalato-bis(ethylenediamine)-cobalt(iii) Hydrogen-d-tartrate dihydrate .... [(-)(76. 31
O)-ethylenediamine-bis(glycinato)-cobalt(iii) Hydrogen-d-tartrate monohydrate [(-)-trans-(76. 30
O)-ethylenediamine-bis(glycinato)-cobalt(iii) Hydrogen-d-tartrate trihydrate [(+)-trans-(76. 29
tris(N-methyl-thiobenzo Hydroxamato)-cobalt(iii) 85. 72
perchlorate] (μ-peroxo)-(μ-*Hydroxo)-bis(bis(ethylenediamine)-cobalt(iii)) 76. 53
(iii)] chloro-bis(α- Hydroxylamino-isobutyrophenone-oximato)-cobalt 83.147
t(ii)] bis(N-methyl Imidazole)-bis(2,4,6-trichlorophenolato)-cobal 84. 81
) tetraphenylborate> bis(1,2-benzoquinone-di- Imidazole)-cobalt 84. 37
<nzoquinone-di-imide)-1,2-benzosemiquinone-di- Imide)-1,2-benzosemiquinone-di-imido-cobalt(ii 83.111
rate monohydrate] bis(2-(3-aminopropyl)- Imido-cobalt(ii) tetraphenylborate pentahydra> 83.111
sodium bis(Imino-butan-3-one-oximato)-cobalt(iii) perchlo 83. 98
(iii)] dichloro-(2,2'-(1,3-di- Imino-diacetato)-cobalt(ii) heptahydrate 82. 15
bis(dimethylglyoximato)-(N- Iminopyridine)-methyl-cobalt 83. 96
-cobalt)] bis(triphenylphosphine)*Immonium nona-μ-carbonyl-nitrido-hexa(carbonyl 64.157
rbonyl-ruthenium)-c> bis((triphenylphosphine)*Immonium) carbonyl-(tri-μ-carbonyl)-tris(trica 64.154
dieneyl)-cobalt)] (μ-1,3-dioxo-*Indan-2-ylidene)-bis(carbonyl-(η(5)-cyclopenta 71.167
phosphine)-hexacarbonyl-di-cobalt-tricarbonyl-*Iron [(μ(3)-phenyl 86. 34
n)-cyclopentadienyl-(cyclopentadienyl-cobalt)-*Iron <lopentadienyl)-bis(dimethylarsenido)-iro 73. 93
)-trihydro-C,C'-dimethyl-dicarba-hexaborato))-*Iron <5)-cyclopentadienyl-cobalt)-(μ-(η(3):η(4 71. 70
)-D-tris(2,2'-bipyridyl)-cobalt(ii) hexacyano- Iron(iii) octahydrate ....[(- 83.197a
omium)-bis(μ-dimethylarsenido)-(tetracarbonyl-*Iron)-cobalt <adienyl-(μ-dimethylarsenido)-chr 73.111
<nyl-(cyclopentadienyl)-bis(dimethylarsenido)-*Iron)-cyclopentadienyl-(cyclopentadienyl-coba> 73. 93
bonyl-cobalt)-bis(dicarbonyl-cyclopentadienyl- Iron(iv) .... [chloro-(tetracar 73. 71
chloro-bis(α-hydroxylamino- Isobutyrophenone-oximato)-cobalt(iii) 83.147
octakis(t-butyl Isocyanide)-di-cobalt 71.263
(tetrachloro-cobalt) trihydr> octakis(1,3-di- Isocyanopropane)-chloro-tetra-rhodium tetrakis 71.261
<)-4-phenyl-1,2,4-triazole-N(1),N(2))-bis(bis(*Isothiocyanato)-(4-phenyl-1,2,4-triazole-N(1)> 83.218
i)] catena-bis(*Isothiocyanato)-bis(μ-1,2,4-triazole)-cobalt(i 83. 21
oboglobin-type complex chloride] bis(*Isothiocyanato)-bis(N-methyl)-(μ(2)-n-hexyl)-c 83.196
bis(3,5-dimethyl-1-phenylpyrazole)-bis(Isothiocyanato)-cobalt(ii) 83.170
t(ii)] bis(Isothiocyanato)-diaquo-bis(nicotinamide)-cobal 83. 94
trans-bis(Isothiocyanato)-tetrakis(pyridine)-cobalt(ii) 83.159
iodoform solvate] trans-bis(Isothiocyanato)-tetrakis(pyridine)-cobalt(ii) 83.160
6-dihydrocobyrinic-pentamethylester-a-amide-c- Lactam hemihydrate ....[6-amino-dicyano-5, 49. 30
is(dicarbonyl-cobalt) di-isopropyl ether so> *Lithium (μ(3)-carbonyl)-tris(μ(2)-carbonyl)-tr 71. 15
))-cobalt(i)] Lithium (N,N'-ethylene-bis(salicylideneiminato 78. 3
lt(iii) chloride trihydrate] Malonato-(2,2',2''-triaminotriethylamine)-coba 76. 54
iii) perchlorate chloride] (2- Mercapto-aniline)-bis(ethylenediamine)-cobalt(76. 63
sodium methyl Mercapto-cobalt(ii) cobalt aluminium silicate 85. 1
trans-di Mesityl-bis(diethylphenylphosphine)-cobalt(ii) 71.256
(η(5)-cyclopentadienyl)-(η(4)-1,3-di Mesityl-2,4-diphenyl-cyclobutadiene)-cobalt.... 73.125
lt benzene solvate] (benzoyl-(1-pyridinio- Methanido-bis(dimethylglyoximato)-methyl-coba 71.182
thylene-2-yl)-bis(pyridine-2-aldimino)))-aquo- Methanol-cobalt(ii) diperchlorate <e-methoxyme 84.109
rate] aqua-chloro-bis(N-(2-pyridyl Methylene)-aniline)-cobalt(ii) nitrate monohyd 83.169
yl-cobalt)] (μ(2)-carbonyl)-(μ(2)-*Methylene)-bis(η(5)-pentamethyl-cyclopentadien 71.184
<ylene-(6,6'-(1,2-dihydrobenzimidazole-methoxy Methylene-2-yl)-bis(pyridine-2-aldimino)))-aq> 84.109
<adienyl)-(μ(3)-(pentacarbonyl-chromium-thio)-*Methylidyne)-(μ(3)-sulfido)-triangulo-tri-cob> 71.169
obalt-(μ-diphenylacetylene)-(cyclopentadienyl-*Nickel)........................ [tricarbonyl-c 72. 39
dicyanato-diaquo-bis(N,N-diethyl Nicotinamide)-cobalt 83.162
bis(isothiocyanato)-diaquo-bis(Nicotinamide)-cobalt(ii) 83. 94
hydrazone))-cobalt(ii) nitrate] aqua- Nitrato-(2,6-diacetylpyridine-bis(benzoic acid 84. 94
ecane)-nitrate] Nitrato-cobalt(iii)-(1,4,7,10-tetraoxacyclodod 84. 16
bis(1H-3,5-diethyl-4-methylpyrazole-N(2))-di Nitrato-O,O-cobalt(ii) 84. 64
s(triphenylphosphine)immonium nona-μ-carbonyl-*Nitrido-hexa(carbonyl-cobalt) ....[bi 64.157
bis(μ-3,5-dimethylpyrazolyl)-bis(di*Nitrosyl-cobalt) 83. 51
μ-peroxo-bis((1,9-bis(2-pyridyl)-2,5,8-triaza*Nonane) cobalt(iii)) tetra-iodide .... [(76.106
chloro-1,9-bis(2-pyridyl)-2,5,8-triaza Nonane-cobalt(iii) tetrachloro-zinc............ 76. 92
acetone solvat> 1,4,7,10,13,16-hexaoxacyclo- Octadecane hexa-aquo-cobalt tetrachloro-cobalt 38. 47
<henyl-1,10-dioxa-4,7,13,16-tetraphosphacyclo- Octadecane)-cobalt(ii) bis(tetraphenylborate) 86.122
<henyl-1,10-dioxa-4,7,13,16-tetraphosphacyclo- Octadecane)-cobalt(ii) bis(tetraphenylborate)> 86.123
hloro-cobalt] (6,6'-(2,5,8,11,14,17,20-hepta Oxa-heneicosane-1,21-diyl)-2,2'-bipyridyl)-dic 84. 99
balt(ii) bi> (α-4,7,13,16-tetraphenyl-1,10-di Oxa-4,7,13,16-tetraphosphacyclo-octadecane)-co 86.122
balt(ii) bi> (β-4,7,13,16-tetraphenyl-1,10-di Oxa-4,7,13,16-tetraphosphacyclo-octadecane)-co 86.123
ro-cobalt acetone solvat> 1,4,7,10,13,16-hexa Oxacyclo-octadecane hexa-aquo-cobalt tetrachlo 38. 47
nitrato-cobalt(iii)-(1,4,7,10-tetra Oxacyclododecane)-nitrate 84. 16
bis(aqua)-(1,4,7,10,13-penta Oxacyclopentadecane)-cobalt dinitrate 84. 26
dodecane-2,11-diamine)-cobalt(iii)> (+)(546)- Oxalato-((2S,4S,9S,11S)-4,9-dimethyl-5,8-diaza 81. 59a
gen-d-tartrate dihydrate] (-)(589)- Oxalato-bis(ethylenediamine)-cobalt(iii) hydro 76. 31
dithionate dihydrate] bis((thio- Oxalato-O,S)-bis(ethylenediamine)-cobalt(iii)) 76. 28

bis(ethylenediamine)-(thio- Oxalato-O,S)-cobalt chloride monohydrate 76. 27
te)]   triethylmethylammonium 8,8'- Oxido-3-cobalta-bis(η(5)-1,2-dicarbadodecabora 62. 5
o-(2,2'-(1,3-di-iminopropane)-bis(butan-3-one- Oximato))-cobalt(iii) .... [dichlor 83. 68
chloro-bis(α-hydroxylamino-isobutyrophenone- Oximato)-cobalt(iii) 83.147
bis(2-(3-aminopropyl)-imino-butan-3-one- Oximato)-cobalt(iii) perchlorate monohydrate 83. 98
Isalicylideneiminato)-cobalt]   di Oxygen-(N-pyridine)-N,N'-ethylene-bis(3-t-buty 76.105
-(4-5-η-1,2-difluoro-1,2-bis(trifluoromethyl)- Pent-4-enyl)-(trimethylphosphite)-cobalt <onyl 71. 38
hydra> (μ(2)-dioxo)-bis(1,5,8,11,15-penta-aza* Pentadecane-cobalt) dithionate dinitrate tetra 76. 97
Pentammine-carbamato-cobalt(iii) nitrate 81. 1
do-ethen-2-ylidene)-bis(tricarbonyl-> (μ-3-n-* Pentyl-but-2-en-3-olid-4-ylidene)-(μ-1,1-dio 71.107
tetrahydrate]   bis(Phenanthroline)-carbonato-cobalt(iii) bromide 83.181
<ta-azacyclopentadeca(3,4,5,6,7,8-aklmn)-1,10- Phenanthroline)-cobalt(ii) bis(tetrafluorobor> 83.164
e]   carbonato-bis(1,10- Phenanthroline)-cobalt(iii) chloride trihydrat 83.182
drate]   dimethylglyoximato-bis(o- Phenanthroline)-cobalt(iii) dithiocyanate dihy 83.192
bis(N-methylimidazole)-bis(2,4,6-trichloro Phenolato-cobalt(ii) 84. 81
oxymethylene-2-yl)-bis(pyridine-2-a> (N,N'-o- Phenylene-(6,6'-(1,2-dihydrobenzimidazole-meth 84.109
<,13,16-tetraphenyl-1,10-dioxa-4,7,13,16-tetra Phosphacyclo-octadecane)-cobalt(ii) bis(tetra> 86.123
<,13,16-tetraphenyl-1,10-dioxa-4,7,13,16-tetra Phosphacyclo-octadecane)-cobalt(ii) bis(tetra> 86.122
bis(η(6)-hexamethylbenzene)-cobalt hexafluoro Phosphate 74. 11
tetradeca-1,3,8,10-tetraene)-cobalt hexafluoro Phosphate <10-diphenyl-1,4,8,11-tetra-azacyclo 83.176
-hexyl)-coboglobin-type complex bis(hexafluoro* Phosphate) .... [bis(N-methyl)-(μ(2)-n 83.188
penta-aquo-(2'-deoxyguanosine-5'- Phosphato)-cobalt(ii) trihydrate 83. 62
bis((μ(2)-carbonyl)-(μ(4)-phenyl* Phosphido))-tetrakis(dicarbonyl-cobalt)....... 86. 77
trimethylphosphine)-cobalt)]   (μ-dimethyl* Phosphido)-(μ-dimethylphosphinomethyl)-bis(di(71.116
tra-cobalt > bis((μ(2)-carbonyl)-(μ(4)-phenyl* Phosphido)-triphenylphosphine-hexacarbonyl-te 86.208
nyl)-arsenyl)-cobalt(iii))-> bis((tri-n-butyl Phosphine-bis(dimethylglyoximato)-(methyl(phe 86.206
trans-dimesityl-bis(diethylphenyl Phosphine-cobalt(ii) 71.256
2-(5-trifluoro-methyltetrazolato)-(tri-n-butyl Phosphine-cobalt(iii) <s(dimethylglyoximato)- 86. 81
(+)(589)-fac-tris(2-aminoethyl-dimethyl Phosphine-cobalt(iii) tribromide trihydrate 86. 28a
)-μ-dimethylphosphinomethyl)-bis(di(trimethyl* Phosphine-cobalt] .... [(μ-dimethylphosphido 71.116
iron]   (μ(3)-phenyl* Phosphine-hexacarbonyl-di-cobalt-tricarbonyl- 86. 34
<2)-carbonyl)-(μ(4)-phenylphosphido)-triphenyl* Phosphine-hexacarbonyl-tetra-cobalt benzene > 86.208
(μ(3)-phenyl* Phosphine-nonacarbonyl-tri-cobalt 86. 35
potassium ethylene-tris(trimethyl Phosphine-semicobaltate 71. 33
a(carbonyl-cobalt)]   bis(triphenyl* Phosphine)immonium nona-μ-carbonyl-nitrido-hex 64.157
tris(tricarbonyl-ruthenium)-c> bis((triphenyl* Phosphine)immonium) carbonyl-(tri-μ-carbonyl)- 64.154
trans-chloro-bis(dimethylglyoximato)-tributyl Phosphine-cobalt(iii) 86. 73
(μ-diarsenic)-bis(dicarbonyl-triphenyl* Phosphine-cobalt) 86.143
yl-cobalt)-(μ-diarsenic)-(dicarbonyl-triphenyl* Phosphine-cobalt) ....[(tricarbon 86. 83
cobalt)-(μ-diphosphorus)-(dicarbonyl-triphenyl* Phosphine-cobalt) .... [(tricarbonyl- 86. 84
N-(phenylcarbenyl> (dicarbonyl-dimethylphenyl* Phosphine-cobalt)-μ-carbonyl-μ-(N,N'-diphenyl- 71.257
η(5)-cyclopentadienyl-trimethyl Phosphine-cyclopentasulfur-cobalt 73. 11
(iii)]   triphenyl Phosphine-nitro-bis(dimethylglyoximato)-cobalt 86. 91
bis(μ(2)-methylthio)-(1,1,1-tris(((diphenyl* Phosphino)methyl)))ethane-cobalt bis(tetraphe> 86.229
ide sol> bis((μ(2)-thio)-(1,1,1-tris(diphenyl* Phosphino)methyl)ethane-cobalt dimethylformam 86.226
ate dim> bis(μ(2)-thio)-(1,1,1-tris((diphenyl* Phosphino)methyl)ethane-cobalt tetraphenylbor 86.227
carbonyl-chromium)-cycl> 1,1,1-tris((diphenyl Phosphino)methyl)ethane-cobalt-η(3)-(bis(penta 86.181
hosphorus)]   1,1,1-tris((diphenyl Phosphino)methyl)ethane-cobalt-(η(3)-cyclotrip 86.149
balt)]   (μ-dimethylphosphido)-(μ-dimethyl* Phosphinomethyl)-bis(di(trimethylphosphine)-co 71.116
(μ(2)-carbon-disulfide)-bis((1,1,1-(diphenyl* Phosphinomethyl)ethane-cobalt) bis(tetraphen> 72. 69
trienenyl)-cobalt perchl> 1,1,1-tris(diphenyl Phosphinomethyl)ethane-(η(4)-cyclohepta-1,3,5- 75. 47
alt-(μ-diphenylacetylene)-(1,1,1-tris(diphenyl* Phosphinomethyl)ethane-carbonyl-cobalt) <l-cob 72. 66
<-η(3)-cyclotriphosphorus)-1,1,1-tris(diphenyl* Phosphinomethyl)ethane-cobalt bis(tetraphenyl> 86.225
iphosphorus)-1,1,1-tris(> 1,1,1-tris(diphenyl* Phosphinomethyl)ethane-rhodium-(μ-η(3)-cyclotr 86.225
2-bis(trifluoromethyl)-pent-4-enyl)-(trimethyl Phosphite)-cobalt <onyl-(4-5-η-1,2-difluoro-1, 71. 38
trans-chloro-bis(dimethylglyoximato)-trimethyl* Phosphite-cobalt(iii) 86. 16
)-ethinyl)-triangulo-tris(dicarbonyl-trimethyl* Phosphite-cobalt) .... [(μ(3 71.112
enyl-η(4)-(1-methoxy-1-oxo-2,3,4,5-tetraphenyl Phosphole)-cobalt .... [η(5)-cyclopentadi 73.121
lphosphino)methyl)ethane-cobalt-(η(3)-cyclotri Phosphorus)........... [1,1,1-tris((dipheny 86.149
alt-η(3)-(bis(pentacarbonyl-chromium)-cyclotri Phosphorus) <phenylphosphino)methyl)ethane-cob 86.181
alt)]   (tricarbonyl-cobalt)-(μ-di* Phosphorus)-(dicarbonyl-triphenylphosphine-cob 86. 84
<osphinomethyl)ethane-rhodium-(μ-η(3)-cyclotri Phosphorus)-1,1,1-tris(diphenylphosphinomethy> 86.225
Phthalocyaninato-cobalt(ii) 49. 1n
bis(4-benzyl- Piperidinium) tetrachloro-cobalt(ii) 33. 46
tetraphenyl Porphyrinato-cobalt(ii)..................... 49. 17
micobaltate]   Potassium ethylene-tris(trimethylphosphine)-se 71. 33
dichloro-(2,2'-(1,3-di-imino Propane)-bis(butan-3-one-oximato))-cobalt(iii) 83. 68
loro-cobalt) trihydr> octakis(1,3-di-isocyano Propane)-chloro-tetra-rhodium tetrakis(tetrach 71.261
cobalt(iii) tetrac> (-)(589)-Δ(S)-cis(1-amino Propane-2-ol-N)-chloro-bis(1,2-diaminoethane)- 76. 45
bis(3,5-dimethyl-1-phenyl Pyrazole)-bis(isothiocyanato)-cobalt(ii) 83.170
bis(1H-3,5-diethyl-4-methyl Pyrazole-N(2))-dinitrato-O,O-cobalt(ii)........ 84. 64
bis-μ-3,5-dimethyl* Pyrazolyl)-bis(dinitrosyl-cobalt) 83. 51
uorophenylthiolato-(hydro-tris(3,5-dimethyl-1- Pyrazolyl)-borato)-cobalt .... [pentafl 85. 67
trans-bis(isothiocyanato)-tetrakis(Pyridine)-cobalt(ii) 83.159
trans-bis(isothiocyanato)-tetrakis(Pyridine)-cobalt(ii) iodoform solvate 83.160

```
           bis(dimethylglyoximato)-(isopropyl)-( Pyridine)-cobalt(iii) ........................ 71.100
hydrate]   ethylenediamine-bis(2-aminomethyl Pyridine)-cobalt(iii) hexacyano-cobalt(iii) di 76. 83
            bis(dimethylglyoximato)-(N-imino Pyridine)-methyl-cobalt......................... 83. 96
eneiminato)-cobalt]            dioxygen-(N- Pyridine)-N,N'-ethylene-bis(3-t-butylsalicylid 76.105
i) nitrate]        aqua-nitrato-(2,6-diacetyl Pyridine-bis(benzoic acid hydrazone))-cobalt(i 84. 94
thoxy-cyclobutyloxy)-dioxosulfuranyl)-phenyl- Pyridine-cobalt <glyoximato-N,N')-(4-((cis-3-e 71.196
<ydrobenzimidazole-methoxymethylene-2-yl)-bis( Pyridine-2-aldimino)))-aquo-methanol-cobalt(i> 84.109
methyl-cobalt benzene solvate]    (benzoyl-(1- Pyridinio)-methanido-bis(dimethylglyoximato)- 71.182
                         dichloro-bis( Pyridinium-2-thiolato)-cobalt(ii) ............ 85. 34
         chloro-(tris(2-aminoethyl)amine)- Pyridino-cobalt(iii) tetrachloro-zinc(ii) .... 76. 68
  diaqua-(7,15-dihydro-7,9,13,15-tetramethyl- Pyrido(2',1',6':12,13,14)(1,2,4,7,9,10,13)-he> 83.164
a-iodide]       (μ-peroxo)-bis((1,9-bis(2-*Pyridyl)-2,5,8-triazanonane) cobalt(iii)) tetr 76.106
hloro-zinc]            chloro-1,9-bis(2- Pyridyl)-2,5,8-triazanonane-cobalt(iii) tetrac 76. 92
trachloro-cobalt(ii) hem>  chloro-(1,11-bis(2- Pyridyl)-2,6,10-triazaundecane)-cobalt(iii) te 83.131
)-cobalt(i) trihydrate]           (N-(2- Pyridylmethyl)-L-aspartato)-(L-phenylalaninato 82. 38
monohydrate]       aqua-chloro-bis(N-(2- Pyridylmethylene)-aniline)-cobalt(ii) nitrate 83.169
<odium carbonato-((2S,2'S)-1,1'-ethylene-di-2- Pyrrolidine-carboxylato)-cobalt(iii) trihydra> 84. 49
  (+)(589)-β(2)-((R)-alaninato)-(1,7-bis(2-(S)- Pyrrolidyl)-2,6-diazaheptane)-cobalt(iii) /bis> 82. 34
<5-di-t-butylcatecholato)-(3,5-di-t-butyl-semi Quinone)-(bipyridyl)-cobalt(iii) toluene solv> 84.123
  octakis(3,5-di-t-butyl-1,2-semi Quinone)-tetra-cobalt(ii) benzene solvate .... 84.139
   bis(1,2-benzoquinone-di-imide)-1,2-benzosemi Quinone-di-imido-cobalt(ii) tetraphenylborate 83.111
  octakis(1,3-di-isocyanopropane)-chloro-tetra- Rhodium tetrakis(tetrachloro-cobalt) trihydro> 71.261
(>   1,1,1-tris(diphenylphosphinomethyl)ethane-*Rhodium-(μ-η(3)-cyclotriphosphorus)-1,1,1-tris 86.225
m) carbonyl-(tri-μ-carbonyl)-tris(tricarbonyl-*Ruthenium)-cobalt <(triphenylphosphine)immoniu 64.154
      lithium (N,N'-ethylene-bis( Salicylideneiminato))-cobalt(i)................ 78.  3
n solvate]      sodium (N,N'-ethylene-bis( Salicylideneiminato))-cobalt(i) tetrahydrofura 78.  4a
xygen-(N-pyridine)-N,N'-ethylene-bis(3-t-butyl Salicylideneiminato)-cobalt....... [dio 76.105
      •   N,N'-ethylene-bis(3-t-butyl Salicylideneiminato)-cobalt(ii)............ 76.101
         (μ-cyclopentadienyl)-(di*Selena-dodecaborane)-cobalt................ 73.  3
ans-dichloro-bis(ethylenediamine)-cobalt(iii) Selenotetrathionate.....................[tr 76. 14
s-dichloro-bis(ethylenediamine)-cobalt(iii) di Selenotetrathionate monohydrate........ [tran 76. 12
s-dichloro-bis(ethylenediamine)-cobalt(iii) di Selenotetrathionate monohydrate........ [tran 76. 13
   potassium ethylene-tris(trimethylphosphine)- Semicobaltate ..........................  71. 33
  (3,5-di-t-butylcatecholato)-(3,5-di-t-butyl- Semiquinone)-(bipyridyl)-cobalt(iii) toluene > 84.123
        octakis(3,5-di-t-butyl-1,2- Semiquinone)-tetra-cobalt(ii) benzene solvate 84.139
ate>  bis(1,2-benzoquinone-di-imide)-1,2-benzo Semiquinone-di-imido-cobalt(ii) tetraphenylbor 83.111
um methylmercapto-cobalt(ii) cobalt aluminium Silicate                        [sodi 85.  1
ine)-cobalt)-(μ-carboxylato-methyl)-thiolato)-*Silver perchlorate .... [bis((bis(ethylenediam 76. 80
obutadiene)-cobalt]        (η(5)-trimethyl Silyl-cyclopentadienyl)-(η(4)-tetraphenyl-cycl 73.122
)-cobalt(i) tetrahydrofuran solvate]    Sodium (N,N'-ethylene-bis(salicylideneiminato) 78.  4a
drate]         Sodium bis(imino-diacetato)-cobalt(ii) heptahy 82. 15
pyrrolidine-carboxylato)-cob>   (+)(546)-cis-α- Sodium carbonato-((2S,2'S)-1,1'-ethylene-di-2- 84. 49
iii) dihydrate]         Sodium dinitro-bis(dimethylglyoximato)-cobalt( 83. 37
ium silicate]         Sodium methylmercapto-cobalt(ii) cobalt alumin 85.  1
balt(iii) trihydrate]         Sodium trans-bis(ethylenediamine)-disulfito-co 76. 17
mine-cobalt(iii) tetrahydrate]     Sodium trans-sulfito-bis(dimethylglyoximato)am 85. 24
<enzothiazoline)-chloro-cobalt)-acetone-(2,2'-*Spiro(8-methoxy-3-methyl-6-nitro-chromene)-3'> 84.120
ethyl-benzothiazolin>  bis(μ-chloro)-bis(2,2'-*Spiro(8-methoxy-3-methyl-6-nitro-chromene-3'-m 84.120
     tris(η-cyclopentadienyl)-tri-cobalt-di Sulfide......................................... 73. 54
     tris(η-cyclopentadienyl)-tri-cobalt-di Sulfide......................................... 73. 53
thane)-cobalt bis(tetraphen>  (μ(2)-carbon-di*Sulfide)-bis((1,1,1-(diphenylphosphinomethyl)e 72. 69
alt]     tris(η(5)-cyclopentadienyl)-(μ(3)-*Sulfido)-(μ(3)-thiocarbonyl)-triangulo-tri-cob 71. 97
<tacarbonyl-chromium-thio)-methylidyne)-(μ(3)-*Sulfido)-triangulo-tri-cobalt tetrahydrofuran> 71.169
ii) tetrahydrate]         sodium trans- Sulfito-bis(dimethylglyoximato)ammine-cobalt(i 85. 24
     trans-chloro-bis(ethylenediamine)- Sulfito-cobalt(iii) monohydrate.................. 76.  9
     trans-ammine-bis(ethylenediamine)- Sulfito-cobalt(iii) perchlorate.................. 76. 21
   sodium trans-bis(ethylenediamine)-di Sulfito-cobalt(iii) trihydrate .............. 76. 17
cyclopentadienyl-trimethylphosphine-cyclopenta Sulfur-cobalt ......................... [η(5)- 73. 11
o-N,N')-(4-((cis-3-ethoxy-cyclobutyloxy)-dioxo Sulfuranyl)-phenyl)-pyridine-cobalt <glyoximat 71.196
o-bis(ethylenediamine)-cobalt(iii) hydrogen-d- Tartrate dihydrate ........ [(-)(589)-oxalat 76. 31
diamine-bis(glycinato)-cobalt(iii) hydrogen-d- Tartrate monohydrate  [(-)-trans-(O)-ethylene 76. 30
diamine-bis(glycinato)-cobalt(iii) hydrogen-d- Tartrate trihydrate....[(+)-trans-(O)-ethylene 76. 29
      tetramethylammonium hexa-cobalt- Tetradecacarbonyl-carbide .................... 3. 19
<bis(dimethylglyoximato)-2-(5-trifluoro-methyl Tetrazolato)-(tri-n-butylphosphine)-cobalt(ii> 86. 81
          penta-ammine-5-methyl Tetrazolato-cobalt(iii) perchlorate............ 83.  7
chloride]    Δ-bis(ethylenediamine)-((3R,4R)- Thiazolidine-4-carboxylato-N,O)-cobalt(iii) di 76. 47
<oxy-3-methyl-6-nitro-chromene-3'-methyl-benzo*Thiazolidine-chloro-cobalt)-acetone-(2,2'-spir 84.120
oxy-3-methyl-6-nitro-chromene)-3'-methyl-benzo*Thiazoline)-cobalt <acetone-(2,2'-spiro(8-meth 84.120
)ethane-cobalt bis(tetraphen>  bis(μ(2)-methyl*Thio)-(1,1,1-tris(((diphenylphosphino)methyl)) 86.229
hane)-cobalt tetraphenylborate dim>  bis(μ(2)-*Thio)-(1,1,1-tris((diphenylphosphino)methyl)et 86.227
ane)-cobalt dimethylformamide sol>  bis((μ(2)-*Thio)-(1,1,1-tris(diphenylphosphino)methyl)eth 86.226
vate]         bis((propyl Thio)-acetato)-cobalt(ii) hexamer p-xylene sol 81. 97
<lopentadienyl)-(μ(3)-(pentacarbonyl-chromium- Thio)-methylidyne)-(μ(3)-sulfido)-triangulo-t> 71.169
           dichloro-bis(2,2'- Thio-diethanol)-cobalt(ii) .................... 85. 25
i-cobalt)-μ-(dithiocarboxylato-carbido)-(μ(3)-*Thio-heptacarbonyl-tri-cobalt) <onacarbonyl-tr 71.118
```

iii)) dithionate dihydrate]   bis((Thio-oxalato-O,S)-bis(ethylenediamine)-cobalt(76. 28
               bis(ethylenediamine)-(Thio-oxalato-O,S)-cobalt chloride monohydrate   76. 27
                          tris(N-methyl- Thiobenzohydroxamato)-cobalt(iii) 85. 72
             tris(N,N-dimethyldi Thiocarbamato)-cobalt(iii) 80. 3
s(η(5)-cyclopentadienyl)-(μ(3)-sulfido)-(μ(3)-*Thiocarbonyl)-triangulo-tri-cobalt .... [tri 71. 97
nyl-tri-coba> (nonacarbonyl-tri-cobalt)-μ-(di*Thiocarboxylato-carbido)-(μ(3)-thio-heptacarbo 71.118
                  tris(S-methylethene-1,2-di Thiolato)-cobalt 85. 32
                 dichloro-bis(pyridinium-2- Thiolato)-cobalt(ii) 85. 34
hylenediamine)-cobalt)-(μ-carboxylato-methyl)-*Thiolato)-silver perchlorate   [bis((bis(et 76. 80
-borato)-cobalt]   pentafluorophenyl Thiolato-(hydro-tris(3,5-dimethyl-1-pyrazolyl) 85. 67
o-bis(ethylenediamine)-cobalt(iii) selenotetra Thionate .... [trans-dichlor 76. 14
lato-O,S)-bis(ethylenediamine)-cobalt(iii)) di Thionate dihydrate .... [bis((thio-oxa 76. 28
is(1,5,8,11,15-penta-azapentadecane-cobalt) di*Thionate dinitrate tetrahydrate <μ(2)-dioxo-b 76. 97
bis(ethylenediamine)-cobalt(iii) diselenotetra Thionate monohydrate   [trans-dichloro- 76. 13
bis(ethylenediamine)-cobalt(iii) diselenotetra Thionate monohydrate .... [trans-dichloro- 76. 12
         hexakis(N-methylurea)-cobalt(ii) Thiosulfate.................................. 79. 9
(ethylenediamine)-cobalt(iii) S-hydroxymethyl- Thiosulfate..............[trans-dichloro-bis 76. 11a
cobalt)-bis(dicarbonyl-cyclopentadienyl-iron)- Tin(iv) ....[chloro-(tetracarbonyl- 73. 71
                   η(6)-Toluene-bis(perfluorophenyl)-cobalt............. 71.133
               dichloro-bis(vinylbenzo Triazole)-cobalt(ii) 83.103
   catena-bis(isothiocyanato)-bis(μ-1,2,4-*Triazole)-cobalt(ii) 83. 21
N(2))-bis(bis(isothiocyanato)-(4-phenyl-1,2,4-*Triazole-N(1))-cobalt(ii)) hydrate <zole-N(1), 83.218
-phenyl-1,2,4-tria> tris(μ(2)-4-phenyl-1,2,4-*Triazole-N(1),N(2))-bis(isothiocyanato)-(4 83.218
                  tris(Tropolonato)-cobalt(iii) 84. 86
kis(trifluoromethyl)-buta-1,3-diene-1,4-diyl)- Tungsten-dicarbonyl-cobalt <I-η-(1,2,3,4-tetra 71.103
em> chloro-(1,11-bis(2-pyridyl)-2,6,10-triaza Undecane)-cobalt(iii) tetrachloro-cobalt(ii) h 83.131
                hexakis(N-methyl Urea)-cobalt(ii) sulfate 79. 10
                hexakis(N-methyl Urea)-cobalt(ii) thiosulfate 79. 9
        dichloro-bis(1- Vinyl-2-hydroxymethylimidazole)-cobalt 84. 37
           dichloro-bis(Vinylbenzotriazole)-cobalt(ii) 83.103
I)-2,5,8-triazanonane-cobalt(iii) tetrachloro- Zinc .... [chloro-1,9-bis(2-pyridy 76. 92
is(1,2-diaminoethane)-cobalt(iii) tetrachloro- Zinc <Δ(S)-cis(1-aminopropane-2-ol-N)-chloro-b 76. 45
ethyl)amine)-pyridino-cobalt(iii) tetrachloro- Zinc(ii) .... [chloro-(tris(2-amino 76. 68

Copper

-copper(ii)]   bis(chloro Acetato)-(N,N,N',N'-tetraethylethylenediamine) 76. 85
te]   tetrakis(μ-2,4-dichlorophenoxy*Acetato)-bis(dioxane-copper(ii)) dioxane solva 81. 92
          tetrakis(pyridine)-bis(trifluoro Acetato)-copper(ii)..................... 81. 77
              (2,2'-bipyridyl)-(imino-di Acetato)-copper(ii) hexahydrate................ 82. 27
        (2-diethylaminoethanolato)-(trichloro Acetato)-copper(ii) tetramer.............. 81. 87
             tetrakis((μ(2)-methoxy)-(μ(2)-*Acetato)-copper(ii)) 84. 38
ylenediamine)-copper-(μ-ethylenediamine-tetra-*Acetato)-nickel tetrahydrate .... [bis(eth 84. 58
0))-(μ-2-diethylam> bis((μ-acetato-(0,0'))-di*Acetato-(μ(3)-2-diethylaminoethanolato-N,μ(3)- 84.122
i))]   tetrakis(chloro*Acetato-(μ-2-dimethylaminoethanolato)-copper(i 81. 82
i))]   tetrakis(dichloro*Acetato-(μ-2-dimethylaminoethanolato)-copper(i 81. 81
ethanolato-N,μ(3)-0))-(μ-2-diethylam> bis((μ-*Acetato-(0,0'))-diacetato-(μ(3)-2-diethylamino 84.122
rate monohydrate]   Acetato-bis(2,2'-bipyridyl)-copper(ii) perchlo 81. 72
uoroborate]   Acetato-bis(2,2'-bipyridyl)-copper(ii) tetrafl 81. 71
is(μ-2-diethylaminoethanolato)-tetrakis(chloro*Acetato-copper(ii) ....[tetrak 81. 89
       aqua-(2,2'-bipyridyl)-nitro Acetato-copper(ii) monohydrate............. 84. 36
        calcium ethylenediamine-tetra- Acetato-copper(ii) tetrahydrate................ 76. 55
hylenediamine)-copper)-(ethylenediamine-tetra- Acetato-nickel) dihydrate <ine)-copper (bis(et 76. 18
)-hexacarbonyl-hexa-rhodium methan> bis(μ(3)-*Acetonitrile-copper)-carbido-ennea-(μ-carbonyl 71.147
o-phenanthroline)-copper(ii)]   (Acetylacetonato)-(hexafluoroacetylacetonato)-(77. 8
afluoroacetylacetone monohydrate]   aqua-(Acetylacetonato)-(o-phenanthroline)-copper hex 77. 6
         (acetylacetonato)-(hexafluoro Acetylacetonato)-(o-phenanthroline)-copper(ii) 77. 8
I-piperidinyl-N-oxy)-copper(i> bis(hexafluoro Acetylacetonato)-(4-hydroxy-2,2,6,6-tetramethy 77. 7
cetonato)-(o-phenanthroline)-copper hexafluoro Acetylacetone monohydrate .... [aqua-(acetyla 77. 6
          N,N'-ethylene-bis(monothio Acetylacetone-iminato)-copper(ii) 76. 71
itrato-(S-methylthiosemicarbazone-8-quinoline- Aldehyde)-copper(ii) .... [din 83. 75
) nitrate]   salicylidene-anthranyl- Aldehyde-4-phenylthiosemicarbazonato-copper(ii 78. 7
per(ii)]   chloro-N,N-diethyl Amide-pyruvate-4-phenylthiosemicarbazonato-cop 79. 12
              aqua-bis(Amidonicotinato)-bis(formato)-copper(ii) 84. 55
        bis(N-isopropyl-2-oxy-1-naphthylidene- Aminato)-copper(ii)...................... 78. 12
utyl-(5-chloro-α-phenyl-2-hydroxybenzylidene)- Aminato)-copper(ii) ....[bis(N-isob 78. 15
odime> bis(N-isopropyl-2-oxy-1-naphthylidene- Aminato)-copper(ii) bis(7,7,8,8-tetracyanoquin 78. 13
         aqua-bis(methyl Amine)-bis(benzoato)-copper(ii) monohydrate.... 81. 61
      ammine-(tris(2-aminoethyl) Amine-copper(ii) diperchlorate.............. 76. 36
ra-bis(2'-benzimidazolyl-methyl)-1,2-ethane-di Amine-copper(ii)   [N,N,N',N'-tet 83.209
perchlorate]   Ammine-(tris(2-aminoethyl)amine-copper(ii) di 76. 36
    bis(5,5-diphenyl-hydantoinato)-di Ammine-copper(ii) 83.199
ii) dihydrate]   N-(2- Ammonioethyl)-piperazinium pentachloro-copper(33. 19
  tetrakis(μ-o-benzoylbenzoato)-bis((p-iodo*Anilino)-copper(ii)) 81. 98
-copper(ii) nitrate]   salicylidene- Anthranyl-aldehyde-4-phenylthiosemicarbazonato 78. 7

                                        L- Asparaginato-L-histidinato-copper(ii) 82. 18
rate]   L- Asparaginato-L-histidinato-copper(ii) tetrahyd 82. 19
<idazole-(2,3,9,10-tetramethyl-1,4,8,11-tetra- Aza-1,3,8,10-cyclotetradecatetraene)-copper(i> 83.132
) bis(> (2,3,9,10-tetramethyl-1,4,8,11-tetra- Aza-1,3,8,10-cyclotetradecatetraene)-copper(ii 83. 97
  bis((methylcyano)-(μ(2)-chloro)-(μ(2)-1,3-di*Aza-2,4-dithietidine)-chloro-copper) 83. 4
 perchlorate methanol s> (3,9-dimethyl-4,8-di Aza-3,8-undecadiene-2,10-dioximato)-copper(ii) 83. 69
ladium)-3-methyl-1,4-bis(methoxyphenyl)-1,4-di Azabutadien-1,4-diyl-dichloro-copper <ine)-pal 86.196
<S>-2,4,4,11,13,-hexamethyl-1,5,10,14-tetra-   Azacyclo-octadeca-1,10-diene)-copper(ii) perc> 83.153
<S>-2,4,4,11,11,13-hexamethyl-1,5,10,14-tetra- Azacyclo-octadeca-1,13-diene)-copper(ii) perc> 83.152
                               (1,5-di Azacyclo-octane)-dinitrato-copper(ii) 84. 11
<-carbonato)-bis(2,4,4,9-tetramethyl-1,5,9-tri*Azacyclododec-1-ene)-di-copper(ii) diperchlor> 83.189
   chloro-(1,4,7,10-tetrabenzyl-1,4,7,10-tetra- Azacyclododecane) copper(ii) nitrate 76.107
                         dibromo-(1,4,7-tri Azacyclononane)-copper(ii) 76. 25
                        dichloro-(1,4,7-tri Azacyclononane)-copper(ii) 76. 26
olyl)-borato)-copper(i) a> potassium (p-nitro Benzenethiolato)-(hydro-tris(3,5-dimethylpyraz 85. 68
er(ii)]   N,N,N',N'-tetra-bis(2'- Benzimidazolyl-methyl)-1,2-ethane-diamine-copp 83.209
  bromo-(2,3,7,8,12,13,17,18-octamethoxy-tetra Benzo(b,e,h,k)(1,4,7,10)tetrathia-cyclododeca> 85. 84
  chloro-(2,3,7,8,12,13,17,18-octamethoxy-tetra Benzo(b,e,h,k)(1,4,7,10)tetrathia-cyclododeca> 85. 85
                   tetrakis(μ-o-benzoyl*Benzoato)-bis((p-iodoanilino)-copper(ii)) 81. 98
       bis(1,3-propanediamine)-bis(m-nitro Benzoato)-copper(ii) 81. 68
       aqua-bis(methylamine)-bis(Benzoato)-copper(ii) monohydrate 81. 61
                    tetrakis(μ(2)-*Benzoato)-dioxane-di-copper dioxane solvate.... 84.113
<-chloro-1,3-dihydro-1-methyl-5-phenyl-3H)-1,4- Benzodiazepin-2-one)-copper(ii) dihydrate chl> 83.204
<(benzothiazole-2-yl)-trisulfide)-(2-mercapto- Benzothiazole)-copper(i) perchlorate chlorofo> 85. 65
nzothiazole)-copper(i) perchlorate chl> (bis(Benzothiazole-2-yl)-trisulfide)-(2-mercapto-be 85. 65
                   tetrakis(μ-o-*Benzoylbenzoato)-bis((p-iodoanilino)-copper(ii 81. 98
))]   bis(1,1-diethyl-3- Benzoylthiourea)-copper(ii).................... 79. 15
                        bis(4- Benzyl-piperidinium) hexachloro-di-copper(ii) 33. 47
(ii) nitrate]   chloro-(1,4,7,10-tetra Benzyl-1,4,7,10-tetra-azacyclododecane) copper 76.107
   bis(N-isobutyl-(5-chloro-α-phenyl-2-hydroxy Benzylidene)-aminato)-copper(ii) 78. 15
drate]   (2,2'- Bipyridyl)-(imino-diacetato)-copper(ii) hexahy 82. 27
copper(ii) tetrahydrate]   bis(2,2'- Bipyridyl)-bis(pyridine-2,6-dicarboxylato)-di- 81. 90
                   bis(6,6'-dimethyl-2,2'- Bipyridyl)-copper(i) tetrafluoroborate 83.171
                        aqua-bis(2,2'- Bipyridyl)-copper(ii) dithionate 83.145
                acetato-bis(2,2'- Bipyridyl)-copper(ii) perchlorate monohydrate 81. 72
                acetato-bis(2,2'- Bipyridyl)-copper(ii) tetrafluoroborate........ 81. 71
                         bis(2,2'- Bipyridyl)-mono-iodo-copper(ii) perchlorate.... 83.141
ate]   bis(2,2'- Bipyridyl)-monobromo-copper(ii) tetrafluoroborob 83.140
ate]   aqua-(2,2'- Bipyridyl)-nitrato-copper(ii) nitrate monohydr 83.142
                         bis(2,2'- Bipyridyl)-nitroacetato-copper(ii) monohydrate 84. 36
rate]   bis(2,2'- Bipyridyl)-peroxo-disulfato-copper(ii) monohyd 84. 82
                      4,4'-*Bipyridylium di-μ-chloro-tetrachloro-di-copper 33. 36
ethiolato)-(hydro-tris(3,5-dimethylpyrazolyl)- Borato)-copper(i) acetone solvate <nitrobenzen 85. 68
                  bis(hydro-tris(1-pyrazolyl)- Borato)-copper(ii) 83.112
ethylene-dinitrilo)-bis(butan-2-one-oximato))- Borato)-copper(ii) perchlorate <ro-(3,3'-(trim 83. 70
ethylene-dinitrilo)-bis(butan-2-one-oximato))- Borato)-copper(ii) [iodo-(difluoro-(3,3'-(trim 83. 66
methylene-dinitrilo)-bis(butan-2-one-oximato)- Borato)-pyridine-copper(ii) perchlorate <-(tri 83.105
ium)-3-methyl-1,4-bis(methoxyphenyl)-1,4-diaza Butadien-1,4-diyl-dichloro-copper <ine)-pallad 86.196
<(difluoro-(3,3'-(trimethylene-dinitrilo)-bis(Butan-2-one-oximato))-borato)-copper(ii) perc> 83. 70
-(difluoro-(3,3'-(trimethylene-dinitrilo)-bis(Butan-2-one-oximato))-borato)-copper(ii) [iodo 83. 66
ono> aquo-(3,3'-(trimethylene-dinitrilo)-bis(Butan-2-one-oximato))-copper(ii) perchlorate m 83. 71
   (difluoro-3,3'-(trimethylene-dinitrilo)-bis(Butan-2-one-oximato)-borato)-pyridine-copper(> 83.105
rc> bis(μ-2-(N,N-dimethyl-2-aminoethyl)imino-*Butan-3-one-oximato)-diaqua-di-copper(ii) dipe 84. 67
-copper(ii) sulfate > tris(m-xylene-α,α'-dia*Butane-2-monoximato-3-imino)-oxo-hydroxo-hexa 83.221
<ium(i) (μ(8)-chloro)-dodecakis(α-mercapto-iso*Butyrato)-octa-copper(i)-hexa-copper(i) dode> 81. 95
i) tetrahydrate]   Calcium ethylenediamine-tetra-acetato-copper(i 76. 55
                         bis(1- Carbamoyl-3,5-dimethyl-pyrazolato)-copper(ii) 83. 78
hodium methan> bis(μ(3)-acetonitrile-copper)-*Carbido-ennea-(μ-carbonyl)-hexacarbonyl-hexa-r 71.147
ne)-copper(ii))]   asym>*Carbonato)-bis(chloro-(tetraethylethylenediami 76. 98
acyclododec-1-ene)-di-copper(ii) diperch> (μ-*Carbonato)-bis(2,4,4,9-tetramethyl-1,5,9-triaz 83.189
nomethyl)-pyridine-copper(ii))]   (μ(3)-*Carbonato)-tris(aqua-2-(2-(2-pyridyl)-ethylimi 84.126
bis(diethylene)-triethylene-tetramine)-bis(per Chlorato)-copper .... [(76. 44
                bis(pyrazine)-bis(per Chlorato)-copper(ii) 83. 34
ethyl)-imidazolato)-di-co> aqua-dinitrato-per Chlorato-(4,5-bis((2-(2-pyridyl)-ethylimino)-m 83.136
              (tris(triethylene)-tetramine)-per Chlorato-copper perchlorate.................... 84. 22
aminoethyl)-1-phenylpropane-2-oximato)-bis(per*Chlorato-copper(ii)) diperchlorate <imethyl-2- 76.104
                di-copper hexyl Cofacial diporphyrin-7 .... 49. 36
<2,4,4,11,13,13-hexamethyl-1,5,10,14-tetra-aza Cyclo-octadeca-1,10-diene)-copper(ii) perchlo> 83.153
<2,4,4,11,11,13-hexamethyl-1,5,10,14-tetra-aza Cyclo-octadeca-1,13-diene)-copper(ii) perchlo> 83.152
                         (1,5-diaza Cyclo-octane)-dinitrato-copper(ii) 84. 11
<rbonato)-bis(2,4,4,9-tetramethyl-1,5,9-triaza*Cyclododec-1-ene)-di-copper(ii) diperchlorate> 83.189
hloro-(1,4,7,10-tetrabenzyl-1,4,7,10-tetra-aza Cyclododecane) copper(ii) nitrate .... [c 76.107
ethoxy-tetrabenzo(b,e,h,k)(1,4,7,10)tetrathia- Cyclododecatetraene)-copper(i) <13,17,18-octam 85. 84
<thoxy-tetrabenzo(b,e,h,k)(1,4,7,10)tetrathia- Cyclododecatetraene)-copper(i) chloroform sol> 85. 85
                      bis(N- Cyclohexyl-salicylaldiminato)-copper(ii) 78. 18

(N,N,N',N'-tetramethylethylenediamine)-bis(Cyclohexylisocyanide)-copper(i) tetraphenylbo> 71.161
dibromo-(1,4,7-triaza Cyclononane)-copper(ii).......................... 76. 25
dichloro-(1,4,7-triaza Cyclononane)-copper(ii)........................ 76. 26
N,N'-trimethylene-bis(methyl-2-amino-1- Cyclopentene-dithiocarboxylato)-copper(ii) 85. 60
<9,10-tetramethyl-1,4,8,11-tetra-aza-1,3,8,10- Cyclotetradecatetraene)-copper(ii) bis(hexafl> 83.132
<9,10-tetramethyl-1,4,8,11-tetra-aza-1,3,8,10- Cyclotetradecatetraene)-copper(ii) bis(tetrap> 83. 97
di Cytosinium tetrachloro-copper 44. 6
copper formate tetra Deuterohydrate 81. 2n
copper formate tetra Deuterohydrate 81. 3n
<ro-1,3-dihydro-1-methyl-5-phenyl-3H-1,4-benzo Diazepin-2-one)-copper(ii) dihydrate chlorofo> 83.204
tetrakis(μ-2,4-dichlorophenoxyacetato)-bis(*Dioxane-copper(ii)) dioxane solvate............ 81. 92
tetrakis(μ(2)-benzoato)-*Dioxane-di-copper dioxane solvate 84.113
<to)-bis(2,4,4,9-tetramethyl-1,5,9-triazacyclo*Dodec-1-ene)-di-copper(ii) diperchlorate dime> 83.189
-(1,4,7,10-tetrabenzyl-1,4,7,10-tetra-azacyclo Dodecane) copper(ii) nitrate ....[chloro 76.107
-aminoethylthiolato)-tetradeca-copper sulfate Eicosahydrate   [chloro-dodeca(1,1-dimethyl-2 85. 91
thane)-platinum(iv) tetrachl> bis(1,2-diamino Ethane)-platinum(ii) dichloro-bis(1,2-diaminoe 76. 20
oethane)-platinum(ii) dichloro-bis(1,2-diamino Ethane)-platinum(iv) tetrachloro-copper(i) <in 76. 20
N',N'-tetra-bis(2'-benzimidazolyl-methyl)-1,2- Ethane-diamine-copper(ii) .... [N,N, 83.209
amer]   (2-diethylamino Ethanolato)-(trichloroacetato)-copper(ii) tetr 81. 87
chloro-(2-(2'-ethanolthio)- Ethanolato)-copper(ii) tetramer.............. 85. 53
tetrakis(chloroacetato-(μ-2-dimethylamino- Ethanolato)-copper(ii)).............. 81. 82
tetrakis(dichloroacetato-(μ-2-dimethylamino*Ethanolato)-copper(ii))............... 81. 81
olvate]   tetrakis(bromo-(2-diethylamino Ethanolato)-copper(ii) carbon tetrachloride s 84.103
bis(cis-bis(2-amino Ethanolato)-copper(ii)) dinitrate 84. 6
tetrakis(μ-2-diethylamino*Ethanolato)-tetrakis(chloroacetato-copper(ii) 81. 89
pper(i) thiocyanate] 2-((3-aminopropyl)amino) Ethanolato-copper(ii) tetra(isothiocyanato)-co 84. 8
<cetato-(0,0'))-diacetato-(μ(3)-2-diethylamino*Ethanolato-N,μ(3)-0))-(μ-2-diethylaminoethano> 84.122
<laminoethanolato-N,μ(3)-0))-(μ-2-diethylamino*Ethanolato-N,μ-0)-(μ(3)-hydroxo)-tri-copper(i> 84.122
chloro-(2-(2'- Ethanolthio)-ethanolato)-copper(ii) tetramer 85. 53
orate]   (tris(tri Ethylene)-tetramine)-perchlorato-copper perchl 84. 22
to)-copper]   (bis(di Ethylene)-triethylene-tetramine)-bis(perchlora 76. 44
pper(ii)]   N,N'- Ethylene-bis(monothioacetylacetone-iminato)-co 76. 71
<xafluoropentane-2,4-dionato)-copper(ii)-N,N'- Ethylene-bis(salicylideneiminato)-copper(ii) 76.102
aqua-((RS)-N,N'- Ethylene-bis(serinato))-copper(ii) 82. 17
er(ii)]   N,N'- Ethylene-bis(1-iminomethyl-2-naphtholato)-copp 76. 99
minato)-copper(ii)]   N,N'- Ethylene-bis(1,1,1-trifluoropentane-2,4-dionei 77. 1
(bis(diethylene)-tri Ethylene-tetramine-bis(perchlorato)-copper.... 76. 44
copper(i)) bis(tetraphenylborate)]   (μ(2)-*Ethylenediamine)-bis(carbonyl-ethylenediamine- 76. 49
per(i) tetraphenylbor> (N,N,N',N'-tetramethyl Ethylenediamine)-bis(cyclohexylisocyanide)-cop 71.161
copper)-(ethylenediamine-tetra-acetato-> bis(Ethylenediamine)-copper (bis(ethylenediamine)- 76. 18
dibromo-(N-(2-hydroxyethyl)- Ethylenediamine)-copper(ii).................. 76. 6
dinitrato-(N,N,N',N'-tetraethyl Ethylenediamine)-copper(ii).................. 76. 65
bis(chloroacetato)-(N,N,N',N'-tetraethyl Ethylenediamine)-copper(ii).................. 76. 85
bis(N-ethyl Ethylenediamine)-copper(ii) diperchlorate 76. 48
bis(N,N-diethyl Ethylenediamine)-copper(ii) perchlorate........ 76. 77
bis(N,N-diethyl Ethylenediamine)-copper(ii) perchlorate........ 76. 78
drate]   (N,N,N',N',-tetra(2-hydroxypropyl)- Ethylenediamine)-copper(ii) perchlorate monohy 76. 86
asym-(μ-carbonato)-bis(chloro-(tetraethyl*Ethylenediamine)-copper(ii)) 76. 98
chloro-bis(dicarbonyl-chloro-(tetramethyl Ethylenediamine)-copper(ii)) tetraphenylborate 76. 87
a-acetato-> bis(ethylenediamine)-copper (bis(Ethylenediamine)-copper)-(ethylenediamine-tetr 76. 18
ra-acetato)-nickel tetrahydrate] bis(*Ethylenediamine)-copper-(μ-ethylenediamine-tet 84. 58
carbonyl- Ethylenediamine-copper(i) tetraphenylborate.... 76. 4
te)]   (μ(2)-ethylenediamine)-bis(carbonyl-*Ethylenediamine-copper(i)) bis(tetraphenylbora 76. 49
vate]   bis((μ(2)-phenoxo)-*Ethylenediamine-phenoxy-copper(ii)) phenol sol 76.103
rate]   bis(ethylenediamine)-copper-(μ-*Ethylenediamine-tetra-acetato)-nickel tetrahyd 84. 58
hydrate]   calcium Ethylenediamine-tetra-acetato-copper(ii) tetra 76. 55
<amine)-copper (bis(ethylenediamine)-copper)-(Ethylenediamine-tetra-acetato-nickel) dihydra> 76. 18
hydrate]   aqua-(N,N-dimethyl Ethylenediamine-N-oxide)-oxalato-copper(ii) di 81. 23
henylborate]   (di Ethylenetriamine)-(hex-1-ene)-copper(i) tetrap 72. 10
bis(di Ethylenetriamine)-dinitrato-copper 76. 50
bis(di Ethylenetriamine)-dinitrato-copper 76. 51
o-iron hexahydrate]   bis((di*Ethylenetriamine-copper)-μ(2)-cyano)-tetracyan 71. 78
per(ii)-bis(μ-chloro)-(chloro-bis-cis(dimethyl*Formamide))-copper(ii) <imethylformamide))-cop 84. 44
is-cis(dimethylform> (chloro-bis-cis(dimethyl*Formamide))-copper(ii)-bis(μ-chloro)-(chloro-b 84. 44
copper Formate tetradeuterohydrate.................... 81. 2n
copper Formate tetradeuterohydrate.................... 81. 3n
diaqua-bis(Formato)-bis(pyridine)-copper 81. 46
bis(Formato)-bis(pyridine)-copper monohydrate 81. 41
aqua-bis(amidonicotinato)-bis(Formato)-copper(ii).......................... 84. 55
bis(Formato)-pyridine-copper 81. 56
bis(2,2,5,5-tetramethylpyrrolidinyl)-3-N-(2- Formyl-5-methyl-salicylaldiminato)-copper 78. 16
i) tetrahydrate]   (glycyl*Glycinato)-copper(i 82. 22a
tetrakis(μ-N-acetyl*Glycinato)-bis(aquo-copper(ii)) 82. 32
Glycyl-L-methioninato-copper(ii) 82. 14
pper(ii) tetrahydrate]   (Glycylglycinato)-(7,9-dimethylhypoxanthine)-co 82. 22a

(diethylenetriamine)-(Hex-1-ene)-copper(i) tetraphenylborate 72. 10
bis(2,5-dithia Hexane)-copper tetrakis(tetrafluoroborate) 85. 27
bis(2,5-dithia Hexane)-copper tetraperchlorate................. 85. 26
di-copper Hexyl cofacial diporphyrin-7 .... 49. 36
dichloro- Histamino-copper(ii) 83. 18
L-asparaginato-L- Histidinato-copper(ii) 82. 18
L-asparaginato-L- Histidinato-copper(ii) tetrahydrate............ 82. 19
bis(5,5-diphenyl- Hydantoinato)-diammine-copper(ii) 83.199
(ii) chloride dihydrate] aquo-bis(2- Hydrazino-4-hydroxy-6-methylpyrimidine)-copper 83. 59
imino)-oxo-hydroxo-hexa-copper(ii) sulfate di Hydrogen phosphate pentahydrate <monoximato-3- 83.221
-0))-(μ-2-diethylaminoethanolato-N,μ-0)-(μ(3)- *Hydroxo)-tri-copper(ii) monohydrate <o-N,μ(3) 84.122
<e-α,α'-bis(butane-2-monoximato-3-imino))-oxo- Hydroxo-hexa-copper(ii) sulfate dihydrogen ph> 83.221
(glycylglycinato)-(7,9-dimethyl Hypoxanthine)-copper(ii) tetrahydrate 82. 22a
o-(4,5-bis((2-(2-pyridyl)-ethylimino)-methyl)- Imidazolato)-di-copper(ii) hydrate <perchlorat 83.136
-aza-1,3,8,10-cyclotetradecatetraen) 1-methyl Imidazole-(2,3,9,10-tetramethyl-1,4,8,11-tetra 83.132
chloro-bis(N-ethyl- Imidazolidine-2-thione)-copper(ii) 79. 8
N,N'-ethylene-bis(monothioacetylacetone- Iminato)-copper(ii) 76. 71
-ethylene-bis(1,1,1-trifluoropentane-2,4-dione Iminato)-copper(ii)..................... [N,N' 77. 1
tris(m-xylene-α,α'-bis(butane-2-monoximato-3- Imino))-oxo-hydroxo-hexa-copper(ii) sulfate d> 83.221
<ato-perchlorato-(4,5-bis((2-(2-pyridyl)-ethyl Imino)-methyl)-imidazolato)-di-copper(ii) hyd> 83.136
) diperc> bis(μ-2-(N,N-dimethyl-2-aminoethyl)* Imino-butan-3-one-oximato)-diaqua-di-copper(ii 84. 67
(2,2'-bipyridyl)-(Imino-diacetato)-copper(ii) hexahydrate........ 82. 27
3)-carbonato)-tris(aqua-2-(2-(2-pyridyl)-ethyl* Iminomethyl)-pyridine-copper(ii)) .... [(μ(84.126
N,N'-ethylene-bis(1- Iminomethyl-2-naphtholato)-copper(ii) 76. 99
hylenetriamine-copper)-μ(2)-cyano)-tetracyano-* Iron hexahydrate .... [bis((diet 71. 78
,α,α-tetra(o-nicotinamidophenyl)-porphyrinato- Iron(iii))-copper(ii) perchlorate dihydrate <α 83.225
ne)-aminato)-copper(ii)] bis(N- Isobutyl-(5-chloro-α-phenyl-2-hydroxybenzylide 78. 15
<allium(i) (μ(8)-chloro)-dodecakis(α-mercapto-* Isobutyrato)-octa-copper(i)-hexa-copper(ii) d> 81. 95
<'-bis(2-(2-pyridyl)ethyl)-2-hydroxy-5-methyl-* Isophthalaldiminato)-(μ-pyrazolato)-di-copper> 84.107
i)] catena-bis(* Isothiocyanato)-bis(μ-1,2,4-triazole)-copper(i 83. 22
aminopropyl)amino)ethanolato-copper(i) tetra Isothiocyanato)-copper(i) thiocyanate [2-((3- 84. 8
er(ii)-copper(i)-pseudo-porphyrin-macrocyclic Ligand complex .... [copp 84. 98
copper(ii)-copper(i)-pseudo-porphyrin- Macrocyclic ligand complex 84. 98
chl> (bis(benzothiazole-2-yl)-trisulfide)-(2- Mercapto-benzothiazole)-copper(i) perchlorate 85. 65
er(ii> thallium(i) (μ(8)-chloro)-dodecakis(α-* Mercapto-isobutyrato)-octa-copper(i)-hexa-copp 81. 95
bis(5-methylthio-8- Mercapto-quinolinato)-copper(ii).............. 85. 64
ophosphor> bis((μ-bis(diphenylthiophosphoryl)* Methane)-chloro-copper(i)) bis(bis(diphenylthi 85. 93
oro-copper(i)) bis(bis(diphenylthiophosphoryl)* Methane-chloro-copper(i)) <phoryl)methane)-chl 85. 93
<2-aminoethyl)-1-phenylpropane-2-oximato)-bis(* Methanol-copper(ii)) bis(μ-(N,N-dimethyl-2-am> 76.104
glycyl-L- Methioninato-copper(ii)...................... 82. 14
thiocarboxylato)-copper(ii)] N,N'-tri Methylene-bis(methyl-2-amino-1-cyclopentene-di 85. 60
-borato)-copper(ii> aquo-(difluoro-(3,3'-(tri Methylene-dinitrilo)-bis(butan-2-one-oximato) 83. 70
-borato)-copper(ii> iodo-(difluoro-(3,3'-(tri Methylene-dinitrilo)-bis(butan-2-one-oximato)) 83. 66
-copper(ii) perchlorate mono> aquo-(3,3'-(tri Methylene-dinitrilo)-bis(butan-2-one-oximato)) 83. 71
borato)-pyridine-copper(> (difluoro-3,3'-(tri Methylene-dinitrilo)-bis(butan-2-one-oximato)- 83.105
dichloro-(tetra Methylenesulfoxide)-copper(ii)] 84. 4
sulfido)-tris(triphenylphosphine-copper)-thio-* Molybdenum .... [(μ(3)-chloro)-tris(μ(3)- 86.202
(μ(3)-chloro)-(trisulfur-oxo-* Molybdenum)-tris(triphenylphosphine-copper).... 86.201
bis(triphenylphosphine)-copper-bis(μ(2)-thio)-* Molybdenum-bis(μ(2)-thio)-triphenylphosphine-> 86.203
i) sulfate > tris(m-xylene-α,α'-bis(butane-2- Monoximato-3-imino))-oxo-hydroxo-hexa-copper(i 83.221
r(ii) methanol solvate) (diacetyl Monoxime)-(salicyloylhydrazinato)-chloro-coppe 84. 31
e) bis((μ(2)-bromo)-bromo-copper)] *Naphthaceno(5,6-cd:11,12-c'd')-bis(1,2-dithiol 39. 66
N,N'-ethylene-bis(1-iminomethyl-2- Naphtholato)-copper(ii) 76. 99
bis(N-isopropyl-2-oxy-1- Naphthylidene-aminato)-copper(ii) 78. 12
tetracyanoquinodime> bis(N-isopropyl-2-oxy-1- Naphthylidene-aminato)-copper(ii) bis(7,7,8,8- 78. 13
bis(1,8- Naphthyridine)-dichloro-copper(ii) 83.102
ine)-copper-(μ-ethylenediamine-tetra-acetato)-* Nickel tetrahydrate........ [bis(ethylenediam 84. 58
amine)-copper)-(ethylenediamine-tetra-acetato- Nickel) dihydrate <ine)-copper (bis(ethylenedi 76. 18
pper(ii) perch> chloro-(meso-α,α,α,α-tetra(o- Nicotinamidophenyl)-porphyrinato-iron(iii))-co 83.225
aqua-bis(amido Nicotinato)-bis(formato)-copper(ii)........... 84. 55
copper(ii)] di Nitrato-(N,N,N',N'-tetraethylethylenediamine)- 76. 65
-aldehyde)-copper(ii)] di Nitrato-(S-methylthiosemicarbazone-8-quinoline 83. 75
bis(diethylenetriamine)-di Nitrato-copper 76. 51
bis(diethylenetriamine)-di Nitrato-copper 76. 50
(1,5-diazacyclo-octane)-di Nitrato-copper(ii) 84. 11
bis(2,2'-bipyridyl)- Nitrato-copper(ii) nitrate monohydrate 83.142
hylimino)-methyl)-imidazolato)-di-co> aqua- Nitrato-perchlorato-(4,5-bis((2-(2-pyridyl)-et 83.136
2-methyl-1,2-propanediamine)-nitro-copper(ii)- Nitrite................ [bis(N-isopropyl- 76. 90
,11,13,13-hexamethyl-1,5,10,14-tetra-azacyclo- Octadeca-1,10-diene)-copper(ii) perchlorate <4 83.153
,11,11,13-hexamethyl-1,5,10,14-tetra-azacyclo- Octadeca-1,13-diene)-copper(ii) perchlorate <4 83.152
(1,5-diazacyclo- Octane)-dinitrato-copper(ii) 84. 11
disodium bis(Oxalato)-copper(ii) dihydrate 81. 9
disodium catena-bis(μ-* Oxalato)-copper(ii) dihydrate 81. 8
aqua-(N,N-dimethylethylenediamine-N-oxide)- Oxalato-copper(ii) dihydrate 81. 23
μ-aqua)-bis(μ-N,N'-bis(2-hydroxyethyl)-dithio-* Oxamidato)-bis(sulfato-copper(ii)) <(ii))-bis(85. 45

3,3'-(trimethylene-dinitrilo)-bis(butan-2-one- Oximato))-borato)-copper(ii) perchlorate ⟨ro-(83. 70
3,3'-(trimethylene-dinitrilo)-bis(butan-2-one- Oximato))-borato)-copper(ii) [iodo-(difluoro-(83. 66
⟨,3'-(trimethylene-dinitrilo)-bis(butan-2-one- Oximato))-copper(ii) perchlorate monohydrate   83. 71
⟨N,N-dimethyl-2-aminoethyl)-1-phenylpropane-2-*Oximato-bis(methanol-copper(ii)) bis(μ-(N,N-⟩   76.104
⟨N,N-dimethyl-2-aminoethyl)-1-phenylpropane-2-*Oximato-bis(perchlorato-copper(ii)) diperchl⟩   76.104
⟨,3'-(trimethylene-dinitrilo)-bis(butan-2-one- Oximato)-borato)-pyridine-copper(ii) perchlor⟩   83.105
⟨,9-dimethyl-4,8-diaza-3,8-undecadiene-2,10-di Oximato-copper(ii) perchlorate methanol solv⟩   83. 69
rate]   bis(2,6-diacetylpyridine-di*Oximato-di-copper(ii) tetrafluoroborate dihyd   84. 72
-(N,N-dimethyl-2-aminoethyl)imino-butan-3-one-*Oximato-diaqua-di-copper(ii) diperchlorate ⟨2   84. 67
-diazabut⟩ 2-(chloro-bis(triphenylphosphine)- Palladium)-3-methyl-1,4-bis(methoxyphenyl)-1,4   86.196
bis(salicylidenei⟩ bis(1,1,1,5,5,5-hexafluoro Pentane-2,4-dionato)-copper(ii)-N,N'-ethylene-   76.102
        N,N'-ethylene-bis(1,1,1-trifluoro Pentane-2,4-dioneiminato)-copper(ii) 77. 1
monohydrate]   aqua-(acetylacetonato)-(o- Phenanthroline)-copper hexafluoroacetylacetone   77.· 6
etylacetonato)-(hexafluoroacetylacetonato)-(o- Phenanthroline-copper(ii)   [(ac   77. 8
        bis((μ-methoxy)-(2,4,6-trichloro*Phenolato)-quinoline-copper(ii)) 84.112
henol solvate]   bis((μ(2)-*Phenoxo)-ethylenediamine-phenoxy-copper(ii)) p   76.103
        bis((μ(2)-phenoxo)-ethylenediamine-*Phenoxy-copper(ii)) phenol solvate 76.103
e solvate]   tetrakis(μ-2,4-dichloro*Phenoxyacetato)-bis(dioxane-copper(ii)) dioxan   81. 92
7,7,8,8-tetracyanoquinodimethane⟩ N,N'-(1,2- Phenylene)-bis(salicylaldiminato)-copper(ii) -   60. 40
xo-hydroxo-hexa-copper(ii) sulfate dihydrogen Phosphate pentahydrate ⟨monoximato-3-imino))-o   83.221
lotetradecatetraene)-copper(ii) bis(hexafluoro Phosphate) ⟨yl-1,4,8,11-tetra-aza-1,3,8,10-cyc   83.132
phenyl-dithiocarboxylato-bis(triphenyl Phosphine)-copper(i) 86.160
        tetrakis((tri-p-tolyl Phosphine)-copper)-dioxo-hexathio-di-tungsten   86.230
s(μ(2)-thio)-triphenylphosphin⟩ bis(triphenyl Phosphine)-copper-bis(μ(2)-thio)-molybdenum-bi   86.203
phenyl)-1,4-diazabut⟩ 2-(chloro-bis(triphenyl Phosphine)-palladium)-3-methyl-1,4-bis(methoxy   86.196
μ(2)-thio)-molybdenum-bis(μ(2)-thio)-triphenyl*Phosphine-copper methylene chloride solvate ⟨(86.203
oro)-(trisulfur-oxo-molybdenum)-tris(triphenyl*Phosphine-copper) .... [(μ(3)-chl   86.201
(3)-sulfido)-bis(oxo-tungsten)-tetra(triphenyl*Phosphine-copper)   [hexakis(μ   86.219
μ(3)-chloro)-tris(μ(3)-sulfido)-tris(triphenyl*Phosphine-copper)-thio-molybdenum .... [(86.202
hydrate]   aquo-(pyridoxylidene-O- Phospho-D,L-threoninato)-copper(ii) dimer mono   82. 39
⟨is(2-(2-pyridyl)ethyl)-2-hydroxy-5-methyl-iso*Phthalaldiminato)-(μ-pyrazolato)-di-copper(i)   84.107
        N-(2-ammonioethyl)- Piperazinium pentachloro-copper(ii) dihydrate   33. 19
                Piperidinium tetrachloro-copper(ii).............   33. 15
        bis(4-benzyl- Piperidinium) hexachloro-di-copper(ii)   33. 47
)]   bis(4-(N-methyl- Piperidinium))-disulfide tetrachloro-copper(ii   33. 50
etylacetonato)-(4-hydroxy-2,2,6,6-tetramethyl- Piperidinyl-N-oxy)-copper(ii) ⟨is(hexafluoroac   77. 7
latinum(iv) tetrachl⟩ bis(1,2-diaminoethane)- Platinum(ii) dichloro-bis(1,2-diaminoethane)-p   76. 20
-platinum(ii) dichloro-bis(1,2-diaminoethane)- Platinum(iv) tetrachloro-copper(i) ⟨inoethane)   76. 20
copper(ii)-copper(i)-pseudo- Porphyrin-macrocyclic ligand complex 84. 98
di-copper hexyl cofacial di Porphyrin-7..........................   49. 36
⟨ro-(meso-α,α,α,α-tetra(o-nicotinamidophenyl)- Porphyrinato-iron(iii))-copper(ii) perchlorat⟩ 83.225
(3,5-dimethylpyrazolyl)-borato)-copper(i) a⟩ Potassium (p-nitrobenzenethiolato)-(hydro-tris   85. 68
s⟩ bis(μ-(N,N-dimethyl-2-aminoethyl)-1-phenyl*Propane-2-oximato-bis(methanol-copper(ii)) bi   76.104
⟨)) bis(μ-(N,N-dimethyl-2-aminoethyl)-1-phenyl*Propane-2-oximato-copper(ii))⟩   76.104
)]   bis(1,3- Propanediamine)-bis(m-nitrobenzoato)-copper(ii   81. 68
        bis(N-isopropyl-2-methyl-1,2- Propanediamine)-nitro-copper(ii)-nitrite 76. 90
        tetrakis((2-amino-2-methyl-1- Propanolato)-chloro-copper(ii)).............   84. 70
        bis(aqua-bis(2-amino-2-methyl Propanolato)-copper(ii)) dinitrate 84. 18
        copper Propionate 1. 9
        bis(Pyrazine)-bis(perchlorato)-copper(ii) 83. 34
        bis(1-carbamoyl-3,5-dimethyl- Pyrazolato)-copper(ii) 83. 78
l)-2-hydroxy-5-methyl-isophthalaldiminato)-(μ-*Pyrazolato)-di-copper(i) ⟨is(2-(2-pyridyl)ethy 84.107
oro)-dichloro-tetrakis(1H-3,5-diethyl-4-methyl*Pyrazole)-di-copper(ii)........ [bis(μ(2)-chl   83.208
⟨itrobenzenethiolato)-(hydro-tris(3,5-dimethyl Pyrazolyl)-borato)-copper(i) acetone solvate   85. 68
        bis(hydro-tris(1- Pyrazolyl)-borato)-copper(ii)   83.112
        bis(2-dimethylamino- Pyridine 1-oxide)-copper(ii) diperchlorate 84. 57
        tetrakis(Pyridine)-bis(trifluoroacetato)-copper(ii) 81. 77
        diaqua-bis(formato)-bis(Pyridine)-copper   81. 46
        bis(formato)-bis(Pyridine)-copper monohydrate 81. 41
        dibromo-bis(3,5-dimethyl Pyridine)-copper(ii) 83. 95
diaqua-bis(N-acetyl-D,L-tryptophanato)-bis(Pyridine)-copper(ii) 82. 41
        bis(formato)- Pyridine-copper...........................   81. 56
e-dinitrilo)-bis(butan-2-one-oximato)-borato)- Pyridine-copper(ii) perchlorate ⟨-(trimethylen   83.105
        bis(μ(2)-bromo)-bis(bromo*Pyridine-copper(ii)) .... [(μ(3)-carbonato)   83. 46
-tris(aqua-2-(2-(2-pyridyl)-ethyliminomethyl)-*Pyridine-copper(ii)) .... [(μ(3)-carbonato)   84.126
orate dihydrate]   bis(μ-2,6-diacetyl*Pyridine-dioximato)-di-copper(ii) tetrafluorob   84. 72
        hexakis(Pyridine-N-oxide)-copper(ii) dinitrate 84.111
ahydrate]   bis(2,2'-bipyridyl)-bis(Pyridine-2,6-dicarboxylato)-di-copper(ii) tetr   81. 90
        diaquo-bis(Pyridine-3-sulfonato)-copper(ii) 84. 23
        tetra-aquo-bis(Pyridine-3-sulfonato)-copper(ii)-zinc(ii) 83. 57
        Pyridinium tris(thiocyanato)-di-copper(i) 33. 7
r(ii) monohydrate]   chloro-(N- Pyridoxylidene-N'-salicyloylhydrazinato)-coppe   78. 2
er(ii) dimer monohydrate]   aquo-(Pyridoxylidene-O-phospho-D,L-threoninato)-copp   82. 39
o⟩ aqua-dinitrato-perchlorato-(4,5-bis((2-(2- Pyridyl)-ethylimino)-methyl)-imidazolato)-di-c   83.136
)]   (μ(3)-carbonato)-tris(aqua-2-(2-(2-*Pyridyl)-ethyliminomethyl)-pyridine-copper(ii)   84.126

iminato)-(μ-pyrazolato)-di> (μ-N,N'-bis(2-(2-*Pyridyl)ethyl)-2-hydroxy-5-methyl-isophthalald   84.107
hydrate]   bis(2,2'-di Pyridylamino)-chloro-copper(ii) chloride tetra   83.144
          bis(3,3'-dimethyl-2,2'-di Pyridylamino)-copper(ii) 83.172a
aquo-bis(2-hydrazino-4-hydroxy-6-methyl Pyrimidine)-copper(ii) chloride dihydrate 83. 59
                chloro-(bis(2- Pyrimidyl)-disulfide)-copper(i) monohydrate....   83. 32
diminato)-copper]   bis(2,2,5,5-tetramethyl Pyrrolidinyl)-3-N-(2-formyl-5-methyl-salicylal   78. 16
)]   chloro-N,N-diethylamide- Pyruvate-4-phenylthiosemicarbazonato-copper(ii   79. 12
cylaldiminato)-copper(ii) - 7,7,8,8-tetracyano Quinodimethane complex <,2-phenylene)-bis(sali   60. 40
ene-aminato)-copper(ii) bis(7,7,8,8-tetracyano Quinodimethane) <-isopropyl-2-oxy-1-naphthylid   78. 13
         bis(5-methylthio-8-mercapto- Quinolinato)-copper(ii)..................   85. 64
         dinitrato-(S-methylthiosemicarbazone-8- Quinoline-aldehyde)-copper(ii) 83. 75
  bis((μ-methoxy)-(2,4,6-trichlorophenolato)-*Quinoline-copper(ii)) 84.112
        di(bis(μ-acetyl)-tetracarbonyl-*Rhenium)-copper 71. 93
-carbido-ennea-(μ-carbonyl)-hexacarbonyl-hexa-*Rhodium methanol solvate <acetonitrile-copper)   71.147
tramethylpyrrolidinyl)-3-N-(2-formyl-5-methyl- Salicylaldiminato)-copper   ....[bis(2,2,5,5-te   78. 16
             bis(N-cyclohexyl- Salicylaldiminato)-copper(ii) 78. 18
yanoquinodimethane> N,N'-(1,2-phenylene)-bis(Salicylaldiminato)-copper(ii) - 7,7,8,8-tetrac   60. 40
micarbazonato-copper(ii) nitrate]   Salicylidene-anthranyl-aldehyde-4-phenylthiose   78. 7
                       (N- Salicylidene-L-valinato)-aquo-copper(ii) 78. 1
           bis(N-isopropyl-3-methoxy- Salicylideneiminato)-copper(ii)..................   78. 9
ane-2,4-dionato)-copper(ii)-N,N'-ethylene-bis(Salicylideneiminato)-copper(ii) <exafluoropent   76.102
nol solvate]   (diacetylmonoxime)-(Salicyloylhydrazinato)-chloro-copper(ii) metha   84. 31
          chloro-(N-pyridoxylidene-N'- Salicyloylhydrazinato)-copper(ii) monohydrate   78. 2
chloro-N,N-diethylamide-pyruvate-4-phenylthio Semicarbazonato-copper(ii) 79. 12
   salicylidene-anthranyl-aldehyde-4-phenylthio Semicarbazonato-copper(ii) nitrate   78. 7
           dinitrato-(S-methylthio Semicarbazone-8-quinoline-aldehyde)-copper(ii)   83. 75
         aqua-((RS)-N,N'-ethylene-bis(Serinato))-copper(ii) 82. 17
                       di Sodium bis(oxalato)-copper(ii) dihydrate   81. 9
ate]   di*Sodium catena-bis(μ-oxalato)-copper(ii) dihydr   81. 8
        bis(2,2'-bipyridyl)-peroxo-di Sulfato-copper(ii) monohydrate 84. 82
,N'-bis(2-hydroxyethyl)-dithio-oxamidato-bis(*Sulfato-copper(ii)) <(ii)-bis(μ-aqua)-bis(μ-N   85. 45
       bis(4-(N-methyl-piperidinium))-di Sulfide tetrachloro-copper(ii) 33. 50
     (bis(benzothiazole-2-yl)-tri Sulfide)-(2-mercapto-benzothiazole)-copper(i)   85. 65
perchlorate chl>   chloro-(bis(2-pyrimidyl)- Sulfide)-copper(i) monohydrate 83. 32
phine-copper)]   hexakis(μ(3)-*Sulfido-bis(oxo-tungsten)-tetra(triphenylphos   86.219
molybdenum]   (μ(3)-chloro)-tris(μ(3)-*Sulfido)-tris(triphenylphosphine-copper)-thio-   86.202
          diaquo-bis(pyridine-3- Sulfonato)-copper(ii) 84. 23
     tetra-aquo-bis(pyridine-3- Sulfonato)-copper(ii)-zinc(ii) 83. 57
              dichloro-(dimethyl Sulfoxide)-copper(ii) 84. 3
       dichloro-(tetramethylene Sulfoxide)-copper(ii) 84. 4
-copper)]   (μ(3)-chloro)-(tri*Sulfur-oxo-molybdenum)-tris(triphenylphosphine   86.201
             copper L- Tartrate trihydrate....................   1. 11
oro-dodeca(1,1-dimethyl-2-aminoethylthiolato)- Tetradeca-copper sulfate eicosahydrate   [chl   85. 91
       (bis(diethylene)-triethylene- Tetramine)-bis(perchlorato)-copper   76. 44
            (tris(triethylene)- Tetramine)-perchlorato-copper perchlorate 84. 22
-isobutyrato)-octa-copper(i)-hexa-copper(ii> *Thallium(i) (μ(8)-chloro)-dodecakis(α-mercapto   81. 95
         diaqua-dichloro- Theophylline-copper(ii)....................   83. 31
octamethoxy-tetrabenzo(b,e,h,k)(1,4,7,10)tetra Thia-cyclododecatetraene)-copper(i) <13,17,18-   85. 84
<ctamethoxy-tetrabenzo(b,e,h,k)(1,4,7,10)tetra Thia-cyclododecatetraene)-copper(i) chlorofor>   85. 85
              bis(2,5-di Thiahexane)-copper tetrakis(tetrafluoroborate)   85. 27
              bis(2,5-di Thiahexane)-copper tetraperchlorate....   85. 26
<othiazole-2-yl)-trisulfide)-(2-mercapto-benzo Thiazole)-copper(i) perchlorate chloroform so>   85. 65
          trans-dichloro-bis(2,4-dimethyl Thiazole)-copper(ii) 83. 49
iazole)-copper(i) perchlorate chl> (bis(benzo Thiazole-2-yl)-trisulfide)-(2-mercapto-benzoth   85. 65
hylcyano)-(μ(2)-chloro)-(μ(2)-1,3-diaza-2,4-di*Thietidine)-chloro-copper)   .... [bis((met   83. 4
hin> bis(triphenylphosphine)-copper-bis(μ(2)-*Thio)-molybdenum-bis(μ(2)-thio)-triphenylphosp   86.203
<e)-copper-bis(μ(2)-thio)-molybdenum-bis(μ(2)-*Thio)-triphenylphosphine-copper methylene chl>   86.203
akis((tri-p-tolylphosphine)-copper)-dioxo-hexa Thio-di-tungsten   ....   [tetr   86.230
μ(3)-sulfido-tris(triphenylphosphine-copper)-*Thio-molybdenum........ [(μ(3)-chloro)-tris(86.202
-bis(μ-aqua)-bis(μ-N,N'-bis(2-hydroxyethyl)-di*Thio-oxamidato)-bis(sulfato-copper(ii)) <(ii))   85. 45
            bis(5-methyl Thio-8-mercapto-quinolinato)-copper(ii)........   85. 64
        N,N'-ethylene-bis(mono Thioacetylacetone-iminato)-copper(ii)   76. 71
       bis(N,N-di-isopropyldi Thiocarbamato)-copper(ii) 80. 12
methylene-bis(methyl-2-amino-1-cyclopentene-di Thiocarboxylato)-copper(ii)........   [N,N'-tri   85. 60
(i)]   phenyl Thiocarboxylato-bis(triphenylphosphine)-copper   86.160
te]   chloro-dodeca(1,1-dimethyl-2-aminoethyl Thiolato)-tetradeca-copper sulfate eicosahydra   85. 91
    naphthaceno(5,6-cd:11,12-c'd')-bis(1,2-di*Thiole) bis((μ(2)-bromo)-bromo-copper) 39. 66
     aqua-bis(2,2'-bipyridyl)-copper(ii) di Thionate 83.145
       chloro-bis(N-ethyl-imidazolidine-2- Thione)-copper(ii) 79. 8
bis(diphenylthiophosphor> bis((μ-bis(diphenyl*Thiophosphoryl)methane)-chloro-copper(i)) bis(85. 93
yl)methane)-chloro-copper>   bis((μ-bis(diphenyl*Thiophosphoryl)methane-chloro-copper(i)) <phor   85. 93
  chloro-N,N-diethylamide-pyruvate-4-phenyl Thiosemicarbazonato-copper(ii) 79. 12
   salicylidene-anthranyl-aldehyde-4-phenyl Thiosemicarbazonato-copper(ii) nitrate 78. 7
(ii)]   dinitrato-(S-methyl Thiosemicarbazone-8-quinoline-aldehyde)-copper   83. 75
       bis(1,1-diethyl-3-benzoyl Thiourea)-copper(ii) 79. 15

aquo'-(pyridoxylidene-O-phospho-D,L- Threoninato)-copper(ii) dimer monohydrate 82. 39
sten] tetrakis((tri-p- Tolylphosphine)-copper)-dioxo-hexathio-di-tung 86.230
catena-bis(isothiocyanato)-bis(μ-1,2,4-*Triazole)-copper(ii) 83. 22
diaqua-bis(N-acetyl-D,L- Tryptophanato)-bis(pyridine)-copper(ii)........ 82. 41
i-p-tolylphosphine)-copper)-dioxo-hexathio-di- Tungsten .... [tetrakis((tr 86.230
hexakis(μ(3)-sulfido)-bis(oxo-*Tungsten)-tetra(triphenylphosphine-copper) 86.219
(N-salicylidene-L- Valinato)-aquo-copper(ii) 78. 1
(glycylglycinato)-(7,9-dimethylhypo Xanthine)-copper(ii) tetrahydrate 82. 22a
oxo-hydroxo-hexa-copper(ii) sulfate > tris(m- Xylene-α,α'-bis(butane-2-monoximato-3-imino))- 83.221
tra-aquo-bis(pyridine-3-sulfonato)-copper(ii)- Zinc(ii) ....[te 83. 57

Dysprosium

tris(di Cyclohexyl-dithiophosphinato)-dysprosium(iii) 85. 87
tris(dicyclohexyl-di Thiophosphinato)-dysprosium(iii) 85. 87

Europium

bis((1,2-dimethoxy Ethane)-bis(trimethylsilyl)amino-europium(ii) 84. 85
-dionato)-europium(iii)] bis(dimethyl Formamide)-tris(2,2,6,6-tetramethylheptane-3,5 77. 10
is(dimethylformamide)-tris(2,2,6,6-tetramethyl Heptane-3,5-dionato)-europium(iii) .... [b 77. 10
bis((1,2-dimethoxyethane)-bis(trimethyl Silyl)amino-europium(ii) 84. 85

Gadolinium

tetrahydrate] bis(μ-*Acetato)-bis(diaquo-diacetato-gadolinium(iii)) 81. 50
bis(μ-acetato)-bis(diaquo-di*Acetato-gadolinium(iii)) tetrahydrate ... 81. 50
tris(Cyclopentadienyl)-tetrahydrofuran-gadolinium 73. 79
.tris(cyclopentadienyl)-tetrahydro Furan-gadolinium 73. 79

Gallium

thyl-gallium] (N,N',N''-trimethyl Acetimidohydrazino-0,N)-dimethyl-gallium-trime 68. 12
2-(N-phenylaminomethylene)-3(2H)- Benzo(b)furan-thionato-dimethyl-gallium........ 68. 19
methoxymethyl)-dicarb> dimethyl-(N,N-dimethyl Ethanolamino)-(1-pyrazolyl)-gallato-(η(2)-thio 72. 22
-dinitrosyl-iron] (dimethyl-(N,N-dimethyl Ethanolamino)-(3,5-dimethylpyrazolyl)-gallato) 84. 32
-nitrosyl-nickel] (dimethyl-(N,N-dimethyl Ethanolamino)-(3,5-dimethylpyrazolyl)-gallato) 84. 33
(η(2)-thiomethoxymeth> dimethyl-(N,N-dimethyl Ethanolamino)-(3,5-dimethylpyrazolyl)-gallato- 68. 16
(dimethyl-(3,5-dimethyl-1-pyrazolyl)-(dimethyl Ethanolamino)-gallato)-nickel(ii) <)-gallato)- 83.166
2-(N-phenylaminomethylene)-3(2H)-benzo(b) Furan-thionato-dimethyl-gallium................ 68. 19
ium] (N,N'-dimethylacetyl Hydrazino-0,N)-dimethyl-gallium-trimethyl-gall 68. 7
)-(3,5-dimethylpyrazolyl)-gallato)-dinitrosyl- Iron [(dimethyl-(N,N-dimethylethanolamino 84. 32
yl-gallium] 2-(N-phenylamino Methylene)-3(2H)-benzo(b)furan-thionato-dimeth 68. 19
)-gallato-(η(2)-thiomethoxymethyl)-dicarbonyl- Molybdenum <dimethylethanolamino)-(1-pyrazolyl 72. 22
)-gallato-(η(2)-thiomethoxymethyl)-dicarbonyl- Molybdenum <hanolamino)-(3,5-dimethylpyrazolyl 68. 16
no)-(3,5-dimethylpyrazolyl)-gallato)-nitrosyl- Nickel .... [(dimethyl-(N,N-dimethylethanolami 84. 33
-1-pyrazolyl)-(dimethylethanolamino)-gallato- Nickel(ii) <)-gallato)-(dimethyl-(3,5-dimethyl 83.166
anolamino)-(3,5-dimethylpyrazolyl)-gallato)-di Nitrosyl-iron ....[(dimethyl-(N,N-dimethyleth 84. 32
thanolamino)-(3,5-dimethylpyrazolyl)-gallato)- Nitrosyl-nickel.... [(dimethyl-(N,N-dimethyle 84. 33
(N,N',N'',N'''-tetramethyl Oxalamidinato)-bis(dimethyl-gallium) 68. 10
bis(dimethyl-gallium)-N,N'-dimethyl-dithio- Oxamide.. 68. 6
<yrazolyl)-gallato)-(dimethyl-(3,5-dimethyl-1- Pyrazolyl)-(dimethylethanolamino)-gallato)-ni> 83.166
pyrazolyl)-(di> (dimethyl-bis(3,5-dimethyl-1- Pyrazolyl)-gallato)-(dimethyl-(3,5-dimethyl-1- 83.166
ethyl-(N,N-dimethylethanolamino)-(3,5-dimethyl Pyrazolyl)-gallato)-dinitrosyl-iron.... [(dim 84. 32
ethyl-(N,N-dimethylethanolamino)-(3,5-dimethyl Pyrazolyl)-gallato)-nitrosyl-nickel.... [(dim 84. 33
<thyl-(N,N-dimethylethanolamino)-(3,5-dimethyl Pyrazolyl)-gallato-(η(2)-thiomethoxymethyl)-d> 68. 16
carb> dimethyl-(N,N-dimethylethanolamino)-(1- Pyrazolyl)-gallato-(η(2)-thiomethoxymethyl)-di 72. 22
N-methyl- Salicylaldiminato-dimethyl-gallium 68. 8
bis(dimethyl-gallium)-N,N'-dimethyl-di Thio-oxamide 68. 6
thylethanolamino)-(1-pyrazolyl)-gallato-(η(2)- Thiomethoxymethyl)-dicarbonyl-molybdenum <dime 72. 22
lamino)-(3,5-dimethylpyrazolyl)-gallato-(η(2)- Thiomethoxymethyl)-dicarbonyl-molybdenum <hano 68. 16
-(N-phenylaminomethylene)-3(2H)-benzo(b)furan- Thionato-dimethyl-gallium .... [2 68. 19

Germanium

Acetylacetonato-iodo-germanium 69. 6
10,10-dimethyl-10-germa-9-thio-9,10-dihydro Anthracene 69. 28
ron] μ-(1-(1,2,3,4-tetraphenyl*Butadienyl)-phenylgermylene)-octacarbonyl-di-i 69. 58
enylgermyl)-(triphenylgermyl-cadmium)-nickel) Cadmium toluene solvate <clopentadienyl-(triph 73.134
entadienyl-(triphenylgermyl)-(triphenylgermyl- Cadmium)-nickel)-cadmium toluene solvate <clop 73.134
tricarbonyl-(triphenylgermyl)-(ethylethoxy Carbene)-cobalt 71.205
rbonyl-(triphenylgermyl)-(ethylethoxycarbene)- Cobalt .... [trica 71.205
dodecaphenyl- Cyclohexagermane benzene solvate 69. 61
5-germyl- Cyclopenta-1,3-diene 69. 7
germyl-mercury)-nickel)-mercury t> bis((η(5)- Cyclopentadienyl)-(triphenylgermyl)-(triphenyl 73.135

ermyl-cadmium)-nickel)-cadmium tol> bis(η(5)- Cyclopentadienyl-(triphenylgermyl)-(triphenylg   73.134
            octaphenyl-thia-tetragermana- Cyclopentane 69. 59
                             (1,1'- Ferrocenediyl)-diphenyl-germanium 73. 91
                      hexaphenyl-di- Germanium 69. 52
            tricarbonyl-(triphenyl Germyl)-(ethylethoxycarbene)-cobalt............   71.205
ium tol> bis(η(5)-cyclopentadienyl-(triphenyl Germyl)-(triphenylgermyl-cadmium)-nickel)-cadm   73.134
ury t> bis((η(5)-cyclopentadienyl)-(triphenyl Germyl)-(triphenylgermyl-mercury)-nickel)-merc   73.135
                    1,8-bis(trimethyl Germyl)-naphthalene.................................   69. 33
<cyclopentadienyl-(triphenylgermyl)-(triphenyl Germyl-cadmium)-nickel)-cadmium toluene solva>   73.134
                              5- Germyl-cyclopenta-1,3-diene....................   69. 7
<yclopentadienyl)-(triphenylgermyl)-(triphenyl Germyl)-(triphenylgermyl-mercury toluene solva>   73.135
                 triphenyl-(triphenyl Germyl-peroxy)-silane...........................   69. 51
    μ-(1-(1,2,3,4-tetraphenylbutadienyl)-phenyl *Germylene)-octacarbonyl-di-iron.............   69. 58
lbutadienyl)-phenylgermylene)-octacarbonyl-di-*Iron .... [μ-(1-(1,2,3,4-tetrapheny   69. 58
enylgermyl)-(triphenylgermyl-mercury)-nickel)- Mercury toluene solvate <lopentadienyl)-(triph   73.135
ntadienyl)-(triphenylgermyl)-(triphenylgermyl)- Mercury)-nickel)-mercury toluene solvate <lope   73.135
                    1,8-bis(trimethylgermyl)- Naphthalene.................................   69. 33
l-(triphenylgermyl)-(triphenylgermyl-cadmium)- Nickel)-cadmium toluene solvate <clopentadieny   73.134
)-(triphenylgermyl)-(triphenylgermyl-mercury)- Nickel)-mercury toluene solvate <lopentadienyl   73.135
               bis(2,6-di-t-butyl-4-methyl Phenoxy)-germanium 69. 49
               bis(2,2,6,6-tetramethyl Piperidinato)-germanium.........................   69. 40
       triphenyl-(triphenylgermyl-peroxy)- Silane................................   69. 51
            bis(triphenyl-germanium)- Sulfide.................................   69. 53
            bis(triphenyl-germanium)- Sulfide.................................   69. 54
                       octaphenyl- Thia-tetragermana-cyclopentane 69. 59
        10,10-dimethyl-10-germa-9- Thio-9,10-dihydroanthracene...................   69. 28

Gold

      hydrogen bis(N,N-dimethyl Acetamide) tetrachloro-gold(iii) 1. 14
      hydrogen bis(N,N-dimethyl Acetamide) tetrachloro-gold(iii) 1. 15n
              tetrakis(dithio Acetato)-tetra-gold(i) 85. 17
ron]   (η(3)- Allyl)-tricarbonyl-(triphenylphosphine-gold)-i   72. 46
      chloro-(bis(2-aminoethyl) Amido)-gold(iii) perchlorate 76. 5
      chloro-bis(2-aminomethyl) Amine-gold(iii) chloride perchlorate 76. 7
                     tetraphenyl Arsonium difulminato-gold 71. 1
                            Auranofin 86. 71
                     2,2'- Bipyridyl potassium dicyano-gold(iii) 33. 34
hosphine)-gold(i) dodecahydro-6-thia-nido-deca Borane .... [tris(triphenylp   86.200
              tetraphenylarsonium di Fulminato-gold 71. 1
         (2,3,4,6-tetra-O-acetyl-1-thio-β-D- Glucopyranosato-S)-(triethylphosphine)-gold....   86. 71
ylphosphine-gold)]   bis(μ(2)-*Hydrido)-undecacarbonyl-tri-osmium-bis(triphen   86.172
o-gold(iii)]   Hydrogen bis(N,N-dimethylacetamide) tetrachlor   1. 15n
o-gold(iii)]   Hydrogen bis(N,N-dimethylacetamide) tetrachlor   1. 14
-allyl)-tricarbonyl-(triphenylphosphine-gold)- Iron .... [(η(3)   72. 46
ta-gold dinitrat> tris(bis(diphenylphosphino) Methane)-(bis(diphenylphosphino)methanido)-pen   71.281
enylphosphino)methane)-(bis(diphenylphosphino) Methanido)-penta-gold dinitrate [tris(bis(diph   71.281
         bis(μ(2)-hydrido)-undecacarbonyl-tri-*Osmium-bis(triphenylphosphine-gold)........   86.172
loride solvate]   (μ(2)-chloro)-bis(triphenyl *Phosphine)-di-gold(i) perchlorate methylene ch   86.118
acetyl-1-thio-β-D-glucopyranosato-S)-(triethyl Phosphine)-gold............ [(2,3,4,6-tetra-O-   86. 71
           (2,6-dimethoxyphenyl)-(triphenyl Phosphine)-gold(i) 71.203
           tris(triphenyl Phosphine)-gold(i) dodecahydro-6-thia-nido-dec   86.200
aborane]   tris(triphenyl Phosphine)-gold(i) dodecahydro-6-thia-nido-dec   86.200
drido)-undecacarbonyl-tri-osmium-bis(triphenyl *Phosphine-gold)................. [bis(μ(2)-hy   86.172
           (η(3)-allyl)-tricarbonyl-(triphenyl Phosphine-gold)-iron 72. 46
anido)-penta-gold dinitrat> tris(bis(diphenyl Phosphino)methane)-(bis(diphenylphosphino)meth   71.281
s(bis(diphenylphosphino)methane)-(bis(diphenyl Phosphino)methanido)-penta-gold dinitrate [tri   71.281
                  2,2'-bipyridyl Potassium dicyano-gold(iii)....................   33. 34
         (2,3,4,6-tetra-O-acetyl-1-thio-β-D-gluco Pyranosato-S)-(triethylphosphine)-gold 86. 71
ris(triphenylphosphine)-gold(i) dodecahydro-6- Thia-nido-decaborane .... [t   86.200
)-gold]   (2,3,4,6-tetra-O-acetyl-1- Thio-β-D-glucopyranosato-S)-(triethylphosphine   86. 71
              tetrakis(di Thioacetato)-tetra-gold(i) 85. 17
              bis(Tricyclohexylphosphine)-gold(i) thiocyanate....   86.127

Hafnium

m]   bis(η(5)- Cyclopentadienyl)-bis(cyclopentadienyl)-hafniu   71.155
         dicarbonyl-bis(Cyclopentadienyl)-hafnium 73. 29
    bis(η(5)-cyclopentadienyl)-bis(Cyclopentadienyl)-hafnium 71.155

·Indium

                                          tris(Acetylacetonato)-indium(iii) 68. 15
zene solvate]   di-μ-(pyridine-2-*Carbaldehyde-oximato)-bis(dimethyl-indium) ben 68. 17
                  (N,N',N'',N'''-tetramethyl- Oxalamidinato)-bis(dimethyl-indium)............ 68. 11
                di-μ-(pyridine-2-carbaldehyde-*Oximato)-bis(dimethyl-indium) benzene solvate   68. 17
                 chloro-(5,10,15,20-tetraphenyl Porphinato)-indium(iii)........................ 49. 16
indium) benzene solvate]   di-μ-(*Pyridine-2-carbaldehyde-oximato)-bis(dimethyl- 68. 17

Iridium

m (dichloro-(dihydrogen-ethylenediamine-tetra- Acetato)-iridium) .... [potassiu 76. 59
m (+)-(chloro-(hydrogen-ethylenediamine-tetra- Acetato)-iridium) monohydrate .... [potassiu 76. 58
-iridium(iii) bis(hexafluorophosphate)] tris(Acetonitrile)-nitrosyl-bis(triphenylphosphine)- 86.152
iridium]   (η(3)-1-phenyl Allyl)-chloro-hydrido-bis(triphenylphosphine)- 72. 63
e)-(1,2-η-penta-1,3-diene)-(2-4-η-1,3-dimethyl Allyl)-iridium(i) <6-η-1,5-cyclo-octa-1,5-dien 72. 34
nyl-chloro-(o-(diphenylphosphino)-N,N-dimethyl Aniline)-iridium .... [carbo 86. 75
                        (o-phenylene-bis(dimethyl Arsine))-decacarbonyl-tetra-iridium............ 86. 64
solv> bis(triphenylphosphine)-(tetrachlorodi Azo-cyclopentadiene)-chloro-iridium(i) toluene 86.148
tri-p-tolylphosphine)-(nido-7,8-dicarba-undeca Borane)-iridium toluene solvate <ihydrido-bis(86.167
           tris(μ-t-butylthiolato)-(μ-hexafluoro-*But-2-ene-2,3-diyl)-tris(dicarbonyl-iridium)   71.181
<5)-cyclopentadienyl-iodo-(methyl(methylthio)- Carbene)-(triphenylphosphine)-iridium(iii) io> 71.206
methylene chloride solvate]   tetrabromo Catecholato-nitroso-triphenylphosphine-iridium 86. 85
(2-4-η-1,3-dimethylallyl)-ir> (1,2,5,6-η-1,5- Cyclo-octa-1,5-diene)-(1,2-η-penta-1,3-diene)- 72. 34
adienediyl)-dodecacarbonyl-hepta-iridium] (Cyclo-octadiene)-(cyclo-octadienyl)-(cyclo-oct 71.246
ro)-bis(mercury-(μ-p-tolyl(ethyl)triazenido)-(*Cyclo-octadiene-iridium)) ....[bis(μ-chlo 75. 40
ium]   (cyclo-octadiene)-(cyclo-octadienyl)-(Cyclo-octadienediyl)-dodecacarbonyl-hepta-irid 71.246
carbonyl-hepta-iridium]   (cyclo-octadiene)-(Cyclo-octadienyl)-(cyclo-octadienediyl)-dodeca 71.246
ium(i)]   (η(4)-1,3- Cycloheptadienyl)-(η(5)-cycloheptadienyl)-irid 75. 8
           (η(4)-1,3-cycloheptadienyl)-(η(5)- Cycloheptadienyl)-iridium(i) 75. 8
v> bis(triphenylphosphine)-(tetrachlorodiazo- Cyclopentadiene)-chloro-iridium(i) toluene sol 86.148
ene)-(triphenylphosphine)-iridium(iii)> η(5)- Cyclopentadienyl-iodo-(methyl(methylthio)-carb 71.206
(1,1,1-tris(diphenylphosphinomethyl)ethane)- Cyclotriphosphorus-iridium 86.150
          (1,1,1-tris(diphenylphosphinomethyl) Ethane)-cyclotriphosphorus-iridium 86.150
disulfido-bis(1,2-bis(diphenylphosphido) Ethane)-iridium chloride N,N'-diphenylurea 86.191
     chloro-bis(bis(dimethylphosphino) Ethane)-methoxycarbonyl-iridium fluorosulfate   71. 79
          potassium (dichloro-(dihydrogen- Ethylenediamine-tetra-acetato)-iridium)........ 76. 59
drate]   potassium (+)-(chloro-(hydrogen- Ethylenediamine-tetra-acetato)-iridium) monohy 76. 58
μ(2)-hydrido)-tris(triethylphosphine)-iridium-*Hydride tetraphenylborate <phine-platinum-bis(86.104
)-di-iridium tetrafluor> dihydrido-tris(μ(2)-*Hydrido)-bis(1,3-bis(diphenylphosphino)propane 86.205
nylphosphino)-propane)-tri> trihydrido-(μ(3)-*Hydrido)-tris(μ(2)-hydrido)-tris(1,3-bis(diphe 86.224
phenyl-triethylphosphine-platinum-bis(μ(2)-*Hydrido)-tris(triethylphosphine)-iridium-hydr> 86.104
ne)-tri> trihydrido-(μ(3)-hydrido)-tris(μ(2)-*Hydrido)-tris(1,3-bis(diphenylphosphino)-propa 86.224
(1,3-bis(diphenylphosphino)-propane)-tri> tri-*Hydrido-(μ(3)-hydrido)-tris(μ(2)-hydrido)-tris 86.224
carba-undecaborane)-iridium toluene s> 3,9-di Hydrido-bis(tri-p-tolylphosphine)-(nido-7,8-di 86.167
         (η(3)-1-phenylallyl)-chloro- Hydrido-bis(triphenylphosphine)-iridium.......... 72. 63
bromo-(2-(diphenylphosphino)phenyl)- Hydrido-bis(triphenylphosphine)-iridium(iii) 71.276
lphenylphosphine)-(o-dimethylphosphinophenyl)- Hydrido-iridium hexafluorophosphate <s(dimethy 71.237
bis(μ-t-butylthiolato)-bis(carbonyl- Hydrido-trimethylphosphite-iridium)............ 86. 42
lphosphino)propane)-di-iridium tetrafluor> di-*Hydrido-tris(μ(2)-hydrido)-bis(1,3-bis(dipheny 86.205
m) monohydrate]   potassium (+)-(chloro-(Hydrogen-ethylenediamine-tetra-acetato)-iridiu 76. 58
m)]   potassium (dichloro-(di Hydrogen-ethylenediamine-tetra-acetato)-iridiu 76. 59
nium-tetra-iridi> bis(bis(triphenylphosphino)*Iminium) heptacarbonyl-μ(2)-hexacarbonyl-ruthe 64.162
nium-tetra-iridi> bis(bis(triphenylphosphine)*Iminium) nonacarbonyl-hexa-μ(2)-carbonyl-ruthe 64.163
                              t-butyl Isocyanide-(undecacarbonyl)-tetra-iridium 71. 92
tadiene-iridium))]   bis(μ-chloro)-bis(*Mercury-(μ-p-tolyl(ethyl)triazenido)-(cyclo-oc 75. 40
osphine)-pyridine-chloro-iridium ace> (2-aci- Nitrato-propan-1-on-1-yl)-bis(dimethylphenylph 71.192
hloride solvate]   tetrabromocatecholato- Nitroso-triphenylphosphine-iridium methylene c 86. 85
osphine)-iridium(i) bis(hexafluorophosphate> Nitrosyl-(1,10-phenanthroline)-bis(triphenylph 86.174
bis(hexafluorophosphate)] tris(acetonitrile)- Nitrosyl-bis(triphenylphosphine)-iridium(iii) 86.152
diyl)-dodecacarbonyl-hepta-iridium]   (cyclo- Octadiene)-(cyclo-octadienyl)-(cyclo-octadiene 71.246
s(mercury-(μ-p-tolyl(ethyl)triazenido)-(cyclo-*Octadiene-iridium)) .... [bis(μ-chloro)-bi 75. 40
  (cyclo-octadiene)-(cyclo-octadienyl)-(cyclo- Octadienediyl)-dodecacarbonyl-hepta-iridium.... 71.246
yl-hepta-iridium]   (cyclo-octadiene)-(cyclo- Octadienyl)-(cyclo-octadienediyl)-dodecacarbon 71.246
m(i) bis(hexafluorophosphate> nitrosyl-(1,10- Phenanthroline)-bis(triphenylphosphine)-iridium 86.174
tra-iridium]   (o- Phenylene-bis(dimethylarsine))-decacarbonyl-te 86. 64
hylphosphinophenyl)-hydrido-iridium hexafluoro Phosphate <s(dimethylphenylphosphine)-(o-dimet 71.237
riphenylphosphine)-iridium(iii) bis(hexafluoro Phosphate) [tris(acetonitrile)-nitrosyl-bis(t 86.152
(triphenylphosphine)-iridium(i) bis(hexafluoro Phosphate) [nitrosyl-(1,10-phenanthroline)-bis 86.174
ylurea]   disulfido-bis(1,2-bis(diphenyl Phosphido)ethane)-iridium chloride N,N'-diphen 86.191
dium toluene s> 3,9-dihydrido-bis(tri-p-tolyl Phosphine)-(nido-7,8-dicarba-undecaborane)-iri 86.167
-iridium hexafluorophosp> tris(dimethylphenyl Phosphine)-(o-dimethylphosphinophenyl)-hydrido 71.237
chloro-iridium(i) toluene solv> bis(triphenyl Phosphine)-(tetrachlorodiazo-cyclopentadiene)- 86.148
3)-1-phenylallyl)-chloro-hydrido-bis(triphenyl Phosphine)-iridium .... [(η(72. 63
bonyl-(1,3-di-p-tolyltriazenido)-bis(triphenyl Phosphine)-iridium(i) .... [trans-car 86.183

nitrosyl-(1,10-phenanthroline)-bis(triphenyl Phosphine)-iridium(i) bis(hexafluorophosphate)   86.174
honio-dithiocarboxylato-carbonyl-bis(triphenyl Phosphine)-iridium(i) tetrafluoroborate <phosp   86.209
iphenylphosphino)phenyl)-hydrido-bis(triphenyl Phosphine)-iridium(iii)........   71.276
e)] tris(acetonitrile)-nitrosyl-bis(triphenyl Phosphine)-iridium(iii) bis(hexafluorophosphat   86.152
l-iodo-(methyl(methylthio)-carbene)-(triphenyl Phosphine)-iridium(iii) iodide <cyclopentadieny 71.206
<hine-platinum-bis(μ(2)-hydrido)-tris(triethyl*Phosphine)-iridium-hydride tetraphenylborate   86.104
<-nitrato-propan-1-on-1-yl)-bis(dimethylphenyl Phosphine)-pyridine-chloro-iridium acetone so>   71.192
onyl-ruthenium-tetra-iridi> bis(triphenyl*Phosphine)iminium) nonacarbonyl-hexa-μ(2)-carb   64.163
tetrabromocatecholato-nitroso-triphenyl Phosphine-iridium methylene chloride solvate   86. 85
μ(2)-t-butylthio)-carbonyl-iodo-dimethylphenyl*Phosphine-iridium(ii)) ....   [bis((86. 92
thylphosphine)-iridium-hydri> phenyl-triethyl*Phosphine-platinum-bis(μ(2)-hydrido)-tris(trie   86.104
<ido)-tris(μ(2)-hydrido)-tris(1,3-bis(diphenyl*Phosphino-propane)-tri-iridium bis(tetrafluo>   86.224
trichloro-(2,2'-bis(o-diphenyl Phosphino)-trans-stilbene)-iridium(iii)........   72. 57
carbonyl-chloro-(o-(diphenyl Phosphino-N,N-dimethylaniline)-iridium   86. 75
rosulfate]   chloro-bis(bis(dimethyl Phosphino)ethane)-methoxycarbonyl-iridium fluo   71. 79
onyl-ruthenium-tetra-iridi> bis(bis(triphenyl*Phosphino)iminium) heptacarbonyl-μ(2)-hexacarb   64.162
ne)-iridium(iii)]   bromo-(2-(diphenyl Phosphino)phenyl)-hydrido-bis(triphenylphosphi   71.276
<drido-tris(μ(2)-hydrido)-bis(1,3-bis(diphenyl*Phosphino)propane)-di-iridium tetrafluorobora>   86.205
dium]   (1,1,1-tris(diphenyl Phosphinomethyl)ethane)-cyclotriphosphorus-iri   86.150
sp> tris(dimethylphenylphosphine)-(o-dimethyl Phosphinophenyl)-hydrido-iridium hexafluoropho   71.237
bis(μ-t-butylthiolato)-bis(carbonyl-trimethyl*Phosphite-iridium) 86. 41
-butylthiolato)-bis(carbonyl-hydrido-trimethyl*Phosphite-iridium) ....   [bis(μ-t   86. 42
henylphosphine)-iridium(i) tetrafl> triphenyl Phosphonio-dithiocarboxylato-carbonyl-bis(trip   86.209
-tris(diphenylphosphinomethyl)ethane)-cyclotri Phosphorus-iridium ....[(1,1,1   86.150
ine)-iridium-hydri> phenyl-triethylphosphine-*Platinum-bis(μ(2)-hydrido)-tris(triethylphosph   86.104
e-tetra-acetato)-iridium) monohydrate]   Potassium (+)-(chloro-(hydrogen-ethylenediamin   76. 58
e-tetra-acetato)-iridium)]   Potassium (dichloro-(dihydrogen-ethylenediamin   76. 59
-pyridine-chloro-iridium ace> (2-aci-nitrato- Propan-1-on-1-yl)-bis(dimethylphenylphosphine)   71.192
s(μ(2)-hydrido)-bis(1,3-bis(diphenylphosphino)-*Propane)-di-iridium tetrafluoroborate <ido-tri 86.205
<(2)-hydrido)-tris(1,3-bis(diphenylphosphino)-*Propane)-tri-iridium bis(tetrafluoroborate) m>   86.224
ropan-1-on-1-yl)-bis(dimethylphenylphosphine)- Pyridine-chloro-iridium acetone solvate <ato-p   71.192
hine)iminium) nonacarbonyl-hexa-μ(2)-carbonyl-*Ruthenium-tetra-iridium <is(bis(triphenylphosp   64.163
hine)iminium) heptacarbonyl-μ(2)-hexacarbonyl-*Ruthenium-tetra-iridium <is(bis(triphenylphosp   64.162
richloro-(2,2'-bis(o-diphenylphosphino)-trans- Stilbene)-iridium(iii) .... [t   72. 57
bis(μ-chloro)-tetrachloro-tetrakis(diethyl*Sulfide)-di-iridium(iii) 85. 56
o)-pentachloro-(μ-diethylsulfide)-tris(diethyl*Sulfide)-di-iridium(iii) ....   [(μ-chlor 85. 55
(μ-chloro)-pentachloro-(μ-diethyl*Sulfide)-tris(diethylsulfide)-di-iridium(iii)   85. 55
iridium chloride N,N'-diphenylurea]   di Sulfido-bis(1,2-bis(diphenylphosphido)ethane)-   86.191
i)> η(5)-cyclopentadienyl-iodo-(methyl(methyl Thio)-carbene)-(triphenylphosphine)-iridium(ii   71.206
idium(ii))]   bis((μ(2)-t-butyl*Thio)-carbonyl-iodo-dimethylphenylphosphine-ir   86. 92
e)-iridium(i) tetrafl> triphenylphosphonio-di Thiocarboxylato-carbonyl-bis(triphenylphosphin   86.209
is(dicarbonyl-iridium)]   tris(μ-t-butyl*Thiolato)-(μ-hexafluoro-but-2-ene-2,3-diyl)-tr   71.181
ite-iridium)]   bis(μ-t-butyl*Thiolato)-bis(carbonyl-hydrido-trimethylphosph   86. 42
ium)]   bis(μ-t-butyl*Thiolato)-bis(carbonyl-trimethylphosphite-irid   86. 41
um))]   bis(μ-chloro)-bis(mercury-(μ-p-*Tolyl(ethyl)triazenido)-(cyclo-octadiene-iridi   75. 40
)-iridium toluene s> 3,9-dihydrido-bis(tri-p- Tolylphosphine)-(nido-7,8-dicarba-undecaborane   86.167
um(i)]   trans-carbonyl-(1,3-di-p- Tolyltriazenido)-bis(triphenylphosphine)-iridi   86.183
bis(μ-chloro)-bis(mercury-(μ-p-tolyl(ethyl)*Triazenido)-(cyclo-octadiene-iridium)) 75. 40
trans-carbonyl-(1,3-di-p-tolyl Triazenido)-bis(triphenylphosphine)-iridium(i)   86.183
osphido)ethane)-iridium chloride N,N'-diphenyl Urea .... [disulfido-bis(1,2-bis(diphenylph   86.191

Iron

tricarbonyl-(methyl α- Acetamido-cinnamate)-iron 72. 23
)]   (μ-*Acetimidoyl)-(μ-hydrido)-tris(tricarbonyl-iron   71. 26
,4,8,11-tetra-aza-1,3,8,10-cyclotetradecate> Acetonitrile(carbonyl)-(2,3,9,10-tetramethyl-1   83.110
hexakis(Acetonitrile)-iron(ii) bis(tetrachloro-iron)   83. 80
hexakis(Acetonitrile)-iron(ii) bis(tetrachloro-iron)   83. 79
S-cyclo-1-(1'-dimethylaminoethyl-ferrocene)-2- Acetylacetonato-palladium .... [(-)-R 73. 80a
2-nitro-3-ferrocenyl- Acrylic acid ethyl ester   73. 55
<η(1),η(3)-(1-methyl-2,3-bis(methoxycarbonyl)-*Allyl))-carbonyl-bis(η(5)-cyclopentadienyl-ir> 72. 38
(η(3)- Allyl)-bromo-tricarbonyl-iron   72. 2
(5)-cyclopentadienyl-bis(μ(2)-carbonyl)-(μ(3)-*Allyl)-hexacarbonyl-tri-iron .... [η   71. 91
ron]   (η(3)- Allyl)-tricarbonyl-(triphenylphosphine-gold)-i   72. 46
2-diphenylphosphino-ferrocenyl)ethyl)-dimethyl Amine ....[(+)-D-((73. 108a
<2-diphenylphosphino-ferrocenyl)ethyl)-dimethyl Amine)-norbornadiene-rhodium hexafluorophosph> 73.120
μ(2)-carbonyl)-nickel-tri-i> benzyl-trimethyl*Ammonium (μ(3)-hydrido)-octacarbonyl-tetrakis(3. 43
is(tricarbonyl-iron)]   tetraethyl*Ammonium (μ(4)-carbomethoxymethylidene)-tetrak 71. 84
(2)-carbonyl)-dicarbonyl-iron)-tr> tetraethyl*Ammonium (μ(5)-carbido)-(dicarbonyl-rhodium-(μ   3. 33
molybdenum-trisulfido-tris(ethylt> tetraethyl*Ammonium (μ-sulfido)-bis(μ-ethylthiolato)-bis(85. 57
2)-sulfido)-tris(ethylthiolat> triethylbenzyl*Ammonium tris(μ-ethylthiolato)-bis(tetrakis(μ(85. 62
)-carbonyl)-dicarbonyl-iron)-> bis(tetraethyl*Ammonium) (μ(6)-carbido)-(dicarbonyl-iron-(μ(2   3. 37
um)-tetrakis(μ-thiola> tetrakis(tetra-n-butyl*Ammonium) (bis(tris(μ-benzylthiolato)-molybden 85. 96
)-tetrakis(μ-thiolato> tetrakis(tetra-n-butyl*Ammonium) (bis(tris(μ-benzylthiolato)-tungsten 85. 97
)-tetrakis(μ-thiolat> tris(N-benzyl-trimethyl*Ammonium) bis(tris(μ-ethylthiolato)-molybdenum 85. 78

n(ii)]   bis(tetramethyl*Ammonium) deca(μ(2)-benzenethiolato)-tetra-iro   85. 95
num>   tetraphenylphosphonium (benzyl-trimethyl*Ammonium) dichloro-iron-bis(μ(2)-thio)-molybde   64.123
m]   bis(trimethylbenzyl Ammonium) hexadecacarbonyl-tetra-iron-palladiu   3. 44
                          bis(trimethylbenzyl Ammonium) hexadecacarbonyl-tetra-iron-platinum   3. 45
-hexa-palladium acetoni> tris(trimethylbenzyl Ammonium) tetracosa-carbonyl-hydrido-hexa-iron   71.190
hiolato-iron)]   tris(tetraethyl*Ammonium) tetrakis(μ(3)-sulfido)-tetra(benzylt   85. 80
3)-thio)-tris(μ(3)-ethylthio> tris(tetraethyl*Ammonium) tris(μ(2)-ethylthio)-bis(tetrakis(μ(3. 39
             2,2',5,5'-tetramethyl-1,1'-di Arsa-ferrocene 73. 38
-(μ-dimethylarsenido)-chromium)-bis(μ-dimethyl*Arsenido)-(tetracarbonyl-iron)-cobalt <adienyl   73.111
on)-manganese tetracarbonyl-chloro-(μ-dimethyl*Arsenido)-(tetracarbonyl-iron)-manganese <l-ir   86. 18
carbonyl-chloro)   tetracarbonyl-bis(μ-dimethyl*Arsenido)-(tetracarbonyl-iron)-manganese tetra 86. 18
<ienyl)-chromium)-cyclopentadienyl-(μ-dimethyl*Arsenido)-chromium)-bis(μ-dimethylarsenido)-(>   73.111
<)-(μ-carbonyl-(cyclopentadienyl)-bis(dimethyl*Arsenido)-iron)-cyclopentadienyl-(cyclopentad>   73. 93
                o-phenylene-bis(dimethyl Arsine)-decacarbonyl-tri-iron 86. 63
<rbonyl)-(2,3,9,10-tetramethyl-1,4,8,11-tetra- Aza-3,8,10-cyclotetradecatetraene)-iron(ii)>   83.110
<henylphosphido)-bis(tricarbonyl-iron) 1,10-di*Aza-4,7,13,16,21,24-hexaoxabicyclo(8.8.8)hexa> 86.101
i) ethylthiolate 4,7,13,16,21-pentaoxa-1,10-di Azabicyclo(8.8.5)tricosane benzene solvate <(i   49. 27
<ato)-iron(ii) 4,7,13,16,21,24-hexaoxa-1,10-di Azabicyclo(8.8.8)hexacosane chlorobenzene sol>   49. 23
bonyl)-bis((1-t-butyl-3-methyl-2-phenyl-η-1,2-*Azaborolinyl)-carbonyl-iron) [cis-bis(μ(2)-car 75. 38
         (1-t-butyl-3-methyl-2-phenyl-η-1,2- Azaborolinyl)-dicarbonyl-iodo-iron 75. 10
<ercapto)-2,3,9,10-tetramethyl-1,4,8,11-tetra- Azacyclotetradeca-1,3,8,10-tetraene-iron(iii)>   83.195
                hexamethyl Benzene-cyclopentadienyl-iron   73. 69
       bis(tetramethylammonium) deca(μ(2)-*Benzenethiolato)-tetra-iron(ii)...............   85. 95
-tricarbonyl)-(c)-phosphepin-iro> 1-phenyl-di Benzo(b,f)-(2,5-diphenyl-ferra-cyclopentadiene   73.124
osphepin-di-iron-hexacarbonyl]   1-phenyl-di Benzo(b,f)-2,5-diphenyl-cyclopentadienone(c)ph   73.128
                tris(N-methyl-thio Benzohydroxamato)-iron(iii) 85. 74
<(tricarbonyl-iron)-bis(μ-diphenylphosphido)-(*Benzoyl-dicarbonyl-iron) tetrahydrofuran solv> 71.245
rbonyl-tetrakis(μ(2)-carbonyl)-nickel-tri-i> *Benzyl-trimethylammonium (μ(3)-hydrido)-octaca   3. 43
lato)-molybdenum)-tetrakis(μ-thiolat>   tris(N-*Benzyl-trimethylammonium) bis(tris(μ-ethylthio 85. 78
2)-thio)-molybdenum>   tetraphenylphosphonium (*Benzyl-trimethylammonium) dichloro-iron-bis(μ(64.123
kis(μ(2)-sulfido)-tris(ethylthiolat>   triethyl*Benzylammonium tris(μ-ethylthiolato)-bis(tetra 85. 62
lladium]   bis(trimethyl Benzylammonium) hexadecacarbonyl-tetra-iron-pa   3. 44
atinum]   bis(trimethyl Benzylammonium) hexadecacarbonyl-tetra-iron-pl   3. 45
a-iron-hexa-palladium acetoni> tris(trimethyl Benzylammonium) tetracosa-carbonyl-hydrido-hex   71.190
on(ii)]   bis(α- Benzyldioximato)-bis(cyclohexyl-isonitrile)-ir   71.264
                tris(N,N-di Benzyldithiocarbamato)-iron(iii) 80. 29
                tris(N,N-di Benzyldithiocarbamato)-iron(iii) 80. 28
tetra-azacyclotetradeca-1,3,8,10-tetrae> bis(Benzylmercapto)-2,3,9,10-tetramethyl-1,4,8,11- 83.195
   tetrakis(tetra-n-butylammonium) (bis(tris(μ-*Benzylthiolato)-molybdenum)-tetrakis(μ-thiola> 85. 96
   tetrakis(tetra-n-butylammonium) (bis(tris(μ-*Benzylthiolato)-tungsten)-tetrakis(μ-thiolato> 85. 97
raethylammonium) tetrakis(μ(3)-sulfido)-tetra(*Benzylthiolato-iron)   85. 80
                                    [tris(tet thiolato)-tungsten)-tetrakis(μ-thiolato)-tris(*Benzylthiolato-iron))-iron <(bis(tris(μ-benzyl 85. 97
iolato)-molybdenum)-tetrakis(μ-thiolato)-tris(*Benzylthiolato-iron))-iron <is(tris(μ-benzylth 85. 96
       η(4)-(2,3,5,6-tetrakis(methylene)-7-oxa Bicyclo(2.2.1)heptane)-endo-tricarbonyl-iron   75. 5
nyl-iron]   ((2,3,6,7-η)- Bicyclo(3.2.2)nona-2,6,8-trien-4-one)-tricarbo   75. 3
           (μ(3)-8-tricarbonyl-7-selena-8-ferra-*Bicyclo(4.3.0)nonen-9-one)-tricarbonyl-iron- 71. 49
ene> bis(carbonyl)-(η(5)-cyclopentadienyl)-(Bicyclo(5.4.0)undeca-1,3,6,8,10-pentaen-5-ylid 71.122
ethylthiolate 4,7,13,16,21-pentaoxa-1,10-diaza Bicyclo(8.8.5)tricosane benzene solvate <(ii)   49. 27
bonyl-iron) 1,10-diaza-4,7,13,16,21,24-hexaoxa*Bicyclo(8.8.8)hexacosane <hosphido)-bis(tricar 86.101
<)-iron(ii) 4,7,13,16,21,24-hexaoxa-1,10-diaza Bicyclo(8.8.8)hexacosane chlorobenzene solvate 49. 23
hydrate]   (-)-D-tris(2,2'- Bipyridyl)-cobalt(ii) hexacyano-iron(iii) octa 83.197a
                     (2,2'- Bipyridyl)-zinc-tetracarbonyl-iron 83.190
(5)-cyclopentadienyl-ferra)-2,3-dicarba-dodeca Borane ....   [8-trifluoroacetoxy-1-(η 71. 13
(η(3):η(4)-trihydro-C,C'-dimethyl-dicarba-hexa*Borato))-iron <5)-cyclopentadienyl-cobalt)-(μ- 71. 70
               bis(hydro-tris(1-pyrazolyl)- Borato)-iron(ii) 83.113
       bis(hydrido-tris(3,5-dimethylpyrazolyl)- Borato)-iron(ii)   83.202
onyl-methyl-bis(trimethylphosphine)-(triphenyl Borato-nitrilo-methyl)-iron(ii)........[dicarb 71.220
uoro-2-isopropyl-4-methyl-1,5-dioxa-3-ferra-6- Borinane <bonyl-3-η(5)-cyclopentadienyl-6-difl 71. 41
2),η(5)-3,4-diethyl-2,5-dimethyl-1,2,5-thia-di*Borolene)-iron <(η(5)-cyclopentadienyl-iron-μ(73.109
yl)-bis((1-t-butyl-3-methyl-2-phenyl-η-1,2-aza*Borolinyl)-carbonyl-iron) [cis-bis(μ(2)-carbon 75. 38
          (1-t-butyl-3-methyl-2-phenyl-η-1,2-aza Borolinyl)-dicarbonyl-iodo-iron............... 75. 10
ocyanide)-(1,4-bis(t-butylimino)-2,3-diphenyl- Buta-1,3-diene)-iron toluene solvate <-butylis 71.259
ron]   μ-(1-(1,2,3,4-tetraphenyl*Butadienyl)-phenylgermylene)-octacarbonyl-di-i 69. 58
dene-2,3-dihydroxy-1,4-bis(diphenylphosphino)- Butane)-iron <lopentadienyl-iodo-(O-isopropyli 73.123
       1,4-dihydroxy-2,3-diethyl- Butane-1,1,4,4-tetrayl-bis(tricarbonyl)-iron 71. 64
(μ-1-diphenylphosphino-2-sulfido-3,3-dimethyl*Butene)-carbonyl-(methoxy-thiocarbonyl)-iron] 72. 49
(1-cyclohexylimino-2-methoxy-3-methoxycarbonyl Butene)-tricarbonyl-iron   [η(4)- 72. 27
yclopentadienyl)-iron)] (μ-carbonyl)-(μ-t-*Butoxy-carbonylmethylene)-bis(carbonyl-(η(5)-c 71.140
                (μ(3)-*Butyronitrile)-tris(tricarbonyl-iron) 72. 19
       nickel-bis(tetracarbonyl-μ-(dimethyl*Carbamoyl)-iron) 71. 63
       cis((μ-carbonyl)-(μ-methyl*Carbene-bis(carbonyl-cyclopentadienyl-iron)) 71. 88
               ferrocenyl-diphenyl Carbenium tetrafluoroborate...................   73. 97
carbonyl-cyclopentadienyl-(5-methyl-2-(phenyl Carbenylamino)-phenyl)-iron................   71.154
rbonyl-iron)-> bis(tetraethylammonium) (μ(6)-*Carbido)-(dicarbonyl-iron-(μ(2)-carbonyl)-dica   3. 37
icarbonyl-iron)-tr> tetraethylammonium (μ(5)-*Carbido)-(dicarbonyl-rhodium-(μ(2)-carbonyl)-d   3. 33

rate]   Cesium tris(ethylnitrosolato)-iron(ii) monohyd 84. 9
                  octaethylporphinato-per Chlorato-iron(iii) 49. 7
<omium)-cyclopentadienyl-(μ-dimethylarsenido)-*Chromium)-bis(μ-dimethylarsenido)-(tetracarbo> 73.111
<yl-(dicarbonyl(tricarbonyl(cyclopentadienyl)-*Chromium)-cyclopentadienyl-(μ-dimethylarsenid> 73.111
            dicarbonyl-(η(4)- Cinnamaldehyde)-(triphenylphosphine)-iron(0)   72. 52
                    (N-methyl- Cinnamaldehyde-imino)-tetracarbonyl-iron 83. 90
         tricarbonyl-(methyl α-acetamido- Cinnamate)-iron.......................   72. 23
-bis(μ-dimethylarsenido)-(tetracarbonyl-iron)-*Cobalt <adienyl-(μ-dimethylarsenido)-chromium) 73.111
            (-)-D-tris(2,2'-bipyridyl)- Cobalt(ii) hexacyano-iron(iii) octahydrate 83.197a
   η(5)-cyclopentadienyl-(η(5)-cyclopentadienyl-*Cobalt)-(μ-(η(3):η(4)-trihydro-C,C'-dimethyl-> 71. 70
              chloro-(tetracarbonyl- Cobalt)-bis(dicarbonyl-cyclopentadienyl-iron)-   73. 71
tin(iv)]
ido)-iron)-cyclopentadienyl-(cyclopentadienyl-*Cobalt)-iron <lopentadienyl)-bis(dimethylarsen 73. 93
        (μ(3)-phenylphosphine)-hexacarbonyl-di-*Cobalt-tricarbonyl-iron..................... 86. 34
o-nicotinamidophenyl)-porphyrinato-iron(iii))- Copper(ii) perchlorate dihydrate <α,α,α-tetra(83.225
e]   bis((diethylenetriamine-*Copper)-μ(2)-cyano)-tetracyano-iron hexahydrat 71. 78
tricarbonyl-(η(4)-1,5-dimethylene-2,6-dimethyl Cyclo-octane)-iron 72. 24
           hexacarbonyl-(μ-(1,2-eta:2-η)-1-*Cyclo-octene-1-selenolato)-di-iron 75. 7
            tris(trimethylphosphite)-(η(3)- Cyclo-octenyl)-iron tetrafluoroborate 75. 19r
            tris(trimethylphosphite)-(η(3)- Cyclo-octenyl)-iron tetrafluoroborate 75. 17
            tris(trimethylphosphite)-(η(3)- Cyclo-octenyl)-iron tetrafluoroborate 75. 18r
)]   (η(3)- Cyclo-octenyl)-tris(trimethylphosphite)-iron(i   75. 16
etylacetonato-palladium]   (-)-RS- Cyclo-1-(1'-dimethylaminoethyl-ferrocene)-2-ac   73. 80a
         η(5)-cyclopentadienyl-(2,4-diphenyl- Cyclobut-1-en-3-on-1-yl)-dicarbonyl-iron 71.188
ne)-iron]   carbonyl- Cyclobutadiene-(1,2-bis(diphenylphosphino)etha 75. 37
-iron hexafluorophosphate]   bis(carbonyl)- Cycloheptatrienylidene-(η(5)-cyclopentadienyl) 71. 62
            tricarbonyl-(1,1-diphenyl-1-sila Cyclohexa-2,4-dienyl)-iron-   75. 24
<carbonyl-(methyl 1-(η-2-5),4-methoxy-1-methyl Cyclohexa-2,4-dienyl)-2-oxo-cyclopentane-carb> 75. 21
-5-η)-exo-1-acetyl-2,4,6-tris(trifluoromethyl Cyclohexadienyl)-(η-cyclopentadienyl)-iron <(1 73. 56
          (2-methyl-3,5,6-tri(methylene)- Cyclohexane-(2-4-η)-ene)-bis(tricarbonyl-iron) 72. 25
        dicarbonyl-cyclopentadienyl-(3-methyl Cyclohexenone)-iron hexafluorophosphate........ 73. 48
n)]   trisila-tetrachloro- Cyclohexyl-bis(dicarbonyl-cyclopentadienyl-iro 73. 66
            trisila-pentachloro- Cyclohexyl-dicarbonyl-cyclopentadienyl-iron.... 73. 20
            bis(α-benzyldioximato)-bis(Cyclohexyl-isonitrile)-iron(ii)................ 71.264
ene)-tricarbonyl-iron]   η(4)-(1- Cyclohexylimino-2-methoxy-3-methoxycarbonylbut 72. 27
            3,4-diethyl-1-tetracarbonyl-ferra Cyclopent-3-ene-2,5-dione   71. 35
o> 1-phenyl-dibenzo(b,f)-(2,5-diphenyl-ferra Cyclopentadiene-tricarbonyl)-(c)-phosphepin-ir 73.124
is(tri> μ-1-5-eta:1'-5'-η-(bis(μ-tetraphenyl-*Cyclopentadienediyl-oxo-O)-bis(aqua-silver))-b 73.133
bonyl] 1-phenyl-dibenzo(b,f)-2,5-diphenyl- Cyclopentadienone(c)phosphepin-di-iron-hexacar 73.128
8,10-pentaen-5-ylidene)> bis(carbonyl)-(η(5)- Cyclopentadienyl)-(bicyclo(5.4.0)undeca-1,3,6, 71.122
osphino)-pent-2-en-3-yl)-iron(ii)]   (η(5)- Cyclopentadienyl)-(2-methyl-4,5-bis(diphenylph 71.243
cyclopentadien> bis(μ-carbonyl)-(μ-carbonyl-(*Cyclopentadienyl)-bis(dimethylarsenido)-iron)- 73. 93
μ-dimet> tricarbonyl-(dicarbonyl(tricarbonyl(*Cyclopentadienyl)-chromium)-cyclopentadienyl-(73.111
,4,6-tris(trifluoromethyl)cyclohexadienyl)-(η- Cyclopentadienyl)-iron <(1-5-η)-exo-1-acetyl-2 73. 56
     bis(carbonyl)-cycloheptatrienylidene-(η(5)- Cyclopentadienyl)-iron hexafluorophosphate 71. 62
-butoxy-carbonylmethylene)-bis(carbonyl-(η(5)-*Cyclopentadienyl)-iron).... [(μ-carbonyl)-(μ-t 71.140
dimethyliminomethylene)-bis(carbonyl-(η-methyl*Cyclopentadienyl)-iron) iodide <-carbonyl)-(μ- 71.128
)-(μ-(η(3):η(4)-trihydro-C,C'-dimethyl> η(5)-*Cyclopentadienyl-(η(5)-cyclopentadienyl-cobalt 71. 70
<onyl(tricarbonyl(cyclopentadienyl)-chromium)-*Cyclopentadienyl-(μ-dimethylarsenido)-chromiu> 73.111
socyanide)-iron)]   trans-anti-bis(η(5)-*Cyclopentadienyl-(μ-phenylisocyanide)-(phenyli 71.253
<yclopentadienyl)-bis(dimethylarsenido)-iron)-*Cyclopentadienyl-(cyclopentadienyl-cobalt)-ir> 73. 93
-on-1-yl)-dicarbonyl-iron]   η(5)- Cyclopentadienyl-(2,4-diphenyl-cyclobut-1-en-3 71.188
hexafluorophosphate]   dicarbonyl- Cyclopentadienyl-(3-methylcyclohexenone)-iron 73. 48
ino)-phenyl)-iron]   carbonyl- Cyclopentadienyl-(5-methyl-2-(phenylcarbenylam 71.154
l)-hexacarbonyl-tri-iron]   η(5)-*Cyclopentadienyl-bis(μ(2)-carbonyl)-(μ(3)-ally 71. 91
oromethyl-1-fluorovinyl)-iron]   dicarbonyl- Cyclopentadienyl-cis-(2-ethoxycarbonyl-2-triflu 71. 51
o-C,C'-dimethyl> η(5)-cyclopentadienyl-(η(5)- Cyclopentadienyl-cobalt)-(μ-(η(3):η(4)-trihydr 71. 70
bis(dimethylarsenido)-iron)-cyclopentadienyl-(*Cyclopentadienyl-cobalt)-iron <lopentadienyl)- 73. 93
onamethyl-cyclopentasilane-dimethylsilyl-η(5)- Cyclopentadienyl-dicarbonyl-iron .... [n 73. 78
yl-iron)-cyclopentasilane)-dimethylsilyl-η(5)- Cyclopentadienyl-dicarbonyl-iron <nyl-dicarbon 73.103
lane)-dimethylsilyl-η(5)-> (octamethyl-(η(5)- Cyclopentadienyl-dicarbonyl-iron)-cyclopentasi 73.103
ne]   8-trifluoroacetoxy-1-(η(5)- Cyclopentadienyl-ferra)-2,3-dicarba-dodecabora 71. 13
hydroxy-1,4-bis(diphenylphosphino)-butane)-> Cyclopentadienyl-iodo-(O-isopropylidene-2,3-di 73.123
        hexamethylbenzene- Cyclopentadienyl-iron 73. 69
     trisila-pentachloro-cyclohexyl-dicarbonyl- Cyclopentadienyl-iron 73. 20
   (μ-carbonyl)-(μ-thiocarbonyl)-bis(carbonyl-*Cyclopentadienyl-iron) 73. 46
trisila-tetrachloro-cyclohexyl-bis(dicarbonyl- Cyclopentadienyl-iron) 73. 66
is(methoxycarbonyl)-allyl)-carbonyl-bis(η(5)-*Cyclopentadienyl-iron) <),η(3)-(1-methyl-2,3-b 72. 38
s((μ-carbonyl)-(μ-methylcarbene)-bis(carbonyl- Cyclopentadienyl-iron)).......................[ci 71. 88
   chloro-(tetracarbonyl-cobalt)-bis(dicarbonyl- Cyclopentadienyl-iron)-tin(iv) 73. 71
5-dimethyl-1,2,5-thia-diborolene)-> bis(η(5)-*Cyclopentadienyl-iron-μ(2),η(5)-3,4-diethyl-2, 73.109
           dicarbonyl- Cyclopentadienyl-isothiocyanato-iron 73. 9
xacarbonyl-di-iron]   (η(5)- Cyclopentadienyl-nickel)-(η-t-butylethynyl)-he 72. 29
       (μ(3)-carbonyl)-bis(η(5)-*Cyclopentadienyl-nickel)-bis(tricarbonyl-iron) 71.105
        (hex-3-ene-3,4-diyl)-bis(η(5)- Cyclopentadienyl-nickel)-bis(tricarbonyl-iron) 71.178
yl-1,5-dioxa-3-ferra-6-bo> 3-carbonyl-3-η(5)- Cyclopentadienyl-6-difluoro-2-isopropyl-4-meth 71. 41

4-methoxy-1-methylcyclohexa-2,4-dienyl)-2-oxo-   Cyclopentane-carboxylato-iron <thyl 1-(η-2-5),   75. 21
<thyl-(η(5)-cyclopentadienyl-dicarbonyl-iron)-   Cyclopentasilane)-dimethylsilyl-η(5)-cyclopen>   73.103
dienyl-dicarbonyl-iron]   nonamethyl-   Cyclopentasilane-dimethylsilyl-η(5)-cyclopenta   73. 78
                          syn-(η(3)-2-vinylyl-   Cyclopentyl-carboxylato)-tricarbonyl-iron 71. 27
                          anti-(η(3)-2-vinylyl-   Cyclopentyl-carboxylato)-tricarbonyl-iron 71. 28
            trans-1-ferrocenyl-2-p-methoxyphenyl-   Cyclopropane 73. 84
              ethyl (Z)-2-ferrocenyl-2-methyl-   Cyclopropane-carboxylate 73. 68
<apto)-2,3,9,10-tetramethyl-1,4,8,11-tetra-aza   Cyclotetradeca-1,3,8,10-tetraene-iron(iii) he>   83.195
<9,10-tetramethyl-1,4,8,11-tetra-aza-1,3,8,10-   Cyclotetradecatetraene)-iron(ii) hexafluoroph>   83.110
            dicarbonyl-(1,4,7,10-tetrathia   Decane)-iron 85. 18
nyl-cyclobutadiene-(1,2-bis(diphenylphosphino)   Ethane)-iron .... [carbo   75. 37
        bis(μ-methylthio)-(μ-tetrafluoro*   Ethane-1,1-diyl)-bis(tricarbonyl-iron) 71. 17
        bis(μ-methylthio)-(μ-tetrafluoro*   Ethane-1,2-diyl)-bis(tricarbonyl-iron) 71. 18
-dinitrosyl-iron]   (dimethyl-(N,N-dimethyl   Ethanolamino)-(3,5-dimethylpyrazolyl)-gallato)   84. 32
          1,1-dicyano-2,2-di(ferrocenyl)-   Ethylene 73. 98
phine-carbonyl-ir> (μ(2)-1,2-bis(carbomethoxy*   Ethylene))-bis((μ(2)-methylthio)-trimethylphos   86. 82
  dichloro-bis(cis-1,2-bis(diphenylphosphino)-   Ethylene)-iron(ii) acetone solvate 86.187
o-iron hexahydrate]   bis((di*   Ethylenetriamine-copper)-μ(2)-cyano)-tetracyan   71. 78
iron)]   (μ(3)-*   Ethylideneimido)-(μ-hydrido)-tris(tricarbonyl-   72. 11
    (η(5)-cyclopentadienyl-nickel)-(η-t-butyl   Ethynyl)-hexacarbonyl-di-iron 72. 29
                                             Ferrichrome 84.108a
                                             Ferrocene-1,1'-dicarboxylic acid 73. 28n
                                             Ferrocene-1,1'-dicarboxylic acid 73. 27
      cis-dichloro-tetrakis(dimethyl   Formamide)-iron tetrachloro-iron 84. 43
ormamide solvate]   carbonyl-(N,N-dimethyl   Formamide)-phthalocyaninato-iron(ii) dimethylf   49. 5
    meso-tetraphenylporphinato-bis(tetrahydro   Furan)-iron(ii)............................   84.136
dimethylethanolamino)-(3,5-dimethylpyrazolyl)-   Gallato)-dinitrosyl-iron   [(dimethyl-(N,N-   84. 32
          (1,1'-ferrocenediyl)-diphenyl-   Germanium 73. 91
    μ-(1-(1,2,3,4-tetraphenylbutadienyl))-phenyl   Germylene)-octacarbonyl-di-iron..............   69. 58
icolinium bis(S-methyldithiocarbazato-dimethyl   Glyoximato)-iron(iii) tetrahydrate 85. 39
  (η(3)-allyl)-tricarbonyl-(triphenylphosphine-   Gold)-iron .... [γ-p   72. 46
,3,5,6-tetrakis(methylene)-7-oxabicyclo(2.2.1)   Heptane)-endo-tricarbonyl-iron .... [η(4)-(2   75. 5
      (3,3,6-trimethyl-1-oxo-2-oxa   Heptyl-(4-6-η)-enyl)-tricarbonyl-iron 71. 39
      (3,3,6-trimethyl-1-oxo-2-oxa   Heptyl-(4-6-η)-enyl)-tricarbonyl-iron 71. 40
nickel)-bis(tricarbonyl-iron)]   (Hex-3-ene-3,4-diyl)-bis(η(5)-cyclopentadienyl-   71.178
10-diaza-4,7,13,16,21,24-hexaoxabicyclo(8.8.8)*   Hexacosane <hosphido)-bis(tricarbonyl-iron) 1,   86.101
7,13,16,21,24-hexaoxa-1,10-diazabicyclo(8.8.8)   Hexacosane chlorobenzene solvate <-iron(ii) 4,   49. 23
        bis(trimethylbenzylammonium)   Hexadecacarbonyl-tetra-iron-palladium 3. 44
        bis(trimethylbenzylammonium)   Hexadecacarbonyl-tetra-iron-platinum 3. 45
<aqua-(2,6-diacetylpyridine)-(2,9-bis(1-methyl   Hydrazone)-1,10-phenanthroline)-iron bis(tetr>   83.165
iron]   (μ(2)-*   Hydrido)-(η-methylidyne)-dodecacarbonyl-tetra-   71. 44
yl-iron)]   bis(μ(2)-*   Hydrido)-(μ(3)-phenylphosphido)-tris(tricarbon   86. 36
ny l-tetra-iron]   (μ(2)-carbonyl)-(μ(2)-*   Hydrido)-(μ(4)-methoxymethylidyne)-undecacarbo   84. 51
ny l-tetra-iron]   (μ(2)-carbonyl)-(μ(2)-*   Hydrido)-(μ(4)-methoxymethylidyne)-undecacarbo   84. 52
<henylphosphine)iminium (μ(2)-carbonyl)-(μ(2)-*   Hydrido)-hexacarbonyl-di-iron-hexacarbonyl-di>   64.156
nickel-tri-i>   benzyl-trimethylammonium (μ(3)-*   Hydrido)-octacarbonyl-tetrakis(μ(2)-carbonyl)-   3. 43
<henylphosphine)iminium (μ(2)-carbonyl)-(μ(2)-*   Hydrido)-tricarbonyl-iron-nonacarbonyl-tri-ru>   64.155n
                (μ-acetimidoyl)-(μ-*   Hydrido)-tris(tricarbonyl-iron)..................   71. 26
                (μ(3)-ethylideneimido)-(μ-*   Hydrido)-tris(tricarbonyl-iron).................   72. 11
<((trimethylbenzylammonium) tetracosa-carbonyl-   Hydrido-hexa-iron-hexa-palladium acetonitrile>   71.190
on(ii)]   bis(Hydrido-tris(3,5-dimethylpyrazolyl)-borato)-ir   83.202
              tris(N-methyl-thiobenzo   Hydroxamato)-iron(iii) 85. 74
amidophenyl)-porphyrinato-iron eth>   (2-methyl   Imidazole)-dioxygen-meso-tetra(α,α,α,α-o-pival   49. 35
          hexakis(1-methyl   Imidazole)-iron(ii) octacarbonyl-di-iron(i)....   83.179
          hexakis(1-ethyl   Imidazole)-iron(ii) tetracarbonyl-iron(i)   83.203
yl)-porphyrinato-iron(ii) ethanol >   (2-methyl   Imidazole)-meso-tetra(α,α,α,α-o-pivalamidophen   49. 34
bonyl-di-iron-hexaca>   bis(triphenylphosphine)*   Iminium (μ(2)-carbonyl)-(μ(2)-hydrido)-hexacar   64.156
onyl-iron-nonacarbon>   bis(triphenylphosphine)*   Iminium (μ(2)-carbonyl)-(μ(2)-hydrido)-tricarb   64.155n
ne)-undecacarbonyl-t>   bis(triphenylphosphine)*   Iminium (μ(2)-carbonyl)-(μ(3)-methoxymethylidy   71. 59
ne)-undecacarbonyl-t>   bis(triphenylphosphine)*   Iminium (μ(2)-carbonyl)-(μ(3)-methoxymethylidy   71. 58
        bis(bis(triphenylphosphine)   Iminium) tridecacarbonyl-tetra-iron............   64.161
          (N-methyl-cinnamaldehyde-   Imino)-tetracarbonyl-iron 83. 90
ne >   tris(t-butylisocyanide)-(1,4-bis(t-butyl   Imino)-2,3-diphenyl-buta-1,3-diene)-iron tolue   71.259
(μ-diphenylphosphino)-(μ-N-methyl-(2-phenyl-2-*   Iminoethyl))-bis(tricarbonyl-iron) 71.211
tadienyl)-iron) iod>   (μ-carbonyl)-(μ-dimethyl   Iminomethylene)-bis(carbonyl-(η-methylcyclopen   71.128
phido)-(acetyl-dicar>   bis(triphenylphosphine)*   Immonium (tricarbonyl-iron)-bis(μ-diphenylphos   71.232
              diferrocenyl-selenium   Iodine tri-iodide methylene chloride solvate   73. 82
)-(μ(4)-N-oxy-ethylamino)-tetrakis(dicarbonyl-*   Iron) .... [tris(μ-carbonyl)-(μ(4)-ethylamino   84. 61
trans-anti-bis(η(5)-cyclopentadienyl-(μ-phenyl*   Isocyanide)-(phenylisocyanide)-iron) 71.253
l-buta-1,3-diene)-iron toluene >   tris(t-butyl   Isocyanide)-(1,4-bis(t-butylimino)-2,3-dipheny   71.259
-cyclopentadienyl-(μ-phenylisocyanide)-(phenyl   Isocyanide)-iron) .... [trans-anti-bis(η(5)   71.253
        bis(α-benzyldioximato)-bis(cyclohexyl-   Isonitrile)-iron(ii) 71.264
hosphino)-butane)->   cyclopentadienyl-iodo-(O-   Isopropylidene-2,3-dihydroxy-1,4-bis(diphenylp   73.123
                            bis(Isothiocyanato)-bis(thiourea)-iron(ii) 79. 5
          bis(N,N-diethyldithiocarbamato)-(Isothiocyanato)-iron(iii) 80. 11

dicarbonyl-cyclopentadienyl- Isothiocyanato-iron............................. 73. 9
hido)-(benzoyl-dicarbonyl-iron) tetrahydrof> *Lithium (tricarbonyl-iron)-bis(μ-diphenylphosp 71.245
oro-(μ-dimethylarsenido)-(tetracarbonyl-iron)-*Manganese <l-iron)-manganese tetracarbonyl-chl 86. 18
<bis(μ-dimethylarsenido)-(tetracarbonyl-iron)-*Manganese tetracarbonyl-chloro-(μ-dimethylars> 86. 18
phenylphosphine-iron(ii) perchlorat> tris(μ-*Mercapto-bis(bis(2-(diphenylphosphino)ethyl)- 86.218
azacyclotetradeca-1,3,8,10-tetrae> bis(benzyl Mercapto-2,3,9,10-tetramethyl-1,4,8,11-tetra- 83.195
a-carbonyl-(μ-meth> (μ-thio-ferrocenyl-methyl*Methane-thiomethylene-C',S(2))-1,1,1,2,2,2-hex 71.153
(R)-ferrocenyl-methyl Methane-S-methyltrithiocarbonate .... 73. 49.
)-iron)]   (μ-carbonyl)-(μ-t-butoxy-carbonyl*Methylene)-bis(carbonyl-(η(5)-cyclopentadienyl 71.140
nyl)-iron) iod> (μ-carbonyl)-(μ-dimethylimino*Methylene)-bis(carbonyl-(η-methylcyclopentadie 71.128
bonyl-iron)]   (2-methyl-3,5,6-tri(Methylene)-cyclohexane-(2-4-η)-ene)-bis(tricar 72. 25
octacarbonyl-(μ-*Methylene)-di-iron 71. 10
octacarbonyl-(μ-*Methylene)-di-iron 71. 9
icarbonyl-iron]   η(4)-(2,3,5,6-tetrakis(Methylene)-7-oxabicyclo(2.2.1)heptane)-endo-tr 75. 5
e)-exo-iron]   tricarbonyl-(6,7- Methylene-exo-3-oxatricyclo(3.2.1.0(2,4))octan 72. 16
tricarbonyl-(exo-2-methoxy-5,6-di Methylene-syn-7-norbornanol)-endo-iron 72. 20
-iron]   (exo-2-chloro-5,6-di Methylene-syn-7-norbornanone)-endo-tricarbonyl 72. 15
μ-meth> (μ-thio-ferrocenyl-methylmethane-thio*Methylene-C',S(2))-1,1,1,2,2,2-hexa-carbonyl-(71.153
tricarbonyl-(η(4)-1,5-di Methylene-2,6-dimethylcyclo-octane)-iron 72. 24
tetraethylammonium (μ(4)-carbomethoxy*Methylidene)-tetrakis(tricarbonyl-iron)........ 71. 84
(μ(2)-hydrido)-(η-*Methylidyne)-dodecacarbonyl-tetra-iron 71. 44
(μ(2)-carbonyl)-(μ(2)-hydrido)-(μ(4)-methoxy*Methylidyne)-undecacarbonyl-tetra-iron 84. 52
(μ(2)-carbonyl)-(μ(2)-hydrido)-(μ(4)-methoxy*Methylidyne)-undecacarbonyl-tetra-iron 84. 51
hosphine)iminium (μ(2)-carbonyl)-(μ(3)-methoxy*Methylidyne)-undecacarbonyl-tetra-iron <henylp 71. 58
hosphine)iminium (μ(2)-carbonyl)-(μ(3)-methoxy*Methylidyne)-undecacarbonyl-tetra-iron <henylp 71. 59
enylphosphonium bis(bis-μ-thio-dichloro-iron)-*Molybdenum .... [tetraph 64.129
trakis(μ(2)-sulfido)-tris(ethylthiolato-iron)-*Molybdenum <nium tris(μ-ethylthiolato)-bis(te 85. 62
-(dicarbonyl-iron-(μ(2)-carbonyl)-tricarbonyl-*Molybdenum)-bis(tricarbonyl-iron) <bonyl-iron) 3. 37
<n-butylammonium) (bis(tris(μ-benzylthiolato)-*Molybdenum)-tetrakis(μ-thiolato)-tris(benzyl 85. 96
<trimethylammonium) bis(tris(μ-ethylthiolato)-*Molybdenum)-tetrakis(μ-thiolato)-tris(ethylth> 85. 78
imethylammonium) dichloro-iron-bis(μ(2)-thio)-*Molybdenum-disulfide <ylphosphonium (benzyl-tr 64.123
<lphosphonium pentasulfur-iron-bis(μ-sulfido)-*Molybdenum-disulfide dimethylformamide solvate 64.130
<mmonium (μ-sulfido)-bis(μ-ethylthiolato)-bis(*Molybdenum-trisulfido-tris(ethylthio)-iron)) 85. 57
n]   (η(5)-cyclopentadienyl- Nickel)-(η-t-butylethynyl)-hexacarbonyl-di-iro 72. 29
(μ(3)-carbonyl)-bis(η(5)-cyclopentadienyl- Nickel)-bis(tricarbonyl-iron) 71.105
hex-3-ene-3,4-diyl)-bis(η(5)-cyclopentadienyl- Nickel)-bis(tricarbonyl-iron) .... [(71.178
-iron)]   *Nickel-bis(tetracarbonyl-μ-(dimethylcarbamoyl) 71. 63
hydrido)-octacarbonyl-tetrakis(μ(2)-carbonyl)-*Nickel-tri-iron <nzyl-trimethylammonium (μ(3)- 3. 43
pper(ii) perch> chloro-(meso-α,α,α,α-tetra(o- Nicotinamidophenyl)-porphyrinato-iron(iii))-co 83.225
cesium tris(ethyl Nitrosolato)-iron(ii) monohydrate 84. 9
anolamino)-(3,5-dimethylpyrazolyl)-gallato)-di Nitrosyl-iron ....[(dimethyl-(N,N-dimethyleth 84. 32
bis(μ-3,5-dimethylpyrazolyl)-bis(di*Nitrosyl-iron) 83. 52
-8-tricarbonyl-7-selena-8-ferra-bicyclo(4.3.0*Nonen-9-one)-tricarbonyl-iron .... [(μ(3) 71. 49
enylphosphino-ferrocenyl)ethyl-dimethylamine) Norbornadiene-rhodium hexafluorophosphate <iph 73.120
carbonyl-(exo-2-methoxy-5,6-dimethylene-syn-7- Norbornanol)-endo-iron .... [tri 72. 20
(exo-2-chloro-5,6-dimethylene-syn-7- Norbornanone)-endo-tricarbonyl-iron............ 72. 15
(6,7-methylene-exo-3-oxatricyclo(3.2.1.0(2,4)) Octane)-exo-iron .... [tricarbonyl- 72. 16
bonyl-(η(4)-1,5-dimethylene-2,6-dimethylcyclo- Octane)-iron ....[tricar 72. 24
hexacarbonyl-(μ-(1,2-eta:2-η)-1-cyclo-*Octene-1-selenolato)-di-iron 75. 7
tris(trimethylphosphite)-(η(3)-cyclo- Octenyl)-iron tetrafluoroborate............... 75. 19
tris(trimethylphosphite)-(η(3)-cyclo- Octenyl)-iron tetrafluoroborate............... 75. 18
tris(trimethylphosphite)-(η(3)-cyclo- Octenyl)-iron tetrafluoroborate............... 75. 17
(η(3)-cyclo- Octenyl)-tris(trimethylphosphite)-iron(i) 75. 16
(3,3,6-trimethyl-1-oxo-2- Oxa-heptyl-(4-6-η)-enyl)-tricarbonyl-iron 71. 40
<to)-iron(ii) ethylthiolate 4,7,13,16,21-penta Oxa-1,10-diazabicyclo(8.8.5)tricosane benzene> 49. 27
<enylporphinato)-iron(ii) 4,7,13,16,21,24-hexa Oxa-1,10-diazabicyclo(8.8.8)hexacosane chloro> 49. 23
adienyl-6-difluoro-2-isopropyl-4-methyl-1,5-di Oxa-3-ferra-6-borinane <bonyl-3-η(5)-cyclopent 71. 41
n]   η(4)-(2,3,5,6-tetrakis(methylene)-7- Oxabicyclo(2.2.1)heptane)-endo-tricarbonyl-iro 75. 5
carbonyl-iron) 1,10-diaza-4,7,13,16,21,24-hexa*Oxabicyclo(8.8.8)hexacosane <hosphido)-bis(tri 86.101
(3,3,6-trimethyl-1-oxo-2- Oxaheptyl-(4-6-η)-enyl)-tricarbonyl-iron 71. 39
tricarbonyl-(6,7-methylene-exo-3- Oxatricyclo(3.2.1.0(2,4))octane)-exo-iron ... 72. 16
bis(α-benzyldi Oximato)-bis(cyclohexyl-isonitrile)-iron(ii) 71.264
porphyrinato-iron eth> (2-methylimidazole)-di Oxygen-meso-tetra(α,α,α,α-o-pivalamidophenyl)- 49. 35
ylbenzylammonium) hexadecacarbonyl-tetra-iron- Palladium .... [bis(trimeth 3. 44
methylaminoethyl-ferrocene)-2-acetacetonato- Palladium .... [(-)-RS-cyclo-1-(1'-di 73. 80
um) tetracosa-carbonyl-hydrido-hexa-iron-hexa- Palladium acetonitrile solvate <ylbenzylammoni 71.190
adienyl)-(2-methyl-4,5-bis(diphenylphosphino)- Pent-2-en-3-yl)-iron(ii)   [(η(5)-cyclopent 71.243
tylpyridine)-(2,9-bis(1-methylhydrazone)-1,10- Phenanthroline)-iron bis(tetrafluoroborate) <e 83.165
-iron]   o- Phenylene-bis(dimethylarsine)-decacarbonyl-tri 86. 63
3,3',4,4'-tetramethyl-1,1'-di Phospha-ferrocene 75. 4
dienyl-(3-methylcyclohexenone)-iron hexafluoro Phosphate .... [dicarbonyl-cyclopenta 73. 48
lidene-(η(5)-cyclopentadienyl)-iron hexafluoro Phosphate ....[bis(carbonyl)-cycloheptatrieny 71. 62
imethylamine)-norbornadiene-rhodium hexafluoro Phosphate <iphenylphosphino-ferrocenyl)ethyl-d 73.120
-1,3,6,8,10-pentaen-5-ylidene)-iron hexafluoro Phosphate <opentadienyl)-(bicyclo(5.4.0)undeca 71.122
10-cyclotetradecatetraene)-iron(ii) hexafluoro Phosphate <tramethyl-1,4,8,11-tetra-aza-1,3,8, 83.110

tradeca-1,3,8,10-tetraene-iron(iii) hexafluoro Phosphate <tramethyl-1,4,8,11-tetra-azacyclote   83.195
-silver))-bis(tricarbonyl-iron) bis(hexafluoro*Phosphate) dichloromethane solvate <)-bis(aqua   73.133
dibenzo(b,f)-2,5-diphenyl-cyclopentadienone(c) Phosphepin-di-iron-hexacarbonyl.... [1-phenyl-   73.128
phenyl-ferra-cyclopentadiene-tricarbonyl)-(c)- Phosphepin-iron-dicarbonyl <benzo(b,f)-(2,5-di   73.124
ine)immonium (tricarbonyl-iron)-bis(μ-diphenyl*Phosphido)-(acetyl-dicarbonyl-iron) <nylphosph   71.232
fur>   sodium (tricarbonyl-iron)-bis(μ-diphenyl*Phosphido)-(acetyl-dicarbonyl-iron) tetrahydro   71.233
of>   lithium (tricarbonyl-iron)-bis(μ-diphenyl*Phosphido)-(benzoyl-dicarbonyl-iron) tetrahydr   71.245
                       (μ-chloro)-(μ-diphenyl*Phosphido)-bis(tricarbonyl-iron) 86. 48a
7,13,16,21,24-hexaoxab> sodium bis(μ-diphenyl*Phosphido)-bis(tricarbonyl-iron) 1,10-diaza-4,   86.101
               bis(μ(2)-hydrido)-(μ(3)-diphenyl*Phosphido)-tris(tricarbonyl-iron) 86. 36
                       (1,1'-ferrocenediyl)-phenyl Phosphine 73. 57
on(ii)]   dicarbonyl-methyl-bis(trimethyl Phosphine)-(triphenylborato-nitrilo-methyl)-ir   71.220
(μ(2)-sulfonyl)-bis((μ-methylthio)-(trimethyl*Phosphine)-dicarbonyl-iron(i)) diethyl ether >   86. 20
iron]   (μ(3)-phenyl*Phosphine)-hexacarbonyl-di-cobalt-tricarbonyl-   86. 34
              tetracarbonyl-(tris(dimethylamino) Phosphine)-iron.....................   86. 12
                 tricarbonyl-bis(tris(dimethylamino) Phosphine)-iron...................   86. 37
      dicarbonyl-(η(4)-cinnamaldehyde)-(triphenyl Phosphine)-iron(0) 72. 52
                            (triphenyl Phosphine)-tetracarbonyl-iron 86. 78
o)-hexacarbonyl-di-iron-hexaca> bis(triphenyl*Phosphine)iminium (μ(2)-carbonyl)-(μ(2)-hydrid   64.156
o)-tricarbonyl-iron-nonacarbon> bis(triphenyl*Phosphine)iminium (μ(2)-carbonyl)-(μ(2)-hydrid   64.155n
ymethylidyne)-undecacarbonyl-t> bis(triphenyl*Phosphine)iminium (μ(2)-carbonyl)-(μ(3)-methox   71. 59
ymethylidyne)-undecacarbonyl-t> bis(triphenyl*Phosphine)iminium (μ(2)-carbonyl)-(μ(3)-methox   71. 58
                    bis(bis(triphenyl Phosphine)iminium) tridecacarbonyl-tetra-iron   64.161
phenylphosphido)-(acetyl-dicar> bis(triphenyl*Phosphine)immonium (tricarbonyl-iron)-bis(μ-di   71.232
hoxyethylene)-bis((μ(2)-methylthio)-trimethyl*Phosphine-carbonyl-iron) tetraphenylborate <et   86. 82
                 (η(3)-allyl)-tricarbonyl-triphenyl Phosphine-gold>-iron 72. 46
  dichloro-bis(2-diphenylphosphinoethyl)-phenyl Phosphine-iron(ii) acetone solvate 86.217
<o)-bis(bis(2-(diphenylphosphino)ethyl)-phenyl*Phosphine-iron(ii)) perchlorate acetone solva>   86.218
dium)]   (μ(2)-dicarbonyl)-(μ(2)-tris(diphenyl*Phosphino))-tricarbonyl-iron-tris(carbonyl-rho   86.161
)-bis(tricarbonyl-iron)]   (μ-diphenyl*Phosphino)-(μ-N-methyl-(2-phenyl-2-iminoethyl)   71.211
-isopropylidene-2,3-dihydroxy-1,4-bis(diphenyl Phosphino)-butane>-iron <lopentadienyl-iodo-(0   73.123
         dichloro-bis(cis-1,2-bis(diphenyl Phosphino)-ethylene)-iron(ii) acetone solvate   86.187
)-cyclopentadienyl)-(2-methyl-4,5-bis(diphenyl Phosphino)-pent-2-en-3-yl)-iron(ii).... [(η(5   71.243
     carbonyl-cyclobutadiene-(1,2-bis(diphenyl Phosphino)ethane)-iron 75. 37
chlorat> tris(μ-mercapto)-bis(2-(diphenyl*Phosphino)ethyl)-phenylphosphine-iron(ii)) per   86.218
                      (+)-D-(1-(2-diphenyl Phosphino-ferrocenyl)ethyl)-dimethylamine 73.108a
ornadiene-rhodium hexafl> ((+-)-α-(2-diphenyl Phosphino-ferrocenyl)ethyl-dimethylamine)-norb   73.120
yl-(methoxy-thiocarbonyl)-trim> (μ-1-diphenyl*Phosphino-2-sulfido-3,3-dimethylbutene)-carbon   72. 49
ne solvate]   dichloro-bis(2-diphenyl Phosphinoethyl)-phenylphosphine-iron(ii) aceto   86.217
roborate]   tris(trimethyl Phosphite)-(η(3)-cyclo-octenyl)-iron tetrafluo   75. 19n
roborate]   tris(trimethyl Phosphite)-(η(3)-cyclo-octenyl)-iron tetrafluo   75. 17
roborate]   tris(trimethyl Phosphite)-(η(3)-cyclo-octenyl)-iron tetrafluo   75. 18n
           dicarbonyl-bis(triphenyl Phosphite)-(sulfur dioxide)-iron 86.135
           dicarbonyl-bis(tri-o-tolyl Phosphite)-(sulfur dioxide)-iron 86.164
       (η(3)-cyclo-octenyl)-tris(trimethyl Phosphite)-iron(i) 75. 16
      dicarbonyl-thiocarbonyl-bis(triphenyl Phosphite)-iron 86.142
ene)-carbonyl-(methoxy-thiocarbonyl)-trimethyl*Phosphite-iron <hino-2-sulfido-3,3-dimethylbut   72. 49
o-iron-bis(μ(2)-thio)-molybdenum> tetraphenyl*Phosphonium (benzyl-trimethylammonium) dichlor   64.123
bdenum]   tetraphenyl*Phosphonium bis(bis-μ-thio-dichloro-iron)-moly   64.129
lybdenum-disulfide dimethylforma> tetraphenyl*Phosphonium pentasulfur-iron-bis(μ-sulfido)-mo   64.130
ngsten-disulfide dimethylformami> tetraphenyl*Phosphonium pentasulfur-iron-bis(μ-sulfido)-tu   64.131
lvate]   carbonyl-(N,N-dimethylformamide)- Phthalocyaninato-iron(ii) dimethylformamide so   49. 5
lglyoximato)-iron(iii) tetrahydrate]   γ- Picolinium bis(S-methyldithiocarbazato-dimethy   85. 39
e]   tris(2- Picolylamine)-iron(ii) chloride ethanol solvat   83.122
e]   tris(2- Picolylamine)-iron(ii) chloride ethanol solvat   83.124
e]   tris(2- Picolylamine)-iron(ii) chloride ethanol solvat   83.125
e]   tris(2- Picolylamine)-iron(ii) chloride ethanol solvat   83.126
e]   tris(2- Picolylamine)-iron(ii) chloride ethanol solvat   83.123
te]   tris(2- Picolylamine)-iron(ii) chloride methanol solva   83.121
te]   tris(2- Picolylamine)-iron(ii) chloride methanol solva   83.119
te]   tris(2- Picolylamine)-iron(ii) chloride methanol solva   83.117
te]   tris(2- Picolylamine)-iron(ii) chloride methanol solva   83.118
te]   tris(2- Picolylamine)-iron(ii) chloride methanol solva   83.120
                                    tris(2- Picolylamine)-iron(ii) di-iodide 83.129
                                    tris(2- Picolylamine)-iron(ii) dichloride dihydrate....   83.127
                                    tris(2- Picolylamine)-iron(ii) dichloride dihydrate....   83.128
<thylimidazole)-dioxygen-meso-tetra(α,α,α,α-o- Pivalamidophenyl)-porphyrinato-iron ethanol s>   49. 35
l >   (2-methylimidazole)-meso-tetra(α,α,α,α-o- Pivalamidophenyl)-porphyrinato-iron(ii) ethano   49. 34
ylbenzylammonium) hexadecacarbonyl-tetra-iron- Platinum ....   [bis(trimeth   3. 45
                                             Poly(trivinyl-ferrocenoato-tin(iv))............   73. 67
      chloro-(N-methyl-5,10,15,20-tetraphenyl Porphinato)-iron(ii) 49. 19
1-pentao> disodium (ethylthio)-(tetra-p-tolyl Porphinato)-iron(ii) ethylthiolate 4,7,13,16,2   49. 27
,10-diazabicyc> sodium ethylthio-(tetraphenyl Porphinato)-iron(ii) 4,7,13,16,21,24-hexaoxa-1   49. 23
        potassium dicyano-(meso-tetraphenyl Porphinato)-iron(iii) acetone solvate   49. 22
                    meso-tetraphenyl Porphinato-bis(tetrahydrofuran)-iron(ii) 84.136

bis(3-chloropyridine)-octaethyl Porphinato-iron(iii) perchlorate 49. 24
octaethyl Porphinato-perchlorato-iron(iii) 49. 7
methoxy-(meso-tetraphenyl Porphyrinato)-iron(iii)...................... 49. 20
oxygen-meso-tetra(α,α,α,α-o-pivalamidophenyl)- Porphyrinato-iron ethanol solvate <idazole)-di 49. 35
azole)-meso-tetra(α,α,α,α-o-pivalamidophenyl)- Porphyrinato-iron(ii) ethanol solvate <hylimid 49. 34
<ro-(meso-α,α,α,α-tetra(o-nicotinamidophenyl)- Porphyrinato-iron(ii)-copper(ii) perchlorat> 83.225
-iron(iii) acetone solvate] Potassium dicyano-(meso-tetraphenylporphinato) 49. 22
tris(1,3-diphenyl-1,3- Propanedionato)-iron(iii) 77. 12
bis(μ-3,5-dimethyl*Pyrazolyl)-bis(dinitrosyl-iron)............... 83. 52
bis(hydro-tris(1- Pyrazolyl)-borato-iron(ii)................ 83.113
bis(hydrido-tris(3,5-dimethyl Pyrazolyl)-borato-iron(ii)................ 83.202
ethyl-(N,N-dimethylethanolamino)-(3,5-dimethyl Pyrazolyl)-gallato)-dinitrosyl-iron.... [(dim 84. 32
nanthroline)-iron bis(t> diaqua-(2,6-diacetyl Pyridine)-(2,9-bis(1-methylhydrazone)-1,10-phe 83.165
orate] bis(3-chloro Pyridine)-octaethylporphinato-iron(iii) perchl 49. 24
3-cyanopyridinium 3-cyano Pyridine-tetrachloro-iron(iii) 33. 17
i)] 3-cyano Pyridinium 3-cyanopyridine-tetrachloro-iron(ii 33. 17
diaqua-tetrakis(perhydro Pyrimidin-2-one)-iron trichloride dihydrate.... 84. 68
thane solvate] tris(Pyrrole-N-carbodithioato)-iron(iii) dichlorome 80. 14
decamethylferrocenium - 7,7,8,8-tetracyano Quinodimethane 60. 42
,1'-dimethylferrocenium bis(7,7,8,8-tetracyano Quinodimethane)....................... [1 60. 20
ferrocenyl)ethyl-dimethylamine)-norbornadiene- Rhodium hexafluorophosphate <iphenylphosphino- 73.120
nylphosphino)-tricarbonyl-iron-tris(carbonyl-*Rhodium   [(μ(2)-dicarbonyl)-(μ(2)-tris(diphe 86.161
tetraethylammonium (μ(5)-carbido)-(dicarbonyl-*Rhodium-(μ(2)-carbonyl)-dicarbonyl-iron)-tris> 3. 33
hydrido)-hexacarbonyl-di-iron-hexacarbonyl-di-*Ruthenium <hine)iminium (μ(2)-carbonyl)-(μ(2)- 64.156
2)-hydrido)-tricarbonyl-iron-nonacarbonyl-tri-*Ruthenium <osphine)iminium (μ(2)-carbonyl)-(μ(64.155
arbonyl-iron]   (μ(3)-8-tricarbonyl-7- Selena-8-ferra-bicyclo(4.3.0)nonen-9-one)-tric 71. 49
solvate] diferrocenyl- Selenium iodine tri-iodide methylene chloride 73. 82
exacarbonyl-(μ-(1,2-eta:2-η)-1-cyclo-octene-1-*Selenolato)-di-iron .... [h 75. 7
ntadienyl-iron] tri Sila-pentachloro-cyclohexyl-dicarbonyl-cyclope 73. 20
lopentadienyl-iron)] tri Sila-tetrachloro-cyclohexyl-bis(dicarbonyl-cyc 73. 66
tricarbonyl-(1,1-diphenyl-1- Silacyclohexa-2,4-dienyl)-iron 75. 24
<-cyclopentadienyl-dicarbonyl-iron)-cyclopenta Silane)-dimethylsilyl-η(5)-cyclopentadienyl-d> 73.103
arbonyl-iron] nonamethyl-cyclopenta Silane-dimethylsilyl-η(5)-cyclopentadienyl-dic 73. 78
<raphenyl-cyclopentadienediyl-oxo-0)-bis(aqua-*Silver))-bis(tricarbonyl-iron) bis(hexafluoro> 73.133
nonamethyl-cyclopentasilane-dimethyl Silyl-η(5)-cyclopentadienyl-dicarbonyl-iron.... 73. 78
yl-dicarbonyl-iron)-cyclopentasilane)-dimethyl Silyl-η(5)-cyclopentadienyl-dicarbonyl-iron <n 73.103
ron(ii) ethylthiolate 4,7,13,16,21-pentao> di Sodium (ethylthio)-(tetra-p-tolylporphinato)-i 49. 27
ido)-(acetyl-dicarbonyl-iron) tetrahydrofur> *Sodium (tricarbonyl-iron)-bis(μ-diphenylphosph 71.233
l-iron) 1,10-diaza-4,7,13,16,21,24-hexaoxab> *Sodium bis(μ-diphenylphosphido)-bis(tricarbony 86.101
ii) 4,7,13,16,21,24-hexaoxa-1,10-diazabicyc> Sodium ethylthio-(tetraphenylporphinato)-iron(49. 23
um) dichloro-iron-bis(μ(2)-thio)-molybdenum-di*Sulfide <ylphosphonium (benzyl-trimethylammoni 64.123
pentasulfur-iron-bis(μ-sulfido)-molybdenum-di*Sulfide dimethylformamide solvate <phosphonium 64.130
um pentasulfur-iron-bis(μ-sulfido)-tungsten-di*Sulfide dimethylformamide solvate <ylphosphoni 64.131
risulfido-tris(ethylt> tetraethylammonium (μ-*Sulfido)-bis(μ-ethylthiolato)-bis(molybdenum-t 85. 57
tetraphenylphosphonium pentasulfur-iron-bis(μ-*Sulfido)-molybdenum-disulfide dimethylformami> 64.130
tris(tetraethylammonium) tetrakis(μ(3)-*Sulfido)-tetra(benzylthiolato-iron)............ 85. 80
<nium tris(μ-ethylthiolato)-bis(tetrakis(μ(3)-*Sulfido)-tris(ethylthiolato-iron)-molybdenum) 85. 62
(μ(3)-carbonyl)-(μ(3)-*Sulfido)-tris(tricarbonyl-iron)............... 71. 16
tetraphenylphosphonium pentasulfur-iron-bis(μ-*Sulfido)-tungsten-disulfide dimethylformamide> 64.131
lfido)-bis(μ-ethylthiolato)-bis(molybdenum-tri*Sulfido-tris(ethylthio-iron)) <ammonium (μ-su 85. 57
thiocarbonyl-trim> (μ-1-diphenylphosphino-2-*Sulfido-3,3-dimethylbutene)-carbonyl-(methoxy- 72. 49
ne)-dicarbonyl-iron(i)) diethyl ether> (μ(2)-*Sulfonyl)-bis((μ-methylthio)-(trimethylphosphi 86. 20
chloro-pentakis(dimethyl Sulfoxide)-iron bis(tetrachloro-iron) 84. 27
cis-dichloro-tetrakis(dimethyl Sulfoxide)-iron chloride 84. 19
n)] chloro-pentakis(dimethyl*Sulfoxide)-iron(iii) (μ-oxo-bis(trichloro-iro 84. 28
dicarbonyl-bis(triphenylphosphite)-(Sulfur dioxide)-iron 86.135
dicarbonyl-bis(tri-o-tolylphosphite)-(Sulfur dioxide)-iron 86.164
e dimethylforma> tetraphenylphosphonium penta*Sulfur-iron-bis(μ-sulfido)-molybdenum-disulfid 64.130
dimethylformami> tetraphenylphosphonium penta*Sulfur-iron-bis(μ-sulfido)-tungsten-disulfide 64.131
adium acetoni> tris(trimethylbenzylammonium) Tetracosa-carbonyl-hydrido-hexa-iron-hexa-pall 71.190
iron-μ(2),η(5)-3,4-diethyl-2,5-dimethyl-1,2,5-*Thia-diborolene)-iron <(η(5)-cyclopentadienyl- 73.109
dicarbonyl-(1,4,7,10-tetra Thiadecane)-iron 85. 18
rbonyl-iron)] bis(μ-methyl*Thio)-(μ-tetrafluoroethane-1,1-diyl)-bis(trica 71. 17
rbonyl-iron)] bis(μ-methyl*Thio)-(μ-tetrafluoroethane-1,2-diyl)-bis(trica 71. 18
thiolate 4,7,13,16,21-pentao> disodium (ethyl Thio)-(tetra-p-tolylporphinato)-iron(ii) ethyl 49. 27
diethyl ether> (μ(2)-sulfonyl)-bis((μ-methyl*Thio)-(trimethylphosphine)-dicarbonyl-iron(i)) 86. 20
hio> tris(tetraethylammonium) tris(μ(2)-ethyl*Thio)-bis(tetrakis(μ(3)-thio)-tris(μ(3)-ethylt 3. 39
e-C',S(2))-1,1,1,2,2,2-hexa-carbonyl-(μ-methyl*Thio)-iron <rocenyl-methylmethane-thiomethylen 71.153
thiolato)-bis(molybdenum-trisulfido-tris(ethyl*Thio)-iron)) <ammonium (μ-sulfido)-bis(μ-ethyl 85. 57
zyl-trimethylammonium) dichloro-iron-bis(μ(2)-*Thio)-molybdenum-disulfide <ylphosphonium (ben 64.123
<2-bis(carbomethoxyethylene))-bis((μ(2)-methyl*Thio)-trimethylphosphine-carbonyl-iron) tetra> 86. 82
<nium) tris(μ(2)-ethylthio)-bis(tetrakis(μ(3)-*Thio)-tris(μ(3)-ethylthio-iron)-di-tungsten) 3. 39
6,21,24-hexaoxa-1,10-diazabicyc> sodium ethyl Thio-(tetraphenylporphinato)-iron(ii) 4,7,13,1 49. 23
tetraphenylphosphonium bis(bis-μ-*Thio-dichloro-iron)-molybdenum 64.129
,S(2))-1,1,1,2,2,2-hexa-carbonyl-(μ-meth> (μ-*Thio-ferrocenyl-methylmethane-thiomethylene-C' 71.153

lthio)-bis(tetrakis(μ(3)-thio)-tris(μ(3)-ethyl *Thio-iron)-di-tungsten) ⟨onium⟩ tris(μ(2)-ethy   3. 39
tris(pyrrole-N-carbodi Thioato)-iron(iii) dichloromethane solvate 80. 14
tris(N-methyl- Thiobenzohydroxamato)-iron(iii)..................   85. 74
bis(N,N-diethyldi Thiocarbamato)-(isothiocyanato)-iron(iii) 80. 11
tris(N,N-dibenzyldi Thiocarbamato)-iron(iii) 80. 29
tris(N,N-dibenzyldi Thiocarbamato)-iron(iii) 80. 28
trahydrate] γ-picolinium bis(S-methyldi Thiocarbazato-dimethylglyoximato)-iron(iii) te   85. 39
(R)-ferrocenyl-methylmethane-S-methyltri Thiocarbonate   73. 49a
on)]   (μ-carbonyl)-(μ- *Thiocarbonyl)-bis(carbonyl-cyclopentadienyl-ir   73. 46
sulfido-3,3-dimethylbutene)-carbonyl-(methoxy- *Thiocarbonyl)-trimethylphosphite-iron ⟨hino-2-   72. 49
dicarbonyl- Thiocarbonyl-bis(triphenylphosphite)-iron(0)   86.142
⟨hio)-(tetra-p-tolylporphinato)-iron(ii) ethyl Thiolate 4,7,13,16,21-pentaoxa-1,10-diazabicy⟩   49. 27
t⟩ tetraethylammonium (μ-sulfido)-bis(μ-ethyl *Thiolato)-bis(molybdenum-trisulfido-tris(ethyl   85. 57
lthiolat⟩ triethylbenzylammonium tris(μ-ethyl *Thiolato-bis(tetrakis(μ(2)-sulfido)-tris(ethy   85. 62
⟨(N-benzyl-trimethylammonium) bis(tris(μ-ethyl *Thiolato-molybdenum)-tetrakis(μ-thiolato)-tr⟩   85. 78
⟨kis(tetra-n-butylammonium) (bis(tris(μ-benzyl *Thiolato-molybdenum)-tetrakis(μ-thiolato)-tr⟩   85. 96
s(tris(μ-benzylthiolato)-tungsten)-tetrakis(μ- *Thiolato)-tris(benzylthiolato-iron))-iron ⟨(bi   85. 97
tris(μ-benzylthiolato)-molybdenum)-tetrakis(μ- *Thiolato)-tris(benzylthiolato-iron))-iron ⟨is(85. 96
(tris(μ-ethylthiolato)-molybdenum)-tetrakis(μ- *Thiolato)-tris(ethylthiolato-iron)-iron ⟨) bis   85. 78
⟨kis(tetra-n-butylammonium) (bis(tris(μ-benzyl *Thiolato)-tungsten)-tetrakis(μ-thiolato)-tris⟩   85. 97
lammonium) tetrakis(μ(3)-sulfido)-tetra(benzyl *Thiolato-iron) ....[tris(tetraethy   85. 80
to)-tungsten)-tetrakis(μ-thiolato)-tris(benzyl *Thiolato-iron))-iron ⟨(bis(tris(μ-benzylthiola   85. 97
)-molybdenum)-tetrakis(μ-thiolato)-tris(benzyl *Thiolato-iron))-iron ⟨is(tris(μ-benzylthiolato   85. 96
o)-molybdenum)-tetrakis(μ-thiolato)-tris(ethyl *Thiolato-iron)-iron ⟨) bis(tris(μ-ethylthiolat   85. 78
hiolato)-bis(tetrakis(μ(2)-sulfido)-tris(ethyl *Thiolato-iron)-molybdenum) ⟨nium tris(μ-ethylt   85. 62
yl-(μ-meth⟩ (μ-thio-ferrocenyl-methylmethane- *Thiomethylene-C',S(2))-1,1,1,2,2,2-hexa-carbon   71.153
bis(isothiocyanato)-bis(Thiourea)-iron(ii) 79. 5
cobalt)-bis(dicarbonyl-cyclopentadienyl-iron)- Tin(iv)...............[chloro-(tetracarbonyl-   73. 71
poly(trivinyl-ferrocenoato- Tin(iv))........................   73. 67
dicarbonyl-bis(tri-o- Tolylphosphite)-(sulfur dioxide)-iron 86.164
,16,21-pentao⟩ disodium (ethylthio)-(tetra-p- Tolylporphinato)-iron(ii) ethylthiolate 4,7,13   49. 27
4,7,13,16,21-pentaoxa-1,10-diazabicyclo(8.8.5) Tricosane benzene solvate ⟨(ii) ethylthiolate   49. 27
tricarbonyl-(6,7-methylene-exo-3-oxa Tricyclo(3.2.1.0(2,4))octane)-exo-iron 72. 16
bis(bis(triphenylphosphine)iminium) Tridecarbonyl-tetra-iron........................   64.161
rakis(μ(3)-thio)-tris(μ(3)-ethylthio-iron)-di- *Tungsten) ⟨onium⟩ tris(μ(2)-ethylthio)-bis(tet   3. 39
⟨n-butylammonium) (bis(tris(μ-benzylthiolato)- *Tungsten)-tetrakis(μ-thiolato)-tris(benzylthi⟩   85. 97
⟨lphosphonium pentasulfur-iron-bis(μ-sulfido)- *Tungsten-disulfide dimethylformamide solvate   64.131
hexakis(diethyl Urea)-iron(ii) perchlorate 79. 16
is(2-ethoxycarbonyl-2-trifluoromethyl-1-fluoro Vinyl)-iron.... [dicarbonyl-cyclopentadienyl-c   71. 51
poly(tri Vinyl-ferrocenoato-tin(iv))....................   73. 67
ron]   syn-(η(3)-2- Vinylyl-cyclopentyl-carboxylato)-tricarbonyl-i   71. 27
ron]   anti-(η(3)-2- Vinylyl-cyclopentyl-carboxylato)-tricarbonyl-i   71. 28
(2,2'-bipyridyl)- Zinc-tetracarbonyl-iron.....................   83.190

Lanthanum

trinitrato-(2,6-diacetylpyridine-bis(Benzoic acid hydrazone))-lanthanum(iii)........   84. 93
octadecane)-lanthanum(iii) perchl⟩ aqua-diper Chlorato-(1,4,10,13-tetraoxa-7,16-dithiacyclo-   85. 42
⟨diperchlorato-(1,4,10,13-tetraoxa-7,16-dithia Cyclo-octadecane)-lanthanum(iii) perchlorate   85. 42
itrato-(2,6-diacetylpyridine-bis(benzoic acid Hydrazone))-lanthanum(iii) .... [trin   84. 93
hydrazone))-lanthanum(iii)]   tri Nitrato-(2,6-diacetylpyridine-bis(benzoic acid   84. 93
chlorato-(1,4,10,13-tetraoxa-7,16-dithiacyclo- Octadecane)-lanthanum(iii) perchlorate ⟨-diper   85. 42
) perchl⟩ aqua-diperchlorato-(1,4,10,13-tetra Oxa-7,16-dithiacyclo-octadecane)-lanthanum(iii   85. 42
m(iii)]   trinitrato-(2,6-diacetyl Pyridine-bis(benzoic acid hydrazone))-lanthanu   84. 93
aqua-diperchlorato-(1,4,10,13-tetraoxa-7,16-di Thiacyclo-octadecane)-lanthanum(iii) perchlor⟩   85. 42

Lead

trimethyl-lead-diazo-ethyl Acetate...   69. 15
bis(o-phenylene)-bis(pyridine-2,6-di Aldimino) lead(ii) diperchlorate dihydrate 34. 11
trimethyl-lead-di Azo-ethylacetate 69. 15
bis(p-amino Benzoato)-lead(ii) 69. 26
nitrato-p-amino Benzoato-lead(ii) 69. 14
Nitrato-p-aminobenzoato-lead(ii) 69. 14
i) diperchlorate dihydrate]   bis(o- Phenylene)-bis(pyridine-2,6-dialdimino) lead(i   34. 11
(5,10,15,20-tetra-n-propyl Porphinato)-lead(ii) 49. 3
e dihydrate] bis(o-phenylene)-bis(Pyridine-2,6-dialdimino) lead(ii) diperchlorat   34. 11
-2,6-dicarboxylic acid solvate ⟩ bis(diaquo-(Pyridine-2,6-dicarboxylato)-lead(ii)) pyridine   69. 29
triphenyl-(Pyridine-4-thiolato)-lead 69. 47
bis(dimethyl Sulfoxide)-di-iodo-lead(ii).................   69. 5
bis(N,N-dimethyldi Thiocarbamato)-lead(ii)...................   69. 11
bis(N,N-di-isopropyldi Thiocarbamato)-lead(ii)...................   69. 30
triphenyl-(pyridine-4- Thiolato)-lead 69. 47

Lutetium

tris(di Cyclohexyl-dithiophosphinato)-lutetium(iii).... 85. 88
tris(dicyclohexyl-di Thiophosphinato)-lutetium(iii) 85. 88

Manganese

anese)]   hexakis(μ(2)-*Acetato)-(μ(3)-oxo)-tris(3-chloropyridine-mang   81. 84
tetra-aqua-(μ(3)-dodeca-oxo)-(μ(2)-hexadeca-*Acetato)-dodeca-manganese acetic acid solvate>   81. 88
potassium ethylenediamine-tetra- Acetato-manganese(iii) dihydrate 76. 56
diaquo- Adipato-manganese 81. 22
ate]   bis(D,L-α- Alanine)-diaqua-manganese(ii) dibromide dihydr   82. 12
opentadienyl-manganese)]   (μ(2)-*Allene)-(μ(2)-carbonyl)-bis(carbonyl-η(5)-cycl   71. 96
manganese-((μ(2)-acetyl)-(μ(2)-bromo)-dibromo-*Aluminium) ....[tetracarbonyl-   71. 3
(tris(2-(5'-chloro-salicylaldiminato)ethyl) Amine)-manganese(iii) trihydrate 78. 11
methyl Ammonium diaqua-trichloro-manganese............   3. 1
1,4-butanediyl-di Ammonium tetrachloro-manganese 3. 28
1,4-butanediyl-di Ammonium tetrachloro-manganese 3. 27
bis(n-propyl Ammonium) tetrachloro-manganese(ii)............   3. 13
o-tris(cyclopentadienyl-dicarbonyl-manganese)- Antimony ....[chlor   73. 89
<rbonyl-(μ-dimethylarsenido)-(μ-tetramethyl-di*Arsenic)-(tricarbonyl-chloro-manganese)-manga>   86. 31
yl-chloro-manganes>   tetracarbonyl-(μ-dimethyl*Arsenido)-(μ-tetramethyl-diarsenic)-(tricarbon   86. 31
on)-manganese tetracarbonyl-chloro-(μ-dimethyl*Arsenido)-(tetracarbonyl-iron)-manganese <l-ir   86. 18
carbonyl-chloro> tetracarbonyl-bis(μ-dimethyl*Arsenido)-(tetracarbonyl-iron)-manganese tetra   86. 18
acarbony>   (2-methylene-1-tetracarbonyl-2,3-di Aza-1-mangana-cyclobutan-4-one)-manganese pent   71. 25
,4-bis(p-chlorobenzyl)-2,5-bis(p-tolyl)-2,5-di Aza-1-mangana-penta-2,4-diene) <-tricarbonyl-3   75. 41
) diperchlor>   (2,12-bis(2-pyridyl)-3,7,11-tri Azatridecane-2,11-diene)-nitrosyl-manganese(ii   83.149
(μ-methanedi*Azo)-(N,N,N')-tris(tetracarbonyl-manganese)....   83. 86
tris(N-methyl-thio Benzohydroxamato)-manganese(ii)...............   85. 75
<l-manganese-η-(1-tricarbonyl-3,4-bis(p-chloro Benzyl)-2,5-bis(p-tolyl)-2,5-diaza-1-mangana->   75. 41
akis(dicarbonyl-cyclopentadienyl-manganese)-di*Bismuth   [bis(μ(2)-chloro)-tetr   73.110
1,4- Butanediyl-diammonium tetrachloro-manganese....   3. 28
1,4- Butanediyl-diammonium tetrachloro-manganese....   3. 27
<lphosphine)-platinum)-(μ-carbonyl)-(μ-p-tolyl*Carbenyl)-(carbonyl-cyclopentadienyl-manganes>   71.173
lorate]   1,4,7,10,13,16-hexaoxa Cyclo-octadecane hexa-aquo-manganese(ii) perch   38. 48
arbonyl-(η(5)-methylcyclopentadienyl)-ma>   (μ-*Cyclobuta(1,2-a:3,4-a')dicyclopentene)-bis(dic   71.202
methylene-1-tetracarbonyl-2,3-diaza-1-mangana- Cyclobutan-4-one)-manganese pentacarbonyl [(2-   71. 25
bis(1,4,7,10-tetraoxa Cyclododecane)-manganese(ii) tribromide........   84. 66
,3,3,-tetracarbonyl-5-phenyl-1-oxa-3-mangana- Cyclopenta-2,4-dione <enylphosphine)immonium 3   71. 45
enyl)->   tetracarbonyl-manganese-(μ-η(5):η(1)-*Cyclopentadienyl)-(carbonyl-(η(5)-cyclopentadi   71. 85
-bis(dimethylphosphane)-bis(dicarbonyl-(η(5)-methyl*Cyclopentadienyl)-manganese(i))........[(μ-oxy   73. 76
(μ-dinitrogen)-bis(dicarbonyl-(η(5)-methyl*Cyclopentadienyl)-manganese)   73. 58
a')dicyclopentene)-bis(dicarbonyl-(η(5)-methyl*Cyclopentadienyl)-manganese) <obuta(1,2-a:3,4-   71.202
μ-η(5):η(1)-cyclopentadienyl)-(carbonyl-(η(5)-*Cyclopentadienyl)-molybdenum) <nyl-manganese-(71. 85
anese(ii)]   iodo-(η(5)-dicyclopentadienyl)-nitrosyl-(thiocarbonyl)-mang   73. 7
y]   chloro-tris(Cyclopentadienyl-dicarbonyl-manganese)-antimon   73. 89
(μ-ethylene-1,1-diyl)-bis(dicarbonyl-*Cyclopentadienyl-manganese)...................   71. 94
(2)-allene)-(μ(2)-carbonyl)-bis(carbonyl-η(5)-*Cyclopentadienyl-manganese).................[(μ   71. 96
<)-(μ-carbonyl)-(μ-p-tolylcarbenyl)-(carbonyl-*Cyclopentadienyl-manganese) tetrafluoroborate>   71.173
etracarbonyl-rhenium)]   (dicarbonyl-*Cyclopentadienyl-manganese)-μ-phenylketenyl-(t   71.136
bis(μ(2)-chloro)-tetrakis(dicarbonyl-*Cyclopentadienyl-manganese)-dibismuth   73.110
tetracarbonyl-manganese diethyl >   dicarbonyl-*Cyclopentadienyl-manganese-(μ-p-tolylketenyl)-   71.149
methoxycarbonyl)-1,3-dithiol->   carbonyl-η(5)- Cyclopentadienyl-trimethoxyphosphine-(4,5-bis(71. 99
pentadienyl)-ma>   (μ-cyclobuta(1,2-a:3,4-a')di*Cyclopentene)-bis(dicarbonyl-(η(5)-methylcyclo   71.202
μ(2)-iodo)-(methyl-di-t-butylphosphine)-(2-oxa*Cyclopentylidene)-platinum-manganese <rbonyl-(71.111
(S)-1-(N,N-dimethylaminomethyl)-2-formyl- Cymantrene   73. 31
bis(1,4,7,10-tetraoxacyclo Dodecane)-manganese(ii) tribromide............   84. 66
bis(Ethanol)-bis(thiocyanato)-manganese(ii)........   84. 10
enyl-manganese)]   (μ-*Ethylene-1,1-diyl)-bis(dicarbonyl-cyclopentadi   71. 94
ihydrate]   potassium Ethylenediamine-tetra-acetato-manganese(iii) d   76. 56
(S)-1-(N,N-dimethylaminomethyl)-2- Formyl-cymantrene   73. 31
bis(μ(2)-chloro)-bis(tetrahydro*Furan)-hexacarbonyl-di-manganese 84. 54
acarbonyl-ma>   μ-2-sigma:2-3-η-(4,5-dihydro-2-*Furyl)-(bis(trimethylphosphine)-platinum)-tetr   71. 76
acarbonyl-ma>   μ-2-sigma:2-3-η-(4,5-dihydro-2-*Furyl)-(bis(trimethylphosphine)-platinum)-tetr   71. 75
catena-diaqua-dichloro-(μ-*Glycine)-manganese(ii) 82. 3
diaqua-bis(Glycollato)-manganese(ii)   84. 5
solvate>   tetra-aqua-(μ(3)-dodeca-oxo)-(μ(2)-*Hexadeca-acetato)-dodeca-manganese acetic acid   81. 88
nonacarbonyl-(μ-(methylene*Hydrazine-carbonyl))-di-manganese   71. 25
se >   (2,6-diacetylpyridine)-(2,9-bis(1-methyl Hydrazone)-1,10-phenanthroline)-chloro-mangane   83.163
tris(N-methyl-thiobenzo Hydroxamato)-manganese(ii) 85. 75
tetra- Imidazolyl-tri-manganese   83.174
3-mangana-cyclopenta>   bis(triphenylphosphine) Immonium 3,3,3,3-tetracarbonyl-5-phenyl-1-oxa-   71. 45
tricarbonyl-(1-4:9-η-1-bromo- Indenyl)-manganese   73. 26
yl-chloro-(μ-dimethylarsenido)-(tetracarbonyl-*Iron)-manganese <l-iron)-manganese tetracarbon   86. 18
<bonyl-bis(μ-dimethylarsenido)-(tetracarbonyl-*Iron)-manganese tetracarbonyl-chloro-(μ-dimet>   86. 18

ricarbonyl)-manganese(i)]   bis(μ-N-*Isopropylidene-S-methyl-dithiocarbazato)-bis(t   85. 51
2,4-triazole)-bis(4-methyl-1,2,4-triazole-bis(*Isothiocyanato)-manganese(ii)) ⟨(μ-4-methyl-1,   83.138
⟨arbonyl-cyclopentadienyl-manganese-(μ-p-tolyl*Ketenyl)-tetracarbonyl-manganese diethyl ethe⟩   71.149
icarbonyl-cyclopentadienyl-manganese)-μ-phenyl*Ketenyl-(tetracarbonyl-rhenium)..............[(d   71.136
                        potassium tetrakis(Malonato)-bis(methanol)-di-manganese(iii) ...   81. 57
onacarbonyl-bis(μ(3)-ethoxy)-1,2-(μ-iodo)-tri-*Manganese   [1,1,1,2,2,2,3,3,3-n   84. 46
acarbonyl-bis(μ(3)-ethoxy)-1,2-(μ-fluoro)-tri-*Manganese .... [1,1,1,2,2,2,3,3,3-non   84. 45
anese)]   (μ-*Methanediazo)-(N,N,N')-tris(tetracarbonyl-mang   83. 86
            potassium tetrakis(malonato)-bis(Methanol)-di-manganese(iii).....................   81. 57
cyclobutan-4-one)-manganese pentacarbony⟩ (2- Methylene-1-tetracarbonyl-2,3-diaza-1-mangana-   71. 25
                      nonacarbonyl-(μ-(*Methylenehydrazine-carbonyl))-di-manganese   71. 25
entadienyl)-(carbonyl-(η(5)-cyclopentadienyl)-*Molybdenum ⟨nyl-manganese-(μ-η(5):η(1)-cyclop   71. 85
          bis(triethylphosphine)-dinitrosyl- Nitrito-manganese 86. 24
adienyl)-manganese)]   (μ-di*Nitrogen)-bis(dicarbonyl-(η(5)-methylcyclopent   73. 58
      iodo-(η(5)-methylcyclopentadienyl)- Nitrosyl-(thiocarbonyl)-manganese(ii) 73. 7
                  triphenylphosphine-tri Nitrosyl-manganese 86. 49
2-pyridyl)-3,7,11-triazatridecane-2,11-diene)- Nitrosyl-manganese(ii) diperchlorate ⟨,12-bis(83.149
              bis(triethylphosphine)-di Nitrosyl-nitrito-manganese 86. 24
nese(i) tetrafluoroborate]   di Nitrosyl-tris(dimethylphenylphosphonito)-manga   86. 87
               1,4,7,10,13,16-hexaoxacyclo- Octadecane hexa-aquo-manganese(ii) perchlorate   38. 48
ine)immonium 3,3,3,3-tetracarbonyl-5-phenyl-1- Oxa-3-mangana-cyclopenta-2,4-dione ⟨enylphosph   71. 45
rchlorate]   1,4,7,10,13,16-hexa Oxacyclo-octadecane hexa-aquo-manganese(ii) pe   38. 48
               bis(1,4,7,10-tetra Oxacyclododecane)-manganese(ii) tribromide 84. 66
l-(μ(2)-iodo)-(methyl-di-t-butylphosphine)-(2-*Oxacyclopentylidene)-platinum-manganese ⟨rbony   71.111
                      potassium tris(Oxalato)-manganese(iii) trihydrate   81. 26
⟨ylpyridine)-(2,9-bis(1-methylhydrazone)-1,10- Phenanthroline)-chloro-manganese tetrafluorob⟩   83.163
cis,trans-(dicarbonyl-bis(trimethylphosphite)- Phenanthroline)-manganese· perchlorate   86. 67
   cis,cis-dicarbonyl-bis(trimethylphosphite)- Phenanthroline-manganese perchlorate 83.148
nyl)-manganese(i))]   (μ-oxy-bis(dimethyl*Phosphane))-bis(dicarbonyl-(η(5)-cyclopentadie   73. 76
   carbonyl-platinum-bis((μ-diphenyl*Phosphido)-(tetracarbonyl-manganese)) 86.111
tetracarbonyl-(μ(2)-iodo)-(methyl-di-t-butyl*Phosphine)-(2-oxacyclopentylidene)-platinum-m   71.111
1,1,1,2,2,2,3,3-octacarbonyl-(dimethylphenyl*Phosphine)-bis(μ(3)-ethoxy)-1,2-(μ-ethoxy)-tr⟩   86. 80
          bis(triethyl Phosphine)-dinitrosyl-nitrito-manganese........   86. 24
rbenyl)-(carbonyl-cyclopentad⟩ (bis(trimethyl*Phosphine)-platinum)-(μ-carbonyl)-(μ-p-tolylca   71.173
⟨ma:2-3-η-(4,5-dihydro-2-furyl)-(bis(trimethyl*Phosphine)-platinum)-tetracarbonyl-manganese   71. 75
⟨ma:2-3-η-(4,5-dihydro-2-furyl)-(bis(trimethyl*Phosphine)-platinum)-tetracarbonyl-manganese   71. 76
nyl-1-oxa-3-mangana-cyclopenta⟩ bis(triphenyl Phosphine)immonium 3,3,3,3-tetracarbonyl-5-phe   71. 45
l-⟩ carbonyl-η(5)-cyclopentadienyl-trimethoxy Phosphine-(4,5-bis(methoxycarbonyl)-1,3-dithio   71. 99
                 triphenyl Phosphine-trinitrosyl-manganese.................   86. 49
         tetracarbonyl-(3-(diphenyl Phosphineoxy)propyl)-manganese 71.139
        tetraphenylphosphonium triphenyl Phosphino-tetracarbonyl-manganese 86. 79
te]   cis,trans-(dicarbonyl-bis(trimethyl Phosphite)-phenanthroline)-manganese perchlora   86. 67
e]   cis,cis-dicarbonyl-bis(trimethyl Phosphite)-phenanthroline-manganese perchlorat   83.148
tris(tricarbonyl-manganese)-nonakis(μ-diethoxy*Phosphite)-tri-manganese 86.170
         dinitrosyl-tris(dimethylphenyl Phosphonito)-manganese(i) tetrafluoroborate....   86. 87
anganese]   tetraphenyl Phosphonium triphenylphosphino-tetracarbonyl-m   86. 79
                             Phthalocyaninato-manganese(ii) 49. 2
                     dichloro-bis(Picolinehydrazide)-manganese 83. 77
rbonyl-cyclopentad⟩ (bis(trimethylphosphine)-*Platinum)-(μ-carbonyl)-(μ-p-tolylcarbenyl)-(ca   71.173
4,5-dihydro-2-furyl)-(bis(trimethylphosphine)-*Platinum)-tetracarbonyl-manganese ⟨gma:2-3-η-(71. 75
4,5-dihydro-2-furyl)-(bis(trimethylphosphine)-*Platinum)-tetracarbonyl-manganese ⟨gma:2-3-η-(71. 76
nyl-manganese)))]   carbonyl-*Platinum-bis((μ-diphenylphosphido)-(tetracarbo   86.111
-di-t-butylphosphine)-(2-oxacyclopentylidene)-*Platinum-manganese ⟨rbonyl-(μ(2)-iodo)-(methyl   71.111
ese(iii) dihydrate]   Potassium ethylenediamine-tetra-acetato-mangan   76. 56
manganese(iii)]   Potassium tetrakis(malonato)-bis(methanol)-di-   81. 57
ate]   Potassium tris(oxalato)-manganese(iii) trihydr   81. 26
nanthroline)-chloro-manganese ⟩ (2,6-diacetyl Pyridine)-(2,9-bis(1-methylhydrazone)-1,10-phe   83.163
hexakis(μ(2)-acetato)-(μ(3)-oxo)-tris(3-chloro*Pyridine-manganese).................   81. 84
trosyl-manganese(ii) diperchlor⟩ (2,12-bis(2- Pyridyl)-3,7,11-triazatridecane-2,11-diene)-ni   83.149
do-manganese]   bis(σ- Pyrrolyl-tricarbonyl-manganese)-tricarbonyl-io   71.104
nyl-manganese)-μ-phenylketenyl-(tetracarbonyl-*Rhenium).............. [(dicarbonyl-cyclopentadie   71.136
trihydrate]   (tris(2-(5'-chloro- Salicylaldiminato)ethyl)amine)-manganese(iii)   78. 11
               tris(N-methyl- Thiobenzohydroxamato)-manganese(ii)............   85. 75
ganese(i) bromide] tricarbonyl-(S-methyldi*Thiocarbazato)-(μ-S-methyldithiocarbazato)-man   85. 16
   bis(μ-N-isopropylidene-S-methyl-di*Thiocarbazato)-bis(tricarbonyl)-manganese(i)   85. 51
rbonyl-(S-methyldithiocarbazato)-(μ-S-methyldi*Thiocarbazato)-manganese(i) bromide.... [trica   85. 16
   iodo-(η(5)-methylcyclopentadienyl)-nitrosyl-(Thiocarbonyl)-manganese(ii)   73. 7
hoxyphosphine-(4,5-bis(methoxycarbonyl)-1,3-di Thiol-2-ylidene)-manganese(i) ⟨tadienyl-trimet   71. 99
tricarbonyl-3,4-bis(p-chlorobenzyl)-2,5-bis(p- Tolyl)-2,5-diaza-1-mangana-penta-2,4-diene) ⟨-   75. 41
⟨methylphosphine)-platinum)-(μ-carbonyl)-(μ-p-*Tolylcarbenyl)-(carbonyl-cyclopentadienyl-man⟩   71.173
     dicarbonyl-cyclopentadienyl-manganese-(μ-p-*Tolylketenyl)-tetracarbonyl-manganese diethyl   71.149
hiocyanato)-manganese⟩ tris(μ-4-methyl-1,2,4-*Triazole)-bis(4-methyl-1,2,4-triazole-bis(isot   83.138
μ-4-methyl-1,2,4-triazole)-bis(4-methyl-1,2,4-*Triazole-bis(isothiocyanato)-manganese(ii)) ⟨(83.138
iperchlor⟩ (2,12-bis(2-pyridyl)-3,7,11-triaza Tridecane-2,11-diene)-nitrosyl-manganese(ii) d   83.149
            dibromo-tris(N,N'-dimethyl Urea)-manganese(ii)..........................   79. 7

Mercury

dibromo-(N-(2-pyridyl) Acetamide)-mercury(ii) 84. 15
phenylpyridine-mercury(ii) trifluoro Acetate........................ 71. 29
imethylaminomethyl)-phenyl)-platin> rac-a-(μ-*Acetato)-b-(O-acetato-mercurio)-cf,de-bis(2-(d 71.183
rcury(ii))] bis(μ-*Acetato)-bis(acetato-(tri-o-tolylphosphine)-me 81. 96
mercury(ii))] bis(μ-*Acetato)-bis(acetato-(tricyclohexylphosphine)- 81. 94
bis(μ-acetato)-bis(*Acetato-(tri-o-tolylphosphine)-mercury(ii)).... 81. 96
bis(μ-acetato)-bis(*Acetato-(tricyclohexylphosphine)-mercury(ii)) 81. 94
) dihydrate] di Acetato-bis(tricyclohexylphosphine)-mercury(ii 81. 93
thyl)-phenyl)-platin> rac-a-(μ-acetato)-b-(O-*Acetato-mercurio)-cf,de-bis(2-(dimethylaminome 71.183
diphospha-dimercura-cyclododecan-6,14-dioxide Acetic acid <k)(1,7,3,9,2,8,4,10)-dioxa-diaza- 83.220
)] bis((μ(2)-bromo)-((μ(2)-1-t-butyl*Acetylene)-nonacarbonyl-tri-ruthenium)-mercury 71.228
rchlorate] (μ-*Adeninato)-3,7,9-tris(methyl-mercury(ii)) dipe 83. 35
1-(methyl-mercury(ii))-9-methyl- Adenine nitrate 71. 5
benzyl Ammonium bis(dichloro-mercury) chloride........ 19. 2
7α-(thio-bromo-mercury)-(17R)-spiro(Androst-4-ene-17,2(3H)-furan) 85. 69
<henoxydibenzo(e,k)(1,7,3,9,2,8,4,10)-dioxa-di Aza-diphospha-dimercura-cyclododecan-6,14-dio> 83.220
ercury(ii)] chloro-(1,4,8,11-tetra- Azacyclotetradecane)-mercury(ii) tetrachloro-m 83. 64
(2'-chlorophenyl-di Azoaminobenzene)-N-(phenyl-mercury)............ 71.124
bis(Benzamido)-mercury(ii) 83. 91
(2'-chlorophenyl-diazoamino Benzene)-N-(phenyl-mercury). 71.124
tri Benzo(b,e,h)(1,4,7)trimercuronin 71.121
<phinyl)-6,7,14,15-tetrahydro-6,14-diphenoxydi Benzo(e,k)(1,7,3,9,2,8,4,10)-dioxa-diaza-diph> 83.220
)] Benzyl-mercury-chloride........................ 71. 4
(N- Benzyl-piperazinium)-pentachloro-di-mercury(ii 83. 65
(2- Benzyl-pyridine)-methyl-mercury(ii) nitrate... 71. 53
Benzylammonium bis(dichloro-mercury) chloride 19. 2
2β-(bromo-mercury-methyl)'-3,4-dioxa Bicyclo(4.4.0)decane 71. 14
bis(2,2'- Bipyridyl)-mercury(ii) dinitrate dihydrate 83.143
)-molybdenum(ii)] (2,2'- Bipyridyl)-tricarbonyl-chloro-(chloro-mercurio 71. 48
α-chloro-mercury- Camphene 71. 22a
diphenyl Carbazone-bis(chloro-mercury) 83. 87
bis(μ-chloro)-bis(per*Chlorato-(tri-o-tolylphosphine)-mercury(ii)) 86.158
ro)-bis(mercury-(μ-p-tolyl(ethyl)triazenido)-(*Cyclo-octadiene-iridium)) ....[bis(μ-chlo 75. 40
,8-octamethyl-2,4,6,8-tetrasila-1,5-dimercura- Cyclo-octane .... [2,2,4,6,6,8 63. 11
3,9,2,8,4,10)-dioxa-diaza-diphospha-dimercura- Cyclododecan-6,14-dioxide acetic acid <k)(1,7, 83.220
trans-1,2-dimercapto- Cyclohexane-bis(methyl-mercury(ii))........... 85. 21
(ii)] (μ-trans-*Cyclohexane-1,2-dithiolato)-bis(methyl-mercury 85. 21
germyl-mercury)-nickel)-mercury t> bis((η(5)- Cyclopentadienyl)-(triphenylgermyl)-(triphenyl 73.135
rcury(ii)] trans-bis(3-dimethylsulfonio-*Cyclopentadienylide)-bis(μ-iodo)-di-iodo-di-me 71. 71
ury(ii)] chloro-(1,4,8,11-tetra-aza Cyclotetradecane)-mercury(ii) tetrachloro-merc 83. 64
hloro-mercury-4-acetoxy-tricyclo(4.2.2.0(2,5)) Dec-7-ene-9,10-cis-endo-dicarboxylate <-endo-c 31. 37
(bromo-mercury-methyl)-3,4-dioxabicyclo(4.4.0) Decane .... [2β- 71. 14
eo-1-bromomercuri-2-t-butylperoxy-1,2-diphenyl Ethane .... [thr 71.129
dichloro-(1-diphenylphosphino-2-diethylamino Ethane)-mercury(ii) 86. 50
o-mercury)-(17R)-spiro(androst-4-ene-17,2(3H)- Furan) .... [7α-(thio-brom 85. 69
ury t> bis((η(5)-cyclopentadienyl)-(triphenyl Germyl)-(triphenylgermyl-mercury)-nickel)-merc 73.135
<yclopentadienyl)-(triphenylgermyl)-(triphenyl Germyl-mercury)-nickel)-mercury toluene solva> 73.135
(9-methyl Guanine)-methyl-mercury(ii) nitrate............ 83. 30
Imidazoline-1-oxy-4-yl)phenyl)-mercury 71.207
bis(o-(2,2,5,5-tetramethyl-Δ(3)- Imidazoline-1-oxy-4-yl)phenyl)-mercury 71.207
(μ-p-tolyl(ethyl)triazenido)-(cyclo-octadiene-*Iridium))........[bis(μ-chloro)-bis(mercury- 75. 40
trans-1,2-di Mercapto-cyclohexane-bis(methyl-mercury(ii)) 85. 21
bis(2,3,4,5-tetrafluorophenyl)- Mercury. 71. 34
dinitrato-(tri Mesitylphosphine)-mercury(ii) , ... 86. 93
tetrakis(acetoxy-mercury) Methane dihydrate 71. 12
methyl-(bis(2-pyridyl) Methane-mercury nitrate 71. 37
pyridyl)-tricarbonyl-chloro-(chloro-mercurio)- Molybdenum(ii) .... [(2,2'-bi 71. 48
)-(triphenylgermyl)-(triphenylgermyl-mercury)- Nickel)-mercury toluene solvate <lopentadienyl 73.135
dichloro Nicotine-mercury(ii) 83. 50
di Nitrato-(trimesitylphosphine)-mercury(ii) 86. 93
s(mercury-(μ-p-tolyl(ethyl)triazenido)-(cyclo-*Octadiene-iridium))........ [bis(μ-chloro)-bi 75. 40
amethyl-2,4,6,8-tetrasila-1,5-dimercura-cyclo-*Octane .... [2,2,4,4,6,6,8,8-oct 63. 11
<14-diphenoxydibenzo(e,k)(1,7,3,9,2,8,4,10)-di Oxa-diaza-diphospha-dimercura-cyclododecan-6,> 83.220
2β-(bromo-mercury-methyl)-3,4-di Oxabicyclo(4.4.0)decane....................... 71. 14
(3,4,7,8-tetramethyl-1,10- Phenanthroline)-(bis(trichlorovinyl))-mercury 71.152
bis(3-chloropropynyl)-(1,10- Phenanthroline)-mercury...................... 71.120
lsulfonate) ethanol solvate] tris(1,10- Phenanthroline)-mercury(ii) bis(trifluoromethy 83.210
<enoxyphosphinyl)-6,7,14,15-tetrahydro-6,14-di Phenoxydibenzo(e,k)(1,7,3,9,2,8,4,10)-dioxa-d> 83.220
iphenoxydibenzo(e,k)(1,7,3,9,2,8> 7,15-bis(di Phenoxyphosphinyl)-6,7,14,15-tetrahydro-6,14-d 83.220
o- Phenylene-mercury trimer 71.121
<dibenzo(e,k)(1,7,3,9,2,8,4,10)-dioxa-diaza-di Phospha-dimercura-cyclododecan-6,14-dioxide a> 83.220
dinitrato-(trimesityl Phosphine)-mercury(ii) 86. 93
bis(thiocyanato)-(triphenyl Phosphine)-mercury(ii) 86. 62
hosphine)-mercury(ii) trichloro-(ethyldimethyl Phosphine)-mercury(ii) <oro-bis(ethyldimethylp 86. 7

Iphosphine)-mercury>   chloro-bis(ethyldimethyl Phosphine)-mercury(ii) trichloro-(ethyldimethy   86. 7
          bis(μ-acetato)-bis(acetato-(tri-o-tolyl*Phosphine)-mercury(ii))........................ 81. 96
          bis(μ-chloro)-bis(perchlorato-(tri-o-tolyl*Phosphine)-mercury(ii)) 86.158
                dichloro-(1-diphenyl Phosphino-2-diethylaminoethane)-mercury(ii).... 86. 50
ydibenzo(e,k)(1,7,3,9,2,8> 7,15-bis(diphenoxy Phosphinyl)-6,7,14,15-tetrahydro-6,14-diphenox 83.220
                    (N-benzyl- Piperazinium)-pentachloro-di-mercury(ii) 83. 65
te]   bis(N-methyl Piperidinium-4-thiolato)-mercury(ii) perchlora 85. 44
io)-cf,de-bis(2-(dimethylaminomethyl)-phenyl)-*Platinum <ac-a-(μ-acetato)-b-(0-acetato-mercur 71.183
        N-tosylamino-octaethyl Porphinato-bis(chloro-mercury)................................. 49. 15
              bis(3-chloro Propynyl)-(1,10-phenanthroline)-mercury........ 71.120
          dibromo-(2,4-dimethyl Pyridine)-mercury(ii) 83. 28
              (2-benzyl- Pyridine)-methyl-mercury(ii) nitrate 71. 53
              phenyl Pyridine-mercury(ii) trifluoroacetate 71. 29
          diaqua-bis(Pyridine-3-sulfonato)-di-mercury(i)............. 83. 48
  methyl-(4,4',4''-triethyl-2,2':6',2''-ter Pyridyl)-mercury nitrate 71.180
              dibromo-(N-(2- Pyridyl)acetamide)-mercury(ii) 84. 15
          methyl-(bis(2- Pyridyl)methane)-mercury nitrate 71. 37
  (1,1-dichloro-1-(2-methyl-5-t-butyl-2H-2- Pyrrolyl))-chloro-mercury 71. 20
  (1,1-dichloro-1-(2-methyl-5-t-butyl-2H-2- Pyrrolyl))-phenyl-mercury 71. 98
)-((μ(2)-1-t-butylacetylene)-nonacarbonyl-tri-*Ruthenium)-mercury)........   [bis((μ(2)-bromo 71.228
  2,2,4,4,6,6,8,8-octamethyl-2,4,6,8-tetra Sila-1,5-dimercura-cyclo-octane 63. 11
          bis(triphenyl Silyl)-mercury 63. 50
        7α-(thio-bromo-mercury)-(17R)- Spiro(androst-4-ene-17,2(3H)-furan)....... 85. 69
henanthroline)-mercury(ii) bis(trifluoromethyl Sulfonate) ethanol solvate .... [tris(1,10-p 83.210
          diaqua-bis(pyridine-3- Sulfonato)-di-mercury(i)................ 83. 48
odo-di-mercury(ii)]   trans-bis(3-dimethyl*Sulfonio-cyclopentadienylide)-bis(μ-iodo)-di-i 71. 71
  methyl-(4,4',4''-triethyl-2,2':6',2''- Terpyridyl)-mercury nitrate.................... 71.180
17,2(3H)-furan)]   7α-(Thio-bromo-mercury)-(17R)-spiro(androst-4-ene- 85. 69
          bis(N,N-di-isopropyl Thiocarbamato)-mercury(ii) 80. 13
catena-(bis(μ-chloro)-tetrakis(μ-N,N-diethyldi*Thiocarbamato)-tri-mercury(ii)). 80. 21
          (1,5-diphenyl Thiocarbazonato-N,S)-methyl-mercury(ii)........ 71. 66
          (1,5-diphenyl Thiocarbazonato-N,S)-phenyl-mercury(ii)........ 71.138
  (μ-trans-cyclohexane-1,2-di*Thiolato)-bis(methyl-mercury(ii)) 85. 21
        bis(N-methylpiperidinium-4- Thiolato)-mercury(ii) perchlorate 85. 44
um))] bis(μ-chloro)-bis(mercury-(μ-p-*Tolyl(ethyl)triazenido)-(cyclo-octadiene-iridi 75. 40
              bis(o- Tolyl)-mercury 71. 65
  bis(μ-acetato)-bis(acetato-(tri-o-*Tolylphosphine)-mercury(ii)) 81. 96
  bis(μ-chloro)-bis(perchlorato-(tri-o-*Tolylphosphine)-mercury(ii)) 86.158
ury)]   N- Tosylamino-octaethylporphinato-bis(chloro-merc 49. 15
  bis(μ-chloro)-bis(mercury-(μ-p-tolyl(ethyl)*Triazenido)-(cyclo-octadiene-iridium)] 75. 40
<methyl-trans-3-endo-chloro-mercury-4-acetoxy- Tricyclo(4.2.2.0(2,5))dec-7-ene-9,10-cis-endo> 31. 37
          diacetato-bis(Tricyclohexylphosphine)-mercury(ii) dihydrate 81. 93
ate]   bis(Tricyclohexylphosphine)-mercury(ii) diperchlor 86.129
          bis(μ-acetato)-bis(acetato-(*Tricyclohexylphosphine)-mercury(ii)) 81. 94
etramethyl-1,10-phenanthroline)-(bis(trichloro Vinyl))-mercury....................[(3,4,7,8-t 71.152
              ethyl Xanthate-mercury(ii) 80. 1

Molybdenum

solvate]   tetrakis(μ(2)-N-2,6-xylyl-*Acetamidato)-di-molybdenum methylene chloride 84.128
              tetrakis(μ-N-(2-pyridyl-*Acetamido))-di-molybdenum 83.194
vate]   tetrakis(μ-*Acetanilido)-di-molybdenum tetrahydrofuran sol 84.115
lylmethyl)-molybdenum(ii))]   bis(μ-*Acetato)-bis((trimethylphosphine)-(trimethylsi 71.132
lybdenum)]   bis(μ-*Acetato)-bis((4-phenylimino-pentan-2-onato)-mo 81. 83
sodium (μ-oxo)-bis(dioxo-(hydrogen-nitrilo-tri*Acetato)-molybdenum(vi)) octahydrate 82. 23
)-oxo)-(μ(2)-thio)-1,2-propylenediamine-tetra-*Acetato-di-molybdenum tetrahydrate <dioxo-(μ(2 81. 39
bis(μ(2)-thio)-(μ-1,2-propylenediamine)-tetra-*Acetato-di-molybdenum trihydrate <odium dioxo 81. 38
-bis(μ(2)-oxo)-(μ-1,2-propylenediamine)-tetra-*Acetato-di-molybdenum trihydrate [sodium dioxo 81. 40
bdenum]   trifluoro Acetato-hydrido-tetra(trimethylphosphite)-moly 81. 60
  bis(N,N-dimethyldithiocarbamato)-(ditoluoyl- Acetylene)-oxo-molybdenum(iv) benzene solvate 72. 47
              tetrakis(1- Adamantoxo)-dimethylamine-molybdenum(iv) 84.130
olybdenum]   trans-bis(1- Adamantylamido)-tetrakis(trimethylsilyl-oxo)-m 84.116
)-pyridine-molybdenum(ii))]   (η- Allyl)-dicarbonyl-(N-phenylsalicylideneiminato 72. 43
lybdenum]   η(3)- Allyl-(η(5)-cyclopentadienyl)-iodo-nitrosyl-mo 72. 4
denum]   η(3)- Allyl-dicarbonyl-(η(5)-cyclopentadienyl)-molyb 72. 9
)-molybdenum(ii))]   1-3-η- Allyl-dicarbonyl-chloro-bis(trimethylphosphite 72. 13
s(octakis(t-butoxy)-tris(μ-fluoro)-(μ-dimethyl*Amido)-tetra-molybdenum <)-tetra-molybdenum bi 84.117
  tetrakis(μ-bis(difluorophosphino)methyl*Amine)-bis(chloro-molybdenum) 86. 1
          tetrakis(1-adamantoxo)-dimethyl Amine-molybdenum(iv) 84.130
hyldiphenylphosphine)-molybdenum)] tetraethyl*Ammonium (μ(2)-hydrido)-bis(tetracarbonyl-(met 86.113
s(dicarbonyl-nitroso-molybdenum)] tetraethyl*Ammonium (μ(3)-methoxy)-tris(μ(2)-methoxy)-tri 84. 24
onyl-molybdenum)] tetraethyl*Ammonium (μ-cyano)-bis(cyclopentadienyl-dicarb 73. 52
,6-dicarboxylato)-molybdenum(vi))] tetraethyl*Ammonium (μ-fluoro)-bis(oxo-peroxo-(pyridine-2 81. 53
ybdenum) dihydrate]   *Ammonium (μ-oxo)-(μ-catecholato)-bis(dioxo-mol 84. 35
uarato-S,S')-oxo-molybdenum(v))] tetrabutyl*Ammonium (μ-oxo)-(μ-sulfido)-bis((1,2-dithiosq 85. 31

molybdenum-trisulfido-tris(ethylt>   tetraethyl*Ammonium (μ-sulfido)-bis(μ-ethylthiolato)-bis(85. 57
                                      tetraethyl Ammonium aquo-tetrabromo-oxo-molybdenum........   3. 34
                                      tetraethyl Ammonium aquo-tetraiodo-oxo-molybdenum 3. 36
oxo-molybdenum(v))]   tetraethyl*Ammonium chloro-tris(thiolato-ethanolato)-bis(85. 9
s(trichloro-molybdenum) aquo-tetr>   tetraethyl*Ammonium hydrogen (μ-hydrido)-bis(μ-chloro)-bi   3. 40
0)-(μ-2-oxoethanethiolato-μ-S,μ-0>   tetraethyl*Ammonium trichloro-(μ-2-oxoethanethiolato-S,μ-   85. 4
2)-sulfido-tris(ethylthiolat>   triethylbenzyl*Ammonium tris(μ-ethylthiolato)-bis(tetrakis(μ(85. 62
)-carbonyl)-dicarbonyl-iron)->   bis(tetraethyl*Ammonium) (μ(6)-carbido)-(dicarbonyl-iron-(μ(2   3. 37
um)-tetrakis(μ-thiola>   tetrakis(tetra-n-butyl*Ammonium) (bis(tris(μ-benzylthiolato)-molybden   85. 96
m)]   bis(tetramethyl*Ammonium) bis((μ(2)-thio)-dithio-oxo-molybdenu   3. 23
)-tetrakis(μ-thiolat>   tris(N-benzyl-trimethyl*Ammonium) bis(tris(μ-ethylthiolato)-molybdenum   85. 78
num>   tetraphenylphosphonium (benzyl-trimethyl*Ammonium) dichloro-iron-bis(μ(2)-thio)-molybde   64.123
                              bis(triethyl Ammonium) dihydronium octacyano-molybdenum(iv)   3. 31
        guanidinium tetramolybdo-dimethyl Arsinate monohydrate 65. 2
        guanidinium hexamolybdo-methyl Arsonate hexahydrate 84. 2
thylphosphine)-molybdenum tri>   (μ(2)-dimethyl*Arsonio)-bis(dicarbonyl-cyclopentadienyl-trime   73. 95
cyclopentadienyl-molybdenum)]   tetraphenyl*Arsonium (μ-dimethylphosphido)-bis(dicarbonyl-   73. 59
enum tetrachloro-oxo-molybde>   bis(tetraphenyl Arsonium) tetrachloro-oxo-methylnitrile-molybd   83. 5
l-(η(5)-cyclopentadienyl)-(1,3-di   Aza-2-phospholidin-2-yl)-molybdenum [dicarbony   73.24
                                tri Azido-nitrido-bipyridyl-molybdenum(vi) 83. 45
yl-molybdenum)]   (η(2)-pentamethylene-di Azirine)-(pentacarbonyl-chromium)-(pentacarbon   83.101
ne))-molybdenum]   bis(di Azo-(diethylphosphine-ethylene-diphenylphosphi   86.124
enzyl)-carbonyl-moly>   cyclopentadienyl-(η(2)- Benzaldehyde)-(α-(2-pyridyl)-α-(methylamino)-b   71.204
cyclopentadienyl-(N-benzyl-N'-(1-phenylethyl)- Benzamidinato)-molybdenum [dicarbonyl-η(5)-   73.114a
-cyclopentadienyl)-(N-benzyl-N'-α-methylbenzyl Benzamidinato)-molybdenum <8)-dicarbonyl-(η(5)   73.115a
e solvate]   tris(dimethyldithiocarbamato)-(Benzenediazenato)-molybdenum methylene chlorid   80. 16
loro-molybdenum]   bis(p-fluoro Benzenediazo)-(η(5)-methylcyclopentadienyl)-ch   73. 72
zolyl)-borato-bis(p-toluenethiolato)-(p-fluoro Benzenediazo-molybdenum [hydro-tris(1-pyra   85. 81
ethyl-1-pyrazolyl)-borato-dicarbonyl-(p-chloro Benzenethiolato)-molybdenum acetone solvate <m   85. 70
        trithio-peroxybenzoato-(η(3)-dithio Benzoato-oxo-molybdenum(iv) 85. 48
iv)]   trithio-peroxy Benzoato-(η(3)-dithiobenzoato)-oxo-molybdenum(85. 48
drazone-triyl)-(N'-isopropylidene-p-chlorothio Benzoyl-hydrazone)-molybdenum <othiobenzoyl-hy   85. 71
ydrazone-triyl)-(N'-isopropyli>   (p-chlorothio Benzoyl-hydrazone-diyl)-(p-chlorothiobenzoyl-h   85. 71
<lorothiobenzoyl-hydrazone-diyl)-(p-chlorothio Benzoyl-hydrazone-triyl)-(N'-isopropylidene-p>   85. 71
-benzaldehyde)-(α-(2-pyridyl)-α-(methylamino)- Benzyl)-carbonyl-molybdenum <pentadienyl-(η(2)   71.204
lato)-molybdenum)-tetrakis(μ-thiolat>   tris(N-*Benzyl-trimethylammonium) bis(tris(μ-ethylthio   85. 78
2)-thio)-molybdenum>   tetraphenylphosphonium (*Benzyl-trimethylammonium) dichloro-iron-bis(μ(64.123
<7)-cycloheptatrienyl)-carbonyl-iodo-(α-methyl Benzyl-N-methylamino-(diphenyl)-phosphine)-mo>   75. 36
<)(578)-dicarbonyl-(η(5)-cyclopentadienyl)-(N- Benzyl-N'-α-methylbenzylbenzamidinato)-molybd>   73.115a
denum]   dicarbonyl-η(5)-cyclopentadienyl-(N- Benzyl-N'-(1-phenylethyl)-benzamidinato)-molyb   73.114a
kis(μ(2)-sulfido)-tris(ethylthiolat>   triethyl*Benzylammonium tris(μ-ethylthiolato)-bis(tetra   85. 62
-(η(5)-cyclopentadienyl)-(N-benzyl-N'-α-methyl Benzylbenzamidinato)-molybdenum <8)-dicarbonyl   73.115a
    tetrakis(tetra-n-butylammonium) (bis(tris(μ-*Benzylthiolato)-molybdenum)-tetrakis(μ-thiola>   85. 96
iolato)-molybdenum)-tetrakis(μ-thiolato)-tris(*Benzylthiolato-iron))-iron <is(tris(μ-benzylth   85. 96
nyl-cyclopentadienyl-molybdenum)]   (μ-*Bicyclo(4.2.2)deca-2,4,7,9-tetraene)-bis(carbo   75. 31
)-molybdenum(ii)]   (2,2'- Bipyridyl)-tricarbonyl-chloro-(chloro-mercurio   71. 48
            triazido-nitrido- Bipyridyl-molybdenum(vi) 83. 45
m(ii)]   hydrido-(tetrahydro Borato)-tetrakis(trimethylphosphine)-molybdenu   86. 30
diazo)-molybdenum]   hydro-tris(1-pyrazolyl)- Borato-bis(p-toluenethiolato)-(p-fluorobenzene   85. 81
lybden>   hydro-tris(3,5-dimethyl-1-pyrazolyl)- Borato-dicarbonyl-(p-chlorobenzenethiolato)-mo   85. 70
hyl-2-hydroxy-3-butanolato)-(2,3-dimethyl-2,3-*Butanediolato)-oxo-molybdenum) <bis((2,3-dimet   84.102
xo-mo>   (μ-oxo)-bis((2,3-dimethyl-2-hydroxy-3-*Butanolato)-(2,3-dimethyl-2,3-butanediolato)-o   84.102
            (μ-carbonyl)-bis(μ-t-*Butoxy)-bis(di-t-butoxy-molybdenum) 84.105
denum]   trans-bis(t-*Butoxy)-cis-dicarbonyl-cis-bis(pyridine)-molyb   84. 84
s(octakis(t-butoxy)-tris(μ-fluoro> octakis(t-*Butoxy)-tetrakis(μ-fluoro)-tetra-molybdenum bi   84.117
<kis(μ-fluoro)-tetra-molybdenum bis(octakis(t-*Butoxy)-tris(μ-fluoro)-(μ-dimethylamido)-tetr>   84.117
    (μ-carbonyl)-bis(μ-t-butoxy)-bis(di-t-*Butoxy-molybdenum) 84.105
    bis(η(5)-cyclopentadienyl)-(thio Camphor)-tetracarbonyl-di-molybdenum 72. 48
rbonyl-iron)-> bis(tetraethylammonium) (μ(6)-*Carbido)-(dicarbonyl-iron-(μ(2)-carbonyl)-dica   3. 37
                ammonium (μ-oxo)-(μ-*Catecholato)-bis(dioxo-molybdenum) dihydrate   84. 35
imethylthiophosphinato)-tetra(oxo-molybdenum) *Chloroform carbon tetrachloride solvate <s(μ-d   85. 29
imethylthiophosphinato)-tetra(oxo-molybdenum) *Chloroform carbon tetrachloride solvate <s(μ-d   85. 28
η(2)-pentamethylene-diazirine)-(pentacarbonyl- Chromium)-(pentacarbonyl-molybdenum) .... [(83.101
-molybdenum-bis(μ(2)-thio)-triphenylphosphine-*Copper methylene chloride solvate <(μ(2)-thio)   86.203
ulfur-oxo-molybdenum)-tris(triphenylphosphine-*Copper)................. [(μ(3)-chloro)-(tris   86.201
o)-tris(μ(3)-sulfido)-tris(triphenylphosphine-*Copper)-thio-molybdenum....... [(μ(3)-chlor   86.202
)-triphenylphosphin> bis(triphenylphosphine)-*Copper-bis(μ(2)-thio)-molybdenum-bis(μ(2)-thio   86.203
7,11,14-tetraene-3,7-diyl-tetracarbony> penta Cyclo(8.6.0.0(2,9).0(5,16).0(6,13))hexadeca-3,   75. 25
m]   bis(μ-η(3),η(3)-*Cyclo-octatetraene)-tetracarbonyl-di-molybdenu   75. 26
            1,5,9-triphospha Cyclododecane-tricarbonyl-molybdenum 86. 19
yl-N-methylamino-(diphenyl)-phosphine> (η(7)- Cycloheptatrienyl)-carbonyl-iodo-(α-methylbenz   75. 36
(μ-methoxy)-((η(7)-cycloheptatrienyl> ((η(3)-*Cycloheptatrienyl)-dicarbonyl-molybdenum)-tris   75. 22
dicarbonyl-molybdenum)-tris(μ-methoxy)-((η(7)-*Cycloheptatrienyl)-molybdenum) <heptatrienyl)-   75. 22
onyl-molybdenum-tris(μ-n-butylthiolato)-(η(7)-*Cycloheptatrienyl-molybdenum) .... [tricarb   75. 32
onyl-molybdenum-tris(μ-t-butylthiolato)-(η(7)-*Cycloheptatrienyl-molybdenum) .... [tricarb   75. 33

```
<ybdenum(iv))-(μ-oxo)-(oxy-(1,5,9,13-tetrathia*Cyclohexadecane)-molybdenum(iv)) trifluoromet>   85. 79
(1,5,9,13-tetrat>  (ethoxy-(1,5,9,13-tetrathia*Cyclohexadecane)-molybdenum(iv))-(μ-oxo)-(oxy-    85. 79
hino)ethane)-molybdenum benzene solv>  iodo-N- Cyclohexyl-diazenido-bis(1,2-bis(diphenylphosp    86.215
enyl)->  tetracarbonyl-manganese-(μ-η(5):η(1)-*Cyclopentadienyl)-(carbonyl-(η(5)-cyclopentadi    71. 85
di-molybdenum]                         bis(η(5)- Cyclopentadienyl)-(thiocamphor)-tetracarbonyl-    72. 48
enzamidinato)-mol>  (+)(578)-dicarbonyl-(η(5)- Cyclopentadienyl)-(N-benzyl-N'-α-methylbenzylb    73.115a
ospholidin-2-yl)-molybdenu>  dicarbonyl-(η(5)- Cyclopentadienyl)-(1,3-dimethyl-1,3-diaza-2-ph    73. 24
denum)]                        bis(μ-sigma:eta-*Cyclopentadienyl)-bis(η-cyclopentadienyl-molyb    71.156
       bis(p-fluorobenzenediazo)-(η(5)-methyl Cyclopentadienyl)-chloro-molybdenum............    73. 72
yl-cyclopentadienyl-molybdenum]  carbonyl-bis( Cyclopentadienyl)-hydrido-molybdenum tricarbon    73. 22
                              η(3)-allyl-(η(5)- Cyclopentadienyl)-iodo-nitrosyl-molybdenum ....    72.  4
              (μ(2)-oxo)-bis((η(5)-*Cyclopentadienyl)-iodo-oxo-molybdenum(v))  ....    73. 15
            η(3)-allyl-dicarbonyl-(η(5)- Cyclopentadienyl)-molybdenum  .................    72.  9
          dicarbonyl-nitrosyl-(η(5)-pentamethyl- Cyclopentadienyl)-molybdenum  .................    73. 36
  trans-bis(μ-thioethyl)-bis(nitrosyl-(η(5)-*Cyclopentadienyl)-molybdenum)  .................    73. 50
    cis-bis(μ-thio-isopropyl)-bis(nitrosyl-(η(5)-*Cyclopentadienyl)-molybdenum)  .................    73. 65
μ-η(5):η(1)-cyclopentadienyl)-(carbonyl-(η(5)-*Cyclopentadienyl)-molybdenum) <nyl-manganese-(    71. 85
η(5)-cyclopentadienyl-carbonyl-rhe>  bis(η(5)-*Cyclopentadienyl)-molybdenum-bis(μ-carbonyl)-(    73. 73
    bis(μ-ethoxycarbonylimido)-bis((η-*Cyclopentadienyl)-oxo-molybdenum)  ............    73. 62
     (μ-ethoxycarbonylimido)-(μ-oxo)-bis((η-*Cyclopentadienyl)-oxo-molybdenum.............    73. 42
idyl)-α-(methylamino)-benzyl)-carbonyl-moly>  Cyclopentadienyl-(η(2)-benzaldehyde)-(α-(2-pyr    71.204
benzamidinato)-molybdenum]  dicarbonyl-η(5)- Cyclopentadienyl-(N-benzyl-N'-(1-phenylethyl)-    73.114a
entan-2-onato)-triphenylphosphine->  carbonyl- Cyclopentadienyl-(1,5α-η(3)-1-methylene-cyclop    72. 54
pentadienyl)-molybdenum-bis(μ-carbonyl)-(η(5)-*Cyclopentadienyl-carbonyl-rhenium) <η(5)-cyclo    73. 73
        tetraethylammonium (μ-cyano)-bis(*Cyclopentadienyl-dicarbonyl-molybdenum).......    73. 52
          (μ-dimethylhydrazo)-bis(*Cyclopentadienyl-iodo-nitrosyl-molybdenum) ....    73. 39
lopentadienyl)-hydrido-molybdenum tricarbonyl- Cyclopentadienyl-molybdenum [carbonyl-bis(cyc    73. 22
yl-trimethylphosphine)-molybdenum tricarbonyl-*Cyclopentadienyl-molybdenum <yl-cyclopentadien    73. 95
    bis(μ-sulfido)-bis(μ-methylthio)-bis(methyl*Cyclopentadienyl-molybdenum(iv))  .............    73. 51
       anti-di-μ-sulfido-bis(sulfido-methyl*Cyclopentadienyl-molybdenum(v)).................    73. 32
      bis(μ-sigma:eta-cyclopentadienyl)-bis(η-*Cyclopentadienyl-molybdenum)  .................    71.156
clo(4.2.2)deca-2,4,7,9-tetraene)-bis(carbonyl-*Cyclopentadienyl-molybdenum)  ...  [(μ-bicy    75. 31
arsonium (μ-dimethylphosphido)-bis(dicarbonyl-*Cyclopentadienyl-molybdenum)  [tetraphenyl    73. 59
    bis(μ-propane-1,2-dithiolato)-bis(*Cyclopentadienyl-molybdenum) tetrafluoroborate    73. 64
m tri>  (μ(2)-dimethylarsonio)-bis(dicarbonyl-*Cyclopentadienyl-trimethylphosphine)-molybdenu    73. 95
<onyl-cyclopentadienyl-(1,5α-η(3)-1-methylene- Cyclopentan-2-onato)-triphenylphosphine-molyb>    72. 54
                             1,5,9-triphosphacyclo Decane-tricarbonyl-molybdenum    86. 19
t>  tris(dimethylthiocarbamato)-(m-nitrophenyl Diazenato)-molybdenum methylene chloride solva    80. 15
)-molybdenum benzene solv>  iodo-N-cyclohexyl- Diazenido-bis(1,2-bis(diphenylphosphino)ethane    86.215
            dicarbonyl-bis(diphenylphosphino Ethane)-fluoro-molybdenum hexafluorophosphate    86.204
hexyl-diazenido-bis(1,2-bis(diphenylphosphino) Ethane-tricarbonyl-di-iodo-molybdenum dichloro    71.224
methane solvate]  1,2-bis(diphenylphosphino) Ethane-molybdenum benzene solvate <do-N-cyclo    86.215
      bis(2-(diphenylphosphino) Ethane-1-thiolato)-oxo-molybdenum(iv)  ........    86. 94
<chloro-(μ-2-oxoethanethiolato-S,μ-O)-(μ-2-oxo*Ethanethiolato-μ-S,μ-O)-bis(oxo-molybdenum(v))    85.  4
-S,μ-O>  tetraethylammonium trichloro-(μ-2-oxo*Ethanethiolato-S,μ-O)-(μ-2-oxoethanethiolato-μ    85.  4
methoxymethyl)-dicarb>  dimethyl-(N,N-dimethyl Ethanolamino)-(1-pyrazolyl)-gallato-(η(2)-thio    72. 22
(η(2)-thiomethoxymeth>  dimethyl-(N,N-dimethyl Ethanolamino)-(3,5-dimethylpyrazolyl)-gallato-    68. 16
lybdenum>  piperidinium (μ-oxo)-(μ-2-mercapto-*Ethanolato)-bis(oxo-(2-mercapto-ethanolato)-mo    85. 10
     tetraethylammonium chloro-tris(thiolato- Ethanolato)-bis(oxo-molybdenum(v))  .............    85.  9
(μ-2-mercapto-ethanolato)-bis(oxo-(2-mercapto-*Ethanolato)-molybdenum) [piperidinium (μ-oxo)-    85. 10
       bis(diazo-(diethylphosphine- Ethylene-diphenylphosphine)-molybdenum........    86.124
              (μ(2)-oxo)-bis(di*Ethylene-trithiolato-oxo-molybdenum)  ........    85. 23
<icarbonyl-P,P-diphenyl-N-methyl-phosphinothio*Formamido)-(μ-P,P-diphenyl-N-methyl-phosphino>    71.280
<amido)-(μ-P,P-diphenyl-N-methyl-phosphinothio*Formamido)-molybdenum) dichloromethane solvate    71.280
       bis(diformato-molybdenum) bis(aqua-di Formato-molybdenum)..................    81. 11
num)]                              bis(di Formato-molybdenum) bis(aqua-diformato-molybde    81. 11
<hyl-(N,N-dimethylethanolamino)-(1-pyrazolyl)- Gallato-(η(2)-thiomethoxymethyl)-dicarbonyl-m>    72. 22
<imethylethanolamino)-(3,5-dimethylpyrazolyl)- Gallato-(η(2)-thiomethoxymethyl)-dicarbonyl-m>    68. 16
ydrate]                        tetrakis(glycyl Glycine)-di-molybdenum(ii) tetrachloride hexah    82. 33
 hexahydrate]                       tetrakis( Glycylglycine)-di-molybdenum(ii) tetrachloride    82. 33
te]                           tetra Guanidinium α-dodecamolybdo-silicate monohydra    8.  7
rate]                               Guanidinium hexamolybdo-methylarsonate hexahyd    84.  2
hydrate]                              Guanidinium tetramolybdo-dimethylarsinate mono    65.  2
tricarbonyl-(1,6-bis(diphenylphosphino)-trans- Hex-3-ene)-molybdenum(0)  .................    72. 55
ony>  pentacyclo(8.6.0.0(2,9).0(5,16).0(6,13)) Hexadeca-3,7,11,14-tetraene-3,7-diyl-tetracarb    75. 25
<dimethylthiocarbamato)-(η-N-methyl-N-phenyl Hydrazido)-(η(2)-N-methyl-N-phenylhydrazido)->    80. 19
molybdenum)]                      (μ-dimethyl*Hydrazido)-bis(cyclopentadienyl-iodo-nitrosyl-    73. 39
       bis(dimethyldithiocarbamato)-bis(diphenyl Hydrazido-molybdenum acetone solvate  .......    80. 25
hyl-N-phenylhydrazido)-(η(2)-N-methyl-N-phenyl Hydrazido-molybdenum tetraphenylborate <N-met    80. 19
triyl)-(N'-isopropylidene-p-chlorothiobenzoyl- Hydrazone-molybdenum <othiobenzoyl-hydrazone-    85. 71
-triyl)-(N'-isopropyli)  (p-chlorothiobenzoyl- Hydrazone-diyl)-(p-chlorothiobenzoyl-hydrazo    85. 71
<benzoyl-hydrazone-diyl)-(p-chlorothiobenzoyl- Hydrazone-triyl)-(N'-isopropylidene-p-chlorot>    85. 71
m) aquo-tetr>  tetraethylammonium hydrogen (μ-*Hydrido)-bis(μ-chloro)-bis(trichloro-molybdenu    3. 40
phine)-molybdenum]  tetraethylammonium (μ(2)-*Hydrido)-bis(tetracarbonyl-(methyldiphenylphos    86.113
     hexakis(trimethylphosphine)-dihydrido-(μ-di*Hydrido)-di-molybdenum(ii) ...................    86. 59
```

hexakis(trimethylphosphine)-di*Hydrido-(μ-dihydrido)-di-molybdenum(ii)........ 86. 59
hosphine)-molybdenum(ii)]   Hydrido-(tetrahydroborato)-tetrakis(trimethylp 86. 30
l-molybdenum] carbonyl-bis(cyclopentadienyl)- Hydrido-molybdenum tricarbonyl-cyclopentadieny 73. 22
trifluoroacetato- Hydrido-tetra(trimethylphosphite)-molybdenum 81. 60
ro-molybdenum) aquo-tetr> tetraethylammonium *Hydrogen (μ-hydrido)-bis(μ-chloro)-bis(trichlo 3. 40
ctahydrate]   sodium (μ-oxo)-bis(dioxo-(*Hydrogen-nitrilo-triacetato)-molybdenum(vi)) o 82. 23
pyridinium tetra-μ-(*Hydrogen-phosphato)-di-molybdenum chloride 33. 10
bis(triethylammonium) di Hydronium octacyano-molybdenum(iv) 3. 31
dioxo-bis(N-methyl-p-tolyl-thio Hydroxamato)-molybdenum(vi)................ 85. 61
)-molybdenum 1,10-phenanthroline di-iodide > Hydroxylamido-nitrosyl-bis(1,10-phenanthroline 83.168
dioxo-bis(N,N-diethyl Hydroxylaminato)-molybdenum(vi)................ 84. 17
bis((μ-N-t-butylimido)-(N-t-butyl*Imido))-tetramethyl-di-molybdenum 83.154
lybdenum)] (μ-ethoxycarbonyl*Imido)-(μ-oxo)-bis((η-cyclopentadienyl)-oxo-mo 73. 42
enum] bis((μ-N-t-butyl*Imido)-(N-t-butylimido))-tetramethyl-di-molybd 83.154
)] bis(μ-acetato)-bis((4-phenyl*Imido)-bis((η-cyclopentadienyl)-oxo-molybdenum 73. 62
bis(μ-acetato)-bis((4-phenyl*Imino-pentan-2-onato)-molybdenum) 81. 83
bis(triphenylphosphine) Immonium tetracyano-oxo-dioxo-molybdenum 64.160
tetrakis(μ-thiolato)-tris(ethylthiolato-iron)-*Iron <> bis(tris(μ-ethylthiolato)-molybdenum)- 85. 78
trakis(μ-thiolato)-tris(benzylthiolato-iron))-*Iron <is(tris(μ-benzylthiolato)-molybdenum)-te 85. 96
onyl)-tricarbonyl-molybdenum)-bis(tricarbonyl-*Iron <bonyl-iron)-(dicarbonyl-iron-(μ(2)-carb 3. 37
to)-bis(molybdenum-trisulfido-tris(ethylthio)-*Iron) <ammonium (μ-sulfido)-bis(μ-ethylthiola 85. 57
num)-tetrakis(μ-thiolato)-tris(benzylthiolato-*Iron))-iron <is(tris(μ-benzylthiolato)-molybde 85. 96
<-(dicarbonyl-iron-(μ(2)-carbonyl)-dicarbonyl-*Iron)-(dicarbonyl-iron-(μ(2)-carbonyl)-tricar> 3. 37
enum)-tetrakis(μ-thiolato)-tris(ethylthiolato)-*Iron)-iron <> bis(tris(μ-ethylthiolato)-molybd 85. 78
etraphenylphosphonium bis(bis-μ-thio-dichloro-*Iron)-molybdenum .... [t 64.129
bis(tetrakis(μ(2)-sulfido)-tris(ethylthiolato-*Iron)-molybdenum) <nium tris(μ-ethylthiolato)- 85. 62
<traethylammonium) (μ(6)-carbido)-(dicarbonyl-*Iron-(μ(2)-carbonyl)-dicarbonyl-iron)-(dicarb> 3. 37
<(μ(2)-carbonyl)-dicarbonyl-iron)-(dicarbonyl-*Iron-(μ(2)-carbonyl)-tricarbonyl-molybdenum)-> 3. 37
osphonium (benzyl-trimethylammonium) dichloro-*Iron-bis(μ(2)-thio)-molybdenum-disulfide <ylph 64.123
hylforma> tetraphenylphosphonium pentasulfur-*Iron-bis(μ-sulfido)-molybdenum-disulfide dimet 64.130
hepta(methyl Isocyanide)-molybdenum(ii) tetrafluoroborate 71. 72
lybdenum] bis(di- Isopropoxyphosphine)methylene-tetracarbonyl-mo 86. 47
<l)-(p-chlorothiobenzoyl-hydrazone-triyl)-(N'- Isopropylidene-p-chlorothiobenzoyl-hydrazone)> 85. 71
bis(hexamethylphosphoramide)-bis(Isothiocyanato)-dioxo-molybdenum(vi) 84. 60
tetrakis(pyridinium) di-μ-oxo-bis(oxo-tris(*Isothiocyanato)-molybdenum(v)) 33. 11
oxo-diperoxo-((S)-N,N-dimethyl- Lactamido)-molybdenum(vi) 84. 7a
luenesulfonate) dihydrate] tetrakis(L- Leucine)-di-molybdenum(ii) dichloride bis(p-to 82. 40
onyl-(η(5)-cyclopentadienyl)-> tetracarbonyl-*Manganese-(μ-η(5):η(1)-cyclopentadienyl)-(carb 71. 85
olato)-molybdenum> piperidinium (μ-oxo)-(μ-2-*Mercapto-ethanolato)-bis(oxo-(2-mercapto-ethan 85. 10
(μ-oxo)-(μ-2-mercapto-ethanolato)-bis(oxo-(2-*Mercapto-ethanolato)-molybdenum) [piperidinium 85. 10
(2,2'-bipyridyl)-tricarbonyl-chloro-(chloro-*Mercurio)-molybdenum(ii) 71. 48
ine> carbonyl-cyclopentadienyl-(1,5α-η(3)-1-Methylene-cyclopentan-2-onato)-triphenylphosph 72. 54
(pentacarbonyl-molybdenum)] (η(2)-penta Methylene-diazirine)-(pentacarbonyl-chromium)- 83.101
bis(di-isopropoxyphosphine) Methylene-tetracarbonyl-molybdenum 86. 47
(v) acetonitrile so> trichloro-oxo-(N-pyridyl Methylene-N'-salicyloylhydrazinato)-molybdenum 84. 47
olybdenum chloroform solvate] phenyl Nitrene-bis(diethyldithiocarbamato)-dichloro-m 80. 18
um] tetrachloro-(pentachloroethyl- Nitrido-(phosphoryl-trichlorido)-oxo-molybden 83. 2
triazido- Nitrido-bipyridyl-molybdenum(vi) 83. 45
)-methoxy)-tris(μ(2)-methoxy)-tris(dicarbonyl-*Nitroso-molybdenum) [tetramethylammonium (μ(3 84. 24
(μ(3)-oxo)-tris(μ(2)-methoxy)-tris(dicarbonyl-*Nitroso-molybdenum) <phosphineiminium) sodium 84. 20
trans-bis(μ-thioethyl)-bis(*Nitrosyl-(η(5)-cyclopentadienyl)-molybdenum) 73. 50
cis-bis(μ-thio-isopropyl)-bis(*Nitrosyl-(η(5)-cyclopentadienyl)-molybdenum) 73. 65
olybdenum] dicarbonyl- Nitrosyl-(η(5)-pentamethyl-cyclopentadienyl)-m 73. 36
,10-phenanthroline di-iodide > hydroxylamido- Nitrosyl-bis(1,10-phenanthroline)-molybdenum 1 83.168
η(3)-allyl-(η(5)-cyclopentadienyl)-iodo- Nitrosyl-molybdenum........................ 72. 4
-dimethylhydrazido)-bis(cyclopentadienyl-iodo-*Nitrosyl-molybdenum........................[(μ 73. 39
osyl-bis(1,10-phenanthroline)-molybdenum 1,10-Phenanthroline di-iodide monohydrate <ido-nitr 83.168
di-iodide > hydroxylamido-nitrosyl-bis(1,10-Phenanthroline)-molybdenum 1,10-phenanthroline 83.168
dichloro-dioxo-(o- Phenanthroline)-molybdenum(vi) 83. 72
1,5,9-tri Phosphacyclodecane-tricarbonyl-molybdenum 86. 19
lphosphinoethane)-fluoro-molybdenum hexafluoro Phosphate ....[dicarbonyl-bis(dipheny 86.204
pyridinium tetra-μ-(hydrogen- Phosphato)-di-molybdenum chloride 33. 10
ybdenum)] tetraphenylarsonium (μ-dimethyl*Phosphido)-bis(dicarbonyl-cyclopentadienyl-mol 73. 59
bis(diazo-(diethylphosphine-ethylene-diphenyl Phosphine))-molybdenum 86.124
i))] bis(μ-acetato)-bis((trimethyl*Phosphine)-(trimethylsilylmethyl)-molybdenum(i 71.132
s(μ(2)-thio)-triphenylphosphin> bis(triphenyl*Phosphine)-copper-bis(μ(2)-thio)-molybdenum-bi 86.203
um(ii)] hexakis(trimethyl*Phosphine)-dihydrido-(μ-dihydrido)-di-molybden 86. 59
iodo-(α-methylbenzyl-N-methylamino-(diphenyl)- Phosphine)-molybdenum <heptatrienyl>-carbonyl- 75. 36
<io)-bis(dicarbonyl-cyclopentadienyl-trimethyl*Phosphine)-molybdenum tricarbonyl-cyclopentad> 73. 95
hydrido-(tetrahydroborato)-tetrakis(trimethyl Phosphine)-molybdenum(ii) 86. 30
(2)-hydrido)-bis(tetracarbonyl-(methyldiphenyl*Phosphine)-molybdenum) [tetraethylammonium (μ 86.113
(bis(2-diphenylphosphinoethyl)-phenyl Phosphine)-tricarbonyl-molybdenum(0) 71.251
enum] bis(triphenyl Phosphine)immonium tetracyano-oxo-dioxo-molybd 64.160
)-> bis(μ-chloro)-bis(bis(carbonyl)-trimethyl*Phosphine-(1-2-η-trimethylsilyl-methylcarbonyl 71.163
μ(2)-thio)-molybdenum-bis(μ(2)-thio)-triphenyl*Phosphine-copper methylene chloride solvate <(86.203

oro)-(trisulfur-oxo-molybdenum)-tris(triphenyl*Phosphine-copper) .... [(μ(3)-chl　86.201
μ(3)-chloro-tris(μ(3)-sulfido)-tris(triphenyl*Phosphine-copper)-thio-molybdenum .... [(　86.202
num]　　　　　　　　　　　　　bis(diazo-(diethyl Phosphine-ethylene-diphenylphosphine))-molybde　86.124
<-bis(η(2)-diphenylphosphinesulfido)-triphenyl Phosphine-molybdenum methylene chloride solva>　86.162
(3)-1-methylene-cyclopentan-2-onato)-triphenyl Phosphine-molybdenum(ii) <opentadienyl-(1,5α-η　72. 54
m(iv)]　　　　　　　　　chloro-trimethyl Phosphine-tris(trimethylsilylmethyl)-molybdenu　71. 89
methoxy)-tris(dicarbonyl-nitr> tris(triphenyl*Phosphineiminium) sodium (μ(3)-oxo)-tris(μ(2)-　84. 20
m methylene chl> dicarbonyl-bis(η(2)-diphenyl Phosphinesulfido)-triphenylphosphine-molybdenum　86.162
num]　　　　　　　　　　1,3-bis(diphenyl Phosphino)-propane-tricarbonyl-di-iodo-molybde　71.229
　　　　　tricarbonyl-(1,6-bis(diphenyl Phosphino)-trans-hex-3-ene)-molybdenum(0) ....　72. 55
<o-N-cyclohexyl-diazenido-bis(1,2-bis(diphenyl Phosphino)ethane)-molybdenum benzene solvate　86.215
m dichloromethane solvate] 1,2-bis(diphenyl Phosphino)ethane-tricarbonyl-di-iodo-molybdenu　71.224
)]　　　　　　　　　bis(2-(diphenyl Phosphino)ethane-1-thiolato)-oxo-molybdenum(iv　86. 94
　　　　tetrakis(μ-bis(difluoro*Phosphino)methylamine)-bis(chloro-molybdenum)　86. 1
hosphate]　　　dicarbonyl-bis(diphenyl Phosphinoethane)-fluoro-molybdenum hexafluorop　86.204
olybdenum(0)]　　　　　(bis(2-diphenyl Phosphinoethyl)-phenylphosphine)-tricarbonyl-m　71.251
yl-ph> bis((dicarbonyl-P,P-diphenyl-N-methyl-*Phosphinothioformamido)-(μ-P,P-diphenyl-N-meth　71.280
<phinothioformamido)-(μ-P,P-diphenyl-N-methyl-*Phosphinothioformamido)-molybdenum) dichlorom>　71.280
　　　trifluoroacetato-hydrido-tetra(trimethyl Phosphite-molybdenum ....　81. 60
1-3-η-allyl-dicarbonyl-chloro-bis(trimethyl Phosphite)-molybdenum(ii) .... 　72. 13
)-cyclopentadienyl)-(1,3-dimethyl-1,3-diaza-2- Phospholidin-2-yl)-molybdenum [dicarbonyl-(η(5　73. 24
cyano-di-molybdenum hexahydrate] tetraphenyl*Phosphonium (μ(2)-sulfonyl)-(μ(4)-dithio)-octa　64.133
o-iron-bis(μ(2)-thio)-molybdenum> tetraphenyl*Phosphonium (benzyl-trimethylammonium) dichlor　64.123
bdenum]　　　　　　　tetraphenyl*Phosphonium bis(bis-μ-thio-dichloro-iron)-moly　64.129
lybdenum-disulfide dimethylforma> tetraphenyl*Phosphonium pentasulfur-iron-bis(μ-sulfido)-mo　64.130
　　　　tetrakis(μ-dimethyl*Phosphonium-dimethylido)-di-molybdenum .... 　71.102
denum(vi)]　　　　　　bis(hexamethyl Phosphoramide)-bis(isothiocyanato)-dioxo-molyb　84. 60
　tetrachloro-(pentachloroethyl-nitrido)-(Phosphoryl-trichloride)-oxo-molybdenum ....　83. 2
　　　　bis(μ-methoxy)-bis(dioxo-(*Pinacolato)-molybdenum(vi)) methanol solvate　84. 59
-bis(oxo-(2-mercapto-ethanolato)-molybdenum> *Piperidinium (μ-oxo)-(μ-2-mercapto-ethanolato)　85. 10
i)]　　　　　　　　bis(Piperidinium) hexabromo-diaqua-di-molybdenum(i　33. 16
　cis-dioxo-(5,10,15,20-tetra-p-tolyl Porphyrinato)-molybdenum(vi) toluene solvate　49. 26
　　　　1,3-bis(diphenylphosphino)- Propane-tricarbonyl-di-iodo-molybdenum .... 　71.229
olybdenum) tetrafluoroborate]　　bis(μ-*Propane-1,2-dithiolato)-bis(cyclopentadienyl-m　73. 64
trihydra> sodium dioxo-bis(μ(2)-thio)-(μ-1,2-*Propylenediamine)-tetra-acetato-di-molybdenum　81. 38
trihydrat> sodium dioxo-bis(μ(2)-oxo)-(μ-1,2-*Propylenediamine)-tetra-acetato-di-molybdenum　81. 40
etr> sodium dioxo-(μ(2)-oxo)-(μ(2)-thio)-1,2-*Propylenediamine-tetra-acetato-di-molybdenum t　81. 39
uorobenzenediazo)-molybdenum] hydro-tris(1- Pyrazolyl)-borato-bis(p-toluenethiolato)-(p-fl　85. 81
hiolato)-molybden> hydro-tris(3,5-dimethyl-1- Pyrazolyl)-borato-dicarbonyl-(p-chlorobenzenet　85. 70
<thyl-(N,N-dimethylethanolamino)-(3,5-dimethyl Pyrazolyl)-gallato-(η(2)-thiomethoxymethyl)-d>　68. 16
carb> dimethyl-(N,N-dimethylethanolamino)-(1- Pyrazolyl)-gallato-(η(2)-thiomethoxymethyl)-di　72. 22
　trans-bis(t-butoxy)-cis-dicarbonyl-cis-bis(Pyridine)-molybdenum .... 　84. 84
　　　　trichloro-tris(4-methyl Pyridine)-molybdenum 4-methylpyridine solvate　83.116
　　　　bis(dibromo-bis(4-methyl Pyridine)-molybdenum(ii))　83.175
　　　　bis(dichloro-bis(4-methyl Pyridine)-molybdenum(ii)) chloroform solvate　83.177
vate]　　　　tribromo-tris(4-methyl Pyridine)-molybdenum(iii) 4-methylpyridine sol　83.115
lyl)-dicarbonyl-(N-phenylsalicylideneiminato)- Pyridine-molybdenum(ii)................. [(η-al　72. 43
tetraethylammonium (μ-fluoro)-bis(oxo-peroxo-(*Pyridine-2,6-dicarboxylato)-molybdenum(vi))....　81. 53
　　　　　4-ethyl *Pyridinium octa-molybdenum ....　33. 26
ybdenum chloride]　　　　*Pyridinium tetra-μ-(hydrogen-phosphato)-di-mol　33. 10
to)-molybdenum(v))]　　tetrakis(*Pyridinium) di-μ-oxo-bis(oxo-tris(isothiocyana　33. 11
　　　tetrakis(2-carboxy Pyridinium) octacyano-molybdenum(iv) ....　33. 18
m)dibromide]　　　　pentakis(Pyridinium) tris(aquo-tetrabromo-oxo-molybdenu　33. 12
　　tetrakis(μ(2)-6-chloro-2-*Pyridinolato)-di-molybdenum ....　84. 77
cyclopentadienyl-(η(2)-benzaldehyde)-(α-(2- Pyridyl)-α-(methylamino)-benzyl)-carbonyl-mol>　71.204
　　　tetrakis(μ-N-(2-*Pyridyl-acetamido))-di-molybdenum ....　83.194
ybdenum(v) acetonitrile so> trichloro-oxo-(N- Pyridylmethylene-N'-salicyloylhydrazinato)-mol　84. 47
s(μ-carbonyl)-(η(5)-cyclopentadienyl-carbonyl-*Rhenium) <η(5)-cyclopentadienyl)-molybdenum-bi　73. 73
(η-allyl)-dicarbonyl-(N-phenyl Salicylideneiminato)-pyridine-molybdenum(ii)　72. 43
ile so> trichloro-oxo-(N-pyridylmethylene-N'- Salicyloylhydrazinato)-molybdenum(v) acetonitr　84. 47
　　　tetraguanidinium α-dodecamolybdo- Silicate monohydrate .... 　8. 7
(carbonyl)-trimethylphosphine-(1-2-η-trimethyl*Silyl-methylcarbonyl)-molybdenum(ii)) <bis(bis　71.163
trans-bis(1-adamantylamido)-tetrakis(trimethyl Silyl-oxo)-molybdenum ....　84.116
μ-acetato)-bis((trimethylphosphine)-(trimethyl*Silylmethyl)-molybdenum(ii)) .... [bis(　71.132
　　　chloro-trimethylphosphine-tris(trimethyl Silylmethyl)-molybdenum(iv)　71. 89
rbonyl-nitr> tris(triphenylphosphineiminium) *Sodium (μ(3)-oxo)-tris(μ(2)-methoxy)-tris(dica　84. 20
acetato)-molybdenum(vi)) octahydrate]　　*Sodium (μ-oxo)-bis(dioxo-(hydrogen-nitrilo-tri　82. 23
enediamine-tetra-acetato-di-molybdenum tetr> *Sodium dioxo-(μ(2)-oxo)-(μ(2)-thio)-1,2-propyl　81. 39
mine)-tetra-acetato-di-molybdenum trihydrat> *Sodium dioxo-bis(μ(2)-oxo)-(μ-1,2-propylenedia　81. 40
amine)-tetra-acetato-di-molybdenum trihydra> *Sodium dioxo-bis(μ(2)-thio)-(μ-1,2-propylenedi　81. 38
ylammonium (μ-oxo)-(μ-sulfido)-bis((1,2-dithio*Squarato-S,S')-oxo-molybdenum(v)) [tetrabut　85. 31
um) dichloro-iron-bis(μ(2)-thio)-molybdenum-di*Sulfide <ylphosphonium (benzyl-trimethylammoni　64.123
pentasulfur-iron-bis(μ-sulfido)-molybdenum-di*Sulfide dimethylformamide solvate <phosphonium　64.130
　　　tetrachloro-tetrakis(diethyl Sulfide)-di-molybdenum(ii) ....　85. 54
　　　bis(bis(2-thiolato-ethyl)- Sulfide)-molybdenum(iv)....................... 　85. 22

ybdenum(v))]   tetrabutylammonium (μ-oxo)-(μ-*Sulfido)-bis((1,2-dithiosquarato-S,S')-oxo-mol   85. 31
risulfido-tris(ethylt>   tetraethylammonium (μ-*Sulfido)-bis(μ-ethylthiolato)-bis(molybdenum-t   85. 57
adienyl-molybdenum(iv))]   bis(μ-*Sulfido)-bis(μ-methylthio)-bis(methylcyclopent   73. 51
tetraphenylphosphonium pentasulfur-iron-bis(μ-*Sulfido)-molybdenum-disulfide dimethylformami>   64.130
<nium tris(μ-ethylthiolato)-bis(tetrakis(μ(2)-*Sulfido)-tris(ethylthiolato-iron)-molybdenum)   85. 62
molybdenum]   (μ(3)-chloro)-tris(μ(3)-*Sulfido)-tris(triphenylphosphine-copper)-thio-   86.202
ybdenum(v))]   anti-di-μ-*Sulfido-bis(sulfido-methylcyclopentadienyl-mol   73. 32
lfido)-bis(μ-ethylthiolato)-bis(molybdenum-tri*Sulfido-tris(ethylthio)-iron)) <ammonium (μ-su   85. 57
clohexadecane)-molybdenum(iv)) trifluoromethyl*Sulfonate monohydrate <y-(1,5,9,13-tetrathiacy   85. 79
m hexahydrate]   tetraphenylphosphonium (μ(2)-*Sulfonyl)-(μ(4)-dithio)-octacyano-di-molybdenu   64.133
e dimethylforma>   tetraphenylphosphonium penta*Sulfur-iron-bis(μ-sulfido)-molybdenum-disulfid   64.130
-copper)]   (μ(3)-chloro)-(tri*Sulfur-oxo-molybdenum)-tris(triphenylphosphine   86.201
<-molybdenum(iv))-(μ-oxo)-(oxy-(1,5,9,13-tetra*Thiacyclohexadecane)-molybdenum(iv)) trifluor>   85. 79
oxy-(1,5,9,13-tetrat>   (ethoxy-(1,5,9,13-tetra*Thiacyclohexadecane)-molybdenum(iv))-(μ-oxo)-(85. 79
i-molybdenum trihydra>   sodium dioxo-bis(μ(2)-*Thio)-(μ-1,2-propylenediamine)-tetra-acetato-d   81. 38
))]   bis(μ-sulfido)-bis(μ-methyl*Thio)-bis(methylcyclopentadienyl-molybdenum(iv   73. 51
bis(tetramethylammonium) bis((μ(2)-*Thio)-dithio-oxo-molybdenum) 3. 23
thiolato)-bis(molybdenum-trisulfido-tris(ethyl*Thio)-iron)) <ammonium (μ-sulfido)-bis(μ-ethyl   85. 57
hin>   bis(triphenylphosphine)-copper-bis(μ(2)-*Thio)-molybdenum-bis(μ(2)-thio)-triphenylphosp   86.203
zyl-trimethylammonium) dichloro-iron-bis(μ(2)-*Thio)-molybdenum-disulfide <ylphosphonium (ben   64.123
etraphenylphosphonium (μ(2)-sulfonyl)-(μ(4)-di*Thio)-octacyano-di-molybdenum hexahydrate   [t   64.133
<e)-copper-bis(μ(2)-thio)-molybdenum-bis(μ(2)-*Thio)-triphenylphosphine-copper methylene chl>   86.203
lybdenum tetr>   sodium dioxo-(μ(2)-oxo)-(μ(2)-*Thio)-1,2-propylenediamine-tetra-acetato-di-mo   81. 39
tetraphenylphosphonium bis(bis-μ-*Thio-dichloro-iron)-molybdenum   64.129
ienyl)-molybdenum)]   cis-bis(μ-*Thio-isopropyl)-bis(nitrosyl-(η(5)-cyclopentad   73. 65
μ(3)-sulfido)-tris(triphenylphosphine-copper)-*Thio-molybdenum........   [(μ(3)-chloro)-tris(86.202
bis(tetramethylammonium) bis((μ(2)-thio)-di*Thio-oxo-molybdenum)   3. 23
molybdenum(iv)]   tri Thio-peroxybenzoato-(η(3)-dithiobenzoato)-oxo-   85. 48
trithio-peroxybenzoato-(η(3)-di Thiobenzoato)-oxo-molybdenum(iv) 85. 48
l-hydrazone-triyl)-(N'-isopropylidene-p-chloro Thiobenzoyl-hydrazone)-molybdenum <othiobenzoy   85. 71
yl-hydrazone-triyl)-(N'-isopropyli>   (p-chloro Thiobenzoyl-hydrazone-diyl)-(p-chlorothiobenzo   85. 71
(p-chlorothiobenzoyl-hydrazone-diyl)-(p-chloro Thiobenzoyl-hydrazone-triyl)-(N'-isopropylide>   85. 71
bis(η(5)-cyclopentadienyl)-(Thiocamphor)-tetracarbonyl-di-molybdenum 72. 48
(η(2)-N-methyl-N-phenylhy>   bis(N,N-dimethyldi Thiocarbamato)-(η-N-methyl-N-phenylhydrazido)-   80. 19
ethylene chloride solvate]   tris(dimethyldi Thiocarbamato)-(benzenediazenato)-molybdenum m   80. 16
denum(iv) benzene solvate]   bis(N,N-dimethyldi Thiocarbamato)-(ditoluoyl-acetylene)-oxo-molyb   72. 47
enum methylene chloride solvat>   tris(dimethyl Thiocarbamato)-(m-nitrophenyldiazenato)-molybd   80. 15
um acetone solvate]   bis(dimethyldi Thiocarbamato)-bis(diphenylhydrazido)-molybden   80. 25
solvate]   phenylnitrene-bis(diethyldi Thiocarbamato)-dichloro-molybdenum chloroform   80. 18
dioxo-(N,N-diethyldi Thiocarbamato)-molybdenum(vi) 80. 7
)-molybdenum)]   trans-bis(μ-*Thioethyl)-bis(nitrosyl-(η(5)-cyclopentadienyl   73. 50
<s((dicarbonyl-P,P-diphenyl-N-methyl-phosphino*Thioformamido)-(μ-P,P-diphenyl-N-methyl-phosp>   71.280
<formamido)-(μ-P,P-diphenyl-N-methyl-phosphino*Thioformamido)-molybdenum) dichloromethane so>   71.280
dioxo-bis(N-methyl-p-tolyl- Thiohydroxamato)-molybdenum(vi)   85. 61
tricarbonyl-molybdenum-tris(μ-n-butyl*Thiolato)-(η(7)-cycloheptatrienyl-molybdenum)   75. 32
tricarbonyl-molybdenum-tris(μ-t-butyl*Thiolato)-(η(7)-cycloheptatrienyl-molybdenum)   75. 33
rafluoroborate]   bis(μ-propane-1,2-di*Thiolato)-bis(cyclopentadienyl-molybdenum) tet   73. 64
t>   tetraethylammonium (μ-sulfido)-bis(μ-ethyl*Thiolato)-bis(molybdenum-trisulfido-tris(ethyl   85. 57
lthiolat>   triethylbenzylammonium tris(μ-ethyl*Thiolato)-bis(tetrakis(μ(2)-sulfido)-tris(ethy   85. 62
bis(1,1-dimethyl-2-N-methylaminoethyl Thiolato)-dioxo-molybdenum   85. 37
<(N-benzyl-trimethylammonium) bis(tris(μ-ethyl*Thiolato)-molybdenum)-tetrakis(μ-thiolato)-tr>   85. 78
<kis(tetra-n-butylammonium) (bis(tris(μ-benzyl*Thiolato)-molybdenum)-tetrakis(μ-thiolato)-tr>   85. 96
bis(2-(diphenylphosphino)ethane-1- Thiolato)-oxo-molybdenum(iv)   86. 94
tris(μ-benzylthiolato)-molybdenum)-tetrakis(μ-*Thiolato)-tris(benzylthiolato-iron)-iron <is(85. 96
(tris(μ-ethylthiolato)-molybdenum)-tetrakis(μ-*Thiolato)-tris(ethylthiolato-iron)-iron <) bis   85. 78
o-(μ-2-oxoethanethiolato-S,μ-O)-(μ-2-oxoethane*Thiolato-μ-S,μ-O)-bis(oxo-molybdenum(v)) <hlor   85. 4
tetraethylammonium chloro-tris(Thiolato-ethanolato)-bis(oxo-molybdenum(v))....   85. 9
bis(bis(2- Thiolato-ethyl)-sulfide)-molybdenum(iv).........   85. 12
)-molybdenum)-tetrakis(μ-thiolato)-tris(benzyl*Thiolato-iron))-iron <is(tris(μ-benzylthiolato   85. 96
o)-molybdenum)-tetrakis(μ-thiolato)-tris(ethyl*Thiolato-iron)-iron <) bis(tris(μ-ethylthiolat   85. 78
hiolato)-bis(tetrakis(μ(2)-sulfido)-tris(ethyl*Thiolato-iron)-molybdenum) <nium tris(μ-ethylt   85. 62
(μ(2)-oxo)-bis(diethylene-tri*Thiolato-oxo-molybdenum) 85. 23
tetraethylammonium trichloro-(μ-2-oxoethane*Thiolato-S,μ-O)-(μ-2-oxoethanethiolato-μ-S,μ->   85. 4
thylethanolamino)-(1-pyrazolyl)-gallato-(η(2)- Thiomethoxymethyl)-dicarbonyl-molybdenum <dime   72. 22
lamino)-(3,5-dimethylpyrazolyl)-gallato-(η(2)- Thiomethoxymethyl)-dicarbonyl-molybdenum <hano   68. 16
orm c>   tetrakis(μ(3)-oxo)-tetrakis(μ-dimethyl*Thiophosphinato)-tetra(oxo-molybdenum) chlorof   85. 28
orm c>   tetrakis(μ(3)-oxo)-tetrakis(μ-dimethyl*Thiophosphinato)-tetra(oxo-molybdenum) chlorof   85. 29
abutylammonium (μ-oxo)-(μ-sulfido)-bis((1,2-di*Thiosquarato-S,S')-oxo-molybdenum(v))   [tetr   85. 31
L-leucine)-di-molybdenum(ii) dichloride bis(p- Toluenesulfonate) dihydrate........ [tetrakis(82. 40
enum]   hydro-tris(1-pyrazolyl)-borato-(p- Toluenethiolato)-(p-fluorobenzenediazo)-molybd   85. 81
solvate]   bis(N,N-dimethyldithiocarbamato)-(di Toluoyl-acetylene)-oxo-molybdenum(iv) benzene   72. 47
dioxo-bis(N-methyl-p- Tolyl-thiohydroxamato)-molybdenum(vi) 85. 61
ate]   cis-dioxo-(5,10,15,20-tetra-p- Tolylporphyrinato)-molybdenum(vi) toluene solv   49. 26
oride solvate]   tetrakis(μ(2)-N-2,6-*Xylyl-acetamidato)-di-molybdenum methylene chl   84.128

Neodymium

trinitrato-(18- Crown-6)-neodymium(iii)........................ 84. 39
trinitrato-(18- Crown-6)-neodymium(iii)........................ 84. 40
trinitrato-(hexaoxa Cyclo-octadecane)-neodymium(iii) 84. 39
trinitrato-(hexaoxa Cyclo-octadecane)-neodymium(iii) 84. 40
i)] tri Nitrato-(hexaoxacyclo-octadecane)-neodymium(ii 84. 39
i)] tri Nitrato-(hexaoxacyclo-octadecane)-neodymium(ii 84. 40
tri Nitrato-(18-crown-6)-neodymium(iii)............ 84. 40
tri Nitrato-(18-crown-6)-neodymium(iii)............ 84. 39
trinitrato-(hexaoxacyclo- Octadecane)-neodymium(iii) 84. 40
trinitrato-(hexaoxacyclo- Octadecane)-neodymium(iii) 84. 39
trinitrato-(hexa Oxacyclo-octadecane)-neodymium(iii)............ 84. 39
trinitrato-(hexa Oxacyclo-octadecane)-neodymium(iii)............ 84. 40
oride solvate] bis(Phthalocyaninato)-neodymium(iii) methylene chl 49. 33

Nickel

hylenediamine-nickel-(μ-ethylenediamine-tetra-*Acetato)-nickel................ [tetra-aqua-et 76. 74
ylenediamine)-copper-(μ-ethylenediamine-tetra-*Acetato)-nickel tetrahydrate .... [bis(eth 84. 58
calcium (ethylenediamine-tetra- Acetato)-nickel(ii) tetrahydrate 76. 57
diaquo-bis(N,N-diethylnicotinamide)-di Acetato-nickel(ii) 81. 80
hylenediamine)-copper)-(ethylenediamine-tetra- Acetato-nickel) dihydrate <ine)-copper (bis(et 76. 18
(N,N'-ethylene-bis(benzoyl Acetoniminato))-nickel(ii) 84. 91
phenyl-2-methylethenyl)-nickel] Acetylacetonato-(triphenylphosphine)-(Z-1,2-di 71.255
tricarbonyl-cobalt-(μ-diphenyl*Acetylene)-(cyclopentadienyl-nickel) 72. 39
nzene solvate] (η(2)-methyl-3-benzoyl Acrylate)-bis(triphenylphosphine)-nickel(0) be 72. 64
tetra-aquo-(Adipato)-nickel(ii)................ 81. 25
bis(D-β-(2-pyridyl)-α- Alaninato)-nickel(ii) dihydrate............... 82. 30
bis(2-(N-benzyl Aldimino)-benzo(b)furan-3-thiolato)-nickel(ii) 85. 83
bis(2-(N-phenyl Aldimino)-benzo(b)furan-3-thiolato)-nickel(ii) 85. 82
bis(η(3)- Allyl)-trimethylphosphine-nickel 72. 8
bis(dimethyl Amido-(methyl)-dithiophosphonato)-nickel(ii) 85. 13
um tris(N,N,N',N'-tetramethyl-but-2-ene-1,4-di Amine)-nickel(ii) .... [di-lithi 72. 51
trans-diaqua-bis(cyclohexane-1,2-di Amine)-nickel(ii) chloride 76. 79
bis(diphenyl-hydantoinato)-diaqua-di Ammine-nickel(ii) 83.201
μ(2)-carbonyl-nickel-tri-i> benzyl-trimethyl*Ammonium (μ(3)-hydrido)-octacarbonyl-tetrakis(3. 43
-bis(nitrosyl-nickel)] tetraethyl*Ammonium (μ-iodo)-bis(μ-3,5-dimethylpyrazolyl) 83. 53
i)] tetrabutyl Ammonium bis(isotrithione-dithiolato)-nickel(i 85. 14
dichloro-bis(2,2'-bis(dimethyl Arsine)-biphenyl)-nickel(ii) 86.110
bis(tri-iodo Arsine)-bis(diethyldithiocarbamato)-nickel(ii) 80. 6
(tris(2-diphenyl Arsinoethyl)-amino)-iodo-nickel(i) 86.156
bis(tetraphenyl Arsonium) dihydrido-dodeca-nickel-carbonyl 65. 16
ne solvate] tris(tetraphenyl Arsonium) hydrido-dodeca-nickel-carbonyl aceto 65. 21n
ne solvate] tris(tetraphenyl Arsonium) hydrido-dodeca-nickel-carbonyl aceto 65. 20
arbonitrile)-nickel(iv)] tris(tetraphenyl Arsonium) tris(2,2-diselenido-1,1-ethylene-dic 85. 47
<-hexamethyl-5,7-dioxa-6-hydra-1,4,8,11-tetra- Aza-13-nitro-cyclotetradeca-3,8,11,13-tetraen> 83. 88
cyanato)-(2,3:11,12)-dibenzo-1,13-dioxa-5,9-di Aza-2,11-cyclopentadecadiene-nickel(ii) <othio 84. 89
itrate] nitrato-(1-oxa-7,10-di Aza-4,13-dithia-cyclopentadecane)-nickel(ii) n 85. 36
)-nickel(ii)> (2,12-dimethyl-3,7,11,17-tetra- Azabicyclo(11.3.1)heptadeca-1(17),13,15-triene 83.100
,12-dimethyl-7,14-di-isopropyl-1,4,8,11-tetra- Azacyclotetradecane-nickel(ii) perchlorate [5 83.135
bis(1,4-bis(3,5-dimethylphenyl)-tetra- Azadiene)-nickel(0)....................... 83.205
(3,10-dihydroxyimino-4,9-dimethyl-5,8-di Azadodeca-4,9-diene-2,11-dionato)-nickel(ii) 76. 70
(2,6-bis(1-methyl-4-methylthio-5-thia-2,3-di Azahexa-1,3-dienyl)-pyridine-dithiocyanato-n> 85. 59
zenato-nickel(ii)] Benzamidoximato-N(1)-oxy-N(2)-benzoylimino-dia 83. 92
bis(2-(N-benzylaldimino)- Benzo(b)furan-3-thiolato)-nickel(ii) 85. 83
bis(2-(N-phenylaldimino)- Benzo(b)furan-3-thiolato)-nickel(ii) 85. 82
diene-nic> bis(isothiocyanato)-(2,3:11,12)-di Benzo-1,13-dioxa-5,9-diaza-2,11-cyclopentadeca 84. 89
bis((m-methyl Benzoato)-(1,3-propanediamine)) nickel(ii) 81. 75
bis((m-nitro Benzoato)-(1,3-propanediamine)) nickel(ii) 81. 69
bis((p-methyl Benzoato)-(1,3-propanediamine)) nickel(ii) 81. 76
dihydrate] diaqua-(2,6-diacetylpyridine-bis(Benzoic acid hydrazone))-nickel(ii) dinitrate 84. 95
at> lithium (μ(2)-benzophenonimino)-tetrakis(*Benzophenonimino)-di-nickel diethyl ether solv 72. 67
i-nickel diethyl ether solvat> lithium (μ(2)-*Benzophenonimino)-tetrakis(benzophenonimino)-d 72. 67
1,4,5,8,9,12,13,16-octamethyl-tetra Benzoporphyrinato-nickel(ii) iodide............ 49. 18
(N,N'-ethylene-bis(Benzoylacetoniminato))-nickel(ii) 84. 91
l(0) benzene solvate] (η(2)-methyl-3- Benzoylacrylate)-bis(triphenylphosphine)-nicke 72. 64
triaquo-bis(N- Benzoylglycinato)-nickel(ii) dihydrate 82. 37
benzamidoximato-N(1)-oxy-N(2)- Benzoylimino-diazenato-nickel(ii) 83. 92
rbonyl-tetrakis(μ(2)-carbonyl)-nickel-tri-i> *Benzyl-trimethylammonium (μ(3)-hydrido)-octaca 3. 43
el(ii)] bis(2-(N- Benzylaldimino)-benzo(b)furan-3-thiolato)-nick 85. 83
ickel(ii)> (2,12-dimethyl-3,7,11,17-tetra-aza Bicyclo(11.3.1)heptadeca-1(17),13,15-triene)-n 83.100
dichloro-bis(2,2'-bis(dimethylarsine)- Biphenyl)-nickel(ii) 86.110
(2,2'- Bipyridyl)-nickela-cyclopentane.............. 71. 67
bis(1,1-dimethyl-4-phenyl-1-sila-4- Bora-cyclohexa-2,5-dienyl)-nickel 75. 34

ickel]   bis(Boranato-bis(dimethylphosphonium-methylide))-n   71. 43
         (tetra-cyclopentadienyl-nickela)-octa Borane 73. 87
                    bis(hydrido-tris(1-pyrazolyl)- Borato)-nickel(ii) 83.114
)-cyclopentadienyl-nickel)-(μ(4)-3,3-dimethyl-*But-1-enyl)-nonacarbonyl-tri-ruthenium   [(η(5   71.151
         di-lithium tris(N,N,N',N'-tetramethyl- But-2-ene-1,4-diamine)-nickel(ii) 72. 51
ydrate]   bis(3-amino-3-methyl- Butan-2-one-oximato)-nickel(ii) chloride monoh   83. 60r
         bromo-(2,2'-(1,3-di-iminopropane)-bis(Butan-3-one oximato)-nickel(ii) 83. 67
         bis(2-(1-iminoethyl)-1-cyano-1,3- Butanedionato)-nickel(ii) 77. 3
enylgermyl)-(triphenylgermyl-cadmium)-nickel)- Cadmium toluene solvate <clopentadienyl-(triph   73.134
entadienyl-(triphenylgermyl)-(triphenylgermyl- Cadmium)-nickel)-cadmium toluene solvate <clop   73.134
(ii) tetrahydrate]   nickel-bis(tetracarbonyl-μ-(dimethyl*Carbamoyl)-iron)-   71. 63
larsonium) tris(2,2-diselenido-1,1-ethylene-di Carbonitrile)-nickel(iv)   ....[bis(tetrapheny   85. 47
-nickel)]   tricarbonyl-*Cobalt-(μ-diphenylacetylene)-(cyclopentadienyl   72. 39
diamine-tetra-acetato-> bis(ethylenediamine)- Copper (bis(ethylenediamine)-copper)-(ethylene   76. 18
<thylenediamine)-copper (bis(ethylenediamine)- Copper)-(ethylenediamine-tetra-acetato)-nickel>   76. 18
l tetrahydrate]   bis(ethylenediamine)-*Copper-(μ-ethylenediamine-tetra-acetato)-nicke   84. 58
)-nickel(ii)]   bis(isoseleno Cyanato-(N,N,N',N'-tetramethylethylenediamine)   76. 89
thylvalyl-D-α-hydroxyisovaleryl-L-methylval> Cyclo(L-methylvalyl-D-α-hydroxyisovaleryl-L-me   50. 22
                    bis(η(3)- Cyclo-octatrienyl)-nickel 75. 9
   bis(1,1-dimethyl-4-phenyl-1-sila-4-bora- Cyclohexa-2,5-dienyl)-nickel   75. 34
                 trans-diaqua-bis(Cyclohexane-1,2-diamine)-nickel(ii) chloride   76. 79
                     cis-dichloro-bis(Cyclohexane-1,2-dione-dioxime)-nickel(ii) 83. 81
(2,3:11,12)-dibenzo-1,13-dioxa-5,9-diaza-2,11- Cyclopentadecadiene-nickel(ii) <othiocyanato)-   84. 89
         nitrato-(1-oxa-7,10-diaza-4,13-dithia- Cyclopentadecane)-nickel(ii) nitrate 85. 36
germyl-mercury)-nickel)-mercury t> bis((η(5)- Cyclopentadienyl)-(triphenylgermyl)-(triphenyl   73.135
ermyl-cadmium)-nickel)-cadmium tol> bis((η(5)- Cyclopentadienyl)-(triphenylgermyl)-(triphenylg   73.134
                 bis(μ-carbonyl)-bis(*Cyclopentadienyl-nickel) 73. 30
            tris(μ(3)-hydrido)-tetra(*Cyclopentadienyl-nickel) 73. 85r
           bis(μ-carbonyl)-bis(methyl*Cyclopentadienyl-nickel) 73. 47
  tricarbonyl-cobalt-(μ-diphenylacetylene)-(*Cyclopentadienyl-nickel) 72. 39
xacarbonyl-di-iron]   (η(5)- Cyclopentadienyl-nickel)-(η-t-butylethynyl)-he   72. 29
t-1-enyl)-nonacarbonyl-tri-ruthenium] (η(5)-*Cyclopentadienyl-nickel)-(μ(4)-3,3-dimethyl-bu   71.151
   (μ(3)-carbonyl)-bis(η(5)-*Cyclopentadienyl-nickel)-bis(tricarbonyl-iron)   71.105
   (hex-3-ene-3,4-diyl)-bis(η(5)- Cyclopentadienyl-nickel)-bis(tricarbonyl-iron)   71.178
              (tetra- Cyclopentadienyl-nickela)-octaborane 73. 87
                 η(5)- Cyclopentadienyl-triphenylphosphine-nickel-eth   73.107
ylxanthate]
                (2,2'-bipyridyl)-nickela- Cyclopentane 71. 67
<,7-dioxa-6-hydra-1,4,8,11-tetra-aza-13-nitro- Cyclotetradeca-3,8,11,13-tetraene)-nickel(ii)   83. 88
-dimethyl-7,14-di-isopropyl-1,4,8,11-tetra-aza Cyclotetradecane-nickel(ii) perchlorate   [5,12   83.135
<phenylphosphinomethyl)ethane-rhodium-(μ-η(3)-*Cyclotriphosphorus)-1,1,1-tris(diphenylphosph>   86.228
   benzamidoximato-N(1)-oxy-N(2)-benzoylimino- Diazenato-nickel(ii) 83. 92
                 L,D,L,D,L,D- Enniatin B sodium nickel nitrate hydrate 50. 22
3,7,9,13-tetramethyl-6-(bis(diphenylphosphino) Ethane)-3,7,9,13-tetracarba-6-nickela-trideca>   71.242
<osphorus)-1,1,1-tris(diphenylphosphinomethyl)*Ethane-nickel bis(tetraphenylborate) tetrahyd>   86.228
,1-tris(> 1,1,1-tris(diphenylphosphinomethyl)*Ethane-rhodium-(μ-η(3)-cyclotriphosphorus)-1,1   86.228
lorate]   bis(2-amino Ethanol)-(2-aminoethanolato)-nickel(iii) perch   84. 14
-nitrosyl-nickel]   (dimethyl-(N,N-dimethyl Ethanolamino)-(3,5-dimethylpyrazolyl)-gallato)   84. 33
(dimethyl-(3,5-dimethyl-1-pyrazolyl)-(dimethyl Ethanolamino)-gallato)-nickel(ii) <)-gallato)-   83.166
         bis(2-aminoethanol)-(2-amino Ethanolato)-nickel(iii) perchlorate.............   84. 14
-(triphenylphosphine)-(Z-1,2-diphenyl-2-methyl Ethenyl)-nickel   [acetylacetonato   71.255
                 (N,N'- Ethylene-bis(benzoylacetoniminato))-nickel(ii)   84. 91
el(ii)]   N,N'- Ethylene-bis(1-iminomethyl-2-naphtholato)-nick   76.100
(tetraphenylarsonium) tris(2,2-diselenido-1,1- Ethylene-dicarbonitrile)-nickel(iv)....   [bis   85. 47
      bis(isothiocyanato-(N,N,N',N'-tetramethyl Ethylenediamine))-nickel(ii) 76. 88
      bis(isoselenocyanato-(N,N,N',N'-tetramethyl Ethylenediamine))-nickel(ii) 76. 89
copper)-(ethylenediamine-tetra-acetato-> bis(Ethylenediamine)-copper (bis(ethylenediamine)-   76. 18
a-acetato-> bis(ethylenediamine)-copper (bis(Ethylenediamine)-copper)-(ethylenediamine-tetr   76. 18
ra acetato)-nickel tetrahydrate]   bis(*Ethylenediamine)-copper-(μ-ethylenediamine-tet   84. 58
                 tris(Ethylenediamine)-nickel(ii) dinitrate 76. 40a
                tetra-aqua Ethylenediamine)-nickel(ii) nitrate............   76. 2
                tetra-aquo- Ethylenediamine-nickel(ii) sulfate dihydrate   76. 3
a-acetato)-nickel]   tetra-aqua-*Ethylenediamine-nickel-(μ-ethylenediamine-tetr   76. 74
    tetra-aqua-ethylenediamine-nickel-(μ-*Ethylenediamine-tetra-acetato)-nickel 76. 74
rate]   bis(ethylenediamine)-copper-(μ-*Ethylenediamine-tetra-acetato)-nickel tetrahyd   84. 58
ahydrate]   calcium (Ethylenediamine-tetra-acetato)-nickel(ii) tetr   76. 57
<amine)-copper (bis(ethylenediamine)-copper)-(Ethylenediamine-tetra-acetato)-nickel) dihydra>   76. 18
ihydrate]   penta-aquo-nickel(ii)-((S,S)- Ethylenediamine-N,N'-disuccinato)-nickel(ii) d   76. 64
          tetrathiafulvalene bis(Ethylenedithiolato)-nickel(i) 60. 5
   (η(5)-cyclopentadienyl-nickel)-(η-t-butyl Ethynyl)-hexacarbonyl-di-iron 72. 29
                 bis(2- Formylpyridine-thiosemicarbazonato)-nickel(ii)   79. 11
                 tetrathia Fulvalene bis(ethylenedithiolato)-nickel(i).   60. 5
thane-nickel bis(tetraphenylborate) tetrahydro*Furan <s)-1,1,1-tris(diphenylphosphinomethyl)e   86.228
         bis(2-(N-benzylaldimino)-benzo(b) Furan-3-thiolato)-nickel(ii) 85. 83
         bis(2-(N-phenylaldimino)-benzo(b) Furan-3-thiolato)-nickel(ii) 85. 82

(di> (dimethyl-bis(3,5-dimethyl-1-pyrazolyl)- Gallato)-(dimethyl-(3,5-dimethyl-1-pyrazolyl)- 83.166
-dimethyl-1-pyrazolyl)-(dimethylethanolamino)- Gallato)-nickel(ii) <)-gallato)-(dimethyl-(3,5 83.166
dimethylethanolamino)-(3,5-dimethylpyrazolyl)- Gallato)-nitrosyl-nickel [(dimethyl-(N,N- 84. 33
ium tol> bis(η(5)-cyclopentadienyl)-(triphenyl Germyl)-(triphenylgermyl-cadmium)-nickel)-cadm 73.134
ury t> bis((η(5)-cyclopentadienyl)-(triphenyl Germyl)-(triphenylgermyl-mercury)-nickel)-merc 73.135
<cyclopentadienyl-(triphenylgermyl)-(triphenyl Germyl-cadmium)-nickel)-cadmium toluene solva> 73.134
<yclopentadienyl)-(triphenylgermyl)-(triphenyl Germyl-mercury)-nickel)-mercury toluene solva> 73.135
triaquo-bis(N-benzoyl Glycinato)-nickel(ii) dihydrate................ 82. 37
penta-aquo-(2'-deoxy Guanosine-5'-phosphato)-nickel(ii) trihydrate 83. 63
<2-dimethyl-3,7,11,17-tetra-azabicyclo(11.3.1) Heptadeca-1(17),13,15-triene)-nickel(ii) perc> 83.100
nickel)-bis(tricarbonyl-iron)] (Hex-3-ene-3,4-diyl)-bis(η(5)-cyclopentadienyl- 71.178
triaquo-bis(Hippurato)-nickel(ii) dihydrate............... 82. 37
bis(diphenyl- Hydantoinato)-diaqua-diammine-nickel(ii) 83.201
ca-3,> (2,2,3,9,10,10-hexamethyl-5,7-dioxa-6- Hydra-1,4,8,11-tetra-aza-13-nitro-cyclotetrade 83. 88
diaqua-(2,6-diacetylpyridine-bis(benzoic acid Hydrazone))-nickel(ii) dinitrate dihydrate 84. 95
bis(triphenylphosphine)iminium (μ-*Hydrido)-bis(tricarbonyl-nickel) 64.152
nickel-tri-i> benzyl-trimethylammonium (μ(3)-*Hydrido)-octacarbonyl-tetrakis(μ(2)-carbonyl)- 3. 43
tris(μ(3)-*Hydrido)-tetra(cyclopentadienyl-nickel)........ 73. 85n
bis(tetraphenylarsonium) di Hydrido-dodeca-nickel-carbonyl 65. 16
bis(tetraphenylphosphonium) di Hydrido-dodeca-nickel-carbonyl 64.128n
bis(tetraphenylphosphonium) di Hydrido-dodeca-nickel-carbonyl 64.127
bis(bis(triphenylphosphine)immonium) di Hydrido-dodeca-nickel-carbonyl 64.164
tris(tetraphenylarsonium) Hydrido-dodeca-nickel-carbonyl acetone solvate 65. 20
tris(tetraphenylarsonium) Hydrido-dodeca-nickel-carbonyl acetone solvate 65. 21n
bis(Hydrido-tris(1-pyrazolyl)-borato)-nickel(ii) 83.114
bis(triphenylphosphine)*Iminium (μ-hydrido)-bis(tricarbonyl-nickel).... 64.152
benzamidoximato-N(1)-oxy-N(2)-benzoyl Imino-diazenato-nickel(ii) 83. 92
,11-dionato)-nickel(ii)] (3,10-dihydroxy Imino-4,9-dimethyl-5,8-diazadodeca-4,9-diene-2 76. 70
ii)] bis(2-(1- Iminoethyl)-1-cyano-1,3-butanedionato)-nickel(77. 3
N,N'-ethylene-bis(1- Iminomethyl-2-naphtholato)-nickel(ii) 76.100
bromo-(2,2'-(1,3-di Iminopropane)-bis(butan-3-one-oximato))-nickel 83. 67
bis(bis(triphenylphosphine) Immonium) dihydrido-dodeca-nickel-carbonyl 64.164
yl-nickel)-(η-t-butylethynyl)-hexacarbonyl-di Iron .... [(η(5)-cyclopentadien 72. 29
tacarbonyl-tetrakis(μ(2)-carbonyl)-nickel-tri-*Iron <nzyl-trimethylammonium (μ(3)-hydrido)-oc 3. 43
ickel-bis(tetracarbonyl-μ-(dimethylcarbamoyl)-*Iron) .... [n 71. 63
η(5)-cyclopentadienyl-nickel)-bis(tricarbonyl-*Iron) .... [(μ(3)-carbonyl)-bis(71.105
η(5)-cyclopentadienyl-nickel)-bis(tricarbonyl-Iron) .... [(hex-3-ene-3,4-diyl)-bis(71.178
ediamine))-nickel(ii)] bis(Isoselenocyanato-(N,N,N',N'-tetramethylethylen 76. 89
-5,9-diaza-2,11-cyclopentadecadiene-nic> bis(Isothiocyanato)-(2,3:11,12)-dibenzo-1,13-dioxa 84. 89
kel(ii) 2-bromonaphthalene clathrate] bis(Isothiocyanato)-tetrakis(4-methylpyridine)-nic 61. 10
kel(ii) 2-methylnaphthalene clathrate] bis(Isothiocyanato)-tetrakis(4-methylpyridine)-nic 61. 11
iamine))-nickel(ii)] bis(Isothiocyanato-(N,N,N',N'-tetramethylethylened 76. 88
tetrabutylammonium bis(Isotrithione-dithiolato)-nickel(ii)............ 85. 14
-α-hydroxyisovaleryl-L-methylvalyl-D-α-hydroxy Isovaleryl) sodium nickel nitrate hydrate <I-D 50. 22
<α-hydroxyisovaleryl-L-methylvalyl-D-α-hydroxy Isovaleryl-L-methylvalyl-D-α-hydroxyisovalery> 50. 22
-L-methylval> cyclo(L-methylvalyl-D-α-hydroxy Isovaleryl-L-methylvalyl-D-α-hydroxyisovaleryl 50. 22
phenonimino)-di-nickel diethyl ether solvat> *Lithium (μ(2)-benzophenonimino)-tetrakis(benzo 72. 67
Lithium bis(dithioxalato)-nickel(ii) dihydrate 81. 15
,4-diamine)-nickel(ii)] di- Lithium tris(N,N,N',N'-tetramethyl-but-2-ene-1 72. 51
enylgermyl)-(triphenylgermyl-mercury)-nickel- Mercury toluene solvate <lopentadienyl-(triph 73.135
ntadienyl)-(triphenylgermyl)-(triphenylgermyl- Mercury)-nickel)-mercury toluene solvate <lope 73.135
)] (ethyl- Methacrylate)-bis(triphenylphosphine)-nickel(0 72. 60
chloro)-bis(chloro-(bis(3,5-dimethylpyrazolyl)*Methane)-nickel(ii)) ....[bis(μ- 83.161
bis(boranato-bis(dimethylphosphonium)- Methylide)-nickel 71. 43
bis(Morpholine-N-carbodithioato)-nickel(ii)....... 80. 4
-tetrakis(4-methylpyridine)-nickel(ii) 2-bromo Naphthalene clathrate [bis(isothiocyanato) 61. 10
tetrakis(4-methylpyridine)-nickel(ii) 2-methyl Naphthalene clathrate [bis(isothiocyanato)- 61. 11
N,N'-ethylene-bis(1-iminomethyl-2- Naphtholato)-nickel(ii)...................... 76.100
)] (-)(579)-bis(S)-(+)-O,O'-(1,1'-bi Naphthyl-2,2'-diyl)-dithiophosphato)-nickel(ii 85. 89a
diaquo-bis(N,N-diethyl Nicotinamide)-diacetato-nickel(ii) 81. 80
tadecane)-nickel(ii) nitrate] Nitrato-(1-oxa-7,10-diaza-4,13-dithia-cyclopen 85. 36
thanolamino)-(3,5-dimethylpyrazolyl)-gallato- Nitrosyl-nickel.... [(dimethyl-(N,N-dimethyle 84. 33
sodium tris(μ-3,5-dimethylpyrazolyl)-bis(*Nitrosyl-nickel(i)) tetrahydrofuran solvate.... 83. 99
bis(bis(3,5-dimethylpyrazolyl)- Nitrosyl-nickel(i))-nickel(ii) 83.150
bis(μ-3,5-dimethylpyrazolyl)-bis(*Nitrosyl-nickel) 83. 54
ium (μ-iodo)-bis(μ-3,5-dimethylpyrazolyl)-bis(*Nitrosyl-nickel) .... [tetraethylammon 83. 53
(1,9-bis(diphenylphosphino)-3,7-dithia Nonane)-nickel(ii) diperchlorate 86.108
iodo-(1,9-bis(diphenylphosphino)-3,7-dithia Nonane)-nickel(ii) tetraphenylborate 86.107
<s(isothiocyanato)-(2,3:11,12)-dibenzo-1,13-di Oxa-5,9-diaza-2,11-cyclopentadecadiene-nickel> 84. 89
etradeca-3,> (2,2,3,9,10,10-hexamethyl-5,7-di Oxa-6-hydra-1,4,8,11-tetra-aza-13-nitro-cyclot 83. 88
ickel(ii) nitrate] nitrato-(1- Oxa-7,10-diaza-4,13-dithia-cyclopentadecane)-n 85. 36
potassium bis(dithio- Oxalato)-nickel(ii)........................[bis(1- 85. 6
sodium bis(dithio- Oxalato)-nickel(ii) dihydrate 85. 7
rate] tris(Oxamide-oximato)-nickel(ii) dichloride hemihyd 83. 24
methyl-3-(2-chloro-6-methylphenyl)-triazine-1- Oxidato)-nickel(ii)....................[bis(1- 84. 62

methyl-3-(2-chloro-6-methylphenyl)-triazine-1- Oxidato)-nickel(ii) benzene solvate....[bis(1-   84. 63
o-(2,2'-(1,3-di-iminopropane)-bis(butan-3-one- Oximato))-nickel(ii)   [brom   83. 67
bis(3-amino-3-methyl-butan-2-one- Oximato)-nickel(ii) chloride monohydrate 83. 60
tris(oxamide- Oximato)-nickel(ii) dichloride hemihydrate 83. 24
cis-dichloro-bis(cyclohexane-1,2-dione-di Oxime)-nickel(ii)   83. 81
ckel(ii)]   chloro-(Pentane-2,4-dithionato)-(triethylphosphine)-ni   86. 15
is(4-methylpyridine)-nickel(ii) bis(hexafluoro Phosphate) ....[tetrak   83.178
penta-aquo-(2'-deoxyguanosine-5'- Phosphato)-nickel(ii) trihydrate   83. 63
ckel]   acetylacetonato-(triphenyl Phosphine)-(Z-1,2-diphenyl-2-methylethenyl)-ni   71.255
di-iodo-(trimethyl Phosphine)-carbonyl-nickel   86. 5
chloro-(pentane-2,4-dithionato)-(triethyl Phosphine)-nickel(ii) 86. 15
(ethyl-methacrylate)-bis(triphenyl Phosphine)-nickel(0) 72. 60
(η(2)-methyl-3-benzoylacrylate)-bis(triphenyl Phosphine)-nickel(0) benzene solvate 72. 64
nickel)]   bis(triphenyl *Phosphine)iminium (μ-hydrido-bis(tricarbonyl-   64.152
rbonyl]   bis(bis(triphenyl Phosphine)immonium) dihydrido-dodeca-nickel-ca   64.164
bis(η(3)-allyl)-trimethyl Phosphine-nickel 72. 8
η(5)-cyclopentadienyl-triphenyl Phosphine-nickel-ethylxanthate 73.107
borate] bis(2-diphenylphosphinoethyl)-phenyl Phosphine-nickel-O-methylsulfinate tetraphenyl   86.115
dicarbonyl-(1,3-bis(phenyl Phosphino)-propane)-nickel(0) 86. 46
hlorate]   (1,9-bis(diphenyl Phosphino)-3,7-dithianonane)-nickel(ii) diperc   86.108
henylborate]   iodo-(1,9-bis(diphenyl Phosphino)-3,7-dithianonane)-nickel(ii) tetrap   86.107
a-tride> 3,7,9,13-tetramethyl-6-(bis(diphenyl Phosphino)ethane)-3,7,9,13-tetracarba-6-nickel   71.242
(tris(2-diphenyl Phosphinoethyl)-amino)-iodo-nickel(i) 86.159
lsulfinate tetraphenylborate] bis(2-diphenyl Phosphinoethyl)-phenylphosphine-nickel-O-methy   86.115
<-η(3)-cyclotriphosphorus)-1,1,1-tris(diphenyl *Phosphinomethyl)ethane-nickel bis(tetraphenyl>   86.228
iphosphorus)-1,1,1-tris(> 1,1,1-tris(diphenyl *Phosphinomethyl)ethane-rhodium-(μ-η(3)-cyclotr   86.228
triaquo-(pyridoxylidene-O- Phospho-D,L-threoninato)-nickel(ii) dihydrate   82. 25
bis(tetraphenyl Phosphonium) dihydrido-dodeca-nickel-carbonyl   64.128
bis(tetraphenyl Phosphonium) dihydrido-dodeca-nickel-carbonyl   64.127
bis(boranato-bis(dimethyl Phosphonium-methylide))-nickel   71. 43
<osphinomethyl)ethane-rhodium-(μ-η(3)-cyclotri *Phosphorus)-1,1,1-tris(diphenylphosphinomethy>   86.228
)]   tris(2- Picolylamine)-nickel(ii) bis(tetrafluoroborate   83.130
ans,cis,trans-octaethyl-1,2,3,7,8,20-hexahydro Porphyrinato)-nickel(ii) ....[(trans,tr   49. 12
s,trans,trans-octaethyl-1,2,3,7,8,20-hexahydro Porphyrinato)-nickel(ii) ....   [(trans,tran   49. 13
tct-2,3,7,8-tetrahydro-octaethyl Porphyrinato)-nickel(ii)..................   49. 8
ttt-2,3,7,8-tetrahydro-octaethyl Porphyrinato)-nickel(ii)..................   49. 9
1,4,5,8,9,12,13,16-octamethyl-tetrabenzo Porphyrinato)-nickel(ii) iodide 49. 18
Potassium bis(dithio-oxalato)-nickel(ii) 85. 6
bromo-(2,2'-(1,3-di-imino Propane)-bis(butan-3-one-oximato))-nickel(ii)   83. 67
dicarbonyl-(1,3-bis(phenylphosphino)- Propane)-nickel(0) 86. 46
bis((m-methylbenzoato)-(1,3- Propanediamine)) nickel(ii)..................   81. 75
bis((m-nitrobenzoato)-(1,3- Propanediamine))-nickel(ii)................   81. 69
bis((p-methylbenzoato)-(1,3- Propanediamine))-nickel(ii)................   81. 76
tetrakis(3-amino-1- Propanethiolato)-tri-nickel(ii) chloride   85. 46
rans-dichloro-tetrakis(1H,3,5-diethyl-4-methyl Pyrazole)-nickel(ii)   [t   83.207
<yrazolyl)-gallato)-(dimethyl-(3,5-dimethyl-1- Pyrazolyl)-(dimethylethanolamino)-gallato)-ni>   83.166
uran solvate]   sodium tris(μ-3,5-dimethyl *Pyrazolyl)-bis(nitrosyl-nickel(i)) tetrahydrof   83. 99
bis(μ-3,5-dimethyl *Pyrazolyl)-bis(nitrosyl-nickel)..............   83. 54
tetraethylammonium (μ-iodo)-bis(μ-3,5-dimethyl *Pyrazolyl)-bis(nitrosyl-nickel)..............   83. 53
bis(hydrido-tris(1- Pyrazolyl)-borato-nickel(ii)   83.114
pyrazolyl)-(di> (dimethyl-bis(3,5-dimethyl-1- Pyrazolyl)-gallato)-(dimethyl-(3,5-dimethyl-1-   83.166
ethyl-(N,N-dimethylethanolamino)-(3,5-dimethyl Pyrazolyl)-gallato)-nitrosyl-nickel.... [(dim   84. 33
bis(bis(3,5-dimethyl Pyrazolyl)-nitrosyl-nickel(i))-nickel(ii) 83.150
bis(μ-chloro-bis(chloro-(3,5-dimethyl *Pyrazolyl)methane)-nickel(ii)-   83.161
4-methylthio-5-thia-2,3-diazahexa-1,3-dienyl)- Pyridine-dithiocyanato-nickel(ii) <(1-methyl-   85. 59
tetrakis(4-methyl Pyridine)-nickel(ii) bis(hexafluorophosphate)   83.178
ate]   bis(isothiocyanato)-tetrakis(4-methyl Pyridine)-nickel(ii) 2-bromonaphthalene clathr   61. 10
rate]   bis(isothiocyanato)-tetrakis(4-methyl Pyridine)-nickel(ii) 2-methylnaphthalene clath   61. 11
i) dinitrate dihydrate]   diaqua-(2,6-diacetyl Pyridine-bis(benzoic acid hydrazone))-nickel(i   84. 95
bis(2-formyl Pyridine-thiosemicarbazonato)-nickel(II)   79. 11
el(ii) dihydrate]   triaquo-(Pyridoxylidene-O-phospho-D,L-threoninato)-nick   82. 25
bis(D-β-(2- Pyridyl)-α-alaninato)-nickel(ii) dihydrate 82. 30
bis(2-isopropyl Quinoline-8-thiolato) nickel   85. 76
(> 1,1,1-tris(diphenylphosphinomethyl)ethane- *Rhodium-(μ-η(3)-cyclotriphosphorus)-1,1,1-tris   86.228
(4)-3,3-dimethyl-but-1-enyl)-nonacarbonyl-tri- *Ruthenium [(η(5)-cyclopentadienyl-nickel)-(μ   71.151
v)]   bis(tetraphenylarsonium) tris(2,2-di Selenido-1,1-ethylene-dicarbonitrile)-nickel(i   85. 47
amine))-nickel(ii)]   bis(iso Selenocyanato-(N,N,N',N'-tetramethylethylenedi   76. 89
(α,β)-bis(S-methylthio Semicarbazido)-nickel(ii) bromide 83. 13
(α,β)-bis(S-methylthio Semicarbazido)-nickel(ii) bromide 83. 14
bis(S-methylthio Semicarbazido)-nickel(ii) iodide 83. 15
bis(2-formylpyridine-thio Semicarbazonato)-nickel(ii)   79. 11
bis(1,1-dimethyl-4-phenyl-1- Sila-4-bora-cyclohexa-2,5-dienyl)-nickel 75. 34
e]   Sodium bis(dithio-oxalato)-nickel(ii) dihydrat   85. 7
L,D,L,D,L,D-enniatin B Sodium nickel nitrate hydrate 50. 22
ovaleryl-L-methylvalyl-D-α-hydroxyisovaleryl) Sodium nickel nitrate hydrate <l-D-α-hydroxyis   50. 22

syl-nickel(i)) tetrahydrofuran solvate]   *Sodium tris(μ-3,5-dimethylpyrazolyl)-bis(nitro   83. 99
aquo-nickel(ii)-((S,S)-ethylenediamine-N,N'-di   Succinato)-nickel(ii) dihydrate........[penta-   76. 64
hosphinoethyl)-phenylphosphine-nickel-0-methyl   Sulfinate tetraphenylborate [bis(2-diphenylp   86.115
                   nitrato-(1-oxa-7,10-diaza-4,13-di   Thia-cyclopentadecane)-nickel(ii) nitrate 85. 36
ocyanato-n> (2,6-bis(1-methyl-4-methylthio-5-   Thia-2,3-diazahexa-1,3-dienyl)-pyridine-dithi   85. 59
)]   tetra   Thiafulvalene bis(ethylenedithiolato)-nickel(i   60.   5
            (1,9-bis(diphenylphosphino)-3,7-di   Thianonane)-nickel(ii) diperchlorate 86.108
      iodo-(1,9-bis(diphenylphosphino)-3,7-di   Thianonane)-nickel(ii) tetraphenylborate 86.107
                          potassium bis(di   Thio-oxalato)-nickel(ii)   85.   6
                            sodium bis(di   Thio-oxalato)-nickel(ii) dihydrate 85.   7
)-dithiocyanato-n> (2,6-bis(1-methyl-4-methyl   Thio-5-thia-2,3-diazahexa-1,3-dienyl)-pyridine   85. 59
                     bis(morpholine-N-carbodi   Thioato)-nickel(ii)........................   80.   4
                   bis(tri-iodoarsine)-bis(diethyldi   Thiocarbamato)-nickel(ii)   80.   6
vate trihydrate]   tris(methyl 2-methyldi   Thiocarbazato)-nickel(ii) chloride ethanol sol   85. 33
                    bis(2-isopropylquinoline-8-   Thiolato) nickel 85. 76
         tetrabutylammonium bis(isotrithione-di   Thiolato)-nickel(ii) 85. 14
      bis(2-(N-benzylaldimino)-benzo(b)furan-3-   Thiolato)-nickel(ii) 85. 83
       bis(2-(N-phenylaldimino)-benzo(b)furan-3-   Thiolato)-nickel(ii) 85. 82
                    chloro-(pentane-2,4-di   Thionato)-(triethylphosphine)-nickel(ii) 86. 15
            tetrabutylammonium bis(isotri   Thione-dithiolato)-nickel(ii) 85. 14
is((S)-(+)-0,0'-(1,1'-binaphthyl-2,2'-diyl)-di   Thiophosphato)-nickel(ii) ....[(-)(579)-b   85. 89a
                   bis(ethylmethyl-di   Thiophosphinato)-nickel(ii)...................   85. 12
            bis(dimethylamido-(methyl)-di   Thiophosphonato)-nickel(ii) 85. 13
              (α,β)-bis(S-methyl   Thiosemicarbazido)-nickel(ii) bromide 83. 14
              (α,β)-bis(S-methyl   Thiosemicarbazido)-nickel(ii) bromide 83. 13
                     bis(S-methyl   Thiosemicarbazido)-nickel(ii) iodide 83. 15
              bis(2-formylpyridine-   Thiosemicarbazonato)-nickel(ii).............   79. 11
                  lithium bis(di   Thioxalato)-nickel(ii) dihydrate 81. 15
            triaquo-(pyridoxylidene-0-phospho-D,L-   Threoninato)-nickel(ii) dihydrate 82. 25
                               η(6)-   Toluene-bis(perfluorophenyl)-nickel...........   71.134
          bis(1-methyl-3-(2-chloro-6-methylphenyl)-   Triazine-1-oxidato)-nickel(ii) 84. 62
          bis(1-methyl-3-(2-chloro-6-methylphenyl)-   Triazine-1-oxidato)-nickel(ii) benzene solvate   84. 63
osphino)ethane)-3,7,9,13-tetracarba-6-nickela-   Tridecaborane <3-tetramethyl-6-(bis(diphenylph   71.242
hydroxyisovaleryl-L-methylvalyl-D-α-hydroxyiso   Valeryl) sodium nickel nitrate hydrate <l-D-α-   50. 22
<ydroxyisovaleryl-L-methylvalyl-D-α-hydroxyiso   Valeryl-L-methylvalyl-D-α-hydroxyisovaleryl) >   50. 22
methylval>' cyclo(L-methylvalyl-D-α-hydroxyiso   Valeryl-L-methylvalyl-D-α-hydroxyisovaleryl-L-   50. 22
<-L-methylvalyl-D-α-hydroxyisovaleryl-L-methyl   Valyl-D-α-hydroxyisovaleryl) sodium nickel ni>   50. 22
hydroxyisovaleryl-L-methylval> cyclo(L-methyl   Valyl-D-α-hydroxyisovaleryl-L-methylvalyl-D-α-   50. 22
<(L-methylvalyl-D-α-hydroxyisovaleryl-L-methyl   Valyl-D-α-hydroxyisovaleryl-L-methylvalyl-D-α>   50. 22
clopentadienyl-triphenylphosphine-nickel-ethyl   Xanthate ....   [η(5)-cy   73.107

Niobium

chloride solvate]   tetraphenyl Arsonium pentachloro-oxo-niobium(v) methylene   65. 17
tadienyl-niobium) benzene so> (bis(tetrahydro*Borato)-zinc)-(μ-hydrido)-(carbonyl-dicyclopen   73. 25
dienyl)-2,3,4,5-tetra(trifluoromethyl)-niobia- Cyclopentadiene.... [1,1-bis(η(5)-cyclopenta   71.119
<2-bis(diphenylphosphino)ethane)-trichloro-(η- Cyclopentadienyl)-niobium(iv) toluene solvate   73.117
           (μ-oxo)-bis(aqua-trichloro-(η-methyl Cyclopentadienyl)-niobium(v)) 73. 40
                 (μ-oxo)-bis(n-butyl-bis(*Cyclopentadienyl)-niobium) 71.221
          bis(η(5)-trimethylsilyl- Cyclopentadienyl)-o-xylidene-niobium(iv) 73.101
                bis(η(5)- Cyclopentadienyl)-peroxochloro-niobium(v) 73. 14
l)-niobia-cyclopentadiene]   1,1-bis(η(5)- Cyclopentadienyl)-2,3,4,5-tetra(trifluoromethy   71.119
iene)-dichloro-niobium]   η(5)- Cyclopentadienyl-(η(4)-3,4-dimethyl-hexa-2,4-d   72. 21
           heptacarbonyl-tris(Cyclopentadienyl-niobium) 73. 90
trahydroborato)-zinc)-(μ-hydrido)-(carbonyl-di*Cyclopentadienyl-niobium) benzene solvate <(te   73. 25
(iv) toluene sol> (1,2-bis(diphenylphosphino) Ethane)-trichloro-(η-cyclopentadienyl)-niobium   73.117
benzene so> (bis(tetrahydroborato)-zinc)-(μ-*Hydrido)-(carbonyl-dicyclopentadienyl   73. 25
ate]   bis(μ-chloro)-bis(trichloro-triphenyl*Phosphinimino-niobium) 1,2-dichloroethane solv   83.211
l)-niobium(iv) toluene sol> (1,2-bis(diphenyl Phosphino)ethane)-trichloro-(η-cyclopentadieny   73.117
            bis(η(5)-trimethyl Silyl-cyclopentadienyl)-o-xylidene-niobium(iv)   73.101
iophene)-di-niobium]   (μ-*Thio)-(μ-dithio)-tetrabromo-tetra(tetrahydroth   85. 52
-niobium]   (μ-di*Thio)-(μ-di*Thio)-tetrabromo-tetra(tetrahydrothiophene)-di   85. 52
μ-thio)-(μ-dithio)-tetrabromo-tetra(tetrahydro*Thiophene)-di-niobium   [(85. 52
   bis(η(5)-trimethylsilyl-cyclopentadienyl)-o- Xylidene-niobium(iv) 73.101
-niobium) benzene so> (bis(tetrahydroborato)-*Zinc)-(μ-hydrido)-(carbonyl-dicyclopentadienyl   73. 25

Osmium

           aqua-(ethylenediamine-tetra- Acetato)-osmium(iv) monohydrate.................   76. 61
ii)]   mer-trichloro-(Acetonitrile)-bis(triphenylphosphine)-osmium(i   86.137
              (μ(4)-*Acetylene)-dodecacarbonyl-tetra-osmium 71. 57
          (μ(4)-ethyl*Acetylene)-dodecacarbonyl-tetra-osmium 71. 90
                N-1- Adamantylimido-trioxo-osmium   83. 55
bis(μ(2)-hydrido)-(η(3)-1-methyl-2,3-diphenyl-*Allyl)-undecacarbonyl-tetra-osmium 71.210
       bis(μ(2)-hydrido)-(μ-N-methylbenzyl*Amido-1-yl)-nonacarbonyl-tri-osmium..........   71.106

trimethyl*Amine-bis(μ-nitrosyl)-nonacarbonyl-tri-osmium   83. 73
tetrakis(tetramethyl Ammonium) chloro-penta(trichloro-tin)-osmium   3. 26
(μ(2)-hydrido)-(μ(3)-2-(μ-N-phenylamino)-*Anilino)-tridecacarbonyl-penta-osmium 83.180
bis(μ(2)-hydrido)-(μ-N-methyl*Benzylamido-1-yl)-nonacarbonyl-tri-osmium 71.106
nyl-tetra-osmium]   (μ(2)-*Benzylmethylene)-bis(μ(2)-hydrido)-dodecacarbo 71.148
triethylphosphine-(μ(2)-1,4-bis(trifluoro)-*But-2-ene)-decacarbonyl-tri-osmium 72. 37
romo-(μ(2)-carbonyl)-(η(2)-1,4-bis(trifluoro)-*But-2-ene)-nonacarbonyl-tri-osmium <)iminium b 71. 56
rido)-(μ(3)-η(4)-1,3-dimethyl-1-(phenylimino)-*Butane)-octacarbonyl-tri-osmium [bis(μ(2)-hyd 71.150
oxo-bis(2,3-dimethyl Butane-2,3-diolato)-osmium(vi) 84. 41
l-tri-osmium]   (μ(2)-hydrido)-(μ(2)-p-tolyl*Carbamoyl)-dimethylphenylphosphine-nonacarbony 71.195
carbonyl-dichloro-dichloro Carbene-bis(triphenylphosphine)-osmium 71.252
bis(triphenylphosphine)immonium (μ(5)-*Carbido)-iodo-pentadecacarbonyl-penta-osmium 3. 56
bis(bis(triphenylphosphine)immonium) Carbido-deca-osmium-carbonyl 64.166
(bis(triphenylphosphine)immonium) octa-osmium- Docosacarbonyl .... [bis 64.165
(3)-η(2)-1-isopropyl-2-methyl-2-(phenylamino)-*Ethylene)-nonacarbonyl-tri-osmium <hydrido)-(μ 71.171
triangulo-> 1,1,1,2,2,2,3,3-nonacarbonyl-2-*Ethylene-1,3-μ-hydrido-1,3-μ-(methylthiolato)- 72. 14
hydrate]   aqua-(Ethylenediamine-tetra-acetato)-osmium(iv) mono 76. 61
dimethylphenylph> (μ(2)-hydrido)-(μ(2)-dithio*Formato)-bis(tricarbonyl-osmium)-tricarbonyl-(81. 64
(μ(2)-hydrido)-(μ(2)-dithio*Formato)-decacarbonyl-tri-osmium 81. 37
ecacarbonyl-tri-osmium-bis(triphenylphosphine-*Gold) ....'[bis(μ(2)-hydrido)-und 86.172
ecacarbonyl-tetra-osmium]   bis(μ(2)-*Hydrido)-(η(3)-1-methyl-2,3-diphenylallyl)-und 71.210
osmium)-tricarbonyl-(dimethylphenylph>   (μ(2)-*Hydrido)-(μ(2)-dithioformato)-bis(tricarbonyl- 81. 64
-osmium]   (μ(2)-*Hydrido)-(μ(2)-dithioformato)-decacarbonyl-tri 81. 37
m]   (μ(2)-*Hydrido)-(μ(2)-methoxy)-decacarbonyl-tri-osmiu 84. 29
bonyl-tri-osmium]   (μ(2)-*Hydrido)-(μ(2)-p-toluenesulfonylamide)-decacar 83.109
lphosphine-nonacarbonyl-tri-osmium]   (μ(2)-*Hydrido)-(μ(2)-p-tolylcarbamoyl)-dimethylpheny 71.195
enylamino)-ethylene)-nonacarbonyl-tri> bis(μ-*Hydrido)-(μ(3)-η(2)-1-isopropyl-2-methyl-2-(ph 71.171
o)-butane)-octacarbonyl-tri-osmium] bis(μ(2)-*Hydrido)-(μ(3)-η(4)-1,3-dimethyl-1-(phenylimin 71.150
idecacarbonyl-penta-osmium]   (μ(2)-*Hydrido)-(μ(3)-2-(μ-N-phenylamino)-anilino)-tr 83.180
onyl-tri-osmium]   bis(μ(2)-*Hydrido)-(μ-N-methylbenzylamido-1-yl)-nonacarb 71.106
(μ(2)-benzylmethylene)-bis(μ(2)-*Hydrido)-dodecacarbonyl-tetra-osmium 71.148
undecacarbonyl-methylisocyanide-tetrakis((μ-*Hydrido)-osmium) 71. 46
ylphosphine-gold)]   bis(μ(2)-*Hydrido)-undecacarbonyl-tri-osmium-bis(triphen 86.172
<,1,2,2,2,3,3,3-nonacarbonyl-2-ethylene-1,3-μ-*Hydrido-1,3-μ-(methylthiolato)-triangulo-tri-> 72. 14
bis(N-t-butyl Imido)-dioxo-osmium.................. 83. 39
ifluoro)-but-2-ene)-> bis(triphenylphosphine)*Iminium bromo-(μ(2)-carbonyl)-(η(2)-1,4-bis(tr 71. 56
(2)-hydrido)-(μ(3)-η(4)-1,3-dimethyl-1-(phenyl*Imino)-butane)-octacarbonyl-tri-osmium [bis(μ 71.150
-penta-osmium]   bis(triphenylphosphine)*Immonium (μ(5)-carbido)-iodo-pentadecacarbonyl 3. 56
bis(bis(triphenylphosphine) Immonium) carbido-deca-osmium-carbonyl 64.166
bis(bis(triphenylphosphine) Immonium) octa-osmium-docosacarbonyl 64.165
undecacarbonyl-methyl*Isocyanide-tetrakis((μ-hydrido)-osmium)........ 71. 46
tra-osmium]   (μ(2)-benzyl*Methylene)-bis(μ(2)-hydrido)-dodecacarbonyl-te 71.148
l-(dimethylphenylphosphine)-tri-osmium)] (μ-*Methylenethiolato)-(μ(3)-sulfido)-(nonacarbony 71.123
l-(dimethylphenylphosphine)-tri-osmiu> (μ(3)-*Methylenethiolato)-(μ(3)-sulfido)-(octacarbony 71.108
trimethylamine-bis(μ-*Nitrosyl)-nonacarbonyl-tri-osmium 83. 73
1,1,1,2,2,2,2,3,3-nonacarbonyl-1,3;1,3-bis(μ-*Nitrosyl)-3-(trimethylphosphite)-triangulo-tr> 86. 17
iphenylphosphine)immonium (μ(5)-carbido)-iodo-*Pentadecacarbonyl-penta-osmium .....[bis(tr 3. 56
ethyldithiocarbamato)-di-osmium(iv) hexafluoro*Phosphate dichloromethane solvate <akis(N,N-di 80. 27
ethyldithiocarbamato)-di-osmium(iv) hexafluoro*Phosphate methylene chloride solvate <s(N,N-di 80. 26
bis(carbonyl)-di-selenium-bis(triphenyl Phosphine)-osmium 86.136
carbonyl-dichloro-dichloro-bis(triphenyl Phosphine)-osmium 86.134
arbonyl-dichloro-dichlorocarbene-bis(triphenyl Phosphine)-osmium .... [c 71.252
ricarbonyl-osmium)-tricarbonyl-(dimethylphenyl*Phosphine)-osmium <-(μ(2)-dithioformato)-bis(t 81. 64
mer-trichloro-(acetonitrile)-bis(triphenyl Phosphine)-osmium(iii) 86.137
)-(μ(3)-sulfido)-(nonacarbonyl-(dimethylphenyl*Phosphine)-tri-osmium) [(μ-methylenethiolato 71.123
)-(μ(3)-sulfido)-(octacarbonyl-(dimethylphenyl*Phosphine)-tri-osmium) <μ(3)-methylenethiolato 71.108
1,4-bis(trifluoro)-but-2-ene)-> bis(triphenyl*Phosphine)iminium bromo-(μ(2)-carbonyl)-(η(2)- 71. 56
cacarbonyl-penta-osmium]   bis(triphenyl*Phosphine)immonium (μ(5)-carbido)-iodo-pentade 3. 56
yl]   bis(bis(triphenyl Phosphine)immonium) carbido-deca-osmium-carbon 64.166
bis(bis(triphenyl Phosphine)immonium) octa-osmium-docosacarbonyl 64.165
decacarbonyl-tri-osmium]   triethyl*Phosphine-(μ(2)-1,4-bis(trifluoro)-but-2-ene)- 72. 37
drido)-undecacarbonyl-tri-osmium-bis(triphenyl*Phosphine-gold) [bis(μ(2)-hy 86.172
ydrido)-(μ(2)-p-tolylcarbamoyl)-dimethylphenyl*Phosphine-nonacarbonyl-tri-osmium [(μ(2)-h 71.195
acarbonyl-1,3;1,3-bis(μ-nitrosyl)-3-(trimethyl*Phosphite)-triangulo-tri-osmium <2,2,2,3,3-non 86. 17
carbonyl-(Phthalocyaninato)-(pyridine)-osmium(ii)........ 49. 14
carbonyl-(phthalocyaninato)-(Pyridine)-osmium(ii) 49. 14
tetrakis(μ-α-*Pyridonato)-bis(chloro-osmium) 84. 80
bis(carbonyl)-di- Selenium-bis(triphenylphosphine)-osmium.... 86.136
e)-tri-osmium)] (μ-methylenethiolato)-(μ(3)-*Sulfido)-(nonacarbonyl-(dimethylphenylphosphin 71.123
e)-tri-osmiu> (μ(3)-methylenethiolato)-(μ(3)-*Sulfido)-(octacarbonyl-(dimethylphenylphosphin 71.108
<iethyldithiocarbamato)-bis(μ-(N,N-(diethyltri*Thiocarbamato))-di-osmium(iii) tetraphenylbor> 80. 23
mato))-di-osmium(iii) tet> tris(N,N-diethyldi*Thiocarbamato)-bis(μ-(N,N-(diethyltrithiocarba 80. 23
<iethyldithiocarbamato)-tetrakis(N,N-diethyldi*Thiocarbamato)-di-osmium(iv) hexafluorophosph> 80. 27
<iethyldithiocarbamato)-tetrakis(N,N-diethyldi*Thiocarbamato)-di-osmium(iv) hexafluorophosph> 80. 26
mato)-di-osmium(iv) h> bis(μ(2)-N,N-diethyldi*Thiocarbamato)-tetrakis(N,N-diethyldithiocarba 80. 27

mato)-di-osmium(iv) h>   bis(μ(2)-N,N-diethyldi*Thiocarbamato)-tetrakis(N,N-diethyldithiocarba   80. 26
                         carbonyl-dichloro- Thiocarbonyl-bis(triphenylphosphine)-osmium....   86.134
yl-(dimethylphenylph>   (μ(2)-hydrido)-(μ(2)-di*Thioformato)-bis(tricarbonyl-osmium)-tricarbon   81. 64
                        (μ(2)-hydrido)-(μ(2)-di*Thioformato)-decacarbonyl-tri-osmium 81. 37
ylphenylphosphine)-tri-osmium)]   (μ-methylene*Thiolato)-(μ(3)-sulfido)-(nonacarbonyl-(dimeth   71.123
ylphenylphosphine)-tri-osmiu>   (μ(3)-methylene*Thiolato)-(μ(3)-sulfido)-(octacarbonyl-(dimeth   71.108
arbonyl-2-ethylene-1,3-μ-hydrido-1,3-μ-(methyl*Thiolato)-triangulo-tri-osmium <,2,3,3,3-nonac   72. 14
s(tetramethylammonium) chloro-penta(trichloro- Tin)-osmium.......................   [tetraki   3. 26
                          (μ(2)-hydrido)-(μ(2)-p-*Toluenesulfonylamide)-decacarbonyl-tri-osmium   83.109
rbonyl-tri-osmium]   (μ(2)-hydrido)-(μ(2)-p-*Tolylcarbamoyl)-dimethylphenylphosphine-nonaca   71.195
)-hydrido)-(μ(3)-2-(μ-N-phenylamino)-anilino)-*Tridecacarbonyl-penta-osmium ....   [(μ(2   83.180

Palladium

ladium bis(dimethylacetamidonium) bis(dimethyl*Acetamide) .... [tetrachloro-μ-dichloro-di-pal   1. 16
trachloro-μ-dichloro-di-palladium bis(dimethyl*Acetamidonium) bis(dimethylacetamide) ....[te   1. 16
ic acid solvate]   tetrakis((μ(2)-*Acetato)-(μ(2)-carbonyl))-tetra-palladium acet   81. 44
henyl)-palladium(ii))]   bis(μ-*Acetato)-bis((2-(2'-benzothiazolyl)-5-methyl-p   81. 86
nyl)-palladium(ii))]   bis(μ-*Acetato)-bis((2-(2'-benzoxazolyl)-5-methyl-phe   71.236
palladium(ii)) chloroform solvate]   bis(μ-*Acetato)-bis(chloro-(dimethylphenylphosphine)-   81. 67
                         bis(diphenylphosphino Acetato-palladium(ii) 81. 85
potassium bromo-(2-hydroxy-ethylimino-di Acetato-palladium(ii) monohydrate 82. 11
dium]   tetrakis(μ-trichloro*Acetato)-tetrakis(μ-t-butylperoxy)-tetra-palla   81. 79
tylacetonato)-bis(diethylamine)-palladium(ii) Acetylacetonate.......................   [(ace   77. 2
i) acetylacetonate]   (Acetylacetonato)-bis(diethylamine)-palladium(i   77. 2
ene]   bis(Acetylacetonato)-palladium(ii) tetrathiafulval   60. 12
azobenzyl-hexafluoro Acetylacetonato-palladium 77. 5
azobenzyl-hexafluoro Acetylacetonato-palladium 77. 4
S-cyclo-1-(1'-dimethylaminoethyl-ferrocene)-2- Acetylacetonato-palladium ....   [(-)-R   73. 80a
N-methylimino)-palladium(ii) perchlorate]   (η- Allyl)-(8-isopropylquinoline-2-carboxaldehyde-   72. 30
                         bis(η(3)- Allyl-triphenylphosphine-palladium)............   72. 61
e))-p>   (+)(589)-(S-dimethyl-(α-methylbenzyl)- Aminato)-(SS-o-phenylene-bis(methylphenylarsin   86.102a
(acetylacetonato)-bis(diethyl Amine)-palladium(ii) acetylacetonate 77. 2
m]   bis(trimethylbenzyl Ammonium) hexadecacarbonyl-tetra-iron-palladiu   3. 44
-hexa-palladium acetoni>   tris(trimethylbenzyl Ammonium) tetracosa-carbonyl-hydrido-hexa-iron   71.190
meso-bis(RR,SS-o-phenylene-bis(methylphenyl Arsine))-palladium(ii) di-iodide 86.146
meso-bis(RR,SS-o-phenylene-bis(methylphenyl Arsine))-palladium(ii) dichloride ethane-1,2->   86.145
phenyl-C(1)N)-(SS-o-phenylene-bis(methylphenyl Arsine))-palladium(ii) hexafluorophosphate <)-   86.102a
zyl)-aminato)-(SS-o-phenylene-bis(methylphenyl Arsine))-palladium(ii) hexafluorophosphate <en   86.102a
dichloro-bis(triphenyl Arsine)-palladium(ii) 86.117
ladium)-3-methyl-1,4-bis(methoxyphenyl)-1,4-di Azabutadien-1,4-diyl-dichloro-copper <ine)-pal   86.196
<oro-(1,4-bis(p-methoxyphenyl)-3-methyl-1,4-di Azabutadien-2-yl)-bis(triphenylphosphine)-pal>   71.274
orobor>   (tetrabenzo(b,f,j,n)(1,5,9,13)-tetra- Azacyclohexadecine)-palladium(ii) bis(tetraflu   83.191
Azobenzyl-hexafluoroacetylacetonato-palladium   77. 4
Azobenzyl-hexafluoroacetylacetonato-palladium   77. 5
ine)-palladium(ii) bis(tetrafluorobor>   (tetra Benzo(b,f,j,n)(1,5,9,13)-tetra-azacyclohexadec   83.191
l-3'-ene)>   1-triphenylphosphine-1-chloro-4,5- Benzo-3-(3'-phenyl-5'-ethylidenyl-1',2'-dithio   71.235
)]   bis(μ-acetato)-bis((2-(2'-*Benzothiazolyl)-5-methyl-phenyl)-palladium(ii)   81. 86
                 bis(μ-acetato)-bis((2-(2'-*Benzoxazolyl)-5-methyl-phenyl)-palladium(ii))   71.236
nylarsine))-p>   (+)(589)-(S-dimethyl-(α-methyl Benzyl)-aminato)-(SS-o-phenylene-bis(methylphe   86.102a
azo Benzyl-hexafluoroacetylacetonato-palladium 77. 5
azo Benzyl-hexafluoroacetylacetonato-palladium 77. 4
m(ii)]   trans-dichloro-bis(1- Benzyl-3,5-di-propyl-4-ethylpyrazole)-palladiu   83.215
lladium]   bis(trimethyl Benzylammonium) hexadecacarbonyl-tetra-iron-pa   3. 44
a-iron-hexa-palladium acetoni>   tris(trimethyl Benzylammonium) tetracosa-carbonyl-hydrido-hex   71.190
ium)-3-methyl-1,4-bis(methoxyphenyl)-1,4-diaza Butadien-1,4-diyl-dichloro-copper <ine)-pallad   86.196
<-(1,4-bis(p-methoxyphenyl)-3-methyl-1,4-diaza Butadien-2-yl)-bis(triphenylphosphine)-pallad>   71.274
o)methane-bis(chloro-palladiu>   (μ-hexafluoro*Butadiene-2,3-diyl)-bis(μ-bis(diphenylphosphin   71.275
rchlorate]   (η-allyl)-(8-isopropylquinoline-2- Carboxaldehyde-N-methylimino)-palladium(ii) pe   72. 30
xyphenyl)-1,4-diazabutadien-1,4-diyl-dichloro- Copper <ine)-palladium)-3-methyl-1,4-bis(metho   86.196
<lphosphonium-(trimethylsilyl)methylide)-(1,5- Cyclo-octadiene)-palladium(ii) hexafluorophos>   71.160
etylacetonato-palladium]   (-)-RS- Cyclo-1-(1'-dimethylaminoethyl-ferrocene)-2-ac   73. 80a
-1,3-bis(diethylamino)-2,4-bis(methoxymethyl)- Cyclobutadiene)-palladium monohydrate <o-(η(4)   75. 14
bor>   (tetrabenzo(b,f,j,n)(1,5,9,13)-tetra-aza Cyclohexadecine)-palladium(ii) bis(tetrafluoro   83.191
dichloro-((S-methyl-L- Cysteine)-sulfoxide)-palladium(ii) monohydrate   82. 8
bis(S-ethyl Ethene-1,2-dithiolato)-palladium(ii) 85. 19
perchlorate]   (di Ethylenetriamine)-(guanosine)-palladium(ii) di   76. 84
<phosphine-1-chloro-4,5-benzo-3-(3'-phenyl-5'- Ethylidenyl-1',2'-dithiol-3'-ene)-1-pallada-2>   71.235
(-)-RS-cyclo-1-(1'-dimethylaminoethyl- Ferrocene)-2-acetylacetonato-palladium 73. 80a
bis(1,3,5-tri-p-tolyl- Formazanyl)-palladium(ii) 83.219
bis(acetylacetonato)-palladium(ii) tetrathia Fulvalene   60. 12
cis-bis(Glycinato)-palladium(ii) trihydrate...........   82. 6
trans-chloro-glycino- Glycinato-palladium(ii) 82. 5
trans-chloro- Glycino-glycinato-palladium(ii)...............   82. 5
(diethylenetriamine)-(Guanosine)-palladium(ii) diperchlorate 76. 84

(dichloro-palladium)] bis(μ-*Heptane-1,7-diyl-bis(di-t-butylphosphine))-bis 86.171
bis(trimethylbenzylammonium) Hexadecacarbonyl-tetra-iron-palladium 3. 44
<(trimethylbenzylammonium) tetracosa-carbonyl- Hydrido-hexa-iron-hexa-palladium acetonitrile> 71.190
bis(bis(diphenylphosphino)-maleic-N-methyl Imide)-palladium(0) 1,2-dichloroethane solvate 86.214
8-isopropylquinoline-2-carboxaldehyde-N-methyl Imino-palladium(ii) perchlorate [(η-allyl)-(72. 30
potassium bromo-(2-hydroxy-ethyl Imino-diacetato)-palladium(ii) monohydrate 82. 11
nzylammonium) tetracosa-carbonyl-hydrido-hexa- Iron-hexa-palladium acetonitrile solvate <ylbe 71.190
imethylbenzylammonium) hexadecacarbonyl-tetra- Iron-palladium ....[bis(tr 3. 44
oethane solvate] bis(bis(diphenylphosphino)- Maleic-N-methylimide)-palladium(0) 1,2-dichlor 86.214
sphino)methane)-bis(chloro-palladium) dichloro*Methane methanol solvate <is(μ-bis(diphenylpho 86.176
tadiene-2,3-diyl)-bis(μ-bis(diphenylphosphino)*Methane)-bis(chloro-palladium) <μ-hexafluorobu 71.275
<-sulfur dioxide)-bis(μ-bis(diphenylphosphino)*Methane)-bis(chloro-palladium) dichloromethan> 86.176
solva> (μ-thio)-bis(μ-bis(diphenylphosphino)*Methane)-bis(chloro-palladium) dichloromethane 86.178a
<o-(dimethylphenylphosphonium-(trimethylsilyl) Methylide)-(1,5-cyclo-octadiene)-palladium(ii> 71.160
<honium-(trimethylsilyl)methylide)-(1,5-cyclo- Octadiene)-palladium(ii) hexafluorophosphate 71.160
i)) sulfate monohydrate] bis((Oxamide-oximato)-(oxamide-oximato)-palladium(i 83. 12
drate] bis((oxamide-oximato)-(Oxamide-oximato)-palladium(ii)) sulfate monohy 83. 12
trans-dichloro-bis(Oxazole)-palladium(ii) 83. 20
ate monohydrate] bis((oxamide- Oximato)-(oxamide-oximato)-palladium(ii)) sulf 83. 12
bis((oxamide-oximato)-(oxamide- Oximato)-palladium(ii)) sulfate monohydrate... 83. 12
ium(ii) toluene solvate] di Oxygen-bis(di(t-butyl)-phenylphosphine-pallad 86. 96
dichloro-(2,2-dimethyl- Pent-(E)-3-enyl-methylsulfide)-palladium(ii) 72. 6
chloro-(syn-4,4-dimethyl-5-methylthio- Pent-(1-3-η)-enyl)-palladium 71. 7
bis(μ-1,5-bis(di-t-butylphosphino)-3-methyl*Pentane)-bis(dichloro-palladium(ii)) 86.168
α-(4-6η)-(3,20-dioxo-pregn-4-enyl) Pentane-2,4-dionato-palladium(ii) 75. 35
<-(S-dimethyl-(α-methylbenzyl)-aminato)-(SS-o- Phenylene-bis(methylphenylarsine))-palladium(> 86.102a
<-(1'-dimethylaminoethyl)-phenyl-C(1)N)-(SS-o- Phenylene-bis(methylphenylarsine))-palladium(> 86.102a
i) di-iodide] meso-bis(RR,SS-o- Phenylene-bis(methylphenylarsine))-palladium(i 86.146
i) dichloride ethane-1,2-d> meso-bis(RR,SS-o- Phenylene-bis(methylphenylarsine))-palladium(i 86.145
(methylphenylarsine))-palladium(ii) hexafluoro Phosphate <)-phenyl-C(1)N)-(SS-o-phenylene-bis 86.102a
(methylphenylarsine))-palladium(ii) hexafluoro Phosphate <enzyl)-aminato)-(SS-o-phenylene-bis 86.102a
(1,5-cyclo-octadiene)-palladium(ii) hexafluoro Phosphate <phonium-(trimethylsilyl)methylide)- 71.160
dichloro-bis(2-methyl-4,4,6,6-tetraphenyl- Phosphazadiene-2-yl)-palladium 86.180
bis(μ-heptane-1,7-diyl-bis(di-t-butyl*Phosphine))-bis(dichloro-palladium)........... 86.171
dichloro-bis(tri-n-propyl Phosphine)-palladium(ii) 86. 52
trans-bromo-(2-pyridyl)-bis(triethyl Phosphine)-palladium(ii) 71.113
trans-bromo-(3-pyridyl)-bis(triethyl Phosphine)-palladium(ii) 71.114
trans-bromo-(4-pyridyl)-bis(triethyl Phosphine)-palladium(ii) 71.115
3-methyl-1,4-diazabutadien-2-yl)-bis(triphenyl Phosphine)-palladium(ii) <is(p-methoxyphenyl)- 71.274
dioxygen-bis(di(t-butyl)-phenyl Phosphine)-palladium(ii) toluene solvate 86. 96
,N)-bis(bromo-(μ(2)-pyridyl-C(2),N)-(triphenyl*Phosphine)-palladium(ii)) [trans(P 71.269
bis(μ-acetato)-bis(chloro-(dimethylphenyl*Phosphine)-palladium(ii)) chloroform solvate 81. 67
phenyl)-1,4-diazabut> 2-(chloro-bis(triphenyl Phosphine)-palladium)-3-methyl-1,4-bis(methoxy 86.196
bis(η(3)-allyl-triphenyl Phosphine-palladium) 72. 61
ntakis(μ(2)-carbonyl)-tetrakis(methyl-diphenyl Phosphine-palladium) ....[pe 86.213
thyl idenyl-1',2'-dithiol-3'-ene)> 1-triphenyl Phosphine-1-chloro-4,5-benzo-3-(3'-phenyl-5'-e 71.235
1,2-dichloroethane solvate] bis(bis(diphenyl Phosphino)-maleic-N-methylimide)-palladium(0) 86.214
dium(ii))] bis(μ-1,5-bis(di-t-butyl*Phosphino)-3-methylpentane)-bis(dichloro-palla 86.168
xafluorobutadiene-2,3-diyl)-bis(μ-bis(diphenyl*Phosphino)methane)-bis(chloro-palladium) <μ-he 71.275
oromet> (μ-sulfur dioxide)-bis(μ-bis(diphenyl*Phosphino)methane)-bis(chloro-palladium) dichl 86.176
oromethane solva> (μ-thio)-bis(μ-bis(diphenyl*Phosphino)methane)-bis(chloro-palladium) dichl 86.178a
bis(diphenyl Phosphinoacetato)-palladium(ii)................. 81. 85
cis-dichloro-bis(1-phenyl-3,4-dimethyl Phosphole)-palladium(ii) 86. 86
cis(thiocyanato)-bis(1,3,4-trimethyl Phosphole)-palladium(ii) 86. 39
clo-octadiene)-pallad> chloro-(dimethylphenyl Phosphonium-(trimethylsilyl)methylide)-(1,5-cy 71.160
o)-palladium(ii) monohydrate] Potassium bromo-(2-hydroxy-ethylimino-diacetat 82. 11
)] α-(4-6η)-(3,20-dioxo- Pregn-4-enyl)-pentane-2,4-dionato-palladium(ii 75. 35
ns-dichloro-bis(1-benzyl-3,5-di-propyl-4-ethyl Pyrazole)-palladium(ii)................. [tra 83.215
dichloro-bis(1,3-dimethyl-1,4,5,6-tetrahydro Pyridazine)-palladium(ii) 83. 82
palladium(ii)] bis(pyridine-N-oxide) bis(*Pyridine-N-hydroxide) tetrabromo-μ-dibromo-di- 33. 6
trabromo-μ-dibromo-di-palladium(ii)] bis(*Pyridine-N-oxide) bis(pyridine-N-hydroxide) te 33. 6
trans-bromo-(2- Pyridyl)-bis(triethylphosphine)-palladium(ii) 71.113
trans-bromo-(3- Pyridyl)-bis(triethylphosphine)-palladium(ii) 71.114
trans-bromo-(4- Pyridyl)-bis(triethylphosphine)-palladium(ii) 71.115
(ii))] trans(P,N)-bis(bromo-(μ(2)-*Pyridyl-C(2),N)-(triphenylphosphine)-palladium 71.269
adium(ii) perchlorate] (η-allyl)-(8-isopropyl Quinoline-2-carboxaldehyde-N-methylimino-pall 72. 30
chloro-(dimethylphenylphosphonium-(trimethyl Silyl)methylide)-(1,5-cyclo-octadiene)-pallad> 71.160
trans-bis(thiocyanato)-bis(di(3-trimethyl Silyl-propyl)-tellurium)-palladium(ii) 70. 18
dichloro-(2,2-dimethyl-pent-(E)-3-enyl-methyl Sulfide)-palladium(ii) 72. 6
dichloro-((S-methyl-L-cysteine)- Sulfoxide)-palladium(ii) monohydrate 82. 8
thane)-bis(chloro-palladium) dichloromet> (μ-*Sulfur dioxide)-bis(μ-bis(diphenylphosphino)me 86.176
(thiocyanato)-bis(di(3-trimethylsilyl-propyl)- Tellurium-palladium(ii) .... [trans- 70. 18
adium acetoni> tris(trimethylbenzylammonium) Tetracosa-carbonyl-hydrido-hexa-iron-hexa-pall 71.190
bis(acetylacetonato)-palladium(ii) tetra Thiafulvalene 60. 12
bis(μ-acetato)-bis((2-(2'-benzo*Thiazolyl)-5-methyl-phenyl)-palladium(ii)) 81. 86

(chloro-palladium) dichloromethane solva> (μ-*Thio)-bis(μ-bis(diphenylphosphino)methane)-bis 86.178a
chloro-(syn-4,4-dimethyl-5-methyl Thio-pent-(1-3-η)-enyl)-palladium 71. 7
trans-dichloro-bis(O-ethyl-N,N-dimethyl Thiocarbamato)-palladium(ii) 80. 9
trans-dibromo-bis(O-ethyl-N,N-dimethyl-mono Thiocarbamato)-palladium(ii) 80. 8
trans-dithiocyanato-bis(O-ethyl-N-methyl-mono Thiocarbamato)-palladium(ii) 80. 5
ethylidenyl-1',2'-dithiol-3'-ene)-1-pallada-2- Thiol-2-ene toluene solvate <-3-(3'-phenyl-5'- 71.235
<,5-benzo-3-(3'-phenyl-5'-ethylidenyl-1',2'-di Thiol-3'-ene)-1-pallada-2-thiol-2-ene toluene> 71.235
bis(S-ethylethene-1,2-di Thiolato)-palladium(ii)........................... 85. 19
bis(1,3,5-tri-p- Tolyl-formazanyl)-palladium(ii)............... 83.219
bis(L- Tyrosinato)-palladium(ii) 82. 35

Platinum

bis(oxamide-oximato)-platinum(ii) sodium Acetate sodium chloride dihydrate 83. 11
cis-dichloro-(dimethylphenylphosphine)-(vinyl Acetate)-platinum(ii) 72. 18
imethylaminomethyl)-phenyl)-platin> rac-a-(μ-*Acetato-b-(0-acetato-mercurio)-cf,de-bis(2-(d 71.183
thyl)-phenyl)-platin> rac-a-(μ-acetato)-b-(0-*Acetato-mercurio)-cf,de-bis(2-(dimethylaminome 71.183
bis(N-acetylimino-benzoyl Acetonato)-platinum(ii)...................... 84. 96
bis(chloro-(O-isopropyl- Acetonium))-platinum 71. 23
henylarsinophenyl)-2-methoxyethyl)-(hexafluoro Acetylacetonato)-platinum(ii) .... [(1-(o-dip 71.200
olvate] cis(N-(o-aminobenzylidene)-anthranil- Aldehydato-0,N,N)-chloro-platinum chloroform s 84. 53
2-hydroxy-N-salicylidene-aniline-(diethyl Amine) platinum(ii)....................... 78. 5
dimethyl-N,N'-bis(α-methylbenzyl)-1,2-ethanedi Amine)-platinum(ii) [dichloroethylene-(N,N'- 72. 41
e)-platinum(iv) tetrabromide] tetrakis(ethyl Amine)-platinum(ii) dibromo-tetrakis(ethylamin 83. 43
dimethyl-N,N'-bis(α-methylbenzyl)-1,2-ethanedi Amine)-platinum(ii) [dichloro-(propene)-(N,N'- 72. 44
thylamine)-platinum(ii) dibromo-tetrakis(ethyl Amine)-platinum(iv) tetrabromide [tetrakis(e 83. 43
)-((Z)-2-chloro-N-(3-methyl-but-1-enyl)-benzen Amine)-triphenylphosphine-platinum(ii) <ro-η(2 72. 53
ter) hydrate] trimethylene-di Amine-bis(guanosine-5'-monophosphate methyl es 83.186
iodo-N,N,N',N'-tetraethyl-diethylene-tri Amine-platinum(ii) iodide 76. 75
num pentachloro-antimony] di Ammine-(N,N,N'-trimethylethylenediamine)-plati 76. 22
(ii) perchlorate trihydrate] di Ammine-(1-methylcytosine)-(thyminato)-platinum 83. 61
tetrahydrate] cesium hydrogen di Ammine-bis(cytidine-3'-phosphato)-platinum(ii) 83.133
rchlorate heptahydrate] cis-di Ammine-bis(guanosine)-platinum(ii) chloride pe 83.151
m chloride trihydrate] Ammine-chloro-diethylenetriamine-nitro-platinu 76. 10
cis- Ammine-chloro-glycinato-platinum(ii) 82. 1
cis- Ammine-dichloro-glycinato-platinum(ii) 82. 2
lpyridine)-platinum(ii))] tetra- Ammine-platinum(ii) bis(trichloro-(2,6-dimethy 83. 29
bis(μ-1-methylthyminato)-bis(cis-di*Ammine-platinum(ii)) dinitrate 84. 42
te] bis(bis(μ-1-methylthyminato)-cis-di*Ammine-platinum(ii)) silver nitrate pentahydra 84.100
bis(oxamide-oximato)-platinum(ii) Ammonium chloride 83. 10
tetramethyl Ammonium hexachloro-platinum 3. 16n
e] tetra-n-butyl Ammonium platinum-carbonyl acetonitrile solvat 3. 52
bis(trimethylbenzyl Ammonium) hexadecacarbonyl-tetra-iron-platinum 3. 45
2-hydroxy-N-salicylidene- Aniline-(diethylamine) platinum(ii)............. 78. 5
2-hydroxy-N-salicylidene- Aniline-(triphenylphosphine) platinum(ii) 78. 14
loroform solvate] cis(N-(o-aminobenzylidene)- Anthranil-aldehydato-0,N,N)-chloro-platinum ch 84. 53
rimethylethylenediamine)-platinum pentachloro- Antimony ....[diammine-(N,N,N'-t 76. 22
dichloro-(diphenyl(o-vinylphenyl) Arsine)-platinum(ii) 72. 36
trans-dichloro-(2,11-bis(diphenyl Arsinomethyl)-benzo(c)phenanthrene)-platinum 65. 26
lacetonato)-platinum(ii)] (1-(o-diphenyl Arsinophenyl)-2-methoxyethyl)-(hexafluoroacety 71.200
nate hydrate] (1,3,6,8,10,13,16,19-octa- Azabicyclo(6.6.6)eicosane)-platinum(iv) dithio 76. 76
thylenediamine> (μ-3,6,6-tetramethyl-3,6-di*Azaoctane-1,8-diyl)-bis(N,N,N',N'-tetramethyle 71.187
<o-η(2)-((Z)-2-chloro-N-(3-methyl-but-1-enyl)- Benzenamine)-triphenylphosphine-platinum(ii) 72. 53
rans-dichloro-(2,11-bis(diphenylarsinomethyl)- Benzo(c)phenanthrene)-platinum .... [t 65. 26
num(ii)] 4-methyl-N-thio Benzohydroximato-bis(triphenylphosphine)-plati 86.163
o-platinum] di Benzotetrathiafulvalene pentabromo-dimethylthi 85. 2
bis(N-acetylimino- Benzoylacetonato)-platinum(ii) 84. 96
hloroethylene-(N,N'-dimethyl-N,N'-bis(α-methyl Benzyl)-1,2-ethanediamine)-platinum(ii) [dic 72. 41
oro-(propene)-(N,N'-dimethyl-N,N'-bis(α-methyl Benzyl)-1,2-ethanediamine)-platinum(ii) [dichl 72. 44
bis(di Benzyl-cis-benzyl-2-yl-phosphine)-platinum(ii) 86.155
bis(dibenzyl-cis- Benzyl-2-yl-phosphine)-platinum(ii)............. 86.155
atinum] bis(trimethyl Benzylammonium) hexadecacarbonyl-tetra-iron-pl 3. 45
o-platinum chloroform solvate] cis(N-(o-amino Benzylidene)-anthranil-aldehydato-0,N,N)-chlor 84. 53
hyl)-platinum)-2-methyl-1,2-dica> cis-1-((tri Benzylphosphine)-(dibenzylphosphino-phenyl-met 71.268
hyl-1,2-dica> cis-1-((tribenzylphosphine)-(di Benzylphosphino-phenyl-methyl)-platinum(ii)-2-met 71.268
ine)-platinum(0)] Bicyclo(4.2.1)non-1(8)-ene-bis(triphenylphosph 75. 46
e hydrate] (1,3,6,8,10,13,16,19-octa-aza Bicyclo(6.6.6)eicosane)-platinum(iv) dithionat 76. 76
-bis(triphenylphosphine)-6-thia-9-platina-deca Borane .... [9,9 86.121
-bis(triphenylphosphine)-6-thia-9-platina-deca Borane .... [8-ethoxy-9,9 86.141
ylphosphine)-2,4-dicarba-1-platina-closo-hepta Borane ....[1,1-bis(trieth 86. 32
riethylphosphine)-7,8-dicarba-9-platina-undeca Borane [9-hydrido-9,9-bis(t 71. 82
riethylphosphine)-7,8-dicarba-9-platina-undeca Borane [9-hydrido-9,10-bis(t 71. 81
ylphosphine)-2,3-dicarba-1-platina-closo-hepta Borane [2,3-dimethyl-1,1-bis(trieth 86. 43
tinum))-(μ(5,6)-hydrido-nido-2,3-dicarba-hexa*Borane <ans-bis(triethylphosphine)-hydrido-pla 86. 33
-methyl)-platinum)-2-methyl-1,2-dicarba-dodeca Borane <ylphosphine)-(dibenzylphosphino-phenyl 71.268

yl)>   chloro-(diphenylphosphino-dicarba-dodeca Borane)-(diphenylphosphino-dicarba-dodecaboran   86. 95
num]   bis(μ-2,3,4-η(3)-nido-hexa*Boranyl)-bis(dimethylphenylphosphine)-di-plati   86. 44
odecaborane)-(diphenylphosphino-dicarba-dodeca Boranyl)-platinum <diphenylphosphino-dicarba-d   86. 95
  cis-dichloro-η(2)-((Z)-2-chloro-N-(3-methyl- But-1-enyl)-benzenamine)-triphenylphosphine-p>   72. 53
(ii) hexafluoroph>   (1,2-bis(methoxycarbonyl)- But-2-en-1-yl)-bis(triethylphosphine)-platinum   71.165
(ii) tetraphenylb>   (1,2-bis(methoxycarbonyl)- But-2-en-1-yl)-bis(triethylphosphine)-platinum   71.164
-dimethylethylenediamine)-platinum bis(α-bromo Camphor-π-sulfonate) [(ethylenediamine)-(N,N'   76. 33
l-trimethylphosphine-tungsten)-(μ-methoxytolyl*Carbene))-bis(trimethylphosphine)-platinum <ny   71.186
yl-(methyl-di-t-butylphos>   (μ-methoxy-phenyl-*Carbene)-(μ-tetracarbonyl)-tungsten-bis(carbon   71.238
entacarbonyl-tungsten)]   (μ-methoxy-phenyl-*Carbene)-(bis(trimethylphosphine)-platinum)-(71.142
is(μ-chloro)-bis(chloro-(neopentyl(isopropoxy)*Carbene)-platinum) .... [b   71.131
<lphosphine)-platinum)-(μ-carbonyl)-(μ-p-tolyl*Carbenyl)-(carbonyl-cyclopentadienyl-manganes>   71.173
hato)-platinum(ii) tetrahydrate]   Cesium hydrogen diamine-bis(cytidine-3'-phosp   83.133
yl)methylen>   (tricarbonyl-trimethylphosphine-*Chromium)-(μ-carbonyl)-(μ-phenyl(methoxycarbon   71.185
s(1,2-diaminoethane)-platinum(iv) tetrachloro- Copper(i) <inoethane)-platinum(ii) dichloro-bi   76. 20
               dichloro-(cis-2-methyl-thia Cyclo-oct-4-ene)-platinum 75.   1
(0)]   (1,2-η- Cyclo-octyne)-bis(triphenylphosphine)-platinum   75. 45
                    trans-dibromo-bis(Cyclohexylamine)-platinum(ii) 83. 83
phoramide solvate]   cis-dichloro-bis(Cyclohexylamine-N)-platinum(ii) hexamethylphos   83. 84
<)-(μ-carbonyl)-(μ-p-tolylcarbenyl)-(carbonyl-*Cyclopentadienyl-manganese) tetrafluoroborate   71.173
μ(2)-iodo)-(methyl-di-t-butylphosphine)-(2-oxa*Cyclopentylidene)-platinum-manganese <rbonyl-(71.111
ii) tetra>   (2,3-bis(methoxycarbonyl)-1-methyl*Cyclopropyl)-bis(triphenylphosphine)-platinum <   71.266
te]   cesium hydrogen diamine-bis(Cytidine-3'-phosphato)-platinum(ii) tetrahydra   83.133
hylthiolato)-2,2',2''-terpyridine-plat>   deoxy Cytidylyl-(3'-5')-deoxyguanosine (2-hydroxy-et   85. 58
trihydrate]   diamine-(1-methyl Cytosine)-(thyminato)-platinum(ii) perchlorate   83. 61
  (1,3,6,8,10,13,16,19-octa-azabicyclo(6.6.6) Eicosane)-platinum(iv) dithionate hydrate 76. 76
thane)-platinum(iv) tetrachl> bis(1,2-diamino Ethane)-platinum(ii) dichloro-bis(1,2-diaminoe   76. 20
oethane)-platinum(ii) dichloro-bis(1,2-diamino Ethane)-platinum(iv) tetrachloro-copper(i) <in   76. 20
e-(N,N'-dimethyl-N,N'-bis(α-methylbenzyl)-1,2- Ethanediamine)-platinum(ii)   [dichloroethylen   72. 41
)-(N,N'-dimethyl-N,N'-bis(α-methylbenzyl)-1,2- Ethanediamine)-platinum(ii) [dichloro-(propene   72. 44
                             bis(S-ethyl Ethene-1,2-dithiolato)-platinum(ii) 85. 20
ine)-platinum)]   (μ-1,2-bis(methoxycarbonyl)*Ethene-1,2-diyl)-bis(carbonyl-(triphenylphosph   71.265
i)]   cis-dichloro-(μ-*Ethylene)-(2,6-dimethyl-piperidine)-platinum(i   72.   7
rifluoromethyl-(cis-1,2-bis(diphenylphosphino) Ethylene)-platinum(ii) .... [chloro-t   71.213
l)-1,2-ethanediamine)-platinum(ii)]   dichloro Ethylene-(N,N'-dimethyl-N,N'-bis(α-methylbenzy   72. 41
            iodo-N,N,N',N'-tetraethyl-di Ethylene-triamine-platinum(ii) iodide 76. 75
hosphine-platinum)]   (μ(2)-N,N'-di-t-butyl-*Ethylenedi-imine)-bis(trans-dichloro-tributylp   86.114
         dichloro-(N,N'-di-t-butyl- Ethylenedi-imine)-styrene-platinum 72. 33
)-platinum bis(α-bromocamphor-π-sulfonate)] (Ethylenediamine)-(N,N'-dimethylethylenediamine   76. 33
-platinum(ii) bis(hexafluorophosphate)]   (Ethylenediamine)-bis(7,9-dimethylhypoxanthine)   76. 93
-sulfonate)]   (ethylenediamine)-(N,N'-dimethyl Ethylenediamine)-platinum bis(α-bromocamphor-π   73. 33
        diammine-(N,N,N'-trimethyl Ethylenediamine)-platinum pentachloro-antimony   76. 22
            chloro-(thyminato)-(Ethylenediamine)-platinum(ii) 76. 42
ylenediamine)-platinum(iv) tetraperchlo>   bis(Ethylenediamine)-platinum(ii) dichloro-bis(eth   76. 19
e dihydrate]   chloro-(uracilato)-(Ethylenediamine)-platinum(ii) hydrogen chlorid   76. 24
                 μ-bromo-bis(*Ethylenediamine)-platinum(ii,iv) perchlorate   76.   8
<s(ethylenediamine)-platinum(ii) dichloro-bis(Ethylenediamine)-platinum(iv) tetraperchlorate   76. 19
<azaoctane-1,8-diyl)-bis(N,N,N',N'-tetramethyl*Ethylenediamine-chloro-platinum(ii)) diperchl>   71.187
latinum(ii) bis(hexafluorophosphate) ses> (di Ethylenetriamine)-(7,9-dimethylhypoxanthine)-p   76. 67
drate]   ammine-chloro-di Ethylenetriamine-nitro-platinum chloride trihy   76. 10
tetrachloride]   bis(di Ethylenetriammonium) tetrachloro-platinum(ii)   3. 30
<inum-(μ-dimethylsilyl-μ-(1-sigma:1-2-η-phenyl*Ethynyl))-(2-phenylethynyl-tricyclohexylphosp   71.277
te]   cis-dichloro-bis(diphenyl(phenyl Ethynyl)phosphine)-platinum acetonitrile solva   86.144
lyl-μ-(1-sigma:1-2-η-phenylethynyl))-(2-phenyl*Ethynyl-tricyclohexylphosphine-platinum) <ylsi   71.277
                trans-dichloro-(dimethyl Formamide)-(2,6-lutidine)-platinum(ii) 84. 25
                    dibenzotetrathia Fulvalene pentabromo-dimethylthio-platinum 85.   2
acarbonyl-ma>   μ-2-sigma:2-3-η-(4,5-dihydro-2-*Furyl)-(bis(trimethylphosphine)-platinum)-tetr   71. 75
acarbonyl-ma>   μ-2-sigma:2-3-η-(4,5-dihydro-2-*Furyl)-(bis(trimethylphosphine)-platinum)-tetr   71. 76
                    cis-bis(Glycinato)-platinum(ii) oxalic acid dihydrate   82.   7
                cis-ammine-chloro- Glycinato-platinum(ii) 82.   1
              cis-ammine-dichloro- Glycinato-platinum(ii) 82.   2
             cis-dibromo-bis(Glycine)-platinum(ii) monohydrate 82.   9
            cis-dichloro-bis(Glycine)-platinum(ii) monohydrate 82. 10
            cis-dichloropyridine- Glycine-platinum(ii) 82. 13
erpyridine-plat>   deoxycytidylyl-(3'-5')-deoxy Guanosine (2-hydroxy-ethylthiolato)-2,2',2''-t   85. 58
eptahydrate]   cis-diammine-bis(Guanosine)-platinum(ii) chloride perchlorate h   83.151
te]   trimethylene-diamine-bis(Guanosine-5'-monophosphate methyl ester) hydra   83.186
            bis(trimethylbenzylammonium) Hexadecacarbonyl-tetra-iron-platinum   3. 45
μ(2)-hydrido)-tris(triethylphosphine)-iridium-*Hydride tetraphenylborate <phine-platinum-bis(86.104
ne)-methyl-platinum) hexafluorophosph>   (μ(2)-*Hydrido)-bis((μ(2)-bis(diphenylphosphino)metha   86.193
hosphine)-di-platinum]   bis(μ-*Hydrido)-bis(dimethylsilyl)-bis(tricyclohexylp   86.147
triethylphosphine)-hydrido-platinum))-(μ(5,6)-*Hydrido-nido-2,3-dicarba-hexaborane <ans-bis(86. 33
phenyl-triethylphosphine-platinum-bis(*Hydrido)-tris(triethylphosphine)-iridium-hydr   86.104
            trans-cyanomethyl- Hydrido-bis(triphenylphosphine)-platinum(ii)   71.254
icarb>   (μ(4,5)-(trans-bis(triethylphosphine)-*Hydrido-platinum))-(μ(5,6)-hydrido)-nido-2,3-d   86. 33

ethane)-(bis(diphenylphosphino)methane)-pla>   *Hydrido-platinum-bis(μ-bis(diphenylphosphino)m   86.223
exafluorophosphate]   Hydrido-tris(triethylphosphine)-platinum(ii) h   86. 55
a-9-platina-undecaborane]   9- Hydrido-9,10-bis(triethylphosphine)-7,8-dicarb   71. 81
-9-platina-undecaborane]   9- Hydrido-9,9-bis(triethylphosphine)-7,8-dicarba   71. 82
         bis(oxamide-oximato)-platinum(ii)   Hydrogen chloride 83. 8
ro-(uracilato)-(ethylenediamine)-platinum(ii)   Hydrogen chloride dihydrate............ [chlo   76. 24
latinum(ii) tetrahydrate]   cesium   Hydrogen diammine-bis(cytidine-3'-phosphato)-p   83.133
i)]   4-methyl-N-thiobenzo Hydroximato-bis(triphenylphosphine)-platinum(i   86.163
hate) ses>   (diethylenetriamine)-(7,9-dimethyl Hydroxymato-bis(triphenylphosphine)-platinum(i   86.163
hate)]   (ethylenediamine)-bis(7,9-dimethyl Hypoxanthine)-platinum(ii) bis(hexafluorophosp   76. 67
)]   tetrakis(N-methyl   Hypoxanthine)-platinum(ii) bis(hexafluorophosp   76. 93
atinum)]   (μ(2)-N,N'-di-t-butyl-ethylenedi-*Imidazole)-platinum(ii) hexachloro-platinum(iv   83.107
         dichloro-(N,N'-di-t-butyl-ethylenedi- *Imine)-bis(trans-dichloro-tributylphosphine-pl   86.114
-(2-methyl-2-nitrosopropane)-(2-(N-oxo-t-butyl   Imine)-styrene-platinum......................   72. 33
2,4,6-trimethylpyridine)-(2-(N-hydroxy-t-butyl   Imino)ethyl)-platinum..................[chloro   71. 24
                            bis(N-acetyl   Imino)ethyl)-platinum ....[dichloro-(71. 77
η(2)-((t-butyl-(trimethylsilyl)amino)-t-butyl   Imino-benzoylacetonato)-platinum(ii) 84. 96
num-bis((μ(2)-hydrido)-tris(triethylphosphine)-*Iminothiophosphorane)-bis(triphenylphosphine)>   72. 65
imethylbenzylammonium) hexadecacarbonyl-tetra- *Iridium-hydride tetraphenylborate <phine-plati   86.104
         bis(μ-chloro)-bis(chloro-(neopentyl(*Isopropoxy)carbene)-platinum)   Iron-platinum.......................[bis(tr   3. 45
trans-dichloro-(dimethylformamide)-(2,6-   *Isopropoxy)carbene)-platinum) 71.131
tinum(ii)]   Lutidine)-platinum(ii) 84. 25
s(trimethylphosphine)-platinum)-tetracarbonyl-*Malonato-(2,2-dimethyl-1,3-diaminopropane)-pla   81. 34
s(trimethylphosphine)-platinum)-tetracarbonyl-*Manganese <gma:2-3-η-(4,5-dihydro-2-furyl)-(bi   71. 75
ylphosphine)-(2-oxacyclopentylidene)-platinum-*Manganese <gma:2-3-η-(4,5-dihydro-2-furyl)-(bi   71. 76
<-p-tolylcarbenyl)-(carbonyl-cyclopentadienyl-*Manganese <rbonyl-(μ(2)-iodo)-(methyl-di-t-but   71.111
inum-bis((μ-diphenylphosphido)-(tetracarbonyl-*Manganese) tetrafluoroborate dichloromethane >   71.173
enyl)-platin>   rac-a-(μ-acetato)-b-(O-acetato-*Manganese)).............. [carbonyl-plat   86.111
m> trimethyl-bis(μ(2)-(bis(diphenylphosphino-*Mercurio)-cf,de-bis(2-(dimethylaminomethyl)-ph   71.183
hydrido-platinum-bis(μ-bis(diphenylphosphino*Methane)))-di-platinum(ii) hexafluorophosphate   86.198
carbonyl-platinum-bis(μ-bis(diphenylphosphino*Methane)-(bis(diphenylphosphino)methane)-plat>   86.223
oro-(μ-methylene)-bis(μ-bis(diphenylphosphino*Methane)-chloro-platinum hexafluorophosphate   86.182
· dicarbonyl-bis(μ-bis(diphenylphosphino*Methane)-di-platinum .... [dichl   86.185
<(2)-hydrido)-bis((μ(2)-bis(diphenylphosphino*Methane)-di-platinum bis(hexafluorophosphate)   86.188
enylphosphino)methane)-(bis(diphenylphosphino*Methane)-methyl-platinum hexafluorophosphate   86.193
)-platinum(ii)]   (bis(diphenylphosphino)*Methane)-platinum hexafluorophosphate <is(diph   86.223
loro-triethylphosphine-bis(diphenylphosphino)- Methanide)-(dimethylphosphonium-bis(methylide)   71.225
<mium)-(μ-carbonyl)-(μ-phenyl(methoxycarbonyl)*Methano-platinum(ii) ....[ch   86.109
)-di-platinum]   dichloro-(μ-*Methylene)-(bis(trimethylphosphine)-platinum)   71.185
platinum benzene sol>   σ-(thiophenyl(sulfinyl) Methylene)-bis(μ-bis(diphenylphosphino)methane   86.185
te methyl ester) hydrate]   tri Methylene)-thiophenyl-bis(triphenylphosphine)-   71.272
phosphino)methanide)-(dimethylphosphonium-bis(Methylene-diamine-bis(guanosine-5'-monophospha   83.186
                        bis(μ-chloro)-bis(chloro-(*Methylide))-platinum(ii) .... [(bis(diphenyl   71.225
platinum]   chloro-(2-methyl-*Neopentyl(isopropoxy)carbene)-platinum).......   71.131
0)]   bicyclo(4.2.1) Nitrosopropane)-(2-(N-oxo-t-butylimino)ethyl)-   71. 24
         dichloro-(cis-2-methyl-thiacyclo-   Non-1(8)-ene-bis(triphenylphosphine)-platinum(75. 46
lenediamine> (μ-3,3,6,6-tetramethyl-3,6-diaza*Oct-4-ene)-platinum.........................   75. 1
                        (1,2-η-cyclo- Octane-1,8-diyl)-bis(N,N,N',N'-tetramethylethy   71.187
I-(μ(2)-iodo)-(methyl-di-t-butylphosphine)-(2-*Octyne)-bis(triphenylphosphine)-platinum(0)....   75. 45
                        potassium bis(Oxacyclopentylidene)-platinum-manganese <rbony   71.111
i) hydrate]   rubidium bis(Oxalato)-platinum hydrate 81. 16
         rubidium bis(oxalato)-platinum(ii) bis(Oxalato)-platinum(ii) bis(oxalato)-platinum(ii   81. 17
         cis-bis(glycinato)-platinum(ii)   Oxalato)-platinum(iii) hydrate 81. 17
e]   bis(Oxalic acid dihydrate 82. 7
e]   bis(Oxamide-oximato)-platinum(ii) ammonium chlorid   83. 10
                                        bis(Oxamide-oximato)-platinum(ii) chloride   83. 9
odium chloride dihydrate]   bis(Oxamide-oximato)-platinum(ii) hydrogen chlorid   83. 8
                        bis(oxamide- Oxamide-oximato)-platinum(ii) sodium acetate s   83. 11
                        bis(oxamide- Oximato)-platinum(ii) ammonium chloride........   83. 10
                        bis(oxamide- Oximato)-platinum(ii) chloride 83. 9
                        bis(oxamide- Oximato)-platinum(ii) hydrogen chloride.......   83. 8
loride dihydrate]   bis(oxamide- Oximato)-platinum(ii) sodium acetate sodium ch   83. 11
um(ii) toluene solvate]   di Oxygen-bis(di(t-butyl)-phenylphosphine)-platin   86. 97
<luoro-5-methoxy-3-(trifluoromethyl)-4-methyl- Pent-3-en-2-yl)-platinum hexafluorophosphate   71.162
                        bis(μ-chloro)-bis(chloro-(neo*Pentyl(isopropoxy)carbene)-platinum)   71.131
hloro-(2,11-bis(diphenylarsinomethyl)-benzo(c) Phenanthrene)-platinum .... [trans-dic   65. 74
rate]   bis(1,10- Phenanthroline)-platinum(ii) dichloride trihyd   83.167
te]   cyano-bis(1,10- Phenanthroline)-platinum(ii) nitrate monohydra   83.183
ris(triethylphosphine)-platinum(ii) hexafluoro Phosphate .... [hydrido-t   86. 55
s(dimethylphenylphosphine)-platinum hexafluoro Phosphate .... [cyanomethyl-tri   71.208
phosphino)methane)-methyl-platinum) hexafluoro*Phosphate <(2)-hydrido)-bis((μ(2)-bis(diphenyl   86.193
)-4-methyl-pent-3-en-2-yl)-platinum hexafluoro Phosphate <fluoro-5-methoxy-3-(trifluoromethyl   71.162
diphenylphosphino)methane)-platinum hexafluoro*Phosphate <is(diphenylphosphino)methane)-(bis(86.223
bis(triethylphosphine)-platinum(ii) hexafluoro Phosphate <is(methoxycarbonyl)-but-2-en-1-yl)-   71.165
trimethylene-diamine-bis(guanosine-5'-mono Phosphate methyl ester) hydrate...............   83.186
phosphinomethane)))-di-platinum(ii) hexafluoro*Phosphate methylene chloride solvate <diphenyl   86.198

Iphosphino)methane)-chloro-platinum hexafluoro*Phosphate [carbonyl-platinum-bis(μ-bis(dipheny   86.182
Iphosphino)methane)-di-platinum bis(hexafluoro*Phosphate) .... [dicarbonyl-bis(μ-bis(dipheny   86.188
ethylhypoxanthine)-platinum(ii) bis(hexafluoro Phosphate) .... [(ethylenediamine)-bis(7,9-dim   76. 93
ethylhypoxanthine)-platinum(ii) bis(hexafluoro Phosphate) sesquihydrate <netriamine>-(7,9-dim   76. 67
cesium hydrogen diammine-bis(cytidine-3'- Phosphato)-platinum(ii) tetrahydrate 83.133
carbonyl-platinum-bis((μ-diphenyl*Phosphido)-(tetracarbonyl-manganese)) 86.111
2-hydroxy-N-salicylidene-aniline-(triphenyl Phosphine) platinum(ii)..................   78. 14
num(ii)]   trans-dichloro-(triethyl Phosphine)-(cis-2,3-dimethylpyrrolidine)-plati   86. 22
num(ii)]   trans-dichloro-(triethyl Phosphine)-(cis-2,4-dimethylpyrrolidine)-plati   86. 21
latinum)-2-methyl-1,2-dica> cis-1-((tribenzyl Phosphine)-(dibenzylphosphino-phenyl-methyl)-p   71.268
cis-dichloro-(dimethylphenyl Phosphine)-(prop-2-en-1-ol)-platinum(ii) 72. 12
cis-dichloro-(dimethylphenyl Phosphine)-(vinylacetate)-platinum(ii) 72. 18
uoromethyl)-4-methyl-pent-3-en-> bis(methyl Phosphine)-(1,1,1-trifluoro-5-methoxy-3-(trifl   71.162
tetracarbonyl-(μ(2)-iodo)-(methyl-di-t-butyl*Phosphine)-(2-oxacyclopentylidene)-platinum-m>   71.111
bis(μ-(3-diphenylphosphinopropyl)phenyl*Phosphine)-bis(methyl-platinum(ii))............   86.166
,3,4-η(3)-nido-hexaboranyl)-bis(dimethylphenyl*Phosphine)-di-platinum .... [bis(μ-2   86. 44
-nido-2,3-dicarb> (μ(4,5)-(trans-bis(triethyl*Phosphine)-hydrido-platinum))-(μ(5,6)-hydrido)   86. 33
<hine-platinum-bis(μ(2)-hydrido)-tris(triethyl*Phosphine)-iridium-hydride tetraphenylborate   86.104
no)-t-butyliminothiophosphorane)-bis(trimethyl*Phosphine)-platinum <butyl-(trimethylsilyl)ami   72. 65
ngsten)-(μ-methoxytolylcarbene)-bis(trimethyl*Phosphine)-platinum <nyl-trimethylphosphine-tu   71.186
cis-dichloro-bis(diphenyl(phenylethynyl) Phosphine)-platinum acetonitrile solvate 86.144
l(sulfinyl)methylene)-thiophenyl-bis(triphenyl Phosphine)-platinum benzene solvate <thiopheny   71.272
chloro-tris(trimethyl Phosphine)-platinum chloride 86. 10
cyanomethyl-tris(dimethylphenyl Phosphine)-platinum hexafluorophosphate........   71.208
bis(dibenzyl-cis-benzyl-2-yl- Phosphine)-platinum(ii)................   86.155
trans-dihydro-bis(tri-t-butyl Phosphine)-platinum(ii)................   86. 89
cis-(ethyl)-chloro-bis(triethyl Phosphine)-platinum(ii)................   71. 80
trans-(methyl)-chloro-bis(triethyl Phosphine)-platinum(ii)................   71. 55
trans-(methyl)-chloro-bis(triphenyl Phosphine)-platinum(ii)................   71.249
trans-di(methylsulfito)-bis(triphenyl Phosphine)-platinum(ii)................   86.139
trans-cyanomethyl-hydrido-bis(triphenyl Phosphine)-platinum(ii)................   71.254
trans-(trifluoromethyl)-chloro-bis(triphenyl Phosphine)-platinum(ii)................   71.248
4-methyl-N-thiobenzohydroximato-bis(triphenyl Phosphine)-platinum(ii)................   86.163
cis-bis(thionylimide-O)-bis(triphenyl Phosphine)-platinum(ii) hemihydrate 86.119
hydrido-tris(triethyl Phosphine)-platinum(ii) hexafluorophosphate....   86. 55
s(methoxycarbonyl)-but-2-en-1-yl)-bis(triethyl Phosphine)-platinum(ii) hexafluorophosphate <i   71.165
chloro-tris(triethyl Phosphine)-platinum(ii) tetrafluoroborate 86. 53
fluoro-tris(triethyl Phosphine)-platinum(ii) tetrafluoroborate 86. 54
xycarbonyl)-1-methylcyclopropyl)-bis(triphenyl Phosphine)-platinum(ii) tetrafluoroborate <tho   71.266
s(methoxycarbonyl)-but-2-en-1-yl)-bis(triethyl Phosphine)-platinum(ii) tetraphenylborate <-bi   71.164
dioxygen-bis(di(t-butyl)-phenyl Phosphine)-platinum(ii) toluene solvate........   86. 97
bis(tri-t-butyl Phosphine)-platinum(0) 86. 88
(1,2-η-cyclo-octyne)-bis(triphenyl Phosphine)-platinum(0) 75. 45
bicyclo(4.2.1)non-1(8)-ene-bis(triphenyl Phosphine)-platinum(0) 75. 46
bonyl)ethene-1,2-diyl)-bis(carbonyl-(triphenyl*Phosphine)-platinum)   [(μ-1,2-bis(methoxycar   71.265
enyl(methoxycarbonyl)methylene)-(bis(trimethyl*Phosphine)-platinum) <mium)-(μ-carbonyl)-(μ-ph   71.185
onyl)-tungsten-bis(carbonyl-(methyl-di-t-butyl*Phosphine)-platinum) <yl-carbene)-(μ-tetracarb   71.238
rbenyl)-(carbonyl-cyclopentad> (bis(trimethyl*Phosphine)-platinum)-(μ-carbonyl)-(μ-p-tolylca   71.173
(μ-methoxy-phenyl-carbene)-bis(trimethyl*Phosphine)-platinum)-(pentacarbonyl-tungsten)   71.142
<ma:2-3-η-(4,5-dihydro-2-furyl)-(bis(trimethyl*Phosphine)-platinum)-tetracarbonyl-manganese   71. 75
<ma:2-3-η-(4,5-dihydro-2-furyl)-(bis(trimethyl*Phosphine)-platinum)-tetracarbonyl-manganese   71. 76
bis(triethyl Phosphine)-platinum-tetrathio-tungsten 86. 25
rane]   2,3-dimethyl-1,1-bis(triethyl Phosphine)-2,3-dicarba-1-platina-closo-heptabo   86. 43
rane]   1,1-bis(triethyl Phosphine)-2,4-dicarba-1-platina-closo-heptabo   86. 32
9,9-bis(triphenyl Phosphine)-6-thia-9-platina-decaborane   86.121
8-ethoxy-9,9-bis(triphenyl Phosphine)-6-thia-9-platina-decaborane 86.141
9-hydrido-9,9-bis(triethyl Phosphine)-7,8-dicarba-9-platina-undecaborane   71. 82
9-hydrido-9,10-bis(triethyl Phosphine)-7,8-dicarba-9-platina-undecaborane   71. 81
num(ii)]   chloro-triethyl Phosphine-bis(diphenylphosphino)-methano-plati   86.109
hoxycarbonyl)methylen> (tricarbonyl-trimethyl*Phosphine-chromium)-(μ-carbonyl)-(μ-phenyl(met   71.185
N-(3-methyl-but-1-enyl)-benzenamine)-triphenyl Phosphine-platinum(ii) <ro-η(2)-((Z)-2-chloro-   72. 53
-ethylenedi-imine)-bis(trans-dichloro-tributyl*Phosphine-platinum).... [(μ(2)-N,N'-di-t-butyl   86.114
loro)-(μ-difluorophosphonato)-(chloro-triethyl*Phosphine-platinum) <phosphine-platinum)-(μ-ch   86. 23
phonato)-(chlo> (difluorophosphonato-triethyl*Phosphine-platinum)-(μ-chloro)-(μ-difluorophos   86. 23
thylphosphine)-iridium-hydri> phenyl-triethyl*Phosphine-platinum-bis(μ(2)-hydrido)-tris(trie   86.104
is(trimethylphosphi> (tetracarbonyl-trimethyl*Phosphine-tungsten)-(μ-methoxytolylcarbene))-b   71.186
bis(μ-(3-diphenylphosphinopropyl)phenyl*Phosphino)-bis(chloro-platinum(ii))............   86.157
chloro-triethylphosphine-bis(diphenyl Phosphino)-methano-platinum(ii)................   86.109
chloro-trifluoromethyl-(cis-1,2-bis(diphenyl Phosphino)ethylene)-platinum(ii) 71.213
ane)-pla> hydrido-platinum-bis(μ-bis(diphenyl*Phosphino)methane)-(bis(diphenylphosphino)meth   86.223
hosphat> carbonyl-platinum-bis(μ-bis(diphenyl*Phosphino)methane)-chloro-platinum hexafluorop   86.182
dichloro-(μ-methylene)-bis(μ-bis(diphenyl*Phosphino)methane)-di-platinum 86.185
hosphate)]   dicarbonyl-bis(μ-bis(diphenyl*Phosphino)methane)-di-platinum bis(hexafluorop   86.188
phosph> (μ(2)-hydrido)-bis((μ(2)-bis(diphenyl*Phosphino)methane)-methyl-platinum) hexafluoro   86.193
<-bis(diphenylphosphino)methane)-(bis(diphenyl*Phosphino)methane)-platinum hexafluorophospha>   86.223

methylide))-platinum(ii)]   (bis(diphenyl Phosphino)methanide)-(dimethylphosphonium-bis(71.225
hino-dicarba-dodecaboranyl)> chloro-(diphenyl Phosphino-dicarba-dodecaborane)-(diphenylphosp   86. 95
henylphosphino-dicarba-dodecaborane)-(diphenyl Phosphino-dicarba-dodecaboranyl)-platinum <dip   86. 95
2-dica> cis-1-((tribenzylphosphine)-(dibenzyl Phosphino-phenyl-methyl)-platinum)-2-methyl-1,   71.268
phosphate m>   trimethyl-bis(μ(2)-(bis(diphenyl*Phosphinomethane)))-di-platinum(ii) hexafluoro 86.198
atinum(ii))]   bis(μ-(3-diphenyl*Phosphinopropyl)phenylphosphino)-bis(methyl-pl   86.166
atinum(ii))]   bis(μ-(3-diphenyl*Phosphinopropyl)phenylphosphino)-bis(chloro-pl   86.157
<hylphosphine-platinum)-(μ-chloro)-(μ-difluoro*Phosphonato)-(chloro-triethylphosphine-platin>   86. 23
oro)-(μ-difluorophosphonato)-(chlo> (difluoro*Phosphonato-triethylphosphine-platinum)-(μ-chl   86. 23
vate]   tetraphenyl Phosphonium platinum-carbonyl acetonitrile sol 64.132
(bis(diphenylphosphino)methanide)-(dimethyl Phosphonium-bis(methylide))-platinum(ii) 71.225
cis-dichloro-(μ-ethylene)-(2,6-dimethyl-*Piperidine)-platinum(ii) 72. 7
cis-dichloro-penta-1,4-diene- Platinum(ii) 72. 1
dichloro-2,5-diphenyl-hexa-1,5-diene- Platinum(ii) 72. 31
Potassium bis(oxalato)-platinum hydrate........   81. 16
cis-dichloro-(dimethylphenylphosphine)-(Prop-2-en-1-ol)-platinum(ii) 72. 12
m]   chloro-(2-methyl-2-nitroso Propane)-(2-(N-oxo-t-butylimino)ethyl)-platinu   71. 24
2-diaminopropane)-platinum(ii) bis(1,2-diamino Propane)-di-iodo-platinum(iv) iodide   [bis(1,   76. 32
malonato-(2,2-dimethyl-1,3-diamino Propane)-platinum(ii) 81. 34
di-iodo-platinum(iv) iodide] bis(1,2-diamino Propane)-platinum(ii) bis(1,2-diaminopropane)-   76. 32
e]   bis((R)-1,2- Propanediamine)-platinum(ii) chloride dihydrat   76. 35
e]   bis((R)-1,2- Propanediamine)-platinum(ii) chloride dihydrat   76. 34
l)-1,2-ethanediamine)-platinum(ii> dichloro-(Propene)-(N,N'-dimethyl-N,N'-bis(α-methylbenzy   72. 44
3-propylenediamine)-platinum(iv) pe> bis(1,3- Propylenediamine)-platinum(ii) dichloro-bis(1,   83. 25
opylenediamine)-platinum(ii) dichloro-bis(1,3- Propylenediamine)-platinum(iv) perchlorate <pr   83. 25
atinum]   dichloro-(2,4,6-trimethyl Pyridine)-(2-(N-hydroxy-t-butylimino)ethyl)-pl   71. 77
trans-dichloro-bis(γ-ethoxycarbonyl Pyridine)-platinum(ii) 83.104
trans-dichloro-(dimethylsulfoxide)-(Pyridine)-platinum(ii) 85. 15
mmine-platinum(ii) bis(trichloro-(2,6-dimethyl Pyridine)-platinum(ii))............   [tetra-a 83. 29
cis-dichloro Pyridine-glycine-platinum(ii) 82. 13
anosine (2-hydroxy-ethylthiolato)-2,2',2''-ter Pyridine-platinum(ii) hydrate <(3'-5')-deoxygu   85. 58
num tetrakis(tetrafluoroborate)]   bis(ter Pyridyl-(2-aminoethylthiolato)-platinum)-plati 85. 86
dichloro-(triethylphosphine)-(cis-2,3-dimethyl Pyrrolidine)-platinum(ii) ....[trans-   86. 22
dichloro-(triethylphosphine)-(cis-2,4-dimethyl Pyrrolidine)-platinum(ii) ....[trans-   86. 21
)-platinum(iii) hydrate] Rubidium bis(oxalato)-platinum(ii) bis(oxalato 81. 17
i)]   2-hydroxy-N- Salicylidene-aniline-(diethylamine) platinum(ii)   78. 5
inum(ii)]   2-hydroxy-N- Salicylidene-aniline-(triphenylphosphine) plat   78. 14
1-methylthyminato)-cis-diammine-platinum(ii)) *Silver nitrate pentahydrate........[bis(bis(μ-   84.100
bis(μ-hydrido)-bis(dimethyl*Silyl)-bis(tricyclohexylphosphine)-di-platinum 86.147
triphenylphosphine> η(2)-((t-butyl-(trimethyl Silyl)amino)-t-butyliminothiophosphorane)-bis(72. 65
tricyclohexylphosphine-platinum-(μ-dimethyl*Silyl-μ-(1-sigma:1-2-η-phenylethynyl))-(2-phe>   71.277
bis(oxamide-oximato)-platinum(ii) Sodium acetate sodium chloride dihydrate ...   83. 11
(oxamide-oximato)-platinum(ii) sodium acetate Sodium chloride dihydrate ....   [bis 83. 11
dichloro-(N,N'-di-t-butyl-ethylenedi-imine)- Styrene-platinum 72. 33
osphine)-platinum benzene sol> σ-(thiophenyl(Sulfinyl)methylene)-thiophenyl-bis(triphenylph   71.272
thylenediamine)-platinum bis(α-bromocamphor-π- Sulfito)-bis(triphenylphosphine)-platinum(ii)   86.139
trans-dichloro-(dimethyl Sulfonate) [(ethylenediamine)-(N,N'-dimethyle   76. 33
yguanosine (2-hydroxy-ethylthiolato)-2,2',2''- Sulfoxide)-(pyridine)-platinum(ii) 85. 15
atinum tetrakis(tetrafluoroborate)]   bis(Terpyridine-platinum(ii) hydrate <(3'-5')-deox   85. 58
9,9-bis(triphenylphosphine)-6- Terpyridyl-(2-aminoethylthiolato)-platinum)-pl   85. 86
8-ethoxy-9,9-bis(triphenylphosphine)-6- Thia-9-platina-decaborane 86.121
dichloro-(cis-2-methyl- Thia-9-platina-decaborane 86.141
dibenzotetra Thiacyclo-oct-4-ene)-platinum 75. 1
dibenzotetrathiafulvalene pentabromo-dimethyl Thiafulvalene pentabromo-dimethylthio-platinum   85. 2
bis(triethylphosphine)-platinum-tetra Thio-platinum 85. 2
latinum(ii)]   4-methyl-N- Thio-tungsten 86. 25
um(ii) cyclohexane solvate]   bis(N-t-butyldi Thiobenzohydroximato-bis(triphenylphosphine)-p   86.163
cis-dichloro-bis(O-ethyl-N,N-dimethyl-mono Thiocarbamato)-(tricyclohexylphosphine)-platin   80. 24
bis(S-ethylethene-1,2-di Thiocarbamato)-platinum(ii)...............   80. 10
oroborate)]   bis(terpyridyl-(2-aminoethyl Thiolato)-platinum(ii) 85. 20
<dylyl-(3'-5')-deoxyguanosine (2-hydroxy-ethyl Thiolato)-platinum)-platinum tetrakis(tetraflu   85. 86
cta-azabicyclo(6.6.6)eicosane)-platinum(iv) di Thiolato)-2,2',2''-terpyridine-platinum(ii) h>   85. 58
um(ii) hemihydrate]   cis-bis(Thionate hydrate ....[(1,3,6,8,10,13,16,19-o   76. 76
triphenylphosphine)-platinum benzene sol> σ-(Thionylimide-O)-bis(triphenylphosphine)-platin   86.119
nzene sol> σ-(thiophenyl(sulfinyl)methylene)- Thiophenyl(sulfinyl)methylene)-thiophenyl-bis(71.272
<((t-butyl-(trimethylsilyl)amino)-t-butylimino Thiophenyl-bis(triphenylphosphine)-platinum be   71.272
chloro-(Thiophosphorane)-bis(triphenylphosphine)-plat> 72. 65
trate]   bis(μ-1-methyl*Thyminato)-(ethylenediamine)-platinum(ii) 76. 42
itrate pentahydrate] bis(bis(μ-1-methyl*Thyminato)-bis(cis-diammine-platinum(ii)) dini   84. 42
diammine-(1-methylcytosine)-(Thyminato)-cis-diammine-platinum(ii)) silver n   84.100
<bonyl-trimethylphosphine-tungsten)-(μ-carbonyl Thyminato)-platinum(ii) perchlorate trihydrate 83. 61
<methylphosphine)-platinum)-(μ-carbonyl)-(μ-p-*Tolylcarbenyl)-bis(trimethylphosphine)-platin   71.186
bis(μ-hydrido)-bis(dimethylsilyl)-bis(*Tolylcarbene)-(carbonyl-cyclopentadienyl-man>   71.173
trans-dichloro-bis(Tricyclohexylphosphine)-di-platinum............   86.147
Tricyclohexylphosphine)-platinum(ii) 86.128

ne solvate]   bis(N-t-butyldithiocarbamato)-(Tricyclohexylphosphine)-platinum(ii) cyclohexa   80. 24
-sigma:1-2-η-phenylethynyl))-(2-phenylethynyl-*Tricyclohexylphosphine-platinum) <ylsilyl-μ-(1   71.277
yl-μ-(1-sigma:1-2-η-phenylethynyl))-(2-phen> *Tricyclohexylphosphine-platinum-(μ-dimethylsil   71.277
   bis(triethylphosphine)-platinum-tetrathio- Tungsten 86. 25
(trimethylphosphine)-platinum)-(pentacarbonyl-*Tungsten)   [(μ-methoxy-phenyl-carbene)-(bis   71.142
ylphosphi>   (tetracarbonyl-trimethylphosphine-*Tungsten)-(μ-methoxytolylcarbene)-bis(trimeth   71.186
 (μ-methoxy-phenyl-carbene)-(μ-tetracarbonyl)-*Tungsten-bis(carbonyl-(methyl-di-t-butylphosp>   71.238
ogen chloride dihydrate]   chloro-(Uracilato)-(ethylenediamine)-platinum(ii) hydr   76. 24
   cis-dichloro-(dimethylphenylphosphine)-(Vinylacetate)-platinum(ii) 72. 18
          dichloro-(diphenyl(o- Vinylphenyl)arsine)-platinum(ii) 72. 36
) ses> (diethylenetriamine)-(7,9-dimethylhypo Xanthine)-platinum(ii) bis(hexafluorophosphate   76. 67
)]   (ethylenediamine)-bis(7,9-dimethylhypo Xanthine)-platinum(ii) bis(hexafluorophosphate   76. 93

Praseodymium

(iii)) urea dihydrate]   bis(μ-*Acetato)-bis(di(urea)-di(acetato)-praseodymium   79. 14
       bis(μ-acetato)-bis(di(urea)-di(*Acetato)-praseodymium(iii)) urea dihydrate 79. 14
)-bis(di(urea)-di(acetato)-praseodymium(iii)) *Urea dihydrate .... [bis(μ-acetato   79. 14
drate]   bis(μ-acetato)-bis(di(*Urea)-di(acetato)-praseodymium(iii)) urea dihy   79. 14

Rhenium

                    bis(μ-*Acetato)-bis(dichloro-rhenium) 81. 12
<iminato))-oxo-rhenium(v))-(N,N'-ethylene-bis(*Acetylacetiminato)) rhenium(v) per-rhenate be>   76.108
lene-bis(> bis(μ-oxo)-bis((N,N'-ethylene-bis(*Acetylacetiminato))-oxo-rhenium(v))-(N,N'-ethy   76.108
 bromo-tricarbonyl-(N,N'-dimethylethane-1,2-di Amine)-rhenium(i) 76. 41
-rhenium]   Ammine-tetracarbonyl-(dimethylthiophosphinito)   86. 2
um(i))]   tetraethyl-*Ammonium tris(μ-methoxy)-bis(tricarbonyl-rheni   84. 21
         bis(o-phenyl Azophenolato-(tricarbonyl)-rhenium(i)) 84.110
   tetrachloro-bis(N,N'-dimethyl Benzamidinato)-di-rhenium 83.193
 solvate]   dichloro-tetrakis(N,N'-dimethyl Benzamidinato)-di-rhenium carbon tetrachloride   83.213
is(acetylacetiminato)) rhenium(v) per-rhenate *Benzene dichloromethane solvate <N'-ethylene-b   76.108
)-rhenium]   (2- Benzoylphenyl)-tricarbonyl-(triphenylphosphine   71.239
    fac-(triacetyl-tricarbonyl-rhenato)- Boron-chloride 71. 11
rbonyl-(η(5)-cyclopentadienyl)-(triphenylsilyl Carbene)-rhenium dichloromethane solvate [dica   71.201
         di(bis(μ-acetyl)-tetracarbonyl-rhenium)-*Copper 71. 93
ium dichloromethane solvat> dicarbonyl-(η(5)- Cyclopentadienyl)-(triphenylsilylcarbene)-rhen   71.201
η(5)-cyclopentadienyl-carbonyl-rhe> bis(η(5)- Cyclopentadienyl)-molybdenum-bis(μ-carbonyl)-(73. 73
pentadienyl)-molybdenum-bis(μ-carbonyl)-(η(5)-*Cyclopentadienyl-carbonyl-rhenium) <η(5)-cyclo   73. 73
etracarbonyl-rhenium)]   (dicarbonyl-*Cyclopentadienyl-manganese)-μ-phenylketenyl-(t   71.136
      bromo-tricarbonyl-(N,N'-dimethyl Ethane-1,2-diamine)-rhenium(i) 76. 41
<is(acetylacetiminato))-oxo-rhenium(v))-(N,N'-*Ethylene-bis(acetylacetiminato)) rhenium(v) p>   76.108
))-(N,N'-ethylene-bis(> bis(μ-oxo)-bis((N,N'-*Ethylene-bis(acetylacetiminato))-oxo-rhenium(v   76.108
e]   tricarbonyl-bis(Ethylenediamine)-rhenium(i) diphenylphosphinat   76. 43
osphino)methane)-bis(tricarbonyl-rhenium> (μ-*Ethylideneimino)-(μ-hydrido)-(μ-bis(diphenylph   86.112
) chloride monohydrate]   dithio-bis(Formamidinium) trans-tetrachloro-oxo-rhenium(v   8. 9
            trichloro-tri(Formato)-di-rhenium 81. 7
de solvate]   di Formato-tetrachloro-di-rhenium diphenylformami   81. 4
imethylphosphinato)-bis(tricarbonyl-tetrahydro*Furan-rhenium) tetrahydrofuran solvate <is(μ-d   84. 73
(tricarbonyl-rhenium> (μ-ethylideneimino)-(μ-*Hydrido)-(μ-bis(diphenylphosphino)methane)-bis   86.112
-bis(tricarbonyl-rhenium)]   bis(μ-*Hydrido)-(μ-1,2-bis(diphenylphosphino)methane)   86.106
            tris(μ-*Hydrido)-decacarbonyl-dipyridine-tri-rhenium   83.139
nonakis(trimethylsilylmethyl)-*Hydrido-(μ(2)-hexachloro)-hexa-rhenium 71.247
<(μ-chloro)-(N-nitroso-N-trimethylsilyl-methyl*Hydroxylaminato)-pentakis(trimethylsilylmethy>   71.194
methylsilyloxy)-(μ-per-rhenato)-bis(di(t-butyl*Imido)-trimethylsilyloxy-rhenium) <oxo>-(μ-tri   84.106
icarbonyl-cyclopentadienyl-manganese)-μ-phenyl*Ketenyl-(tetracarbonyl-rhenium)...........[(d   71.136
ium]   (dicarbonyl-cyclopentadienyl-*Manganese)-μ-phenylketenyl-(tetracarbonyl-rhen   71.136
   bis(μ-hydrido)-(μ-1,2-bis(diphenylphosphino)*Methane)-bis(tricarbonyl-rhenium)   86.106
neimino)-(μ-hydrido)-(μ-bis(diphenylphosphino)*Methane)-bis(tricarbonyl-rhenium) [(μ-ethylide   86.112
i))]   bis(μ-*Methoxo)-(μ-oxo)-bis(di(methoxo)-oxo-rhenium(v   84. 12
   bis(μ-methoxo)-(μ-oxo)-bis(di(*Methoxo)-oxo-rhenium(vi)) 84. 12
-rhenium]   bis(trimethylsilyl Methylidyne)-tetrakis(trimethylsilylmethyl)-di   71.193
nyl-carbonyl-rhe> bis(η(5)-cyclopentadienyl)-*Molybdenum-bis(μ-carbonyl)-(η(5)-cyclopentadie   73. 73
(trichlorophosphine-oxide)-N-pentachloroethyl- Nitrido-rhenium.................. [tetrachloro-   83. 3
)-pentakis(trimethylsilyl> tris(μ-chloro)-(N-*Nitroso-N-trimethylsilyl-methylhydroxylaminato   71.194
3,3-tetracarbonyl-5,5-dimethyl-2,2-diphenyl-1- Oxa-2-phospha-3-rhenacyclohexane .... [3,3,   71.172
         bis(o-phenylazo Phenolato-(tricarbonyl)-rhenium(i))..........   84.110
tracarbonyl-5,5-dimethyl-2,2-diphenyl-1-oxa-2- Phospha-3-rhenacyclohexane ....[3,3,3,3-te   71.172
bonyl-bis(ethylenediamine)-rhenium(i) diphenyl Phosphinate.........................[tricar   76. 43
henium) tetrahydrofuran solva> bis(μ-dimethyl*Phosphinato)-bis(tricarbonyl-tetrahydrofuran-r   84. 73
 (2-benzoylphenyl)-tricarbonyl-(triphenyl Phosphine)-rhenium 71.239
ricarbonyl-(phenyl-bis-(3,5-dimethylpyrazolyl) Phosphine)-rhenium(i) .... [bromo-t   86. 60
m(iii)]   hexamethyl-bis(diethylphenyl*Phosphine)-tris(μ-methyl)-triangulo-tri-rheniu   71.227
enium]   tetrachloro-(trichloro Phosphine-oxide)-N-pentachloroethyl-nitrido-rh   83. 3
   bis(μ-hydrido)-(μ-1,2-bis(diphenyl*Phosphino)methane)-bis(tricarbonyl-rhenium)....   86.106

μ-ethylideneimino)-(μ-hydrido)-(μ-bis(diphenyl*Phosphino)methane)-bis(tricarbonyl-rhenium) [(86.112
bromo-tricarbonyl-(phenyl-bis-(3,5-dimethyl Pyrazolyl)phosphine)-rhenium(i)................ 86. 60
tris(μ-hydrido)-decacarbonyl-di*Pyridine-tri-rhenium 83.139
lsilyl> tris(μ-chloro)-(N-nitroso-N-trimethyl*Silyl-methylhydroxylaminato)-pentakis(trimethy 71.194
dicarbonyl-(η(5)-cyclopentadienyl)-(triphenyl Silylcarbene)-rhenium dichloromethane solvate 71.201
(trimethylsilylmethylidyne)-tetrakis(trimethyl Silylmethyl)-di-rhenium.............. [bis 71.193
enium]   nonakis(trimethyl*Silylmethyl)-hydrido-(μ(2)-hexachloro)-hexa-rh 71.247
ilyl-methylhydroxylaminato)-pentakis(trimethyl*Silylmethyl)-triangulo-tri-rhenium(iii) <thyls 71.194
l)-di-rhenium]   bis(trimethyl Silylmethylidyne)-tetrakis(trimethylsilylmethy 71.193
-trimethylsilyloxy-rhen> (μ-oxo)-(μ-trimethyl*Silyloxy)-(μ-per-rhenato)-bis(di(t-butylimido) 84.106
(μ-per-rhenato)-bis(di(t-butylimido)-trimethyl*Silyloxy-rhenium) <oxo>-(μ-trimethylsilyloxy)- 84.106
rhenium(v) chloride monohydrate]   di Thio-bis(formamidinium) trans-tetrachloro-oxo-   8. 9
ammine-tetracarbonyl-(dimethyl Thiophosphinito)-rhenium 86. 2

Rhodium

o)ethane)-rhodium(i) tetraflu> (methyl-(Z)-α- Acetamido-cinnamate)-(1,2-bis(diphenylphosphin 72. 59
ris((η(5)-cyclopentadienyl)-rhodium) trifluoro*Acetate <-carbonyl)-(μ(3)-methylidyne)-cyclo-t 71.126
tetrakis(μ(2)-*Acetato)-bis(benzylthiolato-rhodium) 81. 74
tetrakis(μ-*Acetato)-bis(caffeine-rhodium) 81. 78
tetrakis(μ-*Acetato)-bis(dimethylsulfoxide-rhodium)........ 81. 48
tetrakis(μ-trifluoro*Acetato)-bis(ethanol-rhodium) 81. 42
tetrakis(μ-*Acetato)-bis(tetrahydrothiophene-rhodium) 81. 62
tetrakis(μ-*Acetato)-bis(theophylline-rhodium) dihydrate 81. 73
bis(dimethylsulfoxide)-tetrakis(trifluoro Acetato)-di-rhodium(ii)................. 81. 43
benzoyl-1,1,1-trifluoro Acetonato-cyclo-octa-1,5-diene-rhodium(i) 75. 20
)-hexacarbonyl-hexa-rhodium methan> bis(μ(3)-*Acetonitrile-copper)-carbido-ennea-(μ-carbonyl 71.147
icarbido-dodecacarbonyl-po> bis(dicarbonyl-di*Acetonitrile-rhodium(i)) uneicosa-μ-carbonyl-d 72. 3
is(diphenylphosphino)methylene)-(μ(2)-dimethyl*Acetylene-carboxylate)-dichloro-di-rhodium <-b 71.279
e dihydrate]   tetrakis(μ-Bη-*Alanine)-bis(aquo-rhodium(ii)) tetraperchlorat 82. 26
ofuran solvate] tetraphenylphosphonium η(3)- Allyl-tetradecacarbonyl-hexa-rhodium tetrahydr 72. 28
<2-diphenylphosphino-ferrocenyl)ethyl-dimethyl Amine)-norbornadiene-rhodium hexafluorophosph> 73.120
etone solvate]   bis(bis(diphenylphosphino) Amine-oxo)-rhodium(iii) hexafluorophosphate ac 86.175
(2)-carbonyl)-dicarbonyl-iron)-tr> tetraethyl*Ammonium (μ(5)-carbido-(dicarbonyl-rhodium-(μ   3. 33
-penta-rhodium]   tetra-n-butyl Ammonium hexakis(μ-carbonyl)-iodo-octacarbonyl-   3. 50
rhodium]   tris(tetraethyl Ammonium) hydrido-pentacosacarbonyl-tetradeca-   3. 5
sphine)-(sulfur dioxide)-rhodium(i) hexafluoro Arsenate <(3-diphenylphosphinopropyl)phenylpho 86.132
> trans-diaqua-bis(μ-hydroxy)-bis((1,4,7-tri*Azacyclononane)-rhodium(iii)) tetraperchlorate 76. 81
ium hexafluorophosphate] (η(6)-hexamethyl Benzene)-(η(4)-hexamethyl-cyclohexadiene)-rhod 74. 12
cta-5,7-diene-rhodium perchlorate] hexamethyl Benzene-(2-3)-tetrafluorobenzo-bicyclo(2.2.2)o 71.191
<amethoxy-10,15-dihydro-5H-5,10,15-trithia-tri Benzo(a,d,g)cyclononene)-trinitrato-rhodium(i> 85. 77
chlorate] hexamethylbenzene-(2-3)-tetrafluoro Benzo-bicyclo(2.2.2)octa-5,7-diene-rhodium per 71.191
ropane)-rhodium]   Benzoyl-dichloro-(1,3-bis(diphenylphosphino)-p 71.240
5-diene-rhodium(i)]   Benzoyl-1,1,1-trifluoroacetonato-cyclo-octa-1, 75. 20
tetrakis(μ(2)-acetato)-bis(*Benzylthiolato-rhodium)........................ 81. 74
te] hexamethylbenzene-(2-3)-tetrafluorobenzo- Bicyclo(2.2.2)octa-5,7-diene-rhodium perchlora 71.191
thyl ether so> 3,3-bis(triphenylphosphine)-3- Bisulfato-3-rhoda-1,2-dicarba-dodecaborane die 86.140
hosphine)-3-nitrato-3-rhoda-1,2-dicarba-dodeca Borane dichloromethane solvate [3-(triphenylp 86. 66
sphine)-3-bisulfato-3-rhoda-1,2-dicarba-dodeca Borane diethyl ether solvate <bis(triphenylpho 86.140
bis(diphenylphosphino)ethane)-(η(6)-tetraphenyl Borato)-rhodium(i) .... [1,2- 74. 15
tetrakis(μ-acetato)-bis(*Caffeine-rhodium)........................ 81. 78
icarbonyl-iron)-tr> tetraethylammonium (μ(5)-*Carbido)-(dicarbonyl-rhodium-(μ(2)-carbonyl)-d   3. 33
<etonitrile-rhodium(i)) uneicosa-μ-carbonyl-di*Carbido-dodecacarbonyl-polyhedro-tetradeca-rh> 72. 3
hodium methan> bis(μ(3)-acetonitrile-copper)-*Carbido-ennea-(μ-carbonyl)-hexacarbonyl-hexa-r 71.147
bis(tetraphenylphosphonium) hexa-μ-carbonyl-di*Carbido-heptacarbonyl-hexa-rhodium 64.126
<s(μ(2)-bis(diphenylphosphino)methylene)-(μ-di*Carbon-tetrasulfide)-chloro-carbonyl-rhodium 71.273
bis(tri-isopropylphosphine)-dihydrido-bi Carbonato-rhodium(iii) 81. 66
)-rhodium(iii) catechol solvate]   Catecholato-(η(5)-pentamethyl-cyclopentadienyl 73. 61
rhodium(i) tetraflu> (methyl-(Z)-α-acetamido- Cinnamate)-(1,2-bis(diphenylphosphino)ethane)- 72. 59
<s)-1,1,1-tris(diphenylphosphinomethyl)ethane)- Cobalt bis(tetraphenylborate) acetone solvate 86.225
ne)-chloro-tetra-rhodium tetrakis(tetrachloro- Cobalt) trihydrogen hexahydrate <isocyanopropa 71.261
l-hexa-rhodium methan> bis(μ(3)-acetonitrile-*Copper)-carbido-ennea-(μ-carbonyl)-hexacarbony 71.147
enylphosphine)-rhodium(i) tetrafluoro> η(4)-(Cyclo-octa-1,5-diene)-bis((R)-menthyl-methylph 75. 44
phenylphosphino-methylpyrrolidine-rhodiu> (η- Cyclo-octa-1,5-diene)-4-diphenylphosphino-2-di 75. 42
benzoyl-1,1,1-trifluoroacetonato- Cyclo-octa-1,5-diene-rhodium(i) 75. 20
lphosphino-2-diphenylphosphino-methy> (η-1,5- Cyclo-octadiene)-((2S,4S)-N-pivaloyl-4-dipheny 75. 43
rhodium(i)]   (1,2:5,6-η-1,5- Cyclo-octadiene)-(1-3-η-1,4-cyclo-octadienyl)- 75. 11
(1,2:5,6-η-1,5-cyclo-octadiene)-(1-3-η-1,4- Cyclo-octadienyl)-rhodium(i) 75. 11
ifluo> bis(μ(2)-carbonyl)-(μ(3)-methylidyne)-*Cyclo-tris((η(5)-cyclopentadienyl)-rhodium) tr 71.126
η(5)-cyclopentadienyl-η(4)-(1,2-diphenyl- Cyclobuta(l)phenanthrene)-rhodium 73.118
propen-2-yl-1,2,3,4-tetrakis(trifluoromethyl)- Cyclohexa-1,3-dienyl)-rhodium <-η(4)-(6-endo- 75. 30
(η(6)-hexamethylbenzene)-(η(4)-hexamethyl- Cyclohexadiene)-rhodium hexafluorophosphate.... 74. 12
ylphosphin> chloro-(bis(3,5-di-t-butyl-4-oxo- Cyclohexadiene-1-ylidene)ethylene)-bis(triphen 72. 68
enylphosphine)-(η(5)D-2,6-di-t-butyl-4-methyl- Cyclohexadienonyl)-rhodium(i) .... [bis(triph 75. 48
entadienyl> (2-6-η-1-exo-dimethoxyphosphoryl- Cyclohexadienyl)-(η(5)-ethyltetramethyl-cyclop 73. 81

<amethyl-cyclopentadienyl-(6-methylimino-1-5η- Cyclohexadienyl)-rhodium hexafluorophosphate   73. 70
 trans-diaqua-bis(μ-hydroxy)-bis((1,4,7-triaza*Cyclononane)-rhodium(iii)) tetraperchlorate t>   76. 81
<15-dihydro-5H-5,10,15-trithia-tribenzo(a,d,g)Cyclononene)-trinitrato-rhodium(iii) N,N-dime>   85. 77
bis(pyridine)-rhoda)-2,3-bis(trifluoromethyl)- Cyclopent-2-ene [1-((di(pivaloyl)-methanato)-   71.216
-platinum(ii) hexafluorop> (η(5)-pentamethyl- Cyclopentadienyl)-(triphenylphosphine)-hydrido   73.130
is(μ-iodo)-di-iodo-bis(η(5)-ethyl-tetramethyl-*Cyclopentadienyl)-di-rhodium(iii) .... [b   73. 96
                       (η(5)-nitro Cyclopentadienyl)-dicarbonyl-rhodium 73. 5
            dinitrato-(η(5)-pentamethyl- Cyclopentadienyl)-rhodium(iii) 73. 21
e]   catecholato-(η(5)-pentamethyl- Cyclopentadienyl)-rhodium(iii) catechol solvat   73. 61
<oryl-cyclohexadienyl)-(η(5)-ethyltetramethyl- Cyclopentadienyl)-rhodium(iii) hexafluorophos>   73. 81
       (μ(2)-methylene)-bis(carbonyl-(η(5)-*Cyclopentadienyl)-rhodium) 71. 52n
       bis(μ-iodo)-bis(iodo-(pentamethyl-*Cyclopentadienyl)-rhodium) toluene solvate 73. 88
carbonyl)-(μ(3)-methylidyne)-cyclo-tris((η(5)-*Cyclopentadienyl)-rhodium) trifluoroacetate <-   71.126
ii)]   dinitrato-(η(5)-pentamethyl- Cyclopentadienyl)-triphenylphosphine-rhodium(i   73.112
s(triphenylphosphine)-rhodium he> (bis(η(5)-*Cyclopentadienyl)-tungsten)-bis(μ-hydrido)-(bi   73.129
l)phenanthrene)-rhodium]   η(5)- Cyclopentadienyl)-η(4)-(1,2-diphenyl-cyclobuta(73.118
       (2,4-dithio-1,3-dithia-1,4-diyl)-η(5)- Cyclopentadienyl)-(trimethylphosphine)-rhodium   71. 21
dienyl)-rhodium hexafluoro η(5)-pentamethyl- Cyclopentadienyl)-(6-methylimino-1-5η-cyclohexa   73. 70
<is(μ(2)-hydrido)-μ(2)-oxo)-tris(pentamethyl-*Cyclopentadienyl-rhodium) hexafluorophosphate>   73.116
       tetrakis(μ-hydrido)-tetrakis(pentamethyl-*Cyclopentadienyl-rhodium) tetrafluoroborate....   73.127
<phenylphosphinomethyl)ethane-rhodium-(μ-η(3)-*Cyclotriphosphorus)-1,1,1-tris(diphenylphosph>   86.228
<phenylphosphinomethyl)ethane-rhodium-(μ-η(3)-*Cyclotriphosphorus)-1,1,1-tris(diphenylphosph>   86.225
       (1,1,1-tris(diphenylphosphinomethyl)ethane- Cyclotriphosphorus-rhodium 86.151
       (1,1,1-tris(diphenylphosphinomethyl) Ethane-cyclotriphosphorus-rhodium 86.151
etamido-cinnamate)-(1,2-bis(diphenylphosphino) Ethane-rhodium(i) tetrafluoroborate <(Z)-α-ac   72. 59
       1,2-bis(diphenylphosphino) Ethane-(η(6)-tetraphenylborato)-rhodium(i) 74. 15
<osphorus)-1,1,1-tris(diphenylphosphinomethyl)*Ethane-cobalt bis(tetraphenylborate) acetone >   86.225
<osphorus)-1,1,1-tris(diphenylphosphinomethyl)*Ethane-nickel bis(tetraphenylborate) tetrahyd>   86.228
,1-tris(>   1,1,1-tris(diphenylphosphinomethyl)*Ethane-rhodium-(μ-η(3)-cyclotriphosphorus)-1,1   86.225
,1-tris(>   1,1,1-tris(diphenylphosphinomethyl)*Ethane-rhodium-(μ-η(3)-cyclotriphosphorus)-1,1   86.228
       tetrakis(μ-trifluoroacetato)-bis(*Ethanol-rhodium) 81. 42
<,5-di-t-butyl-4-oxo-cyclohexadiene-1-ylidene) Ethylene)-bis(triphenylphosphine)-rhodium ace>   72. 68
       cis-dichloro-bis(Ethylenediamine)-rhodium(iii) nitrate 76. 15
       trans-dichloro-bis(Ethylenediamine)-rhodium(iii) nitrate 76. 16
rhodium hexafl>   ((+)-α-(2-diphenylphosphino- Ferrocenyl)ethyl-dimethylamine)-norbornadiene-   73.120
thane-nickel bis(tetraphenylborate) tetrahydro*Furan <s)-1,1,1-tris(diphenylphosphinomethyl)e   86.228
m(i) cis-dicarbonyl-dichloro>   cis-dicarbonyl- Glyoxal-bis(2,4-dimethylpentyl-3-imino)-rhodiu   83.134
       rubidium dichloro-bis(dimethyl Glyoximato)-rhodium(iii) monohydrate 83. 36
e>   tetrakis(μ(2)-2,5-dimethyl-2,5-di-isocyano*Hexane)-di-rhodium bis(hexafluorophosphate) ac   71.262
adienyl-rhodium) hexafluorophosph>   tris(μ(2)-*Hydrido)-(μ(3)-oxo)-tris(pentamethyl-cyclopent   73.116
       tris((μ-*Hydrido)-(bis(trimethylphosphite))-rhodium)....   86. 58n
       (bis(η(5)-cyclopentadienyl)-tungsten)-bis(μ-*Hydrido)-(bis(triphenylphosphine)-rhodium) he>   73.129
e)-dicarbonyl-di-rhod>   (μ(2)-carbonyl)-(μ(2)-*Hydrido)-bis(μ(2)-bis(diphenylphosphino)methan   86.195
-rhodium) tetrafluoroborate]   tetrakis(μ-*Hydrido)-tetrakis(pentamethyl-cyclopentadienyl   73.127
       bis(tri-isopropylphosphine)-di Hydrido-bicarbonato-rhodium(iii) 81. 66
       tris(tetraethylammonium) Hydrido-pentacosacarbonyl-tetradeca-rhodium....   3. 5
methyl-cyclopentadienyl)-(triphenylphosphine)- Hydrido-platinum(ii) hexafluorophosphate <enta   73.130
μ-dinitrogen)-bis(bis(tri-isopropylphosphine)-*Hydrido-rhodium(i))......................... [(86.130
tetra-rhodium tetrakis(tetrachloro-cobalt) tri Hydrate hexahydrate <isocyanopropane)-chloro-   71.261
de)-rhodium bis(dimethylsulfoxide)]   Hydrogen trans-tetrachloro-bis(dimethylsulfoxi   85. 5
hodium]   bis(triphenylphosphine)*Iminium (μ-pentacarbonyl)-decacarbonyl-penta-r   64.158
<-dicarbonyl-glyoxal-bis(2,4-dimethylpentyl-3- Imino)-rhodium(i) cis-dicarbonyl-dichloro-rho>   83.134
       η(5)-pentamethyl-cyclopentadienyl)-(6-methyl Imino-1-5η-cyclohexadienyl)-rhodium hexafluor>   73. 70
,3,4-tetrakis(trifluoromethyl)-cycloh>   (η(5)- Indenyl)-(1-2:3-4-η(4)-(6-endo-propen-2-yl-1,2   75. 30
)-carbonyl)-dicarbonyl-iron)-tris(tricarbonyl-*Iron) <(μ(5)-carbido)-(dicarbonyl-rhodium-(μ(2   3. 33
dicarbonyl-rhodium-(μ(2)-carbonyl)-dicarbonyl-*Iron)-tris(tricarbonyl-iron) <(μ(5)-carbido)-(3. 33
l)-(μ(2)-tris(diphenylphosphino))-tricarbonyl-*Iron-tris(carbonyl-rhodium)   [(μ(2)-dicarbony   86.161
hate) ace>   tetrakis(μ(2)-2,5-dimethyl-2,5-di-*Isocyanohexane)-di-rhodium bis(hexafluorophosp   71.262
octahydrate]   tetrakis(μ-1,3-di-*Isocyanopropane)-bis(chloro-rhodium) chloride   71.157
(tetrachloro-cobalt) trihydr>   octakis(1,3-di-*Isocyanopropane)-chloro-tetra-rhodium tetrakis   71.261
ate) acetonitrile solv>   tetrakis(μ(2)-1,3-di- Isocyanopropane)-di-rhodium bis(tetraphenylbor   71.158
 carbonyl-(triphenylphosphine)-(1-(ethylthio- Maleonitrile-2-thiolato)-rhodium 86. 90
afluoro>   η(4)-(cyclo-octa-1,5-diene)-bis((R)- Menthyl-methylphenylphosphine)-rhodium(i) tetr   75. 44
)-4-methyl-3-(prop-2'-ylidene>   (di(pivaloyl)- Methanato)-(η-1,2,4,5-(1,2-bis(trifluoromethyl   72. 40
oromethyl)-cyclopent-2-ene]   1-((di(pivaloyl)- Methanato)-bis(pyridine-rhoda)-2,3-bis(triflu   71.216
)]   bis(diphenylphosphino)*Methane)-(μ-sulfur dioxide)-bis(chloro-rhodium   86.177
<onyl)-(μ-chloro)-bis(μ-bis(diphenylphosphino)*Methane)-bis(carbonyl-rhodium) dichloro-dicar>   86.194
te]   (μ-chloro)-bis(μ-bis(diphenylphosphino)*Methane)-bis(carbonyl-rhodium) tetrafluorobora   86.186
       bis(bis(diphenylphosphino) Methane)-carbonyl-rhodium(i) tetrafluoroborate   86.184
<μ(2)-hydrido)-bis(μ(2)-bis(diphenylphosphino)*Methane)-dicarbonyl-di-rhodium p-toluenesulfo>   86.195
<(2)-carbonyl)-bis(μ(2)-bis(diphenylphosphino)*Methylene)-(μ(2)-dimethylacetylene-carboxylat>   71.279
chloro-rhodium-bis(μ(2)-bis(diphenylphosphino)*Methylene)-(μ-dicarbon-tetrasulfide)-chloro-c>   71.273
)-rhodium)]   (μ(2)-*Methylene)-bis(carbonyl-(η(5)-cyclopentadienyl   71. 52n
)-rhodium) trifluo>   bis(μ(2)-carbonyl)-(μ(3)-*Methylidyne)-cyclo-tris((η(5)-cyclopentadienyl   71.126
<s)-1,1,1-tris(diphenylphosphinomethyl)ethane-*Nickel bis(tetraphenylborate) tetrahydrofuran   86.228

odium(iii)]   di Nitrato-(η(5)-pentamethyl-cyclopentadienyl)-rh   73. 21
iphenylphosphine-rhodium(iii)]   di Nitrato-(η(5)-pentamethyl-cyclopentadienyl)-tr   73.112
<10,15-trithia-tribenzo(a,d,g)cyclononene)-tri   Nitrato-rhodium(iii) N,N-dimethylacetamide so>   85. 77
vate]   trithiaveratrylene tri Nitrato-rhodium(iii) N,N-dimethylacetamide sol   85. 77
oromethane solvate]   3-(triphenylphosphine)-3-   Nitrato-3-rhoda-1,2-dicarba-dodecaborane dichl   86. 66
ido-rhodium(i))]   (μ-di*Nitrogen)-bis(bis(tri-isopropylphosphine)-hydr   86.130
i)]   trans-chloro-(di Nitrogen)-bis(tri-isopropylphosphine)-rhodium(86. 51
enylphosphino-ferrocenyl)ethyl-dimethylamine)-   Norbornadiene-rhodium hexafluorophosphate <iph   73.120
hino-2-diphenylphosphino-methy>   (η-1,5-cyclo-   Octadiene)-((2S,4S)-N-pivaloyl-4-diphenylphosp   75. 43
m(i)]   (1,2:5,6-η-1,5-cyclo-   Octadiene)-(1-3-η-1,4-cyclo-octadienyl)-rhodiu   75. 11
2:5,6-η-1,5-cyclo-octadiene)-(1-3-η-1,4-cyclo-   Octadienyl)-rhodium(i) ....   [(1,   75. 11
           tris(tetraethylammonium) hydrido-   Pentacosacarbonyl-tetradeca-rhodium...........   3. 5
loro>   cis-dicarbonyl-glyoxal-bis(2,4-dimethyl   Pentyl-3-imino)-rhodium(i) cis-dicarbonyl-dich   83.134
clopentadienyl-η(4)-(1,2-diphenyl-cyclobuta(1)   Phenanthrene)-rhodium .... [η(5)-cy   73.118
-hexamethyl-cyclohexadiene)-rhodium hexafluoro   Phosphate ....[(η(6)-hexamethylbenzene)-(η(4)   74. 12
thyl-cyclopentadienyl)-rhodium(iii) hexafluoro   Phosphate <entadienyl)-tungsten-bis(μ-hydrido   73. 81
)-(bis(triphenylphosphine)-rhodium) hexafluoro*Phosphate <entamethyl-cyclopentadienyl)-(triph   73.129
enylphosphine)-hydrido-platinum(ii) hexafluoro   Phosphate <iphenylphosphino-ferrocenyl)ethyl-d   73.130
imine)-norbornadiene-rhodium hexafluoro   Phosphate <iphenylphosphino-ferrocenyl)ethyl-d   73.120
imino-1-5η-cyclohexadienyl)-rhodium hexafluoro   Phosphate <tamethyl-cyclopentadienyl)-(6-methyl   73. 70
ylphosphino)amine-oxo)-rhodium(iii) hexafluoro   Phosphate acetone solvate ....[bis(diphen   86.175
ntamethyl-cyclopentadienyl-rhodium) hexafluoro*Phosphate monohydrate <ido)-(μ(3)-oxo)-tris(pe   73.116
5-di-isocyanohexane)-di-rhodium bis(hexafluoro*Phosphate) acetonitrile solvate <5-dimethyl-2,   71.262
-tri-rhodium]   tris(μ(2)-diphenyl*Phosphido)-bis(triphenylphosphine)-tricarbonyl   86.221
                         pentakis(μ-diphenyl*Phosphido)-pentacarbonyl-tetra-rhodium 86.216
         trans-carbonyl-chloro-bis(tris(2-pyridyl   Phosphine))-rhodium.........................   86.105
ohexadienonyl)-rhodium(i)]   bis(triphenyl   Phosphine)-(η(5)D-2,6-di-t-butyl-4-methyl-cycl   75. 48
  chloro-(bis(3-diphenylphosphinopropyl)phenyl   Phosphine)-(sulfur dioxide)-rhodium(i) 86.120
carbonyl-(bis(3-diphenylphosphinopropyl)phenyl   Phosphine)-(sulfur dioxide)-rhodium(i) hexafl>   86.132
ato)-rhodium]   carbonyl-(triphenyl Phosphine)-(1-(ethylthio)-maleonitrile-2-thiol   86. 90
                           bis(tri-isopropyl   Phosphine)-dihydrido-bicarbonato-rhodium(iii)   81. 66
(η(5)-pentamethyl-cyclopentadienyl)-(triphenyl   Phosphine)-hydrido-platinum(ii) hexafluoropho>   73.130
         (μ-dinitrogen)-bis(bis(tri-isopropyl*Phosphine)-hydrido-rhodium(i)) 86.130
hia-1,4-diyl)-η(5)-cyclopentadienyl-(trimethyl   Phosphine)-rhodium ....   [(2,4-dithio-1,3-dit   71. 21
clohexadiene-1-ylidene)ethylene)-bis(triphenyl   Phosphine)-rhodium acetone solvate <l-4-oxo-cy   72. 68
               chloro-tris(trimethyl   Phosphine)-rhodium(i) 86. 11
         trans-carbonyl-chloro-bis(trimethyl   Phosphine)-rhodium(i) 86.131
       trans-chloro-(dinitrogen)-bis(tri-isopropyl   Phosphine)-rhodium(i) 86. 51
                     tetrakis(trimethyl   Phosphine)-rhodium(i) chloride 86. 29
o-octa-1,5-diene)-bis((R)-menthyl-methylphenyl   Phosphine)-rhodium(i) tetrafluoroborate <(cycl   75. 44
oro)-(μ-t-butylthio)-bis(carbonyl-(tri-t-butyl*Phosphine)-rhodium)...................[(μ-chl   86.103
ienyl)-tungsten)-bis(μ-hydrido)-bis(carbonyl-(tri-t-butyl*Phosphine)-rhodium) hexafluorophosphate <entad   73.129
          tris(μ(2)-diphenylphosphido)-bis(triphenyl*Phosphine)-tricarbonyl-tri-rhodium 86.221
ecaborane diethyl ether so> 3,3-bis(triphenyl   Phosphine)-3-bisulfato-3-rhoda-1,2-dicarba-dod   86.140
aborane dichloromethane solvate] 3-(triphenyl   Phosphine)-3-nitrato-3-rhoda-1,2-dicarba-dodec   86. 66
yl-penta-rhodium]   bis(triphenyl*Phosphine)iminium (μ-pentacarbonyl)-decacarbon   64.158
-(η(5)-pentamethyl-cyclopentadienyl)-triphenyl   Phosphine-rhodium(i) ....   [dinitrato   73.112
     bis((μ(2)-t-butylthiolato)-(dimethylphenyl*Phosphino))-acetyl-carbonyl-iodo-di-rhodium....   71.217
dium)]   (μ(2)-dicarbonyl)-(μ(2)-tris(diphenyl*Phosphino))-tricarbonyl-iron-tris(carbonyl-rho   86.161
             benzoyl-dichloro-(1,3-bis(diphenyl   Phosphino)-propane)-rhodium.................   71.240
methane solvate]   chloro-(2,2'-bis(o-diphenyl   Phosphino)-trans-stilbene)-rhodium(i) dichloro   72. 56
                 trichloro-(2,2'-bis(o-diphenyl   Phosphino)-trans-stilbene)-rhodium(iii) 72. 58
osphate acetone solvate]   bis(bis(diphenyl   Phosphino)amine-oxo)-rhodium(iii) hexafluoroph   86.175
<-(Z)-α-acetamido-cinnamate)-(1,2-bis(diphenyl   Phosphino)ethane)-rhodium(i) tetrafluoroborate   72. 59
ium(i)]   1,2-bis(diphenyl   Phosphino)ethane-(η(6)-tetraphenylborato)-rhod   74. 15
ro-rhodium)]   bis(μ-bis(diphenyl*Phosphino)methane)-bis(μ-sulfur dioxide)-bis(chlo   86.177
o>   (μ-carbonyl)-(μ-chloro)-bis(μ-bis(diphenyl*Phosphino)methane-bis(carbonyl-rhodium) dichl   86.194
fluoroborate]   (μ-chloro)-bis(μ-bis(diphenyl*Phosphino)methane-bis(carbonyl-rhodium) tetra   86.186
uoroborate]   bis(bis(diphenyl   Phosphino)methane-carbonyl-rhodium(i) tetrafl   86.184
<arbonyl)-(μ(2)-hydrido)-bis(μ(2)-bis(diphenyl*Phosphino)methane-dicarbonyl-di-rhodium p-to>   86.195
arboxy>   (μ(2)-carbonyl)-bis(μ(2)-bis(diphenyl*Phosphino)methylene)-(μ(2)-dimethylacetylene-c   71.279
-chloro>   chloro-rhodium-bis(μ(2)-bis(diphenyl*Phosphino)methylene)-(μ-dicarbon-tetrasulfide)   71.273
ornadiene-rhodium hexafl>   ((+-)-α-(2-diphenyl   Phosphino-ferrocenyl)ethyl-dimethylamine)-norb   73.120
<4S)-N-pivaloyl-4-diphenylphosphino-2-diphenyl   Phosphino-methylpyrrolidine)-rhodium perchlor>   75. 43
<cta-1,5-diene)-4-diphenylphosphino-2-diphenyl   Phosphino-methylpyrrolidine-rhodium(i) perchl>   75. 42
<clo-octadiene)-((2S,4S)-N-pivaloyl-4-diphenyl   Phosphino-2-diphenylphosphino-methylpyrrolidi>   75. 43
e-rhodiu>   (η-cyclo-octa-1,5-diene)-4-diphenyl   Phosphino-2-diphenylphosphino-methylpyrrolidin   75. 42
dium]   (1,1,1-tris(diphenyl   Phosphinomethyl)ethane-cyclotriphosphorus-rho   86.151
<-η(3)-cyclotriphosphorus)-1,1,1-tris(diphenyl*Phosphinomethyl)ethane-cobalt bis(tetraphenyl>   86.225
<-η(3)-cyclotriphosphorus)-1,1,1-tris(diphenyl*Phosphinomethyl)ethane-nickel bis(tetraphenyl>   86.228
iphosphorus)-1,1,1-tris(>   1,1,1-tris(diphenyl*Phosphinomethyl)ethane-rhodium-(μ-η(3)-cyclotr   86.228
iphosphorus)-1,1,1-tris(>   1,1,1-tris(diphenyl*Phosphinomethyl)ethane-rhodium-(μ-η(3)-cyclotri   86.225
de)-rhodium(i) hexa>   carbonyl-(bis(3-diphenyl   Phosphinopropyl)phenylphosphine)-(sulfur dioxi   86.132
de)-rhodium(i)]   chloro-(bis(3-diphenyl   Phosphinopropyl)phenylphosphine)-(sulfur dioxi   86.120
               tris((μ-hydrido)-(bis(trimethyl*Phosphite))-rhodium) 86. 58n

rhodium tetrahydrofuran solvate]   tetraphenyl Phosphonium η(3)-allyl-tetradecacarbonyl-hexa-   72. 28
onyl-hexa-rhodium]   bis(tetraphenyl*Phosphonium) hexa-μ-carbonyl-carbido-heptacarb   64.126
<osphinomethyl)ethane-rhodium-(μ-η(3)-cyclotri*Phosphorus)-1,1,1-tris(diphenylphosphinomethy>   86.225
<osphinomethyl)ethane-rhodium-(μ-η(3)-cyclotri*Phosphorus)-1,1,1-tris(diphenylphosphinomethy>   86.228
-tris(diphenylphosphinomethyl)ethane-cyclotri Phosphorus-rhodium ....[((1,1,1   86.151
thyl-cyclopentadienyl> (2-6-η-1-exo-dimethoxy Phosphoryl-cyclohexadienyl)-(η(5)-ethyltetrame   73. 81
                                  tetrakis(μ-*Pivalato)-bis(aquo-rhodium).....................   81. 70
uoromethyl)-4-methyl-3-(prop-2'-ylidene> (di(Pivaloyl)-methanato)-(η-1,2,4,5-(1,2-bis(trifl   72. 40
bis(trifluoromethyl)-cyclopent-2-ene] 1-((di(Pivaloyl)-methanato)-bis(pyridine)-rhoda)-2,3-   71.216
no-methy> (η-1,5-cyclo-octadiene)-((2S,4S)-N- Pivaloyl-4-diphenylphosphino-2-diphenylphosphi   75. 43
yclopentadienyl)-(triphenylphosphine)-hydrido- Platinum(ii) hexafluorophosphate <entamethyl-c   73.130
 uneicosa-μ-carbonyl-dicarbido-dodecacarbonyl-*Polyhedro-tetradeca-rhodium <trile-rhodium(i))   72. 3
<,2,4,5-(1,2-bis(trifluoromethyl)-4-methyl-3-(Prop-2'-ylidene)-penta-1,4-diene))-rhodium(i)   72. 40
ate]   tetrakis(μ-1,3-di-isocyano*Propane-bis(chloro-rhodium) chloride octahydr   71.157
loro-cobalt) trihydr> octakis(1,3-di-isocyano Propane-chloro-tetra-rhodium tetrakis(tetrach   71.261
tonitrile solv> tetrakis(μ(2)-1,3-di-isocyano*Propane)-di-rhodium bis(tetraphenylborate) ace   71.158
 benzoyl-dichloro-(1,3-bis(diphenylphosphino)- Propane)-rhodium..........................   71.240
cycloh> (η(5)-indenyl)-(1-2:3-4-η(4)-(6-endo- Propen-2-yl-1,2,3,4-tetrakis(trifluoromethyl)-   75. 30
             bis(dimethylsulfoxide)-tetrakis(Propionato)-di-rhodium(ii) 81. 63
                        tetrakis(6-methyl-2-oxo- Pyridinato)-di-rhodium 84. 97
opent-2-ene] 1-((di(pivaloyl)-methanato)-bis(Pyridine)-rhoda)-2,3-bis(trifluoromethyl)-cycl   71.216
           trans-carbonyl-chloro-bis(tris(2- Pyridylphosphine)-rhodium 86.105
4-diphenylphosphino-2-diphenylphosphino-methyl Pyrrolidine)-rhodium perchlorate <-N-pivaloyl-   75. 43
4-diphenylphosphino-2-diphenylphosphino-methyl Pyrrolidine-rhodium(i) perchlorate <,5-diene)-   75. 42
ium(iii) monohydrate]   Rubidium dichloro-bis(dimethylglyoximato)-rhod   83. 36
                                    tris(thio Semicarbazido)-rhodium(iii) chloride 79. 4
  chloro-(2,2'-bis(o-diphenylphosphino)-trans- Stilbene)-rhodium(i) dichloromethane solvate   72. 56
richloro-(2,2'-bis(o-diphenylphosphino)-trans- Stilbene)-rhodium(iii) .... [t   72. 58
diphenylphosphino)methylene)-(μ-dicarbon-tetra*Sulfide)-chloro-carbonyl-rhodium <is(μ(2)-bis(71.273
ro-bis(dimethylsulfoxide)-rhodium bis(dimethyl Sulfoxide) .... [hydrogen trans-tetrachlo   85. 5
   hydrogen trans-tetrachloro-bis(dimethyl Sulfoxide)-rhodium bis(dimethylsulfoxide) 85. 5
             bis(dimethyl Sulfoxide)-tetrakis(propionato)-di-rhodium(ii) ..   81. 63
             bis(dimethyl Sulfoxide)-tetrakis(trifluoroacetato)-di-rhodi   81. 43
um(ii)]
           tetrakis(μ-acetato)-bis(dimethyl*Sulfoxide-rhodium) 81. 48
   bis(μ-bis(diphenylphosphino)methane)-(μ-*Sulfur dioxide)-bis(chloro-rhodium)...........   86.177
s(3-diphenylphosphinopropyl)phenylphosphine)-(Sulfur dioxide)-rhodium(i) ....[chloro-(bi   86.120
<(3-diphenylphosphinopropyl)phenylphosphine)-(Sulfur dioxide)-rhodium(i) hexafluoroarsenate   86.132
tetraethylammonium) hydrido-pentacosacarbonyl- Tetradeca-rhodium .... [tris(3. 5
μ-carbonyl-dicarbido-dodecacarbonyl-polyhedro-*Tetradeca-rhodium <trile-rhodium(i)) uneicosa-   72. 3
solvate]   tetraphenylphosphonium η(3)-allyl- Tetradecacarbonyl-hexa-rhodium tetrahydrofuran   72. 28
            tetrakis(μ-acetato)-bis(*Theophylline-rhodium) dihydrate..............   81. 73
<2,13-hexamethoxy-10,15-dihydro-5H-5,10,15-tri Thia-tribenzo(a,d,g)cyclononene)-trinitrato-r>   85. 77
lphosphine)-rhodium] (2,4-dithio-1,3-di Thia-1,4-diyl)-η(5)-cyclopentadienyl-(trimethy   71. 21
methylacetamide solvate]   tri Thiaveratrylene trinitrato-rhodium(iii) N,N-di   85. 77
ium)]   (μ-chloro)-(μ-t-butyl*Thio)-bis(carbonyl-(tri-t-butylphosphine)-rhod   86.103
     carbonyl-(triphenylphosphine)-(1-(ethyl Thio)-maleonitrile-2-thiolato)-rhodium 86. 90
l-(trimethylphosphine)-rhodium] (2,4-di Thio-1,3-dithia-1,4-diyl)-η(5)-cyclopentadieny   71. 21
rbonyl-iodo-di-rhodium] bis((μ(2)-t-butyl*Thiolato)-(dimethylphenylphosphino)-acetyl-ca   71.217
henylphosphine)-(1-(ethylthio)-maleonitrile-2- Thiolato)-rhodium ....[carbonyl-(trip   86. 90
     tetrakis(μ(2)-acetato)-bis(benzyl*Thiolato-rhodium) 81. 74
     tetrakis(μ-acetato)-bis(tetrahydro*Thiophene-rhodium) 81. 62
                               tris(Thiosemicarbazido)-rhodium(iii) chloride 79. 4
nylphosphino)methane)-dicarbonyl-di-rhodium p-*Toluenesulfonate tetrahydrofuran solvate <iphe   86.195
ne)-rhodium) he> (bis(η(5)-cyclopentadienyl)-*Tungsten)-bis(μ-hydrido)-(bis(triphenylphosphi   73.129
o> bis(dicarbonyl-diacetonitrile-rhodium(i)) *Uneicosa-μ-carbonyl-dicarbido-dodecacarbonyl-p   72. 3
ylacetamide solvate]   trithia Veratrylene trinitrato-rhodium(iii) N,N-dimeth   85. 77

Ruthenium

                        tetrakis(μ-*Acetato)-bis(aquo-ruthenium) tetrafluoroborate 81. 35
                 cesium tetrakis(μ-*Acetato)-bis(chloro-ruthenium) 81. 30
          catena-(μ-chloro)-(tetrakis(μ-*Acetato)-di-ruthenium) dihydrate 81. 29
onyl-bis(triphenylphosphine)-ruthe> trifluoro Acetato-(1,4-diphenyl-but-1-en-3-yn-2-yl)-carb 71.278
lamine))-tetracarbonyl-di-ruthenium] (μ(2)-*Acetylene)-(σ,σ,η(2),η(2)-glyoxal-bis(isopropy 71. 68
)]   bis((μ(2)-bromo)-((μ(2)-1-t-butyl*Acetylene)-nonacarbonyl-tri-ruthenium)-mercury   71.228
   bis(σ,σ,η(2),η(2)-glyoxal-bis(isopropyl Amine))-octacarbonyl-tetra-ruthenium   72. 50
cetylene)-(σ,σ,η(2),η(2)-glyoxal-bis(isopropyl*Amine))-tetracarbonyl-di-ruthenium [(μ(2)-a 71. 68
fluorophosphate]   penta- Amine-(1-methylcytosinato)-ruthenium(iii) hexa   83. 19
ihydrate]   penta- Amine-ruthenium(ii)-fumaric acid dithionate d   71. 2
nium(iii)]   Ammine-trichloro-bis(trimethylphosphite)-ruthe   86. 3
um]   tetramethyl Ammonium carbido-hexadecacarbonyl-hexa-rutheni   71.117
exahydrate]   tetraethyl Ammonium hydrogen dodecachloro-tri-ruthenium h   3. 38
   dichloro-bis(o-(diphenylphosphino) Anisole)-ruthenium(ii) dichloromethane solvate   86.138
thenium)]   tetraphenyl*Arsonium (μ(6)-hydrido)-hexakis(tricarbonyl-ru   65. 13n
-hexakis(dicarbonyl-rutheniu> bis(tetraphenyl*Arsonium) μ(6)-carbido-tetrakis(μ(2)-carbonyl) 65. 15

bis(triphenylphosphine)-(μ(2)-tetrachloro-di*Azacyclopentadienyl)-dicarbonyl-ruthenium dic> 72. 62
pentammine-(3,3-pentamethylene-di Azirine)-ruthenium bis(hexafluorophosphate).... 83. 26
) nonahydrate] Barium tris(dihydrogen-violurato)-ruthenium(ii 84. 34
(1-4-η-cyclo-octatetraene)-(η(6)-hexamethyl Benzene)-ruthenium(0) 74. 8
<-nitrosyl-(2,11-bis(diphenylphosphinomethyl)- Benzo(c)phenanthrene)-ruthenium dideutero-dic> 86.169
<ans-bis(dimethylphenylphosphine)-bis(monothio Benzoato)-(1,10-phenanthroline)-ruthenium(ii) 86.154
<cis-bis(dimethylphenylphosphine)-bis(monothio Benzoato)-(1,10-phenanthroline)-ruthenium(ii) 86.153
arbonyl-iodo-bis(triphenylphosphine)-(p-methyl Benzoyl)-ruthenium .... [c 71.267
(μ-hydrido)-((α-methylene*Benzyl)-acetylido)-tris(tricarbonyl-ruthenium) 71.135
lorate] cis-bis(2,2'- Bipyridyl)-carbonyl-chloro-ruthenium(ii) perch 83.157
<pyridyl)-3-(5-pyridyl)-quinoxaline)-bis(2,2'- Bipyridyl)-ruthenium bis(hexafluorophosphate) 83.216
trans-aquo-hydroxo-bis(Bipyridyl)-ruthenium perchlorate 83.146
te)] tris(2,2'- Bipyridyl)-ruthenium(ii) bis(hexafluorophospha 83.198
phine)-ruthe> trifluoroacetato-(1,4-diphenyl- But-1-en-3-yn-2-yl)-carbonyl-bis(triphenylphos 71.278
)-cyclopentadienyl-nickel)-(μ(4)-3,3-dimethyl- But-1-enyl)-nonacarbonyl-tri-ruthenium [(η(5 71.151
(μ(2)-diphenylphosphido)-(μ(3)-3-methyl*Butynyl)-nonacarbonyl-tri-ruthenium............ 71.198
(μ(2)-diphenylphosphido)-(μ(3)-3,3-dimethyl*Butynyl)-octacarbonyl-tri-ruthenium benzene s> 71.199
dicarbonyl- Carbamoyl-(η(5)-cyclopentadienyl)-ruthenium.... 71. 6
pentacarbonyl-(ethoxy(ruthenocenyl) Carbene)-tungsten 71.125
exa-ruthenium] dodecacarbonyl-(η(5)-*Carbido)-(μ-hydrido)-tris(η(2)-ethylsulfide)-h 71.177
dro-hexa-ruthenium] (μ(3)-1-6-η-bitropyl)-*Carbido-(μ(2)-carbonyl)-tridecacarbonyl-octahe 71.223
tetramethylammonium Carbido-hexadecacarbonyl-hexa-ruthenium........ 71.117
onyl-rutheniu> bis(tetraphenylarsonium) μ(6)-*Carbido-tetrakis(μ(2)-carbonyl)-hexakis(dicarb 65. 15
um) tetrafluoro> (μ(2)-carbonyl)-(μ(2)-methyl*Carbyne)-bis(carbonyl-cyclopentadienyl- rutheni 71. 87
m)] *Cesium tetrakis(μ-acetato)-bis(chloro-rutheniu 81. 30
-(tri-μ-carbonyl)-tris(tricarbonyl-ruthenium)-*Cobalt <(triphenylphosphine)immonium) carbonyl 64.154
<ntacarbonyl-(μ-2',3'-η,7'-8'-eta:4'-6'-η-(4-(*Cyclo-octa-2',5',7'-triene-1',4'-diyl)-1,1,4,> 75. 23
uthenium(0)] (1-4-η- Cyclo-octatetraene)-(η(6)-hexamethylbenzene)-r 74. 8
(dicarbonyl-ruthenium)] (μ(3)-*Cycloheptatrienyl)-(μ(3)-t-butylthiolato)-tris 75. 15
onyl-1,2-diphenylethylene)-carbony> bis(η(5)-*Cyclopentadienyl)-(μ-carbonyl)-(μ-η',η(3)-carb 71.212
<s(triphenylphosphine)-(μ(2)-tetrachloro-diaza*Cyclopentadienyl)-dicarbonyl-ruthenium dichlo> 72. 62
bis(Cyclopentadienyl)-ruthenium.................. 73. 16
bis(Cyclopentadienyl)-ruthenium.................. 73. 17
bis(Cyclopentadienyl)-ruthenium.................. 73. 18
dicarbonyl-carbamoyl-(η(5)- Cyclopentadienyl)-ruthenium.................. 71. 6
t-1-enyl)-nonacarbonyl-tri-ruthenium] (η(5)-*Cyclopentadienyl-nickel)-(μ(4)-3,3-dimethyl-bu 71.151
μ(2)-carbonyl)-(μ(2)-vinylidene)-bis(carbonyl-*Cyclopentadienyl-ruthenium).................. [(71. 86
<-carbonyl)-(μ(2)-methylcarbyne)-bis(carbonyl-*Cyclopentadienyl-ruthenium) tetrafluoroborate 71. 87
penta-amine-(1-methyl Cytosinato)-ruthenium(iii) hexafluorophosphate 83. 19
hinomethyl)-benzo(c)phenanthrene)-ruthenium di Deutero-dichloromethane solvate <diphenylphosp 86.169
rakis(μ-hydrido)-(μ-1,2-bis(diphenylphosphino)*Ethane)-decacarbonyl-tetra-ruthenium [tet 86.116
cis-dihydrido-bis(1,2-bis(diphenylphosphino) Ethane)-ruthenium benzene solvate 86.192
-(μ-carbonyl)-(μ-η',η(3)-carbonyl-1,2-diphenyl*Ethylene)-carbonyl-di-ruthenium <opentadienyl) 71.212
henium] (η(2)-dimethyl*Formamide)-(μ(2)-hydrido)-decacarbonyl-tri-rut 71. 47
chloro-carbonyl-(N-isopropenyl-N'-isopropyl- Formamidinato)-bis(triphenylphosphine)-ruthen> 86.165
potassium tetrakis(μ-*Formato)-bis(chloro-ruthenium) 81. 10
penta-ammine-ruthenium(ii)- Fumaric acid dithionate dihydrate 71. 2
a-ruthenium] bis(σ,σ,η(2),η(2)- Glyoxal-bis(isopropylamine))-octacarbonyl-tetr 72. 50
ruthenium] (μ(2)-acetylene)-(σ,σ,η(2),η(2)-*Glyoxal-bis(isopropylamine))-tetracarbonyl-di- 71. 68
tetramethylammonium carbido- Hexadecacarbonyl-hexa-ruthenium............... 71.117
tricarbonyl-ruthenium)] (μ-*Hydrido)-((α-methylenebenzyl)-acetylido)-tris(71.135
decacarbonyl-tetra-ruthenium] tetrakis(μ-*Hydrido)-(μ-1,2-bis(diphenylphosphino)ethane)- 86.116
(η(2)-dimethylformamide)-(μ(2)-*Hydrido)-decacarbonyl-tri-ruthenium............ 71. 47
<henylphosphine)iminium (μ(2)-carbonyl)-(μ(2)-*Hydrido)-hexacarbonyl-di-iron-hexacarbonyl-di> 64.156
tetraphenylarsonium (μ(6)-*Hydrido)-hexakis(tricarbonyl-ruthenium)...... 65. 13n
<henylphosphine)iminium (μ(2)-carbonyl)-(μ(2)-*Hydrido)-tricarbonyl-iron-nonacarbonyl-tri-ru> 64.155n
m] dodecacarbonyl-(η(5)-carbido)-(μ-*Hydrido)-tris(η(2)-ethylsulfide)-hexa-rutheniu 71.177
phosphit> 1,1,1,2,2,3,3-heptacarbonyl-2,3-(μ-*Hydrido)-2,3-(μ-nitrosyl)-1,2,3-tris(trimethyl 86. 40
ruthenium benzene solvate] cis-di Hydrido-bis(1,2-bis(diphenylphosphino)ethane)- 86.192
henium(ii))] bis(μ-chloro)-bis(chloro-(*Hydrogen bis(diethylphosphonito))-nitrosyl-rut 86. 45
e] tetraethylammonium Hydrogen dodecachloro-tri-ruthenium hexahydrat 3. 38
barium tris(di Hydrogen-violurato)-ruthenium(ii) nonahydrate 84. 34
ium(ii) tetrafluoroborate] tris(μ-*Hydroxo)-hexakis(trimethylphosphine)-di-ruthen 86. 57
trans-aquo- Hydroxo-bis(bipyridyl)-ruthenium perchlorate 83.146
sphine)-ruthenium(ii)] (1,3-bis(4-tolyl)- Imidazolidin-2-ylidene)-chloro-bis(triethylpho 71.226
ylphosphine)-((N-p-tolyl)-(N'-(p-tolyl-o-yl))- Imidazolidin-2-ylidene)-ruthenium <-bis(trieth 71.226
ro-bis(triethylphosphine)-((N-(p-tolyl-o-yl))- Imidazolidin-2-ylidene)-ruthenium(ii) <yl-chlo 71.231
bonyl-di-iron-hexaca> bis(triphenylphosphine)*Iminium (μ(2)-carbonyl)-(μ(2)-hydrido)-hexacar 64.156
onyl-iron-nonacarbon> bis(triphenylphosphine)*Iminium (μ(2)-carbonyl)-(μ(2)-hydrido)-tricarb 64.155n
nium-tetra-iridi> bis(bis(triphenylphosphine)*Iminium) heptacarbonyl-μ(2)-hexacarbonyl-ruthe 64.162
nium-tetra-iridi> bis(bis(triphenylphosphine)*Iminium) nonacarbonyl-hexa-μ(2)-carbonyl-ruthe 64.163
rbonyl-ruthenium)-c> bis((triphenylphosphine)*Immonium) carbonyl-(tri-μ-carbonyl)-tris(trica 64.154
nacarbonyl-hexa-μ(2)-carbonyl-ruthenium-tetra-*Iridium <is(bis(triphenylphosphine)iminium) no 64.163
ptacarbonyl-μ(2)-hexacarbonyl-ruthenium-tetra-*Iridium <is(bis(triphenylphosphino)iminium) he 64.162
μ(2)-carbonyl)-(μ(2)-hydrido)-hexacarbonyl-di-*Iron-hexacarbonyl-di-ruthenium <hine)iminium (64.156

um (μ(2)-carbonyl)-(μ(2)-hydrido)-tricarbonyl-*Iron-nonacarbonyl-tri-ruthenium <osphine)imini   64.155n
(η(2)-dithiomethoxycarbonyl)-carbonyl-(p-tolyl Isocyanide)-bis(triphenylphosphine)-ruthenium>   71.270
(η(2)-dithiomethoxycarbonyl)-carbonyl-(p-tolyl Isocyanide)-bis(triphenylphosphine)-ruthenium>   71.271
iphenylphosphine)-ruthen> chloro-carbonyl-(N- Isopropenyl-N'-isopropyl-formamidinato)-bis(tr   86.165
t-butylacetylene)-nonacarbonyl-tri-ruthenium)-*Mercury) ....[bis((μ(2)-bromo)-((μ(2)-1-   71.228
ium(ii) tetrafluoroborate] (μ-methyl)-bis(μ-*Methylene)-bis(tris(trimethylphosphine)-ruthen   71.176
ium(ii))]   tris(μ-*Methylene)-bis(tris(trimethylphosphine)-ruthen   71.175
ium(iii)) tetrafluoroborate]   bis(μ-*Methylene)-bis(tris(trimethylphosphine)-ruthen   71.166
ium(iii))-ruthenium(iv) bis(te> bis(bis(μ(2)-*Methylene)-tetrakis(trimethylphosphine)-ruthen   71.222
hosphate)]   pentammine-(3,3-penta Methylene-diazirine)-ruthenium bis(hexafluorop   83. 26
uthenium)]   (μ-hydrido)-((α-*Methylenebenzyl)-acetylido)-tris(tricarbonyl-r   71.135
bonyl-tri-ruthenium] (η(5)-cyclopentadienyl-*Nickel)-(μ(4)-3,3-dimethyl-but-1-enyl)-nonacar   71.151
<2,2,3,3-heptacarbonyl-2,3-(μ-hydrido)-2,3-(μ-*Nitrosyl)-1,2,3-tris(trimethylphosphite)-tria>   86. 40
nzo(c)phenanthrene)-rutheni> carbonyl-chloro-Nitrosyl-(2,11-bis(diphenylphosphinomethyl)-be   86.169
bis(chloro-(hydrogen bis(diethylphosphonito))-*Nitrosyl-ruthenium(ii))........[bis(μ-chloro)-   86. 45
-tetracarbonyl-> (1,3,5-tris(trimethylsilyl)- Pentalene)-triangulo(bis(dicarbonyl-ruthenium)   73.104
)-dicarbonyl-ru> (1,3,5-tris(trimethylsilyl)- Pentalene)-triangulo(bis(tricarbonyl-ruthenium   73.105
enium bis(hexafluorophosphate)]   Pentammine-(3,3-pentamethylene-diazirine)-ruth   83. 26
ene-1',4'-diyl)-1,1,4,4-tetramethyl-1,4-disila-*Pentyl))-di-ruthenium <cyclo-octa-2',5',7'-tri   75. 23
<l-(2,11-bis(diphenylphosphinomethyl)-benzo(c) Phenanthrene)-ruthenium dideutero-dichloromet>   86.169
lphenylphosphine)-bis(monothiobenzoato)-(1,10- Phenanthroline)-ruthenium(ii) <ans-bis(dimethy   86.154
lphenylphosphine)-bis(monothiobenzoato)-(1,10- Phenanthroline)-ruthenium(ii) <cis-bis(dimethy   86.153
(1-methylcytosinato)-ruthenium(iii) hexafluoro Phosphate ....   [penta-amine-   83. 19
s(2,2'-bipyridyl)-ruthenium(ii) bis(hexafluoro Phosphate) ....   [tri   83.198
amethylene-diazirine)-ruthenium bis(hexafluoro Phosphate) .... [pentammine-(3,3-pent   83. 26
)-bis(2,2'-bipyridyl)-ruthenium bis(hexafluoro Phosphate) <pyridyl)-3-(5-pyridyl)-quinoxaline   83.216
-tri-ruthenium]   (μ(2)-diphenyl*Phosphido)-(μ(3)-3-methylbutynyl)-nonacarbonyl   71.198
onyl-tri-ruthenium benzene so> (μ(2)-diphenyl*Phosphido)-(μ(3)-3,3-dimethylbutynyl)-octacarb   71.199
ylidene)-ruthen> carbonyl-chloro-bis(triethyl Phosphine)-((N-(p-tolyl-o-yl))-imidazolidin-2-   71.231
idazolidin-2-ylidene)-ru> chloro-bis(triethyl Phosphine)-((N-p-tolyl)-(N'-(p-tolyl-o-yl))-im   71.226
enyl)-dicarbonyl-ruthenium dic> bis(triphenyl*Phosphine)-(μ(2)-tetrachloro-diazacyclopentadi   72. 62
carbonyl-iodo-bis(triphenyl Phosphine)-(p-methylbenzoyl)-ruthenium 71.267
uene solvate] bis((μ-sulfato)-bis(triphenyl*Phosphine)-(sulfur dioxide)-ruthenium(ii)) tol   86.220
carbonyl-iodo-bis(triphenyl Phosphine)-acetyl-ruthenium....................   71.258
hroline)-ru> cis,cis,trans-bis(dimethylphenyl Phosphine)-bis(monothiobenzoato)-(1,10-phenant   86.154
hroline)-ruth> cis,cis,cis-bis(dimethylphenyl Phosphine)-bis(monothiobenzoato)-(1,10-phenant   86.153
tris(μ-hydroxo)-hexakis(trimethyl*Phosphine)-di-ruthenium(ii) tetrafluoroborate   86. 57
nyl-but-1-en-3-yn-2-yl)-carbonyl-bis(triphenyl Phosphine)-ruthenium <fluoroacetato-(1,4-diphe   71.278
bis(μ-bromo)-bis(dicarbonyl-(tri-t-butyl*Phosphine)-ruthenium(i)) 86. 98
l)-imidazolidin-2-ylidene)-chloro-bis(triethyl Phosphine)-ruthenium(ii)   [(1,3-bis(4-toly   71.226
enyl-N'-isopropyl-formamidinato)-bis(triphenyl Phosphine)-ruthenium(ii) <-carbonyl-(N-isoprop   86.165
<l)-carbonyl-(p-tolylisocyanide)-bis(triphenyl Phosphine)-ruthenium(ii) perchlorate chlorofo>   71.271
<l)-carbonyl-(p-tolylisocyanide)-bis(triphenyl Phosphine)-ruthenium(ii) perchlorate chlorofo>   71.270
<thiomethoxycarbonyl)-dicarbonyl-bis(triphenyl Phosphine)-ruthenium(ii) perchlorate cyclohex>   71.260
(μ-methyl)-bis(μ-methylene)-bis(tris(trimethyl*Phosphine)-ruthenium(ii) tetrafluoroborate 71.176
tris(μ-methylene)-bis(tris(trimethyl*Phosphine)-ruthenium(ii))   71.175
bis(μ-methylene)-bis(tris(trimethyl*Phosphine)-ruthenium(iii)) tetrafluoroborate   71.166
e> bis(bis(μ(2)-methylene)-tetrakis(trimethyl*Phosphine)-ruthenium(iii))-ruthenium(iv) bis(t   71.222
tricarbonyl-bis(triphenyl Phosphine)-ruthenium(0).................   86. 9
undecacarbonyl-(triphenyl Phosphine)-triangulo-tri-ruthenium 86. 99
o)-hexacarbonyl-di-iron-hexaca> bis(triphenyl*Phosphine)iminium (μ(2)-carbonyl)-(μ(2)-hydrid   64.156
o)-tricarbonyl-iron-nonacarbon> bis(triphenyl*Phosphine)iminium (μ(2)-carbonyl)-(μ(2)-hydrid   64.155n
onyl-ruthenium-tetra-iridi> bis(bis(triphenyl*Phosphine)iminium) nonacarbonyl-hexa-μ(2)-carb   64.163
tris(tricarbonyl-ruthenium)-c> bis(triphenyl*Phosphine)immonium) carbonyl-(tri-μ-carbonyl)-   64.154
ne solvate] dichloro-bis(o-(diphenyl Phosphino)anisole)-ruthenium(ii) dichlorometha   86.138
tetrakis(μ-hydrido)-(μ-1,2-bis(diphenyl Phosphino)ethane-decacarbonyl-tetra-ruthenium   86.116
cis-dihydrido-bis(1,2-bis(diphenyl Phosphino)ethane)-ruthenium benzene solvate....   86.192
onyl-ruthenium-tetra-iridi> bis(bis(triphenyl*Phosphino)iminium) heptacarbonyl-μ(2)-hexacarb   64.162
carbonyl-chloro-nitrosyl-(2,11-bis(trimethyl Phosphinomethyl)-benzo(c)phenanthrene)-ruthen>   86.169
amine-trichloro-bis(trimethyl*Phosphite-ruthenium(iii) 86. 3
hydrido)-2,3-(μ-nitrosyl)-1,2,3-tris(trimethyl*Phosphite)-triangulo-tri-ruthenium <yl-2,3-(μ-   86. 40
tetracarbonyl-trimethyl Phosphite-ruthenium...................   86. 4
bis(μ-chloro)-bis(chloro-(hydrogen bis(diethyl*Phosphonito))-nitrosyl-ruthenium(ii)) 86. 45
nium)]   *Potassium tetrakis(μ-formato)-bis(chloro-ruthe   81. 10
catena-(μ-chloro)-(tetrakis(μ-*Propionato)-di-ruthenium)   81. 47
hexakis(Pyridine)-ruthenium(ii) bis(tetrafluoroborate)   83.200
enium bis(hexafluoropho> (2-(2-pyridyl)-3-(5- Pyridyl)-quinoxaline)-bis(2,2'-bipyridyl)-ruth   83.216
ipyridyl)-ruthenium bis(hexafluoropho> (2-(2- Pyridyl)-3-(5-pyridyl)-quinoxaline)-bis(2,2'-b   83.216
(hexafluoropho> (2-(2-pyridyl)-3-(5-pyridyl)- Quinoxaline)-bis(2,2'-bipyridyl)-ruthenium bis   83.216
Ruthenocene.................   73. 18
Ruthenocene.................   73. 17
Ruthenocene.................   73. 16
-triene-1',4'-diyl)-1,1,4,4-tetramethyl-1,4-di*Silapentyl))-di-ruthenium <cyclo-octa-2',5',7'   75. 23
octa-1,3,5,7-tetrae> tetracarbonyl-(trimethyl Silyl)-η-(1-4-eta:1-σ,5-8-η-(8-trimethylsilyl-   72. 32
henium)-tetracarbonyl-> (1,3,5-tris(trimethyl Silyl)-pentalene)-triangulo(bis(dicarbonyl-rut   73.104

thenium)-dicarbonyl-ru>   (1,3,5-tris(trimethyl Silyl)-pentalene)-triangulo(bis(tricarbonyl-ru   73.105
methylsilyl)-η-(1-4-eta:1-σ,5-8-η-(8-trimethyl Silyl-octa-1,3,5,7-tetraenyl))-di-ruthenium <i   72. 32
de)-ruthenium(ii)) toluene solvate]   bis((μ-*Sulfato)-bis(triphenylphosphine)-(sulfur dioxi   86.220
nyl-(η(5)-carbido)-(μ-hydrido)-tris(η(2)-ethyl*Sulfide)-hexa-ruthenium........   [dodecacarbo   71.177
    bis((μ-sulfato)-bis(triphenylphosphine)-(*Sulfur dioxide)-ruthenium(ii)) toluene solvate   86.220
<cis,cis-bis(dimethylphenylphosphine)-bis(mono Thiobenzoato)-(1,10-phenanthroline)-ruthenium>   86.153
<s,trans-bis(dimethylphenylphosphine)-bis(mono Thiobenzoato)-(1,10-phenanthroline)-ruthenium>   86.154
)-dicarbonyl-ruthenium(>   (μ-bis(N,N-diethyldi*Thiocarbamato))-bis(N,N-diethyldithiocarbamato   80. 22
<,N-diethyldithiocarbamato))-bis(N,N-diethyldi*Thiocarbamato)-dicarbonyl-ruthenium(ii)-ruthe>   80. 22
    (μ(3)-cycloheptatrienyl)-(μ(3)-t-butyl*Thiolato)-tris(dicarbonyl-ruthenium) 75. 15
de)-bis(triphenylphosphine)-rutheni>   (η(2)-di Thiomethoxycarbonyl)-carbonyl-(p-tolylisocyani   71.271
de)-bis(triphenylphosphine)-rutheni>   (η(2)-di Thiomethoxycarbonyl)-carbonyl-(p-tolylisocyani   71.270
hosphine)-ruthenium(ii) perchlorate>   (η(2)-di Thiomethoxycarbonyl)-dicarbonyl-bis(triphenylp   71.260
    penta-ammine-ruthenium(ii)-fumaric acid di Thionate dihydrate 71. 2
ene)-ru>   chloro-bis(triethylphosphine)-((N-p- Tolyl)-(N'-(p-tolyl-o-yl))-imidazolidin-2-ylid   71.226
thylphosphine)-ruthenium(ii)]   (1,3-bis(4- Tolyl)-imidazolidin-2-ylidene)-chloro-bis(trie   71.226
carbonyl-chloro-bis(triethylphosphine)-((N-p- Tolyl-o-yl))-imidazolidin-2-ylidene)-rutheniu>   71.231
<o-bis(triethylphosphine)-((N-p-tolyl)-(N'-(p- Tolyl-o-yl))-imidazolidin-2-ylidene)-ruthenium   71.226
ni>   (η(2)-dithiomethoxycarbonyl)-carbonyl-(p- Tolylisocyanide)-bis(triphenylphosphine)-ruthe   71.270
ni>   (η(2)-dithiomethoxycarbonyl)-carbonyl-(p- Tolylisocyanide)-bis(triphenylphosphine)-ruthe   71.271
(μ(3)-1-6-η-bitropyl)-carbido-(μ(2)-carbonyl)-*Tridecacarbonyl-octahedro-hexa-ruthenium 71.223
l-octahedro-hexa-ruthenium]   (μ(3)-1-6-η-bi*Tropyl)-carbido-(μ(2)-carbonyl)-tridecacarbony   71.223
    pentacarbonyl-(ethoxy(ruthenocenyl)carbene)- Tungsten 71.125
enium)]   (μ(2)-carbonyl)-(μ(2)-*Vinylidene)-bis(carbonyl-cyclopentadienyl-ruth   71. 86
    barium tris(dihydrogen- Violurato)-ruthenium(ii) nonahydrate 84. 34

Samarium

iii) aquo-p>   (4,7,13,16,21,24-hexaoxa-1,10-di Azabicyclo(8.8.8)hexacosane)-nitrato-samarium(84. 74
c>   tris(perchlorato)-(5,6,14,15-tetrahydro-di Benzo(b,k)(1,4,7,10,13,16)-hexaoxacyclo-octade   84. 83
) aquo-p>   (4,7,13,16,21,24-hexaoxa-1,10-diaza Bicyclo(8.8.8)hexacosane)-nitrato-samarium(iii   84. 74
,4,7,10,13,16)-hexaoxacyclo-octadec>   tris(per Chlorato)-(5,6,14,15-tetrahydro-dibenzo(b,k)   84. 83
etrahydro-dibenzo(b,k)(1,4,7,10,13,16)-hexaoxa Cyclo-octadecin)-samarium(iii) <)-(5,6,14,15-t   84. 83
<,13,16,21,24-hexaoxa-1,10-diazabicyclo(8.8.8) Hexacosane)-nitrato-samarium(iii) aquo-pentan>   84. 74
8)hexacosane)-nitrato-samarium(iii) aquo-penta Nitrato-samarium(iii) <-1,10-diazabicyclo(8.8.   84. 74
<-hexaoxa-1,10-diazabicyclo(8.8.8)hexacosane)- Nitrato-samarium(iii) aquo-pentanitrato-samar>   84. 74
dro-dibenzo(b,k)(1,4,7,10,13,16)-hexaoxacyclo- Octadecin)-samarium(iii) <)-(5,6,14,15-tetrahy   84. 83
o-samarium(iii) aquo-p>   (4,7,13,16,21,24-hexa Oxa-1,10-diazabicyclo(8.8.8)hexacosane)-nitrat   84. 74
5-tetrahydro-dibenzo(b,k)(1,4,7,10,13,16)-hexa Oxacyclo-octadecin)-samarium(iii) <)-(5,6,14,1   84. 83

Scandium

yacetato)-scandium(iii) tetra-aquo-bis(hydroxy Acetato)-scandium(iii)   ....   [tetra(hydrox   81. 31
cetato)-scandium(iii)]   tetra(hydroxy Acetato)-scandium(iii) tetra-aquo-bis(hydroxya   81. 31

Silver

    dipicryl Aminato-silver 15. 3
(4)-bromo)-bis(μ-methyl-bis(diphenylphosphino)*Amine-di-silver)   ....[bis(μ-bromo)-bis(μ   86.179
te]   bis(bis(μ-1-methylthyminato)-cis-di*Ammine-platinum(ii)) silver nitrate pentahydra   84.100
    p-amino Benzoato-silver...............................   81. 27
iphenylphosphine-11-argenta-5,6-dicarba-undeca Borane(11) acetone solvate .... [11-tr   86. 65
r perchlorate]   bis((bis(ethylenediamine)-*Cobalt)-(μ-carboxylato-methyl)-thiolato)-silve   76. 80
    4,7,10,13,16-pentaoxa-1-thia Cyclo-octadecane silver nitrate.................   42. 8
    bis(1,5- Cyclo-octadiene)-silver(i) tetrafluoroborate   75. 12
is(tri>   μ-1-5-eta:1'-5'-η-(bis(μ-tetraphenyl-*Cyclopentadienediyl-oxo-O)-bis(aqua-silver))-b   73.133
)-thiolato)-silver perchlorate]   bis((*Ethylenediamine)-cobalt)-(μ-carboxylato-methyl   76. 80
e]   bis(9-methyl Hypoxanthine)-silver(i) perchlorate monohydrat   83. 74
    hexakis(bis(Imidazole)-silver perchlorate) 83.212
<iyl-oxo-O)-bis(aqua-silver))-bis(tricarbonyl-*Iron) bis(hexafluorophosphate) dichloromethan>   73.133
    bis(μ(3)-bromo)-tris((diphenylphosphino)*Methane-silver) bromide...................   86.222
    4,7,10,13,16-pentaoxa-1-thiacyclo- Octadecane silver nitrate   75. 12
    bis(1,5-cyclo- Octadiene)-silver(i) tetrafluoroborate 75. 12
    4,7,10,13,16-penta Oxa-1-thiacyclo-octadecane silver nitrate 42. 8
-silver))-bis(tricarbonyl-iron) bis(hexafluoro*Phosphate) dichloromethane solvate <)-bis(aqua   73.133
hio)-triphenylphosphine-silver]   bis(triphenyl*Phosphine)-silver-bis(μ-thio)-tungsten-bis(μ-t   86.199
ver-bis(μ-thio)-tungsten-bis(μ-thio)-triphenyl*Phosphine-silver   [bis(triphenylphosphine)-sil   86.199
11) acetone solvate]   11-triphenyl Phosphine-11-argenta-5,6-dicarba-undecaborane(86. 65
omo)-bis(μ(4)-bromo)-bis(μ-methyl-bis(diphenyl*Phosphino)amine-di-silver)   .....   [bis(μ-br   86.179
    bis(μ(3)-bromo)-tris((diphenyl*Phosphino)methane-silver) bromide 86.222
    di Picrylaminato-silver 15. 3
bis(bis(μ-1-methylthyminato)-cis-diammine-*Platinum(ii)) silver nitrate pentahydrate 84.100
    4,7,10,13,16-pentaoxa-1- Thiacyclo-octadecane silver nitrate...........   42. 8
Iphosphine)-silver-bis(μ-thio)-tungsten-bis(μ-*Thio)-triphenylphosphine-silver [bis(tripheny   86.199
silver]   bis(triphenylphosphine)-silver-bis(μ-*Thio)-tungsten-bis(μ-thio)-triphenylphosphine-   86.199

hylenediamine)-cobalt)-(μ-carboxylato-methyl)-*Thiolato)-silver perchlorate   [bis((bis(et   76. 80
itrate pentahydrate]   bis(bis(μ-1-methyl*Thyminato)-cis-diammine-platinum(ii)) silver n   84.100
                                        (1-methyl Thymine)-silver(i) 44. 14
   bis(triphenylphosphine)-silver-bis(μ-thio)-*Tungsten-bis(μ-thio)-triphenylphosphine-silver   86.199
                          bis(9-methylhypo Xanthine)-silver(i) perchlorate monohydrate....   83. 74

Tantalum

rofuran-tantalum)]   (μ-di-t-butyl*Acetylene)-bis((μ(2)-chloro)-dichloro-tetrahyd   72. 35
         trichloro-bis(bis(trimethylsilyl) Amido)-tantalum(v) 83. 85
η(5)-pentamethyl-cyclopentadienyl)-3-tantalla- Bicyclo(3.3.0)octane .... [3,3-dichloro-3-(71.110
tantalum]   dichloro-(η(5)- Cyclopentadienyl)-(2-oxo-4-methyl-pent-3-ene)-   71. 30
<lphosphine-methyl)-bis(η(5)-ethyltetramethyl-*Cyclopentadienyl)-hydrido-bis(dichloro-tantal>   71.209
                  bis(η(5)- Cyclopentadienyl)-methylmethylene-tantalum 71. 42n
   1,1-dichloro-1-(η(5)-pentamethyl- Cyclopentadienyl)-tantalla-cyclopentane........   71. 74
   3,3-dichloro-3-(η(5)-pentamethyl- Cyclopentadienyl)-3-tantalla-bicyclo(3.3.0)oct   71.110
-(η(5)-pentamethyl-cyclopentadienyl)-tantalla- Cyclopentane ....[1,1-dichloro-1   71. 74
     bis((1,2-bis(dimethylphosphino) Ethane)-(diphenylphosphido))-hydrido-tantalum   86.125
etylene)-bis((μ(2)-chloro)-dichloro-tetrahydro*Furan-tantalum)............ [(μ-di-t-butylac   72. 35
)-bis(η(5)-ethyltetramethyl-cyclopentadienyl)-*Hydrido-bis(dichloro-tantalum) <osphine-methyl   71.209
imethylphosphino)ethane)-(diphenylphosphido)- Hydrido-tantalum ....[bis((1,2-bis(d   86.125
     bis(η(5)-cyclopentadienyl)-methyl Methylene-tantalum 71. 42n
yl-cyclopentadienyl)-3-tantalla-bicyclo(3.3.0) Octane ....[3,3-dichloro-3-(η(5)-pentameth   71.110
hloro-(η(5)-cyclopentadienyl)-(2-oxo-4-methyl- Pent-3-ene)-tantalum ....   [dic   71. 30
s((1,2-bis(dimethylphosphino)ethane)-(diphenyl Phosphido))-hydrido-tantalum ....[bi   86.125
clopentadienyl)-hydrido> (μ-oxo)-(μ-trimethyl*Phosphine-methyl)-bis(η(5)-ethyltetramethyl-cy   71.209
-tantalum]   bis((1,2-bis(dimethyl Phosphino)ethane)-(diphenylphosphido))-hydrido   86.125
         pyridinium tetrachloro-(Pyridine)-(tolane)-tantalum....   71.137
talum]   Pyridinium tetrachloro-(pyridine)-(tolane)-tan   71.137
     trichloro-bis(bis(trimethyl Silyl)amido)-tantalum(v) 83. 85
     pyridinium tetrachloro-(pyridine)-(Tolane)-tantalum 71.137

Technetium

     tetrabutylammonium bis(thiomercapto- Acetato)-oxo-technetium....................   85. 3
ium]   tetrabutyl Ammonium bis(thiomercapto-acetato)-oxo-technet   85. 3
ii)]   tris(tetra-n-butyl Ammonium) hexakis(isothiocyanato)-technetium(i   3. 51
   trans-dichloro-bis(o-phenylene-bis(dimethyl Arsine))-technetium(iii) chloride 86. 68
   trans-dichloro-bis(o-phenylene-bis(dimethyl Arsine))-technetium(iii) perchlorate 86. 69
   tetrachloro-bis(o-phenylene-bis(dimethyl Arsine))-technetium(v) hexafluorophosphate 86. 70
     dichloro-(hydro-tris(1-pyrazolyl)- Borato)-oxo-technetium(v) 83. 44
     bis(triphenylphosphine) Immonium oxo-tetrachloro-technetium(v) 64.159
     tris(tetra-n-butylammonium) hexakis(Isothiocyanato)-technetium(iii)............   3. 51
methylene-diphosphonato-technetium tris(aqua)-*Lithium hydrate. [(μ-hydroxy)-   84. 1
     tetrabutylammonium bis(thio Mercapto-acetato)-oxo-technetium   85. 3
lithium hydrate]   (μ-hydroxy)-*Methylene-diphosphonato-technetium tris(aqua)-   84. 1
chloride]   trans-dichloro-bis(o- Phenylene-bis(dimethylarsine))-technetium(iii)   86. 68
perchlorate]   trans-dichloro-bis(o- Phenylene-bis(dimethylarsine))-technetium(iii)   86. 69
exafluorophosphate]   tetrachloro-bis(o- Phenylene-bis(dimethylarsine))-technetium(v) h   86. 70
-bis(dimethylarsine))-technetium(v) hexafluoro Phosphate ....   [tetrachloro-bis(o-phenylene   86. 70
v)]   bis(triphenyl Phosphine)immonium oxo-tetrachloro-technetium(64.159
ate]   (μ-hydroxy)-methylene-di*Phosphonato-technetium tris(aqua)-lithium hydr   84. 1
     dichloro-(hydro-tris(1- Pyrazolyl)-borato)-oxo-technetium(v) 83. 44
     tetrakis(μ-2-oxo-*Pyridinato)-(chloro-di-technetium) 84. 79
     tetrabutylammonium bis(Thiomercapto-acetato)-oxo-technetium 85. 3

Thallium

          thallium(i) tetra- Acetato-thallium(iii) 68. 5
<ium(i) (μ(8)-chloro)-dodecakis(α-mercapto-iso*Butyrato)-octa-copper(i)-hexa-copper(ii) dode>   81. 95
hloro)-dodecakis(α-mercapto-isobutyrato)-octa-*Copper(i)-hexa-copper(ii) dodecahydrate <(8)-c   81. 95
s(α-mercapto-isobutyrato)-octa-copper(i)-hexa-*Copper(ii) dodecahydrate <(8)-chloro)-dodecaki   81. 95
allium(iii))]   di-μ-2-hydroxy*Cyclohepta-2,4,6-trien-1-onato-bis(diphenyl-th   68. 24
     tellurium oxalate di Hydrogen oxalate dihydrate 2. 10
<allium(i) (μ(8)-chloro)-dodecakis(α-mercapto-*Isobutyrato)-octa-copper(i)-hexa-copper(ii) d>   81. 95
er(ii> thallium(i) (μ(8)-chloro)-dodecakis(α-*Mercapto-isobutyrato)-octa-copper(i)-hexa-copp   81. 95
     tellurium oxalate dihydrogen Oxalate dihydrate 2. 10
     tellurium Oxalate dihydrogen oxalate dihydrate 2. 10
-bis(bis(2,3,5,6-tetrafluorophenyl)-(triphenyl*Phosphine-oxide)-thallium(iii))   [di-μ-chloro   68. 25
     trichloro-bis(4-cyano Pyridine-N-oxide)-thallium 68. 13
   di-μ(2)-chloro-tetrakis(trimethyl*Silylmethyl)-di-thallium(iii) 68. 18
          Tellurium oxalate dihydrogen oxalate dihydrate   2. 10
     diethyldi Thiocarbamato-diphenyl-thallium(iii) 68. 20
     diphenyl- Tropolonato-thallium(iii) 68. 24

Thorium

horium] bis(trinitrato-tetrakis(tris(dimethyl Amido)-phosphine-oxide)-thorium) hexanitrato-t 84.104
guanidinium tri(Carbonato)-trifluoro-thorium(iv) 81. 6
an)-thorium]   (η(8)- Cyclo-octatetraene)-dichloro-bis(tetrahydrofur 75. 13
tetrakis(di Cyclohexyl-dithiophosphinato)-thorium(iv) 85. 90
bis(μ-hydrido)-bis(bis(pentamethyl-*Cyclopentadienyl)-thorium) toluene solvate 73.126n
8)-cyclo-octatetraene)-dichloro-bis(tetrahydro Furan)-thorium .... [(η(75. 13
v)]   Guanidinium tri(carbonato)-trifluoro-thorium(i 81. 6
-thorium) toluene solvate] bis(μ-*Hydrido)-bis(bis(pentamethyl-cyclopentadienyl) 73.126n
-oxide)-thorium) hexanitrato-thorium] bis(tri Nitrato-tetrakis(tris(dimethylamido)-phosphine 84.104
(dimethylamido)-phosphine-oxide)-thorium) hexa Nitrato-thorium [bis(trinitrato-tetrakis(tris 84.104
bis(trinitrato-tetrakis(tris(dimethylamido)- Phosphine-oxide)-thorium) hexanitrato-thorium 84.104
tetrachloro-tetrakis(diphenyl Sulfoxide)-thorium(iv) 84.133
tetrakis(dimethyl-di Thiophosphinato)-thorium(iv) 85. 30
tetrakis(dicyclohexyl-di Thiophosphinato)-thorium(iv) 85. 90

Tin

tricyclohexyl-tin-trifluoro Acetate........................... 69. 44
potassium tris(monochloro Acetato)-tin(ii) 69. 9
silyl)-tetra(μ(3)-oxo)-hex> hexakis(μ-t-butyl*Amido)-bis(μ(3)-t-butylamido)-tetra(μ-dimethyl 69. 56
hex> hexakis(μ-t-butylamido)-bis(μ(3)-t-butyl*Amido)-tetra(μ-dimethylsilyl)-tetra(μ(3)-oxo)- 69. 56
bis(β- Amidoethyl)-dichloro-tin(iv) 69. 10
tetrakis(tetramethyl Ammonium) chloro-penta(trichloro-tin)-osmium 3. 26
bis(tetramethyl Ammonium) hexachloro-tin(iv) 3. 22
bis(tetramethyl Ammonium) hexachloro-tin(iv) 3. 21
4-(N-methyl Anilino)-1-phenyl-3-(triphenyl-tin)-2-naphthol 69. 57
N-(trimethyl-tin)-dimethyl Arsino-bis(carbomethoxy)-pyrazole 69. 23
olvate] 1,3-bis(trimethyl-tin)-1,3,5,7-tetra- Aza-2,4,6,8-tetrathiocin-2,2-dioxide benzene s 69. 12
1,3-bis(trimethylsilyl)-4-methyl-di Azastanna-boretidine dimer 63. 17
Azido-trimethyl-tin..................... 69. 3
tris(di Benzotetrathiafulvalene) hexabromo-tin 39. 51
octakis(di Benzotetrathiafulvalene) tris(hexachloro-tin) 39. 52
trimethyl-(N-phenyl-N- Benzoylhydroxylamino)-tin 69. 31
1,3-bis(trimethylsilyl)-4-methyl-diazastanna- Boretidine dimer 63. 17
dichloro-bis(ethyl-3-oxo Butanato)-tin(iv) 69. 22
pentacarbonyl-(diethylamino(triphenyl-tin) Carbene-chromium(0) dichloromethane solvate 71.219
1-trimethyl-tin-methyl-o- Carborane 69. 13
trans-triphenyl-tin-tetracarbonyl-diethylamino Carbyne-chromium 71.215
iphenyl-tin-tetracarbonyl-diethylaminocarbyne- Chromium .... [trans-tr 71.215
carbonyl-(diethylamino(triphenyl-tin)carbene)- Chromium(0) dichloromethane solvate.... [penta 71.219
tin(iv)] chloro-(tetracarbonyl- Cobalt)-bis(dicarbonyl-cyclopentadienyl-iron)- 73. 71
bis(pentamethyl- Cyclopentadienyl)-tin 69. 43
(pentamethyl- Cyclopentadienyl)-tin tetrafluoroborate........ 69. 20
chloro-(tetracarbonyl-cobalt)-bis(dicarbonyl- Cyclopentadienyl-iron)-tin(iv) 73. 71
2,2,4,4,6,6-hexamethyl Cyclotristanna-tellurane 70. 2
chloro-(ethyl-L- Cysteinato-N,S)-dimethyl-tin(iv) 69. 16
is(μ-hydroxo)-bis(aqua-trichloro-tin(iv)) 1,4-*Dioxane.................... [b 38. 4
bis(μ-hydroxo)-bis(aqua-tribromo-tin(iv)) 1,8-*Epoxy-p-menthane 38. 26
is(μ-hydroxo)-bis(aqua-trichloro-tin(iv)) 1,8-*Epoxy-p-menthane .... [b 38. 27
1,2-phenylene-di Ethynyl-bis(trimethyl-tin) 69. 32
poly(trivinyl- Ferrocenoato-tin(iv)) 73. 67
triphenyl-tin-(methyl-dithio Formate) 69. 42
tris(dibenzotetrathia Fulvalene) hexabromo-tin 39. 51
octakis(dibenzotetrathia Fulvalene) tris(hexachloro-tin)........... 39. 52
Glycinato-trimethyl-tin(iv)......... 69. 8
tetrakis(t-butylimino)-di Hydrido-tri-tin..................... 69. 36
p-menthane] bis(μ-*Hydroxo)-bis(aqua-tribromo-tin(iv)) 1,8-epoxy- 38. 26
ne] bis(μ-*Hydroxo)-bis(aqua-trichloro-tin(iv)) 1,4-dioxa 38. 4
-p-menthane] bis(μ-*Hydroxo)-bis(aqua-trichloro-tin(iv)) 1,8-epoxy 38. 27
trimethyl-(N-phenyl-N-benzoyl Hydroxylamino)-tin 69. 31
tetrakis(t-butyl Imino)-dihydrido-tri-tin 69. 36
bonyl-cobalt)-bis(dicarbonyl-cyclopentadienyl- Iron)-tin(iv) .... [chloro-(tetracar 73. 71
droxo)-bis(aqua-tribromo-tin(iv)) 1,8-epoxy-p-*Menthane .... [bis(μ-hy 38. 26
roxo)-bis(aqua-trichloro-tin(iv)) 1,8-epoxy-p-*Menthane .... [bis(μ-hyd 38. 27
1,8-bis(trimethylstannyl)- Naphthalene.................... 69. 34
N-methylanilino)-1-phenyl-3-(triphenyl-tin)-2- Naphthol .... [4-(69. 57
bromo-(8-dimethylaminomethyl-5-methoxy Naphthyl)-methylphenyl-tin(iv) 69. 45
ramethylammonium) chloro-penta(trichloro-tin)- Osmium .... [tetrakis(tet 3. 26
tin(ii) Oxalate........................ 69. 1
(μ-*Oxalato)-tin 69. 2
bis(2,6-di-t-butyl-4-methyl Phenoxy)-tin 69. 50
1,2- Phenylene-diethynyl-bis(trimethyl-tin) 69. 32
tetrachloro-(N-pyridoxylidene-N'- Picolinoylhydrazino)-tin(iv) dihydrate 69. 27

Poly(trivinyl-ferrocenoato-tin(iv)) 73. 67
Potassium tris(monochloroacetato-tin(ii) 69. 9
imethyl-tin)-dimethylarsino-bis(carbomethoxy)- Pyrazole .... [N-(tr 69. 23
bis(5-nitro Pyridine-2-thiolato)-bis(n-butyl)-tin(iv) 69. 38
tetrachloro-(N- Pyridoxylidene-N'-picolinoylhydrazino)-tin(iv) 69. 27
trimethyl-(2- Pyridyl-carboxylato)-tin(iv) monohydrate 69. 19
4-trimethyl-tin- Quinuclidinium perchlorate 69. 21
bis(triphenyl-tin)- Selenide 69. 55
amido)-bis(μ(3)-t-butylamido)-tetra(μ-dimethyl*Silyl)-tetra(μ(3)-oxo)-hexa-tin ⟨kis(μ-t-butyl 69. 56
1,3-bis(trimethyl Silyl)-4-methyl-diazastanna-boretidine dimer 63. 17
2,2,4,4,6,6-hexamethylcyclotristanna- Tellurane 70. 2
tris(dibenzotetra Thiafulvalene) hexabromo-tin 39. 51
octakis(dibenzotetra Thiafulvalene) tris(hexachloro-tin) 39. 52
tris(N,N-diethyldi Thiocarbamato)-(n-butyl)-tin(iv) 69. 41
(N,N-diethyldi Thiocarbamato)-dimethyl-tin(iv) 69. 25
(N,N-diethyldi Thiocarbamato)-dimethyl-tin(iv) 69. 24
tris(N,N-diethyldi Thiocarbamato)-methyl-tin(iv) 69. 35
trimethyl-tin)-1,3,5,7-tetra-aza-2,4,6,8-tetra Thiocin-2,2-dioxide benzene solvate [1,3-bis(69. 12
triphenyl-tin-(methyl-di Thioformate) 69. 42
bis(5-nitropyridine-2- Thiolato)-bis(n-butyl)-tin(iv) 69. 38
bis(O,O'-di-isopropyl-di Thiophosphato)-diphenyl-tin(iv) 69. 48
bis(O,O'-diphenyl-di Thiophosphato)-tin(ii) 69. 60
(O,O'-diethyl-di Thiophosphato)-triphenyl-tin(iv) 69. 46
(β-carbomethoxy-ethyl)-trichloro- Tin(iv)...................................... 69. 4
bis(β-carbomethoxy-ethyl)-dichloro- Tin(iv)...................................... 69. 17
chloro- Tricyclohexyl-tin(iv) 69. 39
Tricyclohexyl-tin-trifluoroacetate 69. 44
poly(tri Vinyl-ferrocenoato-tin(iv))................... 73. 67

Titanium

chloro-tris(dithio Acetato)-titanium(iv) 85. 8
N,N',N'',N'''-tetra-p-tolyl-oxalyl Amidine-bis(di(cyclopentadienyl)-titanium) 73.131
tra-t-butyl-2,2,6,6-tetramethyl-1,3,5,7-tetra- Aza-2,6-disila-4-titana-spiro(3.3)heptane ⟨-te 83.155
⟨clo-octatetraene)-(2,3-dimethyl-1,4-diphenyl Buta-1,3-diene-1,4-diyl-titanium)-titanium te⟩ 71.241
-buta-1,3-diene-1,4-diyl-titanium)⟩ bis(η(8)- Cyclo-octatetraene)-(2,3-dimethyl-1,4-diphenyl 71.241
oxy)-titanium(iii)] bis(Cyclopentadienyl)-(2,6-di-t-butyl-4-methylphen 73.106
bis(Cyclopentadienyl)-chloro-ethoxo-titanium 73. 33
bis(bis(Cyclopentadienyl)-chloro-titanium)-oxide 73. 83
(η(5)-pentamethyl Cyclopentadienyl)-dicarbonyl-titanium ....⟨.. 73. 94
2,4-dithio-pyrimidinato-bis(bis(η(5)-methyl Cyclopentadienyl)-titanium(iii)) 73.113
tetrakis(μ-oxo)-tetrakis(chloro-(η(5)-methyl*Cyclopentadienyl)-titanium) 73.100
',N'',N'''-tetra-p-tolyl-oxalylamidine-bis(di(Cyclopentadienyl)-titanium)............. [N,N 73.131
oride] di Cyclopentadienyl-(trimethylsilyl)-titanium-chl 73. 43
di-η(5)- Cyclopentadienyl-di-η-pyrrolyl-titanium(iv).... 73. 74
(μ(2)-pinacolato)-bis(*Cyclopentadienyl-dichloro-titanium)............ 73. 63
bis(cyclopentadienyl)-chloro- Ethoxo-titanium............................... 73. 33
3,5,7-tetra-aza-2,6-disila-4-titana-spiro(3.3) Heptane ⟨-tetra-t-butyl-2,2,6,6-tetramethyl-1, 83.155
m)] N,N',N'',N'''-tetra-p-tolyl Oxalylamidine-bis(di(cyclopentadienyl)-titaniu 73.131
bis(cyclopentadienyl)-(2,6-di-t-butyl-4-methyl Phenoxy)-titanium(iii) 73.106
nium)] (μ(2)-*Pinacolato)-bis(cyclopentadienyl-dichloro-tita 73. 63
yl)-titanium(iii))] 2,4-dithio- Pyrimidinato-bis(bis(η(5)-methylcyclopentadien 73.113
di-η(5)-cyclopentadienyl-di-η- Pyrrolyl-titanium(iv) 73. 74
l-2,2,6,6-tetramethyl-1,3,5,7-tetra-aza-2,6-di Sila-4-titana-spiro(3.3)heptane ⟨-tetra-t-buty 83.155
dicyclopentadienyl-(trimethyl Silyl)-titanium-chloride 73. 43
amethyl-1,3,5,7-tetra-aza-2,6-disila-4-titana- Spiro(3.3)heptane ⟨-tetra-t-butyl-2,2,6,6-tetr 83.155
adienyl)-titanium(iii))] 2,4-di Thio-pyrimidinato-bis(bis(η(5)-methylcyclopent 73.113
chloro-tris(di Thioacetato)-titanium(iv) 85. 8
chloro-tris(N,N-diethyl-mono Thiocarbamato)-titanium(iv)................... 80. 17
itanium)] N,N',N'',N'''-tetra-p- Tolyl-oxalylamidine-bis(di(cyclopentadienyl)-t 73.131

Tungsten

rimidinato)-di-tungsten⟩ bis(μ-N,N'-diphenyl-*Acetamidinato)-bis(μ-2,4-dimethyl-6-hydroxy-py 84.124
tri-tungsten)] (μ(3)-chloro)-tris(μ-oxo)-(μ-*Acetato)-(tetrachloro-tris(triethylphosphine)- 81. 91
trifluoromethylsulfonate] hexakis(μ-*Acetato)-bis(μ(3)-oxo)-tris(aqua-tungsten(iv)) 81. 49
⟨lopentadienyl)-(μ-η(2),η(2)-carbonyl-dimethyl*Acetylene-carboxylate)-bis(dicarbonyl-tungste⟩ 71.170
bis(η(2)-methyl Acrylate)-tetracarbonyl-tungsten 72. 17
sten] η(3)- Allyl-η(5)-cyclopentadienyl-iodo-nitrosyl-tung 72. 5
ungsten] tetraethylammonium η- Allyl-dicarbonyl-dichloro-triphenylphosphine-t 72. 42
bis(1,3-diphenyltriazenido)-tetrakis(dimethyl Amido-di-tungsten 83.206
phosphine-tungsten] tetraethyl Ammonium η-allyl-dicarbonyl-dichloro-triphenyl 72. 42
pentacarbonyl-tungsten(0))] tetraethyl*Ammonium pentakis(μ-chloro)-thiophenolato-bis(85. 50
)-tetrakis(μ-thiolato⟩ tetrakis(tetra-n-butyl*Ammonium) (bis(tris(μ-benzylthiolato)-tungsten 85. 97
)-oxo-tungsten monohydrate] bis(tetramethyl*Ammonium) bis((dithio-tungsten)-bis(μ(2)-thio) 3. 24

3)-thio)-tris(µ(3)-ethylthio>   tris(tetraethyl*Ammonium) tris(µ(2)-ethylthio)-bis(tetrakis(µ(3. 39
                                  tetraphenyl*Arsonium (decachloro-µ-nitrido-di-tungsten)....   65. 19
ro-tungsten(v))]   tetraphenyl*Arsonium µ-selenido-µ-diselenido-bis(tetrachlo   65. 18
o-tungsten]   tetraphenyl Arsonium pentachloro-N-pentachloroethyl-nitrid   83. 1
ten hexafluorophosphate]   bromo-(dichloro-di Azamethane)-bis(diphenylphosphinoethane)-tungs   86.197
icarbonyl-(η(5)-cyclopentadienyl)-(µ-methanedi Azo)-(pentacarbonyl-chromium)-tungsten .... [d 73. 41
   dicarbonyl-(η(5)-cyclopentadienyl)-methyldi Azo-tungsten 73. 10
<(1,2-bis(diphenylphosphino)ethane-bromo-1-di Azo-1-methylethane-tungsten bromide methanol > 86.207
hane)-tungsten dichloro> bromo-(1-chloro-1-di Azo-2,2-dicyanoethene)-bis(diphenylphosphinoet 86.210
<(1,2-bis(diphenylphosphino)ethane)-bromo-4-di Azobutanol-tungsten hexafluorophosphate ethan) 86.211
  tetrakis(tetra-n-butylammonium) (bis(tris(µ-*Benzylthiolato)-tungsten)-tetrakis(µ-thiolato> 85. 97
thiolato)-tungsten)-tetrakis(µ-thiolato)-tris(*Benzylthiolato-iron))-iron <(bis(tris(µ-benzyl 85. 97
        tetracarbonyl-tungsten-(µ-1-methyl-2-*But-2-ene-di-ylidene)-pentacarbonyl-tungsten   71. 61
                        (µ-3-methyl-*But-2-enylidene)-bis(tetracarbonyl-tungsten)   71. 50
<adienyl-η-(1,2,3,4-tetrakis(trifluoromethyl)- Buta-1,3-diene-1,4-diyl)-tungsten-dicarbonyl-> 71.103
<2-bis(diphenylphosphino)ethane)-bromo-4-diazo Butanol-tungsten hexafluorophosphate ethanol > 86.211
l-trimethylphosphine-tungsten)-(µ-methoxytolyl*Carbene))-bis(trimethylphosphine)-platinum <ny 71.186
yl-(methyl-di-t-butylphos> (µ-methoxy-phenyl-*Carbene)-(µ-tetracarbonyl)-tungsten-bis(carbon 71.238
entacarbonyl-tungsten)]   (µ-methoxy-phenyl-*Carbene)-(bis(trimethylphosphine)-platinum)-(p 71.142
          pentacarbonyl-(ethoxy(ruthenocenyl) Carbene)-tungsten 71.125
n.yl-chloro-((tricarbonyl-chromium)-η(6)-phenyl Carbene)-tungsten   [trans-tetracarbo 71. 60
   trans-tetracarbonyl-chloro-((tricarbonyl- Chromium)-η(6)-phenylcarbene)-tungsten 71. 60
opentadienyl)-(µ-methanediazo)-(pentacarbonyl-*Chromium)-tungsten .... [dicarbonyl-(η(5)-cycl 73. 41
-buta-1,3-diene-1,4-diyl)-tungsten-dicarbonyl- Cobalt <l-η-(1,2,3,4-tetrakis(trifluoromethyl) 71.103
o)-bis(oxo-tungsten)-tetra(triphenylphosphine-*Copper).............. [hexakis(µ(3)-sulfid 86.219
          tetrakis((tri-p-tolylphosphine)-   Copper)-dioxo-hexathio-di-tungsten 86.230
)-dicarbonyl-tungsten)-tris(µ-methoxy)-((η(4)-*Cycloheptatrienyl)-dicarbonyl-tungsten <ienyl 75. 29
-methoxy)-((η(4)-cycloheptatrienyl)-> ((η(3)-*Cycloheptatrienyl)-dicarbonyl-tungsten)-tris(µ 75. 29
  dicarbonyl-(η(5)-cyclopentadienyl)-(1-3-η-(4- Cyclopenta-1',3'-dienyl)-5-(cyclopenta-1'',4'> 73. 92
<ienyl)-(1-3-η-(4-cyclopenta-1',3'-dienyl)-5-(Cyclopenta-1'',4''-dienyl)-cyclopentyl)-tungs> 73. 92
ylacetylene-carboxylate)-bis(dicar> bis(η(5)-*Cyclopentadienyl)-(µ-η(2),η(2)-carbonyl-dimeth 71.170
nyl-chromium)-tungsten   dicarbonyl-(η(5)- Cyclopentadienyl)-(µ-methanediazo)-(pentacarbo 73. 41
ienyl)-5-(cyclopenta-1'',4> dicarbonyl-(η(5)- Cyclopentadienyl)-(1-3-η-(4-cyclopenta-1',3'-d 73. 92
                            (η(5)- Cyclopentadienyl)-chloro-dinitrosyl-tungsten   73. 2
            dicarbonyl-(η(5)- Cyclopentadienyl)-methyldiazo-tungsten 73. 10
         bis(η(5)- Cyclopentadienyl)-phenylhydrazido-tungsten tet 73. 60
  carbonyl-(phenylthiocarbyne)-(η(5)- Cyclopentadienyl)-triphenylphosphine-tungsten 71.234
        dicarbonyl-nitrosyl-(η(5)-pentamethyl- Cyclopentadienyl)-tungsten 73. 37
s(triphenylphosphine)-rhodium) he> (bis(η(5)-*Cyclopentadienyl)-tungsten)-bis(µ-hydrido)-(bi 73.129
methyl)-buta-1,3-diene-1,4-> dicarbonyl-η(5)- Cyclopentadienyl-η-(1,2,3,4-tetrakis(trifluoro 71.103
                         η(3)-allyl-η(5)- Cyclopentadienyl-iodo-nitrosyl-tungsten.......... 72. 5
solvate]   (µ-hydrido)-bis(hydrido-di*Cyclopentadienyl-tungsten) perchlorate acetone 73. 86
          pentacarbonyl-(3-(1,1-diphenylvinyl)- Cyclopentyl)-ethoxymethylene-tungsten(0) 71.214
a-1',3'-dienyl)-5-(cyclopenta-1'',4''-dienyl)- Cyclopentyl)-tungsten <yl)-(1-3-η-(4-cyclopent 73. 92
pentyl)-tungsten>   (1,2-bis(dimethylphosphino) Ethane)-(neopentylidyne)-(neopentylidene)-(neo 71.174
           bis(µ-1,2-bis(diphenylphosphino)*Ethane)-bis(dichloro-tungsten)   86.189
           bis(µ-1,2-bis(diphenylphosphino)*Ethane)-bis(dichloro-tungsten) hemihydrate 86.190
bromide metha> bis(1,2-bis(diphenylphosphino) Ethane)-bromo-1-diazo-1-methylethane-tungsten 86.207
rophosphate e> bis(1,2-bis(diphenylphosphino) Ethane)-bromo-4-diazobutanol-tungsten hexafluo 86.211
diazo-2,2-dicyanoethene)-bis(diphenylphosphino) Ethane)-tungsten dichloromethane solvate <o-1- 86.210
-(dichloro-diazamethane)-bis(diphenylphosphino) Ethane)-tungsten hexafluorophosphate   [bromo 86.197
       bis(dichloro-(1,2-bis(dimethylphosphino) Ethane)-tungsten) toluene solvate 86. 26
phenylphosphino)ethane)-bromo-1-diazo-1-methyl Ethane-tungsten bromide methanol solvate <s(di 86.207
dichloro> bromo-(1-chloro-1-diazo-2,2-dicyano Ethene)-bis(diphenylphosphinoethane)-tungsten 86.210
             bis(η(5)-cyclopentadienyl)-phenyl Hydrazido-tungsten tetrafluoroborate 73. 60
is(phenyl-di-isopropylphosphine)-tungsten-hexa Hydride..............................[tr 86.126
  (bis(η(5)-cyclopentadienyl)-tungsten)-bis(µ-*Hydrido)-(bis(triphenylphosphine)-rhodium) he> 73.129
en) perchlorate acetone solvate]   (µ-*Hydrido)-bis(hydrido-dicyclopentadienyl-tungst 73. 86
te acetone solvate]   (µ-hydrido)-bis(*Hydrido-dicyclopentadienyl-tungsten) perchlora 73. 86
ten]   tetra Hydrido-tetrakis(diphenylethylphosphino)-tungs 86.212
trakis(µ-thiolato)-tris(benzylthiolato-iron))-*Iron <(bis(tris(µ-benzylthiolato)-tungsten)-te 85. 97
ten)-tetrakis(µ-thiolato)-tris(benzylthiolato-*Iron)-iron <(bis(tris(µ-benzylthiolato)-tungs 85. 97
)-bis(tetrakis(µ(3)-thio)-tris(µ(3)-ethylthio-*Iron)-di-tungsten <onium) tris(µ(2)-ethylthio 3. 39
lformami> tetraphenylphosphonium pentasulfur-*Iron-bis(µ-sulfido)-tungsten-disulfide dimethy 64.131
               heptakis(t-butyl Isocyanide)-tungsten(ii) hexatungstate 71.244
nyl-tungsten-(µ(2)-3-(trimethylphosphonium)-1-*Isopentenyl)-tetracarbonyl-tungsten <entacarbo 71.109
   hexafluorophosphate]   bromo-(dichloro-diaza Methane)-tricarbonyl-di-iodo-tungsten 86.197
             bis(diphenylphosphino) Methane-tricarbonyl-di-iodo-tungsten 71.218
n]   dicarbonyl-(η(5)-cyclopentadienyl)-(µ-*Methanediazo)-(pentacarbonyl-chromium)-tungste 73. 41
nyl-(3-(1,1-diphenylvinyl)-cyclopentyl)-ethoxy Methylene-tungsten(0) ....[pentacarbo 71.214
ophosphine)))-bis> (µ(2)-bromo)-(µ(2)-p-tolyl*Methylidene)-(µ(2)-bis(methylamino-bis(difluor 71. 36
no)ethane)-(neopentylidyne)-(neopentylidene)-(Neopentyl)-tungsten(vi) <2-bis(dimethylphosphi 71.174
s(dimethylphosphino)ethane)-(neopentylidyne)-(Neopentylidene)-(neopentyl)-tungsten(vi) <2-bi 71.174
              dichloro-triethylphosphine- Neopentylidene-oxo-tungsten..................... 71. 32
ungsten> (1,2-bis(dimethylphosphino)ethane)-(Neopentylidyne)-(neopentylidene)-(neopentyl)-t 71.174

```
                tetraphenylarsonium (decachloro-μ-*Nitrido-di-tungsten)  ......................  65. 19
phenylarsonium pentachloro-N-pentachloroethyl- Nitrido-tungsten  ....................  [tetra  83.  1
ungsten]                        dicarbonyl- Nitrosyl-(η(5)-pentamethyl-cyclopentadienyl)-t  73. 37
      (η(5)-cyclopentadienyl)-chloro-di Nitrosyl-tungsten  ...........................  73.  2
      η(3)-allyl-η(5)-cyclopentadienyl-iodo- Nitrosyl-tungsten  ......................  72.  5
-tungsten-(μ(2)-3-(trimethylphosphonium)-1-iso*Pentenyl)-tetracarbonyl-tungsten <entacarbonyl  71.109
ethane)-(neopentylidyne)-(neopentylidene)-(neo Pentyl)-tungsten(vi) <2-bis(dimethylphosphino)  71.174
imethylphosphino)ethane)-(neopentylidyne)-(neo Pentylidene)-(neopentyl)-tungsten(vi) <2-bis(d  71.174
        dichloro-triethylphosphine-neo Pentylidene-oxo-tungsten  ....................  71. 32
sten>  (1,2-bis(dimethylphosphino)ethane)-(neo Pentylidyne)-(neopentylidene)-(neopentyl)-tung  71.174
s(diphenylphosphinoethane)-tungsten hexafluoro Phosphate  [bromo-(dichloro-diazamethane)-bi  86.197
)-(bis(triphenylphosphine)-rhodium) hexafluoro Phosphate <entadienyl)-tungsten)-bis(μ-hydrido  73.129
hane)-bromo-4-diazobutanol-tungsten hexafluoro Phosphate ethanol solvate <iphenylphosphino)et  86.211
        trichloro-oxo-bis(triphenyl Phosphine oxide)-tungsten(v)  ................  84.119
ethylidene)-(μ(2)-bis(methylamino-bis(difluoro*Phosphine)))-bis(bromo-carbonyl-tungsten) <ylm  71. 36
        tetrakis((tri-p-tolyl Phosphine)-copper)-dioxo-hexathio-di-tungsten  86.230
ngsten)-(μ-methoxytolylcarbene))-bis(trimethyl*Phosphine)-platinum <nyl-trimethylphosphine-tu  71.186
onyl)-tungsten-bis(carbonyl-(methyl-di-t-butyl*Phosphine)-platinum) <yl-carbene)-(μ-tetracarb  71.238
  (μ-methoxy-phenyl-carbene)-(bis(trimethyl*Phosphine)-platinum)-(pentacarbonyl-tungsten)  71.142
        bis(triethyl Phosphine)-platinum-tetrathio-tungsten  ........  86. 25
ienyl)-tungsten)-bis(μ-hydrido)-(bis(triphenyl*Phosphine)-rhodium) hexafluorophosphate <entad  73.129
hio)-triphenylphosphine-silver]  bis(triphenyl*Phosphine)-silver-bis(μ-thio)-tungsten-bis(μ-t  86.199
(μ-oxo)-(μ-acetato)-(tetrachloro-tris(triethyl*Phosphine)-tri-tungsten)  [(μ(3)-chloro)-tris  81. 91
        dimethyl-tetrakis(trimethyl Phosphine)-tungsten(ii)................  71. 83
,7-dichloro-8-quinolinolato-N,O)-bis(triphenyl Phosphine)-tungsten(ii)  [dicarbonyl-chloro-(5  86.173
tricarbonyl-(phenyl-bis(3,5-dimethylpyrazolyl) Phosphine)-tungsten(0)  ...............  86. 61
        bis(dichloro-bis(trimethyl Phosphine)-tungsten)  ...........  86. 27
        tris(phenyl-di-isopropyl Phosphine)-tungsten-hexahydride................  86.126
(3)-sulfido)-bis(oxo-tungsten)-tetra(triphenyl*Phosphine-copper)  ..............  [hexakis(μ  86.219
        dichloro-triethyl Phosphine-neopentylidene-oxo-tungsten  .........  71. 32
ver-bis(μ-thio)-tungsten-bis(μ-thio)-triphenyl*Phosphine-silver  [bis(triphenylphosphine)-sil  86.199
        ammonium η-allyl-dicarbonyl-dichloro-triphenyl Phosphine-tungsten  ................[tetraethyl  72. 42
thiocarbyne)-(η(5)-cyclopentadienyl)-triphenyl Phosphine-tungsten  [carbonyl-(phenyl  71.234
o-(5,7-dichloro-8-quinolinolato-N,O)-triphenyl Phosphine-tungsten(ii)  ...  [tricarbonyl-chlor  86.100
<s(5,7-dichloro-8-quinolinolato-N,O)-triphenyl Phosphine-tungsten(ii) dichloromethane solvate  86.133
is(trimethylphosphi>  (tetracarbonyl-trimethyl*Phosphine-tungsten)-(μ-methoxytolylcarbene))-b  71.186
        tetrahydrido-tetrakis(diphenylethyl Phosphino)-tungsten  ..................  86.212
dene)-(neopentyl)-tungsten>  (1,2-bis(dimethyl Phosphino)ethane)-(neopentylidyne)-(neopentyli  71.174
        bis(μ-1,2-bis(diphenyl*Phosphino)ethane)-bis(dichloro-tungsten)  ....  86.189
ydrate]        bis(μ-1,2-bis(diphenyl*Phosphino)ethane)-bis(dichloro-tungsten) hemih  86.190
-tungsten bromide metha>  bis(1,2-bis(diphenyl Phosphino)ethane)-bromo-1-diazo-1-methylethane  86.207
n hexafluorophosphate e>  bis(1,2-bis(diphenyl Phosphino)ethane)-bromo-4-diazobutanol-tungste  86.211
        bis(dichloro-(1,2-bis(dimethyl Phosphino)ethane)-tungsten) toluene solvate....  86. 26
        bis(diphenyl Phosphino)methane-tricarbonyl-di-iodo-tungsten  71.218
<hloro-1-diazo-2,2-dicyanoethene)-bis(diphenyl Phosphinoethane)-tungsten dichloromethane sol>  86.210
bromo-(dichloro-diazamethane)-bis(diphenyl Phosphinoethane)-tungsten hexafluorophosphate  86.197
ngsten-disulfide dimethylformami>  tetraphenyl*Phosphonium pentasulfur-iron-bis(μ-sulfido)-tu  64.131
st>  pentacarbonyl-tungsten-(μ(2)-3-(trimethyl*Phosphonium)-1-isopentenyl)-tetracarbonyl-tung  71.109
        tetracarbonyl-(S,S-dimethylsulfonium-2- Picolinylmethylide)-tungsten(0)................  84. 48
methoxytolylcarbene))-bis(trimethylphosphine)-*Platinum <nyl-trimethylphosphine-tungsten)-(μ-  71.186
ten-bis(carbonyl-(methyl-di-t-butylphosphine)-*Platinum) <yl-carbene)-(μ-tetracarbonyl)-tungs  71.238
hoxy-phenyl-carbene)-(bis(trimethylphosphine)-*Platinum)-(pentacarbonyl-tungsten)  ....[(μ-met  71.142
        bis(triethylphosphine)- Platinum-tetrathio-tungsten................  86. 25
        tricarbonyl-(phenyl-bis(3,5-dimethyl Pyrazolyl)phosphine)-tungsten(0)  ...........  86. 61
        bis(pyridinium) tetrabromo-bis( Pyridine)-tungsten(iii)................  83. 47
iii)]        bis( Pyridinium) tetrabromo-bis(pyridine)-tungsten(  83. 47
        tetrakis(μ(2)-6-chloro-2-*Pyridinolato)-di-tungsten  84. 78
<-acetamidinato)-bis(μ-2,4-dimethyl-6-hydroxy-*Pyrimidinato)-di-tungsten tetrahydrofuran sol>  84.124
ten tetrahydrofuran s>  bis(2,4-dimethyl-6-oxo Pyrimidine)-bis(1,3-diphenyltriazino)-di-tungs  84.121
gsten(ii)] dicarbonyl-chloro-(5,7-dichloro-8- Quinolinolato-N,O)-bis(triphenylphosphine)-tun  86.173
(ii) dichloro>  dicarbonyl-bis(5,7-dichloro-8- Quinolinolato-N,O)-triphenylphosphine-tungsten  86.133
(ii)]        tricarbonyl-chloro-(5,7-dichloro-8- Quinolinolato-N,O)-triphenylphosphine-tungsten  86.100
sten)-bis(μ-hydrido)-(bis(triphenylphosphine)-*Rhodium) hexafluorophosphate <entadienyl)-tung  73.129
        pentacarbonyl-(ethoxy( Ruthenocenyl)carbene)-tungsten  .................  71.125
(v))]        tetraphenylarsonium μ-*Selenido-μ-diselenido-bis(tetrachloro-tungsten  65. 18
        tetraphenylarsonium μ-selenido-μ-di*Selenido-bis(tetrachloro-tungsten(v))  ........  65. 18
thio)-tungsten-bis(μ-thio)-triphenylphosphine-*Silver  [bis(triphenylphosphine)-silver-bis(μ-  86.199
nylphosphine-silver]  bis(triphenylphosphine)-*Silver-bis(μ-thio)-tungsten-bis(μ-thio)-triphe  86.199
um pentasulfur-iron-bis(μ-sulfido)-tungsten-di*Sulfide dimethylformamide solvate <ylphosphoni  64.131
phine-copper)]        hexakis(μ(3)-*Sulfido)-bis(oxo-tungsten)-tetra(triphenylphos  86.219
tetraphenylphosphonium pentasulfur-iron-bis(μ-*Sulfido)-tungsten-disulfide dimethylformamide>  64.131
)-oxo)-tris(aqua-tungsten(iv)) trifluoromethyl*Sulfonate  ........[hexakis(μ-acetato)-bis(μ(3  81. 49
        tetracarbonyl-(S,S-dimethyl Sulfonium-2-picolinylmethylide)-tungsten(0)....  84. 48
dimethylformami>  tetraphenylphosphonium penta*Sulfur-iron-bis(μ-sulfido)-tungsten-disulfide  64.131
s((μ(2)-ethylthiolato)-dichloro-(tetrahydro-1-*Thienyl)-tungsten(iv))  ........[(μ(2)-thio)-bi  85. 43
```

ethylammonium) bis((dithio-tungsten)-bis(μ(2)-*Thio))-oxo-tungsten monohydrate....[bis(tetram   3. 24
hydro-1-thienyl)-tungsten(iv))]   (μ(2)-*Thio)-bis((μ(2)-ethylthiolato)-dichloro-(tetra   85. 43
hio> η(7)-tropylium-tungsten-tris(μ(2)-n-butyl*Thio)-carbonyl-tungsten-bis(μ(2)-n-butylthio>   3. 39
-butylthio)-carbonyl-tungsten-bis(μ(2)-n-butyl*Thio)-tetracarbonyl-tungsten <sten-tris(μ(2)-n   75. 39
lphosphine)-silver-bis(μ-thio)-tungsten-bis(μ-*Thio)-triphenylphosphine-silver [bis(tripheny   75. 39
<nium) tris(μ(2)-ethylthio)-bis(tetrakis(μ(3)-*Thio)-tris(μ(3)-ethylthio-iron)-di-tungsten)   86.199
silver] bis(triphenylphosphine)-silver-bis(μ-*Thio)-tungsten-bis(μ-thio)-triphenylphosphine-   3. 39
akis((tri-p-tolylphosphine)-copper)-dioxo-hexa Thio-di-tungsten ....[tetr   86.199
lthio)-bis(tetrakis(μ(3)-thio)-tris(μ(3)-ethyl*Thio-iron)-di-tungsten) <onium) tris(μ(2)-ethy   86.230
bis(triethylphosphine)-platinum-tetra Thio-tungsten 3. 39
nohydrate] bis(tetramethylammonium) bis((di*Thio-tungsten)-bis(μ(2)-thio))-oxo-tungsten mo   86. 25
tricarbonyl-bis(N,N-dimethyldi Thiocarbamato)-tungsten(ii)...................   3. 24
phosphine-tungsten]   carbonyl-(phenyl Thiocarbyne)-(η(5)-cyclopentadienyl)-triphenyl   80. 2
sten(iii,iv))]   tris(μ(2)-ethyl*Thiolato)-bis(dichloro-(dimethylthiolato)-tung   71.234
sten(iv))]   (μ(2)-thio)-bis((μ(2)-ethyl*Thiolato)-dichloro-(tetrahydro-1-thienyl)-tung   85. 38
s(tris(μ-benzylthiolato)-tungsten)-tetrakis(μ-*Thiolato)-tris(benzylthiolato-iron))-iron <(bi   85. 43
ris(μ(2)-ethylthiolato)-bis(dichloro-(dimethyl*Thiolato)-tungsten(iii,iv))................[t   85. 97
<kis(tetra-n-butylammonium) (bis(tris(μ-benzyl*Thiolato)-tungsten)-tetrakis(μ-thiolato)-tris>   85. 38
to)-tungsten)-tetrakis(μ-thiolato)-tris(benzyl*Thiolato-iron)-iron <(bis(tris(μ-benzylthiola   85. 97
tetraethylammonium pentakis(μ-chloro)-*Thiophenolato-bis(pentacarbonyl-tungsten(0))   85. 97
<bonyl-trimethylphosphine-tungsten)-(μ-methoxy*Tolylcarbene))-bis(trimethylphosphine)-platin>   85. 50
fluorophosphine)))-bis> (μ(2)-bromo)-(μ(2)-p-*Tolylmethylidene)-(μ(2)-bis(methylamino-bis(di   71.186
sten]   tetrakis((tri-p- Tolylphosphine)-copper)-dioxo-hexathio-di-tung   71. 36
n]   bis(1,3-diphenyl Triazenido)-tetrakis(dimethylamido)-di-tungste   86.230
<,4-dimethyl-6-oxopyrimidine)-bis(1,3-diphenyl Triazino)-di-tungsten tetrahydrofuran solvate   83.206
onyl-tungsten-bis(μ(2)-n-butylthio)-te> η(7)-*Tropylium-tungsten-tris(μ(2)-n-butylthio)-carb   84.121
)]   pentacarbonyl-(3-(1,1-diphenyl Vinyl)-cyclopentyl)-ethoxymethylene-tungsten(0   75. 39
                                                                                              71.214

Uranium

imino-di Acetato-dioxo-uranium(vi) 82. 4
lphosphine-oxide)-uranium(vi)]   Acetone-bis(isothiocyanato)-dioxo-bis(tripheny   84.129
uoro)-dioxofluoro-uranium monohyd> tetraethyl*Ammonium aqua-dioxofluoro-uranium-(tri-μ(2)-fl   3. 35
-dioxo-uranium(vi)]   (N,N'-(3- Azapentane-1,5-diyl)-bis(salicylideneiminato))   76. 95
Barium uranyl-dimalonate trihydrate............   81. 19
nium]   bis(Benzo-15-crown-5) sodium tetrachloro-dioxo-ura   38. 68
(vi)]   chloro-dioxo-(N-phenyl- Benzohydroxamato)-bis(tetrahydrofuran)-uranium   84. 90
(2,6-diacetylpyridine-bis(4-methoxy Benzoylhydrazone))-dioxo-uranium 83.185
(2,6-diacetylpyridine-bis(4-methoxy Benzoylhydrazone))-dioxo-uranium 83.184
tetrakis(4,4,4-trifluoro-1-phenyl-1,3- Butanedionato)-uranium 77. 11
bis(benzo-15- Crown-5) sodium tetrachloro-dioxo-uranium 38. 68
tris(μ-chloro)-tris(bis(pentamethyl-*Cyclopentadienyl)-uranium) 73.132
aquo-hydroxo- Formato-dioxo-uranium(vi) 81. 5
rate]   aqua-tetrakis(N- Formyl-piperidine)-dioxo-uranium(vi) diperchlo   84.101
oxo-(N-phenyl-benzohydroxamato)-bis(tetrahydro Furan)-uranium(vi) .... [chloro-di   84. 90
diaqua-tetrakis(Glycollato)-uranium(iv)....................   81. 36
(2,6-diacetylpyridine-bis(4-methoxybenzoyl Hydrazone))-dioxo-uranium   83.185
(2,6-diacetylpyridine-bis(4-methoxybenzoyl Hydrazone))-dioxo-uranium 83.184
<bis(dioxo-2,6-diacetylpyridine-bis(2'-pyridyl Hydrazone))-uranium dioxo-uranium-tetranitrate   84. 75
chloro-dioxo-(N-phenyl-benzo Hydroxamato)-bis(tetrahydrofuran)-uranium(vi)   84. 90
aquo- Hydroxo-formato-dioxo-uranium(vi) 81. 5
di-μ-aquo-bis(dioxo-dinitrato-uranium(vi)) di-*Imidazole 32. 6
Imidazolium tetrachloro-dioxo-uranium(vi) 32. 7
Imino-diacetato-dioxo-uranium(vi) 82. 4
xide)-uranium(vi)]   acetone-bis(Isothiocyanato)-dioxo-bis(triphenylphosphine-o   84.129
ide)-uranium(iv)]   tetrakis(Isothiocyanato)-tetrakis(trimethylphosphine-ox   84. 69
barium uranyl-di Malonate trihydrate....................   81. 19
strontium uranyl-di Malonate trihydrate....................   81. 20
pyridylhydrazone))-uranium dioxo-uranium-te> Nitrato-bis(dioxo-2,6-diacetylpyridine-bis(2'-   84. 75
di-μ-aquo-bis(dioxo-di*Nitrato-uranium(vi)) di-imidazole   32. 6
te] (N,N'-bis(salicylidene)-1,5-diamino-3- Oxapentane)-dioxo-uranium(vi) chloroform solva   78. 6
(N,N'-bis(salicylidene)-1,5-diamino-3-oxa Pentane)-dioxo-uranium(vi) chloroform solvate   78. 6
oxo-uranium(vi)]   (N,N'-(3-aza Pentane-1,5-diyl)-bis(salicylideneiminato))-di   76. 95
tetrathiocyanato-tetrakis(triphenyl Phosphine-oxide)-uranium(iv) 84.138
tetrakis(isothiocyanato)-tetrakis(trimethyl Phosphine-oxide)-uranium(iv) 84. 69
cetone-bis(isothiocyanato)-dioxo-bis(triphenyl Phosphine-oxide)-uranium(vi) .... [a   84.129
bis(urea)- Phosphonito-dioxo-uranium 79. 2
aquo-urea- Phosphonito-dioxo-uranium monohydrate 79. 1
aqua-tetrakis(N-formyl- Piperidine)-dioxo-uranium(vi) diperchlorate....   84.101
xo-uranium-te> nitrato-bis(dioxo-2,6-diacetyl Pyridine-bis(2'-pyridylhydrazone))-uranium dio   84. 75
-uranium]   (2,6-diacetyl Pyridine-bis(4-methoxybenzoylhydrazone))-dioxo   83.185
-uranium]   (2,6-diacetyl Pyridine-bis(4-methoxybenzoylhydrazone))-dioxo   83.184
nitrato-bis(dioxo-2,6-diacetylpyridine-bis(2'- Pyridylhydrazone))-uranium dioxo-uranium-tetr>   84. 75
uranium(vi) chloroform solvate]   (N,N'-bis(Salicylidene)-1,5-diamino-3-oxapentane)-dioxo-   78. 6

(N,N'-(3-azapentane-1,5-diyl)-bis(Salicylideneiminato))-dioxo-uranium(vi)........ 76. 95
bis(benzo-15-crown-5) Sodium tetrachloro-dioxo-uranium 38. 68
Strontium uranyl-dimalonate trihydrate 81. 20
tris(Urea)-dioxo-uranium(vi) sulfate................. 79. 3
bis(Urea)-phosphonito-dioxo-uranium.............. 79. 2
aquo- Urea-phosphonito-dioxo-uranium monohydrate 79. 1

Vanadium

iammonium bis((hydrogen ethylenediamine-tetra- Acetato)-vanadium) dihydrate   .... [ethaned 76. 60
trichloro-oxo-(phenyl- Acetonitrile)-vanadium(v) 83. 33
e acetonitrile solvate] bis(μ-fulvalene)-bis(*Acetonitrile-vanadium(iii)) hexafluorophosphat 73. 99
enyl)-vanadium] (η(2)-dimethyl Acetylene-dicarboxylate)-bis(η(5)-cyclopentadi 72. 26
η(5)-cyclopentadienyl)-bis(pentafluorophenyl)- Acetylene-vanadium .... [bis(72. 45
cetato)-vanadium) dihydrate] ethanedi Ammonium bis((hydrogen ethylenediamine-tetra-a 76. 60
o)-di-vanadium(iv) octahydrate] tetraethyl*Ammonium dioxo-(μ-(+)-tartrato)-(μ-(-)-tartrat 81. 28
tetraphenyl Arsonium tetrachloro-oxo-vanadium(v) 65. 14
20-tetrahydro-tribenzo(e,i,m)(1,4,8,11)-tetra- Azacyclotetradecinato)-oxo-vanadium(iv) <8,19, 83.158
tetraphenylphosphonium dioxo-(4-(2-pyridyl Azo)-resorcinolato)-oxo-vanadium(v)........... 84. 30
nato)-oxo-vanadi> (17,18,19,20-tetrahydro-tri Benzo(e,i,m)(1,4,8,11)-tetra-azacyclotetradeci 83.158
bis(Benzonitrile)-trichloro-oxo-vanadium(v)...... 83. 89
Benzonitrile-trichloro-oxo-vanadium(v) 83. 27
lene-vanadium] bis(η(5)- Cyclopentadienyl)-bis(pentafluorophenyl)-acety 72. 45
bis(η(5)- Cyclopentadienyl)-vanadium 73. 19
trimethylsilylnitrene-bis(η(5)- Cyclopentadienyl)-vanadium 73. 44
(2)-dimethylacetylene-dicarboxylate)-bis(η(5)- Cyclopentadienyl)-vanadium ....[(η 72. 26
tetrahydro-tribenzo(e,i,m)(1,4,8,11)-tetra-aza Cyclotetradecinato)-oxo-vanadium(iv) <8,19,20- 83.158
-tetra-acetato)-vanadium) dihydrate] Ethanediammonium bis((hydrogen ethylenediamine 76. 60
(iv) tetraphenylborate] sodium N,N'- Ethylene-bis(salicylideneiminato)-oxo-vanadium 76. 91
rate] ethanediammonium bis((hydrogen Ethylenediamine-tetra-acetato)-vanadium) dihyd 76. 60
afluorophosphate acetonitrile solvate] bis(μ-*Fulvalene)-bis(acetonitrile-vanadium(iii)) hex 73. 99
tetrakis(4-ethylpyridinium) di Hydrido-octacosa-oxygen-deca-vanadium 33. 25
um) dihydrate] ethanediammonium bis((Hydrogen ethylenediamine-tetra-acetato)-vanadi 76. 60
(μ-oxo)-bis(bis(N,N-diethyl*Hydroxylaminato)-oxo-vanadium(v)) 84. 71
trimethylsilyl Nitrene-bis(η(5)-cyclopentadienyl)-vanadium.... 73. 44
tetrakis(4-ethylpyridinium) dihydrido- Octacosa-oxygen-deca-vanadium 33. 25
etrakis(4-ethylpyridinium) dihydrido-octacosa- Oxygen-deca-vanadium .... [t 33. 25
ne)-bis(acetonitrile-vanadium(iii)) hexafluoro*Phosphate acetonitrile solvate [bis(μ-fulvale 73. 99
dichloro-oxo-bis(triphenyl Phosphine-oxide)-vanadium(iv) 84.118
ato)-vanadium(v)] tetraphenyl Phosphonium dioxo-(4-(2-pyridylazo)-resorcinol 84. 30
adium] tetrakis(4-ethyl Pyridinium) dihydrido-octacosa-oxygen-deca-van 33. 25
tetraphenylphosphonium dioxo-(4-(2- Pyridylazo)-resorcinolato)-vanadium(v) 84. 30
etraphenylphosphonium dioxo-(4-(2-pyridylazo)- Resorcinolato)-vanadium(v) .... [t 84. 30
e] dichloro-bis(N-n-butyl- Salicylaldiminato)-vanadium(iv) benzene solvat 78. 8
nylborate] sodium N,N'-ethylene-bis(Salicylideneiminato)-oxo-vanadium(iv) tetraphe 76. 91
um] trimethyl Silylnitrene-bis(η(5)-cyclopentadienyl)-vanadi 73. 44
oxo-vanadium(iv) tetraphenylborate] Sodium N,N'-ethylene-bis(salicylideneiminato)- 76. 91
ahydrate] tetraethylammonium dioxo-(μ-(+)-*Tartrato)-(μ-(-)-tartrato)-di-vanadium(iv) oct 81. 28
raethylammonium dioxo-(μ-(+)-tartrato)-(μ-(-)-*Tartrato)-di-vanadium(iv) octahydrate   [tet 81. 28
Vanadocene 73. 19

Ytterbium

tetrakis(tri(N,N-di-isopropyl- Carbamato)-ytterbium(iii)) heptane solvate 81. 99
) toluene solvate] bis(pentamethyl- Cyclopentadienyl)-tetrahydrofuran-ytterbium(ii 73.102
bis(pentamethyl-cyclopentadienyl)-tetrahydro Furan-ytterbium(ii) toluene solvate........... 73.102

Yttrium

drate] bis(μ-*Acetato)-bis(diaquo-diacetato-yttrium) tetrahy 81. 51
bis(μ-acetato)-bis(diaquo-di*Acetato-yttrium) tetrahydrate 81. 51

Zinc

zinc(ii) Acetate....................................... 81. 13
sodium nitrilo-tri Acetato-zinc monohydrate 81. 21
(N,N'-ethylene-bis(monothio Acetylacetone-iminato))-zinc(ii) 85. 40
drate] 2,2'-bipyridyl-(dihydrogen Adenosine-triphosphato)-zinc(ii) dimer tetrahy 84.125
i)] chloro-(tris(2-aminoethyl) Amine)-pyridino-cobalt(iii) tetrachloro-zinc(i 76. 68
bis(2-aminoethyl) Amine-(di-2-pyridylamino)-zinc(ii) nitrate 76. 82
<ipyrido(c,d,i,j,l,m)(1,4,7,8,10,13,15)-hepta- Aza-1,2,6,7,8,15-hexahydro-cyclopentadecine)-> 83.137
chloro-1,9-bis(2-pyridyl)-2,5,8-tri Azanonane-cobalt(iii) tetrachloro-zinc 76. 92
thiolato)-(methanol-benzenethiolato-hexakis(μ-*Benzenethiolato))-tetra-zinc(ii) <a-(μ-benzene 85. 92
akis(μ-benzenethiolato))-tetra-zi> catena-(μ-*Benzenethiolato)-(methanol-benzenethiolato-hex 85. 92
tra-zi> catena-(μ-benzenethiolato)-(methanol-*Benzenethiolato-hexakis(μ-benzenethiolato))-te 85. 92

bis(p-amino Benzoato)-zinc(ii) hydrate 81. 54
bis(1,2,3,7,8,12,13,17,18,19-decamethyl- Biladiene-a,c)-di-zinc(ii) 83.224
(2,2'- Bipyridyl)-zinc-tetracarbonyl-iron 83.190
2,2'- Bipyridyl-(dihydrogen adenosine-triphosphato)- 84.125
tadienyl-niobium) benzene so> (bis(tetrahydro*Borato)-zinc)-(μ-hydrido)-(carbonyl-dicyclopen 73. 25
i)] aqua-(3-ethoxy-2-oxo Butyraldehyde-bis(thiosemicarbazonato))-zinc(i 79. 6
chloro-1,9-bis(2-pyridyl)-2,5,8-triazanonane- Cobalt(iii) tetrachloro-zinc 76. 92
propane-2-ol-N)-chloro-bis(1,2-diaminoethane)- Cobalt(iii) tetrachloro-zinc <Δ(S)-cis(1-amino 76. 45
chloro-(tris(2-aminoethyl)amine)-pyridino- Cobalt(iii) tetrachloro-zinc(ii) 76. 68
tetra-aquo-bis(pyridine-3-sulfonato)- Copper(ii)-zinc(ii) 83. 57
,8,10,13,15)-hepta-aza-1,2,6,7,8,15-hexahydro- Cyclopentadecine)-zinc(ii) nitrate <l,m)(1,4,7 83.137
trahydroborato)-zinc)-(μ-hydrido)-(carbonyl-di*Cyclopentadienyl-niobium) benzene solvate <(te 73. 25
di Cytosinium tetrachloro-zinc................... 44. 7
(1-aminopropane-2-ol-N)-chloro-bis(1,2-diamino Ethane)-cobalt(iii) tetrachloro-zinc <Δ(S)-cis 76. 45
inc(ii)] (N,N'- Ethylene-bis(monothioacetylacetone-iminato))-z 85. 40
chloro-(N,N,N',N'-tetramethyl Ethylenediamine)-penta-2,4-dienyl-zinc 71. 31
bis(di Ethylenetriamine)-dinitrato-zinc 76. 52
diaqua-di Formato-bis(nicotinamide)-zinc 84. 56
bis(2-dimethylaminoethyl-methyl-amino-zinc- Hydride) 76. 66
benzene so> (bis(tetrahydroborato)-zinc)-(μ-*Hydrido)-(carbonyl-dicyclopentadienyl-niobium) 73. 25
r tetrahydrate] 2,2'-bipyridyl-(di Hydrogen adenosine-triphosphato)-zinc(ii) dime 84.125
te] tetrakis(1-methyl-2(3H)- Imidazoline-thione)-zinc(ii) nitrate monohydra 79. 13
bis(di Imidazolyl zinc) 83. 76
(N,N'-ethylene-bis(monothioacetylacetone- Iminato))-zinc(ii) 85. 40
(2,2'-bipyridyl)-zinc-tetracarbonyl- Iron 83.190
catena-bis(*Isothiocyanato)-bis(μ-1,2,4-triazole)-zinc(ii) 83. 23
de)-zinc] trans-bis(Isothiocyanato)-tetrakis(N,N-diethylnicotinami 83.217
lato))-tetra-zi> catena-(μ-benzenethiolato)-(*Methanol-benzenethiolato-hexakis(μ-benzenethio 85. 92
methyl zinc Methoxide tetramer 71. 8
diaqua-diformato-bis(Nicotinamide)-zinc 84. 56
trans-bis(isothiocyanato)-tetrakis(N,N-diethyl Nicotinamide)-zinc 83.217
inc)-(μ-hydrido)-(carbonyl-dicyclopentadienyl-*Niobium) benzene solvate <(tetrahydroborato)-z 73. 25
(c,d.i,j.l,m)(1,4,7,8,10,13,15)-hepta-aza-1> Nitrato-(2,6-dihydroxy-8,15-dimethyl-tripyrido 83.137
bis(diethylenetriamine)-di Nitrato-zinc 76. 52
chloro-1,9-bis(2-pyridyl)-2,5,8-triaza Nonane-cobalt(iii) tetrachloro-zinc............ 76. 92
di-iodo-bis(2-isopropylamino- Pent-2-en-4-one)-zinc 84. 65
2,2'-bipyridyl-(dihydrogen adenosine-tri Phosphato)-zinc(ii) dimer tetrahydrate 84.125
cobalt(iii) tetrac> (-)(589)-Δ(S)-cis(1-amino Propane-2-ol-N)-chloro-bis(1,2-diaminoethane)- 76. 45
inc) (bromo-dichloro-pyridine-zinc)] Pyridine pyridinium (dibromo-chloro-pyridine-z 83. 16
inc) (bromo-dichloro-pyridine-zinc)] Pyridine pyridinium (dibromo-chloro-pyridine-z 83. 17n
dichloro-bis(4-vinyl Pyridine)-zinc(ii) 83. 93
dibromo-chloro-pyridine-zinc) (bromo-dichloro- Pyridine-zinc) .... [pyridine pyridinium (83. 17n
dibromo-chloro-pyridine-zinc) (bromo-dichloro- Pyridine-zinc) .... [pyridine pyridinium (83. 16
pyridine pyridinium (dibromo-chloro- Pyridine-zinc) (bromo-dichloro-pyridine-zinc) 83. 17n
pyridine pyridinium (dibromo-chloro- Pyridine-zinc) (bromo-dichloro-pyridine-zinc) 83. 16
tetra-aquo-bis(Pyridine-3-sulfonato)-copper(ii)-zinc(ii) 83. 57
tetra-aquo-bis(Pyridine-3-sulfonato)-zinc(ii) 83. 58
mo-dichloro-pyridine-zinc)] pyridine Pyridinium (dibromo-chloro-pyridine-zinc) (bro 83. 16
mo-dichloro-pyridine-zinc)] pyridine Pyridinium (dibromo-chloro-pyridine-zinc) (bro 83. 17n
chloro-(tris(2-aminoethyl)amine)- Pyridino-cobalt(iii) tetrachloro-zinc(ii) 76. 68
a-1> nitrato-(2,6-dihydroxy-8,15-dimethyl-tri Pyrido(c,d.i,j.l,m)(1,4,7,8,10,13,15)-hepta-az 83.137
hloro-zinc] chloro-1,9-bis(2- Pyridyl)-2,5,8-triazanonane-cobalt(iii) tetrac 76. 92
bis(2-aminoethyl)amine-(di-2- Pyridylamino)-zinc(ii) nitrate 76. 82
bis(Pyrrolidine-carbodithioato)-zinc(ii) dimer 80. 20
aqua-(3-ethoxy-2-oxobutyraldehyde-bis(thio Semicarbazonato))-zinc(ii) 79. 6
Sodium nitrilo-triacetato-zinc monohydrate 81. 21
tetra-aqua-(2,2,3,3-tetrafluoro Succinato)-zinc(ii) 81. 14
tetra-aquo-bis(pyridine-3- Sulfonato)-copper(ii)-zinc(ii) 83. 57
tetra-aquo-bis(pyridine-3- Sulfonato)-zinc(ii)........................ 83. 58
(N,N'-ethylene-bis(mono Thioacetylacetone-iminato))-zinc(ii) 85. 40
bis(pyrrolidine-carbodi Thioato)-zinc(ii) dimer. 80. 20
tetrakis(1-methyl-2(3H)-imidazoline- Thione)-zinc(ii) nitrate monohydrate 79. 13
aqua-(3-ethoxy-2-oxobutyraldehyde-bis(Thiosemicarbazonato))-zinc(ii) 79. 6
catena-bis(isothiocyanato)-bis(μ-1,2,4-*Triazole)-zinc(ii) 83. 23
dichloro-bis(4- Vinylpyridine)-zinc(ii)..................... 83. 93

Zirconium

clopentadienyl-zirconium(iv))-2,2-bis(diethyl-*Aluminium)ethane ....[(μ-chloro)-1-(dicy 71.159
bis(η5)-cyclopentadienyl)-2,2-bis(diethyl- Alumino)-ethyl-zirconium-cyclopentadienide 71.197
-zirconium(iv)] bis(diethyl Ammonium) tris(tetraphenyl-disiloxane-diolato) 84.137
tra-t-butyl-2,2,6,6-tetramethyl-1,3,5,7-tetra- Aza-2,6-disila-4-zirconia-spiro(3.3)heptane <e 83.156
bis(η5)-cyclopentadienyl)-zirconia- Benzocyclopentene 71.127
)ethane)-zirconium] hydrido-(η5)- Cyclo-octadiene)-bis(1,2-bis(dimethylphosphino 75. 27
nyl)-2,2-bis(diethyl-alumino)-ethyl-zirconium- Cyclopentadienide ....[bis(η5)-cyclopentadie 71.197

bis(η(5)- Cyclopentadienyl)-zirconia-benzocyclopentene   71.127
do)-(μ-2-η(1):1-2-η(2)-naphthyl)-bis(bis(η(5)-*Cyclopentadienyl)-zirconium)   .... [(μ-hydri 71.230
yl-zirconium-cyclopentadienide]   bis(η(5)- Cyclopentadienyl)-2,2-bis(diethyl-alumino)-eth 71.197
conium]   chloro-di Cyclopentadienyl-(diphenylphosphinomethyl)-zir 71.189
di-η(5)- Cyclopentadienyl-di-η-pyrrolyl-zirconium(iv)   73. 75
l-aluminium)ethane]   (μ-chloro)-1-(di*Cyclopentadienyl-zirconium(iv))-2,2-bis(diethy 71.159
bis(η(5)-cyclopentadienyl)-zirconia-benzo Cyclopentene 71.127
enyl-zirconium(iv))-2,2-bis(diethyl-aluminium)*Ethane .... [((μ-chloro)-1-(dicyclopentadi 71.159
yclo-octadiene)-bis(1,2-bis(dimethylphosphino) Ethane)-zirconium .... [hydrido-(η(5)-c 75. 27
5,7-tetra-aza-2,6-disila-4-zirconia-spiro(3.3) Heptane <etra-t-butyl-2,2,6,6-tetramethyl-1,3, 83.156
η(5)-cyclopentadienyl)-zirconium)]   (μ-*Hydrido)-(μ-2-η(1):1-2-η(2)-naphthyl)-bis(bis(71.230
ethylphosphino)ethane)-zirconium]   Hydrido-(η(5)-cyclo-octadiene)-bis(1,2-bis(dim 75. 27
nium)]   (μ-hydrido)-(μ-2-η(1):1-2-η(2)-*Naphthyl)-bis(bis(η(5)-cyclopentadienyl)-zirco 71.230
e)-zirconium]   hydrido-(η(5)-cyclo- Octadiene)-bis(1,2-bis(dimethylphosphino)ethan 75. 27
e]   bis(N,N'-disalicylidene-1,2- Phenylenediamino)-zirconium(iv) benzene solvat 78. 17
do-(η(5)-cyclo-octadiene)-bis(1,2-bis(dimethyl Phosphino)ethane)-zirconium............ [hydri 75. 27
chloro-dicyclopentadienyl-(diphenyl Phosphinomethyl)-zirconium 71.189
tetrakis(1,3-diphenyl-1,3- Propanedionato)-zirconium(iv) 77. 13
di-η(5)-cyclopentadienyl-di-η- Pyrrolyl-zirconium(iv) 73. 75
disodium hexa Pyrrolyl-zirconium(iv) tetrahydrofuran solvate 83.173
v) benzene solvate]   bis(N,N'-di Salicylidene-1,2-phenylenediamino)-zirconium(i 78. 17
l-2,2,6,6-tetramethyl-1,3,5,7-tetra-aza-2,6-di Sila-4-zirconia-spiro(3.3)heptane <etra-t-buty 83.156
bis(diethylammonium) tris(tetraphenyl-di Siloxane-diolato)-zirconium(iv)................ 84.137
ran solvate]   di Sodium hexapyrrolyl-zirconium(iv) tetrahydrofu 83.173
ethyl-1,3,5,7-tetra-aza-2,6-disila-4-zirconia- Spiro(3.3)heptane <etra-t-butyl-2,2,6,6-tetram 83.156

FORMULA INDEX

C₁

61.2 $3CCl_4$, $C_{12}H_{24}N_9P_3$

64.1 $CCl_9N_3P_3^+$, Cl_6Sb^-

5.1 CD_4

32.4 CD_4N_2O, $C_3D_2N_2O_3$

11.1 $CF_3O_3S^-$, $N_2S_3^+$, $0.5C_2H_3N$

2.1 CHO_2^-, K^+

2.2 $2CHO_2^-$, Ca^{2+}

2.3 $2CHO_2^-$, Ca^{2+}

2.4 $2CHO_2^-$, Ca^{2+}

11.2 $CHO_9S_3^{3-}$, $3K^+$, H_2O

41.1 CH_2N_4S

85.1 $3CH_3CoS^{2+}$, $4Na^+$, $Al_{12}CoO_{48}Si_{12}I_0^-$

2.5 $CH_3N_2OS^-$, K^+

32.1 CH_3N_5, H_2O

35.13 $CH_3O_3S^-$, $C_9H_{14}N_5O^+$, H_2O

58.76 $CH_3O_3S^-$, $C_{32}H_{41}BrN_5O_5^+$, $0.5C_3H_8O$

84.1 $(CH_3O_7P_2Tc^-)_n$, $nH_6LiIO_3^+$, $0.33nH_2O$

8.1 CH_4N_2O

1.8 CH_4N_2O, $C_3H_4O_4$

8.2 CH_4N_2O, F_3Sb

8.3 CH_4N_2O, H_2O_2

8.4 $2CH_4N_2O$, H_6O_6Te

8.5 $3CH_4N_2O$, $2Li^+$, O_4S^{2-}

8.6 $4CH_4N_2O$, Mg^{2+}, $2Br^-$, $2H_2O$

17.40 CH_4N_2S, $C_{26}H_{28}N_2O_5$

3.1 $(CH_6N^+)_n$, $(H_4Cl_3MnO_2^-)_n$

3.2 $2CH_6N^+$, Br_4Cd^{2-}

2.9 $CH_6N_3^+$, $C_2HO_4^-$, H_2O

64.13 $0.5CH_6N_3^+$, $C_5H_{14}N_3O_3P$, $0.5Cl^-$

84.2 $2CH_6N_3^+$, $C_{15}AsMo_6O_{27}^{2-}$, $6H_2O$

8.7 $4CH_6N_3^+$, $Mo_{12}O_{40}Si^{4-}$, H_2O

81.6 $5CH_6N_3^+$, $C_3F_3O_9Th^{5-}$

79.1 CH_7N_2O, PU, H_2O

84.2 $CH_{15}AsMo_6O_{27}^{2-}$, $2CH_6N_3^+$, $6H_2O$

81.1 $CH_{17}CoN_6O_2^{2+}$, $2NO_3^-$

11.3 $3CS_2$, $Cl_{10}O_2S_{12}Sb_2$

C₂

71.1 $C_2AuN_2O_2^-$, $C_{24}H_{20}As^+$

62.1 $C_2BF_8^-$, Cs^+

5.2 C_2Cl_6

5.3 C_2Cl_6

83.1 $C_2Cl_{10}NW^-$, $C_{24}H_{20}As^+$

83.2 $C_2Cl_{12}MoNOP$

83.3 $C_2Cl_{12}NOPRe$

81.2 $(C_2CuD_6O_6)_n$, $2nD_2O$

81.3 $(C_2CuD_6O_6)_n$, $2nD_2O$

3.3 $C_2D_6N_2O_2$

66.1 $C_2F_{12}O_2Sb_2$

2.6 $C_2HBr_2O_2^-$, $C_2H_2Br_2O_2$, Rb^+

2.7 $C_2HCl_2DO_2$, $C_2HCl_2O_2^-$, K^+

2.7 $C_2HCl_2O_2^-$, $C_2HCl_2DO_2$, K^+

2.8 $C_2HCl_2O_2^-$, $C_2H_2Cl_2O_2$, K^+

2.9 $C_2HO_4^-$, $CH_6N_3^+$, H_2O

2.10 $C_2HO_4^-$, $C_2H_2O_4$, Tl^+, $2H_2O$

5.4 C_2H_2

5.5 $0.5C_2H_2$, $AlNaO_4Si$

2.6 $C_2H_2Br_2O_2$, $C_2H_2Br_2O_2^-$, Rb^+

2.11 $C_2H_2Br_2O_2$, $C_2H_2Br_2O_2^-$, K^+

32.2 $C_2H_2ClN_3$

2.8 $C_2H_2Cl_2O_2$, $C_2HCl_2O_2^-$, K^+

81.4 $C_2H_2Cl_4O_4Re_2$, $2C_{13}H_{11}NO$

41.2 $C_2H_2N_6O_3S$

1.1 $C_2H_2O_4$, $2H_2O$

83.4 $(C_2H_3Cl_2CuN_2S)_n$

83.5 $C_2H_3Cl_4MoNO^-$, $2C_{24}H_{20}As^+$, Cl_4MoO^-

1.2 $C_2H_3NO_4$

2.12 $C_2H_3O_2^-$, Na^+, $3H_2O$

5.6 C_2H_4

65.1 $C_2H_4As_4O_4$

61.8 $2C_2H_4Cl_2$, $2C_{25}H_{22}P^+$, $CdCl_4^{2-}$

1.3 $C_2H_4N_2O_3$

8.8 $C_2H_4N_4$

3.4 $C_2H_4N_4$, CHN_2^-, K^+

10.1 $C_2H_4N_4O_2$

61.4 $C_2H_4O_2$, $C_{13}H_{17}N_4O_6^+$, Cl^-

81.44 $2C_2H_4O_2$, $C_{12}H_{12}O_{12}Pd_4$

61.24 $4C_2H_4O_2$, $C_{60}H_{66}O_{12}S_6$

3.5 $3C_2H_5N^+$, $C_{25}HO_{25}Rh_{14}^{3-}$

1.4 C_2H_5NO, $C_2H_6NO^+$, Cl^-

48.1 $C_2H_5NO_2$

48.2 $C_2H_5NO_2$

48.3 $3C_2H_5NO_2$, Ca^{2+}, $2Br^-$

48.4 $4C_2H_5NO_2$, K^+, I_3^-

32.3 $C_2H_5N_5$

68.1 $C_2H_6AlI_2^-$, Cs^+, C_8H_{10}

85.2 $C_2H_6Br_5PtS^-$, $C_{14}H_8S_4^+$

84.3 $(C_2H_6Cl_2CuOS)_n$

64.2 $C_2H_6Cl_3N_4O_2PS_2$

1.5 $C_2H_6NO^+$, NO_3^-

48.5 $C_2H_6NO_2^+$, $H_2O_4P^-$

48.6 $2C_2H_6NO_2^+$, $C_2H_5NO_2$, BeF_4^{2-}

48.7 $2C_2H_6NO_2^+$, $C_2H_5NO_2$, BeF_4^{2-}

1.6 $C_2H_6N_2OS$

1.7 $C_2H_6N_2OS$

3.10 $C_2H_6N_2O_2$

3.6 $C_2H_6N_2O_2$

3.7 $C_2H_6N_2O_2$

3.8 $C_2H_6N_2O_2$

3.9 $C_2H_6N_2O_2$

11.4 $C_2H_6N_2S_2$

4.1 $C_2H_6N_6S_5$

21.1 $C_2H_6O_4P^-$, $C_6H_{14}N^+$

65.2 $C_2H_7AsMo_4O_{15}^{2-}$, $2CH_6N_3^+$, H_2O

82.1 $C_2H_7ClN_2O_2Pt$

82.2 $C_2H_8Cl_2N_2O_2Pt$

64.22 $C_2H_8N^+$, $C_6H_{14}O_2PS_2^-$

8.9 $C_2H_8N_4S_2^{2-}$, $H_2Cl_4O_2Re^-$, Cl^-, H_2O

81.5 $(C_2H_8O_{12}U_2)_n$

82.3 $(C_2H_9Cl_2MnNO_4)_n$

79.2 $C_2H_9N_4O_7PU$

2.26 $C_2H_{10}N_2^{2+}$, $C_6H_6O_8^{2-}$, H_2O

2.25 $C_2H_{10}N_2^{2+}$, $C_6H_6O_8^{2-}$

76.1 $C_2H_{14}CoN_6O_4^+$, NO_3^-

83.6 $C_2H_{15}CoN_{10}$, Br_2

76.2 $C_2H_{16}N_2NiO_4^{2+}$, $2NO_3^-$

76.3 $C_2H_{16}N_2NiO_4^{2+}$, O_4S^{2-}, $2H_2O$

83.7 $C_2H_{18}CoN_9^{2+}$, $2ClO_4^-$

2.13 $C_2O_4^{2-}$, $2K^+$, D_2O_2

2.14 $C_2O_4^{2-}$, $2Rb^+$, D_2O_2

2.15 $C_2O_4^{2-}$, $2Rb^+$, D_2O_2

69.1 $(C_2O_4Sn)_n$

69.2 $(C_2O_4Sn)_n$

C₃

32.4 $C_3D_2N_2O_3$, CD_4N_2O

81.6 $C_3F_3O_9Th^{5-}$, $5CH_6N_3^+$

41.3 $C_3H_2N_3O_5S^-$, Na^+, H_2O

38.1 $C_3H_2O_4$

2.16 $C_3H_2O_4^{2-}$, Ca^{2+}, $2H_2O$

81.7 $C_3H_3Cl_3O_6Re_2$

32.5 $C_3H_3D_2N_2^+$, $C_4H_2DO_4^-$

41.4 $C_3H_3N_3O_5S$

2.17 $C_3H_4FO_4^-$, Na^+

32.6 $2C_3H_4N_2$, $H_4N_4O_{18}U_2$

1.8 $C_3H_4O_4$, CH_4N_2O

64.8 $C_3H_5N_2^+$, $C_4H_8O_3PS^-$, H_2O

32.7 $2C_3H_5N_2^+$, $Cl_4O_2U^{2-}$

9.1 $C_3H_5N_3O_2$

41.5 $C_3H_5N_3O_2S$

1.9 $2C_3H_5O_2^-$, Cu^{2+}

41.6 $C_3H_6N_5OS_3^+$, AsF_6^-

61.5 $C_3H_6O_2$, $C_{13}H_{17}N_4O_6^+$, Cl^-

61.19 C_3H_7NO, $C_{36}H_{60}O_{30}$, $5H_2O$

48.8 $C_3H_7NO_3$, $C_6H_8O_6$

46.1 $C_3H_7O_6P^{2-}$, $2Na^+$, $6H_2O$

64.3 $C_3H_8Cl_4N_3P_3$

76.4 $C_3H_8CuN_2O^+$, $C_{24}H_{20}B^-$

48.9 $C_3H_8NO_2^+$, $C_3H_7NO_2$, Br^-

64.4 $C_3H_8NO_5P$

61.22 C_3H_8O, $C_{48}H_{80}O_{40}$, xH_2O

68.2 $C_3H_9AlI^-$, $C_4H_{12}N^+$

4.2 $C_3H_9NO_2S_2$

69.3 $(C_3H_9N_3Sn)_n$

64.5 C_3H_9PSe

63.1 $6C_3H_9Si^-$, $6Li^+$

62.2 $4C_3H_{10}B^-$, $4Na^+$, $C_4H_{10}O$

3.11 $C_3H_{10}N^+$, $C_{12}H_4N_4$, $0.33I_3^-$

3.12 $C_3H_{10}N^+$, $CdCl_3^-$

3.13 $2C_3H_{10}N^+$, Cl_4Mn^2

3.14 $3C_3H_{10}N^+$, Br_4Cd^{2-}, Br_3Cd^-

64.6 $C_3H_{10}NO_3P$

3.15 $C_3H_{12}N_2O^{2+}$, O_4S^{2-}, H_2O

79.3 $C_3H_{12}N_6O_5U^{2+}$, O_4S^{2-}

62.3 $C_3H_{14}B_{10}O_2S$

79.4 $C_3H_{15}N_9RhS_3^{3+}$, $3Cl^-$

7.1 $C_3N_3O^-$, K^+

C₄

39.1	$C_4Br_2O_2S$
81.8	$C_4CuO_8{}^{2-}$, $2Na^+$, $2H_2O$
81.9	$C_4CuO_8{}^{2-}$, $2Na^+$, $2H_2O$
3.16	$2C_4D_{12}N^+$, Cl_6Pt^{2-}
66.2	$C_4FO_9Sb^{4-}$, $3Na^+$, H_3O^+, H_2O
2.18	$C_4F_4O_4{}^{2-}$, $2Na^+$, $6H_2O$
66.3	$C_4F_{12}O_5Sb_2$
2.19	$C_4HCl_2O_4{}^-$, $C_{16}H_{36}N^+$
2.20	$C_4HCl_2O_4{}^-$, $C_{16}H_{36}N^+$
32.8	$C_4H_2ClNO_2$
32.5	$C_4H_2DO_4{}^-$, $C_3H_3D_2N_2{}^+$
44.1	$C_4H_2D_3N_3O$, D_2O
41.7	$C_4H_2N_3O_3S^-$, Na^+, $2H_2O$
32.9	$C_4H_2N_8O_4$, $2H_2O$
2.21	$C_4H_2O_4{}^{2-}$, Sr^{2+}, $4H_2O$
44.2	C_4H_3ClN
41.8	$C_4H_3N_3O_3S$
33.70	$C_4H_3O_4{}^-$, $C_{16}H_{20}ClN_2O^+$
29.9	$C_4H_3O_4{}^-$, $C_{22}H_{25}ClN^+$
81.10	$C_4H_3Cl_2O_8Ru_2{}^-$, K^+
64.7	$C_4H_3Cl_5N_2P$
81.11	$0.5C_4H_4Mo_2O_8$, $0.5C_4H_8Mo_2O_{10}$
44.3	$C_4H_4N_2$
60.5	$3C_4H_4NiS_4$, $2C_6H_4S_4$
38.2	$C_4H_4O_3$
85.3	$C_4H_4O_3S_4Tc^-$, $C_{16}H_{36}N^+$
1.10	$C_4H_4O_4{}^{2-}$, Ca^{2+}, $3H_2O$
1.11	$C_4H_4O_6{}^{2-}$, Cu^{2+}, $3H_2O$
2.22	$C_4H_4O_6{}^{2-}$, Na^+, H_4N^+, $4H_2O$
82.4	$(C_4H_5NO_6U)_n$
44.4	$C_4H_5N_3$
44.5	$C_4H_5N_3O$, Ca^{2+}, $2Cl^-$, H_2O
32.10	$C_4H_5N_3O_2$
2.23	$C_4H_5O_5{}^-$, Li^+
2.24	$2C_4H_5O_5{}^-$, Ca^{2+}, $6H_2O$
81.12	$C_4H_6Cl_4O_4Re_2$
11.5	$C_4H_6Cl_4O_4S_4$
43.1	$C_4H_6N_2$, $C_8H_{12}N_2O_3$
44.6	$2C_4H_6N_3O^+$, Cl_4Cu^{2-}
44.7	$2C_4H_6N_3O^+$, Cl_4Zn^{2-}
1.12	$C_4H_6O_4$, K^+, F^-
81.13	$(C_4H_6O_4Zn)_n$
82.5	$C_4H_7ClN_2O_4Pd$
69.4	$C_4H_7Cl_3O_2Sn$
61.20	C_4H_7NO, $C_{36}H_{60}O_{30}$, $5H_2O$
42.1	$C_4H_7NO_5S$
32.11	$C_4H_7N_3S$
66.4	$C_4H_8ClS_3Sb$
84.4	$(C_4H_8Cl_2CuOS)_n$
85.4	$C_4H_8Cl_3Mo_2O_4S_2{}^-$, $C_8H_{20}N^+$
81.14	$(C_4H_8F_4O_8Zn)_n$
79.5	$(C_4H_8FeN_6S_4)_n$
33.1	$C_4H_8N_2O$
1.13	$C_4H_8N_2O_2$
10.2	$C_4H_8N_2O_2$
3.17	$C_4H_8N_2O_2$
32.12	$C_4H_8N_2O_2$, $3H_2O$

48.10	$C_4H_8N_2O_3$
82.6	$C_4H_8N_2O_4Pd$, $3H_2O$
82.7	$C_4H_8N_2O_4Pt$, $C_2H_2O_4$, $2H_2O$
33.2	$C_4H_8N_4O_2$
41.9	$C_4H_8N_6S_5$
61.1	$C_4H_8O_2$, $3C_{11}H_{20}O_2$
61.23	$C_4H_8O_2$, $C_{54}H_{54}S_6$
61.25	$C_4H_8O_2$, $C_{80}H_{66}S_6$
38.3	$C_4H_8O_2$, Cl_2OSe
38.4	$3C_4H_8O_2$, $H_6Cl_6O_4Sn_2$
64.8	$C_4H_8O_3PS^-$, $C_3H_5N_2{}^+$, H_2O
61.15	C_4H_8S, $2C_{33}H_{36}O_6$
68.3	$C_4H_9AlNS^-$, Cs^+
61.16	C_4H_9Br, $2C_{33}H_{36}O_6$
82.8	$C_4H_9Cl_2NO_3PdS$, H_2O
48.11	$C_4H_9NO_2$
5.7	$4C_4H_9O^-$, $3Be^{2+}$, $2H_4B^-$
82.9	$C_4H_{10}Br_2N_2O_4Pt$, H_2O
82.10	$C_4H_{10}Cl_2N_2O_4Pt$, H_2O
84.5	$C_4H_{10}MnO_8$
1.14	$C_4H_{10}NO^+$, C_4H_9NO, $AuCl_4{}^-$
1.15	$C_4H_{10}NO^+$, C_4H_9NO, $AuCl_4{}^-$
1.16	$C_4H_{10}NO^+$, C_4H_9NO, $Cl_6Pd_2{}^-$
1.17	$C_4H_{10}NO^+$, Cl^-
83.8	$C_4H_{10}N_6O_4Pt$, $2HCl$
83.9	$C_4H_{10}N_6O_4Pt$, $2HCl$
83.10	$C_4H_{10}N_6O_4Pt$, H_4N^+, Cl^-
83.11	$C_4H_{10}N_6O_4Pt$, $0.666Na^+$, $0.333C_2H_3O_2{}^-$, $0.333Cl^-$, $2H_2O$
64.62	$C_4H_{10}O_2$, $C_{14}H_{10}O_2P^-$, Li^+
11.6	$C_4H_{10}O_4S_2$
86.7	$C_4H_{11}Cl_3HgP^-$, $C_8H_{22}ClHgP_2{}^+$
83.12	$2C_4H_{11}N_8O_4Pd^+$, O_4S^{2-}, H_2O
76.5	$C_4H_{12}AuClN_3{}^+$, $ClO_4{}^-$
76.6	$C_4H_{12}Br_2CuN_2O$
86.1	$C_4H_{12}Cl_2F_{16}Mo_2N_4P_8$
64.10	$C_4H_{12}Cl_4N_5P_3$
64.9	$C_4H_{12}Cl_4N_5P_3$
85.5	$C_4H_{12}Cl_4O_2RhS_2{}^-$, $2C_2H_6OS$, H^+
66.5	$C_4H_{12}Cl_8O_8P_2Sb_2$
3.18	$C_4H_{12}F_2N_2S$
69.5	$C_4H_{12}I_2O_2PbS_2$
67.1	$C_4H_{12}MgO_9$, H_2O
61.13	$C_4H_{12}N^+$, $C_{32}H_{36}O_6$, $ClO_4{}^-$, C_6H_6
84.24	$C_4H_{12}N^+$, $C_{10}H_{12}Mo_3N_3O_{13}{}^-$
3.19	$C_4H_{12}N^+$, $C_{15}Co_9O_{14}{}^-$
62.12	$C_4H_{12}N^+$, $C_{20}H_{18}BN_2O_2{}^-$
3.20	$C_4H_{12}N^+$, F_6P^-
85.94	$2C_4H_{12}N^+$, $C_{60}H_{50}Co_4S_{10}{}^{2-}$
85.95	$2C_4H_{12}N^+$, $C_{60}H_{50}Fe_4S_{10}{}^{2-}$
84.137	$2C_4H_{12}N^+$, $C_{72}H_{60}O_9Si_8Zr^{2-}$
3.21	$2C_4H_{12}N^+$, Cl_6Sn^{2-}
3.22	$2C_4H_{12}N^+$, Cl_6Sn^{2-}
3.23	$2C_4H_{12}N^+$, $Mo_2O_2S_6{}^{2-}$
3.24	$2C_4H_{12}N^+$, $OS_8W_3{}^{2-}$, H_2O
3.25	$3C_4H_{12}N^+$, $Bi_2Br_9{}^{3-}$
3.26	$4C_4H_{12}N^+$, $Cl_{16}OsSn_5{}^{4-}$
13.7	$2C_4H_{12}NO_2{}^+$, $C_{13}H_4I_4O_4{}^{2-}$
76.7	$C_4H_{13}AuClN_3{}^{2+}$, Cl^-, $ClO_4{}^-$

84.6	$2C_4H_{13}CuN_2O_2{}^+$, $2NO_3{}^-$
64.11	$C_4H_{14}BP$
62.4	$C_4H_{14}B_9BrO_2$
3.27	$(C_4H_{14}N_2{}^+)_n$, nCl_4Mn^-
3.28	$(C_4H_{14}N_2{}^{2+})_n$, nCl_4Mn^{2-}
3.29	$C_4H_{14}N_2S_2{}^{2+}$, $2Cl^-$
83.13	$C_4H_{14}N_6NiS_2{}^{2+}$, $2Br^-$
83.14	$C_4H_{14}N_6NiS_2{}^{2+}$, $2Br^-$
83.15	$C_4H_{14}N_6NiS_2{}^{2+}$, $2I^-$
76.8	$(C_4H_{16}BrN_4Pt^{2+})_n$, $2nClO_4{}^-$
76.9	$C_4H_{16}ClCoN_4O_3S$, H_2O
76.10	$C_4H_{16}ClN_5O_2Pt^{2+}$, $2Cl^-$, $3H_2O$
76.11	$C_4H_{16}Cl_2CoN^+$, $CH_3O_4S_2{}^-$
76.12	$2C_4H_{16}Cl_2CoN_4{}^+$, $O_6S_2Se_2{}^{2-}$, H_2O
76.13	$2C_4H_{16}Cl_2CoN_4{}^+$, $O_6S_2Se_2{}^{2-}$, H_2O
76.14	$2C_4H_{16}Cl_2CoN_4{}^+$, $O_6S_3Se^{2-}$
76.19	$C_4H_{16}Cl_2N_4Pt^{2+}$, $C_4H_{16}N_4Pt^{2+}$, $4ClO_4{}^-$
76.20	$3C_4H_{16}Cl_2N_4Pt^{2+}$, $3C_4H_{16}N_4Pt^{2+}$, $4Cl_4Cu^{3-}$
76.15	$C_4H_{16}Cl_2N_4Rh^+$, $NO_3{}^-$
76.16	$C_4H_{16}Cl_2N_4Rh^+$, $NO_3{}^-$
76.17	$C_4H_{16}CoN_4O_6S_2{}^-$, Na^+, $3H_2O$
76.18	$C_4H_{16}CuN_4{}^{2+}$, $C_{24}H_{40}CuN_8Ni_2O_{16}{}^{2-}$, $2H_2O$
3.30	$2C_4H_{16}N_3{}^{3+}$, Cl_4Pt^{2-}, $4Cl^-$
76.19	$C_4H_{16}N_4Pt^{2+}$, $C_4H_{16}Cl_2N_4Pt^{2+}$, $4ClO_4{}^-$
76.20	$3C_4H_{16}N_4Pt^{2+}$, $3C_4H_{16}Cl_2N_4Pt^{2+}$, $4Cl_4Cu^{3-}$
76.21	$C_4H_{19}CoN_5O_3S^+$, $ClO_4{}^-$
71.2	$C_4H_{19}N_5O_4Ru^{2+}$, $O_6S_2{}^{2-}$, $2H_2O$
62.5	$C_4H_{20}B_{18}CoO^-$, $C_7H_{18}N^+$
39.2	$C_4I_2O_2S$
85.6	$C_4NiO_4S_4{}^{2-}$, $2K^+$
81.15	$C_4NiO_4S_4{}^{2-}$, $2Li^+$, $2H_2O$
85.7	$C_4NiO_4S_4{}^{2-}$, $2Na^+$, $2H_2O$
39.3	$C_4O_4S_{10}$
81.16	$0.4C_4O_8Pt^-$, $0.6C_4O_8Pt^{2-}$, $1.6K^+$, $1.2H_2O$
81.17	$0.67C_4O_8Pt^{2-}$, $0.33C_4O_8Pt^-$, $1.67Rb^+$, $1.5H_2O$

C₅

41.10	$C_5H_3N_3OS$
20.1	$C_5H_5{}^-$, $C_6H_{16}N_2$, Na^+
83.16	$C_5H_5BrCl_2NZn^-$, $C_5H_5Br_2ClNZn^-$, $2C_5H_5N$, $2C_5H_6N^+$
83.17	$C_5H_5BrCl_2NZn^-$, $C_5H_5Br_2ClNZn^-$, $2C_5H_5N$, $2C_5H_6N^+$
33.3	$C_5H_5BrN^+$, $ClO_4{}^-$, $0.5H_2O$
73.1	$C_5H_5ClCrN_2O_2$
73.2	$C_5H_5ClN_2O_2W$
33.4	C_5H_5N
33.5	C_5H_5N, $3H_2O$
33.6	$2C_5H_5NO$, $2C_5H_6NO^+$, $Br_6Pd_2{}^{2-}$
44.8	$C_5H_5N_2O_2{}^-$, K^+, $3H_2O$
83.47	$C_5H_6N^+$, $C_{10}H_{10}Br_4N_2W^-$, C_5H_5N
60.1	$C_5H_6N^+$, $C_{10}H_9N$, $C_6H_2N_3O_7{}^-$

71.137 $C_5H_6N^+$, $C_{19}H_{15}Cl_4NTa^-$
33.7 $(C_5H_6N^+)_n$, $nC_3Cu_2N_3S_3^-$
33.8 $2C_5H_6N^+$, $Cl_6OSb_2^{2-}$
33.9 $2C_5H_6N^+$, Cl_6Te^{2-}
33.10 $3C_5H_6N^+$, $H_4Mo_2O_{16}P_4^{2-}$, Cl^-
33.11 $4C_5H_6N^+$, $C_6Mo_2N_6O_4S_6^{4-}$
33.12 $5C_5H_6N^+$, $3H_2Br_4MoO_2^-$, $2Br^-$
33.6 $2C_5H_6NO^+$, $2C_5H_5NO$, $Br_6Pd_2^{2-}$
44.9 $C_5H_6N_2$
40.1 $C_5H_6N_2O_2$
41.11 $C_5H_6N_2O_2S$
41.12 $C_5H_6N_2O_2S$
32.13 $C_5H_6N_4O_5$, $0.5H_2O$
44.10 $2C_5H_6N_5^+$, $2NO_3^-$, H_2O
39.4 $C_5H_6S_3$
69.6 $C_5H_7GeIO_2$
50.1 $C_5H_7NO_2$
44.11 $C_5H_7N_3O$, $C_5H_8N_3O^+$, I^-, H_2O
44.12 $C_5H_7N_3O$, $0.5H_2O$
41.13 $C_5H_7N_3O_2S$
72.1 $C_5H_8Cl_2Pt$
69.7 C_5H_8Ge
32.14 $C_5H_8NS_2^-$, Na^+, $2H_2O$
32.15 $C_5H_8NS_2^-$, Na^+, $2H_2O$
32.16 $C_5H_8NS_2^-$, Na^+, $2H_2O$
20.2 $C_5H_8N_2O_2$
44.13 $C_5H_8N_3O^+$, ClO_4^-
64.12 $C_5H_8N_3O_5P^{2-}$, $2Li^+$, $2H_2O$
39.5 $C_5H_8O_4S_2$
45.1 $C_5H_8O_5$
83.18 $C_5H_9Cl_2CuN_3$
48.12 $C_5H_9NO_4$
48.13 $C_5H_9NO_4$
1.18 $C_5H_9OS_2^-$, K^+
1.19 $2C_5H_{10}O_2$, C_6H_6
1.20 $C_5H_{10}O_5$
45.2 $C_5H_{10}O_5$
45.3 $C_5H_{10}O_5$
84.7 $C_5H_{11}MoNO_7$
48.14 $C_5H_{11}NO_2S$
48.15 $C_5H_{11}NO_2S$
46.2 $C_5H_{11}O_5P$
39.6 $C_5H_{11}S^+$, I^-
81.18 $(C_5H_{12}CdO_7)_n$, $2nH_2O$
33.13 $C_5H_{12}N^+$, $C_6H_{10}NS_2^-$
33.14 $C_5H_{12}N^+$, $C_6H_{10}NS_2^-$
85.10 $2C_5H_{12}N^+$, $C_6H_{12}Mo_2O_6S_3^{2-}$
33.15 $2C_5H_{12}N^+$, Cl_4Cu^{2-}
33.16 $2C_5H_{12}N^+$, $H_4Br_6Mo_2O_2^{2-}$
45.4 $C_5H_{12}O_4$
45.5 $C_5H_{12}O_4$
84.8 $4C_5H_{13}CuN_2O^+$, $C_4CuN_4S_4^{3-}$, CNS^-
69.8 $(C_5H_{13}NO_2Sn)_n$
73.3 $C_5H_{14}B_9CoSe_2$
64.13 $C_5H_{14}N_3O_3P$, $0.5CH_6N_3^+$, $0.5Cl^-$
63.2 $C_5H_{15}N_7S_5Si$
62.6 $C_5H_{16}B_{10}S_2$
62.7 $C_5H_{18}B_9^-$, Cs^+
64.14 $C_5H_{19}B_2P_2^-$, Li^+
76.22 $C_5H_{20}N_4Pt^{2+}$, Cl_5Sb^{2-}

83.19 $C_5H_{21}N_8ORu^{2+}$, $2F_6P^-$
65.3 $C_5H_{23}AsB_{20}$

C_6

60.10 $C_6Br_4O_2$, $C_{10}H_{12}S_4$
60.3 $C_6Cl_4O_2$, $C_6H_4S_4$
18.1 $C_6Cl_4O_2^-$, K^+, C_3H_6O
60.4 $C_6F_4O_2$, $C_6H_4S_4$
60.34 $C_6F_4O_2$, $C_{16}H_{10}$
60.13 C_6F_6, $C_{10}H_{16}N_2$
66.6 $C_6F_9O_6Sb$
11.7 C_6HCl_5S
39.7 $C_6H_2Cl_2N_4S_3$
60.43 $C_6H_2N_3O_7^-$, $C_{36}H_{60}N_4O_{12}$, K^+
60.1 $C_6H_2N_3O_7^-$, $C_5H_6N^+$, $C_{10}H_9N$
17.1 $2C_6H_2N_3O_7^-$, $C_8H_{18}O_5$, Ca^{2+}, H_2O
39.16 $C_6H_2N_3O_9S^-$, $C_7H_{15}S^+$
71.3 $C_6H_3AlBr_3MnO_5$
17.2 $C_6H_4Cl_2O$
17.3 $C_6H_4Cl_2O$
60.2 $C_6H_4Cl_2O$, $C_8H_{10}O$
40.2 $C_6H_4N_2O_2$
16.1 $C_6H_4N_2O_3S$, $C_6H_6N_2O_2$
15.1 $C_6H_4N_2O_4$
11.8 $C_6H_4N_2O_4S^{2-}$, $2Na^+$, $3H_2O$
81.19 $C_6H_4O_{10}U^{2-}$, Ba^{2+}, $3H_2O$
81.20 $C_6H_4O_{10}U^{2-}$, Sr^{2+}, $3H_2O$
60.3 $C_6H_4S_4$, $C_6Cl_4O_2$
60.4 $C_6H_4S_4$, $C_6F_4O_2$
60.12 $C_6H_4S_4$, $C_{10}H_{14}O_4Pd$
39.8 $C_6H_4S_4^+$, ClO_4^-
39.9 $C_6H_4S_4^+$, I_3^-
60.5 $2C_6H_4S_4$, $3C_4H_4NiS_4$
72.2 $C_6H_5BrFeO_3$
19.1 $C_6H_5IO_2$
33.17 $C_6H_5N_2^+$, $C_6H_4N_2$, Cl_4Fe^-
35.1 $C_6H_5N_3O$
70.1 $C_6H_5Te^{3+}$, $1.3Br^-$, $1.7Cl^-$
1.19 C_6H_6, $2C_5H_{10}O_2$
61.3 C_6H_6, $2C_{12}H_{24}N_9P_3$
61.12 $0.5C_6H_6$, $C_{27}H_{30}O_6$, H_2O
61.7 $2C_6H_6$, $C_{23}H_{24}N_4O_4S_2$
83.20 $C_6H_6Cl_2N_2O_2Pd$
69.9 $C_6H_6Cl_3O_6Sn^-$, K^+
83.21 $(C_6H_6CoN_8S_2)_n$
83.22 $(C_6H_6CuN_8S_2)_n$
33.18 $4C_6H_6NO_2^+$, $C_8MoN_8^{4-}$
48.16 $C_6H_6NO_6^{3-}$, Na^+, Ca^{2+}
81.21 $(C_6H_6NO_8Zn^-)_n$, nNa^+, nH_2O
16.1 $C_6H_6N_2O_2$, $C_6H_4N_2O_3S$
32.17 $C_6H_6N_2O_2$, $2H_2O$
72.3 $2C_6H_6N_2O_2Rh^+$, $C_{35}O_{33}Rh_{14}^{2-}$
41.14 $C_6H_6N_2O_2S$
17.4 $C_6H_6N_2O_3$
40.3 $C_6H_6N_2O_4$
64.15 $C_6H_6N_3P$
83.23 $(C_6H_6N_8S_2Zn)_n$
17.5 $C_6H_6O_2$
39.10 $C_6H_6O_2S$

2.25 $C_6H_6O_6^{2-}$, $C_2H_{10}N_2^{2+}$
2.26 $C_6H_6O_6^{2-}$, $C_2H_{10}N_2^{2+}$, H_2O
46.3 $C_6H_6O_{24}P_6^{12-}$, $12Na^+$, $38H_2O$
38.5 $2C_6H_7ClO_5$, CH_3NO_2
2.27 $C_6H_7NO_6^{2-}$, $2Na^+$
44.14 $C_6H_7N_2O_2^-$, Ag^+
41.15 $C_6H_7N_3S$
44.15 $C_6H_7N_5$
48.20 $C_6H_7O_6^-$, $C_6H_{15}N_4O_2^+$
33.84 $C_6H_7O_7^-$, $C_{22}H_{29}N_2O^+$, C_7H_8
76.23 $(C_6H_8Cd_2N_6)_n$
85.39 $C_6H_8N^+$, $C_{12}H_{18}FeN_6O_2S_4^-$, $4H_2O$
34.1 $C_6H_8N_2$
34.2 $C_6H_8N_2$
41.16 $C_6H_8N_3S^+$, Br^-, H_2O
44.16 $C_6H_8N_5^+$, Cl^-
48.8 $C_6H_8O_6$, $C_3H_7NO_3$
82.11 $C_6H_9BrNO_5Pd^-$, K^+, H_2O
85.8 $C_6H_9ClS_6Ti$
84.9 $C_6H_9FeN_6O_6^-$, Cs^+, H_2O
48.17 $C_6H_9NO_2$, H_2O
41.17 $C_6H_9NO_3S$
48.18 $C_6H_9NO_4$, H_2O
86.2 $C_6H_9NO_4PReS$
66.7 $C_6H_9O_3S_3Sb$
64.16 $C_6H_9O_5P$
46.4 $C_6H_9O_5PS$
66.8 $C_6H_9O_6Sb$
80.1 $(C_6H_{10}HgO_2S_4)_n$
33.13 $C_6H_{10}NS_2^-$, $C_5H_{12}N^+$
33.14 $C_6H_{10}NS_2^-$, $C_5H_{12}N^+$
45.6 $C_6H_{10}O_4$
45.7 $C_6H_{10}O_4$
45.8 $C_6H_{10}O_5$
64.17 $C_6H_{10}O_5P$
76.24 $C_6H_{11}ClN_4O_2Pt$, H^+, Cl^-, $2H_2O$
64.18 $C_6H_{11}N_4OPS$
85.9 $C_6H_{12}ClMo_2O_5S_3^-$, $C_8H_{20}N^+$
69.10 $C_6H_{12}Cl_2N_2O_2Sn$
84.10 $(C_6H_{12}MnN_2O_2S_2)_n$
81.22 $C_6H_{12}MnO_6$
85.10 $C_6H_{12}Mo_2O_6S_3^{2-}$, $2C_5H_{12}N^+$
48.19 $C_6H_{12}N_2O_4$, $2H_2O$
69.11 $C_6H_{12}N_2PbS_4$
64.19 $C_6H_{12}N_3PS$
45.10 $C_6H_{12}O_5$
45.9 $C_6H_{12}O_5$
64.20 $C_6H_{12}P_2S_2$
39.11 $C_6H_{12}S_3$
64.21 $C_6H_{13}O_4P$
85.11 $C_6H_{14}CdN_4O_4S_2$
81.23 $C_6H_{14}CuN_2O_6$, $2H_2O$
84.11 $C_6H_{14}CuN_4O_6$
21.1 $C_6H_{14}N^+$, $C_2H_6O_4P^-$
32.18 $C_6H_{14}N^+$, $C_6H_7O_3^-$
64.22 $C_6H_{14}O_2PS_2^-$, $C_2H_8N^+$
68.4 $C_6H_{15}AlCl_2N_2$
65.4 $C_6H_{15}AsS$
76.25 $C_6H_{15}Br_2CuN_3$
76.26 $C_6H_{15}Cl_2CuN_3$

48.20 $C_6H_{15}N_4O_2^-$, $C_6H_7O_6^-$
73.4 $C_6H_{16}B_{10}Co^-$, $C_4H_{12}N^+$
76.27 $C_6H_{16}CoN_4O_3S^+$, Cl^-, H_2O
76.28 $2C_6H_{16}CoN_4O_3S^+$, $O_6S_2^{2-}$, $2H_2O$
76.30 $C_6H_{16}CoN_4O_4^-$, $C_4H_5O_6^-$, H_2O
76.31 $C_6H_{16}CoN_4O_4^+$, $C_4H_5O_6^-$, $2H_2O$
76.29 $C_6H_{16}CoN_4O_4^-$, $C_4H_5O_6^-$, $3H_2O$
81.24 $C_6H_{16}CoO_8$
3.31 $2C_6H_{16}N^+$, $C_8MoN_8^{4-}$, $2H_3O^+$
63.23 $2C_6H_{16}N^+$, $C_{18}H_{12}O_6Si^{2-}$
3.32 $2C_6H_{18}N^+$, $H_{12}B_{12}^{2-}$
20.1 $C_6H_{16}N_2$, $C_5H_5^-$, Na^+
81.25 $(C_6H_{16}NiO_8)_n$
85.12 $C_6H_{16}NiP_2S_4$
63.3 $C_6H_{18}AlCl_4N_3Si$
63.4 $C_6H_{18}B_2N_2S_2Si$
66.9 $C_6H_{18}Cl_8N_2O_6P_2Sb_2$
82.12 $(C_6H_{18}MnN_2O_6^{2+})_n$, $2nBr^-$, $2nH_2O$
85.13 $C_6H_{18}N_2NiP_2S_4$
33.19 $C_6H_{18}N_3^{3+}$, Cl_5Cu^{3-}, $2H_2O$
69.12 $C_6H_{18}N_4O_2S_4Sn_2$, $0.5C_6H_6$
83.24 $C_6H_{18}N_{12}NiO_6^{2+}$, $2Cl^-$, $0.5H_2O$
84.12 $C_6H_{18}O_9Re_2$
70.2 $C_6H_{18}Sn_3Te_3$
84.13 $C_6H_{19}CoN_3O_3^+$, $C_6H_{20}CoN_3O_3^{2+}$, $3ClO_4^-$, $0.5H_2O$
62.8 $C_6H_{20}B_{10}$
83.25 $C_6H_{20}Cl_2N_4Pt^{2+}$, $C_6H_{20}N_4Pt^{2+}$, $4ClO_4^-$
76.32 $C_6H_{20}I_2N_4Pt^{2+}$, $C_6H_{20}N_4Pt^{2+}$, $4I^-$
84.14 $C_6H_{20}N_3NiO_3^+$, ClO_4^-
83.25 $C_6H_{20}N_4Pt^{2+}$, $C_6H_{20}Cl_2N_4Pt^{2+}$, $4ClO_4^-$
76.32 $C_6H_{20}N_4Pt^{2+}$, $C_6H_{20}I_2N_4Pt^{2+}$, $4I^-$
76.33 $C_6H_{20}N_4Pt^{2+}$, $2C_{10}H_{14}BrO_4S^-$
76.34 $C_6H_{20}N_4Pt^{2+}$, $2Cl^-$, $2H_2O$
76.35 $C_6H_{20}N_4Pt^{2+}$, $2Cl^-$, $2H_2O$
86.3 $C_6H_{21}Cl_3NO_6P_2Ru$
76.36 $C_6H_{21}CuN_5^{2+}$, $2ClO_4^-$
63.5 $C_6H_{22}B_{10}Si$
69.13 $C_6H_{22}B_{10}Sn$
76.37 $C_6H_{24}CoN_6^{3+}$, $C_6H_{24}CrN_6^{3+}$, $6CNS^-$
76.38 $C_6H_{24}CoN_6^{3+}$, $3I^-$, H_2O
76.39 $C_6H_{24}CrN_6^{3+}$, $3Br^-$, $0.6H_2O$
76.40 $C_6H_{24}N_6Ni^{2+}$, $2NO_3^-$
64.23 $C_6H_{25}B_{20}P$
83.26 $C_6H_{26}N_7Ru^{2+}$, $2F_6P^-$
81.26 $C_6MnO_{12}^{3-}$, $3K^+$, $3H_2O$
60.9 C_6N_4, C_9H_9N
60.15 C_6N_4, $2C_{12}H_8N_2$
85.14 $C_6NiS_{10}^{2-}$, $2C_{16}H_{36}N^+$

C_7

33.20 $C_7H_3D_2NO_4$
33.21 $C_7H_3D_2NO_4$
33.22 $C_7H_3D_2NO_4$
7.2 C_7H_4ClN
73.5 $C_7H_4NO_4Rh$
38.6 $C_7H_4N_2O_3S$

41.18 $C_7H_5ClN_2S$
35.2 $C_7H_5Cl_2N_3$
83.27 $C_7H_5Cl_3NOV$
17.6 $C_7H_5Cl_3O_2$
73.6 $C_7H_5CrNO_2S$
13.1 $C_7H_5NO_4$
41.19 $C_7H_5N_2S_2$
14.1 $2C_7H_5O_3^-$, Ca^{2+}, $2H_2O$
14.2 $2C_7H_5O_3^-$, Ca^{2+}, $2H_2O$
14.3 $2C_7H_5O_3^-$, Sr^{2+}, $2H_2O$
81.27 $(C_7H_6AgNO_2)_n$
10.3 C_7H_6ClNO
13.2 C_7H_6ClNO
69.14 $(C_7H_6N_2O_5Pb)_n$
44.17 $C_7H_6N_5O_2^-$, $C_{10}H_{13}N_2^+$, $0.5H_2O$
13.3 $C_7H_6O_2$
38.7 $C_7H_6O_3$
71.4 C_7H_7ClHg
64.24 $C_7H_7Cl_3OP^+$, Cl^-, HCl
73.7 $C_7H_7IMnNOS$
60.6 $C_7H_7N_2O_4^-$, $C_8H_{12}NO^+$, H_2O
61.21 C_7H_8, $C_{44}H_{56}O_4$
84.15 $(C_7H_8Br_2HgN_2O)_n$
38.8 $C_7H_8Br_2O$
13.4 $C_7H_8NO_2^+$, Cl^-
33.23 $C_7H_8N_4O_4$, H_2O
31.1 C_7H_8O
39.12 $C_7H_8OS_2$
27.1 $C_7H_9BrO_2$
38.9 $C_7H_9BrO_3$
83.28 $(C_7H_9Br_2HgN)_n$
27.2 $C_7H_9ClO_2$
83.29 $2C_7H_9Cl_3NPt^-$, $H_{12}N_4Pt^{2+}$
33.24 C_7H_9NO, $C_7H_{10}NO^+$, ClO_4^-
39.13 $C_7H_9NO_2S_2$
39.14 $C_7H_9NO_2S_2$
40.4 $C_7H_9NO_3$
44.18 $C_7H_9N_4O_2^+$, I_3^-, $0.5I_2$
44.19 $C_7H_9N_5$
86.4 $C_7H_9O_7PRu$
50.2 $C_7H_{10}ClN_2O^+$, Cl^-
82.13 $C_7H_{10}Cl_2N_2O_2Pt$
5.8 $C_7H_{10}Cl_6$
71.5 $C_7H_{10}HgN_5^+$, NO_3^-
83.30 $C_7H_{10}HgN_5O^+$, NO_3^-
61.6 $C_7H_{10}N^+$, $2C_{18}H_{38}N_2O_6$, CNS^-
19.2 $C_7H_{10}N^+$, Cl^-, $2Cl_2Hg$
33.25 $4C_7H_{10}N^+$, $H_2O_{28}V_{10}^{4-}$
33.26 $4C_7H_{10}N^+$, $Mo_8O_{26}^{4-}$
32.19 $C_7H_{10}NO_5^-$, $C_8H_{11}BrN^+$
35.3 $C_7H_{10}N_2O_2$
40.5 $C_7H_{10}N_2O_4$, H_2O
85.15 $C_7H_{11}Cl_2NOPtS$
33.27 $C_7H_{11}NO_2$
41.20 $C_7H_{11}NO_3S$
48.21 $C_7H_{11}NO_4$, H_2O
45.11 $C_7H_{11}N_3O_4S$
76.41 $C_7H_{12}BrN_2O_3Re$
83.31 $C_7H_{12}Cl_2CuN_4O_4$
82.14 $C_7H_{12}CuN_2O_3S$

85.16 $C_7H_{12}MnN_4O_3S_4^+$, Br^-
48.22 $C_7H_{12}N_2O_4$
38.10 $C_7H_{12}O_4$
76.42 $C_7H_{13}ClN_4O_2Pt$
37.1 $C_7H_{13}N$
34.3 $C_7H_{13}N_2^+$, ClO_4^-
32.20 $C_7H_{13}N_7OS_2$
64.25 $C_7H_{14}ClO_3P$
64.26 $C_7H_{14}Cl_2N_2O_2P$
69.15 $C_7H_{14}N_2O_2Pb$
42.2 $C_7H_{14}O_3S$
63.6 $C_7H_{15}ClN_2O_2Si$
37.2 $2C_7H_{15}N_4^+$, $2I_3^-$, I_2
39.15 $C_7H_{15}S^+$, Br^-
39.16 $C_7H_{15}S^+$, $C_6H_2N_3O_9S^-$
39.17 $C_7H_{15}S^+$, I^-
69.16 $C_7H_{16}ClNO_2SSn$
62.14 $C_7H_{16}NO_2^+$, $C_{24}H_{20}B^-$
76.43 $C_7H_{16}N_4O_3Re^+$, $C_{12}H_{10}O_2P^-$
86.5 $C_7H_{18}I_2NiOP_2$
85.66 $C_7H_{18}N^+$, $C_{21}H_{15}CrN_3O_3S_3^{3-}$, $2.5Na^+$, $0.5HO^-$, $18.5H_2O$
62.5 $C_7H_{18}N^+$, $C_4H_{20}B_{18}CoO^-$
76.44 $C_7H_{20}Cl_2CuN_4O_8$
76.45 $C_7H_{25}ClCoN_5O^{2+}$, Cl_4Zn^{2-}

C_8

18.2 $C_8Cl_2N_2O_2$
35.4 $C_8Cl_6N_2$
33.28 $C_8D_5N_3$
33.29 $C_8D_5N_3$
39.18 $C_8H_4N_2O_4S_2$
40.6 $C_8H_4N_4O_2$
14.4 $C_8H_4O_4^{2-}$, $2K^+$, H_2O
81.28 $C_8H_4O_{14}V_2^{4-}$, $4C_8H_{20}N^+$, $8H_2O$
73.8 $C_8H_5CrNO_4$
73.9 $C_8H_5FeNO_2S$
73.22 $C_8H_5MoO_3^-$, $C_{11}H_{11}MoO^+$
73.95 $C_8H_5MoO_3^-$, $C_{22}H_{34}AsMo_2O_4P_2^+$
63.7 $C_8H_6BrF_3O_2Si$
83.32 $(C_8H_6ClCuN_4S_2)_n$, nH_2O
17.7 $C_8H_6Cl_4O_2$
35.5 $C_8H_6N_2$
44.20 $C_8H_6N_4S_2$
39.19 $C_8H_6S_2$, $2Cl_3Sb$
83.33 $C_8H_7Cl_3NOV$
44.21 $C_8H_7Cl_3N_4$
64.27 $C_8H_7F_{12}O_3P$
71.6 $C_8H_7NO_3Ru$
39.20 $C_8H_7NO_3S$
39.21 $C_8H_7NO_3S$
17.8 $C_8H_7NO_5$
40.7 $C_8H_7N_3O_2$
41.21 $C_8H_7N_3O_3S$
32.18 $C_8H_7O_3$, $C_6H_{14}N^+$
2.28 $C_8H_7O_3$, $C_8H_{12}N^+$
83.34 $(C_8H_8Cl_2CuN_4O_8)_n$
33.30 $C_8H_8Cl_6N_2O_2$
70.3 $C_8H_8I_2Te$

35.6 $C_8H_8N_2O_2$

35.7 $C_8H_8N_2O_2$

73.10 $C_8H_8N_2O_2W$

45.66 $C_8H_8N_2O_3$, $C_{42}H_{70}O_{35}$, xH_2O

35.8 $C_8H_8N_3^+$, Br^-, H_2O

50.3 $C_8H_9NO_4$

61.9 C_8H_{10}, $3C_{26}H_{24}P_2Se_2$

39.22 $C_8H_{10}ClN_3S$

5.9 $C_8H_{10}Cl_8$

82.15 $C_8H_{10}CoN_2O_8^{2-}$, $2Na^+$, $7H_2O$

82.16 $C_8H_{10}CrN_2O_8^-$, K^+, $3H_2O$

72.4 $C_8H_{10}IMoNO$

72.5 $C_8H_{10}INOW$

38.11 $C_8H_{10}N_2O_3$

38.12 $C_8H_{10}N_2O_3$

60.2 $C_8H_{10}O$, $C_6H_4Cl_2O$

45.12 $C_8H_{10}O_5$

32.19 $C_8H_{11}BrN^+$, $C_7H_{10}NO_5^-$

86.6 $(C_8H_{11}CdCl_2P)_n$

45.13 $C_8H_{11}IO_5S$

36.11 $C_8H_{11}NO$, $C_{14}H_{13}N_4O_4$, $4H_2O$

40.8 $C_8H_{11}NO_4$

85.17 $C_8H_{12}Au_4S_8$

81.29 $(C_8H_{12}ClO_8Ru_2)_n$, $2nH_2O$

81.30 $C_8H_{12}Cl_2O_8Ru_2^-$, Cs^+

64.28 $C_8H_{12}F_{12}N_2P_2$

85.18 $C_8H_{12}FeO_2S_4$

2.28 $C_8H_{12}N^+$, $C_8H_7O_3^-$

2.32 $C_8H_{12}N^+$, $C_{10}H_{11}O_2^-$

60.6 $C_8H_{12}NO^+$, $C_7H_7N_2O_4^-$, H_2O

59.1 $C_8H_{12}NO_2^+$, Cl^-

33.31 $C_8H_{12}NO_3^+$, Cl^-

37.3 $C_8H_{12}N_2$

35.9 $C_8H_{12}N_2O$

48.23 $C_8H_{12}N_2O_2$

32.21 $C_8H_{12}N_2O_3$

43.1 $C_8H_{12}N_2O_3$, $C_4H_6N_2$

34.4 $C_8H_{12}N_2O_4S_2$

45.14 $C_8H_{12}N_2O_6S$

41.22 $C_8H_{12}N_2S_2$

47.1 $C_8H_{12}N_3O_9P$, $3H_2O$

22.1 $C_8H_{12}O_2$

39.23 $C_8H_{12}O_2S$

38.13 $C_8H_{12}O_4$

39.24 $C_8H_{12}O_4S$

68.5 $C_8H_{12}O_8Tl^-$, Tl^+

81.31 $C_8H_{12}O_{12}Sc^-$, $C_4H_{14}O_{10}Sc^+$

83.35 $C_8H_{13}Hg_3N_5^{2+}$, $2ClO_4^-$

32.22 $C_8H_{13}NO_2S_2$

33.32 $C_8H_{13}NO_3$

83.36 $C_8H_{14}Cl_2N_4O_4Rh^-$, Rb^+, H_2O

69.17 $C_8H_{14}Cl_2O_4Sn$

75.1 $C_8H_{14}Cl_2PtS$

83.37 $C_8H_{14}CoN_6O_8^-$, Na^+, $2H_2O$

73.11 $C_8H_{14}CoPS_5$

22.2 $C_8H_{14}N_2O_2$

48.24 $C_8H_{14}N_2O_3$, H_2O

45.15 $C_8H_{14}O_5$

85.19 $C_8H_{14}PdS_4$

85.20 $C_8H_{14}PtS_4$

64.29 $C_8H_{15}B_{10}Cl_5N_3P_3$

71.7 $C_8H_{15}ClPdS$

21.2 C_8H_{16}

72.6 $C_8H_{16}Cl_2PdS$

84.16 $C_8H_{16}CoN_2O_{10}$

83.38 $C_8H_{16}CoN_5O_7$

81.32 $C_8H_{16}Cr_2O_{10}$

81.33 $C_8H_{16}Cr_2O_{10}$

82.17 $C_8H_{16}CuN_2O_7$

85.21 $C_8H_{16}Hg_2S_2$

85.22 $C_8H_{16}MoS_6$

85.23 $C_8H_{16}Mo_2O_3S_6$

48.25 $C_8H_{16}NO_2^+$, Br^-, H_2O

41.23 $C_8H_{16}N_2O_2S$

40.9 $C_8H_{16}N_2O_2S_4$

40.10 $C_8H_{16}N_2O_2Se_2$

40.11 $C_8H_{16}N_2O_2Se_3$

40.12 $C_8H_{16}N_2O_2Se_4$

81.34 $C_8H_{16}N_2O_4Pt$

1.21 $C_8H_{16}N_2S_2$

79.6 $C_8H_{16}N_6O_2S_2Zn$

21.3 $C_8H_{16}O_2$

22.3 $C_8H_{16}O_2$

81.35 $C_8H_{16}O_{10}Ru_2^+$, BF_4^-

81.36 $C_8H_{16}O_{14}U$

85.24 $C_8H_{17}CoN_5O_7S^-$, Na^+, $4H_2O$

42.3 $C_8H_{17}NO_2S$

40.13 $C_8H_{17}N_2O_2S^+$, I^-

5.10 C_8H_{18}

5.11 C_8H_{18}

62.9 $C_8H_{18}B_2N_2$

68.6 $C_8H_{18}Ga_2N_2S_2$

83.39 $C_8H_{18}N_2O_2Os$

50.4 $C_8H_{18}N_2O_4$

63.8 $C_8H_{19}NSi$

71.179 $C_8H_{20}Al^-$, $C_{22}H_{22}CoN_4^+$

85.25 $C_8H_{20}Cl_2CoO_4S_2$

83.40 $C_8H_{20}CoN_6O_4^+$, Br^-

83.41 $C_8H_{20}CoN_6O_4^+$, CNS^-, H_2O

83.42 $C_8H_{20}CoN_6O_4^+$, NO_3^-

85.27 $2C_8H_{20}CuS_4^+$, $C_8H_{20}CuS_4^{2+}$, $4BF_4^-$

85.26 $2C_8H_{20}CuS_4^+$, $C_8H_{20}CuS_4^{2+}$, $4ClO_4^-$

85.27 $C_8H_{20}CuS_4^{2+}$, $2C_8H_{20}CuS_4^+$, $4BF_4^-$

85.26 $C_8H_{20}CuS_4^{2+}$, $2C_8H_{20}CuS_4^+$, $4ClO_4^-$

84.17 $C_8H_{20}MoN_2O_4$

85.4 $C_8H_{20}N^+$, $C_4H_8Cl_3Mo_2O_4S_2^-$

85.9 $C_8H_{20}N^+$, $C_8H_{12}ClMo_2O_5S_3^-$

84.21 $C_8H_{20}N^+$, $C_9H_9O_9Re_2^-$

83.53 $C_8H_{20}N^+$, $C_{10}H_{14}IN_6Ni_2O_2^-$

60.7 $C_8H_{20}N^+$, $C_{12}H_4N_4^-$, $C_{12}H_4N_4$

81.53 $C_8H_{20}N^+$, $C_{14}H_6FMo_2N_2O_{14}^-$

3.33 $C_8H_{20}N^+$, $C_{15}Fe_4O_{14}Rh^-$

73.52 $C_8H_{20}N^+$, $C_{15}H_{10}Mo_2NO_4^-$

85.50 $C_8H_{20}N^+$, $C_{16}Cl_5O_{10}SW_2^-$

86.113 $C_8H_{20}N^+$, $C_{34}H_{27}Mo_2O_8P_2^-$

3.34 $C_8H_{20}N^+$, $H_2Br_4MoO_2^-$

3.35 $C_8H_{20}N^+$, $H_2F_5O_5U_2^-$, H_2O

3.36 $C_8H_{20}N^+$, $H_2I_4MoO_2^-$

3.37 $2C_8H_{20}N^+$, $C_{18}Fe_5MoO_{17}^{2-}$

3.38 $2C_8H_{20}N^+$, $Cl_{12}Ru_3^{4-}$, $6H_2O$, $2H^+$

85.57 $3C_8H_{20}N^+$, $C_{16}H_{40}Fe_6Mo_2S_{17}^{3-}$

3.39 $3C_8H_{20}N^+$, $C_{18}H_{45}Fe_6S_{17}W_2^{3-}$

85.80 $3C_8H_{20}N^+$, $C_{28}H_{28}Fe_4S_8^{3-}$

3.40 $3C_8H_{20}N^+$, $HCl_4Mo_2^{3-}$, $H_2Cl_4MoO_2^-$, H^+, $2H_2O$

81.28 $4C_8H_{20}N^+$, $C_8H_4O_{14}V_2^{4-}$, $8H_2O$

86.7 $C_8H_{22}ClHgP_2^+$, $C_4H_{11}Cl_3HgP^-$

76.46 $C_8H_{22}CoN_5O_2^+$, $2I^-$, $0.84H_2O$

76.47 $C_8H_{22}CoN_5O_2S^{2+}$, $2Cl^-$

67.2 $C_8H_{22}MgN_2$

84.18 $2C_8H_{23}CuN_2O_3^+$, $2NO_3^-$

3.41 $C_8H_{24}B_4N_4S_2$

84.19 $C_8H_{24}Cl_2FeO_4S_4^+$, Cl^-

69.18 $(C_8H_{24}Cl_4O_2Sn_4)_n$

76.48 $C_8H_{24}CuN_4^{2+}$, $2ClO_4^-$

76.49 $C_8H_{24}Cu_2N_6O_2^{2+}$, $2C_{24}H_{20}B^-$

85.28 $C_8H_{24}Mo_4O_{12}P_4S_4$, $xCHCl_3$, $yCCl_4$

85.29 $C_8H_{24}Mo_4O_{12}P_4S_4$, $xCHCl_3$, $yCCl_4$

64.30 $C_8H_{24}N_4P_4S_3$

71.8 $C_8H_{24}O_4Zn_4$

85.30 $C_8H_{24}P_4S_8Th$

76.50 $C_8H_{26}CuN_6^{2+}$, $2NO_3^-$

76.51 $C_8H_{26}CuN_6^{2+}$, $2NO_3^-$

76.52 $C_8H_{26}N_6Zn^{2+}$, $2NO_3^-$

62.10 $C_8H_{28}B_{10}S$

83.43 $C_8H_{28}Br_2N_4Pt^{2+}$, $C_8H_{28}N_4Pt^{2+}$, $4Br^-$

83.43 $C_8H_{28}N_4Pt^{2+}$, $C_8H_{28}Br_2N_4Pt^{2+}$, $4Br^-$

76.53 $C_8H_{33}Co_2N_8O_3^{3+}$, $3ClO_4^-$

85.31 $C_8Mo_2O_7S_5^{2-}$, $2C_{16}H_{36}N^+$

C₉

71.10 $C_9H_2Fe_2O_8$

71.9 $C_9H_2Fe_2O_8$

38.14 $C_9H_5NO_4$

35.10 $C_9H_6NO_3S_2^-$, Rb^+

35.11 $C_9H_6NS^-$, Na^+, $2H_2O$

42.4 $C_9H_6O_5S$

60.8 $C_9H_6S_2Se$, $C_{12}H_4N_4$

86.8 $C_9H_7CrN_2O_5P$

35.12 C_9H_7NO

41.24 $C_9H_7NO_2S$

40.14 $C_9H_7N_3O_2$

70.4 $C_9H_8Cl_3NO_2Te$

41.25 $2C_9H_8NOS_2^-$, Cl_6Te^{2-}, $C_4H_8O_2$

71.11 $C_9H_9BClO_6Re$

32.23 $C_9H_9ClN_4O$, CH_4O

84.20 $2C_9H_9Mo_3N_3O_{13}^{2-}$, $3C_{36}H_{30}NP_2^+$, Na^+

60.9 C_9H_9N, C_6N_4

13.5 $C_9H_9NO_3$

32.24 $C_9H_9N_3S$

32.25 $C_9H_9N_3S$

32.26 $C_9H_9N_3S$

32.27 $C_9H_9N_3S$

2.29 $C_9H_9O_3^-$, $C_9H_{10}O_3$, Na^+

2.30 $C_9H_9O_3^-$, $C_9H_{10}O_3$, Rb^+

84.21 $C_9H_9O_9Re_2^-$, $C_8H_{20}N^+$

83.44 $C_9H_{10}BCl_2N_6OTc$

17.9 $C_9H_{10}BrNO_2$

31.2 $C_9H_{10}Br_2O$

8.10 $C_9H_{10}Cl_2N_2O$

32.28 $C_9H_{10}Cl_2N_3{}^+$, Cl$^-$

47.2 $C_9H_{10}N_2O_4S$

32.29 $C_9H_{10}N_4O_2$

2.29 $C_9H_{10}O_3$, $C_9H_9O_3{}^-$, Na$^+$

2.30 $C_9H_{10}O_3$, $C_9H_9O_3{}^-$, Rb$^+$

13.6 $C_9H_{10}O_3S$

18.3 $C_9H_{10}O_4$

38.15 $C_9H_{10}O_4$

8.11 $C_9H_{11}ClN_2O$

1.22 $C_9H_{11}NO_4S$

47.3 $C_9H_{11}N_2O_8P^{2-}$, 2Na$^+$, 5H$_2$O

47.4 $C_9H_{11}N_2O_9P^{2-}$, 2Na$^+$, 7H$_2$O

29.1 C_9H_{12}

38.16 $C_9H_{12}Br_2O$

71.12 $C_9H_{12}Hg_4O_8$, 2H$_2$O

47.5 $C_9H_{12}IN_3O_4$

47.6 $C_9H_{12}IN_3O_5$

80.2 $C_9H_{12}N_2O_3S_4W$

47.7 $C_9H_{12}N_2O_6$

47.8 $C_9H_{12}N_3O_7P$, 3H$_2$O

47.9 $C_9H_{12}N_3O_8P^{2-}$, Ba^{2+}, 8.5H$_2$O

47.10 $C_9H_{12}N_3O_{10}P$, H$_2$O

21.4 $C_9H_{12}O_2$

21.5 $C_9H_{12}O_2$

11.9 $C_9H_{12}O_4S_2$

45.16 $C_9H_{12}O_6$

37.4 $C_9H_{13}NO$

69.19 $C_9H_{13}NO_2Sn$, H$_2$O

32.30 $C_9H_{13}NO_3$

50.5 $C_9H_{13}NO_4$

17.10 $C_9H_{13}NO_5S$, 0.5H$_2$O

17.11 $C_9H_{13}NO_5S$, 0.5H$_2$O

64.31 $C_9H_{13}O_2PS$

64.32 $C_9H_{13}O_6P$

47.11 $C_9H_{14}N_2O_9P^-$, K$^+$, 0.5H$_2$O

35.13 $C_9H_{14}N_5O^+$, CH$_3$O$_3$S$^-$, H$_2$O

31.3 $C_9H_{14}O_2$

31.4 $C_9H_{14}O_2S$

31.5 $C_9H_{14}O_2S$

11.10 $C_9H_{14}O_3S$

71.13 $C_9H_{15}B_9F_3FeO_2$

71.14 $C_9H_{15}BrHgO_2$

85.32 $C_9H_{15}CoS_6$

2.31 $C_9H_{16}NO_5{}^-$, Ca^{2+}, Br$^-$

64.33 $C_9H_{16}NO_5P$

32.31 $C_9H_{16}N_4S_2$

40.15 $C_9H_{16}N_6O_3$, H$_2$O

38.17 $C_9H_{16}O_3$

45.17 $C_9H_{16}O_6$

70.5 $C_9H_{17}Br_3OTe$

33.33 $C_9H_{17}NO$

63.9 $C_9H_{18}ClNO_3Si$

80.3 $C_9H_{18}CoN_3S_6$

3.42 $C_9H_{18}N^-$, Li$^+$

86.9 $C_9H_{18}O_3P_2Ru$

72.7 $C_9H_{19}Cl_2NPt$

72.8 $C_9H_{19}NiP$

64.34 $C_9H_{19}O_4P$

1.23 $C_9H_{20}ClN_2O^-$, BF$_4{}^-$

76.54 $C_9H_{20}CoN_4O_4{}^+$, Cl$^-$, 3H$_2$O

63.10 $C_9H_{21}NSi$

73.12 $C_9H_{23}B_9CoN$

79.7 $C_9H_{24}Br_2MnN_6O_3$

84.22 $C_9H_{24}ClCuN_4O_4{}^+$, ClO$_4{}^-$

68.7 $C_9H_{24}Ga_2N_2O$

85.33 $C_9H_{24}N_6NiS_6{}^{2+}$, 2Cl$^-$, 0.5C$_2$H$_6$O, 3H$_2$O

73.13 $C_9H_{25}B_{16}Co_2{}^-$, C$_8$H$_{20}$N$^+$

65.5 $C_9H_{27}As_7Si_3$

86.10 $C_9H_{27}ClP_3Pt^+$, Cl$^-$

86.11 $C_9H_{27}ClP_3Rh$

64.35 $C_9H_{27}P_{11}Si_3$

64.36 $C_9H_{33}N_6PSSi_3$

C_{10}

71.15 $C_{10}Co_3O_{10}{}^-$, Li$^+$, C$_8$H$_{14}$O

60.19 $C_{10}F_8$, C$_{12}$H$_{12}$

71.16 $C_{10}Fe_3O_{10}S$

60.25 $C_{10}H_2N_4$, C$_{14}$H$_{10}$

60.26 $C_{10}H_2N_4$, C$_{14}$H$_{10}$

60.27 $C_{10}H_2N_4$, C$_{14}$H$_{10}$

60.28 $C_{10}H_2N_4$, C$_{14}$H$_{10}$

60.29 $C_{10}H_2N_4$, C$_{14}$H$_{10}$

60.30 $C_{10}H_2N_4$, C$_{14}$H$_{10}$

60.31 $C_{10}H_2N_4$, C$_{14}$H$_{10}$

38.18 $C_{10}H_2O_6$

60.17 $C_{10}H_2O_6$, C$_{12}$H$_9$NS

24.1 $C_{10}H_4Br_4$

41.26 $C_{10}H_4N_4S_2$

42.5 $C_{10}H_4N_4Se_2$

86.18 $C_{10}H_6AsClFeMnO_8{}^-$, C$_{12}$H$_{12}$As$_2$FeMnO$_8{}^+$

27.3 $C_{10}H_6Cl_6$

71.17 $C_{10}H_8F_4Fe_2O_6S_2$

71.18 $C_{10}H_8F_4Fe_2O_6S_2$

20.3 $C_{10}H_8F_{12}$

25.1 $C_{10}H_8O_4$

61.10 2C$_{10}$H$_7$Br, C$_{26}$H$_{28}$N$_6$NiS$_2$

24.2 $C_{10}H_7Cl$

41.27 $C_{10}H_7N_3OS$

35.14 $C_{10}H_7N_5$

41.28 $C_{10}H_8BrNOS_2$

66.10 $C_{10}H_8Cl_3N_2Sb$

83.45 $C_{10}H_8MoN_{12}$

33.34 $C_{10}H_8N_2$, K$^+$, C$_2$AuN$_2{}^-$

35.15 $C_{10}H_8N_2O_4$

33.35 $C_{10}H_8N_4$

24.3 $C_{10}H_8O_2$

38.113 $C_{10}H_8O_2$, C$_{22}$H$_{26}$O$_7$, H$_2$O

38.19 $C_{10}H_8O_4$

59.2 $C_{10}H_8O_4$

31.6 $C_{10}H_9Br_2NO_3S$

60.1 $C_{10}H_9N$, C$_6$H$_2$N$_3$O$_7{}^-$, C$_5$H$_6$N$^+$

38.20 $C_{10}H_9NO_3$

31.7 $C_{10}H_{10}BrClO_2$

46.5 $C_{10}H_{10}BrCl_2O_4P$

83.46 $C_{10}H_{10}Br_4Cu_2N_2$

83.47 $C_{10}H_{10}Br_4N_2W^-$, C$_5$H$_6$N$^+$, C$_5$H$_5$N

73.14 $C_{10}H_{10}ClNbO_2$

85.34 $C_{10}H_{10}Cl_2CoN_2S_2$

71.19 $C_{10}H_{10}CrNO_5{}^+$, BF$_4{}^-$

73.15 $C_{10}H_{10}I_2Mo_2O_3$

72.9 $C_{10}H_{10}MoO_2$

33.36 $C_{10}H_{10}N_2{}^{2+}$, Cl$_6$Cu$_2{}^{2-}$

16.2 $C_{10}H_{10}N_2OS$

41.29 $C_{10}H_{10}N_2OS$

37.5 $C_{10}H_{10}N_2O_2$

41.30 $C_{10}H_{10}N_2O_2S$

41.31 $C_{10}H_{10}N_2O_4S$

47.12 $C_{10}H_{10}N_2O_7$

64.37 $C_{10}H_{10}N_3P$

19.3 $C_{10}H_{10}O_2$

39.25 $C_{10}H_{10}O_2S$

39.26 $C_{10}H_{10}O_2S$

73.16 $C_{10}H_{10}Ru$

73.17 $C_{10}H_{10}Ru$

73.18 $C_{10}H_{10}Ru$

73.19 $C_{10}H_{10}V$

64.38 $C_{10}H_{11}Br_2OP$

17.12 $C_{10}H_{11}ClO_3$

73.20 $C_{10}H_{11}Cl_5FeO_2Si_3$

17.13 $C_{10}H_{11}NO_4$

44.30 $C_{10}H_{11}N_3O_3S$, C$_{14}$H$_{18}$N$_4$O$_3$

32.32 $C_{10}H_{11}N_3S$

47.13 $C_{10}H_{11}N_5O_6P^-$, Na$^+$, 4H$_2$O

2.32 $C_{10}H_{11}O_2{}^-$, C$_8$H$_{12}$N$^+$

64.39 $C_{10}H_{11}O_2P$

85.35 $C_{10}H_{12}Cd_2Cl_4N_8O_2S_2$

21.6 $C_{10}H_{12}Cl_2O$

76.55 $C_{10}H_{12}CuN_2O_8{}^{2-}$, Ca^{2+}, 4H$_2$O

84.23 $(C_{10}H_{12}CuN_2O_8S_2)_n$

83.48 $C_{10}H_{12}Hg_2N_2O_8S_2$

76.56 $C_{10}H_{12}MnN_2O_8{}^-$, K$^+$, 2H$_2$O

84.24 $C_{10}H_{12}Mo_3N_3O_{13}{}^-$, C$_4H_{12}N^+$

76.57 $C_{10}H_{12}N_2NiO_8{}^{2-}$, Ca^{2+}, 4H$_2$O

35.16 $C_{10}H_{12}N_2O$

35.17 $C_{10}H_{12}N_2O_2$

35.18 $C_{10}H_{12}N_2O_2$

34.5 $C_{10}H_{12}N_2O_4$

38.21 $C_{10}H_{12}N_2O_4$

48.26 $C_{10}H_{12}N_2O_8{}^{4-}$, 2Ba^{2+}, 2.5H$_2$O

48.27 $C_{10}H_{12}N_2O_8{}^{4-}$, 2Ca^{2+}, 7H$_2$O

40.16 $C_{10}H_{12}N_3O^+$, BF$_4{}^-$

35.19 $C_{10}H_{12}N_4O_3$

47.14 $C_{10}H_{12}N_4O_5$

60.10 $C_{10}H_{12}S_4$, C$_6$Br$_4$O$_2$

60.11 $C_{10}H_{12}S_4$, C$_{10}$H$_{12}$S$_4{}^+$, I$^-$

39.27 $C_{10}H_{12}S_4{}^+$, C$_{10}$H$_{12}$S$_4$, CNS$^-$

60.35 3C$_{10}$H$_{12}$S$_4$, 2C$_{16}$H$_{12}$N$_4$

52.1 $C_{10}H_{13}Br_2Cl_3$

76.58 $C_{10}H_{13}ClIrN_2O_8{}^-$, K$^+$, H$_2$O

32.33 $C_{10}H_{13}ClN_3{}^+$, NO$_3{}^-$

19.4 $C_{10}H_{13}NO$

38.22 $C_{10}H_{13}NO_2$

1.24 $C_{10}H_{13}NO_4S$

44.17 $C_{10}H_{13}N_2{}^+$, C$_7$H$_6$N$_5$O$_2{}^-$, 0.5H$_2$O

47.15 $C_{10}H_{13}N_5O_3$

47.16 $C_{10}H_{13}N_5O_4$

47.17 $C_{10}H_{13}N_5O_4S$, H_2O
39.28 $2C_{10}H_{13}OS^+$, $C_{10}H_{12}OS$, $2ClO_4^-$
61.18 $C_{10}H_{14}$, $C_{34}H_{44}N_4^{4+}$, $4Cl^-$, $4H_2O$
76.33 $2C_{10}H_{14}BrO_4S^-$, $C_6H_{20}N_4Pt^{2+}$
5.12 $C_{10}H_{14}Br_3Cl_3$
83.49 $C_{10}H_{14}Cl_2CuN_2S_2$
83.50 $C_{10}H_{14}Cl_2HgN_2$
76.59 $C_{10}H_{14}Cl_2IrN_2O_8^-$, K^+
71.20 $C_{10}H_{14}Cl_3HgN$
66.11 $C_{10}H_{14}Cl_6O_5Sb_2$
83.51 $C_{10}H_{14}Co_2N_8O_4$
83.52 $C_{10}H_{14}Fe_2N_8O_4$
68.8 $C_{10}H_{14}GaNO$
83.53 $C_{10}H_{14}IN_6Ni_2O_2^-$, $C_8H_{20}N^+$
31.8 $C_{10}H_{14}I_2$
64.170 $C_{10}H_{14}NO_2^+$, $C_{40}H_{36}O_5P_2$, $CF_3O_3S^-$
17.14 $C_{10}H_{14}NO_2^+$, $C_{10}H_{13}NO_2$, I_3^-, I_2
48.28 $C_{10}H_{14}N_2O_3$, H_2O
47.18 $C_{10}H_{14}N_2O_5$
47.19 $C_{10}H_{14}N_2O_6$
76.60 $2C_{10}H_{14}N_2O_6V^-$, $C_2H_{10}N_2^{2+}$, $2H_2O$
76.61 $C_{10}H_{14}N_2O_9Os$, H_2O
47.20 $C_{10}H_{14}N_4O_6$, H_2O
47.21 $2C_{10}H_{14}N_5O_4^+$, $C_{13}H_{12}N_3^+$, $1.5O_4S^{2-}$, $6.5H_2O$
47.22 $C_{10}H_{14}N_5O_8P$, $3H_2O$
47.23 $C_{10}H_{14}N_5O_{10}P_2^-$, K^+, $2H_2O$
83.54 $C_{10}H_{14}N_6Ni_2O_2$
17.15 $C_{10}H_{14}O$
31.9 $C_{10}H_{14}O_2$
38.23 $C_{10}H_{14}O_3$
17.16 $C_{10}H_{14}O_4$
60.12 $C_{10}H_{14}O_4Pd$, $C_6H_4S_4$
71.21 $C_{10}H_{14}PRhS_4$
39.29 $C_{10}H_{14}S_2$
71.22 $C_{10}H_{15}ClHg$
82.18 $C_{10}H_{15}CuN_5O_5$
82.19 $C_{10}H_{15}CuN_5O_5$, $4H_2O$
83.55 $C_{10}H_{15}NO_3Os$
73.21 $C_{10}H_{15}N_2O_6Rh$
37.6 $C_{10}H_{15}N_3OS$
35.20 $C_{10}H_{15}N_3O_2$
47.24 $C_{10}H_{15}N_5O_{10}P_2$, $3H_2O$
8.12 $C_{10}H_{15}N_5S$
69.20 $C_{10}H_{15}Sn^+$, BF_4^-
82.20 $C_{10}H_{16}CdN_2O_9^{2-}$, $2Na^+$
83.56 $C_{10}H_{16}Cd_2N_{16}O_{20}$
84.25 $C_{10}H_{16}Cl_2N_2OPt$
82.21 $C_{10}H_{16}CoN_2O_4S_2^+$, ClO_4^-
76.62 $C_{10}H_{16}CrN_2O_4$
83.57 $0.5C_{10}H_{16}CuN_2O_{10}S_2$, $0.5C_{10}H_{16}N_2O_{10}S_2Zn$
3.43 $C_{10}H_{16}N^+$, $C_{12}HFe_3NiO_{12}^-$
3.44 $2C_{10}H_{16}N^+$, $C_{16}Fe_4O_{16}Pd^{2-}$
3.45 $2C_{10}H_{16}N^+$, $C_{16}Fe_4O_{16}Pt^{2-}$
85.78 $3C_{10}H_{16}N^+$, $C_{24}H_{60}Fe_7Mo_2S_{20}^{3-}$
64.40 $C_{10}H_{16}NOP$, $C_{10}H_{17}NOP^+$, Cl^-
60.13 $C_{10}H_{16}N_2$, C_6F_6
80.4 $C_{10}H_{16}N_2NiO_2S_4$
31.10 $C_{10}H_{16}N_2O_2$

35.21 $C_{10}H_{16}N_2O_2$
41.32 $C_{10}H_{16}N_2O_2S$
39.30 $C_{10}H_{16}N_2O_3Se$
38.24 $C_{10}H_{16}N_2O_4$
48.29 $C_{10}H_{16}N_2O_4$, H_2O
39.31 $C_{10}H_{16}N_2O_4S$
48.30 $C_{10}H_{16}N_2O_5$, $3H_2O$
39.32 $C_{10}H_{16}N_2O_5S$
83.58 $C_{10}H_{16}N_2O_{10}S_2Zn$
83.57 $0.5C_{10}H_{16}N_2O_{10}S_2Zn$, $0.5C_{10}H_{16}CuN_2O_{10}S_2$
34.6 $C_{10}H_{16}N_4O_4$
32.34 $C_{10}H_{16}N_6S$, H_2O
21.7 $C_{10}H_{16}O_2$
31.11 $C_{10}H_{16}O_2$
31.12 $C_{10}H_{16}O_2$
52.2 $C_{10}H_{16}O_2$
52.3 $C_{10}H_{16}O_2$
38.25 $C_{10}H_{16}O_5$
70.6 $C_{10}H_{17}Cl_3OTe$
45.18 $C_{10}H_{17}NO_6$
83.59 $C_{10}H_{18}CuN_6O_3^{2+}$, $2Cl^-$, $2H_2O$
83.60 $C_{10}H_{18}D_5N_4NiO_2^+$, Cl^-, D_2O
86.12 $C_{10}H_{18}FeN_3O_4P$
41.33 $C_{10}H_{18}N_2O_2S$
40.17 $C_{10}H_{18}N_2O_4$
80.5 $C_{10}H_{18}N_4O_2PdS_4$
83.61 $C_{10}H_{18}N_7O_3Pt^+$, ClO_4^-, $3H_2O$
38.26 $4C_{10}H_{18}O$, $H_8Br_6O_4Sn_2$
38.27 $4C_{10}H_{18}O$, $H_8Cl_6O_4Sn_2$
27.4 $C_{10}H_{18}S_2$
10.4 $C_{10}H_{19}NO$
23.1 $C_{10}H_{19}NO$
58.1 $C_{10}H_{19}NO$
48.31 $C_{10}H_{19}NO_4$
32.35 $C_{10}H_{19}N_3S$, $CHCl_3$
80.6 $C_{10}H_{20}As_2I_6N_2NiS_4$
79.8 $C_{10}H_{20}ClCuN_4S_2$
71.23 $C_{10}H_{20}Cl_2O_2Pt$
80.7 $C_{10}H_{20}MoN_2O_2S_4$
58.2 $C_{10}H_{20}NO^+$, Cl^-
46.6 $C_{10}H_{20}NO_5P$
33.37 $C_{10}H_{20}N_2Se_4$
32.36 $C_{10}H_{20}N_6O$
64.41 $C_{10}H_{20}O_4P_2S_2$
71.24 $C_{10}H_{21}ClN_2O_2Pt$
80.8 $C_{10}H_{22}Br_2N_2O_2PdS_2$
80.9 $C_{10}H_{22}Cl_2N_2O_2PdS_2$
80.10 $C_{10}H_{22}Cl_2N_2O_2PtS_2$
83.62 $C_{10}H_{22}CoN_5O_{12}P$, $3H_2O$
76.63 $C_{10}H_{22}CoN_5S^{2+}$, ClO_4^-, Cl^-
69.21 $C_{10}H_{22}NSn^+$, ClO_4^-
76.64 $C_{10}H_{22}N_2Ni_2O_{13}$, $2H_2O$
40.18 $C_{10}H_{22}N_2O_2$, $C_7H_{10}N^+$, CNS^-
85.36 $C_{10}H_{22}N_3NiO_4S^+$, NO_3^-
83.63 $C_{10}H_{22}N_5NiO_{12}P$, $3H_2O$
68.9 $C_{10}H_{24}Al_2N_4$
83.64 $2C_{10}H_{24}ClHgN_4^+$, Cl_4Hg^{2-}
84.26 $C_{10}H_{24}CoO_7^{2+}$, $2NO_3^-$
76.65 $C_{10}H_{24}CuN_4O_6$

68.10 $C_{10}H_{24}Ga_2N_4$
68.11 $C_{10}H_{24}In_2N_4$
85.37 $C_{10}H_{24}MoN_2O_2S_2$
3.46 $2C_{10}H_{24}N^+$, $CdCl_4^{2-}$
3.47 $2C_{10}H_{24}N^+$, $CdCl_4^{2-}$
40.19 $C_{10}H_{24}N_2O_2^{2+}$, $2ClO_4^-$
11.11 $C_{10}H_{24}N_4S_2^{2+}$, Cl_6Te^{2-}
72.10 $C_{10}H_{25}CuN_3^+$, $C_{24}H_{20}B^-$
64.42 $C_{10}H_{26}N_4O_4P_2S_2$, C_2H_3N
85.38 $C_{10}H_{27}Cl_4S_5W_2$
68.12 $C_{10}H_{27}Ga_2N_3$
64.43 $C_{10}H_{27}P_3^{2+}$, $2I^-$
63.11 $C_{10}H_{28}Hg_2Si_4$
76.66 $C_{10}H_{28}N_4Zn_2$
84.27 $C_{10}H_{30}ClFeO_5S_5^{2+}$, $2Cl_4Fe^-$
84.28 $C_{10}H_{30}ClFeO_5S_5^{2+}$, $Cl_6Fe_2O^{2-}$

C_{11}

71.25 $C_{11}H_2Mn_2N_2O_{10}$
81.37 $C_{11}H_2O_{10}Os_3S_2$
84.29 $C_{11}H_4O_{11}Os_3$
38.28 $C_{11}H_5BrCl_3NO_6S$
71.26 $C_{11}H_5Fe_3NO_9$
72.11 $C_{11}H_5Fe_3NO_9$
38.29 $C_{11}H_6O_3$
38.30 $C_{11}H_6O_3$
24.4 $C_{11}H_7N_2O_2^+$, I^-, H_2O
84.30 $C_{11}H_7N_3O_4V^-$, $C_{24}H_{20}P^+$
39.33 $C_{11}H_7S_2^+$, BF_4^-
33.38 $C_{11}H_8Br_2N_2O_2$
32.37 $C_{11}H_8ClNO$
41.34 $C_{11}H_8ClN_3S$
27.5 $C_{11}H_8O_3$
38.31 $C_{11}H_9BrO_3$
32.38 $C_{11}H_9NO$
33.39 $C_{11}H_9NO_2S$
33.40 $C_{11}H_9N_3$
36.1 $C_{11}H_9N_3$
61.11 $2C_{11}H_{10}$, $C_{26}H_{28}N_6NiS_2$
71.27 $C_{11}H_{10}FeO_5$
71.28 $C_{11}H_{10}FeO_5$
71.29 $C_{11}H_{10}HgN^+$, $C_2F_3O_2^-$
1.25 $C_{11}H_{10}O_2$
42.6 $C_{11}H_{10}O_3SSe$
38.32 $C_{11}H_{10}O_4$
73.22 $C_{11}H_{11}MoO^+$, $C_8H_5MoO_3^-$
39.34 $C_{11}H_{11}NOS$
39.35 $C_{11}H_{11}NOS$
41.35 $C_{11}H_{11}NO_2S_3$
27.6 $C_{11}H_{12}Cl_2N_2$
32.39 $C_{11}H_{12}Cl_2N_2O$
41.36 $C_{11}H_{12}Cl_2N_2S$
31.13 $C_{11}H_{12}N_2O$
44.22 $C_{11}H_{12}N_3O^+$, NO_3^-
45.19 $C_{11}H_{12}N_8O_{14}$
27.7 $C_{11}H_{12}O_3$
84.31 $C_{11}H_{13}ClCuN_3O_3$, CH_4O
44.23 $C_{11}H_{13}ClN_4O_3$
32.40 $C_{11}H_{13}Cl_2N_2O^+$, Cl^-

64.44 $C_{11}H_{13}Cl_2O_5P$

40.20 $C_{11}H_{13}NOS$

40.21 $C_{11}H_{13}NOS$

33.41 $C_{11}H_{13}NO_2$

40.22 $C_{11}H_{13}NO_2S$

1.26 $C_{11}H_{13}NO_4S$

41.37 $C_{11}H_{13}NO_5S$

38.33 $C_{11}H_{13}NO_9$

32.41 $C_{11}H_{13}N_3O_2$

35.22 $C_{11}H_{13}N_3O_3$

40.23 $C_{11}H_{13}N_3O_4$

64.45 $C_{11}H_{13}OPS$

64.46 $C_{11}H_{13}O_2P$

86.13 $C_{11}H_{14}CrNO_6P$

82.22 $C_{11}H_{14}CuN_6O_4$, $4H_2O$

81.38 $C_{11}H_{14}Mo_2N_2O_{10}S_2{}^{2-}$, $2Na^+$, $3H_2O$

81.39 $C_{11}H_{14}Mo_2N_2O_{11}S^{2-}$, $2Na^+$, $4H_2O$

81.40 $C_{11}H_{14}Mo_2N_2O_{12}{}^{2-}$, $2Na^+$, $3H_2O$

47.25 $C_{11}H_{14}N_2O_6$

47.26 $C_{11}H_{14}N_2O_6$

20.4 $C_{11}H_{14}O_2S$

59.3 $C_{11}H_{14}O_5$

73.23 $C_{11}H_{15}B_2Co$

71.30 $C_{11}H_{15}Cl_2OTa$

73.24 $C_{11}H_{15}MoN_2O_2P$

31.14 $C_{11}H_{15}N$

10.5 $C_{11}H_{15}NO$

45.20 $C_{11}H_{15}NO_5S$, H_2O

44.24 $C_{11}H_{15}N_5O_3$

39.36 $C_{11}H_{15}OS^+$, $C_{11}H_{14}OS$, $ClO_4{}^-$

39.37 $C_{11}H_{15}OS^+$, $ClO_4{}^-$

64.47 $C_{11}H_{15}O_2P$

45.21 $C_{11}H_{16}O_6$

72.12 $C_{11}H_{17}Cl_2OPPt$

83.65 $C_{11}H_{17}Cl_5Hg_2N_2$

45.22 $C_{11}H_{17}NO_8$

64.48 $C_{11}H_{17}N_4O_2PS$

47.27 $C_{11}H_{17}N_6O_7P$, $2H_2O$

83.66 $C_{11}H_{18}BCuF_2IN_4O_2$

75.2 $C_{11}H_{18}B_2CrN_2O_3$

86.14 $C_{11}H_{18}CrO_5P_4Si_3$

17.17 $C_{11}H_{18}NO^+$, Br^-

35.23 $C_{11}H_{18}N_2O_3$

39.38 $C_{11}H_{18}N_2O_3S$

37.7 $C_{11}H_{18}N_4O_2$

27.8 $C_{11}H_{18}O$

27.9 $C_{11}H_{18}O_2$

38.34 $C_{11}H_{18}O_7$

73.25 $C_{11}H_{19}B_2NbOZn$, $0.5C_6H_6$

83.67 $C_{11}H_{19}BrN_4NiO_2$

83.68 $C_{11}H_{19}Cl_2CoN_4O_2$

83.69 $C_{11}H_{19}CuN_4O_2{}^+$, ClO_4^- $0.5CH_4O$

48.32 $C_{11}H_{19}NO_3$

48.33 $C_{11}H_{19}NO_4$

83.70 $C_{11}H_{20}BCuF_2N_4O_3{}^-$, ClO_4^-

80.11 $C_{11}H_{20}FeN_3S_5$

48.34 $C_{11}H_{20}N_2O_5$

61.1 $3C_{11}H_{20}O_2$, $C_4H_8O_2$

83.71 $C_{11}H_{21}CuN_4O_3{}^-$, ClO_4^- H_2O

23.2 $C_{11}H_{21}NO$

39.39 $C_{11}H_{21}NO_2S$, $0.5H_2O$

76.67 $C_{11}H_{21}N_7OPt^{2+}$, $2F_6P^-$, $1.5H_2O$

39.40 $C_{11}H_{21}S^+$, BF_4^-

86.15 $C_{11}H_{22}ClNiPS_2$

58.3 $C_{11}H_{22}NO^+$, I^-

64.49 $C_{11}H_{22}NO_2P$

48.35 $C_{11}H_{22}N_2O_2S$

86.16 $C_{11}H_{23}ClCoN_4O_7P$

76.68 $C_{11}H_{23}ClCoN_5$, Cl_4Zn

72.13 $C_{11}H_{23}ClMoO_8P_2$

71.31 $C_{11}H_{23}ClN_2Zn$

84.32 $C_{11}H_{23}FeGaN_5O_3$

84.33 $C_{11}H_{23}GaN_4NiO_2$

71.32 $C_{11}H_{25}Cl_2OPW$

76.69 $C_{11}H_{25}CoN_4O_2{}^{2+}$, $2Cl^-$, $2H_2O$

71.33 $C_{11}H_{31}CoP_3{}^-$, $C_{11}H_{31}CoP_3$, K^+

C_{12}

19.5 $C_{12}Br_2F_8$

38.35 $C_{12}F_8O_2$

39.41 $C_{12}F_8S_2$

11.12 $C_{12}F_8Se_2$

60.16 $C_{12}F_{10}$, $C_{12}H_9Br$

60.18 $C_{12}F_{10}$, $C_{12}H_{10}$

60.23 $C_{12}F_{10}$, $C_{13}H_{12}$

15.2 $C_{12}H_2F_6N_2O_4S$

71.34 $C_{12}H_2F_8Hg$

14.5 $C_{12}H_2O_{12}{}^{4-}$, $2Ca^{2+}$, $9H_2O$

26.1 $C_{12}H_4Cl_4O_2$, $C_{14}H_{10}$

3.11 $C_{12}H_4N_4$, $C_3H_{10}N^+$, $0.33I_3{}^{3-}$

60.8 $C_{12}H_4N_4$, $C_9H_6S_2Se$

36.2 $C_{12}H_4N_4$, $0.4C_{12}H_8N_2$, $0.6C_{14}H_{14}N_2$

60.14 $C_{12}H_4N_4$, $C_{13}H_9N$

60.32 $C_{12}H_4N_4$, $C_{14}H_{12}$

60.33 $C_{12}H_4N_4$, $C_{16}H_8S_3$

60.38 $C_{12}H_4N_4$, $C_{18}H_{20}O_4$

60.40 $C_{12}H_4N_4$, $2C_{20}H_{14}CuN_2O_2$

60.7 $C_{12}H_4N_4{}^-$, $C_8H_{20}N^+$, $C_{12}H_4N_4$

60.20 $C_{12}H_4N_4{}^-$, $C_{12}H_{14}Fe^+$, $C_{12}H_4N_4$

60.41 $C_{12}H_4N_4{}^-$, $C_{20}H_{20}P^+$, $C_{12}H_4N_4$

60.42 $C_{12}H_4N_4{}^-$, $C_{20}H_{30}Fe^+$

60.21 $2C_{12}H_4N_4{}^-$, $C_{12}H_{30}N_2{}^{2+}$, $2C_{12}H_4N_4$

78.13 $2C_{12}H_4N_4$, $C_{28}H_{28}CuN_2O_2$

15.3 $C_{12}H_4N_7O_{12}{}^-$, Ag^+

16.3 $C_{12}H_4N_7O_{12}{}^-$, K^+

73.26 $C_{12}H_6BrMnO_3$

17.18 $C_{12}H_6Cl_2INO_3$

17.19 $C_{12}H_6Cl_4O$

41.38 $C_{12}H_6N_2S$

17.20 $C_{12}H_6N_4O_{10}$

84.34 $2C_{12}H_6N_9O_{12}Ru^-$, Ba^{2+}, $9H_2O$

35.24 $C_{12}H_7N_5O_4$

38.36 $C_{12}H_8ClNO_5S$

83.72 $C_{12}H_8Cl_2MoN_2O_2$

68.13 $C_{12}H_8Cl_3N_4O_2Tl$

37.8 $C_{12}H_8Cl_5N$

31.15 $C_{12}H_8F_6$

75.3 $C_{12}H_8FeO_4$

84.35 $C_{12}H_8Mo_2O_9{}^{2-}$, $2H_4N^-$, $2H_2O$

36.2 $0.4C_{12}H_8N_2$, $0.6C_{14}H_{14}N_2$, $C_{12}H_4N_4$

60.15 $2C_{12}H_8N_2$, C_6N_4

36.3 $C_{12}H_8N_2O_4$, $2C_2H_4O_2$

70.7 $C_{12}H_8O_2S_2Te$

38.37 $C_{12}H_8O_3$

72.14 $C_{12}H_8O_9Os_3S$

19.6 $C_{12}H_9Br$

60.16 $C_{12}H_9Br$, $C_{12}F_{10}$

72.15 $C_{12}H_9ClFeO_4$

60.17 $C_{12}H_9NS$, $C_{10}H_2O_6$

86.17 $C_{12}H_9N_2O_{14}Os_3P$

83.73 $C_{12}H_9N_3O_{11}Os_3$

60.18 $C_{12}H_{10}$, $C_{12}F_{10}$

65.6 $C_{12}H_{10}As_2N_4S_2$

72.16 $C_{12}H_{10}FeO_4$

73.27 $C_{12}H_{10}FeO_4$

73.28 $C_{12}H_{10}FeO_4$

71.35 $C_{12}H_{10}FeO_6$

73.29 $C_{12}H_{10}HfO_2$

33.42 $C_{12}H_{10}N_2$

33.43 $C_{12}H_{10}N_2$

9.2 $C_{12}H_{10}N_2$

33.44 $C_{12}H_{10}N_2O$

41.39 $C_{12}H_{10}N_2S_3$

64.50 $C_{12}H_{10}N_3PS_2$

73.30 $C_{12}H_{10}Ni_2O_2$

17.21 $C_{12}H_{10}O_2$

76.43 $C_{12}H_{10}O_2P^-$, $C_7H_{16}N_4O_3Re^+$

84.36 $C_{12}H_{11}CuN_3O_5$, H_2O

16.4 $C_{12}H_{11}N$, Cl_3Sb

32.42 $C_{12}H_{11}NO_2$

39.42 $C_{12}H_{11}NO_2S$

60.19 $C_{12}H_{12}$, $C_{10}F_8$

83.74 $C_{12}H_{12}AgN_8O_2{}^+$, ClO_4^-, H_2O

86.18 $C_{12}H_{12}As_2FeMnO_8{}^+$, $C_{10}H_6AsClFeMnO_8^-$

32.43 $C_{12}H_{12}BrNO_3$

31.16 $C_{12}H_{12}BrNO_5S$

77.1 $C_{12}H_{12}CuF_6N_2O_2$

81.41 $C_{12}H_{12}CuN_2O_4$, H_2O

83.75 $C_{12}H_{12}CuN_6O_6S$

81.42 $C_{12}H_{12}F_{12}O_{10}Rh_2$

81.43 $C_{12}H_{12}F_{12}O_{10}Rh_2S_2$

73.31 $C_{12}H_{12}MnNO_4$

36.4 $C_{12}H_{12}N_2O_2$, H_2O

35.25 $C_{12}H_{12}N_2O_3$

41.40 $C_{12}H_{12}N_4S_2$

83.76 $(C_{12}H_{12}N_8Zn_2)_n$

38.38 $C_{12}H_{12}O_3$

38.39 $C_{12}H_{12}O_4$

38.40 $C_{12}H_{12}O_5$

72.17 $C_{12}H_{12}O_8W$

81.44 $C_{12}H_{12}O_{12}Pd_4$, $2C_2H_4O_2$

71.36 $C_{12}H_{13}Br_3F_8N_2O_2P_4W_2$

71.37 $C_{12}H_{13}HgN_2{}^+$, NO_3

35.26 $C_{12}H_{13}NO_2$

16.5 $C_{12}H_{13}N_4{}^+$, Cl^- $2H_2O$

40.24 $C_{12}H_{14}BrN_2O^-$, Br^-

83.77 $C_{12}H_{14}Cl_2MnN_6O_2$

71.38 $C_{12}H_{14}CoF_8O_5P$

70.8 $C_{12}H_{14}CrN_2O_5Te$

74.1 $C_{12}H_{14}CrO_3Si$

63.12 $C_{12}H_{14}FNO_4Si$

60.20 $C_{12}H_{14}Fe^-$, $C_{12}H_4N_4^-$, $C_{12}H_4N_4$

71.39 $C_{12}H_{14}FeO_5$

71.40 $C_{12}H_{14}FeO_5$

82.23 $C_{12}H_{14}Mo_2N_2O_{17}{}^{2-}$, $2Na^+$, $8H_2O$

73.32 $C_{12}H_{14}Mo_2S_4$

47.28 $C_{12}H_{14}N_2O_4S$

47.29 $C_{12}H_{14}N_2O_6$

38.41 $C_{12}H_{14}O_6$

71.41 $C_{12}H_{15}BF_2FeO_3$

73.33 $C_{12}H_{15}ClOTi$

73.34 $C_{12}H_{15}CoO_2$

73.35 $C_{12}H_{15}CrNO_3$

78.1 $C_{12}H_{15}CuNO_4$

73.36 $C_{12}H_{15}MoNO_3$

73.37 $C_{12}H_{15}NO_3W$

38.42 $C_{12}H_{15}NO_7$

71.42 $C_{12}H_{15}Ta$

30.1 $C_{12}H_{16}$

73.38 $C_{12}H_{16}As_2Fe$

32.44 $C_{12}H_{16}BrNO_2S$

11.13 $C_{12}H_{16}ClNO_4S$

64.51 $C_{12}H_{16}ClO_2P$

84.37 $C_{12}H_{16}Cl_2CoN_4O_2$

33.45 $C_{12}H_{16}Cl_2N_5O^+$, Cl^-, $0.29H_2O$

82.24 $C_{12}H_{16}CoN_6O_4{}^+$, $ClO_4{}^-$, $2H_2O$

81.45 $(C_{12}H_{16}Cr_2N_2O_8)_n$

81.46 $C_{12}H_{16}CuN_2O_6$

83.78 $C_{12}H_{16}CuN_6O_2$

75.4 $C_{12}H_{16}FeP_2$

73.39 $C_{12}H_{16}I_2Mo_2N_4O_2$

41.41 $C_{12}H_{16}N_2S$

76.70 $C_{12}H_{16}N_4NiO_4$

40.25 $C_{12}H_{16}N_4OS$

27.10 $C_{12}H_{16}O_2$

39.43 $C_{12}H_{16}S_3$

72.18 $C_{12}H_{17}Cl_2O_2PPt$

1.27 $C_{12}H_{17}NO$

1.28 $C_{12}H_{17}NO_4$

1.29 $C_{12}H_{17}N_3O_2$

64.52 $C_{12}H_{17}O_6P$

19.7 $C_{12}H_{18}$

69.22 $C_{12}H_{18}Cl_2O_6Sn$

11.14 $C_{12}H_{18}Cl_4O_2S_3$

73.40 $C_{12}H_{18}Cl_6Nb_2O_3$

76.71 $C_{12}H_{18}CuN_2S_2$

83.79 $C_{12}H_{18}FeN_6{}^{2+}$, $2Cl_4Fe^-$

83.80 $C_{12}H_{18}FeN_6{}^{2+}$, $2Cl_4Fe^-$

85.39 $C_{12}H_{18}FeN_6O_2S_4{}^-$, $C_6H_8N^+$, $4H_2O$

33.46 $2C_{12}H_{18}N^+$, Cl_4Co^{2-}

33.47 $2C_{12}H_{18}N^+$, $Cl_6Cu_2{}^{2-}$

1.30 $C_{12}H_{18}N_2O_2S$

41.42 $C_{12}H_{18}N_2O_3S$

48.36 $C_{12}H_{18}N_2O_5$

85.40 $C_{12}H_{18}N_2S_2Zn$

44.25 $C_{12}H_{18}N_4OS^{2+}$, $2Cl^-$, $0.5H_2O$

34.7 $C_{12}H_{18}N_4O_2$

38.43 $C_{12}H_{18}O_2$

45.23 $C_{12}H_{18}O_6$

38.44 $C_{12}H_{18}O_8$

58.4 $C_{12}H_{19}NO_5$

81.47 $(C_{12}H_{20}ClO_6Ru_2)_n$

83.81 $C_{12}H_{20}Cl_2N_4NiO_4$

70.9 $C_{12}H_{20}Cl_4Te$

82.25 $C_{12}H_{20}N_2NiO_{11}P$, $2H_2O$

1.31 $C_{12}H_{20}N_2O_2$

39.44 $C_{12}H_{20}N_2O_3S$

32.45 $C_{12}H_{20}N_4$

34.8 $C_{12}H_{20}N_4O_4$

39.45 $C_{12}H_{20}O_2S_2$

69.23 $C_{12}H_{21}AsN_2O_4Sn$

86.19 $C_{12}H_{21}MoO_3P_3$

76.72 $C_{12}H_{22}Cr_2N_4O_{10}$, $4H_2O$

45.24 $C_{12}H_{22}O_{11}$

45.25 $C_{12}H_{22}O_{11}$

45.26 $C_{12}H_{22}O_{11}$

23.3 $C_{12}H_{23}NO$

64.53 $C_{12}H_{23}O_2P$

64.54 $C_{12}H_{23}O_4P$

65.7 $C_{13}H_{24}AsN_3O_3$

70.10 $C_{12}H_{24}BrN_3O_3S_6Te$

83.82 $C_{12}H_{24}Cl_2N_4Pd$

84.38 $C_{12}H_{24}Cu_4O_{12}$

86.20 $C_{12}H_{24}Fe_2O_6P_2S_3$, $0.5C_4H_{10}O$

1.32 $C_{12}H_{24}N_2O_4$

63.13 $C_{12}H_{24}N_2Si_2$

84.39 $C_{12}H_{24}N_3NdO_{15}$

84.40 $C_{12}H_{24}N_3NdO_{15}$

70.11 $C_{12}H_{24}N_3O_3PTe$

61.2 $C_{12}H_{24}N_9P_3$, $3CCl_4$

61.3 $2C_{12}H_{24}N_9P_3$, C_6H_6

84.41 $C_{12}H_{24}O_5Os$

42.7 $C_{12}H_{24}O_5S$

42.8 $C_{12}H_{24}O_5S$, Ag^+, $NO_3{}^-$

42.9 $C_{12}H_{24}O_5S$, CNS^-, K^+

42.10 $C_{12}H_{24}O_5S$, CNS^-, Na^+

42.11 $C_{12}H_{24}O_5S$, CNS^-, Rb^+

38.45 $C_{12}H_{24}O_6$

38.46 $C_{12}H_{24}O_6$, $C_2N_2P^-$, Na^+, C_4H_8O

38.47 $C_{12}H_{24}O_6$, $H_{12}CoO_6{}^{2+}$, Cl_4Co^{2-}, C_3H_6O

38.48 $C_{12}H_{24}O_6$, $H_{12}MnO_6{}^{2+}$, $2ClO_4{}^-$

81.48 $C_{12}H_{24}O_{10}Rh_2S_2$

81.49 $C_{12}H_{24}O_{17}W_3{}^{2+}$, $2CF_3O_3S^-$

33.48 $C_{12}H_{25}N_2O^+$, $ClO_4{}^-$

83.83 $C_{12}H_{26}Br_2N_2Pt$

85.41 $(C_{12}H_{26}CdN_2S_2{}^{2+})_n$, $2nClO_4{}^-$, $2nH_2O$

76.73 $(C_{12}H_{26}Cd_2N_{10}S_4)_n$

85.42 $C_{12}H_{26}Cl_2LaO_{13}S_2{}^+$, $ClO_4{}^-$

83.84 $C_{12}H_{26}Cl_2N_2Pt$, $2C_6H_{18}N_3OP$

85.43 $C_{12}H_{26}Cl_4S_5W_2$

81.50 $C_{12}H_{26}Gd_2O_{16}$, $4H_2O$

85.44 $C_{12}H_{26}HgN_2S_2{}^{2+}$, $2ClO_4{}^-$

33.49 $C_{12}H_{26}N_2O_2{}^{2+}$, $2ClO_4{}^-$

33.50 $C_{12}H_{26}N_2S_2{}^{2+}$, Cl_4Cu^{2-}

69.24 $C_{12}H_{26}N_2S_4Sn$

69.25 $C_{12}H_{26}N_2S_4Sn$

84.42 $C_{12}H_{26}N_6O_4Pt_2{}^{2+}$, $2NO_3{}^-$

81.51 $C_{12}H_{26}O_{16}Y_2$, $4H_2O$

68.14 $C_{12}H_{28}Al_3Cl_5O_4$

84.43 $C_{12}H_{28}Cl_2FeN_4O_4{}^+$, Cl_4Fe^-

86.21 $C_{12}H_{28}Cl_2NPPt$

86.22 $C_{12}H_{28}Cl_2NPPt$

84.44 $C_{12}H_{28}Cl_4Cu_2N_4O_4$

85.45 $C_{12}H_{28}Cu_2N_4O_{16}S_6$

76.74 $C_{12}H_{28}N_4Ni_2O_{12}$

76.75 $C_{12}H_{29}IN_3Pt^+$, I^-

86.23 $C_{12}H_{30}Cl_2F_4O_2P_4Pt_2$

86.24 $C_{12}H_{30}MnN_3O_4P_2$

60.21 $C_{12}H_{30}N_2{}^{2+}$, $2C_{12}H_4N_4^-$, $2C_{12}H_4N_4$

76.76 $C_{12}H_{30}N_8Pt^{4+}$, $2O_6S_2{}^{2-}$, $2.5H_2O$

66.12 $C_{12}H_{30}O_6P_3S_6Sb$

86.25 $C_{12}H_{30}P_2PtS_4W$

86.26 $C_{12}H_{32}Cl_4P_4W_2$, C_7H_8

76.77 $C_{12}H_{32}CuN_4{}^{2+}$, $2ClO_4{}^-$

76.78 $C_{12}H_{32}CuN_4{}^{2+}$, $2ClO_4{}^-$

76.79 $C_{12}H_{32}N_4NiO_2{}^{2+}$, $2Cl^-$

85.46 $C_{12}H_{32}N_4Ni_3S_4{}^{2+}$, $2Cl^-$

82.26 $C_{12}H_{32}N_4O_{10}Rh_2{}^{4+}$, $4ClO_4{}^-$, $2H_2O$

70.12 $C_{12}H_{32}N_8S_4Te^{2+}$, $2Cl^-$

76.80 $C_{12}H_{36}AgCo_2N_8O_4S_2{}^{3+}$, $3ClO_4{}^-$

65.8 $C_{12}H_{36}As_4Si_6$

71.43 $C_{12}H_{36}B_2NiP_4$

83.85 $C_{12}H_{36}Cl_3N_2Si_4Ta$

86.27 $C_{12}H_{36}Cl_4P_4W_2$

86.28 $C_{12}H_{36}CoN_3P_3{}^{3+}$, $3Br^-$, $3H_2O$

79.9 $C_{12}H_{36}CoN_{12}O_6{}^{2+}$, $O_3S_2{}^{2-}$

79.10 $C_{12}H_{36}CoN_{12}O_6{}^{2+}$, O_4S^{2-}

76.81 $C_{12}H_{36}N_6O_4Rh_2{}^{4+}$, $4ClO_4{}^-$, $4H_2O$

64.55 $C_{12}H_{36}N_7O_3P_3$, H_2O

86.29 $C_{12}H_{36}P_4Rh^+$, Cl^-

63.14 $C_{12}H_{39}N_5Si_6$

86.30 $C_{12}H_{41}BMoP_4$

7.3 $C_{12}N_6$

85.47 $C_{12}N_6NiSe_6{}^{2-}$, $2C_{24}H_{20}As^+$

C_{13}

71.44 $C_{13}H_2Fe_4O_{12}$

83.86 $C_{13}H_3Mn_3N_2O_{12}$

13.7 $C_{13}H_4I_4O_4{}^{2-}$, $2C_4H_{12}NO_2{}^+$

71.45 $C_{13}H_6MnO_7{}^-$, $C_{36}H_{30}NP_2{}^+$

35.27 $C_{13}H_7Br_2N_3O$

72.19 $C_{13}H_7Fe_3NO_9$

71.46 $C_{13}H_7NO_{11}Os_4$

71.47 $C_{13}H_7NO_{11}Ru_3$

71.48 $C_{13}H_8Cl_2HgMoN_2O_3$

39.46 $C_{13}H_8Cl_2O_4S$

73.41 $C_{13}H_8CrN_2O_7W$

30.2 $C_{13}H_8F_6$

71.49 $C_{13}H_8Fe_2O_7Se$

28.1 $C_{13}H_8O_2$

71.50 $C_{13}H_8O_8W_2$

16.6 $C_{13}H_9BrClN$

16.7 $C_{13}H_9BrClN$

41.43 $C_{13}H_9BrNS^+$, $BF_4{}^-$

60.14 $C_{13}H_9N$, $C_{12}H_4N_4$

36.5 $C_{13}H_9N_3$

64.56 $C_{13}H_{10}BrCl_2O_2PS$

83.87 $C_{13}H_{10}Cl_2Hg_2N_4O$

84.45 $C_{13}H_{10}FMn_3O_{11}$

71.51 $C_{13}H_{10}F_4FeO_4$

75.5 $C_{13}H_{10}FeO_4$

84.46 $C_{13}H_{10}IMn_3O_{11}$

32.46 $C_{13}H_{10}N_4O$

32.47 $C_{13}H_{10}N_4S$

38.49 $C_{13}H_{10}O_3$

40.26 $C_{13}H_{11}ClN_2O_2$

40.27 $C_{13}H_{11}ClN_2O_2$

31.17 $C_{13}H_{11}ClN_2O_4S$

84.47 $C_{13}H_{11}Cl_3MoN_3O_3$, C_2H_3N

36.6 $C_{13}H_{11}N$

39.47 $C_{13}H_{11}NO_3S_2$

38.50 $C_{13}H_{11}NO_5S$

84.48 $C_{13}H_{11}NO_5SW$

36.7 $C_{13}H_{11}N_2^+$, Br^-

36.8 $C_{13}H_{11}N_2^+$, Cl^-. $2H_2O$

60.22 $4C_{13}H_{11}N_2^+$, $4C_{19}H_{23}IN_8O_{12}P^-$, $21H_2O$

60.39 $C_{13}H_{11}N_3$, $C_{19}H_{26}N_8O_{11}P$. $10H_2O$

47.38 $2C_{13}H_{11}N_3$, $2C_{19}H_{25}N_8O_{10}P$, xH_2O

44.26 $C_{13}H_{11}N_5O_2$

60.23 $C_{13}H_{12}$, $C_{12}F_{10}$

31.18 $C_{13}H_{12}ClNO_2S$

64.57 $C_{13}H_{12}ClOP$

33.51 $C_{13}H_{12}N_2O_2$

47.21 $C_{13}H_{12}N_3^+$, $2C_{10}H_{14}N_5O_4^+$, $1.5O_4S^{2-}$, $6.5H_2O$

60.24 $1.5C_{13}H_{12}N_3^+$, $C_{19}H_{24}N_8O_{12}P^-$, $0.5O_4S^-$, $11.5H_2O$

47.36 $2C_{13}H_{12}N_3^+$, $2C_{19}H_{23}IN_8O_{12}P^-$, $15H_2O$, CH_4O

8.13 $C_{13}H_{12}N_4O$

38.51 $C_{13}H_{12}O_2$

71.52 $C_{13}H_{12}O_2Rh_2$

25.2 $C_{13}H_{12}O_6$

65.9 $C_{13}H_{13}AsN_2O_2$

75.6 $C_{13}H_{13}CrO_3S_2^-$, Li^+, $3C_4H_8O_2$

33.52 $C_{13}H_{13}N_2^+$, I^-

8.14 $C_{13}H_{13}N_3$

38.52 $C_{13}H_{13}N_3O_3$

33.53 $C_{13}H_{13}N_5S$

64.58 $C_{13}H_{13}O_2PS$

40.28 $C_{13}H_{14}ClNO_2$

63.15 $C_{13}H_{14}F_3NO_4Si$

72.20 $C_{13}H_{14}FeO_5$

71.53 $C_{13}H_{14}HgN^+$, NO_3^-

73.42 $C_{13}H_{15}Mo_2NO_5$

20.5 $C_{13}H_{15}NO_4S$

48.37 $C_{13}H_{15}N_3O_3$

48.38 $C_{13}H_{15}N_3O_3$, $0.5C_6H_6$

64.59 $C_{13}H_{15}OP$

32.48 $C_{13}H_{16}BrN_3O$

44.27 $C_{13}H_{16}N_2O_3S$

44.28 $C_{13}H_{16}N_2O_4$

44.29 $C_{13}H_{16}N_2O_4$

45.27 $C_{13}H_{16}N_2O_5$

41.44 $C_{13}H_{16}N_2S_2$

41.45 $C_{13}H_{16}N_4O_2S$

40.29 $C_{13}H_{16}N_4O_6$

37.9 $C_{13}H_{17}BrCl_3F_3O_4$

70.13 $C_{13}H_{17}Cl_3Te$

32.49 $C_{13}H_{17}N_2O_2$

61.4 $C_{13}H_{17}N_4O_6^+$, $C_2H_4O_2$. Cl^-

61.5 $C_{13}H_{17}N_4O_6^+$, $C_3H_6O_2$, Cl^-

40.30 $C_{13}H_{17}N_4O_6^+$, Cl^-. $2H_2O$

86.31 $C_{13}H_{18}As_3ClMn_2O_7$

33.54 $C_{13}H_{18}ClN_2O^+$, Cl^-

33.55 $C_{13}H_{18}ClN_2O^+$, Cl^-

33.56 $C_{13}H_{18}ClN_2O^+$, Cl^-

33.57 $C_{13}H_{18}ClN_2O^+$, Cl^-, H_2O

84.49 $C_{13}H_{18}CoN_2O_7^-$, Na^+, $3H_2O$

41.46 $C_{13}H_{18}N_2S$

41.47 $C_{13}H_{18}N_2S$

31.19 $C_{13}H_{18}O_2$

38.53 $C_{13}H_{18}O_3$

38.54 $C_{13}H_{18}O_3$

48.39 $C_{13}H_{19}ClNO_2^+$, Cl^-, H_2O

73.43 $C_{13}H_{19}ClSiTi$

64.60 $C_{13}H_{19}Cl_2N_2O_2P$

72.21 $C_{13}H_{19}Cl_2Nb$

41.48 $C_{13}H_{19}NO_2S_2$

33.58 $C_{13}H_{19}NO_4$

73.44 $C_{13}H_{19}NSiV$

48.40 $C_{13}H_{19}N_7O_5S$, $2H_2O$

31.20 $C_{13}H_{20}BrCl_3O_4$

59.4 $C_{13}H_{20}N_2O_5$, H_2O

39.48 $C_{13}H_{20}N_2O_5S$

32.50 $C_{13}H_{20}N_4O_2$

35.28 $C_{13}H_{20}N_6O_2$

38.55 $C_{13}H_{20}O_4$

31.21 $C_{13}H_{21}NOS$

63.16 $C_{13}H_{21}NO_2Si$

83.88 $C_{13}H_{21}N_5NiO_4$

48.41 $C_{13}H_{22}N_2O_5$

13.8 $C_{13}H_{22}N_3O^+$, Cl^-

48.42 $C_{13}H_{23}N_3O_4$, H_2O

71.54 $C_{13}H_{24}B_7Co$

84.50 $C_{13}H_{24}CrN_2O_4^+$, I^-, H_2O

72.22 $C_{13}H_{24}GaMoN_3O_3S$

45.28 $C_{13}H_{24}O_6$

45.29 $C_{13}H_{24}O_{11}$

23.4 $C_{13}H_{25}NO$

77.2 $C_{13}H_{29}N_2O_2Pd^+$, $C_5H_7O_2^-$

71.55 $C_{13}H_{33}ClP_2Pt$

64.61 $C_{13}H_{36}N_3PSi_3$

C_{14}

71.56 $C_{14}H_2BrF_6O_{10}Os_3^-$, $C_{36}H_{30}NP_2^+$

71.57 $C_{14}H_2O_{12}Os_4$

71.58 $C_{14}H_3Fe_4O_{13}^-$, $C_{36}H_{30}NP_2^+$

71.59 $C_{14}H_3Fe_4O_{13}^-$, $C_{36}H_{30}NP_2^+$

84.51 $C_{14}H_4Fe_4O_{13}$

84.52 $C_{14}H_4Fe_4O_{13}$

71.60 $C_{14}H_5ClCrO_7W$

81.52 $C_{14}H_6CrN_2O_8^-$, Rb^+

81.53 $C_{14}H_6FMo_2N_2O_{14}^-$, $C_8H_{20}N^+$

50.6 $C_{14}H_7Cl_5N_2O_6$

39.49 $C_{14}H_7N_4O_3S^-$, Rb^+

13.9 $C_{14}H_8BrIO_4$

40.31 $C_{14}H_8Cl_5NO$

58.5 $C_{14}H_8N_2O_2$, H_2O

20.6 $C_{14}H_8N_4O$

71.61 $C_{14}H_8O_9W_2$

85.2 $C_{14}H_8S_4^+$, $C_2H_6Br_5PtS^-$

39.50 $C_{14}H_8S_4^+$, I_3^-

39.51 $2C_{14}H_8S_4^+$, $C_{14}H_8S_4$, Br_6Sn^{2-}

39.52 $6C_{14}H_8S_4^+$, $2C_{14}H_8S_4$, $3Cl_6Sn^{2-}$

24.5 $C_{14}H_9Cl_3O_4$

40.32 $C_{14}H_9Cl_4NO$

28.2 $C_{14}H_{10}$

28.3 $C_{14}H_{10}$

28.4 $C_{14}H_{10}$

28.5 $C_{14}H_{10}$

60.25 $C_{14}H_{10}$, $C_{10}H_2N_4$

60.26 $C_{14}H_{10}$, $C_{10}H_2N_4$

60.27 $C_{14}H_{10}$, $C_{10}H_2N_4$

60.28 $C_{14}H_{10}$, $C_{10}H_2N_4$

60.29 $C_{14}H_{10}$, $C_{10}H_2N_4$

60.30 $C_{14}H_{10}$, $C_{10}H_2N_4$

60.31 $C_{14}H_{10}$, $C_{10}H_2N_4$

26.1 $C_{14}H_{10}$, $C_{12}H_4Cl_4O_2$

17.22 $C_{14}H_{10}Cl_2N_2O_6$

83.89 $C_{14}H_{10}Cl_3N_2OV$

73.45 $C_{14}H_{10}Cr_2O_4S$

73.46 $C_{14}H_{10}Fe_2O_3S$

85.48 $C_{14}H_{10}MoOS_5$

40.33 $C_{14}H_{10}N_2O_2$, $0.25C_6H_6$

64.62 $C_{14}H_{10}O_2P^-$, $C_4H_{10}O_2$, Li^+

26.2 $C_{14}H_{10}O_3$

28.6 $C_{14}H_{10}O_4$

84.53 $C_{14}H_{11}ClN_2OPt$, $CHCl_3$

83.90 $C_{14}H_{11}FeNO_4$

71.62 $C_{14}H_{11}FeO_2^+$, F_6P^-

36.9 $C_{14}H_{11}NO$

13.10 $C_{14}H_{11}NO_5$

35.29 $C_{14}H_{11}N_3O$

41.49 $C_{14}H_{11}N_3OS$

29.2 $C_{14}H_{12}$

60.32 $C_{14}H_{12}$, $C_{12}H_4N_4$

61.14 $C_{14}H_{12}$, $C_{33}H_{36}O_6$

61.17 $C_{14}H_{12}$, $2C_{33}H_{36}O_6$

16.8 $C_{14}H_{12}BrN$

16.9 $C_{14}H_{12}BrN$

19.8 $C_{14}H_{12}Br_2O$

16.10 $C_{14}H_{12}ClN$

16.11 $C_{14}H_{12}ClN$

29.3 $C_{14}H_{12}Cl_4$

24.6 $C_{14}H_{12}F_3NO$

71.63 $C_{14}H_{12}Fe_2N_2NiO_{10}$

75.7 $C_{14}H_{12}Fe_2O_6Se$

71.64 $C_{14}H_{12}Fe_2O_8$

83.91 $(C_{14}H_{12}HgN_2O_2)_n$

33.59 $C_{14}H_{12}N_2O$

33.60 $C_{14}H_{12}N_2O$

69.26 $C_{14}H_{12}N_2O_4Pb$

81.54 $(C_{14}H_{12}N_2O_4Zn)_n$, $1.5nH_2O$

41.50 $C_{14}H_{12}N_2O_5S_2$

59.5 $C_{14}H_{12}O_5$

31.22	$C_{14}H_{13}ClN_2O_4S$
82.27	$C_{14}H_{13}CuN_3O_4$, $6H_2O$
36.10	$C_{14}H_{13}N$
38.56	$C_{14}H_{13}N_3O_4$
36.11	$C_{14}H_{13}N_4O_4$, $C_8H_{11}NO$, $4H_2O$
83.92	$C_{14}H_{13}N_5NiO_2$
66.13	$C_{14}H_{13}O_2Sb$
81.55	$(C_{14}H_{14}CdN_2O_5)_n$, $2nH_2O$
36.12	$C_{14}H_{14}ClN_3$, H^+, Cl^-, $2H_2O$
15.4	$C_{14}H_{14}ClN_3O_6$
83.93	$C_{14}H_{14}Cl_2N_2Zn$
69.27	$C_{14}H_{14}Cl_4N_4O_3Sn$, $2H_2O$
81.56	$C_{14}H_{14}Cu_2N_2O_8$
69.28	$C_{14}H_{14}GeS$
71.65	$C_{14}H_{14}Hg$
71.66	$C_{14}H_{14}HgN_4S$
64.63	$C_{14}H_{14}NO_4PS$
36.13	$C_{14}H_{14}N_2$
36.2	$0.6C_{14}H_{14}N_2$, $C_{12}H_4N_4$, $0.4C_{12}H_8N_2$
69.29	$C_{14}H_{14}N_2O_{12}Pb_2$, $2C_7H_5NO_4$, H_2O
77.3	$C_{14}H_{14}N_4NiO_4$
79.11	$C_{14}H_{14}N_8NiS_2$
73.47	$C_{14}H_{14}Ni_2O_2$
27.11	$C_{14}H_{14}O_3$
38.57	$C_{14}H_{14}O_3$
11.15	$C_{14}H_{14}O_4S_2$
73.48	$C_{14}H_{15}FeO_3^+$, F_6P^-
24.7	$C_{14}H_{15}NO$
24.8	$C_{14}H_{15}NO$
33.61	$C_{14}H_{15}NO_2S$
64.64	$C_{14}H_{15}O_2P$
64.65	$C_{14}H_{15}O_4P$
84.54	$C_{14}H_{16}Cl_2Mn_2O_8$
83.94	$C_{14}H_{16}CoN_6O_4S_2$
84.55	$C_{14}H_{16}CuN_4O_7$
73.49	$C_{14}H_{16}FeS_3$
81.57	$(C_{14}H_{16}Mn_2O_{16}{}^{2-})_n$, $2nK^+$
33.62	$C_{14}H_{16}N_2{}^{2+}$, $2I^-$
33.63	$C_{14}H_{16}N_2{}^{2+}$, $2I^-$
71.67	$C_{14}H_{16}N_2Ni$
35.30	$C_{14}H_{16}N_2O$
35.31	$C_{14}H_{16}N_2O$
41.51	$C_{14}H_{16}N_2OS_2$
33.64	$C_{14}H_{16}N_2O_3$, $2H_2O$
20.7	$C_{14}H_{16}O_2S$
38.58	$C_{14}H_{16}O_3$
38.59	$C_{14}H_{16}O_3$
31.23	$C_{14}H_{16}O_6$
22.4	$C_{14}H_{17}BrO_2$
31.24	$C_{14}H_{17}Cl$
85.49	$C_{14}H_{17}CrNO_6S_2$
31.25	$C_{14}H_{17}F$
35.32	$C_{14}H_{17}NO_3$
36.14	$C_{14}H_{17}N_3O$
31.26	$C_{14}H_{18}$
83.95	$C_{14}H_{18}Br_2CuN_2$
40.34	$C_{14}H_{18}ClNO$
32.51	$C_{14}H_{18}N_2O_2$
59.6	$C_{14}H_{18}N_2O_3$
71.68	$C_{14}H_{18}N_2O_4Ru_2$
35.33	$C_{14}H_{18}N_2O_5$
44.30	$C_{14}H_{18}N_4O_3$, $C_{10}H_{11}N_3O_3S$
84.56	$C_{14}H_{18}N_4O_6Zn$
31.27	$C_{14}H_{18}O_6$
31.28	$C_{14}H_{18}O_6$
45.30	$C_{14}H_{18}O_9$
71.69	$C_{14}H_{19}B_3Co_2$
79.12	$C_{14}H_{19}ClCuN_4OS$
38.60	$C_{14}H_{19}ClO_3$
75.8	$C_{14}H_{19}Ir$
31.29	$C_{14}H_{19}NO$
31.30	$C_{14}H_{19}NO$
48.43	$C_{14}H_{19}NO_4$
71.70	$C_{14}H_{20}B_3CoFe$
64.66	$C_{14}H_{20}ClO_3P$
81.58	$C_{14}H_{20}CoN_2O_8$
84.57	$C_{14}H_{20}CuN_4O_2{}^{2+}$, $2ClO_4^-$
71.71	$C_{14}H_{20}Hg_2I_4S_2$
73.50	$C_{14}H_{20}Mo_2N_2O_2S_2$
73.51	$C_{14}H_{20}Mo_2S_4$
32.52	$C_{14}H_{20}N_3O^+$, I^-
58.6	$C_{14}H_{20}N_4O_7$, H_2O
47.30	$C_{14}H_{20}N_6O_5$, $2H_2O$
48.44	$C_{14}H_{20}N_6O_5S$, $2.5H_2O$
27.12	$C_{14}H_{20}O_2$
28.7	$C_{14}H_{20}O_2$
31.31	$C_{14}H_{20}O_2$
38.61	$C_{14}H_{20}O_2$
38.62	$C_{14}H_{20}O_2$
38.63	$C_{14}H_{20}O_4$
38.64	$C_{14}H_{20}O_4$
38.65	$C_{14}H_{20}O_5$, $C_6H_2N_3O_7^-$, Li^+, $2H_2O$
38.66	$C_{14}H_{20}O_5$, Na^+, ClO_4^-
62.15	$2C_{14}H_{20}O_5$, $C_{24}H_{20}B^-$, Na^+
38.67	$2C_{14}H_{20}O_5$, Na^+, ClO_4^-
38.68	$2C_{14}H_{20}O_5$, $2Na^+$, $Cl_4O_2U^{2-}$
45.31	$C_{14}H_{20}O_9$
39.53	$C_{14}H_{20}S_3$
45.32	$C_{14}H_{21}ClO_6$
71.72	$C_{14}H_{21}MoN_7{}^{2+}$, $2BF_4^-$
64.67	$C_{14}H_{21}NO_2P$
58.7	$C_{14}H_{21}N_2O^+$, ClO_4^-
64.68	$C_{14}H_{21}O_4P$, $0.5C_6H_{14}$
82.28	$C_{14}H_{22}Cd_2Cl_2N_8O_6$
71.73	$C_{14}H_{22}CoN_5O_4$
82.29	$C_{14}H_{22}CrN_2O_8^-$, Na^+, $2H_2O$
47.31	$C_{14}H_{22}N_6O_4$
76.82	$C_{14}H_{22}N_6Zn^{2+}$, $2NO_3^-$
38.69	$C_{14}H_{22}O_4$
13.11	$C_{14}H_{23}ClN_3O_2^+$, Cl^-, H_2O
71.74	$C_{14}H_{23}Cl_2Ta$
83.96	$C_{14}H_{23}CoN_6O_4$
71.75	$C_{14}H_{23}MnO_5P_2Pt$
71.76	$C_{14}H_{23}MnO_5P_2Pt$
40.35	$C_{14}H_{23}NO$
71.77	$C_{14}H_{24}Cl_2N_2OPt$
76.83	$C_{14}H_{24}CoN_6{}^{3-}$, $C_6CoN_6{}^{3-}$, $2H_2O$
83.97	$C_{14}H_{24}CuN_4{}^{2-}$, $2C_{24}H_{20}B^-$
39.54	$C_{14}H_{24}O_4S_2$
39.55	$C_{14}H_{24}S_2$
71.78	$(C_{14}H_{26}Cu_2FeN_{12})_n$, $6nH_2O$
76.84	$C_{14}H_{26}N_8O_5Pd^{2+}$, $2ClO_4^-$
23.5	$C_{14}H_{27}NO$
5.13	$C_{14}H_{28}$
76.85	$C_{14}H_{28}Cl_2CuN_2O_4$
83.98	$C_{14}H_{28}CoN_6O_2^+$, ClO_4^-, H_2O
80.12	$C_{14}H_{28}CuN_2S_4$
84.58	$C_{14}H_{28}CuN_6NiO_8$, $4H_2O$
80.13	$C_{14}H_{28}HgN_2S_4$
69.30	$C_{14}H_{28}N_2PbS_4$
81.59	$C_{14}H_{30}CoN_4O_4^+$, Br^-, $3H_2O$
76.86	$C_{14}H_{31}CuN_2O_4^+$, ClO_4^-, H_2O
76.87	$C_{14}H_{32}ClCu_2N_4O_2^+$, $C_{24}H_{20}B^-$
84.59	$C_{14}H_{32}Mo_2O_{10}$, $2CH_4O$
76.88	$C_{14}H_{32}N_6NiS_2$
76.89	$C_{14}H_{32}N_6NiSe_2$
71.79	$C_{14}H_{35}ClIrO_2P_4^+$, FO_3S^-
71.80	$C_{14}H_{35}ClP_2Pt$
86.32	$C_{14}H_{36}B_4P_2Pt$
76.90	$C_{14}H_{36}CuN_5U_2^+$, NO_2^-
84.60	$C_{14}H_{36}MoN_8O_4P_2S_2$
64.69	$C_{14}H_{36}N_4O_2P_2Si_2$
81.60	$C_{14}H_{37}F_3MoO_{14}P_4$
86.33	$C_{14}H_{38}B_4P_2Pt$
71.81	$C_{14}H_{40}B_8P_2Pt$
71.82	$C_{14}H_{41}B_8P_2Pt$
63.17	$C_{14}H_{42}B_2N_4Si_4Sn_2$
71.83	$C_{14}H_{42}P_4W$

C_{15}

71.84	$C_{15}H_3Fe_4O_{14}^-$, $C_8H_{20}N^+$
86.34	$C_{15}H_5Co_2FeO_9P$
86.35	$C_{15}H_5Co_3O_9P$
32.53	$C_{15}H_6F_{12}N_6$
86.36	$C_{15}H_7Fe_3O_9P$
71.85	$C_{15}H_9MnMoO_5$
35.34	$C_{15}H_{10}BrClN_2O$
84.61	$C_{15}H_{10}Fe_4N_2O_{12}$
73.52	$C_{15}H_{10}Mo_2NO_4^-$, $C_8H_{20}N^+$
41.52	$C_{15}H_{10}N_2S_2$
26.3	$C_{15}H_{11}Br$
26.4	$C_{15}H_{11}Br$
16.12	$C_{15}H_{12}ClNO_2$
17.23	$C_{15}H_{12}Cl_2N_2O_6$
80.14	$C_{15}H_{12}FeN_3S_6$, $0.5CH_2Cl_2$
71.86	$C_{15}H_{12}O_3Ru_2$
5.14	$C_{15}H_{12}O_4$
72.23	$C_{15}H_{13}FeNO_6$
28.8	$C_{15}H_{13}NO$
71.87	$C_{15}H_{13}O_3Ru_2^+$, BF_4^-
78.2	$C_{15}H_{14}ClCuN_3O_4$, H_2O
71.88	$C_{15}H_{14}Fe_2O_3$
44.31	$C_{15}H_{14}N_4$, CH_4O
41.53	$C_{15}H_{14}N_4O_2S$
41.54	$C_{15}H_{14}N_4S$
41.55	$C_{15}H_{14}N_4S$
38.70	$C_{15}H_{14}O_5Se$
50.7	$C_{15}H_{15}BrO_3$
50.8	$C_{15}H_{15}BrO_4$

36.15 $C_{15}H_{15}Br_2N$
73.53 $C_{15}H_{15}Co_3S_2$
73.54 $C_{15}H_{15}Co_3S_2$
73.55 $C_{15}H_{15}FeNO_4$
31.32 $C_{15}H_{15}NO_4S$
13.12 $C_{15}H_{15}N_3O_2$
44.32 $C_{15}H_{15}N_3O_3$
64.70 $C_{15}H_{16}ClN_2OP$
41.56 $C_{15}H_{16}N_2O_2S_4$
27.13 $C_{15}H_{16}N_2O_6$
15.5 $C_{15}H_{16}N_4O_2, 0.25C_6H_6$
40.36 $C_{15}H_{16}N_4O_5$
21.8 $C_{15}H_{17}Cl_3O_3S$
64.71 $C_{15}H_{17}NO_2P_2S_3$
64.72 $C_{15}H_{17}O_2P$
53.1 $C_{15}H_{18}BrClO_2$
33.65 $C_{15}H_{18}BrN_3O_2$
40.37 $C_{15}H_{18}ClNO_2$
33.66 $C_{15}H_{18}ClN_4O_3S^+, HO_4S^-$
40.38 $C_{15}H_{18}N_2O_3$
50.9 $C_{15}H_{18}N_4O_5, 2H_2O$
53.2 $C_{15}H_{18}O_3$
53.3 $C_{15}H_{18}O_3$
53.4 $C_{15}H_{18}O_3$
53.5 $C_{15}H_{18}O_3$
40.39 $C_{15}H_{19}NO_7, K^+, CNS^-$
64.73 $C_{15}H_{19}N_6PS$
30.3 $C_{15}H_{20}$
59.7 $C_{15}H_{20}BrClO_2$
59.8 $C_{15}H_{20}BrClO_3$
53.6 $C_{15}H_{20}Br_2O_3$
72.24 $C_{15}H_{20}FeO_3$
64.74 $C_{15}H_{20}N_3O_4P_3$
31.33 $C_{15}H_{20}O$
31.34 $C_{15}H_{20}O$
59.9 $C_{15}H_{20}O_2$
21.9 $C_{15}H_{20}O_3$
59.10 $C_{15}H_{20}O_3$
45.33 $C_{15}H_{20}O_{10}$
53.7 $C_{15}H_{21}BrO_3$
17.24 $C_{15}H_{21}IO_4$
68.15 $C_{15}H_{21}InO_6$
21.10 $C_{15}H_{21}NO$
58.8 $C_{15}H_{21}N_2O^+, Br^-$
58.9 $C_{15}H_{21}N_2O^+, ClO_4^-$
48.45 $C_{15}H_{21}N_3O_3$
83.99 $C_{15}H_{21}N_8Ni_2O_2^-, Na^+, 2C_4H_8O$
63.18 $C_{15}H_{21}O_6Si^-, ClO_4^-$
27.14 $C_{15}H_{22}Br_3Cl$
80.15 $2C_{15}H_{22}MoN_6O_2S_6, 0.5CH_2Cl_2, 0.5H_2O$
33.67 $C_{15}H_{22}N_2$
31.35 $C_{15}H_{22}O$
39.56 $C_{15}H_{22}OS$
31.36 $C_{15}H_{22}O_2$
38.71 $C_{15}H_{22}O_2$
38.72 $C_{15}H_{22}O_4$
53.8 $C_{15}H_{22}O_4$
45.34 $C_{15}H_{22}O_{10}$
80.16 $C_{15}H_{23}MoN_5S_6, CH_2Cl_2$

40.40 $C_{15}H_{23}NO_5, C_4H_{12}N^+, ClO_4^-$
58.10 $C_{15}H_{23}N_2O^+, ClO_4^-, 0.5H_2O$
63.19 $C_{15}H_{23}N_3O_3Si$
44.33 $C_{15}H_{23}N_4^+, C_2H_5O_3S^-$
10.6 $C_{15}H_{24}ClN_2O_2^+, Cl^-$
58.11 $C_{15}H_{24}N_2O$
58.12 $C_{15}H_{24}N_2O$
58.13 $C_{15}H_{24}N_2O$
58.14 $C_{15}H_{24}N_2O$
58.15 $C_{15}H_{24}N_2O_2$
38.73 $C_{15}H_{24}O_2$
53.9 $C_{15}H_{24}O_2$
53.10 $C_{15}H_{24}O_4$
58.16 $C_{15}H_{25}N_2O^+, ClO_4^-$
83.100 $C_{15}H_{26}N_4Ni^{2+}, 2ClO_4^-, H_2O$
53.11 $C_{15}H_{26}O, 0.5C_5H_5N$
53.12 $C_{15}H_{26}O_2$
53.13 $C_{15}H_{26}O_2$
53.14 $C_{15}H_{26}O_2$
68.16 $C_{15}H_{28}GaMoN_3O_3S$
53.15 $C_{15}H_{28}O_2$
53.16 $C_{15}H_{28}O_2$
80.17 $C_{15}H_{30}ClN_3O_3S_3Ti$
86.37 $C_{15}H_{36}FeN_6O_3P_2$
71.89 $C_{15}H_{42}ClMoPSi_3$
64.75 $C_{15}H_{64}B_{50}P_2, C_6H_6$

C_{16}

85.50 $C_{16}Cl_5O_{10}SW_2^-, C_8H_{20}N^+$
17.25 $C_{16}D_{18}N_2O_4$
64.76 $C_{16}H_4F_{18}P_2$
71.90 $C_{16}H_6O_{12}Os_4$
59.11 $C_{16}H_8Br_2N_2O_2$
59.12 $C_{16}H_8Br_2N_2O_2$
71.91 $C_{16}H_8Fe_3O_8$
60.33 $C_{16}H_8S_3, C_{12}H_4N_4$
71.92 $C_{16}H_9Ir_4NO_{11}$
60.34 $C_{16}H_{10}, C_6F_4O_2$
83.101 $C_{16}H_{10}CrMoN_2O_{10}$
72.25 $C_{16}H_{10}Fe_2O_7$
39.57 $C_{16}H_{11}BrO_2S_2$
27.15 $C_{16}H_{11}Cl_5$
86.38 $C_{16}H_{11}CrO_7P$
73.56 $C_{16}H_{11}F_9FeO$
41.57 $C_{16}H_{12}ClNO_2S_2$
83.102 $C_{16}H_{12}Cl_2CuN_4$
71.93 $C_{16}H_{12}CuO_{12}Re_2$
71.94 $C_{16}H_{12}Mn_2O_4$
27.16 $C_{16}H_{12}NO_3S^-, H_4N^+, H_2O$
32.54 $C_{16}H_{12}N_2O$
41.58 $C_{16}H_{12}N_2O_3S$
60.35 $2C_{16}H_{12}N_4, 3C_{10}H_{12}S_4$
27.17 $C_{16}H_{12}O$
39.58 $C_{16}H_{12}S_8$
35.35 $C_{16}H_{13}ClN_2O_2$
32.55 $C_{16}H_{13}Cl_2N_3O, CH_4O$
73.57 $C_{16}H_{13}FeP$
38.74 $C_{16}H_{13}N_3O_8$
35.36 $C_{16}H_{14}BrNO_2$

32.56 $C_{16}H_{14}BrN_3O_2$
35.37 $C_{16}H_{14}BrN_3O_3$
35.38 $C_{16}H_{14}ClN_3O$
35.39 $C_{16}H_{14}ClN_3O$
83.103 $C_{16}H_{14}Cl_2CoN_6$
78.3 $C_{16}H_{14}CoN_2O_2^-, Li^+, 1.5C_4H_8O$
78.4 $C_{16}H_{14}CoN_2O_2^-, Na^+, C_4H_8O$
71.95 $C_{16}H_{14}CrO_3$
73.58 $C_{16}H_{14}Mn_2N_2O_4$
71.96 $C_{16}H_{14}Mn_2O_3$
35.40 $C_{16}H_{14}N_2O_2$
76.91 $2nC_{16}H_{14}N_2O_3V, nNa^+, nC_{24}H_{20}B^-$
22.5 $C_{16}H_{14}N_2O_6$
35.41 $C_{16}H_{14}N_3O_3^+, Br^-$
20.8 $C_{16}H_{14}O_2$
38.75 $C_{16}H_{14}O_4$
13.13 $C_{16}H_{14}O_4S$
13.14 $C_{16}H_{14}O_4S_2$
39.59 $C_{16}H_{14}S_2$
43.2 $C_{16}H_{15}BrN_2O_3$
71.97 $C_{16}H_{15}Co_3S_2$
36.16 $C_{16}H_{15}N_3O_2$
38.76 $C_{16}H_{15}O^+, ClO_4^-$
58.17 $C_{16}H_{16}ClNO_3$
35.42 $C_{16}H_{16}ClN_2^+, Cl^-, H_2O$
17.26 $C_{16}H_{16}Cl_2O_2$
73.59 $C_{16}H_{16}Mo_2O_4P^-, C_{24}H_{20}As^+$
13.15 $C_{16}H_{16}N_2O_2$
47.32 $C_{16}H_{16}N_6O_8PS^-, C_6H_{16}N^+$
39.60 $C_{16}H_{16}OS$
39.61 $C_{16}H_{16}OS$
59.13 $C_{16}H_{16}O_3$
39.62 $C_{16}H_{16}O_4S_6$
72.26 $C_{16}H_{16}O_4V$
59.14 $C_{16}H_{16}O_5$
39.63 $C_{16}H_{16}S_2$
39.64 $C_{16}H_{16}S_4^+, C_{16}H_{16}S_4, ClO_4^-$
32.57 $C_{16}H_{17}ClN_2OS$
73.60 $C_{16}H_{17}N_2W^+, BF_4^-$
44.34 $C_{16}H_{17}N_3O_2$
47.33 $C_{16}H_{17}N_5O_4Se$
83.104 $C_{16}H_{18}Cl_2N_2O_4Pt$
84.62 $C_{16}H_{18}Cl_2N_6NiO_2$
84.63 $C_{16}H_{18}Cl_2N_6NiO_2, C_6H_6$
63.20 $C_{16}H_{18}Cl_2Si_2$
85.51 $C_{16}H_{18}Mn_2N_4O_6S_4$
36.17 $C_{16}H_{18}N_2^+, C_{16}H_{18}N_2, I_3^-, 0.1I_2$
48.46 $C_{16}H_{18}N_2O_3$
17.27 $C_{16}H_{18}N_2O_4$
82.30 $C_{16}H_{18}N_4NiO_4, 2H_2O$
36.18 $C_{16}H_{18}N_4O_4S_2$
11.16 $C_{16}H_{18}N_4S$
75.9 $C_{16}H_{18}Ni$
75.10 $C_{16}H_{19}BFeINO_2$
31.37 $C_{16}H_{19}ClHgO_6$
71.98 $C_{16}H_{19}Cl_2HgN$
72.27 $C_{16}H_{19}FeNO_6$
69.31 $C_{16}H_{19}NO_2Sn$
33.68 $C_{16}H_{19}NO_3$
33.69 $C_{16}H_{19}N_2^+, I^-$

73.61 $C_{16}H_{19}O_2Rh$, $2C_6H_6O_2$
33.70 $C_{16}H_{20}ClN_2O^+$, $C_4H_3O_4^-$
71.99 $C_{16}H_{20}MnO_6PS_2$
73.62 $C_{16}H_{20}Mo_2N_2O_6$
58.18 $C_{16}H_{20}NO_3^-$, Cl^-, $2H_2O$
40.41 $C_{16}H_{20}N_2O_4$
50.10 $C_{16}H_{20}N_2O_4$
38.77 $C_{16}H_{20}N_2O_7$
53.17 $C_{16}H_{20}O_3$
45.35 $C_{16}H_{20}O_5S_2$
45.36 $C_{16}H_{20}O_5S_2$
45.37 $C_{16}H_{20}O_5S_2$
45.38 $C_{16}H_{20}O_6$
27.18 $C_{16}H_{20}O_8$
31.38 $C_{16}H_{21}NO$
1.33 $C_{16}H_{21}NO_5$
45.39 $C_{16}H_{21}NO_9$
73.63 $C_{16}H_{22}Cl_4O_2Ti_2$
81.61 $C_{16}H_{22}CuN_2O_5$, H_2O
68.17 $C_{16}H_{22}In_2N_4O_2$, $0.5C_6H_6$
73.64 $C_{16}H_{22}Mo_2S_4^+$, BF_4^-
32.58 $C_{16}H_{22}N_2O_2$
33.71 $C_{16}H_{22}N_2O_2$
86.39 $C_{16}H_{22}N_2P_2PdS_2$
25.3 $C_{16}H_{22}O_2$
31.39 $C_{16}H_{22}O_2$
31.40 $C_{16}H_{22}O_2$
31.41 $C_{16}H_{22}O_4$
45.40 $C_{16}H_{22}O_6$
45.41 $C_{16}H_{22}O_6$
69.32 $C_{16}H_{22}Sn_2$
83.105 $C_{16}H_{23}BCuF_2N_5O_2^+$, ClO_4^-
76.92 $C_{16}H_{23}ClCoN_5$, Cl_4Zn
58.19 $C_{16}H_{23}NO_6$
58.20 $C_{16}H_{23}N_3O$
48.47 $C_{16}H_{23}N_6O_4^+$, $C_4H_5O_6^-$, H_2O
75.11 $C_{16}H_{23}Rh$
75.12 $C_{16}H_{24}Ag^+$, BF_4^-
75.13 $C_{16}H_{24}Cl_2O_2Th$
83.106 $C_{16}H_{24}CrF_2N_8^+$, BF_4^-
69.33 $C_{16}H_{24}Ge_2$
73.65 $C_{16}H_{24}Mo_2N_2O_2S_2$
83.107 $C_{16}H_{24}N_8Pt^{2+}$, Cl_6Pt^{2-}
79.13 $C_{16}H_{24}N_8S_4Zn^{2+}$, $2NO_3^-$, H_2O
76.93 $C_{16}H_{24}N_{10}O_2Pt^{2+}$, $2F_6P^-$
27.19 $C_{16}H_{24}O_2$
27.20 $C_{16}H_{24}O_2$
38.78 $C_{16}H_{24}O_2$
63.21 $C_{16}H_{24}O_{12}Si_6$
69.34 $C_{16}H_{24}Sn_2$
80.18 $C_{16}H_{25}Cl_2MoN_3S_4$, $CHCl_3$
53.18 $C_{16}H_{25}N$
58.21 $C_{16}H_{25}NO_5$
64.77 $C_{16}H_{25}N_2O_2PS_2$
58.22 $C_{16}H_{25}N_3O_5$
71.100 $C_{16}H_{26}CoN_5O_4$
82.31 $C_{16}H_{26}CrN_2O_8^-$, K^+, $4H_2O$
27.21 $C_{16}H_{26}NO^+$, Cl^-
32.59 $C_{16}H_{26}N_2$
40.42 $C_{16}H_{26}N_2O_8$

84.64 $C_{16}H_{28}CoN_6O_6$
82.32 $C_{16}H_{28}Cu_2N_4O_{14}$
86.40 $C_{16}H_{28}NO_{17}P_3Ru_3$
81.62 $C_{16}H_{28}O_8Rh_2S_2$
58.23 $C_{16}H_{29}N_2^+$, ClO_4^-
58.24 $C_{16}H_{29}N_2^+$, I^-
48.48 $C_{16}H_{29}N_3O_3$, H_2O
75.14 $C_{16}H_{30}Cl_2N_2O_2Pd$, H_2O
84.65 $C_{16}H_{30}I_2N_2O_2Zn$
58.25 $C_{16}H_{30}NO^+$, Cl^-
45.42 $C_{16}H_{30}O_5S_2$
85.52 $C_{16}H_{32}Br_4Nb_2S_7$
84.66 $C_{16}H_{32}MnO_6^{2+}$, $2Br^-$
82.33 $C_{16}H_{32}Mo_2N_8O_{12}^{4+}$, $4Cl^-$, $6H_2O$
70.14 $C_{16}H_{32}N_4O_4S_8Te$
64.78 $C_{16}H_{32}N_8P_2$
51.55 $0.5C_{16}H_{32}O_2$, $4C_{24}H_{40}O_4$, $0.5C_2H_6O$
1.34 $C_{16}H_{32}O_4$
81.63 $C_{16}H_{32}O_{10}Rh_2S_2$
69.35 $C_{16}H_{33}N_3S_6Sn$
82.34 $C_{16}H_{34}CoN_5O_2^{2+}$, $2ClO_4^-$, $2H_2O$
79.14 $C_{16}H_{34}N_8O_{16}Pr_2$, $2CH_4N_2O$, $2H_2O$
85.53 $C_{16}H_{36}Cl_4Cu_4O_8S_4$
84.67 $C_{16}H_{36}Cu_2N_6O_4^{2+}$, $2ClO_4^-$
84.68 $C_{16}H_{36}FeN_8O_6^{3+}$, $3Cl^-$, $2H_2O$
86.41 $C_{16}H_{36}Ir_2O_8P_2S_2$
2.19 $C_{16}H_{36}N^+$, $C_4HCl_2O_4^-$
2.20 $C_{16}H_{36}N^+$, $C_4HCl_2O_4^-$
85.3 $C_{16}H_{36}N^+$, $C_4H_4O_3S_4Tc^-$
3.48 $C_{16}H_{36}N^+$, I_3^-
3.49 $C_{16}H_{36}N^+$, $N_3S_3^-$
85.14 $2C_{16}H_{36}N^+$, $C_6NiS_{10}^{2-}$
85.31 $2C_{16}H_{36}N^+$, $C_8Mo_2O_7S_5^{2-}$
3.50 $2C_{16}H_{36}N^+$, $C_{14}IO_{14}Rh_5^{2-}$
3.51 $3C_{16}H_{36}N^+$, $C_6N_6S_6Tc^{3-}$
3.52 $4C_{16}H_{36}N^+$, $8C_2H_3N$, $C_{22}O_{22}Pt_{19}^{4-}$
85.96 $4C_{16}H_{36}N^+$, $C_{84}H_{84}Fe_7Mo_2S_{20}^{4-}$
85.97 $4C_{16}H_{36}N^+$, $C_{84}H_{84}Fe_7S_{20}W_2^{4-}$
34.9 $C_{16}H_{36}N_4$, $2H_2O$
84.69 $C_{16}H_{36}N_4O_4P_4S_4U$
64.79 $C_{16}H_{36}P^+$, BF_4^-
83.108 $C_{16}H_{38}BrCoN_4O^{2+}$, $2Br^-$
86.42 $C_{16}H_{38}Ir_2O_8P_2S_2$
69.36 $C_{16}H_{38}N_4Sn_3$
62.11 $C_{16}H_{40}B_4N_{10}$
86.43 $C_{16}H_{40}B_4P_2Pt$
86.44 $C_{16}H_{40}B_{12}P_2Pt_2$
84.70 $C_{16}H_{40}Cl_4Cu_4N_4O_4$
85.54 $C_{16}H_{40}Cl_4Mo_2S_4$
69.37 $(C_{16}H_{40}Cl_4O_2Sn_4)_n$
85.55 $C_{16}H_{40}Cl_6Ir_2S_4$
85.56 $C_{16}H_{40}Cl_6Ir_2S_4$
71.101 $C_{16}H_{40}Cr_2P_4$
85.57 $C_{16}H_{40}Fe_6Mo_2S_{17}^{3-}$, $3C_8H_{20}N^+$
71.102 $C_{16}H_{40}Mo_2P_4$
84.71 $C_{16}H_{40}N_4O_7V_2$
86.45 $C_{16}H_{42}Cl_4N_2O_{14}P_4Ru_2$
68.18 $C_{16}H_{44}Cl_2Si_4Tl_2$
76.94 $C_{16}H_{45}Co_2N_8O_3^{3+}$, $3ClO_4^-$, $2H_2O$

C_{17}

71.103 $C_{17}H_5CoF_{12}O_4W$
72.28 $C_{17}H_5O_{14}Rh_6^-$, $C_{24}H_{20}P^+$, C_4H_8O
16.13 $C_{17}H_7F_{12}N_3O$
71.104 $C_{17}H_8IMn_3N_2O_9$
83.109 $C_{17}H_9NO_{12}Os_3S$
38.79 $C_{17}H_{10}Cl_2O_3$
77.4 $C_{17}H_{10}F_6N_2O_2Pd$
77.5 $C_{17}H_{10}F_6N_2O_2Pd$
71.105 $C_{17}H_{10}Fe_2Ni_2O_7$
71.106 $C_{17}H_{11}NO_9Os_3$
71.107 $C_{17}H_{12}Co_2I_2O_8$
36.19 $C_{17}H_{12}N_2O_9$, $2H_2O$
31.42 $C_{17}H_{12}O$
71.108 $C_{17}H_{13}O_9Os_3PS_2$
72.29 $C_{17}H_{14}Fe_2NiO_6$
19.9 $C_{17}H_{14}O$
39.65 $C_{17}H_{14}OS$
59.15 $C_{17}H_{14}O_6$, H_2O
35.43 $C_{17}H_{15}BrN_2O$
35.44 $C_{17}H_{15}ClN_2O$
38.80 $C_{17}H_{15}NO_6$
60.36 $2C_{17}H_{15}N_2^+$, $2C_{19}H_{23}IN_6O_{12}P^-$, $20H_2O$, $2CH_4O$
41.59 $C_{17}H_{16}BrNO_5S$
35.45 $C_{17}H_{16}ClFN_2O$, H_2O
73.66 $C_{17}H_{16}Cl_4Fe_2O_4Si_3$
68.19 $C_{17}H_{16}GaNOS$
85.58 $2C_{17}H_{16}N_3OPtS^+$, $2C_{19}H_{24}N_8O_{10}P^-$, xH_2O
38.81 $C_{17}H_{16}O_2$
1.35 $C_{17}H_{16}O_3$
38.82 $C_{17}H_{16}O_4$
38.83 $C_{17}H_{16}O_5$
75.15 $C_{17}H_{16}O_6Ru_3S$
32.60 $C_{17}H_{17}BrN_2O_2$
50.11 $C_{17}H_{17}ClO_6$, C_6H_6
77.6 $C_{17}H_{17}CuN_2O_3^+$, $C_5HF_6O_2^-$, H_2O
32.61 $C_{17}H_{17}NO_2$
40.43 $C_{17}H_{17}NO_2$
38.84 $C_{17}H_{17}NO_3$
33.72 $C_{17}H_{17}N_2^+$, Br^-, $0.5H_2O$
41.60 $C_{17}H_{17}N_2O_2S_3$
41.61 $C_{17}H_{17}N_3OS$
41.62 $C_{17}H_{17}N_3O_3S_2$
13.16 $C_{17}H_{17}N_5O_6$
71.109 $C_{17}H_{17}O_9PW_2$
19.10 $C_{17}H_{18}Br_2$
4.3 $C_{17}H_{18}ClNO_3S$
73.67 $(C_{17}H_{18}FeO_2Sn)_n$
16.14 $C_{17}H_{18}N_2O_4$
41.63 $C_{17}H_{18}N_2O_5S$
86.46 $C_{17}H_{18}NiO_2P_2$
64.80 $C_{17}H_{19}Cl_2O_5P$
13.17 $C_{17}H_{19}NO_3$
60.37 $2C_{17}H_{19}N_2^+$, $2C_{19}H_{23}IN_8O_{12}P^-$, $17H_2O$, $2CH_4O$
50.12 $C_{17}H_{19}N_3O_4$, H_2O

73.68 $C_{17}H_{20}FeO_2$
68.20 $C_{17}H_{20}NS_2Tl$
32.62 $C_{17}H_{20}N_2$, $2HBr$
78.5 $C_{17}H_{20}N_2O_2Pt$
48.49 $C_{17}H_{20}N_2O_3S$
32.63 $C_{17}H_{20}N_2O_8$
47.35 $C_{17}H_{20}N_3^+$, $C_{19}H_{23}IN_8O_{12}P^-$, $12H_2O$
11.17 $C_{17}H_{20}O_2S_2$
59.16 $C_{17}H_{20}O_6$
64.81 $C_{17}H_{20}P^+$, I^-
22.6 $C_{17}H_{21}Cl_3O_3S$
45.43 $C_{17}H_{21}NO_5$
72.30 $C_{17}H_{21}N_2Pd^+$, ClO_4^-
85.59 $C_{17}H_{21}N_7NiS_6$
64.82 $C_{17}H_{22}NO_4P$
32.64 $C_{17}H_{22}N_2O_2$
37.10 $C_{17}H_{22}N_2O_2S$
31.43 $C_{17}H_{22}N_2O_4S$
33.73 $C_{17}H_{22}N_4O_3S$
48.50 $C_{17}H_{22}N_4O_4S$
59.17 $C_{17}H_{23}BrCl_2O_3$
73.69 $C_{17}H_{23}Fe$
85.60 $C_{17}H_{24}CuN_2S_4$
73.70 $C_{17}H_{24}NRh^{2+}$, $2F_6P^-$
3.53 $C_{17}H_{24}N_4O_2S$
47.34 $C_{17}H_{24}N_5O_{10}P$, $3.5H_2O$
48.51 $C_{17}H_{24}N_6O_4S$, $1.5H_2O$
38.85 $C_{17}H_{24}O_5$
38.86 $C_{17}H_{24}O_5$
53.19 $C_{17}H_{25}Br_2ClO_2$
38.87 $C_{17}H_{26}O_3$
63.22 $C_{17}H_{26}O_4Si$
52.4 $C_{17}H_{26}O_{10}$
71.110 $C_{17}H_{27}Cl_2Ta$
83.110 $C_{17}H_{27}FeN_5O^{2+}$, $2F_6P^-$
71.111 $C_{17}H_{27}IMnO_5PPt$
64.83 $C_{17}H_{27}OP$
64.84 $C_{17}H_{27}OP$
40.44 $C_{17}H_{28}N_2O_7$
53.20 $C_{17}H_{28}O_4$
64.85 $C_{17}H_{29}NO_3P^+$, I^-
48.52 $C_{17}H_{29}N_3O_6$, $0.5H_2O$
71.112 $C_{17}H_{30}Co_3O_{15}P_3$
86.47 $C_{17}H_{30}MoO_8P_2$
64.86 $C_{17}H_{30}O_2P^+$, Br^-
71.113 $C_{17}H_{34}BrNP_2Pd$
71.114 $C_{17}H_{34}BrNP_2Pd$
71.115 $C_{17}H_{34}BrNP_2Pd$
75.16 $C_{17}H_{40}FeO_9P_3$
75.17 $C_{17}H_{40}FeO_9P_3^+$, BF_4^-
75.18 $C_{17}H_{40}FeO_9P_3^+$, BF_4^-
75.19 $C_{17}H_{40}FeO_9P_3^+$, BF_4^-
71.116 $C_{17}H_{50}Co_2P_6$
71.117 $C_{17}O_{16}Ru_6^{2-}$, $2C_4H_{12}N^+$

C_{18}

16.15 $C_{18}Cl_{15}N$
71.118 $C_{18}Co_6O_{16}S_3$
19.11 $C_{18}H_8F_{14}$

39.66 $2C_{18}H_8S_4$, Br_4Cu_2
73.71 $C_{18}H_{10}ClCoFe_2O_8Sn$
86.48 $C_{18}H_{10}ClFe_2O_6P$
35.46 $C_{18}H_{10}Cl_2N_2O_2$
35.47 $C_{18}H_{10}Cl_2N_2O_2$
71.119 $C_{18}H_{10}F_{12}Nb$
71.120 $C_{18}H_{12}Cl_2HgN_2$
71.121 $C_{18}H_{12}Hg_3$
35.48 $C_{18}H_{12}N_2S_2$
38.88 $C_{18}H_{12}O_3$
38.89 $C_{18}H_{12}O_5$
63.23 $C_{18}H_{12}O_6Si^{2-}$, $2C_6H_{16}N^+$
39.67 $C_{18}H_{12}S_4$
71.122 $C_{18}H_{13}FeO_2^+$, F_6P^-
39.68 $C_{18}H_{13}N_3O_2S_2$
71.123 $C_{18}H_{13}O_9Os_3PS_2$
81.64 $C_{18}H_{13}O_9Os_3PS_2$
27.22 $C_{18}H_{14}$
71.124 $C_{18}H_{14}ClHgN_3$
32.65 $C_{18}H_{14}Cl_4N_2O$, $0.5H_2O$
36.20 $C_{18}H_{14}N_2$
29.4 $C_{18}H_{14}O_2$
29.5 $C_{18}H_{14}O_2$
71.125 $C_{18}H_{14}O_6RuW$
73.72 $C_{18}H_{15}ClF_2MoN_4$
35.49 $C_{18}H_{15}Cl_2N_3O_3$
35.50 $C_{18}H_{15}Cl_2N_3O_3$
86.49 $C_{18}H_{15}MnN_3O_3P$
73.73 $C_{18}H_{15}MoO_3Re$
16.16 $C_{18}H_{15}N$, $2Cl_3Sb$
27.23 $C_{18}H_{15}NO_2$
64.87 $C_{18}H_{15}N_2PS_3$
64.88 $C_{18}H_{15}OP$, $C_{18}H_{16}OP^+$, ClO_4^-
64.89 $C_{18}H_{15}OP$, $C_{18}H_{16}OP^+$, ClO_4^-
64.90 $C_{18}H_{15}OP$, $0.5H_2O_2$
64.91 $2C_{18}H_{15}OP$, H_2O, HBr
11.18 $C_{18}H_{15}Se^+$, Cl^-, H_2O
30.4 $C_{18}H_{16}$
35.51 $C_{18}H_{16}ClN_3O_2$
1.36 $C_{18}H_{16}F_5NO_2$
41.64 $C_{18}H_{16}N_2OS$
36.21 $C_{18}H_{16}N_2O_2$
36.22 $C_{18}H_{16}N_2O_2$
59.18 $C_{18}H_{16}N_2O_4$
41.65 $C_{18}H_{16}N_2O_4S$
71.126 $C_{18}H_{16}O_2Rh^{3+}$, $C_2F_3O_2^-$
63.24 $C_{18}H_{16}Si$
41.66 $C_{18}H_{17}Cl_2N_3O_2S$
36.23 $C_{18}H_{17}NO_3$
72.31 $C_{18}H_{18}Cl_2Pt$
83.111 $C_{18}H_{18}CoN_6^+$, $C_{24}H_{20}B^-$, $5H_2O$
75.20 $C_{18}H_{18}F_3O_2Rh$
41.67 $C_{18}H_{18}N_2O_2S$
37.11 $C_{18}H_{18}N_2O_4$
78.6 $C_{18}H_{18}N_2O_5U$, $CHCl_3$
73.74 $C_{18}H_{18}N_2Ti$
73.75 $C_{18}H_{18}N_2Zr$
64.92 $C_{18}H_{18}N_3P$
9.3 $C_{18}H_{18}N_4$
20.9 $C_{18}H_{18}O_2S_2$

38.90 $C_{18}H_{18}O_4$
39.69 $C_{18}H_{18}S_4$
71.127 $C_{18}H_{18}Zr$
36.24 $C_{18}H_{19}Br_3N_2O_3$
31.44 $C_{18}H_{19}Cl$
31.45 $C_{18}H_{19}Cl$
76.95 $C_{18}H_{19}N_3O_4U$
83.112 $C_{18}H_{20}B_2CuN_{12}$
83.113 $C_{18}H_{20}B_2FeN_{12}$
83.114 $C_{18}H_{20}B_2N_{12}Ni$
74.2 $C_{18}H_{20}Cr^+$, F_6P^-
74.3 $C_{18}H_{20}Cr^+$, I_3^-
84.72 $C_{18}H_{20}Cu_2N_6O_4^{2+}$, $2BF_4^-$, $2H_2O$
75.21 $C_{18}H_{20}FeO_7$
71.128 $C_{18}H_{20}Fe_2NO_3^+$, I^-
85.61 $C_{18}H_{20}MoN_2O_4S_2$
82.35 $C_{18}H_{20}N_2O_6Pd$
31.46 $C_{18}H_{20}O$
60.38 $C_{18}H_{20}O_4$, $C_{12}H_4N_4$
39.70 $C_{18}H_{20}S_6$
71.129 $C_{18}H_{21}BrHgO_2$
83.115 $C_{18}H_{21}Br_3MoN_3$, $0.5C_6H_7N$
83.116 $C_{18}H_{21}Cl_3MoN_3$, $0.5C_6H_7N$
27.24 $C_{18}H_{21}NO_2$
37.12 $C_{18}H_{21}NO_3$
58.26 $C_{18}H_{21}NO_4$
58.27 $C_{18}H_{21}NO_4$
64.93 $C_{18}H_{21}PS_2$
65.10 $C_{18}H_{22}As_2N_4S_2$
82.36 $(C_{18}H_{22}CoN_2O_9)_n$, $2nH_2O$
81.65 $C_{18}H_{22}Cr_2N_2O_6$
73.76 $C_{18}H_{22}Mn_2O_5P_2$
82.37 $(C_{18}H_{22}N_2NiO_9)_n$, $2nH_2O$
58.28 $C_{18}H_{22}N_2O$
48.53 $C_{18}H_{22}N_2O_5$
39.71 $C_{18}H_{22}N_2O_6S_2$
40.45 $C_{18}H_{22}N_2O_7$
36.25 $C_{18}H_{22}N_4O_4S$
51.1 $C_{18}H_{22}O_2$
51.2 $C_{18}H_{22}O_2$, $0.5H_2O$
51.3 $C_{18}H_{22}O_4$
38.91 $C_{18}H_{22}O_6$
38.92 $C_{18}H_{22}O_6$
38.93 $C_{18}H_{22}O_6$
64.94 $C_{18}H_{22}P^+$, Br^-
64.95 $C_{18}H_{22}P^+$, Br^-
64.96 $C_{18}H_{22}P^+$, I^-
33.74 $C_{18}H_{23}N_2^+$, Cl^-
37.13 $C_{18}H_{23}N_3O_2$
30.5 $C_{18}H_{24}$
38.94 $C_{18}H_{24}BrNO_7$
38.95 $C_{18}H_{24}BrNO_7$
86.50 $C_{18}H_{24}Cl_2HgNP$
83.117 $C_{18}H_{24}FeN_6^{2+}$, $2Cl^-$, CH_4O
83.118 $C_{18}H_{24}FeN_6^{2+}$, $2Cl^-$, CH_4O
83.119 $C_{18}H_{24}FeN_6^{2+}$, $2Cl^-$, CH_4O
83.120 $C_{18}H_{24}FeN_6^{2+}$, $2Cl^-$, CH_4O
83.121 $C_{18}H_{24}FeN_6^{2+}$, $2Cl^-$, CH_4O
83.122 $C_{18}H_{24}FeN_6^{2+}$, $2Cl^-$, C_2H_6O
83.123 $C_{18}H_{24}FeN_6^{2+}$, $2Cl^-$, C_2H_6O

83.124 $C_{18}H_{24}FeN_6^{2+}$, $2Cl^-$, C_2H_6O
83.125 $C_{18}H_{24}FeN_6^{2+}$, $2Cl^-$, C_2H_6O
83.126 $C_{18}H_{24}FeN_6^{2+}$, $2Cl^-$, C_2H_6O
83.127 $C_{18}H_{24}FeN_6^{2+}$, $2Cl^-$, $2H_2O$
83.128 $C_{18}H_{24}FeN_6^{2+}$, $2Cl^-$, $2H_2O$
83.129 $C_{18}H_{24}FeN_6^{2+}$, $2I^-$
39.72 $C_{18}H_{24}N_2O_4S$
69.38 $C_{18}H_{24}N_4O_4S_2Sn$
83.130 $C_{18}H_{24}N_6Ni^{2+}$, $2BF_4$
64.97 $C_{18}H_{24}OP^+$, Br^-
31.47 $C_{18}H_{24}O_2$
31.48 $C_{18}H_{24}O_2$
31.49 $C_{18}H_{24}O_2$
58.29 $C_{18}H_{25}NO_5$
45.44 $C_{18}H_{25}NO_{11}$
58.30 $C_{18}H_{26}NO_6^+$, Br^-, C_2H_6O
32.66 $C_{18}H_{26}N_6O_2S_2$
48.54 $C_{18}H_{26}N_6O_6$, $4H_2O$
63.25 $C_{18}H_{26}O_2Si_2$
72.32 $C_{18}H_{26}O_4Ru_2Si_2$
45.45 $C_{18}H_{26}O_{11}$
45.46 $C_{18}H_{26}O_{11}$
83.131 $C_{18}H_{27}ClCoN_5^{2+}$, Cl_4Co^{2-}, $0.5H_2O$
58.31 $C_{18}H_{27}NO_5$, C_6H_6
58.32 $C_{18}H_{27}NO_6$
72.33 $C_{18}H_{28}Cl_2N_2Pt$
58.33 $C_{18}H_{28}NO_2^+$, Br^-
58.34 $C_{18}H_{28}NO_5^+$, I^-
33.75 $C_{18}H_{28}N_2$
38.96 $C_{18}H_{28}O_8$
45.47 $C_{18}H_{28}O_{10}$
84.73 $C_{18}H_{28}O_{12}P_2Re_2$, $2C_4H_8O$
73.77 $C_{18}H_{29}B_8Co$
72.34 $C_{18}H_{29}Ir$
16.17 $C_{18}H_{29}NOS$
41.68 $C_{18}H_{29}NOS_2$
41.69 $C_{18}H_{29}NS_2$
48.55 $C_{18}H_{29}N_3O_3$
71.130 $C_{18}H_{30}CdN_2Si_2$, $0.5C_{10}H_8N_2$
83.132 $C_{18}H_{30}CuN_6^{2+}$, $2F_6P^-$
83.133 $C_{18}H_{30}N_8O_{16}P_2Pt^{2-}$, $0.5Cs^+$, $1.5H^+$, $4H_2O$
21.11 $C_{18}H_{30}O_2$
54.1 $C_{18}H_{30}O_2$
38.97 $C_{18}H_{30}O_4$
48.56 $C_{18}H_{31}N_3O_6$, $0.25H_2O$
83.134 $C_{18}H_{32}N_2O_2Rh^+$, $C_2Cl_2O_2Rh^-$
40.46 $C_{18}H_{32}N_4O_4$, $2C_3H_2N_2$
69.39 $C_{18}H_{33}ClSn$
72.35 $C_{18}H_{34}Cl_6O_2Ta_2$
36.26 $C_{18}H_{34}N_4$
36.27 $C_{18}H_{34}N_4$
36.28 $C_{18}H_{34}N_4$
21.12 $C_{18}H_{34}O_2$
21.13 $C_{18}H_{34}O_2$
71.131 $C_{18}H_{36}Cl_4O_2Pt_2$
69.40 $C_{18}H_{36}GeN_2$
1.37 $C_{18}H_{36}N_2O_2$
86.101 $2C_{18}H_{36}N_2O_6$, $C_{30}H_{20}Fe_2O_6P_2^{2-}$, $2Na^+$
84.74 $C_{18}H_{36}N_3O_9Sm^{2+}$, $H_2N_5O_{16}Sm^{2-}$

36.29 $C_{18}H_{36}N_4$, $0.25H_2O$
70.15 $C_{18}H_{36}N_{12}S_6Te_2^{4+}$, $4ClO_4^-$
73.78 $C_{18}H_{38}FeO_2Si_6$
61.6 $2C_{18}H_{38}N_2O_6$, $C_7H_{10}N^-$, CNS^-
83.135 $C_{18}H_{40}N_4Ni^{2+}$, $2ClO_4^-$
86.51 $C_{18}H_{42}ClN_2P_2Rh$
86.52 $C_{18}H_{42}Cl_2P_2Pd$
64.98 $C_{18}H_{42}N_3O_3P_3$
64.99 $C_{18}H_{42}N_4O_4P_2Si_2$
86.53 $C_{18}H_{45}ClP_3Pt^+$, BF_4^-
86.54 $C_{18}H_{45}FP_3Pt^+$, BF_4^-
85.62 $C_{18}H_{45}Fe_6Mo_2S_{17}^{3-}$, $3C_{13}H_{22}N^+$
71.132 $C_{18}H_{46}Mo_2O_4P_2Si_2$
86.55 $C_{18}H_{46}P_3Pt^+$, F_6P^-
86.56 $C_{18}H_{48}CrP_6$
76.96 $C_{18}H_{54}Co_2N_{12}O_2^{4+}$, $4ClO_4^-$, $2H_2O$
63.26 $C_{18}H_{54}N_2O_8Si_{10}$
86.57 $C_{18}H_{57}O_3P_6Ru_2^+$, BF_4^-
86.58 $C_{18}H_{57}O_{18}P_6Rh_3$
86.59 $C_{18}H_{58}Mo_2P_6$

C_{19}

28.9 $C_{19}Cl_{12}O$
71.133 $C_{19}H_8CoF_{10}$
71.134 $C_{19}H_8F_{10}Ni$
71.135 $C_{19}H_8O_9Ru_3$
71.136 $C_{19}H_{10}MnO_7Re$
13.18 $C_{19}H_{11}Cl_2I_2NO_3$
65.11 $C_{19}H_{13}As$
64.100 $C_{19}H_{13}Cl_2O_3P$
64.101 $C_{19}H_{13}P$
29.6 $C_{19}H_{14}$
39.73 $C_{19}H_{14}O_2S_2$
22.7 $C_{19}H_{15}^+$, BF_4^-
71.137 $C_{19}H_{15}Cl_4NTa^-$, $C_5H_6N^+$
28.10 $C_{19}H_{15}NO$
35.52 $C_{19}H_{15}N_3O_2$
64.102 $C_{19}H_{15}N_3O_9P^+$, $C_6H_5NO_3$, $0.5C_6H_6$, Cl^-
64.103 $C_{19}H_{15}OPS_2$
71.138 $C_{19}H_{16}HgN_4S$
71.139 $C_{19}H_{16}MnO_5P$
64.104 $C_{19}H_{16}NOP$, C_6H_6O
17.28 $C_{19}H_{16}O_2$
17.29 $C_{19}H_{16}O_2$
20.10 $C_{19}H_{18}O$
59.19 $C_{19}H_{18}O_5$
86.60 $C_{19}H_{19}BrN_4O_3PRe$
32.67 $C_{19}H_{19}Cl_3N_2O_4$
31.50 $C_{19}H_{19}NO_4$
86.61 $C_{19}H_{19}N_4O_3PW$
48.57 $C_{19}H_{19}N_7O_6$, $2H_2O$
84.75 $2C_{19}H_{19}N_8O_5U^+$, $N_4O_{14}U^{2-}$
82.38 $C_{19}H_{20}CoN_3O_6$, $3H_2O$
77.7 $(C_{19}H_{20}CuF_{12}NO_6)_n$
71.140 $C_{19}H_{20}Fe_2O_5$
58.35 $C_{19}H_{20}N_2O$, CH_4O
58.36 $C_{19}H_{20}N_2O_7$, $0.67C_4H_8O_2$
38.98 $C_{19}H_{21}BrN_2O_3S$
38.99 $C_{19}H_{21}BrN_2O_3S$

83.136 $C_{19}H_{21}ClCu_2N_8O_{11}$, H_2O
36.30 $C_{19}H_{21}N$
7.4 $C_{19}H_{21}N$
83.137 $C_{19}H_{21}N_8O_5Zn^+$, NO_3^-
71.141 $C_{19}H_{22}CrN_2O_4$
58.37 $C_{19}H_{22}NO_3^+$, Br^-
58.38 $C_{19}H_{22}N_2O$
39.74 $C_{19}H_{22}OS$
51.4 $C_{19}H_{22}O_2$
38.100 $C_{19}H_{22}O_4$
53.21 $C_{19}H_{22}O_7$
51.5 $C_{19}H_{23}BrO_3$
73.79 $C_{19}H_{23}GdO$
47.35 $C_{19}H_{23}IN_8O_{12}P^-$, $C_{17}H_{20}N_3^+$, $12H_2O$
60.36 $2C_{19}H_{23}IN_8O_{12}P^-$, $2C_{17}H_{15}N_2^+$, $20H_2O$, $2CH_4O$
60.37 $2C_{19}H_{23}IN_8O_{12}P^-$, $2C_{17}H_{19}N_2^+$, $17H_2O$, $2CH_4O$
47.36 $2C_{19}H_{23}IN_8O_{12}P^-$, $2C_{13}H_{12}N_3^+$, $15H_2O$, CH_4O
60.22 $4C_{19}H_{23}IN_8O_{12}P^-$, $4C_{13}H_{11}N_2^+$, $21H_2O$
75.22 $C_{19}H_{23}Mo_2O_5$
28.11 $C_{19}H_{23}N$
17.30 $C_{19}H_{23}NO$
17.31 $C_{19}H_{23}NO_2$
28.12 $C_{19}H_{23}NO_2$
40.47 $C_{19}H_{23}NO_3$
35.53 $C_{19}H_{23}N_2O^+$, Cl^-
32.68 $C_{19}H_{23}N_3O$
48.58 $C_{19}H_{23}N_3O_3$
44.35 $C_{19}H_{23}N_3O_3S^{2+}$, $2Cl^-$, $3H_2O$
64.105 $C_{19}H_{23}O_3P$
58.39 $C_{19}H_{24}BrNO_3$
35.54 $C_{19}H_{24}NO_5^+$, Cl^-, H_2O
41.70 $C_{19}H_{24}N_2OS$
40.48 $C_{19}H_{24}N_2O_4$
36.31 $C_{19}H_{24}N_4O_4S_2$
85.58 $2C_{19}H_{24}N_8O_{10}P^-$, $2C_{17}H_{16}N_3OPtS^+$, xH_2O
60.24 $C_{19}H_{24}N_8O_{12}P^-$, $1.5C_{13}H_{12}N_3^+$, $0.5O_4S^-$, $11.5H_2O$
47.37 $C_{19}H_{24}N_8O_{12}P^-$, $C_{21}H_{20}N_3^+$
17.32 $C_{19}H_{24}O_2$
51.6 $C_{19}H_{24}O_2$
51.7 $C_{19}H_{24}O_2$
75.23 $C_{19}H_{24}O_5Ru_2Si_2$
53.22 $C_{19}H_{24}O_6$
53.23 $C_{19}H_{24}O_6$
51.10 $C_{19}H_{25}BrO_2$
51.8 $C_{19}H_{25}BrO_2$
51.9 $C_{19}H_{25}BrO_2$
73.80 $C_{19}H_{25}FeNO_2Pd$
83.138 $C_{19}H_{25}Mn_2N_{19}S_4$
63.27 $C_{19}H_{25}NOSi$
63.28 $C_{19}H_{25}NO_2Si$
48.59 $C_{19}H_{25}N_7O_5$, $4H_2O$
47.38 $2C_{19}H_{25}N_8O_{10}P$, $2C_{13}H_{11}N_3$, xH_2O
74.4 $C_{19}H_{26}CrN_2O_3$
74.5 $C_{19}H_{26}CrN_2O_3$
74.6 $C_{19}H_{26}CrN_2O_3$

58.40 $C_{19}H_{26}N_2O_2$, $0.5H_2O$
40.49 $C_{19}H_{26}N_2O_7$
60.39 $C_{19}H_{26}N_8O_{11}P$, $C_{13}H_{11}N_3$, $10H_2O$
51.11 $C_{19}H_{26}O_2$
51.12 $C_{19}H_{26}O_2$
51.13 $C_{19}H_{26}O_2$, C_2H_6O
51.14 $C_{19}H_{26}O_2$, H_2O
51.15 $C_{19}H_{26}O_2$, $0.5H_2O$
51.16 $C_{19}H_{26}O_3$
71.142 $C_{19}H_{26}O_6P_2PtW$
31.51 $C_{19}H_{26}O_7$
53.24 $C_{19}H_{26}O_7$
27.25 $C_{19}H_{27}BrO_7$
64.106 $C_{19}H_{27}ClNOPSi_2$
58.41 $C_{19}H_{27}NO_2$
48.60 $C_{19}H_{28}N_2O_5S$
63.29 $C_{19}H_{28}Si_3$
63.30 $C_{19}H_{28}Si_3$
71.143 $C_{19}H_{29}CoN_6O_4$
71.144 $C_{19}H_{29}CoN_6O_4$
71.145 $C_{19}H_{29}CoN_6O_4$
71.146 $C_{19}H_{29}CoN_6O_4$
73.81 $C_{19}H_{29}O_3PRh^+$, F_6P^-
58.42 $C_{19}H_{30}NO_2^+$, Br^-
59.20 $C_{19}H_{30}O_2$
45.48 $C_{19}H_{31}N_2O_7^+$, Cl^-, H_2O
48.61 $C_{19}H_{31}N_5O_5$, $3H_2O$
48.62 $C_{19}H_{32}N_4O_4$
51.17 $C_{19}H_{32}O_2$
38.101 $C_{19}H_{38}O_3$
69.41 $C_{19}H_{39}N_3S_6Sn$
3.54 $C_{19}H_{42}N^+$, Cl_2I^-
81.66 $C_{19}H_{45}O_3P_2Rh$
64.107 $C_{19}H_{57}N_7P_2Si_4$

C_{20}

71.147 $C_{20}H_6Cu_2N_2O_{15}Rh_6$, $0.5CH_4O$
71.148 $C_{20}H_{10}O_{12}Os_4$
84.76 $C_{20}H_{12}Cl_4Cr_2N_4O_4$
84.77 $C_{20}H_{12}Cl_4Mo_2N_4O_4$
84.78 $C_{20}H_{12}Cl_4N_4O_4W_2$
71.149 $C_{20}H_{12}Mn_2O_7$, $0.5C_4H_{10}O$
40.50 $C_{20}H_{12}N_2O_2$
41.71 $C_{20}H_{12}N_2S_2$
83.139 $C_{20}H_{13}N_2O_{10}Re_3$
17.33 $C_{20}H_{14}Br_4O_6S_2$
60.40 $2C_{20}H_{14}CuN_2O_2$, $C_{12}H_4N_4$
35.55 $C_{20}H_{14}NO_2$
35.56 $C_{20}H_{14}N_2$
35.57 $C_{20}H_{14}N_2$
38.102 $C_{20}H_{14}O_4$
86.62 $(C_{20}H_{15}HgN_2PS_2)_n$
41.72 $C_{20}H_{15}NO_3S_2$
71.150 $C_{20}H_{15}NO_8Os_3$
71.151 $C_{20}H_{15}NiO_9Ru_3$
65.12 $C_{20}H_{16}As^+$, Cl^-, H_2O
86.63 $C_{20}H_{16}As_2Fe_3O_{10}$
86.64 $C_{20}H_{16}As_2Ir_4O_{10}$
67.3 $C_{20}H_{16}BeN_2O_2$, $2H_2O$

83.140 $C_{20}H_{16}BrCuN_4^-$, BF_4^-
85.63 $C_{20}H_{16}CdN_2S_2$
58.43 $C_{20}H_{16}ClNO$
84.79 $(C_{20}H_{16}ClN_4O_4Tc_2)_n$
84.80 $C_{20}H_{16}Cl_2N_4O_4Os_2$
28.13 $C_{20}H_{16}Cl_2O_7$
84.81 $C_{20}H_{16}Cl_6CoN_4O_2$
71.152 $C_{20}H_{16}Cl_6HgN_2$
83.141 $C_{20}H_{16}CuIN_4^+$, ClO_4^-
85.64 $C_{20}H_{16}CuN_2S_4$
84.82 $C_{20}H_{16}CuN_4O_8S_2$, H_2O
83.142 $C_{20}H_{16}CuN_5O_3^-$, NO_3^-, H_2O
31.52 $C_{20}H_{16}F_6N_2O_2$
75.24 $C_{20}H_{16}FeO_3Si$
71.153 $C_{20}H_{16}Fe_3O_6S_3$
83.143 $C_{20}H_{16}HgN_4^{2+}$, $2NO_3^-$, $2H_2O$
75.25 $C_{20}H_{16}MoO_4$
75.26 $C_{20}H_{16}Mo_2O_4$
33.76 $0.4C_{20}H_{16}N_2$, $0.6C_{20}H_{16}N_2$
33.76 $0.6C_{20}H_{16}N_2$, $0.4C_{20}H_{16}N_2$
35.58 $C_{20}H_{16}N_2O_2$
39.75 $C_{20}H_{16}N_2O_4S_2$
33.77 $C_{20}H_{16}N_8$
39.76 $C_{20}H_{16}O_2S_2$
29.7 $C_{20}H_{16}O_5$
72.36 $C_{20}H_{17}AsCl_2Pt$
72.37 $C_{20}H_{17}F_6O_{10}Os_3P$
71.154 $C_{20}H_{17}FeNO$
35.59 $C_{20}H_{17}NO$
27.26 $C_{20}H_{17}NO_3$
58.44 $C_{20}H_{17}NO_6$
64.108 $C_{20}H_{17}OP$
27.27 $C_{20}H_{18}$
62.12 $C_{20}H_{18}BN_2O_2^-$, $C_4H_{12}N^+$
62.13 $C_{20}H_{18}B_2F_{10}N_2$
83.144 $C_{20}H_{18}ClCuN_6^+$, Cl^-, $4H_2O$
83.145 $C_{20}H_{18}CuN_4O^{2+}$, $O_6S_2^{2-}$
73.82 $C_{20}H_{18}Fe_2Se^+$, I_3^-, I_2, $0.5CH_2Cl_2$
71.155 $C_{20}H_{18}Hf$
67.4 $C_{20}H_{18}MgN_8O_{10}$
71.156 $C_{20}H_{18}Mo_2$
32.69 $C_{20}H_{18}N_4S_2$
69.42 $C_{20}H_{18}S_2Sn$
36.32 $C_{20}H_{19}ClN_2O_3S$
37.14 $C_{20}H_{19}ClN_2O_3S$
37.15 $C_{20}H_{19}I_2NO_2$
41.73 $C_{20}H_{19}N_7O_2S_2$
83.146 $C_{20}H_{19}N_4O_2Ru^{2+}$, $2ClO_4^-$
37.16 $C_{20}H_{20}ClN_3O_4S$, C_5H_5N
73.83 $C_{20}H_{20}Cl_2OTi_2$
73.84 $C_{20}H_{20}FeO$
72.38 $C_{20}H_{20}Fe_2O_6$
50.13 $C_{20}H_{20}N_2O_7S$
50.14 $C_{20}H_{20}N_2O_7S$, C_3H_7NO
39.77 $C_{20}H_{20}OS_2$
38.103 $C_{20}H_{20}O_4$
60.41 $C_{20}H_{20}P^+$, $C_{12}H_4N_4^-$, $C_{12}H_4N_4$
64.109 $C_{20}H_{20}P_2$
64.110 $C_{20}H_{21}P_2^+$, BF_4^-
35.60 $C_{20}H_{22}N_4O_5S_2$, $3.5H_2O$

31.53 $C_{20}H_{22}O_4$
59.21 $C_{20}H_{22}O_6$
39.78 $C_{20}H_{32}S_4$
59.22 $C_{20}H_{23}IN_2O_4$
64.111 $C_{20}H_{23}NO_2P^+$, Br^-
37.17 $C_{20}H_{23}N_3O_2$
36.33 $C_{20}H_{23}N_7O_6^{2+}$, $2Br^-$, $2H_2O$
48.63 $C_{20}H_{23}N_7O_6^{2+}$, $2Br^-$, $2H_2O$
73.85 $C_{20}H_{23}Ni_4$
73.86 $C_{20}H_{33}W_2^+$, ClO_4^-, C_3H_6O
73.87 $C_{20}H_{24}B_4Ni_4$
41.74 $C_{20}H_{24}ClN_3S$, $2CH_4O_3S$
71.157 $C_{20}H_{24}Cl_2N_8Rh_2^{2+}$, $2Cl^-$, $8H_2O$
84.83 $C_{20}H_{24}Cl_3O_{18}Sm$
17.34 $C_{20}H_{24}N_2O_4$
17.35 $C_{20}H_{24}N_2O_9$, K^+, CNS^-
71.158 $C_{20}H_{24}N_8Rh_2^{2+}$, $2C_{24}H_{20}B^-$, C_2H_3N
51.18 $C_{20}H_{24}O_2$
51.19 $C_{20}H_{24}O_5$
39.79 $C_{20}H_{24}S$
39.80 $C_{20}H_{24}S$
33.78 $C_{20}H_{25}NO$
17.36 $C_{20}H_{25}NO_2$
58.45 $C_{20}H_{25}NO_5$
59.23 $C_{20}H_{25}NO_6$
86.65 $2C_{20}H_{26}AgB_8P$, C_3H_6O
86.66 $C_{20}H_{26}B_9NO_3PRh$, $3CH_2Cl_2$
83.147 $C_{20}H_{26}ClCoN_4O_4$
74.7 $C_{20}H_{26}CrO_5Si_2$
83.148 $C_{20}H_{26}MnN_2O_8P_2^+$, ClO_4^-
86.67 $C_{20}H_{26}MnN_2O_8P_2^+$, ClO_4^-
1.38 $C_{20}H_{26}NO_2^+$, Br^-
40.51 $C_{20}H_{26}N_2O_6$
59.24 $C_{20}H_{26}O_4$
74.8 $C_{20}H_{26}Ru$
83.149 $C_{20}H_{27}MnN_6O^{2+}$, $2ClO_4^-$
28.14 $C_{20}H_{27}N$
58.46 $C_{20}H_{27}NO$
63.31 $C_{20}H_{27}NOSi$
81.67 $C_{20}H_{28}Cl_2O_4P_2Pd_2$, $0.5CHCl_3$
81.68 $C_{20}H_{28}CuN_6O_8$
84.84 $C_{20}H_{28}MoN_2O_4$
58.47 $C_{20}H_{28}NO^+$, Br^-
58.48 $C_{20}H_{28}N_2O$
48.64 $C_{20}H_{28}N_2O_5$
41.75 $C_{20}H_{28}N_4O_2S$
41.76 $C_{20}H_{28}N_4O_2S$
81.69 $C_{20}H_{28}N_6NiO_8$
83.150 $C_{20}H_{28}N_{10}Ni_3O_2$
54.2 $C_{20}H_{28}O$
51.20 $C_{20}H_{28}O_2$
51.21 $C_{20}H_{28}O_2$
54.3 $C_{20}H_{28}O_6$
54.4 $C_{20}H_{28}O_7$, H_2O
53.25 $C_{20}H_{28}O_7S$
27.28 $C_{20}H_{29}BrN_2O_2S$
51.22 $C_{20}H_{29}FO_3$
80.19 $C_{20}H_{29}MoN_6S_4^+$, $C_{24}H_{20}B^-$
48.65 $C_{20}H_{29}N_3O_6$
60.42 $C_{20}H_{30}Fe^+$, $C_{12}H_4N_4^-$

73.88 $C_{20}H_{30}I_4Rh_2$, $2C_7H_8$
54.5 $C_{20}H_{30}O_2$
54.6 $C_{20}H_{30}O_3$
54.7 $C_{20}H_{30}O_4$
54.8 $C_{20}H_{30}O_4$
53.26 $C_{20}H_{30}O_5$
31.54 $C_{20}H_{30}S_2$
69.43 $C_{20}H_{30}Sn$
86.68 $C_{20}H_{32}As_4Cl_2Tc^+$, Cl^-
86.69 $C_{20}H_{32}As_4Cl_2Tc^+$, ClO_4^-
86.70 $C_{20}H_{32}As_4Cl_4Tc^+$, F_6P^-
58.49 $C_{20}H_{32}NO^+$, Br^-
80.20 $C_{20}H_{32}N_4S_8Zn_2$
83.151 $C_{20}H_{32}N_{12}O_{10}Pt^{2+}$, $1.5Cl^-$, $0.5ClO_4^-$, $7H_2O$
54.9 $C_{20}H_{32}O_3$
54.10 $C_{20}H_{32}O_4$
59.25 $C_{20}H_{32}O_5$
71.159 $C_{20}H_{33}Al_2ClZr$
71.160 $C_{20}H_{33}ClPPdSi^+$, F_6P^-
69.44 $C_{20}H_{33}F_3O_2Sn$
54.11 $C_{20}H_{34}$
86.71 $C_{20}H_{34}AuO_9PS$
54.12 $C_{20}H_{34}O_2$
59.26 $C_{20}H_{34}O_5$
59.27 $C_{20}H_{34}O_6$
63.32 $C_{20}H_{34}O_7Si_6$
45.49 $C_{20}H_{35}NO_{16}$
58.50 $C_{20}H_{35}N_3$, $C_{20}H_{33}N_3$
20.11 $C_{20}H_{36}$
86.72 $C_{20}H_{36}Cr_2O_8P_2Si_6$
58.51 $2C_{20}H_{36}N_3^{3+}$, $6Br^-$, $3.5H_2O$
71.161 $C_{20}H_{38}CuN_4^+$, $C_{24}H_{20}B^-$
67.5 $C_{20}H_{38}MgN_2$
71.162 $C_{20}H_{39}F_6OP_2Pt^+$, F_6P^-
80.21 $(C_{20}H_{40}Cl_2Hg_3N_4S_8)_n$
71.163 $C_{20}H_{40}Cl_2Mo_2O_6P_2Si_2$
83.152 $C_{20}H_{40}CuN_4^{2+}$, $2ClO_4^-$
83.153 $C_{20}H_{40}CuN_4^{2+}$, $2ClO_4^-$
81.70 $C_{20}H_{40}O_{10}Rh_2$
86.73 $C_{20}H_{41}ClCoN_4O_4P$
71.164 $C_{20}H_{41}O_4P_2Pt^+$, $C_{24}H_{20}B^-$
71.165 $C_{20}H_{41}O_4P_2Pt^+$, F_6P^-
3.55 $C_{20}H_{42}NO_6P$, H_2O
64.112 $C_{20}H_{42}N_2P_2S_2$
75.27 $C_{20}H_{44}P_4Zr$
83.154 $C_{20}H_{48}Mo_2N_4$
83.155 $C_{20}H_{48}N_4Si_2Ti$
83.156 $C_{20}H_{48}N_4Si_2Zr$
76.97 $C_{20}H_{54}Co_2N_{10}O_2^{4+}$, $O_6S_2^{2-}$, $4H_2O$, $2NO_3^-$
84.85 $C_{20}H_{56}EuN_2O_4Si_4$
71.166 $C_{20}H_{58}P_6Ru_2^{2+}$, $2BF_4^-$

C_{21}

85.65 $C_{21}H_{13}CuN_3S_7^+$, ClO_4^-, $2CHCl_3$
71.167 $C_{21}H_{14}Co_2O_4$
71.168 $C_{21}H_{14}Co_2O_7$
73.89 $C_{21}H_{15}ClMn_3O_6Sb$

84.86 $C_{21}H_{15}CoO_6$
71.169 $C_{21}H_{15}Co_3CrO_5S_2$, $0.5C_4H_8O$
85.66 $C_{21}H_{15}CrN_3O_3S_3^{3-}$, $C_7H_{18}N^+$, $2.5Na^+$, $0.5HO^-$, $18.5H_2O$
84.87 $C_{21}H_{15}CrN_3O_6^{3-}$, $3Na^+$, $8H_2O$, C_2H_6O
84.88 $C_{21}H_{15}CrN_3O_6^{3-}$, $5Na^+$, I^-, HO^-, $9H_2O$, $3CH_4O$, C_2H_6O
41.77 $C_{21}H_{15}N_3OS$
83.157 $C_{21}H_{16}ClN_4ORu^+$, ClO_4^-
35.61 $C_{21}H_{16}NO$
35.62 $C_{21}H_{16}N_2O_2$
40.52 $C_{21}H_{16}N_2O_4S$
71.170 $C_{21}H_{16}O_9W_2$
78.7 $C_{21}H_{17}CuN_4OS^+$, NO_3^-
71.171 $C_{21}H_{17}NO_9Os_3$
13.19 $C_{21}H_{17}N_3OS$
36.34 $C_{21}H_{17}N_3O_2$
40.53 $C_{21}H_{17}N_3O_5$
28.15 $C_{21}H_{18}$
25.4 $C_{21}H_{18}ClNO_4$
35.63 $C_{21}H_{18}N_2O$
41.78 $C_{21}H_{18}N_2O_5S_2$
38.104 $C_{21}H_{18}N_2O_9$
21.14 $C_{21}H_{18}O$
21.15 $C_{21}H_{18}O$
21.16 $C_{21}H_{18}O$
59.28 $C_{21}H_{19}BrN_2O_2$
86.74 $C_{21}H_{19}CrO_9P$
33.79 $C_{21}H_{19}N_2^+$, Br^-
86.75 $C_{21}H_{20}ClIrNOP$
25.5 $C_{21}H_{20}ClNO_6S$
47.37 $C_{21}H_{20}N_3^+$, $C_{19}H_{24}N_8O_{12}P^-$
17.37 $C_{21}H_{20}O_2$
71.172 $C_{21}H_{20}O_5PRe$
75.28 $C_{21}H_{21}CrNO_6$
27.29 $C_{21}H_{21}NO_4$
66.14 $C_{21}H_{21}Sb$
85.67 $C_{21}H_{22}BCoF_5N_6S$
36.35 $C_{21}H_{22}BrN_5$
35.64 $C_{21}H_{22}ClFN_4O_2$
84.120 $C_{21}H_{22}Cl_2CoN_2O_5S$, $C_{36}H_{32}Cl_4Co_2N_4O_8S_2$
64.113 $C_{21}H_{22}NO_4P$
32.70 $C_{21}H_{22}N_2O$
59.29 $C_{21}H_{22}O_2$
58.52 $C_{21}H_{23}NO_5$
58.53 $C_{21}H_{23}NO_6$, $C_4H_8O_2$, H_2O
69.45 $C_{21}H_{24}BrNOSn$
84.89 $C_{21}H_{24}N_4NiO_2S_2$
40.54 $C_{21}H_{24}N_4O_3$
38.105 $C_{21}H_{24}O_4$
75.29 $C_{21}H_{24}O_7W_2$
53.27 $C_{21}H_{24}O_9$
35.65 $C_{21}H_{25}NO$
13.20 $C_{21}H_{25}NO_5$
85.68 $C_{21}H_{26}BCuN_7O_2S^-$, K^+, $2C_3H_6O$
84.90 $C_{21}H_{26}ClNO_6U$
41.79 $C_{21}H_{26}F_3N_3S^{2+}$, $2Cl^-$
35.66 $C_{21}H_{26}N_4$
51.23 $C_{21}H_{26}O_2$

51.24 $C_{21}H_{26}O_3$
54.13 $C_{21}H_{26}O_4$
51.25 $C_{21}H_{26}O_5$
53.28 $C_{21}H_{27}BrN_2$
37.18 $C_{21}H_{27}N_3O_2$
31.55 $C_{21}H_{28}$
74.9 $C_{21}H_{28}CrN_2O_4$
38.106 $C_{21}H_{28}NO_4^+$, Br^-
58.54 $C_{21}H_{28}N_2O_2$
51.26 $C_{21}H_{28}O_2$, $0.5H_2O$
63.33 $C_{21}H_{29}N_2Si^+$, $C_4H_3O_4^-$
21.17 $C_{21}H_{30}BrNO_2$
71.173 $C_{21}H_{30}MnO_2P_2Pt^+$, BF_4^-, CH_2Cl_2
48.66 $C_{21}H_{30}N_6O_6$
48.67 $C_{21}H_{30}N_6O_6$, $0.5Ca^{2+}$, ClO_4^-, $1.5H_2O$
48.68 $C_{21}H_{30}N_6O_6$, $1.5H_2O$
48.69 $C_{21}H_{30}N_6O_6$, Mg^{2+}, $2ClO_4^-$, $3H_2O$
48.70 $2C_{21}H_{30}N_6O_6$, Ca^{2+}, $2ClO_4^-$
54.14 $C_{21}H_{30}O_2$
51.27 $C_{21}H_{31}BrO_3$
54.15 $C_{21}H_{31}N$
54.16 $C_{21}H_{31}N$
27.30 $C_{21}H_{32}N_2O_2S$
63.34 $C_{21}H_{32}N_2Si_2$
38.107 $C_{21}H_{32}O_3$
38.108 $C_{21}H_{32}O_4$
58.55 $C_{21}H_{33}NO_3$
58.56 $C_{21}H_{33}NO_4$
51.28 $C_{21}H_{34}O_2$
38.109 $C_{21}H_{36}O_7$
86.76 $C_{21}H_{45}CrO_3P_3$
71.174 $C_{21}H_{46}P_2W$
76.98 $C_{21}H_{48}Cl_2Cu_2N_4O_3$
71.175 $C_{21}H_{60}P_6Ru_2$
71.176 $C_{21}H_{61}P_6Ru_2^+$, BF_4^-

C_{22}

86.77 $C_{22}H_{10}Co_4O_{10}P_2$
64.114 $C_{22}H_{14}F_6O_6P^-$, $C_6H_{16}N^+$
64.115 $C_{22}H_{15}ClN_3P$
72.39 $C_{22}H_{15}CoNiO_3$
75.30 $C_{22}H_{15}F_{12}Rh$
86.78 $C_{22}H_{15}FeO_4P$
86.79 $C_{22}H_{15}MnO_4P^-$, $C_{24}H_{20}P^+$
38.110 $C_{22}H_{15}NO_6$
73.90 $C_{22}H_{15}Nb_3O_7$
27.31 $C_{22}H_{16}$
77.8 $C_{22}H_{16}CuF_6N_2O_4$
67.6 $C_{22}H_{16}MgN_6O_{10}$
30.6 $C_{22}H_{16}O_2$
30.7 $C_{22}H_{16}O_2$
71.177 $C_{22}H_{16}O_{15}Ru_6S_3$
34.10 $C_{22}H_{18}BrNO_4$
73.91 $C_{22}H_{18}FeGe$
83.158 $C_{22}H_{18}N_4OV$
33.80 $C_{22}H_{18}N_6$
33.81 $C_{22}H_{18}N_6$
38.111 $C_{22}H_{18}O_2$
36.36 $C_{22}H_{19}Br_3N_2O_3$

81.71 $C_{22}H_{19}CuN_4O_2^-$, BF_4^-

81.72 $C_{22}H_{19}CuN_4O_2^-$, ClO_4^-, H_2O

19.12 $C_{22}H_{19}NO$

64.116 $C_{22}H_{19}N_2P$

35.67 $C_{22}H_{19}N_5O_4$

64.117 $C_{22}H_{19}O_4P$

20.12 $C_{22}H_{20}Cl_2N_4$

20.13 $C_{22}H_{20}Cl_2N_4$

83.159 $C_{22}H_{20}CoN_6S_2$

83.160 $C_{22}H_{20}CoN_6S_2$, $2CHI_3$

71.178 $C_{22}H_{20}Fe_2Ni_2O_6$

75.31 $C_{22}H_{20}Mo_2O_2$

1.39 $C_{22}H_{20}N_2O_4$

30.10 $C_{22}H_{20}O_2$

30.11 $C_{22}H_{20}O_2$

30.12 $C_{22}H_{20}O_2$

30.8 $C_{22}H_{20}O_2$

30.9 $C_{22}H_{20}O_2$

73.92 $C_{22}H_{20}O_2W$

59.30 $C_{22}H_{21}BrO_3$

31.56 $C_{22}H_{21}NO_6S$

41.80 $C_{22}H_{21}N_3O_5S_2$

71.179 $C_{22}H_{22}CoN_4^+$, $C_8H_{20}Al^-$

84.91 $C_{22}H_{22}N_2NiO_2$

11.19 $C_{22}H_{22}O_6S_2$

59.31 $C_{22}H_{22}O_8$

36.37 $C_{22}H_{23}BrN_4O_4$

29.8 $C_{22}H_{23}ClN^+$, Cl^-

38.112 $C_{22}H_{23}NO_5$

50.15 $C_{22}H_{24}N_2O_9$, $4H_2O$, CH_2Cl_2

64.118 $C_{22}H_{24}P^+$, BF_4^-

29.9 $C_{22}H_{25}ClN^+$, $C_4H_3O_4^-$

74.10 $C_{22}H_{25}CrO_2P$

58.57 $C_{22}H_{25}NO_7$

33.82 $C_{22}H_{25}N_2^+$, $C_7H_7O_3S^-$

69.46 $C_{22}H_{25}O_2PS_2Sn$

64.119 $C_{22}H_{25}O_3P$

58.58 $C_{22}H_{26}Cl_2N_2O_4$

71.180 $C_{22}H_{26}HgN_3^+$, NO_3^-

86.80 $C_{22}H_{26}Mn_3O_{11}P$

38.113 $C_{22}H_{26}O_7$, $C_{10}H_8O_2$, H_2O

73.93 $C_{22}H_{27}As_2CoFe_2O_3$

71.181 $C_{22}H_{27}F_6Ir_3O_6S_3$

78.8 $C_{22}H_{28}Cl_2N_2O_2V$, C_6H_6

71.182 $C_{22}H_{28}CoN_5O_5$, C_6H_6

78.9 $C_{22}H_{28}CuN_2O_4$

58.59 $C_{22}H_{28}NO^+$, Br^-, $0.5H_2O$

33.83 $C_{22}H_{28}N_2O_2$

81.73 $C_{22}H_{28}N_6O_{12}Rh_2$, $2H_2O$

54.17 $C_{22}H_{28}O_5$

53.29 $C_{22}H_{28}O_7$

53.30 $C_{22}H_{28}O_7$, H_2O

53.31 $C_{22}H_{28}O_8$

81.74 $C_{22}H_{28}O_8Rh_2S_2$

53.32 $C_{22}H_{29}BrO_2$

33.84 $C_{22}H_{29}N_2O^+$, $C_6H_7O_7^-$, C_7H_8

19.13 $C_{22}H_{30}$

19.14 $C_{22}H_{30}$

68.21 $C_{22}H_{30}Al_2$

84.92 $C_{22}H_{30}CeN_{14}O_4^{3+}$, $3ClO_4^-$, $3H_2O$

71.183 $C_{22}H_{30}HgN_2O_4Pt$

35.68 $C_{22}H_{30}NO_2^+$, I^-

63.35 $C_{22}H_{30}N_2Si$

51.29 $C_{22}H_{30}O_2$

51.30 $C_{22}H_{30}O_2$

73.94 $C_{22}H_{30}O_2Ti$

54.18 $C_{22}H_{30}O_7$

85.69 $(C_{22}H_{31}BrHgO_2S)_n$

53.33 $C_{22}H_{31}BrO_4$

72.40 $C_{22}H_{31}F_6O_2Rh$

72.41 $C_{22}H_{32}Cl_2N_2Pt$

83.161 $C_{22}H_{32}Cl_4N_6Ni_2$

83.162 $C_{22}H_{32}CoN_6O_6$

71.184 $C_{22}H_{32}Co_2O$

54.19 $C_{22}H_{32}N_2$

33.85 $C_{22}H_{32}N_4OS_4$

51.31 $C_{22}H_{32}O_3$

59.32 $C_{22}H_{32}O_7$

51.32 $C_{22}H_{33}D_3O_2$, CH_4O

51.33 $C_{22}H_{33}IO_3$

58.60 $C_{22}H_{33}NO_2$

59.33 $C_{22}H_{33}NO_5$

58.61 $C_{22}H_{33}NO_8$

73.95 $C_{22}H_{34}AsMo_2O_4P_2^+$, $C_8H_5MoO_3^-$

73.96 $C_{22}H_{34}I_4Rh_2$

75.32 $C_{22}H_{34}Mo_2O_3S_3$

75.33 $C_{22}H_{34}Mo_2O_3S_3$

81.75 $C_{22}H_{34}N_4NiO_4$

81.76 $C_{22}H_{34}N_4NiO_4$

51.34 $C_{22}H_{34}O_3$

63.36 $C_{22}H_{34}O_3Si_2$

51.35 $C_{22}H_{34}O_4$, H_2O

71.185 $C_{22}H_{35}CrO_6P_3Pt$

23.6 $C_{22}H_{35}N_3O$

68.22 $C_{22}H_{36}Al_2Cl_2$

38.114 $C_{22}H_{36}O_2$

51.36 $C_{22}H_{36}O_2$, CH_4O

51.37 $C_{22}H_{36}O_2$, CH_4O

54.20 $C_{22}H_{36}O_3$

54.21 $C_{22}H_{36}O_4$

54.22 $C_{22}H_{36}O_4$

63.37 $C_{22}H_{36}Si_4$

71.186 $C_{22}H_{37}O_5P_3PtW$

48.71 $C_{22}H_{38}N_2O_6$

48.72 $C_{22}H_{38}N_2O_6$

80.22 $C_{22}H_{40}N_4O_2Ru_2S_8^+$, BF_4^-

63.38 $C_{22}H_{40}O_{12}Si_4$

86.81 $C_{22}H_{41}CoF_3N_8O_4P$

1.40 $C_{22}H_{42}NO_2^+$, I^-

86.82 $C_{22}H_{49}Fe_2O_6P_4S_2^+$, $C_{24}H_{20}B^-$

67.7 $C_{22}H_{50}Li_2N_4S_4$

71.187 $C_{22}H_{56}Cl_2N_6Pt_2^{2+}$, $2ClO_4^-$

63.39 $C_{22}H_{60}N_4Si_6$

C_{23}

86.83 $C_{23}H_{15}As_2Co_2O_5P$

86.84 $C_{23}H_{15}Co_2O_5P_3$

71.188 $C_{23}H_{16}FeO_3$

37.19 $C_{23}H_{18}N_2O_4$

24.9 $C_{23}H_{18}O$

24.10 $C_{23}H_{18}O_2$

73.97 $C_{23}H_{19}Fe^+$, BF_4^-

69.47 $C_{23}H_{19}NPbS$

72.42 $C_{23}H_{20}Cl_2O_2PW^-$, $C_8H_{20}N^+$

72.43 $C_{23}H_{20}MoN_2O_3$

83.163 $C_{23}H_{21}ClMnN_7^+$, BF_4^-

84.93 $C_{23}H_{21}LaN_8O_{11}$

19.15 $C_{23}H_{21}NO$

35.69 $C_{23}H_{21}N_3O_3S$

64.120 $C_{23}H_{21}P$

71.189 $C_{23}H_{22}ClPZr$

50.16 $C_{23}H_{22}N_2O_3S$

38.115 $C_{23}H_{22}O_{10}$

84.94 $C_{23}H_{23}CoN_6O_6^+$, NO_3^-

58.62 $C_{23}H_{23}NO_7$

64.121 $C_{23}H_{23}N_2OP$

61.7 $C_{23}H_{24}N_4O_4S_2$, $2C_6H_6$

37.20 $C_{23}H_{24}N_4O_5$

39.81 $C_{23}H_{25}ClOS_3$

83.164 $C_{23}H_{25}CoN_7O_2^{2+}$, $2BF_4^-$

83.165 $C_{23}H_{25}FeN_7O_2^{2+}$, $2BF_4^-$

21.18 $C_{23}H_{25}NO_4$

48.73 $C_{23}H_{25}N_3O_3$, H_2O

84.95 $C_{23}H_{25}N_5NiO_4^{2+}$, $2NO_3^-$, $2H_2O$

31.57 $C_{23}H_{26}$

85.70 $C_{23}H_{26}BClMoN_6O_2S$, $0.2C_3H_6O$

39.82 $C_{23}H_{26}OS_3$

59.34 $C_{23}H_{26}O_6$

39.83 $C_{23}H_{26}S_4$

58.63 $C_{23}H_{28}NO_5^+$, I^-

36.38 $C_{23}H_{30}N_2$

33.86 $C_{23}H_{30}N_2O$, HNO_3, C_3H_8O

53.34 $C_{23}H_{31}BrO_3$

35.70 $C_{23}H_{31}N_3O_2$

51.38 $C_{23}H_{32}O_2$

51.39 $C_{23}H_{32}O_5$, H_2O

54.23 $C_{23}H_{32}O_6$

51.40 $C_{23}H_{32}O_7$, $C_4H_8O_2$

54.24 $C_{23}H_{32}O_8$

59.35 $C_{23}H_{33}NO_5S$

72.44 $C_{23}H_{34}Cl_2N_2Pt$

51.41 $C_{23}H_{34}O_5$, $2H_2O$

50.17 $C_{23}H_{34}O_7$

51.42 $C_{23}H_{35}IO_3$

51.43 $C_{23}H_{35}NO_3$

51.44 $C_{23}H_{36}O_4$

83.166 $C_{23}H_{43}Ga_2N_7NiO$

C_{24}

71.190 $C_{24}HFe_6O_{24}Pd_6^{3-}$, $3C_{10}H_{16}N^+$, $2C_2H_3N$

72.45 $C_{24}H_{10}F_{10}V$

77.9 $C_{24}H_{12}CdF_9O_9^-$, H_4N^+

38.116 $C_{24}H_{12}O_3$

64.122 $C_{24}H_{14}F_{11}N_4O_2P$

36.39 $C_{24}H_{14}N_6$

86.85 $C_{24}H_{15}Br_4IrNO_3P$, CH_2Cl_2

36.40 $C_{24}H_{16}N_2O_2$

83.167	$C_{24}H_{16}N_4Pt^{2-}$, $2Cl^-$, $3H_2O$
31.58	$C_{24}H_{18}$
31.59	$C_{24}H_{18}$
76.99	$C_{24}H_{18}CuN_2O_2$
73.98	$C_{24}H_{18}Fe_2N_2$
83.168	$C_{24}H_{18}MoN_6O_2^{2-}$, $C_{12}H_8N_2$, $2I^-$, H_2O
76.100	$C_{24}H_{18}N_2NiO_2$
36.41	$C_{24}H_{18}N_4O_3$
38.117	$C_{24}H_{18}O_7$
85.71	$C_{24}H_{19}Cl_3MoN_6S_3$
71.1	$C_{24}H_{20}As^+$, $C_2AuN_2O_2^-$
83.1	$C_{24}H_{20}As^+$, $C_2Cl_{10}NW^-$
73.59	$C_{24}H_{20}As^+$, $C_{16}H_{16}Mo_2O_4P^-$
65.13	$C_{24}H_{20}As^+$, $C_{18}HO_{18}Ru_6^-$
65.14	$C_{24}H_{20}As^+$, Cl_4OV^-
83.5	$2C_{24}H_{20}As^+$, $C_2H_3Cl_4MoNO^-$, Cl_4MoO^-
85.47	$2C_{24}H_{20}As^+$, $C_{12}N_6NiSe_6^{2-}$
65.15	$2C_{24}H_{20}As^+$, $C_{17}O_{16}Ru_6^{2-}$
65.16	$2C_{24}H_{20}As^+$, $C_{21}H_2Ni_{12}O_{21}^{2-}$
65.17	$2C_{24}H_{20}As^+$, Cl_5NbO^{2-}, $2CH_2Cl_2$
65.18	$2C_{24}H_{20}As^+$, $Cl_8Se_3W_2^{2-}$
65.19	$2C_{24}H_{20}As^+$, $Cl_{10}NW_2^{2-}$
65.20	$3C_{24}H_{20}As^+$, $C_{21}HNi_{12}O_{21}^{3-}$, $2C_3H_6O$
65.21	$3C_{24}H_{20}As^+$, $C_{21}HNi_{12}O_{21}^{3-}$, $2C_3H_6O$
65.22	$C_{24}H_{20}As_4S_4$
72.46	$C_{24}H_{20}AuFeO_3P$
76.4	$C_{24}H_{20}B^-$, $C_3H_8CuN_2O^+$
62.14	$C_{24}H_{20}B^-$, $C_7H_{16}NO_2^+$
72.10	$C_{24}H_{20}B^-$, $C_{10}H_{25}CuN_3^+$
62.15	$C_{24}H_{20}B^-$, $2C_{14}H_{20}O_5$, Na^+
71.161	$C_{24}H_{20}B^-$, $C_{20}H_{38}CuN_4^+$
71.164	$C_{24}H_{20}B^-$, $C_{20}H_{41}O_4P_2Pt^+$
86.82	$C_{24}H_{20}B^-$, $C_{22}H_{49}Fe_2O_6P_4S_2^+$
80.23	$C_{24}H_{20}B^-$, $C_{25}H_{50}N_5Os_2S_{12}^+$
86.107	$C_{24}H_{20}B^-$, $C_{31}H_{34}INiP_2S_2^+$
86.115	$C_{24}H_{20}B^-$, $C_{35}H_{36}NiO_2P_3S^+$
62.16	$C_{24}H_{20}B^-$, H_4N^+
62.17	$C_{24}H_{20}B^-$, H_4N^+
71.158	$2C_{24}H_{20}B^-$, $C_{20}H_{24}N_8Rh_2^{2+}$, C_2H_3N
86.123	$2C_{24}H_{20}B^-$, $C_{36}H_{44}CoO_2P_4^{2+}$, $2C_3H_7NO$
83.97	$2C_{24}H_{20}B^-$, $C_{14}H_{24}CuN_4^{2+}$
86.122	$2C_{24}H_{20}B^-$, $C_{36}H_{44}CoO_2P_4^{2+}$
62.18	$C_{24}H_{20}BNO$
38.118	$C_{24}H_{20}Cl_4O_6$
38.119	$C_{24}H_{20}Cl_4O_6$
81.77	$C_{24}H_{20}CuF_6N_4O_4$
33.87	$C_{24}H_{20}N_4$
33.88	$C_{24}H_{20}N_4$
31.60	$C_{24}H_{20}O_2$
64.123	$C_{24}H_{20}P^+$, $C_{10}H_{16}N^+$, $Cl_2FeMoS_4^{2-}$
84.30	$C_{24}H_{20}P^+$, $C_{11}H_7N_3O_4V^-$
86.79	$C_{24}H_{20}P^+$, $C_{22}H_{15}MnO_4P^-$
64.124	$C_{24}H_{20}P^+$, IN_6^-
64.125	$2C_{24}H_{20}P^+$, $Br_{10}Te_2^{2-}$
64.126	$2C_{24}H_{20}P^+$, $C_{14}O_{13}Rh_6^{2-}$
64.127	$2C_{24}H_{20}P^+$, $C_{21}H_2Ni_{12}O_{21}^{2-}$
64.128	$2C_{24}H_{20}P^+$, $C_{21}H_2Ni_{12}O_{21}^{2-}$
64.129	$2C_{24}H_{20}P^+$, $Cl_4Fe_2MoS_4^{2-}$
64.130	$2C_{24}H_{20}P^-$, $FeMoS_9^{2-}$, $0.5C_3H_7NO$
64.131	$2C_{24}H_{20}P^-$, FeS_9W^{2-}, $0.5C_3H_7NO$
64.132	$4C_{24}H_{20}P^-$, $4C_2H_3N$, $C_{22}O_{22}Pt_{10}^{4-}$
64.133	$4C_{24}H_{20}P^-$, $C_8Mo_2N_8O_2S_3^{4-}$, $6H_2O$
66.15	$C_{24}H_{20}Sb_2$
37.21	$C_{24}H_{21}N_3O_3$
66.16	$C_{24}H_{21}O_3Sb$, $C_{24}H_{19}O_2Sb$
83.169	$C_{24}H_{22}ClCoN_4O^-$, NO_3^-, H_2O
36.42	$C_{24}H_{22}N_2O_4$
73.99	$C_{24}H_{22}N_2V_2^{2+}$, $2F_6P^-$, C_2H_3N
27.32	$C_{24}H_{22}O$
63.40	$C_{24}H_{22}Si_2$
85.72	$C_{24}H_{24}CoN_3O_3S_3$
83.170	$C_{24}H_{24}CoN_6S_2$
85.73	$C_{24}H_{24}CrN_3O_3S_3$
83.171	$C_{24}H_{24}CuN_4^+$, BF_4^-
83.172	$C_{24}H_{24}CuN_6$
71.191	$C_{24}H_{24}F_4Rh^+$, ClO_4^-
85.74	$C_{24}H_{24}FeN_3O_3S_3$
85.75	$C_{24}H_{24}MnN_3O_3S_3$
85.76	$C_{24}H_{24}N_2NiS_2$
84.96	$C_{24}H_{24}N_2O_4Pt$
85.77	$C_{24}H_{24}N_3O_{15}RhS_3$, $3C_4H_9NO$
84.97	$C_{24}H_{24}N_4O_4Rh_2$
83.173	$C_{24}H_{24}N_6Zr^{2-}$, $12C_4H_8O$, $2Na^+$
64.134	$C_{24}H_{24}O_3P^+$, $C_{24}H_{23}O_3P$, BF_4^-
39.84	$C_{24}H_{24}O_6S_3$, $2CHCl_3$
64.135	$C_{24}H_{25}F_3NO_5P$
32.71	$C_{24}H_{25}N_2^+$, Br^-
86.86	$C_{24}H_{26}Cl_2P_2Pd$
84.98	$C_{24}H_{26}Cu_2N_4O_2^+$, ClO_4^-, $0.5CH_4O$
83.174	$(C_{24}H_{26}Mn_3N_{16})_n$
72.47	$C_{24}H_{28}MoN_2O_3S_4$, C_6H_6
72.48	$C_{24}H_{26}Mo_2O_4S$
59.36	$C_{24}H_{26}O_4$
83.175	$C_{24}H_{28}Br_2Mo_2N_4$
83.176	$C_{24}H_{28}Cl_2CoN_4^+$, F_6P^-
83.177	$C_{24}H_{28}Cl_4Mo_2N_4$, $CHCl_3$
73.100	$C_{24}H_{28}Cl_4O_4Ti_4$
35.71	$C_{24}H_{28}N_3^+$, BF_4^-
83.178	$C_{24}H_{28}N_4Ni^{2+}$, $2F_6P^-$
51.45	$C_{24}H_{28}OS_2$
54.25	$C_{24}H_{28}O_9$
28.16	$C_{24}H_{29}NO_5$, $0.5C_2H_3N$
75.34	$C_{24}H_{30}B_2NiSi_2$
71.192	$C_{24}H_{30}ClIrN_2O_3P_2$, C_3H_6O
76.101	$C_{24}H_{30}CoN_2O_2$
79.15	$C_{24}H_{30}CuN_4O_2S_2$
33.89	$C_{24}H_{30}N_2O_3$
41.81	$C_{24}H_{30}N_2O_4S_4$
13.21	$C_{24}H_{30}N_2O_6$
38.120	$C_{24}H_{30}O_7$
54.26	$C_{24}H_{30}O_9$
51.46	$C_{24}H_{31}FO_6$, $0.67CH_4O$
45.50	$C_{24}H_{32}Br_2O_{15}$, $CHCl_3$
72.49	$C_{24}H_{32}FeO_5P_2S_2$
58.64	$C_{24}H_{32}NO^+$, Br^-
17.38	$C_{24}H_{32}N_2O_7$
17.39	$2C_{24}H_{32}N_2O_7$, $2K^+$, $2CNS^-$, H_2O
72.50	$C_{24}H_{32}N_4O_8Ru_4$
33.90	$C_{24}H_{32}N_6O_2$, Ba^{2-}, $2ClO_4^-$
81.78	$C_{24}H_{32}N_8O_{12}Rh_2$
54.27	$C_{24}H_{32}O_6$
51.47	$C_{24}H_{32}O_7$, CH_4O
86.87	$C_{24}H_{33}MnN_2O_8P_3^+$, BF_4^-
84.99	$C_{24}H_{34}Cl_2CoN_2O_7$
82.39	$C_{24}H_{34}Cu_2N_4O_{16}P_2$, $2H_2O$
48.74	$C_{24}H_{34}N_2O_9$
73.101	$C_{24}H_{34}NbSi_2$
51.48	$C_{24}H_{34}O_2$
55.1	$C_{24}H_{34}O_2$
58.65	$C_{24}H_{35}NO_6$
31.61	$C_{24}H_{36}$
81.79	$C_{24}H_{36}Cl_{12}O_{16}Pd_4$
74.11	$C_{24}H_{36}Co^+$, F_6P^-
83.179	$C_{24}H_{36}FeN_{12}^{2+}$, $C_8Fe_2O_8^{2-}$
31.62	$C_{24}H_{36}N_2O_2$
51.49	$C_{24}H_{36}O_3$
81.80	$C_{24}H_{38}N_4NiO_8$
73.102	$C_{24}H_{38}OYb$, $0.5C_7H_8$
69.48	$C_{24}H_{38}O_4P_2S_4Sn$
74.12	$C_{24}H_{38}Rh^+$, F_6P^-
51.50	$C_{24}H_{39}O_4^-$, Rb^+, H_2O
51.51	$C_{24}H_{39}O_5^-$, Na^+, H_2O
76.18	$C_{24}H_{40}CuN_8Ni_2O_{16}^{2-}$, $C_4H_{16}CuN_4^{2+}$, $2H_2O$
73.103	$C_{24}H_{40}Fe_2O_4Si_6$
51.52	$C_{24}H_{40}O_4$
51.53	$C_{24}H_{40}O_4$, $1.5H_2O$
51.54	$2C_{24}H_{40}O_4$, C_2H_6O, H_2O
51.55	$4C_{24}H_{40}O_4$, $0.5C_{16}H_{32}O_2$, $0.5C_2H_6O$
36.43	$C_{24}H_{42}N_2$
84.100	$C_{24}H_{44}AgN_{12}O_8Pt_2^+$, NO_3^-, $5H_2O$
81.81	$C_{24}H_{44}Cl_6Cu_4N_4O_{12}$
84.101	$C_{24}H_{46}N_4O_7U^{2+}$, $2ClO_4^-$
63.41	$C_{24}H_{46}Si_2$
81.82	$C_{24}H_{48}Cl_4Cu_4N_4O_{12}$
82.40	$C_{24}H_{48}Mo_2N_4O_8^{4+}$, $2Cl^-$, $2C_7H_7O_3S^-$, $2H_2O$
70.16	$C_{24}H_{48}N_{12}S_6Te_2^{4+}$, $4ClO_4^-$
70.17	$C_{24}H_{48}N_{12}S_6Te_2^{4+}$, $4ClO_4^-$
84.102	$C_{24}H_{50}Mo_2O_{11}$
64.136	$C_{24}H_{51}N_3O_6P_2$
72.51	$C_{24}H_{54}N_6Ni^{2-}$, $2Li^+$
86.88	$C_{24}H_{54}P_2Pt$
84.103	$C_{24}H_{56}Br_4Cu_4N_4O_4$, $4CCl_4$
86.89	$C_{24}H_{56}P_2Pt$
85.78	$C_{24}H_{60}Fe_7Mo_2S_{20}^{3-}$, $3C_{10}H_{16}N^+$
71.193	$C_{24}H_{62}Re_2Si_6$
71.194	$C_{24}H_{66}Cl_3N_2O_2Re_3Si_6$
84.104	$2C_{24}H_{72}N_{15}O_{13}P_4Th^+$, $N_6O_{18}Th^{2-}$

C_{25}

83.180	$C_{25}H_{10}N_2O_{13}Os_5$
83.181	$C_{25}H_{16}CoN_4O_3^+$, Br^-, $4H_2O$
83.182	$C_{25}H_{16}CoN_4O_3^+$, Cl^-, $3H_2O$
83.183	$C_{25}H_{16}N_5Pt^+$, NO_3^-, H_2O
65.23	$C_{25}H_{17}As$
64.137	$C_{25}H_{19}BrOP^+$, Br^-

64.138 $C_{25}H_{20}NOP$

71.195 $C_{25}H_{20}NO_{10}Os_3P$

86.90 $C_{25}H_{20}N_2OPRhS_2$

63.42 $C_{25}H_{20}OSi$

38.121 $C_{25}H_{20}O_2$

24.11 $C_{25}H_{22}O_2$

61.8 $2C_{25}H_{22}P^+$, $2C_2H_4Cl_2 \cdot CdCl_4{}^{2-}$

15.6 $C_{25}H_{23}N_5O_6$

83.184 $C_{25}H_{23}N_5O_6U$

83.185 $C_{25}H_{23}N_5O_6U$

64.139 $C_{25}H_{25}BrOP^-$, Br_3^-

40.55 $C_{25}H_{25}N_3O_7$

36.44 $C_{25}H_{26}BrN_3O_5$

36.45 $C_{25}H_{26}N_2$

36.46 $C_{25}H_{27}N_3O_5$

64.140 $C_{25}H_{29}NP_2$

73.104 $C_{25}H_{30}O_8Ru_3Si_3$

73.105 $C_{25}H_{30}O_8Ru_3Si_3$

36.47 $C_{25}H_{31}N$

33.91 $C_{25}H_{31}N_2{}^+$, $C_7H_7O_3S^-$

64.141 $C_{25}H_{31}O_2P$

33.92 $C_{25}H_{32}N_2O_3$, $C_2H_2O_4$

38.122 $C_{25}H_{32}O_8$

73.106 $C_{25}H_{33}OTi$

71.196 $C_{25}H_{34}CoN_5O_6S$

51.56 $C_{25}H_{35}NO_6$

51.57 $C_{25}H_{35}NO_6$

51.58 $C_{25}H_{36}O_2$

55.2 $C_{25}H_{37}O_2^-$, $C_{12}H_{24}N^+$

71.197 $C_{25}H_{38}Al_2Zr$

58.66 $C_{25}H_{40}NO_6{}^+$, I^-

83.186 $C_{25}H_{44}N_{12}O_{16}P_2Pt$, $11H_2O$

80.23 $C_{25}H_{50}N_5Os_2S_{12}{}^+$, $C_{24}H_{20}B^-$

84.105 $C_{25}H_{54}Mo_2O_7$

84.106 $C_{25}H_{63}N_4O_8Re_3Si_3$

C_{26}

28.17 $C_{26}HCl_{17}$

28.18 $C_{26}H_2Cl_{16}$

30.13 $C_{26}H_{16}O_2$

26.5 $C_{26}H_{17}Br$

36.48 $C_{26}H_{17}N_3O$

71.198 $C_{26}H_{17}O_9PRu_3$

76.102 $C_{26}H_{18}Cu_2F_{12}N_2O_6$

24.12 $C_{26}H_{18}N_2$

34.11 $C_{26}H_{18}N_6$, Pb^{2+}, $2ClO_4^-$, $2H_2O$

59.37 $C_{26}H_{19}ClO_{10}$, $2CH_4O$

71.199 $C_{26}H_{19}O_8PRu_3$, $0.5C_6H_6$

83.187 $C_{26}H_{20}CdClN_6O_5{}^+$, ClO_4^-, CH_4O

33.93 $C_{26}H_{20}Cl_2N_2$

45.51 $C_{26}H_{20}Cl_2O_7$

64.142 $C_{26}H_{20}P_2S_2$

64.143 $C_{26}H_{20}P_2Se_2$

65.24 $C_{26}H_{21}As$

71.200 $C_{26}H_{21}AsF_6O_3Pt$

71.201 $C_{26}H_{21}O_2ReSi$, CH_2Cl_2

71.202 $C_{26}H_{22}Mn_2O_4$

45.52 $C_{26}H_{22}O_7$

45.53 $C_{26}H_{22}O_7$

42.12 $C_{26}H_{23}Cl_2N_3O_4S_2$

63.43 $C_{26}H_{23}NSi$

71.203 $C_{26}H_{24}AuO_2P$

71.204 $C_{26}H_{24}MoN_2O_2$

61.9 $3C_{26}H_{24}P_2Se_2$, C_8H_{10}

71.205 $C_{26}H_{25}CoGeO_4$

73.107 $C_{26}H_{25}NiOPS_2$

84.107 $C_{26}H_{26}Cu_2N_6O$

71.206 $C_{26}H_{26}IIrPS^+$, I^-

45.54 $C_{26}H_{26}O_8$

73.108 $C_{26}H_{28}FeNP$

16.18 $C_{26}H_{28}N_2O_4$

35.72 $C_{26}H_{28}N_2O_4$

17.40 $C_{26}H_{28}N_2O_5$, CH_4N_2S

61.10 $C_{26}H_{28}N_6NiS_2$, $2C_{10}H_7Br$

61.11 $C_{26}H_{28}N_6NiS_2$, $2C_{11}H_{10}$

30.14 $C_{26}H_{28}O_2$

86.91 $C_{26}H_{29}CoN_5O_6P$

17.41 $C_{26}H_{29}NO$

81.83 $C_{26}H_{30}Mo_2N_2O_6$

71.207 $C_{26}H_{32}HgN_4O_2$

59.38 $C_{26}H_{32}O_9$

66.17 $C_{26}H_{33}O_4Sb$

71.208 $C_{26}H_{35}NP_3Pt^+$, F_6P^-

75.35 $C_{26}H_{36}O_4Pd$

54.28 $C_{26}H_{36}O_8$

48.75 $C_{26}H_{38}N_2O_9$

51.59 $C_{26}H_{38}O_2$

86.92 $C_{26}H_{40}I_2Ir_2O_2P_2S_2$

51.60 $C_{26}H_{40}N_8$

73.109 $C_{26}H_{42}B_4Fe_3S_2$

58.67 $C_{26}H_{42}NO_6{}^+$, I^-

58.68 $C_{26}H_{42}N_2O_2$

58.69 $C_{26}H_{43}N_2{}^+$, I^-, $2C_2H_6O$

83.188 $C_{26}H_{44}CoN_6{}^{2+}$, $2F_6P^-$

71.209 $C_{26}H_{45}Cl_4OPTa_2$

21.19 $C_{26}H_{46}$

63.44 $C_{26}H_{46}N_4Si_4$

85.79 $C_{26}H_{53}Mo_2O_3S_8{}^{3+}$, $3CF_3O_3S^-$, H_2O

70.18 $C_{26}H_{60}N_2PdS_2Si_4Te_2$

C_{27}

30.15 $C_{27}H_{16}O$

71.210 $C_{27}H_{16}O_{11}Os_4$

30.16 $C_{27}H_{18}$

68.23 $C_{27}H_{18}AlN_3O_3$, CH_4O

64.144 $C_{27}H_{18}Cl_4NO_2P$

71.211 $C_{27}H_{20}Fe_2NO_6P$

71.212 $C_{27}H_{20}O_3Ru_2$

42.13 $C_{27}H_{20}O_3S$

65.25 $C_{27}H_{22}AsN_3O$, $0.5C_6H_6$

71.213 $C_{27}H_{22}ClF_3P_2Pt$

17.42 $C_{27}H_{22}N_2O_6$

78.10 $C_{27}H_{24}Cl_3CrN_4O_3$, $3H_2O$

78.11 $C_{27}H_{24}Cl_3MnN_4O_3$, $3H_2O$

38.123 $C_{27}H_{24}O_2$

71.214 $C_{27}H_{24}O_6W$

45.55 $C_{27}H_{24}O_8$

50.18 $C_{27}H_{24}O_{10}$

71.215 $C_{27}H_{25}CrNO_4Sn$

36.49 $C_{27}H_{25}F_3N_2O$

38.124 $C_{27}H_{28}N_2O_9$

38.125 $C_{27}H_{28}N_2O_9$

38.126 $C_{27}H_{29}NO_7$

81.84 $C_{27}H_{30}Cl_3Mn_3N_3O_{13}$

61.12 $C_{27}H_{30}O_6$, $0.5C_6H_6 \cdot H_2O$

59.39 $C_{27}H_{31}NO_3$

54.29 $C_{27}H_{31}NO_4$

38.127 $C_{27}H_{32}N_2O_9$

59.40 $C_{27}H_{32}O_9$

71.216 $C_{27}H_{33}F_6N_2O_2Rh$

86.93 $C_{27}H_{33}HgN_2O_6P$

37.22 $C_{27}H_{33}N$

58.70 $C_{27}H_{33}NO_{10}S$, $2C_2H_6O \cdot H_2O$

58.71 $C_{27}H_{33}N_3O_5$, H_2O

54.30 $C_{27}H_{37}BrO_2$

54.31 $C_{27}H_{37}BrO_3$

63.45 $C_{27}H_{37}BrO_3Si_2$

54.32 $C_{27}H_{38}O_2$

51.61 $C_{27}H_{38}O_5$

51.62 $C_{27}H_{40}O_9$

84.108 $C_{27}H_{42}FeN_9O_{12}$

71.217 $C_{27}H_{43}IO_2P_2Rh_2S_2$

58.72 $C_{27}H_{44}NO^+$, Cl^-

51.63 $C_{27}H_{44}O$

51.64 $C_{27}H_{44}O$

51.65 $C_{27}H_{45}Br$

51.66 $C_{27}H_{45}Cl$

51.67 $C_{27}H_{46}O_2$

51.68 $C_{27}H_{46}O_5$

51.69 $C_{27}H_{48}O_4$

83.189 $C_{27}H_{52}Cu_2N_6O_3{}^{2+}$, $2ClO_4^-$, C_3H_7NO

C_{28}

29.10 $C_{28}Cl_{18}$, $0.5CCl_4$

83.190 $C_{28}H_{16}Fe_2N_4O_8Zn_2$

36.50 $C_{28}H_{16}N_2$

36.51 $C_{28}H_{18}N_6$, $CHCl_3$

30.17 $C_{28}H_{18}O$

73.110 $C_{28}H_{20}Bi_2Cl_2Mn_4O_8$

20.14 $C_{28}H_{20}N_4$

83.191 $C_{28}H_{20}N_4Pd^{2+}$, $2BF_4^-$

66.18 $C_{28}H_{20}O_3Sb_2$

13.22 $C_{28}H_{20}O_4$

32.72 $C_{28}H_{21}NO$

41.82 $C_{28}H_{21}N_3O_2S$

71.218 $C_{28}H_{22}I_2O_3P_2W$

38.128 $C_{28}H_{22}O_9$

83.192 $C_{28}H_{23}CoN_6O_2{}^{2+}$, $2CNS^-$, $2H_2O$

62.19 $C_{28}H_{24}B_2O$

35.73 $C_{28}H_{24}N_4S^{2+}$, $2BF_4^-$

64.145 $C_{28}H_{24}O_2P^+$, BF_4^-

81.85 $C_{28}H_{24}O_4P_2Pd$

71.219 $C_{28}H_{25}CrNO_5Sn$, $0.5CH_2Cl_2$

83.193 $C_{28}H_{26}Cl_4N_4Re_2$

64.146 $C_{28}H_{26}P_2$

37.23 $C_{28}H_{27}N_5O_3$

73.111 $C_{28}H_{28}As_3CoCr_2FeO_{12}$

84.109 $C_{28}H_{28}CoN_6O_3^{2-}$, $2ClO_4^-$
78.12 $C_{28}H_{28}CuN_2O_2$
78.13 $C_{28}H_{28}CuN_2O_2$, $2C_{12}H_4N_4$
85.80 $C_{28}H_{28}Fe_4S_8^{3-}$, $3C_8H_{20}N^-$
86.94 $C_{28}H_{28}MoOP_2S_2$
83.194 $C_{28}H_{28}Mo_2N_8O_4$
31.63 $C_{28}H_{30}$
73.112 $C_{28}H_{30}N_2O_6PRh$
73.113 $C_{28}H_{30}N_2S_2Ti_2$
50.19 $2C_{28}H_{31}NO_{10}$, C_2H_6O
19.16 $C_{28}H_{32}N_3O^+$, ClO_4^-
38.129 $C_{28}H_{32}O_6$
23.7 $C_{28}H_{32}O_8$
63.46 $C_{28}H_{32}O_8Si_6$
53.35 $C_{28}H_{35}N_3O_2$
58.73 $C_{28}H_{35}N_3O_4$
71.220 $C_{28}H_{36}BFeNO_2P_2$
76.103 $C_{28}H_{36}Cu_2N_4O_4$, $2C_6H_6O$
63.47 $C_{28}H_{36}N_4Si_4$
54.33 $C_{28}H_{36}O_5$
54.34 $C_{28}H_{37}BrO_8S$, H_2O
83.195 $C_{28}H_{38}FeN_4S_2^+$, F_6P^-
71.221 $C_{28}H_{38}Nb_2O$
51.70 $C_{28}H_{38}O_7$
45.56 $C_{28}H_{38}O_{19}$
54.35 $C_{28}H_{39}NO$, $C_4H_8O_2$
59.41 $C_{28}H_{39}NO_2$, $0.5CH_4O$
38.130 $C_{28}H_{40}O_{10}$, CNS^-, K^+
38.131 $C_{28}H_{40}O_{10}$, $2Na^+$, $2CNS^-$, H_2O
86.95 $C_{28}H_{41}B_{20}ClP_2Pt$
83.196 $C_{28}H_{44}CoN_8S_2^+$, Cl^-
76.104 $C_{28}H_{44}Cu_2N_6O_4^{2+}$, $C_{26}H_{36}Cl_2Cu_2N_6O_{10}$, $2ClO_4^-$
86.96 $C_{28}H_{46}O_2P_2Pd$, C_7H_8
86.97 $C_{28}H_{46}O_2P_2Pt$, C_7H_8
51.71 $C_{28}H_{48}O_6$
51.72 $C_{28}H_{48}O_6$, H_2O
80.24 $C_{28}H_{53}N_2PPtS_4$, C_6H_{12}
86.98 $C_{28}H_{54}Br_2O_4P_2Ru_2$
63.48 $C_{28}H_{72}O_2Si_8$
71.222 $C_{28}H_{80}P_8Ru_3^{2+}$, $2BF_4^-$

C_{29}

71.223 $C_{29}H_{14}O_{14}Ru_6$
86.99 $C_{29}H_{15}O_{11}PRu_3$
33.94 $C_{29}H_{21}NO_2$
32.73 $C_{29}H_{22}N_2O_2$
72.52 $C_{29}H_{23}FeO_3P$
71.224 $C_{29}H_{24}I_2MoO_3P_2$, CH_2Cl_2
35.74 $C_{29}H_{25}N_5O_5$
73.114 $C_{29}H_{26}MoN_2O_2$
73.115 $C_{29}H_{26}MoN_2O_2$
85.81 $C_{29}H_{28}BFMoN_8S_2$
72.53 $C_{29}H_{29}Cl_3NPPt$
75.36 $C_{29}H_{29}IMoNOP$
71.225 $C_{29}H_{31}P_3Pt$
50.20 $C_{29}H_{34}N_3O_9^+$, Br^-, H_2O
76.105 $C_{29}H_{35}CoN_3O_4$
58.74 $C_{29}H_{35}NO_5$

50.21 $C_{29}H_{37}BrO_8S$, C_2H_6OS
35.75 $C_{29}H_{37}N_2^-$, BF_4^-
48.76 $C_{29}H_{37}N_6O_6S^+$, Cl^-. CH_4O. $0.5C_4H_{10}O$
32.74 $C_{29}H_{38}N_4^{2+}$, $2Br^-$. $2CHCl_3$
38.132 $C_{29}H_{43}NO_4$
51.73 $C_{29}H_{44}O_9$
35.76 $C_{29}H_{45}NO_2$
34.12 $C_{29}H_{45}NO_2$, C_2H_6O
71.226 $C_{29}H_{47}ClN_2P_2Ru$
51.74 $C_{29}H_{48}N_2O_4$
51.75 $C_{29}H_{48}O_3$
51.76 $C_{29}H_{50}O_6$
51.77 $C_{29}H_{50}O_6$
71.227 $C_{29}H_{57}P_2Re_3$

$C_{30}-C_{34}$

71.228 $C_{30}H_{18}Br_2Hg_2O_{18}Ru_6$
84.110 $C_{30}H_{18}N_4O_8Re_2$
28.19 $C_{30}H_{18}O_2$
86.100 $C_{30}H_{19}Cl_3NO_4PW$
31.64 $C_{30}H_{20}$
86.101 $C_{30}H_{20}Fe_2O_6P_2^{2-}$, $2C_{18}H_{36}N_2O_6$, $2Na^+$
85.82 $C_{30}H_{20}N_2NiO_2S_2$
38.133 $C_{30}H_{20}O$
41.83 $C_{30}H_{22}N_2O_2S_2$, CH_4O
13.23 $C_{30}H_{22}O_7$
13.24 $C_{30}H_{22}O_7$, C_6H_6
74.13 $C_{30}H_{23}CrO_5P$
48.77 $C_{30}H_{23}NO_2$
48.78 $C_{30}H_{23}NO_2$
48.79 $C_{30}H_{23}NO_2$
83.197 $C_{30}H_{24}CoN_6^{3+}$, $C_6FeN_6^{3-}$, $8H_2O$
83.198 $C_{30}H_{24}N_6Ru^{2+}$, $2F_6P^-$
31.65 $C_{30}H_{24}O_4$
74.14 $C_{30}H_{25}CrO_2PS$
71.229 $C_{30}H_{26}I_2MoO_3P_2$
72.54 $C_{30}H_{27}MoO_2P$
83.199 $C_{30}H_{28}CuN_6O_4$
71.230 $C_{30}H_{28}Zr_2$
84.111 $C_{30}H_{30}CuN_6O_6^{2+}$, $2NO_3^-$
40.56 $C_{30}H_{30}N_2O_4$
83.200 $C_{30}H_{30}N_6Ru^{2+}$, $2BF_4^-$
64.147 $C_{30}H_{31}NP_2$, C_3H_6O
64.148 $C_{30}H_{31}P_3$
80.25 $C_{30}H_{32}MoN_6S_4$, C_3H_6O
83.201 $C_{30}H_{32}N_6NiO_6$
64.149 $C_{30}H_{32}P_2^{2+}$, $2Br^-$
86.102 $C_{30}H_{34}As_2NPd^+$, F_6P^-
35.77 $C_{30}H_{35}N_3O_3$
56.1 $C_{30}H_{38}O_9$
35.78 $C_{30}H_{40}AsN_2^+$, BF_4^-, $0.5CH_2Cl_2$
58.75 $C_{30}H_{40}N_2O_4$
35.79 $C_{30}H_{40}N_2P^+$, BF_4^-
35.80 $C_{30}H_{40}N_2P^+$, BF_4^-, $0.5CH_2Cl_2$
35.81 $C_{30}H_{40}N_3^+$, BF_4^-
45.57 $C_{30}H_{40}O_{10}$
55.3 $C_{30}H_{41}BrO_4S$
83.202 $C_{30}H_{44}B_2FeN_{12}$
69.49 $C_{30}H_{46}GeO_2$

69.50 $C_{30}H_{46}O_2Sn$
71.231 $C_{30}H_{47}ClN_2OP_2Ru$
83.203 $C_{30}H_{48}FeN_{12}^{2+}$, $2C_4FeO_4^-$
73.116 $C_{30}H_{48}ORh_3^+$, F_6P^-. H_2O
56.2 $C_{30}H_{48}O_2$
56.3 $C_{30}H_{48}O_4$
56.4 $C_{30}H_{50}$
61.26 $0.5C_{30}H_{50}$, $C_{72}H_{90}S_6$
56.5 $C_{30}H_{50}O$
56.6 $C_{30}H_{50}O$
63.49 $C_{30}H_{60}N_4Si_6$
80.26 $C_{30}H_{60}N_6Os_2S_{12}^{2+}$, $2F_6P^-$. CH_2Cl_2
80.27 $C_{30}H_{60}N_6Os_2S_{12}^{2+}$, $2F_6P^-$. $2CH_2Cl_2$
86.103 $C_{30}H_{63}ClO_2P_2Rh_3S$
86.104 $C_{30}H_{68}IrP_4Pt^+$, $C_{24}H_{20}B^-$
79.16 $C_{30}H_{72}FeN_{12}O_6^{2+}$, $2ClO_4^-$
40.57 $C_{31}H_{22}ClNO_2$
71.232 $C_{31}H_{23}Fe_2O_6P_2^-$, $C_{36}H_{30}NP_2^+$
71.233 $C_{31}H_{23}Fe_2O_6P_2^-$, Na^+, $2C_4H_6O$
86.105 $C_{31}H_{24}ClN_6OP_2Rh$
78.14 $C_{31}H_{24}NO_2PPt$
86.106 $C_{31}H_{24}O_6P_2Re_2$
71.234 $C_{31}H_{25}OPSW$
19.17 $C_{31}H_{26}$
71.235 $C_{31}H_{26}ClPPdS_3$, $2C_7H_8$
75.37 $C_{31}H_{28}FeOP_2$
73.117 $C_{31}H_{29}Cl_3NbP_2$, $2C_7H_8$
27.33 $C_{31}H_{33}N_3O_9$
86.107 $C_{31}H_{34}INiP_2S^{2+}$, $C_{24}H_{20}B^-$
86.108 $C_{31}H_{34}NiP_2S_2^{2+}$, $2ClO_4^-$
86.109 $C_{31}H_{36}ClP_3Pt$
35.82 $C_{31}H_{41}N_2^+$, ClO_4^-
56.7 $C_{31}H_{48}O_4$
49.1 $C_{32}H_{16}CoN_8$
49.2 $C_{32}H_{16}MnN_8$
38.134 $C_{32}H_{18}O_6$
39.85 $C_{32}H_{22}S_4$
40.58 $C_{32}H_{23}N_3O_7S$
84.112 $C_{32}H_{24}Cl_6Cu_2N_2O_4$
85.83 $C_{32}H_{24}N_2NiO_2S_2$
83.204 $C_{32}H_{26}Cl_4CuN_4O_2$, $2H_2O$, $0.33CHCl_3$
81.86 $C_{32}H_{26}N_2O_4Pd_2S_2$
71.236 $C_{32}H_{26}N_2O_6Pd_2$
21.20 $C_{32}H_{26}O$
64.150 $C_{32}H_{27}P_2^-$, Na^+, C_4H_8O
84.113 $(C_{32}H_{28}Cu_2O_{10})_n$, $2nC_4H_8O_2$
21.21 $C_{32}H_{28}O_2$
85.84 $C_{32}H_{32}BrCuO_8S_4$
85.85 $C_{32}H_{32}ClCuO_8S_4$, $3CHCl_3$
84.114 $C_{32}H_{32}Cr_2N_4O_4$
84.115 $C_{32}H_{32}Mo_2N_4O_4$, $2C_4H_8O$
39.86 $C_{32}H_{32}O_8S_4$, $2CHCl_3$
39.87 $C_{32}H_{32}S_8$
49.3 $C_{32}H_{36}N_4Pb$
83.205 $C_{32}H_{36}N_8Ni$
38.135 $C_{32}H_{36}O_6$
38.136 $C_{32}H_{36}O_6$
61.13 $C_{32}H_{36}O_6$, $C_4H_{12}N^+$, ClO_4^-. C_6H_6
75.38 $C_{32}H_{38}B_2Fe_2N_2O_4$
29.11 $C_{32}H_{38}NO_{12}^+$, Cl^-

86.110 $C_{32}H_{40}As_4Cl_2Ni$
35.83 $C_{32}H_{40}N_2O_8$, Rb^+, I^-
38.137 $C_{32}H_{40}O_8$
58.76 $C_{32}H_{41}BrN_5O_5{}^-$, $CH_3O_3S^-$, $0.5C_3H_8O$
71.237 $C_{32}H_{44}IrP_4{}^+$, F_6P^-
83.206 $C_{32}H_{44}N_{10}W_2$
76.106 $C_{32}H_{48}Co_2N_{10}O_2{}^{4-}$, $4I^-$
71.238 $C_{32}H_{50}O_7P_2Pt_2W$
56.8 $C_{32}H_{52}O_2$
75.39 $C_{32}H_{52}O_5S_5W_3$
83.207 $C_{32}H_{56}Cl_2N_8Ni$
83.208 $C_{32}H_{56}Cl_4Cu_2N_8$
81.87 $C_{32}H_{56}Cl_2Cu_4N_4O_{12}$
81.88 $C_{32}H_{58}Mn_{12}O_{48}$, $2C_2H_4O_2$, $4H_2O$
81.89 $C_{32}H_{64}Cl_4Cu_4N_4O_{12}$
84.116 $C_{32}H_{68}MoN_2O_4Si_4$
84.117 $C_{32}H_{72}F_4Mo_4O_8$, $2C_{34}H_{78}F_3Mo_4NO_8$
86.111 $C_{33}H_{20}Mn_2O_9P_2Pt$
73.118 $C_{33}H_{23}Rh$
73.119 $C_{33}H_{25}Co$
45.58 $C_{33}H_{26}O_9$
86.112 $C_{33}H_{27}NO_6P_2Re_2$
54.36 $C_{33}H_{30}Br_2O_9$
72.55 $C_{33}H_{30}MoO_3P_2$
36.52 $C_{33}H_{33}N_3$
73.120 $C_{33}H_{36}FeNPRh^+$, F_6P^-
61.14 $C_{33}H_{36}O_6$, $C_{14}H_{12}$
61.15 $2C_{33}H_{36}O_6$, C_4H_8S
61.16 $2C_{33}H_{36}O_6$, C_4H_9Br
61.17 $2C_{33}H_{36}O_6$, $C_{14}H_{12}$
58.77 $C_{33}H_{38}N_5O_5{}^+$, $CH_3O_3S^-$, H_2O
45.59 $C_{33}H_{38}O_{14}$
32.75 $C_{33}H_{41}N_4O_4{}^+$, Br^-, C_2H_6O
32.76 $C_{33}H_{42}N_4O_4{}^{2+}$, $2Br^-$
38.138 $C_{33}H_{42}O_6$
55.4 $C_{33}H_{43}BrO_3$
50.22 $C_{33}H_{57}N_3O_9$, Na^+, $6.25H_2O$, $N_3NiO_9{}^-$
81.90 $C_{34}H_{22}Cu_2N_6O_8$, $4H_2O$
71.239 $C_{34}H_{24}O_4PRe$
39.88 $C_{34}H_{24}S_2$
39.89 $C_{34}H_{24}S_2$, $0.36I_3$, $0.4I_5$
39.90 $C_{34}H_{24}S_2$, $0.72I_3$
64.151 $C_{34}H_{27}ClNOP$
86.113 $C_{34}H_{27}Mo_2O_8P_2{}^-$, $C_8H_{20}N^+$
26.6 $C_{34}H_{28}$
73.121 $C_{34}H_{28}CoO_2P$
83.209 $C_{34}H_{28}CuN_{10}{}^{2+}$, $2BF_4{}^-$
71.240 $C_{34}H_{31}Cl_2OP_2Rh$
50.23 $C_{34}H_{32}BrNO_9$
71.241 $C_{34}H_{32}Ti_2$
78.15 $C_{34}H_{34}Cl_2CuN_2O_2$
85.86 $C_{34}H_{34}N_8Pt_3S_2{}^{4+}$, $4BF_4{}^-$
58.78 $C_{34}H_{40}N_2O_5$, C_3H_6O
31.66 $C_{34}H_{42}N_2$
49.4 $C_{34}H_{43}N_4{}^+$, $I_3{}^-$
71.242 $C_{34}H_{44}B_8NiP_2$
78.16 $C_{34}H_{44}Cu_2N_4O_6$
61.18 $C_{34}H_{44}N_4{}^{4+}$, $C_{10}H_{14}$, $4Cl^-$, $4H_2O$
48.80 $C_{34}H_{46}N_6O_6$, $2H_2O$
75.40 $C_{34}H_{48}Cl_2Hg_2Ir_2N_6$

51.78 $C_{34}H_{54}N_4O_2S$
38.139 $C_{34}H_{54}O_8$
51.79 $C_{34}H_{55}NO_3S$
86.114 $C_{34}H_{74}Cl_4N_2P_2Pt_2$
84.117 $2C_{34}H_{78}F_3Mo_4NO_8$, $C_{32}H_{72}F_4Mo_4O_8$

$C_{35}-C_{39}$

26.7 $C_{35}H_{30}$
71.243 $C_{35}H_{34}FeP_2$
46.7 $C_{35}H_{35}O_4P$
86.115 $C_{35}H_{36}NiO_2P_3S^+$, $C_{24}H_{20}B^-$
62.20 $C_{35}H_{45}BF_2N_4O_2$, $CHCl_3$
51.80 $C_{35}H_{48}O_4$
71.244 $C_{35}H_{63}N_7W^{2+}$, $O_{19}W_6{}^{2-}$
49.5 $C_{36}H_{23}FeN_9O_2$, C_3H_7NO
83.210 $C_{36}H_{24}HgN_8{}^{2+}$, $2CF_3O_3S^-$, C_2H_6O
71.245 $C_{36}H_{25}Fe_2O_6P_2{}^-$, Li^+, $3C_4H_8O$
75.41 $C_{36}H_{26}Cl_2Mn_2N_2O_6$
31.67 $C_{36}H_{28}$
41.84 $C_{36}H_{28}N_2O_2S_2$
86.116 $C_{36}H_{28}O_{10}P_2Ru_4$
86.117 $C_{36}H_{30}As_2Cl_2Pd$
86.118 $C_{36}H_{30}Au_2ClP_2{}^+$, $ClO_4{}^-$, CH_2Cl_2
84.118 $C_{36}H_{30}Cl_2O_3P_2V$
84.119 $C_{36}H_{30}Cl_3O_3P_2W$
83.211 $C_{36}H_{30}Cl_8N_2Nb_2P_2$, $C_2H_4Cl_2$
69.51 $C_{36}H_{30}GeO_2Si$
69.52 $C_{36}H_{30}Ge_2$
69.53 $C_{36}H_{30}Ge_2S$
69.54 $C_{36}H_{30}Ge_2S$
63.50 $C_{36}H_{30}HgSi_2$
64.152 $C_{36}H_{30}NP_2{}^+$, $C_6HNi_2O_6{}^-$
64.153 $C_{36}H_{30}NP_2{}^+$, $C_{10}Cr_2DO_{10}{}^-$
64.154 $C_{36}H_{30}NP_2{}^+$, $C_{13}CoO_{13}Ru_3{}^-$
64.155 $C_{36}H_{30}NP_2{}^+$, $C_{13}HFeO_{13}Ru_3{}^-$
64.156 $C_{36}H_{30}NP_2{}^+$, $C_{13}HFe_2O_{13}Ru_2{}^-$
71.45 $C_{36}H_{30}NP_2{}^+$, $C_{13}H_6MnO_7{}^-$
64.157 $C_{36}H_{30}NP_2{}^+$, $C_{15}Co_8NO_{15}{}^-$
64.158 $C_{36}H_{30}NP_2{}^+$, $C_{15}O_{15}Rh_5{}^-$
3.56 $C_{36}H_{30}NP_2{}^+$, $C_{16}IO_{15}Os_5{}^-$
71.232 $C_{36}H_{30}NP_2{}^+$, $C_{31}H_{23}Fe_2O_6P_2{}^-$
64.159 $C_{36}H_{30}NP_2{}^+$, Cl_4OTc^-
64.160 $2C_{36}H_{30}NP_2{}^+$, $C_4MoN_4O_3{}^{2-}$
64.161 $2C_{36}H_{30}NP_2{}^+$, $C_{13}Fe_4O_{13}{}^{2-}$
64.162 $2C_{36}H_{30}NP_2{}^+$, $C_{13}Ir_4O_{13}Ru^{2-}$
64.163 $2C_{36}H_{30}NP_2{}^+$, $C_{15}Ir_4O_{15}Ru^{2-}$
64.164 $2C_{36}H_{30}NP_2{}^+$, $C_{21}H_2Ni_{12}O_{21}{}^{2-}$
64.165 $2C_{36}H_{30}NP_2{}^+$, $C_{22}O_{22}Os_8{}^{2-}$
64.166 $2C_{36}H_{30}NP_2{}^+$, $C_{25}O_{24}Os_{10}{}^{2-}$
84.20 $3C_{36}H_{30}NP_2{}^+$, $2C_9H_9Mo_3N_3O_{13}{}^{2-}$, Na^+
64.167 $C_{36}H_{30}N_6P_2S_4$
30.18 $C_{36}H_{30}O_5$, $0.93CHCl_3$
59.42 $C_{36}H_{30}O_7$, C_6H_6
66.19 $C_{36}H_{30}P_3S_6Sb$
69.55 $C_{36}H_{30}SeSn_2$
84.120 $C_{36}H_{32}Cl_4Co_2N_4O_8S_2$, $C_{21}H_{22}Cl_2CoN_2O_5S$
86.119 $C_{36}H_{32}N_2O_2P_2PtS_2$, $0.5H_2O$
73.122 $C_{36}H_{33}CoSi$

71.246 $C_{36}H_{33}Ir_7O_{12}$
84.121 $C_{36}H_{34}N_{10}O_2W_2$, $2C_4H_8O$
83.212 $C_{36}H_{36}Ag_6N_{24}{}^{6+}$, $6ClO_4{}^-$
35.84 $C_{36}H_{36}N_4O_3$, $2H_2O$
35.85 $C_{36}H_{36}N_4O_3$, Rb^+, I^-
86.120 $C_{36}H_{37}ClO_2P_3RhS$
73.123 $C_{36}H_{37}FeIO_2P_2$
49.6 $C_{36}H_{37}N_5O_6$, $C_2H_4Cl_2$
86.121 $C_{36}H_{40}B_8P_2PtS$
82.41 $C_{36}H_{40}CuN_6O_8$
17.43 $C_{36}H_{40}N_2O_{13}$, $2K^+$, $2CNS^-$
76.107 $C_{36}H_{44}ClCuN_4{}^+$, $NO_3{}^-$
49.7 $C_{36}H_{44}ClFeN_4O_4$
83.213 $C_{36}H_{44}Cl_2N_8Re_2$, CCl_4
86.122 $C_{36}H_{44}CoO_2P_4{}^{2+}$, $2C_{24}H_{20}B^-$
86.123 $C_{36}H_{44}CoO_2P_4{}^{2+}$, $2C_{24}H_{20}B^-$, $2C_3H_7NO$
83.214 $C_{36}H_{44}Cr_2N_8$
38.140 $C_{36}H_{44}O_{16}$
63.51 $C_{36}H_{46}Si_2$
54.37 $C_{36}H_{47}BrO_{11}$
86.124 $C_{36}H_{48}MoN_4P_4$
49.8 $C_{36}H_{48}N_4Ni$
49.9 $C_{36}H_{48}N_4Ni$
49.10 $C_{36}H_{50}N_4$
49.11 $C_{36}H_{50}N_4$
83.215 $C_{36}H_{52}Cl_2N_4Pd$
49.12 $C_{36}H_{52}N_4Ni$
49.13 $C_{36}H_{52}N_4Ni$
86.125 $C_{36}H_{53}P_8Ta$
76.108 $C_{36}H_{54}N_6O_{10}Re_3{}^+$, O_4Re^-, $1.5C_6H_6$, xCH_2Cl_2
51.81 $C_{36}H_{54}O_2$
64.168 $C_{36}H_{58}ClOP$
60.43 $C_{36}H_{60}N_4O_{12}$, $C_8H_9N_3O_7{}^-$, K^+
45.60 $C_{36}H_{60}O_{30}$, C_2H_6O, $8H_2O$
61.19 $C_{36}H_{60}O_{30}$, C_3H_7NO, $5H_2O$
61.20 $C_{36}H_{60}O_{30}$, C_4H_7NO, $5H_2O$
45.61 $C_{36}H_{60}O_{30}$, $6H_2O$
45.62 $C_{36}H_{60}O_{30}$, $6H_2O$
45.63 $2C_{36}H_{60}O_{30}$, $0.5Cd^{2+}$, $I_5{}^-$, $27H_2O$
45.64 $2C_{36}H_{60}O_{30}$, Li^+, $I_3{}^-$, I_2, $8H_2O$
51.82 $C_{36}H_{62}O_2$
86.126 $C_{36}H_{63}P_3W$
86.127 $C_{36}H_{66}AuP_2{}^+$, CNS^-
86.128 $C_{36}H_{66}Cl_2P_2Pt$
85.87 $C_{36}H_{66}DyP_3S_6$
86.129 $C_{36}H_{66}HgP_2{}^{2+}$, $2ClO_4{}^-$
85.88 $C_{36}H_{66}LuP_3S_6$
84.122 $C_{36}H_{76}Cu_6N_4O_{18}$, H_2O
86.130 $C_{36}H_{86}N_2P_4Rh_2$
67.8 $C_{36}H_{90}Mg_6N_{10}$
71.247 $C_{36}H_{100}Cl_6Re_6Si_9$
31.68 $C_{37}H_{25}BrO$, C_6H_6, CH_4O
71.248 $C_{37}H_{30}ClF_3P_2Pt$
86.131 $C_{37}H_{30}ClOP_2Rh$
71.249 $C_{37}H_{33}ClP_2Pt$
71.250 $C_{37}H_{33}CrO_3P_3$
71.251 $C_{37}H_{33}MoO_3P_3$
59.43 $C_{37}H_{33}N_7O_8$, $C_{36}H_{33}N_7O_8$, H_2O, $3CH_4O$

86.132 $C_{37}H_{37}O_3P_3RhS^-$, AsF_6
75.42 $C_{37}H_{41}NP_2Rh^+$, ClO_4^-
59.44 $C_{37}H_{42}O_{10}$
56.9 $C_{37}H_{54}O_2$
51.83 $C_{37}H_{64}O_2$
49.14 $C_{38}H_{21}N_9OOs$
86.133 $C_{38}H_{23}Cl_4N_2O_4PW$, CH_2Cl_2
22.8 $C_{38}H_{24}Cl_4$
83.216 $C_{38}H_{28}N_8Ru^{2+}$, $2F_6P^-$
37.24 $C_{38}H_{29}NO_3$
21.22 $C_{38}H_{30}$
72.56 $C_{38}H_{30}ClP_2Rh$, CH_2Cl_2
86.134 $C_{38}H_{30}Cl_2OOsP_2S$
72.57 $C_{38}H_{30}Cl_3IrP_2$
72.58 $C_{38}H_{30}Cl_3P_2Rh$
71.252 $C_{38}H_{30}Cl_4OOsP_2$
86.135 $C_{38}H_{30}FeO_{10}P_2S$
71.253 $C_{38}H_{30}Fe_2N_4$
86.136 $C_{38}H_{30}O_2OsP_2Se_2$
68.24 $C_{38}H_{30}O_4Tl_2$
11.20 $C_{38}H_{30}S$
30.19 $C_{38}H_{32}O_4$, $0.5C_6H_6$
86.137 $C_{38}H_{33}Cl_3NOsP_2$
71.254 $C_{38}H_{33}NP_2Pt$
86.138 $C_{38}H_{34}Cl_2O_2P_2Ru$, CH_2Cl_2
71.255 $C_{38}H_{35}NiO_2P$
20.15 $C_{38}H_{36}O_2$
86.139 $C_{38}H_{36}O_6P_2PtS_2$
72.59 $C_{38}H_{37}NO_3P_2Rh^+$, BF_4^-
86.140 $C_{38}H_{42}B_9O_4P_2RhS$, $C_4H_{10}O$
86.141 $C_{38}H_{44}B_8OP_2PtS$
84.123 $C_{38}H_{48}CoN_2O_4$, $0.5C_7H_8$
63.52 $C_{38}H_{48}N_2OSi_2$
71.256 $C_{38}H_{52}CoP_2$
67.9 $C_{38}H_{66}Li_2O_4$
50.24 $C_{38}H_{66}N_6O_{10}$
51.84 $C_{38}H_{66}O_2$
81.91 $C_{38}H_{84}Cl_5O_5P_3W_3$
73.124 $C_{39}H_{23}Fe_2O_5P$
86.142 $C_{39}H_{30}FeO_8P_2S$
71.257 $C_{39}H_{31}Co_2N_2O_5P$
71.258 $C_{39}H_{33}IO_2P_2Ru$
73.125 $C_{39}H_{37}Co$
50.25 $C_{39}H_{47}NO_{14}$, H_2O
71.259 $C_{39}H_{55}FeN_5$, C_7H_8
77.10 $C_{39}H_{71}EuN_2O_8$

$C_{40}-C_{49}$

77.11 $C_{40}H_{24}F_{12}O_6U$
85.89 $C_{40}H_{24}NiO_4P_2S_4$
78.17 $C_{40}H_{28}N_4O_4Zr$, $2.5C_6H_6$
86.143 $C_{40}H_{30}As_2Co_2O_4P_2$
86.144 $C_{40}H_{30}Cl_2P_2Pt$, $2C_2H_3N$
64.169 $C_{40}H_{30}P_2$
71.260 $C_{40}H_{33}O_2P_2RuS_2^+$, ClO_4^-, C_6H_{12}
81.92 $C_{40}H_{36}Cl_8Cu_2O_{16}$, $3C_4H_8O_2$
64.170 $C_{40}H_{36}O_5P_2$, $C_{10}H_{14}NO_2^-$, $CF_3O_3S^-$
38.141 $C_{40}H_{36}O_6$
86.145 $C_{40}H_{40}As_4Pd^{2+}$, $2Cl^-$, $2C_2H_6O_2$

86.146 $C_{40}H_{40}As_4Pd^{2-}$, $2I^-$
84.124 $C_{40}H_{40}N_8O_2W_2$, $2C_4H_8O$
84.125 $C_{40}H_{44}N_{14}O_{26}P_6Zn_2$, $4H_2O$
84.126 $C_{40}H_{45}Cu_3N_9O_6^{4-}$, $4NO_3^-$
51.85 $C_{40}H_{47}BrO_{14}S$
47.39 $C_{40}H_{47}N_{14}O_{25}P_4^{5-}$, $5H_4N^+$, $57H_2O$
71.261 $C_{40}H_{48}ClN_{16}Rh_4^{5-}$, $4Cl_4Co^{2-}$, $3H^+$, $6H_2O$
84.127 $C_{40}H_{48}Cr_2N_4O_4$, $2CH_2Cl_2$
84.128 $C_{40}H_{48}Mo_2N_4O_4$, $2CH_2Cl_2$
64.171 $C_{40}H_{56}N_{20}P_4$
73.126 $C_{40}H_{62}Th_2$, C_7H_8
71.262 $C_{40}H_{64}N_8Rh_2^{2+}$, $2F_6P^-$, $2C_2H_3N$
73.127 $C_{40}H_{64}Rh_4^{2+}$, $2BF_4^-$
48.81 $C_{40}H_{66}N_6O_{10}$, $2H_2O$
50.26 $C_{40}H_{67}O_{11}^-$, Na^+
71.263 $C_{40}H_{72}Co_2N_8$
81.93 $C_{40}H_{72}HgO_4P_2$, $2H_2O$
86.147 $C_{40}H_{80}P_2PtSi_2$
69.56 $C_{40}H_{96}N_8O_4Si_4Sn_6$
73.128 $C_{41}H_{23}Fe_2O_7P$
86.148 $C_{41}H_{30}Cl_5IrN_2P_2$, C_7H_8
69.57 $C_{41}H_{33}NOSn$
84.129 $C_{41}H_{36}N_2O_5P_2S_2U$
86.149 $C_{41}H_{39}CoP_6$
86.150 $C_{41}H_{39}IrP_6$
86.151 $C_{41}H_{39}P_6Rh$
50.27 $C_{41}H_{50}N_2O_{11}$
59.45 $C_{41}H_{52}N_4O_6$
51.86 $C_{41}H_{64}O_{14}$
51.87 $C_{41}H_{64}O_{14}$
38.142 $0.3C_{41}H_{66}O_{12}$, $0.7C_{42}H_{68}O_{12}$
50.28 $C_{41}H_{70}O_8^{2-}$, Ca^{2+}, $0.5C_7H_{16}$
50.29 $C_{41}H_{70}O_8^{2-}$, Cd^{2+}, $0.5C_7H_{16}$
30.20 $C_{42}H_{22}$
69.58 $C_{42}H_{26}Fe_2GeO_8$
32.77 $C_{42}H_{30}N_2O$
60.44 $C_{42}H_{38}$, $C_{42}H_{38}O_4$
60.44 $C_{42}H_{38}O_4$, $C_{42}H_{38}$
86.152 $C_{42}H_{39}IrN_4OP_2^{2+}$, $2F_6P^-$
86.153 $C_{42}H_{40}N_2O_2P_2RuS_2$
86.154 $C_{42}H_{40}N_2O_2P_2RuS_2$
72.60 $C_{42}H_{40}NiO_2P_2$
72.61 $C_{42}H_{40}P_2Pd$
86.155 $C_{42}H_{40}P_2Pt$
86.156 $C_{42}H_{42}As_3INNi$
86.157 $C_{42}H_{42}Cl_2P_4Pt_2$
86.158 $C_{42}H_{42}Cl_4Hg_2O_8P_2$
86.159 $C_{42}H_{42}INNiP_3$
58.79 $C_{42}H_{42}N_4O_2$, C_3H_6O
71.264 $C_{42}H_{44}FeN_6O_4$
75.43 $C_{42}H_{49}NOP_2Rh^+$, ClO_4^-
83.217 $C_{42}H_{56}N_{10}O_4S_2Zn$
57.1 $C_{42}H_{58}O_6$
75.44 $C_{42}H_{66}P_2Rh^+$, BF_4^-
84.130 $C_{42}H_{87}MoNO_4$
38.142 $0.7C_{42}H_{68}O_{12}$, $0.3C_{41}H_{66}O_{12}$
45.65 $C_{42}H_{70}O_{35}$, $2C_7H_8O$, $10H_2O$
45.66 $C_{42}H_{70}O_{35}$, $C_8H_8NO_3$, xH_2O
45.67 $C_{42}H_{70}O_{35}$, $12H_2O$

45.68 $C_{42}H_{70}O_{35}$, Na^+, I_3^-, $8H_2O$
72.62 $C_{43}H_{30}Cl_4N_2O_2P_2Ru$, CH_2Cl_2
59.46 $C_{43}H_{32}O_8$
86.160 $C_{43}H_{35}CuP_2S_2$
49.15 $C_{43}H_{51}Cl_2Hg_2N_5O_2S$
51.88 $C_{43}H_{56}Br_2O_4$
51.89 $C_{43}H_{66}O_4$
49.16 $C_{44}H_{28}ClInN_4$
49.17 $C_{44}H_{28}CoN_4$
86.161 $C_{44}H_{30}FeO_8P_3Rh_3$
64.172 $C_{44}H_{31}N_4O_3P$, $2H_2O$
65.26 $C_{44}H_{34}As_2Cl_2Pt$
83.218 $C_{44}H_{35}Co_2N_{19}S_4$, $2.7H_2O$
86.162 $C_{44}H_{35}MoO_2P_3S_2$, CH_2Cl_2
49.18 $C_{44}H_{36}N_4Ni^+$, I^-, $0.08I$
71.265 $C_{44}H_{36}O_6P_2Pt_2$
86.163 $C_{44}H_{37}NOP_2PtS$
45.69 $C_{44}H_{40}O_5$, H_2O
71.266 $C_{44}H_{41}O_4P_2Pt^+$, BF_4^-
86.164 $C_{44}H_{42}FeO_{10}P_2S$
83.219 $C_{44}H_{42}N_8Pd$
75.45 $C_{44}H_{42}P_2Pt$
86.165 $C_{44}H_{43}ClN_2O_2P_2Ru$
86.166 $C_{44}H_{48}P_4Pt_2$
86.167 $C_{44}H_{56}B_9IrP_2$, C_7H_8
84.131 $C_{44}H_{56}Cr_2N_4O_5$, C_7H_8
61.21 $C_{44}H_{56}O_4$, C_7H_8
48.82 $C_{44}H_{59}N_{13}O_{10}S$
84.132 $C_{44}H_{60}Cr_2N_8O_5$
81.94 $C_{44}H_{78}Hg_2O_8P_2$
86.168 $C_{44}H_{96}Cl_4P_4Pd_2$
49.19 $C_{45}H_{31}ClFeN_4$
49.20 $C_{45}H_{31}FeN_4O$
77.12 $C_{45}H_{33}FeO_6$
86.169 $C_{45}H_{34}ClNO_2P_2Ru$, CCl_2D_2
71.267 $C_{45}H_{37}IO_2P_2Ru$
72.63 $C_{45}H_{40}ClIrP_2$
80.28 $C_{45}H_{42}FeN_3S_6$
80.29 $C_{45}H_{42}FeN_3S_6$
75.46 $C_{45}H_{44}P_2Pt$
49.21 $C_{45}H_{47}N_5O_8$, $1.5CH_2Cl_2$
50.30 $C_{45}H_{51}IN_2O_{14}$, C_2H_6OS, H_2O
50.31 $C_{45}H_{53}IN_2O_{13}$, C_3H_6O, $5H_2O$
71.268 $C_{45}H_{54}B_{10}P_2Pt$
50.32 $C_{45}H_{57}N_3O_9$, $2C_6H_2N_3O_7^-$, Ba^{2+}, $2C_7H_8$
86.170 $C_{45}H_{90}Mn_6O_{36}P_9$
30.21 $C_{46}H_{26}$
49.22 $C_{46}H_{28}FeN_6^-$, K^+, $2C_3H_6O$
49.23 $C_{46}H_{33}FeN_4S^-$, $C_{18}H_{36}N_2O_6$, Na^+, $2C_6H_5Cl$
71.269 $C_{46}H_{38}Br_2N_2P_2Pd_2$
73.129 $C_{46}H_{42}P_2RhW^+$, F_6P^-
73.130 $C_{46}H_{46}P_2Rh^+$, F_6P^-
49.24 $C_{46}H_{52}Cl_2FeN_6^+$, ClO_4^-
50.33 $C_{46}H_{54}N_{12}O_{12}S_2$, $2CH_4O$, xH_2O
50.34 $C_{46}H_{54}N_{12}O_{12}S_2$, $12H_2O$
86.171 $C_{46}H_{100}Cl_4P_4Pd_2$
86.172 $C_{47}H_{32}Au_2O_{11}Os_3P_2$
86.173 $C_{47}H_{34}Cl_3NO_3P_2W$

71.270 $C_{47}H_{40}NOP_2RuS_2^-$, ClO_4^-. 0.5CHCl$_3$. 0.5H$_2$O

71.271 $C_{47}H_{40}NOP_2RuS_2^-$, ClO_4^-. 0.5CHCl$_3$. 0.5H$_2$O

72.64 $C_{47}H_{40}NiO_3P_2$, C_6H_6

72.65 $C_{47}H_{57}N_2P_3PtSSi$

49.25 $C_{47}H_{61}CoN_8O_{12}$, 3H$_2$O

30.22 $C_{48}H_{24}$

49.26 $C_{48}H_{36}MoN_4O_2$, 1.5C$_7$H$_8$

83.220 $C_{48}H_{38}Hg_2N_2O_{12}P_4$, 2C$_2H_4O_2$

86.174 $C_{48}H_{38}IrN_3OP_2^{2-}$, 2F$_6P^-$

31.69 $C_{48}H_{38}N_4$

84.133 $C_{48}H_{40}Cl_4O_4S_4Th$

69.59 $C_{48}H_{40}Ge_4S$

63.53 $C_{48}H_{40}O_4Si_4$

69.60 $C_{48}H_{40}O_8P_4S_8Sn_2$

86.175 $C_{48}H_{42}N_2O_2P_4Rh^+$, F$_6P^-$. 0.5C$_3H_6$O

75.47 $C_{48}H_{47}CoP_3^+$, ClO$_4^-$. 0.5CH$_2$Cl$_2$

83.221 $C_{48}H_{61}Cu_6N_{12}O_8^{3+}$, O$_4$S^{2-}. H$_2O_4P^-$. 5H$_2$O

84.134 $C_{48}H_{64}Cr_2N_4O_6$, C$_4$H$_8$O

81.95 $C_{48}H_{72}ClCu_1O_{24}S_{12}^{5-}$, 5Tl$^+$, 12H$_2$O

61.22 $C_{48}H_{80}O_{40}$, C$_3$H$_8$O, xH$_2$O

45.70 $C_{48}H_{80}O_{40}$, 17H$_2$O

50.35 $C_{48}H_{82}O_{20}$

85.90 $C_{48}H_{88}P_8S_8Th$

85.91 $C_{48}H_{120}ClCu_{14}N_{12}S_{12}^{7+}$, 3.5O$_4$S^{2-}. 20H$_2$O

71.272 $C_{49}H_{40}OP_2PtS_3$, C$_6$H$_6$

85.92 $(C_{49}H_{44}OS_8Zn_4)_n$

C_{50}–C_{74}

74.15 $C_{50}H_{44}BP_2Rh$

85.93 $C_{50}H_{44}Cl_2Cu_2P_4S_4$, 2C$_{25}H_{22}$ClCuP$_2S_2$

86.176 $C_{50}H_{44}Cl_2O_2P_4Pd_2S$, 0.5CH$_2Cl_2$. CH$_4$O

86.177 $C_{50}H_{44}Cl_2O_2P_4Rh_2S$

86.178 $C_{50}H_{44}Cl_2P_4Pd_2S$, 3CH$_2$Cl$_2$

86.179 $C_{50}H_{46}Ag_4Br_4N_2P_4$

86.180 $C_{50}H_{48}Cl_2N_6P_6Pd$

73.131 $C_{50}H_{48}N_4Ti_2$

32.78 $C_{50}H_{50}N_4$

34.13 $C_{50}H_{50}N_4$

34.14 $C_{50}H_{50}N_4$

81.96 $C_{50}H_{54}Hg_2O_8P_2$

84.135 $C_{50}H_{58}Cr_2N_6O_4$, C$_5$H$_5$N

86.181 $C_{51}H_{39}CoCr_2O_{10}P_6$

49.27 $C_{51}H_{41}FeN_4OS^-$, C$_2$H$_5$S$^-$. 2C$_{16}$H$_{32}$N$_2$O$_5$. 1.5C$_6$H$_6$. 2Na$^+$

86.182 $C_{51}H_{44}ClOP_2Pt_2^+$, F$_6$P$^-$

86.183 $C_{51}H_{44}IrN_3OP_2$

86.184 $C_{51}H_{44}OP_4Rh^+$, BF$_4^-$

86.185 $C_{51}H_{46}Cl_2P_4Pt_2$

75.48 $C_{51}H_{53}OP_2Rh$, 2C$_6$H$_6$

49.28 $C_{52}H_{44}CdN_4O_4$

86.186 $C_{52}H_{44}ClO_2P_4Rh_2^-$, BF$_4^-$

86.187 $C_{52}H_{44}Cl_2FeP_4$, 2C$_3$H$_6$O

84.136 $C_{52}H_{44}FeN_4O_2$

86.188 $C_{52}H_{44}O_2P_4Pt_2^{2+}$, 2F$_6P^-$

86.189 $C_{52}H_{48}Cl_4P_4W_2$

86.190 $C_{52}H_{48}Cl_4P_4W_2$, 0.5H$_2$O

86.191 $C_{52}H_{48}IrP_4S_2^-$, Cl$^-$. 2C$_{13}$H$_{12}$N$_2$O

86.192 $C_{52}H_{50}P_4Ru$, C$_6$H$_6$

86.193 $C_{52}H_{51}P_4Pt_2^-$, F$_6$P$^-$

78.18 $C_{52}H_{64}Cu_2N_4O_4$

49.29 $C_{52}H_{69}CoN_6O_{15}$, H$_2$O

49.30 $C_{52}H_{71}CoN_8O_{12}$, 0.5H$_2$O

86.194 $C_{53}H_{44}ClO_3P_4Rh_2^+$, C$_2Cl_2O_2Rh^-$. CH$_2Cl_2$

71.273 $C_{53}H_{44}Cl_2OP_4Rh_2S_4$

86.195 $C_{53}H_{45}O_3P_4Rh_2^+$, C$_7$H$_7$O$_3$S$^-$. 2C$_4$H$_8$O

71.274 $C_{53}H_{47}ClN_2O_2P_2Pd$

86.196 $C_{53}H_{47}Cl_3CuN_2O_2P_2Pd$

86.197 $C_{53}H_{48}BrCl_2N_2P_4W^+$, F$_6P^-$

86.198 $C_{53}H_{53}P_4Pt_2^+$, F$_6$P$^-$. 3CH$_2$Cl$_2$

83.222 $C_{54}H_{38}CrN_6$, C$_7$H$_8$

71.275 $C_{54}H_{44}Cl_2F_6P_4Pd_2$

86.199 $C_{54}H_{45}Ag_2P_3S_4W$

86.200 $C_{54}H_{45}AuP_3^+$, H$_{12}$B$_9$S$^-$

71.276 $C_{54}H_{45}BrIrP_3$

86.201 $C_{54}H_{45}ClCu_3MoOP_3S_3$

86.202 $C_{54}H_{45}ClCu_3MoP_3S_4$

86.203 $C_{54}H_{45}Cu_2MoP_3S_4$, 0.8CH$_2$Cl$_2$

86.204 $C_{54}H_{48}FMoO_2P_4^+$, F$_6$P$^-$

49.31 $C_{54}H_{52}N_4O_8$, 0.5CH$_2$Cl$_2$

61.23 $C_{54}H_{54}S_6$, C$_4$H$_8$O$_2$

86.205 $C_{54}H_{57}Ir_2P_4^+$, BF$_4^-$

83.223 $C_{54}H_{73}CoN_6O_{14}$, xH$_2$O

71.277 $C_{54}H_{82}P_2Pt_2Si$

50.36 $C_{54}H_{90}N_6O_{18}$, 2Ba^{2+}, 4ClO$_4^-$, 4H$_2$O

86.206 $C_{54}H_{96}As_2Co_3N_8O_{10}P_2$, 2H$_2$O

71.278 $C_{55}H_{41}F_3O_3P_2Ru$

86.207 $C_{55}H_{54}BrN_2P_2W^+$, Br$^-$. 0.5CH$_4$O

50.37 $C_{55}H_{93}N_7O_{17}$

86.208 $C_{56}H_{40}Co_4O_8P_4$, C$_6$H$_6$

86.209 $C_{56}H_{45}IrOP_3S_2^+$, BF$_4^-$

86.210 $C_{56}H_{48}BrClN_4P_4W$, CH$_2$Cl$_2$

86.211 $C_{56}H_{56}BrN_2OP_4W^+$, F$_6$P$^-$. 0.5C$_2$H$_6$O

86.212 $C_{56}H_{64}P_4W$

71.279 $C_{57}H_{50}Cl_2O_5P_4Rh_2$

86.213 $C_{57}H_{52}O_5P_4Pd_4$

48.83 $C_{57}H_{93}N_{15}O_{15}$, H$_2$O, C$_2$H$_6$OS

29.12 $C_{58}H_{40}$

86.214 $C_{58}H_{46}N_2O_4P_4Pd$, 5C$_2$H$_4$Cl$_2$

86.215 $C_{58}H_{59}IMoN_2P_4$, 0.5C$_6$H$_6$

83.224 $C_{58}H_{68}N_8Zn_2$

56.10 $C_{58}H_{84}O_{18}$

72.66 $C_{59}H_{49}Co_2O_4P_3$

68.25 $C_{60}H_{34}Cl_2F_{16}O_2P_2Tl_2$

77.13 $C_{60}H_{44}O_8Zr$

85.94 $C_{60}H_{50}Co_4S_{10}^{2-}$, 2C$_4H_{12}N^+$

85.95 $C_{60}H_{50}Fe_4S_{10}^{2-}$, 2C$_4H_{12}N^+$

71.280 $C_{60}H_{52}Mo_2N_4O_4P_4S_4$, 4CH$_2Cl_2$

79.17 $C_{60}H_{60}Cr_2N_8O_6$, C$_6$H$_{12}$

61.24 $C_{60}H_{66}O_{12}S_6$, 4C$_2$H$_4$O$_2$

61.25 $C_{60}H_{66}S_6$, C$_4$H$_8$O$_2$

17.44 $C_{60}H_{86}O_4$, 0.5C$_4$H$_{10}$O

73.132 $C_{60}H_{90}Cl_3U_3$

50.38 $2C_{60}H_{96}N_{12}O_{12}$, 2Rb$^+$, 2C$_6H_2N_3O_7^-$. 3C$_7H_8$, 2CHCl$_3$

50.39 $C_{60}H_{102}N_6O_{18}$

81.97 $C_{60}H_{108}Co_6O_{24}S_{12}$, C$_8H_{10}$

49.32 $C_{62}H_{44}N_4O_{12}$, 5CHCl$_3$, CH$_4$O

49.33 $C_{64}H_{33}N_{16}Nd$, CH$_2$Cl$_2$

73.133 $C_{64}H_{44}Ag_2Fe_2O_{10}^{2+}$, 2F$_6P^-$

50.40 $C_{64}H_{96}N_{12}O_{12}$

72.67 $C_{65}H_{50}N_5Ni_2^{3-}$, 3Li$^+$, 2C$_4$H$_{10}$O

86.216 $C_{65}H_{50}O_5P_5Rh_4^-$, Li$^+$

72.68 $C_{66}H_{70}ClO_2P_2Rh$, 1.5C$_3$H$_6$O

50.41 $C_{66}H_{74}Cl_2N_8O_{25}$, xH$_2$O

83.225 $C_{68}H_{44}ClCuFeN_{12}O_4^{2+}$, 2ClO$_4^-$. 2H$_2$O

81.98 $C_{68}H_{48}Cu_2I_2N_2O_{12}$

86.217 $C_{68}H_{66}Cl_2FeP_6$, 2C$_3$H$_6$O

86.218 $C_{68}H_{69}Fe_2P_6S_3^+$, ClO$_4^-$. C$_3$H$_6$O

49.34 $C_{68}H_{70}FeN_{10}O_4$, C$_2$H$_6$O

49.35 $C_{68}H_{70}FeN_{10}O_6$, C$_2$H$_6$O

86.219 $C_{72}H_{60}Cu_4O_2P_4S_6W_2$

69.61 $C_{72}H_{60}Ge_6$, 7C$_6$H$_6$

84.137 $C_{72}H_{60}O_9Si_6Zr^{2-}$, 2C$_4H_{12}N^+$

86.220 $C_{72}H_{60}O_{12}P_4Ru_3S_4$, C$_7H_8$

61.26 $C_{72}H_{90}S_6$, 0.5C$_{30}$H$_{50}$

C_{75}–C_{99}

86.221 $C_{75}H_{60}O_3P_5Rh_3$

86.222 $C_{75}H_{66}Ag_3Br_2P_6^+$, Br$^-$

86.223 $C_{75}H_{67}P_6Pt_2^+$, F$_6$P$^-$

84.138 $C_{76}H_{60}N_4O_4P_4S_4U$

86.224 $C_{81}H_{85}Ir_3P_6^{2+}$, 2BF$_4^-$, 4CH$_2$Cl$_2$

73.134 $C_{82}H_{70}Cd_3Ge_4Ni_2$, C$_7$H$_8$

73.135 $C_{82}H_{70}Ge_4Hg_3Ni_2$, C$_7$H$_8$

86.225 $C_{82}H_{78}CoP_9Rh^{2+}$, 2C$_{24}$H$_{20}$B$^-$. 2C$_3$H$_6$O

86.226 $C_{82}H_{78}Co_2P_6S_2^+$, C$_3$H$_7$NO$^-$. 0.4C$_3$H$_7$NO

86.227 $C_{82}H_{78}Co_2P_6S_2^+$, C$_{24}$H$_{20}$B$^-$. 0.5C$_3$H$_7$NO

86.228 $C_{82}H_{78}NiP_9Rh^{2+}$, 2C$_{24}$H$_{20}$B$^-$. C$_4$H$_8$O

72.69 $C_{83}H_{78}Co_2P_6S_2^{2+}$, 2C$_{24}H_{20}B^-$. 2C$_3H_6$O

86.229 $C_{84}H_{84}Co_2P_6S_2^{2+}$, 2C$_{24}H_{20}B^-$. 2C$_3H_6$O

86.230 $C_{84}H_{84}Cu_4O_2P_4S_6W_2$

85.96 $C_{84}H_{84}Fe_7Mo_2S_{20}^{4-}$, 4C$_{16}H_{36}N^+$

85.97 $C_{84}H_{84}Fe_7S_{20}W_2^{4-}$, 4C$_{16}H_{36}N^+$

81.99 $C_{84}H_{168}N_{12}O_{24}Yb_4$, 2C$_7H_{16}$

49.36 $C_{92}H_{124}Cu_2N_{10}O_2$

C_{100}–

71.281 $C_{100}H_{87}Au_5P_6^{2+}$, 2NO$_3^-$

84.139 $C_{112}H_{160}Co_4O_{16}$, 2C$_6$H$_6$

PERMUTED FORMULA INDEX

Ag

81.27	(Ag	$C_7H_6NO_2)_n$
83.74	Ag	$C_{12}H_{12}N_8O_2^+$, ClO_4^- , H_2O
76.80	Co_2 Ag	$C_{12}H_{36}N_8O_4S_2^{3+}$, $3ClO_4^-$
75.12	Ag	$C_{16}H_{24}^+$, BF_4^-
86.65	$2B_8$ Ag	$C_{20}H_{28}P$, C_3H_6O
84.100	Pt_2 Ag	$C_{24}H_{44}N_{12}O_8^+$, NO_3^- , $5H_2O$
86.199	W Ag_2	$C_{54}H_{45}P_3S_4$
73.133	Fe_2 Ag_2	$C_{64}H_{44}O_{10}^{2+}$, $2F_6P^-$
86.222	Ag_3	$C_{75}H_{66}Br_2P_6^+$, Br^-
86.179	Ag_4	$C_{50}H_{46}Br_4N_2P_4$
83.212	Ag_6	$C_{36}H_{36}N_{24}^{6+}$, $6ClO_4^-$

Al

68.1	Al	$C_2H_6I_2^-$, Cs^+ , C_6H_{10}
68.2	Al	$C_3H_9I^-$, $C_4H_{12}N^+$
68.3	Al	$C_4H_9NS^-$, Cs^+
71.3	Mn Al	$C_6H_3Br_3O_5$
68.4	Al	$C_6H_{15}Cl_2N_2$
63.3	Si Al	$C_6H_{18}Cl_4N_3$
71.179	Al	$C_8H_{20}^-$, $C_{22}H_{22}CoN_4^+$
68.23	Al	$C_{27}H_{18}N_3O_3$, CH_4O
68.9	Al_2	$C_{10}H_{24}N_4$
71.159	Zr Al_2	$C_{20}H_{33}Cl$
68.21	Al_2	$C_{22}H_{30}$
68.22	Al_2	$C_{22}H_{36}Cl_2$
71.197	Zr Al_2	$C_{25}H_{38}$
68.14	Al_3	$C_{12}H_{28}Cl_5O_4$

As

84.2	Mo_6 As	$CH_{15}O_{27}^{2-}$, $2CH_6N_3^+$, $6H_2O$
65.2	Mo_4 As	$C_2H_7O_{15}^{2-}$, $2CH_6N_3^+$, H_2O
65.3	B_{20} As	C_5H_{23}
65.4	As	$C_6H_{15}S$
86.18	MnFe As	$C_{10}H_6ClO_8^-$, $C_{12}H_{12}As_2FeMnO_8^+$
69.23	Sn As	$C_{12}H_{21}N_2O_4$
65.7	As	$C_{12}H_{24}N_3O_3$
65.9	As	$C_{13}H_{13}N_2O_2$
65.11	As	$C_{19}H_{13}$
65.12	As	$C_{20}H_{16}^+$, Cl^- , H_2O
72.36	Pt As	$C_{20}H_{17}Cl_2$
73.95	Mo_2 As	$C_{22}H_{34}O_4P_2^+$, $C_8H_5MoO_3^-$
71.1	As	$C_{24}H_{20}^+$, $C_2AuN_2O_2^-$
83.1	As	$C_{24}H_{20}^+$, $C_2Cl_{10}NW^-$
73.59	As	$C_{24}H_{20}^+$, $C_{16}H_{16}Mo_2O_4P^-$
65.13	As	$C_{24}H_{20}^+$, $C_{18}HO_{18}Ru_6^-$
65.14	As	$C_{24}H_{20}^+$, Cl_4OV^-

83.5	2 As	$C_{24}H_{20}^+$
85.47	2 As	$C_{24}H_{20}^+$, $C_{12}N_6NiSe_6^{2-}$
65.15	2 As	$C_{24}H_{20}^+$, $C_{17}O_{16}Ru_6^{2-}$
65.16	2 As	$C_{24}H_{20}^+$, $C_{21}H_2Ni_{12}O_{21}^{2-}$
65.17	2 As	$C_{24}H_{20}^+$, Cl_5NbO^{2-} , $2CH_2Cl_2$
65.18	2 As	$C_{24}H_{20}^+$, $Cl_8Se_3W_2^{2-}$
65.19	2 As	$C_{24}H_{20}^+$, $Cl_{10}NW_2^{2-}$
65.20	3 As	$C_{24}H_{20}^+$, $C_{21}HNi_{12}O_{21}^{3-}$, $2C_3H_6O$
65.21	3 As	$C_{24}H_{20}^+$, $C_{21}HNi_{12}O_{21}^{3-}$, $2C_3H_6O$
65.23	As	$C_{25}H_{17}$
65.24	As	$C_{26}H_{21}$
71.200	PtF_6 As	$C_{26}H_{21}O_3$
65.25	As	$C_{27}H_{22}N_3O$, $0.5C_6H_6$
35.78	As	$C_{30}H_{40}N_2^+$, BF_4^- , $0.5CH_2Cl_2$
80.6	Ni As_2	$C_{10}H_{20}I_6N_2S_4$
65.6	As_2	$C_{12}H_{10}N_4S_2$
86.18	MnFe As_2	$C_{12}H_{12}O_8^+$, $C_{10}H_6AsClFeMnO_8^-$
73.38	Fe As_2	$C_{12}H_{16}$
65.10	As_2	$C_{18}H_{22}N_4S_2$
86.64	Ir_4 As_2	$C_{20}H_{16}O_{10}$
86.63	Fe_3 As_2	$C_{20}H_{16}O_{10}$
73.93	Fe_2Co As_2	$C_{22}H_{27}O_3$
86.83	Co_2 As_2	$C_{23}H_{15}O_5P$
86.102	Pd As_2	$C_{30}H_{34}N^+$, F_6P^-
86.117	Pd As_2	$C_{36}H_{30}Cl_2$
86.143	Co_2 As_2	$C_{40}H_{30}O_4P_2$
65.26	Pt As_2	$C_{44}H_{34}Cl_2$
86.206	Co_3 As_2	$C_{54}H_{96}N_8O_{10}P_2$, $2H_2O$
86.31	Mn_2 As_3	$C_{13}H_{18}ClO_7$
73.111	$FeCr_2Co$ As_3	$C_{28}H_{28}O_{12}$
86.156	Ni As_3	$C_{42}H_{42}IN$
65.1	As_4	$C_2H_4O_4$
65.8	Si_6 As_4	$C_{12}H_{36}$
86.68	Tc As_4	$C_{20}H_{32}Cl_2^+$, Cl^-
86.69	Tc As_4	$C_{20}H_{32}Cl_2^+$, ClO_4^-
86.70	Tc As_4	$C_{20}H_{32}Cl_4^+$, F_6P^-
65.22	As_4	$C_{24}H_{20}S_4$
86.110	Ni As_4	$C_{32}H_{40}Cl_2$
86.145	Pd As_4	$C_{40}H_{40}^{2+}$, $2Cl^-$, $2C_2H_6O_2$
86.146	Pd As_4	$C_{40}H_{40}^{2+}$, $2I^-$
65.5	Si_3 As_7	C_9H_{27}

Au

71.1	Au	$C_2N_2O_2^-$, $C_{24}H_{20}As^+$
76.5	Au	$C_4H_{12}ClN_3^+$, ClO_4^-
76.7	Au	$C_4H_{13}ClN_3^{2+}$, Cl^- , ClO_4^-
86.71	Au	$C_{20}H_{34}O_9PS$
72.46	Fe Au	$C_{24}H_{20}O_3P$
71.203	Au	$C_{26}H_{24}O_2P$
86.127	Au	$C_{36}H_{66}P_2^+$, CNS^-
86.200	Au	$C_{54}H_{45}P_3^+$, $H_{12}B_9S^-$
86.118	Au_2	$C_{36}H_{30}ClP_2^+$, ClO_4^- , CH_2Cl_2
86.172	Os_3 Au_2	$C_{47}H_{32}O_{11}P_2$
85.17	Au_4	$C_8H_{12}S_8$
71.281	Au_5	$C_{100}H_{87}P_8^{2+}$, $2NO_3^-$

B

62.1	F_8	**B**	C_2^- . Cs^+
62.2	4	**B**	$C_3H_{10}^-$. $4Na^+$. $C_4H_{10}O$
64.11		**B**	$C_4H_{14}P$
71.11	Re	**B**	$C_9H_9ClO_6$
83.44	Tc	**B**	$C_9H_{10}Cl_2N_6O$
83.66	F_2Cu	**B**	$C_{11}H_{18}IN_4O_2$
83.70	F_2Cu	**B**	$C_{11}H_{20}N_4O_3^+$. ClO_4^-
71.41	FeF_2	**B**	$C_{12}H_{15}O_3$
86.30	Mo	**B**	$C_{12}H_{41}P_4$
75.10	Fe	**B**	$C_{16}H_{19}INO_2$
83.105	F_2Cu	**B**	$C_{16}H_{23}N_5O_2^+$. ClO_4^-
62.12		**B**	$C_{20}H_{16}N_2O_2^-$. $C_4H_{12}N^+$
85.67	F_5Co	**B**	$C_{21}H_{22}N_6S$
85.68	Cu	**B**	$C_{21}H_{26}N_7O_2S^-$. K^+ . $2C_3H_6O$
85.70	Mo	**B**	$C_{23}H_{26}ClN_6O_2S$. $0.2C_3H_6O$
76.4		**B**	$C_{24}H_{20}^-$. $C_3H_8CuN_2O^+$
62.14		**B**	$C_{24}H_{20}^-$. $C_7H_{16}NO_2^+$
72.10		**B**	$C_{24}H_{20}^-$. $C_{10}H_{25}CuN_3^+$
62.15		**B**	$C_{24}H_{20}^-$. $2C_{14}H_{20}O_5$. Na^+
71.161		**B**	$C_{24}H_{20}^-$. $C_{20}H_{38}CuN_4^+$
71.164		**B**	$C_{24}H_{20}^-$. $C_{20}H_{41}O_4P_2Pt^+$
86.82		**B**	$C_{24}H_{20}^-$. $C_{22}H_{49}Fe_2O_6P_4S_2^+$
80.23		**B**	$C_{24}H_{20}^-$. $C_{25}H_{50}N_5Os_2S_{12}^+$
86.107		**B**	$C_{24}H_{20}^-$. $C_{31}H_{34}INiP_2S_2^+$
86.115		**B**	$C_{24}H_{20}^-$. $C_{35}H_{36}NiO_2P_3S^+$
62.16		**B**	$C_{24}H_{20}^-$. H_4N^+
62.17		**B**	$C_{24}H_{20}^-$. H_4N^+
71.158	2	**B**	$C_{24}H_{20}^-$
86.123	2	**B**	$C_{24}H_{20}^-$
83.97	2	**B**	$C_{24}H_{20}^-$. $C_{14}H_{24}CuN_4^{2+}$
86.122	2	**B**	$C_{24}H_{20}^-$. $C_{36}H_{44}CoO_2P_4^{2+}$
62.18		**B**	$C_{24}H_{20}NO$
71.220	Fe	**B**	$C_{28}H_{36}NO_2P_2$
85.81	MoF	**B**	$C_{29}H_{28}N_8S_2$
62.20	F_2	**B**	$C_{35}H_{45}N_4O_2$. $CHCl_3$
74.15	Rh	**B**	$C_{50}H_{44}P_2$
64.14		**B₂**	$C_5H_{19}P_2^-$. Li^+
63.4	Si	**B₂**	$C_6H_{18}N_2S_2$
62.9		**B₂**	$C_8H_{18}N_2$
73.23	Co	**B₂**	$C_{11}H_{15}$
75.2	Cr	**B₂**	$C_{11}H_{18}N_2O_3$
73.25	ZnNb	**B₂**	$C_{11}H_{19}O$. $0.5C_6H_6$
71.43	Ni	**B₂**	$C_{12}H_{36}P_4$
63.17	Sn_2Si_4	**B₂**	$C_{14}H_{42}N_4$
83.112	Cu	**B₂**	$C_{18}H_{20}N_{12}$
83.113	Fe	**B₂**	$C_{18}H_{20}N_{12}$
83.114	Ni	**B₂**	$C_{18}H_{20}N_{12}$
62.13	F_{10}	**B₂**	$C_{20}H_{18}N_2$
75.34	Si_3Ni	**B₂**	$C_{24}H_{30}$
62.19		**B₂**	$C_{28}H_{24}O$
83.202	Fe	**B₂**	$C_{30}H_{44}N_{12}$
75.38	Fe_2	**B₂**	$C_{32}H_{38}N_2O_4$
71.69	Co_2	**B₃**	$C_{14}H_{19}$
71.70	FeCo	**B₃**	$C_{14}H_{20}$
3.41		**B₄**	$C_8H_{24}N_4S_2$

86.32	Pt	**B₄**	$C_{14}H_{36}P_2$
86.33	Pt	**B₄**	$C_{14}H_{38}P_2$
62.11		**B₄**	$C_{16}H_{40}N_{10}$
86.43	Pt	**B₄**	$C_{16}H_{40}P_2$
73.87	Ni_4	**B₄**	$C_{20}H_{24}$
73.109	Fe_3	**B₄**	$C_{26}H_{42}S_2$
71.54	Co	**B₇**	$C_{13}H_{24}$
71.81	Pt	**B₈**	$C_{14}H_{40}P_2$
71.82	Pt	**B₈**	$C_{14}H_{41}P_2$
73.77	Co	**B₈**	$C_{18}H_{29}$
86.65	2Ag	**B₈**	$C_{20}H_{26}P$. C_3H_6O
71.242	Ni	**B₈**	$C_{34}H_{44}P_2$
86.121	Pt	**B₈**	$C_{36}H_{40}P_2S$
86.141	Pt	**B₈**	$C_{38}H_{44}OP_2S$
62.4		**B₉**	$C_4H_{14}BrO_2$
73.3	Se_2Co	**B₉**	C_5H_{14}
62.7		**B₉**	$C_5H_{18}^-$. Cs^+
71.13	FeF_3	**B₉**	$C_9H_{15}O_2$
73.12	Co	**B₉**	$C_9H_{23}N$
86.66	Rh	**B₉**	$C_{20}H_{26}NO_3P$. $3CH_2Cl_2$
86.140	Rh	**B₉**	$C_{38}H_{42}O_4P_2S$. $C_4H_{10}O$
86.167	Ir	**B₉**	$C_{44}H_{56}P_2$. C_7H_8
62.3		**B₁₀**	$C_3H_{14}O_2S$
62.6		**B₁₀**	$C_5H_{16}S_2$
73.4	Co	**B₁₀**	$C_6H_{16}^-$. $C_4H_{12}N^+$
62.8		**B₁₀**	C_6H_{20}
63.5	Si	**B₁₀**	C_6H_{22}
69.13	Sn	**B₁₀**	C_6H_{22}
64.29		**B₁₀**	$C_8H_{15}Cl_5N_3P_3$
62.10		**B₁₀**	$C_8H_{26}S$
71.268	Pt	**B₁₀**	$C_{45}H_{54}P_2$
86.44	Pt_2	**B₁₂**	$C_{16}H_{40}P_2$
73.13	Co_2	**B₁₆**	$C_9H_{25}^-$. $C_8H_{20}N^+$
62.5	Co	**B₁₈**	$C_4H_{20}O^-$. $C_7H_{18}N^+$
65.3	As	**B₂₀**	C_5H_{23}
64.23		**B₂₀**	$C_6H_{25}P$
86.95	Pt	**B₂₀**	$C_{28}H_{41}ClP_2$
64.75		**B₅₀**	$C_{15}H_{64}P_2$. C_6H_6

Be

67.3	**Be**	$C_{20}H_{16}N_2O_2$. $2H_2O$

Bi

73.110	Mn_4	**Bi₂**	$C_{28}H_{20}Cl_2O_8$

Cd

81.18		(**Cd**	$C_5H_{12}O_7)_n$. $2H_2O$
85.11		**Cd**	$C_6H_{14}N_4O_4S_2$
86.6		(**Cd**	$C_8H_{11}Cl_2P)_n$
82.20		**Cd**	$C_{10}H_{16}N_2O_9^{2-}$. $2Na^+$
85.41		(**Cd**	$C_{12}H_{26}N_2S_2^{2+})_n$. $2nClO_4^-$. $2nH_2O$
81.55		(**Cd**	$C_{14}H_{14}N_2O_5)_n$. $2nH_2O$
71.130	Si_2	**Cd**	$C_{18}H_{30}N_2$. $0.5C_{10}H_8N_2$
85.63		**Cd**	$C_{20}H_{16}N_2S_2$
77.9	F_9	**Cd**	$C_{24}H_{12}O_9^-$. H_4N^+

83.187		**Cd**	$C_{26}H_{20}ClN_6O_5{}^+$, $ClO_4{}^-$. CH_4O
49.28		**Cd**	$C_{52}H_{44}N_4O_4$
76.23		(**Cd$_2$** $C_6H_8N_6)_n$	
85.35		**Cd$_2$**	$C_{10}H_{12}Cl_4N_8O_2S_2$
83.56		**Cd$_2$**	$C_{10}H_{16}N_{16}O_{20}$
76.73		(**Cd$_2$** $C_{12}H_{26}N_{10}S_4)_n$	
82.28		**Cd$_2$**	$C_{14}H_{22}Cl_2N_8O_6$
73.134	Ni$_2$Ge$_4$	**Cd$_3$** $C_{82}H_{70}$. C_7H_8	

Ce

84.92		**Ce**	$C_{22}H_{30}N_{14}O_4{}^{3+}$, $3ClO_4{}^-$, $3H_2O$

Co

85.1		3 **Co**	CH_3S^{2+} , $4Na^+$, $Al_{12}CoO_{48}Si_{12}{}^{10-}$
81.1		**Co**	$CH_{17}N_6O_2{}^{2+}$, $2NO_3{}^-$
76.1		**Co**	$C_2H_{14}N_6O_4{}^+$, $NO_3{}^-$
83.6		**Co**	$C_2H_{15}N_{10}$, Br_2
83.7		**Co**	$C_2H_{18}N_9{}^{2+}$, $2ClO_4{}^-$
76.9		**Co**	$C_4H_{16}ClN_4O_3S$, H_2O
76.11		**Co**	$C_4H_{16}Cl_2N_4{}^+$, $CH_3O_4S_2{}^-$
76.12		2 **Co**	$C_4H_{16}Cl_2N_4{}^+$, $O_6S_2Se_2{}^{2-}$, H_2O
76.13		2 **Co**	$C_4H_{16}Cl_2N_4{}^+$, $O_6S_2Se_2{}^{2-}$, H_2O
76.14		2 **Co**	$C_4H_{16}Cl_2N_4{}^+$, $O_6S_3Se^{2-}$
76.17		**Co**	$C_4H_{16}N_4O_6S_2{}^-$, Na^+ , $3H_2O$
76.21		**Co**	$C_4H_{19}N_5O_3S^+$, $ClO_4{}^-$
62.5	B$_{18}$	**Co**	$C_4H_{20}O^-$, $C_7H_{18}N^+$
73.3	Se$_2$B$_9$	**Co**	C_5H_{14}
83.21		(**Co** $C_6H_6N_8S_2)_n$	
73.4	B$_{10}$	**Co**	$C_6H_{16}{}^-$, $C_4H_{12}N^+$
76.27		**Co**	$C_6H_{16}N_4O_3S^+$, Cl^- , H_2O
76.28		2 **Co**	$C_6H_{16}N_4O_3S^+$, O_6S^{2-} , $2H_2O$
76.30		**Co**	$C_6H_{16}N_4O_4{}^+$, $C_4H_5O_6{}^-$, H_2O
76.31		**Co**	$C_6H_{16}N_4O_4{}^+$, $C_4H_5O_6{}^-$, $2H_2O$
76.29		**Co**	$C_6H_{16}N_4O_4{}^+$, $C_4H_5O_6{}^-$, $3H_2O$
81.24		**Co**	$C_6H_{16}O_8$
84.13		**Co**	$C_6H_{19}N_3O_3{}^+$, $C_6H_{20}CoN_3O_3{}^{2+}$, $3ClO_4{}^-$, $0.5H_2O$
76.37		**Co**	$C_6H_{24}N_6{}^{3+}$, $C_6H_{24}CrN_6{}^{3+}$, $6CNS^-$
76.38		**Co**	$C_6H_{24}N_6{}^{3+}$, $3I^-$, H_2O
76.45		**Co**	$C_7H_{25}ClN_5O^{2+}$, Cl_4Zn^{2-}
82.15		**Co**	$C_8H_{10}N_2O_8{}^{2-}$, $2Na^+$, $7H_2O$
83.37		**Co**	$C_8H_{14}N_6O_8{}^-$, Na^+ , $2H_2O$
73.11		**Co**	$C_8H_{14}PS_5$
84.16		**Co**	$C_8H_{16}N_2O_{10}$
83.38		**Co**	$C_8H_{16}N_5O_7$
85.24		**Co**	$C_8H_{17}N_5O_7S^-$, Na^+ , $4H_2O$
85.25		**Co**	$C_8H_{20}Cl_2O_4S_2$
83.40		**Co**	$C_8H_{20}N_6O_4{}^+$, Br^-
83.41		**Co**	$C_8H_{20}N_6O_4{}^+$, CNS^- , H_2O
83.42		**Co**	$C_8H_{20}N_6O_4{}^+$, $NO_3{}^-$
76.46		**Co**	$C_8H_{22}N_5O_2{}^{2+}$, $2I^-$, $0.84H_2O$
76.47		**Co**	$C_8H_{22}N_5O_2S^{2+}$, $2Cl^-$
85.32		**Co**	$C_9H_{15}S_6$
80.3		**Co**	$C_9H_{18}N_3S_6$
76.54		**Co**	$C_9H_{20}N_4O_4{}^+$, Cl^- , $3H_2O$
73.12	B$_9$	**Co**	$C_9H_{23}N$
85.34		**Co**	$C_{10}H_{10}Cl_2N_2S_2$

82.21		**Co**	$C_{10}H_{16}N_2O_4S_2{}^+$, $ClO_4{}^-$
83.62		**Co**	$C_{10}H_{22}N_5O_{12}P$, $3H_2O$
76.63		**Co**	$C_{10}H_{22}N_5S^{2+}$, $ClO_4{}^-$, Cl^-
84.26		**Co**	$C_{10}H_{24}O_7{}^{2+}$, $2NO_3{}^-$
73.23	B$_2$	**Co**	$C_{11}H_{15}$
83.68		**Co**	$C_{11}H_{19}Cl_2N_4O_2$
86.16		**Co**	$C_{11}H_{23}ClN_4O_7P$
76.68		**Co**	$C_{11}H_{23}ClN_5$, Cl_4Zn
76.69		**Co**	$C_{11}H_{25}N_4O_2{}^{2+}$, $2Cl^-$, $2H_2O$
71.33		**Co**	$C_{11}H_{31}P_3{}^-$, $C_{11}H_{31}CoP_3$, K^+
71.38	F$_8$	**Co**	$C_{12}H_{14}O_5P$
73.34		**Co**	$C_{12}H_{15}O_2$
84.37		**Co**	$C_{12}H_{16}Cl_2N_4O_2$
82.24		**Co**	$C_{12}H_{16}N_6O_4{}^+$, $ClO_4{}^-$, $2H_2O$
86.28		**Co**	$C_{12}H_{36}N_3P_3{}^{3+}$, $3Br^-$, $3H_2O$
79.9		**Co**	$C_{12}H_{36}N_{12}O_6{}^{2+}$, $O_3S_2{}^{2-}$
79.10		**Co**	$C_{12}H_{36}N_{12}O_6{}^{2+}$, O_4S^{2-}
84.49		**Co**	$C_{13}H_{18}N_2O_7{}^-$, Na^+ , $3H_2O$
71.54	B$_7$	**Co**	$C_{13}H_{24}$
83.94		**Co**	$C_{14}H_{16}N_6O_4S_2$
71.70	FeB$_3$	**Co**	$C_{14}H_{20}$
81.58		**Co**	$C_{14}H_{20}N_2O_8$
71.73		**Co**	$C_{14}H_{22}N_5O_4$
83.96		**Co**	$C_{14}H_{23}N_6O_4$
76.83		**Co**	$C_{14}H_{24}N_6{}^{3+}$, $C_6CoN_6{}^{3-}$, $2H_2O$
83.98		**Co**	$C_{14}H_{28}N_6O_2{}^+$, $ClO_4{}^-$, H_2O
81.59		**Co**	$C_{14}H_{30}N_4O_4{}^+$, Br^- , $3H_2O$
83.103		**Co**	$C_{16}H_{14}Cl_2N_6$
78.3		**Co**	$C_{16}H_{14}N_2O_2{}^-$, Li^+ , $1.5C_4H_8O$
78.4		**Co**	$C_{16}H_{14}N_2O_2{}^-$, Na^+ , C_4H_8O
76.92		**Co**	$C_{16}H_{23}ClN_5$, Cl_4Zn
71.100		**Co**	$C_{16}H_{26}N_5O_4$
84.64		**Co**	$C_{16}H_{28}N_6O_6$
82.34		**Co**	$C_{16}H_{34}N_5O_2{}^{2+}$, $2ClO_4{}^-$, $2H_2O$
83.108		**Co**	$C_{16}H_{38}BrN_4O^{2+}$, $2Br^-$
71.103	WF$_{12}$	**Co**	$C_{17}H_5O_4$
73.71	SnFe$_2$	**Co**	$C_{18}H_{10}ClO_8$
83.111		**Co**	$C_{18}H_{18}N_6{}^+$, $C_{24}H_{20}B^-$, $5H_2O$
82.36		(**Co** $C_{18}H_{22}N_2O_9)_n$, $2nH_2O$	
83.131		**Co**	$C_{18}H_{27}ClN_5{}^{2+}$, Cl_4Co^{2-} , $0.5H_2O$
73.77	B$_8$	**Co**	$C_{18}H_{29}$
71.133	F$_{10}$	**Co**	$C_{19}H_8$
82.38		**Co**	$C_{19}H_{20}N_3O_6$, $3H_2O$
71.143		**Co**	$C_{19}H_{29}N_6O_4$
71.144		**Co**	$C_{19}H_{29}N_6O_4$
71.145		**Co**	$C_{19}H_{29}N_6O_4$
71.146		**Co**	$C_{19}H_{29}N_6O_4$
84.81		**Co**	$C_{20}H_{16}Cl_6N_4O_2$
83.147		**Co**	$C_{20}H_{26}ClN_4O$
86.73		**Co**	$C_{20}H_{41}ClN_4O_4P$
84.86		**Co**	$C_{21}H_{15}O_6$
84.120		**Co**	$C_{21}H_{22}Cl_2N_2O_5S$, $C_{36}H_{32}Cl_4Co_2N_4O_8S_2$
85.67	F$_5$B	**Co**	$C_{21}H_{22}N_6S$
72.39	Ni	**Co**	$C_{22}H_{15}O_3$
83.159		**Co**	$C_{22}H_{20}N_6S_2$
83.160		**Co**	$C_{22}H_{20}N_6S_2$, $2CHI_3$
71.179		**Co**	$C_{22}H_{22}N_4{}^+$, $C_8H_{20}Al^-$
73.93	Fe$_2$As$_2$	**Co**	$C_{22}H_{27}O_3$

71.182   Co $C_{22}H_{28}N_5O_5$, C_6H_6

83.162   Co $C_{22}H_{32}N_6O_6$

86.81 F_3 Co $C_{22}H_{41}N_8O_4P$

84.94   Co $C_{23}H_{23}N_6O_6{}^+$, $NO_3{}^-$

83.164   Co $C_{23}H_{25}N_7O_2{}^{2+}$, $2BF_4{}^-$

83.169   Co $C_{24}H_{22}ClN_4O^+$, $NO_3{}^-$, H_2O

85.72   Co $C_{24}H_{24}N_3O_3S_3$

83.170   Co $C_{24}H_{24}N_6S_2$

83.176   Co $C_{24}H_{28}Cl_2N_4{}^+$, F_6P^-

76.101   Co $C_{24}H_{30}N_2O_2$

84.99   Co $C_{24}H_{34}Cl_2N_2O_7$

74.11   Co $C_{24}H_{36}{}^+$, F_6P^-

83.181   Co $C_{25}H_{16}N_4O_3{}^+$, Br^- , $4H_2O$

83.182   Co $C_{25}H_{16}N_4O_3{}^+$, Cl^- , $3H_2O$

71.196   Co $C_{25}H_{34}N_5O_8S$

71.205   Ge Co $C_{26}H_{25}O_4$

86.91   Co $C_{26}H_{29}N_5O_6P$

83.188   Co $C_{26}H_{44}N_6{}^{2+}$, $2F_6P^-$

83.192   Co $C_{28}H_{23}N_6O_2{}^{2+}$, $2CNS^-$, $2H_2O$

84.109   Co $C_{28}H_{28}N_6O_3{}^{2+}$, $2ClO_4{}^-$

73.111 FeCr$_2$As$_3$ Co $C_{28}H_{28}O_{12}$

83.196   Co $C_{28}H_{44}N_8S_2{}^+$, Cl^-

76.105   Co $C_{29}H_{35}N_3O_4$

83.197   Co $C_{30}H_{24}N_6{}^{3+}$, $C_6FeN_6{}^{3-}$, $8H_2O$

49.1   Co $C_{32}H_{16}N_8$

73.119   Co $C_{33}H_{25}$

73.121   Co $C_{34}H_{28}O_2P$

73.122   Si Co $C_{36}H_{33}$

86.122   Co $C_{36}H_{44}O_2P_4{}^{2+}$, $2C_{24}H_{20}B^-$

86.123   Co $C_{36}H_{44}O_2P_4{}^{2+}$, $2C_{24}H_{20}B^-$, $2C_3H_7NO$

84.123   Co $C_{38}H_{48}N_2O_4$, $0.5C_7H_8$

71.256   Co $C_{38}H_{52}P_2$

73.125   Co $C_{39}H_{37}$

86.149   Co $C_{41}H_{39}P_6$

49.17   Co $C_{44}H_{28}N_4$

49.25   Co $C_{47}H_{61}N_8O_{12}$, $3H_2O$

75.47   Co $C_{48}H_{47}P_3{}^+$, $ClO_4{}^-$, $0.5CH_2Cl_2$

86.181   Cr$_2$ Co $C_{51}H_{39}O_{10}P_6$

49.29   Co $C_{52}H_{69}N_6O_{15}$, H_2O

49.30   Co $C_{52}H_{71}N_8O_{12}$, $0.5H_2O$

83.223   Co $C_{54}H_{73}N_6O_{14}$, xH_2O

86.225   Rh Co $C_{82}H_{78}P_6{}^{2+}$, $2C_{24}H_{20}B^-$, $2C_3H_6O$

76.53   Co$_2$ $C_8H_{33}N_8O_3{}^{3+}$, $3ClO_4{}^-$

73.13   B$_{16}$ Co$_2$ $C_9H_{25}{}^-$, $C_8H_{20}N^+$

83.51   Co$_2$ $C_{10}H_{14}N_8O_4$

76.80   Ag Co$_2$ $C_{12}H_{36}N_8O_4S_2{}^{3+}$, $3ClO_4{}^-$

71.69   B$_3$ Co$_2$ $C_{14}H_{19}$

86.34   Fe Co$_2$ $C_{15}H_5O_9P$

76.94   Co$_2$ $C_{16}H_{45}N_8O_3{}^{3+}$, $3ClO_4{}^-$, $2H_2O$

71.107   Co$_2$ $C_{17}H_{12}I_2O_8$

71.116   Co$_2$ $C_{17}H_{50}P_6$

76.96   Co$_2$ $C_{18}H_{54}N_{12}O_2{}^{4+}$, $4ClO_4{}^-$, $2H_2O$

76.97   Co$_2$ $C_{20}H_{54}N_{10}O_2{}^{4+}$, $O_6S_2{}^{2-}$, $4H_2O$, $2NO_3{}^-$

71.167   Co$_2$ $C_{21}H_{14}O_4$

71.168   Co$_2$ $C_{21}H_{14}O_7$

71.184   Co$_2$ $C_{22}H_{32}O$

86.83   As$_2$ Co$_2$ $C_{23}H_{15}O_5P$

86.84   Co$_2$ $C_{23}H_{15}O_5P_3$

76.106   Co$_2$ $C_{32}H_{46}N_{10}O_2{}^{4+}$, $4I^-$

84.120   Co$_2$ $C_{36}H_{32}Cl_4N_4O_8S_2$, $C_{21}H_{22}Cl_2CoN_2O_5S$

71.257   Co$_2$ $C_{39}H_{31}N_2O_5P$

86.143   As$_2$ Co$_2$ $C_{40}H_{30}O_4P_2$

71.263   Co$_2$ $C_{40}H_{72}N_8$

83.218   Co$_2$ $C_{44}H_{35}N_{19}S_4$, $2.7H_2O$

72.66   Co$_2$ $C_{59}H_{49}O_4P_3$

86.226   Co$_2$ $C_{82}H_{78}P_6S_2{}^+$, $C_3H_7NO^-$, $0.4C_3H_7NO$

86.227   Co$_2$ $C_{82}H_{78}P_6S_2{}^+$, $C_{24}H_{20}B^-$, $0.5C_3H_7NO$

72.69   Co$_2$ $C_{83}H_{78}P_6S_2{}^{2+}$, $2C_{24}H_{20}B^-$, $2C_3H_6O$

86.229   Co$_2$ $C_{84}H_{84}P_6S_2{}^{2+}$, $2C_{24}H_{20}B^-$, $2C_3H_6O$

71.15   Co$_3$ $C_{10}O_{10}{}^-$, Li^+ , $C_6H_{14}O$

86.35   Co$_3$ $C_{15}H_5O_9P$

73.53   Co$_3$ $C_{15}H_{15}S_2$

73.54   Co$_3$ $C_{15}H_{15}S_2$

71.97   Co$_3$ $C_{16}H_{15}S_2$

71.112   Co$_3$ $C_{17}H_{30}O_{15}P_3$

71.169   Cr Co$_3$ $C_{21}H_{15}O_5S_2$, $0.5C_4H_8O$

86.206   As$_2$ Co$_3$ $C_{54}H_{96}N_8O_{10}P_2$, $2H_2O$

86.77   Co$_4$ $C_{22}H_{10}O_{10}P_2$

86.208   Co$_4$ $C_{56}H_{40}O_8P_4$, C_6H_6

85.94   Co$_4$ $C_{60}H_{50}S_{10}{}^{2-}$, $2C_4H_{12}N^+$

84.139   Co$_4$ $C_{112}H_{160}O_{16}$, $2C_6H_6$

71.118   Co$_6$ $C_{18}O_{16}S_3$

81.97   Co$_6$ $C_{60}H_{108}O_{24}S_{12}$, C_8H_{10}

Cr

73.1   Cr $C_5H_5ClN_2O_2$

76.39   Cr $C_6H_{24}N_6{}^{3+}$, $3Br^-$, $0.6H_2O$

73.6   Cr $C_7H_5NO_2S$

73.8   Cr $C_8H_5NO_4$

82.16   Cr $C_8H_{10}N_2O_8{}^-$, K^+ , $3H_2O$

86.8   Cr $C_9H_7N_2O_5P$

71.19   Cr $C_{10}H_{10}NO_5{}^+$, $BF_4{}^-$

76.62   Cr $C_{10}H_{16}N_2O_4$

86.13   Cr $C_{11}H_{14}NO_6P$

75.2   B$_2$ Cr $C_{11}H_{18}N_2O_3$

86.14   Si$_3$ Cr $C_{11}H_{18}O_5P_4$

70.8   Te Cr $C_{12}H_{14}N_2O_5$

74.1   Si Cr $C_{12}H_{14}O_3$

73.35   Cr $C_{12}H_{15}NO_3$

73.41   W Cr $C_{13}H_8N_2O_7$

75.6   Cr $C_{13}H_{13}O_3S_2{}^-$, Li^+ , $3C_4H_8O_2$

84.50   Cr $C_{13}H_{24}N_2O_4{}^+$, I^- , H_2O

71.60   W Cr $C_{14}H_5ClO_7$

81.52   Cr $C_{14}H_6N_2O_8{}^-$, Rb^+

85.49   Cr $C_{14}H_{17}NO_6S_2$

82.29   Cr $C_{14}H_{22}N_2O_8{}^-$, Na^+ , $2H_2O$

83.101   Mo Cr $C_{16}H_{10}N_2O_{10}$

86.38   Cr $C_{16}H_{11}O_7P$

71.95   Cr $C_{16}H_{14}O_3$

83.106   F$_2$ Cr $C_{16}H_{24}N_8{}^+$, $BF_4{}^-$

82.31   Cr $C_{16}H_{26}N_2O_8{}^-$, K^+ , $4H_2O$

74.2   Cr $C_{18}H_{20}{}^+$, F_6P^-

74.3   Cr $C_{18}H_{20}{}^+$, $I_3{}^-$

86.56   Cr $C_{18}H_{48}P_6$

71.141   Cr $C_{19}H_{22}N_2O_4$

74.6		**Cr**	$C_{19}H_{26}N_2O_3$
74.5		**Cr**	$C_{19}H_{26}N_2O_3$
74.4		**Cr**	$C_{19}H_{26}N_2O_3$
74.7	Si$_2$	**Cr**	$C_{20}H_{26}O_5$
85.66		**Cr**	$C_{21}H_{15}N_3O_3S_3{}^{3-}$, $C_7H_{18}N^+$, 2.5Na$^-$, 0.5HO$^-$, 18.5H$_2$O
84.87		**Cr**	$C_{21}H_{15}N_3O_6{}^{3-}$, 3Na$^+$, 8H$_2$O, C_2H_6O
84.88		**Cr**	$C_{21}H_{15}N_3O_6{}^{3-}$, 5Na$^+$, I$^-$, HO$^-$, 9H$_2$O, 3CH$_4$O, C_2H_6O
71.169	Co$_3$	**Cr**	$C_{21}H_{15}O_5S_2$, 0.5C_4H_8O
86.74		**Cr**	$C_{21}H_{19}O_9P$
75.28		**Cr**	$C_{21}H_{21}NO_6$
74.9		**Cr**	$C_{21}H_{28}N_2O_4$
86.76		**Cr**	$C_{21}H_{45}O_3P_3$
74.10		**Cr**	$C_{22}H_{25}O_2P$
71.185	Pt	**Cr**	$C_{22}H_{35}O_6P_3$
85.73		**Cr**	$C_{24}H_{24}N_3O_3S_3$
78.10		**Cr**	$C_{27}H_{24}Cl_3N_4O_3$, 3H$_2$O
71.215	Sn	**Cr**	$C_{27}H_{25}NO_4$
71.219	Sn	**Cr**	$C_{28}H_{25}NO_5$, 0.5CH$_2$Cl$_2$
74.13		**Cr**	$C_{30}H_{23}O_5P$
74.14		**Cr**	$C_{30}H_{25}O_2PS$
71.250		**Cr**	$C_{37}H_{33}O_3P_3$
83.222		**Cr**	$C_{54}H_{38}N_6$, C_7H_8
81.32		**Cr$_2$**	$C_8H_{16}O_{10}$
81.33		**Cr$_2$**	$C_8H_{16}O_{10}$
81.45	(	**Cr$_2$**	$C_{12}H_{16}N_2O_8)_n$
76.72		**Cr$_2$**	$C_{12}H_{22}N_4O_{10}$, 4H$_2$O
73.45		**Cr$_2$**	$C_{14}H_{10}O_4S$
71.101		**Cr$_2$**	$C_{16}H_{40}P_4$
81.65		**Cr$_2$**	$C_{18}H_{22}N_2O_8$
84.76		**Cr$_2$**	$C_{20}H_{12}Cl_4N_4O_4$
86.72	Si$_6$	**Cr$_2$**	$C_{20}H_{36}O_8P_8$
73.111	FeCoAs$_3$	**Cr$_2$**	$C_{28}H_{28}O_{12}$
84.114		**Cr$_2$**	$C_{32}H_{32}N_4O_4$
83.214		**Cr$_2$**	$C_{36}H_{44}N_8$
84.127		**Cr$_2$**	$C_{40}H_{48}N_4O_4$, 2CH$_2$Cl$_2$
84.131		**Cr$_2$**	$C_{44}H_{56}N_4O_5$, C_7H_8
84.132		**Cr$_2$**	$C_{44}H_{60}N_8O_5$
84.134		**Cr$_2$**	$C_{48}H_{64}N_4O_6$, C_4H_8O
84.135		**Cr$_2$**	$C_{50}H_{58}N_6O_4$, C_5H_5N
86.181	Co	**Cr$_2$**	$C_{51}H_{39}O_{10}P_6$
79.17		**Cr$_2$**	$C_{60}H_{60}N_6O_6$, C_6H_{12}

Cu

83.4	(	**Cu**	$C_2H_3Cl_2N_2S)_n$
84.3	(	**Cu**	$C_2H_6Cl_2OS)_n$
81.2	(D$_6$	**Cu**	$C_2O_6)_n$, 2nD$_2$O
81.3	(D$_6$	**Cu**	$C_2O_6)_n$, 2nD$_2$O
76.4		**Cu**	$C_3H_8N_2O^+$, $C_{24}H_{20}B^-$
84.4	(	**Cu**	$C_4H_8Cl_2OS)_n$
76.6		**Cu**	$C_4H_{12}Br_2N_2O$
84.6	2	**Cu**	$C_4H_{13}N_2O_2{}^+$, 2NO$_3{}^-$
76.18		**Cu**	$C_4H_{16}N_4{}^{2+}$, $C_{24}H_{40}CuN_8Ni_2O_{16}{}^{2-}$, 2H$_2$O
81.8		**Cu**	$C_4O_8{}^{2-}$, 2Na$^+$, 2H$_2$O
81.9		**Cu**	$C_4O_8{}^{2-}$, 2Na$^+$, 2H$_2$O
83.18		**Cu**	$C_5H_9Cl_2N_3$

84.8	4	**Cu**	$C_5H_{13}N_2O^+$, $C_4CuN_4S_4{}^{3-}$, CNS$^-$
83.22	(	**Cu**	$C_6H_6N_8S_2)_n$
81.23		**Cu**	$C_6H_{14}N_2O_6$, 2H$_2$O
84.11		**Cu**	$C_6H_{14}N_4O_6$
76.25		**Cu**	$C_6H_{15}Br_2N_3$
76.26		**Cu**	$C_6H_{15}Cl_2N_3$
76.36		**Cu**	$C_6H_{21}N_5{}^{2+}$, 2ClO$_4{}^-$
83.31		**Cu**	$C_7H_{12}Cl_2N_4O_4$
82.14		**Cu**	$C_7H_{12}N_2O_3S$
76.44		**Cu**	$C_7H_{20}Cl_2N_4O_8$
83.32	(	**Cu**	$C_8H_6ClN_4S_2)_n$, nH$_2$O
83.34	(	**Cu**	$C_8H_8Cl_2N_4O_8)_n$
82.17		**Cu**	$C_8H_{16}N_2O_7$
85.27	2	**Cu**	$C_8H_{20}S_4{}^+$, $C_8H_{20}CuS_4{}^{2+}$, 4BF$_4{}^-$
85.26	2	**Cu**	$C_8H_{20}S_4{}^+$, $C_8H_{20}CuS_4{}^{2+}$, 4ClO$_4{}^-$
85.27		**Cu**	$C_8H_{20}S_4{}^{2+}$
85.26		**Cu**	$C_8H_{20}S_4{}^{2+}$
84.18	2	**Cu**	$C_8H_{23}N_2O_3{}^+$, 2NO$_3{}^-$
76.48		**Cu**	$C_8H_{24}N_4{}^{2+}$, 2ClO$_4{}^-$
76.50		**Cu**	$C_8H_{26}N_6{}^{2+}$, 2NO$_3{}^-$
76.51		**Cu**	$C_8H_{26}N_6{}^{2+}$, 2NO$_3{}^-$
84.22		**Cu**	$C_9H_{24}ClN_4O_4{}^+$, ClO$_4{}^-$
76.55		**Cu**	$C_{10}H_{12}N_2O_8{}^{2-}$, Ca^{2+}, 4H$_2$O
84.23	(	**Cu**	$C_{10}H_{12}N_2O_8S_2)_n$
83.49		**Cu**	$C_{10}H_{14}Cl_2N_2S_2$
82.18		**Cu**	$C_{10}H_{15}N_5O_5$
82.19		**Cu**	$C_{10}H_{15}N_5O_5$, 4H$_2$O
83.57	0.5	**Cu**	$C_{10}H_{16}N_2O_{10}S_2$, 0.5$C_{10}H_{16}N_2O_{10}S_2Zn$
83.59		**Cu**	$C_{10}H_{18}N_8O_3{}^{2+}$, 2Cl$^-$, 2H$_2$O
79.8		**Cu**	$C_{10}H_{20}ClN_4S_2$
76.65		**Cu**	$C_{10}H_{24}N_4O_6$
72.10		**Cu**	$C_{10}H_{25}N_3{}^+$, $C_{24}H_{20}B^-$
84.31		**Cu**	$C_{11}H_{13}ClN_3O_3$, CH$_4$O
82.22		**Cu**	$C_{11}H_{14}N_6O_4$, 4H$_2$O
83.66	F$_2$B	**Cu**	$C_{11}H_{18}N_4O_2$
83.69		**Cu**	$C_{11}H_{19}N_4O_2{}^+$, ClO$_4{}^-$, 0.5CH$_4$O
83.70	F$_2$B	**Cu**	$C_{11}H_{20}N_4O_3{}^+$, ClO$_4{}^-$
83.71		**Cu**	$C_{11}H_{21}N_4O_3{}^+$, ClO$_4{}^-$, H$_2$O
84.36		**Cu**	$C_{12}H_{11}N_3O_5$, H$_2$O
77.1	F$_6$	**Cu**	$C_{12}H_{12}N_2O_2$
81.41		**Cu**	$C_{12}H_{12}N_2O_4$, H$_2$O
83.75		**Cu**	$C_{12}H_{12}N_6O_6S$
78.1		**Cu**	$C_{12}H_{15}NO_4$
81.46		**Cu**	$C_{12}H_{16}N_2O_6$
83.78		**Cu**	$C_{12}H_{16}N_6O_2$
76.71		**Cu**	$C_{12}H_{18}N_2S_2$
76.77		**Cu**	$C_{12}H_{32}N_4{}^{2+}$, 2ClO$_4{}^-$
76.78		**Cu**	$C_{12}H_{32}N_4{}^{2+}$, 2ClO$_4{}^-$
82.27		**Cu**	$C_{14}H_{13}N_3O_4$, 6H$_2$O
84.55		**Cu**	$C_{14}H_{16}N_4O_7$
83.95		**Cu**	$C_{14}H_{18}Br_2N_2$
79.12		**Cu**	$C_{14}H_{19}ClN_4OS$
84.57		**Cu**	$C_{14}H_{20}N_4O_2{}^{2+}$, 2ClO$_4{}^-$
83.97		**Cu**	$C_{14}H_{24}N_4{}^{2+}$, 2$C_{24}H_{20}B^-$
76.85		**Cu**	$C_{14}H_{28}Cl_2N_2O_4$
80.12		**Cu**	$C_{14}H_{28}N_2S_4$
84.58	Ni	**Cu**	$C_{14}H_{28}N_6O_8$, 4H$_2$O
76.86		**Cu**	$C_{14}H_{31}N_2O_4{}^+$, ClO$_4{}^-$, H$_2$O

76.90	**Cu**	$C_{14}H_{36}N_5O_2^+ \cdot NO_2^-$
78.2	**Cu**	$C_{15}H_{14}ClN_3O_4 \cdot H_2O$
83.102	**Cu**	$C_{16}H_{12}Cl_2N_4$
71.93	Re$_2$ **Cu**	$C_{16}H_{12}O_{12}$
81.61	**Cu**	$C_{16}H_{22}N_2O_5 \cdot H_2O$
83.105	F$_2$B **Cu**	$C_{16}H_{23}N_5O_2^+ \cdot ClO_4^-$
77.6	**Cu**	$C_{17}H_{17}N_2O_3^+ \cdot C_5HF_6O_2^- \cdot H_2O$
85.60	**Cu**	$C_{17}H_{24}N_2S_4$
83.112	B$_2$ **Cu**	$C_{18}H_{20}N_{12}$
83.132	**Cu**	$C_{18}H_{30}N_6^{2+} \cdot 2F_6P^-$
77.7	(F$_{12}$ **Cu**	$C_{19}H_{20}NO_6)_n$
60.40	2 **Cu**	$C_{20}H_{14}N_2O_2 \cdot C_{12}H_4N_4$
83.140	**Cu**	$C_{20}H_{16}BrN_4^+ \cdot BF_4^-$
83.141	**Cu**	$C_{20}H_{16}IN_4^+ \cdot ClO_4^-$
85.64	**Cu**	$C_{20}H_{16}N_2S_4$
84.82	**Cu**	$C_{20}H_{16}N_4O_8S_2 \cdot H_2O$
83.142	**Cu**	$C_{20}H_{16}N_5O_3^+ \cdot NO_3^- \cdot H_2O$
83.144	**Cu**	$C_{20}H_{18}ClN_6^+ \cdot Cl^- \cdot 4H_2O$
83.145	**Cu**	$C_{20}H_{18}N_4O^{2+} \cdot O_6S_2^{2-}$
81.68	**Cu**	$C_{20}H_{28}N_6O_8$
71.161	**Cu**	$C_{20}H_{38}N_4^+ \cdot C_{24}H_{20}B^-$
83.153	**Cu**	$C_{20}H_{40}N_4^{2+} \cdot 2ClO_4^-$
83.152	**Cu**	$C_{20}H_{40}N_4^{2+} \cdot 2ClO_4^-$
85.65	**Cu**	$C_{21}H_{13}N_3S_7^+ \cdot ClO_4^- \cdot 2CHCl_3$
78.7	**Cu**	$C_{21}H_{17}N_4OS^+ \cdot NO_3^-$
85.68	B **Cu**	$C_{21}H_{26}N_7O_2S^- \cdot K^+ \cdot 2C_3H_6O$
77.8	F$_6$ **Cu**	$C_{22}H_{16}N_2O_4$
81.71	**Cu**	$C_{22}H_{19}N_4O_2^+ \cdot BF_4^-$
81.72	**Cu**	$C_{22}H_{19}N_4O_2^+ \cdot ClO_4^- \cdot H_2O$
78.9	**Cu**	$C_{22}H_{28}N_2O_4$
76.99	**Cu**	$C_{24}H_{18}N_2O_2$
81.77	F$_6$ **Cu**	$C_{24}H_{20}N_4O_4$
83.171	**Cu**	$C_{24}H_{24}N_4^+ \cdot BF_4^-$
83.172	**Cu**	$C_{24}H_{24}N_6$
79.15	**Cu**	$C_{24}H_{30}N_4O_2S_2$
76.18	Ni$_2$ **Cu**	$C_{24}H_{40}N_8O_{16}^{2-}$
78.12	**Cu**	$C_{26}H_{28}N_2O_2$
78.13	**Cu**	$C_{26}H_{28}N_2O_2 \cdot 2C_{12}H_4N_4$
83.199	**Cu**	$C_{30}H_{28}N_6O_4$
84.111	**Cu**	$C_{30}H_{30}N_6O_8^{2+} \cdot 2NO_3^-$
83.204	**Cu**	$C_{32}H_{26}Cl_4N_4O_2 \cdot 2H_2O \cdot 0.33CHCl_3$
85.84	**Cu**	$C_{32}H_{32}BrO_8S_4$
85.85	**Cu**	$C_{32}H_{32}ClO_8S_4 \cdot 3CHCl_3$
83.209	**Cu**	$C_{34}H_{28}N_{10}^{2+} \cdot 2BF_4^-$
78.15	**Cu**	$C_{34}H_{34}Cl_2N_2O_2$
82.41	**Cu**	$C_{36}H_{40}N_6O_8$
76.107	**Cu**	$C_{36}H_{44}ClN_4^+ \cdot NO_3^-$
86.160	**Cu**	$C_{43}H_{35}P_2S_2$
86.196	Pd **Cu**	$C_{53}H_{47}Cl_3N_2O_2P_2$
83.225	Fe **Cu**	$C_{68}H_{44}ClN_{12}O_4^{2+} \cdot 2ClO_4^- \cdot 2H_2O$
76.49	**Cu$_2$**	$C_8H_{24}N_6O_2^{2+} \cdot 2C_{24}H_{20}B^-$
83.46	**Cu$_2$**	$C_{10}H_{10}Br_4N_2$
84.44	**Cu$_2$**	$C_{12}H_{28}Cl_4N_4O_4$
81.56	**Cu$_2$**	$C_{14}H_{14}N_2O_8$
71.78	(Fe **Cu$_2$**	$C_{14}H_{26}N_{12})_n \cdot 6nH_2O$
76.87	**Cu$_2$**	$C_{14}H_{32}ClN_4O_2^+ \cdot C_{24}H_{20}B^-$
82.32	**Cu$_2$**	$C_{16}H_{28}N_4O_{14}$
84.67	**Cu$_2$**	$C_{16}H_{36}N_6O_4^{2-} \cdot 2ClO_4^-$
84.72	**Cu$_2$**	$C_{18}H_{20}N_6O_4^{2+} \cdot 2BF_4^- \cdot 2H_2O$
83.136	**Cu$_2$**	$C_{19}H_{21}ClN_8O_{11} \cdot H_2O$
71.147	Rh$_8$ **Cu$_2$**	$C_{20}H_8N_2O_{15} \cdot 0.5CH_4O$
76.98	**Cu$_2$**	$C_{21}H_{48}Cl_2N_4O_3$
84.98	**Cu$_2$**	$C_{24}H_{26}N_4O_2^+ \cdot ClO_4^- \cdot 0.5CH_4O$
82.39	**Cu$_2$**	$C_{24}H_{34}N_4O_{18}P_2 \cdot 2H_2O$
76.102	F$_{12}$ **Cu$_2$**	$C_{26}H_{18}N_2O_6$
84.107	**Cu$_2$**	$C_{26}H_{26}N_6O$
83.189	**Cu$_2$**	$C_{27}H_{52}N_6O_3^{2+} \cdot 2ClO_4^- \cdot C_3H_7NO$
76.103	**Cu$_2$**	$C_{28}H_{36}N_4O_4 \cdot 2C_6H_6O$
76.104	**Cu$_2$**	$C_{28}H_{44}N_6O_4^{2+} \cdot C_{26}H_{36}Cl_2Cu_2N_6O_{10} \cdot 2ClO_4^-$
84.112	**Cu$_2$**	$C_{32}H_{24}Cl_6N_2O_4$
84.113	(**Cu$_2$**	$C_{32}H_{28}O_{10})_n \cdot 2nC_4H_8O_2$
83.208	**Cu$_2$**	$C_{32}H_{56}Cl_4N_8$
81.90	**Cu$_2$**	$C_{34}H_{22}N_6O_8 \cdot 4H_2O$
78.16	**Cu$_2$**	$C_{34}H_{44}N_4O_6$
81.92	**Cu$_2$**	$C_{40}H_{36}Cl_8O_{16} \cdot 3C_4H_8O_2$
85.93	**Cu$_2$**	$C_{50}H_{44}Cl_2P_4S_4 \cdot 2C_{25}H_{22}ClCuP_2S_2$
78.18	**Cu$_2$**	$C_{52}H_{64}N_4O_4$
86.203	Mo **Cu$_2$**	$C_{54}H_{45}P_3S_4 \cdot 0.8CH_2Cl_2$
81.98	**Cu$_2$**	$C_{68}H_{48}I_2N_2O_{12}$
49.36	**Cu$_2$**	$C_{92}H_{124}N_{10}O_2$
84.126	**Cu$_3$**	$C_{40}H_{45}N_9O_6^{4+} \cdot 4NO_3^-$
86.201	Mo **Cu$_3$**	$C_{54}H_{45}ClOP_3S_3$
86.202	Mo **Cu$_3$**	$C_{54}H_{45}ClP_3S_4$
84.38	**Cu$_4$**	$C_{12}H_{24}O_{12}$
85.45	**Cu$_4$**	$C_{12}H_{28}N_4O_{16}S_6$
85.53	**Cu$_4$**	$C_{16}H_{36}Cl_4O_8S_4$
84.70	**Cu$_4$**	$C_{16}H_{40}Cl_4N_4O_4$
81.81	**Cu$_4$**	$C_{24}H_{44}Cl_8N_4O_{12}$
81.82	**Cu$_4$**	$C_{24}H_{48}Cl_4N_4O_{12}$
84.103	**Cu$_4$**	$C_{24}H_{56}Br_4N_4O_4 \cdot 4CCl_4$
81.87	**Cu$_4$**	$C_{32}H_{56}Cl_{12}N_4O_{12}$
81.89	**Cu$_4$**	$C_{32}H_{64}Cl_4N_4O_{12}$
86.219	W$_2$ **Cu$_4$**	$C_{72}H_{60}O_2P_4S_6$
86.230	W$_2$ **Cu$_4$**	$C_{84}H_{84}O_2P_4S_6$
84.122	**Cu$_6$**	$C_{36}H_{76}N_4O_{18} \cdot H_2O$
83.221	**Cu$_6$**	$C_{48}H_{61}N_{12}O_8^{3+} \cdot O_4S^{2-} \cdot H_2O_4P^- \cdot 5H_2O$
81.95	**Cu$_{14}$**	$C_{48}H_{72}ClO_{24}S_{12}^{5-} \cdot 5Tl^+ \cdot 12H_2O$
85.91	**Cu$_{14}$**	$C_{48}H_{120}ClN_{12}S_{12}^{7+} \cdot 3.5O_4S^{2-} \cdot 20H_2O$

D

2.7	**D**	$C_2HCl_2O_2$
32.5	**D**	$C_4H_2O_4^- \cdot C_3H_3D_2N_2^+$
32.5	**D$_2$**	$C_3H_3N_2^+ \cdot C_4H_2DO_4^-$
32.4	**D$_2$**	$C_3N_2O_3 \cdot CD_4N_2O$
33.20	**D$_2$**	$C_7H_3NO_4$
33.21	**D$_2$**	$C_7H_3NO_4$
33.22	**D$_2$**	$C_7H_3NO_4$
44.1	**D$_3$**	$C_4H_2N_3O \cdot D_2O$
51.32	**D$_3$**	$C_{22}H_{33}O_2 \cdot CH_4O$
5.1	**D$_4$**	C
32.4	**D$_4$**	$CN_2O \cdot C_3D_2N_2O_3$
33.29	**D$_5$**	C_8N_3
33.28	**D$_5$**	C_8N_3
83.60	Ni **D$_5$**	$C_{10}H_{18}N_4O_2^+ \cdot Cl^- \cdot D_2O$
3.3	**D$_6$**	$C_2N_2O_2$

81.2	(Cu D_6	$C_2O_6)_n$. $2nD_2O$
81.3	(Cu D_6	$C_2O_6)_n$. $2nD_2O$
3.16	2 D_{12}	C_4N^+ . Cl_6Pt^{2-}
17.25	D_{18}	$C_{16}N_2O_4$

Dy

85.87	Dy	$C_{36}H_{66}P_3S_6$

Eu

84.85	Si_4 Eu	$C_{20}H_{56}N_2O_4$
77.10	Eu	$C_{39}H_{71}N_2O_8$

F

2.17	F	$C_3H_4O_4^-$. Na^+
66.2	Sb F	$C_4O_9^{4-}$. $3Na^+$. H_3O^+ . H_2O
63.12	Si F	$C_{12}H_{14}NO_4$
84.45	Mn_3 F	$C_{13}H_{10}O_{11}$
81.53	Mo_2 F	$C_{14}H_6N_2O_{14}^-$. $C_8H_{20}N^+$
31.25	F	$C_{14}H_{17}$
35.45	F	$C_{17}H_{16}ClN_2O$. H_2O
86.54	Pt F	$C_{18}H_{45}P_3^+$. BF_4^-
51.22	F	$C_{20}H_{29}O_3$
35.64	F	$C_{21}H_{22}ClN_4O_2$
51.46	F	$C_{24}H_{31}O_6$. $0.67CH_4O$
85.81	MoB F	$C_{29}H_{28}N_8S_2$
86.204	Mo F	$C_{54}H_{48}O_2P_4^+$. F_6P^-
3.18	F_2	$C_4H_{12}N_2S$
83.66	CuB F_2	$C_{11}H_{18}IN_4O_2$
83.70	CuB F_2	$C_{11}H_{20}N_4O_3^+$. ClO_4^-
71.41	FeB F_2	$C_{12}H_{15}O_3$
83.105	CuB F_2	$C_{16}H_{23}N_5O_2^+$. ClO_4^-
83.106	Cr F_2	$C_{16}H_{24}N_8^+$. BF_4^-
73.72	Mo F_2	$C_{18}H_{15}ClN_4$
62.20	B F_2	$C_{35}H_{45}N_4O_2$. $CHCl_3$
11.1	F_3	CO_3S^- . $N_2S_3^+$. $0.5C_2H_3N$
81.6	Th F_3	$C_3O_9^{5-}$. $5CH_6N_3^+$
63.7	Si F_3	$C_8H_6BrO_2$
71.13	FeB_9 F_3	$C_9H_{15}O_2$
63.15	Si F_3	$C_{13}H_{14}NO_4$
37.9	F_3	$C_{13}H_{17}BrCl_3O_4$
24.6	F_3	$C_{14}H_{12}NO$
81.60	Mo F_3	$C_{14}H_{37}O_{14}P_4$
75.20	Rh F_3	$C_{18}H_{18}O_2$
69.44	Sn F_3	$C_{20}H_{33}O_2$
41.79	F_3	$C_{21}H_{26}N_3S^{2+}$. $2Cl^-$
86.81	Co F_3	$C_{22}H_{41}N_8O_4P$
64.135	F_3	$C_{24}H_{25}NO_5P$
71.213	Pt F_3	$C_{27}H_{22}ClP_2$
36.49	F_3	$C_{27}H_{25}N_2O$
84.117	$2Mo_4$ F_3	$C_{34}H_{78}NO_8$. $C_{32}H_{72}F_4Mo_4O_8$
71.248	Pt F_3	$C_{37}H_{30}ClP_2$
71.278	Ru F_3	$C_{55}H_{41}O_3P_2$
81.14	(Zn F_4	$C_4H_8O_8)_n$
2.18	F_4	$C_4O_4^{2-}$. $2Na^+$. $6H_2O$
60.4	F_4	C_6O_2 . $C_6H_4S_4$
60.34	F_4	C_6O_2 . $C_{16}H_{10}$
71.17	Fe_2 F_4	$C_{10}H_6O_6S_2$
71.18	Fe_2 F_4	$C_{10}H_6O_6S_2$
86.23	Pt_2 F_4	$C_{12}H_{30}Cl_2O_2P_4$
71.51	Fe F_4	$C_{13}H_{10}O_4$
71.191	Rh F_4	$C_{24}H_{24}^+$. ClO_4^-
84.117	Mo_4 F_4	$C_{32}H_{72}O_8$. $2C_{34}H_{78}F_3Mo_4NO_8$
1.36	F_5	$C_{18}H_{16}NO_2$
85.67	CoB F_5	$C_{21}H_{22}N_6S$
60.13	F_6	C_6 . $C_{10}H_{16}N_2$
15.2	F_6	$C_{12}H_2N_2O_4S$
31.15	F_6	$C_{12}H_8$
77.1	Cu F_6	$C_{12}H_{12}N_2O_2$
30.2	F_6	$C_{13}H_8$
71.56	Os_3 F_6	$C_{14}H_2BrO_{10}^-$. $C_{36}H_{30}NP_2^+$
77.4	Pd F_6	$C_{17}H_{10}N_2O_2$
77.5	Pd F_6	$C_{17}H_{10}N_2O_2$
31.52	F_6	$C_{20}H_{16}N_2O_2$
72.37	Os_3 F_6	$C_{20}H_{17}O_{10}P$
71.162	Pt F_6	$C_{20}H_{39}OP_2^+$. F_6P^-
64.114	F_6	$C_{22}H_{14}O_6P^-$. $C_6H_{16}N^+$
77.8	Cu F_6	$C_{22}H_{16}N_2O_4$
71.181	Ir_3 F_6	$C_{22}H_{27}O_6S_3$
72.40	Rh F_6	$C_{22}H_{31}O_2$
81.77	Cu F_6	$C_{24}H_{20}N_4O_4$
71.200	PtAs F_6	$C_{26}H_{21}O_3$
71.216	Rh F_6	$C_{27}H_{33}N_2O_2$
71.275	Pd_2 F_6	$C_{54}H_{44}Cl_2P_4$
62.1	B F_8	C_2^- . Cs^+
60.19	F_8	C_{10} . $C_{12}H_{12}$
11.12	Se_2 F_8	C_{12}
19.5	F_8	$C_{12}Br_2$
71.34	Hg F_8	$C_{12}H_2$
71.36	W_2 F_8	$C_{12}H_{13}Br_3N_2O_2P_4$
71.38	Co F_8	$C_{12}H_{14}O_5P$
38.35	F_8	$C_{12}O_2$
39.41	F_8	$C_{12}S_2$
66.6	Sb F_9	C_6O_6
73.56	Fe F_9	$C_{16}H_{11}O$
77.9	Cd F_9	$C_{24}H_{12}O_9^-$. H_4N^+
60.16	F_{10}	C_{12} . $C_{12}H_9Br$
60.18	F_{10}	C_{12} . $C_{12}H_{10}$
60.23	F_{10}	C_{12} . $C_{13}H_{12}$
71.134	Ni F_{10}	$C_{19}H_8$
71.133	Co F_{10}	$C_{19}H_8$
62.13	B_2 F_{10}	$C_{20}H_{18}N_2$
72.45	V F_{10}	$C_{24}H_{10}$
64.122	F_{11}	$C_{24}H_{14}N_4O_2P$
66.1	Sb_2 F_{12}	C_2O_2
66.3	Sb_2 F_{12}	C_4O_5
64.27	F_{12}	$C_8H_7O_3P$
64.28	F_{12}	$C_8H_{12}N_2P_2$
20.3	F_{12}	$C_{10}H_6$
81.42	Rh_2 F_{12}	$C_{12}H_{12}O_{10}$
81.43	Rh_2 F_{12}	$C_{12}H_{12}O_{10}S_2$
32.53	F_{12}	$C_{15}H_6N_6$
71.103	WCo F_{12}	$C_{17}H_5O_4$
16.13	F_{12}	$C_{17}H_7N_3O$

71.119	Nb	F_{12}	$C_{18}H_{10}$
77.7	(Cu	F_{12}	$C_{19}H_{20}NO_6)_n$
75.30	Rh	F_{12}	$C_{22}H_{15}$
76.102	Cu_2	F_{12}	$C_{26}H_{18}N_2O_6$
77.11	U	F_{12}	$C_{40}H_{24}O_8$
19.11		F_{14}	$C_{18}H_8$
86.1	Mo_2	F_{16}	$C_4H_{12}Cl_2N_4P_8$
68.25	Tl_2	F_{16}	$C_{60}H_{34}Cl_2O_2P_2$
64.76		F_{18}	$C_{16}H_4P_2$

Fe

79.5	(	Fe	$C_4H_8N_6S_4)_n$
72.2		Fe	$C_8H_5BrO_3$
84.9		Fe	$C_8H_9N_6O_6^-$, Cs^+ , H_2O
73.9		Fe	$C_8H_5NO_2S$
85.18		Fe	$C_8H_{12}O_2S_4$
84.19		Fe	$C_8H_{24}Cl_2O_4S_4^+$, Cl^-
71.13	F_3B_9	Fe	$C_9H_{15}O_2$
86.18	MnAs	Fe	$C_{10}H_6ClO_6^-$, $C_{12}H_{12}As_2FeMnO_8^+$
73.20	Si_3	Fe	$C_{10}H_{11}Cl_5O_2$
86.12		Fe	$C_{10}H_{18}N_3O_4P$
84.27		Fe	$C_{10}H_{30}ClO_5S_5^{2+}$, $2Cl_4Fe^-$
84.28		Fe	$C_{10}H_{30}ClO_5S_5^{2+}$, $Cl_6Fe_2O^{2-}$
71.28		Fe	$C_{11}H_{10}O_5$
71.27		Fe	$C_{11}H_{10}O_5$
80.11		Fe	$C_{11}H_{20}N_3S_5$
84.32	Ga	Fe	$C_{11}H_{23}N_5O_3$
75.3		Fe	$C_{12}H_8O_4$
72.15		Fe	$C_{12}H_9ClO_4$
73.27		Fe	$C_{12}H_{10}O_4$
73.28		Fe	$C_{12}H_{10}O_4$
72.16		Fe	$C_{12}H_{10}O_4$
71.35		Fe	$C_{12}H_{10}O_6$
86.18	$MnAs_2$	Fe	$C_{12}H_{12}O_8^+$, $C_{10}H_6AsClFeMnO_8^-$
60.20		Fe	$C_{12}H_{14}^+$, $C_{12}H_4N_4^-$, $C_{12}H_4N_4$
71.39		Fe	$C_{12}H_{14}O_5$
71.40		Fe	$C_{12}H_{14}O_5$
71.41	F_2B	Fe	$C_{12}H_{15}O_3$
73.38	As_2	Fe	$C_{12}H_{16}$
75.4		Fe	$C_{12}H_{16}P_2$
83.79		Fe	$C_{12}H_{18}N_6^{2+}$, $2Cl_4Fe^-$
83.80		Fe	$C_{12}H_{18}N_6^{2+}$, $2Cl_4Fe^-$
85.39		Fe	$C_{12}H_{18}N_6O_2S_4^-$, $C_6H_8N^+$, $4H_2O$
84.43		Fe	$C_{12}H_{28}Cl_2N_4O_4^+$, Cl_4Fe^-
71.51	F_4	Fe	$C_{13}H_{10}O_4$
75.5		Fe	$C_{13}H_{10}O_4$
72.20		Fe	$C_{13}H_{14}O_5$
83.90		Fe	$C_{14}H_{11}NO_4$
71.62		Fe	$C_{14}H_{11}O_2^+$, F_6P^-
73.48		Fe	$C_{14}H_{15}O_3^+$, F_6P^-
73.49		Fe	$C_{14}H_{16}S_3$
71.70	CoB_3	Fe	$C_{14}H_{20}$
71.78	$(Cu_2$	Fe	$C_{14}H_{26}N_{12})_n$, $6nH_2O$
86.34	Co_2	Fe	$C_{15}H_5O_9P$
80.14		Fe	$C_{15}H_{12}N_3S_6$. $0.5CH_2Cl_2$
72.23		Fe	$C_{15}H_{13}NO_6$
73.55		Fe	$C_{15}H_{15}NO_4$
72.24		Fe	$C_{15}H_{20}O_3$
86.37		Fe	$C_{15}H_{36}N_6O_3P_2$
73.56	F_9	Fe	$C_{16}H_{11}O$
73.57		Fe	$C_{16}H_{13}P$
75.10	B	Fe	$C_{16}H_{19}INO_2$
72.27		Fe	$C_{16}H_{19}NO_6$
84.68		Fe	$C_{16}H_{36}N_6O_6^{3+}$, $3Cl^-$, $2H_2O$
73.67	(Sn	Fe	$C_{17}H_{18}O_2)_n$
73.68		Fe	$C_{17}H_{20}O_2$
73.69		Fe	$C_{17}H_{23}$
83.110		Fe	$C_{17}H_{27}N_5O^{2+}$, $2F_6P^-$
75.16		Fe	$C_{17}H_{40}O_9P_3$
75.17		Fe	$C_{17}H_{40}O_9P_3^+$, BF_4^-
75.18		Fe	$C_{17}H_{40}O_9P_3^+$, BF_4^-
75.19		Fe	$C_{17}H_{40}O_9P_3^+$, BF_4^-
71.122		Fe	$C_{18}H_{13}O_2^+$, F_6P^-
83.113	B_2	Fe	$C_{18}H_{20}N_{12}$
75.21		Fe	$C_{18}H_{20}O_7$
83.117		Fe	$C_{18}H_{24}N_6^{2+}$, $2Cl^-$, CH_4O
83.118		Fe	$C_{18}H_{24}N_6^{2+}$, $2Cl^-$, CH_4O
83.119		Fe	$C_{18}H_{24}N_6^{2+}$, $2Cl^-$, CH_4O
83.120		Fe	$C_{18}H_{24}N_6^{2+}$, $2Cl^-$, CH_4O
83.121		Fe	$C_{18}H_{24}N_6^{2+}$, $2Cl^-$, CH_4O
83.122		Fe	$C_{18}H_{24}N_6^{2+}$, $2Cl^-$, C_2H_6O
83.123		Fe	$C_{18}H_{24}N_6^{2+}$, $2Cl^-$, C_2H_6O
83.124		Fe	$C_{18}H_{24}N_6^{2+}$, $2Cl^-$, C_2H_6O
83.125		Fe	$C_{18}H_{24}N_6^{2+}$, $2Cl^-$, C_2H_6O
83.126		Fe	$C_{18}H_{24}N_6^{2+}$, $2Cl^-$, C_2H_6O
83.127		Fe	$C_{18}H_{24}N_6^{2+}$, $2Cl^-$, $2H_2O$
83.128		Fe	$C_{18}H_{24}N_6^{2+}$, $2Cl^-$, $2H_2O$
83.129		Fe	$C_{18}H_{24}N_6^{2+}$, $2I^-$
73.78	Si_6	Fe	$C_{18}H_{38}O_2$
73.80	Pd	Fe	$C_{19}H_{25}NO_2$
75.24	Si	Fe	$C_{20}H_{16}O_3$
71.154		Fe	$C_{20}H_{17}NO$
73.84		Fe	$C_{20}H_{20}O$
60.42		Fe	$C_{20}H_{30}^+$, $C_{12}H_4N_4^-$
86.78		Fe	$C_{22}H_{15}O_4P$
73.91	Ge	Fe	$C_{22}H_{18}$
71.188		Fe	$C_{23}H_{16}O_3$
73.97		Fe	$C_{23}H_{19}^+$, BF_4^-
83.165		Fe	$C_{23}H_{25}N_7O_2^{2+}$, $2BF_4^-$
72.46	Au	Fe	$C_{24}H_{20}O_3P$
85.74		Fe	$C_{24}H_{24}N_3O_3S_3$
72.49		Fe	$C_{24}H_{32}O_5P_2S_2$
83.179		Fe	$C_{24}H_{36}N_{12}^{2+}$, $C_8Fe_2O_8^{2-}$
73.108		Fe	$C_{26}H_{28}NP$
84.108		Fe	$C_{27}H_{42}N_9O_{12}$
73.111	Cr_2CoAs_3	Fe	$C_{28}H_{28}O_{12}$
71.220	B	Fe	$C_{28}H_{36}NO_2P_2$
83.195		Fe	$C_{28}H_{38}N_4S_2^+$, F_6P^-
72.52		Fe	$C_{29}H_{23}O_3P$
83.202	B_2	Fe	$C_{30}H_{44}N_{12}$
83.203		Fe	$C_{30}H_{48}N_{12}^{2+}$, $2C_4FeO_4^-$
79.16		Fe	$C_{30}H_{72}N_{12}O_6^{2+}$, $2ClO_4^-$
75.37		Fe	$C_{31}H_{28}OP_2$
73.120	Rh	Fe	$C_{33}H_{36}NP^+$, F_6P^-
71.243		Fe	$C_{35}H_{34}P_2$

49.5		Fe	$C_{36}H_{23}N_9O_2 \cdot C_3H_7NO$
73.123		Fe	$C_{36}H_{37}IO_2P_2$
49.7		Fe	$C_{36}H_{44}ClN_4O_4$
86.135		Fe	$C_{38}H_{30}O_{10}P_2S$
86.142		Fe	$C_{39}H_{30}O_8P_2S$
71.259		Fe	$C_{39}H_{55}N_5 \cdot C_7H_8$
71.264		Fe	$C_{42}H_{44}N_6O_4$
86.161	Rh$_3$	Fe	$C_{44}H_{30}O_8P_3$
86.164		Fe	$C_{44}H_{42}O_{10}P_2S$
49.19		Fe	$C_{45}H_{31}ClN_4$
49.20		Fe	$C_{45}H_{31}N_4O$
77.12		Fe	$C_{45}H_{33}O_6$
80.28		Fe	$C_{45}H_{42}N_3S_6$
80.29		Fe	$C_{45}H_{42}N_3S_6$
49.22		Fe	$C_{46}H_{28}N_6{}^- \cdot K^+ \cdot 2C_3H_6O$
49.23		Fe	$C_{46}H_{33}N_4S^- \cdot C_{18}H_{36}N_2O_6 \cdot Na^+ \cdot 2C_6H_5Cl$
49.24		Fe	$C_{46}H_{52}Cl_2N_6{}^+ \cdot ClO_4{}^-$
49.27		Fe	$C_{51}H_{41}N_4OS^- \cdot C_2H_5S^- \cdot 2C_{16}H_{32}N_2O_5 \cdot 1.5C_6H_6 \cdot 2Na^+$
86.187		Fe	$C_{52}H_{44}Cl_2P_4 \cdot 2C_3H_6O$
84.136		Fe	$C_{52}H_{44}N_4O_2$
83.225	Cu	Fe	$C_{66}H_{44}ClN_{12}O_4{}^{2+} \cdot 2ClO_4{}^- \cdot 2H_2O$
86.217		Fe	$C_{66}H_{66}Cl_2P_6 \cdot 2C_3H_6O$
49.34		Fe	$C_{66}H_{70}N_{10}O_4 \cdot C_2H_6O$
49.35		Fe	$C_{66}H_{70}N_{10}O_6 \cdot C_2H_6O$
71.10		Fe$_2$	$C_9H_2O_8$
71.9		Fe$_2$	$C_9H_2O_8$
71.17	F$_4$	Fe$_2$	$C_{10}H_8O_6S_2$
71.18	F$_4$	Fe$_2$	$C_{10}H_8O_6S_2$
83.52		Fe$_2$	$C_{10}H_{14}N_8O_4$
86.20		Fe$_2$	$C_{12}H_{24}O_6P_2S_3 \cdot 0.5C_4H_{10}O$
71.49	Se	Fe$_2$	$C_{13}H_8O_7$
73.46		Fe$_2$	$C_{14}H_{10}O_3S$
71.63	Ni	Fe$_2$	$C_{14}H_{12}N_2O_{10}$
75.7	Se	Fe$_2$	$C_{14}H_{12}O_6$
71.64		Fe$_2$	$C_{14}H_{12}O_8$
71.88		Fe$_2$	$C_{15}H_{14}O_3$
72.25		Fe$_2$	$C_{16}H_{10}O_7$
71.105	Ni$_2$	Fe$_2$	$C_{17}H_{10}O_7$
72.29	Ni	Fe$_2$	$C_{17}H_{14}O_6$
73.66	Si$_3$	Fe$_2$	$C_{17}H_{16}Cl_4O_4$
86.48		Fe$_2$	$C_{18}H_{10}ClO_6P$
73.71	SnCo	Fe$_2$	$C_{18}H_{10}ClO_8$
71.128		Fe$_2$	$C_{18}H_{20}NO_3{}^+ \cdot I^-$
71.140		Fe$_2$	$C_{19}H_{20}O_5$
73.82	Se	Fe$_2$	$C_{20}H_{18}{}^+ \cdot I_3{}^- \cdot I_2 \cdot 0.5CH_2Cl_2$
72.38		Fe$_2$	$C_{20}H_{20}O_6$
71.178	Ni$_2$	Fe$_2$	$C_{22}H_{20}O_6$
73.93	CoAs$_2$	Fe$_2$	$C_{22}H_{27}O_3$
86.82		Fe$_2$	$C_{22}H_{49}O_6P_4S_2{}^+ \cdot C_{24}H_{20}B^-$
73.98		Fe$_2$	$C_{24}H_{18}N_2$
73.103	Si$_6$	Fe$_2$	$C_{24}H_{40}O_4$
71.211		Fe$_2$	$C_{27}H_{20}NO_6P$
83.190	Zn$_2$	Fe$_2$	$C_{28}H_{16}N_4O_8$
86.101		Fe$_2$	$C_{30}H_{20}O_6P_2{}^{2-} \cdot 2C_{18}H_{36}N_2O_6 \cdot 2Na^+$
71.232		Fe$_2$	$C_{31}H_{23}O_6P_2{}^- \cdot C_{36}H_{30}NP_2{}^-$
71.233		Fe$_2$	$C_{31}H_{23}O_6P_2{}^- \cdot Na^+ \cdot 2C_4H_8O$
75.38	B$_2$	Fe$_2$	$C_{32}H_{38}N_2O_4$

71.245		Fe$_2$	$C_{36}H_{25}O_6P_2{}^- \cdot Li^+ \cdot 3C_4H_8O$
71.253		Fe$_2$	$C_{38}H_{30}N_4$
73.124		Fe$_2$	$C_{39}H_{23}O_5P$
73.128		Fe$_2$	$C_{41}H_{23}O_7P$
69.58	Ge	Fe$_2$	$C_{42}H_{26}O_8$
73.133	Ag$_2$	Fe$_2$	$C_{64}H_{44}O_{10}{}^{2+} \cdot 2F_6P^-$
86.218		Fe$_2$	$C_{68}H_{69}P_6S_3{}^+ \cdot ClO_4{}^- \cdot C_3H_6O$
71.16		Fe$_3$	$C_{10}O_{10}S$
71.26		Fe$_3$	$C_{11}H_5NO_9$
72.11		Fe$_3$	$C_{11}H_5NO_9$
72.19		Fe$_3$	$C_{13}H_7NO_9$
86.36		Fe$_3$	$C_{15}H_7O_9P$
71.91		Fe$_3$	$C_{16}H_8O_8$
71.153		Fe$_3$	$C_{20}H_{16}O_6S_3$
86.63	As$_2$	Fe$_3$	$C_{20}H_{16}O_{10}$
73.109	B$_4$	Fe$_3$	$C_{26}H_{42}S_2$
71.44		Fe$_4$	$C_{13}H_2O_{12}$
71.58		Fe$_4$	$C_{14}H_3O_{13}{}^- \cdot C_{36}H_{30}NP_2{}^+$
71.59		Fe$_4$	$C_{14}H_3O_{13}{}^- \cdot C_{36}H_{30}NP_2{}^+$
84.51		Fe$_4$	$C_{14}H_4O_{13}$
84.52		Fe$_4$	$C_{14}H_4O_{13}$
71.84		Fe$_4$	$C_{15}H_3O_{14}{}^- \cdot C_8H_{20}N^+$
84.61		Fe$_4$	$C_{15}H_{10}N_2O_{12}$
85.80		Fe$_4$	$C_{28}H_{28}S_8{}^{3-} \cdot 3C_8H_{20}N^+$
85.95		Fe$_4$	$C_{60}H_{50}S_{10}{}^{2-} \cdot 2C_4H_{12}N^+$
85.57	Mo$_2$	Fe$_6$	$C_{16}H_{40}S_{17}{}^{3-} \cdot 3C_8H_{20}N^+$
85.62	Mo$_2$	Fe$_6$	$C_{18}H_{45}S_{17}{}^{3-} \cdot 3C_{13}H_{22}N^+$
71.190	Pd$_6$	Fe$_6$	$C_{24}HO_{24}{}^{3-} \cdot 3C_{10}H_{16}N^+ \cdot 2C_2H_3N$
85.78	Mo$_2$	Fe$_7$	$C_{24}H_{60}S_{20}{}^{3-} \cdot 3C_{10}H_{16}N^+$
85.97	W$_2$	Fe$_7$	$C_{84}H_{84}S_{20}{}^{4-} \cdot 4C_{16}H_{36}N^+$
85.96	Mo$_2$	Fe$_7$	$C_{84}H_{84}S_{20}{}^{4-} \cdot 4C_{16}H_{36}N^+$

Ga

68.8		Ga	$C_{10}H_{14}NO$
84.33	Ni	Ga	$C_{11}H_{23}N_4O_2$
84.32	Fe	Ga	$C_{11}H_{23}N_5O_3$
72.22	Mo	Ga	$C_{13}H_{24}N_3O_3S$
68.16	Mo	Ga	$C_{15}H_{28}N_3O_3S$
68.19		Ga	$C_{17}H_{18}NOS$
68.6		Ga$_2$	$C_8H_{18}N_2S_2$
68.7		Ga$_2$	$C_9H_{24}N_2O$
68.10		Ga$_2$	$C_{10}H_{24}N_4$
68.12		Ga$_2$	$C_{10}H_{27}N_3$
83.166	Ni	Ga$_2$	$C_{23}H_{43}N_7O$

Gd

73.79		Gd	$C_{19}H_{23}O$
81.50		Gd$_2$	$C_{12}H_{26}O_{16} \cdot 4H_2O$

Ge

69.6		Ge	$C_5H_7IO_2$
69.7		Ge	C_5H_8
69.28		Ge	$C_{14}H_{14}S$
69.40		Ge	$C_{18}H_{36}N_2$
73.91	Fe	Ge	$C_{22}H_{18}$

71.205	Co **Ge**	$C_{26}H_{25}O_4$
69.49	**Ge**	$C_{30}H_{46}O_2$
69.51	Si **Ge**	$C_{36}H_{30}O_2$
69.58	Fe$_2$ **Ge**	$C_{42}H_{26}O_8$
69.33	**Ge$_2$**	$C_{16}H_{24}$
69.52	**Ge$_2$**	$C_{36}H_{30}$
69.53	**Ge$_2$**	$C_{36}H_{30}S$
69.54	**Ge$_2$**	$C_{36}H_{30}S$
69.59	**Ge$_4$**	$C_{48}H_{40}S$
73.135	Ni$_2$Hg$_3$ **Ge$_4$**	$C_{82}H_{70}$, C_7H_8
73.134	Ni$_2$Cd$_3$ **Ge$_4$**	$C_{82}H_{70}$, C_7H_8
69.61	**Ge$_6$**	$C_{72}H_{60}$, $7C_6H_6$

Hf

73.29	**Hf**	$C_{12}H_{10}O_2$
71.155	**Hf**	$C_{20}H_{18}$

Hg

86.7	**Hg**	$C_4H_{11}Cl_3P^-$, $C_8H_{22}ClHgP_2^+$
80.1	(**Hg**	$C_6H_{10}O_2S_4)_n$
71.4	**Hg**	C_7H_7Cl
84.15	(**Hg**	$C_7H_8Br_2N_2O)_n$
83.28	(**Hg**	$C_7H_9Br_2N)_n$
71.5	**Hg**	$C_7H_{10}N_5^+$, NO_3^-
83.30	**Hg**	$C_7H_{10}N_5O^+$, NO_3^-
86.7	**Hg**	$C_8H_{22}ClP_2^+$, $C_4H_{11}Cl_3HgP^-$
71.14	**Hg**	$C_9H_{15}BrO_2$
83.50	**Hg**	$C_{10}H_{14}Cl_2N_2$
71.20	**Hg**	$C_{10}H_{14}Cl_3N$
71.22	**Hg**	$C_{10}H_{15}Cl$
83.64	2 **Hg**	$C_{10}H_{24}ClN_4^+$, $Cl_4Hg_2^-$
71.29	**Hg**	$C_{11}H_{10}N^+$, $C_2F_3O_2^-$
71.34	F$_8$ **Hg**	$C_{12}H_2$
71.37	**Hg**	$C_{12}H_{13}N_2^+$, NO_3^-
85.44	**Hg**	$C_{12}H_{26}N_2S_2^{2+}$, $2ClO_4^-$
71.48	Mo **Hg**	$C_{13}H_8Cl_2N_2O_3$
71.53	**Hg**	$C_{13}H_{14}N^+$, NO_3^-
83.91	(**Hg**	$C_{14}H_{12}N_2O_2)_n$
71.65	**Hg**	$C_{14}H_{14}$
71.66	**Hg**	$C_{14}H_{14}N_4S$
80.13	**Hg**	$C_{14}H_{28}N_2S_4$
31.37	**Hg**	$C_{16}H_{19}ClO_6$
71.98	**Hg**	$C_{16}H_{19}Cl_2N$
71.120	**Hg**	$C_{18}H_{12}Cl_2N_2$
71.124	**Hg**	$C_{18}H_{14}ClN_3$
71.129	**Hg**	$C_{18}H_{21}BrO_2$
86.50	**Hg**	$C_{18}H_{24}Cl_2NP$
71.138	**Hg**	$C_{19}H_{16}N_4S$
86.62	(**Hg**	$C_{20}H_{15}N_2PS_2)_n$
71.152	**Hg**	$C_{20}H_{16}Cl_6N_2$
83.143	**Hg**	$C_{20}H_{16}N_4^{2+}$, $2NO_3^-$, $2H_2O$
71.180	**Hg**	$C_{22}H_{28}N_3^+$, NO_3^-
71.183	Pt **Hg**	$C_{22}H_{30}N_2O_4$
85.69	(**Hg**	$C_{22}H_{31}BrO_2S)_n$
71.207	**Hg**	$C_{26}H_{32}N_4O_2$
86.93	**Hg**	$C_{27}H_{33}N_2O_6P$
83.210	**Hg**	$C_{36}H_{24}N_6^{2-}$, $2CF_3O_3S^-$, C_2H_6O
63.50	Si$_2$ **Hg**	$C_{36}H_{30}$
86.129	**Hg**	$C_{36}H_{66}P_2^{2+}$, $2ClO_4^-$
81.93	**Hg**	$C_{40}H_{72}O_4P_2$. $2H_2O$
85.21	**Hg$_2$**	$C_8H_{16}S_2$
83.48	**Hg$_2$**	$C_{10}H_{12}N_2O_8S_2$
63.11	Si$_4$ **Hg$_2$**	$C_{10}H_{28}$
83.65	**Hg$_2$**	$C_{11}H_{17}Cl_5N_2$
83.87	**Hg$_2$**	$C_{13}H_{10}Cl_2N_4O$
71.71	**Hg$_2$**	$C_{14}H_{20}I_4S_2$
71.228	Ru$_6$ **Hg$_2$**	$C_{30}H_{18}Br_2O_{18}$
75.40	Ir$_2$ **Hg$_2$**	$C_{34}H_{48}Cl_2N_6$
86.158	**Hg$_2$**	$C_{42}H_{42}Cl_4O_8P_2$
49.15	**Hg$_2$**	$C_{43}H_{51}Cl_2N_5O_2S$
81.94	**Hg$_2$**	$C_{44}H_{78}O_8P_2$
83.220	**Hg$_2$**	$C_{48}H_{38}N_2O_{12}P_4$, $2C_2H_4O_2$
81.96	**Hg$_2$**	$C_{50}H_{54}O_8P_2$
83.35	**Hg$_3$**	$C_8H_{13}N_5^{2+}$, $2ClO_4^-$
71.121	**Hg$_3$**	$C_{18}H_{12}$
80.21	(**Hg$_3$**	$C_{20}H_{40}Cl_2N_4S_8)_n$
73.135	Ni$_2$Ge$_4$ **Hg$_3$**	$C_{82}H_{70}$, C_7H_8
71.12	**Hg$_4$**	$C_9H_{12}O_8$, $2H_2O$

In

68.15	**In**	$C_{15}H_{21}O_6$
49.16	**In**	$C_{44}H_{28}ClN_4$
68.11	**In$_2$**	$C_{10}H_{24}N_4$
68.17	**In$_2$**	$C_{16}H_{22}N_4O_2$, $0.5C_6H_6$

Ir

76.58	**Ir**	$C_{10}H_{13}ClN_2O_8^-$, K^+ , H_2O
76.59	**Ir**	$C_{10}H_{14}Cl_2N_2O_8^-$, K^+
75.8	**Ir**	$C_{14}H_{19}$
71.79	**Ir**	$C_{14}H_{35}ClO_2P_4^+$, FO_3S^-
72.34	**Ir**	$C_{18}H_{29}$
86.75	**Ir**	$C_{21}H_{20}ClNOP$
86.85	**Ir**	$C_{24}H_{15}Br_4NO_3P$, CH_2Cl_2
71.192	**Ir**	$C_{24}H_{30}ClN_2O_3P_2$, C_3H_6O
71.206	**Ir**	$C_{26}H_{26}IPS^+$, I^-
86.104	Pt **Ir**	$C_{30}H_{68}P_4^+$, $C_{24}H_{20}B^-$
71.237	**Ir**	$C_{32}H_{44}P_4^+$, F_6P^-
72.57	**Ir**	$C_{38}H_{30}Cl_3P_2$
86.148	**Ir**	$C_{41}H_{30}Cl_5N_2P_2$, C_7H_8
86.160	**Ir**	$C_{41}H_{36}P_6$
86.152	**Ir**	$C_{42}H_{39}N_4OP_2^{2+}$, $2F_6P^-$
86.167	B$_9$ **Ir**	$C_{44}H_{56}P_2$, C_7H_8
72.63	**Ir**	$C_{45}H_{40}ClP_2$
86.174	**Ir**	$C_{48}H_{38}N_3OP_2^{2+}$, $2F_6P^-$
86.183	**Ir**	$C_{51}H_{44}N_3OP_2$
86.191	**Ir**	$C_{52}H_{48}P_4S_2^+$, Cl^- , $2C_{13}H_{12}N_2O$
71.276	**Ir**	$C_{54}H_{45}BrP_3$
86.209	**Ir**	$C_{56}H_{45}OP_3S_2^+$, BF_4^-
86.41	**Ir$_2$**	$C_{16}H_{36}O_8P_2S_2$
86.42	**Ir$_2$**	$C_{16}H_{38}O_8P_2S_2$
85.55	**Ir$_2$**	$C_{16}H_{40}Cl_6S_4$
85.56	**Ir$_2$**	$C_{16}H_{40}Cl_6S_4$

86.92   **Ir**$_2$ $C_{26}H_{40}I_2O_2P_2S_2$

75.40   **Hg**$_2$ **Ir**$_2$ $C_{34}H_{48}Cl_2N_6$

86.205   **Ir**$_2$ $C_{54}H_{57}P_4{}^-$. $BF_4{}^-$

71.181   **F**$_6$ **Ir**$_3$ $C_{22}H_{27}O_6S_3$

86.224   **Ir**$_3$ $C_{81}H_{85}P_6{}^{2-}$. $2BF_4{}^-$. $4CH_2Cl_2$

71.92   **Ir**$_4$ $C_{16}H_9NO_{11}$

86.64   **As**$_2$ **Ir**$_4$ $C_{20}H_{16}O_{10}$

71.246   **Ir**$_7$ $C_{36}H_{33}O_{12}$

La

85.42   **La** $C_{12}H_{26}Cl_2O_{13}S_2{}^+$. $ClO_4{}^-$

84.93   **La** $C_{23}H_{21}N_8O_{11}$

Li

67.7   **Li**$_2$ $C_{22}H_{50}N_4S_4$

67.9   **Li**$_2$ $C_{38}H_{66}O_4$

Lu

85.88   **Lu** $C_{36}H_{66}P_3S_6$

Mg

67.1   **Mg** $C_4H_{12}O_9$. H_2O

67.2   **Mg** $C_8H_{22}N_2$

67.4   **Mg** $C_{20}H_{18}N_8O_{10}$

67.5   **Mg** $C_{20}H_{38}N_2$

67.6   **Mg** $C_{22}H_{16}N_6O_{10}$

67.8   **Mg**$_6$ $C_{36}H_{90}N_{10}$

Mn

82.3   (**Mn** $C_2H_9Cl_2NO_4)_n$

84.5   **Mn** $C_4H_{10}O_8$

71.3   **Al** **Mn** $C_6H_3Br_3O_5$

84.10   (**Mn** $C_6H_{12}N_2O_2S_2)_n$

81.22   **Mn** $C_6H_{12}O_6$

82.12   (**Mn** $C_6H_{18}N_2O_6{}^{2+})_n$. $2nBr^-$. $2nH_2O$

81.26   **Mn** $C_6O_{12}{}^{3-}$. $3K^+$. $3H_2O$

73.7   **Mn** C_7H_7INOS

85.16   **Mn** $C_7H_{12}N_4O_3S_4{}^+$. Br^-

79.7   **Mn** $C_9H_{24}Br_2N_6O_3$

86.18   **FeAs** **Mn** $C_{10}H_6ClO_8{}^-$. $C_{12}H_{12}As_2FeMnO_8{}^+$

76.56   **Mn** $C_{10}H_{12}N_2O_8{}^-$. K^+ . $2H_2O$

73.26   **Mn** $C_{12}H_6BrO_3$

73.31   **Mn** $C_{12}H_{12}NO_4$

86.18   **FeAs**$_2$ **Mn** $C_{12}H_{12}O_8{}^+$. $C_{10}H_6AsClFeMnO_8{}^-$

83.77   **Mn** $C_{12}H_{14}Cl_2N_8O_2$

86.24   **Mn** $C_{12}H_{30}N_3O_4P_2$

71.45   **Mn** $C_{13}H_6O_7{}^-$. $C_{36}H_{30}NP_2{}^+$

71.75   **Pt** **Mn** $C_{14}H_{23}O_5P_2$

71.76   **Pt** **Mn** $C_{14}H_{23}O_5P_2$

71.85   **Mo** **Mn** $C_{15}H_9O_5$

71.99   **Mn** $C_{16}H_{20}O_8PS_2$

84.66   **Mn** $C_{16}H_{32}O_8{}^{2+}$. $2Br_3{}^-$

71.111   **Pt** **Mn** $C_{17}H_{27}IO_5P$

86.49   **Mn** $C_{18}H_{15}N_3O_3P$

71.136   **Re** **Mn** $C_{19}H_{10}O_7$

71.139   **Mn** $C_{19}H_{16}O_5P$

86.67   **Mn** $C_{20}H_{26}N_2O_8P_2{}^+$. $ClO_4{}^-$

83.148   **Mn** $C_{20}H_{26}N_2O_8P_2{}^+$. $ClO_4{}^-$

83.149   **Mn** $C_{20}H_{27}N_6O^{2+}$. $2ClO_4{}^-$

71.173   **Pt** **Mn** $C_{21}H_{30}O_2P_2{}^-$. $BF_4{}^-$. CH_2Cl_2

86.79   **Mn** $C_{22}H_{15}O_4P^-$. $C_{24}H_{20}P^+$

83.163   **Mn** $C_{23}H_{21}ClN_7{}^+$. $BF_4{}^-$

85.75   **Mn** $C_{24}H_{24}N_3O_3S_3$

86.87   **Mn** $C_{24}H_{33}N_2O_8P_3{}^+$. $BF_4{}^-$

78.11   **Mn** $C_{27}H_{24}Cl_3N_4O_3$. $3H_2O$

49.2   **Mn** $C_{32}H_{16}N_8$

71.25   **Mn**$_2$ $C_{11}H_2N_2O_{10}$

86.31   **As**$_3$ **Mn**$_2$ $C_{13}H_{18}ClO_7$

84.54   **Mn**$_2$ $C_{14}H_{16}Cl_2O_8$

81.57   (**Mn**$_2$ $C_{14}H_{16}O_{18}{}^{2-})_n$. $2nK^+$

71.94   **Mn**$_2$ $C_{16}H_{12}O_4$

73.58   **Mn**$_2$ $C_{16}H_{14}N_2O_4$

71.96   **Mn**$_2$ $C_{16}H_{14}O_3$

85.51   **Mn**$_2$ $C_{16}H_{18}N_4O_8S_4$

73.76   **Mn**$_2$ $C_{18}H_{22}O_5P_2$

83.138   **Mn**$_2$ $C_{19}H_{25}N_{19}S_4$

71.149   **Mn**$_2$ $C_{20}H_{12}O_7$. $0.5C_4H_{10}O$

71.202   **Mn**$_2$ $C_{26}H_{22}O_4$

86.111   **Pt** **Mn**$_2$ $C_{33}H_{20}O_9P_2$

75.41   **Mn**$_2$ $C_{36}H_{26}Cl_2N_2O_6$

83.86   **Mn**$_3$ $C_{13}H_3N_2O_{12}$

84.46   **Mn**$_3$ $C_{13}H_{10}IO_{11}$

84.45   **F** **Mn**$_3$ $C_{13}H_{10}O_{11}$

71.104   **Mn**$_3$ $C_{17}H_8IN_2O_9$

73.89   **Sb** **Mn**$_3$ $C_{21}H_{15}ClO_6$

86.80   **Mn**$_3$ $C_{22}H_{26}O_{11}P$

83.174   (**Mn**$_3$ $C_{24}H_{26}N_{16})_n$

81.84   **Mn**$_3$ $C_{27}H_{30}Cl_3N_3O_{13}$

73.110   **Bi**$_2$ **Mn**$_4$ $C_{28}H_{20}Cl_2O_8$

86.170   **Mn**$_6$ $C_{45}H_{90}O_{36}P_9$

81.88   **Mn**$_{12}$ $C_{32}H_{56}O_{48}$. $2C_2H_4O_2$. $4H_2O$

Mo

83.2   **Mo** $C_2Cl_{12}NOP$

83.5   **Mo** $C_2H_3Cl_4NO^-$. $2C_{24}H_{20}As^+$. Cl_4MoO^-

84.7   **Mo** $C_5H_{11}NO_7$

73.22   **Mo** $C_8H_5O_3{}^-$. $C_{11}H_{11}MoO^+$

73.95   **Mo** $C_8H_5O_3{}^-$. $C_{22}H_{34}AsMo_2O_4P_2{}^+$

72.4   **Mo** $C_8H_{10}INO$

85.22   **Mo** $C_8H_{16}S_6$

84.17   **Mo** $C_8H_{20}N_2O_4$

83.45   **Mo** $C_{10}H_8N_{12}$

72.9   **Mo** $C_{10}H_{10}O_2$

80.7   **Mo** $C_{10}H_{20}N_2O_2S_4$

85.37   **Mo** $C_{10}H_{24}N_2O_2S_2$

73.22   **Mo** $C_{11}H_{11}O^+$. $C_8H_5MoO_3{}^-$

73.24   **Mo** $C_{11}H_{15}N_2O_2P$

72.13   **Mo** $C_{11}H_{23}ClO_8P_2$

83.72   **Mo** $C_{12}H_6Cl_2N_2O_2$

73.36   **Mo** $C_{12}H_{15}NO_3$

86.19		**Mo**	$C_{12}H_{21}O_3P_3$
86.30	B	**Mo**	$C_{12}H_{41}P_4$
71.48	Hg	**Mo**	$C_{13}H_8Cl_2N_2O_3$
84.47		**Mo**	$C_{13}H_{11}Cl_3N_3O_3 . C_2H_3N$
72.22	Ga	**Mo**	$C_{13}H_{24}N_3O_3S$
85.48		**Mo**	$C_{14}H_{10}OS_5$
71.72		**Mo**	$C_{14}H_{21}N_7{}^{2-} . 2BF_4{}^-$
84.60		**Mo**	$C_{14}H_{36}N_8O_4P_2S_2$
81.60	F_3	**Mo**	$C_{14}H_{37}O_{14}P_4$
71.85	Mn	**Mo**	$C_{15}H_9O_5$
80.15	2	**Mo**	$C_{15}H_{22}N_6O_2S_6 . 0.5CH_2Cl_2 . 0.5H_2O$
80.16		**Mo**	$C_{15}H_{23}N_5S_6 . CH_2Cl_2$
68.16	Ga	**Mo**	$C_{15}H_{28}N_3O_3S$
71.89	Si_3	**Mo**	$C_{15}H_{42}ClP$
83.101	Cr	**Mo**	$C_{16}H_{10}N_2O_{10}$
80.18		**Mo**	$C_{16}H_{25}Cl_2N_3S_4 . CHCl_3$
86.47		**Mo**	$C_{17}H_{30}O_8P_2$
73.72	F_2	**Mo**	$C_{18}H_{15}ClN_4$
73.73	Re	**Mo**	$C_{18}H_{15}O_3$
85.61		**Mo**	$C_{18}H_{20}N_2O_4S_2$
83.115		**Mo**	$C_{18}H_{21}Br_3N_3 . 0.5C_6H_7N$
83.116		**Mo**	$C_{18}H_{21}Cl_3N_3 . 0.5C_6H_7N$
75.25		**Mo**	$C_{20}H_{16}O_4$
84.84		**Mo**	$C_{20}H_{28}N_2O_4$
80.19		**Mo**	$C_{20}H_{29}N_6S_4{}^+ . C_{24}H_{20}B^-$
72.43		**Mo**	$C_{23}H_{20}N_2O_3$
85.70	B	**Mo**	$C_{23}H_{26}ClN_6O_2S . 0.2C_3H_6O$
83.168		**Mo**	$C_{24}H_{18}N_6O_2{}^{2+} . C_{12}H_8N_2 . 2I^- . H_2O$
85.71		**Mo**	$C_{24}H_{19}Cl_3N_6S_3$
72.47		**Mo**	$C_{24}H_{26}N_2O_3S_4 . C_6H_6$
71.204		**Mo**	$C_{26}H_{24}N_2O_2$
86.94		**Mo**	$C_{28}H_{28}OP_2S_2$
71.224		**Mo**	$C_{29}H_{24}I_2O_3P_2 . CH_2Cl_2$
73.115		**Mo**	$C_{29}H_{26}N_2O_2$
73.114		**Mo**	$C_{29}H_{26}N_2O_2$
85.81	FB	**Mo**	$C_{29}H_{28}N_8S_2$
75.36		**Mo**	$C_{29}H_{29}INOP$
71.229		**Mo**	$C_{30}H_{26}I_2O_3P_2$
72.54		**Mo**	$C_{30}H_{27}O_2P$
80.25		**Mo**	$C_{30}H_{32}N_6S_4 . C_3H_6O$
84.116	Si_4	**Mo**	$C_{32}H_{68}N_2O_4$
72.55		**Mo**	$C_{33}H_{30}O_3P_2$
86.124		**Mo**	$C_{36}H_{48}N_4P_4$
71.251		**Mo**	$C_{37}H_{33}O_3P_3$
84.130		**Mo**	$C_{42}H_{67}NO_4$
86.162		**Mo**	$C_{44}H_{35}O_2P_3S_2 . CH_2Cl_2$
49.26		**Mo**	$C_{48}H_{36}N_4O_2 . 1.5C_7H_8$
86.201	Cu_3	**Mo**	$C_{54}H_{45}ClOP_3S_3$
86.202	Cu_3	**Mo**	$C_{54}H_{45}ClP_3S_4$
86.203	Cu_2	**Mo**	$C_{54}H_{45}P_3S_4 . 0.8CH_2Cl_2$
86.204	F	**Mo**	$C_{54}H_{48}O_2P_4{}^+ . F_6P^-$
86.215		**Mo**	$C_{58}H_{59}IN_2P_4 . 0.5C_6H_6$
81.11	0.5	**Mo$_2$**	$C_4H_4O_8 . 0.5C_4H_8Mo_2O_{10}$
85.4		**Mo$_2$**	$C_4H_8Cl_3O_4S_2{}^- . C_8H_{20}N^+$
86.1	F_{16}	**Mo$_2$**	$C_4H_{12}Cl_2N_4P_8$
85.9		**Mo$_2$**	$C_6H_{12}ClO_5S_3{}^- . C_8H_{20}N^+$
85.10		**Mo$_2$**	$C_6H_{12}O_6S_3{}^{2-} . 2C_5H_{12}N^+$
85.23		**Mo$_2$**	$C_8H_{16}O_3S_6$
85.31		**Mo$_2$**	$C_8O_7S_5{}^{2-} . 2C_{16}H_{36}N^+$
73.15		**Mo$_2$**	$C_{10}H_{10}I_2O_3$
81.38		**Mo$_2$**	$C_{11}H_{14}N_2O_{10}S_2{}^{2-} . 2Na^+ . 3H_2O$
81.39		**Mo$_2$**	$C_{11}H_{14}N_2O_{11}S^{2-} . 2Na^- . 4H_2O$
81.40		**Mo$_2$**	$C_{11}H_{14}N_2O_{12}{}^{2-} . 2Na^+ . 3H_2O$
84.35		**Mo$_2$**	$C_{12}H_8O_9{}^{2-} . 2H_4N^+ . 2H_2O$
82.23		**Mo$_2$**	$C_{12}H_{14}N_2O_{17}{}^{2-} . 2Na^+ . 8H_2O$
73.32		**Mo$_2$**	$C_{12}H_{14}S_4$
73.39		**Mo$_2$**	$C_{12}H_{16}I_2N_4O_2$
73.42		**Mo$_2$**	$C_{13}H_{15}NO_5$
81.53	F	**Mo$_2$**	$C_{14}H_6N_2O_{14}{}^- . C_8H_{20}N^+$
73.50		**Mo$_2$**	$C_{14}H_{20}N_2O_2S_2$
73.51		**Mo$_2$**	$C_{14}H_{20}S_4$
84.59		**Mo$_2$**	$C_{14}H_{32}O_{10} . 2CH_4O$
73.52		**Mo$_2$**	$C_{15}H_{10}NO_4{}^- . C_8H_{20}N^+$
73.59		**Mo$_2$**	$C_{16}H_{16}O_4P^- . C_{24}H_{20}As^+$
73.62		**Mo$_2$**	$C_{16}H_{20}N_2O_6$
73.64		**Mo$_2$**	$C_{16}H_{22}S_4{}^+ . BF_4{}^-$
73.65		**Mo$_2$**	$C_{16}H_{24}N_2O_2S_2$
82.33		**Mo$_2$**	$C_{16}H_{32}N_8O_{12}{}^{4+} . 4Cl^- . 6H_2O$
85.54		**Mo$_2$**	$C_{16}H_{40}Cl_4S_4$
71.102		**Mo$_2$**	$C_{16}H_{40}P_4$
85.57	Fe_6	**Mo$_2$**	$C_{16}H_{40}S_{17}{}^{3-} . 3C_8H_{20}N^+$
85.62	Fe_6	**Mo$_2$**	$C_{18}H_{45}S_{17}{}^{3-} . 3C_{13}H_{22}N^+$
71.132	Si_2	**Mo$_2$**	$C_{18}H_{46}O_4P_2$
86.59		**Mo$_2$**	$C_{18}H_{58}P_6$
75.22		**Mo$_2$**	$C_{19}H_{23}O_5$
84.77		**Mo$_2$**	$C_{20}H_{12}Cl_4N_4O_4$
75.26		**Mo$_2$**	$C_{20}H_{16}O_4$
71.156		**Mo$_2$**	$C_{20}H_{18}$
71.163	Si_2	**Mo$_2$**	$C_{20}H_{40}Cl_2O_6P_2$
83.154		**Mo$_2$**	$C_{20}H_{48}N_4$
75.31		**Mo$_2$**	$C_{22}H_{20}O_2$
75.32		**Mo$_2$**	$C_{22}H_{34}O_3S_3$
75.33		**Mo$_2$**	$C_{22}H_{34}O_3S_3$
73.95	As	**Mo$_2$**	$C_{22}H_{34}O_4P_2{}^+ . C_8H_5MoO_3{}^-$
72.48		**Mo$_2$**	$C_{24}H_{26}O_4S$
83.175		**Mo$_2$**	$C_{24}H_{28}Br_4N_4$
83.177		**Mo$_2$**	$C_{24}H_{28}Cl_4N_4 . CHCl_3$
82.40		**Mo$_2$**	$C_{24}H_{48}N_4O_8{}^{4+} . 2Cl^- . 2C_7H_7O_3S^- . 2H_2O$
84.102		**Mo$_2$**	$C_{24}H_{50}O_{11}$
85.78	Fe_7	**Mo$_2$**	$C_{24}H_{60}S_{20}{}^{3-} . 3C_{10}H_{16}N^+$
84.105		**Mo$_2$**	$C_{25}H_{54}O_7$
81.83		**Mo$_2$**	$C_{26}H_{30}N_2O_6$
85.79		**Mo$_2$**	$C_{26}H_{53}O_3S_8{}^{3+} . 3CF_3O_3S^- . H_2O$
83.194		**Mo$_2$**	$C_{28}H_{28}N_8O_4$
84.115		**Mo$_2$**	$C_{32}H_{32}N_4O_4 . 2C_4H_8O$
86.113		**Mo$_2$**	$C_{34}H_{27}O_8P_2{}^- . C_8H_{20}N^+$
84.128		**Mo$_2$**	$C_{40}H_{48}N_4O_4 . 2CH_2Cl_2$
71.280		**Mo$_2$**	$C_{60}H_{52}N_4O_4P_4S_4 . 4CH_2Cl_2$
85.96	Fe_7	**Mo$_2$**	$C_{84}H_{84}S_{20}{}^{4-} . 4C_{16}H_{36}N^+$
84.20	2	**Mo$_3$**	$C_9H_9N_3O_{13}{}^{2-} . 3C_{36}H_{30}NP_2{}^+ . Na^+$
84.24		**Mo$_3$**	$C_{10}H_{12}N_3O_{13}{}^- . C_4H_{12}N^+$
65.2	As	**Mo$_4$**	$C_2H_7O_{15}{}^{2-} . 2CH_6N_3{}^+ . H_2O$
85.28		**Mo$_4$**	$C_8H_{24}O_{12}P_4S_4 . xCHCl_3 . yCCl_4$
85.29		**Mo$_4$**	$C_8H_{24}O_{12}P_4S_4 . xCHCl_3 . yCCl_4$
84.117	F_4	**Mo$_4$**	$C_{32}H_{72}O_8 . 2C_{34}H_{78}F_3Mo_4NO_8$
84.117	$2F_3$	**Mo$_4$**	$C_{34}H_{78}NO_8 . C_{32}H_{72}F_4Mo_4O_8$

84.2	As **Mo₆**	$CH_{15}O_{27}^{2-} . 2CH_6N_3^- . 6H_2O$

Nb

73.14	**Nb**	$C_{10}H_{10}ClO_2$
73.25	ZnB₂ **Nb**	$C_{11}H_{19}O . 0.5C_6H_6$
72.21	**Nb**	$C_{13}H_{19}Cl_2$
71.119	F₁₂ **Nb**	$C_{18}H_{10}$
73.101	Si₂ **Nb**	$C_{24}H_{34}$
73.117	**Nb**	$C_{31}H_{29}Cl_3P_2 . 2C_7H_8$
73.40	**Nb₂**	$C_{12}H_{18}Cl_6O_3$
85.52	**Nb₂**	$C_{16}H_{32}Br_4S_7$
71.221	**Nb₂**	$C_{28}H_{38}O$
83.211	**Nb₂**	$C_{36}H_{30}Cl_8N_2P_2 . C_2H_4Cl_2$
73.90	**Nb₃**	$C_{22}H_{15}O_7$

Nd

84.39	**Nd**	$C_{12}H_{24}N_3O_{15}$
84.40	**Nd**	$C_{12}H_{24}N_3O_{15}$
49.33	**Nd**	$C_{64}H_{33}N_{16} . CH_2Cl_2$

Ni

76.2	**Ni**	$C_2H_{16}N_2O_4^{2+} . 2NO_3^-$
76.3	**Ni**	$C_2H_{16}N_2O_4^{2+} . O_4S^{2-} . 2H_2O$
60.5	3 **Ni**	$C_4H_4S_4 . 2C_6H_4S_4$
83.13	**Ni**	$C_4H_{14}N_6S_2^{2+} . 2Br^-$
83.14	**Ni**	$C_4H_{14}N_6S_2^{2+} . 2Br^-$
83.15	**Ni**	$C_4H_{14}N_6S_2^{2+} . 2I^-$
85.6	**Ni**	$C_4O_4S_4^{2-} . 2K^+$
81.15	**Ni**	$C_4O_4S_4^{2-} . 2Li^+ . 2H_2O$
85.7	**Ni**	$C_4O_4S_4^{2-} . 2Na^+ . 2H_2O$
81.25	(**Ni**	$C_6H_{16}O_6)_n$
85.12	**Ni**	$C_6H_{16}P_2S_4$
85.13	**Ni**	$C_6H_{18}N_2P_2S_4$
83.24	**Ni**	$C_6H_{18}N_{12}O_6^{2+} . 2Cl^- . 0.5H_2O$
84.14	**Ni**	$C_6H_{20}N_3O_3^+ . ClO_4^-$
76.40	**Ni**	$C_6H_{24}N_6^{2+} . 2NO_3^-$
85.14	**Ni**	$C_8S_{10}^{2-} . 2C_{16}H_{36}N^+$
86.5	**Ni**	$C_7H_{18}I_2OP_2$
72.8	**Ni**	$C_9H_{19}P$
85.33	**Ni**	$C_9H_{24}N_6S_6^{2+} . 2Cl^- . 0.5C_2H_6O . 3H_2O$
76.57	**Ni**	$C_{10}H_{12}N_2O_8^{2-} . Ca^{2+} . 4H_2O$
80.4	**Ni**	$C_{10}H_{16}N_2O_2S_4$
83.60	D₅ **Ni**	$C_{10}H_{18}N_4O_2^+ . Cl^- . D_2O$
80.6	As₂ **Ni**	$C_{10}H_{20}I_6N_2S_4$
85.36	**Ni**	$C_{10}H_{22}N_3O_4S_2^+ . NO_3^-$
83.63	**Ni**	$C_{10}H_{22}N_5O_{12}P . 3H_2O$
83.67	**Ni**	$C_{11}H_{19}BrN_4O_2$
86.15	**Ni**	$C_{11}H_{22}ClPS_2$
84.33	Ga **Ni**	$C_{11}H_{23}N_4O_2$
76.70	**Ni**	$C_{12}H_{16}N_4O_4$
83.81	**Ni**	$C_{12}H_{20}Cl_2N_4O_4$
82.25	**Ni**	$C_{12}H_{20}N_2O_{11}P . 2H_2O$
76.79	**Ni**	$C_{12}H_{32}N_4O_2^{2-} . 2Cl^-$
71.43	B₂ **Ni**	$C_{12}H_{36}P_4$
85.47	Se₆ **Ni**	$C_{12}N_6^{2-} . 2C_{24}H_{20}As^-$
83.88	**Ni**	$C_{13}H_{21}N_5O_4$
71.63	Fe₂ **Ni**	$C_{14}H_{12}N_2O_{10}$
83.92	**Ni**	$C_{14}H_{13}N_5O_2$
77.3	**Ni**	$C_{14}H_{14}N_4O_4$
79.11	**Ni**	$C_{14}H_{14}N_8S_2$
71.67	**Ni**	$C_{14}H_{16}N_2$
84.58	Cu **Ni**	$C_{14}H_{28}N_6O_8 . 4H_2O$
76.89	Se₂ **Ni**	$C_{14}H_{32}N_6$
76.88	**Ni**	$C_{14}H_{32}N_6S_2$
83.100	**Ni**	$C_{15}H_{26}N_4^{2+} . 2ClO_4^- . H_2O$
75.9	**Ni**	$C_{16}H_{18}$
84.62	**Ni**	$C_{16}H_{18}Cl_2N_6O_2$
84.63	**Ni**	$C_{16}H_{18}Cl_2N_6O_2 . C_6H_6$
82.30	**Ni**	$C_{16}H_{18}N_4O_4 . 2H_2O$
72.29	Fe₂ **Ni**	$C_{17}H_{14}O_6$
86.46	**Ni**	$C_{17}H_{18}O_2P_2$
85.59	**Ni**	$C_{17}H_{21}N_7S_6$
83.114	B₂ **Ni**	$C_{18}H_{20}N_{12}$
82.37	(**Ni**	$C_{18}H_{22}N_2O_9)_n . 2nH_2O$
83.130	**Ni**	$C_{18}H_{24}N_6^{2+} . 2BF_4^-$
83.135	**Ni**	$C_{18}H_{40}N_4^{2+} . 2ClO_4^-$
71.134	F₁₀ **Ni**	$C_{19}H_8$
71.151	Ru₃ **Ni**	$C_{20}H_{15}O_9$
81.69	**Ni**	$C_{20}H_{28}N_6O_8$
84.89,	**Ni**	$C_{21}H_{24}N_4O_2S_2$
72.39	Co **Ni**	$C_{22}H_{15}O_3$
84.91	**Ni**	$C_{22}H_{22}N_2O_2$
81.75	**Ni**	$C_{22}H_{34}N_4O_4$
81.76	**Ni**	$C_{22}H_{34}N_4O_4$
84.95	**Ni**	$C_{23}H_{25}N_5O_4^{2+} . 2NO_3^- . 2H_2O$
83.166	Ga₂ **Ni**	$C_{23}H_{43}N_7O$
76.100	**Ni**	$C_{24}H_{18}N_2O_2$
85.76	**Ni**	$C_{24}H_{24}N_8S_2$
83.178	**Ni**	$C_{24}H_{28}N_4^{2+} . 2F_6P^-$
75.34	Si₂B₂ **Ni**	$C_{24}H_{30}$
81.80	**Ni**	$C_{24}H_{38}N_4O_8$
72.51	**Ni**	$C_{24}H_{54}N_6^{2-} . 2Li^+$
73.107	**Ni**	$C_{26}H_{25}OPS_2$
61.10	**Ni**	$C_{26}H_{28}N_6S_2 . 2C_{10}H_7Br$
61.11	**Ni**	$C_{26}H_{28}N_6S_2 . 2C_{11}H_{10}$
85.82	**Ni**	$C_{30}H_{20}N_2O_2S_2$
83.201	**Ni**	$C_{30}H_{32}N_6O_6$
86.107	**Ni**	$C_{31}H_{34}IP_2S_2^+ . C_{24}H_{20}B^-$
86.108	**Ni**	$C_{31}H_{34}P_2S_2^{2+} . 2ClO_4^-$
85.83	**Ni**	$C_{32}H_{24}N_2O_2S_2$
83.205	**Ni**	$C_{32}H_{36}N_8$
86.110	As₄ **Ni**	$C_{32}H_{40}Cl_2$
83.207	**Ni**	$C_{32}H_{56}Cl_2N_8$
71.242	B₈ **Ni**	$C_{34}H_{44}P_2$
86.115	**Ni**	$C_{35}H_{36}O_2P_3S^+ . C_{24}H_{20}B^-$
49.9	**Ni**	$C_{36}H_{48}N_4$
49.8	**Ni**	$C_{36}H_{48}N_4$
49.13	**Ni**	$C_{36}H_{52}N_4$
49.12	**Ni**	$C_{36}H_{52}N_4$
71.255	**Ni**	$C_{38}H_{35}O_2P$
85.89	**Ni**	$C_{40}H_{24}O_4P_2S_4$
72.60	**Ni**	$C_{42}H_{40}O_2P_2$
86.156	As₃ **Ni**	$C_{42}H_{42}IN$

86.159	**Ni**	$C_{42}H_{42}INP_3$
49.18	**Ni**	$C_{44}H_{36}N_4^+$, I^- . 0.08I
72.64	**Ni**	$C_{47}H_{40}O_3P_2$, C_6H_6
86.228	**Rh Ni**	$C_{82}H_{78}P_9^{2+}$. $2C_{24}H_{20}B^-$. C_4H_8O
83.53	**Ni₂**	$C_{10}H_{14}IN_8O_2^-$, $C_8H_{20}N^+$
83.54	**Ni₂**	$C_{10}H_{14}N_6O_2$
76.64	**Ni₂**	$C_{10}H_{22}N_2O_{13}$. $2H_2O$
73.30	**Ni₂**	$C_{12}H_{10}O_2$
76.74	**Ni₂**	$C_{12}H_{28}N_4O_{12}$
73.47	**Ni₂**	$C_{14}H_{14}O_2$
83.99	**Ni₂**	$C_{15}H_{21}N_8O_2^-$, Na^+ . $2C_4H_8O$
71.105	**Fe₂ Ni₂**	$C_{17}H_{10}O_7$
71.178	**Fe₂ Ni₂**	$C_{22}H_{20}O_6$
83.161	**Ni₂**	$C_{22}H_{32}Cl_4N_8$
76.18	**Cu Ni₂**	$C_{24}H_{40}N_8O_{16}^{2-}$
72.67	**Ni₂**	$C_{65}H_{50}N_5^{3-}$, $3Li^+$, $2C_4H_{10}O$
73.135	**Hg₃Ge₄ Ni₂**	$C_{82}H_{70}$, C_7H_8
73.134	**Ge₄Cd₃ Ni₂**	$C_{82}H_{70}$, C_7H_8
85.46	**Ni₃**	$C_{12}H_{32}N_4S_4^{2+}$, $2Cl^-$
83.150	**Ni₃**	$C_{20}H_{28}N_{10}O_2$
73.85	**Ni₄**	$C_{20}H_{23}$
73.87	**B₄ Ni₄**	$C_{20}H_{24}$

Os

83.39	**Os**	$C_8H_{18}N_2O_2$
76.61	**Os**	$C_{10}H_{14}N_2O_9$, H_2O
83.55	**Os**	$C_{10}H_{15}NO_3$
84.41	**Os**	$C_{12}H_{24}O_5$
49.14	**Os**	$C_{38}H_{21}N_9O$
86.134	**Os**	$C_{38}H_{30}Cl_2OP_2S$
71.252	**Os**	$C_{38}H_{30}Cl_4OP_2$
86.136	**Se₂ Os**	$C_{38}H_{30}O_2P_2$
86.137	**Os**	$C_{38}H_{33}Cl_3NP_2$
84.80	**Os₂**	$C_{20}H_{16}Cl_2N_4O_4$
80.23	**Os₂**	$C_{25}H_{50}N_5S_{12}^+$, $C_{24}H_{20}B^-$
80.26	**Os₂**	$C_{30}H_{60}N_6S_{12}^{2+}$, $2F_6P^-$, CH_2Cl_2
80.27	**Os₂**	$C_{30}H_{60}N_6S_{12}^{2+}$, $2F_6P^-$, $2CH_2Cl_2$
81.37	**Os₃**	$C_{11}H_2O_{10}S_2$
84.29	**Os₃**	$C_{11}H_4O_{11}$
72.14	**Os₃**	$C_{12}H_8O_9S$
86.17	**Os₃**	$C_{12}H_9N_2O_{14}P$
83.73	**Os₃**	$C_{12}H_9N_3O_{11}$
71.56	**F₆ Os₃**	$C_{14}H_2BrO_{10}^-$, $C_{36}H_{30}NP_2^+$
83.109	**Os₃**	$C_{17}H_9NO_{12}S$
71.106	**Os₃**	$C_{17}H_{11}NO_9$
71.108	**Os₃**	$C_{17}H_{13}O_8PS_2$
81.64	**Os₃**	$C_{18}H_{13}O_9PS_2$
71.123	**Os₃**	$C_{18}H_{13}O_9PS_2$
71.150	**Os₃**	$C_{20}H_{15}NO_8$
72.37	**F₆ Os₃**	$C_{20}H_{17}O_{10}P$
71.171	**Os₃**	$C_{21}H_{17}NO_9$
71.195	**Os₃**	$C_{25}H_{20}NO_{10}P$
86.172	**Au₂ Os₃**	$C_{47}H_{32}O_{11}P_2$
71.46	**Os₄**	$C_{13}H_7NO_{11}$
71.57	**Os₄**	$C_{14}H_2O_{12}$
71.90	**Os₄**	$C_{16}H_6O_{12}$
71.148	**Os₄**	$C_{20}H_{10}O_{12}$
71.210	**Os₄**	$C_{27}H_{16}O_{11}$
83.180	**Os₅**	$C_{25}H_{10}N_2O_{13}$

Pb

69.5	**Pb**	$C_4H_{12}I_2O_2S_2$
69.11	**Pb**	$C_6H_{12}N_2S_4$
69.14	**(Pb**	$C_7H_6N_2O_5)_n$
69.15	**Pb**	$C_7H_{14}N_2O_2$
69.26	**Pb**	$C_{14}H_{12}N_2O_4$
69.30	**Pb**	$C_{14}H_{28}N_2S_4$
69.47	**Pb**	$C_{23}H_{19}NS$
49.3	**Pb**	$C_{32}H_{36}N_4$
69.29	**Pb₂**	$C_{14}H_{14}N_2O_{12}$, $2C_7H_5NO_4$, H_2O

Pd

82.5	**Pd**	$C_4H_7ClN_2O_4$
82.6	**Pd**	$C_4H_8N_2O_4$, $3H_2O$
82.8	**Pd**	$C_4H_9Cl_2NO_3S$, H_2O
83.12	**2 Pd**	$C_4H_{11}N_8O_4^+$, O_4S^{2-} , H_2O
83.20	**Pd**	$C_6H_8Cl_2N_2O_2$
82.11	**Pd**	$C_6H_9BrNO_5^-$, K^+ , H_2O
85.19	**Pd**	$C_8H_{14}S_4$
71.7	**Pd**	$C_8H_{15}ClS$
72.6	**Pd**	$C_8H_{16}Cl_2S$
60.12	**Pd**	$C_{10}H_{14}O_4$, $C_6H_4S_4$
80.5	**Pd**	$C_{10}H_{18}N_4O_2S_4$
80.8	**Pd**	$C_{10}H_{22}Br_2N_2O_2S_2$
80.9	**Pd**	$C_{10}H_{22}Cl_2N_2O_2S_2$
83.82	**Pd**	$C_{12}H_{24}Cl_2N_4$
77.2	**Pd**	$C_{13}H_{29}N_2O_2^+$, $C_5H_7O_2^-$
76.84	**Pd**	$C_{14}H_{28}N_8O_5^{2+}$, $2ClO_4^-$
86.39	**Pd**	$C_{16}H_{22}N_2P_2S_2$
75.14	**Pd**	$C_{16}H_{30}Cl_2N_2O_2$, H_2O
77.4	**F₆ Pd**	$C_{17}H_{10}N_2O_2$
77.5	**F₆ Pd**	$C_{17}H_{10}N_2O_2$
72.30	**Pd**	$C_{17}H_{21}N_2^+$, ClO_4^-
71.113	**Pd**	$C_{17}H_{34}BrNP_2$
71.114	**Pd**	$C_{17}H_{34}BrNP_2$
71.115	**Pd**	$C_{17}H_{34}BrNP_2$
82.35	**Pd**	$C_{18}H_{20}N_2O_8$
86.52	**Pd**	$C_{18}H_{42}Cl_2P_2$
73.80	**Fe Pd**	$C_{19}H_{25}NO_2$
71.160	**Si Pd**	$C_{20}H_{33}ClP^+$, F_6P^-
86.86	**Pd**	$C_{24}H_{26}Cl_2P_2$
75.35	**Pd**	$C_{26}H_{36}O_4$
70.18	**Te₂Si₄ Pd**	$C_{26}H_{60}N_2S_2$
83.191	**Pd**	$C_{28}H_{20}N_4^{2+}$, $2BF_4^-$
81.85	**Pd**	$C_{28}H_{24}O_4P_2$
86.96	**Pd**	$C_{28}H_{46}O_2P_2$, C_7H_8
86.102	**As₂ Pd**	$C_{30}H_{34}N^+$, F_6P^-
71.235	**Pd**	$C_{31}H_{26}ClPS_3$, $2C_7H_8$
86.117	**As₂ Pd**	$C_{36}H_{30}Cl_2$
83.215	**Pd**	$C_{36}H_{52}Cl_2N_4$
86.145	**As₄ Pd**	$C_{40}H_{40}^{2+}$, $2Cl^-$, $2C_2H_6O_2$
86.146	**As₄ Pd**	$C_{40}H_{40}^{2+}$, $2I^-$
83.219	**Pd**	$C_{44}H_{42}N_8$

86.180	Pd	$C_{50}H_{48}Cl_2N_6P_6$
71.274	Pd	$C_{53}H_{47}ClN_2O_2P_2$
86.196	Cu Pd	$C_{53}H_{47}Cl_3N_2O_2P_2$
86.214	Pd	$C_{58}H_{46}N_2O_4P_4$. $5C_2H_4Cl_2$
81.67	Pd$_2$	$C_{20}H_{28}Cl_2O_4P_2$. $0.5CHCl_3$
81.86	Pd$_2$	$C_{32}H_{26}N_2O_4S_2$
71.236	Pd$_2$	$C_{32}H_{26}N_2O_6$
72.61	Pd$_2$	$C_{42}H_{40}P_2$
86.168	Pd$_2$	$C_{44}H_{96}Cl_4P_4$
71.269	Pd$_2$	$C_{46}H_{38}Br_2N_2P_2$
86.171	Pd$_2$	$C_{46}H_{100}Cl_4P_4$
86.176	Pd$_2$	$C_{50}H_{44}Cl_2O_2P_4S$. $0.5CH_2Cl_2$, CH_4O
86.178	Pd$_2$	$C_{50}H_{44}Cl_2P_4S$. $3CH_2Cl_2$
71.275	F$_6$ Pd$_2$	$C_{54}H_{44}Cl_2P_4$
81.44	Pd$_4$	$C_{12}H_{12}O_{12}$. $2C_2H_4O_2$
81.79	Pd$_4$	$C_{24}H_{36}Cl_{12}O_{16}$
86.213	Pd$_4$	$C_{57}H_{52}O_5P_4$
71.190	Fe$_6$ Pd$_6$	$C_{24}HO_{24}^{3-}$, $3C_{10}H_{16}N^+$, $2C_2H_3N$

Pr

79.14	Pr$_2$	$C_{16}H_{34}N_8O_{16}$, $2CH_4N_2O$, $2H_2O$

Pt

85.2	Pt	$C_2H_8Br_5S^-$, $C_{14}H_8S_4^+$
82.1	Pt	$C_2H_7ClN_2O_2$
82.2	Pt	$C_2H_8Cl_2N_2O_2$
82.7	Pt	$C_4H_8N_2O_4$, $C_2H_2O_4$, $2H_2O$
82.9	Pt	$C_4H_{10}Br_2N_2O_4$, H_2O
82.10	Pt	$C_4H_{10}Cl_2N_2O_4$, H_2O
83.9	Pt	$C_4H_{10}N_8O_4$, $2HCl$
83.8	Pt	$C_4H_{10}N_8O_4$, $2HCl$
83.10	Pt	$C_4H_{10}N_8O_4$, H_4N^+ , Cl^-
83.11	Pt	$C_4H_{10}N_8O_4$, $0.666Na^+$, $0.333C_2H_3O_2^-$, $0.333Cl^-$, $2H_2O$
76.8	(Pt	$C_4H_{16}BrN_4^{2+})_n$, $2nClO_4^-$
76.10	Pt	$C_4H_{16}ClN_5O_2^{2+}$, $2Cl^-$, $3H_2O$
76.19	Pt	$C_4H_{16}Cl_2N_4^{2+}$
76.20	3 Pt	$C_4H_{16}Cl_2N_4^{2+}$
76.19	Pt	$C_4H_{16}N_4^{2+}$, $C_4H_{16}Cl_2N_4Pt^{2+}$, $4ClO_4^-$
76.20	3 Pt	$C_4H_{16}N_4^{2+}$, $3C_4H_{16}Cl_2N_4Pt^{2+}$, $4Cl_4Cu^{3-}$
81.16	0.4 Pt	$C_4O_8^-$, $0.6C_4O_8Pt^{2-}$, $1.6K^+$, $1.2H_2O$
81.17	0.67 Pt	$C_4O_8^{2-}$, $0.33C_4O_8Pt^-$, $1.67Rb^+$, $1.5H_2O$
72.1	Pt	$C_5H_8Cl_2$
76.22	Pt	$C_5H_{20}N_4^{2+}$, Cl_5Sb^{2-}
76.24	Pt	$C_6H_{11}ClN_4O_2$, H^+ , Cl^- , $2H_2O$
83.25	Pt	$C_6H_{20}Cl_2N_4^{2+}$
76.32	Pt	$C_6H_{20}I_2N_4^{2+}$
83.25	Pt	$C_6H_{20}N_4^{2+}$, $C_6H_{20}Cl_2N_4Pt^{2+}$, $4ClO_4^-$
76.32	Pt	$C_6H_{20}N_4^{2+}$, $C_6H_{20}I_2N_4Pt^{2+}$, $4I^-$
76.33	Pt	$C_6H_{20}N_4^{2+}$, $2C_{10}H_{14}BrO_4S^-$
76.34	Pt	$C_6H_{20}N_4^{2+}$, $2Cl^-$, $2H_2O$
76.35	Pt	$C_6H_{20}N_4^{2+}$, $2Cl^-$, $2H_2O$
83.29	2 Pt	$C_7H_9Cl_3N^-$, $H_{12}N_4Pt^{2+}$
82.13	Pt	$C_7H_{10}Cl_2N_2O_2$
85.15	Pt	$C_7H_{11}Cl_2NOS$
76.42	Pt	$C_7H_{13}ClN_4O_2$

75.1	Pt	$C_8H_{14}Cl_2S$
85.20	Pt	$C_8H_{14}S_4$
81.34	Pt	$C_8H_{16}N_2O_4$
83.43	Pt	$C_8H_{28}Br_2N_4^{2-}$
83.43	Pt	$C_8H_{28}N_4^{2+}$. $C_8H_{28}Br_2N_4Pt^{2+}$. $4Br^-$
72.7	Pt	$C_9H_{19}Cl_2N$
86.10	Pt	$C_9H_{27}ClP_3^+$. Cl^-
84.25	Pt	$C_{10}H_{16}Cl_2N_2O$
83.61	Pt	$C_{10}H_{18}N_7O_3^+$. ClO_4^- . $3H_2O$
71.23	Pt	$C_{10}H_{20}Cl_2O_2$
71.24	Pt	$C_{10}H_{21}ClN_2O_2$
80.10	Pt	$C_{10}H_{22}Cl_2N_2O_2S_2$
72.12	Pt	$C_{11}H_{17}Cl_2OP$
76.67	Pt	$C_{11}H_{21}N_7O^{2+}$. $2F_6P^-$. $1.5H_2O$
72.18	Pt	$C_{12}H_{17}Cl_2O_2P$
83.83	Pt	$C_{12}H_{26}Br_2N_2$
83.84	Pt	$C_{12}H_{26}Cl_2N_2$, $2C_6H_{18}N_3OP$
86.21	Pt	$C_{12}H_{28}Cl_2NP$
86.22	Pt	$C_{12}H_{28}Cl_2NP$
76.75	Pt	$C_{12}H_{29}IN_3^+$, I^-
76.76	Pt	$C_{12}H_{30}N_8^{4+}$, $2O_6S_2^{2-}$, $2.5H_2O$
86.25	W Pt	$C_{12}H_{30}P_2S_4$
71.55	Pt	$C_{13}H_{33}ClP_2$
84.53	Pt	$C_{14}H_{11}ClN_2O$, $CHCl_3$
71.75	Mn Pt	$C_{14}H_{23}O_5P_2$
71.76	Mn Pt	$C_{14}H_{23}O_5P_2$
71.77	Pt	$C_{14}H_{24}Cl_2N_2O$
71.80	Pt	$C_{14}H_{35}ClP_2$
86.32	B$_4$ Pt	$C_{14}H_{36}P_2$
86.33	B$_4$ Pt	$C_{14}H_{38}P_2$
71.81	B$_8$ Pt	$C_{14}H_{40}P_2$
71.82	B$_8$ Pt	$C_{14}H_{41}P_2$
83.104	Pt	$C_{16}H_{18}Cl_2N_2O_4$
83.107	Pt	$C_{16}H_{24}N_8^{2+}$, Cl_6Pt^{2-}
76.93	Pt	$C_{16}H_{24}N_{10}O_2^{2+}$, $2F_6P^-$
86.43	B$_4$ Pt	$C_{16}H_{40}P_2$
85.58	2 Pt	$C_{17}H_{16}N_3OS^+$, $2C_{19}H_{24}N_9O_{10}P^-$, xH_2O
78.5	Pt	$C_{17}H_{20}N_2O_2$
71.111	Mn Pt	$C_{17}H_{27}IO_5P$
72.31	Pt	$C_{18}H_{18}Cl_2$
72.33	Pt	$C_{18}H_{28}Cl_2N_2$
83.133	Pt	$C_{18}H_{30}N_8O_{16}P_2^{2-}$, $0.5Cs^+$, $1.5H^+$, $4H_2O$
86.53	Pt	$C_{18}H_{45}ClP_3^+$, BF_4^-
86.54	F Pt	$C_{18}H_{45}P_3^+$, BF_4^-
86.55	Pt	$C_{18}H_{46}P_3^+$, F_6P^-
71.142	W Pt	$C_{19}H_{26}O_6P_2$
72.36	As Pt	$C_{20}H_{17}Cl_2$
83.151	Pt	$C_{20}H_{32}N_{12}O_{10}^{2+}$, $1.5Cl^-$, $0.5ClO_4^-$, $7H_2O$
71.162	F$_6$ Pt	$C_{20}H_{39}OP_2^+$, F_6P^-
71.164	Pt	$C_{20}H_{41}O_4P_2^+$, $C_{24}H_{20}B^-$
71.165	Pt	$C_{20}H_{41}O_4P_2^+$, F_6P^-
71.173	Mn Pt	$C_{21}H_{30}O_2P_2^+$, BF_4^- , CH_2Cl_2
71.183	Hg Pt	$C_{22}H_{30}N_2O_4$
72.41	Pt	$C_{22}H_{32}Cl_2N_2$
71.185	Cr Pt	$C_{22}H_{35}O_6P_3$
71.186	W Pt	$C_{22}H_{37}O_5P_3$
72.44	Pt	$C_{23}H_{34}Cl_2N_2$
83.167	Pt	$C_{24}H_{16}N_4^{2+}$, $2Cl^-$, $3H_2O$

84.96		Pt	$C_{24}H_{24}N_2O_4$
86.88		Pt	$C_{24}H_{54}P_2$
86.89		Pt	$C_{24}H_{56}P_2$
83.183		Pt	$C_{25}H_{16}N_5^-$. NO_3^- . H_2O
83.186		Pt	$C_{25}H_{44}N_{12}O_{16}P_2$. $11H_2O$
71.200	F_6As	Pt	$C_{26}H_{21}O_3$
71.208		Pt	$C_{26}H_{35}NP_3^-$. F_6P^-
71.213	F_3	Pt	$C_{27}H_{22}ClP_2$
86.95	B_{20}	Pt	$C_{28}H_{41}ClP_2$
86.97		Pt	$C_{28}H_{46}O_2P_2$. C_7H_8
80.24		Pt	$C_{28}H_{53}N_2PS_4$. C_6H_{12}
72.53		Pt	$C_{29}H_{29}Cl_3NP$
71.225		Pt	$C_{29}H_{31}P_3$
86.104	Ir	Pt	$C_{30}H_{68}P_4^+$. $C_{24}H_{20}B^-$
78.14		Pt	$C_{31}H_{24}NO_2P$
86.109		Pt	$C_{31}H_{36}ClP_3$
86.111	Mn_2	Pt	$C_{33}H_{20}O_9P_2$
86.119		Pt	$C_{36}H_{32}N_2O_2P_2S_2$. $0.5H_2O$
86.121	B_8	Pt	$C_{36}H_{40}P_2S$
86.128		Pt	$C_{36}H_{66}Cl_2P_2$
71.248	F_3	Pt	$C_{37}H_{30}ClP_2$
71.249		Pt	$C_{37}H_{33}ClP_2$
71.254		Pt	$C_{38}H_{33}NP_2$
86.139		Pt	$C_{38}H_{36}O_6P_2S_2$
86.141	B_8	Pt	$C_{38}H_{44}OP_2S$
86.144		Pt	$C_{40}H_{30}Cl_2P_2$. $2C_2H_3N$
86.155		Pt	$C_{42}H_{40}P_2$
65.26	As_2	Pt	$C_{44}H_{34}Cl_2$
86.163		Pt	$C_{44}H_{37}NOP_2S$
71.266		Pt	$C_{44}H_{41}O_4P_2^+$. BF_4^-
75.45		Pt	$C_{44}H_{42}P_2$
75.46		Pt	$C_{45}H_{44}P_2$
71.268	B_{10}	Pt	$C_{45}H_{54}P_2$
72.65	Si	Pt	$C_{47}H_{57}N_2P_3S$
71.272		Pt	$C_{49}H_{40}OP_2S_3$. C_6H_6
84.42		Pt_2	$C_{12}H_{26}N_8O_4^{2+}$. $2NO_3^-$
86.23	F_4	Pt_2	$C_{12}H_{30}Cl_2O_2P_4$
86.44	B_{12}	Pt_2	$C_{16}H_{40}P_2$
71.131		Pt_2	$C_{18}H_{36}Cl_4O_2$
71.187		Pt_2	$C_{22}H_{56}Cl_2N_6^{2+}$. $2ClO_4^-$
84.100	Ag	Pt_2	$C_{24}H_{44}N_{12}O_8^+$. NO_3^- . $5H_2O$
71.238	W	Pt_2	$C_{32}H_{50}O_7P_2$
86.114		Pt_2	$C_{34}H_{74}Cl_4N_2P_2$
86.147	Si_2	Pt_2	$C_{40}H_{80}P_2$
86.157		Pt_2	$C_{42}H_{42}Cl_2P_4$
71.265		Pt_2	$C_{44}H_{36}O_6P_2$
86.166		Pt_2	$C_{44}H_{48}P_4$
86.182		Pt_2	$C_{51}H_{44}ClOP_4^+$. F_6P^-
86.185		Pt_2	$C_{51}H_{46}Cl_2P_4$
86.188		Pt_2	$C_{52}H_{44}O_2P_4^{2+}$. $2F_6P^-$
86.193		Pt_2	$C_{52}H_{51}P_4^+$. F_6P^-
86.198		Pt_2	$C_{53}H_{53}P_4^+$. F_6P^- . $3CH_2Cl_2$
71.277	Si	Pt_2	$C_{54}H_{82}P_2$
86.223		Pt_2	$C_{75}H_{67}P_6^+$. F_6P^-
85.86		Pt_3	$C_{34}H_{34}N_8S_2^{4+}$. $4BF_4^-$

Re

83.3		Re	$C_2Cl_{12}NOP$
86.2		Re	$C_6H_9NO_4PS$
76.41		Re	$C_7H_{12}BrN_2O_3$
76.43		Re	$C_7H_{16}N_4O_3^-$. $C_{12}H_{10}O_2P^-$
71.11	B	Re	$C_9H_9ClO_6$
73.73	Mo	Re	$C_{18}H_{15}O_3$
71.136	Mn	Re	$C_{19}H_{10}O_7$
86.60		Re	$C_{19}H_{19}BrN_4O_3P$
71.172		Re	$C_{21}H_{20}O_5P$
71.201	Si	Re	$C_{26}H_{21}O_2$. CH_2Cl_2
71.239		Re	$C_{34}H_{24}O_4P$
81.4		Re_2	$C_2H_2Cl_4O_4$. $2C_{13}H_{11}NO$
81.7		Re_2	$C_3H_3Cl_3O_6$
81.12		Re_2	$C_4H_6Cl_4O_4$
84.12		Re_2	$C_6H_{18}O_9$
84.21		Re_2	$C_9H_9O_9^-$. $C_8H_{20}N^+$
71.93	Cu	Re_2	$C_{16}H_{12}O_{12}$
84.73		Re_2	$C_{18}H_{28}O_{12}P_2$. $2C_4H_8O$
71.193	Si_2	Re_2	$C_{24}H_{62}$
83.193		Re_2	$C_{28}H_{26}Cl_4N_4$
84.110		Re_2	$C_{30}H_{18}N_4O_8$
86.106		Re_2	$C_{31}H_{24}O_6P_2$
86.112		Re_2	$C_{33}H_{27}NO_6P_2$
83.213		Re_2	$C_{36}H_{44}Cl_2N_8$. CCl_4
83.139		Re_3	$C_{20}H_{13}N_2O_{10}$
71.194	Si_6	Re_3	$C_{24}H_{66}Cl_3N_2O_2$
84.106	Si_3	Re_3	$C_{25}H_{63}N_4O_8$
71.227		Re_3	$C_{29}H_{57}P_2$
76.108		Re_3	$C_{36}H_{54}N_6O_{10}^+$. O_4Re^- . $1.5C_6H_6$. xCH_2Cl_2
71.247	Si_9	Re_6	$C_{36}H_{100}Cl_6$

Rh

79.4		Rh	$C_3H_{15}N_9S_3^{3+}$. $3Cl^-$
85.5		Rh	$C_4H_{12}Cl_4O_2S_2^-$. $2C_2H_6OS$. H^+
76.15		Rh	$C_4H_{16}Cl_2N_4^+$. NO_3^-
76.16		Rh	$C_4H_{16}Cl_2N_4^+$. NO_3^-
72.3	2	Rh	$C_6H_6N_2O_2^+$. $C_{35}O_{33}Rh_{14}^{2-}$
73.5		Rh	$C_7H_4NO_4$
83.36		Rh	$C_8H_{14}Cl_2N_4O_4^-$. Rb^+ . H_2O
86.11		Rh	$C_9H_{27}ClP_3$
71.21		Rh	$C_{10}H_{14}PS_4$
73.21		Rh	$C_{10}H_{15}N_2O_6$
86.29		Rh	$C_{12}H_{36}P_4^+$. Cl^-
73.61		Rh	$C_{16}H_{19}O_2$. $2C_6H_6O_2$
75.11		Rh	$C_{16}H_{23}$
73.70		Rh	$C_{17}H_{24}N^{2+}$. $2F_6P^-$
75.20	F_3	Rh	$C_{18}H_{18}O_2$
83.134		Rh	$C_{18}H_{32}N_2O_2^+$. $C_2Cl_2O_2Rh^-$
86.51		Rh	$C_{18}H_{42}ClN_2P_2$
73.81		Rh	$C_{19}H_{29}O_3P^+$. F_6P^-
81.66		Rh	$C_{19}H_{45}O_3P_2$
86.66	B_9	Rh	$C_{20}H_{28}NO_3P$. $3CH_2Cl_2$
75.30	F_{12}	Rh	$C_{22}H_{15}$
72.40	F_6	Rh	$C_{22}H_{31}O_2$
71.191	F_4	Rh	$C_{24}H_{24}^+$. ClO_4^-
85.77		Rh	$C_{24}H_{24}N_3O_{15}S_3$. $3C_4H_9NO$
74.12		Rh	$C_{24}H_{38}^+$. $\bar{F}_6P^-$

86.90	**Rh**	$C_{25}H_{20}N_2OPS_2$
71.216	F_6 **Rh**	$C_{27}H_{33}N_2O_2$
73.112	**Rh**	$C_{28}H_{30}N_2O_6P$
86.105	**Rh**	$C_{31}H_{24}ClN_6OP_2$
73.118	**Rh**	$C_{33}H_{23}$
73.120	Fe **Rh**	$C_{33}H_{36}NP^+$, F_6P^-
71.240	**Rh**	$C_{34}H_{31}Cl_2OP_2$
86.120	**Rh**	$C_{36}H_{37}ClO_2P_3S$
86.131	**Rh**	$C_{37}H_{30}ClOP_2$
86.132	**Rh**	$C_{37}H_{37}O_3P_3S^+$, AsF_6^-
75.42	**Rh**	$C_{37}H_{41}NP_2^+$, ClO_4^-
72.56	**Rh**	$C_{38}H_{30}ClP_2$, CH_2Cl_2
72.58	**Rh**	$C_{38}H_{30}Cl_3P_2$
72.59	**Rh**	$C_{38}H_{37}NO_3P_2^+$, BF_4^-
86.140	B_9 **Rh**	$C_{38}H_{42}O_4P_2S$, $C_4H_{10}O$
86.151	**Rh**	$C_{41}H_{39}P_6$
75.43	**Rh**	$C_{42}H_{49}NOP_2^+$, ClO_4^-
75.44	**Rh**	$C_{42}H_{66}P_2^+$, BF_4^-
73.129	W **Rh**	$C_{46}H_{42}P_2^+$, F_6P^-
73.130	**Rh**	$C_{46}H_{46}P_2^+$, F_6P^-
86.175	**Rh**	$C_{46}H_{42}N_2O_2P_4^+$, F_6P^- , $0.5C_3H_6O$
74.15	B **Rh**	$C_{50}H_{44}P_2$
86.184	**Rh**	$C_{51}H_{44}OP_4^+$, BF_4^-
75.48	**Rh**	$C_{51}H_{53}OP_2$, $2C_6H_6$
72.68	**Rh**	$C_{66}H_{70}ClO_2P_2$, $1.5C_3H_6O$
86.225	Co **Rh**	$C_{82}H_{79}P_9^{2+}$, $2C_{24}H_{20}B^-$, $2C_3H_6O$
86.228	Ni **Rh**	$C_{82}H_{79}P_9^{2+}$, $2C_{24}H_{20}B^-$, C_4H_8O
81.42	F_{12} **Rh₂**	$C_{12}H_{12}O_{10}$
81.43	F_{12} **Rh₂**	$C_{12}H_{12}O_{10}S_2$
81.48	**Rh₂**	$C_{12}H_{24}O_{10}S_2$
82.26	**Rh₂**	$C_{12}H_{32}N_4O_{10}^{4+}$, $4ClO_4^-$, $2H_2O$
76.81	**Rh₂**	$C_{12}H_{36}N_6O_4^{4+}$, $4ClO_4^-$, $4H_2O$
71.52	**Rh₂**	$C_{13}H_{12}O_2$
81.62	**Rh₂**	$C_{16}H_{28}O_8S_2$
81.63	**Rh₂**	$C_{16}H_{32}O_{10}S_2$
71.157	**Rh₂**	$C_{20}H_{24}Cl_2N_8^{2+}$, $2Cl^-$, $8H_2O$
71.158	**Rh₂**	$C_{20}H_{24}N_8^{2+}$, $2C_{24}H_{20}B^-$, C_2H_3N
73.88	**Rh₂**	$C_{20}H_{30}I_4$, $2C_7H_8$
81.70	**Rh₂**	$C_{20}H_{40}O_{10}$
81.73	**Rh₂**	$C_{22}H_{28}N_8O_{12}$, $2H_2O$
81.74	**Rh₂**	$C_{22}H_{28}O_8S_2$
73.96	**Rh₂**	$C_{22}H_{34}I_4$
84.97	**Rh₂**	$C_{24}H_{24}N_4O_4$
81.78	**Rh₂**	$C_{24}H_{32}N_6O_{12}$
71.217	**Rh₂**	$C_{27}H_{43}IO_2P_2S_2$
86.103	**Rh₂**	$C_{30}H_{63}ClO_2P_2S$
86.130	**Rh₂**	$C_{36}H_{66}N_2P_4$
71.262	**Rh₂**	$C_{40}H_{64}N_8^{2+}$, $2F_6P^-$, $2C_2H_3N$
86.177	**Rh₂**	$C_{50}H_{44}Cl_2O_2P_4S$
86.186	**Rh₂**	$C_{52}H_{44}ClO_2P_4^+$, BF_4^-
86.194	**Rh₂**	$C_{53}H_{44}ClO_3P_4^+$, $C_2Cl_2O_2Rh^-$, CH_2Cl_2
71.273	**Rh₂**	$C_{53}H_{44}Cl_2OP_4S_4$
86.195	**Rh₂**	$C_{53}H_{45}O_3P_4^+$, $C_7H_7O_3S^-$, $2C_4H_8O$
71.279	**Rh₂**	$C_{57}H_{50}Cl_2O_5P_4$
71.126	**Rh₃**	$C_{18}H_{16}O_2^+$, $C_2F_3O_2^-$
86.58	**Rh₃**	$C_{18}H_{57}O_{18}P_6$
73.116	**Rh₃**	$C_{30}H_{48}O^+$, F_6P^- , H_2O
86.161	Fe **Rh₃**	$C_{44}H_{30}O_8P_3$

86.221	**Rh₃**	$C_{75}H_{60}O_3P_5$
71.261	**Rh₄**	$C_{40}H_{48}ClN_{16}^{5+}$, $4Cl_4Co^{2-}$, $3H^+$, $6H_2O$
73.127	**Rh₄**	$C_{40}H_{64}^{2+}$, $2BF_4^-$
86.216	**Rh₄**	$C_{65}H_{50}O_5P_5^-$, Li^+
72.28	**Rh₆**	$C_{17}H_5O_{14}^-$, $C_{24}H_{20}P^+$, C_4H_8O
71.147	Cu_2 **Rh₆**	$C_{20}H_6N_2O_{15}$, $0.5CH_4O$

Ru

71.2	**Ru**	$C_4H_{19}N_5O_4^{2+}$, $O_6S_2^{2-}$, $2H_2O$
83.19	**Ru**	$C_5H_{21}N_6O^{2+}$, $2F_6P^-$
86.3	**Ru**	$C_6H_{21}Cl_3NO_6P_2$
83.26	**Ru**	$C_6H_{25}N_7^{2+}$, $2F_6P^-$
86.4	**Ru**	$C_7H_9O_7P$
71.6	**Ru**	$C_8H_7NO_3$
86.9	**Ru**	$C_9H_{18}O_3P_2$
73.18	**Ru**	$C_{10}H_{10}$
73.17	**Ru**	$C_{10}H_{10}$
73.16	**Ru**	$C_{10}H_{10}$
84.34	2 **Ru**	$C_{12}H_6N_9O_{12}^-$, Ba^{2+} , $9H_2O$
71.125	W **Ru**	$C_{18}H_{14}O_6$
83.146	**Ru**	$C_{20}H_{19}N_4O_2^{2+}$, $2ClO_4^-$
74.8	**Ru**	$C_{20}H_{26}$
83.157	**Ru**	$C_{21}H_{16}ClN_4O^+$, ClO_4^-
71.226	**Ru**	$C_{29}H_{47}ClN_2P_2$
83.198	**Ru**	$C_{30}H_{24}N_6^{2+}$, $2F_6P^-$
83.200	**Ru**	$C_{30}H_{30}N_6^{2+}$, $2BF_4^-$
71.231	**Ru**	$C_{30}H_{47}ClN_2OP_2$
83.216	**Ru**	$C_{36}H_{26}N_8^{2+}$, $2F_6P^-$
86.138	**Ru**	$C_{38}H_{34}Cl_2O_2P_2$, CH_2Cl_2
71.258	**Ru**	$C_{39}H_{33}IO_2P_2$
71.260	**Ru**	$C_{40}H_{33}O_2P_2S_2^+$, ClO_4^- , C_6H_{12}
86.154	**Ru**	$C_{42}H_{40}N_2O_2P_2S_2$
86.153	**Ru**	$C_{42}H_{40}N_2O_2P_2S_2$
72.62	**Ru**	$C_{43}H_{30}Cl_4N_2O_2P_2$, CH_2Cl_2
86.165	**Ru**	$C_{44}H_{43}ClN_2OP_2$
86.169	**Ru**	$C_{45}H_{34}ClNO_2P_2$, CCl_2D_2
71.267	**Ru**	$C_{45}H_{37}IO_2P_2$
71.270	**Ru**	$C_{47}H_{40}NOP_2S_2^+$, ClO_4^- , $0.5CHCl_3$, $0.5H_2O$
71.271	**Ru**	$C_{47}H_{40}NOP_2S_2^+$, ClO_4^- , $0.5CHCl_3$, $0.5H_2O$
86.192	**Ru**	$C_{52}H_{50}P_4$, C_6H_6
71.278	F_3 **Ru**	$C_{55}H_{41}O_3P_2$
81.10	**Ru₂**	$C_4H_4Cl_2O_8^-$, K^+
81.29	(**Ru₂**	$C_8H_{12}ClO_8)_n$, $2nH_2O$
81.30	**Ru₂**	$C_8H_{12}Cl_2O_8^-$, Cs^+
81.35	**Ru₂**	$C_8H_{16}O_{10}^+$, BF_4^-
81.47	(**Ru₂**	$C_{12}H_{20}ClO_8)_n$
71.68	**Ru₂**	$C_{14}H_{18}N_2O_4$
71.86	**Ru₂**	$C_{15}H_{12}O_3$
71.87	**Ru₂**	$C_{15}H_{13}O_3^+$, BF_4^-
86.45	**Ru₂**	$C_{16}H_{42}Cl_4N_2O_{14}P_4$
72.32	Si_2 **Ru₂**	$C_{18}H_{26}O_4$
86.57	**Ru₂**	$C_{18}H_{57}O_3P_6^+$, BF_4^-
75.23	Si_2 **Ru₂**	$C_{19}H_{24}O_5$
71.166	**Ru₂**	$C_{20}H_{58}P_6^{2+}$, $2BF_4^-$
71.175	**Ru₂**	$C_{21}H_{60}P_6$
71.176	**Ru₂**	$C_{21}H_{61}P_6^+$, BF_4^-
80.22	**Ru₂**	$C_{22}H_{40}N_4O_2S_8^+$, BF_4^-

71.212   **Ru**$_2$ $C_{27}H_{20}O_3$

86.98   **Ru**$_2$ $C_{28}H_{54}Br_2O_4P_2$

86.220   **Ru**$_2$ $C_{72}H_{60}O_{12}P_4S_4$. C_7H_8

71.47   **Ru**$_3$ $C_{13}H_7NO_{11}$

86.40   **Ru**$_3$ $C_{16}H_{28}NO_{17}P_3$

75.15   **Ru**$_3$ $C_{17}H_{16}O_6S$

71.135   **Ru**$_3$ $C_{19}H_8O_9$

71.151   Ni **Ru**$_3$ $C_{20}H_{15}O_9$

73.104   Si$_3$ **Ru**$_3$ $C_{25}H_{30}O_8$

73.105   Si$_3$ **Ru**$_3$ $C_{25}H_{30}O_8$

71.198   **Ru**$_3$ $C_{26}H_{17}O_9P$

71.199   **Ru**$_3$ $C_{26}H_{19}O_8P$. $0.5C_6H_6$

71.222   **Ru**$_3$ $C_{28}H_{60}P_8{}^{2+}$. $2BF_4{}^-$

86.99   **Ru**$_3$ $C_{29}H_{15}O_{11}P$

72.50   **Ru**$_4$ $C_{24}H_{32}N_4O_8$

86.116   **Ru**$_4$ $C_{36}H_{28}O_{10}P_2$

71.117   **Ru**$_6$ $C_{17}O_{16}{}^{2-}$. $2C_4H_{12}N^+$

71.177   **Ru**$_6$ $C_{22}H_{16}O_{15}S_3$

71.223   **Ru**$_6$ $C_{29}H_{14}O_{14}$

71.228   Hg$_2$ **Ru**$_6$ $C_{30}H_{18}Br_2O_{18}$

Sb

66.4   **Sb** $C_4H_8ClS_3$

66.2   F **Sb** $C_4O_9{}^{4-}$. $3Na^+$. H_3O^+ . H_2O

66.7   **Sb** $C_6H_9O_3S_3$

66.8   **Sb** $C_6H_9O_6$

66.6   F$_9$ **Sb** C_6O_6

66.10   **Sb** $C_{10}H_8Cl_3N_2$

66.12   **Sb** $C_{12}H_{30}O_6P_3S_6$

66.13   **Sb** $C_{14}H_{13}O_2$

73.89   Mn$_3$ **Sb** $C_{21}H_{15}ClO_6$

66.14   **Sb** $C_{21}H_{21}$

66.16   **Sb** $C_{24}H_{21}O_3$. $C_{24}H_{19}O_2Sb$

66.17   **Sb** $C_{26}H_{33}O_4$

66.19   **Sb** $C_{36}H_{30}P_3S_6$

66.1   F$_{12}$ **Sb**$_2$ C_2O_2

66.5   **Sb**$_2$ $C_4H_{12}Cl_6O_6P_2$

66.3   F$_{12}$ **Sb**$_2$ C_4O_5

66.9   **Sb**$_2$ $C_6H_{18}Cl_6N_2O_6P_2$

66.11   **Sb**$_2$ $C_{10}H_{14}Cl_6O_5$

66.15   **Sb**$_2$ $C_{24}H_{20}$

66.18   **Sb**$_2$ $C_{28}H_{20}O_3$

Sc

81.31   **Sc** $C_8H_{12}O_{12}{}^-$. $C_4H_{14}O_{10}Sc^+$

Se

64.5   **Se** C_3H_9P

60.8   **Se** $C_9H_6S_2$. $C_{12}H_4N_4$

39.30   **Se** $C_{10}H_{16}N_2O_3$

42.6   **Se** $C_{11}H_{10}O_3S$

71.49   Fe$_2$ **Se** $C_{13}H_8O_7$

75.7   Fe$_2$ **Se** $C_{14}H_{12}O_6$

38.70   **Se** $C_{15}H_{14}O_5$

47.33   **Se** $C_{16}H_{17}N_5O_4$

11.18   **Se** $C_{18}H_{15}{}^+$. Cl^- . H_2O

73.82   Fe$_2$ **Se** $C_{20}H_{18}{}^+$. $I_3{}^-$. I_2 . $0.5CH_2Cl_2$

69.55   Sn$_2$ **Se** $C_{36}H_{30}$

73.3   CoB$_9$ **Se**$_2$ C_5H_{14}

40.10   **Se**$_2$ $C_8H_{16}N_2O_2$

42.5   **Se**$_2$ $C_{10}H_4N_4$

11.12   F$_8$ **Se**$_2$ C_{12}

76.89   Ni **Se**$_2$ $C_{14}H_{32}N_6$

64.143   **Se**$_2$ $C_{26}H_{20}P_2$

61.9   3 **Se**$_2$ $C_{26}H_{24}P_2$. C_8H_{10}

86.136   Os **Se**$_2$ $C_{38}H_{30}O_2P_2$

40.11   **Se**$_3$ $C_8H_{16}N_2O_2$

40.12   **Se**$_4$ $C_8H_{16}N_2O_2$

33.37   **Se**$_4$ $C_{10}H_{20}N_2$

85.47   Ni **Se**$_6$ $C_{12}N_6{}^{2-}$. $2C_{24}H_{20}As^+$

Si

63.1   6 **Si** $C_3H_9{}^-$. $6Li^+$

63.2   **Si** $C_5H_{15}N_7S_5$

63.3   Al **Si** $C_6H_{18}Cl_4N_3$

63.4   B$_2$ **Si** $C_6H_{18}N_2S_2$

63.5   B$_{10}$ **Si** C_6H_{22}

63.6   **Si** $C_7H_{15}ClN_2O_2$

63.7   F$_3$ **Si** $C_8H_6BrO_2$

63.8   **Si** $C_8H_{19}N$

63.9   **Si** $C_9H_{18}ClNO_3$

63.10   **Si** $C_9H_{21}N$

63.12   F **Si** $C_{12}H_{14}NO_4$

74.1   Cr **Si** $C_{12}H_{14}O_3$

63.15   F$_3$ **Si** $C_{13}H_{14}NO_4$

73.43   Ti **Si** $C_{13}H_{19}Cl$

73.44   V **Si** $C_{13}H_{19}N$

63.16   **Si** $C_{13}H_{21}NO_2$

63.18   **Si** $C_{15}H_{21}O_6{}^+$. $ClO_4{}^-$

63.19   **Si** $C_{15}H_{23}N_3O_3$

63.22   **Si** $C_{17}H_{26}O_4$

63.23   **Si** $C_{18}H_{12}O_6{}^{2-}$. $2C_6H_{16}N^+$

63.24   **Si** $C_{18}H_{16}$

63.27   **Si** $C_{19}H_{25}NO$

63.28   **Si** $C_{19}H_{25}NO_2$

75.24   Fe **Si** $C_{20}H_{16}O_3$

63.31   **Si** $C_{20}H_{27}NO$

71.160   Pd **Si** $C_{20}H_{33}ClP^+$. F_6P^-

63.33   **Si** $C_{21}H_{29}N_2{}^+$. $C_4H_3O_4{}^-$

63.35   **Si** $C_{22}H_{30}N_2$

63.42   **Si** $C_{25}H_{60}O$

71.201   Re **Si** $C_{26}H_{21}O_2$. CH_2Cl_2

63.43   **Si** $C_{26}H_{23}N$

69.51   Ge **Si** $C_{36}H_{30}O_2$

73.122   Co **Si** $C_{36}H_{33}$

72.65   Pt **Si** $C_{47}H_{57}N_2P_3S$

71.277   Pt$_2$ **Si** $C_{54}H_{82}P_2$

63.13   **Si**$_2$ $C_{12}H_{24}N_2$

64.69   **Si**$_2$ $C_{14}H_{36}N_4O_2P_2$

63.20   **Si**$_2$ $C_{16}H_{18}Cl_2$

63.25   **Si**$_2$ $C_{18}H_{26}O_2$

72.32   Ru$_2$ **Si**$_2$ $C_{19}H_{26}O_4$

71.130	Cd	Si_2	$C_{18}H_{30}N_2$. $0.5C_{10}H_8N_2$
64.99		Si_2	$C_{18}H_{42}N_4O_4P_2$
71.132	Mo_2	Si_2	$C_{18}H_{46}O_4P_2$
75.23	Ru_2	Si_2	$C_{19}H_{24}O_5$
64.106		Si_2	$C_{19}H_{27}ClNOP$
74.7	Cr	Si_2	$C_{20}H_{26}O_5$
71.163	Mo_2	Si_2	$C_{20}H_{40}Cl_2O_6P_2$
83.155	Ti	Si_2	$C_{20}H_{48}N_4$
83.156	Zr	Si_2	$C_{20}H_{48}N_4$
63.34		Si_2	$C_{21}H_{32}N_2$
63.36		Si_2	$C_{22}H_{34}O_3$
63.40		Si_2	$C_{24}H_{22}$
75.34	NiB_2	Si_2	$C_{24}H_{30}$
73.101	Nb	Si_2	$C_{24}H_{34}$
63.41		Si_2	$C_{24}H_{46}$
63.45		Si_2	$C_{27}H_{37}BrO_3$
63.50	Hg	Si_2	$C_{36}H_{30}$
63.51		Si_2	$C_{36}H_{46}$
63.52		Si_2	$C_{38}H_{48}N_2O$
86.147	Pt_2	Si_2	$C_{40}H_{80}P_2$
65.5	As_7	Si_3	C_9H_{27}
64.35		Si_3	$C_9H_{27}P_{11}$
64.36		Si_3	$C_9H_{33}N_6PS$
73.20	Fe	Si_3	$C_{10}H_{11}Cl_5O_2$
86.14	Cr	Si_3	$C_{11}H_{18}O_5P_4$
64.61		Si_3	$C_{13}H_{36}N_3P$
71.89	Mo	Si_3	$C_{15}H_{42}ClP$
73.66	Fe_2	Si_3	$C_{17}H_{16}Cl_4O_4$
63.29		Si_3	$C_{19}H_{28}$
63.30		Si_3	$C_{19}H_{28}$
73.104	Ru_3	Si_3	$C_{25}H_{30}O_8$
73.105	Ru_3	Si_3	$C_{25}H_{30}O_8$
84.106	Re_3	Si_3	$C_{25}H_{63}N_4O_8$
63.11	Hg_2	Si_4	$C_{10}H_{28}$
83.85	Ta	Si_4	$C_{12}H_{36}Cl_3N_2$
63.17	Sn_2B_2	Si_4	$C_{14}H_{42}N_4$
68.18	Tl_2	Si_4	$C_{16}H_{44}Cl_2$
64.107		Si_4	$C_{19}H_{57}N_7P_2$
84.85	Eu	Si_4	$C_{20}H_{56}N_2O_4$
63.37		Si_4	$C_{22}H_{36}$
63.38		Si_4	$C_{22}H_{40}O_{12}$
63.44		Si_4	$C_{26}H_{46}N_4$
70.18	Te_2Pd	Si_4	$C_{26}H_{60}N_2S_2$
63.47		Si_4	$C_{28}H_{36}N_4$
84.116	Mo	Si_4	$C_{32}H_{68}N_2O_4$
69.56	Sn_6	Si_4	$C_{40}H_{96}N_8O_4$
63.53		Si_4	$C_{48}H_{40}O_4$
65.8	As_4	Si_6	$C_{12}H_{36}$
63.14		Si_6	$C_{12}H_{39}N_5$
73.78	Fe	Si_6	$C_{18}H_{38}O_2$
63.32		Si_6	$C_{20}H_{34}O_7$
86.72	Cr_2	Si_6	$C_{20}H_{36}O_8P_8$
63.39		Si_6	$C_{22}H_{60}N_4$
73.103	Fe_2	Si_6	$C_{24}H_{40}O_4$
71.193	Re_2	Si_6	$C_{24}H_{62}$
71.194	Re_3	Si_6	$C_{24}H_{66}Cl_3N_2O_2$
63.46		Si_6	$C_{28}H_{32}O_8$
63.49		Si_6	$C_{30}H_{60}N_4$

84.137	Zr	Si_6	$C_{72}H_{60}O_9^{2-}$. $2C_4H_{12}N^+$
63.21		Si_8	$C_{16}H_{24}O_{12}$
63.48		Si_8	$C_{28}H_{72}O_2$
71.247	Re_6	Si_9	$C_{36}H_{100}Cl_6$
63.26		Si_{10}	$C_{18}H_{54}N_2O_8$

Sm

84.74	Sm	$C_{18}H_{36}N_3O_9^{2+}$, $H_2N_5O_{16}Sm^{2-}$
84.83	Sm	$C_{20}H_{24}Cl_3O_{18}$

Sn

69.1		(Sn	$C_2O_4)_n$
69.2		(Sn	$C_2O_4)_n$
69.3		(Sn	$C_3H_9N_3)_n$
69.4		Sn	$C_4H_7Cl_3O_2$
69.8		(Sn	$C_5H_{13}NO_2)_n$
69.9		Sn	$C_6H_6Cl_3O_6^-$, K^+
69.10		Sn	$C_6H_{12}Cl_2N_2O_2$
69.13	B_{10}	Sn	C_6H_{22}
69.16		Sn	$C_7H_{16}ClNO_2S$
69.17		Sn	$C_8H_{14}Cl_2O_4$
69.19		Sn	$C_9H_{13}NO_2$, H_2O
69.20		Sn	$C_{10}H_{15}^+$, BF_4^-
69.21		Sn	$C_{10}H_{22}N^+$, ClO_4^-
69.22		Sn	$C_{12}H_{18}Cl_2O_6$
69.23	As	Sn	$C_{12}H_{21}N_2O_4$
69.24		Sn	$C_{12}H_{26}N_2S_4$
69.25		Sn	$C_{12}H_{26}N_2S_4$
69.27		Sn	$C_{14}H_{14}Cl_4N_4O_3$, $2H_2O$
69.31		Sn	$C_{16}H_{19}NO_2$
69.35		Sn	$C_{16}H_{33}N_3S_6$
73.67		(Fe Sn	$C_{17}H_{18}O_2)_n$
73.71	Fe_2Co	Sn	$C_{18}H_{10}ClO_8$
69.38		Sn	$C_{18}H_{24}N_4O_4S_2$
69.39		Sn	$C_{18}H_{33}Cl$
69.41		Sn	$C_{19}H_{39}N_3S_6$
69.42		Sn	$C_{20}H_{18}S_2$
69.43		Sn	$C_{20}H_{30}$
69.44	F_3	Sn	$C_{20}H_{33}O_2$
69.45		Sn	$C_{21}H_{24}BrNO$
69.46		Sn	$C_{22}H_{25}O_2PS_2$
69.48		Sn	$C_{24}H_{38}O_4P_2S_4$
71.215	Cr	Sn	$C_{27}H_{25}NO_4$
71.219	Cr	Sn	$C_{28}H_{25}NO_5$. $0.5CH_2Cl_2$
69.50		Sn	$C_{30}H_{46}O_2$
69.57		Sn	$C_{41}H_{33}NO$
69.12		Sn_2	$C_6H_{18}N_4O_2S_4$. $0.5C_6H_6$
63.17	Si_4B_2	Sn_2	$C_{14}H_{42}N_4$
69.32		Sn_2	$C_{16}H_{22}$
69.34		Sn_2	$C_{16}H_{24}$
69.55	Se	Sn_2	$C_{36}H_{30}$
69.60		Sn_2	$C_{48}H_{40}O_8P_4S_8$
70.2	Te_3	Sn_3	C_6H_{18}
69.36		Sn_3	$C_{16}H_{38}N_4$
69.18		(Sn_4	$C_8H_{24}Cl_4O_2)_n$
69.37		(Sn_4	$C_{16}H_{40}Cl_4O_2)_n$

69.56	Si$_4$ Sn$_6$	C$_{40}$H$_{96}$N$_8$O$_4$

Ta

71.30	Ta	C$_{11}$H$_{15}$Cl$_2$O
71.42	Ta	C$_{12}$H$_{15}$
83.85	Si$_4$ Ta	C$_{12}$H$_{36}$Cl$_3$N$_2$
71.74	Ta	C$_{14}$H$_{23}$Cl$_2$
71.110	Ta	C$_{17}$H$_{27}$Cl$_2$
71.137	Ta	C$_{19}$H$_{15}$Cl$_4$N$^-$, C$_5$H$_6$N$^+$
86.125	Ta	C$_{36}$H$_{53}$P$_6$
72.35	Ta$_2$	C$_{18}$H$_{34}$Cl$_6$O$_2$
71.209	Ta$_2$	C$_{26}$H$_{45}$Cl$_4$OP

Tc

84.1	(Tc	CH$_3$O$_7$P$_2$$^-$)$_n$, nH$_6LiO_3$$^+$, 0.33nH$_2$O
85.3	Tc	C$_4$H$_4$O$_3$S$_4$$^-$. C$_{16}H_{36}N^+$
83.44	B Tc	C$_9$H$_{10}$Cl$_2$N$_8$O
86.68	As$_4$ Tc	C$_{20}$H$_{32}$Cl$_2$$^+$, Cl$^-$
86.69	As$_4$ Tc	C$_{20}$H$_{32}$Cl$_2$$^+$, ClO$_4$$^-$
86.70	As$_4$ Tc	C$_{20}$H$_{32}$Cl$_4$$^+$, F$_6P^-$
84.79	(Tc$_2$	C$_{20}$H$_{16}$ClN$_4$O$_4$)$_n$

Te

70.1	Te	C$_6$H$_5$$^{3+}$, 1.3Br$^-$, 1.7Cl$^-$
70.3	Te	C$_8$H$_8$I$_2$
70.4	Te	C$_9$H$_8$Cl$_3$NO$_2$
70.5	Te	C$_9$H$_{17}$Br$_3$O
70.6	Te	C$_{10}$H$_{17}$Cl$_3$O
70.7	Te	C$_{12}$H$_8$O$_2$S$_2$
70.8	Cr Te	C$_{12}$H$_{14}$N$_2$O$_5$
70.9	Te	C$_{12}$H$_{20}$Cl$_4$
70.10	Te	C$_{12}$H$_{24}$BrN$_3$O$_3$S$_6$
70.11	Te	C$_{12}$H$_{24}$N$_3$O$_3$P
70.12	Te	C$_{12}$H$_{32}$N$_8$S$_4$$^{2+}$, 2Cl$^-$
70.13	Te	C$_{13}$H$_{17}$Cl$_3$
70.14	Te	C$_{16}$H$_{32}$N$_4$O$_4$S$_8$
70.15	Te$_2$	C$_{18}$H$_{36}$N$_{12}$S$_6$$^{4+}$. 4ClO$_4$$^-$
70.16	Te$_2$	C$_{24}$H$_{48}$N$_{12}$S$_6$$^{4+}$, 4ClO$_4$$^-$
70.17	Te$_2$	C$_{24}$H$_{48}$N$_{12}$S$_6$$^{4+}$, 4ClO$_4$$^-$
70.18	Si$_4$Pd Te$_2$	C$_{26}$H$_{60}$N$_2$S$_2$
70.2	Sn$_3$ Te$_3$	C$_6$H$_{18}$

Th

81.6	F$_3$ Th	C$_3$O$_9$$^{5-}$. 5CH$_6N_3$$^+$
85.30	Th	C$_8$H$_{24}$P$_4$S$_8$
75.13	Th	C$_{16}$H$_{24}$Cl$_2$O$_2$
84.104	2 Th	C$_{24}$H$_{72}$N$_{15}$O$_{13}$P$_4$$^-$. N$_6O_{18}$Th^{2-}
84.133	Th	C$_{48}$H$_{40}$Cl$_4$O$_4$S$_4$
85.90	Th	C$_{48}$H$_{88}$P$_4$S$_8$
73.126	Th$_2$	C$_{40}$H$_{62}$. C$_7$H$_8$

Ti

85.8	Ti	C$_6$H$_9$ClS$_6$

73.33	Ti	C$_{12}$H$_{15}$ClO
73.43	Si Ti	C$_{13}$H$_{19}$Cl
80.17	Ti	C$_{15}$H$_{30}$ClN$_3$O$_3$S$_3$
73.74	Ti	C$_{18}$H$_{18}$N$_2$
83.155	Si$_2$ Ti	C$_{20}$H$_{48}$N$_4$
73.94	Ti	C$_{22}$H$_{30}$O$_2$
73.106	Ti	C$_{25}$H$_{33}$O
73.63	Ti$_2$	C$_{16}$H$_{22}$Cl$_4$O$_2$
73.83	Ti$_2$	C$_{20}$H$_{20}$Cl$_2$O
73.113	Ti$_2$	C$_{28}$H$_{30}$N$_2$S$_2$
71.241	Ti$_2$	C$_{34}$H$_{32}$
73.131	Ti$_2$	C$_{50}$H$_{48}$N$_4$
73.100	Ti$_4$	C$_{24}$H$_{28}$Cl$_4$O$_4$

Tl

68.5	Tl	C$_8$H$_{12}$O$_8$$^-$, Tl$^+$
68.13	Tl	C$_{12}$H$_8$Cl$_3$N$_4$O$_2$
68.20	Tl	C$_{17}$H$_{20}$NS$_2$
68.18	Si$_4$ Tl$_2$	C$_{16}$H$_{44}$Cl$_2$
68.24	Tl$_2$	C$_{38}$H$_{30}$O$_4$
68.25	F$_{16}$ Tl$_2$	C$_{60}$H$_{34}$Cl$_2$O$_2$P$_2$

U

79.1	U	CH$_7$N$_2$O$_7$P , H$_2$O
79.2	U	C$_2$H$_9$N$_4$O$_7$P
79.3	U	C$_3$H$_{12}$N$_6$O$_5$$^{2+}$, O$_4$S^{2-}
82.4	(U	C$_4$H$_5$NO$_6$)$_n$
81.19	U	C$_6$H$_4$O$_{10}$$^{2-}$, Ba^{2+} , 3H$_2$O
81.20	U	C$_6$H$_4$O$_{10}$$^{2-}$, Sr^{2+} , 3H$_2$O
81.36	U	C$_8$H$_{16}$O$_{14}$
84.69	U	C$_{16}$H$_{36}$N$_4$O$_4$P$_4$S$_4$
78.6	U	C$_{18}$H$_{18}$N$_2$O$_5$, CHCl$_3$
76.95	U	C$_{18}$H$_{19}$N$_3$O$_4$
84.75	2 U	C$_{19}$H$_{19}$N$_8$O$_5$$^+$, N$_4O_{14}$U^{2-}
84.90	U	C$_{21}$H$_{26}$ClNO$_6$
84.101	U	C$_{24}$H$_{46}$N$_4$O$_7$$^{2+}$, 2ClO$_4$$^-$
83.184	U	C$_{25}$H$_{23}$N$_5$O$_6$
83.185	U	C$_{25}$H$_{23}$N$_5$O$_6$
77.11	F$_{12}$ U	C$_{40}$H$_{24}$O$_8$
84.129	U	C$_{41}$H$_{36}$N$_2$O$_5$P$_2$S$_2$
84.138	U	C$_{76}$H$_{60}$N$_4$O$_4$P$_4$S$_4$
81.5	(U$_2$	C$_2$H$_8$O$_{12}$)$_n$
73.132	U$_3$	C$_{60}$H$_{90}$Cl$_3$

V

83.27	V	C$_7$H$_5$Cl$_3$NO
83.33	V	C$_8$H$_7$Cl$_3$NO
73.19	V	C$_{10}$H$_{10}$
76.60	2 V	C$_{10}$H$_{14}$N$_2$O$_8$$^-$. C$_2H_{10}N_2$$^{2+}$, 2H$_2$O
84.30	V	C$_{11}$H$_7$N$_3$O$_4$$^-$. C$_{24}H_{20}P^-$
73.44	Si V	C$_{13}$H$_{19}$N
83.89	V	C$_{14}$H$_{10}$Cl$_3$N$_2$O
76.91	2n V	C$_{16}$H$_{14}$N$_2$O$_3$. nNa$^-$. nC$_{24}$H$_{20}$B$^-$
72.26	V	C$_{16}$H$_{16}$O$_4$
83.158	V	C$_{22}$H$_{18}$N$_4$O

78.8		V	$C_{22}H_{28}Cl_2N_2O_2 \cdot C_6H_6$
72.45	F_{10}	V	$C_{24}H_{10}$
84.118		V	$C_{36}H_{30}Cl_2O_3P_2$
81.28		V_2	$C_8H_4O_{14}{}^{4-} \cdot 4C_8H_{20}N^+ \cdot 8H_2O$
84.71		V_2	$C_{16}H_{40}N_4O_7$
73.99		V_2	$C_{24}H_{22}N_2{}^{2+} \cdot 2F_6P^- \cdot C_2H_3N$

W

83.1		W	$C_2Cl_{10}N^- \cdot C_{24}H_{20}As^+$
73.2		W	$C_5H_5ClN_2O_2$
73.10		W	$C_8H_8N_2O_2$
72.5		W	$C_8H_{10}INO$
80.2		W	$C_9H_{12}N_2O_3S_4$
83.47		W	$C_{10}H_{10}Br_4N_2{}^- \cdot C_5H_6N^+ \cdot C_5H_5N$
71.32		W	$C_{11}H_{25}Cl_2OP$
72.17		W	$C_{12}H_{12}O_8$
73.37		W	$C_{12}H_{15}NO_3$
86.25	Pt	W	$C_{12}H_{30}P_2S_4$
73.41	Cr	W	$C_{13}H_8N_2O_7$
84.48		W	$C_{13}H_{11}NO_5S$
71.60	Cr	W	$C_{14}H_5ClO_7$
71.83		W	$C_{14}H_{42}P_4$
73.60		W	$C_{16}H_{17}N_2{}^+ \cdot BF_4{}^-$
71.103	$F_{12}Co$	W	$C_{17}H_5O_4$
71.125	Ru	W	$C_{18}H_{14}O_6$
86.61		W	$C_{19}H_{19}N_4O_3P$
71.142	Pt	W	$C_{19}H_{26}O_6P_2$
71.174		W	$C_{21}H_{46}P_2$
73.92		W	$C_{22}H_{20}O_2$
71.186	Pt	W	$C_{22}H_{37}O_5P_3$
72.42		W	$C_{23}H_{20}Cl_2O_2P^- \cdot C_8H_{20}N^+$
71.214		W	$C_{27}H_{24}O_6$
71.218		W	$C_{28}H_{22}I_2O_3P_2$
86.100		W	$C_{30}H_{13}Cl_3NO_4P$
71.234		W	$C_{31}H_{25}OPS$
71.238	Pt_2	W	$C_{32}H_{50}O_7P_2$
71.244		W	$C_{35}H_{63}N_7{}^{2+} \cdot O_{19}W_6{}^{2-}$
84.119		W	$C_{36}H_{30}Cl_3O_3P_2$
86.126		W	$C_{36}H_{63}P_3$
86.133		W	$C_{38}H_{23}Cl_4N_2O_4P \cdot CH_2Cl_2$
73.129	Rh	W	$C_{46}H_{42}P_2{}^+ \cdot F_6P^-$
86.173		W	$C_{47}H_{34}Cl_3NO_3P_2$
86.197		W	$C_{53}H_{48}BrCl_2N_2P_4{}^+ \cdot F_6P^-$
86.199	Ag_2	W	$C_{54}H_{45}P_3S_4$
86.207		W	$C_{55}H_{54}BrN_2P_4{}^+ \cdot Br^- \cdot 0.5CH_4O$
86.210		W	$C_{56}H_{48}BrClN_4P_4 \cdot CH_2Cl_2$
86.211		W	$C_{56}H_{56}BrN_2OP_4{}^+ \cdot F_6P^- \cdot 0.5C_2H_6O$
86.212		W	$C_{56}H_{64}P_4$
85.38		W_2	$C_{10}H_{27}Cl_4S_5$
71.36	F_8	W_2	$C_{12}H_{13}Br_3N_2O_2P_4$
85.43		W_2	$C_{12}H_{26}Cl_4S_5$
86.26		W_2	$C_{12}H_{32}Cl_4P_4 \cdot C_7H_8$
86.27		W_2	$C_{12}H_{36}Cl_4P_4$
71.50		W_2	$C_{13}H_8O_8$
71.61		W_2	$C_{14}H_8O_9$
85.50		W_2	$C_{16}Cl_5O_{10}S^- \cdot C_8H_{20}N^+$
71.109		W_2	$C_{17}H_{17}O_9P$
84.78		W_2	$C_{20}H_{12}Cl_4N_4O_4$
73.86		W_2	$C_{20}H_{23}{}^+ \cdot ClO_4{}^- \cdot C_3H_6O$
71.170		W_2	$C_{21}H_{16}O_9$
75.29		W_2	$C_{21}H_{24}O_7$
83.206		W_2	$C_{32}H_{44}N_{10}$
84.121		W_2	$C_{36}H_{34}N_{10}O_2 \cdot 2C_4H_8O$
84.124		W_2	$C_{40}H_{40}N_8O_2 \cdot 2C_4H_8O$
86.189		W_2	$C_{52}H_{48}Cl_4P_4$
86.190		W_2	$C_{52}H_{48}Cl_4P_4 \cdot 0.5H_2O$
86.219	Cu_4	W_2	$C_{72}H_{60}O_2P_4S_6$
86.230	Cu_4	W_2	$C_{84}H_{84}O_2P_4S_6$
85.97	Fe_7	W_2	$C_{84}H_{84}S_{20}{}^{4-} \cdot 4C_{16}H_{36}N^+$
81.49		W_3	$C_{12}H_{24}O_{17}{}^{2+} \cdot 2CF_3O_3S^-$
75.39		W_3	$C_{32}H_{52}O_5S_5$
81.91		W_3	$C_{38}H_{84}Cl_5O_5P_3$

Y

81.51	Y_2	$C_{12}H_{26}O_{16} \cdot 4H_2O$

Yb

73.102	Yb	$C_{24}H_{38}O \cdot 0.5C_7H_8$
81.99	Yb_4	$C_{84}H_{168}N_{12}O_{24} \cdot 2C_7H_{16}$

Zn

81.13		(Zn	$C_4H_6O_4)_n$
81.14		(F_4 Zn	$C_4H_8O_8)_n$
83.16		Zn	$C_5H_5BrCl_2N^- \cdot C_5H_5Br_2ClNZn^- \cdot 2C_5H_5N \cdot 2C_5H_6N^+$
83.17		Zn	$C_5H_5BrCl_2N^- \cdot C_5H_5Br_2ClNZn^- \cdot 2C_5H_5N \cdot 2C_5H_6N^+$
81.21		(Zn	$C_6H_6NO_6{}^-)_n \cdot nNa^+ \cdot nH_2O$
83.23		(Zn	$C_6H_6N_8S_2)_n$
79.6		Zn	$C_8H_{16}N_6O_2S_2$
76.52		Zn	$C_8H_{26}N_6{}^{2+} \cdot 2NO_3{}^-$
83.58		Zn	$C_{10}H_{16}N_2O_{10}S_2$
83.57	0.5	Zn	$C_{10}H_{16}N_2O_{10}S_2 \cdot 0.5C_{10}H_{16}CuN_2O_{10}S_2$
73.25	NbB_2	Zn	$C_{11}H_{19}O \cdot 0.5C_6H_6$
71.31		Zn	$C_{11}H_{23}ClN_2$
85.40		Zn	$C_{12}H_{18}N_2S_2$
81.54		(Zn	$C_{14}H_{12}N_2O_4)_n \cdot 1.5nH_2O$
83.93		Zn	$C_{14}H_{14}Cl_2N_2$
84.56		Zn	$C_{14}H_{18}N_4O_8$
76.82		Zn	$C_{14}H_{22}N_6{}^{2+} \cdot 2NO_3{}^-$
79.13		Zn	$C_{16}H_{24}N_8S_4{}^{2+} \cdot 2NO_3{}^- \cdot H_2O$
84.65		Zn	$C_{16}H_{30}I_2N_2O_2$
83.137		Zn	$C_{19}H_{21}N_8O_5{}^+ \cdot NO_3{}^-$
83.217		Zn	$C_{42}H_{56}N_{10}O_4S_2$
76.66		Zn_2	$C_{10}H_{28}N_4$
83.76		(Zn_2	$C_{12}H_{12}N_8)_n$
80.20		Zn_2	$C_{20}H_{32}N_4S_8$
83.190	Fe_2	Zn_2	$C_{28}H_{16}N_4O_8$
84.125		Zn_2	$C_{40}H_{44}N_{14}O_{26}P_6 \cdot 4H_2O$
83.224		Zn_2	$C_{58}H_{68}N_8$
71.8		Zn_4	$C_8H_{24}O_4$
85.92		(Zn_4	$C_{49}H_{44}OS_8)_n$

Zr

71.127		Zr	$C_{18}H_{18}$
73.75		Zr	$C_{18}H_{18}N_2$
71.159	Al_2	Zr	$C_{20}H_{33}Cl$
75.27		Zr	$C_{20}H_{44}P_4$
83.156	Si_2	Zr	$C_{20}H_{48}N_4$
71.189		Zr	$C_{23}H_{22}ClP$
83.173		Zr	$C_{24}H_{24}N_6{}^{2-}$, $12C_4H_8O$, $2Na^+$
71.197	Al_2	Zr	$C_{25}H_{38}$
78.17		Zr	$C_{40}H_{28}N_4O_4$, $2.5C_6H_6$
77.13		Zr	$C_{60}H_{44}O_8$
84.137	Si_6	Zr	$C_{72}H_{60}O_9{}^{2-}$, $2C_4H_{12}N^+$
71.230		Zr_2	$C_{30}H_{28}$

AUTHOR INDEX

Abe, F. 45.57
Abe, M. 45.60
Abel, E.W. 76.41, 84.45, 84.46, 86.80
Abello, L. 82.14
Abeysekera, A.M. 49.4
Ablov, A.V. 1.9, 76.1, 84.112
Abraham, F. 84.34
Abrahams, S.C. 28.1
Abu−Dari, K. 84.87, 84.88, 85.66, 85.72,
  85.73, 85.74, 85.75
Abuaan, M.M. 45.22
Achari, A. 60.24
Acharya, K.R. 35.25
Acheson, R.M. 32.69
Achiwa, K. 75.42
Ackermann, E. 38.141
Acklin, W. 54.29
Ad'yaasuren, P. 63.26
Adam, G. 59.32
Adam, T. 63.18, 63.23
Adamiak, D.A. 64.26
Adams, H.−N. 71.15
Adams, J.M. 2.13, 2.14, 2.15
Adams, M.A. 73.22, 73.107
Adams, R.D. 46.2, 71.106, 71.108, 71.123,
  71.150, 71.154, 71.171, 71.195, 71.243,
  71.257, 81.37, 81.64
Adiwidjaja, G. 11.14, 32.35, 42.4, 64.54,
  69.32
Adler, A.D. 49.3
Admiraal, G. 48.82
Adrian, H.W.W. 81.5
Aeppli, L. 32.62
Aerts, J. 41.45, 41.62
Afanasova, O.B. 75.24
Agarwal, S.K. 59.36
Agnus, Y. 83.221, 85.36
Agre, V.M. 36.35, 76.18, 76.74, 83.207,
  83.208, 83.215, 84.58, 84.64
Agrifoglio, G. 73.114, 73.115
Ahlgren, M. 76.85, 76.88, 76.89, 81.81,
  81.82, 81.87, 81.89, 84.122
Ahmed, F.R. 26.2, 28.12, 51.44, 58.22,
  58.46, 58.47, 58.65, 64.9, 64.10, 72.48
Ahmed, K. 51.73
Ahmed, N.A. 24.3
Aiba, K. 50.20
Aime, S. 71.35, 71.64
Aimoto, S. 48.66, 48.67, 48.68, 48.69, 48.70
Ainscough, E.W. 85.93
Airoldi, C. 84.15
Ajioka, M. 81.59
Aken, D.van 46.4, 64.16
Akhmedov, E.L. 81.11
Akhmedova, R.Sh. 31.17, 31.18, 31.22,
  31.32
Akhmedova, S. 16.2, 41.29
Akimoto, T. 9.3
Akimov, V.M. 85.11
Akiyama, T. 56.3, 58.71

Akrigg, D. 17.17
Al−Salem, N.A. 86.168
Aladzheva, I.M. 64.134, 64.145
Albano, P. 74.15
Albano, V.G. 3.19, 64.126, 71.147, 72.3
Albert, A. 41.40
Albert, W. 73.95
Alberts, A.H. 64.170
Albertsson, J. 80.28, 80.29
Albinati, A. 59.3, 75.12, 83.65, 84.21
Albrecht, H.A. 41.5
Albright, J.D. 35.14
Albright, M.J. 63.11, 63.50
Alcock, N.W. 11.8, 19.1, 36.26, 36.27, 36.29,
  38.3, 40.16, 65.6, 65.10, 71.3, 73.129,
  78.10, 78.11, 81.36, 83.64, 85.21, 86.39,
  86.52, 86.81, 86.119
Alderweireldt, F.C. 33.58
Aldoshin, S.M. 85.82, 85.83
Aldoshina, M.Z. 39.50, 60.8
Aleksandrov, G.G. 1.38, 39.36, 39.37,
  39.68, 39.79, 39.80, 64.68, 71.23, 71.91,
  71.131, 73.31, 83.26, 84.110
Alemany, A. 45.31, 45.47
Alexander, J. 75.41
Alexeev, N.V. 63.6, 63.19
Alexeev, V.I. 63.21
Ali, M.A. 85.59
Ali, S. 53.4
Alkire, R.W. 2.12
Allain, A. 82.8
Allcock, H.R. 64.3, 64.29
Alleaume, M. 50.26
Allemand, J. 63.24
Allen, D.W. 64.108, 64.137, 86.110
Allen, R.M. 58.36
Allen Junior, H.C. 77.1
Allibon, J. 65.13
Allison, N.T. 71.62, 71.122
Allmann, R. 35.71, 35.75, 35.78, 35.79,
  35.80, 35.81, 35.82, 69.3
Almeida, V.N.de 13.5
Alper, H. 72.48
Altenbach, H.−J. 34.1, 34.4, 63.13
Altermatt, D. 3.2
Altheff, G. 73.69
Altmeppen, D. 85.31
Altnau, G. 73.43
Alvarado, S. 53.3
Alvarez, A.G. 32.21
Alvarez, R.L. 64.160
Aly, A.A.M. 86.179, 86.222
Alyea, E.C. 71.7, 72.6, 81.93, 81.94, 81.96,
  86.93, 86.129, 86.158
Amanov, A.Z. 81.51
Amaro, A.A. 5.5
Amaudrut, J. 71.119
Ambrosius, H.P.M.M. 71.280, 86.162
Ambrosius, K. 64.122
Ambuehl, J. 86.21, 86.22
Amero, R.B. 70.5, 70.9, 70.13
Amiell, J. 60.21
Amigo, J.M. 33.25
Amiraslanov, I.R. 69.14, 69.26, 81.27,
  81.54, 81.55, 81.58, 81.98, 83.94
Amirzadeh−Asl, D. 64.42
Amma, E.L. 49.28, 85.35
Ammon, H.L. 33.45, 33.72, 33.79, 37.15,
  38.133, 41.82
Amoureux, J.P. 31.14

Amstutz, R. 67.7
Ananchenko, S.N. 51.13, 51.15
Andersen, R.A. 73.102, 84.85
Anderson, B.B. 71.133, 71.134
Anderson, B.F. 31.19
Anderson, J.B. 17.11, 21.9, 59.46
Anderson, L. 45.9
Anderson, O.P. 49.19, 77.7, 83.66, 83.70,
  83.71, 83.105
Anderson, T.J. 63.11, 63.50
Andersson, J.−E. 39.33
Ando, D.J. 11.19
Andreetti, G. 71.107
Andreetti, G.D. 35.55, 35.59, 35.61, 38.22,
  41.15, 41.16, 41.24, 41.27, 41.34, 51.45,
  52.2, 52.3, 61.10, 61.11, 61.21, 69.28,
  72.64
Andrews, G.C. 45.17
Andrews, L.C. 2.9
Andrews, M.A. 71.26, 72.11, 72.19
Andrianov, K.A. 63.14, 63.47
Andrianov, V.G. 8.14, 31.17, 32.43, 35.34,
  64.22, 71.22, 71.51, 71.104, 71.188, 73.84
Andrianov, V.I. 38.10, 64.17, 64.32, 64.52
Andrieu, J.M. 32.21
Andronati, S.A. 35.34, 35.43, 35.44, 35.69
Angelici, R.J. 31.10, 71.234, 82.30, 82.38
Angenault, J. 69.22
Angenot, L. 58.79
Angst, C. 49.12, 49.13
Ania, F. 69.12
Anker, W. 39.63
Annesley, T.M. 64.12
Annibale, G. 76.5, 76.7
Annis, G.D. 71.27, 71.28
Ansell, G.B. 71.84, 71.117, 71.223
Anteunis, M. 31.20
Anthonj, R. 60.17
Antipin, M.I. 83.192
Antipin, M.Yu. 16.13, 19.7, 19.11, 39.18,
  39.20, 39.21, 51.81, 62.7, 63.8, 63.10,
  64.49, 64.88, 64.89, 64.115, 64.134,
  64.145, 71.4, 73.19
Antolini, L. 33.19, 33.46
Antonov, P.G. 3.26
Antsans, Yu.E. 1.29
Antsyshkina, A.S. 73.25, 81.18, 81.31,
  81.51, 84.55
Anzai, H. 83.59
Aoe, K. 33.68
Aoki, K. 73.121, 81.73, 81.78, 82.25, 82.39
Aoyagi, T. 20.1
Apinitis, S.K. 35.11
Appel, R. 64.107
Apreda, M.C. 35.58, 36.3, 36.21, 36.22
Arabi, M.S. 86.20
Arad−Yellin, R. 61.14, 61.15, 61.16, 61.17
Arai, H. 9.3
Arai, T. 50.20
Arakelova, E.R. 1.38
Aranda, R.A. 76.79, 81.13
Aravamudhan, S. 38.18
Arbuzov, B.A. 64.116, 64.121, 65.25
Arcamone, F. 50.18
Archer, R.D. 78.17, 86.100, 86.133, 86.173
Arend, H. 3.2, 3.27, 3.28, 3.46, 3.47
Arene, F. 45.24, 45.29
Aresta, M. 74.15

AUTHOR INDEX

Argay, G. 32.52, 35.30, 35.31, 35.68, 36.14.
40.28. 40.37, 41.36, 41.41, 41.46, 41.47.
41.66
Ariaans, G.J.A. 86.162
Arigoni, D. 54.29
Aripov, T.F. 58.11, 58.12, 58.13, 58.14,
58.15
Arison, B.H. 27.13
Arkhangel'skii, I.V. 81.31, 81.51
Armand, J. 10.1
Arnold, E. 53.34
Arnold, E.V. 1.34, 38.85, 48.46, 55.2, 59.40
Arnold, W. 38.5, 45.19
Arnoux, B. 48.22, 48.28, 48.29, 48.36,
50.35
Arora, S.K. 50.10, 50.12
Arrese, F. 17.34
Arrieta, J.M. 33.25
Arte, E. 10.3, 20.7, 41.12
Aruffo, A. 76.54, 76.69, 83.195
Arutyunyan, L.D. 83.79, 83.80, 84.19,
84.27, 84.28, 84.43, 84.44
Arzeno, H.B. 75.28
Arzoumanian, H. 64.160
Ashaks, Ya.V. 35.10
Ashby, C.I.H. 82.28
Ashcroft, P.L. 40.1
Ashida, T. 48.24, 48.52, 48.56, 48.60, 48.64
Ashworth, T.V. 71.142, 71.238
Aslanov, L.A. 38.10, 46.6, 64.17, 64.32,
64.33, 64.52, 64.53
Astapov, B.A. 63.47
Astier, J.P. 36.10
Astruc, D. 73.69
Atkinson, A. 40.18, 40.19
Atkinson, A.M. 71.27, 71.28
Atovmyan, L.O. 1.16, 17.25, 17.27, 17.31,
17.36, 32.19, 33.6, 40.52, 63.43, 64.31,
64.51, 64.66, 64.72, 64.85, 71.207, 83.79,
83.80, 84.19, 84.27, 84.28, 84.35, 84.41,
84.43, 84.44, 85.82, 85.83
Atsumi, K. 53.21
Attig, T.G. 71.164, 71.165, 71.266
Atwood, J.D. 86.76, 86.170
Atwood, J.L. 2.18, 37.8, 41.1, 67.9, 68.1,
68.2, 68.3, 68.4, 69.40, 69.49, 69.50,
71.89, 71.127, 71.155, 71.163, 72.45, 73.5,
73.8, 73.26, 73.29, 73.35, 73.36, 73.37,
73.74, 73.75, 73.79, 73.92, 73.94, 73.101,
73.106, 75.48, 83.173, 86.30, 86.113,
86.170
Au, G.von 71.172
Aubry, A. 1.27, 19.4, 45.32, 48.35, 48.48
Auburn, M. 86.147
Audisio, G. 32.64
Aumann, R. 71.39, 71.40
Aurivillius, K. 86.107, 86.108
Avduevskaya, K.A. 79.1, 79.2
Avenel, D. 45.29
Aycard, J.-P. 27.6
Ayres, B.E. 51.27, 51.34, 51.35
Azzolina, R. 59.17

Baacke, M. 86.46
Baba, I.B. 85.52
Babin, D. 53.35
Baccon, M.le 41.65, 84.120

Bachechi, F. 65.26, 78.5. 78.14, 86.21.
86.22, 86.169
Bachet, B. 41.53
Bacon, B.E. 45.17
Bacon, G.E. 33.31
Baenziger, N.C. 7.4, 64.129, 64.130, 64.131.
71.71
Baert, F. 33.28, 33.29, 44.32, 45.49, 59.16
Bagirov, Sh.A. 81.7
Baidina, I.A. 63.21, 76.15, 76.16, 82.1, 82.2,
82.5, 82.6, 82.9, 82.10, 82.13
Baikalova, L.V. 84.37
Baikie, A.R.E. 81.84
Bailey, A.S. 36.32, 37.14, 37.16, 42.12
Bailey, N.A. 33.23, 73.81, 77.6, 77.8
Bailey, P.M. 73.61, 73.70, 73.116, 73.127
Baine, P. 1.17
Bajaj, A.G. 53.32
Bakakin, V.V. 76.15, 76.16, 79.14, 82.1,
82.2, 82.6, 82.13
Baker, J.R. 64.45, 64.46, 84.108
Baker, P.K. 73.133
Baker, R.J. 41.42
Baker, S.R. 38.115
Bakhmutov, V.I. 19.11
Balagot, L.L. 84.10
Balaram, P. 48.62, 48.65
Balasubramaniam, M. 17.42
Balavoine, G. 73.123
Balch, A.L. 71.275, 86.176, 86.178, 86.194
Balimann, G. 65.26
Balitskii, Yu.V. 64.49
Balitz, D.M. 59.6
Ball, R.G. 49.16
Baloniak, S. 33.38, 33.65
Ban, T. 37.21
Bandaranayake, W.M. 59.29
Bandoli, G. 1.8, 3.54, 33.39, 71.274, 82.4,
83.114, 83.184, 83.185, 84.75, 86.196
Banek, M. 64.122
Banerji, J. 36.45
Banfield, J.E. 59.29
Banister, A.J. 41.19
Bannier, G. 33.55
Bansal, R.K. 54.9
Bar, I. 16.6, 16.7, 16.8, 16.9, 16.10, 16.11,
64.138
Baran, J. 2.6, 2.11
Baranovskii, I.B. 81.42
Barborak, J.C. 75.3
Bardi, R. 18.2, 39.12, 71.55, 71.80, 71.248,
71.249, 80.5, 80.8, 80.9, 80.10
Barelli, N. 13.5
Bargar, T. 75.6
Bargouth, M.O. 2.2
Barikina, L.R. 39.28, 39.36, 39.37
Barker, G.K. 71.81, 71.82, 86.32, 86.33,
86.43
Barkhanova, N.N. 76.18
Barkigia, K.M. 49.3
Barnes, J.C. 38.4, 38.26. 38.27, 45.40, 45.41
Barnes, J.F. 40.1
Barnett, B.L. 2.27, 48.16, 48.27, 81.21, 84.1
Barnett, K.W. 73.9
Barnier, J.-P. 29.1, 30.1, 30.3, 30.5
Barr, P.J. 47.26
Barrans, Y. 50.26, 51.1, 51.21
Barreda, A.de A.G.de 13.21
Barrera, H. 85.46
Barrick, J. 58.52

Barrick, J.C. 71.73, 83.41, 83.42, 85.24
Barriola, A. 73.3
Barrow, M.J. 40.1, 69.7
Bars, O. 86.111
Bart, J.C.J. 5.8, 5.9, 17.26, 20.3, 61.8
Bartczak, T.J. 29.3
Bartlett, P.D. 31.64. 38.57, 38.58, 38.59
Barton, D.H.R. 75.28
Bass, L. 38.85
Bassett, J.-M. 71.33, 71.225, 71.259
Bassi, I.W. 5.8, 5.9, 17.26, 20.3, 61.8
Bassinet, P. 10.1
Basson, S.S. 3.31, 33.18, 75.20
Bastani, B. 63.16, 63.28
Batail, P. 73.69
Batcho, A.D. 15.4, 32.63
Bates, P. 5.12
Bates, R.B. 39.5
Batir, D.G. 71.264, 83.81
Bats, J.W. 2.1, 19.2, 39.3, 48.43, 48.45,
48.58, 48.73, 69.12
Batsanov, A.S. 83.90
Batschelet, W.H. 86.100, 86.133, 86.173
Batta, A.K. 51.68
Battaglia, L.P. 33.47, 79.8, 82.41, 84.89
Battaglia, R. 83.101
Battaile, J. 54.5, 54.12
Battersby, A.R. 49.21, 49.31
Battiston, G.A. 82.4
Bau, R. 27.29, 73.85, 83.62, 83.63, 83.133,
84.20, 84.24, 86.24, 86.49
Bauer, W.R. 85.86
Baughman, R.G. 8.10, 8.11
Baukova, T.V. 71.221
Bavoux, C. 17.2, 17.3, 60.2
Baxter, C.E. 83.172
Baxter, S.G. 21.19, 63.40, 63.41, 63.51
Bayon, J.C. 85.41, 85.44
Beale, J.P. 64.149
Bear, J.L. 82.26
Beauchamp, A.L. 44.14, 71.5, 83.35, 83.74,
86.62
Becher, H.J. 66.18
Bechgaard, K. 60.10
Bechtel, D.W. 35.12, 68.23
Bechtel, F. 48.55
Bechtold, R.A. 86.75
Beck, A.K. 63.16, 63.28
Beck, W. 71.1, 86.163
Becker, G. 64.62, 64.106
Becker, H.-D. 28.19, 31.23, 31.27, 31.28,
34.12, 35.76, 38.91, 38.92, 38.93, 38.132
Becker, K.B. 75.46
Becker, W. 49.29, 49.30, 59.45
Beckhaus, H. 20.9
Beckhaus, H.-D. 19.13, 19.14
Beckingsale, P.G. 38.97
Beckman, D.E. 73.46
Bee, M. 31.14
Beecham, A.F. 59.20
Begley, M.J. 3.4, 35.32, 38.115, 40.23, 66.19,
69.18, 69.37
Behr, D. 54.22
Behr, H. 32.35
Behrens, C.B. 71.133, 71.134
Behrens, H. 71.6, 73.71
Behrens, U. 29.1, 30.1, 30.3, 30.5, 45.12,
73.97
Beindorf, G. 65.14
Bel'skii, V.K. 36.9, 63.19, 64.104, 65.4

A 2

Belanger-Gariepy, F. 83.74
Belew, J.S. 38.110
Belichuk, N.I. 84.31
Belikov, V.M. 73.31
Belin, C. 33.3
Bell, E.A. 48.17, 48.18
Bell, N.A. 62.2, 71.152, 76.66, 83.28, 86.6, 86.7
Bell, R.A. 71.157, 71.158, 71.262
Bellaart, A.C. 46.4, 64.16
Bellan, J. 1.33
Bellard, S. 86.17
Bellesia, F. 52.2, 52.3
Belluco, U. 71.213
Belmonte, P. 71.209
Belokon, Yu.N. 73.31
Belov, N.V. 69.51, 79.4, 85.11
Belsky, V.K. 35.40, 66.14
Belyaev, A.V. 76.15, 76.16
Benard, M. 81.32, 81.33
Benassi, R. 35.55, 35.61
Benci, P. 44.15
Bendall, M.R. 84.86
Benechie, M. 21.6
Benedetti, E. 1.17, 48.33, 48.34, 48.41, 48.74
Benedix, R. 83.181
Benes, J. 3.27, 3.28
Benetollo, F. 76.68, 76.92, 76.95, 78.6, 81.19, 81.20, 84.39
Benkovic, P.A. 40.15
Benkovic, S.J. 40.15
Benn, M.H. 58.67
Benn, R. 72.8, 72.17, 75.9
Benner, L.S. 86.176, 86.178
Bennett, M.A. 74.8
Beno, M.A. 38.8, 38.16, 53.23, 71.44
Benoit, A. 71.99, 71.153, 73.49
Bensmann, W. 86.214
Bentley, P.H. 50.16
Benton, W.H. 40.53, 40.55
Bereman, R.D. 76.25, 80.14, 85.60
Berestevich, B.K. 36.35
Berg, J.-E. 38.102, 54.22
Berg, J.M. 71.232, 71.233, 71.245, 80.7, 85.37, 85.57, 85.62, 85.78, 85.80, 85.95, 85.96, 85.97
Berg, K.P. 38.5
Berg, O. 11.6, 11.9, 11.15
Berg, R.W. 3.16, 3.21, 3.22
Bergamaschi, E. 51.45
Berger, S. 19.3
Bergerhoff, R.G. 65.22
Bergman, J. 38.134, 70.6
Bergman, R.G. 71.255
Berimzhanov, B.A. 1.5
Berlin, K.D. 36.4, 39.56, 39.74, 64.45, 64.46
Berman, H.M. 47.38, 48.40, 48.44, 60.24
Bernabe, M. 45.31, 45.47
Bernal, I. 27.5, 27.11, 29.2, 49.14, 53.27, 71.140, 71.167, 71.204, 73.13, 73.26, 73.114, 73.115, 73.118, 73.119, 73.125, 74.14, 75.36, 76.40, 82.26
Bernard, M.A. 81.1, 81.41, 81.46, 81.56
Bernardinelli, G. 64.57
Bernassau, J.-M. 51.63, 51.64, 51.67
Bernath, G. 35.30, 35.31, 35.68, 40.28, 40.37
Bernauer, K. 32.62
Berndt, A.F. 19.8, 73.9

Berndt, B. 33.78, 63.27, 63.31, 63.52
Bernier, F. 45.27
Bernier, J.L. 44.32
Bernstein, J. 16.6, 16.7, 16.8, 16.9, 16.10, 16.11, 35.17, 35.18, 35.46, 35.47, 45.54, 60.1, 60.34, 64.138
Bernstein, J.L. 28.1
Berry, K.J. 83.225
Berry, M. 71.75, 71.76, 71.111, 71.238, 84.97
Berti, C. 35.59
Bertinsson, G.-I. 86.107, 86.108
Bertolasi, V. 33.67, 33.70, 33.74, 33.75, 35.39, 41.75, 41.76, 72.52
Bertrand, J.A. 84.6, 84.13, 84.14, 84.18
Besenhard, J.O. 71.125
Bestmann, H.J. 20.14
Beswetherick, S. 73.23
Bettinetti, G.P. 44.30
Bettolo, G.B.M. 59.34
Beurich, H. 86.34, 86.35
Beurskens, G. 41.56
Beurskens, P.T. 41.56, 71.281
Beveridge, K.A. 69.29
Beyer, L. 79.15
Bezler, H. 83.211
Bezmenova, T.E. 39.25, 39.26
Bhadbhade, M.M. 38.120
Bhagwat, V.W. 38.65
Bhandary, K.K. 60.36, 60.37
Bharadwaj, M.M. 11.5
Bharara, P.C. 62.11
Bhat, T.N. 48.8
Bhattacharyya, A.A. 86.119
Bhattacharyya, R.G. 3.24
Bhattacharyya, S.C. 48.9
Bhatti, M.M. 76.41
Bianchini, C. 72.66, 72.69, 75.47, 86.150, 86.151, 86.225, 86.228
Biavati, A. 72.64
Bick, I.R.C. 58.48
Bideau, J.-P. 38.29, 38.30, 38.37, 38.49, 48.55
Bierner, M.W. 53.24, 53.26
Biersack, H. 71.30, 73.90
Bill, E. 83.179, 83.203
Billig, E. 86.221
Biname, R. 20.7
Binamira-Soriaga, E. 83.20, 85.34
Binger, P. 71.67
Binnewies, M. 64.35, 65.5
Bino, A. 3.34, 3.36, 3.38, 3.40, 33.10, 33.12, 81.10, 81.29, 81.30, 81.35, 81.47, 81.49, 82.33, 82.40, 83.214, 84.114, 84.115, 86.63
Bird, P. 84.109
Birker, P.J.M.W.L. 81.95, 83.209
Birkhahn, M. 64.62, 69.15, 69.23
Birnbaum, G.I. 10.5, 47.5, 47.7, 47.19, 72.48
Birtle, S.L. 60.19
Bisacchi, G.S. 38.72
Bishop, M.M. 83.165
Bishop, R. 61.1
Bistram, S.A. 73.41
Bjelica, L. 83.13, 83.14, 83.15
Bkouche-Waksman, I. 73.14
Black, D.St.C. 35.36, 59.29
Blackmer, G.L. 84.66
Blackwell, J. 16.14

Blaha, K. 37.4
Blaicher, W. 64.144
Blakeney, A.J. 38.57, 38.58, 38.59
Blanck, K. 86.180
Blank, N.E. 74.2, 74.3
Blann, W.G. 25.1
Blaschke, G. 64.79
Blasio, B.di 1.17, 48.33, 48.34, 48.41, 48.74, 72.41, 72.44
Blaton, N.M. 8.13, 13.8, 13.11, 32.65, 33.84, 33.86, 33.92, 35.41, 58.33, 58.41, 58.42, 83.87
Blecher, A. 70.2
Bleidelis, Ya.Ya. 1.29, 25.4, 35.11, 38.31, 38.33, 44.23, 44.24, 45.21, 45.23, 85.63
Blessing, R.H. 35.23, 38.24, 39.30, 39.31, 39.32, 39.38, 51.2, 51.4, 51.6, 51.7
Blinc, R. 3.46, 3.47
Blinn, E.L. 76.107
Blinova, V.A. 71.22
Blom, N.S. 21.22
Bloor, D. 11.19
Blosl, S. 66.11
Blount, J.F. 2.25, 2.26, 15.4, 16.15, 32.23, 32.29, 32.63, 37.17, 37.18, 38.94, 38.95, 38.101, 40.34, 41.5, 48.77, 48.78, 48.79, 50.8, 50.23, 51.47, 51.68, 53.2, 53.8, 53.24, 53.25, 53.26, 53.29, 53.31, 54.3, 54.7, 54.15, 54.16, 54.19, 54.24, 63.40, 63.41, 63.51, 64.83, 64.84, 69.33, 69.34, 75.44
Blount, J.R. 35.72
Bluhm, T.L. 45.25
Blum, D.M. 55.2
Blunt, E.H. 73.81
Blunt, J.W. 5.12
Bobek, M. 45.15
Bocelli, G. 33.27, 35.16, 35.20, 35.33, 35.37, 35.55, 35.59, 35.61, 35.63, 38.22, 38.32, 41.15, 41.16, 41.27, 41.34, 51.45, 52.2, 52.3, 69.28, 72.64
Bochmann, M. 71.193, 84.130
Boczon, W. 58.23
Bodalski, R. 64.105
Bodot, H. 27.6
Bodrikov, I.V. 31.56
Boeckman Junior, R.K. 29.7, 55.2
Boer, E.de 84.111
Boese, R. 75.10, 75.38
Bogatskii, A.V. 34.6, 34.8, 35.34, 35.43, 35.44, 35.69, 38.96, 40.13, 40.42, 40.44, 40.49
Bogdanovic, B. 71.235
Bogge, H. 64.123, 86.199, 86.201, 86.202, 86.203, 86.219
Boggs, J.E. 39.10
Bogomolova, G.S. 39.68
Bogucka-Ledochowska, M. 50.11
Bohle, C. 73.109
Bohme, R. 38.141, 73.71
Bois, C. 10.1, 73.14
Bojes, J. 3.49, 64.167
Bok, L.D.C. 3.31, 75.20
Boldeskul, I.E. 64.49
Boles, M.O. 41.63, 50.16
Bolhuis, F.van 5.4
Boll, P.M. 38.15
Bologa, O.A. 83.36, 83.37, 83.38, 86.91
Bolotin, B.M. 40.52
Bolsman, T.A.B.M. 71.192

Bolz, P.-R. 69.42
Bombieri, G. 27.15, 71.206, 71.254, 76.68, 76.92, 76.95, 78.6, 81.19, 81.20, 83.131, 84.39, 84.129, 84.138, 86.183
Bondar, V.I. 85.11
Bondarenko, N.A. 64.72
Bone, S.P. 66.13
Boniak, L. 81.61, 84.113
Boniface, S.M. 71.260, 71.270, 71.271, 86.209
Bonnet, J.-J. 64.136, 71.181, 71.217, 86.92
Bonnet, J.J. 71.17, 71.18, 83.204, 86.41, 86.42, 86.82
Bonnet, M.C. 49.26
Bonnett, R. 49.6, 62.20
Bonny, A. 73.87
Bonomo, R.P. 81.90, 82.27
Bonora, G.M. 48.33, 48.34, 48.41
Book, L. 80.21
Boom, J.H.van 85.58
Boorman, P.M. 85.38, 85.43
Bor, G. 71.118
Bordner, J. 17.42, 18.3, 36.49, 45.17, 58.8
Borea, P.A. 33.70, 33.74
Borel, M.M. 81.1, 81.41, 81.46, 81.56, 81.61, 84.113
Borisenko, A.A. 46.6
Borishok, V.P. 63.9
Borisov, A.P. 86.212
Borisov, S.V. 53.11, 54.1, 63.21, 82.1, 82.2, 82.5, 82.9, 82.10, 82.13
Borowiak, T. 43.2, 58.9, 58.10
Borowitz, I.J. 29.12
Borowski, E. 50.11
Bos, H.J.T. 39.77, 39.81, 39.82, 39.83
Boschelli, D. 59.19
Boschi, T. 71.208, 72.15, 72.16, 72.20, 72.25, 75.5
Bosman, W.P. 71.280, 80.6
Botoshanskii, M.M. 34.6, 34.8, 38.96, 40.42, 40.44, 83.81, 83.192, 84.38
Bottino, F. 39.87
Bottrill, M. 71.38, 73.56
Bouachir, F. 86.2
Boucherle, A. 33.61
Bouffard, F.A. 40.45
Bour, J.J. 71.281
Bourgault, M. 8.2
Bourne, P.E. 31.24, 31.25, 31.26, 44.6, 44.7, 79.6
Bourshteyn, I.F. 86.117
Boussard, G. 48.48
Bovill, M.J. 61.6
Bowden, B.F. 54.8
Bowen, D.V. 38.90
Bowser, J.R. 73.87
Boyd, I.W. 85.4, 85.9
Boyd, V.L. 64.60
Boyer, B. 10.4, 21.10
Boyer, M. 72.42
Bozopoulos, A.P. 40.57
Bradaczek, H. 50.11
Braden, B. 50.32
Bradley, D.C. 83.85
Bradley, J.S. 71.84, 71.117, 71.223
Brady, F. 68.18
Brady, K. 13.16
Braekman, J.C. 54.20, 54.28, 54.33
Braga, D. 63.53, 64.126, 71.147, 72.3
Brant, P. 71.72

Brassy, C. 41.53
Bratek-Wiewiorowska, M.D. 58.20
Brattsev, V.A. 62.7
Brauer, D.J. 62.1, 68.21, 72.51, 83.155, 83.156
Brauman, J.I. 49.34, 49.35
Braund, N.C. 73.116
Brauninger, M. 66.5, 66.9
Braunstein, P. 86.111
Braverman, S. 39.29
Bravic, G. 3.3, 3.6, 3.7, 3.8, 3.9, 3.10, 38.29, 38.30, 38.37, 38.49
Bray, D.D. 32.1
Bregadze, V.I. 64.75, 68.8, 68.19
Bregeault, J.-M. 64.133
Breining, T. 35.16, 35.20, 35.33, 35.37
Breitenstein, W. 59.40
Bremard, C. 84.34
Bremner, J.B. 40.47
Bren, V.A. 62.18, 68.8, 68.19, 85.82, 85.83
Bren, Zh.V. 68.8, 68.19, 85.82, 85.83
Brencic, J.V. 33.16, 83.47, 83.115, 83.116, 83.175, 83.177
Brennan, T.F. 17.10, 35.54, 83.182
Brenner, L.S. 86.194
Bresadola, S. 71.268
Bresciani-Pahor, N. 1.31, 32.34, 39.62, 39.70, 39.87, 71.100, 71.268, 76.102, 84.65, 86.16, 86.73
Breuninger, M. 34.2, 34.5, 34.7
Brewer, A.D. 39.5
Brewer, D. 38.20
Brianso, J.L. 33.50, 85.41, 85.44, 85.46
Brianso, M.C. 2.28, 2.32, 85.41
Briant, C.E. 30.6, 30.7
Briggs, J.R. 72.12, 72.18
Bright, W.M. 31.9
Brimacombe, J.S. 45.22, 45.38, 45.40, 45.41, 45.43
Brindle, I.D. 40.24
Brinegar, A.C. 83.222
Brinker, W.den 83.21, 83.22, 83.23, 83.218
Brint, P. 76.36
Brisdon, B.J. 72.13
Brisse, F. 13.15, 45.50
Britten, J. 3.30
Britton, D. 7.2, 40.2
Broach, R.W. 64.127, 64.128, 64.164, 65.16, 65.20, 65.21, 73.126
Broadley, K. 73.133
Brock, C.P. 19.6, 71.164, 71.165, 71.266
Brodersen, K. 83.48
Brodie, A.M. 85.93
Brodie, A.M.H. 51.16
Brodie, H.J. 51.16
Brook, A.G. 63.48
Brookhart, M. 60.20
Brossi, A. 37.12, 58.57
Brotherton, P.D. 71.48
Broude, V.L. 17.25, 17.27
Brouty, C. 76.37, 76.38, 76.39
Brouwer, A.C. 39.77, 39.81, 39.82, 39.83
Brouwers, A.M.F. 71.272
Brown, A. 27.1, 27.2, 38.9
Brown, D. 84.129, 84.138
Brown, D.H. 61.9
Brown, D.S. 71.34, 71.121
Brown, E. 59.31
Brown, G.M. 73.126
Brown, I.D. 68.5

Brown, J.N. 15.5, 33.53, 84.57
Brown, K.L. 85.93
Brown, M.P. 86.193, 86.223
Brown, R.K. 64.153, 75.17, 75.18, 75.19, 86.58
Brown, S.B. 39.43, 64.93, 64.140, 64.147, 64.148
Brown, T.L. 82.28
Browne, A.R. 37.17
Browne, E.J. 40.47
Browning, J. 86.109
Brownstein, S.K. 3.20
Bruce, A.E. 85.37
Bruce, M.I. 86.3
Bruckner, S. 41.13, 59.3
Brudgam, I. 83.159
Brufani, M. 50.25, 50.30, 50.31
Bruice, T.C. 36.37
Brun, G. 39.27, 39.64
Bruncks, N. 73.43
Brunie, S. 61.14, 61.17, 73.123
Bruniquel, M.F. 85.6
Brunner, H. 71.204, 73.114, 73.115, 75.36
Bruno, G. 13.3, 27.15, 71.55, 71.80, 71.206, 71.248, 71.249
Brustolon, M. 1.8
Bruvo, M. 11.17
Bruylants, A. 21.20, 21.21
Bryan, D.B. 40.36
Bryan, R.F. 41.28
Bubnov, N.N. 19.11
Buchanan, G.W. 10.5
Buchanan, R.M. 71.160, 84.123, 84.139, 86.85
Buck, H.M. 46.4, 64.16
Buckingham, D.A. 83.225
Buckley, T.F. 40.36
Budarina, Z.N. 22.5, 38.74
Bugg, C.E. 2.31, 36.33, 44.16, 44.19, 48.61, 48.63, 48.83
Buhl, H. 39.53
Bui, X.-H. 21.11, 21.12, 21.13
Bukowska-Strzyzewska, M. 33.36, 36.15, 39.13, 39.14, 39.47
Bulc, N. 2.10
Buldt, E. 38.118, 38.119
Bulgak, I.I. 71.264, 83.81
Bulgarovskaya, I.V. 38.88
Bullivant, D.P. 66.1, 66.3, 66.6
Bunker, M.J. 72.21
Bunzli, J.-C.G. 84.40
Buranbaev, M.Zh. 1.5
Burch Junior, R.R. 86.113
Burford, N. 64.50
Burger, H. 62.1, 83.155, 83.156
Burger, N. 81.2, 81.3
Burgers, P.M.J. 47.32
Burke, P.J. 83.171
Burlitch, J.M. 84.54
Burns, J.H. 84.74
Burns, K. 61.25
Burreson, B.J. 53.18, 54.23
Burrows, E.P. 44.27, 44.28, 44.29
Burschka, C. 73.11
Burton, G.W. 27.3
Busch, D.H. 83.188, 83.196
Buscher, K. 64.125
Buschmann, J.F. 39.59
Busetta, B. 36.8, 51.14, 51.21
Busetti, V. 33.93, 39.12

Busetto, L. 75.1
Bushnell, G.W. 39.63, 69.29, 71.130, 86.109
Buskirk, G.van 64.161, 71.26, 72.11
Busman, S.C. 30.18
Busnot, F. 81.41, 81.46, 81.56, 81.61, 84.113
Butcher, R.J. 76.104
Butler, D.N. 31.67
Butler, W.M. 31.8, 69.58, 73.52, 86.121, 86.141
Butman, L.A. 84.37, 84.56
Buttafava, A. 84.111
Butts, S.B. 71.3
Bychkov, V.T. 73.134, 73.135
Byers, L.R. 73.30, 73.34, 73.47
Byushkin, V.N. 59.10, 76.1, 78.7, 79.12, 84.31

Cabral, J.de O. 34.11, 83.187
Cabral, M.F. 34.11, 83.187
Caccamese, S. 59.17
Cachia, P. 31.36
Caddy, P. 75.30
Cady, M.A. 21.17
Cagle Junior, F.W. 83.190
Caira, M.R. 17.24, 84.118
Cais, M. 74.10, 74.13
Calabrese, J. 72.68
Calabrese, J.C. 73.78, 73.103
Calcaterra, M. 5.8, 5.9, 17.26, 61.8
Calderazzo, F. 49.5, 76.103, 81.99
Calleja, F.J.B. 69.12
Calligaris, M. 1.31, 39.62, 39.70, 39.87, 71.100, 73.122, 76.102, 86.16, 86.73
Callot, H.J. 49.15
Calogero, S. 11.11, 33.1, 33.39, 69.39, 69.44, 70.12, 79.5, 79.16, 84.68
Calvert, L.D. 3.20
Calvino, R. 40.14
Calzada, J. 53.3
Camalli, M. 72.7, 83.104
Camarda, L. 59.3
Cambie, R.C. 59.20
Camellini, M.T. 71.105, 71.151, 71.178, 71.187, 71.228, 72.29, 83.31, 86.174
Camerman, A. 48.57, 58.76
Camerman, N. 48.57, 58.6, 58.76
Cameron, T.S. 30.13, 31.6, 31.16, 31.65, 61.3, 64.70, 64.74, 70.5, 70.9, 70.13, 84.45, 84.46, 86.80
Campaigne, E. 41.21
Campana, C.F. 68.22, 73.24, 81.28, 83.7, 86.83, 86.84, 86.143
Campbell, M.L. 42.7, 42.8, 42.9, 42.10, 42.11
Campsteyn, H. 32.71, 51.49
Camus, A. 86.160
Canada, L.G. 37.8
Canfield, D. 58.52
Canning, W.M. 48.1, 48.2
Cannon, J.F. 31.9
Cannon, J.R. 25.2
Cano, F.H. 35.2, 35.58, 35.60, 36.3, 36.21, 36.22, 41.84, 45.56, 71.191
Cantrell, T.S. 13.22
Canty, A.J. 71.37, 71.53, 71.180, 83.30
Cape, T. 60.12
Cape, T.W. 60.42

Capilla, A.V. 76.79, 81.13
Caplar, V. 40.43
Caradonna, J.P. 44.11, 44.13, 44.22
Cardellina, J.H. 1.34, 58.43
Cardellini, L. 35.63
Cardin, C.J. 83.157
Cariati, F. 83.65
Caric, S. 83.75
Carmona-Guzman, E. 86.30
Caron, C. 49.23, 49.27
Carpy, A. 10.6, 32.33, 32.39, 32.40, 33.61, 39.22, 39.46, 59.5
Carre, F. 71.205
Carrell, H.L. 2.17, 36.37, 60.24
Carroll, J.A. 73.60
Carroll, W.E. 71.263
Carrondo, M.A.A.F.de C.T. 76.108, 82.7
Carte, B. 54.10
Carturan, G. 75.12
Carty, A.J. 64.169, 71.198, 71.199, 71.211, 72.49, 86.48, 86.144
Caruso, F. 72.7, 83.104, 85.15
Casalnuovo, A.L. 86.184
Casalone, G. 5.13
Casellato, U. 73.67
Cassels, B.K. 17.13, 58.62
Cassinelli, G. 50.18
Cassoux, P. 60.40, 85.6
Castellano, E.E. 32.21, 60.9
Castellano, G.de 51.70
Castensson, S. 48.50
Castro, A.L. 35.53
Castro, B.de 76.67, 76.93
Catsoulakos, P. 41.59
Caubere, P. 38.73
Caulton, K.G. 71.94, 71.96, 73.22, 73.33, 73.63, 73.107
Cavagna, F. 64.25
Cavell, R.G. 64.28
Cazaux, L. 42.3
Ceccarelli, C. 45.8
Cefola, M. 77.9
Ceh, B. 83.47
Celikel, R. 17.17
Cellai, L. 50.25
Ceriotti, A. 3.52, 64.132, 72.28
Cerrini, S. 36.44, 36.46, 40.48, 48.49, 50.25, 50.30, 50.31, 59.34, 61.12, 71.20, 71.98
Cerutti, M. 71.119
Cerveau, G. 71.205
Cesari, M. 67.8, 76.91
Cetinkaya, B. 67.9, 69.49, 69.50, 73.106, 75.48
Chadaeva, N.A. 65.25
Chadwick, D.J. 61.6
Chadwick, D.L. 49.20
Chaichit, N. 58.48, 71.37, 71.53, 71.180, 83.30
Chalii, V.P. 15.3
Chalilpoyil, P. 83.186
Chambers, R. 32.57
Chamot, E. 37.17
Chan, A.S.C. 72.59
Chan, C. 30.13, 31.6, 31.16, 31.65, 61.3, 64.74, 70.5
Chan, L.T. 76.56
Chan, L.Y.Y. 58.6, 58.76
Chan, T.-L. 61.7

Chananont, P. 35.38, 35.42, 35.45, 35.51, 47.26, 81.8
Chand, P. 2.21, 81.22
Chandler, T. 86.204
Chandler, W.D. 13.10, 13.17
Chandrasekhar, K. 35.26
Chang, C.-M. 86.124
Chang, C.H. 10.2
Chang, C.Y. 59.24
Chang, H.W. 41.80
Chang, K.H. 13.9
Chang, M.N.T. 58.45
Chang, Y.-M. 63.48
Chang, Z. 39.42
Chanh, N.B. 24.2
Chanton, J.P. 84.20, 84.24
Chao, C.H. 27.29
Chao, G. 39.42
Chapman, O.L. 30.18
Chapuis, G. 3.12, 3.46, 3.47, 71.2, 72.25, 75.5
Charbonneau, G.P. 11.7
Charpentier, R. 81.79
Charpin, P. 77.11
Charumilind, P. 42.13
Chasseau, D. 60.21, 60.35
Chatt, J. 80.19, 80.25
Chatterjee, A. 36.45
Chattopadhyaya, J.B. 47.30, 47.31
Chauvel, C. 85.45
Chauvin, M.M. 66.12
Chauvin, Y. 72.23
Cheer, C.J. 39.45
Cheetham, A.K. 19.17, 21.4, 21.5
Chel'tsov, P.A. 76.58, 76.59
Chen, C.-M. 53.17
Chen, H.W. 81.77
Chen, J.P. 51.86
Chen, J.S. 21.3
Chen, S.-M. 38.53, 38.54
Chen, Y.J. 38.111
Cheng, C.-H. 86.90
Cheng, P.-T. 58.50, 81.67
Cherepinskii-Malov, V.D. 64.116, 64.121
Chernyshev, E.A. 75.24
Cherpillod, O. 72.25, 75.5
Cherry, P.C. 41.78
Chetkina, L.A. 24.12, 36.6
Chevrier, B. 49.15
Chexal, K.K. 59.37
Chheda, G.B. 44.26, 47.25
Chiang, C.C. 76.67, 76.93, 82.22, 83.186
Chiaroni, A. 51.89
Chibber, S.S. 48.5
Chiche, L. 73.38
Chidester, C. 38.60
Chidester, C.G. 50.17, 59.43
Chieh, C. 80.1, 80.21, 86.88
Chini, P. 3.19, 3.43, 3.52, 64.126, 64.127, 64.128, 64.132, 64.158, 64.162, 64.163, 64.164, 65.16, 65.20, 65.21, 71.147, 72.3, 72.28
Chiou, D.M. 31.44, 31.45, 31.46
Chipens, G.I. 1.29
Chisholm, M.H. 83.206, 84.84, 84.105, 84.117
Chiu, K.W. 86.59
Chiu, S. 41.42
Chiusoli, G.P. 51.45, 72.64
Chivers, T. 3.49, 64.50, 64.87, 64.167

Chizhov, O.S. 1.28

Chodosh, D.F. 46.2, 71.154, 71.257, 72.4, 72.9

Chogovadze, T.V. 63.36

Chollet, A. 72.15, 72.16, 72.20

Chong, K.S. 68.16, 72.22, 83.51, 83.52, 83.53, 83.54, 83.99, 83.150, 83.166, 84.32, 84.33

Choong, W. 45.34, 85.59

Chottard, J.-C. 71.24, 71.77

Christensen, A.T. 40.26, 40.27

Christensen, B.G. 40.45

Christidis, P.C. 80.24

Christie, A.D. 69.1

Christoph, G.G. 38.8, 38.16, 53.23, 81.74, 83.188, 83.196, 86.216

Christou, G. 3.39

Chu, S.S.C. 39.60, 39.61

Chumakov, Yu.M. 84.31

Chun, H.K. 77.13

Churchill, M.R. 69.6, 71.32, 71.46, 71.74, 71.92, 71.110, 71.174, 71.209, 71.236, 73.88, 76.25, 76.98, 80.14, 81.86, 83.109, 84.29, 85.60, 86.64, 86.76, 86.116

Churkin, Yu.D. 39.18, 39.20, 39.21

Chwang, A. 47.12

Ciajolo, A. 48.33, 48.34, 48.41

Ciampolini, M. 84.83, 85.42, 86.122, 86.123

Cian, A.de 72.23

Ciani, G. 3.5, 3.19, 3.50, 64.157, 72.28, 83.139, 84.21

Ciccio, J.F. 53.3

Ciechanowicz-Rutkowska, M. 48.6, 48.7

Cielusek, G. 86.103

Ciereszko, L.S. 54.21, 59.24

Cimarusti, C.M. 41.80

Cingi, M.B. 83.31, 83.107, 84.81

Cini, R. 76.71, 84.83, 84.125, 85.40

Cinquantini, A. 76.71, 85.40

Ciriano, M. 71.277, 86.147

Ciunik, Z. 82.3, 82.12

Civis, S. 81.85

Clardy, J. 1.34, 21.6, 38.41, 38.70, 38.85, 48.17, 48.18, 48.46, 50.18, 53.1, 53.18, 53.32, 53.34, 54.5, 54.10, 54.12, 54.23, 54.29, 54.30, 54.35, 55.2, 59.6, 59.7, 59.8, 59.37, 59.39, 59.40, 59.41, 59.42, 75.6

Clark, G. 75.6

Clark, G.R. 71.252, 71.260, 71.270, 71.271, 72.55, 73.50, 73.65, 79.11, 86.134, 86.209

Clark, H.C. 71.162

Clark, J.I. 58.70

Clark, P.E. 83.225

Clark, P.W. 72.55

Clark, S.J. 69.9

Clark-Lewis, J.W. 36.41

Clarke, P.J. 58.5

Clausen, T. 58.27

Clay, R. 84.11

Clayton, J.P. 35.19, 35.22, 41.10, 50.16

Clear, J.M. 83.157

Clearfield, A. 73.118, 73.119, 73.125, 76.106, 83.152, 83.153

Clegg, W. 3.23, 33.34, 38.78, 63.39, 63.44, 63.49, 84.97

Clemente, D.A. 1.8, 3.54, 71.274, 82.4, 83.114, 83.184, 83.185, 84.75, 86.196

Clery, F. 85.6

Cliff, C.A. 85.61

Cline, S.J. 47.16

Cloke, F.G.N. 86.56

Cobbledick, R.E. 51.51, 86.4, 86.60, 86.61

Cochrane, C.M. 72.55

Codding, P.W. 58.66, 58.67, 85.38, 85.43

Cody, V. 13.7, 32.28, 44.31, 44.33

Cogne, A. 64.5

Cohen, A.I. 50.28, 50.29

Cohen, N. 40.34

Coiro, V.M. 51.50, 51.54, 51.55

Colapietro, M. 13.4

Cole, R.J. 59.38

Coll, J.C. 54.8

Collamati, I. 49.5

Colligiani, A. 1.8, 76.103

Collins, D.J. 75.35

Collman, J.P. 49.34, 49.35, 71.232, 71.233, 71.245, 86.101

Colomer, E. 71.205

Colquhoun, H.M. 86.65, 86.197, 86.210

Colvin, E.W. 63.16, 63.28

Combremont, J.J. 39.15, 39.16, 39.17

Comin, F. 35.77

Commereuc, D. 72.23

Compere Junior, E.L. 2.29, 2.30

Conde, A. 38.42, 45.16, 45.27

Confalone, D.L. 39.44, 39.72

Confalone, P.N. 39.44, 39.72

Conia, J.-M. 29.1, 30.1, 30.3, 30.5

Connelly, N.G. 73.133

Connop, A.H. 75.25, 75.26

Consiglio, G. 86.21, 86.22

Conway, P. 86.135, 86.142, 86.164

Cook, D.F. 78.10, 78.11

Cook, W.J. 48.61, 48.83

Cook Junior, J.C. 51.72

Cooke, R.G. 59.14, 59.28

Cooper, D.J. 83.149

Cooper, M. 85.50

Cooper, M.K. 71.200, 72.36

Cooper, S.J. 86.193

Coppens, P. 1.1, 81.32, 81.33

Coppernolle, A.van 38.13, 38.39

Coppola, J.C. 21.1, 64.13

Corain, B. 77.3

Corbin, D.R. 73.113

Corbin, J.L. 72.47, 85.37

Cordell, F.R. 39.10

Cordell, G.A. 53.22, 54.27

Cordes, A.W. 64.50, 64.87, 64.167

Cordes, R.E. 64.70, 70.5, 70.9, 70.13, 84.45, 84.46, 86.80

Cordt, F. 31.61

Corey, E.J. 38.109

Corey, E.R. 63.33

Corey, J.Y. 63.33

Corradi, A.B. 33.47, 79.8, 82.41, 84.89

Corral, R.A. 32.21

Corriu, R.J.P. 71.205

Corvaja, C. 1.8

Corvan, P.J. 60.20

Costa, M. 72.64

Costa, S.R. 83.157

Costa, T. 64.146

Cotrait, M. 32.2, 48.23, 48.55

Cottier, L. 45.45, 45.46

Cotton, F.A. 3.34, 3.36, 3.38, 3.40, 33.10, 33.12, 64.159, 71.72, 71.101, 71.102, 71.137, 72.35, 79.17, 81.10, 81.29, 81.30, 81.35, 81.43, 81.45, 81.47, 81.48, 81.62, 81.63, 81.65, 81.70, 81.83, 81.91, 82.33, 82.40, 83.193, 83.194, 83.213, 83.214, 84.76, 84.77, 84.78, 84.79, 84.80, 84.105, 84.114, 84.115, 84.121, 84.124, 84.127, 84.128, 84.131, 84.132, 84.134, 84.135, 85.54, 86.1, 86.26, 86.27, 86.63, 86.189, 86.190

Coucouvanis, D. 64.129, 64.130, 64.131

Coughlin, D.J. 38.81

Couldwell, C. 36.32, 37.14, 37.16, 42.12, 72.21, 73.15

Coulombe, B. 76.84

Courseille, C. 17.41, 21.2, 28.15, 36.8, 48.21

Courtois, A. 38.73

Cowie, M. 71.273, 71.279, 73.60, 86.177, 86.186

Cowley, A.H. 3.18, 63.3, 86.12, 86.37

Cox, A.G. 79.13

Cox, R.H. 58.78, 59.38

Coyle, C.L. 71.158, 71.262

Cozak, D. 73.69

Cozzi, F. 69.33, 69.34

Crabtree, R.H. 71.237

Cragel Junior, J. 85.79

Cragg, G.M.L. 51.78, 51.79

Cragoe Junior, E.J. 32.36, 41.2

Craig, D.C. 64.58

Cramer, R.E. 83.151

Cras, J.A. 71.280, 80.6

Craven, B.M. 10.2, 32.4, 43.1, 44.1, 44.15, 51.82, 51.83, 51.84

Cremin, D.J. 35.32

Creswick, M. 71.140, 71.167, 71.204, 73.13, 73.115

Creswick, M.W. 73.114

Crichton, B.A.L. 80.25

Criegern, T.von 64.113, 64.135

Cristau, B. 36.10

Critchley, S.R. 32.69

Crociani, B. 71.274, 86.196

Crocker, C. 72.12, 72.18

Crombie, L. 38.115

Crosby, J. 40.1

Cross, R.J. 61.9

Crowley, J.I. 60.3, 60.4

Crumley, F.G. 58.78

Cruse, W.B. 49.21, 49.31

Cruse, W.B.T. 31.13, 36.19, 48.76, 50.34

Csoregh, I. 45.6, 45.7, 45.10, 45.69

Cucinella, S. 67.8

Cudd, M.A. 71.14

Cullen, D. 64.172

Cullen, D.L. 49.14

Cullen, W.R. 73.120, 86.206

Cung, M.T. 1.27, 19.4, 48.35

Cunningham, J.A. 77.10

Curran, D.P. 21.6

Curson, E.H. 83.64

Curtin, D.Y. 1.39, 17.28, 17.29, 17.37, 21.14, 21.15, 21.16, 24.10, 25.1

Curtis, M.D. 69.58, 73.52

Cushner, M.C. 63.3

Cuthbertson, A.F. 41.49

Cuthbertson, E. 41.67

Cutler, H.G. 58.78

Cybulski, J. 38.22
Cygler, M. 64.67, 73.68
Cymbaluk, T.H. 83.190
Cynkier, I. 41.60, 41.81, 62.19
Czugler, M. 35.3, 45.6, 45.7, 45.10, 45.69

D'Alfonso, G. 83.139
D'Andrea, A. 51.54, 51.55
D'Angelo, J. 75.14
D'yachenko, O.A. 32.19, 63.43, 85.82, 85.83
D'yachenko, S.A. 83.147
D'yakov, V.M. 63.7, 63.9
Dabovic, M. 51.63, 51.64, 51.67
Dabrow, M. 47.38
Dabrowski, Z. 38.22
Dahan, F. 71.50, 71.61
Dahl, L.F. 3.52, 64.127, 64.128, 64.132,
    64.164, 65.16, 65.20, 65.21, 71.232,
    71.233, 71.245, 73.30, 73.34, 73.47,
    86.77, 86.83, 86.84, 86.101, 86.143,
    86.208
Dahl, O. 64.15
Dahl, T. 60.13
Dahlenburg, L. 71.276
Dahlstrom, P. 80.25, 85.71
Dahlstrom, P.L. 80.19, 83.151
Dalley, N.K. 38.34, 40.39, 42.7, 42.8, 42.9,
    42.10, 42.11
Daloze, D. 54.20, 54.28, 54.33
Daluga, P. 73.3
Daly, J.J. 34.2, 34.5, 34.7, 53.28, 83.223
Daly, J.W. 58.25
Damewood Junior, J.R. 69.33, 69.34
Damm, L. 40.35
Dana, G. 27.8
Dance, I. 61.1
Dance, I.G. 85.4, 85.9, 85.10, 85.92, 85.94
Daniel, H. 36.12
Daniels, P.H. 37.8
Danilin, V.V. 35.43, 35.44, 35.69
Danishefsky, S. 27.25, 31.51, 38.41, 38.70
Daoud, A. 3.14, 19.2
Dapporto, P. 35.5, 72.66, 75.47, 83.106,
    86.122, 86.123, 86.156, 86.159
Daran, J.-C. 33.17, 71.109, 71.214, 72.42,
    73.40, 73.117, 83.89
Darby, N. 31.36
Darensbourg, M.Y. 86.113
Darias, J. 27.14
Darriet, J. 83.34
Dartiguenave, M. 86.5
Dartiguenave, Y. 86.5
Dartmann, M. 64.123, 86.201, 86.203
Darzynkiewicz, E. 47.7
Das, B.C. 38.20
Date, T. 31.41, 33.68
Datta, N. 62.14
Dauner, H. 71.196
Dauter, Z. 50.11
David, L. 50.26
Davidovich, R.L. 3.35
Davidson, J.L. 71.103
Davidson, T.A. 21.9
Davies, D.L. 71.86, 71.87
Davies, G. 76.98
Davies, P.E. 40.47
Davies, S.G. 75.28
Davis, A.R. 83.189

Davis, M.A. 3.51, 85.3
Davis, P.D. 35.72
Davis, R.E. 32.18, 32.44, 39.43, 64.93,
    64.140, 64.147, 64.148, 71.9, 71.10,
    71.62, 71.122, 75.37, 86.12, 86.37, 86.78,
    86.79
Davison, A. 3.51, 64.159, 71.243, 85.3
Dawoodi, Z. 71.56, 72.37, 83.180
Dawson, B.A. 10.5
Dawson, P.A. 71.59, 71.112, 84.51
Day, C.S. 3.33, 3.37, 73.132, 74.11, 74.12,
    86.215
Day, R.O. 31.57, 64.78, 64.96, 66.12, 78.17,
    86.100, 86.133, 86.173
Day, V.W. 3.33, 3.37, 64.159, 73.99, 73.132,
    74.11, 74.12, 81.60, 86.215
De Clercq, P. 38.64, 38.69, 38.86
De Clercq, P.J. 38.63
De Micheli, C. 40.31, 40.32
De Paoli, G. 84.129
De Ranter, C.J. 8.13, 13.8, 13.11, 32.65,
    33.84, 33.86, 33.92, 35.41, 58.33, 58.37,
    58.41, 58.42, 58.49, 58.59, 58.64, 83.87
De Renzi, A. 72.41, 72.44, 72.53
Deacon, G.B. 68.25, 83.210
Dean, C.E. 71.216, 72.40
Debaerdemaeker, T. 33.25, 36.38, 38.118,
    38.119, 63.18, 63.23, 63.34, 63.35
Debuyst, R. 14.2, 14.3
Declercq, J.P. 1.23, 1.30, 1.33, 3.53, 10.3,
    13.12, 13.23, 14.2, 14.3, 16.5, 19.16, 20.7,
    21.11, 21.12, 21.13, 21.20, 21.21, 28.8,
    28.9, 28.17, 28.18, 29.10, 31.21, 31.29,
    31.30, 32.45, 32.54, 32.66, 32.70, 32.78,
    33.2, 33.35, 33.40, 33.42, 33.43, 33.52,
    33.62, 33.63, 33.77, 33.80, 33.81, 33.87,
    33.88, 34.13, 34.14, 35.28, 35.56, 35.57,
    35.66, 37.22, 37.23, 38.13, 38.39, 38.64,
    38.69, 38.86, 40.25, 41.11, 41.12, 41.17,
    41.22, 41.44, 41.73, 44.21, 53.10, 54.20,
    54.28, 54.33, 54.37, 58.18, 64.77
Dederer, B. 16.1
Dee, T.D. 86.6
Deeming, A.J. 72.30
Deger, H.M. 31.60
Degtyarev, A.N. 64.75
Dehnicke, K. 64.124, 65.14, 65.19, 83.1,
    83.2, 83.3, 83.5
Deigin, V.I. 48.53
Deiters, J.A. 64.96
Dejehet, F. 14.2, 14.3
Dekandelaer, M.-C. 14.2, 14.3
Dekaprilevich, M.O. 1.28
Del Piero, G. 67.8
Del Pra, A. 18.1, 18.2, 33.1, 71.213, 71.254,
    76.68, 77.3, 81.19, 81.20, 83.131, 84.68,
    86.128, 86.131
Delarge, J. 33.66
Delaunay, J. 79.7
Delbaere, L.T.J. 45.28
Delettre, J. 51.18, 51.23, 51.30
Delhaes, P. 60.21
Delker, G. 85.26, 85.27
Dell'Amico, D.B. 81.99
Dell'Amico, M. 76.103
Delton, M.H. 29.7
DeLucas, L. 2.31
DeLucia, M.L. 81.32, 81.33
Delugeard, Y. 11.7
Demaret, A. 82.14

DeMartinis, C. 59.15
Demidov, M.P. 63.9
Demidowicz, Z. 86.66
Demir, T. 64.70
DeMunk, P. 50.6
Demuynck, M. 38.86
Denisoff, O. 13.8, 13.11
Dennis, A.M. 82.26
Denny, G.H. 32.36, 41.2
DePamphilis, B.V. 85.3
Depmeier, W. 3.1, 3.13
Dereppe, J.M. 38.13, 38.39
Derkach, L.G. 51.81
Deroski, B.R. 2.9
Derricott, C. 32.12
Derunov, V.V. 84.110
Desai, H.K. 54.11, 58.60
Desauvage, S. 21.20, 21.21
Deschamps, B. 75.4
Deschler, U. 64.150
Descotes, G. 45.45, 45.46
DeSimone, R.E. 76.107, 85.79
Desimoni, G. 35.77
Deslauriers, R. 47.19
Dessy, G. 71.78, 85.33
Destro, R. 39.40
Desvergne, J.P. 38.29
DeTar, M.B. 31.47, 31.48, 31.49, 38.98,
    38.99
DeTitta, G.T. 32.28, 35.23, 38.24, 38.81,
    39.30, 39.31, 39.32, 39.38, 39.48, 59.25,
    59.26, 59.27
Deuten, K.von 1.2, 17.16, 39.84, 39.86,
    40.46, 41.25, 62.12, 65.1, 66.15, 70.7,
    71.276, 84.36, 85.77, 85.84, 85.85
Deutsch, E. 76.28, 76.63, 76.80, 83.44, 84.1,
    86.68, 86.69, 86.70
Dev, S. 53.32
Devarajan, S. 50.36
Devillanova, F.A. 79.8
Devillers, J. 64.136, 71.181
Devis, R. 21.11, 21.12, 21.13
Devos, L. 33.28, 33.29
Devyatkina, E.T. 3.26
Dewan, J.C. 63.22, 69.9, 83.136, 85.86
Dhaneshwar, N.N. 56.1
Dhathathreyan, K.S. 64.74
Di Cesare, P. 45.32
Dianova, E.N. 64.116, 64.121, 65.25
Dias, S. 86.158
Dias, S.A. 71.7, 72.6, 81.93, 81.94, 81.96,
    86.93, 86.129
Dias-Rodrigues, A.M.G. 83.170
Dickman, M.H. 76.63
Dickson, R.S. 71.273, 71.279
Dideberg, O. 32.71, 33.66, 33.73, 33.85,
    45.48, 51.49, 56.10, 58.58, 58.79
Dieck, H.tom 83.134
Diederich, F. 30.22
Diel, B.N. 86.19
Dietz, R. 27.32, 74.9
Dietzsch, W. 85.47
Dignam, K.J. 40.23
Dikareva, L.M. 76.58, 76.59, 81.31, 81.42
Dilworth, J.R. 80.19, 80.25, 85.71
DiMatteo, F. 50.18
DiPierro, M.J. 71.261
Dirkx, J. 47.1
Dirscherl, A. 38.5
Dive, G. 33.85

Divjakovic, V. 83.13, 83.14, 83.15
Dixit, N.S. 76.70
Dixneuf, P. 71.99
Dixneuf, P.H. 72.49, 86.45
Dixon, K.R. 86.109
Djermanovic, M. 51.61
Djura, P. 53.34
Dobler, M. 38.44
Dobrowolska, W. 39.13, 39.14, 39.47
Dobson, A. 71.278
Doddrell, D.M. 84.86
Dodge, J. 82.36, 82.37
Dodge, J.A. 84.107
Doedens, R.J. 21.18, 32.57, 39.57, 76.63
Doehner Junior, R.F. 37.18
Doesburg, H.M. 11.13, 30.16
Doherty, G.O.P. 59.22
Doherty, R. 86.230
Doherty, R.M. 86.25
Dohling, A. 31.53
Doi, J.A. 86.167
Dolak, T.M. 29.7
Dolcetti, G. 86.85, 86.152, 86.174
Dolganov, V.K. 17.25, 17.27
Dolling, R. 83.48
Dolphin, D. 86.206
Domaille, P.J. 86.125
Domazetis, G. 69.16, 69.38
Domenicano, A. 13.4, 15.1
Domiano, P. 69.27, 69.28, 78.2, 84.47
Dommisse, R.A. 33.58
Domnina, E.S. 84.37
Domrachev, G.A. 73.134, 73.135
Donaldson, J.D. 69.9
Donati, D. 84.125
Dondoni, A. 41.75, 41.76
Donohue, J. 73.98
Doorn, J.A.van 71.192
Doretti, L. 83.114
Dorn, H. 35.49, 35.50
Dorner, J.W. 59.38
Doskotch, R.W. 53.23
Dotz, K.H. 27.32, 71.141, 74.7, 74.9
Dougherty, D.A. 19.13, 19.14, 63.41
Doughty, D.H. 71.240, 86.184
Dove, M.F.A. 66.1, 66.3, 66.6
Dowdell, L.R.J. 86.60, 86.61
Downey, R.F. 73.70
Doyle, M.J. 71.67
Doyle, T.W. 59.6
Dozzi, G. 67.8
Drager, M. 66.4, 69.52, 69.59, 69.61, 70.2
Drahnak, T.J. 73.78, 73.103
Dregeris, Ya.Ya. 25.4
Dreissig, W. 50.11
Dreme, M. 71.24, 71.77
Dresow, B. 49.25, 49.30
Drew, M.G.B. 33.90, 34.11, 65.18, 72.13,
    72.43, 83.100, 83.187, 84.109, 85.52
Drews, R.E. 46.2
Dromzee, Y. 71.109
Drozd, V.N. 39.68
Drozdov, Yu.N. 69.51
Drozdz, B. 53.30
Drumheller, J.E. 33.15
Duarte, P.G. 85.41, 85.44
Duax, W.L. 48.81, 51.2, 51.4, 51.5, 51.6,
    51.7, 51.8, 51.11, 51.12, 51.13, 51.15,
    51.16, 51.17, 51.20, 51.22, 51.26, 51.31,
    51.36, 51.43, 59.25, 59.26

Dubchak, I.L. 63.38
Dubgen, R. 64.124
DuBois, D.L. 73.32, 73.51
DuBois, M.R. 73.32, 73.51, 73.64, 86.191
Dubourg, A. 38.23
Dubovenko, Zh.V. 53.12, 53.13, 53.15,
    53.16, 54.1
Dubrawski, J. 86.213
Duchamp, D.J. 59.43
Duchaussoy, P. 45.32
Duchek, J.R. 31.4, 31.5
Duckworth, P.A. 85.50
Ducourant, B. 8.2, 66.2
Duesler, E.N. 1.39, 17.29, 21.16, 24.10, 25.1,
    59.17, 83.34
Duggan, M. 76.36
Duisenberg, A.J.M. 71.183
Dumas, Y. 35.24
Dunaj–Jurco, M. 83.78, 84.55
Dunand, A. 26.6, 26.7, 38.43, 41.40
Dunitz, J.D. 30.21, 38.45, 40.35, 63.16,
    63.28, 67.7
Dunkelblum, E. 30.8, 30.9, 30.10, 30.11,
    30.12
Dupont, A. 54.20
Dupont, L. 32.71, 33.66, 33.73, 33.85,
    45.48, 51.49, 56.10, 58.58, 58.79
Durant, F. 4.3, 33.83, 33.89, 35.64, 42.6
Durgeat, M. 41.53
Durham, B. 83.146
Durley, R.C.E. 76.75
Dusausoy, Y. 3.14
Dutky, S.R. 51.71
Dutt, S.K. 48.5
Dutta, S.P. 44.26, 47.25
Duyne, R.P.van 60.12, 60.42
Dvorkin, A.A. 34.6, 34.8, 35.43, 35.44,
    35.69, 38.96, 40.13, 40.42, 40.44, 40.49,
    71.264, 83.147, 84.112
Dwight, S.K. 71.273, 71.279, 86.177, 86.186
Dyke, A.F. 71.86, 71.87, 71.88, 71.212,
    72.38
Dymock, K.R. 68.15
Dyong, I. 39.54
Dzhafarov, N.Kh. 41.20, 69.14, 69.26
Dzhavakhishvili, Z.O. 83.77, 83.94
Dzhemilev, U.M. 39.23
Dziomko, V.M. 36.35, 83.208
Dzyabchenko, A.V. 35.40, 36.9

Eaborn, C. 63.29, 63.30
Ealick, S.E. 36.4, 54.21, 64.45, 64.46,
    76.101, 76.105
Eastland, G.W. 83.212
Easwaran, K.R.K. 50.36
Ebby, N.D. 27.6
Eberle, H.–J. 71.139
Ebina, F. 76.61, 81.97
Ebner, S.R. 82.30
Ebsworth, E.A.V. 69.7
Eckle, E. 29.11, 48.51, 48.59, 50.19
Eckstein, F. 47.32
Edmonds, J.W. 59.25, 59.26
Edwards, A.J. 81.53, 83.157
Edwards, D.A. 72.13
Edwards, J.M. 59.46
Edwards, P. 71.194, 71.227, 84.12
Edwards, P.G. 71.247

Efraty, A. 73.7
Efremov, V.A. 76.18
Egert, E. 45.51, 48.76
Egestad, B. 38.134
Eggleston, D.S. 83.176
Egorov, Yu.P. 64.115
Egorova, V.V. 51.13, 51.15
Eguren, L. 54.25
Eichelberger, H.R. 82.36, 82.37
Eicher, T. 32.77
Eichhorn, K. 39.75
Eilerman, D. 1.20, 45.4
Einck, J.J. 51.69
Einspahr, H. 2.31, 48.61, 48.83
Einstein, F.W.B. 51.51, 73.108, 73.120,
    83.40, 83.41, 83.42, 83.44, 85.24, 86.68,
    83.189, 85.5, 86.4, 86.60, 86.61, 86.206
Einstein, J.R. 44.10
Eisenberg, R. 86.90, 86.195
Ekiel, I. 47.7
Ekstrand, J.D. 33.33
El–Sayed, M.A. 76.98
El–Shazly, M.F. 76.98
Elder, R.C. 37.10, 71.73, 76.28, 76.80,
    86.69, 86.70
Elding, I. 80.28, 80.29
Elepina, L.G. 64.53
Elepina, L.T. 38.10, 46.6, 64.17, 64.33
Elia, A. 83.97, 83.132
Eller, P.G. 84.13, 84.14, 86.120, 86.132
Ellermann, J. 86.175
Ellis, P.D. 49.28
Elvers, B. 45.35, 45.36, 45.37
Emerson, J. 47.11, 47.22
Emerson, K. 33.15, 83.102
Emoto, S. 41.48
Emsley, J. 1.12
Endesfelder, A. 71.86, 71.87
Endres, H. 36.2, 36.17, 36.39, 76.8, 76.20,
    76.32, 83.8, 83.9, 83.10, 83.11, 83.12,
    83.24, 83.43
Enemark, J.H. 86.204
Eng–Wilmot, D.L. 84.108
Engberts, J.B.F.N. 1.22, 1.24, 1.26, 20.5
Engel, J. 32.74, 32.75, 32.76, 83.224
Engel, P. 27.22, 31.55, 75.46
Engelen, B. 76.99, 76.100
Engelfriet, D.W. 83.21, 83.22, 83.23,
    83.138, 83.218
Engelhardt, U. 64.18, 64.36
Engelmann, M. 86.13
Engler, R. 66.4
English, R.B. 32.37, 32.38, 32.42
English, R.P. 84.104
Engman, L. 70.6
Enjalbert, R. 61.2
Enkelmann, V. 16.18
Ensuque, A. 82.14
Enwall, E.L. 83.102
Enzell, C.R. 54.22
Epe, B. 58.27
Epstein, A.J. 60.20
Epstein, J.M. 71.48
Epstein, J.W. 36.42
Erb, W. 73.43
Erhardt, W.D. 33.72, 33.79
Eriks, K. 62.10
Erman, M.G. 38.81, 59.27
Ermer, O. 75.46
Ermer, S. 71.135

Ernst, J. 56.4
Ernst, L. 33.78, 63.27, 63.52
Ernst, R.D. 83.190
Escande, A. 35.24, 39.67
Escande, P. 66.2
Eschenmoser, A. 40.35, 49.12, 49.13
Esho, F.S. 34.11, 83.187, 84.109
Espinet, P. 73.61, 73.70, 73.127
Espiritu, A.A. 29.12
Essig, E. 83.155, 83.156
Esteban-Calderon, C. 41.3, 41.4, 41.7, 41.8
Estes, G.W. 83.44
Estienne, J. 36.10
Etter, M.C. 77.4, 77.5
Eudy, N.H. 41.57
Evans, D.N. 41.78
Evrard, G. 33.83, 33.89, 35.64, 73.13
Extine, M.W. 84.105

Fabre, J.M. 39.27, 39.64, 60.35
Fachinetti, G. 71.15, 78.3, 78.4
Fackler Junior, J.P. 76.56, 81.77, 86.15
Fagan, P.J. 73.126, 73.132
Faggiani, R. 68.5, 76.24, 76.42
Fahey, J.A. 22.8
Fahmy, R. 71.228
Fair, C.K. 32.5, 83.60
Fairchild, E.H. 53.23
Fairhurst, G. 73.81, 73.96
Fajer, J. 49.3
Falk, M. 62.16, 62.17
Fallab, S. 76.53, 76.94, 76.96
Faller, J.W. 72.4, 72.9
Fallon, G.D. 59.29, 76.17, 85.61
Falvello, L. 47.39, 71.256
Fanshawe, W.J. 35.70
Fanwick, P.E. 82.40, 84.79, 85.54
Faraone, F. 71.206
Faraone, G. 71.55, 71.80, 71.206, 71.248, 71.249
Fares, V. 71.78, 85.33
Farina, R. 75.6
Farnsworth, N.R. 53.22, 54.27
Farrar, D.H. 86.136
Farrugia, L. 86.172
Faught, J.B. 11.5, 85.69
Faulkner, D.J. 53.34, 54.10
Faure, R. 45.45, 45.46
Favas, M.C. 71.250, 71.251, 81.50
Favez, R. 71.208, 86.10
Fawcett, J.K. 21.1
Fawcett, T.G. 32.31, 76.26, 76.44, 84.22
Fay, R.C. 77.13, 80.17
Fayos, J. 1.37, 17.34, 27.14, 53.19, 54.9, 54.25, 54.26, 64.2, 83.148, 86.67
Fedeli, W. 36.44, 36.46, 40.48, 48.49, 50.25, 50.30, 50.31, 59.34, 71.20, 71.98
Fedotov, N.S. 63.6, 63.19
Fee, J.A. 76.56
Feher, O. 36.14
Fehlhammer, W.P. 71.253
Felcht, U.-H. 64.25
Felkin, H. 71.237
Feltham, R.D. 86.213

Felthouse, T.R. 81.10, 81.29, 81.30, 81.35, 81.43, 81.45, 81.47, 81.48, 81.62, 81.63, 81.65, 81.70, 81.91, 85.91, 86.26, 86.27, 86.189, 86.190
Fenical, W. 53.1, 54.5, 54.12, 59.7, 59.8
Fenske, D. 64.35, 65.5, 66.18, 83.1, 83.2, 83.3, 86.214
Fenton, D.E. 76.102, 77.6, 77.8, 78.6
Ferguson, G. 13.16, 27.33, 30.20, 34.3, 51.27, 51.28, 51.33, 51.34, 51.35, 51.42, 59.22, 71.7, 72.6, 73.72, 81.93, 81.94, 81.96, 85.70, 85.81, 86.89, 86.93, 86.129, 86.158
Ferguson, J. 26.6, 26.7
Ferrari, F. 59.34
Ferraro, J.R. 83.191
Ferreira, M.A. 53.4
Ferrino, S. 38.107
Ferris, M.J. 32.69
Ferruti, P. 32.64
Feser, R. 71.21
Ficini, J. 75.14
Fielding, P.E. 49.1
Figgis, B.N. 49.1, 49.2, 79.9, 79.10
Fild, M. 86.47
Filhol, A. 3.3, 3.6, 3.7, 3.8, 3.9, 3.10, 3.11
Filipenko, O.S. 17.31, 17.36, 40.52, 83.79, 83.80, 84.19, 84.27, 84.28, 84.43, 84.44
Filipic, C. 3.46, 3.47
Filippakis, S.E. 41.37, 41.59
Filippo Junior, J.S. 71.244
Filira, F. 77.3
Fillebeen-Khan, T. 71.237
Finer, J. 54.5, 54.12, 54.29
Finer, J.S. 53.18
Finer-Moore, J. 54.11, 54.23, 58.60, 58.68
Fink, R. 24.9, 24.11
Finke, R.G. 86.101
Finnimore, S.R. 71.170
Fischer, E.O. 71.19, 71.36, 71.60, 71.125, 71.136, 71.201, 71.219
Fischer, J. 75.4, 81.79
Fischer, K. 22.4, 39.75
Fischer, M.B. 75.27
Fischer, N.H. 53.27
Fischli, A. 83.223
Fisher, J. 84.17, 84.71
Fisher, J.R. 86.223
Fitzgerald, B.J. 71.265, 84.139
Flack, H.D. 85.55, 85.56
Flandrois, S. 60.21
Flemal, J. 64.77
Fleming, I. 31.13
Fleming, R.J. 60.41
Flippen-Anderson, J.L. 38.105, 44.27, 44.28, 44.29, 51.71, 51.72, 51.76, 51.77
Floch, L. 41.1
Florencio, F. 37.6, 37.7, 37.13, 38.103
Floriani, C. 71.161, 72.10, 73.131, 76.4, 76.49, 76.87, 76.91, 78.3, 78.4, 78.8
Floss, H.G. 48.46
Flowerday, P. 36.1, 36.5, 41.38
Fluck, E. 83.220
Fludzinski, P. 21.6
Flynn, R.M. 71.71
Foces-Foces, C. 35.2, 35.58, 35.60, 36.3, 36.21, 36.22, 41.84, 45.31, 45.47, 45.56, 71.191
Folting, K. 71.94, 73.22
Fonseca, I. 19.12, 19.15

Font-Altaba, M. 28.9, 28.17, 28.18
Fontecilla-Camps, J.C. 36.33, 48.63
Forbes, E.J. 86.99
Forenza, S. 50.18
Forlani, L. 41.75, 41.76
Forsellini, E. 71.254, 76.68, 76.92, 78.6, 83.131, 84.129, 84.138
Forster, H.-J. 36.12
Fortier, S. 10.5, 36.47, 51.44, 59.27, 64.9, 64.10
Foss, O. 33.37, 40.9, 40.10, 40.11, 40.12
Foster, C. 86.172
Foulon, M. 44.32, 59.16
Fourcade, R. 8.2, 66.2
Fouret, R. 33.28, 33.29, 45.49
Fourme, R. 37.1
Fourneron, J.-D. 53.35
Fournet, B. 45.49
Foust, A.S. 70.15, 70.16, 70.17, 76.11, 76.12, 76.13, 76.14, 86.83, 86.143
Foust, D.F. 72.45
Fowles, G.W.A. 65.18
Fox, J.R. 64.154
Foxman, B.M. 58.45, 73.48, 86.125
Foy, R.M. 71.218, 71.224, 71.229
Francesconi, L.C. 73.113
Francetic, V. 80.20
Francisco, R.H.P. 84.15
Franck, R.W. 40.8
Frank, A. 71.125, 75.8
Frank, R.M. 31.61
Frank-Kamenetskaya, O.V. 32.3, 32.8, 32.9
Franke, L.A. 31.57
Frankel, R.B. 85.57, 85.62
Franklin, M.V. 77.6, 77.8
Fratini, A.V. 40.50, 41.71, 64.7
Frediani, S. 49.5
Fredrich, M.L. 73.99
Freeland, B.H. 72.39
Freer, A. 61.24, 61.26
Freer, A.A. 54.27
Freiburg, C. 76.99, 76.100
Freijee, F.J.M. 64.101, 65.11
Freimanis, Ya.F. 25.4
Freni, M. 83.139
Freund, M. 39.29
Frew, A.A. 86.185, 86.188, 86.193, 86.198
Frew, A.J. 38.80
Freyberg, D.P. 32.7, 85.66, 85.72, 85.73, 85.74, 85.75
Fric, I. 37.4
Fridey, S.M. 45.15
Fridman, A.Ya. 76.18
Friedrichsen, W. 38.118, 38.119
Friesen, G.D. 73.3
Fristad, W.E. 38.8, 38.16
Fritchie Junior, C.J. 8.3
Fritsch, N. 38.41, 38.70
Fritz, H. 19.13, 19.14, 21.19, 34.2, 34.5, 34.7
Frolov, F. 41.60
Frolov, Yu.L. 63.7, 63.9
Frolova, N.A. 1.18
Frolow, F. 40.17
From, A.H.L. 51.73
Fronckowiak, M.D. 51.32, 51.37
Fronczek, F. 40.51, 40.53, 40.55, 84.99
Fronza, G. 41.13
Fu, P.P. 29.4, 29.5

Fu, Y.L. 45.15
Fuchs, B. 36.24, 36.36
Fueno, T. 86.96, 86.97
Fuente, G.de la 58.55, 58.56
Fuess, H. 2.1, 19.2, 38.79, 39.3, 48.43, 48.45, 48.58, 48.73, 69.12, 81.2, 81.3
Fugen-Koster, B. 71.141
Fuhrer, H. 51.75
Fuhrhop, J. 64.160
Fuhrhop, J.-H. 56.4
Fujii, S. 47.2, 47.28
Fujii, T. 33.68
Fujimoto, H. 58.71
Fujinami, S. 76.46
Fujino, M. 48.47
Fujino, S. 64.76
Fujioka, H. 38.55
Fujita, E. 41.28, 84.6, 84.13, 84.14, 84.18
Fujita, J. 86.28
Fujita, R. 59.30
Fujiwara, T. 45.60, 47.2, 47.28, 50.9, 58.25
Fukazawa, Y. 59.9
Fukutani, Y. 13.19, 41.61, 41.77
Fukuyama, K. 13.19, 36.16, 36.23, 36.34, 38.122, 41.61, 41.77
Full, R. 62.9, 75.2
Fullerton, D.S. 51.41, 51.73
Fumagalli, A. 3.43, 64.158, 64.162, 64.163, 71.42, 71.52
Funke, P.T. 50.28, 50.29
Furberg, S. 44.2, 44.3, 44.4, 44.9
Furey Junior, W. 85.91
Furmanova, N. 62.19
Furmanova, N.G. 15.6, 22.5, 38.74, 62.18, 64.23, 64.75, 65.3, 68.8, 68.14, 68.19, 69.47, 71.4, 71.13
Furst, W. 81.52
Furuhata, K. 83.59
Furukawa, J. 53.20
Furusaki, A. 48.30, 58.40
Fusstetter, H. 3.41, 63.4, 63.17
Fyfe, C.A. 25.1
Fyne, P.J. 86.56

Gaasdal, A. 11.6, 11.9, 11.15
Gabe, E.J. 9.2, 38.20
Gadret, M. 10.6
Gadwood, R.C. 27.25, 31.51
Gaetani, E. 41.52
Gaetani, M. 38.106
Gaetani-Manfredotti, A. 76.4, 76.49
Gaete, W. 33.50
Gafner, G. 71.66, 71.138, 76.62, 85.49
Gage, L.D. 64.159, 84.79
Gagne, R.R. 84.98, 84.107
Gainsford, G.J. 58.61, 76.27, 76.47
Gal, M. 36.14
Galas, A.M.R. 71.83, 71.193, 71.222, 71.263, 86.9, 86.11, 86.29, 86.57, 86.59
Galdecki, Z. 29.3, 32.20, 35.35, 46.5, 64.39, 64.41, 64.47, 64.105, 64.119, 64.141, 83.18
Galesic, N. 84.30
Galf, S. 28.9, 28.17, 28.18
Galiazzo, G. 33.93
Galigne, J.L. 39.27, 39.64, 60.11
Galitskii, N.M. 48.31, 48.53, 48.81, 50.39
Gall, J.H. 36.52

Gallagher, M.J. 64.58, 64.149
Gallagher, P.T. 31.13
Gallagher, R.T. 54.29, 54.35
Gallazi, A. 38.106
Gallenkamp, B. 34.2, 34.5, 34.7
Galli, B. 83.82
Gallicano, K.D. 64.171
Galloy, J. 32.45, 35.66
Gallucci, J.C. 64.94, 64.95
Galy, J. 33.26, 61.2, 64.48, 64.112, 69.2, 71.181, 71.217, 73.38, 81.9, 86.5, 86.41, 86.42
Gambacorta, A. 71.20, 71.98
Gambaryan, N.P. 16.13
Gammel, F.J. 71.60, 71.125
Gammill, R.B. 38.70
Gandlovich, M. 84.55
Gandolfi, R. 40.31, 40.32
Gane, P.A.C. 41.63
Gange, D. 38.85
Ganin, Yu.G. 40.13
Ganis, P. 69.39, 69.44, 72.53
Ganter, C. 31.50
Ganyushkin, A.V. 63.25, 69.51
Garafalo, A.R. 17.10, 35.54
Garay, R.J. 35.53
Garbay-Jaureguiberry, C. 48.22, 48.28, 48.29, 48.36
Garcia-Blanco, S. 13.21, 19.12, 19.15, 24.5, 35.2, 35.58, 35.60, 36.3, 36.21, 36.22, 37.6, 37.7, 37.13, 38.103, 41.3, 41.4, 41.7, 41.8, 41.84, 45.56, 54.13, 68.13, 71.120, 71.191, 83.169
Garcia-Mina, M.C. 17.34
Gardner, K.H. 16.14
Gardner, S.A. 70.4, 73.118
Garner, C.D. 3.39, 84.97
Garrigou-Lagrange, C. 32.2
Gasparrini, F. 83.82
Gassman, P.G. 27.4, 31.52
Gastaldi, L. 85.8
Gatehouse, B.M. 58.48, 59.29, 71.37, 71.53, 71.180, 83.30, 85.61
Gatehouse, B.M.K. 75.35
Gatehouse, B.M.K.C. 35.36
Gatilov, Yu.V. 53.11, 53.12, 53.13, 53.15, 53.16, 54.1
Gaudemer, A. 47.33
Gaultier, J. 3.11, 38.37, 60.21
Gauthier, M.D. 73.60
Gavel, D.P. 83.49
Gavuzzo, E. 46.3, 59.34
Gavuzzo, G. 71.20, 71.98
Gawron, M. 38.76
Gazo, J. 83.78
Gdaniec, M. 33.24, 58.1, 58.2, 58.10, 64.26
Gebert, E. 60.20, 60.42
Gebler, W. 64.37
Geddes, A.J. 17.17
Geiger, D.K. 49.24
Geise, H.J. 31.20, 37.9, 40.4, 42.1, 42.2, 64.21
Geita, L.S. 27.26
Gelders, Y.G. 58.37, 58.49, 58.59, 58.64
Gellatly, B.J. 84.118
Gellert, R.W. 83.62, 83.63
Geoffre, S. 17.41, 36.8, 51.21
Geoffroy, G.L. 64.154, 64.155, 64.156
George, T.A. 86.215
Gerasimov, V.I. 78.7, 79.12

Gerdil, R. 38.43, 39.15, 39.16, 39.17, 63.24, 64.57
Gerloch, M. 71.256
Gerlt, J.A. 46.2
Germain, G. 1.23, 1.30, 1.33, 3.53, 10.3, 13.12, 13.23, 16.5, 20.7, 21.11, 21.12, 21.13, 21.20, 21.21, 28.8, 28.9, 28.17, 28.18, 29.10, 31.21, 31.29, 31.30, 32.54, 32.66, 32.70, 32.78, 33.2, 33.35, 33.40, 33.42, 33.43, 33.52, 33.62, 33.63, 33.77, 33.80, 33.81, 33.87, 33.88, 34.13, 34.14, 35.28, 35.56, 35.57, 37.22, 37.23, 38.13, 38.39, 38.64, 38.69, 38.86, 40.25, 41.11, 41.12, 41.17, 41.22, 41.44, 41.73, 44.21, 48.74, 53.10, 54.20, 54.28, 54.33, 54.37, 58.18, 64.77
Gerr, R.G. 51.81, 71.4
Gerstner, F. 68.7, 68.9, 68.10, 68.11, 68.12
Gervasio, G. 71.118, 84.61
Geschwandtner, W. 83.155, 83.156
Ghali, N.I. 26.5
Ghatak, I. 86.220
Ghedini, M. 86.85, 86.152, 86.174
Ghilardi, C.A. 35.5, 86.149, 86.181, 86.226, 86.227, 86.229
Ghosez, L. 20.7
Ghosh, A.C. 51.26
Ghosh, D. 10.2, 76.60
Ghosh, S.K. 16.3
Ghouse, K.M. 41.67
Gieren, A. 16.1, 31.61, 32.53, 36.1, 36.5, 41.26, 41.38, 42.5, 71.196
Giesecke, J. 27.21, 59.1
Giessen, B.C. 58.52
Gifkins, M.R. 64.63
Giglio, E. 51.50, 51.54, 51.55, 61.12
Gilardi, R.D. 38.105, 51.76, 51.77
Gilchrist, T.L. 28.6
Giles, R.G.F. 20.8, 32.51, 33.71
Gill, G.B. 21.8, 22.6
Gill, T.P. 71.261
Gilli, G. 33.67, 33.70, 33.74, 33.75, 35.39, 72.52
Gillier-Pandraud, H. 45.29
Gillies, D.G. 68.18
Gilmore, C.J. 36.52, 54.27, 58.36, 61.24, 61.25, 61.26
Ginderow, D. 71.220
Ginsburg, R.E. 71.232, 71.233, 71.245, 86.101
Ginzburg, S.L. 27.26
Giordano, F. 44.30
Giordano, T.J. 84.94, 84.95
Giral, L. 39.27, 39.64, 60.35
Girault, J.-P. 71.24, 71.77
Girerd, J.J. 85.45
Girven, R.J. 50.16
Giunchi, L.M. 41.13
Giuseppetti, G. 38.106, 44.30
Gladfelter, W.L. 64.154
Gladkaya, A.Sh. 76.10
Gladysz, J.A. 71.45
Glaser, R. 48.77, 48.78, 48.79
Glass, R.S. 31.4, 31.5, 39.11
Glatzmaier, G. 73.64
Glavan, K.A. 86.68, 86.69, 86.70
Gleason, J.G. 40.36
Glehn, M.von 36.18, 36.25, 36.31
Gleizes, A. 60.40, 69.2, 81.9, 81.15, 85.6, 85.7, 86.5

Glen, R. 27.1, 27.2, 38.9
Glen, R.C. 38.7, 38.52
Glick, M.D. 29.7, 63.1, 63.11, 63.33, 63.50, 71.65, 85.79
Glidewell, C. 32.50, 41.39, 41.49
Glikberg, S. 76.48, 76.77, 76.78
Glowiak, T. 2.6, 2.11, 29.3, 64.6, 82.3, 82.8, 82.12
Glowka, M.L. 32.20, 35.35, 46.5, 64.39, 64.47, 64.105, 64.119, 64.141, 83.18
Glozbach, E. 69.15
Glusenkamp, K.-H. 52.4
Glusker, J.P. 2.17, 2.25, 2.26, 29.4, 29.5, 36.37, 60.24
Gluzinski, P. 34.9, 35.1, 36.28, 38.31, 45.21, 45.23, 83.135
Go, K. 51.86, 51.87
Goaman, L.C.G. 39.49
Goddard, R. 71.38, 71.39, 71.40, 71.63, 72.8, 72.17, 72.24, 72.32, 72.67, 75.23
Godovikov, N.N. 64.31, 64.51, 64.66, 64.75, 64.85
Godtfredsen, W.O. 50.3
Goel, A.B. 86.89
Goel, R.G. 86.88
Gol'dshtein, I.P. 65.4
Goldberg, I. 35.6, 35.7, 35.17, 35.18, 35.46, 35.47, 37.5, 38.135, 38.136, 39.29, 40.29, 40.30, 41.31, 61.4, 61.5, 61.13
Goldsmith, L. 38.72
Goldstein, E. 35.17, 35.18, 35.46, 35.47
Goldstein, M. 83.28, 86.6, 86.7
Golembeski, N.M. 71.108, 71.123, 71.154, 71.257
Golic, L. 2.10, 2.19, 2.20, 13.20, 30.2, 31.15, 31.38, 41.18, 83.175, 83.177, 85.47
Gololobov, Yu.G. 64.49
Golubinskaya, L.M. 68.8, 68.19
Golubovskaya, E.V. 82.5, 82.6
Gomez-Beltran, F. 76.79
Goncharova, I.N. 44.23, 44.24
Gonschorek, W. 39.1, 39.2
Gonzalez, A.G. 27.14, 58.55, 58.56
Gonzalez-Martinez, J. 17.34
Good, M.L. 82.36, 82.37
Goodfellow, R.J. 73.116
Goodgame, D.M.L. 82.7
Goodhand, N. 15.2, 86.99
Goodman, M. 48.74
Goodman, P.D. 11.8
Gopal, R. 13.10, 13.17
Gordon, E.M. 41.80
Gorenysheva, O.N. 59.10
Gorewit, B.V. 85.26, 85.27
Gorter, S. 83.21, 83.22, 83.23, 83.209
Gorting, K. 73.71
Gosselink, J.W. 71.272
Goswami, K.N. 58.34
Gotoh, H. 13.19, 41.61, 41.77
Gottlieb, H.E. 54.17
Gotz, V. 72.67
Goubitz, K. 33.84, 59.36
Gougoutas, J.Z. 13.9, 14.4, 24.4, 41.58, 41.80, 50.28, 50.29
Gould, R.O. 45.26, 86.153, 86.154
Gourdon, A. 83.27, 83.33, 83.89, 86.56
Goursolle, M. 39.46
Gouzerh, P. 81.52, 84.9
Govindan, S.V. 53.24, 53.26, 53.29, 53.31, 54.7

Gowda, D.S.S. 35.25
Graafland, T. 1.22, 1.24, 1.26, 20.5
Grabarczyk, H. 53.30
Grabley, S. 37.23
Graeber, E.J. 83.6
Graf, C. 41.9
Graffeuil, M. 61.3
Grampel, J.C.van de 64.2
Grand, A. 64.5, 64.73
Grandin, A. 81.1
Grandjean, D. 71.99, 71.153, 73.49, 86.45, 86.111
Grant, S.M. 86.135, 86.164
Graves, B.J. 83.19
Gray, H.B. 71.157, 71.158, 71.262
Graziani, R. 73.67
Grdenic, D. 71.12, 83.143
Greaves, W.W. 71.234
Greci, L. 35.55, 35.59, 35.61, 41.15, 41.16
Green, B.S. 45.54, 61.14, 61.15, 61.16, 61.17
Green, M. 71.38, 71.81, 71.82, 71.259, 71.263, 73.56, 75.30, 86.32, 86.33, 86.43
Green, M.L.H. 72.21, 86.56
Greenaway, A.M. 83.122, 86.204
Greenhough, T.J. 25.3, 27.10, 27.12, 27.19, 27.20, 31.39, 31.40, 38.62, 38.87, 40.16, 72.5, 73.1, 73.2, 73.6, 73.45, 86.65
Greenwood, A.J. 83.158
Greenwood, N.N. 86.44
Gregson, D. 86.126
Greiser, T. 67.2
Grenier-Loustalot, M.F. 33.27, 38.32
Grenthe, I. 76.48, 76.77, 76.78
Grevels, F.-W. 72.17, 72.24
Grevtsev, A.M. 82.11
Grieco, P.A. 38.107
Griend, L.V. 64.28
Grierson, D.S. 58.28
Griffin, G.F. 72.43
Griffin, J.F. 51.16, 58.74
Griffin, R.T. 68.20, 68.24
Griffith, E.A.H. 85.35
Griffith, E.H. 49.28
Griffith, L. 72.26
Griffith, R.C. 21.9
Grigg, R. 49.4
Grimes, R.N. 71.54, 71.69, 71.70, 71.242, 73.77, 73.87
Grochowski, J. 38.82
Grogaard, J. 44.2, 44.3, 44.4, 44.9
Gromek, J.M. 73.98
Gronowitz, S. 41.81
Gross, B. 45.32
Gross, H. 16.18
Gross, J. 71.33
Gross Junior, R.A. 54.21
Grossel, M.C. 19.17, 21.4, 21.5
Grossenbacher, L. 31.58, 31.59, 40.40
Grossert, J.S. 11.5
Groth, P. 13.1, 23.1, 23.2, 23.3, 23.4, 23.5, 39.73
Grouhi-Witte, G. 69.32
Grove, M.D. 51.72
Groweiss, A. 59.35
Grubmuller, P. 31.55
Grulich, R.E. 59.6
Grundy, K.R. 86.136
Grynpas, M. 56.8, 56.9
Grzesiak, E. 73.55

Guastini, C. 20.12, 20.13, 40.54, 73.131
Guay, F. 44.14
Gubanova, L.I. 63.7
Gubin, A.I. 1.5
Guerchais, J.E. 71.119, 73.14, 81.53
Guerina, N.G. 51.82
Guerney, P.J. 71.200, 72.36
Guggenberger, L.J. 86.200
Guggolz, E. 71.30, 71.126
Guglielmetti, R. 41.65, 84.120
Guild, B.C. 83.182
Guilhem, J. 47.33, 71.24, 71.77
Gulya, A.P. 76.1
Gumrukcu, I. 67.9, 69.49, 69.50
Gumz, J.-P. 73.76
Gunter, M.J. 83.225
Gunther, H. 11.14
Gupta, I. 31.67
Gupta, M.P. 2.21, 14.1, 27.7, 81.22, 81.24, 81.25
Gupta, P.K.S. 86.50
Gupta, R.K. 2.17
Gurkova, S.N. 63.6, 63.19, 64.104
Gurst, J.E. 31.31
Guryanova, E.N. 66.14
Gusakovskaya, I.G. 83.79, 83.80, 84.19, 84.27, 84.28, 84.43, 84.44
Gusejnova, M.K. 41.20, 78.16, 81.98
Gusev, A.I. 63.6, 63.19, 63.36, 64.104, 64.116, 64.121, 75.24
Gustafsson, K. 34.12, 35.76, 38.132
Gutierrez-Puebla, E. 68.13, 71.120, 83.169
Gutkoska, R. 80.25
Guy, E. 38.23
Guyot, M. 41.53
Guzman, E.C. 71.89, 71.163
Gysling, H.J. 70.4, 70.18

Ha, F.C. 76.45
Haack, J.L. 26.5
Haase, W. 8.4, 84.103
Haddad, M.S. 83.34, 85.91
Haddon, R.C. 28.1, 36.1, 36.5, 41.26, 41.38, 42.5
Hadjioannou, C.R. 82.7
Hadjoudis, E. 33.44, 33.51
Hadzi, D. 2.7, 2.8
Haeberlen, U. 38.18
Haenel, M.W. 74.2, 74.3
Hafner, K. 28.11, 28.14, 31.66, 36.13
Hagadone, M.R. 53.18
Hagelee, L. 72.68
Hagen, A.P. 74.1
Hagen, K.S. 85.95
Haget, Y. 24.2
Haginiwa, J. 58.8
Hagiwara, H. 40.26, 40.27
Hahnlein, W. 72.34
Hai, M.A. 58.48
Haiduc, I. 66.19, 69.46, 69.48
Hails, M.J. 86.44
Haines, R.J. 86.161
Haire, M.J. 51.74
Haisa, M. 1.25, 17.4, 19.9
Hajek, M. 31.29, 31.30
Hajos, G. 41.43
Hakoshima, T. 38.55

Halbert, T.R. 71.184
Halder, T. 68.6
Haley, M.J. 66.1, 66.3, 66.6
Halfpenny, J. 71.29, 71.129, 83.91
Hall, D. 73.50, 73.65
Hall, G. 76.45
Hall, J.R. 11.2
Hall, M. 33.8, 66.7, 66.8, 66.16, 77.6, 77.8
Hall, S.R. 17.44
Hall, W.T. 71.137, 72.35
Hall Junior, H.T. 75.6
Haller, K.J. 49.22
Halpern, J. 72.59
Halstenberg, M. 62.9, 64.107, 75.2
Haltiwanger, R.C. 38.107, 64.92, 73.32, 73.51, 73.64, 84.66, 86.19, 86.191
Halton, B. 22.8
Hamada, A. 73.121
Hamalainen, R. 76.85, 76.88, 76.89, 78.1, 81.81, 81.82, 81.87, 81.89
Hamana, H. 64.76
Hamelin, M. 1.35
Hamid, M.A. 35.27
Hamilton, A.D. 49.21, 49.31
Hamilton, J.A. 50.32, 50.38, 60.43
Hamon, J.-R. 73.69
Hamor, M.J. 19.5
Hamor, T.A. 1.40, 15.2, 19.5, 35.38, 35.42, 35.45, 35.51, 47.26, 86.99
Hampton, M. 84.66
Han, D.-S. 56.3
Han, K.R. 73.52
Hanaoka, M. 35.65
Hancock, R.D. 76.2
Handke, W. 86.47
Hanessian, S. 45.50
Hanisch, P. 72.55
Hanlon, D.J. 74.1
Hanna, R. 45.38
Hanson, A.W. 38.20
Hanson, B.E. 71.101, 71.102
Hanton, L.R. 83.164
Happer, D.A.R. 63.29, 63.30
Haque, Z.P. 83.137
Hara, H. 58.63
Harada, S. 28.13
Haran, R. 83.204
Harano, K. 31.68, 37.21, 37.24, 40.26, 40.27
Harata, K. 61.19, 61.20
Harding, A. 39.43, 64.93, 64.140, 64.147, 64.148
Harding, M.M. 40.1, 45.66, 50.40, 69.7
Hardy, A.D.U. 61.23
Haring, H.-W. 73.44
Harkema, S. 32.58, 40.38
Harley, A.D. 64.154
Harlow, R.L. 11.10, 32.72, 32.73, 33.94, 35.62, 36.40, 51.74, 71.79, 75.16, 75.17, 75.18, 75.19, 83.39, 83.55, 83.154, 84.106, 84.116
Harms, K. 38.78
Harms, R. 36.17
Harper, A. 3.4
Harpp, D.N. 39.45
Harris, A.D. 86.165
Harris, A.R. 31.63
Harris, M. 58.28
Harris, P.J. 64.3

Harrison, P.G. 69.4, 69.10, 69.17, 69.18, 69.19, 69.31, 69.37
Harrison, W.D. 83.145, 84.82
Harrison Junior, E.A. 38.133
Hart, D.W. 27.29, 73.85
Hart, H. 30.8, 30.9, 30.10, 30.11, 30.12, 38.53, 38.54
Hart, W.P. 73.5
Hartl, H. 83.159, 83.160
Hartley, P. 41.28
Hartley, S.G. 49.21, 49.31
Hartshorn, M.P. 5.12
Hartstock, F. 72.49
Hartung, H. 64.20
Harvey, R.G. 29.4, 29.5
Hasek, J. 38.130, 62.6
Hata, G. 59.23
Hatada, M. 49.36
Hatano, K. 49.22
Hatfield, W.E. 60.20
Hathaway, B. 76.36, 76.50, 76.51, 76.52, 76.82
Hathaway, B.J. 81.71, 81.72, 83.57, 83.58, 83.112, 83.140, 83.141, 83.145, 84.23, 84.82
Haudegond, J.-P. 72.23
Haupt, H.-J. 71.239
Hausen, B.M. 59.2
Hausen, H.-D. 68.6, 68.7, 68.9, 68.10, 68.11, 68.12
Hauser, H. 3.55
Hauw, C. 60.21
Hawke, D. 51.26
Hawthorne, M.F. 86.66, 86.140, 86.167
Hawthorne, S.L. 80.17
Hay, D.G. 17.22, 17.23, 50.6
Hayashi, S. 53.21, 54.31, 54.32
Hayashi, T. 13.2, 38.142, 41.50
Hayashi, Y. 54.36
Hayden, T. 83.182
Hayes, K.S. 16.15
Hayhurst, G. 71.37, 71.180
Haymore, B.L. 73.10, 80.18
Haynes, N.B. 59.22
Hayward, R.C. 59.20
Hazell, A. 3.17, 28.2, 28.3, 28.4, 28.5, 83.167, 83.183
Hazell, R.G. 28.2, 28.3, 28.4, 28.5
Hazra, B.G. 51.26
Head, R.A. 86.207, 86.211
Healy, M.A. 69.4, 69.10, 69.17
Healy, P.C. 84.86
Heatley, F. 35.13
Heaton, B.T. 3.43, 64.157, 64.158, 64.162, 64.163
Hebert, H. 58.77
Hecht, H.J. 36.12, 56.7
Heck, R.F. 27.29
Hedge, V.R. 38.14
Heeg, E. 33.78, 63.27, 63.31, 63.52
Heeg, M.J. 71.73, 76.80, 83.40, 83.41, 83.42, 85.24
Hegarty, A.F. 13.16, 35.32, 40.23
Heiker, F.R. 45.52, 45.53, 45.58
Heitz, S. 41.53
Helland, B.J. 8.10, 8.11, 71.234, 82.30
Hellmann, G. 21.19
Hellwinkel, D. 64.144

Helm, D.van der 24.9, 24.11, 33.33, 36.4, 39.56, 39.74, 48.80, 50.33, 54.21, 59.24, 64.45, 64.46, 69.46, 69.48, 69.60, 74.1, 84.108, 86.50
Hembre, R.I. 8.11
Hempel, A. 50.11
Henc, B. 72.8, 75.9
Hencsei, P. 63.12, 63.15
Hendrickson, D.N. 3.48, 48.4, 73.82, 73.113, 83.34, 85.91
Hendriks, H.J.M. 83.209
Hengartner, U. 15.4, 32.63
Henichart, J.P. 44.32
Henjes, H. 64.71
Henkel, G. 11.1, 41.9, 64.98, 86.13
Henkel, J.G. 17.11
Henling, L.M. 84.98
Henn, L. 69.57
Hennessee, G.L.A. 45.17
Hennig, H. 83.181
Henrick, K. 49.4, 64.86, 64.97, 68.18, 68.20, 68.24, 68.25, 71.56, 72.37, 83.158
Henriet, M. 32.70, 41.22, 41.44
Henry, R.A. 86.81
Herak, R. 76.72
Herberich, G.E. 73.23, 75.34
Herbert, J.A. 38.113
Herbstein, F.H. 3.48, 17.14, 44.18, 48.4, 60.1, 73.82
Hergold-Brundic, A. 83.143
Herk, A.van 46.4, 64.16
Herk, A.M.van 33.84
Hermann, R. 39.54
Hermecz, I. 35.16, 35.20, 35.33, 35.37
Herring, F.G. 80.4
Herrmann, W.A. 71.30, 71.42, 71.52, 71.126, 71.140, 71.167, 71.202, 73.10, 73.26, 73.41, 73.58, 73.90
Herron, N. 83.64
Herskovitz, T. 71.79
Hervat, A.W. 2.15
Herz, C.P. 31.53
Herz, W. 53.2, 53.8, 53.24, 53.25, 53.26, 53.29, 53.31, 54.7
Hesbain-Frisque, A.M. 20.7
Hess, H. 41.43, 64.30, 64.69, 64.71, 64.99, 86.180
Hesse, K.-F. 81.49
Hessner, B. 73.23
Hewat, A.W. 2.13, 2.14
Hewgill, F.R. 38.138
Hey, R.G. 41.19
Heyding, R.D. 73.83
Hickel, D. 32.33, 32.39, 32.40, 39.22, 59.5
Hickel, S. 64.20
Hidaka, J. 82.21
Highet, R.J. 58.25
Higuchi, T. 54.36, 64.168
Hill, D.T. 86.71
Hill, E.W. 71.84
Hill, L.H. 84.119
Hill, N.L. 85.26, 85.27
Hillhouse, G.L. 77.1
Hillier, I.H. 35.13, 84.97
Hilty, T.K. 86.121, 86.141
Himbert, G. 69.57
Himes, V.L. 17.18, 64.60
Hingerty, B. 45.61
Hingerty, B.E. 44.10
Hirama, M. 38.41

Hirano, A. 58.40
Hirao, K. 31.41
Hirata, Y. 59.4
Hirayama, N. 38.137, 48.13, 50.4
Hirobe, M. 58.26
Hirose, T. 31.35
Hirotsu, K. 53.1, 54.35, 54.36, 59.37, 59.42, 64.168, 75.6
Hirshfeld, F.L. 8.8
Hirshfield, J.M. 32.36, 41.2
Hirst, D.M. 40.16
Hisano, T. 40.26, 40.27
Hiskey, R.G. 48.75
Hitchcock, P.B. 63.29, 63.30, 71.226, 71.231, 73.106, 75.48, 86.207, 86.211
Hite, G. 17.11, 21.9, 59.46
Hixson, S.S. 31.57
Hlavata, D. 38.130
Hluchy, H. 63.44
Ho, B.Y.K. 69.8
Ho, P. 48.19
Hoberg, H. 72.67
Hobi, R. 40.35
Hodgson, D.J. 38.100, 47.16, 48.75, 60.20, 77.1, 83.19, 83.49, 83.146
Hodgson, K.O. 71.232, 71.233, 71.245, 80.7, 85.37, 85.57, 85.62, 85.78, 85.80, 85.96, 85.97
Hodgson, P.G. 17.12
Hoehne, S. 73.76, 76.43, 84.73, 86.2
Hoek, W.G.M.van den 23.6
Hoeksema, H. 45.18
Hoffman, B.M. 49.18
Hoffmann, E.G. 75.9
Hofmann, P. 69.20, 69.43
Hoge, R. 22.4, 39.75, 64.139, 69.57
Hoggard, R. 71.49
Hogle, J. 47.9
Hohenberger, E.F. 86.175
Hohlfeld, R. 69.3, 69.23
Hohlwein, D. 5.2, 5.3
Hohne, E. 16.12
Holden, H.D. 32.50, 41.39, 41.49
Holden, K.G. 40.36
Holderegger, R. 86.169
Holladay, A. 86.76
Hollander, F.J. 69.6, 71.46, 83.109
Hollis, S. 83.100
Holm, R.H. 85.57, 85.62, 85.78, 85.80, 85.95, 85.96, 85.97
Holt, E.M. 38.47, 38.48, 65.6, 65.10, 71.3, 71.58, 84.16, 84.26, 84.52, 86.39, 86.81, 86.119
Holt, S.L. 38.47, 38.48
Holub, M. 53.30
Holy, N.L. 71.71
Holzbach, W. 83.168
Hon, P.K. 37.2
Honan, G.J. 84.101
Honan, N.B. 73.26
Honle, W. 64.35, 65.5, 65.8, 73.20, 73.66, 86.14, 86.72
Honneger, H. 64.64
Honore, T. 40.5
Hope, D.A.O. 19.17
Hope, H. 8.8, 41.81, 71.189, 85.26, 85.27
Hopkins, P.B. 38.109
Hopper, S.P. 63.29, 63.30
Hori, H. 53.6

Hori, M. 39.65
Horiguchi, Y. 34.10
Horne, W. 64.59, 64.111
Horvath, A. 35.16, 35.20, 35.33, 35.37
Horvath, G. 36.14
Horvath, I.T. 71.107
Hoser, A. 27.17, 58.7, 58.16, 58.23, 73.55
Hoshino, O. 58.63
Hospital, M. 17.41, 21.2, 48.21, 51.1
Hossain, M.A. 84.137
Hossain, M.B. 48.80, 50.33, 59.24, 69.46, 69.48, 69.60, 84.108, 86.50
Hosur, M.V. 47.24
Houalla, D. 64.136
Houghton, T.J. 17.21
Houk, L.W. 86.50
House, D.A. 76.45
House, H.O. 26.5, 27.28, 27.30, 31.47, 31.48, 31.49, 38.98, 38.99
Howard, B.M. 53.1, 59.7, 59.8
Howard, J.A.K. 17.30, 71.75, 71.76, 71.111, 71.142, 71.149, 71.173, 71.185, 71.186, 71.238, 71.259, 71.277, 73.23, 73.104, 73.105, 75.15, 86.126, 86.147, 86.172
Howard, R.A. 82.26
Howarth, O.W. 73.129
Howatson, J. 5.7
Howie, R.A. 69.1
Howlader, N.C. 84.119
Hoxmeier, R.J. 71.85
Hoyano, J.K. 86.60, 86.61
Hoyer, E. 85.14
Hseu, T.H. 38.111
Hsu, B. 83.60
Huang, C.-H. 73.120
Huang, C.-T. 53.23
Huang, S.-P. 39.71
Hubbard, C.R. 17.18, 86.25, 86.230
Hubberstey, P. 3.4
Huber, C.P. 35.25
Huber, F. 64.90, 64.91
Hubert, A.J. 32.71
Hubert, J. 83.35
Hubner, R. 71.196
Hudson, C.W. 39.43, 64.93, 64.140, 64.147, 64.148
Hudson, H.R. 64.86, 64.97
Huffman, J.C. 38.61, 38.71, 41.21, 71.94, 71.96, 73.3, 73.12, 73.22, 73.33, 73.63, 73.86, 73.107, 83.206, 84.84, 84.117
Hufford, C.D. 58.57, 59.42
Hugel, H.M. 22.8
Hugel, R.P. 79.7
Huggins, J.M. 71.255
Hughes, B.B. 84.66
Hughes, G.R. 86.139
Huie, B.T. 76.101, 76.105
Huiszoon, C. 35.4
Hull, W.E. 40.18, 40.19
Hulot, G. 54.33
Humber, L.G. 36.47
Humblet, C. 33.83, 33.89
Huml, K. 38.130
Hummel, G.J.van 40.38
Hummel, J.P. 63.41
Hummel, L. 62.3, 62.5, 73.4
Hunger, K. 35.74
Hunter, B.K. 73.83

Hunter, W.E. 71.89, 71.127, 71.163, 72.45, 73.74, 73.75, 73.92, 73.101, 73.106, 75.48, 83.173, 86.30, 86.113
Hunziker, M. 36.39
Hurley, T.J. 2.17
Hursthouse, M.B. 11.19, 20.10, 20.15, 49.6, 62.20, 64.118, 71.83, 71.132, 71.166, 71.175, 71.176, 71.193, 71.194, 71.222, 71.227, 71.247, 71.278, 72.30, 73.21, 73.112, 73.130, 76.41, 81.84, 83.85, 84.12, 84.119, 84.130, 84.137, 86.9, 86.11, 86.29, 86.57, 86.59, 86.165, 86.220
Husain, J. 36.43, 51.60, 58.69
Husain, S.A. 55.4
Husebye, S. 64.96, 70.10, 70.14
Hussain, M.S. 1.14, 1.15, 32.5, 83.88
Husson, H.-P. 58.28
Hutchins, L.D. 73.24
Hutchinson, B.B. 83.113, 83.202
Hutchinson, J.P. 71.92, 76.98, 86.64
Huttner, G. 62.9, 64.107, 73.89, 73.110, 75.2, 75.8, 85.18, 86.36, 86.38, 86.74
Hutton, A.T. 11.16, 41.54, 41.55, 71.66, 71.138
Hux, J.W. 72.39
Huybrechts, L. 3.53, 32.66
Hvastijova, M. 83.78
Hwang, T.K. 86.219
Hyde, J. 85.22, 85.23, 86.94

Ibers, J.A. 49.18, 49.32, 49.34, 49.35, 71.17, 71.18, 72.62, 72.63, 81.66, 85.67, 85.68, 86.51, 86.82, 86.130, 86.148
Ibragimov, B.T. 58.11, 58.12, 58.13, 58.14, 58.15, 81.80, 85.64
Ichida, H. 8.7
Ichikawa, M. 40.26, 40.27
Ida, D. 33.39
Ida, Y. 40.26, 40.27
Idrissi, M.S. 32.2
Igeta, H. 9.3
Ihara, M. 39.71
Iitaka, Y. 9.3, 31.35, 34.10, 36.20, 36.48, 38.142, 39.65, 47.34, 50.2, 51.80, 53.6, 53.14, 53.17, 53.20, 55.1, 55.3, 55.4, 56.3, 56.5, 58.17, 58.63, 58.71, 61.18, 75.42, 83.59
Ikehara, M. 47.17
Ikemoto, I. 39.8
Il'ina, I.G. 38.88
Ilinskii, A.I. 83.77
Illingsworth, M.L. 78.17
Ilsley, W.H. 63.1, 63.11, 63.50, 71.65, 71.101, 71.102, 79.17, 81.83, 83.193, 83.194, 83.213, 84.76, 84.77, 84.78, 84.121, 84.124, 84.127, 84.128, 84.131, 84.132, 84.134, 84.135, 85.79, 86.1
Ilvonen, P. 76.6
Ilyukhin, V.V. 85.11
Immirzi, A. 86.104, 86.183
Inamoto, N. 64.168
Inayama, S. 31.35, 53.6
Inoue, K. 59.30
Inoue, M. 36.11, 44.17
Inouye, H. 59.30
Institoris, L. 45.6, 45.7, 45.10, 45.69
Interrante, L.V. 60.5
Ionov, B.M. 46.6

Iorio, M.A. 58.57
Ippen, J. 39.58, 39.69, 39.78
Ireland, R.E. 28.7
Irngartinger, H. 20.11, 31.1, 31.42, 37.11, 38.18
Iroff, L.D. 69.33, 69.34
Irving, H.M.N.H. 11.16, 41.54, 41.55, 67.3, 71.66, 71.138
Isago, T. 82.21
Isakina, A.P. 5.1
Ishida, T. 36.11, 44.17, 60.6
Ishido, H. 38.122
Ishii, H. 28.13
Ishikawa, F. 33.59, 33.60
Ishimaru, T. 41.70
Ishizu, J. 72.60
Ishley, J.N. 64.7
Isobe, K. 71.113, 71.114, 71.115, 71.269
Isomura, K. 32.61
Israilov, I.A. 58.44
Itai, A. 51.80, 53.20, 55.1, 55.3, 56.5, 61.18
Ito, M. 69.30, 80.13
Ito, S. 59.9
Ito, T. 50.1
Itoh, H. 48.24, 48.64
Itokawa, H. 53.14
Ittel, S.D. 75.16, 75.17, 75.18, 75.19
Ivanov, O.I. 83.207
Ivanov, O.V. 83.208, 83.215
Ivanov, S.B. 3.35
Ivanov, V.T. 50.39
Ivanova, A.I. 9.1
Iversen, A.J. 70.11
Iwakuma, T. 31.41
Iwamoto, T. 76.23, 76.61
Iwasaki, F. 16.17, 41.68, 41.69
Iwasaki, H. 69.11, 69.30, 80.3, 80.12, 80.13
Iwasaki, S. 55.4
Iwata, M. 2.7, 2.8, 36.37
Izawa, T. 50.21

Jackels, S.C. 83.176
Jackson, P.F. 3.56, 64.165, 64.166, 65.13
Jackson, P.J. 83.188, 83.196
Jackson, R. 71.57, 71.90
Jackson, W.G. 76.27, 76.47
Jackson, W.R. 75.35
Jacobsen, H.-J. 69.55
Jacobson, A.E. 37.12
Jacobson, K.B. 83.56
Jacobson, R.A. 2.3, 2.4, 8.10, 8.11, 38.2, 46.7, 64.19, 64.56, 64.63, 71.234, 73.46, 82.30, 82.38, 83.144
Jacques, J. 2.28
Jahnig, W. 81.92
Jahnisch, K. 83.26
Jain, S.C. 47.37, 60.36, 60.37
Jamerson, J.D. 86.221
James, B.D. 69.38
James, B.R. 85.5
James, D. 21.4, 21.5
James, E.J. 75.27
Jameson, G.B. 49.32, 49.34, 49.35
Jamoulle, J.C. 33.85
Jandacek, R.J. 14.5
Janickis, V. 33.37, 40.9, 40.10, 40.11, 40.12, 76.11, 76.12, 76.13, 76.14
Janik, T.S. 86.170

Jannack, T. 83.24
Jansen, J.C. 83.161
Jansen, M. 41.83
Janta, R. 73.95
Janzen, D.H. 48.17, 48.18
Jarchow, O. 1.13, 1.21, 32.77, 38.114, 45.59
Jarvis, B.B. 41.82
Jaskolski, M. 33.24, 33.48, 33.49, 58.9
Jastrzebski, J.T.B.H. 69.45, 71.203
Jaud, J. 33.26, 42.3, 64.48
Jeannin, S. 27.24, 72.54, 85.65
Jeannin, Y. 1.35, 27.24, 33.17, 71.50, 71.61, 71.109, 71.214, 72.42, 72.54, 81.52, 83.27, 83.33, 83.89, 84.9, 85.45, 85.65
Jecny, J. 81.85
Jefferies, P.R. 38.124, 38.125, 38.126, 38.127
Jeffery, J.C. 71.173, 71.185, 86.138
Jeffrey, G.A. 11.20, 45.2, 45.3, 45.8, 45.20
Jeffreys, J.A.D. 71.168
Jenkins, J.A. 37.18
Jensen, W.P. 83.144
Jerkovich, G. 36.14
Jernow, J. 50.7, 50.8
Jewers, K. 58.5
Jezowska, M. 82.35
Jezowska-Trzebiatowska, B. 82.8
Jircitano, A.J. 83.191
Jogun, K.H. 45.65
Johansen, J.E. 49.12, 49.13
Johansen, O.H. 39.73, 39.76
Johansson, A.A. 84.110
John, A.M. 39.43, 64.93, 64.140, 64.147, 64.148
Johnson, B.F.G. 3.56, 64.165, 64.166, 65.13, 65.15, 71.57, 71.59, 71.90, 71.148, 71.177, 71.210, 72.14, 83.73, 84.51, 86.40
Johnson, C.K. 39.9
Johnson, D.K. 64.169, 86.144
Johnson, J.F. 86.68, 86.69, 86.70
Johnson, M.R. 61.6
Johnson, R. 73.7
Johnson, R.A. 11.2
Johnston, D.B.R. 40.45
Jolly, P.W. 72.8, 72.61, 75.9
Jones, A.G. 3.51, 85.3
Jones, A.J. 5.12, 58.67, 72.55
Jones, A.S. 47.26
Jones, D.J. 1.12
Jones, D.L. 86.99
Jones, D.S. 83.198, 83.216
Jones, D.W. 29.6, 30.6, 30.7, 30.19
Jones, G. 40.18, 40.19
Jones, G.P. 59.13
Jones, H.F. 58.5
Jones, N.F. 61.6
Jones, P.G. 33.34, 47.39, 50.34, 50.41, 52.4, 86.118
Jones, R.A. 64.118, 71.83, 71.166, 71.175, 71.176, 71.222, 86.9, 86.11, 86.29, 86.57, 86.59
Jones, R.D.G. 26.3, 26.4
Jones, T. 83.28, 86.7
Jones, W. 20.10, 20.15, 33.76
Jones, W.A. 51.52
Jones, W.M. 71.62, 71.122
Jonk, H.-E. 85.18
Jordaan, J.H. 45.33
Jorgensen, F.S. 31.54
Joshi, B.S. 38.14

Joshi, K.C. 54.9
Joss, S. 27.22
Jostes, R. 64.123, 86.201, 86.203
Joyce, C.J. 40.1
Julia, M. 53.35
Julis, S.A. 73.88
Jung, C.Y. 58.74
Jungbauer, A. 71.6
Jungmann, H. 72.65
Juranic, I. 51.63, 51.64, 51.67
Jurd, L. 38.117, 38.128
Jutzi, P. 69.20, 69.43

Kabachnik, M.I. 64.31, 64.51, 64.66, 64.72, 64.75, 64.85, 64.88, 64.89, 64.134, 64.145
Kabuto, C. 27.31
Kaesz, H.D. 64.161, 71.26, 71.47, 71.85, 72.11, 72.19
Kaftanat, V.N. 85.39
Kaftory, M. 3.48, 17.38, 17.39, 37.19, 37.20, 38.38, 38.40, 38.56, 38.75, 38.84, 38.89, 48.4, 74.10, 74.13
Kagan, H.B. 73.123
Kahn, O. 85.45
Kai, E. 71.113, 71.114, 71.115
Kai, Y. 28.10, 31.33, 31.34, 63.16, 63.28, 67.5, 71.31, 77.2
Kaim, W. 79.17, 81.83, 83.193, 83.194, 83.213, 84.76, 84.77, 84.78, 84.114, 84.115, 84.121, 84.124, 84.127, 84.128, 84.131, 84.132, 84.134, 84.135, 86.1
Kaimierczak, W. 83.18
Kaise, H. 50.21
Kaiser, J. 13.20, 64.20, 79.15, 83.181, 85.19, 85.20, 85.32, 85.47
Kaisin, M. 53.10
Kaitner, B. 77.12
Kaizu, Y. 83.197
Kajfez, F. 40.43
Kajiwara, A. 38.55
Kajtar, M. 35.3
Kakos, G.A. 75.35
Kakudo, M. 28.13, 36.23
Kalb, W.C. 86.140
Kalck, P. 71.217, 86.92
Kalinin, A.E. 39.28, 39.36, 39.37, 39.68, 64.22, 64.88, 64.89, 64.115, 64.134, 64.145
Kalman, A. 4.2, 13.6, 13.13, 13.14, 32.10, 32.52, 35.3, 35.30, 35.31, 35.68, 36.14, 40.28, 40.37, 41.36, 41.41, 41.47, 41.66
Kaluski, Z. 27.17, 58.7, 58.16, 58.23, 73.55, 73.68
Kalvoda, J. 51.75
Kalyanaraman, B. 2.18
Kambu, K. 58.79
Kamdar, B.V. 38.60
Kamenar, B. 33.11, 77.12, 83.143
Kametani, T. 39.71
Kamijo, N. 2.3, 2.4, 54.31, 73.53, 73.54
Kaminskii, V.F. 39.66, 60.7, 60.8
Kaminsky, W. 71.159, 71.197
Kamiya, K. 39.39, 48.47
Kanda, H. 71.269
Kane, M.J. 51.26
Kaneda, M. 34.10, 58.17
Kanehisa, N. 32.61

Kaneko, K. 58.72
Kaneko, T. 36.48
Kanematsu, K. 31.68, 37.21, 37.24
Kanev, A.N. 63.21
Kanoktanaporn, S. 36.7
Kanters, J.A. 21.22, 33.82, 33.91
Kanunnikova, E.N. 66.17
Kaplan, M.L. 36.1, 36.5, 41.26, 41.38, 42.5
Kapon, M. 3.48, 17.14, 44.18, 48.4
Kapor, A. 35.30, 35.31, 41.36, 41.46, 41.47
Kapovits, I. 13.6, 13.13, 13.14
Kappenstein, C. 27.32, 74.9, 79.7
Kapundu, M. 56.10
Karanovic, L. 51.61
Karapetyan, A.A. 32.43, 35.34
Karaulov, A.I. 48.72, 50.22, 50.24, 50.37
Karaulova, E.N. 39.28, 39.36, 39.37
Karban, J. 38.110
Karcher, B.A. 64.19
Kardanov, N.A. 64.31, 64.51, 64.66, 64.85
Karhu, M. 64.142, 64.143
Karimov, Z. 48.71
Karipides, A. 1.10, 67.1, 81.14
Karl, N. 60.17
Karle, I.L. 48.37, 48.38, 58.25
Karlsson, B. 28.19, 38.102, 38.116
Karolak-Wojciechowska, J. 41.64, 64.41
Karpelus, R. 71.135
Karraker, D. 82.36, 82.37
Karsch, H.H. 64.43
Kartha, G. 47.4, 47.13, 48.66, 48.67, 48.68,
  48.69, 48.70, 51.86, 51.87
Karup-Nielsen, I. 51.26
Kasai, H. 32.61, 47.34
Kasai, M. 38.137
Kasai, N. 28.10, 31.33, 31.34, 32.61, 41.70,
  67.5, 71.31, 77.2
Kashaev, A.A. 1.18, 63.7
Kashino, S. 1.25, 17.4, 19.9
Kashiwabara, K. 86.28
Kashiwada, Y. 54.34
Kashiwagi, M. 83.151
Kashman, Y. 59.35
Kashutina, E.A. 63.38
Kasper, J.S. 60.5
Kasuga, K. 49.33
Katagiri, K. 50.5
Katahira, D. 72.4, 72.9
Kataoka, H. 53.21
Kataoka, T. 39.65
Katayama, C. 53.5, 53.9, 59.4
Kato, H. 35.65
Kato, M. 44.12
Kato, T. 59.23
Katritzky, A.R. 32.72, 32.73, 33.94, 35.62,
  36.40
Katrusiak, A. 58.7, 58.23
Katsube, Y. 13.19, 36.16, 36.34, 38.122,
  41.61, 41.77
Katti, S.K. 47.23
Katz, B.A. 83.117, 83.118, 83.119, 83.120,
  83.121, 83.123, 83.127, 83.128, 83.129
Kausch, E. 8.12
Kawafune, I. 84.48
Kawaguchi, S. 71.113, 71.114, 71.115,
  71.269, 84.96
Kawamata, T. 31.35
Kawamura, N. 58.72
Kawano, S. 78.12, 78.13
Kawasaki, T. 45.57, 51.62, 51.85, 54.4, 54.6

Kawashima, T. 17.4
Kawato, T. 40.51
Kawazu, K. 59.9
Kawazura, H. 60.33
Kazakova, L.Z. 1.18
Kazakova, V.V. 63.14
Kazlauskas, R. 54.15, 54.16, 54.19
Keeping, J.W. 38.20
Keese, R. 75.46
Kehr, W. 71.253, 86.175
Keijzers, C.P. 84.111
Keister, J.B. 83.109
Keith, A.N. 71.103
Keks, B. 3.47
Kelland, J.W. 71.148, 71.210
Keller, C.D. 83.216
Keller, E. 73.111
Keller, H.J. 36.2, 36.17, 76.8, 76.20, 76.32,
  83.43
Keller-Schierlein, W. 38.44
Kellerer, W. 71.36
Kelly, B.A. 73.133
Kelly, D.A. 86.3
Kelly, D.P. 22.8
Kelly, J.M. 83.157
Kelly, P. 17.21
Kelly, R.L. 83.206, 84.84, 84.105, 84.117
Kemme, A. 38.31, 45.21, 45.23
Kemme, A.A. 35.11, 85.63
Kemmitt, R.D.W. 71.216, 72.40
Kemp, T.J. 11.8, 40.16, 81.36, 86.52
Kende, A.S. 21.6
Kennard, C.H.L. 11.2, 17.12, 17.32, 84.86
Kennard, O. 21.1, 31.13, 36.19, 47.39,
  48.76, 49.21, 49.31, 50.34, 50.41, 64.13
Kennedy, D. 81.71, 81.72
Kennedy, D.A. 86.110
Kennedy, F.G. 75.15, 75.25, 75.26
Kennedy, J.D. 86.44
Kepert, D.L. 71.218, 71.224, 71.229, 71.250,
  71.251, 81.50, 84.101
Keppler, B. 83.43
Kergoat, R. 81.53
Kerr, I.S. 40.18, 40.19
Kerr, K.A. 21.1, 40.20, 40.21, 40.22, 58.66,
  58.67, 64.13, 85.38, 85.43
Kessissoglou, D.P. 73.107
Kessler, H. 27.27, 48.43
Kessler, R.M. 85.26, 85.27
Kettmann, V. 76.64
Keubler, M. 86.163
Keuchler, T.C. 77.7
Khan, M. 86.158
Khan, M.A. 27.33, 51.28, 73.72, 81.93,
  81.94, 81.96, 85.70, 85.81, 86.129
Khan, V.A. 53.12, 53.13, 53.15, 53.16
Khananashvili, L.M. 63.36
Kharchenko, V.G. 39.79, 39.80
Kharitonova, R.I. 81.18
Khidekel, M.L. 1.16, 33.6, 39.50, 60.8,
  83.79, 83.80, 84.19, 84.27, 84.28, 84.43,
  84.44
Khodadad, P. 33.9, 84.60
Khodashova, T.S. 3.26, 83.162, 84.56
Khomutov, M.A. 71.51
Kida, S. 76.19, 78.12, 78.13, 83.25
Kido, M. 54.34
Kido, T. 48.24, 48.64
Kiec-Kononowicz, K. 41.64
Kiedrowski, G.von 38.78

Kierkegaard, P. 36.18, 36.25, 36.31
Kim, W. 83.214
Kimura, M. 20.4, 20.9, 31.64, 38.19, 38.57,
  38.58, 38.59, 38.83
Kind, R. 3.27, 3.28, 3.46, 3.47
King, G.S.D. 4.3, 41.45, 41.62, 47.10, 47.29
King, K. 71.228
King, T.J. 3.39, 32.46, 32.47, 35.19, 35.22,
  36.7, 40.6, 40.7, 40.33, 41.10, 41.72, 50.3,
  53.4, 56.2, 59.33, 69.4, 69.10, 69.17,
  69.31, 83.112, 86.197, 86.210
Kinney, J.B. 71.79
Kinoshita, I. 86.28
Kinoshita, K. 71.113, 71.114, 71.115,
  71.269, 72.27
Kiosse, G.A. 2.22
Kipalova, N. 8.5
Kirby, A.J. 1.22, 1.24, 1.26, 20.5
Kirby, G.W. 58.36
Kirfel, A. 38.25, 58.29
Kirillova, N.I. 62.7, 62.8, 63.5, 69.13,
  71.221
Kirin, V.N. 31.37, 31.56
Kirkup, M. 54.5, 54.12
Kirkup, M.P. 58.43
Kirner, J.F. 81.60, 83.222
Kirtley, S.W. 84.20, 84.24
Kisch, H. 83.101
Kiseleva, I.N. 79.4
Kisliuk, R.L. 36.33, 48.63
Kispert, L.D. 2.18
Kiss, J. 38.5
Kistenmacher, T.J. 44.11, 44.13, 44.22,
  60.12, 76.67, 76.93, 82.22, 83.186, 84.98
Kitagawa, I. 38.55
Kitajima, K. 32.55
Kitamura, K. 38.55, 47.17
Kitano, Y. 59.23
Kitchen, M.D. 73.133
Kito, Y. 40.3
Kitschke, B. 19.13, 19.14, 21.19, 28.11,
  28.14, 30.15, 30.17, 31.66, 36.51
Kivekas, R. 3.15, 38.108
Klaar, M. 56.7
Klabunde, K.J. 71.133, 71.134
Klar, B. 45.61
Klar, G. 1.2, 17.16, 39.84, 39.86, 41.25,
  62.12, 65.1, 70.7, 84.36, 85.77, 85.84,
  85.85
Klaska, K.-H. 1.13, 1.21, 3.1, 32.77, 38.114,
  45.59
Klaus, J. 83.134
Klein, B. 84.40
Klein, C. 41.42
Klein, H.-F. 71.33, 71.116
Klemarczyk, P.T. 73.48
Klement'ev, I.Yu. 63.46
Klima, R. 71.234
Klimova, E.I. 73.84
Klimova, T.V. 62.8, 63.5, 69.13
Klimusheva, G.V. 39.25, 39.26
Klinga, M. 81.68, 81.69, 81.75, 81.76
Klingebiel, U. 63.39, 63.44, 63.49
Klingler, R.J. 73.86
Klunder, A.J.H. 31.2
Knachel, H.C. 64.7
Knappenberg, M. 47.1
Knapstein, C.-M. 33.78, 63.27, 63.52
Knapstein, K.M. 63.31
Knobler, C. 70.3, 82.23, 84.59

Knobler, C.B. 64.161, 71.26, 71.85, 72.11, 72.19, 76.45
Knochel, A. 40.46
Knop, O. 62.16, 62.17
Knossow, M. 61.14, 61.15, 61.16, 61.17
Knott, R.B. 64.64
Knox, C.V. 33.90
Knox, S.A.R. 71.86, 71.87, 71.88, 71.170, 71.212, 72.38, 75.15, 75.25, 75.26, 75.31
Knunyants, I.L. 19.11
Knuuttila, H. 64.4
Knuuttila, P. 64.4
Knyazev, S.P. 62.7
Kobak, V.V. 71.13
Kobayashi, A. 8.7, 81.17
Kobayashi, H. 39.85, 81.17, 83.197
Kobayashi, K. 39.85, 80.3, 80.12
Kobayashi, Y. 64.76
Kober, J. 35.68
Kochi, J.K. 73.86
Kocman, V. 81.67
Kodama, M. 59.9
Kodama, Y. 50.1
Koetzle, T.F. 3.43, 33.20, 33.21, 33.22, 48.1, 48.2, 48.10, 64.158, 64.162, 64.163, 71.42, 71.52, 73.16, 73.17, 73.18, 73.27, 73.28, 73.85
Koga, K. 61.18
Kogure, T. 75.43
Kohl, F. 69.20, 69.43
Kohli, D.K. 84.99
Kohn, D.H. 74.10, 74.13
Kohn, H. 42.13
Kohout, J. 83.78
Koie, Y. 71.265
Koike, K. 54.27
Kojic-Prodic, B. 20.9, 31.38, 32.34, 32.67, 45.39, 45.44, 64.82
Kojima, T. 48.24, 48.60, 48.64
Kok, A.J.de 48.25, 64.34, 64.55
Kok, P. 38.63, 38.64, 38.69
Kokel, B. 35.28
Kokkes, M. 58.59, 58.64, 72.33, 75.40, 86.114
Kokkou, S.C. 40.57
Koknat, F.W. 81.77
Kolakowski, A. 17.9
Kolb, O. 71.21, 71.97, 71.169
Kolbe, R. 85.19, 85.20
Kolesnikov, A.I. 17.25, 17.27
Kolks, G. 84.126
Koll, P. 45.12
Kolobova, N.E. 71.91
Kolthammer, B.W.S. 73.1, 73.2, 73.6, 73.45
Komalenkova, N.G. 75.24
Komarov, V.N. 9.1
Komatsu, M. 41.73
Komiya, S. 71.179, 72.60
Komiya, Y. 72.27
Komori, T. 40.26, 40.27, 51.85
Kompan, O.E. 15.6, 22.5, 38.74, 62.18
Kompan, O.Y. 68.8, 68.19
Kondo, H. 59.23
Kong, P.C. 76.84, 84.25
Koniger-Ahlborn, E. 3.24, 86.199
Koningsveld, H.van 58.39, 59.36, 83.161
Konnert, J. 7.2
Konno, M. 60.33, 76.76, 83.124, 83.125, 83.126
Konnov, V.I. 3.26

Konzelmann, F.M. 41.5
Kopelove, A.B. 49.19
Kopf, J. 39.84, 39.86, 40.46, 65.1, 67.2, 71.159, 71.197, 83.134, 85.77, 85.84, 85.85
Kopilov, V.V. 24.12
Kopp, M. 72.1, 72.31
Koppe, H. 36.12
Koppenhagen, V.B. 49.25, 49.30
Korableva, L.G. 83.79, 83.80, 84.19, 84.27, 84.28, 84.43, 84.44
Koray, A.R. 74.2, 74.3
Korhonen, K. 78.1
Koroteev, M.P. 38.10, 46.6, 64.17, 64.32, 64.52
Korotenko, T.I. 35.34
Korp, J.D. 27.5, 27.11, 29.2, 38.110, 74.14, 76.40, 82.26
Korswagen, R. 73.42, 73.62
Korte, H.-J. 69.53, 69.54
Korte, L. 16.16, 39.19, 66.10
Korvenranta, J. 83.67, 83.68, 83.98, 84.67
Koshibe, S. 31.33, 31.34
Koshinaka, E. 35.65
Kostansek, E.C. 48.54
Kostikas, A. 64.129, 64.130, 64.131
Kosturkiewicz, Z. 33.24, 33.48, 33.49, 58.1, 58.2, 58.3, 58.24
Kostyanowskii, R.G. 32.19
Kotake, S. 77.2
Kotel'nikova, A.S. 81.4, 81.7, 81.11
Koten, G.van 69.45, 71.183, 71.203, 71.272, 72.33, 75.40, 83.205
Kotera, K. 31.41
Kothe, G. 45.52, 45.53, 45.58
Kotrelev, G.V. 63.14, 63.26
Kouba, J.K. 81.60
Koussini, R. 28.15
Kovach, Ya. 38.6, 38.11, 38.12, 38.21, 38.36, 38.50
Kovnat, L. 60.42
Kovredov, A.I. 71.13
Kow, A. 64.86, 64.97
Koyama, H. 32.68
Koyano, K. 41.32
Koz'min, A.S. 31.37, 31.56
Koz'min, P.A. 81.4, 81.7, 81.11, 81.12
Kozikowski, A.P. 25.5
Kozinski, V.A. 41.13
Koziol, A.E. 58.1, 58.2, 58.3, 58.24
Kozlova, N.P. 83.215, 84.64
Kozlowski, H. 82.8, 82.35
Kozunov, V.A. 68.14
Krajewski, J.W. 34.9, 35.1, 36.28, 38.31, 45.21, 45.23, 83.135
Kralingen, C.G.van 81.34, 83.107
Kramarenko, F.G. 15.3, 82.15
Kramer, J.A. 73.82
Kramer, R. 64.48
Krampe, C. 59.11, 63.39, 63.49
Krapp, W. 64.144
Krasan, Yu.P. 15.3
Krasavin, I.A. 35.29, 35.52
Krasnova, N.F. 83.192
Krasochka, O.N. 1.16, 33.6, 71.207
Kratky, C. 49.8, 49.9, 49.10, 49.11, 49.12, 49.13
Krauth, L.R. 72.1, 72.31
Krauth-Siegel, L.R. 75.39
Kravers, M.A. 19.7, 63.8, 63.10

Kravtsov, D.N. 69.47, 71.124
Krebs, A. 38.114
Krebs, B. 11.1, 41.9, 64.98, 64.125, 69.53, 69.54, 69.55, 82.32, 83.50, 85.25, 85.53, 86.13
Kreh, R.P. 84.107
Kreissl, F.R. 71.136
Kress, R.B. 17.29, 24.10
Kreter Junior, P.E. 86.216
Kriege-Simondsen, J.C. 86.213
Kretschmer, R.G. 33.56, 33.57, 35.49
Krieger, C. 30.22, 31.53, 38.18, 39.69, 39.78
Kriek, G.R. 39.5, 86.204
Krilova, S.P. 17.25, 17.27
Krische, B. 42.4
Krishnamurthy, S.S. 64.74
Krishnanachari, H. 46.7
Kristine, F.J. 76.60
Kritskaya, I.I. 71.22
Krohnke, C. 16.18
Krol, I.A. 83.207, 83.208, 84.64
Kroner, J. 38.46
Kroon, J. 23.6, 39.77, 39.81, 39.82, 39.83, 69.45
Krstanovic, I. 51.61
Krueger, W.C. 59.43
Kruger, C. 36.24, 36.36, 68.21, 69.20, 69.43, 71.39, 71.40, 71.63, 71.67, 71.235, 72.8, 72.17, 72.24, 72.51, 72.61, 72.67, 73.109, 75.9, 83.101, 86.38, 86.74
Kruger, G.J. 54.18, 76.62, 85.49
Krylova, L.F. 82.1, 82.2, 82.13
Kubiak, C.P. 86.195
Kubiak, M. 82.8
Kubo, A. 50.20
Kubo, S. 35.65
Kucharski, E.S. 49.2
Kuchkarov, S. 58.14
Kuchta, B. 7.3
Kucsman, A. 4.2, 13.6, 13.13, 13.14
Kukina, G.A. 76.10
Kuklina, Y.F. 82.9, 82.10
Kukushkin, Yu.N. 3.26
Kulishov, V.I. 41.35, 51.81
Kulpe, S. 64.44, 64.80, 64.100, 64.103
Kumadaki, I. 64.76
Kumamoto, F. 51.62
Kumar, N. 53.2, 53.8, 53.25
Kumar, S.V. 17.8
Kumihashi, M. 44.5
Kumin, A. 40.35
Kunakova, R.V. 39.23
Kundu, M.L. 16.3
Kung, W.-J.H. 23.7, 38.53, 38.54, 38.129
Kunze, U. 69.42
Kuo, C.-S. 60.12
Kuo, K.T. 33.69
Kupfer, E. 38.44
Kuppers, H. 81.49
Kuramoto, M. 76.29, 76.30, 76.31
Kurilla, M.G. 76.101, 76.105
Kuriyama, K. 59.30
Kuroda, H. 39.8
Kuroda, R. 47.31
Kursanov, D.N. 71.104, 73.31
Kurskov, S.N. 1.16
Kurvyakov, P.A. 53.12
Kusa, A. 38.6, 38.11, 38.12, 38.21, 38.28, 38.36, 38.50

Kushch, N.D. 39.66
Kushi, Y. 53.21, 54.31, 76.31
Kushmuradov, Yu.K. 58.11, 58.12, 58.13, 58.14, 58.15
Kustyn, M. 85.71
Kutchan, T.M. 15.5
Kutovaya, T.M. 78.16
Kutschabsky, L. 35.49, 35.50, 59.32
Kuyper, J. 65.6, 65.10
Kuz'min, E.A. 69.51
Kuz'mina, L.G. 39.23, 58.44, 71.124, 73.80, 81.44
Kuzmin, A.M. 32.8, 32.9
Kuznetsov, V.G. 3.35
Kuzuhara, H. 41.48
Kvick, A. 38.1, 48.1, 48.2, 48.10
Kvitashvili, A.I. 76.3
Kvitashvili, A.N. 83.94
Kyba, E.P. 39.43, 64.93, 64.140, 64.147, 64.148

L'abbe, G. 3.53, 32.66, 40.25, 41.73, 64.77
La Monica, G. 83.92
Labarre, J.-F. 61.2, 61.3
Ladd, M.F.C. 45.5
Laguna, A. 86.118
Laguna, M. 71.142, 71.173, 71.185, 71.238
Lahajnar, G. 3.46, 3.47
Lahav, M. 51.53
Lahuerta, P. 86.63
Lai, R. 64.160
Lai, T.F. 48.19
Laidlow, W.G. 3.49
Laing, M. 86.87, 86.161
Lalancette, R.A. 32.31, 38.90, 76.26, 76.44, 84.22, 85.91
Lalitha, V. 17.42, 18.3
Lalor, F.J. 73.72, 85.70, 85.81
Lam, K.P. 19.17
Lam, S. 7.2
Lamaty, G. 10.4, 21.10
Lamazouere, A.M. 31.21
Lamm, V. 31.61, 36.1, 36.5, 41.26, 41.38, 42.5, 71.196
Lamotte, J. 32.71, 33.73, 33.85, 45.48, 51.49, 58.58, 58.79
Lampe, P.A. 85.21
Lamprecht, G.J. 75.20
Landee, C.P. 84.4
Landee, D.D. 84.3
Landers, A.E. 85.4, 85.9, 85.10
Landis, M.E. 32.60
Landschoot, R.van 84.81
Lane, M.D. 39.48
Lanfranchi, M. 86.152
Lanfredi, A.M.M. 71.35, 71.64, 71.135, 83.31, 83.82, 83.107, 84.81
Lang, G. 84.136
Lang Junior, S.A. 32.41, 32.56
Langenbach, H.-J. 73.95, 86.18
Langenbach, H.J. 73.111
Langs, D.A. 17.13, 31.9, 38.81, 48.81, 59.25, 59.26, 59.27
Lantratova, O.B. 17.20, 26.1
Lanzilotti, A.E. 35.70
Lapasset, J. 10.4, 21.10, 27.6, 39.67
Lapidaire, J.C.P.M. 64.55
Lapluye, G. 82.14

Lapouyade, R. 28.15
Lapp, R.L. 2.3, 2.4, 64.56
Lappert, M.F. 67.9, 69.40, 69.49, 69.50, 70.8, 71.127, 71.226, 71.231, 73.101, 73.106, 75.48
Larina, T.B. 81.4, 81.7, 81.11, 81.12
Larkina, T.I. 83.79, 83.80, 84.19, 84.27, 84.28, 84.43, 84.44
Larscheid, M.E. 15.4, 32.63
Larsen, E. 31.62
Larsen, F.K. 28.2, 28.3, 28.4, 28.5
Larsen, I.K. 1.3
Larsen, K. 39.43, 64.93, 64.140, 64.147, 64.148
Larsen, S. 31.62, 59.12, 59.18, 64.15
Larson, S.B. 38.34, 40.39
Larson, S.L. 42.7, 42.8, 42.9, 42.10, 42.11
LaRue, W.A. 71.244
Lashewycz, R.A. 86.116
Lasswell Junior, W.L. 59.42
Lattman, M. 86.12, 86.37
Laugier, J. 64.5
Lauher, J.W. 72.2, 72.46
Lauridsen, J. 40.5
Lauzon, G.de 64.58, 64.64, 75.4
Lavallee, D.K. 49.19
Lavallee, P. 45.50
Lavie, D. 54.17
Lavigne, G. 85.45, 85.65
Lavrent'ev, I.P. 1.16, 33.6, 83.79, 83.80, 84.19, 84.27, 84.28, 84.43, 84.44
Lavrent'eva, E.A. 83.79, 83.80, 84.19, 84.27, 84.28, 84.43, 84.44
Lay, D.G. 81.91, 86.26, 86.27, 86.189, 86.190
Lazar, D. 35.68, 40.28, 40.37
Lazarini, F. 3.25
Le Bozec, H. 71.99, 72.49
Le Marouille, J.Y. 71.99, 71.153, 73.49, 86.45
Le Page, Y. 9.2, 73.83
Leach, M. 38.94, 38.95
Leadbetter, A.J. 17.30
Leban, I. 2.7, 2.8, 2.19, 2.20, 30.2, 31.15, 41.18, 80.20, 83.47, 83.115, 83.116, 83.175, 83.177
Lebedev, V.A. 69.51
Lebioda, L. 8.6, 19.10, 35.1, 38.51, 38.82, 38.123, 39.55
Lechat, J.R. 13.5, 84.15
Leclaire, A. 81.1, 81.41, 81.46, 81.56, 81.61, 84.113
Leclerc, G. 10.6
Leclercq, M. 2.28
Ledet, M.T. 73.8
Ledon, H.J. 49.26
Ledoux, M.J. 86.56
Lee, B. 11.18
Lee, C.-L. 71.275
Lee, J.-S. 63.48
Lee, J.D. 35.73
Lee, K.M. 49.16
Lee, S. 38.53, 38.54
Lee, T.J. 33.69
Lee, V.Y. 60.10
Lee, Y.N. 45.25
Leeis, Ya.E. 85.64
Lefferts, J.L. 69.60
Leger, J.M. 10.6, 32.33, 32.39, 32.40, 33.61, 39.22, 39.46, 59.5

Legros, J.P. 64.112
Legzdins, P. 72.5, 73.1, 73.2, 73.6, 73.45
Lehmann, H. 71.2
Lehmann, M.S. 48.12
Lehnert, R. 83.76, 83.174, 83.179, 83.203
Leibnitz, P. 33.54
Leidl, E. 86.163
Leimert, J. 39.24
Leimgruber, W. 15.4, 32.63
Leipoldt, J.G. 3.31, 33.18, 75.20
Leiserowitz, L. 24.6, 24.7, 24.8, 51.53
Lelay, C. 71.99
Lemenovskii, D.A. 71.221
Lemke, G. 13.20
Lemke, P. 48.17, 48.18
Lemke, T.L. 27.5, 27.11
Lenhert, P.G. 71.41, 71.93
LeNoble, W.J. 31.44, 31.45
Lenoir, D. 31.61, 71.196
Lenstra, A.T.H. 2.23, 2.24, 31.20, 33.41, 33.58, 35.67, 37.9, 40.4, 42.1, 42.2, 64.21
Leonhard, K. 71.97, 71.169, 73.11, 86.191
Leoni, M. 35.77
Leonov, S.B. 1.18
Leonowicz, M.E. 62.4, 71.184
Leovac, V. 83.13, 83.14, 83.15, 83.75
Lepicard, G. 51.18, 51.23, 51.30
Leroy, F. 21.2, 48.21, 48.23, 51.14
Letuchij, Ya.A. 33.6
Leutwiler, A. 54.29
LeVanda, C. 75.13
Levi, H.A. 39.57
Levisalles, J. 71.50, 71.61, 71.109
Levitan, P. 32.23, 32.29
Levstik, A. 3.46, 3.47
Levy, H.A. 83.198, 83.216
Levy, L.A. 24.1
Lewin, G. 58.58
Lewinski, K. 8.6
Lewis, J. 3.56, 64.165, 64.166, 65.13, 65.15, 71.57, 71.59, 71.90, 71.148, 71.177, 71.210, 72.14, 83.73, 83.163, 83.165, 84.
Lewis, L.N. 71.94, 71.96, 73.107
Lewis, T.W. 17.28, 17.29, 17.37, 21.14, 21.15, 21.16, 24.10
Lex, J. 31.60, 34.1, 34.4, 63.13
Ley, S.V. 71.27, 71.28
Li, G. 39.42
Li, R. 51.40, 51.47
Liautard, B. 39.27, 39.64
Libertini, E. 84.65
Libson, K. 84.1
Lichtenthaler, F.W. 45.51
Lidert, Z. 41.81
Lidor, R. 36.24, 36.36
Liebelt, W. 65.14, 65.19
Liehr, G. 73.71, 83.48
Liepin'sh, E.E. 44.23
Liles, D.C. 32.50, 41.39, 41.49
Lillhouse, G.L. 73.10
Liminga, R. 38.1
Limn, W. 47.27
Lin, G.H.Y. 47.9
Lin, S.Y. 31.44, 31.45
Lin, T.-S. 47.5, 47.19
Lin, Y. 32.41, 32.56
Lincoln, S.F. 84.101
Lind, H. 41.23
Lindeman, S.V. 31.17
Lindemann, M. 72.17

Lindley, P.F. 51.52, 56.8, 56.9
Lindner, E. 71.139, 71.172, 73.76, 76.43, 84.73, 86.2
Lindner, H.J. 19.13, 19.14, 21.19, 27.27, 28.11, 28.14, 30.15, 30.17, 31.66, 36.13, 36.51
Lindner, K. 61.22
Lindqvist, O. 83.181, 85.14
Lindsay, C.H. 71.275, 86.194
Lindsey, R.L. 32.60
Lindstrom, K. 17.6
Linek, A. 62.3, 62.5, 62.6, 73.4
Lingafelter, E.C. 76.54, 76.69, 83.97, 83.110, 83.132, 83.195
Liopo, V.A. 1.18
Lipka, A. 16.4, 16.16, 39.19, 66.10
Lipkin, A.E. 39.18
Lipkowski, J. 61.10, 61.11
Lippard, S.J. 3.51, 83.136, 84.126, 85.86
Lippert, B. 44.8, 76.24, 76.42, 83.61, 84.42, 84.100
Lipscomb, W.N. 3.32, 38.109, 48.54
Liptak, D. 71.65
Liptrot, R.E. 73.48
Lis, T. 8.9, 81.26, 81.57, 81.88, 84.5, 86.105
Lisini, A. 40.43
Littell, R. 35.70
Liu, A. 59.14
Liu, A.T. 71.244
Liu, L. 37.15
Liu, L.-K. 32.18, 32.44, 36.24, 36.36, 83.101
Liu, R.S.H. 54.2, 54.14
Lively, M.O. 84.13, 84.14
Lloyd, D. 34.3
Lobkovskaya, R.M. 39.51, 39.52
Lobkovskii, E.B. 73.19, 86.212
Locatelli, P. 71.235
Lock, C.J.L. 3.30, 44.8, 76.24, 76.42, 83.83, 83.84
Lodge, B.A. 51.33, 51.42
Loghry, R.A. 39.56, 39.74, 74.1, 84.108
Logvinenko, V.A. 76.55, 76.57
Loim, N.M. 73.31
Loiseleur, H. 45.45, 45.46
Lojanapiwatna, V. 25.2
Loliger, H. 83.111
Longato, B. 71.268
Longchambon, F. 45.24
Longoni, G. 3.43, 3.44, 3.45, 3.52, 64.127, 64.128, 64.132, 64.152, 64.162, 64.163, 64.164, 65.16, 65.20, 65.21, 71.190
Lopez de Lerma, J. 54.13
Lopirev, V.A. 32.3
Lopresti, R.J. 40.34
Lorberth, J. 69.3, 69.15, 69.23
Lorca, E. 86.127
Lorenc, L. 51.63, 51.64, 51.67, 51.75
Lorenz, I. 65.17
Los, M.G. 63.19
Lotter, H. 53.22
Loub, J. 8.4, 81.85
Loubinoux, B. 38.73
Louis, R. 83.221, 85.36
Love, R.A. 84.20, 84.24
Lovell, F.M. 35.14, 35.70, 41.57
Lovell, M.F. 32.41, 32.56, 36.42
Low, J.N. 45.22
Lozar, B. 3.46, 3.47
Lu, C.T. 47.13

Lu, T.H. 33.69
Lucente, G. 36.44, 36.46, 40.48, 48.49
Ludi, A. 71.2
Luger, P. 45.30, 45.35, 45.36, 45.37, 45.52, 45.53, 45.55, 45.58
Luk'yanenko, N.G. 34.6, 34.8, 38.96, 40.13, 40.42, 40.44, 40.49
Luk, K.S. 11.10
Lukas, K.L. 31.1, 31.42, 37.11
Lukaszewicz, K. 48.6, 48.7
Lukehart, C.M. 71.11, 71.41, 71.93
Lukevits, E.Ya. 63.8, 63.10
Lundeen, M. 44.20, 83.20, 83.32, 85.34
Lunel, J. 50.35
Lusby, W. 51.71
Lusinchi, Z. 75.28
Luss, H.R. 39.4, 39.88, 39.89, 39.90, 70.4, 70.18
Lustenberger, P. 27.22
Luth, H. 81.67
Luthy, C.L. 40.15
Lux, D. 64.69, 64.71, 64.99
Lydon, J.D. 76.28
Lyubarskaya, A.E. 68.8, 68.19
Lyubovskaya, R.N. 39.50, 60.8

Ma, J.-L. 86.124
Ma, M.S. 31.10
Maas, G. 64.139
Maatta, E.A. 80.18
Mabbs, F.E. 84.119
Macaulay, E.W. 51.33, 51.42
MacBride, J.A.H. 36.7
MacDougall, J.J. 86.39, 86.86
MacDowell, A.A. 84.97
Mackay, M.F. 13.18, 17.22, 17.23, 50.6, 58.51, 59.15, 59.21, 69.16, 69.38, 84.49
Macke, H. 76.94
Mackenzie, G.A. 3.16
Mackenzie, R.K. 41.67
MacLean, G. 64.167
Maclennan, J.M. 45.65, 45.66, 45.67, 45.70
MacNicol, D.D. 36.52, 41.67, 61.9, 61.23, 61.24, 61.25, 61.26
Madach, T. 71.16, 86.34, 86.35
Madesclaire, M. 33.61
Maebayashi, Y. 38.122
Maeda, C. 76.34, 76.35
Maehashi, H. 53.17
Maehr, H. 38.94, 38.95
Magee, R.J. 69.38
Magerramov, A.I. 81.55
Maggi, L.D. 38.112
Magin, L. 85.22, 85.23, 86.94
Magnus, P. 38.85
Magomedova, N.S. 27.23, 27.26, 35.40
Mahaim, C. 72.25, 75.5
Mahato, S.B. 54.6
Mahe, C. 71.153, 73.49
Maheu, L.J. 80.23
Mahmoud, M.M. 51.52
Maier, G. 20.11
Maisonnat, A. 86.41, 86.42
Maitlis, P.M. 73.61, 73.70, 73.116, 73.127
Majer, J. 76.64, 82.17
Majeste, R. 41.42, 82.36, 82.37
Majoral, J.-P. 64.48, 64.73

Mak, T.C.W. 31.36, 37.2, 58.6, 61.7, 74.4, 74.5, 74.6
Makarenko, G.V. 9.1
Makarevich, L.G. 83.215
Makarova, N.N. 63.32
Makarskii, V.V. 32.3
Makhaev, V.D. 86.212
Makhija, R.C. 86.62
Makitova, D.D. 1.16, 33.6
Mal'chikov, G.D. 82.6, 82.9, 82.10
Malatesta, L. 83.92
Malatesta, V. 84.91
Maleki, L. 82.36, 82.37
Malik, K.M.A. 64.118, 71.132, 71.166, 71.175, 71.176, 71.193, 71.194, 71.222, 71.227, 71.247, 72.30, 73.21, 73.112, 73.130, 76.41, 83.85, 84.12, 84.119, 84.130, 86.9, 86.11, 86.29, 86.57, 86.220
Malik, M.A. 85.59
Malinovskii, S.T. 34.6, 34.8, 38.96, 40.13, 40.42, 40.44, 40.49, 86.117
Malinovskii, T.I. 2.22, 34.6, 34.8, 35.43, 35.44, 35.69, 38.96, 40.13, 40.42, 40.44, 40.49, 59.10, 76.1, 78.7, 79.12, 83.37, 83.38, 83.81, 83.147, 83.192, 84.31, 84.112, 86.91, 86.117
Malisch, W. 73.95
Malito, J.T. 73.35, 73.36, 73.37
Mallinson, P.R. 61.9, 61.25, 73.39
Malmanis, A.Ya. 25.4
Malon, P. 37.4
Malsch, K.-D. 20.11
Maluszynska, H. 27.17
Maly, K. 62.3, 62.5, 73.4
Mamedov, Kh.S. 69.14, 69.26, 78.16, 81.27, 81.54, 81.55, 81.58, 81.98
Mamedova, A.Kh. 41.20
Manassero, M. 3.44, 3.45, 3.52, 64.132, 64.152, 71.190, 74.15, 83.92
Manchand, P.S. 54.3, 54.24
Mandava, N. 51.71, 51.72
Mandel, G.S. 28.7
Mandel, N.S. 28.7
Mander, L.N. 31.19, 83.225
Mandl, J.R. 71.225
Manfredotti, A.G. 20.12, 20.13, 40.54, 71.161, 72.10, 76.87
Mangani, S. 64.172, 84.83, 84.125
Mangia, A. 84.89
Mangion, M. 59.46
Mani, F. 83.106
Mann, A.L. 71.148, 71.210
Mann, K.R. 71.157, 71.158, 71.261, 71.262
Manna, A.la 44.30
Manna, S. 36.45
Manning, A.R. 71.128, 86.135, 86.142, 86.164
Manohar, H. 38.65, 76.70
Manoharan, P.T. 84.62, 84.63
Manojlovic-Muir, L. 71.103, 86.95, 86.182, 86.185, 86.188, 86.193, 86.198, 86.223
Manoli, J.-M. 64.133
Manoussakis, G.E. 73.107
Manriquez, J.M. 73.126, 73.132
Mansuy, D. 71.24, 71.77
Marangoni, G. 83.184, 83.185, 84.75
Marchetti, F. 76.91, 76.103, 78.8, 81.99
Marchetti, L. 35.55, 35.61, 41.15, 41.16
Marcotrigiano, G. 33.19, 33.46, 33.47, 82.41, 83.65

Mare, P.B.D.de la 27.3
Marecek, J.F. 64.114, 67.4, 67.6
Marel, G.van der 85.58
Maresca, L. 71.187
Margulis, T.N. 58.70
Mari, A. 86.5
Marinetti, A. 71.105, 71.178, 72.29
Marini, G. 76.87
Marini–Bettolo, G.B. 56.2
Marinone, F.A. 38.112
Mariot, J.–P. 73.69
Mark, C. 84.7
Markham, R. 86.168
Marko, L. 71.16, 71.107, 86.84
Marks, T.J. 39.9, 73.126, 73.132, 85.67,
    85.68
Maroy, K. 76.12, 76.13, 76.14
Marquez, R. 35.53, 38.42, 45.16, 45.27
Marraud, M. 1.27, 19.4, 48.35, 48.48
Marrison, K. 21.8, 22.6
Marsau, P. 51.14
Marsden, K. 71.252, 73.50, 73.65, 86.134
Marsella, J.A. 73.33, 73.63
Marsh, R.E. 2.16, 48.19, 48.42, 76.73, 78.15,
    83.69, 83.93
Marsh, W. 30.21
Marsh, W.C. 34.3
Marshall, A.G. 49.16
Marshall, D.R. 34.3
Marsich, N. 86.160
Martell, A.E. 76.106, 83.152, 83.153
Martens, C. 41.73
Martin, D.G. 59.43
Martin, D.T. 72.5
Martin, I.L. 35.38, 35.42, 35.45, 35.51
Martin, J.D. 27.14
Martin, J.W.L. 83.152, 83.153
Martin, L.M. 33.17
Martin, R. 76.8, 76.20, 76.32, 83.43
Martin, T.R. 70.8, 71.127, 73.101
Martin, V.S. 27.14
Martin–Frere, J. 27.24
Martin–Gil, J. 71.111, 71.149
Martin–Lomas, M. 45.31, 45.47
Martin–Smith, M. 31.63
Martinengo, S. 3.5, 3.19, 3.43, 3.50, 64.126,
    64.157, 64.158, 64.162, 64.163, 71.147,
    72.3, 72.28
Martinez–Carrera, S. 19.12, 19.15
Martinez–Ripoll, M. 13.21, 17.34, 33.26,
    41.3, 41.4, 41.7, 41.8, 53.19, 54.9, 54.26,
    83.169
Martinis, C.de 59.21
Martinon, S. 75.14
Martynova, T.N. 63.21
Maruyama, K. 28.13
Marzilli, L.G. 44.11, 44.13, 44.22, 71.100,
    76.67, 76.93, 82.22, 83.186
Mascarenhas, Y.P. 13.5, 83.170
Mascherpa, G. 8.2, 66.2
Mashiko, T. 84.136
Maskill, H. 31.43
Maslinski, C. 83.18
Mason, J. 64.157
Mason, R. 49.1
Mason, S.A. 49.1, 65.13
Massa, W. 64.62, 69.15, 69.23
Massey, A.G. 71.34, 71.121
Masters, A.F. 86.15
Mastropaolo, D. 48.57

Mastryukova, T.A. 64.134, 64.145
Masuda, H. 49.7
Mather, A.M. 45.38
Matheson, A.J. 76.45, 84.102
Matheson, T.W. 74.8
Mathey, F. 64.58, 64.64, 73.38, 75.4, 86.39,
    86.86
Mathieson, D.W. 59.22
Mathieu, R. 71.17, 71.18, 86.20, 86.82
Matrosov, E.I. 64.72, 64.88, 64.89
Matsubayashi, G. 84.48, 85.48
Matsui, M. 2.3, 2.4, 39.6
Matsumoto, H. 54.2, 54.14
Matsumoto, K. 76.22, 76.33, 76.34, 76.35,
    84.50, 84.96, 86.28
Matsumoto, K.Y. 65.2, 84.2
Matsumoto, M. 86.96, 86.97
Matsumoto, N. 76.19, 78.12, 78.13, 83.25
Matsumoto, T. 48.30, 58.40
Matsuo, A. 53.21, 54.31, 54.32
Matsuo, M. 36.48
Matsuoka, T. 40.26, 40.27
Matsuura, H. 28.10
Matsuura, T. 36.16, 36.34
Matsuura, Y. 28.13
Matsuzaki, K. 32.55
Matt, D. 86.111
Mattes, R. 1.6, 1.7, 2.5, 11.4, 39.54, 85.16,
    85.28, 85.29, 85.31, 85.51
Matthews, R.W. 68.18, 68.20, 68.24, 68.25
Mattice, W. 40.51
Mattschei, P.K. 40.53, 40.55
Matuszewski, J. 81.26, 81.57
Matuzenko, G.S. 84.38
Matyukhina, O.G. 35.48, 85.64
Maurin, M. 39.27, 39.64
Maury, F. 81.9, 81.15, 85.7
Maverick, A.W. 35.9
Maverick, E. 38.45, 40.35, 40.40
Maverick, E.F. 31.58, 31.59, 35.9
Mavridis, A. 33.4, 33.51
Maxfield, P. 59.34
Mayanza, A. 71.217
Maydonovitch, D.J. 71.184
Mayer, A.J. 71.49, 75.7
Mayerle, J.J. 60.3, 60.4, 60.10, 65.6, 65.10,
    86.86
Maynard, R.B. 71.69, 71.70
Mayorzik, H. 60.15
Mayr, A. 71.253
Mays, M.J. 71.56, 72.37, 83.180, 86.106,
    86.112
Mazanec, T.J. 86.157, 86.166
Mazid, M.A. 76.41, 83.212, 86.53, 86.54,
    86.55
Mazurov, A.A. 35.44, 35.69
Mazus, M.D. 84.31
Mazza, F. 36.44, 36.46, 40.48, 48.49, 61.12
Mazzi, F. 38.106
Mazzocchi, P.H. 37.15
Mazzola, E.P. 17.18
McAlees, A.J. 71.7, 72.6
McAlister, J. 46.1, 47.20
McAtee Junior, J.L. 38.110
McBride, J.M. 21.7, 21.17, 37.3
McBride, S.S. 71.162
McCabe, T. 50.18, 54.35, 59.6
McCall, J.M. 38.60
McCallum, P.A. 41.54, 41.55, 51.78, 51.79
McCandlish, L.E. 83.110

McCauley, J.A. 32.36, 41.2
McClune, G.J. 76.56
McCollor, D.P. 71.133, 71.134
McConnell, J.F. 45.34, 45.42
McCormack, M.T. 13.16
McCormick, B.J. 84.72
McCowan, J.D. 73.83
McCrindle, R. 71.7, 72.6
McCullough, K.J. 19.13, 19.14
McCurry Junior, P.M. 38.70
McDonald, J.W. 72.47
McDonald, W.S. 30.19, 72.12, 72.18, 86.44,
    86.168, 86.171
McDougall, D.J. 58.36
McDougall, G.J. 76.2
McDowell, J.J.H. 41.74, 41.79
McElearney, J.N. 84.10
McEwen, W.E. 66.12
McGillivray, G. 32.37, 32.38, 32.42
McGuiggan, M.F. 71.240
McKee, M. 86.12, 86.37
McKeever, B. 64.114, 67.4, 67.6
McKenzie, E.D. 78.10, 78.11
McKenzie, T.C. 35.70, 36.42
McKinney, J.D. 17.19, 24.1
McKinney, R.J. 75.16, 83.39, 83.55
McLaughlin, G.M. 70.8, 83.225, 86.3
McLaughlin, S.A. 71.198, 71.199
McLean, S. 58.50
McMillin, D.R. 83.171
McMullan, R.K. 8.3, 32.4, 44.1, 44.15, 45.2,
    45.3, 51.32, 51.37, 64.153
McNeese, T.J. 75.27, 86.125
McPartlin, M. 3.56, 64.166, 65.13, 65.15,
    68.20, 68.24, 68.25, 71.177, 71.200,
    72.36, 83.137, 85.50
McPhail, A.T. 22.1, 22.2, 22.3, 31.31, 35.21,
    39.24, 51.63, 51.64, 51.67, 58.75, 71.14,
    75.3
McPhaul, M.J. 39.43, 64.93, 64.140, 64.147,
    64.148
McSharry, W.O. 77.9
Mead, K.A. 71.186
Meade, L. 40.51
Mealli, C. 35.5, 72.69, 85.42, 86.115, 86.150,
    86.151, 86.226, 86.227, 86.229
Medici, A. 41.75, 41.76
Medzhidov, A.A. 78.16
Meek, D.W. 86.157, 86.166, 86.216
Meerssche, M.van 1.23, 1.30, 3.53, 10.3,
    14.2, 14.3, 19.16, 20.7, 21.11, 21.12,
    21.13, 21.20, 21.21, 28.8, 31.29, 31.30,
    32.45, 32.54, 32.66, 32.70, 32.78, 33.35,
    33.40, 33.42, 33.43, 33.52, 33.62, 33.63,
    33.77, 33.80, 33.81, 33.87, 33.88, 34.13,
    34.14, 35.28, 35.56, 35.57, 35.66, 37.23,
    38.13, 38.39, 38.64, 38.69, 38.86, 40.25,
    41.11, 41.12, 41.17, 41.22, 41.44, 41.73,
    44.21, 53.10, 54.20, 54.28, 54.33, 64.77
Meester, P.de 86.39
Mehrotra, S.K. 64.42
Mehta, G. 31.6, 31.16, 31.65
Meicheng, S. 48.26
Meier, E. 72.25, 75.5
Meier, H. 39.53
Meijer, H.J.de L. 71.241
Meijere, A.de 29.1, 30.1, 30.3, 30.5
Meille, S.V. 75.12, 83.65
Meinwald, J. 54.30
Meiske, L.A. 82.38

Melanson, R. 76.84, 83.29, 84.25
Melchers, M. 76.99, 76.100
Meli, A. 72.66, 72.69, 75.47, 86.150, 86.151, 86.225, 86.228
Mello, J.F.de 56.2
Menabue, L. 33.19, 33.46, 33.47, 82.41, 83.65
Menig, H. 72.34, 75.8
Mentzen, B.F. 49.26
Merbach, P. 73.71
Mercier, R. 71.119
Meresse, A. 24.2
Mergehenn, R. 8.4, 84.103
Merrien, M.-A. 58.73
Mertes, K.B. 11.18, 83.191, 84.53
Mertis, K. 71.194, 71.227, 71.247
Merz, L. 84.103
Meszaros, Z. 35.16, 35.20, 35.33, 35.37
Metcalfe, J.C. 40.18, 40.19
Metter, H.-P. 64.36
Metzger, J. 64.160
Metzger, R.M. 60.12
Meunier, B. 38.104, 71.156, 75.28
Meurs, F.van 32.10
Meyer, B. 64.54
Meyer, B.N. 64.7
Meyer, E. 64.172
Meyer, T.J. 83.146
Meyer Junior, E.F. 49.14, 49.33
Meyers, E.A. 11.12, 38.35, 39.41, 70.1
Michalska, M. 32.73, 35.62, 36.40
Michaud, P. 73.69
Michel, A. 4.3, 33.83, 33.89, 42.6
Michel, P. 20.7
Michelin, R.A. 71.208, 71.254
Middlebrook, R.E. 54.21
Middlemas, E.D. 39.24
Middlemiss, D. 36.30
Middleton, A.R. 76.108
Midgley, J.M. 51.27, 51.33, 51.34, 51.35, 51.42
Midollini, S. 86.149, 86.181, 86.187, 86.217, 86.218, 86.226, 86.227, 86.229
Mieroop, W.van de 37.9
Mighell, A.D. 17.18, 64.60, 86.25, 86.230
Mignani, G. 71.153, 73.49
Mihailovic, M.L. 51.63, 51.64, 51.67, 51.75
Mihashi, S. 53.14
Mihichuk, L.M. 86.206
Mikami, M. 76.76, 83.124, 83.125, 83.126
Mikeladze, E.A. 76.3
Mikelashvili, Z.V. 81.80, 83.217
Mikhailov, I.E. 15.6
Mikhailov, Yu.N. 3.35, 79.1, 79.2
Mikhaleva, I.I. 48.31
Miki, K. 41.70, 77.2
Mikolajczyk, M. 41.64, 64.8
Milani, G. 41.15, 41.16
Milano, E. 17.10
Miler-Srenger, E. 41.65, 84.120
Miles, H.T. 47.27
Milewski-Mahrla, B. 64.11, 64.146
Milia, F. 3.46, 3.47
Miller, D.J. 73.64
Miller, J.A. 14.4
Miller, J.S. 60.20, 60.42
Miller, K.F. 85.37
Miller, R.M. 3.39

Miller, R.W. 22.1, 22.2, 22.3, 31.31, 35.21, 39.24, 51.63, 51.64, 51.67, 58.75, 71.14, 75.3
Miller, W. 75.41
Milligan, W.O. 38.110, 83.113, 83.202
Mills, H.H. 41.67
Mills, K. 31.63
Mills, R.M. 75.25
Milne, C.R.C. 73.101
Milone, L. 71.35, 71.64
Miminoshvili, E.B. 76.3
Mimoun, H. 81.79, 84.17, 84.71
Minacheva, L.Kh. 84.56
Minale, L. 54.5, 54.12
Minamikawa, J.-I. 37.12
Mingiardi, M.R. 41.52
Mingos, D.M.P. 73.21, 73.112, 73.130, 86.139, 86.220
Mink, R.I. 73.73, 74.11, 74.12
Minkin, V.I. 15.6, 22.5, 38.74, 62.18, 68.8, 68.19, 85.82, 85.83
Minkina, L.S. 85.82, 85.83
Minshall, P.C. 73.130, 86.139
Mintz, E. 73.119, 73.125
Miravitlles, C. 13.12, 28.9, 28.17, 28.18, 29.10, 33.2, 54.37
Mirsky, K. 31.58, 31.59
Misailova, T.V. 81.7
Mishnev, A.F. 1.29, 38.33, 44.23, 44.24
Mislow, K. 16.15, 19.13, 19.14, 21.19, 48.77, 48.78, 48.79, 63.40, 63.41, 63.51, 69.33, 69.34
Misra, R. 58.50
Missert, J.R. 71.32
Mitani, K. 35.65
Mitcham, R.V. 11.18
Mitchell, P.R.K. 20.8
Mitchell, R.H. 39.63
Mitchell, S.J. 54.8
Mitschler, A. 49.23, 49.27, 75.4, 81.79, 84.17, 84.71
Mitsudo, T.-A. 72.27
Mitsuhashi, H. 58.72
Mitsuhashi, Y. 83.197
Miura, S. 71.135
Miwa, T. 71.113, 71.114, 71.115
Miyagawa, Y. 72.27
Miyagi, Y. 28.13
Miyahara, K. 45.57, 51.62, 54.4, 54.6
Miyamae, H. 69.5
Miyao, A. 38.55, 47.17
Miyata, N. 58.26
Miyazaki, S. 36.16, 36.34
Miyazaki, W. 50.21
Miyazawa, T. 47.34
Miyoshi, F. 39.6
Mizsak, S.A. 45.18, 50.17
Mizuno, H. 47.17
Mizuno, S. 28.13
Mizusawa, E. 62.10
Mo, F. 11.6, 11.9, 11.15
Moad, G. 40.15
Modro, T.A. 64.40
Mody, N.V. 54.11, 58.60, 58.67, 58.68
Moedritzer, K. 64.38
Moews, P. 59.46
Mohan, N. 3.23
Mohr, G. 86.36
Mohr, S. 28.16, 40.56, 58.27
Mok, K.F. 36.26, 36.27, 36.29

Molin-Norris, J.A. 2.29, 2.30
Molinaro, F.S. 49.34, 49.35
Moll, M. 73.71
Molle, L. 13.8, 13.11
Molloy, K.C. 69.8, 69.18, 69.31, 69.37, 69.46, 69.48, 69.60
Molodkin, A.K. 85.11
Moloy, K.G. 73.33, 73.63
Monache, F.D. 56.2, 59.34
Mondal, P. 62.14
Mondi, C. 69.22
Mondon, A. 58.27
Money, T. 31.36
Monge, A. 83.169
Monger, D.J. 51.88
Montauzon, D.de 71.181
Montgomery, J.A. 36.33, 48.63
Monti, C.T. 39.48
Montreuil, J. 45.49
Moody, D.C. 38.68
Moore, D.S. 71.278
Moore, P. 36.26, 36.27, 36.29, 73.129, 83.64, 85.21
Moore, P.B. 2.29, 2.30
Moore, R.E. 1.34, 54.5, 54.12, 58.43
Moore, R.H. 51.40, 51.47
Mootz, D. 16.16, 33.4, 33.5, 39.19, 64.24, 66.10, 82.16, 82.29, 82.31
Moreau, S. 59.16
Moreiras, D. 13.12, 16.5
Morelock, M.M. 82.36, 82.37
Moreno, E. 38.42, 45.16
Morisaki, N. 53.20
Moriyama, Y. 56.5
Mornon, J.-P. 51.18, 51.23, 51.30
Moroni, W. 36.2, 36.17
Morosetti, S. 51.50
Morosin, B. 5.7, 83.6, 83.7
Morpurgo, G.O. 71.78
Morris, D.G. 30.13
Morris, G.E. 73.129
Morris, J.S. 69.24, 69.25, 69.35, 69.41
Morris, R.H. 85.5
Morrison, R.M. 83.178
Morton, S. 86.46
Moseley, J.E. 59.6
Moseley, P.T. 76.66
Moser, W. 69.1
Mosina, L.V. 84.112
Mosini, V. 71.78
Moss, G.P. 57.1
Mosset, A. 83.204
Mossini, F. 41.52
Motevalli, M. 11.19, 20.10, 20.15
Motschi, H. 78.5, 78.14
Mott, G.N. 71.211, 86.48
Moustakali-Mavridis, I. 33.44, 33.51
Movsumov, E.M. 69.14, 69.26, 81.27, 81.54, 81.55, 81.58
Moynihan, K.J. 80.1, 86.88
Mucker, K.F. 76.107
Muetterties, E.L. 3.33, 3.37, 71.44, 74.11, 74.12, 81.60, 86.58
Mugnoli, A. 84.91
Muhlsiepen, K. 85.28, 85.29
Muhonen, H. 84.70
Muir, J.A. 84.10, 86.127
Muir, K.W. 1.4, 71.103, 86.95, 86.182, 86.185, 86.188, 86.193, 86.198, 86.223
Muir, M.M. 86.127

Mukhopadhyay, A. 3.17, 83.167
Mullane, J. 76.50, 76.51, 76.52
Mullen, D. 8.1
Muller, A. 3.23, 3.24, 64.123, 86.199, 86.201, 86.202, 86.203, 86.219
Muller, B. 83.4
Muller, F.A. 38.2
Muller, G. 64.11, 71.43
Muller, J. 72.34, 73.85, 75.8
Muller, K.-H. 34.2, 34.5, 34.7
Muller, R. 73.93, 86.31
Muller, T. 41.6
Muller, U. 64.1, 64.124, 65.17, 83.5
Mullica, D.F. 38.110, 83.113, 83.202
Mulligan, K.J. 76.28
Mullins, M.J. 27.4
Munakata, T. 41.70
Munari, M. 80.9
Mundt, O. 64.106
Munro, M.H.G. 5.12
Mura, P. 78.5, 78.14, 86.169
Muradyan, L.A. 32.49
Murakami, T. 53.17
Murakoshi, I. 58.8
Muraoka, K. 40.26, 40.27
Murata, I. 22.7
Muratova, N.M. 82.15
Murmann, R.K. 83.88
Murphey, M.B. 85.26, 85.27
Murphy, A. 76.50, 76.51, 76.52, 83.112, 83.140, 83.141
Murphy, B. 81.71, 81.72
Murphy, P.T. 54.15, 54.16, 54.19
Murray, K.S. 83.225, 85.61
Murray, M. 71.263, 86.147
Murray-Rust, J. 27.1, 27.2, 31.3, 31.11, 31.63, 35.13, 36.30, 38.9, 41.78, 84.11
Murray-Rust, P. 27.1, 27.2, 31.3, 31.11, 31.43, 31.63, 35.13, 36.30, 38.7, 38.9, 38.52, 41.78, 84.11
Mursaliev, A.A. 81.27
Murtaza, N. 55.4
Murthy, A.R.V. 64.74
Musaev, F.N. 81.54, 81.55, 81.58, 81.98
Musatti, A. 35.63, 69.27, 78.2, 84.47
Muschiol, M. 75.36
Musco, A. 86.104
Musker, W.K. 85.26, 85.27
Mutai, K. 39.85
Mynderse, J.S. 58.43
Mynott, R. 72.8, 75.9

Naae, D.G. 60.16, 60.18, 60.19, 60.23
Nabeya, A. 32.68
Nader, B. 40.8
Nadig, H. 50.27
Nadzhafov, G.N. 69.14, 69.26, 81.27, 81.54, 81.55, 81.58
Naegeli, P. 53.28
Nagao, Y. 41.28
Nagaraj, R. 48.65
Nagata, K. 83.197
Nagata, M. 69.5
Nagel, U. 71.1, 86.163
Nagele, W. 5.2, 5.3
Nagumo, M. 16.15
Naik, V.M. 76.70
Nair, C.M. 48.62

Nair, C.M.K. 40.41, 50.36
Naish, P.J. 71.86, 71.87, 71.88, 71.212, 72.38
Nakahara, A. 82.18, 82.19
Nakahara, S. 50.20
Nakai, H. 50.5, 59.44, 83.142
Nakaidze, L.I. 63.36
Nakajima, S. 59.9
Nakamachi, H. 44.25
Nakamura, A. 67.5, 71.31, 86.96, 86.97
Nakamura, H. 50.2
Nakamura, Y. 71.113, 71.114, 71.115, 71.269, 84.96
Nakanish, H. 20.10, 20.15
Nakanishi, H. 33.76
Nakano, Y. 50.21
Nakao, K. 22.7
Nakata, K. 13.2, 48.11
Nakatsu, K. 22.7, 71.113, 71.114, 71.115, 71.269, 72.27, 84.48, 85.48, 86.96, 86.97
Nakayama, E. 41.50
Nakayama, J. 31.64
Nakayama, K. 51.80
Nakayama, M. 53.21, 54.31, 54.32
Nakazawa, T. 22.7
Nalewajek, D. 80.14
Namgung, H. 65.22
Namiki, M. 40.3
Nanjan, M.J. 17.42
Narayana, S.V.L. 53.7
Narayanan, C.R. 56.1
Narayanan, P. 67.4, 67.6
Nardelli, M. 41.52, 69.27, 78.2, 84.47
Nardi, N. 84.83, 85.42, 86.122, 86.123
Nardin, G. 1.31, 76.5, 76.7, 76.102, 81.90, 82.27, 86.160
Narita, K. 39.65
Nasakkala, E. 76.6, 83.68, 83.98
Nasakkala, M. 76.65, 81.23, 83.68, 83.98, 84.67
Nasini, G. 59.3
Nasirov, S.-M. 58.44
Nassimbeni, L.R. 11.16, 20.8, 32.51, 33.71, 41.54, 41.55, 51.78, 51.79, 64.65, 64.117, 67.3, 71.66, 71.138, 83.130, 84.104
Natarajan, S. 48.3
Nathans, J.N. 85.58
Natile, G. 71.187, 76.5, 76.7, 83.82
Navarro, R. 71.149, 71.173, 71.185, 71.186
Navaza, A. 77.11
Navech, J. 64.48, 64.73
Nawata, Y. 38.142
Nayak, A. 40.51, 40.53, 40.55
Neagley, C.H. 85.26, 85.27
Neckers, D.C. 35.72
Nefedov, V.I. 86.226, 86.227, 86.229
Neidle, S. 47.31, 47.38, 60.24
Neigauz, M.G. 27.23
Nelson, J.H. 86.39, 86.81, 86.86
Nelson, S.M. 33.90, 34.11, 83.187, 84.109
Nelson, W.J.H. 3.56, 64.166, 65.13, 65.15
Nemchinova, L.A. 83.36
Nerone, A. 71.73, 83.41, 83.42, 85.24
Nersisyan, A.M. 31.17, 31.22
Nes, G.J.H.van 5.4, 5.6
Nesmeyanov, A.N. 73.84, 83.90
Nesterov, D.Yu. 63.36
Nesterova, Ya.M. 76.55, 76.57, 82.15
Nettleton, D.E. 59.6
Neuenschwander, M. 27.22

Neugebauer, D. 27.32, 64.43, 64.150, 71.60, 71.125, 71.141, 71.201, 74.9, 83.61, 84.42, 84.100, 86.38, 86.74, 86.179, 86.222
Neuman, A. 45.24, 45.29
Neumann, A. 3.23, 3.24
Neumann, S. 86.23
Neupert-Laves, K. 38.44
Neustadt, R.J. 83.190
Neverov, V.A. 76.1, 78.7, 79.12
Nevestveit, O. 32.58
New, L. 71.278, 81.84
Newell, C.E. 41.78
Newkome, G.R. 40.51, 40.53, 40.55, 84.99
Newman, P.J. 85.61
Newsam, J.M. 21.4, 21.5
Newton, C.G. 33.23
Newton, R.F. 27.1, 27.2, 38.9, 61.6
Newton, W.E. 72.47
Ng, W.W. 31.67, 64.40
Ni, C.-C. 86.124
Nicholls, J.N. 3.56, 86.126
Nichols, B.H. 45.40, 45.41
Nichols, S. 64.45, 64.46
Nicholson, G.A. 84.72
Nicholson, J. 54.35
Nicoletti, R. 71.20, 71.98
Nicolini, M. 3.54
Niecke, E. 64.98, 86.13
Niedner, R. 33.78, 63.27, 63.52
Niedzwiecki, S. 39.43, 64.93, 64.140, 64.147, 64.148
Niekerk, J.C.van 32.51, 33.71, 51.78, 51.79, 67.3
Nielsen, K. 3.21, 3.22, 32.11, 32.24, 32.25, 32.26, 32.27, 32.32
Nieminen, K. 64.143, 84.8
Nieuwenhuis, J.J. 45.33
Nifant'ev, E.E. 38.10, 46.6, 64.17, 64.32, 64.33, 64.52, 64.53
Nifontova, G.A. 83.79, 83.80, 84.19, 84.27, 84.28, 84.43, 84.44
Niggli, A. 3.2
Nikitina, E.V. 32.8, 32.9
Nikolaev, V.P. 84.37
Nishida, T. 54.22
Nishigaki, S. 84.48, 85.48
Nishikawa, J. 50.5
Nishikawa, M. 48.47
Nishikiori, S. 76.23
Nishimoto, K. 71.113, 71.114, 71.115
Nishimura, S. 39.8, 47.34
Nishimura, T. 32.55
Nishioka, I. 54.4, 54.34
Niswander, R.H. 76.106
Niven, M.L. 83.130
Nixon, P.E. 81.8
Noble, W.J.le 31.46
Nogaideli, A.I. 63.36
Nohara, T. 51.85, 54.4, 54.34
Nolte, M.J. 32.30
Noltemeyer, M. 45.63, 45.64, 45.68
Noltes, J.G. 64.170, 69.45, 71.203
Nomura, A. 50.9
Nonaka, Y. 78.12, 78.13
Noorani, R. 55.4
Noordik, J.H. 5.14, 11.13, 31.2, 31.7, 41.56, 71.280, 86.162
Nordenson, S. 45.20
Norman, A.D. 64.92, 86.19

Norman, J. 72.68
Norte, M. 27.14
North, H. 63.17
Norton, M.C. 86.168
Norton, T. 83.151
Noth, H. 3.41, 62.11, 63.4
Nothe, D. 36.2, 36.17
Nourmamode, A. 28.15
Novakovskaya, L.A. 27.23
Novotny, M. 76.8, 76.32
Nowacki, W. 31.55
Nowell, I.W. 64.108, 64.137, 71.152, 73.96, 79.13, 83.28, 86.6, 86.7, 86.110
Nowicki, S.C. 86.184
Nowogrocki, G. 84.34
Nowoswiat, E.F. 45.11
Nozaki, H. 53.21, 53.33, 54.31, 54.32
Nozoe, S. 53.20, 55.1, 55.3
Nuber, B. 76.81, 83.168
Nuffel, P.van 42.2
Nugent, W.A. 83.39, 83.55, 83.154, 84.106, 84.116
Numahata, Y. 69.5
Numazawa, M. 51.5, 51.8, 51.43
Nunes, A.C. 48.12
Nurakhmetov, N.N. 1.5
Nutton, A. 73.116
Nyburg, S.C. 31.67, 58.50, 63.48, 64.40, 75.27, 81.67

O'Brien, E. 73.56, 75.30
O'Brien, J.P. 64.29
O'Brien, N. 81.71, 81.72
O'Connell, C.M. 83.157
O'Connor, C.J. 76.104, 83.122
O'Connor, J.P. 86.77, 86.208
O'Connor, M.J. 84.49
O'Donnell, M. 20.7
O'Donoghue, T.D. 83.163, 83.165
O'Hanlon, P.J. 41.10
O'Leary, M.K. 47.26
Oakley, R.T. 64.50, 64.87, 64.167
Oberhansli, W.E. 32.62, 38.5
Oberti, R. 35.77, 38.112, 40.31, 40.32
Ochi, M. 59.44
Oda, M. 27.31
Odashima, K. 61.18
Odou, G. 59.16
Oehler, J. 40.46
Oesch, N. 27.22
Oetjen, H.-H. 84.73
Ogawa, K. 44.5, 50.9, 60.6, 78.9, 78.18
Ogawa, N. 35.65
Ogoshi, H. 49.7
Ogura, H. 32.55, 39.65, 58.40, 83.59
Ohashi, T. 39.6
Ohashi, Y. 38.137, 48.13, 50.4, 71.143, 71.144, 71.145, 71.146, 83.197
Ohba, M. 33.68
Ohba, S. 76.46
Ohga, Y. 75.42
Ohgo, Y. 71.143, 71.144, 71.145, 71.146
Ohki, E. 41.50
Ohmasa, M. 82.21
Ohmiya, S. 58.8
Ohno, S. 39.65
Ohnuma, Y. 71.31
Ohsaki, K. 58.72

Ohsawa, A. 9.3, 64.76
Oida, S. 41.50
Ojala, W.H. 14.4
Ojima, I. 75.43
Okamoto, K.-I. 82.21
Okamoto, T. 36.20
Okano, T. 81.66, 86.130
Okaya, Y. 31.44, 31.45, 31.46
Okuda, S. 55.4
Okuda, Y. 13.19, 41.61, 41.77
Olejnik, S. 35.71, 35.75, 35.78, 35.79, 35.80, 35.81, 35.82, 48.6, 48.7
Olekhnovich, L.P. 15.6, 22.5, 38.74
Oleksyn, B. 35.3, 58.38
Olieman, C. 58.39
Oliver, J.D. 2.27, 81.21, 83.113, 83.202
Oliver, J.P. 63.1, 63.11, 63.50, 71.65
Olivier, M.J. 71.5
Ollis, W.D. 50.3
Olmstead, M.M. 71.275, 85.26, 85.27, 86.176, 86.178, 86.194
Olsen, H. 35.9
Olsen, R.K. 50.33, 50.34
Olson, J.M. 40.2
Olthof, G.J. 59.36
Omiya, S. 49.14
Omura, S. 58.40
Onak, T.P. 86.33
Ono, T. 82.18, 82.19
Ooi, S. 76.22, 76.33, 76.34, 76.35, 81.38, 81.39, 81.40, 84.50, 84.96, 86.28
Ooijen, J.A.C.van 83.95, 83.161
Oonk, H.A.J. 23.6
Opdenbosch, N.van 49.33
Opitz, J. 86.98
Orama, M. 76.86
Orama, O. 64.142, 64.143, 71.43, 71.136, 76.86
Orazi, O.O. 32.21
Orbeladze, F.V. 83.77
Orel, B. 2.7, 2.8
Orioli, P. 76.71, 84.83, 84.125, 85.40
Orlandini, A. 35.5, 72.69, 86.149, 86.181, 86.226, 86.227, 86.229
Orlov, V.D. 27.17
Orpen, A.G. 71.86, 71.87, 71.88
Ortega, R. 83.7
Ortega, R.B. 81.28
Osaki, K. 49.7, 59.30
Osawa, T. 40.3
Osawa, Y. 51.5, 51.8, 51.32, 51.36, 51.43
Osborn, R.S. 1.12
Osborne, A.G. 73.57, 73.91
Osipov, O.A. 85.82, 85.83
Oskarsson, A. 32.14, 32.15, 32.16, 80.28, 80.29
Osmanov, N.S. 81.4, 81.7
Osterberg, F. 17.6
Osterle, F. 36.38, 63.34, 63.35
Ostrikova, V.N. 73.25, 81.18, 81.31, 81.51
Otani, S. 36.50
Otomasu, H. 58.8
Otsuka, H. 48.46
Otsuka, S. 81.66, 86.96, 86.97, 86.130
Ott, W. 27.27
Ottersen, T. 39.34, 39.35, 39.76
Ou, C.-C. 85.91
Ouchi, A. 76.61, 81.97
Ovchinnikov, Yu.A. 50.39
Overbeek, O. 83.205

Overbosch, P. 83.205
Overman, L.E. 21.18
Owen, J.D. 38.66, 38.67, 38.131, 62.15
Owston, P.G. 83.158
Oya, A. 36.50
Ozol, L.D. 83.81
Ozola, A.D. 35.10
Ozols, Ya.K. 35.10, 35.48, 85.63, 85.64, 85.76

Pacheco, P. 38.83
Packard, A.B. 37.10, 83.66, 83.70, 83.71, 83.105
Paddick, K.E. 72.13
Paddock, N.L. 64.171
Paetzold, P. 62.13
Page, E.M. 65.18
Page, S.W. 17.18
Pagnoni, U.M. 52.2, 52.3
Paine, R.T. 68.22, 73.24
Pajunen, A. 76.6, 76.65, 76.90, 81.23
Pajunen, S. 76.90
Pal, B.C. 54.6
Palenik, G.J. 68.15, 79.11, 84.92, 84.93, 84.94, 84.95
Palenik, R.C. 84.93, 84.94, 84.95
Palleschi, A. 51.50
Palm, J. 31.60
Palm, T.-B. 17.32
Palmer, A. 13.15
Palmer, R. 58.69
Palmer, R.A. 36.43, 51.60, 76.40, 84.16, 84.26
Palyi, G. 71.107, 86.84
Pampaloni, G. 49.5
Panagiotopoulos, N.C. 41.59
Panagiotopoulos, N.G. 41.37
Panfilova, L.V. 39.18, 39.20, 39.21
Pankiewicz, K. 64.26
Pannell, K.H. 71.49, 75.7
Pant, L.M. 56.1
Panunzi, A. 72.53
Paoletti, P. 76.48, 76.77, 76.78
Paoli, G.de 84.39, 84.138
Paolucci, G. 83.114, 83.184, 83.185, 84.75
Papaefthymiou, V. 64.129, 64.130, 64.131
Pappalardo, G.C. 33.39
Pappalardo, S. 39.87
Paquette, L.A. 37.17, 37.18, 38.8, 38.16
Pardy, R.B.A. 71.219
Parish, E.J. 51.88
Park, J.M. 80.4
Parkanyi, L. 4.2, 13.6, 13.13, 13.14, 13.24, 35.16, 35.20, 35.33, 35.37, 41.43, 63.12, 63.15
Parker, W.C. 31.3, 31.11
Parker, W.L. 50.28, 50.29
Parkes, R.L. 86.137
Parkinson, G.M. 33.76
Parnes, Z.N. 73.31
Parrott, S.J. 21.8, 22.6
Parry, R.J. 58.45
Parthasarathy, R. 35.23, 38.24, 39.30, 39.31, 39.32, 39.38, 44.26, 45.14, 45.15, 47.8, 47.25
Parusnikov, B.V. 35.29, 35.52
Parvez, M. 30.20, 51.27, 51.34, 51.35, 71.7, 72.6, 73.72, 85.70, 85.81, 86.93

Pascard, C. 29.1, 30.1, 30.3, 30.5, 36.45, 38.104, 38.140, 48.22, 48.28, 48.29, 48.36, 50.35, 58.73, 59.31, 71.237, 75.28
Pascher, I. 3.55
Pasquali, M. 71.161, 72.10, 73.131, 76.4, 76.49, 76.87, 76.91, 78.8
Pasteels, J. 54.20
Pasynskii, A.A. 73.25
Patel, C.C. 76.70
Patel, G.N. 1.39
Patel, V.D. 85.38, 85.43
Paternostro, M. 54.26
Patin, H. 71.153, 73.49
Patni, R. 54.9
Paton, R.M. 40.1, 45.66
Paton, W.F. 63.33, 82.28
Patrick, J.M. 84.101
Patroni, J.J. 45.13
Pattabhi, V. 35.26, 51.83
Paul, I.C. 1.39, 17.28, 17.29, 17.37, 21.14, 21.15, 21.16, 24.10, 25.1, 38.139, 59.17
Paul, W. 38.141
Pauling, P. 62.14
Pauling, P.J. 59.13
Paulsen, H. 45.30, 45.35, 45.36, 45.37, 45.52, 45.53, 45.55, 45.58
Paulus, E.F. 32.13, 35.74, 64.25
Pavel, N.V. 61.12
Pavelcik, F. 76.64, 82.17
Pavkovic, S.F. 84.57
Pavlovic, V. 51.75
Pavone, V. 48.33, 48.34, 48.41, 48.74
Pawelke, G. 62.1
Pawley, G.S. 3.16
Paxton, L.De C. 39.49
Payne, D.R. 40.16
Payne, N.C. 71.162, 72.39, 86.136, 86.137
Pearson, A.J. 75.21
Pech, L.Ya. 85.63, 85.76
Peckler, S. 41.28
Pedersen, B.F. 56.6
Pedone, C. 48.33, 48.34, 48.41, 48.74
Peeters, O.M. 8.13, 13.8, 13.11, 32.65, 33.84, 33.86, 33.92, 35.41, 58.33, 58.41, 58.42, 83.87
Pelin, K. 76.36
Pelister, Y. 38.61, 38.71
Pelizzi, C. 69.27, 78.2, 84.47
Pelizzi, G. 41.52, 49.5, 76.103
Pellacani, G.C. 33.19, 33.46, 82.41, 83.65
Pelletier, S.W. 54.11, 58.60, 58.67, 58.68
Penavic, M. 33.11
Penfold, B.R. 82.23, 84.59, 84.102
Pennings, M.L.M. 32.58
Pentegova, V.A. 53.12
Pentilla, R.E. 73.101
Pepe, G. 49.33
Perales, A. 27.14, 53.19, 54.9, 54.25, 64.2
Perego, G. 81.99
Perez, C. 27.14
Perkins, M.J. 19.17, 36.1, 36.5, 41.38
Perkinson, N.A. 32.41, 32.56, 35.14, 36.42, 41.57
Perkowska, A. 58.16
Perlmutter, H.D. 38.90
Perret, R. 3.14
Perrin, D.R. 59.15
Perrin, M. 17.2, 17.3, 17.15, 60.2
Perry, D.L. 32.6, 32.7
Pertici, P. 86.192

Perucaud, M.C. 33.50, 85.44, 85.46
Peruzzini, M. 86.115
Peruzzo, V. 69.39, 69.44
Peters, E.-M. 27.27
Peters, K. 27.27, 32.59, 64.35, 65.5, 71.1
Petersen, J.L. 64.153, 72.26, 73.59, 73.100, 84.72
Peterson, L.K. 86.60, 86.61
Peterson, S.W. 81.16
Petit, G.H. 33.41, 33.58, 40.4, 42.1, 42.2
Petricek, V. 62.3, 62.5, 73.4
Petrignani, J.-F. 64.160
Petrina, A. 62.3, 73.4
Petropavlov, N.N. 51.25, 51.39
Petrouleas, V. 64.129, 64.130, 64.131
Petrovic, D. 83.75
Pett, V.B. 48.40, 48.44
Petter, W. 3.2
Pettersen, R.C. 49.33, 71.49, 75.7
Pettit, G.R. 51.69
Pettit, R. 71.9, 71.10
Petty, C.B. 21.18
Petz, W. 71.63
Pevzner, M.S. 32.8, 32.9
Peytavin, S. 39.27, 39.64
Pez, G.P. 71.230
Pfab, R. 83.61
Pfeil, H.-R. 85.18
Pfoertner, K.-H. 32.62
Pfund, R.A. 31.50
Philips, R.C. 69.19
Phillips, D.J. 85.59
Phillips, G.H. 51.27, 51.34, 51.35
Phillips, L.R. 38.61, 38.71
Phillips, R.J. 68.25
Phillips, T.E. 49.18
Phillips Junior, G.N. 64.12
Piazzesi, A. 71.213
Piazzesi, A.M. 39.12, 71.55, 71.80, 71.248, 71.249, 80.5, 80.8, 80.9, 80.10
Picard, J.-P. 63.48
Piccardi, P. 17.26, 20.3
Pickardt, J. 11.3, 72.34, 73.43, 75.11, 86.98, 86.103
Pickering, M. 1.19
Pierpont, C.G. 38.47, 71.160, 71.246, 71.265, 84.66, 84.123, 84.139, 86.85
Pietrusiewicz, K.M. 64.105
Pignolet, L.H. 71.240, 80.22, 80.23, 80.26, 80.27, 83.219, 86.184, 86.205, 86.224
Pilati, T. 5.13, 39.40
Pilon, P. 44.8
Pilotti, A.-M. 28.19, 38.102, 38.116, 54.22
Pinal, J.L.B. 13.21
Pinkerton, A.A. 72.15, 72.16, 72.20, 72.25, 75.5, 85.30, 85.87, 85.88, 85.90, 86.10
Pinnen, F. 40.48, 48.49
Pinsky, B.L. 73.99
Pinzino, C. 1.8
Piovesana, O. 85.17
Piozzi, F. 54.26
Pipal, J.R. 71.54, 71.242, 73.77, 73.87
Pippard, D. 72.14
Piraino, P. 73.127
Pirtle, S.L. 60.16, 60.23
Pitteri, B. 76.5, 76.7
Pittman Junior, C.U. 86.77, 86.208
Pivovarova, M.G. 81.18
Piwinski, J.J. 21.7
Pizzala, L. 27.6

Pizzolato, G. 39.44, 39.72
Plaas, D. 71.86, 71.87
Plana, F. 28.9, 28.17, 28.18
Plank, J. 71.126, 71.140
Plant, J.S. 33.31
Plas, H.C.van der 33.38, 33.65
Plastas, L.A. 33.45
Plati, J.T. 41.5
Platte, C. 64.42
Plattner, J.J. 36.49
Plazzogna, G. 73.67
Plesko, S. 3.46, 3.47
Pletcher, J. 44.35
Pletnev, V.Z. 48.31, 48.53, 48.81, 50.39
Ploeg, A.F.M.J.van der 71.183
Ploeger, F. 71.68, 72.50
Plute, K. 86.191
Pluth, J.J. 2.29, 2.30, 72.59
Pocar, D. 20.12, 20.13
Pochini, A. 61.21
Podberezskaya, N.V. 63.21, 76.15, 76.16, 79.14, 82.1, 82.2, 82.5, 82.6, 82.9, 82.10, 82.13
Podjarny, A.D. 60.9
Podlahova, J. 81.85
Poel, H.van der 72.33
Pohl, L.R. 1.36
Pohl, S. 11.1, 64.61, 86.8
Poilblanc, R. 71.17, 71.18, 71.181, 71.217, 86.41, 86.42, 86.92
Pointer, D.J. 35.73
Poje, M. 11.17, 32.13
Pokrovskaya, I.E. 17.20, 26.1, 66.17
Polishchuk, A.A. 51.81
Polishchuk, V.R. 19.11
Poll, W. 64.24, 85.89
Pollard, D.R. 59.22
Polm, L.H. 71.68, 72.50
Poloni, M. 35.59
Polonsky, J. 38.140, 58.73
Polyakova, I.N. 35.29, 35.52, 82.11
Polyakova, M.V. 63.6
Polynova, T.N. 38.6, 38.11, 38.12, 38.21, 38.28, 38.36, 38.50, 82.11, 82.15
Pomeroy, R.K. 86.4
Pommerening, H. 3.41, 63.4
Pomponi, M. 56.2
Ponomarev, V.I. 17.25, 17.27, 40.52, 83.79, 83.80, 84.19, 84.27, 84.28, 84.43, 84.44, 85.39
Pook, K.-H. 36.12
Poon, Y.C. 63.48
Poonia, N.S. 17.1, 38.65
Poonian, M.S. 45.11
Popa, E.V. 76.1
Popkov, Yu.A. 34.6, 34.8, 38.96, 40.42, 40.44
Popova, E.G. 36.6
Popovitz-Biro, R. 51.53
Popowski, E. 63.12, 63.15
Poppleton, B.J. 58.51, 59.20
Porai-Koshits, M.A. 3.26, 7.1, 38.6, 38.11, 38.12, 38.21, 38.28, 38.36, 38.50, 73.25, 76.10, 76.55, 76.57, 76.58, 76.59, 81.18, 81.31, 81.42, 81.51, 82.11, 82.15, 83.103, 83.162, 84.37, 84.55, 84.56
Porta, P. 71.78
Portalone, G. 13.4
Porter, N.A. 71.14
Porzio, W. 32.64, 86.155, 86.183, 86.192

Posin, B. 75.27
Post, M.L. 47.3
Postnov, V.N. 73.84
Potekhin, K.A. 19.7
Potenza, J.A. 32.31, 73.7, 76.26, 76.44, 84.22, 85.91
Potvin, C. 64.133
Poucke, M.van 35.67
Pougny, J.-R. 45.29
Pouskoulelis, G. 85.69
Povet'eva, Z.P. 24.12
Powell, H.M. 17.5
Power, P.P. 85.78, 85.96, 85.97
Prabhu, U.D.G. 31.4, 31.5
Pradillas, J. 81.77
Prandl, W. 5.2, 5.3
Prange, T. 29.1, 30.1, 30.3, 30.5, 36.45, 38.140, 48.22, 48.28, 48.29, 48.36, 53.35, 58.73
Prasad, B.V.V. 48.65
Prasad, V.A.V. 64.114
Precigoux, G. 17.41, 51.1, 51.14, 51.21
Predieri, G. 69.27, 78.2, 84.47
Preez, J.G.H.du 84.104
Pregosin, P.S. 71.208, 78.5, 78.14, 86.21, 86.22, 86.104
Press, J.B. 41.57
Prest, D.W. 86.106, 86.112
Preston, N.W. 58.48
Preston, P.N. 32.46, 32.47, 40.6, 40.7, 40.33
Prestwich, G.D. 54.30
Pretorius, J.A. 45.33
Preut, H. 32.77, 64.90, 71.239
Prewo, R. 50.13, 50.14, 50.15
Prick, P.A.J. 31.7
Prikner, B. 76.81
Prinzbach, H. 34.2, 34.5, 34.7
Proctor, G.R. 38.80
Prokhvatilov, A.I. 5.1
Prokscha, H. 66.18
Protas, J. 1.27, 19.4, 38.73, 45.32, 48.35, 48.48
Prout, C.K. 86.56
Prout, K. 36.32, 37.14, 37.16, 42.12, 71.156, 72.21, 73.15, 73.40, 73.117
Pruchnik, F. 86.105
Pruett, R.L. 86.221
Prusoff, W.H. 47.5, 47.19
Pryce, R.J. 48.17, 48.18
Przybylska, M. 28.12, 36.47, 58.22, 58.65
Pschigoda, L.M. 45.18
Puddephatt, R.J. 86.193, 86.223
Puebla, P. 86.63
Pugh, N.J. 86.147
Pukhnarevich, V.B. 71.23, 71.131
Purzer, A. 86.175
Putnik, C.F. 71.230
Putten, N.van der 64.120
Puza, M. 26.6, 26.7
Pye, P.L. 71.226, 71.231
Pyshnograeva, N.I. 71.104
Pyzalska, D. 43.2, 58.9, 58.10
Pyzalski, R. 43.2

Quast, H. 32.59
Quin, L.D. 39.24
Quiocho, F.A. 51.88, 64.12

Quirk, J.M. 71.237
Qureshi, I.H. 55.4
Qureshi, M.Y. 48.17, 48.18

Rabinovich, D. 40.17
Radanovic, D.J. 76.72
Radonovich, L.J. 71.133, 71.134
Radwan, M.M. 47.6, 47.15
Rae, A.D. 64.64, 64.149
Rae, I.D. 33.32
Rafikov, S.R. 39.23
Ragatz, P. 73.3
Ragulina, N.B. 79.1, 79.2
Rainville, D. 11.12, 38.35, 39.41, 70.1
Raithby, P.R. 64.165, 65.13, 71.56, 71.57, 71.59, 71.90, 71.148, 71.210, 72.14, 72.37, 75.21, 83.73, 83.163, 83.164, 83.165, 83.180, 84.51, 86.17, 86.40, 86.106, 86.112
Rajapaksa, D. 38.134
Rajasekharan, M.V. 84.62, 84.63
Raju, M.S. 51.63, 51.64, 51.67
Ram, P. 27.7
Ramakumar, S. 37.4
Ramalingam, K. 36.4, 39.56, 39.74
Ramdas, V. 2.13, 2.14, 2.15
Ramirez, F. 64.114, 67.4, 67.6
Rampal, A.L. 58.74
Ramsden, C.A. 32.72, 33.94
Ramsden, J.N. 83.165
Ramuz, H. 32.23, 32.29
Ramzaeva, N.P. 44.23, 44.24
Randaccio, L. 1.31, 13.3, 32.34, 39.62, 39.70, 39.87, 71.100, 76.5, 76.7, 76.102, 81.90, 82.27, 84.65, 86.16, 86.73
Rango, C.de 77.11
Rankin, D.W.H. 69.7
Rao, G.S.R.S. 38.120
Rao, J.K.M. 48.3, 48.5
Rao, K.S. 31.6, 31.16
Rao, L.M. 17.8
Rao, M.N.S. 4.1, 41.9, 63.2
Rao, P.N. 51.11
Rao, V.R. 31.57
Raper, E.S. 79.13
Rapposch, M. 17.11, 59.46
Rasmusson, G.H. 27.13
Rastogi, R.P. 59.36
Raston, C.L. 17.44, 25.2, 33.7, 33.32, 34.12, 35.76, 38.132, 38.138, 40.47, 54.8, 59.14, 59.28, 71.218, 71.224, 71.229, 76.9, 76.17, 76.21, 80.11, 83.210, 84.86
Rastrup-Andersen, N. 50.3
Ratajczak, H. 2.6, 2.11
Ratka, R. 72.1, 72.31
Rau, T.F. 31.18, 31.32
Rau, V.G. 31.18, 31.32, 31.37, 31.56, 85.11
Raubenheimer, H.G. 76.62, 85.49
Rauchfuss, T.B. 86.138
Rausch, M.D. 72.45, 73.5, 73.29, 73.118, 73.119, 73.125
Ravenscroft, M.D. 83.149
Ray, N. 76.36, 76.82, 81.71, 81.72
Raymond, K.N. 1.32, 84.87, 84.88, 84.90, 85.66, 85.72, 85.73, 85.74, 85.75
Raynaud, L. 50.35
Razenberg, H. 33.82, 33.91
Razuvaev, G.A. 73.134, 73.135

Razzouk, C. 28.8
Real, F.M. 86.11, 86.29
Reboul, J.P. 36.10
Reck, G. 33.54, 33.55, 33.56, 33.57, 59.32, 81.92
Recktenwald, O. 69.56
Reddy, B.S. 47.35, 47.36, 60.22
Reed, A.T. 1.10
Reed, C.A. 83.222, 84.136
Reedijk, J. 81.34, 83.95, 83.107, 83.161, 83.209, 84.81
Rees, C.W. 28.6
Reese, C.B. 47.30, 47.31
Reffstrup, T. 38.15
Regev, H. 60.1, 60.34
Rehder, D. 66.15
Reichert, W. 76.99, 76.100
Reichsfeld, V.O. 63.43
Reidinger, F. 69.8
Reimann, R.H. 86.87
Reina, M. 58.55, 58.56
Reinhardt, R. 17.1
Reinhoudt, D.N. 32.58, 40.38
Reis Junior, A.H. 60.20, 60.42, 81.16
Reisenauer, H.P. 36.51
Reisner, G.M. 73.118, 73.119, 73.125, 75.36
Rejto, M. 45.54
Rekers, J.W. 37.10
Remadna, K. 86.12, 86.37
Remy, G. 45.45, 45.46
Rengstl, A. 63.37, 63.42
Rentzeperis, P.J. 40.57, 80.24
Renz, W. 66.18
Rettig, S.J. 64.171, 68.16, 72.22, 80.4, 83.51, 83.52, 83.53, 83.54, 83.99, 83.150, 83.166, 84.32, 84.33
Reumov, O.A. 73.80
Rey, M. 50.3
Rey-Lafon, M. 3.3, 3.6, 3.7, 3.8, 3.9, 3.10
Reymann, J. 38.73
Reynolds, G.F. 27.13
Reynolds, P.A. 5.10, 5.11
Reynolds, W.F. 63.48
Ribakov, V.B. 38.10, 51.25, 51.39, 64.17, 64.32, 64.33, 64.52, 64.53
Ribar, B. 32.52, 35.30, 35.31, 35.68, 40.28, 40.37, 41.36, 41.46, 41.47, 41.66, 83.75
Ricard, L. 49.23, 49.27, 72.47
Ricci, J. 75.41
Ricci, J.S. 2.9
Riccio, R. 54.5, 54.12
Rice, D.A. 65.18, 85.52
Rice, G.W. 71.101, 71.102
Rice, K.C. 37.12
Rich, A. 85.58
Richard, L. 71.205
Richardson, M.F. 40.24
Riche, C. 27.8, 51.89, 58.28, 58.35, 58.54, 75.14
Richman, J.E. 64.78
Richter, A. 62.13
Richter, F. 86.34, 86.35
Richter, H. 83.220
Richter, R. 13.20, 64.20, 79.15, 85.19, 85.20, 85.32, 85.47
Richter, S.I. 86.116
Rickard, C.E.F. 84.69, 84.133
Riding, G.H. 75.25, 75.26
Ried, W. 38.79
Riegler, N. 20.11

Riemann, U. 32.53
Rienzo, F.di 15.1
Riffel, H. 86.180
Righetti, P. 35.77, 38.112
Rihs, G. 41.23, 51.75
Riley, J.P. 35.13
Riley, P.E. 3.18, 63.3, 71.9, 71.10, 71.62, 71.122, 75.37, 86.78, 86.79
Rillema, D.P. 83.198, 83.216
Rimsky, A. 69.22
Rinehart Junior, K.L. 59.17
Ring, H. 71.39, 71.40
Ripamonti, M.C. 50.18
Rist, G. 41.23, 51.75
Rist, P.E. 15.5
Ritsko, J.J. 60.42
Rittner, W. 3.23, 3.24
Riva di Sanseverino, L. 15.1
Rivero, B.E. 60.9
Rivest, R. 86.62
Rivett, D.E.A. 54.18
Riviere, G. 49.23, 49.27
Rivola, G. 50.18
Rizzardi, G. 71.187
Rizzarelli, E. 81.90, 82.27
Robbers, J.E. 48.46
Robbins, A. 11.20, 45.2, 45.3
Robbins, W.E. 51.71
Roberfroid, M. 28.8
Robert, J.B. 64.5, 64.73
Robert, P. 72.49
Robertiello, A. 38.90
Roberts, J.C. 59.33
Roberts, N.K. 64.109, 64.110
Roberts, P.J. 13.16, 81.94, 81.96
Robertson, B.E. 13.10, 13.17, 60.17, 76.75
Robertson, G.B. 26.6, 26.7, 72.56, 72.57, 72.58, 74.8, 75.45, 86.3
Robertson, J.M. 51.33, 51.42, 59.22
Robin, J.-P. 59.31
Robinson, B.H. 71.112
Robinson, J.C. 76.40
Robinson, S.D. 71.278, 86.165
Robinson, W.R. 83.171
Robinson, W.T. 5.12, 76.45, 82.23, 84.59
Rocchiccioli-Deltcheff, C. 84.9
Rochon, F.D. 76.84, 83.29, 84.25
Rocic, B. 32.13
Rode, G.G. 39.25, 39.26
Rodek, E. 41.6
Rodesiler, P.F. 49.28
Rodgers, A.L. 51.78, 51.79
Rodier, N. 33.9, 83.72, 84.60
Rodig, O.R. 83.172
Rodin, N.P. 78.16
Rodrigues, M.M. 83.170
Rodriguez, B. 54.9, 54.26
Rodriguez, J.G. 24.5, 54.13
Roeder, E. 38.25, 58.29
Roelofsen, G. 21.22
Roesky, H. 11.1
Roesky, H.W. 4.1, 39.3, 41.6, 41.9, 63.2, 64.42, 64.122, 69.12
Roey, P.van 85.38, 85.43
Roey, P.M.A.O.van 40.20, 40.21, 40.22
Rogan, P.K. 60.20
Rogers, D. 38.14, 46.3, 71.27, 71.28
Rogers, N.H. 35.19, 35.22

Rogers, R.D. 37.8, 68.1, 68.2, 69.49, 69.50, 71.89, 71.155, 71.163, 73.29, 73.74, 73.75, 73.79, 73.92, 73.94, 83.173
Rogerson, K. 17.21
Rogic, V. 45.39, 45.44
Rohly, W.G. 84.53
Rohr, M. 53.28
Rohrer, D.C. 45.25, 51.2, 51.4, 51.6, 51.7, 51.11, 51.12, 51.17, 51.20, 51.41, 51.73
Rohwedder, W.K. 51.72
Rojas, R.M. 81.19, 81.20
Rokhlina, E.M. 69.47
Roman, E. 73.69
Roman, P. 33.26
Romanenko, G.V. 79.14
Romeo, A. 36.44, 36.46
Romers, C. 48.25, 64.34
Romiti, P. 83.139
Romm, I.P. 66.14
Romming, C. 65.7, 70.11
Rooijen-Reiss, C.van 58.37, 65.23
Rooney, C.S. 32.36, 41.2
Roper, W.R. 71.252, 71.258, 71.267, 86.134, 86.136
Roques, B.P. 48.21, 48.22, 48.28, 48.29, 48.36
Roques, R. 1.33, 31.21, 37.22, 38.23, 58.18
Ros, J. 33.50
Ros, R. 71.208, 71.213, 71.254
Rosch, L. 73.43
Rose, E. 49.34, 49.35
Rose, J.D. 36.33, 48.63
Rose, N.J. 83.110
Rose-Munch, F. 71.109
Roselli, M.E. 60.9
Rosen, P. 50.7, 50.8
Rosenberg, E. 71.135, 71.228
Rosenblum, M. 73.48
Rosenstein, R.D. 39.60, 39.61
Rosner, M. 58.57
Ross, F.K. 2.12
Ross, L. 69.52, 69.59, 69.61
Rossetti, R. 71.118, 84.61
Rossi, J.C. 1.33, 58.18
Rossi, M. 44.11, 44.13, 44.22
Rossmanith, E. 8.12
Rossner, H. 73.95
Roth, B. 64.42
Rothrock, R.K. 71.232, 71.233, 71.245, 86.101
Rothwell, I.P. 72.30
Rottinger, E. 71.97, 71.169, 73.93, 73.95, 86.18, 86.31
Roulet, R. 71.208, 72.15, 72.16, 72.20, 72.25, 75.5, 86.10
Roundhill, D.M. 86.75
Roundhill, S.G.N. 86.75
Rouse, K.D. 65.13
Roush, P.B. 85.26, 85.27
Roux, S.D.le 81.5
Rozanov, I.A. 79.1, 79.2
Rozantsev, E.G. 71.207
Rozenberg, L.P. 39.50, 39.51, 39.52, 85.2
Rozhkova, N.K. 16.2, 41.29
Roziere, J. 33.3
Ruban, G. 39.59, 60.32, 60.38
Ruben, H. 32.6, 73.102, 79.3, 84.85
Rubinchik, B.Ya. 83.162
Ruble, J.R. 32.4, 45.8, 76.60
Ruchardt, C. 19.13, 19.14, 21.19

Rudchenko, V.F. 32.19
Rudich, S. 73.7
Rudich, S.M. 76.44, 84.22
Rudler, H. 71.50, 71.61, 71.109
Rudman, R. 1.20, 45.4
Rudnick, S.E. 62.10
Rudolf, P. 83.152, 83.153
Rudolph, G. 40.46
Rudolph, R.W. 86.121, 86.141
Rueda, D.R. 1.37
Ruge, B. 31.23, 31.27, 31.28, 38.91, 38.92, 38.93
Ruiter, B.de 64.2
Rupich, M.W. 76.98
Rusanovskii, M.E. 79.4
Russell, D.R. 71.216, 72.40, 83.212, 86.53, 86.54, 86.55
Russo, U. 11.11, 33.1, 70.12, 79.5, 79.16, 84.68
Rustemeyer, P. 71.201
Rutsch, W. 27.22
Ruzic-Toros, Z. 31.38, 32.34, 32.67, 45.39, 45.44, 64.82
Ryabova, N.A. 85.39
Ryan, R.C. 86.77, 86.208
Ryan, R.R. 38.68, 86.120, 86.132
Ryang, H.-S. 44.27, 44.28, 44.29
Ryasin, G.V. 63.6, 63.19
Rybakov, V.B. 31.17
Rybin, L.V. 83.90
Rybinskaya, M.I. 83.90
Rychlewska, U. 41.14, 53.30, 58.20

Saarinen, H. 83.67, 83.68, 83.98, 84.67
Sabat, M. 76.71, 82.35, 85.40
Sabesan, M.N. 50.32, 50.38, 60.43
Saburi, M. 82.34
Sacconi, L. 72.69, 75.47, 86.149, 86.150, 86.151, 86.156, 86.159, 86.181, 86.187, 86.217, 86.218, 86.225, 86.226, 86.227, 86.228, 86.229
Sacerdoti, M. 33.74, 35.39, 72.52
Sadikov, A.S. 58.12, 58.13, 58.15
Sadikov, G.G. 7.1, 73.25, 76.58, 76.59, 81.42, 84.55, 84.56
Sadovaya, N.K. 31.17, 31.18, 31.22, 31.32
Sadurski, E.A. 63.11, 63.50
Saeed, M.S. 45.43
Saenger, W. 3.48, 17.35, 17.38, 17.39, 17.40, 17.43, 33.64, 35.83, 45.60, 45.61, 45.62, 45.63, 45.64, 45.68, 47.1, 47.32, 48.4, 48.53, 61.22
Safa, K.D. 63.29, 63.30
Safina, Z.Sh. 17.31, 17.36, 38.10, 64.17, 64.32, 64.52
Safronova, Z.V. 16.13
Saha, A.P. 14.1
Saha, B.N. 81.25
Saha, N.N. 48.9
Sahajpal, A. 86.165
Saiki, Y. 53.17
Saint-Joly, C. 86.5
Saito, I. 36.16, 36.34
Saito, J. 32.68
Saito, K. 81.38, 81.39, 81.40
Saito, M. 76.61
Saito, S. 53.17
Saito, T. 71.182, 83.96

Saito, Y. 60.33, 71.265, 76.46, 76.76, 83.124, 83.125, 83.126
Sakaguchi, H. 83.59
Sakakibara, T. 45.51
Sakharova, V.G. 84.56
Sakore, T.D. 47.35, 47.36, 60.22
Sakuda, Y. 59.9
Sakurai, K. 13.2, 48.11, 48.14, 48.15
Sakurai, T. 41.48, 82.18, 82.19, 83.108
Sala-Pala, J. 71.119, 73.14
Salaneck, W.R. 60.42
Salazar, A.P. 41.19
Saleh, A.A. 53.22
Salen, G. 51.68
Salerno, G. 51.45
Salisbury, S. 47.39
Salisbury, S.A. 36.19
Sallfrank, R.W. 38.141
Salomon, R.G. 38.81
Salz, R. 72.8, 75.9
Samek, Z. 53.30
Sammes, M.P. 11.10
Sampson, H.A. 38.4, 38.26, 38.27
Sams, D.D. 8.10
Samuel, E. 72.45
Samus, I.D. 79.4
Samus, N.M. 83.192
Sanderson, M.R. 47.31
Sandros, K. 28.19
Sandstrom, M. 76.48, 76.77, 76.78
Sankawa, U. 56.3, 58.71
Sankey, S.W. 65.15, 71.57, 71.90
Sano, T. 34.10, 58.17
Sansoni, M. 3.19, 3.44, 3.45, 3.52, 64.132, 64.152, 71.147, 71.190, 83.92
Santarsiero, B.D. 76.54, 76.69, 83.110, 83.195
Saporito, A. 72.41, 72.44
Saporta, M. 85.50
Sappa, E. 71.35, 71.64, 71.105, 71.151, 71.178, 72.29
Sargeson, A.M. 76.27, 76.47
Sarko, A. 45.25
Sarma, R. 64.114, 67.4, 67.6
Sasada, Y. 31.69, 35.8, 38.137, 41.32, 44.12, 48.13, 50.4, 60.44, 71.143, 71.144, 71.145, 71.146, 71.179, 72.60, 82.18, 82.19, 83.197
Sasaki, K. 50.21
Sasaki, Y. 8.7, 71.182, 81.17, 81.38, 81.39, 81.40, 83.96
Sasvari, K. 13.23, 13.24, 41.43
Sathyanarayana, B.K. 47.32
Sato, M. 36.50, 41.70
Sato, S. 41.50, 50.20, 76.46
Satyamurthy, N. 39.56, 39.74
Sauer, J.D. 40.53, 40.55
Saussine, L. 84.17, 84.71
Sauvaitre, H. 32.2
Savona, G. 54.26
Sawka-Dobrowolska, W. 64.6
Sawyer, J.F. 19.1, 38.3
Sawzik, P. 51.84
Sax, M. 44.35
Sazonova, V.A. 73.84
Sbrignadello, G. 82.4
Scalone, M. 72.41, 72.44
Scaramuzza, L. 85.8
Scaringe, R.P. 49.18
Scarlata, G. 33.39

Schaaf, T.F. 63.1, 63.11, 63.50
Schaap, C.A. 71.203
Schaber, P.M. 76.25
Schaefer, W.P. 76.101, 76.105
Schafer, P. 32.59
Scham, W. 1.13, 1.21
Schammel, W.P. 83.188, 83.196
Schappacher, M. 49.23, 49.27
Schaumann, E. 8.12, 32.35, 37.23
Scheffer, J.R. 38.87
Scheidt, W.R. 49.22, 49.24, 83.222, 84.136
Schenetti, M.L. 41.15, 41.16
Schenk, K.J. 71.2
Schepper, L. 32.32
Scherer, O.J. 72.65
Scherer, P. 63.45
Scherz, M. 41.33
Scheuer, P.J. 53.18, 54.23
Schick, K.-P. 72.61
Schilling, B. 86.2
Schilling, M.D. 71.216, 72.40
Schimanski, U. 64.123, 86.201, 86.202, 86.203
Schinzel, H. 38.79
Schlabrendorff, C.von 62.12
Schlatzer, R.K. 83.172
Schlebusch, J.J.J. 75.20
Schlemper, E.O. 1.14, 1.15, 2.12, 32.5, 69.24, 69.25, 69.35, 69.41, 83.16, 83.17, 83.60, 83.88
Schleyer, P.von R. 31.55
Schlicksupp, L. 83.8, 83.9
Schlingmann, G. 49.25, 49.30
Schmalle, H.W. 1.13, 1.21, 59.2
Schmalstieg, H. 38.114
Schmid, G. 20.14, 75.10, 75.38
Schmidbaur, H. 64.11, 64.14, 64.79, 64.146, 64.150, 71.43, 71.225
Schmidpeter, A. 38.46, 64.37, 64.113, 64.135, 86.8, 86.180
Schmidt, A. 66.5, 66.9, 66.11
Schmidt, A.H. 38.79
Schmidt, H. 62.9, 75.2
Schmidt, R. 69.23
Schmidt, W. 76.81
Schmitz, D. 75.34
Schmitz, E. 83.26
Schmutzler, R. 64.27, 86.23
Schnabel, W. 41.25, 70.7
Schneider, F. 32.62
Schneider, J. 86.36
Schneider, K. 72.24
Schneider, K.-A. 20.11
Schnering, H.G.von 27.27, 32.59, 64.35, 65.5, 65.8, 71.1, 73.20, 73.66, 86.14, 86.72
Scholer, F.R. 62.4
Scholten, K. 1.6, 1.7, 2.5
Schomaker, V. 2.16, 76.54, 76.69, 76.73, 78.15, 83.69, 83.93, 83.97, 83.132, 83.195
Schomburg, D. 3.32, 38.109, 48.54, 64.27, 64.102, 64.113, 64.135, 86.23
Schonberg, P.R. 68.22
Schonholzer, P. 34.2, 34.5, 34.7, 45.19
Schore, N.E. 71.189
Schramm, K.D. 72.62, 86.148
Schreurs, A.M.M. 39.81, 39.82, 39.83
Schrock, A. 83.122
Schrock, R.R. 71.32, 71.42, 71.52, 71.209
Schroeder, A.C. 45.14
Schroepfer Junior, G.J. 51.88

Schroth, G. 75.9
Schubert, U. 27.32, 63.20, 63.37, 63.42, 64.11, 64.43, 64.79, 64.146, 64.150, 71.19, 71.21, 71.33, 71.36, 71.43, 71.95, 71.116, 71.136, 71.215, 71.219, 73.44, 74.7, 74.9, 76.86, 84.42, 86.179, 86.222
Schuck, R. 48.43
Schuckmann, W. 38.79
Schuep, W. 50.23
Schugar, H.J. 32.31, 76.26, 76.44, 84.22, 85.91
Schulte, G.R. 59.7, 59.8
Schultz, A.J. 64.127, 64.128, 64.164, 65.16, 65.20, 65.21, 73.126, 75.17, 75.18, 75.19
Schultz, R.V. 73.12
Schulz, J. 45.12
Schulze, J. 75.10, 75.38
Schulze, W. 75.39
Schuman, C.A. 38.8, 38.16
Schumann, H. 86.98, 86.103
Schurig, V. 84.7
Schwab, J.M. 58.45
Schwartz, A. 45.42
Schwartz, J. 2.17
Schwarz, W. 41.43, 64.30, 64.69, 64.71, 64.99, 66.5, 66.9, 66.11, 68.9, 68.10, 68.11, 83.220
Schwarzenbach, D. 72.15, 72.16, 72.20, 72.25, 75.5, 85.30, 85.87, 85.88, 85.90, 86.10
Schweda, E. 83.45
Schweitzer, D. 30.22, 39.58
Schweizer, B. 67.7
Schweizer, W.B. 31.50, 38.45, 63.45
Schwindinger, W. 32.31
Schwindinger, W.F. 76.26
Scibelli, J. 69.58
Scopelianos, A.G. 64.29
Scott, F.L. 21.9
Scott, J.W. 15.4, 32.63
Scourides, P.A. 49.6
Secor, H.V. 64.60
Seddon, K.R. 86.223
Sedov, B.B. 31.18, 31.32, 31.37, 31.56
Seebach, D. 63.16, 63.28, 67.7
Seel, F. 83.76, 83.174, 83.179, 83.203, 85.13
Seevogel, K. 72.8, 75.9
Seff, K. 5.5, 44.20, 54.2, 54.14, 58.43, 83.20, 83.32, 85.1, 85.34
Sega, A. 40.43
Segala, P. 86.131
Segaloff, A. 51.2, 51.4, 51.6, 51.7, 51.12
Segedin, P. 33.16, 83.175, 83.177
Sei, T. 67.5, 77.2
Seidel, I. 16.12, 51.9, 64.44, 64.80, 64.100, 64.103
Seiders, R.P. 60.20
Seiler, P. 38.45, 40.35, 67.7
Sekikawa, I. 48.30
Sekizaki, M. 76.83
Sekutowski, J.C. 71.72, 72.51, 75.9
Selby, T.P. 41.21
Selegue, J.P. 71.106, 71.108, 71.123, 71.150, 71.171, 71.195, 71.243, 81.37, 81.64
Seliger, J. 3.46, 3.47
Sellmann, D. 85.18
Selover, J.C. 71.45
Semenenko, K.N. 73.19, 86.212
Semenov, A.A. 59.10

Semmelhack, M.F. 75.6
Senda, S. 44.17
Senechal, M. 32.2
Sengier, L. 47.10, 47.29
Seno, K. 41.28
Sequin, U. 50.27
Serafino, A. 40.14
Sergienko, V.S. 8.5, 76.10, 76.58, 76.59, 83.162, 83.217
Serra, A.M. 49.5
Seshadri, T.P. 47.3, 47.4, 47.35, 47.36
Setkina, V.N. 71.104
Setzer, W.N. 31.4, 31.5, 39.11
Seu, M.J.T. 83.151
Sevin, A. 75.14
Seyer, U. 85.25
Seyerl, J.von 62.9, 64.107, 73.89, 73.110, 75.2, 85.18, 86.36, 86.38, 86.74
Sgarabotto, P. 35.55, 35.59, 35.61, 38.22, 41.15, 41.16, 41.27, 41.34, 51.45, 52.2, 52.3, 61.10, 61.11, 69.28, 72.64
Shaanan, B. 60.14
Shaikh, M.A. 60.41
Shakir, R. 2.18, 67.9, 68.3, 73.8, 73.35, 73.36, 73.37, 86.170
Shakked, Z. 47.39
Shamala, N. 48.65
Shanzer, A. 40.17
Shapiro, A.B. 71.207
Shapkin, V.A. 34.6, 34.8, 38.96, 40.42, 40.44
Shapley, J.R. 83.109, 86.64, 86.116
Sharapov, V.A. 64.116, 64.121, 75.24
Sharipova, F.V. 39.23
Sharma, R.P. 48.5
Sharma, S.D. 58.34
Sharp, P.R. 71.42, 71.52
Sharrock, P. 83.204
Shaw, B.L. 72.12, 72.18, 86.168
Shaw, R.A. 64.70, 64.74
Shchedrin, B.M. 76.3
Shchegoleva, T.M. 17.20, 26.1, 66.17
Shchegolikhina, O.I. 63.38
Shcherbakov, A.A. 39.36, 39.37, 39.68, 39.79, 39.80
Shchurkina, V.N. 83.162
Shearer, H.M.M. 3.42, 20.1, 62.2, 68.17, 71.8, 76.66
Shefer, S. 51.68
Sheka, E.F. 17.25, 17.27
Sheldon, R.I. 84.53
Sheldrick, B. 17.17
Sheldrick, G.M. 3.23, 33.34, 35.84, 35.85, 38.78, 47.39, 49.21, 49.31, 50.34, 50.41, 52.4, 63.39, 63.44, 63.49, 86.118
Sheldrick, W.S. 4.1, 33.78, 38.46, 49.25, 49.29, 49.30, 59.45, 63.2, 63.27, 63.31, 63.52, 64.27, 64.37, 64.113, 64.122, 64.135, 64.144, 83.224, 86.46, 86.47
Shemyakin, N.F. 64.23, 65.3
Shepherd, A.G. 30.13
Shepherd, R.E. 76.60
Sherman, E.O. 86.137
Sherwood, M. 17.30
Shestakova, N.A. 82.6, 82.9, 82.10
Shevchenko, L.L. 15.3
Shevyrev, A.A. 32.49
Shia, Z.-S. 86.124
Shiau, G.T. 47.5, 47.19
Shiau, W.-I. 25.1

Shiba, J.K. 83.62, 83.63
Shibaeva, R.P. 39.50, 39.51, 39.52, 39.66, 60.7, 60.8, 85.2
Shibanova, E.F. 32.3
Shibata, M. 36.11, 76.46
Shibata, T. 53.6
Shibuya, H. 38.55
Shie, S. 39.42
Shieh, H.-S. 47.38, 48.40, 48.44
Shields, G. 85.60
Shima, I. 64.168
Shimanouchi, H. 31.69, 35.8, 41.32, 60.44, 82.18, 82.19
Shimba, S. 76.46
Shimizu, H. 39.65
Shimizu, N. 32.48, 53.6, 83.199, 83.201
Shimoi, M. 81.97
Shimozono, K. 36.16, 36.34
Shin, W. 44.35
Shinoda, S. 71.265
Shinohara, M. 50.21
Shirahata, K. 38.137, 48.13, 50.4
Shiro, M. 50.5, 59.44
Shirotake, S. 44.5
Shiying, S. 48.26
Shklover, V.E. 63.9, 63.14, 63.25, 63.26, 63.32, 63.38, 63.46, 63.47
Shkuratova, I.G. 2.22
Shkurpelo, A.I. 83.37, 83.38, 86.91
Shmidt, E.N. 54.1
Shmueli, U. 59.35, 60.14, 60.15
Sho, M. 39.42
Shochet, N. 40.17
Shoham, G. 3.32
Shoja, M. 29.12
Shoolery, J.N. 36.45
Shorthill, W.B. 86.85
Shou-Dao, W. 58.21
Shrivastava, H.N. 53.7
Shriver, D.F. 71.3, 71.58, 84.52
Shudo, K. 36.20
Shugar, D. 47.7
Shur, V.B. 83.26
Shvelashvili, A.E. 76.3
Shvets, A.E. 25.4
Sibley, L.D. 27.3
Siebert, W. 62.9, 73.109, 75.2
Siedle, A.R. 60.12, 77.4, 77.5, 83.219, 86.25, 86.230
Sieger, H. 17.38, 17.39
Sieler, J. 83.181, 85.14, 85.19, 85.20, 85.32
Sievers, R.E. 77.10
Sievert, A.C. 3.33, 3.37
Siew, P.Y. 86.89
Siftar, J. 2.10, 85.47
Sikirica, M. 11.17, 35.15, 40.43, 71.12
Sikora, D.J. 73.29
Silavwe, N.D. 72.48
Siler, P. 40.36
Silva, M. 38.83
Silver, J. 69.9
Silverton, J.V. 13.22, 27.9, 31.9, 37.12, 38.80, 47.27, 58.53, 58.57
Sim, G.A. 54.27, 73.39
Simhon, E.D. 64.129, 64.130, 64.131
Simmons, C.J. 44.20, 54.2, 54.14, 58.43, 83.32
Simon, D. 69.61
Simon, F.E. 72.2, 72.46

Simon, K. 35.16, 35.20, 35.33, 35.37, 45.6, 45.7, 45.10, 45.69
Simonen, T. 38.108
Simonetta, M. 5.13, 39.40
Simonov, M.A. 60.7
Simonov, V.I. 32.49, 51.25, 51.29, 51.38, 51.39, 51.48, 51.58, 51.59
Simonov, Yu.A. 1.9, 34.6, 34.8, 35.43, 35.44, 35.69, 38.96, 40.13, 40.42, 40.44, 40.49, 71.264, 83.36, 83.37, 83.38, 83.147, 83.192, 84.38, 84.112, 86.91
Simonov, Yu.M. 83.81
Simonsen, O. 38.15
Simonsen, S.H. 32.72, 32.73, 33.94, 35.12, 35.62, 36.40, 39.10, 42.13, 68.23
Simonyan, L.A. 16.13
Simopoulos, A. 64.129, 64.130, 64.131
Simpson, J. 71.112
Simpson, T.H. 59.22
Sims, J.J. 52.1
Sims, M. 36.32, 37.14, 37.16, 42.12
Sinay, P. 45.29
Sinchai, W. 25.2
Sinclair, J.D. 86.83, 86.143
Sindellari, L. 80.5, 80.8, 80.10
Sindt, A.C. 13.18
Singh, A.K. 31.35
Singh, B. 35.8
Singh, P. 17.19, 24.1
Singh, R.K. 38.70
Singh, T.P. 17.1
Singh, V. 31.65
Singleton, E. 86.87
Sinha, S.K. 81.24
Sinn, E. 71.54, 71.69, 71.70, 71.242, 73.77, 76.104, 83.122, 83.172
Siroki, M. 84.30
Sironi, A. 3.5, 3.50, 64.157, 72.28, 83.139, 84.21
Sisoeva, T.F. 76.18, 76.74, 84.58
Sitkina, L.M. 62.18, 68.8, 68.19
Sivak, A.J. 86.58
Sivakumar, R. 39.56, 39.74
Sivaramakrishnan, R. 71.27, 71.28
Siwapanyoyos, G. 86.33
Sixl, H. 16.18
Sjolin, L. 83.181, 85.14
Sjostrand, U. 69.33, 69.34
Skapski, A.C. 76.108, 82.7
Skelton, B.W. 31.23, 31.27, 31.28, 38.91, 38.92, 38.93, 38.124, 38.125, 38.126, 38.127, 45.13, 58.4, 60.41, 64.109, 64.110, 71.250, 71.251, 79.9, 79.10, 81.50, 86.102, 86.145, 86.146
Skinnemoen, K. 39.34, 39.35
Skopenko, V.V. 7.1
Skoweranda, J. 36.15
Skripkin, V.V. 71.91
Skripkin, Yu.V. 73.25
Skrzypczak-Jankun, E. 58.16, 73.55, 73.68
Skvortsova, G.G. 84.37
Slade, M.J. 69.40
Slak, J. 3.46, 3.47
Slavjanov, E.V. 78.7, 79.12
Slim, D.R. 81.53
Slokar, M. 83.115
Slot, H.J.B. 39.81, 39.82, 39.83
Slovokhotov, Yu.L. 71.188
Slutsky, J. 31.55

Sly, W.G. 80.11
Smal, E. 32.37, 32.38, 32.42
Small, R.W.H. 1.19, 71.29, 71.129, 83.91
Smart, J.C. 73.99
Smart, L.E. 73.56, 75.30
Smedsrud, B. 44.2, 44.3, 44.4, 44.9
Smelyanskaya, E.M. 27.26
Smets, G. 33.35, 33.40, 33.42, 33.43, 33.52,
   33.62, 33.63, 33.77, 33.80, 33.81, 33.87,
   33.88, 35.56, 35.57
Smirnova, V.I. 50.22, 50.24, 50.37
Smit, F. 65.12
Smith, A.B. 59.19
Smith, A.J. 76.95
Smith, A.K. 74.8
Smith, A.R.P. 80.15, 80.16
Smith, D.L. 39.4, 39.88, 39.89, 39.90, 70.18
Smith, G. 11.2, 17.12, 17.32, 84.86
Smith, G.A. 50.41
Smith, G.D. 51.13, 51.15
Smith, J.L. 47.12
Smith, V.J. 35.19, 35.22
Smith, W.L. 1.32, 84.90
Smith-Verdier, P. 24.5, 37.7, 37.13,
   38.103
Smits, J.M.M. 5.14, 71.281
Smolander, K. 81.87, 81.89, 84.122
Snyder, J.P. 31.54
Sobell, H.M. 47.35, 47.36, 60.22, 60.36,
   60.37
Sobolev, A.N. 66.14
Soderholm, A.-C. 38.102, 38.116
Sokol, V.I. 83.103, 84.37
Sokolov, V.I. 73.80
Sokolov, V.V. 63.6
Sokolova, Yu.A. 84.41
Solans, J. 13.12, 16.5
Solans, X. 13.12, 28.9, 28.17, 28.18, 29.10,
   33.2, 54.37
Solar, J.P. 75.13
Solheim, B. 53.1, 59.7, 59.8
Solheim, B.A. 54.30
Solodovnikov, S.P. 19.11
Solomun, T. 86.95, 86.182
Soloveichik, G.L. 73.19
Sombroek, J. 31.60
Sommerville, P. 86.161
Songstad, J. 65.7, 70.11
Sonneveld, E.J. 83.95
Soos, Z.G. 36.17
Soroka, G.M. 39.25, 39.26
Sorrell, T. 85.67
Sorrell, T.N. 84.20, 84.24
Sorum, H. 32.17, 45.1
Sostero, S. 81.36
Sotiropoulos, J. 31.21, 37.22
Sotman, S.S. 38.10, 46.6, 64.17, 64.32,
   64.33, 64.52, 64.53
Sotofte, I. 32.11, 32.24, 32.25, 32.26, 32.27,
   32.32
Southern, T.G. 71.153, 71.273, 71.279,
   86.45
Sowden, J.M. 29.6
Sowerby, D.B. 33.8, 66.7, 66.8, 66.13, 66.16,
   66.19
Soylu, H. 1.11, 82.20
Spagna, R. 85.15
Spartalian, K. 84.136
Speakman, J.C. 1.4
Spek, A.L. 64.170, 69.45, 71.183, 81.34

Spence, R.D. 84.10
Spencer, B. 73.102, 79.3, 84.85
Spencer, C.B. 62.2, 71.8, 76.66
Spencer, G.F. 51.72
Spencer, J.L. 71.277, 86.126, 86.147
Spencer, S.A. 31.12
Speranzini, R.A. 83.84
Spetch, E.R. 86.4
Spik, G. 45.49
Spinat, P. 76.37, 76.38, 76.39
Spirlet, M.R. 56.10
Springer, J.P. 25.5, 27.13, 27.25, 31.51,
   32.36, 41.2, 58.78, 59.38, 59.39, 59.41
Srdanov, G. 76.72
Srere, P.A. 2.25, 2.26
Srikrishna, A. 31.65
Srikrishnan, T. 44.26, 45.14, 47.8, 47.25
Staab, H.A. 30.22, 31.53, 39.58, 39.69,
   39.78
Staal, L.H. 71.68, 72.50
Stadnicka, K. 19.10, 35.1, 38.51, 38.82,
   38.121, 38.123, 39.55
Stahl, K. 32.14, 32.15, 32.16
Stahle, H. 36.12
Stahly, G.P. 41.82
Stalick, J.K. 64.60
Stallings, W. 35.23, 38.24, 39.30, 39.31,
   39.32, 39.38, 60.24
Stallings, W.C. 2.25, 2.26, 39.48
Stam, C. 33.38, 33.65
Stam, C.H. 33.30, 46.4, 64.16, 64.101,
   64.120, 65.11, 65.12, 65.23, 65.24, 71.68,
   72.33, 72.50, 86.114
Stamm, E. 75.46
Stanghellini, P.L. 71.118, 84.61
Stanko, V.I. 62.7, 62.8, 63.5, 69.13
Stankovic, S. 41.24, 41.41, 41.66
Stanovnik, B. 41.18
Stansfield, R.F.D. 73.104, 73.105, 75.31
Starikova, Z.A. 17.20, 26.1, 35.29, 35.52,
   66.17
Starish, M.P. 71.264
Starker, B. 39.58, 39.69, 39.78
Starova, G.L. 32.3, 32.8, 32.9
Stec, W.J. 64.26
Steen, N.D.C.T. 86.161
Stefanovic, M. 51.61
Steffen, W.L. 77.13
Steffens, G.L. 51.72
Steffl, I. 71.167
Stegelmeier, H. 34.1, 34.4, 63.13
Steggerda, J.J. 80.6
Steglich, W. 36.12, 56.7
Steidel, J. 11.3
Steidl, S. 83.160
Steiger, W. 83.43
Steimecke, G. 85.14
Stein, J. 76.56
Steinberg, N.G. 27.13
Steinhardt, P.C. 64.154
Steinrauf, L.K. 50.32, 50.38, 60.43
Steliou, K. 39.45
Stelzer, N.A. 83.90
Stelzer, O. 64.27, 86.46
Stempel, A. 50.23
Stensland, B. 36.18, 36.25, 36.31, 48.50
Stephens, F.S. 71.128, 86.135, 86.142
Stephenson, N.C. 45.34, 85.59
Stephenson, T.A. 86.153, 86.154
Sternglanz, H. 44.16, 44.19

Stetsenko, A.I. 83.147
Steudel, R. 11.3
Stevens, E.D. 1.1, 48.10, 49.17, 49.20,
   81.32, 81.33
Stevens, J.C. 83.188, 83.196
Stevens, J.D. 45.34, 45.42
Stevens, R.V. 38.72
Stevenson, R. 35.19, 35.22
Stewart, J. 86.230
Stewart, J.M. 86.25
Stewart Junior, R.P. 73.59
Steyn, P.S. 32.30
Stezowski, J.J. 29.11, 45.65, 45.67, 45.70,
   48.51, 48.59, 50.13, 50.14, 50.15, 50.19,
   60.17, 60.25, 60.26, 60.27, 60.28, 60.29,
   60.30, 60.31
Stick, R.V. 45.13
Stiefel, E.I. 85.37
Stierle, D.B. 52.1, 53.34, 54.10
Stimson, R.E. 71.3
Stobart, S.R. 69.6, 71.130
Stobbe, S. 72.8
Stoddart, J.F. 40.18, 40.19
Stoeckli-Evans, H. 58.30, 73.57, 73.91
Stokie, G.J. 54.8
Stone, F.G.A. 71.75, 71.76, 71.81, 71.82,
   71.111, 71.142, 71.149, 71.173, 71.185,
   71.186, 71.238, 71.259, 71.277, 75.31,
   86.32, 86.33, 86.43, 86.147, 86.172
Stone, L.B. 68.2
Stoppioni, P. 86.115
Storr, A. 68.16, 72.22, 83.51, 83.52, 83.53,
   83.54, 83.99, 83.150, 83.166, 84.32, 84.33
Stotter, D.A. 83.149
Stout, C.D. 76.60
Strahle, J. 65.14, 71.15, 83.45, 83.211
Strauss, S.H. 71.3
Strecker, G. 45.49
Strecker, M. 33.78, 63.27, 63.31, 63.52
Street, G.B. 65.6, 65.10
Streitweiser Junior, A. 75.13
Stremple, P. 64.129, 64.130, 64.131
Strong, P.D. 13.7, 51.2, 51.4, 51.6, 51.7,
   51.12, 51.13, 51.15, 51.16, 51.20, 51.22,
   51.31
Strouse, C.E. 38.72, 71.45, 71.47, 83.117,
   83.118, 83.119, 83.120, 83.121, 83.123,
   83.127, 83.128, 83.129
Strub, K. 32.62
Struchkov, Yu.T. 1.5, 1.38, 8.14, 15.6,
   16.13, 19.7, 19.11, 22.5, 31.17, 31.18,
   31.22, 31.32, 32.43, 35.34, 38.74, 39.18,
   39.20, 39.21, 39.23, 39.28, 39.36, 39.37,
   39.68, 39.79, 39.80, 41.35, 51.81, 58.44,
   62.7, 62.8, 62.18, 63.5, 63.8, 63.10, 63.14,
   63.25, 63.26, 63.32, 63.38, 63.46, 63.47,
   64.22, 64.23, 64.49, 64.68, 64.75, 64.88,
   64.89, 64.115, 64.134, 64.145, 65.3,
   65.25, 65.8, 68.14, 68.19, 69.13, 69.47,
   71.4, 71.13, 71.22, 71.23, 71.51, 71.91,
   71.104, 71.124, 71.131, 71.188, 71.221,
   73.19, 73.31, 73.80, 73.84, 73.134, 73.135,
   81.44, 83.26, 83.90, 84.110, 85.11
Struckmeier, G. 32.74, 32.75, 32.76, 36.38,
   63.34, 63.35
Strumolo, D. 71.147, 72.3
Stucky, G.D. 71.230, 73.73, 73.113, 75.17,
   75.18, 75.19
Stuhler, H.-O. 75.11
Stults, B.R. 64.38

Stuntz, G.F. 86.64
Sturis, A.P. 35.48, 85.63, 85.76
Sturkovic, R.Ya. 63.8, 63.10
Suades, J. 85.46
Subbiah, A. 47.31
Subramanian, E. 17.42, 18.3, 41.51, 47.14
Subramanian, V. 85.1
Subrtova, V. 62.5, 62.6
Suck, D. 47.1
Sudhakar, V. 48.8, 48.20
Sueur, S. 84.34
Suffolk, J.S. 32.46, 32.47, 40.6, 40.7, 40.33
Sugano, T. 39.8
Sugawara, M. 50.21
Sugimoto, H. 49.7
Sugiura, K. 50.21
Sugiyama, K. 25.5
Suh, I.–H. 17.35, 17.38, 17.39
Suib, S.L. 71.230
Sulaimankulov, K. 8.5
Suleimanov, A.S. 41.20
Suleimanov, Kh. 8.5
Sullivan, B. 53.34
Sullivan, D.A. 84.94, 84.95
Sumner Junior, C.E. 71.9, 71.10
Sundaralingam, M. 45.9, 46.1, 47.9, 47.11, 47.12, 47.20, 47.21, 47.22, 47.30, 60.39
Sundell, S. 3.55
Sung, C.-K. 56.3
Sung, T.V. 59.32
Sunjic, V. 40.43
Surazhskaya, M.D. 81.4, 81.7, 81.11, 81.12
Surcouf, E. 51.24, 51.46
Suri, O.P. 58.34
Suri, S.C. 31.6, 31.16
Sushchinskaya, S.P. 71.23, 71.131
Suslick, K.S. 49.34, 49.35
Susse, P. 59.11
Sutherland, I.O. 61.6
Sutton, B.M. 86.71
Sutton, D. 73.60
Suwinska, K. 39.55
Suzuki, K. 58.71
Suzuki, K.Z. 81.38, 81.39, 81.40
Suzuki, Y. 39.71
Svensson, C. 28.1, 32.14, 32.15, 32.16
Svinning, T. 32.17, 45.1
Svistunov, V.S. 63.26
Swaminathan, P. 45.9, 47.20, 47.21
Swaminathan, V. 47.30
Swanepoel, H. 76.62, 85.49
Swank, D.D. 83.46, 84.3, 84.4
Swanson, K. 85.26, 85.27
Swanson, S. 61.23, 61.24, 61.25
Swenson, D. 64.129, 64.130, 64.131
Swenson, D.C. 7.4, 51.5, 51.8, 51.13, 51.15, 51.43
Swenson, W. 21.6
Swepston, P. 42.1, 64.87
Swepston, P.N. 64.50
Swiatkiewicz, J. 7.3
Swindells, D.C.N. 51.3, 51.19
Symons, M.C.R. 83.212
Syrchina, A.I. 59.10
Sysoeva, T.F. 36.35
Szafran, M. 33.24
Szalda, D.J. 3.51
Szobota, J.S. 3.18
Szostak, R. 71.47
Szulzewsky, K. 51.9, 51.10

Tacconi, G. 35.77, 38.112
Tachikawa, M. 3.33, 3.37, 71.44
Tacke, R. 33.78, 63.27, 63.31, 63.52
Taddei, F. 35.55, 35.61, 41.15, 41.16
Tadini, C. 38.106, 44.30
Taeger, T. 3.41, 63.4
Tafeenko, V.A. 36.35
Taga, T. 49.7, 59.30
Tagle, B. 53.32, 54.10
Tagliaferri, E. 36.51
Tagliavini, G. 69.39, 69.44
Tago, K. 60.6
Tahara, T. 59.9
Takach, N.E. 86.81
Takaegami, Y. 72.27
Takagi, K. 84.96
Takagi, S. 45.2, 45.3
Takagi, Y. 59.23
Takahashi, K. 27.18, 50.20, 54.36
Takahashi, T. 56.5
Takaki, Y. 13.2, 48.11, 48.14, 48.15
Takamoto, M. 39.39, 48.47
Takani, M. 27.18
Takaoka, D. 54.31, 54.32
Takata, T. 40.58
Takayama, H. 51.80
Takayanagi, H. 32.55, 39.65, 58.40
Takenaka, A. 44.12, 71.179, 72.60
Takeuchi, A. 78.9, 78.18
Takeuchi, S. 71.143, 71.144, 71.145, 71.146
Takeuchi, T. 81.97
Takusagawa, F. 3.43, 33.20, 33.21, 33.22, 64.158, 64.162, 64.163, 71.42, 71.52, 73.16, 73.17, 73.18, 73.27, 73.28
Talipov, S.A. 58.11, 58.14
Talley, J.J. 31.52
Tamm, C. 59.37, 59.40
Tamura, C. 50.20
Tamura, H. 78.9, 78.18
Tanaka, I. 48.52, 48.56
Tanaka, J. 53.5, 53.9
Tanaka, K. 60.44
Tanaka, N. 28.13, 36.23, 41.70, 53.17
Tanaka, O. 54.6
Tanaka, T. 84.48, 85.48
Tang, C.P. 51.53
Taniguchi, M. 31.41
Taniguchi, T. 48.14, 48.15
Tao-pen, C. 39.58, 39.69, 39.78
Tapscott, R.E. 81.28
Tarassoli, A. 64.92
Tarkhova, T.N. 1.9
Tarnopol'skii, B.L. 17.31, 17.36
Tasaki, S. 44.25
Tashkhodzhaev, B. 16.2, 41.29, 58.31, 58.32
Tasker, P.A. 83.137, 83.158
Tatarsky, D. 74.10, 74.13
Tatsumi, K. 49.33, 86.96, 86.97
Tatsumisago, M. 85.48
Tau, K.D. 86.157, 86.166
Tautz, H. 86.8
Tautz, W. 50.7, 50.8
Tavale, S.S. 56.1
Tavanaiepour, I. 81.60, 86.215
Tavberidze, M.G. 76.3

Taylor, A. 38.20
Taylor, G. 60.24
Taylor, G.E. 71.86, 71.87, 71.170, 71.212, 71.258, 71.267, 72.38
Taylor, H.C.R. 40.51
Taylor, H.L. 58.75
Taylor, M.R. 31.24, 31.25, 31.26, 36.41, 44.6, 44.7, 79.6
Taylor, N.J. 64.169, 71.198, 71.199, 71.211, 72.49, 86.20, 86.48, 86.144
Teichert, H. 66.18
Teitelbaum, R.C. 39.9
Telezhenetskaya, M.V. 58.31
Teller, R.G. 86.66, 86.140, 86.167
Telser, J. 53.32
Temple Junior, C. 36.33, 48.63
Templeton, D.H. 32.6, 73.102, 79.3, 84.85
Templeton, J.L. 80.2, 83.200
TenBrink, R.E. 38.60
Tench, A.W. 47.18
Terenghi, G. 72.64
Ternay Junior, A.L. 39.60
Terzis, A. 11.5, 85.69
Thaller, V. 38.20
Theobald, F. 35.36
Thetaz, C. 36.51
Thewalt, U. 36.38, 63.18, 63.23, 63.34, 63.35, 76.53, 76.94, 76.96, 76.97, 83.4
Thich, J.A. 71.158, 71.262, 85.91
Thiele, G. 71.6, 86.175
Thiem, J. 45.59, 64.54
Thierbach, D. 64.90, 64.91
Thiessen, W.E. 48.54
Thijssen, T. 62.13
Thiollet, G. 73.38
Thoennes, D. 67.2
Thomas, J.E. 84.92, 84.93
Thomas, J.M. 20.10, 20.15
Thomas, M. 3.3, 3.6, 3.7, 3.8, 3.9, 3.10, 33.28, 33.29
Thomas, R.W. 83.44
Thomas, S. 71.226, 71.231
Thomas, T.L. 21.9
Thomas Junior, J.W. 64.12
Thomas-David, G. 45.45, 45.46
Thompson, D.A. 86.121, 86.141
Thompson, D.J. 59.33
Thompson, D.M. 40.24
Thompson, J.B. 58.75
Thompson, J.L. 84.80
Thompson, J.S. 85.67, 85.68
Thompson, J.T. 31.43
Thompson, M.J. 51.71
Thompson, M.R. 3.33, 3.37, 74.11, 74.12
Thompson, R.C. 83.178
Thompson, R.S. 73.116
Thompson, W.J. 28.7
Thomson, M.A. 86.153, 86.154, 86.193, 86.223
Thomson, R.H. 41.72, 53.4, 56.2
Thonnessen, M. 75.34
Thorez, A. 86.41, 86.42
Thorkildsen, G. 11.6, 11.9, 11.15
Thorn, D.L. 81.66, 86.51, 86.130
Thornton, P. 81.84
Thorup, N. 82.24
Thozet, A. 17.3, 17.15, 60.2
Thulasidhass, R. 48.5
Thuy, N.H. 72.15, 72.16, 72.20
Tichy, K. 3.27, 3.28

Tichy, M. 37.4
Tickle, I. 58.69
Tietze, L.F. 38.78, 52.4
Tihanyi, E. 36.14
Tikhonova, I.A. 83.26
Tilley, J.W. 32.23, 32.29
Tilley, T.D. 73.102, 84.85
Timakov, I.A. 78.16
Timberlake, J.W. 41.42
Timken, M.D. 83.191, 84.53
Timmer, K. 64.170
Timmons, J.H. 76.106, 83.152, 83.153
Timms, R.N. 75.35
Timofeeva, T.V. 19.11
Tinant, B. 1.23, 19.16, 32.54, 41.11, 41.12, 41.17, 44.21
Tipton, D.L. 73.85, 84.20, 84.24
Tiripicchio, A. 71.35, 71.64, 71.105, 71.135, 71.151, 71.178, 71.187, 71.228, 72.29, 83.31, 83.82, 83.107, 84.81, 86.152, 86.174
Tishchenko, G.N. 48.32, 48.71, 48.72, 50.22, 50.24, 50.37, 58.11, 58.12, 58.13, 58.14, 58.15
Tisler, M. 41.18
Tisnes, P. 42.3
Titova, S.N. 73.134, 73.135
Tkachev, V.V. 64.31, 64.51, 64.66, 64.72, 64.85, 84.35
Tobe, M.L. 76.68, 76.92, 83.131
Tobias, R.S. 83.30
Toby, B.H. 76.44, 84.22
Toda, F. 60.44
Toda, J. 58.17
Todd, L.J. 73.3, 73.12
Toder, B.H. 59.19
Toeplitz, B. 41.80
Toeplitz, B.K. 50.28, 50.29
Toffoli, P. 33.9
Tokubuchi, N. 54.4, 54.34
Tokuno, K. 39.6
Tokuoka, R. 45.60
Tokuyama, T. 58.25
Tolbert, M. 81.74
Tolle, H.-G. 64.123, 86.201, 86.203
Tolstikov, G.A. 39.23
Tomat, G. 82.4
Tomimatsu, T. 54.34
Tomita, K. 38.55, 47.17, 50.9
Tomita, K.-I. 44.5, 44.17, 45.60, 47.2, 47.28, 60.6
Tomkins, I.B. 86.3
Tomlinson, G. 76.36
Tong, E. 39.42
Tongchang, L. 48.26
Toniolo, C. 48.33, 48.34, 48.41
Toniolo, L. 86.183
Toppet, S. 3.53, 32.66, 33.77, 33.80, 33.81, 33.87, 33.88, 41.73
Tori, K. 50.5
Torrance, J.B. 60.3, 60.4, 60.10
Torreilles, E. 39.27, 39.64, 60.35
Torroni, S. 73.106, 75.48
Toscano, P.J. 71.100
Tosi, G. 35.63
Tosik, A. 33.36
Toth, J. 32.10
Toth, J.A. 54.23
Toth, K. 64.83, 64.84, 75.44
Touchard, D. 71.99

Touillaux, R. 21.12, 21.13, 28.8, 32.70, 37.23, 41.22, 41.44
Toupet, L. 11.7
Toure, S. 10.4, 21.10
Toussaint, J. 33.66
Towle, I.D.H. 84.45, 84.46, 86.80
Trad, S. 76.43, 84.73
Traeger, U. 76.8, 76.20, 76.32, 83.43
Trager, W.F. 1.36
Tranter, R.L. 31.3, 31.11
Trautwein, A. 83.179, 83.203
Trave, R. 52.2, 52.3
Traverso, O. 81.36, 86.52
Trefonas, L.M. 41.42, 82.36, 82.37
Trenkle, A. 73.93, 86.31
Trimarco, P. 40.54
Trocha-Grimshaw, J. 49.4
Troitskaya, L.L. 73.80
Trompenaars, W.P. 32.58, 40.38
Trop, H.S. 3.51, 64.159
Trotter, J. 25.3, 27.10, 27.12, 27.19, 27.20, 31.12, 31.36, 31.39, 31.40, 37.2, 38.62, 38.87, 49.6, 49.16, 61.7, 62.20, 64.171, 68.16, 72.5, 72.22, 73.1, 73.2, 73.6, 73.45, 74.4, 74.5, 74.6, 80.4, 83.51, 83.52, 83.53, 83.54, 83.99, 83.149, 83.150, 83.166, 83.178, 84.32, 84.33
Troup, J.M. 73.119, 73.125
Troy, D. 64.112
Trsic, M. 3.49
Trueblood, K. 31.58, 31.59
Trueblood, K.N. 30.18, 40.40
Trullinger, D.P. 40.34
Trunov, V.K. 17.20, 26.1, 35.29, 35.52, 36.35, 66.17, 76.18, 76.74, 83.207, 83.208, 83.215, 84.58, 84.64
Trust, R.I. 35.14
Truter, M.R. 38.113, 38.131
Tsay, Y.-H. 69.20, 69.43, 71.67, 86.38, 86.74
Tseikinskii, V.M. 51.25, 51.29, 51.38, 51.39, 51.48, 51.58, 51.59
Tsetlina, E.O. 71.23, 71.131
Tsintsadze, G.V. 81.80, 83.77, 83.94, 83.162, 83.217, 84.55, 84.56
Tsipis, C. 73.107
Tsiryapkin, V.A. 73.31
Tsivtsivadze, T.I. 81.80, 83.77, 83.217
Tskhovrebashvili, V.S. 63.36
Tso, T.C. 84.129
Tsoucaris, G. 61.14, 61.15, 61.16, 61.17
Tsuboyama, K. 83.108
Tsuboyama, S. 83.108
Tsuda, Y. 34.10, 58.17
Tsuge, O. 40.58
Tsuji, A. 9.3
Tsuji, K. 40.3
Tsujino, I. 48.30
Tsukada, H. 31.69, 60.44
Tsukihara, K. 13.19, 41.61, 41.77
Tsurita, Y. 83.96
Tsutsui, M. 49.14, 49.33
Tsuyuki, T. 56.5
Tsuzuki, K. 27.25, 31.51
Tsvetkov, E.N. 64.72
Tuchagues, J.P. 83.204
Tucker, P.A. 72.56, 72.57, 72.58, 74.8, 86.53, 86.54, 86.55
Tuddenham, D. 28.6
Tulinsky, A. 27.16, 49.36

Tulip, T.H. 72.63, 86.51, 86.130
Tumanskii, B.L. 19.11
Tunaley, D. 83.210
Turnbull, K. 32.46, 32.47, 40.6, 40.7, 40.33
Turner, A.G. 86.119
Turner, D.G. 86.126
Turner, J.V. 31.19
Turney, T.W. 71.263
Turova, N.Ya. 68.14
Turpeinen, U. 76.85, 76.88, 76.89, 81.81, 81.82, 81.87, 81.89, 84.122
Tursch, B. 53.10, 54.20, 54.28, 54.33
Turte, K.I. 84.112
Twiss, J. 68.17
Tyers, K.G. 72.39

Uba, K. 58.26
Uchiyama, T. 84.96
Uchtman, V.A. 14.5, 48.16, 48.27
Ude, W. 28.11, 28.14, 31.66
Udel'nov, A.I. 85.39
Udupa, M.R. 82.32, 83.50, 85.25, 85.53
Ueda, H. 53.5, 53.9
Ueda, I. 40.58, 76.19, 78.12, 78.13, 83.25
Ueda, T. 47.2, 47.28
Uehiro, T. 76.61
Uemura, D. 59.4
Uggla, R. 64.142, 64.143
Ugi, I. 71.196
Uhl, W. 64.62
Ul-Haque, M.- 32.22, 64.59, 64.81, 64.111
Ulibarri, M. 83.148, 86.67
Umemura, T. 48.60
Umezawa, B. 58.63
Umezawa, H. 50.2
Undheim, K. 39.73, 39.76
Ungaro, R. 61.21
Uno, T. 32.48, 83.199, 83.201
Urabe, H. 71.182
Urbanczyk-Lipkowska, Z. 34.9, 35.1, 36.28, 38.31, 45.21, 45.23, 83.135
Urgast, K. 22.4, 39.75
Uriarte, R. 86.157, 86.166
Urzua, A. 58.62
Usha, R. 20.2, 20.6, 37.4
Uskokovic, M.R. 39.44, 39.72
Uson, R. 86.63, 86.118
Usov, O.A. 32.8, 32.9
Usubaliev, B.T. 69.14, 69.26
Usubillaga, A. 51.70
Utsuno, S. 76.83

Vaciago, A. 50.25, 50.30, 50.31, 59.34
Vahrenkamp, H. 71.16, 71.97, 71.169, 73.93, 73.95, 73.111, 86.18, 86.31, 86.34, 86.35
Vainshtein, B.K. 48.72
Vaira, M.di 86.187, 86.217, 86.218, 86.225, 86.228
Valach, F. 83.78
Valente, E.J. 1.36, 38.100, 48.75
Valentine Junior, D. 64.83, 64.84, 75.44
Valentini, F. 84.9
Valkonen, J. 3.15
Valle, G. 11.11, 33.93, 70.12, 72.53, 79.5, 79.16

Van Bree, J. 31.20
Van de Mieroop, W. 31.20
Van der Poel, H. 86.114
Van Havere, W. 2.23, 2.24
Van Koten, G. 86.114
Van Nuffel, P. 64.21
Van Opdenbosch, N. 35.64
Van Tets, A. 81.5
Vance Junior, T.B. 38.47, 38.48, 84.16, 84.26
Vanderbilt, J.J. 31.8
VanDerveer, D. 26.5, 31.47, 31.48, 31.49, 38.98, 38.99
VanDerveer, D.G. 27.28, 27.30, 84.6, 84.13, 84.14, 84.18
VanDerveer, M.C. 73.32, 73.51, 84.54
Vandewalle, M. 38.64, 38.69, 38.86
Vandewalle, M.E. 38.63
Vandrevala, M. 36.32, 37.14, 37.16, 42.12
Vangedal, S. 50.3
Vangehr, K. 45.30, 45.52, 45.53, 45.55, 45.58
Vani, G.V. 51.65, 51.66
Vanier, N. 40.35
Vankan, J.M.J. 31.2
Vann Bynum, R. 71.155, 73.74, 73.75, 73.79, 83.173
Vanoni, P.C. 38.106
Vansant, J. 33.35, 33.40, 33.42, 33.43, 33.52, 33.62, 33.63, 33.77, 33.80, 33.81, 33.87, 33.88, 35.56, 35.57
Varenne, J. 38.140
Varie, D.L. 38.48
Varnek, V.A. 3.26
Varret, F. 73.69
Vartanian, M.H. 36.17
Varughese, K.I. 47.13, 48.66, 48.67, 48.68, 48.69, 48.70, 84.62, 84.63
Vary, M.W. 21.7, 21.17, 37.3
Vashakidze, R.M. 76.3
Vaughan, K. 27.25, 31.51
Vedavathi, B.M. 3.29
Veenstra, S.J. 71.241
Vega, R. 35.53
Vegas, A. 68.13, 71.120
Vegas-Molina, A. 41.19
Vegkh, D. 38.6, 38.11, 38.12, 38.21, 38.36, 38.50
Veith, M. 69.36, 69.56
Velden, J.W.A.van der 71.281
Veldman, M.E.E. 71.241
Vella, P. 85.23, 86.94
Venanzi, L.M. 65.26, 86.21, 86.22, 86.104, 86.169
Venier, C.G. 20.4, 20.9
Venkatachalapathi, Y.V. 48.62
Venkataramu, S.D. 64.45, 64.46
Venkatasubramanian, K. 73.122
Venkatesan, K. 20.2, 20.6, 37.4, 38.120
Venkov, A.P. 58.68
Venters, K.K. 38.33
Venturoli, C. 41.75, 41.76
Verani, G. 79.8
Verhoeckx, G.J. 39.77, 39.81, 39.82, 39.83, 69.45
Verkade, J.G. 64.19
Vermeulen, A.J.W.A. 35.4
Vermin, W.J. 83.138
Verschoor, G.C. 83.21, 83.22, 83.23, 83.138, 83.218

Verweij, A.B. 48.82
Veselinovic, D.S. 76.72
Vichnewski, W. 53.2
Vickovic, I. 11.17, 35.15, 40.43
Vidari, G. 38.107
Videnova, V. 2.6, 2.11
Vidra, I. 45.6, 45.7, 45.10, 45.69
Viehe, H.G. 35.28
Vigevani, A. 50.18
Vigroux, M. 60.35
Vijayan, K. 3.29, 51.65, 51.66
Vijayan, M. 40.41, 48.8, 48.20, 48.62, 50.36
Vilches, J. 24.5, 37.6, 37.7
Villa, A.C. 20.12, 20.13, 40.54, 71.161, 72.10, 73.131
Villarreal-Salinas, B.E. 83.16, 83.17
Villiers, J.P.R.de 76.62, 85.49
Vincent, M.G. 85.55, 85.56
Viossat, B. 33.9, 83.72, 84.60
Visser, J.W. 83.95
Viswamitra, M.A. 47.3, 47.4, 47.23, 47.24, 47.39, 48.76, 50.34
Vitagliano, A. 72.41, 72.44, 72.53
Vitali, D. 49.5
Viterbo, D. 1.31, 40.14
Vitulli, G. 86.192
Vizi-Orosz, A. 86.84
Vladimirov, S. 32.52
Vliet, P.I.van 75.40
Voet, D. 44.33
Vogel, E. 31.60, 34.1, 34.4, 63.13
Vogel, P. 72.15, 72.16, 72.20, 72.25, 75.5
Vogtle, F. 17.38, 17.39
Vol'pin, M.E. 83.26
Voliotis, S. 81.6
Volka, K. 31.29, 31.30
Vollenbroek, F.A. 71.281
Vollmer, H.-J. 71.159, 71.197
Vollmer, S.H. 73.132
Volodarskii, L.B. 71.207, 83.147
Voronkov, M.G. 32.3, 63.7, 63.9, 71.23, 71.131
Voronkova, V.K. 84.112
Vorontsova, L.G. 1.28, 51.56, 51.57
Vos, A. 5.6, 48.82
Vos, A.M.de 39.81, 39.82, 39.83
Vos, J.G. 83.157
Voss, B. 34.1, 34.4, 63.13
Voss, J. 1.13, 1.21, 11.14
Vrieze, K. 71.68, 71.183, 71.272, 72.33, 72.50, 75.40, 86.114
Vuru, G.B.C. 83.85
Vuuren, C.P.J.van 84.104

Wachter, J. 71.204
Wada, A. 59.4
Wada, Y. 27.18, 39.39, 44.25, 48.47
Wade, K. 3.42, 20.1, 68.17
Wadepohl, H. 71.277
Wagenaar, A. 1.22, 1.24, 1.26, 20.5
Wagner, A. 29.8, 29.9, 31.60, 48.39
Wagner, H. 53.22, 71.6
Wahlberg, A. 33.13, 33.14
Wahlberg, I. 54.22
Wajda, K. 86.105
Wakahara, A. 47.17
Wakelin, L. 50.34
Wal, H.R.van der 71.241

Walba, D.M. 38.17
Walker, A. 86.136
Walker, M.L. 3.18
Walker, P.E. 64.108, 64.137
Walker, R.T. 47.26
Walkinshaw, M.D. 45.26
Wall, M.E. 58.75
Wallace, B. 21.8, 22.6
Wallbridge, M.G.H. 86.65
Wallis, R.C. 86.3
Wallwork, S.C. 17.5, 17.21
Walsh, B. 83.57, 83.58, 84.23
Walsh, L. 38.87
Walter, B. 33.7, 38.124, 38.125, 38.126, 38.127
Walter, W. 42.4
Walton, D.R.M. 63.29, 63.30
Walton, R.A. 71.72
Waltz, W.L. 76.75
Wampler, D.L. 64.13
Wand, M.D. 38.17
Wang, A. 43.1
Wang, A.H.J. 85.58
Wang, C.J. 38.111
Wang, H.H. 86.205, 86.224
Wang, S.Y. 44.27, 44.28, 44.29
Wang, Sh.-T. 58.19
Wang, Y. 3.20
Wani, M.C. 58.75
Ward, B.C. 80.2
Ward, D.L. 23.7, 30.4, 30.14, 38.53, 38.54, 38.129
Ward, M.A. 20.4
Warfield, L.T. 71.11, 71.41, 71.93
Warin, V. 44.32, 45.49
Waring, M. 50.34
Warren, R.G. 38.101
Warthen Junior, J.D. 51.72
Washburne, S.S. 63.29, 63.30
Washecheck, D.M. 3.52, 64.132
Waskowska, A. 35.71, 35.75, 35.78, 35.79, 35.80, 35.81, 35.82, 48.6, 48.7, 69.3
Wassem, W. 36.13
Wasserman, H.J. 71.236, 81.86, 84.29
Waszczak, J.V. 84.126
Watanabe, A. 44.25
Watanabe, H. 72.27
Watanabe, K. 53.14, 58.26
Watanabe, T. 2.3, 2.4, 29.7, 39.6, 73.53, 73.54
Watanabe, Y. 72.27
Waters, J.M. 27.3, 38.97, 71.258, 71.267, 81.8
Waters, T.N. 38.97, 81.8
Watjen, F. 59.12, 59.18
Watkin, D. 21.2, 36.32, 37.14, 37.16, 42.12
Watkin, D.J. 32.69
Watkins, S.F. 40.53, 40.55, 53.27
Watson, A.D. 76.47
Watson, D.G. 21.1
Watson, F.E. 51.52
Watson, N.S. 41.78
Watson, S.L. 75.3
Watson, W.H. 17.13, 20.4, 20.9, 21.3, 31.64, 32.60, 38.19, 38.57, 38.58, 38.59, 38.83, 51.70, 53.3, 58.55, 58.56, 58.62
Watt, C.I.F. 31.3, 31.11
Wawzonek, S. 7.4
Weakley, T.J.R. 38.4, 38.26, 38.27, 45.38, 45.40, 45.41, 45.43

Weber, G. 17.35, 17.38, 17.39, 17.40, 17.43, 33.64, 35.83, 35.84, 35.85, 64.151
Weber, H. 1.6, 1.7, 2.5, 11.4, 85.16, 85.51
Weber, H.P. 32.4, 44.1
Weber, L.D. 27.16
Wedd, A.G. 85.4, 85.9
Weeks, C.M. 51.13, 51.15, 51.26, 51.40
Weferling, N. 64.27
Wehner, E. 41.6
Wehri-Altenburger, S. 48.22, 48.28, 48.29, 48.36
Wei, C.H. 44.10, 83.56
Wei, K.-T. 30.14
Weibel, F. 54.29
Weidenhammer, K. 71.25, 71.202, 72.1, 72.31, 73.42, 73.58, 73.62, 73.90, 74.2, 74.3, 75.22, 75.29, 75.32, 75.33, 83.86
Weidlein, J. 68.6, 68.7, 68.9, 68.10, 68.11, 68.12
Weiher, U. 83.1, 83.2, 83.3, 83.5
Weiner, S. 21.19
Weinheimer, A.J. 54.21
Weinmaier, J.H. 86.8
Weinreb, S.M. 40.8
Weinstein, D.A. 2.29, 2.30
Weinstein, S. 24.6, 24.7, 24.8
Weisbuch, F. 27.8
Weishaupt, M. 86.2
Weiss, E. 64.14, 67.2
Weiss, J. 41.33, 76.81, 83.168
Weiss, R. 49.15, 49.23, 49.27, 71.205, 72.23, 72.47, 81.79, 83.221, 85.36
Weissman, E.C. 71.154, 71.257
Welberry, T.R. 26.3, 26.4
Welch, A.J. 71.81, 71.82, 71.263, 86.32, 86.33, 86.43
Welch, W.M. 36.49
Weller, F. 65.19, 83.5
Wells, R.J. 38.101, 54.15, 54.16, 54.19
Welter, J.J. 73.73
Wenger, J. 72.15, 72.16, 72.20
Wengrovius, J.H. 71.32
Wenkert, E. 51.63, 51.64, 51.67
Wenninger, J. 71.116
Wentrup, C. 36.51
Wentworth, R.A.D. 80.18
Wermuth, C.G. 10.6
Wernberg, O. 83.183
Werner, H. 71.21, 71.97, 71.169, 73.11
Wessels, P.L. 32.30
Wessner, D. 84.40
West, A.P. 76.108
West, D.X. 84.57
West, R. 72.68, 73.78, 73.103
Westerhaus, W.J. 62.16, 62.17
Westhof, E. 47.21, 60.39
Westover, G.F. 73.119, 73.125
Westphalen, J. 36.41
Wet, J.F.de 17.24
Wetter, H. 63.45
Weyland, M. 35.64
Whaley, H.A. 50.17
Whalley, W.B. 27.33, 51.27, 51.28, 51.33, 51.34, 51.35, 51.42, 59.22
Wheeler, S.H. 80.22, 80.26, 80.27
Whimp, P.O. 72.56, 72.57, 72.58, 75.45

White, A.H. 17.12, 17.44, 25.2, 31.23, 31.27, 31.28, 33.7, 33.32, 34.12, 35.76, 38.91, 38.92, 38.93, 38.124, 38.125, 38.126, 38.127, 38.132, 38.138, 40.47, 45.13, 54.8, 58.4, 59.14, 59.28, 60.41, 64.109, 64.110, 71.48, 71.218, 71.224, 71.229, 71.250, 71.251, 76.9, 76.17, 76.21, 79.9, 79.10, 80.11, 81.50, 83.210, 84.86, 84.101, 86.102, 86.145, 86.146
White, C. 73.81, 73.96
White, D.W. 64.19
White, J.G. 29.12, 32.1, 77.9
White, P.S. 51.3, 51.19
White, R.G. 81.84
Whitehead, G. 3.42, 20.1
Whiteley, R.H. 73.57, 73.91
Whitmire, K. 71.58, 84.52
Whittle, R. 71.73, 83.41, 83.42, 85.24, 86.68, 86.69, 86.70
Whuler, A. 76.37, 76.38, 76.39
Wiberg, N. 73.44
Wickens, D.A. 71.34, 71.121
Widjaja, H. 1.13, 1.21
Wieczorek, M. 64.41
Wieczorek, M.W. 17.7, 17.33, 41.64, 64.8
Wiedenfeld, H. 38.25, 58.29
Wieghardt, K. 76.81, 83.168
Wiemer, D.F. 54.30
Wiesenfeld, L. 64.5
Wiewiorowski, M. 58.20
Wild, S.B. 71.48
Wiley, P.F. 50.19
Wilford, J.B. 35.73
Wilhelm, E. 20.14
Wilhelm, M. 34.1, 34.4, 63.13
Wilke, G. 72.8, 72.61, 75.9
Wilkes, M.C. 38.17
Wilkins, C.J. 82.23, 84.59
Wilkinson, G. 64.118, 71.83, 71.89, 71.163, 71.166, 71.175, 71.176, 71.193, 71.194, 71.222, 71.227, 71.247, 76.108, 84.12, 84.130, 86.9, 86.11, 86.29, 86.30, 86.57, 86.59
Wilkowski, K. 76.67, 76.93, 82.22
Will, G. 2.2, 38.25, 58.29
Willems, G.J. 8.13
Willemse, J. 80.6
Willett, R.D. 83.46, 84.3, 84.4
Williams, A.F. 85.55, 85.56
Williams, D.A. 17.10, 35.54
Williams, D.H. 50.41
Williams, D.J. 28.6, 38.14, 40.18, 40.19, 46.3, 71.27, 71.28
Williams, D.M. 85.52
Williams, G. 45.50
Williams, G.A. 49.1, 49.2, 80.15, 80.10
Williams, G.J.B. 22.8, 48.1, 48.2, 49.3
Williams, J. 75.41
Williams, J.M. 2.7, 2.8, 64.127, 64.128, 64.153, 64.164, 65.16, 65.20, 65.21, 71.44, 73.126, 75.17, 75.18, 75.19, 86.58
Williams, J.R. 51.40, 51.47
Williams, R.L. 11.19
Williams, T.H. 37.12, 38.94, 38.95, 50.7, 50.23
Williamson, D.G. 41.67
Willis, A. 85.5
Willis, A.C. 73.108, 73.120, 86.206
Willis, S. 71.128
Willocx, A. 40.25

Willoughby, S.D. 73.21, 73.112, 73.130
Wilson, A.A. 31.43
Wilson, B.J. 54.35
Wilson, D.R. 61.23, 61.26
Wilson, F.B. 41.67
Wilson, G.S. 31.4, 31.5, 39.11
Wilson, H.R. 47.6, 47.15
Wilson, R.D. 86.24, 86.49
Wilson, R.M. 37.10
Wilson, S.R. 38.61, 38.71, 60.20, 83.146
Wilton, J.H. 53.23
Wimmer, F.L. 86.52
Wing, R.M. 52.1
Winkler, H. 38.141
Winkler, M.E. 76.25
Winter, H.-W. 36.51
Winter, M.J. 75.31
Winter, W. 19.3, 39.53, 69.42, 73.124, 73.128, 84.7
Wiseman, J.R. 31.8
Wit, A.D.de 32.58, 40.38
Witczak, M. 64.8
Witt, M. 69.12
Wittmann, D. 71.19
Wnuk, R.J. 50.17
Wohlfahrt, L. 73.89
Wojcik, G. 11.7
Wolf, R. 64.136
Wolff, C. 58.27
Wolff, M.E. 51.17, 51.22
Wolff, T.E. 85.57, 85.62, 85.78, 85.96, 85.97
Wolford, T.L. 85.26, 85.27
Wolinska-Mocydlarz, J. 58.7, 58.9, 58.10
Wolsey, W.C. 71.81, 71.82
Wong, A. 86.76
Wong, C. 33.69
Wong, C.S. 71.162
Wong, J.L. 37.8
Wong, K. 65.15, 71.177
Wong, L.C.H. 35.36
Wong, R.Y. 38.117, 38.128
Wong-Ng, W. 58.50, 75.27, 81.67
Wood, J.S. 84.111
Wood, T.E. 71.72
Woodhouse, D.I. 73.39
Woods, M. 64.74
Woodward, P. 71.38, 71.263, 72.32, 73.23, 73.56, 73.104, 73.105, 73.133, 75.23, 75.25, 75.30, 75.31, 86.147, 86.172
Woollard, D.C. 84.69, 84.133
Woon, T.C. 84.49
Worley, J.F. 51.71, 51.72
Worthington, J.M. 78.10, 78.11
Worthington, R.D. 41.72
Wreford, S.S. 75.27, 81.60, 86.125
Wright, L.J. 71.252, 71.258, 71.267, 86.134
Wright, M.R.W. 20.8, 64.65, 64.117
Wrobel, J.T. 38.22
Wu, K.K. 36.4
Wu, M.T. 38.111
Wu, S.-M. 83.133
Wucherer, E.J. 3.52, 64.132
Wudl, F. 39.7
Wunderlich, H. 64.24, 65.9, 82.16, 82.29, 82.31, 85.12, 85.89
Wurtenberg, S. 62.13
Wussow, H.-G. 33.4, 33.5, 64.24
Wyk, A.J.van 33.18
Wyss, P.C. 45.19

Yaba, S. 81.59
Yablokov, V.A. 63.25, 69.51
Yablokov, Yu.V. 84.112
Yagodina, L.A. 64.104
Yaguadaev, M.R. 58.32
Yagubskii, E.B. 39.66
Yakovlev, I.P. 63.43
Yakushi, K. 39.8
Yamada, K. 60.33
Yamada, S. 51.80, 78.9, 78.18
Yamada, Y. 48.56, 48.60
Yamagata, Y. 47.2, 47.17, 47.28
Yamagishi, K. 35.65
Yamaguchi, K. 36.20
Yamaguchi, M. 82.34
Yamaizumi, Z. 47.34
Yamamoto, A. 71.179, 72.60
Yamamoto, I. 13.19, 41.61, 41.77
Yamamoto, T. 71.179, 72.60
Yamamoto, Y. 38.55
Yamamura, K. 22.7
Yamane, T. 48.24, 48.60, 48.64
Yamasaki, K. 54.6
Yamashita, M. 76.19, 83.25
Yamauchi, M. 67.5
Yamauchi, O. 82.18, 82.19
Yamauchi, T. 45.57
Yamazaki, H. 73.121, 81.73, 81.78, 82.25, 82.39
Yamazaki, M. 38.122, 58.71
Yampol'skaya, M.A. 84.38, 84.112
Yan, C.-F. 38.70
Yanagi, K. 83.197
Yandell, J.K. 76.9, 76.17, 76.21
Yang, C.H. 33.53
Yano, S. 81.59, 82.34
Yanovskii, A.I. 1.5, 31.22, 64.23, 64.75, 65.3, 65.25, 68.14, 71.188
Yasuda, H. 67.5, 71.31
Yasuda, M. 31.68, 37.24
Yasufuku, K. 73.121
Yasuoka, N. 28.10, 31.33, 31.34, 32.61, 67.5, 71.31, 77.2
Yau, C.-C. 27.28, 27.30
Yeh, E.-S. 73.120
Yelon, W.B. 2.12
Yeung, H.W. 58.6
Ymen, I. 32.14, 32.15, 32.16
Yokohama, S. 39.71
Yokoyama, S. 47.34
Yokoyama, Y. 56.5
Yoneda, H. 76.31
Yonemitsu, O. 31.41
Yong, S.H. 82.23, 84.59
Yorke, S.C. 5.12
Yoshida, A. 41.50
Yoshida, T. 81.66, 86.96, 86.97, 86.130
Yoshida, Z.-I. 49.7
Yoshifuji, M. 64.168, 75.6
Yoshikawa, S. 81.59, 82.34
Yoshimura, J. 47.2
Yoshimura, Y. 50.5
Yoshino, Y. 76.23, 76.61, 81.97
Yoshioka, K. 51.73
Yougi, T. 48.26
Young, G.B. 84.130
Young, G.J. 71.236, 81.86

Young, J.C. 71.205
Young, P.R. 73.73
Youngs, W.J. 69.6, 71.32, 71.74, 71.110, 71.174, 71.209
Yunusov, M.S. 58.44
Yunusov, S.Yu. 58.31, 58.32, 58.44
Yurashek, A. 38.28
Yurchenko, E.N. 3.26
Yurchenko, V.M. 64.49

Zabel, V. 17.13, 32.60, 51.70, 53.3, 58.55, 58.56, 58.62
Zacharias, D.E. 29.4, 29.5
Zagar, V. 3.46, 3.47
Zaitseva, E.S. 84.64
Zakaria, Z. 32.72, 33.94
Zakharkin, L.I. 64.23, 65.3, 71.13
Zakharov, L.N. 8.14, 41.35, 73.134, 73.135
Zakrzewski, S.F. 44.31, 44.33
Zalkin, A. 32.6, 32.7, 73.102, 79.3, 84.85
Zamankhan, H. 39.3
Zamanov, A.Z. 81.31
Zambonelli, L. 65.26, 72.7, 78.5, 78.14, 83.104, 85.15, 86.21, 86.22, 86.169
Zanari, A.R. 78.3, 78.4
Zanazzi, P.F. 38.77, 78.3, 78.4, 85.17
Zanobini, F. 35.5, 86.122, 86.123
Zanotti, G. 18.1, 18.2, 33.93, 40.48, 48.49, 63.53, 71.213, 77.3, 86.128, 86.131
Zatorski, A. 41.64
Zavodnik, V.E. 35.40, 36.9, 63.19, 64.104, 65.4
Zaworotko, M.J. 41.1, 68.3, 68.4, 69.40, 69.49, 69.50, 71.89, 71.163, 73.5, 73.106, 75.48
Zechlin, L. 56.7
Zefirov, N.S. 31.17, 31.18, 31.22, 31.32, 31.37, 31.56
Zefirov, Yu.V. 83.103
Zehnder, M. 50.27, 69.21, 76.53, 76.94, 76.96, 76.97, 83.111
Zeiss, W. 64.30, 64.69, 64.71, 64.99
Zejc, A. 41.64
Zeks, B. 3.46
Zel'bst, E.A. 63.7
Zelenskaja, O.V. 41.13
Zelentsov, V.V. 85.39
Zeller, K.-P. 19.3
Zellers, E.T. 39.7
Zellweger, J.-M. 85.30, 85.90
Zelnik, R. 54.17
Zeltzer, I.E. 73.31
Zengren, T. 48.26
Zetta, L. 32.64
Zhdankin, V.V. 31.56
Zhdanov, A.A. 63.32, 63.38
Zhdanova, E.A. 63.26
Zhukhlistova, N.E. 48.32, 50.22, 50.24, 50.37
Ziegler, F.E. 21.7, 21.17
Ziegler, M.L. 71.25, 71.30, 71.126, 71.202, 72.1, 72.31, 73.42, 73.58, 73.62, 73.90, 74.2, 74.3, 75.22, 75.29, 75.32, 75.33, 75.39, 83.86
Ziegler, R.J. 71.266
Ziffer, H. 27.9
Ziffer, M. 27.9

Zimmer-Gasser, B. 64.14, 64.43, 64.79, 64.150, 71.36
Zinczuk, J. 32.21
Zindler, G. 85.13
Zingaro, R.A. 11.12, 38.35, 39.41, 70.1
Ziolo, R.F. 11.18, 70.3
Zivanov-Stakic, D. 32.52
Zlotina, I.B. 71.51
Zobel, D. 60.32, 60.38
Zocchi, M. 86.192
Zolotoi, A.B. 32.19, 63.43
Zon, G. 64.60
Zorn, H. 64.98, 86.13
Zub, Yu.L. 7.1
Zubieta, J. 80.19, 85.22, 85.23, 85.71, 86.94
Zubieta, J.A. 69.8, 80.25
Zuccaro, C. 83.73, 86.40
Zuckerman, J.J. 69.8, 69.46, 69.48, 69.60
Zuniga, F.J. 3.12
Zvagulis, M. 83.83, 83.84
Zvarikina, A.V. 39.66
Zvonkova, Z.V. 27.23, 27.26, 38.88
Zwanenburg, B. 31.2
Zwaschka, F. 38.46, 64.37
Zwierzak, A. 46.5
Zwinselman, J.J. 33.38, 33.65
Zylber, J. 47.33
Zylber, N. 47.33

List of Classes

1 Aliphatic Carboxylic Acids and their Derivatives
2 Aliphatic Carboxylic Acid Salts (Ammonium, IA, IIA Metals)
3 Aliphatic Amines
4 Aliphatic (N and S) Compounds
5 Aliphatic Miscellaneous
6 Enolates (Aliphatic and Aromatic)
7 Nitriles (Aliphatic and Aromatic)
8 Urea Compounds (Aliphatic and Aromatic)
9 Nitrogen-Nitrogen Compounds (Aliphatic and Aromatic)
10 Nitrogen-Oxygen Compounds (Aliphatic and Aromatic)
11 Sulfur and Selenium Compounds
12 Carbonium Ions, Carbanions, Radicals
13 Benzoic Acid Derivatives
14 Benzoic Acid Salts (Ammonium, IA, IIA Metals)
15 Benzene Nitro Compounds
16 Anilines
17 Phenols and Ethers
18 Benzoquinones
19 Benzene Miscellaneous
20 Monocyclic Hydrocarbons (3, 4, 5-Membered Rings)
21 Monocyclic Hydrocarbons (6-Membered Rings)
22 Monocyclic Hydrocarbons (7, 8-Membered Rings)
23 Monocyclic Hydrocarbons (9- and Higher-Membered Rings)
24 Naphthalene Compounds
25 Naphthoquinones
26 Anthracene Compounds
27 Polycyclic Hydrocarbons (2 Fused Rings)
28 Polycyclic Hydrocarbons (3 Fused Rings)
29 Polycyclic Hydrocarbons (4 Fused Rings)
30 Polycyclic Hydrocarbons (5 or More Fused Rings)
31 Bridged Ring Hydrocarbons
32 Hetero-Nitrogen (3, 4, 5-Membered Monocyclic)
33 Hetero-Nitrogen (6-Membered Monocyclic)
34 Hetero-Nitrogen (7- and Higher-Membered Monocyclic)
35 Hetero-Nitrogen (2 Fused Rings)
36 Hetero-Nitrogen (More than 2 Fused Rings)
37 Hetero-Nitrogen (Bridged Ring Systems)
38 Hetero-Oxygen
39 Hetero-Sulfur and Hetero-Selenium
40 Hetero-(Nitrogen and Oxygen)
41 Hetero-(Nitrogen and Sulfur)
42 Miscellaneous Heterocycles
43 Barbiturates
44 Pyrimidines and Purines
45 Carbohydrates
46 Phosphates
47 Nucleosides and Nucleotides
48 Alpha-Amino-Acids and Peptides
49 Porphyrins and Corrins
50 Antibiotics
51 Steroids
52 Monoterpenes
53 Sesquiterpenes
54 Diterpenes
55 Sesterterpenes
56 Triterpenes
57 Tetraterpenes
58 Alkaloids
59 Miscellaneous Natural Products
60 Molecular Complexes
61 Clathrates
62 Boron Compounds
63 Silicon Compounds
64 Phosphorus Compounds
65 Arsenic Compounds
66 Antimony and Bismuth Compounds
67 Groups IA and IIA Compounds
68 Group III Compounds
69 Germanium, Tin, Lead Compounds
70 Tellurium Compounds
71 Transition Metal-C Compounds
72 Metal π-Complexes (Open-Chain)
73 Metal π-Complexes (Cyclopentadiene)
74 Metal π-Complexes (Arene)
75 Metal π-Complexes (Miscellaneous Ring Systems)
76 Metal Complexes (Ethylenediamine)
77 Metal Complexes (Acetylacetone)
78 Metal Complexes (Salicylic Derivatives)
79 Metal Complexes (Thiourea)
80 Metal Complexes (Thiocarbamate or Xanthate)
81 Metal Complexes (Carboxylic Acid)
82 Metal Complexes (Amino-Acid)
83 Metal Complexes (Nitrogen Ligand)
84 Metal Complexes (Oxygen Ligand)
85 Metal Complexes (Sulfur or Selenium Ligand)
86 Metal Complexes (P, As, Sb Ligand)